Encyclopedia of Physics

Edited by
Rita G. Lerner and George L. Trigg

Volume 1

Encyclopedia of Physics

Third, Completely Revised and Enlarged Edition

Edited by
Rita G. Lerner and George L. Trigg

Volume 1

WILEY-
VCH

WILEY-VCH Verlag GmbH & Co. KGaA

Editors:

Rita G. Lerner †

George L. Trigg
New Paltz, New York

All books published by Wiley-VCH are carefully produced. Nevertheless, authors, editors, and publisher do not warrant the information contained in these books, including this book, to be free of errors. Readers are advised to keep in mind that statements, data, illustrations, procedural details or other items may inadvertently be inaccurate.

Library of Congress Card No.: applied for

British Library Cataloguing-in-Publication Data
A catalogue record for this book is available from the British Library.

Bibliographic information published by
Die Deutsche Bibliothek
Die Deutsche Bibliothek lists this publication in the Deutsche Nationalbibliografie; detailed bibliographic data is available in the Internet at <http://dnb.ddb.de>.

© 2005 WILEY-VCH Verlag GmbH & Co. KGaA, Weinheim

Typesetting   Michael Bär
Printing and Binding AALEXX, Großburgwedel
Coverdesign 4t Matthes + Traut Werbeagentur GmbH, Darmstadt

Printed in the Federal Republic of Germany
Printed on acid-free paper

ISBN-13: 978-3-527-40554-1
ISBN-10: 3-527-40554-2

Table of Contents

Table of Contents, v
Preface, xiii
List of Contributors, xv

Volume 1

Absorption Coefficient, 1
Accelerators, Linear, 2
Accelerators, Potential-Drop Linear, 10
Acoustical Measurements, 21
Acoustics, 29
Acoustics, Architectural, 32
Acoustics, Linear and Nonlinear, 35
Acoustics, Physiological, 50
Acoustoelectric Effect, 54
Adsorption, 57
Aerosols, 61
Allotropy and Polymorphism, 63
Alloys, 65
Alpha Decay, 67
Ampère's Law, 71
Anelasticity, 72
Angular Correlation of Nuclear
    Radiation, 76
Antimatter, 81
Arcs and Sparks, 84
Astronomy, High-Energy Neutrino, 89
Astronomy, Optical, 100
Astronomy, Radio, 103
Astronomy, X-Ray, 110
Astrophysics, 122
Atmospheric Physics, 126
Atomic Spectroscopy, 130
Atomic Structure Calculations, Electronic
    Correlation, 138
Atomic Structure Calculations, One-Electron
    Models, 141
Atomic Structure Calculations, Relativistic
    Atoms, 149
Atomic Trapping and Cooling, 157
Atoms, 161
Auger Effect, 169

Aurora, 173
Balmer Formula, 177
Baryons, 179
Beams, Atomic and Molecular, 182
Beta Decay, 186
Betatron, 190
Bethe–Salpeter Equation, 192
Binding Energy, 197
Biophysics, 198
Black Holes, 203
Blackbody Radiation, 208
Bohr Theory of Atomic Structure, 212
Bose–Einstein Condensation, 218
Bose–Einstein Statistics, 221
Boundary Layers, 223
Bremsstrahlung, 226
Brillouin Scattering, 230
Brownian Motion, 233
Calorimetry, 235
Capillary Flow, 238
Carnot Cycle, 241
Casimir Effect, 244
Catalysis, 246
Catastrophe Theory, 250
Cellular Automata, 258
Center-of-Mass System, 262
Ceramics, 263
Čerenkov Radiation, 267
Channeling, 268
Chaos, 272
Charge-Density Waves, 277
Charged-Particle Optics, 281
Charged-Particle Spectroscopy, 286
Chemical Bonding, 295
Chemiluminescence, 304
Circuits, Integrated, 306

Encyclopedia of Physics, Third Edition. Edited by George L. Trigg and Rita G. Lerner
Copyright ©2005 WILEY-VCH Verlag GmbH & Co. KGaA, Weinheim
ISBN: 3-527-40554-2

Clocks, Atomic and Molecular, 310
Cloud and Bubble Chambers, 312
Cold Atoms and Molecules, 316
Collisions, Atomic and Molecular, 317
Color Centers, 332
Combustion and Flames, 349
Complementarity, 354
Complex Systems, 356
Compton Effect, 359
Conduction, 362
Conservation Laws, 364
Constants, Fundamental, 371
Coriolis Acceleration, 393
Corona Discharge, 394
Cosmic Rays, Astrophysical Effects, 397
Cosmic Rays, Solar System Effects, 410
Cosmic Strings, 418
Cosmology, 421
Counting Tubes, 428
CPT Theorem, 438
Critical Points, 441
Cryogenics, 444
Crystal and Ligand Fields, 447
Crystal Binding, 449
Crystal Defects, 451
Crystal Growth, 455
Crystal Symmetry, 458
Crystallography, X-Ray, 474
Currents in Particle Theory, 487
Cyclotron, 490
Cyclotron Resonance, 497
Deformation of Crystalline Materials, 499
de Haas–van Alphen Effect, 502
Demineralization, 507
Diamagnetism and Superconductivity, 508
Dielectric Properties, 512
Diffraction, 516
Diffusion, 526
Dispersion Theory, 528
Doppler Effect, 532
Dynamic Critical Phenomena, 534
Dynamics, Analytical, 535
Eigenfunctions, 547
Elasticity, 548
Electric Charge, 553

Electric Moments, 554
Electrochemical Conversion and
    Storage, 557
Electrochemistry, 563
Electrodynamics, Classical, 567
Electroluminescence, 574
Electromagnetic Interaction, 579
Electromagnetic Radiation, 583
Electromagnets, 589
Electron, 591
Electron and Ion Beams, Intense, 595
Electron and Ion Impact Phenomena, 600
Electron Beam Technology, 605
Electron Bombardment of Atoms and
    Molecules, 611
Electron Diffraction, 616
Electron Energy States in Solids and
    Liquids, 621
Electron–Hole Droplets in
    Semiconductors, 631
Electron Microscopy, 635
Electron Spin Resonance, 637
Electron Tubes, 644
Electronics, 651
Electronic Noses, 659
Electrophoresis, 663
Electrophotography, 665
Electrostatics, 669
Elementary Particles in Physics, 671
Elements, 713
Ellipsometry, 718
Energy and Work, 720
Entropy, 723
Equations of State, 725
Ergodic Theory, 728
Error Analysis, 730
Excitons, 734
Far-Infrared Spectra, 737
Faraday Effect, 741
Faraday's Law of Electromagnetic
    Induction, 742
Fatigue, 744
Fermi–Dirac Statistics, 746
Fermi Surface, 748
Ferrimagnetism, 753

Ferroelasticity, 761
Ferroelectricity, 763
Ferromagnetism, 768
Feynman Diagrams, 777
Fiber Optics, 779
Field Emission, 782
Field-Ion Microscopy, 785
Field Theory, Axiomatic, 791
Field Theory, Classical, 797
Field Theory, Unified, 803
Fields, 805
Fine and Hyperfine Spectra and
    Interactions, 807
Fluctuation Phenomena, 813
Fluid Physics, 819
Formation of Stars and Planets, 835
Fourier Transforms, 840
Fractals, 847
Franck–Condon Principle, 850
Fraunhofer Lines, 851
Free Energy, 853
Friction, 854
Fullerenes, 858
Galaxies, 863
Galvanomagnetic and Related Effects, 866
Gamma Decay, 871
Gamma-Ray Spectrometers, 875
Gauge Theories, 877
Gauss's Law, 882
Geochronology, 884
Geomagnetism, 888
Geometric Quantum Phase, 891
Geophysics, 893
Glass, 902
Glassy Metals, 904
Grand Unified Theories, 909
Gratings, Diffraction, 916
Gravitation, 918
Gravitational Lenses, 931
Gravitational Waves, 934
Gravity, Earth's, 937
Group Theory in Physics, 941
Gyromagnetic Ratio, 952
H Theorem, 955
Hadrons, 956

Hadron Colliders at High Energy, 958
Hall Effect, 965
Hall Effect, Quantum, 969
Hamiltonian Function, 972
Heat, 975
Heat Capacity, 976
Heat Engines, 978
Heat Transfer, 982
Heavy-Fermion Materials, 988
Helium, Liquid, 992
Helium, Solid, 1001
Hidden Variables, 1010
High-Field Atomic States, 1013
High Temperature, 1015
History of Physics, 1024
Holography, 1044
Hot Atom Chemistry, 1052
Hot Cells and Remote Handling
    Equipment, 1054
Hubble Effect, 1057
Hydrodynamics, 1060
Hydrogen Bond, 1065
Hypernuclear Physics and Hypernuclear
    Interactions, 1072
Hyperons, 1076
Hysteresis, 1078
Ice, 1081
Inclusive Reactions, 1087
Inertial Fusion, 1090
Infrared Spectroscopy, 1096
Insulators, 1102
Interatomic and Intermolecular Forces, 1106
Interferometers and Interferometry, 1110
Intermediate Valence Compounds, 1114
Internal Friction in Crystals, 1120
Interstellar Medium, 1123
Invariance Principles, 1127
Inversion and Internal Rotation, 1132
Ionization, 1139
Ionosphere, 1143
Ising Model, 1145
Isobaric Analog States, 1147
Isomeric Nuclei, 1150
Isospin, 1155
Isotope Effects, 1160

Isotope Separation, 1163
Isotopes, 1176
Jahn–Teller Effect, 1197
Josephson Effects, 1198
Kepler's Laws, 1205
Kerr Effect, Electro-Optical, 1208
Kerr Effect, Magneto-Optical, 1209
Kinematics and Kinetics, 1212
Kinetic Theory, 1218
Kinetics, Chemical, 1226
Klystrons and Traveling-Wave Tubes, 1228
Kondo Effect, 1232
Laser Spectroscopy, 1239
Laser Cooling, 1246
Lasers, 1254
Lattice Dynamics, 1284

Lattice Gauge Theory, 1294
Leptons, 1297
Levitation, Electromagnetic, 1299
Lie Groups, 1308
Light, 1310
Light Scattering, 1316
Light-Sensitive Materials, 1319
Lightning, 1321
Liquid Crystals, 1325
Liquid Metals, 1334
Liquid Structure, 1338
Lorentz Transformations, 1344
Low-Energy Electron Diffraction
    (LEED), 1345
Luminescence (Fluorescence and
    Phosphorescence), 1349

Volume 2

Mach's Principle, 1355
Magnetic Circular Dichroism, 1356
Magnetic Cooling, 1359
Magnetic Domains and Bubbles, 1366
Magnetic Fields, High, 1372
Magnetic Materials, 1379
Magnetic Moments, 1385
Magnetic Monopoles, 1389
Magnetic Ordering in Solids, 1392
Magnetoacoustic Effect, 1396
Magnetoelastic Phenomena, 1398
Magnetohydrodynamics, 1401
Magnetoresistance, 1412
Magnetosphere, 1415
Magnetostriction, 1421
Magnets (Permanent) and
    Magnetostatics, 1425
Many-Body Theory, 1428
Masers, 1440
Mass, 1447
Mass Spectroscopy, 1448
Matrices, 1454
Maxwell–Boltzmann Statistics, 1463
Maxwell's Equations, 1464
Mechanical Properties of Matter, 1467

Mesons, 1473
Mesoscopic Physics, 1474
Metal–Insulator Transitions, 1477
Metallurgy, 1482
Metals, 1485
Meteorology, 1486
Metrology, 1490
Michelson–Morley Experiment, 1493
Microscopy, Optical, 1496
Microwave Spectroscopy, 1508
Microwaves and Microwave Circuitry, 1512
Milky Way, 1520
Molecular Spectroscopy, 1522
Molecular Structure Calculations, 1600
Molecules, 1615
Molten Salts, 1622
Moment of Inertia, 1626
Momentum, 1633
Monte Carlo Techniques, 1635
Mössbauer Effect, 1642
Multipole Fields, 1659
Muonic, Mesonic, and Other Exotic
    Atoms, 1662
Muonium, 1667
Musical Instruments, 1671

Nanobionics, 1677
Nanocatalysis, 1681
Network Theory: Analysis and
    Synthesis, 1686
Neutrinos, 1700
Neutron Diffraction and Scattering, 1705
Neutron Spectroscopy, 1714
Neutron Stars, 1721
Newton's Laws, 1725
Noise, Acoustical, 1728
Nonlinear Wave Propagation, 1731
Novel Particle Acceleration Methods, 1734
Nuclear Fission, 1739
Nuclear Forces, 1751
Nuclear Fusion, 1756
Nuclear Magnetic Resonance, 1766
Nuclear Moments, 1771
Nuclear Polarization, 1774
Nuclear Properties, 1778
Nuclear Quadrupole Resonance, 1788
Nuclear Reactions, 1792
Nuclear Reactors, 1798
Nuclear Scattering, 1804
Nuclear States, 1807
Nuclear Structure, 1810
Nucleon, 1820
Nucleosynthesis, 1822
Operators, 1829
Optical Activity, 1832
Optical Pumping, 1834
Optics, Geometrical, 1838
Optics, Nonlinear, 1841
Optics, Physical, 1846
Optics, Statistical, 1852
Order–Disorder Phenomena, 1857
Organic Conductors and
    Superconductors, 1861
Organic Semiconductors, 1866
Oscilloscopes, 1877
Paramagnetism, 1881
Parity, 1882
Partial Waves, 1895
Partons, 1899
Phase Transitions, 1901
Philosophy of Physics, 1921

Phonons, 1926
Photoconductivity, 1930
Photoelastic Effect, 1931
Photoelectron Spectroscopy, 1933
Photoionization, 1942
Photon, 1944
Photonic Crystals, 1948
Photonuclear Reactions, 1953
Photosphere, 1956
Photovoltaic Effect, 1958
Piezoelectric Effect, 1959
Plasma Confinement Devices, 1961
Plasmas, 1975
Plasma Waves, 1984
Plasmons, 1991
Polarizability, 1995
Polarization, 1998
Polarized Light, 2000
Polaron, 2004
Polymers, 2028
Positron, 2045
Positron Annihilation in Condensed
    Matter, 2047
Positron–Electron Colliding Beams, 2050
Positronium, 2055
Precession, 2059
Probability, 2061
Proton, 2067
Pulsars, 2068
Pyroelectricity, 2072
Quantum Information, 2077
Quantum Electrodynamics, 2083
Quantum Field Theory, 2095
Quantum Fluids, 2105
Quantum Mechanics, 2111
Quantum Optics, 2128
Quantum Statistical Mechanics, 2134
Quantum Structures in
    Semiconductors, 2138
Quantum Theory of Measurement, 2144
Quarkonium, 2152
Quarks, 2158
Quasars, 2163
Quasiparticles, 2168
Radar, 2171

Radiation Belts, 2174
Radiation Chemistry, 2177
Radiation Damage in Solids, 2181
Radiation Detection, 2187
Radiation Interaction with Matter, 2192
Radioactivity, 2196
Radiochemistry, 2200
Radiological Physics, 2205
Radiometry, 2217
Raman Spectroscopy, 2221
Rare Earths, 2227
Rare Gases and Rare-Gas Compounds, 2231
Rayleigh Scattering, 2235
Reflection, 2236
Reflection High-Energy Electron Diffraction
    (RHEED), 2240
Refraction, 2241
Regge Poles, 2247
Relativity, General, 2249
Relativity, Special Theory, 2257
Relaxation Phenomena, 2274
Renormalization, 2278
Resistance, 2283
Resonance Phenomena, 2285
Resonances, Giant, 2291
Rheology, 2298
Rotation and Angular Momentum, 2310
S-Matrix Theory, 2333
Scanning Tunneling Microscopy, 2337
Scattering Theory, 2339
Schrödinger Equation, 2347
Scintillation and Čerenkov Counters, 2348
Second Sound, 2353
Secondary Electron Emission, 2354
Sedimentation and Centrifugation, 2357
Seismology, 2361
Semiconductor Radiation Detectors, 2369
Semiconductors, Amorphous, 2377
Semiconductors, Crystalline, 2393
Servomechanism, 2410
Shock Waves and Detonations, 2411
Soil Physics, 2418
Solar Energy, 2421
Solar Neutrinos, 2442
Solar System, 2451

Solar Wind, 2456
Solid-State Physics, 2459
Solid-State Switching, 2472
Solitons, 2483
Sound, Underwater, 2485
Space Science and Technology, 2488
Spacetime, 2494
Spectrophotometry, 2500
Spin, 2502
Statics, 2509
Statistical Mechanics, 2511
Statistics, 2519
Stellar Energy Sources and Evolution, 2524
Stochastic Processes, 2531
String Theory, 2539
Strong Interactions, 2551
Sum Rules, 2556
Sun, 2560
Superconducting Materials, 2565
Superconductive Devices, 2571
Superconductivity Theory, 2580
Superheavy Elements, 2591
Supersymmetry and Supergravity, 2598
Surface Tension, 2605
Surface Waves on Fluids, 2607
Surfaces and Interfaces, 2609
SU(3) and Symmetry Groups, 2613
Symbols, Units, and Nomenclature,
    2619
Symmetry Breaking, Spontaneous, 2642
Synchrotron, 2650
Synchrotron Radiation, 2659
Tachyons, 2667
Temperature, 2668
Thermal Analysis, 2671
Thermal Expansion, 2674
Thermionic Emission, 2678
Thermodynamic Data, 2682
Thermodynamics, Equilibrium, 2684
Thermodynamics, Nonequilibrium, 2689
Thermoelectric Effects, 2701
Thermoluminescence, 2705
Thermometry, 2720
Thin Films, 2727
Three-Body Problem, Gravitational, 2734

Three-Body Problem, Quantum
      Mechanical, 2737
Time, 2741
Transducers, 2744
Transistors, 2746
Transition Elements, 2756
Transmission Lines and Antennas, 2760
Transport Properties, 2766
Transport Theory, 2771
Transuranium Elements, 2774
Tribology, 2780
Tunneling, 2783
Turbulence, 2789
Twin Paradox, 2796
Ultracold Quantum Gases, 2799
Ultrahigh-Pressure Techniques, 2803
Ultrashort Optical Pulses, 2816
Ultrasonic Biophysics, 2820
Ultrasonics, 2822

Uncertainty Principle, 2829
Universe, 2832
Vacuums and Vacuum Technology, 2837
Vapor Pressure, 2846
Vector and Tensor Analysis, 2848
Vibrations, Mechanical, 2854
Viscosity, 2856
Visible and Ultraviolet Spectroscopy, 2860
Vision and Color, 2866
Vortices, 2884
Water, 2891
Waves, 2893
Weak Interactions, 2900
Weak Neutral Currents, 2908
Whiskers, 2913
Work Function, 2914
X-Ray Spectra and X-Ray
      Spectroscopy, 2917
Zeeman and Stark Effects, 2927

Preface to the Third Edition

For more than 20 years now the Wiley-VCH Encyclopedia of Physics has been recognized as one of the leading international encyclopedias in physics. Worldwide, this work has served the continued reference needs of scientists, students, librarians, and the general public interested in modern physics. With this third, completely revised, updated and enlarged edition of the Encyclopedia, the publisher intends to meet the demand for current, precise information on both classical and recent concepts in physics. The Encyclopedia is intended to be the starting point for everybody who either works in or studies physics, in particular for everyone who looks for information on subjects outside her or his field of expertise.

In their preface to the previous edition of this Encyclopedia, the editors Rita G. Lerner and George L. Trigg emphasized that this Encyclopedia "describes physics only as of a particular moment in time". In the almost two decades since the first edition appeared, there have been significant changes in many areas of physics.

In recent years, scientific research has been driven predominantly by the approach to nanometer scale systems and devices. Scanning probe microscopy was a pre-requisite for our present knowledge of the structure and dynamics of surfaces and interfaces on an atomic scale. The experimental realization of the trapping and cooling of neutral atoms attracted wide interest among the scientific community and was the starting point for a new branch of physics. As a consequence of this development, for example, the first experimental evidence of Bose–Einstein condensation was achieved. Possible approaches for the realization of quantum entanglement and quantum computation have been realized at the border between the physics of the twentieth century and the present decade. Among others, these results of modern basic research in physics will possibly have an impact on future developments in technology and science. The present development in physics, as well as in other scientific and technical disciplines, requires more and more knowledge of other, adjacent areas of modern research. This tendency is reflected in particular by the interdisciplinary approach of a number of novel and further 'booming' subject areas such as biophysics or econophysics.

All of the more than 500 articles on the most important areas of physics in this Encyclopedia have been written or revised by renowned authorities in science and physical technology. A number of new entries have been included to cover the most recent terminology and techniques in physics. As in the previous editions, articles are arranged in alphabetical order. At the end of most articles, cross-references are given to related articles in this work. Since it was not our intention to further expand the size of this volume and to leave the concept of a concise, easy-to-access approach to modern physics behind, most authors have provided references for further reading. These bibliographies have been completely updated for the existing articles.

The publisher is grateful to the many scientific colleagues and friends who have cooperated and contributed to this new edition and thanks them for their suggestions and comments.

Alexander Grossmann
Editorial Director Wiley-VCH

Encyclopedia of Physics, Third Edition. Edited by George L. Trigg and Rita G. Lerner
Copyright © 2005 WILEY-VCH Verlag GmbH & Co. KGaA, Weinheim
ISBN: 3-527-40554-2

List of Contributors

George O. Abell[†]
Late of Department of Astronomy, University of
California, Los Angeles, CA, USA
Universe

Mark J. Ablowitz
Department of Applied Mathematics, University
of Colorado, Boulder, CO, USA
Nonlinear Wave Propagation

S. C. Abrahams
Physics Department, Southern Oregon University,
Ashland, OR, USA
Ferroelasticity
Pyroelectricity

R. K. Adair
Physics Department, Yale University, New Haven,
CT, USA
Quarks

Charles S. Adams
Department of Physics, University of Durham,
Durham, United Kingdom
Cold Atoms and Molecules

Gail D. Adams
Pagosa Springs, CO, USA
Radiological Physics

W. M. Adams
Department of Geology, Western Washington
University, Bellingham, WA, USA
Seismology

Irshad Ahmad
Argonne National Laboratory, Argonne, Illinois,
USA
Alpha Decay

Thomas J. Ahrens
Department of Geophysics, California Institute of
Technology, Pasadena, CA, USA
Geophysics

Hashem Akhavan-Tafti
Lumigen, Inc., Southfield, MI, USA
Chemiluminescence

D. E. Alburger
Brookhaven National Laboratory, Upton, NY,
USA
Beta Decay

B. Alder
Lawrence Livermore National Laboratory,
University of California, Livermore, CA, USA
Liquid Structure

George A. Alers
Retired from Magnasonics, Inc., Albuquerque,
NM, USA
Mechanical Properties of Matter

R. Casanova Alig
Sigma Xi, Princeton Chapter, Princeton, NJ, USA
Electrostatics

Douglas R. Allen
US Naval Research Lab, Washington, DC, USA
Meteorology

Philip B. Allen
Department of Physics and Astronomy, Stony
Brook University, Stony Brook, NY, USA
Superconductivity Theory

D. Ambrose
Surbiton, Surrey, United Kingdom
Vapor Pressure

Betsy Ancker-Johnson
Austin, TX, USA
Plasmons

A. C. Anderson
Department of Physics, University of Missouri,
Kansas City, MO, USA
Cryogenics

James L. Anderson
Department of Physics, Stevens Institute of
Technology, Hoboken, NJ, USA
Field Theory, Unified
Relativity, General

P. W. Anderson
Department of Physics, Princeton University,
Princeton, NJ, USA
Quasiparticles

[†]deceased

Encyclopedia of Physics, Third Edition. Edited by George L. Trigg and Rita G. Lerner
Copyright © 2005 WILEY-VCH Verlag GmbH & Co. KGaA, Weinheim
ISBN: 3-527-40554-2

Robert A. Anderson
Sandia National Laboratories Albuquerque, NM,
USA
Insulators

J.-P. Antoine
Institute of Theoretical Physics, Catholic
University of Louvain, Louvain-la-Neuve,
Belgium
Group Theory in Physics

Donald G. Archer
National Institute of Standards, Gaithersburg,
MD, USA
Thermal Expansion

J. A. Arnaud
Université des Sciences et Techniques du
Languedoc, Montpellier, France
Klystrons and Traveling-Wave Tubes

A. S. Arrott
Department of Physics, Simon Fraser University,
Burnaby, British Columbia, Canada
Ferromagnetism

J. D. Axe
Retired from Brookhaven National Laboratory,
Upton, NY, USA
Ferroelectricity

A. D. Baer
U.S. Naval Weapons Center, China Lake, CA,
USA
Work Function

John N. Bahcall
Department of Astrophysical Science, Princeton
University, Princeton, NJ, USA
Solar Neutrinos

K. M. Baird
Applied Physics Division, National Research
Laboratory, Ottawa, Canada
Interferometers and Interferometry

Stanley S. Ballard[†]
Late of Department of Physics, University of
Florida, Gainesville, FL, USA
Visible and Ultraviolet Spectroscopy

C. J. Ballhausen
Retired from Kemisk Laboratorium, University of
Copenhagen, Denmark
Crystal and Ligand Fields

R. C. Barber
Department of Physics, University of Manitoba,
Winnipeg, Canada
Isotopes

J. A. Barker
Retired from IBM Alamaden Research Center,
San Jose, CA, USA
Interatomic and Intermolecular Forces

James A. Barnes[†]
National Institute of Standards, Gaithersburg,
MD, USA
Clocks, Atomic and Molecular

Richard G. Barnes
Retired from Department of Physics, Iowa State
University, Ames, IA, USA
Nuclear Quadrupole Resonance

Eric Baron
GANIL, Caen, France
Cyclotron

Ernesto Barreto
Retired from Atmospheric Sciences Research
Center, State University of New York, Albany,
NY, USA
Corona Discharge

Stephen M. Barr
Bartol Research Institute, University of Delaware,
Newark, DE, USA
Grand Unified Theories

Matthias Bartelmann
Institute of Theoretical Astrophysics, University
of Heidelberg, Heidelberg, Germany
Astrophysics

Neil Bartlett
Department of Chemistry, University of
California, Berkeley, CA, USA
Rare Gases and Rare-Gas Compounds

Uwe H. Bauder
Retired from Physics Department, Technical
University of Munich, Germany
Arcs and Sparks

[†]deceased

James E. Bayfield
Retired from Department of Physics, University
of Pittsburgh, Pittsburgh, PA, USA
High-Field Atomic States

Gordon Baym
Department of Physics, University of Illinois,
Urbana, IL, USA
Neutron Stars

Earl C. Beaty
Joint Institute for Laboratory Astrophysics,
University of Colorado, Boulder, CO, USA
Electron Bombardment of Atoms and Molecules

R. E. Bedford
National Research Council, Ottawa, Canada
Blackbody Radiation

A. H. Benade[†]
Late of Department of Physics, Case Western
Reserve University, Cleveland, OH, USA
Musical Instruments

James N. Benford
Beam and Plasma Research Group, Physics
International Company, San Leandro, CA, USA
Electron and Ion Beams, Intense

F. F. Bentley
Retired from Department of Chemistry, U.S.
Wright-Patterson Air Force Base, OH, USA
Raman Spectroscopy

Henry A. Bent
Chemistry Department, University of Pittsburgh,
Pittsburgh, PA, USA
Chemical Bonding

Robert Berg
National Institute of Standards and Technology,
Gaithersburg, MD, USA
Viscosity

Karl Berkelman
Newman Laboratory, Cornell University, Ithaca,
NY, USA
Compton Effect

Stephan Berko[†]
Late of Department of Physics, Brandeis
University, Waltham, MA, USA
Positronium

A. E. Berkowitz
Center for Magnetic Recording Research,
University of California, San Diego, CA, USA
Magnets (Permanent) and Magnetostatics

Joan B. Berkowitz
The Weinberg Group, Washington, D.C., USA
High Temperature

Robert A. Bernheim
Department of Chemistry, Pennsylvania State
University, University Park, PA, USA
Optical Pumping

D. W. Berreman
Retired from Bell Laboratories, Murray Hill, NJ,
USA
Rayleigh Scattering

Daniel Bershader[†]
Late of Department of Aeronautics and
Astronautics, Stanford University, Stanford, CA,
USA
Fluid Physics

F. Besenbacher
Interdisciplinary Nanoscience Center, University
of Aarhus, Aarhus, Denmark
Nanocatalysis

P. E. Best
Institute of Materials Science, University of
Connecticut, Storrs, CT, USA
Secondary Electron Emission

Robert T. Beyer
Department of Physics, Brown University,
Providence, RI, USA
Acoustics

Dieter Bimberg
Physics Department, Technical University of
Berlin, Germany
Quantum Structures in Semiconductors

Yvon G. Biraud
Astronomie Infrarouge LAM, Observatoire de
Meudon, France
Fourier Transforms

James D. Bjorken
Retired from Stanford Linear Accelerator Center,
Stanford, CA, USA
Partons

[†] deceased

R. Blankenbecler
Stanford Linear Accelerator Center, Stanford
University, Stanford, CA, USA
Regge Poles

George E. Blomgren
Blomgren Consulting Services, Lakewood, OH,
USA
Electrochemical Conversion and Storage

Frank A. Blood, Jr.
Retired from Department of Physics, Rutgers
University, Newark, NJ, USA
Brownian Motion

Elliott D. Bloom
Stanford Linear Accelerator Center, Stanford
University, Stanford, CA, USA
Quarkonium

A. R. Bodmer
Argonne National Laboratory, Argonne, IL, USA
*Hypernuclear Physics and Hypernuclear
Interactions*

Manfred Böhm
Physics Department, University of Giessen,
Giessen, Germany
Crystal Symmetry

Arno Bohm
Department of Physics, University of Texas,
Austin, TX, USA
Momentum

B. J. Bok[†]
Late of Seward Observatory, University of
Arizona, Tucson, AZ, USA
Milky Way

John J. Bollinger
National Institute of Standards, Gaithersburg,
MD, USA
Clocks, Atomic and Molecular

David W. Bonnell
Retired from National Institute of Standards,
Gaithersburg, MD, USA
High Temperature

Nicholas Bottka
Retired from Department of Physics, University
of Virginia, USA
Dielectric Properties

Edward A. Boudreaux
Retired from Chemistry Department, University
of New Orleans, LA, USA
Diamagnetism and Superconductivity

B. N. Brockhouse[†]
Late of Department of Physics, Mcmaster
University, Hamilton, Ontario, Canada
Neutron Spectroscopy

W. E. Bron
Retired from Department of Physics, University
of California, Irvine, CA, USA
Elasticity

Edgardo Browne
Lawrence Berkeley National Laboratory,
University of California, Berkeley, CA, USA
Radioactivity

Robert W. Brown
Department of Physics, Case Western Reserve
University, Cleveland, OH, USA
Gauss's Law

Ronald A. Bryan
Department of Physics, Texas A&M University,
College Station, TX, USA
Nuclear Forces

Olof Bryngdahl
Retired from Department of Physics, University
of Essen, Essen, Germany
Optics, Physical

Wolfgang Brütting
Institute of Physics, University of Augsburg,
Germany
Organic Semiconductors

Jeffrey Bub
Philosophy Department, University of Maryland,
College Park, MD, USA
Complementarity

E. Bucher
Retired from Physics Department, University of
Constance, Germany
Transition Elements

R. C. Budhani
Indian Institute of Technology, Kanpur, India
Thin Films

[†]deceased

Martin J. Buerger[†]
Late of Department of Earth and Planetary
Sciences, Massachusetts Institute of Technology,
Cambridge, MA, USA
Crystal Symmetry
Crystallography, X-Ray

R. Bullough
United Kingdom Atomic Energy Authority,
Oxfordshire, United Kingdom
Deformation of Crystalline Materials

E. Margaret Burbidge
Center for Astrophysics and Space Sciences,
University of California, San Diego, CA, USA
Quasars

C. P. Burgess
Department of Physics, Mcgill University,
Montreal, Quebec, Canada
Supersymmetry and Supergravity

Edward A. Burke[†]
Late of Department of Physics and Astronomy,
Adelphi University, Garden City, NY, USA
Bohr Theory of Atomic Structure

Kurt Busch
Department of Physics, Karlsruhe University,
Karlsruhe, Germany, and Department of Physics,
University of Central Florida, Orlando, FL, USA
Neutron Stars

W. H. Butler
Martin Marietta Systems, Inc., Oak Ridge, TN,
USA
Transport Properties

David O. Caldwell
Department of Physics, University of California,
Santa Barbara, CA, USA
Hyperons

Joseph Callaway[†]
Late of Department of Physics and Astronomy,
Louisiana State University, Baton Rouge, LA,
USA
Electron Energy States in Solids and Liquids

Karl F. Canter
Department of Physics, Brandeis University,
Waltham, MA, USA
Positronium

[†]deceased

Manuel Cardona
Max Planck Institute of Condensed Matter
Research, Stuttgart, Germany
Synchrotron Radiation

Michael P. Carpenter
Argonne National Laboratory, Argonne, IL, USA
Gamma-Ray Spectrometers

Peter A. Carruthers[†]
Late of Department of Physics, University of
Arizona, Tenmpe, AZ, USA
Dispersion Theory

Francesco Cataliotti
European Laboratory for Nonlinear Spectroscopy,
University of Florence, Florence, Italy
Laser Spectroscopy

Peter Caws
Department of Philosophy, George Washington
University, Washington, D.C., USA
Philosophy of Physics

P. M. Chaikin
Department of Physics, Princeton University,
Princeton, NJ, USA
Organic Conductors and Superconductors

Bruce Chalmers[†]
Late of Division of Engineering and Applied
Physics, Harvard University, Cambridge, MA,
USA
Metallurgy

Owen Chamberlain
Retired from Department of Physics, University
of California, Berkeley, CA, USA
Nucleon

B. S. Chandrasekhar
Walther–Meissner Institute for Low-Temperature
Research, Bavarian Academy of Sciences,
Garching, Germany
Magnetostriction

David B. Chang
Tustin, CA, USA
Plasmons

Alan J. Chapman
Department of Mechanical Engineering and
Materials Science, Rice University, Houston, TX,
USA
Heat Transfer

Georges Charpak
Retired from European Organization for Nuclear
Research, Geneva, Switzerland
Counting Tubes

Geoffrey F. Chew
Retired from Lawrence Berkeley National
Laboratory, University of California, Berkeley,
CA, USA
S-Matrix Theory

Hong Yee Chiu
Goddard Space Flight Center, Greenbelt, MD,
USA
Pulsars

Andreas Chrambach
Retired from National Institute of Child Health
and Human Development, Bethesda, MD, USA
Electrophoresis

Charles C. Church
National Center for Physical Acoustics,
University of Mississippi, University, MS, USA
Ultrasonic Biophysics

Steven Chu
Department of Physics, Stanford University,
Stanford, CA, USA
Laser Spectroscopy

John Clarke
Department of Physics, University of California,
Berkeley, CA, USA
Superconductive Devices

E. Richard Cohen
Encino, CA, USA
Symbols, Units, and Nomenclature

R. V. Coleman
Department of Physics, University of Virginia,
Charlottesville, VA, USA
Whiskers

Stirling A. Colgate
Los Alamos National Laboratory, Los Alamos,
NM, USA
Cosmic Rays, Astrophysical Effects

M. B. Colket
United Technologies Research Center, East
Hartford, CT, USA
Combustion and Flames

Esther M. Conwell
Retired from Xerox Webster Research Center,
Webster, New York, USA
Acoustoelectric Effect

Homer E. Conzett
Retired from Lawrence Berkeley National
Laboratory, University of California, Berkeley,
CA, USA
Polarization

Bernard R. Cooper
Retired from Department of Physics, West
Virginia University, Morgantown, VA, USA
Fermi Surface

F. V. Coroniti
Department of Physics, University of California,
Los Angeles, CA, USA
Magnetosphere

Bernd Crasemann
Retired from Physics Department, University of
Oregon, Eugene, OR, USA
Auger Effect

J. H. Crawford, Jr.
Retired from Oak Ridge National Laboratory, Oak
Ridge, TN, USA
Radiation Interaction with Matter

Gerard M. Crawley
Department of Physics and Astronomy, Michigan
State University, East Lansing, MI, USA
Charged-Particle Spectroscopy

Michael Creutz
Brookhaven National Laboratory, Upton, NY,
USA
Lattice Gauge Theory

Clarence R. Crowell
Retired from Department of Materials Science,
University of Southern California, Los Angeles,
CA, USA
Photoconductivity

Paul L. Csonka
Institute of Theoretical Science, University of
Oregon, Eugene,OR, USA
Partial Waves

N. E. Cusack
Retired from School of Mathematics and Physics,
University of East Anglia, Norwich, United
Kingdom
Liquid Metals

F. Cyrot-Lackmann
Laboratoire des Propriétés Electronique des
Solides, Centre National de la Recherche
Scientifique, Grenoble, France
Surfaces and Interfaces

I. J. D'Haenens
Hughes Research Laboratories, Malibu, CA, USA
Thermionic Emission

Oliver Dalton[†]
Late of Tektronix, Inc., Beaverton, OR, USA
Oscilloscopes

Michael Danos
Retired from National Institute of Standards and
Technology, Gaithersburg, Boulder, CO, USA
Čerenkov Radiation

J. G. Dash
Department of Physics, University of Washington,
Seattle, WA, USA
Adsorption

J. P. Davidson
Department of Physics and Astronomy, University
of Kansas, Lawrence, KS, USA
Nuclear Structure

Michael W. Davidson
National High Magnetic Field Laboratory, Florida
State University, Tallahassee, FL, USA
Microscopy, Optical

P. C. W. Davies
Department of Theoretical Physics, The
University, Newcastle-upon-Tyne, United
Kingdom
Time

J. M. Davis
Rockwell Hanford Operations, Richland, WA,
USA
Hot Cells and Remote Handling Equipment

Morris H. DeGroot[†]
Department of Physics, University of Nijmegen,
The Netherlands
Statistics

Wolfgang Demtröder
Department of Physics, University of
Kaiserslautern, Kaiserslautern, Germany
Collisions, Atomic and Molecular

Malcolm Derrick
Argonne National Laboratory, Argonne, IL, USA
Mesons

M. C. Desjonquères
Commissariat a lEnergie Atomique, Paris, France
Surfaces and Interfaces

R. W. Detenbeck
Department of Physics, University of Vermont,
Burlington, VT, USA
Light Scattering

Jozef T. Devreese
Department of Physics, University of Antwerpen,
Antwerpen, Belgium
Polaron

F. J. DiSalvo
Department of Chemistry, Cornell University,
Ithaca, NY, USA
Allotropy and Polymorphism

J. F. Dillon, Jr.
Retired from Bell Laboratories, Murray Hill, NJ,
USA
Kerr Effect, Magneto-Optical

Volker Dohm
Department of Physics, Aachen University,
Aachen, Germany
Phase Transitions

John. F. Donoghue
Department of Physics, University of
Massachusetts, Amherst, MA, USA
SU(3) and Symmetry Groups

T. M. Donovan
U.S. Naval Weapons Center, China Lake, CA,
USA
Work Function

[†] deceased

J. Robert Dorfman
Institute for Physical Science and Technology,
University of Maryland, College Park, MD, USA
Statistical Mechanics

David A. Dows
Department of Chemistry, University of Southern
California, Los Angeles, CA, USA
Infrared Spectroscopy

Charles W. Drake
Department of Physics, Oregon State University,
Corvallis, OH, USA
Polarizability

G. W. F. Drake
Department of Physics, University of Windsor,
Windsor, Ontario, Canada
Fine and Hyperfine Spectra and Interactions

Joseph Dreitlein
Retired from Department of Physics, University
of Colorado, Boulder, CO, USA
Twin Paradox

G. Dresselhaus
Francis Bitter National Magnet Laboratory,
Massachusetts Institute of Technology,
Cambridge, MA, USA
de Haas–van Alphen Effect

M. S. Dresselhaus
Departments of Electrical Engineering and
Physics, Massachusetts Institute of Technology,
Cambridge, MA, USA
de Haas–van Alphen Effect

Alan Dressler
The Observatories, Carnegie Institution of
Washington, Pasadena, CA, USA
Galaxies

Charles B. Duke
Wilson Center for Research and Technology,
Xerox Innovation Group, Webster, NY, USA
Low-Energy Electron Diffraction (LEED)

R. A. Dunlap
Department of Physics and Atmospheric Science,
Dalhousie University, Halifax, Canada
Hysteresis

Floyd Dunn
Retired from Department of Physics, University
of Illinois, Urbana-Champaign, IL, USA
Ultrasonic Biophysics

James R. Durig
Department of Chemistry, University of Missouri,
Kansas City, MO, USA
Microwave Spectroscopy

Pol Duwez
Department of Materials Science and Engineering,
University of Pennsylvania, Pittsburgh, PA, USA
Glassy Metals

Dean E. Eastman
Department of Physics, University of Chicago,
Chicago, IL, USA
Photoelectron Spectroscopy

Philippe Eberhard
Lawrence Berkeley National Laboratory,
University of California, Berkeley, CA, USA
Center-of-Mass System

Alan S. Edelstein
U.S. Army Research Laboratory, Adelphi, MD,
USA
Kondo Effect

David L. Ederer
Department of Physics, Tulane University, New
Orleans, LA, USA
Photoionization

Takeshi Egami
Department of Materials Science and Engineering,
University of Pennsylvania, Pittsburgh, PA, USA
Glassy Metals

F. R. Eirich
Department of Chemistry, Polytechnic Institute of
New York, New York, USA
Rheology

Leonard Eisenbud
Retired from Department of Physics, State
University of New York at Stony Brook, Stony
Brook, NY, USA
Eigenfunctions

T. Embleton
Division of Physics, National Research Council,
Ottawa, Ontario, Canada
Noise, Acoustical

V. J. Emery[†]
Late of Brookhaven National Laboratory, Upton,
NY, USA
Quantum Statistical Mechanics

P. R. Emtage
Westinghouse Electric Corporation, Research and
Development Center, Pittsburgh, PA, USA
Hall Effect

John E. Enderby
The Royal Society, London, United Kingdom
Paramagnetism

Hermann Engelhardt
Division of Geological and Planetary Sciences,
California Institute of Technology, Pasadena, CA,
USA
Ice

Stanley Engelsberg
Department of Physics and Astronomy, University
of Massachusetts, Amherst, MA, USA
Metals

Gerhard Ertl
Fritz Haber Institute, Max Planck Society, Berlin,
Germany
Catalysis

Allen E. Everett
Department of Physics and Astronomy, Tufts
University, Medford, MA, USA
Lorentz Transformations

Richard Eykholt
Department of Physics, Colorado State University,
Fort Collins, CO, USA
Sedimentation and Centrifugation

Henry A. Fairbank
Retired from Department of Physics, Duke
University, Durham, NC, USA
Second Sound

Fereydoon Family
Department of Physics, Emory University,
Atlanta, GA, USA
Fractals

W. G. Fateley[†]
Late of Department of Chemistry, Kansas State
University, Manhattan, KS, USA
Raman Spectroscopy

Norman Feather[†]
Late of University of Edinburgh, Scotland, United
Kingdom
Statics

G. Feinberg
Department of Physics, Columbia University,
New York, USA
Tachyons

A. J. Fennelly
Teledyne Brown Engineering, Huntsville, AL,
USA
Photon

Thomas Ferbel
Department of Physics, University of Rochester,
Rochester, NY, USA
Inclusive Reactions

Herman Feshbach[†]
Late of Department of Physics, Massachusetts
Institute of Technology, Cambdridge, MA, USA
Nuclear Scattering

Alexander L. Fetter
Department of Physics, Stanford University,
Stanford, CA, USA
Quantum Fluids

Douglas K. Finnemore
Ames Laboratory, Iowa State University, Ames,
IA, USA
Superconducting Materials

Val L. Fitch
Department of Physics, Princeton University,
Princeton, NJ, USA
Weak Interactions

Carl L. Foiles[†]
Department of Physics, Michigan State
University, East Lansing, MI, USA
Thermoelectric Effects

Leslie L. Foldy[†]
Late of Department of Physics, Case Western
Reserve University, Cleveland, OH, USA
Faraday's Law of Electromagnetic Induction

[†]deceased

Kenneth J. Foley
Brookhaven National Laboratory, Upton, NY, USA
Scintillation and Čerenkov Counters

Simon Foner
National Magnet Laboratory, Massachusetts Institute of Technology, Cambridge, MA, USA
Magnetic Fields, High

Jerry W. Forbes
University of Maryland, Port Tobacco, MD, USA
Shock Waves and Detonations

Joseph Ford[†]
Late of School of Physics, Georgia Institute of Technology, Atlanta, GA, USA
Ergodic Theory

G. W. Ford
Department of Physics, University of Michigan, Ann Arbor, MI, USA
H Theorem

F. M. Fowkes
Retired from Department of Chemistry, Lehigh University, Bethlehem , PA, USA
Surface Tension

Susan E. Fox[†]
Late of Department of Physics, Yale University, New Haven, CT, USA
Quantum Mechanics

Donald R. Franceschetti
Department of Physics, University of Memphis, Memphis, TN, USA
Electrochemistry

Jack H. Freed
Department of Chemistry and Chemical Biology, Cornell University, Ithaca, NY, USA
Gyromagnetic Ratio

Robert D. Freeman
Enody Unlimited, Stillwater, OK, USA
Thermodynamic Data

A. P. French
Department of Physics, Massachusetts Institute of Technology, Cambridge, MA, USA
Atoms

J. B. French
Retired from Department of Physics and Astronomy, University of Rochester, Rochester, NY, USA
Sum Rules

J. P. Friedberg
Plasma Fusion Center, Massachusetts Institute of Technology, Cambridge, MA, USA
Nuclear Fusion

F. J. Friedlaender
School of Chemical Engineering, Purdue University, West Lafayette, IN, USA
Electromagnets

John Friedman
Department of Physics, University of Wisconsin, Madison, WI, USA
Gravitation

Josef W. Fried
Max Planck Institute for Astronomy, Heidelberg, Germany
Astronomy, Optical

Thomas Fulton
Department of Physics, The Johns Hopkins University, Baltimore, MD, USA
Feynman Diagrams

H. P. Furth
Plasma Physics Laboratory, Princeton University, Princeton, NJ, USA
Plasma Confinement Devices

F. L. Galeener[†]
Late of Department of Physics, Colorado State University, Fort Collins, CO, USA
Absorption Coefficient

Serge Gambarelli
Department of Fundamental Research on Condensed Matter, Commissariat à l'Energie Atomique, Grenoble, France
Electron Spin Resonance

M. Garbuny[†]
Late of Westinghouse Research and Development Center, Pittsburgh, PA, USA
Heat Engines

[†]deceased

Elsa Garmire
Thayer School of Engineering, Dartmouth
College, Hanover, NH, USA
Reflection

Paul D. Garn
Department of Chemistry, University of Akron,
Akron, OH, USA
Thermal Analysis

S. Gasiorowicz
School of Physics and Astronomy, University of
Minnesota, Minneapolis, MN, USA
Elementary Particles in Physics

Peter P. Gaspar
Department of Chemistry, Washington University,
St. Louis, MO, USA
Hot Atom Chemistry

B. D. Gaulin
Department of Physics, Mcmaster University,
Hamilton, Ontario, Canada
Neutron Spectroscopy

J. D. Gavenda
Department of Physics, University of Texas,
Austin, TX, USA
Magnetoacoustic Effect

P. G. de Gennes
College de France, Paris, France
Liquid Crystals

B. F. Gibson
Los Alamos National Laboratory, Los Alamos,
NM, USA
Three-Body Problem, Quantum Mechanical

Carl H. Gibson
Scripps Institution of Oceanography, University
of California, San Diego, CA, USA
Turbulence

J. A. Giordmaine
NEC Research Institute, Princeton, NJ, USA
Optics, Nonlinear

P. Glansdorff
University of Brussels, Belgium
Thermodynamics, Nonequilibrium

Henry R. Glyde
Department of Computer and Information
Science, University of Delaware, Newark, DE,
USA
Helium, Solid

Hubert F. Goenner
Institute of Theoretical Physics, University of
Göttingen, Göttingen, Germany
Mach's Principle

Joshua N. Goldberg
Department of Physics, Syracuse University,
Syracuse, NY, USA
Spacetime

Walter I. Goldburg
Department of Physics and Astronomy, University
of Pittsburgh, Pittsburgh, PA, USA
Dynamic Critical Phenomena

Alfred Scharff Goldhaber
C. N. Yang Institute for Theoretical Physics, State
University of New York, Stony Brook, NY, USA
Magnetic Monopoles

B. Golding
Department of Physics and Astronomy, Michigan
State University, East Lansing, MI, USA
Glass

Herbert Goldstein
Retired from Department of Mechanical and
Nuclear Engineering, Columbia University, New
York, USA
Dynamics, Analytical

R. H. Good, Jr.
Department of Physics, Pennsylvania State
University, University Park, PA, USA
Hamiltonian Function

A. R. H. Goodwin
Schlumberger, Sugar Land, TX, USA
Vapor Pressure

James P. Gordon
Bell Laboratories, Holmdel, NJ, USA
Masers

Paul Gorenstein
Smithsonian Astrophysical Observatory,
Cambridge, MA, USA
Astronomy, X-Ray

N. B. Gove
Retired from Oak Ridge National Laboratory, Oak Ridge, TN, USA
Binding Energy

C. D. Graham, Jr.
Department of Materials, University of Pennsylvania, Philadelphia, PA, USA
Magnetic Materials

Andrew V. Granato
Department of Physics, University of Illinois, Urbana-Champaign, IL, USA
Internal Friction in Crystals

Victor L. Granatstein
Institute for Research in Electronics and Applied Physics, University of Maryland, College Park, MD, USA
Light

O. W. Greenberg
Department of Physics, University of Maryland, College Park, MD, USA
Bose–Einstein Statistics

Joshua E. Greenspon
Retired form J. G. Engineering Research Associates, Baltimore, MD, USA
Acoustics, Linear and Nonlinear

Sandra C. Greer
Department of Chemical Engineering, and Department of Chemistry and Biochemistry, University of Maryland, College Park, MD, USA
Order–Disorder Phenomena

Walter Greiner
Frankfurt Institute of Advanced Studies, University of Frankfurt, Frankfurt, Germany
Superheavy Elements

Thomas J. Greytak
Department of Physics, Massachusetts Institute of Technology, Cambridge, MA, USA
Fluctuation Phenomena

Robert B. Griffiths
Department of Physics, Carnegie-Mellon University, Pittsburgh, PA, USA
Thermodynamics, Equilibrium

C. C. Grimes
Bell Laboratories, Murray Hill, NJ, USA
Cyclotron Resonance

Franz Gross
Thomas Jefferson National Accelerator Facility, Newport News, VA, USA
Bethe–Salpeter Equation

K. A. Gschneidner, Jr.
Institute for Physical Research and Technology, and Department of Materials Science and Engineering, Iowa State University, Ames, IA, USA
Rare Earths

David Gubbins
School of Earth & Environment, University of Leeds, Leeds, United Kingdom
Geomagnetism

R. A. Guyer
Department of Physics, University of Massachusetts, Amherst, MA, USA
Helium, Liquid

Feza Gürsey[†]
Late of Physics Department, Yale University, New Haven, CT, USA
Invariance Principles

Steven W. Haan
Lawrence Livermore National Laboratory, University of California, Livermore, CA, USA
Inertial Fusion

M. H. Hablanian
Vacuum Products Division, Varian Associates, Lexington, MA, USA
Vacuums and Vacuum Technology

W. Haeberli
Department of Physics, University of Wisconsin, Madison, WI, USA
Nuclear Polarization

Richard L. Hahn
Brookhaven National Laboratory, Upton, NY, USA
Transuranium Elements

[†]deceased

Yukap Hahn
Retired from Department of Physics, University
of Connecticut, Storrs, CT, USA
Multipole Fields

Astrid Haibel
Hahn–Meitner Institute, Berlin, Germany
Tunneling

John J. Hall
Shearwater Company, Brooklyn, NY, USA
Photovoltaic Effect

Arthur M. Halpern
Department of Chemistry, Indiana State
University, Terre Haute, IN, USA
*Luminescence (Fluorescence and
Phosphorescence)*

M. Hamermesh[†]
Late of School of Physics and Astronomy,
University of Minnesota, Minneapolis, USA
Lie Groups

Robert J. Hamers
Department of Chemistry, University of
Wisconsin, Madison, WI, USA
Scanning Tunneling Microscopy

J. M. Hammersley
Retired from Mathematical Institute, Oxford
University, Oxford, United Kingdom
Probability

Richard S. Handley
Lumigen, Inc., Southfield, MI, USA
Chemiluminescence

John H. Hannay
H. H. Wills Physical Laboratory, University of
Bristol, United Kingdom
Geometric Quantum Phase

William Happer
Department of Physics, Princeton University,
Princeton, NJ, USA
Resonance Phenomena

P. Hariharan
Csiro Division of Applied Physics, National
Measurement Laboratory, Lindfield, Australia
Holography

Hanns L. Harney
Max Planck Institute for Nuclear Physics,
Heidelberg, Germany
*Isobaric Analog States
Probability*

Franklin S. Harris, Jr.[†]
Late of Rockville, UT, USA
Aerosols

J. C. Harrison
Geodynamics Corporations, Santa Barbara, CA,
USA
Gravity, Earth's

Frank E. Harris
Department of Physics, University of Florida,
Gainesville, FL, USA
Molecules

F. Joachim Hartmann
Department of Physics, Technical University of
Munich, Garching, Germany
Muonic, Mesonic, and Other Exotic Atoms

Bernard G. Harvey
Lawrence Berkeley National Laboratory,
University of California, Berkeley, CA, USA
Radioactivity

John W. Hastie
Retired from National Institute of Standards,
Gaithersburg, MD, USA
High Temperature

Stephen W. Hawking
Department of Applied Mathematics and
Theoretical Physics, University of Cambridge,
Cambridge, United Kingdom
Black Holes

Sabih I. Hayek
Department of Engineering Science and
Mechanics, Pennsylvania State University,
University Park, PA, USA
Vibrations, Mechanical

Raymond W. Hayward[†]
Late of National Institute of Standards and
Technology, Gaithersburg, MD, USA
CPT Theorem

[†]deceased

Evans Hayward
National Institute of Standards, Gaithersburg,
MD, USA
Photonuclear Reactions

Robert Hebner
Center for Electromechanics, University of Texas
at Austin, TX, USA
Kerr Effect, Electro-Optical
Kerr Effect, Magneto-Optical

Dennis J. Hegyi
Department of Physics, University of Michigan,
Ann Arbor, MI, USA
Interstellar Medium

Thomas Heinzel
Department of Physics, University of Freiburg,
Freiburg, Germany
Mesoscopic Physics

Ernest M. Henley
Department of Physics, University of Washington,
Seattle, WA, USA
Parity

R. G. Herb[†]
Late of Department of Physics, University of
Wisconsin, Madison, WI, USA
Accelerators, Potential-Drop Linear

Gerd Hermann
Institute of Physics, University of Gießen, Gießen,
Germany
Polarization
Polarized Light

Frank Herman
Research Staff Member Emeritus, IBM Almaden
Research Center, San Jose, CA, USA
*Atomic Structure Calculations, One-Electron
Models*

J. C. Herrera
Retired from Brookhaven National Laboratory,
Upton, NY, USA
Electromagnetic Radiation

Conyers Herring
Retired from Department of Applied Physics,
Stanford University, Stanford, CA, USA
Solid-State Physics

Lloyd O. Herwig
U.S. Department of Energy, Washington, D.C.,
USA
Solar Energy

Jacqueline N. Hewitt
Department of Physics, Massachusetts Institute of
Technology, Cambridge, MA, USA
Gravitational Lenses

Ken-ichi Hikasa
Department of Physics, University of Tohoku,
Sendai, Japan
CPT Theorem

Franz J. Himpsel
Department of Physics, University of Wisconsin,
Madison, WI, USA
Photoelectron Spectroscopy

David G. Hitlin
California Institute of Technology, Pasadena, CA,
USA
Baryons

Joseph V. Hollweg
Department of Physics and Space Science Center,
University of New Hampshire, Durham, NH, USA
Cosmic Rays, Solar System Effects

Michael Horne
Department of Physics, Stonehill College, Easton,
MA, USA
Hidden Variables

Robert Horton
Department of Agronomy, Iowa State University,
Ames, IA, USA
Soil Physics

Ottó Horváth
Department of General and Inorganic Chemistry,
University of Veszprem, Veszprem, Hungary
Light-Sensitive Materials

R. P. Hudson
National Institute of Standards and Technology,
Gaithersburg, MD, USA
Thermometry

Vernon W. Hughes[†]
Late of Physics Department, Yale University, New
Haven, CT, USA
Muonium

[†] deceased

Andrew P. Hull[†]
Brookhaven National Laboratory, Upton, NY, USA
Radiation Detection

H. B. Huntington
Retired from Department of Physics and Astronomy, Rensselaer Polytechnic Institute, Troy, NY, USA
Diffusion

James L. Hunt
Department of Physics, University of Guelph, Ontario, Canada
Brillouin Scattering

Theodor W. Hänsch
Max Planck Institute of Quantum Optics, Garching, Germany
Doppler Effect

O. Häusser
Triumf, Vancouver, Canada
Nuclear Moments

Takanobu Ishida
Department of Chemistry, State University of New York, Stony Brook, NY, USA
Isotope Separation

Werner Israel
Department of Physics, University of Alberta, Edmonton, Alberta, Canada
Black Holes

J. D. Jackson
Lawrence Berkeley National Laboratory, University of Berkeley, CA, USA
Electrodynamics, Classical

John Arthur Jacobs[†]
Late of Department of Geology, University of Wales, Aberystwyth, United Kingdom
Geomagnetism

Robert V. F. Janssens
Argonne National Laboratory, Argonne, IL, USA
Gamma-Ray Spectrometers

George J. Janz
Molten Salts Data Center, Rensselaer Polytechnic Institute, Troy, NY, USA
Molten Salts

[†]deceased

Carson Jeffries[†]
Late of Department of Physics, University of California, Berkeley, CA, USA
Excitons

C. K. Jen
Silver Spring, MD, USA
Zeeman and Stark Effects

Robert L. Johnson
Department of Physics, University of Hamburg, Hamburg, Germany
Synchrotron Radiation

J. R. Jokipii
Committee on Theoretical Astrophysics, University of Arizona, Tucson, AZ, USA
Solar Wind

David L. Judd[†]
Late of Lawrence Berkeley National Laboratory, University of California, Berkeley, CA, USA
Cyclotron

B. R. Judd
Department of Physics and Astronomy, The Johns Hopkins University, Baltimore, MD, USA
Franck–Condon Principle
Jahn–Teller Effect

Pierre Y. Julien
Department of Civil Engineering, Colorado State University, Fort Collins, CO, USA
Sedimentation and Centrifugation

M. N. Kabler
Retired from U.S. Naval Research Laboratory, Washington, D.C., USA
Electroluminescence

Mark Kac[†]
Late of Rockefeller University, New York, USA
H Theorem

Claude Kacser
Department of Physics, University of Maryland, MD, USA
Relativity, Special Theory

Rafael Kalish
Department of Physics, Technion University, Haifa, Israel
Nuclear Properties

M. H. Kalos
Lawrence Livermore National Laboratory,
Berkeley, CA, USA
Monte Carlo Techniques

Isabella Karle
Laboratory for the Structure of Matter, Naval
Research Laboratory, Washington, D.C, USA
Crystallography, X-Ray

Jerome Karle
Laboratory for the Structure of Matter, Naval
Research Laboratory, Washington, D.C, USA
Crystallography, X-Ray

Henry Kasha
Physics Department, Yale University, New Haven,
CT, USA
Proton

O. Lewin Keller, Jr.
Retired from Oak Ridge National Laboratory, Oak
Ridge, TN, USA
Elements

Paul L. Kelley
Department of Physics, Tufts University,
Medford, MA, USA
Lasers

Walter Kellner
Retired from Siemens AG, Munich, Germany
Semiconductors, Crystalline

Gary L. Kellogg
Sandia National Laboratories, Albuquerque, NM,
USA
Field-Ion Microscopy

B. T. Kelly
United Kingdom Atomic Energy Authority,
Salwick, United Kingdom
Radiation Interaction with Matter

Francis E. Kennedy, Jr.
Thayer School of Engineering, Dartmouth
College, Hanover, NH, USA
Friction

George C. Kennedy[†]
Late of Institute of Geophysics and Planetary
Physics, University of California, Los Angeles,
CA, USA
Ultrahigh-Pressure Techniques

Edward H. Kerner[†]
Late of Department of Physics, University of
Delaware, Newark, DE, USA
Kepler's Laws

Dieter P. Kern
Department of Physics, Tübingen University,
Tübingen, Germany
Electron Beam Technology

Donald W. Kerst[†]
Late of Department of Physics, University of
Wisconsin, Madison, WI, USA
Betatron

R. Kingslake
Rochester, NY, USA
Optics, Geometrical

R. W. P. King
Division of Applied Sciences, Harvard University,
Cambridge, MA, USA
Transmission Lines and Antennas

Toichiro Kinoshita
Laboratory for Nuclear Studies, Cornell
University, Ithaca, NY, USA
Electromagnetic Interaction

Gordon S. Kino
Retired from Department of Physics, Stanford
University, Stanford, CA, USA
Microscopy, Optical

Arthur F. Kip[†]
Late of University of California, Berkeley, CA,
USA
Resistance

S. Kirkpatrick
Retired from IBM Research Center, Yorktown
Heights, New York, USA
Conduction

Louis T. Klauder, Jr.
Retired from Bell Laboratories, Murray Hill, NJ,
USA
Ampère's Law

John R. Klauder
Department of Physics, University of Florida,
Gainesville, FL, USA
Quantum Optics

[†] deceased

P. G. Klemens
Department of Physics, University of Connecticut,
Storrs, CT, USA
Phonons

Andreas Kling
Instituto Tecnológico e Nuclear, Sacavém,
Portugal
Channeling

Klaus von Klitzing
Max Planck Institute of Condensed-Matter
Research, Stuttgart, Germany
Hall Effect, Quantum

G. M. Klody
Retired from National Electrostatics Corporation,
Middleton, WI, USA
Accelerators, Potential-Drop Linear

Ulrich Kogelschatz
Retired from ABB Corporate Research,
Baden-Dättwil, Switzerland
Corona Discharge

Dieter Kohl
Department of Physics, University of Gießen,
Gießen, Germany
Electronic Noses

Michal Kopcewicz
Institute of Electronic Materials Technology,
Warsaw, Poland
Mössbauer Effect

H. W. Kraner
Brookhaven National Laboratory, Upton, NY,
USA
Semiconductor Radiation Detectors

Reinhard Krause-Rehberg
Department of Physics, University of Halle, Halle,
Germany
Positron Annihilation in Condensed Matter

Edgar A. Kraut
Retired from Rockwell Scientific Company, LLC.,
USA
Matrices

Siegfried Krewald
Department of Physics, University of Essen,
Essen, Germany
Resonances, Giant

Lawrence C. Krisher
Institute for Physical Science and Technology,
University of Maryland, College Park, MD, USA
Inversion and Internal Rotation

Wolfgang Krätschmer
Max Planck Institute of Nuclear Physics,
Heidelberg, Germany
Fullerenes

D. Kurath
Argonne National Laboratory, Argonne, IL, USA
Nuclear States

Frederick K. Lamb
Department of Physics, University of Illinois,
Urbana, IL, USA
Neutron Stars

Paul Langacker
Department of Physics and Astronomy, University
of Pennsylvania, Philadelphia, PA, USA
Elementary Particles in Physics

Kenneth R. Lang
Department of Physics, Tufts University,
Medford, MA, USA
Hubble Effect

Marvin A. Lanphere
U.S. Geological Survey, United States
Department of the Interior, Menlo Park, CA, USA
Geochronology

Jeppe V. Lauritsen
Interdisciplinary Nanoscience Center, University
of Aarhus, Aarhus, Denmark
Nanocatalysis

Jon M. Lawrence
Department of Physics, University of California,
Irvine, CA, USA
Intermediate Valence Compounds

J. D. Lawson
Retired from Rutherford Appleton Laboratory,
Chilton, United Kingdom
Charged-Particle Optics

Paul L. Leath
Department of Physics and Astronomy, Rutgers
University, Piscataway, NJ, USA
Alloys

Hartmut S. Leipner
Interdisciplinary Center for Materials Science,
Martin Luther University Halle-Wittenberg, Halle,
Germany
Positron Annihilation in Condensed Matter

Alfred Leitner
Department of Physics, Rensselaer Polytechnic
Institute, Troy, NY, USA
Fraunhofer Lines

Lionel M. Levinson
Vartek Associates LLC, Schenectady, NY, USA
Ceramics

Eugene H. Levy
Department of Planetary Sciences, University of
Arizona, Tucson, AZ, USA
Solar System

Paul W. Levy
Retired from Brookhaven National Laboratory,
Upton, NY, USA
Color Centers
Thermoluminescence

P. F. Liao
Retired from Bell Communications Research, Red
Bank, NJ, USA
Polarized Light

David R. Lide, Jr.
CRC Handbook of Chemistry and Physics,
Gaithersburg, MD, USA
Molecular Spectroscopy

Elliott H. Lieb
Department of Physics and Mathematics,
Princeton University, Princeton, NJ, USA
Uncertainty Principle

M. H. Lietzke
Oak Ridge National Laboratory, Oak Ridge, TN,
USA
Demineralization

M. E. Lines
Retired from Bell Laboratories, Murray Hill, NJ,
USA
Ferroelectricity

J. D. Litster
Department of Physics, Massachusetts Institute of
Technology, Cambridge, MA, USA
Critical Points

P. B. Littlewood
Department of Physics, University of Cambridge,
Cambridge, United Kongdom
Charge-Density Waves

Ling-Fong Li
Department of Physics, Carnegie-Mellon
University, Pittsburgh, PA, USA
Symmetry Breaking, Spontaneous

C. J. Lobb
Department of Physics, Harvard University,
Cambridge, MA, USA
Conduction

E. G. Loewen
Analytical Products Division, Milton R Company,
Rochester, NY, USA
Gratings, Diffraction

Maurice W. Long
Radar Consultant, Atlanta, Georgia, USA
Radar

A. E. Lord
Department of Physics and Atmospheric Science,
Drexel University, Philadelphia, PA, USA
Levitation, Electromagnetic

J. D. Louck
Los Alamos National Laboratory, Los Alamos,
NM, USA
Rotation and Angular Momentum

O. V. Lounasmaa[†]
Late of Low Temperature Laboratory, Helsinki
University of Technology, Helsinki, Finland
Magnetic Cooling

Walter D. Loveland
Department of Chemistry, Oregon State
University, Corvallis, OR, USA
Radiochemistry

Sudarshan K. Loyalka
Department of Nuclear Engineering, University of
Missouri, Columbia, MO, USA
Transport Theory

[†]deceased

G. Lucovsky
Department of Physics, North Carolina State
University, Raleigh, NC, USA
Semiconductors, Amorphous

Allen Lurio[†]
Late of IBM Research Laboratory, Yorktown
Heights, New York, USA
Atomic Spectroscopy

Per-Olov Löwdin[†]
Late of Florida Quantum Theory Project,
University of Florida, Gainesville, FL, USA
Molecular Structure Calculations

H. Lüth
Institute for Semiconductor Thin Films and
Devices, Jülich Research Centre, Jülich, Germany
Ellipsometry

David L. MacAdam
Institute of Optics, University of Rochester,
Rochester, NY, USA
Spectrophotometry

Douglas E. MacLaughlin
Department of Physics, University of California,
Riverside, CA, USA
Heavy-Fermion Materials

J. Ross Macdonald
Department of Physics and Astronomy, University
of North Carolina, Chapel Hill, NC, USA
Electrochemistry

Joseph Macek
Department of Physics, University of Tennessee,
Knoxville, TN, USA
Electric Moments

Hartmut Machner
Department of Physics, University of Essen,
Essen, Germany
Resonances, Giant

Alfred K. Mann
Department of Physics, University of
Pennsylvania, Philadelphia, PA, USA
Weak Neutral Currents

Robert H. March
Department of Physics, University of Wisconsin,
Madison, WI, USA
Coriolis Acceleration

William J. Marciano
Brookhaven National Laboratory, Upton, NY,
USA
Gauge Theories

H. Mark[†]
Late of Polytechnic Institute of New York, New
York, USA
Polymers

L. Marton[†]
Late of Advances in Eletronics and Electron
Physics, Washington, D.C., USA
Electron Diffraction

Warren P. Mason[†]
Late of Columbia University, New York, USA
Ultrasonics

E. der Mateosioan
Retired from Brookhaven National Laboratory,
Upton, NY, USA
Gamma Decay

H. Matsumoto
Institute of Physics, University of Tsukuba,
Ibaraki, Japan
Bose–Einstein Condensation

J. R. Matthews
United Kingdom Atomic Energy Authority,
Oxfordshire, United Kingdom
Deformation of Crystalline Materials

D. C. Mattis
Department of Physics, University of Utah, Salt
Lake City, UT, USA
Many-Body Theory

Leonard C. Maximon
National Institute of Standards and Technology,
Gaithersburg, MD, USA
Bremsstrahlung

James A. McCray
Department of Physics and Atmospheric Science,
Drexel University, Philadelphia, PA, USA
Electronics

R. Bruce McKibben
Department of Physics and Space Science Center,
University of New Hampshire, Durham, NH, USA
Cosmic Rays, Solar System Effects

[†]deceased

E. M. McMillan
Department of Physics, University of California,
Berkeley, CA, USA
Synchrotron

John P. McTague
IBM Almaden Research Center, San Jose, CA,
USA
Interatomic and Intermolecular Forces

R. L. Melcher
Retired from IBM Research Center, Yorktown
Heights, New York, USA
Magnetoelastic Phenomena

J. C. Melrose
Retired from Department of Petroleum
Engineering, Stanford University, Stanford, CA,
USA
Capillary Flow

Robert J. Meltzer
The RJM Consultancy, Kirkland, WA, USA
Optical Activity

W. Kendall Melville
Scripps Institution of Oceanography, University
of California San Diego, La Jolla, California, CA,
USA
Surface Waves on Fluids

N. D. Mermin
Department of Physics, Cornell University, Ithaca,
NY, USA
Maxwell–Boltzmann Statistics

Harold Metcalf
Physics Department, State University of New
York, Stony Brook, NY, USA
Atomic Trapping and Cooling

H. O. Meyer
Cyclotron Facility, Indiana University,
Bloomington, IN, USA
Nuclear Polarization

D. L. Miller
Retired from Department of Physics, University
of Pennsylvania, Philadelphia, PA, USA
Thin Films

A. P. Mills, Jr.
Department of Physics, University of California,
Riverside, CA, USA
Positronium

Eugene Mittelmann
Consulting Engineer, Chicago, IL, USA
Transducers

K. D. Moeller
Department of Physics, Fairleigh Dickinson
University, Madison, NJ, USA
Far-Infrared Spectra

Peter J. Mohr
National Institute of Standards and Technology,
Gaithersburg, MD, USA
Constants, Fundamental

J. W. Morris, Jr.
Department of Materials Science and Mineral
Engineering, University of California, Berkeley,
CA, USA
Heat

R. L. Morse
Department of Physics, University of New
Mexico, Albuquerque, NM, USA
Plasmas

N. F. Mott[†]
Late of Department of Physics, University of
Cambridge, United Kingdom
Metal–Insulator Transitions

Raymond D. Mountain
National Institute of Standards and Technology,
Gaithersburg, MD, USA
Equations of State

R. G. Munro
Data and Standards Technology Group, National
Institute of Standards and Technology,
Gaithersburg, MD, USA
Tribology

Volker Müller
Astrophysical Institute, Potsdam, Germany
Universe

Wolfgang Nadler
Department of Physics and Atmospheric Science,
Drexel University, Philadelphia, PA, USA
Electronics

[†]deceased

Lorenzo M. Narducci
Department of Physics and Atmospheric Science,
Drexel University, Philadelphia, PA, USA
Optics, Statistical

Reinhard B. Neder
Department of Physics, University of Würzburg,
Würzburg, Germany
Diffraction

Mark Nelkin
Department of Applied Physics, Cornell
University, Ithaca, NY, USA
Hydrodynamics

David R. Nelson
Department of Physics, Harvard University,
Cambridge, MA, USA
Renormalization

Donald F. Nelson
Department of Physics, Worcester Polytechnic
Institute, Worcester, MA, USA
Photoelastic Effect
Piezoelectric Effect

R. K. Nesbet
Retired from IBM Almaden Research Center, San
Jose, CA, USA
*Atomic Structure Calculations, Electronic
Correlation*
Magnetic Moments

Alfred O. Nier[†]
Late of School of Physics and Astronomy,
University of Minnesota, Minneapolis, MN, USA
Mass Spectroscopy

Günter Nimtz
Institute of Physics, Köln University, Köln,
Germany
Tunneling

Frank Noll
Institute of Physical Chemistry, Marburg
University, Marburg, Germany
Nanobionics

Thomas D. Northwood
Retired from Division of Building Research,
National Research Council, Ottawa, Ontario,
Canada
Acoustics, Architectural

A. S. Nowick
Department of Chemical Engineering and
Materials Science, University of California,
Irvine, CA, USA
Anelasticity

Richard M. Noyes[†]
Late of Department of Chemistry, University of
Oregon, Eugene, OR, USA
Kinetics, Chemical

M. A. O'Keefe
Lawrence Berkeley National Laboratory,
Berkeley, CA, USA
Electron Diffraction

G. Davis O'Kelley
Oak Ridge National Laboratory, Oak Ridge, TN,
USA
Radiation Detection

Donald C. O'Shea
Retired from School of Physics, Georgia Institute
of Technology, Atlanta, GA, USA
Visible and Ultraviolet Spectroscopy

Robert J. Oakes
Department of Physics, Northwestern University,
Evanston, IL, USA
Currents in Particle Theory

Lothar Oberauer
Department of Physics, Munich Technical
University, Munich, Germany
Neutrinos

Y. Ohashi
Institute of Physics, University of Tsukuba,
Ibaraki, Japan
Bose–Einstein Condensation

Paul S. Olmstead[†]
Late of Winter Park, FL, USA
Error Analysis

Irwin Oppenheim
Department of Chemistry, Massachusetts Institute
of Technology, Cambridge, MA, USA
Stochastic Processes

R. Orbach
Retired from Department of Physics, University
of California, Los Angeles, CA, USA
Relaxation Phenomena

[†] deceased

Brian J. Orr
Department of Physics, Macquarie University,
Sydney, NSW, Australia
Visible and Ultraviolet Spectroscopy

Colin G. Orton
Retired from Department of Radiation Oncology,
Wayne State University, Detroit, MI, USA
Radiological Physics

Konrad Osterwalder
Department of Mathematics, ETH Zürich, Zürich,
Switzerland
Operators

Edward Ott
Laboratory for Plasma Research, University of
Maryland, College Park, MD, USA
Chaos

Thornton Page[†]
Late of Nassau Bay, TX, USA
Newton's Laws

Abraham Pais[†]
Late of Department of Physics, Rockefeller
University, New York, USA
Electron

R. L. Parker
Consultant, Washington, D.C., USA
Crystal Growth

Philip M. Pearle
Physics Department, Hamilton College, Clinton,
NY, USA
Quantum Theory of Measurement

Norman Pearlman
Department of Physics, Purdue University, West
Lafayette, IN, USA
Heat Capacity

P. J. E. Peebles
Department of Physics, Princeton University,
Princeton, NJ, USA
Cosmology

A. A. Penzias
Retired from Bell Laboratories, Murray Hill, NJ,
USA
Astronomy, Radio

G. J. Perlow
Retired from Argonne National Laboratory,
Argonne, IL, USA
Mössbauer Effect

Martin L. Perl
Stanford Linear Accelerator Center, Menlo Park,
CA, USA
Hadrons
Leptons

Michael E. Peskin
Stanford Linear Accelerator Center, Menlo Park,
CA, USA
Quantum Field Theory

J. M. Peterson
Universities Research Association, Berkeley, CA,
USA
Synchrotron

Christopher J. Pethick
Nordic Institute for Theoretical Physics,
Copenhagen, Denmark
Quantum Fluids

Melba Phillips[†]
Late of New York, USA
Electric Charge

Jim C. Phillips
Bell Laboratories, Murray Hill, NJ, USA
Crystal Binding

Nan Phinney
Stanford Linear Accelerator Center, Stanford, CA,
USA
Positron–Electron Colliding Beams

George C. Pimentel[†]
Late of Department of Chemistry, University of
California, Berkeley, CA, USA
Hydrogen Bond

F. Plasil
Oak Ridge National Laboratory, Oak Ridge, TN,
USA
Nuclear Fission

P. M. Platzman
Retired from Bell Laboratories, Murray Hill, NJ,
USA
Fermi–Dirac Statistics

[†] deceased

Udo W. Pohl
Physics Department, Technical University of
Berlin, Germany
Quantum Structures in Semiconductors

J. Preskill
Department of Physics, California Institute of
Technology, Pasadena, CA, USA
Cosmic Strings

I. Prigogine[†]
Late of University of Brussels, Belgium
Thermodynamics, Nonequilibrium

Morris Pripstein
Lawrence Berkeley National Laboratory,
University of California, Berkeley, CA, USA
Center-of-Mass System

J. Prost
Groupe de Physico Chimie Théorique, Ecole de
Physique et Chimie, Paris, France
Liquid Crystals

Lewis Pyenson
Center for Louisiana Studies, University of
Louisiana, Lafayette, USA
History of Physics

Elisabeth Rachlew
Physics Department, The Royal Institute of
Technology, Stockholm, Sweden
X-Ray Spectra and X-Ray Spectroscopy

Norman F. Ramsey
Department of Physics, Harvard University,
Cambridge, MA, USA
Precession

André Rassat
Chemistry Laboratory, Ecole Normale Supérieure,
Paris, France
Electron Spin Resonance

Tor O. Raubenheimer
Stanford Linear Accelerator Center, Stanford, CA,
USA
Positron–Electron Colliding Beams

R. Ronald Rau
Brookhaven National Laboratory, Upton, NY,
USA
Hadron Colliders at High Energy

W. A. Reed
Bell Laboratories, Murray Hill, NJ, USA
Magnetoresistance

John D. Reichert
Retired from Department of Electrical
Engineering and Computer Engineering,
University of New Mexico, Albuquerque, NM,
USA
Refraction

Frederick Reines[†]
Late of Department of Physics, University of
California, Irvine, CA, USA
Neutrinos

A. Rich[†]
Late of Department of Physics, University of
Michigan, Ann Arbor, MI, USA
Positron
Quantum Electrodynamics

Burton Richter
Department of Physics, Stanford University,
Stanford, CA, USA
Positron–Electron Colliding Beams

Thomas A. Rijken
Institute for Theoretical Physics, University of
Nijmegen, Nijmegen, The Netherlands
Nuclear Forces

P. H. Roberts
Department of Mathematics, University of
California, Los Angeles, CA, USA
Magnetohydrodynamics

Gert Roepstorff
Department of Physics, Aachen University,
Aachen, Germany
Casimir Effect
Quantum Electrodynamics

Fritz Rohrlich
Retired from Department of Physics, Syracuse
University, Syracuse, NY, USA
Quantum Electrodynamics

Fred J. Rosenbaum
Department of Electrical Engineering,
Washington University, Seattle, WA, USA
Microwaves and Microwave Circuitry

[†]deceased

Carol Z. Rosen
Plessey Electronic Systems Corporation, Totawa, NJ, USA
Transducers

Marvin Ross
Lawrence Livermore National Laboratory, University of California, Livermore, CA, USA
Shock Waves and Detonations

Walter G. Rothschild
Scientific Laboratories, Ford Motor Company, Dearborn, MI, USA
Far-Infrared Spectra

Laura M. Roth
Department of Physics, State University of New York, Albany, NY, USA
Faraday Effect

John M. Rowell
Retired from Conductors, Inc., Sunnyvale, CA, USA
Tunneling

M. W. Ruckman
Fusion UV Systems, Gaithersburg, MD, USA
Thin Films

M. E. Rudd
Department of Physics and Astronomy, University of Nebraska, Lincoln, NV, USA
Ionization

Arnold Russek
Retired from Department of Physics, University of Connecticut, Storrs, CT, USA
Collisions, Atomic and Molecular

John D. Ryder[†]
Ocala, FL, USA
Electron Tubes

R. M. Ryder
Summit, NJ, USA
Transistors

P. G. Saffman
Department of Applied Mathematics, California Institute of Technology, Pasadena, CA, USA
Vortices

Richard H. Sands
Retired from Department of Physics, University of Michigan, Ann Arbor, MI, USA
Spin

Antonio Sasso
Department of Physical Science, University of Naples, Italy
Laser Cooling

Todd Satogata
Brookhaven National Laboratory, Upton, NY, USA
Hadron Colliders at High Energy

Peter R. Saulson
Department of Physics, Syracuse University, Syracuse, NY, USA
Gravitational Waves

D. J. Scalapino
Physics Department, University of California, Santa Barbara, CA, USA
Josephson Effects

A. Paul Schaap
Lumigen, Inc., Southfield, MI, USA
Chemiluminescence

Gertrude Scharff-Goldhaber[†]
Late of Brookhaven National Laboratory, Upton, NY, USA
Isomeric Nuclei

L. B. Schein
Independent Consultant, San Jose, CA, USA
Electrophotography

John P. Schiffer
Argonne National Laboratory, Argonne, IL, USA
Nuclear Reactions

K. E. Schmidt
Department of Physics, Arizona State University, Tucson, AZ, USA
Monte Carlo Techniques

H. W. Schmitt
Atomic Science, Inc., Oak Ridge, TN, USA
Nuclear Fission

John Schroeder
Department of Physics, Rensselaer Polytechnic Institute, Troy, NY, USA
Ultrahigh-Pressure Techniques

[†]deceased

Heinz-Georg Schuster
Department of Physics, University of Kiel, Kiel,
Germany
Complex Systems

Harold A. Schwarz
Brookhaven National Laboratory, Upton, NY,
USA
Radiation Chemistry

John H. Schwarz
California Institute of Technology, Pasadena, CA,
USA
String Theory

Richard L. Sears
Department of Astronomy, University of
Michigan, Ann Arbor, MI, USA
Stellar Energy Sources and Evolution

D. J. Seery
United Technologies Research Center, East
Hartford, CT, USA
Combustion and Flames

B. G. Segal[†]
Late of Barnard College, Columbia University,
New York, USA
Balmer Formula

Ernest A. Seglie
Office of the Secretary of Defense, Washington,
D.C., USA
Quantum Mechanics

Harvey Segur
Department of Applied Mathematics, University
of Colorado, Boulder, CO, USA
Solitons

D. J. Sellmyer
Center for Materials Research and Analysis,
University of Nebraska, Lincoln, NE, USA
Galvanomagnetic and Related Effects

R. B. Setlow
Biology Department, Brookhaven National
Laboratory, Upton, NY, USA
Biophysics

Jagdeep Shah
Microsystems Technology Office, Defense
Advanced Research Projects Agency, Arlington,
VA, USA
Electron–Hole Droplets in Semiconductors

Paul Shaman
Department of Statistics, University of
Pennsylvania, Philadelphia, PA, USA
Statistics

S. M. Shapiro
Brookhaven National Laboratory, Upton, NY,
USA
Neutron Diffraction and Scattering

Melvin P. Shaw
Retired from Department of Electrical and
Computer Engineering, Wayne State University,
USA
Solid-State Switching

D. Sherrington
Department of Physics, Imperial College of
Science and Technology, London, United
Kingdom
Magnetic Ordering in Solids

Eric P. Shettle
Naval Research Laboratory, Washington, D.C.,
USA
Atmospheric Physics

G. Shirane
Brookhaven National Laboratory, Upton, NY,
USA
Neutron Diffraction and Scattering

Ferdinand J. Shore
Department of Physics, Queens College of the
City University of New York, New York, USA
Nuclear Reactors

Howard A. Shugart
Department of Physics, University of California,
Berkeley, CA, USA
Beams, Atomic and Molecular

Kurt E. Shuler
Department of Chemistry, University of
California, San Diego, CA, USA
Stochastic Processes

[†]deceased

R. P. Shutt[†]
Late of Brookhaven National Laboratory, Upton, NY, USA
Cloud and Bubble Chambers

Walter P. Siegmund
Schott Fiber Optics, Inc., Southbridge, MA, USA
Fiber Optics

R. H. Siemann
Stanford Linear Accelerator Center, Stanford University, Stanford, CA, USA
Novel Particle Acceleration Methods

John A. Simpson
Retired from National Institute of Standards and Technology, Gaithersburg, MD, USA
Metrology

David J. Smith
John M. Cowley Center for High-Resolution Electron Microscopy, Arizona State University, Tempe, AZ, USA
Electron Microscopy

Henrik Smith
The Niels Bohr Institute, University of Copenhagen, Copenhagen, Denmark
Quantum Fluids

Domina Eberle Spencer
Department of Mathematics, University of Connecticut, Storrs, CT, USA
Vector and Tensor Analysis

H. E. Spencer
Retired from Department of Chemistry, Oberlin College, Oberlin, OH, USA
Light-Sensitive Materials

Steve W. Stahler
Department of Astronomy, University of California, Berkeley, CA, USA
Formation of Stars and Planets

Anthony F. Starace
Department of Physics and Astronomy, University of Nebraska, Lincoln, NE, USA
Atomic Spectroscopy

Rolf M. Steffen[†]
Late of Department of Physics, Purdue University, West Lafayette, IN, USA
Angular Correlation of Nuclear Radiation

Philip Stehle
Department of Physics and Astronomy, University of Pittsburgh, Pittsburgh, PA, USA
Maxwell's Equations

Gary Steigman
Department of Physics, Ohio State University, Columbus, OH, USA
Antimatter

Philip J. Stephens
Department of Chemistry, University of Southern California, Los Angeles, CA, USA
Magnetic Circular Dichroism

Frank H. Stillinger
Department of Physics, Princeton University, Princeton, NJ, USA
Water

Thomas H. Stix[†]
Late of Plasma Physics Laboratory, Princeton University, Princeton, NJ, USA
Plasma Waves

M. Strasberg
Acoustical Society of America, Washington, D.C., USA
Acoustical Measurements

K. G. Strassmeier
Astrophysical Institute, Potsdam, Germany
Milky Way

Myron Strongin
Retired from Brookhaven National Laboratory, Upton, NY, USA
Thin Films

E. C. G. Sudarshan
Department of Physics, University of Texas, Austin, TX, USA
Kinematics and Kinetics

D. Gary Swanson
Department of Physics, Auburn University, Auburn, AL, USA
Plasma Waves

L. W. Swanson
Department of Applied Physics, Oregon Graduate Center, Beaverton, OR, USA
Field Emission

[†]deceased

L. S. Swenson, Jr.
Department of History, University of Houston, Houston, TX, USA
Michelson–Morley Experiment

Victor Szebehely[†]
Late of Department of Aerospace Engineering and Engineering Mechanics, University of Texas, Austin, TX, USA
Three-Body Problem, Gravitational

Michael Tabor
Department of Mathematics, University of Arizona, Tucson, AZ, USA
Dynamics, Analytical

Frank R. Tangherlini
Department of Physics, College of the Holy Cross, Worcester, MA, USA
Moment of Inertia

C. L. Tang
School of Electrical Engineering, Cornell University, Ithaca, NY, USA
Ultrashort Optical Pulses

Barry N. Taylor
National Institute of Standards and Technology, Gaithersburg, MD, USA
Constants, Fundamental

Cyrus Taylor
Institute for Technology Innovation, Commercialization and Entrepreneurship, Case Western Reserve University, Cleveland, OH, USA
Fields

Edwin F. Taylor
Department of Physics, Massachusetts Institute of Technology, Cambridge, MA, USA
Schrödinger Equation

P. C. Taylor
Department of Physics, University of Utah, Salt Lake City, UT, USA
Semiconductors, Amorphous

G. J. Thaler
United States Naval Postgraduate School, Monterey, CA, USA
Servomechanism

Edward W. Thomas
School of Physics, Georgia Institute of Technology, Atlanta, GA, USA
Electron and Ion Impact Phenomena

R. N. Thurston
Bell Communications Research, Red Bank, NJ, USA
Ultrasonics

D. R. Tilley
Retired from Department of Physics, University of Essex, Colchester, United Kingdom
Waves

Laszlo Tisza
Department of Physics, Massachusetts Institute of Technology, Cambridge, MA, USA
Entropy

A. M. Title
Lockheed Palo Alto Research Laboratory, Palo Alto, CA, USA
Photosphere

Tommaso Toffoli
Department of Electrical and Computer Engineering, Boston University, Boston, MA, USA
Cellular Automata

Juergen Tonndorf[†]
Late of College of Physicians and Surgeons, Columbia University, New York, USA
Acoustics, Physiological

Gérard Toulouse
Ecole Normale Superieure, Paris, France
Catastrophe Theory

Sol Triebwasser
Retired from IBM, Thornwood, New York, USA
Circuits, Integrated

James W. Truran
Department of Astronomy and Astrophysics, University of Chicago, Chicago, IL, USA
Nucleosynthesis

E. Y. Tsymbal
Center for Materials Research and Analysis, University of Nebraska, Lincoln, NE, USA
Galvanomagnetic and Related Effects

[†] deceased

Wallace Tucker
Department of Physics, University of California,
Irvine, CA, USA
Astronomy, X-Ray

B. E. Turner
National Radio Astronomy Observatory,
Charlottesville, VA, USA
Astronomy, Radio

J. Anthony Tyson
Retired from Bell Laboratories, Murray Hill, NJ,
USA
Gravitational Waves

Martin A. Uman
Department of Electrical and Computer
Engineering, University of South Florida,
Gainesville, FL, USA
Lightning

R. J. Urick
Silver Spring, MD, USA
Sound, Underwater

James A. Van Allen
Department of Physics and Astronomy, University
of Iowa, Iowa City, IA, USA
Radiation Belts

C. W. Van Atta
Retired from Department of Applied Mechanics
and Engineering Sciences, University of
California, San Diego, CA, USA
Boundary Layers
Vacuums and Vacuum Technology

H. Van Dam
Department of Physics, University of North
Carolina, Chapel Hill, NC, USA
Rotation and Angular Momentum

W. Alexander Van Hook
Retired from Department of Chemistry,
University of Tennessee, Knoxville, TN, USA
Isotope Effects

Michel A. Van Hove
Lawrence Berkeley National Laboratory,
Berkeley, CA, USA
Electron Diffraction

Johannes A. Van den Akker[†]
Late of Appleton, WI, USA
Mass

Peter Van der Straten
Department of Physics, Utrecht University,
Utrecht, The Netherlands
Atomic Trapping and Cooling

R. N. Varney
Palo Alto, CA, USA
Kinetic Theory

G. P. Vella-Coleiro
Bell Laboratories, Murray Hill, NJ, USA
Magnetic Domains and Bubbles

Richard W. Vook
Department of Physics, Syracuse University,
Syracuse, NY, USA
*Reflection High-Energy Electron Diffraction
(RHEED)*

James T. Waber
Retired from Department of Physics, Michigan
Technical University, Houghton, MI, USA
Atomic Structure Calculations, Relativistic Atoms

Kameshwar C. Wali
Department of Physics, Syracuse University, New
York, USA
Strong Interactions

Jearl D. Walker
Retired from Department of Physics, Cleveland
State University, Cleveland, OH, USA
Energy and Work

Richard F. Wallis
Department of Physics, University of California,
Irvine, CA, USA
Lattice Dynamics

R. E. Walstedt
Advanced Science Research Center, JAERI,
Tokai-mura, Ibaraki-ken, Japan
Nuclear Magnetic Resonance
Nuclear Quadrupole Resonance

Ulrich Walter
Institute of Astronautics, Munich Technical
University, Germany
Space Science and Technology

[†] deceased

Thomas P. Wangler
Retired from Los Alamos National Laboratory,
Los Alamos, NM, USA
Accelerators, Linear

George D. Watkins
Department of Physics, Lehigh University,
Bethlehem, PA, USA
Crystal Defects

Kenneth M. Watson
Scripps Institution of Oceanography, University
of California, San Diego, CA, USA
Scattering Theory

Eli Waxman
Physics Department, Weizmann Institute of
Science, Rehovot, Israel
Astronomy, High-Energy Neutrino

Alfons Weber
Optical Technology Division, National Institute of
Standards and Technology, Gaithersburg, MD,
USA
Molecular Spectroscopy

Michael A. Webster
Department of Psychology, University of Nevada,
Reno, NV, USA
Vision and Color

Johannes Weertman
Department of Materials Science, Northwestern
University, Evanston, IL, USA
Fatigue

M. Weidemüller
Physics Institute, Freiburg University, Freiburg,
Germany
Ultracold Quantum Gases

Louis Weinberg
Department of Electrical Engineering, The City
College of the City University of New York, New
York, USA
Network Theory: Analysis and Synthesis

Don Weingarten
Retired from IBM Research Center, Yorktown
Heights, NY, USA
Field Theory, Classical

Gabriel Weinreich
Retired from Department of Physics, University
of Michigan, Ann Arbor, MI, USA
Free Energy
Musical Instruments

J. Weinstock
National Oceanic and Atmospheric
Administration, Boulder, CO, USA
Ionosphere

Robert Weinstock
Department of Physics, Oberlin College, Oberlin,
OH, USA
Temperature

George H. Weiss
National Institutes of Health, Bethesda, MD, USA
Stochastic Processes

David O. Welch
Brookhaven National Laboratory, Upton, NY,
USA
Radiation Damage in Solids

Heinrich J. Welker
Retired from Siemens AG, Erlangen, Germany
Semiconductors, Crystalline

Edgar F. Westrum, Jr.
Retired from Department of Chemistry,
University of Michigan, Ann Arbor, MI, USA
Calorimetry

A. S. Wightman
Department of Physics, Princeton University,
Princeton, MA, USA
Field Theory, Axiomatic

Sir Denys Wilkinson
University of Sussex, Falmer, Brighton, United
Kingdom
Isospin

Lincoln Wolfenstein
Retired from Department of Physics, Carnegie
Mellon University, Pittsburgh, PA, USA
Conservation Laws

William L. Wolfe
Optical Sciences Center, University of Arizona,
Tucson, AZ, USA
Radiometry

W. P. Wolf
Retired from Department of Physics, Yale
University, New Haven, CT, USA
Ferrimagnetism

Gary L. Womack
Zetetic Institute, Tucson, AZ, USA
Sun

G. Wouch
Lisle Technical Center, R. R. Connelly & Sons,
Lisle, IL, USA
Levitation, Electromagnetic

Francis J. Wright
School of Mathematical Sciences, Queen Mary,
University of London, London, United Kingdom
Catastrophe Theory

Fa Yueh Wu
Department of Physics, Northeastern University,
Boston, MA, USA
Ising Model

Jonathan S. Wurtele
Lawrence Berkeley National Laboratory,
University of California, Berkeley, CA, USA
Novel Particle Acceleration Methods

John J. Zayhowski
Lincoln Laboratory, Massachusetts Institute of
Technology, Cambridge, MA, USA
Lasers

Anton Zeilinger
Institute of Experimental Physics, University of
Vienna, and Institute of Quantum Optics and
Quantum Information, Austrian Academy of
Sciences, Vienna, Austria
Quantum Information

Mark W. Zemansky[†]
Late of The City College of the City University of
New York, New York, USA
Carnot Cycle

Da-Ming Zhu
Department of Modern Physics, University of
Science and Technology of China, Hefei, China
Cryogenics

E. C. Zipf
Department of Physics and Astronomy, University
of Pittsburgh, Pittsburgh, PA, USA
Aurora

[†]deceased

Absorption Coefficient

F. L. Galeener[†]

The absorption coefficient α measures the spatial decrease in intensity of a propagating beam of waves or particles due to progressive conversion of the beam into different forms of energy or matter. Absorption usually implies the creation of some form of internal energy in the traversed medium, e. g., the production of heat; however, it may also be associated with other inelastic scattering events, such as the ultimate conversion of incident particles into new types, or the change in frequency of waves from their incident values. Removal of intensity from the beam merely by diversion into new directions is called elastic scattering, and this process is not properly included in the absorption coefficient.

The extinction coefficient α_e measures the reduction in beam intensity due to *all* contributing processes and is often represented as a sum $\alpha_e = \alpha + \alpha_s$ where α_s is the coefficient associated with elastic scattering. Additional subdivisions of processes that remove intensity from the beam are possible and sometimes used.

These coefficients appear in the Bouguer or Lambert–Beer Law in the form $I(x) = I(0)e^{-\alpha_e x}$, where $I(x)$ is the beam intensity after it has traveled a distance x in the medium, while α_e, α and α_s have units of inverse length, often written cm^{-1}.

The absorption coefficient α appears frequently in discussions of the optical properties of homogeneous solids, liquids, and gases, where α may be a strong function of the wavelength of the light involved, the temperature, and various sample parameters. The theory of electromagnetic waves relates α to the complex permittivity ε and permeability μ of the medium.

See also: Electromagnetic Radiation

Bibliography
M. Born and E. Wolf, *Principles of Optics*, 7th ed. Pergamon, New York, 1999. (A)
F. A. Jenkins and H. E. White, *Fundamentals of Optics*, 4th ed. McGraw-Hill, New York, 2001. (E)
R. B. Leighton, *Principles of Modern Physics*, Chapter 12. McGraw-Hill, New York, 1959. (I)

[†]deceased

Encyclopedia of Physics, Third Edition. Edited by George L. Trigg and Rita G. Lerner
Copyright © 2005 WILEY-VCH Verlag GmbH & Co. KGaA, Weinheim
ISBN: 3-527-40554-2

Accelerators, Linear

T. P. Wangler

Overview

It is common in the particle-accelerator field to use the term linear accelerator (linac) for a device in which a beam of charged subatomic particles moves on a straight path and is accelerated by time-varying electric fields. In a radiofrequency (RF) linac, a beam of electrons, protons, or heavier ions is accelerated by an RF electric field with a harmonic time dependence using a linear array of waveguides or cavities excited in a resonant electromagnetic mode. Another type of linear accelerator, related to the betatron and called a linear induction accelerator or an induction linac, uses pulsed nonharmonic electric fields for acceleration in a nonresonant device called an induction module. For both the RF and the induction linac, the accelerating voltages are related to a changing magnetic field through Faraday's law. In both cases the beam is synchronized with the applied time-varying accelerating voltages.

Since the end of World War II, the linac has undergone a remarkable development. Its technological base is a consequence of the science of both the nineteenth and twentieth centuries. Included are the discoveries of electromagnetism by Faraday, Maxwell, and Hertz in the nineteenth century and the discovery of superconductivity in the twentieth century. The linear accelerator has developed as a powerful tool for basic research. It provides beams of high quality and high energy, sufficient to resolve the internal structure of the nucleus and its constituent subnuclear particles. Like a microscope, it has been used to probe the internal structure of the nuclear constituents, the proton and neutron, and has given us our present picture that these constituents are themselves made of point-like particles called quarks. Electron linacs are used in hospitals throughout the world as a source of X-rays for radiation therapy to treat cancer, an application that may represent the most significant spin-off of high energy and nuclear physics research.

The main advantage of the linac relative to other accelerator types is its capability for producing high-energy high-intensity charged particle beams of high quality with small beam diameter and small energy spread. Other attractive characteristics of the linac include the following: (1) Strong focusing can be provided to confine a high-intensity beam. (2) The beam traverses the linac in a single pass eliminating conditions that cause destructive beam resonances in circular accelerators. (3) Injection and extraction are simple compared with circular accelerators, because the natural orbit of the linac is open at each end. (4) Because the beam travels in a straight line, there is no power loss from synchrotron radiation, which is a major limitation for high-energy electron beams in circular accelerators. (5) The linac can operate in a pulsed mode at any duty factor, up to and including 100%. High duty factor provides the capability for higher average beam currents and beam powers. (6) The linac is capable of simultaneous acceleration of multiple charge states of a given mass species resulting in higher intensities for heavy-ion linacs. The principal disadvantage of the linac is the multiplicity and cost of accelerating structures, RF equipment, and operating power needed to achieve a given final beam energy. The increasing application of RF superconductivity to linacs within recent years, promises to reduce costs and make linacs even more attractive for many new applications.

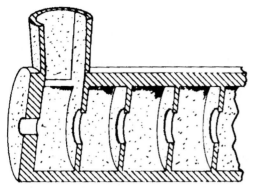

Fig. 1: Schematic drawing of the iris-loaded traveling-wave structure, showing the input waveguide through which the electromagnetic wave is injected into the structure at the end cell. The beam propagates along the central axis and is accelerated by the electric field of the traveling wave.

History

The first published proposal for a linac was made by Gustav Ising [1] in 1924. The first concept and experimental demonstration of an RF linac was due to Rolf Wideröe [2], who in 1928 showed that by applying a high-frequency voltage to a drift tube between two grounded electrodes, phased so that the beam experienced an accelerating voltage in both gaps, one could deliver an effective voltage gain to the beam of twice the applied voltage. This would not have been possible using a time-independent voltage, and it was clear that a multi-gap sequence of properly positioned drift tubes would allow repeated acceleration of the beam to arbitrarily high energies. The Wideröe result established the basis not only of RF linacs, but of the entire class of resonance RF accelerators that includes the cyclotron, and the synchrotron. But, linear accelerators that were useful for physics research were not feasible until after the developments of microwave technology stimulated by radar development during WWII.

Accelerating Structures

The modern RF linac uses high-Q resonant cavities or waveguides, either of which can be excited to high RF electromagnetic field levels at frequencies typically in the very high frequency (VHF) or ultrahigh frequency (UHF) ranges. These accelerating structures are tuned to resonance and are driven by external, high-power RF tubes such as klystrons, magnetrons, or gridded vacuum tubes. The accelerating structures are designed, through an optimized configuration of the internal geometry, for efficient transfer of electromagnetic energy to beam kinetic energy.

One can excite in a waveguide, a TM_{01}-like transverse magnetic traveling wave (having an axial electric accelerating field) that co-propagates with the beam. The waveguide uses a periodic array of conducting irises that reduces the phase velocity of the traveling wave to the velocity of the beam. Major developments of the iris-loaded traveling-wave linac (Fig. 1) for acceleration of relativistic electron beams began shortly after World War II at Stanford. These developments over two decades led to the design and construction of the 3-km-long Stanford Linear Accelerator (SLAC) linac [3].

Another method is to excite a standing wave in a multicell cavity, where the beam is accelerated by the longitudinal electric field in each of the cells. Various types of multicell structures have been invented for efficient acceleration over different ranges of beam velocity. One such structure is the Alvarez drift-tube linac (DTL), which is used to accelerate medium-velocity protons or other ions in the velocity range from about 0.04 to 0.4 times the speed of light. The first such accelerator was a 200-MHz DTL, proposed and built in 1946 at University of California by Luis Alvarez and co-workers [4]. The concept uses a quasi-periodic array of copper drift tubes enclosed in a high-Q cylindrical cavity. Figure 2 shows a DTL structure at the SNS facility at Oak Ridge National Laboratory. A TM_{010}-like transverse-magnetic mode is excited, which has an axial electric field in the gaps between drift tubes and zero field inside the drift tubes to shield the beam from deceleration when the field reverses. Unlike the original Wideröe structure, the fields in adjacent gaps are in phase, and the spacing between the centers of the gaps is equal to the distance the beam travels in a single RF period. Beam-focusing, necessary to confine the beam radially, is usually provided by installing magnetic-quadrupole lenses within the drift tubes. The DTL structure is not needed for electrons because electrons are so light that their velocity, as supplied from the electron source, is already above the applicable velocity range for a DTL.

Various types of standing-wave, coupled-cavity linac structures are used for acceleration of protons or electrons in the velocity range from about 0.4 to 1.0 times the speed of light. For example, a coupled-cavity linac operating at 805-MHz, called the side-coupled linac [5] (Fig. 3), invented at Los Alamos in the 1960s, is used at the Los Alamos Neutron Science Center (LANSCE) linac to accelerate a beam of protons or negative hydrogen ions from 100 to 800 MeV. The same structure type is used at the Spallation Neutron Source (SNS) now under construction at Oak Ridge.

The radiofrequency quadrupole (RFQ) linac, invented by I. M. Kapchinskiy and V. A. Tepliakov [6], is used for bunching and acceleration of low-velocity ion beams in the velocity range from about 0.01 to 0.1 times the speed of light. The RFQ bunches and captures most of the continuous beam injected from the ion source, and provides strong electric focusing to confine and accelerate high-current ion beams. The RFQ is a typically a few meters in length and is used as the first accelerating structure in a modern ion linac. A transverse electric-quadrupole mode is excited in a resonant RFQ cavity loaded with four equally-spaced conducting rods or vanes that are parallel to the beam axis (Fig. 4). The transverse focusing, provided by the electric-quadrupole field, is superior to magnetic focusing for these low-velocity ions. A longitudinal electric field for acceleration is obtained by machining a longitudinal modulation pattern onto the four vanetips, producing an array of accelerating cells.

Copper, because of its high electrical and thermal conductivity, is the most commonly used metal for the RF surfaces of room-temperature accelerating structures. The RF equipment and ac power for accelerator operation are major costs for a room-temperature linac. Reducing the fields to reduce the RF power dissipation means that more real estate is needed for the linac. For high-duty-factor room-temperature linacs, cooling the structures also becomes an important engineering requirement. The application of superconductivity to RF linacs has long been recognized as an important step toward better performance and lower costs. The use of superconducting niobium cavities reduces the Ohmic power dissipation to roughly 10^{-5} that of room-temperature copper. Even after the cryogenic refrigeration requirements are included, the net operating power for the superconducting linac is reduced typically by about two orders

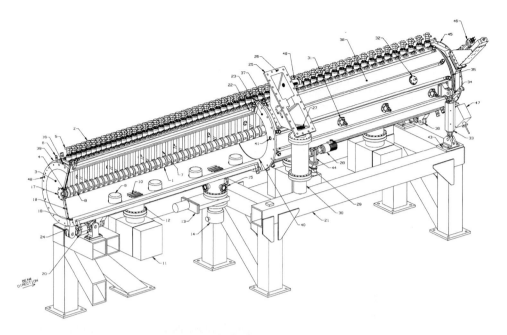

Fig. 2: Cut-away drawing of a 402.5-MHz 4.2-m 60-cell drift-tube-linac structure at the Spallation Neutron Source (SNS) at Oak Ridge National Laboratory. The photograph shows copper drift tubes and the drift-tube support stems.

of magnitude. For many applications the superconducting option is now superior to the room-temperature linac, resulting in reduced capital and operating costs, and a number of additional advantages including higher accelerating gradients and shorter linacs, and larger bore radii to reduce beam losses and to reduce wakefield effects in electron accelerators.

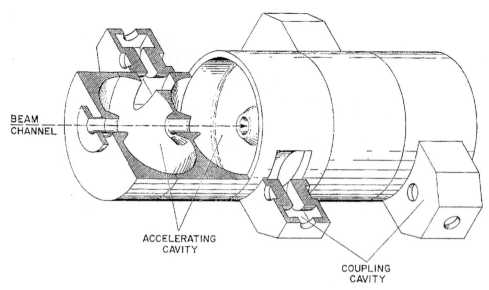

BEAM
CHANNEL

ACCELERATING
CAVITY

COUPLING
CAVITY

Fig. 3: The side-coupled linac structure was invented at Los Alamos in the 1960s. The cavities on the beam axis are the accelerating cavities. The coupling cavities on the side are nominally unexcited and stabilize the accelerating-cavity fields against perturbations from fabrication errors and beam loading.

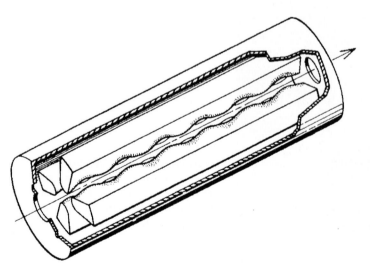

Fig. 4: Schematic drawing of a 4-vane RFQ accelerator. The four conducting vanes are excited with electric-quadrupole-mode RF voltages to focus the beam. The vanetips are modulated in the axial direction to produce longitudinal electric fields that bunch and accelerate low-velocity ions.

Beam Dynamics

A linac is designed for acceleration of a single on-axis particle, called the synchronous particle, which maintains exact synchronism with the accelerating field. Focusing forces must be provided to ensure stability for particles that deviate from the synchronous particle trajectory.

Longitudinal focusing is obtained when the synchronous particle is accelerated by an electric field that is increasing in time, i.e., before the peak of the sinusoidal waveform. Then, early particles experience a smaller field, which reduces their speed relative to the synchronous particle, whereas late particles experience a larger field, which increases their relative speed. This is known as the principle called phase stability, which results in stable phase and energy oscillations about the synchronous particle. Longitudinal stability is not a concern for relativistic electron linacs, where all particles travel at nearly the speed of light.

When off-axis particles experience an accelerating field that increases with time, as is required for longitudinal focusing, Maxwell's equations require that the particles also experience a radial RF defocusing force. In the case of relativistic electrons, the net radial force, comprised of oppositely directed electric and magnetic forces, is nearly zero, so that focusing may be unnecessary. For nonrelativistic ion beams, radial focusing must be provided to compensate for the RF defocusing force. With the exception of the RFQ, which provides its own transverse quadrupole RF electric focusing, the most effective solution to radial defocusing has been to include magnetic focusing lenses, usually magnetic quadrupoles, installed either between accelerating structures, or within the drift tubes, as is done in the DTL.

Controlling high-intensity effects can significantly influence the choices for the accelerator design, particularly for the beam-focusing requirements [7]. The main intensity limitation in nonrelativistic ion linacs is caused by the mutually repulsive Coulomb electric forces between the beam particles, commonly referred to as space charge. The space-charge force is important for lower velocity beams where the beam density is high, and where the attractive magnetic force from the moving beam is small. The nonlinearity of the space-charge force can also produce an extended halo surrounding the main core of the beam. This is of concern at high intensities, because halo particles may impact the accelerating structure, and induce undesired radioactivity.

In relativistic electron linacs the electric-space-charge force is nearly canceled by the oppositely directed magnetic force of the moving beam. But, Lorentz-compressed electromagnetic fields from the moving charges produce scattered radiation known as wakefields at discontinuities in the surrounding walls of the accelerator. Wakefields exert forces on trailing charges in both the same and later bunches, causing radial defocusing, energy loss, and energy spread in the beam. Another important intensity-limiting effect in electron linacs is known as the beam-breakup (BBU) instability. In this case, when a bunch travels off axis in an accelerating structure, it can excite a resonant mode that deflects trailing bunches. BBU is an instability that grows in time and distance along the accelerator, eventually leading to loss of the beam.

Linac Applications And Some Major Linac Facilities

A worldwide compendium of existing and planned scientific linacs published in 1996, listed 176 linacs distributed over the Americas, Europe, and Asia [8]. Contemporary electron linac applications include (1) a future TeV electron-positron international linear collider (ILC) proposed for high-energy physics research, (2) linacs for free-electron lasers (FEL), and (3) compact electron linacs as X-ray sources for cancer therapy. The high intensity and excellent beam quality provided by the linac are important for achieving high luminosity in the collider, and high brilliance in the FEL. An important and very successful commercial application of RF linacs is the use of compact 10- to 20-MeV electron linacs for cancer therapy. Several thousand electron linacs are used worldwide for medical irradiations and this number is growing. In addition compact electron linacs are used for industrial applications including X-ray radiography, materials processing, and food sterilization.

Light-ion-linac applications include: (1) injectors to high-energy synchrotrons for physics research, (2) high-power proton linacs as drivers for spallation-neutron sources for condensed matter and material science research, nuclear material production for national defense, transmutation of nuclear wastes, and accelerator-driven fission reactors, (3) short-lived radioisotope production for medical diagnostics, (4) deuteron-linac-driven neutron sources for materials irradiation studies for fusion reactors, (5) multi-GeV linacs for heavy-ion inertial-confinement fusion, and (6) proton RFQ linacs for boron-neutron capture therapy. Important applications for heavy-ion linacs include (1) superconducting accelerators for a rare-isotope accelerator (RIA) facility to be used for nuclear-physics research with radioactive ion beams, and (2) ion implantation for commercial semiconductor fabrication.

The largest electron linac is the 3-km 50-GeV room-temperature traveling-wave structure at the Stanford Linear Accelerator Center (SLAC). This linac, which operates at 2856 MHz, has had a very productive physics history, beginning operation as a fixed-target accelerator facility in 1966, and as an electron-positron collider (SLC) beginning in 1989. The newest application of the SLAC linac is the Linac Coherent Light Source (LCLS), which will use electrons accelerated in the last kilometer of the linac for the world's first hard-X-ray free-electron laser, when it becomes operational in 2009.

At the Deutsches Electronen Synchrotron (DESY) laboratory in Hamburg, Germany, a superconducting standing-wave electron linac serves both as a state-of-the-art FEL facility, and as a pilot facility for the TESLA electron-positron-collider development project. The planned coherent X-ray FEL will use a 50-GeV electron linac built from superconducting niobium cavities operating at 1.3 GHz. The ILC collider concept, for which the site is not yet determined at the time this article is being written, uses an approximately 33-kilometer-long tunnel that will house two superconducting linear accelerators in which electrons and positrons will be accelerated and made to collide at a center-of-mass energy of 0.5 TeV. The linacs will use 21 000 superconducting niobium cavities (Fig. 5), cooled with liquid helium to an operating temperature of 2 K. The TESLA project has made significant advances in the performance of the superconducting accelerating cavities, achieving accelerating gradients larger than 25 MV/m.

The Continuous Electron Beam Accelerator Facility (CEBAF) at the Thomas Jefferson National Accelerator Facility (TJNAF) in Virginia is a 5-pass recirculating linear-accelerator facility that uses two superconducting electron linacs (320 5-cell superconducting cavities in 40 cryomodules) joined by two 180-degree magnet arcs through which the beam is recirculated and accelerated to a final energy of 6 GeV. CEBAF is used for nuclear physics research and began operation in 1994.

The first proton linac with beam energy near 1-GeV kinetic energy is the 800-m-long LANSCE linac at Los Alamos. This high-intensity standing-wave proton linac began operation in 1972 as a pion factory (Los Alamos Meson Physics Facility) for nuclear physics research,

Fig. 5: Two 1.3-m 9-cell 1.3-GHz superconducting niobium elliptical cavities used for the X-ray FEL at the TESLA Test Facility and for the proposed linacs of the 0.5-TeV center-of-mass e^+–e^- International Linear Collider.

and is now a multipurpose facility used as a driver for a spallation-neutron source and for proton radiography. LANSCE can simultaneously accelerate H^+ and H^- ions to an energy of 800-MeV. The linac uses normal-conducting copper technology, and is comprised of a 201.25-MHz DTL (0.75 to 100 MeV) followed by an 805-MHz side-coupled linac (100 to 800 MeV).

The 1-GeV H^- Spallation Neutron Source (SNS) linac, now under construction at Oak Ridge, will incorporate the world's first proton superconducting linac. The room-temperature part of the linac consists of a 402.5-MHz RFQ and DTL, which accelerates the beam to 87 MeV, followed by an 805-MHz side-coupled linac that accelerates the beam to 186 MeV. The 805-MHz superconducting accelerator, comprised of 81 niobium superconducting cavities, continues the acceleration to the final beam kinetic energy of 1 GeV.

Hydrodynamic testing of mockups of nuclear-weapons implosion systems is an important application of induction linacs. At Los Alamos, a pair of electron induction linacs at the Dual Axis Radiographic Hydrodynamic Test (DARHT) facility is used for X-ray radiography. Energetic (20 MeV), intense (2 kA), short (60 ns) bursts of electrons produce X-rays in a tungsten target. The pair of linacs, oriented at 90°, provides time-resolved tomographic reconstructions of implosions driven by high explosives.

Heavy-ion linacs are used for study of the atomic nucleus. At Argonne National Laboratory the approximately 150-m-long superconducting ATLAS linac is capable of accelerating ions of any element to energies as high as 17 MeV per nucleon. ATLAS is the world's first heavy-ion accelerator to use superconducting cavities. ATLAS contains 62 independently phased superconducting cavities; the independent phasing of these cavities provides adjusta-bility needed for efficient acceleration over the wide range of ion species.

Heavy-ion superconducting linacs are the basis of the planned nuclear-physics facility called the Rare Isotope Accelerator (RIA), which is being designed for acceleration of ra-dioactive ion beams. A powerful 1.4-GV RIA driver linac is used for acceleration of sta-ble heavy-ion beams that bombard a target in which rare-isotopes are produced. The driver linac can simultaneously accelerate multiple charge states of ions of a given mass, yielding higher intensity beams than any other type of accelerator for this application. The rare iso-tope species produced in the target are then selected magnetically for acceleration by a second superconducting linac, providing energetic radioactive beams for a variety of nuclear-physics and astrophysics studies.

See also: Accelerators, Potential-Drop Linear; Microwaves and Microwave Circuitry

References

[1]   G. Ising, *Ark. Mat. Fys.* **18** (no. 30), 1–4 (1924).
[2]   R. Wideröe, *Arch. Electrotech.* **21**, 387 (1928).
[3]   R. B. Neal, W. A. Benjamin (eds.), *The Stanford Two-Mile Accelerator.* New York, 1968.
[4]   L. W. Alvarez *et al.*, *Rev. Sci. Instrum.* **26**, 111–133 (1955).
[5]   E. A. Knapp, B. C. Knapp, and J. M. Potter, *Rev. Sci. Instrum.* **39**, 979–991 (1968); D. E. Nagel,
       E. A. Knapp, and B. C. Knapp, *Rev. Sci. Instrum.* **38**, 1583–1587 (1967).
[6]   I. M. Kapchinskiy and V. A. Tepliakov, *Prib. Tekh. Eksp.* **2**, 19–22 (1970).
[7]   Thomas P. Wangler, *Principles of RF Linear Accelerators*, Wiley, New York, 1998.
[8]   J. Clendenin *et al.*, *Compendium of Scientific Linacs*. CERN report CERN/PS 96-32 (DI),
       Nov. 1996.

Accelerators, Potential-Drop Linear

R. G. Herb[†] *and G. M. Klody*

Introduction

Potential-drop accelerators were the first to open up nuclear physics to extensive experimen-tation. Despite the higher-energy beams that were soon available with the cyclotron, betatron, and later accelerators that provide very high energies through many small-energy increments, potential-drop accelerators have continued to play an important, and often dominant, role in nuclear physics. In addition, potential-drop accelerators have a broad range of other applica-tions in many, diverse fields in industry, research, and medicine.

Potential-drop accelerators consist of a high-voltage terminal supported by an insulating column, a means of generating the high voltage, and an acceleration tube. To reduce their size, many potential-drop accelerators use high-pressure gas for electrical insulation. The

---

[†]deceased

electrostatic potential drop between the high-voltage terminal and ground accelerates charged particles to an energy equal to the charge times the terminal voltage ($E = qV$). A variety of ion and electron sources are available to produce the beams of charged particles for acceleration.

The acceleration tube is an insulating assembly, mounted between the high-voltage terminal and ground potential, through which the beam is accelerated. High vacuum inside the tube minimizes beam losses. Mechanical strength is also important for tubes in accelerators insulated with high-pressure gas. The tube design that maximizes voltage-holding capability and voltage stability under these conditions is a laminated series of insulating rings and metal electrodes. The electrodes are connected to a resistive voltage divider from the terminal to ground for a uniform voltage gradient (linear potential drop) along the tube.

Single-stage potential-drop accelerators can accelerate ions or electrons. Two-stage accelerators, called tandems, are for ion acceleration. Tandems accelerate negative ions from ground potential to a positively charged high-voltage terminal, change the ionic charge from negative to positive in the terminal, and accelerate the ions through a second acceleration tube, back to ground potential. For a given terminal voltage, tandems produce higher-energy ions than single-stage accelerators, but usually with less beam current.

Potential-drop accelerators are generally distinguished by the type of generator used to charge the high-voltage terminal. Each high-voltage generator design has advantages for specific applications. For precise measurements of nuclear-energy-level characteristics and many analytical techniques, the electrostatic accelerator is far superior. For certain neutron-induced reactions the Cockcroft–Walton-type accelerator may be most convenient. Needs for intense positive-ion or electron beams of a few MeV energy are most easily met by the Dynamitron.

Modern potential-drop accelerators are very reliable and simple to operate. This has greatly expanded their range of applicability in many fields. They are routinely used for ion implantation in the manufacture of semiconductor devices, nondestructive analysis of materials, surface hardening to reduce corrosion and wear, polymerization, sterilization, pollution monitoring, age determination of samples for archaeology, cosmology, and oceanography, and analysis of biomedical specimens.

Each of the accelerators described below is serving an important need in science and technology, and each is playing an a expanding role.

Cockcroft–Walton Voltage Multiplier Accelerator

This accelerator was the first to be used successfully for nuclear transmutation and gained wide recognition when results were published in 1932. It employs a circuit developed by H. Greinacher in 1920, as illustrated in Fig. 1, which utilizes two stacks of series connected capacitors. One capacitor stack is fixed in voltage except for voltage ripple with one terminal connected to ground and the other to the load which, in this case, is an evacuated accelerating tube.

One terminal of the second capacitor stack is connected to a transformer giving peak voltages of $\pm V$, and voltages at all points along the second capacitor stack oscillate over a voltage range of $2\,V$. The two capacitor stacks are linked by series-connected rectifiers. As voltage on the second stack oscillates, charge is transferred stepwise from ground to the high-voltage terminal. Voltage here is steady except for ripple caused by power drain and stray capacitance. Its value is approximately $2\,VN$, where N is the number of stages. The power supply furnishes

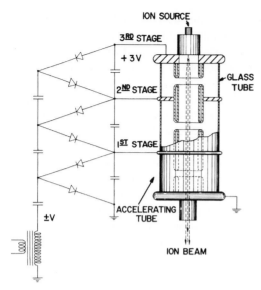

Fig. 1: Schematic drawing of a Cockcroft–Walton accelerator utilizing three stages.

power to an evacuated accelerating tube equipped with an ion source. Usually the tube and ion source are continually pumped and the multi-section tube may have one tube section per accelerator stage.

Cockcroft–Walton accelerators operating in open air at 1 million volts are large in size and must be housed in a very large room to avoid voltage flashover. Figure 2 shows an 850-keV accelerator of this type built by Emile Haefely & Co. Ltd. It serves to inject pulses of hydrogen ions into a high-intensity linear accelerator at the Los Alamos National Laboratory for production of intense meson beams.

Size and space requirements increase rapidly as voltage is extended above 1 million volts in open air and the practical upper voltage limit for these accelerators appears to be about 1.5 MV.

G. Reinhold of Emile Haefely & Co. Ltd. has shown that the terminal voltage developed by multipliers, using the circuit of Fig. 1, does not depart greatly from $2VN$ for multipliers going to a few hundred kilovolts. However, above about 500 kV, voltages achieved fall substantially below values given by this simple expression because of stray capacitances and other effects. He has developed another circuit called a symmetrical cascade rectifier in which the shortcomings of the simple circuit are largely eliminated. It employs two transformers and two capacitor stacks that oscillate in voltage. Both feed one fixed capacitor stack. Using this symmetric system, Emile Haefely & Co. Ltd. has built an open air rectifier without an accelerating tube going to 2.5 million volts.

The Philips Gloeilampenfabrieken has also manufactured accelerators with Cockcroft–Walton charging. Emile Haefely & Co. Ltd. has built accelerators utilizing voltage multipliers housed in tanks containing insulating gases such as SF_6 or a mixture of N_2 and CO_2. These machines have ranged in voltage from about 1 million up to 4 million volts. Nisshin-High

Fig. 2: 850-keV Haefely accelerator which serves to inject pulses of hydrogen
ions into a linear accelerator at the Los Alamos National Laboratory.

Voltage Company also builds SF_6-insulated Cockcroft–Walton accelerators from 500 kV to
3 million volts for their industrial electron processing systems. The use of insulating gases at
high pressure permits great savings in the size of equipment for a given voltage.

Dynamitron Accelerator

M. R. Cleland invented a cascaded rectifier system termed the Dynamitron in which series
connected rectifiers are driven in parallel. The circuit is shown schematically in Fig. 3, and
Fig. 4 is a photograph of a 4-million-volt positive ion Dynamitron accelerator.

Rectifiers connected in series between ground and the high-voltage terminal are positioned
in two columns on opposite sides of the accelerating tube of the Dynamitron and the high-
voltage column is enclosed by half rings that have a smooth exterior surface to inhibit corona
and spark discharge. The half-rings serve as capacitor plates coupled capacitively to the large
semicylindrical rf electrodes positioned between the walls of the tank and the high-voltage
column of the accelerator. The rf electrodes form the tuning capacitance of an *LC* resonant
circuit which is driven by a separate power supply through an oscillator tube.

Since ac power is fed in parallel to each of the series-connected rectifiers, the relatively
large storage capacitors that are connected between successive stages of other cascaded rectifier systems are not required. Stored energy in the Dynamitron is low and does not differ
greatly from that in electrostatic machines. This feature is important since damage due to
discharge can be a serious problem in multimillion volt accelerators, especially for discharge
in the accelerating tubes.

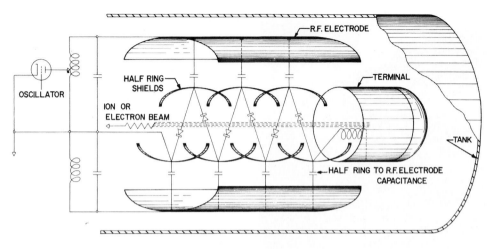

Fig. 3: Schematic diagram of a Dynamitron accelerator in a pressure tank.

Fig. 4: A 4-MeV Dynamitron positive-ion accelerator with pressure tank rolled away from the high-voltage column.

These accelerators are enclosed in pressure tanks containing high-pressure SF_6 gas. At a pressure of 1 atm, this gas has a dielectric strength about 2.7 times that of air at the same pressure. Its dielectric strength rises approximately linearly with pressure up to a few atmospheres and at a pressure of about 7 atm it will sustain fields of about 200 kV/cm.

Radiation Dynamics manufactured many single-stage and double-stage (tandem) positive-ion Dynamitrons operating at terminal potentials up to 4 MV. They are especially advantageous for applications requiring high currents.

A larger proportion of Dynamitrons manufactured have been electron accelerators for industrial applications such as polymerization of plastics and sterilization of disposable medical products. These applications require electron energies up to a few MeV and from 10 to 100 kilowatts of electron beam power.

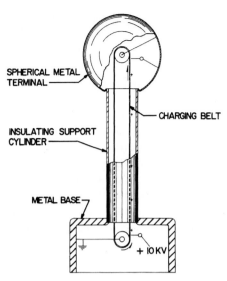

Fig. 5: Schematic drawing of a Van de Graaff generator for operation in atmospheric air.

The most powerful machines built by Radiation Dynamics include a 0.5-MV machine with an electron beam power of 50 kW, a 1-MV machine giving 100 kW of electron beam power, a 1.5-MV machine providing 75 kW, a 3-MV machine providing 150 kW, and a 5-MV machine providing electron beam power of 200 kW.

Van de Graaff Accelerator

In this accelerator, high voltage is generated by means of an insulating belt which carries charge from ground to the high-voltage terminal. Robert Van de Graaff built the first successful belt-charged high-voltage generator in 1929.

In this device, which is illustrated in Fig. 5, electrical charge is deposited on an insulating, motor-driven belt and is carried into a smooth, well-rounded metal shell which is shown in the figure as a sphere. Here charge is removed from the belt and passes to the sphere which rises in voltage until the sphere is discharged by a spark or until the charging current is balanced by a load current.

To charge the belt, a corona discharge is maintained between a series of points or a fine wire on one side of the belt and the grounded lower pulley or a well-rounded, grounded, inductor plate on the other side of the belt. If the corona needles are at a positive potential, the belt intercepts positive ions as they move from needle points toward the grounded pulley. Charge is carried into the sphere where it is removed by an array of needle points and passes to the outer surface of the sphere. The generator can provide a high negative voltage if the corona needles are operated at a negative voltage with respect to the grounded pulley.

Belt charging electrodes must be well shielded from the field of the high-voltage terminal and charge must be carried well within the sphere before removal is attempted. Charging current is then completely independent of voltage on the terminal and voltage will rise until limited by corona or spark-over to ground, by leakage current along insulators, or by a load such as ion current through an accelerating tube.

The Van de Graaff belt-type generator was first used to accelerate ions in 1932 at the Department of Terrestrial Magnetism of the Carnegie Institution of Washington, D.C. A machine completed in 1934 at this institution was used extensively for research in nuclear physics and is now on display at the Smithsonian Institute of Washington, D.C.

Open-air belt-type accelerators for 1 million volts or more are large and require a very large enclosure. At Wisconsin this accelerator was adapted to a pressure tank and by 1940 a belt-charged accelerator was operated successfully up to 4.5 million volts.

These machines insulated by high-pressure gas were manufactured from 1946 by the High Voltage Engineering Corporation for accelerating of electrons and for positive ions. In 1988, Vivirad High Voltage Corporation purchased the Accelerator Division of High Voltage Engineering Corporation to build accelerators, for research and materials analysis. High Voltage Engineering–Europa also manufactures accelerators with Van de Graaff belt-charging to about 2 million volts, primarily for ion-implantation applications.

Two-Stage Accelerators

In 1958, the High Voltage Engineering Corporation completed a machine for acceleration of negative ions as illustrated schematically in Fig. 6. Negative hydrogen ions from a source developed for this purpose are accelerated as they pass from ground to the terminal which is

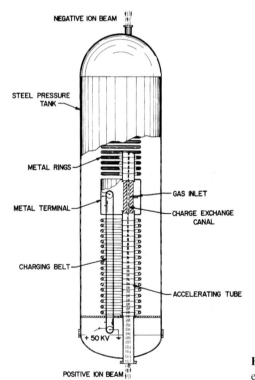

Fig. 6: Schematic drawing of a tandem electrostatic accelerator in a pressure tank.

at a high positive voltage V. Here they are stripped of both electrons as they pass through a very thin foil or through a small-diameter tube containing adequate gas. The protons are again accelerated as they pass from the terminal to ground and they emerge from the accelerator with an energy of $2Ve$. These machines, which give two stages of acceleration, are called tandems.

Atoms of a large proportion of the elements form stable negative ions. A negative oxygen ion gains an energy of 10 MeV as it goes from ground to the terminal of an accelerator operating at 10 million volts. Here the oxygen atoms may be stripped of all of their eight electrons. The oxygen nuclei gain 80 MeV as they pass to ground and they emerge from the machine with an energy of 90 MeV.

The High Voltage Engineering Corporation manufactured a large number of two-stage accelerators, which are in use in laboratories throughout the world. The largest reach terminal voltages of over 16 million volts. Many are used for acceleration of heavy ions for research and materials analysis applications.

Nisshin-High Voltage Company also manufactures tandem accelerators with Van de Graaff belt-charging to 3 million volts for basic research and materials modification and analysis.

Tandetron Accelerators

In 1980, K. H. Purser began development of the Tandetron, a line of compact tandem accelerators designed to meet the requirements for applications in materials analysis, accelerator mass spectrometry, high-energy implantation, and basic research. An MeV implanter which uses a Tandetron accelerator is shown schematically in Fig. 7.

The high-voltage generator in the Tandetron is a power supply with a parallel-driven cascaded rectifier circuit similar to that of the Dynamitron. Using silicon rectifiers and driven at high frequency (about 50 kHz), the power supply delivers several milliamperes of charging current at up to 3 million volts with high stability and negligible terminal voltage ripple. For accessibility, the high-voltage stack is mounted at right angles to the accelerator column, rather than being built into the column (compare Fig. 3). The accelerator tank contains SF_6 gas at about 500 kPa (5 atm) to insulate the accelerator and the power supply.

Through 1988, General Ionex Corporation manufactured a large number of Tandetron systems, many with specialized accessories that they developed for materials analysis and modification applications. Genus Corporation now manufactures Tandetron systems.

Pelletron^{TM} Accelerators

Pelletron accelerators use a charging chain, rather than a belt or power supply, to generate high voltage on the accelerator terminal. The chain consists of steel cylinders (pellets) joined by links of solid insulating material such as nylon (Fig. 8). The chain is intrinsically spark-protected (undamaged in test sparks at over 30 million volts). Pellet-charging current is adjustable and highly uniform, so that terminal voltages are easily maintained at very precise values with very little voltage ripple. This is advantageous for many measurements in nuclear physics and is usually required in other applications. Many belt-charged machines have been converted to Pelletron charging.

The pellets are charged inductively, so there is no contact with the charging electrodes (inductors). For a positive terminal voltage, positive charge is induced on the pellets at a

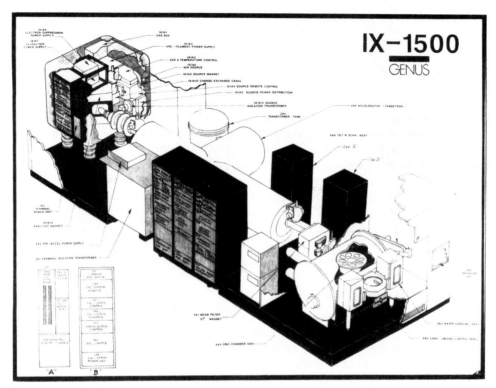

Fig. 7: Schematic diagram of a Tandetron accelerator in an ion-implant
system. The high-voltage solid-state multiplier stack is at right angles to the
tandem accelerator column, and the high-frequency driver is outside the tank
for accessibility.

motor-driven pulley at ground potential. This charge is removed at a pulley in the terminal,
and the pellets are then negatively charged. Thus, the up-going and down-going runs of the
chain contribute equally to charging. Reversing the inductor voltage polarities gives a negative
terminal voltage.

Pelletron accelerators are manufactured by the National Electrostatics Corporation. Single-
stage Pelletrons have either an electron source or a positive-ion source in the terminal. Pelletrons manufactured by this company for industries and laboratories around the world range
in voltage from 1 to 25 million volts, including the largest potential-drop accelerator, the 25-
million-volt tandem at the Oak Ridge National Laboratory (described below).

Most Pelletrons above 4 million volts are vertically oriented for simplicity of support. The
high-voltage column, which supports the terminal, charging chain, and acceleration tubes,
is an assembly of standard column modules, each module holding 1 million volts. Most
smaller Pelletrons, built for applications requiring only 4 million volts or less, use a simple,
inexpensive cast acrylic support column.

Acceleration tubes in Pelletrons have ceramic insulators and titanium electrodes bonded
together with a metal for organic-free, ultra-high-vacuum operation. This is important for

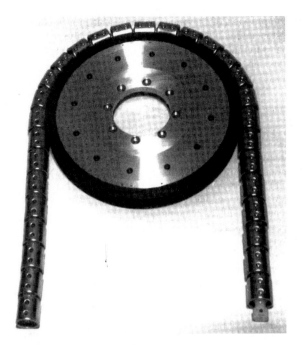

Fig. 8: Charging chain and pulley.

reliability at high voltage and for contaminant-free vacuum. Earlier tube designs have glass insulators which are bonded to the metal electrodes with organic glues.

Large Pelletrons (8 to 25 million volts) are in operation on five continents. They are used principally for basic research with heavy ions. Most Pelletrons are smaller, to 5 million volts, and are used primarily for applications such as production ion implantation, industrial materials analysis, accelerator mass spectroscopy, and biomedical analysis.

Folded Tandems

Folded tandems are like straight-through tandems (see Two-Stage Accelerators, above), except that both acceleration tubes are in a single insulating column with a 180° magnet in the terminal to steer the beam from one tube into the other. Figure 9 shows the folded 25-million-volt tandem at the Oak Ridge National Laboratory. The steel tank, which contains about 700 kPa (7 atm) of SF_6 gas for high-voltage insulation, is large (30 m high and 10 m in diameter). The straight-through design, however, would be much taller, because it has insulating columns above and below the high-voltage terminal plus an ion-beam injector above the tank (compare designs in Figs. 6 and 9).

The folded design, first successfully used by Naylor in New Zealand, reduces the costs for the tank, SF_6, and building for very high-voltage tandems. For higher-energy ion beams, Oxford University converted their single-stage accelerator to a folded tandem. General Ionex Corporation manufactured several 660-kilovolt folded tandem systems for materials analysis by Rutherford backscattering. Not only does the folded design give a compact system, but it also locates both the ion source and sample chamber near the system control panel.

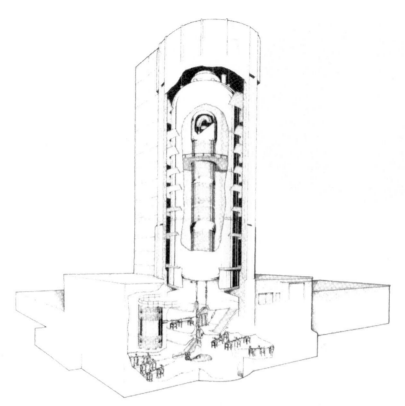

Fig. 9: Drawing of 25-MV folded tandem. Tank is approximately 30 m high and 10 m in diameter. Service platforms are shown in use inside and outside the column.

With +25 million volts on the terminal, negative ions injected from below the accelerator (see Fig. 9) are accelerated up to the terminal to an energy of 25 MeV, where they are stripped of some or all of their electrons. Stripping produces ions in a range of different charge states, and the energy gained in the second acceleration depends on which charge state is selected. If the magnet in the terminal is set for ions of charge +15, these ions gain $15 \times 25 = 375$ MeV in the second acceleration for a total energy of 400 MeV. A second stripper, one-third of the way down the second acceleration tube, can increase the charge to +34 to give ions with a final energy of over 700 MeV. The terminal magnet completely filters out all undesired components from the beam before the second acceleration, so there is no extra beam loading, and ion transmission is straightforward.

The 25-million-volt accelerator column consists of 1-million-volt modules which use the same components as in smaller Pelletrons. Six charging chains generate over 600 µA of current to the terminal. Service platforms inside and outside the column give complete access for maintenance without column disassembly. The computer-based control system uses light link telemetry to transmit over 200 control and monitoring signals to components at high-voltage locations in the accelerator column and terminal.

The ion beam injector at Oak Ridge (cylinder in lower left of Fig. 9) operates at $-500\,\text{kV}$ in air to preaccelerate ions from the negative-ion sources for injection into the accelerator. The 90° magnets below the accelerator provide high resolution of the injected ion mass and high resolution of the accelerated beam energy.

The other large folded tandem is the 20-million-volt Pelletron at the Japanese Atomic Energy Research Institute in Tokai-Mura.

Large Straight-Through Tandems

Argentina's Comisión Nacíonal de Energia Atómica operates a 20-million-volt Pelletron in their Tandar facility in Buenos Aires.

Daresbury Laboratory in England built a large tandem for scientific research. This accelerator is charged by a Laddertron, which looks like two parallel Pelletron chains with metal bars connecting the pellets of one chain to the adjacent pellets of the other chain.

At Strasbourg, M. Letournel is developing the Vivitron, a design for a very high-voltage belt-charged tandem accelerator. The design uses metal electrodes, called porticos, to modify the electric-field distribution between the accelerator column and the grounded steel tank for higher terminal voltages.

Bibliography

M. R. Cleland and P. Farrell, "Dynamitrons of the Future," *IEEE Trans. Nucl. Sci.* **NS-12**, 227 (1965).

Large Electrostatic Accelerators (D. Allan Bromley, ed.) [Reprinted from *Nuclear Instruments and Methods* **122** (1974)].

M. S. Livingston and J. P. Blewett, *Particle Accelerators*. McGraw-Hill, New York, 1962.

C. C. Thompson and M. R. Cleland, "Design Equations for Dynamitron Type Power Supplies in the Megavolt Range," *IEEE Trans. Nucl. Sci.* **NS-16**, 124 (1969).

Acoustical Measurements

M. Strasberg

Acoustics is concerned with fluctuations in the value of various mechanical quantities characterizing the state of matter – fluctuations of the pressure or other components of stress, or fluctuations of density, temperature, and the position of individual particles of matter. Primary acoustical measurements determine the magnitude and wave form of the oscillations of one of these quantities at one or more positions in space, whereas secondary measurements characterize the wavelike propagation of these oscillations through space by determining the speed of propagation, the intensity or rate of propagation of acoustic energy, and the absorption or rate of dissipation of acoustic energy.

Most of the early primary acoustical measurements used mechanical devices that would determine only the magnitude of the oscillatory particle displacement or velocity. The advent of electronic amplifiers led to development of electromechanical transducers which convert the oscillating mechanical quantities into an emf. Nowadays, primary acoustical measurements usually utilize linear electromechanical transducers to generate an oscillating emf that is instantaneously proportional to the oscillating mechanical quantity. The magnitude and wave form of the oscillating mechanical quantity are deduced from measurements of the corresponding characteristics of the emf.

Several textbooks survey acoustical measurements. Wood [1], Stephens and Bate [2], and Meyer and Neuman [3], for example, describe mechanical techniques used before the present electronic age, and the latter two texts also cover contemporary instruments. The four-volume *Encyclopedia of Acoustics* has 150 pages discussing measurements [24]. The textbook *Acoustical measurements* [4] is devoted entirely to the subject, albeit primarily to audible sound. Various ultrasonic measurement techniques are described by Fry and Dunn [5] and are discussed in several chapters in the 18-volume collection edited by Mason [6]. Handbooks published by several manufacturers discuss the use of their instruments for sound measurements [7–9]. Standard procedures and apparatus for performing certain conventional measurements have been published by organizations concerned with standardization [10].

Measurement of Oscillating Pressure

Most present-day acoustic measurements in fluids utilize electromechanical transducers sensing the oscillating pressure. These are called *microphones* when used in gases and *hydrophones* when waterproofed for use in liquids.

Various physical effects are utilized for generating the alternating emf; a comprehensive survey is given in Chapter 8 of Olson [11]. The type chosen for a particular application depends on a compromise among conflicting desirable characteristics, e. g., small size, high sensitivity, constant sensitivity over the frequency range of interest, stability to varying temperature and humidity, and low cost.

Whatever type of transducer is used, the magnitude, wave form, and spectral characteristics of the generated alternating emf are measured with electronic instruments such as voltmeters, oscilloscopes, wave analyzers, and spectrum analyzers. The corresponding characteristics of the oscillating pressure are determined from the measured characteristics of the alternating emf by dividing the electrical magnitudes by a proportionality factor (called a sensitivity or calibration factor) that is equal to the generated emf per unit sound pressure, including the amplification in the electronic system. Ideally this factor is independent of the magnitude and frequency of the oscillating quantity, but in practical systems there is usually some variation in sensitivity with frequency within the range of interest that must be taken into account.

Carbon Microphone

The carbon microphone, an early pressure-sensing transducer still used in some telephones, depends on the piezoresistive property of compressed carbon granules. The oscillating sound pressure causes fluctuations in the compression of the granules that result in a fluctuating resistance that is detected electrically as an alternating potential difference developed by a dc current through the granules. The device has a high, albeit unstable, sensitivity; its useful frequency range covers from about 250 to 4000 Hz.

Electromagnetic Microphone

Sometimes called *dynamic* or *moving-coil* microphones, electromagnetic microphones comprise a light coil of wire located in the field of a permanent magnet and attached to a thin diaphragm that vibrates in response to the sound pressure, so that an alternating emf is generated by electromagnetic induction. The vibrating system is designed so that the generated emf is instantaneously proportional to the oscillating pressure. The frequency range of relatively constant sensitivity can extend from perhaps 100 to 10000 Hz.

Another type of electrodynamic microphone utilizes a limp metallic ribbon instead of the coil. If both sides of the ribbon are exposed to the sound, the ribbon tends to generate an alternating emf proportional to the particle velocity of the gas, and for this reason it is sometimes called a velocity microphone, but it is really responding to the gradient of the oscillating pressure.

Piezoelectric Microphones and Hydrophones

The piezoelectric microphones and hydrophones contain piezoelectric elements that develop an emf in response to mechanical stress. In the early days, thin disks cut from natural or synthetic single crystals were used. Rochelle salt crystals were used because of their high sensitivity, but they deteriorated easily. Lithium sulfate crystals were more stable but less sensitive. Disks cut from natural quartz and tourmaline were used if stability and ruggedness were important, particularly for underwater applications, but they were relatively insensitive. Most present-day piezoelectric microphones and hydrophones utilize polarized ceramic piezoelectric elements of barium titanate, lead zirconate, or others, which have relatively high sensitivity and are available in various shapes, e. g., cylinders and spheres as well as disks (see Chapter 3 in Vol. 1A of the Mason collection [6]). Thin sheets of polarized piezoelectric plastics, such as polyvinyladine fluoride, which can be adhered to curved surfaces, are also available.

Most hydrophones use piezoelectric elements that can withstand the high static pressures existing at deep submergences. The hydrophone dimensions can vary from small 1-mm cylinders, used as "probe" hydrophones, to 10-cm (or larger) units. Depending on the design, the useful frequency range can extend down to a few hertz and up to 1 MHz. Bobber [12] presents a detailed discussion of underwater acoustic measurements.

Capacitance or Electrostatic Microphones

Capacitance or electrostatic microphones comprise a small variable capacitance consisting of a metallic stretched membrane or thin diaphragm, separated from a rigid back plate by an air gap perhaps 10^{-3} cm thick. The oscillating sound pressure results in oscillating deflections of the diaphragm, which change the width of the air gap and the associated capacitance. The variations in capacitance are usually detected by placing a fixed electric charge onto the electrodes of the capacitance through a high-resistance leak from a source of several hundred volts dc, so that variations in capacitance result in a varying potential difference across the capacitor. For very low-frequency measurements, the capacitance can be placed in one arm of an ac bridge so that the capacitance variations result in an amplitude-modulated carrier output from the bridge; alternatively, the varying capacitance can be used to control the frequency generated by an oscillator and thus provide a frequency-modulated carrier.

To eliminate the need for dc polarization, electret capacitance microphones have been developed utilizing polarized electrets holding a permanent charge on the metal electrodes.

Calibration of Microphones and Hydrophones

The ratio of the generated emf to the oscillating pressure generating the emf is called the *sensitivity* of the microphone. The sensitivity is determined by a procedure called *calibration*. Most microphones are calibrated by comparing their generated emf to that of another primary standard microphone subjected to the same sound pressure at various frequencies.

Microphones used as primary standards can be calibrated by one of several absolute methods. The method usually used nowadays is called a *reciprocity calibration*. This procedure involves two sets of measurements, the first being a comparison of the emf generated by the microphone with that of a second microphone when both are responding to the same sound source, and the second being a measurement of the emf of the microphone when the second microphone is itself used as a source (see Chapter 4.2 of Beranek [4]).

Another absolute calibration uses a Rayleigh disk, to be described subsequently, to determine the oscillating particle velocity. A third method uses an oscillating piston driven by a rotating cam and forming one wall of a small gas cavity; the known volume change is used to calculate the oscillating pressure, taking into account departure from the adiabatic gas law because of heat conduction at the walls of the cavity. Microphones having flat diaphragms can be calibrated by placing an auxiliary electrode close to the diaphragm and applying an alternating potential difference between this electrode and the diaphragm so as to develop a calculable alternating electrostatic pressure on the diaphragm.

Hydrophones are calibrated by methods generally the same as those used for microphones. An absolute method suitable only for hydrophones is to hold the hydrophone in a vibrating container of liquid; the vibration results in an oscillating pressure that can be calculated in terms of the measured acceleration of the container [13].

Measurement of Density and Temperature Oscillations

Measurements of fluctuating density are usually done optically, utilizing variations in index of refraction associated with the density variations. The Debye–Sears apparatus is suitable for plane waves of sound, the spatially periodic density variations acting as an optical diffraction grating. The diffraction angles depend on the wavelength of the sound, and the intensity of the diffracted light depends on the magnitude of the oscillating refraction index and density (see Sect. 6.81 of Meyer and Neumann [3]). Optical holographic techniques with laser light sources are now being developed to indicate density oscillations. All these optical methods require relatively large-amplitude sounds and are useful only at high frequencies.

Temperature fluctuations can be observed with small probe wires whose electrical resistance fluctuates with the temperature fluctuations [14].

Measurements of Oscillating Particle Motion

Direct Visual Observation

The oscillating motion of fluid particles associated with ordinary sound is usually too small to be observed visually, even with a microscope. For example, a sound in air having a frequency of 1000 Hz and an oscillating rms sound pressure of $1 \, N/m^2$ (94 dB re 20 µPa) may be loud enough to cause some damage to hearing, but the oscillatory particle displacement is only about 3.7×10^{-5} cm. However, the oscillatory motion can be observed if the sound is strong enough and the frequency low enough. Photographs through a microscope showing streaks of smoke particles oscillating in an intense sound field are shown on p. 292 of Stephens and Bate [2].

Rayleigh Disk

The Rayleigh disk is an early device used for absolute measurement of the oscillating particle velocity associated with sound in gases. A Rayleigh disk consists of a small thin disk, perhaps 1 cm in diameter, suspended by a fine fiber (see [1–3] or Sec. 4.3A of Beranek [4]). The oscillating particle velocity results in a steady torque on the disk, proportional to the mean-square velocity, tending to turn the disk broadside to the direction of the oscillating velocity. The suspension fiber is used as a balance to indicate the magnitude of the steady torque. Since the torque has been calculated theoretically, the Rayleigh disk permits an absolute measurement of the particle velocity. It is used nowadays mainly to provide an absolute calibration of microphones. A difficulty is that the disc must be shielded from even the slightest steady wind.

Electronic Measurement of Fluid Particle Velocity

The oscillating particle velocity can be measured using the small-amplitude relation between the instantaneous pressure gradient and particle acceleration, viz., the particle acceleration is equal to the negative of the pressure gradient divided by the fluid density. The pressure gradient can be determined using two closely spaced pressure-sensing microphones or hydrophones, spaced much more closely than the wavelength of sound at the highest frequency of interest. The pressure gradient is equal to the instantaneous difference in the two sensed pressures divided by their separation (see Sec. 5.2.2 of Fahy [21]). Since both individual pressures usually are much larger than their difference, the two sensors must be accurately matched in both amplitude and phase response.

In a liquid, the particle velocity can be determined with a single vibration sensor attached to the interior wall of a small, rigid, and waterproof spherical or cylindrical shell immersed in the medium. If the shell is neutrally buoyant and much smaller than the wavelength of sound at the highest frequency of interest, its instantaneous vibration duplicates the vibration of the liquid it displaces (see Sec. 5.12.1 of Bobber [12]). The shell vibration is sensed by the attached vibration pickup. If an accelerometer is used, its electrical output can be converted from instantaneous acceleration to velocity by passing its electrical signal through an integrating circuit; if the acceleration spectral density is calculated digitally, the acceleration density can be converted to velocity by dividing each value by the square of its radian frequency.

Electromechanical Pickup of Surface Vibration

The most common method for measuring the vibration of a solid is to mount an electromechanical transducer on the surface to convert the vibratory motion into an alternating emf. These transducers are called displacement pickups, velocity pickups, or accelerometers, depending on whether they generate emfs instantaneously proportional to the vibratory displacement or to its first or second time derivatives [15, 16]. Their sizes range from units smaller than 1 cm and with masses of only a few grams, to units having dimensions of 5 cm or more. Although the larger units are more sensitive, their large mass can perturb the vibration they are intended to measure.

Displacement pickups are used for measuring relatively low-frequency vibrations covering the frequency range from zero to a few hundred hertz. Velocity pickups are usually electromagnetic and cover the frequency range from perhaps 5 to 1000 Hz. Accelerometers for measuring very low-frequency accelerations utilize active transducers, whereas measurements in the frequency range 10 to 10000 Hz and higher utilize piezoelectric elements. The sensitivity of vibration pickups can be determined in various ways, including an absolute reciprocity calibration (see Chapter 18 of Harris and Crede (15)).

Variable magnetic reluctance pickups are available which sense the vibratory velocity of a magnetic surface without any part of the pickup being in contact with the surface. Conventional phonograph pickups can be used as low-mass vibration pickups. If the two stereophonic outputs are combined in phase, the pickup is sensitive to vibrations parallel to the axis of the pickup stylus, whereas if combined out of phase, the pickup is sensitive to vibrations perpendicular to the stylus axis. The sensitivity of phonograph pickups can be determined as is done with other vibration pickups, or with a phonograph test record as described in Sec. 10.5 of Olson [11].

Piezoresistive Strain Gauges

The development of semiconductor strain gauges, which are some 50 times more sensitive than the older resistance-wire gauges, has made it possible to measure oscillating strains at the surface of a solid associated with sounds of ordinary magnitude. The fluctuating resistance associated with the oscillating strain can be measured electrically in various ways (see Vaughan [17] or Chapter 17 of Harris and Crede [15]).

Fiber Optics

Fine optical fibers can be used to measure oscillatory strain and displacement. Minute oscillations in the optical path length through the fiber can be sensed as oscillations in the phase of monochromatic light transmitted through the fiber. Various methods for converting the oscillatory phase into an electrical signal are available. To measure acoustic pressure, the optical fiber is bonded to a diaphragm so as to sense the strain in the diaphragm caused by acoustic pressure acting on it (see Chapter 7 of Vol. 16 of Mason [6]).

Optical Measurements of Surface Vibration

Optical interferometry has been used to determine the magnitude of the vibratory displacement of the surface of a solid or liquid. The laser vibrometer splits a beam of monochromatic light into two beams, one of which is reflected off the vibrating surface and then recombined with the other beam. Motion of the reflecting surface changes the path length and causes the phase of the light beams to change relative to each other, so that the intensity of the recombined beam oscillates in response to vibrations of the reflecting surface. The method can be used to measure the vibration of a small region of a surface [18] as well as to display the distribution of vibration amplitude over a large surface vibrating in a complicated vibration pattern. If a laser is used as the light source, holographic reconstructions of the vibration pattern are possible (see Sec. 6.3.1 of Meyer and Neumann [3], also Brown [25]). A laser-Doppler vibrometer senses the Doppler shift in the frequency of the light reflected by the vibrating surface to indicate its velocity of vibration.

Miscellaneous

Other methods for observing and measuring oscillatory displacements are used occasionally. Sensitive flames provided one of the early methods for detecting the oscillating particle velocity associated with sound in gases (see p. 409B of Wood [1]). Oscillating strains in the interior of a photoelastic solid can be measured using polarized light [19]. The hot wire, involving an electrically heated fine wire whose temperature and resistance fluctuate in response to the cooling effect of a fluctuating flow past the wire, is the conventional device for measuring the fluctuating velocity in turbulent flows in gases; it can also be used to measure the oscillatory particle velocity associated with sound (see p. 440 of Wood [1]). Fluctuating electrochemical potentials detected by a pair of probes inserted in an electrolyte have been used to measure the oscillating particle velocity associated with sound (see p. 439 of Wood [1]). Finally, mention should be made of the use of sand grains or small ball bearings to indicate when the vertical oscillatory acceleration of a solid surface exceeds $1g$ (see p. 210 of Meyer and Neumann [3]); or of the use of sand to provide a visual display of a complicated pattern of surface vibration (see discussion of Chladni figures on p. 172 of Wood [1]). 1

Secondary Measurements

The velocity of propagation of sound can be determined by direct measurement of the transit time of a transient sound, or by observing various phenomena that depend on the wavelength of continuous sounds of constant frequency. These methods are discussed in Chapters V and IX of Herzfeld and Litovitz [20]; also methods for measuring acoustic absorption.

Acoustic Intensity

Measurements of the distribution of acoustic intensity around a sound source have become increasingly commonplace since the advent of digital spectrum analyzers. The intensity is a vector quantity whose component in a specified direction is the acoustic power passing through unit cross section perpendicular to the specified direction, this being equal to the time average of the product of the instantaneous sound pressure and the instantaneous particle velocity component in that direction.

For frequencies below 10 kHz, the intensity can be measured using an intensity probe consisting of two small microphones spaced a small fraction of an acoustic wavelength apart. The intensity can be shown to be proportional to the time average of the instantaneous product of the sum of the sound pressures at the two microphone positions multiplied by the difference between the two pressures. An alternative procedure, especially convenient if a two-channel digital spectrum analyzer is available, is to measure the complex cross-spectral density of the sound pressures at the two-microphone positions; the spectral density of the intensity at any frequency can be shown to be proportional to the imaginary part of the cross-spectral density of the two pressures at that frequency (see Fahy [21]). Although the sum-and-difference method may be used for measurements covering any band of frequencies, the cross-spectral technique is suitable only for narrow bands.

Since the two microphones of intensity probes must be much less than a wavelength apart, they are not suitable for measurements at frequencies above about 10 kHz. For ultrasonic frequencies, radiometers are used which sense the radiation pressure on a small disc in the sound field; the force on the disc is proportional to the acoustic intensity (see Fry and Dunn [5], Sec. 6.7 of Beyer [22], or Sec. 2.8 of Bobber [12]).

Acoustical Holography

Acoustical holography, as distinct from acoustical imaging, is used to determine the distribution of sound pressure or vibration velocity over the surface of a sound source. The procedure involves measurements of the magnitude and phase of the sound pressure at many positions in the sound field and summation of these pressures with appropriate weighting functions to reconstruct the pressure or velocity distribution over the source surface. Only the radiating or nonevanescent portion of the pressure or velocity distribution can be determined from measurements made several wavelengths removed from the source; but the entire distribution can be reconstructed from measurements in the "near field" (see, e. g., Maynard [23]).

See also: Acoustics; Piezoelectric Effect; Sound, Underwater; Transducers; Waves.

References

[1]   A. B. Wood, *A Textbook of Sound*. G. Bell and Sons Ltd., London, 1957.

[2]   R. W. B. Stephens and A. E. Bate, *Acoustics and Vibrational Physics*. Edward Arnold Ltd., London, 1966.

[3]   E. Meyer and E. G. Neumann, *Physical and Applied Acoustics*. Academic Press, New York, 1972.

[4]   L. Beranek, *Acoustical Measurements*. Acoustical Society of America, New York, 1989.

[5]   W. J. Fry and F. Dunn, "Ultrasound: Analytical and Experimental Methods in Biological Research," in *Physical Techniques in Biological Research* (W. L. Nastuk, ed.), Vol. 4, Chapter 6. Academic Press, New York, 1962.

[6]   W. P. Mason, ed., *Physical Acoustics: Principles and Methods*, 18 vols. Academic Press, New York, 1964–1988.

[7]   A. P. G. Peterson and E. E. Gross, Jr., *Handbook of Noise Measurement*. General Radio Co., Concord, MA, 1972.

[8]   J. T. Brock, *Application of B&K Equipment to Acoustic Noise Measurements*. Bruel and Kjaer, Naerum, Denmark, 1971.

[9]   *Acoustic Handbook*, Application Note 100. Hewlett Packard Co., Palo Alto, CA.

[10]   *ASA National Standards Catalog 26-2004*. Acoustical Society of America, New York, 2004.

[11]   H. F. Olson, *Acoustical Engineering*. Van Nostrand-Reinhold, Princeton, NJ, 1957.

[12]   R. J. Bobber, *Underwater Electroacoustic Measurements*. U.S. Government Printing Office, Washington, DC, 1970.

[13]   F. Schloss and M. Strasberg, *J. Acoust. Soc. Am.* **34**, 1958 (1962).

[14]   J. Hojstrup, K. Rasmussen, and S. E. Larsen, "Dynamic Calibration of Temperature Wires in Still Air," published in *DISA Information*, No. 20, DISA Electronics, Franklin Lakes, NJ, 1975.

[15]   C. M. Harris and C. F. Crede, eds., *Shock and Vibration Handbook*. McGraw-Hill, New York, 1976.

[16]   J. T. Brock, *Application of B&K Equipment to Mechanical Vibration and Shock Measurement*. Bruel and Kjaer, Denmark, 1972.

[17]   J. Vaughan, *Application of B&K Equipment to Strain Measurement*. Bruel and Kjaer, Denmark, 1975.

[18]   F. J. Eberhart and F. A. Andrews, *J. Acoust. Soc. Am.* **48**, 603 (1970).

[19]   R. C. Dove and P. H. Adams, *Experimental Stress Analysis and Motion Measurement*. Charles E. Merrill, Columbus, Ohio, 1964.

[20]   K. F. Herzfeld and T. A. Litovitz, *Absorption and Dispersion of Ultrasonic Waves*. Academic Press, New York, 1959.

[21]   F. J. Fahy, *Sound Intensity*. Elsevier, New York, 1989.

[22]   R. T. Beyer, *Nonlinear Acoustics*. U.S. Government Printing Office, Washington, 1974.

[23]   J. D. Maynard, E. G. Williams, and Y. Lee, *J. Acoust. Soc. Am.* **78**, 1395 (1985).

[24]   M. J. Crocker (ed.), "Acoustical Measurements" in *Encyclopedia of Acoustics*, Vol. 4, Part XVII. Wiley, New York, 1997.

[25]   G. M. Brown *et al.*, *J. Acoust. Soc. Am.* **45**, 1166 (1969).

Acoustics

R. T. Beyer

Acoustics is the science of sound. What is sound? When you open your mouth and utter speech you are said to produce a sound. Under normal conditions a nearby person with so-called normal hearing says that he hears the sound. In its study of the production and reception of sound and its transmission through material media, acoustics is a branch of physics, though speech and hearing obviously involve biological elements.

Let us examine the physics of what happens when a person speaks. A disturbance is produced in the air in front of the mouth, involving a slight compression of the air or, alternatively, an increase in the air pressure of the order of $0.1 \, \text{N}/\text{m}^2$. This amount is about one-millionth of the normal atmospheric pressure. Since air is an elastic medium, it does not stay compressed but tends to expand again and hence produces a disturbance in the neighboring air. This in turn passes on the disturbance to the air adjoining it, and the result is a pressure fluctuation that moves through the air in the form of a sound wave. When the wave reaches the ear of an observer, it produces a motion of the eardrum, which in turn moves the little bones in the middle ear, and this movement communicates motion to the hair cells in the cochlea in the inner ear. The ultimate result, by a rather complicated biophysical process that is not yet completely understood, is the hearing of the sound.

While acoustics, as defined above, applies only to audible sound in air, its scope has been gradually expanded until today it encompasses mechanical waves and vibrations in all material media – solids, liquids, and gases, and includes waves of any frequency, as well as aperiodic disturbances, such as shocks and noise. A perusal of the pages of the *Journal of the Acoustical Society of America* will easily demonstrate that this is the case.

It is convenient to divide this discussion of acoustics into three parts: the production, the transmission, and the reception of sound.

Production

Any change in stress or pressure leading to a local change in density or displacement from equilibrium in an elastic material medium can serve as a source of sound. We have already mentioned the human vocal mechanism as an example. All sound sources in practical use involve the vibrations of solids, liquids, or gases. Such a vibration is an oscillation of stress or pressure with a definite frequency. The unit of frequency used in acoustics is the hertz

(Hz), which is one complete cycle per second. The standard musical instruments are common sources of sound of more or less definite frequency. In sophisticated scientific and technological applications the sound source is called a transducer. Those in standard use are electroacoustic in character, that is, they depend on electrical action to produce mechanical vibrations. An example is the electrodynamic loudspeaker, in which an alternating electric current in a coil of wire placed in a magnetic field produces oscillatory motion in the coil. This motion is communicated to a membrane, whose resulting vibrations are radiated as sound. Another commonly used electroacoustic transducer is based on the piezoelectric effect, according to which certain crystals (e. g., quartz) can be made to vibrate when placed in an oscillating electric field.

Certain ceramic materials, such as barium titanate and lead zirconate, can be polarized by the use of an applied electric field and can thereafter serve as transducers in the same way as that described for quartz. In addition, the change in shape and dimensions of a piece of magnetic material, like nickel, when placed in a magnetic field, can also be used in the construction of transducers. This is known as the magnetostrictive effect. All these types of transducers are useful, not only in the production of sound but also in its reception.

The rapid flow of air over a rough surface or through a nozzle produces sound that can vary over a wide range of frequency and intensity. An example is the aerodynamic sound from a jet engine on an airplane. Sounds of this character are commonly known as noise, one of the chief problems of environmental acoustics in the late twentieth century.

An exciting new field has been the optoacoustic generation of sound by the thermoacoustic effect of absorption of the energy of a pulsed light source (laser beam) by a medium, which thereupon emits acoustic pulses. It is of interest to note that this effect was first observed by Alexander Graham Bell in 1881 (without the use of a laser!).

Transmission

Sound demands a material medium for its transmission from place to place. As previously mentioned, the propagation takes place by means of wave motion, a good visual example of which is provided by a wave on the surface of water. A sound wave travels with a definite velocity, depending on the type of wave, the physical nature of the medium, and the temperature. Through air at standard room temperature (20 °C) sound travels with the velocity 344 m/s, a value that increases as the temperature is raised. Under similar conditions, sound travels through water with a velocity somewhat more than four times as great. The velocity of sound in a highly elastic solid like steel is even greater, being as high as 6000 m/s.

A sound wave represents the transmission of mechanical energy through the medium in which the sound travels. The measure of this transmission is called the intensity of the sound wave. It is defined as the average flow of energy per unit time through a unit area of the surface through which the sound passes (with direction normal to the surface). The strictly scientific unit of sound intensity is the watt per square meter. In practice, however, this unit is replaced by a system in which 10 times the logarithm to the base 10 of the ratio of the given intensity to that corresponding to minimum audibility is taken as the number of decibels (abbreviated dB) represented by the sound in question. Ordinary conversational speech at a distance of 1 m from the speaker has an intensity of about 60 dB, whereas the intensity in the neighborhood of a jet airplane with the engine running can be as high as 140 dB. Sound of intensity in excess of

90 dB at the human ear can be harmful to that delicate and valuable organ of hearing. Sounds of very high intensity are called macrosonic and form the subject of what is termed nonlinear acoustics.

As sound spreads out from a source, its intensity decreases with distance. For a highly localized source from which the sound travels in all directions, the intensity varies inversely as the square of the distance from the source. Sound intensity also falls off with distance traveled through a process known as absorption, a dissipative action in which the energy being transmitted by the sound wave is gradually converted into heat. When a sound wave strikes an obstacle, it undergoes reflection and refraction, following the same laws as those that hold for light waves. The acoustic echo is a well-known phenomenon.

The simplest type of sound wave is the periodic one, in which at any given point in the medium being traversed by the wave the disturbance (e. g., the excess pressure for a wave in air) varies periodically between a minimum and a maximum value and back again a certain number of times a second. As indicated earlier in the case of sound source vibrations, this number is called the frequency of the wave and is measured in hertz. Another important quantity characterizing a periodic sound wave is the wavelength, or the distance between successive points in a wave train at which the disturbance has the same magnitude and is doing the same thing (i. e., is either increasing or decreasing in magnitude). The wavelength is related to the frequency by the simple but fundamental relation that the wavelength is equal to the velocity divided by the frequency. For a given velocity, a long wavelength means a low frequency and vice versa.

Reception

The frequency of a periodic sound wave determines, in a rather complicated way, the pitch at which it is heard. High pitch corresponds to high frequency. Sounds of frequency below 20 Hz are not normally heard by human beings even at very high intensity. These sound waves are called infrasonic. The upper frequency limit of audible sound varies markedly. For young people, this upper limit is at about 20 000 Hz. For older adults, this figure drops to 10 000 Hz or even lower. This phenomenon is known as presbycusis. Frequencies above the audible range are known as ultrasonic; they can be generated and detected into the gigahertz range (10^9 Hz). Ultrasound has many practical applications in such areas as sound signaling, metallurgy, and medicine.

The most important receiver of audible sound in human experience is, of course, the ear, a marvelously sensitive mechanism that can normally detect sound of intensity as low as 10^{-12} W/m^2 and can stand intensity as high as 1 W/m^2 before pain ensues and possible ear damage develops. The normal ear is most sensitive at around 2000 Hz. Loudness, though of course related to intensity in such a way that it increases with intensity, is a subjective quantity connected with the biological response of the listener. The unit of loudness developed by the psychophysicists is the sone, defined as the loudness produced by a tone of 1000 Hz at 40 dB above the minimum audible threshold. A loudness scale in sones has been established by the statistical study of the hearing of a large number of individuals.

The principal artificial sound receiver in common use is the microphone, of which there are many varieties, mainly based on the electroacoustic effects used for transducers in general, as mentioned earlier. Their widespread employment in the audio industry – radio, television, public address systems, sound recording and reproduction, hearing aids, etc. – is well known.

More detailed information about the topics mentioned here involving the application of acoustics to fields like architectural acoustics, ultrasonics, vibration problems, and underwater sound, can be found in the relevant articles in the encyclopedia.

See also: Acoustical Measurements; Acoustics, Architectural; Acoustics, Linear and Nonlinear; Acoustics, Physiological; Acoustoelectric Effect; Musical Instruments; Noise, Acoustical; Sound, Underwater; Ultrasonic Biophysics; Ultrasonics; Vibrations, Mechanical; Waves.

Bibliography

L. L. Beranek, *Acoustics*. Reprint of the 1954 edition, with revisions. Acoustical Society of America, New York, 1986. (I)

F. V. Hunt, *Acoustics*. Reprint of the 1954 edition. Acoustical Society of America, New York, 1982. (I)

L. E. Kinsler, A. R. Frey, A. B. Coppens, and J. V. Sanders, *Fundamentals of Acoustics*, 3rd ed., Wiley, New York, 1982. (I)

James Lighthill, *Waves in Fluids*. Cambridge University Press, Cambridge, 1978. (A)

R. B. Lindsay, ed., *Acoustics: Historical and Philosophical Development*. Dowden, Hutchinson k Ross, Stroudsburg, PA., 1973. (E)

Iain G. Main, *Vibrations and Waves in Physics*. 2nd ed. Cambridge University Press, Cambridge, 1984. (I)

P. M. Morse and K. U. Ingard, *Theoretical Acoustics*. Reprint of the 1968 edition. Princeton University Press, Princeton, 1986. (A)

A. D. Pierce, *Acoustics, An Introduction to its Physical Principles and Applications*. 1981 text reprinted by the Acoustical Society of America, New York, 1989. (I)

J. R. Pierce, *Almost All About Waves*. MIT Press. Cambridge, MA, 1974. (E)

R. W. B. Stephens and A. E. Bate, *Acoustics and Vibrational Physics*, 2nd ed. St. Martin's Press, New York, 1966. (I)

J. W. Strutt (Lord Rayleigh). *The Theory of Sound*. 1877 (reprinted by Dover Publications, New York, 1945). (A)

For information on current progress in acoustical research the *Journal of the Acoustical Society of America* may be consulted. This is published monthly by the American Institute of Physics for the Acoustical Society of America.

Acoustics, Architectural

T. D. Northwood

Architectural acoustics may be defined as the science of acoustics applied to the design of buildings. It thus derives from such diverse disciplines as physics, psychology, the arts, architecture, and engineering, the techniques needed to produce acoustical environments acceptable to the building occupants. The objective is twofold: to bring to the occupants the sounds they wish to hear, and to protect them from the sounds they do not wish to hear. The first of these tasks is the more attractive and creative one, but the exercise will be successful only if the second task is handled with equal care.

The profession of architectural acoustics may be said to have been inaugurated by Wallace Clement Sabine (1868–1919), a professor of physics at Harvard University, who in 1895 was assigned the task of curing the acoustical defects of a new lecture theater at the university. In solving that problem he went on to evolve the first quantitative theory of reverberation processes in rooms, and applied it to the solution of problems in many theaters and concert halls, the most famous of which was Symphony Hall in Boston.

For studies of these rooms the instrumentation available to Sabine consisted of a set of organ pipes, a stop watch, and his two ears. He used these with great ingenuity to determine what is now known as the "reverberation time" of a room and, by extension, the "sound absorption coefficients" of typical room surfaces. From his collected data for many rooms he was able to develop the Sabine reverberation theory that is still used by acousticians everywhere.

Another technique first developed by Sabine was the use of two-dimensional models of complex surfaces, utilizing a spark technique and schlieren photography to observe the progression of waves in such models. Today, 80 years later, the same topics are dealt with in more sophisticated ways, but there is scarcely any problem in architectural acoustics on which Sabine did not make a useful contribution.

Most of the sounds of interest in architectural acoustics, such as speech or music, consist of a sequence of transient impulses, and these brief transients constitute the "message." Following the progression of one of these sounds in a room, we can identify three phases: first, the direct transmission of sound through the air from source to listener; then the first reflections from the various room surfaces; and finally, the reverberant field, composed of multiple reflections from the room surfaces.

The acoustical design of a room involves a consideration of these three phases. The direct transmission does not carry much energy very far, but it is an important reference, establishing the location of the source and the timing of the sequence of sounds. The first-order reflections, if they arrive soon enough, provide useful reinforcement of the direct transmission. Strong delayed reflections (echoes) tend to garble the sequence of transient sounds and thus interfere with the perception of speech. Even without forming discrete echoes, the ensemble of multiple reflections forms a persistent reverberation that also can interfere with perceptions of the original sound sequences.

In halls used for speech the design objective is to shape the room surfaces so as to provide short-delay reflections that reinforce the direct sounds and thus extend the range for perception of intelligible speech. For one-way communication the range can be further extended by electroacoustic reinforcement. At the same time the reverberation time must be limited so that it does not blur the sequence of sounds. For musical performances, on the other hand, the hall plays a more active role. A certain optimal amount of reverberation provides a desirable blending and smoothing of musical sounds. In addition, the presence of early reflections, reaching the listeners especially from lateral directions, are essential to the listeners' impression of the space enveloping them and the performers. For the performers an additional requirement is that each should hear his own part in relation to the whole and to what is heard in the hall. To achieve this sort of ambience for every listener and performer and every kind of music is a delicate task, which becomes increasingly difficult as halls and audiences become larger.

The reduction of unwanted or disturbing noises in an enclosure involves the opposite sort of techniques: surfaces are designed to absorb rather than reflect sounds, so that the first-order reflections and the reverberant portions of the sound are made negligible. The direct sound is reduced by interposing partitions or screens between source and listener, or by a partial or complete enclosure around the source. These principles apply with minor variations to all large spaces containing noise sources: factory areas, open-plan offices, restaurants, air terminals, and so on. In many instances the noise may be intrusive speech, for which the annoyance increases rapidly with intelligibility. Reduction of intelligibility can be accomplished either by reducing the level of the transmitted speech or by increasing the level of "background" noise. The latter sometimes takes the form of background music, which is deemed a lesser evil than the intrusive speech. This question of detection of intrusive noises in the presence of acceptable "background" noise is implicit in most noise control problems.

Sound insulation between rooms is mainly a function of the separating wall or floor. In the design of such partition elements it is usual to distinguish between airborne sounds, which travel through the air to reach the partition, and structure-borne sounds, which begin as vibrations in the structure itself. With respect to airborne sounds the most important virtues of a simple partition are that it be impermeable and heavy. For a given total weight, however, it is found more effective to use a multiplicity of relatively independent layers.

The sound transmission loss of a partition increases systematically with frequency in the incident sound, and this must be taken into account in arriving at a representative performance rating for partitions. For sounds such as speech and music (live or by way of radio or television) the customary figure of merit is the sound transmission class (STC), which emphasizes the importance of middle- and high-frequency components.

With respect to structure-borne sounds, such as footsteps and machinery vibrations, the first requirement is structural discontinuity between the vibrating surface and the contiguous structure. A floor, for example, may be composed of a finished floor panel separated from the main structural slab by a soft layer. In the case of machinery it is usual to provide an individually designed mounting tuned to filter out the driving frequency of the machine. Other problems characteristic of modern buildings, especially office buildings, are noise propagation in ventilation ducts and in continuous plenum spaces over suspended acoustical ceilings, and noise produced by plumbing appliances. These are but special cases of the noise mechanisms already described.

See also: Noise, Acoustical.

Bibliography

L. L. Beranek, *Music, Acoustics and Architecture*. Wiley, New York, 1962. (I)

L. Cremer, H. A. Muller, and T. J. Schultz (translator, English edition), *Principles and Applications of Room Acoustics*. Applied Science Publishers, 1982.

Vern O. Knudsen, and Cyril M. Harris, *Acoustical Designing in Architecture*. Wiley, New York, 1950. (E)

Heinrich Kuttruff, *Room Acoustics*. Taylor & Francis, New York, 2000. (A)

Anita Lawrence, *Architectural Acoustics*. Elsevier, Amsterdam, 1970. (I)

T. D. Northwood, *Architectural Acoustics, Benchmark Papers in Acoustics, Vol. 10*. Dowden, Hutchinson & Ross, Stroudsburg, PA, 1977.

W. C. Sabine, *Collected Papers on Acoustics*. Peninsula, 1993. (E)

Acoustics, Linear and Nonlinear

J. E. Greenspon

Definitions and Nomenclature

Linear acoustics is the study of sounds of relatively small amplitude. Nonlinear acoustics is the study of sounds of relatively large amplitude. The physical phenomena associated with what is called relatively small and what is termed relatively large are, in part, the topics to be discussed in this article. The categories of linear and nonlinear acoustics have to be discussed independently of architectural acoustics, underwater acoustics, physical acoustics, and musical acoustics since linear and nonlinear refers to an amplitude characteristic of the sound whereas the other categories are associated with the acoustic environment and the mechanisms causing the sound. The study of linear acoustics is sometimes referred to as the study of infinitesimally small amplitude waves or just the study of ordinary sound waves.

Nonlinear acoustics is sometimes referred to as finite-amplitude acoustics, high-intensity acoustics, or macrosonics. There are many problems in fluid mechanics which are nonlinear because of the nature of the mathematics involved in their solution. Those physical problems in fluid mechanics which are primarily concerned with the sounds produced are acoustic problems. From a mathematical standpoint the problems involving the classical linear partial differential equation of sound waves (which will be discussed later in this article) are termed linear acoustic problems whereas the ones involving nonlinear partial differential equations (some examples being discussed later) are nonlinear acoustic problems.

Physical Phenomena in Acoustics

Sound Propagation in Air, Water, and Solids

Many practical problems are associated with the propagation of sound waves in air or water. Sound does not propagate in free space but must have a dense medium to propagate. Thus, for example, when a sound wave is produced by a voice, the air particles in front of the mouth are vibrated, and this vibration, in turn, produces a disturbance in the adjacent air particles, and so on.

If the wave travels in the same direction as the particles are being moved, it is called a longitudinal wave. This same phenomenon occurs whether the medium is air, water, or a solid. If the wave is moving perpendicular to the moving particles, it is called a transverse wave.

The rate at which a sound wave thins out, or attenuates, depends to a large extent on the medium through which it is propagating. For example, sound attenuates more rapidly in air than in water, which is the reason that sonar is used more extensively under water than in air. Conversely, radar (electromagnetic energy) attenuates much less in air than in water, so that it is more useful as a communication tool in air.

Sound waves travel in solid or fluid materials by elastic deformation of the material, which is called an elastic wave. In air (below a frequency of 20 kHz) and in water, a sound wave travels at constant speed without its shape being distorted. In solid material, the velocity of the wave changes, and the disturbance changes shape as it travels. This phenomenon in solids is called dispersion. Air and water are for the most part nondispersive media, whereas most solids are dispersive media.

Reflection, Refraction, Diffraction, Interference, and Scattering

Sound propagates undisturbed in a nondispersive medium until it reaches some obstacle. The obstacle, which can be a density change in the medium or a physical object, distorts the sound wave in various ways. (It is interesting to note that sound and light have many propagation characteristics in common: The phenomena of reflection, refraction, diffraction, interference, and scattering for sound are very similar to the phenomena for light.)

Reflection. When sound impinges on a rigid or elastic obstacle, part of it bounces off the obstacle, a characteristic that is called reflection. The reflection of sound back toward its source is called an echo. Echoes are used in sonar to locate objects under water. Most people have experienced echoes in air by calling out in an empty hall and hearing their words repeated as the sound bounces off the walls.

Refraction and Transmission. Refraction is the change of direction of a wave when it travels from a medium in which it has one velocity to a medium in which it has a different velocity. Refraction of sound occurs in the ocean because the temperature of the water changes with depth, which causes the velocity of sound also to change with depth. For simple ocean models, the layers of water at different temperatures act as though they are layers of different media. The following example explains refraction: Imagine a sound wave that is constant over a plane (i. e., a plane wave) in a given medium and a line drawn perpendicular to this plane (i. e., the normal to the plane) which indicates the travel direction of the wave. When the wave travels to a different medium, the normal bends, thus changing the direction of the sound wave. This normal line is called a ray.

When a sound wave impinges on a plate, part of the wave reflects and part goes through the plate. The part that goes through the plate is the transmitted wave. Reflection and transmission are related phenomena that are used extensively to describe the characteristics of sound baffles and absorbers.

Diffraction. Diffraction is associated with the bending of sound waves around or over barriers. A sound wave can often be heard on the other side of a barrier even if the listener cannot see the source of the sound. However, the barrier projects a shadow, called the shadow zone, within which the sound cannot be heard. This phenomenon is similar to that of a light that is blocked by a barrier.

Interference. Interference is the phenomenon that occurs when two sound waves converge. In linear acoustics the sound waves can be superimposed. When this occurs, the waves interfere with each other, and the resultant sound is the sum of the two waves, taking into consideration the magnitude and the phase of each wave.

Scattering. Sound scattering is related closely to reflection and transmission. It is the phenomenon that occurs when a sound wave envelops an obstacle and breaks up, producing a sound pattern around the obstacle. The sound travels off in all directions around the obstacle. The sound that travels back toward the source is called the backscattered sound, and the sound that travels away from the source is known as the forward-scattered field.

Standing Waves, Propagating Waves, and Reverberation

When a sound wave travels freely in a medium without obstacles, it continues to propagate unless it is attenuated by some characteristic of the medium, such as absorption. When sound waves propagate in an enclosed space, they reflect from the walls of the enclosure and travel in a different direction until they hit another wall. In a regular enclosure, such as a rectangular room, the waves reflect back and forth between the sound source and the wall, setting up a constant wave pattern that no longer shows the characteristics of a traveling wave. This wave pattern, called a standing wave, results from the superposition of two traveling waves propagating in opposite directions. The standing wave pattern exists as long as the source continues to emit sound waves. The continuous rebounding of the sound waves causes a reverberant field to be set up in the enclosure. If the walls of the enclosure are absorbent, the reverberant field is decreased. If the sound source stops emitting the waves, the reverberant standing wave field dies out because of the absorptive character of the walls. The time it takes for the reverberant field to decay is sometimes called the time constant of the room.

Sound Radiation

The interaction of a vibrating structure with a medium produces disturbances in the medium that propagate out from the structure. The sound field set up by these propagating disturbances is known as the sound radiation field. Whenever there is a disturbance in a sound medium, the waves propagate out from the disturbance, forming a radiation field.

Coupling and Interaction between Structures and the Surrounding Medium

A structure vibrating in air produces sound waves, which propagate out into the air. If the same vibrating structure is put into a vacuum, no sound is produced. However, whether the vibrating body is in a vacuum or air makes little difference in the vibration patterns, and the reaction of the structure to the medium is small. If the same vibrating body is put into water, the high density of water compared with air produces marked changes in the vibration and consequent radiation from the structure. The water, or any heavy liquid, produces two main effects on the structure. The first is an added mass effect, and the second is a damping effect known as radiation damping. The same type of phenomenon also occurs in air, but to a much smaller degree unless the body is traveling at high speed. The coupling phenomenon in air at these speeds is associated with flutter.

Deterministic (Single-Frequency) Versus Random Linear Acoustics

When the vibrations are not single frequency but are random, new concepts must be introduced. Instead of dealing with ordinary parameters such as pressure and velocity, it is necessary to use statistical concepts such as autocorrelation and cross-correlation of pressure in the time domain and auto- and cross-spectrum of pressure in the frequency domain. Frequency is a continuous variable in random systems, as opposed to a discrete variable in single-frequency systems. In some acoustic problems there is randomness in both space and time. Thus statistical concepts have to be applied to both time and spatial variables.

Some of the Practical Problems in Linear and Nonlinear Acoustics

The majority of problems in architectural and musical acoustics involve small-amplitude sounds and therefore come under the category of linear acoustics. In fact, as will be seen later, all sounds which are below the threshold of pain[1] are well within the linear acoustic region. Most problems in submarine and surface-ship sound radiation which involve interaction of a vibrating structure with water are also linear acoustic problems since only very small motions of both the structure and the medium in contact with it, are involved. The problem of propagation of explosive waves is a large-amplitude, nonlinear acoustic problem. The transition from subsonic to supersonic flow and the associated production of shock waves is also a problem in nonlinear acoustics.

References to Fundamental Developments in Linear and Nonlinear Acoustics

There are a number of books on acoustics which contain various degrees of mathematical sophistication. The only presentations on acoustics that the writer would classify as elementary are the accounts given in standard physics texts such as Duff *et al.* [1] and Sears and Zemansky [2]. There are many of these texts, and they usually give good elementary discussions of acoustics. The reader can almost pick any college physics text at random, and it will usually give a reasonably good presentation of elementary acoustics. Of the books which are devoted entirely to acoustics, the writer would rate as number one the treatise of Lord Rayleigh [3]. The writer would classify Rayleigh's two volumes as advanced. The reader who is reasonably familiar with elementary differential equations would do well to go through such books as Kinsler and Frey [4], Beranek [5], and Morse [6] before going to Rayleigh's work. Two very fine books of recent vintage which the writer would classify as advanced are the one written by Skudrzyk [7] and the treatise by Morse and Ingard [8]. The most recent reference is Beyer's excellent work on nonlinear acoustics [13].

The most complete mathematical treatment to date on linear and nonlinear wave motion is contained in a very recent book by G. B. Whitham [9] which is devoted entirely to this subject. A most easily readable account of the mathematics associated with both linear and nonlinear acoustics is given by R. B. Lindsay [10]. In this book Lindsay devotes an entire chapter to discussing sound waves in fluids, and he considers both small- and large-amplitude motions in a general way. Short but readily readable accounts of finite-amplitude waves in acoustics and their comparison to linear waves are given in both Rayleigh's [3] and Lamb's [11] texts on sound. Finally, the *Journal of the Acoustical Society of America*, which is a monthly technical publication published by the American Institute of Physics, has two of its sections devoted entirely to work in general linear acoustics and nonlinear acoustics or macrosonics. Finally, a review of the important aspects of linear acoustics is contained in a recent article [14].

Basic Governing Equations of Acoustics

The fundamental concepts of both linear and nonlinear acoustics can be covered for the one-dimensional case in a simple but general way. From these notions the reader will be able to

---

[1]The threshold of pain is the intensity at which an average person starts to feel pain in his ears and at which permanent damage to the ears could result if exposure to the sound is sustained.

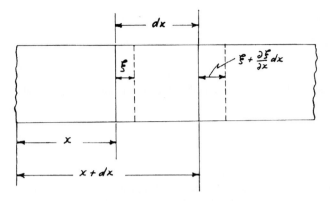

Fig. 1: Propagation of plane one-dimensional sound wave.

grasp a physical feeling concerning sounds produced in both the linear and nonlinear acoustic regimes. The extension to two and three dimensions is much more mathematically complicated, but involves no new physical concepts. For this discussion the writer will follow a combination of the presentation on plane sound waves by Kinsler and Frey [4] and Lamb [11]. Let

x = equilibrium coordinate of a given particle of the medium from some given origin (see Fig. 1);

ξ = particle displacement along the x axis from the equilibrium position;

u = velocity of the particle = $\partial \xi / \partial t$;

ρ = instantaneous density at any point;

ρ_0 = equilibrium density of the medium;

S = condensation at any point, which is defined as $S = (\rho - \rho_0)/\rho_0$ so that $\rho = \rho_0(1+S)$;

$\bar{p}$ = instantaneous pressure at any point;

p_0 = equilibrium pressure in the medium;

p = excess pressure (which is the acoustic pressure) at any point, thus $p = \bar{p} - p_0$;

C = velocity of propagation of the wave.

It will be assumed that the wave is plane, i. e., all particles on the plane at x have the same displacement, and that this displacement is a function only of space coordinate x and time t. We employ three basic concepts to derive three independent equations and then we will combine these equations into a single equation. The first of these concepts is the conservation of mass. We apply the principle of conservation of mass to a cross-sectional area $\bar{S}$ of the undisturbed fluid contained between planes positioned at x and $x + dx$. The mass of this undisturbed fluid is $\rho_0 \bar{S} dx$. Assume that upon passage of a sound wave the plane at x is displaced a distance ξ to the right (see Fig. 1) and that the plane at $x + dx$ is displaced a distance $(\xi + d\xi$ [$d\xi = (\partial \xi / \partial x) dx$]. The volume enclosed is therefore changed to $\bar{S} dx(1 + \partial \xi / \partial x)$ and its mass is $\rho \bar{S} dx(1 + \partial \xi / \partial x)$. Equating the original mass to the new mass (i. e., employing the conservation of mass) gives

$$\rho \bar{S} \, dx = \left(1 + \frac{\partial \xi}{\partial x}\right) = \rho_0 \bar{S} \, dx \; ; \tag{1}$$

noting that $\rho = \rho_0(1+S)$ we obtain

$$(1+S)\left(1+\frac{\partial\xi}{\partial x}\right) = 1 . \tag{2}$$

If both S and $\partial\xi/\partial x$ are *small* (i. e., $\ll 1$), then we can neglect the product of S and $\partial\xi/\partial x$ and Eq. (2) reduces to

$$S = -\frac{\partial\xi}{\partial x} . \tag{3}$$

This equation is known as the equation of continuity.

The second basic concept that shall be employed is Newton's equation of motion of the element. The resultant pressures on the two faces of the volume element $\bar{S}\,dx$ will be slightly different from each other producing a net force which will accelerate the element. The external force acting on each face is equal to the product of the pressure and the area of the face. The net force acting upon $\bar{S}\,dx$ in the positive x direction is

$$dF_x = \left[p-\left(p+\frac{\partial p}{\partial x}\,dx\right)\right]\bar{S} = -\frac{\partial p}{\partial x}\,dx\bar{S} . \tag{4}$$

This net force is equal, by Newton's second law of motion, to the product of the element's mass and its acceleration, thus

$$-\frac{\partial p}{\partial x} = \rho_0\frac{\partial^2\xi}{\partial t^2} . \tag{5}$$

One other relation is necessary to combine the equation of continuity (2) or (3) with the equation of motion (5), and this is a relation between the pressure and the density. If we assume that the process is isothermal, then the temperature does not change during the passage of the sound wave and the pressure density relation follows Boyle's law, i. e.,

$$\frac{\bar{p}}{p_0} = \frac{\rho}{\rho_0} . \tag{6}$$

However, in ordinary sound waves the condensation S changes sign so frequently and the temperature, consequently, rises and falls so rapidly, that there is no time for transfer of heat between adjacent portions of the fluid. The flow of heat has hardly gone from one element to another before its direction is reversed; therefore the conditions are close to being adiabatic, i. e., no heat transfer occurring, and the pressure density relation becomes

$$\frac{\bar{p}}{p_0} = \left(\frac{\rho}{\rho_0}\right)^{\gamma} , \tag{7}$$

where γ is the adiabatic constant having a value of about 1.4 for air.

For large-amplitude adiabatic waves we combine (2), (5), and (7) as follows:

From before we had, by definition

$$\rho = \rho_0(1+S) ; \tag{8}$$

Eq. (2) gives

$$(1+S) = \frac{1}{1+\partial\xi/\partial x} \ .$$

Thus

$$\rho = \frac{\rho_0}{1+\partial\xi/\partial x} \ ; \tag{9}$$

Eq. (7) gives

$$\frac{\bar{p}}{p_0} = \left(\frac{\rho}{\rho_0}\right)^{\gamma} \tag{10}$$

or

$$\frac{\bar{p}}{p_0} = \left(\frac{\rho_0}{\rho_0(1+\partial\xi/\partial x)}\right)^{\gamma} = \left(\frac{1}{1+\partial\xi/\partial x}\right)^{\gamma} \ . \tag{11}$$

but

$$\bar{p} = p_0 + p \ . \tag{12}$$

So

$$\bar{p} = \frac{p_0}{(1+\partial\xi/\partial x)^{\gamma}} \ . \tag{13}$$

So

$$\frac{\partial p}{\partial x} = \frac{\partial \bar{p}}{\partial x} = -p_0\gamma\left(1+\frac{\partial\xi}{\partial x}\right)^{-\gamma-1}\frac{\partial^2\xi}{\partial x^2} \ . \tag{14}$$

Thus (5) gives

$$\frac{p_0\gamma}{\rho_0} \frac{\partial^2\xi/\partial x^2}{(1+\partial\xi/\partial x)^{\gamma+1}} = \frac{\partial^2\xi}{\partial t^2} \ . \tag{15}$$

Let

$$C^2 = p_0\gamma/\rho_0 \ . \tag{16}$$

Then the final adiabatic equation of motion for large amplitude nonlinear plane waves is

$$C^2\frac{\partial^2\xi/\partial x^2}{(1+\partial\xi/\partial x)^{\gamma+1}} = \frac{\partial^2\xi}{\partial t^2} \ . \tag{17}$$

If the process is isothermal, $\gamma = 1$.

For very small motions the condensation S is very small. When we employ this assumption and eq. (3) the equation of motion (17) reduces to the linear equation of sound waves

$$C^2 \frac{\partial^2 \xi}{\partial x^2} = \frac{\partial^2 \xi}{\partial t^2} . \tag{18}$$

The general solution of Eq. (18) can be written in the form

$$\xi = f_1(Ct - x) + f_2(Ct + x) . \tag{19}$$

This is easily verified by substituting (19) into (18) and performing the indicated differentiations. The solution (19) states that small displacements propagate with velocity C without change of shape. If we neglect any losses in the medium, a sound disturbance which is started at a given point will propagate with the velocity C without being distorted as it propagates. This is not true of large-amplitude waves which satisfy Eq. (17). To see this consider the transitional region between small- and large-amplitude waves in a long straight tube in which a piston (at $x = 0$) is made to move in some arbitrary manner described by

$$\xi = f(t) \tag{20}$$

If we expand the denominator of Eq. (17) and neglect the terms greater than the second derivative of ξ, we obtain the following approximate equation for the transition region:

$$\frac{\partial^2 \xi}{\partial t^2} = C^2 \frac{\partial^2 \xi}{\partial x^2} - (\gamma + 1)C^2 \frac{\partial \xi}{\partial x} \frac{\partial^2 \xi}{\partial x^2} . \tag{21}$$

By means of a procedure adopted by Airy [11] an approximate solution can be constructed. First we note that the solution of (18) is

$$\xi = f(t - x/C) . \tag{22}$$

Substituting this value of ξ in the last term on the right-hand side of (21) we obtain

$$\frac{\partial^2 \xi}{\partial t^2} = C^2 \frac{\partial^2 \xi}{\partial x^2} - \frac{1}{2}(\gamma + 1) \frac{\partial}{\partial x} \left\{ f' \left(t - \frac{x}{C} \right) \right\}^2 . \tag{23}$$

The solution of this equation is

$$\xi = f \left(t - \frac{x}{C} \right) + \frac{\gamma + 1}{4C^2} x \left\{ f' \left(t - \frac{x}{C} \right) \right\}^2 . \tag{24}$$

Now assume that the motion of the piston at one end of the tube is simple harmonic, i. e.,

$$f(t) = a \cos \omega t . \tag{25}$$

Formula (24) then gives

$$\xi = a \cos \omega \left(t - \frac{x}{C} \right) + \frac{(\gamma + 1)\omega^2 a^2}{8C^2} x \left\{ 1 - \cos 2\omega \left(t - \frac{x}{C} \right) \right\} . \tag{26}$$

It is thus seen that the displacement of any particle is no longer simple harmonic at $x > 0$ but is distorted from the original wave shape.

Sound Intensity and Pressure in Linear and Nonlinear Acoustics – Sine Waves (Siren-Type Sounds)

In order to give the reader a physical feeling of the relative sounds coming from various processes in both linear and nonlinear acoustics let us first compute the intensity of sine waves that might come from any physical source such as a siren or transducer of some sort. The sound intensity is defined as the time average of the power flow per unit area (or rate at which energy is transmitted per unit area [6]) as follows:

$$I = \frac{1}{T} \int_0^T p\dot{\xi}\, dt \tag{27}$$

where I is the intensity, p is the excess pressure, and $\dot{\xi}$ is the velocity of the medium particle. The value of T is taken as some arbitrary value. In the case of sine waves we will take T to be a number of periods of the wave. In the last section it was found that the pressure p was (for adiabatic processes)

$$\bar{p} = p_0 (\rho/\rho_0)^\gamma , \tag{28}$$

(for an isothermal process $\gamma = 1$) and that the density ρ was connected to the condensation S by the relation

$$S = \rho/\rho_0 - 1 . \tag{29}$$

It has been found [11] that the velocity $\dot{\xi}$ has the following values for a general nonlinear wave:

$$\dot{\xi} = \pm C \log(1+S) \qquad \text{(for an isothermal process);} \tag{30}$$

$$\dot{\xi} = \pm \frac{2C}{\gamma-1}[1 - (1+S)^{(\gamma-1)/2}] \qquad \text{(for an adiabatic process).} \tag{31}$$

For both processes it is easily verified that for the linear case ($S \ll 1$)

$$\dot{\xi} = \pm CS . \tag{32}$$

For sine waves of small amplitude

$$I = \frac{1}{T} \int_0^T p_0 \gamma C S^2\, dt , \tag{33}$$

so

$$\frac{I}{p_0 C} = \gamma \frac{1}{T} \int_0^T S^2\, dt ,$$

$$S = S_0 \cos \omega t ,$$

therefore

$$\frac{I}{p_0 C} \quad = \quad \gamma \frac{S_0^2}{2} \qquad \text{(for the adiabatic case),} \tag{34}$$

$$\quad = \quad S_0^2/2 \qquad \text{(for the isothermal case).}$$

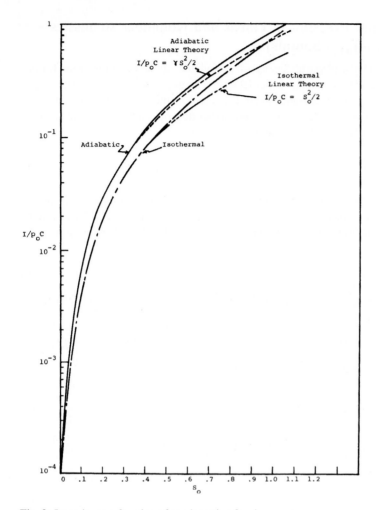

Fig. 2: Intensity as a function of condensation for sine waves.

For the nonlinear case we just substitute (28), (29), (30), or (31) into (27) and integrate numerically. The result is given in Fig. 2 along with the linear approximation. It is seen that the linear theory holds for values of S_0 less than 0.4.

The sound pressure generated for either the linear or nonlinear adiabatic case is

$$p = p_0(1+S)^\gamma - p_0 \ .$$

Assuming that the ambient pressure is sea level pressure (i. e., 1 atmosphere) the pressure level in dB relative to $0.0002 \, \text{dyn}/\text{cm}^2$, which is a standard measure for pressure levels in air, is as follows:

Sound Pressure Level $= 2\log_{10} p + 74\,\text{dB}$ relative to $0.0002 \, \text{dyn}/\text{cm}^2$ where p is expressed in dynes/cm^2. Table 1 gives the sound pressure levels for the adiabatic case as a function of the condensation S.

Table 1: Sound pressure level (in dB relative to $0.0002\,\text{dynes/cm}^2$)[a] as a function of condensation.

Condensation S	Sound Pressure Level(SPL)	Condensation S	Sound Pressure Level(SPL)
10^{-10}	-3	0.9	196
10^{-9}	17	1.0	197
10^{-8}	37	2.0	203
10^{-7}	57	3	207
10^{-6}	77	4	209
10^{-5}	97	5	211
10^{-4}	117	6	213
10^{-3}	137	7	214
10^{-2}	157	8	215
10^{-1}	177	9	216
0.2	183	10	217
0.3	186	12	219
0.4	189	14	220
0.5	191	16	221
0.6	193	18	222
0.7	194	20	223
0.8	195		

[a] $1\,\text{dyn/cm}^2$ is called a microbar, written µbar.

In order to get a physical feeling of the order of magnitude of the sound as a function of condensation S_0, Fig. 3 illustrates the intensity as a function of frequency showing the threshold of hearing and the threshold of pain [6]. The vertical lines, which are not frequency dependent, illustrate the intensity values for various values of condensation amplitude S_0. The intensity level is defined as follows:

$$\text{Intensity Level} = 10\log_{10}(10^9 I) = 90 + 10\log_{10} I , \tag{35}$$

where I is expressed in ergs per square centimeter per second. Thus the intensity level is in dB relative to 10^{-10} microwatts per square centimeter per second. Note that at 3000 Hz (i. e., 3000 cyles per second) the threshold of hearing is about $-6\,\text{dB}$ relative to 10^{-10} microwatts/cm^2. This corresponds to a sine wave condensation of about 10^{-10} in air and would even correspond to a much smaller S_0 in water. This means that an average person could hear a sound in air at a frequency of 3000 Hz if the condensation amplitude was as small as 10^{-10}. The threshold of pain, which Fig. 3 shows is almost frequency independent, corresponds to condensations of the order of 5×10^{-4} in air. Thus the entire auditory area (i. e., from threshold of hearing to threshold of pain) is well within the region of linear acoustics. Comparing Table 1 with Fig. 3 it is seen that the sound pressure level corresponding to a condensation of 10^{-10}, which corresponds to the intensity at threshold of hearing, is about $-3\,\text{dB}$ relative to 0.0002 µbar and that the sound pressure level at threshold of pain is of the order of 125 dB relative to 0.0002 µbar.

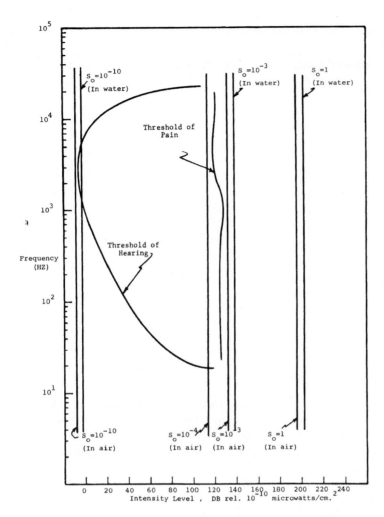

Fig. 3: Intensity of sine waves as a function of frequency.

Linearity and Nonlinearity as Related to Elasticity of the Medium

In any physical system which involves forces and displacements it is the usual understanding that if the force is a linear function of displacement, then the system is linear and when the force becomes a nonlinear function of displacement, the system becomes nonlinear. The same concept holds true in acoustics. In order to see this use Eqns. (28), (29), (30), (31) and write the pressure in terms of velocity as follows:

$$\frac{\bar{p}}{p_0} = \left(1 + \frac{\gamma-1}{2C}\dot{\xi}\right)^{2\gamma/(\gamma-1)} \quad \text{(Adiabatic Case);} \tag{36}$$

$$\frac{\bar{p}}{p_0} = e^{\xi/C} \quad \text{(Isothermal Case).}$$

Table 2: Pressure and condensation for blast waves.

$\bar{R}$	P_1/P_0	dB relative to 0.0002 μbar P_1	S_1	P_2/P_0	dB relative to 0.0002 μbar P_2	S_1
0.10	67.9	231	6.0	585	249	32.2
0.2	20.4	220	3.9	146	237	17.1
0.3	7.3	211	2.6	37.7	225	9.0
0.4	3.5	205	1.7	15.3	218	5.1
0.5	2.1	200	1.1	9.4	213	3.2
0.6	1.4	197	0.8	6.1	210	2.1
0.8	0.77	192	0.5	2.6	202	1.1
1	0.51	188	0.33	1.3	196	0.66
2	0.16	178	0.11	0.36	185	0.22
3	0.089	173	0.063	0.19	180	0.12
4	0.062	170	0.044	0.13	176	0.087
5	0.047	167	0.033	0.095	174	0.066
6	0.037	165	0.027	0.077	172	0.053
8	0.026	162	0.019	0.054	169	0.039
10	0.020	160	0.014	0.040	166	0.028
20	0.0087	153	0.0062	0.018	159	0.012
30	0.0054	149	0.0039	0.011	155	0.0077
40	0.0039	146	0.0028	0.0079	152	0.0056
50	0.0030	144	0.0022	0.0061	150	0.0043
60	0.0025	142	0.0018	0.0050	148	0.0035
80	0.0018	139	0.0010	0.0036	145	0.0021
100	0.0014	137	0.00082	0.0028	143	0.0016
500	0.00024	122	0.00017	0.00049	128	0.00033
1000	0.00012	116	0.000082	0.00023	121	0.00017

For sine waves $\xi = \xi_0 \sin \omega t$, so

$$\left(\frac{\bar{p}}{p_0}\right)_{max} = \left(1 + \frac{\gamma-1}{2}\frac{\omega\xi_0}{C}\dot{\xi}\right)^{2\gamma/(\gamma-1)} \quad \text{(Adiabatic Case)}; \tag{37}$$

$$\left(\frac{\bar{p}}{p_0}\right)_{max} = e^{\omega\xi_0/c} \quad \text{(Isothermal Case)}.$$

but $p = \bar{p} - p_0$. In Fig. 4 the dimensionless pressure ratio, p/p_0 is plotted as a function of the dimensionless deformation parameter $\omega\xi_0/C$. It is seen in Fig. 4 that when the displacements are small the pressure is a linear function of displacement. As the displacements become larger the medium becomes stiffer and a small change in displacement gives a proportionally larger increase in pressure. The nonlinear region starts at $\omega\xi_0/C \approx 0.1$. This corresponds to an $S \approx 0.1$, which is a much better criterion for linearity than Fig. 2.

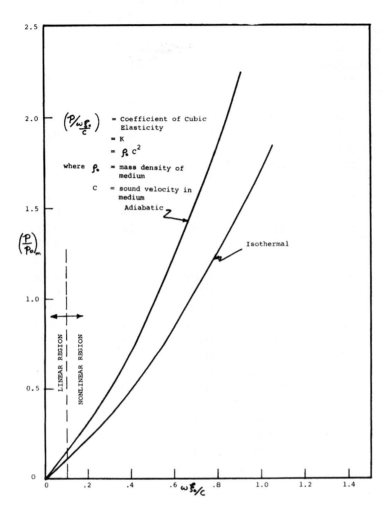

Fig. 4: Sound pressure as a function of displacement for sine waves.

Explosive Waves

Consider next the sounds generated by explosive-type waves. Unless the charge weight of the explosive is exceptionally small, the sounds generated by the explosion are in the nonlinear acoustic region at the point of the explosion, but as they die out away from the explosion, they become smaller and enter the linear region. In his book *Explosions in Air* [12], Baker gives the necessary information needed to estimate the sounds obtained from the explosions. The waves produced by the explosions are shock waves and have steep wave fronts. Therefore they have to be analyzed in a different manner from the relatively simple way that has been explained previously in this article. However, the results for shock waves will be given here in order to give the reader a feeling of the sounds produced by these waves as a function of the explosive weight, the distance from the explosion, and the type of explosive. Table 2 gives the pressure levels and condensation values for explosive waves.

Table 3: Characteristics of explosives.

Explosive	Specific Energy E/M (in./lb$_m$)
Pentolite	20.50×10^6
TNT	18.13×10^6
RDX	21.5×10^6

In Table 2 $\bar{R} = Rp_0^{1/3}/E^{1/3}$, where R is the distance from the explosion, p_0 is the ambient pressure, and E is the total energy in the explosive charge. P_1, S_1 correspond to the pressure and condensation in the incident blast wave (i. e., the wave coming directly from the explosion to the point at which the pressure is being measured) and P_2, S_2, correspond to the pressure and condensation in the reflected wave. The reflected wave parameters are determined from the assumption that the reflection occurs from a rigid wall.

Table 3 contains the energy characteristics for several of the more important explosives [12]. In Table 3 the symbol # represents a pound of force (the weight) while lb$_m$ represents a pound of mass. In order to obtain the energy for a given weight of explosive we use the following relation:

$$E = (E/M)(W/g) .$$

Thus the energy contained in 1000 # of TNT is

$$E = 18.13 \times \frac{1000}{386} = 47 \times 10^6 \text{# in.}$$

The value of R at 1 mile from an explosion of 1000 # of TNT would be

$$\bar{R} = 5280 \times 12 \times \left(\frac{14.7}{47 \times 10^6} \right) = 430$$

Examination of Table 2 indicates that for $\bar{R} = 430$ the incident blast pressure is of the order of 125 dB relative to 0.0002 μbar, and the reflected pressure of the order of 130 dB relative to 0.0002 μbar, both being around the threshold of pain for hearing. However, since the blast wave acts only for a very short time, the pain in the ear will undoubtedly not have time to develop for such an explosion.

The values of condensation S contained in Tables 1 and 2 compare very favorably for sound pressure levels less than 200 dB relative to 0.0002 μbar, i. e., for S values less than 1. For the larger S values the characteristics of the shock wave front enter the problem and there is no longer any correlation between the value contained in the two tables.

See also: Acoustics, Architectural; Fluid Physics; Nonlinear Wave Propagation; Shock Waves And Detonations; Sound, Underwater.

References

[1]   A. W. Duff (ed.), *Physics for Students of Science and Engineering*. Blakiston Co., Philadelphia, PA, 1937. (E)
[2]   F. W. Sears and M. W. Zemansky, *College Physics*. AddisonWesley, Reading, MA, 1960. (E)
[3]   Lord Rayleigh, *The Theory of Sound*. Dover, New York, 1945. (A)
[4]   L. E. Kinsler and A. R. Frey, *Fundamentals of Acoustics*. Wiley, New York, 1962. (I)
[5]   L. L. Beranek, *Acoustics*. McGraw-Hill, New York, 1954. (I)
[6]   P. M. Morse, *Vibration and Sound*. McGraw-Hill, New York, 1948. (I)
[7]   E. Skudrzyk, *The Foundations of Acoustics*. Springer Verlag, Wein, 1971. (A)
[8]   P. M. Morse and K. U. Ingard, *Theoretical Acoustics*. McGrawHill, New York, 1968. (A)
[9]   G. B. Whitham, *Linear and Nonlinear Waves*. Wiley, New York, 1974. (A)
[10] R. B. Lindsay, *Mechanical Radiation*. McGraw-Hill, New York, 1960. (I)
[11] H. Lamb, *The Dynamical Theory of Sound*. Dover, New York, 1960. (I)
[12] W. E. Baker, *Explosions in Air*. University of Texas Press, Austin Texas, 1973. (I)
[13] R. T. Beyer, *Nonlinear Acoustics*. Naval Sea Systems Command, 1974. (I)
[14] J. E. Greenspon "Acoustics, Linear", *Encyclopedia of Physical Science and Technology, Vol. I*, p. 135. Academic Press, San Diego, California 1987.

Acoustics, Physiological

J. Tonndorf[†]

Physiological acoustics, an expression coined after Helmholtz's *Physiological Optics* (1856), concerns itself with analytical assessments of the reception of sound by the ear and of the further processing of the signals thus received at the various levels of the central auditory nervous system.

This work requires close cooperation between physiologists and anatomists. Formal analyses are made possible by inputs from fluid mechanics (inner-ear dynamics); biochemistry (chemical events underlying the responses of the sense organ); systems analysis, including electrical network theory, and advanced statistics (electrical responses in both the organ and the central nervous system, equivalent network analysis); and many others.

The following brief description of the current state of the art in physiological acoustics must include some anatomical remarks.

The ear is traditionally divided into three parts: the outer, middle, and inner ears (Fig. 1). Outer and middle ears help shape the acoustic signal for optimal reception by the inner ear and its receptor cells.

The outer ear consists of the pinna and ear canal. The ear canal, a continuation of the funnel-shaped pinna, is an open tube terminated at its inner end by the tympanic membrane. The latter seals off the middle ear from the outside. The middle ear, an air-filled cavity (with a volume of approximately $2\,cm^3$), can be aerated via the Eustachian tube, a connection with the upper pharynx. The tympanic membrane, a thin elastic structure, vibrates in response to sound. It is connected by a mechanical transmission chain, consisting of a series of three small

---

[†]deceased

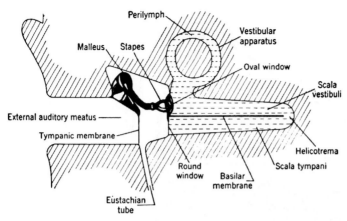

Fig. 1: Highly schematic outline of the ear (from van Békésy and Rosenblith, 1951).

leverlike ossicles, to one of the two "windows" of the inner ear, the oval window. The other one, the round window, looks likewise into the middle ear, but it is simply closed over by a membrane. The inner ear is deeply hidden in hard dense bone. Like that of all vertebrates, it is a fluid-filled cavity, very complex in shape, whence its classical name, otic labyrinth. In addition to the auditory receptor organ, the inner ear houses five other organs that have to do with spatial orientation and maintenance of equilibrium, the so-called vestibular system. (The "vestibule" is part of the inner ear.)

The functions of outer and middle ears are (a) protection of the inner ear (foremost by virtue of their position, which shields the inner ear against physical insults; then there are middle-ear nonlinearities occurring at high amplitudes; and finally there are the two small middle-ear muscles; their reflex contraction on sound exposure attenuates middle-ear transmission); (b) optimization of transmission of acoustic energy into the inner ear (the impedance of the fluid-filled inner ear is much higher than that of the air on the outside; thus, impedance matching is needed; this task is accomplished by a series of mechanical transformers that are completely integrated with one another, involving both the outer and middle ears).

In addition to the route just described, i. e., via tympanic membrane–ossicular chain–oval window (so-called air conduction), mechanoacoustic energy may also be brought into the ear when the bones of the head are set in vibration by contact with a vibrating object. This bone-conduction mode plays an important role in the clinical diagnosis of hearing disorders.

The auditory receptor organ is located in the lower, or cochlear portion of the inner ear. (The vestibular system occupies the upper portion.) The cochlea is a long (35 mm) but narrow bony chamber, coiled up $2\frac{1}{2}$ times like a snail shell (which is what cochlea means in Latin). It is subdivided by a number of membranes into a system of three compartments ("scalae") (Fig. 2). The receptor organ, the organ of Corti, lies in the middle scala and stretches over the whole length of the cochlea. It consists of about 16 000 to 20 000 receptor cells, the hair cells, distributed in a characteristic four-row pattern and held in place by a supporting-cell structure. Each hair cell carries a tuft of 80–100 sensory hairs on its top surface. Sensory cells of this kind are also found in other receptor organs; all of them are stimulated by a mechanical, sideways deflection of their hairs. In the cochlea, the necessary mechanism is provided

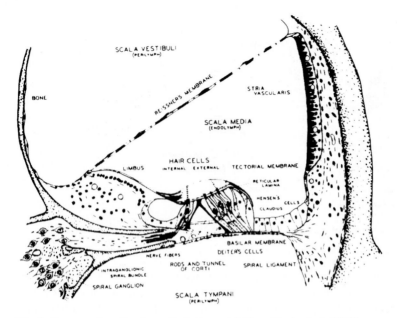

Fig. 2: Cross section of scala media (guinea pig) (from Davis *et al.*, 1953).

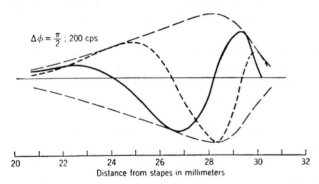

Fig. 3: Schematic outline of cochlear traveling wave at two instances, 90° apart
in phase. Waves are moving from base to apex, i. e., left to right in the figure;
amplitudes overstated (from von Békésy and Rosenblith, 1951).

by the connection of the sensory hairs with a membrane that covers the organ of Corti from
side to side for its full length. This tectorial membrane (Fig. 2) executes a sliding ("shearing")
movement across the organ that leads to the deflection of the sensory hairs. This constitutes
the ultimate mechanical input to the hair cells; it is the final one in a series of interlinked mechanical events that are elicited when mechanoacoustic energy enters the cochlea, usually via
the oval window. Such signals set up a series of displacements of the cochlear membranes,
including the basilar membrane, on which the organ of Corti is situated (Fig. 2). These displacements progress along the basilar membrane in the manner of traveling waves (Fig. 3),
invariably in the direction from the cochlear base to its apex. In this respect, the cochlea acts

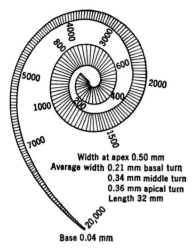

Width at apex 0.50 mm
Average width 0.21 mm basal turn
0.34 mm middle turn
0.36 mm apical turn
Length 32 mm

Base 0.04 mm

Fig. 4: Outline of the human basilar membrane. Note its width increasing with distance. Places of frequency maxima as indicated (from Stuhlmann, 1943).

like a mechanical delay line. For a given sine-wave input, the traveling-wave mechanism creates a displacement maximum at a distinct, frequency-dependent place along the membrane ("place principle"). For high frequencies, the maxima are formed near the cochlear base, and as frequency goes lower this place moves toward the apex in a systematic manner (Fig. 4). Therefore, given sine-wave signals stimulate only limited regions along the basilar membrane and hence distinct groups of hair cells. This tonotopic relation is maintained throughout the entire central auditory system. It enables the latter to process frequencies, by substituting place for frequency, up to approximately 20 kHz, while single fibers of the auditory nerve are capable of responding to frequencies not higher than 3 – 4 kHz.

At its bottom end, each hair cell is supplied by a fiber (or fibers) of the cochlear (sensory) nerve. There are some 25 000 to 30 000 individual fibers, and their pattern of distribution to the hair cells is complex but systematic. The mechanoacoustic signals received by the hair cells elicit – in a multiple-step operation that bridges the hair-cell–nerve junction – nerve-action potentials (bursts of negative electrical impulses) of essentially the same kind as those observed in fibers of all other nerves. Their energy source is chemoelectric and inherent to the nervous system. These potentials represent a signal code particularly suited for neural transmission and processing.

After entering the part of the brain known as the brain stem, the auditory fibers run in well-defined tracts that go sequentially from one central station ("nucleus") to the next higher one (Fig. 5). This chain of nuclei finally terminates in the auditory portion of the cortex located in the temporal lobe of the brain. In each nucleus, the signal carried by the incoming fibers is switched onto a new set of outgoing fibers. The underlying networks are structurally very intricate, allowing for complex signal processing. Most, but by no means all, fibers in each tract cross over to the opposite side, so that the left brain receives primarily signals from the right ear, and vice versa. The left auditory cortex appears to handle mainly signals for which analytical processing is of importance (e. g., speech signals), while the right one is primarily concerned with signals of emotional importance (e. g., music). In the region where the two

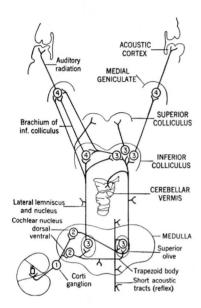

Fig. 5: Schematic outline of the afferent auditory system (from Davis, 1951).

tracts first cross over, signals received by the two ears are brought together in a set of special nuclei (superior olive; Fig. 5) initiating neural processing that concerns spatial hearing. These ascending ("afferent") fiber tracts are paralleled by a similar system of descending ("efferent") tracts that appear to exert a central (feedback) control upon the input at various levels, mainly at the hair-cell–nerve junction.

See also: Fluid Physics; Network Theory, Analysis and Synthesis; Statistics.

Bibliography

G. van Békésy, *Experiments in Hearing*. Krieger, 1980.

G. van Békésy and W. Rosenblith, in *Experimental Psychology*, (S. S. Stevens, ed.), Chapter 27. Wiley, New York, 1951.

P. Dallos, *The Auditory Periphery*. Academic, New York, 1974.

H. Davis, in *Experimental Psychology* (S. S. Stevens, ed.) Chapter 28. Wiley, New York, 1951.

O. Stuhlmann, Jr., *Introduction to Biophysics*, Wiley, New York, 1943.

Acoustoelectric Effect

E. M. Conwell

The acoustoelectric effect is the appearance of a dc electric field when an acoustic wave propagates in a medium containing mobile charges. It was first named and discussed, theoretically, by R. H. Parmenter in 1953. As pointed out by G. Weinreich, who first detected it experimentally, it is an example of the general phenomenon of "wave–particle drag," of which the operation of a linear accelerator and the motion of driftwood toward a beach are other examples. Although first proposed for metals, it is only significant in semiconductors. It can be quite strong in piezoelectric semiconductors, such as CdS, ZnO and GaAs, but is also seen in nonpiezoelectric semiconductors. In the piezoelectric case, the wave and mobile charges interact through the electric field arising from the strain associated with the wave. This interaction is significant when the electric field is longitudinal, i.e., parallel to the wave propagation direction, and in what follows we assume that the type of wave (i.e., longitudinal or shear) and sample orientation have been chosen to make this the case. In nonpiezoelectric media the interaction is through the shift of carrier energy induced by the strain, the so-called deformation potential. At the frequencies ordinarily used, $\ll 1\,\text{GHz}$, the latter interaction is much smaller than that in piezoelectric materials. In either case, however, because of the interaction the passage of an acoustic wave through the medium causes a periodic spatial variation of the potential energy of the charge carriers. If the mean free path l of the carriers is small compared to the acoustic wavelength λ, as is the case for most of the usual (ultrasonic) frequency range, this results in a bunching of the carriers in the potential-energy troughs. Since the wave is propagating, it drags the bunches along with it. This is the origin of the acoustoelectric field and clearly also causes attenuation of the wave. The effect is stronger, the stronger the bunching. Thus it is very weak in ordinary metals, where space-charge effects prevent appreciable bunching. It is enhanced in those nonpiezoelectric semiconductors where space-charge effects are minimized by either (1) the presence of both positively and negatively charged carriers or (2) the existence of different groups of carriers (many-valley band structure) whose energy is affected differently by the strain so that they bunch in different phases of the wave. Being sensitive to the number of free carriers, acoustoelectric interaction has proven to be a powerful tool in nondestructive testing of semiconductors, testing that does not even require contacts.

The foregoing discussion suggests that if an external electric field were applied to the sample to give the carriers a drift velocity, v_d, greater than the wave velocity, v_s, the carriers should drag the wave, i.e., the wave should be amplified. This conjecture was verified experimentally on CdS samples at frequencies of 15 and 45 MHz by Hutson, McFee, and White in 1962. They found acoustic gain for shear waves at fields greater than $700\,\text{V/cm}$, at which field v_d equals the shear wave velocity. For not too large acoustic wave amplitudes it was possible to account quite well for the size of the gain and its variation with frequency, etc., with a linear phenomenological theory taking into account the currents and space charge produced by the piezoelectric fields that accompany the acoustic wave. The gain, called *acoustoelectric gain*, is found to be low at low frequencies (less than the conductivity relaxation frequency σ/ε), where the carriers can redistribute themselves quickly enough to essentially cancel out the piezoelectric field. It peaks at the frequency for which the acoustic wavelength is of the order of the Debye length, where the bunching is optimum. In a fairly strong piezoelectric like CdS acoustoelectric gains as high as $40\,\text{dB/cm}$ have been found, leading to consideration of this effect for practical use as an amplifier (*acoustoelectric amplifier*). This type of amplification has been found particularly useful for amplification of surface acoustic waves (SAWs). Because the amplitude of a SAW decays exponentially with distance below a free surface, the surface acts as a waveguide for such a wave. SAWs at microwave frequencies are easily introduced into a piezoelectric material such as LiNbO_3 by coupling in microwaves through a suitable transducer. A similar transducer can reconvert the SAWs into microwaves. The SAW velocity being smaller by a factor of 10^5 than electromagnetic wave velocity, a short length of a SAW-propagating material is useful as a delay line and for various types of signal processing. For maximum utility the losses of the SAWs in the guide are conveniently overcome by incorporating acoustoelectric gain. If the piezoelectric material is insulating, this may be accomplished by providing a conducting layer, e. g., Si, in contact with it. Alternatively, a conducting piezoelectric material, e. g., GaAs, may be used to support both the SAW and the electrons drifting in the electric field.

The theory and effects considered so far are linear in the sound amplitude provided it is small, i. e., small enough to bunch only a small fraction of the carriers. At large sound amplitude new effects appear. When two acoustic waves are present, the interaction of the piezoelectric field of one with the bunched carriers of the other may result in the generation of the difference or sum frequency of the two. As a special case of this, for a single large wave the interaction of its piezoelectric field with its own carrier bunches results in the generation of a dc current, called the *acoustoelectric current*, flowing in a direction opposite to the usual or Ohmic current, and may result in the generation of the second harmonic.

When an outside acoustic wave is not introduced, application to a highly conducting sample of CdS or ZnO, for example, of a field high enough to make $v_d > v_s$ causes a large amplification of the thermal equilibrium acoustic waves or flux present in the sample. This gives rise to unusual behavior, the exact nature of which depends on the details of sample inhomogeneity and contacts. One possibility is that immediately after the high field is applied a dc current will flow of the expected magnitude for the Ohmic resistance, i. e., the resistance displayed for $v_d < v_s$, but in a short time the current will drop to a much smaller value and remain there. The smaller value is due to the opposing acoustoelectric current arising from the flux amplification. Another frequently seen possibility is the onset of strong current oscillations with a period equal to the length of the sample divided by v_s. These oscillations are due to the creation close to the cathode of a domain or narrow region of high acoustic flux density, which typically moves down the sample with velocity v_s. When it exits at the anode, the current rises to its Ohmic value, strong flux generation begins again at the cathode, and the process is repeated.

The above discussion has been couched in terms appropriate for $l \ll \lambda$, since this is the case for most of the experimental situations studied. However, both linear and nonlinear processes have been studied with a microscopic theory that does not make this restriction. In this theory acoustic gain, for example, may be thought of as due to an excess of stimulated phonon emission by the carriers over absorption.

See also: Piezoelectric Effect; Semiconductors.

Bibliography

N. G. Einspruch, "Ultrasonic Effects in Semiconductors", in *Solid State Physics* (F. Seitz and D. Turnbull, eds.), Vol. 17, p. 217. Academic Press, New York, 1965.

J. H. McFee, "Transmission and Amplification of Acoustic Waves in Piezoelectric Semiconductors," in *Physical Acoustics* (W. Mason, ed.), Vol. IV, part A, p. 1. Academic Press, New York, 1966.

H. N. Spector "Interaction of Acoustic Waves and Conduction Electrons", in *Solid State Physics* (F. Seitz and D. Turnbull, eds.) Vol. 19, p. 291. Academic Press, New York, 1966.

R. Bray, "A Perspective on Acoustoelectric Instabilities," *IBM J. Res. Devel.* **13**, 487 (1969). See also other articles in this volume., pp. 494–510.

N. I. Meyer and M. H. Jorgensen, "Acoustoelectric Effects in Piezoelectric Semiconductors with Main Emphasis on CdS and ZnO", in *Festkörper Probleme X* (0. Madelung, ed.), p. 21. Vieweg, Braunschweig, Germany, 1970.

E. M. Conwell and A. K. Ganguly, "Mixing of Acoustic Waves in Piezoelectric Semiconductors", *Phys. Rev.* **B4**, 2535 (1971).

Adsorption

J. G. Dash

All surfaces are typically coated with films of foreign molecules that are either specifically applied or unintentionally drawn from their environment, and these films can affect most of the properties of the interface. The very active field of surface science [1] is driven by interest in the fundamental properties of surfaces and films, and by their technical importance, for they are prime factors in many industrial areas, including adhesion, catalysis, corrosion, fracture, lubrication, and solid state electronics. The films are broadly classified by the nature of the forces binding them to the substrate; *chemisorption* when the bonding is primarily chemical, and *physisorption*, or simply *adsorption*, when there is little electron transfer. Most substrates are structurally and chemically heterogeneous, so that their films are highly disordered and difficult to analyze on a fundamental level. Steele [2] reviews many studies of heterogeneous adsorption, and Rudzinski and Everett [3] describe the characterization of heterogeneous adsorbents by vapor pressure isotherms. More uniform substrates, which are essential in modern electronic and optical devices, began to be produced and studied in the latter part of the 20th Century. Accounts of some of the crucial steps in the development of surface science are described in the collections edited by Duke [1], and Duke and Plummer [4]. In what follows we limit the discussion to films adsorbed on uniform solid surfaces, which have both contributed to and benefited from, modern surface science.

The forces of attraction that cause adsorption are the relatively weak and long-range interactions that exist between neutral atoms, molecules, and macroscopic objects. These *dispersion* forces are due to the attractions between the electric dipole moments induced in each body by the fluctuating fields of their neighbors [5]. Dispersion forces between neutral atoms and molecules are responsible for the condensation of vapors to liquid or solid phases at sufficiently low temperature.

The states of surface films depend on their thickness, temperature, and composition, as well as the structure, uniformity, and constitution of the substrate. As a result of this interplay films display a great variety of distinctive regimes. Monolayer films exhibit two-dimensional analogs of the familiar vapor, liquid, and solid phases of bulk matter, as well as others that have no three-dimensional equivalents [5–7]. In thicker films some layers may behave with distinctly different thermodynamic character while the remainder is diffuse. Experimental techniques for film studies include calorimetry, nuclear resonance, and electron, xray, neutron, gamma ray, molecular and optical spectroscopies. Special methods developed rapidly since the 1960s and are developing still.

Considerable attention has been focused on the properties of films in the context of general questions involving the physics of two-dimensional (2D) systems. Theories had indicated that there can be no perfectly ordered states or structures in 1D and 2D matter; that lower-dimensional crystals, magnets, superconductors, and superfluids cannot exist above absolute zero. However, more recent theories have shown that certain types of 2D long-range order may persist at finite temperatures, and several experiments bear out these newer ideas [8].

A crystal has long range *positional* order if it has periodicity of unlimited extent. In principle, a 3D crystal has long-range positional order, barring dislocations that span the entire structure. Peierls [9] showed that thermal excitations destroy the long range positional order of a 1D chain. At $T = 0$ the position of the nth atom, in a chain with interparticle spacing d, is predictably at nd even as n diverges, but at finite T the uncertainty in position becomes greater than d for $n \to \infty$. Mermin [10] extended Peierls' model to 2D, but went on to show that long-range *directional* order in a 2D crystal is much more robust. Directional order (a more general term is *topological order*) is preserved if a large closed loop can be traced stepwise from atom to atom, and as long as the bonds remain unbroken such a closed path will be intact at all temperatures. The distinction between positional and topological order is that of elasticity and rupture; an elastic net when undisturbed has both forms of order; stretching it may destroy positional order, but the net retains topologically order as long as it is not torn apart. Kosterlitz and Thoules [11] described how topological order in a 2D crystal could be destroyed by thermally excited dislocations, in a continuous melting transition.

Subsequent elaboration of the theory by several investigators [8] predicted that the development of complete liquid-like disorder may take place via two successive continuous transitions. However, up to the present time, there have been no unambiguous confirmations of continuous 2D melting in any experimental monolayer films; in contrast, several experimental films undergo first order melting. It is possible that the predicted continuous transitions would occur if they were not preempted by first-order melting. The theory neglects the effects of grain boundaries and the edges of films of finite extent, where the weakness of the solid allows premelting at relatively low temperature, and enables the first order phase change to proceed at a lower temperature than the theoretical transitions [12].

The new ideas have been much more successfully applied to superfluidity [8]. Helium films provide the simplest example of a topological phase transition. The order parameter, which is the condensate wave function, is a complex function of the 2D position, so that the system is equivalent to a 2D planar spin model. In this system the singularities that destroy long-range order are vortices. Kosterlitz and Thouless [11] predicted a continuous transition to superfluidity, and that the transition temperature would vary with the film coverage as a power law. The theory was confirmed in a very sensitive experiment by Bishop and Reppy [13]. Their method was based on the changes of period and dissipation of the torsional oscillations of a spiral of Mylar plastic, covered by a thin film of adsorbed helium. The experiments showed that the superfluid transition temperature followed a power law in film coverage, with the theoretical exponent of 1/2, over more than a decade in the temperature.

Several international conferences have explored the great variety of phases and phase transitions displayed by monolayer and multilayer films [14]. When a uniform surface is sparsely covered, the adsorbed atoms can act like a 2D gas [6]. The essential characteristics are adhesion to the substrate, which at sufficiently low temperature leads to a "freezing out" of higher states of motion normal to the surface, and low surface density. The atoms' surface mobility depends on the substrate's lattice size and the amplitude of the corrugation of binding energy. On strongly corrugated surfaces the adatoms tend to be immobilized at surface sites for long dwell times. As T rises the dwell time decreases, due to more rapid hopping between sites. Surface mobility may be appreciable even at low T for low-mass adatoms on relatively smooth surfaces, due to quantum-mechanical tunneling between sites. Low-density helium and hydrogen films adsorbed on graphite exhibit 2D mobile gas-like heat capacities at low T, evidently as a result of rapid quantum tunneling. Interactions between the adsorbed atoms become important at low temperature and high surface density. Corrections to the equation of state can be in the form of a series expansion in the surface density, similar to the virial expansion for a 3D gas. The second virial coefficients of several films have been deduced from calorimetric measurements or vapor-pressure isotherms, and they agree well with theoretical coefficients calculated from atomic pair potentials [15].

Phase condensation of low coverage films to 2D liquid or solid phases occurs when they are cooled to low temperature. Critical temperatures of the 2D gas-liquid transition are somewhat less than half of the critical temperatures of the bulk phases for most of the noble gases and other simple molecular gases. The gas–liquid critical point is especially interesting, since it belongs to the universality class of the 2D Ising model. Measurements on several monolayer films show critical behavior in good agreement with theory [16].

Typical monolayer systems have dense spatially ordered phases at high density and low temperature, where the structure of the film is incommensurate with that of the substrate [6, 7, 14, 17]. Such "floating solid" monolayers are effectively 2D solids; exemplary systems are helium and hydrogen isotopes, neon and xenon on basal-plane graphite. Their heat capacities exhibit temperature dependence varying as T^2, the 2D analog of the well-known Debye T^3 law, and they melt at sharp triple points. In some films the substrate structure imposes a regularity on the atomic arrangement in the monolayer. In the simplest cases the adsorbed atoms have the same regularity as the substrate atoms, but they may have more complex structures due to the competition between substrate-atom and atom-atom interactions. Particular interest focuses on the order-disorder transition of the registered phases of helium and hydrogen isotopes on graphite, which exhibit strong heat capacity peaks with power-law temperature dependence. The transitions belong to the universality class of the three-state Potts model, and the experimental exponent is in excellent agreement with the theory [18].

Films of several layers thickness display a variety of habits. In some examples the one or two layers closest to the substrate behave as relatively distinct 2D solids while a topmost third layer is effectively a two-dimensional gas. In many systems there is a *wetting transition* [19] between layer formation and cluster growth, generally occurring at a thickness of several atomic layers. The instability of layer formation *vis a vis* cluster formation is an important phenomenon in all types of films. It can occur in systems that have strongly cohesive interactions as well as those having relatively weak dispersion forces. In strongly adsorbed solid films the substrate attraction tends to produce strained layers next to the substrate, which prevents the formation of very thick uniform deposits. This effect belongs to a complex of wetting phenomena, which are of fundamental interest and practical importance. Adsorbed multilayer films are also valuable as test systems for the study of phenomena that can occur on typical surfaces of bulk solid materials. Recent examples are surface roughening and surface melting, which have been observed in multilayer noble gas and light molecular films on carbon nanotubes [20].

See also: Catalysis; Ising Model; Order–Disorder Phenomena; Phase Transitions; Surfaces and Interfaces; Thin Films.

References

[1]   C. B. Duke (ed.), *Surface Science, The First Thirty Years*. North-Holland, 1994.

[2]   W. A. Steele, *The Interaction of Gases with Solid Surfaces*. Pergamon, Oxford, 1974.

[3]   W. Rudzinski and D. H. Everett, *Adsorption of Gases on Heterogeneous Surfaces*. Academic Press, New York, 1992.

[4]   C. B. Duke and E. W. Plummer (eds.), *Frontiers in Surface and Interface Science*. North-Holland, 2002.

[5]   L. Bruch, M. W. Cole and E. Zaremba, *Adsorption: Thermodynamics and Structure*. Wiley, New Yrok, 1998.

[6]   J. G. Dash, *Films on Solid Surfaces*. Academic Press, New York, 1975.

[7]   I. Lyuksyutov, A. G. Naumovets and V. Pokrovsky, *Two-Dimensional Crystals*. Academic Press, New York, 1992.

[8]   D. J. Thouless, *Topological Phase Transitions*. World Scientific, Singapore, 2000.

[9]   R. E. Peierls, *Ann. Inst. H. Poincaré* **5**, 177 (1935).

[10] N. D. Mermin, *Phys. Rev.* **176**, 250 (1968).

[11] M. Kosterlitz and D. J. Thouless, *J. Phys. C* **5**, L124 (1971) and **6**, 1181 (1973).

[12] J. G. Dash, "Melting From One to Two to Three Dimensions", *Contemp. Phys.* **43**, 427 (2002).

[13] D. J. Bishop and J. D. Reppy, *Phys. Rev. Lett.* **40**, 1727 (1978).

[14] Coll. Int. du CNRS, "Phases Bidimensionelles Adsorbées", *J. Phys. (Paris)* **38**, C-4 (1977); S. K. Sinha (ed.), *Ordering in Two Dimensions*, North-Holland 1980; J. G. Dash and J. Ruvalds (eds.), *Phase Transitions in Surface Films*, Plenum, New York, 1980; H. Taub, G. Torzo, H. J. Lauter and S. C. Fain, Jr., *Phase Transitions in Surface Films 2*, Plenum, New York, 1991; M. Michailov and I. Gutzow, *Thin Films and Phase Transitions on Surfaces*, Inst. Phys. Chem. Bulgarian Acad. Sci., 1994.

[15] R. L. Siddon and M. Schick, *Phys. Rev. A* **9**, 907 (1974).

[16] R. B. Griffiths, in C. Domb and M. S. Green (eds.) *Phase Transitions and Critical Phenomena*, Vol. 1. Academic Press, New York 1972.

[17] E. Domany, M. Schick, J. S. Walker and R. B. Griffiths, *Phys. Rev. B* **18**, 2209 (1978).

[18] J. G. Dash, M. Schick and O. E. Vilches, *Surf. Sci.* **299/300**, 405 (1994).

[19] S. Dietrich, in C. Domb and J. Lebowitz (eds.) *Phase Transitions and Critical Phenomena*, Vol. 12, Academic Press, New York, 1987.

[20] M. M. Calbi, M. W. Cole, S. M. Gatica, M. J. Bojan and G. Stan, *Rev. Mod. Phys.* **73**, 857 (2000); T. Wilson, A. Tyburski, M. R. DePies, O. E. Vilches, D. Becquet and M. Bienfait, *J. Low Temp. Phys.* **126**, 403 (2002).

Aerosols

F. S. Harris, Jr.[†]

An aerosol is a suspension of liquid, solid, or mixed particles in a gas, usually air. The size of the particles ranges from about 10^{-9} m, just larger than molecules, to a radius of about 25 μm, as in cloud droplets and dusts with short-time stability due to gravitational settling. Examples are hazes, mists, fogs, clouds, smokes, and dusts, as well as living bacteria, viruses, and molds. Aerosols are important in atmospheric electricity, cloud formation, precipitation processes, atmospheric chemistry, air pollution, visibility, radiation transfer, and hence climate.

The smallest particles, called Aitken nuclei, are from molecular sizes up to about 0.05 μm in radius. They vary in concentration from a few particles per cubic centimeter over the South Pole Plateau to 300 in clean continental air, up to hundreds of thousands in a polluted city or downwind from a combustion source. The condensation nuclei serve as centers upon which cloud and fog droplets form. The large droplets, as in fags and clouds, ordinarily range from 1 to 25 μm with a concentration of 20–500/cm^3 and a liquid water content up to 1 g/m^3.

The tropospheric aerosol sources are (1) inorganic gas-to-particle conversion, primarily SO_2, NH_3, NO_x both natural and man-made; (2) mineral dust, primarily from arid zones and deserts; (3) sea salt; and (4) organic matter of apparently complex but still unidentified origins, but thought to have heavy contributions from plant-derived terpene compounds, forest fires, and oxygenated hydrocarbons. The residence time in the lower troposphere is about 3–6 days for the particles. Above the ocean the maritime aerosol is found only at the lower altitudes; at higher altitudes there is continental aerosol such as the Sahara Desert dust over the North Atlantic Ocean. The wind (aeolian) transport of aerosols is sometimes as far as 10 000 km. By using elemental tracers regional pollution aerosols of both North America and Europe have been followed for several thousand kilometers downwind. Dusts from the central and eastern Asian deserts have been carried to Hawaii and the Marshall Islands in the Pacific Ocean. The continental atmospheric aerosol has approximately 60% of the total mass water-soluble material and 25–30% organic matter; 25% of the material is volatile at temperatures below 150 °C. The hazes over remote areas may be due to photochemical transformation of terpenes from vegetation. Recent work has shown the importance of sulfate regionally distributed sources. Clean air background in remote areas of the earth is about 10 μm/m^3. The standard mass loading established by the U.S. Environmental Protection Agency, not to be exceeded appreciably, is 75 μm/m^3. The mass loading sometimes reaches to about 2000 μg/m^3. The number concentration varies from 10^2 to 10^7/cm^3. In the stratosphere the total number above about 0.01 μm is about 10 particles/m^3, primarily sulfates. There is a maximum at an altitude of about 20 km, the amount depending on the length of time since a major volcanic eruption.

Often a simple function has been found to represent the natural aerosol from 0.1 to 20 μm when in equilibrium, the Junge power law, $dn/d\log r = Cr^{-b}$, where dn is the number of particles in a logarithmic size interval, C a constant, r the radius, and b usually a value of about 3. No simple size model can represent the wide variety of sources and complex interactions in the atmosphere. With differing sources, often a log normal or modified gamma distribution can be used for each, or a combination of distributions with one for each source, such as one for small particles from combustion and one for larger dust particles. The particle sizes

---

[†]deceased

important for health through retention in the human body are those retained in the breathing system after passing through the nose, in the range 4.5 μm down to 0.25 μm radius. The particle size distribution in the atmosphere is affected by the type of source, the changes due to gas-to-particle conversion, condensation, and removal through aggregation, precipitation formation and washout, and gravitational settling.

Experiments have shown that at 75–95% relative humidity (RH) 0.3–0.9 of the submicron aerosol mass can be liquid water, and that even at 50% RH 0.1–0.2 may be liquid water. Actual dry maritime aerosol particles collected over the Atlantic Ocean may increase in volume from 5 to 15 times when the RH is increased to 96%. In air pollution from combustion sources the particles are originally small or are formed by gas-to-particle conversion, and in the Los Angeles, California, basin, 50–80% of the submicron aerosol mass is volatile at 220 °C, with primarily sulfate, nitrate, noncarbonate carbon, and liquid water.

The optical behavior of the particles is determined by the complex refractive index (which includes the wavelength-dependent real and absorption parts), the shape, the size relative to the wavelength of the radiation, and the size distribution. For particles small compared to the wavelength, the scattering intensity is proportional to the inverse fourth power of the wavelength. For spherical particles (many particles are not), in the range of the radiation wavelength size, the Lorenz–Mie theory must be used in which complicated functions describe the polarization parameters as a function of scattering angle, refractive index, and size distribution. The maximum scattering per unit volume for visible light is for particles 0.5 μm in radius. The amount of solar energy absorbed is comparable in amount with the absorption by atmospheric gases. The absorption part of the refractive index is the critical parameter in determining whether such particles on a world-wide basis will tend to cause cooling or warming of the earth and hence climatic change. For a variety of purposes aerosols are often produced by using a gas under pressure to disperse liquids or solids into the atmosphere. One of the propellants commonly used is a group of chlorofluoromethanes which are chemically quite stable and nontoxic. Currently, however, there is serious investigation of the possible accumulation in the stratosphere and by complex processes reducing the ozone, letting more solar ultraviolet reach the earth's surface.

See also: Atmospheric Physics.

Bibliography

G. Bouesbet and G. Brehan, eds. *Optical Particle Sizing*. Plenum, New York, 1988. (A)

Ardash Deepak, ed. *Atmospheric Aerosols, Their Formation, Optical Properties and Effects*. Deepak, Hampton, VA, 1982. (A)

S. K. Friedlander, *Smokes, Dust, and Hazes, Fundamentals of Aerosol Behavior*. Oxford University Press, New York, 2000. (I)

Peter V. Hobbs and M. P. McCormick, eds. *Aerosols and Climate*. Wiley, 1988. (A)

Kenneth Pye, *Aeolian Dust and Dust Deposits*. Academic Press, New York, 1987. (I)

Parker C. Reist, *Introduction to Aerosol Science*. Macmillan, New York, 2000. (E)

Allotropy and Polymorphism
F. J. DiSalvo

The equilibrium crystal structure of some solids changes when the external conditions, such as pressure or temperature, are varied. In addition, the structure of some compounds depends upon the preparation conditions. One of these structures may be the thermodynamically stable structure, while the remainder are metastable phases. This phenomenon is called allotropy when it occurs in an element, and polymorphism when it occurs in a compound. Three common examples of allotropy are presented below.

Sulfur

Solid sulfur consists of nearly flat S_8 molecular rings that are stacked on top of one another. When heated to 95 °C the molecules change orientation, forming a differently stacked structure. Sulfur at room temperature is called rhombic sulfur, and above 95 °C, monoclinic sulfur (after the shapes of their respective crystallographic unit cells). Monoclinic sulfur melts at 120 °C.

Sulfur can also exist in an amorphous form. When liquid sulfur is heated to several hundred degrees centigrade, most of the S_8 molecules break open and join with others to form long sulfur chains. If this liquid is rapidly cooled to room temperature, the chains remain intact and are randomly packed together to form a rubbery solid. At room temperature, amorphous sulfur will very slowly change back into rhombic sulfur. Rhombic sulfur is the stable, or equilibrium, form of sulfur at room temperature. By other preparation methods a number of other metastable forms of sulfur can be obtained at room temperature. Consequently, sulfur has a large number of allotropes; however, rhombic and monoclinic sulfur are the only equilibrium forms (in their respective temperature ranges of stability and at atmospheric pressure).

Iron

At room temperature, iron has a body-centered cubic (bcc) structure; the unit cell is shown in Fig. 1a. (The structure can be visualized by imagining space to be filled with closely packed cubes. At each corner, where eight cubes come together, place an iron atom and then put another in the center of each cube.) When iron is heated to 910 °C its structure changes to face-centered cubic (fcc); a unit cell is shown in Fig. 1b. (In this structure an iron atom is placed at each cube corner and one iron atom on each face of the cube, where two cubes touch.) Iron changes back to the bcc structure at 1390 °C and melts at 1536 °C.

The allotropy of iron is very important for the production of steels. Carbon is moderately soluble in fcc iron, the carbon atoms occupying some of the holes between the iron atoms in this structure (at the center in Fig. 1b). However, the solubility of carbon is much lower in bcc iron. If iron containing several weight percent of carbon is cooled from 1100 °C (fcc phase) to room temperature, the carbon not soluble in bcc iron forms a compound, Fe_3C. The Fe_3C exists in small plate-like regions dispersed in bcc iron. Fe_3C is called cementite, since it makes the iron much stronger. Iron prepared in this manner is called carbon steel.

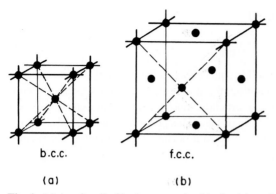

b.c.c.   f.c.c.

(a)   (b)

Fig. 1: (a) A unit cell of body-centered cubic (bcc) iron. Iron atoms are spheres that touch along the dotted lines (body center to cell edge). (b) A unit cell of face-centered cubic (fcc) iron. The cubic unit cell is larger than the bcc cell because the iron atoms now touch from the face center to the cell edge.

Carbon

Carbon exists in two structural forms: graphite and diamond. Graphite has a layered structure with weak interlayer bonding. Because of the weak interlayer bonds the layers slide easily over each other. Consequently graphite is used as a lubricant and in pencil lead. Graphite becomes diamond under high pressure (greater than 10000 bar). With Ni as a catalyst, small diamonds can be manufactured at 1200 °C and high pressure. These diamonds are used in making grinding wheels and cutting tools, since diamond is a very hard material. Recently, diamond films up to millimeters thick have been prepared in the laboratory by a low-pressure plasma deposition technique using methane as a source gas.

Conclusion

Obviously the properties of some materials are quite affected by a change in their structure, and a knowledge of the allotropic or polymorphic forms of materials is important to the development of many technologies.

See also: Crystal Symmetry; Metallurgy.

Bibliography
B. Meyer, *Elemental Sulfur*. Wiley (Interscience), New York, 1965. (I)
A. L. Ruoff, *Introduction to Materials Science*. Prentice-Hall, Englewood Cliffs, NJ, 1972. (I)
W. J. Moore, *Physical Chemistry*, 3rd ed. Prentice-Hall, Englewood Cliffs, NJ 1963. (A)

Alloys

P. L. Leath

An alloy is a macroscopically homogeneous mixture (solution or compound) of metals or, as in the case of carbon steel, a metallic mixture of metals and nonmetals. If they are macroscopically inhomogeneous they would often be called *composites*. Most, but not all, alloys are metallic (for one exception, indium antimonide is a semiconductor). Most pairs of metals are miscible (i. e., form *binary alloys*) at some concentrations, although there are many notable exceptions (e. g., indium is insoluble in gallium). Since there are 70 elemental metals, the subject of alloys is immense, and an enormous variety of electronic and other physical properties is possible. The subject has now expanded even further with recent interest in ternary (three-component) alloys, tertiary or quaternary (four-component) alloys, etc. Several examples of alloys include cast iron, steel, stainless steel, brass, bronze, pewter, solder, intermetallics, aluminum alloys, stellite, chromel, mu-metals, nichrome, constantan, invar, alnico, sterling silver, electrum, and type metal.

Alloys are often classified into ordered (or stoichiometric) and disordered alloys. The ordered alloys have the symmetry of a Bravais lattice with a multiatomic unit cell. Their structure is specified by giving the location of each atom in the unit cell. Some alloys exist essentially only as ordered alloys over the corresponding very narrow ranges of composition necessary for stoichiometry; these alloys are called *intermetallic compounds*. Other alloys (such as β-brass) have ordered phases at stoichiometric concentrations when the temperature is below a phase transition temperature but are disordered otherwise. (That is, they undergo an order–disorder phase transition .) More recently *quasiperiodic* alloys or *quasicrystals* (most notably Al_4Mn) have been discovered which display sharp diffraction peaks that form three-dimensional icosahedral patterns with 5-fold symmetry axes and which thus are not Bravais lattices and do not have the translational symmetry of crystals but do have point symmetries.

Disordered alloys, also called *solid solutions*, occur usually over appreciable ranges of composition. Most common are substitutional alloys, where the various types of atoms randomly occupy the normal sites of a lattice. But there are also *interstitial* alloys (such as carbon in γ-iron), where the solute atoms are small enough to occupy randomly the interstices between the normal lattice sites of the host metal; and there are *amorphous* alloys, where the atoms are not on sites of a regular lattice but are randomly placed, as in a liquid or glass. The occupations of the sites in a disordered alloy by the atomic types may be purely random but generally there is some degree of short-range order; that is, the species occupation of a particular site may be dependent on the occupation of the neighboring sites (e. g., in the disordered phases of brass, the copper atoms are more likely to have zinc than copper nearest neighbors).

Generally, alloys do not have a single melting point, but a solidus temperature at which melting begins, and a liquidus at which melting is complete. In specially designed *eutectic* mixtures, these two temperatures merge into a single melting point.

Certain principal variables that qualitatively give the alloy structures and phases were pointed out in the classic work of Hume-Rothery and Jones. It is, however, only rarely possible to use these few variables to predict detailed behavior of alloy phases. Clearly the relative sizes of the atoms constitute a vital factor in alloys because the volume-dependent potentials in the cohesive energy are an order of magnitude larger than the interatomic rearrangement potentials. This size factor is especially important in interstitial alloys and certain intermetallic compounds (e. g., interstitial alloys are generally not formed when the ratio of the radii of the solute atoms to those of the host atoms exceeds about 0.6). Electrochemical differences are such that generally we find only intermetallic compounds or very restricted ranges of solubility for elements widely separated in the electrochemical series. Particularly interesting are the interstitial alloys of hydrogen in metals.

When size and electrochemical factors allow solid solutions, the alloy structure can in some cases be directly related to the electron density or electron-to-atom ratio. According to the Hume-Rothery rules, in nearly free-electron alloys the stable crystal structure at a particular electron-to-atom ratio will be that which minimizes the energies of the electrons in the crystal potential; thus the position of the Fermi surface relative to the Brillouin zone faces is an essential ingredient. These rules seem to work qualitatively well for the *d*-band transition metal alloys (especially copper, silver, and gold alloys), but they fail for the more nearly free-electron alkali metal alloys because of electrochemical differences. Clearly the *d* bands play an important role.

Only recently have basic calculational methods been developed to predict the physical behavior of alloys accurately from first principles. The pseudopotential, orthogonalized plane wave (OPW), augmented plane wave (APW), and Korringa–Kohn–Rostoker (KKR) Green's function methods of calculating electronic energy-band structure of metals beginning with the atomic potentials (the atomic potentials in alloys look very much the same as those atomic potentials do in the respective pure metals) are now in many cases being directly applied successfully in the calculation of such physical properties of alloys as energy-band structure, crystal structure, lattice vibration spectra, electrical and thermal resistivity, and magnetic and superconducting properties. In those cases where the potentials of the constituent atoms do not vary greatly the average potential may be used; this is called the *virtual crystal* (or rigid band, or common band) *approximation*. In the cases of strong disorder, when the potentials differ greatly, the *average t-matrix approximation* (ATA) and the *coherent potential approximation* (CPA), which are capable of producing the separate energy bands for each atomic species, are used. Although these calculations have been somewhat successful, such effects as charge transfer between atoms and atomic cluster effects are often important but are not included in the simple approximations just mentioned. A fine review of the experimental electronic properties of alloys is given by Sellmyer (1978).

Disordered alloys are dramatically different from ordered alloys and pure metals in their electrical resistance at low temperatures. In ordered metals there is a striking decrease in resistance with decreasing temperature that is absent in disordered alloys. For example, in very pure disordered brass at liquid-helium temperatures the electrical resistance is about half its room-temperature value, in contrast to drops by factors of about 10^{-4} in comparable ordered metals. This phenomenon is caused by electronic scattering off of the disorder or of those regions where the periodicity is destroyed.

Finally, the physical properties of alloys are often greatly affected by heat and mechanical treatment, which may introduce or eliminate such defects as vacancies, dislocations, or grain boundaries. For example, wrought alloys, which have been hot or cold worked and hence are generally very anisotropic and fibrous in contrast to cast alloys, which are generally crystalline, are generally more ductile. And there are *shape memory alloys* (notably Ti–Ni) which under certain treatment will return to an original shape. The effect of such defects is only understood qualitatively, although progress is rapidly being made.

See also: Electron Energy States in Solids and Liquids; Metals.

Bibliography

G. Alefeld and J. Volkl, eds. *Hydrogen in Metals*, Vols. I & II. Springer-Verlag, Berlin, 1978.

R. Banks, *Shape Memory Effects in Alloys*. Plenum, New York, 1975.

C. S. Barrett and T. B. Massalski, *Structure of Metals*, 3rd ed. McGraw-Hill, New York, 1966. (I)

R. J. Elliott, J. A. Krumhansl, and P. L. Leath, "The Theory and Properties of Randomly Disordered Crystals and Related Physical Systems," *Rev. Mod. Phys.* **46**, 465–543 (1974). (A)

M. Hansen and K. Anderko, *Constitution of Binary Alloys*, 2nd ed. (1958); R. P. Elliott, 1st suppl. (1965); F. A. Shunk, 2nd suppl. (1969). McGraw-Hill, New York. (A compendium of data on specific alloys.)

V. Heine and D. Weaire, "Pseudopotential Theory of Cohesion and Structure," in *Solid State Physics* (F. Seitz, D. Turnbull, and H. Ehrenreich, eds.), Vol. 24, pp. 249–463. Academic, New York, 1970. (A)

J. Janssen, M. Fallon, and L. Delacy, *Strength of Metals and Alloys* (P. Haasen, ed.). Pergamon, London, 1979.

F.E. Luborsky, ed, *Amorphous Metallic Alloys*, Butterworths, London and Boston, 1983.

N. F. Mott and H. Jones, *The Theory and Properties of Metals and Allloys*. Oxford, London and New York, 1936. (E)

P. S. Rudman, J. Stringer, and R. I. Jaffee, eds., Phase Stability in Metals and Alloys. McGraw-Hill, New York, 1967. (I)

D. J. Sellmyer, "Electronic Structure of Metallic Compounds and Alloys," in *Solid State Physics* (H. Ehrenreich, F. Seitz, and D. Turnbull, eds.), Vol. 33, pp. 83–248. Academic, New York, 1978.

W.F. Smith, "Structure and Properties of Engineering Alloys", 2nd ed., McGraw-Hill, New York, 1993.

P. J. Steinhardt and S. Ostlund, *The Physics of Quasicrystals*. World Scientific, Singapore, 1987.

K. Tien and G.S. Ansell, eds., "Alloy and Microstructural Design", Academic, New York, 1976.

Alpha Decay

I. Ahmad

Soon after the discovery of radioactivity by Becquerel in 1896, it was established that three types of radiations are emitted by radioactive substances. The most easily absorbed radiations were named alpha (α) rays. In 1909, Rutherford and Royds obtained a direct experimental proof that α particles are doubly ionized helium atoms. Since α particles are electrically charged they are deflected in electric and magnetic fields and produce intense ionization in matter. The thickness of material required to stop an α particle is called its range and it depends on the kinetic energy of the α particle and on the nature of the stopping medium. The range of an α particle with the typical energy of 6.0 MeV is ~ 5 cm in normal air and ~ 0.05 mm in aluminum.

At present more than 500 α-emitting nuclides are known and most of these are produced artificially. The kinetic energies of α particles range from 1.83 MeV for ^{144}Nd to 11.65 MeV for ^{212m}Po; these energies correspond to α particle velocities of $(1–3)\times10^9$ cm/s. The measured half-lives of known α emitters vary from 3.0×10^{-7} s for ^{212}Po decay to 2.1×10^{15} years for ^{144}Nd decay. Normally α decay occurs from the ground state of the parent nucleus and several groups of α particles (each group contains monoenergetic α particles) are emitted leaving the daughter nucleus in its ground state or in an excited state. In a few cases α particles are also emitted from an excited state of the parent nucleus. These α particles have kinetic energies of 9–12 MeV and are called long-range α particles. Examples of such α emitters are ^{212}Po and ^{214}Po.

Alpha decay has recently been used to characterize newly produced transactinide elements. By following the decay chain down to a known nuclide, it has been possible to determine the atomic number of a new element. This procedure has been used to identify elements with $Z = 107$–112.

For a nucleus to be unstable toward α decay, its mass must be greater than the sum of the masses of the daughter nucleus and the α particle. If we write the α decay of a nucleus with mass number A and atomic number Z as

$$^{A}Z \rightarrow {}^{(A-4)}(Z-2) + {}^{4}_{2}\text{He} ,$$ (1)

then the α decay energy, also called Q value, is given by

$$Q = (M_A - M_{A-4} - M_{\text{He}})c^2 .$$ (2)

In the above equation, M represents the atomic mass and c is the velocity of light. Q values have been calculated from known atomic masses. Calculations show that Q values are positive for all β-stable nuclei with $A >\sim 150$. Although such nuclei are thus unstable with regard to α emission, in many cases the half-life is too long for the α decay to have been detected. Experimentally α radioactivity has been detected in most translead and some rare-earth nuclei.

In order to conserve linear momentum, the decay energy Q is divided between the α particle and the daughter nucleus in inverse proportion to their masses. The energy imparted to the daughter nucleus is called recoil energy. The α particle energy E_α and the recoil energy E_R are given by the equations

$$E_\alpha = (M_{A-4}/M_A)Q \quad \text{and} \quad E_R = (M_\alpha/M_A)Q .$$ (3)

The laws of conservation of angular momentum and of parity (even or odd character of the state wave function) plus the fact that the α particle has no intrinsic spin and even parity lead to simple selection rules for α decay. The orbital angular momentum L of the emitted α particle is restricted to integral values between the sum and the difference of the total spins of the initial and final nuclear states. If the parent and the daughter nuclear states have the same parity, only even values of L are permitted; if their parities are opposite, only odd L values are allowed.

The energies of α particles and the intensities of α groups are measured with gas ionization counters, solid-state detectors, or magnetic spectrographs. At present, Passivated Implanted Planar Silicon (PIPS) detectors are widely used in spectroscopic measurements. These silicon detectors, under the best conditions, have resolutions [full width at half-maximum (FWHM) of the α peak] of 9.0 keV and efficiencies of ~30%. Magnetic spectrographs, on the other hand, have low transmission (~0.1%) but can achieve resolution (FWHM) of less than 3.0 keV for

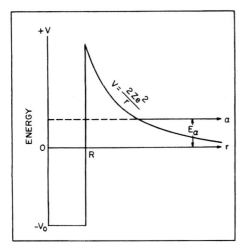

Fig. 1: Schematic representation of the potential energy of an α particle in the vicinity of a heavy nucleus. The potential energy is plotted against the distance r between the centers of the α particle and the residual nucleus.

6.0-MeV α particles. Very thin, essentially massless, sources are used in these measurements. The absolute energies of the α particles emitted by a few nuclides have been measured with high precision by Rytz using a magnetic spectrograph; energies of other α groups are measured relative to these standards. Because of the monoenergetic character of α groups plus the fact that their energies and intensities can be measured with high precision, α particle spectroscopy has been extensively used in nuclear-structure studies of heavy elements.

The systematic relationship between the α decay half-life and the decay energy was first discovered by Geiger and Nuttall in 1911. According to a modified version of their rule, when the logarithm of the α decay half-life is plotted against the inverse square root of Q, a straight line is obtained for each element; i. e.,

$$\log t_{1/2} = A/Q^{1/2} + B \, , \tag{4}$$

where A and B are constants and depend on the atomic number Z. This relationship applies only to α transitions between ground states of even–even nuclei. Because of the strong dependence of the α decay rate on the Q value, only states up to a few hundred kilovolts excitation are measurably populated.

The mechanism of α decay and the Geiger–Nuttall rule were first explained by Gamow and independently by Gurney and Condon in 1927. In the potential-energy diagram (see Fig. 1) the maximum occurs at R, where R is equal to the sum of the radii of the α particle and the residual nucleus. At distances $r < R$ the potential is attractive because of the short-range nuclear force and for $r > R$ electrostatic repulsion gives a positive potential which decreases with increasing r according to Coulomb's law. The typical energy of an α particle within the nucleus, with respect to the zero of energy at $r = \infty$, is 6.0 MeV and the height of the potential barrier at R (called the Coulomb barrier) for heavy elements is ~ 20 MeV. Since the kinetic energy of the α particle is less than the barrier height, according to classical mechanics the α particle can never leave the nucleus. However, the wave nature of matter permits a

6.0-MeV α particle occasionally, on the nuclear time scale, to "tunnel" through the 20-MeV potential barrier. Using a simplified shape for the potential and assuming that the α particle pre-exists as a clustered entity inside the nucleus and is constantly impinging on the barrier, Gamow derived an expression for the α decay rate which explains the observed exponential dependence of transition rates on Q values.

The measured partial half-lives of α groups in the decay of odd-mass and odd–odd nuclei and α transitions to the excited states of even-even nuclei are found to be longer than the partial half-life $(t_{1/2})_{e-e}$ of an α group of the same energy emitted in the decay between ground states of even–even nuclei. The relative retardation of the former decay is called its hindrance factor and its reciprocal gives the relative reduced α transition probability. The values of $(t_{1/2})_{e-e}$ are calculated either by Eq. (4) or by some other theory. In most recent publications the values of $(t_{1/2})_{e-e}$ are computed with the one-body α decay theory of Preston. In this theory, as in Gamow's, the α particle is assumed to preexist inside the nucleus and is ejected with no orbital angular momentum. Radius parameters of even–even nuclei are obtained by normalizing to the measured transition rates between their ground states and the radius parameters of odd-mass nuclei are determined by interpolation between the values of adjacent even–even nuclei.

Hindrance factors for α transitions of odd-mass nuclei vary from unity to several thousands and these yield significant information on the parent and daughter states involved in the decay. Alpha transitions of odd-mass and odd–odd nuclei with hindrance factors of 1–4 are called favored transitions; in these decays the parent and the daughter states have similar wave functions. Since the α transitions are only mildly inhibited by angular momentum changes L, high hindrance factors give a clear indication that the α particle does not exist as a clustered entity all the time in the corresponding nucleus. Instead, the α particle is formed from four nucleons (two protons and two neutrons) at the nuclear surface at the time of its ejection. The probability for the formation of an α particle from four nucleons can be calculated theoretically. Alpha-decay rates for spherical nuclei in the lead region and spheroidal actinide nuclei have been calculated by Mang and Rasmussen and these reproduce the general trend in the observed α decay rates. Although in most cases the calculated and measured rates agree within a factor of 2, there are several unfavored transitions for which the calculations and measurements differ by a factor of ~10. Despite these deficiencies, these calculations are extremely useful in nuclear structure studies.

See also: Nuclear Properties; Radioactivity.

Bibliography

R. D. Evans, in *McGraw-Hill Encyclopedia of Science and Technology*, Vol. 1, p. 305. McGraw-Hill, New York, 1971. (E)

I. Perlman and J. O. Rasmussen, in *Handbuch der Physik*, Vol. 42, p. 109. Springer-Verlag, Berlin, 1957. (I)

J. O. Rasmussen, in *Alpha-, Beta-, and Gamma-Ray Spectroscopy* (K. Siegbahn, ed.), Vol. 1, p. 701. North-Holland, Amsterdam, 1965. (A)

Ampère's Law

L. T. Klauder, Jr.

The name Ampère's law has been applied to several of the formulas that give magnetic effects of time-independent electric currents. (Formulas given here assume rationalized mks units.) The equation

$$\oint_C \mathbf{H} \cdot d\mathbf{I} = \mathbf{I} \tag{1}$$

relating the magnetic intensity along a closed curve C and the current I linking C is commonly referred to as Ampère's law or Ampère's circuital law. This nomenclature has become popular in recent years because it associates a useful elementary formula with the most important of the original investigators. Historically, this law was discovered by Gauss with help from the theorem of Ampére stating that the field produced by a magnetic shell is the same as that due to a current flowing around the boundary of the shell. The modern form distinguishing between B and H was first given by Maxwell.

A few authors apply the term Ampère's law to both the integral relationship (1) and the corresponding differential equation

$$\nabla \times \mathbf{H} = \mathbf{j} \tag{2}$$

where $\mathbf{j}$ is the electric current density. (For the generalization to cases in which fields are time dependent, *see* Maxwell's equations.)

A number of authors apply the term Ampère's law to the formula for the force exerted by a current element $I_2\,d\mathbf{l}_2$ on another current element $I_1\,d\mathbf{l}_1$:

$$\begin{aligned} d\mathbf{F}_{12} &= \frac{\mu_0}{4\pi} \frac{I_1 I_2}{r_{12}^3} d\mathbf{l}_1 \times (d\mathbf{l}_2 \times \mathbf{r}_{12}) \\ &= \frac{\mu_0}{4\pi} \frac{I_1 I_2}{r_{12}^3} [(d\mathbf{l}_1 \cdot \mathbf{r}_{12})d\mathbf{l}_2 - (d\mathbf{l}_1 \cdot d\mathbf{l}_2)\mathbf{r}_{12}] \end{aligned} \tag{3}$$

where $\mu_0 = 4\pi \times 10^{-7}$ is the permeability of free space and $\mathbf{r}_{12}$ is the vector from current path element $d\mathbf{l}_1$ to $d\mathbf{l}_2$. This formula is the basis for the SI unit of electric current referred to as the absolute Ampère.

The existence of the interaction between electric currents was discovered by Ampère in 1820, and he subsequently carried out a remarkable program of experiments and analysis that led him to a formula related to Eq. (3). The reason for the difference is itself interesting. Ampère shared the general view that electrostatic and gravitational forces were cases of action at a distance and assumed that the same was true of the force between currents. Thus, in the interest of conservation of momentum, he assumed that the force between two current elements would have to be directed along the line between them. Accordingly, his result lacked the first term in the second line of Eq. (3) but included another term directed along $\mathbf{r}_{12}$ and causing the entire expression to conform to his experimental result that the force exerted by a closed electric circuit on a current element is perpendicular to the current element. When applied to complete circuits, the formula deduced by Ampère gives the same results as Eq. (3). It can be shown that Eq. (3) is consistent with conservation of momentum as long as the momentum of the electromagnetic field is taken into account.

Following a suggestion by Heaviside, some authors have applied the term Ampère's law to the related formula

$$d\mathbf{F} = I \, d\mathbf{l} \times \mathbf{B} \,, \tag{4}$$

giving the force exerted by the magnetic field $\mathbf{B}$ on the current element $I \, d\mathbf{l}$.

Finally, a number of authors apply the term Ampère's law to the formula

$$d\mathbf{H}_1 = \frac{1}{4\pi} \frac{I_2}{r_{12}^3} \, d\mathbf{l}_2 \times \mathbf{r}_{12} \,, \tag{5}$$

giving the contribution of a current element $I_2 \, d\mathbf{l}_2$ to the magnetic intensity at location 1. However, this equation is a little more frequently referred to as the Biot–Savart law. By their experiments, Biot and Savart established the r^{-1} dependence of the force on a magnetic pole due to current in a long straight wire, and Biot credited Laplace with having inferred from their result that the field contribution from a current element must have the r^{-2} behavior exhibited in formula (5).

See also: Maxwell's Equations; Electrodynamics, Classical; Electromagnets.

Bibliography

For physical explanations of the formulas in this article see any college physics text.

For a discussion of Ampère's work from the point of view of the history of ideas, see the article on Ampère in the *Dictionary of Scientific Biography*, C. C. Gillespie (ed.). Scribners, New York, 1970.

Historical references for Eqns. (1) and (2) are C. F. Gauss's article "Allgemeine Theorie des Erdmagnetismus" in *Carl Friederich Gauss, Werke*, Vol. 5, pp. 170, 171; (Göttingen, 1867) and J. C. Maxwell's article "On Faraday's Lines of Force" in *Trans. Camb. Phil. Soc.* **10**, 27 (1856) [reprinted in Vol. 1 of *The Scientific Papers of J. C. Maxwell*. Cambridge, 1890].

Ampère's counterpart to Eq. (3) is discussed in E. T. Whittaker's *A History of the Theories of Aether and Electricity*, 2nd ed., Vol. 1, pp. 85–87, (London, 1951; reprinted by Harper, New York, 1960) and in J. C. Maxwell's *A Treatise on Electricity and Magnetism*. 3rd ed., Vol. 2, pp. 163–174 (Oxford, 1892). Translations of most of Ampère's papers are available in R. A. R. Tricker's *Early Electrodynamics: The First Law of Circulation* (Pergamon, New York, 1965).

For a demonstration that formula (3) does not violate conservation of momentum when the role of the electromagnetic field is included, see the article by L. Page and N. E. Adams in *Am. J. Phys.* **13**, 141 (1945). For an extended treatment see F. Rohrlich, *Classical Charged Particles* (Addison-Wesley, Reading, Mass., 1965).

Anelasticity

A. S. Nowick

The term anelasticity, although once used loosely to refer to nonelastic behavior, was given a more specific meaning by C. Zener in 1946; this meaning has since been widely adopted. Anelasticity is a generalization of Hooke's law of elasticity, which allows for time-dependent

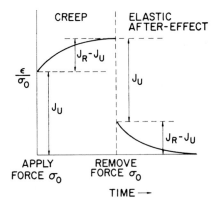

Fig. 1: Behavior of the standard anelastic solid upon application of a static stress σ_0 (creep), and upon the subsequent release of this stress (elastic aftereffect).

effects. Hooke's law may be stated as $\varepsilon = J\sigma$, where ε is strain, σ is stress, and J is the compliance constant. In anelasticity, the instantaneous response and single-valuedness inherent in Hooke's law are discarded. However, two restrictions are retained: (a) *linearity*, in the sense that doubling the stress doubles the strain at each instant of time; and (b) a *unique equilibrium relationship*, which means that to every value of stress there corresponds a unique value of strain that is attained if sufficient time is allowed. The simplest relation between stress and strain and their time derivatives that obeys these conditions is

$$J_R\sigma + \tau J_U\dot{\sigma} = \varepsilon + \tau\dot{\varepsilon} , \tag{1}$$

involving three constants σ, J_R, and J_U. Any material that obeys Eq. (1) is called a *standard anelastic solid*. Equation (1) can be solved under conditions of constant stress, say σ_0, (which constitutes a "creep" experiment), to give

$$\frac{\varepsilon(t)}{\sigma_0} = J_U + (J_R - J_U)\left[1 - \exp\left(\frac{-t}{\tau}\right)\right] . \tag{2}$$

From this equation the meaning of the constants becomes clear: J_U, called the unrelaxed compliance, corresponds to the instantaneous response, $\varepsilon(0)/\sigma_0$, at $t = 0$; J_R, the relaxed compliance, is ε/σ_0 as $t \to \infty$; τ is the relaxation time (at constant stress). Figure 1 shows this creep behavior, as well as the time-dependent recovery, or "elastic aftereffect," which takes place after the stress is removed. Equation (1) can also be solved under conditions of constant strain to obtain an exponentially decreasing stress, describing a "stress-relaxation experiment." The most important manifestation of anelasticity, however, occurs in the dynamical case, where stress and strain are both periodic, with the strain lagging behind the stress by a phase angle ϕ. Then we can express a complex compliance by $J^* = \varepsilon/\sigma \equiv J_1 - iJ_2$, where J_1 is the real part, which is in phase with the applied stress, and J_2 the imaginary part, which lags σ by $\pi/2$. Using Eq. (1), we can express J_1 and J_2 as functions of the angular frequency ω:

$$J_1(\omega) = J_U + (J_R - J_U)/(1 + \omega^2\tau^2) , \tag{3}$$
$$J_2(\omega) = (J_R - J_U)\omega\tau/(1 + \omega^2\tau^2) , \tag{4}$$

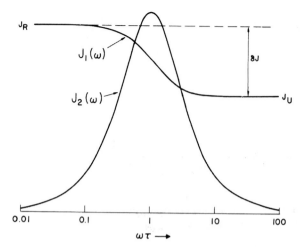

Fig. 2: Dependence of the dynamical functions $J_1(\omega)$ and $J_2(\omega)$ on $\omega\tau$ for the standard anelastic solid.

Equations (3) and (4) are the celebrated Debye equations. The function $J_2(\omega)$ when plotted versus $\log(\omega\tau)$ gives a symmetrical peak centered about $\log(\omega\tau) = 0$ (i. e., $\omega\tau = 1$), which is called a Debye peak. Figure 2 shows the variation of both J_1 and J_2 with the variable $\log(\omega\tau)$. The phase angle ϕ by which ε lags behind a is given by $\tan\phi = J_2/J_1$. This quantity, which also takes the form of a Debye peak, is often called the internal friction, since it is a measure of the energy dissipated per cycle.

Since the Debye peak depends on the variable $\omega\tau$, it can be traced out either by varying the frequency, ω, or by changing τ. Although a continuous variation of the vibration frequency is sometimes possible, the usual experimental methods make it preferable to work at one frequency. It is therefore quite valuable to have a method for tracing out a Debye peak by varying τ while keeping ω constant. This is possible when τ is controlled by a thermally activated process involving an Arrhenius-type relation

$$\tau = \tau_0 \exp(Q/kT) \tag{5}$$

where Q is the activation energy for the process, τ_0 is the preexponential constant, k is Boltzmann's constant, and T is the absolute temperature. In fact, it then turns out that a plot of $\tan\phi$ (or J_2) versus T^{-1} gives a symmetrical peak like that in Fig. 2 except for a change in scale factor. If the peak is then obtained at two or more different frequencies, the activation energy is readily obtained from the shift of the peak location with frequency. Figure 3 shows an example of this type of plot, which is most important in studying anelastic phenomena.

Many phenomena in solids are describable in terms of the equations of the standard anelastic solid. Often, however, this simple model is insufficient to describe the behavior of a material. For example, the internal friction may show a superposition of two or more Debye peaks instead of a single one, or in other cases a single peak may be obtained that is broader than a Debye peak. To treat such cases, it is necessary to introduce a spectrum of relaxation times in place of the single relaxation time of the standard anelastic solid. Thus, instead of a single exponential in the creep function of Eq. (2) there may be a summation of terms with different

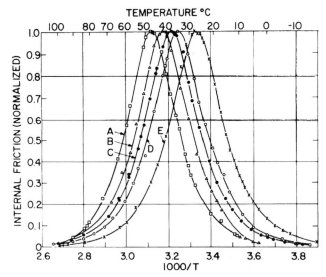

Fig. 3: A series of internal friction peaks for an Fe–C alloy as function of $1/T$ for five different frequencies: A, 2.1; B, 1.17; C, 0.86; D, 0.63; and E, 0.27 Hz. From C. Wert and C. Zener, *Phys. Rev.* **76**, 1169 (1949).

τ values and weighting factors. Correspondingly, Eq. (4) becomes a sum of Debye peaks. Such behavior is called a discrete relaxation spectrum. In more complex cases there may be a continuous variation in τ, a continuous spectrum, described by an appropriate distribution function. In either of these situations, the display of data in the form of a plot of internal friction ($\tan\phi$) versus T^{-1} is widely used and interpreted.

The physical origins of anelasticity are very varied and encompass almost all aspects of solid-state physics. In crystalline materials, anelastic behavior can result from any of the following mechanisms:

1. Point-defect relaxations: redistribution of point defects (whose symmetry is lower than that of the crystal) into sites that become preferential in the presence of a stress field.

2. Dislocation relaxations: motion of dislocation segments, present either from growth or from plastic deformation, in a variety of ways with the aid of jogs, kinks, and impurity atoms on the dislocation lines.

3. Grain-boundary relaxation: viscous sliding of one grain over another in a polycrystalline material.

4. Phonon relaxation: change in the frequency distribution of phonons (lattice vibrations) due to stress.

5. Magnetic relaxations: magnetoelastic coupling via magnetostriction of a ferromagnetic material, giving rise to a number of different relaxation processes.

6. Electronic relaxations: change in the energetics of the electronic configuration produced by stress leading to a redistribution of both free and bound electrons in various materials.

Glasses (including amorphous alloys) are also capable of showing a variety of relaxations, many of which are similar in origin to those found in crystalline materials. In amorphous polymers a major relaxation is associated with the glass transition and attributed to large-scale rearrangements of the main polymer chain. Secondary relaxations, at lower temperatures, are due to side groups that are capable of independent hindered rotations.

See also: Elasticity; Relaxation Phenomena; Rheology.

Bibliography

W. Benoit and G. Gremaud, eds., "Internal Friction and Ultrasonic Attenuation in Solids." *J. Phys. (Paris)* **42**, Colloque No. 5 (1981).

R. De Batist, *Internal Friction of Structural Defects in Crystalline Solids.* North-Holland, Amsterdam, 1972.

R. De Batist and J. Van Humbeeck, eds., "Internal Friction and Ultrasonic Attenuation in Solids." *J. Phys. (Paris)* **48**, Colloque C8 (1987).

J. D. Ferry, *Viscoelastic Properties of Polymers.* Wiley, New York, 1961.

A. V. Granato, G. Mozurkewich, and C. A. Wert (eds.), "Internal Friction and Ultrasonic Attenuation in Solids." *J. Phys. (Paris)* **46**, Colloque C10 (1985).

R. R. Hasiguti and N. Mikoshiba, eds., *Internal Friction and Ultrasonic Attenuation in Solids.* Univ. Tokyo Press, Tokyo, 1977.

D. Lenz and K. Lücke, eds., *Internal Friction and Ultrasonic Attenuation in Crystalline Solids*, Vols. I and II. Springer, Berlin and New York, 1975.

W. P. Mason and R. N. Thurston (eds.), *Physical Acoustics*, Vols. 1–18. Academic Press, New York, 1964–1988.

N. G. McCrum, B. E. Read, and G. Williams, *Anelastic and Dielectric Effects in Polymeric Solids.* Wiley, New York, 1967.

A. S. Nowick and B. S. Berry, *Anelastic Relaxation in Crystalline Solids.* Academic Press, New York, 1972.

R. Truell, C. Elbaum, and B. B. Chick, *Ultrasonic Methods in Solid State Physics.* Academic Press, New York, 1969.

C. Zener, *Elasticity and Anelasticity of Metals.* Univ. of Chicago Press, Chicago, 1948.

Angular Correlation of Nuclear Radiation
R. M. Steffen[†]

The probability of emission of a particle or a quantum by a decaying nucleus depends, in general, on the angle between the nuclear spin axis I and the direction of emission k. In most cases (e. g., ordinary radioactive sources) the total radiation is isotropic, because the nuclear spin axes are randomly oriented in space. An anisotropic intensity distribution of the radiation is only observed if it is emitted from an ensemble of nuclei that is *not* randomly oriented, i. e., in which the spin axes of the decaying nuclei show some preferred direction in space. Such an ensemble is called an *oriented ensemble*.

[†]deceased

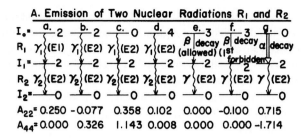

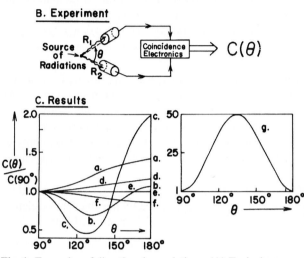

Fig. 1: Examples of directional correlations. (A) Typical gamma–gamma radiation cascades [(a)–(d)], beta–gamma radiation cascades [(e)–(f)], and an alpha–gamma cascade (g). Below each cascade are given the values of the directional correlation coefficients. (B) Experimental arrangements for measurement of directional correlations. (C) Directional correlations of the radiation cascades (a)–(g).

Oriented ensembles of nuclei can be prepared, e. g., by placing a radioactive sample at a very low temperature in strong magnetic or electrostatic gradient fields, thereby polarizing or aligning the nuclei by virtue of the interaction of the magnetic and electric moments of the nuclei with the external fields. The angular distribution of the radiation emitted by such an oriented source is then, in general, anisotropic with respect to the direction of the applied fields. Another method of preparing an oriented subensemble of nuclei is based on selecting only those nuclei whose spin axes happen to be in a preferred direction. Nuclear reactions or decay processes that lead to the formation of nuclei in a particular excited state of spin I_1 (the "intermediate" state) can be used in such a selection process.

Many nuclei decay through the *successive emissions of two radiations* R_1 and R_2 via a short-lived (lifetime $\tau \lesssim 10^{-9}$ s) intermediate nuclear state of spin I_1. Some examples of such cascade decays are depicted in Fig. 1A. The observation of RE in a fixed direction $\mathbf{k}_1$ selects from the originally random ensemble of nuclei with spin I_0 a subensemble of nuclei in the intermediate state with spin I_1. This subensemble has, in general, a preferred direction of the spin axes $\mathbf{I}_1$ with respect to the observation direction $\mathbf{k}_1$ of R_1, because the radiation emission probability depends on the angle between $\mathbf{k}_1$ and $\mathbf{I}_0$. Since the second radiation R_2 is now emitted from this *oriented* subensemble of spin I_1, the intensity of R_2 observed in a direction $\mathbf{k}_2$, depends, in general, on the angle θ between $\mathbf{k}_1$ and $\mathbf{k}_2$, i.e., the second radiation R_2 has an anisotropic angular distribution with respect to the direction $\mathbf{k}_1$ in which R_1 has been observed. The angular distribution of R_2 with respect to $\mathbf{k}_1$ (or of R_1 with respect to $\mathbf{k}_2$) is called the *angular correlation* of the radiations R_1 and R_2.

If only the propagation directions (no polarization phenomena) of the two radiations are measured, the *directional* correlation is observed. If the linear or circular polarization of one or of both of the radiations is measured, a *polarization–directional correlation* or a *polarization–polarization correlation*, respectively, is observed. The term angular correlation comprises all three cases.

The observation of an angular correlation requires a coincidence experiment, i.e., the two radiations R_1 and R_2 must be recorded, each in one of two detectors that respond only if R_1 and R_2 strike the detectors simultaneously (actually within a very short time interval $\tau_0 \approx 10^{-9} - 10^{-8}$ s) in order to maximize the probability that the observed radiations R_1 and R_2 are emitted from the same nucleus. A directional correlation experiment consists thus simply of measuring the coincidence rate $C(\theta)$ of R_1 and R_2 as a function of the angle θ between the axes of the two detectors (Fig. 1B).

The relative probability $W(\theta)\, d\Omega$ that R_2 is emitted into the solid angle $d\Omega$ at an angle θ with respect to the propagation direction $\mathbf{k}_1$ of R_1 is characterized by the angular correlation function $W(\theta)$. For an *ordinary directional correlation*, $W(\theta)$ can be expressed in the general form

$$W(\theta) = 1 + A_{22}P_2(\cos\theta) + A_{44}P_4(\cos\theta) . \tag{1}$$

The angular functions $P_i(\cos\theta)$ are Legendre polynomials, i.e., $P_2(\cos\theta) = (3\cos^2\theta - 1)/2$ and $P4(\cos\theta) = (35\cos^4\theta - 30\cos^2\theta + 3)/8$. The directional correlation coefficients A_{ii} ($i = 2, 4$) can be expressed as the product, $A_{ii} = A_i(R_1, I_1; I_0) \cdot A_i(R_2, I_1; I_2)$, of two directional distribution coefficients $A_i(R_1, I_1; I_0)$ and $A_i(R_2, I_1; I_2)$, each being characteristic of one of the two emission processes R_1 and R_2 that make up the radiation cascade. The directional distribution coefficients $A_i(R, I; I')$ depend on the properties of the radiation R that is emitted in the transition $I \to I'$ and on the spins I and I' of the initial and final nuclear states, respectively. In particular, the distribution coefficients depend on the so-called multipolarity L of the emitted radiation R. A 2^L-pole radiation carries away an angular momentum of $L\hbar$ with respect to the center of the emitting nuclei. The directional distribution coefficients, however, do not depend on the reflection symmetry of the emitted radiation. For emission of gamma radiation, e.g., the directional distribution coefficients do not distinguish between electric 2^L-pole (EL) and magnetic 2^L-pole (ML) radiation. The observation of the linear polarization of the gamma radiation is required to distinguish between EL and ML radiation.

The directional distribution coefficients $A_i(\gamma, I; I')$ for gamma transitions do not depend on the energy of the gamma transitions. For alpha-particle emission the $A_i(\alpha, I; I')$ depend on the energy of the alpha particles only if two (or more) alpha-particle waves of different L interfere with each other. In beta emission two particles are emitted simultaneously, an electron (or positron) and an antineutrino (or neutrino), of which only the electron (or positron) is, in general, observed in an angular correlation observation. The electrons (or positrons) have a continuous energy spectrum up to a maximum energy E_0 and the directional distribution coefficients for these electrons (or positrons) depend on the energy of the observed particle.

Theoretical expressions for the directional distribution coefficients (and for polarization distribution coefficients) are available for all types of radiations and for all cases of interest. For details see Refs. [1]–[4]. Four illustrative examples of various multipole gamma–gamma directional correlations are shown in Figs. 1A and 1C, (a)–(d). Beta–gamma directional correlations involving so-called allowed beta transitions are isotropic (e), first-forbidden beta–gamma directional correlations (f) are, in general, nonisotropic. Alpha–gamma directional correlations can show very large anisotropies (g).

Angular correlations are, in general, observed with the initial nuclear state of spin I_0, which emits R_1, randomly oriented (ordinary angular correlation). If R_1 itself is emitted from an oriented state, e. g., from a state produced in a nuclear reaction, the angular correlation from an oriented state (ACO) or the directional correlation from an oriented state (DCO) is observed.

An equivalent situation prevails in triple angular correlations where the angular correlation of the radiations R_1 and R_2 is observed with respect to the observation direction $\mathbf{k}_0$ of a preceding radiation R_0 that is emitted from a random ensemble I_{00} resulting in an oriented nuclear ensemble I_0 from which the radiation R_1 is emitted.

Directional correlations of two successively emitted gamma radiations R_1 and R_2 *emitted by an oriented nuclear ensemble* I_0 that is axially symmetric with respect to a direction $\mathbf{k}_0$ are characterized by a correlation function of the general form

$$W(\theta_1, \tag{2}$$

$$theta_2; \varphi) = \sum_{i,k,l} B_l(I_0) A_l^{ki}(R_1, I_0, I_1)$$

$$\times A_{ll}(R_2, I_1, I_2) H_{ikl}(\theta_1, \theta_2, \varphi) \tag{3}$$

$$(i, k, l) = 0, 2, 4$$

where θ_1 and θ_2 are the polar angles, with respect to the orientation axis $\mathbf{k}_0$, of the directions $\mathbf{k}_1$ and $\mathbf{k}_2$ in which the radiations R_1 and R_2, respectively, are observed and the azimuthal angle φ is the angle between the planes determined by $\mathbf{k}_0\mathbf{k}_2$ and $\mathbf{k}_0\mathbf{k}_1$. The parameter $B_1(I_0)$ describes the state of orientation of the nuclear ensemble I_0 and $A_l^{ki}(R_1, I_0, I_1)$ is a generalized directional distribution coefficient. Expressions for the latter and for the angular function $H_{ikl}(\theta_1, \theta_2, \varphi)$ can be found in Ref. 5. DCO measurements are particularly useful in assigning spins to nuclear states that are produced in nuclear reactions and in exploring the multipole character of gamma radiations between such states.

In many experimental situations the time t elapsed between the formation of the intermediate oriented state I_1 by the radiation R_1 and the time moment of emission of the second radiation R_2 is long enough ($\sim 10^{-9} - 10^{-6}$ s) to cause an appreciable change of the orientation of the nuclear ensemble through the interactions of the electromagnetic nuclear moments (magnetic-dipole moment μ, electric-quadrupole moment Q) of the individual nuclei with external fields. In such cases the angular correlation can be influenced by the external fields and a perturbed angular correlation (PAC) is observed (see Refs. [6]–[8]).

A strong external (or internal atomic) magnetic field B, e. g., causes a precession of the magnetic moment μ of the nucleus in the intermediate state about the direction of B as axis (Fig. 2A) with a frequency ω_B that is proportional to B (Larmor precession). The angular

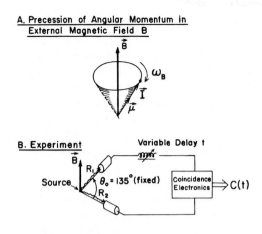

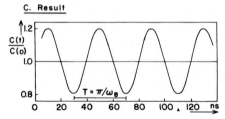

Fig. 2: Extranuclear perturbations by an external magnetic field. (A) Precession of spin about a magnetic field B (Larmor precession). (B) Delayed-coincidence observation. (C) Periodic variation with time t of the relative coincidence rate C(t)/C(0) reflecting the spin precession in the intermediate nuclear state.

distribution pattern of R_2 is then rotated about the direction of B by an angle $\Delta\theta = \omega_B t$. By observation of the angular shift $\Delta\theta$ of the angular correlation pattern, ω_B can be determined and thus either μ or B can be measured. For larger values of $\omega_B \tau$ (i.e., $\omega_B \tau \gtrsim 1$) and if $\tau \gg \tau_0$, the precession of the angular distribution pattern of R_2, i.e., the precession of nuclei in the intermediate state, can be directly observed by measuring the coincidence rate $C(t)$ in two fixed detectors as a function of the time during which the intermediate nuclear state is exposed to the magnetic field B. In practice this is done through delaying (electronically) the detector signal caused by R_1 by a time t before it reaches the coincidence circuit (Fig. 2B). The oscillating behavior of the observed coincidence rate as a function of t represents the spin precession of the nuclei in the intermediate state I_1 (Fig. 2C).

Angular correlation observations are a very important tool in nuclear spectroscopy for the determination of angular momenta and electromagnetic moments of excited nuclear states and for precise measurements of the multipolarities of nuclear radiations. Perturbed angular correlation experiments are also used to explore the magnetic and electric field gradients at the site of nuclei and thus can be applied to atomic, solid-state, and liquid-state problems.

See also: Alpha Decay; Beta Decay; Gamma Decay; Multipole Fields; Nuclear Polarization; Polarization.

References

[1] H. Frauenfelder and R. M. Steffen, "Angular Correlations," in *Alpha, Beta and Gamma Ray Spectroscopy* (K. Siegbahn, ed.), pp. 997–1198. North-Holland, Amsterdam, 1965. (I)

[2] R. M. Steffen and K. Alder, "Angular Distributions and Correlations of Gamma Radiation: Theoretical Basis," in *The Electromagnetic Interaction in Nuclear Spectroscopy* (W. D. Hamilton, ed.), pp. 505–581. North-Holland, Amsterdam, 1975. (A)

[3] S. Devons and L. J. B. Goldfarb, in *Handbuch der Physik* (S. Flügge, ed.), Vol. 42, p. 362. Springer-Verlag, Berlin and New York, 1957. (A)

[4] A. J. Ferguson, *Angular Correlation Methods in Gamma Ray Spectroscopy*. North-Holland, Amsterdam, 1965. (A)

[5] K. S. Krane, R. M. Steffen, and R. M. Wheeler, "Directional Correlations of Gamma Radiations Emitted from Nuclear States Oriented by Nuclear Reactions or Cryogenic Methods," *Atomic and Nuclear Data Tables, Vol. 11*, pp. 351–405. Academic Press, New York, 1975. (A)

[6] R. M. Steffen, "Extranuclear Effects on Angular Correlations of Nuclear Radiation," *Adv. Phys. (Phil. Mag. Suppl.)*, **4**, 293–362 (1955). (E)

[7] R. M. Steffen and H. Frauenfelder, "The Influence of Extranuclear Perturbations in Angular Correlations," in *Perturbed Angular Correlations* (E. Karlsson, E. Mathias, and U. Siegbahn, eds.), pp. 1–89. North-Holland, Amsterdam, 1964. (A)

[8] R. M. Steffen and K. Alder, "Extranuclear Perturbations and Angular Distributions and Correlations," in *The Electromagnetic Interaction in Nuclear Spectroscopy* (W. D. Hamilton, ed.), pp. 583–643. North-Holland, Amsterdam, 1975. (A)

Antimatter

G. Steigman

All quantum theories consistent with the special theory of relativity and the requirement of causality require that particles exist in pairs. Particles and their antiparticles have the same masses and lifetimes; electrically charged particles (e. g., electron, proton) have antiparticles (e. g., positron, antiproton) with equal but opposite electric charges; some electrically neutral particles (e. g., photon) are their own antiparticles (self-conjugate). Following the discovery of the positron (Anderson 1933) there was a hiatus of some 22 years before the antiproton was produced and detected at an accelerator (Chamberlain *et al.* 1955). Subsequent accelerator experiments have provided strong confirmation that all particles do, indeed, exist in pairs and, further, that particles carry certain quantum numbers (baryon number, lepton number, etc.) which *seem* to be conserved in all reactions. If these conservation "laws," inferred from experimental data, are exact, then matter is restricted to appear (creation) or disappear (annihilation) only as particle–antiparticle pairs. This apparent symmetry (about which, more later) in the laws of physics has stimulated serious speculation on the antimatter content of the Universe (Goldhaber 1956) and the possible astrophysical consequences of macroscopic amounts of antimatter (Burbidge and Hoyle 1956). In approaching the issue of the matter–antimatter symmetry (or, asymmetry) of the Universe, it is valuable to distinguish between two distinct questions: Is the Universe symmetric? Must the Universe be symmetric? The first question will be considered before a "modern" (a la "Grand Unified Theory") answer is given to the second question. For further details and references see Steigman (1976).

Searching for Antimatter in the Universe

Antimatter is, in principle, trivially easy to detect. You place your sample in a detector (the most rudimentary device will suffice) and, if the detector disappears (annihilates), the sample was made of antimatter. Unfortunately, only the solar system and the cosmic rays provide a sample of the Universe which may be subjected to such a direct test. Lunar landings and the Venus probes establish that the Moon and Venus are made of ordinary matter. The absence of annihilation gamma rays when the solar wind sweeps through the solar system establishes that the Solar System consists only of (ordinary) matter; were any of the planets made of antimatter, annihilation of the solar wind particles which strike their surfaces would have made them the strongest gamma ray sources in the sky.

Cosmic rays provide the only direct sample of extrasolar system material in the Universe. The cosmic rays, perhaps the debris of exploding stars (supernovae) or the accelerated nuclei of interstellar gas, bring information about the material in our Galaxy. As the cosmic rays traverse the Galaxy they collide with interstellar gas nuclei, occasionally producing (secondary) antiprotons. Therefore, antiprotons in the cosmic rays do not provide an unambiguous signal for the presence of "primary" sources of antimatter (e. g., antistars) in the Galaxy. In contrast, virtually no antihelium (antialpha) nuclei would be present as secondaries in the cosmic rays. The discovery of even one antialpha particle in the cosmic rays would provide compelling evidence for the existence of antimatter in the Galaxy. None has ever been found. In contrast, antiprotons have been observed in the cosmic rays (Golden *et al.*, 1979; Bogomolov *et al.*, 1979). However, the upper limits to the $\bar{p}$ flux at low energies (Ahlen *et al.*, 1988) are completely consistent with a secondary origin; these latest results are in conflict with – and cast doubt on the reality of – an earlier claim (Buffington *et al.*, 1981) of a positive detection.

Astronomy is an observational science, often relying on the interpretation of indirect evidence. When matter and antimatter meet, they annihilate producing, among the debris, gamma rays of energy from several tens to several hundred MeV. The annihilation gamma rays can provide an indirect probe for antimatter in the Universe. However, since gamma rays may be produced by other astrophysical processes (synchrotron radiation, Compton scattering, etc.), they are not an unambiguous signal; the best approach is to use the observed gamma ray flux to place upper limits on annihilation and, hence, on the amount of mixed matter and antimatter at various astrophysical sites (Steigman 1976).

Gamma ray observations of the Galaxy limit the antimatter fraction in the interstellar gas to less than 10^{-15}. Such a small limit is not surprising once it is realized that the lifetime – against annihilation – of an antiparticle in the interstellar gas is less than 300 years (the Galaxy is at least 10 billion years old). Our galaxy is clearly made entirely of ordinary matter. What of other galaxies?

Clusters of galaxies are the largest astrophysical entities which can be probed by gamma rays for a possible antimatter component. Many rich clusters shine in the x-ray part of the spectrum; the x-rays are the thermal bremsstrahlung emission from a hot intracluster gas. From the absence of gamma rays, less than one part in 10^5 of that hot gas could be made of antimatter. Thus, the observational data shows no evidence for astrophysically interesting amounts of antimatter in the Universe. If antimatter were present, it would have to be separated from ordinary matter on scales at least as large as clusters of galaxies ($\sim 10^{15}\,M_\odot \sim 10^{48}\,\text{gm} \sim 10^{72}$ protons). Could matter and antimatter be separated in the Universe on such large scales?

Symmetric Cosmologies

In the context of the hot big bang model, particle–antiparticle pairs were present in great abundance during the early ($\lesssim 10\,\mu$s), hot ($T \gtrsim$ few hundred MeV) epochs in the evolution of the Universe. As the Universe expanded and cooled, however, these pairs annihilated. In a completely symmetric, perfectly mixed Universe the annihilation is so efficient that less than one nucleon–antinucleon pair remains for each 10^{18} microwave background photons in the Universe; this is less matter than is present in our Galaxy alone! Either the Universe at these early times ($\lesssim 10\,\mu$s) was asymmetric or, possibly, matter and antimatter were separated. However, although the Universe was very dense, it was also very small and only ~ 1 g of nucleons (antinucleons) could have been separated by any causal process (none is known which would effect such a separation). Inevitably, then, astrophysicists have led the way to the conclusion that the early Universe ($\lesssim 10\,\mu$s) was asymmetric.

Baryon Asymmetry and GUT

Inspired by the (then recent) discovery of *CP* violation in the $K^0 - -\bar{K}^0$ system, Sakharov (1967) outlined the recipe for generating a baryon asymmetry in an initially (matter–antimatter) symmetric Universe. First, there must be interactions which violate conservation of baryon number (matter–antimatter symmetry is a "broken" symmetry). Next, he noted a technical – but crucial – requirement that *CP* conservation be violated (e. g., the branching ratios for decays into certain channels for particles and antiparticles differ; note that the total lifetimes are still required to be equal). Finally, Sakharov (1967) pointed out that these *B*- and *CP*- violating processes must occur "out of equilibrium." The offspring of the marriage of particle physics (Grand Unified Theories) and cosmology (the expanding, hot big bang model) is endowed with the requisite properties (Yoshimura 1978; Dimopoulos and Susskind 1978; Ellis, Gaillard and Nanopoulos 1979; Toussaint *et al.*, 1979; Weinberg 1979). Baryon-number- and *CP*-violating interactions occurring very early ($\lesssim 10^{-35}$ s) in the evolution of the Universe would have, due to the expansion and cooling of the Universe, dropped out of equilibrium and left behind a small matter–antimatter asymmetry (the net baryon number is $\sim 10^{-10} - 10^{-9}$ of the photon number). This tiny relic, the legacy of the earliest epochs in the evolution of the Universe, is responsible for the presently observed, matter–antimatter asymmetric Universe.

See also: Cosmology; Elementary Particles; Positron.

Bibliography

S. P. Ahlen *et al.*, *Phys. Rev. Lett.* **61**, 145 (1988).

C. D. Anderson, *Phys. Rev.* **43**, 491; 44, 406 (1933).

E. A. Bogomolov *et al.*, *Proc. of the 16th Int. Cosmic Ray Conf.* **1**, 330 (1979).

A. Buffington, S. M. Schindler, and C. R. Pennypacker, *Astrophys. J.* **248**, 1179 (1981).

G. R. Burbidge and F. Hoyle, *Nuovo Cimento* **1**, 558 (1956).

O. Chamberlain, E. Segre, C. Wiegand, and T. Ypsilantis, *Phys. Rev.* **100**, 947 (1955).

S. Dimopoulos and L. Susskind, *Phys. Rev.* **D18**, 4500 (1978).

J. Ellis, M. K. Gaillard, and D. V. Nanopoulos, *Phys. Lett.* **B80**, 360 (1979).

R. L. Golden *et al.*, *Phys. Rev. Lett.* **43**, 1196 (1979).

M. Goldhaber, *Science* **124**, 218 (1956).
A. Sakharov, *JETP Len.* **5**, 24 (1967).
G. Steigman, *Ann. Rev. Astron. Astrophys.* **14**, 339 (1976).
D. Toussaint, S. Treiman, F. Wilczek, and A. Zee, *Phys. Rev.* **D19**, 1036 (1979).
S. Weinberg, *Phys. Rev. Lett.* **42**, 850 (1979).
M. Yoshimura, *Phys. Rev. Lett.* **41**, 381 (1978).

Arcs and Sparks

U. H. Bauder

General Properties

The electric arc is characterized by high current densities, low potential differences between the electrodes, and small differences compared to other discharges between the temperatures of the different particle species present in the column. In equilibrium temperatures exceeding 50 000 K have been reached [1]. Arcs can be operated over a wide pressure range: from several millibars gas pressure to 1000 bars and in all gaseous media. The initiation of an arc discharge may be achieved either by contacting the electrodes or by preionizing the gas in the discharge channel by breakdown processes following Paschen's law [2]. The unstationary discharge preceding a sustained arc discharge is the spark; besides duration time the main difference between arcs and sparks relates to electrode effects. Most stationary arc discharges operate with a combined thermal and field emission of electrons at a hot cathode [3], whereas duration times of sparks are too small to allow for a substantial increase of the bulk temperature of electrodes.

Depending on the external conditions under which the arc is operated one distinguishes between electrode-stabilized (short) arcs, flow-stabilized arcs (longitudinal or swirl flow), wall-stabilized arcs (operated in a segmented cascaded tube if high power levels have to be achieved), high pressure arcs, and vacuum arcs.

The High-Pressure Arc Column

Arcs operated at atmospheric pressure or above are high-pressure arcs. Once established by one of the ignition processes mentioned above, the arc column is sustained by ionizing processes whose energy is supplied by the electric field generated by the power supply. Free electrons coming from the cathode are accelerated in the cathode fall region and in the column during a free path length between two collisions. High-pressure arc columns are characterized by mean free path lengths which are much smaller than the geometrical dimensions of the arc column. Due to their small mass, electrons reach higher velocities in the field than ions; mainly collisions of electrons with molecules, atoms, and ions lead to further ionization. Ionization is possible from the ground state as well as starting from excited states. Depending on the gas species as well as on the plasma density and temperature, photoionization processes may also play a role. Carriers are lost by recombination processes. In some application plasmas such as SF_6 plasmas, electron attachment with the formation of negative ions may also reduce the number density of free electrons in the column. The large difference between the masses of electrons and heavy particles frequently leads to thermal nonequilibrium in the plasma. It takes approximately 10^3–10^4 elastic collisions of electrons with atoms or ions until thermalization is achieved. Due to this fact local thermodynamic equilibrium (LTE) is only achieved in high-temperature arc plasmas with high collision rates (elevated pressures); in other arc columns partial local thermodynamic equilibrium (PLTE) prevails. In the case of PLTE only the higher excited states are populated in equilibrium with the electron temperature, the ground state being largely overpopulated.

Considering the differential energy balance of the cylindrical arc column without convection losses [4] as it is realized experimentally in the cascade arc [5],

$$\sigma E^2 + \frac{1}{r}\frac{d}{dr}\left(r\kappa\frac{dT}{dr}\right) - e + a = 0 . \tag{1}$$

it can be noted that the energy input to the volume element of the plasma column is mainly due to Ohmic heating (σ being the electrical conductivity and E the electric field strength). Absorption processes may also play a role; the quantity a is the volumetric absorption coefficient integrated over all frequencies. The energy is lost from the column by radiation ($e =$ volumetric emission coefficient) and by thermal conduction. The thermal conductivity κ is composed of the contact conductivities of all species; in the temperature regions of dissociation of molecules and that of ionization a large contribution to κ may be due to the radial transport of dissociation and ionization energies. Due to this transport, temperature profiles of arc discharges operating in molecular gases exhibit a "shoulder" as shown in Fig. 1 [6] for the case of an atmospheric pressure nitrogen arc operated in a cascade channel of 5 mm diameter. By contrast, the $T(r)$ profiles of noble gas cascade arc columns at elevated pressures have a flat central portion and very steep gradients at the wall (Fig. 2 [7]). The energy balance of the central portion of such arcs is largely radiation dominated. It has been shown [8] that up to 80% of the energy of a high-pressure argon arc may be lost by radiation; this property of the column is used in high-energy radiation source applications.

The field-strength/current (E–I) characteristic of the high-pressure arc column exhibits a falling portion at small currents and a rising part in the higher current range. Stable arc operation at small currents (point C in Fig. 3 [9]) therefore requires an Ohmic resistance in series with the arc; thus the resulting total characteristic can be made to increase at the current of interest and stable operation is achieved. Introducing the heat flux potential $S = \int \kappa \, dT$ into the transport function $\sigma(T) = \sigma(S)$ and neglecting radiation Eq. (1) may be rewritten

$$\sigma E^2 + \frac{1}{r}\frac{\partial}{\partial r}\left(r\frac{\partial S}{\partial r}\right) = 0 . \tag{2}$$

Defining the Ohmic heating per unit arc length as $L = IE$ and normalizing the arc radius $v = r/R$ (R being the total radius) Eq. (2) yields

$$\frac{\sigma L}{\pi \int_0^1 \sigma \, d(v^2)} + \frac{1}{v}\frac{\partial}{\partial v}\left(v\frac{\partial S}{\partial v}\right) = 0 . \tag{3}$$

Since Eq. (3) is independent of the total radius R of the column, equal values of L lead to identical $T(r)$-distributions. This independence on radius may be used to predict electrical arc data (E, I) for arcs of different diameters (similarity laws).

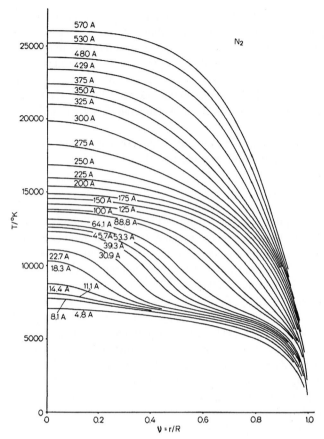

Fig. 1: Temperature distributions in a 5-mm nitrogen cascade arc (from [6], with permission).

Due to its importance for applications the interaction of flow fields and magnetic fields with the arc column has been studied extensively [10–12]. By solving the MHD conservation laws with the appropriate boundary conditions, predictions of arc motion, gas flow in the arc and arc stability may be obtained. The energy Eq. (1) has to be used in its more general form:

$$\rho\frac{dh}{dt} = \rho\frac{\partial h}{\partial t} + \rho\mathbf{v}\cdot\nabla h = \sigma E^2 + \nabla\cdot\kappa\nabla T - e + a \,, \tag{4}$$

where $\mathbf{v}$ is the velocity of the gas, ρ is its mass density, and h the enthalpy. In addition, the momentum and the mass balance equations (5) and (6) have to be considered:

$$\rho\frac{d\mathbf{v}}{dt} = \rho\frac{\partial \mathbf{v}}{\partial t} + \rho\mathbf{v}\cdot\nabla\mathbf{v} = \mathbf{j}\times\mathbf{B} - \nabla p + \nabla\cdot\tau \,, \tag{5}$$

$$\frac{\partial \rho}{\partial t} + \nabla\cdot(\rho\mathbf{v}) = 0 \tag{6}$$

with $\mathbf{v}$ being the friction tensor.

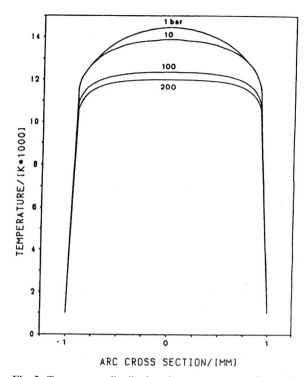

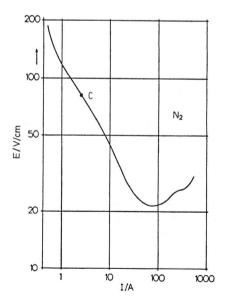

Fig. 2: Temperature distributions in an argon arc at elevated pressures. Arc current 60 A [7].

Fig. 3: Electrical arc characteristic (from [9], with permission).

Vacuum Arcs

This arc discharge is initiated in an evacuated environment. Since plasma can only be formed if ionizable gas is present, the vacuum arc has to generate its own gaseous environment. Vaporization of the electrodes, mainly that of the cathode material, provides the necessary atoms. Vacuum arcs therefore operate typically in a metal vapor atmosphere. Since the crater formation process which is necessary for the evaporation at the cathode ends after a certain crater size has been reached [13], vacuum arcs are not stationary in nature. The cathode spot only remains at a given crater for times shorter than 50 ns; thus a single (undivided) short vacuum arc is very similar in nature to a spark.

If magnetic fields are present, the vacuum arc moves against the $\mathbf{j} \times \mathbf{B}$ direction; this "retrograde motion" is due to cathode jet phenomena. At higher currents (above $50 - 100$ A) the single vacuum arc splits up; a multitude of cathode or anode spots may coexist in such arcs.

Applications

Arc discharges are used in a host of scientific and technical applications. Due to the temperature range which is achievable in the column ($T \leq 50000$ K) different ionization stages of the gas atoms are reached. Fundamental data such as collision cross sections, transition probabilities, index of refraction, and spectral line broadening parameters can be determined from the investigation of arc plasmas.

Technical applications also make use of the high plasma temperatures which lead to an excellent electrical conductivity, to high radiative power losses, to high enthalpies in the gas – to name only the most important properties. The high electrical conductivity of an arc column, together with the short recovery time of dielectric strength after current zero, is used in AC-circuit breakers. Very high voltage levels can be handled with gas filled-gas breakers (SF_6) where the interaction of flow fields and magnetic fields with the arc column plays an important role. Vacuum circuit breakers are much more compact, however, their voltage handling capability is reduced. At present, voltages of 50–60 kV can be handled. The high specific radiation of noble gases and rare earths is used in radiation source applications, while the high enthalpies reached in arc heaters allow for special chemical reactions and for surface treatment processes, such as spark etching and plasma coating.

See also: Corona Discharge; Ionization; Lightning; Photoionization.

Bibliography

H. Maecker and W. Finkelnburg, "Elektrische Bögen und thermisches Plasma", *Handbuch der Physik*, Bd. XXII. Springer-Verlag, Berlin, 1957.

J. M. Lafferty, ed., *Vacuum Arcs, Theory and Application*. Wiley, New York, 1980.

H. Raether, *Electron Avalanches and Breakdown in Gases*. Butterworths, London, 1964.

K. Günther and R. Radtke, *Electric Properties of Weakly Nonideal Plasmas*. Birkhaeuser, Basel, 1984.

K. Ragaller, ed., *Current Interruption in High-Voltage Networks*. Plenum, New York, 1978.

References

[1]   F. Burhorn, H. Maecker, and Th. Peters, *Z. Phys.* **131**, 28 (1951).

[2]   F. Paschen, *Ann. Phys. (Leipzig)* **37**, 69 (1889).

[3]   G. Burkhard, PhD Thesis, Technische Hochschule Ilmenau (1971).
[4]   W. Elenbaas, *Physica* **3**, 947 (1936).
[5]   H. Maecker and S. Steinberger, *Z. Angew. Phys.* **23**, 456 (1967).
[6]   E. Schade, *Z. Phys.* **233**, 53 (1970).
[7]   H. Poisel, F. J. Landers, P. Höß, and U. H. Bauder, *IEEE Trans. Plasma Sci.* **PS-14**, 306 (1986).
[8]   U. H. Bauder and P. Schreiber, *Proc. IEEE* **59**, 633 (1971).
[9]   U. Plantikow, *Z. Phys.* **237**, 388 (1970).
[10] H. Maecker and H. G. Stablein, *IEEE Trans. Plasma Sci.* **PS-14**, 291 (1986).
[11] N. Sebald, *Proc. XIIth ICPIG*, North Holland/American Elsevier Eindhoven, p. 187 (1975).
[12] J. Blass and U. H. Bauder, *Proc. IIW Asian Pacific Regional Welding Congress*, p. 528. Hobart, Australia, 1988.
[13] J. Prock, *IEEE Trans. Plasma Sci.* **PS-14**, 482 (1986).

Astronomy, High-Energy Neutrino

E. Waxman

Most of the information we have on astronomical objects and systems is obtained by observing the electromagnetic radiation, i. e., the photons, they emit. In the past four decades a new type of astronomy has emerged, where information is carried to us from astronomical objects not by photons but rather by a different type of particle, the neutrino. Neutrinos are nearly massless particles traveling at essentially the speed of light (*see* Neutrino). The great strength of neutrino astronomy is related to the fact that neutrinos interact very weakly with matter. Looking at the Sun, for example, we can only observe photons emitted from the Sun's surface. Photons produced within the Sun can not reach us, since they can not propagate much through the dense solar plasma. Neutrinos produced in the depth of the Sun, on the other hand, can propagate almost unhindered through the Sun and reach our detectors on Earth, carrying information on the physical processes taking place in the core of the Sun.

The strength of neutrino astronomy is also its challenge. Since neutrinos interact only weakly with matter, large detector mass is required in order to detect them. The detection of solar neutrinos became possible with the construction of detectors with kilo-tons of detecting medium, *see* Solar Neutrinos [3]. The probability that a neutrino passing through kilo-tons of matter would be "captured", i. e., would interact within the detector, is still very small. However, the large flux of neutrinos from the Sun, some 100 billion neutrinos per square centimeter per second, allows hundreds of them to be detected every year. The detection of solar neutrinos enabled direct observations of nuclear reactions in the core of the Sun, confirming the hypothesis that the energy source of the Sun is nuclear fusion and demonstrating the validity of solar structure models. It has also taught us that the standard model of particle physics is incomplete. Neutrinos come in three types, or "flavors": electron-type, muon-type and tau-type. Solar neutrino detection demonstrated that neutrinos can change their flavor as they propagate in matter or in vacuum, a phenomenon not accounted for by the standard model. This flavor change also indicates that neutrinos are not massless, as assumed in the standard model, but rather have finite, albeit small, masses (*see* Neutrino; Solar Neutrinos).

The characteristic energy of neutrinos produced in the Sun is Mega (million) electron-Volt, MeV. An eV is the energy typically required to detach an electron from an atom, while MeV is the characteristic energy released in the fusion or fission of atomic nuclei. Solar neutrino detectors, "telescopes", are capable of detecting MeV neutrinos from supernova explosions in our "local" Galactic neighborhood, at distances smaller than 100 000 light years. Since the rate of such explosions is one per few decades, only one supernova, the famous SN 1987A, has so far been detected. Much like the detection of solar neutrinos, the detection of neutrinos from SN 1987A provided a direct observation of the physical process powering the explosions, the collapse of the core of a massive star to a neutron star, as well as constraints on fundamental neutrino properties [14].

The detection of MeV neutrinos from sources well outside our local neighborhood, lying at distances ranging from several million light years, the typical distance between galaxies, to several billion light years, the size of the observable universe, is impossible using present techniques. In order to extend the distance accessible to neutrino astronomy to the edge of the observable universe, several high energy neutrino telescopes are currently being constructed [10]. These telescopes are designed for the detection of neutrinos with energies exceeding Terra-electron-Volt (TeV, equal to million-MeV).

The detection of astrophysical high energy neutrinos will allow to answer some of the most important open questions of high energy astrophysics [19], e. g., the identity and physics of the most powerful accelerators in the universe and the mechanisms for energy extraction from black-holes. It will also allow to study fundamental neutrino physics issues, e. g., neutrino coupling to gravity and the existence of weakly interacting massive particles (WIMPs). These issues are discussed at some length below. It should be kept in mind, however, that as the construction of high energy neutrino telescopes opens a new, unexplored window of observations on the universe, one should be ready for surprises. It may well be that the most important things that we will learn would be related to such surprises, rather than to the open questions discussed below.

High Energies for Large Distances

The neutrino flux from cosmologically distant sources is too low to be detectable by kiloton detectors of MeV neutrinos. The construction of orders of magnitude larger detectors, that would be required for the detection of extragalactic sources at this energy, is currently unfeasible. This situation changes at higher neutrino energy, due to two reasons. First, the interaction cross section increases with energy, that is, higher energy neutrinos are more likely to interact with matter than lower energy ones. This implies that smaller fluxes of higher energy neutrinos may be detectable with a given detector mass. Second, at TeV neutrino energy the construction of gigaton, rather than kiloton, telescopes becomes feasible.

Interactions of high energy, > 1 TeV, muon-type neutrinos with atomic nuclei on Earth produce muons (charged particles about hundred times heavier than the electron), which propagate at a straight line and at nearly the speed of light over more than a kilometer through rock, water or ice. While propagating at nearly speed of light, the muon emits visible light, "Čerenkov radiation". Thus, if the muon propagates through transparent water or ice, its "track" may be identified by detecting the light it emits. Since the muon track is colinear to within one degree with the initial neutrino trajectory, the direction to the neutrino source may be determined.

The feasibility of detection of high energy muons in deep sea or lake water has been demonstrated by the DUMAND experiment off the coast of Hawaii and by the Lake Baikal experiment. The AMANDA collaboration has demonstrated that the deep ice of the south pole is also a suitable medium, and that construction of a cubic kilometer ice detector, with a gigaton of ice as detecting medium, is feasible. A schematic description of the AMANDA detector is presented in Fig. 1.

The reason for constructing neutrino telescopes in deep water or ice is twofold. First, the ice and water properties improve with depth. The scattering of light is weaker at large depth, thus allowing to place the photomultipliers, the detectors of the muon Čerenkov light, at larger spacing. This implies that at larger depth a smaller number of photomultipliers is required to instrument a given detector volume, and allows the instrumentation of a km^3 of ice or water at an acceptable cost. Second, the atmospheric muon background to the neutrino signal is reduced at larger depth. High energy muons are produced in the atmosphere of the Earth by interaction of cosmic rays with air. The cosmic rays which hit the atmosphere are mostly high energy nuclei, believed to be produced in Galactic supernovas. The muons produced by their interactions in the atmosphere constitute a background to the signal of neutrino-induced muons, that is, to the signal of muons produced by neutrino interactions. Since the muons penetrate a distance of order a km in ice or water, putting the detector several kilometers deep under water or ice strongly suppresses the atmospheric muon background. The large depth does not affect the neutrino induced muon signal, since neutrinos can easily penetrate the Earth and interact near or within the detector.

Figure 2 shows a neutrino event detected by the AMANDA telescope. The fact that the muon track is "upgoing", crossing the detector from bottom to top, allows to confidently identify it as a as neutrino-induced muon: Only neutrinos can cross the Earth to approach the detector from below and produce upgoing muons. This neutrino is, most likely, an "atmospheric neutrino". The interaction of cosmic rays in the atmosphere produce both muons and neutrinos. While the atmospheric muon flux is suppressed by going to large depth, the atmospheric neutrino flux is not. Atmospheric neutrinos constitute therefore an unavoidable background, which sets a lower limit to the flux of astronomical sources which are detectable (for a given detector size).

The AMANDA detector, which is roughly 0.1 gigaton in mass, is currently expanded in the IceCube project to a km^3, 1 gigaton, detector. In addition, several efforts are currently underway in the Mediterranean to construct gigaton-scale underwater detectors (the ANATRES, NESTOR and NEMO projects).

Cosmic Accelerators as Neutrino Sources

Nuclear fusion in stars generally does not lead to production of neutrinos at energy much higher than MeV. Should we expect, therefore, any sources of TeV neutrinos to be out there? If so, what is the mechanism by which they produce neutrinos, and what is the detector size that is required to detect them? The answers to these questions rely largely on observations of high energy cosmic rays.

The cosmic-ray spectrum extends to energies of 10^{20} eV, 100 million TeV. We have strong indications that ultrahigh-energy cosmic rays (UHECRs), cosmic rays of energy exceeding 10 million TeV, are produced by extragalactic sources, and that they are light nuclei, most

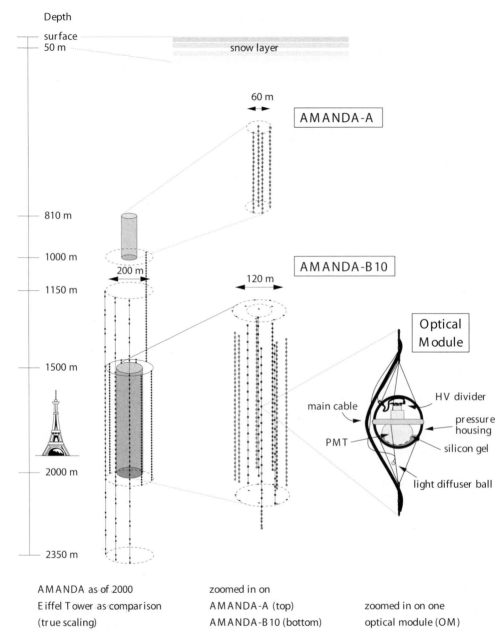

Depth

surface

50 m

snow layer

60 m

AMANDA-A

810 m

1000 m

200 m

AMANDA-B10

1150 m

120 m

Optical
Module

1500 m

main cable

HV divider

pressure
housing

PMT

silicon gel

2000 m

light diffuser ball

2350 m

AMANDA as of 2000	zoomed in on	
Eiffel Tower as comparison	AMANDA-A (top)	zoomed in on one
(true scaling)	AMANDA-B10 (bottom)	optical module (OM)

Fig. 1: The AMANDA experiment. $> 1\,\text{TeV}$ muons, produced by high-energy
neutrinos interacting with atomic nuclei near or within the detector, are
identified by an array of photomultipliers (PMTs) deployed 2 km deep under
the South-Pole surface. Detection of the muon emission of visible, Čerenkov,
radiation allows to reconstruct the > 1 kilometer-long muon track. A neutrino
event recorded in AMANDA is presented in Fig. 2.

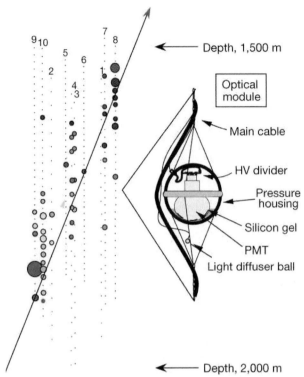

Fig. 2: A high-energy muon track reconstructed by the AMANDA detector [E. Andres *et al.*, *Nature* **410**, 441 (2001)]. Each dot represents an optical module, shown in detail on the right. The gray circles show pulses from the PMTs: The size of the circle indicates the pulse amplitude. The arrow indicates the upward-moving muon track.

likely protons [13]. The identity of the sources is yet unknown: The high energies, 100 million times larger than the highest energy achieved by man made accelerators, challenge all models proposed for particle acceleration. Moreover, the cosmic-ray arrival directions do not necessarily point back to their sources, since protons (being charged particles, unlike photons and neutrinos) are deflected by Galactic and extragalactic magnetic fields, and do not propagate along straight lines.

The essence of the challenge of accelerating to 10^{20} eV, or 100 million TeV, can be understood from Fig. 3. Most models involve the acceleration of charged particles, like protons, which are confined to the accelerator by magnetic fields. Magnetic confinement requires the product of accelerator magnetic field strength and accelerator size to exceed a value, which increases with particle energy. Only two types of astrophysical sources are known to be large enough and to contain magnetic fields strong enough to possibly allow proton acceleration to 10^{20} eV: Gamma-Ray Bursts (GRBs) [17] and Active Galactic Nuclei (AGN) [9, 12]. GRBs are the brightest transient sources known in the universe. Lying at cosmological distances, they produce short (typically 1 to 100s long) flashes of γ-rays with luminosity exceeding that of the Sun by 19 orders of magnitude. AGN are the brightest known steady sources, with

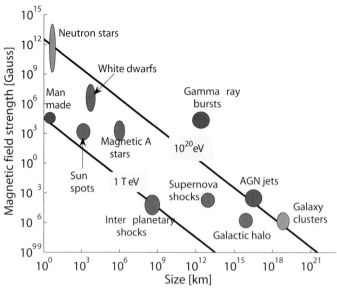

Fig. 3: Size and magnetic field strength of possible sites of particle acceleration (following [9]). Proton acceleration to $1\,\mathrm{TeV}$ or $10^{20}\,\mathrm{eV}$ is possible only for sources lying above the appropriately marked lines. This is a necessary, but not sufficient requirement: Proton acceleration to $10^{20}\,\mathrm{eV}$ is impossible in galaxy clusters, since the acceleration time in these objects is larger than the age of the universe, and unlikely in highly magnetized neutron stars, due to severe energy losses.

luminosity exceeding that of the Sun by 12 orders of magnitude. While GRBs and AGN are plausible candidates for UHECR production, we have no direct evidence for proton acceleration in these sources despite many years of photon observations. Furthermore, our theoretical models describing these sources are incomplete, a point to which we return below.

Irrespective of the nature of the UHECR sources, some fraction of their energy output is bound to be carried by high energy neutrinos. Protons, p's, of sufficiently high energy may interact with photons, γ's, to produce charged pions, π^+'s, particles that decay to muons and neutrinos. This interaction is represented symbolically as

$$p + \gamma \rightarrow n + \pi^+ , \tag{1}$$

where n stands for a neutron. The subsequent decay of the pion produces neutrinos:

$$\pi^+ \rightarrow \nu_\mu + \mu^+ \rightarrow \nu_\mu + e^+ + \nu_e + \bar{\nu}_\mu . \tag{2}$$

The positively charged pion decays to a positively charged muon (μ^+) and to a muon-type neutrino (ν_μ), and the positively charged muon decays to a positron (e^+, the anti-particle of the negatively charged electron), electron-type neutrino (ν_e) and anti-muon-type neutrino ($\bar{\nu}_\mu$, the antiparticle of ν_μ). Charged pions may be produced also in collisions of high energy protons with other, low energy, nucleons (protons and neutrons). In this case, both positively and negatively charged pions (π^+ and π^-) may be produced. The decay of the negatively charged pions produces neutrinos in a manner similar to that described by Eq. (2),

$$\pi^- \to \bar{\nu}_\mu + \mu^- \to \bar{\nu}_\mu + e^- + \bar{\nu}_e + \nu_\mu \ . \tag{3}$$

The neutrinos produced by the decay of the pions carry a significant fraction of the energy of the parent proton. Neutrinos produced, for example, by the decay of pions produced by interaction with photons, Eq. (1), typically carry 5% of the proton energy. UHECR sources are expected therefore to be sources also of high energy neutrinos. The detection of high energy neutrinos emitted by extragalactic sources will provide the first direct evidence for acceleration of protons in such sources, and may resolve the mystery of the UHECR source identity.

The Waxman–Bahcall Bound and Gigaton Neutrino Telescopes

UHECR observations provide guidance to estimating the expected high energy neutrino signal and the detector size required to detect it. Assuming that UHECRs are protons produced by extragalactic sources, the observed flux of UHECRs determines the average rate, per unit time and volume, at which such high energy protons are produced in the universe. The inferred rate at which energy is injected into the universe in the form of protons with energies in the range of 10^{19} eV to 10^{20} eV is [20]

$$\dot{\varepsilon} = 1.5 \times 10^{44} \,\mathrm{erg\,Mpc}^{-3}\,\mathrm{yr}^{-1} \ . \tag{4}$$

The distance unit used here, megaparsec (Mpc), equals approximately to 3-million light years. Observations are consistent with the rate of energy injection being independent of the proton energy decade, i. e., the rate of energy production in protons in the energy range of, e. g., 10^{18} eV to 10^{19} eV is also given by Eq. (4).

Waxman and Bahcall have shown that the observed UHECR production rate sets an upper bound to the neutrino flux produced by extragalactic sources. High energy protons produced in candidate UHECR accelerators, such as AGN and GRBs, are likely to escape the source with only few pion production interactions. This implies that protons do not lose a large fraction of their energy to pion production. For sources of this type, the energy generation rate of neutrinos must be smaller than the proton energy generation rate, given by Eq. (4). Assuming that neutrinos are produced at this rate over the age of the universe, the resulting upper bound (for muon and anti-muon neutrinos, neglecting propagation flavor changes) is [21]

$$E_\nu^2 \Phi_\nu < 5 \times 10^{-8} \,\mathrm{GeV\,cm}^{-2}\,\mathrm{s}^{-1}\,\mathrm{sr}^{-1} \ . \tag{5}$$

Here Φ_ν is the number flux of neutrinos (number of neutrinos per unit area, time and solid angle) per unit neutrino energy, E_ν is the neutrino energy, and GeV stands for Giga-eV (1000 MeV). $E_\nu^2 \Phi_\nu$ describes the energy flux of neutrinos (energy per unit area, time and solid angle) carried by neutrinos with energies spread over (approximately) half a decade. The upper bound, which came to be known as the "Waxman–Bahcall" (WB) bound, is compared in Fig. 4 with current experimental limits, and with the expected sensitivity of planned neutrino telescopes.

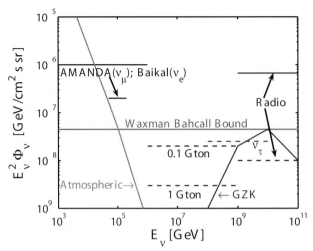

Fig. 4: The upper bound imposed by UHECR observations on the extragalactic high energy muon neutrino intensity, compared with the atmospheric neutrino background and with the experimental upper bounds (solid lines) of optical Čerenkov experiments, BAIKAL [4] and AMANDA [1, 10], and of coherent Čerenkov radio experiments (RICE [11], GLUE [7, 16]). The curve labelled "GZK" shows the intensity due to interaction with micro-wave background photons. Dashed curves show the expected sensitivity of 0.1 Gton (AMANDA, ANTARES, NESTOR) and 1 Gton (IceCube, NEMO) optical Čerenkov detectors [10], of the coherent radio Čerenkov (balloon) experiment ANITA [16] and of the Auger air-shower detector (sensitivity to ν_τ) [15]. Space air-shower detectors (OWL-AIRWATCH) may also achieve the sensitivity required to detect fluxes lower than the WB bound at energies $> 10^{18}$ eV [15].

The figure indicates that gigaton neutrino telescopes are needed to detect the expected extragalactic flux in the energy range of ~ 1 TeV to ~ 1000 TeV, and that much larger effective mass is required to detect the flux at higher energy. Few tens of ~ 100 TeV events per year are expected in a gigaton telescope if GRBs are the sources of ultra-high energy protons [18, 21]. These events will be correlated in time and direction with GRB photons, allowing for an essentially background free experiment. A lower rate is expected if AGN are the UHECR proton sources [2].

"GZK" Neutrinos

Protons of sufficiently high energy, exceeding $\sim 5 \times 10^{19}$ eV, may interact with photons of the cosmic microwave background, the big bang "relic" of 2.7 K radiation permeating the universe, to produce pions as described by Eq. (1) [8]. Protons of sufficiently high energy, $> 10^{20}$ eV, lose most of their energy over less than 300 million years, a time much shorter than the age of the universe, which is ~ 10 billion years. All the energy injected into the universe in the form of such protons is thus converted to pions which decay to neutrinos, producing a background neutrino intensity similar to the WB bound [5]. The expected neutrino intensity is schematically shown in Fig. 4. It is denoted "GZK neutrinos", where "GZK" stands for Greisen, Zatsepin and Kuszmin, who were the first to point out the rapid energy loss of high energy protons due to interaction with the microwave background. The GZK neutrino flux peaks at $\sim 5 \times 10^{18}$ eV, since neutrinos produced by $p\gamma$ interactions typically carry 5% of the $\sim 10^{20}$ eV proton energy (The GZK intensity in Fig. 4 decreases at the highest energies since it was assumed that the maximum energy of protons produced by UHECR sources is 10^{21} eV).

The detection of GZK neutrinos will be a milestone in neutrino astronomy. Most important, neutrino detectors with sensitivity better than the WB bound at energies $> 10^{18}$ eV will test the hypothesis that the UHECR are protons (possibly somewhat heavier nuclei) of extragalactic origin. The large effective detector mass, much larger than gigaton, required to achieve this sensitivity may be obtained by detectors searching for radio, rather than optical, Čerenkov emission. "Air shower" detectors, which detect the "shower" of high-energy particles produced in the atmosphere following the interaction of a high-energy neutrino in the atmosphere, may also achieve sufficiently large effective mass at ultrahigh energy.

The challenge posed by the existence of UHECRs to models of particle acceleration, and the lack of direct evidence for proton acceleration in any extragalactic source, have led many to speculate that modifications of the basic laws of physics are required in order to account for the existence of UHECRs. Such "new physics" models commonly postulate the existence of very massive particles, the decay of which produces the observed UHECRs, and generally predict large fluxes of $\sim 10^{20}$ eV neutrinos [6], well above the WB bound. Measurements of the neutrino flux above $\sim 10^{19}$ eV would therefore allow to discriminate between "new physics" models for UHECR production and models where UHECRs are produced by "standard physics" acceleration in astrophysical objects, like GRBs and AGN.

Probing Astrophysical Accelerators with High-Energy Neutrino Telescopes

GRBs and AGN are believed to be powered by the accretion of mass onto black holes. GRBs are most likely powered by the accretion of a fraction of a Solar mass on second time scale onto a newly born Solar mass black hole. Recent observations strongly suggest that the formation of the black hole is associated with the collapse of the core of a very massive star. AGN are believed to be powered by accretion of mass onto massive, million to billion Solar mass, black holes residing at the centers of distant galaxies. As illustrated in Fig. 5, the gravitational energy released by the accretion of mass onto the black hole is assumed in both cases to drive a relativistic jet, which travels at nearly the speed of light and produces the observed radiation at a large distance away from the central black hole. The models describing the physics responsible for powering these objects, though successful in explaining most observations, are largely phenomenological. In particular, the answer to the question of whether or not the out flowing jet carries protons, which has major implications to our understanding of the mechanism by which gravitational energy is harnessed to power the jet, is not known despite many years of photon observations. This situation is common also to our understanding of Galactic micro-Quasars, which may be considered as a scaled down versions of AGN, with ~ 1 Solar mass black hole (or neutron star) "engines". Neutrino observations of GRBs, AGN and micro-Quasars will provide new information that can not be obtained using photon observations, and that may allow to answer the underlying open questions.

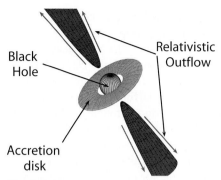

Fig. 5: GRBs and AGN are believed to be powered by black holes. The accretion of mass onto the black hole, through an accretion disk, releases large amounts of gravitational energy. If the black hole is rotating rapidly, another energy source becomes available: The rotational energy may be released by slowing the black hole down through interaction with the accretion disk. The energy released drives a jet-like relativistic outflow. The observed radiation is produced as part of the energy carried by the jets is converted, at large distance from the central black hole, to electromagnetic radiation.

Fundamental Neutrino Properties

Since neutrinos are expected to be produced in astrophysical sources via the decay of charged pions, production of high-energy muon and electron neutrinos with a 2:1 ratio, as described in Eqs. (2) and 3), is expected. Because of neutrino flavor changes during propagation, usually termed "neutrino oscillations", neutrinos that get to Earth are expected to be almost equally distributed between flavors. This implies that one should detect equal numbers of muon-type and tau-type neutrinos. Upgoing taus, rather than muons, would be a distinctive signature of such oscillations. Although the Čerenkov emission along the track of a high energy tau is similar to that produced by a muon, it may be possible to distinguish between taus and muons in a km^3 ice or water detector, since at 1000 TeV the tau decays after propagating $\sim 1\,km$. This will allow a "tau appearance experiment". At present, the oscillation of muon-type neutrinos to tau-type neutrinos is inferred from the "disappearance" of muon-type neutrinos produced by cosmic-ray interactions in the atmosphere, without detecting the tau-type neutrinos that should be produced by such oscillation. The detection of taus in a neutrino telescope will provide direct confirmation of the oscillation hypothesis.

Detection of neutrinos from GRBs could be used to test the simultaneity of neutrino and photon arrival to an accuracy of $\sim 1\,s$ ($\sim 1\,ms$ for short bursts), checking the assumption of special relativity that photons and neutrinos have the same limiting speed (The time delay due to the neutrino mass is negligible: for a neutrino of energy $\sim 1\,TeV$ with mass $\sim 1\,eV$ traveling 1000 Mpc, the delay is $\sim 0.1\,s$.) These observations would also test the weak equivalence principle, according to which photons and neutrinos should suffer the same time delay as they pass through a gravitational potential. With 1 s accuracy, a burst at a distance of 1000 Mpc would reveal a fractional difference in limiting speed of 1 part in 10^{17}, and a fractional difference in gravitational time delay of order 1 in 10^6 (considering the Galactic potential alone). Previous applications of these ideas to supernova 1987A, where simultaneity could be checked only to an accuracy of order several hours, yielded much weaker upper limits: of order 10^{-8} and 10^{-2} for fractional differences in the limiting speed and time delay, respectively.

Weakly Interacting Massive Particles

Most of the mass in the universe is currently believed to be in the form of "dark matter", composed of particles which were not detected in laboratories on Earth, and which interact with the normal matter that we know essentially only through gravitational forces. Weakly Interacting Massive Particles (WIMPs) are the leading dark matter particle candidates. If WIMPS populate the halo of our galaxy, the Sun or Earth would capture them, where they would annihilate occasionally into high-energy neutrinos. The annihilation rate depends on the details of the model. A widely discussed WIMP candidate is the lightest neutralino in minimal super-symmetric models. Gigaton neutrino telescopes currently under construction complement direct search detectors by reaching good sensitivity at high neutralino masses, typically in excess of a few hundred GeV, and by allowing us to probe regions of parameter space with large branching fractions to W and Z bosons.

References

[1]   J. Ahrens *et al.*, *Phys. Rev. Lett.* **90**, 251101 (2003). The upper bound on high energy neutrino flux derived by the AMANDA experiment (A).

[2]   A. Atoyan and C. D. Dermer, *Phys. Rev. Lett.* **87**, 221102 (2001); J. Alvarez-Muñiz and P. Meszaros, *Phys. Rev.* **D70**, 123001 (2004); J. K. Becker, P. L. Biermann and W. Rhode, astro-ph/0502089. Model predictions for neutrino emission from AGN (A).

[3]   J. N. Bahcall, *Neutrino Astrophysics*. Cambridge University Press, New York, 1989. Textbook review of neutrino astrophysics, with focus on Solar neutrinos (I,A).

[4]   V. Balkanov *et al.*, *Nucl. Phys. B (Proc. Suppl.)* **110**, 504 (2002). Description of the first operating optical Čerenkov detector of high energy neutrinos, the lake Baikal experiment (A).

[5]   V. S. Berezinsky and G. T. Zatsepin, *Phys. Lett.* **28B**, 423 (1969). Prediction of the exitence of GZK neutrinos (A); R. Engel, D. Seckel and T. Stanev, *Phys. Rev.* **D64**, 093010 (2001). A calculation of the expected GZK neutrino flux (A).

[6]   P. Bhattacharjee and G. Sigl, *Phys. Rep.* **327**, 109 (2000). Review of models of sources of ultra-high energy cosmic-rays, with detailed discussion of "new physics" models (A).

[7]   P. W. Gorham *et al.*, *Phys. Rev. Lett.* **93**, 041101 (2004). Description of the upper bound on ultra-high energy neutrino flux derived by radio observations of the moon (A).

[8]   K. Greisen, *Phys. Rev. Lett.* **16**, 748 (1966); G. T. Zatsepin, V. A. Kuzmin, *JETP* **4**, 78 (1966). The first papers to point out the rapid energy loss of high energy protons due to interaction with cosmic microwave background photons (A).

[9]   A. M. Hillas, *ARA&A* **22**, 425 (1984). A review of the phenomenology of ultra high energy cosmic ray observations, and of the theoretical challenges facing models for particle acceleration (I).

[10] F. Halzen, *Proc. Nobel Symp. 129 "Neutrino Physics"* to appear in *Physica Scripta*, L. Bergstrom, O. Botner, P. Carlson, P. O. Hulth, and T. Ohlsson (eds.), astro-ph/0501593. Description of the current status of gigaton-scale optical Čerenkov neutrino detectors (I); F. Halzen and D. Hooper, *Rep. Prog. Phys.* **65**, 1025 (2002). Review of high energy neutrino astronomy (A).

[11] I. Kravchenko *et al.*, *Astropar. Phys.* **19**, 15 (2003). Description of the south pole coherent Čerenkov radio detector of high energy neutrinos (A).

[12] R. V. E. Lovelace, *Nature* **262**, 649 (1976). Phenomenological arguments suggesting that the brightest active galactic nuclei jets may accelerate protons to ultra high energy (I).

[13] M. Nagano and A. A. Watson, *Rev. Mod. Phys.* **72**, 689 (2000). Review of high energy cosmic-ray observations (I,A).

[14] G. Raffelt, *Stars as Laboratories for Fundamental Physics: The Astrophysics of Neutrinos, Axions, and Other Weakly Interacting Particles*. University of Chicago Press, Chicago, 1996. Textbook review of constraints imposed by stellar astrophysics on fundumental particle properties (I,A).

[15] D. Saltzberg, *Proc. Nobel Symp. 129 "Neutrino Physics"*, to appear in *Physica Scripta*, L. Bergstrom, O. Botner, P. Carlson, P. O. Hulth, and T. Ohlsson (eds.), astro-ph/0501364. Analysis of the sensitivity of large air-shower arrays as detectors of ultra-high energy neutrinos (A).

[16] A. Silvestri *et al.*, astro-ph/0411007. Description of the south-pole ballon radio experiment designed for detection of ultra-high energy neutrinos (A).

[17] E. Waxman, *Phys. Rev. Lett.* **75**, 386 (1995); M. Vietri, *Astrophys. J.* **453**, 883 (1995); M. Milgrom, and V. Usov, *Astrophys. J.* **449**, L37 (1995). First papers suggesting gamma-ray bursts to be ultrahigh energy cosmic-ray sources (A).

[18] E. Waxman and J. N. Bahcall, *Phys. Rev. Lett.* **78**, 2292 (1997). First calculations of high energy neutrino emission from gamma-ray bursts (A).

[19] E. Waxman, *Proc. Nobel Symp. 129 "Neutrino Physics"*, to appear in *Physica Scripta*, L. Bergstrom, O. Botner, P. Carlson, P. O. Hulth, and T. Ohlsson (eds.), astro-ph/0502159. Description of some major open questions of theoretical high energy astrophyscis, that may be addressed by high energy neutrino telescopes (I).

[20] E. Waxman, *Astrophys. J.* **452**, L1 (1995); J. N. Bahcall and E. Waxman, *Phys. Lett.* **B556**, 1 (2003). Derivation of the high energy cosmic-ray energy generation rate in the universe (A).

[21] E. Waxman and J. N. Bahcall, *Phys. Rev.* **D59**, 023002 (1999); J. N. Bahcall and E. Waxman, *Phys. Rev.* **D64**, 023002 (2001). Derivation of the upper bound on the extragalactic neutrino flux, implied by high-energy cosmic ray observations (A).

Astronomy, Optical

J. W. Fried

Scope

The purpose of astronomical observations is to provide information that can be used to explain the composition, structure, formation, dynamics and evolution of these objects by applying physical laws. The list of celestial objects includes the solar system, stars, star clusters, galaxies, clusters of galaxies, and the universe as a whole. Virtually all celestial objects can be studied by means of optical observations. In this article, optical astronomy includes observations using ultraviolet, visible, or near infrared radiation, since the technology used is nearly identical in these spectral regimes.

Telescopes

The purpose of a telescope is to collect the light from an object onto a detector. Obviously, the larger the telescope, the more light is collected and hence fainter and so more distant objects can be studied. All large telescopes use mirrors rather than lenses because mirrors are much easier to mount in such a way that they will not distort and degrade the images as the telescope points in different directions. Telescopes are usually used with a small convex "secondary" mirror placed in the converging beam from the "primary" just before it comes to a focus. The beam is reflected by the secondary through a hole in the center of the primary and comes to a focus just behind the primary's mount. This is the "Cassegrain" focus and is a much more convenient place to mount instruments than the "prime" focus of the primary. The convex secondary magnifies the image, that is, it increases the effective focal length of the telescope by typically a factor of 3 or more. This design provides a long focal length in a compact size.

Telescopes are specified by the diameter of their primary. Many telescopes with diameters of 4–6 m existed already until the 1980's, and at the end of the last century, several new telescopes with diameters of 8–10 m have been built. Their mirrors are thinner and lighter than the older smaller ones, but since they are also more flexible their mounting requires actively moving supports to compensate changes in the gravitational force as the telescope moves. There are two ways to build large mirrors, a single monolithic design (used for example in the VLT of ESO) or a segmented mirror, assembled from many small segments. The 10-m mirrors of the Keck telescopes on Mauna Kea, Hawaii, for example, consist of 36 hexagonal segments. The segments have the surface shape appropriate to their location in the mirror which is approximately a parabola.

Instruments

Optical instruments range from simple cameras to complex computer-controlled multi purpose instruments. In the simplest case, a cameras just consists of a box to hold a 2-dimensional detector at the telescope's focus. Formerly this was generally a photographic plate, but since the 1980's the detector is almost exclusively a CCD (see below). Several of these detectors can be mounted side by side in the focal plane to cover a large field of view. Filters placed in front of the detector allow only one color band to pass. Images taken in several filters already give information on the physical state of the objects detected. For example, a hot star will appear brighter in a blue pass band than in a red pass band, i. e., measuring the brightness difference between blue and red gives the temperature of the star. An image of the sky is the most basic observation to be done, yielding position, brightness and form of the objects.

More information about the objects can be derived if several pass bands are used. Modern multi color surveys, carried out on telescopes of the 2–4 m class, use up to ~ 20 filters; from these data spectral types and red-shifts of the objects detected can be derived, albeit with moderate precision. These surveys detect objects which are up to 100 times fainter than those found in the famous "Palomar Sky Survey" which used the 48-in. Schmidt telescope on Mt. Palomar to photograph the entire northern sky on 935 pairs of 14-in. square plates (6×6 degrees each) in the blue and red pass bands.

In order to get more detailed information about an object, its spectrum – that is its intensity versus wavelength – has to be measured at much higher wavelength resolution. A diffraction-grating spectrograph indexspectrograph is normally used for these measurements. With such a spectrum from a star one can measure the star's temperature, composition, and surface gravity. From this information and a liberal amount of theory one can deduce the entire structure of the star right down to its center. Another important use is as a speedometer. An object's relative speed toward or away from the observer can be found by measuring the Doppler shift of its entire spectrum; the amount of shift is proportional to the speed. Among other things this is how the cosmological red-shift is measured.

There are many other, more special purpose instruments such as Fourier-transform spectrometer, Fabry–Perot interferometers, spectroheliographs, occultation photometers, speckle interferometers, etc.

Instruments in the infrared are very similar to those used at visible wavelengths. Observations from the ground are limited to a few spectral regions, since the Earth's atmosphere is not transparent outside these regions. At ultraviolet wavelengths, the Earth's atmosphere is opaque, too. Observations in this regime must be done from space.

Detectors

Photographic plates were the first recording detectors. The spatial resolution of a plate is 10–20 µm and so a 4×5-in. plate contains over 10 million independent measurements. Their main problem is that they are not very sensitive to light (low efficiency), and it is very difficult to obtain accurate intensities from them. The other main class of detectors are the photoelectric devices that convert incident light to an electrical signal. The photomultiplier tube is an ultraviolet and visible light detector. It uses the 'outer' photoelectric effect to convert an incident photon into an electron which is then amplified a million-fold within the tube.

While the photomultiplier makes only one measurement at a time, charged coupled devices (CCDs) contain up to 16 million individual detectors (pixel) and so can be used to take images. Making use of the 'inner' photoelectrical effect, an optical image is converted into an electrical one which is transmitted to a computer. CCDs are linear, record up to 90% of the incoming light, and have very little noise. Therefore, they are practically ideal detectors and are nowadays used almost exclusively in optical astronomy.

Infrared detectors are generally also semiconductor devices. These detectors can have over 80% efficiency but they are fairly noisy in most cases unless they are cooled down to under 4 K. The advent of these detectors opened up the field of infrared astronomy to the point where it is now a major branch of observational astronomy. The reason for this is that dust between the object under study and the observer blocks light at optical wavelengths, but much less at infrared wavelengths: infrared observations penetrate dust. One active field of infrared astronomy is the study of formation of stars which is usually going on at dusty places.

Observatories

There are many small and large observatories throughout the world. They are owned or operated by colleges, universities, research organizations and nations. Major observatories are located on Kitt Peak near Tucson, Arizona, Mauna Kea on Hawaii, La Silla, Cerro Tololo and Cerro Paranal in Chile. Recent site testing campaigns have shown that at least one location in Antarctica has outstanding characteristics and is possibly inferior only to space. Observatories in space are not hampered by the Earth's atmosphere and so are not restricted to certain atmospheric windows; the ultraviolet region is observable from space (or at least high flying balloons or rockets) only. Space observatories such as the Hubble Space Telescope also do not suffer from 'seeing', i. e., image smearing due to atmospheric turbulences, which degrades the resolution of ground-based telescopes. Seeing effects can be reduced in ground-based observations if a deformable mirror is placed in the light path and deformed in such a way that atmospheric distortions of the incoming wave front are compensated ('adaptive optics').

See also: Cosmology; Galaxies; Milky Way; Solar System; Sun.

Bibliography

M. A. Seeds, *Horizons: Exploring the Universe*. Belmont, Calif., Wadsworth, 1987. (E)

J. H. Robinson, *Astronomy Data Book*. Wiley, New York, 1972.

Astronomy (Astromedia Corp., Milwaukee, 1972). (E)

B. V. Barlow, *The Astronomical Telescope*. Wykeham Publ, Ltd., 1975. (I)

C. R. Kitchin, *Astrophysical Techniques*. Adam Hilger Ltd., 1984. (A)

W. H. Steel, *Interferometry*, 2nd ed. Cambridge University Press, Cambridge, 1983. (A)

R. N. Wilson, *Reflecting Telescope Optics I*. Springer, Berlin, Heidelberg, New York, 1996. (A)

R. N. Wilson, *Reflecting Telescope Optics II*. Springer, Berlin, Heidelberg, New York, 1996. (A)

P. L'ena, F. Lebrun, F. Mignard, *Observational Astrophysics*. Springer, Berlin, Heidelberg, New York, 1998. (A)

J. N. Hardy, *Adaptive Optics for Astronomical Telescopes*. Oxford University Press, 1998. (A)

Astronomy, Radio

A. A. Penzias and B. E. Turner

Radio astronomy differs from the other observational branches of astronomy in that the incident energy is coherently amplified. Unlike film, bolometers, or particle counters, the phase of the incident wave is preserved in the process. The terminals of the receiver are located at the focal point of the *antenna*, where the field in the aperture is coherently added. The most generally used antenna and the one which most nearly resembles its optical counterpart is the parabolic reflector (Fig. 1). The angular distribution of the antenna response, the "antenna pattern," has an angular width at half-intensity of $\sim \lambda/a$ radians, where a is the width of the antenna aperture.

Mechanical limitations on antenna size and precision limit the resolution (minimum beam width) of the largest single antennas to about 10 arc seconds. To obtain higher resolution, an array of antennas is connected together to form an *interferometer* (Fig. 2).

When the signals obtained by a pair of antennas are multiplied together, the product is proportional to the cosine of the difference between the phases of the received signals. The angular dependence of the response of such a system to a distant source of monochromatic radiation will have a one-dimensional periodicity in the direction which is coplanar with the line of sight between the two antennas. The angular frequency of the response is given by the projected distance, in wavelengths, between the antennas as viewed from the source.

By observing the source with a number of differently spaced antenna pairs one can obtain enough angular frequency components to construct a Fourier synthesis of the angular distribution of the source intensity. These different spacings can be obtained by use of a number of antennas with different spacing, by moving one or more of the antennas between observations, and by making use of the earth's rotation to change the projected spacing. Two-dimensional information can be obtained by arranging the antenna array in the form of a cross or "Y"

Fig. 1: The 15-m submillimeter telescope of the Swedish–European Southern Observatory consortium, at an elevation of 2350 m at La Silla, Chile. The surface accuracy (deviation from a perfect paraboloid) is 35 μm rms, allowing operation to wavelengths as short as 550 μm (Onsala Space Observatory photo).

(Fig. 2) or making use of the change in orientation of the array axis with respect to the source caused by the rotation of the earth. The resulting intensity map of the source extending over the entire area covered by the relatively broad beams of the individual antennas can have the resolution equivalent to that of a single antenna whose aperture encompasses the entire array; hence the name, *aperture synthesis*.

Radiometry

The intensity scale used in radio astronomy is antenna temperature. Consider a warm uniform opaque cloud extending over an angular area much larger than the antenna beam. The power radiated per unit area by this cloud is given by the usual blackbody formula

$$B = \frac{4\pi h \nu^3}{c^2 (e^{h\nu/kT} - 1)} \Delta\nu \, ,$$

and, in the limit $h\nu/kT \ll 1$, the power radiated per unit solid angle becomes the familiar Rayleigh–Jeans formula

$$b = \frac{2kT}{\lambda^2} \Delta\nu \, .$$

Fig. 2: The Very Large Array, consisting of 27 antennas, each 25 m in diameter, in a Y configuration. Shown is the most compact configuration (arm length 0.6 km). At its shortest wavelength, 1.3 cm, and largest configuration (arm length 21 km), the VLA provides a spatial resolution of 0.08 arcsec (NRAO photo).

At some distance R from the cloud we put an antenna with an aperture of width a pointed toward it. The antenna will receive energy only from that portion of the cloud intercepted by its beam, i. e., $\sim (\lambda/a)^2 R^2$. Furthermore, we must multiply by the solid angle subtended by the antenna at the cloud surface, $\sim (a/R)^2$. Thus the power incident upon the antenna terminals in one plane of polarization is

$$\frac{kT\Delta\nu}{\lambda^2} \times \frac{\lambda^2 R^2}{a^2} \times \frac{a^2}{R^2} = kT\Delta\nu \;.$$

This relation leads to the definition of antenna temperature as the power per unit bandwidth received at the antenna terminal divided by Boltzmann's constant. Note, therefore, that antenna temperature is an intensity which only corresponds to a thermodynamic temperature in the Rayleigh–Jeans limit.

Radio astronomy observations are simply the measurements of antenna temperature as a function of angle and frequency. To relate the observed quantity to the property of the emitting medium we introduce the concept of equivalent brightness temperature of an astronomical object. This is the physical temperature of a perfect absorber with the same angular dimensions as the object which would produce the observed antenna temperature. To obtain the equivalent brightness temperature from the antenna temperature we must take the response pattern of the antenna into account. If the object is small compared to the antenna beam, we must determine its angular size by other means. Failing that, we can only assign a flux (incident power density per unit bandwidth per unit frequency interval) to the object by dividing the observed antenna temperature by Boltzmann's constant and the effective area of the antenna, and multiplying the result by 2. (The multiplication by 2 reflects the fact that a coherent receiver responds to only one polarization of the incident field.)

Antenna temperature is measured by means of a radiometer, which in its simplest form consists of an amplifier, frequency-selective filter, rectifier–detector, and an integration and voltage-measurement circuit. Incident power levels from astronomical sources upon even the largest antennas are extremely small and must be measured in the presence of the generally much larger noise power of the radiometer itself.

Only the portion of this power contained within the particular frequency interval selected is detected; it is thus useful to characterize the radiometer's sensitivity by a quantity proportional to its noise power per unit frequency interval, its *equivalent noise temperature*, T_N, defined as the temperature of a perfect absorber placed at the input which delivers the same output noise power for an equivalent noiseless amplifier.

The fluctuation in output due to receiver noise is expressed in terms of the equivalent change in noise power at its input in units of antenna temperature by the radiometer formula

$$\Delta T_{RMS} = T_N / \sqrt{B\tau} \,,$$

where B is the bandwidth and τ is the postdetection integration time. (A heuristic understanding of the origin of this relation may be obtained by thinking of $B\tau$ as the number of independent measurements one may make of the noise temperature within the integration time.)

Practical radiometry normally employs comparisons in making measurements. The most common such comparisons are made between: the antenna and a reference termination; two positions in the sky; or two different radiometer frequencies. A common method of periodic comparison used is called synchronous detection. The output of the receiver is inverted in synchronism with the switching of the input from the signal to reference condition. Thus, over a number of cycles the integration accumulates a voltage proportional to the difference between the two inputs. The great virtue of this arrangement can be illustrated by considering the effect of a small change in gain ΔG. In the unswitched case this would cause a change in output of ΔG times the total system temperature, whereas in the switched case the fluctuation in output is ΔG times the difference of two temperatures, which can be made as small as desired by proper selection of the reference temperature. The penalty paid is a decrease in the signal-to-noise ratio by approximately a factor of 2. An array of such radiometers, usually sharing a common amplifier and with their filters spaced adjacently in frequency, is the most generally used system for line studies, the multichannel line receiver. An alternative to the multichannel radiometer is the autocorrelation receiver, which uses the fact that the autocorrelation function of the time variation of the input signal is the Fourier transform of the power spectrum in frequency.

It is useful to compare the sensitivity of radiometers with that of incoherent devices as detectors of astronomical radiation. The radiometer contains active elements which amplify both the incident power as well as its own noise to levels much higher than the noise associated with its detector. Thus, it is the noise in the amplifier which limits the sensitivity of the device.

Since this noise is proportional to bandwidth, whereas that of the incoherent detector is not, a meaningful comparison of the two types of devices can only be made when the bandwidth of the observation is specified. We may relate the noise equivalent power, NEP, of an incoherent detector to the minimum detectable increment in antenna temperature:

$$\Delta T_{RMS} = NEP/kB\sqrt{\tau}\,.$$

Astrophysical Sources of Radio Emission

Free–free emission, often called "thermal emission" for historical reasons, arises from the interaction (acceleration) between unbound charged particles in an ionized gas (HII region). It has a characteristic spectrum whose brightness temperature decreases roughly as the inverse square of the frequency above a certain "turnover" frequency, below which it is equal to the temperature of the ionized gas. The turnover frequency marks where the gas becomes opaque to radiation and ranges from 100 MHz for large, diffuse HII regions to as high as 30 GHz for small, dense ones. Several hundred HII regions are known in the Milky Way; they serve as signposts for massive hot stars since the gas is ionized by uv photons at wavelengths shorter than 912 Å from the central stars. The balance between uv heating rate and cooling by emission from trace elements (O, N) endows most HII regions with a temperature of 10 000 K. Superimposed upon the broadband spectrum of the ionized gas are *recombination lines* caused by electronic transitions in the constituent atoms (mostly H, He) as given by the Rydberg formula ($n \sim 50$ to 500). The line-to-continuum intensity ratio is a sensitive indicator of physical conditions in HII regions.

Synchrotron emission is generated by high-energy electrons moving in a magnetic field. It is called "nonthermal emission" because its intensity is related not to the temperature of the emitter but to the strength of the field and the number and energy distribution of the electrons. It produces much of the radiation from supernova remnants, radio galaxies, and quasars. Radio galaxies and quasars comprise extragalactic radio astronomy. Their emission falls into two categories: the extended structure (which is transparent) and the compact structure (which is opaque to its own radiation). The extended emission is typically associated with galaxies, but in many cases with quasars with no visible optical extent. Most compact sources are identified with quasars or with active galactic nuclei. In less powerful radio galaxies, the radio emission is often confined to the region of optical emission (about 30 000 light-years in size), but in more powerful radio galaxies the emission comes from two well-separated regions hundreds of thousands of light-years across. Figure 3 shows a radio image of the powerful radio galaxy 3C175, made with the VLA at 6 cm wavelength at a spatial resolution of 0.35 arcsec. The unresolved bright spot near the center is a compact source coincident with a quasar at a redshift of 0.77. The long, narrow, one-sided radio jet is typical of powerful double-lobed (extended) sources. There are prominent hot spots in both lobes suggesting they have both been recently supplied with relativistic particles despite the appearance of only a single jet.

Our own galaxy radiates largely by synchrotron emission. An all-sky map of the galaxy's emission at 408 MHz is shown in Fig. 4; it was made at a single resolution of 0.85 degrees using three of the world's largest parabolic reflectors. The galactic center serves as the center of symmetry for the entire sky, with the low galactic latitude intensity dropping away steeply on each side out to longitudes ~ 60 and 280 degrees. Large-scale prominences, rising from low galactic latitudes and extending nearly to the poles in both hemispheres, distort the symmetry

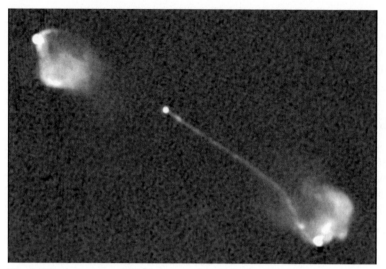

Fig. 3: The radio galaxy 3C175, observed with the VLA at a resolution of
0.35 arcsec at 6 cm wavelength (photo courtesy A. Bridle).

of the high-latitude emission. These loops and spurs trace large corresponding features in the
magnetic field of the galaxy.

Line emission provides a powerful method for the study of neutral (and some ionized)
interstellar matter. Hydrogen, the most abundant element, is studied in our own galaxy, in
external galaxies, and on cosmological scales by means of its ground-state magnetic hyperfine
transition at 1420 MHz (21 cm wavelength). Together with optical redshift data, HI data has
established the existence of large-scale "voids" over cosmological distances, regions that are
underpopulated at least in matter concentrated in normal galaxies. Current cosmological HI
research is aimed at determining whether significant amounts of baryonic matter, capable of
contributing importantly to the mass density distribution and therefore to the question whether
the universe is closed or open, may reside in noncollapsed intergalactic clouds of HI.

HI studies of clusters of galaxies have shown gas deficiencies in cluster spiral galaxies –
evidence that gas has been swept from at least their outer regions by passage through the center
of the cluster. Several mechanisms may be at work (galaxy collisions, tidal interactions, ram
pressure sweeping by the intracluster medium, evaporation) and may remove up to 90% of the
initial HI mass. Reduced star formation rates are evident.

HI studies of noncluster galaxies, especially nearby ones for which good spatial resolution
exists, serve to relate the structure of the interstellar medium with that of the stellar populations. Clearly defined HI disks correspond to optical disks but extend much further out, allowing galaxy mass determinations. Spiral arms seen in HI often reveal important differences
from their optical counterparts that relate to star formation mechanisms. Strong warps are
often seen in the outer regions of HI disks, which in many cases appear to be self-maintaining
rather than resulting from perturbations such as tidal interactions. HI studies of velocity distributions in galaxies show in the case of spirals that the dynamical masses greatly exceed
the total mass (by factors of 2 to 3) seen within the optical radius. HI studies also show that
low-luminosity galaxies contain the same large mass-to-light ratio as luminous galaxies do,

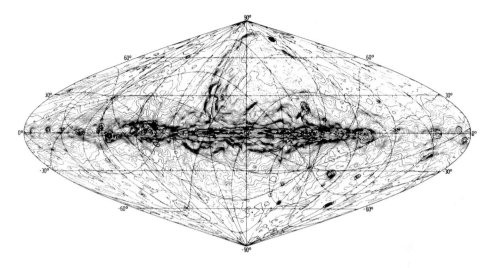

Fig. 4: An all-sky map of the Milky Way at 408 MHz (photo courtesy G. Haslam).

and that some types also contain dark halos. Since low-luminosity galaxies are by far the most numerous, these HI results are important to the question of the "missing mass" needed to close the universe.

Line emission is also observed from a large number of interstellar molecules, which reside in the denser regions of the interstellar medium. Together with studies of HI in the Milky Way, these lines have delineated the spiral structure and have established the existence of several thousand giant, massive molecular clouds (10^5 solar masses or greater) as well as many more smaller ones. These "GMCs" are the most massive entities in our galaxy and have been established as the sites of massive star formation. In other galaxies as well as our own, the GMCs are nor, known to trace the spiral arms. Molecular CO has now been observed at redshifts up to 0.16, and promises to become a cosmological tool as important as HI.

Around 130 interstellar molecular species have now been identified in dense molecular clouds, forming the current dicipline of astrochemistry. The interstellar chemistry is carbon rich, as on earth, but produces many exotic species not found on earth, as well as many familiar in the laboratory. These molecules play a major role in how stars form from the interstellar gas. The identification rate of new species is still about 2 per year. Recent discoveries have focused on larger organic molecules such as vinyl alcohol, propenal and propanal, the latter two being "sugars". Glycine, the smallest biologically important amino acid, has possibly been identified, but is highly controversial. Equally intriguing is the discovery of highly deuterated species such as doubly deuterated ammonia NHD_2, and formaldehyde (D_2CO), and most important triply deuterated protonated hydrogen (H_3^+), which does not occur on Earth but along with H_3^+ is the most basic progenitor of astrochemistry. At a cold 10 Kelvin deep inside thick molecular clouds, H_3^+ and D_3^+ are the only species not frozen out onto icy grains.

Blackbody radiation is characteristic of solar system objects. At long radio wavelengths the brightness temperature of the sun is very large, $\gtrsim 10^6$ K, because the observed emission comes from the corona and material ejected during solar flares. This ionized matter becomes essentially transparent at shorter wavelengths, and the millimeter wavelength brightness temperature of the sun is essentially that of the photosphere, $\sim 6000\,\mathrm{K}$. Conversely, in the case of a *planet*, the long wavelengths are able to penetrate un-ionized atmosphere better than shorter wavelengths. Thus the long-wavelength brightness temperature is that of the surface, whereas at shorter wavelengths the brightness temperature corresponds more closely to that of the cooler atmosphere.

Blackbody radiation fills the entire sky and is the remnant of the big-bang fireball expanded and cooled to $3\,\mathrm{K}$ at the present epoch. It is unique in that it has the same brightness temperature, $3\,\mathrm{K}$, at all wavelengths. The radio spectrum of "empty" sky is a superposition of galactic radiation, unresolved distant radio sources, and the cosmic background radiation.

See also: Astrophysics; Blackbody Radiation; Galaxies; Interferometers and
Interferometry; Photosphere; Radiometry; Sun; Synchrotron Radiation;
Transmission Lines and Antennas

Bibliography
J. D. Kraus, *Radio Astronomy.* Cygnus-Quasar Books, 1986. An intermediate level book emphasizing technical aspects, with advanced references.
G. L. Verschuur and K. I. Kellermann (eds.), *Galactic and Extragalactic Radio Astronomy.* Springer, New York, 1988. A graduate-level text covering 15 major areas of radio astrophysical research.

Astronomy, X-Ray

P. Gorenstein and W. Tucker

Introduction

X-ray astronomy involves photons of cosmic origin in the energy band 0.2 to 30 keV. The low-energy limit is determined by the opacity of the interstellar medium, the higher limit by the falling spectra of sources. Observations take place above the absorption of the Earth's atmosphere. X-ray production in a cosmic setting is associated with thermal radiation from plasmas with temperatures from 10^6 to $10^8\,\mathrm{K}$, synchrotron radiation from highly energetic electrons in a magnetic field and inverse Compton scattering where electrons elevate lower energy photons to the x-ray band. Solar x-ray emission was detected in 1948 and a rocket flight in 1962 made the first positive detection of sources outside the solar system. A series of spinning satellites, initially bearing proportional counters beginning with UHURU in 1970 and continuing through ROSAT in 1990's with an imaging telescope, surveyed the sky and cataloged about 80 000 sources. Even more have been found serendipitously in the fields of pointed observations. Their celestial distribution shows two components: one distributed along the plane of the galaxy and another having the isotropy characteristic of extragalactic objects. In addition, aside from an irregular soft component associated with the local interstellar medium, there is an intense isotropic background which now appears to be fully explained as the sum of distant extragalactic point-like sources.

X-ray astronomy was transformed by the findings of two facility-class observatories with focusing telescopes launched in 1999 plus results from the large area counters of the Rossi X-Ray Timing Explorer (1995). NASA's Chandra X-Ray Observatory, with the ability to make images with sub-arc-second resolution is well complemented by the lower-resolution, but higher throughput European Space Agency's observatory, XMM-Newton. Both observatories are equipped with dispersive spectrometers that allow high- resolution spectra ($E/\delta E \sim 300$–1000). Chandra and XMM have studied planetary atmospheres, normal stars of every type, brown dwarfs, white dwarfs, neutron stars and black holes, the remnants of exploded stars, galaxies of every size, shape, and stage of evolution, galaxy clusters with thousands of galaxies embedded in vast clouds of hot gas and dark matter millions of light years across, and even larger streamers of hot gas between the clusters. These studies plus the fast temporal intensity variations measured by the Rossi X-Ray Timing Explorer have led to advances on such fundamental questions as the geometry of spacetime around a black hole, the ability of accreting black holes to produce highly collimated, relativistic jets a million light years in length, the distribution and properties of dark matter, and the existence of dark energy. Their findings have placed x-ray astronomy in realm of fundamental physics where several basic principles can be tested in domains that are not accessible on Earth.

X Rays from Planets and Comets

While the temperature of planets, satellites, asteroids and comets, is typically well below 1000 K they produce x rays through mechanisms that involve the Sun directly or indirectly. The x-ray power is weak, ranging from a few megawatts in the Martian atmosphere to a few gigawatts for Jupiter but provides information about chemical composition, mass, and radiation belts that are difficult to obtain with measurements at other wavelengths.

Charge exchange, primarily between heavy neutral atoms such as oxygen in the atmosphere of a comet or a planet, and fast ions in the solar wind, operates throughout the solar system. It is especially important for comets, which have extended atmospheres. By observing x rays from comets, it is possible to study the elements present in the solar wind, the structure of the comet's atmosphere, and cometary rotation.

Planetary atmospheres and in the absence of an atmosphere, the bare surface, emit fluorescence x rays when illuminated by solar x radiation. Fluorescent radiation from oxygen and other atoms has been detected in the Venusian atmosphere between 120 and 140 kilometers above the surface of the planet. In contrast, the optical light from Venus is caused by the reflection of sunlight from clouds at altitudes 50 to 70 kilometers.

Fluorescent x rays from oxygen atoms in the Martian atmosphere probe similar heights. The lack of variation of Martian x radiation during a dust storm on the surface shows that the storm did not reach into the upper atmosphere. The detection of a faint halo of x rays, presumably due to the solar wind charge–exchange process operating in the tenuous extreme upper atmosphere of Mars some 7 000 kilometers above the surface indicates that Mars is still losing its atmosphere into deep space.

Jupiter produces x rays as a result of its substantial magnetic field. X rays are emitted when high-energy particles from the Sun get trapped in its magnetic field and are accelerated toward the polar regions where they collide with atoms in Jupiter's atmosphere. Chandra's image of Jupiter shows strong concentrations of x rays near the north and south magnetic poles. Weaker x-ray signals have been detected from two of Jupiter's moons, Io and Europa, and from the Io Plasma Torus, a doughnut-shaped ring of energetic particles that circles Jupiter. Gases, such as sulfur dioxide, are produced by Io's volcanoes, escape from Io and become trapped in an orbit around Jupiter, where they are accelerated to high energies. Collisions between the particles within the torus, and with the surfaces of Io and Europa can account for the observed x rays.

Chandra's observation of Saturn, which has a weaker magnetic field than Jupiter, revealed an increased x-ray brightness in the equatorial region. Furthermore, Saturn's x-ray spectrum, or the distribution of its x rays according to energy, was consistent with scattered solar x rays. The same process may be responsible for the weak equatorial x radiation observed from Jupiter.

Galactic X-Ray Sources

The bright galactic sources are associated with the final phases of stellar evolution. They are identified with remnants of supernova explosions or binary star systems containing a compact object such as a neutron star or a black hole. They are among the most luminous objects in the galaxy, radiating 10^{36} to 10^{38} erg/s in the x-ray band. In comparison, the Sun, an average star, radiates 4×10^{33} erg/s, principally at optical wavelengths. A much lower level of x-ray emission, 10^{28}–10^{32} erg/s, has been detected from many relatively nearby stars. The x-ray emission from normal stars is produced by processes that are analogous to solar coronal and flare activity, or shock waves in the winds of massive stars or colliding stellar winds of binary stars. Binary systems containing white dwarf stars, and isolated neutron stars with surface temperatures of the order of a million degrees have also been detected at these x-ray luminosities.

Young Star Clusters

About a thousand faint x-ray-emitting stars have been detected in the Orion star cluster by the Chandra X-Ray Observatory. Since flaring activity is more pronounced at x rays than at optical wavelengths, long-term monitoring of this sample of stars, all at the same distance, and approximately the same age is indicative of the rate and magnitude of flaring in young stars. This information on the high-energy activity of young stars is relevant to understanding the environment in which the planets were formed and evolved.

In a related discovery, non-thermal x rays have been detected from a cloud of high-energy electrons enveloping the young star cluster RCW 38. These extremely high-energy particles could cause dramatic changes in the chemistry of the disks that will eventually form planets around stars in the cluster.

Compact X-Ray Binaries

All of the bright luminous galactic x-ray sources that have not been identified with supernova remnants fall into a class designated as "compact x-ray binaries." The essential features of these sources are (i) a 1–10-keV luminosity in the range 10^{36}–10^{38} erg/s; (ii) membership in binary systems, as evidenced by eclipses or periodic variations on a time scale of days in the radiation of the compact object or its companion star; (iii) a spectrum similar to that produced by radiation from a hot gas having a temperature in the range 50–500 million degrees; (iv) fast, and in some cases quasi-periodic oscillations on a time scale of seconds or less.

The model consists of matter lost from the primary star in a close binary system accreting onto a neutron star or black hole companion. Gravitational potential energy heats a gas which radiates the input energy as x rays. If the secondary has a mass M_x and a radius R, the gravitational energy released per gram would be on the order of GM_x/R. For a mass accretion rate $\dot{m}$, the energy released per second is

$$L \sim \frac{GM_x\dot{m}}{R} \sim 10^{41} \left(\frac{M_x}{M_\odot}\right) \left(\frac{R_\odot}{R}\right) \dot{m} \quad \text{erg/s} \tag{1}$$

where $M_\odot$ and $R_\odot$ are the mass and radius of the sun and $\dot{m}$ is the accretion rate in units of solar masses per year. For both neutron stars and black holes $R_\odot/R \sim 10^5$, we have, therefore,

$$L \sim 10^{46}\dot{m} \quad \text{erg/s} \tag{2}$$

Equation (2) shows that mass accretion rates in the range 10^{-10}–10^{-8} solar masses per year can produce x-ray luminosities in the range 10^{36}–10^{38} erg/s.

For comparable accretion rates, the expected luminosity for white dwarf companion stars $(R/R_\odot \sim 10^{-2})$ is three orders of magnitude smaller. Accreting white dwarf binaries are perhaps the most common type of binary x-ray source in the galaxy.

Compact binary systems also differ with respect to the nature of the noncompact primary star. In young binary systems the primary is a giant blue star that has a mass more than 10 times that of the sun and whose age is less than 10 million years. They are found in regions of active star formation, such as the galactic spiral arms. In old or low-mass x-ray binaries, the noncompact star is often less massive than the sun. These systems have existed for at least a hundred million years and show no preference for spiral arms.

Young neutron stars have intense magnetic fields which modify the accretion flow and produce an asymmetric radiation pattern that, when coupled with the rotation of the neutron star, appears to a distant observer as a series of pulses with eclipses and Doppler variations from the binary motion. For old x-ray binaries, two physical effects change the nature of their radiation. First, the accreting plasma transfers angular momentum to the neutron star, causing it to spin faster over time. Second, the magnetic field of the neutron star has weakened to the point where it can no longer effectively channel the accreting plasma. Consequently, these sources are characterized, not by stable periodic pulses on a time scale of seconds, but by quasi-periodic oscillations (QPO's) down to a scale of milliseconds. QPO's were discovered in neutron star binaries in 1996 by investigators analyzing data from the large area counters of the Rossi X-Ray Timing Explorer. Relativistic dragging of inertial frames, known as the Lense–Thirring effect, around fast rotating collapsed stars, has been proposed by some theoreticians as a mechanism responsible for QPO's.

Similar behavior is observed from an accreting black hole. Strong magnetic fields cannot exist in the vicinity of black holes and the period of the last stable orbit around a black hole having the mass of a few solar masses is on the order of milliseconds. The similarity in the radiation patterns of accreting black holes and old neutron stars has made it impossible to identify black holes conclusively on the basis of x-ray data alone. X-ray novae are a clue. Some of these sources exhibit a faint x-ray luminosity in quiescence, whereas others are undetectable. The absence of x radiation in quiescence has been interpreted as evidence for an event horizon in these sources, since the presence of a solid surface would lead to low-level x-ray emission, or x-ray bursts due to thermonuclear burning of the accreted matter.

In general, by far the strongest evidence for a black hole consists of optical observations of the primary star indicating that the primary has an invisible companion with a mass greater than $3M_\odot$ the theoretical upper limit for the mass of a neutron star. To date, about 20 such systems have been discovered, with estimated black hole masses ranging from 4 to 16 solar masses.

With most physicists and astrophysicists in agreement that black holes do indeed exist, intense theoretical and observational efforts are underway to understand the detailed properties of these systems. The K-α fluorescent line of iron provides an especially useful probe of the region within a few gravitational radii of the event horizon of a black hole. The gravitational, or Schwarzschild radius is defined as

$$R_S = \frac{2GM}{c^2} \tag{3}$$

Detailed x-ray spectroscopy of broadened iron line features has been used to study Doppler and gravitational redshifts, thereby providing key information on the location and kinematics of the cold material. Observations of both stellar mass black holes and supermassive black holes have provided intriguing and impressive evidence for the gravitational red-shift and the effects of black hole spin on the spacetime around black holes.

The orbit of a particle near a black hole depends on the curvature of space around the black hole, which also depends on how fast the black hole is spinning. A spinning black hole drags space around with it and allows atoms to orbit closer to the black hole than is possible for a non-spinning black hole. For example, x-ray observations of the K-α line from the stellar black hole Cygnus X-1 show that the profile of the iron line is skewed to lower energies in a manner consistent with a slowly rotating or non-rotating black hole, whereas data from the black hole, XTE J1650-500, show a much larger skewing to low energies, consistent with a rapidly spinning black hole. Previous observations of some supermassive black holes by Japan's ASCA satellite, XMM-Newton and Chandra have indicated that they may also be rotating rapidly.

In recent years, evidence has been found for a third class of black holes intermediate in mass between stellar mass black holes and supermassive black holes at the center of galaxies. If their emission is not beamed towards us, preferentially, these x-ray sources would have a significantly higher luminosity than normal stellar mass black holes which suggests that they may be black holes with masses in the range of a few hundred solar masses. Current possible explanations for the formation of intermediate mass black holes include the mergers of scores of stellar black holes, or the collapse of a extremely massive star. If their x-ray emission is beamed toward the Earth it would reduce the overall output of x rays from the source, and reduce the estimate of their mass to a value consistent with stellar mass black holes but their higher flux at the Earth would still place them in a separate class.

Supernova Remnants & Pulsars

Over eighty sources in our galaxy and numerous sources in nearby galaxies have been identified with supernova remnants (SNR). Nearly all are characterized by a fragmentary shell with a diameter that gets larger and a spectrum that gets softer with age. The evolution of lumi-

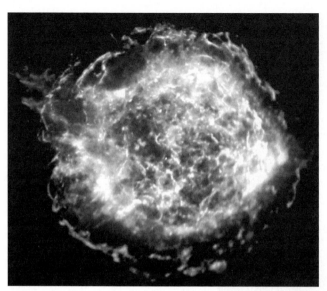

Fig. 1: X-ray image of the supernova remnant Cas A taken with the Chandra
X-ray Observatory. There is evidence for a point source at the center, which
could be a hot neutron star. The image scale is 7 arcminutes on a side.

nosity is more complex. Some twenty SNR show evidence for the existence of a neutron star. Four of these are fast pulsars that are losing rotational energy at a rate sufficient or more than sufficient to explain their luminosity. Figures 1 and 2 show x-ray images of two relatively young SNR: Cas A, and the Crab Nebula which contains a rotationally powered pulsar. Cas A was born in 1680 and the Crab in 1054. The distance to both is about 2 kpc. Their x-ray spectra are shown in Figure 3. Cas A contains x-ray lines of highly ionized silicon, sulfur, calcium, and argon, indicative of radiation from a hot plasma ($\sim 10^7$ K) with enriched elemental abundances expected to be associated with supernova ejecta. Two shock waves are visible: the outward-moving shock that is moving into the interstellar medium, and the reverse shock that is moving into the stellar ejecta. In a high quality image a point source is seen near the center which may be a neutron star that appeared when the core of a massive star collapsed to initiate the supernova explosion.

The Crab Nebula presents a dramatically different picture. It is atypical in that there is no indication of a shell of ejecta. Its appearance is dominated by the effects of the rapidly rotating neutron star, or pulsar at its center. The spectrum of the continuum radiation of the Crab Nebula is due to synchrotron radiation from relativistic electrons in the magnetic field of the nebula. It is intrinsically featureless but absorption in the interstellar medium produces a low energy cutoff and elemental absorption edges. The rapidly rotating neutron star is seen directly as a point source twice during each 33-ms period when the radiation is beamed in our direction. Enormous electrical voltages generated by the rotating, highly magnetized neutron star accelerate particles outward along its equator to produce the pulsar wind. These pulsar voltages also produce the polar jets seen spewing x-ray emitting matter and antimatter electrons perpendicular to the rings.

Fig. 2: Chandra's x-ray image of the Crab Nebula (SN 1054) with a 33-ms period pulsar at the center of the rings. The pulsar brightens twice during each period when its rotating beams are pointing toward the Earth. The Crab Nebula is unusual among supernova remnants in that there is no shell produced by supernova ejecta interacting with interstellar material. The image scale is 1.7 arcminutes on a side

As this relativistic wind of matter and antimatter particles from the pulsar plows into the surrounding nebula, it creates a shock wave and forms the inner ring. Energetic shocked particles move outward to brighten the outer ring and produce an extended x-ray glow called a pulsar wind nebula. Observations with the Chandra X-Ray Observatory have led to the discovery of about 50 pulsar wind nebulae in various stages of evolution, both inside supernova remnants and outside. Some of them show spectacular bow shock waves due to their supersonic motion through the interstellar gas.

Extragalactic Sources

Galaxies, Starburst Galaxies

The x-ray emission from galaxies consists of three basic components: (i) individual sources of the type discussed above, (ii) x rays from a hot interstellar medium, and (iii) for at least some galaxies x rays connected with a supermassive black hole in the center. Their neutron star and black hole x-ray binary populations provide clues to their history. For example, the Chandra image of the elliptical galaxy NGC 4261 reveals dozens of black holes and neutron stars strung out across tens of thousands of light years like beads on a necklace. The structure, which is not apparent from the optical image of the galaxy, is thought to be the remains of a collision between galaxies a few billion years ago. According to this interpretation, a smaller galaxy was captured and pulled apart by the gravitational tidal forces of NGC 4261. As it fell into the larger galaxy, large streams of gas were pulled out into long tidal tails.

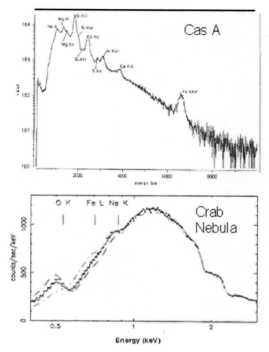

Fig. 3: X-ray spectra of Cas A and the Crab nebula. Strong elemental lines in Cas A's spectrum are indicative of thermal radiation from a hot plasma containing enriched material released from the exploded star and shocked heated matter from the circumstellar medium. The absence of lines in the Crab's spectrum and the presence of polarization is consistent with synchrotron radiation. The low energy cutoff and the absorption edges are caused by the detector response and absorption in the interstellar medium

Shock waves induced by these tidal tails triggered the formation of many massive stars causing it to become a "starburst" galaxy. The shock waves push on giant clouds of gas and dust, causing them to collapse and form a few hundred stars. The massive stars use up their fuel quickly and explode as supernovas, which produce more shock waves and more star formation. The process continues until the gas is consumed or blown away by the explosions to end the starburst phase. During a starburst, stars can form at tens, even hundreds of times greater rates than the star formation rate in normal galaxies. Over the course of a few million years, these stars evolved into neutron stars or black holes. A few of these collapsed stars had companion stars, and became bright x-ray sources as gas from the companions was captured by their intense gravitational fields. Starburst activity typically lasts for ten million years or more, a small fraction of the ten billion year age of the galaxy. The optical evidence for starburst activity fades rather quickly into the stellar background of the galaxy, whereas the x-ray signature such as Chandra's image of the many point sources in NGC 4261 lingers for hundreds of millions of years and may be the best means of identifying the ancient remains of mergers between galaxies.

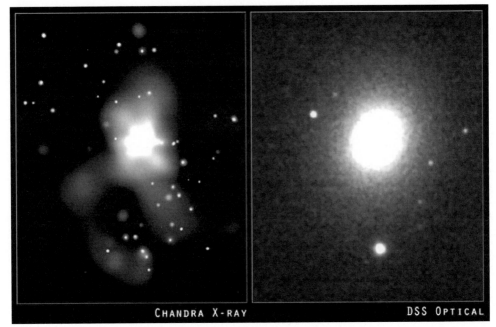

Fig. 4: X-ray image obtained by Chandra (left panel) and optical image of the galaxy NGC 4261. The x-ray image contains many neutron star and black hole binary systems, the residue of an intense period of star formation that was probably induced by the collision between and the merger of a large and smaller galaxy. Each panel is 4×5 arcminutes.

Active Galactic Nuclei

Thousands of x-ray sources have been identified with quasars, Seyfert galaxies, radio galaxies, and BL Lacertae objects, known collectively as active galactic nuclei (AGN). Some differences among them may be due to our observing direction relative to their galactic plane or a jet emanating from a supermassive black hole at the center. The x-ray luminosity of AGN is an appreciable fraction of their total radiative output, in some cases accounting for most of it. Because most supermassive black holes produce x radiation that can penetrate the clouds of dust and gas that surround them, Chandra and XMM have been able to show that most galaxies harbor a supermassive black hole at their centers.

The Chandra Deep Sky surveys showed that the x-ray background radiation is not diffuse. Rather, it resolves into numerous point-like sources due to supermassive black holes in active galactic nuclei. An important, as yet unanswered, question is whether the first supermassive black holes were formed before, at the same time as, or after the first galaxies were formed. At present we know only that some supermassive black holes already existed about a billion years after the Big Bang, about the same time as the most distant observed galaxies.

Larger black holes and/or larger supplies of gas produce higher luminosities, which suggests that the typical galactic nucleus was once a quasar and that the typical quasar will eventually become the nucleus of a normal galaxy when all the matter susceptible to accretion is consumed. This process might take a few million years, or a few hundredths of a percent of the age of a galaxy, consistent with the observed ratio of quasars to normal galaxies and their ages.

The estimated masses of supermassive black holes range from about a million solar masses to more than a billion solar masses. For example, the Milky Way has a small bulge and a supermassive black hole with a mass of only about 3 million solar masses, whereas the giant elliptical galaxy M87 contains a black hole with a mass in excess of a billion solar masses. An empirical relation between the mass of the black hole, and the mass of the spherical bulge containing the black hole (as deduced from the magnitude of the average random velocity of the stars in the bulge) suggests that the initial rotation rate, or collisions with other galaxies may play a role.

X-ray and radio observations have shown that the influence of supermassive black hole continues well beyond the central regions of the galaxy. High-energy jets extend away from some supermassive black holes moving at nearly the speed of light in tight beams that travel hundreds of thousands of light years. These jets are thought to originate in the accretion disk around the black hole, where the twisting of magnetic field lines creates large electromagnetic fields that launch and collimate the jet.

X-radiation from the jets near the nucleus of the galaxy is produced most likely by synchrotron radiation from relativistic electrons in the jet, or possibly by Compton upscattering of optical, infrared or radio synchrotron photons by high energy electrons in the jet. Outside the galaxy, the X-radiation of the jets is probably due to Compton upscattering of cosmic microwave background photons by the electrons in the jet. The number and energy density of the cosmic microwave background photons increases with increasing redshift (or look-back time) and offsets the effect of increasing distance, so the large jets should be detectable at very large distances, and can provide a useful probe of the environment of young galaxies.

The energy content of the large x-ray jets is enormous, suggesting that the conversion of the jet's kinetic energy into heat could play a role in regulating the growth of galaxies. A striking example of episodic outbursts from the central regions of a galaxy is provided by the Chandra x-ray image of the galaxy Perseus A, which shows wavelike features produced by explosions occurring at intervals of about ten million years.

Clusters of Galaxies

Clusters of galaxies contain hundreds to thousands of galaxies immersed in an enormous cloud of hot gas held together by dark matter. They are the largest gravitationally bound systems in the universe. The hot gas clouds in clusters are detected as x-ray sources extending over a region several million light years in diameter. The emission is characteristic of radiation from a hot, optically thin plasma with temperatures in the range 10–100 MK, central densities of $\sim 10^{-3}$ cm^{-3}, and relative elemental abundances, based mainly upon the measurement of x-ray lines from Fe ions with one and two electrons, that are between one-quarter and one-half of the solar abundance (Fe/H $\sim 3 \times 10^{-4}$).

The mass of the intracluster gas exceeds the mass of all the stars, gas and dust in all the galaxies in the cluster. The gas was likely heated primarily by shock waves generated as the system collapsed. Large N-body computer simulations indicate that clusters are formed gradually over the eons through the mergers of groups and subclusters of galaxies. X-ray images of clusters provide dramatic confirmation of this picture, showing many examples of clusters with rich substructure due to merging subclusters.

Warm-hot Intergalactic Medium and Missing Baryons

However, the mass of the intracluster gas plus all the stars and dust in galaxies is insufficient by a factor of two to account for all the baryons that must have existed in the early universe. Therefore theorists have predicted that the missing baryons must reside in the intergalactic medium where the atoms are too hot and therefore too highly ionized to be detectable by absorbing the radio or visible light emissions of objects along the line of sight but not so highly ionized that they cannot absorb x rays. Indeed measurements of x-ray absorption lines in the spectrum of an AGN have confirmed the hypothesis that there exists a diaphanous filamentary warm-hot intergalactic medium (WHIM) outside of galaxies and clusters. If the points sampled are typical then the number of baryons that exists in the WHIM is sufficient to resolve the factor of two discrepancy.

Dark Matter

Because the cluster gas is in hydrostatic equilibrium in the cluster's gravitational potential, the morphology of the x-ray source reflects the distribution of mass in the cluster. Hence, the underlying mass of subcluster components and extended dark halos around individual galaxies can be studied by measuring the distribution of gas density and temperature. X-ray studies of numerous clusters have shown that 70 to 90 percent of the mass of a typical cluster consists of dark matter, i. e., matter which does emit any radiation - mysterious particles left over from the dense early universe that interact with each other and "normal" matter only through gravity. The favored type of dark matter is cold dark matter, which gets its name from the assumption that its constituent particles were moving slowly when galaxies and galaxy clusters began to form.

The exact nature of cold dark matter remains a mystery, but x-ray observations have provided a constraint. For example, the Chandra image of the galaxy Abell 2029 shows a smooth increase in the intensity of x rays all the way into the central galaxy of the cluster. By precisely measuring the temperature and intensity distribution of the x rays, astronomers were able to make the best map yet of the distribution of dark matter in the inner region of the galaxy cluster. The x-ray data imply that the density of dark matter increases smoothly all the way into the central galaxy of the cluster. This discovery agrees with the predictions of cold dark matter models, and is contrary to other dark matter models, such as self-interacting dark matter, that predict a leveling off of the amount of dark matter in the center of the cluster.

Dark Energy

While an explanation for dark matter is still lacking an effect that is even more enigmatic has been discovered. Optical astronomers have observed that the visible light from Type-1a supernovas, which act as standard candles, is fainter than expected in distant galaxies. The best explanation is that they are more distant than originally thought, which implies that the expansion of the universe must be accelerating. Chandra's measurements of the dark matter content of clusters of galaxies corroborates this astounding result by a method that is both independent of and complementary to the Type 1a supernova findings.

X-ray observations have determined the gas fraction of the total mass, i. e., the ratio of the mass of the hot gas and the mass of the dark matter for a number of cluster of galaxies. The observed values of the gas fraction depend on the distance scale adopted, which in turn depends on the expansion rate of the universe. Because galaxy clusters are the largest bound structures in the Universe, they are thought to represent a fair sample of the matter content in the universe. If so, the ratio of hot gas and dark matter should be the same for every cluster. Using this assumption, the parameters in the distance scale can be adjusted to determine which one fits the data best. The best fit parameters are consistent with a model in which the expansion of the Universe was first decelerating until about six billion years ago, and then began to accelerate.

The driving force behind cosmic acceleration is being attributed to a new entity known as dark energy. Accounting for the existence of dark energy requires either a refinement of Einstein's theory of general relativity or a major revision of some other area of fundamental physics. Assuming that dark energy is responsible for the acceleration, combining the x-ray results with observations of Type 1a supernovas and the cosmic microwave background radiation indicates that dark energy makes up about 75% of the Universe, dark matter about 21%, and visible matter about 4%.

Concluding Remarks

The success of the x-ray astronomy observatories launched at the turn of the century has placed x-ray astronomy on a par with optical astronomy as an area of research into the structure and evolution of the universe. Furthermore, the observation of strong gravity effects around black holes plus the constraints placed on dark matter and dark energy have demonstrated that x-ray astronomy has an important role to play in fundamental physics research. For the future we require much larger area telescopes to provide more photons for spectroscopy and to extend our reach further back in time. International collaboration, already a prominent feature of the current generation of observatories will be even more essential for the larger instruments of the next generation.

See also: Black Holes; Galaxies; Interstellar Medium; Neutron Stars; Pulsars; Quasars; Synchrotron Radiation.

Bibliography

R. Giacconi, H. Gursky, F. Paolini, and B. Rossi, *Phys. Rev. Lett.* **9**, 439 (1962).

P. A. Charles and F. D. Seward, *Exploring the X-Ray Universe*. Cambridge University Press, Cambridge, 1995. (New edition in few years.)

M. C. Weisskopf, B. Brinkman, C. Canizares, G. Garmire, S. Murray, and L. P. van Spreybroeck, *PASP* **1**, 114 (2002).

W. H. Lewin and M. van der Klis (eds.), *Compact Stellar X-Ray Sources*. Cambridge University Press, Cambridge, 2005.

Web Sites that are resources for current information

Chandra X-Ray Observatory: `chandra.harvard.edu`

NASA's High Energy Astrophysics Science Archive Research Center: `heasarc.gsfc.gov`

Astrophysics

M. Bartelmann

Astrophysics is unique among the physical disciplines for two reasons. First, experiments with astrophysical objects are generally impossible, except perhaps for the rare occasions when extraterrestrial material like lunar rocks or meteorites can be studied under laboratory conditions. Instead, astrophysics relies on information transmitted mostly by light from radio waves to γ rays, but also elementary particles (e. g., neutrinos) or gravitational waves (in future). Second, there is hardly any area of physics which does not play a role in astrophysics, from quantum field theory to the general theory of relativity.

Physics was first applied to astronomy by Galileo, Kepler and Newton, who succeeded in explaining the motion of bodies in the solar system. Modern astrophysics emerged from the study of stars, and the discovery that stellar spectra exhibit characteristic absorption lines. Questions as to the origin of these spectra, the internal constitution of the stars and their energy source led to the vigorous subsequent development of astrophysics.

Of the four fundamental interactions known in physics, only gravity is relevant on the largest scales, because the strong and weak interactions are restricted to subatomic scales, and the electromagnetic interaction can be shielded by opposite charges. A possible exception are magnetic fields, which can be important even on cosmological scales. The current theory of gravity is Einstein's general theory of relativity, which allows simple, symmetric models for the universe as a whole to be constructed. Perhaps one of the most surprising recent successes of astrophysics was the discovery that these models indeed seem to describe our universe extremely well. According to them, the universe originated 14 billion years ago in a hot, dense state, called the Big Bang, from which it subsequently expanded and cooled.

At times very close to the Big Bang, quantum physics must surely be used, but it is yet unknown how general relativity and quantum theory are to be combined. It appears that a particular class of quantum fields, so-called scalar fields, are crucially important for the appearance of the universe and for the origin of the structures contained in it. The cooling of the universe is described by the thermodynamics of (relativistic) quantum gases. When the universe was about three minutes old, helium and a few light elements could be formed from hydrogen by nuclear fusion. About 400 000 years after the Big Bang, atoms formed from nuclei and electrons. The sudden disappearance of charged particles allowed the ubiquitous background of electromagnetic radiation to decouple from matter and stream almost freely through the universe since.

Cosmic structures as we see them today were already laid out at that time, and they left their imprint in the cosmic background radiation, which we can observe today. The corresponding fluctuation patterns in the radiation density can be predicted using kinetic theory and Compton scattering. Although they are at the level of 10 parts per million only, they have been observed and accurately mapped, thereby confirming and tightly constraining the generally-relativistic cosmological models.

Three questions immediately arise from there. First, what was the origin of the structures we see in the density of the cosmic background radiation? Second, starting from these tiny fluctuations, how could the cosmic structures be built that we see today? And third, how could the rich diversity of cosmic objects form?

The speculative answer to the first question, which has become plausible recently, is that vacuum fluctuations of quantum fields very early in the universe were stretched to cosmic scales in a process called cosmological inflation. Structures as large as galaxies and galaxy clusters could then be traced back to quantum fluctuations which are an inevitable consequence of Heisenberg's uncertainty principle.

The second question leads us to assume that the majority of matter in the universe does not interact with light, because otherwise the fluctuations in the cosmic radiation background would have to be much larger. Such a hypothetic form of "dark matter" has not been observed yet, but apparently we have to accept that ordinary matter composed of protons, neutrons and electrons is the exception rather than the rule. Assuming that there is dark matter, and that it consists of weakly-interacting, massive elementary particles, the appearance of cosmic structures from scales smaller than galaxies to much larger than galaxy clusters can well be explained. According to general relativity, any matter inhomogeneities deflect light. This gives rise to the gravitational lensing effect, which allows even dark structures to be traced.

The third question is extremely complicated in detail. Accepting that the skeleton of cosmic structures is provided by dark matter, we have to study when and how visible entities like stars, galaxies and others could originate. Almost all ordinary gas in the universe became neutral 400 000 years after the Big Bang when the cosmic radiation background was released. In the spectra of the bright, distant quasars, we see that the gas must have been ionized again when the universe was about a billion years old. The picture is plausible that gas fell into the structures provided by the dark matter, heated up, and formed the first stars, whose energetic radiation could re-ionize the cosmic gas. Much of the cosmic gas is not bound in stars and galaxies, but forms the diffusely distributed intergalactic medium.

It is equally plausible that the first galaxies formed at about the same time from gas falling into "halos" of dark matter. The most obvious observational facts on galaxies which need to be explained are that they broadly fall into two morphological classes, ellipticals and spirals, whose parameters exhibit tight correlations. How galaxies form and develop, and how they change in response to the density of their environment, is an area of active research. Galaxy clusters, which are assemblies of hundreds or even thousands of galaxies held together by the gravity of dark matter, formed much later in cosmic history than the galaxies, when the universe had approximately reached half of its present age.

We have no consistent theory of how stars form. Necessary conditions are that sufficient amounts of gas are concentrated sufficiently so that they can cool, contract and reach central densities and temperatures necessary for nuclear fusion to set in. For the first generation of stars, the cooling seems to be the most eminent problem. Efficient cooling agents are traces of heavy elements, broadly called "metals" in astrophysics, which can radiate energy away by line emission. Primordial gas, as it was available shortly after the Big Bang, did not contain any elements heavier than Lithium in appreciable quantities. All heavier elements had to be produced by nuclear fusion in stellar interiors, and then released by stellar winds and supernova explosions. That way, the gas could be enriched with metals, which allowed for progressively more efficient cooling, and the formation of new generations of stars.

It marked one of the most fundamental achievements of astrophysics when it was realized that stars produce their energy by nuclear fusion. A firm proof of nuclear energy production in the sun was given by the detection of solar neutrinos, which are emitted in the β decay accompanying many nuclear reactions. The energy produced in stellar cores has to travel a long way until it is finally released, mostly as visible light, at the stellar surfaces. The energy transport depends on the macroscopic constitution of the stars, as described by their density, temperature, and pressure profiles, and on microscopic quantities such as the opacity of the stellar material. While the stellar structure is determined by hydrodynamics and energy-transport mechanisms, opacities have to be calculated from the quantum mechanics of atoms. The radiation that we receive from stellar "surfaces", or photospheres, shows spectral lines caused by atomic absorption. Those "Fraunhofer lines" allow the chemical composition of outer stellar layers and many physical properties of the stars to be deduced. Their detection marks the onset of modern astrophysics.

Observations show that stars are not randomly distributed in the plane spanned by temperature and luminosity, but fall within well-defined, sharp regions on that plane. It was one of the breakthroughs of astrophysics when the theory of stellar structure, based on the assumption of nuclear fusion in stellar cores, was able to explain those patterns. The most prominent of those are the "main sequence", on which stars fall while they produce helium from hydrogen in their cores, and the "giant branch", to which stars move when their central hydrogen supply is exhausted and they proceed to burning helium.

As isolated, gravitationally-bound systems, stars can oscillate in a multitude of modes. Such oscillations can be inferred from velocity patterns on the solar surface, and their wave lengths and frequencies can be compared to those predicted from stellar models. The overall excellent agreement between theory and observations of solar oscillations provides further support for the theory of stellar structure and energy production.

It was thus perceived as a fundamental physical problem when neutrino detectors kept finding substantially fewer solar neutrinos than predicted by the otherwise well-established solar model. This problem was solved when it was proven that neutrinos can change flavour in a process called neutrino oscillations, which requires the neutrinos to have a small, but non-vanishing mass. This is a prototypical example for astrophysics driving theoretical physics beyond the standard model of elementary-particle theory.

Stars end in different ways depending on their mass. When their nuclear fuel is exhausted, low-mass stars cool and contract to form white dwarfs, which keep radiating until their internal energy is lost. White dwarfs are stabilized by the electron degeneracy pressure in their interiors, which follows from Pauli's exclusion principle. Masses higher than 1.4 solar masses (the Chandrasekhar limit) cannot be stabilized against gravity that way. They collapse further until electrons and protons are converted to neutrons by inverse β decay. Thus neutron stars are formed, which are stabilized by the degeneracy pressure of the neutrons. Yet more massive objects can collapse to form black holes, i. e., singularities in spacetime from which even light cannot escape. Stellar collapse is accompanied by explosions giving rise to supernovae. Despite intense research, the physics of core-collapse supernovae is not yet fully understood. Neutrino transport seems to be crucially important. It is likely that the mysterious γ-ray bursts are also related to the collapse of (very) massive stars.

Another type of supernova (type Ia) occurs in binary systems in which a white dwarf accretes mass from an overflowing companion star. When the Chandrasekhar mass limit is reached, the white dwarf collapses and explodes. Since the exploding mass is approximately fixed, so is the luminosity. Thus, supernovae of type Ia can be used as "standard candles" in cosmology. From their apparent brightness, compared to their known luminosity, their distance can be inferred. Since looking at large distances means looking back in time, observations of supernovae of type Ia allow the expansion history of the universe to be reconstructed, yielding the surprising result that the universe turned over from decelerated to accelerated expansion when it was about half of its present age. This is interpreted as evidence for a "dark energy" which can drive accelerated cosmic expansion due to its negative pressure. Besides dark matter, dark energy is one of the most fundamental enigmas in current astrophysical research, which may find its solution in the physics of quantum fields.

Black holes are not only possible end products of stellar evolution. Detailed measurements of stellar dynamics in the vicinity of the center of our Galaxy has revealed the presence of a black hole with about a million solar masses. Similar, but necessarily much less well-resolved observations indicate that black holes exist in the cores of most, if not all, galaxies. A surprising recent discovery showed that the masses of galactic black holes are correlated with the masses of their host galaxies, which indicates that their formation processes may be closely linked. Black holes in galactic cores are believed to be the central machines of the so-called quasars, objects which appear point-like as stars, but have luminosities well exceeding those of ordinary galaxies. It was realized soon after quasars were first discovered that nuclear fusion is insufficient to release such amounts of energy. The only viable energy-production mechanism is the conversion of gravitational potential energy into radiation in an accretion process. When gas streams into a black hole, its angular momentum forces it to orbit around the black hole in a disk. The friction in the disk allows the gas to lose its angular momentum and to gradually flow inward. At the same time, it heats up the gas and causes it to radiate in a broad wavelength range. Quasars were abundant in the young universe and apparently ceased being effective when the gas supply was exhausted.

A widespread phenomenon in astrophysical objects are magnetic fields. They may be produced in stars and accretion disks by battery or dynamo mechanisms, but possibly also during phase transitions in the early universe. Stellar magnetic fields can be blown into the ambient medium by winds or supernova explosions. Contraction processes like those leading to the formation of white dwarfs, neutron stars, or accretion disks can then produce ordered magnetic fields of considerable strength. Even galaxy clusters are found to be permeated by large-scale magnetic fields, whose origin is yet unclear. Magnetic fields can collimate jets streaming away along the symmetry axes of accretion disks. Relativistic particles, possibly accelerated in shock fronts, emit synchrotron radiation in the radio waveband when gyrating along magnetic-field lines.

On the smallest scales, accretion disks are also crucially important for the formation of planets around stars. The material in dusty disks can efficiently cool, fragment, and coagulate to form planetesimals and progressively larger objects. A large number of extra-solar planets was recently discovered, almost exclusively through their gravitational pull on their host stars. The distribution of these planets in mass and orbital parameters is much different from that seen in our solar system, illustrating that the theory of planet formation is in an early state.

Astrophysical observations started in the visible light, then expanded to the radio, infrared, X-ray, γ-ray and ultraviolet wave bands. The last remaining wide gap in the observed electromagnetic spectrum will be closed by submillimeter observations which have already begun and will be intensified in the near future. Gravitational-wave detectors are being built which promise to provide insight into a broad variety of astrophysical processes, from supernova explosions to the physics of the early universe.

See also: Black Holes; Cosmology; Interstellar Medium; Neutron Stars; Nucleosynthesis; Pulsars; Quasars; Stellar Energy Sources And Evolution; Sun; Universe.

Bibliography
The best references on astrophysics are current journals, review volumes, and reports of symposia.

Journals
Astrophysical Journal, published by the University of Chicago Press, Chicago, Ill.
Astrophysics and Space Science, published by Reidel Publishers, Dordrecht, The Netherlands.
Astronomy and Astrophysics, published by Springer-Verlag, Berlin.
Monthly Notices of the Royal Astronomical Society, published by Blackwell Scientific Publications, Oxford, England.

Review Volumes
Annual Review of Astronomy and Astrophysics. Annual Reviews Inc., Palo Alto, Calif.

Reports of Symposia
International Astronomical Union Symposia Proceedings, published by Reidel Publishers, Dordrecht, The Netherlands. Examples pertaining to the *Sun*: Vols. 35, 43, 56, 68, 71, 86; pertaining to *galactic structure, galaxies, quasars*, etc.: Vols. 38, 44, 58, 60, 63, 64, 69, 74, 79, 84, 92, 97, 104, 106, 108, 116, 117, 119, 124, 126, 127, 130; pertaining to *planets and meteors*: Vols. 33, 40, 47, 48, 62, 65, 89, 90; pertaining to *gaseous nebulae and the interstellar medium*: Vols. 34, 39, 46, 52, 76, 87, 103, 120, 131; pertaining to *stars, stellar spectra, and stellar evolution*: Vols. 42, 50, 52, 54, 55, 59, 66, 67, 70, 72, 75, 83, 95, 98, 99, 101, 105, 113, 115, 122, 123, 125.

Atmospheric Physics
Eric P. Shettle

The atmosphere is the gaseous shell that surrounds the earth and atmospheric physics is the discipline describing the physical processes occurring in the atmosphere. The atmosphere is hundreds of kilometers thick, providing the air we breathe. It presence is most notable through the phenomena that we know as weather.

Atmospheric Composition

The composition of the earth's atmosphere is dominated by two gases, nitrogen and oxygen which together make up about 99% of clean dry air, with most of the remaining 1% argon. These are nearly homogeneous up to altitudes of 80 to 90 km. Water vapor is highly variable; values near the surface can range from 3% in the tropics to a few tenths of a percent at mid-latitudes during the winter. The concentration of water vapor decreases rapidly above the surface and only makes up a few parts per million of the upper atmosphere. The concentrations of these and the other principal atmospheric gases are summarized in Table 1.

Table 1: Composition of the atmosphere.

Substance	Vol. % in dry air	Mol. weight
Total atmosphere		
Dry air	100.000	28.97
Nitrogen	78.083	28.02
Oxygen	20.946	32.00
Argon	0.934	39.88
Carbon dioxide	0.037	44.00
Neon	0.0018	20.0
Helium	0.00052	4.00
Ozone	Variable	48.00
Water vapor	0 to 3.0	18.02

Vertical Structure of the Atmosphere

As noted in the previous section the primary atmospheric species are uniformly mixed up to about 80 to 90 kilometers. At higher altitudes molecular diffusion begins to separate the different gases by their molecular weight. So the concentration of oxygen decreases relative to nitrogen. Also at these altitudes the solar ultraviolet radiation dissociates the oxygen molecules providing a source of atomic oxygen, and the photodissociation of water vapor at lower altitudes produces atomic hydrogen. The atmosphere is nearly in hydrostatic equilibrium which means that the pressure at any altitude is equal to the weight of the atmosphere above that altitude. The result of this and the ideal gas law is that both the atmospheric pressure and the density decrease exponentially with height. Half of the total mass of the atmosphere is below about 5.5 km and nearly 90% is below 16 km. By 100 km the density of the atmosphere has dropped by a factor of three million. Satellite orbits are generally above at least 300 km to minimize atmospheric drag. Satellite orbits at 700 to 800 km are still modified by drag effects over several years.

The atmospheric temperature tends to decrease up to an altitude of about 8 km in the polar regions to about 16 km in the tropics. This region of generally decreasing temperatures is known as the *troposphere* and the local temperature minimum the *tropopause*. The tropopause temperatures range from 180 to 220 K. The *lapse rate*, which means the decrease of temperature with altitude, is about 6.5 K/km on a global average, but can vary considerably and there can be regions where there is a temperature inversion where the temperature increases with altitude over a shallow layer. The troposphere is the part of the atmosphere where most of our weather occurs. Large-scale turbulence and mixing play a significant role in the distribution of the atmospheric properties. Above the tropopause there is a marked change in the lapse rate in the region known as the *stratosphere*. In the first few kilometers of the stratosphere the temperature is nearly constant and then the temperature increases with altitude up to a height of about 50 km where the *stratopause* is reached. This increase in temperature is caused by the absorption of the incident solar radiation by ozone which reaches in maximum concentrations in the stratosphere. Above stratopause the temperature again decreases with height, in the region known as the *mesosphere*. This region ends with the *mesopause*, where the lowest temperature in the atmosphere is about 180 K and can be as cold as 120 to 130 K during the polar summers. Above the mesopause the atmospheric temperatures once again increase due the absorption the solar ultraviolet radiation for wavelengths less than 185 nm. This region is known as the *thermosphere*.

The incident ultraviolet radiation and energetic particles from the sun produces dissociation and ionization of the atmospheric constituents, in an atmospheric region known as the *ionosphere*, which overlaps with the mesosphere and thermosphere. This ionization produces a free electron and a positively charged ion. The ionosphere is divided into three regions or layers based on the number of free electrons and the dominant ion. The D layer extends from about 50 to 90 km, the E layer from roughly 90 to 150 km, and the F layer above about 150 km.

Atmospheric Radiation

The atmospheric structure discussed in the previous section is largely controlled by the absorption of the incident solar radiation and the long wave radiation emitted by the earth. This absorbed radiation provides the energy required to drive the atmospheric motions. The intensity of the solar radiation incident on a perpendicular plane at the top of the atmosphere is about 1370 watt/m^2. The daily solar flux per unit horizontal incident on the top of the atmosphere depends on the angle of incidence and the length of daylight. It ranges a maximum at poles at summer equinox, decreasing slowly towards the equator and then more rapidly in the winter hemisphere going to zero at latitudes where there are 24 hours of darkness. About half (51%) of the sunlight is absorbed at the earth's surface. Another 30% is reflected back into space by the surface, clouds, or atmospheric scattering. The remaining 19% is absorbed by the atmosphere. Essentially all of the ultraviolet radiation at wavelengths less than 300 nm is absorbed before it reaches the surface. By contrast much of the visible and some of the infrared solar radiation reaches the earth's surface. Significant portions of the infrared solar radiation are absorbed by water vapor and carbon dioxide, as well other gases.

The earth emits radiation with a characteristic blackbody temperature of about 270 K which peaks around 10 μm. However much of these is absorbed in the atmosphere predominately by water vapor and carbon dioxide plus ozone and other species. The atmosphere re-emits the radiation, and some of the upwelling radiation is absorbed in turn at higher levels in the atmosphere where it is again re-emitted. Much of the surface radiation in the *atmospheric window* between about 8 to 12 μm reaches space without being absorbed. The characteristic blackbody temperature of the long wavelength radiation emitted into space is a weighted average temperature of where in the earth/atmosphere system the radiation was emitted from. This is called the *terrestrial* (or *earth*) *radiation*.

On an annual basis the global average the solar radiation absorbed by the earth/atmosphere system equals the terrestrial radiation emitted into space. However on a regional basis this is not true. On an annual basis the equatorial latitudes out to about 35° absorb more solar radiation than terrestrial radiation is emitted back into space, and at higher latitudes the reverse ins true. Over most of the winter hemisphere, the terrestrial radiation lost to space exceeds the absorbed solar radiation with the reverse holding in much of the summer hemisphere. These energy imbalances drive the large scale atmospheric circulation.

The Greenhouse Effect

The atmospheric radiation emitted downward heats the lower atmosphere and the earth's surface causing them to be warmer than they would be in the absence of an atmosphere. This is called the *greenhouse effect* by analogy with how greenhouses were thought to be heated. The glass transmits most of the incident solar radiation but absorbs most of the outgoing infrared radiation from inside the greenhouse and radiating half of it back down. However this is a misnomer, since R. W. Wood demonstrated that primary mechanism heating greenhouses is that the glass blocks the removal of heat by convection. He compared the temperature inside two small greenhouses one of which used rock salt, which is transparent in the infrared instead of glass.

There is a concern that the increase in carbon dioxide (and other gases such as methane) since the industrial revolution could increase the atmospheric absorption of the outgoing terrestrial radiation. Carbon dioxide has increased from less than 290 parts per million by volume (ppmv) of the atmosphere before 1900, to 375 ppmv in 2003. The combustion of fossil fuels such as coal or oil produces carbon dioxide. The global average temperature has increased over the last century by about half a degree Centigrade, with geographic and seasonal variations. This is consistent with the predictions of global climates on the impact of such an increase of the greenhouse gases.

Remote Sensing

The radiation emitted, scattered or transmitted by the atmosphere is used increasingly to remotely measure the properties of the atmosphere from satellites, ground based, or airborne instruments. Measurements of the radiation emitted by carbon dioxide at different wavelengths where the strength of the absorption varies can be used to measure the atmospheric temperature as function of altitude. Similar measurements as function of wavelength across the water vapor absorption bands can be used to determine the vertical distribution of water vapor. Satellite measurements of the amount of sunlight backscattered at different ultraviolet wavelengths can be used to determine the vertical distribution of ozone.

Moisture in the Atmosphere

Water plays a unique role in the atmosphere because at the range of atmospheric temperatures and pressures it can be present as a solid, as a liquid, or as gas. The phase changes between these different states require that heat be absorbed or released. The absorbed heat energy is known as latent heat and is released back into the atmosphere when the phase transition is reversed. This provides one of the mechanisms for the energy transport required to compensate for the regional imbalances between the absorbed solar radiation and emitted terrestrial radiation. Water which evaporates from the tropical oceans absorbs latent heat which is when the air containing that water vapor is cooled sufficiently for the water vapor to condense forming cloud droplets. If it cools further so that the cloud droplets freeze, forming ice particles additional latent heat is released.

See also: Aurora; Corona Discharge; Ionosphere; Lightning; Magnetosphere; Meteorology; Rayleigh Scattering; Refraction.

Bibliography

R. G. Fleagle and J. A. Businger, *An Introduction to Atmospheric Physics*. Academic Press, San Diego, CA, 1980.

A. S. Jursa (ed.), *Handbook of Geophysics and the Space Environment*. National Technical Information Service, Springfield, Virginia, 1985.

F. K. Lutgens and E. J. Tarbuck, *The Atmosphere – An Introduction to Meteorology*. Prentice Hall, Englewood Cliffs, NJ, 1995.

M. L. Salby, *Fundamentals of Atmospheric Physics*. Academic Press, San Diego, CA, 1996.

Atomic Spectroscopy

A. Lurio[†] and A. F. Starace

Introduction

The progressively more detailed understanding of the emission and absorption spectra of atoms has led from the Bohr theory of the hydrogen atom (1913), to the discovery of electron spin by Uhlenbeck and Goudsmit (1925), to the Pauli exclusion principle (1926), and ultimately to the nonrelativistic quantum-mechanical Schrödinger equation (1926).

From these developments we find the following principles apply to the interpretation of all atomic spectra.

1. An atomic system can exist only in discrete stationary states corresponding to a well-defined energy E (within the Heisenberg uncertainty limit $\Delta E \Delta t \sim \hbar$, where Δt is the lifetime of the state). Transitions between these states, including the emission and absorption of radiation, require the complete transfer of an amount of energy equal to the difference in energy between these states.

2. The frequency of the emitted or absorbed radiation in going from state 2 to state 1 is given by $\omega = (E_2 - E_1)/\hbar$ (ω negative is absorption).

3. Each stationary state has associated with it a definite quantized angular momentum J and a definite parity (defined later). The projection of the angular momentum on any chosen direction in space is quantized with allowed values $m_J = J, J-1, \ldots, -J$.

Each atom has its own unique spectrum. The interpretation of this spectrum has led to the classification of many of the stationary-state energy levels of neutral and several-times-ionized atoms. These results are tabulated in a classic three-volume NBS publication by Charlotte Moore. We shall attempt here to give a simplified treatment of the physical basis of this classification.

[†] deceased

Table 1: Designation of electron states.

	$l = 0$	$l = 1$	$l = 2$	$l = 3$	$l = 4$
	s	p	d	f	g
$n = 1$	1s				
$n = 2$	2s	2p			
$n = 3$	3s	3p	3d		
$n = 4$	4s	4p	4d	4f	
$n = 5$	5s	5p	5d	5f	5g

Electronic Configurations

The specification of the stationary states of an N-electron atom is given by the solution of the time-independent Schrödinger equation

$$H\Psi = E\Psi ,\tag{1}$$

where the dominant terms contributing to the Hamiltonian H are

$$H = \sum_{i=1}^{N} \frac{p_i^2}{2m} - \sum_{i=1}^{N} \frac{Ze^2}{r_i} + \sum_{i>j=1}^{N} \frac{e^2}{r_{ij}} .\tag{2}$$

The first term is the kinetic energy of the electrons, and the other terms are the potential energies of Coulomb interaction of the electrons with the nucleus and with each other. A good starting approximation to H is the "central-field approximation" in which one replaces the second and third terms of Eq. (2) by $\sum_{i=1}^{N} V(r_i)$, where $V(r_i)$ is the spherically symmetric average potential seen by the ith electron due to all other electrons. Equation (1) is now solvable with a wave function which is the product of N single-electron wave functions. The Schrödinger equation for each of these single-electron wave functions is like that for hydrogen (see Burke) except that the hydrogenic potential-energy term e^2/r is replaced by $V(r)$. Consequently, the same set of quantum numbers n, l, s, m_l, m_s, used to describe the hydrogenic electron apply here. These are respectively, the principal quantum number n, the orbital and spin angular momentum quantum numbers, and their projections on the quantization axis.

Complete specification of an N-electron state requires N sets of these quantum numbers with the Pauli restriction that no two sets can be identical. For a given n and l if all possible m_l and m_s states are occupied, we have a closed subshell; if all $2n^2$ states ($l = 0, 1, \ldots, n-1$) are filled, we have a closed shell. Closed shells and subshells have exactly spherically symmetric charge distributions and zero net angular momentum.

The standard notation for designating the individual electron configurations or orbitals (n, l values) which are combined to form an N-electron product state is shown in Table 1. To illustrate, the ground-state configuration of sodium ($Z = 11$) is written $(1s)^2(2s)^2(2p)^63s$, where the superscripts indicate the number of electrons of a given n,l type.

Spectroscopic Notation

To completely specify an atomic state, besides listing as above the individual electron orbitals, we must also specify the coupling of the angular momenta for all unfilled subshells. The closed shells couple to give zero angular momentum.

The Hamiltonian of Eq. (2) commutes with $\mathbf{L} = \sum_{i=1}^{N} \mathbf{l}_i$, the total orbital angular momentum, and with $\mathbf{S} = \sum_{i=1}^{N} \mathbf{s}_i$, the total spin angular momentum, and thus with $\mathbf{J} = \mathbf{L} + \mathbf{S}$, the total electronic angular momentum of the atom. Within a given configuration, therefore, one may take linear combinations of products of the central-field-approximation orbitals to form states having exact values of the quantum numbers L^2, S^2, J^2, and M_J. All these states are degenerate in energy in the central-field approximation. When we use perturbation theory to take into account the difference between the central-field-approximation potential energy and the potential-energy terms of Eq. (2) we find that states with different L and S have significantly different energies. This treatment, which works well for many atoms (especially the light ones), is called the LS or Russell–Saunders coupling scheme. In the LS coupling scheme we couple vectorially all open-shell electron spins to obtain a number of different S values and couple all the open shell L_i to obtain a number of different L values. In early atomic spectroscopy the *vector model* was used to visualize these different coupling schemes. White gives an extensive discussion of this model. The number of permitted L and S values is discussed in detail by Condon and Shortley. $\mathbf{L}$ and $\mathbf{S}$ are now coupled vectorially to form the total angular momentum $\mathbf{J} = \mathbf{L} + \mathbf{S}$. J takes values from $|L+S|$ to $|L-S|$.

At this point we add the spin–orbit interaction $H_{so} = \xi \mathbf{L} \cdot \mathbf{S}$ (*see* Fine and Hyperfine Spectra and Interactions) which removes the degeneracy between the different J values of the same L and S. The spectroscopic notation to designate these LS coupled states is $^{2S+1}L_{2J+1}$, where, similar to Table 1, for $L = 0, 1, 2, 3$ we use the capital letters S, P, D, F, etc. To illustrate, if $S = 1$, $L = 2$, and $J = 1$ we would write $^3D_1$. The levels of a given J have $2J + 1$ magnetic sublevels which in the presence of an external magnetic field split apart (*see* Zeeman and Stark Effects).

Transition Rates

Any isolated atom in an excited state will decay spontaneously to lower energy states emitting radiation and ultimately ending in the ground state. Also, in the presence of an external radiation field of the proper frequency, an atom can absorb radiation and make a transition to an excited state.

We will discuss only allowed electric dipole transitions. Classically the time-averaged power radiated per unit solid angle in direction $\mathbf{n}$ by a harmonically varying charge distribution $\rho(r)$ with an electric dipole moment $\mathbf{P} = \int_0^{\infty} \mathbf{r}\rho(\mathbf{r})\, dv$ is (in Gaussian units)

$$\frac{dw}{dt} = \frac{ck^4}{8\pi} |\mathbf{n} \times (\mathbf{n} \times \mathbf{P})|^2 \,,$$

which for a linearly oscillating dipole moment parallel to the z axis reduces to $dw/dt = (ck^4/8\pi)|\mathbf{P}|^2 \sin\theta$, where $k = \omega/c$ and θ is the angle between the z axis and the direction of radiation n. For a charge distribution rotating about the z axis, $dw/dt = (ck^4/8\pi)|\mathbf{P}|^2(1+\cos\theta)$. We make the connection with the quantum-mechanical description of the atomic system by $\mathbf{P} = 2\mathbf{P}_{ij}$ where $\mathbf{P}_{ij} = \int \bar{\Psi}_i \mathbf{P} \Psi_j\, dv$, and by setting the harmonic frequency ω equal to $\omega = (E_i - E_j)/\hbar$, where i and j refer to the initial and final states of the transition.

Electric-Dipole Selection Rules

Electric dipole transitions are possible only between certain atomic energy levels. Rules that tell us which transitions are allowed are called selection rules. We will consider only electric dipole transitions, i. e., "allowed" transitions. If electric dipole transitions are forbidden, transitions can still occur by other radiation processes such as higher-order multipole radiation (see Garstang) but these transitions are much weaker.

The parity operation $P\mathbf{r_i} = -\mathbf{r_i}$ commutes with H (i.e., with the complete atomic Hamiltonian, not just our approximation) and for a product wave function yields $P\Psi(\mathbf{r_i}) = (-1)^{\Sigma l_i}\Psi(\mathbf{r_i})$. If Σ_{l_i} is (even/odd) we say the state is an (even/odd) parity state. In spectroscopic notation an upper righthand superscript "o" is sometimes used to indicate explicitly states having odd parity. Electronic dipole transitions *only* take place between states of *different* parity so that for the common case of a one-electron jump, $\Delta l = \pm 1$. No condition on Δn is required.

By arguments similar to those given to explain hydrogenic selection rules, we also find

$$\Delta S = 0 \text{ (no spin change)}, \quad \Delta J = 0, \pm 1 \ (0 \rightarrow 0 \text{ forbidden}),$$
$$\Delta L = 0, \pm 1, \quad\quad\quad \Delta M_J = 0, \pm 1.$$

Radiation from a $\Delta M_J = 0$ transition is linearly polarized ($\sin^2\theta$ dependence); radiation from a $\Delta M_j = \pm 1$ transition is circularly polarized [$(1 + \cos^2\theta)$ dependence] when viewed along the axis of space quantization.

The strength of a transition depends on the magnitude of the radial part of the $\mathbf{P}_{ij}$ integral, which is quite sensitive to the approximations used in finding ψ and is difficult to calculate accurately.

Examples of Simple Spectra

Alkali-like Spectra

Alkalis and ions with a single electron outside of closed subshells have energy levels and spectra very similar to hydrogen because the core electrons, not taking any role in optical transitions, act principally to screen the nuclear charge. The larger the n of the valence electron, the larger its orbit and so the more complete is the screening. The energy level diagram of sodium is shown in Fig. 1. The configurations responsible for these levels are:

$$(1s)^2(2s)^2(2p)^6 ns\ ^2S_{1/2} \quad\quad n = 3, 4, \ldots,$$
$$(1s)^2(2s)^2(2p)^6 np\ ^2P_{1/2, 3/2} \quad n = 3, 4, \ldots,$$
$$(1s)^2(2s)^2(2p)^6 nd\ ^2D_{3/2, 5/2} \quad n = 3, 4, \ldots,$$

Early spectroscopists called the $nd\ ^2D \rightarrow 3p\ ^2P$ transitions the diffuse series because in low resolution the three allowed lines, $^2D_{5/2} \rightarrow\ ^2P_{3/2}$ and $^2D_{3/2} \rightarrow\ ^2P_{3/2, 1/2}$ appeared as a blend.

Two-Electron Spectra

The alkaline-earth elements Be, Mg, Ca, Sr, and Ba are representative of atoms with two electrons outside of closed shells. We shall consider Be, whose term diagram is shown in Fig. 2. The low-lying excited states arise from excitation of one of the ground state $(2s)^2$ electrons. The low-lying excited configurations are

$$(1s)^2 2s\ ns\ ^1S_0\ ^3S_1, \quad\quad n = 3, 4, \ldots,$$
$$(1s)^2 2s\ np\ ^1P_0\ ^3P_{2, 1, 0}, \quad n = 2, 3, 4,$$
$$(1s)^2 2s\ nd\ ^1D_0\ ^3D_{3, 2, 1}, \quad n = 3, 4, 5, \ldots$$

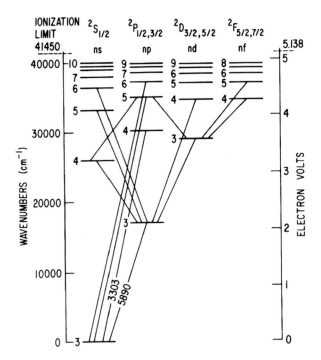

Fig. 1: Term diagram of sodium.

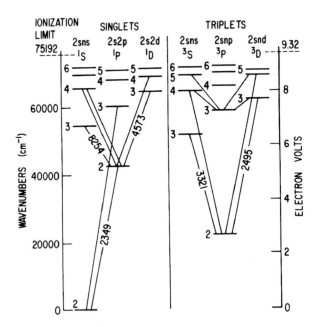

Fig. 2: Term diagram of beryllium.

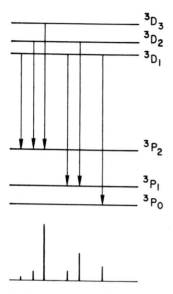

Fig. 3: Allowed ^{3}D–^{3}P transitions and relative line strengths.

In each configuration the $s = \frac{1}{2}$ spins of the two unpaired electrons are coupled to form the resultant total spin $S = 0, 1$. The L value is that of the excited electron. To illustrate: for the D states, S and L are combined to form the resultant J as follows: (for the singlet) $L = 2, S = 0$, $J = 2$; (for the triplets) $L = 2, S = 1, J = L+S, L+S-1, L-S = 3, 2, 1$. The $\Delta S = 0$ selection rule prohibits singlet to triplet transitions. The allowed ^{3}D to ^{3}P triplet transitions are shown in Fig. 3. The same methods can be applied to more complicated spectra (see White and Kuhn).

Quantum Defect Theory

A key feature of the attractive Coulomb field in which atomic electrons move is that it supports an infinite number of bound states which converge in energy to a particular ionization threshold. These states may be grouped in series. In simple cases each series may be identified by the term level of the ion to whose threshold the series converges, the orbital angular momentum of the excited electron, and the coupling of the electron to the atomic core. In general, the various series of states may interact with each other, thereby complicating the analysis in the very region where the number of levels is becoming infinite.

The quantum defect theory (QDT) is a method of using the analytically known properties of excited electrons moving in a pure Coulomb field to describe such atomic spectra in terms of a few parameters. These parameters may be determined either from experimental data or from *ab initio* theoretical calculations. In addition, they are usually nearly independent of energy in the threshold energy region (i. e., within a few electron volts of the atomic ionization threshold). Thus the determination of these parameters at any *single* energy suffices to predict the *variation with energy* of numerous atomic properties in the threshold energy region such as total and partial photoionization cross sections, photoelectron angular distributions, discrete line strengths, autoionization profiles, etc. These properties are often very strongly energy dependent and difficult to measure or to calculate by other methods. Yet all these phenomena, according to the QDT, depend on only a few essential parameters which represent the proper interface between theory and experiment.

The QDT assumes that the configuration space for an excited atomic electron can be divided into two regions: an *inner region*, $0 \leq r \leq r_0$, where electron correlations are strong and difficult to treat, and an *outer region*, $r_0 \leq r \leq \infty$, where the electron–ion interaction potential is assumed to be purely Coulombic and where the form of the electron wave function is known analytically. The boundary radius r_0 between the two regions is typically of the order of the atomic radius.

Consider the simple problem of an excited electron of angular momentum l in an alkali atom: the electron sees a Coulomb field for $r \geq r_0$, where r_0 is roughly the ionic radius. We measure the energy ε of the excited electron relative to the ionization threshold as $\varepsilon = -0.5\nu^{-2}$, where the parameter ν is our measure of energy. The Schrödinger equation for $r \geq r_0$ has two solutions, one regular and one irregular for small values of r:

$$f(\nu, r) \sim r^{l+1} \quad \text{as} \quad r \to 0 \,, \tag{3a}$$

$$g(\nu, r) \sim r^{-l} \quad \text{as} \quad r \to 0 \,. \tag{3b}$$

A general solution of the Schrödinger equation for $r \geq R_0$ is a linear combination of $f(\nu, r)$ and $g(\nu, r)$ with coefficients to be determined by application of boundary conditions at infinity and at r_0. This general solution may be written as

$$\psi(\nu, r) = N_\nu \{f(\nu, r) \cos \pi\mu - g(\nu, r) \sin \pi\mu\} \quad \text{for } r \geq r_0 \,, \tag{4}$$

where N_ν is a normalization factor that is determined by the behavior of $\psi(\nu, r)$ at large r. μ on the other hand, is the relative phase with which the regular and irregular solutions are superimposed. Its value is determined by the behavior of $\psi(\nu, r)$ in the core region, $0 \leq r \leq r_0$, where the effective potential is non-Coulombic: i. e., μ has that value which allows the analytically determined $\psi(\nu, r)$ given by Eq. (4) for $r \geq r_0$ to be joined smoothly at $r = r_0$ onto the numerically determined portion of $\psi(\nu, r)$ that obtains in the inner core region, $0 \leq r \leq r_0$.

Alternatively, μ may be determined semiempirically from atomic spectral data on energy levels, as we show here. Consider the asymptotic behavior of $\psi(\nu, r)$ in the case of excited electron energies below threshold, i. e., $\psi(\nu, r)$ must tend toward zero. The asymptotic forms of the regular and irregular Coulomb functions are

$$f(\nu, r) \to u(\nu, r) \sin \pi\nu - v(\nu, r) \exp[i\pi\nu] \quad \text{as} \quad r \to \infty \,, \tag{5a}$$

$$g(\nu, r) \to -u(\nu, r) \cos \pi\nu + v(\nu, r) \exp\left[i\pi\left(\nu + \tfrac{1}{2}\right)\right] \quad \text{as} \quad r \to \infty \,, \tag{5b}$$

where $u(\nu, r)$ is an exponentially increasing function of r and $v(\nu, r)$ is an exponentially decreasing function of r. Substituting Eq. (5) in Eq. (4) gives

$$\psi(\nu, r) \to N_\nu \{u(\nu, r) \sin \pi(\nu + \mu) - v(\nu, r) \exp[i\pi(\nu + \mu)]\} \quad \text{as} \quad r \to \infty \,. \tag{6}$$

In order that $\psi(\nu, r)$ tend toward zero at large values of r, the coefficient of $u(\nu, r)$ must be zero; i. e., $\sin \pi(\nu + \mu) = 0$ or $\nu + \mu = n$, where n is an integer. Substituting $\nu = n - \mu$, in the expression for the electron's energy gives (in atomic units)

$$\varepsilon_n = -\frac{1}{2\nu^2} = \frac{1}{2(n - \mu)^2} \,. \tag{7}$$

μ is thus the quantum defect of spectroscopy and may be determined directly from Rydberg energy level data for the alkalis.

It is well known empirically in atomic spectroscopy that the quantum defect μ is a nearly constant function of energy near the ionization threshold. Theoretically, μ *should* be only a slowly varying function of energy since it is determined from the wave function in the inner core region, where the electron's large instantaneous kinetic energy makes it insensitive to the relatively small energy differences between the energy levels near threshold. Knowledge of the parameter μ therefore enables one to predict the energies of a whole series of atomic energy levels according to Eq. (7), thereby illustrating the ability of QDT to describe large amounts of atomic spectral information in a very compact way.

QDT may also be used to understand the variation with energy of the intensities of atomic spectral levels. From Eq. (4) we can see directly that for small $r \geq r_0$, the wavefunction $\psi(v, r)$ depends on energy mainly through the normalization factor N_v since μ is weakly energy dependent and so are $f(v, r)$ and $g(v, r)$ at small r [cf. Eq. (3)]. The normalization factor N_v is determined by the asymptotic behavior of $\psi(v, r)$ and may be very energy dependent. The point of this discussion thus is that at small radii $\psi(v, r)/N_v$ is likely to be quite insensitive to energy in the threshold energy region. Yet it is in this small r region that spectral transitions between the ground and the excited states occur. We conclude that the intensities of these transitions along a series of lines converging to the ionization threshold should depend on the energies of the excited levels in proportion to N_v^2. QDT shows that N_v^2 is proportional to $v^{-3} = (n - \mu)^3$ for discrete (i. e., negative) electron energies. This implies that multiplication of the measured intensity of the nth level by $(n - \mu)^3$ will produce a renormalized intensity that is only a slowly varying function of n. Hence, accurate measurements of only a few level intensities allows one to determine this slowly varying function and therefore to predict the intensities of all other levels in the series.

The QDT may also be used to describe atomic spectra more complicated than those of the alkali metals as well as to relate an atom's discrete spectrum to collision processes occurring at energies above the atom's ionization threshold. These topics, however, are beyond the scope of this article. The interested reader is referred to review articles on QDT by Fano and by Seaton.

See also: Bohr Theory Of Atomic Structure; Hamiltonian Function; Rotation And Angular Momentum; Schrödinger Equation; Zeeman And Stark Effects.

Bibliography

E. U. Condon and G. H. Shortley, *Theory of Atomic Spectra*. Cambridge University Press, Cambridge, 1953. (A)

U. Fano, "Unified Treatment of Perturbed Series, Continuous Spectra, and Collisions," *J. Opt. Soc. Am.* **65**, 979 (1975).

R. H. Garstang, "Forbidden Transitions," in *Atomic and Molecular Processes*. Academic Press, New York, 1962.

G. Herzberg, *Atomic Spectra and Atomic Structure*, 2nd ed. Dover, New York, 1944. (E)

H. G. Kuhn, *Atomic Spectra*. Academic Press, New York and London, 1962. (E)

C. E. Moore, *Atomic Energy Levels*, Circular 467, Vols. I, II, and III. National Bureau of Standards, Washington, DC, 1949, 1952, 1958.

M. J. Seaton, "Quantum Defect Theory," *Rept. Prog. Phys.* **46**, 167 (1983).

B. W. Shore and D. H. Menzel, *Principles of Atomic Spectra.* Wiley, New York, 1968. (A)

J. C. Slater, *Quantum Theory of Atomic Structure*, Vols. 1 and 2. McGraw-Hill, New York and London, 1960. (Vol. 1, I; Vol. 2, A)

I. I. Sobelman, *Introduction to the Theory of Atomic Spectra.* Pergamon Press, Oxford, 1972. (I)

H. E. White, *Introduction to Atomic Spectra.* McGraw-Hill, New York and London, 1934. (E)

Atomic Structure Calculations, Electronic Correlation

R. K. Nesbet

The electronic wave function for noninteracting electrons would be a single Slater determinant, an antisymmetrized product of one-electron "orbital" wave functions. The Hartree–Fock model of interacting electrons [1] optimizes this model function by solving variational equations for the occupied orbital functions. A simple linear combination of Slater determinants is used when the state in question is not invariant under rotation. The Hartree–Fock equations are analogous to noninteracting Schrödinger equations, but contain a nonlocal "self-consistent-field" potential, due to Coulomb and exchange interactions between electrons. This is an independent-particle model, in which electron quasiparticles interact only indirectly through a self-consistent mean field. The electrostatic energy depends on correlations between the locations of the electrons and cannot be computed exactly in the Hartree–Fock approximation [2].

The expression "electronic correlation" refers generally to corrections to the Hartree–Fock approximation. For nonspherical atomic states, the self-consistent potential function is usually spherically averaged, resulting in one-electron "correlation" associated with spin or rotational symmetry-breaking [3]. Otherwise the principal correlation effect in atoms is short-range Coulomb-cusp relaxation due to the electronic Coulomb repulsion. In molecules and solids this is supplemented by the long-range correlation effect of multipole polarization response.

One-electron mean-value properties computed in the closed-shell Hartree–Fock approximation are subject only to second-order correlation corrections [2, 3]. The practical effect of this is that the total electronic density distribution is well described for many purposes in the Hartree–Fock approximation, but reliable and consistent theoretical results for atomic properties sensitive to open-shell structure or to the response to external perturbations require a quantitative treatment of electronic correlation. Such properties include hyperfine structure, polarizabilities, oscillator strengths, and electron scattering cross sections [4, 5].

While some physical properties are described more or less accurately in the Hartree–Fock approximation, others are not described at all. For example, the van der Waals or dispersion potential energy of two spatially separated electronic systems is simply the long-range limit of the correlation energy of spatially separated electrons [6]. Orbital functional theory (OFT) [7] extends Hartree–Fock to a mean-field model that incorporates correlation. The polarization potential that dominates low-energy electron scattering is the asymptotic form of the nonlocal OFT correlation potential. In the usual Hartree–Fock approximation, the magnetic hyperfine structure constant is zero by symmetry in the ground states of nitrogen and phosphorus. Observed values differ from zero because of a combination of spin symmetry-breaking and Coulombic correlation effects [4].

Theoretical methods for the computation of correlation effects are often based on a preliminary Hartree–Fock calculation, or on a mean-field model such as density functional theory (DFT) [8], which includes an approximation to both exchange and correlation. OFT [7] optimizes the model or reference state for exact exchange and correlation within some many-body formalism. The optimized reference state [7] defines a "vacuum state" for formal perturbation theory [9]. A shell model is appropriate for atomic electrons because the nuclear attraction dominates the self-consistent radial potential. Occupied orbitals for a particular state in the Hartree–Fock approximation are labeled by quantum numbers n, l, m_l, m_s (or n, j, m) appropriate to the one-electron Schrödinger equation for a central potential. The conventional open-shell Hartree–Fock equations [1] are spherically averaged so that for each value of orbital angular momentum l there is a set of orthonormal radial functions $R_{nl}(r)$, not dependent on the axial quantum numbers m_l and m_s. A "configuration" is defined by a set of occupation numbers $d_{nl} \leq 2(l+1)$ which assign occupied orbital functions to subshells of given n, l but arbitrary m_l, m_s, subject to

$$m_l = -l, -l+1, \ldots, l; \qquad m_s = -\frac{1}{2}, \frac{1}{2}. \tag{1}$$

Conventional Hartree–Fock theory is formulated in terms of eigenfunctions of total orbital angular momentum and spin constructed from the Slater determinants of a specified configuration.

The set of radial functions R_{nl} for occupied orbitals in a reference configuration can be extended to a complete orthonormal set. This generates a complete basis for the atomic N-electron wave function as a hierarchy of virtual excitations, defined by substitution of "unoccupied" orbitals (from the extended set) for orbitals occupied in a particular configuration. An n-electron virtual excitation is defined by n such substitutions [7].

Since the electronic Hamiltonian contains only one- and two-electron operators, only one- and two-electron virtual excitations contribute to the first-order wave function of perturbation theory. The closed-shell Hartree–Fock approximation causes one-electron matrix elements with the reference state to vanish [2], but one-electron virtual excitations cannot be neglected for open-shell states. The total energy or correlation energy can be expressed exactly in terms of the coefficients of all one- and two-electron virtual excitations of the reference state [7]. Valid estimates of these coefficients can be made either by formal perturbation theory [9], or by approximate solution of the matrix eigenvalue problem defined in the basis of all virtual excitations of all orders $n \leq N$, for N electrons. The latter method, superposition of configurations or "configuration interaction," has been widely applied and highly developed in its computational and data-handling aspects [10]. The direct use of relative coordinates for pairs of electrons is not computationally feasible for more than two or three electrons, although it has given very accurate results for two-electron atoms and ions [11].

Because of the great complexity of such calculations for virtual excitations with $n > 2$, several methods have been introduced for approximate incorporation of terms of higher order [10]. These methods either treat electron pairs as uncoupled from each other [12] or modify the variational equations for such separated pairs to allow for the higher-order virtual excitations implied by a cluster expansion of the N-electron wave function [13]. This level of approximation appears to be the most natural extension of theory beyond Hartree–Fock to include electronic pair correlation. In the multiconfiguration Hartree–Fock method, configuration interaction is incorporated into an iterative variational calculation. A numerical version of this method has been used for accurate calculations of atomic oscillator strengths and photoionization cross sections [14].

These methods have been tested by numerous calculations of total atomic energies or of excitation energies [5], primarily for atoms in the first third of the periodic table. Their most important application, however, has been to the calculation of physical atomic properties that are difficult or impossible to measure experimentally. Theoretical calculations have helped to resolve discrepancies between conflicting experimental data on oscillator strengths [5], and have provided values of polarizabilities, particularly for atomic excited states, that have not been measured.

Hyperfine-structure calculations of high accuracy have helped to establish the fact that electronic correlation affects the three tensorially distinct magnetic hyperfine-structure interaction operators differently, so that three independent parameters must be used to fit experimental data. Theoretical calculations of the electric field gradient at a nucleus, which cannot be measured directly, have made it possible to obtain accurate nuclear quadrupole moments from measured quadrupole hyperfine coupling constants [4, 15].

Low-energy electron scattering by neutral atoms is dominated by the electric dipole polarization potential, essentially a correlation effect. Theoretical calculations by methods that can describe this effect quantitatively have been carried out for electron scattering by hydrogen, helium, alkali metals, and several other atoms [16]. These theoretical results have helped to elucidate observed structural features (resonance and threshold structures) and to establish absolute values of cross sections.

See also: Atomic Structure Calculations, One-Electron Models; Atomic Structure Calculations, Relativistic Atoms; Fine and Hyperfine Spectra and Interactions.

References

[1] D. R. Hartree, *The Calculation of Atomic Structures*. Wiley, New York, 1957. C. Froese Fischer, *The Hartree–Fock Method for Atoms*. Wiley-Interscience, New York, 1977.

[2] C. Møller and M. S. Plesset, *Phys. Rev.* **46**, 618 (1934); L. Brillouin, *Les Champs "Self-Consistent" de Hartree et de Fock*. Hermann et Cie, Paris, 1934.

[3] R. K. Nesbet, *Proc. Roy. Soc. (London)* **A230**, 312 (1955).

[4] N. C. Dutta, C. Matsubara, R. T. Pu, and T. P. Das, *Phys. Rev. Lett.* **21**, 1139 (1968); *Phys. Rev.* **177**, 33 (1969).

[5] A. Hibbert, *Rept. Prog. Phys.* **38**, 1217 (1975).

[6] F. London, *Z. Phys.* **63**, 245 (1930).

[7] R. K. Nesbet, *Adv. Chem. Phys.* **9**, 321 (1965); R. K. Nesbet, *Variational Principles and Methods in Theoretical Physics and Chemistry*. Cambridge Univ. Press, New York, 2003.

[8] W. Kohn and L. J. Sham, *Phys. Rev.* **140**, A1133 (1965); R. G. Parr and W. Yang, *Density-Functional Theory of Atoms and Molecules*. Oxford Univ. Press, New York, 1989.

[9] H. P. Kelly, *Adv. Chem. Phys.* **14**, 129 (1969); I. Lindgren and J. Morrison, *Atomic Many-Body Theory*. 2nd ed., Springer-Verlag, Berlin, 1986.

[10] I. Shavitt, in *Methods of Electronic Structure Theory*, H. F. Schaefer III, ed., p. 189. Plenum, New York, 1977; A. Szabo and N. S. Ostlund, *Modern Quantum Chemistry*. McGraw-Hill, New York, 1989.

[11] C. L. Pekeris, *Phys. Rev.* **115**, 1216 (1959); **126**, 1470 (1962); K. Frankowski and C. L. Pekeris, *Phys. Rev.* **146**, 46 (1966).

[12] O. Sinanoglu, *Adv. Chem. Phys.* **6**, 315 (1964); **14**, 237 (1969); R. K. Nesbet, *Adv. Chem. Phys.* **14**, 1 (1969).

[13] J. Cizek, *Adv. Chem. Phys.* **14**, 35 (1969); J. Cizek and J. Paldus, *Int. J. Quantum Chem.* **5**, 359 (1971); W. Meyer, *J. Chem. Phys.* **58**, 1017 (1973); W. Kutzelnigg, in *Methods of Electronic Structure Theory*, H. F. Schaefer III, ed., p. 127. Plenum, New York, 1977; W. Meyer, in *Methods of Electronic Structure Theory*, H. F. Schaefer III, ed., p. 413. Plenum, New York, 1977; R. J. Bartlett, *Ann. Rev. Phys. Chem.* **32**, 359 (1981).

[14] C. Froese Fischer, *Comput. Phys. Commun.* **14**, 145 (1978); H. P. Saha and C. Froese Fischer, *Phys. Rev.* **A35**, 5240 (1987); H. P. Saha, C. Froese Fischer, and P. W. Langhoff, *Phys. Rev.* **A38**, 1279 (1988).

[15] J. D. Lyons, R. T. Pu, and T. P. Das, *Phys. Rev.* **178**, 103 (1969); R. K. Nesbet, *Phys. Rev.* **A2**, 661 (1970); J. D. Lyons and T. P. Das, *Phys. Rev.* **A2**, 2250 (1970); R. K. Nesbet, *Phys. Rev. Lett.* **24**, 1155 (1970).

[16] B. L. Moiseiwitsch, *Rept. Prog. Phys.* **40**, 843 (1977); R. K. Nesbet, *Variational Methods in Electron-Atom Scattering Theory*. Plenum, New York, 1980.

Atomic Structure Calculations, One-Electron Models

F. Herman

The one-electron theory of atoms, molecules, and solids [1, 2] has enjoyed wide success in many branches of physics and chemistry. This theory postulates that the exact wave function for a many-electron system can be represented accurately by an approximate many-electron wave function constructed from one-electron wave functions or spin orbitals. Emphasis shifts from a consideration of complex many-electron wave functions depending on the coordinates of all the electrons, to one-electron wave functions. These are easier to treat because they depend only on the spatial and spin coordinates of single electrons. The ground state of the many-electron system can then be described in terms of the occupied spin orbitals, and excitations in terms of transitions between occupied and unoccupied spin orbitals.

The one-electron theory describes the electronic structure and related physical and chemical properties of many-electron systems. The theory is widely used in atomic spectroscopy [3] and many other applications as a conceptual tool. In addition, the theory offers a convenient and systematic framework for carrying out detailed numerical calculations for atoms [4–6]. In such calculations, it is usually necessary to introduce many simplifying assumptions, both physical and mathematical, to make progress.

For some applications, where gross electronic properties are of primary interest, very crude phenomenological models can be adopted. In other instances, where the effects are subtle, it is necessary to employ highly sophisticated theoretical models [7, 8], even though their use leads to extensive numerical computation. Fortunately, large-capacity high-speed electronic digital computers are widely available, as are atomic structure computer codes [5, 9, 10]. Extensive tabulations of the results of atomic structure calculations are also readily at hand [5, 10, 11].

Considerable effort has been devoted to perfecting atomic structure calculations, not only with a view to studying a wide variety of atomic properties, but also as a starting point for computer modeling and simulation in physics, chemistry, biology, and materials science. Such investigations have become increasingly important in industrial and government settings as well as in academe.

A notable characteristic of atomic structure calculations is the wide spectrum of available methods and points of view: different approximations are advantageous for different applications. In this introductory sketch we can hardly do justice to the many ingenious techniques that are currently in use for treating many-electron systems starting with atoms. We will provide a general perspective and some useful references.

Hartree and Hartree–Fock Methods

In orbital theories of many-electron systems, one derives a set of one-electron wave equations by using the variational method. The form of these equations is determined by the manner in which the exact many-electron wave function Φ is represented by one-electron wave functions or spin orbitals ϕ_q where q denotes all the relevant quantum numbers. By using the variational method, one guarantees that the solutions of the one-electron wave equations, the ϕ_q, are the best possible consistent with the assumed form of Φ.

In the Hartree (H) approximation, Φ is represented by a simple product of spin orbitals, ϕ_q, each factor corresponding to a different occupied state q. In the Hartree–Fock (HF) approximation, Φ is represented by an antisymmetrized product of the ϕ_q also known as a Slater determinant and a determinantal wave function. Both treatments lead to an independent-particle model in which each of the electrons moves independently of all the others in a time-averaged potential field produced by all the other electrons and the nucleus. As a result of this time-averaging, spatial correlations in the motions of pairs of electrons produced by their instantaneous Coulomb repulsion are neglected.

In the HF approximation, Φ is represented by a determinantal wave function to take into account the fact that a many-electron wave function must be an antisymmetric function of the electron coordinates. This feature is closely related to the requirement that electrons satisfy the Fermi–Dirac statistics and the Pauli exclusion principle. In contrast to the H approach, the use of determinantal wave functions in the HF method leads to additional terms in the one-electron wave equations, the exchange terms. These tend to keep electrons of like spin out of each other's way, as required by the Pauli exclusion principle. Thus, the HF approximation includes a certain type of spatial correlation between like-spin electrons. This is called exchange and also statistical correlation because it arises from the Fermi–Dirac statistics.

Spatial correlation over and above statistical correlation is usually described simply as correlation. Thus, it can be said that the H approximation neglects both exchange and correlation, while the HF approximation includes exchange but neglects correlation. HF calculations were originally favored over H calculations because they included exchange effects. However, we now realize that exchange effects are to some degree offset by correlation effects. Accordingly, it is best to include exchange and correlation effects together, especially for determining delicate electronic properties. [7, 8].

In the following, we first discuss the essentials of atomic structure calculations within the context of the traditional H and HF methods and the simplified HFS method. Next, we turn to the density functional method that allows simultaneous incorporation of exchange and correlation effects and greatly improves the physical and chemical accuracy. Finally, we examine the pseudopotential method that provides further simplifications leading to significant reductions in computational effort and the possibility of treating very large systems.

Self-Consistent Iteration and Central-Field Approximation

Since the potential terms appearing in the H and HF one-electron wave equations depend on the wave functions solving these equations, it is necessary to treat these equations by iterative techniques. One chooses an initial set of wave functions and inserts these into the potential terms. Solving the wave equations, one obtains a final set of wave functions for this cycle. The next cycle begins when we use a new set of starting wave functions constructed from a suitable average of the initial and final wave functions of the preceding cycle. This averaging represents a tradeoff between rapid convergence and numerical stability. The process continues until the initial and final wave functions in a given cycle agree with one another to some specified accuracy. In this way we obtain a self-consistent solution in the sense that the wave functions are generated by wave equations whose potential terms are determined by these very wave functions.

One can simplify the self-consistent field equations for an atomic system by taking advantage of the spherical symmetry of the atomic field. One introduces the central field approximation by considering only the spherically averaged component of the atomic field. The complete three-dimensional wave equation can then be separated into an angular wave equation and a radial wave equation by writing the spin-orbital as the product of a radial function $R(r)$, an angular function $Y(\theta, \varphi)$, and a spin function. The angular functions are the well-known spherical harmonics $Y_{lm}(\theta, \varphi)$, where l and m are the azimuthal and magnetic quantum numbers characteristic of a central field. The radial functions, $R_{nls}(r)$, in general depend on n, l, and s, where n and s are the principal and spin quantum numbers. The $R_{nls}(r)$ can be determined by solving the radial wave equations numerically [4, 5] or by the expansion method [6]. In the expansion method, the $R_{nls}(r)$ and all the potential terms are expanded in terms of a suitably chosen set of analytic functions, while the wave equations are solved by matrix methods.

In closed-shell atoms, it is possible to represent Φ by a single determinantal wave function. For the more general case of open shell atoms, it is usually necessary to represent Φ by a linear combination of determinantal wave functions. Each corresponds to a different assignment of the one-electron quantum numbers ($q = n, l, m, s$) compatible with the assumed overall symmetry of the many-electron atom. This leads to interactions over and above the central field approximation, and to multiplet structure that is of great importance in atomic spectroscopy [3].

Correlation effects can be included by expanding the many-electron wave function Φ as a linear combination of determinantal wave functions, each representing a different electronic configuration. In this approach, known as configuration interaction (CI), it is usually necessary to include large numbers of configurations to insure an accurate description of correlation effects. Apart from dealing with the intricacies of multi-configuration calculations, much of the underlying theoretical effort in quantum chemistry has been devoted to the development of efficient basis sets for representing atomic and molecular orbitals [12, 13]. The same is true in solid-state physics.

CI with optimized basis sets is very popular with quantum chemists. It is widely used and has had considerable success for atoms and moderately small molecules [12, 13]. However, CI is not practical for large molecules and crystals because of the prohibitive amount of computational effort required.

Hartree–Fock–Slater Equations

The HF equations can be simplified by averaging the HF exchange potentials over all occupied states and then using a free-electron model to determine the averaged exchange potential [14]. The exchange potential for any many-electron system can then be approximated at any point $\mathbf{r}$ by the value of the exchange potential $V_{exch}(\mathbf{r})$ of a free-electron gas having an electronic charge density ρ equal to $\rho(\mathbf{r})$. This is known as the free-electron exchange approximation. This procedure avoids the non-local potentials $V_{exch}(\mathbf{r}, \mathbf{r}')$ inherent in the HF method, at the same time eliminating the need to solve the HF equations for each orbital separately. Attention now focuses on a one-electron wave equation, the Hartree–Fock–Slater (HFS) equation, that contains the averaged exchange potential [1, 5, 14] and is the same for all orbitals.

For atomic systems containing equal numbers of electrons with spin up and spin down (balanced spins), the radial HFS equations take the form

$$\left(-\frac{d^2}{dr^2} - \frac{2Z}{r} + V_{coul}(r) + V_{exch}(r) + \frac{l(l+1)}{r^2} \right) P_{nl}(r) = E_{nl} P_{nl}(r) , \tag{1}$$

where the energy eigenvalues are denoted by E_{nl} and the radial eigenfunctions by $P_{nl}(r) = rR_{nl}(r)$. We will measure distances in Bohr units (1 Bohr = 0.529 Å) and energies in Rydberg units (1 Ry = 13.6 eV). The first term on the LHS is the kinetic energy operator. The second is the nuclear Coulomb potential, with Z denoting the nuclear charge. Next, $V_{coul}(r)$ is the spherically averaged electronic Coulomb potential, $V_{exch}(r)$ the free-electron exchange potential, and the last term the centrifugal potential.

The radial wave functions $P_{nl}(r)$ must vanish at the origin and at infinity. They have $n - l - 1$ nodes between the origin and infinity. The azimuthal quantum number l ranges from 0 to $n - 1$, and the magnetic quantum number m from $-l$ to $+l$. With the radial wave functions normalized, $\int [(P_{nl}(r)]^2 \, dr = 1$, the spherically averaged electronic charge density $\rho(r)$ and Coulomb potential $V_{coul}(r)$ are

$$\rho(r) = -(4\pi r^2)^{-1} \sum_{nl} \omega_{nl} [P_{nl}(r)]^2 \tag{2}$$

and

$$V_{coul}(r) = -(8\pi/r) \int_0^r \rho(t) t^2 \, dt - 8\pi \int_r^\infty \rho(t) t \, dt , \tag{3}$$

where $\omega_{nl}(r)$ is the occupation number for the orbital nl (both spins). In the special case of a closed shell, $\omega_{nl} = 2(2l + 1)$. The total number of electrons in the atom is $N = \sum_{nl} \omega_{nl}$, and the ionicity is $Z - N$.

Slater's exchange potential based on his free-electron model can be written as

$$V_{exch}^{\alpha}(r) = -6\alpha [(3/8\pi)|\rho(r)|]^{1/3} , \tag{4}$$

where α is a parameter whose value is 1 according to Slater's intuitive variational derivation [14].

In HF theory, the energy eigenvalues E_{nl} represent one-electron ionization energies (Koopmans' theorem). This is not the case for HFS theory, where ionization and excitation energies are determined by the transition state method [15].

The self-interaction correction, describing the Coulomb interaction of an electron with itself, is taken into account properly in the HF, but not in the HFS method. Consequently, HFS potentials are flawed at large distances from the nucleus. In the interior of non-magnetic solids, where atoms lie close together, this feature is less important than in atoms and molecules. Magnetic solids require special treatment. In spite of its simplified exchange model and the neglect of self-interaction corrections, the HFS method gained favor over the HF method because it usually led to improved ionization and excitation energies and reduced computational effort.

In an early study, the HFS equations were solved with $\alpha = 1$ for all normal neutral atoms in the periodic table [5]. Calculated energy eigenvalues E_{nl} and radial eigenfunctions $P_{nl}(r)$ were tabulated and the computer codes listed. Because of their simplicity and the favorable agreement with experiment obtained, the Herman–Skillman atomic structure codes were and still are widely used for atomic, molecular, and solid state problems [10, 16], and for pedagogical purposes as well [17]. Over time the codes have remained substantially the same except for the incorporation of improved exchange-correlation potentials [10, 21].

Kohn and Sham [18] took issue with Slater's use of $\alpha = 1$ because their variational calculations led to a value of $\alpha = 2/3$. (Note that Eq. 4 with $\alpha = 2/3$ is known as the Kohn–Sham equation.) However, atomic results for $\alpha = 2/3$ often departed more from experiment than those for $\alpha = 1$. This led to the Xα method, where X stands for exchange, and the value of α is optimized for each atom [19]. This pragmatic approach for atoms led to conceptual difficulties in applications to molecules and solids, where it is more reasonable to use the same value of α in all regions of space than different values in different atomic regions.

The significance of the Xα approximation was clarified by HFS atomic calculations that included a charge-density-gradient correction to the exchange potential [20]. The gradient-corrected exchange potential was represented by the expression

$$V_{\mathrm{exch}}^{\alpha\beta}(r) = -6\left[\alpha + \beta G(\rho)\right]\left[(3/8\pi)|\rho(r)|\right]^{1/3}, \tag{5}$$

where the dimensionless function $G(\rho)$ is defined as

$$G(\rho) = \frac{1}{\rho^{2/3}}\left[\frac{4}{3}\left(\frac{\nabla\rho}{\rho}\right)^2 - \frac{2\nabla^2\rho}{\rho}\right]. \tag{6}$$

This form was determined by dimensional analysis. Here α and β are variational parameters that were determined by minimizing the total energy for each of a large set of representative atoms using Slater's average-over-configurations method [1, 2]. In this so-called X$\alpha\beta$ approximation, the optimum value of α was found to be equal to 2/3 for all atoms investigated, demonstrating that the Z-dependence of α in the Xα method is due to the neglect of gradient corrections. (The optimal value of β was nearly independent of Z.)

Since this study was based on minimizing the total energy, and most of the total energy resides in the inner shells and very little in the outer shells, the treatment of inner shell electrons was improved, but that for outer shell electrons was not. These early calculations left much to be desired, but they stimulated many subsequent efforts to improve the treatment of gradient corrections in atoms.

The good agreement between theory and experiment found earlier using the HFS method with $\alpha = 1$ [5] is due in part to the fortuitous cancelation of neglected correlation effects and enhanced free-electron exchange arising from the use of $\alpha = 1$ instead 2/3 (cf. Ref. [21], p. 165).

Density Functional Theory and the Kohn–Sham Equations

The underlying idea of density functional theory [18, 22–25] is that the total energy of a many-electron system is determined by a functional of the charge density $\rho(r)$, $E[\rho(r)]$. From this functional it is possible to derive a potential function $V(r)$ that determines the one-electron wave functions and energies of the system. The theory does not provide an algorithm for determining this functional, but we are assured that the use of improved functionals will result in more accurate descriptions of electronic structure.

The density functional approach to many-electron systems leads to one-electron wave functions and methods of solution analogous to those for the HFS method, with the important exception that the Kohn–Sham exchange potential $V_{exch}^{\alpha=2/3}(r)$ is replaced by a local potential $V_{exch}^{corr}(r)$ representing exchange and correlation effects. This potential can be derived from many-electron theories of the free-electron gas [26, 27]. A sophisticated version of the local density approximation (LDA), the screened-exchange plus Coulomb hole approximation [26], forms the basis of the GW method that is used in solid-state physics to determine energy band structure features with great accuracy [28].

The search for improved density-functional exchange-correlation potentials has been an active field of research for many years and continues to thrive. Some notable advances include generalized gradient approximations (GGA) [29]; semi-empirical schemes where parameters are adjusted to experiment [30]; and orbital-dependent potentials (OEP) [31]. Treatments of self-interaction corrections [32], relativistic effects [33], and spin polarization [34] have also been implemented. These and many other developments have improved the accuracy of density functional calculations considerably [21] (cf. Part II). However, the goal of achieving "chemical accuracy" of about 1 kcal/mole ($= 0.0434$ eV) for ionization, excitation, and relative energies remains elusive.

Pseudopotential Theory

Although the idea of pseudopotentials is very old, dating back to Fermi, recent developments began with a seminal paper by Herring [35] dealing with crystals, though his ideas also apply to atoms and molecules. Herring showed that orthogonalizing the wave functions of the outer shell electrons to those of the inner shells introduces a repulsive term in the wave equation that partially cancels the remaining electronic terms. This process leaves a residual effective potential or pseudopotential that is considerably smoother and weaker than the original "all-electron" potential. Morever, orthogonalization removes most of the nodal structure in outer shell electron wave functions, so that they can be expanded in relatively compact basis sets (orthogonalized plane waves). The use of pseudopotentials greatly increases the number of atoms that can be included in a molecular cluster and the range of problems that can be treated. Combined with density functional theory, the pseudopotential method paves the way for realistic computer simulation and modeling in many fields of science and technology. [36]

The trail from Herring's work [35] to present-day pseudopotential theory runs as follows: In 1952, the first realistic calculation of the band structure of the diamond crystal [37] was carried out using Herring's OPW method. Seeing the detailed numerical results, Phillips noted the remarkable cancelation brought about by orthogonalization in semiconductors. This led Phillips and Kleinman [38] to devise a practical method for replacing the complicated orthogonality terms by simpler, more easily managed analytical forms. Independently, Harrison [39] pioneered the use of pseudopotentials in the study of the Fermi surface of metals. Further progress included understanding pseudopotential theory at a deeper level [25, 40], and developing efficient semi-empirical [41] and first-principles pseudopotentials [42]. For a comprehensive account of the current state of the theory, see Ref. [21], Chap. 11.

Concluding Remarks

Although we have emphasized atomic structure calculations here, appreciable cross-fertilization has taken place among atomic, molecular, and solid-state calculational studies. Methods for dealing with central field problems in atoms have naturally found their way into studies of molecules and crystals. Extensive investigations of efficient basis sets in molecular calculations provided valuable experience for the subsequent development of sophisticated basis sets for solids. Concepts and methods such as pseudopotentials and density functionals that originated in solid state physics now become increasingly important in atomic and molecular investigations.

Purely theoretical calculations continue to shed insight on complex problems. At the same time, numerical studies have expanded almost exponentially owing to the availability of more powerful computers and the pressing need for detailed results for practical problems, as opposed to analytical results for model problems. Perhaps this trend reflects John C. Slater's point of view (private communication), namely, that you don't really understand physical or chemical properties unless you can actually calculate them. Others like Charles Kittel and Sir Nevill Mott (private communications) took the opposite view, arguing that "back-of-the-envelope" calculations giving order-of-magnitude estimates and essential insights are preferable. Apart from this dichotomy, computational physicists and chemists should bear in mind that their calculations will in all likelihood be improved by future investigators using more powerful algorithms and computational tools, while mathematical theorems proved today will remain true forever [43].

See also: Atomic Structure Calculations, Electronic Correlation; Fine and Hyperfine Spectra and Interactions, Solid-State Physics.

References

[1] J. C. Slater, *Quantum Theory of Atomic Structure*, Vols. 1 and 2. McGraw–Hill, New York, 1960.
[2] J. C. Slater, *Quantum Theory of Molecules and Solids*, Vols. 1 to 4. McGraw–Hill, New York, 1963, 1965, 1967, 1974.
[3] I. I. Sobelman, *Introduction to the Theory of Atomic Spectra*. Pergamon Press, Oxford, 1972; R. D. Cowan, *The Theory of Atomic Structure and Spectra*. University of California Press, Berkeley, 1981; B. R. Judd, *Rept. Prog. Phys.* **48**, 907 (1985).

[4]   D. R. Hartree, *The Calculation of Atomic Structures*. Wiley, New York, 1957. For a personal account of Hartree's work in atomic physics, see B. S. Jeffreys, *Comments Atom. Molec. Phys.* **20**, 189 (1987); C. Froese Fischer, *Douglas Rayner Hartree: His Life in Science and Computing*. World Scientific, Singapore, 2004.

[5]   F. Herman and S. Skillman, *Atomic Structure Calculations*. Prentice-Hall, Englewood Cliffs, NJ, 1963.

[6]   C. Froese-Fischer, *The Hartree-Fock Method for Atoms*. Wiley-Interscience, New York, 1977.

[7]   A. Hibbert, Rept. *Prog. Phys.* **38**, 1217 (1975); I. Lindgren and J. Morrison, *Atomic Many-Body Theory*, 2nd ed. Springer-Verlag, Berlin, 1986.

[8]   R. K. Nesbet, "Atomic Structure Calculations, Electronic Correlation" (this volume).

[9]   Many atomic structure codes are available through *Computer Physics Communications Program Library*, www.cpc.cs.qub.ac.uk/cpc and *Quantum Chemistry Program Exchange*, www.qcpe.indiana.edu.

[10]   S. Kotochigova, Z. Levine, E. Shirley, M. Stiles, and C. Clark, *Phys. Rev. A* **55**, 191 (1997); *ibid*, 5191E. Computer codes are available from these authors: physics.nist.gov/PhysRefData/DFTdata/contents/html.

[11]   For atomic tabulations, see the internet URL in the previous reference. A wide variety of specialized tabulations appear in *Atomic Data and Nuclear Data Tables*.

[12]   J. A. Pople, Nobel Prize lecture, *Rev. Mod. Phys.* **71**, 1267 (1999).

[13]   P.-O. Löwdin, "Molecular Structure Calculations" (this volume).

[14]   J. C. Slater, *Phys. Rev.* **81**, 385 (1951).

[15]   J. C. Slater, *Adv. Quantum Chem.* **6**, 1 (1972); see also Ref. [2], Vol. IV, Sec. 2–5.

[16]   K. H. Johnson, *Adv. Quantum Chem.* **7**, 143 (1973); *Ann. Rev. Phys. Chem.* **26**, 39 (1975); F. Herman, A. R. Williams, and K. H. Johnson, *J. Chem. Phys.* **61**, 3508 (1974); see also Ref. [2], Vol. IV.

[17]   C. M. Quinn, *Computational Quantum Chemistry – An Interactive Guide to Basis Set Theory*. Academic Press, San Diego, 2002.

[18]   W. Kohn and L. J. Sham, *Phys. Rev.* **140**, A1133 (1964).

[19]   K. Schwarz, *Phys. Rev.* **B5**, 2466 (1972); see also Ref. [2], Vol. IV.

[20]   F. Herman, I. B. Ortenberger, and J. P. Van Dyke, *Phys. Rev. Lett.* **22**, 807 (1969); *Intern. J. Quant. Chem.* **3S**, 827 (1970).

[21]   R. M. Martin, *Electronic Structure: Basic Theory and Practical Methods*. Cambridge University Press, Cambridge, 2004. For supplementary material, see http://mcc.uiuc.edu/estructure.

[22]   P. Hohenberg and W. Kohn, *Phys. Rev.* **136**, B864 (1963); N. D. Mermin, *Phys. Rev.* **137**, A1441 (1965).

[23]   R. G. Parr and W. Yang, *Density-Functional Theory of Atoms and Molecules*. Oxford University Press, Oxford, 1989; R. O. Jones and O. Gunnarsson, *Rev. Mod. Phys.* **61**, 689 (1989); W. Koch and M. C. Holthausen, *Chemist's Guide to Density Functional Theory*, 2nd ed. John Wiley, New York, 2004.

[24]   W. Kohn, Nobel Prize lecture, *Rev. Mod. Phys.* **71**, 1253 (1999).

[25]   C. Herring, "Solid State Physics" (this volume).

[26]   L. Hedin, *Phys. Rev.* **39**, A796 (1965); L. Hedin and S. Lundqvist, *Solid State Phys.* **23**, 1 (1969).

[27]   L. Hedin and B. I. Lundqvist, *J. Phys. C* **4**, 2064 (1971); U. von Barth and L. Hedin, *J. Phys. C* **5**, 1629 (1972).

[28]   M. S. Hybertsen and S. G. Louie, *Phys. Rev.* **34**, 5390 (1986); **35**, 5585, 5602 (1987).

[29]   J. P. Perdew and Y. Wang, *Phys. Rev. B* **45**, 13244 (1992); J. P. Perdew, K. Burke, and M. Ernzerhof, *Phys. Rev. Lett.* **77**, 3865 (1996).

[30]   A. D. Becke, *Phys. Rev. A* **38**, 3098 (1988); C. Lee, W. Yang, and R. G. Parr, *Phys. Rev. B* **37**, 785 (1988).

[31] D. M. Bylander and L. Kleinman, *Intern. J. Mod. Phys.* **10**, 399 (1996).
[32] J. P. Perdew and A. Zunger, *Phys. Rev. B* **23**, 5048 (1981); A. Svane and O. Gunnarsson, *Phys. Rev. B* **37**, 9919 (1988).
[33] J. T. Waber, "Atomic Structure Calculations, Relativistic Atoms" (this volume).
[34] O. Gunnarsson and B. I. Lundqvist, *Phys. Rev. B* **13**, 4274 (1976).
[35] C. Herring, *Phys. Rev.* **57**, 1169 (1940).
[36] M. L. Cohen, *Annu. Rev. Mater. Sci.* **30**, 1 (2000).
[37] F. Herman, *Phys. Rev.* **88**, 1210 (1952).
[38] J. C. Phillips and L. Kleinman, *Phys. Rev.* **116**, 287 (1959).
[39] W. A. Harrison, *Pseudopotentials in the Theory of Metals*. W. A. Benjamin, New York, 1966.
[40] M. L. Cohen and V. Heine, *Solid State Physics* **24**, 1 (1970).
[41] M. L. Cohen and T. K. Bergstresser, *Phys. Rev.* **141**, 1979 (1966).
[42] D. R. Hamann, M. Schlüter, and C. Chiang, *Phys. Rev. Lett.* **43**, 1494 (1979); G. B. Bachelet, D. R. Hamann, and M. Schlüter, *Phys. Rev. B* **26**, 4199 (1982).
[43] See comments by George E. Kimball to the author in F. Herman, *Phys. Today* **37**, 56 (June, 1984).

Atomic Structure Calculations, Relativistic Atoms

J. T. Waber

Relativistic atomic calculations have become available for all of the atoms of the periodic table in the last decade and have extended our understanding in many areas of physics and chemistry. Some of these will be reviewed briefly below.

Apparently, the first such calculation [1] was made for the $_{29}Cu^{+1}$ ion in 1940. It passed relatively unnoticed until Mayers [2] did $_{80}Hg$ in 1957. These were formidable tasks done without the aid of modern computers. Only a limited number of atomic calculations of any kind existed before 1963 when Herman and Skillman published their book on *Atomic Structure Calculations* [3]. Within a few years [4–7] a large number of nonrelativistic calculations started to appear with increasing array of complications. With these in hand, the importance of the various improvements could be assessed. Excellent agreement with experiment for energy values was demonstrated in 1975.

Several differences between the nonrelativistic Hartree–Fock–Slater (discussed elsewhere by Herman [8]) and the relativistic Dirac–Slater wavefunctions [9] are important. Instead of a single orbital $P(\mathbf{r})$ for each electron with the quantum numbers (nlm), one writes

$$\Psi_k(\mathbf{r}) = \frac{1}{r} \begin{pmatrix} i^l F(r)\Omega(jlm(\theta,\varphi) \\ i^{l'} G(r)\Omega(jl'm(\theta,\varphi) \end{pmatrix} \tag{1}$$

as the orbital for the kth electron, where $F(r)$ stands for the major radial component and $G(r)$ for the minor. Instead of one spherical harmonic or associated Legendre polynomial $Y_l^m(\theta,\varphi)$ to represent the angular dependence, there are several. The angular dependence for a given component is defined as

$$\Omega_{lj\mu} = A\{B(lj)Y_l^{\mu-\sigma}(\theta,\varphi) \begin{bmatrix} 0 \\ 1 \end{bmatrix} + C(lj)Y_l^{\mu+\sigma}(\theta,\varphi) \begin{bmatrix} 1 \\ 0 \end{bmatrix} \}, \tag{2}$$

Table 1: Quantum numbers for relativistic atoms.

		Symbol		s	p	d	f	g
	a	j	l	0	1	2	3	4
Number of	-1	$l-\frac{1}{2}$		0	2	4	6	8
electrons	$+1$	$l+\frac{1}{2}$		2	4	6	8	10
Sum				2	6	10	14	18
Kappa	-1			0	1	2	3	4
value	$+1$			-1	-2	-3	-4	-5

where A is a normalizing constant, $B(lj)$ and $C(lj)$ are coefficients dependent on the angular momenta, and the two functions $Y_l^{\mu\pm\sigma}$ are the same type of spherical harmonic as occurs in the nonrelativistic case. Finally $\begin{bmatrix}0\\1\end{bmatrix}$ and $\begin{bmatrix}1\\0\end{bmatrix}$ represent the Pauli spin matrices which represent up and down spin, respectively. Because the angular momentum l' differs from l by the quantity a (which may be either plus or minus 1), the angular dependence of the major and minor components depend either on l or l' or vice versa. Hence one involves an odd and the other an even power of $\cos\theta$.

The angular momentum j has the definition $j=l+\frac{1}{2}a$. Another aspect of a will be taken up below, and the μ values differ by 1 since $|\sigma|=\frac{1}{2}$. Concerning the two $Y_l^{\mu\pm\sigma}$ functions in the formula for $\Omega_{il\mu}$, the quantity μ is the resolved component of j. For these reasons $\Psi(\mathbf{r})$ is a four-component wave function. Details of the formulation are given by Grant [10] but his notation differs slightly from these formulas. The reader is cautioned that phase factors may differ in the various representations which have been published.

A simple table (Table 1) will indicate these quantum numbers as well as Dirac quantum number κ, and how many electrons can occupy a complete subshell. Thus, for example, the p shell is divided into two subshells; one is called $2p_{1/2}$ and the other $2p_{3/2}$. Each contains $2j+1$ electrons. This is one additional way in which the relativistic and nonrelativistic wave functions and atomic calculations differ.

The phenomenon just mentioned is called spin–orbit splitting and the energy separation between the two subshells varies roughly as Z^4, where Z is the atomic number. This fact will indicate why relativistic effects have only limited importance when dealing with the electrons of atoms and ions with low Z values but become very important for heavy elements.

Some of the details of making a self-consistent-field calculation have been indicated in this volume [8]. For greater detail, the reader is referred to recent papers [11–13]. The probability of finding the ith electron in a spherical shell of radius r reduces to the simple formula

$$\rho_i(r) = [F_i(r)]^2 + [G_i(r)]^2 \tag{3}$$

after integration over the angles θ and φ. The total charge density $\rho(r)$ is obtained by summing this over all the occupied orbitals. Slater's approximation [14] to exchange potential is related to the diameter of a Fermi hole, and hence is given by

$$V_{ex}(r) = \alpha_s \left[\frac{3}{8\pi}\sum\rho_i(r)\right]^{1/3} \tag{4}$$

where α_s is an adjustable constant in the range of $\frac{2}{3}$ to 1 and he summation runs over all of the occupied orbitals. Mann [15] has discussed the slightly more complicated equations which arise when exchange is handled in the Hartree–Fock manner.

A convenient way [7] to write the radial equations is as a matrix for a given set of quantum numbers:

$$\frac{d}{dr}\begin{pmatrix} F(r) \\ G(r) \end{pmatrix} = \begin{pmatrix} \dfrac{-\kappa}{r} & \dfrac{(V - E_0 - W)}{r} \\ -\dfrac{(V + +E_0 - W)}{r} & \dfrac{\kappa}{r} \end{pmatrix} \tag{5}$$

so that the derivative of F is equal to the sum of radial factors times both F and G – a similar situation applies for the minor component G. That is, the two radial functions are coupled together by first-order differential equations.

In Eq. (5), V is the Coulomb potential Z/r, E_0 is the rest mass of the electron, and W is the energy eigenvalue $E(nlj)$. The relation of the quantum number κ to the more familiar ones, l and j, is listed above.

By differentiating G with respect to r, and substituting, one can get from (5) a second-order differential equation in the major component. This is similar to the Schrödinger equation but contains additional components. One of these additional terms gives rise to the spin–orbit splitting. By assuming that such terms are only a small perturbation, one can estimate the relativistic effects with single-component nonrelativistic wave functions. This was the traditional approach found in textbooks before the relativistic calculations became available.

Relativistic atomic calculations are important for treating phenomena which involve the interaction between the nucleus and electrons. References to typical areas are: Mössbauer spectroscopy [12], hyperfine interaction [13], and beta capture [17].

Another area where relativistic effects are important is in connection with the valence electrons of heavy elements. This can be illustrated by the two diagrams which contrast the "order" of filling electrons in the lowest-energy state of a given shell nl. In a nonrelativistic treatment of the electrons in a rare earth (lanthanide) or actinide element, one first arranges the spin of the f electrons parallel according to LS or Russell–Saunders coupling scheme (Fig. 1). Progressively one spin is occupied after another and the energy increases due to electron–electron interaction. With seven electrons, the resultant is $J = \frac{7}{2}$, corresponding to the seven unpaired spins. Completion of the shell by adding the other $2l + 1$ electrons brings the total spin to $J = 0$.

In Fig. 2, there are the two subshells with six and eight electrons separated by a large spin–orbit splitting. According to the modified Hund's rule [18], the first three electrons go in a parallel arrangement hence $J = -\frac{5}{2} - \frac{3}{2} - \frac{1}{2} = -\frac{9}{2}$. However, the next three are opposite in sign so that the sum of the six individual J values, namely J, becomes zero. The seventh electron would lead to a J value of $-\frac{7}{2}$ just as was found for LS coupling. In fact, in many cases, the result found by jj coupling (which is characteristic of a relativistic treatment) is the same as one would obtain by LS coupling; in others, the two results differ. For example, for six electrons, LS coupling would give -3 for J, where as the better answer is 0.

Another area which involves the outer electrons is the anomalous behavior of the angular distribution of photoelectrons. The jj coupling between the bound and the emergent electron in the continuum state leads to observed dependence and explains the occurrence of the Cooper minimum [19] in the asymmetry coefficient β as well as its energy dependence. In contrast,

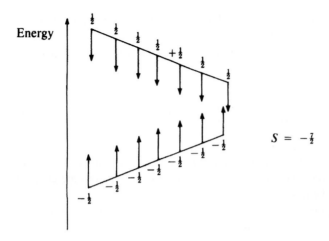

Fig. 1: Schematic of energy involved in adding electrons in *LS* coupling.

LS coupling gives a constant value [20, 21] for β.

The accuracy of the relativistic atomic calculations is indicated by the calculated theoretical binding energy of the $1s_{1/2}$ electron [22]:

Energy eigenvalue	−142.929 keV
Magnetic contribution	+0.715
Retardation effect	−0.041
Vacuum fluctuation	+0.457
Vacuum polarization	−0.155
Total binding energy (theoretical)	141.953 (±0.053) keV
Experimental value [23]	141.963 (±0.013) keV

The agreement is very good with the experimental value of Dittner *et al.* [21] and with the independent theoretical value (also based on a Dirac–Fock calculation) of Freedman *et al.* [23].

This listing also serves to indicate the magnitude of other relativistic effects which cannot be discussed fully here. Both the magnetic and the retardation corrections are involved in the Breit interaction – a coupling of two electrons by means of a virtual photon. The vacuum fluctuation is also called Zitterbewegung. Polarization of the vacuum results from the strong Coulomb field of the nucleus. Its estimation is discussed by Pyykkö [24].

Most of these calculations have assumed that the field experienced is spherically symmetric. While this is guaranteed by Unsold's theorem for a closed shell, it is not true for open shells and a more complex treatment is required for dealing with even relatively simple atoms. The reader is referred to the article in this volume by Nesbet [25].

The other important use of such relativistic wave functions is in the construction of a reasonable molecular or crystal potential when heavy elements are concerned. Two types of relativistic molecular calculations have been done. The "discrete variational method" was used in connection with a linear combination of numerical (relativistic) orbitals by Rosen and Ellis [26–28]. At nearly the same time, Yang and his collaborators developed a relativistic

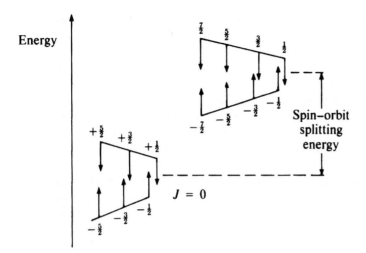

Fig. 2: Energy sketch of adding electrons with various spins in jj coupling.

version of the scattered wave formalism [29–32]. A review article by Pyykkö [24] covers a number of aspects of relativistic quantum chemistry. The earliest relativistic band calculations were made by Loucks [33, 34] who employed the augmented plane wave (APW) method. Sommers [35] developed a relativistic Kohn–Korringa–Rostoker method and Soven [37] a relativistic orthogonalized plane wave. More recently, Koelling and Freeman have published a series of relativistic APW calculations [38–41].

With the ready availability of even multiconfigurational atomic Dirac–Fock MCDF programs of Desclaux *et al.* [42] and Grant *et al.* [43], progress has been made in the last 10 years on the fronts of molecular and atomic structure calculations. Despite this fact, experimentalists continue to use a collection of approximation and *Ansatz* corrections to NR calculations to bring about improved comparisons with their data. These corrections are of questionable validity and may pertain to only portions of the periodic table. The MCDF calculations are "exact" and now with current computers are easy to perform.

While the word "exact" has been used, some difficulties remain; for example, the question of how to treat the interaction between two electrons is not resolved. In the comprehensive review in 1970 by Grant [44] of relativistic atomic calculations, two problems had not emerged, even though Brown and Ravenhall [45] had pointed out that the simple Dirac–Coulomb equation did not yield normalizable solutions. The success of obtaining good numerical eigenfunctions had diverted attention from such fundamental problems.

When finite basis sets were employed, unrealistically negative solutions were obtained and the "Brown–Ravenhall disease" and "variational collapse" became of significance. One cause [46] of the latter was that the same basis set was used to expand the major and minor components. The reader is referred to the studies of Kutzelnigg [47] and Goldman [48], Grant has made a number of comments [49] on these types of problem in recent conference volumes. These difficulties were attributable to the fact that the single-electron relativistic Hamiltonian is not bounded from below, i. e., there exists a negative energy continuum which has features in common with positive energy Rydberg states, and spurious roots with $E \sim -2c^2$ were obtained [50]. The recommended procedure is to project out properly behaved solutions and constrain the solutions to pertain only to a fixed number of electrons, as was effectively done with numerical orbitals [51].

The use of basis sets was motivated by interest in following the NR procedures of Roothan [52] for molecules. Goldman and Dalgarno implemented a variational procedure for two-electron atoms which avoided spurious roots, variational collapse, and continuum dissolution. Results with $Z \leq 10$ were published [53]. It should be mentioned that the proper nodal behavior is complex. That is, in representing the orbital by such basis functions, the minor component has one more node than the major [54], i.e., the significance increases rapidly with Z.

Concerning the importance of utilizing relativistic orbitals, Torboem, Fricke, and Rosen [55] discuss isomer shifts for low-lying states of IIa and IIb elements and compare nonrelativistic NF, DF, and MCDF results. The inadequacy using the contact term $|\psi_{ns}|^2$ rather than the wave functions integrated over the range of the nuclear charge distribution is illustrated. They note that $np_{1/2}$ orbitals contribute very little to the overall electronics charge density in this region. However, they note that the $ns(n-1)d$ states make an important contribution in the multiconfigurational treatment.

Kim, Huang, Cheng and Desclaux [56] emphasize that care should be taken when comparing DF calculations and experimental spin–orbit splittings $^2P_{1/2}$–$^2P_{3/2}$. They presented results for two negative ions, B^- and F^-. Problems arose because $^2P_{1/2}$ and $^2P_{3/2}$ orbitals do not converge to the same nonrelativistic results and can yield spurious nonrelativistic contributions when one attempts to obtain good agreement with experimental data.

Pitzer and colleagues [57] have investigated the use of an effective potential obtained from DF calculations for molecular problems.

While retardation interaction between two electrons is not very large when treating an atom [54], it will become significant in treating molecules because of larger interelectronic distances. The necessity of treating atomic cases with nonorthogonal orbitals also has not been evaluated for molecular problems.

Zangwill and Liberman [58] have made some improvements of the Liberman–Waber–Cromer program to take in account time-dependent optical response.

Relativistic continuum orbitals have been discussed in the texts by Berestetskii, Lifschitz, and Pitaerskii [59] and Rose [60]. A computer code for such orbitals has been accepted by Computer Physics Communications [61] by Perger and a study involving both discrete and positive energy continuum orbitals expressed in an *ab initio* basis set obtained from the Dirac equation is near completion.

One way of summarizing this article on atomic calculations is to indicate the typical references to various kinds of relativistic atomic calculations which are available as well as some differences (Table 2). Most of these are available as numerical tabulations. The results of Fraga, Saxena, etc. are as coefficients of algebraic basis functions. The maximum atomic number studied for each type is indicated.

Pyykkö [62] organized an exhaustive bibliography of several hundred references covering the period 1916 to 1985. The reader is directed to it for locating the literature on various relativistic studies. The recent paper by Grant and Quiney [63] is an excellent review of the theoretical and computational situation.

Table 2: Some typical relativistic atomic calculations.

Name of type	Author	Type of exchange	max. Z calculated	Year	Footnote
Hartree–Fock	Cohen	None	92	1951	*a*
Dirac–Slater	Carlson *et al.*	Statistical	126	1966	*b*
numerical	Liberman *et al.*	Statistical	126	1966	*c, q, p*
	Schofield	Statistical	104	1974	*d*
	Rosin, Lindgren	Modified Slater	95	1968	*e*
	Band *et al.*	Statistical	95	1977	*f*
	Huang *et al.*	Statistical	106	1965	*g*
Analytic expansion	Fraga *et al.*	Fock integrals	102	1965	*h*
	Kim		40	1967	*i*
	Kagawa		50	1975	*j*
Dirac–Fock	Mann, Waber	Fock integrals plus	131	1970	*k*
		Breit and correlation corrections	126	1970	*l*
Multiconfiguration	Desclaux	Fock integrals plus	120	1973	*m*
Dirac-Fock	Desclaux	Breit and	126	1976	*n*
	Fricke, Soff	correlation corrections	173	1977	*o*

a S. Cohen, *Phys. Rev.* **118**, 489 (1960).

b C. C. Lu, T. A. Carlson, F. B. Malik, T. C. Tucker, and C. W. Nestor, Jr., *At. Data* **3**, 1 (1971); (erratum) **14**, 89 (1974); **2**, 63 (1970).

c D. Liberman, J. T. Waber, and D. T. Cromer, *Phys. Rev.* **137**, A27 (1965).

d J. H. Schofield, *At. Nucl. Data Tab.* **14**, 121 (1974); *Phys. Rev.* **9**, 1041 (1974).

e A. Rosin and I. Lindgren, *Phys. Rev.* **176**, 114 (1968); see also I. Lindgren and A. Rosin, *Atom. Phys.* **4**, 93 (1974).

f I. M. Band, M. A. Listengarten, M. B. Trzhaskovskaya, and V. I. Fomichev, *Leningrad. Inst. Idernov. Fiz. Report* **298** (April 1977).

g K. H. Huang, M. Aoyagi, M. H. Chen, B. Craseman, and H. Mark, *At. Data* **18**, 243 (1976).

h S. Fraga and K. M. S. Saxena. *At. Data* **3**, 323 (1971); **4**, 255 (1972); **4**, 269 (1972); **5**, 467 (1973).

i Y. K. Kim, *Phys. Rev.* **154**, 17 (1967); (erratum) **159**, 190 (1967).

j T. Kagawa, *Phys. Rev. A* **12**, 2245 (1975).

k J. B. Mann and J. T. Waber, *J. Chem. Phys.* **53**, 2397 (1970).

l J. B. Mann, *J. Chem. Phys.* **51**, 841 (1969).

m J. P. Desclaux, *At. Data* **2**, 311 (1973).

n J. P. Desclaux, *At. Data* **18**, 243 (1976).

o B. Fricke and G. Soff, *At. Nucl. Data Tab.* **19**, 83 (1977).

p D. A. Liberman, *Phys. Rev. B* **2**, 244 (1970).

q W. Kohn and L. T. Sham. *Phys. Rev.* **140**, A1133 (1965).

See also: Atomic Structure Calculations, One-Electron Models.

References

[1]   A. O. Williams, *Phys. Rev.* **58**, 723 (1970).

[2]   D. F. Mayers, *Proc. Roy. Soc. (London)* **A241**, 93 (1957). J. P. Desclaux, D. F. Mayers, and F. O'Brien, *Phys. Rev. B* **4**, 631 (1971).

[3]   F. Herman and S. Skillman, *Atomic Structure Calculations* (Prentice-Hall, Englewood Cliffs, N.J., 1963).

[4]   M. A. Coulthard, *Proc. Roy. Soc. (London)* **91**, 44 (1967); **91**, 421 (1967).

[5]   F. C. Smith and W. R. Johnson, *Phys. Rev.* **160**, f36 (1967).

[6]   C. C. Lu, T. A. Carlson, F. B. Malik, T. C. Tucker, and C. W. Nestor, Jr., *At. Data* **3**, 1 (1971); (erratum) **14**, 89 (1974); **2**, 63 (1970).

[7]   D. Liberman, J. T. Waber, and D. T. Cromer, *Phys. Rev.* **137**, A27 (1965).

[8]   F. Herman, this Encyclopedia.

[9]   V. M. Burke and I. P. Grant, *Proc. Phys. Soc. (London)* **90**, 297 (1967).

[10]   I. P. Grant, *Adv. Phys.* **19**, 747 (1970). A slightly different notation was used in earlier papers [*Proc. Roy. Soc. A* **262**, 555 (1961); *Proc. Phys. Soc. (London)* **86**, 523 (1965)].

[11]   D. Liberman, J. T. Waber, and D. T. Cromer, *Comp. Phys. Commun.* **2**, 107 (1971).

[12]   B. Fricke and J. T. Waber, *Actinide Rev.* **1**, 433 (1971); *Theo. Chim. Acta* **21**, 235 (1971); *Phys. Rev. B* **5**, 3445 (1972).

[13]   J.-P. Desclaux, *Int. J. Quantum Chem.* **6**, 25 (1972); J.-P. Desclaux and N. Bessis, *Phys. Rev. A* **2**, 1623 (1970).

[14]   J. C. Slater, *Phys. Rev.* **81**, 385 (1951).

[15]   J. B. Mann, *J. Chem. Phys.* **51**, 841 (1969).

[16]   J. B. Mann and J. T. Waber. *At. Data* **5**, 201 (1973); *J. Chem. Phys.* **53**, 2397 (1970).

[17]   T. A. Carlson, C. W. Nestor, Jr., F. B. Malik, and T. C. Tucker, *Nucl. Phys. A* **135**, 57 (1969).

[18]   T. E. H. Walker and J. T. Waber, *Phys. Rev. A* **7**, 1218 (1973).

[19]   T. E. H. Walker and J. T. Waber, *Phys. Rev. Lett.* **30**, 307 (1973).

[20]   T. E. H. Walker and J. T. Waber, *J. Phys. B* **6**, 1 165 (1973).

[21]   T. E. H. Walker and J. T. Waber, *J. Phys. B* **7**, 674 (1974).

[22]   B. Fricke, J.-P. Desclaux, and J. T. Waber, *Phys. Rev. Lett.* **28**, 714 (1972).

[23]   M. S. Freedman, F. T. Porter, and J. B. Mann, *Phys. Rev. Lett.* **28**, 711 (1972). P. F. Dittmer, C. E. Bemis, D. C. Hansley, R. J. Silva, and D. C. Goodmen *Phys. Rev. Lett.* **26**, 1037 (1971).

[24]   P. Pyykko, *Adv. Quantum Chem.* (to be published).

[25]   R. Nesbet, this Encyclopedia.

[26]   A. Rosen and D. E. Ellis, *J. Chem. Phys.* **62**, 3039 (1975).

[27]   D. E. Ellis, A. Rosin, and P. F. Walch, *Int. J. Quantum Chem.* **S9**, 351 (1975).

[28]   P. F. Walch and D. E. Ellis, *J. Chem. Phys.* **65**, 2387 (1976).

[29]   C. Y. Yang and S. Rabii, *Phys. Rev. A* **12**, 362 (1975).

[30]   C. Y. Yang, K. H. Johnson, and J. A. Horsley, *Bull. Am. Phys. Soc.* **21**, 382 (1976).

[31]   C. Y. Yang, *Chem. Phys. Lett.* **41**, 588 (1976).

[32]   C. Y. Yang and S. Rabii, *J. Chem. Phys.* **78**, 6S (1978).

[33]   T. Loucks, *Phys. Rev.* **139**, A f333 (1965).

[34]   T. Loucks, *Augmented Plane Wave Method* (Benjamin, New York, 1966).

[35]   C. Sommers and H. Amar, *Phys. Rev.* **188**, 1117 (1969).

[36]   C. Sommers, *J. Phys. C* **3**, 39 (1972).

[37]   P. Soven, *Phys. Rev.* **137**, A1706 (1965).

[38]   D. Koelling, *Phys. Rev.* **188**, 1049 (1969).

[39]   D. Koelling and A. J. Freeman, *Phys. Rev. B* **7**, 4454 (1973).

[40]   D. Koelling and A. J. Freeman, *Phys. Rev. B* **12**, 5622 (1975).

[41]   D. Koelling and A. J. Freeman, *Plutonium and Other Actinides*, p. 2911 (H. Blank and R. Lindner, eds.). (North Holland, Amsterdam, 1976).

[42]   J.-P. Desclaux, *Comp. Phys. Commun.* **9**, 31 (1975).

[43]   I. P. Grant, *Comp. Phys. Commun.* **21**, 207 (1980).

[44]   Lan P. Grant, *Adv. Phys.* **19**, 747–811 (1970).

[45]   G. E. Brown and D. G. Ravenhall, *Proc. R. Soc. (London) A* **208**, 552 (1951).

[46]   Y. Ishikawa, R. C. Binning, and K. M. Sand, *Chem. Phys. Lett.* **105**, 189 (1984); **101**, 111 (1983). J. Mark and P. Rozickey, *Chem. Phys. Lett.* **74**, 562 (1980).

[47]   W. Kutzelnigg, *Int. J. Quantum Chem.* **25**, 107 (1984).

[48]   S. P. Goldman, *Phys. Rev. A* **30**, 1219 (1984); **31**, 354 (1985); **37**, 16–30 (1988).

[49]   I. P. Grant, in *Atom Theory Workshop on Relativistic and QED Effect in Heavy Atoms* (H. Kelly and Y.-Ki Kim, eds., pp. 17–19, 200–203, 299–301, American Institute of Physics, New York, 1985). See also *Phys. Rev. A* **25**, 1230 (1982).

[50]   See J. Sucher, in *Proceedings of the NATO Advance Study Institute on Relativistic Effects in Atoms, Molecules and Solids* (G. Malli, ed.), Plenum, New York, 1982; also *Proceedings of Argonne Workshop on the Relativistic Theory of Atomic Structure* (H. G. Berry, K. T. Chem, W. K. Johnson, and Y.-Ki Kim, eds.), ANL-80-116, Argonne National Laboratory, Argonne, IL, 1980. See also M. Mittleman, *Phys. Rev. A* **4**, 893 (1971); *A* **15**, 2395 (1972).

[51]   W. D. Sepp and B. Fricke, in *Atomic Theory Workshop on Relativistic and QED Effects in Heavy Atoms*, AIP Conf. 136 (H. Kelly and Y.-Ki Kim, eds.), pp. 20–25, American Institute of Physics, New York, 1985.

[52]   C. C. J. Roothan, *Rev. Mod. Phys.* **32**, 179 (1960).

[53]   S. P. Goldman and A. Dalgarno, *Phys. Rev. Lett.* **57**, 408 (1988).

[54]   P. J. C. Airts and W. C. Nieuwport, *Chem. Phys. Lett.* **113**, 165 (1985).

[55]   G. Torboem, B. Fricke, and A. Rosin, *Phys. Rev. A* **31**, 2038–2053 (1985).

[56]   K. N. Huang, Y.-Ki Kim, K. T. Cheng, and J.-P. Desclaux, *Phys. Rev. Lett.* **48**, 1245 (1982).

[57]   Y. S. Lee, W. C. Krumler, and K. S. Pitzer, *J. Chem. Phys.* **67**, 5861 (1977); **69**, 976 (1978); **70**, 288–293 (1979); **73**, 360 (1980); **74**, 1162 (1981).

[58]   A. Zangwill and D. Liberman, *Comp. Phys. Commun.* **37**, 75–82 (1984).

[59]   V. D. Berestetskii, E. M. Lifschitz, and L. P. Pitaerskii, *Relativistic Quantum Theory*, Vol. 1, pp. 113–115. Pergamon Press, New York, 1971.

[60]   M. E. Rose, *Relativistic Electron Theory*, p. 82ff, Wiley, New York, 1961.

[61]   W. Perger, private communication, Michigan Technological University. Accepted by *J. Comp. Phys.* (1990).

[62]   Pekka Pyykko, in *Relativistic Theory of Atoms and Molecules*, Lecture Notes in Chemistry **41**, Springer-Verlag, New York, 1986.

[63]   Ian Grant and H. M. Quiney, *Adv. At. Mol. Phys.* **23**, 37–86 (1988).

Atomic Trapping and Cooling

P. van der Straten and H. Metcalf

Optical Forces

The notion of optical forces goes back to Maxwell, but their modern implementation for laser cooling is most commonly described in terms of the momentum of light when it is absorbed by an atom making a discrete transition between states whose energy difference is ΔE. The magnitude of this momentum exchange is related to the energy through the relativistic formula $p = \Delta E/c = h\nu/c = \hbar k$ where $k \equiv 2\pi/\lambda$ and λ is the wavelength of the light. In order for the momentum exchange between the atom and the light field to be efficient, the light must drive a resonant transition between atomic states, and so must have the right frequency. This leads immediately to a model description that involves only two atomic states, one ground and one excited.

Both absorption and emission exchange nearly the same magnitude of momentum between the atoms and the light, so any net momentum exchange must arise from directionality. At low light intensity I the dominant return to the ground state is through spontaneous emission, and directional exchange is thus implemented because it occurs in random directions. Atoms can only undergo spontaneous emission from their excited states whose lifetime is $\tau \equiv 1/\gamma$, and even under the strongest excitation, they spend no more than 50% of the time in the excited state, so the maximum average rate of spontaneous emission from optically excited atoms is $\gamma/2$ and the maximum force is $\hbar k\gamma/2$. At high intensities the momentum exchange is limited by stimulated emission because the absorption and stimulated emission are both parallel to the laser beam. High-intensity forces are usually produced in the presence of multiple beams so that absorbtion from one can be followed by stimulated emission into the other. The momentum difference between these is imparted to the atoms.

In the low intensity domain the spontaneous emission rate is $\gamma_p = (\gamma s/2)/(1 + s + \Delta^2)$. Here $s \equiv (3\lambda^3\tau)/(\pi hc)I$, and Δ is related to the detuning of the light ω_ℓ from exact atomic resonance ω_{atom} by $\Delta = 2(\omega_\ell - \omega_{\text{atom}})/\gamma \equiv 2\delta/\gamma$. Because spontaneous emission is in random directions its average vanishes so the direction of the force is the same as the direction of the light. This "radiative" force is usually written as $\mathbf{F} = \hbar \mathbf{k}\gamma_p$. The optical frequency is measured in the reference frame of the lab which is different from that of the atoms moving at velocity $\mathbf{v}$, and so the Doppler shift $\omega_D \equiv -\mathbf{k} \cdot \mathbf{v}$ must be included in the detuning δ. The negative sign arises because the frequency is increased when $\mathbf{k}$ and $\mathbf{v}$ are in opposite directions. In the atomic rest frame the optical frequency is $\omega_\ell - \mathbf{k} \cdot \mathbf{v}$.

The most common form of laser cooling uses counter-propagating beams of light tuned just below atomic resonance. The formula $\mathbf{F} = \hbar \mathbf{k}\gamma_p$ is applied for each of the two beams, but the force has opposite directions and the Doppler shifts are different. The result is a total force $\mathbf{F}_{\text{tot}} = -\beta\mathbf{v}$ which damps atomic motion in either direction. Here β is a constant that depends on the atomic and laser parameters. The notion of this damping from the radiative force is readily extended to three dimensions, and such a cooling configuration is called "Optical Molasses".

A pure damping force would bring the atoms to rest and thus to the impossible temperature of $T = 0$. Therefore it is necessary to consider the discreteness of each momentum exchange near the cooling limit. The result is an ultimate low temperature of $T_D = \hbar\gamma/2k_B \sim 10^{-4}$ K, where k_B is Boltzmann's constant and the subscript 'D' refers to the Doppler-shift dependence of the mechanism of cooling. Experiments have shown the inadequacy of the two-level atom model that leads to T_D, and in fact, both theory and experiment involving real atoms that have multiple energy levels show the ultimate low temperature obtainable by optical cooling in the presence of spontaneous emission is a few times $T_r = \hbar k^2/2Mk_B$ where M is the atomic mass and the subscript 'r' refers to recoil. At $T_r \sim$ few μK the atomic de Broglie wavelength is comparable to λ, and this has important consequences for further cooling.

When atoms interact with nearly-resonant light, they not only absorb its energy and momentum, but they experience also shifts of their energy levels given by $\Delta E_{LS} = \frac{\hbar}{2}\{\sqrt{\Omega^2 + \delta^2} - \delta\}$ where the Rabi frequency Ω characterizes the strength of interaction between the atoms and the light such that $\Omega^2 = s\gamma^2/2$. In the limit of $\delta \gg \Omega$ that characterizes many low intensity experiments, $\Delta E_{LS} \approx \Omega^2/4\delta$ is proportional to the light intensity and is therefore called the light shift. This light shift can result in forces on atoms when the light intensity varies in space, such as between the nodes and anti-nodes of a standing wave, because the spatial energy dependence can be viewed as a potential. In some sense, it derives from multiple sequences of absorption and stimulated emission, and is therefore conservative and cannot be used for cooling, just as is the case for any force derived from a potential. Such interactions are labeled the "dipole force" in analogy to the static case, because the light induces an atomic dipole moment, and the atom is in an inhomogeneous field.

Thus optical forces on atoms arise from absorption followed by either spontaneous or stimulated emission, and in general, both processes take place. Atoms can be confined or steered by the dipole force, and cooled by the radiative force, thereby providing physicists with enormously powerful and flexible tools for controlling atomic motion.

Confinement – Light Beams and Magnetic Fields

The sign of the the light shift depends on the detuning. For $\delta < 0$, called "red" detuning, the energy of ground state atoms is lower in more intense light and thus atoms in an inhomogeneous light field are attracted to regions of high intensity. The region near the focus of a single laser beam therefore provides a radial attraction for atoms via the dipole force. If the focus is sufficiently sharp, meaning that the light intensity decreases strongly in either longitudinal direction moving away from the focus, this dipole force can exceed the radiative force that tends to expel the atoms from the focal region longitudinally, especially if the detuning is large enough. Thus a single focussed laser beam having sufficiently large Ω and δ can confine atoms to a quite small region of space. Changing the position of the focus by steering the beam allows atoms to be manipulated as if they were held by tweezers. Such "optical tweezers" have been used on single atoms and on Bose–Einstein condensates, as well as on macroscopic objects.

By contrast, ground state atoms are repelled from the region of high light intensity if the detuning is "blue", $\delta > 0$. Atomic mirrors have been made by focussing a blue detuned laser beam into a sheet of light with cylindrical optics, and a trap has been made with an array of such sheets of light. Atomic mirrors have also been made using the very thin film of light produced by a blue detuned evanescent wave near the surface of a flat piece of glass illuminated from the inside with a beam of light near the critical angle. Finally, blue detuned Laguerre–Gaussian beams with a hollow center have been used to confine atoms in two dimensions and thus guide their motion along the path of the light beam. Two orthogonal, separated sheets of light can cap such an elongated region making a quasi one dimensional trap.

Many new phenomena appear in multiple laser beams because there can be absorption from one beam and stimulated emission into the other. Optical molasses described above is achieved in counter-propagating beams having $\delta \sim \gamma$ and $s \sim 1$ so that spontaneous emission is more likely than stimulated emission. Under these conditions, an atomic sample is cooled. By contrast, in the parameter range $\delta \gg \gamma$ and $s \gg 1$ satisfying $\delta \gg \Omega$, the dominant interaction is the dipole force. In a standing wave, this dipole force oscillates in space on the wavelength scale thereby forming a periodic potential, and atoms are said to be subject to an optical lattice. (Optical crystal is an inappropriate name except under conditions where all the sites can be filled with exactly one atom.) Atomic motion in such a periodic potential is subject to the well-known conditions described by the Bloch theorem and Bloch wave functions. By careful choices of the mean kinetic energy (temperature) and well depth of the lattice, very many phenomena of condensed matter can studied.

Most atoms have a non-zero magnetic moment in their ground state, and hence can be confined by an inhomogeneous magnetic field. The simplest imaginable case uses a pair of coaxial coils carrying opposite currents to form a quadrupole field. (A single coil cannot work because a local field maximum cannot exist.) The B field at their geometric center is zero, but is non-zero everywhere else in space. Thus atoms whose magnetic moments $\boldsymbol{\mu}$ are properly oriented are attracted to the field zero, and can be confined there if their kinetic energy is less than $\boldsymbol{\mu} \cdot \mathbf{B}_{max}$ where $\mathbf{B}_{max}$ is the smallest magnetic field in the vicinity of the coils that constitutes a local maximum. Such magnetic trapping of atoms is free from the disturbing effects of light beams, and can therefore confine atoms with far less disturbance than optical traps. Magnetic traps are widely used for the containment of ultra-cold samples of gas, and particularly Bose–Einstein condensates.

Perhaps the most widely used atomic confinement method is the magneto-optical trap (MOT). It exploits the selection rules associated with the polarization of light and the magnetic sublevels of atoms by making the absorption of light from multiple beams depends upon atomic position in an inhomogeneous magnetic field. The field and light beams are carefully arranged so that atoms not in the center of the trap preferentially absorb light from beams that tend to push them back toward the trap center through the radiative force. The MOT has an extremely large velocity capture range and depth, much larger than either pure optical or magnetic traps.

A much more important property, however, is that the force in a MOT is velocity dependent through the Doppler shift, similar to the force in optical molasses. Both magnetic and optical traps are conservative, that is, atoms entering from one side will readily pass through them and escape from the other side. They can be loaded only by applying them to an already cooled sample of atoms, or by applying a cooling force to atoms traveling in them. By contrast, a MOT tends to slow atoms traversing it so that they cannot escape, and therefore it can capture atoms without auxiliary cooling. Moreover, their capture range can be chosen by the laser and magnetic parameters, and can be changed after a sample is captured to enable further cooling and/or compression. Finally, the magnetic field configuration is the same as that of the quadrupole magnetic trap, so that a MOT can be converted to a magnetic trap simply by shutting off the laser beams, thereby transferring a captured sample from the MOT into a dark and cold purely magnetic trap.

Applications

Although laser cooling has first been discussed in relation with high-resolution atomic spectroscopy, today it is used in many areas of atomic physics. One of the most prominent applications is the research of atomic collisions, where the researchers have been able to study atomic interactions at very low energies. This has led to photoassociation spectroscopy, where during the interaction between two ultracold atoms a photon is absorbed and the total system is bound in a transient molecular state. Since the translational energy in the initial state is very low, this offers a novel way of studying molecular states with very high resolution. One of the results of these studies is the measurement of the scattering length in the ground state, which can be measured for with unprecedented resolution. This information is of crucial importance for the achievement of Bose–Einstein condensation in such systems.

For sufficient low energy and high densities atomic gases can make a transition to a new state, the so-called Bose–Einstein condensation. Although this has been predicted already in 1925 by S. N. Bose and A. Einstein, its experimental realization in dilute atomic gases was first realized in 1995 using laser cooling and trapping in combination with evaporative cooling. If the atomic gas makes a transition to a condensate, all atoms are in an identical state and have to be described by one macroscopic wavefunction. The coherence properties of such a condensate can be exploited for many novel experiments, like superfluidity, quantized vortices, parametric down-conversion, collapsing condensates and matter–wave interferometry.

The study of Bose–Einstein condensation forms a bridge between the research in atomic physics on laser cooled atoms and many other fields in physics, like low-temperature physics, condensed-matter physics and statistical physics. For example, loading a Bose–Einstein condensate in an optical lattice make it possible to study in this atomic system the quantum phase transition between a superfluid and Mott insulator state. In analogy with electron conduction in condensed matter, the conduction of atoms in an optical lattice can be tuned by increasing or decreasing the optical potential for the atoms, leading to an "conduction" state, where the atoms are distributed over the lattice in a random way, or an "insulator" state, where the atoms are evenly distributed over the lattice sites, thus creating an integer filling of all sites. It is shown, that this transition can be crossed many times without loosing the coherence of the atoms in the superfluid state.

See also: Bose–Einstein Condensation; Cold Atoms and Molecules; Quantum Optics; Ultracold Quantum Gases.

Bibliography

H. J. Metcalf and P. van der Straten, *Laser Cooling and Trapping*. Springer, New York, 1999.
C. J. Pethick and H. Smith, *Bose–Einstein Condensation in Dilute Gases*. Cambridge University Press, Cambridge, 2002.

Atoms

A. P. French

Introduction

Despite all the discoveries that have been made since the beginning of the twentieth century in the fields of nuclear and subnuclear physics, the atom remains the most important type of unitary system in our picture of the physical world. This is in part because the electrically neutral, stable atom is the basic building block in the structure of condensed matter as we are most familiar with it. Under more severe conditions, e. g., at the enormously high temperatures and densities characteristic of stellar interiors, the individual atom ceases to be an identifiable unit of the structure. Nevertheless, whenever conditions permit, the atom will establish its existence because of its property of being the system of least energy and greatest stability that can be formed from its constituent particles. This property guarantees the atom a permanent place in our description of nature.

Early History

The birth of the theory that the basic structure of matter is discrete, not continuous, is usually attributed to the ancient Greeks, in particular Democritus (ca. 420 B.C.). (The name "atom" comes directly from the Greek atomos – indivisible.) However, it was not until about 1800 that the quantitative study of chemical reactions provided evidence that the behavior of bulk matter might indeed be governed by individual processes on a submicroscopic level. In the latter part of the eighteenth century, A. L. Lavoisier found that the total mass was conserved in chemical reactions, whatever changes in form and appearance took place among the reactants, and J. L. Proust showed that every pure chemical compound contains fixed and constant proportions (by weight) of its constituent elements. (The modern concept of element was propounded by Robert Boyle in 1661.) Building on these results, John Dalton in 1808 enunciated a detailed theory of chemical combination, based on the picture that each element is made up of a host of identical atoms and that the formation of a chemical compound from its elements takes place through the formation of "compound atoms" containing a definite (and small) number of atoms of each element.

From the known mass ratios for many different reactions, Dalton was able to suggest values of mass ratios of individual kinds of atoms. However, his scheme led to certain ambiguities and inconsistencies, which were not resolved until it came to be realized that the basic units of a pure element were not necessarily single atoms, but were often compound atoms in the form of diatomic molecules – a hypothesis first put forward by A. Avogadro in 1811, but which met resistance (despite its success in removing internal contradictions from Dalton's theory) because there was no obvious reason why such well-defined compounds of identical atoms should exist.

Relative Atomic Masses

In 1860 an international conference on atomic weights officially adopted the Dalton–Avogadro scheme, and during the succeeding decades a highly accurate tabulation of the relative atomic masses of the elements was built up from the analysis of thousands of different compounds. A natural unit for this scheme of relative atomic masses was the mass of hydrogen, the lightest atom. This was Dalton's choice, but in 1902 the basis was changed to a slightly different value, namely, one sixteenth of the atomic weight of oxygen. (A prime reason for this change was the practical one that oxygen, in contrast to hydrogen, forms stable and tractable compounds with most elements.) However, the discovery of isotopy (that a given element may have atoms of several different characteristic masses) led to a decision, in 1961, to redefine the atomic mass unit as one twelfth of the mass of the particular isotope carbon-12, and to express chemical atomic weights and the masses of individual isotopes as multiples of this unit.

Avogadro's Number and the Atomic Mass Unit

The atomic-molecular theory was developed in the absence of any direct evidence for the granular structure of matter (although the small irregular movements of microscopic particles in liquid suspension, discovered by the botanist Robert Brown in 1827, were at least suggestive). However, if the theory is assumed to be correct, a definite value must exist for the number of atoms in a given mass of material of known composition. In particular, the *Avogadro number*, N_A, is defined as the number of atoms in a mass of elementary substance equal in grams to its (relative) atomic weight: The mass of an individual atom is thus equal to its atomic weight (in grams) divided by N_A, and the atomic mass unit (1 amu) is numerically equal to $1/N_A$.

One early source of information leading to quantitative estimates of N_A was the study of the properties of gases. The atomic-molecular kinetic theory of gases led to a picture of a gas as made up of small particles traveling at high speeds (hundreds of meters per second). The slowness of the processes of diffusion and mixing in gases implied, however, that individual molecules travel only very short distances before suffering changes of direction through collisions. If there are n molecules of diameter d per unit volume, the mean free path (which can be related directly to the observed diffusion rate) is of the order of $1/nd^2$. To obtain separately the values of n and d, we can use the fact that the total volume of n molecules is about nd^3, and will represent the volume occupied by these molecules if they are condensed to the almost incompressible liquid phase. Analysis along these lines indicated that N_A must be of the order of 10^{24}.

Much more precise values of N_A were obtained later by quite different methods. In 1900, Max Planck inferred a value within 3% of the currently accepted figure as a result of his quantum analysis of the continuous radiation spectrum of hot bodies. In 1916, R. A. Millikan's experimental proof that electric charge exists only in integral multiples of a unit, e, led to a value of N_A as given by the ratio F/e, where F, the faraday, is the amount of electric charge associated with the transport of 1 gram-equivalent of a substance in electrolysis. Later, in about 1930, J. A. Bearden and others made determinations of interatomic distances in crystals, through the process of diffraction of x-rays of known wavelength (measured with a ruled grating); from this the number of atoms in a known mass of crystal could be inferred directly.

The currently accepted value of the Avogadro number is 6.02214×10^{23}, from which follows the result

$$1 \, \text{amu} = 1.66054 \times 10^{-24} \, \text{g}$$

Masses and Sizes of Individual Atoms

The relative atomic weights of the known atoms range from about 1 (hydrogen) to 250 (highly unstable transuranic elements). The corresponding absolute masses range from 1.67×10^{-24} g to about 4.1×10^{-22} g. In this range are about 100 different elements (as characterized by the atomic numbers Z in the periodic table) comprising about 300 naturally occurring distinct atomic species – a given atomic species being characterized by (besides its Z value) its *mass number*, A, the integer closest to its relative atomic weight.

The large range of atomic masses is not accompanied by a correspondingly large or systematic variation in size. Atomic radii all lie between about 0.5 and 2.5 Å (10^{-10} m), with no marked increase from the lightest to the heaviest. One of the simplest sources of this informa-

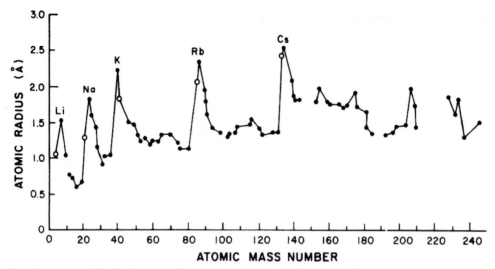

Fig. 1: Atomic radii. Black dots: radius based on experimental analyses of crystals and/or molecules; open circles: radius based on mean-free-path experiments.

tion is a knowledge of the densities of the elements in their solid state; another source is the measurement of cross sections for interatomic collisions. Figure 1 shows how the radii vary with Z; particularly noteworthy are the relatively small radii for the atoms of the noble gases and the especially large radii for the alkali metal atoms that immediately follow them in the periodic table. The general features of Fig. 1 can be understood from the standpoint of the internal structure of atoms (see later).

The Nuclear Atom

The discovery of the electron by J. J. Thomson in 1897 led quickly to the conclusion that all atoms contain some number of electrons. It followed that atoms, being electrically neutral, must be composed of some combination of electrons and positively charged material. Then, during the period 1911–1913, Ernest Rutherford, with H. Geiger and E. Marsden, carried out the alpha-particle scattering experiments that proved that the positive charge and most of the mass of an atom are concentrated in a minute volume, with a radius of less than 10^{-14} m (i. e., less than 10^{-4} of the outer radius of the atom).

Hard on the heels of Rutherford's discovery came the quantum model of the hydrogen atom, published by Niels Bohr in 1913. Using only the known constants of nature (including Planck's constant h), and without the use of any adjustable parameters, Bohr developed an accurate and quantitative model of the hydrogen atom as a nucleus of charge $+e$ (associated with 99.95% of the mass of the whole atom) with a single electron in orbit around it. For the atom in its ground (lowest-energy) state, Bohr's theory gave a radius of 0.53 Å and an ionization energy of 13.6 eV, in good agreement with observation. Furthermore, and perhaps even more striking, his theory accounted naturally for the systematic series of lines in the visible spectrum (Balmer spectrum) of atomic hydrogen. The basis was, according to the theory, that the possible (quantized) energies of an electron bound to a central charge of magnitude Qe are given by

$$E(n) = -\frac{2\pi^2 me^4 Q^2}{h^2} \times \frac{1}{n^2} \qquad (n = 1, 2, 3, \ldots) \tag{1}$$

and that the wavelengths λ of possible spectral lines are defined by the amounts of energy carried away by photons in quantum jumps between levels, using the relations

$$E_{photon} = E(n_1) - E(N_2)$$

and

$$E_{photon} = h\nu = hc/\lambda .$$

X-Rays, Atomic Number, and Nuclear Charge

Bohr's theory had little success in accounting for optical spectra other than those of hydrogen or equally simple systems (e. g., singly ionized helium). However, it was possible to apply the theory to the so-called characteristic x-rays emitted by many elements when bombarded with energetic electrons. By 1913 it was established that these x-rays are electromagnetic radiations similar to visible light but of much shorter wavelength (typically of the order of 1 Å). In 1913 H. G.-J. Moseley made a systematic study of the characteristic x-rays and found that their frequencies ν were just what would be expected, according to the Bohr energy-level formula, for a single electron falling to the lowest ($n = 1$) level from a level with $n = 2$ or $n = 3$. A graph of $\sqrt{\nu}$ against atomic number Z was a pair of straight lines (Fig. 2). Moseley concluded that the characteristic x-rays were produced when an atom returned to its ground state after an electron in its lowest possible orbit, close to the nucleus, had been knocked out. He concluded further that the chemical atomic number could be identified with the number of units of positive charge on the nucleus.

Quantum Mechanics and Atomic Structure

With the development of wave mechanics by E. Schrödinger in 1926, the picture of an electron in the field of a nucleus was drastically changed. In place of well-defined quantized orbits it became necessary to think in terms of smoothly varying probability distributions that permitted the electron to be found at any distance from the nucleus, or even inside it. The quantized energies of the possible bound states duplicated the results of the Bohr theory, but in every other way the description was quite different.

The spatial state of an individual electron could be characterized with the help of three quantum numbers, n, l, and m, all integers. The value of n, called the *principal quantum number*, defined the electron energy, just as in the Bohr theory, but the identification of a state also required the two quantum numbers l and m that defined the orbital angular momentum {of magnitude $[l(l+1)]^{1/2}h/2\pi$} of the electron and the projection, of magnitude $mh/2\pi$, of this orbital angular momentum along a specified axis. The quantum analysis required $0 \leq l << n - 1$, and $-l \leq m \leq +l$.

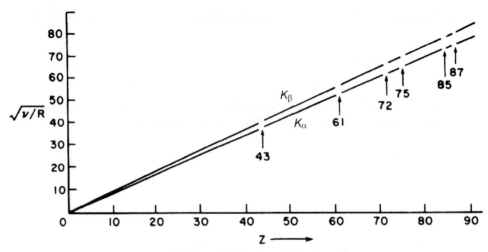

Fig. 2: Linear dependence of $\sqrt{\nu}$ on atomic number for characteristic x-rays. The line marked K_α corresponds to the transitions $n = 2$ to $n = 1$; the line marked K_β corresponds to $n = 3$ to $n = 1$. The graphs show gaps at values of Z corresponding to elements not discovered until later. (The unit of measurement for ν is the rydberg, R, equal to $2\pi^2 me^4/h^3$.)

Added to this was a property first recognized by W. Pauli in 1924: states defined by particular values of n, l, and m are in general split into two components of slightly different energy (as manifested, e. g., in the close doublet structure of many spectral lines). This property, subsequently interpreted as the consequence of an intrinsic spin and associated magnetic moment of the electron (G. Uhlenbeck and S. A. Goudsmit, 1925), meant that the full characterization of the quantum state of an electron in an atom required a total of *four* quantum numbers, the last of which simply took on the values $\pm\frac{1}{2}$ to correspond to the two possible quantized projections ($\pm h/4\pi$) of the spin angular momentum. This allows a total of $2(2l+1)$ different quantum states for given values of n and l.

An essential further ingredient had to be supplied, however. This was the *exclusion principle* of W. Pauli (1924), according to which no two electrons can have the same set of quantum numbers. This principle is the fundamental key to the internal structure of atoms, because it means that the electrons in a many-electron atom cannot all congregate in the lowest energy state ($n = 1$), but are forced to occupy "shells" of progressively increasing energy. We might expect that a given shell would be defined by a particular value of n, since for a *single* electron in the field of a central charge the energy depends only on n. In a many-electron atom, however, the situation is drastically modified by the partial screening of the nuclear charge by the electrons close to it. The innermost electrons, in states with $n = 1$, "see" almost the full nuclear charge Ze, but electrons in states of higher n are exposed to an effective central charge that is less than Ze and depends on l as well as n. Small l, in the quantum-mechanical picture, corresponds to a greater probability for the electron to be near the nucleus and therefore to be exposed to the full nuclear charge. The result of these considerations is to give rise to a fairly well-defined sequence of energy shells, each able to accommodate a certain maximum number of electrons, and corresponding to increasing energy and increasing mean distance from

Table 1: Electron shell structure.

Values of (n, l)	Shell capacity	Cumulative total
(1,0)	2	2
(2,0)+ (2,1)	8	10
(3,0) + (3,1)	8	18
(3,2) + (4,0) + (4,1)	18	36
(4,2) + (5,0) + (5,1)	18	54
(4,3) + (5,2) + (6,0) + (6,1)	32	86

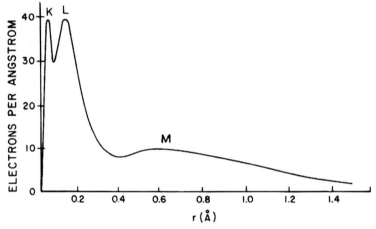

Fig. 3: The radial density distribution for the electron charge cloud in argon atoms, exhibiting the spatial shell structure as inferred from electron scattering experiments. The peaks marked K, L, M correspond to the approximate shell radii for $n = 1$, 2, and 3, respectively. [After L. S. Bartell and L. O. Brockway, *Phys. Rev.* **90**, 833 (1953).]

the nucleus. A partial tabulation showing the shell structure up to $Z = 86$ is shown in Table 1. Completion of a shell leads to a particularly stable and compact structure. The next electron added is relatively weakly bound and can stray quite far from the nucleus. This theoretical model is in good accord with the empirical features of the periodic table of the elements. The atomic numbers $Z = 2$, 10, 18, 36, 54, and 86 are those of the noble gases, He, Ne, A, Kr, Xe, and Rn. The atoms immediately following these ($Z = 3$, 11, 19, etc.) are the alkali metals, which have large atomic radii (cf. Fig. 1) and single, weakly bound valence electrons. Many other less prominent features of the atomic sequence can be understood within the framework of this theoretical model.

The reality of the shell structure has been directly demonstrated in a few cases by probing the atomic structure with x-rays or electrons. Figure 3 shows some experimental results on the radial variation of electron charge density in argon.

General Structure of a Massive Atom

To bring together the results discussed thus far, it may be helpful to consider the complete picture of a particular massive atom with many electrons. Let us take the most abundant isotope of tin, with $Z = 50$, $A = 120$.

The nucleus of this atom has a charge of $+50e$; it contains 50 protons and a number of neutrons equal to $A - Z$, i. e., 70 (but the nuclear structure is not the concern of this article). Closest to the nucleus (on the average) are two electrons in states with $n = 1$; these electrons are exposed to almost the full force of the nuclear electric field. According to Eq. (1), with $n = 1$, $Q = 50$, each electron would be bound with a negative energy equal to almost 2500 times the binding energy of the electron in the hydrogen atom (13.6 eV); this would be of the order of 30 keV. Direct measurements show that it takes x-rays with a quantum energy of 29.2 keV to dislodge an electron from this innermost shell. The average distance of these electrons from the nucleus is comparable to the orbit radius calculated from the Bohr theory for $n = 1$, $Q = 50$; this is about 0.01 Å, or 10^{-12} m – about 150 times the nuclear radius of tin.

Going to the next electron shell (cf. Table 1), we have eight electrons for which $n = 2$ and for which the nuclear charge is significantly shielded, first by the two innermost electrons; second, by the other electrons in this same shell; and third, by penetrating electrons from shells still farther out. The effective central charge is not easy to estimate in this case, but experiment shows that x-rays of about 4 keV can eject electrons from this shell. (Actually, this and subsequent shells have a substructure, based on the involvement of two or more values of the quantum number l, which we shall not go into.) Putting $n = 2$ in Eq. (1), and using the observed electron binding energy of about 4 keV, we would infer an effective central charge of about $35e$ and a mean shell radius of about 0.06 Å.

Proceeding in the same way, we find that the third shell (eight electrons, $n = 3$) has a critical x-ray absorption energy of about 0.9 keV, corresponding to $Q \approx 24$, $r \approx 0.2$ Å; and the fourth shell (18 electrons) has a critical x-ray absorption energy of about 120 eV, corresponding to $Q \approx 12$, $r \approx 0.7$ Å.

This accounts for 36 out of the 50 electrons in the atom. The remaining 14 belong to the fifth electron shell, which can accommodate up to 18 electrons. The situation in this region of the atom is complicated, but we know that these electrons are the main contributors to the outer parts of the atomic charge cloud, which extends to a radius of about 1.5 Å (see Fig. 1). To remove one of these electrons requires an energy of 7.3 eV (the first ionization potential of tin). Thus we see that the energy-level structure within the Sn atom ranges all the way from weakly bound electrons (less than 10 eV) to very tightly bound electrons (tens of keV).

The chemical and spectroscopic characteristics of an element depend on the details of the quantum states of its outermost electrons. In particular, the chemical valence depends on the extent to which the total number of electrons in an atom represents an excess or a deficit with respect to the more stable configuration of a completed shell. In the case of tin, for example, the total of 50 electrons is four short of completing the fifth shell of Table 1. In these terms we can understand why tin is quadrivalent (although it takes a study of finer details to understand in physical terms why it also exhibits divalency). In such matters as these, however, although the properties certainly have their complete basis in electric forces and quantum theory, the theoretical analysis is at best semiempirical.

See also: Atomic Spectroscopy; Atomic Structure Calculations, Electronic Correlation; Atomic Structure Calculations, One-Electron Models; Bohr Theory of Atomic Structure; Elements; Isotopes.

Bibliography

H. A. Boorse and Lloyd Motz, *The World of the Atom*. Basic Books, New York, 1966.

Max Born (tr. J. Dougal), *Atomic Physics*, 8th. rev. ed. Dover Publications, New York, 1989.

A. P. French and Edwin F. Taylor, *Introduction to Quantum Physics*. Norton, New York, 1978.

H. Haken and H. C. Wolf (tr. W. D. Brewer), *The Physics of Atoms and Quanta*, 6th. ed. Springer, New York, 2000.

G. P. Harnwell and W. E. Stephens, *Atomic Physics*. McGraw–Hill, New York, 1955.

G. Herzberg, *Atomic Spectra and Atomic Structure*. Dover, New York, 1945.

Alan Holden, *The Nature of Atoms*. Oxford (Clarendon Press), 1971.

H. G. Kuhn, *Atomic Spectra*. Academic, New York, 1962.

F. K. Richtmeyer, E. H. Kennard, and T. Lauritsen, *Introduction to Modern Physics*, 6th ed. McGraw–Hill, New York, 1969.

Auger Effect

B. Crasemann

An atom that contains an inner-shell vacancy becomes deexcited through a cascade of transitions that are due to two kinds of competing processes: x-ray emission and radiationless , or Auger, transitions. In either process, the original vacancy is filled by an electron from a higher-energy level. In radiative transitions, the released energy is carried off by a photon; in radiationless transitions, this energy is instead transferred through the Coulomb interaction to another atomic electron, which is ejected. The emitted electron is called a K–LL Auger electron, for example, if a K-shell vacancy is filled by an L-shell electron, and another L electron is ejected.

Direct experimental evidence for the existence of radiationless transitions was gained by Pierre Auger through cloud-chamber experiments reported in 1923. X-rays traversing the chamber produced photoelectrons; the tracks of these photoelectrons increased if more energetic x-rays were used. In addition to the photoelectron tracks, Auger observed numerous short tracks, each of which originated at the same point as a photoelectron track (Fig. 1). The length of the short tracks did not change with x-ray energy, but depended only on the kind of gas that was placed in the cloud chamber. Auger was able to show that the length of many of the short tracks corresponded to the energy that would be released in K–LL radiationless transitions. In some cases, Auger found additional tracks, caused by electrons emitted in a second (L–MM or L–MN) step of the radiationless deexcitation cascade of an atom.

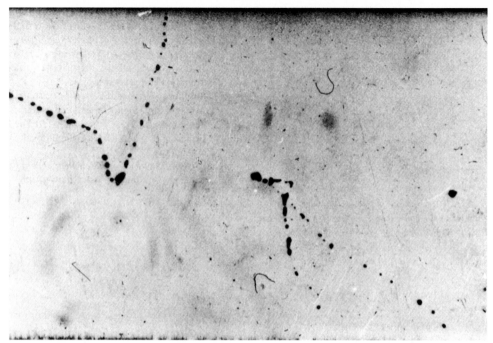

Fig. 1: Photoelectrons and Auger electrons from krypton ionized with 60-keV x rays, photographed in a cloud chamber by Pierre Auger *ca.* 1923. (Courtesy P. Auger).

The quantum-mechanical theory of radiationless transitions was formulated by G. Wentzel in 1927. From perturbation theory, the nonrelativistic matrix element for a direct Auger transition is

$$D = \in \int \psi^*_{n''l''j''}(1)\psi^*_{\infty l_A j_A}(2)\left|\frac{e^2}{\mathbf{r}^2_{12}}\right|\psi_{nlj}(1)\psi_{n'l'j'}(2)\,d\tau_1 d\tau_2 \;,$$

where the quantum numbers n, l, j characterize electrons that are identified schematically in Fig. 2. The state of the continuum (Auger) electron is labeled by $\infty l_A j_A$. In the physically indistinguishable exchange process, described by a matrix element E, the roles of electrons nlj and $n'l'j'$ are interchanged (Fig. 2). The total radiationless transition probability per unit time is

$$w_{fi} = \hbar^{-2}|D-E|^2 \;,$$

if the continuum-electron wave function is normalized so as to correspond to one electron emitted per unit time. The matrix elements D and E can be separated into radial and angular factors. Evaluation of the angular factors depends on a choice of the appropriate angular-momentum coupling scheme. If spin–orbit interaction is neglected, the initial and final two-electron (or two-hole) states can be expressed for different values of the total angular momentum J in the $(LSJM)$ representation of Russell–Saunders coupling. For the heavier atoms, inner-shell states are expressed more realistically in j–j coupling.

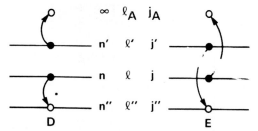

Fig. 2: Energy levels involved in the direct (*D*) and exchange (*E*) Auger processes, and notation for the principal, orbital-angular-momentum, and total-angular-momentum quantum numbers that characterize the pertinent electron states.

The selection rules governing radiationless transitions require that the total angular momentum and the parity of the final-state system (ion plus emitted electron) must be the same as of the initial-state ion. A large number of different transitions is generally possible from any given initial state; for example, 2784 matrix elements are required to describe the radiationless decay of an L_3 vacancy in a high-Z atom.

The relative probability that a K-shell hole is filled by an Auger process, rather than by x-ray emission, ranges from 0.999 for the lightest elements ($Z \leq 5$) to 0.02 for uranium. For vacancies in other shells, the Auger transition rate is generally several orders of magnitude greater than the radiative rate, and thus, essentially determines the lifetime of the vacancy. Auger rates can be very fast. Particularly intense are *Coster–Kronig transitions* that shift a hole to a higher subshell within one major shell. Thus, N_1–$N_{4.5}N_{4.5}$ "super-Coster–Kronig" transitions produce a 4s level width $\Gamma > 60\,\text{eV}$ at $Z = 50$, which corresponds to an N_1 vacancy mean life of $< 10^{-17}\,\text{s}$. The limits of validity of perturbation theory are strained when transition rates of such magnitude are calculated.

The Auger-electron energy is

$$E_{\infty l_A} = E_{n''l''} - E_{nl} - E_{n'l'} \ .$$

The subscripts pertain to the states indicated in Fig. 2; $E_{n''l''}$ and E_{nl} are (absolute values of) neutral-atom binding energies, while $E_{n'l'}$ is the binding energy of $n'l'$ electron *in the presence of an nl vacancy*. The energy of electrons from solid samples can additionally include a contribution from extra-atomic relaxation, typically of the order of $10\,\text{eV}$.

Measurements of electron spectra from radiationless transitions (Fig. 3) and the theory of Auger processes constitute an active field of research, with relevance to fundamental atomic and solid-state theory as well as to surface physics, chemistry, and materials science.

New impetus has been given to Auger spectrometry in recent years with the advent of tunable synchrotron radiation. Selective photoexcitation of atomic subshells near threshold leads to Auger spectra that reveal details of electron rearrangement process, including correlation effects which produce a multifaceted many-electron response to inner-shell ionization. Important new insights into atomic structure and dynamics are being generated through such studies.

See also: Electron Energy States in Solids and Liquids.

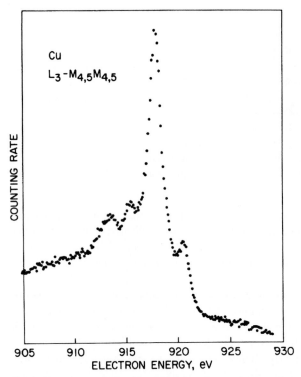

Fig. 3: Energy spectrum of L_3–$M_{4.5}M_{4.5}$ Auger electrons from metallic Cu. The separate peaks result from multiplet splitting due to various couplings of the two final-state holes. (Courtesy Dr. Lo I Yin, NASA Goddard Space Flight Center.)

Bibliography

P. Auger, "The Auger Effect," *Surf. Sci.* **48**, 1 (1975). (E) A first-hand account of the discovery of radiationless transitions.

For up-to-date atomic data see *Photon Interaction Data* (EPDL97); *Electron Interaction Data* (EEDL); *Atomic Relaxation Data* (EADL). Contact for users within the USA: National Nuclear Data Center (NNDC), Brookhaven National Laboratory, Vicki McLane (`services@bnlnd2.dne.bnl.gov`). Contact for users outside the USA: Nuclear Data Section, (NDS), International Atomic Energy Agency (IAEA), Vienna, Austria, Vladimir Pronyaev (`v.pronyaev@iaea.org`).

V. Schmidt, *Electron Spectrometry of Atoms using Synchrotron Radiation*. Cambridge University Press, Cambridge, 1997. A comprehensive treatise.

B. Crasemann, *Can. J. Phys.* **76**, 251 (1998). An overview.

R. H. Pratt, "Some Frontiers of X-Ray/Atom Interactions", in *X-Ray and Inner-Shell Processes*, R.W. Dunford *et al.* (eds.). American Institute of Physics, Melville, 2000; W. Mehlhorn, "Atomic Auger Spectroscopy: Historical Perspective and Recent Highlights", *ibid.*

G. Materlik, C. J. Sparks and K. Fischer (eds.), *Resonant Anmalous X-Ray Scattering – Theory and Applications*. North-Holland, Amsterdam, 1994. Summaries of recent research.

Aurora

E. C. Zipf

An aurora is an often spectacular optical phenomenon that occurs at altitudes above 95 km in the polar regions of the north and south hemispheres. Auroral displays occur most frequently in relatively narrow doughnut-shaped regions which encircle the geomagnetic poles. The Earth rotates underneath these patterns which are fixed with respect to the sun, resulting in a characteristic diurnal vibration in the morphology and temporal behavior of aurora at a particular geographic location. In addition to these geometrical effects, the physical thickness and diameter of the auroral ovals varies considerably with the degree of geomagnetic activity. Enhancements in the magnitude of the solar wind due to solar flares or to a general increase in particle flow from the sun during the normal solar cycle result in the equatorward expansion of the auroral zone which is also accompanied by an increase in the frequency of auroral displays and in their average intensity. The instantaneous position of the auroral zone is determined primarily by the precipitation of electrons with energies in the range 1–20 keV which are guided into the polar regions from the plasma sheet by the Earth's magnetic field lines. As these charged particles descend into the denser regions of the atmosphere, they lose their energy in inelastic collisions with atmospheric atoms and molecules. Most of this energy is consumed in ionizing, dissociating, and heating the atmosphere in the altitude range 95–300 km. Less than 5% of the energy of the primary particles is used to produce the visible radiation for which the aurora is so noteworthy. The plasma densities created by the precipitating electrons are comparable in magnitude to those produced by solar radiation in the normal ionosphere. When auroral activity is unusually intense, still larger ionospheric plasmas are created that cause interruptions in global radio communications (polar blackout).

The most common auroral form is the arc or band which is striking because of its extreme length (hundreds or thousands of kilometers) compared with its width (typically 1–10 km). Auroras are frequently classified in terms of their internal structure: homogeneous or rayed; their apparent motion: active or quiet; their brightness: on a scale from I through IV corresponding to an intensity variation from the brightness of the Milky Way to that of the full moon, respectively; and their visual color: the ordinary green or whitish aurora is designated Type C, while the dramatic veil auroras that fill the entire sky with red light are classified as Type-A aurora.

Three other types of auroral forms deserve special mention. The first is the evening hydrogen arc which is produced as the result of proton and electron precipitation and appears in the form of a broad diffuse arc equatorward of the main portion of the auroral oval during the evening hours, The second is the Polar Cap Absorption event (PCA) which is produced by very energetic solar cosmic rays (1–100 MeV) that enter the atmosphere over the entire polar cap down to a geomagnetic latitude of 60° where the cosmic ray cutoff limits further penetration. PCA events are associated with major flare activity on the sun and give rise to a bright uniform glow over the poles that is often accompanied by a complete radio blackout in the polar region. The third unusual auroral form is the midlatitude red arc (M-arc) which is a subvisual, broad arc elongated in the geomagnetic east–west direction and located generally between geomagnetic latitudes 41° and 60°. M-arcs are approximately 600 km wide in north–south extent and are found at altitudes above 300 km. Their east–west extent is for thousands of kilometers, possibly circling the entire globe. The light emitted from these arcs is essentially monochromatic consisting of two atomic lines with wavelengths of 630.0 and 636.6 nm emitted by metastable oxygen atoms in the ^{2}D state. This unusual spectrum can be contrasted with the variety of features found in normal auroral radiation. These include many molecular bands emitted by N_2, N_2^+, O_2, and O_2^+ as well as more than 100 spectral lines radiated by atomic O and N and their ions.

Our knowledge of auroral morphology and spectroscopy has been enhanced significantly by recent sounding rocket and satellite experiments. The auroral spectrum has been measured quantitatively from the x-ray region (0.1 nm) to the deep infrared (100 µm), and excitation models for the principal emission features have been developed. Satellites with vacuum ultraviolet imagers have also discovered a new type of auroral form: the theta aurora. Theta arcs are aligned in north–south direction and bisect the auroral oval giving it the appearance of the Greek letter "θ" when viewed from afar.

Perhaps the most dramatic auroral phenomenon is the substorm or breakup event. This highly dynamic display develops from a quiet arc and is characterized by intense swirls, surges, and eddies of multicolored light that expand poleward from the original position of the are and, in a matter of minutes, cover most of the sky. These substorms will frequently last 20 min and are accompanied by disturbances in the local magnetic field that indicate the presence of large currents flowing in the auroral ionosphere. The field-aligned electric currents, which flow into the auroral zone from the magnetosphere and are the source of the magnetic effects associated with aurora, are called Birkeland currents. When these generally vertical currents enter the auroral ionosphere, they change direction and begin flowing horizontally in an east-west sense at altitudes near 100 km until they ultimately find their way along other field lines back to the plasma sheet. The horizontal portion of the current, which is often as much as 1000 km long, is called the auroral electrojet, and it carries currents as high as 10^7 A on some occasions. There is some evidence that the auroral substorms are triggered by plasma instabilities in the electrojet that develop and grow dramatically when the current density in the electrojet exceeds a threshold value. A variety of plasma waves are also generated in auroras. These include Alfvén waves as well as electrostatic drift waves. The excitation of these waves provides a mechanism for heating electrons locally to very high temperatures. As these energetic electrons cool, they produce vibrationally excited molecules that can modify the ion and neutral chemistry of the aurora.

Large electric fields (~ 50 mV/m) are occasionally observed along the borders of auroral arcs. These fields produce kinetically energetic ions and contribute to the thermal economy of the upper atmosphere through Joule heating. Some auroras also generate substantial x-ray fluxes with energies of 20 keV or above. These x-rays penetrate down into the stratosphere where they contribute to the formation of nitric oxide. This mechanism is one example of how energy deposited high in the atmosphere (> 90 km) can be coupled into the stratosphere where it can affect the polar ozone budget and may influence long-term climate patterns.

See also: Arcs and Sparks; Atmospheric Physics: Ionosphere; Magnetosphere; Solar Wind.

Bibliography

Joseph W. Chamberlain, *Physics of the Auroras and Airglow*. American Geophysical Union, 1995. A still useful comprehensive text.

A. Valiance Jones, *Auroras*. D. Reidel, Dordrecht, 1974. A comprehensive contemporary text on the physics of auroras.

B. M. McCormac (ed.), *The Radiating Atmosphere*. D. Reidel, Dordrecht, 1971. Emphasis on auroral morphology, particle precipitation, and the aurora as a visual manifestation of large-scale magnetospheric processes.

A. Omholt, *The Optical Aurora*. Springer, New York, 1971. Text emphasizes the emission spectroscopy of aurora.

Balmer Formula

B. G. Segal[†]

When an electric discharge (a spark) is passed through a tube containing hydrogen gas at a few millimeters pressure, it dissociates some of the H_2 molecules into H atoms and the tube is observed to glow with a red light. If the radiation from the discharge tube is passed through a slit, dispersed by a prism, detected by means of a photographic plate, and then analyzed, it is found that there are only four wavelengths emitted in the visible region and many more in the near ultraviolet. These emitted wavelengths, called "lines," since they are images of the slit, appear at different positions on the photographic plate. The four lines in the visible region, arranged in order of decreasing intensity, are observed to have the wavelengths (λ) shown in Table 1. These four wavelengths were measured by the Swedish physicist A. J. Ångstrøm during the second half of the nineteenth century; in addition, several lines in the near ultraviolet were measured by Vogel and by Huggins.

Johann Jacob Balmer, a Swiss physicist, was intrigued by the four discrete lines in the visible region whose wavelengths had been very carefully measured by Ångstrøm. He studied these four values and looked for a pattern, for some relation between them. In 1885 Balmer showed that the wavelengths of the first four lines of the emission spectrum of the H atom are related by the equation

$$\lambda = 3645.6 \left(\frac{n^2}{n^2 - 4} \right)$$

where $n = 3, 4, 5$, and 6, and the result is in angstroms. On the basis of this equation, Balmer predicted that there should be a fifth line, whose wavelength he obtained by setting $n = 7$. He did not know that Vogel and Huggins had measured wavelengths in the near ultraviolet, but on being informed of their measurements, he showed that the wavelengths of the first nine lines of the H atom spectrum all agreed, within experimental error, with his formula, by setting n at all integral values between 3 and 11.

[†] deceased

Encyclopedia of Physics, Third Edition. Edited by George L. Trigg and Rita G. Lerner
Copyright ©2005 WILEY-VCH Verlag GmbH & Co. KGaA, Weinheim
ISBN: 3-527-40554-2

Table 1: Lines in the visible region of the Balmer series.

Name	Color	λ (nm)
H_α	Red	656.28
H_β	Blue-green	486.13
H_γ	Blue-violet	434.05
H_δ	Violet	410.17

Table 2: Line series in the spectrum of atomic hydrogen.

Series	Spectral region	n_L	n_H
Lyman	Far ultraviolet	1	2, 3, 4, …
Balmer	Visible and near ultraviolet	2	3, 4, 5, …
Paschen	Infrared	3	4, 5, 6, …
Brackett	Infrared	4	5, 6, 7, …
Pfund	Infrared	5	6, 7, 8, …

The original Balmer formula is now more usually given as an expression for the wave number, $1/\lambda$, and is

$$\frac{1}{\lambda} = \frac{10^8}{3645.6} \left(\frac{n^2 - 2^2}{n^2} \right) = \frac{4 \cdot 10^5}{3.6456} \left(\frac{1}{2^2} - \frac{1}{n^2} \right) \text{cm}^{-1}$$

or

$$\frac{1}{\lambda} = \mathcal{R} \left(\frac{1}{2^2} - \frac{1}{n^2} \right)$$

where $\mathcal{R}$ is called the Rydberg constant after the Swedish spectroscopist, J. R. Rydberg. The Rydberg constant is one of the most accurately known physical constants and has the value $109\,677.576 \pm 0.012\,\text{cm}^{-1}$.

With the more sensitive emission spectrometers available since the time of Balmer's calculations, other spectral regions have been investigated and many more lines have been detected. All of the observed spectral lines for the H atom fit the relation

$$\frac{1}{\lambda} = \mathcal{R} \left(\frac{1}{n_L^2} - \frac{1}{n_H^2} \right)$$

where n_L and n_H are both integers, with n_H having a value higher than that of n_L. The set of lines for a given value of n_L constitutes a "series" named after its discoverer. Within each series there is a striking pattern: as the wavelength decreases, the spacing between adjacent lines also decreases and the intensity decreases as well. The principal series in the spectrum of atomic hydrogen are shown in Table 2. Each series converges to a limit: as $n_H \to \infty$, $1/\lambda \to \mathcal{R}/n_L^2$.

Balmer's work showed that the emission spectrum of hydrogen has an underlying unifying simplicity, but the reason why the atomic spectrum of H (and of all other atoms) should consist of a discrete set of lines instead of being a continuum, as predicted by classical mechanics, remained a great puzzle until the pioneering theoretical work of Niels Bohr in 1913. By introducing the concept of quantization of the energies allowed for the hydrogen atom, Bohr was able to derive an expression for the Rydberg constant in terms of fundamental constants:

$$\mathcal{R} = 2\pi^2 \mu e^4 / c h^3$$

where e is the charge on the electron; c the speed of light; h Planck's constant; μ the reduced mass of the H atom, $m_e m_p / (m_e + m_p)$; m_e the mass of the electron; and m_p the mass of the proton. The calculated value of the Rydberg constant is $109\,678\,\mathrm{cm}^{-1}$ and the agreement between the theoretical and experimental values was such a triumph for the Bohr theory that it hastened the acceptance of the revolutionary concepts of the quantum theory.

Bibliography

J. J. Balmer, *Ann. Phys. Chem.* **25**, 80 (1885); translated in *The World of the Atom* (H. Boorse and L. Motz, eds.), Vol. 1, p. 365. Basic Books, New York, 1966.

Gerhard Herzberg, *Atomic Spectra and Atomic Structure*, Chapter 1. Dover, New York, 1944.

Gerald W. King, *Spectroscopy and Molecular Structure*, Chapters 1 and 2. Holt, Rinehart & Winston, New York, 1964.

Baryons

D. G. Hitlin

All known elementary (that is, subatomic) particles may be classified into four families: *gauge bosons*, *leptons*, *mesons*, and *baryons*. The gauge bosons are the carriers of the forces which govern the structure of matter; the leptons, mesons, and baryons, names derived from the Greek for light, medium, and heavy, comprise both stable and unstable varieties, some of which are the constituents of atoms and nuclei, while others are produced in high-energy collisions at accelerators or in cosmic ray interactions. The mesons and baryons are further grouped together as *hadrons*, denoting that they interact through the nuclear or strong force. The original mass-derived nomenclature has long since been rendered obsolete, as leptons and mesons more massive than many baryons have been discovered; the basic distinction is now one of internal structure.

Leptons, which are fermions, having half-integral spin, are thought to be truly elementary; that is, they have no known substructure. Mesons, which are bosons, having integral spin, and baryons, which are fermions, have a substructure; they are composed of particular combinations of quarks. It is thus the quarks which are the truly elementary constituents. Mesons are made up of particular combinations of a quark and an antiquark; baryons are composed of three quarks (their antiparticles, antibaryons, are composed of three antiquarks). There are three known *generations* of quarks: (*up, down*), (*charm, strange*), and (*top, bottom*), a total of six *flavors* of quarks. The question of whether there are additional quark generations is under active investigation; a measurement which could settle this question is possible within a few years. All possible combinations of quarks of the same and different flavors can produce a baryon.

The lightest baryon is the proton, the nucleus of the hydrogen atom, which has a mass of 938 MeV. It has traditionally been thought to be stable, but grand unified theories (GUTs) of elementary particles, which seek to find a common origin of the known forces, predict that protons, in fact, decay. Such predictions set off a flurry of experimental activity, which to date has yielded no evidence for proton decay; the current lifetime is known to be greater than 3×10^{32} years, which rules out the simplest GUTs. The next heaviest baryon is the neutron, 1.293 MeV heavier than the proton, discovered by Chadwick in 1932. A free neutron decays via the weak interaction to a proton, an electron, and a neutrino with a lifetime of 896 s. The atomic properties and the bulk of the nuclear properties of matter can be accounted for in detail by the strong and electroweak interactions of protons, neutrons, electrons, and neutrinos. The similarities of the proton and neutron led to the concept of *isospin*, the proton and neutron being viewed as two different states of a single object, the *nucleon*. With the discovery of more massive baryons in cosmic rays by Rochester and Butler and later at accelerators, this concept was eventually generalized, leading to the current picture of the quark substructure of hadrons.

These more massive baryons, the Λ^0, $\Sigma^\pm$, Ξ^-, and Ξ^0, are identified by the characteristic "V" signatures they produce upon their decay. They all have spin $\frac{1}{2}$, as do the proton and neutron, and their predominant decays, $\Lambda^0 \rightarrow p\pi^-$, $\Sigma^\pm \rightarrow n\pi^+$, $\Sigma^0 \rightarrow \Lambda^0\gamma$, and $\Xi^{-,0} \rightarrow \Lambda^0\pi^{-,0}$, with subsequent Λ^0 decay, all result in production of the baryon ground state, a proton or a neutron, accompanied by π mesons, the lightest mesons. These heavier baryons, produced in high-energy strong interactions, decay with lifetimes of the order of 10^{-10} s, which are characteristic of weak processes, instead of lifetimes of $\sim 10^{-23}$ s which are characteristic of strong interactions. The solution to this puzzle was provided by Gell-Mann and Nishijima, who postulated the existence of a new quantum number, *strangeness*, possessed by the heavier baryons but not by the proton and neutron. The strangeness quantum number is conserved in the strong interaction production process, but not in the weak decay process. The observed decay chains also provide for yet another quantum number, *baryon number*, which is conserved both in production and decay. It is the postulated violation of baryon number, a common feature of GUTs, which has engendered the intense interest in searches for proton decay, as the decay of the lowest known baryon state would manifestly be evidence for nonconservation of baryon number.

There is another class of baryons, which decay with lifetimes characteristic of strong interactions, called the *baryon resonances*. The first of these, the $\Delta(1232)$, was originally called the "3,3" resonance, signifying that it simultaneously possessed the novel properties of having both spin and isospin of $\frac{3}{2}$. There are thus four charge states of the $\Delta(1238)$: ++, +, 0, and −. This first baryon resonance has been joined by a host of others. The lowest-mass baryon resonances are bound states of light quarks with nonzero orbital angular momentum. There are also the Y^*, baryon resonances containing strange quarks.

The modern picture of the structure of baryons is the quark model. All known baryons, whether they are stable(?), namely, the proton, or decay via the weak interactions (n, Λ^0, $\Sigma^\pm$, $\Xi^{-,0}$,...), the electromagnetic interaction (the Σ^0), or the strong interaction [$\Delta(1232)$, $N(1520)$, $\Lambda(1405)$] are composed of specific combinations of three quarks, arranged in *multiplets*. This regularity was first identified by Gell-Mann and Ne'eman, using techniques which involved the eight generators of the Lie group $SU(3)$, and was called "the eightfold way." This approach was very successful in organizing the then-known baryon states, allowing the accurate calculation of the masses and decay rates of baryon states. It received striking confirmation with the discovery of the Ω^- baryon by Samios and co-workers. This unusual particle, possessing three units of strangeness and existing in only a single (negative) charge state, had been predicted to exist as the tenth member of the spin-$\frac{3}{2}$ baryon multiplet.

The eightfold way predicted baryon multiplets with up to 27 members, while only smaller groupings, with 8 or 10 members, have been identified. The answer to this puzzle was implicit in the quark model, devised independently by Gell-Mann and Zweig. By postulating that the baryons are composed of three quarks with noninteger electric charge [initially the up (u), down (d), and strange (s) quark, but now extended to include the charmed (c), bottom (b) and the (undiscovered) top (t) quark], only those multiplets with 1, 8, and 10 members, which have actually been found experimentally, appear. The larger multiplets, which contain baryonic states with "exotic" combinations of quantum numbers, do not appear in the quark model, and none have been identified in the more than 25 years since the genesis of the idea. The u, c, and t quarks have charge $+\frac{2}{3}$ (in units of the electron charge, e), while the d, s, and b quarks have charge $-\frac{1}{3}$. Thus baryons composed of three quarks can have charges as large as $+2e$ or as small as $-e$, in accord with observation. Antibaryons, composed of three antiquarks, can have charge states ranging from $-2e$ to $+e$. The proton is composed of (uud) quarks, the neutron of (udd), the λ^0 of (uds) quarks, and so forth. The Ω^-, whose prediction and discovery firmly established the role of $SU(3)$ symmetry in the structure of hadrons, is composed of (sss) quarks, and thus has strangeness -3 and charge $-e$.

There are several known examples of families of baryons which include the heavier quark species. The first of these to be discovered is the lightest of them, the Λ_c^+, which is composed of a (udc) quark combination and which decays via the weak interactions. Several other members of the lowest *charmed baryon* multiplet are also firmly established. No firm evidence has as yet appeared for baryons containing b quarks. The rich spectroscopy of baryons containing heavier quarks has just begun to be explored; much interesting experimental work remains.

While the success of the quark model in explaining the properties of the known mesons and baryons has firmly established the "reality" of quarks, no evidence for free quarks has yet been found. The signature would be striking; a particle with charge $\pm\frac{2}{3}$ or $\pm\frac{1}{3}$. Searches in cosmic rays, in seawater, and in rocks, both terrestrial and from the moon, have yielded no evidence for free quarks. An explanation is to be found in quantum chromodynamics (QCD), which successfully describes the strong interaction as a gauge theory of quarks and *gluons*. The detailed properties of the QCD interaction cause quarks to be bound tightly into hadrons, and do not allow a single quark to be removed.

The culmination of the quark model of baryons is found in its extension to $SU(6)$ symmetry, which describes baryon states in terms of products of an $SU(3)$ portion (the quark content) and an $SU(2)$ portion (the spin degrees of freedom). Those combinations of quark species and spin alignment which produce states which are totally antisymmetric under the interchange of quarks, that is, those combinations which are *fermions*, are identified with physical baryon states. The $SU(6)$ model incorporates the successes of the simple quark model, but in addition, allows the successful prediction of more detailed properties of baryons, such as their magnetic moments.

See also: Elementary Particles in Physics; Hadrons, Hyperons; Quarks; $SU(3)$ and Higher Symmetries.

Beams, Atomic and Molecular

H. A. Shugart

An atomic or molecular beam is a narrow, collision-free stream of electrically neutral atoms or molecules. In experimental practice a high-vacuum system provides the collision-free environment for the beam, while apertures shape its cross section. Since 1911 when it was demonstrated that a stream of atoms remained collimated in a sufficiently good vacuum, beam techniques have contributed to basic understanding in physics, chemistry, and engineering. This success in elucidating fundamental properties of nature results because each atom or molecule in the beam is essentially isolated from the others in the beam, from residual background gas, and from the containment apparatus. Such a "free" state is ideal for studying the properties of an individual particle, and its interactions with other particles, with electric and magnetic fields, or with a surface.

A beam apparatus usually consists of the following components: (1) vacuum enclosure, (2) beam source(s), (3) interaction region(s), and (4) beam detector(s). *The vacuum enclosure* provides the collisionless environment for the beam. In a gas the average distance an atom or molecule will travel between collisions (mean free path) is inversely proportional to the pressure. Whereas this distance is about 4×10^{-7} m in standard atmospheric air, it increases to about 300 m if the pressure is reduced to 10^{-6} mm Hg, a typical maximum operating pressure in beam experiments. Vacuum techniques currently achieve pressures in the range 10^{-8} to 10^{-10} mm Hg so that scattering is negligible in most applications. The *beam source* shown in Fig. 1 consists of an "oven" having a small orifice from which the gas of atoms or molecules emerge. The oven may be the end of a tube for gases or high-vapor-pressure liquids, a heated tantalum or ceramic crucible for metals requiring high-temperature evaporation, a molten spot on a refractory surface heated by electron bombardment, a microwave discharge for dissociating gaseous molecules or for exciting metastable states, or one of the numerous other devices which evaporate and/or excite a substance. At low oven temperatures (mean free path $\geq$ exit slit dimensions) particles effuse from the oven with a modified Maxwell–Boltzmann velocity distribution

$$N(v)\,\mathrm{d}v \approx (v^3/\alpha^4)\exp(-v^2/\alpha^2)\,\mathrm{d}v ,$$

where $N(v)\,\mathrm{d}v$ is the number of atoms with velocity between v and $v + \mathrm{d}v$, $\mathrm{d}v$ is a small increment in velocity, $\alpha = (2kT/m)^{1/2}$ is the most probable velocity, k is Boltzmann's constant, m

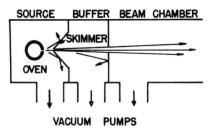

Fig. 1: Diagram of an atomic or molecular beam source. After emerging from a small orifice in the oven, atoms or molecules pass through regions of successively lower pressure until the beam is collision-free.

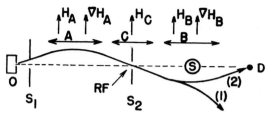

Fig. 2: Schematic of an atomic-beam, magnetic resonance apparatus. An atom leaving the oven O and having an effective magnetic moment $-\mu$, follows path (1) through the inhomogeneous magnetic deflecting fields in regions A and B. If a transition to a state with $+\mu$ is induced by a radio-frequency field (rf) in region C, the atom follows path (2) and reaches the detector D. Stop S intercepts fast atoms which experience small deflections, and apertures S_1 and S_2 collimate the beam.

is the molecular mass of the particles, and T is the absolute temperature. Depending on the values of m and T, the most probable velocity can range from 100 to 10000 m/s. At much higher pressure in the oven the mean free path decreases until collisions between atoms become important and gaseous expansion occurs beyond the nozzle. Under these supersonic jet conditions the beam is much more intense, travels faster, and has a smaller velocity spread than the Maxwell–Boltzmann beam from a low-pressure effusion oven at the same temperature. Because of the larger gas flow from a jet, a well-designed skimmer and buffer stages of differential pumping are needed to isolate the high source-chamber pressure from the low beam-chamber pressure.

Detection of weak beams requires methods which are simultaneously sensitive, efficient, and selective of particular atoms or molecules. (a) The earliest detectors condensed the beam on a cooled surface. The optical density and pattern of the deposit gave both intensity and spatial distribution information. Deposits have also been assayed by neutron activation or by direct counting of nuclear decay when the beam contained a radioactive isotope. (b) For beams consisting of alkali atoms (sodium, potassium rubidium, and cesium) or others with low ionization potential, the detector consists of a hot filament of high work function (tungsten) that produces ions which are counted or measured as a current. (c) A universal detector uses ionization of a beam by electron bombardment with subsequent mass analysis for discrimination against background gas ions. Although such systems have an efficiency of only 10^{-4} to 10^{-5}, they excel in fast response, and are capable of selecting a particular mass species in the beam. (d) Single beam atoms or molecules are selectively detected by fluorescence using monochromatic laser light. This detection method resolves not only different types of atoms or molecules but also the particular quantum state of the particle. (e) Other beam detectors include radiometers, pressure manometers (Pirani gauge), thermopiles, bolometers, and space-charge-limited diodes.

Figure 2 snows the basic elements of a beam apparatus for magnetic resonance detection in atoms, molecules, or clusters of atoms. After preparation in the source chamber, the beam passes through a polarization region A, a resonance region C, and an analysis region B before encountering a detector D. The polarization and analysis regions contain an inhomogeneous magnetic field ($\partial \mathbf{H}/\partial z \neq 0$) which deflects the particles by virtue of the force on their effective magnetic moment $F_z = \mu_{\text{eff}}(\partial H/\partial z))$, where $\mu_{\text{eff}} = -\partial W/\partial H$ is the negative slope of energy W versus magnetic field H. This force produces a deflection which spatially separates atoms with different magnetic moments. In the C region an oscillating or rotating radio-frequency magnetic field of frequency ν induces transitions between levels of energy E_2 and E_1 when the resonant condition $h\nu = E_2 - E_1$ is satisfied (h is Planck's constant). In the B region spatial deflection once again discriminates levels and the detector senses any change of magnetic moment and hence of path induced by the application of the radio-frequency field.

For example, suppose in a magnetic field H an atom has two energy levels, $W = -\mu H$ and $+\mu H$, where the effective magnetic moments are $+\mu$ and $-\mu$, respectively. If the gradients, $\partial H/\partial z$, in the A and B regions are parallel, a typical trajectory of the system remaining in the state $\mu_{\text{eff}} = -\mu$ is shown by (1) in Fig. 2. If a transition to the other state with $\mu_{\text{eff}} = +\mu$ is induced by radio-frequency field ($\nu = 2\mu H/h$, amplitude, polarization, and transit time must be correctly related), the atom will now follow path (2) and reach the detector. Thus an increase in beam flux at the detector is an indication that a change of state occurred in the C region. By suitable arrangement of beam stops and of directions and amplitudes of H_A, H_B, H_C, $\partial H_A/\partial z$, and $\partial H_B/\partial z$, a variety of experimental arrangements permit flexible measurements of various transitions in atoms and molecules. Enormous versatility results when the A and/or B deflecting magnets are replaced by intense light from tunable single-mode lasers. This monochromatic light may deplete or fill (pump) a single energy level in the beam, act as a shutter for time-of-flight or beam modulation, detect a single state through fluorescence, and produce short-lived excited electronic states for study.

The history of atomic beams reveals a succession of important accomplishments. By 1911 vacuum pumps and technology had improved to the point that Dunoyer was able to produce the first collision-free collimated beam of sodium atoms. Beginning in 1919 Otto Stern and collaborators commenced a series of experiments which demonstrated the power, simplicity, and directness of the "molecular-ray method." Using macroscopic apparatus to study isolated neutral atoms or molecules, they experimentally confirmed several fundamental assumptions of contemporary theory. These early experiments (1920), for example, established the Maxwell–Boltzmann velocity distribution predicted by kinetic theory. Using deflection of the silver atom by the force on its magnetic moment in an inhomogeneous magnetic field, Stern and Gerlach (1924) demonstrated the validity of the space quantization of angular momentum and established the electron spin as 1/2. This remarkable experiment supported a basic hypothesis of quantum mechanics and led to studies of electromagnetic properties of other atoms, molecules, and nuclei. In 1929 Knauer and Stern observed the de Broglie wave character of neutral matter [λ(wavelength) $= h$(Planck's constant)$/p$(momentum)] by diffracting a helium beam from a cleaved NaCl crystal surface. Frisch and Stern (1933) and Estermann and Stern (1933) measured the proton magnetic moment to be about 2.5 nuclear magnetons rather than the 1 nuclear magneton expected if it were a simple Dirac particle. In the same year Frisch and Stern observed the radiation-reaction deflection of a beam atom when a photon of light is emitted or absorbed.

High-precision radio-frequency spectroscopy began in 1937 when I. I. Rabi combined the magnetic resonance proposal (Gorter, 1936) with the state selection and analysis capability of an atomic beam apparatus. This development permitted very accurate measurement of differences between energy levels of a system. The frequency ν at which transitions occur between states of energy E_2 and E_1 is given by the resonance condition $h\nu = E_2 - E_1$ mentioned previously. For transitions between states of long lifetime, the frequency width, Δv, of the resonance is determined by the time, Δt, that the system interacts with the radio-frequency field, according to the Heisenberg uncertainty principle $\Delta v \Delta t \geq (4\pi)^{-1}$. Subjecting the beam to a longer interaction region produces narrower resonances for which the peak can be more accurately determined. Using radio-frequency methods Kellogg, Rabi, Ramsey, and Zacharias (1939) established a nonzero quadrupole moment for the deuteron, a discovery which indicated the tensor character of strong forces between nuclear particles. Kusch and Foley (1949) found the anomalous electron magnetic moment. Lamb and Retherford (1947) observed the finite Lamb shift in the atomic hydrogen fine structure. Subsequent refinements in quantum electrodynamic theory have explained both these effects to the accuracy of the experimental measurements. During the 1950s and 1960s a large effort went into the atomic beam study of nuclear spins (I) and of magnetic-dipole (μ) and electric-quadrupole (Q) moments of both stable and radioactive nuclei. This information proved valuable in testing the shell model (1949) and collective model (1953) of nuclei. At the same time, measurements of atomic hyperfine structure constants, which describe the electron–nucleus interaction, and of internal molecular interaction constants contributed to improving the theory of atomic and molecular structure. The most widespread practical application of beam techniques has been in frequency standards employing the cesium "clock" (1952). Presently the hyperfine structure frequency of ^{133}Cs in zero magnetic field is detained to be exactly 9192.631770 MHz, and its resonant detection in an atomic beam provides the current operational standard of time and frequency. Although frequency comparisons are now commonly made to 13 decimal places, it is anticipated that measurements to a few parts in 10^{15} will be soon possible. The ammonia maser (Gordon, Zeiger, and Townes, 1954) and the hydrogen maser (Kleppner, Crampton, and Ramsey, 1960) employ beam techniques for quantum state selection and function as secondary frequency standards.

During the 1970s beam techniques in conjunction with lasers produced significant spectroscopic and chemical reaction data. The decade witnessed developments in isotope separation, in frequency standards, in sources of spin-polarized particles for nuclear accelerators, and in neutral beam injectors for fueling fusion plasmas.

The slowing or cooling of atomic beams with lasers was accomplished in the decade of the 1980s and led to trapping of single neutral atoms in a small volume of space. Velocities corresponding to equivalent temperatures in the few microkelvin range have been obtained. In earlier experiments charged particles in electromagnetic traps were cooled with laser light. Laser cooling occurs when the interaction light is tuned to the "red" or low-frequency side of a resonant absorption transition of the beam atoms. Atoms heading into the laser beam encounter Doppler-shifted radiation nearer the peak of the absorption curve and have a higher probability of interacting than those atoms traveling in the same direction as the laser light, since the latter encounter light Doppler shifted away from the absorption peak. Each time an atom absorbs a photon head-on, its forward momentum is decreased slightly. The subsequent reemission of a photon takes place isotropically (averaged over many absorption–emission cycles) and does not change the average momentum of the atom. After many thousand such absorption–emission cycles the atom's velocity can be reduced to a few centimeters per second. The resulting slow particles have a very small second-order Doppler shift and are amenable to the highest precision spectroscopy and to implementing atomic frequency standards.

In recognition of important advances utilizing atomic or molecular beam methods, Nobel Prizes have been awarded to several scientists: Otto Stern (1943) "for his contribution to the development of the molecular-ray method and for his discovery of the magnetic moment of the proton"; Isidor I. Rabi (1944) "for his application of the resonance method to the measurement of the magnetic properties of atomic nuclei"; Polykarp Kusch (1955) "for his precision determination of the magnetic moment of the electron"; Willis E. Lamb, Jr., (1955) "for his discoveries concerning the fine structure of the hydrogen spectrum"; Charles H. Townes, Nikolai G. Basov, and Alexander M. Prochorov (1964) "for fundamental work in the field of quantum electronics, which has led to the construction of oscillators and amplifiers based on the maser-laser principle"; and Dudley R. Herschbach, Yuan T. Lee, and John C. Polanyi (1986) "for their contributions concerning the dynamics of chemical elementary processes."

See also: Clocks, Atomic and Molecular; Fine and Hyperfine Spectra and Interactions; Kinetic Theory; Kinetics, Chemical; Magnetic Moments; Masers; Microwave Spectroscopy; Optical Pumping; Vacuums and Vacuum Technology; Zeeman and Stark Effects.

Bibliography

W. J. Childs, "Hyperfine and Zeeman Studies of Metastable Atomic States by Atomic-Beam, Magnetic-Resonance," in *Case Studies in Atomic Physics*, Vol. 3, No. 4. North-Holland, Amsterdam, 1973.

T. C. English and J. C. Zorn, "Molecular Beam Spectroscopy," in *Methods of Experimental Physics*, 2nd ed., Vol. 3. Academic Press, New York, 1973.

M. A. D. Fluendy and K. P. Lawley, *Chemical Applications of Molecular Beam Scattering*. Chapman and Hall, London, 1973.

P. Kusch and V. W. Hughes, in *Handbuch der Physik*, Vol. 37/1. Springer, Berlin, 1959.

W. A. Nierenberg, "Nuclear Spins and Static Moments of Radioactive Isotopes," *Ann. Rev. Nucl. Sci.* **7**, 349 (1957).

Laser-cooled and trapped atoms/Workshop on Spectroscopic Applications of Slow Atomic Beams, National Bureau of Standards, Gaithersburg, MD; W. D. Phillips (ed.). Oxford; New York, Pergamon Press, 1984. Series title: *Progress in Quantum Electronics* Vol. 8, No. 3/4.

N. F. Ramsey, *Molecular Beams*. Oxford University Press, London, 1963 (reprinted).

K. F. Smith, *Molecular Beams*. Methuen, London, 1955.

Beta Decay

D. E. Alburger

In a chart of the elements the naturally occuring atoms lie along the so-called valley of nuclear stability. If a nucleus, either existing naturally or produced artificially, has too many or too few neutrons, it will have an excess energy with respect to neighboring nuclei in the valley. Such a nucleus is said to be radioactive and it rids itself of the excess energy by undergoing a spontaneous transformation in which either alpha particles or beta rays are generally emitted. Beta decay is the process in which the radioactive nucleus changes into a neighboring element differing in atomic number, or nuclear electrical charge, by +1 or −1, but with no change in atomic weight. (There is actually a small decrease in the mass because of the energy released.) The relationship is referred to as a mother-to-daughter decay.

Before their nature was understood, beta rays were so called because their power of penetrating through matter was intermediate as compared with alpha rays (now called alpha particles and known to be helium nuclei), which penetrated through very little matter, and gamma rays (now known to be electromagnetic radiation or high-energy photons), which penetrated through large thicknesses of matter. It was shown that beta rays were the most strongly deflected by a magnetic field and that the sign of their electric charge was negative. From their measured charge and their ratio of charge to mass it was demonstrated that beta rays, designated by β^-, were not rays at all but were identical to ordinary electrons. In spite of this the term beta ray is still commonly used.

The simple picture might have been that beta decay occurs when an electron residing inside the nucleus comes out, with a resulting change of +1 in the charge on the residual nucleus. It is now known, however, that nuclei are composed only of protons and neutrons. Furthermore, it has been shown by calorimetric methods that only about 40% of the energy that is given off by a radioactive beta-decaying sample appears as beta-ray energy, and it has been further demonstrated that an individual beta ray can emerge with any energy from nearly zero up to the maximum possible allowed by the mother–daughter mass difference. This continuous intensity distribution has a maximum at somewhat less than half the maximum energy. Another feature is that the beta ray has a spin of $\frac{1}{2}$ unit, whereas the nuclear states connected by beta decay always differ from each other by either zero or an integral number of units of angular momentum.

The puzzle as to where the major portion of the beta-decay energy was going and why angular momentum was apparently not being conserved was solved by Pauli, who postulated that another particle is emitted simultaneously but is undetected in the beta-decay process. Not knowing its properties, he called this particle the neutrino, or little neutron (as discussed in the following, it is actually an antineutrino $\bar{\nu}$ that accompanies β^- emission). The neutrino was presumed to have a spin of $\frac{1}{2}$ unit so as to conserve the total angular momentum of the system and to take away, on the average, the missing 60% of the decay energy. After the neutrino had been shown experimentally to have no measurable charge or mass and an extremely high penetrating power, it was eventually detected by means of a large and complex apparatus.

Beta decay, as described theoretically by Fermi and as now understood, occurs when one of the neutrons in a radioactive nucleus transforms into a proton with the creation and simultaneous emission of a negative beta ray (moving with a speed approaching the velocity of light) and an antineutrino (moving at the velocity of light). In analogy to the atomic case, where a photon or light wave does not exist before an excited atomic electron makes an orbital transition, thereby creating the photon, the beta ray and antineutrino do not exist prior to the beta-decay event. Thus the beta ray and antineutrino emerge with their spins oppositely directed and the sum of their energies is exactly equal to the difference in the mother–daughter energies. They can share the energy in all possible divisions, resulting in the bell-shaped distribution of intensity versus energy for the beta rays.

The discovery of the positron (or positively charged beta ray) in 1932 confirmed Dirac's theory, according to which the electron can exist in an antimatter form. The first observation of a positron, or β^+, decay came about through the formation of the isotope ^{13}N, a positron activity produced in a nuclear reaction in which elemental boron was exposed to a radioactive source emitting alpha particles. It was not until particle accelerators were invented that the means became available to easily produce neutron-deficient radioactive nuclei on the other side of the valley of stability. These nuclei are the positron emitters that, on decaying, change the atomic number of the nucleus by -1. The theory requires that positrons be accompanied by the simultaneous emission of a neutrino and that all the same requirements for conservation of energy and angular momentum as in β^- decay be fulfilled.

An alternative process to positron–neutrino emission, one that also changes the atomic number by -1, is the capture of an orbital atomic electron by the nucleus (electron capture, EC) in the course of the internal transformation of a proton into a neutron. In this case the orbital electron disappears but a neutrino is still emitted. We now include in the term beta decay all three of the processes described thus far: β^-–$\bar{\nu}$, β^+–ν, and EC–ν.

When a β^- particle is emitted it is found experimentally that its spin vector points backward with respect to the direction of motion, a situation referred to as negative helicity; β^+ particles, on the other hand, emerge with positive helicity. This has led to the conclusion that in the weak beta-decay interaction occurring in the nucleus, parity (see below) is not conserved, since otherwise we would expect β^- and β^+ emissions each to result in particles of both positive and negative helicity.

As with all types of radioactivity, the most striking feature of beta decay is the exponential rate of decay, characterized by a half-life that can vary all the way from 0.009 s for ^{13}O to 6×10^4 y for ^{115}In. The free neutron, which is the simplest beta-ray emitter of all, decays into a proton and has a half-life of 12 min. For a given radioactivity the decay constant λ, a number inversely proportional to the half-life $t_{1/2}$, is a precisely fixed quantity that, except in a few very unusual cases of electron-capture decay, cannot be altered at all by laboratory measuring conditions. The value for the half-life is fixed mainly by the energy available for the decay (the higher the energy, the shorter the half-life) and by the relationship between the properties of the mother and daughter states. In Fermi's theory and in later theoretical work by Gamow and Teller, selection rules were developed according to which a beta decay can take place most easily if the spins of the mother and daughter nuclei do not differ by more than one unit, and if the parities are the same. Parity is a symmetry property of a nucleus and is always described as either even ($+$) or odd ($-$). Beta-ray transitions of this type are called *allowed*. If the spin difference is greater than one unit and/or the parity changes, the transition is called *forbidden*, which is really a misnomer and should be taken to mean inhibited. There are various degrees of forbiddenness, depending on the amount of spin change: first-forbidden, second-forbidden, etc., and for each higher degree of forbiddenness, other things being equal, the half-life is longer by a factor of ~ 1000.

Beta decay can take place not only to the lowest-energy or ground state of the daughter nucleus but to excited states as well. Each such decay is characterized by a branching ratio and a consequent partial half-life, or the half-life that the mother nucleus would have if only that particular decay mode were to take place. Evidently the observed half-life results from the composite effect of all possible partial halflives, or in other words, the overall decay constant λ is the sum of the decay constants of the individual branches, $\lambda_1 + \lambda_2 + \ldots$.

All beta decays, whether they are branches in the decay of a given radioactive nucleus or from radioisotopes in different regions of the table of elements, can be reduced to a common denominator by the use of a function f, which is found in tables and effectively removes the energy and atomic-number dependence of the decay. The product $ft_{1/2}$, is called the comparative half-life and is expressed in seconds. Since this number can vary from $\sim 10^3$ to $\sim 10^4$ s for most radioisotopes, it is more convenient to discuss the decay in terms of the $\log_{10} ft_{1/2}$.

The study of beta decay is made partly to learn about the properties of nuclear states. Thus it is known that the $\log ft$ value depends on the degree of forbiddenness, which in turn places restrictions on the spin and parity changes in the decay. The shape of the beta-ray intensity versus energy distribution can also be measurably different in forbidden decay, and the departure from the "allowed" shape can sometimes be predicted theoretically. Measurements of the shapes of beta-ray spectra can best be carried out with several types of magnetic spectrometer. In order to make the shape analysis easier to see, the bell-shaped intensity distribution can be converted, by use of the Fermi function F, into a plot, the so-called Fermi–Kurie plot, which for allowed beta-ray spectra is a linearly decreasing function of energy. Not only can departures from the allowed shape be seen more clearly in such a plot, but the extrapolation of the plot to the point of zero intensity is an accurate measure of the maximum energy of the beta rays. The total decay energy obtained from the end point of a ground-state decay, or for a beta decay to an excited state with the subsequent gamma-ray energy added, is essentially a measure of the mother–daughter mass difference.

Although the general features of beta decay are fairly well understood, there are many important details that are continuing to occupy the attention of both theorists and experimentalists. For example, attempts to predict accurately the $\log ft$ values for beta decay are being made on the basis of theories such as the shell model of the nucleus, but these calculations are thus far only partially successful. Intensive efforts are still being made to understand the exact nature of the weak interaction responsible for beta decay, and experiments of increasingly greater precision are being carried out to aid in these studies.

See also: Neutrino; Parity; Radioactivity; Weak Interactions.

Bibliography

E. J. Konopinski and M. E. Rose, in *Alpha-, Beta- and Gamma-Ray Spectroscopy* (Kai Siegbahn, ed.), Vol. 2, Chap. 23. North-Holland, Amsterdam, 1965. (A)

O. Kofoed-Hansen and C. J. Christensen, in *Handbuch der Physik* (S. Flügge, ed.), Vol. 41/2. Springer-Verlag, Berlin and New York, 1962. (I)

J. M. Soper, in *Encyclopedic Dictionary of Physics*, Vol. 1, p. 394. Pergamon, New York, 1961. (E)

R. D. Evans, in *McGraw–Hill Encyclopedia of Science and Technology*, Vol. 11, p. 286. McGraw–Hill, New York, 1971. (E)

Encyclopaedia Britannica, Vol. I, p. 1028. Encyclopaedia Britannica, Chicago, 1973. (E)

Van Nostrand's Scientific Encyclopedia, 4th ed., p. 198. Van Nostrand, Princeton, 1968. (E)

Betatron

D. W. Kerst[†]

The name betatron is applied to a device for accelerating charged particles by continuously increasing the magnetic flux within the particle's orbit. Since the method seemed especially suitable for the acceleration of electrons, the word beta was incorporated in the name of the accelerator. The electron starts executing an orbit in the magnetic field while a time-varying flux within the orbit provides a voltage gain (*see* Faraday's Law) for each revolution which the electron makes in its orbit. The action is the same as that in an electrical transformer for which the secondary voltage is proportional to the number of turns of wire.

To enable the electron to circulate around the enclosed flux approximately a million times a typical electron must closely follow a path in a well-evacuated vessel. For betatrons operating with a time-varying magnetic field of 180 Hz a vacuum better than 10^{-5} mbar is required, and a scheme of stably confining the particle to its lengthy orbit within the vacuum vessel is provided by shaped magnetic fields which produce so-called betatron or orbital oscillations.

In betatrons used for practical applications there are typically 100 eV of energy gained per turn of the electron around its orbit and in terms of the magnetic bending field at the orbit the particle's energy is $300Br$ MeV, where B is the magnetic field in teslas and r is the orbit radius in meters.

There were numerous attempts to achieve induction acceleration of electrons prior to the first successful achievement of acceleration at the University of Illinois. These attempts have been described in some detail [1]. The necessary theory for injecting particles into a time-varying magnetic field indicated that if electrons were injected from an internal gun with a speed sufficient to avoid serious scattering on background gas molecules, and if the magnetic field varied rapidly enough in time, the electron orbit executed a transient oscillation of decreasing amplitude about an equilibrium orbit. The decreasing amplitude allowed the electron to avoid striking the injection structure thereby allowing the beam to be captured [2] for its long journey.

The axial and vertical focusing forces which cause the electrons to oscillate around the equilibrium orbit are produced by shaping the guide field so that it decreases less rapidly in magnitude than $1/r$. Since the centripetal force required to hold the particle in a circular orbit varies as $1/r$, the magnetic force from the shaped field will be too large to hold the electron at a radius larger than the equilibrium radius and too small to hold it in a circle smaller than the equilibrium radius. The particles will then oscillate across the equilibrium orbit with an ever decreasing amplitude as the magnetic restoring forces continue to rise.

For axial focusing, or oscillation about the equilibrium orbit, the decrease of magnetic field with increasing radius provides a curvature of the lines of force or a bulging of the field which supplies a little focusing force directed back to the plane of the orbit. This same slightly bulging field was also used in the cyclotron to provide axial focusing of particles.

A further condition on the magnetic field is necessary if the orbit is to be prevented from continually spiralling into smaller radius as the guide field rises. The particles can be held at a large fixed radius so they enclose a maximum flux change by supplying extra flux to the center of the orbit. This can be seen from the equations for the equilibrium orbit position. We have

---

[†]deceased

the momentum $p = eBr$, while the electrical field is $d\phi/dt$ divided by $2\pi r$, provided that the particle is following in the orbit of radius r. The rate of gain in momentum is then

$$\frac{dp}{dt} = \frac{e}{2\pi r}\frac{d\phi}{dt} .$$

Integrating this we find that the final momentum is

$$p = \frac{e}{2\pi r}(\phi_2 - \phi_1) .$$

This gives the condition on the change in flux

$$\phi_2 - \phi_1 = 2\pi r^2 B ,$$

and thus the change in flux linkage during acceleration should be twice that which would occur if the magnetic field were uniform within the orbit. The extra flux thus provides the momentum to keep the orbit at large radius as the bending field increases. It is evident that this central flux could be biased in a reverse direction at the outset so that the iron transformer core providing this flux change could be operated between a negative value close to saturation and a positive value again close to saturation. This enables the iron core to be efficiently used.

In addition, the application of alternating-gradient focusing, in place of the simple shaped field previously described, can allow dc guide magnets to contain both the initial low-energy orbit and the high-energy orbit without a great change in orbital radius. Then the rising central accelerating flux can bring continuously injected electrons from the low-energy orbit to the high-energy orbit in a stream lasting almost as long as the rising flux. This so-called fixed-field alternating-gradient betatron would thus not have the single accelerated pulse for every magnetic field cycle but rather it would provide a high-powered stream of electrons for perhaps 25% of each cycle.

Typical medical betatrons between 25 and 35 MeV generate x-rays having an intensity of about 100 R/min at 1 m. They can have the accelerated electron beam extracted for therapeutic purposes giving similar dosage rates for similar areas of treatment with about 2 MeV of their energy being absorbed per centimeter of penetration into tissue. X-rays from a 25-MeV electron beam have the property of being the most penetrating in iron for industrial radiography. Photons of higher energy are more strongly absorbed because of increasing electron pair production, while photons of less energy are more strongly absorbed because of Compton scattering.

For nuclear physics research the photons and the electrons have been useful, since with energies of several million electron volts they can photodisintegrate the nucleus. Radiation from the largest betatron for 320 MeV can produce mesons.

The time-average currents of pulses from a medical betatron is very small and approaches only a microampere. However, the circulating electron beam current in the chamber is about 3 A. Numerous attempts to raise this circulating current have included the use of plasma effects to neutralize the space-charge limitation due to beam charge; and subsequently the addition of a toroidal magnetic field parallel to the orbit in the so-called modified betatron. This greatly increases the space charge limit into the 1000-A region if high-energy injection is used and if the limitations from instabilities can be surmounted. Alternating gradient focussing in addition to the longitudinal field has been successful in achieving 1000 A circulating at 20 MeV with injection filling assisted by plasma effects. The control of instabilities is a major problem with the intense currents.

References

[1]   D. W. Kerst, *Nature* **157**, 90 (1946).
[2]   D. W. Kerst and R. Serber, *Phys. Rev.* **60**, 54 (1941); D. W. Kerst, *Phys. Rev.* **60**, 47 (1941).

Bibliography

A. Arnold, P. Bailey, and J. S. Laughlin, *Neurology* **4**, 165 (1953). (I)

H. Gartner, *Strahlentherapie* **96**, 201 (1955); **96**, 378 (1955). (I)

D. Greene, *Br. J. Radiol.* **34**, 129 (1961). (I)

D. Harder, G. Harigel, and K. Schultze, *Strahlentherapie* **115**, 1 (1961). (I)

H. E. Johns, E. K. Darby, R. N. M. Harlam, L. Katz, and E. L. Harrington, *Am. J. Roentgenol. Rad. Ther.* **62**, 257 (1949). (I)

D. W. Kerst, *Rev. Sci. Instru.* **13**, 387 (1942). (I)

D. W. Kerst, *Handbuch der Physik*, Vol. XLIV. Springer, Berlin, 1959.

C. A. Kapetanakos, P. S. Sprangle, D. P. Chemin, S. J. Marsh, and I. Haber, *Phys. Fluids* **26**, 1034 (1983).

Ishizuka, Leslie, Mandelbaum, Fisher, and Rostoker, *IEEE Trans. Nucl. Sci.* **NS-32**, 2727–2729 (1985).

D. Lamarque, *J. Radiol. Electrol.* **32**, 491 (1951). (I)

J. S. Laughlin, *Nucleonics* **8**, 5 (1951). (I)

J. S. Laughlin, J. W. Beattie, J. E. Lindsay, and R. A. Harvey, *Am. J. Roentgenol.* **65**, 787 (1951). (I)

J. S. Laughlin, J. Ovadia, J. W. Beattie, J. W. Henderson, W. J. Harvey, and L. L. Haas, *Radiology* **60**, 165 (1953). (I)

N. Maikoff and M. Sempert, *Acta Radiol. Suppl.* **313**, 95 (1972). (I)

B. Markus, *Strahlentherapie* **97**, 376 (1955). (I)

M. Sempert, *Radiology* **74**, 105 (1960). (I)

F. Wachsmann, *Nippon Acta Radiol.* **23**, 375 (1963). (I)

R. Wideroe, *Arch. Elektrotech.* **21**, 387 (f928). (1)

R. Wideroe, *Strahlentherapie* **35**, 266 (1956). (A)

Bethe–Salpeter Equation

F. Gross

Definition: The Bethe–Salpeter equation [1] is a linear relativistic equation describing the scattering (or bound states) of two off mass-shell particles with a conserved total four-momentum. In more detail, if $M(p, p'; P)$ is the two-body scattering amplitude (or four-point function), P is the total four-momentum of the system, and p and p' are the relative four-momenta of the final and initial particle pair (see Fig. 1), then the Bethe–Salpeter (BS) equation for M is

$$M(p, p'; P) = V(p, p'; P) - \int \frac{d^4 k}{(2\pi)^4} V(p, k; P) \, G(k, P) \, M(k, p'; P) \,, \tag{1}$$

with V the kernel of the equation and G the full two-particle propagator

$$G(k, P) = D_1(k_1) D_2(k_2) \,. \tag{2}$$

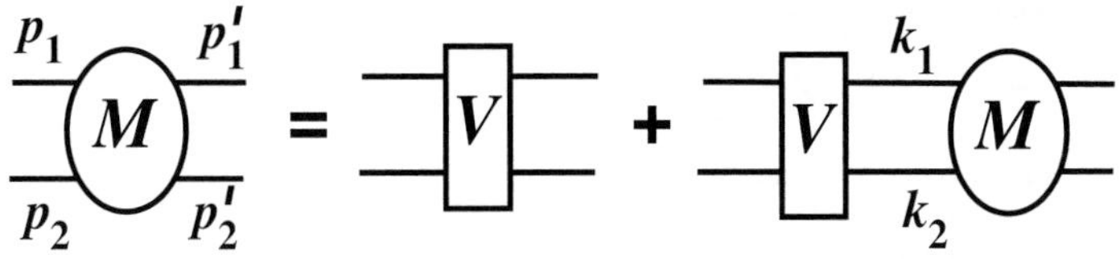

Fig. 1: Bethe–Salpeter equation in diagrammatic form. The single-particle momenta are $p_1 = \frac{1}{2}P + p$ and $p_2 = \frac{1}{2}P - p$ (similarly for p' and k), M is the four-point function, and V is the kernel.

Here $D_i(k_i)$ is the complete one-particle propagator for particle i (with mass m_i) including all self-energy contributions. At all but exceptional points in the region of integration, $k_i^2 \neq m_i^2$, which means the particles are off mass-shell. The essential features which define Eq. (1) as a BS equation are (i) integration over all four components of the relative momentum k of the internal particle pair, (ii) conservation of the total four-momentum P, and (iii) linear form of the equation. Equations for the bound state vertex function and its normalization can be derived directly from (1) without further assumptions [2].

The BS equation gives the exact solution to the scattering problem only if the kernel V is the *exact* irreducible four-point function (i. e. the part of the four-point function that cannot be split into two disconnected diagrams by cutting only two lines) and D is the *exact* dressed propagator. If the kernel is small so that the Born series of iterated irreducible kernels converges, the equation can be viewed as summing this series, and when it is large so that the series diverges, the solution to the equation is the analytic continuation of the Born series.

In general, the BS equation is useful whenever it is essential to evaluate an infinite sum of irreducible kernels. The study of bound states, one of the most important questions in contemporary physics, is such a case. A bound state produces a pole in the scattering matrix in the $s = P^2$ channel. If the bound state is truly composite, no such pole exists in any Feynman diagram (or any finite sum); it can only be generated by an infinite sum. The situation is similar to the generation of the pole at $z = 1$ that arises when the geometric series is summed and analytically continued beyond its region of convergence $|z| < 1$:

$$S = 1 + z + z^2 + z^3 + \cdots = \frac{1}{1 - z}. \tag{3}$$

A unitary description of elastic scattering near threshold also requires an infinite number of diagrams [3].

The ladder approximation

Because the exact irreducible kernel is as difficult to calculate as the amplitude M itself, the kernel is usually approximated by taking the first term of its perturbation expansion [diagram (a) in Fig. 2]. In this case the BS equation sums the ladder diagrams illustrated in Fig. 2. This may be a good approximation when one-particle exchange is believed to describe the important physics, such as photon exchange in atomic physics, gluon exchange in perturbative quantum chromodynamics (QCD), or meson (in particular pion) exchange in low-energy nuclear physics.

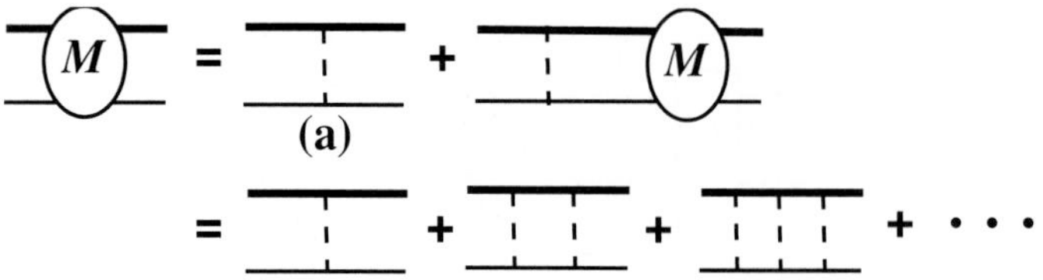

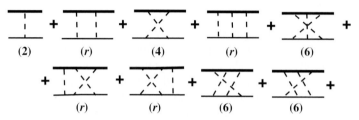

Fig. 2: The ladder approximation to the BS equation. The iteration of the one-boson exchange kernel (a) gives the infinite ladder sum.

Fig. 3: Ladder and crossed ladder diagrams up to sixth order. Diagrams labeled (2), (4), and (6) are second-, fourth-, and sixth-order irreducible contributions to the kernel. Those labeled (*r*) are reducible, and are generated by iterating the kernel.

But, the ladder sum is only an infinitely small subclass of all the Feynman diagrams that contribute to the exact solution. Care must be taken in reading the literature because some investigators give the impression that the ladder approximation to the BS equation is an exact solution to the problem. While this is not the case, the ladder approximation is nevertheless a useful tool.

One technique for obtaining solutions to the BS equation in the ladder approximation is to rotate the k_0 integration contour from the real to the imaginary axis [4] (referred to as a Wick rotation). This converts the relativistic four-dimensional space with indefinite metric into a space with a positive definite metric, allowing the equation to be solved using standard techniques. Exact analytic solutions for the case of spinless particles interacting through the exchange of a massless scalar particle have been obtained [4, 5] by exploiting the symmetry of such a system. While this method is good for calculating bound state masses, the wave functions determined using a Wick rotation have imaginary values of the relative energies p_0 and p'_0, and hence have limited use.

Three-dimensional reductions

Unfortunately, the ladder approximation to the BS equation has some serious limitations. In situations where the physics is dominated by the ladder diagrams of Fig. 2, the crossed ladders (see Fig. 3) are also important. In some theories (ϕ^3 and scalar QED for example) it can be shown [6] that when one of the particle masses becomes very large (let m_1 be the larger mass), the crossed ladders continue to give contributions of the same order in $1/m_1$ as the ladders, and are required in order to obtain the correct one-body equation for the lighter particle (m_2) in the limit as m_1 approaches infinity (referred to as the one-body limit). A similar situation obtains for peripheral interactions at high energy; the eikonal limit cannot be obtained without consideration of crossed ladders [7]. Hence the BS equation will give good nonrelativistic and one-body limits only if its kernel includes irreducible crossed ladder diagrams to all orders, a requirement that is next to impossible to satisfy.

Ironically, this problem can be solved by simplifying the structure of the equation. To maintain relativistic invariance it is not necessary to keep the full four-dimensional integration d^4k. A number of equations, sometimes referred to as quasipotential equations, have been introduced which reduce the internal integration to three dimensions. These equations have an integration volume of the form

$$d^4k \, \delta_+ [f(P,k)] \,, \tag{4}$$

where the function f is a scalar and the δ function therefore eliminates the energy component, k_0, in a covariant fashion. Well-known choices for f are (i) $f = k \cdot P$, which in the rest system fixes $k_0 = 0$ [8], and (ii) $f = m_1^2 - k_1^2$, which restricts the heavier particle to its positive energy mass shell [9]. Alternatively, one may fix the energies k_0 and p_0 in the kernel (usually by setting them to zero) allowing the integration to be reduced to three dimensions. This latter approach gives the Salpeter [10] or equal-time [11] equations. While these equations are new equations with different properties, they are occasionally referred to in the literature as "Bethe–Salpeter" or "Bethe–Salpeter-like" equations, and in this sense the term "Bethe–Salpeter equation" is sometimes used to refer (incorrectly) to any relativistic quantum-mechanical equation.

These simpler quasipotential equations have the remarkable feature that the crossed ladders are suppressed, so that the ladder approximation to these equations already gives the correct one-body and nonrelativistic limits, and this, together with the ease with which they can be solved and interpreted (arising from their three-dimensional nature), makes them better suited for many applications [12, 13].

Applications

The BS equation has been very useful in obtaining exact asymptotic results for bound systems in QCD [14], and in classifying and studying the small-distance and scaling behavior of current-correlation functions, as measured in deep inelastic electron scattering (DIS) [15, 16]. Also it has enjoyed some modest success in developing our understanding of Feynman graph models for the asymptotic behavior of scattering processes at fixed momentum transfer [17]. In the study of electromagnetically bound systems, the equation has been useful in the calculation of relativistic corrections to binding energies and magnetic moments [18, 19]. The equation has been applied to the description of nucleon–nucleon scattering, where it has been solved numerically [20], and to the study of the quark structure of hadrons [21,22].

For example, DIS can be studied by applying the BS equation to the calculation of the forward Compton scattering amplitude (fCSA) for a virtual photon (produced by the electron) with spacelike momentum q ($q^2 < 0$) scattering from a particle of four-momentum p and mass m. The total DIS cross section can be obtained from the imaginary part of the fCSA which in turn can be computed from the BS equation. In this application the BS equation is "crossed" so that four-momentum transfer $\Delta = p - p = 0$ plays the role of P. The DIS cross section depends on only two variables, q^2 and $x = -q^2/2p \cdot q$, and it is found experimentally to "scale" (i. e. to depend only on x) as $-q^2$ becomes large. In most field theory models studied, the scaling law is found [15] to be violated by powers of q^{-2}. In Yang–Mills theories the scaling law is violated [16] by powers of $\ln q^2$. In a somewhat different context, the Bethe–Salpeter equation and the closely related multiperipheral model were used to study the Regge behavior of high energy scattering. Here Eq. (1) is again "crossed," with momentum transfer $\Delta \neq 0$ playing the role of the total momentum P and the energy $s = (p_1 - p_1')^2$ playing the role of the momentum transfer. In ϕ^3 theory, Regge pole behavior s^α, where α is the Regge trajectory, emerges in certain situations [17].

Recently, the BS equation has been used to study quark models of hadrons and the dynamical breaking of chiral symmetry [22]. The dressed quark propagator and the quark antiquark bound states are calculated together using the same kernel. The Dyson–Schwinger equation is used for the propagator and the BS equation for the bound states. This provides a self consistent calculation of both the dynamical quark and bound state masses and shows how the breaking of chiral symmetry can lead simultaneously to a large dynamical quark mass and a very small pion mass.

A complete review of this subject prior to 1969 has been given by Nakanishi [23].

See also: Feynman Diagrams; Quantum Field Theory; Quantum Mechanics.

References

[1]   Y. Nambu, *Prog. Theoret. Phys. (Kyoto)* **5**, 614 (1950); E. E. Salpeter and H. A. Bethe, *Phys. Rev.* **84**, 1232 (1951); M. Gell-Mann and F. E. Low, *Phys. Rev.* **84**, 350 (1951); S. Mandelstam, *Proc. Roy. Soc. (London)*, **A233**, 248 (1955).

[2]   J. Adam, Jr., F. Gross, C. Savkli, and J. W. Van Orden, *Phys. Rev. C* **56**, 641 (1997).

[3]   See the texts by F. Gross, *Relativistic Quantum Mechanics and Field Theory*, Wiley Interscience, New York, 1993, and J. Bjorken and S. Drell, *Relativistic Quantum Mechanics*, Vol. 1, and *Relativistic Quantum Fields*, Vol. 2. McGraw–Hill, New York, 1965.

[4]   G. C. Wick, *Phys. Rev.* **96**, 1124 (1954).

[5]   R. E. Cutkosky, *Phys. Rev.* **96**, 1135 (1954).

[6]   S. Deser, *Phys. Rev.* **99**, 325 (1955); F. Gross, *Phys. Rev. C* **26**, 2203 (1982).

[7]   M. Levy and J. Sucher, *Phys. Rev.* **182**, 1852 (1969); H. D. I. Abarbanel and C. Itzykson, *Phys. Rev. Lett.* **23**, 53 (1969); S. J. Wallace and J. A. McNeil, *Phys. Rev. D* **16**, 3565 (1977).

[8]   A. A. Logunov and A. N. Tavkhelidze, *Nuovo Cim.* **29**, 380 (1963); R. Blankenbecler and R. Sugar, *Phys. Rev.* **142**, 1051 (1966); I. T. Todorov, *Phys. Rev. D* **3**, 2331 (1971).

[9]   F. Gross, *Phys. Rev.* **186**, 1448 (1969); *Phys. Rev. D* **10**, 223 (1974); *Phys. Rev. C* **26**, 2203 (1982).

[10]   E. E. Salpeter, *Phys. Rev.* **87**, 328 (1952).

[11]   D. R. Phillips, S. J. Wallace and N. K. Devine, *Phys. Rev. C* **58**, 2261 (1998); V. Pascalutsa and J. A. Tjon, *Phys. Lett. B* **435**, 245 (1998).

[12]   T. Nieuwenhuis and J. A. Tjon, *Phys. Rev. Lett.* **77**, 814 (1996); C. Savkli, F. Gross, and J. Tjon, *Phys. Lett. B* **531**, 161 (2002),

[13]   See, for example, G. P. Lepage, *Phys. Rev. A* **16**, 863 (1977); G. T. Bodwin, D. R. Yennie, and M. A. Gregorio, *Phys. Rev. Lett.* **41**, 1088 (1978); G. T. Bodwin and D. R. Yennie, *Phys. Rep.* **43**, 267 (1978).

[14]   G. R. Farrar and D. R. Jackson, *Phys. Rev. Lett.* **43**, 246 (1979); S. J. Brodsky and G. P. Lepage, *Phys. Rev. D* **22**, 2157 (1980).

[15]   A. Mueller, *Phys. Rev. D* **9**, 963 (1974).

[16]   D. Gross and F. Wilczek, *Phys. Rev. Lett.* **30**, 1343 (1973); H. Politzer, *Phys. Rev. Lett.* **30**, 1346 (1973).

[17]   M. Baker and I. J. Muzinich, *Phys. Rev.* **132**, 2291 (1963); B. W. Lee and R. F. Sawyer, *Phys. Rev.* **127**, 2266 (1962); D. Amati, A. Stanghellini, and S. Fubini, *Nuovo Cim.* **26**, 896 (1962); D. Z. Freedman and J.-M. Wang, *Phys. Rev.* **153**, 1596 (1967).

[18]   T. Fulton and R. Karplus, *Phys. Rev.* **93**, 1109 (1954); T. Fulton and P. C. Martin, *Phys. Rev.* **95**, 811 (1954).

[19]   S. J. Brodsky and J. R. Primack, *Ann. Phys. (N. Y.)* **52**, 315 (1969).

[20]   J. Fleischer and J. A. Tjon, *Nucl. Phys. B* **84**, 375 (1975); *Phys. Rev. D* **15**, 2537 (1977); **21**, 87 (1980).

[21]   See, for example, A. Le Yaouanc, L. Oliver, S. Ono, O. Pene, and J.-C. Raynal, *Phys. Rev. D* **31**, 137 (1985).

[22]   P. Maris and C. D. Roberts, *Phys. Rev. C* **56**, 3369 (1997); P. Maris and P. C. Tandy, *Phys. Rev. C* **60**, 055214 (1999); H. Munczek and P. Jain, *Phys. Rev. D* **46**, 438 (1992).

[23]   N. Nakanishi, *Prog. Theoret. Phys. (Kyoto)* **43**, 1 (1969).

Binding Energy

N. B. Gove

The energy required to separate an atom of atomic number Z and mass number A into Z atoms of hydrogen (^{1}H) and N neutrons (^{1}n), where $A = Z + N$, is called the binding energy, or BE, for that atom. The BE is thus also the energy equivalent of the mass difference between the hydrogen atoms plus neutrons and the atom:

$$BE(A, Z) = Zm(^1H) + Nm(^1n) - m(A, Z)$$

(nuclear and atomic ground states are assumed here). Since the hydrogen mass, $m(^1H)$, and the neutron mass, $m(^1n)$, are by now well known, the BE can be computed if the atomic mass $m(A, Z)$ is known. The atomic mass is known for about 1700 (A, Z)-species through combinations of experimental data, for example, mass spectrometer doublets, reaction Q values, radioactive decay energies. Table 1 shows a few binding energies (BE) and binding energy per nucleon (BE/A) based on a least-squares adjustment of experimental results.

The binding energy per nucleon shows some small peaks at magic numbers but generally rises to a maximum of 8.79 at $A = 60$ and drops gradually to 8.6 at $A = 100$ and 7.6 for the heavy nuclei. Thus, about 1 MeV/nucleon is released in fission of heavy nuclei.

Two related terms are proton binding energy, $B(p)$, and neutron binding energy, $B(n)$. The former is the net energy required to remove a proton; the latter is the energy required to remove a neutron. (It is generally not possible to extract a proton with zero velocity. The term "net energy" means here the minimum of externally supplied energy that must be converted into internal energy in a proton extraction process.) Thus:

$$B(p) \;=\; m(A-1, Z-1) + m(^1H) - m(A, Z)$$
$$B(n) \;=\; m(A-1, Z) + m(^1n) - m(A, Z)$$

For deuterium, the binding energy, the proton binding energy, and the neutron binding energy are the same. More generally, the binding energy of any bound system is the energy equivalent

Table 1: Binding energies and binding energy per nucleon.

Nucleus	BE (MeV)			BE/A
^{1}n		0		0
^{1}H		0		0
^{2}H	2.224574	$\pm$	0.000006	1.112287
^{3}H	8.481855	$\pm$	0.000013	2.827285
^{4}H	5.66	$\pm$	0.38	1.4
^{4}He	28.295875	$\pm$	0.000026	7.074
^{12}C	92.16239	$\pm$	0.00014	7.680
^{14}C	105.28522	$\pm$	0.00014	7.520
^{16}O	127.62022	$\pm$	0.00015	7.976
^{40}Ca	342.0549	$\pm$	0.0013	8.551
^{58}F	509.9508	$\pm$	0.0016	8.792
^{62}Ni	545.2650	$\pm$	0.0016	8.795
^{100}Ru	861.9338	$\pm$	0.0024	8.619
^{200}Hg	1581.1215	$\pm$	0.005	7.906
^{235}U	1783.881	$\pm$	0.003	7.591

of the difference between the mass of the bound system and the sum of the masses of its free constituents. The expression above for BE(A,Z) treats hydrogen atoms and neutrons as the constituents. Other definitions of constituents are possible. The binding energy of a "bare" nucleus (i.e., fully ionized, no orbital electrons) would differ from the above because of electron energies:

$$\mathrm{BE}_{\mathrm{bare}}(A,Z) = Zm(\mathrm{p}) + Nm(^1\mathrm{n}) - m_{\mathrm{bare}}(A,Z) \, ,$$

where $m(\mathrm{p})$ is the proton mass.

See also: Nuclear Properties; Nuclear Structure; Nucleon.

Bibliography

D. Halliday, *Introductory Nuclear Physics*, 2nd ed., pp. 261–263. Wiley, New York, 1960.
A. H. Wapstra and N. B. Gove, "The 1971 Atomic Mass Evaluation," *Nuclear Data Tables* **9**, 267 (1971).
A. H. Wapstra and G. Audi, "The 1983 Atomic Mass Evaluation," *Nuclear Physics* **A431**, 1 (1985).

Biophysics

R. B. Setlow

Biophysics may be defined as the application of physical techniques and ideas to biological problems. Like any field which borders several disciplines, it is difficult to determine precisely where it begins and ends. On the one hand, it overlaps molecular biology and physical biochemistry, and, on the other hand, cell biology and physiology. An excellent indication of

Table 1: Subject areas in the *Biophysical Journal* (October 2004).

Biophysical Theory and Modeling
Channels, Receptors and Electrical Signaling
Membranes
Nucleic Acids
Photobiophysics
Proteins
Szupramolecular Assemblies
Spectroscopy, Imaging and Other Techniques
Cell Biophysics
Bioenergetics

the breadth of the subject of biophysics is a listing of subject areas in the *Biophysical Journal* (Table 1). Biophysics blossomed as a separate discipline in the 1950s with the rapid influx of physicists, and their techniques, into biology. The early biophysicists, the natural scientists of their day, also worked at the interface between biology and quantitative science. Such early biophysicists were, for example, Galvani, who started the field of electrophysiology; Mayer, who was first to enunciate the law of conservation of energy; and D'Arcy Thompson, who in his momentous book *Growth and Form* beautifully laid out the application of dimensional analysis to biological problems, such as why all animals, including fleas and horses, can jump to approximately the same height. At the present time biophysics – what biophysicists do – can be divided into molecular, cellular, and organismal biophysics. We concentrate in this article on the first two. However, it is worth emphasizing that advances in instrumentation designed to elucidate molecular structures have had important applications in medical physics. General references will be found at the end.

The most important molecules in cells are DNA, RNA, and protein. They are large polymers of subunits called deoxyribonucleotides, ribonucleotides, and amino acids, respectively. The genetic information of cells is in their genes which are represented by sequences in DNA of four possible deoxyribonucleotide subunits containing adenine (A), guanine (G), thymine (T), or cytosine (C). The sequence of these four so-called bases is transcribed by the cellular machinery into RNA which, in turn, is translated with the aid of a number of complex cellular molecules, structures, and organelles to give the sequence of amino acids in proteins. Thus, although the major information of cells is carried by DNA, the day-to-day reactions are carried out by RNA and protein. If we are to understand these reactions and the others catalyzed by enzymes – proteins with specific catalytic activities – it is important that we be able to relate structure of the macromolecules to their function. All of them are large; for example, a typical protein molecular weight may be 6×10^4 (~ 600 amino acids), whereas DNA may range from 10^6 for a small virus to 10^9 or greater for bacteria. Most DNA molecules are found to be double-stranded helical structures (see below) in which the helices are held together by complementary base pairs, A pairing with T and G pairing with C. A DNA of mass 10^9 dalton would contain $\sim 1.5\,M$ base pairs. Since the two DNA strands are complementary to one another, specification of the sequence of one strand sets the sequence of the second. This redundancy in the information carried by DNA is essential if DNA damaged by environmental, physical, or chemical agents, or by endogenous chemical reactions is to be repaired efficiently so that the genetic information is relatively stable. The haploid DNA content of the human cell is $\sim 3\,\mathrm{B}$ base pairs and if all this material contained in the 23 chromosomes were laid end to end, it would extend for $\sim 1\,\mathrm{m}$.

The physical techniques for handling and analyzing large molecules are different from those used to analyze atoms and nuclei. For example, DNA is a long very thin molecule (diameter $\sim 2\,\mathrm{nm}$) and is readily subject to shear degradation by hydrodynamic forces. Thus, it took many years to realize that DNA was such a large molecule and to develop techniques appropriate to handling and carrying out reproducible measurements on it. Large molecules may be considered to have an overall shape (the tertiary structure), the folding of the polymer subunits of which they are composed (the secondary structure), and the order of the subunits themselves (the primary structure). The classical way to analyze the tertiary structure is by measurements of sedimentation in centrifugal fields and by the rate of diffusion. However, by the conventional techniques the latter measurement is laborious and almost impossible for large molecules and because the diffusion constant is small and the experiment times are long, there is a very low signal-to-noise ratio. A solution to the problem is to use laser scattering and measure, in times of the order of minutes, the frequency shift of the laser light resulting from molecular Brownian motion. Here is a perfect example of the application of a new physical technique to solve a biological problem.

The local conformation of these long-chain polymers is also important. For example, DNA is a double helical molecule (see below), but its conformation could be linear, circular, or twisted circular. Such possibilities are not as esoteric as they may seem because the replication of a linear molecule often proceeds by way of a circular one. Moreover, a twisted molecule can be caused to open to a circular molecule by a single nick in one of the two chains. Such a change is easy to detect as an alteration in the frictional coefficient of the molecule, observed by sedimentation or electrophoresis, and hence such twisted circular molecules are often used to measure the deleterious physical effects of radiation or environmental chemicals on DNA. The local conformation of molecules may also be assessed by their ultraviolet absorption (molecules with stacked units tend to absorb less) or by circular dichroism, and the interaction between neighboring groups can be measured by nuclear magnetic resonance by observing the fine structure that results from such interactions. These estimates of local conformation are important because to understand how macromolecules work it is necessary to know both inter- and intramolecular interactions among subgroups. For example, it is desirable to discover how the structure of hemoglobin facilitates its interaction with oxygen, and how subtle changes in such structure affect this interaction, as in the hemoglobin of individuals with sickle-cell anemia. Such structural information with a high degree of resolution is necessary to understand the enormous specificity of the various interactions between large molecules and small (such as enzymes and substrates), and between large and large (such as antibodies and antigens).

X-ray diffraction techniques have been used to determine the detailed three-dimensional structure of a large number of proteins and to demonstrate that DNA is normally a double-stranded helical molecule. In the latter case, however, the x-ray diffraction data were not sufficient, and chemical information was necessary to infer that the molecule was helical with complementary base pairs. The application of x-ray diffraction techniques to macromolecules was more than just an extension of the ideas for small molecules. Not only is it more difficult to crystallize large ones, but new schemes for determining experimentally the phases of the diffracted waves had to be developed in order to analyze the structures. The advent of dedicated synchrotron light sources with high-intensity beam lines for biological research has not only speeded up the rate of data collection from protein crystals by orders of magnitude but, by reducing radiation damage to the crystals, improved resolution. The ability to change wavelength makes it possible to phase the reflections by virtue of the anomalous dispersion of heavy-metal atoms. The use of insertion devices, wigglers and undulators, results in intensity increases suitable to record a Laue diffraction pattern in times of the order of milliseconds, making it possible to investigate dynamic changes in macromolecular conformation in muscle contractility and in the interactions between enzymes and substrates or between controlling proteins and DNA. A big challenge to the effective use of synchrotron radiation is the design and construction of two-dimensional detectors capable of recording data at about 10^8 counts/min.

A potentially important spin-off of synchrotron technology is the possibility of performing coronary angiography with little risk by imaging the heart with monochromatic radiation at the K edge of a contrast agent and subtracting from this image the absorption by background radiation off the K edge. The resulting greatly increased contrast is sufficient to enable the reconstruction of the coronary arteries after the injection of the contrast agent into a peripheral vein.

A complementary technique to x-ray diffraction is neutron diffraction. The technique is especially useful because hydrogen has a large negative scattering power and scatters strongly whereas it cannot be visualized by x-ray diffraction because of its very small scattering cross section. Moreover, by substituting deuterium for hydrogen, the magnitude and sign of the scattering power is changed making it possible to examine a structure in which the scattering has been changed without a change in conformation. The labeling of parts of molecular and cellular organelles with deuterium permits one to detect their positions and conformation readily. The visualization of macromolecules has been enhanced by the use of scanning transmission electron microscopy in which the object is scanned by a small, 2.5-Å diameter, electron beam. The electrons transmitted and scattered by the object are collected quantitatively and permit the estimation of the scattering cross section and hence, with high precision, the atomic mass of the object or a portion of the object from molecular weights of 20 000 on up. It is possible to detect clusters of heavy-metal atoms bound covalently to specific macromolecular binding sites so as to locate these sites on proteins or on nucleic acids within < 8 Å. The techniques discussed above are only a small part of the extensive ones applied to studies of macromolecules. Table 2 lists these techniques.

In addition to the large number of techniques used to determine molecular structures, a number of others are used to alter the characteristics of a molecule, its structure and its function, to gain insight into the way such molecules work. For example, there are a number of so-called relaxation methods in which the temperature or the pH in a molecular or cellular system is changed suddenly. The rate at which the system approaches a new steady state is then measured by one of the techniques of Table 2 and, from this rate, information is gained about the stability of the new and the old molecular conformations. This is one approach used in attempts to determine how proteins – polyamino acids – manage to fold to a unique functional structure rather than just a random nonfunctional one.

A second group of techniques involves measurements of fluorescence and phosphorescence of molecular or cellular systems. From the excitation and the emission spectra, both the absorbers and the emitting subunits may be identified. Measurements of the lifetimes, the

Table 2: Physical techniques used for studying macromolecules.

Technique	Information obtained
Osmotic pressure	M
Diffusion	(L, V, H_2O)
Sedimentation, equilibrium	M, ρ
velocity	(L, V, H_2O), M_{DNA}
Viscosity	(L, H_2O)
viscoelastometry	M_{DNA}
Birefringence, electric	(L, H_2O)
flow	(L, H_2O)
Dispersion, dielectric	(L, H_2O)
optical rotatory	S, H
Light scattering	M, anisotropy, diffusion constant
Light absorption	Composition, S
Dichroism linear	O, C
circular	S, H
magnetic circular	C, anisotropy
Luminescence	Composition, S, O, local environment, energy transfer
Electrophoresis	(charge, L, V, H_2O), M
X-ray scattering	$(L, V), M, O$
diffraction	C, O, atomic coordinates
Neutron scattering	$(L, V), O$
diffraction	C, O, atomic coordinates
Electron microscopy	L, V, M
Electron paramagnetic resonance	Free radicals, reaction mechanisms
Nuclear magnetic resonance	O, S, C, H_2O
Irradiation, ionizing	Sensitive volumes and their location in cells (radiobiology)
nonionizing	Sensitive components in cells, damage to DNA, photosynthesis (photobiology)

M: molecular weight; L: shape; V: volume; H_2O: bound H_2O; ρ: buoyant density; S: stacking of subunits; H: helical content; O: orientation of subunits; C: conformation.

yields, and the fluorescence depolarization make it possible to estimate the proximity of the absorber and the emitter, and to infer any special aspects of the local environment such as hydrophobic or hydrophilic characteristics. Such measurements also give information about excited states, and by inference reaction mechanisms of interest in photobiology. Fluorescent depolarization of reporter groups (fluorescent subunits) has been used to measure the fluidity of membranes.

The effects of ionizations or excitations on biological systems are also a part of biophysics. Studies of ionizing radiation are necessary for quantitative evaluation of its hazards, and they also may yield target volumes of molecules, which can lead to some of the same types of information given by other techniques of Table 2. However, this particular technique is unique in that it can be used not only with impure preparations, but also with complexes of molecules, or even whole cells. Photobiologists not only investigate basic mechanisms in photosynthesis, but also the killing and mutation of cells by ultraviolet radiation of different wavelengths. Such action spectra were one of the first lines of experimental evidence that nucleic acids are the important component of the genetic material. They have also led to the discovery of cellular repair mechanisms that serve to ameliorate most of the deleterious changes made in cellular DNA by endogenous reactions or by radiation or chemicals. The examples given above are some of the clearest ones showing the impact of physical techniques and physical methods of thought on the study of biological systems.

Acknowledgement

This work is supported by the Office of Health and Environmental Research of the U.S. Department of Energy.

See also: Crystallography, X-ray; Diffraction; Fluctuation Phenomena; Luminescence; Neutron Diffraction and Scattering; Nuclear Magnetic Resonance; Optical Activity; Polymers; Radiological Physics; Relaxation Phenomena; Sedimentation and Centrifugation.

Bibliography

Annual Review of Biochemistry **1** (1932) to present.
Annual Review of Biophysics and Bioengineering **1** (1972–1984).
Annual Review of Biophysics and Biophysical Chemistry **14** (1985–1991).
Annual Review of Biophysics and Biomolecular Structure **21** (1992) to present.
Annual Review of Physical Chemistry **1** (1950) to present.
Biophysical Journal **1** (1960) to present.
Progress in Biophysics (and Molecular Biology) **1** (1950) to present.
Quarterly Review of Biophysics **1** (1968) to present.

Black Holes

S. W. Hawking and W. Israel

Gravity is by far the weakest interaction known to physics: The gravitational attraction between two electrons is completely dominated by the electrical repulsion between them, which is about 10^{43} times stronger. Despite their weakness, gravitational forces are much more a matter of everyday experience than electrical forces. The reason is that gravity makes up for its extreme weakness by having a property that no other interaction has, the property of *universality* or *equivalence*, which means that gravity affects the trajectories of all freely moving particles in the same way, no matter what their internal constitution. This characteristic has been verified to an extremely high accuracy both in laboratory experiments and by laser ranging on the moon. Mathematically, the property of universality or equivalence is expressed by saying that gravity couples to the energy-momentum tensor of matter. Together with the fact that quantum mechanics seems to require that the energy density be nonnegative (otherwise the vacuum could spontaneously decay into infinite numbers of positive- and negative-energy particles), this implies that gravity is always attractive. The gravitational fields of all the particles in a large body like the earth therefore add up to produce a significant field at the surface. For the earth this field is still weak compared to the electrical and exclusion-principle forces between atoms, so that the density of the terrestrial material is not much higher than it would be in the absence of gravity. For more massive bodies, however, gravity becomes more and more important, until it can dominate all other forces and give rise to catastrophic collapse inward.

The best-understood way in which such collapse can occur is for stars that are more than a few times the mass of the sun. A star is thought to be formed by condensation out of a cloud of interstellar gas. At first it would contract under its own gravity. As it did so, the temperature at the center would rise until it became high enough to start the thermonuclear reaction that converts hydrogen into helium. The heat generated by this process would create enough thermal pressure to prevent the star from contracting any further. The star would spend the next thousand million years or so in a quasi-stationary state burning hydrogen into helium and radiating the heat into space. Eventually, however, the nuclear fuel would all be used up and the star would begin to contract again. If the star were less than about one and a half times the mass of the sun, the collapse could be halted by degeneracy pressure of electrons or neutrons. In the first case the resultant body is called a white dwarf and has a radius of a few thousand kilometers, whereas in the second case it is called a neutron star and has a radius of the order of 10 km. For stars of more than a couple of solar masses, however, there is no final equilibrium state. Some stars may manage to reduce their masses to below this limit by throwing off material, but it seems virtually certain that this will not occur in all cases. It seems that such stars must continue to shrink until their density becomes infinite, creating what is called a singularity of space-time, a place where the notion of space-time as a continuum or manifold breaks down, as do all the laws of physics because they are formulated on a space-time background.

As the star contracts, the escape velocity from the surface will rise. In a spherical collapse, the escape velocity will become equal to the speed of light when the star reaches the Schwarzschild radius $r_S = 2GM/c^2$ (about $3 \times M/M_\odot$ km). After this, any further light emitted from the star cannot escape to infinity, but is dragged inward by the intense gravitational field. The region of space-time from which light cannot escape to infinity is called a black hole. Its boundary is called the event horizon and is formed by the wavefront of light that just fails to escape to infinity but remains hovering at the Schwarzschild radius. It acts as a sort of one-way membrane: objects can fall into the black hole through the event horizon but nothing can come out of the black hole because the event horizon (as viewed by a local observer in free fall) is moving outward at the speed of light and, according to special relativity, nothing can travel faster than light.

The collapse of a rotating, aspherical star is expected to be qualitatively similar, leading again to formation of a space-time singularity enclosed within a black hole. According to theorems of Penrose and Hawking, a singularity inevitably forms whenever outgoing light waves can no longer expand because of the growing force of gravity. (Essentially the only assumption required is that the energy density of the contracting material remains positive and larger than the sum of the three principal pressures.) That such a singularity will be surrounded by an event horizon which permanently screens it from outside view is a widely accepted, though still unproven, hypothesis, known as "cosmic censorship," whose plausibility is based chiefly on a persistent failure to find persuasive counterexamples. If this is true, then in the wake of the collapse, the horizon and the exterior gravitational field will settle, like a newly formed soap bubble, into the simplest configuration compatible with the external constraints. The end state, known as a Kerr–Newman black hole, is accordingly characterized by just three parameters: mass, angular momentum, and (in principle) charge. Other characteristics originally anchored in the stellar precursor are precipitated into the hole or radiated to infinity. In the words of John Wheeler, "a black hole has no hair."

The "stationary limit" of a black hole is the surface which marks out the innermost set of points at which a physical particle can be at rest with respect to a distant stationary observer while remaining subluminal (i. e., not exceeding the speed of light with respect to a local observer in free fall). It generally lies outside the horizon, but coincides with it for a spherical (nonspinning) black hole. In the intervening layer, called the ergosphere, subluminal orbiting observers, able to communicate with the outside, can still exist but are now forced into partial corotation with the hole. On the horizon itself only a single subluminal observer can exist at each point. He orbits with the speed of light, and his angular speed (the same at all latitudes on the horizon) is called the angular velocity of the hole.

The negative gravitational potential is so large in the ergosphere that a weight can in principle be lowered into this region in such a way as to recover more than its rest-mass energy as work; the extra energy is derived from the rotational energy of the hole. In a realistic astrophysical setting, a black hole's rotational energy can be tapped through hydromagnetic interactions with an ambient plasma. In such interactions, the event horizon behaves much like the surface of a conductor. (For instance, electric field lines threading the hole must cross the horizon orthogonally in the frame of a corotating observer.) Thus a black hole spinning in a uniform paraxial magnetic field anchored in a surrounding accretion disk will behave like a unipolar inductor. Processes of this kind may be responsible for jets and other forms of activity observed in active galactic nuclei.

Astronomical evidence for the existence of black holes is not yet overwhelming, but is steadily growing. A handful of binary x-ray sources in our own galaxy and in the Large Magellanic Cloud are thought (on the basis of more-or-less trustworthy lower bounds for the mass of the unseen component) to contain a black hole of a few solar masses in orbit around a somewhat more massive normal star (*see* Astronomy, X-Ray). On a larger scale, the motion of gas and stars in the central regions of our own and other galaxies, and the detection of rapidly variable γ- or x-ray emission from "hot spots" in these regions, point to the presence of central condensations of a million or so solar masses, which may be black holes. The most plausible source for the highly luminous quasars is accretion onto black holes of up to a billion solar masses.

The event horizon, the boundary of a black hole, has the property that its surface area always increases when more matter or radiation falls into the black hole. If two black holes collide and merge to form a single black hole, the area of the event horizon around the final black hole is greater than the sum of the areas of the two original black holes. There is thus a similarity between the area of an event horizon and the concept of entropy in thermodynamics, with the law of the area increase being the analog of the second law of thermodynamics. There is also a black-hole analog of the first law of thermodynamics, which relates the change in energy (or mass) of a system to the change in entropy (or area). In the thermodynamic case the constant of proportionality is called the temperature. In the black-hole case it turns out to be a quantity called the *surface gravity*, which is a (redshifted) measure of the strength of the gravitational field at the horizon, equal to GM/r_S^2 for an uncharged spherical black hole. An analog of the third law states that the surface gravity cannot be reduced to zero in a finite (advanced) time by injecting material with bounded, positive density and bounded stresses into the hole. (The surface gravity of an uncharged black hole becomes zero when its angular momentum reaches GM^2/c, the maximum possible for a black hole of mass M.)

Although the area of the event horizon and the surface gravity of a black hole have a strong resemblance to entropy and temperature, respectively, it is not possible, in a purely classical theory, to regard some multiples of the area and the surface gravity as the actual entropy and temperature of the black hole. For according to classical theory, a black hole can absorb radiation but cannot emit anything. It therefore could not be in equilibrium with blackbody radiation at any nonzero temperature. It was, nevertheless, suggested by Bekenstein that the entropy S of a black hole is a universal multiple of its area A. This idea was vindicated when Hawking, taking quantum theory into account, proved in 1974 that one must, after all, assign a finite temperature T to a black hole, because it emits particles and radiation by a quantum tunneling process. The Bekenstein–Hawking relations are

$$S = \frac{1}{4}\frac{kc^3}{G\hbar}A\,, \qquad T = \frac{1}{2\pi}\frac{\hbar}{kc}\kappa\,,$$

where κ is the surface gravity. The temperature of a spherical black hole of mass M is $10^{-7}(M/M_\odot)^{-1}\,\mathrm{K}$.

A black hole radiating into empty space will slowly evaporate. Its area, and therefore its entropy, will decrease. However, this decrease is more than compensated by the entropy of the emergent radiation. In the quantum domain the second laws of black-hole dynamics and thermodynamics are subsumed and sublimated into the statement that the total entropy of a black hole and its surroundings can never decrease.

One way in which the quantum-mechanical emission process can be understood is as follows. The uncertainty principle implies that "empty" space is filled with virtual particle–antiparticle pairs that come into existence together at some point of space-time, move apart, and then come together again and annihilate each other. They are called "virtual" because unlike "real" particles they cannot be observed by a particle detector but their indirect effects can be measured in cases such as the Lamb shift in the spectrum of atomic hydrogen and the Casimir effect between two parallel conducting plates. When a black hole is present, one member of a virtual pair may fall into the hole, leaving the other without a partner with which to annihilate. The forsaken particle or antiparticle may follow its mate into the black hole but it may also escape to infinity where it will appear to be radiation emitted by the black hole. An alternative way of looking at the process is to regard the member of the pair that falls into the hole (say an antiparticle) as a particle that is traveling backward in time. It would come out of the black hole and then, when it reached the point where the particle–antiparticle pair first appeared, it would be scattered by the gravitational field into a particle traveling forward in time. In effect, the particle would have quantum mechanically tunneled through the potential barrier represented by the event horizon, a barrier that it would be impossible for it to surmount classically.

The reason this tunneling process gives a thermal spectrum for the emitted particles is that these particles come from the region inside the event horizon, which someone outside the black hole cannot observe directly. According to the "no hair" property, all such a person could measure about this region would be the three quantities mass, angular momentum, and charge. It implies that a priori probabilities to emit all different configurations of particles with the same total energy, angular momentum, and charge are the same. A black hole could indeed emit a television set or a belly dancer but the number of configurations represented by these possibilities is very small. The overwhelming probability is that the radiation will be very nearly thermal because this represents by far the largest number of possible configurations.

That the particles come from a region about which an external observer has only a very limited knowledge means that the emission from black holes has an extra degree of unpredictability or uncertainty over and above that normally associated with quantum mechanics. In classical physics we can predict with certainty both the position and velocity of a particle. In ordinary quantum mechanics we can make definite predictions of either the position or the velocity or some combination of the two. In this case of particles emitted from a black hole, however, we cannot definitely predict any observable quantity. All we can do is deduce the probabilities that particles will be emitted in certain modes. Thus there is associated with black holes an additional degree of randomness that seems to arise from their containing singularities of space-time at which the laws of physics break down. According to some recent speculations, a conceivable way in which this extra randomness might arise is through loss of information channeled through "wormholes" into microscopic regions ("baby universes"), which form by quantum fluctuations of the gravitational field near the singularity, and which subsequently detach themselves from our universe. But here we trespass beyond the borders of present knowledge and understanding.

See also: Astrophysics; Spacetime.

Bibliography

Popular

S. W. Hawking, *A Brief History of Time: From the Big Bang to Black Holes*. Bantam, New York, 1987.

J. P. Luminet, *Black Holes*. Cambridge University Press, Cambridge, 1990.

Elementary

W. Israel, *Sci. Prog. (Oxford)* **68**, 333–363 (1983).

R. D. Blandford, "Astrophysical Black Holes," in *300 Years of Gravitation* (S. W. Hawking and W. Israel, ed.), pp. 277–329. Cambridge University Press, Cambridge, 1987.

Advanced

K. S. Thorne, R. H. Price, and D. A. MacDonald, *Black Holes: The Membrane Paradigm*. Yale University Press, New Haven, CT, 1986.

I. D. Novikov and V. P. Frolov, *Physics of Black Holes*. Kluwer, Boston, 1989.

Blackbody Radiation
R. E. Bedford

Every object in the universe continuously emits and absorbs energy in the form of electro-magnetic radiation. This kind of energy transport differs from the processes of conduction and convection in requiring neither the presence of a medium nor the transport of matter. The radiated energy originates from the internal energy associated with atomic and molecu-lar motion and the accompanying accelerations of electrical charges within the object. When the atoms or molecules are only loosely coupled, as in gases at ordinary temperatures and pressures, they can radiate semi-independently with the emitted energy appearing in discrete spectral lines or bands characteristic of the internal energy-level structures. In opaque solids and liquids, on the other hand, the atoms and molecules interact so strongly with one another that they cannot radiate independently and the emitted energy becomes entirely determined by the object's temperature and surface structure. For this reason it is called *thermal radiation*. Thermal radiation is distributed continuously over the entire electromagnetic spectrum but conventionally is taken to lie within the spectral range from roughly 0.2 to 500 μm. When the object's temperature is between 10 and 5000 K, the radiation is concentrated in the infrared, the region commonly associated with the production of heat.

For an object to maintain thermal equilibrium its rates of emission and absorption of en-ergy must be equal. This equality leads to *Kirchhoff's Law* (established in 1859) that, if $M_\lambda(T)$ (variously called spectral radiant emittance, spectral self-exitance, or spectral emissive power) is the power emitted per unit area per unit of wavelength at the wavelength λ, and $\alpha_\lambda(T)$ (spectral absorptance) is the fraction of the incident power per unit area per unit of wavelength that is absorbed, then for every particular wavelength, all objects in thermal equi-librium at the same absolute temperature T have the same ratio of $M_\lambda(T)/\alpha_\lambda(T)$. It follows that an object having the maximum possible absorptance [$\alpha_\lambda(T) = 1$] also has the maximum possible spectral radiance emittance [$M_\lambda^b(T)$]. Such a perfect absorber and perfect emitter is called a *blackbody* and the electromagnetic energy it emits is called *blackbody radiation*. No true blackbody exists, but a small hole through which radiation escapes from an isothermal enclosure is an excellent approximation. Any real substance is characterized by a spectral emissivity $\varepsilon_\lambda(T)$ [$0 < \varepsilon_\lambda(T) < 1$] expressing its spectral radiant emittance as some fraction of that of a blackbody at the same temperature, $M_\lambda(T) = \varepsilon_\lambda(T)M_\lambda^b(T)$. With this, Kirchhoff's Law becomes $\varepsilon_\lambda(T) = \alpha_\lambda(T)$. This is an important result – a good emitter of thermal radi-ation is also a good absorber. It follows also that a good reflector is a poor emitter because, for opaque substances, $\rho_\lambda(T) + \alpha_\lambda(T) = 1$ [where $\rho_\lambda(T)$ is the spectral reflectance], and so $\rho_\lambda(T) = 1 - \varepsilon_\lambda(T)$. The incident radiation is usually absorbed in a thin surface layer whose properties then determine the value of $\varepsilon_\lambda(T)$. As a rule of thumb, smooth or polished surfaces have lower emissivities than rough, grainy, or oxidized surfaces; metals have lower values than nonmetals. In general $\varepsilon_\lambda(T)$ varies only slowly with temperature, but rather more with wavelength. For a large number of materials, however, $\varepsilon_\lambda(T)$ is approximately constant over a fairly wide wavelength range. These are called *graybodies*.

Materials are also characterized by a total emissivity $\varepsilon(T)$ relating the total radiant emit-tance $M(T)$ (total power emitted per unit area) to that of a blackbody $M^b(T)$ at the same tem-perature by $M(T) = \varepsilon(T)M^b(T)$. $\varepsilon(T)$ is an integrated average of weighted values of $\varepsilon_\lambda(T)$

Table 1: Representative values of emissivities of solids.

Material	$\varepsilon_\lambda(T)$ $(k = 0.65 \times 10^{-4}\,\mathrm{cm})$	$\varepsilon(T)$ $(T = 300\,\mathrm{K})$
Aluminum foil	0.15	0.02
Copper, polished	0.1	0.03
Copper, oxidized	0.8	0.5
Iron, polished	0.4	0.08
Iron, oxidized	0.9	0.8
Carbon	0.9	0.8
Red brick	0.7	0.9
Concrete	0.6	0.94
Soot	0.97	0.95
Flat black paint	0.98	0.94
Flat white paint	0.2	0.87

using the blackbody spectral radiant emittance as the weighting function. Typical values of these emissivities for a few solids are given in Table 1.

The theoretical foundations of blackbody radiation have played an important role in the history of physics. In 1879 Josef Stefan deduced the empirical relation that the total radiant emittance varies directly with the fourth power of the absolute temperature, i. e.,

$$M^b(T) = \sigma T^4 \,, \tag{1}$$

where σ is a constant. Ludwig Boltzmann established the theoretical proof of Eq. (1) (the Stefan–Boltzmann law) in 1884 from an analysis of the thermodynamics of the pressure exerted by radiation. In searching for a specific relation between the spectral radiant emittance of a blackbody and temperature, Wilhelm Wien showed (1893), on the basis of classical thermodynamics, that $M_\lambda^b(T)$ must have the form

$$M_\lambda^b(T) = c_1 \lambda^{-5} f(kT) \tag{2}$$

where c_1 is a constant and $f(kT)$ is some function of the product kT only. Although thermodynamics could not provide the specific form of $f(kT)$, Wien was able to prove that the product $\lambda_m T$ (where λ_m is the wavelength at which $M_\lambda^b(T)$ has a maximum value) is a universal constant having the value 0.2898 cm K. It follows from this *displacement law* that (a) the maximum of $M_\lambda^b(T)$ and, correspondingly, most of the radiated energy shift to lower wavelengths as the temperature increases; (b) the value of $M_{\lambda_m}^b(T)$ increases with the fifth power of the temperature. In 1896 Wien suggested that the function $f(kT)$ should be set equal to $e^{-c_2/\lambda T}$ (where c_2 is a constant), but this was found to depart increasingly from experimental observations as λT increased (a difference of 1% when $\lambda T = 0.3124$ cm K). For these studies Wien received the Nobel Prize in 1911. In 1900 Lord Rayleigh deduced from classical statistical mechanics that $f(kT) = \lambda T/c_2$. Although a good approximation for large values of λT, this result is clearly wrong for small λ because it predicts the emission of an infinite amount of energy.

The dilemma was finally resolved in 1900 by Max Planck who introduced the revolutionary idea that energy can be emitted (or absorbed) only in discrete amounts that are integral multiples of a fundamental *quantum* of energy hc/λ (where c is the speed of electromagnetic radiation in vacuum and h is a universal constant known as Planck's constant). With this assumption Planck derived the correct formula

$$M_\lambda^b(T) = 2\pi hc^2\lambda^{-5}(e^{hc/\lambda kT} - 1)^{-1} , \tag{3}$$

where k is the Boltzmann constant. Equation (3) may be written in equivalent forms in terms of frequency (ν) or numbers of photons (n) instead of wavelength (λ). This formula satisfies the demands of Eq. (2) and also, by equating to zero the first derivative with respect to λ, gives $\lambda_m T = hc/4.9651k$, which is Wien's displacement law. When $\lambda T \gg hc/k$, Eq. (3) reduces to Wien's approximation, and when $\lambda T \ll hc/k$ it reduces to Rayleigh's. Integration of Eq. (3) over all wavelengths leads to the Stefan–Boltzmann Law

$$M^b(T) = \frac{2\pi^5 k^4}{15h^3c^2}T^4 \tag{4}$$

which relates σ to the other universal physical constants. For this spectacular success, which marked the beginnings of the quantum theory, Planck received the Nobel Prize in 1918. Equation (3) was subsequently derived by others (notably Einstein and Bose) from completely different lines of reasoning.

The Planck distribution [Eq. (3)] is illustrated in Fig. 1 for some representative temperatures, for one of which the Wien and Rayleigh approximations are also shown. The variation of λ_m with temperature is also indicated. Both Eqns. (3) and (4) are valid only when the blackbody radiates into a vacuum; otherwise $M_\lambda^b(T)$ and $M^b(T)$ must be multiplied by n^2, where n is the refractive index of the surrounding medium and λ is the wavelength in vacuum. For most gases (including air), n is very near to unity so the factor n^2 may be neglected.

In the past two decades blackbody simulators (experimental approximations to true blackbodies) have become increasingly important. These simulators almost always take the form of isothermal cavities, some of fairly simple, and others of highly complicated, design. They have been used over a wide temperature range from 4 K to above 2000 K. The material of the radiating surface of the simulator must be suited to the temperature range of use: black paint, oxidized or anodized metal below about 500 K; graphite (suitably protected from oxidation) to about 1600 K; ceramics above. In all cases the spectral (or total) emissivity of this radiating surface is made as high as possible and the cavity geometry is designed so as to increase the simulator's "effective emissivity" to be very close to unity. Sometimes this is attained by a judicious combination of black diffuse surfaces and highly reflective surfaces. Depending upon the application, it is required to know the value of the effective emissivity to within 1 part in 10^{-4} to 10^{-6}. Measurement of the effective emissivity to this accuracy is not possible so recourse to calculated values is necessary. Many methods for these calculations have been described. Bedford [1] has given a detailed discussion of these.

Blackbody simulators are widely used in precision metrology. They serve, for example, as calibration sources for the realization of the radiation portion of the International Temperature Scale. They are similarly used as reference sources in the measurement of the spectral emissivities of materials. They form the heart of cryogenic radiometers that underpin the

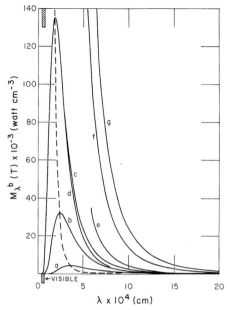

Fig. 1: Spectral radiant emittance of a blackbody: (a) 800 K, (b) 1200 K, (c) 1600 K, (d) 1600 K (Wien), (e) 1600 K (Rayleigh), (f) 6000 K, (g) 10 000 K. The variation of λ_m with temperature is shown by the dashed curve.

calibrations of photodetectors for fundamental spectro-radiometric measurements. They have also been used in the measurement of fundamental physical constants. Using a complex combination of two simulators, one as a radiator and one as a thermal detector held at cryogenic temperatures, Quinn and Martin [2] measured the Stefan–Boltzmann constant. With the same apparatus they measured thermodynamic temperatures over the range 230 K to 375 K. Their instrumentation was the forerunner of the commercially-supplied cryogenic radiometers that are now fundamental to precision spectro-radiometry. Many firms now supply simulators that are widely used in industrial measurements. Simulators of special design have been flown in space for measurements of the solar constant. Thermographic measurements of sea and land temperatures depend for their calibrations on simulators. More esoterically, blackbody radiation is important in cosmology, where it is a powerful probe for elucidating the early history of the universe.

See also: Heat.

References

[1]   R. E. Bedford, "Calculation of Effective Emissivities of Cavity Sources of Thermal Radiation," Ch. 12 in *Theory and Practice of Radiation Thermometry* (D. P. DeWitt and Gene D. Nutter, eds.). Wiley, New York, 1988.

[2]   T. J. Quinn and J. E. Martin, "A Radiometric Determination of the Stefan–Boltzmann Constant and Thermodynamic Temperatures between −40 °C and +100 °C," *Phil. Trans. Roy. Soc.* **316**, 85 (1985).

Bibliography

Blackbody radiation is discussed in most textbooks that deal with heat, heat transfer, infrared radiation and detection, or radiometry. Examples of good, detailed treatments are:

G. R. Noakes, *A Text-Book of Heat*, 3rd ed., pp. 368–409. Macmillan, London, 1953. (E)

R. Siegel and J. R. Howell, *Thermal Radiation Heat Transfer*, 4th ed. Taylor & Francis, New York, 2001. (I)

F. Reif, *Fundamentals of Statistical and Thermal Physics*, pp. 373–388. McGraw–Hill, New York, 1965. (I)

H. P. Baltes and F. K. Kneubühl, "Thermal Radiation in Finite Cavities," *Helv. Phys. Acta* **45**, 481–529 (1972). (A)

Bohr Theory of Atomic Structure

E. A. Burke[†]

The idea that all matter is composed of basic indivisible parts was first suggested by the Greek philosopher Democritus (who coined the word atom in the fourth century) and subsequently alluded to in medieval writings. Early in the 19th century, John Dalton proposed an atomic theory of matter. This theory assumes that (1) all matter is composed of atoms, (2) all atoms of a given element have a distinctive mass, and (3) in chemical reactions whole atoms, never fractions of atoms, are involved. The theory proved invaluable in understanding chemical reactions. Later on in the 19th century, the works of Rudolf Clausius, James Clerk Maxwell, and Josef Loschmidt on the kinetic theory of gases, climaxing the work done by Daniel Bernoulli in the middle of the 18th century, firmly established the concept of an atom as the fundamental constituent of matter which differentiated one element from another.

However, it soon became apparent that the atom has a structure. In particular, the electrolysis research performed by Michael Faraday indicated the existence of electrically charged constituent parts, and in 1891 G. Johnstone Stoney proposed the name electron for the basic unit of charge in an atom. In 1895, the experiments of Jean-Baptiste Perrin showed that the cathode rays emitted from electrical discharge through gases consisted of negative charges, and in 1897 Joseph J. Thomson performed his now famous experiments on these cathode rays, passing them through combined electric and magnetic fields. This established the ratio of charge to mass, e/m, of this electrical constituent of atoms. The concept of an electron as a negatively charged fundamental constituent of an atom was further reinforced through an analysis of the Zeeman effect, in which spectral lines were observed to broaden in a strong magnetic field. The magnitude of the electron charge was determined by Robert A. Millikan in his classical oil drop experiments begun in 1906.

Since atoms are normally uncharged electrically, the discovery that constituent parts were negatively charged (and by inference that an equal amount of positive charge was in the atom) required a theory of atomic structure to accommodate charged particles, such that the total atomic charge would be zero. By assuming that an atom consists of an equal number of positive and negative ions that obey a force law that is slightly different from an inverse square law,

---

[†]deceased

James H. Jeans in 1901 was able to calculate some of the observed features of atomic spectra. However, this model could not account for the observed periodicity of the chemical properties of the elements. In 1904, Thomson proposed an atomic structure consisting of a uniformly and positively charged sphere embedded with negatively charged electrons. The total charge was taken to be zero and the electrons confined to positions on rings concentrically located about the center of the sphere. For equilibrium, the numbers of electrons on each ring for a given atom were assumed identical. Thomson calculated the distribution of electrons on the rings for various elements. This model proved effective in describing the periodicity of the elements, but was not very useful in explaining the spectral lines that were observed in atomic spectra. In the same issue of the *Philosophical Magazine* in which Thomson proposed his model, H. Nagaoka, relying on earlier (1890) calculations of Maxwell on the rings of Saturn, first proposed a nuclear atom with extranuclear electrons. He recognized the difficulties associated with such a model in that the electrons could not remain stationary for otherwise they would be drawn into the nucleus by the Coulomb force. Although mechanically stable if they moved about the nucleus, they should radiate energy by virtue of their accelerated motion then slow down, and ultimately collapse. These objections were left unanswered. Radiation was accounted for by the vibrations of the electrons within their orbits. This model was unable to predict more than a few of the observed spectral lines. More importantly, this model did not gain general acceptance because of the inherent instability of the proposed physical system. In order to improve Thomson's model so that spectroscopic phenomena could also be explained, Lord Rayleigh (1906) assumed that the number of electrons could be infinite. This proposal preceded calculations published by Thomson on the basis of an analysis of several different experimental observations that the number of electrons (referred to by Thomson as corpuscles) was the same order as the atomic weight. In 1911 Barkla more accurately ascertained that the number of electrons was more nearly half the atomic weight, except for the lightest elements. And so Rayleigh's modification of Thomson's atomic model was found to be untenable. Also, in 1906, G. A. Schott proposed other modifications of Thomson's model to account for more details of atomic spectra. At this time the accepted model of atomic structure was that due to Thomson. In this model, radiation, as observed in spectral lines, was accounted for by the mechanism of electronic vibration.

However, the experiments of Hans Geiger and Ernest Marsden begun in 1909 at the suggestion of Ernest Rutherford, and the subsequent analysis of the results by Rutherford, first published in 1911, indicated that an atom must consist of a small but massive and positively charged central nucleus surrounded by extranuclear electrons. Following the discovery of radioactivity by Antoine Henri Becquerel in 1896, Rutherford had identified α rays as ionized helium atoms moving very rapidly. Geiger and Marsden observed the scattering of α rays through thin metallic foils. The surprising result of these experiments was an unexpectedly large amount of backscattering. In the words of Rutherford: "It was almost as incredible as if you fired a 15-inch shell at a piece of tissue paper and it came back and hit you." It was not possible with the Thomson model to predict any significant backscattering, even if one assumed multiple (from many atoms) scattering. On the other hand, Rutherford predicted, and Geiger and Marsden confirmed, that the scattering pattern observed could be completely understood by assuming a nuclear model for atoms. Rutherford was able to conclude that the dimension of a nucleus is of the order of 10^{-13} cm. It can be deduced from kinetic theory that an atom has dimensions of the order of 10^{-8} cm. Hence an atom is mostly empty space. In

1906 Thomson's experiments indicated most probably that the smallest known atom, hydrogen, contained just a single electron.

In 1913, Niels Bohr (a student of Rutherford), in a paper entitled, "On the Constitution of Atoms and Molecules," which appeared in Vol. 26 of the *Philosophical Magazine*, first proposed his atomic theory. This theory is known as the Bohr theory of atomic structure. The principal success of the theory was in explaining the spectra of atomic hydrogen.

If we assume a circular orbit of an electron about a stationary nucleus, the total energy of the system W is given by

$$W = mv^2 - eQ/r \, , \tag{1}$$

where m is the electron's mass, v is its speed, e and Q are, respectively, the magnitudes of the electron and nuclear charge, and r is the radius of the orbit. Following Bohr, cgs units are employed. In order for the electron to remain in a circular orbit, the centripetal force must be supplied by the Coulomb force between the electron and the nucleus, and hence

$$\frac{mv^2}{r} = \frac{eQ}{r^2} \, . \tag{2}$$

Combining Eqns. (1) and (2) we obtain

$$W = -\tfrac{1}{2}eQ/r = -\tfrac{1}{2}mv^2 \tag{3}$$

so that the energy of the bound system is less than 0, whereas for the free electron, $r = \infty$, the energy is taken to be zero. If the frequency of orbital motion is f, then

$$v = 2\pi r f \, . \tag{4}$$

By eliminating r and v from the preceding relations, we obtain

$$f = \frac{\sqrt{2}}{\pi} \frac{(-W)^{3/2}}{eQ\sqrt{m}} \, . \tag{5}$$

In describing blackbody radiation, Max Planck (1900) assumed that radiation must be emitted in separate emissions (quanta) of $h\nu$, where h is Planck's constant and ν is the frequency of the radiation. (This concept of quantized radiation was successfully employed later by Albert Einstein in 1905 in describing the photoelectric effect.) Following his predecessors in relating radiation frequency to the mechanical frequencies of the electrons, Bohr assumed that the radiation frequency emitted by an atom when an electron goes from an unbound to a bound state would be the average of the mechanical (in this case the orbital) frequencies of the initial and final state. Since the orbital frequency of the unbound electron is zero and its bound frequency is f, then the relationship between f and the radiation frequency ν is

$$\nu = f/2 \tag{6}$$

and, following Planck's hypothesis, the energy W radiated would be

$$\mathcal{W} = nh\nu = nh(f/2) \, , \tag{7}$$

where n is an integer. Combining Eq. (7) with Eq. (5), noting that $\mathcal{W} = -W$, we obtain

$$W = -\frac{2\pi^2 e^2 Q^2 m}{n^2 h^2} \tag{8}$$

and

$$f = \frac{4\pi^2 m e^2 Q^2}{n^3 h^3} \tag{9}$$

with

$$r = \frac{n^2 h^2}{4\pi^2 m e Q} . \tag{10}$$

We note that the most stable state, the state with the lowest energy, occurs for $n = 1$. For the hydrogen atom $Q = e$, and if we use the experimental values known to Bohr of

$$
\begin{aligned}
e &= 4.7 \times 10^{-10} \text{ statcoulombs} , \\
e/m &= 5.31 \times 10^{17} \text{ statcoul/q} ,
\end{aligned}
\tag{11}
$$

and

$$h = 6.5 \times 10^{-27} \text{ erg s} ,$$

we get, to the accuracy of these data,

$$W = -13 \text{eV}, \quad f = 6.2 \times 10^{15} \text{ Hz}, \quad r = 0.55 \times 10^{-8} \text{ cm} .$$

These values are at least the same order of magnitude as the ionization potential of hydrogen, the observed optical frequencies, and atomic dimensions.

Larger integer values of n correspond to higher-energy stationary states of the system. We note, from Eq. (8), (9), and (10) that, as $n \to \infty$, W and $f \to 0$, and $r \to \infty$. We approach the unbound state. All the energy of the atom is emitted as radiation when the electron proceeds from an initial unbound state to the bound state, and from Eq. (8) this positive amount of energy is

$$W_n = \frac{2\pi^2 e^4 m}{n^2 h^2} \tag{12}$$

for the hydrogen atom. The amount of energy emitted by the passing of the system from a state corresponding to $n = n_1$, to one corresponding to $n = n_2 < n_1$ is, consequently,

$$W_{n_2} - W_{n_1} = \frac{2\pi^2 e^4 m}{h^2} \left(\frac{1}{n_2^2} - \frac{1}{n_1^2} \right) . \tag{13}$$

If the frequency of the radiation is ν, then according to Planck this energy must be $h\nu$ so that

$$W_{n_2} - W_{n_1} = h\nu \tag{14}$$

and hence

$$v = \frac{2\pi^2 e^4 m}{h^3} \left(\frac{1}{n_2^2} - \frac{1}{n_1^2} \right). \tag{15}$$

By selecting a value of n_2 and letting n_1 take on all integral values larger than n_2, we obtain a formula for a sequence of spectral lines. In particular, by selecting $n_2 = 2$, this formula gives results of the series discovered by Johann Balmer in 1884. If we rewrite Eq. (14) as

$$v = Rc \left(\frac{1}{2^2} - \frac{1}{k^2} \right), \tag{16}$$

we then have the same form as written by Balmer. R is the Rydberg constant, which, by the Bohr theory, is

$$R = \frac{2\pi^2 m e^4}{ch^3}. \tag{17}$$

(More properly we should replace m by μ, the reduced mass of the electron–nucleus system. In 1932 this refinement led to the discovery of deuterium.) Putting $n_2 = 3$ we obtain the series first observed by Friederich Paschen in 1908. For $n_2 = 1$ we get the Lyman series. Theodore Lyman discovered the first term in this ultraviolet region in 1906 but did not attribute this to hydrogen until after the publication of Bohr's theory. For $n_2 = 4$ and $n_2 = 5$ we have series in the infrared. These series, referred to as the Brackett and Pfund series, were predicted by Bohr's theory and were subsequently observed in the 1920s.

Thus the Bohr theory proved spectacularly successful in accounting for known spectral series and predicting future series of hydrogen spectra. With the Bohr theory it also becomes possible to explain the greater number of spectral lines observed from interstellar space. The additional lines can be considered to arise from transitions from very high values of n_1 in Eq. (14). This, in turn, implies orbits of very large radii which could only be possible in the rare atmosphere of outer space.

One can show (as was done by Bohr) that the choice of $\mathcal{W} = nhf/2$ in Eq. (7) is consistent with the known Balmer series expansion and leads to a correspondence of radiation frequency with orbital frequency for transitions between two adjacent highly excited states.

We can state Eq. (7) in an alternative way. We see from Eq. (3) that $\mathcal{W} = -W = T$, where T is the kinetic energy of the electron and for circular motion $T = mr^2 f^2 2\pi^2$, hence

$$2\pi^2 mr^2 f^2 = nhf/2. \tag{18}$$

But the angular momentum p of the system is $2\pi mr^2 f$. From this we obtain

$$p = n\hbar \tag{19}$$

where $\hbar = h/2\pi$. The importance of angular momentum had been indicated earlier (1912) by J. W. Nicholson. The condition represented by Eq. (18) is referred to as the quantization of orbital angular momentum. Implicitly, the condition includes all orbitals, i. e., stationary states of the system, for which the angular momentum is an integral multiple of $\hbar$ and precludes all other possibilities. Furthermore, the existence of stationary states implies, by definition, that the atom cannot radiate when in any of its allowed states. Finally, all transitions between stationary states result in emission or absorption of radiation in quanta of energy equal to $h\nu$ with ν the frequency of the radiation. Transitions to lower-energy states result in emission while transitions to higher-energy states are required for absorption.

One can derive the Bohr theory of atomic structure starting with the quantization of angular momentum, Eq. (18), and proceed from Eqns. (1) through (5) above.

The quantization of the angular momentum condition may be thought of as a necessary result of the wave nature of matter, as proposed by Louis de Broglie in 1924. In this theory, we ascribe a wavelength λ to a particle with linear momentum P by $\lambda = h/P$. For circular orbits in stationary states, we would thus require that the allowed orbits would be integral multiples of the particle wavelength λ. Thus we would require

$$2\pi r = n\lambda = nh/P . \tag{20}$$

Angular momentum p is related to linear momentum P in this case by $p = rP$. Hence Eq. (19) reduces to Eq. (18), which represents the quantization of angular momentum.

That the electron can remain in stationary states can be understood somewhat from quantum mechanics. Quantum mechanically, the electron is spread out in space and if the electrostatic attraction of the nucleus were to confine it within a small region near the nucleus, the indeterminancy principle would yield a great uncertainty in momentum. But this, in turn, would mean a higher expected energy, thereby enabling the electron to move away from the nucleus. As a result, there is an electrical equilibrium allowing for the possibility of stationary states.

Despite the success of the Bohr model, it fails in several important ways, viz.,

1. It does not account for the intensities of the observed spectral lines.

2. Despite some moderate success in describing the spectra of alkali-like atoms, the model is insufficient for atoms of more than one electron.

3. It is unable to explain chemical binding.

4. It is unable to explain fine details observed in the hydrogen spectra.

Even with its failures, the Bohr theory represents a significant advance in thought toward understanding atomic structure. It represents a model which bridges the gap between classical and quantum mechanics. Some of the more important, permanent contributions to our understanding of atomic structure that come from the Bohr theory of atomic structure are

1. the existence of stationary atomic states,

2. energy and momentum quantization,

3. the explanation of emission and absorption of radiation as transitions between stationary states.

See also: Atoms; Atomic Spectroscopy; Atomic Structure Calculations, One-Electron Models; Balmer Formula.

Bibliography

For an historical survey see Robert M. Besanqon (ed.), *The Encyclopedia of Physics*. Reinhold, New York, 1966; and Kenneth W. Ford, *Basic Physics*. Ginn/Blaisdell, Waltham, Mass., 1968.

For a collection of early publications see G. K. T. Conn and H. D. Turner, *The Evolution of the Nuclear Atom*. American Elsevier, New York, 1965.

For an elementary discussion see Richard T. Weidner and Robert L. Sells, *Elementary Modern Physics*. Allyn and Bacon, Boston, 1965.

Bose–Einstein Condensation

H. Matsumoto and Y. Ohashi

Bose–Einstein condensation(BEC) is a quantum many-body phenomenon in systems of bose particles, in which the bosons condensate into the lowest-energy state below certain critical temperature T_c. The condensate forms a macroscopically ordered quantum state. This phenomenon was predicted theoretically by A. Einstein in 1925 [1], extending the work of S. N. Bose on the statistical-mechanical treatment of photons [2]. He circumvented the breakdown of quantum statistical mechanics in the free massive bose gas below T_c by allowing a fraction of particles to condensate into the lowest-energy state. In 1938, F. London [3] suggested that the superfluidity of ^{4}He is the result of BEC. Later, many physical ideas followed, such as the coherent ordered state and the spontaneous breakdown of symmetry.

Although it is widely accepted that the superfluidity of ^{4}He is the result of BEC, it is still no direct proof of BEC for a free bose gas or a bose gas with repulsive interaction, since ^{4}He atoms have a weak attractive interaction outside of the hard-core repulsive interaction. The realization of BEC in atomic gases needs an ultra-low temperature of μK (10^{-6} K) or even nK (10^{-9} K) due to the large masses of atoms.

BEC in Alkaline Atomic Bose Gases [4, 5]

In 1995, BEC was experimentally realized in dilute alkaline atomic bose gases, using laser cooling, magnetic traps and evaporation. Wieman and Cornell used ^{87}Rb at $T = 170$ nK with 2000 particles [6], Ketterle used ^{23}Na at $T = 2\mu$K with 5×10^5 particles [7].

In a magnetic trap, neutral alkaline atoms experience Zeeman splitting and a certain hyperfine spin state becomes the lowest-energy state in the trap. The bose gas is then described by a one-component complex boson field $\Psi(\mathbf{x},t)$. The field Ψ is decomposed as the sum of the condensate part $\phi(\mathbf{x},t)$ and the quantum excitation part $\psi(\mathbf{x},t)$, $\Psi(\mathbf{x},t) = \phi(\mathbf{x},t) + \psi(\mathbf{x},t)$. The condensate part ϕ, which describes the quantum macroscopic behavior of the system, satisfies the following approximate equation called Gross–Pitaevskii equation,

$$i\hbar \frac{\partial}{\partial t}\phi(\mathbf{x},t) = \left[-\frac{\hbar^2}{2m}\nabla^2 - \mu + U(\mathbf{x}) \right]\phi(\mathbf{x},t) + V\phi^\dagger(\mathbf{x},t)\phi(\mathbf{x},t)^2 .$$

The first term of the right hand side is the kinetic energy, μ is the chemical potential, and the trap potential $U(\mathbf{x})$ is given approximately by the harmonic potential $U(\mathbf{x}) = \frac{1}{2}m\omega_h^2(x^2 + y^2 +$

z^2) (in the spherical case). The last term arises from the s-wave atom–atom scattering and is given by the s-wave scattering length a_s as $V = (4\pi\hbar^2 a_s)/m$. When the typical length scale is taken as $a_h = [\hbar/(m\omega_h)]^{1/2}$, a_s/a_h becomes very small, of the order $10^{-3}\cdots 10^{-4}$. The present alkaline atomic gas seems close to the ideal free bose gas.

In reality, the situation is not so simple. The shape of the condensate is affected by the total number of atoms, $N = \int d^3x \langle \Psi^\dagger(\mathbf{x},t)\Psi(\mathbf{x},t)\rangle$. The scaled effective coupling constant is proportional to $(a_s/a_h)N$, indicating that a shift from weak to strong coupling can be realized by changing the particle number. In fact, for smaller N, a sharp Gaussian distribution of the density profile above the thermal Gaussian distribution is observed, reflecting the shape of the wave function for the lowest energy in the harmonic potential. For larger N, it is changed to the bell shape of the parabolic distribution.

An experiment to observe the phase coherence of the BEC state was performed by Andrews *et al.* [8]. Two bose gases were prepared in two separate trapped potentials and were cooled down to form the BEC states and then released into free space. The expected interference pattern was observed, indicating that the two condensates are in the phase-coherent state. By using the phase-ordered state property, an atomic laser could also be realized. The zero sound, which is expected from the phase-ordered state (or superfluidity), could also be observed by exciting the density wave in an elongated arrangement of the condensate [9].

Another experiment relating to the phase-ordered state is the formation of a vortex lattice in the condensate. A triangular vortex lattice is formed up to the Thomas–Fermi edge of the condensate [10].

The BEC in alkaline dilute gas is unique compared with the superfluidity of ^{4}He in that a variety of external controls are possible; one can control not only the external environment such as number of atoms, temperature and applied potential but also the self interaction itself. The following are a number of unique new features:

1. Formation of condensate: Due to the weakness of the self interaction, the relaxation time is long; it ranges from a few ten milliseconds to a hundred milliseconds. Therefore the time evolution is observable experimentally. Thermalization processes and the time evolution of the condensate after evaporation were observed [11], which makes the condensates a good laboratory example of a nonequilibrium process.

2. Control of the interaction strength: The self interaction among atoms is given by the effective s-wave scattering of atoms. Using the hyperfine structure of atomic levels, one can control the resonant scattering by changing the applied magnetic fields. A resonating state of atom–atom scattering is called Feshbach resonance, and by using this Feshbach resonance, one can vary the effective atom–atom interaction from repulsive to attractive in an arbitrary strength [12].

3. By preparing the BEC, and by changing the interaction from repulsive to attractive, the time evolution of the bose gas can be observed [13]. As in the gravitational collapse, the system first collapses and then explodes, like a supernova in the universe; the effect is therefore named "bosenova". This kind of time-dependent observation is possible in BEC systems because of their "millisecond" time-scale of evolution and the possibility to control the interaction strength.

4. Optical lattice and superfluid–insulator transition: Recently another unique feature of the atomic gas has been explored, namely atomic gases trapped optically in a lattice. When an atomic gas is located in a standing wave of a laser, it feels an effective periodic potential due to electric polarization of atoms induced by the electromagnetic field (the Stark effect), $U(\mathbf{x}) = V_0[\sin^2(x/\lambda) + \sin^2(y/\lambda) + \sin^2(z/\lambda)]$ with λ the wave length and V_0 the potential energy, which is proportional to the intensity of the laser. A superfluid–insulator transition is observed at certain V_0, when the intensity of the laser is increased [14].

Fermion Superfluidity and BCS–BEC Crossover

In contrast to bosons, more than two fermions cannot occupy the same quantum state due to the Pauli exclusion principle. However, in metallic superconductivity and in superfluid ^{3}He, when fermions (electrons and ^{3}He atoms) form Cooper pairs, these molecular "bosons" can be Bose-condensed. In 2004, fermion superfluidity was also discovered in ultracold gases of ^{40}K [15] and ^{6}Li Fermi atoms [16, 17].

In fermion superfluidity, an attractive pairing interaction between particles is essential for the formation of Cooper pairs. When the strength of the pairing interaction becomes strong, it is known that the BCS–BEC crossover phenomenon [18, 19] occurs, where the character of superfluidity continuously changes from the weak-coupling BCS (Bardeen–Cooper–Schrieffer) type to the BEC type of tightly-bound Cooper-pair molecules. In the strong-coupling regime, Cooper pairs have already appeared above the superfluid phase transition temperature, which is quite different from the weak-coupling BCS regime, where the formation of Cooper pairs and their condensation occur at the same temperature. From the viewpoint of the BCS–BEC crossover, Fermi BCS superfluids and the BEC of bosons can be understood in a unified manner.

To observe the BCS–BEC crossover experimentally, the key is how to tune the strength of a pairing interaction between fermions. In metallic superconductivity and superfluid ^{3}He, such a tuning is actually difficult. On the other hand, a tunable pairing interaction can be realized in a trapped Fermi atom gas with a Feshbach resonance [19, 20]. In the Feshbach resonance, two atoms (open channel) form a quasi-molecular boson (closed channel), and this molecule dissociates into two Fermi atoms. The pairing interaction associated with the Feshbach resonance can be adjusted by an external magnetic field, using different atomic hyperfine states between atoms in the closed channel and the open channel.

The BCS–BEC crossover in an ultracold Fermi gas with a Feshbach resonance was first predicted theoretically [19], and was experimentally observed in superfluid ^{40}K Fermi gas [15], as well as in ^{6}Li Fermi gas [16, 17]. In ^{40}K, Cooper pairs are formed between atoms in the two hyperfine states, $|F, F_z\rangle = |\frac{9}{2}, -\frac{9}{2}\rangle$ and $|\frac{9}{2}, -\frac{7}{2}\rangle$. In ^{6}Li, $|F, F_z\rangle = |\frac{1}{2}, \frac{1}{2}\rangle$ and $|\frac{1}{2}, -\frac{1}{2}\rangle$ are used. The energy gap in the BCS–BEC crossover region has been observed in superfluid ^{6}Li, using the rf-tunneling current spectroscopy [21]. Some collective modes associated with surface oscillations of a trapped superfluid gas have been also observed in ^{6}Li [17, 22]. So far, only the s-wave superfluidity has been discovered in trapped Fermi gases [15–17]. The realization of a p-wave Fermi gas superfluid is still an open problem. The tunable p-wave pairing interaction has been observed in ^{40}K and ^{6}Li [23].

See also: Atomic Trapping and Cooling; Cold Atoms and Molecules; Helium, Liquid; Laser Cooling; Superconductivity Theory.

References

[1]   A. Einstein, *Sitzungsberichte der Preußischen Akademie der Wissenschaften, Physikalisch-ma-thematische Klasse*, 261 (1924); 3 (1925).

[2]   S. N. Bose, *Z. Phys.* **26**, 178 (1924).

[3]   F. London, *Nature* **141**, 643 (1938); *Phys. Rev.* **54**, 947 (1938).

[4]   See the following review papers and references cited therein: A. S. Parkins and D. F. Walls, *Phys. Rep.* **303**, 1 (1998); F. Dalfovo, S. Giorgini, L. P. Pitaevskii and S. Stringari, *Rev. Mod. Phys.* **71**, 463 (1999); A. J. Legget, *Rev. Mod. Phys.* **73**, 307 (2001).

[5]   See also the following books: C. J. Pethick and H. Smith, *Bose-Einstein Condensation in Dilute Gases.* Cambridge University Press, Cambridge, 2002; L. P. Pitaevskii and S. Stringari, *Bose–Einstein Condensation.* Oxford University Press, Oxford 2003.

[6]   M. H. Anderson *et al.*, *Science* **269**, 198 (1995).

[7]   K. B. Davis *et al.*, *Phys. Rev. Lett.* **75**, 3969 (1995).

[8]   M. R. Andrews *et al.*, *Nature* **275**, 637 (1997).

[9]   M. R. Andrews, *et al.*, *Phys. Rev. Lett.* **79**, 533 (1977).

[10]   J. R. Abo-Shaeer, *et al.*, *Phys. Rev. Lett.* **88**, 070409 (2002).

[11]   H.-J. Miesner, *et al.*, *Science* **279**, 1005 (1998).

[12]   S. Inouye, *et al.*, *Nature* **392**, 151 (1998).

[13]   S. L. Cornish, *et al.*, *Phys. Rev. Lett.* **85**, 1795 (2000).

[14]   M. Greiner, *et al.*, *Nature* **415**, 39 (2002).

[15]   C. A. Regal *et al.*, *Phys. Rev. Lett.* **92**, 040403 (2004).

[16]   M. Bartenstein *et al.*, *Phys. Rev. Lett.* **92**, 120401 (2004); M. W. Zwierlein *et al.*, *Phys. Rev. Lett.* **92**, 120403 (2004).

[17]   J. Kinast *et al.*, *Phys. Rev. Lett.* **92**, 150402 (2004).

[18]   D. M. Eagles, *Phys. Rev.* **186**, 456 (1969); A. J. Leggett, in *Modern Trend in the Theory of Condensed Matter*, A. Pekalski and J. Przystawa (eds.), p. 14. Springer, Berlin, 1980; Nozières and S. Schmitt-Rink, *J. Low Temp. Phys.* **59**, 195 (1985).

[19]   Y. Ohashi and A. Griffin, *Phys. Rev. Lett.* **89**, 130402 (2002).

[20]   E. Timmermans *et al.*, *Phys. Lett. A* **285**, 228 (2001); M. Holland *et al.*, *Phys. Rev. Lett.* **87**, 120406 (2001).

[21]   C. Chin *et al.*, *Science* **305**, 1128 (2004).

[22]   M. Bartenstein *et al.*, *Phys. Rev. Lett.* **92**, 203201 (2004).

[23]   C. Regal *et al.*, *Phys. Rev. Lett.* **90**, 053201 (2003); J. Zhang, *et al.*, *Phys. Rev. A* **70**, 030720 (2004).

Bose–Einstein Statistics

O. W. Greenberg

Identical atomic, nuclear, and subnuclear quantum-mechanical particles and systems of particles are indistinguishable. Experimental evidence confirms the relativistic quantum field theory prediction that the wave functions that describe several identical particles with integral spin must be symmetric under joint permutations of the coordinates and spins associated with the identical particles. This property of the wave functions leads to a counting, or "statistics" (called *Bose–Einstein statistics*), of the probability that a given set of identical integral-spin particles will be distributed in the accessible energy levels in a given way. Bose–Einstein statistics leads to characteristic statistical-mechanical (partition, etc.) and thermodynamic (energy distribution, etc.) functions. Composite systems obey Bose–Einstein statistics unless they contain an odd number of constituents that obey Fermi–Dirac statistics; in this case they obey Fermi–Dirac statistics.

Among elementary particles that obey Bose–Einstein statistics are the photon, the pions, the kaons, and mesons generally. Among composite systems that obey Bose–Einstein statistics are the hydrogen atom, the deuteron, the ^{4}He atom, and even atomic number nuclei generally. Particles that obey Bose–Einstein statistics are called *bosons*.

Bose derived the Planck frequency distribution for blackbody radiation by treating the photon as a massless particle that obeys Bose–Einstein statistics. Bose–Einstein statistics produces a tendency for identical particles to cluster in the same quantum state. For example, consider two particles, 1 and 2, each of which can be in either of two states, A or B. If these particles are classical distinguishable particles that obey Maxwell–Boltzmann statistics, then at high temperatures the following four configurations each have probability $\frac{1}{4}$: 1 and 2 in A, 1 in A and 2 in B, 1 in B and 2 in A, and 1 and 2 in B. If the particles 1 and 2 are identical and obey Bose–Einstein statistics, then there are three possible symmetric wave functions: $A(1)A(2)$, $2^{-1/2}[A(1)B(2)+B(1)A(2)]$, and $B(1)B(2)$, and the probabilities are $\frac{1}{3}$ for the state corresponding to each wave function. Thus the particles that obey Bose–Einstein statistics are in the same state (either both in A or both in B) two thirds of the time, while the particles that obey Maxwell–Boltzmann statistics are in the same state only one half of the time. Identical particles that obey Fermi–Dirac statistics are never in the same state (Pauli exclusion principle). This clustering tendency, which increases with the number of particles, acts in the absence of forces between the particles; however, forces between the particles can modify this tendency. The clustering tendency plays an important role in processes, such as superradiant states and the action of the laser and maser, that involve coherent electromagnetic radiation. Einstein pointed out an extreme case of such clustering: A gas of identical particles obeying Bose–Einstein statistics will undergo a phase transition to a state in which a finite fraction of the particles occupies the lowest-energy quantum state when the temperature of the gas is brought below a critical temperature. This fraction goes to unity when the temperature goes to absolute zero. Liquid ^{4}He, which obeys Bose–Einstein statistics, undergoes such a transition (the λ transition) when it is cooled below 2.18 K. The detailed properties of the transition and of the superfluid state below the λ point depend on the Bose–Einstein statistics obeyed by ^{4}He and on the spectrum of excitations in the ^{4}He liquid. The λ transition does not occur in ^{3}He, which obeys Fermi–Dirac statistics.

See also: Fermi–Dirac Statistics; Maxwell–Boltzmann Statistics; Quantum Statistical Mechanics; Statistical Mechanics.

Bibliography

R. P. Feynman, R. B. Leighton, and M. Sands, *The Feynman Lectures in Physics*, Vol. III. Addison-Wesley, Reading, Mass., 1965. (E)

K. Huang, *Statistical Mechanics*, 2nd ed. Wiley, New York, 1987. (A)

L. D. Landau and E. M. Lifshitz, *Statistical Physics*. Addison-Wesley, Reading, Mass., 1958. (A)

J. E. Mayer and M. G. Mayer, *Statistical Mechanics*. Wiley, New York, 1940. (A)

R. K. Pathria, *Statistical Mechanics*. Pergamon, Oxford, 1985. (A)

F. Reif, *Fundamentals of Statistical and Thermal Physics*. McGraw–Hill, New York, 1965. (I)

E. Schrödinger, *Statistical Thermodynamics*. Cambridge, London and New York, 1962. (A)

Boundary Layers

C. W. Van Atta

Boundary layers are a ubiquitous feature of natural flow systems and of fluid flows in scientific and technological applications. For most boundary layers of geophysical or technological interest, the Reynolds number of the layer $R = U\delta/v$, where U is the mean velocity, δ the layer thickness, and v the kinematic viscosity, is sufficiently large that the flow is fully turbulent. Geophysical cases include the atmospheric surface layer, oceanic surface and benthic layers, western inertial boundary currents like the Gulf Stream and Kuroshio, and turbidity currents in submarine canyons and alluvial fans. Technological examples include external boundary layers on aircraft, ships, land vehicles, and buildings, and internal boundary layers in pipes, ducts, turbomachinery, shock tubes, wind tunnels, chimneys, and sewer outfalls. Biological examples include flow in the boundary layers on swimming fish (it has been suggested that the unusual speed of some fish may be due to reduction of the viscous boundary layer drag by compliant boundaries or secretion of drag-reducing additives), and blood flow in large arteries, for which the flow properties in the boundary layer may be important for determining the formation and degree of clotting.

Boundary layers are formed whenever a flowing fluid encounters a solid boundary or free surface, or is subject to other dynamically constraining forces over some restricted domain of the flow. The thickness of the layer is generally small compared with a characteristic dimension in the main flow direction. In some boundary layer problems, the external "free stream" conditions far from the boundary and conditions on the boundary are given, and we are interested in the variation of the flow variables (e. g., velocity, temperature, salinity, species concentration) as a function of distance from and along the boundary, or in the flux of momentum, energy, or other transferable property across the layer and the consequent shear stress on the boundary, rate of heat transfer between fluid and boundary, rate of ablation or evaporation of the boundary, etc.

In a viscous boundary layer mean pressure gradients produce a flux of mean vorticity across the fluid–solid boundary. When the Reynolds number of the flow is large enough, certain spatial derivatives of flow variables with respect to the principal flow direction along the boundary are small compared with derivatives in the transverse direction across the boundary layer, leading to justifiable neglect of certain terms in the Navier–Stokes equations for the fluid motion compared with other terms, an approximation first introduced by Ludwig Prandtl. When the Reynolds number of the flow is small enough for the motion to remain laminar, this leads to a well-posed theoretical problem whose solution is obtained by solving the boundary layer momentum and energy equations along with appropriate boundary conditions. When the flow is turbulent, fluctuating vorticity associated with coherent flow structures is produced by local, highly intermittent, flow instabilities. The result is much greater momentum and scalar transport, and hence substantially increased drag and heat transfer compared with laminar flows. The Reynolds-averaged Navier–Stokes equations do not lead to a closed set of equations for the statistical quantities describing the turbulent motion, and no theoretical solutions can be found without invoking further assumptions about correlations between fluctuating variables. These assumptions have taken many forms, from simple models for turbulent stresses patterned after molecular diffusion to elaborate modeling of higher-order correlations occurring in the hierarchy of statistical equations associated with the "closure" problem of turbulence theory. Some types of modeling have been found to yield computation schemes that are satisfactory for engineering calculations and for simulation of certain aspects of geophysical boundary layers.

Direct numerical simulations of the unaveraged Navier–Stokes equations for the case of a turbulent boundary layer on a smooth flow surface accurately predict the mean flow structure and turbulence statistics, and the computed unaveraged instantaneous flow exhibits many of the coherent flow structures observed in experiments.

In addition to the viscosity, thermal conductivity, and other physical properties of the fluid, the parameters of importance in describing the characteristics of any given boundary layer flow depend on the scale of the flow involved and on the various dominant physical mechanisms operating in the flow. The characteristic length scales of all of the technological applications mentioned earlier are sufficiently small that the effect of the rotation of the earth, and hence the Coriolis parameter f, can be neglected, but this parameter is of dominant influence in the planetary boundary layer. The thickness of a nonrotating turbulent boundary layer scales roughly like $\delta \sim (v/U)^{0.2} x^{0.8}$, where x is the distance along the boundary. In contrast, for the rotationally influenced atmospheric surface layer or upper mixed layer in the ocean, the thickness of the "Ekman layer" is independent of x and v and is proportional to $\tau^{1/2}/f$, where τ is the surface skin friction drag. In the nonrotating case the unopposed tendency of the turbulent eddies to mix momentum downward increases the vertical extent of the boundary layer with downstream distance. In the rotating case the vertical growth of the layer is apparently limited by the spontaneous formation of inertial oscillations when the layer depth exceeds a certain value. These instability waves then radiate energy away from the boundary layer, removing part of the source that would otherwise be available for vertical diffusion and growth. In the ocean, the stratification of the underlying thermocline plays an important role in determining the depth of the turbulent boundary layer. The dimensionless parameter describing the degree of stability of a stratified boundary layer is the Richardson number,

$$R_i = -g \frac{d\rho}{dz} \bigg/ \rho \left(\frac{dU}{dz} \right)^2$$

where ρ is the fluid density and z the vertical coordinate. The structure of the layer, as well as the heat, momentum, and density differences transported by it, is a sensitive function of R_i. R_i is approximately proportional to z/L, where z is the vertical height in the boundary layer and L is the Monin–Obukhov length, $L = u^3/kN$, where $u = (\tau/\rho)^{1/2}$, k is von Kármán's constant, and N is the net flux of buoyancy across the interface, proportional to the heat flux for the atmospheric boundary layer. Experimental results are then cast in the form of universal functions for dimensionless turbulent and mean quantities as functions of R_i or z/L. The Monin–Obukhov similarity theory has been successful in describing available data for the surface layer, and numerical models have developed to a state wherein some of the universal functions can be predicted for comparison with measurements.

The Coriolis forces are not important in the near surface layers ($z \leq 100$ m or so) of the atmosphere and ocean, but become increasingly important with increasing height or depth. The balance of Coriolis forces and turbulent stresses outside the inner surface layer produces the turbulent Ekman boundary layer, in which the direction of the mean velocity vector changes continuously throughout the depth of the layer from its external geostrophic direction to that of the surface stress at the boundary. The net fluid transport in the Ekman boundary layer is responsible for the wind-driven ocean circulation patterns of the oceans through the Sverdrup relation $\tau_N = - \operatorname{curl} \tau/(\partial f/\partial y)$, which states that the total northward fluid transport is equal to the inverse of the northward gradient of the Coriolis parameter times the curl of the wind stress on the ocean surface. The invalidity of this relation at a western boundary is related to the formation of inertial western boundary layers in the oceanic circulation. A westward-directed current having small relative vorticity tends to intensify into a thin jet as it is forced toward high latitudes by the eastern continental margins. The thickness of these major features, such as the Gulf Stream and the Kuroshio, is proportional to $(\bar{T}/\bar{f})^{1/2}$, where $\bar{T}$ is the total volume transport of the jet and $\bar{f}$ is a Coriolis parameter characteristic of the entire jet.

The relatively thin boundary layers formed on the vertical boundaries of turbulent convection cells for high Rayleigh numbers are thought to be important in convection in stars, in the earth's molten core, and in the earth's crust. The crustal flow is highly viscous and nonturbulent. At the vertical boundaries of the mantle convection cells, relatively thin, hot ascending jets drive spreading centers at oceanic ridges and descending jets form subduction trenches at opposite margins of the global tectonic plate system. The experimental and theoretical evidence indicates that boundary layers are the preferred natural modes for the transport of momentum, heat, and other forms of energy on the macroscopic level.

See also: Hydrodynamics; Rheology; Turbulence; Viscosity; Vortices.

Bibliography

G. K. Batchelor, *An Introduction to Fluid Mechanics*. Cambridge University Press, London and New York, 2000. (E)

T. Cebeci and A. M. O. Smith, *Analysis of Turbulent Boundary Layers*. Academic Press, New York, 1974. (A)

E. A. Eichelbrenner, "Three-Dimensional Boundary Layers," *Annu. Rev. Fluid Mech.* **5**, 339–360 (1973). (A)

H. Greenspan, *The Theory of Rotating Fluids*. Breukelen Press, 1990. (A)

J. O. Hinze, *Turbulence*. McGraw–Hill, New York, 1975. (I)

J. Kim, P. Moin, and R. D. Moser, "Turbulence Statistics in Fully Developed Channel Flow at Low Reynolds Number," *J. Fluid Mech.* **177**, 133–166 (1987).

J. L. Lumley, "Drag Reduction by Additives," *Annu. Rev. Fluid Mech.* **1**, 367–384 (1969). (I)

A. S. Monin and A. M. Yaglom, *Statistical Fluid Mechanics: Mechanics of Turbulence*, Vols. 1 and 2. MIT Press, Cambridge, MA, 1971, 1975. (I–A)

E. Reshotko, "Boundary-Layer Stability and Transition," *Annu. Rev. Fluid Mech.* **8**, 311–349 (1976). (A)

W. C. Reynolds, "Computation of Turbulent Flows," *Annu. Rev. Fluid Mech.* **8**, 183–208 (1976). (A)

A. R. Robinson, "Boundary Layers in Ocean Circulation Models," *Annu. Rev. Fluid Mech.* **2**, 293–312 (1970). (A)

L. Rosenhead, *Laminar Boundary Layers*. Dover, New York, 1988. (I–A)

H. Schlichting, *Boundary Layer Theory*, 8th ed. Springer, New York, 1999. (I)

E. A. Spiegel, "Convection in Stars: I. Basic Boussinesq Convection," *Annu. Rev. Astron. Astrophys.* **9**, 323–352 (1971); "II. Special Effects," **10**, 261–304 (1972). (A)

H. M. Stommel, *The Gulf Stream*, 2nd ed. University of California Press, 1977. (I)

I. Tani, "Boundary-Layer Transition," *Annu. Rev. Fluid Mech.* **1**, 169–196 (1969). (I-A)

A. A. Townsend, *The Structure of Turbulent Shear Flow*, 2nd ed. Cambridge University Press, London and New York, 1980. (I – A)

D. L. Turcotte and E. R. Oxburgh, "Mantle Convection and the New Global Tectonics." *Annu. Rev. Fluid Mech.* **4**, 33–68 (1972). (I – A)

J. S. Turner, *Buoyancy Effects in Fluids*. Cambridge University Press, London and New York, 1979. (I–A)

W. W. Willmarth, "Structure of Turbulence in Boundary Layers," *Adv. Appl. Mech.* **15**, 159–254 (1975). (A)

Bremsstrahlung

L. C. Maximon

One of the straightforward consequences of the classical theory of electrodynamics is that a charged particle, when accelerated, radiates energy in the form of an electromagnetic wave. The large radio broadcasting antennas are a familiar example of this phenomenon on a macroscopic scale. At the other end of the spectrum is an equally familiar example, used for medical diagnosis and therapy: the tubes in which an electron beam is accelerated by a high potential, focused, and directed at a target, with the emission of x rays. It was in describing the continuous spectrum of these x rays that the term bremsstrahlung (literally, braking radiation) was first coined by Sommerfeld in 1909. Here, however, the force acting on the electron is that of the screened nuclear charge, so that the acceleration takes place over atomic distances, necessitating a quantum-mechanical treatment. This was first given by Sommerfeld in 1931, for nonrelativistic electrons ($v/c \ll 1$), for the idealized problem in which the atom is represented by a pure Coulomb potential; the screening of the nucleus by the atomic electrons is neglected. With this simplification he was able to give the intensity, angular distribution, and polarization of the emitted x rays in a closed, analytic form. A fully relativistic calculation of the bremsstrahlung process, in which the electron wave functions are solutions of the Dirac equation, first appeared in 1934, given independently by Racah, by Sauter, and by Bethe and Heitler, the last two names generally being associated with the cross section for this process. These calculations require, for their validity, that the nuclear charge be small and the electron velocities high – specifically $2\pi(Z/137)/(v/c) \ll 1$ – in that they include the interaction between electron and nucleus only to first order (the Born approximation). Of greatest significance, both for analysis of experiments and for an understanding of the details of the process, is, however, the work of Bethe (1934) in which, for the experimentally important case of high electron energies (greater than a few MeV), he obtained the bremsstrahlung spectrum (the differential cross section integrated over the angles of the final particles) in a very simple form as an integral over the atomic form factor, accounting in this way for the screening of the nucleus by the atomic electrons. This analysis not only permitted comparison with experimental measurements, but also brought out very clearly the significant theoretical details of the process. Since then, and particularly in the last 25 years, when high-energy electron accelerators have become a most important tool for physics research, the increasing accuracy of experiments has provided an important incentive for the calculation of corrections to these theoretical expressions. Nonetheless, the articles just mentioned have remained an essential basis for most of the theoretical work that has been done on this subject.

There are a variety of reasons why the bremsstrahlung process occupies such an important place in physics. They are related to the fundamentals of the theory, to experiment, and to applications – to the nature of the process itself, in that bremsstrahlung is in fact present in every actual scattering of a charged particle: Since any detection apparatus has perforce a finite energy resolution, particles that have scattered elastically cannot be distinguished from those that have radiated a small amount of energy in the form of bremsstrahlung. This fact has important consequences for the analysis of almost all experiments using electron scattering as a probe to investigate nuclear structure, since the bremsstrahlung that accompanies the scattering (the radiative tail) must be subtracted in order to extract the purely nuclear information. Within the framework of the theoretical formalism, bremsstrahlung is very closely related to a number of other fundamental quantum-electrodynamical processes: inverse bremsstrahlung, pair production, radiative capture, and the photoelectric effect. The amplitude for each of these processes is closely related to that for bremsstrahlung; they differ formally only in that the particles are described by either outgoing or ingoing waves, positive or negative energy states, continuum or bound wave functions.

Thus far we have spoken only of the radiation from electrons. Experimentally they have been the most important source of bremsstrahlung, the cross section for the emission of bremsstrahlung from nonrelativistic charged particles being inversely proportional to the square of the mass of the particle. As the velocity of the particle approaches that of light, however, its mass becomes of decreasing importance. The radiation from heavier particles is therefore not easily detected unless the particles have relativistic velocities. With the advent of high-energy accelerators and the production of beams of heavier particles, the bremsstrahlung produced by the scattering of muons, pions, and protons has also been studied. In particular, the bremsstrahlung produced in proton–proton collisions has been studied extensively, both theoretically and experimentally, as a possible means of gaining greater understanding of the nucleon–nucleon force than is obtained solely from the analysis of elastic scattering experiments.

Another extension of the original work on bremsstrahlung concerns the variety of possible scatterers of the incident particle. In the aforementioned theoretical considerations the atomic electrons served only to screen the Coulomb potential of the nucleus, the recoil momentum being taken up by the atom as a whole, which remains in its ground state. However, the bremsstrahlung process may also take place in the field of the atomic electrons; they then absorb the recoil momentum and are either ejected or raised to an excited state. This effect was first calculated by Wheeler and Lamb in 1939. Subsequent improvements of their calculation have appeared, some quite recently, but in all theoretical work on this topic, both the screening of the scatterer by the other atomic electrons and its binding to the nucleus is neglected; the atomic electrons are treated as free electrons at rest. The closely related process of pair production in the field of the electron (triplet production) has also been the subject of recent experimental and theoretical research.

The field that serves to scatter the incident particle may, however, be magnetic rather than electric. The process is then known as magnetic bremsstrahlung (also referred to as cyclotron radiation when the emission is from nonrelativistic electrons, or as synchrotron radiation when it is from relativistic electrons). It was predicted on the basis of classical electromagnetic theory at the end of the nineteenth century, but the first observation was not made until 1947, and a detailed theoretical calculation of its properties (angular distribution, polarization, total energy loss) was first given by Schwinger in 1949. Since then, and particularly in the last few years, synchrotron radiation has become an increasingly important research tool. As a source of photons for research in other fields of physics, synchrotron radiation complements the more traditional bremsstrahlung, which is produced by scattering electrons from high-energy accelerators in targets placed in the beam. The latter has served for over 20 years as the primary source of photons for photonuclear research in the energy range 1 MeV to 1 GeV. Synchrotron radiation now provides an important source of photons for research in solid-state physics in the ultraviolet and soft x-ray regions and covers the energy range 5 eV to 50 keV (1–10000 Å).

Most recently, the bremsstrahlung process has become the subject of numerous studies concerned with the design of very high-energy (TeV) electron–positron linear colliders. In order to obtain the luminosity necessary for high-energy physics experiments, the beams in such colliders must have very small transverse dimensions ($\sim 10^{-4}$–10^{-8} cm) and the high density of charged particles provides strong electromagnetic fields as viewed by the particles of the opposite beam. The particles in each beam thus emit intense synchrotron radiation, termed *beamstrahlung*, in the collision region. These studies aim at providing a quantitative description of beam–beam interactions in linear colliders and investigating the new physics they make possible.

The foregoing discussion of bremsstrahlung has referred primarily to laboratory-produced phenomena. However, some of the most interesting research of the last few years has been in a number of branches of astrophysics, where bremsstrahlung and synchrotron radiation occur naturally, and as essential mechanisms in the physical processes. In solar physics, cyclotron and synchrotron radiation are largely responsible for the radio and optical emission from flares, while the emission of x rays and gamma rays is due mainly to bremsstrahlung arising in electron–electron and electron–ion (hydrogen and helium) collisions. [We note here the analogy to laboratory plasmas, for the operation of which both bremsstrahlung and cyclotron radiation are crucial processes. Their measurement serves as an important diagnostic tool; from it can be determined the plasma density, the electron temperature, the impurities (ions other than hydrogen isotopes), and the presence of high-energy particles in the plasma.] Farther away, but within the galaxy, synchrotron radiation is the primary source of radio waves, and both the continuum gamma-ray sources and the diffuse soft x-ray background are due to bremsstrahlung. Finally, there are the very important extragalactic sources of radiation. The x-ray emission from clusters of galaxies is believed to be due largely to bremsstrahlung, while the radio waves from the largest known energy source, the quasars, are due to synchrotron radiation.

See also: Electrodynamics, Classical; Quantum Electrodynamics; Synchrotron Radiation.

Bibliography
Classical Radiation
J. D. Jackson, *Classical Electrodynamics*, 2nd ed. Wiley, New York, 1975.(1)

Quantum-Mechanical Bremsstrahlung
V. B. Berestetskii, E. M. Lifshitz, and L. P. Pitaevskii, *Relativistic Quantum Theory*, Vol. 4 of *Course of Theoretical Physics*, Part I. Pergamon, New York, 1971.

H. A. Bethe and E. E. Salpeter, "Quantum Mechanics of One- and Two-Electron Systems," *Handbuch der Physik* (S. Flügge, ed.), Vol. 35, Atoms I. Springer-Verlag, Berlin and New York, 1957.

G. Diambrini Palazzi, "High-Energy Bremsstrahlung and Electron Pair Production in Thin Crystals," *Rev. Mod. Phys.* **40**, 611 (1968).

W. Heitler, *The Quantum Theory of Radiation*, 3rd ed. Oxford, London and New York, 1954.

H. W. Koch and J. W. Motz, "Bremsstrahlung Cross-Section Formulas and Related Data," *Rev. Mod. Phys.* **31**, 920 (1959).

J. W. Motz, Haakon A. Olsen, and H. W. Koch, "Pair Production by Photons," *Rev. Mod. Phys.* **41**, 581 (1969).

Haakon A. Olsen, *Applications of Quantum Electrodynamics*. Springer Tracts in Modern Physics, Vol. 44. Springer-Verlag, Berlin and New York, 1968.

A. Sommerfeld, *Atombau und Spektrallinien*, 3rd ed., Vol. II. F. Vieweg, Braunschweig, 1960.

Synchrotron Radiation
G. N. Kulipanov and A. N. Skrinskil, "Utilization of synchrotron radiation: current status and prospects," *Usp. Fiz. Nauk* **122**, 369 (1977). English translation *Sov. Phys. Vsp.* **20** (7), 559 (1977).

C. Kunz, ed. *Synchrotron Radiation, Techniques and Applications* (Springer Topics in Current Physics, Vol. 10). Springer-Verlag, Berlin, Heidelberg, New York, 1979.

E. M. Rowe and J. H. Weaver, "The Uses of Synchrotron Radiation," *Sci. Am.* **236**, 32 (1977). (I)

H. Winick and A. Bienenstock, "Synchrotron Radiation Research," *Annu. Rev. Nucl. Part. Sci.* **28**, 33 (1978).

H. Winick and S. Doniach, eds., *Synchrotron Radiation Research*. Plenum Press, New York, 1980.

Plasma Physics
G. Bekefi, *Radiation Processes in Plasmas*. Wiley, New York, 1966.

Beamstrahlung
M. Bell and J. S. Bell, "Quantum Beamstrahlung," *Particle Accelerators* **22**, 301 (1988); "End Effects in Quantum Beamstrahlung", *Particle Accelerators* **24**, 1 (1988).

R. Blankenbecler and S. D. Drell, "Quantum Treatment of Beamstrahlung", *Phys. Rev. D* **36**, 277 (1987); "Quantum Beamstrahlung from Ribbon Pulses," *Phys. Rev. D* **37**, 3308 (1988); "Quantum Beamstrahlung: Prospects for a Photon–Photon Collider," *Phys. Rev. Lett.* **61**, 2324 (1988).

P. Chen, "An Introduction to Beamstrahlung and Disruption," in *Frontiers of Particle Beams*, M. Month and S. Turner, eds. Lecture Notes in Physics, Vol. 296, pp. 495–532. Springer-Verlag, Berlin, Heidelberg, New York, 1988; "Review of Linear Collider Beam–Beam Interaction," SLAC-PUB-4823 (Jan. 1989), SLAC, Stanford, CA, also contributed to U.S. Particle Accelerator School, Batavia, IL, July 20–August 14, 1987; "Disruption, Beamstrahlung, and Beamstrahlung Pair Creation," SLAC-PUB-4822 (Dec. 1988), SLAC, Stanford, CA, also contributed paper at the DPF Summer Study Snowmass '88. High Energy Physics in the 1990's, Snowmass, Col., June 27–July 15, 1988.

M. Jacob and T. T. Wu, "Quantum Approach to Beamstrahlung," *Phys. Lett.* **B197**, 253 (1987); "Quantum Calculation of Beamstrahlung," *Nucl. Phys.* **B303**, 373 (1988), **B303**, 389 (1988). B. Richter, "Very High Energy Colliders," *IEEE Trans. Nucl. Sci.* **NS-32**, 3828 (1985).

P. B. Wilson, "Future e^+e^- Linear Colliders and Beam–Beam Effects," SLAC-PVB-3985 (May 1986) SLAC, Stanford, CA.

Solar and Astrophysics

T. Bai, "Studies on solar hard x-rays and gamma-rays; Compton backscatter, anisotropy, polarization, and evidence for two phases of acceleration," Ph.D. thesis, Univ. of Maryland, Goddard Space Flight Center Report X-660-77-85, April 1977. See also references in this work.

R. Giacconi and H. Gursky, eds., *X-ray Astronomy* (Astrophysics and Space Science Library, Vol. 43). Reidel, Dordrecht, 1974.

V. L. Ginzburg, *Elementary Processes for Cosmic Ray Astrophysics*. Gordon & Breach, New York, 1969.

V. L. Ginzburg and S. I. Syrovatskii, *The Origin of Cosmic Rays*. Pergamon, New York, 1964.

F. B. McDonald and C. E. Fichtel, eds., *High-Energy Particles and Quanta in Astrophysics*. MIT Press, Cambridge, Mass., 1974. A. G. Pacholczyk, *Radio Astrophysics*. W. H. Freeman, San Francisco, 1970.

G. L. Verschuur and K. I. Kellermann, eds., *Galactic and Extra-Galactic Radio Astronomy*. Springer-Verlag, Berlin and New York, 1974.

T. C. Weeks, *High-Energy Astrophysics*. Chapman & Hall, London, 1969. (I)

R. J. Weymann, T. L. Swihart, R. E. Williams, W. J. Cocke, A. G. Pacholczyk, and J. E. Felton, *Lecture Notes on Introductory Theoretical Astrophysics*. Pachart, Tucson, Arizona, 1976.

Brillouin Scattering

J. L. Hunt

Brillouin scattering in a pure fluid is the scattering of light by transitory and localized thermally driven fluctuations in the density of the scattering medium. The effect was first predicted by Brillouin, later by Mandel'shtam (in Russia it is known as "Mandel'shtam–Brillouin scattering"), and was first observed by Gross. The history of the subject before the advent of the laser is reviewed by Rank [1].

The thermally driven density fluctuations in a fluid are of two types: the adiabatic (pressure) fluctuations and the isobaric (entropy) fluctuations. The former are collective modes of motion of the molecules and propagate in the fluid with the velocity of sound, and, thus, light scattered from them experiences a Doppler shift. The latter do not propagate and so do not alter the frequency of the scattered light; they correspond to a diffusional motion of the molecules.

The propagating fluctuations may be viewed as a set of Debye waves traveling in all directions with speed v in the fluid (see Fig. 1). The choice of a direction *IX* for the input beam of light and an observing direction *XO* selects two sets of plane waves which satisfy the law of reflection. The components of the sound speed of these two sets in the direction of observation are $\pm v\sin(\theta/2)$ which yield equal and opposite Doppler shifts (Δf) whose magnitude relative to the frequency (f) of the incident light is

$$\left|\frac{\Delta f}{f}\right| = 2n\frac{v}{c}\sin\frac{\theta}{2} \ . \tag{1}$$

In Eq. (1), n is the index of refraction and c is the velocity of light in vacuum.

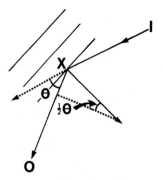

Fig. 1: Reflection of light from plane sound waves.

The expected "Brillouin spectrum," i.e., the distribution of scattered light intensity as a function of frequency, should thus consist of three components: one at frequency f (scattered from the nonpropagating fluctuations), one at $f + |\Delta f|$, and one at $f - |\Delta f|$. The whole spectrum is sometimes called the Rayleigh–Brillouin triplet. In all of the foregoing, it is assumed that the light is polarized with the electric vector perpendicular to the plane of *IXO* and that the scattered light is also observed in this polarization.

The recording of Brillouin spectra requires special techniques since the frequency shifts, Δf, are small ($\Delta f/f = 10^{-6}$), and high-resolution optics are required, e.g., Fabry–Perot etalons. The use of the laser as a light source has resulted in greatly increased activity in Brillouin scattering research. Polarized Brillouin spectra have been observed in almost all isotropic media, e.g., organic and inorganic liquids, amorphous solids, crystals, and gases. An example of the Rayleigh–Brillouin triplet is shown for benzene in Fig. 2.

The ratio of the intensity, I_c, of the central component to the intensity, $2I_B$ of the components shifted in frequency is called the "Landau–Placzek ratio":

$$\frac{I_c}{2I_B} = \frac{c_p - c_V}{c_V} = \gamma - 1 , \qquad (2)$$

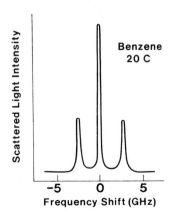

Fig. 2: The Rayleigh–Brillouin triplet for benzene at 20 °C.

where c_p and c_V, are the specific heats of the fluid at, respectively, constant pressure and constant volume, and $c_p = \gamma c_V$. The formula is only approximately correct and many refinements to it have been made for various media.

Even if the incident light were strictly monochromatic and the spectrum were infinitely well resolved, the shifted components in the spectrum would have a finite width due to the damping of the sound waves in the medium. Thus the width of the shifted components is related to the viscosity of the fluid. The width of the unshifted component is related to the thermal diffusivity.

Another broad non-Lorentzian component of the scattered spectrum with its central frequency at the unshifted position is called the "Mountain mode". As the temperature of the scattering sample is lowered, it narrows and appears as a contribution to the Rayleigh line. This contribution arises because of the dispersion of the shear and bulk viscosities and the resulting distribution of relaxation times.

A different spectrum results when the incident light is polarized with the electric vector perpendicular to the plane *IXO* but observations are made on light polarized in the plane. This depolarized light is in part due to the reorientation of anisotropic molecules; this scattering is not usually called "Brillouin scattering".

Brillouin scattering has also been observed in binary mixtures of fluids. In this case light scattering may also occur as a result of concentration fluctuations; thus information on diffusion coefficients and reaction rates may be obtained. A further type of scattering is "stimulated Brillouin scattering". In this type of scattering the intensity of the scattered light increases nonlinearly with the dimensions of the scattering volume. The effect is observed when very intense sources of light, such as the pulsed ruby laser, are used. In this case, the electric field of the light is sufficiently high to alter (through electrostriction) the nature of the sound waves. Thus the incident and scattered light now alters the motion of the medium whereas in normal Brillouin scattering the motion of the medium is completely described by the Debye waves.

Brillouin scattering is a widely used analytical tool for investigating properties of matter at the molecular level (e. g., relaxations) and bulk properties (e. g., sound velocity and moduli).

A thorough discussion and bibliography of Brillouin scattering up to 1967 can be found in the book by Fabelinskii [2]. More recent reviews are those of Patterson and Carroll [3] dealing with fluids and Dil [4] on condensed matter including solids. The book by Damzen *et al.* [5] deals with theoretical and experimental aspects of the stimulated effects.

See also: Doppler Effect, Light Scattering.

References
[1]   D. H. Rank, *J. Acoust. Soc. Am.* **49**, 937 (1971).
[2]   I. L. Fabelinskii, *Molecular Scattering of Light*. Plenum Press, New York, 1968.
[3]   G. D. Patterson and P. J. Carroll, *J. Phys. Chem.* **89**, 1344 (1985).
[4]   J. G. Dil, Rept. *Prog. Phys.* **45**, 285 (1982).
[5]   M. J. Damzen, V. I. Vlad, V. Babin and A. Mocofonescu, *Stimulated Brillouin Scattering*. IOP Publishing, Philadelphia, 2003.

Brownian Motion

Frank A. Blood, Jr.

Brownian motion is a random motion of microscopic particles suspended in a fluid. The Scottish botanist Robert Brown, although not the first to see this phenomena, observed it in 1827 for many different types of materials – particles within various pollen grains, powdered coal, glass, and metals – suspended in liquids [1]. Smoke particles in air also show this random motion. A satisfactory explanation of these observations was given in 1906 by Einstein (and separately by Smoluchowski) and constituted the most direct proof up to that time of the existence of atoms. The small random motions occur because atoms bombard the smoke particle in a haphazard manner, sometimes applying more force on one side than another. This random variation in the force causes the particle to move in a random fashion, sometimes called a random walk [2]. An analogy is that of BBs being shot from all directions toward a billiard ball. Slight fluctuations in the number of BBs hitting one side will cause the billiard ball to move.

Einstein derived a formula [3–5] which can be compared with observation. Suppose you observe one particle and note how far it is from where it started after 1 min, 2 min, etc. Then you make the same measurements for many particles and take the average distance traveled after 1 min, 2 min, etc. Einstein's formula says that the average distance traveled is proportional to the square root of the time. That is, a particle, on the average, is 1.414 times farther from the origin after 2 min than it was after 1 min. Furthermore, the average distance traveled depends on the temperature of the fluid the particle is in. The reason for this is that the atoms move faster at higher temperatures and therefore kick the particle harder and farther. The average distance moved also becomes smaller as the size of the particle becomes larger, and, in fact, the Brownian motion is not even observable for large smoke particles.

Another method of observing Brownian motion is to look at the angular movements of a very small mirror suspended from fine quartz fibers. An accurate determination of Boltzmann's constant can be made from such observations [6–7]. An analogous phenomenon, Johnson noise [8–10], occurs in electrical circuits. The density of electrons in a wire will fluctuate, again because of random thermal motions. Too many electrons (or too few) in a particular region will produce an electric field and therefore a voltage. The average size of the random fluctuation in voltage in a resistor depends on the resistance and, again, on the temperature. As an example, a 400 000-Ω resistor at room temperature will have an average spurious voltage fluctuation of about 10 μV when fluctuations faster than 20 000 Hz are discounted.

See also: Fluctuation Phenomena.

References

[1]   *Dictionary of Scientific Biography*, Vol. II, p. 516. Scribners, New York, 1970.

[2]   R. P. Feynman, R. B. Leighton, and M. Sands, *The Feynman Lectures on Physics*, Vol. 1, Sec. 6, p. 5. Addison-Wesley, Reading, Mass., 1963. (E)

[3]   F. Reif, *Fundamentals of Statistical and Thermal Physics*, pp. 560–564. McGraw–Hill, New York, 1965. (I)

[4]   G. Wannier, *Statistical Physics*, pp. 475–478. Wiley, New York, 1966.

[5]   A. Sommerfeld, *Thermodynamics and Statistical Mechanics*, pp. 180–183. Academic Press, New York, 1965. (I)

[6]   R. P. Feynman *et al.*, in Ref. 1, Sec. 41, p. 2.

[7]   A. Sommerfeld, in Ref. 4, pp. 486–489.

[8]   R. P. Feynman *et al.*, in Ref. 1, Sec. 41, p. 2.

[9]   G. Wannier, in Ref. 3, pp. 486–489.

[10]   A. L. King, *Thermophysics*, pp. 212–214. Freeman, San Francisco, 1962. (I)

Calorimetry

E. F. Westrum, Jr.

A calorimeter is a device for determining energetic quantities typically by measurement of temperature changes produced upon input of electrical work to a system or by heat released by some reaction or process. Typical examples are:

Heat capacities, defined by $C_x = (dq/dT)_x$. Here dq is the heat involved (e. g., as laser radiation, the electrical supplied work, etc.) and x represents the constraints under which the experiment is performed such as constant pressure, constant volume, constant magnetic field, etc.

Enthalpies of phase transitions (e. g., solid/solid, fusion, vaporization, sublimation) or enthalpy increments associated with reactions, mixing, solution processes, or with transitions involving order/disorder, magnetic, ferroelectric, electronic (e. g., Schottky contributions associated with electronic energy levels) phenomena, etc.

Transport properties of materials such as the thermal diffusivity.

Such experimental quantities provide precise characterization of the energetic states of matter. At very low temperatures, the heat capacity assumes a primary role in such matters. At increasingly higher temperatures – but still cryogenic – one enters a thermal chaos in the complexity of the energy spectrum of the thermal properties, such that heat-capacity studies in this region are of considerably less interest to physicists than are those at extremely low temperature. On the other hand, the region above 10 K to – and above – ambient temperatures is of great interest to chemical thermodynamicists concerned with thermophysical properties at or above ambient temperatures. The region from 30 to 300 K is the region over which most of the entropy and the enthalpy are developed. Since the energy interpretation becomes largely statistical in nature, the entropy – particularly the entropy increments associated with the various transformations of matter – often become a more relevant parameter or criterion than does the heat capacity from which they were derived. Hence, this temperature region also

Encyclopedia of Physics, Third Edition. Edited by George L. Trigg and Rita G. Lerner
Copyright © 2005 WILEY-VCH Verlag GmbH & Co. KGaA, Weinheim
ISBN: 3-527-40554-2

assumes a crucial role in the critical evaluation of ambient-temperature chemical thermodynamic values for science (physics, chemistry, materials science, biology, geology, mineralogy,
and petrology).

Thus, entropy S in a heat-capacity or gradual transformation region provides a definitive
thermodynamic criterion:

$$S_x = \int_0^T \frac{C_p(T)\,dT}{T} + S_0$$

in which upon evaluation of the entropy at zero Kelvin (S_0), practical "absolute" values of
the entropy can be ascertained at higher temperatures for thermodynamic utilization. For an
isothermal phase transition,

$$\Delta_{trs}S = \Delta_{trs}H/T_{trs} \ .$$

Such thermophysical and thermochemical data are invaluable in establishing thermodynamic
properties which in turn are needed in the design of physical apparatus or of technological
chemical operations. The design of calorimetric apparatus ranges over many types since a
great variety of substances, changes of state, and thermodynamic processes may be studied.
In physics, quantum effects are often studied with low-temperature heat-capacity cryostats
employing solid or liquid nitrogen, liquid hydrogen, liquid helium-4, or liquid helium-3 as
refrigerants.

A typical aneroid cryostat is depicted in Figure 1. The sample under investigation, usually
a massive, crystalline, or even a finely divided specimen, is contained in a calorimeter (**D**),
sealed so that a few kPa of helium may be added to ensure thermal diffusivity between sample, heater, and thermometer. An electrical "heater" (i.e., a coil of resistance wire through
which electrical work may be provided) and a capsule-type resistance thermometer (**E**) are
often inserted in an axial entrant well with a film of grease to provide adequate thermal contact. A surrounding jacket or shield surrounds the calorimeter to reduce the heat exchange
with the surroundings to a negligible value. This is achieved by controlling the temperature
of the shield (**C**) or the principal portions of it dynamically to follow the temperature of the
calorimeter within the ability of thermocouples (**F**) to monitor the difference. Separate channels of an analog microprocessor control the adiabatic shield. To enhance the quality of this
control two other channels of such control monitor the temperature of the surrounding guard
shield (**B**) as well as that of the ring (**A**) to "temper" the temperature of the bundle of electrical
lead wires – a main source of thermal leak at low temperatures.

Usually the determination of the temperature increment (10^{-6} to $10\,K$) is the limiting factor
in the accuracy of the heat-capacity measurement as the energy increment can be readily
ascertained. Temperature determination involves not only the measurement of the resistance
of the thermometer, but the stability and calibration of the thermometer as well. Moreover, the
establishment of the temperature scale is relevant.

A series of certified heat-capacity standards established by the (U.S.) Calorimetry Conference have enabled calorimetrists to test the reliability of their calorimetric systems.

Because thermometry often limits the accuracy of the calorimetric results, it should be
noted that any temperature-dependent property such as electrical resistance, magnetic susceptibility, vapor pressure, or PVT relations of gases can be used as the thermometric property. Electrical resistance thermometers, capsule-type with a fine helically wound coil of wire

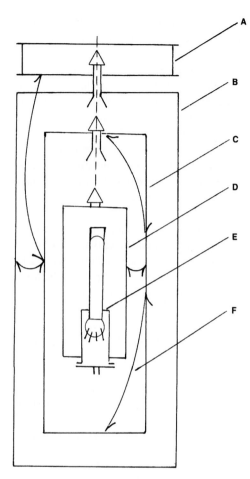

Fig. 1: Schematic cryogenic, adiabatic, heat-capacity cryostat (a superambient thermostat would be similar except in the provision of additional heated and/or floating guard shields). **A**, ring for tempering lead bundle; **B**, guard shield (heated); **C**, adiabatic shield (three separate controlled temperature regions); **D**, calorimeter containing sample; **E**, resistance thermometer (within "heater" sleeve); **F**, one of five differential thermals (multijunction thermocouples).

wound in a strain-free and noninductive configuration on a notched mica cross-like frame, are commonly employed in cryogenic calorimetry. These are typically sealed within a platinum capsule sealed with a glass head to permit entry of the leads and to provide for the sealing of a small amount of helium gas to assist thermal diffusion. Thermal contact with calorimeter vessel or the "heater" sleeve is achieved with vacuum grease or low-melting solder. The sensitivity of platinum resistance thermometers diminishes rapidly below about 20 K and they become too insensitive for accurate thermometry. The use of an iron-rhodium thermometric alloy avoids this problem to below the temperatures of liquid helium (4 K). At sufficiently low temperatures semiconducting resistors with negative temperature coefficients of resistance – even ordinary carbon resistors – are often used. Encapsulated doped germanium resistance thermometers provide a more accurate thermometer which may be cycled over the temperature range without effect.

More exotic thermometers are used below 0.1 K including magnetic moment and γ- and x-ray emission. (Such thermometers are described in Ref. [2].)

The utilization of thermocouples and thermals in ascertaining temperature differentials in the control of adiabatic shields and in relating the temperature of the calorimeter to a more remotely located (more massive) accurate temperature sensor is almost universal.

Although modes of calorimeter operation other than the adiabatic exist – or the quasiadiabatic where a slight offset in the control temperature relative to that of the calorimeter provides effective cancelation of the energy supplied by thermometric current – adiabatic working does permit measurements at *equilibrium*. Isoperibol operation – with a shield at a more or less fixed temperature – does have some convenience at very low temperatures.

At higher temperatures, calorimetry by the method of mixtures or drop calorimetry is often useful. Inverse temperature drop calorimetry has certain advantages in eliminating slow equilibrium and hysteresis effects. Because of the availability of computerized commercial units requiring only milligram size samples, differential scanning calorimetry (dsc) is often useful above 100 K.

By computerized automation in calorimetric operation and data acquisition even equilibrium, adiabatic measurements do not require much more operator endeavor than does dsc.

A recent comprehensive monograph on thermophysical calorimetric techniques [1] updates the earlier treatise [2] published by The International Union of Pure and Applied Chemistry as well as the excellent discussion of cryogenic and thermometric techniques at extremely low temperature [3]. A more general treatise is also available [4].

See also: Heat Capacity; Magnetic Cooling; Temperature; Thermometry

Bibliography
C. Y. Ho (ed.) *Specific Heat of Solids, Cindas Data Series on Materials Properties, Volume I–II*. Hemisphere, New York, 1988. (I)
J. P. McCullough and D. W. Scott, *Experimental Thermodynamics, Vol. I: Calorimetry of Non-reacting Systems*. Butterworth, London, 1968. (A)
O. V. Lounasmaa, *Experimental Principles and Methods below* 1 K. Academic, London, 1974. (A)
W. Hemminger and G. Hohne, *Calorimetry – Fundamentals and Practice*. VCH Publishers, New York, 1984. (Also available in a German edition.) (E)

Capillary Flow

J. C. Melrose

The term capillary flow refers to the motion of two or more fluid phases within the interstices of a finely porous solid. Viscous flows of this type are encountered in a wide variety of contexts in natural science and technology. The transport of water in soils and the recovery of oil and gas from reservoir rocks are examples of particular importance. Other well-known examples are found in such diverse fields as paper and textile manufacturing, food processing, and powder metallurgy.

The driving forces responsible for capillary flow involve the fluid pressure differences associated with the highly curved interfaces between the several fluid phases (menisci). For most cases in which capillary flow is of significance, the average diameter of the pores will be of the order of 10^{-5}–10^{-1} mm. Consequently, the effect of gravity on the curvature of the interfaces is either negligible or can be accounted for by small corrections. Also, the flows in question will usually be sufficiently slow that hydrostatic conditions are approached in the vicinity of the interfacial regions separating any pair of fluid phases.

In these circumstances, the pressure differences that give rise to the driving forces for capillary flow are determined, at least approximately, by the classical hydrostatic principles applicable to fluid interfaces. The first of these is the Young–Laplace equation, $P_c = \gamma J$. Here, P_c is the static pressure difference (capillary pressure) between a pair of contiguous bulk fluids, γ is the corresponding fluid/fluid interfacial tension, and J is the (constant) mean curvature of the interface separating the two phases. A rigorous thermodynamic analysis of the Young–Laplace equation, as developed initially by Gibbs, indicates that the dependence of γ on curvature is of significance only when the reciprocal of the curvature, J^{-1}, approaches molecular dimensions.

For nonaxisymmetric interfaces the precise description of a surface of constant mean curvature involves the solution of a nonlinear partial differential equation of the elliptic type. However, in the simple case of a very small cylindrical capillary tube with uniform wetting characteristics, such a surface is spherical in shape, and the mathematical difficulties arising in the general case are avoided.

A second principle is the Young equation, $\gamma_{12} = \gamma_{13} + \gamma_{23}\cos\theta$. Here, subscript 1 refers to the solid phase, while subscripts 2 and 3 refer to the two fluid phases in contact with the solid. The Young equation thus interrelates the properties (tensions) of three interfaces that meet at a three-phase contact line. Included in this relationship is the geometrical parameter characterizing the confluence of phase boundaries (the contact angle, θ). This principle and the Young–Laplace equation represent the line and surface analogs of the condition for hydrostatic equilibrium applicable to bulk fluid phases. Whereas the Young–Laplace equation corresponds to a differential equation, the contact angle (as defined by the Young equation), together with the configuration of the solid surface comprising an individual pore, provides the boundary condition required for its solution.

Interfacial properties and the physical principles that determine static interfacial configurations are thus of central importance in capillary flow. The dynamical process by which one fluid displaces another in a cylindrical tube of small diameter is a simple and well-known example of such a flow. In this special case the curvature of the meniscus separating the two fluids under static conditions is given by twice the ratio of the cosine of the contact angle (assumed to be uniform) to the radius of the tube. Under dynamic conditions, however, the flows that occur in the vicinity of the moving fluid interface are quite complex, and it is observed that the so-called dynamic contact angle is generally rate dependent. When this complicating factor is ignored, the analysis of the capillary flow in a small cylindrical tube is straightforward, yielding the well-known Washburn equation.

More generally, typical porous solids possess very complex systems of interconnecting pores, any one of which has a varying cross-sectional area available for flow. Regular or irregular packings of spherical solid particles provide examples of such complex pore systems. The pores in this more general type of porous solid constitute channels that are alternately convergent and divergent. Consequently, the capillary pressure and the meniscus curvature, for a particular static configuration of the meniscus, no longer depend simply on the cosine of the contact angle. This more complex dependence on contact angle is frequently overlooked in analyzing capillary flow behavior in real porous solids. The structural feature associated with pore channels which are alternately convergent and divergent also plays an important role in determining which configurations of the fluid/fluid interfaces are actually stable under hydrostatic conditions. A third physical principle, also governing static capillary phenomena, must therefore be considered in the case of typical porous solids.

This principle is a stability condition that must be satisfied by the various individual fluid–fluid interfaces within the pores. It is found that the quasi-static motion of an interface through a given pore involves interfacial configurations that are alternately stable and unstable. A rather sudden hydrodynamic event (Haines jump or rheon) will occur when the configuration of an interface within a particular pore becomes unstable. Displacement within those pores that are occupied by interfaces is therefore initiated by a slow flow process, which is then followed by a rapid burst of flow. The process is completed when the interface achieves a stable configuration in a neighboring pore.

Sudden flow, or rheons, resulting from unstable fluid/fluid interfacial configurations, are thus a microscopic feature of the highly irregular displacement front characterizing displacement processes in typical porous solids. Also associated with such hydrodynamic events is the trapping of segments of the displaced fluid in certain pores or in certain portions of typical pores. The interfacial configurations and the relative volume of the displaced phase that is trapped depend on the degree to which the pores are interconnected. These factors are also dependent on the prevailing curvature (convex or concave with respect to the displaced phase) of the fluid interfaces, i. e., on whether the displacement is the so-called drainage or imbibition type of process.

It should be noted in this connection that the distinction between drainage and imbibition is difficult to make in the case of intermediate wetting conditions, i. e., when the contact angle is more than about 40° or less than about 140°. Under strongly wetting conditions, however, the term drainage unambiguously refers to the displacement of the wetting phase (contact angle less than 40°) by the nonwetting phase, while imbibition refers to the reverse process. If these processes are each carried out under quasi-static conditions, two different relationships between the interfacial curvature (proportional to the static pressure difference between the two bulk fluid phases) and the relative volume of one of the fluids are defined. These relationships therefore characterize the hysteresis associated with the two different quasi-static displacement paths. This type of hysteresis is different from and should be distinguished from the hysteresis that is frequently observed in the value of the contact angle itself.

The term capillary flow often refers to a displacement process of the imbibition type, since external driving forces are required in drainage processes. Consequently, the interfacial curvature is concave with respect to the displaced phase. Thus, it is the nonwetting phase that is subjected to the fluid-phase entrapment associated with this type of capillary flow. It is found that the relative volume of trapped fluid varies from about 10% to as much as 50%, depending on the structure of the porous solid.

In many applications, capillary flow processes of the drainage type are also encountered. Often, one of the fluid phases is a gas, while the other is a volatile liquid. In this case, adsorption at the gas/solid interface of those components of the liquid which are volatile must be taken into account. First, it is found that the adsorption is usually sufficient to establish a zero value of the contact angle. Second, the total volume or mass of the wetting liquid retained within the solid phase at a particular value of the capillary pressure depends on the thickness of the adsorbed (wetting) film, as well as on the principles of static capillary phenomena already considered.

This extension of the physical principles involved in capillary flow clearly requires a consideration of diffusional transport in the gas phase, as well as viscous flow in the liquid phase. The classical hydrostatic principles applicable to fluid interfaces must be supplemented by the Gibbs principle of chemical equilibrium between the gas and liquid phases. This principle is expressed by the well-known Kelvin equation (even though not originally so conceived).

See also: Hydrodynamics; Surface Tension.

Bibliography

F. P. Buff, "The Theory of Capillarity," in *Handbuch der Physik*, S. Flügge, ed., Vol. 10, pp. 281–304. Springer-Verlag, Berlin and New York, 1960.

F. A. L. Dullien, *Porous Media*, 2nd ed. Academic Press, New York, 1992.

E. B. Dussan V., "On the Spreading of Liquids on Solid Surfaces: Static and Dynamic Contact Lines," *Ann. Rev. Fluid Mech.* **11**, 371–400 (1979).

D. H. Everett, "Pore Systems and their Characteristics," in *Characterization of Porous Solids*, K. K. Unger, J. Rouquerol, and K. S. W. Sing (eds.), pp. 1–22. Elsevier, Amsterdam, 1988.

R. Finn, *Equilibrium Capillary Surfaces*. Springer-Verlag, New York, 1986.

P. G. De Gennes, "Wetting: Statics and Dynamics," *Rev. Mod. Phys.* **57**, 827–863 (1985). J. C. Melrose, "Thermodynamics of Surface Phenomena," *Pure Appl. Chem.* **22**, 273–286 (1970).

J. R. Philip, "Flow in Porous Media," *Ann. Rev. Fluid Mech.* **2**, 177–204 (1970).

G. F. Teletzke, H. T. Davis, and L. E. Scriven, "How Liquids Spread on Solids," *Chem. Eng. Comm.* **55**, 41–81 (1987).

R. A. Wooding and H. J. Morel-Seytoux, "Multiphase Fluid Flow through Porous Media," *Ann. Rev. Fluid Mech.* **8**, 233–274 (1976).

Carnot Cycle

M. W. Zemansky[†]

A heat engine, such as an internal combustion engine or a steam engine, is a device for converting heat into work. To this end, a substance, such as a mixture of fuel and air or a mixture of water and steam, called the *working substance*, is caused to undergo a series of processes, called a *cycle*, in which the working substance returns periodically to its initial state. During a part of a heat engine cycle, some heat Q_H is absorbed from a chemically reacting mixture of fuel and air, known as the *hot reservoir*, and during another part of the cycle, a smaller amount of heat Q_C is rejected to a *cold reservoir*. The engine is therefore said to operate between these

---

[†]deceased

two reservoirs. The difference between the heat Q_H and the smaller quantity Q_C is converted to work W, so that

$$W = Q_H - Q_C .$$

The work W is the output of the engine and the heat Q_H is the input, so that the efficiency E is

$$E = \frac{W}{Q_H} = \frac{Q_H - Q_C}{Q_H} = 1 - \frac{Q_C}{Q_H} .$$

Since it is a fact of experience that some heat is always rejected to the cooler reservoir, the efficiency of an actual engine is never 100%. If we assume that we have at our disposal two reservoirs at given temperatures, it is important to answer the following questions: (1) What is the maximum efficiency that can be achieved by an engine operating between these two reservoirs? (2) What are the characteristics of such an engine? (3) Of what effect is the nature of the substance undergoing the cycle? The importance of these questions was recognized by Nicolas Léonard Sadi Carnot, a brilliant young French engineer, who in the year 1824, before the first law of thermodynamics was firmly established, described, in a paper entitled "Reflexions sur la puissance motrice du feu," an ideal engine operating in a particularly simple cycle known today as the *Carnot cycle*.

A Carnot cycle is a set of processes that can be performed by any thermodynamic system whatever, whether chemical, electrical, magnetic, or other. The working substance is imagined first to be in thermal equilibrium with a cold reservoir at the temperature T_C. Four processes are then performed, in the following order:

1. A reversible adiabatic process is performed in such a direction that the temperature rises to that of the hotter reservoir, T_H.

2. The working substance is maintained in contact with the reservoir at T_H and a reversible isothermal process is performed in such a direction and to such an extent that heat Q_H is absorbed from the reservoir.

3. A reversible adiabatic process is performed in a direction opposite to (1) until the temperature drops to that of the cooler reservoir, T_C.

4. The working substance is maintained in contact with the reservoir at T_C and a reversible isothermal process is performed in a direction opposite to (2) until the working substance is in its initial state. During this process, heat Q_C,is rejected to the cold reservoir.

An engine operating in a Carnot cycle is called a *Carnot engine*. A Carnot engine operates between two reservoirs in a particularly simple way. All the heat that is absorbed is absorbed at a constant high temperature, namely, that of the hot reservoir. Also, all the heat that is rejected is rejected at a constant lower temperature, that of the cold reservoir. Since all four processes are reversible, the Carnot cycle is a reversible cycle. The expression "Carnot engine," therefore, means "a reversible engine operating between *only* two reservoirs."

It was shown by Kelvin that an absolute temperature scale could be defined in terms of the operation of a Carnot engine. If T_H and $T Q_C$ are the Kelvin temperatures of the hot and cold reservoirs, respectively, then Kelvin showed that

$$\frac{Q_C}{Q_H} = \frac{T_C}{T_H} ,$$

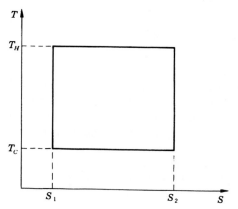

Fig. 1: A Carnot cycle of any system when represented on a *T–S* diagram is a rectangle.

and the Carnot efficiency becomes

$$E(\text{Carnot}) = 1 - \frac{T_C}{T_H} \, .$$

For a Carnot engine to have an efficiency of 100% T_C must be zero. Since nature does not provide us with a reservoir at absolute zero, a heat engine with 100% efficiency is a practical impossibility.

A temperature–entropy diagram is particularly suited to display the characteristics of a Carnot cycle. The two reversible adiabatic processes are vertical lines, and the two reversible isothermal processes are horizontal lines lying between the two vertical lines, so that the Carnot cycle is represented by a rectangle, as shown in Fig. 1. This is true regardless of the nature of the system.

In describing and explaining the behavior of this ideal engine, Carnot made use of three terms: *feu, chaleur,* and *calorique*. By feu he meant fire or flame, and when the word is so translated no misconceptions arise. Carnot gave, however, no definitions for chaleur and calorique, but in a footnote stated that they had the same meaning. If both of these words are translated as heat, then Carnot's reasoning is contrary to the first law of thermodynamics. There is, however, some evidence that, in spite of the unfortunate footnote, Carnot did not mean the same thing by chaleur and calorique. Carnot used chaleur when referring to heat in general, but when referring to the motive power of heat that is brought about when heat enters at high temperature and leaves at low temperature, he used the expression "chute de calorique," never "chute de chaleur." It is the opinion of a few scientists that Carnot had in the back of his mind the concept of entropy, for which he reserved the term calorique. This seems incredible, and yet it is a remarkable circumstance that, if the expression chute de calorique is translated "fall of entropy," many of the objections to Carnot's work raised by Kelvin, Clapeyron, Clausius, and others are no longer valid. In spite of possible mistranslations, Kelvin recognized the importance of Carnot's ideas and put them in the form in which they appear today.

See also: Thermodynamics, Equilibrium; Heat Engines.

Casimir Effect

G. Roepstorff

Overview

In 1948 the Dutch physicist Hendric Casimir published a famous paper [1] in which he offered an unusual explanation for the origin of the van der Waals force between molecules. The novel understanding of its nature was achieved by shifting the emphasis from *action at a distance* to *local action* of the quantized photon field: the short-distance attraction was thus thought of as a manifestation of the zero-point field energy. It was much later that the paper received proper attention, and at present the influence of the basic ideas behind the calculation on our concept of the physical vacuum and the Casimir force, as it is called now, is immeasurable.

It was predicted and later on experimentally verified – first measurements were carried out by M. J. Sparnaay in 1957 – that two uncharged perfectly conducting parallel plates attract each other when separated by a very small distance a of the order $1\,\mu m$. As found by Casimir, the force per unit area is

$$F = -\frac{\pi^2}{240}\frac{\hbar c}{a^4} \,. \tag{1}$$

But the essence of the paper may be phrased differently: the cavity between the two plates exerts a negative pressure, F, on its boundary. The smallness of the effect is immediately seen from

$$a^4|F| = \frac{\pi^2}{240}\hbar c = 1.30014 \cdot 10^{-27}\,\text{Newton m}^2 \,. \tag{2}$$

The pressure follows from the vacuum energy per unit area,

$$E = -\frac{\pi^2}{720}\frac{\hbar c}{a^3} \,, \tag{3}$$

by taking the negative derivative with respect to a. The startling fact is that the Casimir cavity, an empty space, carries a negative energy density, $\rho = E/a$, whereas typical ground state energies of quantum oscillators such as $\frac{1}{2}\hbar\omega$ are all positive. The assertion here is that by removing a divergent *inefficient* part of the zero-point energy the *efficient* part comes out negative but finite.

At first sight the claim that the vacuum energy should have such an observable effect seems paradoxical since the ground state energy of any quantized field turns out to be infinite and, when renormalized, will be zero. Therefore, according to a widespread belief at the time, vacuum energies had no real physical significance. However, with recent impressive experiments [2, 3] (using a gold coated sphere and flat plate rather than two flat plates) we have come to realize that the vacuum energy in a restricted domain is neither infinite nor zero and demands understanding. The appearance of Planck's constant in Eq. (1) clearly shows the quantum nature of the effect: it disappears when $\hbar \to 0$. The other surprising fact is that no QED interaction seems to be involved: as it stands, the Casimir force shows no dependence on α, the fine structure constant, and in its calculation one never uses Feynman diagrams from quantum electrodynamics. Therefore the frequent claim that the Casimir effect is caused by virtual particle–antiparticle pairs filling the vacuum misses the point.

It should be emphasized that the theoretical results depend on the boundary conditions, the geometry of the macroscopic objects, the spatial dimension and whether we are dealing with a scalar or vector field. For instance, a change of the boundary conditions and the geometry may reverse the sign of the Casimir force. There are also corrections due to finite temperature and limited conductivity, as well as radiative corrections that have to be taken into account in precision experiments [4, 5, 6]. Last not least J. Schwinger succeeded in deriving Eq. (1) from source theory, his version of quantum field theory, without ever referring to zero-point energy, a concept foreign to his framework.

Cosmological Implications

At present it is believed that the Casimir effect operates at all levels from from subnuclear to cosmological scales. Evidence is growing that 70% of the energy density in the universe exists in an unknown form dubbed *dark energy* but postulated as somehow related to some kind of vacuum energy, perhaps of the type leading to the Casimir effect. In Einstein's theory of gravitation, it is the energy density (more generally, the components of energy–momentum tensor) which determines the geometry of space-time. Quite the other way round, we have come to think that energies may be generated by space-time itself. This interplay between gravitational and quantum theory is on the forefront of present-day research.

Spaces with non-trivial topology such as the Friedman universe have the same effect as boundary conditions. There is also the possibility of a non-trivial vacuum effect in higher-dimensional spaces (as in theories of Kaluza–Klein type) where the Casimir energy–momentum tensor can lead to a spontaneous compactification of the extra dimensions, that is, they are curled up on the scale of the Planck length. The most prominent proposal, however, is that the Casimir energy density ought to be identified with the famous *cosmological constant*, previously considered by Einstein to arrive at a static universe. As for today, the reasons for introducing this constant are different.

Let us recall Einstein's equation with the cosmological constant Λ,

$$R_{\mu\nu} - \frac{1}{2} g_{\mu\nu} R - \Lambda g_{\mu\nu} = \frac{8\pi G}{c^4} T_{\mu\nu} , \tag{4}$$

where $R_{\mu\nu}$ and $g_{\mu\nu}$ are the Ricci and the metric tensor, R is the Ricci scalar and G is Newton's gravitational constant, and observe that the introduction of the Λ term on the left hand side amounts to a change of the energy–momentum tensor:

$$T_{\mu\nu} \rightarrow T_{\mu\nu} + \frac{\Lambda c^4}{8\pi G} g_{\mu\nu} . \tag{5}$$

Equivalently, the energy density has changed by the amount

$$\rho_0 = \frac{\Lambda c^2}{8\pi G} , \tag{6}$$

and the modern interpretation of Λ is as follows: ρ_0 represents the vacuum density.

By definition, dark energy is a hypothetical form of energy which permeates all space leading to a negative pressure (similar to what one observes in the Casimir cavity) with the result of an effective *repulsive gravitational force* which causes expansion of the universe to accelerate. In astronomy, the expansion rate is estimated by comparing the redshifts of distant galaxies with the apparent brightness of Type 1a supernovae found in them. Present measurements suggest that the expansion rate is growing.

The introduction of a cosmological constant, however, is but one possibility to explain the growth of the expansion rate. Various other types of dark energy have been proposed, including a cosmic field called "quintessence", normally associated with inflation. Whereas a cosmological constant would be static, i. e., not dependent on space-time, the proposed cosmic field is dynamic and, hence, may give rise to an acceleration rate which is time dependent. Sad to say, naive attempts to calculate the cosmological constant failed in that the results came out 100 orders of magnitude bigger that what has been observed by astronomers. We will soon see the next stage in that development.

See also: Cosmology; Photon; Quantum Electrodynamics; Quantum Field Theory.

Bibliography

H. B. G. Casimir, "On the attraction between two perfectly conducting plates", *Proc. Kon. Ned. Akad. Wetensch.*, **51**, 793 (1948).

S. K. Lamoreaux, "Demonstration of the Casimir force in the 0.6 to 6 μm range", *Phys. Rev. Lett.*, **78**, 5 (1997).

B. W. Harris, F. Chen, U. Mohideen, "Precision measurement of the Casimir force using gold surfaces", *Phys. Rev.*, **A 62**, 052109 (2000).

M. Bordag, U. Mohideen, V. M. Mostepanenko: "New Developments in the Casimir Effect", arXiv: quant-ph/0106045 (2001).

V. M. Mostepanenko, N. N. Trunov, R. L. Znajek, *The Casimir Effect and its Applications.* Oxford Science Publications, 1997.

K. A. Milton, *The Casimir Effect*, World Scientific, Singapore, 2001.

Catalysis

G. Ertl

If a chemical reaction is favored thermodynamically, but proceeds too slowly, its rate may be enhanced by the addition of a catalyst. The term "catalysis" (derived from the Greek and meaning "break down") was introduced in 1835 by the Swedish chemist J. Berzelius and received its final definition in 1895 by W. Ostwald: A catalyst is a substance which affects the rate of a chemical reaction without appearing in its end products. Although in principle this phenomenon can refer to either an increase or diminution in the rate, one in practice refers to the agent which accelerates a reaction as a catalyst and to an agent which decreases its rate as an inhibitor. In short, this effect is caused by the ability of the catalyst to form intermediate bonds with the molecules involved in the reaction, such that an alternate reaction path with an enhanced overall rate is offered. Let us consider a general chemical reaction of the type $A + B \rightarrow AB$ (1), whereby the back-reaction ($AB \rightarrow A + B$) may be neglected for the moment; then a catalyst C may form an intermediate species $A + C \rightarrow AC$ (2) reacting further to $AC + B \rightarrow AB + C$ (3). The reaction will be catalyzed by C if the sum of the rates (2) and (3) exceeds that of the direct reaction (1).

The efficiency of a catalyst is characterized by its *activity* and *selectivity*. The activity may be identified with the number of product molecules formed per "active site" of the catalyst per second ("turnover frequency"). In general from a given set of interacting molecules, the reactants, not only a single but several reactions may occur, yielding different reaction products. Selectivity is then defined as the ratio of the rate of the desired reaction over the sum of the rates of all possible reactions.

Catalysis is a very widespread phenomenon which is of greatest importance for living organisms as well as technical processes: The molecules catalyzing the very complex chemical reactions in biological systems are called enzymes and are unique with respect to their extreme selectivity. Solid catalysts form, on the other hand, the basis of chemical and petroleum industries as well as of processes for controlling atmospheric pollution.

Nonbiological catalysis is usually classified into *homogeneous* and *heterogeneous* catalysis. In the former the molecules involved in the reaction, including the catalyst, are within a single (gaseous or liquid) phase, while in the latter case the catalyst is in a different state of aggregation (often solid) from the reacting species (gas or liquid), and the catalytic action occurs at the interface between these phases. In the following, a few examples will serve to illustrate the basic principles underlying the various forms of catalysis.

Homogeneous Catalysis

A simple example for a reaction sequence as outlined above is offered by the reaction $2\,SO_2 + O_2 \rightarrow 2\,SO_3$, which is catalyzed by nitric oxide, NO, through the intermediate steps

$$O_2 + 2\,NO \rightarrow 2\,NO_2\ ,$$

$$NO_2 + SO_2 \rightarrow SO_3 + NO\ .$$

This is the basis of an old and important technical process for the formation of sulfuric acid.

NO as a catalyst may, on the other hand, have fatal consequences in the destruction of ozone (O_3) in the stratosphere, which reaction may be schematically formulated as

$$NO + O_3 \quad \rightarrow \quad NO_2 + O_2\ ,$$
$$O_3 + h\nu \quad \rightarrow \quad O_2 + O\ ,$$
$$NO_2 + O \quad \rightarrow \quad NO + O_2\ .$$

Obviously, the catalyst NO is not consumed in the overall reaction $2\,O_3 \rightarrow 3\,O_2$, whose rate is, however, considerably enhanced by the presence of the atmospheric pollutant.

Acid–base catalysis is the most important form of homogeneous catalysis in liquid phase. The reaction

$$N_2CHCOOC_2H_5 + H_2O \rightarrow HOCH_2COOC_2H_5 + N_2\ ,$$

is, for example, catalyzed by hydrogen ions through the intermediate step

$$N_2CHCOOC_2H_5 + H^+ \rightleftharpoons {}^+N_2\text{--}CH_2COOC_2H_5\ ,$$

whereby the product decomposes readily into

$${}^+N_2\text{--}CH_2COOC_2H_5 + H_2O \rightarrow HOCH_2COOC_2H_5 + N_2 + H^+\ ,$$

and the catalyst (H^+) is formed back. The overall rate will in this case obviously be determined by the pH ($= H^+$ concentration) or, more generally, by the strength of the acid available for the hydrogen ion transfer process.

A variety of technical processes is based on homogeneous catalysis by transition metal compounds, such as the Wacker process for converting ethylene (C_2H_4) to acetaldehyde (CH_3CHO) in which the species $PdCl_4^{2-}$ plays the role of the catalytic agent. Separation of the catalyst from the product after completion of the reaction is, however, a general problem in processes of this type occurring in homogeneous phase.

Heterogeneous Catalysis

The just-mentioned difficulty is almost nonexistent if, e. g., the catalyst is in the form of a solid which is in contact with the reacting species from the gaseous phase. The formation of chemical bonds between these species and the surface of the solid (chemisorption) is obviously of fundamental importance for this phenomenon of heterogeneous catalysis. The development of surface physical methods in recent years was of tremendous impact toward elucidation of the underlying elementary steps.

The activity of a given amount of a heterogeneous catalyst will depend on its surface area, and hence usually finely dispersed materials with a high specific surface area (typically up to $100\,m^2/g$) are applied. This may be achieved by depositing the catalytically active material (e. g., a noble metal) as small particles on a high surface area support such as alumina (Al_2O_3) or silica (SiO_2), by forming small catalyst particles during the chemical pretreatment (e. g., reduction) which may be stabilized by structural "promoters," or by using substances with a large *internal* surface area such as the important class of zeolites (crystalline aluminosilicates).

In general, the surface of a heterogeneous catalyst will be structurally and chemically nonuniform which renders this field so complex: The catalyst particle will expose various crystal planes exhibiting, in addition, defects such as steps and kinks. Moreover, the chemical composition of the surface may differ considerably from the (nominal) bulk composition; purposely added "electronic" promoters such as alkali atoms may enhance its specific activity, while the latter may be sensitively suppressed by other species acting as "poisons."

Access to the atomic processes underlying heterogeneous catalysis can be sought via the "surface science" approach: Well-defined single crystal surfaces under ultrahigh vacuum conditions serve as model systems for which the interaction with gaseous particles can be studied in detail. The "gap" to "real" catalysis can, on the other hand, be bridged by analyzing the surface properties of the practical catalysts and by systematically varying the properties of the model system (e. g., defect or impurity concentrations).

The reaction

$$2\,CO + O_2 \rightarrow 2\,CO_2$$

is readily catalyzed by the platinum group metals and is of practical relevance in the control of car exhaust gases. Its mechanism is basically quite simple and may be formulated as

$$CO + * \rightleftharpoons CO_{ad}$$
$$O_2 + 2* \rightarrow 2\,O_{ad}$$
$$CO_{ad} + O_{ad} \rightarrow CO_2 + 2* \,;$$

$*$ denotes schematically a bare site on the catalyst surface to which either a CO molecule or an O atom is chemisorbed. The adsorbed oxygen atoms are formed through dissociation of O_2

molecules which process proceeds with high efficiency on the platinum surface, but which, on the other hand, is suppressed by a too high concentration (coverage) of chemisorbed CO. That is why under steady-state conditions the temperature has to be high enough in order to drive part of the adsorbed CO back into the gas phase (desorption).

The synthesis of ammonia (NH_3) from nitrogen (N_2) and hydrogen (H_2) is a large-scale industrial process (Haber–Bosch) which forms, inter alia, the basis for the production of fertilizers. This reaction proceeds with a promoted iron catalyst. Its mechanism, i. e., the sequence of elementary steps, could recently be elucidated following the surface science strategy, whereby the dissociative chemisorption of nitrogen, viz., $N_2 + 2* \rightarrow 2\,N_2$, was identified as rate-limiting. In addition, based on the kinetic parameters derived from low-pressure studies with single crystal surfaces, even the yields under industrial conditions could be successfully modelled theoretically.

Progress in this area is rapid, and the complex area of heterogeneous catalysis is currently developing from a "black art" to a physical science.

Enzyme Catalysis

Biological reactions are catalyzed by macromolecules which belong to the class of proteins and which are called enzymes. This word is derived from the Greek "yeast," a material which played a key role in the development of this branch of science: E. Buchner showed in 1897 that a cell-free extract of yeast is able to catalyze the conversion of sugar into alcohol and carbon dioxide. The classification into "homogeneous" and "heterogeneous" catalysis breaks down in this case: The enzyme molecules are in solution in the form of colloidal dispersions, while interaction with the reacting molecules can, on the other hand, be considered as bond formation at the "surface" of such a macromolecule.

The key for understanding the catalytic properties of enzymes, and in particular of their fascinating selectivities towards specific reactions, lies in their complex structures. In 1894, E. Fischer compared an enzyme molecule with a lock which fits only with a single key, namely, the "correct" reacting molecule – a view which has essentially been confirmed by application of modern x-ray crystallography. Since the basic types of reactions of nonbiological catalysis are also present with enzyme catalysis, several formal analogies between both fields may be found. However, the high structural complexity of biological molecules introduces also cooperative phenomena and novel qualitative aspects.

See also: Adsorption; Kinetics, Chemical; Surfaces And Interfaces.

Bibliography

Advances in Catalysis. Academic Press, New York, 19XX.

J. R. Anderson and M. Boudart, eds. *Catalysis – Science and Technology.* Springer-Verlag, Berlin, 1981–96, 11 Volumes.

M. L. Bender and L. J. Brubaker, *Catalysis and Enzyme Action.* McGraw–Hill, New York, 1973.

G. C. Bond, *Heterogeneous Catalysis – Principles and Applications.* Clarendon Press, Oxford, 1974.

M. Boudart and G. Djega-Mariadassou, *Kinetics of Heterogeneous Catalytic Reactions.* Princeton University Press, Princeton, NJ, 1984.

G. Ertl, H. Knözinger, J. Weitkamp, eds. *Handbook of Heterogenous Catalysis*, 5 Vols. VCH, Weinheim 1997.

Catastrophe Theory
F. J. Wright and G. Toulouse

The theory of catastrophes is one product of the study of singularities. For a long time, mathematics has been concerned with the study of regular objects (functions, mappings, differential equations, etc.) and theorems have been so phrased as to eliminate the singularities. But actually, it quite often appears that the singularities contain a great deal of qualitative information, and a strong trend has developed in mathematics to make a systematic study of them. A general approach consists in building a classification of archetypical singularities in order of increasing complexity. That is the content of Thom's classification of elementary catastrophes.

Elementary catastrophes are singularities associated with the bifurcation of extrema of smooth real functions $f : \mathbb{R}^n \to \mathbb{R}$. As such, they are also singularities of differentiable mappings $Df : \mathbb{R}^n \to \mathbb{R}^n$, and singularities of differential equations of gradient type, $dx_i/dt = X_i$, with $X_i = \partial f/\partial x_i$. Within the theory of differential equations (also called the theory of dynamical systems), gradient systems are a simple restricted class (e. g., the dynamical systems, often used in physics, with limit cycles or strange attractors are clearly outside the gradient class). The term *catastrophe* is sometimes applied outside the elementary framework to describe a breakdown of stability of any equilibrium causing a system to *jump* into another state, or the transition of a dynamical system from a stable to a chaotic orbit. But usually *catastrophe* is used to mean *elementary catastrophe* as defined above, and it will be mainly used that way here.

Before giving specific details about the nature and classification of elementary catastrophes, it is appropriate to discuss where such a mathematical theory may be relevant in physics. Very many physical laws are expressed as variational principles: in optics (Fermat's principle), mechanics (principle of least action), thermodynamics (maximal entropy or minimal free energy), etc. In all these cases, the physical state (or trajectory) is obtained as an extremum of a real-valued function. When a bifurcation of extrema occurs, a sudden qualitative change of the physical state may follow; such sudden changes happen in optics (light caustics), mechanics (instabilities), hydrodynamics (shock waves), thermodynamics (phase transitions), etc. Catastrophe theory may be used as a unifying language in these various domains, and as a way of treating general rather than special cases.

Definitions and Theorem

What follows is a presentation of the concepts that are necessary for a first approach and are immediately accessible to a non-mathematician. Many mathematical expositions, at various levels of abstraction, may be found in the literature, with the help of the references in the Bibliography.

Let $f(a_1,\ldots,a_i,\ldots,a_r;x_1,\ldots,x_j,\ldots,x_n)$ be a smooth real-valued function of $r+n$ real variables:

$$f : \mathbb{R}^r \times \mathbb{R}^n \to \mathbb{R}.$$

The variables a_i $(i = 1,\ldots,r)$ will be called *control* parameters; the variables x_j $(j = 1,\ldots,n)$ will be called *state* variables. The function f is then seen as an r-parameter family of functions of n variables.

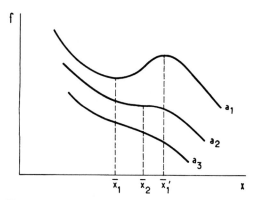

Fig. 1: Three representative curves of a one-parameter family of functions of one variable $f(a;x)$; a is the parameter, x the variable. The stationary values of f have abscissas $\bar{x}$.

Consider then the stationary values (local extrema) of f with respect to variation of the state variables x_j. The set of points in $\mathbb{R}^r \times \mathbb{R}^n$ for which f is stationary,

$$\frac{\partial f}{\partial x_j} = 0 \qquad (j = 1, \ldots, n),$$

is called M.

At this stage in the definitions, a simple example facilitates understanding and the introduction of other concepts. Consider a one-parameter family of functions of one variable $f(a;x)$. Figure 1 shows, on a graph of f versus x, three typical curves corresponding to three values of the parameter a. The upper curve, parametrized by a_1, shows two extrema (one minimum of abscissa $\bar{x}_1$, one maximum of abscissa $\bar{x}'_1$); the lower curve, parametrized by a_3, has no extremum; the intermediate curve, parametrized by a_2, has an inflection point at the abscissa value $\bar{x}_2$. It is then useful, and this is done in Fig. 2, to draw the set M (as defined above) as a graph of x versus a: for each value of a are plotted the values $\bar{x}$ of x that make f stationary. In this example, M is seen to be a curve, whose projection on the parameter axis has a singularity at a_2 (vertical slope) corresponding to the coalescence of two extrema (as visible in Fig. 1). On the control-parameter axis there is one point a_2, called a catastrophe or bifurcation point, separating two domains ($a > a_2$ and $a < a_2$) where the shapes of f are qualitatively different, with different numbers of extrema.

More generally, instead of a one-dimensional control-parameter axis, there will be an r-dimensional control-parameter space Σ; the set M will be a manifold; the points of Σ where the projection of M on Σ is singular (vertical slope) will constitute a set, called the catastrophe or bifurcation set.

Obviously, it is possible to imagine functions f that have very complicated singularities. Clarification occurs if we decide to look for the singularities that occur *generically* (generic means typical). This concept is not unfamiliar to physicists: terms that are not forbidden by symmetry are (generically) nonzero, but they may become (accidentally) zero if some parameter is varied (such is the distinction between systematic and accidental degeneracies of levels in quantum mechanics; cf. also the discussion on the order of a transition in the

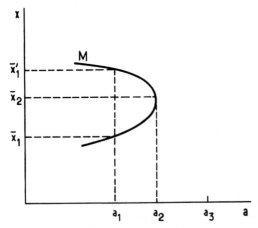

Fig. 2: The fold catastrophe. For a one-parameter family of functions of one variable $f(a;x)$, the set M of values of the variable making f stationary is plotted as a function of the parameter a. The fold singularity occurs at point $(a_2, \bar{x}_2)$ where the slope is vertical.

Landau theory of phase transitions). A precise definition of genericity is clearly crucial for the mathematical proofs (see the mathematical references in the Bibliography). With a hypothesis of genericity for f, it is then possible to give a classification of the singularities, as follows.

If f is generic, then M is an r-dimensional manifold and the singularities of the projection of M onto the control-parameter space Σ are equivalent to a number of types called elementary catastrophes. The number of elementary catastrophes in Thom's classification depends only on r (not on n, provided $n > 1$) and is finite for $r \leq 5$, infinite for $r \geq 6$.

The Elementary Catastrophes

For $r = 1$, there is only one elementary catastrophe (illustrated in Fig. 2) called, rather naturally, the fold catastrophe. In the vicinity of a fold point the function f has, generally after a smooth change of coordinates, the canonical local form

$$x_1^3 + ax_1 + \cdots,$$

where $\cdots$ represents a diagonal quadratic form in the remaining state variables, in this case x_j $(j = 2, \ldots, n)$.

For $r = 2$, there are fold curves (imagine Fig. 2 continued smoothly with another horizontal axis) and another elementary catastrophe called the cusp catastrophe: the catastrophe set (in the control-parameter space Σ) is made up of fold arcs joining in cusps (Fig. 3). The canonical form of f near a cusp catastrophe is

$$x_1^4 + a_2 x_1^2 + a_1 x_1 + \cdots.$$

For $r = 3$, three new elementary catastrophes appear; their names, together with their associated canonical local forms, follow:

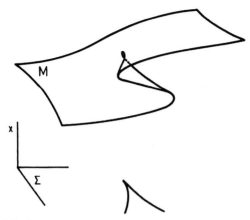

Fig. 3: The cusp catastrophe. The set M is a surface; its projection on the control-parameter plane Σ is singular along the catastrophe set, which exhibits a characteristic cusp shape.

1. the swallowtail,

$$x_1^5 + a_3 x_1^3 + a_2 x_1^2 + a_1 x_1 + \cdots ;$$

2. the elliptic umbilic,

$$x_1^2 x_2 - x_2^3 + a_3 (x_1^2 + x_2^2) + a_2 x_2 + a_1 x_1 + \cdots ;$$

3. the hyperbolic umbilic,

$$x_1^2 x_2 + x_2^3 + a_3 (x_1^2 + x_2^2) + a_2 x_2 + a_1 x_1 + \cdots .$$

Sketches of the catastrophe sets (in the control space Σ) are shown in Fig. 4. Note that for the two umbilics, the canonical forms necessarily involve two state variables; canonical forms involving two or more state variables cannot in general be uniquely defined, and slightly different choices are sometimes used.

For $r = 4$, two new elementary catastrophes appear: the butterfly and the parabolic umbilic. In many applications, the control parameters will be space and time parameters and therefore $r = 4$. If we add up all the elementary catastrophes that can appear when $r = 4$, we get the famous total of seven elementary catastrophes.

For $r = 5$, four new catastrophes appear. For $r \geq 6$, the classification becomes infinite because there are equivalence classes depending on continuous parameters called *modal* parameters. The simplest such class involving two state variables is the *uni-modal* catastrophe called (by Arnol'd) X_9, which is important for example in optical systems that are nearly rotation-symmetric about their optical axis. Part of X_9 contains two cusp catastrophes and is therefore sometimes called the *double cusp*.

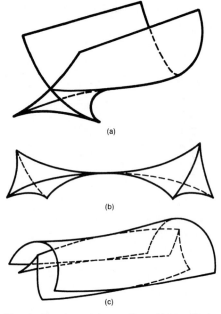

(a)

(b)

(c)

Fig. 4: Sketches of the swallowtail (a), elliptic umbilic (b), and hyperbolic umbilic (c) catastrophe sets in three-dimensional control space.

Applications in Physics

Elementary catastrophe theory (ECT) has been applied to a wide range of topics, including most kinds of wave propagation, stability of mechanical structures, stability of ships and aircraft, laser action, singularities of phonon spectra, fluid flow, electric and magnetic phenomena, various phase transitions, nuclear and high-energy physics, semiclassical quantum mechanics, stellar evolution, relativity, geology, meteorology and climatology. Here are sketches of two application areas.

The canonical form for a cusp catastrophe is identical to the expression used to describe an ordinary critical point in the Landau theory of phase transitions. The meanings of x_1, a_1, a_2, and f are now (respectively) order parameter, field coupled to the order parameter, temperature measured from the critical temperature, and free energy density. Similarly, the canonical form for a butterfly catastrophe has its analogy in the expression used to describe tricritical points. However, symmetry considerations are of paramount importance in phase-transition problems and usual symmetry restrictions eliminate many of the elementary catastrophes listed above. Also, the predictions that can be obtained from such expressions for the free energy density ignore spatial fluctuations, which play a dominant role near the critical point. To include these effects, we have to add gradient terms to the free energy density and to sum over contributions due to the spatial fluctuations of the order parameter. This program is very remote from catastrophe theory.

Improvements on the results of simple catastrophe models have been achieved, but only at the expense of some *ad hoc* additions, such as coordinate changes outside those allowed by ECT, or use of a fractal lattice of dimension 3.09. However, the very successful renormalization group technique may be viewed as a catastrophe-theoretic model. A contribution of catastrophe theory to the field of phase transitions might be in a rational vocabulary for higher-order critical points.

A very neat application of catastrophe theory has been made in the theory of caustics. Caustics are envelopes of geometrical rays of a wave field. When the wavelength becomes small, the geometrical approximation of the wave theory becomes appropriate and most of the intensity of the wave field is concentrated on the caustics, which become the dominant features of the wave pattern. In this semigeometrical limit, not only does ECT produce a useful classification of caustics, but also of the wave fields – called *diffraction catastrophes* – that surround the caustics.

The geometrical rays are obtained via a variational principle (this is in essence the content of Fermat's principle), the main control parameters are space coordinates of points where the wave amplitude is to be measured, and the state variables are space coordinates of the intersections of the rays with some arbitrary wave surface. Typically, in a three-dimensional geometry, the number of control parameters is three and, from the classification above, it is possible to predict that the caustic pattern will generically consist of fold surfaces joining along cusp lines meeting in points which may be swallowtails, elliptic umbilics, or hyperbolic umbilics. The wave field around a caustic described by the catastrophe with canonical form f is modelled by a canonical diffraction integral of the form

$$\int e^{\mathrm{i} f(a;x)} \, \mathrm{d}x.$$

These integrals have been evaluated numerically for all the caustics that can occur generically in three dimensions and used to produce simulated diffraction catastrophes, which agree spectacularly well with observations.

On a caustic, the amplitude A of the wave diverges as a power of the wavelength λ:

$$A \sim \left(\frac{1}{\lambda}\right)^{\sigma}, \qquad \lambda \to 0.$$

By scaling out the wavelength from the physical diffraction integral to produce the above canonical form, the value of the index σ is obtained from the canonical form of the elementary catastrophe associated with a given caustic:

$$\text{fold,} \quad \sigma = \tfrac{1}{6};$$
$$\text{cusp,} \quad \sigma = \tfrac{1}{4};$$
$$\text{swallowtail,} \quad \sigma = \tfrac{3}{10};$$
$$\text{elliptic or hyperbolic umbilic,} \quad \sigma = \tfrac{1}{3}.$$

By a technique of averaging over caustics, Berry successfully predicted the "twinkling" statistics of starlight, using a second set of scaling indices describing the wavelength dependence of the spatial size of caustics. The amplitude and size indices are quantitative results, which can be measured by varying the wavelength λ.

Perspectives and References

Catastrophe theory is now sufficiently mature that there are a number of articles reviewing its applications in the physical sciences; those by Stewart and Gilmore give good introductions to the main areas. The classic book by René Thom (1923–2002) gives marvellous insight into the motivation for this branch of mathematics, although it was Christopher Zeeman who coined the name "catastrophe theory" while developing many of its applications, some of which appear in his book of selected papers. The MacTutor History of Mathematics web site gives brief biographies of Thom and Zeeman, and Weisstein's MathWorld web site has some nice pictures of elementary catastrophes, especially the swallowtail. The Russian mathematician V. I. Arnol'd vastly extended Thom's original classification of elementary catastrophes, using a complex rather than real setting, and contributed significantly to the more general version known as "singularity theory."

Since the initial surge of interest in catastrophe theory in the 1970s there has been a steady stream of applications in physics (and many other disciplines). Applications to phase transitions and wave propagation have continued to be particularly successful; for a recent example of the former see the paper by Bogdan and Wales. The background to the latter is set out in the review by Berry and Upstill and the subject is greatly developed in the monograph by Nye. The review by Dangelmayr and Güttinger discusses the inverse wave problem.

Optical applications of catastrophe theory have become very diverse. For example, it has been used to classify higher-order resonances in the spectroscopic Hamiltonian for two coupled vibrational modes and polarization transverse-pattern dynamics in lasers, and the cusp catastrophe has been used to explain the multiple peaks produced by light scattering from very rough dielectric surfaces in the geometrical limit.

Other major application areas include quantum mechanics and gravitation. The time evolution of a kicked quantum rotor shows spectacular semiclassical catastrophes; the cusp catastrophe describes the spontaneous spatial symmetry breaking occurring in the ground state of two-component Bose-Einstein condensates; and quantum morphogenesis has been described as a variation on Thom's catastrophe theory. In Fermi's wave mechanics of collision processes, the composite wave function for a particle and its analogous rotator can be expressed in terms of Hankel functions for absorbing and radiating waves whose index is related to the caustic circle, near which catastrophe theory can be applied. Catastrophe theory has been used to show the stability of black holes and nonuniform black strings.

More elaborate mechanical systems have been considered than those introduced by Zeeman, such as multiple-parameter non-linear buckling, a cylinder acted upon by a torque rolling on a slope and the stationary states of an inverted double-link pendulum. In an alternative paradigm for physical computing, information processing results from cycling the parameters of a system, such as magnetic memory, through a path containing a cusp-type catastrophe. Catastrophe theory has been used to derive a non-dimensional elastocapillary number for the adhesion criterion between a centre-anchored circular plate and its underlying substrate caused by strong capillary forces.

It is perhaps an indication of the success of catastrophe theory that the term "catastrophe" is now being applied to phenomena that have some analogy, but no obvious direct mathematical connection, with those considered by Thom, Zeeman *et al.* For example, at the event horizon of a black hole, time stands still and light behaves geometrically. However, quantum effects leading to Hawking radiation soften this ray singularity in the same way that diffraction softens caustics for light of nonzero wavelength. Leonhardt calls this phenomenon a "quantum catastrophe" and has proposed a laboratory analogue using "slow light", which is light in a nonlinear atomic medium that is coerced to have zero group velocity.

ECT grew partly out of the bifurcation theory of solutions of differential equations, which has fed back into that area as "imperfect" bifurcation theory. This considers the effects of perturbations on an idealized model, and the perturbation parameters play an analogous role to the control parameters in ECT. Thus some of the relative computational ease of ECT can be brought to bear on "conventional" bifurcation theory, and this has been particularly successful for including symmetry and symmetry-breaking effects, as described in the two volumes by Golubitsky *et al.* For example, it has been applied to "mode jumping" in the buckling of plates and to the way that fluid flows can display highly symmetric cell structures (e. g., planar Bénard convection and cylindrical Taylor–Couette flow). The range of physical systems to which ECT can be applied directly is important but limited, and the influence that it has had back on its origins – dynamical systems and bifurcation theory, which both have broader applicability – may yet prove to be its greatest contribution.

See also: Chaos; Diffraction; Dynamics; Fluid Physics; Hydrodynamics; Optics; Phase Transitions; Renormalization; Shock Waves; Thermodynamics.

Bibliography

V. I. Arnol'd, *Singularity Theory: Selected Papers*. Cambridge University Press, London, 1981.

V. I. Arnol'd, *Catastrophe Theory*. Springer-Verlag, New York, 1984.

M. V. Berry and C. Upstill, "Catastrophe Optics: Morphologies of Caustics and Their Diffraction Patterns," *Prog. Opt.* **18**, 257–346 (1980).

T. V. Bogdan and D. J. Wales, "New Results for Phase Transitions from Catastrophe Theory," *J. Chem. Phys.* **120**(23), 11090–9 (2004).

G. Dangelmayr and W. Güttinger, "Topological Approach to Remote Sensing," *Geophys. J. R. Astr. Soc.* **71**, 79–126 (1982).

R. Gilmore, "Catastrophe Theory: What it is – Why it exists – How it works," *Chaos and the Changing Nature of Science and Medicine: An Introduction.* American Institute of Physics Conference Proceedings **376**, 35–53 (1996).

M. Golubitsky and D. G. Schaeffer, *Singularities and Groups in Bifurcation Theory*, Vol. 1. Springer-Verlag, New York, 1985.

M. Golubitsky, I. N. Stewart, and D. G. Schaeffer, *Singularities and Groups in Bifurcation Theory*, Vol. 2. Springer-Verlag, New York, 1988.

U. Leonhardt, "A Laboratory Analogue of the Event Horizon using Slow Light in an Atomic Medium," *Nature* **415**, 406–409 (2002)

U. Leonhardt, "Theory of a Slow-Light Catastrophe," *Phys. Rev. A,* **65**(4), 1–15 (2002).

MacTutor History of Mathematics archive. www-history.mcs.st-andrews.ac.uk/.

J. F. Nye, *Natural Focusing and Fine Structure of Light: Caustics and Wave Dislocations.* Institute of Physics Publishing, Bristol and Philadelphia, 1999.

T. Poston and I. N. Stewart, *Catastrophe Theory and its Applications*. Pitman, London, 1978.

I. N. Stewart, "Catastrophe Theory in Physics," *Rept. Prog. Phys.* **45**, 185–221 (1982).

I. N. Stewart, "Applications of Nonelementary Catastrophe Theory," *IEEE Trans. Circuits Syst.* **31**, 165–74 (1984).

R. Thom, *Structural Stability and Morphogenesis*. Benjamin, Reading, Mass., 1975.

E. W. Weisstein, "Catastrophe". From MathWorld – A Wolfram Web Resource.
　　`mathworld.wolfram.com/Catastrophe.html`.

E. C. Zeeman, *Catastrophe Theory: Selected Papers*, 1972–1977. Addison-Wesley, Reading, Mass., 1977.

E. C. Zeeman, "Bifurcation and Catastrophe Theory," *Contemp. Math.* **9**, 207–272 (1982).

Cellular Automata

T. Toffoli

Cellular automata are abstract dynamical systems that play in discrete mathematics a role analogous to partial differential equations in the continuum. It is not surprising that they have been reinvented innumerable times; the canonical attribution is to von Neumann, circa 1950. The latter, following a suggestion by Ulam, used them as an elaborate thought experiment to show that a particular toy world – a game played with tokens on a board and governed by simple local rules – supported arbitrary complex structures that could make copies of themselves, and was thus in principle capable of hosting evolution based on Darwinian mechanisms. Von Neumann's cellular automaton provided a plausible reductionistic model for "life" just as the Turing's machine had for "thought."

For a surprisingly small entry fee in terms of required scientific expertise, cellular automata offer access to experimentation with the world of complex systems and the ideas behind it: emergent structures and behavior; nonlinear dynamics; parallel computation, construction, growth, reproduction, competition, and evolution; particles and fields; the interplay between dissipative tendencies and conservation laws. Perhaps because of this bridging of reductionism and holism, they have tended to achieve a cult status among successive generations of amateur scientists – as when Martin Gardner presented Conway's 'game of life' in *Scientific American* (1970). Yet, besides being entertaining and instructive, cellular automata constitute a fully professional mathematical tool.

In a cellular automaton, space is represented by an infinite array; to each site (or "cell") of this array there is associated a state variable ranging over a finite set – typically just a few bits' worth of data. Time advances in discrete steps, and the dynamics is given by an explicit *rule* – say, a lookup table – through which at every step a cell determines its next state from the current state of its neighbors. Thus, the system's laws are *local* (no action-at-a-distance) and *uniform* (the same rule applies to all sites); in this respect, they reflect fundamental aspects of physics. Moreover, they are *finitary*: even though the state space of a cellular automaton has the cardinality of the *continuum* much as a differential equation, yet the evolution over a finite time of a finite portion of the system can be computed not just with arbitrary precision (as in the case with differential equations), but *exactly*, much as in digital logic; this is what sets cellular automata apart from more conventional dynamical systems. A formal characterization of cellular automata may be given in terms of topological dynamics on the Cantor set.

Fig. 1: *Left:* Excitation patterns ("forest fire") in a two-dimensional array of coupled relaxation oscillator. *Right:* Diffusion-limited aggregation of particles undergoing Brownian motion on a lattice, leading to fractal dendritic growth.

In spite of their structural simplicity, cellular automata can support complex behavior. They supply useful models for investigations in natural sciences, combinatorial mathematics, computer science, and theoretical physics. One of their most natural and successful applications is the modeling of *mesoscopic-scale* phenomena – surface tension, fractal growth, annealing, flow through porous media, mixing of fluids, phase transitions, activation patterns in excitable media – by reduction to simple fine-grained mechanisms (Fig. 1). With current progress in computer technology, reductionistic models of this kind – which are in any case conceptually very attractive – are also becoming widely accessible as effective *computational* models.

As a simple example, consider an array where each cell can be in one of two states – black and white – and can see as a *neighborhood* the entire 3×3 window (9 sites) centered on the cell itself. At each step, let the new state of the cell follow the *majority* of its nine neighbors, i.e., become white if at least five are white, and black if fewer. With this rule, the system, started from a random state (Fig. 2a), in a few steps converges to a stable majority pattern (b).

Injecting a small amount of randomness into this dynamics allows the *annealing* of the two phases to proceed indefinitely, yielding ever larger and more rounded domains. A similar effect, as shown in Fig. 2c–d, can be obtained by slightly "frustrating" the original rule (e. g., a bare majority of 5 votes is now made to *lose* rather than win, and conversely a minority of 4 to *win* rather than lose). Note that with the latter approach we can have each part of the system act as a source of (pseudo-)random disturbances for the adjacent parts *without forgoing determinism*. In either case, this rule provides a rudimentary mechanism for phase separation; *surface tension* emerges as a "statistical force" from the underlying combinatorics. A slightly more sophisticated rule, the well-known *Ising spin model*, gives a physically more accurate account of this phenomenology.

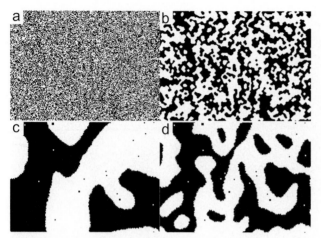

Fig. 2: With a plain-majority rule, from a random initial state (a) black and white domains emerge and reach a stable state (b). With a frustrated variant of this rule, annealing proceeds indefinitely: "bays" are gradually filled and "capes" eroded (c, d).

For robust and enlightening models, a good approach is to devise rules that capture, albeit in a stylized way, essential aspects of the underlying physics – such as symmetries and conservation laws. By devising microscopically reversible rules one insures the conservation of fine-grained entropy, and thus obtains models that obey the second principle of thermodynamics. Recent advances in the theory of cellular automata have made it straightforward to build such "detailed balance" constraints into a rule, as well as custom-tailor many other desirable properties. Most of these advances rely on a variant computational scheme called *lattice-gas automata*, in which discrete tokens travel on the arcs of a uniform graph ("lattice") and interact at its nodes, much as a gas whose particles were constrained to travel on discrete tracks and could only collide at the intersections.

Figure 3 illustrates structure and behavior of a simple lattice gas called *tm* (see, e. g., the book by Toffoli and Margolus). Its interaction rule f, applied at each node, basically specifies that particles travel straight through a node except when two of them collide head-on, in which case they are both scattered at 90°. The result is a rudimentary fluid: pressure is rapidly equalized in all directions. Note the emergence of *circular isotropy* in wave propagation, in spite of the fact that the microscopic dynamics takes place on a square lattice.

Though cellular automata and lattice gases are alternative algorithmic implementations of essentially the same family of abstract dynamical systems, they differ in what properties they can more easily express. Cellular automata are preferred for modeling macroscopic, dissipative phenomenology (biology, economics, traffic), while lattice gases are best for modeling fundamental physics. Lattice-gas fluid dynamics is a well-established discipline aimed at both theoretical understanding and applications; because of their computational efficiency, the related *lattice-Boltzmann* algorithms lend themselves to industrial-scale modeling.

Having proven to be natural tools for investigations in statistical mechanics, cellular automata are now beginning to be used in an exploratory role for quantum and relativistic modeling.

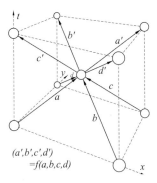 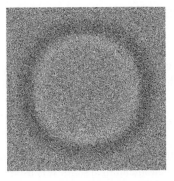

Fig. 3: *Left:* Spacetime graph showing tracks and intersections for a simple lattice gas. *Right:* In the gas at equilibrium, a localized disturbance generates a circular wave.

The cellular-automaton/lattice-gas scheme of computation is unique in the way it lends itself to extremely efficient concrete implementations. We have seen that the volume elements of the physical system to be modeled are directly mapped into volume elements of a cellular automaton. Conversely, when the time comes to concretely run such a model, one can directly map volume elements of the cellular automaton into volume elements of an appropriate physical computer – a *cellular automata machine*. Such a massively parallel architecture is *indefinitely scalable*, i. e., the number of sites in the array can be made as large as desired without having to slow down the rate of updating of individual sites.

The Turing-machine paradigm characterizes, according to Church's thesis, *what* can in principle be computed by physical means. A stronger version of this thesis is gaining credence today, namely, that the cellular-automaton paradigm characterizes *how efficiently* anything can in principle be computed by physical means. Thus, cellular automata are relevant not only when one asks how much physical simulation one can get out of computers, but also how much computing power one can get out of physics.

The *Programmable Matter* site pm.bu.edu maintains a general list of publications, tools, and links on the subject of cellular and lattice-gas automata.

Bibliography

Arthur Burks, *Essays on Cellular Automata*. University of Illinois Press, 1970.

Gary Doolen, *et al.* (ed.), *Lattice Gas Methods for Partial Differential Equations*. Addison–Wesley, New York, 1990.

David Griffeath and Cristopher Moore (eds.), *New Constructions in Cellular Automata*. Oxford University Press, Oxford, 2002.

Andrew Ilachinski, *Cellular Automata: A Discrete Universe*. World Scientific, Singapore, 2001.

Barry Simon, *The Statistical Mechanics of Lattice Gases*, vol. 1. Princeton University Press, 1993.

Tommaso Toffoli, "Cellular automata as an alternative to (rather than an approximation of) differential equations in modeling physics," *Physica* **10D** 117–127 (1984).

Tommaso Toffoli and Norman Margolus, *Cellular Automata Machines – A new environment for modeling*. MIT Press, 1987.

Jörg Weimar, *Simulation with Cellular Automata*. Logos-Verlag, Berlin, 1997.

Stephen Wolfram, *A New Kind of Science*. Wolfram Media, 2002.

Center-of-Mass System

P. Eberhard and M. Pripstein

For an ensemble of bodies the center-of-mass system is that inertial system of reference axes in which the vectorial sum of the momenta of the bodies is zero:

$$\sum_i \mathbf{p}_i^* = 0 \tag{1}$$

where $\mathbf{p}_i^*$ is the momentum of the ith body. Because of the definition [Eq. (1)], that system is often referred to as the center-of-momentum system. It is used because it generally makes the kinematics simpler. An asterisk is added to a symbol to designate its value in that restframe.

That system is named the "center of mass system" because the center of mass of the ensemble, i. e., of the point defined nonrelativistically in any inertial rest frame by its vectorial coordinate,

$$\mathbf{R} = \frac{\sum_i m_i \mathbf{r}_i}{M} \tag{2}$$

where $\mathbf{r}_i$ is the position of the ith body, m_i is its mass, and where

$$M = \sum_i m_i . \tag{3}$$

is at rest.

Nonrelativistically, the position $\mathbf{r}_k^*$ and the velocity $\mathbf{v}_k^*$ of the kth body in the center-of-mass system are defined as

$$\mathbf{r}_k^* = \mathbf{r}_k - \mathbf{R} \tag{4}$$

and

$$\mathbf{v}_k^* = \mathbf{v}_k - \mathbf{V} \tag{5}$$

where $\mathbf{v}_k$ and $\mathbf{V}$ are the derivatives of $\mathbf{r}_k$ and $\mathbf{R}$, respectively, with respect to time.

Whenever relativistic corrections cannot be neglected, Eqs. (2), (4) and (5) are not valid because a Lorentz transformation has to be applied instead to ensure that Eq. (1) is satisfied. Instead the position of the 'center of mass', i. e., of the point at rest in the 'center of mass system', should be defined as

$$\mathbf{R} = \frac{\sum_i e_i \mathbf{r}_i}{E} \tag{6}$$

where e_i is the position of the ith body and where

$$E = \sum_i e_i . \tag{7}$$

The momentum $\mathbf{p}_k^*$ and energy e_k^* of the kth body in the center-of-mass system can be defined from the momentum $\mathbf{p}_k$ and energy e_k of the body in the initial reference frame by the equations

$$\mathbf{p}_k^* = \mathbf{p}_k - \left[\frac{\gamma e_k}{c} - \frac{\gamma^2}{\gamma+1} (\boldsymbol{\beta} \cdot \mathbf{p}_k) \right] \boldsymbol{\beta} \ , \tag{8}$$

$$e_k^* = \gamma e_k - \gamma (\boldsymbol{\beta} \cdot \mathbf{p}_k) c \tag{9}$$

where c is the velocity of light, and

$$e_k = \frac{m_k c^2}{[1 - (\mathbf{v}_k/c)^2]^{1/2}} \ , \tag{10}$$

$$\boldsymbol{\beta} = \frac{\sum_i \mathbf{p}_i c}{\sum_i e_i} \ , \tag{11}$$

$$\gamma = \frac{\sum_i e_i}{E^*} \ , \tag{12}$$

and

$$E^* = \left[\left(\sum_i e_i \right)^2 - \left(\sum_i \mathbf{p}_i \right)^2 c^2 \right]^{1/2} \ . \tag{13}$$

As in any Lorentz transformation, the component of the momentum $\mathbf{p}_k$ that is orthogonal to $\boldsymbol{\beta}$ is conserved while that which is parallel to $\boldsymbol{\beta}$, and the energy e_k, are modified. The position of the center of mass in the center-of-mass system is now defined by Eqns. (2) and (3) but with m_i and M replaced by e_i^* and E^*, respectively.

Because of Eq. (1), the number of parameters necessary to describe the ensemble in the center-of-mass system can in general be easily reduced, and the description of the motion of the individual bodies is made simpler. For example, in the center-of-mass system of two stars revolving around each other, the trajectory of each one is an ellipse. In particle physics, the scattering of two particles can be described by only one parameter, the scattering angle θ^*, because the momenta of the two particles of the ensemble are always opposite to one another in the initial and in the final states.

See also: Kinematics and Kinetics; Dynamics, Analytical.

Ceramics

L. M. Levinson

The conventional understanding of ceramics encompasses the production and use of clays and silicate materials for pottery, bricks, and refractories. Modern ceramics also include a wide variety of inorganic nonmetallic materials with precisely controlled properties which play critical roles in the function of numerous technological devices. The need for materials with closely specified characteristics has transformed the ancient ceramic art into a modern-day branch of

Table 1: Modern technical ceramic materials.

Material	Outstanding property	Use
Alumina, Al_2O_3	High-temperature strength	Furnace ware, thermocouples
	Low electrical loss	IC multilayer package; microcircuit substrates
Beryllia, BeO	High thermal conductivity	Electronic component heat sink
Aluminum nitride, AlN	High thermal conductivity and low coefficient of thermal expansion	Si IC substrates
Magnesia, MgO	High-temperature resistivity	Heater element insulation
Ferrite, MFe_2O_4 (M= Mn, Fe, Ni, Zn, Cu, Mg, etc.)	Ferrimagnetic insulator	Computer memories, high-frequency transformers
Garnet, $R_3Fe_5O_{12}$ (R = rare earth)	Low microwave loss Ferrimagnetic insulator	Microwave devices
Zinc oxide, ZnO	Nonlinear conductor	Overvoltage transient suppression
Insulating barium titanate, $BaTiO_3$	High dielectric constant	Capacitors
Conducting barium titanate, $BaTiO_3$	Positive temperature coefficient of resistivity	Relay, sensor
Lead zirconate/titanate (PZT), $Pb(Zr, Ti)O_3$	Piezoelectric	Ultrasonic transducers
	High-temperature superconductivity	
$YBa_2Cu_3O_{7-x}$	$T_c = 92\,K$	Magnetic imaging (MRI), wireless
$Bi_2Sr_2Ca_2Cu_3O_{8-x}$	$T_c = 110\,K$	communication filters, power
$HgCa_2Ba_2Cu_3O_{8+x}$	$T_c = 133\,K$	transmission, current limiters

materials science and engineering [1]. We will briefly list some of the modern technological ceramics and then examine more closely two particular ceramic systems of interest. In Table 1 we present a selection of technical ceramic materials along with some special features and uses of the ceramic. It should be emphasized that the table makes no pretence at completeness and that the use of a particular material in a given application will usually depend on a spectrum of material properties in addition to the particular feature listed. Of the various polycrystalline ceramics in Table 1 it is clear that the utility of some lies in the ease of fabrication and low cost inherent in ceramic processing. For example, single-crystal alumina could provide superior microcircuit substrates and the use of single-crystal lead zirconate titanate would result in the fabrication of better ultrasonic transducers. There are, however, some ceramic devices in which the ceramic has properties essentially not feasible in the single-crystal analog of the ceramic. In this case the ceramic microstructure produces the new effect. We shall examine two such examples: (1) the high-frequency ferrite and (2) the ZnO varistor.

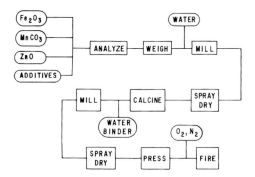

Fig. 1: Typical flowchart for processing of manganese zinc ferrite.

Manganese Zinc Ferrite

The double oxide spinel $MO \cdot Fe_2O_3$, where M can be (among others) Mn, Zn, or ferrous iron, is a magnetic ceramic material which combines the properties of high magnetic permeability with high electrical resistivity. The latter property results in low eddy-current loss so that the material can be used at high (tens of kilohertz) frequencies. A typical application of such ferrites is in the tuned circuit used for frequency-synthesis Touch-Tone telephones [2]. The ferrite must have high quality factor Q, low sensitivity to variations in the ambient temperature, and good stability with time. Since Q is the ratio of energy stored ($\propto$ permeability μ) to energy dissipated ($\propto$ conductivity σ), the highest possible value of μ/σ is desired. To maximize μ the Mn:Zn:Fe^{2+}:Fe^{3+} ratio is chosen so that the magnetic anisotropy and magnetostriction are very small. Ideal properties require close control of the purity of the raw materials and processing techniques such as milling and firing atmosphere. A typical processing flowchart is shown in Fig. 1.

The necessary presence of both Fe^{2+} and Fe^{3+} in manganese zinc ferrite produces an increase in the ferrite conductivity by allowing easy electron transfer from Fe^{2+} to Fe^{3+} sites. This in turn produces an unwanted high-frequency loss by eddy-current effects. This loss, while unavoidable in single-crystal material, can be controlled in fine-grain ceramics by the addition of controlled impurities. CaO and SiO_2 have been found [3] to segregate to the grain boundaries to form a low-conductivity phase, thereby substantially decreasing eddy-current losses. Figure 2 gives the effect of small amounts of CaO on μ and σ. Since μ decreases only slightly with CaO content while σ drops precipitously, a few tenths mol-% CaO addition substantially improves the performance of the ceramic ferrite.

ZnO Varistors

Zinc oxide is an ohmic semiconductor. However, it has been found [4] that ZnO ceramics containing a few mol-% of other oxide additives (for example, Bi_2O_3, Co_2O_3, MnO_2 and Sb_2O_3) exhibit a marked electrical nonlinearity in their current–voltage characteristic. Figure 3 gives a typical current–voltage characteristic of one of these devices. At voltages below a critical value the device acts as an insulator; above that voltage it acts as a conductor. In the vicinity of the turn-on voltage a 5% increase in voltage will typically cause the current to increase by an order of magnitude!

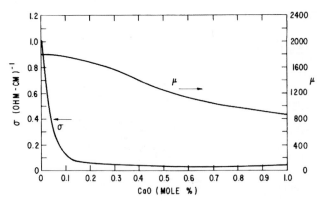

Fig. 2: Conductivity σ and permeability μ as a function of CaO addition to manganese zinc ferrite.

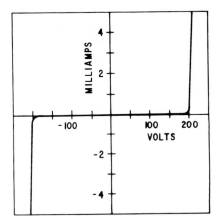

Fig. 3: Current-voltage characteristic of a ZnO varistor.

Figure 4 indicates how the ZnO varistor is used to protect electrical and electronic equipment from overvoltage transients due to switching surges or lightning strokes. The varistor is connected in parallel with the load and its turn-on voltage chosen so that it is insulating in normal use. The device conducts when a high-voltage surge in excess of the turn-on voltage appears on the line. This shunts the current and prevents the surge from reaching the protected equipment.

The origin of the varistor effect lies in the ceramic microstructure of this device [5, 6]. The ZnO grains are conducting, but during sintering they become surrounded with an adsorbed layer of aliovalent dopants (e. g., Bi) which segregate to the grain boundary region. The dopants are derived from oxides specifically added to produce the nonlinear electrical behavior in the ZnO ceramic. The dopant segregation results in the creation of electrical depletion layers at the grain boundaries which inhibit electron flow at low applied voltages. When increasing voltage is applied to the varistor, very high fields appear across the thin insulating depletion layers since the interior of the ZnO grains is conducting and will not sustain

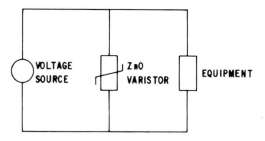

Fig. 4: Use of ZnO varistor to protect electrical equipment.

a voltage. At a critical applied voltage [6] (about 3 volts per grain boundary) it becomes possible for electrons to transfer through the depletion layers from one conducting ZnO grain to a neighboring grain, thereby giving rise to the observed high nonlinearity in the electrical characteristics.

See also: Conduction; Magnetic Materials.

References

[1]   W. D. Kingery, H. K. Bowen, and D. R. Uhlmann, *Introduction to Ceramics*. Wiley, New York, 1976; M. Barsoum, *Fundamentals of Ceramics*, Institute of Physics, London, 2003.

[2]   F. J. Schnettler, "Microstructure and Processing of Ferrites," in *Physics of Electronic Ceramics*, Part B, p. 833, L. L. Hench and D. B. Dove (eds.). Marcel Dekker, New York, 1972.

[3]   T. Akashi, *Trans. Jpn. Inst. Metals* **2**, 171 (1961).

[4]   M. Matsuoka, *Jpn. J. Appl. Phys.* **10**, 736 (1971).

[5]   L. M. Levinson and H. R. Philipp, *J. Appl. Phys.* **46**, 1332 (1975).

[6]   L. M. Levinson, in *Ceramic Materials for Electronics*, p. 431, R. C. Buchanan (ed.). Marcel Dekker, New York, 2004.

Čerenkov Radiation

M. Danos

In 1934 P. A. Čerenkov reported that light is emitted by relativistic particles traversing a non-scintillating liquid; he also described the properties of this radiation. (Čerenkov radiation had been observed, but not understood, by Marie Curie around the turn of the century.) Soon thereafter the nature of this effect was completely explained by I. M. Frank and I. E. Tamm as being the electrodynamic analog to the Mach waves of hydrodynamics: A charged particle traveling with velocity v through a nonconducting medium (or close to such a medium) having an index of refraction $n(v)$ emits electromagnetic radiation at all frequencies v for which the phase velocity c/n is less than v. In principle, a neutral particle having a magnetic moment (e. g., a neutron) also will emit Čerenkov radiation. Because of the smallness of the magnetic moment, this radiation will be negligibly small. The radiation is emitted at one angle, the Čerenkov angle $\theta = \arccos(c/nv)$, with respect to the momentum vector of the particle. It is 100% linearly polarized, the electric vector being coplanar with the particle momentum vector. The emitted power is given by $dE/dz = (4\pi^2 e^2/c^2) \int v\, dv[1 - (c/nv)^2]$ where the integration is limited to those frequencies for which the Čerenkov condition $(c/nv) < 1$ is fulfilled. Typically, relativistic electrons in water radiate about 250 visible light photons per centimeter of travel. The radiated energy integrated over all frequencies is rather small, since even for $v = c$ the Čerenkov condition is fulfilled only over a limited frequency interval; it is less than $1\,\mathrm{keV/cm}$ (i. e., its contribution to the stopping power is negligible).

On the other hand, the Čerenkov condition is fulfilled by the vacuum itself for all frequencies for particles that travel at $v > c$, i. e., for tachyons. This case was treated in terms of classical electrodynamics in prerelativity times (1892) by A. Sommerfeld, who thus in effect anticipated the Čerenkov effect. He found that a charged tachyon would radiate an energy equal to its rest mass while traveling a distance of the order of its classical radius, i. e., $d \approx e^2/mc^2$. This result appears to eliminate the possibility that charged tachyons may exist in nature.

The principal applications of the Čerenkov effect are in detectors for high-energy particle physics. We utilize here the existence of a threshold velocity: Only charged particles that fulfill the Čerenkov condition emit radiation, which then is detected, e. g., by photomultipliers. Such counters can be used either to select particles above a certain energy (e. g., the Čerenkov threshold energy for protons in air of normal density is about 55 GeV), or to discriminate against the heavier particles in a beam of given momentum containing a mixture of particles. Alternatively, Čerenkov counters have been built that respond to particles of a particular velocity by detecting only light emitted at a predetermined angle with respect to the beam direction. Such counters can be used to select particles of a given rest mass from a fixed-momentum beam.

See also: Electrodynamics, Classical; Tachyons.

Channeling

A. Kling

When beams of swift charged particles are incident on monocrystalline targets a strong directional dependence is observed for the yield of reactions requiring small impact parameters (e. g., Rutherford backscattering, nuclear reactions, particle-induced x-ray emission). Reductions of up to two orders of magnitude are found for the reaction rates in experiments with light ions impinging along low-index axial directions.

In the case of positively charged particles the effect can be understood qualitatively by the following model: Crystal atoms are highly ordered, forming chains or planes along axial and planar directions, respectively, which enclose channels completely void of atoms. Incident particles experience small-angle scattering induced by the repelling Coulomb potential of these target atom arrangements (see Fig. 1). Therefore the particles are steered in these channels, hence the term *channeling*. Since virtually no small impact parameters occur in the collisions with target atoms, the probabilities of nuclear reactions and inner shell vacancy production are drastically reduced.

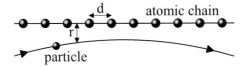

Fig. 1: Trajectory of a positively charged axially channeled particle during a large impact parameter collision with an atomic chain.

More quantitative results can be obtained using the continuum model of channeling developed by Lindhard in 1965. It assumes that a positively charged particle undergoes a series of correlated elastic collisions with the crystal atoms. In the case of axial channeling a continuous cylindrical potential can be derived by averaging over the individual screened Coulomb potentials of all atoms in a chain. The overall potential $U(r)$ for a channel is obtained by summing up the potentials of all strings surrounding the channel. The transversal energy $E_\perp$ of a charged particle in the crystal is composed of $U(r)$ and the transversal component of its total kinetic energy E_0,

$$E_\perp = E_0 \Psi^2 + U(r) , \tag{1}$$

where Ψ is the angle between the flight direction of the particle and the nearest atomic row. If particles approach an atomic chain closer than r_{min} they will experience the influence of its discrete character and thermal vibrations. Therefore it is a good approximation to set $r_{min} = \rho$, where ρ is the transversal vibrational amplitude. A maximum angle of incidence Ψ_c for successful particle steering can be calculated by setting the critical $E_\perp$ equal to the one at the point of closest approach ($\Psi = 0$) yielding

$$\Psi_c = \sqrt{\frac{U(\rho)}{E_0}} . \tag{2}$$

Values for critical angles in the case of protons and alpha particles with energies of a few MeV incident along a low-index axial direction are typically in the order of $1°$.

Due to the steering effect exerted by the continuum potential the charged particles are forced towards the center of the channel. If the particles are incident parallel to the crystal axis this leads to a flux peaking in the middle of the channel. At larger incidence angles (but below Ψ_c) the flux distribution varies in a more complex way. After traveling a few micrometers in a crystalline solid the flux distribution has reached equilibrium. Particles with a certain transversal energy are confined to an area of the channel enclosed by the equipotential line for which $U = E_\perp$.

The analytical approach used by the continuum theory has been quite successful in describing the channeling effect in simple crystals but it fails in more complex cases. Therefore various simulation codes based on the Monte Carlo method have been developed to describe different channeling phenomena.

The main applications of channeling are related to the use of swift light ions (H^+, He^+) obtained from accelerators in the analysis of crystalline materials. Typical experimental channeling studies measure the dependence of reaction yields, χ, on the tilt angle (angle of incidence relative to a major crystallographic direction). Two important characteristic parameters

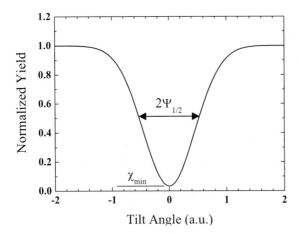

Fig. 2: The reaction yield dependence on the tilt angle for an angular scan and the two main characteristic parameters.

of such angular scans are the angular width at half minimum, $\Psi_{1/2}$, and the minimum yield, χ_{min} (see Fig. 2). A good estimate of $\Psi_{1/2}$ is given by

$$\Psi_{1/2} = \alpha(\rho/a)\sqrt{\frac{2Z_1 Z_2 e^2}{E_0 d}} , \tag{3}$$

where the function $\alpha(\rho/a)$ contains the information on the screening with a being the Thomas–Fermi screening radius. Z_1 and Z_2 are the atomic numbers of incident particle and target atom, respectively, and d is the distance between atoms in a row.

The minimum yield is mainly equivalent to the geometrical fraction of the channel area in which no channeling trajectories can exist, i. e., where the initial transversal energy of the particle exceeds the critical value. The continuum theory predicts that for a perfect single crystal

$$\chi_{min} = Nd\pi(2\rho^2 + a^2) , \tag{4}$$

with N being the atomic density of the target.

Although channeling cannot compete with x-ray diffraction in the structural analysis of perfect single crystals it presents excellent and versatile features in the detection and characterization of intrinsic and extrinsic defects.

In the preceding paragraphs the implicit assumption was made that a particle traveling on a channeled trajectory will never be removed from it. In fact such *dechanneling* occurs even in perfect crystals due to scattering at electrons or thermally displaced atoms. Defect-containing crystals provide additional scattering centers, e. g., displaced atoms, dislocations, stacking faults etc., that increase the dechanneling leading to an increase of χ_{min}. Channeling measurements using Rutherford backscattering and similar methods yield detailed information on the density and depth distribution of defects in crystalline materials. In addition, studies of the energy dependence of χ_{min} can be used to identify the defect type and size.

Another important application of ion channeling is lattice site location of impurities. A foreign atom incorporated substitutionally does not influence the channeling and therefore shows almost identical characteristic parameters in an experiment. In contrast to this, a dopant located on an interstitial site will be, at least in one crystallographic direction, located within the open channel area. Due to the flux-peaking effect the reaction yields for the dopant in an angular scan differ significantly from those of the host lattice atoms. Each lattice site has its characteristic angular scan. Measurements along three different crystallographic directions are sufficient to determine unambiguously the exact position of the impurity in the lattice.

Channeling studies reveal also many properties of epitaxial layers. Strained layers show slight angular differences with regard to the substrate for identically indexed crystallographic axes. High-resolution depth profiles can be extracted by analyzing angular scans for both. The dechanneling occurring in the layer–substrate interfaces or in the epitaxial layer itself yields additional information on mismatch and lattice disorder.

An alternative technique for the study of single-crystalline materials is *emission channeling*. In this case a radioactive nuclide is introduced into the target, e. g., by ion implantation. Electrons, positrons or α particles, released in the radioactive decay, experience channeling effects on their path to the crystal surface. It is important to note that a positively charged particle emitted by an atom on a substitutional site in a crystallographic direction will actually be blocked by the corresponding atomic chain (blocking effect). On the other hand, particles emitted from interstitial locations will be channeled. For electrons the effects are reversed due to the attractive force of the atomic chains. Measuring the anisotropic emission distributions along different axes allows the direct determination of impurity lattice sites. An advantage of this technique is that only very low dopant concentrations are necessary.

Related with the emission channeling technique is the determination of nuclear lifetimes by the blocking technique. A compound nucleus formed by irradiating the target with neutrons or charged particles will recoil from the original lattice site. If the decay of the compound nucleus occurs before it moved more than about 10 pm the emission of the decay product will be blocked along axial directions. If the lifetime is long enough (10^{-18}–10^{-16} s) the decay of the recoiled nucleus occurs in the open channel and the blocking effect is drastically reduced.

Relativistic electrons and positrons impinging on a monocrystalline target along a major axis or plane perform a transversal oscillatory motion. In a classical view the occurring accelerations force the emission of electromagnetic radiation, the *channeling radiation*. In a quantum-mechanical approach the emission is related to spontaneous transitions between bound eigenstates of the transverse Hamiltonian. The quasi-monochromatic photons are emitted in a forward-directed cone and have energies in the keV or MeV range for MeV and GeV particle energies, respectively. The main application of channeling radiation is as a tunable linear polarized radiation source with picosecond pulse lengths. The channeling radiation itself can provide information on the interaction of electrons with solids, e. g., incoherent scattering.

See also: Scattering Theory.

Bibliography

J. Lindhard, *Mat.-Fys. Medd. Danske Vid. Selsk.* **34**, No. 14 (1965). The fundamental theoretical treatment of channeling by the continuum theory.

D. S. Gemmell, *Rev. Mod. Phys.* **46**, 129 (1974). The classic review article on fundamentals and applications of channeling.

L. C. Feldman, J. W. Mayer, S. T. Picraux, *Materials Analysis by Ion Channeling*. Academic Press, New York, NY, 1982. Comprehensive overview on the application of channeling in materials science.

H. Hofsäß, *Hyperfine Interactions* **97**–**98**, 247 (1996). Detailed review on emission channeling.

J. U. Andersen, E. Bonderup, R. H. Pantell, *Ann. Rev. Nucl. Sci.* **33**, 453 (1983). Overview on basics and applications of channeling radiation.

Chaos

E. Ott

The word chaos, as used in this article, describes a type of behavior of dynamical systems in which the time evolution can be very complicated. (A definition of chaos is given later.) In particular, chaotic systems, although strictly deterministic, nevertheless display many apparent attributes which are reminiscent of randomness. Indeed when viewing the time evolution of such systems one often has the feeling that a statistical description is called for. Surprisingly such complicated chaotic dynamics can be present even in deceptively simple systems.

The existence of chaotic dynamics has been discussed in the mathematical literature for many decades starting with Poincaré at the turn of this century, with subsequent important early contributions made by the mathematicians Birkhoff, Cartwright, Littlewood, Levinson, Smale, and Kolmogorov and his students, among others. Nevertheless, it is only comparatively recently that the wide-ranging impact of chaos in physics has been recognized. Specific examples where chaotic dynamics arise are celestial mechanics, convection in fluids, oscillations in lasers, heating of plasmas by electromagnetic waves, the determination of the limits to weather forecasting, stirred chemical reactor systems, and an ever-growing host of other examples.

In what follows we discuss some of the concepts, definitions and results basic to the field of chaotic dynamics.

Dynamical Systems

This is a system of equations that allows one to predict the future given the past. One example is an autonomous system of first-order ordinary differential equations in time, $d\mathbf{x}(t)/dt = \mathbf{F}(x)$, where $\mathbf{x}(t)$ is a D-dimensional vector and $\mathbf{F}$ is a D-dimensional vector function of $\mathbf{x}$. Given $\mathbf{x}(0)$, the differential equation determines $\mathbf{x}(t)$ for $t \geq 0$. Another example is a *map* which is an equation of the form $\mathbf{x}_{t+1} = \mathbf{G}(\mathbf{x}_t)$ where here the "time" t is discrete and integer-valued. Thus given $\mathbf{x}_0$, the map gives $\mathbf{x}_1$, which when inserted in $\mathbf{G}$ then gives $\mathbf{x}_2$, and so on. The importance of maps derives from the fact that continuous time systems such as ordinary differential equations can often be reduced to maps by the Poincaré surface-of-section technique: pick some $(D-1)$-dimensional surface (the surface of section) in the D-dimensional phase space of a continuous-time dynamical system, and consider the orbits of the system and where they pierce the surface of section. Since the location $\mathbf{x}_n$ of the nth piercing uniquely determines the location of the $(n+1)$th piercing $\mathbf{x}_{n+1}$ (via integration of the ordinary differential equations), there is, in principal, a functional relation $\mathbf{x}_{n+1} = \mathbf{G}(\mathbf{x}_n)$ relating $\mathbf{x}_{n+1}$ to $\mathbf{x}_n$. That is, there is a map.

Conservative and Nonconservative Systems

For the purposes of the following discussion it is useful to distinguish between "nonconservative" dynamical systems and "conservative" dynamical systems. By a conservative system we mean one under which volumes in phase space are preserved as time evolves (or else, if they are not preserved, they can be made to be preserved by a smooth change of variables). Thus, if we consider all the points on the surface of a subset of phase space as initial conditions and evolve these initial conditions in time, then the surface continuously contorts its shape, but the total volume enclosed by it does not change. Hamiltonian systems are conservative in this sense. The most important difference between conservative and nonconservative dynamical systems is that the latter typically have attractors, while the former do not.

Attractor

If one considers a system and its phase space, then the initial conditions in some region B may be asymptotic as $t \to \infty$ to some smaller set A contained in B. In such a case A is said to be an attractor. For example, the phase-space variables specifying the state of a damped harmonic oscillator, $d^2x/dt^2 + v \, dx/dt + \omega^2 x = 0$, are the position x and velocity $v = dx/dt$. Any initial condition eventually comes to rest at the point $(x, v) = (0, 0)$, and this is the attractor for the system. Thus, here the attractor is very simple, a single point, which is a set of dimension $d = 0$. It is often the case, however, that for chaotic systems the attracting set can be geometrically much more complicated. In fact, it can have fractal geometry with a dimension d which is not an integer. In such cases the attractor is often called a strange attractor.

Dimension

There are various ways to define the dimension of a set. Perhaps the simplest is the box-counting dimension (also called the capacity dimension) which is given by the formula

$$d = \lim_{\varepsilon \to 0} \frac{\ln N(\varepsilon)}{\ln(1/\varepsilon)} , \tag{1}$$

where we imagine the set to be covered by small D-dimensional cubes of edge length ε (D denotes the Euclidian dimension of the phase space) and $N(\varepsilon)$ is the minimum number of such cubes needed to cover the set. For example, if the set were a single point, then $N(\varepsilon) = 1$ independent of ε, and Eq. (1) yields $d = 0$, as it should. If the set is a smooth curve, then $N(\varepsilon) \sim l/\varepsilon$, where l is the length of the curve, and Eq. (1) yields $d = 1$, again as it should. A more interesting example is the case of a Cantor set: Take the interval on the real line from 0 to 1; divide it in thirds; discard the middle third; take the two remaining thirds; divide each of them in thirds; discard the two middle thirds from these; and continue this process. In the limit that the process is applied an infinite number of times the remaining set is a Cantor set. This set is uncountable, and application of Eq. (1) shows that its box-counting dimension is $d = (\ln 2)/(\ln 3)$. Thus d is a number between 0 and 1 and the set is fractal. This type of geometric structure is typical of strange attractors.

Chaos

The dynamics on an attractor for a nonconservative dynamical system is said to be chaotic if typical orbits on the attractor display sensitivity to initial conditions. That is, consider an orbit $\mathbf{x}(t)$ evolving from an initial condition $\mathbf{x}(0)$ in the basin B of a chaotic attractor A (the basin of an attractor is the set of all initial conditions leading to that attractor). Now give the initial condition an infinitesimal perturbation $\delta\mathbf{x}(0)$, and consider the orbits $\mathbf{x}(t)$ and $\mathbf{x}(t) + \delta\mathbf{x}(t)$ which, respectively, evolve from the initial conditions $\mathbf{x}(0)$ and $\mathbf{x}(0) + \delta\mathbf{x}(0)$. If, for typical choices of $\delta\mathbf{x}(0)$, the distance between the two orbits grows exponentially with time, $|\delta\mathbf{x}(t)| \sim \exp(ht)$ with $h > 0$, then we say that the dynamics on A is chaotic. The quantity h defined by

$$h = \lim_{t \to \infty} \frac{1}{t} \ln \frac{|\delta\mathbf{x}(t)|}{|\delta\mathbf{x}(0)|}$$

is called the Lyapunov exponent. For Hamiltonian systems there are no attractors but chaos is said to be present if typical orbits yield $h > 0$. The extreme (i. e., exponential) sensitivity to initial conditions displayed by chaotic systems has the practical importance that small errors (such as computer round-off) eventually make it impossible to obtain the exact long-time behavior of the system. This was originally pointed out in the context of weather prediction in the seminal 1961 paper of Edward Lorenz. The condition $h > 0$ gives what is perhaps the most common definition of chaos. This definition, however, is not universally accepted. For example, other attributes of chaos which might be taken as its defining property are the infinity of unstable periodic orbits embedded in chaotic regions of phase space, and the property of positive topological entropy. (The topological entropy is a quantitative measure of how rapidly the number of distinct system orbits one can discern under finite resolution grows with the length of observation time.) These latter properties are more relevant when discussing unstable chaotic sets.

Unstable Chaotic Sets

Attractors refer to sets which "attract" orbits and hence determine typical long-term behavior. It is also possible to have sets in phase space on which the dynamics can be exceedingly complicated, but which are not attracting. In such cases orbits placed exactly on the set stay there forever, but typical neighboring orbits eventually leave the neighborhood of the set, never to return. One indication of the possibility of complex behavior on such nonattracting (unstable) sets is the presence on the set of an infinite number of unstable periodic orbits whose number increases exponentially with their period, as well as the presence of the uncountable number of nonperiodic orbits. Even though nonattracting, unstable chaotic sets can have important observable macroscopic consequences. Three such consequences are the phenomena of chaotic transients, fractal basin boundaries, and chaotic scattering.

Chaotic Transients

In chaotic transients one observes that typical initial conditions initially behave in an apparently chaotic manner for a possibly long time, but, after a while, then rapidly move off to some other region of phase space, perhaps asymptotically approaching a nonchaotic attractor.

The length of such chaotic transients depends sensitively on initial conditions and exhibits a characteristic Poisson distribution for randomly chosen initial conditions.

Fractal Basin Boundaries

Basin boundaries arise in dissipative dynamical systems when two, or more, attractors are present. In such situations each attractor has a basin of initial conditions which lead asymptotically to that attractor. The basin boundaries are the sets which separate different basins. It is very common for basin boundaries to contain unstable chaotic sets. In such cases the basin boundaries can have very complicated fractal structure. Because of this complicated very fine-scaled structure, fractal basin boundaries can pose an impediment to predicting long-term behavior. In particular, if an initial condition is specified with only finite precision, it may be very difficult *a priori* to determine in which basin it lies if the boundaries are fractal.

Chaotic Scattering

In the classical dynamics potential scattering problem one considers a Hamiltonian $H = p^2/2m + V(\mathbf{r})$, where the potential V approaches zero for large $|\mathbf{r}|$. One then asks how outgoing orbits at large $|\mathbf{r}|$ depend on incoming orbits. For example, one might plot scattering angle as a function of impact parameter. In typical cases, such functions can have exceedingly complex behavior, where the function is singular on a fractal (uncountable) set of impact parameter values. This type of behavior is indicative of the presence of an unstable chaotic set in the dynamics.

Routes to Chaos

It is a common procedure in experiments to examine the observed behavior as some parameter of the system is varied. One can then attempt to observe transitions between regions of parameter space where qualitatively different properties occur (e. g., phase transitions and critical phenomena in condensed matter physics). In nonlinear dynamics, particular attention attaches to the study of transitions to chaos in which one observes nonchaotic behavior (e. g., periodic motion) for some range of the parameter, but then observes a chaotic attractor as the parameter is varied. The question then is *how* does the chaotic attractor come into being as the parameter is varied. Generally, it is found that chaotic attractors come about in a limited number of often-observed characteristic ways. These include period doubling cascades, crises, intermittency, and quasiperiodic transitions. In the following we discuss the first two of these routes to chaos.

Period Doubling

In some range of the parameter, we might have time periodic behavior of a relevant dynamical variable, $x(t) = x(t+T)$, where T is the period. As the parameter, call it p, increases through a value p_1, the period of $x(t)$ can double to $2T$ in the following way. For $p < p_1$ the periodic signal has a peak (maximum) which repeats every period T. For $p > p_1$, the signal bifurcates so that peaks separated by T are now unequal but repeat after every $2T$. (For $p \to p_1$ from above, the difference between the two adjacent maxima approaches zero.) This is called a period-doubling bifurcation. It is often observed that period doublings occur in cascades. That is, one finds that as p is increased, there is a succession of period doublings at $p_1 < p_2 < p_3 \ldots < p_\infty$, where p_n accumulates geometrically on some finite value p_∞. The rate of geometric accumulation is universal and has been determined by the renormalization group technique by Feigenbaum. He obtains $\lim_{n \to \infty} (p_n - p_{n+1})/(p_{n+1} - p_n) = 4.669201 \ldots$ For $p > p_\infty$ there is typically attracting chaotic behavior.

Crises

Another type of transition to a chaotic attractor is the crisis. Basically, what happens in this case is that the unstable chaotic set responsible for a chaotic transient becomes stable as the parameter p is increased through a critical crisis value p_c. When it becomes stable, the chaotic set formerly responsible for the chaotic transient becomes a chaotic attractor. For parameter values in the transient range, $p < p_c$ there is typically a characteristic dependence of the mean duration of chaotic transients on p. Namely, $\tau \sim (p_c - p)^{-\gamma}$ where the critical exponent γ can be obtained from a knowledge of the instability properties of certain unstable periodic orbits on the chaotic set. This dependence of τ on γ makes clear the nature of the transition: τ increases to infinity as p approaches p_c from below, thus converting the transient to long-term time-asymptotic behavior.

KAM Surfaces

Dynamics in conservative systems, and, in particular, Hamiltonian systems, can differ qualitatively from that in nonconservative systems. In particular, Hamiltonian systems are characterized by the absence of attractors and by the typical occurrence of KAM (for Kolmogorov, Arnol'd, and Moser) surfaces. KAM surfaces are N-dimensional toroidal surfaces in the $2N$-dimensional phase space of coordinates and momenta $(q_1, q_2, \ldots, q_N; p_1, p_2, \ldots, p_N)$ on which orbits execute N-frequency quasiperiodic motion. (In such quasiperiodic motion the orbit winds around each of the N possible angular paths on the torus with N frequencies which are incommensurate.) For completely integrable systems, such surfaces permeate all of the phase space. Small perturbations from perfect integrability typically lead to chaotic motions, but only in a relatively small volume of the phase space, the remainder of the phase-space volume (Lebesgue measure) being occupied by KAM surfaces. In such cases the KAM surfaces lie on a Cantor set of positive volume; in particular, an arbitrarily small neighborhood of a KAM surface will typically contain chaotic orbits. As the perturbation from exact integrability is increased, the relative volume of chaos increases and can eventually occupy an order-one fraction of the phase-space volume.

Quantum Chaos

According to the correspondence principle, there is a limit where classical behavior as described by Hamilton's equations becomes similar, in some suitable sense, to quantum behavior as described by the appropriate wave equation. Formally, one can take this limit to be $h \to 0$, where h is Planck's constant; alternatively, one can look at successively higher energy levels, etc. Such limits are referred to as "semiclassical." It has been found that the semiclassical limit can be highly nontrivial when the classical problem is chaotic. The study of how quantum systems, whose classical counterparts are chaotic, behave in the semiclassical limit has been called quantum chaos. More generally, these considerations also apply to elliptic partial differential equations that are physically unrelated to quantum considerations. For example, the same questions arise in relating classical acoustic waves to their corresponding ray equations. Among recent results in quantum chaos is a prediction relating the chaos in the classical problem to the statistics of energy-level spacings in the semiclassical quantum regime. Other notable work has concerned Anderson localization phenomena and microwave ionization of atoms in high-energy states.

The Future

Much remains to be done in applying chaotic dynamics to specific physical systems. In addition, many important fundamental questions in chaotic dynamics remain unanswered. Some of these are the following: characterization of universal properties of dynamics in the chaotic regime; the interaction of spatial patterns and temporal chaos; the understanding of long chaotic transients in higher-dimensional dynamical systems; bifurcations involving chaotic behavior in higher-dimensional dynamical systems; how to extract information from experimental data obtained from chaotic processes; how to find the minimum dimensionality of a dynamical system needed to describe given chaotic data; how best to characterize and determine the properties of fractal sets arising in chaotic dynamics; how to use small external perturbations to control chaotic processes; the development of new computer methods for the study of chaos; the behavior of random (or noisy) dynamical systems. This is but a partial list. What seems certain, however, is that research in chaotic dynamics will be both important and exciting for a long time to come.

See also: Dynamic Critical Phenomena; Fractals.

Bibliography
H.-G. Schuster and W. Just, *Deterministic Chaos*, 4th ed. Wiley-VCH, Weinheim, 2005.

Charge-Density Waves

P. B. Littlewood

A charge-density wave (CDW) in a solid refers to a weak periodic perturbation of the valence-band electronic charge density, accompanied by a periodic lattice distortion (PLD) of the crystal, sometimes also called a Peierls distortion. The appellation CDW is usually reserved for small perturbations in the charge density of a material which would be otherwise a metal, in distinction to periodic structural distortions in insulators (e. g., $NaNO_2$) where the driving mechanism is different.

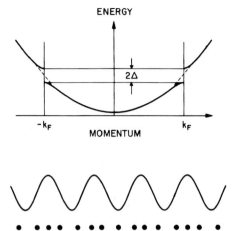

Fig. 1: Electronic energy spectrum (top) in the presence of a periodic lattice distortion (bottom) with wavelength π/k_F. The dashed curve is the metallic spectrum in the absence of the CDW. The periodic CDW is also shown.

The physical reason for the distortion is most easily pictured in a model of a one-dimensional chain of atoms; the Fermi energy corresponds to partial filling of the band up to the Fermi wave vector k_F (see Fig. 1). If now a small lattice distortion is introduced with a wavelength $2\pi/Q$, the band structure will have a gap at a wave vector $Q/2$ because of scattering of electronic Bloch waves by the new periodicity in the potential. States of wave vector $|k| < \frac{1}{2}Q$ have their energy lowered, while those of larger momenta have their energy raised; consequently the single-particle energy will be a minimum when $Q = 2k_F$, so that the gap lies exactly at the Fermi level. Because the electronic energy is lowered by the presence of the gap, the undistorted metallic state is unstable to the periodically deformed CDW.

In a three-dimensional material the new gaps introduced by the distortion will lie along planes in momentum space (the new Brillouin zone boundaries); a significant lowering of the total energy will require that much of the Fermi surface lies in the new gaps, which in turn requires that the metallic Fermi surface must have flat pieces translated by the CDW wave vector **Q**. This "nesting" condition explains why CDW are prevalent in materials which are electronically low dimensional, typically having either chain-like structures [e. g., NbSe$_3$, (TaSe$_4$)$_2$I, and some organic conductors of which tetrathiafulvalene-tetracyanoquinodimethane (TTF–TCNQ) is the prototype] or layer structures (e. g., NbSe$_2$, TaSe$_2$). If a gap is opened over the whole Fermi surface, the CDW will be semiconducting; in many cases the Fermi surface is not completely gapped, and the material is then a semimetal.

A closely related concept is that of the spin-density wave (SDW). A SDW can be regarded as a combination of two CDW, one for each spin polarization, arranged 180° out of phase; thus the spin density is periodic while the charge density is unchanged from that of the metal. The driving force for the transition is the exchange interaction between electrons rather than the direct Coulomb interaction (mediated by the lattice) in the CDW case.

In the simple model picture given above, the period of the density wave is related to the "nesting" wave vector, and may not be related to the period of the underlying lattice. In such

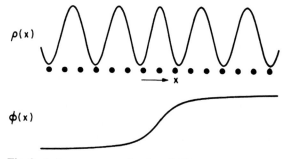

Fig. 2: A discommensuration in a CDW whose wavelength is just less than three lattice constants. At the top is the CDW charge density, and the bottom curve shows the spatial dependence of the CDW phase.

a case the CDW is termed incommensurate. Also commonly observed is a commensurate CDW, where the period of the CDW is a multiple of the lattice constant. A further complexity is that the lattice symmetry may be such that different directions of the CDW wave vector Q_i may be related by symmetry. In this case interactions between CDW in different directions may give rise to a single-Q CDW, or a multiple-Q CDW where all periods coexist. Different combinations of these situations can exist even in the same material at different temperatures. A well-studied example is the layer compound $2H$-TaSe$_2$, where the individual layers have triangular symmetry, and there are three equivalent CDW under $120°$ rotation. The low-temperature state is commensurate, with a periodicity of three times the lattice constant. With increasing temperature, there is a phase transition where two of the three Q vectors become incommensurate (the "striped" phase); at a higher temperature all three become incommensurate and equivalent before the final phase transition to the normal metallic phase when the CDW amplitude vanishes.

Such complex situations are not uncommon, and are best understood within a phenomenological Ginzburg–Landau expansion of the free energy in terms of the CDW order parameter. For a CDW, one might use as an order parameter the lattice distortion itself, or the periodic components of the charge density; the latter is conventional, and one may write the charge density as $\rho(\mathbf{r}) = \rho_0(\mathbf{r})[1 + \alpha(\mathbf{r})]$. Here ρ_0 is the charge density in the normal state, and $\alpha(\mathbf{r}) = \mathrm{Re}\,\Psi(\mathbf{r})$ the real order parameter. In the case of a simple incommensurate CDW we shall have $\Psi(\mathbf{r}) = \Psi_0 \exp[i(\mathbf{Q} \cdot \mathbf{r} + \phi)]$.

The value of ϕ measures the position of the CDW relative to the underlying lattice; clearly if $\mathbf{Q}$ is incommensurate, the value of ϕ is arbitrary, if constant. More generally, there will be some interaction with the lattice which will attempt to "lock" the $\mathbf{Q}$ vector to some nearby commensurate value, say $\mathbf{Q}_c$. In this case, we shall obtain local regions of the CDW which are commensurate separated by periodic jumps in phase ("discommensurations"); the phase ϕ is now a periodic function of position. The situation is illustrated in Figure 2; it bears close similarity with models for atoms adsorbed on surfaces (e. g., Kr on graphite) as well as dislocations in solids.

If the interaction with the underlying lattice is unimportant, then the freedom in choosing ϕ suggests that an incommensurate CDW may be free to slide and conduct a current. Such an idea was proposed by Fröhlich in 1954 as a possible mechanism for superconductivity. In fact, incommensurate CDW are always pinned by impurities (via local fluctuations in the phase ϕ); furthermore a sliding CDW dissipates energy quite rapidly so this is not a plausible mechanism for superconductivity. In recent years, a number of sliding CDW have been discovered, principally among inorganic linear chain compounds (e. g., $NbSe_3$, $K_{0.3}MoO_3$) but also reported in TTF–TCNQ, and possibly also a sliding SDW in tetramethyltetraselenafulvalinium nitrate, $(TMTSF)_2NO_3$.

In all of these materials a finite electric field (the threshold field E_T, typically $\sim 100\,mV/cm$) must be applied for the CDW to break free from impurities and begin to slide. In the pinned state at an electric field $E < E_T$, the CDW behaves like a collection of damped harmonic oscillators, with a distribution of oscillator frequencies reflecting the disorder induced by the pinning. As in all disordered systems, the pinned configuration is not unique, leading to hysteresis in the electrical polarization of the pinned CDW. In the sliding state there is a net current transported by the CDW; in many systems, this is found not to be uniform in time, but has both a nearly periodic component ("narrow band noise," NBN) and a low-frequency broadband noise with an approximately $1/f$ power spectrum. The frequency of the NBN varies proportionately to the average velocity $\mathbf{v}$ of the CDW, and matches the "washboard" frequency $\omega_0 = \mathbf{Q} \cdot \mathbf{v}$ of the CDW. The appearance of this internal frequency naturally reflects the periodicity of the CDW interacting with underlying pinning sites.

If, in addition to a DC electric field, a fairly large AC component of the electric field is introduced there develop interference features in the response whenever the washboard frequency is a harmonic or subharmonic of the driving frequency. In extreme cases the CDW velocity can mode-lock to the AC driving frequency ω so that $\omega_0 = (p/q)\omega$, with p and q integers; at the same time, the broadband noise vanishes. This appears as plateaus in the current–voltage characteristics, often called Shapiro steps after a similar phenomenon observed in flux flow in type-II superconductors. A remarkable phenomenon has been discovered under repetitive pulsed driving, whereby the phase of the transient oscillations at the washboard frequency becomes entrained to the *end* of the pulse. Such experiments, and the models to explain them, have given rise to ideas concerning dynamical selection of states which may have general application to the dynamics of driven nonlinear systems.

In semiconducting CDW at low temperatures, additional phenomena are observed, especially an abrupt rise in conductivity of many orders of magnitude at a second threshold field higher than E_T. This behavior may reflect the Coulomb self-interaction of the CDW, unscreened at low temperatures, and producing a rigid CDW; thus we have the possibility of true superconductivity as originally envisioned by Fröhlich, although only at the absolute zero of temperature.

Bibliography

"Charge Density Waves in Solids", in L. P. Gor'kov and G. Grüner (eds.), *Modern Problems in Condensed Matter Sciences*. North-Holland, Amsterdam, 1989. (A)

G. Grüner, "The Dynamics of Charge-Density Waves," *Rev. Mod. Phys.* **60**, 1129 (1988). (A)

"Low-Dimensional Conductors and Superconductors", in D. Jerome and L. G. Caron (eds.), *NATO ASI Series B*, Vol. 155. Plenum, New York, 1987. (I,A)

K. Bechgaard and D. Jerome, "Organic Superconductors," *Sci. Am.* **247(1)**, 52, 1987.

Charged-Particle Optics

J. D. Lawson

Beams of electrons, protons, or heavy ions have many applications in basic science, technology, and industry. The formation of these beams and their transportation and focusing are accomplished with the aid of the principles of charged-particle optics. This is a very wide field, with many specialized branches, relevant to such varied devices as electron microscopes, spectrometers, particle accelerators, cathode ray tubes, beam probe analysis and isotope separators. Nevertheless, despite much specialized technical elaboration, all depend on relatively few, rather general, and basically simple principles.

In the broadest sense, charged-particle optics is concerned with the motion of particles in external electric and magnetic fields. Once these fields are specified, the trajectory of a particle is uniquely determined in terms of its charge q, rest mass m_0 and the three components of velocity at some given point. The equation of motion has the simple form

$$\mathbf{F} = \frac{d\mathbf{p}}{dt} = q(\mathbf{E} + \mathbf{v} \times \mathbf{B}) \, , \tag{1}$$

where the momentum $\mathbf{p}$ is equal to $\gamma m_0 \mathbf{v}$, and γ is the relativistic factor $(1 - v^2/c^2)^{-1/2}$. The essential physics is contained in this equation. In order to find appropriate and convenient solutions of practical value, concepts originally developed in light optics, such as "focusing," "dispersion," and "aberrations," are introduced. At the most fundamental level the curvature of particle trajectories in electric and magnetic fields may be compared with the curvature of light rays in a nonuniform medium. Starting with equivalence between Fermat's principle in optics and the principle of least action in mechanics, a suitable refractive index for charged particle optics may be defined. Although formally elegant, this method is not the most convenient for practical purposes; it is simpler to apply the laws of mechanics more directly. Most applications are concerned with an ensemble of particles in the form of a beam. At any point along the beam these particles are moving in roughly the same direction and have a small spread in energy.

Lenses and prisms in charged-particle optics consist of suitable localized configurations of electric and magnetic fields, and the design of electrodes, coils, and iron magnets to produce fields of appropriate shape for different applications is a highly developed art. In a "perfect" system, the image of a planar object is sharp, and geometrically similar to the object. This idealization is, in charged-particle optics, characterized by the paraxial ray equation, which specifies the particle motion in terms of the electrostatic potential and magnetic field on the axis of the beam. (The axis may be the natural symmetry axis, or, more generally, the trajectory of a particular particle.) It is assumed that particle trajectories (or "rays") make a small angle with the axis, and that the deflecting fields can be expressed as first-order expansions of the fields on the axis.

The detailed form of this equation, of which several derivations may be found in the standard texts, depends on the symmetry and geometrical form of the axis. For axial symmetry the assumptions imply that field components directed along the axis are independent of the distance from the axis, and radial components are proportional to it. The nonrelativistic form of the paraxial equation is then

$$r'' + \frac{\phi' r'}{2\phi} + \left(\frac{\phi''}{4\phi} - \frac{qB_z^2}{8m_0\phi} \right) r + \frac{q\Psi_0^2}{8\pi^2 m_0\phi} \frac{1}{rs} = 0 , \tag{2}$$

where primes denote d/dz, ϕ is the electrostatic potential on the axis such that $-q\phi$ is equal to the kinetic energy of the particle at that point, and B is the magnetic field. The quantity Ψ_0 will be explained after consideration of the subsidiary angular equation

$$\theta - \theta_0 = - \int_0^z \frac{q}{2m_0 v} \left(B - \frac{\Psi_0}{\pi r^2} \right) dz , \tag{3}$$

which, together with Eq. (2), is necessary to specify both r and θ when a magnetic field is present. The magnetic field introduces a feature absent in light optics: the image can be *rotated* as well as scaled in size from the object. It will be seen from the form of Eq. (3) that Ψ_0 is related to the initial angular velocity of a particle about the axis; if, at some radius r_0, $d\theta/dz = 0$, then Ψ_0 is the magnetic flux through a circle of radius r_0 where $d\theta/dz = 0$. Alternatively, $\Psi_0 = 2\pi P_\theta/q$, where $P_\theta = p_\theta + qAr$, the conserved canonical angular momentum of the particle about the axis.

Since ϕ and B are known functions of z and r, trajectories can be calculated. By following a trajectory with initial conditions $r = 1$, $r' = 0$ through the lens, both the focal length and image rotation (for magnetic lenses) may readily be found. For a "thin" lens, in which the radial position of the particle remains essentially constant during its passage through the lens, the focal lengths of electrostatic and magnetic lenses in nonrelativistic approximation are, respectively,

$$\frac{1}{f_e} = \frac{3}{16} \left(\frac{\phi_1}{\phi_2} \right)^{1/4} \int \left(\frac{\phi'}{\phi} \right)^2 dz , \tag{4}$$

$$\frac{1}{f_m} = \int \left(\frac{qB_z}{2m_0 c} \right)^2 dz . \tag{5}$$

For an electrostatic lens ϕ_1 and ϕ_2 represent potentials at entry and exit. Two features of these lenses are evident from the form of these expressions. First, f is always positive, so that all lenses are focusing, and second, the focusing strength depends on the *square* of the fields. These are "second-order" lenses, as a detailed physical consideration of just how they work will reveal.

The paraxial Eq. (2) may be put into linear form by introducing the Larmor transformation, which decouples the radial and angular motion. By observing the orbits in a reference frame which rotates about the axis with frequency $Q_L = -\frac{1}{2}qB_z/m_0$ the force on the particles appears purely radial, and it becomes possible to specify the motion in rotating rectangular coordinates, with equations of the form

$$x'' + \alpha_x(z)x' + \kappa_x(z)x = 0 ,$$
$$y'' + \alpha_y(z)y' + \kappa_y(z)y = 0 , \tag{6}$$

where $\alpha_x = \alpha_y$, and $\kappa_x = \kappa_y$. These are now linear, and from this fact many useful relations can be established.

Although introduced here in a particular context, equations of the same form apply in all paraxial situations when there is no coupling between motion in the x and y directions. They describe oscillations about orbits in cyclic particle accelerators, for example, where $B_z = 0$. On the other hand, the focusing is different in the x and y planes, so that the coefficients are not equal. If the accelerating field is negligible or absent (as in a beam transport system), then $\alpha = 0$. When κ is independent of z, the orbits are evidently sinusoidal or exponential in shape.

A useful property which follows from this linear form is that of matrix representation. The values of x and x' at any point z_2 are related to those at an "upstream" point z_1 by the equation

$$\begin{pmatrix} x_2 \\ x_2' \end{pmatrix} = M \begin{pmatrix} x_1 \\ x_1' \end{pmatrix} , \tag{7}$$

where M is a 2×2 matrix. Furthermore, when $\alpha = 0$, $|M| = 1$. Once the appropriate matrices for various elements are known, groups of particles may readily be traced through the system. Particles lying on an ellipse in $x - x'$ space remain on an ellipse. The shape and orientation of the ellipse may change, but its area remains inversely proportional to the momentum of the particles. This fact forms the basis of the emittance concept, which leads to useful design procedures for systems of lenses and accelerator magnet lattices.

The lenses discussed earlier possess axial symmetry. An important lens which does not have this symmetry is the quadrupole, in which the radial and circumferential fields vary as $\cos\theta$ and $\sin\theta$, respectively. (Both types are illustrated in Fig. 1(a)–(d), showing some commonly used elements in electron and ion optics. Solid and dashed lines denote orbits and fields, respectively.) In the two orthogonal symmetry planes the field directions are such that there is focusing in one plane and equal defocusing in the other. For a magnetic quadrupole, with field gradient $\partial B_\theta / \partial x$, the paraxial equation in the two planes is

$$\begin{aligned} x'' + \kappa_x x &= 0 , \\ y'' + \kappa_y y &= 0 , \end{aligned} \tag{8}$$

where $\kappa = q(\partial B_\theta / \partial x)/\gamma m_0 v$. The abandonment of axial symmetry permits first-order focusing in one plane at the expense of defocusing in the other. If, however, two quadrupoles are arranged as a pair with opposite polarity, a net focusing effect is produced. Focusing of this type is much stronger than that obtainable by lenses with axial symmetry. Known as alternating gradient focusing, it forms the basis of the focusing in large particle accelerators where economy of power and size is essential.

In Eq. (2) the coefficients are functions of particle energy, so that energy spread in the beam results in the formation of "blurred" images, a phenomenon known by analogy with light optics as "chromatic aberration." Likewise, if the optical axis is curved, particles with identical initial conditions but different energies follow different trajectories, and a unique "axis" can only be defined with reference to a particle of particular energy. This property, sometimes made use of and sometimes an embarrassment, is known as "dispersion." It may be handled analytically by using 3×3 matrices, with a third row for $\Delta p/p$, the fractional excess of momentum.

Many of the basic ideas of *linear* or paraxial optics have now been introduced. Once the appropriate matrix elements for various devices have been determined, straightforward design procedures, making use of simple computer programs, can be developed. When nonlinear

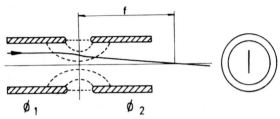

Fig. 1(a): Electrostatic lens (across which there is a difference of potential $\phi_2 - \phi_1$). Hollow disks or cones can be used in place of cylinders. A cylinder with two gaps of opposite polarity, so that the total difference of potential across the pair is zero, is known as an "einzel lens."

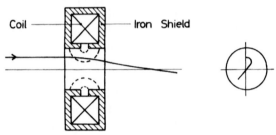

Fig. 1(b): Magnetic lens. The coil is clad in an iron shield containing a gap. Radial and azimuthal components of velocity interact with azimuthal and radial components of field, and this produces a rotation of the image.

Fig. 1(c): Magnetic quadrupole pair, in which defocusing is followed by focusing. Pole pieces approximate the ideal hyperbolic shape which gives rise to a pure quadrupole field in the absence of edge effects and saturation.

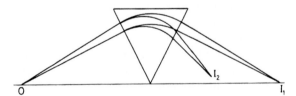

Fig. 1(d): Magnetic prism, showing bending, focusing, and dispersion. Object, apex, and image are collinear for a magnet with uniform field, normal incidence and exit, and "hard" edges.

features are introduced, the elegant simplicity disappears; so also does the generality. Aberration problems encountered in different fields tend to be special and different, and need to be considered individually. Small nonlinearities in the focusing systems of accelerators and storage rings, especially those which couple the motion in the two planes, again introduce a wide range of new phenomena. Sophisticated techniques for dealing with these aberrations and nonlinearities have been developed. Recent methods employing Lie algebra may be noted. These are now used particularly in the design of very large accelerators and storage rings.

So far, attention has been confined to the characteristics of individual particle orbits. Many practical problems are concerned with ensembles of orbits, arising perhaps from a Maxwellian energy distribution of electrons or ions in a cathode or source plasma. The subject can be developed further to investigate these collective properties, using the ideas of statistical mechanics, and particularly the theorem of Liouville. According to this theorem the density of noninteracting particles in phase space remains invariant, and this property is made use of in defining the beam emittance, $\pi\varepsilon$. It is often possible to decouple the longitudinal and transverse motion, and assign an emittance to each. Then, by dividing the transverse momentum by $\beta\gamma m_0 c$, the emittance is defined as $1/\pi$ times the projected area on the $x - \mathrm{d}x/\mathrm{d}z$ or $y - \mathrm{d}y/\mathrm{d}z$ planes. The meaning of the area when the density of points tapers gradually to zero requires further consideration, and the rms value of the distribution is often used. A more general quantity is the normalized emittance, equal to $\beta\gamma\varepsilon$, and the rms value of this quantity is invariant in a linear focusing system even in the presence of acceleration along the axis. The emittance (un-normalized) corresponds to the Helmholtz–Lagrange invariant in light optics, and means physically that at a waist or position of maximum radius the angular spread of the particles in the beam multiplied by the beam diameter remains constant. This may also be thought of as a "gas law" for a beam considered as a drifting gas; the temperature multiplied by the two-dimensional volume remains constant. In intense beams, effects arising from self-fields and scattering may become significant. Here again, however, we leave the realm of optics and begin to impinge on the discipline of plasma physics. Indeed, the study of charged-particle beams in full depth and breadth embraces the three separate disciplines of charged-particle optics, statistical mechanics, and plasma physics.

See also: Charged-Particle Spectroscopy; Electron and Ion Beams, Intense.

Bibliography

P. Dahl, *Introduction to Electron and Ion Optics*. Academic Press, New York, 1973. Clear introduction to basic principles. (E)

P. Grivet and A. Septier, *Electron Optics*, 2nd ed. Pergamon Press, Oxford, 1972. Basic electron optics, with application to a wide range of practical devices. Good bibliography. (I)

S. Humphries, Jr., *Principles of Charged Particle Acceleration*. John Wiley, New York, 1986. Broad coverage, with good introduction to basic principles. (I)

J. D. Lawson, *The Physics of Charged Particle Beams*. Clarendon Press, Oxford, 1977. Synoptic view of basic physical ideas. (I)

J. J. Livingood, *Principles of Cyclic Particle Accelerators*. Van Nostrand, Princeton, 1969. Rather old, but clear elementary treatment of theory. (E)

P. Sturrock, *Static and Dynamic Electron Optics*. Cambridge University Press, Cambridge, England, 1955. Elegant but formal treatment making use of variational methods. (A)

H. Wollnik, *Optics of Charged Particles*. Academic Press, New York, 1987. Comprehensive theoretical
treatment. (I)

The advanced theory of particle accelerators may be found in the proceedings of a number of Accelerator
Schools in the USA and Europe from 1982 onwards. These are published in the *American Institute
of Physics Conference Series*, or as *CERN reports*, CERN, Geneva. In particular, an account of
the techniques making use of Lie algebra is given by A. J. Dragt (1982) in *American Institute of
Physics Conference Proceedings* No. 87, p. 145.

Charged-Particle Spectroscopy

G. M. Crawley

The study of the properties of nuclear quantum states by bombarding nuclei with charged-particle beams and detecting the emitted charged particles is called charged-particle spectroscopy. Information sought by charged-particle spectroscopy includes the mass (energy) of the nuclear states and their spin (J), parity (π), and isospin (T), the ultimate goal being to obtain the complete wave functions of individual nuclear states.

A typical experimental arrangement is shown in Fig. 1. A beam of charged particles from an accelerator impinges on a "target" in an evacuated chamber. The particles in the beam that pass through the foil without interacting are collected in an insulated Faraday cup, the charge

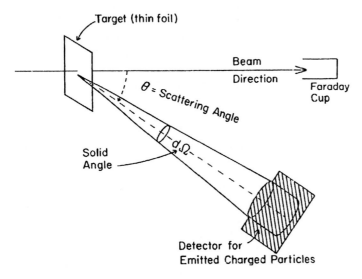

Fig. 1: Schematic diagram of standard experimental arrangement for charged-particle spectroscopy.

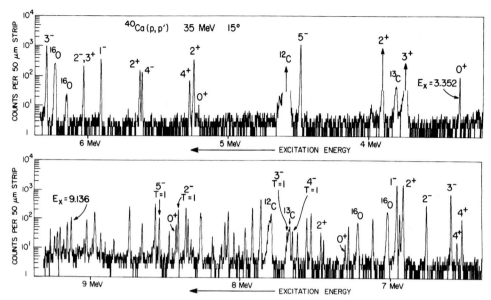

Fig. 2: ^{40}Ca (p,p') ^{40}Ca spectrum taken at $\theta_{lab} = 31.2°$ for $E_p = 34.78$ MeV (from [3]).

thereby collected giving a measure of the number of beam particles. The relatively few particles that actually interact with nuclei in the target can transfer energy, angular momentum, and nucleons between the beam and the target nucleus. The emitted particles are collected by a particle detector that subtends a solid angle $d\Omega$ at a mean scattering angle θ with respect to the incident beam. The probability of particle emission into solid angle $d\Omega$ is termed the *differential cross section*, $\sigma(\theta)$, which is measured in units of square centimeters per steradian or, commonly, millibarns per steradian (mb/sr), where 1 mb $= 10^{-27}$ cm^2. An *angular distribution* refers to the measurement of $\sigma(\theta)$ at a number of different scattering angles θ.

The plot of the number of emitted particles of a particular kind as a function of their energy is called a spectrum, by analogy with an optical spectrum, in which intensity is plotted as a function of wavelength or frequency. An example of a spectrum of inelastically scattered protons from ^{40}Ca using a 35-MeV incident beam is shown in Fig. 2. The spectrum consists of a number of peaks, each denoting the excitation of a particular nuclear state. As increasing amounts of energy are transferred to the nucleus, states of higher excitation energy in the nucleus are produced and the emitted particles have lower energies. The spectrum shown has an experimental energy resolution of 4.5 keV, which is the full width at half-maximum height of the peaks from ^{40}Ca (p,p'). This spectrum illustrates the need for good energy resolution in order to distinguish peaks from the isotope under study from contaminants in the thin foil targets and, more particularly, to resolve close-lying states.

Excellent energy resolution can also be obtained at lower energies in resonance reactions where compound-nucleus states are studied. This kind of experiment has been pioneered at Duke University where widths of nuclear states as small as 130 eV have been measured. (See, e. g. [2].)

Thus the mass (energy) of a particular nuclear state can be obtained from the energy spectrum. The spin and parity (J^π) are obtained by comparing the measured angular distribution with theoretical predictions. In the case of resonance reactions, the shape of the yield versus bombarding energy curve also indicates the J^π of the state.

The experimental challenges of charged-particle spectroscopy also include the preparation of incident beams of particles with accurately known energies and with small energy spread, the preparation of clean uniform isotopic targets, and particularly the use of detector systems with both good energy resolution and the ability to discriminate between different particle types emitted from the target. Each of these aspects is discussed in what follows.

Accelerators and Beam Preparation

Since charged particles must have sufficient energy to overcome the Coulomb repulsion of the positively charged target nuclei, special particle accelerators have been constructed to produce the charged-particle beams. The accelerators most commonly used today are Van de Graaff accelerators (including the tandem type), linear accelerators, cyclotrons, and synchrotrons, each having certain advantages as well as limitations. For example, tandem Van de Graaff accelerators produce beams with good energy resolution whose energy can be readily varied, making them particularly suitable for studies of resonance reactions. However, the maximum energy of beams of particles with charge Q from even the largest of such accelerators is presently less than about $25(1+Q)\,\text{MeV}$.

Modern cyclotrons with azimuthally varying magnetic fields can readily accelerate protons to much higher energies (up to a few hundred MeV) and the beam quality from these machines is now comparable to that obtained from tandem Van de Graaffs. Using dispersion-matching techniques, in which beam energy spread is canceled by the dispersion of a magnetic spectrograph, final-state energy resolution of 1 part in 10^4 in energy has been obtained. Because of their higher energy capability, cyclotrons are generally used for direct reaction studies where the beam energy remains constant during the experiment.

Linear accelerators (linacs) have been used to produce large-intensity beams of very high-energy protons. The beam quality from linacs is generally poorer than that from cyclotrons or Van de Graaff accelerators. However, since the beam intensity is high, excellent energy resolution can be obtained even at high energy by selecting a small fraction of the beam and using dispersion matching. For example, an energy resolution of better than 50 keV has been obtained at the 800-MeV proton linac (LAMPF) at Los Alamos National Laboratory.

The most efficient scheme to reach the highest energies is by the use of synchrotrons, where the ions travel along a circular path and are accelerated by a series of electromagnetic kicks. At the energy of the ions increases the magnetic field strength is increased to keep the same circular path. High-energy heavy-ion synchrotrons are in use at the Relativistic Heavy-Ion Collider, RHIC, at Brookhaven National Laboratory and the Center for Heavy Ion Research, GSI in Darmstadt Germany. The RHIC facility can accelerate gold ions to 250 GeV/nucleon or 99.999% the speed of light. In a few years heavy ions will be accelerated by the synchrotrons at CERN as part of the Large Hadron Collider project to 6 TeV/nucleon.

Superconducting technology has had a significant impact on accelerator design. For example, superconducting radio-frequency cavities have been used to provide post acceleration at several Van de Graaff laboratories, the first being at Argonne National Laboratory in 1978. Superconducting technology has also been applied to cyclotron magnets allowing much higher field strengths at comparatively modest cost. This in turn has allowed the acceleration of more massive charged particles (heavy ions). Heavy ion cyclotrons are in operation at Michigan State University, Texas ARM University, RIKEN in Japan, JINR in Russia, and at GANIL at Caen, France. The first two cyclotrons mentioned use superconducting magnets. The development of electron cyclotron resonance (ECR) ion sources, which produce intense beams of highly charged ions, has also contributed significantly to the more extensive use of heavy ions for charged particle spectroscopy. Another very recent development in accelerator technology has been the production of beams with very small intrinsic energy spread which are stored in a ring of magnetic elements. The beams are "cooled" by repeated interaction with electron beams having very small energy spread. Such a system was demonstrated in 1988 at Indiana University and similar systems have been used in Heidelberg and GSI in Darmstadt, Germany and Uppsala, Sweden. The use of these very high-quality beams for charged-particle spectroscopy holds the promise of dramatic improvements in energy resolution.

Beam preparation systems consist of a series of magnetic-quadrupole lenses and dipoles to focus and switch the charged-particle beams to different target positions. A dipole magnet can also provide energy dispersion of the beam, and combined with a slit system can improve the energy resolution of the incident beam. More sophisticated systems produce dispersed as well as focused beams at the target, so that dispersion matching can be used to provide good energy resolution without loss of beam intensity.

At low energies, below about 20 MeV, collimator systems are generally used to constrain geometrically the beam position and direction. At higher energies, scattering from the collimator slits makes this method unsatisfactory, so that alternative methods of monitoring the beam position with retractable detectors (scintillators or wire chambers) are usually employed.

Scattering Chambers and Beam Monitoring

Since charged-particle beams lose energy and scatter when passing through air, the beam lines and experimental apparatus through which the beam or reaction products pass must be evacuated. Oil diffusion pumps, which were once used extensively in vacuum systems, have largely been replaced by oil-free pumping systems to eliminate carbon and silicon impurities from the system since these can contaminate targets. Modern systems generally employ cryogenic pumping either with a molecular-sieve adsorber at liquid nitrogen temperatures or with surfaces held at liquid helium temperatures. Turbo-molecular pumps, carbon vane pumps, titanium sublimation pumps, and ion pumps are also used to obtain clear vacua.

Scattering chambers, which house the target and often the charged-particle detectors, come in many shapes and sizes. The principal design criteria for a general-purpose chamber are (1) accurate setting of the geometry of the experiment, particularly the laboratory scattering angle, (2) the availability of multiple detector systems that can be moved independently; (3) the ability to use a number of targets without breaking vacuum; and (4) the accurate recording of the total charge of the beam that passes through the target in an insulated Faraday cup.

Beam current, target thickness, and beam alignment can be measured by one or more monitor detectors, which are placed at some fixed angles and which can record a prolific reaction such as elastic scattering. The measurement of cross section at a series of angles relative to the number of counts recorded by the monitor provides an accurate measure of relative cross section.

Targets

The success of many charged-particle reaction experiments depends on the preparation of suitable targets. Most targets are in the form of a thin foil of the isotope of interest, although in some cases compounds containing the isotope are more convenient. In some cases thin backing foils of carbon or various plastics like Formvar are used as a substrate for the isotopic foil. Gas targets are also used and differential pumping is sometimes employed to enable the entrance or exit window of the gas cell to be eliminated. At another extreme, for higher energy experiments done at facilities such as GSI in Germany with high energy ions, the targets can be several centimeters thick.

The choice of target thickness is usually a compromise between the small counting rate from a thin target and the large energy loss and consequent degradation of the energy resolution from a thick target.

The elimination of unwanted impurities, particularly carbon and oxygen, can, however, be an important element in target preparation. This often involves the preparation and transfer of targets to the scattering chamber in vacuo, and the availability of clean high-vacuum systems. In some cases a liquid-nitrogen-cooled shroud may be used around the target. The use of cooled beams which repeatedly interact with a target poses special problems since ultrathin targets are required. Gas jet and powder targets are under development for these applications.

Detectors

Perhaps the single most important factor in a charged-particle reaction experiment is the detector for the emitted charged particles. Just as accelerator technology has produced higher-energy and better-quality beams, recent detector technology has led to much improved energy resolution and better particle discrimination for charged particles. The use of heavy-ion beams has placed even more demands on detectors and has helped to drive the improvement of the technology. There are four general types of particle detectors in current use: (1) scintillation counters, (2) gas counters, (3) solid-state detectors of silicon or germanium, and (4) magnetic spectrometers plus a particle detector in the focal plane.

Scintillation counters consist of inorganic crystals such as NaI, CsI, YAP, GSO, BaF_2, or CaF_2 as well as many kinds of plastic scintillators, bonded to a photomultiplier tube (PMT). Bismuth germanate (BGO), which has even higher stopping power, has also been used as a charged-particle detector. These detectors can be sufficiently thick to stop high-energy charged particles and can also be used when a large solid angle is required. However, the energy resolution obtainable with these detectors is not very good (e. g., about 2% for 30-MeV protons) and they are quite sensitive to gamma rays and neutrons, which are often present as unwanted background in charged-particle experiments. In cases where timing information is needed, a plastic scintillator mounted on a fast photomultiplier tube can be used. The pulses produced in plastic scintillators have rapid rise times, and timing resolution better than 0.5×10^{-9} s can readily be obtained.

A gas counter consists of a gas volume containing an electric field. The primary electrons, produced by the passage of a charged particle through the detector, drift to the anode, often a thin wire, to produce an electrical signal. If the field is low enough, the size of the electrical signal is proportional to the energy loss of the charged particle (ionization and proportional counters). A typical gas mixture used is argon (90%) plus methane (10%). Gas ionization chambers are particularly useful as detectors for heavy ions or low energy light ions because of the small mass of material traversed by the particle. Other useful features are the large areas which can be produced, the lack of radiation damage and the fact that they can be made position sensitive. Gas detectors come in many configurations and have many applications. One of the most common applications is their use as a thin front energy loss detector in a multi-element telescope used for particle identification. This is illustrated in Fig. 3 where charge and even isotope identification is obtained using a gas-filled ionization chamber. Another common use is as a position sensitive detector in the focal plane of a magnetic spectrograph.

Solid-state ionization chambers made from reverse-biased diodes of semiconducting crystals of silicon or germanium are also very common detectors for charged particles. These detectors have far better energy resolution than gas ionization chambers, since the energy required to produce an ion pair is only about 3 eV in Si compared to about 30 eV in a gas. Thus about 10 times more ion pairs are produced in a silicon detector, with consequently better statistical accuracy in the energy determination. An energy resolution of about 0.1% can now be obtained for such devices.

Solid-state detectors are limited in size because of the difficulty of producing large pure crystals of silicon or germanium. However, thicker detectors can be produced by drifting lithium through the material to compensate exactly for impurities. These lithium-drifted detectors [Si(Li) or Ge(Li)] can be produced in thicknesses up to about 7 mm. Through the use of thin entrance and exit windows, all solid-state detectors can be stacked to produce a thicker amount of material to stop higher-energy particles. Germanium detectors have greater stopping power per unit thickness but have the disadvantage that they must be operated at the temperature of liquid nitrogen. Silicon detectors can also be made position sensitive by evaporating a resistive coating along the back of the detector and taking signals from both ends. However, the maximum length of such devices is only about 10 cm and their position resolution is at best about 1% of their length.

One further advantage of solid-state detectors is their use in multi-element systems to differentiate between different particles emitted from a reaction. In a typical two-detector telescope, the thin front detector gives a signal proportional to the rate of energy loss (dE/dx) of the traversing particle, and the second, thicker detector measures the total energy loss (E). (More accurately, the sum of the energy signals from the two detectors gives the total energy loss.) Since the rate of energy loss (dE/dx) of a charged particle is approximately proportional to MZ^2/E, the product (dE/dx) $\cdot E$ is proportional to MZ^2 and so can be used to differentiate particles of different mass and charge. This is the basis for particle identification using a series of solid-state detectors where the logic is carried out with electronic hardware, or more commonly in a computer.

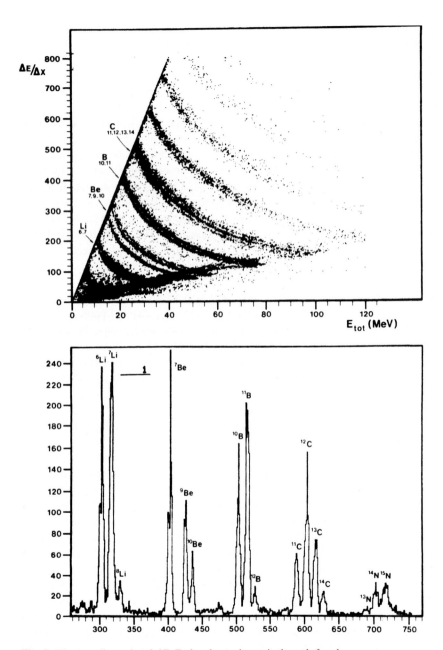

Fig. 3: The two-dimensional dE, E plot shows the typical result for charge resolution in an ionization chamber. The reaction is 84 MeV/u ^{12}C + Au. If the particle energy is above 3 MeV/u, the resolution is good enough to separate the isotopes of these light ions. This is seen in the lower part of the figure, where the projection on the particle identification (charge) axis is shown (from [4]).

A similar arrangement of a pair of scintillation detectors connected to a single PMT is called a phoswich detector. Either a CaF_2 and plastic scintillator or two plastic scintillators such as NE102 (decay time 2.5 ns) and NE115 (decay time 225 ns) are used. The total light pulse is sampled with both a short and a long gate to distinguish the signals from the fast and slow scintillators. Much larger area two-element detector systems are possible using phoswich scintillators than can be obtained with solid state Si or Ge detectors.

Various kinds of magnetic spectrometers are used to detect the charged products following nuclear reactions. Most modern types include dipole, quadrupole, and higher-multipole elements, and focus particles both radially and vertically (double focusing) to obtain a large solid angle. Some spectrographs designs, e. g., Big Karl at Jülich and the K600 at Indiana University allow variation of the dispersion in the focal plane. The radius of curvature gives a measure of the momentum and therefore the energy of the emitted charged particle, and is normally obtained by measuring the position of the particle in the focal plane of the spectrometer. Originally the preferred device to measure position was a nuclear emulsion. This is still the most precise technique but the inconvenience of emulsions, their lack of discrimination of particle type, and, particularly, the improvement in position sensitive gas and solid-state detectors has almost eliminated the use of nuclear emulsions.

Many modern spectrographs, such at the S800 at Michigan State University in the US and the VAMOS spectrograph at GANIL in France use software techniques to obtain high position and angular resolution. Particle tracks are measured at the focal plane of the device and the information coupled with a precise knowledge of the magnetic fields of the device are used to reconstruct the momenta and angles of the particles from the interaction target area.

The three main types of position sensitive gas detectors are multiwire, drift, and single wire proportional counters. Multiwire proportional counters (MWPC) have a position resolution which depends on the wire spacing but is typically ≤ 0.5 mm. They also have high count rate capability. Single-wire proportional counters (SWPC) use various methods to measure the position of the incident particle, including charge division of the signals from either end of the counter or measurement of the transit time of the signal along a delay line. By such methods a position resolution of between 0.1 and 0.2 mm can be obtained. Drift chamber use the drift time of electrons through the gas and a start signal to determine the location the particle enters the detector. A new type of gas detector is the GEM, or gas electron multiplication, detector. Where high electric fields generated by precisely fabricated electrodes with small distances are used to generate high multiplication of electrons released when the ions pass through the gas.

Apart from large solid angles, another advantage of magnetic spectrometers is that unwanted reaction products with high count rates can often be "bent off" the detector. Very good final-state energy resolution can also be obtained with a magnetic spectrometer, as was illustrated in Fig. 2. This spectrum was measured using a dispersion-matching technique where the energy dispersion of the incident beam at the target position is canceled by the dispersion of the spectrometer, so that only the uncorrelated energy spread gives a contribution to the energy resolution (Fig. 4). By using multiple detector elements in the focal plane of a spectrometer, particle identification can be carried out very reliably. A typical setup would consist of a series of gas proportional counters to give position and energy-loss information followed by either a solid-state detector or a plastic scintillator to record total energy or time of flight. Such combinations allow very small cross sections to be measured in the presence of prolific backgrounds of other particles. An example of a focal-plane detector used at Berkeley is shown in Fig. 5.

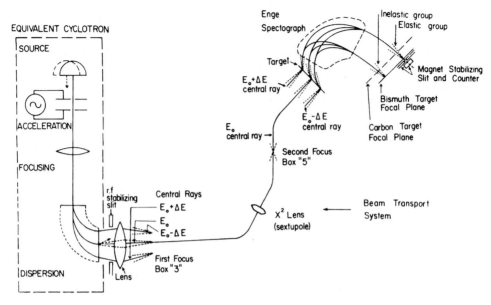

Fig. 4: "Equivalent circuit" of cyclotron and high-resolution beam line showing dispersion matching. The correlated energy dispersion of the beam on target is canceled by the dispersion of the spectrograph (from [1]).

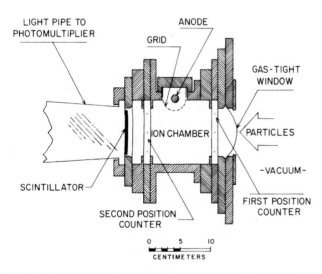

Fig. 5: Spectrometer focal-plane detector system. Gas is usually 200 Torr propane (from *Lawrence Berkeley Lab. Ann. Rep.* 1975, p. 355).

See also: Angular Correlation of Nuclear Radiation; Cyclotron; Nuclear Reactions; Nuclear Scattering; Radiation Detection; Semiconductor Radiation Detectors.

Bibliography
Experimental Aspects and Resonance Reactions
J. Cerny (ed.), *Nuclear Spectroscopy and Reactions*, Vols. 1–4. Academic Press, New York, 1974–1975. (Updates and expands upon the material in the following book.) (I)

W. W. Buechner, "The Measurement of the Spectra of Charged Nuclear Particles"; H. T. Richards, "Charged-Particle Reactions"; N. S. Wall, "Charged-Particle Detectors," in *Nuclear Spectroscopy*, Part A (F. Ajzenberg-Selove, ed.). Academic Press, New York, 1960.

Experimental Methods
B. L. Cohen, *Concepts of Nuclear Physics*, Chapter 9. McGraw–Hill, New York, 1971. (I)

H. Enge, *Introduction to Nuclear Physics*, Chapters 7 and 12; Section 3 of Chapter 13. Addison-Wesley, Reading, Mass., 1966.

Instrumentation for Heavy Ion Nuclear Research, Vol. 7 of *Nuclear Science Research Conference Series*, D. Shapira (ed.). Harwood Academic Publishers, New York, 1985.

References

[1] H. G. Blosser, G. M. Crawley, R. deForest, E. Kashy, and B. H. Wildenthal, *Nucl. Instr. Methods* **91**, 61 (1971).

[2] G. A. Keyworth, G. C. Kyker, E. G. Bilpuch, and H. W. Newson, *Nucl. Phys.* **89**, 590 (1966).

[3] J. A. Nolen and R. S. Gleitsmann, *Phys. Rev. C* **11**, 1159 (1975).

[4] H. Sann, *Instrumentation for Heavy Ion Nuclear Research*, Vol. 7, p. 27, D. Shapira (ed.). Harwood Academic Publishers, 1985.

Chemical Bonding

H. A. Bent

Introduction

That chemical combination is essentially an electrical phenomenon Faraday reasoned must be so from the fact that passage of an electrical current between inert electrodes immersed in aqueous solutions produces at the electrodes chemical decomposition ("electro-lysis"). Mathematical development of Faraday's views had to await, however, the introduction into physical theory of the electron, electron spin, the nuclear atom, the wave equation, and the exclusion principle. Chemical theory, meanwhile, developed along purely phenomenological lines, graphically illustrated in the structural theory of organic chemistry and the doctrine of coordination. In recent decades a union of physical and chemical theory has been achieved through the use of the principle of indistinguishability and the creation of localized molecular orbitals that correspond closely to the graphic formulas of classical structural theory.

Physical Models of the Chemical Bond

The concept of a chemical *bond* arose from the use in chemistry of graphic formulas to illustrate the rule that carbon atoms in stable compounds, such as CH_4, CH_2Cl_2, CO_2 and HCN, are "tetravalent" – hydrogen and chlorine (in, e. g., HCl) being taken as "univalent"; oxygen (as in H_2O) "divalent"; nitrogen (as in NH_3) "trivalent":

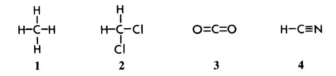

For half a century no connection existed between the "bonds" or "valence strokes" in graphic formulas such as **1–4** and physical theory. Then, soon after Moseley's determination of atomic numbers and, thereby, the number of electrons in compounds (usually an even number, for compounds of the nontransition metals), Lewis suggested that in such graphic formulas as **1–4**, (i) each valence stroke represents, improbable as it might seem, two "valence-shell" electrons and that, correspondingly, (ii) the symbols for the elements, H, C, N, O, represent the corresponding atoms' positively charged "kernels" or "atomic cores" [after Rutherford and Bohr, the atoms' nuclei and inner-shell (here two K-shell or $1s$) electrons]; e. g., in **1–4**: H^+, C^{4+}, N^{5+}, O^{6+}.

A valence stroke (–) may be viewed as the union of a pair of oppositely directed Faraday lines of force (in the organic chemist's convention: $\longleftarrow\longrightarrow$) that stretch between a valence-shell electron pair and the two adjacent atomic cores that that pair helps to bond together. In accordance with the principle of local electrical neutrality and Gauss's law (suitably modified, numerically), the number of lines of force that terminate on an atomic core generally equals that core's positive charge.

In, for example, HCN, the nitrogen atom's kernel, N^{5+}, has, according to Lewis's theory, like C^{4+}, four valence-shell electron pairs, only three of which, however, are simultaneously in the valence shell of another atom (carbon) and shown in the molecule's graphic formula by conventional, straight, bonding valence strokes. The fourth pair is "unshared." It may be shown as

Like lines of force, valence strokes never cross each other. Valence-shell electrons obey a principle of spatial exclusion: Two but no more than two electrons may be in the same place at the same time. About each spatially coincident electron pair is a Fermi hole, a van der Waals-like domain of diameter approximately equal to the electrons' de Broglie wavelengths, into which other valence-shell electrons cannot easily penetrate.

Because of the decrease in a particle's kinetic energy with increasing de Broglie wavelength, the van der Waals-like domains of electron pairs tend to expand to fill the available space. In a full axiomatization of structural theory, there is associated with valence strokes, as with Faraday's lines of force, a stereochemically significant property of mutual repulsion.

Stereochemical Models of Chemical Bonding

Together, the exclusion principle and the kinetic-energy operator in the molecular Hamiltonian, reinforced by Coulombic repulsions, cause the four electron pairs about the carbon atom of methane (CH_4) and its derivatives [such as CH_2Cl_2, (2)] to spread out tetrahedrally in three dimensions **7** rather than to remain crowded together in one plane. [Graphic formula **2**, for example, implies, contrary to observation, that there should be two isomers of dichloromethane: a cis isomer, depicted, and a trans isomer.] Similarly, ammonia is pyramidal **8** not planar, and water is bent **9** not linear.

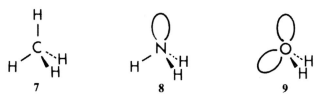

In thought, methane can be transformed into ammonia via an *al*chemical movement of the atomic core of, for example, the upper hydrogen atom in **7**, a bare proton, together with its attached line of force, to the central, heavy-atom nucleus, thereby converting C^{4+} to N^{5+} ("proton capture") and changing the electron pair of the C–H bond of methane into the unshared valence-shell electron pair of ammonia (electron "capture" by the heavy-atom's core). Ammonia and water are similarly related. The three molecules, **7, 8, 9**, have similar bond angles (HCH 109.5°, HNH 107°, HOH 104.5°) and, by inference, similar electronic structures. They are said to be "isoelectronic." Likewise, H–C≡N (hydrogen cyanide) and H–C≡C–H (acetylene), both linear molecules, are isoelectronic with N≡N (N_2). Another isoelectronic family of molecules is H_3C–CH_3 (ethane), H_3C–OH (methyl alcohol), H_3N–NH_3 (hydrazine), HO–OH (hydrogen peroxide), and F–F (fluorine).

The tetrahedral model of directional chemical affinities applies chiefly to C, N, O, F, and other electronegative, "octet-rule"-satisfying atoms (ones that in chemical combination have eight, or four pairs, of valence-shell electrons) in so-called "covalent" compounds (ones that contain no large-atomic-core, electropositive, metallic-like elements).

The tetrahedral model accounts for the nonexistence of quadruply bonded C, and, more importantly, for most of the facts of organic stereochemistry, particularly the stereochemistry of many displacement reactions (as described in the last section) and the planar and linear geometries about double and triple bonds, Fig. 1.

Single, double, and triple bonds may be viewed, c. f. Fig. 1, as the sharing, respectively, of a corner, an edge, and a face of two tetrahedral polyhedra of electron pairs.

Similar terminology is used to describe the structures of Born-type "ion compounds." Calcium oxide, for example, may be viewed as Ca^{2+} cations (a cation being the atomic core of a metallic, large-core element) surrounded by a coordinated polyhedron of ("valence-shell") O^{2-} anions. When such compounds are "anion deficient" (number of anions per cation less than the cation's usual coordination number), the polyhedra of anions about the cations generally share corners, edges, or faces, Fig. 2.

With the exception of (possibly) He and Ne, all uncombined atoms are "electron deficient": they contain fewer than the maximum number of valence-rule-allowed electrons. They are, so to speak, "coordinately unsaturated." Particularly electron deficient are atoms to the left of

(a)
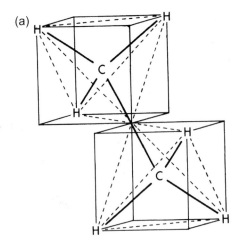

(b)
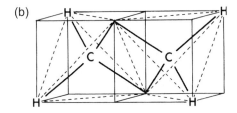

(c)
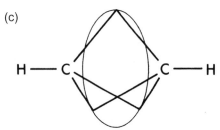

Fig. 1: Sharing of, respectively, a corner, an edge, and a face by a pair of tetrahedra of electron pairs about two atomic cores, C^{4+}, in (a) ethane ($H_3C–CH_3$), (b) ethylene ($H_2C=CH_2$), and (c) acetylene (HC=CH). Heavy lines represent valence-shell electron pairs. The two carbon atoms of a double bond (b) and their four substituents (H) lie in a plane. The carbon atoms of a triple bond (c) and their substituents lie on a straight line.

carbon in the periodic table. To utilize fully the low-potential-energy space for electrons about atomic cores of elements from Groups I, II, and III of the Periodic Table, electrons, or pairs of electrons, must often be shared by more than two atomic cores.

B_2H_6, which is isoelectronic with C_2H_4 (**10**), has the electronic structure depicted in **11**.

10   **11**

The boron–boron bond in "electron deficient" B_2H_6 may be described variously as: a "protonated double bond"; as two B–H–B "three-center bonds"; as two electron pairs simultaneously in the valence shells of three atoms; as two "bridging" hydride ions (H^-); or as two tetrahedral BH_4 ions sharing an edge.

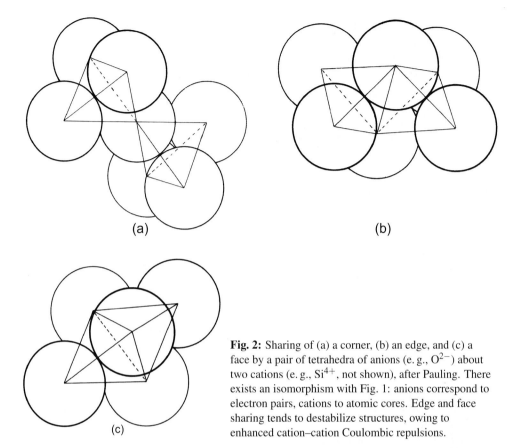

(a)

(b)

(c)

Fig. 2: Sharing of (a) a corner, (b) an edge, and (c) a face by a pair of tetrahedra of anions (e. g., O^{2-}) about two cations (e. g., Si^{4+}, not shown), after Pauling. There exists an isomorphism with Fig. 1: anions correspond to electron pairs, cations to atomic cores. Edge and face sharing tends to destabilize structures, owing to enhanced cation–cation Coulombic repulsions.

Atoms with large atomic cores may contain in their valence shells more than eight electrons. Whereas oxygen forms only one mono-oxygen fluoride, OF_2, sulfur forms, in addition to SF_2, SF_4 (**12**) and coordinatively saturated, extremely inert SF_6 (**13**).

12   13

Preferred geometries for coordination of five and six anions, or electron pairs, are the capped tetrahedron or trigonal bipyramid (**12**) and the octahedron (**13**).

To achieve stability, highly electron-deficient, large-core elements – the metals – may adopt simultaneously the bonding strategies of boron and sulfur. Potassium hydride, for example, a saltlike substance iso-structural with potassium and sodium chloride, may be viewed as a Born-type, ion compound, K^+H^-, in which each potassium atomic core is surrounded by six protonated electron pairs (H^- ions), each of which is in the valence shell of six potassium atoms, forming thereby "seven-center" bonds.

Isoelectronic with K^+H^- is the structure $Ca^{2+}E_2^{2-}$, a model similar to one first proposed by Thompson for the face-centered-cubic modification of calcium metal. E_2^{2-} stands for the deprotonated hydride ion, i. e., in the present instance, for a six-center electron-pair bond.

Historically, bonds have been termed covalent, ionic, or metallic. Those three bond types correspond to electrons shared between or among, respectively, solely small atomic cores (radius $< 0.05\,nm$), small and large atomic cores, and exclusively large atomic cores.

Orbital Models of Chemical Bonding

Mathematical theories of chemical bonding require a quantitative expression for the wavelike behavior of electrons. Usually used is Schrödinger's equation. Approximations to molecular wave functions for use in Schrödinger's equation generally are constructed from atom-centered, hydrogen-like orbitals.

To mimic the directional properties of chemical bonding, linear combinations of the spectroscopic, nondirectional, doughnut-shaped p orbitals of the hydrogen-atom problem were first used to produce bond orbitals p_x, p_y, p_z that point along the coordinate axes. Formation of additional linear combinations with the s orbital of the same principal quantum number (a process called "hybridization") yields a set of tetrahedrally directed orbitals.

To produce from component orbitals a wave function that does not distinguish between indistinguishable electrons and that does satisfy the exclusion principle requires antisymmetrization with respect to electron labels. Antisymmetrization yields a sum of orbital products that, for closed-shell molecules, may be written as a determinant, each column of which refers to a different orbital.

Since a determinant is unchanged when one column is added to or subtracted from another column, the component orbitals of a determinantal wave function may be described in several ways. In conventional molecular-orbital descriptions of a molecule, the component orbitals extend over the entire molecule. From linear combinations of those orbitals can be created localized molecular orbitals that occupy nearly mutually exclusive domains.

Localized orbitals correspond closely to Lewis's bonding electron pairs, unshared valence-shell pairs, and inner-shell electrons. They constitute a bridge between quantum-mechanical theories of chemical bonding and Lewis's electronic interpretation of classical structural theory, when taken with the discovery of the "classically nondescribable two-valuedness of the electron" called "spin" and the rule that two spin-opposed electrons may share the same spatial orbital.

Calculated energies of many-electron systems are always improved, however, if different spatial orbitals are used for electrons of different spins. Molecular oxygen, for example, is in fact paramagnetic in its ground state. It has five valence-shell electrons of one spin, seven of the other. One may picture the five-membered spin set as having a configuration of maximum probability (CMP) that is triple-bond-like (two tetrahedra sharing a face), with the seven-membered spin set having a single-bond-like CMP (two tetrahedra sharing a corner). Electrons of opposite spin are, thus, somewhat spatially anticoincident. In the bonding region are placed altogether $3 + 1 = 4$ electrons, as in a conventional double bond.

The Virial Theorem and Chemical Bonding

As a result of electron–electron repulsions, individual electron, or electron-pair, orbitals cannot be rigorously defined for many-electron systems. Applicable, however, to all systems (for which the Born–Oppenheimer approximation is valid) is the virial theorem.

The virial theorem states that the average kinetic and potential energies, $\bar{T}$ and $\bar{V}$, of, for example, a diatomic molecule are related to the molecule's total energy E and internuclear separation R according to the relations

$$\bar{T} = -E - R\frac{\mathrm{d}E}{\mathrm{d}R} \; ; \qquad \bar{V} = 2E + R\frac{\mathrm{d}E}{\mathrm{d}R} \; .$$

Analysis of the exact E-vs-R curve for the hydrogen molecule ion, H_2^+, reveals that as R approaches from above that value for which E is a minimum, the decrease in E that produces the minimum, and chemical bonding, arises from a decrease in potential energy, owing to contraction of the wave function into the internuclear, bonding region; concomitantly, the kinetic energy rises. As R decreases still further, the fall of potential energy owing to a continued enhancement of nuclear–electron attraction is offset by its rise owing to nuclear–nuclear repulsion, which causes the potential energy, and total energy, to approach infinity as R approaches zero.

When E is a minimum, $\mathrm{d}E/\mathrm{d}R = 0$ and $E = \frac{1}{2}\bar{V}$. In polyatomic, many-electron systems $\bar{V}$ receives contributions from nuclear–electron attractions, nuclear–nuclear repulsions, and electron–electron repulsions:

$$E_{\min} = \frac{1}{2}\bar{V} = \frac{1}{2}(\bar{V}_{\mathrm{ne}} + \bar{V}_{\mathrm{nn}} + \bar{V}_{\mathrm{ee}}) \; .$$

Only the first term, $\bar{V}_{\mathrm{ne}}$ is negative (with respect to infinite separation of the parts) and contributes to chemical bonding. Its central role in theories of chemical bonding is reflected in such words and phrases as coordination, valence saturation, the octet rule, inner-shell electrons, atom-centered orbitals, and spatial pairing of electrons (in, especially, bonding regions).

The role of nuclear–nuclear repulsion term, $\bar{V}_{\mathrm{nn}}$, in chemical bonding is reflected in the rule that atomic cores (or cations) with large charges tend not to share electrons (or anions) with each other. F–F, HO–F, and HO–OH are thermodynamically highly reactive molecules. The structure of nitrous oxide is NNO (not NON), of nitrosyl fluoride FNO (not NOF), of stable cyanate ion NCO$^-$ (not CNO$^-$, explosive fulminate).

The role of the electron–electron repulsion term, $\bar{V}_{\mathrm{ee}}$ is reflected in the use of different orbitals for electrons of different spin.

Intermolecular Forces and Reaction Mechanisms

Nuclear–nuclear repulsion between the proton of a hydrogen atom and a highly charged atomic core of an electronegative atom to which it is chemically bonded, as, e. g., in hydrogen iodide, H–I, produces departures from local electrical neutrality, dipole moments, and a positive patch, "electrophilic center," or "acidic site" on the molecule's surface that, via stray feeler lines of force, may interact with a negative patch, "nucleophilic center," or "basic site" on an adjacent molecule or ion, usually an unshared valence-shell electron-pair, as,

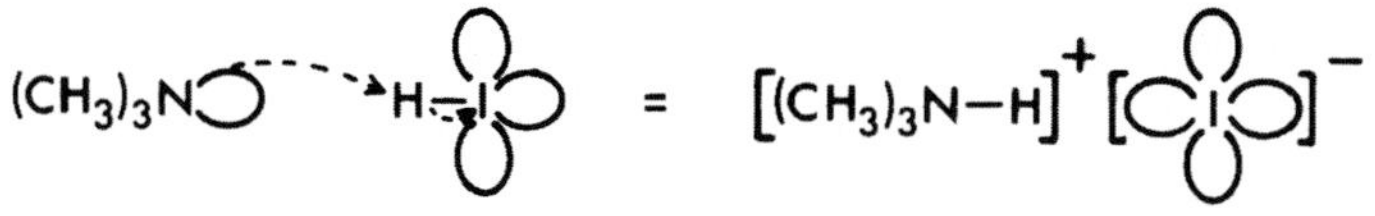

$(CH_3)_3N\overset{\cdot\cdot}{\underset{\cdot\cdot}{\colon}} \text{- - - -} \overset{+}{H}\text{---}I \quad = \quad [(CH_3)_3N\text{---}H]^+ \, [I]^-$

Fig. 3: Graphic representation of diversions of lines of force in the creation of a chemical bond between nitrogen and hydrogen and the simultaneous annihilation of a chemical bond between hydrogen and iodine in the reaction of the base trimethyl amine, $(CH_3)_3N$, with the acid hydrogen iodide, HI. Valence strokes, straight and curved, represent, respectively, localized, doubly-occupied bonding and nonbonding (or unshared) valence-shell molecular orbitals. Symbols H, C, N, I stand for the atomic cores H^+, C^{4+}, N^{5+}, I^-. Formed in the first step of the proton transfer is a hydrogen-bond, N–H–I, in which the bond angle NHI is $180°$.

e. g., in the trimethyl derivative of ammonia, $(CH_3)_3N$. Formed thereby is an intermolecular "hydrogen bond," the first step in a proton-transfer, "Brønsted acid–base reaction", Fig. 3.

    $(CH_3)_3N$ interacts similarly with the electrophilic C^{4+} center of methyl iodide, H_3C–I, through "backside attack." The base $(CH_3)_3N$ approaches with its nucleophilic, unshared electron pair that face of the tetrahedron of substituents surrounding the acidic, C^{4+} cation opposite to the eventual "leaving group," I^-. Formed thereby is an intermolecular "charge-transfer complex," the first step in a methyl-group transfer via a "Walden inversion" at carbon, Fig. 4.

    Caveats concerning the notation used in Figs. 3 and 4 have been cogently expressed by Michael Faraday (*Experimental Researches in Electricity*, 14th Series, 1838, par. 1684 [Added remarks appear in brackets]).

> "The terms free charges and dissimulated electricity convey erroneous notions if they are meant to imply any differences as to the mode or kind of action. The charge upon an insulator [e. g., immobile $(CH_3)_3N^+$] in the middle of a room [or ionic crystal] is [with respect to lines of force] in the same relation to the walls of that room [and surrounding ions, I^-] as the charge upon the inner coating of a Leyden jar [or the carbon kernel of CH_3I] is to the outer coating [the surrounding "ions": H^- and I^-] of the same jar [or molecule]. The one is not more free or dissimulated [or, with respect to lines of force, more detached] than the other; and when sometimes we make electricity appear [as, e. g., in the formation of $(CH_3)_3N^+I^-$] where it was not evident before, as upon the outside of a charged jar [or molecule] when, after insulating it [$(CH_3)_3I$] [by placing it in solution], we touch the inner coating [with $(CH_3)_3N°_o$] it is only because we divert more or less of the inductive force from one direction into another [from C to N rather than to I; from I to H rather than to C; etc.]."

    In a full graphical representation of a reaction mechanism, the path traced by the curly arrows employed in Figs. 3 and 4 would not be left open. The diverted lines of force would form a closed, Gauss circuit. Therein lies the major role of the solvent in many chemical reactions.

See also: Molecular Structure Calculations; Molecules.

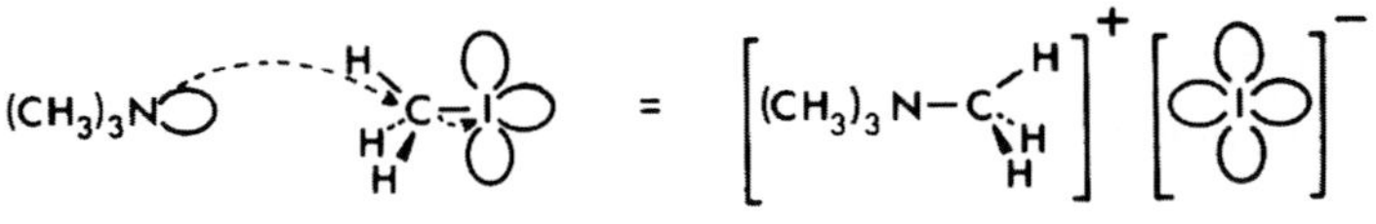

Fig. 4: Reaction of(CH$_3$)$_3$N with methyl iodide, CH$_3$I (rather than, as in Fig. 3, hydrogen iodide, HI). Formed in the first step of the transfer of the methyl cation CH$_3^+$ from I to N is a "charge-transfer complex" or "face-centered bond," N–C–I, in which the bond angle NCI is 180°. As the carbon core, C^{4+}, leaves I^{7+} and approaches the unshared pair about N^{5+}, it passes through the plane defined by its three hydrogen atoms. The methyl group undergoes, it is said, "Walden inversion." It turns inside out, like an umbrella: H$_3$C– → –CH$_3$.

Bibliography

H. A. Bent, "Isoelectronic Systems," *J. Chem. Ed.* **43**, 170 (1966) (E); "The Tetrahedral Atom," *Chemistry* **39**, 8 (1966) (E); **40**, 8 (1967) (E); "Ion-Packing Models of Covalent Compounds," *J. Chem. Ed.* **45**, 768 (1968). (E-I) Written for high school and college students and teachers.

H. A. Bent, *Isoelectronic Molecules in Molecular Structure and Energetics: Chemical Bonding Models*, Vol. 1, pp. 17–50 (Joel S. Liebman and Arthur Greenberg, eds.). VCH Publishers, New York, 1986. (I)

H. A. Bent, The Isoelectronic Principle and the Periodic Table, in *Molecular Structure and Energetics: From Atoms to Polymers: Isoelectronic Analogies*, pp. ix–xii (Joel S. Liebman and Arthur Greenberg, eds.). VCH Publishers, New York, 1988. (E)

C. A. Coulson, *Valence*. Oxford, London, 1961. (I) A readable, relatively nonmathematical introduction to quantum-mechanical theories of valence by a major contributor to the field.

C. Edmiston and K. Ruedenberg, "Localized Atomic Molecular Orbitals," *Rev. Mod. Phys.* **35**, 457 (1963). (A) A landmark paper that has stimulated much research on the creation of quantum-mechanically based, chemically interpretable, transferable molecular orbitals.

R. J. Gillespie, *Molecular Geometry*. Van Nostrand Reinhold, New York, 1972. (E–I) A nonmathematical account of numerous applications of the author's "valence-shell-electron-pair-repulsion" model for allowing in structural theory for the effects of electron correlation between electrons of parallel spin.

G. N. Lewis, *Valence and the Structure of Atoms and Molecules*. Dover, New York, 1966 (reprint). (E) A lively, personal, historical, and still-provocative account of the introduction and uses of the concept of the electron-pair bond. A deservedly everpopular classic.

E. H. Lieb, "The Stability of Matter," *Rev. Mod. Phys.* **48**, 553 (1976). (A) An explanation in terms of the exclusion principle and electrostatic screening of the fundamental paradox of classical physics as to why matter, which is held together by Coulombic forces, neither implodes nor explodes.

J. W. Linnett, *The Electronic Structure of Molecules: A New Approach*. Wiley, New York, 1964. (E–I) A largely nonmathematical account of the author's "doublet-quartet" model for allowing in structural theory for the effects of electron correlation between electrons of opposite spin.

L. Pauling, *The Nature of the Chemical Bond*. Cornell University Press, Ithaca, New York, 1960. (I) The author has introduced more concepts into structural theory than any other living scientist. Although his theory of resonance has, in many minds, been superseded (though not the classification of molecules based upon it), this classic remains a mine of interesting ideas and information.

J. C. Slater, *Quantum Theory of Molecules and Solids*, Vol. 1, *Electronic Structure of Molecules*. McGraw–Hill, New York, 1963. (A) A lucid introduction to the mathematical theory of chemical bonding by an early and long-active contributor to the field. Includes an expanded account of the author's classic discussion of applications of the virial theorem to H$_2^+$.

Chemiluminescence

A. P. Schaap, H. Akhavan-Tafti, and R. S. Handley

In the 1880s Eilhard Wiedernann was investigating various phenomena which resulted in the emission of light. He was the first to use the term "Chemiluminescenz" for chemical reactions which produced light. In 1888, he wrote: *"Das bei chemischen Processen auftretende Leuchten würde Chemiluminescenz..."*. Chemiluminescence is produced as a result of the generation of electronically excited products of a chemical reaction which subsequently emit photons. Such reactions are relatively uncommon as most exothermic chemical processes release energy as heat. The initial step is termed *chemiexcitation*, and involves conversion of chemical energy into electronic excitation energy with an efficiency denoted as Φ_{CE}. If the excited state species is fluorescent, the process is *direct* chemiluminescence and occurs with an overall efficiency (Φ_{CL}) that is a product of the efficiencies for the two steps: $\Phi_{CL} = \Phi_{CE} \times \Phi_{F}$. If the initial excited state species transfers energy to a molecule which then emits light, the process is *indirect* chemiluminescence and occurs with an efficiency that is a product of the efficiencies for the three steps (Fig. 1).

  The thermochemical requirement for chemiluminescence is that the total energy available from the reaction, i.e. the enthalpy of activation, $\Delta H^{\ddagger}$, and the enthalpy of reaction, ΔH_R, be at least as great as the energy of the lowest excited state of one of the products: $\Delta H^{\ddagger} - \Delta H_R \geq E_{ex}$. The most widely studied type of chemiluminescent reaction which meets this criterion is the decomposition of 1,2-dioxetanes to produce two carbonyl-containing products. In general there is only sufficient energy available to produce one of the products in the excited state.

  The thermal stability of 1,2-dioxetanes varies widely, with half-lives at room temperature ranging from minutes for most simple dioxetanes prepared in the laboratory to several years for dioxetanes with bulky acyclic or polycyclic alkyl groups such as adamantylideneadamantane dioxetane shown below in Eq. (1). In addition, these dioxetanes are inefficient producers of chemiluminescence generating predominantly triplet state products so the yield of direct chemiluminescence is low.

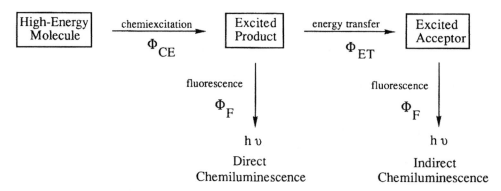

Fig. 1: Direct and indirect chemiluminescence.

$$ (1) $$

In contrast, model dioxetanes containing electron-rich aromatic groups are markedly less stable and can produce direct chemiluminescence with high efficiencies. This type of dioxetane more closely resembles the behavior of intermediates in biological processes such as the familiar firefly bioluminescence. These properties have been utilized to design dioxetanes which are stable indefinitely at room temperature but which can be "triggered" to undergo efficient chemiluminescent decomposition on demand. Removal of a protecting group from the dioxetanes shown below in Eq. (2) converts them to the unstable aryloxide form which rapidly decomposes with emission of light.

$$ (2) $$

The activating agent may be a simple chemical reagent or an enzyme. In the case of the t-butyldimethylsiloxy-substituted dioxetane shown above, reaction of the dioxetane with fluoride ion in DMSO produces brilliant bluish chemiluminescence with an efficiency of 25%. The phosphate-substituted dioxetane is triggered in aqueous solution by alkaline phosphatase. This and other stabilized dioxetanes have been used in extremely sensitive assays with a detection limit of 10^{-21} moles of enzyme. These types of dioxetanes are in wide use as reporters for enzyme-linked immunoassays and nucleic acid hybridization assays in medical testing and research applications.

Numerous other chemiluminescent organic reactions have been put to practical use. Reaction of oxalate esters with hydrogen peroxide in the presence of a fluorescer produces a long lived emission which finds use in novelty items. Chemiluminescent oxidation of luminol, catalyzed by iron-containing species, provides a forensic test for detecting trace amounts of blood. Acridinium esters, like the dioxetanes, are useful in medical testing applications functioning as labels which produce a brief burst of light when treated with hydrogen peroxide in alkaline solution.

See also: Luminescence.

Bibliography

A. K. Campbell, *Chemiluminescence: Principles and Applications in Biology and Medicine*. Ellis Horwood Ltd., Chichester (England), 1988.

K. D. Gundermann and F. McCapra, *Chemiluminescence in Organic Chemistry*. Springer-Verlag, New York, 1987.

A. P. Schaap, *Photochem. Photobiol.* **47S, 50S**, 1988.

R. Schreiner, M. E. Testen, B. Z. Shakashiri, G. E. Dirreen, and L. G. Williams, "Chemiluminescence," Chap. 2 in *Chemical Demonstrations* (B. Z. Shakashiri, ed.). University of Wisconsin Press, Madison, Wisconsin, 1983.

T. Wilson, "Chemiluminescence in the Liquid Phase: Thermal Cleavage of Dioxetanes," Chap. 7 in *International Review of Science, Physical Chemistry, Series Two*, Vol. 9 (D. R. Herschbach, ed.). Butterworth, Boston, 1976.

K. Van Dyke, Ed., *Bioluminescence and Chemiluminescence: Instruments and Applications*, Volumes I and II. CRC Press, Inc., Boca Raton, Florida, 1985.

L. J. Kricka, *Analytica Chimica Acta* **500(1-2)**, 279–286 2003.

Circuits, Integrated

S. Triebwasser

The invention of the transistor in 1948 opened the way to the rapid development of a series of increasingly sophisticated and useful device structures. Within 10 years, transistors had become the basis of a major industry which has revolutionized the way we live. Integrated circuit chips, consisting of millions of devices for applications in computer memories, have been fabricated, as well as chips that perform the basic functions of today's sophisticated personal computers.

At the other end of the spectrum, we have seen the displacement of the slide rule by powerful hand-held calculators and remarkably accurate, inexpensive timepieces.

Silicon integrated circuits use active devices of two basic types, the field-effect transistor (FET) and the bipolar transistor.

The FET operates on a simple and easily understood principle: application of a voltage to a capacitively coupled electrode creates an electric field which alters the number of charge carriers in a semiconductor, thus modulating its conductivity. The electrode has been realized as a *p-n* junction (junction FET, J-FET), as a Schottky barrier (metal–silicon FET, MESFET), or as a plate separated from the semiconductor by an insulating layer (insulated gate FET, IGFET). This last structure is usually a metal plate insulated from the surface of the semiconductor by silicon dioxide (metaloxide–semiconductor FET, MOSFET). The MOSFET, or, more commonly, the MOS device, is of two forms, the *n* channel and *p* channel, distinguished by the nature of the current-carrying species, electrons and holes respectively. In addition, the metal gate has been replaced by heavily doped (for enhanced conductivity) polycrystalline silicon or, in some cases, alloys of Si with certain metals. Today's integrated circuit industry is dominated by the Complementary Metal Oxide Semiconductor (CMOS) technology in which the basic circuit configuration consists of an *n*-channel MOS device in series with a

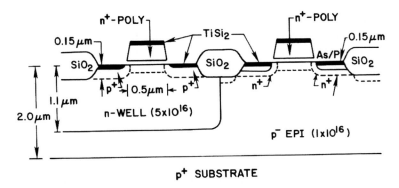

Fig. 1: Cross section of complementary MOS device structures.

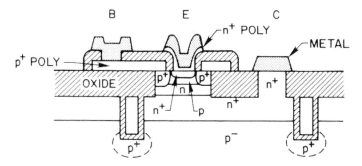

Fig. 2: Cross section of a bipolar transistor. Minimum dimensions are similar to those shown in Fig. 1.

p-channel device. This circuit has virtues of being simple to design, but, more important, it draws virtually no current except while being switched. Hence, it represents a very low-power technology, a dominant consideration as the number of circuits being fabricated in very small areas continues to increase almost geometrically with time. Figure 1 shows a cross section of an advanced CMOS device configuration.

The bipolar transistor, unlike the FET, does not rely on capacitive coupling. For an n-p-n transistor, it consists of a thin p region between two n regions. These regions correspond to the emitter, base, and collector of the transistor. In operation, the emitter–base junction is forward-biased such that electrons flow from the emitter region into the base. The base–collector junction is usually reverse-biased, that is, the n region is positive with respect to the base p region. The collector junction therefore collects the electrons that do not recombine in the base region. Current and voltage gain between base input and collector output can be obtained by using the base electrode as the input.

A p-n-p transistor operates similarly except that the emitter injects holes into the base region. Again, because electrons are more mobile than holes, n-p-n transistors with the same dimensions perform at higher frequencies than p-n-p transistors. Figure 2 shows a cross section of a modern n-p-n bipolar transistor.

Until about 1959, all semiconductor circuits were fabricated with discrete bipolar devices such as those just discussed. These were used, much like vacuum tubes, in circuits that also contained other, passive components such as diodes (simple *p–n* junctions), resistors, and capacitors. The introduction of transistors made it possible to operate the circuits at much lower voltage and power levels and offered significant improvements in reliability. Integrated electronics as we know it today had its start in 1958–59, when government-sponsored and industrial research sparked the development of semiconductor integrated circuits. Whereas germanium had dominated the technology in the 1950s, silicon now wins out, since it grows a natural oxide that serves as an insulator, as a passivation layer, and, most importantly, as a diffusion mask.

Integrated circuits are fabricated by a series of processes that form interconnected active and passive circuit components on a single piece of monocrystalline material, usually a thin silicon wafer, perhaps 125 mm in diameter and $\frac{1}{3}$ mm thick. Millions of devices can be fabricated and interconnected simultaneously. It is this batch fabrication capability that makes integrated circuits so inexpensive. The following briefly describes the principal steps in fabricating the typical integrated device shown in Figs. 1 and 2.

The first step is the production of single-crystal material, most commonly by the Czochralski method. Wafers are cut from single crystals, and lapped and polished to a mirror finish. An epitaxial film, that is, a film whose lattice structure is an extension of the substrate crystal, is grown on the surface to a thickness of several micrometers. This film is generally made to have conductivity opposite to that of the substrate, so that a *p–n* junction occurs at the interface; this is desirable because it provides an isolating junction. In other applications, the film has the same conductivity type as the substrate but a different doping level. An oxide layer (SiO_2), several thousand Ångstrøms thick, is grown on the Si surface by heating the wafer in an oxidizing atmosphere. The oxide layer is then coated with a film of photosensitive material called photoresist. The resist has the characteristics that it can be polymerized by radiation; originally optical radiation, but, by today, electrons, ions and x-ray radiation are being used as well. The resist is patterned either through a mask or by directed electron or ion beams. The wafer is then rinsed in a developer, which removes the nonpolymerized unexposed film. After the remaining resist is baked, the wafer is etched. Where the resist has been removed, the etchant leaves openings in the oxide layer. The wafer is now ready for ion implantation or diffusion. With the oxide layer acting as a mask, materials such as phosphorus or boron are ion implanted or diffused (mostly implanted today) through the openings in the oxide layer. Such processes of oxidation, photolithography and implantation are repeated to create the multiple-layer structures shown in Figs. 1 and 2 and to form resistors and diodes, isolation regions, and, finally, several layers of metal conductors to complete the highly functional integrated circuit chips that are on the market. Today's advanced chips require 15 or more lithographic steps and hundreds of additional processing steps.

Once processed, the wafers are cut into chips which may measure 10 mm or more on a side. A chip may contain millions of components interconnected to form an electronic function such as a microprocessor, or millions of bits of an electronic memory. Figure 3 shows a chip containing 4×10^6 bits of memory and all the peripheral circuitry required to address the memory locations, write and read the information in this random access memory (RAM). This chip is less than 80 mm^2 in area and contains enough storage capacity to store 400 pages of double-spaced typewritten pages.

Fig. 3: Photograph of a 4×10^6-bit integrated circuit dynamic random access memory chip. This chip was fabricated using $0.7\,\mu$m minimum dimension.

The chips are attached to a protective package which provides the interconnecting leads for signal and power external connections. Chips can be attached to packages with wire bonds, beam leads, beam tape, or solder pads. At present, the minimum line width and line spacings used by industry are about $0.5\,\mu$m, a dimension that is taxing the capability of optical systems. Innovative approaches will be required to meet the lithographic challenges of the decade of the '90s.

Integrated circuits have revolutionized computers, television, radio, and electronic products, and have made possible lower costs, higher reliability, savings in size, weight, power, and better performance and have made a major impact on how we live.

See also: Semiconductors, Crystalline; Transistors.

Bibliography

S. M. Sze, *Semiconductor Devices, Physics and Technology*. Wiley, New York, 1985.
D. A. Hodges and H. G. Jackson, *Analysis and Design of Integrated Circuits*, 2nd ed. McGraw–Hill, New York, 1988.
C. Mead and L. Conway, *Introduction to VLSI Systems*. Addison-Wesley, Reading, MA, 1980.
I. Brodie and J. J. Murray, *The Physics of Microfabrication*. Plenum, New York, 1982.
S. K. Ghandi, *VLSI Fabrication Principles*. Wiley, New York, 1983.

Clocks, Atomic and Molecular

J. A. Barnes[†] and J. J. Bollinger

Almost any clock consists of three main parts: (1) a pendulum or other nearly periodic device, which determines the rate of the clock; (2) a counting mechanism, which accumulates the number of cycles of the periodic phenomenon; and (3) a display mechanism to indicate the accumulated count (i. e., time).

An atomic clock makes use of an atomic resonance to control the periodic phenomenon. Similarly, a resonance in a molecule could be used to control the periodic phenomenon. The atomic or molecular clock is a very good clock because these resonances are determined by the atom's properties rather than by the man-made dimensions of an artifact; they are among the most stable and accurately measured phenomena known to man. Since all clocks are just devices that count and display the total of a series of periodic events (such as swings of a pendulum, passages of the sun overhead, or oscillations of an atom), the accuracy of the clock depends directly on the stability and accuracy of the periodic phenomenon used to establish the rate of the clock.

The most accurate clocks today make use of a microwave resonance in the ground state of cesium. In fact, the unit of time, the second, is defined in terms of this microwave resonance in cesium, and the national standards of time and frequency for the United States and other countries are cesium clocks. Time (and frequency) can be measured with the smallest uncertainty of any physical quantity. Current estimates of possible errors in various national standards laboratories are of the order of a few parts in 10^{14}. This can be expressed by saying that independent cesium clocks can maintain synchronism with one another to better than one-millionth of a second after one year's operation. This is more than 100 000 times more predictable than the earth's rotation on its axis.

The operation of a cesium atomic clock depends on the observation of a particular resonance in cesium atoms. The atoms are not radioactive, and radioactive decay processes play no part in the scheme. Neutral atoms boiled off from a quantity of liquid cesium are allowed to escape through narrow holes in a small oven and form a beam, which traverses an evacuated chamber. To prevent the cesium atoms from colliding with air molecules and being scattered out of the beam, a good vacuum must be maintained in the chamber. After passing through the strong, inhomogeneous magnetic field of a Stern–Gerlach magnet, the beam of atoms is separated into two beams with opposite magnetic polarizations. In many cesium-beam devices one of the polarized beams is absorbed in graphite and is of no further interest, while the other continues down the chamber.

Farther down the chamber is another strong, inhomogeneous magnetic field nearly identical to the first. At the end of the chamber there is a detector (which is sensitive to cesium atoms) placed in just such a position as to detect only those atoms that somehow change their polarization while traveling between the two magnetic field regions. Thus, the detector would not detect cesium atoms unless something happened to the atoms between the two strong magnetic field regions to change their polarity.

What happens is that the atoms are exposed to microwave radiation at a frequency of about 9 GHz. If this frequency is adjusted very precisely to the proper resonance frequency of cesium

---

[†] deceased

(9 192 631 770 Hz), the magnetic polarization of the atoms reverses, and the beam is deflected by the second magnetic field toward the detector. The detector indicates the presence of cesium atoms by means of an electric current. In actual operation the frequency of the microwave signal is controlled electronically to maximize the detector current, so the resonance condition of the microwave signal with the cesium atoms is ensured. A clock is obtained by counting the cycles of the microwave radiation.

Other kinds of atomic and molecular clocks use similar principles to extract frequency information from the atoms or molecules. The first atomic or molecular clock ever developed (completed in 1949 by Harold Lyons of the U.S. National Bureau of Standards) used the absorption of a microwave signal in ammonia to control the frequency, while hydrogen maser clocks use the stimulated emission of microwave radiation, and rubidium gas cell clocks use absorption of microwave radiation.

Several manufacturers produce atomic clocks commercially. The most accurate commercial devices are based on a resonance technique using a beam of cesium atoms like the various national standards. Somewhat less expensive atomic clocks are based on rubidium vapor. There is a tradeoff between cost and stability or accuracy. Most clocks in use today are based on resonances in atoms rather than molecules. At present there are a few tens of thousands of atomic clocks in routine use in many areas. For example, atomic clocks are used to control Loran-C, Omega, and GPS navigational systems, to do very long baseline interferometry and measure continental drift, to control network television signals, and to define an internationally accepted time-of-day system that is the time reference for most of the world.

Recent experimental advances and techniques are improving the accuracy of atomic clocks. The spectral width or possible frequency fluctuations of the cesium resonance used in the cesium atomic clock is approximately equal to the inverse of the transit time of a cesium atom down the beam. More narrow spectral widths have been obtained on atomic resonances with ions stored in traps. Ion traps use the force of electric and magnetic fields on ions (which have a net charge) to confine the ions to a small region in a good vaccum. Confinement times of several hours can be obtained routinely. Two types of traps have been used in the development of new atomic clocks: the rf or Paul trap, and the Penning trap. The rf or Paul trap uses spatially inhomogeneous rf fields to confine the ions, much as an rf quadrupole mass filter works in a mass spectrometer. The Penning trap uses static magnetic and electric fields to confine the ions. In general there is a trade-off between the number of trapped ions which gives good stability and the accuracy in a stored-ion clock. One of the first atomic clocks using ion storage is based on a microwave resonance in the mercury ion with about one million ions stored in an rf trap.

Lasers are starting to be incorporated into present atomic clocks with anticipated improvements in performance. The Stern–Gerlach magnets used in conventional cesium atomic clocks are being replaced with diode lasers. Through the technique of optical pumping, the polarization of the laser light is used to give the cesium atoms a magnetic polarization. The diode lasers are also used at the end of the beam line to detect the polarization state of the cesium atoms. The diode lasers enable all of the atoms in the beam to be used in measuring the atomic resonance.

Laser cooling or the use of radiation pressure from lasers to slow atoms or ions may provide further improvements for atomic clocks. According to Einstein's special theory of relativity, moving clocks tick slower than clocks at rest. Laser cooling enables both the spread in the atomic velocities and the mean atomic velocity to be reduced. This could improve the accuracy of the cesium clock. With the use of laser cooling, stored-ion frequency standards based on microwave resonances in ions are expected to have accuracies and stabilities better than one part in 10^{15}.

In general, atomic clocks are based on resonances in atoms at microwave frequencies, that is, frequencies less than 100 GHz. This is because of the technical difficulty of counting the cycles of higher frequencies. In addition, there is a lack of readily available, stable, narrow-band sources at high frequencies, especially at infrared and optical frequencies. Current research is making progress on both these problems, and in the future atomic clocks may be based on optical resonances in atoms or ions. An advantage of optical resonances is that the ratio of the optical frequency to the spectral width of the resonance can be very high. In a stored-ion clock, this means a good stability can be obtained with only a single ion in the trap. The projected accuracy of an atomic clock based on a single ion in an rf trap is on the order of one part in 10^{18}.

See also: Beams, Atomic and Molecular; Lasers; Masers; Optical Pumping; Time.

Bibliography

J. Jespersen and J. Fitz-Randolph, *From Sundials to Atomic Clocks*. Dover, New York, 1982.

N. F. Ramsey, "History of Atomic Clocks," *J. Res. Natl. Bur. Stand.* **88**(5), 301–320 (1983).

N. F. Ramsey, "Precise Measurement of Time," *Am. Sci.* **76**(1), 42–49 (1988).

J. Vanier and C. Audoin, *The Quantum Physics of Atomic Frequency Standards, Vol. I, II*. Adam Hilger, Bristol and Philadelphia, 1989.

D. J. Wineland, "Trapped Ions, Laser Cooling, and Better Clocks," *Science* **226**, 395–400 (1984). *Frequency Standards and Metrology*. Proc. 4th Symposium, Ancona, Italy. (A. DeMarchi, ed.). Springer-Verlag, Berlin, Heidelberg, 1989.

Cloud and Bubble Chambers

R. P. Shutt[†]

In order to detect nuclear interactions, for instance in radioactivity, cosmic radiation, or at high-energy accelerators, one can use the trails of ions and of more energetic "knock-on" electrons produced by charged particles colliding with atoms when passing through matter. In cloud chambers, first built by C. T. R. Wilson in 1912, tracks consisting of drops of liquid are formed along these trails, whereas in bubble chambers, invented by D. A. Glaser in 1952, the tracks consist of bubbles in a liquid. Both techniques have many important discoveries to their credit, but have been almost completely displaced by electronic methods. Electronic detectors are now able to distinguish between passages of two charged particles in space as well as the older techniques, in addition to their superior ability to distinguish in time between successively passing particles. Among the last major applications of bubble chambers were

---

[†]deceased

studies of neutrino interactions. Neutrinos do not leave ion trails and have very small nuclear interaction cross sections, so that intense, partially collimated neutrino beams could be passed through large bubble chambers without producing much background radiation, but producing weak-interaction events in the dense liquid, which acted as a target as well as a particle-detector medium.

In cloud chambers a gas containing a saturated vapor is expanded adiabatically, lowering the temperature, so that the vapor becomes supersaturated. The liquid surface tension prevents spontaneous drop formation without the presence of some kind of condensation nuclei. The electrostatic field of ions opposes the effect of the surface tension and permits drops to form and grow beyond a critical radius if the vapor is sufficiently supersaturated. Surface tension also prevents spontaneous bubble formation in a liquid in a superheated state, which in a bubble chamber is accomplished by letting the liquid expand elastically. But here the ionization energy due to a passing particle is very quickly converted to molecular kinetic energy, thus producing "heat spikes," locally raising the vapor pressure, which now can overcome the contracting effect of the surface tension, enabling bubbles to grow beyond a critical radius. In gases ions diffuse, but do so sufficiently slowly that cloud chamber expansions can be triggered for interaction events of special interest by means of electronic detectors. This is not possible for bubble chambers because the heat spikes responsible for bubble formation diffuse in about 10^{-12} s.

Chamber vessels usually must withstand pressure differences of less than 1 atm for most cloud chambers, up to 25 atm for bubble chambers. Excellent temperature control must be provided to obtain uniform drop or bubble size and to minimize convection currents, which produce track distortions. Expansions are provided by adequately sealed diaphragms, bellows, or pistons, or by valves letting gas pass through heat regenerators. Volume expansions range from $< 1\%$ for liquid hydrogen to 30% for air saturated with water vapor and must take only 10–20 ms to reduce convective distortions and to allow cloud chamber expansions to be triggered. Recompression must be equally fast in order to restore temperature equilibrium as soon as possible, so that the "dead time" during which a chamber is not ready for further expansions can be short. Dead times as short as 20 ms have been achieved with small bubble chambers, several seconds with small cloud chambers, but have lasted up to 30 min in high-pressure cloud chambers operated with argon at 300 atm. Time resolution between successive particles is of the order of 1 ms, sensitive time somewhat longer.

A special kind of continuously sensitive cloud chamber, the diffusion chamber, was built by A. Langsdorf in 1939, and for several years before the advent of bubble chambers it was used for particle physics research, mostly filled with about 20 atm of hydrogen. Here a temperature gradient is impressed on the gas volume, usually by cooling the bottom of the pressure vessel and heating the top, where a source of vapor of a liquid is also provided. As the vapor diffuses downward, it becomes supersaturated at some level, again resulting in track formation along charged-particle trails. Such chambers are relatively simple, since no expansion system is needed, but track-sensitive regions are only up to 8 cm high, and intense radiation can flood them with background tracks. Nevertheless, a chamber simply cooled with dry ice and filled with air or argon at atmospheric pressure, and with methyl alcohol as a vapor supply, serves well for demonstration of cosmic rays or other radiation and can be used to explore background near extended radiation sources. Just as in expansion cloud chambers, an electrostatic field is needed to sweep out unwanted ions.

During the short time while tracks are present in a chamber, they must be illuminated and photographed through suitable glass or quartz windows. Large bubble chambers now employ relatively small spherically curved windows, so that tracks are photographed through a layer of the operating liquid. Illumination is usually provided by xenon in a tube discharged at energies of 20–300 J during less than or about 1 ms. Windows and camera lenses must be of excellent optical quality, often also accommodating wide angles. The depth of the chamber volume, up to 3 m, requires large depth of focus and therefore small aperture as well as small magnification. Photography is thus diffraction limited. Of course, the film grain size, contrast, and speed are also of great importance. Usually, 35- to 70-mm films have been employed. To avoid distortion, the film must be held consistently flat. Illumination at about 90° to the lens axis has been sufficient for liquids with high refractive index, resulting in "dark-field illumination." The low index of, for instance, hydrogen requires light scattered at small angles, $\sim 1°$, and large chambers universally employ a specially developed Scotchlite, which reflects light from a source placed around the camera lens back into the latter, resulting in bright-field illumination. Three or more stereo views are usually photographed for redundancy and accuracy for all track orientations.

Expansion cloud chambers were most frequently used with argon and a water–propyl-alcohol mixture requiring expansion ratios as low as 1.08. In bubble chambers various fluids have been in use, which are operated at one half to two thirds their critical pressure. Propane, operated near 20 atm and 60 °C, has a density of 0.43 g/cm^3 and a radiation length of 108 cm, and contains more hydrogen per unit volume than liquid hydrogen. It is therefore suited for studies of interactions with individual protons (whose recoil tracks can also act as neutron detectors) as targets and of neutral interaction products decaying into γ rays, which have a reasonably high probability for conversion into electron pairs in propane chambers of suffi-cient size. The latter advantage can be reinforced by mixing propane with 30% by volume of methyl iodide, which results in a radiation length of 12 cm. Xenon has a radiation length of only 3.5 cm but, of course, contains no hydrogen and is extremely expensive. Freons, with density of 1.5 and radiation length of 10 cm, are very suitable for bubble chamber operation but contain no hydrogen. When electron-pair conversion is not needed and the large number of interactions with the heavier nuclei that occur in the mentioned liquids cannot be toler-ated, pure liquid hydrogen (or deuterium, for interactions with the extra neutron) is used. The density of liquid hydrogen is 0.06, its operating pressure only 5 atm, but its operating temper-ature is 25–27 K, requiring cryogenic techniques. Window and flange seals are usually made with indium–silver alloy wire often pushed against seal surfaces by means of helium-inflated bellows. For insulation, chambers are almost surrounded by multilayers of aluminum-coated Mylar in a vacuum chamber with windows for photography. Temperature control is provided by hydrogen- or helium-operated refrigerators.

Solid plates can be mounted in the chambers to aid in particle identification – for instance, to distinguish muons from electrons or from pions – but this can create optical problems and promote turbulence. Cryogenic chambers can also be used with liquid-neon–hydrogen mix-tures. Pure neon has a radiation length of 25 cm, so that its admixture results in increased pair conversion. "Track-sensitive targets" are relatively small transparent chambers with flexible walls, filled with liquid hydrogen or deuterium. When one is inserted into a neon–hydrogen-filled chamber, interactions can be produced in pure hydrogen or deuterium, and pair conver-sion can be observed outside the target. Finally, helium has also been a liquid of interest in

K-capture experiments, since the helium nucleus has zero spin. It is operated at 3.2 K and about 1 atm.

Most chambers have been positioned in magnetic fields, which cause a curvature in the tracks that is proportional to the field strength and inversely proportional to the momentum of a charged particle. Superconducting magnets employing copper-stabilized niobium–titanium alloy, cooled to liquid-helium temperatures, have been used, providing fields up to almost 4 Tesla. These magnets require very much less electric power than conventional electromagnets but require an additional cryogenic installation. Large cryogenic bubble chamber installations were found at the Fermi National Accelerator Laboratory, Brookhaven and Argonne National Laboratories, and CERN, the European Organization for Nuclear Research, where 15-ft, 7-ft, 12-ft, and 3.7-m chambers, respectively, were located, the largest of which contained about 30 m^3 of liquid. Installation costs have ranged between $7 and $20 million, and around-the-clock operation required 4–6 persons per shift. Another very large chamber, with a conventional magnet, was located at Serpukhov, USSR.

At particle accelerators, chambers can be exposed to beams of protons, antiprotons, electrons, or charged pi and *K* mesons, selected by arrays of magnets and electrostatic or radio-frequency beam separators from a spray of particles emerging from a target exposed to the primary circulating beam. Exposures to γ rays, neutrons, and, particularly, neutrinos are made possible by means of different arrangements. Exposures can amount to millions of pictures per experiment and can be distributed to many research groups at various locations where the photographs are scanned for events of interest and analyzed.

Locations of bubbles on the stereo views of event tracks are measured accurately with respect to images of fiducial marks located in the chamber. Angles between tracks and their curvatures can then be determined. Bubbles (or drops) along the tracks or gaps between bubbles can also be counted. The bubble density is inversely proportional to the square of the particle velocity, thus allowing determination of particle masses at not too relativistic energies. Energies of particles stopping in the chamber liquid can be measured from their range. The geometry and particle momenta in an event are fitted to hypotheses that are consistent with conservation of energy, momentum, and all relevant quantum numbers. Goodness of fit is tested with well-established statistical methods, such as the χ^2 test, and best fits are selected for inclusion in final results. Occasionally, a single event has proved the existence of a long-lived new particle or of a new quantum number. Usually large statistics are required, for instance, to show the existence of a short-lived resonant state occurring in strong interactions. For meaningful results, event selection must be as definite as possible, and therefore uncertainties in the tracks and their measurements must be as small as possible. Overall errors in bubble position, due to chamber turbulence, photography, and measuring, can be made as small as 50 μm, but multiple Coulomb scattering on atoms produces spurious curvatures, especially in the heavier liquids. Great care is required in evaluation of all possible errors in cloud and bubble chamber experiments.

High-statistics experiments involving hundreds of thousands of events have only been possible with the help of automatic measuring machines, guided by digital computers, and of elaborate pattern-recognition-type programs for event reconstruction, recognition, and compilation.

An example of an automatic measuring machine has been the flying-image digitizer at Brookhaven National Laboratory, where rotating mirrors sweep a projected image over small light-sensitive elements that move in a direction perpendicular to the motion of the image. By proper digitization and synchronization of motions, a picture can here be measured in a few seconds. Previous examples were the spiral reader, where a fine slit spirals over a photograph, producing light signals when passing over bubble images; and the flying-spot digitizer, where a small light spot, 10–15 μm in size, scans pictures. Events that do not result in satisfactory fits can often be saved by injecting into the analysis procedure additional information derived by human inspection.

See also: Radiation Detection.

Bibliography
J. G. Wilson, *The Principles of Cloud-Chamber Technique*. Cambridge, London and New York, 1951.
C. M. York, "Cloud Chambers," in *Handbuch der Physik* (S. Flügge, ed.) Vol. 45, 260. Springer-Verlag, Berlin and New York, 1958.
D. A. Glaser, in Handbuch der Physik (S. Fliigge, ed.), Vol. 45, 314. Springer-Verlag, Berlin and New York, 1958.
D. V. Bugg, "The Bubble Chamber," *Progr. Nucl. Phys.* **7**, 1 (1959).
H. Bradner, "Bubble Chambers," *Ann. Rev. Nucl. Sci.* **10**, 109 (1960).
R. P. Shutt, ed., *Bubble and Spark Chambers, Principles and Use*. Academic Press, New York, 1967.
Yu. A. Aleksandrov, G. S. Voronov, V. M. Gorbunkov, N. B. Delone, and Yu. I. Nechayev, in *Bubble Chambers* (W. R. Frisken, ed.), Indiana Univ. Press, Bloomington, IN, 1967.

Cold Atoms and Molecules

C. S. Adams

Cold atoms and molecules refers to gaseous systems of atoms and molecules at very low temperature, typically less than a few degrees Kelvin, i. e., within a few degrees of absolute zero −273.15 °C. The term ultra-cold is often used to describe temperatures of less than 1 milli-Kelvin. At such temperatures, the kinetic energies of the atoms and molecules are sufficiently small that they can be manipulated or trapped using relatively weak conservative potentials produced by magnetic fields, far-off resonant laser light or electrostatic fields in the case of polar molecules. However, even in the trapped case, the gases are typically sufficiently dilute that only two-body collisions are important, which prevents the metastable gas collapsing into a cluster or solid.

A variety of cooling methods are used to create cold atoms and molecules. For atoms, laser cooling is commonly used to produce atomic clouds with temperatures as low as one micro-Kelvin. Lower temperatures are achieved using evaporative cooling. For very low temperature, in the nano-Kelvin range, and sufficiently high density, the spatial extent of the atomic wavefunction becomes macroscopic and the sample behaves as a quantum many-body system (a Bose–Einstein condensate or degenerate Fermi gas). Buffer gas cooling or sympathetic cooling is also used on atomic species where there are no convenient closed transitions for efficient laser cooling.

In general for molecules laser cooling is not effective due to their more complex energy level structure. Cold molecular beams can be produced by supersonic expansion of the desired molecular species in inert buffer gas. Polar molecular beams can be slowed using switched electrostatic fields. Cold molecules can also be produced by buffer gas cooling, and ultra-cold cold molecules can be created from cold atoms using photo-association spectroscopy or in a high density system by driving an external field across a Feshbach resonance to form a long-range bound state.

Currently, the main applications of cold atoms and molecules are in precision measurements (atomic clocks), quantum information processing (quantum computing) and in the development of our understanding of quantum many-body systems such as dilute Bose–Einstein condensation or degenerate Fermi gases, and emergence of cooperative phenomena such as superfluidity.

See also: Bose–Einstein Condensation; Laser Cooling; Quantum Fluids; Ultracold Quantum Gases.

Bibliography

C. S. Adams and I. G. Hughes, "Laser cooling and trapping", in *Handbook of Laser Technology and Applications*, C. E. Webb and J. D. C. Jones (eds.). Institute of Physics Publishing, Bristol, 2003.

J. Doyle, B. Friedrich, R. V. Krems, F. Masnou-Seeuws, "Quo vadis, cold molecules?", *Eur. Phys. J. D* **31**, 149–164 (2004).

Collisions, Atomic and Molecular

A. Russek and W. Demtröder

Introduction

One of the fundamental tools available to explore the submicroscopic domain of quantum physics is the performance and analysis of collision experiments. The present body of knowledge of the structure of, and the interactions between, atomic and molecular systems stems in large measure from the interpretation of the results of scattering experiments. Indeed, it was the analysis by Sir Ernest Rutherford in 1911 of the α-particle scattering experiments of Geiger and Marsden that established the structure of the atom as a small, massive, positively charged nucleus about which the light, negatively charged electrons move. Scattering by a point Coulomb field is to this day referred to as Rutherford scattering.

Experimental research in atomic collisions continued on a very modest scale throughout the 1930s. Theoretically, however, even with the fragmentary evidence then available, L. Landau, C. Zener, and E. C. G. Stückelberg in 1932 outlined the conceptual understanding of molecular processes in collisional excitation. In the same year, W. Weisel and O. Beeck, using the newly created molecular model of F. Hund and R. S. Mullikan, fully described the role of particular orbitals in excitation processes.

There are different areas of atomic and molecular collision physics, depending on the kinetic energy of the collision partners. At sufficiently high energies (in the keV range) outer shell electronic excitation of projectile or target particles and impact ionization can take place. At even higher energies inner shell excitation with subsequent x-ray emission or Auger electron production may be observed. At thermal energies mainly elastic collisions happen. If molecules are involved rotational or vibrational excitation is possible. Recently ultra-cold collisions have been studied between atoms in optically cooled gases at temperatures below $1\,\mu K$, where three-body or radiative recombination become major processes [1].

In recent years, as a result of massive technological advances in vacuum systems, electronics, and particle detectors, a flowering of experimental research has taken place. Beginning in the late 1950s with the pioneering experiments of E. Everhart and his collaborators at the University of Connecticut and of N. V. Fedorenko and his collaborators at the Ioffe Physical-Technical Institute in Leningrad, the analysis of moderate-energy collisions (in the energy range of 1–100 keV) between atoms has provided a significant increase in the understanding of the energy states of diatomic molecules, of the limitations and breakdown of adiabatic behavior, and of the nonadiabatic couplings responsible for excitation processes. Although the experiments are carried out in a much higher kinetic energy range than that at which most chemical reactions take place, the knowledge gleaned from these experiments and their interpretations bears directly on that subject. In this connection, the important parameter is the ratio of the relative translational velocity v_r, to the orbital velocity v_0 of the electron involved. If v_r/v_0 is very much less than unity, a molecular description is in order. The orbital velocity of an electron in a hydrogen atom, which is essentially characteristic of all outer-shell electrons, is equal to the velocity of light divided by 137, or roughly $2 \times 10^6\,\mathrm{m/s}$. For a hydrogen atom of kinetic energy 25 keV, the ratio $v_r/v_0 = 1$, while an argon atom must have a kinetic energy of 1 MeV before this ratio is equal to unity for its outer-shell electrons, and of 18 MeV for inner-shell electrons. At higher energies, up to several MeV, collisional excitation of inner-shell electrons has been explored. Much of the theoretical understanding of the molecular processes observed in the experiments from the early 1960s on is due to the interpretations by D. R. Bates, U. Fano, W. Lichten, and M. Barat.

Besides these collisions with high and medium kinetic energies of the collision partners low energy collisions at thermal energies (0.03–0.1 eV) have been intensively studied in recent years. This has become possible through the technical development of well collimated molecular beams, where the velocity distribution of the molecules in the beam follows either a Maxwell–Boltzmann distribution (effusive beams) or is much narrower (supersonic molecular beams) [2].

Most of these collisions are elastic and they give detailed information on the long range of the interaction potential between the collision partners [3]. If collisions between molecules are considered, also inelastic collisions are observed resulting in rotational or vibrational transitions [4] or in deactivation of electronically excited states. Collisions between different molecules might also result in reactive processes such as

$$
\begin{aligned}
AB + CD \quad &\rightarrow \quad AC + BD \\
&\rightarrow \quad ABC + D \\
&\rightarrow \quad ACD + B
\end{aligned}
$$

where the relative probability for the different reaction channel depends on the internal states of the reactants and on their relative velocity [5, 6].

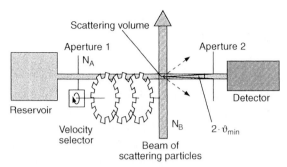

Fig. 1: Schematic diagram for measuring integral collision cross sections as a function of the relative kinetic energy of the collision partners.

A different approach to collision physics is based on measurements of spectral line profiles, broadened by collisions, because this profile depends on the interaction between the collision partners. Inelastic collisions result in broadening of the spectral lines, while elastic collisions lead to a shift of the line center and a broadening of the line profile [7]. For some time the two techniques were applied separately and the spectroscopists had little communication with the collision physicists. This has very much changed with the introduction of lasers to collision physics, which has allowed more detailed studies of elastic, inelastic and reactive collisions and has brought about a closer cooperation between the two communities [8].

The combination of laser spectroscopy with molecular beam techniques now enables the realization of ideal scattering experiments, where the initial and the final states of the collision partners before and after the collision, the scattering angle and even the orientation of the molecular collision partner can be determined [9].

Collision Cross Sections

In the analysis of any collision experiment the confrontation between theory and experiment is to be found in the cross section for some particular process (e. g., the excitation of a given electronic state of the target particle, or the deflection of the projectile by a certain angle). The collision cross section σ describes the probability for such a process. It is a quantity with the dimensions of an area, which can be both experimentally determined and calculated from the theory being tested [10]. It gives the probability that a scattering event will occur per target particle when one projectile is incident onto this area s around the target particle. We distinguish between integral (or total) and differential cross sections, which can be defined as follows (Fig. 1):

- The *integral cross section* σ_{int} for elastic collisions is that area around a target particle, through which a projectile has to pass in order to be deflected. Assume a flux of N_p projectiles per sec flying in a molecular beam with area A pass in the x-direction through a target volume with n_t target particles per cm^3. If L is the path length of the projectiles through the target volume, the number of projectile particles, scattered per s within the path length dx out of the incident direction is

$$dN_p = -N_p n_t \sigma_{int} \, dx = -(N_p/A) \, n_t A \, dx \cdot \sigma , \qquad (1)$$

where $A \, dx$ is the volume of the target intercepted by the projectile beam along the path length dx and $n_t A \, dx$ is the number of target particles within this volume.

Integration over the path through the scattering volume between $x = 0$ and $x = L$ gives the flux of transmitted particles, reaching the detector:

$$N(L) = N_p(0) e^{-n\sigma L} . \tag{2}$$

Note that the value of the integral cross section depends on the acceptance angle α of the detector, because all projectiles which are scattered by an angle $\theta > \alpha$ cannot reach the detector. The integral cross section can be represented by a circular area with radius b_{max} (called the maximum impact parameter) around a target particle. Its value depends on the interaction potential between the collision partner, on their relative velocity and on the acceptance angle $\alpha = \theta_{min}(b_{max})$ of the detector.

- The *differential cross section* σ_{diff} is that annular ring area $2\pi b \, db$ with radius b and width db through which a projectile has to pass in order to be deflected by an angle within the range from $\theta - d\theta/2$ to $\theta + d\theta/2$. Its measurement relates the impact parameter b with the scattering angle θ.

The number of projectile particles reaching the detector per s under an angle θ against the incident direction is

$$dN_p = N_p n_t \sigma_{diff}(\theta) L \, d\theta \, d\Omega , \tag{3}$$

where $d\Omega$ is the solid angle accepted by the detector. The differential cross section is therefore proportional to the number of events per unit solid angle around the scattering angle θ per target particle for each incident projectile. Measurements of differential cross sections single out only those collisions that give rise to the process under consideration caused by collisions within a small impact parameter range, where the projectile particles are scattered into a definite small angular range $d\theta$ around θ. Integration of the differential cross section over all scattering angles θ and ϕ, i. e., over all solid angles $d\Omega = \sin\theta \, d\theta \, d\phi$ gives the integral cross section. Therefore the differential cross section is written as

$$\sigma_{diff} = \frac{d(\sigma_{int})}{d\Omega} . \tag{4}$$

In a similar way cross sections for inelastic or reactive collisions can be defined as that area through which a projectile has to pass in order to cause the particular excitation of the target particle or to induce the wanted reaction.

Only for monotonic potentials $E_{pot}(r)$ is the relation between scattering angle θ and impact parameter b unambiguous. For potentials with a minimum there are generally at least two different impact parameters resulting in the same scattering angle. The superposition of these different contributions to the same scattering angle gives rise to interference phenomena, which shows the wave nature of the scattered particles. Their measurement is a sensitive tool for probing the interaction potential, because the interference structure depends on the different phases a particle experiences when traveling through the interaction region.

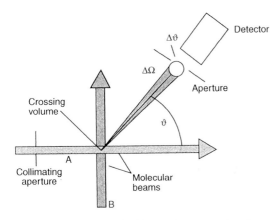

Fig. 2: Measurement of differential cross sections in crossed molecular beams.

Experimentally, it is important that the product of target density and path length in the target be kept sufficiently small so that the probability of making any collision at all (not just the ith process) is very much less than unity. This ensures the "single-collision regime" wherein the likelihood of making more than one collision is negligible. This is usually accomplished by reducing the target gas density in successive measurements and verifying that the number of scattering events N_i varies linearly with target density. Typical pressures are 10^{-4} mbar.

Instead of using a gas cell as target, two crossed collimated beams allow a better definition of the collision region and therefore a better angular resolution for measurements of differential cross sections (Fig. 2). The relative velocity between the collision partners can be varied within certain limits by velocity selectors in the two beams.

Information can be gained from observation of the incident projectiles, scattered targets, and photons and electrons generated in the collision. Basically, collision experiments fall into one of two general categories. Total cross-section measurements lump together all collisions that give rise to a given process at a given energy, independent of impact parameter. They are dominated by small-angle scatterings that come from large-impact-parameter collisions.

On the other hand, differential cross-section measurements, $d\sigma_i/d\Omega$, single out only those collisions that give rise to the process under consideration and, at the same time, result in scattering into a small angular region at a prescribed angle θ of scattering.

Experimental Methods

Atomic and molecular collision experiments require a source and accelerator to produce a collimated beam of the desired projectile at the desired energy, a scattering chamber containing the target gas, and a detector able to detect the projectile and/or recoiling target and/or secondary particles with known efficiency. Projectiles can be atoms or molecules or their ions; targets can only be atoms or molecules. Figure 3 schematically illustrates a typical (perhaps it should be called ideal) experimental arrangement for scattering experiments with higher energies of the projectiles. Ions are extracted from an ion source and electrostatically accelerated to the desired energy. Removal of unwanted ions coming out of the ion source (and there will always be some) is accomplished by passing the ion beam through a magnetic field, which

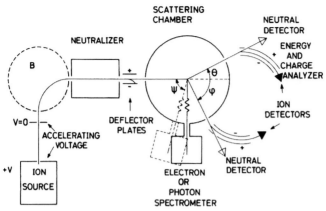

Fig. 3: Principal components of an ideal collision experiment for high-energy collisions. The vacuum system, gas-handling system, and electronics for counting detected particles are not shown. As depicted here, the apparatus is set up for neutral projectiles. The neutralizer and deflector plates are not used if the projectiles are ions. In this illustration, scattered projectiles, recoiling targets, and secondary electrons or photons are all detected and analyzed. Commonly, however, only one or two of these are encompassed in a collision experiment.

separates components with different values of charge-to-mass ratio Q/M. If neutral projectiles (which cannot, of course, be electrostatically accelerated) are to be used, these must be produced by passing the ion beam, after acceleration, through a neutralization chamber containing a gas for which there is substantial probability of electron capture by the ion. Very little energy is exchanged in this process, so that a neutral atom at the energy of the accelerated ion passes on to the scattering chamber. Ions that have not been neutralized must be deflected out of the beam by means of deflection plates.

Since the target is a gas or vapor, it must be confined in a scattering chamber with small apertures to allow the incoming beam to enter and the scattered particles to leave. The same, of course, is true for the neutralization chamber. These apertures are generally integrated into the collimation systems that define the respective beams. The target gas (and neutralizer gas, if used) will inevitably leak out through the apertures, so that the regions between the components must all be kept evacuated by constant pumping of these regions. At the same time, target and neutralizer gases must be constantly replenished. In a dynamic situation such as this, the pressure of the target gas (from which its density is determined) must be measured within the scattering chamber itself. Before detection, the energy and charge state can be determined by electrostatic means if an ion, or by time-of-flight means if a neutral. In the latter case, the beam must be pulsed, in order to establish a zero time. The experimental arrangement in Fig. 3 shows a detector for scattered projectile and one for recoiling target. Since atoms in collision behave somewhat like billiard balls, the motion will take place in a plane. It is not too difficult to detect both scattered and recoil atoms *in coincidence* (more precisely, with a slight, but exactly known, time difference, due to the different velocities of projectile and recoiling target).

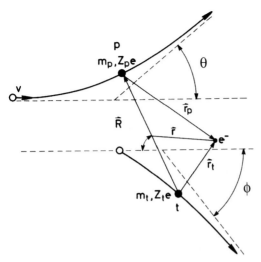

Fig. 4: Collision trajectory in the laboratory frame of reference. **R** is the vector separation between target and projectile; **r** locates an electron in the time-varying quasimolecule formed during the collision process. In general, of course, there will be many electrons. The angles θ and ϕ represent the angle of scattering of the incident projectile and the angle made by the recoiling target with the incident-beam direction, respectively. Mathematically, the quasimolecule is quantum mechanically described in terms of the non-Newtonian rotating coordinate system, called the *body-fixed system*, for which **R** is the polar axis. The change in the angle θ with time generates the rotational coupling term in Eq. (14).

As illustrated in Fig. 4, the collision is described by the angle θ through which the projectile is scattered and the angle ϕ made by the recoiling target with the incident-beam direction. These angles, as well as the final energies E_p and E_t of projectile and target, can all be determined in a coincidence experiment. Since the incident-projectile energy E_0 is set by the accelerator voltage, and the initial energy of the target can safely be taken equal to zero (thermal energies are $\sim 1/40\,\mathrm{eV}$), the inelastic energy transferred from translational kinetic energy to internal energy (i. e., the excitation energy) can be determined simply by conservation of energy and conservation of momentum. Letting Q denote the excitation energy, we have

$$
\begin{aligned}
E_0 &= E_p + E_t + Q, \\
(2M_pE_0)^{1/2} &= (2M_pE_p)^{1/2}\cos\theta + (2M_tE_t)^{1/2}\cos\phi, \\
0 &= (2M_pE_p)^{1/2}\sin\theta - (2M_tE_t)^{1/2}\sin\phi.
\end{aligned}
\tag{5}
$$

With five final quantities, Q, θ, ϕ, E_p, and E_t and with three relations existing between these five, it is clear that measurement of any two automatically determines the others. Thus, the inelastic energy Q can be determined just by measuring the angles θ and ϕ, or alternatively, just by measuring E_p and θ. A coincidence measurement is not essential to determine the inelastic energy.

Ideally, the experiment should measure the differential cross section for a collision with specified final states of both projectile and target. Most often, however, it is understandably

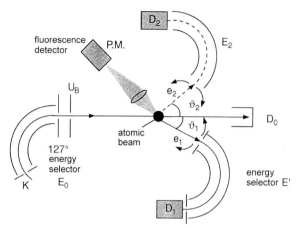

Fig. 5: Experimental arrangement for electron-molecule scattering and electron impact ionization of atoms with coincidence measurements of the two outgoing electrons with selectable angles θ_1 and θ_2. The energies of the incident electron and of the two outgoing electrons are selected by 127° electrostatic analyzers.

necessary to settle for less. The situation is made more difficult by the fact that, except for special cases of long-lived states, transit times through the scattering region are much longer than electronic rearrangement times in the atoms themselves. These rearrangements, generally termed deexcitation processes, occur by photon emission or electron ejection. The latter is called Auger emission, or autoionization, and involves an energy exchange between two electrons in the atoms. One drops down to fill a vacancy in a lower energy state, giving up its energy to the second, which then has enough energy to escape, and does so with a characteristic energy. As a consequence of these rearrangements, it is often the case that the states of excitation produced by the collision must be inferred, rather than measured directly. For this reason, many experiments also detect and energy-analyze electrons or photons coming from the collision region. Often, the angular distributions of these secondary particles are measured. As an example, collisional excitation of an inner-shell electron in an atom leaves a vacancy in that inner shell immediately after the collision. Before the atom leaves the scattering region, either an Auger electron or an x ray will be emitted with an energy characteristic of the shell excited. Much can be learned about the process if the angular distribution of Auger electrons or x rays is determined.

Up to now we have dealt only with high collisions where the energy of the projectiles is much higher than thermal energies. By scattering low energy electrons on atoms or molecules much can be learned about the electron correlation in atoms or molecules. If the energy of the incident electron is sufficiently high to allow ionization of the target particle (This is still below 50 eV), two electrons with different energies and scattering angles are present after the collision. They can be measured in coincidence with the setup shown in Fig. 5. The incident electrons are energy-selected by a 127° electrostatic energy analyzer, hit the target atoms or molecules in a collimated beam perpendicular to the drawing plane. The two electrons emerging from the ionizing collision are measured in coincidence by two movable

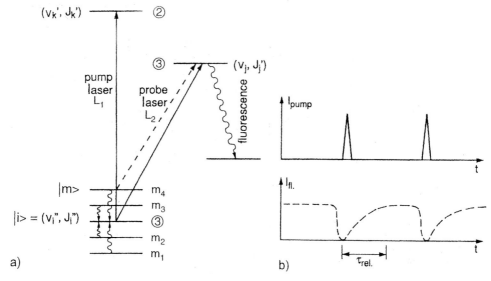

Fig. 6: Measurements of total and state selective integral cross sections using time-dependent saturation of the population in selected levels and pump–probe techniques with pulsed lasers.

detectors under two selectable angles θ_1 and θ_2 behind energy analyzers. This allows the determination of threefold-differential cross sections $d^3\sigma/dE_1\,dE_2\,d\Omega$ for varying energies of the incident electron [11].

The analysis of the experimental results gives information on the ionization energy, the correlation between the atomic electrons and the post-collision interaction between the two outgoing electrons.

Application of Laser Spectroscopy to Collision Processes

Laser spectroscopy has considerably increased the information derived from scattering experiments. First of all the projectile particles can be prepared in a specific quantum state by selective absorption of photons from a "pump-laser" before the collision takes place. Secondly their final state after the collision can be monitored by laser-induced fluorescence techniques [12].

For measurements of integral cross sections for inelastic collisions between molecules or atoms at thermal energies cell experiments are sufficient. The basic principle is explained in Fig. 6. A pulsed pump laser is tuned to a specific molecular transition $|i\rangle \rightarrow |k\rangle$. Due to the high laser intensity the lower level $|i\rangle$ of the transition is depleted and the upper level population is increased. Collisions refill the depleted level. This can be monitored by a weak pulsed probe laser, which is tuned to a transition $|i\rangle \rightarrow |j\rangle$. The intensity of the laser-induced fluorescence, induced by the probe laser, which is a measure for the population n_i is monitored as a function of the delay time between pump and probe. This time resolved measurements give the total refilling rate of level $|i\rangle$ due to collision-induced transitions from all other levels. If the density of the molecular gas is known the total cross section for the sum of all possible transitions can be determined.

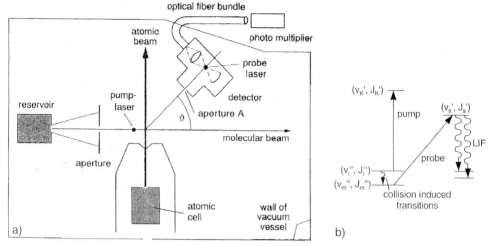

Fig. 7: Measurement of differential cross sections for inelastic collisions with selection of initial and final states by pump- and probe-lasers.

Cross sections for selected transitions can be measured, if the probe laser is tuned to a transition $|m\rangle \rightarrow |j\rangle$, starting from a level other than $|i\rangle$. Any collision-induced transition $|m\rangle \rightarrow |i\rangle$ will result in a partial depletion of level $|m\rangle$ which can again be measured as a function of delay time.

Selective laser excitation of individual upper levels is in particular suited to measure collisions between electronically excited atoms or molecules. The decay time of the population in the upper level, measured as a function of pressure in the gas cell, gives the sum of collision rates from level $|k\rangle$ to all other levels. If τ_k^{rad} is the radiative lifetime of $|k\rangle$, determined by the total radiative decay rate $A_k = \sum A_{kj} = 1/\tau_k$, the total decay rate is

$$R = A_k + n\sigma\langle v \rangle \tag{6}$$

and the effective lifetime becomes

$$\frac{1}{\tau_k^{\mathrm{eff}}} = R = \frac{1}{\tau_k^{\mathrm{rad}}} + n\sigma\langle v \rangle , \tag{7}$$

where $\langle v \rangle$ is the mean relative velocity between the collision partners and σ the total deactivation (= quenching) cross section. The absolute collision cross sections for individual transitions between selected levels in the excited electronic state can be obtained by measuring the relative intensities of the fluorescence emitted from these levels as a function of pressure.

Differential cross sections for inelastic collisions can be obtained by crossed beam experiments depicted in Fig. 7. A collimated beam of molecules A is intersected by the beam of a pump laser just before the collision region, which is defined by the crossing volume with a second atomic or molecular beam which contains the target molecules B. The molecules A scattered by the angle θ are monitored by laser-induced fluorescence behind a narrow aperture which defines the angular resolution. The pump laser is tuned to a selected transition and can nearly completely deplete the lower level $|i\rangle$. The probe laser is tuned to another transition $|m\rangle \rightarrow |s\rangle$ and the laser-induced fluorescence is a measure for the population of level

$|m\rangle$ after the collision. If the pump laser is periodically chopped, the population n_i will be modulated correspondingly. Due to collisions the population n_m will also show a (smaller) modulation. A lock-in, tuned to the chopping frequency measures the difference of the populations nm with and without the pump laser. This difference signal represents that part of inelastic collisions that causes transitions $|i\rangle \rightarrow |m\rangle$ at an impact parameter b resulting in a scattering angle θ. The technique has been applied to measurements of rotational transitions in diatomic molecules induced by collisions with noble gas atoms [13]. The experimental results allowed the determination of the anisotropic interaction potential between the molecule and the atom [14].

Reactive collisions represent the microscopic basis for every chemical reaction. In order to understand the reaction and its dependence on the internal energy of the reactands measurements of state to state collisions are necessary. Meanwhile many examples of such measurements have been published for different reactions [15, 16], such as

$$Ba + HF(v = 0) \rightarrow BaF(v = 0 \ldots 12) + H \, ,$$

$$Ba + CO_2 \rightarrow BaO + CO \, .$$

The simplest exchange reaction

$$H + D_2 \rightarrow HD + D$$

has been controversely discussed, because there is a potential barrier for this reaction which has to be overcome by the reactants. Tunneling through this barrier can considerably enhance the cross section for the reaction. Recently unambiguous experimental results have determined the heights of the barrier and the collision cross section for different vibration excitation of the D_2 molecule [17]. Also reactive collisions involving polyatomic molecules have been studied [18]

A good survey on the historical development of scattering experiments can be found in [19]

Theoretical Methods

Theoretically, the cross sections are most often discussed within the framework of one or the other of two very important approximations which can only be briefly described here [20–22].

The Born Approximation

The Born approximation is generally applied to high-energy collisions. It is based on the physical assumption that the collision takes place so rapidly that there is little time for the systems to adjust to the changing internuclear separation. The wave function describing the state of the overall system is therefore expanded in terms of the states of the isolated projectile and target. The Hamiltonian H for the system is written in the form

$$H = T_R + H_p + H_t + V_{pt} = H_0 + V_{pt} \tag{8}$$

Here, T_R represents the kinetic energy of translational motion, H_p and H_t are the Hamiltonians describing the isolated projectile and target, and V_{pt} is the sum of all potential energy terms between pairs of components one of which is in the projectile and one in the target. The wave function for the system is expanded in terms of the eigenstates of H_0, which are of the form $\phi_p \phi_t \times \exp(i\mathbf{K} \cdot \mathbf{R})$, $\mathbf{R}$ is the internuclear separation vector, with energy $\varepsilon_p + \varepsilon_t + \hbar^2 K^2/2m$, where ϕ_p and ϕ_t are the internal states of projectile and target, m is the reduced mass, and $\hbar^2 K^2/2m$ is the translational kinetic energy in the center-of-mass reference frame. The differential cross section for scattering $\mathbf{K}_i \rightarrow \mathbf{K}_f$ together with electronic transition $\phi_p \rightarrow \phi_{p'}$ and $\phi_t \rightarrow \phi_{t'}$ can be equally well derived from either time-dependent or time-independent perturbation theory (where the final state is in the continuum). The differential cross section is given by the expression

$$\frac{d\sigma}{d\omega} = \left(\frac{m}{2\pi\hbar^2}\right)^2 \frac{K_f}{K_i} |M(p,t \rightarrow p',t')|^2 \, \delta(E_i - E_f) \tag{9}$$

where the δ function ensures that energy is conserved between initial and final states of the overall system. The matrix element M is

$$M(p,t \rightarrow p',t') = \int d^3\mathbf{R} \, e^{i(\mathbf{K}_i - \mathbf{K}_f) \cdot \mathbf{R}} \times \int d^3\mathbf{r} \, \phi_{p'}^* \phi_{t'}^* \, V_{pt} \, \phi_p \phi_p \tag{10}$$

where $\mathbf{r}$ stands for all electronic coordinates.

A variant of the above procedure that is sometimes used treats the nuclear motion classically, with well-defined trajectory as a function of time, $\mathbf{R}(t)$, yet continues to expand the electronic states in terms of those of the isolated systems H_p and H_t.

The Molecular Approximation

The molecular approximation (also called the "adiabatic" or "quasi-adiabatic" approximation) is generally applied to slow collisions. It is based on the physical assumption that the collision process evolves sufficiently slowly in time that the electrons are able not only to adjust continuously to the changing internuclear separation, but also to take energy from, and give energy to, the kinetic energy of nuclear motion. In this way, the electronic motion acts as a potential for the nuclear motion which supplements the repulsive internuclear Coulomb potential. The adiabatic electronic state is defined as the eigenfunction of the electronic Hamiltonian, $\mathcal{H}$:

$$\mathcal{H} = \sum_i \left(\frac{p_i^2}{2m_e} - \frac{Z_p e^2}{r_{ip}} - \frac{Z_t e^2}{r_{it}} + \sum_{j<i} \frac{e^2}{r_{ij}} \right) \tag{11}$$

where r_{ip} and r_{it} are the respective distances of the ith electron to projectile and target nuclei, and m_e is the electronic mass. $\mathcal{H}$ includes the kinetic energy of each electron along with its potential energy of attraction to both nuclei and, finally, the interelectron repulsive term between each pair of electrons. Thus, $\mathcal{H}$ contains the internuclear separation $\mathbf{R}$ as a parameter; it is implicitly contained in r_{ip} and r_{it}. As a consequence, both the eigenstate and eigenenergy of $\mathcal{H}$ depend on $\mathbf{R}$ as a parameter:

$$\mathcal{H}\Phi_n(\mathbf{r};\mathbf{R}) = E_n(R)\Phi_n(\mathbf{r};\mathbf{R}) \tag{12}$$

where again $\mathbf{r}$ stands for all electronic coordinates.

In the adiabatic approximation, the electronic energy $\mathcal{E}(R)$ is one of the potentials influencing the nuclear motion, so that the wave function χ describing the nuclear motion is a solution to the eigenvalue equation:

$$\left[-(\hbar^2/2m)\nabla_R^2 + Z_p Z_t e^2/R + \mathcal{E}(R)\right]\chi(\mathbf{R}) = E_\chi(\mathbf{R}) . \tag{13}$$

The wave function Ψ describing the overall system in the adiabatic approximation is the product of the two functions:

$$\Psi(\mathbf{r};\mathbf{R}) = \Phi(\mathbf{r};\mathbf{R})\chi(\mathbf{R}) . \tag{14}$$

It is not an exact eigenfunction of the total Hamiltonian H for the overall collision system. To see this, H, given by (18), is regrouped as

$$H = T_R + Z_p Z_t e^2/R + \mathcal{H} . \tag{15}$$

From (12) and (13) it follows that

$$\begin{aligned} H\Psi &= (T_R + Z_p Z_t e^2/R + \mathcal{H})\Phi(\mathbf{r};\mathbf{R})\chi(\mathbf{R}) \\ &= E\Psi - (\hbar^2/2m)[2(\nabla_R\Phi)\cdot(\nabla_R\chi) + \chi\nabla_r^2\Phi] . \end{aligned} \tag{16}$$

If the last two terms on the right-hand side of (16) are neglected, then Ψ is indeed an eigenfunction of H, and this is the adiabatic approximation. These terms are, in fact, quite negligible for vibrational states of diatomic molecules and for collisions in the eV energy range. However, for increasing collision energy, the second term on the right-hand side of (16) becomes increasingly important, since $\nabla_r\chi/m$ is essentially the relative velocity of collisions. This gradient term couples different adiabatic states and is responsible for deviations from adiabatic behavior and, therefore, for electronic excitation. In order more easily to describe these couplings mathematically, several variants of the adiabatic states defined by (12) have been introduced. These include "diabatic" states and "traveling orbitals," two subject areas of recent activity. Adiabatic electronic states are calculated at fixed nuclear positions, a built-in contradiction for a theory describing a collision process. Traveling orbitals allow for the nuclear motion by incorporating traveling wave factors $\exp^{iv_N r}$ into the electronic states, where the velocity v_N is that of the nucleus about which the basis function is centered. This subject has been reviewed by Delos in 1981 [23].

Several formulations of diabatic states have been advanced, all of which have in common an attempt to maintain for each diabatic state a well-defined topological "character" that remains invariant throughout the collision. This subject was reviewed by Russek and Furlan in 1989 [24]. By contrast, adiabatic states of the same symmetry rapidly interchange their characters in near-degeneracy situations, and the adiabatic energy levels do not intersect. This avoided crossing phenomenon was first elucidated by von Neumann and Wigner and is known as the "noncrossing rule." The characters of diabatic states can be formulated in terms of topologically defined projection operators which select out specified nodal structures. The diabatic states are eigenfunctions of a diabatic Hamiltonian, $\mathcal{H}_D$:

$$\mathcal{H}_D = P\mathcal{H}P + Q\mathcal{H}Q . \tag{17a}$$

It is not hard to show that a pseudosymmetry operator $\pi = P - Q$ commutes with $\mathcal{H}_D$, so that crossings of the diabatic energy levels do not constitute a violation of the noncrossing rule.

The actual electronic Hamiltonian, $\mathcal{H}$, is the sum of $\mathcal{H}_D$ and an interaction term $\mathcal{H}'$, which includes the remaining terms of $\mathcal{H}$ not included in $\mathcal{H}_D$:

$$\mathcal{H} = \mathcal{H}_D + P\mathcal{H}P + Q\mathcal{H}Q = \mathcal{H}_D + \mathcal{H}' . \tag{17b}$$

Writing the state describing the collision system in the form

$$\Phi = \sum_{n=1} c_n(t)\phi_n(\mathbf{r};\mathbf{R}(t)) \exp\left(-i\int^t \mathcal{E}_n \, d\tau/h\right) , \tag{18}$$

the time-dependent Schrödinger equation yields a set of coupled first-order differential equations for the coefficients:

$$\dot{c}_m = -\sum_n \left[i\langle\phi_m|\mathcal{H}'|\phi_n\rangle + \langle\phi_m|\dot{\phi}_n\rangle\right] c_n \exp\left[i\int^t (\mathcal{E}_m - \mathcal{E}_n) \, d\tau/\hbar\right] . \tag{19}$$

Except at very small angles of scattering, the nuclear motion can be treated in terms of a classical trajectory $\mathbf{R}(t)$ because of the very small de Broglie wavelength (when compared with atomic dimensions) associated with this motion. The quantum description must, however, be retained for electronic behavior.

Denoting by v_R the radial component of internuclear velocity and by $\dot{\Theta}$ the angular rotation of the line joining the two nuclei, we can show that

$$\dot{\phi}_n = v_R \frac{\partial \phi_n}{\partial R} + \frac{\dot{\Theta}}{2\hbar}(L_+ - L_-)\phi_n , \tag{20}$$

where L_+ and L_- are the raising and lowering operators for the component of angular momentum along the internuclear axis. The first term on the right-hand side of (14) is called radial coupling and leaves unaltered the component of angular momentum along the internuclear axis, whereas the second term, called rotational coupling, changes this quantity by $\pm\hbar$.

Figure 8 illustrates an example of energies of a set of diabatic molecular orbitals plotted as functions of internuclear distance. It was used by U. Fano and W. Lichten to analyze inner-shell excitation produced in collisions of argon atoms (or ions) on argon atoms, which have filled K, L, and M shells. For example, L-shell excitation can be explained as follows: The $4f\sigma$ orbital, which contains two electrons, rises rapidly as R decreases below 0.5 a.u., crossing many vacant orbitals. These two electrons have a substantial probability of being coupled to one of the orbitals crossed, leaving one or two vacancies in the $4f\sigma$ orbital, which separates, after the collision, into a 2p vacancy in the argon L-shell.

See also: Atoms; Auger Effect; Beams, Atomic and Molecular; Scattering Theory.

References

[1]   J. Weiner, "Advances in Ultracold Collisions", *Adv. At. Mol. Opt. Phys.* **35**, 45 (1995).
[2]   G. Scoles (ed.), *Atomic and Molecular Beam Methods*, Vols. I and II. Oxford University Press, Oxford, 1988 and 1992.
[3]   J. P. Toennies, "Low Energy Atomic and Molecular Collisions", in *Semiclassical Description of Atomic and Nuclear Collisions*, J. Bang and J. deBoer (eds.). North Holland, Amsterdam, 1985.

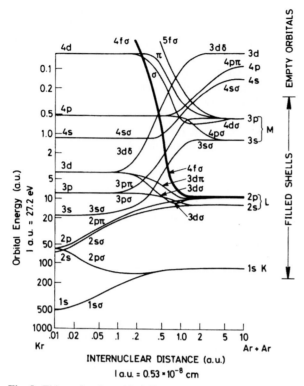

Fig. 8: This molecular orbital diagram, taken from the work of U. Fano and W. Lichten, has been used to explain the detailed behavior of inner-shell excitations in argon–argon collisions. Energies of various orbitals are plotted as functions of internuclear separation R. To use this diagram, start from the right-hand side ($R = \infty$) and proceed to the left down to the minimum internuclear separation for the collision under consideration, and then back out to $R = \infty$. The filled orbitals are indicated on the right-hand side. The two electrons in the 4fσ orbital, which rises rapidly as internuclear separation decreases, are excited into normally vacant states provided that the minimum internuclear separation is less than approximately 0.5 a.u. (0.26×10^{-8} cm), where the first level crossing with unoccupied orbitals (e. g., 3sσ) occurs.

[4] M. Faubel, "Vibrational and rotational excitation in molecular collisions", *Adv. At. Mol. Phys.* **19**, 345 (1983).

[5] M. A. D. Fluendy and K. P. Lawley, *Chemical Applications of Molecular Beam Scattering*. Chapman and Hall, London, 1973.

[6] R. E. Wyatt and J. Z. Zhang, *Dynamics of Molecules and Chemical Reactions*. Marcel Dekker, New York, 1996.

[7] See, for instance: *Proc. Intern. Conf. on Spectral Line Shapes*, Vols. 1–12. Library of Congress Catalogue.

[8] See the Proceedings on the ICPEAC conferences.

[9] M. Innokuti (ed.), "Cross Section Data", *Adv. At. Mol. Opt. Phys.*, **33**, (1994).

[10] N. Andersen and K. Bartschat, *Polarisation, Alignment and Orientation in Atomic Collisions*. Springer, Berlin, Heidelberg, 2001.

[11] P. Schlemmer, M. K. Srivastava, T. Rösel and H. Ehrhardt, "Electron impact ionization of helium at intermediate collision energies", *J. Phys. B* **24**, 977 (1971).

[12] P. R. Berman, "Studies of Collisions by Laser Spectroscopy", *Adv. At. Mol. Phys.* **13**, 57 (1977).

[13] K. Bergmann, "State Selection via Optical Methods", in *Atomic and Molecular Beam Methods*, G. Scoles (ed.). Oxford University Press, Oxford, 1989.

[14] R. Schinke, "Theory of Rotational Transitions in Molecules". *Int. Conf. Phys. El. At. Collisions* (ICPEAC), Vol. XIII, p. 429. North Holland, Amsterdam, 1984.

[15] S. R. Leone, "State-Resolved Molecular Reaction Dynamics", *Ann. Rev. Phys. Chem.* **35**, 109 (1984).

[16] K. Liu, "Crossed-Beam Studies of Neutral Reactions", *Ann. Rev. Phys. Chem.* **52**, 139 (2001).

[17] K. D. Rinnen, D. A. V. Kliner, R. N. Zare, "The $H + D_2$ Reaction", *J. Chem. Phys.* **91**, 7514 (1989).

[18] J. J Valentin, "State-to-state chemical reaction dynamics in polyatomic systems", *Ann. Rev. Phys. Chem.* **52**, 41 (2001).

[19] G. Boato, G. G. Volpi, "Experiments on the Dynamics of Molecular Processes: A Chronicle of 50 Years", *Ann. Rev. Phys. Chem.* **50**, 23 (1999).

[20] M. Kimura and N. F. Lane, "The Low-Energy Heavy Particle Collisions – A Close-Coupling Treatment", *Adv. At. Mol. Opt. Phys.* **26** (1989).

[21] P. W. Atkins, *Molecular Quantum Mechanics*, Parts I, II, and III. Oxford (Clarendon Press), London and New York, 1970 (E); James C. Slater, *Quantum Theory of Molecules and Solids*, Vol. 1. McGraw–Hill, New York, 1963 (I); Ta-You Wu and Takashi Ohmura, *Quantum Theory of Scattering*. Prentice-Hall, Englewood Cliffs, NJ, 1962 (I).

[22] Bernd Crasemann, ed., *Atomic Inner-Shell Processes*, Vols. I and II. Academic Press, New York, 1975. (A)

[23] J. B. Delos, *Rev. Mod. Phys.* **53**, 287 (1981).

[24] A. Russek and R. J. Furlan, *Phys. Rev.* **39**, 5034 (1989).

Color Centers

P. W. Levy

Normally transparent crystals and glasses often appear colored because they contain color centers. More specifically, their colored appearance is attributable to absorption bands associated with defects, impurities, or more complex imperfections. Many different types of color centers are formed when the charge or valence state of these imperfections is changed. The defects or imperfections may be incorporated into the lattice in one charge state and not produce observable absorption bands. Subsequently, if the charge state is altered, most often by exposure to ionizing radiation, absorption bands are formed. Alternatively, color centers are often formed by impurities that were introduced into the lattice in their normal valence state. Familiar examples of solids that appear colored because they contain color centers include:

- Solids colored by radiation – e. g., crystals and glasses colored by exposure to nuclear radiation, to x rays, to high-energy particles, or to sunlight for a long period of time (solarization). In fact, almost all transparent solids will develop color centers when exposed to any radiation that will create electron–hole ionization pairs in the interior of the solid.

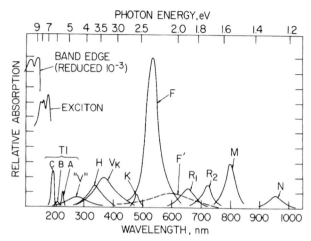

Fig. 1: Principal color-center absorption bands in a "typical" alkali halide. The energies and intensities of bands other than the exciton bands have been chosen to approximate the spectra observed after x-ray irradiation near liquid-nitrogen temperature, roughly 88 K. Exciton bands are observed only near 0 K.

- Colored minerals – e. g., ruby is Al_2O_3 with Cr impurity; smoky quartz usually contains Al and alkalis and the formation of the color requires exposure to ionizing radiation. The many colored fluorites are due to the presence of different rare-earth ions and other impurities.

- Colored glasses – e. g., "stained" glass windows are made of different glasses with a variety of impurities.

Color-Center Properties

Almost all of the properties of color centers are included in recent books [1–4] and may be illustrated by describing F centers in alkali halides. Historically, this was the first center investigated in any detail and has been studied more extensively than any other center. All of the alkali halides, such as NaCl, KCl, and LiF, become colored when exposed to sufficiently large doses of x-rays, nuclear radiation, or any other radiation producing ionization. The most commonly occurring bands, induced in a typical alkali halide by x-ray irradiation at liquid-nitrogen temperature, are shown in Fig. 1. In many alkali halides the most prominent band is in the visible. Much of the original research on these bands was done in Germany, and they were labeled F bands after the German word for color, *Farbe*. Likewise, the center associated with the F band was labeled the F center. The nature of the F center was established by studying additively colored crystals that were prepared by exposing pristine samples to alkali-metal vapor. For example, KCl crystals that have been heated in potassium vapor and rapidly cooled to room temperature contain a prominent absorption band that is identical to the largest band induced by irradiation. During exposure to the potassium vapor potassium atoms attach themselves to the KCl lattice. To maintain the lattice-ion and defect concentrations appropriate for the crystal and vapor temperature, each potassium ion incorporated into the lattice surface is accompanied by the introduction of a Cl^- ion vacancy into the lattice. To maintain charge neutrality, the electron released from each potassium atom, as it becomes a lattice ion, is transported into the crystal (it "wanders about" in the conduction band) until it is trapped by a Cl^- ion vacancy.

The defect (i. e., the configuration) formed when a single electron is trapped on an isolated monovalent negative-ion vacancy – such as a Cl^- vacancy in KCl – is called an F center. It, and most other color centers, contain a ground state and at least one excited state. Associated with the F center is the F-center optical absorption band which is due to a transition between the ground state and the next highest electronic state. In most crystals the next highest state lies in the band "gap" a few tenths of an electron volt below the conduction band. The lifetime of the excited state is usually quite short and radiative transitions from the upper to the lower state are often observed as luminescence. Also, and increasingly so with increasing temperature, the center can return to the ground state by nonradiative transitions, or the electron in the upper level, i. e., in the excited state, can thermally untrap into the conduction band.

Additional features of F centers and other defect-related color centers are illustrated by measurements made on additively colored crystals. If a crystal is cooled rapidly to room temperature after exposure to a hot alkali-vapor environment and then cleaved into thin sections, it is found that the F-center concentration is a maximum at the surface and diminishes toward the interior. A uniform distribution of F centers is obtained by heating an additively colored crystal for a comparatively longer time period in a vacuum. This clearly illustrates the introduction of F centers at the surface and the relatively rapid diffusion of centers in the crystal. Additively colored crystals that are slowly cooled, in contrast to the rapidly cooled ones just described, exhibit additional absorption bands that occur on the long-wavelength side of the F band and with one-tenth, or less, of its intensity. One of these bands is called the M band and it has been identified with a center formed by two adjacent F centers along $\langle 110 \rangle$ directions. Its properties were established by optical absorption measurements on strained and unstrained crystals that utilized light polarized in different directions with respect to the crystal axes. In a similar way other bands have been attributed to R centers, which consist of three adjacent F centers in an equilateral-triangle arrangement with the sides along $\langle 110 \rangle$ directions. This arrangement is energetically favored over a linear three F-center configuration along a single $\langle 110 \rangle$ direction. The appearance of multidefect M and R centers in slowly cooled additively colored crystals is attributable to the coagulation of F centers by diffusion processes. These multiple-defect centers form only at temperatures high enough to ensure F-center mobility but low enough to prevent the complete dissociation of previously formed multiple defects. Pure additively colored alkali halides contain very weak absorption bands on the short-wavelength side of the F band. These are called L bands and usually are observed only at liquid-nitrogen or lower temperatures. It has been shown that they are coupled in some manner to the F center. However, they are not well understood.

Additional electronic properties of F centers can be demonstrated by exposing additively colored crystals to light that is absorbed by the F band (termed F light). As the exposure to light increases, the F band diminishes and an additional broad band, lying principally to the long-wavelength side of the F band, appears. This broad band is called the F′ band (F-prime band) and is formed by the capture of a second electron by existing F centers. During exposure to F light a fraction of the F center absorbs photons, which are excited to an excited state lying slightly below the conduction band. The lifetime of this state is sufficiently long to allow the electron to be excited thermally into the conduction band. Once in the conduction band the electron is mobile and "wanders about" until it is captured by a negative-ion vacancy to recreate an F center, captured by an F center to become an F$'$ center, or removed from the conduction band in some other way such as by electron–hole recombination. The F$'$ center formation process may be reversed by exposing the crystal to light that is absorbed by the F$'$ center or by raising the crystal temperature. Both processes release electrons from the F$'$ center into the conduction band, which leads to the reestablishment of F centers. Bleaching experiments such as these have provided much of the existing information on the nature of electron-trap centers and their interaction with mobile charges in the crystal. Bleaching measurements also have been very useful in studies of the electrical antimorphs of the electron centers, namely, the (electronic) hole centers.

The properties of the alkali-halide centers described in this article, and other more commonly occurring centers, are included in Table 1.

The electronic nature of F, M, and R centers, or more specifically the fact that they are formed by trapping of electrons on defects, is readily understood for additively colored crystals. As described earlier, the electron released by each metal atom as it is incorporated into the surface of the lattice is trapped on the negative-ion vacancy concomitantly formed to maintain electronic neutrality. These and other considerations suggested that there should be centers that could be described as the electronic and ionic antimorphs of the electron centers, i. e., centers formed by trapping (electronic) holes on positive-ion vacancies. Historically, answers to this and related questions were more readily obtained by studies on crystals colored by exposure to radiation, for reasons that are difficult to describe briefly, than by investigating additively colored crystals prepared by heating them in halogen vapor. In the latter case we would expect the halogen atom to join the lattice and release an electronic hole (capture an electron), which in turn is trapped (by the release of an electron) on a positive-ion vacancy. The resulting configuration (i. e., a hole trapped on an isolated positive-ion vacancy) is often referred to as the V center. However, it has not been established that the simple V center described here actually exists. As described later, the hole centers that have been identified are usually more complicated.

To understand the formation of color centers by radiation it is essential to describe the more important processes that occur in solids during exposure to energetic radiation, such as ultraviolet and visible light, x rays, gamma rays, electrons, alpha particles, neutrons, and high-energy particles. Most important is the formation of electron–hole ionization pairs, which are often referred to simply as ion pairs. Electron–hole ionization pairs are created whenever the incident radiation is sufficiently energetic to excite lattice electrons, i. e., valence-band electrons, into the conduction band. Once formed, the electron member of the electron-hole ionization pair can "wander about" independently of the concomitantly formed hole. Almost all of the ionization charges enter into recombination. Usually ionization holes are trapped on defects, impurities, or other lattice imperfections more rapidly than ionization electrons are trapped on "opposite-sign" lattice defects. Hole centers usually have large cross sections for electron-hole recombination processes. Thus recombination accounts for a very large fraction, usually more than 99%, of all of the ionization pairs produced during irradiation. The charges that escape recombination during irradiation remain trapped (e. g., to form various electron and hole centers) when the irradiation is terminated.

Table 1: Principal alkali-halide color-center absorption bands[a].

Classical[b] notation	Defect associated with given band, properties of band, etc.
α	Isolated negative lattice-ion vacancy
F	One electron on a single isolated negative lattice-ion vacancy
F'	Two electrons on a single isolated monovalent negative-ion vacancy
K	Transition(s) to level(s) above the F-band absorption level
$L_{1,2,3}$	Very weak absorption associated with F centers; origin not known
M	Two adjacent F centers in $\langle 110 \rangle$ directions
N	Four adjacent F centers: arrangement not certain
$R_{1,2}$	Three adjacent F centers in "triangle" arrangement
M^+	One electron on two adjacent negative lattice-ion vacancies (ionized M center)
F_A	F center adjacent to a single substitutional, usually alkali-ion, impurity
F_B	F center adjacent to two substitutional, usually alkali-ion, impurities
$F_A(I)$	F_A center that relaxes to a "single-well" configuration after excitation by photon absorption
$F_A(II)$	F_A center that relaxes to a "double-well" configuration after excitation by photon absorption
M_A^+, M_B^+	M^+ center adjacent to single (M_A^+) or two (M_B^+) substitutional impurities
Z_1	One electron on coupled substitutional divalent impurity and adjacent positive-ion vacancy
β	Isolated positive lattice-ion vacancy
V (or V_1)	The postulated antimorph of the F center, consisting of one hole on a single isolated positive lattice-ion vacancy (or an adjacent negative ion); it is not certain that this center has been observed
$V_{1,2,...}$	Bands in V region on the short-wavelength side of the F center; often assumed to be produced by hole centers; may be identical with "identified" hole centers
V_K	One hole trapped on a pair of adjacent negative-ion lattice ions to produce a single ionized two-atom molecule, e. g., Cl_2^-, aptly called the "self-trapped hole"
I	Isolated ionized lattice atom in an interstitial site
H	One hole trapped on a pair of negative lattice atoms centered on a single lattice-atom site, i. e., one hole, one lattice atom, and one interstitial; often described as a "split interstitial molecular ion on a single lattice atom site," e. g., a Cl_2^- molecular ion on a Cl^- site
H_A	H center adjacent to a substitutional positive-ion impurity
V_A	V center adjacent to a substitutional negative-ion impurity such as Br^- or OH^- on a Cl^- site.
Exciton bands	Transitions between the "Bohr-atom" like levels of bound electron–hole pairs, i. e., excitons. Since exciton stability decreases rapidly with temperature stable, or nontransient, exciton bands are observed only near zero Kelvin

Table 1 *(continued)*: Principal alkali-halide color-center absorption bands[a].

Classical[b] notation	Defect associated with given band, properties of band, etc.
Colloid or "X" bands	Bands arising from the combined absorption and scattering, called extinction, of colloid particles; usually consisting of regions, in the lattice, containing 5–15 or more metallic ions and few, if any, negative ions

[a] Unless otherwise stated, the absorption bands given in this table refer to ground-state to first-excited-state transitions.

[b] Originally the letter designations, such as F in F band, referred to specific absorption bands. More recently, it has been common to use the letter designation to refer to the center responsible for a given absorption band. This often causes confusion, e. g., there is a K band but not a K center. This multiple usage, and the use of other systems for designating absorption bands and/or defects that are often similarly ambiguous, require the reader of color-center literature to pay particular attention to nomenclature.

The ionization-producing radiation, other than photons, is always a charged particle and interacts with (excites) the lattice electrons through a Coulombic (charge–charge) interaction. As a result, almost all outer-shell electrons on atoms along the path of the incident charged particle are excited to some extent. Those that receive a small amount of excitation lose excitation energy by contributing phonons to the lattice, i. e., heating the lattice. Electrons that receive sufficient excitation, usually an amount equal to or greater than the band "gap," become ionization electrons; the ionization electron and concomitantly formed hole constitute an electron–hole ion pair. To create one ion pair, the average energy given up by the lattice traversing charged particle is very roughly twice the band-gap energy in low- (average) Z materials and increases to roughly three times the gap energy in higher- (average) Z materials.

To retain electrical neutrality the total number of trapped electrons and holes must remain very nearly equal. However, large local charge imbalance may exist in certain circumstances. For example, when samples are subjected to low-energy electron bombardment from one direction, the trapped charges are unequally distributed and discharges can occur between regions of opposite charge. Also, excess charge can discharge out of the target. The resulting treelike discharge patterns are known as Lichtenberg figures. However, for radiation-induced color-center formation the total number of electrons trapped in electron-type color centers is approximately equal to the number of holes trapped in hole centers.

The nature of the more simple identified hole-trapping centers is more complex than the nature of the antimorph electron centers. To achieve local charge neutrality the missing positive lattice-ion charge must be compensated by a trapped hole. To a large extent the electronic nature of hole centers can be understood if it is remembered that the electronic hole is a convenient fiction and that holes are in reality "missing electrons." A hole trapped on a positive-ion vacancy is shared by, i. e., moves between, each of the adjacent negative ions. Thus in the portion of the crystal containing a hole on an isolated positive-ion vacancy there would be local charge neutrality. However, as mentioned earlier it is not certain that this conceptually simple center actually exists. A discussion of reasons why it does not exist or has not been observed is too long to include in this article. The more commonly occurring and established centers formed by trapped holes are described in what follows.

Among the various kinds of V centers that definitely do occur in alkali halides, the V_K center has been studied most extensively. This center assumes an unexpected and surprising configuration. It can enter into a number of color-center formation processes and is comparable to the F center in controlling radiation-induced color-center formation at all temperatures. It is stable, or nearly so, at temperatures below roughly 100 K. At temperatures up to the 550-K region it is stable long enough to enter into a number of reactions, as described below. The V_K center is aptly described and referred to as a "self-trapped hole." The center is formed when an ionization-created hole attaches itself to a pair of adjacent halogen atoms to create a halogen negative ion. For example, in KCl, the V_K centers are Cl_2^- ions oriented in the six equivalent $\langle 110 \rangle$ directions. The V_K centers are unique: The available evidence indicates that they are formed in the initially unperturbed or defect-free regions of the lattice. In contrast, other centers are formed at one or more existing lattice defects. The processes controlling the formation of these centers, i.e., the factors determining exactly where in the crystal a V_K center will be formed, or if it is formed in a mobile state where it will localize, are not understood in detail.

The properties of self-trapped holes have been very extensively studied – especially in alkali halides – by optical absorption using polarized and unpolarized light, electron spin resonance, luminescence, etc. The characteristics of this center are illustrated by the numerous processes that take place at different temperatures. At temperatures appreciably less than 100 K, for most alkali halides, the V_K centers are stable and oriented in $\langle 110 \rangle$ directions. At higher temperatures, specific to each type of crystal, the defect axis may change from the initial $\langle 110 \rangle$ to any one of the $\langle 110 \rangle$ directions at an angle of 60° to the original; reorientation to the $\langle 110 \rangle$ directions at 90° from the original direction apparently does not occur. At even higher temperatures, where the orientation occurs very rapidly, the defect becomes quite mobile and usually enters into several different kinds of interactions. It is likely to form any number of other hole centers, such as H or V_A centers. Also it can interact with electron centers to undergo electron–hole center neutralization or recombination. During recombination, light (i.e., luminescence) may be emitted, or energy may be released as heat or phonon emission and, as described below, provide the energy that results in the formation of a negative ion vacancy and a negative ion interstitial, the process called *ionization damage*.

A second type of commonly occurring hole center is the H center. It should be carefully contrasted with the V_K center. The V_K center is a negative molecular ion, such as Cl_2^- formed by hole trapping in the normal or unperturbed lattice. The V_K center, i.e., the Cl_2^- ion, occupies the volume normally containing two Cl^- lattice ions. In contrast, the H center can be regarded as a similar Cl_2^- molecular ion formed by hole trapping on an interstitial negative atom and an associated lattice atom. In its stable configuration the H center is a negative ion pair, e.g., Cl_2^- ion occupying the volume normally containing a single Cl^- lattice ion. The lattice surrounding the H and V_K centers – in fact the lattice surrounding practically all centers – distorts, or relaxes, to accommodate the atoms (ions) in the center.

An additional commonly occurring hole center is formed when H centers couple with other lattice defects, such as substitutional impurities, to form centers usually labeled H_A centers. Often the A atom is identified, e.g., an H center coupled to a Li atom is labeled H_{Li}.

Next, the *exciton* bands will be described. These bands are stable, i.e., have lifetimes long enough to permit their being easily observed, only near 0 K. They normally occur at wavelengths near, but longer than, the normal band "gap" valence-band to conduction-band transitions (see Fig. 1). To understand exciton bands, recall first that electron-hole ionization pairs formed at high temperatures, e. g., room temperature, appear to dissociate immediately and the charges move independently of each other. At low temperatures, e. g., near 0 K, some of the ionization pairs remain coupled. More precisely, some (ionization) electrons remain coupled to ionized atoms. The coupled pair has many of the properties of a "Bohr-like" atom. The observed energies of exciton absorption bands are, in fact, in reasonable agreement with the Bohr atom model. Near 0 K excitons appear to have little tendency to dissociate but there is considerable evidence they are quite mobile. They can be observed well after being formed and are localized or trapped on lattice defects such as V_K centers. There is increasing evidence that the dominant process for exciton formation is the trapping of mobile electrons by trapped holes, particularly electron trapping by V_K centers. The fact that ionization damage (see below) is observed in NaCl at 250–350 °C, and that excitons are essential for this process, indicates that excitons have appreciable lifetimes – in terms of lattice vibration frequencies, $\sim 10^{13}$ /s – at these temperatures.

Exciton bands observed in crystals at low temperatures disappear from the absorption spectra as the crystal temperature is raised. This disappearance can be attributed to the dissociation of the excitons with the individual charges entering into the processes usually associated with ionization. Alternatively, they could have become mobile and interacted with lattice defects, surface states, etc. Such interactions produce luminescence or, alternatively, result in the release of an amount of energy, as phonons or heat, slightly less than the band-gap energy. Also, as described below, the recombination energy can contribute to defect formation.

Centers Formed by Optically Absorbing Impurities

A large variety of light-absorbing impurities may be incorporated into alkali halides, oxides, and other crystals. The more common examples include:

- Alkali impurities in alkali halides, e. g., Na in KCl, KBr, etc.

- Alkaline earths in alkali halides, e. g., Ca and Sr in KCl, LiF, etc.

- Metals] such as Pb, Cu, Tl.

- Molecular ions like SO_4^{2-}, CN^-, OH^-

- Nonmetals, including S^{2-}, O^{2-}.

In fact, a very large fraction of the conceivable ions can be incorporated in the alkali halides and simple oxides. Crystals prepared with specific impurities are described by the *colon notation*. Thus NaCl with lead impurity is specified as NaCl:Pb. This is ambiguous if the Pb was introduced as a compound, e. g., $PbCl_2$. In terms of this example, it has become increasingly common to specify NaCl with lead introduced as a metal by the notation NaCl:Pb, and when the lead is incorporated as $PbCl_2$ by the notation NaCl:$PbCl_2$.

In the development of color-center physics, measurements on alkali halides containing thallium substituted for the alkali metals played a pivotal role. The Tl^+ ion produces three easily observed absorption bands, usually called A, B, and C bands. These three bands occur on the long-wavelength side of the valence-band to conduction-band transition. There is appreciable evidence for a fourth band, the D band, which lies very close to or on the band-gap absorption.

The properties of the A, B, and C bands have been deduced theoretically by considering how the electronic energy levels of the free Tl^+ ions are modified by embedding the ion in the host alkali-halide crystal. The A and C bands are attributable to transitions that are allowed in the free Tl^+ atom. The B band is associated with a normally forbidden electronic transition in which the forbiddance is removed by interaction with lattice phonon modes. Inasmuch as this interaction increases with increasing temperature, the B-band absorption increases with increasing temperature. This is a unique property, observed only with these and similar atomic impurity bands. The origin of the D band is not unequivocally established. It has been attributed to transitions between the ground state and levels above the A-, B-, and C-band levels and, alternatively, to transitions from the ground state to the conduction band. A large fraction of the properties of the impurity-related absorption bands are very similar to the defect-related bands. Both types are described well by the configuration coordinate formalism described below.

Many transparent but colored minerals and various kinds of colored glasses are examples of materials colored by the presence of specific light-absorbing atoms. Ruby is Al_2O_3 (corundum) containing Cr^{3+} atoms substitutionally replacing Al^{3+} atoms. Emerald is $Be_3Al_2Si_6O_{18}$ (beryl) containing Cr^{3+} in Al sites. Aquamarine is beryl containing Fe^{2+} and Fe^{3+} impurities in Al sites. Citrine is quartz given a yellow color by the presence of Fe^{3+} ions, most likely in Si^{4+} sites. Examples of colored glasses include the common blue cobalt glass that obviously includes Co, and pink-tinted spectacle lenses, made from didymium glass, which contains a mixture of rare-earth elements. The ubiquitous beer bottle's brown color is due to Fe^{3+} ions, and the light green of wine bottles is ascribable to Fe^{2+} ions.

Almost all colored crystals, in contrast to host crystals that do not contain other than negligible quantities of impurities, exhibit absorption bands in the visible. A large fraction of these bands can be attributed to transitions between atomic energy levels in the impurities, which are often split by crystal-field effects. Ruby and emerald are in this category. The absorption bands in another large group, which includes aquamarine, are due to molecular-orbital and/or charge-transfer transitions. A small number of crystals are colored because they possess valence-band to conduction-band transitions. Cinnabar (HgS) is a good example in this category. Finally, there are a number of crystals that appear colored because they contain defect-related color centers of the type described earlier for the alkali halides, i.e., they contain color centers formed by charge trapping on lattice defects. Smoky quartz and some, but not all, colored fluorites are in this group.

To reduce the existing confusion, it is necessary to mention three types of crystals that exhibit color not attributable to any of the processes mentioned so far, i.e., not caused by defect and impurity color centers. First, halite, which is the natural form of NaCl and is usually called rock salt, occasionally exhibits an intense color ranging from very light blue to a deep blue-black purple. Often the color is layered along $\langle 110 \rangle$ planes. These blue crystals exhibit a *de facto* absorption band in the red that is attributable to a combination of absorption and scattering from colloid Na metal particles. This blue coloring can be induced by exposing NaCl crystals to large doses of ionizing radiation, particularly above room temperature, and/or exposing crystals to room light after irradiation. Second, crystals can appear colored if they contain numerous microinclusions of a second mineral, e. g., the color in blue quartz is attributed to microscopic rutile (TiO_2) inclusions. Third, crystals that are composites of regularly arranged microcrystals exhibit colors because they diffract light. The classic example

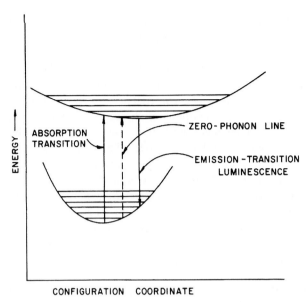

Fig. 2: A typical color-center configuration-coordinate diagram. Many quantitative optical properties of color centers can be deduced from such diagrams. Typical absorption. luminescence, and a "zero-phonon" absorption transition are shown.

is opal. This mineral consists of nearly equal-size spheres of a hydrated silica located on a uniformly spaced three-dimensional lattice.

Optical Properties of Color Centers and the Configuration Coordinate Diagram

A large fraction of the optical properties of color centers can be described by utilizing a configuration-coordinate diagram such as Fig. 2. Such diagrams describe the electron and vibrational, or thermal energy, levels of simple electron or hole centers. To a moderately good approximation such centers can be described by quantum-mechanical calculations based on the Born–Oppenheimer approximation. This procedure separates the Schrödinger equation solution into an electronic part and a vibrational part. The solution of the vibrational part is obtained by using a "spring-constant" potential which leads to harmonic-oscillator wavefunction solutions. The solution of the electronic part, for the lowest two electronic states, is represented by the curved envelopes, properly called "adiabatic potential-energy curves." Superimposed on the electronic states are the vibrational or phonon states, shown as closely spaced lines. Usually the lowest electronic level is an s state and the next highest level a p state. At temperatures approaching $0\,\mathrm{K}$ the system can be described by the s-state envelope, the p-state envelope, and the lowest-lying phonon levels in each state. At low temperatures the absorption transitions will originate from the lowest-lying levels at the minimum in the ground state and terminate on the phonon levels of the excited-state envelope which lie directly above in the configuration-coordinate diagram. Such transitions are indicated schematically in Fig. 2

as the absorption transition. The resulting absorption band will be quite narrow, since all of the absorbing transitions have nearly the same energy. At higher temperatures the transitions originate and terminate from wider distributions of phonon states, and as the temperature increases the absorption bands become wider. Optical transitions from upper to lower electronic states result in the emission of light, i.e., luminescence. Similarly, the luminescent emission is narrow at low temperatures and increases with increasing temperature. This approach to the optical properties of color centers provides a good description of absorption, luminescence, and many other properties of color centers. The phonon contribution is well approximated by a Boltzmann distribution for all temperatures except those near $0\,\mathrm{K}$. Near $0\,\mathrm{K}$ accurate calculations require the inclusion of discrete phonon levels and the zero-point energy. Also, for most calculations the electronic envelope can be regarded as parabolic; in the pure harmonic-oscillator approximation it is precisely parabolic. Calculations made with this model lead to the following expressions for the temperature dependence of the absorption band peak E_0 and the full width of the band at half maximum U (usually both E_0 and U are given in electron volts):

$$E_0(T) \;=\; C_1 - C_2 \coth(\hbar\omega/2kT)\,, \tag{1}$$
$$U(T) \;=\; 8\ln 2 (\hbar\omega)^2 S \coth(\hbar\omega/2kT)\,, \tag{2}$$

where T is the temperature in kelvins, C_1 and C_2 are constants, $\omega/2\pi$ is the effective vibration frequency in s^{-1}, $\hbar = h/2\pi$ where h is the Planck constant, k is the Boltzmann constant, and S is the Huang–Rhys factor. The last-mentioned factor is difficult to explain briefly [5]. It is a coupling constant that is a measure of the relative horizontal displacement of the adiabatic potential energy curves along the configuration-coordinate axis and is approximately equal to the mean vibrational quantum number in the transition.

The optical absorption of a crystal containing color centers can be calculated by considering it as an assemblage of a large number of (harmonic oscillator) absorbing centers. The result is known as *Smakula's formula*, which relates the measured optical absorption coefficient α (in cm^{-1}) to the concentration of color centers N:

$$Nf = 0.87 \times 10^{17} \frac{n}{(n^2+2)^2} U \alpha_\mathrm{m}\,, \tag{3}$$

where f is the oscillator strength, which must be determined independently for each color center and is approximately 0.9 for most alkali-halide centers; n is the index of refraction of the host crystal; U is the full width at half maximum of the absorption band (in eV), and α_m is the absorption coefficient (in cm^{-1}) at the peak of the band. This formula is based on the assumption that the absorption band is precisely Gaussian shaped. The numerical coefficient changes by roughly 10% for other common shapes. Precisely Gaussian-shaped bands are observed in many cases and in crystalline materials represent cases where the lower adiabatic potential-energy curve for the centers can be approximated by parabolas and simple harmonic oscillator wave functions are obtained. In many cases the actual potential cannot be represented by simple parabolic curves and the observed absorption and emission bands are not well described by Gaussian-shaped bands; in fact, the observed bands may be quite asymmetric. In glasses and amorphous solids the absorption (luminescence) bands may be Gaussian shaped because they represent a superposition of absorption (emission) by centers randomly influenced by their surroundings, e. g., the randomness superimposed on centers in glasses by the random atomic separation and band angles.

The usefulness of the configuration-coordinate diagram approach and the behavior of many color centers at low temperatures is illustrated by zero-phonon line phenomena. A good example is the R-center absorption in LiF at liquid-helium temperatures, i. e., near 4 K. The absorption spectrum consists of a number of very sharp lines superimposed on a broad underlying continuum. The continuum resembles the spectrum observed at higher temperatures but is slightly narrower and shifted in accord with the peak-energy and bandwidth formulas given earlier. The sharp lines are not observed at higher temperatures. They arise from transitions between individual low-lying phonon states in the upper and lower levels. The most prominent sharp-line transition is between the lowest-lying states, and since these do not include lattice vibration, i. e., phonon modes (the vibrational quantum number is zero), it is called the "zero-phonon line." This transition is illustrated in Fig. 2. It is often observed when the other sharp lines are obscured.

Radiation-Induced Color-Center Formation

Most studies on color centers were undertaken to determine the properties of the different types of centers. More specifically, they were designed to identify the atomic and electronic configuration of each center and to determine its properties. A less emphasized but equally interesting aspect of color-center physics, one that has numerous applications, is the study of radiation-induced color-center formation. Curves of color-center concentration versus irradiation time or dose depend on a large number of physical parameters. The more important are:

1. The nature of the material being irradiated. Some aspects of the color-center formation kinetics are common to most materials. Other aspects depend on the material type. There are features peculiar to each of the following: alkali halides, oxides, mixed ionic and covalent crystals (such as nitrates and bromates), covalent crystals, semiconductors, glasses, glass-ceramics, and organic glasses.

2. Sample temperature.

3. The type of radiation producing the coloring, which includes infrared, visible, ultraviolet, x-ray, and gamma-ray *photons* and the various charged particles such as electrons or beta rays, protons, alpha particles, fission fragments, and the entire list of charged high-energy particles. Low-energy photons can create ionization by optical absorption processes such as band "edge" transitions. Higher-energy photons efficiently create ionization almost entirely by the formation of charged recoils resulting from the photoelectric, Compton, and pair-production processes. Energetic neutral particles, such as reactor neutrons, produce copious ionization and radiation damage, by creating charged lattice atom recoils from elastic or inelastic (nuclear event) collisions.

4. Most importantly, the incident particle energy, the total energy, i. e., *dose*, imparted to the material and the rate of energy deposition, i. e., the *dose rate*. Often the irradiation conditions are given in terms of the total number of particles incident on the sample,

i. e., the *fluence*, or the irradiation time and the flux, or particles per unit time (fluence = integration of flux over irradiation time). Color-center formation often depends on details of the irradiation conditions such as continuous versus pulse irradiation, pulse length and interval between pulses, etc.

5. The presence of impurities and defects in the lattice. Impurities control the coloring of many materials.

6. Strain state and/or plastic deformation. In the alkali halides the coloring curves are strongly influenced by dislocations and dislocation-related defects. Plastically deformed alkali halides color much more readily than unstrained ones.

Early color-center formation studies were made by irradiating a crystal with x rays, or other radiation, and then making measurements at a later time. If the crystal temperature was low enough, e. g., liquid-nitrogen temperature, to suppress all temperature-dependent decay occurring during and after irradiation, most of the color-center versus dose or irradiation-time curves are linear or monotonic increasing and are true representations of the color-center formation kinetics. However, at higher temperatures and particularly at room temperature, the center concentrations change both during and after irradiation and the measured coloring curves do not accurately describe the color-center growth kinetics. In many crystals, particularly the alkali halides, the coloring curves contain three distinct stages. The initial part, stage I, increases monotonically at a continuously decreasing rate until a low- or zero-slope plateau is reached. The plateau region is called stage II. The irradiation time included in stage II varies from close to zero to very large values. The following region, stage III, when it is observed, usually is concave upward and continues until the irradiation is terminated. At very high doses some materials exhibit growth beyond stage III which flattens out to a plateau, or occasionally reaches a maximum and then decreases. Sometimes this last region is referred to as stage IV.

First- and second-stage coloring curves obtained by making measurements at some controlled time after irradiation and at a fixed temperature (e. g., at room temperature) are crudely approximated by an expression consisting of the sum of a saturating exponential and a linear component. However, the stage-I and -II color-center versus irradiation-time curves *obtained by making measurements during irradiation* [6, 7] are very accurately described by the expression

$$a(t) = \sum_{i=1}^{n} A_i(1 - e^{-a_i t}) + \alpha_L t . \tag{4}$$

The number of saturating exponential terms, n, usually does not exceed four. Also, the possibility that this expression is merely a good empirical curve-fitting formula is almost certainly eliminated by the occurrence, in different materials, of stage-I curves containing one exponential component or curves with two or three unambiguously resolvable exponential components.

An understanding of the various coloring curves that are obtained during irradiation and the changes occurring after irradiation can be obtained by considering the coloring of alkali halides (LiF is often an exception). While Eq. (4) can be regarded as an empirical equation, it can be readily derived from relatively simple considerations. The primary coloring processes is the conversion of traps, often called precursors, into color centers by charge capture. For example, vacancies are precursors in the sense that they can be converted to F centers by electron capture. Usually, prior to irradiation a crystal will contain a given concentration of precursors for each type of color center. Also, it is likely that more than one type of precursor may produce the same color center. For example, in the alkali halides F-center precursors may exist as (1) isolated defects, (2) vacancies coupled to one or more different kinds of impurities, (3) coupled vacancies with trapped charges, such as M or R centers, that separate during irradiation and become F centers or vacancies, and (4) additional vacancies may be introduced by radiation-damage processes. In other words, during irradiation the precursor concentration may be increased (or in rare cases decreased) by radiation-induced processes. Let N_0 be the preirradiation concentration of traps – that are converted to color centers by capturing ionization-produced electrons – K the rate that traps are introduced during irradiation (the assumption of a linear defect formation rate appears to apply to almost all material at low total doses [6,7]), f the fraction of empty traps converted to color centers per unit time, ϕ a measure of the radiation-induced ionization electron concentration (usually dose rate), and n the color center concentration at time t. During irradiation at time t the number of empty traps is $N_0 + Kt - n$ and the rate these empty traps are converted to color centers is

$$\mathrm{d}n/\mathrm{d}t = \phi f (N_0 + Kt - n) \tag{5}$$

which has the solution, with $n = 0$ at $t = 0$,

$$n = (N_0 - K/\phi f)(1 - \mathrm{e}^{-\phi f t}) + Kt . \tag{6}$$

If the same center is formed by independent processes, e. g., from two different types of precursors, the growth is described by the superposition of a number of equations like (6), i. e., Eq. (4).

For long, or very long, irradiations this simple treatment must often be modified to include other processes such as electron–hole recombination. In other words, a color center may be converted back into a precursor by recombination with opposite-sign charges; e. g., during irradiation F centers can be converted to vacancies by recombination with holes.

Measurements made after irradiations are terminated show several different features. Such measurements are particularly meaningful if they are made with apparatus for studying color-center formation during irradiation [6, 7]. First, in most materials, when the irradiation is terminated, the measured color-center concentrations undergo abrupt changes. In most crystals and glasses the coloring decreases rapidly. In others, one or more centers increase and in some they increase initially and then decrease. Often the changes occurring after irradiation can be resolved unambiguously into decreasing exponential components or into combinations of increasing saturating exponential and decreasing exponential components. Second, at room temperature the radiation-induced color centers usually do not completely disappear after irradiation. The coloring remaining a week or longer after irradiation ranges from a few percent to 98–99% of that present at the end of irradiation. In most materials the decrease is in the 20–60% range. Third, the removal of color after irradiation is almost always increased by raising the sample temperature. The changes occurring after irradiation indicate that one or more parameters in Eq. (6) include both growth and decay process [7]. Also, a large fraction of the radiation-induced coloring often can be removed by exposure to strong visible and/or ultraviolet light.

The formation of precursors that become color centers during irradiation, as mentioned above, is usually attributable to either the *displacement damage* [1] or the *ionization damage* [6–8] process. The latter is particularly important in some alkali halides. *Displacement damage* occurs when an incident particle makes an elastic (elastic event) or occasionally an inelastic (nuclear event) collision with a lattice atom that transfers sufficient energy, usually about 25 eV, to the struck atom to move it from its normal lattice position to a nearby interstitial position. Often the newly created nearby interstitial undergoes immediate recombination with the vacancy at its original lattice position. If the conditions are "right," e. g., if the interstitial is sufficiently removed, a stable vacancy–interstitial pair is formed. The vacancy and/or the interstitial can capture ionization-induced charges to become color centers. The incident bombarding particle, that initiates the process, is usually a neutral or charged particle. It must be emphasized that gamma rays also produce appreciable displacement damage; but by a two-step process. The first step is the formation of a recoil electron by photoelectric, Compton, or pair-formation processes. Once formed, the recoil electron gives up energy – primarily by ion-pair formation – as it moves through material. If sufficiently energetic, it can transfer enough energy to a lattice atom, by making an elastic collision, to create a vacancy–interstitial pair. Once it has degraded below the threshold energy, which is primarily a function of the initial recoil energy and lattice atom mass, it cannot transfer enough energy to displace atoms. Yet, this is a surprisingly efficient process. Very roughly, 1 out of 10^4–10^5 incident 1-MeV gamma rays will create a single displacement damage pair in most low-Z materials. In contrast, only 10–100 incident 1-MeV electrons are required to create a single displacement damage pair in the same material.

The importance of the *ionization damage* process has become widely accepted only in the last decade [6–8]. This process utilizes the energy released as a trapped exciton, i. e., a trapped hole–electron "Bohr-atom" like ion pair, undergoes recombination. Although this process is not completely understood, it is best described as it is presumed to occur in a typical alkali halide, namely, NaCl. The first step is the formation of a V_K center (a Cl_2^- ion in the space normally occupied by two adjacent Cl^- lattice ions) by the "self-trapping" of an ionization-produced hole. This step is accompanied by the relaxation, or distortion, of the nearest and next nearest lattice atoms. The V_K center is stable, or decays very slowly, at temperatures around 80 K and its lifetime decreases as the lattice temperature increases. In NaCl at 250–350 °C it is long enough for the ionization damage process to proceed. The next step is the capture of an ionization electron by the V_K center to form a trapped exciton. It is likely that additional rearrangement of the atoms in and surrounding the V_K center occurs as the exciton is formed. The lifetime of the trapped-exciton V_K-center complex is quite short. It is likely that the next step, the recombination step, can occur in two distinct ways. First, it can decay radiatively by the emission of luminescence, i. e., the emission of a photon. In this case all of the energy available in the exciton – and perhaps some contained in lattice distortion – is emitted as luminescence; or, a large part of the available energy is emitted as light and the rest nonradiatively as lattice phonons, i. e., heat. This sequence will restore the lattice to its original unaltered condition. Second, the decay step can occur by the available energy being transferred to one of the Cl atoms, in the Cl_2^- ion, in a way that provides enough energy for the atom to move to an interstitial position. The probability of this occurring is increased by the possibility that the Cl atom movement is assisted by the formation of a focused series of successive Cl atom displacements, in favored crystallographic directions, to produce an interstitial Cl^- ion four, five, or more Cl separations from the site of the original V_K center. Both theoretical and experimental evidence support the existence of such focused collision sequences. This second decay process creates a Cl^- vacancy and an interstitial Cl at a sufficiently distant site to create a stable vacancy–interstitial pair. The interstitial Cl atom is usually neutral, the valence electron having been trapped on the (or an equivalent) Cl^- vacancy to convert it to an F center. It is correct to say that the *ionization damage* process creates (well-separated, uncorrelated) F-center Cl^0 interstitial pairs.

The average energy required to produce a single F-center interstitial pair by the ionization damage process is extremely low compared to the energy required to produce the same defect by displacement damage. As mentioned above, the average energy required to produce an electron–hole ion pair ranges from twice the band-gap energy, in low-Z materials, to three times the gap energy in high-Z materials. Only a fraction of the ion pairs formed by incident radiation produce stable ionization damage: a large fraction of the electron–hole ion pairs formed undergo recombination without forming excitons, a fraction of the excitons radiatively recombine, a fraction of the vacancy–interstitial pairs immediately recombine, and other processes occur that compete with the ionization-damage process. An average of 400–500 eV is required to produce a single vacancy–interstitial pair in NaCl by ionization damage. In NaCl this process is 4 000 to 10 000 times more efficient than displacement damage. If the figure of merit for ionization damage in alkali halides is 4 000, it is roughly 2 000 for alkaline earth halides and very roughly 1 for fused silica. A technically important consequence of ionization damage is included in the Applications section.

Color-Center Applications

Color-center research, because it was the first aspect of "defect solid-state physics" studied in detail, played a pivotal role in the development of a number of applications of solid-state concepts. Examples include luminescence, e. g., fluorescent lights; video display tubes; semiconductors; lasers; etc. Yet, only a few applications based directly on defect color centers can be cited. One application, which is commercially viable and directly utilizes defect color centers – in contrast to substitutional impurities or to molecules – is the *color-center laser* [10, 11]. Color-center lasers are based on the same principles that apply to many other laser active materials, particularly dye lasers. These principles are described in laser articles. Quite a number of different color centers, in a variety of host crystals, have been shown to be laser active. These centers and the corresponding host crystals include (but are not limited to) the following: M^+, LiF; M^+, KF; M_A^+, KCl:Na; M_A^+, KCl:Li; $M^+:O^{2-}$, NaCl:O_2; $F_B(II)$, KCl:Na; $F_A(II)$, KCl:Li and F^+, CaO. Most color-center lasers operate in the infrared, particularly in the 1–3-μm region, some are useful in the visible, a few operate in the ultraviolet, and at least one in the vacuum ultraviolet.

Two other direct applications involving color centers are photochromic (spectacle) glasses that darken in strong ultraviolet, such as sunlight, and radiation dosimeters, especially for high total dose applications, that are based on the radiation-induced color-center formation kinetics (coloring) described above.

The development of optical devices that operate in intense light or in non-negligible x-ray and/or nuclear radiation fields, particularly if they include light-transmitting elements such as lenses and/or windows, must always include steps to demonstrate that they will not be rendered inoperable by radiation damage; more specifically, if they are susceptible to non-negligible radiation-induced color-center formation [6]. The number of devices subject to this kind of radiation damage is surprisingly large. Included are satellites and space probes (cosmic rays and Van Allen belt radiation), optical viewers for reactors and hot cells, fiber optics (some fibers are exposed to radioactivity from soil and the seabed), and particle detectors for high-energy physics research – especially in colliding beam experiments where the detectors must be located close to the intersecting accelerator beams.

A most surprising application of color centers has emerged in the search for a permanent way to dispose safely of radioactive waste [7, 11]. About three-fourths of the initial disposal effort assumed repositories would be located in the worldwide natural rock salt formations, occurring 1–3 km below the surface and often more than 1 km thick, that cover much of Europe and the southwest United States. Salt deposits were considered resistant to radiation damage. However, recent studies show that both natural and synthetic rock salt is particularly susceptible to radiation induced F-center and colloid center, i.e., X-band, formation because of the ionization damage process. Since one Cl vacancy and interstitial atom pair are formed for each 400–500 eV (*not* MeV) deposited in the salt by radiation from the radioactive waste, the damage formation rate is very high. Curves of F-center concentration versus dose increase monotonically to a plateau that "levels off" at 10^6–10^7 rad. As the F centers reach saturation a sodium metal colloid absorption band appears that is accurately described by classical nucleation and growth curves with induction period and power law growth. The colloid particle nucleation is related to dislocations; the induction period is reduced markedly by straining samples prior to irradiation. Colloid particles grow by the diffusion of F centers and Cl^- vacancies to particle surfaces. Because of the high defect formation rate, typical waste doses of 10^{10} rad will convert 1–10% of the irradiated salt to colloid in about 400 years and 10^{11} rad will convert 10–50%. When salt irradiated to 10^{10} rad is dissolved in brine, pH values of 12 to 14 are obtained. Such solutions are highly corrosive and make it difficult to find acceptable corrosion-resistant materials to contain the waste.

This color-center and colloid particle formation process is only partially understood [7, 11]. The colloid formation rate is low at 90 °C, increases to a maximum at 150–170 °C, and decreases to a negligible level at 250–350 °C. The colloid formation rate increases as the dose rate decreases. Measurements made during irradiation show that decreasing (increasing) the dose rate, after an initial irradiation, decreases (increases) the F-center concentration and increases (decreases) the colloid formation rate. After irradiation, large changes occur in the F-center and colloid band intensities that depend primarily on irradiation temperature. This is one of many examples of research pursued for purely basic reasons, in this case the study of ionization damage, that played an important role in a practical and pressing national and international problem, the disposal of radioactive waste. The studies provided information on the levels of damage expected in natural salt and explained why they were so high; effects that had not been included in the original plans for radioactive waste repositories in salt.

Additional Information on Color Centers

The information on color centers given in this article is a very much abbreviated overview of the subject. Many aspects of color-center physics have been mentioned but not described and others are not included. Examples are: electron spin resonance (ESR); electromechanical and magnetooptical properties; luminescence, which includes fluorescence, phosphorescence, and thermoluminescence; ionic, electronic, and photoconductivity; diffusion of defects producing color centers; thermal annealing; dielectric properties; and strain-induced reorientation. Only trivial color-center theory is included. The bibliography includes books that contain advanced level information on almost all aspects of color-center physics.

See also: Crystal Defects; Electron Energy States in Solids and Liquids; Electron–Hole Droplets in Semiconductors; Excitons; Lasers; Luminescence; Radiation Damage in Solids; Radiation Interaction with Matter; Thermoluminescence.

References

[1]   J. H. Crawford, Jr. and L. M. Slifkin (eds.), *Point Defects in Solids*, Vols. I, II, III. Kluwer, 1975.
[2]   W. B. Fowler (ed.), *Physics of Color Centers*. Academic, New York, 1968.
[3]   J. J. Markham, *F Centers in Alkali Halides*. Academic, New York, 1966.
[4]   J. H. Schulman and W. D. Compton, *Color Centers in Solids*. Macmillan, New York, 1962.
[5]   A. M. Stoneham, *Theory of Defects in Solids*. Oxford University Press, Oxford, 2001.
[6]   P. W. Levy, *Radiation Effects in Optical Materials, SPIE Proc. 541*, P. W. Levy (ed.), pp. 2–24. SPIE, Bellingham, WA, 1985.
[7]   P. W. Levy, *Defect and Impurity Center in Ionic Crystals – I, Optical and Magnetic Properties*, P. W. M. Jacobs (ed.), *J. Phys. Chem. Solids*: Special Topics Issue, in press.
[8]   F. C. Brown and N. Itoh (eds.), "Recombination Induced Defect Formation in Crystals", *Semiconductors and Insulators: Special Topics Issue* **5**, 1983.
[9]   L. F. Mollenauer, *The Laser Handbook*, M. Stich and M. Bass (eds.), Chap. III. North-Holland, Amsterdam, 1985.
[10] L. F. Mollenauer, *Topics in Applied Physics*, Vol. 59, L. F. Mollenauer and J. C. White (eds.). Springer-Verlag, Berlin, 1987.
[11] P. W. Levy, J. M. Loman, K. J. Swyler, and R. W. Klaffky, *The Technology of High-Level Nuclear Waste Disposal, I*, P. L. Hofmann (ed.), pp. 136–167. Technical Information Center, U.S. Dept. of Energy, Oak Ridge, TN, 1981.

Combustion and Flames

D. J. Seery and M. B. Colket

Combustion is a self-sustaining exothermic process involving the reaction of a fuel with an oxidizer. When visible light is also produced, the term flame is applied. The type of fuel and the conditions under which the fuel and oxidizing air are brought together vary depending upon the intended application of the combustion process, and may include domestic and industrial heating, electric power generation, or transportation. Other manifestations of combustion include incineration, explosives, rocket propulsion, unwanted fires, and some illuminating devices. The manner in which the heat release from combustion is transformed into useful kinetic energy will differ with end use, such as, in gas turbines, boilers, piston engines, or other applications.

Combustion processes have been the major source of energy for our civilization. While this situation is likely to continue for the foreseeable future, we must also contend with the polluting exhausts from these combustion systems (even with extended use to areas such as incineration of hazardous wastes). In addition, the infrared-active combustion products, CO_2 and NO_2, contribute to "greenhouse" warming. These emissions must be included with the pollution of the lower atmosphere as major deleterious effects of combustion. Resolving these concerns related to exhaust products of combustion requires a quantitative understanding of the details of combustion. Many of the present-day combustion problems often relate to species in very low concentrations (ca. $< 100\,ppm$); hence, control strategies require detailed information on chemical reactions and species not normally important to bulk flame structure.

Both chemical and physical processes occur in combustion and either may dominate in different applications. For example, most practical combustion involves non-premixed flames: the reactants are initially unmixed and the progress of the reaction is dominated by the rate of mixing. For liquid fuels, mixing also depends on atomization and vaporization, which may vary greatly. In most combustion applications the overall progress of the reaction is determined by a combination of mass transport, heat transport, and chemical reaction rates.

To illustrate how competition between these latter two processes affects the start of combustion, consider the ignition of a gaseous mixture of fuel and oxidizer. If the temperature of this mixture is sufficiently high, chemical reactions will occur, some of which result in the heating of the mixture. If the temperature of the gases rises to a value, T, higher than the wall temperature, T_W, then the rate of heat loss by conduction to the walls, Q_L, is proportional to the difference between them:

$$Q_L = hS(T - T_W) \,,$$

where h is the heat transfer coefficient and S is the surface area of the vessel. The rate of heat production, Q_G, is proportional to the reactant concentrations and the reaction rate, which has an exponential dependence on temperature

$$Q_G = f(N)H_0A\,e^{-E/RT} \,,$$

where H_0 is the heat of reaction, $f(N)$ is an appropriate function of concentration, and $Ae^{-E/RT}$ is the kinetic rate constant (*see* Kinetics, Chemical). The energy conservation equation can be written

$$\rho C_V \frac{dT}{dt} = f(N)H_0A\,e^{-E/RT} - hS(T - T_W) \,,$$

where the left-hand side represents the energy accumulation in the gases with density ρ and heat capacity C_V.

When the rate of heat production from chemical reaction exceeds the rate of heat loss, the temperature rises. Rising temperature increases the reaction rate exponentially and combustion of the remaining reactants occurs rapidly or 'explosively'. Alternatively, if the heat loss term dominates, temperature decreases, slowing down the reaction rate and quenching the combustion. Such an energy balance helps to explain the critical limits observed for flame phenomena. For example, for mixtures of fuels with air or oxygen there are limits of composition for which flame propagation can occur but outside of these limits no self-sustaining

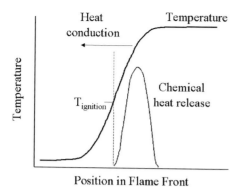

Fig. 1: Heat release, heat conduction and temperature in a premixed, laminar Flame.

flame can be initiated. It appears that for many flames these composition limits correspond to some minimum final flame temperature achieved during combustion. For methane (CH_4) this temperature is about 1400 K. Outside the composition limits ("flammability limits"), the flame is not hot enough to be self-sustaining since radiative and conductive losses successfully compete with chemical heat release rates.

A flame can be understood as a thin reaction zone that propagates through a mixture of reactants; within this reaction zone, reactants are converted to products and the accompanying energy release results in a temperature rise across the flame front. At ambient conditions, premixed laminar flame speeds can vary from ~ 10 cm/s for fuel lean hydrocarbon/air flames to ~ 300 cm/s for fuel-rich hydrogen/air flames. At the elementary level, both chemistry (of oxidation reactions) and transport (of energy and species) control ignition as well as the rate at which a flame moves through a combustible mixture. A schematic structure of a flame is shown in Fig. 1 in the frame of reference of the flame. Once the flame gases have ignited and post-flame gases have been heated to elevated temperatures, energy is conducted upstream of the reaction front towards the cool reactants due to the high spatial temperature gradient. This transferred energy preheats the reactant gases to their ignition conditions. Once ignited, rapid (explosive) reaction leads to rapid heat release and the process is self-sustaining. Accenting the importance of both chemical and transport processes to flame propagation, the laminar flame speed (SL) can be shown to be proportional to $\sqrt{\alpha f(N)A\,e^{-E/RT}}$, where α is the diffusivity and $f(N)A\,e^{-E/RT}$ is the reaction rate.

A simple arrangement for studying a propagating flame is a premixed flat flame often operated at low pressures (~ 5 kPa). The low pressure decreases the concentration of reactants thus slowing down the overall combustion process and stretching out the reaction zone. In this configuration the temperature profile and the species concentration profiles delineate the entire combustion process, including chemical reactions, molecular diffusion, and thermal conduction. An example of the concentration profiles for major species in a methane (17.9%)–oxygen (31.6%)–argon (50.5%) flame burning at 4.67 kPa (35 Torr) is shown in Fig. 2. The flame is slightly rich; that is, it contains excess fuel. Stated differently, the oxygen concentration is inadequate to burn all the fuel to CO_2 and H_2O. These data were obtained using a mass spectrometer coupled to a molecular beam sampling system. An examination of the profiles

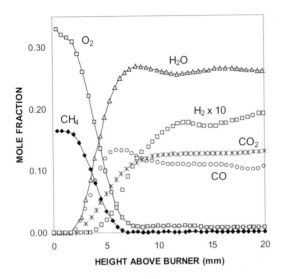

Fig. 2: Major species concentration profiles of fuel-rich methane flame.

reveals the fairly smooth decline of the reactants CH_4 and O_2 and the rise of the products H_2O and CO_2. The intermediate species CO and H_2 are also produced in quantity and continue downstream relatively unchanged because of the lack of O_2.

The two most important reactions in a hydrocarbon flame front are

$$H + O_2 \rightleftharpoons OH + O$$

and

$$CO + OH \rightleftharpoons CO_2 + H \,.$$

The first of these is a chain branching reaction, primarily responsible for the chemically 'explosive' characteristic of a flame. For this reaction, two reactive radicals (O and OH) are formed while only one (H) is lost. Furthermore, one of the formed species (O) is a biradical. The second reaction results in the formation of the thermodynamically stable CO_2 molecule, and hence is a primary contributor to heat release in a flame.

In contrast to the premixed flames discussed above, combustion processes in non-premixed flames are largely controlled by mixing (or diffusion) rates. A well-known example of a non-premixed flame is a candle flame with fuel vaporizing from the wick along the centerline while air convects and diffuses to the outside of the flame. Alternatively, consider gaseous fuel flowing upward through a central tube while an oxidizer (e. g., air) is flowing upward through a concentric tube; under these conditions, a laminar diffusion flame can be stabilized at the end of the tube where the reactants begin mixing. A simplified diagram of the radial concentration profiles at some distance above the burner tubes is shown in Fig. 3. In these flames, the mixing of the reactants occurs by molecular diffusion and the reactants are all consumed at the interface, i. e., flame front. Hence, in such non-premixed flames, heat release is governed by the mixing/diffusion rates. Note, that because of the significant temperature

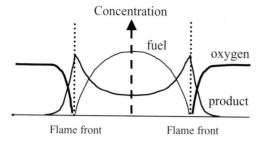

Fig. 3: Concentration profiles through a laminar non-premixed flame.

Fig. 4: Contrasting flame structure for candle flames at normal gravity (A) and at microgravity (B) (Courtesy NASA/Glenn Research Center).

increase from combustion and the accompanying density decrease, buoyancy plays a dominant role in defining the structure of these flames. Dramatic changes in flame structure occur in microgravity environments, and the conical shape of a candle flame, as occurs on the earth surface, becomes spherical (see Fig. 4). For both of these flames, the fuel is vaporized from the candlewick following heat transfer from the flame. At normal gravity, the buoyancy-induced upward convective flow collapses the flame into a vertical column. At microgravity there is no buoyant acceleration and the fuel and oxidant diffuse uniformly (from opposite directions) to the flame front where they are consumed. The grayish portion of the visible flames (shown in the black and white figure) is normally blue because of chemiluminescence from low molecular weight gas-phase species. The whitish portion is continuum radiation from soot particulates formed in the fuel-rich portion of the flame. This radiation from soot is the yellow-orange light that has made candle flames so useful. There is little formation of soot in the microgravity flame.

In most practical applications, combustion is altered through aerodynamic turbulence (*see* Turbulence). Turbulent enhancements to mixing dramatically increase burning rates and volumetric heat release rates. At very low levels of turbulence (that is when characteristic chemical time scales are much less than turbulent time scales), a flame front is wrinkled by the turbulence, effectively increasing the surface area of the flame and hence reaction rates. With increasing turbulence, the wrinkling of the flame increases until the turbulent scales start to be comparable with the flame scales creating a distributed reaction region. As the turbulence is increased further, a theoretical limit of well-mixed reactions (i. e., mixing between reactants and products) is reached. However, depending on the characteristics of the flow, extinction (blow-off) of the flame can occur and the combustion process is no longer self-sustaining. For a fixed turbulent environment, reductions in the chemical kinetic rates (occurring, for example, at low power conditions) will have a similar quenching effect. Design of robust systems with low emissions, high efficiencies, and wide operability range remains a challenge to engineers.

Non-premixed, turbulent flames are found in most practical applications including the combustion of gaseous, solid, and liquid fuels. Use of liquid and solid fuels further complicates combustion systems since vaporization, heat and mass transport, thermal decomposition, or surface reactions may assume key (i. e., rate-limiting) roles in the combustion process.

See also: Chemiluminescence; High Temperature; Kinetics, Chemical; Turbulence.

Bibliography
A. G. Gaydon and H. G. Wolfhard, *Flames: Their Structure, Radiation and Temperature*, 4th ed. Wiley, New York, 1979. (E)
B. Lewis and G. von Elbe, *Combustion, Flames and Explosion of Gases*, 3rd ed. Academic Press, New York, 1987. (I)
I. Glassman, *Combustion*, 2nd ed. Academic Press, New York, 1987. (I)
R. A. Strehlow, *Combustion Fundamentals*. McGraw-Hill, New York, 1984. (I)
F. A. Williams, *Combustion Theory*, 2nd ed. Benjamin/Cummings, Menlo Park, 1985. (A)
S. Turns, *An Introduction to Combustion; Concepts and Applications*. McGraw-Hill, New York, 1999.

Complementarity

J. Bub

The concept of complementarity lies at the heart of Niels Bohr's interpretation of quantum mechanics. Early versions of the quantum theory emphasized a wave-particle duality: microsystems seem to manifest wave aspects under certain experimental conditions and particle aspects under other conditions. The uncertainty principle, derivable from an analysis of a wave packet as a representation of a relatively localized system and as a theorem in quantum mechanics, may be regarded as a formal expression of this duality. It says, in effect, that there are certain quantities, like position and momentum, that necessarily satisfy a reciprocal relationship with respect to the precision with which these quantities can be specified in any possible experimental arrangement. More precisely, if a measuring instrument fixes the position q of a system to within a certain latitude Δq, then the momentum p can be fixed only to within a latitude Δp, where Δp is proportional to $1/\Delta q$ (the proportionality constant being a multiple of Planck's constant $\hbar$, which has the dimensions of an 'action': momentum $\times$ length, or energy $\times$ time). Now, this reciprocal relation between position and momentum might reflect a necessary disturbance of position values in any momentum measurement (and conversely), as required by the existence of a quantum of action that is non-negligible at the microlevel, or it might concern the very definition of these quantities. On Bohr's view, it follows from the finite value of $\hbar$ that the conditions for the precise applicability of the concept "position" exclude the conditions for the precise applicability of the concept "momentum." He terms the relationship between quantities like position and momentum "complementary," since precise values for both quantities are required for a complete specification of the classical state of the system.

For Bohr, the wave and particle pictures of the early quantum theory are complementary, in the sense that they involve different aspects of a single physical system revealed under mutually incompatible experimental conditions. Bohr argued that macrosystems *qua* measuring instruments are necessarily characterized in terms of the concepts of classical physics (like position and momentum). It follows that an interaction between a microsystem and a measuring instrument cannot be sharply separated from the undisturbed or independent behavior of the microsystem itself, because the classical description of the functioning of the measuring instrument entails that the unavoidably finite interaction of the microsystem with the measuring instrument cannot be controlled more accurately than is compatible with the uncertainty principle. For this reason, the interaction between a microsystem and a macroscopic measuring instrument forms an integrable part of the behavior of the microsystem as revealed in the interaction. Bohr used the term "phenomenon" to refer to an observation obtained under a complete specification of classically described experimental conditions, including an account of the whole experimental arrangement. A quantum phenomenon cannot be "subdivided" by a further experimental arrangement, because every such attempt would yield a new phenomenon incompatible with the original phenomenon. In this sense, a quantum phenomenon exhibits a feature of "wholeness" or "individuality." The concept of complementarity characterizes phenomena observed under mutually exclusive experimental conditions, such as the conditions for space-time coordination and the conditions for the applicability of dynamical conservation laws, both of which are required for the complete specification of a classical mechanical state.

Bohr regarded the concept of complementarity as a generalization of causality appropriate for the description of systems whose undisturbed behavior cannot be sharply separated from the interactions with the measuring instruments that reveal this behavior. From this standpoint, the deterministic description of classical physics characterizes the behavior of systems for which $\hbar$ can be neglected in measurement interactions. The irreducibly probabilistic character of the quantum-mechanical description reflects the fact that the behavior of microsystems can only be revealed in complementary phenomena constrained by the uncertainty principle . The principle of complementarity forms the core of Bohr's explanation for the origin of quantum probabilities as irreducible, on the basis of his analysis of quantum measurements. This "complementarity interpretation" influenced – but should be distinguished from – a loosely related variety of views known as the "Copenhagen interpretation," usually taken as the standard or orthodox interpretation of the theory.

See also: History of Physics; Philosophy of Physics; Quantum Mechanics; Uncertainty
    Principle.

Bibliography

M. Beller, *Quantum Dialogue: The Making of a Revolution.* Chicago University Press, Chicago, 1999.
    (I)

D. Bohm, "On Bohr's Views Concerning the Quantum Theory," in *Quantum Theory and Beyond* (T.
    Bastin, ed.). Cambridge, London and New York, 1971. (I)

N. Bohr, (a) *Atomic Theory and the Description of Nature.* Cambridge, London and New York, 1934.
    (I) (b) *Atomic Physics and Human Knowledge.* Wiley, New York, 1958. (I) (c) *Essays 1958–1962
    on Atomic Physics and Human Knowledge.* Wiley (Interscience), New York, 1963. (I)

J. Faye, *Niels Bohr: His Heritage and Legacy.* Kluwer, Dordrecht, 1991. (I)

J. Faye and H.J. Folse (eds.), *Niels Bohr and Contemporary Philosophy.* Kluwer, Dordrecht, 1994. (A)

P. K. Feyerabend, (a) "Complementarity," *Aristotelian Soc. Suppl.* **32**, 75–104, (1958). (I) (b) "Niels
    Bohr's Interpretation of the Quantum Theory," in *Current Issues in the Philosophy of Science* (H.
    Feigl and G. Maxwell, eds.). Holt, Rinehart & Winston, New York, 1961. (E) (c) "On a Recent
    Critique of Complementarity," Parts I and II, *Philos. Sci.* **35**, 309–331 (1968); **36**, 82–105 (1969).
    (A)

H. J. Folse, *The Philosophy of Niels Bohr.* North-Holland, Amsterdam, 1985. (I)

W. Heisenberg, "Quantum Theory and Its Interpretation," in *Niels Bohr* (S. Rozental, ed.). North-
    Holland, Amsterdam, 1967. (E)

D. Howard, "Who Invented the Copenhagen Interpretation? A Study in Mythology," in *PSA 2002* Part
    II, *Symposium Papers.* Proceeding of the 2002 Biennial Meeting of the Philosophy of Science
    Association, Milwaukee, Wisconsin, November 7–9, 2002. Forthcoming in a special issue of
    Philosophy of Science **71**. (I)

M. Jammer, *The Conceptual Development of Quantum Mechanics.* McGraw–Hill, New York, 1966. (I)

A. Petersen, *Quantum Physics and the Philosophical Tradition.* M.I.T. Press, Cambridge, Mass., 1968.
    (E)

E. Scheibe, *The Logical Analysis of Quantum Mechanics.* Pergamon, New York, 1973. (A)

Complex Systems

H.-G. Schuster

The word complex derives from the Latin "complexus" and means a whole made up of many interacting interwoven parts. This describes for example our intuitive picture of the human brain. The parts are the neurons which are connected in an intricate interwoven way by synapses. So the brain is certainly the most complex system of human scales. Some other examples of complex systems are the interwoven chemical reactions involved in a cell, the many different reaction schemes of the immune system, economical systems with their mutual interdependencies and the information transport in the world wide web (see Table 1).

Table 1: Examples of complex adaptive systems.

Prebiotic evolution	Eigen [5]
Darwinian evolution	Maynard Smith and Szamary [10]
Chemical networks	Kauffman [9]
Insect colonies	Bonabeau, Dorigo and Thoreau [4]
Immune system	Segel [14]
Nervous system	Kandel and Squire [12]
Economic networks	Brian Arthur, Durlauf and Lane [3]
Social networks	Frank [6]
Communication networks	Albert, Jeong and Barabasi [1]
Transportation networks	Narguney [11]
Evolutionary games	Hofbauer and Sigmund [8]

All complex systems have a basic common structure. They are made up of a network of elements or agents such as molecules, genes, neurons or players in a game which interact with each other and with their environment in a nonlinear fashion. These nonlinear interactions imply that a complex system is more than the sum of its constituents. It can display emergent properties like waves which are not a property of a single particle but of the network of water molecules. The structural and dynamical properties of complex systems are typically hard to predict because there are many connected agents which influence each other by feedback loops [13].

A point in case is the traveling-salesman problem. Here, the positions of N cities in a plane are given and the task is to predict, i. e., to compute, the shortest path which starts at a given city, passes through all other cities and returns to the starting point. There are of order $N!$ paths from which the shortest one has to be selected. A sequential check of all these paths would require a time which which increases like $N!$, i. e., the traveling salesman problem has a high computational complexity. More about the notion of computational complexity, including many examples of hard computational problems, can be found in [7].

If we say that a system is complex we intuitively mean that its number of basic constituents is large. A system which consists of one neuron can be considered as "simple" as long as we describe it on the functional level as a switching element which fires an impulse if it receives a sufficiently large input current. A closer look shows that it contains many constituents such as different ion channels. Therefore, measuring complexity by the number of constituents depends on the level of coarse graining that we adopt when we are defining these elements.

Many properties of complex systems like their ability to transport information or their resistance to damage depend of the architecture of the underlying network. It has recently been found that many real networks are scale-free, which means that the probability $P(k)$ that an arbitrary chosen node is connected to k others decays with a power law $P(k) \sim k^{-\gamma}$ where $\gamma \lesssim 4$ [1]. This is due to the fact that these nets grow with preferential attachment, e. g., we link with a higher probability to a more connected document in the world wide web. Whereas in randomly connected networks most nodes have approximately the same number of links, scale-free nets are inhomogeneous displaying "hubs", i. e., nodes with a very large number of links (see Fig. 1). These multiply connected hubs make the transport in these networks less

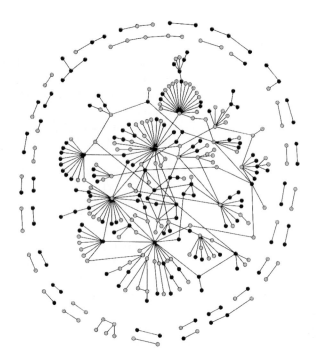

Fig. 1: Network of physical interactions between nuclear proteins in yeast. Note the highly connected "hubs" (S. Maslov, K. Sneppen and U. Alon in [2].

vulnerable to damage because removal of a few nodes or links still preserves longe-ranged connections. Such a resistance against damage is most important for natural networks such as biological or ecological nets.

A better understanding of complex systems implies better and new possibilities for their control, prediction and improvement, i. e., very practical aspects whose use and far reaching consequences for the understanding of biological and economical systems, for the stable flow of traffic or the reliable transport of information in large-scale communication networks, are becoming more and more realized [2].

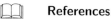

References
[1]   R. Albert, H. Jeong and A. L. Barabasi, "Error and Attack Tolerance of Complex Networks", *Nature* **406**, 378 (2000).

[2]   S. Bornholdt and H. G. Schuster (eds.), *Handbook of Graphs and Networks*. Wiley-VCH, Weinheim, 2003.

[3]   W. Brian Arthur, S. N. Durlauf and D. Lane (eds.) *The Economy as an Evolving Complex System II*. Addison Wesley, 1997.

[4]   E. Bonabeau, M. Dorigo and G. Theraulaz, *Swarm Intelligence: From Natural to Artificial Systems*. Oxford University Press, Oxford, 1999.

[5]   M. Eigen, "The Physics of Molecular Evolution", in *Molecular Evolution of Life*, H. Baltscheffsky, H. Jörnvall and R. Rigler (eds.), p. 13. Published in *Chemica Scripta* **26B** (1986).

[6]   S. A. Frank, *Foundations of Social Evolution*. Princeton University Press, Princeton, 1998.

[7]   M. R. Garey and D. S. Johnson, *Computers and Intractability: A Guide to the Theory of NP-Completeness*. W. H. Freeman, San Francisco, 1979.

[8]   J. Hofbauer and K. Sigmund, *Evolutionary Games and Population Dynamics*. Cambridge University Press, Cambridge, 1998.

[9]   S. A. Kauffman, *The Origins of Order*. Oxford University Press, New York, 1993.

[10] J. Maynard Smith and E. Szathmary, *The Major Transitions in Evolution*. W. H. Freeman, Oxford, New York, Heidelberg, 1995.

[11] A. Narguney, *Sustainable Transportation Networks*. Edward Elgar Publishing, Cheltenham, 2000.

[12] E. R. Kandel and L. R. Squire, "Neuroscience: Breaking down Scientific Barriers to the Study of Brain and Mind", *Science* **290**, 1113 (2000).

[13] H. G. Schuster, *Complex Adaptive Systems*. Scator Verlag, Saarbruecken, 1992.

[14] L. A. Segel and I. Cohen (eds.), *Design Principles for the Immune System and Other Distributed Autonomous Systems*, Oxford University Press, Oxford, 2000.

[15] K. Sigmund, *Games of Life*. Oxford University Press, Oxford, 1993.

Compton Effect

K. Berkelman

The elastic scattering of a photon by an electron is called the Compton effect when the interaction can be considered as the collision of two otherwise free particles. This is true when the energy $h\nu$ of the photon is comparable to or higher than the rest energy $m_e c^2$ of the electron (0.51 MeV).

The scattering of electromagnetic radiation of lower quantum energies by electrons was explained by J. J. Thomson in terms of the classical oscillatory motion of the electron in the incident electromagnetic wave field and the consequent reradiation of the absorbed energy. In this picture the scattered radiation has the same frequency as that of the motion of the electron, which in turn is equal to the frequency of the incident wave.

However, in 1923 A. H. Compton observed x rays scattered at various angles in thin targets of light elements and noted that the scattered wave lengths were longer than the incident wavelength and increasing with scattering angle. This was in fact one of the decisive early demonstrations of the quantum theory. For if we treat the x rays as particles (called photons) having energy $h\nu$ and momentum $h\nu/c$, and apply energy and momentum conservation to the photon–electron collision (using relativistic kinematics), we can solve for the final photon energy:

$$h\nu' = \frac{m_e c^2}{1 + \cos\theta + (m_e c^2 / h\nu)} \ ,$$

which is reduced from the incident energy by the energy of electron recoil. The corresponding wavelength shift

$$\lambda' - \lambda = \left(\frac{h}{m_e c}\right)(1 - \cos\theta)$$

reproduces Compton's observed dependence on scattering angle. The quantity $h/m_e c$ has been named the "Compton wavelength of the electron." Note that in the low-energy limit, in which

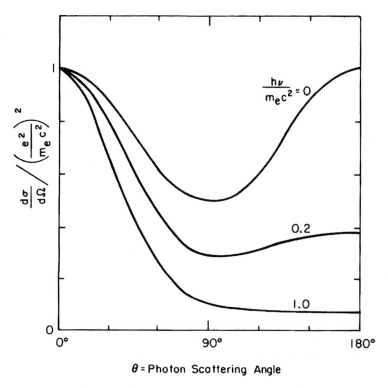

Fig. 1: The differential cross section for Compton scattering of unpolarized photons, expressed in terms of $(e^2/m_e c^2)^2 = 7.95 \times 10^{-26} \, \text{cm}^2$, plotted as a function of the photon scattering angle for several values of the incident photon energy $h\nu$ divided by the electron rest energy $m_e c^2 = 0.51 \, \text{MeV}$.

$h\nu \ll m_e c^2$ (or $\lambda = c/\nu \gg h/m_e c$ in terms of wavelength), the classical Thomson relation between incident and scattered frequencies (or wavelengths) holds. Compton's experiment is a beautiful example of the wave–particle duality in quantum mechanics. The particle nature of electromagnetic radiation is demonstrated by wavelength measurements.

In one of the earliest applications of relativistic quantum mechanics, Klein and Nishina calculated in 1928 the differential cross section (i. e., the probability of scattering through an angle θ into unit solid angle, in a target of unit thickness containing one electron per unit volume) for Compton scattering:

$$\frac{d\sigma}{d\Omega} = \frac{r_e^2}{2} \left(\frac{\nu'}{\nu}\right)^2 \left[\left(\frac{\nu}{\nu'} + \frac{\nu'}{\nu} - \sin^2\theta\right) \pm \left(\frac{\nu}{\nu'} - \frac{\nu'}{\nu}\right) \cos\theta \right] .$$

Here $r_e = e^2/m_e c^2 = 2.82 \times 10^{-13} \, \text{cm}^2$, and is called the classical radius of the electron. The $\pm$ on the last term apply if the initial photons are right or left circularly polarized, respectively. The differential cross section for scattering of unpolarized photons (plotted in Fig. 1) is obtained by omitting the $\pm$ term. At low energies $h\nu$ it becomes the classical Thomson cross

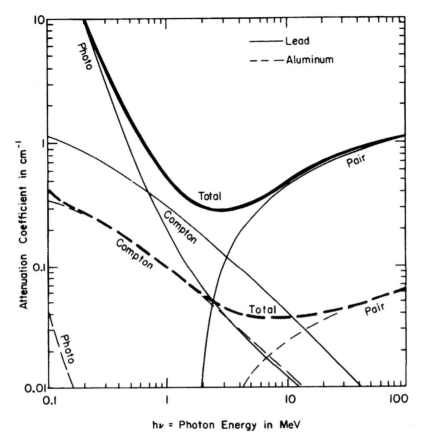

Fig. 2: The attenuation coefficient μ for a typical light and heavy element, plotted as a function of incident photon energy $h\nu$.

section $\frac{1}{2}r_e^2(1+\cos^2\theta)$, while at higher energies the scattering is strongly peaked forward. For plane-polarized photons the differential cross section is

$$\frac{d\sigma}{d\Omega} = \frac{r_e^2}{2}\left(\frac{\nu'}{\nu}\right)^2\left(\frac{\nu}{\nu'}+\frac{\nu'}{\nu}-2\sin^2\theta\cos^2\phi\right)$$

where ϕ is the angle between the plane of the incident electric vector and the scattering plane. The yield is a maximum when the photon scatters in the plane normal to the polarization plane. The total cross section, integrated over all angles, decreases from the classical Thomson value, $(8\pi/3)r_e^2 = 0.84 \times 10^{-24}\,\mathrm{cm}^2$, monotonically as the photon energy is increased. The probability that a photon will survive without interaction after passing through x cm of material is $e^{-\mu x}$ where the attenuation coefficient μ is the product of the total cross section and the number of electrons per unit volume of material. In Fig. 2 is plotted the total attenuation coefficient for aluminum and lead, including the effects of the photoelectric effect, which dominates at low energies $h\nu$, and of electron–positron pair production, important at high energies. Compton scattering is the dominant photon attenuation process only in materials of low atomic number Z and for energies $h\nu$ near 1 MeV. Since the scattering probability per atom is proportional to number of electrons Z, it becomes overshadowed by photoelectric absorption, proportional to Z^5, and pair production, proportional to Z^2, as Z becomes large.

The elastic scattering of photons from charged particles other than the electron is often called Compton scattering also. The kinematic relations between incident and scattered photon energies and scattering angle are changed only by substituting the appropriate particle mass for the electron mass. For photons of low energy the proton Compton scattering cross section is again the Thomson cross section, reduced, however, by the square of the electron–proton mass ratio. At energies above the pion photoproduction threshold (about 150 MeV) proton Compton scattering occurs mainly through the excitation and radiative decay of the various nucleon resonance states observed in pion-proton scattering and pion photoproduction. Above 1 GeV, proton Compton scattering becomes diffractive and is understood in terms of intermediate states involving vector mesons, such as the ρ^0, ω, and ϕ.

See also: Scattering Theory.

Bibliography

An elementary discussion of the Compton effect is given in every introductory modern physics text; see, for example, H. Semat and J. R. Albright, *Introduction to Atomic and Nuclear Physics*, 5th ed., pp. 143–149. Holt, Rinehart & Winston, New York, 1972. (E)

R. D. Evans, *The Atomic Nucleus*, Chapter 23. McGraw–Hill, New York, 1955. (I)

The relativistic quantum-mechanical description of the Compton effect is discussed in many advanced texts; see, for example, J. D. Bjorken and S. D. Drell, *Relativistic Quantum Mechanics*, pp. 127–132. McGraw–Hill, New York, 1964. (A)

R. D. Evans, "Compton Effect," in *Encyclopedia of Physics* (S. Flügge, ed.), Vol. 34, pp. 234–298. Springer-Verlag, Berlin and New York 1958. (A)

Conduction

S. Kirkpatrick and C. J. Lobb

Conduction processes in condensed matter consist of the transport of heat, electric charge, mass, or some combination of the three, in response to an imposed temperature gradient, electric field, or density gradient. (Convection, distinguished among conduction processes by the fact that macroscopic motion of all the atoms or molecules occurs, also contributes to transport in fluids, but will not be discussed here.) Conductivities vary more widely among materials than do other physical properties. While the densities of the liquids and solids which form at room temperature vary over roughly one decade, electrical conductivities vary by 20 decadees or more from metals to insulators, and may be infinite in superconductors at sufficiently low temperatures. At cryogenic temperatures, thermal conductivities also exhibit a range of many decades, with metals offering the highest thermal conductivities, and insulators the lowest.

The usefulness of specific materials is very often due to their high or low conductivities. Discovery of new materials with extreme conductivities frequently opens novel applications, e. g., superconducting magnets for producing very large magnetic fields. But here we shall be concerned with conduction, especially at low temperatures where the largest variations are observed, as a probe of the microscopic particle-like excitations which carry the heat, charge, or mass currents. These particles are most commonly phonons (long-wavelength mechanical vibrations) or electrons, but ions may contribute (*see* Electrochemistry) and Cooper pairs carry the lossless current in superconductors (*see* Superconductivity Theory).

A semiclassical description, first employed in the kinetic theory of gases, is convenient for discussing conduction by electrons or ions. A mobility μ is defined as the mean carrier drift velocity induced by a unit applied force. For electrons,

$$\mu \equiv \langle v \rangle / e\mathcal{E} = \tau / m^* ,$$

where τ is a relaxation time (10^{-14} s is typical for electrons in metals at room temperature), and m^* the effective mass of an electron moving in the attractive potential of the ions in the material. The conductivity σ is obtained from the associated electric current $j^{(e)}$ by

$$\sigma \equiv j^{(e)} / \mathcal{E} = ne^2 \tau / m ,$$

with n being the density of carriers and e being the electronic charge. Diffusion processes, in which a particle current $j^{(d)}$ is induced by a density gradient, are characterized by a diffusion coefficient D:

$$j^{(d)} = -D\nabla n .$$

The two types of process are intimately connected. The Einstein relation,

$$\mu = eD / k_{\mathrm{B}} T ,$$

holds for any particles which obey classical Maxwell–Boltzmann statistics, for example, electrons and holes in semiconductors, or ions in electrolytes.

The thermal conductivity κ is defined from the heat current j^q which flows in response to a temperature gradient:

$$j^{(q)} \equiv -\kappa \nabla T .$$

A useful expression for κ is

$$\kappa = \frac{1}{3} C_V v^2 \tau ,$$

where C_V is the specific heat at constant volume, and v is the instantaneous velocity of the particle. For electrons, v is the Fermi velocity; for phonons, the speed of sound is appropriate in this expression.

Both n and τ can vary widely with temperature and from one material to another. In insulators, n vanishes as the temperature tends to zero. If E is the excitation energy required to create a free carrier, then n is proportional to $\exp(-E/k_{\mathrm{B}}T)$. "Arrhenius plots" of $\ln(\sigma)$ against $1/T$ yield straight lines for most semiconductors, and provide a means of determining E. Heat is transported principally by the phonons in insulators. These scatter only from defects or sample boundaries at low temperatures, so in this limit $\kappa \propto C_V \propto T^3$ is observed.

In metals, n and v are not sensitive to temperature, but τ can be. Phonon scattering alone gives $\tau \propto T^{-5}$, while scattering from impurities gives a constant contribution to τ. As a result, in a nonsuperconducting metal σ tends to a constant at low temperatures while $\kappa \propto T$.

In noncrystalline solids, such as insulating or semiconducting glasses, localized electronic and vibrational excitations introduce novel transport properties. The dominant electrical conduction mechanism at low temperatures may involve "hopping" of carriers between spatially separated long-lived states, assisted by the absorption of phonons. The characteristic distance and phonon energy involved in hopping change in a complicated way with decreasing temperature. The result, first explained by N. F. Mott, is

$$\ln \sigma \propto T^{-1/4} \, .$$

Localized vibrations give rise to a phonon scattering time $\propto T^{-1}$ in glasses. As a result, κ is proportional to T^2 at the lowest temperatures.

One additional idea should be mentioned which is not included in the classical theory of electronic conduction. At cryogenic temperatures, electrons maintain their quantum-mechanical phase coherence over longer and longer distances as inelastic scattering becomes less frequent. The resulting quantum-mechanical interference causes a number of interesting effects. One of these effects is localization: the backscattering rate is effectively enhanced due to interference, leading to a reduced o as the temperature is lowered. Another is the occurrence of quantum conductance fluctuations: when the magnetic field applied to a sample is changed, the phases of the wave functions are shifted, altering the amount of interference. This leads to noisy, seemingly random variations in σ as the field is varied, except that the noise reproduces itself as the magnetic field is swept back and forth.

See also: Diffusion; Electrochemistry; Insulators; Metal–Insulator Transitions; Metals; Photoconductivity; Semiconductors, Crystalline; Superconductivity Theory; Transport Properties; Tunneling.

Bibliography

A good short discussion of transport is given by C. Kittel and H. Kroemer in *Thermal Physics*, 2nd ed., pp. 397–406, W. H. Freeman and Company, New York, 1980. A recent review on localization is Gerd Bergman, "Weak Localization in Thin Films, a Time of Flight Experiment with Conduction Electrons," *Phys. Rep.* **107**, 58 (1984). Quantum conductance fluctuations are reviewed in Sean Washburn and Richard A. Webb, "Aharonov–Bohm Effect in Normal Metal Quantum Coherence and Transport," *Adv. Phys.* **35**, 375 (1986).

Conservation Laws

L. Wolfenstein

A conservation law equates the value of a physical quantity in the initial state to its subsequent values, in particular to its value in the final state, for some process. The great importance of conservation laws is that they provide significant constraints on complicated processes for which a detailed mathematical description may be practically impossible.

The simplest form of a conservation law involves scalar quantities. The conservation of the sum of the kinetic energies and potential energies of a set of interacting particles provides an extremely useful integral of the motion in classical mechanics. The basic discovery leading to modern chemistry in the 1800s was the discovery of the conservation of mass, equating the total mass of the reagents to that of the products for a chemical reaction. The theory of special relativity requires a modification of these two conservation principles because of the possible interconversion of mass and energy. Although this modification is quantitatively insignificant for chemical reactions, it is of essential importance in nuclear and elementary-particle reactions. The law appropriate for these reactions is the *conservation of the relativistic energy*, sometimes called mass-energy:

$$\sum_i (T_i + m_i c^2) = \sum_f (T_f + m_f c^2) \,, \tag{1}$$

where T_i and m_i are the kinetic energy and rest mass of an initial particle and T_f and m_f are for a final particle.

As a simple application, consider the photodisintegration of the deuteron into a neutron and a proton,

$$\gamma + d \rightarrow n + p \,,$$

where Eq. (1) states (for a deuteron at rest) that

$$E_\gamma + M_d c^2 = M_n c^2 + M_p c^2 + T_n + T_p \,. \tag{2}$$

The minimum γ-ray energy E_γ for this reaction is 2.3 MeV, which corresponds to the fact that the mass of the neutron plus that of the proton is greater than that of the deuteron. Note that the potential energy between the neutron and proton and their kinetic energy inside the deuteron are not included in this equation because the deuteron is considered a single particle. An alternative approach is to consider the deuteron a composite system bound by nuclear forces. In this approach we can write

$$M_d c^2 = M_n c^2 + M_p c^2 + \langle V_{np} \rangle + \langle T_n \rangle + \langle T_p \rangle \,, \tag{3}$$

where $\langle V_{np} \rangle$ is the average value of the potential energy associated with the nuclear force inside the deuteron and $\langle T_n \rangle$, $\langle T_p \rangle$ are average values of the kinetic energies inside the deuteron.

In elementary-particle physics there are much more dramatic examples of the conversion of mass into energy. In the annihilation of a positron (e^+) by an electron (e^-), the total mass is converted into the energy of gamma rays:

$$e^+ + e^- \rightarrow \gamma + \gamma \,,$$

so that for annihilation at rest each γ-ray is emitted with the energy $m_e c^2 = 0.51$ MeV.

Every elementary particle, and therefore every system of particles, can be characterized by a set of numbers, often referred to as quantum numbers. The most familiar of these are charge Q, and baryon number B. For a set of particles the value of one of these numbers is the sum of the values for the members of the set. In any physical process each of these numbers is conserved. Since antiparticles can combine with particles to produce photons (which have a zero value for each of these numbers), these quantum numbers for an antiparticle must be the negative of their values for a particle.

The constituents of ordinary matter, proton, neutron, and electron, have electric charges Q of $+1$, 0, and -1. The conservation of electric charge states that the algebraic sum of the charges of all the particles does not change during a reaction. A very intriguing aspect of charge conservation is that it is the electric charge that determines the strength of the interaction of the particle with the electromagnetic field. In terms of Q this strength is given by Qe where $e = 1.5 \times 10^{-19}$ C. This suggests a fundamental relation between a conserved quantity and a particular form of interaction. This relation is embodied in the concept of gauge invariance in field theory.

The reason that only a small fraction ($< 1\%$) of the mass is converted into energy in nuclear reactions is the conservation of the number of nucleons (neutrons plus protons) in interactions involving normal matter. The more general form of this law is the *conservation of baryon number B*; this law states that the number of baryons minus the number of antibaryons is conserved in any reactions. A baryon is an elementary particle with $B = 1$, such as the neutron or proton or the strange baryons Λ, Σ, Ξ and Ω. An antibaryon is the antiparticle of one of these, such as the antineutron or antiproton. According to the quark model the elementary constituents of baryons are quarks with $B = \frac{1}{3}$ and a baryon is made of three quarks. As an example, consider a possible annihilation of an antiproton ($\bar{\text{p}}$) by a deuteron

$$\bar{\text{p}} + \text{d} \rightarrow \bar{\Lambda} + \Sigma^{+} + \text{n} + \pi^{-}\ .$$

The total baryon number B on both sides is $+1$. Note that $\bar{\Lambda}$ and $\bar{\text{p}}$ are antibaryons, the deuteron has baryon number $B = 2$, and the π^{-} is a meson with $B = 0$. It is the conservation of baryon number that is responsible for the stability of normal matter.

Many extensions of the standard model of weak interactions, particularly grand unified theories, allow for nonconservation of baryon number. This has led to an extensive unsuccessful search for proton and neutron decay. Lower limits on $\tau_{\text{p}}/(\text{BR})$ of the order 10^{32} years exist for a number of possible decay modes ($\text{e}^{+}\pi^{0}$, $\mu^{+}\pi^{0}$, $\text{e}^{+}K^{0}$) where τ_{p} is the proton lifetime and (BR) is the branching ratio to these modes.

In ordinary reactions involving normal matter the number of electrons is also conserved. The conservation of the number of electrons plus the number of nucleons may be considered as the basis for the conservation of mass in chemistry.

Analogous to the conservation of baryon number there exists a conservation of electron number L_{e}. In electrodynamic processes this law states that electrons cannot be created or destroyed except as e^{+}–e^{-} pairs. When weak interactions are included, however, this law must be generalized to include the electron-type neutrino ν_{e}, and antineutrino $\bar{\nu}_{\text{e}}$. Then ν_{e} and e^{-} have $L_{\text{e}} = +1$; $\bar{\nu}_{\text{e}}$ and e^{+} have $L_{\text{e}} = -1$. In nuclear beta decay, for example, either an e^{-}–$\bar{\nu}_{\text{e}}$ or e^{+}–ν_{e} pair is emitted. A similar law is the *conservation of muon number* L_{μ}, for which the set (e^{-} e^{+} ν_{e} $\bar{\nu}_{\text{e}}$) is replaced by (μ^{-} μ^{+} ν_{μ} $\bar{\nu}_{\mu}$). This is illustrated in the decay of the muon:

$$\mu^{-} \longrightarrow \text{e}^{-} + \bar{\nu}_{\text{e}} + \nu_{\mu}.$$

This law can also be extended to the τ lepton and its neutrinos ν_{τ} giving the *conservation of* L_{τ}. The conservation of $L_{\text{e}} + L_{\mu} + L_{\tau}$ is called the *conservation of lepton number*. If we label the lepton number L the conservation of $B - L$ may be called *fermion number conservation*.

While L_e, L_μ, L_τ are separately conserved in weak interactions there is now strong evidence for their violation in the propagation of neutrinos, a phenomenon known as *neutrino oscillations*. The ν_e emitted by the sun appear to transform, or oscillate, partially into ν_μ and ν_τ as they propagate out from the center of the sun. Almost half of the ν_μ in cosmic rays disappear after propagating thousands of kilometers and are believed to oscillate into ν_τ. These oscillations are believed to be associated with the small neutrino mass and it is the neutrino mass matrix that violates the separate lepton numbers. No evidence for such violations has been found for charged leptons, although there are continuing searches for decays such as $\mu \longrightarrow e + \gamma$.

The best test for the conservation of the total lepton number L is the search for neutrinoless double beta decay, a process in which a nucleus emits two electrons but no neutrinos which would indicate a change of two units in lepton number ($\Delta L = 2$). Experiments provide a lower limit of about 10^{25} years for the half-life of ^{76}Ge for this decay mode. If $\Delta L = 2$ is not exactly forbidden, it is expected that neutrinos would acquire a nonzero mass of the Majorana type. This is the most popular explanation of the non-zero masses implied by neutrino oscillations; however, there is no way known to prove this other than the observation of neutrinoless double beta decay.

The fact that K mesons and Λ and Σ baryons were produced in strong interactions but decayed only weakly led to their designation as strange particles. In the strong interactions these are produced in pairs, typically K^+ or K^0 mesons together with Λ or Σ. This leads to the law of *conservation of strangeness*. Each particle is assigned an integer, analogous to charge, called strangeness S: K^+ and K^0 have strangeness $S = +1$, Λ and Σ have $S = -1$, while "normal" particles like nucleons and pions have $S = 0$. The sum of the strangeness numbers is conserved; thus in a reaction such as $\pi^- + p \rightarrow K^0 + \Lambda$ the sum equals zero at the beginning and at the end. Strangeness is conserved in both the strong and electromagnetic interactions but not in weak interactions. As a consequence, strange particles decay into normal particles by means of the weak interaction. This is an example of a conservation law that is not exact but holds only for a class of interactions. Another example is isospin (q.v.) conservation that holds only for the strong interactions but not for electromagnetic or weak. Instead of strangeness S we often use hypercharge Y defined as $S + B$, where B is the baryon number, which is separately conserved.

Each conserved quantity discussed so far is *additive* in the sense that its value for a set of particles equals the sum of the values for the individual particles. In contrast, parity (q.v.) is *multiplicative*: the parity of a set of particles is the product of the orbital parity times the intrinsic parities of each of the particles. The parity is plus or minus one, depending on the behavior of the quantum-mechanical state under inversion of the spatial coordinates. For a pair of particles the orbital parity is $(-1)^l$ where l is the relative orbital angular momentum in units of $\hbar$. For a composite system the intrinsic parity may simply represent the orbital parity of its components; for example, the parity of the ground state of ^{7}Li is negative because the valence neutron is in a p state ($l = 1$). On the other hand, for elementary particles that can be created or destroyed, the intrinsic parity is a basic characteristic of the particle. The intrinsic parity of the antiparticle of a fermion is opposite to that of the particle. Mesons may have either positive or negative intrinsic parity; the lightest mesons (π and K) have been found to have negative parity. This is understood in the quark model because π and K consist of a quark and an antiquark in an $l = 0$ state. Similarly the ground state of positronium has odd parity.

Parity conservation does not hold for weak-interaction processes, such as beta decay. In a weak decay of a particle like the neutron, which has a well-defined parity, the final state is a mixture of both even and odd parities. This shows up in observations sensitive to interference between the final states of even and odd parity. Such interference effects show up as nonzero expectation values of pseudoscalar quantities. For example, in beta decay the electrons emerge predominantly left-handed, corresponding to a nonzero expectation value of $\boldsymbol{\sigma} \cdot \mathbf{p}$, where $\boldsymbol{\sigma}$ is the spin vector and $\mathbf{p}$ the momentum vector of the electron. Since $\boldsymbol{\sigma}$ is an axial vector and $\mathbf{p}$ is a polar vector, $\boldsymbol{\sigma} \cdot \mathbf{p}$ is a pseudoscalar. The parity nonconservation in the weak interactions is not a small effect but appears to be as large as possible.

It is believed that parity conservation holds for the strong and electromagnetic interactions. However, processes governed by these interactions may demonstrate a small amount of parity nonconservation as a result of the perturbation by the weak interactions. This has been demonstrated by observations of admixtures of electric dipole (odd parity) with magnetic dipole (even parity) in certain nuclear gamma-ray transitions (^{181}Ta, ^{175}Lu). The admixture results in a small net circular polarization of the emitted gamma ray.

For every system there exists a system related to it by changing every particle into its antiparticle. For some systems the resulting system is identical to the original system except possibly for a multiplicative factor C equal to plus or minus one. For example, the positronium atom made up of a positron and an electron has the value $C = +1$ for the $^1S_0$ state and $C = -1$ for the $^3S_1$ state. Some particles are also their own antiparticles with definite C values: for example, $C = -1$ for a photon and $C = +1$ for a neutral pion π^0. C is called the *charge conjugation* (or the particle–antiparticle conjugation) *quantum number*. It is believed that the conservation of C holds for strong and electromagnetic interactions, although the evidence is much less precise than for parity conservation. For example, the state of n photons has $C = (-1)^n$, so that the $^1S_0$ state of positronium decays into an even number of photons but the $^3S_1$ state must decay into an odd number.

The weak interactions that do not conserve parity P also do not conserve charge conjugation C. For example, the decays $\pi^+ \rightarrow \mu^+ \nu_\mu$ and $\pi^- \rightarrow \mu^- \bar{\nu}_\mu$ are related by C, but in these decays the μ^+ is emitted with left-handed polarization and the μ^- with right. The V–A theory developed in 1957 yielded the result that both C and P are not conserved in weak interactions but that the product CP is conserved. The failure of CP conservation was discovered in 1964 from a study of K^0 decays. The antiparticle of the K^0 meson ($S = 1$) is $\bar{K}^0$ ($S = -1$). Since strangeness is not exactly conserved the states with definite mass and lifetimes are linear combinations of K^0 and $\bar{K}^0$, called K_S and K_L. If CP is conserved, then K_S and K_L should be CP eigenstates: $K_S = (K^0 - \bar{K}^0)/\sqrt{2}$ ($CP = +1$) and $K_L = (K^0 + \bar{K}^0)/\sqrt{2}$ ($CP = -1$). We find, in accordance with this expectation, that the major decay modes of K_S are $\pi^+\pi^-$ and $2\pi^0$ ($C = +1$, $CP = +1$), whereas an important decay mode of K_L is $3\pi^0$ ($C = +1$, $CP = -1$). These decays are consistent with CP conservation and also demonstrate nonconservation of C. However, it was discovered that K_L has as a rare decay mode the 2π state with $CP = +1$. Thus the conservation of CP is not an exact law, although it appears to be much better than conservation of C or P separately.

The interaction responsible for the nonconservation of CP has not been identified. Within the standard Weinberg–Salam gauge theory of weak interactions it is possible to incorporate CP violation by means of a mechanism suggested by Kobayashi and Maskawa.

A consequence of this theory is that there should be a large *CP* violation in the decay of *B* mesons. This prediction has now been verified, but further experiments are needed to test this theory.

It follows from the *CPT* theorem (q.v.) that the interaction responsible for *CP* noninvariance must also violate time-reversal invariance. The most sensitive search for a failure of time reversal invariance outside of the *K* system are experiments attempting to find an electric dipole moment of the neutron.

Certain conservation laws relate vector quantities. The *conservation of momentum* requires that the vector sum of the momenta of a set of interacting particles remain constant during any process. In the theory of special relativity, momentum and energy form a four-vector so that the *conservation of the energy–momentum four-vector* defines one covariant law. The *conservation of angular momentum* (q.v.) requires that the vector sum of the angular momenta of a set of particles interacting by means of a rotationally invariant force law be conserved. In quantum mechanics it is in general impossible to specify all three components of the angular momentum of a system since the different components are represented by noncommuting operators. Thus effectively in quantum mechanics we use the conservation of the square of the angular momentum J^2 and one component J_z. In calculating the angular momentum of a set of particles it is necessary to add vectorially the orbital angular momenta plus the spin angular momenta of all the particles.

Conservation laws are closely related to invariance principles (q.v.). In quantum mechanics invariance principles are stated in terms of a unitary transform U, which leaves the Hamiltonian invariant:

$$UHU^{-1} = H .\tag{4}$$

If U is a continuous transformation, it is convenient to consider an infinitesimal form of U

$$U = 1 + i\varepsilon F ,$$

where ε is an infinitesimal parameter and F is a Hermitian operator. From Eq. (4) to order ε it is required that

$$FH - HF = 0 .\tag{5}$$

The fact that F commutes with H guarantees that it is conserved, since it is H that governs the time development of a quantum-mechanical system. The eigenvalues of the operator F thus are conserved quantities. In the case of a discrete transformation such as space reflection, we can define an operator $\mathcal{P}$, which is both unitary and Hermitian, so that $\mathcal{P}$ plays the role of both U in Eq. (4) and F in Eq. (5). The eigenvalue of $\mathcal{P}$ is the conserved quantity called parity P. It is possible to define a kind of gauge invariance [often referred to as a $U(1)$ group] to be associated with each of the conserved additive numbers Q, B, L_e, L_μ L_τ and S. However, although in the case of the original gauge invariance associated with Q there is a profound relation between this invariance and the law of electromagnetic interactions, the significance of the other gauge invariances is not clear.

Table 1 contains a list of conserved quantities. For each there is indicated whether it is additive or multiplicative, the range of validity, and the invariance principle with which it is associated.

Table 1: Summary of conserved quantities.

Quantity	Symbol	Type	Interactions for which it is not conserved	Associated invariance principle
Energy	E	Additive	None	Time translation
Momentum	**p**	Additive vector	None	Space translation
Angular momentum	**J**	Additive vector	None	Space rotation
Parity	P	Multiplicative	Weak	Space inversion
Charge conjugation	C	Multiplicative	Weak	Charge conjugation
CP	CP	Multiplicative	Weak	CP; time reversal
Charge	Q	Additive	None	Gauge invariance
Baryon number	B	Additive	None	$U(1)$
Electron number	L_e	Additive	Neutrino Mass	$U(1)$
Muon number	L_μ	Additive	Neutrino Mass	$U(1)$
Strangeness	S	Additive	Weak	$U(1)$
Hypercharge	Y	Additive	Weak	$U(1)$
Isospin	I	Additive vector	Electromagnetic; weak	$SU(2)$

See also: *CPT* Theorem; Dynamics, Analytical; Invariance Principles; Isospin; Kinematics and Kinetics; Momentum; Parity.

Bibliography
Classical Conservation Laws
G. Holton and S. G. Brush, *Physics, The Human Adventure*, pp. 201–261. Rutgers, New Brunswick, NJ, 2001. (E)

L. D. Landau and E. M. Lifschitz, *Mechanics*. Pergamon, New York, 1960. (I)

Conservation Laws in Quantum Physics
L. I. Schiff, *Quantum Mechanics*, 4th ed., Chapter 7. McGraw–Hill, New York, 1968. (A)

R. P. Feynman, R. B. Leighton, and M. Sands, *The Feynman Lectures on Physics*, Vol. III, Chapters 17, 18, and 20. Addison-Wesley, Reading, Mass., 1965. (I)

Conservation Laws in Elementary-Particle Physics
R. P. Feynman, *The Character of Physical Law*, Chapters 3 and 4. MIT Press, Cambridge, Mass., 1965. (E)

H. Frauenfelder and E. M. Henley, *Subatomic Physics*, Part III. Prentice-Hall, Englewood Cliffs, NJ, 1991. (I)

A. Das and T. Ferbel, *Introduction to Nuclear and Particle Physics*. John Wiley, New York, 1994. (I)

D. H. Perkins, *Introduction to High Energy Physics*. Addison-Wesley, Reading, Mass., 1987. (I)

Constants, Fundamental

B. N. Taylor and P. J. Mohr

Introduction

This article touches upon three main topics: (1) the motivation for "the romance of the next decimal place", or why the fundamental physical constants are important and why their determination with ever smaller uncertainties can have a profound effect on physics; (2) how a self-consistent set of "best values" of the fundamental constants is obtained, with emphasis on the 2002 least-squares adjustment of the values of the constants (the most recent comprehensive study carried out); and (3) future trends – where the field is heading over the next 5 to 10 years.

Origins

Three distinct "sources" of fundamental constants may be identified. The first is physical theory and its application to the real world, such as Maxwell's theory of electromagnetism, Einstein's theories of relativity, quantum mechanics, and quantum electrodynamics (QED). The speed of light in vacuum c, the Planck constant h, which relates the energy E of a photon to its frequency ν via the relation $E = h\nu$, the fine-structure constant α, which is the coupling constant of one of the four fundamental forces of nature – the electromagnetic force – and the Rydberg constant R_∞, which sets the scale of the energy levels in atoms, fall within this category.

The second is the fundamental particles such as the electron e, proton p, and neutron n. The elementary charge e, the rest mass of the electron and of the proton m_e, and m_p, and the magnetic moment of the proton in units of the nuclear magneton μ_p/μ_N, where $\mu_N = h/4\pi m_p$, are all examples of quantities that characterize a basic property of nature's elementary building blocks.

The third is conversion factors. Although not true fundamental constants like those in the first two categories, these quantities nevertheless have played an important role in the past in the fundamental-constants field, and one or two continue to do so today. (The present article was completed in early 2005.) This is because knowledge of their values is or was essential to many fundamental-constant determinations. The best current example in this category is the local value of the gravitational acceleration g, which relates force to mass; past examples include the ratio of the old x-ray unit of length, the kilo-x-unit, symbol kxu, to the ångström, symbol Å; and the ratio of various so-called as-maintained electrical units to the corresponding units of the International System of Units (SI), for example, Ω_{NIST}/Ω, where Ω_{NIST} was the National Institute of Standards and Technology (NIST) ohm defined in terms of the mean resistance of a particular group of wire-wound standard resistors, and Ω is the SI ohm.

It should also be noted that $\alpha = \mu_0 c e^2/2h$ and $R_\infty = m_e c \alpha^2/2h$ are examples of constants that are actually combinations of other constants, but are considered fundamental constants in their own right since the combination always appears in theoretical equations in the same way; and that c and m_e are examples of constants used as basic measurement units, for example, the mass of the muon m_μ is expressed as (approximately) 207 electron masses, or $m_\mu = 207 m_e$. Further, in the above expression for α, $\mu_0 = 4\pi \times 10^{-7}\,\text{N/A}^2$ exactly is the magnetic constant, also called the permeability of vacuum. It is an example of a constant that has its origins in the way certain laws of physics are formulated and certain units are defined, in this case Maxwell's equations of electromagnetism and the SI unit of current, the ampere.

Importance

There are at least four reasons why the values of the fundamental physical constants play a critical role in science and technology, and thus must be known with the smallest possible uncertainties. First, accurate values of the constants are required for the critical comparison of theory with experiment, and it is only such comparisons that enable our understanding of the physical world to advance. A closely related idea is that by comparing the numerical values of the same fundamental constants obtained from experiments in different fields of physics, the self-consistency of the basic theories of physics themselves can be tested.

Second, determining the fundamental constants to ever-greater levels of accuracy fosters the development of state-of-the-art measurement methods that may have wide application. Determining the next decimal place is never trivial and usually requires an entirely new measurement technology. An example is the determination of the Avogadro constant, N_A, which necessitated the development of techniques to measure the lattice spacing of a pure, single crystal of silicon in meters with a relative uncertainty of less than one part in 10^7. The end result has been the extension of our length scale, with this impressively small uncertainty, to the picometer range, which has had a number of important practical applications in the field of dimensional measurements.

Third, the fundamental constants are the obvious key to the development of a readily reproducible and invariant system of measurement units, a major goal of measurement science or metrology. If our measurement system could be completely based on fundamental constants, there would be no need for artifacts such as the 90% platinum–10% iridium (by weight) cylinder approximately 39 mm in height and diameter kept in a vault at the International Bureau of Weights and Measures (BIPM), Sévres, near Paris. Called the international prototype of the kilogram, the mass of this precious piece of metal defines the SI unit of mass, the kilogram. A prime example of our progress towards this long-term metrological goal of a unit system based solely on invariants of nature is the 1983 redefinition of the SI unit of length, the meter, in terms of the speed of light in vacuum c. Indeed, this new definition has the effect of fixing c to be 299 792 458 m/s exactly, which allows any laser radiation of known frequency ν to be used to realize the meter using the value of its wavelength λ obtained from the relation $\lambda = c/\nu$.

Finally, values of the fundamental constants are required for computations and measurements throughout science and technology, for example, the calculation of the properties of atoms and molecules important in the fields of air pollution, nuclear fusion, and astrophysics.

As an example of the significant role improved measurements of fundamental constants can play in increasing our understanding of physical theory, we consider the now historical determination of the Josephson constant $K_J = 2e/h$ using the Josephson effect in weakly coupled superconductors, which was first reported in 1967, nearly 40 years ago. (This phenomenon was predicted by Josephson in 1962 and the constant K_J is named after him.) When a Josephson device, for example, two thin films of lead separated by a 1 nm thick thermally-grown oxide layer, is cooled below the transition temperature of the superconductors of which it is composed and irradiated with microwave radiation of frequency f, its current vs. voltage curve exhibits current steps at quantized Josephson voltages U_J. The voltage of the nth step $U_J(n)$, where n is an integer, is related to f by

$$U_J(n) = \frac{nf}{K_J} = \frac{nf}{2e/h} .$$ (1)

A measurement of f and U_J thus yields $K_J = 2e/h$. The value obtained in the late 1960s using X-band microwaves (8 GHz to 12 GHz) was

$$K_J = \frac{2e}{h} = 4.835\,976(12) \times 10^{14}\,\text{Hz}/V_{\text{NIST}} \quad [2.4 \times 10^{-6}] ,$$ (2)

where V_{NIST} is the unit of voltage maintained at the National Institute of Standards and Technology (the national metrology institute or NMI of the United States), which at the time was based on the mean emf of a group of electrochemical standard cells. In Eq. (2), the number in parentheses is the standard uncertainty u, that is, the one-standard-deviation estimate, of the last digits of the quoted value. The number in brackets is its fractional equivalent, that is, the relative standard uncertainty u_r, which is defined according to $u_r(y) = u(y)/|y|, y \neq 0$. (When these experiments were performed at the University of Pennsylvania in Philadelphia, NIST was called the National Bureau of Standards; its name was changed in 1988.) The difference between this value when expressed in SI units and the previous best value, that resulting from the 1963 compilation of recommended values of the fundamental constants, far exceeded the uncertainty of the difference. As we shall see, the cause of the discrepancy was the use in the 1963 compilation of a value of the fine-structure constant α derived from the early 1950s measurement of the deuterium fine-structure splitting, which subsequently turned out to be incorrect. (Deuterium D, or ^{2}H, is similar to hydrogen, ^{1}H, but its nucleus consists of a proton and neutron instead of just a proton; this nucleus is called the deuteron, d. Helium 3, ^{3}He, has a nucleus consisting of two protons and one neutron; it is called the helion, h.)

Because QED, which describes the interaction between leptons (e. g., electron, e, muon, μ, tau, τ) and electromagnetic radiation, is one of the most precise and important theories of modern physics, it is essential to see how well it can withstand experimental tests. The Josephson $2e/h$ determination was therefore significant mainly because a reliable indirect value of α independent of quantum electrodynamic theory could be derived from it, and this value could in turn be used to compare QED theory and experiment critically and unambiguously. This was in marked contrast to the situation that existed prior to 1967 when no such value was available and tests of QED were mainly checks of internal consistency.

In the late 1960s, the equation relating α to $2e/h$ was written as

$$\alpha = \left[\frac{4R_\infty(\Omega_{\text{NIST}}/\Omega)\gamma_p'(\text{low})_{\text{NIST}}}{c(\mu_p'/\mu_B)(2e/h)_{\text{NIST}}} \right]^{1/2} ,$$ (3)

where as before $\Omega_{\text{NIST}}/\Omega$ is the ratio of the NIST as-maintained ohm to the SI ohm; μ_p'/μ_B is the magnetic moment of the proton in units of the Bohr magneton $\mu_B = h/4\pi m_e$ (throughout, the prime means for protons in a spherical sample of pure H_2O at 25 °C); and $\gamma_p'(\text{low})_{\text{NIST}}$ is the gyromagnetic ratio of the proton obtained by the so-called low-(magnetic) field method measured in terms of the NIST as-maintained unit of current $A_{NIST} = V_{NIST}/\Omega_{NIST}$. (It should be noted that the subscript NIST may be replaced by the more general Lab; and that $\gamma'_p = \omega'_p/B = 4\pi\mu'_p/h$, where ω'_p is the proton nuclear magnetic resonance or spin-flip angular frequency in the magnetic flux density B. In the low-field method, B is established via a precision solenoid of known dimensions carrying a current known in terms of A_{Lab}.) The value of α derived from Eq. (3) in 1968 using the best data available at the time was

$$\frac{1}{\alpha} = 137.03608(26) \quad [1.9 \times 10^{-6}] . \tag{4}$$

Included among the quantities that require an accurate value of α for comparing theory and experiment are the g-factors of the electron and muon; the energy levels in hydrogen-like atoms, especially the $n = 2$ Lamb shift ($2\,^2P_{1/2}-2\,^2S_{1/2}$ splitting); and the ground-state hyperfine splitting in hydrogen, muonium (an electron bound to a positive muon or μ^+e^- atom), and positronium (a bound electron–positron pair or e^+e^- atom). Of particular interest in the late 1960s was the hydrogen hyperfine splitting (hfs) transition frequency. This quantity, which corresponds to the energy difference between a hydrogen atom in which the electron and proton spins are aligned and one in which they are in opposite directions, can be measured with the extraordinarily small uncertainty of $u_r < 1 \times 10^{-12}$ using the hydrogen maser. In contrast, the theoretical QED equation for the hydrogen hfs, which mainly involves well-known constants and α, is limited to a relative uncertainty of a few times 10^{-6} because of the difficulty in calculating some of the terms in the equation from theory.

The most uncertain term is δ_{pp}, the proton polarizability contribution, which arises from the various excited states or internal structure of the proton. In the late 1960s, theoretical calculations predicted that δ_{pp} was in the range 0 to 2×10^{-6}. This was in conflict with what was implied by the value $1/\alpha = 137.0388(6)$ $[4.4 \times 10^{-6}]$ accepted at that time, which, as noted earlier, was derived from a measurement of the fine-structure splitting in deuterium. When this value of α was used to calculate a theoretical value of the hydrogen hfs, and when the difference between this calculated value and the hydrogen maser value was assumed to arise solely from the existence of a polarizability term, it was found that $\delta_{pp} = 42(9) \times 10^{-6}$. This meant that the probability for δ_{pp} to be as small as predicted by direct calculation was only about 1 in 20 000, an obvious inconsistency. In contrast, when the value of α derived from the Josephson-effect measurement of $2e/h$ [Eq. (4)] was used in place of the deuterium value, it was found that $\delta_{pp} = 2.5(4.0) \times 10^{-6}$, in agreement with the theoretical calculations. Hence the Josephson-effect value of α removed a discrepancy that during the 1960s was termed one of the major unsolved problems of QED. This example further illustrates the overall unity of science – a low-temperature, solid-state physics experiment provided information about the excited states of the proton – as well as how precise measurements of the fundamental constants can illuminate apparent inconsistencies in our physical description of nature.

Another example of a critical test of QED, but one that is still under active experimental and theoretical investigation, involves the g-factor of the electron $g_e = -2(1 + a_e) = 2\mu_e/\mu_B$, where a_e is the electron magnetic moment anomaly and μ_e is the magnetic moment of the electron. The deviation of g_e from 2, that is, the finite value of a_e ($\approx 1.16 \times 10^{-3}$), arises mainly from the electron's virtual emission and absorption of photons and the polarization of the vacuum with electron–positron pairs – called radiative corrections.

An experimental value of the electron g-factor, g_e(expt), with $u_r = 4.2 \times 10^{-12}$, has been obtained from a direct experimental determination of the anomaly, a_e(expt), with $u_r = 3.7 \times 10^{-9}$, based on measurements on either a single electron or positron stored in a Penning trap. The theoretical expression for the g-factor obtained from QED is g_e(theor) $= -2[1 + a_e$(theor)], where

$$a_e(\text{theor}) = C_1 \left(\frac{\alpha}{\pi}\right) + C_2 \left(\frac{\alpha}{\pi}\right)^2 + C_3 \left(\frac{\alpha}{\pi}\right)^3 + C_4 \left(\frac{\alpha}{\pi}\right)^4 + C_5 \left(\frac{\alpha}{\pi}\right)^5 + \cdots + \delta a_e . \quad (5)$$

The coefficient C_1 is exactly known, the coefficients C_2 and C_3 have negligible uncertainties, the coefficient C_4 is currently believed to have an uncertainty (arising from numerical integrations) that contributes an uncertainty to a_e(theor) of $0.96 \times 10^{-9} a_e$, the coefficient C_5 is unknown, but because of the small size of the term $(\alpha/\pi)^5$, the lack of knowledge of C_5 is estimated to contribute a nearly negligible uncertainty to a_e(theor), and δa_e, which accounts for comparatively small effects that are not purely QED, has an uncertainty that is very nearly negligible. Indeed, if α were exactly known, then a_e(theor) would have an uncertainty of $0.99 \times 10^{-9} a_e$, which is only slightly larger than that due to C_4 itself, and g_e(theor) would have an uncertainty of $1.1 \times 10^{-12} g_e$. To test QED by comparing g_e(theor) with g_e(expt) to the full accuracy of a_e(expt) requires, therefore, a value of α with u_r rather less than the 3.8×10^{-9} relative standard uncertainty of the value of α obtained by equating a_e(expt) with Eq. (5), which is

$$\frac{1}{\alpha} = 137.035\,998\,80(52) \quad [3.8 \times 10^{-9}] . \quad (6)$$

[A comparison of Eqs. (4) and (6) shows how far physics has advanced in the last 40 years or so.] Although an independent value of α with u_r this small is not yet available, there is one value with $u_r = 7.7 \times 10^{-9}$ and another with $u_r = 18 \times 10^{-9}$. The following paragraph discusses the latter and its implications for QED; it results from measurements of the von Klitzing constant, which is the basic constant of the quantum Hall effect or QHE.

The QHE is characteristic of certain high-mobility semiconductor devices of standard Hall-bar geometry when in an applied magnetic flux density of order 10 T and cooled to about 1 K. Hence, like the Josephson effect, it is a low-temperature, solid-state physics phenomenon. For a fixed current I through a particular type of QHE device, the curve of Hall voltage, U_H, vs. magnetic field displays regions of constant Hall voltage termed Hall plateaus. The Hall resistance of the ith plateau, $R_H(i)$, defined as the quotient of the Hall voltage of the ith plateau, $U_H(i)$, to the current I, is quantized and given by

$$R_H(i) = \frac{U_H(i)}{I} = \frac{R_K}{i} , \quad (7)$$

where $R_K = h/e^2$ is the von Klitzing constant and is named after Klaus von Klitzing, who discovered the QHE in 1980. (We consider only the integral QHE where i is an integer.) Since

$$R_K = \frac{\mu_0 c}{2\alpha} , \quad (8)$$

measurement of the quantized Hall resistance of the ith plateau $R_H(i)$ in SI units, and hence of $R_K = h/e^2$ in SI units, with a given u_r will yield a value of the fine-structure constant α

with the same u_r. In practice, $R_H(i)$ is measured in terms of a reference resistance R_{ref}, so the conversion factor R_{ref}/Ω must be determined in a separate experiment using an apparatus known as a calculable capacitor. Nevertheless, the weighted mean of the results of five independent measurements of the von Klitzing constant carried out in five different laboratories has yielded the value

$$\frac{1}{\alpha} = 137.036\,0030\,(25) \quad [18 \times 10^{-9}], \tag{9}$$

which has an uncertainty less than five times larger than that of the value in Eq. (6). If the value of α in Eq. (9) is now used to evaluate Eq. (5), thereby obtaining a value of $g_e(\text{theor})$, and if the latter is then compared to $g_e(\text{expt})$, one finds agreement within statistically acceptable limits. Since theory predicts the entire g-factor, not just the anomaly, the end result is the confirmation of theory at the extraordinary level of uncertainty of about 4 parts in 10^{11}.

Relationships

In addition to showing how accurate values of the constants can provide critical tests of the fundamental theories of physics, the above discussion shows that a value of a particular constant, in this case the fine-structure constant α, can be obtained from different experiments, and that complex relationships can exist among groups of fundamental constants. Some other examples of such relationships are

$$e = \left(\frac{2\alpha h}{\mu_0 c}\right)^{1/2} , \quad m_e = \frac{2R_\infty h}{c\alpha^2} , \quad N_A = \frac{cA_r(e)M_u\alpha^2}{2R_\infty h} , \quad K_J^2 R_K = \frac{4}{h} , \quad F = N_A e , \tag{10}$$

where N_A is the Avogadro constant, $A_r(e)$ is the relative atomic mass of the electron, $M_u = 10^{-3}\,\text{kg/mol}$ exactly is the molar mass constant, and F is the Faraday constant. [The relative atomic mass of a particle is its mass relative to 1/12th of the mass of the carbon-12 atom $m(^{12}C)$, which implies that $A_r(^{12}C) = 12$ exactly.] Such relationships also point up the fact that a particular constant may be obtained either by direct measurement or indirectly by appropriately combining other directly measured constants. If the direct and indirect values have comparable uncertainties, both must be considered in order to arrive at a "best value" for that constant. However, each of the various routes that can be followed to a particular constant, both direct and indirect, provides a somewhat different numerical value. The best way to handle this situation is by the mathematical technique known as least squares.

Least Squares Adjustments

The least-squares method furnishes a well-defined procedure for calculating best "compromise" values of the constants from all of the available data. It is a mathematical technique that automatically takes into account all possible routes to a particular constant, and yields a single value for each constant by weighting the different routes according to their uncertainties. The weights themselves are obtained from the *a priori* standard uncertainties u assigned to the individual measurements or calculated values that constitute the original set of data.

Least-squares adjustments (or studies) of the fundamental constants provide the scientific and technological communities with a self-consistent set of internationally recommended values of the basic constants and conversion factors of physics and chemistry based on all of the relevant data available at a given point in time. Equally important, however, is the critical data review that must necessarily accompany such studies: besides requiring the reviewer to summarize in one place a vast amount of rather diverse information available only after much searching, it forces him or her to reassign uncertainties to individual experiments on the same basis, thus making possible the ready identification of discrepancies among different measurements and calculations. This identification, in turn, can stimulate new experimental and theoretical work.

Comprehensive least-squares (or its equivalent) studies of the constants were pioneered by R. T. Birge starting in the late 1920s and continued by others, including DuMond and Cohen, and Bearden and Thomson. The last six adjustments (see Bibliography) are those of

1. Cohen and DuMond, dated 1963;

2. Taylor, Parker, and Langenberg, dated 1969 and based on their Josephson-effect determination of $K_J = 2e/h$;

3. Cohen and Taylor, dated 1973, which was carried out under the auspices of the Committee on Data for Science and Technology (CODATA) Task Group on Fundamental Constants and recommended for international use by CODATA;

4. Cohen and Taylor, dated 1986, again carried out under CODATA auspices;

5. Mohr and Taylor (the authors of the present article), dated 1998, also carried out under CODATA auspices; and

6. Mohr and Taylor, dated 2002, done once again under the auspices of CODATA. The latter study, which is referred to as the 2002 CODATA adjustment, is the most recent, and the set of recommended constants resulting from it are those currently adopted by CODATA for international use; values of selected constants from this adjustment are given in Table 1, and the complete set is available in the first reference of the bibliography as well as electronically on the Web at `physics.nist.gov/constants`. This site also provides the means to convert online the value of an energy expressed in one unit to the equivalent value expressed in another unit, for example, to convert a value expressed in joules to the corresponding value expressed in electron volts.

That CODATA now plays an important role in the fundamental constants field is a direct consequence of its mission. CODATA was established in 1966 as an interdisciplinary committee of the International Council for Science (ICSU, formerly the International Council of Scientific Unions), with the aim of improving the quality, reliability, processing, management, and accessibility of data of importance to science and technology. In keeping with this aim, the CODATA Task Group on Fundamental Constants was established in 1969 with the express purpose of providing from time to time a set of recommended values of the constants for worldwide use. The 1973 adjustment was the first carried out under its sponsorship.

The 2002 CODATA Adjustment: Basic Approach and Principles

The approach used to carry out the 2002 adjustment was the same as that employed to carry out the 1998 adjustment and is conceptually straightforward. The foundation of both is the method of least squares, which enables all of the information available relevant to the constants

Table 1: A highly abbreviated list of the 2002 CODATA recommended values of the fundamental physical constants of physics and chemistry. The complete set may be found in P. J. Mohr and B. N. Taylor, *Rev. Mod. Phys.* **77**, 1 (2005), and at physics.nist.gov/constants.

Quantity	Symbol	Numerical value	Unit	Relative std. uncert. u_r
speed of light in vacuum	c, c_0	299 792 458	$m\,s^{-1}$	(exact)
magnetic constant	μ_0	$4\pi \times 10^{-7}$	$N\,A^{-2}$	
		$= 12.566370614...\times 10^{-7}$	$N\,A^{-2}$	(exact)
electric constant $1/\mu_0 c^2$	ϵ_0	$8.854187817...\times 10^{-12}$	$F\,m^{-1}$	(exact)
Newtonian constant				
of gravitation	G	$6.6742(10)\times 10^{-11}$	$m^3\,kg^{-1}\,s^{-2}$	1.5×10^{-4}
Planck constant	h	$6.6260693(11)\times 10^{-34}$	$J\,s$	1.7×10^{-7}
$h/2\pi$	$\hbar$	$1.05457168(18)\times 10^{-34}$	$J\,s$	1.7×10^{-7}
elementary charge	e	$1.60217653(14)\times 10^{-19}$	C	8.5×10^{-8}
magnetic flux quantum $h/2e$	Φ_0	$2.06783372(18)\times 10^{-15}$	Wb	8.5×10^{-8}
conductance quantum $2e^2/h$	G_0	$7.748091733(26)\times 10^{-5}$	S	3.3×10^{-9}
electron mass	m_e	$9.1093826(16)\times 10^{-31}$	kg	1.7×10^{-7}
proton mass	m_p	$1.67262171(29)\times 10^{-27}$	kg	1.7×10^{-7}
proton–electron mass ratio	m_p/m_e	$1836.15267261(85)$		4.6×10^{-10}
fine-structure const. $e^2/4\pi\epsilon_0\hbar c$	α	$7.297352568(24)\times 10^{-3}$		3.3×10^{-9}
inverse fine-structure const.	α^{-1}	$137.03599911(46)$		3.3×10^{-9}
Rydberg constant $\alpha^2 m_e c/2h$	R_∞	$10973731.568525(73)$	m^{-1}	6.6×10^{-12}
Avogadro constant	N_A, L	$6.0221415(10)\times 10^{23}$	mol^{-1}	1.7×10^{-7}
Faraday constant $N_A e$	F	$96485.3383(83)$	$C\,mol^{-1}$	8.6×10^{-8}
molar gas constant	R	$8.314472(15)$	$J\,mol^{-1}\,K^{-1}$	1.7×10^{-6}
Boltzmann constant R/N_A	k, k_B	$1.3806505(24)\times 10^{-23}$	$J\,K^{-1}$	1.8×10^{-6}
Stefan-Boltzmann constant				
$(\pi^2/60)k^4/\hbar^3 c^2$	σ	$5.670400(40)\times 10^{-8}$	$W\,m^{-2}\,K^{-4}$	7.0×10^{-6}

Non-SI units accepted for use with the SI:

electron volt: (e/C) J	eV	$1.60217653(14)\times 10^{-19}$	J	8.5×10^{-8}
(unified) atomic mass unit				
$1\,u = m_u = \frac{1}{12}m(^{12}C)$	u	$1.66053886(28)\times 10^{-27}$	kg	1.7×10^{-7}
$= 10^{-3}\,kg\,mol^{-1}/N_A$				

to be taken into account in a consistent way, including correlations among the data. The basic steps were to

1. identify and critically review all possible input data, both experimental and theoretical, with special attention to uncertainty assignment;

2. express the initial N *input data* in terms of a subset of M quantities called *adjusted constants*, which are really the variables or "unknowns" of the least-squares calculation, by means of a set of relations called *observational equations*;

3. investigate the compatibility of the input data and adjust *a priori* assigned uncertainties and/or eliminate data as deemed appropriate;

4. obtain "best values" in the least-squares sense of the M adjusted constants by solving the observational equations using the input data finally selected; and

5. calculate all other constants of interest from the "best values" of the M adjusted constants, taking into account their uncertainties and correlations, that is, their covariances.

A number of guiding principles aided this process:

1. the input data had to be available by 31 December 2002, except in a few special cases;

2. ordinary physics was assumed to be valid, for example, special relativity, QED, the Standard Model of particle physics, including combined charge conjugation, parity inversion, and time-reversal (*CPT*) invariance;

3. the validity of the fundamental Josephson and quantum Hall effect relations, $K_J = 2e/h$ and $R_K = h/e^2$, was assumed, although as discussed later, tests of this assumption were performed as part of the examination of the consistency of the data;

4. uncertainties were evaluated by the method currently accepted internationally, that is, all components were estimated at the level of one standard deviation (a standard uncertainty), even those arising from systematic effects, and then combined according to the law of propagation of uncertainty (commonly called the "root-sum-of squares", or "rss" method);

5. each input datum considered for initial inclusion in the adjustment had to have a sufficiently small standard uncertainty u so that its weight in the adjustment $w = 1/u^2$ was non-trivial compared with the weight of other directly measured values of the same quantity. This leads to the "factor of five" rule: for the same quantity, if the uncertainty u_1 of result 1 is greater than about 5 times the uncertainty u_2 of result 2, then result 1 is not considered for inclusion. Such a rule is consistent with the fact that experiments with uncertainties that differ by a factor of 5 or more are qualitatively different;

6. the latest result from a given laboratory for a particular quantity, which usually had the smallest uncertainty, was normally viewed as superseding an earlier result from the same laboratory for the same quantity, because they were not independent; and

7. all input data were treated on an equal footing – data were not classified as "auxiliary constants" (data with uncertainties considered small enough to be neglected) or "stochastic input data" (data with significantly larger uncertainties), as was usually the case prior to the 1998 adjustment. This allowed all components of uncertainty and all correlations among the input data to be properly taken into account. It also eliminated the somewhat arbitrary division of the data into two categories and hence the possible shift in category of a particular quantity from one adjustment to the next.

It is important to recognize that the underlying philosophy of most adjustments of the values of the constants since the pioneering work of Birge over 75 years ago has been to provide the user community with values of the constants having the smallest possible uncertainties consistent with the information available at the time. The motivation for this approach is that it gives the most critical users of the values of the constants the best possible tools for their work based on the current state of knowledge. The downside is that the information available may include an error or oversight. Nevertheless, The CODATA Task Group on Fundamental Constants rejects the idea of making uncertainties sufficiently large that any future change in the recommended value of a constant will likely be less than its uncertainty – simply put, the Task Group eschews the use of a "safety factor" as employed in some data compilations.

The reader may have noticed that for the four adjustments carried out under CODATA auspices, which are dated 1973, 1986, 1998, and 2002, the 4-year time interval between the third and fourth is much shorter than the 12 to 13-year time interval between the first and second and second and third. Because new experimental and theoretical results that influence our knowledge of the values of the constants appear nearly continuously, because the World Wide Web allows new sets of recommended values of the constants to be rapidly and widely distributed, and because the Web has engendered new modes of work and thought – its users expect to find the latest information available electronically only a mouse-click away – the CODATA Task Group on Fundamental Constants decided at the time of the 1998 adjustment to take advantage of the high degree of computerization that had been incorporated in that effort to provide a new set of recommended values every 4 years. The 2002 set is the first from the new schedule.

The 2002 CODATA Adjustment: Data, Adjusted Constants, and Observational Equations

The N input data initially considered for inclusion in the 2002 CODATA adjustment consisted of 112 separate items, with 170 associated and distinct correlation coefficients that varied from -0.375 to 0.991. These 112 items are summarized in the following paragraphs, where for convenience the data are divided into two main categories or groups: the principal data that contribute to the determination of the 2002 recommended value of the Rydberg constant R_∞ (50 items); and the principal data that contribute to the determination of the 2002 recommended values of the "other" constants (62 items, not counting 8 measured values of the Newtonian constant of Gravitation G). Although all 112 data, which are in fact treated in the adjustment as a single group, are expressed in SI units, a number of data in the "other" category were actually measured in terms of "conventional" electrical units based on the Josephson and quantum Hall effects and conventional values of the Josephson and von Klitzing constants, namely, $K_{J-90} = 483\,597.9\,\mathrm{GHz/V}$ exactly and $R_{K-90} = 25\,812.807\,\Omega$ exactly. Initially requested by the General Conference on Weights and Measures (CGPM), these values were adopted by the International Committee for Weights and Measures (CIPM) for international use starting 1 January 1990 in order to ensure the worldwide consistency of electrical measurements.

Data: Group I – Rydberg constant data

- 23 frequencies corresponding to transitions between two different energy levels of the hydrogen atom ^{1}H (9 transition frequencies), of the deuterium atom ^{2}H (5 transition frequencies), differences between different transition frequencies (6 transition frequency differences for ^{1}H and 2 for ^{2}H), and the difference in frequency between a transition in ^{1}H and the same transition in ^{2}H (1 such difference);

- 1 value of the bound-state root-mean-square (rms) charge radius R_p of the proton and 1 value of the similar radius R_d of the deuteron; and

- 25 additive corrections δ_i that arise as follows: In order to properly take into account the uncertainty of the theoretical expression used to relate any one of the 23 frequencies to the adjusted constants, and which in fact may contain up to four distinct theoretical expressions, an additive correction δ_i was introduced for each distinct expression. The δs were then included among the variables or adjusted constants of the least-squares adjustment, and their estimated values were taken as input data. The best *a priori* estimate of each δ_i was taken to be zero, but with a standard uncertainty u equal to that of the theoretical expression. This approach enabled the uncertainties of the theory to be taken into account in a rigorous way.

Data: Group II – "Other" constants data

- 8 relative atomic masses of various atoms and particles, for example, $A_r(^1\text{H})$, $A_r(^{16}\text{O})$ and $A_r(e)$;

- 1 value of the electron magnetic moment anomaly a_e and 1 frequency ratio related to the determination of the muon magnetic moment anomaly a_μ;

- 1 frequency ratio related to the determination of the g-factor of the electron bound in a hydrogen-like, or hydrogenic, atom of ^{12}C (five electrons removed), and 1 similar ratio for an electron bound in a hydrogenic atom of ^{16}O (seven electrons removed);

- 5 magnetic moment ratios involving the electron, proton, neutron, deuteron, and helion;

- 2 values of the ground-state hyperfine splitting $\Delta\nu_{\text{Mu}}$ in muonium and 2 values of a related frequency difference that provides information about the electron–muon mass ratio m_e/m_μ;

- 5 additive corrections δ_i related to the theoretical expressions for a_e, a_μ, $\Delta\nu_{\text{Mu}}$, and the g-factor of the electron bound in hydrogenic ^{12}C and in hydrogenic ^{16}O;

- 6 values of the gyromagnetic ratios of the proton and helion determined by both the low and high-field methods in conventional electrical units based on K_{J-90} and R_{K-90};

– 2 values of K_J, 5 of R_K, 2 of $K_J^2 R_K$, and 1 of the Faraday constant, the latter determined in conventional electrical units based on K_{J-90} and R_{K-90};

– 2 values of the ratio of h to the mass of an atomic particle, in one case the mass of the cesium-133 atom and in the other case the mass of the neutron times the {220} lattice spacing d_{220} of a particular sample of silicon (all measurements involving silicon were carried out on highly pure, nearly crystallographically perfect single crystals and, where appropriate, the results were converted to the same reference temperature and pressure);

– 1 value of the ratio of the wavelength of the 2.2 MeV capture γ ray emitted in the reaction $n + p \rightarrow d + \gamma$ to the d_{220} lattice spacing of a particular single crystal of silicon, 9 fractional differences between the d_{220} lattice spacings of a number of silicon single crystal samples, 1 such difference between d_{220} of one of these samples used as a reference and that of ideal silicon, 4 ratios of the wavelengths of three different x-ray radiations to the d_{220} lattice of two different silicon crystals, 1 value of d_{220} in meters for a particular silicon sample, and 1 value of the molar volume of ideal silicon $V_m(Si)$ [that is, the volume of one mole of silicon atoms in a perfect crystal of naturally occurring silicon]; and

– 2 values of the molar gas constant R.

The 50 Rydberg constant input data had relative standard uncertainties u_r that ranged from 2.0×10^{-2} for R_p to 1.9×10^{-14} for the $1 S_{1/2}$–$2 S_{1/2}$ transition frequency in ^1H. The 62 "other" data (but also considering the 8 measured values of the Newtonian constant of gravitation G discussed below) had u_r that ranged from 1.0×10^{-4} for G to 1.0×10^{-11} for $A_r(^{16}O)$. Although space does not permit a detailed discussion of how these 112 input data were obtained, it should be recognized that the determination of many required an enormous amount of hard work by many researchers extending over a number of years, in some cases more than 30. One need only look at the original papers to gain an appreciation of the extreme difficulty of experimentally determining the value of a quantity such as the von Klitzing constant with $u_r \approx 1 \times 10^{-8}$ (one part in 100 million), or of calculating from theory a fractional contribution of 1×10^{-8} to a theoretical expression such as $a_e(\text{theor})$.

Adjusted constants

As discussed above, a least-squares adjustment is carried out by first expressing the input data in terms of a set of quantities called adjusted constants. The latter are the variables or "unknowns" of the adjustment and the resulting equations, which are the theoretical relations between the measured and calculated input data and the adjusted constants, are called observational equations. The set of adjusted constants are to some extent arbitrary – the two principal requirements that they must meet are that all of the input data must be expressible in terms of them, and no adjusted constant can be eliminated by expressing it in terms of other adjusted constants. In the 2002 CODATA adjustment, 61 quantities were taken as adjusted constants, 28 primarily for the Rydberg-constant data and 33 mainly for the "other" data. The 28 included R_∞, R_p, R_d, and the 25 δs related to the theoretical expressions of various energy levels in hydrogen and deuterium; the 33 included 8 relative atomic masses such as $A_r(e)$, $A_r(p)$, and $A_r(n)$, α, h, and R, 5 magnetic moment ratios such as that of the electron to that of the proton μ_e/μ_p, the mass ratio m_e/m_μ, 5 δs such as δ_e, the additive correction to $a_e(\text{theor})$,

3 specialized units in the field of x rays of historic interest, and 8 d_{220} lattice spacings of different silicon crystals, including that of a crystal of ideal silicon.

Observational equations

Forty-nine different types of observational equation were required for the 50 Rydberg-constant data, because two of the 23 frequencies were for the same hydrogen transition. Fifty one different types of observational equation were required for the 62 "other" constant data, because there were 2 measurements each of $\Delta\nu_{Mu}$, of the gyromagnetic ratio of the proton obtained by the low field method, of the same quantity obtained by the high-field method, of the gyromagnetic ratio of the helion obtained by the low-field method, of K_J, of $K_J^2 R_K$, and of R, and 5 measurements of R_K.

Space limitations do not allow us to give the 100 observational equations used to analyze the 112 input data initially considered in the 2002 adjustment. However, the following examples, which are the observational equations for the experimentally measured values of K_J, R_K, $K_J^2 R_K$, and $\mathcal{F}_{90}$, where the latter is the Faraday constant determined in terms of the conventional electrical units based on K_{J-90} and R_{K-90}, should provide a sense of what they are about:

$$K_J \doteq \left(\frac{8\alpha}{\mu_0 ch}\right)^{1/2} \, , \qquad R_K \doteq \frac{\mu_0 c}{2\alpha} \, , \tag{11a}$$

$$K_J^2 R_K \doteq \frac{4}{h} \, , \qquad \mathcal{F}_{90} \doteq \frac{cM_u A_r(e)\alpha^2}{K_{J-90}R_{K-90}R_\infty h} \, . \tag{11b}$$

In each of these observational equations, the quantities on the left-hand side of the $\doteq$ sign are the measured values of the indicated quantities. The symbol $\doteq$ is used rather than the ordinary equals sign to indicate the fact that the two sides of the equations are equal in principle but not numerically, because the set of observational equations is over determined. The right-hand sign contains only quantities with values that are exactly known in SI units such as μ_0, c, and M_u, and adjusted constants – α, h, and $A_r(e)$ in these four examples. Although they are among the simplest observational equations, some are even simpler. For example, the observational equation for δ_e – and for that matter, all of the other δs – is simply of the form $\delta_e \doteq \delta_e$, because δ_e cannot be expressed in terms of any other adjusted constant. The observational equation for R is of the same form: $R \doteq R$.

The 2002 CODATA Adjustment: Analysis of Data, Final Data Selection, and Final Adjustment

The focus of the data analysis was to identify discrepancies among the data and to understand the degree to which a particular datum would contribute to the determination of the 2002 set of recommended values. It proceeded in three stages. First, directly measured values of the same quantity were compared, that is, data of the same type, usually by calculating their weighted mean, which is actually a single variable (as compared to a many variable or multivariate) least-squares adjustment. An example of this case was the comparison of the five measured values of the von Klitzing constant. (For the case where there are only two values of the same quantity, simply calculating the difference between the values and comparing that difference to its uncertainty is an equivalent procedure.)

Second, directly measured values of different quantities, that is, data of different types, were compared through a third quantity that could be inferred from the values of the directly measured quantities. The most important of these inferred values were the fine-structure constant α and the Planck constant h. Indeed, some 15 different items of input data were compared through their inferred values of α and 8 through their inferred values of h.

Finally, a multivariate analysis of the data was carried out using the method of least-squares for correlated input data, which was developed in about the first third of the 20th century. (The method of least squares has its origins in the work of the famous mathematicians Legendre, Gauss, and Laplace in the first quarter of the 19th century.) In carrying out either a single variable or multivariate adjustment, which often involved sequentially deleting various input data to see what impact these data might have on the weighted mean or on the values of the adjusted constants, several statistical tools were used to decide if the data were compatible and if a datum contributed in a meaningful way to the determination of the mean or the adjusted constants. These included

- the well-known statistic chi-square, symbol χ^2, and the closely associated *Birge ratio* $R_B = (\chi^2/\nu)^{1/2}$, where $\nu = N - M$ is the number of *degrees of freedom* of the adjustment, with N the number of input data and M the number of adjusted constants;

- the *normalized residual* r_i of a given input datum, which is the ratio of the difference between the value of the input datum and the best estimated value of that datum resulting from the adjustment, to the *a priori* uncertainty assigned the datum (for uncorrelated input data, χ^2 is equal to the sum of the squares of the normalized residuals); and

- the *self sensitivity coefficient* S_c of a particular datum, which is normally between 0 and 1 and is a measure of the influence of that datum on the best estimated value of the corresponding quantity.

A value of R_B much larger than about 1 usually indicates some inconsistencies in the data, while a normalized residual much larger than about 2 for a particular datum usually indicates that the datum is inconsistent with the other input data. A value of S_c of only about 0.01 (1%) for a particular datum indicates that the datum plays a rather inconsequential role in determining the best value of the corresponding quantity and can be omitted with little effect (recall the "factor-of-five" rule discussed above).

As a consequence of these analyses, two significant inconsistencies among the data were identified, as were a number of data whose values of S_c were below 0.01. This led to the final least-squares adjustment on which the 2002 CODATA set of recommended values was based, which may be summarized as follows: from the initially considered set of 112 potential input data, seven were deleted, one because of its significant disagreement with the other data and because of its extremely small influence on the best estimated value of the corresponding quantity, and six solely for the latter reason. The first of these data was a measurement of the proton gyromagnetic ratio by the low-field method, while the remaining six were two measurements of the helion gyromagnetic ratio by the low field method, and four measurements of the von Klitzing constant.

In addition, the *a priori* uncertainties of five input data were weighted by the multiplicative factor 2.325 in order to reduce their inconsistency to an acceptable level. These five data, all of which contributed to the determination of the Planck constant h, were two measurements of the Josephson constant K_J, two moving-coil watt balance measurements of the product $K_J^2 R_K$, and one value of the molar volume of silicon $V_m(Si)$ [the factor 2.325 reduced the absolute value of the normalized residual $|r_i|$ of $V_m(Si)$ from $|-3.18|$ to $|-1.50|$].

The $N = 105$ final input data were expressed in terms of $M = 61$ adjusted constants, corresponding to $\nu = N - M = 44$ degrees of freedom. For this final adjustment, $\chi^2 = 31.2$, $R_B = 0.84$, and the probability p that this observed value of χ^2 for $\nu = 44$ degrees of freedom would have exceeded that observed value is the quite acceptable $p = 0.93$. Each input datum included in the final adjustment had a value of $S_c > 0.01$, or was a subset of the data of an experiment that provided a datum with such a value of S_c. The final input data with the four largest $|r_i|$ had values of r_i of 2.20, -1.50, 1.43, and 1.39; three other input data had values of r_i of 1.11, -1.11, and 1.05, and all other values of $|r_i|$ were less than 1.

It is worth noting that the disagreement discussed above involving the measurements of K_J, $K_J^2 R_K$, and $V_m(Si)$ led the authors to consider whether relaxing the assumptions $K_J = 2e/h$ and $R_K = h/e^2$ would reduce or possibly even eliminate the inconsistency, even though these fundamental relations are strongly supported by both experiment and theory. To this end, various adjustments were carried out in which K_J and/or R_K were treated simply as phenomenological constants unrelated to e and h. However, no statistically significant evidence was found that indicated the above relations are not exact.

The 2002 CODATA Adjustment: Calculation of Other Constants from the Adjusted Constants

The direct output of the final adjustment was best estimated values in the least-squares sense of the 61 adjusted constants, together with their covariances; all of the 300-plus 2002 CODATA recommended values and their uncertainties were obtained from these and, as appropriate,

1. those constants that have defined values such as c, μ_0 and M_u;

2. the value of the Newtonian constant of gravitation G resulting from the weighted mean of eight independent results, but taking into account the historical difficulty of measuring G in assigning its uncertainty (because G has no known relationship with any other constant, for convenience, the eight values were treated independently of the main adjustment); and

3. the values of the tau mass m_τ, Fermi coupling constant G_F, and $\sin^2 \theta_W$, where θ_W is the weak mixing angle, all as given in the 2002 biennial Review of Particle Physics by the Particle Data Group.

Of course, a number of the 300-plus CODATA recommended values are the adjusted constants themselves, for example, α, h, R_∞, $A_r(e)$, R, and the mass ratio m_e/m_μ, and thus are a direct consequence of the final adjustment. However, most must be calculated from the adjusted constants using the relations that exist among the constants, such as the first three relationships in Eq. (10). Some others are as follows:

$$k = \frac{2R_\infty hR}{cA_r(e)M_u\alpha^2}, \quad \mu_B = \left(\frac{c\alpha^5 h}{32\pi^2\mu_0 R_\infty^2}\right)^{1/2}, \quad m_\mu = \frac{2R_\infty h}{c\alpha^2}\left(\frac{m_e}{m_\mu}\right)^{-1}, \quad R_K = \frac{\mu_0 c}{2\alpha}, \quad (12)$$

Table 2a: Comparison of the 2002 and 1998 CODATA recommended values of a representative group of constants. Each number in the third column is the factor by which the standard uncertainty of the 2002 value of the indicated constant has decreased (increased) relative to the standard uncertainty of the 1998 value. Each number in the fourth column is obtained by subtracting from the 2002 value of the indicated constant the 1998 value and dividing by the standard uncertainty of the 1998 value (that is, each is the change in the value of the indicated constant from 1998 to 2002 relative to its 1998 standard uncertainty). The 2002 and 1998 adjustments are described in detail in the first and third references of the bibliography, respectively (Mohr and Taylor, authors).

Constant	2002 rel. std. uncert. u_r	Factor by which 2002 std. uncert. has decreased (increased)	Change in 1998 value relative to 1998 std. uncert.
α	3.3×10^{-9}	1.1	1.3
h	1.7×10^{-7}	(2.2)	1.1
e	8.5×10^{-8}	(2.2)	1.1
G	1.5×10^{-4}	10	0.2
R	1.7×10^{-6}	1.0	0.0
N_A	1.7×10^{-7}	(2.2)	-1.0
m_p/m_e	4.6×10^{-10}	4.6	1.3
R_∞	6.6×10^{-12}	1.1	-0.3

where k is the Boltzmann constant. In the first three expressions in Eq. (10) and in all of those of Eq. (12), the quantities on the right-hand side of the equals sign are either adjusted constants or exactly defined constants.

Although evaluating such equations is a matter of simple substitution, calculating the uncertainty of the resulting constant requires some care, because the covariances of the constants entering the equations must be properly taken into account in order to obtain the correct uncertainty.

Comparison of Selected Recommended Values of the Constants from Different Adjustments

It is of interest to compare the recommended values of the constants resulting from various adjustments in order to see how our knowledge of the values of the constants has evolved over the years. This is done in Tables 2a to 2e for a small but representative group of recommended values resulting from the last six adjustments, starting with the CODATA 2002 adjustment and going back in time over 40 years to the 1963 adjustment of Cohen and DuMond, which was published in 1965.

Even a casual perusal of Tables 2a to 2e tells a number of interesting tales. First and foremost is how far our knowledge of the values of the constants has advanced during this period as measured by their uncertainties – one need only compare the second column of Table 2a to the last column of Table 2e. This is due to advances in both experiment and theory, the former aided by the discovery of new phenomena such as the Josephson and quantum Hall effects and the development of new and better instrumentation such as the laser, Penning trap,

Table 2b: As in Table 2a but for the 1998 and 1986 CODATA recommended values. The 1986 adjustment is described in detail in the fourth reference of the bibliography (Cohen and Taylor, authors).

Constant	1998 rel. std. uncert. u_r	Factor by which 1998 std. uncert. has decreased (increased)	Change in 1986 value relative to 1986 std. uncert.
α	3.7×10^{-9}	12	-1.7
h	7.8×10^{-8}	7.7	-1.7
e	3.9×10^{-8}	7.8	-1.8
G	1.5×10^{-3}	(12)	0.0
R	1.7×10^{-6}	4.8	-0.5
N_A	7.9×10^{-8}	7.5	1.5
m_p/m_e	2.1×10^{-9}	9.5	-0.9
R_∞	7.6×10^{-12}	157	2.7

Table 2c: As in Table 2a but for the 1986 and 1973 CODATA recommended values. The 1973 adjustment is described in detail in the fifth reference of the bibliography (Cohen and Taylor, authors).

Constant	1986 rel. std. uncert. u_r	Factor by which 1986 std. uncert. has decreased (increased)	Change in 1973 value relative to 1973 std. uncert.
α	4.5×10^{-8}	18	0.4
h	6.0×10^{-7}	9.0	-2.8
e	3.0×10^{-7}	9.4	-2.6
G	1.3×10^{-4}	4.8	0.1
R	8.4×10^{-6}	3.7	0.4
N_A	5.9×10^{-7}	8.6	3.0
m_p/m_e	2.0×10^{-8}	19	1.7
R_∞	1.2×10^{-9}	64	-0.3

and desktop computer, and the latter by new calculational techniques as well as very high speed computer workstations.

Second, mistakes have been made: changes in the recommended values of the constants from one adjustment to the next larger than one would expect from their uncertainties have occurred all too frequently. On the other hand, it could be argued that such changes are inevitable when one provides recommended values with uncertainties based only on current knowledge – a policy whose purpose is to provide users with the sharpest knife possible for doing their work. Although incorporating some sort of "safety factor" in the evaluation of uncertainties would reduce the occurrence of such changes, in the opinion of the Task Group (see above), doing so would likely restrict the usefulness of the recommended values – it would diminish their role in identifying discrepancies among different results, which quite often fosters new experimental and theoretical work aimed at understanding and eliminating them.

Table 2d: As in Table 2a but for the 1973 and 1969 recommended values. The 1969 adjustment is described in detail in the sixth reference of the bibliography (Taylor, Parker, and Langenberg, authors).

Constant	1973 rel. std. uncert. u_r	Factor by which 1973 std. uncert. has decreased (increased)	Change in 1969 value relative to 1969 std. uncert.
α	8.2×10^{-7}	1.8	0.0
h	5.4×10^{-6}	1.4	-0.4
e	2.9×10^{-6}	1.5	-0.4
G	6.2×10^{-4}	(1.3)	-0.4
R	3.1×10^{-5}	1.3	0.2
N_A	5.1×10^{-6}	1.3	-3.1
m_p/m_e	3.8×10^{-7}	16	3.9
R_∞	7.5×10^{-8}	1.3	0.5

Table 2e: As in Table 2a but for the 1969 and 1963 recommended values. The 1963 adjustment is described in detail in the seventh reference of the bibliography (Cohen and DuMond, authors). (It should be noted that because the uncertainties of many of the 1963 recommended values were expressed in the form of a single digit, some of the values in the last three columns must be viewed as approximate.)

Constant	1969 rel. std. uncert. u_r	Factor by which 1969 std. uncert. has decreased (increased)	Change in 1963 value relative to 1963 std. uncert.	1963 rel. std. uncert. u_r
α	1.5×10^{-6}	2.9	4.6	4.4×10^{-6}
h	7.6×10^{-6}	3.2	3.8	2.4×10^{-5}
e	4.4×10^{-6}	2.9	4.6	1.2×10^{-5}
G	4.6×10^{-4}	1.6	0.6	7.5×10^{-4}
R	4.2×10^{-5}	1.0	0.0	4.2×10^{-5}
N_A	6.6×10^{-6}	2.3	-3.9	1.5×10^{-5}
m_p/m_e	6.2×10^{-6}	(1.1)	1.3	5.5×10^{-6}
R_∞	1.0×10^{-7}	1.0	0.2	1.0×10^{-7}

Finally, although new data usually lead to smaller uncertainties for the constants from one adjustment to the next, the comparison of the 2002 recommended values with their 1998 counterparts in Table 2a shows quite clearly that this is not always the case. As discussed above, the value of the molar volume of silicon $V_m(Si)$ that became available for the 2002 adjustment turned out to be inconsistent with four previously available and quite consistent measurements, two of K_J and two of $K_J^2 R_K$, and led the Task Group to decide to weight the *a priori* assigned uncertainties of all five data by the multiplicative factor 2.325 in the final adjustment on which the 2002 CODATA recommended values are based. The end result of this decision was an increase of $u_r(h)$ and the u_r of those constants that depend strongly on h by a factor of about 2.2 compared to their 1998 values. Nevertheless, one can argue that because the larger uncertainties to which the new data gave rise are presumably closer to the truth, our knowledge of the values of the constants has actually advanced, even though the uncertainties of some constants have increased.

A related case is the 1986, 1998, and 2002 recommended values of the Newtonian constant of gravitation G. The uncertainty of the 1998 value is about 12 times larger than that of 1986, because of the availability in the 1998 adjustment of a credible result for G obtained in an apparently careful experiment carried out at a major NMI, but which was in gross disagreement with all other values. Because of this credible but discrepant result, the Task Group decided to retain the 1986 recommended value as the 1998 recommended value, but with a significantly increased uncertainty to reflect the discrepant value's existence. Then, between 1998 and 2002, additional new results for G became available, and three researchers at the NMI, two of whom were involved in the original work that led to the questionable value of G, carried out experimental investigations of several critical aspects of the experiment that led them to conclude that the original result and the uncertainty assigned to it could no longer be considered correct. Based on this new information, the Task Group was able to include in the 2002 CODATA set of recommended values a new value of G with a reduced uncertainty.

A Brief Look Into the Future

If one looks at how our knowledge of the values of the constants has improved over the last 75 years – orders of magnitude reductions in uncertainty for many constants – one can only wonder how such progress can continue. Can ever more ingenious methods of overcoming the limitations of ever lower levels of electrical and mechanical noise be devised by experimentalists, and can ever more sophisticated techniques for calculating the contributions from an ever increasing number of higher-order Feynman diagrams be devised by QED theorists? An unequivocal answer to this question is obviously not possible, but the ingenuity continually demonstrated by both experimentalists as well as theoreticians over this period provides a good basis for optimism about the future.

If one looks at the current structure of the fundamental constants, that is, the relationships that exist among the different constants and the uncertainties of those constants that play a key role in those relationships, one finds that our knowledge of the values of many of the constants can be significantly advanced if the uncertainties of the fine-structure constant α, Planck constant h, and molar gas constant R can be significantly reduced. Fortunately, an experiment to measure the electron magnetic moment anomaly a_e and the necessary theoretical calculations to reduce the uncertainty of a_e(theor) [see Eq. (5)] are currently well underway and should lead to a value of α with $u_r < 1 \times 10^{-9}$. This would correspond to about a factor of four reduction in the uncertainty of the most accurate value of α currently available, that obtained from the current experimental value and theoretical expression for a_e and which is given in Eq. (6).

Similarly, there are a number of moving-coil watt-balance (MCWB) experiments currently under way that should provide values of h through the relation $K_J^2 R_K = 4/h$ with $u_r \approx 1 \times 10^{-8}$ (note that the MCWB determines this product directly). This would correspond to about a factor of nine reduction in the uncertainty of the most accurate value of h currently available, which was obtained using an MCWB at the authors' institution in the late 1990s and which has an assigned uncertainty of $u_r = 8.7 \times 10^{-8}$. A value of h with an uncertainty comparable to that expected from the several new MCWB experiments should also be available from an experiment being carried out by an international collaboration of researchers working at a number of different laboratories to determine the Avogadro constant by means of the x-ray-crystal-density (XRCD) method using a sample of silicon that has been highly enriched, that is, a sample that is 99.985% ^{28}Si (naturally occurring silicon contains about 92% ^{28}Si).

While achieving either of these advances in the determination of α or h would lead to significant reductions in the uncertainties of a number of constants, achieving both would lead to improvements in our knowledge of the values of a broad spectrum of constants. For example, u_r of m_e, e, K_J, and γ'_p would be reduced by about a factor of nine, while u_r of R_K, the Bohr radius a_0, the electron Compton wavelength λ_C, and the classical electron radius r_e would be reduced by about a factor of four. Unfortunately, however, no experiment is currently underway that would lead to a value of R with u_r smaller than 1.7×10^{-6}, which is the uncertainty of the CODATA 2002 recommended value of R. Thus, until such an experiment can be devised and carried out, our knowledge of the values of a number of physicochemical constants that depend on R, including the Boltzmann constant k and Stefan–Boltzmann constant σ, will remain unchanged.

Another area of quite significant current activity is the improved determination of the frequencies of various transitions in atomic hydrogen and deuterium, ^{1}H and ^{2}H, and the improved calculation of the theory of the energy levels of these atoms. Together, these advances will lead to a more accurate value of the Rydberg constant R_∞, which, with its current uncertainty of $u_r = 6.6 \times 10^{-12}$, is already among the most accurately known constants. They will also lead to improved values of the bound-state rms charge radii R_p and R_d of the proton and deuteron.

As pointed out above in our discussion of the importance of the fundamental constants, a major goal of metrology is the development of a reproducible and invariant system of measurement units, and the fundamental constants are critical to this endeavor. Of the seven base units of the SI – the meter, kilogram, second, ampere, kelvin, mole, and candela – only the kilogram is still defined in terms of a material artifact. Its definition is "The kilogram is the unit of mass; it is equal to the mass of the international prototype of the kilogram", which has the effect of fixing the mass $m(\mathcal{K})$ of the international prototype to be one kilogram exactly.

Although the international prototype has served science and technology well since it was adopted in 1889 as the unit of mass in the metric system – the forerunner of the SI adopted in 1960 – it has a number of limitations, not the least of which is that its mass may be changing with time. Thus, in recent years, serious thought has been given to redefining the kilogram in such a way as to fix the value of either the Planck constant h or Avogadro constant N_A. One such definition might be "The kilogram is the mass of a body at rest such that the value of the Planck constant h is exactly $6.6260693 \times 10^{-34}$ joule second". Although a u_r of about 1×10^{-8} in relating $m(\mathcal{K})$ to a fundamental constant has often been stated as being a desirable goal to achieve before the definition of the kilogram is changed, arguments have recently been put forward that would seem to justify doing so even before this level of uncertainty is reached. One of the great advantages of doing so now rather than later is that our knowledge of the values of many constants would be immediately improved. For example, if the above definition for the kilogram were to be adopted without delay, not only would the value of h be exactly known immediately, but the uncertainties of the values of N_A, m_e, and the masses of other fundamental particles would at the same time be reduced by a factor of 25, and the uncertainties of e, K_J, and the magnetic flux quantum Φ_0 would be reduced by a factor of 50.

In this scenario, $m(\mathcal{K})$ would be assigned the "conventional" value "one kilogram" and the international prototype would serve as the basis for a conventional mass measurement system in much the same way that the Josephson and quantum Hall effects, together with the conventional values of the Josephson and von Klitzing constants K_{J-90} and R_{K-90}, serve as the basis for the current worldwide conventional electrical measurement system. The redefinition of any base unit requires the approval of the Consultative Committee on Units (CCU) of the CIPM, the CIPM itself, and, most importantly, the CGPM. The CGPM meets every 4 years and next convenes in October 2007. It will be most interesting to see if in the coming few years the desired uncertainty goal for relating either h or N_A to $m(\mathcal{K})$ of $u_r \approx 1 \times 10^{-8}$ is reached, or if the responsible international bodies decide to proceed with the redefinition prior to it being reached.

"Numerology" and Time Variation

No article on the fundamental physical constants would be complete without at least a brief discussion of why the constants have the values they do, and whether they are in fact really constant. For example, why does the ratio of the rest mass of the proton to that of the electron m_p/m_e happen to be 1836.15...? Why does $1/\alpha$ happen to be 137.03...? Why not other, perhaps simpler, values? Three rather different points of view have evolved about the constants, especially their dimensionless combinations.

The first viewpoint is that the values of the constants are not at all arbitrary but can be calculated from some as yet unknown basic theory (or theories) in much the same way as the g-factor of the electron can be calculated from QED, or even in the way one can calculate the ratio of the circumference of a circle to its diameter (the constant π need not be experimentally determined, but can be calculated to arbitrary precision from pure mathematics). For example, in 1969 Wyler derived an expression for α involving only simple integers and π based on the volumes of certain bounded spaces associated with the invariance group of a relativistic quantum wave equation, and which yielded a value of α in excellent agreement with the recommended value at the time. However, the physical basis for the derivation was not at all clear, and it is safe to say that there does not yet exist a physically meaningful "derivation" of an accurate value for a fundamental constant, a dimensionless ratio of constants, or a particular combination of constants.

The second viewpoint is not aimed at explaining the values of the physical constants but simply notes that if they were terribly different from their currently observed values, we would not be here to measure them – life as we know it on earth, and perhaps the existence of the universe itself, depends on complex physical processes that require the constants to have their observed values. This is the idea behind the so-called weak anthropic principle. The strong anthropic principle goes even further by speculating that the fundamental constants and laws of physics must be such that life and the universe as we know it can exist. Nevertheless, if the constants were different, life might still be able to exist, but in a form that we cannot imagine.

An extreme statement of the third point of view is that the fundamental constants change with time as the universe evolves and therefore the currently observed values are not particularly significant. This idea had its origins in the 1937–1938 papers of Dirac. He noted that the ratio of the electrical to gravitational forces F_e and F_g between the electron and proton is of the order 10^{39}, very nearly equal to the age of the universe t expressed in an appropriate atomic unit of time, for example, the time required for light to travel a distance equal to the classical electron radius, which is a unit that is a combination of the constants e, m_e, and c. He further noted that the ratio of the mass of the universe to the proton mass m_p was approximately 10^{78}. This led him to speculate that the fact that such large numbers were of the form $a10^{39n}$, $n = 1,2\ldots$ and a of order unity, was not mere coincidence, but indicated a "deep connection in Nature between cosmology and atomic theory". He thus argued that if one adopts a system of units based on atomic constants and requires that the relations between such large numbers be invariant with time – that is, when the age of the universe in an atomic unit of time reaches, say 10^{50}, the F_e to F_g force-ratio must be 10^{50} and the universe to m_p mass-ratio must be 10^{100} – then the Newtonian constant of gravitation G must vary inversely with the age of the universe: $G \sim t^{-1}$.

Such speculations have stimulated a considerable amount of theoretical and experimental work over the last 70 years or so, the latter often involving astronomical or geophysical observations. For example, it was recently claimed by a group of researchers that their measurements of the absorption line-spectra of optical radiation reaching earth from distant quasars after traversing gas surrounding an intervening galaxy show that approximately 10^{10} years ago, the fine-structure constant was smaller than it is today by about the fractional amount 6×10^{-6}. However, these claims have not been substantiated by similar measurements of other workers. It is therefore quite safe to say that there is no widely accepted experimental evidence indicating that the value of a fundamental constant varies with time. Indeed, various experiments and data analyses have placed stringent limits (albeit frequently model dependent) on the possible time dependence of a number of physical constants. For example, in our epoch and region of space, it has been shown from lunar ranging data that G changes less than 1 part in 10^{13} per year, and from the laboratory comparison of different atomic transition frequencies over time that α changes less than 3 parts in 10^{15} per year. Clearly, for a long time to come, the changes in the values of the fundamental constants caused by man's inability to determine their exact values (see Tables 2a to 2e) will dominate any changes that might be caused by Nature's laws!

Conclusion

The following quotation by a famous physicist of the early decades of the 20th century seems to be a quite fitting conclusion to this article:

> *Why would one wish to make measurements with ever increasing precision?*
> *Because the whole history of physics proves that a new discovery is quite likely to be found lurking in the next decimal place.*

> F. K. Richtmyer, 1932

Bibliography

P. J. Mohr and B. N. Taylor, "CODATA recommended values of the fundamental physical constants: 2002", *Rev. Mod. Phys.* **77**, 1 (2005).

T. J. Quinn, S. Leschiutta, and P. Tavella (eds.), *Advances in Metrology and Fundamental Constants*, Proceedings of the International School of Physics "Enrico Fermi", Course CXLVI. IOS Press, Amsterdam, 2001.

P. J. Mohr and B. N. Taylor, "CODATA recommended values of the fundamental physical constants: 1998", *Rev. Mod. Phys.* **72**, 351 (2000).

E. R. Cohen and B. N. Taylor, "The 1986 adjustment of the fundamental physical constants", *Rev. Mod. Phys.* **59**, 1121 (1987).

E. R. Cohen and B. N. Taylor, "The 1973 Least-Squares Adjustment of the Fundamental Constants", *J. Phys. Chem. Ref. Data* **2**, 663 (1973).

B. N. Taylor, W. H. Parker, and D. N. Langenberg, "Determination of e/h Using Macroscopic Quantum Phase Coherence in Superconductors: Implications for Quantum Electrodynamics and the Fundamental Constants", *Rev. Mod. Phys.* **41**, 375 (1969).

E. R. Cohen and J. W. M. DuMond, "Our Knowledge of the Fundamental Constants of Physics and Chemistry in 1965", *Rev. Mod. Phys.* **37**, 537 (1965).

I. M. Mills, P. J. Mohr, T. J. Quinn, B. N. Taylor, and E. R. Williams, "Redefinition of the kilogram: a decision whose time has come", *Metrologia* **42**, 71 (2005).

Coriolis Acceleration

R. H. March

The apparent or "fictitious" acceleration attributed to a body in motion with respect to a rotating coordinate system (e. g., the frame of reference fixed with respect to the earth's surface) is known as the Coriolis acceleration. It is fictitious in the sense that if the same motion is described in a nonrotating (inertial Newtonian) reference frame, no such acceleration is present. It is merely an apparent deviation from inertial motion due to the choice of a noninertial reference frame.

Coriolis acceleration is experienced *only* by bodies with nonzero velocity in the rotating frame, unlike *centrifugal* acceleration, which is independent of the state of motion or rest.

If the body has linear velocity **v** with respect to a frame rotating with angular velocity as, the Coriolis acceleration **a** is given by the pseudovector expression

$$\mathbf{a} = -2\boldsymbol{\omega} \times \mathbf{v} \ .$$

In certain cases, the origin of Coriolis effects is intuitively obvious if the motion is described in an inertial frame. The simplest examples are those where the motion brings the object closer to or farther from the axis of rotation, changing the linear speed of motion due to the frame's rotation. For example, a falling body released from rest will be "deflected" to the east of a vertical line, because its point of release is moving faster than points that are below it and hence closer to the axis of rotation. By a similar argument, a projectile in the northern hemisphere will be deflected to the right, and one in the southern hemisphere to the left. At the equator, or for a trajectory that is due east or west, **a** is vertical and hence affects the range rather than the direction of the projectile.

Terrestrial applications of Coriolis acceleration include corrections to long-range artillery tables and the analysis of the precession of Foucault's pendulum. It is responsible for establishing the sense of rotation of large-scale weather systems, so that the wind pattern of a major storm is always "cyclonic" (counterclockwise) in the northern hemisphere and anticyclonic (clockwise) in the southern.

The phenomenon is named after Gaspard Gustave de Coriolis (1792–1843), who in 1835 published the first complete analysis of motion in rotating coordinate systems. Some Coriolis effects, however, were known much earlier. The eastward deflection of a falling body, for example, was cited by Robert Hooke and Isaac Newton as a possible experimental proof of the earth's rotation. In general relativity, certain terms that appear in the metric tensor of rotating coordinate systems are known as Coriolis terms, because for velocities much less than the speed of light and in weak gravitational fields they reduce to the classical expression for Coriolis acceleration.

See also: Dynamics. Analytical.

Bibliography
Keith R. Symon, *Mechanics*, 3rd ed., Chap. 7. Addison-Wesley, Reading, Mass., 1971. (I)
Herbert Goldstein, *Classical Mechanics*, 2nd ed., pp. 135ff. Addison-Wesley, Reading. Mass., 1980.
(A)

Corona Discharge

E. Barreto and U. Kogelschatz

A protruding convex region of radius a in an electric field E concentrates the electric flux by inducing excess charge on its surface. Thus an isolated region of high field intensity is produced when the gap d, over which a voltage difference exists, is much larger than the convex radius of curvature considered (e. g., when $d \gg a$). In a gas, at sufficiently high voltage, this condition produces local ionization that does not propagate across the whole gap d because the Laplacian field fades away in the low field region ($r \gg a$) that represents most of the space being considered. Instead, weak luminosity may be observed that can be either diffuse or attached to the stressed surface in one or more bright spots that move or become localized. This is a corona discharge. It might require dark-adapted eyes or even photon multiplication to detect its luminosity onset. As the voltage is increased across the gap d, the region of luminosity grows and eventually produces thin filaments that are called streamers. Positive streamers from a thin needle in air are particularly conspicuous and can be emitted at different angles to the axis of symmetry faster than the eye can resolve sequential events. They are responsible for the name corona. On the mast of a ship, sailors observe bunches of streamers and call them St. Elmo's fire. A pressure decrease causes the luminous corona region to expand because ionization is a function of E/n, with n the neutral number density. However, at a critical pressure (in air $\sim 10\,\text{mm Hg}$) coronas become impossible because of the onset of a new, diffusion-dominated, glow discharge. This is characterized by a weakly ionized glowing plasma ($T_e > T_n$) that fills most of the region between electrodes and can follow the contours of a discharge tube (e. g., a neon sign). In two dimensions, coronas occur when a wire or thin rod is surrounded by a much larger cylinder or placed parallel to a plane surface. In three dimensions they occur whenever an object protrudes into a uniform field configuration. Examples are an asperity in a metal surface, a needle facing a plane electrode, a tree high above others, or an airplane. Subsistence of a corona depends on the ability of the electrodes to remove the charge that reaches them. If isolated or nonconducting objects are involved, charge of polarity opposite to that originally on their surface will reach them. The discharge stops but a net charge transfer has taken place. This may charge or neutralize the objects involved. For instance, sharp wires are placed on the tips of airplane wings to promote local ionization that maintains the aircraft close to the potential of the region where it flies. Conversely, a nonconducting surface is purposely charged by coronas in electrophotography. Whenever a charged dust, powder, or spray is produced, induced coronas on the walls of the containing surface provide free charge that limits the space-charge density. These induced coronas indicate the onset of a dangerous condition in an explosive atmosphere and are detrimental for the purposes of electrostatic spraying in painting or agricultural fumigation. Coronas from liquid surfaces are associated with the production of filaments, charged aerosols, and surface oscillations.

For the sake of concreteness, consider only unipolar metal electrodes at atmospheric pressure. Avalanching electrons in the corona region are accelerated between collisions in the field direction (eE/m_e) but collide primarily and at high frequency with neutral molecules ($v_{en} \sim 10^{12}\,\mathrm{s}^{-1}$). Very rapidly, electrons obtain an equilibrium velocity distribution ($nt \sim 10^{15}\,\mathrm{m}^{-3}\mathrm{s}$; or $4 \times 10^{-11}\,\mathrm{s}$ at $n = 2.5 \times 10^{25}\,\mathrm{m}^{-3}$). The electron average energy is limited by inelastic collisions and is, therefore, always significantly smaller than the thresholds for optical excitation and ionization of the gas particles. Ionization is produced only by a small number of electrons in the high-energy tail of a primarily isotropic electron velocity distribution with an average speed, c_e that is much larger than the electron drift velocity, v_d, in the field direction: $c_e \gg v_d(E)$. Consequently, for ionization, it makes little difference whether the electrons drift into a positive electrode or away from a negative one. By contrast, the positive ions stay practically in place during the time it takes electrons to get into equilibrium with the electric field. Their space charge plays a dominant but very different role in positive and negative coronas.

Near a negative point, positive ions left by avalanches enhance the electric field at the electrode surface but reduce the extent of a high-field region from which free electrons have just been expelled at speeds thousands of times the drift velocity of ions ($\sim 100\,\mathrm{m/s}$). If the electrons attach, they become slow negative ions. These interact with the positive ions and the discharge stops until the negative ions can disperse far into the low-field region. A Trichel pulse is formed. In dry air, for point radii above $20\,\mu\mathrm{m}$, a series of equally spaced Trichel pulses are produced at their onset ($\sim 1.0\,\mathrm{kHz}$, $1.5\,\mathrm{ns}$ rise time, milliamperes peak current and $\sim 50\,\mathrm{ns}$ decay time). An increase in voltage increases their frequency but does not change their shape; this depends on the gas used. Each pulse maximum has been identified with the cathode region of a miniature glow discharge. This glow discharge is fully developed with Crooke's dark space, negative glow, Faraday dark space, and a positive column. This last one starts only about $100\,\mu\mathrm{m}$ from the surface and fades away in another $100\,\mu\mathrm{m}$ in the shape of a luminous, sharply divergent cone or brush. Trichel pulses start at frequencies of kHz, increase to MHz, and disappear to make the current steady (at about $120\,\mu\mathrm{A}$) with only a small jump in the current–voltage characteristic. At the high-frequency limit, pulses might become irregularly timed because two or more cathode spots may act simultaneously. This phenomenon can also be observed at the low-frequency limit for a pin with a large radius of curvature.

Near a positive electrode, positive ions decrease the geometrical field intensification. The discharge stops because the self-sustainment criterion is no longer fulfilled. It will start again when the positive ions disperse into the low-field region, or when a free electron is produced by photoionization in a nearby region that is, nevertheless, sufficiently far not to be affected by the positive space charge. A burst pulse is produced. These have 10^6–10^7 ions, durations of 10–100 μs, and average currents of 10^{-12}–10^{-10} A. Bursts traveling along a wire in an easily photoionizable mixture of gases were fundamental to the development of Geiger counters. If the burst has 10^8–10^9 ions the localized positive space charge facing the drift region provides a local field sufficient to ionize. Filamentary preonset streamers (~ 20 μm in diameter) are produced that travel a few millimeters at speeds of 10^5–10^6 m/s (larger than the electron drift velocity) into the low-field region.

It must be emphasized that coronas depend on localized fields due to a geometrical configuration. Characteristic times and lengths in the ionization region are nanoseconds and micrometers, while in the drift region they become milliseconds and meters. In air, coronas start at about 10 kV if a rounded rod 0.1 cm in radius is used in a 1.0-cm point-to-plane gap. However, if a rod 0.02 cm in radius is used in the same gap, a corona starts at about 5.0 kV. Atmospheric pressure, humidity, and polarity affect the actual onset value. In all cases, the electric field near the stressed electrode is many times the magnitude of the breakdown value in a uniform field. An increase in voltage causes ions in the drift region to disperse faster and the corona pulses (Trichel, or preonset streamers) to become closer. Eventually steady current and luminosity are established around the stressed electrode. The current density distribution in a point-to-plane drift region is given by $j(\theta) = j(0) \cos^m \theta$; θ is the angle from the axis of symmetry. For positive coronas $m = 4.82$, for negative ones $m = 4.65$. $j(\theta) = 0$ for $\theta > 60$ degrees. This distribution is known as Warburg's $\cos^5 \theta$ law. The average corona current was calculated in 1914 by Townsend to be proportional to $V_{gap}(V_{gap} - V_{onset})$.

As the current increases, the corona luminous region also grows and the field becomes less divergent. Eventually, a breakdown streamer about 50 μm in diameter will be emitted into the low-field region ($r \gg a$) at speeds between 10^4 and 10^6 m/s depending on a, d, and the gas composition. In uniform fields, streamers propagate simultaneously in both directions when an avalanche is purposely started in midgap and electron amplification reaches about 10^8 in a distance of 3–5 mm. Negative streamers (anode directed) are timed to be faster than the positive ones. In all cases, streamers have speeds larger than the corresponding electron drift velocity in the given electric field and may exhibit changes in direction that do not follow the Laplacian field configuration. This shows that they are driven by local space-charge accumulation that provides the field which pushes them into unionized gas. In molecular gases such as air (usually "laboratory air": 79% N_2, 21% O_2) the electrons in a streamer have a small average energy (2–4 eV) fixed by a high peak in the cross section of N_2 due to a vibrational excitation peak at 2.5 eV. Only molecular radiation is observed and indicates that the gas is not heated. The electron densities are 10^{18}–10^{21} m^{-3}. This picture of a streamer as a fairly conducting filament where net charge is limited to very small distances contrasts with an earlier, alternative, model where a positive streamer has a head with net positive space charge that results because of the difference in mobility between electrons and ions but without multiple interactions or Coulomb shielding. Positive streamers can be injected through a small aperture into a uniform field and propagate at field strengths between 4×10^5 and 5×10^5 V/m for distance at least, but probably exceeding, 1 m. When streamers cross a discharge gap, the current either significantly decreases or saturates before a spark is produced.

Corona discharges have found many industrial applications including electrostatic precipitators, electrophotography, copying machines, printers, liquid spray guns, and powder coating processes.

See also: Arcs and Sparks; Electrophotography; Lightning.

Bibliography

Yu. S. Akishev, I. V. Kochetov, A. I. Loboiko and A. P. Napartovich, "Numerical Simulations of Trichel Pulses in a Negative Corona in Air", *Plasma Phys. Rep.* **28**, 1049–1059 (2002).

N. Yu. Babaeva and G. V. Naidis, "Modeling of Streamer Propagation", in *Electrical Discharges for Environmental Purposes*, E. M. van Veldhuizen (ed.), Ch. 3, pp. 21–48. Nova Science Publishers, New York, 2000.

J. S. Chang, P. A. Lawless and T. Yamamoto, " Corona Discharge Processes", *IEEE Trans. Plasma Sci.* **19**, 1152–1166 (1991).

J. A. Cross, *Electrostatics: Principles, Problems and Applications*. Adam Hilger, Bristol, 1987.

J. M. Crowley, "Electrophotography", in *Wiley Encyclopedia of Electrical and Electronic Engineering*, J. G. Webster (ed.), Vol. 6, pp. 719–734. Wiley, New York, 1999.

W. Egli, U. Kogelschatz, E. A. Gerteisen and R. Gruber, "3D Computation of Corona, Ion-Induced Secondary Flows, and Particle Motion in Technical ESP Configurations", *J. Electrostat.* **40/41**, 425–439 (1997).

E. E. Kunhardt and L. H. Luessen, eds. *Electrical Breakdown and Discharge in Gases*, NATO ASI Series Vols. 89a and 89b. Plenum Press, New York, 1983.

L. B. Loeb, *Electrical Coronas: Their Basic Physical Mechanisms*. University of California Press, Berkeley, 1965.

M. K. Mazumder, "Electrostatic Processes", in *Wiley Encyclopedia of Electrical and Electronic Engineering*, J. G. Webster (ed.), Vol. 7, pp. 15–39. Wiley, New York, 1999.

K. R. Parker (ed.), *Applied Electrostatic Precipitation*. Blackie, London, 1997.

R. S. Sigmond, "Corona Discharges", in *Electrical Breakdown of Gases* (J. M. Meek and J. D. Craggs, eds.). Wiley, New York, 1978; also, *J. Electrostatics* **18**, 249–272 (1986).

Cosmic Rays, Astrophysical Effects

S. A. Colgate

Overview of Cosmic Rays

Cosmic rays are energetic particles incident upon earth from sources that include the sun, the galaxy, and other galaxies. The most important particles are nuclei, but an electron flux of a few percent is also present. In addition to high-energy gamma rays and secondary products, neutrinos and antiprotons are also now considered as cosmic rays.

Energetic Particles that are not Cosmic Rays

The extraordinary pulse of neutrinos, ($\sim 50\,\mathrm{MeV}$), detected from the neutron star collapse, initiating the supernova event 1987a in the Large Magellanic Cloud, might also be considered as cosmic rays, but for an important distinction: namely these neutrino particles are emitted from a thermal source. One would like to define cosmic rays as energetic cosmic particles originating from nonthermal sources – if for no other reason than we understand thermal sources and the origin of nonthermal sources is still open to debate. Gamma ray bursts are somewhat ambiguous in their definition since the origin of gamma bursts is still open to argument.

The Astrophysical and Intellectual Significance

The characteristics of cosmic rays that are significant astrophysically are (1) the energy spectrum; (2) the composition or number frequency of various nuclei or positron–electron ratio; and (3) the angular distribution, commonly measured as the degree of anisotropy. The intellectual content is the question of the origin of cosmic rays, the reasons for their distributions in energy, angle, and space, and finally the effects of cosmic rays and the implications of their origin on the structure of the universe.

Measured Quantities

The Spectrum

The energy spectrum describes the most surprising feature of cosmic rays, namely, the extraordinary energy of individual particles and the large total energy flux. The kinetic energy of individual particles is measured variously in eV (electron volts) per nucleon, MeV (10^6 eV), GeV (10^9 eV), TeV (10^{12} eV), and EUV (10^{19} eV). When cosmic rays were first discovered in the early 1920s as residual ionization in an ion chamber and later as pulses in a Geiger counter and later as tracks in a cloud chamber and nuclear emulsion, i. e., photographic emulsion, the implied energy necessary to penetrate the earth's atmosphere, because of ionization loss alone, was necessarily greater than a GeV, i. e., the rest-mass of the proton. The flux was like rain, of the order of $1\,\mathrm{cm}^{-2}\,\mathrm{s}^{-1}$.

Figure 1 shows a composite of of the presently most likely spectrum of nonsolar cosmic rays. The most striking feature is the extension of a nearly constant power law, $\mathrm{d}\log n/\mathrm{d}\log E = -\Gamma \simeq -2.7$ or the differential spectrum $\mathrm{d}n/\mathrm{d}E = n_0 E^{-\Gamma}$ over a range of 10^{11} in energy, from 10^9 to 10^{20} eV, and then also the extraordinary upper energy limit of $\sim 3 \times 10^{20}$ eV. However, the small deviations from a constant power law, Γ are important to the interpretation of galactic versus extra galactic origin. A particle of this extraordinary energy, 10^{20} eV or ~ 10 joules is only slightly deflected by the magnetic field of the whole galaxy and at 10^{19} eV the galaxy would contain less than one Larmor orbit. This compels the conclusion that some particles must reach us from the universe outside our galaxy. Hence the term galactic cosmic rays is too parochial and we must consider extra galactic as truly the "cosmic" rays of equal or greater importance. As we discuss later the implied total energy of extra galactic cosmic rays within a galaxy spacing volume (the volume between galaxies) is $\sim 10^6$ greater than within the galaxy itself. Of equal importance is the composition of cosmic rays in A, atomic number and Z, charge. The near constancy of composition in the low energy range implies a near charge independent acceleration mechanism, a significant theoretical challenge.

Composition

The composition of cosmic rays surprisingly mimics the abundance distribution of the average matter of galaxies with a major exception of the light elements, Li, Be, and Boron. These elements are highly deficient, by a factor of $\sim 10^{-3}$, within stars and galaxies, compared to

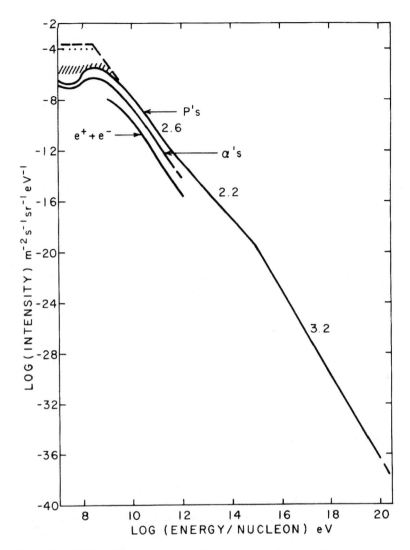

Fig. 1: The differential energy spectra of cosmic rays as log intensity versus log energy. The numbers attached to the curve are the negative exponent of the corresponding power law that approximates the measurements. Between 10^{15} eV and 10^{18} eV the spectrum is slightly steeper and the primary particles are progressively enriched in atomic number, most likely nuclei up to iron in charge, rather than protons (p's) as indicated at lower energy. A likely explanation of this is discussed in the text. At lower energies the composition ratios are nearly constant. The low-energy spectra of helium nuclei (α's) are shown and the curves for the heavier nuclei are similar but of lower intensity as discussed in the text. The electrons (e^+, e^-) have similar spectra. The shaded curve is the estimated galactic flux as interpreted (with some questions) from satellite measurements. The dashed curve is the flux required for interstellar cloud heating and the dotted is that required for pressure support of the galactic disk.

the average of other slightly heavier elements, C, N, O, because of the ease by which they are converted by thermonuclear reactions to heavier elements in stellar nuclear synthesis. Their highly enriched presence in cosmic rays is interpreted as due to the partial spallation (proton–nuclear collisions) of heavier ($A \geq 12$), more common nuclei, which make up the bulk of heavy cosmic rays. The fractional spallation interpreted from the composition ratios, $\sim 50\%$ implies that heavy cosmic rays diffuse out of the galaxy just about as frequently as they are destroyed by collisions while traversing galactic matter with a density of $\sim 5\,\mathrm{g/cm^2}$ before escaping from the galaxy. This leads to an estimated life time of cosmic rays within the galaxy of $\sim 10^6$ years. An independent estimate of the life time of cosmic rays has been made from the ratios of the isotopes produced in the radioactive decay of ^{7}Be. This gives a life time of ten times longer than the spallation time. This factor of ten difference is reconciled by assuming that the magnetic field of the galaxy is extended as a magnetized halo filled with cosmic rays, with a thickness about 1 kpc, about ten times the thickness ($\sim 100\,\mathrm{pc}$) of the stellar disk as observed for stars and gaseous matter, which in turn is the spallation "target". The difficulty with this picture is that the observed ratio of the spalled versus non spalled composition is nearly independent of energy and yet cosmic ray loss from the galaxy is believed to be by particle diffusion in the magnetic field. This diffusion may be composed of a random walk of Larmor orbits, or alternately as particles following a turbulently distorted field with a correlation length much larger than their Larmor orbit radius and therefore independent of energy. On the other hand, shock acceleration, the favorite acceleration model depends upon particle scattering by Alfvén waves and therefore Larmor orbit scattering. If this were the case for all scattering, the time for diffusion of a particle out of the galaxy should be $\propto R_L^{-1} \propto E^{-1}$, leading to a highly energy-dependent ratio of spallation products (light nuclei) to primaries, contrary to what is observed.

The Exclusion of Solar Cosmic Rays

Finally, if one excludes the highly variable solar flux of cosmic rays, then this allows one to consider galactic and extra galactic cosmic rays as a unified subject where the measurements have been exceedingly numerous and self consistent over long periods of time.

Solar cosmic rays, on the other hand, have been recognized as correlated with solar activity and contribute a highly variable composition and spectrum which is almost exclusively less than $\sim 1\,\mathrm{GeV}$ and is mostly in the energy region of 10 to $\sim 100\,\mathrm{MeV}$. The lack of constancy of solar-cosmic-ray measurements has made their interpretation of origin and flux difficult. In addition this variability, reflected in the solar wind, leads to a modulation in flux of the observed galactic cosmic rays up to energies of several GeV.

Isotropy of Cosmic Rays

In Fig. 2 we show the compiled data of some years ago of the surprising high degree of isotropy of galactic cosmic rays. As one goes to higher energy both the statistical accuracy decreases because of a smaller flux and also an anisotropy emerges. However, this general picture has not significantly changed as more data has accumulated. Instead its interpretation and extension to ultra high energies has lead to a continuing confrontation with theories. These will be discussed later in terms of acceleration and sources.

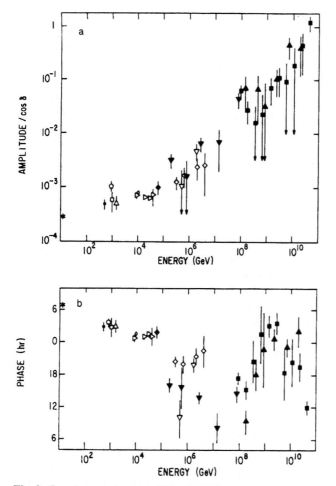

Fig. 2: Cosmic ray anisotropy. The panels show measurements of the equatorial projected amplitude and phase of the first harmonic of count rate versus energy (δ being the latitude of the experiment), compiled by J. Linsley, *Proc. 18th International Cosmic Ray Conf. (Bangalore)* **12**, 135 (1983).

Time Variability

Galactic cosmic rays exclusive of gamma ray bursts, but including high-energy gamma rays as secondary products of pion decay, are constant in time, in all measured quantities, as one would expect of charged particles diffusing in the galactic magnetic fields. The total flux has evidently not changed significantly in 10^8 years as measured by fossil tracks in lunar rocks.

Astrophysical Effects of Cosmic Rays

Pressure and Heating by Cosmic Rays

The pressure and heating by cosmic rays gives rise to the principal astrophysical effects of cosmic rays both within and external to our galaxy. The bulk of the energy and hence pressure is calculated by integrating E times the spectrum with the result that the total energy, $W_T \propto E^{(2-\Gamma)}$. This result is sensitive to the low-energy cutoff, which we assumed is $m_p c^2$. The heating rate per particle, for $E < mc^2$, is $\propto N_{CR} E^{-0.8}$, and so since $N_{CR} \propto W_T$, both are a negative power of the energy per particle, and so heating and pressure are concentrated in the low-energy part of the spectrum, i. e., $< 1\,\mathrm{GeV}$.

In Fig. 1 the flattening of the spectrum below $\sim 1\,\mathrm{GeV}$ and hence reduction in the total energy of the CRs is variously attributed to the properties of the source(s), energy loss in propagation, and partial exclusion from the solar-system magnetosphere of cosmic rays originating from elsewhere in the galaxy. The actual galactic flux at low energy ($E \lesssim 300\,\mathrm{MeV}$) is not known accurately although several secondary indicators are considered. The satellite measurements out to $\sim 3\,\mathrm{A.U.}$ would indicate a galactic flux within the shaded area of Fig. 1. This is less than the flux required for the heating of the interstellar medium by low-energy cosmic rays, which would require a flux indicated by the dashed line. Finally, the support of the gas of the galactic disk by cosmic-ray pressure would require an intermediate flux indicated by the dotted curve. The heating and pressure support of the galactic gas by cosmic rays are each major theories of galactic structure. The fact that both these lines lie above the extrapolation of the satellite measurements is taken as an indication that these major theories of galactic structure may have to be revised. The expected degree of ionization of the interstellar gas from cosmic ray heating is also not observed. On the other hand, two theories of the origin of cosmic rays – the shock acceleration by supernovae and the quasi-continuous reconnection acceleration both lead to spectra at least as steep as that required to meet galactic structure theories. (A steep spectrum implies that if one extrapolates a given cosmic ray flux at say $m_p c^2 = 10^9\,\mathrm{eV}$ with a steeper slope to lower energy, the flux, heating, and total energy will be greater than with a flatter slope.)

Energy Density and Spacial Distribution

A series of satellite measurements of the 100-MeV gamma-ray distribution in space has been interpreted as due to the creation of π^0 mesons by cosmic rays of energy $\simeq 1\,\mathrm{GeV}$ colliding with hydrogen atoms of the interstellar medium and where the mesons subsequently decay to two gamma rays. The spacial distribution of the gamma rays is consistent with cosmic rays being distributed uniformly with constant flux within the galaxy, $\simeq 1\,\mathrm{cm}^2/\mathrm{s}$ or an energy density $1\,\mathrm{eV/cm}^3 = 1.6 \times 10^{-12}\,\mathrm{ergs/cm}^3$. This flux in turn is consistent with the radio synchrotron emission flux from electrons at $\simeq 3\,\mathrm{GeV}$ and a density of 1% of the proton density. This does, however, conflict with two theories of cosmic-ray heating and cosmic-ray pressure support of the galactic gas. On the other hand, the natural tendency for cosmic rays to fill the low-density holes or tunnels in the interstellar medium leaves these latter conclusions possibly uncertain.

Cosmic-Ray Acceleration

Adiabatic Energy Loss from Expansion: Discrete Event versus Uniform Acceleration Theory

There are two primary theories of cosmic ray acceleration the one, shock acceleration primarily in shocks produced by supernova in the inter stellar region, ISM, and the second is the acceleration by parallel electric fields produced by reconnection of force-free fields. (A force-free field is one where the current is parallel to the field so that $\mathbf{J} \times \mathbf{B} = 0$. In shock acceleration

$\mathbf{J} \times \mathbf{B} = \nabla P_{\text{shock}}$, the jump in pressure across the shock.) Shock acceleration has had the major concentration of theoretical endeavor and is to be considered a mature theory that is nearly universally accepted. However there are a few points that challenge this acceptance. The first point is that there is a lack of experimental evidence of shock acceleration in the laboratory. By contrast the reconnection acceleration process has been demonstrated countless times in plasma laboratory experiments. Heliospheric shocks in the plasma within the planetary system are sometimes quoted as proof of shock acceleration, but the observed multiplication of particle energy has been no more than ten times the thermal energy, $E < 10kT$, far less than what is needed to test a scale-independent high energy acceleration process.

A second problem with the shock acceleration theory, in addition to the lack of experimental proof, is the consequences of the adiabatic energy loss due to the expansion of the over-pressurized nebula of cosmic rays produced in a singular acceleration event such as a supernova. This over-pressurized nebula blows a bubble in the ISM and the necessary $P\,dV$ work to create the bubble by the cosmic-ray pressure greatly reduces the total cosmic ray energy content by adiabatic expansion. Only if cosmic rays are accelerated fairly uniformly in space throughout the whole galaxy and at the near uniform pressure of the galaxy can the difficulty of adiabatic energy loss be totally avoided; on the other hand, if the interstellar medium is as inhomogeneous as present measurements would indicate with minimum densities less than 10^{-2} of the mean, then single-origin events like supernovae or pulsars might allow cosmic rays to escape into low-density regions non adiabatically. However, quasi-spherical confinement bubbles such as the Crab nebula, illuminated by GeV electrons, leads one to believe that confinement and adiabatic loss should significantly modify any discrete event explosion type source of cosmic rays such as a supernova shock.

Once adiabatic expansion terminates when pressure equilibrium is attained, then cosmic rays should propagate preferentially in low-density regions which are naturally interconnected and have been described as "tunnels" in the interstellar medium. The modification of the spectra of the low-energy cosmic rays as they propagate in such an inhomogeneous region is unknown so that the source spectrum of low-energy cosmic rays below $\sim 1\,\text{GeV}$, although of major importance to galactic processes, must be accepted as uncertain. On the other hand the average expansion of a source "bubble" to the lower density of equilibrium pressure will lead to a fractional loss directly proportional to the pressure or linearly with radius of a spherical volume V since cosmic rays collectively behave as a relativistic gas where $P \propto V^{1/3}$.

Total Energy Requirements

The total volume of the galaxy including the halo out to a radius of 10 kpc and thickness 1 kpc is $\sim 300\,\text{kpc}^3 =\sim 10^{67}\,\text{cm}^3$. When this volume is filled with a mean energy density of one half eV, half the local flux, the total energy in cosmic rays within the galaxy becomes $\simeq 10^{55}$ ergs. Since the mean lifetime is 10^7 years from the Be measurements, the power necessary to sustain the galactic cosmic ray flux becomes $\simeq 3 \times 10^{40}\,\text{ergs/s}$. If we had taken the spallation life time of 2×10^6 years, then we would have to use the matter thickness of the galaxy or 1/5 the halo thickness and thus 1/5 the volume with the resulting same power requirement. One might rightfully compare this energy requirement to several other galactic processes:

1. The binding energy of the galaxy divided by a Hubble time is $\simeq 10^{41}\,\text{ergs/s}$, barely greater.

2. Supernova kinetic energy times supernova rate (one in 30 years) or 10^{51} erg / 30 years $\simeq$ 10^{42} erg/s. This is ample ample energy, ~ 30 times the requirement, but requires 100% acceleration efficiency and is subject to expansion loss.

3. Neutron star spin-down energy, 100 Hz, or $\simeq 3 \times 10^{50}$ erg per supernova or 3×10^{41} erg/s, is sufficient but the magnetic coupling of angular momentum by flux winding and reconnection acceleration is uncertain.

4. Star formation binding energy can be converted to magnetic winding energy. Here a star formation rate of ~ 10 stars/year, and a binding energy of $\sim 10^{48}$ erg/star (T-tauri stars, the earliest stars, have 1/10 of this energy) or 3×10^{40} erg/s. This energy is just sufficient, but this source requires 100% of the stellar binding energy to be transformed into magnetic energy. Since this energy source of stellar winding appears marginal, we compare just supernova shocks as the energy source (the accepted theory of shock acceleration) to the energy of neutron star spin energy.

Available Energies: SN Shocks and Neutron Star Spin

Neutron star spin energy is adequate to account for galactic cosmic rays provided there is an efficient mechanism to convert the free energy of spin to accelerated cosmic rays. It is already known that the torque necessary to spin down neutron stars by the magnetic field of the neutron star, $\sim 10^{12}$ G, is sufficient. Several mechanisms have been suggested for cosmic ray acceleration in pulsars, the escape of relativistic particles associated with the light cylinder modulation by the pulsed phenomena, and a second, the formation of extensive helical force-free magnetic fields by the rotational twisting of a polar cap flux tube that subsequently undergoes reconnection and acceleration. In the Crab nebula both mechanisms appear feasible.

In shock acceleration theory a given particle diffuses ahead of the acceleration zone, the shock, (by Alvén wave scattering) a distance x, at steady state, exactly as fast as the shock progress, v_s. Hence over larger dimensions, the diffusion velocity is $x/t \propto D/x$, where D is the diffusion coefficient and x is the growing diffusion zone width. The velocity of growth of the diffusion zone becomes slow compared to the expansion velocity and the recently shock accelerated particles are confined as a diffusing cloud. Thus they are subject to adiabatic energy loss where the pressure of the recently accelerated cosmic rays does $P\,dV$ work on the inter stellar medium, the ISM. The initial supernova shock pressure where the maximum $P\,dV$ work can be performed on accelerating cosmic rays is $P_{SN} \simeq \frac{1}{3}\rho_{ISM}E_{SN}/M_{SN} \simeq 3 \times 10^{-7}$ dynes/cm^2 when 10^{51} ergs is deposited in an equal mass of the ISM, or in $10M_\odot$. This is to be compared to the pressure of the ISM, which at equipartition with the cosmic rays is the magnetic pressure due to 6μG $\simeq 1.6 \times 10^{-12}$ dynes/cm^2. This results in a pressure ratio of $\simeq 10^5$. Such a large adiabatic loss would be prohibitive for cosmic ray acceleration. However, as the shock propagates in the ISM, its velocity decreases as in an inelastic collision as $v_s \propto 1/M_{ISM}$ where M_{ISM} is the mass of the ISM that has been over-taken by the shock. In a strong shock the energy is divided equally between heat behind the shock, presumably the accelerated cosmic rays, and kinetic energy of motion. Hence the available energy in the shock, the heated fraction is $W_{shock} \propto M_{ISM}^{-1}$ and $P_{shock} \propto M_{ISM}^{-2}$. The result is that the effectiveness of the shock for acceleration with adiabatic losses included decreases as $W_{shock} \propto$

$P_{shock} \propto 1/M_{ISM}$. An integration leads to a useful accelerated fraction as small as 10^{-4}, which is far too small to account for cosmic ray acceleration. Corrections to this estimate such as using the Sedov solutions for an expanding shock in a uniform medium is not likely to alter this estimate of efficiency by a significant factor. When one also recognizes various inefficiencies in shock acceleration, the energy problem with supernova shock acceleration becomes still more difficult.

Shock Acceleration

One possible circumvention of adiabatic losses of a singular large injection event becomes some unknown form of non adiabatic loss. However, such a non adiabatic loss confronts the very basis of shock acceleration theory. In this theory relativistic particles, cosmic rays, scatter very many times back and forth across the shock gaining an increment of energy on each traversal. The scattering takes place adiabatically from large, $\sim 10\%$, amplitude Alfvén waves, of roughly several Larmor radii in wave length. This scattering statistically reverses the direction of particles on each side of the shock so that the particles cross the shock many times thereby gaining many increments of energy. However, each crossing requires a further large number of scatterings in order to return across the shock. Since the fractional increment in energy gained per shock traversal is $\Delta E/E \simeq V_{shock}/c$, the number of crossings necessary to double in energy becomes $n_{cross} \simeq c/v_{shock}$ or very many times. Similarly the number of scatterings per crossing is the same ratio, so that the number of scatterings per doubling in energy is $n_{scatter} \simeq (c/v_{shock})^2 \sim 10^5$. Since each doubling in energy requires this very large number of scatterings, even the smallest nonadiabatic loss, $\leq 10^{-5}$ per scattering, necessary to allow escape from scattering confinement would likely greatly interfere with the acceleration process. Therefore adiabatic losses are inevitable with shock acceleration unless the diffusion coefficient increases sufficiently rapidly with expansion. Such a model will be discussed in connection with the "leaky box" model.

The Leaky Box Model

A standard picture of cosmic rays in the Galaxy has been the "Leaky Box Model". In this model cosmic rays both "leak" from the galaxy into the "metagalaxy", the intergalactic space external to galaxies and collide with the nuclei of atoms or molecules of inter stellar space, the ISM. In this fashion there are then two loss mechanisms for cosmic rays: one by collisions and the other by diffusing out of the galaxy. This model was invented when only a small portion of the spectrum of the light elements Lithium, Berylium, and Boron had been observed in cosmic rays. These elements are under-abundant in nature, $\sim 10^{-3}$ to 10^{-4}, because, as pointed out before, they are destroyed easily in thermonuclear burning in stars, leading to the abundant elements, C, N, and O. However, both groups of elements were present in cosmic rays in comparable numbers, implying that some of the higher energy C, N, O had been spalled in nuclear collisions with the primarily protons of the ISM and in turn some had not collided. In order to achieve steady state with a mean life time of $\sim 10^7$ years and a galactic life time of a Hubble time of $\sim 10^{10}$ years, required that some fraction of the cosmic rays must be continuously lost. This was explained as diffusion out of the galaxy, the "leaky box" model.

This model has lasted despite the observation of a constant ratio of light to heavy nuclei over a much larger range of energy of $E \leq 1\,TeV$. The difficulty with this picture is that if the diffusion time and spallation time are roughly equal at one energy then at a higher (or lower) particle energy diffusion by scattering from Alfvén waves (proportional to the respective Larmor radii, R_L, or energy) will be different by $t \propto R_L^{-1} \propto E^{-1}$. Hence, the fractional spallation during escape from the galaxy would be vastly different as a function of particle energy. Two models can save the leaky box model.

In the shock acceleration theory the scattering of particles back and forth across the shock depends upon the excitation of large amplitude Alfvén waves by the steep gradient in energetic particle density ahead of the shock. The gradient in number density is created by the diffusion gradient of the particles by the wave scattering itself. The excitation of the waves and consequently the diffusion coefficient is inversely proportional to the particle gradient, and so one obtains an energy-independent diffusion.

Let the maximum wave amplitude be limited by non-linear effects (i. e., wave breaking) such that the initial scattering mean free path in the shock is $\lambda_0 = 10 R_L$ where R_L is the Larmor orbit of a particle of energy . In the 6 μG field of the Galaxy, $R_L \simeq 10^{12}\,\mathrm{cm} = 3 \times 10^{-7}\,\mathrm{pc}$ for $E > m_p c^2$. The initial shock thickness, δ_0 must be such that for a given particle energy, this thickness must remain near constant during propagation at a constant velocity, v_{shock} so that $\delta_0 = \sqrt{(\lambda_0 (c/3) t)}$ where, in order to be near constant, $t = (\delta_0 / v_{\mathrm{shock}})$. Then $\delta_0 = \lambda_0 c / (3 v_{\mathrm{shock}})$. Then with the assumption that the wave excitation and hence scattering mean free path is proportional to the diffusion gradient, δ / δ_0, one then obtains $\delta = \sqrt{(\lambda_0 c/3)(\delta/\delta_0) t} = v_{\mathrm{shock}} t$. In other words, when the scattering length is proportional to the diffusion gradient, which is also the width, the diffusion velocity becomes a constant, the shock velocity, v_{shock}. This then saves the leaky box model and partially reduces the adiabatic loss in the supernova shock acceleration model.

In this model at a typical SN shock velocity of $3 \times 10^8\,\mathrm{cm/s}$ in $10^6\,\mathrm{years} = 3 \times 10^{13}\,\mathrm{s}$, the diffusing particles have reached 3 kpc, well beyond the bounds of the galactic halo. However, in order for diffusion to transport the particles to this distance in this time, the scattering length has become larger than the reversal length of the fields in the ISM. This length has been measured to be $\lambda_{\mathrm{revers}} \simeq 0.5\,\mathrm{pc}$ depending upon inter-cloud turbulence. This then limits the scattering length of the gradient dependent diffusion to no greater than this constant length independent of energy up E_{crit} where $R_L = 0.5\,\mathrm{pc}$, or $E_{\mathrm{crit}} = 10^{15}\,\mathrm{eV}$. The diffusion coefficient with this scattering length $D = \lambda_{\mathrm{revers}} c / 3 = 1.5 \times 10^{28}\,\mathrm{cm^2/s}$, resulting in a diffusion distance of $\sim 1\,\mathrm{kpc}$ in a time of 10^7 years. Thus must particles diffuse from the shock as a gradient dependent diffusion until reaching a scattering length of $\lambda_{\mathrm{revers}}$ and thereafter diffuse out of the halo in 10^7 years. The attractive feature of this theory is that at $E > 10^{15}\,\mathrm{eV}$ particles will diffuse out of the galaxy more rapidly than all lower energy ones, which have a constant loss time. The consequence is the knee where a steeper slope of the spectrum implies just such a loss.

Another possible explanation of the near equal spallation and loss of the leaky box model is that cosmic rays are continuously re-accelerated as well as new ones added. In this picture the spallation products are accelerated the same as the primaries and their near equal numbers is a result of their near equal probability for acceleration and spallation. The protons, on the other hand accumulate relative to higher-Z nuclei because of their smaller cross section for loss. The loss that was presumably "leaking out of the box" is in reality the loss by further acceleration. Additional analysis and modeling is needed to decide among these possibilities, but regardless, a lifetime of cosmic rays of 10^7 years and the isotropy measurements have implications for the likely sources of acceleration.

Isotropy and the Discrete Energetic Events

In the energy range $10^{11} \le E \le 10^{14}$ eV, the measurements of isotropy and spectral charge dependence lead to information concerning the properties of the source of cosmic rays. There are two measurements of dominant concern:

1. The near-independence of the energy spectra in eV per nucleon regardless of the wide range of energy and of nuclear charge is a major challenge to any acceleration mechanism.

2. The incredible isotropy of the flux independent of arrival direction, is as low as 5×10^{-4} at energies great enough, $E > 10^{12}$ eV, to be unaffected by terrestrial and solar-system magnetic fields, Fig. 2. This extraordinary isotropy continues to $E < 10^{14}$ eV before larger anisotropies become evident at higher energies. However, the extraordinary isotropy extends all the way to the highest energies observed $\sim 10^{20}$ eV. The increasing error bars as a function of energy are a reflection of the limiting statistics of a decreasing number of higher energy particles. Before considering possible origins of this directional independence, the opposite extreme of the small anisotropy at lower energies will be considered.

In the leaky box model half the particles diffuse out of the galaxy halo in $t = 10^7$ years independent of energy, and therefore, in turn, would not diffuse a larger distance laterally, within the disk, than the half thickness of the halo, or 0.5 kpc. The fluctuation in number and spacing of supernova events within this volume is determined by the expected number of supernova events in 10^7 years. Since a supernova occurs roughly every 30 years within the Galaxy, the number per cosmic ray lifetime is $N_{\text{galaxy}} = 3 \times 10^5$ and the number per diffusion volume, $N_{\text{diff}} = N_{\text{galaxy}}(A_{\text{diff}}/A_{\text{galaxy}}) \simeq N_{\text{galaxy}}/400 = 7.5 \times 10^2$ where the Galaxy area, $A_{\text{galaxy}} = 100\pi \text{kpc}^2$ and the area of diffusion is $A_{\text{diff}} \simeq 0.25\pi \text{kpc}^2$. The fluctuation in number between any two hemispheres in a direction centered on us is $\sqrt{N_{\text{diff}}} \simeq 0.03$. This expected fluctuation in number will give rise to the same expected anisotropy, but the minimum anisotropy measured over $10^{12} < E < 10^{14}$ eV is smaller by a factor of 1/50. The chance or likelihood of this isotropy with a statistical error so much larger is negligible and so we conclude that either a more uniform acceleration mechanism is operating, or that the interpretation of the leaky box model needs revision. We note that if gradient diffusion is not invoked to give an energy independent diffusion, and acceleration loss is invoked instead, then the isotropy problem becomes still more difficult, because the diffusion sphere becomes smaller.

The spectrum from the Knee to the Ankle

The steepening of the spectrum above $E > 10^{15}$ eV is expected on very general grounds, because at this energy and at still greater energies the diffusion time out of the galaxy/halo becomes less than the confinement time at lower energies independent of gradient diffusion or turbulent distortion of the field, and so an enhanced or more rapid loss is expected. If cosmic rays were only accelerated within the Galaxy and there were only protons, then we would expect a rapid fall off in intensity above the energy of confinement to the value corresponding to no confinement at all. This ratio of predicted confined to unconfined fluxes is the transit time of the galaxy thickness, $t_{\text{transit}} = 1 \text{kpc}/c = 3000$ years versus $t_{\text{diff}} = 10^7$ years for the confined flux. This corresponds to a decrease in flux of 3×10^{-4}. In Fig. 3 the ratio of the low

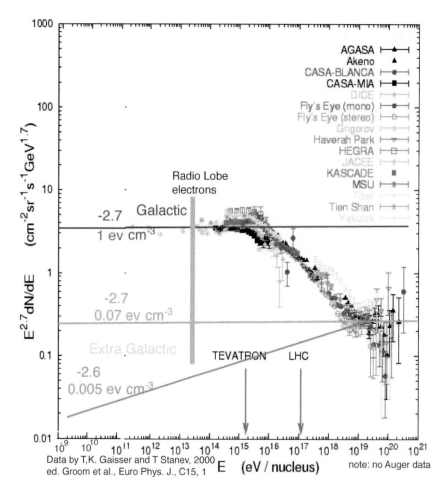

Data by T.K. Gaisser and T Stanev, 2000
ed. Groom et al., Euro Phys. J., C15, 1

Fig. 3: The many measurements of cosmic rays at high energy as compiled by Gaisser and Stanev (2000). The data are plotted as $E^{2.7} \, dN/dE$ vs. E and thus with the slope $\Gamma = -2.7$ removed. The change in slope to $\Gamma \simeq -3.0$ in the energy interval $E \le 10^{15} \, \text{eV} \le E \le 10^{18} \, \text{eV}$ is evident between the "knee" and "ankle". This change in slope is interpreted as a transition from galactic confinement to extragalactic CRs. The width of the transition region is interpreted as progressive loss of nuclei of progressively higher $\langle A \rangle / Z$ with the highest being iron nuclei. Particles above $10^{18} \, \text{eV}$ are interpreted as extragalactic UHECRs, primarily protons. Several lines are drawn at different slopes tangent to the UHECR spectrum and extrapolated back in energy to $E_{CR} = m_p c^2$ where the maximum in total CR energy resides. The slope $\Gamma = -2.7$ is the galactic slope and results in an energy density of extragalactic CRs is $\sim 7\%$ of the galactic value. The slope for CRs in the IGM for the same acceleration mechanism should be slightly flatter because of the lack of spallation. Spallation attenuates galactic CRs by $\sim \frac{1}{2}$ in 6 decades in energy and so we estimate that without spallation $\Gamma \simeq -2.6$. This results in the energy density of $\simeq 0.005$ of galactic CRs.

energy flux the high energy flux beyond the ankle is inferred to be $N_{\text{extra}}/N_{\text{galaxy}} \sim 0.03$, where the spectral dependence has been removed by multiplying the flux by E^{Γ} where $\Gamma = 2.7$ for the galaxy. This predicted flux, if the galaxy were transparent, and if produced in the galaxy alone, is therefore orders of magnitude less than the observed flux. This very much larger flux of ultra high energy cosmic rays is explained by an extra galactic source of cosmic rays that is $\sim 3\%$ in number density of the galactic component. The existence of ultra high energy cosmic rays, $E > 10^{18}$ eV also argues for an extra galactic origin as well, because there is insufficient magnetic flux within the galaxy to confine such particles during their acceleration. If these "ultra-high energy" (UHE) cosmic rays are extra galactic in origin, that is, accelerated from matter of the IGM rather than "born" from the decay of unknown primordial particles, then, on very general grounds, one would expect lower energy cosmic rays from the same origin for two reasons.

1. All accelerators have losses during acceleration, i.e., they contain lower-energy particles. In galactic acceleration the fractional loss in number per fractional gain in energy is $(dN/N)/(dE/E) = 1 - \Gamma$ so that a perfect accelerator with no losses like Fermilab or CERN corresponds to $\Gamma = 1$. A measure of the imperfection or loss of the natural accelerator of the Galaxy is then $\Gamma - 1 = 1.7$. If there were any mechanism within the Galaxy that could produce a spectrum of flatter slope, smaller Γ, then there would be more particles at high energy from such an accelerator and they would be observed as a change to a flatter slope. One concludes that the most likely accelerator is one with $\Gamma \sim 2.7$.

2. Lower energy extra galactic cosmic rays are observed in an amount that is comparable to or exceeds the extrapolated extra galactic flux using a standard $\Gamma = 2.7$. These cosmic rays are the electrons of the giant radio lobes associated with the quasar or active galactic nuclei, AGN phenomena. The luminosity, spectrum, and size of the lobes leads to an estimate of a minimum energy in a combined relativistic electrons, 10 to 100 GeV, and magnetic field in order to account for the synchrotron radiation. The calculated magnetic field at this minimum, $\sim 5\,\mu G$ has now been observed in Faraday rotation so that the existence of immense energy in relativistic electrons in radio lobes is more likely and by inference so also are cosmic rays.

In Fig. 3, the cosmic ray spectrum is multiplied by $E^{2.7}$ and therefore shows the low energy part of the spectrum as a horizontal line. If one assumes the same acceleration mechanism for galactic as well as extragalactic cosmic rays, the slope of the UHCR spectrum above 10^{18} eV is extrapolated backwards to $E = mc^2$ at a slope of -2.7 for galactic and at a slope -2.6 for an extragalactic acceleration. The small difference would account for the additional attenuation expected within the Galaxy due to losses from spallation. The total energy in this inferred spectrum of extra galactic cosmic rays is then $W_{\text{CR}} \sim 10^{60}$ ergs per volume of galaxy spacing. This is then 10^6 times greater than the energy in Galactic cosmic rays again inferring an extra galactic mechanism. The only known energy source this larger is the AGN phenomena associated with the birth of nearly every galaxy.

In this discussion, total energy has been emphasized as the most astonishing aspect of UHE extra galactic cosmic rays, but comparably important is the isotropy with a lack of any clearly indentifiable nearby sources. (Several close pairs and one triplet have been detected, but at a frequency below a clear identification.) Finally the upper energy limit of UHE extra galactic cosmic rays of 3×10^{20} eV is comparably challenging to acceleration mechanisms as well as its very existence challenges the expectation that it and all UHE extragalactic cosmic rays should be attenuated by the GZK cutoff due to the production of pions by the interaction of high-energy protons ($\gtrsim 10^{19}$ eV) with the cosmic background radiation. This observation alone limits possible sources to $\lesssim 30$ Mpc distance.

The Ingredients of a Solution

One possible example of an explanation of these many disparate observations and interpretations is that in the formation of the massive black holes of the AGN phase of galaxy formation, that the free energy of the massive black hole formation of nearly every galaxy is converted into partially stable magnetic field energy, $\sim 1\,\mu$G that during a Hubble time, is converted in a space filling fashion by reconnection, into cosmic rays. These cosmic rays in turn, diffuse from the filamentary cosmic structure of galaxies and are lost in the lower density voids, thereby suggesting the necessary source of the total energy, the containment of the maximum particle energy, isotropy, and lack of an observed GZK cutoff by instead more rapid loss from the filaments to the voids . Only a continuing interaction between new experiments and new theories will give us a consistent and satisfactory picture of our universe.

See also: Cosmic Rays, Galaxies; Interstellar Medium; Nucleosynthesis.

Bibliography

The most comprehensive treatment of cosmic rays in earlier years is still the authoritative book by V. L. Ginzburg and S. I. Syrovatskii, *The Origin of Cosmic Rays*; Pergamon Press, New York, 1964. An updated version of this authoritative text is *The Astrophysics of Cosmic Rays* by Berezinskii, Bulanov, Dogiel, Ginzburg, and Ptuskin; North-Holland, Amsterdam, 1990. The current scientific changes are reviewed in the "Rapporteur" papers of each International Cosmic Ray Conference (every two years) and the details appear in the published papers.

Cosmic Rays, Solar System Effects

R. B. McKibben and J. V. Hollweg

Introduction

Cosmic rays were discovered early in the 20th century when scientists investigating the electrical conductivity of air found a low and nearly constant level of ionization even in heavily shielded chambers. They suggested an unknown and very penetrating type of radiation as the cause. In a series of balloon flights in 1912 the Austrian physicist Viktor Hess showed that this radiation increased with height above the Earth's surface, suggesting an extra-terrestrial source. In 1926, mistakenly believing that the radiation consisted mainly of γ rays, two American scientists, R.A. Millikan and G.H. Cameron, referred to the radiation as "cosmic rays", and the term became generally used. Later observations of the effects of the Earth's magnetic field on radiation intensities at various latitudes (*latitude effect*) and from different directions (*east-west effect*) showed that the cosmic rays are mainly high-energy positively charged particles that impinge on the Earth's atmosphere from space. Measurements in space have now shown that almost all cosmic rays are relativistic atomic nuclei, stripped of all electrons, with all naturally occurring elements represented. Energetic electrons and positrons are also present but make up only about 1% of the cosmic rays. The observed cosmic ray energies extend from a few million electron Volts (MeV) up to energies $> 10^{20}$ eV, the highest energy ever measured for a nuclear particle. Direct and indirect observations show that cosmic rays are present not only near Earth, but throughout the solar system, and indeed throughout the galaxy. Because the particles with the highest energies cannot be contained within the galaxy, very high energy cosmic rays must also be present in intergalactic space. In this article, we concentrate on effects involving cosmic rays in the solar system. For more detailed discussions of the astrophysical importance of cosmic rays, *see* Cosmic Rays, Astrophysical Effects)

The Heliosphere and Cosmic Rays

To discuss cosmic rays in the solar system, we must first discuss the heliosphere, which is the region of space dominated by the solar wind (*see* Solar Wind). The solar wind is a supersonic flow of hot plasma continuously blowing off from the Sun at speeds of several hundred km/s. As the wind accelerates in the Sun's million degree upper atmosphere (the solar Corona) it draws magnetic fields from the top of the corona into interplanetary space. In interplanetary space the Sun's rotation and the radial outflow of the wind combine to stretch the fields into an Archimedean spiral pattern, much like the spiral pattern of water from the nozzles of a spinning garden sprinkler. As the wind expands outward, it becomes more tenuous, the magnetic field weakens and, because of its spiral pattern, the field becomes nearly perpendicular to the wind flow. Eventually the wind becomes so tenuous that pressure from the local interstellar medium exerts a significant influence. At a distance from the Sun of about 90–100 Astronomical Units (AU; 1 AU = 150 000 000 km), most scientists believe this external pressure causes the wind to undergo a shock transition to a hotter and more dense subsonic flow. Recent observations from the Voyager 1 spacecraft, first reported in 2003, suggest that, since 2002 when the spacecraft was near 85 AU, Voyager 1 has been very near and may even have encountered this termination shock. Beyond the termination shock, the magnetic field is expected to strengthen as a result of compression of the solar wind, and the subsonic flow is expected to continue outward for many AU until it is finally deflected by pressure from the interstellar medium. Since the Sun is moving through the local medium, the wind will be deflected back around the termination shock to form a tail, much like a wake in the interstellar medium. The interface between the interstellar medium and the solar wind is called the heliopause, and everything inside the heliopause is part of the heliosphere.

The most important feature of the heliosphere for cosmic rays is the magnetic field carried by the solar wind. Since the cosmic rays are charged particles, magnetic fields strongly influence their motion. Charged particles propagate easily along magnetic fields, but if their velocities are perpendicular to the field direction their paths are bent into circles whose radii, called gyro-radii, are proportional to the particles' momentum and inversely proportional to the strength of the field. As a result, particles have difficulty crossing a magnetic field, and the cosmic rays are carried outward with the solar wind magnetic field. They can still reach the inner heliosphere by propagating along the spiral field, or by drifts (discussed below) driven by the non-uniformity of the field, but irregularities of various sizes in the field scatter the cosmic rays, resulting in diffusive motion along the field rather than simple streaming, and also disrupting the pattern of drift motions. The cosmic rays also lose energy as a result of the expansion and weakening of the magnetic field as it is carried outward, an effect called adiabatic deceleration.

Much of what we know about cosmic rays in the heliosphere is the result of observations from several space missions. Particularly important have been the now complete Pioneer 10 and 11 missions and the continuing Voyager 1 and 2 missions, which have returned the first measurements from the far outer heliosphere beyond the orbit of Pluto. More recently, the NASA/ESA Ulysses mission in the inner heliosphere has made the first observations in regions over the Sun's polar latitudes.

Components of the Cosmic Radiation

The cosmic rays observed above the atmosphere are usually divided into three groups, depending on whether they enter the solar system from the galaxy (galactic cosmic rays), are associated with solar activity (solar cosmic rays), or originate as interstellar neutral atoms ionized and then locally accelerated in the heliosphere (anomalous component cosmic rays). In addition, the radiation in the atmosphere that Hess observed, while commonly called cosmic radiation, is really secondary radiation produced by interaction of the primary cosmic rays from space with nuclei in the atmosphere. The particles that make up this atmospheric radiation are therefore sometimes called secondary cosmic rays.

Importance of Cosmic Ray Observations

The study of cosmic rays in the solar system touches on a wide variety of areas. For example:

1. Solar cosmic rays carry information about the solar atmosphere and about the particle-acceleration processes that take place both in solar flares and in association with strong shock waves in the corona and solar wind.

2. Within the heliosphere the spectra and intensities of galactic and anomalous component cosmic rays are modified, or modulated, as they propagate through the solar wind, and these changes yield information about conditions in distant regions of the heliosphere.

3. Study of the interaction of cosmic rays with the plasmas and magnetic fields in the solar wind provides a laboratory where we can directly observe astrophysical processes that are important in many other environments throughout the galaxy, for example acceleration of energetic charged particles by shocks in a thin magnetized plasma.

4. Cosmic rays have effects on the atmosphere, including modifications of ionospheric structure and of the ozone layer. They may also influence weather and climate by producing ions which can serve as nucleation points for condensation of water vapor, thus contributing to cloud formation.

5. Since the galactic cosmic rays include particles with energies very much higher than can be produced by accelerators, they can provide information important for high-energy particle physics.

6. Radioactive isotopes produced by cosmic ray interactions with the atmosphere are important for many dating methods. For example, ^{14}C, produced by interactions with atmospheric nitrogen atoms, is important for dating archaeological objects. Other cosmic-ray-produced radioactive isotopes are used to date ocean sediments and layers of glacial and polar ice.

7. Heavy cosmic ray nuclei can cause temporary malfunctions, known as single event upsets and latchups, in sensitive micro-electronic circuits used in spacecraft.

8. Finally, cosmic rays account for an important part of the radiation dose accumulated by travelers at high altitudes or in space.

Galactic Cosmic Rays

Galactic cosmic rays consist of protons (85%), helium nuclei (13%), heavier nuclei (1%), and electrons and positrons (1%). For a number of reasons, including their low flux compared to nuclei, the electrons and positrons are difficult to observe. Therefore far more is known about the nuclear components in cosmic rays. An important feature of cosmic rays is that they provide a sample of matter that was not part of the cloud from which the Sun and everything else in the solar system condensed.

For energies above about 1 GeV (10^9 eV) the flux of galactic cosmic rays decreases rapidly with increasing energy. From 10^{10} eV up to about 10^{15} eV the integral intensity spectrum, $J(E)$, defined as the intensity of all cosmic rays with energy greater than E, decreases by a factor of about 50 for each factor of 10 increase in E. Above a few times 10^{15} eV, which is the maximum energy that can be confined by galactic magnetic fields, the rate of decrease becomes slightly steeper, a bend referred to as the knee in the cosmic ray spectrum. The measured spectrum covers about 19 decades of intensity, ranging from about $3\,\mathrm{cm^{-2}\,s^{-1}}$ at energies around 1 GeV to about $6\,\mathrm{km^{-2}\,century^{-1}}$ at 10^{20} eV. Inside the heliosphere and below $\sim 10\,\mathrm{GeV/n}$ (GeV per nucleon, or per proton and neutron in the nucleus) the intensities are significantly reduced by the difficulty the cosmic rays experience in propagating inward through the solar wind, an effect called solar modulation. The modulated intensity is controlled by the competition between the inward propagation by diffusion along the fields and by gradient and curvature drifts (discussed below), and the outward convection and adiabatic deceleration. Intensity levels of the modulated cosmic rays vary during the 11-year solar activity (or sunspot) cycle roughly in inverse correlation with the sunspot number. This variation is a result of changes in the structure of the Sun's magnetic field and of the solar wind source region (the corona) during the activity cycle, and of the more frequent occurrence near sunspot maximum of large interplanetary disturbances produced by eruptions on the Sun. All of these effects greatly reduce the rates of inward drift and diffusion of cosmic rays in the solar wind. Even near sunspot minimum, when the Sun is comparatively quiet, these effects significantly reduce the intensity of low energy cosmic rays near Earth compared to that outside the heliosphere. In fact, the modulation effectively excludes galactic cosmic rays with energies below a few hundred MeV/n from the inner heliosphere. Those observed at these energies near Earth have been decelerated from higher energies outside the heliosphere by adiabatic deceleration.

Recent models of the solar modulation have emphasized the role of gradient and curvature drifts, which are generated by the non-uniform strength and the spiral curvature of the interplanetary magnetic field. Gradient drifts arise from the fact that the gyro-radius of a particle in a strong field is smaller than in a weak field. If a particle's orbit samples regions with different field strengths, the periodic change in the radius of its orbit leads to a drift motion perpendicular to the direction of the gradient in the field. For somewhat more complicated reasons, curvature of the field leads to a similar effect. Drifts are most significant when the global structure of field is well organized. Near solar minimum the Sun's magnetic field approximates a dipole and the interplanetary field is well described to first order by an Archimedean spiral both in the ecliptic and, as Ulysses has shown, up to solar latitudes as high as 80°. Because of the dipole nature of the Sun's field, the prevailing field directions in the heliosphere are opposite in the northern and southern hemispheres, and the polarities are separated by a near-equatorial sheet of electric current in the solar wind plasma. In such a field the drift patterns result in particle flow either from the polar regions toward the equator, or vice-versa, depending on the sign of the solar dipole. Since the solar dipole reverses its sign near solar maximum in every solar cycle, over two cycles (22 years) the effects of both signs of the solar dipole are observable.

As a result of the solar modulation, the cosmic ray intensity in the heliosphere varies with both radius and latitude. As the Pioneer and Voyager spacecraft traveled outwards, they measured rates of increase in the radiation intensities (or gradients) of the order of a few percent per AU, largest for low energy (< 100 MeV) cosmic rays, and decreasing gradually with increasing distance from the Sun. Recent theoretical results suggest that gradients persist beyond the termination shock into the compressed subsonic solar wind and that the full interstellar intensity is reached only after crossing the heliopause.

Latitudinal gradients are controlled primarily by the drift motions. Therefore they reverse in alternate solar activity cycles. Latitudinal gradients near solar minimum were measured first by Pioneer 11 in 1974/75, then by Voyager 1 in 1985, and most recently by Ulysses in 1994/95. Pioneer 11 and Ulysses both observed intensities to increase from the equator towards the poles, whereas Voyager 1 found the highest intensity near the equator. Thus, as expected, the latitude gradients reversed in successive solar minima. The observed gradients are small, of the order of 1% per degree, but the values are not consistent between the various measurements, most likely because the measurements were all made at different radii and over different latitudinal ranges. In the most recent and complete observations by Ulysses, which extend all the way to 80° latitude over a radial range from 1.3 AU to about 3 AU, the cosmic ray intensity in the polar regions was at most 50% greater than the equatorial intensity. When the very complex and disturbed state of the solar and interplanetary magnetic fields at solar maximum disrupts drift patterns, latitudinal gradients become very small. In the recent measurements performed by Ulysses near solar maximum in 2000–2001, no measurable latitude gradients were observed.

Solar Cosmic Rays

Solar cosmic rays, or solar energetic particles, are often observed near Earth following sudden localized brightenings on the Sun known as solar flares, and other large disturbances on the Sun. All such events are believed to be the result of explosive release of energy stored in tangled magnetic fields in the lower corona, most often in complex sunspot groups. We now believe that there are two different forms of acceleration that produce the solar cosmic rays. In most cases one or the other dominates, but both often contribute.

The first form of acceleration is associated primarily with very small events that do not produce major interplanetary disturbances. In these events the acceleration probably occurs in direct association with the release of stored magnetic energy. Such events, called impulsive events because of their rapid time-intensity profile, produce low intensity, low-energy (up to tens of MeV) particles that are injected very near the flare site and nearly simultaneously with the flare. They seem to travel to Earth almost without scattering along interplanetary magnetic field lines. Because the particles are injected over a narrow region and are strongly guided by the interplanetary magnetic field, they are observed over only a very narrow range of solar latitudes and longitudes, and the events typically last only a few hours, at most a day. The particles from these events have unusual elemental and isotopic compositions. Heavy elements such as Iron are much more abundant than they are in the solar atmosphere, and ^{3}He/^{4}He ratios, normally less than 1/10000, can exceed 1. The abundance anomalies are most likely produced by still poorly understood details of the injection and acceleration processes in the dense lower corona.

The other form of acceleration, characteristic of the largest events, is believed to take place at the shock front ahead of a large and fast Coronal Mass Ejection (CME) produced by the energy release associated with the flare. A CME is a sudden ejection of coronal material into interplanetary space, sometimes at very high velocity. A fast CME (e. g., $> 1000 \, \text{km/s}$) drives a shock wave ahead of it in the solar wind, and this shock can accelerate solar cosmic rays.

The highest energy particles, which can reach into the GeV range, seem to be injected almost immediately after the initiation of the CME, when the shock is at its strongest in the lower corona. However acceleration across the broad front of the shock continues as the CME expands outwards into the heliosphere, continuing in some cases even until the CME passes the orbit of Earth. While details are not fully worked out, the particles accelerated are believed to be drawn from pre-existing seed populations of superthermal particles in the solar wind. In these events, the elemental and isotopic abundances of the accelerated particles are similar to general solar system abundances. As a result of the broad front of acceleration in longitude and latitude, the extended acceleration time, and perhaps the fact that such events tend to occur mainly at solar maximum when the interplanetary medium is in a highly disturbed state, these events tend to have a slow rise to maximum intensity (hours to a day or more at low energies) after which the enhanced intensities decay over a period that may last days or even weeks. Thus these events are called gradual events. Simultaneous observations from multiple spacecraft have shown that energetic particles from a single gradual event can often spread throughout the inner heliosphere, affecting the radiation environment everywhere in the inner solar system.

The shock associated with a CME or gradual event affects galactic cosmic rays as well. It acts as a propagating barrier that sweeps cosmic rays out of the inner solar system, decreasing their intensity by several percent for periods of up to several days. Such abrupt but short-term intensity decreases are called Forbush decreases after the scientist who discovered and described them.

While impulsive and gradual events represent extremes of the solar cosmic ray acceleration process, contributions from both can be found in many events; for example, there may be a gradual time profile, but the abundances may show some anomalies characteristic of impulsive events. For producing effects at Earth, such as ionospheric disturbances and errors in spacecraft circuitry, the gradual events are by far the more significant.

Anomalous Components

The anomalous components of the cosmic radiation were discovered in the early 1970s when sophisticated sensors on spacecraft first studied the composition of low energy cosmic rays near solar minimum. The first to be discovered was anomalous Helium, which appeared as an additional flux of Helium nuclei at energies below about 30 MeV/n. At these energies the flux of Helium even exceeded that of Hydrogen, normally almost 10 times more abundant. Soon similar enhancements of Nitrogen, Oxygen, Neon, and Argon were found, and much later an anomalous component of Hydrogen itself was identified in the outer heliosphere. No other elements showed similar enhancements. The composition and spectra were so different from galactic cosmic rays that these enhancements acquired the name "anomalous components". Even though we now understand their origin, the name persists.

The anomalous components are derived from neutral gas in the local interstellar medium just outside the heliosphere. Much of the interstellar gas is ionized by the ultraviolet (UV) radiation in starlight. Hydrogen is by far the most abundant element in the gas, but it requires a higher energy UV photon to become ionized than most elements. Ionization of Hydrogen in the interstellar gas absorbs a large fraction of the UV photons with energies above its threshold for ionization. Thus the few elements with higher ionization thresholds than Hydrogen (namely Helium, Nitrogen, Oxygen, Neon, and Argon, the same elements as make up the anomalous components) are exposed to a much reduced flux of UV that can convert them to ions, and a large fraction remain as neutral atoms. For elements with lower ionization thresholds, no element is abundant enough to significantly reduce the flux of UV photons capable of ionizing them, and so most of these atoms are ionized. As charged particles, ions are excluded from the heliosphere by the heliospheric magnetic field. Neutral atoms, however, are not affected by magnetic fields. Therefore, as the Sun moves through the interstellar gas they drift into the inner heliosphere where they are directly exposed to intense solar UV. If they approach close enough to the Sun, most are converted to ions when the UV removes one (and occasionally more than one) electron. The new ions are immediately picked up by the interplanetary magnetic field and convected outward with the solar wind. They form a separate and very hot component of the wind called pick-up ions. On reaching the solar wind termination shock they are more effectively accelerated (because of their higher thermal energies) than other components of the solar wind, and can reach energies typical of low energy cosmic rays. Some then diffuse back into the inner heliosphere in the same manner as galactic cosmic rays, except that, because most are missing only one electron they are much less affected by solar modulation than are the fully stripped galactic cosmic ray nuclei. Still, because of their relatively low energies and the stronger modulation at solar maximum, they reach the inner heliosphere in appreciable numbers only when the Sun's activity level is very low.

The significance of the anomalous components lies in their origin as part of the very local interstellar medium, which was not part of the original cloud from which the Sun and solar system condensed, and from their unusual response to solar modulation as a result of their low ionization state compared to galactic cosmic rays. Thus they provide not only another sample of extra-solar matter, but also a crucial test of our understanding of the transport of cosmic rays through the solar wind.

Secondary cosmic rays

Ground-based detectors almost never observe a primary cosmic ray before it is destroyed by interactions with atoms high in the Earth's atmosphere. These interactions produce neutral pions, charged pions, and nucleons. The neutral pions decay into γ rays, which in turn produce electron-positron pairs in a cascading process called an electromagnetic shower. The charged pions decay almost immediately into muons and neutrinos. The neutrinos, since they interact hardly at all with matter, easily penetrate the whole Earth. The muons have a lifetime of only about 2 microseconds before they decay into electrons and neutrinos. However the muons in cosmic ray showers are of such high energy that relativistic effects allow most of them to reach ground-level. The nucleons also interact further with the atmosphere to produce more pions as well as neutron-proton cascades. All of the particles produced by these processes form an extensive air shower, which may have a radius of 100 m or more at ground level if produced by a very high energy primary.

Muons are the most penetrating ionizing component of the secondary cosmic radiation. For experiments that require very low levels of background radiation, laboratories have been established deep underground where very few muons penetrate. The neutrinos have no charge and are very difficult to observe. Even so, observations in 1998 of a difference in flux between neutrinos coming down from the atmosphere above and those coming up from air showers on the other side of the Earth has been interpreted as evidence for oscillation between types of neutrinos, a very important observation for high energy particle physics.

The secondary cosmic radiation does not have the same intensity everywhere on Earth. Because the cosmic rays consist of charged particles, the Earth's magnetic field bends their trajectories. This has two main effects. First, away from the magnetic poles there is an energy below which the trajectories of particles are so strongly bent that the particles cannot reach the atmosphere. This energy, called the cutoff energy, varies from about 15 GeV near the magnetic equator to near zero over the poles. This produces a pole-to-equator variation in the intensity of the secondary radiation called the *latitude effect*, which provided crucial early evidence that the cosmic rays were charged particles. A second effect arises from the fact that cosmic rays are primarily positively charged nuclei, so that the cutoff energy for particles arriving from the west is lower than for particles arriving from the east. As a result more showers arrive from the west than the east. Observation of this *East–West effect* provided the first evidence for the positive charge of cosmic rays.

Today the most important practical effect of the deflection of cosmic rays by the Earth's field is the protection it affords to manned and unmanned spacecraft in low-Earth quasi-equatorial orbits. Even the most energetic solar cosmic rays are effectively blocked from reaching spacecraft in such orbits. This provides important protection during the largest solar energetic particle events, which can produce levels of radiation hazardous to both electronic components and astronauts. Shielding against such dramatic increases in radiation from solar cosmic rays, and even from galactic cosmic rays, is a still-unsolved problem for long-duration interplanetary manned missions.

Bibliography

A. Balogh, R.G. Marsden, and E. J. Smith (Eds.), *The Heliosphere Near Solar Minimum: The Ulysses Perspective*, Springer-Praxis, Chichester, 2001.

L. A. Fisk, J. R. Jokipii, G. M. Simnett, R. von Steiger, and K.-P. Wenzel (Eds.), *Cosmic Rays in the Heliosphere*, Kluwer, Dordrect/Boston/London, 1998.

M. A. Forman and G. M. Webb, "Acceleration of Energetic Particles", in *Collisionless Shocks in the Heliosphere: A Tutorial Review*, R. G. Stone and B. Tsurutani, eds. American Geophysical Union Monograph 34, Washington, D.C., 1985.

M. W. Friedlander, *A Thin Cosmic Rain* Harvard University Press, Cambridge, MA, 2000

L.I. Miroshnichenko, *Solar Cosmic Rays*, Kluwer Academic Publishers, Dordrecht/Boston/London, 2001.

Cosmic Strings

J. Preskill

Cosmic strings are hypothetical line defects that arise in some relativistic quantum field theories as a consequence of the spontaneous breakdown of gauge or global symmetries. These defects are one-dimensional objects with a width of order m^{-1} and an energy per unit length of order m^2 (in units with $\hbar = c = 1$), where m is a characteristic mass scale of the symmetry breakdown. No cosmic strings are expected in the "standard model" that describes the electroweak and strong interactions at energies of order a few hundred GeV and below. Thus, if cosmic strings can actually occur in nature, they must be associated with very short-distance physics that has not yet been directly explored in accelerator experiments.

Although there is at present no direct observational evidence indicating that cosmic strings exist, it has been proposed that cosmic strings play a significant role in the physics of the very early universe. This proposal is the origin of the use of the adjective "cosmic" as applied to these objects. In current usage, the term "cosmic string" distinguishes the composite strings that arise in relativistic field theories from fundamental strings (such as the superstring) that are truly elementary dynamical entities.

That strings may arise in relativistic quantum field theory was emphasized in 1973 by Nielsen and Olesen, though the mathematics underlying this suggestion had been worked out in 1957 by Abrikosov and applied by him to the theory of superconductivity. These strings were considered in the context of cosmology in 1976 by Kibble, who observed that strings could have been produced in a phase transition in the very early universe. That such cosmic strings might have stimulated the formation of galaxies and other types of large- scale structure in the universe was proposed in 1980 by Zeldovich, and, in a more specific form, by Vilenkin.

Cosmic strings may be classified into two general types, gauge strings and global strings, according to the nature of the spontaneously broken symmetry with which they are associated. In fact, one-dimensional defects that are analogous to cosmic strings occur in some condensed-matter systems, and for each type there is a corresponding condensed-matter analog that serves as a prototypical example. The concept of a gauge string is nicely illustrated by a magnetic flux tube in a type-II superconductor. Magnetic fields are expelled from a superconductor by the Meissner effect, but a type-II superconductor may be penetrated by filaments that carry the quantum of magnetic flux. These filaments have a characteristic finite thickness that is known as the penetration depth of the superconductor. In this example, the spontaneously broken symmetry associated with the string is the $U(1)$ gauge symmetry of electrodynamics; this symmetry is spontaneously broken in a superconductor, and the photon acquires a mass m via the Higgs mechanism, where m^{-1} is the penetration depth.

Analogously, one can construct a relativistic quantum field theory in which the $U(1)$ gauge symmetry of electrodynamics is spontaneously broken in the vacuum state. Strings appear in such a theory as stable solutions to the classical field equations. Similar string solutions occur in other theories with various patterns of gauge symmetry breaking. In general, a field theory contains cosmic strings whenever there is a conserved magnetic flux that is carried only by gauge fields that have acquired mass via the Higgs mechanism; then the conserved flux becomes confined by a generalized Meissner effect to a tube of finite width. Thus, the existence of gauge strings is predicted by some "grand unified" theories, in which the gauge group of the standard model is embedded in a larger group that undergoes the Higgs mechanism at a large mass scale.

The concept of a global string is illustrated by a vortex line in superfluid helium. In the superfluid, there is a conserved particle number, and the associated $U(1)$ global symmetry is spontaneously broken by Bose condensation. The vortex is a thin tube of normal fluid that is trapped by the surrounding superfluid. A characteristic property that distinguishes gauge and global strings is that global strings have a long-range interaction mediated by the Goldstone bosons to which they couple, while gauge strings typically have no long-range interaction other than gravity.

Cosmic strings have no ends; they either form closed loops or extend to infinity. (Actually, other topologies are also possible, depending on the details of the associated pattern of symmetry breakdown. But strings that form closed loops seem to be of the greatest cosmological interest.) The tension in a string is equal to its energy per unit length, so that its motion is highly relativistic; transverse oscillations propagate along the string with velocity c, and a closed loop of length L executes a periodic motion with a period comparable to L/c.

When two strings cross one another, they are apt to break at their point of intersection and change partners before rejoining, a process known as intercommuting. A typical closed loop may intersect itself at some point during its motion; then intercommuting may cause the loop to fission into two "offspring" loops. An oscillating closed loop that does not self-intersect can decay by emission of gravitational radiation.

It is not known whether the actual pattern of symmetry breakdown that governs particle physics at very short distances is such that cosmic strings are expected to exist in nature. But if so, it is likely that a phase transition occurred in the very early universe in which strings were copiously produced. Symmetries that are spontaneously broken at low temperature tend to be restored when the temperature exceeds a critical value T_c, of order m, where m is the symmetry-breaking mass scale. As the rapidly expanding universe cools down through such a critical temperature, defects inevitably freeze in.

The cosmic strings thus formed comprise a complicated network of vibrating open strings and oscillating closed loops. As the network evolves, strings frequently cross and intercommute. It is believed that the strings eventually attain a self-similar equilibrium configuration that is only weakly dependent on the details of the structure of the network when it was initially formed.

The string network generates inhomogeneous perturbations in the primordial distribution of matter in the universe. Although initially small, these perturbations grow because of gravitational instabilities. They may thus account for the origin of galaxies, clusters of galaxies, and other large-scale inhomogeneities that we observe in the universe today. Because the dynamics of the string network is quite complicated, it is not easy to extract from this scenario detailed predictions about large-scale structure. But one can infer that the primordial mass-density perturbations have the important property of being independent of length scale, as a consequence of the self-similarity of the equilibrium string network. The predicted scale independence is in qualitative agreement with some observed features of large-scale structure. One can also infer that, if cosmic strings did play a significant role in the origin of galaxies and other large structures, then they must have an energy per unit length of order $(10^{16}\,\mathrm{GeV})^2$, in units with $\hbar = c = 1$. Such strings, with a width of order 10^{-30} cm, would be a relic of the first 10^{-38} s after the big bang, and would be associated with particle physics at an energy far beyond the reach of foreseeable accelerator experiments.

Even though precise predictions concerning large-scale structure are not easily extracted from the cosmic string scenario for galaxy formation, this scenario may nevertheless be subject to observational confirmation or refutation. Decaying loops of cosmic string generate a stochastic background of gravitational radiation that should soon be detectable in pulsar-timing measurements. Also, strings might be detected as gravitational lenses, or through the characteristic way that they distort the cosmic microwave background radiation.

In some field theory models, cosmic strings have the property that electric charge carriers are bound to the core of the string, and the string therefore behaves like a superconducting wire. (This was pointed out by Witten in 1985.) Such superconducting strings are capable of carrying very large currents and can be enormously powerful sources of electromagnetic radiation. If superconducting strings occur in nature, the characteristic radiation that they emit might be detected by radio telescopes.

For now, the proposal that cosmic strings stimulated the formation of galaxies and other large-scale structures must be regarded as highly speculative. But this speculation may be confirmed or refuted. Confirmation would establish a truly remarkable connection between particle physics at very short-distance scales and cosmology at very long-distance scales.

See also: Cosmology; Galaxies; Gauge Theories; Symmetry Breaking, Spontaneous.

Bibliography

R. H. Brandenberger, "Inflation and Cosmic Strings: Two Mechanisms for Producing Structure in the Universe," *Int. J. Mod. Phys.* **A2**, 77–131 (1987).

N. D. Mermin, "The Topological Theory of Defects in Ordered Media," *Rev. Mod. Phys.* **51**, 591–648 (1979).

J. Preskill, "Vortices and Monopoles," in *Architecture of Fundamental Interactions at Short Distance* (P. Ramond and R. Stora, ed.), pp. 235–335. North-Holland, Amsterdam, 1987.

A. Vilenkin, "Cosmic Strings and Domain Walls," *Phys. Rep.* **121**, 263–315 (1985).

A Vilenkin, "Cosmic Strings," *Sci. Am.* **257** (No. 6), 94–102 (1987).

E. Witten, "Superconducting Cosmic Strings," *Nucl. Phys.* **B249**, 557–592 (1985).

Cosmology
P. J. E. Peebles

Homogeneity, Hubble's Law, and the Expansion of the Universe

In the standard and observationally successful cosmology the universe is homogeneous and isotropic in the large-scale average, which means the universe appears much the same, apart from local fluctuations, viewed from any galaxy. Also, space is uniformly expanding, which means that well-separated bits of matter are moving apart, the mean recession velocity at separation **r** being

$$\mathbf{v} = H\mathbf{r} , \quad h = 100h\,\mathrm{km\,s^{-1}\,Mpc^{-1}} . \tag{1}$$

This is Hubble's law. The functional form is dictated by homogeneity. The length unit is $1\,\mathrm{Mpc} = 3.086 \times 10^{24}4\,\mathrm{cm}$. The value of Hubble's constant, H, is uncertain because it is hard to fix absolute distances of galaxies; the dimensionless factor h is thought to be in the range $0.5 < h < 1$.

The cosmological redshift, z, of a distant object is defined by the ratio of the observed wavelength λ_o to emitted wavelength λ_e of spectral features in the object:

$$1 + z = \frac{\lambda_o}{\lambda_e} = \frac{a_o}{a_e} . \tag{2}$$

In the relativistic theory, this ratio also is equal to the factor by which the universe has expanded between now and the epoch of emission of the radiation, as indicated in the last part of the equation. It is standard practice to label epochs in the early universe by the redshift factor. The present distance to an object on our light cone and at very high redshift is on the order of the Hubble length, $cH^{-1} = 3000h^{-1}\,\mathrm{Mpc}$.

The first large-scale survey of the galaxy distribution was the Lick catalog based on visual counts of galaxy angular positions by C. D. Shane and C. A. Wirtanen, completed in 1967. Recently, improved automated large-scale counts have been obtained by groups at the Institute of Astronomy in Cambridge and the Royal Observatory in Edinburgh. All these surveys extend to depths $\sim 10\%$ of the Hubble length.

The surveys are consistent with large-scale homogeneity and a very clumpy small-scale distribution. One measure of this is the reduced two-point correlation function, $\xi(r)$, defined by the joint probability of finding a galaxy in each of the volume elements dV_1 and dV_2 at separation r,

$$dP = n^2 dV_1\, dV_2 [1 + \xi(r)] , \tag{3}$$

where n is the mean space number density. At $r < 10h\,\mathrm{Mpc}$, $\xi(r)$ is close to a power law, $\xi \propto r^{-1.77}$. This with similar power law behavior of the higher-order correlation functions indicates that the small-scale galaxy distribution approximates a nested clustering hierarchy, or fractal, with dimension $D = 1.23$. The hierarchy terminates on the small-scale end at about the size of an individual galaxy.

Redshifts are much more powerful tracers of structure than angular positions alone because the redshift reduces the ambiguity in distance. However, redshifts require much more observing time so redshift surveys still are relatively small. They have revealed in the small-scale galaxy distribution a curious pattern, that has been variously called frothy, bubbly, and cellular, in which rather linear structures surround empty regions, or voids.

The coherence length in the galaxy distribution, where $\xi = 1$, is $hr_o \sim 5\,\text{Mpc}$. Fluctuations on scales larger than this are relatively small. For example, the rms fluctuation in galaxy counts within a sphere of diameter $100h^{-1}\,\text{Mpc} \sim 3\%$ of the Hubble length is about 20% of the mean. One can pick out apparently coherent structures in these fluctuations, the largest well-accepted ones being $\sim 50h^{-1}\,\text{Mpc}$ across.

Large-scale fluctuations in the mass distribution are indicated by the gravitational production of departures from the pure Hubble flow of Eq. (1). In linear perturbation theory, the peculiar velocity, or departure from Hubble flow, caused by a spherical density fluctuation $\delta\rho/\rho$ extending over radius R is

$$v \sim \frac{1}{3}\frac{\delta\rho}{\rho}HR . \qquad (4)$$

The dipole ($\propto \cos\theta$) anisotropy of the microwave background radiation discussed below is most reasonably interpreted as the result of the peculiar motion of our neighborhood at about $600\,\text{km s}^{-1}$. The nearby galaxies at distances $\sim 10h^{-1}\,\text{Mpc}$ are moving at about the same peculiar velocity.) There are indications that the coherence length of this peculiar velocity field may be as large as $\sim 50h^{-1}\,\text{Mpc}$, the local motion roughly converging on the "Great Attractor" in the direction of Centaurus. With $R = 50h^{-1}\,\text{Mpc}$ and the above-mentioned $\delta\rho/\rho \sim 0.2$, Eq. (4) gives $v \sim 300\,\text{km s}^{-1}$ on the order of what is observed.

Most measures of structure on still larger scales yield only upper bounds. Apart from the dipole term, the microwave background is isotropic to better than one part in 10^4 on angular scales down to $\sim 10\,\text{arc min}$. Since this radiation is not now appreciably interacting with matter, the isotropy says nothing directly about the present matter distribution, but it does show that the initial entropy density and the cosmological redshift in different parts of space must have been very close to uniform. The extragalactic x-ray background comes from some combination of clusters of galaxies, exploding galaxies, quasars, and hot intercluster gas. This background is isotropic to better than 5% on scales greater than a few degrees. Thus the surface density of these sources integrated to the Hubble length must be isotropic to like accuracy.

In some versions of the inflation scenario mentioned below, we observe at the Hubble distance part of an island outside of which the universe is very different. However, there is no known way to substantiate this. To the limits of our observations, the universe is very close to isotropic on large scales.

The Cosmic Background Radiation

The microwave-submillimeter background radiation (also called the primeval fireball, relict, or cosmic background radiation, or CBR for short) has spectrum close to blackbody at temperature $T = 2.76 \pm 0.03\,\text{K}$. There is no natural way to produce a blackbody spectrum in the universe as it is now, for the universe is optically thin (radio sources are observed at redshifts $z > 1$ without appreciable attenuation), so this radiation must have been thermalized when the universe was denser and hotter than it is now. Isotropic and homogeneous expansion of the universe automatically preserves the thermal character of the radiation, with no need for a thermalizing agent: expansion simply lowers the temperature.

We have two candidate theories for the epoch of origin of this radiation. It is possible that dust made the universe optically thick at redshift $z \sim 100$, and that enough starlight was present to be thermalized by the dust to produce the CBR. However, the demands on the dust and the energy production by stars at $z \sim 100$ are so tight that most workers do not consider this picture very likely. Also, we have evidence for the other possibility, that the CBR originated at extremely high redshift.

The evidence comes from the production of light elements. Knowing the present CBR temperature, and assuming entropy was nearly conserved back to high redshifts, we can trace the thermal history of the universe back in time to temperatures high enough to have driven thermonuclear reactions. Computations of the nuclear reactions, under the assumptions (1) the early universe is accurately homogeneous and isotropic, (2) the conventional laws of physics apply, and (3) the universe is not filled with a sea of degenerate neutrinos, predict that most matter comes out of this "hot big bang" as hydrogen, with about 20% by mass helium, and significant amounts of deuterium, ^{3}He, and ^{7}Li. The computation depends on the mean baryon number density, n_b. It is remarkable that a choice of n_b consistent with what is observed to be present in and around galaxies yields computed abundances concordant with what is seen in old stars. This is evidence of reality of the hot big bang, though a fuller test should be possible with improved precision in the measurement of abundances in the oldest stars.

The Relativistic Cosmological Model

The relativistic cosmological model involves the following parameters:

1. The present rate of expansion, H [Eq. (1)]. This fixes the scales of length, time and mass.

2. The mean mass density and pressure, ρ and p. By homogeneity, ρ and p are functions of the world time t kept by any observer at rest relative to the local matter distribution. The order of magnitude of ρ is set by the combination H^2/G, where G is Newton's constant. In the standard model, the prefactor is on the order of unity, and $\rho \sim 1$ proton per cubic meter, which is within an order of magnitude of what is observed.

3. The ages of the oldest objects. In classical general relativity theory, the expansion of the universe traces back to a singularity at a finite time in the past. The world time since the singularity is $\sim H^{-1} \sim 10^{10}$ years, the prefactor again depending on the cosmological parameters. The singularity presumably reflects a failure of the classical theory, but since no object we could hope to date could have survived from preclassical epochs the ages of the oldest objects must be less than the model age. A test of this is limited by the uncertainties in H and the evolutionary ages of stars and the elements; we only know that predicted and observed ages agree to a factor of about 3.

4. The rate of change of the expansion rate. This is measured by the "acceleration parameter," q:

$$\frac{dH(t)}{dt} = -(q+1)H(t)^2 \ . \tag{5}$$

This fixes the first-order relativistic corrections to angular sizes and counts of galaxies as a function of redshift, and so in principle is measurable. Until recently, it has not been possible to correct the measurements for systematic errors: distant galaxies are seen at an earlier world time, when evolution may have made the mean properties of galaxies significantly different from now. The situation has been improving with advances in infrared detectors, where evolutionary corrections are smaller, and in the understanding of constraints on evolution from the spectra of galaxies and their luminosity distributions, so a significant test may be within reach.

5. The cosmological constant, Λ. This acts as constant contributions ρ_Λ and p_Λ to the net mass density and pressure, with $\rho_\Lambda = -p_\Lambda$, a Lorentz invariant arrangement. (Here and below units are chosen so the velocity of light is unity.) Einstein introduced this term in his gravitational field equations so as to allow a static homogeneous universe with non-negative pressure in ordinary matter. When the expansion of the universe was discovered he proposed that Λ be abandoned; opinion since then has been mixed. A Λ term might be expected from zero-point energy in a Lorentz-invariant theory, but a "reasonable" value would be much larger than would be acceptable to cosmology. Such a large effective Λ is present during inflation in the scenario discussed below.

6. Space curvature, $R(t)^{-2}$. If $R(t)^{-2} > 0$, space sections at fixed cosmic time are curved and closed like a sphere, with radius of curvature $R(t)$. The volume of space is finite [and proportional to $R(t)^3$] even though there is not a spatial boundary. If $R(t)^{-2} < 0$, space curvature resembles that of a saddle, and there is not direct closure of space. If $R(t)^{-2}$ is negligibly small, the model is said to be cosmologically flat (though spacetime is curved).

The gravitational field equations yield the relations

$$(1-\Omega)H^2 = \Lambda/3 - R^{-2} ,$$
$$H^2[\Omega(1-3p/\rho) - 2q] = 2\Lambda/3 , \tag{6}$$

where the density parameter is

$$\Omega = 8\pi G\rho/3H^2 . \tag{7}$$

The factor $1+3p/\rho$ ranges from 1 for nonrelativistic matter to 2 for electromagnetic radiation or zero-mass particles, an unimportant spread compared to the observational uncertainties in the other parameters.

In the Einstein–de Sitter model, Λ and R^{-2} both are negligibly small, so Eq. (6) says $\Omega = 1$. This is a particularly attractive case because the universe presents us with no fixed characteristic scales. (In all other observationally possible cases, the model defines a characteristic world time that must coincidentally be the same order of magnitude as the epoch of life on earth.) However, direct measurements of ρ based on galaxy masses derived from the dynamics of systems of galaxies fall short of the Einstein–de Sitter prediction [Eq. (7) with $\Omega = 1$] by a factor ~ 5. Opinion is divided on whether this is because the dynamical studies miss mass not in systems of galaxies (the "biasing" hypothesis) or whether it might mean the model fails.

In the more general two-parameter set of models with $\Lambda = 0$, it is easy to categorize the time histories. Here Eq. (6) may be rewritten as

$$\Omega H^2 - H^2 = R^{-2} . \tag{8}$$

The mean mass density varies with time as $\rho \propto R(t)^{-3}$ (if $p \ll \rho$; if pressure is large ρ varies still more rapidly with R because of $p\,dV$ work). Therefore Eq. (7) says ΩH^2 is decreasing more rapidly than $R(t)^{-2}$. Thus in a closed model, where $R^{-2} > 0$, Eq. (8) indicates there must come a time when $\Omega H^2 = R^{-2}$ and H therefore vanishes, which means the universe stops expanding. That is, a closed model with $\Lambda = 0$ expands from a "big bang" that has singular space curvature (in the classical theory) through the present epoch to a point of maximum expansion, after which it contracts back to a singular "hot crunch."

An open model expands forever after the big bang, the deceleration parameter q approaching zero, which means the gravitational slowing of the expansion becomes negligible. In the Einstein–de Sitter case, $R^{-2} = 0$, expansion also continues forever but with $q = \frac{1}{2}$. Unlimited expansion has been characterized as a heat death, because stars and the elements eventually exhaust the energy supply needed to power observers (though there would be isolated bursts of action for an exceedingly long time from relativistic gravitational collapse of galaxies and clusters of galaxies). The debate on whether the universe ends with a heat death or a hot crunch has the virtue that it concentrates our attention, but better physics is likely to come out of attempts to understand the past history of the universe, because that may have left clues.

Dark Mass

Masses of galaxies and systems of galaxies can be estimated from dynamics, through the gravitational field needed to account for observed motions. In compact stable-looking systems, one assumes gravity is strong enough to bind the system against its tendency to fly apart; in extended systems, one assumes gravity is the cause of perturbations from Hubble flow, as in Eq. (4).

In the bright central parts of galaxies, the dynamical mass estimates are consistent with what is seen to be present in the stars. In the outer parts of galaxies, the dynamical mass is significantly greater than what is seen: if our understanding of gravity physics is adequate, many bright galaxies have massive dark halos with density that varies with radius, r, as $\rho \propto r^{-2}$. It is interesting that the constant of proportionality has rather a hard upper bound, reflected in the observation that the circular velocity in a galaxy seldom is much in excess of $300\,\mathrm{km\,s}^{-1}$.

The discrepancy between what is seen and what is inferred from dynamics increases with increasing size of the system, reaching a ratio of inferred to seen ~ 10 in rich clusters. As noted above, if the Einstein–de Sitter model were valid the discrepancy would increase by another order of magnitude for the universe as a whole.

What is the nature of the dark mass? The most conservative guess is starlike objects, "brown dwarfs," with low luminosities because their masses are below the limit $\sim 0.1\,M_\odot$ for hydrogen burning. Most of the stars in our neighborhood have masses close to this limit. There could not be a large number of less massive brown dwarfs, because we do not have a large local mass problem, but it is easy to imagine that the ratio of brown dwarfs to visible stars increases with increasing distance from the center of the galaxy, matching the observed increase of mass per unit of starlight. The best hope for probing this is improved measures of the frequency distribution of star masses at $M > 0.1 M_\odot$ as a function of position in the galaxy.

The dark mass need not be baryonic. Among exotic dark mass candidates, massive neutrinos are the most conservative guess, because neutrinos are known to exist. Thermal production of neutrinos in the hot big bang leaves a present residue of about $100\,\mathrm{cm}^{-3}$ per neutrino type. If one type had mass $\sim 30\,\mathrm{eV}$, the mass density would be about right for the Einstein–de Sitter model, and there would be the pleasant coincidence that the wanted density of neutrinos in galaxy halos would just saturate the phase space bound. Modern particle physics suggests a host of other exotic candidates, and the possibility of their detection in the halo of our galaxy, through the occasional particle that scatters in a cold detector as it passes through the earth. It would be hard to overstate the impact on cosmology of a detection of exotic halo particles, but even if the dark mass is not exotic the identification of its nature is a great discovery waiting to be made.

The Physics Of The Very Early Universe: Inflation

In the classical cosmological models, the expansion of the universe traces back to a singularity. This means our theory is physically incomplete; worse, it means we have no way to predict the large-scale structure of the universe. A particular problem is the particle horizon: in the classical models, material now inside our light cone back to the singularity was outside our light cone in the past. The curious consequence is that we can see distant galaxies that at the time we see them were not in causal connection with each other subsequent to the singularity. How then can we understand why well-separated parts of space look so similar to each other, much less understand the small systematic differences that we call large-scale structures? The problem is compounded by the fact that the cosmological models are gravitationally unstable: inhomogeneities in mass density tend to grow with time. Thus if causal connection were established in the past, how could it happen that this promotes the observed homogeneity rather than clustering?

An elegant possible solution is the inflation scenario pioneered by A. Guth, A. Linde, P. Steinhardt, and others. It is based on the idea mentioned above that quantum physics encourages us to think space may contain something approximating the effect of a cosmological constant, Λ. The present value of this effective Λ must be small, but there may have been a time in the very early universe when it was large. In a patch of space-time where this term dominated the effect of matter and curvature, the first of Eqns. (6) gives $H \sim (\Lambda/3)^{1/2}$, which is to say the patch is expanding at a large and nearly constant rate. Where this persisted for a long time, the expansion would make the density of any remaining matter negligibly small, which would eliminate inhomogeneities, and it would make the volume of the patch very large. This "inflationary" epoch of nearly constant expansion would have to terminate in a phase transition in which the large energy density associated with the Λ-like term is converted into the entropy we see as the CBR, and the entropy would have to make the baryons out of which we are made.

A patch that contains observers like us would have to last a long time, because we required a long time to appear. A long-lived patch has to be big, so we might not expect to see the gradients across it. Thus the universe could appear homogeneous to us, even though it truly is chaotic.

It is to be emphasized that there is no very strong reason to believe any of this really happened; this is a scenario rather than a theory. Nonetheless the idea has been influential because it is the only route by which we have been able to see how we might have made *ab initio* computations of the origin of the observed large-scale structure of the universe.

There is one generic testable prediction. Primeval inhomogeneities are supposed to be reduced by the large expansion during inflation, which made the length scale over which any quantity varies much greater than the Hubble length. The same surely would apply also to $R(t)$. Thus we would predict that the first of Eqns. (6) reduces to

$$(1 - \Omega)H^2 = \Lambda/3 .$$

This is testable, though the observational uncertainties are still somewhat too large. If in addition the present value of Λ is negligibly small, $\Omega = 1$. As noted above, the status of this prediction is controversial.

Cosmogony

The origin of planets and stars generally is taken to be a problem for astronomy and astrophysics, while galaxies and the structure in the galaxy distribution is left to cosmology.

The division is historical and not necessarily rational. What are the main effects involved in the formation of galaxies and clusters of galaxies? The list generally is thought to include gravity, because that is what holds these objects together; dissipative gas dynamics, because that certainly is needed to produce the stars that dominate the mass in the central parts of bright galaxies; and perhaps also explosions, because the effect of supernovae is very visible in the interstellar medium. Light pressure ("mock gravity") from early generations of stars may also play a role. Primeval magnetic fields could induce galaxy formation by inhomogeneous magnetic stress and could also account for the observations of cosmic magnetic fields. Cosmic strings, that act like magnetic flux tubes in a superconductor, could enter through gravity or the pressure generated by annihilation. Which played the central roles in galaxy formation is a matter of considerable debate in the cosmology community. A few general aspects of the considerations are worth listing.

A partial answer to the problem almost certainly is to be found in the gravitational instability of the relativistic cosmological models. That means a mass concentration such as a cluster of galaxies has to be growing with time; that is, the large-scale structure in the galaxy distribution is growing by gravitational instability.

For a more complete answer, we need the character of the primeval density fluctuations out of which the gravitational clustering of mass grew. A popular possibility is the Zeldovich spectrum, in which the space curvature caused by the mass fluctuations varies as the logarithm of the length scale, so we get convergence by cutting off the spectrum at some exceedingly small and large scales. This has the virtue of simplicity (we need not specify the values of the cutoffs) and is predicted in some versions of the inflation scenario. Numerical solutions for the growth of structure out of this spectrum do seem to resemble the observed "bubbly" appearance of the galaxy distribution. The most serious challenge is that this spectrum may be unable to account for the largest known coherent structures.

Did the same instability operating at an earlier epoch produce compact groups and clusters of galaxies, and galaxies themselves? One might see evidence for it in the fact that the mean value of the mass density at distance r from a bright galaxy, averaged over a fair sample of galaxies, varies as r^{-2}. This applies within a galaxy, at scales $1\,\mathrm{kpc} < r < 30\,\mathrm{kpc}$, and it applies with about the same constant of proportionality to the mean mass density in the pattern of clustering around a galaxy, for another factor ~ 30 in radius. This scaling property might mean that galaxies and clusters formed by scaled versions of the same process, which would be gravitational instability.

Scaling cannot be the whole story, however, for a galaxy is a distinct and coherent mass distribution, while the small-scale galaxy distribution resembles a fractal. Perhaps gas dynamics redistributed mass on the scale of galaxies; perhaps galaxies did not form by gravitational instability. Essential hints surely will come from the recent progress in ability to observe galaxies at high redshifts by more efficient detectors, larger telescopes, and space-based observations. All this promises to give us an increasingly better view of galaxies as they were in the distant past. The details of what these young galaxies are like will be followed with great interest.

See also: Galaxies; Hubble Effect; Universe; Astronomy, X-ray.

Bibliography
For a longer account see J. Silk's *The Big Bang*. Freeman, New York, 1989.
For more detailed accounts see P. J. E. Peebles, *Physical Cosmology*, Princeton University Press, Princeton, NJ, 1971; S. Weinberg, *Gravitation and Cosmology: Principles and Applications of the General Theory of Relativity*, Wiley, New York, 1972; and P. J. E. Peebles, *The Large-Scale Structure of the Universe*, Princeton University Press, Princeton, NJ, 1980.
Recent research can be traced through the following: for the cosmological tests see the references in P. J. E. Peebles, *Publ. Astron. Soc. Pacific* **100**, 670, 1988; for large-scale velocity fields, see *Large-Scale Motions in the Universe*, V. C. Rubin and G. V. Coyne, eds., Princeton University Press, Princeton, NJ, 1989; for element production in the hot big bang see J. Yang *et al.*, *Astrophys. J.* **281**, 493, 1984; for the dark mass problem see *Dark Matter in the Universe*, J. Kormendy and G. R. Knapp, eds., Reidel, Dordrecht, 1987; papers on the inflationary scenario are collected in *Inflationary Cosmology*, L. F. Abbott and So-Young Pi, eds., World Scientific, Singapore, 1986; for recent progress in theories of galaxy formation see *Large Scale Structures of the Universe*, J. Audouze *et al.*, eds., Reidel, Dordrecht, 1988.

Counting Tubes

G. Charpak

The term counting tubes is applied to instruments detecting particles, charged or neutral, such as electrons, protons, gammas, and neutrons. Originally the most common shape was a tube with a central wire and an appropriate gas filling. Now the shape is much more diversified and the structure quite different. Since 1968 important properties of multiwire structures have been brought to light. The counters are used not only for counting, but more for localizing the position of the radiations. This has extended the application to new domains in nuclear physics, biophysics, astrophysics, and medicine. The considerable reduction in the cost of the electronics has played a great role in the evolution of the counters. The correlation between the time of detection of a signal and the ionization localization has led to an important new class of detectors, the drift chambers, which permit the construction of detecting surfaces of several square meters, with localization accuracies of better than $100\,\mu m$.

Some Basic Phenomena in Counting Tubes

All counting tubes are filled with gases. Liquefied noble gases are used in very rare cases. The particles to be detected and counted liberate free ions in the gas, either by direct interaction or by extracting charged particles from the walls of the tube.

An electric field, applied in the gas, drifts the ions to different electrodes, according to their sign. The ions are either collected without multiplication (ionization counter mode) or various types of multiplication may occur. In the proportional counter, each electron gives rise to a single, well-localized avalanche in the intense field surrounding a thin wire. In the Geiger–Müller counter each avalanche is followed by a series of avalanches propagating along the wire. In the limited streamer mode a succession of avalanches along the electric field lines gives rise to a much stronger current pulse than in the proportional mode. It has the advantage, over the preceding mode, of a very localized dead region along the wire with the same advantage of an almost constant pulse height but much faster. It is now used in very large detectors, in high-energy physics, because of the simplification it brings to the electronics. In the spark counter, the avalanches produce a conductive channel of avalanches bridging the anode–cathode gap, with a subsequent spark. The propagation is mediated by ultraviolet photons. Figure 1 illustrates the different modes of operation and Fig. 2 shows the current amplification factor in a cylindrical counter operating in the different modes.

Among the many factors influencing the choice of the filling gases are the following.

1. Ability to operate in the desired mode: Although all gases are suitable for ionization chambers, those working in stable conditions in a multiplicative mode are rare.

   Noble gases mixed with CO_2 or various organics such as alcohols, isobutane, and methane are often chosen. In these gases electrons liberated by the radiations are not captured by the atoms and move freely under the influence of electric fields.

   The additives to noble gases are chosen for their action on the electron drift velocity or their ability to control the mean free path of the ultraviolet radiations emitted by the atoms excited by electronic collisions.

2. Interaction properties with the radiations to be detected: For low-energy (up to $30\,keV$) x rays xenon is the best filling gas. Above these energies, the gaseous efficiency drops so drastically that the efficiency is dominated by the electrons extracted from the walls. For slow neutrons, ^3He or BF_3 are the best suited, since they have a considerable cross section for slow-neutron absorption.

   For charged particles, argon is the main component because of its low cost.

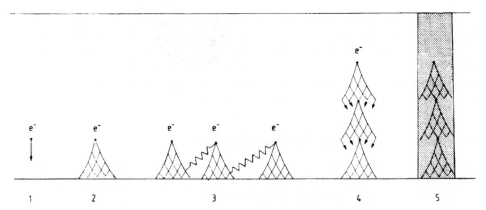

Fig. 1: The different mechanisms of electron multiplication around a wire.
(1) Ionization chamber. The electrons are attracted by the anode wire. No charge gain.
(2) Proportional counter. The field close to the wire is more intense, and by inelastic
collisions each electron gives rise to an avalanche. Maximum charge gain $\sim 10^6$.
(3) The ultraviolet light emitted by the excited atoms left behind by an avalanche
produces secondary electrons in the close vicinity of the avalanche. This gives rise to a
succession of nearby avalanches along the wire. Propagation speed 200 ns/cm:
maximum charge gain $\sim 10^{10}$. (4) Limited streamer mode. A succession of avalanches
propagates along the electric field lines but does not bridge the gap. (5) A conductive
channel is produced after the first avalanches and a spark short-circuits the electrodes.

The Different Gaseous Detectors

Ionization Counters

Any electrode shape is suitable for a counting tube that is to be used as an ionization counter.

In intense radiation fields the average collected current can be measured with a high accuracy. It is the ideal instrument for monitoring slowly varying radiation intensities. An example of this application is the measurement of the density of the products flowing in a pipeline, which can easily be determined, to accuracies of better than 10^{-3}, from the average current collected in a high-pressure ionization chamber detecting the 0.66-MeV γ rays from a ^{131}Cs source on the opposite side of the pipe.

For heavily ionizing particles, such as $\alpha >$ particles or fission fragments, the charge purses collected from every single track are detectable. The progress in the cost and noise of amplifiers progressively pushes down the minimum detectable energy lost in the gas. The limit is around 1 keV for very expensive amplifiers and around 100 keV for very cheap amplifiers, whereas in proportional counters one single electron liberated in the gas can easily be detected. This is why detectors with gaseous amplification have a wider range of applications.

Proportional Wire Counters

Proportional counters usually consist of a single thin wire in the axis of a cylinder. The wire is at a positive potential with respect to the cylinder. The electric field, as a function of the distance r to the axis, is equal to $2q/r$ where q is the charge per unit length. Close to the wire, which is usually 10–50 μm in diameter, the electric field is so high that the free electrons

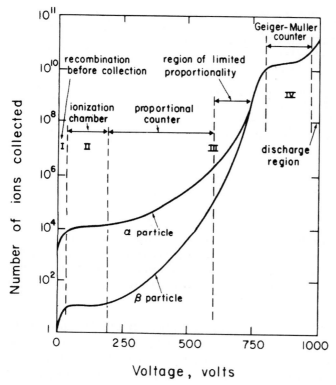

Fig. 2: Charges collected in a cylindrical counter as a function of applied voltage.

acquire, between two successive collisions, a sufficient energy to ionize the gas atoms and liberate a new electron. The mean free path at atmospheric pressure for an ionizing collision is close to $1\,\mu m$. Since the maximum gain is around 10^6, a maximum length of about $20\,\mu m$ is required for this amplification.

The majority of the electrons, which are produced in the last few microns, are collected in times less than 10^{-10} s and give rise to an undetectable pulse: it is too fast and too small; its size is proportional to the potential drop experienced by the electrons in the last few microns close to the wire.

The useful pulse is produced by the motion of the positive ions away from the wire. They have to undergo the total potential drop between cathode and anode, and despite their large mass, the field is so intense that the risetime of the induced pulse can be of the order of $20\,\text{ns}$. For a point charge the pulse rises like $\log(1 + t/t_0)$ where t_0 is close to $2\,\text{ns}$ for most common counter parameters.

The important properties are listed below.

1. *Energy response.* Since the mechanism of pulse formation is close to a wire, all the electrons freed in the gaseous volume give rise to the same pulse amplitude, and the charge induced in the cathode is proportional to the number of electrons and hence to the energy lost in the gas.

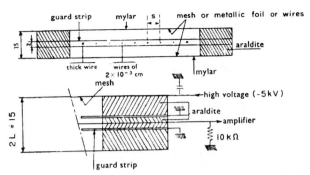

Fig. 3: Example of a multiwire proportional counter. Anode wires 20 μm in diameter, regularly spaced at distances of 2 mm between two cathode planes. Gas filling: argon + isobutane + methylal. High voltage 5 kV.

2. *Tune resolution.* When an electron is freed in the counter it will produce no pulse as long as it has not reached the anode. The delay in its response is thus a function of its distance to the axis. For particles irradiating a counter uniformly, this results in a time jitter that is equal at most to the maximum drift time of electrons in the volume of the cylinder. It is of the order of 1 μs for a counter 1 cm in diameter.

3. *Dead time.* The electric field between the positive-ion sheath left behind by an avalanche and the anode is reduced. For high amplification factors the anode may be locally dead for times as long as 100 μs, but over a region extending over only 0.2 mm. High counting rates, of the order of 10^4 counts s^{-1} mm^{-1} of wire, are tolerated by proportional counters.

Multiwire Proportional Chambers

These consist of a layer of thin anode wires, of diameter d, separated by equal distances s, sandwiched between two cathode planes at a distance L. The usual diameter ranges from 3 to 100 μm; the wire distance s is from 0.5 to a few millimeters (Fig. 3 shows an example of construction). The charge per unit length is

$$q = \frac{V_0}{2(\ln \sinh \pi L/s - \ln \sinh \pi d/2s)}$$

where V_0 is the cathode–anode potential difference. The field varies like $1/r$ close to the wire and is uniform at distances larger than $s/2$. Since all the multiplication processes occur within about 20 μm from the wire, the amplification processes around the wire are exactly the same as in a single-wire cylindrical counter.

Two essential properties characterize such a structure:

1. The electric field is sharply separated into two regions: Far from the wires the field is nearly uniform, whereas close to the wire it is similar to the radial field produced in a cylindrical tube with a central wire (Fig. 4). When electrons are freed in the gas they drift in the uniform field and then experience inelastic collisions close to the wire, as in a cylindrical counter.

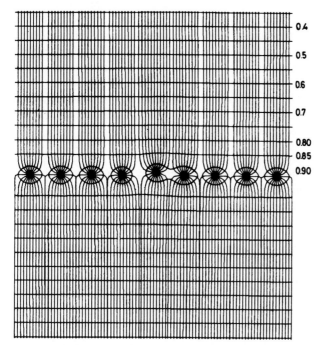

0.4
0.5
0.6
0.7
0.80
0.85
0.90

Fig. 4: Equipotentials in a multiwire chamber. Potential 1 on the anode wires, 0 in outer cathodes. $s = 1$ mm, $L = 8$ mm. Two regions of field: around the wire the field is cylindrical; in most of the volume it is uniform.

2. On a wire that is collecting the electrons from an avalanche, the pulse-producing mechanism is the same as in a cylindrical counter, the positive-ion sheath moving away in the intense field inducing the fast negative pulses. But the essential point is that it induces, at the same time, positive charges on the neighboring wires. Despite the strong capacitive coupling between wires, this ensures the localization of the detecting wire.

While large surfaces can be constructed, it is easy to localize with precision the position of the trajectory of the ionizing particle by detecting the wire carrying the pulse. The most remarkable properties of such structures, explaining their expanding use since 1968, are the following:

1. *Time resolution.* The time resolution is determined by the maximum distance between a trajectory and the nearest wire. With 2-mm wire spacing it is 25 ns, considerably better than in the cylindrical tube where the freed electrons had to drift in low fields over 1 cm.

2. *Dead time.* After the detection of an avalanche, the dead region is localized to about 0.2 mm along a wire and lasts for tens of microseconds. Since every wire acts as an independent detector, large counting rates can be reached; 10^7 counts/s over surfaces of 100 cm^2 are possible.

3. *Two-dimensional localization and imaging properties.* The motion of the positive ions, responsible for this localization of the detecting wire, induces positive pulses on the

cathode. If the cathode is made of wires or strips, orthogonal or at an angle to the wire, the determination of the centroid of the induced pulses yields the position of the avalanche along the wire with an accuracy of the order of 0.1 mm. This is an essential property for the two-dimensional localization of neutral radiations such as xrays or neutrons. Since the secondary ionizing radiation that they produce is usually of a very small range and cannot cross several chambers, the coordinates in a plane have to be obtained from a single gap. Several methods have been designed to obtain the two-dimensional readout: delay line methods, current division methods, etc.

4. *Examples of applications.* Chambers several meters in length, with a total of 20 000–100 000 wires, are common in high-energy physics, where they are used to localize trajectories. The imaging of slow neutrons or x-ray distributions with proportional chambers has growing applications in medicine and biology.

Drift Chambers

Detecting Drift Spaces

Detection volumes can be attached to wire chambers. They are used to extend the useful volume in case of gaseous detection, or they can be filled with various solid converters for an efficient conversion of γ rays or slow or fast neutrons. The electrons are transferred to the multiwire chamber by appropriate electric fields.

Localization Drift Chambers

The interval of time separating the production of ionization in a gas and the detection of a pulse at a wire is strictly correlated to the distance between the trajectory of the ionizing event and the wire. If the drift velocity of electrons is known, the measurement of time gives the trajectory position.

A typical drift velocity is 5×10^6 cm/s in argon–isobutane mixtures at fields of 1000 V/cm. A great variety of drift-chamber shapes are possible; an example of a widely used chamber is given in Fig. 5. The sense wires are spaced by 5 cm. The cathode planes, 6 mm apart, are made of wires or strips at a linearly increasing potential, producing an electric field parallel to the chamber plane.

The accuracy in position determination can reach 50 μm. Detectors of several square meters are common. It is the cheapest accurate detector so far built.

Geiger–Müller Counters

Geiger–Müller counters are constructed like proportional counters. When the field around a wire goes beyond some limit, the positive ions produced close to the wire are so dense that they produce a local electrostatic shielding that will prevent further development of avalanches on the same spot. On the sides of the avalanche, the electric field is high, photoelectrons produced by the ultraviolet photons emitted by the excited atoms will give rise to rapidly developing new avalanches, and the discharge will spread along the wire at a velocity depending on many factors but close to 10^7 cm/s. The pulse shape is governed by this propagation mechanism. It consists of a constant current as the discharge propagates to both sides of the wire; it then drops by a factor of two when one end has been reached, the total pulse width being a function

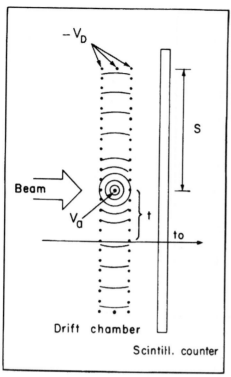

Fig. 5: Example of a drift chamber. The cathode is made of wires or strips at potentials growing linearly from 0 to $-V_D$. The drift field V_D/s is about 1000 V/cm. The electrons are detected on the anode wire at potential V_a. The length varies from 1 cm to 100 cm. The time t is a linear function of position. The zero time is given by a scintillator; the accuracy is better than 100 μm.

of the position of the initial avalanche. The total charge, however, is constant and easy to detect, since it is much greater than in a proportional counter.

Time Resolution

Time resolution is the same as in proportional counters; the time jitter is governed by the same mechanism, the time taken by the free electrons to reach the anode wire.

Dead Time

After a discharge the electric field between the ion sheath and the wire is reduced all along the wire, and the chamber is paralyzed for times of the order of 100 μs. For this reason Geiger counters have lost much of their interest, since it is now extremely easy and cheap to amplify the purses from proportional counters to the same level without paralyzing the counter for such long times. Geiger counters can be used in multiwire structures. The propagation of the discharge along the wires permits extremely simple methods of localization, although these methods are much less accurate than in the proportional mode.

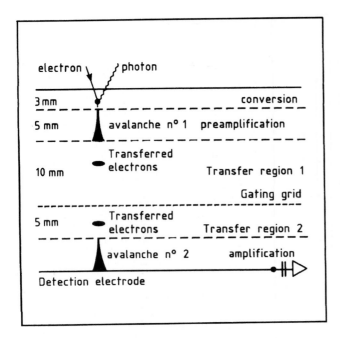

Fig. 6: Example of a multistep structure.

Limited Streamer Counters

A new mode of operation of wire counters has been discovered which is adopted for many large size detectors used in high-energy physics. With a proper choice of gases it is possible to have a succession of avalanches propagating along the field lines and stopping before reaching the cathode. This gives rise to large pulses of current, easy to handle by the electronics. The dead region is limited to a small region of the wire near the avalanches, while in a Geiger counter all the wire is shielded by the slow moving positive ions.

Proportional Counters With Parallel-Plane Electrodes

Electron avalanches between parallel electrodes can be used for the detection of the electrons liberated by charged particles. It is more difficult than with wires to reach large amplification. The limiting factor is the occurrence of sparks at some points because of inhomogeneities in the electric field, or because of some contamination by α emitters, which are often present in most gases. With wires the decrease of the electric field far from the wire is unfavorable for spark propagation.

A great extension in the use of the amplification between parallel electrodes can be foreseen with the introduction of multistep structures (Fig. 6). It has been shown that with electrodes made of grids transparent to electrons it is possible, by having a transfer gap for electrons between successive gaps, to reach amplifications even larger than with wire counters permitting for instance the detection and localization of single electrons liberated by the photoionization in the gap by ultraviolet photons.

Multistep Avalanche Chambers

The electrons liberated in a conversion gap by ionizing events drift in an electric field to a preamplification gap made of two wire meshes where they are multiplied by a factor 10^4 by an adequate electric field and a fraction of them is transferred to a region of lower field and then to a second amplification gap where the gain can be 10^3. Taking into account all the transfer losses gains of 10^6 are easy to achieve. An intermediate grid permits the gating, between the transfer regions 1 and 2, within times of few tens of nanoseconds, of the swarm of drifting electrons. This detector is ideal for single VUV photon detection and localization.

Counters Sensitive to UV or VUV Photons

An important class of new gaseous detectors has been introduced with the use of vapors which can be ionized by vacuum ultraviolet photons. For instance TMAE (tetrakis dimethylamine ethylene) can be ionized by radiations of a wavelength shorter than 230 nm, thus permitting, by introducing these vapors in wire counters or multistep parallel-plate counters, detection and imaging of VUV photons. The discovery of heavy scintillators, like BaF_2, emitting radiations at wavelength short enough to be detected by the gaseous detectors, permits one in some cases to replace photomultipliers or photodiodes and opens the way for a new wide range of applications.

Scintillating Proportional Counters

The scintillating proportional counter relies on a mechanism of gaseous amplification that is quite different from the one in operation in the preceding counters.

If electrons are drifted in some gases (such as pure xenon) in appropriate electric fields, they excite light-emitting atomic levels without ionizing the gas. The photons emitted by these atomic levels are detected by photomultipliers. The advantage is a better energy resolution. Typically, a full width of 8.5% is achieved with x rays of 6 keV, as against 14% with a proportional counter.

See also: Radiation Detection.

Bibliography
G. Charpak, R. Bouclier, T. Bressani, J. Favier, and C. Zupancic, *Nucl. Instr. Meth.* **62**, 235 (1968).
G. Charpak, D. Rahm, and H. Steiner, *Nucl. Instr. Meth.* **80**, 13 (1970).
G. Charpak and F. Sauli, *Phys. Lett.* **78B**, 523 (1978).
P. Rice Evans, Spark, Streamer, *Proportional and Drift Chambers*. Richelieu Press, London, 1974.
    (Contains a detailed description of the basic processes and the operation modes of most gaseous detectors in use prior to 1974.)
T. Ferbel, *Experimental Techniques in High Energy Physics*. Addison-Wesley, Reading, Mass., 1987.

CPT Theorem

R. W. Hayward[†] and K. Hikasa

The *CPT* theorem is one of the most far-reaching theorems in quantum field theory. This theorem states that any local Lagrangian theory which is invariant under proper Lorentz transformations is also invariant with respect to the combined operations of charge conjugation, *C*, space inversion, *P*, and time reversal, *T*, taken in any order, although the theory may not be separately invariant under the individual operations *C*, *P*, or *T*. One important corollary of this theorem is that if a physical process is not invariant under one of the operations *C*, *P*, or *T*, it is not invariant under at least one of the remaining two processes. A second corollary is that if a physical process is invariant under one of the operations *C*, *P*, or *T*, it is invariant under the product of the other two.

The validity of the *CPT* theorem is based on the following assumptions:

(a) The theory is invariant under proper orthochronous Lorentz transformations. No further assumptions concerning invariance under space and time reflections need be made.

(b) The Lagrangian density is local: Each term in the Lagrangian density is constructed out of field operators and derivatives of field operators of at most a finite order.

(c) Any two fields at spacelike separation commute or anticommute with each other. It can be proven that fields with an integral spin commute and those with a half-integral spin anticommute (*spin-statistics theorem*).

(d) The Lagrangian density is Hermitian as a whole.

A homogeneous Lorentz transformation from one coordinate frame, *S*, to another, *S'*, is given by

$$x'^\mu = \Lambda^\mu{}_\nu x^\nu,$$

where $x^\mu = (x^0, x^1, x^2, x^3) \equiv (ct, x, y, z)$ is the space-time coordinate four vector and summation over a repeated index is implied. The $\Lambda^\mu{}_\nu$ are real parameters independent of the space-time coordinates such that the relativistic interval remains constant, i. e.,

$$g_{\mu\nu} x'^\mu x'^\nu = g_{\mu\nu} x^\mu x^\nu,$$

where $g_{\mu\nu}$ is the Lorentzian metric: $g_{00} = 1$, $g_{11} = g_{22} = g_{33} = -1$, $g_{\mu\nu} = 0$ $(\mu \neq \nu)$. The invariance requires that

$$g_{\mu\nu} \Lambda^\mu{}_\rho \Lambda^\nu{}_\sigma = g_{\rho\sigma} .$$

A general (inhomogeneous) Lorentz transformation is given by

$$x'^\mu = \Lambda^\mu{}_\nu x^\nu + a^\nu,$$

where a^ν are real parameters corresponding to the displacement of the origin of the coordinates. These transformations form a ten-parameter continuous group called the Poincaré (or

---

[†]deceased

inhomogeneous Lorentz) group. Those with $a^\nu = 0$ form a six-parameter subgroup of the Poincaré group called the (homogeneous) Lorentz group.

These six parameters correspond to the space rotation (three of the six parameters) describing the change of the orientation of the space coordinates, and "boosts" (other three parameters) connecting coordinates having different velocities. The four parameters of the inhomogeneous part describe space-time translations, change of the origin of the coordinates. The requirement of equivalence of different Lorentz frames leads directly to the conservation of energy, momentum, and angular momentum.

The Lorentz group consists of four disjoint classes of transformations depending on signs of $\det\Lambda$ and $\Lambda^0{}_0$. A transformation satisfying $\det\Lambda = 1$ is called proper, and one satisfying $\Lambda^0{}_0 \geq 1$ is called orthochronous. Proper orthochronous Lorentz transformations constitute one of the four classes and can be obtained from a series of infinitesimal transformations from unity. These transformations do not change either the handedness of the space coordinates nor the direction of time and form a subgroup of the homogeneous Lorentz group.

Other three classes are not continuously connected to unity and do not form subgroups. Representative elements of these classes are:

(a) Space inversion (parity): $(x^0, \mathbf{x}) \rightarrow (x^0, -\mathbf{x})$; $\Lambda^\mu{}_\nu = \mathrm{diag}\,(1, -1, -1, -1)$, class $\det\Lambda = -1$, $\Lambda^0{}_0 > 1$.

(b) Time reversal: $(x^0, \mathbf{x}) \rightarrow (-x^0, \mathbf{x})$; $\Lambda^\mu{}_\nu = \mathrm{diag}\,(-1, 1, 1, 1)$, class $\det\Lambda = +1$, $\Lambda^0{}_0 < 1$.

(c) Total inversion: $(x^0, \mathbf{x}) \rightarrow (-x^0, -\mathbf{x})$; $\Lambda^\mu{}_\nu = \mathrm{diag}\,(-1, -1, -1, -1)$, class $\det\Lambda = -1$, $\Lambda^0{}_0 < 1$.

Any transformation in each class can be written as the product of one of these representative elements and a proper orthochronous Lorentz transformation.

The last class of the transformations containing total inversion become continuously connected to the unity, if the Lorentz transformation is extended to complex values, thus not maintaining the reality of the space-time coordinates in general. For example, total reflection corresponds to a real rotation through an angle π in the x^1–x^2 plane followed by an imaginary "rotation" through an angle π in the x^0–x^3 plane. This fact is crucial in axiomatic proof of the *CPT* theorem.

Under the parity transformation, a general relativistic field $\Phi(\mathbf{x}, t)$ undergoes a transformation of the form

$$\Phi(\mathbf{x},t) \rightarrow \mathcal{P}\Phi(\mathbf{x},t)\mathcal{P}^{-1} = \eta_P S_P \Phi(-\mathbf{x},t).$$

Here S_P is a matrix which acts on the spin components of Φ, and is common to all fields of the same spin. η_P is a phase factor which may vary for each field. The matrix S_P for lower spin fields are given in Table 1.

Similar transformation law applies for time reversal, except that the operator $\mathcal{T}$ is antiunitary and any c number is transformed to its complex conjugate:

$$\Phi(\mathbf{x},t) \rightarrow \mathcal{T}\Phi(\mathbf{x},t)\mathcal{T}^{-1} = \eta_T S_T \Phi(\mathbf{x},-t).$$

The matrix S_T is found in Table 1 and the η_T is a field-dependent phase factor.

Table 1: Transformation matrix of fields under space inversion, charge conjugation, time reversal, and *CPT*. Here P is the matrix diag $(1, -1, -1, -1)$. See text for the definition of C.

Field	$\Phi(\mathbf{x},t)$	S_P	S_T	S_C	S_{CPT}
Scalar	$\varphi(\mathbf{x},t)$	1	1	1	1
Dirac spinor	$\psi(\mathbf{x},t)$	γ^0	$B = C^\dagger \gamma_5$	$C\gamma^{0T}$	γ_5^*
Vector	$A^\mu(\mathbf{x},t)$	$-P$	P	1	-1

Under charge conjugation, the field Φ transforms as

$$\Phi(\mathbf{x},t) \rightarrow C\Phi(\mathbf{x},t)C^{-1} = \eta_C S_C \Phi^*(\mathbf{x},t).$$

Unlike P and T, the field transforms into its conjugate. This operation exchanges the role of the particle and its antiparticle. Again the matrix S_C is listed in Table 1 and η_C is a field-dependent phase. The matrix C for spin 1/2 Dirac field is a unitary matrix satisfying $C\gamma^{\mu T}C^\dagger = -\gamma^\mu$. In the Dirac representation of the gamma matrices, it can be chosen as $C = i\gamma^2\gamma^0$.

Self-conjugate fields may be defined by the condition $\Phi = S_C\Phi^*$. This is nothing but the reality condition for the scalar and vector fields and Majorana condition for the spin 1/2 field. Self-conjugate fields describe a particle which is identical to its antiparticle. The phase factors η_X ($X = P, T, C$) for these fields are restricted to $\eta_X = \pm 1$ for integer-spin fields or $\eta_X = \pm i$ for half-integer-spin fields. Spin-0 fields with $\eta_P = +1$ is called scalar (in the narrow sense) and those with $\eta_P = -1$ pseudoscalar. Likewise, spin-1 fields with $\eta_P = +1$ is called (polar) vector and those with $\eta_P = -1$ axial vector.

In order that each transformation P, T, or C is a symmetry of the theory, the Lagrangian density has to be invariant under the transformation, by appropriate choices of the phase factors η_X associated to the fields in the Lagrangian. This is not always the case for all three transformations.

The *CPT* transformation is a combined operation of the three transformations. By an appropriate choice of phases, it can be written as

$$\Phi(\mathbf{x},t) \rightarrow C\mathcal{P}\mathcal{T}\Phi(\mathbf{x},t)(C\mathcal{P}\mathcal{T})^{-1} = S_{CPT}\Phi^*(-\mathbf{x}, -t),$$

with the matrix S_{CPT} listed in Table 1. Under *CPT* transformation, an integer-spin operator with spin J transforms to its conjugate with the phase factor $(-1)^J$. This is also true for operators containing an even number of fermion fields. Since Lorentz invariant field theories have Lorentz scalar Lagrangian density which is also hermitian, the Lagrangian density is found to be invariant under *CPT* transformation.

More rigorous proof of the *CPT* theorem is made in axiomatic field theory. The proof utilizes the invariance of the theory under proper orthochronous Lorentz group, which can be extended to complex Lorentz group to include the total reflection. The vacuum expectation values of products of fields can be analytically continued to a region where the space-time coordinates take complex values. Existence of such a region is ensured by the energy condition that the energy eigenvalue of any physical state is nonnegative. The relation between a matrix element and its *CPT* conjugate can then be established using the (anti)commutativity of the fields in spacelike separations.

Three of the four known fundamental interactions in Nature, gravitation, electromagnetic interaction, and the strong interaction, are known to preserve C, P, and T symmetry separately. The weak interaction, however, violates these individual symmetries. The W boson, the mediator of the charged weak processes, interacts only with left-handed leptons and quarks and with right-handed antileptons and antiquarks. Parity and charge conjugation symmetries are thus maximally broken. The product CP is conserved in a large class of weak interaction processes.

The study of neutral kaon decays in 1964 revealed a small violation of CP invariance in the weak interactions. One of the neutral kaon states K_L^0, which mainly decays to three pions and was supposed to have $CP = -1$, is found to decay to two pions (even CP) with a branching fraction of 0.3%. This had been the only observed CP noninvariance for 35 years. In 2000, large CP violating asymmetries were discovered in a class of B meson decays. The observed effect is in agreement with the origin of CP violation proposed by Kobayashi and Maskawa in 1972, which now constitutes a part of the Standard Model of elementary particle interactions.

There is no experimental indication that CPT invariance is violated. The most stringent test is provided by the measurement of the mass difference of neutral kaon and its antiparticle. Its upper limit is 10^{-18} times the kaon mass itself.

See also: Bose–Einstein Statistics; Elementary Particles in Physics; Fermi–Dirac Statistics; Invariance Principles; Lorentz Transformations; Matrices; Operators; Parity; Quantum Field Theory.

Bibliography

G. Lüders, "Proof of the TCP Theorem," *Ann. Phys. (New York)* **2**, 1 (1957).

W. Pauli, "Exclusion principle, Lorentz group, and reflection of space-time and charge," in *Niels Bohr and the Development of Physics*, W. Pauli (ed.). Pergamon Press, London, 1955. (I)

J. J. Sakurai, *Invariance Principles and Elementary Particles*. Princeton University Press, Princeton, N.J., 1964. (I)

R. F. Streater and A. S. Wightman, *PCT, Spin and Statistics, and All That*. Benjamin, New York, 1964. (A)

S. Weinberg, *The Quantum Theory of Fields*, Vol. 1: *Foundations*, especially Ch. 5. Cambridge University Press, Cambridge, 1996.

Particle Data Group, "Review of Particle Physics," *Phys. Lett. B* **592**, 1 (2004). See pdg.lbl.gov for the newest version.

Critical Points

J. D. Litster

If the pressure on a fluid, such as water, is increased, the boiling temperature will also increase. At the same time, the density difference between the liquid and vapor phases becomes smaller, and vanishes continuously as the liquid–gas critical point (at temperature T_c and pressure p_c) is approached. This is commonly called a second-order phase transition, because a material property known as the order parameter (here, the density difference between the liquid and vapor phases) vanishes continuously. Near the critical point anomalous behavior (divergence) is observed in the range over which order parameter fluctuations are correlated as well as in a number of thermodynamic derivatives, such as the compressibility, and the specific heat. Nearly all materials near a second-order phase transition have been found to show behavior similar to that of a pure fluid near its critical point. This is because properties near a critical point are determined primarily by the correlation length for fluctuations in the order parameter; when this length exceeds the range of the interactions responsible for the phase transition, the behavior becomes insensitive to the detailed nature of the interactions. With the order parameter given in parentheses, some examples of critical points are: ferromagnets (magnetization), antiferromagnets (sublattice magnetization), separating binary fluid mixtures (concentration of components in the two phases), ferroelectrics (spontaneous polarization), and superconductors and superfluids (condensate wave function). While there are quantitative differences in the critical behavior near these different second-order phase transitions, it is the similarity that is most striking.

The classical models for these phase transitions all predicted the same behavior in the vicinity of the critical point as was given by Landau's 1937 theory [1] of phase transitions. The anomalous quantities diverged according to powers of $T - T_c$, (e. g., the compressibility diverged as $[T - T_c]^{-1}$) and the specific heat was predicted to have a discontinuity rather than to diverge. Some materials, such as ferroelectrics and superconductors, were found to follow the classical theory closely, but most showed power-law divergences with different exponents than predicted as well as specific heats that diverged approximately logarithmically.

The Landau model failed because it does not consider properly the divergent order-parameter fluctuations near the critical point. The renormalization group (RG) method [2] can be used for statistical mechanical calculations in the presence of strong fluctuations, a discovery for which Kenneth Wilson was awarded the Nobel Prize in Physics in 1982. As a result of this breakthrough, we have now a quantitative understanding of the role of critical fluctuations and an appreciation of the importance of the effects of spatial dimensionality and symmetry of the order parameter on critical behavior. Critical points fall into different "universality classes," which should show identical behavior, according to the symmetry, and hence the number of degrees of freedom, of the order parameter [3]. The effect of thermally induced fluctuations of the order parameter diminishes as the spatial dimension increases. There is an upper "marginal" dimension, which is four for most critical points, where the classical Landau theory becomes valid. There is also a lower marginal dimension, two for most critical points; below this dimension the effect of the fluctuations is so strong the phase change does not occur. (Rare gases adsorbed on the surface of crystals are a realizable example of a two-dimensional system.)

The RG methods have been used to explain theoretically a wide variety of second-order phase transitions. This incudes multicritical points where two or more lines of critical points join. An example is a tricritical point, where three lines of critical points meet and the transition becomes first order; this commonly occurs when two order parameters compete in a system. When order parameters are of different universality classes, RG methods can be used to calculate the "crossover" behavior as one dominates the other, and also to calculate the interplay between two- and three-dimensional physics in a wide variety of materials. For example, RG theories of multicritical phenomena in anisotropic magnets have been applied to a detailed model for the growth of two- and three-dimensional liquid crystal phases [4].

The discussion above applies to systems in which the fluctuations are driven by temperature and are described by classical statistical mechanics. Of recent interest are quantum critical points [5]. Quantum critical behavior arises in magnetic-nonmagnetic phase changes at very low temperatures that occur because an interaction is varied (by alloying or changes in pressure, for example) rather than because the temperature is changed, as would be the case for a conventional critical point. The fluctuations are not driven by temperature but rather by quantum mechanical effects (i. e., the "zero point" fluctuations that are a consequence of the Heisenberg uncertainty principle). At a conventional critical point the static and dynamic critical behavior may be considered independently, but at a quantum critical point they are intertwined and theoretical models must consider both equilibrium and nonequilibrium properties together [6]. A consequence is that static (equilibrium) behavior at a quantum critical point is like that at a conventional critical point in a higher spatial dimension; the dimension change is determined by a dynamic critical exponent z that can vary from 1 to 4, depending on the system. Quantum critical behavior often occurs in systems where antiferromagnetism and superconductivity interact; thus there is intellectual overlap with the problem of understanding high T_c superconductivity.

The experimental elucidation of the properties of fluids near critical points and their theoretical explanation, which were among the most challenging problems in physics during the 1960s, have enabled us to understand in detail such diverse physical systems as rare earth magnets, thin film substrates on solid substrates, lamellar (high T_c,) superconductors, liquid crystals, and biological membranes. The experimental and theoretical tools that were developed are relevant to important current investigations in physics.

See also: Ferroelectricity; Ferromagnetism; Fluctuation Phenomena; Helium, Liquid; Liquid Crystals; Phase Transitions; Renormalization; Superconductivity Theory.

References

[1] L. D. Landau, "On the Theory of Phase Transitions" (English translation), in *The Collected Papers of L. D. Landau* (D. Ter Haar, ed.). Gordon & Breach, New York, 1965. (E)

[2] K. G. Wilson, *Phys. Rev. B* **4**, 3174, 3184 (1971). (A) M. E. Fisher, *Rev. Mod. Phys.* **46**, 597 (1974). (I)

[3] See the extensive discussions of various theoretical approaches in the multivolume series *Phase Transitions and Critical Phenomena* (C. Domb and M. S. Green, eds.). Academic, New York, 1972, 1974. (A)

[4] J. D. Brock, R. J. Birgeneau, J. D. Litster, and A. Aharony, "Liquids, Crystals, and Liquid Crystals," *Phys. Today* (July 1989), and references therein. (E)

[5] W. Montfrooij, M. C. Aronson, B. D. Rainford, J. A. Mydosh, A. P. Murani, P. Haen, and T. Fukuhara, "Extended versus Local Fluctuations in Quantum Critical Ce$(Ru_{1-x}Fe_x)_2Ge_2$ ($x = x_c = 0.76$)," *Phys. Rev. Lett.* **91**, 087202 (2003). (A)

[6] John A. Hertz, "Quantum Critical Phenomena," *Phys. Rev. B* **14**, 1165 (1976). (A) A. J. Millis, "Effect of a nonzero temperature on quantum critical points in itinerant fermion systems." *Phys. Rev. B* **48**, 7183 (1993). (A)

Cryogenics

D.-M. Zhu and A. C. Anderson

Cryogenics is the science of producing and maintaining low temperatures, studying the properties of matter and systems at low temperatures, and safely and efficiently utilizing low temperatures for various applications. The field of cryogenics is very broad and is generally associated with the use of various gases and liquefied gases. In the industrial sector cryogenics includes the preparation and storage of food and biological materials, the production and delivery of liquid oxygen to medical and metallurgical facilities, the transport and storage of liquefied natural gases, the liquefaction of hydrogen for the space program, and development of new and efficient refrigerators and cryogenic techniques for various applications. Cryogenics provides techniques for cooling various detectors used in space, medical, and defense applications. One current area of active research and developmental work is directed toward the application of low temperature techniques to the generation, transmission, storage, and utilization of electrical energy. At sufficiently low temperatures, almost all metals and some oxides become superconductors which have zero electrical resistance and hence provide a considerable reduction in power loss when used in electrical power and electronics industries. Cryogenic techniques provide platforms for searching for new superconducting materials and developing superconducting devices and instruments.

In measuring low temperatures, although various temperature scales such as Celsius and Fahrenheit scales are used, the Kelvin scale of temperature is generally used in physics. On this scale the freezing and boiling points of water are approximately 273 and 373 K, respectively. The zero degree on the Kelvin scale is the absolute zero of temperature which can only be approached but can not be reached. To appreciate the Kelvin scale and in fact to understand the importance of cryogenics to the physicist, it is necessary to recognize that it is often the ratio of two temperatures, not the difference, that controls a physical or chemical process. Therefore the temperature interval 0.001–0.01 K may contain as much interesting physics as the much "larger" interval from 100 to 1000 K. The calibration of a low-temperature thermometer is based on a variety of phase transitions, such as the temperature of the triple point of hydrogen or the equilibrium vapor pressure of liquid helium. The temperatures of the phase transitions have been estimated through much exacting experimental work; the accuracy of the determinations is constantly being improved. A thermometer generally utilizes a physical property of matter, such as an electrical or magnetic property, as the thermometric parameter. Examples are the electrical resistivity of metals and semiconductors or the weak magnetic susceptibility of certain salts and metals. Each type of thermometer has been carefully studied to be certain that impurities, strain, and other variables do not influence the calibration. A variety of thermometers, some precalibrated, are available commercially for use down to 0.01 K or lower.

Providing a research laboratory with appropriate refrigeration facilities has been greatly simplified in recent years. Liquid cryogens such as liquid nitrogen and liquid helium are widely used. Liquid nitrogen at 77 K and liquid helium at 4 K are available from commercial sources or may be liquefied on site using commercial liquefiers. Gaseous ^{4}He is extracted from natural gas wells in several countries and therefore is sufficiently abundant, at present, that many laboratories vent the evaporated ^{4}He to the atmosphere. Temperatures between 4 K and 300 K can be conveniently obtained by partial thermal isolation of the experimental samples from the liquid cryogen. Another widely used method to generate refrigeration is to circulate pure ^{4}He gas in a gas expansion cycle. ^{4}He gas is supplied at room temperature to a refrigerator by a compressor. The compressed gas is cooled by expansion, providing cooling at the refrigeration station. After cooling the refrigerator, the gas is returned to the compressor to repeat the cycle. Typical two-stage closed cycle ^{4}He refrigerators can produce temperatures down to about 10 K. Multiple stage version of closed cycle refrigerators which employ special heat exchanging materials have been developed to extend the temperature down to about 4 K. Closed-cycle ^{4}He refrigerators are simple to operate and maintain, and are especially useful in applications where liquid nitrogen and liquid helium are either impractical or unavailable. In some applications, using closed cycle refrigerators can be more cost effective than using those employing a liquid cryogen.

Temperatures below 4 K are produced with liquid ^{4}He by rapidly pumping away the evaporating gas. Since helium, unlike other liquids, does not solidify during this process, a liquid bath at a temperature of about 1 K can be produced. Below 2 K, liquid ^{4}He is a superfluid having an extremely large effective thermal conductivity. As a result, the helium bath has the desirable property of being nearly isothermal. On the other hand, a superfluid film creeps up the walls of the cryostat, a process that is often the most important contribution to the liquid loss rate.

The lighter isotope of helium, ^{3}He, has become available as a by-product in the manufacture of nuclear devices. As a liquid, ^{3}He has a higher vapor pressure than ^{4}He and is not a superfluid above 0.003 K. The liquid may be pumped to 0.25 K. The price of ^{3}He dictates that it be used in a closed system. The evaporated gas is either collected for reuse or returned to the cryostat to form a closed-cycle, continuously operating refrigerator. Pumped ^{4}He at 1–2 K is required to absorb the heat as the ^{3}He gas is reliquefied. A more important closed ^{3}He cycle has recently been developed in which the liquid ^{3}He is dissolved in a small stationary bath of liquid ^{4}He. This process absorbs heat and therefore provides refrigeration. Commercial ^{3}He–^{4}He dilution refrigerators are available that provide temperatures down to 0.01 K continuously. Dilution refrigerators can be constructed in a variety of configurations to match the experimental problem.

Refrigeration may be provided for temperatures below 0.01 K but not, at present, in a continuous fashion. The ^{3}He–^{4}He dilution process may be used to reach roughly 0.002 K by not returning the ^{3}He gas to the cryostat. Alternatively, pure ^{3}He can be solidified by the application of a pressure of about 2.9×10^6 Pa (29 atm). Since the freezing characteristics of pure ^{3}He below 0.3 K are such that it cools with the application of pressure, compression and solidification of liquid ^{3}He can provide temperatures to about 0.002 K. Still lower temperatures can be obtained by the use of slightly paramagnetic materials. Application of a large magnetic field causes such a material to warm. The evolved heat may be absorbed by a dilution refrigerator. Removal of the magnetic field then causes the paramagnetic material to cool below the temperature of the refrigerator. Techniques combining dilution refrigerators with multiple nuclear cooling stages in cascade and using superconducting magnets have been developed. Refrigerators utilizing these techniques have achieved temperatures of nearly 1 μK (one-millionth of a degree above absolute zero) in metals and nuclear temperatures (when the temperature is sufficiently low, electrons and nucleus in matter can stay at two different temperatures) of about 0.25 nK (one quarter of one billionth of a degree above absolute zero).

It is to be emphasized that the temperatures mentioned in the foregoing refer to the sample or device that is to be cooled. The temperature of the refrigerator or refrigerating material may be much lower, since the transport of heat to the refrigerator becomes an increasingly severe problem with decreasing temperature. It is therefore important to minimize any heat influx to the sample. This is done by providing several stages of stepwise reduction in temperature between 77 and 0.02 K to reduce heat transport by thermal conduction. In addition, and especially at temperatures below 1 K, heat influx from vibration and radiation must also be eliminated. For radiation this includes the electromagnetic spectrum from light waves to radio waves, and many low-temperature laboratories are therefore constructed inside copper-shielded rooms.

In experimental physics, cryogenic techniques are employed either because a phenomenon of interest occurs only at low temperature, or because a physical property being studied becomes less complex and more amenable to theoretical interpretation. Phenomena that occur only at low temperatures include superconductivity, the superfluid phases of liquid ^{3}He and liquid ^{4}He, and the effects of the motion of atoms and molecules that persist after thermal motion has been suppressed through a reduction in temperature. Since these phenomena can be understood only in terms of quantum mechanics, cryogenics provides the opportunity for a rigorous test of the modern theory of quantum mechanics applied to condensed matter. A well-documented example of the use of cryogenics to isolate and simplify physical properties, and hence facilitate theoretical interpretation, has been the study of nonmetallic crystals. Thermal expansion, thermal conductivity, and specific heat can be described in terms of collective atomic vibrations or lattice waves called phonons. At low temperatures those phonons that are of greatest importance have the velocity of a sound wave and a wavelength large compared to the atomic spacing in the solid. The phonon behavior is then rather independent of the detailed atomic arrangement and bonding, and the phonon may be treated as a classical sound wave. This permits the development of a simple theory that still contains quantization, and that may be tested quantitatively against experimental data. Historically this procedure has been of great importance in the development of our knowledge of the physics of solids. Cryogenics continues in this manner to provide the physicist with a highly effective experimental tool.

Finally, it should be mentioned that an entirely different technique, laser cooling, developed outside of the field of cryogenics, has been successfully developed and used to cool gaseous phases of matter to very low temperatures. In laser cooling, laser beams are used to slow down the motion of gaseous atoms and thus reduce their temperature. The technique has produced a record low temperature of half-a-billionth of a degree above absolute zero for gaseous atoms. Using this technique, a fundamentally intriguing phenomenon, Bose–Einstein condensation, has been finally observed in recent years.

See also: Heat Capacity; Heat Transfer; Helium, Liquid; Magnetic Cooling; Laser Cooling; Paramagnetism; Temperature; Thermal Expansion.

Bibliography
Bulletin of the International Institute of Refrigeration. (A)

E. M. Codlin, *Cryogenics and Refrigeration (A Bibliographical Guide).* Mac-Donald, London, 1968; Part 2, IFI/Plenum, New York, 1970.

F. E. Hoare, L. C. Jackson, and N. Kurti, *Experimental Cryophysics.* Butterworth, London, 1961. (A)

O. V. Lounasmaa, *Experimental Principles and Methods Below* 1 K. Academic, New York, 1974. (A)

K. Mendelssohn, *The Quest for Absolute Zero*. McGraw-Hill, New York, 1977. (E)

H. M. Rosenberg, *Low Temperature Solid State Physics*. Oxford, London and New York, 1965. (A)

M.W. Zemansky, *Temperature Very Low and Very High*. Van Nostrand, Princeton, NJ, 1981. (E)

G. K. White and P. J. Meeson, *Experimental Techniques in Low-Temperature Physics*, 4th ed. Oxford University Press, New York, 2002.

Crystal and Ligand Fields

C. J. Ballhausen

The basic idea of crystal field theory is that the electrons located upon a transition-metal ion which is embedded in an ionic lattice are subjected to an electric field originating from the surrounding electric charges. The "lattice" can also be taken to consist of the groups or ions directly associated with the metal ion, called the ligands. For instance in the CrF_6^{3-} complex the $(3d)^3$ electrons, taken to be located upon the Cr^{3+} center, experience an electric intramolecular field of octahedral symmetry coming from the six F ligands. In $Ni(H_2O)_6^{++}$ the $(3d)^8$ electrons located on the Ni^{++} ion are subjected to a field originating from the six water ligands. Historically this notion was first used by Van Vleck in the early 1930s to explain the paramagnetism of the complexes of the rare earths, $(4f)^n$, and the $(3d)^n$ transition series.

It was known that for a free ion in vacuum the magnetic moment would have a contribution both from the total orbital angular momentum L and from the total spin angular momentum S of the electrons. For complexes of the 3d series, in contrast to those of the 4f series, it was, however, found that the orbital angular-momentum contribution was quenched. Van Vleck explained this by pointing out that the degeneracy of the orbital angular momentum would be lifted by the crystal field – an intramolecular Stark effect occurs. In the Cr^{+++} ion, for instance, the atomic ground state is a ^4F state. This has sevenfold orbital degeneracy. An octahedral crystal field will now split this state into three levels, being respectively three- three-, and onefold degenerate (Fig. 1). The ground state is onefold degenerate, and separated from the first excited state by several thousand wave numbers. It is therefore only the ground state which is populated at room temperature, and since an orbital contribution to the magnetic susceptibility requires degeneracy, there will be none such in octahedral Cr^{3+} complexes.

On the other hand, for the complexes of the $(4f)^n$ series, the f electrons are so well "shielded" from the electric charges of the ligands that they experience only a small crystal field. The degeneracies are indeed broken up, but the splittings are small compared to kT. Consequently it is only at low temperatures that deviations from free-ion behavior are observed.

The splittings of the atomic levels due to the crystal field had been worked out by Hans Bethe in 1929, using point-group symmetry. Applying Bethe's theory, Van Vleck and his co-workers then performed quantitative calculations of the magnetic susceptibilities of transition-metal complexes. Using only the symmetry of the complex – octahedral, tetrahedral, or a distorted arrangement of the ligands – and with the "strength" of the crystal field as a parameter, they succeeded in giving a complete explanation of the experimental features.

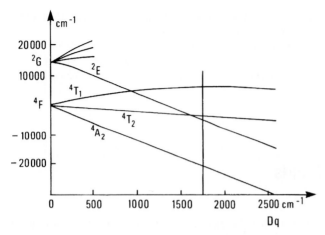

Fig. 1: Partial orbital splitting diagram for $(3d)^3$ in an octahedral field. For $Cr(H_2O)_6^{+++}$, $Dq = 1740\,cm^{-1}$ with the first spin-allowed band $^4A_2 \rightarrow {}^4T_2$ placed at $17400\,cm^{-1}$. The $^4A_2 \rightarrow {}^2E$ lines are found at approximately $15000\,cm^{-1}$. The second spin-allowed band $^4A_2 \rightarrow {}^4T_1$ is placed at $24500\,cm^{-1}$

The idea of an operative "crystal field" was, however, not accepted at the time by the chemists. The valence bond treatment of the complexes, as taught by Linus Pauling in his famous book *The Nature of the Chemical Bond*, had a strong grip on the chemical community. It was therefore a major breakthrough when Van Vleck showed in 1935 that there are no contradictions between the two theories: The "crystal field splitting" is indeed the unavoidable consequence of the chemical bonding between the ligands and the metal ion. Taking this into account, the name "crystal field" has been supplanted by the name "ligand field" to acknowledge the role played by the "covalent" ligand-metal bond.

An important consequence of the theory which was not appreciated until much later is the impossibility of calculating the crystal field splittings using an electrostatic approach as originally done by Van Vleck. These splittings can only be calculated by doing a full molecular orbital calculation, including all the many important "exchange integrals." Such a calculation was first performed in 1956 by Tanabe and Sugano. The difficulties of calculating the splittings from first principles are, however, not of any practical importance, since the strength of the crystal field, measured in units of Dq, is always extracted from experiments. Dq is thus a semiempirical parameter analogous with the β parameter in Hückel theory.

In addition to accounting for the magnetic features, crystal or ligand field theory can also explain the often beautiful colors of the transition-metal complexes as being due to transitions from the ground state to the excited atomic levels as split by the crystal field. The first absorption band which was identified is a transition in octahedral Cr^{+++} complexes. This was done by Finkelstein and Van Vleck in 1940. The atomic configuration of $(3d)^3$ gives rise to (among others) ^4F and ^2G states, the latter being placed some $15000\,cm^{-1}$ above ^4F (Fig. 1). When a cubic crystal field is added, the ^4F splits as indicated, while ^2G gives rise to (among others) a ^2E state. In the diagram, the levels are calculated as a function of Dq, and with a value of $Dq = 1820\,cm^{-1}$ Finkelstein and Van Vleck could place the $^4A_2 \rightarrow {}^2E$ transition at $14900\,cm^{-1}$.

The weakness of the intensity of this transition is explained by its being both spin- and parity-forbidden. The transitions which take place between the split components of (for example) ^4F in $(3d)^3$ are also parity-forbidden, taking place inside the $(3d)^3$ configuration, but being spin-allowed, they have a somewhat larger intensity. The bands are broad and structureless. The identification of such bands had, however, to wait until after World War II – the first such identification being made by Ilse and Hartmann for $Ti(H_2O)_6^{+++}$ in 1951. Extensive correlation diagrams which enable one to identify the excited states of all $(3d)^n$ complexes, have been published by many authors.

The crystal or ligand field theory is exceptional among semiempirical electronic theories in that it permits one to perform identifications of excited states solely from a knowledge of the symmetry of the complex ion. However, the lower the point-group symmetry of the complex, the more splitting parameters have to be introduced. Only in octahedral and tetrahedral symmetries is there only one one-electron parameter, Dq. Fortunately, the Dq parameter is rather insensitive to lower fields, and a gross identification is rarely obscured by the presence of such fields. The order of magnitude of Dq is $1000\,\mathrm{cm}^{-1}$ for doubly ionized metal ions of the first transition $(3d)^n$ series. For triply ionized ions, it is approximately $2000\,\mathrm{cm}^{-1}$.

The aim of the so-called "angular overlap models" is to establish a correlation between the orbital splittings caused by the lower symmetries and chemical behavior. Unfortunately, the experimental observations rarely permit an unambiguous assignment of the splitting parameters, and ingenious "chemical" assumptions have to be introduced. The inherent postulate that all "effects" are solely related to the one-electron parameters is also a necessity. Indeed, over-refinements of the crystal and ligand field theories obscure their main merits: the deep understanding they have given us of the ground and excited states of many electronic systems.

See also: Chemical Bonding; Group Theory in Physics; Magnetic Moments.

Bibliography

C. J. Ballhausen, *Introduction to Ligand Field Theory*. McGraw–Hill, New York, 1962.
C. J. Ballhausen, *Molecular Electronic Structures of Transition Metal Complexes*. McGraw–Hill, New York, 1979.
F. A. Cotton, *Chemical Applications of Group Theory*. Wiley, New York, 1979.
T, M. Dunn, D. S. McClure, and R. G. Pearson, *Crystal Field Theory*. Harper & Row, New York, 1965.
J. P. Fackler (ed.), *Symmetry in Chemical Theory*. Dowden, Hutchinson & Ross, Allentown, PA, 1972.
B. N. Figgis, *Introduction to Ligand Fields*. Wiley, New York, 1966.
J. S. Griffith, *The Theory of Transition-Metal Ions*. Cambridge University Press, Cambridge, 1961.

Crystal Binding

J. C. Phillips

Solids are often classified according to the nature of their binding. Four broad types of binding are covalent, ionic, metallic, and molecular.

The simplest of these is ionic binding, which arises when there is substantial charge transfer from one or more cations to one or more anions. As an example, consider NaCl (table salt). In the crystal the electronic configuration is Na^+Cl^-. The electrostatic interaction energy of point charges centered on the ions is called the Madelung energy; in NaCl this energy is about $-9\,\text{eV}$ per atom pair. The cohesive energy of NaCl is $-8\,\text{eV}$ per atom pair; about $+1\,\text{eV}$ per atom pair arises from closed-shell repulsion (exclusion principle). Many naturally occurring minerals have ionic binding with complex crystal structures, e. g., mica [$KAl_3Si_3O_{10}(OH)_2$]. Ions are rather incompressible, so that ionic radii empirically determined from interatomic distances in simple ionic crystals can be used to rationalize complex ionic structures.

Metallic binding ideally occurs when the valence electrons of metallic (cationic) atoms such as Na or Al are weakly bound to the ion cores. In the solid these electrons then form a nearly free electron gas, with a cohesive energy of about $1\,\text{eV}$ per valence electron. In this case there is no classical electrostatic interaction (as in ionic crystals) that can account for the observed cohesion. Instead, quantum-mechanical exchange forces (between electrons of parallel spin) or dynamical correlation (avoidance) between electrons of antiparallel spin takes place; these interaction energies are large in metals just because the electrons are spread out away from the ionic centers.

Three kinds of metals should be distinguished: "normal" metals, with s and p valence electrons only; transition metals (with d electrons); and rare-earth metals (unfilled f shell). Only the s and p valence electrons can become free and contribute fully to cohesion, but the d electrons contribute substantially and the f contribution in the rare earths is small. Because there are in some elements more d electrons than s and p electrons in normal metals, the elements with the largest cohesive energies are those in the middle of the transition series. The 3d, 4d, and 5d electrons contribute with increasing effectiveness to cohesion. Tungsten, in the middle of the 5d series, has the greatest cohesion of any element ($9\,\text{eV}$ per atom).

Covalent bonding describes sharing of valence electrons in pairs, as in classical chemical theories. Atoms with $N = 4$ to $N = 7$ s–p valence electrons often form covalent bonds, and in the elemental crystals the coordination configurations are such that each atom tends to have $8 - N$ nearest neighbors. The classical covalent crystals are those with $N = 4$ (tetrahedral bonding with sp^3 hybrids, in chemical terms), including diamond and Si. With the exception of elements from the first period, covalent binding energies are about $(8 - N)\,\text{eV}$ per atom.

Molecular binding arises because of van der Waals interactions between, e. g., benzene molecules in solid benzene. It is weak and is typically of the order of $0.1\,\text{eV}$ per molecule or less.

In most crystals the binding is best described as a mixture of ionic, covalent, and metallic interactions. We can distinguish between ionic and metallic binding on the one hand and covalent binding on the other because covalent crystals (such as grey Sn, with the diamond structure) are about 20% less dense than their metallic morphotropes (white Sn, in this example). When there is more than one kind of atom in the crystal there will always be some ionic contribution to the binding. For example, there are many tetrahedrally coordinated compounds with eight valence electrons per atom pair, such as III–V and II–VI compounds; here the bonding is predominantly covalent (tetrahedral coordination) but it contains an ionic component, which chemists associate with the heat of formation of the compound from the elements. An example of mixed metallic–ionic bonding is provided by transition-metal carbides (e. g., WC), which contain solids with the highest known melting points.

See also: Chemical Bonding; Interatomic and Intermolecular Forces; Metals; Rare Earths; Transition Elements.

Bibliography

A. Navrotsky and J. C. Phillips, "Ionicity and Phase Transitions," *Phys. Rev.* **11**B, 1583 (1975). (A)

J. C. Phillips, "Covalent-Ionic and Covalent-Metallic Transitions," *Phys. Rev. Lett.* **27**, 1196 (1971) (A)

J. C. Phillips, "Chemical Bonds in Solids," in *Treatise on Solid State Chemistry* (N. B. Hannay, ed.). Plenum, New York, 1974. (I)

E. P. Wigner and F. Seitz, "Qualitative Analysis of the Cohesion in Metals," *Sol. State Phys.*, **1**, 97 (1955). (E)

Crystal Defects

G. D. Watkins

A crystalline solid with no defects would consist of a perfectly periodic three-dimensional array of unit cells, each containing the identical arrangement of atoms. Such an idealized crystal, however, cannot exist. In a real crystal, there are always structural imperfections.

If a perfect crystal could exist, it would be very strong mechanically. Aside from this desirable feature, however, it might be rather dull. Indeed, many of the important properties of crystalline solids – electrical, optical, mechanical – are determined not so much by the properties of the perfect crystal, but rather by its imperfections: The semiconductor industry is based on the ability to dissolve trace amounts of chemical impurities in silicon or germanium to control the conductivity level and type. The optical absorption and luminescence of solids is controlled primarily by impurities, and other simple lattice imperfections. The ductility and strength of structural materials are determined by imperfections and, as a result, can be modified and controlled by the addition of impurities, heat treatment, cold work, etc.

In many respects, therefore, it is often instructive to consider the host crystal as simply the matrix (or solvent) for defects, which, in turn, play the dominant role in determining the physical properties of the material. The basic structural defects can be classified as follows:

1. Point defects. These include *lattice vacancies* (lattice sites which are missing an atom), *interstitials* (extra atoms not on regular lattice sites), and *foreign atoms* (chemical impurities).

2. Line defects, such as *dislocations*.

3. Surface defects, such as *grain boundaries* and *stacking faults*.

Lattice Vacancies and Interstitials

In any real crystal, there must always be a finite number of lattice vacancies and interstitials (see Fig. 1). This can be seen simply as follows: The external surface of a crystal provides an inexhaustible source of vacancies (the empty space beyond) and interstitial atoms (the last atomic layer.) The equilibrium vacancy and interstitial concentrations in the interior of the crystal therefore represent the finite solubility of each of these point defects in the lattice matrix (solvent) as they dissolve in from the surface. For a monatomic metal, for instance, simple

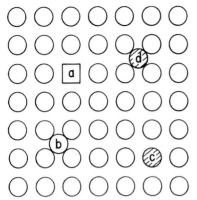

Fig. 1: Simple point defects: (a) vacancy; (b) interstitial; (c) substitutional impurity; (d) interstitial impurity.

thermodynamic arguments lead to the expression for the equilibrium solubility concentration of each species,

$$c_i \sim \exp\left(-\frac{E_i}{kT}\right) , \tag{1}$$

where E_i is the energy required to transfer each defect into the crystal (its formation energy). At any finite temperature T, there is therefore always a finite stable concentration, which increases exponentially as the temperature is raised. For most materials, E_I the energy to form an interstitial, is greater than E_V that for the vacancy, and the vacancy tends to be the dominant defect. Typically, near the melting point, a crystal may have a vacancy concentration of 0.01–0.1%.

In insulators and compound materials (more than one atom species per unit cell), the thermodynamics can be somewhat more complicated in that defects can carry charge, and chemical reactions with crystal components in the gas phase must also be considered (e. g., an oxide in an oxygen atmosphere). However, a proper treatment still leads to equilibrium concentrations of the form of Eq. (1), but where E_i may require a more general definition.

The presence of vacancies and interstitial atoms in a crystal provides a mechanism by which mass transport (diffusion) can occur in the lattice. The vacancy provides a missing site into which a neighboring atom can jump. When the atom jumps, the vacancy has moved, now occupying the original site of the neighbor. Similarly, the interstitial may move by jumping to a nearby empty interstitial site, or by pushing a neighbor into an adjacent interstitial site and taking its place. Both defects therefore stir up the lattice atoms as they randomly jump from one lattice position to another. The jump rate for a defect will have a temperature dependence of the form

$$v_i \sim \exp\left(-\frac{U_i}{kT}\right) , \tag{2}$$

where U_i represents an energy barrier that the atoms must overcome in order to accomplish the jump. The contribution of a defect to atomic diffusion in the crystal is therefore proportional to its jump rate (2) times the number of the defects (1) leading to a temperature dependence for the diffusion coefficient:

$$D \sim \exp\left[-\frac{E_i + U_i}{kT}\right].$$

$$(3)$$

In nonmetals, vacancies and interstitials may also be charged. In this case, defect motion produces electrical (*ionic*) conductivity which is also proportional to (3). This is often the dominant source of electrical conductivity in ionic-compound insulators such as alkali halides and ceramic oxides.

At temperatures well below the melting temperature, the *equilibrium* concentration of vacancies and interstitials may be small, because of the exponential temperature dependence, Eq. (1). However, vacancies and interstitials are often present greatly in excess of this amount as a result of (a) quenching from a high-temperature equilibrium, (b) mechanical distortion of the crystal, and (c) radiation damage, where the incoming particle displaces atoms. In these cases, return to equilibrium (annealing) may be extremely slow, being governed by the migration rate (2) which is also exponentially dependent on temperature.

The presence of vacancies and interstitials in semiconductors can alter the *electronic* properties of the material in that they serve to trap and scatter free electrons and holes. Electronic transitions of the trapped charges may also give rise to optical absorption and luminescence bands in semiconductors and insulators.

Foreign Atoms

Chemical impurities may dissolve in a crystal as extra atoms (interstitial) or by replacing normal lattice atoms (substitutional). Even the purest crystals inevitably contain trace amounts of impurities. In addition, impurities may be introduced intentionally by addition during crystal growth, by diffusion into the crystal from the surface at elevated temperatures, or by ion bombardment, where a high-energy ion of the impurity is driven in through the surface of the material. Transmutation doping has also been used, where a nuclear reaction converts a host atom into an impurity. Like vacancies and interstitials, impurities may trap and donate electrons or holes and alter the electronic and optical properties of the crystal. In addition, they may introduce some of their characteristic atomic optical absorption or luminescent spectra to the crystal. Their elastic interactions with dislocations are important in determining the strength of materials.

Dislocations

A perfect crystal should be very strong, the critical shear stress necessary to slide one perfect atom plane past another being comparable to the shear modulus G of the crystal. Contrary to this, however, it is found experimentally that the purer and more "perfect" the crystal is made, the softer it usually tends to become. The critical shear stress in a well-annealed single crystal may be only 10^{-5}–$10^{-4}G$. The reason for this is the presence of *dislocations*.

Two basic types of dislocations are illustrated in Fig. 2. In Fig. 2a, the *edge* dislocation can be viewed as having been formed by slicing the crystal with a knife partway in from the top and inserting an extra half plane of atoms in the cut. In Fig. 2b, the *screw* dislocation

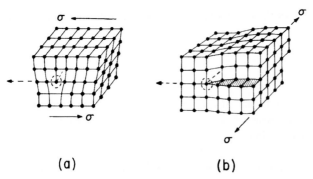

(a)   **(b)**

Fig. 2: (a) An edge dislocation; (b) a screw dislocation (after B. Henderson, *Defects in Crystalline Solids*. Edward Arnold, London, 1972). Applied shear stress a causes the upper half of the crystal to slip one atom spacing as the dislocation line moves through the crystal to the left.

could be produced by slicing part way in from the right and shearing the crystal parallel to the slice by one atomic spacing. For both, the imperfection is the line which terminates the cut, the surrounding material being a strained but otherwise perfect crystal. As shown in the figure, applied shear stress o allows the upper part of the crystal to slip one atomic spacing as the dislocation moves to the left. A reduction in the critical shear stress arises because now only a single line of atoms is required to move at a time, rather than the whole plane of atoms together.

Dislocations are not thermodynamically stable, the energy to form one in a bulk crystal being too great. However, they are almost always present in crystals as a result of accidents in growth. Typically, densities for an as-grown crystal may be $\sim 10^{12}/m^2$. The rapid and chaotic manner in which atoms fall into place during growth makes the formation of lattice misfits of this type likely. Once formed on a small scale, they can propagate as the crystal grows. Also vacancies and interstitials, present in high concentrations during growth from the melt, can nucleate dislocation loops as they condense out upon cooling.

Dislocation motion can be impeded by the presence of point defects, precipitates, or other dislocations in or near its slip plane. Interaction with point defects and precipitates is an important mechanism for the strengthening of constructional metals such as steel, where controlled additions of impurities and subsequent heat treatments to disperse the impurities or allow precipitation are used to tie up the dislocations to the desired degree. *Work hardening* is another mechanism for strengthening materials. Here dislocation movement under stress generates more dislocations, ultimately leading to a massive tangle of "pinned" dislocations.

Surface Defects

The interface between two regions of different crystal orientation is called a grain boundary. These can originate in growth as two separately growing regions merge. Grain boundaries can also form by regrouping of dislocations during anneal. Stacking faults are the result of an error in layer growth. A stacking fault for a crystal normally growing in the layer order *ABCABCABC* might be *ABCBCABC*.

See also: Color Centers; Crystal Growth; Diffusion; Electron Energy States in Solids and Liquids; Luminescence.

Bibliography

A. H. Cottrell, *Dislocations and Plastic Flow in Crystals*. Oxford University Press, Oxford, 1953. (I–A)

J. H. Crawford, Jr. and L. M. Slifkin (eds.), *Point Defects in Solids*, Vols. 1–3. Plenum Press, New York, 1972. (I)

B. Henderson, *Defects in Crystalline Solids*. Edward Arnold, London, 1972. (I)

D. Hull, *Introduction to Dislocations*, 2nd ed. Pergamon Press, Oxford, 1975. (E-I)

F. A. Kröger, *The Chemistry of Imperfect Crystals*, 2nd ed., Vols. 1–3. North-Holland, Amsterdam, 1973. (A)

F. Seitz, in *Imperfections in Nearly Perfect Crystals*, Chap. 1 (W. Shockley, J. H. Hollomon, R. Maurer, and F. Seitz, eds.). Wiley, New York, 1952. (E, I)

Crystal Growth

R. L. Parker

Among the many techniques for growing crystals, the following may be mentioned as representative examples: growth from the melt (solidification), growth from the vapor phase, growth from solution (aqueous, molten salts, other solvents), growth by precipitation from solid solution, growth by chemical (vapor) transport reaction, epitaxial growth from vapor or solution, electrocrystallization, and growth from the solid phase (strain annealing). It is useful, in classifying the various growth processes, to consider the thermodynamic phase diagram for the system; it specifies the regions of equilibrium between the desired solid and the mother medium from which the crystal grows, which may be liquid, vapor, solution, or also solid. Thus, in melt growth of a single-component system at constant pressure, the temperature of the liquid is gradually lowered so that a portion of it is at the freezing point; further cooling may then cause nucleation of the new solid phase, which, if properly controlled, may form a seed upon which further precipitation of the liquid can take place in an orderly way, yielding a single crystal.

Probably the most characteristic and fundamental problem in crystal growth is understanding, predicting, and controlling the crystal morphology or form. By morphology is meant not only the external forms of crystals, important and fascinating as these can be (as, for example, in snow crystals), but more generally the complete history of the solid–liquid, solid–vapor, or solid–solid interface morphology during the entire growth process. It is this morphology that determines the microstructure, the distribution of impurities, and the dislocation and imperfection substructure of the crystal, which in turn control the useful properties of the crystalline material. By no means is the prediction of morphology an easy matter. The interface morphology is controlled simultaneously and in varying proportions by bulk transport effects (flow of heat and/or solute) and by interface kinetic effects.

Heat flow is clearly important in, for example, the freezing of a pure liquid, since the latent heat must be removed. Solute flow is clearly important in, for example, the precipitation from a solid solution. Interface kinetics, which refers to the structure, on a molecular level, of the surface and to the relative ease or difficulty that a molecule may have in moving from the mother medium to a relatively final position in the crystal surface, may be controlling for particular materials and for particular molecular and crystal structures. Thus, both macroscopic and microscopic theory are involved in understanding crystal growth mechanisms.

Unless a crystal seed is present, the new phase must start by a nucleation process. Nucleation is the formation of very small particles of new phase in the mother medium and may take place when the appropriate thermodynamic variable (temperature, concentration, pressure, etc.) has departed sufficiently from the equilibrium value. It is generally assumed to take place by a random fluctuation process (heterophase fluctuation) analogous to the better-known density fluctuations in a gas; dimers, trimers, and higher species are assumed to form, with the larger sizes having short lifetimes and dilute concentrations. There is found, by an appropriate balancing of the free energy of formation of a particle against its surface energy, to be a critical embryo size for nucleation; particles greater than this size have a tendency to grow, and those below it tend to dissolve. The dependence of the number of these critical nuclei on temperature or on solute concentration is extremely sensitive, so that only a few degrees change in undercooling can cause many orders of magnitude change in nucleation rate. While the preceding discussion is for "three-dimensional" nucleation, "two-dimensional" nucleation of a monolayer disk may take place from the random fluctuation of adsorbed particles on an otherwise plane crystal surface, and provides one kind of interface kinetic mechanism.

Model calculations of nucleation, interface structure, and interface kinetics have been made for many years. Perhaps the best-known model for crystals is the BCF or Burton–Cabrera–Frank model of growth at the emerging step of a screw dislocation at the crystal surface. Modern supercomputers now permit the calculation, using the methods of molecular dynamics, of the actual three-dimensional trajectory of each atom in entire systems of thousands of atoms interacting with each other with a given interatomic potential. As the temperature or average kinetic energy is lowered, both condensation of vapor to liquid or solid, and freezing of liquid to solid, can be observed and studied. Atom-by-atom and layer-by-layer growth, as for example in epitaxy, is thus calculated at intervals of 10^{-12} s, from first principles. This is an improvement over the Monte Carlo technique. In this technique, atoms are deposited at random sites on an initially flat crystal, but their evaporation depends on the number of their first neighbors. Not only interface structure, but growth rates can be calculated. It is possible to confirm and make more precise predictions of surface roughness that were made previously by statistical mechanics (Ising model).

It is often observed that crystal forms are not stable shapes. An example is the quite common dendritic or tree-shaped form, which exhibits periodic branching and subbranching, as in snow crystals. This results from the unstabilizing point effect of diffusion (of heat or solute) acting on a small protuberance. This effect can be counteracted by the additional surface energy thereby produced, and it can for particular materials be overcome by the size and/or anisotropy of the interface kinetic effect (sluggish kinetics). Morphological stability theory has been applied to this phenomenon, using the transport equations and appropriate boundary conditions, and has been highly successful in such areas as the breakdown of planar to cellular growth of alloy crystals in the presence of undercooling caused by solute rejection, and in such other areas as ice cylinder growth from supercooled water. It seems likely that the ultimate calculation of crystal form will use this macroscopic theory, with boundary conditions determined by computer modeling on the atomic scale.

When the mother medium is a fluid (liquid or gas), fluid flow and convection can greatly influence crystal growth rates and morphology. Although often overlooked in the past, these effects are due to the breakup of relatively long-range diffusion fields of heat or solute. These fields are translated away and are replaced by relatively short-range diffusion fields. The result is that the fluid of bulk concentration or temperature is now quite close to the interface, and is only separated from it by a thin hydrodynamic boundary layer; this will tend to increase growth rates. Thus, modern crystal growth technology (for example, large Czochralski machines pulling rotating dislocation-free silicon crystals from the convecting melt) must concern itself with hydrodynamics. In addition, there are many deep and unexpected effects on interface morphological stability calculations when the fluid flows are included, as, for example, in double-diffusive convection. These simulations also require extensive computer calculations, particularly those in which the diffusive transport and convective transport are strongly coupled, and indeed the two fields, formerly separate, of morphological stability and hydrodynamic stability are now joined.

Crystalline perfection or freedom from defects is not solely due to the growth process, for subsequent processing, annealing, handling, and the like often alter the concentration and types of defects. However, the growth process is certainly the key aspect. Among the defects that may be introduced are point defects (vacancies, interstitials, foreign atoms), line defects (dislocations), planar defects (stacking faults, twins, low-angle boundaries), and three-dimensional defects (striations, cellular growth, voids, inclusions). An example of controlling defects is the use of substantial temperature gradients in directional solidification to remove constitutional supercooling and consequently to prevent cellular growth by morphological instability. Another is to control dislocation density by use of a bottlenecked seed at which dislocations may emerge on the crystal surfaces. Of the many crystal characterization techniques, x-ray tomography is rapidly growing as a major tool for structural defects such as dislocations, including real-time *in situ* observation, using such powerful x-ray sources as coherent synchrotron radiation, during the melt–growth process. Electrically or optically active defects are often detected by measuring appropriate physical properties such as resistivity or absorption.

Finally, while the physicist may be interested in crystals as good specimens in which to measure physical properties or because the mechanisms of growth are complex and fascinating, it cannot be ignored that substantial recent advances in the technology of crystal growth are due to immense demand for production of industrial and consumer products. These demands exist in the electronics, optical, computer and communications industries for such materials as silicon for transistors, integrated circuit chips, random access memory chips and solar cells; GaAs for higher-speed and higher-frequency ICs including microwave and lightwave chips; and metallorganic chemical vapor deposition epitaxial growth of heterostructures for laser diodes. In a typical year's issues of the *Journal of Crystal Growth*, more than 100 different compositions of new or improved crystals may be found.

See also: Crystal Defects; Crystal Symmetry.

Bibliography

A. A. Chernov, *Crystal Growth*, Modern Crystallography 3, Springer-Verlag, New York, 1984.

M. E. Glicksman, S. R. Coriell, and G. B. McFadden, "Interaction of Flows with the Crystal-Melt Interface," *Ann. Rev. Fluid Mech.* **18**, 307–335 (1986).

F. Rosenberger, *Fundamentals of Crystal Growth I.* Springer-Verlag, New York, 1979.

D. Elwell and H. J. Scheel, *Crystal Growth from High Temperature Solutions.* Academic Press, New York, 1975.

Journal of Crystal Growth, North-Holland, Amsterdam, 1967–present.

R. L. Parker, "Crystal Growth Mechanisms," in *Solid State Physics* (H. Ehrenreich, F. Seitz, and D. Turnbull, eds.), 25. Academic Press, New York, pp. 151–299, 1970.

Crystal Symmetry

M. J. Buerger[†] and M. Böhm

Introduction

Homogeneity

Crystals are anisotropic homogeneous solids. The homogeneity is on the scale of a cluster of a small number of atoms. It implies that the solid consists of clusters related to each other by direct or opposite isometrics, and that the environments of homologous points of all clusters are the same.

Coincidence Operations

Every cluster is related to any other cluster by one of four simple operations: translation, rotation about an axis (proper rotation), reflection across a plane (improper rotation), or inversion through a point (improper rotation); or by acceptable combinations of these four. (The condition for two different operations to form an acceptable combination is that each of the component operations transforms the other into itself.)

In every crystal the clusters are related to each other at least by the operations of translations. Accordingly, any other *symmetry operations* must be consistent with translations. It is readily demonstrated that this requires every rotation operation to be through an angle of $2\pi/n$, where $n = 1, 2, 3, 4,$ or 6.

Each of the operations relating the crystal clusters generates a group. As an aid to describing the operations it is convenient to designate the locus that remains unmoved under the operations of the cyclical group as the *symmetry element* of that group. The symmetry element of a group of rotations is the axis A of the rotation; of a group of reflections is the mirror plane m of the reflection; of an inversion is the point through which the operation inverts; etc.

Any of the permissible operations that relate a pair of atomic clusters in a particular homogeneous solid must also relate all clusters in the solid. If the solid contains operations beyond the generally required translations, the solid is said to be *symmetrical* with respect to each of the symmetry elements corresponding to the groups of operations relating its atomic clusters.

[†]deceased

All *symmetry transformations* are operators in the 3-dimensional Euclidean space acting on vectors **r** as the elements of the vector space. They can be expressed by

$$\mathbf{r}' = \hat{\mathbf{d}}\mathbf{r} + \mathbf{t} \equiv \{\hat{\mathbf{d}}|\mathbf{t}\}\mathbf{r}, \tag{1}$$

where $\hat{\mathbf{d}}$ is a general rotation (proper or improper) and **t** is a translation. From the requirement that the distance of two points remains unchanged the transformations are found to be orthogonal. The set of symmetry transformations constitute a group in the mathematical sense. As a consequence, the results of representation theory can be used as most effective tools in discussing physical problems.

Crystallographic Groups

Two kinds of symmetries are especially important in crystallography: *point groups* and *space groups*. Point groups are so named because in each group one point remains unmoved under the operations of the group; they contain no translations. Space groups are so named because in each group all three-dimensional space remains invariant under the operations of the group; they contain pure translations and may also contain operations with translation components.

Since all crystals are ideally homogeneous, every crystal must conform to one of the space-group symmetries. Nevertheless it is useful to consider the point groups because many of the properties of crystals are not dependent on its translations. For example, the appearance of a crystal conforms to its point-group symmetry because the eye, even when aided by the microscope, cannot see the tiny translations. Furthermore, the point groups, which are easy to construct, lead to a simple way of enumerating all space groups.

The Crystallographic Point Groups

Symbolism for Symmetry Elements

Crystallographic point groups G that leave the lattice invariant are finite subgroups of the set of all rotations. This forms a Lie group that is isomorphic to the orthogonal group $O(3, \mathbb{R})$ in three dimensions. The symbols universally used by crystallographers to designate symmetry elements are the Hermann–Mauguin, or international, symbols. These are shown in Table 1 for both point groups and space groups. These symbols replace the 1890 point-groups symbols of Schoenflies, which cannot be suitably extended to space groups. The equivalences are shown in Table 2.

Chirality

Any operation or group of operations that relates objects of the same chirality (i. e., objects related by direct isometry, which can be regarded arbitrarily as all right-handed or all left-handed) is said to be of the first sort. Any operation or group that relates objects of opposite chirality is said to be of the second sort. Pure rotations relate objects of the same chirality; they are called proper rotations. The set of all proper rotations forms a Lie group that is isomorphic to the special orthogonal group $SO(3, \mathbb{R})$ in three dimensions consisting of the set of all rotation matrices $\hat{\mathbf{d}}$ with det $\hat{\mathbf{d}} = +1$.

Rotoinversions and rotoreflections are called improper rotations (Fig. 1); they relate a sequence of objects that are alternately right-handed and left-handed. The set of all improper rotations is isomorphic to the set of all orthogonal matrices $\hat{\mathbf{d}}$ with det $\hat{\mathbf{d}} = -1$. The objects of homogeneous patterns are either all of the same chirality or they are half right-handed and half left-handed.

Table 1: Hermann–Mauguin symbols for the symmetry elements of crystallographic groups

Point-group symbols		Isogonal space-group symbols				
1		1				
2		2	2_1			
3		3	3_1	3_2		
4		4	4_1	4_2	4_3	
6		6	6_1	6_2	6_3	6_4 6_5
$\bar{1}$ $(=\tilde{2})$	$= i$, an inversion center	$\bar{1} = i$				
$\bar{2}$ $(=\tilde{1})$	$= m$, a mirror	Various glide planes				
$\bar{3}$ $(=\tilde{6})$	$= 3+i$	$\bar{3} = 3+i$				
$\bar{4}$ $(=\tilde{4})$	Not decomposable	$\bar{4}$				
$\bar{6}$ $(=\tilde{3})$	$= \frac{3}{m}$, i.e., $3 \perp m$	$\bar{6} = \frac{3}{m}$				

Glide planes	
Symbol	Translation component
m	0
a	$\frac{1}{2}a$
b	$\frac{1}{2}b$
c	$\frac{1}{2}c$
n	$\frac{1}{2}(a+b)$ or $\frac{1}{2}(b+c)$ or $\frac{1}{2}(a+c)$
d	$\frac{1}{4}(a+b)$ or $\frac{1}{4}(b+c)$ or $\frac{1}{4}(a+c)$ or $\frac{1}{4}(a+b+c)$

Monaxial Groups with Proper Rotations

In Hermann–Mauguin notation a proper rotation axis is labeled by the numerical value of its n. Accordingly there are five such crystallographic point groups, namely, 1, 2, 3, 4, and 6, shown in Fig. 1.

Groups with Proper Rotations about Intersecting Axes

It can readily be shown by Euler's construction, or by multiplying together matrices $\hat{d}$ that represent rotations, that a rotation $\hat{d}_A(\alpha)$ through an angle α about an axis A, followed by a rotation $\hat{d}_B(\beta)$ through β about B, is equivalent to a rotation $\hat{d}_C(-\gamma)$ through $-\gamma$ about an axis C. In the product notation for combining the elements of a group, this can be expressed as

$$\hat{d}_A(\alpha)\hat{d}_B(\beta) = \hat{d}_C(-\gamma) = \hat{d}_C^{-1}(\gamma) . \tag{2}$$

This implies that

$$\hat{d}_A(\alpha)\hat{d}_B(\beta)\hat{d}_A C(\gamma) = \hat{1} , \tag{3}$$

Table 2: The distribution of point-group symmetries over the seven crystal systems and their possible 14 lattice types.

Point-group symbols		Crystal system		
Schoenflies	Hermann–Mauguin[a]	Name	Geometry	Bravais lattice
C_1	1	Triclinic	$a \neq b \neq c$	P
C_i	$\bar{1}$		$\alpha \neq \beta \neq \gamma$	
C_2	2	Monoclinic	$a \neq b \neq c$	P, A (or B)
C_s	m		$\pi/2 = \alpha$	
C_{2h}	$\dfrac{2}{m}$		$= \beta \neq \gamma$	
C_3	3	Trigonal	$a = b = c$	R
D_3	32		$\alpha = \beta = \gamma$	
C_{3v}	$3m$		$\neq \pi/2$	
C_{3i}	$\bar{3}$			
D_{3d}	$\bar{3}\dfrac{2}{m}$ $(3m)$			
C_6	6	Hexagonal	$a = b \neq c$	P
D_6	622		$\pi/2 = \alpha = \beta$	
C_{6h}	$\dfrac{6}{m}$		$\neq \gamma = 2\pi/3$	
C_{6v}	$6mm$			
C_{3h}	$\bar{6}$			
D_{3h}	$\bar{6}m2$ or $\bar{6}2m$			
D_{6h}	$\dfrac{6\ 2\ 2}{m\ m\ m}$ $(6/mmm)$			
D_2	222	Orthorhombic	$a \neq b \neq c$	P, I, A
C_{2v}	$mm2$		$\pi/2 = \alpha = \beta$	(or B or C), F
$D_{2h} = V$	$\dfrac{2\ 2\ 2}{m\ m\ m}$ $(2/mmm)$		$= \gamma$	
C_4	4	Tetragonal	$a = b \neq c$	P, I
D_4	422		$\alpha = \beta = \gamma$	
C_{4h}	$\dfrac{4}{m}$		$= \pi/2$	
C_{4v}	$4mm$			
S_4	$\bar{4}$			
D_{2d}	$\bar{4}m2$ or $\bar{4}2m$			
D_{4h}	$\dfrac{4\ 2\ 2}{m\ m\ m}$ $(4/mmm)$			
T	23	Cubic	$a = b = c$	P, I, F
O	432		$\alpha = \beta = \gamma$	
T_h	$\dfrac{2}{m}\bar{3}$		$= \pi/2$	
T_i	$\bar{4}3m$			
O_h	$\dfrac{4}{m}\bar{3}\dfrac{2}{m}$ $(m\bar{3}m)$			

[a] Hermann–Mauguin symbols in parentheses are abbreviated forms in common use.

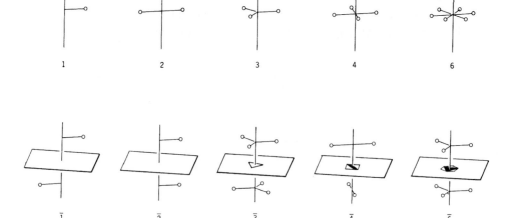

Fig. 1: Top: The symmetry axes of the five crystallographic point groups with proper rotations. Bottom: The symmetry axes of the five crystallographic point groups with improper rotations.

where the right-hand side of (3) the identity operation is represented by the unit matrix $\hat{1}$, indicating no net motion. Since

$$\alpha = \frac{2\pi}{n_1}, \quad \beta = \frac{2\pi}{n_2}, \quad \gamma = \frac{2\pi}{n_3} \tag{4}$$

in crystal symmetry, it is obvious that if there are two rotations about intersecting axes, a rotation about a third intersecting axis is implied. (Exceptionally, the third rotation is one of the other two with an opposite rotation.)

Euler's construction allows the angular relations between rotation axes to be obtained from a spherical triangle in which axes A, B and C occur in a clockwise sense at distances $A \wedge B = c$, $B \wedge C = a$, $C \wedge A = b$, and with angles $\alpha/2$, $\beta/2$, $\gamma/2$. For this spherical triangle, the sum of its angles is

$$\frac{\alpha}{2} + \frac{\beta}{2} + \frac{\gamma}{2} \geq \pi . \tag{5}$$

With the aid of (4) this can be expressed as

$$\frac{1}{n_1} + \frac{1}{n_2} + \frac{1}{n_3} \geq 1 . \tag{6}$$

The only distinct solutions of (6) are

$$2,2,n_3 \quad 2,3,3 \quad 2,3,4 \quad 2,3,5 .$$

The crystallographic possibilities for these solutions are the following.

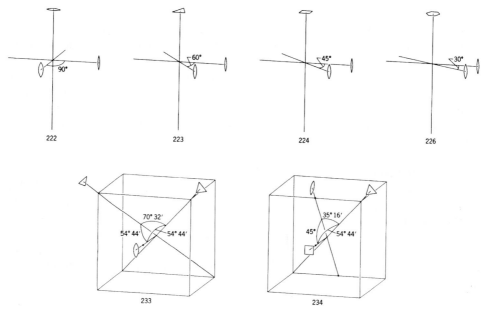

Fig. 2: The six combinations of crystallographic proper rotation axes. [From M. J. Buerger, *Elementary Crystallography* (Wiley, New York, 1963).]

4 "dihedral" sets: 2,2,2   2,2,3   2,2,4   2,2,6
1 "tetrahedral" set: 2,3,3
1 "octahedral" set: 2,3,4

The angles $\tilde{\alpha}$, $\tilde{\beta}$, and $\tilde{\gamma}$ between the rotation axes A, B, and C can be found from relations like

$$\cos\tilde{\alpha} = \frac{\cos(\alpha/2) + \cos(\beta/2) + \cos(\gamma/2)}{\sin(\beta/2)\sin(\gamma/2)}. \tag{7}$$

The arrangements of these six axial combinations are shown in Fig. 2.

The Axial Frames of Crystallography

The five monaxial point groups and the six sets of polyaxial point groups just demonstrated are the only crystallographic point groups with proper rotations only referred to as *proper point groups* . These 11 arrangements of axes provide axial frames on whose geometries all other angular relationships in other crystallographic groups are based.

Groups with Improper Rotations

Corresponding to the five monaxial groups with proper rotations are five with improper rotations, symbolized by the rotoinversion axes $\bar{1}$, $\bar{2}$, $\bar{3}$, $\bar{4}$, and $\bar{6}$, as listed in Table 1 and shown in Fig. 1, all but four can also be represented by other symmetry elements.

A proper and an improper rotation axis of the same n can occur along the same line. When these n and $\bar{n}$ axes are written as a fraction $n/\bar{n}$ and reduced, there arise three new groups: $2/\bar{2} \rightarrow 2/m$; $[3/\bar{3} \rightarrow 3/(3+i) \equiv \bar{3}]$; $4/\bar{4} \rightarrow 4/m$; $6/\bar{6} \rightarrow 6/(3/m) \rightarrow 6/m$.

Intersecting axes can be combined as were the three proper rotations, provided either that proper and improper axes coincide, as just noted, or that there are two improper axes and one proper axis. These intersecting combinations generate the other point groups having improper rotations listed in Table 2.

The groups with improper rotations called *improper point groups* are isomorphic to subgroups of the orthogonal group $O(3, \mathbb{R})$ resulting from the direct product between the special orthogonal group $SO(3, \mathbb{R})$ and the inversion group $C_i = \{\hat{1}, -\hat{1}\}$. As a consequence, besides the proper rotations we expect improper rotations such as the inversion, the rotary inversion, and the rotary reflection with det $\hat{d} = -1$.

The 32 Point Groups

The arrangements of symmetry elements in the 32 point groups are illustrated in Fig. 3. The "point" of the point group is assumed to be at the center of the sphere whose outline is the dotted circle; if there are intersecting symmetry elements, the point of intersection is at the center of the circle. A reflection plane is represented by the line where it intersects the sphere's surface. An *n*-fold axis is shown as a polygon with n vertices at the point where the axis intersects the sphere's surface. A tiny circle in the center of a diagram indicates that the point group has a center of symmetry.

Translation Groups

Lattices

A crystal is characterized by its regular, periodically repeated structure. As a consequence, the lattice is invariant to translations over distances which are integer multiples of the lattice period. This is only true, of course, for an ideal, infinite crystal or for a crystal which has been subjected to periodic boundary conditions. In three-dimensional space a translation can be represented by its components $\mathbf{a}_1$, $\mathbf{a}_2$ and $\mathbf{a}_3$ of three non-coplanar vectors. Different lattices may be distinguished according to the relative length of these basic vectors and the angles between them. For crystallographic applications, let $\mathbf{a}_1$ be the shortest translation in some direction. A set of points equivalent by translation $\mathbf{a}_1$ constitute a line parallel to $\mathbf{a}_1$. Let $\mathbf{a}_2$ be chosen so that it translates the points of this line to the points of a parallel line that is next adjacent to it in some direction; then let $\mathbf{a}_3$ be chosen so that it translates a point of the plane parallel to $\mathbf{a}_1$ and $\mathbf{a}_2$ to a point in the next parallel plane. When $\mathbf{a}_1$, $\mathbf{a}_2$ and $\mathbf{a}_3$ are chosen in this way they are called *conjugate translations*. There are an infinite number of sets of such translations in a homogeneous solid. Any translation of the solid which transforms the empty lattice into its self can then be described by

$$\mathbf{R_n} = n_1\mathbf{a}_1 + n_2\mathbf{a}_2 + n_3\mathbf{a}_3 \equiv \{\hat{1}|\mathbf{R_n}\} \tag{8}$$

with

$$\mathbf{n} = (n_1, n_2, n_3) \qquad n_1, n_2, n_3 : \text{integers} .$$

The set of all *primitive translations* $\mathbf{R_n}$ forms the discrete Abelian *translation group* $\mathcal{T}$ which is an invariant subgroup of the space group $\mathcal{R}$. If the elements of this group operate on an arbitrary point, the resulting collection of points is called the *point lattice* of the crystal. (This is the only correct meaning of "lattice" in crystallography.)

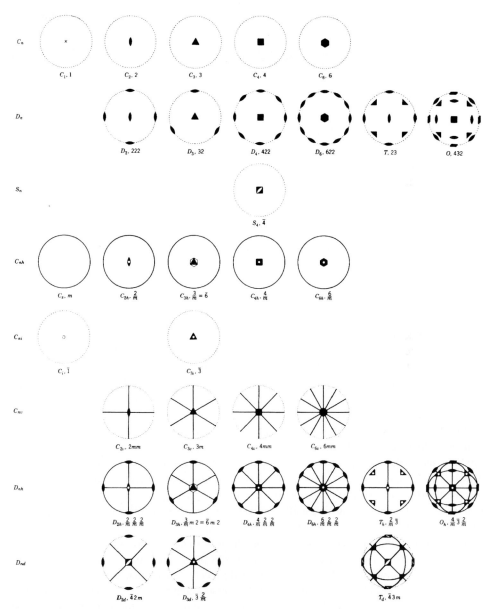

Fig. 3: The arrangements of symmetry elements in the 32 crystallographic point groups. [From M. J. Buerger, *Elementary Crystallography* (Wiley, New York, 1963).]

Table 3: Crystallographic cells with respect to different types of point lattices; P: primitive, I: body-centered, A (B, C): face-centered on face A (B, C), R: rhombohedral, F: face-centered on all faces.

Lattice type symbol	Coordinates of lattice points (in fraction of cell edges a, b, c)	Cell multiplicity
P	$(0,0,0)$	1
I	$(0,0,0)$; $(\frac{1}{2},\frac{1}{2},\frac{1}{2})$	2
A	$(0,0,0)$; $(0,\frac{1}{2},\frac{1}{2})$	2
B	$(0,0,0)$; $(\frac{1}{2},0,\frac{1}{2})$	2
C	$(0,0,0)$; $(\frac{1}{2},\frac{1}{2},0)$	2
R	$(0,0,0)$; $(\frac{2}{3},\frac{1}{3},\frac{1}{3})$; $(\frac{1}{3},\frac{2}{3},\frac{2}{3})$	3
F	$(0,0,0)$; $(0,\frac{1}{2},\frac{1}{2})$; $(\frac{1}{2},0,\frac{1}{2})$; $(\frac{1}{2},\frac{1}{2},0)$	4

Cells

The parallelepiped, three of whose adjacent edges are conjugate translations, is known as the *primitive (elementary, fundamental) unit cell*. The adjective "primitive" implies that to each cell of this kind there corresponds one point of the lattice. The infinite number of primitive cells of a given crystal all have the same volume. A particular kind of primitive cell, known as the *reduced cell*, is characterized by having the three shortest translations.

Except as noted below, the cell edges of the reduced cell offer a natural coordinate system to which the geometrical features of a crystal can be referred. When a crystal has point-group symmetry other than 1 or $\bar{1}$, however, there are certain cases for which it is mathematically simpler to use as the coordinate system the edges of a more symmetrical cell known as *crystallographic cell* that contains more than one set of translationally equivalent atoms. These cells are called *multiple cells*, specifically *double*, *triple*, and *quadruple* cells, according as the cell contains 2, 3, or 4 sets of translationally equivalent atoms. Such cells have, in addition to a lattice point at the cell origin, one or more additional lattice points at various special positions in the cell. The characteristics of these cells and their designations are given in Table 3.

A third possible choice of the unit cell is obtained by placing the coordinate system origin at one of the lattice points and drawing the planes that perpendicularly bisect the lines joining the origin to the nearest (and sometimes to the next-nearest) neighbours. The smallest polyhedron defined by these planes is the *Wigner–Seitz cell*. This unit cell has the same volume as the primitive cell, contains only one lattice point and also exhibits the point symmetry of the lattice, just as the crystallographic unit cell does.

When an appropriate cell has been selected, it is customary to label its three adjacent edges a, b, and c, in such a way as to define a right-handed system of coordinates. The angles between cell axes are designated $\alpha = b \wedge c$, $\beta = c \wedge a$, $\gamma = a \wedge b$.

Crystal Systems

The set of maximal point group symmetries $\hat{d}$ which leave the point lattice invariant form the *holosymmetric point group* G_0. The periodicity of the lattice implies that the inversion is a necessary symmetry element. Thus only improper point groups are to be considered. As a result, there are seven holosymmetric point groups which define seven *crystal systems* listed in Table 2. These seven primitive (simple) lattices can be constructed by seven different triple of basis vectors.

The extension of the primitive cell to a multiple cell so as to include the symmetry elements $\hat{d}$ of the crystallographic point group G with lower symmetry $(G \subseteq G_0)$ gives rise to multiple primitive lattices of which there are seven types, too. As a result, we find 14 possible lattices (*Bravais lattice*) shown in Figure 4 which are compatible with the point group G. The distribution of the point groups over the crystal systems is given in Table 2.

Rational Lines and Directions

An aspect of crystal geometry is said to be rational if it can be expressed in integral numbers of lattice translations. A *direction* is rational if it has the form

$$\mathbf{r} = u\mathbf{a} + v\mathbf{b} + w\mathbf{c} \qquad u, v, w : \text{integers} . \tag{9}$$

The three integers u, v, w, are known as the indices of the direction. A given direction in a crystal is specified by the smallest set of integers having the same ratio as the components or direction cosines of a vector $\mathbf{r}$ drawn in that direction. Since sets of three integers are used for various indices, to indicate that they refer to a direction they are traditionally written between square brackets $[uvw]$. *Equivalent directions* are directions that are crystallographically equivalent. The set of all equivalent directions is denoted by $\langle uvw \rangle$.

Rational Planes

A *plane* is rational if it contains three lattice points (in which case it contains an infinite number). The translations of the lattice require such a plane to be one of an infinite stack of equally spaced parallel planes whose spacing is designated d. The equation of the nth plane from the origin is

$$hx + ky + lz = n , \tag{10}$$

where h, k, and l are integers and x, y, and z are fractions based on a, b, and c, respectively, as units. A rearrangement of (10) for $n = 1$ provides the intercept form for the plane of the stack nearest the origin

$$\frac{x}{1/h} + \frac{y}{1/k} + \frac{z}{1/l} = 1 . \tag{11}$$

This plane has intercepts on the a, b, and c axes of $1/h$, $1/k$, and $1/l$, respectively, so the stack cuts these units into h, k, and l parts, respectively. These three integers, placed between parentheses, are the *indices of the stack of planes* known as *Miller indices* (hkl). A set of planes which is crystallographically equivalent is defined by Miller indices in curly brackets $\{hkl\}$. The Miller indices notation is a reciprocal lattice representation. Thus, it is quite useful for the discussion of interference phenomena.

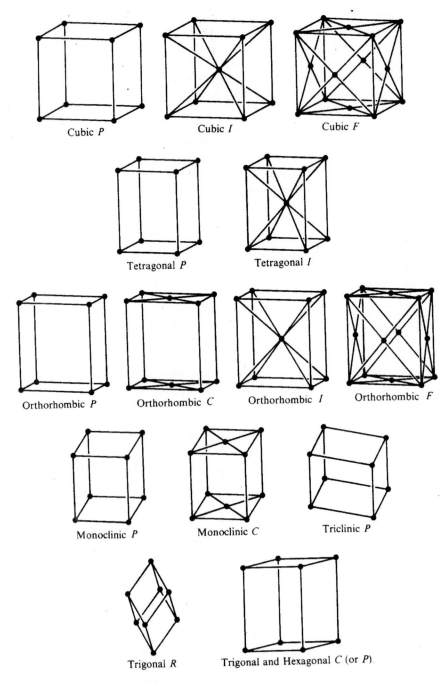

Cubic *P*   Cubic *I*   Cubic *F*

Tetragonal *P*   Tetragonal *I*

Orthorhombic *P*   Orthorhombic *C*   Orthorhombic *I*   Orthorhombic *F*

Monoclinic *P*   Monoclinic *C*   Triclinic *P*

Trigonal *R*   Trigonal and Hexagonal *C* (or *P*)

Fig. 4: Crystallographic unit cells for the 14 point lattices (Bravais lattices) arranged according to the seven crystal systems.

Space Groups

General Features

Every crystal has a group of translations that are consistent with the symmetry of its point group, specifically one of the Bravais-lattice types of Table 2. Its symmetry may include rotational operations with translation components parallel to the axis of rotation, and its reflectional operations may have translation components parallel to the reflection plane, from Table 1. The angles of the rotational component of the screws, and the angles between various axes and planes, are not affected by the translations, however, so that the resulting angular geometry of every combination of symmetry elements is *isogonal* with the angular geometry of the corresponding translation-free symmetry elements in some possible crystallographic point group.

From the point of view of group theory the set of symmetry elements $\{\hat{d}|t\}$ constitute the space group $\mathcal{R}$. It is isomorphic to a subgroup of the Euclidean group $E(3,\mathbb{R})$ in three dimensions which includes all continuous transformations in the Euclidean space such as translations $\{\hat{1}|t\}$ and orthogonal transformations $\{\hat{d}|0\}$.

The set of all rotational parts $\{\hat{d}|0\}$ of the space group operations form the point group G of the space group $\mathcal{R}$ which is a subgroup of the holosymmetric group G_0. Space groups having the same point group G are said to belong to the same *crystal class*, so there are 32 different crystal classes.

The correlation of elements of the point group G to those of the translation group T gives rise to a splitting of the 32 point groups into 230 space groups. Each of these space groups is *isomorphic* to one of the 32 point groups. This relation makes it possible to derive every space group by considering, in turn, the combinations of permissible symmetry operations that are isogonal with the symmetry operations of the appropriate point groups.

Any symmetry element of the space group $\mathcal{R}$ can be expressed by

$$\{\hat{d}|t\} = \{\hat{d}|v+R_m\} = \{\hat{d}|v\}\{\hat{1}|R_n\} , \tag{12}$$

where the non-primitive translation v is unique up to a lattice vector R_n. If $v = 0$ for every rotational part $\{\hat{d}|0\}$, then the space group is said to be a *symmorphic space group*. In this case the point group G_0 is a subgroup of the space group $\mathcal{R}$. Only 73 of the 230 crystallographic space groups are symmorphic. If the space group $\mathcal{R}$ is non-symmorphic, then for some rotational elements $\{\hat{d}|0\}$ of the point group G_0 there exists a nonvanishing translation $v(\hat{d})$. In this case the point group G_0 is not a subgroup of the space group $\mathcal{R}$.

Groups with Only Operations of the First Sort

For each rotation operation that generates a symmetry element in point-group symmetry, there are several isogonal operations in space-group symmetry with different translation components. The rotation $\hat{d}$ is accompanied by a non-primitive translation component v parallel to the axis A such that n translations v are equal to an integral number p of one of the primitive lattice translations $\{a_i|i = 1,2,3\}$ along the axis

$$nv = pa_i \quad n, p: \text{integers} \quad i = 1,2,3 . \tag{13}$$

Thus the space-group operation is characterized by two parameters: n, which is the same as in the isogonal point group; and p, which may be zero. Since these are, in general, helicoidal motions, the symmetry elements are known as *screw axes*. They are designated by the general symbol n_p; the various possibilities are shown in the upper right-hand part of Table 1. Some of them are enantiomorphic that is, they differ only in the sense of the screw (right-left). When $p/n < \frac{1}{2}$, the screw is right-handed; when $p/n > \frac{1}{2}$, it is left-handed; when $p/n = \frac{1}{2}$, it is neutral. The number of threads in the screw is p for $p/n \geq \frac{1}{2}$, but $n-p$ for $p/n \leq \frac{1}{2}$.

A translation $\mathbf{t}'$ normal to the axis A of the screw produces a translation-equivalent operation A' as well as a nonequivalent but equal screw operation about parallel axes B and C located along the perpendicular bisector of AA' and at a distance of $t' \cot \alpha/2$ from AA'. These screw operations accumulate along axes B and C, producing screw axes parallel with A. The resulting sets of symmetry elements, repeated by the lattice translations, constitute infinite sets of parallel axes, which have been called "flocks of axes." An interesting way of appreciating the values of the n's for the A, B, and C axes is as follows: If the sphere and its spherical triangle on which the reasoning of (2) through (6) were based is allowed to expand to infinite radius, the three axes may be allowed to become parallel so that the spherical triangle becomes a plane triangle. Then inequalities (5) and (6) become equalities. The solutions for the n's of the three parallel axes and their separations become

$$
\begin{array}{lll}
2,2,[\infty] & \text{with} & a:b:c = [\infty:\infty]:1, \\
2,3,6 & \text{with} & a:b:c = 2:\sqrt{3}:1, \\
2,4,4 & \text{with} & a:b:c = \sqrt{2}:1:1, \\
3,3,3 & \text{with} & a:b:c = 1:1:l.
\end{array}
$$

The possible space groups isogonal with a particular point group differ from each other by various characteristic translations. Such a set of space groups can be formed by combining the translations of the several Bravais-lattice types permitted to the point group with the different symmetry elements isogonal with each symmetry element of the point group.

Groups with Operations of the Second Sort

The space-group elements of the second sort are listed in the lower right-hand part of Table 1. Only the symmetry elements isogonal with point-group symmetry element m call for special comment. In space groups a reflection may have a non-primitive translation $\mathbf{v}$ in the reflecting plane, since the translation and reflection each transform the other into itself. The resulting glide reflection operation generates a glide plane as a symmetry element. A sequence of two glide operations is the same as a primitive translation, so the non-primitive translation can only be parallel to a lattice translation.

$$
\mathbf{v} = \frac{\mathbf{a}_i}{2} \quad \text{or} \quad \frac{\mathbf{a}_i + \mathbf{a}_j}{2} \quad \text{or} \quad \frac{\mathbf{a}_i + \mathbf{a}_j}{4} \qquad i,j = 1,2,3 . \tag{14}
$$

The Hermann–Mauguin symbol for a glide plane reveals its translation component (*n*: face-diagonal, *d*: diagonal); the six possibilities are shown at the lower right in Table 1.

Symbolism

Just as in point groups both Schönflies and Hermann-Mauguin (international) symbols are used for denoting space groups. In the Schönflies notation the space group is given the isogonal point group and different space groups having the same isogonal point group are given an arbitrary distinguishing superscript. There are, for example, 10 space groups isogonal to the octahedral point group O_h ($m3m$) and these are numbered $O_h^1, O_h^2, \ldots, O_h^{10}$. The 10 groups are distributed among the three cubic Bravais lattices, four primitive (1–4), four face-centered (5–8) and two body-centered (9, 10). This notation has two disadvantages. First, it is not possible to tell at a glance on which Bravais lattice the space group is based. Secondly, there is no indication of wether the space group is symmorphic or non-symmorphic.

The two disadvantages of the Schönflies notation are overcome by using the Hermann–Mauguin (international) notation, which not only determines the two features fist mentioned but also gives some idea of the nature and orientation of the symmetry elements present in the space group. Starting with the international point group symbol, for example $m3m$, this is prefixed by the letter describing the Bravais lattice type that specifies the lattice transformations of the space group and is one of the several symbols listed in the last column of Table 2. Thus, we obtain the symbols $Pm3m$, $Fm3m$, and $Im3m$. These symbols denote the three symmorphic space groups isogonal to the point group $m3m$ (O_h). When a point group can be situated on a Bravais lattice in more than one orientation, as for instance $\bar{4}2m$ (D_{2d}) in the tetragonal system, this is allowed for by attaching significance to the ordering of the characters in the international space group symbol; thus $P\bar{4}2m$ (D_{2d}^1) and $P\bar{4}m2$ (D_{2d}^5) are distinct space groups.

For non-symmorphic space groups the symbol is modified as follows. In a glide operation, the non-primitive translation must be parallel to the glide plane, and its direction specifies the type (a, b, c, n, or d) of the glide. The symbols a, b, and c stand for glides with non-primitive translations $\{\mathbf{a}_i | i = 1, 2, 3\}$, while n stands for face-diagonal translations $\{(\mathbf{a}_i + \mathbf{a}_j)/2 | i, j = 1, 2, 3\}$; and d stands for the glide operation with none-primitive diagonal translations $\{(\mathbf{a}_i + \mathbf{a}_j)/4 | i, j = 1, 2, 3\}$ (Table 1). In a screw operation, the non-primitive translation component must be directed along the screw axis n_p (Table 1).

To deduce the geometrical arrangements of the symmetry elements requires a knowledge of the results of combining the various symmetry operations of space groups with the translations of the lattice and with each other. This process is easily carried out with the aid of the algebra of operations. After some of the simpler space groups have been derived, the more complicated ones can be deduced from them by a stepwise augmentation.

Equivalent Positions

A space group $\mathcal{R}$ can be characterized by listening the coordinates of those points that are equivalent to a chosen point $\mathbf{r}$. Such analytical description is based on the fact that for a given space group $\mathcal{R}$ all points of the Euclidean space can be subdivided into sets of symmetrically equivalent points (*equipoints*) called *crystallographic orbits*.

The points $\{\mathbf{r}_i | i = 1, \ldots, s\}$ of a given crystallographic orbit may be obtained from a generating orbit point $\mathbf{r}$ arbitrarily chosen by applying to the latter all the operations $\{\hat{d} | \mathbf{t}\}$ of the space group $\mathcal{R}$

$$\mathbf{r}_i = \{\hat{d}_i | \mathbf{t}_i\}\mathbf{r} = \{\hat{d}_i | \mathbf{v}_i + \mathbf{R_n}\}\mathbf{r} \qquad \{\hat{d}_i | \mathbf{t}_i\} \in \mathcal{R} \qquad i = 1, \ldots, s(\mathbf{r}) \,. \tag{15}$$

The infinite number of translations gives rise to an infinite number of points in each space group crystallographic orbit. Any one of the orbit points may represent the whole orbit, i. e., may be a generating point.

The set of symmetry transformations $\{\{\hat{d}_j|\mathbf{v}_j\}|\, j=1,\ldots,m(\mathbf{r})\}$ of the space group $\mathcal{R}$ that leaves the point $\mathbf{r}$ invariant

$$\mathbf{r} = \{\hat{d}_j|\mathbf{v}_j\}\mathbf{r} \qquad j=1,\ldots,m(\mathbf{r}) \tag{16}$$

constitute a finite group $G(\mathbf{r})$ of the *site symmetry (group of the* $\mathbf{r}$-*vector* which is isomorphic to one of the crystallographic point groups. The site symmetry group $G(\mathbf{r}_i)$ of different points $\{\mathbf{r}_i|i=1,\ldots,s(\mathbf{r})\}$ of the same crystallographic orbit is a conjugate subgroup of $G(\mathbf{r})$

$$G(\mathbf{r}_i) = \{\hat{d}_i|\mathbf{t}_i\}\,G(\mathbf{r})\{\hat{d}_i|\mathbf{t}_i\}^{-1} \qquad i=1,\ldots,s(\mathbf{r}) \tag{17}$$

the elements of which leave the points $\mathbf{r}_i$ invariant according to (16). For a point $\mathbf{r}$ at a general position the site symmetry group $G(\mathbf{r})$ consists only of the identity operation $\{\hat{1}|\}$; the site symmetry group of a point at a special position includes at least one other symmetry element in addition to the identity operation.

An infinite number of crystallographic orbits for a given space group $\mathcal{R}$ can be subdivided into sets of so-called *Wyckoff positions* of $\mathcal{R}$. These positions are labelled by small Roman letters arranged alphabetically according to the degree of symmetry. All the crystallographic orbits which have the same site symmetry group belong to the same Wyckoff position. If coordinates of the generating point of an orbit do not contain variables parameters the corresponding Wyckoff position consists of only one crystallographic orbit; in the other case an infinite number of orbits belongs to the same Wyckoff position with variable parameters. The various possible sets of such positions for all space groups are given in the *International Tables for Crystallography*.

As an example Table 4 lists those Wyckoff positions for the rutil structure with the space group $P\frac{4_2}{m}\frac{2_1}{n}\frac{2}{m}$ (D_{4h}^{14}). There are 11 different Wyckoff positions denoted by letters from a to k. The order s of an orbit which means the number of equivalent points of the orbit in the primitive cell is

$$s(\mathbf{r}) = \frac{\text{ord } G}{\text{ord } G(\mathbf{r})} = \frac{g}{m}, \tag{18}$$

where the order of the point group G $(=\frac{4}{m}\frac{2}{m}\frac{2}{m})$ is $g=16$. Only generating points are given in Table 4 for the orbits of special positions h to j. The number of points in a Wyckoff position and their coordinates are given with respect to the crystallographic unit cell of the lattice which coincides with the primitive unit cell. All the crystallographic orbit points in the unit cell are given for a point k of general Wyckoff position $\mathbf{r}=(x,y,z)$. The coordinates of Wyckoff positions e to j contain at least one variable parameter. This means that an infinite number of Wyckoff sets e to j exists in the crystal but the sets a to d consist of only one crystallographic orbit. Pairs of Wyckoff positions (a,b), (f,g), and (i,j) show isomorphic site symmetry groups.

Crystal Structures

The pattern of matter in a crystal consists of sets of integral numbers of atoms in each cell distributed over the available eguipoints of its space group. In this way the integral subscripts of a Daltonian compound are accommodated by the integral number of available locations in

Table 4: Wyckoff positions and site symmetries for the rutil structure with the space group $P\frac{4_2}{m}\frac{2_1}{n}\frac{2}{m}$ (D_{4h}^{14}).

Order of the orbit s	Wyckoff position $\mathbf{r}$	Site symmetry $\mathcal{G}(\mathbf{r})$	Wyckoff position in the unit cell
2	a	$m.mm(D_{2h})$	$(0,0,0)$, $(\frac{1}{2},\frac{1}{2},\frac{1}{2})$
2	b	$m.mm(D_{2h})$	$(0,0,\frac{1}{2})$, $(\frac{1}{2},\frac{1}{2},0)$
4	c	$2/m..(C_{2h})$	$(0,\frac{1}{2},0)$, $(0,\frac{1}{2},\frac{1}{2})$, $(\frac{1}{2},0,\frac{1}{2})$, $(\frac{1}{2},0,0)$
4	d	$\bar{4}..(S_4)$	$(0,\frac{1}{2},1/4)$, $(0,\frac{1}{2},3/4)$, $(\frac{1}{2},0,1/4)$, $(\frac{1}{2},0,3/4)$
4	e	$2.mm(C_{2v})$	$(0,0,z)$, $(\frac{1}{2},\frac{1}{2},z+\frac{1}{2})$, $(\frac{1}{2},\frac{1}{2},\frac{1}{2}-z)$, $(0,0,-z)$
4	f	$m.2m(C_{2v})$	$(x,x,0)$, $(-x,-x,0)$, $(\frac{1}{2}-x,\frac{1}{2}+x,\frac{1}{2})$, $(\frac{1}{2}+x,\frac{1}{2}-x,\frac{1}{2})$
4	g	$m.2m(C_{2v})$	$(x,-x,0)$, $(-x,x,0)$, $(\frac{1}{2}+x,\frac{1}{2}+x,\frac{1}{2})$, $(\frac{1}{2}-x,\frac{1}{2}-x,\frac{1}{2})$
8	h	$2..(C_2)$	$(0,\frac{1}{2},z)$
8	i	$m..(C_s)$	$(x,y,0)$
8	j	$..m(C_s)$	(x,x,z)
16	k	$1(C_1)$	(x,y,z), $(-x,-y,z)$, $(x,y,-z)$, $(-y,-x,-z)$, $(-x,-y,-z)$, $(x,y,-z)$, $(-y,-x,z)$, (y,x,z) $(-\frac{1}{2}-y,\frac{1}{2}+x,\frac{1}{2}+z)$, $(\frac{1}{2}+y,\frac{1}{2}-x,\frac{1}{2}+z)$, $(\frac{1}{2}-x,\frac{1}{2}+y,\frac{1}{2}-z)$, $(\frac{1}{2}+x,\frac{1}{2}-y,\frac{1}{2}-z)$, $(\frac{1}{2}+y,\frac{1}{2}-x,\frac{1}{2}-z)$, $(\frac{1}{2}-y,\frac{1}{2}+x,\frac{1}{2}-z)$, $(\frac{1}{2}-x,\frac{1}{2}+y,\frac{1}{2}+z)$, $(\frac{1}{2}+x,\frac{1}{2}-y,\frac{1}{2}+z)$

the cell. Ideally, the translations of the lattice provide that all regions of the crystal have the same composition. If the conditions of growth, such as temperature, pressure, and composition of the nutrient material, vary with time, however, the composition of the crystal may vary from cell to cell due to the vicarious replacement of one atomic species by another, or of several atomic species by several others, in such a way that the replacing set has certain properties similar to the replaced set.

See also: Group Theory in Physics; Crystallography, X-ray.

Bibliography

M. Böhm, *Symmetrien in Festkörpern*. Wiley-VCH, Berlin, 2002.

W. Borchardt-Ott, *Crystallography*. Springer, Berlin, Heidelberg, New York, 1995.

M. J. Buerger, *Introduction to Crystal Geometry*. McGraw–Hill, New York, 1971. (E)

M. J. Buerger, *Elementary Crystallography*. MIT Press, Cambridge, MA, 1977. (I)

E. S. Federow, *Symmetry of Crystals* (transl. by David and Katherine Harker). ACA Monograph 7, 1971. (A)

T. Hahn (ed.), *International Tables for Crystallography*, Vol. A. Kluwer Academic Publishers, Dordrecht, 2002.

I. Janssen, *Crystallographic Groups*. North-Holland, Amsterdam, 1973. (A)

F. C. Philips, *An Introduction to Crystallography*, 4th ed. Wiley, New York, 1971. (E)

J.-J. Rousseau, *Basic Crystallography*. John Wiley a. Sons, New York, 1998.

A. V. Shubnikov and V. A. Koptsik, *Symmetry in Science and Art* (David Harker, ed.). Plenum, New York, 1974. (I)

Crystallography, X-Ray

M. J. Buerger[†], Isabella Karle, and Jerome Karle

Introduction

Background

Until about the second decade of this century, experimental crystallography was carried out largely by the observation of crystal faces and forms with the aid of the reflecting goniometer and by the study of crystal optics with the aid of the polarizing microscope. Chiefly with these tools the point-group symmetries, crystal systems, and axial ratios of many crystals were established.

The discovery of the diffraction of x rays by crystals afforded the opportunity to obtain much more information concerning the atomic arrangements in matter. Not only is the geometry of the unit cell in a crystal readily determined, but also the locations of all the atoms in crystals of metals, alloys, minerals, salts, organic compounds, natural products, nucleic acids, proteins, and, even, to a certain extent, viruses are determinable. Beams of x rays, electrons, and neutrons are used for diffraction experiments, but x radiation is used primarily because

---

[†]deceased

of readily available instrumentation for the individual laboratory, fewer experimental complications, and well-established techniques. Electron and neutron beams have special purposes. For example, electron beams are suitable for very thin crystalline samples and neutron beams are particularly useful for locating hydrogen atoms. Very intense beams of x rays are available from synchrotron sources and high-intensity neutron beams are available from spallation sources.

Max van Laue's and Sir Lawrence Bragg's descriptions of the diffraction of radiation by three-dimensional gratings are treated in most elementary physics texts. Therefore they are not discussed here.

Space Groups

The morphological study of crystals of different symmetry indicated that they could be referred to seven different sets of crystallographic axes of reference known as triclinic, monoclinic, orthorhombic, tetragonal, trigonal, hexagonal, and cubic. Each of these systems possesses characteristic symmetry. The internal symmetry operations in a crystal may consist of an inversion center, reflection across a plane, 2-, 3-, 4-, or 6-fold rotation, translation, or combinations of these operations called screw axes and glide planes. The seven crystal systems combined with the possible symmetry operators lead to 230 unique space-group types. The characteristics of the 230 space groups are detailed in the International Tables of X-Ray Crystallography, Vol. A.

Space-Group Extinctions

Reflections in a diffraction experiment arise from sets of imaginary parallel planes cutting through a crystal. The planes are indexed with integral numbers that are the reciprocals of their intercepts with the crystallographic axes. The reflected waves are indexed with the integers h, k, and l, the same indices as those of the corresponding planes from which they appear to be reflected. *Extinctions* are defined as certain classes of reflections that are missing in a diffraction pattern owing to the presence of certain symmetry elements, as shown by Paul Niggli in his 1919 book *Geometrische Kristallographie des Discontinuums*. Translation components of glide planes give rise to missing reflections of the type $hk0$, $h0l$, $0kl$, or hhl, and translation components of screw axes cause missing reflections of the type $h00$, $0k0$, $00l$, or $hh0$ (with odd-numbered h, k, or l). Space-group identification is facilitated by the missing reflections.

Friedel's Law

Georges Friedel in 1913 stated that the intensities of reflections from planes $(\bar{h}\bar{k}\bar{l})$ are the same as those from planes (hkl) and consequently it was deemed impossible to determine by means of an x-ray photograph whether a crystal possesses a center of symmetry. However, the statistical distribution of intensities of all the reflections or of certain groups of reflections were shown by A. J. C. Wilson to be different for centric and acentric crystals, generally allowing an unequivocal distinction to be made. Another way of distinguishing centrosymmetric from noncentrosymmetric crystals arises when the wavelength of the incident beam is sufficiently close to an absorption edge of one or more of the atoms present. In that case, Friedel's law is violated by the intensities for noncentrosymmetric crystals, i. e., $I(hkl) \neq I(\bar{h}\bar{k}\bar{l})$, but for centrosymmetric crystals, $I(hkl) = I(\bar{h}\bar{k}\bar{l})$.

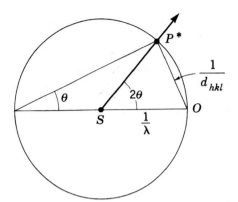

Fig. 1: Ewald's interpretation of Bragg's law with the aid of the reciprocal lattice. [From M. J. Buerger. *Contemporary Crystallography*. (McGraw–Hill, New York, 1970).]

Reciprocal Lattice

To every crystal lattice there corresponds a related lattice, called the reciprocal lattice, that has properties useful in experimental diffraction by crystals. It has the same symmetry as the real space lattice and, for the special cases (orthorhombic, tetragonal, and cubic) when the angles $\alpha = \beta = \gamma = 90°$, the angles of the reciprocal lattice $\alpha^* = \beta^* = \gamma^* = 90°$ and the cell edges are related by $a^* = K/a$, $b^* = K/b$, and $c^* = K/c$, where K can be related to an instrumental constant (see the Precession Method). For the crystal systems with any angle $\neq 90°$, the angle and edge relationships are somewhat more complex.

Application of the Reciprocal Lattice

Bragg's law can be written

$$\sin\theta = \frac{\lambda}{2d(hkl)} = \frac{1/d(hkl)}{2\lambda} , \tag{1}$$

where 2θ is the angle between the incident beam of wavelength λ and the beam reflected by the planes whose indices are the integers h, k, and l. The distance between such parallel planes is $d(hkl)$.

A geometrical interpretation of the right-hand part of (1), due to Ewald, is illustrated in Fig. 1. The primary x-ray beam is along the direction SO and the origin of the reciprocal lattice is O. Whenever a crystal is so oriented that any reciprocal-lattice point P^*, at the end of a vector of length $/d$, comes to lie on the surface of a sphere (known as the *Ewald sphere*), Bragg's law is satisfied and a reflected beam develops in the direction SP^* that makes an angle 2θ with the x-ray beam.

The concept of the reciprocal lattice has proven to be of considerable value in the interpretation of the geometrical aspects of diffraction data and the progress of many aspects of diffraction physics.

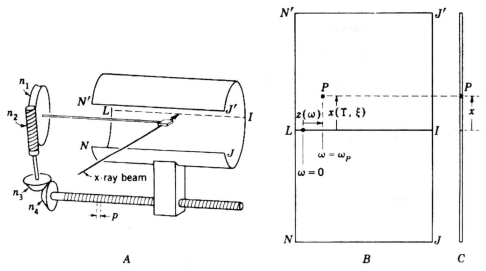

Fig. 2: Diagram of a Weissenberg instrument showing (A) the coupling of the camera translation to the crystal rotation; (B) the resulting position of a reflection on the unrolled film; (C) the position of the corresponding spot along a layer line of a rotating-crystal photograph. [From M. J. Buerger, *X-Ray Crystallography* (Wiley, New York, 1942)].

Experimental Methods of X-Ray Crystallography

The Weissenberg Method

If a crystal is rotated about a rational axis, the diffraction directions lie along generators of a nest of Laue cones coaxial with the axis. In the Weissenberg instrument, each cone intersects a coaxial cylindrical film in a circle that, when the film is unrolled, becomes a straight line known as a *layer line*. With the cone containing the incoming x-ray beam counted as the zero cone, the nth cone contains reflections with indices hkn if the rotation axis is the crystallographic c axis. In the Weissenberg method the desired cone is allowed to reach the film through a circular slot in a cylindrical metal screen called the layer-line screen. The cylindrical camera can be translated parallel to its axis, and this translation is coupled to the rotation of the crystal about the camera axis (Fig. 2). In this process the diffraction spots are spread out on the two-dimensional surface of the film so that their two variable indices h and k are determinable. The coordinates x and y of a spot are readily transformed into indices h and k.

The photographs taken with a Weissenberg camera have been almost universally recorded by *equi-inclination*, a technique in which the primary beam enters along a generator of the cone whose reflections hkn are being photographed. This technique has the advantage that the shapes of rows of the reciprocal lattice as recorded on the Weissenberg film are the same regardless of the value of n.

With the introduction of the Weissenberg method in 1924, it became possible for the first time to determine with one instrument not only information about the symmetry of the crystal, but also the indices and intensities of all reflections hkl.

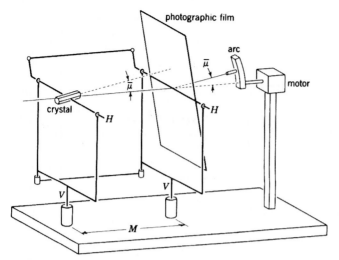

Fig. 3: Diagram of a precession instrument. The motor causes the normal to the photographic film to precess around the direction of the x-ray beam; the coupling of the crystal to the film causes a chosen rational direction of the crystal to precess around the x-ray beam also. [From M. J. Buerger. *Contemporary Crystallography* (McGraw–Hill, New York, 1970)].

The Weissenberg method has gained new attention for macromolecular structure analysis. With a film cylinder of larger diameter, a reusable flexible phosphor "plate" for recording the reflections, and automated and computerized scanners for indexing the reflections and reading their intensities from the plate, this method is a viable alternative for collecting the thousands of data for large cells as occur with macromolecules.

The Precession Method

In the precession instrument, a crystal is held in one gimbal and a flat film in another gimbal, as shown in Fig. 3. The unmoved points of these gimbals are aligned along the incoming primary x-ray beam, so that the beam encounters the crystal first. The vertical and horizontal axes of the two gimbals are coupled and a rational axis of the crystal is set parallel with the normal to the film surface. These axes are then inclined at an angle μ, to the x-ray beam and caused to precess about it. The nest of Laue cones coaxial with the rational direction of the crystal then also precesses about the x-ray beam. If the rational axis is the crystallographic c axis, the reflections of the nth cone are hkn. The reflections of any chosen cone are allowed to reach the film through an annular opening in a flat layer-line screen that is attached to and precesses with the crystal.

Precession photographs are undistorted pictures of the reciprocal lattice, level by level. Accordingly they not only reveal the symmetry at each level, but the indices of all reflections are obvious by mere inspection. Indeed, indexing for space-group determination is unnecessary since, for example, a glide plane parallel to a level can be detected by simple comparison of the appearances of the zero level and an nth level; if a glide plane is present, alternate lines of spots on the zero level are missing in a direction along the translation component of the

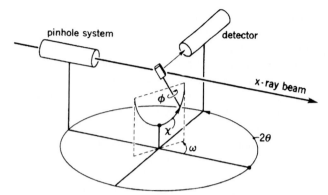

Fig. 4: Diagram of the four-circle diffractometer. [From M. J. Buerger. *Contemporary Crystallography* (McGraw–Hill, New York, 1970).]

glide plane. The limitation of the method is that only reflection arising at relatively small scattering angles can be recorded on the flat film (as compared with the cylindrical film in the Weissenberg method).

Diffractometer Methods

Several instruments are currently used to record the intensities of x-ray reflections with the aid of photon counters. Two versions have come into common use, one with Weissenberg geometry and the other known as a four-circle diffractometer, shown in Fig. 4. The photons are usually counted by means of proportional counters or scintillation counters.

Diffractometers furnish more precise measurements of the diffraction intensities than the photographic methods and so are preferred for this reason. The operation is automated by a computerized program that sets four angles that are associated with each reflection, scans the diffraction spot, corrects the raw data for various experimental factors, and presents a data record of all the measured reflections in a form suitable for deriving the atomic arrangement in the crystal. The mode of operation is a sequential one in that each reflection is scanned separately.

Other Methods

The Laue, rotating- or oscillating-crystal, and powder methods are among the earliest methods used for recording x-ray diffraction diagrams. Over the years, the facility with which these methods could be used has increased dramatically. This was achieved by the design of more sophisticated instrumentation, the use of high-intensity neutron and x-ray beams, and the advent of convenient advanced computing facilities and associated software packages.

The powder method uses a sample of tiny crystals that are glued or tamped into a pellet or inserted into a thin-walled glass capillary. It usually consists of materials for which larger crystals are not available. Since the powder sample has crystallites oriented in all directions, the diffraction pattern is in the form of concentric circles rather than individual spots. Present technology can analyze a trace of a complex powder pattern (even with overlapped lines) for space-group information and individual *hkl* assignments. Of importance is the least-squares refinement of an approximately known structure by the Rietveld procedure.

In the oscillation method, a variant of the rotating crystal method, the crystal is oscillated through a small-angle range about a particular axis, the x-ray data recorded, the procedure repeated for the next angle range, until the entire pattern for a complete rotation is recorded in segments. In this method, the overlap of spots experienced in the rotation method is avoided. Computerized procedures are used to index the spots and measure their intensities. This method is often used for macromolecules with high-intensity synchrotron radiation that permits very short exposures for each oscillation pattern recorded.

The Laue method uses *white radiation*, that is, radiation consisting of a large range of wavelengths, rather than monochromatic radiation which is used in all other diffraction methods. With white radiation, it is not necessary to continually reorient a crystal by, for example, rotation or precession. A single Laue photograph of a protein crystal, taken at a carefully chosen crystal orientation, with the use of high-intensity synchrotron radiation, can record more than 75% of the useful intensity data in less than a millisecond. This emerging technology has the promise of developing into a method for recording ongoing chemical reactions, for example, between a macromolecule and its substrate.

The Raw Data and their Treatment

The Integrated Reflection
It is usual to measure the radiation reflected by a crystal by the integrated reflection, which is given by

$$\frac{Q(hkl)\omega}{I_0} = \frac{e^4}{m^2 c^4} \lambda^3 N^2 \Delta V Lp |F(hkl)|^2 . \tag{2}$$

Here $Q(hkl)$ is the energy reflected in the orders hkl by a tiny fragment of crystal composed of N cells and volume ΔV, so small that absorption and extinction can be ignored as the crystal is rotated with angular velocity ω in a beam of radiation whose intensity is I_0 and whose wavelength is λ. L and p are the Lorentz and polarization factors, respectively, while e, m, and c are universal constants. The $F(hkl)$ are the amplitudes of the reflected x-ray waves, also called *structure factors*. The left-hand side of (2) has dimensions of power times radians and represents the power of the reflection hkl integrated over its angular range. Since the terms before L are fixed for the entire experiment, (2) can be written as

$$Q(hkl)\omega/I_0 = KLp|F(hkl)|^2 , \tag{3}$$

where K is a constant for the experimental setup.

Measurement of Intensities
Integrated reflections can be obtained with a single-crystal diffractometer. The reflections can be automatically integrated by counting while the crystal is uniformly rotated through the region of the Bragg reflection; alternatively, the crystal may be held fixed at equally spaced angular intervals for equal times while counting, and the results summed. In either case the background is measured and subtracted.

Correction of Raw Data

To transform the integrated reflections into the $|F(hkl)|^2$, they must be corrected for the Lorentz factor, polarization, absorption, and extinction. Convenient software is available for making the corrections.

Fourier Transformations

A crystal structure and its diffraction amplitudes are Fourier mates, so that each can be found if the other is known. The diffraction amplitude $F(hkl)$ is given in terms of the *electron density* $\rho(xyz)$ by

$$F(hkl) = V \int_0^1 \int_0^1 \int_0^1 \rho(xyz) \, \exp[i2\pi(hx + ky + lz)] \, dx \, dy \, dz \,. \tag{4}$$

Here V is the volume of the unit cell and x, y, and z are coordinates for the *electron density* expressed in terms of fractions of the cell edges a, b, and c, respectively. The electron density of a crystal cell is the sum of the contributions from a set of N discrete atoms. The scattering power of the jth atom is denoted by f_j, expressed in units of the scattering power of a free electron. The amplitude is then given by

$$F(hkl) = \sum_{j=1}^{N} f_j(hkl) \, \exp[i2\pi(hx_j + ky_j + lz_j)]T_j \,. \tag{5}$$

Here x_j, y_j, and z_j are coordinates of the atoms in fractions of the cell edge, and T_j is a factor that represents the effects of the thermal motion.

To recover the *electron density* $\rho(xyz)$ from the $F(hkl)$'s the relation is

$$\rho(xyz) = V^{-1} \sum_h \sum_k \sum_l F(hkl) \, \exp[-i2\pi(hx + ky + lz)] \,. \tag{6}$$

Any crystal can be solved for its electron density by making use of (6) if the diffraction amplitudes $F(hkl)$ are fully known. As seen in (4) and (5), however, these are complex quantities, and all that can be determined by experimental methods are the magnitudes $|F(hkl)|^2$. Some progress has been made in deriving phase information rather directly from experiment, by making use of multiple-scattering phenomena. This technique is in its early stages and does not afford the large amount of phase information required for current structure determination. It would appear then that (6) cannot be calculated from the experimental data that are obtained in a standard diffraction experiment. It is quite a remarkable fact, however, that the required phase information is subtly contained in the measured intensities from which, it would seem, only the values of the corresponding $|F(hkl)<^2$ are directly obtainable, i. e., by use of (2). This fact has led to the "direct methods" described below. Other ways to overcome the phase problem will also be noted.

Deduction of the Correct Arrangement of Atoms

The Patterson Function

In 1934 A. L. Patterson showed that the centrosymmetrical function

$$P(uvw) = V^{-2} \sum_h \sum_k \sum_l |F(hkl)|^2 \, \cos[2\pi(hu + kv + lw)] \tag{7}$$

has high values at coordinates uvw whenever the vector from the origin to uvw is the same as a vector between two atoms p and q in the crystal cell: i. e., when

$$u = \pm(x_q - x_p), \quad v = \pm(y_q - y_p), \quad w = \pm(z_q - z_p), \tag{8}$$

where only all the plus or all the minus signs are taken together. The magnitude of $P(uvw)$ is proportional to $f_p f_q$ so that $P(uvw)$ is large when p and q are both heavy atoms. Thus, the vector distance between heavy atoms is easy to determine, and if the atoms are related by symmetry they can be easily located in the cell.

While the Patterson function provides the interatomic distances in a crystal structure, in practice it does not generally, in itself, solve the structure in a direct way. The function suffers from the fact that if there are n atoms in the crystal cell, there are $n(n-1)$ high values of $P(uvw)$ for $p \neq q$. The clutter due to this overlap increases approximately as n^2, so the usefulness of the function decreases as the number of atoms in the cell increases.

The Patterson function has been particularly useful in applications requiring the location of heavy atoms in structures. This has not only permitted the solution of very small structures, it also has made the solution of macromolecular structures feasible when used in combination with data from the isomorphous replacement and anomalous dispersion techniques. Another worthwhile use of the Patterson function has been to fix the orientation and position of a partially known or sometimes completely known structure. In the case of a partial structure, once its position and orientation are known, there are several procedures for completing the structure.

The Heavy-Atom Method

Crystals that are particularly suited to the application of this method are characterized as consisting of a small number of heavy atoms among many lighter ones. It is often possible to find the locations of the heavy atoms by use of the Patterson function because the interatomic vectors associated with the heavy atoms predominate in this function. The phases and amplitudes of many of the $F(hkl)$'s of (5) are dominated by the heavy atoms. These phases can then be used as preliminary phases of the coefficients in (6), from which a fair approximation of the electron density can be obtained. This reveals not only the heavy atoms, but some of the light ones as well. From this improved set of atoms an improved set of phases can be computed by (5), which can then be used in another Fourier summation (6). This iteration eventually comes to an end because no phases computed by (5) need to be changed. The heavy-atom method has been responsible for the solution of a great many structures.

Direct Methods in Fourier Space

It is possible to show by use of (5), corrected for the effects of T_j, that because the atoms in a unit cell are discrete and their scattering powers are known to very good approximation, the structure problem is highly overdetermined by the number of independent intensity data obtained from the radiation of an x-ray tube with a Cu target. This applies to the x_j, y_j, z_j coordinates of the atoms as well as the phases. In principle, it should be possible to determine the coordinates directly. It has, however, been found to be much more feasible to determine *phase values* first and then obtain the atomic coordinates from (6).

With strong clues from relations among the $F(hkl)$ that arise from the *non-negativity* of the *electron density* distributions in crystals, phase-determining formulas were obtained by

Herbert Hauptman and Jerome Karle in a probabilistic context that has served well in the direct determination of crystal structures containing no heavy atoms. Structures with up to 100 nonhydrogen atoms to be placed can be solved rather readily. With increasing complexity beyond 100 atoms, determinations become more difficult to carry out although a number of the more complex structures have been solved.

The main phase-determining formulas are

$$\phi_{hkl} \sim \langle \phi_{h'k'l'} - \phi_{h-h',k-k',l-l'} \rangle_{h'k'l'} \tag{9}$$

and

$$\tan\phi_{hkl} \sim \frac{\sum_{h'k'l'} |E_{h'k'l'} E_{h-h',k-k',l-l'}| \times \sin(\phi_{h'k'l'} + \phi_{h-h',k-k',l-l'})}{\sum_{h'k'l'} |E_{h'k'l'} E_{h-h',k-k',l-l'}| \times \cos(\phi_{h'k'l'} + \phi_{h-h',k-k',l-l'})}, \tag{10}$$

where the E are values of F rescaled in such a way that $\langle E_{hkl}^2 \rangle_{hkl} = 1$. Because of the latter property, they are called *normalized structure factors*. The E corresponds to atoms for which the source of scattering is concentrated in a point at the atomic center ("point atoms"). These formulas are used either in a stepwise fashion or with a random set of phase values followed by further refinement with (10). For the stepwise procedure, it is also necessary to have some initial phase values. These are obtained from permitted assignments of some phase values in order to specify the origin in a crystal and the additional use of symbols whose values are determined as the phase-determination proceeds. Other procedures use alternative numerical values for the same phase instead of symbols. The particular phase assignments for a given space group have been tabulated and arise from a theory of phase invariants and structure semi-invariants. The initial phase assignments are used to define the values of additional ones and in a series of steps the phase determination proceeds. The steps are chosen so as to maximize the probability of correctness by use of probability measures associated with (9) and (10). Correct structures are recognized by their good agreement with the data as computed from (5) and by confirming that the bond lengths, bond angles, and connectedness make good chemical sense. It should be noted that the direct methods of phase determination are particularly effective when applied to structures that contain heavy atoms, because of an enhancement of probability, and when the positions of the heavy atoms are known, development of the remainder of the structure by use of (10) is generally considerably more efficient than the heavy-atom method.

Macromolecular Structure Determination

Heavy atoms have played an indispensable role in the determination of the structures of macro-molecules in the crystalline state. The names of the techniques used are *isomorphous replace-ment* and *anomalous dispersion* . Crystals that have the same unit-cell geometry but differ in chemical composition are called isomorphous. Isomorphous protein crystals are usually comprised of the native protein and the native protein to which some heavy atoms have been added. The information contained in the different intensities of scattering from isomorphous crystals permits the determination of their structures.

If heavy atoms are added to an unsubstituted crystal, the relationship between wave amplitudes is

$$F_R(hkl) + F_Y(hkl) = F_{R+Y}(hkl), \tag{11}$$

where R represents the native substance, Y the added atoms, and $R + Y$ the substituted substance. Analysis of (11) is benefited from the determination of the positions of the atoms Y, for example, from Patterson maps or direct methods which may be applied with enhanced utility to isomorphous replacement data. From this information, $F_Y(hkl)$ may be computed with the use of (5). With knowledge of $F_Y(hkl)$ and the experimentally measured magnitudes in (11), it is possible to derive phase information. If the isomorphous pair of crystals are both centrosymmetric, the phase information will be unique. If the isomorphous pair of crystals are both noncentrosymmetric, the phase information will be obtained with a twofold ambiguity. Many methods have been developed to resolve the ambiguity, e. g., additional substitutions, anomalous dispersion information, probabilistic and algebraic analyses, and a filtering technique.

It has been customary to make several isomorphous replacements when possible in order to compensate for experimental error and lack of precise isomorphy. The method based on several isomorphous replacements is called the *multiple isomorphous replacement* method. It has so far played a predominant role in the progress of protein crystallography.

The second technique, *anomalous dispersion*, is based on the modification of the intensities of scattering as a consequence of absorption processes that take place in some of the atoms that comprise the crystal of interest. The role of the heavy atoms in macromolecules derives from their relatively strong anomalous scattering. The modifications of the scattered intensities provide the information from which the structures can be elucidated.

When atoms are present in a structure that scatter anomalously to a significant extent, the wave amplitude may be written

$$F_\lambda(hkl) = F^n(hkl) + F_\lambda^a(hkl) ,\qquad(12)$$

where $F_\lambda(hkl)$ is the amplitude of the scattered wave whose magnitude may be measured by experiment, $F^n(hkl)$ is the wave amplitude that would be obtained if all atoms scattered without any anomalous dispersion, and $F_\lambda^a(hkl)$ is the wave amplitude that arises from the anomalous corrections to the normal atomic scattering. In noncentrosymmetric crystals, the Friedel mates are no longer equal when there is significant anomalous scattering so that $F_\lambda(hkl) \neq F_\lambda(\bar{h}\bar{k}\bar{l})$ and a second independent equation follows:

$$F_\lambda(\bar{h}\bar{k}\bar{l}) = F^n(\bar{h}\bar{k}\bar{l}) + F_\lambda^a(\bar{h}\bar{k}\bar{l}) .\qquad(13)$$

Identification of the anomalous scatterers and their location permits $F_\lambda^a(hkl)$ and $F_\lambda^a(\bar{h}\bar{k}\bar{l})$ to be computed from (5) in which the atomic scattering factors are replaced by the real and imaginary corrections to the atomic scattering factors. Patterson and direct methods techniques can be used to locate the anomalous scatterers. It was found to be worthwhile to manipulate the right sides of (12) and (13) to express the quantities in somewhat different form. This resulted, with the use of various wavelengths, in providing a useful set of linear simultaneous equations. Phase determination can be made with these equations. If there is only a one-wavelength experiment, the result for noncentrosymmetric crystals will involve a twofold ambiguity. To overcome the ambiguity, it is possible to employ, for example, more wavelengths, isomorphous replacement, probabilistic calculations, and a filtering technique. In Fig. 5 the combination of single isomorphous replacement and one-wavelength anomalous dispersion are illustrated in a manner that emphasizes how their individual ambiguities

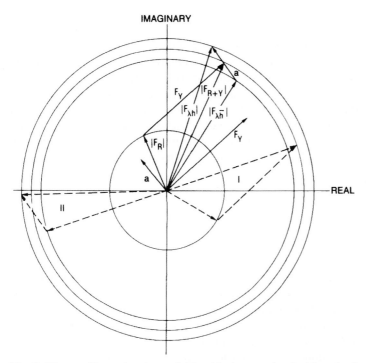

Fig. 5: Diagram illustrating the twofold ambiguity associated with each of single isomorphous replacement and one-wavelength anomalous dispersion and the manner in which the combination of the two techniques resolves the ambiguity. Here we have set $F_{\mathbf{h},R+Y} = 0.5(F_{\lambda\bar{\mathbf{h}}} + F^*_{\lambda\bar{\mathbf{h}}})$ to good approximation, $\mathbf{a} = F^a_{\lambda\mathbf{h}} - F^a_{\lambda\bar{\mathbf{h}}}$ and the vector triangles I and II are the incorrect alternatives for the isomorphous replacement and the anomalous dispersion data, respectively $[\mathbf{h} = (hkl)]$.

are mutually resolved. The anomalous dispersion technique is especially facilitated by the use of x-ray radiation from synchrotron sources. Its high-intensity, tunable radiation is well suited to multiple-wavelength anomalous dispersion experiments. Up to the present, anomalous dispersion has been applied to a lesser extent than isomorphous replacement and usually in combination with the latter. The value of each technique is enhanced when used in combination. Anomalous dispersion is beginning to be used more often, however, as the sole method for the investigation of macromolecular structure. It is particularly useful when it appears to not be possible to form isomorphous crystals.

Another approach to macromolecular structure determination arises from the occurrence of symmetry within the unit cell of a crystal in addition to the space-group symmetry. For example, there may be additional rotational symmetry among the atomic arrangements within the unit cell. A methodology for facilitating the analysis of macromolecules based on this so-called noncrystallographic symmetry has been particularly stimulated by the work of Michael G. Rossmann and it has been quite effective in applications to complex viral structures.

Refinement

An approximately solved structure becomes more exactly defined by successive Fourier syntheses, but the final synthesis suffers from errors due to the arbitrary termination of the infinite Fourier series. It also suffers from lack of knowledge of the exact temperature factor. Corrections can be applied by several devices, such as the *difference Fourier synthesis*. This uses coefficients $(F_{obs} - F_{cal})$ for each reflection in (6), where obs and cal signify observed and calculated, respectively. F_{obs} and F_{cal} differ in magnitude. The same phases are used for both. The result of this synthesis is the difference between the observed and calculated electron densities, which shows the magnitudes and directions in which the model must be changed to bring it into conformity with the actual structure.

The progress of refinement is followed by computing at each stage the discrepancy index, a residual summed over the data set,

$$R = \frac{\sum ||F_{obs}| - |F_{cal}||}{\sum |F_{obs}|} . \tag{14}$$

The value of R is usually between 5 and 10% for an acceptably refined structure, and may be as small as 2% for a very well-refined structure. The acceptable values of R for data from macromolecular structure determinations are generally somewhat higher. The R value is 82.8% for a completely wrong model if centrosymmetrical, and 58.6% if noncentrosymmetrical. A preliminary model that has $R = 45\%$ is regarded as promising.

The use of least squares in fitting an approximately correct model to the observed amplitudes was first suggested by E. W. Hughes in 1941 and applied to refinement of the structure of melamine. Least-squares refinement became popular with the rise of electronic digital computers, and currently is the standard way of finishing the details of the structure. Because the structures of most inorganic and the simpler organic structures are overdetermined by the large numbers of their diffraction-amplitude data and because modern computers have large capacities, it is usual not only to refine the three coordinates of each atom, but also to determine the three principal axes of the ellipsoids of thermal motion of each atom and their orientations with respect to the cell axes.

Data Files

Compendia of structural information are available on separate computer files for inorganic, organic, and macromolecular substances. An extensive file is also available concerning information obtained by powder diffraction. For details concerning these files, contact International Union of Crystallography, 5 Abbey Square, Chester CH1 2HU, England.

See also: Crystal Symmetry; Fourier Transforms; X-Ray Spectra and X-Ray Spectroscopy.

Bibliography

F. R. Ahmed, K. Huml, and B. Sedláček, (eds.), *Crystallographic Computing Techniques*. Munksgaard, Copenhagen, 1976. (I)

D. M. Blow, "How Bijvoet Made the Difference: The Growing Power of Anomalous Scattering," in *Methods in Enzymology*, Vol. 374, pp. 1–22, Elsevier, 2003.

M. J. Buerger, *X-ray Crystallography*. Wiley, New York, 1942. (E)

M. J. Buerger, *The Precession Method*. Wiley, New York, 1964. (E)

M. J. Buerger, *Contemporary Crystallography*. McGraw–Hill, New York, 1970. (E)

M. A. Carrondo and G. A. Jeffrey (eds.), *Chemical Crystallography with Pulsed Neutrons and Synchrotron X-Rays*. D. Reidel, Dordrecht, 1988. (A)

C. W. Carter, Jr., and R. M. Sweet (eds.), *Methods in Enzymology*, Vols. 276 and 277, Academic Press, New York, 1997; Vols. 368 and 374, Academic Press, New York, 2003.

T. Hahn (ed.), *International Tables for Crystallography*. D. Reidel, Dordrecht, 1987. (A)

S. R. Hall and T. Ashida (eds.), *Methods and Applications in Crystallographic Computing*. Clarendon Press, Oxford, 1984. (I)

R. W. James, *The Optical Principles of the Diffraction of X-Rays*. G. Bell and Sons, London, 1948. (A)

J. Karle, *Angew. Chem. (Int. Ed. Eng.)* **25**, 614 (1986). (I)

M. F. C. Ladd and R. A. Palmer (eds.), *Theory and Practice of Direct Methods in Crystallography*. Plenum, New York, 1980. (I)

R. Newnham, *Structure–Property Relations*, Vol. 2 of *Crystal Chemistry of Non-Metallic Materials Series*. Springer-Verlag, Berlin, 1975. (I)

D. Sayre (ed.), *Computational Crystallography*. Clarendon Press, Oxford, 1982. (A)

T. C. Terwilliger, "Automated Structure Solution and Density Modification," in *Methods in Enzymology*, Vol. 374, pp. 22–37, Elsevier, 2003.

A. F. Wells, *Structural Inorganic Chemistry*. Clarendon Press, Oxford, 1984. (I)

H. Wyckoff (ed.), *Diffraction Methods in Biological Macromolecules*. Academic Press, New York, 1985. (A)

Currents in Particle Theory

R. J. Oakes

Currents have played a fundamental role throughout the development of present theoretical ideas about elementary particles. The concept of charges and currents is intimately related to the notion that matter is made up of fundamental constituents and that all the properties of matter are ultimately determined by the nature of these elementary parts and their interactions. For example, the electric charge and electric current densities of matter are superpositions of the individual constituent densities and fluxes weighted by the electric charge carried on each constituent. Although discoveries of new particles and new phenomena have caused drastic revisions in views about which particles are elementary and what laws govern their interactions, the basic concept of charges and currents has remained essentially unchanged.

In nonrelativistic quantum mechanics, where the position of a particle is described by a probability distribution, the electric charge and electric current densities at a point $\mathbf{x}$ at time t due to a particle of charge e are $e\rho(\mathbf{x},t)$ and $e\mathbf{j}(\mathbf{x},t)$ where $\rho(\mathbf{x},t)$ and $\mathbf{j}(\mathbf{x},t)$ are, respectively, the position probability and the probability current density at $\mathbf{x}$ and t. The density $\rho(\mathbf{x},t)$ and current $\mathbf{j}(\mathbf{x},t)$ depend on the Schrödinger wave function $\psi(\mathbf{x},t)$. The local conservation law

$$\frac{\partial \rho}{\partial t}(\mathbf{x},t) + \mathbf{\nabla} \cdot \mathbf{j}(\mathbf{x},t) = 0$$

is insured by virtue of $\psi(\mathbf{x},t)$ being a solution to the Schrödinger equation of motion for the particle.

In relativistic quantum field theory, which incorporates the infinite number of degrees of freedom necessary to describe an indefinite number of particles, for each kind of elementary particle there is a canonical field operator that plays a role analogous to the classical canonical coordinates. The Lagrangian describing the entire system of particles is a function of all these fields and their first derivatives. In general, there will be several currents in the theory, corresponding to the charges carried by the particles (e. g., electric charge, isospin, and color). Since these currents are functions of the canonical fields, they can be thought of as analogs of classical (noncanonical) coordinates of the system. The currents and their associated charges have proven to be of enormous practical value in studying the interactions of the elementary particles, since the Euler–Lagrange equations of motion, whose complete solutions would determine all the canonical fields, are exceedingly complex. For each current $J_\mu(\mathbf{x}, t)$ there is a charge

$$Q = \int d^3x \, J_0(\mathbf{x}, t) \, ,$$

which will be a constant of the motion if the current is conserved, thus providing dynamical information about the system. Nonconserved currents have also proven to be very important in the phenomenological studies of elementary-particle interactions.

Currents are essential in describing the interactions among the elementary particles, which at present include at least quarks, leptons, and intermediate vector bosons. While quarks cannot be directly observed, since they carry color charges which are confined to very small regions of space, there is substantial indirect evidence that the observed strongly interacting hadrons, e. g., protons, neutrons, and mesons, are composed of these spin-$\frac{1}{2}$ quark constituents. The leptons, which so far include the electron, the muon, and the tau, as well as their associated neutrinos, do not participate in strong interactions directly but do have electromagnetic, weak, and, of course, gravitational interactions. The photon (γ), which mediates electromagnetic interactions, and the vector bosons ($W^\pm$, Z^0) that mediate the weak interactions have all been directly observed, experimentally. The gluons (g), which mediate the strong interactions, also carry color and, like the quarks, are confined to very small regions of space and cannot, therefore, be directly observed. Nevertheless, there is clear evidence that gluons do exist together with the quarks inside the hadrons, and, indeed, can be emitted or radiated from these hadrons, resulting in observed jets of more hadrons. It is not a settled question, at present, how many quarks, leptons, and intermediate bosons there really are, or what all their properties are, or even if there *must* also be other elementary particles, particularly, elementary scalar bosons.

The electromagnetic interactions of the elementary particles are described by the coupling of the electromagnetic current J_μ to the electromagnetic field A_μ. That is, the interaction Lagrangian is

$$\mathcal{L}_{\text{EM}} = e J_\mu A^\mu$$

and therefore the quantum of the electromagnetic field, the photon (γ), which is a massless vector boson, mediates the electromagnetic interactions of all the elementary particles with a strength equal to the electric charge of the particle. Very precise experimental tests of quantum electrodynamics, e. g., the Lamb shift, the electron and muon anomalous magnetic moments, and the hyperfine splitting, have confirmed this theory down to very small distances ($\sim 10^{-15}$ cm) or equivalently, to very high energies (~ 20 GeV).

It has long been known that weak interactions at low energies, e. g., beta decay, muon decay, and muon capture, can phenomenologically be described by an effective coupling of currents to currents. It is now known that this is only an approximation, valid in the low-energy regime, to a theory which unifies these weak interactions with the electromagnetic interaction. In this electroweak theory the weak interactions are mediated by the massive vector bosons ($W^{\pm}$, Z^0) coupled to weak currents, just as the electromagnetic interaction is mediated by the massless vector boson, the photon (γ), being coupled to the electromagnetic current. The weak-interaction Lagrangian then has a form analogous to the electromagnetic interaction; that is,

$$\mathcal{L}_{\text{WK}} = g_C C_\mu W^\mu + g_N N_\mu Z^\mu \, ,$$

where C_μ is a charged current which changes the electric charge by one unit and N_μ is a neutral current which conserves the electric charge. The charged and neutral intermediate vector bosons $W^\pm$ and Z^0 are quite massive: approximately, $M_W = 80$ GeV/c^2 and $M_Z = 90$ GeV/c^2.

The strong interactions are mediated by massless vector bosons, the gluons (g), which are coupled to the strong currents composed of the particles which carry color: the quarks and the gluons, themselves. These strong interactions are responsible for binding the quarks and gluons together to form the observed hadrons. There are three colors (red, blue, and green) and the unbroken symmetry of the theory among the three colors implies invariance under a symmetry group of transformations, viz., $SU(3)$. As, a consequence, the $3 \times 3 - 1 = 8$ currents corresponding to the possible non-neutral color combinations are conserved, like the electromagnetic current.

It is entirely possible that ultimately all the interactions among the elementary particles can be unified into a grand theory in which the electromagnetic, weak, strong, and even gravitational interactions are all mediated by bosons coupled to currents which are conserved by virtue of the symmetry of the complete theory under a group of transformations. Such theories are called gauge theories, or Yang–Mills theories, and are based on some gauge group of symmetry transformations.

Certainly the electromagnetic interaction is a gauge theory of this type. The conservation of the electromagnetic current and the electric charge is a consequence of invariance of the Lagrangian under a one-dimensional (Abelian) group $U(1)$ of gauge transformations. The weak and electromagnetic interactions have been unified in a generalized (non-Abelian) gauge theory based on the group $SU(2) \otimes U(1)$. The strong interactions are based on the non-Abelian gauge group $SU(3)$, the color symmetry group. The search is now under way for a grand symmetry group, containing as subgroups both the color symmetry group $SU(3)$ and the electroweak symmetry group $SU(2) \otimes U(1)$. If this program succeeds, all the interactions could be understood in terms of intermediate bosons being exchanged between elementary particles in a way prescribed by their gauge-invariant couplings to the elementary-particle currents.

Independent of this ambitious program, currents and their associated charges, whether conserved or not, play an important role in the phenomenology of elementary particles. Measurements of the hadron matrix elements of the electromagnetic and weak currents probe the quark and gluon structure of the hadrons. From elastic electron–nucleon scattering experiments the nucleon form factors of the electromagnetic current have been extracted, while deep inelastic electron-scattering experiments have yielded the proton and neutron structure functions, which naively can be related to constituent quark wave functions. Analogous matrix elements of the weak currents have been measured in neutrino-scattering experiments and, in fact, it was in these neutrino experiments that the existence of *neutral* weak currents in addition to the charged weak currents was first observed. Electron–positron annihilation into hadrons has also been used to probe the matrix elements of the electromagnetic current between the vacuum and various hadronic states yielding many surprising results; particularly the production of new quark–antiquark pairs, which then undergo strong interactions to form the more familiar hadrons.

Currents also play an important role in the study of approximate symmetries of the hadrons. The space integrals of the time components of the currents give charges which close under equal-time commutation to form algebras. The eight vector charges in the algebra associated with the lightest quarks (up, down, and strange) generate an approximate $SU(3)$ symmetry of strong interactions, which contains the isotopic spin and electromagnetic symmetries as subgroups, and has enjoyed enormous success in classifying the hadrons and relating their properties.

Even though this hadron symmetry is only approximate, and therefore not all the associated currents are exactly conserved, the concept of partial conservation of certain currents has proven very useful. In particular, partial conservation of axial vector currents (PCAC) together with the $SU(2) \otimes SU(2)$ current algebra leads to sum rules, notably the Adler–Weisberger sum rule, which tests the underlying approximate symmetries and how they are broken.

Clearly currents have been exceedingly important in the study of the elementary particles and their interactions and no doubt they will continue to play an essential role in this area of research.

See also: Elementary Particles in Physics; Gauge Theories; Quantum Field Theory; *SU(3)* and Higher Symmetries.

Cyclotron

D. L. Judd[†] and E. Baron

The cyclotron is a particle accelerator conceived by Ernest O. Lawrence in 1929 and developed, with his colleagues and students, at the University of California in the 1930s. His goal was to produce beams of high-energy ions, without using high electrostatic voltages, to study their reactions with atomic nuclei. Since then, over 200 cyclotrons of greatly differing sizes, energies, and other properties have been built for a continually growing variety of uses. Three successive generations of cyclotrons ("classic", frequency-modulated, and sector-focused) are described here. In addition, the classic cyclotron is the prototype not only of the later types but also of a wider class of *magnetic resonance accelerators* which include microtrons and various kinds of synchrotrons that accelerate electrons, protons, and heavier ions.

[†]deceased

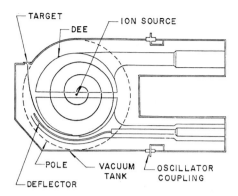

TARGET

DEE

ION SOURCE

POLE

VACUUM
TANK

OSCILLATOR
COUPLING

DEFLECTOR

Fig. 1: Some components of a conventional cyclotron.

By 1940 "classic" cyclotrons had been built and used at many laboratories throughout the world. Exemplified by the 60-in. cyclotron at Berkeley, their principal components were a dc electromagnet having circular pole pieces with a magnetic gap small compared to their diameter, producing a nearly uniform, axially symmetric magnetic field; two hollow flat D-shaped copper electrodes (dees) placed in the magnet gap, open toward each other, with a small space (dee gap) between them along a diameter; an electric field produced across the dee gap, oscillating at a constant radiofrequency (rf), driven by an external oscillator; a vacuum tank within the magnet gap, enclosing the dees and their supporting structures (dee stems) used as resonant lines; an ion source (electric discharge) between the dees at the center, supplied with a small flow of neutral gas; internal targets to be bombarded at large radii; and a deflection system to bring ion beams outside the accelerator (Fig. 1).

The operation depends on the *cyclotron resonance condition*: a particle of charge Qe (with e the electric charge of the electron and Q a positive or negative integer) and mass m moving in a circle perpendicular to a magnetic field B circulates with frequency f and angular frequency ω given by $\omega = 2\pi f = QeB/m$ (SI units) which is the same for all energies, velocities, and radii for constant B and m. This can be seen from Newton's force law $F = ma$, with $F = QeBv$ the centripetal Lorentz force, v the velocity, $a = v^2/r$ the centripetal acceleration, r the radius, and $\omega = v/r$. The kinetic energy at r is $T = \frac{1}{2}mv^2 = \frac{1}{2}(QeBr)^2/m$. An ion making a circle around the source may gain energy, if the oscillator has this *cyclotron frequency*, on each crossing of the dee gap. The electric field there will reverse in each half-turn, having the tangential direction to speed up an ion each time the ion "sees" it. Between crossings the ions are shielded, inside a dee, from the electric field. Because of the resonance condition an ion will stay in step; its energy and radius will grow on every turn until it strikes a target or enters a deflector. (The *cyclotron frequency* of a charged particle in a magnetic field is also important in physics of the solid state, plasmas, the ionosphere, and astrophysics.)

It is not hard to get ions started at the center. They are drawn into a dee each time it is negatively charged (for positive ions), start to circulate at once, and tend to bunch at the correct phase relative to the rf within a few turns. However, they will soon drift away from the magnet midplane and strike the inside surface of a dee unless a vertical focusing force directed toward this plane is provided. To produce it the magnetic field strength B is made to decrease slightly with increasing radius, which causes the magnetic field lines to curve as

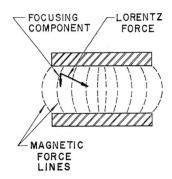

FOCUSING
COMPONENT

LORENTZ
FORCE

MAGNETIC
FORCE
LINES

Fig. 2: Curved magnetic field lines showing vertical focusing force.

shown in Fig. 2. The Lorentz force, perpendicular to these lines, then has a small component directed toward the midplane, causing the ions to oscillate slowly up and down across it as they circulate. This small variation of B also serves to define well-centered circular orbits, about which oscillations in radius may occur. Their frequency is determined by the difference between the centripetal acceleration and the inward radial Lorentz force for ions which depart slightly in radius from the proper circle. These vertical and radial motions are called betatron oscillations because they were analyzed in connection with *betatron accelerators* in which the focusing actions are similar. The radial variation of B is described by the *field index* $n = -(r/B)(dB/dr)$; the angular frequencies of the vertical and radial betatron oscillations are, respectively, $\omega_v = \omega n^{1/2}$, $\omega_r = \omega(1-n)^{1/2}$ with ω as above. For $0 < n < 1$ both frequencies are real, giving stable motions. In these "classic" (flat pole tips) cyclotrons n is small, rising from zero at the center to ≤ 0.2 near the outer edge. A resonant effect ($\omega_r/\omega_v = 2$) can cause beam loss at $n = 0.2$ by transferring energy of radial oscillations, for which there is plenty of room, into vertical oscillations that have limited clearance within the dees.

With radial and vertical motions thus controlled, only an unwanted variation in azimuthal position, or phase relative to the rf oscillator, remains. This effect limits the energy of a classic cyclotron, in which there is no phase stability. The small decrease of B and (particularly for light ions) the relativistic increase of mass ($m = m_0 + T/c^2$, with m_0 the rest mass and c the speed of light) both act to decrease an ion's frequency $\omega = QeB/m$ as its energy and radius increase. Its times of gap crossings will lag more and more behind the phase of maximum electric field, and could eventually lead to crossings at times of reversed field, causing energy loss. To reach a high energy in a limited number of turns, before excessive phase lag can accumulate, the largest possible dee voltage is needed. The highest proton energy reached with such a cyclotron was 22 MeV; this required about 500 kV on the dees. Currents of 100 μA were typical, with a maximum of ~ 1 mA. Deuterons, alpha particles, and heavier ions in higher charge states were also accelerated. Such cyclotrons quickly became the leading tools of nuclear physics research. They were also used to produce radioisotopes for a rapidly growing number of applications in other fields. The first transuranium elements, neptunium and plutonium, were discovered through cyclotron bombardment of heavy targets. Nevertheless, the classic cyclotron is nowadays abandoned for the sector-focused type.

Following World War II new methods were applied which extended the energy and other capabilities of circular magnetic accelerators. The *principle of phase stability* (discovered independently by E. M. McMillan and V. Veksler) led to several new types of accelerators, of which the frequency-modulated cyclotron (synchrocyclotron), electron synchrotron, and proton synchrotron have played important roles in research. They differ from the classic cyclotron by producing particle beams in pulses separated in time (with pulse repetition frequencies ranging from 0.1 to 1000 Hz) rather than as a steady current modulated only at the radiofrequency. In synchrotrons the magnetic field is made to vary periodically at the pulse repetition frequency, but the synchrocyclotron, or FM cyclotron, has a static magnetic field. Its radiofrequency is smoothly lowered during each cycle of acceleration and is followed by that of the ions locked stably to it. Frequency modulation is done with rotating capacitors or vibrating blades. Lawrence's pioneering FM cyclotron at Berkeley accelerated protons to 340 MeV and produced the first man-made pi mesons in 1948. After extensive improvements in 1956 it produced 720-MeV protons, which move at over four-fifths the speed of light, with a mass increase of more than 75%. Deuterons and alpha particles were also accelerated. Protons in the range 600–700 MeV have been produced by similar machines at Dubna, U.S.S.R., and Geneva, Switzerland. These and others producing somewhat lesser energy protons employ magnetic fields on the order of 2 T. Pulse repetition frequencies range from 50 to 1000 Hz and typical currents are 1 μA. Work with these machines greatly enlarged our knowledge of nuclear and meson physics, particularly of pion and muon properties and nucleon–nucleon interactions. Also, large numbers of patients have been treated for certain conditions, particularly pituitary tumors, employing techniques developed at Berkeley.

A different method of overcoming the energy limitation of the cyclotron is based on work by L. H. Thomas in 1938, but the developments needed to put his concept and its later elaborations into practice did not begin until 1949. Because many characteristics of the classic cyclotron (static magnetic field, constant rf frequency, and steady, non-pulsed beam) are retained by the accelerators that have evolved along this line they are called *sector-focused, isochronous, spiral-ridge,* or *azimuthally-varying-field (AVF) cyclotrons.* In these machines the magnetic field (averaged along a full turn of an ion's orbit) *increases* with increasing radius just enough to offset the relativistic mass increase, matching the cyclotron frequency to that of the oscillator at all ion energies. The pole tips are carefully shaped, and correcting fields produced by adjustable currents in pole-face windings called trim coils provide fine-tuning so that every turn takes the same time; the orbits are *isochronous.* However, the radially increasing field tends to make the vertical motions unstable. Field line curvature is reversed from that shown in Fig. 2, so the averaged field index n is negative. The focusing force required to achieve net vertical stability by overriding this vertical defocusing effect is obtained by making the orbits noncircular. This is done by introducing *azimuthal variations* in the magnetic field. Threefold or fourfold symmetry is common on smaller machines, sixfold and eightfold on the largest. In some machines the gradual wave-like variations proposed by Thomas are used, but in others the magnet is divided into *sectors* with field-free spaces between them, in which rf accelerating structures (descendants of the classic dees) are located. In either case, vertical focusing forces result from interaction of the radial component of velocity with the azimuthal component of magnetic field (and from additional interactions for spiraled ridges as in Fig. 4(b) below). The focusing effect is not steady, but varies along an orbit; it becomes a succession of impulses if the field changes at sector edges are steep. Such an impulse, which occurs whenever an ion obliquely crosses a fringing field at a magnet edge in a magnetic spectrometer or other device, is known as *edge focusing.*

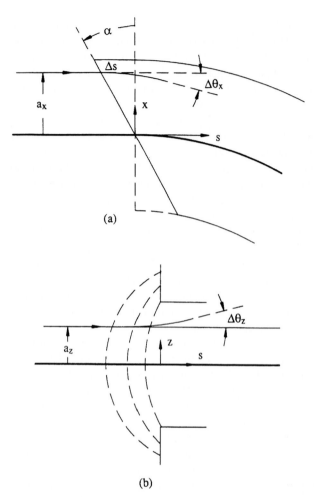

Fig. 3: Orbits experiencing edge focusing from non-normal magnet entry:
(a) in bending plane, (b) in vertical plane.

To see how edge focusing operates, consider particles of identical charge Qe and momentum p entering a magnetic field region at an angle α from the normal to its boundary, as shown in Fig. 3(a). An ion entering with displacement a_x, will encounter "excess" magnetic field (compared with normal entry) along the path Δs, producing excess deflection $\Delta\theta_x < 0$. This deflection represents the focusing effect of a thin converging lens. Similarly, a particle entering above the plane of bend with displacement a_z as shown in Fig. 3(b), will experience excess deflection $\Delta\theta > 0$ that represents the defocusing effect of a thin diverging lens. These focal effects also occur at non-normal exits. In sector-focused cyclotrons the angles α are on average negative, providing the requisite net vertical focusing.

The first experimental tests of this concept at Berkeley used electron model cyclotrons (1949–1952) to simulate a 150-MeV proton cyclotron. "Thomas shims" to improve the fo-

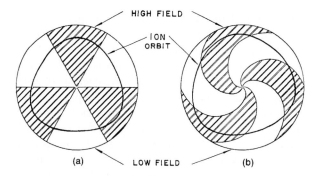

Fig. 4: Sectors of high and low fields; (a) radial, (b) spiral.

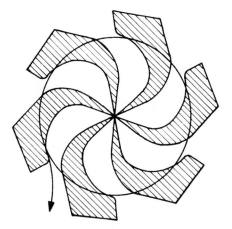

Fig. 5: Separated sectors of TRIUMF H⁻ cyclotron (Canada).

cusing were first inserted into a cyclotron at Los Alamos, New Mexico (~1956). The first isochronous proton cyclotron (12 MeV) was completed at Delft, Netherlands, in 1958. At many laboratories in North America, Europe, the U.S.S.R., and Japan new ideas were explored and programs were started to improve existing machines and to build new ones. In the early 1950s the designs were based on sectors or ridges whose center lines were straight along radii, as proposed by Thomas [Fig. 4(a)]. In the mid-1950s studies at the Midwestern Universities Research Association (MURA) in Michigan and Wisconsin, aimed at developing new types of high-energy accelerators, resulted in the development of spiral-ridge field geometries. It was found that this concept was applicable to cyclotrons, and that by spiraling the ridges or sectors (with larger spiral angle at larger radius) the focusing strength could be greatly increased and higher energies attained [Fig. 4(b)]. (*see* Synchrotron, where alternating-gradient focusing is described.) The advent of powerful digital computers greatly improved the accurate design of the required magnetic field configurations.

The highest proton energies have been attained by the PSI (Zürich, Switzerland, 590 MeV) and TRIUMF (Vancouver, Canada, 520 MeV) machines (Fig. 5). These were first called meson factories because of their copious pion production by high-current (100–200-μA) proton beams. They are now also used as intense spallation neutron sources or for production of radioactive ion beams. The Canadian machine accelerates H^- ions, requiring a low magnetic field ($< 0.6\,T$) to avoid stripping an electron by the motional ($\mathbf{v} \times \mathbf{B}$) electric field, and therefore a large orbit radius ($\sim 8\,m$), but beam extraction at any energy is easily accomplished by stripping in a thin foil to H^+ ions which come directly out of the magnet. An intensity of $2\,mA$ is now extracted from the Swiss accelerator, therefore corresponding to a $1\,MW$ proton beam.

In addition to such large, special-purpose machines, many other third-generation cyclotrons of smaller size and energy but greater versatility are now operating, and others are under construction or design. Prototypes for some of these are the ORIC (Oak Ridge, Tennessee) and 88-in. (Berkeley) machines. However, since the kinetic energy at extraction is proportional to the square of the product Br, and B is limited to about $2\,T$, the race to higher energies would lead to larger and larger radii and consequently to gigantic magnets. A method to overcome this barrier is the use of superconducting coils producing B values up to 5 teslas, thus reducing the extraction radius to 1 meter or less. Several such cyclotrons are presently in operation, like at NSCL (East Lansing, USA), KVI (Groningen, The Netherlands) and LNS-NFN (Catania, Italy). Concerning acceleration of heavy ions, since the energy is proportional to Q^2/m, another mutation came from the development of ion sources with a high degree of ionization: instead of placing the source at the center of the magnet, the beam is injected through a channel drilled along the axis of symmetry of the poles and driven on the proper first orbit in the median plane by an electrostatic device called *inflector*. This method allows taking full advantage of the properties of various types of external sources, both in terms of beam quality and high charge state Q.

Among the capabilities of sector-focused cyclotrons are variation in extracted beam energy by changing magnetic field and oscillator frequency, injection and extraction of polarized beams, and acceleration of a wide variety of heavy ions. They can be put in cascade (two or three cyclotrons in a row, one injecting into the other), receive input beams from other types of accelerator or used as injectors into beam storage and cooling rings. This makes the AVF cyclotron a perfect tool for research in nuclear and particle physics as well as studies in chemistry, biology, medicine, solidstate physics, and other fields. The further elaboration of unique types of cyclotrons for specialized purposes continues. These include compact, simply controlled cyclotrons producing radioisotopes, either short-lived for medical diagnostics or proton-rich for hundreds of applications, and cyclotrons for proton therapy.

A recent tabulation of presently active cyclotrons shows more than 200 machines in 23 countries. The best collected accounts of these diverse accomplishments, activities, and prospects are the 16 volumes mentioned in the last reference in the Bibliography.

See also: Betatron; Cyclotron Resonance; Synchrotron.

Bibliography

A. A. Kolomensky and A. N. Lebedev, *Theory of Cyclic Accelerators*. Wiley, New York, 1966.

John J. Livingood, *Principles of Cyclic Particle Accelerators*. Van Nostrand, New York, 1961.

M. Stanley Livingston and John P. Blewett, *Particle Accelerators*, McGraw–Hill, New York, 1962.

Edwin M. McMillan, "Particle Accelerators," in *Experimental Nuclear Physics*, Vol. III (E. Segre, ed.). Wiley, New York, 1959.

F. Marti (ed.), *Proceedings of the Sixteenth International Conference on Cyclotrons and their Applications*, East Lansing, MI, USA, 13–17 May 2001. American Institute of Physics, New York, 2001. This volume contains references (p. XIX) to the Proceedings of the 15 preceding international conferences in this series starting in 1959. Like them, it includes comprehensive discussions of all aspects of cyclotron technology and applications in many fields. It also contains "data sheets" with much detailed information on the cyclotrons of the world.

Cyclotron Resonance
C. C. Grimes

Cyclotron resonance denotes the resonant absorption of power from an alternating electromagnetic field by charged particles orbiting about a steady magnetic field. The resonance occurs when the frequency of the alternating field is equal to the orbital frequency of the particles. This phenomenon has been extensively employed to study the properties of electrons and holes in solids, and electrons and ions in plasmas. It forms the basic principle utilized in cyclotron particle accelerators.

When a charged particle moves in a magnetic field, it is acted on by the Lorentz force, which continuously rotates the component of the particle's velocity that is perpendicular to the magnetic field. Consequently, the particle moves along a helical orbit with the axis of the helix parallel to the field. The particle's orbital motion is then periodic, having an angular frequency $\omega_c = qB/m^*c$, which is called the cyclotron frequency. Here q/m^* is the charge-to-mass ratio for the particle, B is the magnetic flux density, and c is the velocity of light. If an electromagnetic field having angular frequency equal to ω_c acts on an orbiting particle in a direction that accelerates it, then the particle continually gains energy from the electromagnetic field. The transfer of energy from an electromagnetic field to charged particles is employed in cyclotrons and leads to the resonant absorption of power observed in cyclotron resonance. For a particle to experience a well-defined resonant interaction, it must spend enough time in an orbit to complete a substantial fraction of a revolution about the magnetic field. This condition is expressed by $\omega_c\tau \geq 1$ where τ is the characteristic time between scattering events. The scattering time is limited in solids by the density of impurities, crystal defects, and lattice vibrations, and in plasmas by the density of ions, neutral atoms, and molecules. Examples of typical scattering times, magnetic fields, and frequencies are given below.

Historically, the first observation of cyclotron resonance was in the ionosphere where it was observed that the absorption of radio waves had a peak at a frequency of 1500 kHz. This absorption peak was identified as due to cyclotron resonance of electrons moving in the earth's magnetic field, which has an average flux density of 0.5 G in the ionosphere. Resonance at such a low frequency is possible because the density of scatterers is very low and the electron scattering time is relatively long ($\tau \approx 10^{-5}$ s). Similarly, in gaseous plasmas it is relatively easy to achieve suitable conditions to study cyclotron resonance of electrons or even ions. Since the charge-to-mass ratio of ions is much smaller than for electrons, ion cyclotron resonance occurs at proportionately lower frequencies for the same magnetic field. Cyclotron resonance of electrons and ions in gaseous plasmas constitutes a useful diagnostic tool and provides a means by which energy can be coupled into a plasma to heat it.

Cyclotron resonance has been widely applied to the study of the energy band structure of solids. Its utility lies in the fact that the shapes of the constant-energy surfaces in momentum space (Fermi surfaces) for holes and electrons can be deduced from the variations of the effective masses m* with magnetic field direction relative to the crystal axes. Typical scattering times in crystals of high purity and perfection cooled to a few degrees Kelvin are $\tau \approx 10^{-11}$ s. In semiconductors m^* is typically $0.1m$ where m is the electron mass. To satisfy the condition $\omega_c \tau \geq 1$, microwave frequencies of 10^{10}–10^{11} Hz are commonly employed in conjunction with magnetic fields of 1–10 kG. In ionic crystals, electrons interact strongly with optical phonons to form polarons, which have frequency-dependent effective masses. Polarons have been studied by cyclotron resonance at frequencies from the microwave region to the infrared region in conjunction with magnetic fields ranging from a few kilogauss to well above 100 kG. Infrared frequencies and high magnetic fields have also been utilized in many recent cyclotron resonance studies on two-dimensional systems such as inversion and accumulation layers at semiconductor–insulator interfaces and electrons in semiconductor heterojunctions and super-lattices.

In semimetals and metals the skin effect restricts high-frequency electromagnetic wave penetration to a very thin layer at the surface of a specimen. To achieve cyclotron resonance, the static magnetic field is applied parallel to the specimen surface, so that orbiting electrons and holes near the surface will repeatedly pass through the skin layer and be accelerated by the electromagnetic field for a brief interval during each orbital period. Cyclotron resonance studies of metals and semimetals have yielded a wealth of information on Fermi surface shapes, the density of states at the Fermi surface, the contribution of electron–phonon interactions to the density of states, and the Fermi velocities, as well as some information on the anisotropy of electron scattering times.

See also: Fermi Surface; Resonance Phenomena.

Bibliography

C. Kittel, *Introduction to Solid State Physics*, 6th ed., p. 196. Wiley, New York, 1986.

B. Lax and J. G. Mavroides, in *Solid State Physics* (F. Seitz and D. Turnbull, eds.), Vol. 11, p. 261. Academic Press, New York, 1960. (I, A)

Deformation of Crystalline Materials

R. Bullough and J. R. Matthews

When a crystalline body is subjected to applied forces it will, in general, undergo a change of shape. If the forces are small this shape deformation can be entirely elastic, in which case the atomic structure also suffers a homogeneous deformation (referred to as a homogeneous lattice deformation) that is identical to the shape deformation; removal of the forces after such a deformation results in the body reverting to its original shape. Such a deformation may be termed a simple elastic deformation. On the other hand, if the applied forces are sufficiently large then the formation and movement of extended lattice defects, such as dislocations and twin lamellae, will occur, which ensures that a lattice-invariant or plastic deformation occurs and contributes to the total shape deformation of the body. Removal of the forces now results in the body only partly recovering its original shape. Such a deformation may be termed an elastic-plastic deformation. A complete survey of the physical processes that contribute to the onset and development of the plastic deformation and its detailed contribution to the total shape deformation in, generally, a polycrystalline body is not feasible in this article; it will therefore suffice merely to highlight some of the more important factors involved and refer the reader to the general references for appropriate amplification.

The onset of observable bulk plastic deformation is defined by the *yield strength* of the material, which is simply the applied stress required to produce some specified plastic deformation, usually a few tenths of a percent. This essentially macroscopic concept of yielding implies the *simultaneous sustained* movement of large numbers of dislocations. A single dislocation will glide under an applied (resolved) shear stress, referred to as the Peierls stress for a pure crystal or as the dislocation flow stress, that is much lower than the bulk yield stress and, in fact, if all the preexisting dislocations were simply driven out of the crystal by the applied stress, the total plastic strain would be quite small for annealed materials. It follows that some mechanism for the continuous generation of dislocations is essential and that the macroscopic yield stress is that stress required to generate the dislocations rather than simply to move individual dislocations. Several dislocation sources have been suggested, of which the

Encyclopedia of Physics, Third Edition. Edited by George L. Trigg and Rita G. Lerner
Copyright © 2005 WILEY-VCH Verlag GmbH & Co. KGaA, Weinheim
ISBN: 3-527-40554-2

Frank–Read source is probably the most important, since it clearly demonstrates how certain arrangements of dislocations can interact under the applied stress to generate, purely geometrically, a large number of dislocations and thereby help to define, in dislocation terms, the observed slip lines often associated with plastic deformation.

However, it is certain that the Frank–Read source is not the only athermal source of dislocations, and it is probably not even the most prevalent. Dislocations can be rather easily emitted from various internal stress concentrations, such as surface imperfections and grain boundary irregularities; the precise atomic mechanisms involved have not, as yet, been clearly identified. Since the free energy per atom spacing along a dislocation is about 7 eV, it is quite impossible for thermal fluctuations to generate dislocations spontaneously. Special configurations of mobile dislocations have been identified with observable macroscopic features of the plastic deformation. For example, when the mobile dislocations, which may be causing slip on two or more intersecting crystallographic planes, are confined to a broad band, such a band is referred to as a *deformation band*. If the dislocation motion is confined to a single set of parallel crystallographic planes (single slip) and moreover occurs in a band approximately normal to the slip planes we have a *kink band*. In polycrystalline materials the strain associated with the region of the body where plastic flow begins will often lead to effective softening adjacent to the region, with the result that the plastic zone propagates as a band through the entire crystal. Such deformation bands are often easily visible and are referred to as *Lu"ders bands*. The plastic deformation of a polycrystalline aggregate involves each grain plastically deforming subject to compatibility with its adjoining grains. Such successful accommodation requires, in general, at least five operative slip systems in each grain to ensure the requisite arbitrary shape changes. Large plastic deformation will often severely distort the grains out of their original shape and tend to rotate the operating slip planes toward the direction of deformation, producing preferred orientations of the grains. In addition to dislocation motion causing the plastic deformation, a further contribution to the lattice-invariant deformation, which is usually conceptually subsumed in the total plastic deformation, can arise by the formation of mechanical twin lamellae. The essential relation between the deformation arising from the dislocation movement and that arising by twinning is perhaps clarified by regarding the twin boundary as a special "surface" dislocation whereby it separates *volumes* of crystal that differ only by a lattice *rotation*, in contrast to the slip dislocation, which is a line that separate regions of different lattice *translation*.

As deformation proceeds, the dislocations multiply and interact with each other. This interaction combined with the overall increase in dislocation density ensures that, in general, the *flow stress* (stress required to maintain plastic deformation) will increase – the material will work-harden. In tensile deformation, however, the load-bearing capacity of the body can reach a maximum, since the effect of the reduction in the cross-sectional area is eventually greater than the increase of flow stress due to work-hardening. At greater loads the specimen will "neck" and finally fracture; the *ductility* of the material is defined by the strain or elongation to fracture.

Although dislocations cannot be formed by thermal fluctuations, temperature will assist their movement. At low temperatures, where the temperature is less than half the absolute melting point, the effect is limited to thermally activated reduction of the flow stress and the enabling of slip on new glide planes, cross slip. At higher temperatures, dislocations can move nonconservatively by a process known as climb involving the absorption and emission of point defects. The freeing of constraints on dislocation motion enables low-energy config-uration and annihilation processes to take place, collectively known as recovery. Under these conditions the deformation will continue to increase with time, i. e., the material creeps. On first application of a load at elevated temperature the deformation rate is initially high but continuously decreases as a dislocation substructure is established. If the temperature is high enough, a dynamic balance is eventually established between the deforming dislocation net-work and its simultaneous recovery, producing a constant creep rate. During this type of creep a stable subgrain structure is frequently observed, with a scale that is inversely proportional to the applied load. Steady-state creep also occurs for small loads and very high temperatures by a purely diffusional process that may involve grain boundaries or dislocations as sources and sinks for the point defects. Under a tensile load, a third stage of creep is observed at large accumulated strains, where the creep rate accelerates. Cavities are generated within the material and failure quickly follows.

Crystalline materials are often used as structural components in the cores of nuclear reac-tors. In such an environment they will be bombarded by neutrons and other radiation which can strongly affect the response of such materials to any applied forces that are present. Thus, for example, fast neutrons can produce copious numbers of displacement defects (interstitials and vacancies) which at temperatures prevailing in such reactor cores are often very mobile. These displaced atoms can then precipitate into dislocation loops which can harden the mate-rial, but under sustained loads displacement damage can produce creep at temperatures well below those expected for conventional creep. The main source of irradiation creep is thought to be the preferred absorption of displaced atoms by dislocations of different orientation, the preference being induced by the applied stress. The resulting dislocation climb may directly produce the creep deformation or the accompanying evolving dislocation substructure may permit dislocation glide. Nuclear fuels undergoing fission damage often exhibit large irradi-ation creep effects. Materials with anisotropic crystal structures show shape changes without load, known as growth, which are brought about by similar processes.

Engineering structural materials are not usually pure metals or even simple alloys but multi-component alloys with complex microstructures. A common means of hardening a metal is to add a solute with an atomic size different from the host crystal. These solutes act as obstacles to dislocations increasing the flow stress. Fine precipitates or finely layered eutectic structures also impede dislocation movement, but often at the expense of ductility. Phase transitions within an alloy system may be used to produce hard structures with internal strains, a classic example being the iron–carbon system that may have a wide range of yield strength and duc-tility according to heat treatment. More recently new classes of alloys have been developed, resistant to creep, for applications at high temperature such as gas turbines. The super-alloys rely on a duplex structure in the alloy with regions of an Ni–Al intermetallic compound al-ternating with the Ni–Fe matrix. The coherence of the interface between the relatively soft matrix and hard second phase gives great strength with acceptable ductility.

A complete microscopic theory of plastic deformation of crystalline solids does not yet exist. There is a reasonable understanding of the mechanics of single dislocations and very simple dislocation configurations, and a mathematically sophisticated continuum theory of plasticity has been developed; there is, however, no formal bridge between the two. To achieve such a connection we require a mathematical framework for the deformation behavior of a complex dislocation network. Some progress using the continuous distribution theory of dislocations with its associated non-Riemannian representation of the imperfect crystal has been made, but the complete bridge is certainly not yet built to permit a full microscopic understanding of deformation.

See also: Anelasticity; Crystal Defects; Rheology.

Bibliography

A. S. Argon, ed., *Physics of Strength and Plasticity*. MIT Press, Cambridge, Mass., 1969. (I,A)

B. A. Bilby, *Prog. Solid Mech.* **1**, 329 (1960). (A)

R. Bullough, "Microplasticity," in *Surface Effects in Crystal Plasticity*, pp. 5–14. NATO Adv. Study Inst., Sept. 1975, Hohegeiss, Germany. (I)

R. Bullough and J. A. Simmons, "On the Deformation of an Imperfect Solid," in *Physics of Strength and Plasticity*, p. 47. MIT Press, Cambridge, Mass., 1969. (A)

A. H. Cottrell, *Dislocations and Plastic Flow in Crystals*. Oxford (Clarendon Press), London and New York, 1953. (E,I)

R. de Wit, *Solid State Phys.* **10**, 249 (1960). (E)

F. C. Frank and W. T. Read, *Phys. Rev.* **79**, 772 (1950). (I)

J. Friedel, *Dislocations*. Addison-Wesley, Reading, Mass., 1964. (I)

R. Hill, *The Mathematical Theory of Plasticity*. Oxford University Press, New York, 1998. (A)

J. P. Hirth and J. Lothe, *Theory of Dislocations*. Krieger, 1992. (I)

U. F. Kocks, A. S. Argon, and M. F. Ashby, *Thermodynamics and Kinetics of Slip*. Pergamon, New York, 1975. (I,A)

K. Kondo, RAAG Memoirs of the Unifying Study of Basic Problems in *Engineering and Physical Sciences by means of Geometry* **1**, 458 (1955). (A)

E. Kroner, *Arch. Rat. Mech. Anal.* **4**, 273 (1960). (A)

N. H. Loretto, ed., Dislocations and properties of real materials, *Proc. Conf. Royal Society*, London, Dec. 1984, pub. Inst. of Metals, 1985. (E,I,A)

F. A. McClintock and A. S. Argon, eds., *Mechanical Behaviour of Materials*. Addison-Wesley, Reading Mass., 1965. (E)

W. T. Read, *Dislocations in Crystals*. McGraw–Hill, New York, 1953. (E)

S. Takeuchi and A. S. Argon, *J. Mat. Sci.* **11**, 1542 (1976). (I)

de Haas–van Alphen Effect

M. S. Dresselhaus and G. Dresselhaus

It was in the year 1930 that an oscillatory magnetic field dependence was first observed in the electrical resistance of bismuth by Shubnikov and de Haas and in the magnetization by de Haas and van Alphen. It was not long before Peierls showed how these effects could be understood in principle. Landau had implicitly predicted oscillatory behavior even before the experimental discovery on the basis of his theory of the quantization of the magnetic energy levels normal to the magnetic field direction. Nevertheless these magneto-oscillatory effects remained somewhat of a scientific curiosity for more than 20 years after their first observation. It was only in the 1950s with the observation of magnetic oscillations in many metals other than bismuth and the advent of improved theoretical understanding that it began to be realized that the de Haas–van Alphen effect provided a powerful tool for understanding the electronic structure of metals. Many of the most sensitive experimental techniques were developed by the pioneering work of Shoenberg, who dominated the experimental developments of this field for several decades. The theoretical breakthrough was Onsager's observation that the extremal Fermi-surface cross-sectional area $_0$ is related to the de Haas–van Alphen period.

During the 1960s this effect was widely exploited as researchers joined the "band wagon," and with ever-improving experimental and theoretical techniques, an immense amount of detailed information was accumulated about the "Fermiology" of individual metals, including magnetic materials, intermetallic compounds, and alloys. During the 1970s the pace slackened, but new areas emerged in the late 1970s and the 1980s with applications to quantum well structures and to heavy fermion systems.

The de Haas–van Alphen effect refers to a very small magnetic field-dependent oscillatory term in the diamagnetic susceptibility χ. This oscillatory term can be observed under suitable conditions at high magnetic fields B. The de Haas–van Alphen phenomenon is important because the period P of the oscillatory effect is proportional to $1/B$, and P can be very accurately related to extremal cross sections of Fermi surfaces. By varying the direction of the externally applied magnetic field, the Fermi surface of each carrier pocket can be mapped out independently. The de Haas–van Alphen effect can also be used to determine other electronic parameters, such as the cyclotron effective mass and the effective g factor for electrons and holes.

Application of the Bohr–Sommerfield quantization condition to the k-space orbit of electrons perpendicular to the magnetic field leads to the relation for the extremal cross-sectional area

$$S_0 = \frac{2\pi e B_n (n+\gamma)}{c\hbar} \tag{1}$$

from which the de Haas–van Alphen period is defined as

$$P = \left| \frac{1}{B_n} - \frac{1}{B_{n+1}} \right| = \frac{2\pi e}{c\hbar S_0}, \tag{2}$$

where B is a resonant magnetic field corresponding to quantum number n. The observed periods in metals normally are between $\sim 10^{-5}$ and $10^{-9} \, \mathrm{G}^{-1}$, corresponding to Fermi surface cross sections ranging from $\sim 10^{13}$ to $10^{17} \, \mathrm{cm}^{-2}$. Magneto-oscillatory effects are often reported in terms of the de Haas–van Alphen frequency $\nu = 1/P$.

The significance of this resonant magnetic field can be understood in terms of the energy levels of an electron in a magnetic field. In the effective mass approximation, the magnetic energy levels are given by

$$E_n(k_H) = \frac{\hbar^2 k_H^2}{2m_H^2} + \frac{\hbar e B}{m_c^* c} \tag{3}$$

in which k_H is the wave vector along the magnetic field **B**, while m_H^* is the effective mass component projected along **B**, and m_c^* is the cyclotron effective mass corresponding to the motion perpendicular to **B**. For a three-dimensional electron system in a magnetic field, each magnetic subband n has an energy minimum at $k_H = 0$ where the density of the states becomes infinite. According to Eq. (3) the energy of each magnetic subband extremum increases linearly with B. Because of the singularity in the density of the states in a magnetic field, a series of resonant responses is observed, each resonance corresponding to the passage of a magnetic energy subband extremum through the Fermi level:

$$E_F = \frac{\hbar e B_n}{m_c^* c} \left(n + \frac{1}{2} \right) \tag{4}$$

so that the B_n specifies the magnetic field where subband n becomes unoccupied (occupied) with increasing (decreasing) B. From this point of view, for every value of B_n, the density of states at the Fermi level becomes infinite, thereby introducing an oscillatory magnetic-field-dependent term in the density of states. Thus, any observable depending on the density of states also exhibits an oscillatory magnetic field dependence with a period $1/B$. We refer to such oscillations in the electrical conductivity as the Shubnikov–de Haas effect, in honor of its discovery in 1930 by Shubnikov and de Haas. In the intervening time, de Haas–van Alphen oscillations have been found in a large number of observables, including sample temperature (called magnetothermal oscillations), the Hall coefficient, the sound attenuation coefficient, the velocity of sound, the thermoelectric power, and the optical reflectivity.

In order that the de Haas–van Alphen effect (and the magneto-oscillatory phenomena related to this effect) be observable, it is necessary for the electronic motion to be dominated by the magnetic field. This is necessary to exploit the singularity in the magnetic-field-dependent density of states at each magnetic subband extremum. In contrast, there are no singularities in the zero-field density of states for a 3D system; in this case, only a single nonresonant threshold is present at the band extremum. For lower-dimensional systems the functional form of the density of states changes, giving increasing emphasis to the subband extrema as the dimensionality decreases to 2 (quantum wells), to 1 (quantum wires) and finally to 0 (quantum dots).

The magnetic field dominates the electronic motion when B is large enough for a carrier to complete an electron orbit before scattering. This condition is expressed mathematically as

$$\omega_c \tau \gg 1 \tag{5}$$

where the cyclotron frequency ω_c is related to B according to

$$\omega_c = \frac{eB}{m_c^* c} \tag{6}$$

and τ is the mean time between collisions. Thus to observe the de Haas–van Alphen effect, it is necessary to use high magnetic fields to maximize ω_c, and to carry out the measurement at low temperatures to minimize the collision probability and thereby maximize τ. From the Lorentz force equation,

$$\dot{\mathbf{p}} = \frac{e}{c} \mathbf{v} \times \mathbf{B} \tag{7}$$

it follows that the electron orbit in real space has the same shape as in reciprocal space but the real-space orbit is changed by a scale factor (c/eH) and rotated by $\pi/2$.

Since the condition $\omega_c \tau \gg 1$ can be satisfied for many materials, the de Haas–van Alphen and related effects can be used to determine Fermi-surface parameters for a large variety of materials. With ever-improving materials, synthesis techniques, and the improving sensitivity of the magneto-oscillatory measurements, these techniques have been applied to increasing numbers of materials, including new classes of materials. Of particular interest to developments in the 1980s has been the application of magneto-oscillatory techniques to study the Fermi surfaces of heavy fermions and of quantum wells and superlattices in semiconducting and metallic systems.

The detailed interpretation of the de Haas–van Alphen data to yield electronic parameters is made on the basis of the Lifshitz–Kosevich theory, which gives the oscillatory component of the electrodynamic potential as

$$
\Omega_{\rm osc} = 2V k_B T \left(\frac{eB}{\hbar c}\right)^{3/2} \left(\frac{\partial^2 S}{\partial k_N^2}\right)_{k_m}^{-1/2}
$$

$$
\times \sum_{j=1}^{\infty} \frac{\exp(-2\pi^2 jk_B T_D/\beta^* B)}{j^{3/2} \sinh(2\pi^2 jk_B T/\beta^* B)} \times \cos\left[\frac{j\hbar e S_m}{eB} - 2\pi j\gamma \pm \frac{\pi}{4}\right] \cos\left[\frac{j\pi g m_c^*}{2m_0}\right],
$$

(8)

in which $k_H = k_m$ defines the location of the extremal Fermi-surface cross-sectional area S_m, β^* is the effective double Bohr magneton $\beta^* = e\hbar/m_c^* c$, and m_c^* is the cyclotron effective mass $m_c^* = \hbar^2(\partial S/2\pi\partial E)_{E_F}$, on the constant energy surface at E_F and T_D is the Dingle temperature related to the width $\hbar/\tau$ of the Landau levels given by $k_B T_D = \hbar/\pi\tau$. Also, g is the spin-splitting factor, m_s^* is the spin effective mass, and m_0 is the free-electron mass. The upper or lower signs in the phase correspond to a maximum or minimum, respectively, in the external cross-sectional area of the Fermi surface. The parameter γ enters the theory from the quantization condition given in Eq. (1). The magnetic susceptibility is obtained from Eq. (8) through differentiation $\chi = -(\partial^2\Omega_{\rm osc}/\partial B^2)_\gamma$. The oscillations in χ arise from the argument of the magnetic-field-dependent cosine function in Eq. (8). Furthermore, the argument of the sinh function gives rise to an exponential temperature dependence of χ, the exponential terms depending also on the parameters m_c^* and T_D. Careful temperature-dependent measurements of χ thus can be analyzed to yield explicit values for m_c and T_D. Values of m_c can also be determined by other techniques, the most important being cyclotron resonance experiments. When comparisons can be made, values of m_c determined by the de Haas–van Alphen effect are in good agreement with m_c values obtained by analysis of cyclotron resonance data. However, the effective mass determinations using the de Haas–van Alphen effect are not nearly as accurate as the accurate measurements of Fermi-surface cross-sectional areas.

Introduction of electron spin into the electronic problem results in a spin splitting of the magnetic subbands. In this case, whenever a magnetic subband extremum associated with either spin orientation (parallel or antiparallel to the applied field), crosses the Fermi level, a resonant response will be achieved. Measurement of the de Haas–van Alphen amplitudes as a function of B also can be used to obtain the effective g factor for conduction electrons and holes at the Fermi energy. From analysis of de Haas–van Alphen line shapes, information can be obtained on relaxation times and relaxation processes in the materials under investigation.

"Many-body effects" in an electron gas have been studied on the alkali metals yielding Landau Fermi-liquid parameters for the spin correlation or interaction function. This determination is possible because in Fermi-liquid theory, the effective g factor as measured by a de Haas–van Alphen experiment differs from the corresponding quantity as determined by a microwave electron-spin resonance measurement.

Since about 1980, particular attention has been given to magneto-oscillatory effects in the two-dimensional electron gas, largely related to the discovery of the quantum Hall effect, which is a 2D manifestation of the quantum oscillatory behavior when the magnetic field is normal to the 2D electron gas. Since the zero-field density of states in two dimensions is a constant independent of energy, a δ-function singularity is found for the density of states in a magnetic field for each magnetic subband for the ideal 2D electron gas. Magneto-oscillatory behavior is usually studied in the Hall resistivity ρ_{xy} (appearing as steps in ρ_{xy}), the plateau following each step being associated with a zero in both the transverse magnetoresistance ρ_{xx} and the transverse magnetoconductivity σ_{xx}; these relations are valid when $\rho_{xy} \gg \rho_{xx}$. Physical realizations of the 2D electron gas are the bound states in semiconductor quantum wells and superlattices and the inversion layer in a MOSFET. In these systems, magneto-oscillatory measurements are used to determine the effective mass of the carriers of the 2D electron gas as the geometry and composition of the quantum wells are varied. Of special interest has been the use of the Shubnikov–de Haas effect to investigate the semiconductor–semimetal transition in InAs/GaSb quantum wells (type II superlattice) as a function of the width of the quantum well.

The de Haas–van Alphen and related magneto-oscillatory effects today represent the most accurate and widely applicable means for the determination of Fermi-surface properties of metals, semimetals, and semiconductors described by a 2D electron gas. Though mature, the field remains active.

See also: Cyclotron Resonance; Diamagnetism and Superconductivity; Fermi Surface; Heavy Fermion Materials.

Bibliography

N. W. Ashcroft and N. D. Mermin, *Solid State Physics*, p. 264. Holt, Rinehart and Winston, New York, 1976. (E)

W. J. de Haas and P. M. van Alphen, *Leiden Commun.* **208a**, **212a** (1930) (of historical interest).

C. Kittel, *Introduction to Solid State Physics*, 6th ed., p. 241. Wiley, New York, 1986. (E)

I. M. Lifshitz and A. M. Kosevich, *Zh. Eksp. Fiz.* **29**, 730 (1955) [English translation: *Sov. Phys. JETP* **2**, 636 (1956)] (a classic paper in the field). (A)

L. Onsager, *Philos. Mag.* **43**, 1006 (1952) (of historical interest).

Proceedings of the 6th International Conference on "Crystal-Field Effects and Heavy Fermion Physics," *J. Magn. Mag. Mater.* **76/77**, 1–42 (1988) (reports on a recent application of de Haas–van Alphen effect in heavy-fermion systems).

D. Shoenberg, *Magnetic Oscillations in Metals*. Cambridge University Press, Cambridge, 1984 (complete book devoted to de Haas–van Alphen effect by one of the pioneers).

D. Shoenberg, *Philos. Trans. R. Soc. London A* **245**, 210 (1952) (a classic paper in the field).

W. Shubnikov and W. J. de Haas, *Leiden Commun.* **207**, 210 (1930) (of historical interest).

Demineralization

M. H. Lietzke

Demineralization is the process of producing water free of dissolved minerals. This may be accomplished either by removing the minerals from the water or by separating the water from the dissolved minerals. The most important methods for demineralizing water on a large scale are distillation and electrodialysis. On a smaller scale ion exchange, hyperfiltration, or the freezing process may be used. Applications of demineralization include the production of potable water from brackish or other mineralized water, such as seawater, and the removal of dissolved salts from water to be used in some industrial processes or in steam boilers, where scaling must be avoided.

The most promising method for producing fresh water from mineralized water on a large scale is distillation. There are several large distillation plants in operation in various parts of the world for recovering fresh water from seawater. Some of these plants utilize heat from electrical power plants and produce over three million gallons ($\sim 10^7$ liters) of fresh water per day. Projected distillation plants with capacities of around 100 million gallons (3.8×10^8 liters) per day would have to be coupled with nuclear reactors to provide the power. On a much smaller scale, solar energy may also be used to provide the heat required for distillation of fresh water from mineralized water. For example, a solar still producing 4000 gallons per day of fresh water from seawater has been constructed on the Greek island of Symi.

Demineralization by electrodialysis is accomplished by passing the stream of water to be demineralized through a cell containing a large number of alternating anion- and cation-selective membranes separated from each other by electrically resistive spacers. The spacers have a central area cut out to permit solution and current to flow between the membranes. At each end of the stack are located electrodes connected to a source of direct current. Two inlet streams and two outlet streams are manifolded to the alternate compartments formed by each pair of membranes and separating gaskets. As flow proceeds through the system the cations in all the cells migrate toward the cathode and the anions migrate toward the anode. In a diluting cell, the cations pass out of the process stream through the cation membrane into the adjacent concentrating cell, while the anions migrate in the opposite direction through the anion-selective membrane into the other adjacent concentrating cell. Because of the high ion selectivity of the membranes, ions are restrained from migrating from the concentrating cells back into the diluting cells. The end result of these ion transfers is that the process stream in the diluting cells is demineralized and the transfer stream in the concentrating cells becomes enriched in electrolyte. Charge balance in the transfer streams is maintained since cations migrate in from one adjacent diluting cell, while anions migrate in from the other. Charge balance at the ends of the stack is maintained by the oxidation–reduction reactions occurring at the electrodes. A number of large-scale electrodialysis plants have been constructed and operated in various parts of the world. Several of these produce over one million gallons (3.8×10^6 liters) of demineralized water per day. Electrodialysis is now considered to be the most economical process available for demineralizing mildly brackish water.

Another promising method for demineralizing brackish water is the process of hyperfiltration or reverse osmosis. The principle of this method can be explained as follows: Suppose we have a membrane through which water can pass but through which salt cannot pass. If pure water is put on one side of the membrane and water containing dissolved minerals on the other side, then water will pass through the membrane from the region of pure water, where its effective concentration is higher, into the water containing the dissolved minerals, where its effective concentration is lower. However, if sufficient pressure is applied to the mineral-ized water, the flow of water through the membrane may be reversed and water forced out of the salt solution. One of the major problems in designing hyperfiltration plants has been in the development of suitable membranes. Cellulose acetate, which has excellent salt rejection properties, has been most widely used for this purpose. A number of experimental units have been built which produce several thousand gallons ($\sim 10^4$ liters) of fresh water per day from brackish water.

In the ion-exchange process for demineralization, metallic cations are removed by a hydro-gen cation exchanger, such as Dowex 50, which is a polystyrene divinyl benzene derivative containing sulfonic acid groups. As the metallic ions are adsorbed an equivalent amount of hy-drogen ions is released. Anions are removed by an anion exchanger, such as Dowex 1, which is a polystyrene divinyl benzene derivative containing quaternary amine groups. In this step an equivalent amount of hydroxyl ions is released. The hydrogen and hydroxyl ions set free in the exchange processes then combine to form water. In practice the water to be demineralized may pass through the cation and anion exchangers sequentially, or may pass through a mixed bed of cation and anion exchangers. Carbon dioxide formed from bicarbonates in the ex-change process may be removed mechanically by an aerator, degasifier, or vacuum deaerator, or chemically by a strongly basic anion exchanger. The ion-exchange process for demineral-ization is usually carried out on a much smaller scale than either electrodialysis or distillation.

The principle of the freezing process is that ice frozen from water containing dissolved salts is essentially free of mineral matter. The ice is skimmed off and melted to produce fresh water. A large-scale freezing plant, constructed and operated in Israel in the 1960s and 1970s, has since been shut down.

Bibliography

K. S. Spiegler, *Principles of Desalination*, 2nd ed. Academic Press, New York, 1980. (E)
F. Helfferich, *Ion Exchange*. Dover, New York, 1995. (I)
S. Hwang and K. Kammermeyer, *Membranes in Separations*. Krieger, 1984. (E)

Diamagnetism and Superconductivity

E. A. Boudreaux

The phenomenon of diamagnetism is a consequence of induction, whereby a substance experi-ences polarization when placed in a magnetic field. In this respect it is similar to the analogous phenomenon of paramagnetism, except that the induced magnetic polarization in diamag-netism is negative and is some hundred to a thousand times smaller than in paramagnetism. Furthermore, diamagnetic substances contain no permanent magnetic dipoles, whereas nor-mal paramagnetic materials do. Also, there is neither a normal temperature dependence nor a magnetic field dependence associated with diamagnetism.

In discussing inductive magnetic effects it is customary to invoke the definition $K = I/H$, where K is the magnetic susceptibility per unit volume of substance, I is the intensity of magnetization of the substance, and H is the strength of the magnetic field. For diamagnetism I is negative and smaller than H, and thus the susceptibility is negative and small (i. e., $\sim 10^{-6}$ emu). The measurement of diamagnetic susceptibilities is attained through the observation of an apparent mass change which the substances exhibit when placed in a sufficiently strong magnetic field. In a diamagnetic material this is observed as a net mass loss resulting from the magnetic force tending to repel the substance from the field.

According to the atomic theory of matter, diamagnetism is a consequence of the orbital motion of electrons, and hence is a universal property of all matter. In fact the majority of substances are observed to be diamagnetic, while, relatively speaking, the net nondiamagnetic behavior of matter is rather rare in nature. Common substances such as wood, plastics, water, sand (SiO_2), nitrogen, CO_2, NaCl, $CaCO_3$, Al_2O_3 etc., and the millions of organic compounds and most biological substances are all diamagnetic. In graphite and certain metals an anomalous large diamagnetism is observed, which is also found to be somewhat temperature and magnetic field dependent. This is called "Landau diamagnetism," after the famous physicist, and results from the large delocalization of the orbital motion of electrons in such systems.

Much of the fundamental theory of diamagnetism and its applications were established by noted scientists such as Langevin, Oxley, Stoner, Pascal, Van Vleck, and others, within the first third of the 20th century. Langevin, for example, is credited with formulating the theory of atomic diamagnetism, while Pascal derived a systematic empirical means for calculating diamagnetic susceptibilities of organic compounds. Van Vleck applied quantum mechanics in extending the theory of diamagnetism from atoms to molecules. Thus he was able to show that in diamagnetic molecules a "temperature-independent paramagnetism" (TIP) must also be inherently present.

The sum total of these effects are outlined in the following flow diagram:

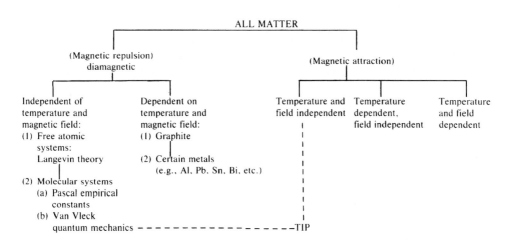

Although the application of static diamagnetic susceptibilities is not currently utilized to any significant extent, diamagnetism still remains an important diagnostic tool to the researcher.

Superdiamagnetism

According to the general theory of superconductivity, a superconductor generates inherently large diamagnetic supercurrents which persist up to the maximum superconducting critical temperature, T_c. At this point, the material enters into a state in which its diamagnetic susceptibility χ_d approaches a maximum value of $-\frac{1}{4}\pi \sim 8 \times 10^{-2}$ emu/g. The magnitude of this diamagnetism (which is some four orders of magnitude larger than of normal non-superconductors) is sufficient to cancel the paramagnetic component of the susceptibility. Hence, the sample is levitated in the magnetic field, which is called the Meissner effect.

The question as to whether superdiamagnetism is possible in non-superconductors was recently addressed theoretically by Ginzburg and co-workers [1]. They showed that for superdiamagnetism to exist in non-superconducting condensed media, there must be a *toroidal moment* **T**, giving rise to a macroscopic electric current **I**. Such a situation is possible in systems capable of being activated to a state of stable *excitonic* behavior; i.e., pairwise movement of an electron and its hole through the crystal lattice upon photoexcitation. Reasonable candidates for such behavior are insulators like CuCl and CdS, for example.

However, prior to addressing superdiamagnetism *per se*, it is essential to consider the fact that in condensed solids there are several contributions to the magnetic susceptibility. If there are no electron spins producing paramagnetism or ferromagnetism, the total magnetic susceptibility, χ_t, contains contributions from five terms, all having orders of magnitude in the 10^{-7} to 10^{-6} emu/g range,

$$\chi_t = \chi_P + \chi_{Ld} + \chi_{LP} + \chi_{Lv} + \chi_{VV} \; . \tag{1}$$

Here, χ_P is the Pauli paramagnetic susceptibility due to the density of states at the Fermi level. It is given by

$$\chi_P = \frac{1}{2} g^2 \mu_B^2 N(0) \tag{2}$$

with g the electron g factor, μ_B the electron Bohr magneton, $e/2m_e c$, and $N(0)$ the density of states at the Fermi level.

χ_{Ld} is the Landau diamagnetic term, also associated with the density of states at the Fermi level,

$$\chi_{Ld} = -\frac{e\mu_B N(0)}{3\pi m^*} \; , \tag{3}$$

where m^* is the electron effective mass in the medium.

χ_{LP} is the Landau–Peierls term derived from a combination of eqs. (2) and (3). This diamagnetic term is the dominating contribution if the valence band is nearly filled or nearly empty. The resulting expression is

$$\chi_{LP} = -\frac{2}{3}\left(\frac{m}{m^*}\right)\mu_B^2 N(0) \; . \tag{4}$$

χ_{Lv} is the regular Langevin diamagnetism due to closed-shell electrons in the atoms. In condensed media, this contribution has been shown to be reasonably well accounted for by the relation [2]

$$\chi_{Lv} = -\frac{\sqrt{3}\mu_B e}{8cR},\tag{5}$$

where e and c are electron charge and speed of light, respectively, and R is the internuclear distance. Of course, eq. (5) will be a sum of terms in cases having variable R.

Finally, χ_{VV} is the van Vleck paramagnetic susceptibility resulting from electron orbital angular momenta interactions between bonded atoms. This is a second-order effect having essentially the same magnitude as χ_{Lv}. An acceptable expression for this term in condensed media is [2]

$$\chi_{VV} = \frac{\sqrt{3}\mu_B e a_c^3}{4.59cR},\tag{6}$$

where a_c is the covalency bonding parameter (from the normalized wave function).

Returning to the question of superdiamagnetism in a non-superconductor, as already stated, if the requirement of a toroidal state producing a macroscopic inhomogeneous electric current is met, the susceptibility is given by the equation [1]

$$\chi_D = \chi_{Ld}\left(\frac{\lambda_l}{R}\right)^2,\tag{7}$$

where χ_{Ld} is given in eq. (3) and λ_l is an effective radius of the current correlations. This latter quantity can be some two orders of magnitude greater than the interatomic distance. Consequently, χ_D may have a magnitude in the range of 10^{-2}–10^{-1} emu/g.

So far, there do not appear to be any further experimental or theoretical data validating the predictions of eq. (7).

See also: Paramagnetism.

References
[1]   V. I. Ginzburg, A. A. Gorbatsevich, Yu. V. Kopayuev and B. A. Valkov, *Sol. State Commun.* **50**, 339–343 (1984).
[2]   W. A. Harrison, *Elementary Electronic Structure*, pp. 158–161. World Scientific, New Jersey, 1999.

Bibliography
L. N. Mulay and E. A. Boudreaux, *Theory and Applications of Molecular Diamagnetism.* Wiley-Interscience, New York, 1976.

L. N. Mulay, "Techniques of Magnetic Susceptibility." in *Physical Methods of Chemistry*, A. Weiss-berger and B. W. Rossiter (eds.). Wiley-Interscience, New York, 1972; Part 1, Vol. IV.

L. N. Mulay, *Magnetic Susceptibility* (reprint monograph). Wiley, New York, 1966.

Ya. G. Dorfman, *Diamagnetism and the Chemical Bond*. Edward Arnold, London. 1965.

P. W. Selwood, *Magnetochemistry*, 2nd ed. Interscience, New York, 1956.

J. H. Van Vleck, *The Theory of Electric and Magnetic Susceptibilities*. Oxford, London, 1932.

E. C. Stoner, *Magnetism and Atomic Structure*. Methuen, London, 1926.

Dielectric Properties
N. Bottka

The term *dielectric* applies to the material properties governing the interaction between matter and an electromagnetic field. Induced or permanent electric polarization or magnetization of matter as a function of a static or an alternating electric, magnetic, or electromagnetic field constitutes the *dielectric properties* of the material.

The macroscopic Maxwell equations describe the response of a system to an external electromagnetic field, or probe, described in space and time by the complex field vectors $\mathbf{E}(\mathbf{r}, t)$ and $\mathbf{H}(\mathbf{r}, t)$. The electric displacement $\mathbf{D}(\mathbf{r}, t)$ and the magnetic flux density $\mathbf{B}(\mathbf{r}, t)$ are material-related parameters and, in general, are not proportional to $\mathbf{E}(t)$ and $\mathbf{H}(t)$ at arbitrary times, i. e., the response of the system depends on the past history of the source $\mathbf{E}(t)$. This causality is expressed by the Fourier transform of the Maxwell relation (in cgs electrostatic units)

$$\mathbf{D}(\mathbf{k}, \omega) = \mathbf{E}(\mathbf{k}, \omega) + 4\pi i \mathbf{J}(\mathbf{k}, \omega)/\omega \tag{1}$$

where $\mathbf{k}$, the wave vector (or propagation direction) of the electromagnetic probe, will have complex components if a wave of angular frequency ω is absorbed by the medium. $\mathbf{J}$ is the current density; it describes all the material properties.

The electromagnetic field can be a time-varying *longitudinal* or *transverse* field; its interaction with a system must conserve both energy and momentum. The electromagnetic wave in optical reflectivity is an example of a transverse probe, coupling directly to the transverse current-density fluctuation of the electrons. Slow neutrons and fast electrons are examples of longitudinal probes, in that they couple directly to elementary excitations in the medium, e. g., density fluctuations in a solid such as phonons and plasmons.

Ideally the external probe is weakly coupled to the material so that the system response can be represented in terms of the properties of the excitations in the absence of the probe. One then speaks of a linear response of the system (as opposed to nonlinear interactions, such as discussed elsewhere, *see* Nonlinear Wave Propagation), and Eq. (1) is redefined as

$$\begin{aligned} \mathbf{D}(\mathbf{k}, \omega) &= \mathbf{E}(k, \omega) + 4\pi i \boldsymbol{\sigma}(\mathbf{k}, \omega) \cdot \mathbf{E}(\mathbf{k}, \omega)/\omega \\ &= \boldsymbol{\varepsilon}(\mathbf{k}, \omega) \cdot \mathbf{E}(\mathbf{k}, \omega) , \end{aligned} \tag{2}$$

where $\boldsymbol{\varepsilon}$ and $\boldsymbol{\sigma}$ are the complex *dielectric function* and *conductivity* of the material, respectively. Analogously one can define a complex magnetic *permeability* $\boldsymbol{\mu}$ relating $\mathbf{B}$ to $\mathbf{H}$. In general $\boldsymbol{\varepsilon}$ is a tensor and represents the response of the system to a perturbing field.

For a transverse probe $\mathbf{E} = \mathbf{E}_0 \exp(i\mathbf{k} \cdot \mathbf{r} - i\omega t)$ in an isotropic nonmagnetic medium, the Maxwell relations for the linear case in Eq. (2) yield the dispersion relation

$$\mathbf{k} \cdot \mathbf{k} = \omega^2 \boldsymbol{\varepsilon}(\mathbf{k}, \omega)/c^2 . \tag{3}$$

When $\boldsymbol{\varepsilon}$ is complex, Eq. (3) can be satisfied only by complex $\mathbf{k}$; all solutions correspond to damped waves in a passive medium. The converse is not necessarily true; i. e., even if $\boldsymbol{\varepsilon}$ is real there is a solution with complex $\mathbf{k}$ corresponding to an evanescent wave (this wave is damped, but there is no dissipation of energy since $\boldsymbol{\varepsilon}$ is real). If $\boldsymbol{\varepsilon}$ and $\mathbf{k}$ are real, Eq. (3) reduces to the *Maxwell relation* $n^2 = \boldsymbol{\varepsilon}$, relating the refractive index n to the dielectric constant $\boldsymbol{\varepsilon}$ of the nonabsorbing medium. For optical frequencies, $\boldsymbol{\varepsilon}$ is usually assumed to be independent of $\mathbf{k}$.

The **k** dependence of **ε** gives rise to such phenomena as the *anomalous skin effect* in metals and the interaction of light with excitons.

Historically, one of the greatest scientific achievements of our times has been the understanding of the correlation between the experimentally observed *macroscopic* quantities **ε**, **μ**, and **σ** and the underlying *microscopic* nature of matter. A class of interactions known as *dipole interactions* contributed substantially to this understanding. For an isotropic atomic system consisting of N elementary dipoles, the macroscopic polarization **P** (electric dipole moment per unit volume) is related to the microscopic average dipole moment $\langle \mathbf{p} \rangle = \langle q\mathbf{r} \rangle$ (in general, represented as a sum of bound electronic, atomic, and permanent dipole moments) by

$$\mathbf{P} = N\langle \mathbf{p} \rangle = N\alpha\mathbf{E}' = (\boldsymbol{\varepsilon} - 1)\mathbf{E}/4\pi = \boldsymbol{\chi}\mathbf{E}/4\pi , \tag{4}$$

where q is the charge and **r** is the displacement from equilibrium of the elementary particle; α is the *polarizability* and $\boldsymbol{\chi}$ is the *susceptibility!magnetic* of the system. $\mathbf{E}'$ is the *local electric-field strength*; in addition to the external field it includes the contribution from induced and permanent dipoles. An expression analogous to Eq. (4) exists for magnetic materials relating the macroscopic magnetization **M** to the magnetic susceptibility $\boldsymbol{\chi}_M$ and the magnetic field.

For a system of noninteracting oscillating dipoles, the classical model for polarizability yields

$$\boldsymbol{\varepsilon} - 1 = 4\pi q^2 \sum_s (\omega_s^2 - \omega^2 + i\gamma_s\omega)^{-1} N_s/M_s , \tag{5}$$

where N_s is the number of particles per unit volume of mass M_s in the system, ω_s is the natural resonant frequency, and γ_s is the damping constant of the dipole oscillator. The sum over s indicates the possibility of various dipoles. Figure 1 shows the behavior of the real and imaginary parts of $\boldsymbol{\varepsilon} - 1$ as a function of frequency in the vicinity of ω_0, for a system of N identical noninteracting oscillators. Strong absorption is indicated by the peak in the imaginary part of **ε**. The frequency dependence of the real part of the dielectric function describes the classical dispersion characteristics of the dielectric medium. Far from ω_0, the real part of **ε** rises with increased frequency; this behavior is called *normal dispersion* in contrast to the *anomalous dispersion* in the halfwidth region of the spectral line where the real part of **ε** falls with increased frequency.

In the case of electronic polarizability where $\omega_0 \simeq 10^{15}\,\mathrm{s}^{-1}$ the strong absorption takes place in the visible region of the spectrum. In the case of ionic polarizability [q and M in Eq. (5) refer to an ion] where $\omega_0 \simeq 10^{13}\,\mathrm{s}^{-1}$ resonant absorption takes place in the infrared region.

Although the dispersion formula of classical physics correctly predicts the general appearance of a spectral line, neither the intensity nor the resonance frequency can be derived from such classical considerations. Moreover, the classical picture leads to a number of unexplained physical catastrophes. The more realistic quantum-mechanical calculation leads to the dispersion formula

$$\boldsymbol{\varepsilon} - 1 = (8\pi\hbar^{-1}) \sum_{f \neq j} \omega_{jf} |\langle \boldsymbol{\mu}_{jf} \rangle|^2 (\omega_{jf}^2 - \omega^2 + i\gamma\omega)^{-1} (N_j - N_f) , \tag{6}$$

where the resonance frequency now corresponds to a transition between two stationary quantum-mechanical energy states E_j and E_f, and $\langle \boldsymbol{\mu}_{jf} \rangle$ is the matrix element of the transition

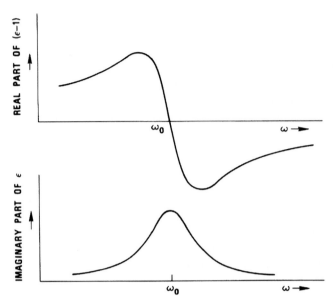

Fig. 1: The real and the imaginary parts of the dielectric function ε as a function of frequency near the resonance frequency ω_0.

coupling the electromagnetic field to the initial and final quantum-mechanical states j and f; N_j and N_f represent the number of atoms in the lower- and the higher-energy state, respectively.

In addition to the bound electronic and atomic polarizability described above, there are cases where free conduction electrons also contribute to absorption, such as in metals and semiconductors. This occurs at low frequencies, anywhere from microwave to infrared. In such a regime the dielectric function takes the form

$$\varepsilon_{\text{free}} = 1 - 4\pi q^2 (N/M)/(\omega^2 - i\gamma\omega) . \tag{7}$$

In conductors and semiconductors, the damping constant γ is inversely proportional to the mobility of the free carrier type (which for intrinsic semiconductors can be both electrons and holes). At these frequencies the optical absorption coefficient is proportional to the imaginary part of ε, and one speaks of a "λ^2" dependence of the free-carrier absorption. For regions of low absorption, the real part of ε in Eq. (7) goes through zero at $\omega_p = (4\pi N q^2/M)^{1/2}$. This is the frequency where the free-electron gas undergoes collective density fluctuations known as *plasma oscillations*. The quanta of energy associated with these oscillations is called the *plasmon*. For most insulators and semiconductors, and many metals, the free-electron model is approximately valid in the frequency range $\omega_v < \omega < \omega_c$ where ω_v is the maximum interband transition of the valence electrons and ω_c is the threshold for interband transitions of the core electrons.

The many excitations discussed above can be put into perspective by considering the absorption spectrum of a hypothetical solid shown in Fig. 2. The strength and the frequency (or photon energy) of the excitation give an insight into the nature of matter and its dielectric properties.

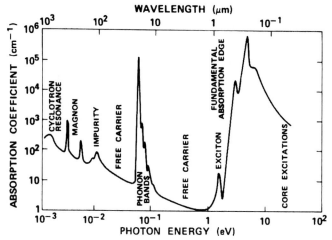

Fig. 2: The absorption spectrum of a hypothetical solid showing the frequency regime where the various excitations give rise to structure in the dielectric function.

The dielectric properties of a material, as defined by the real and imaginary parts of the dielectric function $\boldsymbol{\varepsilon}(\omega)$ over the frequency range $(0,\infty)$ and as exemplified by some of the absorption processes shown in Fig. 2, constitute a unique "fingerprint" of that material. However, despite their great diversity, the basic dielectric properties of solids are rigorously limited by nature. These limitations take the form of *sum rules* which arise from the requirement of causality mentioned before and reflect the physical laws governing the dynamics of interaction between radiation and matter. These sum rules may be viewed as the ω-space equivalent of the dynamic laws of motion in time-space. They give insight into the mathematical structure of the dielectric function, provide a means of relating different physical properties of the solid without model fits to spectra, and most importantly, serve as self-consistency tests. The best-known sum rule in optics (also known as f sum rule) may be written in a variety of forms useful in optical analysis:

$$\int_0^\infty \omega \varepsilon_2(\omega)\, d\omega = \frac{\pi}{2}\omega_p^2 \tag{8}$$

and

$$\int_0^\infty \omega \operatorname{Im}[\varepsilon^{-1}(\omega)]\, d\omega = -\frac{\pi}{2}\omega_p^2 . \tag{9}$$

These sum rules involve an infinite frequency interval $(0,\infty)$ and include all absorptive processes. In practice the absorption spectra is only known over limited frequency intervals, i. e., regions corresponding to a single class of absorption processes such as due to electronic or impurity transitions, In order to treat these individual processes, finite-energy sum rules and dispersion relations have been developed. In order for these finite sum rules to be meaningful, the key requirement is that the absorption in question be isolated and not overlap with other absorption spectra. The absorption singularities can be viewed as taking place in a transparent medium with real dielectric function arising from the dispersion of absorptive processes in all other spectral ranges. Sum rules provide a useful means of testing optical measurements, particularly wide-range composite data both against theoretical and experimental constraints

In summary, the concept of dielectric has been very useful in understanding the physical nature of matter. The numerous dielectric properties of solid or liquid materials have given rise to many useful electronic and/or magnetic devices. Their study and understanding have also given birth to new fields in physics. Recent research on "artificial materials" such as superlattices and photonic crystals pose new questions as to the nature of their "dielectric properties". Photonic crystals, for example, are made by periodically varying the dielectric constant of the material either in one, two, or three dimensions. Periodicity in the dielectric constant gives rise to "forbidden frequencies" called photonic band gap (analogous to a crystal semiconductor band gap where the atomic lattice presents a periodic potential to an electron propagating through it). Patterned dielectric material will block light with wavelength in the photonic band gap, while allowing other wavelengths to pass freely. Such "engineered materials" have great application potential and will continue challenging physicists and engineers.

See also: Dispersion Theory; Excitons; Maxwell's Equations; Phonons; Photonic Crystals; Plasmons; Sum Rules.

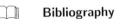

Bibliography

C. J. F. Böttcher, *Theory of Electric Polarization*. Elsevier, Amsterdam, 1952. (I).

K. Busch, S. Lolkes, R. B. Wehrsprohn, and H. Foll (eds.), *Photonic Crystals: Advances in Design, Fabrication, and Characterization*. Wiley-VCH, Weinheim, 2004.

A. R. von Hippel, *Dielectrics and Waves*. Wiley, New York, 1954. (I).

J. N. Hodgson, *Optical Absorption and Dispersion in Solids*. Chapman and Hall, London, 1970. (A).

D. Pines, *Elementary Excitations in Solids*. Benjamin, Advanced Book Program, Reading, Mass. 1963. (A).

D. Y. Smith, "Dispersion Theory, Sum Rules, and their Application to the Analysis of Optical Data," in Edward D. Palik (ed.), *Handbook of Optical Constants of Solids*. Academic Press, New York, 1985.

Diffraction

R. B. Neder

Introduction

The term diffraction refers to the superposition of waves that have passed through a medium or obstacle whose dimensions are comparable to the wavelength. Diffraction occurs independent of the particular radiation and is observed for the whole spectrum of electromagnetic radiation, waves on the surface of a liquid, sound waves, as well as for particle beams like neutrons or electrons, which may be understood as waves. The interaction between the particular wave and matter is very different, yet the geometrical description of diffraction is independent of the nature of the particular wave.

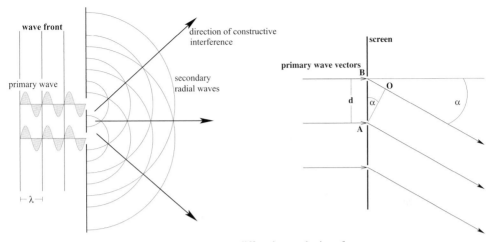

Fig. 1: Diffraction grating. The left image shows diffraction as the interference of secondary waves. Each circle represents the loci of a fixed phase value at a snapshot in time. The right image shows schematically the direction of rays.

The superposition of the waves behind the object is essentially the interference of the waves originating from different parts of the object. For diffraction to occur, there must be a well defined phase-relationship between the incident wave and the diffracted wave. Given such a defined phase-relationship, the diffraction pattern of an object can be determined if the location and nature of all source points within the object is known. A complete treatment of diffraction requires the calculation of the wave propagation within the object, taking into full account changes of the initial and the secondary wave functions. For many cases, however, substantial simplifications can be made, especially when trying to understand the geometrical effects of diffraction rather than the intensity of the diffracted wave.

While many applications exist, prominent applications include the diffraction of visible light by periodic arrangements of obstacles [1, 2], holography [3], and in particular structural studies of matter by x-ray [4], neutron [5, 6] and electron diffraction [7].

An important distinction has to be made in regards to the distance between the object and the observer. In Fraunhofer diffraction, the observer is at a large distance to the object. Under these conditions, the incident wave may be considered to be a plane wave. All wave fronts that originate from different parts of the object are effectively seen by the observer under the same angle to the incident wave vector and may therfore be considered parallel. In Fresnel diffraction, the observer is close to the object and waves originating from different parts of the object are no longer parallel.

Diffraction Grating

The simplest example of diffraction is that of diffraction of a plane wave by a series of parallel equidistant very thin slits in an otherwise opaque screen.

Consider the plane wave hitting the screen as depicted in Fig. 1. According to the Huygens principle, each slit may be considered a source of a new spherical wave. Here we assume the slits to be of infinitesimal width so that they serve as the source of a single spherical wave. At

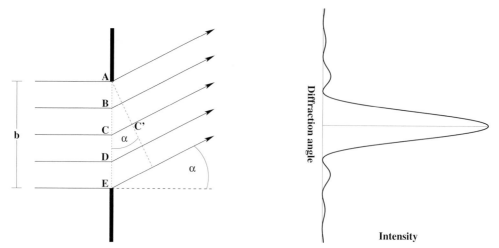

Fig. 2: Diffraction by a single wide slit. *Left:* Direction of rays and the resulting path difference. *Right:* Resulting intensity distribution.

a large distance, all wavelets which originate from the individual slits and are observed under an angle α to the incident wave direction, can be considered parallel. When observing the diffracted wave at different angles α, one finds directions of maximum intensity for values of α, which are defined by the relationship

$$d \sin \alpha = n\lambda \qquad (1)$$

where d is the distance between slits, λ the wavelength and n an integer number. This result can be understood by analyzing the path difference between rays originating from adjacent slits, which is $\overline{BO} = d \sin \alpha$. If this path difference is an integer multiple of the wavelength, constructive interference occurs for all rays and a sharp intensity maximum is observed. If we observe the diffraction pattern at a slightly different angle α, such that for example $d \sin \alpha = 1.1\lambda$, the path difference between two rays from pairs of slits separated by 5 times d will be 1.5λ and these two will interfere destructively, as will all other rays originating from a pair of slits separated by 5 times d, resulting in zero intensity in this direction. Generalizing this to a grating with a very large number of equidistant slits, the corresponding diffraction pattern will consist of sharp lines that fulfil Eq. (1) with zero intensity for all other angles α.

Diffraction by a Single Wide Slit

A single wide slit, as shown in Fig. 2, will also produce a diffraction pattern. This time, each point along the line AE serves as a source of a new spherical wave. The path difference between a ray originating at A and one from the center of the slit at C is $CC' = AC \sin \alpha = b/2 \sin \alpha = n(\lambda/2)$. If this path shift is equal to $\lambda/2$, i.e., $n = \pm 1, \pm 3, \pm 5, \ldots$, destructive interference occurs. Similarly all other pairs of rays originating from points separated by $d/2$ will interfere destructively. Since we have a pairwise extinction of all rays, the net intensity is zero. Correspondingly, pairwise destructive interference will occur for even n from points separated by $b/4$. Thus, a single wide slit will produce a diffraction pattern with intensity minima given by

$$b \sin \alpha = n\lambda . \tag{2}$$

The intensity distribution of the diffraction pattern of a single wide slit can be calculated by dividing the slit into very narrow strips of width δx, each serving as a source of waves of amplitude $d\zeta$. The phase difference of a wavelet originating at a distance x from point A to the wave originating at A is

$$\delta = \frac{2\pi x \sin \alpha}{\lambda} , \tag{3}$$

where x is the distance from A along line AE.

The integration along line AE results in the amplitude A, relative to the amplitude A_0, observed at $\alpha = 0$ as

$$A = A_0 \frac{\sin(\pi b \sin \alpha / \lambda)}{(\pi b \sin \alpha / \lambda)} , \tag{4}$$

and since the intensity is proportional to the square of the amplitude it follows that

$$I = I_0 \left(\frac{\sin(\pi b \sin \alpha / \lambda)}{\pi b \sin \alpha / \lambda} \right)^2 = I_0 \left(\frac{\sin u}{u} \right)^2 \tag{5}$$

with $u = \pi b \sin \alpha / \lambda$.

Diffraction as Fourier Transform

Compare the intensity of the single wide slit from Eq. (5) with the Fourier transform of a box-shaped function,

$$f(x) = \begin{cases} 1 & |x| \leq b/2 \\ 0 & |x| > b/2 \end{cases} , \tag{6}$$

which is

$$F(u) = \frac{\sin \pi b u}{\pi b u} . \tag{7}$$

It is obvious that the intensity of the diffraction pattern of a single wide slit is proportional to the square of the Fourier transform of the slit function. This property of a diffraction pattern is of great help in understanding the relationship between an object and its corresponding diffraction pattern. Several important assumptions have to hold for the intensity to be proportional to the square of the Fourier transformation:

- The incident intensity is constant throughout the object

- A secondary, i. e., diffracted wave, does not interfere with the incident wave.

- The secondary waves do not interfere with each other inside the object.

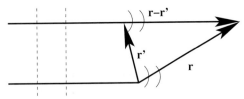

plane incident wave

Fig. 3: Vector relationships in a diffraction experiment.

A more rigorous development of diffraction principles [8] starts from the wave equation

$$\left[\nabla^2 + k_0^2 + \mu \phi(\mathbf{r}) \right] \psi = 0 , \tag{8}$$

where k_0 represents the incident wave number $2\pi/\lambda$, μ the strength of the interaction between the wave and the potential field $\psi(\mathbf{r})$. Using Greens's function $G(\mathbf{r},\mathbf{r}')$, which represents the amplitude at point of observation $\mathbf{r}$, due to a point of unit scattering strength at $\mathbf{r}'$, the wave may be written as

$$\psi(\mathbf{r}) = \psi^0(\mathbf{r}) + \mu \int G(\mathbf{r},\mathbf{r}')\phi(\mathbf{r}')\psi(\mathbf{r}')\, d\mathbf{r}' . \tag{9}$$

An appropriate form of Green's function is

$$G(\mathbf{r},\mathbf{r}') = \frac{e^{(ik|\mathbf{r}-\mathbf{r}'|)}}{4\pi|\mathbf{r}-\mathbf{r}'|} , \tag{10}$$

which is the amplitude of a spherical wave emitted from point $\mathbf{r}'$. Thus Eq. (9) is analogous to the Kirchhoff integral. Each point $\mathbf{r}'$ emits a spherical wave whose amplitude depends on the potential $\phi(\mathbf{r}')$. The diffracted wave $\psi(\mathbf{r})$ in its most general form cannot be directly calculated from Eq. (9), since it forms part of the wave $\psi(\mathbf{r}')$. If the amplitude of the diffracted wave is much smaller than the incident wave amplitude, the wave function $\psi(\mathbf{r}')$ in the integral in Eq. (9) may be replaced by the incident wave $\psi^0(\mathbf{r})$. This is known as the first Born approximation. If we take the usual condition that the point of observation is far from the object, i. e., the Fraunhofer condition, we have $\mathbf{r} = \mathbf{R} = \mathbf{r} - \mathbf{r}'$ (Fig. 3). This will eventually lead to the wave equation

$$\psi^0(\mathbf{r}) + \psi^1(\mathbf{r}) = e^{-ik_0 R} + \frac{\mu}{4\pi} \frac{e^{-ik_0 R}}{R} \int \phi(\mathbf{r}')e^{-i\mathbf{q}\mathbf{r}'}\, d\mathbf{r}' . \tag{11}$$

Thus, the amplitude of the diffracted wave is proportional to the integral in Eq. (11). This integral, however, is the Fourier transform of the potential $\phi(\mathbf{r})$. This shows in a more general form that the first Born approximation implies that the diffracted amplitude under Fraunhofer conditions is proportional to the Fourier transform of the object.

Fortunately, the first Born approximation holds for most applications, including optical gratings as well, as x-ray and neutron diffraction. For electron diffraction [7, 8] and x-ray diffraction by very perfect crystals [9], the first Born approximation is no longer valid, and the intensity has to be calculated through a more involved procedure. The treatment of this dynamical diffraction is beyond this short chapter, see [9] for further reading.

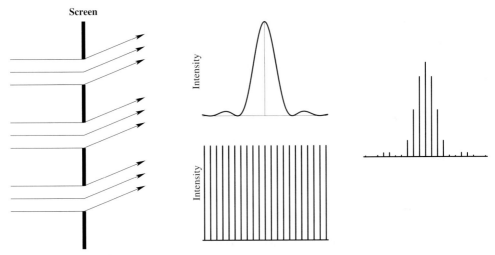

Fig. 4: Diffraction by a periodic wide slits. *Left:* Schematic display of the rays originating from different points in the screen. *Top center:* Intensity of a diffraction pattern of a single wide slit. *Bottom center:* Diffraction pattern of a periodic arrangement of thin slits. These two intensity distributions are multiplied to give the intensity of a diffraction pattern of periodic wide slits (*right*).

Diffraction by Periodic Wide Slits

In Fig. 4 the periodic arrangement of very narrow slits (see Fig. 1) has been replaced by a periodic arrangement of wide slits. The resulting diffraction pattern is the superposition of a diffraction pattern of the periodic very narrow slits with that of a single wide slit. In addition to the phase difference between rays from adjacent slits we also have the phase difference between rays originating from within an individual slit. The resulting diffraction pattern is the product of the diffraction pattern of the periodic thin slits and that of a single wide slit. The explanation follows from the Fourier transformation relationship between an object and its diffraction pattern. The periodic assembly of wide slits can be thought of as the convolution of a single wide slit with a periodic arrangement of delta distributions. The delta distributions are a good description of periodic infinitely narrow slits. Now we will use the convolution theorem to derive the diffraction pattern of the periodic arrangement of wide slits. The convolution theorem states that the Fourier transform of a convolution of two functions g, h is the product of the Fourier transforms G, H of the two individual functions: $F(g \otimes h) = F(g) \cdot F(h) = G \cdot H$. The Fourier transformation of a periodic arrangement of delta distributions is again a periodic arrangement of delta distributions with distance proportional to $1/d$, where d is the distance between the original delta distributions. The Fourier transformation of the single wide slit is function (7). The final diffraction pattern is thus a set of sharp intensity maxima, whose intensity is modulated by Eq. (7). The distance of the maxima is given by the distance between the slits, while the intensity distribution reflects the width of the individual slits.

The convolution theorem also helps to understand the diffraction by other objects. A finite set of very narrow slits can be described as an infinite periodic arrangement of narrow slits that has been multiplied by a box-shaped function, which is zero outside the set of slits.

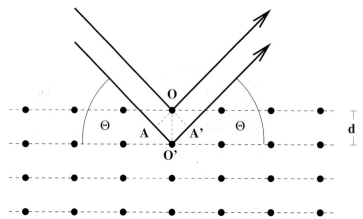

Fig. 5: Diffraction by a single crystal. The points represent the atom positions in a simple crystal structure, the broken lines one set of lattice planes. The path difference between the incident and diffracted wave correspond to the line $A - O' - A'$. The distance between lattice planes is denoted by d.

The corresponding diffracted amplitude is therefore the Fourier transform of a product, which is the convolution of the individual Fourier transforms. The Fourier transform of the periodic arrangement of narrow slits is again the periodic distribution of delta distributions, the Fourier transform of the box shaped function is given by Eq. (7). The convolution of these two functions gives a periodic arrangement of amplitude maxima each of which is shaped like Eq. (7).

These conclusions can be generalized to the following guidelines:

- The existence of regularly spaced maxima in a diffraction pattern indicates a periodic object, their absence indicates an object that lacks periodicity.

- The distance between maxima is inversely proportional to the repeat distance of the object.

- The width of intensity maxima is inversely related to the size of the periodic pattern.

- The intensity modulation of the maxima reflects the Fourier transform of the individual repeat unit. It thus allows a determination of the internal structure of an object.

Diffraction by Crystals

A crystal consists of a three dimensional periodic arrangement of atoms or molecules [4]. Interaction by radiation with wavelengths around 1 Å produces diffraction patterns that relate to the lattice dimensions and the crystal structure.

Figure 5 shows a cross section through a simple crystal structure. The incident radiation falls onto the horizontal lattice plane under an angle of Θ, and is *reflected* under an exit angle Θ. The path difference between radiation *reflected* at point O and point O' follows from the triangles OAO' and $OA'O'$,

$$n\lambda = 2d \sin \Theta, \tag{12}$$

$$\lambda = 2d/n \sin \Theta. \tag{13}$$

where d is the inter-planar spacing. This equation is known as Bragg's equation. Due to the large number of parallel equidistant planes in a crystal, the intensity of a *reflected* wave is limited to a very small angular range. The term *reflection* is commonly used yet misleading. The diffraction by crystals results from the interference of secondary spherical waves emitted by the atoms within the crystal. Calculating the phase difference between different atoms one finds that there is no phase difference for all atoms within one plane. The phase difference depends only on the distance of an atom from a reference plane. Constructive interference occurs only if the primary wave impinges on a plane under the angle Θ as given by Bragg's equation for an integer n and is observed under the identical exit angle Θ. Thus, a ray tracing of the incident wave vector and the diffracted wave vector corresponds to a specular reflection of visible light on a mirrors surface. It is common to use the second form of the Bragg equation, where higher orders of n are instead interpreted as first order diffraction by virtual planes with inter-planar spacing of d/n.

A crystal can be described by convoluting the three-dimensional lattice, i.e., a three-dimensional periodic arrangement of delta distributions, with the content of one unit cell. Consequently, the diffraction pattern of a crystal is the product of the Fourier transforms of the lattice and the unit cell, respectively. The diffraction pattern of a crystal consists of a three dimensional arrangement of sharp intensity maxima, which are commonly called Bragg reflections. The angular distribution of the Bragg reflections reflects the geometrical shape of the unit cell, while their intensities reflect the content of the unit cell, meaning the type and position of the atoms within the unit cell. By interpreting the intensities, the full crystal structure can be determined (*see* Crystallography, X-ray for details).

Although the basic diffraction geometry is similar, x-rays, electrons and neutrons interact differently with matter and thus often yield complementary results. Since x-rays and electrons interact with the electron distribution of a crystal, an x-ray or electron diffraction pattern yields information on the electron distribution including information on chemical bonds. A strong dependence of the interaction to the atomic number exists. Elements with higher atomic number diffract x rays and electrons much more strongly.

Neutrons on the other hand, interact with the nucleus and a neutron diffraction experiment yields the distribution of the nuclei. Compared to x rays and electrons, light elements can be detected much easier by neutron diffraction experiments. The interaction of neutrons with matter is fairly weak. Thus, fairly large crystals can be examined. Another special property of neutron diffraction is the sensitivity to magnetic order within a crystal.

Diffraction by crystals has developed into a a very predominant application. Using the very intense x rays generated by a synchrotron one can even determine the crystal structure of protein or virus crystals which contain many thousands of atoms per unit cell. Also, crystals as small as one micrometer in diameter can be studied. On the other hand fast changes in a crystal may be studied [10, 11]. Deviations from the periodic order produce diffuse scattering located between the Bragg reflections. The interpretation of the diffuse diffraction will in turn yield information on the local structure of a crystal. Diffraction by high energy x rays has picked up many of the neutron diffraction fields. Since the absorption of high-energy x rays is very small, large samples, like engines under running conditions, can be studied. The interaction between x rays and atoms depends on the x-ray wavelength, i. e., the energy. At several energies, the x-ray absorption is enhanced, these are called absorption edges. See the entry X-ray Spectra and X-ray Spectroscopy for further details. Near the absorption edge, the x-ray diffraction strength changes correspondingly. The is used to obtain element specific diffraction data.

Diffraction is not limited to periodic structures. The surface of crystals can be studied by grazing incidence x-ray diffraction. Here the x rays impinge on the surface under an angle of some tenths of a degree. This enhances the signal due to the surface structure compared to that from the bulk crystal. Complementary information is gained from Low-Energy Electron Diffraction (LEED) and Reflection High-Energy Electron Diffraction (RHEED), see the corresponding separate articles in this Encyclopedia.

Glasses, liquids, and gases do not possess a periodic order of the atoms. A corresponding diffraction pattern shows broad maxima only. They provide information on the short range order in the material.

Inelastic Scattering

In most diffraction experiments the wave length of the incident and diffracted waves are identical and this is referred to as elastic diffraction or elastic scattering. This condition holds for the diffraction of visible light by a diffraction grating. The interaction between the incident wave and the object may, however, also involve an energy transfer. In a diffraction experiment with a crystal, for example, the incident wave may excite a phonon or dampen an existing phonon. Each of these situations changes the diffracted wave energy correspondingly. Although the energy transfer is small, it can be observed as a change of wavelength, especially for neutrons with a wavelength around $1\,\text{Å}$, which have an energy of about $80\,\text{meV}$, comparable to phonon energies. Correspondingly, the wavelength shift is a significant fraction of the incident wavelength, which makes neutrons an ideal probe for phonon phenomena in matter. Since x rays with wavelength $1\,\text{Å}$ have an energy of $12.36\,\text{keV}$, an energy change by a few meV causes only a very small wavelength change. To detect this change a very high resolution setup is required, which has now become available at synchrotron sources such as the European Synchrotron Radiation Facility (ESRF) in Grenoble, France.

Fresnel Diffraction

Fresnel diffraction takes place when either a point source of incident waves or the observation plane are at finite distance to the object. The full calculation of Fresnel diffraction is much more involved than Fraunhofer diffraction, but the underlying principle – the interference of secondary waves – remains the same. To calculate the resulting diffraction pattern, the Huygens-Kirchhoff integral has to be calculated. This integral gives the amplitude $u(P)$ at point P due to a wave originating at a point source Q diffracted through an aperture s,

$$u(P) = \frac{-i}{2\lambda} \iint \frac{e^{i(r+r_0)/\lambda}}{rr_0} [\cos(\mathbf{n},\mathbf{r}) - \cos(\mathbf{n},\mathbf{r}_0)] \, ds \,. \tag{14}$$

Here $\mathbf{r}_0$ is the vector from the point source to the aperture s, $\mathbf{r}$ the vector from the aperture to P and $\mathbf{n}$ a vector of unit length normal to the aperture s on the side toward Q.

The calculation can often be simplified by dividing the aperture into zones, whose path lengths $(r + r_0)$ differ by $\lambda/2$. With the exception of the phase factor, all terms in the integral vary slowly. Thus adjacent zones will have almost identical amplitudes, yet opposite phase values and thus tend to cancel each other. The integral in Eq. (14) is thus replaced by a sum which often can be evaluated by inspection.

An outstanding example of Fresnel diffraction is observed in the near region of the shadow of a circular disc. A bright spot, the Arago or Poisson spot is observed along the line of the incident beam and is a convincing proof of the wave nature of light.

Shadow Scattering, Babinet's Principle

Consider two complementary objects such as an aperture in an opaque screen and a small circular disk with diameter identical to the aperture. For diffraction of visible light by these two objects one can set the potential function $\phi(\mathbf{r})$ in Eq. (11) to be 1 inside the aperture and outside the disk, respectively. Outside the aperture and within the disk one can set the potential function to 0. Thus these functions are complementary to each other and we can write

$$\phi_1(\mathbf{r}) = 1 - \phi_2(\mathbf{r}) . \tag{15}$$

Under Fraunhofer conditions, the diffracted amplitude is proportional to the Fourier transforms of each respective function, which are related to each other as

$$\Phi_1(\mathbf{q}) = \delta(\mathbf{q}) - \Phi_2(\mathbf{q}) . \tag{16}$$

The intensity in the diffraction pattern is proportional to $|\Phi_1(\mathbf{q})|^2$, which is identical to $|\Phi_2(\mathbf{q})|^2$ except at the origin. Thus we have the result that the diffraction pattern of these two complementary objects are identical. The disk does not cast a simple shadow but produces a diffraction pattern identical to the aperture. This correspondence is known as Babinet's principle.

Given the brief introduction into diffraction, it has not been possible to describe in detail all the applications of diffraction processes. The reader is referred to the references, as well as to other parts of this volume.

See also: Crystallography, X-Ray; Electron Diffraction; Electron Microscopy; Fourier Transforms; Gratings, Diffraction; Holography; Low-Energy Electron Diffraction (LEED); Microscopy, Optical; Neutron Diffraction and Scattering; Optics, Physical; Polarized Light; Reflection High-Energy Electron Diffraction (RHEED); Waves.

References

[1]   S.A. Akhmanov and S.Yu. Nikitin, *Physical Optics*. Clarendon Press, Oxford, 1997.
[2]   E. Hecht, *Optics*, 3rd ed. Addison-Wesley, Reading, 1998.
[3]   P. Hariharan, *Optical Holography*, 2nd ed. Cambridge University Press, 1996.
[4]   C. Giacovazzo, H. L. Monaco, D. Viterbo, F. Scordari, G. Gilli, G. Zanotti and M. Catti, *Fundamental Crystallography*. Oxford University Press, 2002.
[5]   V.F. Sears, *Neutron Optics*. Oxford University Press, 1989.
[6]   G. E. Bacon, *X-Ray and Neutron Diffraction*. Pergamon, Oxford, New York, 1966.
[7]   D.L. Dorset, *Structural Electron Crystallography*. Plenum Press, 1995.

[8]   J.M. Cowley, *Diffraction Physics*, 2nd ed. North-Holland, Amsterdam, 1981.

[9]   A. Authier, *Dynamical Theory of X-ray Diffraction*. Oxford University Press, 2003.

[10]   J.R. Helliwell and P.M. Rentzepis (eds.), *Time-resolved Diffraction*. Clarendon Press, Oxford, 1997.

[11]   P. Coppens, I. I. Vorontsov, T. Graber, M. Gembicky and A. Y. Kovalevsky, "The structure of short lived excited states of molecular complexes by time-resolved X-ray diffraction", *Acta Cryst. A* **61**, 162–172 (2005).

Diffusion

H. B. Huntington

Diffusion is basically the process by which a concentrated quantity is spread out over a wider extent. As applied to matter in three dimensions it is the homogenization that occurs from random (or nearly random) motion in the microcosmos, apart from convective flow.

The basic quantity in diffusion, D, is called the *diffusivity* or the *diffusion constant* and is defined empirically by Fick's first law

$$\mathbf{F}_i = -D\nabla c_i \tag{1}$$

where c_i and $\mathbf{F}_i$ are, respectively, the concentration and flux of the ith constituent. (The more basic scientific approach is to recognize that the flux is proportional to the gradient of the chemical potential rather than of the concentration.) Fick's second law is a heat-flow-type equation that results from combining the first law with the continuity equation:

$$\frac{\partial c}{\partial t} = \nabla \cdot D\nabla c . \tag{2}$$

It is applied to determine D from measurements on specimens of simple geometry.

Diffusion occurs in all phases of matter [1]. In gases the kinetic theory analysis shows that D varies as the square of the mean free path. In crystalline solids atoms move by jumps of fixed length related to the lattice constant. As a first approximation the random walk analysis gives for D

$$D = \frac{1}{6}\sum \lambda_i^2 \nu_i , \tag{3}$$

where the summation is over various possible mechanisms, λ_i are the lengths of the respective jumps, and ν_i are the respective frequencies of jumping. Liquids and amorphous solids present an intermediate situation that is less easily modeled but does appear to depend strongly on the "free-volume" character of the structure. In crystalline phases the diffusional jump is almost always associated with the presence of some sort of a structural defect, such as a vacancy or an interstitial. The population of these defects therefore enters as an important factor in Eq. (3). The presence of these defects keeps the jumping process from being completely random and introduces a correlation factor, f, usually less than 1, into the expression for D in Eq. (3).

The study of diffusion in solids [2–11] has both technological and scientific importance. In the field of materials science the term has become synonymous with atom movements. As such it plays a role in all heat treatments: homogenization, aging, grain coarsening, annealing, precipitation, sintering, segregation, tempering, and the like. From the scientific side the study of diffusion provides much basic information on the energies of formation and motion of lattice defects, since diffusion data frequently show an Arrhenius-type dependence on temperature, indicating an activated process for the atom motions. For this reason data on diffusion are almost invariably presented as

$$D(T) = D_0 e^{-Q/RT} , \qquad (4)$$

where Q plays the role of activation energy.

The magnitude of Q is directly dependent on the nature of the transport. It is largest for diffusion through the crystalline bulk, smaller for grain boundary diffusions, and least for surface diffusion. This last is quite difficult to determine unequivocally. The best measurements are perhaps those determined by field ion microscopy (FIM). Depending on the crystalline face the diffusivity may be anisotropic even for cubic crystals for which the bulk diffusivity is always isotropic.

The grain-boundary diffusivity is a field which has been attracting ever-increasing attention of late. A complication in its study is the simultaneous presence of bulk diffusion in nearly every situation. For diffusion from the surface in the usual planar geometry three ranges can be distinguished: (A) close to the surface where bulk diffusion predominates, (B) where material first moves out along the grain boundaries and then spreads laterally by bulk diffusion, and (C) which involves diffusion primarily down the grain boundaries. It is the B range which is of most interest in the study of grain-boundary diffusion [12]. The analysis is complicated and has been attacked under various simplifying assumptions by several investigators, e. g., Fisher [13], Whipple [14], and Suzuoka [15]. While bulk diffusion gives a concentration falling off as $\exp[-(x^2/4DT)]$ with penetration x, the concentration in the B range appears to be fit best by $\exp(-Ax^{6/5})$.

The techniques for measuring the diffusivity are numerous and depend on a wide variety of material properties. The use of radioisotopes followed by sequential sectioning has on the whole proved to be the most direct, precise, and reliable method. There are, however, several interesting techniques, such as internal friction, nuclear magnetic resonance, and Mössbauer studies, that give atom movement data by determining the jump time of the atoms rather than the overall distances that they travel.

Basic information on the mechanisms for atom motion can be obtained by measurements other than temperature dependence. The influence of pressure on diffusion shows the volume of the activated complex. The effect of impurity enhancement on diffusion can rule out certain mechanisms. Comparative studies of the motion of different radioactive isotopes of the same element can be used to help determine the extent to which a diffusing atom moves alone or as a partner in a cooperative process. Comparison between diffusion measurements and those of ionic conductivity in the appropriate materials can be used to determine, through the application of the Nernst–Einstein equation, the correlation coefficient f involved.

See also: Kinetic Theory; Crystal Defects.

References
[1]   W. Jost, *Diffusion in Solids, Liquids and Gases*. Academic, New York, 1960. (I)
[2]   A. S. Novick and J. J. Burton (eds.), *Diffusion in Solids: Recent Developments*. Academic, New York, 1975. (I)
[3]   H. I. Aaronson, ed., *Diffusion* (ASM Seminar for 1972). American Society for Metals, Metals Park, Ohio, 1973. (I)
[4]   L. A. Girifalco, *Atomic Migration in Crystals*. Blaisdell, New York, 1964. (E)
[5]   P. G. Shewman, *Diffusion in Solids*, 2nd ed. Minerals, Metals & Materials Society, 1998. (I)
[6]   David Lazarus, "Diffusion in Metals," in *Solid State Physics* (F. Seitz, D. Turnbull, and H. Ehrenreich, eds.), Vol. 10 (1960); N. L. Peterson, "Diffusion in Metals," Vol. 22 (1968). Academic, New York. (I)
[7]   J. R. Manning, *Diffusion Kinetics for Atoms in Crystals*. Van Nostrand, Princeton, NJ, 1968. (A)
[8]   Y. Adda and J. Philibert, *La Diffusion dans les Solides*. Bibliotheque des Sciences et Techniques Nucléaires, Saclay, 1966. (I)
[9]   C. P. Flynn, *Point Defects and Diffusion*. Oxford University Press, London and New York, 1972. (A)
[10] G. E. Murch and A. S. Nowick, *Diffusion in Crystalline Solids*. Academic Press, New York, 1984. (I)
[11] R. J. Borg and G. J. Dienes, *An Introduction to Solid State Diffusion*. Academic Press, New York, 1988. (E)
[12] See Ref. 8, Chapter 12.
[13] J. C. Fisher, *J. Appl. Phys.* **22**, 74 (1951).
[14] R. T. Whipple, *Phil. Mag.* **45**, 1225 (1954).
[15] T. Suzuoka, *Trans. Jap. Inst. Met.* **2**, 25 (1961).

Dispersion Theory

P. A. Carruthers[†]

The original "dispersion relation" of Kramers [1] and Kronig [2] expresses the real part of the frequency-dependent dielectric constant $n(\omega)$ in terms of the absorption coefficient $\alpha(\omega)$ [the attenuation factor is $\exp(-\alpha x)$] according to (P means principal value)

$$n_\mathrm{r}(\omega) = 1 + \frac{c\mathrm{P}}{\pi} \int_0^\infty \frac{\alpha(\omega')\,\mathrm{d}\omega'}{\omega'^2 - \omega^2} \tag{1}$$

where c is the velocity of light. Alternatively, Eq. (1) is a relation between the real and imaginary parts of the complex index of refraction $n(\omega) = n_\mathrm{r}(\omega) + \mathrm{i}(c/2\omega)\alpha(\omega)$. The structure of Eq. (1) is also a consequence of analyticity in the complex frequency z of the function $n(z)$. This analyticity is in turn related to the physical requirement of causality, i. e., that cause must precede effect. The interesting character of the causal properties of the propagation of light in dispersive media was first clarified by Sommerfeld and Brillouin [3].

[†]deceased

   The simplest example [4] exhibiting the pertinent analyticity properties is the classical damped harmonic oscillator model of a bound electron subject to an external electric field E of frequency ω. The equation of motion is

$$m(\ddot{x} + \gamma\dot{x} + \omega_0^2 x) = -eEe^{-i\omega t} .$$ (2)

Here m and e are, respectively, the electron mass and charge, and γ and ω_0 the oscillator damping constant and frequency. The dielectric constant $\varepsilon(\omega)$ for a medium of N electrons per unit volume is computed from the relation $P = (\varepsilon - 1)E/4\pi$ and $P = -Nex$:

$$\varepsilon(\omega) - 1 = \frac{4\pi nE^2}{m} \frac{1}{\omega_0^2 - \omega^2 - i\gamma\omega} .$$ (3)

The right-hand side of (3) has poles in the lower complex ω plane at $\omega = \frac{1}{2}i\gamma \pm (\omega_0^2 - \frac{1}{4}\gamma^2)^{1/2}$ for $\omega_0 > \gamma/2$ (i. e., "weak" damping). As a consequence of Cauchy's integral theorem we have directly

$$
\begin{aligned}
\mathrm{Re}(\varepsilon(\omega)) &= \frac{P}{\pi} \int_{-\infty}^{\infty} \frac{d\omega' \, \mathrm{Im}\varepsilon(\omega')}{\omega' - \omega} \\
&= \frac{2}{\pi} P \int_0^{\infty} \frac{d\omega' \, \omega' \, \mathrm{Im}\varepsilon(\omega')}{\omega'^2 - \omega^2}
\end{aligned}
$$

since $\mathrm{Re}\varepsilon$ and $\mathrm{Im}\varepsilon$ are even and odd functions of ω. To make connection with (1) we use $\varepsilon(\omega) = n^2(\omega)$ and note that for weak absorption, $n \gg n_1$, we have $\mathrm{Re}(\varepsilon - 1) \cong 2n$ and $\mathrm{Im}\varepsilon \cong 2n_1$. Actually (1) is an exact consequence of (4), as explained by Nussenzweig [5]. In the foregoing model the positivity of the decay constant γ expresses the causal nature of the process. Clearly the first of Eqns. (4) follows [61 from the less specific assumption that there exists a function $\varepsilon(z)$ analytic in the upper half-plane satisfying $\varepsilon(z) \to 1$ as $|z| \to \infty$. The functions $\mathrm{Re}\varepsilon(\omega)$ and $\mathrm{Im}\varepsilon(\omega)$ are to be interpreted as the boundary values as $z = \omega + i\eta$ approaches the real axis from above, e. g., $\mathrm{Re}\varepsilon(\omega) == \lim_{\eta \to 0} \mathrm{Re}\varepsilon(\omega + i\eta)$ with $\eta > 0$. The simplification due to the second equality in (4) depends on the symmetry properties of $\mathrm{Re}\varepsilon$ and $\mathrm{Im}\varepsilon$.

   The required analyticity is indeed a consequence of a rather general causality argument, as now shown. Consider the time dependence of a physical quantity $F(t)$ driven by a source $s(t)$ (e. g., electric or magnetic field, current) with a linear response function $R(t - t')$:

$$F(t) = \int_{-\infty}^{\infty} R(t - t')s(t') \, dt'$$ (4)

Causality implies that the behavior of F at time t depends only on the source values at times $t' \leq t$, i. e.

$$R(t) = 0 , \qquad t < 0 .$$ (5)

In terms of Fourier transforms

$$R(\omega) = \int_{-\infty}^{\infty} dt \, e^{i\omega t} R(t) ,$$ (6)

Eq. (5) simplifies to $F(\omega) = R(\omega)s(\omega)$, in analogy to the classical oscillator example. Since $R(t)$ vanishes for $t < 0$ we can immediately continue (7) into the complex $(z = w_1 + iw_2)$ upper half-plane:

$$R(\omega_1 + i\omega_2) = \int_0^\infty dt \, e^{i\omega_1 t - \omega_2 t} R(t) , \qquad (7)$$

since the integral converges still better than before (convergence assumed!) for $\omega_2 < 0$.

It is now possible to assert that $R(z)$ is analytic in the upper half-plane (Titchmarsh theorem [7]) under various conditions. As a simple case, assume (7) to be absolutely convergent for $\omega = 0$. Then if $R(t)$ is piecewise continuous, (8) is uniformly convergent ($\omega_2 > 0$), which is adequate to establish the analyticity of (8) in the upper half-plane.

If $R(z)$ vanishes as $|z| \to \infty$ in the upper half-plane, we can write

$$R(\omega + i\eta) = \frac{1}{2\pi i} \int_{-\infty}^\infty \frac{d\omega' R(\omega')}{\omega' - \omega} , \qquad (8)$$

which reduces to

$$\operatorname{Re} R(\omega) = \frac{P}{\pi} \int_{-\infty}^\infty \frac{d\omega' \, \operatorname{Im} R(\omega')}{\omega' - \omega} , \qquad (9)$$

$$\operatorname{Im} R(\omega) = -\frac{P}{\pi} \int_{-\infty}^\infty \frac{d\omega' \, \operatorname{Re} F(\omega')}{\omega' - \omega} . \qquad (10)$$

Such relations are known as Hilbert transforms. They are useful not only in physics but also in electrical engineering. When $R(z)$ does not vanish as $|z| \to \infty$, a modification of the foregoing argument is needed. Suppose $R(z)$ becomes a constant for large z (as was the case for the dielectric constant). Then the function $[R(z) - R(z_0)]/(z - z_0)$ is analytic as before but vanishes as $1/z$ for large z. Choosing z_0 to be real leads to

$$\operatorname{Re} R(\omega) = \operatorname{Re} R(\omega_0) + \frac{\omega - \omega_0}{\pi} P \int_{-\infty}^\infty \frac{d\omega' \, \operatorname{Im} R(\omega')}{(\omega' - \omega)(\omega' - \omega_0)} . \qquad (11)$$

Equation (11) is known as a "subtracted" dispersion relation since it results if we formally subtract from (10) the same "equation" evaluated for $\omega = \omega_0$.

The Kramers–Kronig relation may be reexpressed as a connection between the real and imaginary parts of the forward scattering amplitude $f(\omega)$ for light using the Lorentz relation [5]

$$n(\omega) = 1 + \frac{2\pi c^2}{\omega^2} N f(\omega) . \qquad (12)$$

If we use in addition the optical theorem

$$\operatorname{Im} f(\omega) = \frac{\omega \sigma(\omega)}{4\pi c} , \qquad (13)$$

where $\sigma(\omega)$ is the cross section, (1) can be written as

$$\operatorname{Re} f(\omega) = \frac{\omega^2}{2\pi^2 c} P \int_0^\infty \frac{d\omega' \sigma(\omega')}{\omega'^2 - \omega^2} . \qquad (14)$$

Relations similar to (13) exist for the nonrelativistic and relativistic scattering amplitudes of particles, although considerable care has to be given to the question of subtractions. For example, Eq. (13) fails for scattering off *free* electrons, for which $f(0) \neq 0$ but rather $f(0) = -e^2/mc^2$ (see [5, p. 51]). Dispersion relations also exist for partial-wave amplitudes. These matters are discussed in detail by Goldberger and Watson [8].

The connection between causality and the analyticity of the S matrix has been much studied in scattering theory [6, 8, 9]. Thus far we have mentioned only one-variable functions, such as the total forward scattering amplitude and the partial-wave amplitude, which obey certain dispersion relations. However, it has been found possible to establish analytic properties not only in the energy variable but also in the momentum transfer variable. In certain nonrelativistic potential scattering problems the full amplitude can be obtained from the associated dispersion relation supplemented by unitarity and plausible assumptions on asymptotic behavior. Efforts to extend this principle to high-energy physics have not succeeded because of the essentially many-body nature of the problem. Nevertheless dispersion relations constitute one of the principal tools of elementary-particle physics and have led to many significant insights. The use of dispersion relations as a dynamical tool is explained in Refs. [8]–[11].

The expression of the causality condition in nonrelativistic quantum mechanics requires some care since there is no limiting velocity and because it is not possible to construct propagating wave fronts having a sharp edge in time. Various studies of this problem, associated with the names of Schutzer, Tiomno, van Kampen, and Wigner, are discussed in Ref. [5]. In relativistic quantum field theory the condition of local commutativity of the field variables [called $\phi(x)$ where $x \equiv (x, y, z, t)$],

$$[\phi(x), \phi(x')] = 0 , \qquad x - x' \text{ spacelike,} \tag{15}$$

for spacelike separations gives a precise but operationally obscure definition of causality. The "microscopic causality" expressed in (14) plays the same role as (6) in establishing the analytic properties necessary to write dispersion relations for the scattering amplitude. Since the condition (14) is a very precise geometric constraint it could be concluded that any violation of a dispersion relation might be due to the violation of the local commutativity condition characteristic of quantum field theory. Precise tests of (rigorously valid) dispersion relations for forward pion–nucleon scattering have confirmed [12] rather than called into question the general theoretical structure [including (14)] on which the dispersion relations are based.

See also: Absorption Coefficients; *S*-Matrix Theory; Scattering Theory.

References

[1] H. A. Kramers, *Atti Congr. Int. Fis. Como* **2**, 545 (1927).
[2] R. de L. Kronig, *J. Opt. Soc. Am.* **12**, 547 (1926).
[3] L. Brillouin, *Wave Propagation and Group Velocity*. Academic Press, New York, 1960.
[4] L. Rosenfeld, *Theory of Electrons*. North-Holland, Amsterdam, 1951.
[5] H. M. Nussenzweig, *Causality and Dispersion Relations*. Academic Press, New York, 1972.
[6] L. D. Landau and E. M. Lifshitz, *Electrodynamics of Continuous Media*. Butterworth-Heinemann, 1984.

[7]   E. C. Titchmarsh, *Introduction to the Theory of Fourier Integrals*, 3rd ed. Chelsea Publishing, 1986.

[8]   M. L. Goldberger and K. M. Watson, *Collision Theory*. Dover, Mineola, 2004.

[9]   R. J. Eden, P. V. Landshoff, D. I. Olive, and J. C. Polkinghorne, *The Analytic S-Matrix*. Cambridge University Press, London and New York, 2002.

[10] H. Burkhardt, *Dispersion Relation Dynamics*. North-Holland, Amsterdam, 1969.

[11] G. F. Chew, *S-Matrix Theory of Strong Interactions*. Benjamin, New York, 1962.

[12] S. S. Lindenbaum, *Particle Physics Interactions at High Energies*, Chapter 5. Oxford, London and New York, 1973.

Doppler Effect

T. W. Hänsch

A change in the observed frequency of light, sound, and other waves, caused by a relative motion between source and observer is known as the Doppler effect. A familiar example is the change in the pitch of a train whistle as the train approaches and passes. The observed frequency v' is higher than the source frequency v if the distance between source and observer is diminishing, and vice versa.

For sound waves that propagate with a characteristic velocity u relative to a medium (air, water), the Doppler shift depends on the velocities of source and listener relative to this medium. The number of waves per second arriving at the observer can be calculated by simply counting the waves emitted per second by the source, and any change per second in the number of waves "in flight" traveling from source to observer. We obtain

$$v' = v \left(1 - \frac{v_o}{u}\right) \left(1 - \frac{v_s}{u}\right) \tag{1}$$

where v_o and v_s are the velocities of observer and source relative to the medium along the direction from source to observer.

The Doppler effect for light is of particular importance for spectroscopy and astronomy. According to the special theory of relativity, the velocity of light has the same value c in all inertial frames. Consequently, the optical Doppler effect, unlike its acoustical counterpart, depends only on the relative velocity v between source and observer. With the help of the familiar Lorentz transformation we calculate a Doppler-shifted frequency

$$v' = v \frac{1 - \frac{v \cos \phi}{c}}{\left(1 - \frac{v^2}{c^2}\right)^{1/2}} \tag{2}$$

where ϕ is the angle between the relative velocity and the line of sight between source and observer. For small velocities, this result is essentially the same as expected classically for a source at rest and a moving observer: a red shift proportional to the velocity if the observer is moving away from the source, and a blue shift if the observer is approaching (*linear* Doppler effect).

For higher velocities, the denominator $(1 - v^2/c^2)^{1/2}$ in Eq. (2) predicts an additional purely relativistic effect: a red shift independent of the direction of the relative velocity between source and observer. This shift persists even if source or observer is moving in transverse direction relative to the line of sight. Though observations of this *transverse* Doppler effect are difficult, they have been carried out in the laboratory with fast-moving atoms, and they give direct evidence for the relativistic "time dilation" (moving clocks seems to oscillate more slowly).

The linear optical Doppler effect manifests itself much more readily. In astronomy, Doppler shifts of spectral lines are observed which can be used to determine the rotation of planets and stars or to identify binary stars. The red shift of the light from remote galaxies gives proof of the expansion of the universe.

In the laboratory we commonly observe a Doppler *broadening* of spectral lines in gases. The absorption or emission lines appear blurred because atoms or molecules of different velocities contribute with different Doppler shifts. For a gas of temperature T (in kelvins) and mass number M, the relative Doppler width (full width at half maximum) is given by

$$\frac{\Delta v}{v} = 7.16 \times 10^{-7} \left(\frac{T}{M}\right)^{1/2} . \tag{3}$$

For high-resolution spectroscopy and precision measurements it is often important to reduce or eliminate Doppler broadening. This can be accomplished by restricting the range of atomic velocities along the direction of light propagation, as in the transverse observation of a collimated atomic beam. "Doppler-free" spectroscopy of a gas is possible by selectively exciting the absorbing atoms in a narrow velocity interval by a monochromatic saturating laser beam, and by probing the resulting bleaching or "spectral hole burning" with a second, counterpropagating probe laser beam. Resolutions approaching 1 part in 10^{11} have been achieved by this method of saturation spectroscopy or Lamb-dip spectroscopy. Doppler-free spectroscopy without velocity selection is possible by two-photon excitation of a gas with two counterpropagating laser beams, whose linear Doppler shifts cancel. The Mössbauer effect provides a means to eliminate the Doppler shift due to recoil in nuclear spectroscopy.

See also: Hubble Effect; Laser Spectroscopy; Mössbauer Effect; Relativity, Special Theory; Waves.

Bibliography

V. P. Chebotaev and V. S. Letokhov, "Nonlinear Optical Resonances Induced by Laser Radiation," *Prog. Quantum Electron.* **4**(2), 111 (1975). (I–A)

R. W. Ditchburn, *Light.* 1963. (E)

T. P. Gill, *The Doppler Effect.* 1965. (I)

W. C. Michels, "Phase Shifts and the Doppler Effect," *Am. J. Phys.* **24**, 51 (1956). (E–I)

J. J. Snyder and J. L. Hall, "A New Measurement of the Relativistic Doppler Shift," *Lecture Notes in Physics*, Vol. 43 (1975). (I–A)

Dynamic Critical Phenomena
W. I. Goldburg

The equilibrium properties of systems near the critical point are now quite well understood. We know, for example, that the isothermal compressibility of fluids and the magnetic susceptibility of ferromagnets diverge in the same universal fashion as this point is approached. What is more, the critical exponents that characterize this universal behavior can be calculated, starting with the partition function of statistical mechanics and using the well-justified renormalization group approximations.

Here we are concerned, not with the time-independent properties of critical systems, but (a) with fluctuations about the equilibrium state and (b) with the rate at which the equilibrium state is approached when the system is perturbed. A typical perturbation might be a temperature change, which causes the system to evolve from one equilibrium state to another. Of special interest is the case where the initial state is a "disordered" one, for example, a ferromagnetic material in its paramagnetic state, and the final state is the "ordered" ferromagnetic state. For a fluid in the ordered state, both the liquid and vapor phases are generally present. To study the dynamics of such systems, one might monitor the time dependence of the magnetization or, in fluid systems, the radius or number density of the emerging droplets. In magnetic systems, fluids, and alloys, it is especially interesting to probe the evolving system by scattering electromagnetic radiation or neutrons from it, as discussed below.

Relaxation of Fluctuations About the Equilibrium State

Starting from the known interactions between molecules in a system and using the principles of statistical mechanics, one can, in principle, predict the mean square deviation of thermodynamic variables from their equilibrium values, as well as the equilibrium values themselves. However, it is not possible to determine the *rate* at which the fluctuations occur from a knowledge of the interactions alone. Additional assumptions about the dynamical features of the system are required. Near the critical point simple and compelling assumptions about the dynamics can satisfactorily account for the relaxation properties of entire classes of systems. Just as with the equilibrium properties of systems near the critical point, their dynamical properties also display universality and self-similarity.

To illustrate, consider the temporal fluctuations in the density of a fluid $\rho(\mathbf{r},t)$ at the point $\mathbf{r}$ and at the time t. It is actually more useful to focus on spatial Fourier components of this quantity, namely, $\rho(\mathbf{k},t)$. Here $\mathbf{k}$ is the wave number of the fluctuation, its wavelength being $2\pi/k$. Near the critical point, the relaxation rate $\Gamma(k)$ of the long-wavelength fluctuations will go to zero as its critical temperature T_c is approached. Quite generally $\Gamma(k)$ will vary with temperature as $|T - T_c|^\theta$, with the "critical exponent" θ having the same value for entire classes of systems (one-component fluids and fluid mixtures all lie in the same class).

The Dynamics of Phase Separation

As an example, consider a fluid, such as CO_2, in a sealed container, its temperature being sufficiently high and its mean density ρ low enough that the system is in the (disordered) vapor phase. Assume, now, that the fluid is abruptly cooled through its coexistence temperature $T_{cx}(\rho)$, so that droplets or domains of liquid form. If ρ differs appreciably from the critical density (or the quench is very shallow), the developing droplets will occupy a fractionally small volume of the container. This type of phase separation is called nucleation. In the special case of systems prepared at the critical density, a quench will give rise to two phases having almost equal volumes. This type of phase separation is very different from nucleation in that there is no surface-energy barrier associated with the creation of the domains of the newly developing phase. This barrier-free phase separation is called spinodal decomposition.

Both nucleation and spinodal decomposition exhibit universal features near the critical point. For example, in nucleation, the emerging phase takes the form of spherical droplets whose mean radius grows as some characteristic power of the time t after the quench. In spinodal decomposition, the two phases form an interconnected structure that has a worm-like appearance in a microscope. Again the mean domain size increases as some characteristic power of time. What is more, critically quenched systems exhibit self-similarity: a micrograph of the structure taken at late times after the quench will look the same as an early time picture, except that it will appear to have been magnified.

The two phases that appear in nucleation (or spinodal decomposition) do not have identical densities or compositions, and hence will scatter neutrons as well as light (if the system is transparent). One can obtain detailed information about the dynamics of phase separation by measuring the scattering cross section to determine the structure factor $S(k,t)$, which is proportional to the mean square modulus of $\rho(k,t)$. In spinodal decomposition the structure factor is found to exhibit a very interesting type of self-similarity that is being actively studied experimentally and theoretically. Spinodal decomposition is observed in diverse types of systems, including fluids, fluid mixtures, polymers, metallic alloys, glasses, liquid crystals, gels, and colloidal systems.

See also: Fluctuation Phenomena; Phase Transitions; Renormalization Group; Transport Properties.

Bibliography

P. C. Hohenberg and B. I. Halperin, *Rev. Mod. Phys.* **49**, 435 (1977).

W. I. Goldburg, in *Light Scattering Near Phase Transitions* (H. Z. Cummins and L. P. Levanyuk, eds.), p. 53. North-Holland, New York, 1983.

J. D. Gunton, M. San Miguel, and P. S. Sahni, *Phase Transitions and Critical Phenomena*, Vol. 8 (C. Domb and J. L. Lebowitz, eds.). Academic, London, 1983.

Dynamics, Analytical

H. Goldstein and M. Tabor

Analytical dynamics is concerned with the classical motion of mass points (called particles), caused either by the forces of mutual interaction or by forces externally imposed on the system. By the adjective "classical" is meant nonquantum, nonrelativistic. The least number of independent coordinates needed to describe the motion of the system of particles is spoken of as *the number of degrees of freedom* of the system. In practice, the field of analytical dynamics is usually restricted to systems where the number of degrees of freedom is small, either because only a few particles are involved or because of constraints on the system (e. g., when the system is a rigid body). Where many independent particles are present, the motion forms the subject of statistical mechanics.

Many of the techniques of analytical dynamics were originally devised to facilitate solving problems about the motion of material, macroscopic bodies. In present day mathematics and physics (as distinguished from, say, engineering or celestial mechanics) the interest rather is in the formulations themselves of analytical dynamics, particularly as they may serve as models or springboards for the construction of theories in other areas. What will be summarized here, very briefly and omitting proofs, is the skeleton structure of the various formulations of analytical dynamics. The starting point will be the Newtonian form – the most basic and generally applicable – followed successively by formulations of increasing abstraction and, often, of reduced physical content. Running like a thread throughout all the formulations is the search for constants of the motion, and the connection between the symmetry properties of the system and the existence of meaningful constants of motion.

However, in the end we shall see that most Hamiltonians do not possess sufficient constants of motion to render, in a certain sense, exactly soluble equations of motion. In these situations the classical motion can be chaotic.

Newtonian Formulation

If $\mathbf{r}_i$ and $\mathbf{p}_i = m_i \dot{\mathbf{r}}_i$ are the position vector and linear momentum, respectively, of the ith particle, then Newton's equations of motion for the system are

$$\frac{d\mathbf{p}_i}{dt} = \mathbf{F}_i^{(e)} + \sum_{i \neq j} \mathbf{F}_{ji} \, . \tag{1}$$

Here $\mathbf{F}_i^{(e)}$ represents the external force (including constraint forces, if any) acting on the ith particle, and $\mathbf{F}_{ji}$ stands for the interaction force of the jth particle on the ith particle. The equations of motion described by Eq. (1) are of second order and the solutions thus involve two initial conditions for each particle, e. g., of initial position and velocity. First integrals of the equations of motion are constant functions of positions and velocities. Where the first integrals are algebraic expressions of these variables, they are spoken of as *constants of the motion*, or *conserved quantities*. The conditions for the existence of conserved quantities are described in *conservation theorems*.

Depending on the nature of the forces acting on the particles, various conservation theorems can be derived from Newton's equations of motion. For example, if $\mathbf{F}_{ji} = -\mathbf{F}_{ij}$ (the weak form of the law of action and reaction) and the external forces on the particles sum to zero, then the total linear momentum of the system,

$$\mathbf{P} = \sum_i \mathbf{p}_i \, , \tag{2}$$

is a constant of the motion. When the strong form of the law of action and reaction holds, i. e., $\mathbf{F}_{ji} = -\mathbf{F}_{ij}$ and is along the direction of $\mathbf{r}_i - \mathbf{r}_j$, and the net torque of the external forces about the arbitrary origin vanishes, then the total angular momentum of the system about the given origin,

$$\mathbf{L} = \sum_i \mathbf{r}_i \times \mathbf{p}_i \, , \tag{3}$$

is a constant of the motion. Conservation of linear momentum can be shown to hold if the system is symmetric under linear translation, i. e., the conditions operating on the system are unaffected if the system is displaced linearly as if it were a rigid body. Similarly, conservation of angular momentum about a point is equivalent to symmetry of rotation about the point. Another conservation theorem can be obtained if all of the forces acting on the ith particle can be obtained as the negative gradient (with respect to the coordinates of the ith particle) of a scalar V, a function of all the coordinates. Under these conditions the *total energy* of the system

$$E = T + V \tag{4}$$

is a constant of the motion, where T is the kinetic energy,

$$T = \frac{1}{2} \sum_i m_i (\dot{\mathbf{r}}_i)^2 \, , \tag{5}$$

and V is known as the *potential energy*.

Lagrangian Formulation

In the presence of constraints, the number of degrees of freedom of a system of k particles is less than the $3k$ position coordinates. If constraints can be expressed by the equations of the form

$$f_i(\mathbf{r}_1, \ldots \mathbf{r}_k, t) = 0 \, , \quad i = 1, \ldots 3k - n \, , \tag{6}$$

then the motion of the system can be entirely described by n independent coordinates, which may either be an appropriate subset of the $\mathbf{r}_i$, or new, generalized, coordinates defined by equations of transformation

$$q_j = q_j(\vec{r}_1, \ldots \mathbf{r}_k, t) \, , \quad j = 1, \ldots, n \, . \tag{7}$$

When the constraint equations are of the form of Eq. (6), the constraints (and the system) are said to be *holonomic*. The Lagrangian formulation has its simplest and most useful expression when the mechanical systems are holonomic, although partial extensions to some types of nonholonomic constraints are possible. The constraints defining a rigid body are simple examples of holonomic constraints, and reduce the number of degrees of freedom of a rigid body, no matter how many particles make up the body, to no more than six.

In the Lagrangian formulation it is also assumed that the forces which arise by virtue of the constraint do no work in an infinitesimal displacement of the system coordinates at a given moment (a *virtual displacement*). Almost all constraints in the absence of friction or other dissipative phenomena are workless in this sense, as is also the constraint described by "rolling" even though it requires (rolling) friction. Under these conditions the equations of motion of the system reduce to n independent equations of the form

$$\frac{\mathrm{d}}{\mathrm{d}t}\frac{\partial T}{\partial \dot{q}_i} - \frac{\partial T}{\partial q_i} = Q_i \equiv \sum_{j=1}^{k} \mathbf{F}_j \cdot \frac{\partial \mathbf{r}}{\partial q_i}, \quad i = 1,\ldots,n, \tag{8}$$

where the Q_i are called *generalized forces*, evaluated in terms of all of the forces on the particles except the constraint forces. When the Q_i can be obtained from a generalized potential function $U(q,\dot{q},t)$ according to the prescription

$$Q_i = -\frac{\partial U}{\partial q_i} + \frac{\mathrm{d}}{\mathrm{d}t}\frac{\partial U}{\partial \dot{q}_i}, \tag{9}$$

the equations of motion clearly simplify to

$$\frac{\mathrm{d}}{\mathrm{d}t}\frac{\partial L}{\partial \dot{q}_i} - \frac{\partial L}{\partial q_i} = 0, \quad i = 1,\ldots,n, \tag{10}$$

where L, the Lagrangian function is

$$L = T - U. \tag{11}$$

Most classical interaction forces on a microscopic level, including electromagnetic forces, are derivable from a generalized potential in the form given by Eq. (9). Equations (10) are customarily called the Lagrangian equations of motion, although Eqns. (8) are often also so designated.

The Lagrangian formulation, as summarized by Eqns. (10), is thus restricted to holonomic systems where the constraints are workless, and where the forces are derivable from a generalized potential. Under such assumptions, an alternative expression of the Lagrangian formulation can be given as a variational principle. The generalized coordinates q_i are considered as the axes in an n-dimensional *configuration space*, in which any given positional configuration of the system corresponds to a point. As the system evolves in time the system point describes a trajectory in configuration space. Between any two fixed points 1 and 2 in configuration space the equations of motion of the system determine a trajectory, and the manner in which the system point traverses it, in such a fashion that

$$\delta \int_1^2 L(q,\dot{q},t)\,\mathrm{d}t = 0; \tag{12}$$

that is to say, the value of the integral in Eq. (12) (known as the action integral) evaluated on the actual trajectory has an extremum value relative to all neighboring trajectories with the same end points. Equation (12), Hamilton's principle, is a variational principle for which the Euler–Lagrange variational equations are exactly the Lagrange equations of motion of the system, Eqns. (10).

The Lagrangians satisfying Hamilton's principle, and therefore determining the equations of motion, are not uniquely given by the recipe of Eq. (11). Indeed it is clear that if L in any Lagrangian function for which Eq. (12) holds, then an equally suitable Lagrangian is

$$L' = L + \frac{\mathrm{d}F(q,t)}{\mathrm{d}t}, \tag{13}$$

where F is any (twice differentiable) function of the q's and t. In addition, valid Lagrangians can be found that are not related in this simple manner. Thus, for a single particle moving in one dimension under the influence of a potential V not explicitly dependent on time, then Eq. (11) prescribes the Lagrangian

$$L = \frac{1}{2}m\dot{q}^2 - V \ . \tag{14}$$

As may be easily verified, however, the Lagrangian

$$L = \frac{1}{12}m^2\dot{q}^4 + m\dot{q}^2V - V^2 \tag{15}$$

will lead to the same equations of motion, and, consequently, also satisfies Hamilton's principle. The statement of the Lagrangian formulation of mechanics via a variational principle thus leads naturally to extension of the concept of a Lagrangian beyond its initial prescription.

Conservation theorems are easily established in the Lagrangian formulation. If a given coordinate q_i does not appear in L (a so-called *cyclic* and *ignorable* coordinate), then Eqns. (10) show that there exists a first-order constant of the motion,

$$p_i \equiv \frac{\partial L}{\partial \dot{q}_i} \ . \tag{16}$$

The physical nature of the quantities p_i (known as *generalized* or *conjugate momenta*) depends on what sort of coordinates the q_i are. If q_i is a Cartesian, or translation, coordinate, it can be shown that p_i has the dimensions of a linear momentum; if q_i measures a rotation, p_i is the corresponding angular momentum, etc. Very often, p_i will have no interpretation as a mechanical momentum. The two momentum conservation theorems of Newtonian mechanics can be recovered if q_i is a coordinate measuring either the rigid displacement or rigid rotation, respectively, of the system as a whole.

An additional conserved quantity can be found if L does not contain time explicitly. It can then be shown that a first-order constant of the motion h exists, where

$$h(q,\dot{q}) = \sum_i \dot{q}_i \frac{\partial L}{\partial \dot{q}_i} - L \ . \tag{17}$$

It very frequently happens that the Lagrangian prescribed by Eq. (11), or otherwise suitably determined, can be written as the sum of three functions, one of which is independent of $\dot{q}_i$, and the other two are homogeneous functions of $\dot{q}_i$ of first and second order, respectively:

$$L = L_0 + L_1 + L_2 \ . \tag{18}$$

While it is not universally required that the Lagrangian conform to this pattern, many of the further developments in classical mechanics are constructed with a view to Lagrangians of the form of Eq. (18). The condition that Eq. (18) holds at least serves to reduce possible ambiguity in the form of the Lagrangian, *vide* Eqns. (14) and (15). If L is of the type described by Eq. (18), application of Euler's theorem on homogeneous functions shows that h, the so-called *Jacobi integral*, is given by

$$h = L_2 - L_0 \ . \tag{19}$$

When further, the Lagrangian is constructed according to Eq. (11) *and* the potential is V independent of $\dot{q}_i$ *and* the defining Eqns. (7) for q_i are not explicit functions of the time, then $L_2 = T$ and $L_0 = -V$, and h is the total energy of the system. Conservation of h then is equivalent to conservation of energy.

To solve problems in analytical dynamics appearing in most engineering situations it is rarely necessary to go beyond the Lagrangian formulation. It has the advantage that the existence of holonomic (and certain types of nonholonomic) constraints can be made visible in the formulation and the forces of constraint obtained by the method of Lagrange multipliers. Finally, extensions can be made to systems with some forms of dissipative forces, through a so-called Rayleigh dissipation function.

Hamiltonian Formulation

The Hamiltonian formulation is intended for the same restricted class of mechanical systems considered in the Lagrangian formulation, but starts from a radically different viewpoint. It seeks to describe the motion of a holonomic system of n degrees of freedom by $2n$ independent coordinates. Consequently, it must lead to $2n$ first-order equations of motion, which prescribe the trajectory of a system point in the $2n$-dimensional space of the independent variables, known as phase space. The $2n$ quantities are divided into two sets, n coordinates q_i and n variables p_i, the conjugate momenta. A function $H(q_i, p_i, t)$, the Hamiltonian, is constructed such that the equations of motion are given in terms of the two sets, as,

$$\dot{q}_i = \frac{\partial H}{\partial p_i} , \dot{p}_i = -\frac{\partial H}{\partial q_i} , \quad i = 1, 2, \ldots, n ; \tag{20}$$

Eqns. (20) are clearly first-order differential equations in time. The procedures for the initial selection of the $2n$ variables and the construction of a suitable H are not in principle part of the Hamiltonian formulation. It is usual, however, to start on the basis of some particular generalized coordinates and a corresponding Lagrangian. The conjugate momenta, defined by Eqns. (16), are then selected to play the role of the other half of the independent Hamiltonian coordinates. It can then be shown that Hamiltonian's equations, Eqns. (20), describe the motion if the Hamiltonian is obtained as

$$H(q, p, t) = \sum_i p_i \dot{q}_i - L . \tag{21}$$

In carrying out the prescription implied in Eq. (21) the generalized velocities $\dot{q}_i$ are expressed as functions of q and p by inverting Eqns. (16) so that H is a function only of the Hamiltonian variables and (possibly) the time.

Comparison of Eqns. (17) and (21) shows that h and H have the same values and magnitudes. Functionally (aside from time dependence), however, they are different, for h is a function of the n q_i's and their time derivatives while H is a function of the $2n$ independent variables q_i, p_i. Physically H can be identified as the total energy of the system under the same conditions specified above for h.

Although the Lagrangian and Hamiltonian descriptions of a mechanical system have the same *physical* content their mathematical structure is profoundly different. Velocities transform as *contravariant* variables and the combined set of generalized coordinates and velocities in the Lagrangian description form a $2n$-dimensional manifold known as a *tangent bundle*

(TM). The Lagrangian thus provides the mapping from the tangent bundle space to a scalar field, namely, $L : \mathrm{TM} \rightarrow R$. By contrast the generalized momenta transform as *covariant* variables and the Hamiltonian phase space of generalized coordinates and momenta is a $2n$-dimensional *symplectic manifold* with a very different geometric structure from the tangent bundle space of the Lagrangian description.

It is important to keep in mind the distinction between the functional bases of the Lagrangian and Hamiltonian formulations, a distinction that is half-hidden by the customary manner of constructing the Hamiltonian from the Lagrangian. In the Lagrangian formulation the conjugate momenta p_i are *defined* by Eqns. (16) as specific functions of q, $\dot{q}$ (and t). As part of the Hamiltonian formulation they are independent variables, whose relationships with q and $\dot{q}$ are *consequences* of the $\dot{q}_i$ half of the equations of motion.

The prescription of Eq. (21) for constructing the Hamiltonian on the basis of the Lagrangian formulation does not always work. If the velocity-dependent part of the Lagrangian is entirely of type L_1, namely a homogeneous function of the $\dot{q}$'s in the first degree, then Eqns. (16) cannot be inverted to find the $\dot{q}$'s as functions of the p's. [For the limiting case in which the $\dot{q}$'s appear only linearly in L, Eqns. (16) do not contain $\dot{q}$ at all!] In effect, then, not all the p's are independent and some of Eqns. (16) act as constraint equations. What one does in this case is not yet clear or universally agreed to. Occasionally (as in relativistic analytical dynamics) one can avoid the problem by seeking an alternative, nonpathologic, Lagrangian which still gives the correct equations of motion.

Hamilton's equation of motion can also be derived from a modified Hamilton's variational principle that says that the equations of motion describe the trajectory in phase space, between fixed end points, which makes the integral

$$\int_1^2 \left(H(q,p,t) - \sum_i p_i \dot{q}_i \right) dt \tag{22}$$

an extremum. If the integrand in Eq. (22) is considered as a function of $q, \dot{q}, p, \dot{p}$, then the corresponding Euler–Lagrange variational equations are exactly Hamilton's equations of motion. The modified Hamilton's principle contains no physical limitations on the system beyond those applying to Hamilton's principle in the Lagrangian formulation.

Conservation theorems in the Hamiltonian formulation flow directly from the equations of motion. Clearly, from Eqns. (20) the momentum variable p_i is conserved if the conjugate coordinate q_i does not appear in the Hamiltonian. It can be shown easily that a coordinate cyclic in L will also be cyclic in the Hamiltonian constructed from it. From Eqns. (20) it also follows that

$$\frac{dH}{dt} = \sum_i \left(\frac{\partial H}{\partial q_i} \dot{q}_i + \frac{\partial H}{\partial p_i} \dot{p}_i \right) + \frac{\partial H}{\partial t} = \frac{\partial H}{\partial t} . \tag{23}$$

Hence H will be conserved if it is not an explicit function of time. Frequently H is physically the total energy of the system and Eq. (23) then implies a conservation theorem for energy. But it can happen that Eqns. (7) defining q are such that H is conserved when the energy is not conserved, and vice versa, or both H and E can be conserved but may be different constants of the motion.

Canonical Transformation Theory

In the Lagrangian formulation, any set of independent generalized coordinates are suitable variables. The Lagrangian has the same significance, $T - U$, in all such sets. In other words, all transformations of configuration space lead to another configuration space with a Lagrangian of the same magnitude, and with the same *form* of the equation of motion. Invariance of this kind does not exist for the Hamiltonian phase space. An arbitrary transformation $(q, p) \rightarrow (Q, P)$ will not necessarily lead to variables satisfying equations of motion of the Hamiltonian type. Conditions for transformations leading to Hamilton's equations of motion in the new variables can be expressed in terms of the modified Hamilton's principle. If the transformed set of variables (Q, P) obey equations of motion with a Hamiltonian $K(Q, P, t)$, then in the transformed space the trajectory must make the integral

$$\int_1^2 \left(K - \sum_i P_i \dot{Q}_i \right) dt \tag{24}$$

an extremum. Transformations for which the integrands of (22) and (24), at equivalent points, differ by only the time derivative of an arbitrary point function F are known as canonical transformations. It can be shown that if a canonical transformation $(q, p) \rightarrow (Q, P)$ does not contain the time explicitly, then $K = H$; otherwise the original and transformed Hamiltonians are related by

$$K = H + \frac{\partial F}{\partial t} . \tag{25}$$

When F is suitably expressed in terms of the old and the new set of variables, it is known as the generating function for the canonical transformation; the equations of transformation can be expressed in terms of the partial derivatives of F.

Necessary and sufficient conditions for a canonical transformation can be expressed in a variety of forms; one of the most useful is in terms of *Poisson brackets*. If u and v are two functions of the canonical variables, then the Poisson bracket of u, v with respect to the set (q, p) is the bilinear differential form

$$[u, v]_{q,p} = \sum_i \left(\frac{\partial u}{\partial q_i} \frac{\partial v}{\partial p_i} - \frac{\partial v}{\partial q_i} \frac{\partial u}{\partial p_i} \right) . \tag{26}$$

A transformation $(q, p) \rightarrow (Q, P)$ is canonical if

$$[Q_i, Q_j]_{q,p} = 0 \quad = \quad [P_i, P_j]_{q,p} \tag{27}$$
$$[Q_i, P_j]_{q,p} \quad = \quad \delta_{ij}$$

The Poisson brackets appearing in Eq. (27) are called the fundamental Poisson brackets. In words Eq. (27) says that the principal Poisson brackets are invariant under a canonical transformation. It then follows, rather easily, that *all* Poisson brackets are invariant under canonical transformations. Poisson brackets are the classical analogs of the quantum-mechanical commutator, and they share similar algebraic and physical properties.

A more geometric definition of canonical transformations is provided by the use of differential forms. A transformation is canonical if the differential 2-form is preserved under the
change of variables, namely,

$$\sum_{i=1}^{n} \mathrm{d}p_i \wedge q_i = \sum_{i=1}^{n} \mathrm{d}P_i \wedge \mathrm{d}Q_i \, ,$$

where $\wedge$ denotes the so-called wedge product.

The solution of the mechanical problem can be looked on as a particular canonical transformation. If the new Hamiltonian is identically zero, the transformed variables are all constants
in time. By Eq. (25) such a transformation must necessarily be time dependent. The initial
values (q_0, p_0) of the canonical set (q, p) form a suitable set of canonical variables. Hence the
transformation $(q_0, p_0) \leftrightarrow (q, p)$ is a canonical time-dependent transformation, for which the
equations of transformation give (q, p) as functions of time and the initial values, i. e., the desired solution. In principle, at least, solving the mechanical problem can be reduced to finding
the necessary canonical transformation for which $K \equiv 0$. Note that in this view the canonical
transformation transforms one system point in phase space (the initial point) to another point
in the same space (the system point at time t). This *active* interpretation of a canonical transformation is to be distinguished from the *passive* view in which the transformation relates the
coordinates of a given system point in the one phase space to the coordinates in another phase
space.

Poisson Bracket Formulation

From the definition of the Poisson bracket, Eq. (26), and Hamilton's equations, Eqns. (20), it
follows that the time derivative of any general function $u(q, p, t)$ is

$$\frac{\mathrm{d}u}{\mathrm{d}t} = [u, H] + \frac{\partial u}{\partial t} \, . \tag{28}$$

Equation (28) is a general form for the equation of motion of quantities related to the mechanical system. With $u = q_i$ or p_i, it reduces to Hamilton's equations of motion. If $u = H$
itself, Eq. (28) leads to Eq. (23). Furthermore, Eq. (28) says that constants of the motion not
involving the time explicitly must have vanishing Poisson brackets with H (corresponding to
the quantum condition of commuting with H).

The relation between the constants of the motion and the symmetry properties of the system
can be derived in an elegant fashion in terms of Poisson bracket behavior. It has been noted
previously that the motion of the system point in time from its initial position corresponds to
the evolution of a canonical transformation with time as a parameter. There are other canonical
transformations depending continuously on some parameter, e. g., spatial rotation. For these
transformations one can introduce the concept of an infinitesimal canonical transformation
(ICT) corresponding to an infinitesimal change of the parameter from the initial conditions.
The change in the value of a function u as the ICT moves the system (on the active viewpoint)
from one point to an infinitesimally near point can be shown to be given by

$$\partial u = \varepsilon [u, G] \, , \tag{29}$$

where ε is the infinitesimal change in the parameter and G is the generating function of the
ICT. (It will not come as a surprise to note that $G = H$ for the ICT of motion in time, i. e.,

where $\varepsilon = dt$.) If the system is symmetric under the operation implied by a given ICT, then the Hamiltonian (among other functions of the system) will be unchanged in value by the ICT. Equation (29) then says that the Poisson bracket of H with the generating function G of the given ICT vanishes, which by Eq. (28) immediately shows that G is a constant of the motion, at least where G is not an explicit function of time. More detailed considerations remove even this last restriction, leading to the general theorem: *the constants of motion of a system are the generators of the symmetry transformations of the system, i. e., of ICT's which leave the Hamiltonian invariant.*

The Poisson bracket formulation of the equations of motion is of special consequence in statistical mechanics, where Eq. (28) leads directly to Liouville's theorem concerning the density of system points in phase space.

Completely Integrable Hamiltonians and Action-Angle Variables

A Hamiltonian system is said to be completely integrable if there exist n (the number of degrees of freedom) single-valued, analytic integrals of the motion $F_1, \ldots, F_n$ with say $F_1 = H$, which are in *involution*. This means that all the F_i commute with each other, that is,

$$[F_i, F_j] = 0, \quad i, j = 1, \ldots, n .$$

The existence of the n integrals F_i implies that the phase-space trajectories are confined to some n-dimensional manifold in the $2n$-dimensional phase space. It may be shown that this manifold has the topology of an *n-dimensional torus*. This topology imparts a naturally periodic structure to the motion and provides the basis for defining a particularly important set of canonical variables known as the *action-angle* variables.

On an n-dimensional torus one can define n topologically independent closed paths C_k, $k = 1, \ldots, n$, which cannot be continuously deformed into each other or shrunk to zero. Using these paths the set of action variables is defined as

$$I_k = \oint_{C_k} \sum_{i=1}^{n} p_i \, dq_i , \quad k = 1, \ldots, n .$$

The set of conjugate angle variables θ_k are defined via a generating function (found by solving the appropriate Hamilton–Jacobi equation, as discussed below). It is possible to effect a canonical transformation in which the action variables play the role of constant conjugate variables. This results in the transformed Hamiltonian being a function of the action variables only, i. e., $K = K(I_1, \ldots, I_n)$ with the consequence that Hamilton's equation take the particularly simple form

$$\dot{I}_k = -\frac{\partial K}{\partial \theta_k} , \quad \dot{\theta}_k = -\frac{\partial K}{\partial I_k} = \omega_k ,$$

where the ω_k denote the set of frequencies associated with a given torus. The equations are trivially integrated leading to the result

$$\theta_k = \omega_k t + \delta_k ,$$

where the δ_k are a set of initial phases. The trajectories of completely integrable systems are thus seen to be *multiply periodic* functions of time.

Hamilton–Jacobi Equation and Perturbation Theory

The observation that the motion of the system can be described by a canonical transformation leads to a reduction of the problem to finding a complete solution of a first-order partial differential equation in $n+1$ variables. Generating functions for finite canonical transformations can be used to provide the equations of transformation if they are functions of a mixture of original and the transformed variables. To find the canonical transformation leading to an identically vanishing K, the generating function, denoted by S, is a function of the old coordinates and the new, constant, momenta. One-half of the equations of transformation are then given by the relations

$$p_i = \frac{\partial S}{\partial q_i} \,. \tag{30}$$

Equation (25) then says that for $K = 0$, S must satisfy the partial differential equation

$$H\left(q_i, \frac{\partial S}{\partial q_i}, t\right) + \frac{\partial S}{\partial t} = 0 \,, \tag{31}$$

known as the Hamilton–Jacobi equation. A complete solution to Eq. (31) will have n nontrivial constants of integration and they, or any n independent functions of them, can serve as the new momenta which we may denote by α_i. The theory of canonical transformations says that the new coordinates, say β_i, are then related to S, known as Hamilton's principal function, by

$$\beta_i = \frac{\partial S(q_i, \alpha_i, t)}{\partial \alpha_i} \,. \tag{32}$$

Equations (30) and (32), between them, determine (q, p) as functions of the time and $2n$ constants, i. e., solve the problem of the motion in time. A related approach, applicable when H is conserved, is to obtain the solution in terms of the generating function W (Hamilton's characteristic function) of a canonical transformation in which all the new coordinates are cyclic. For conservative systems S in fact is $W - Ht$.

The Hamilton–Jacobi approach has applications to fields outside analytical dynamics. In geometrical optics rays are obtained as normals to the constant-W surfaces. Quantum wave mechanics appears also as an extension of the Hamilton–Jacobi method, in which classical mechanics is to quantum mechanics as geometrical optics is to the wave theory of optics.

One of the main uses of the Hamilton–Jacobi equation is in the formulation of perturbation theory for analytical dynamics. A most important example concerns the case of an integrable Hamiltonian $H_0 = H_0(I)$, with associated action-angle variables (I, θ), perturbed by an additional small Hamiltonian term, $H_1 = H_1(I, \theta)$. The (I, θ) variables are still canonical but the equations of motion are now

$$\dot{I}_k = -\frac{\partial H_1}{\partial \theta_k} \,, \quad \dot{\theta}_k = \omega_k + \frac{\partial H_1}{\partial I_k} \,,$$

The Hamilton–Jacobi equation can be used to attempt to find a new set of action-angle variables (J, ϕ), where the J are the new actions corresponding to perturbed tori on which the trajectories still execute (modified) multiply periodic motion. In principle, perturbation expansions may be generated to arbitrary order and, for small perturbations, can be useful for calculating the shifts in the frequencies of the motion. Unfortunately, except for special cases, the perturbation expansions generated by the Hamilton–Jacobi formalism are *divergent*. This divergence is more than a mathematical artifact and has extremely important dynamical implications.

The KAM Theorem and the Onset of Chaos

The problems associated with Hamilton–Jacobi perturbation theory were manifested in celestial mechanics as the problem of *small divisors* encountered in the famous three-body problem which is concerned with, for example, the effect of Jupiter on the orbit of the earth about the sun. The divergences found in this problem suggested the impossibility of making statements about the longtime stability of planetary orbits. An alternative point of view is that these divergences raise serious doubts about the existence of phase-space tori under generic perturbation and hence the fundamental issue of the extent to which phase space is explored by typical trajectories. This fundamental problem was only resolved in the early 1960s by the Kolomogorov–Arnold–Moser (KAM) theorem which proved, for the first time, that for sufficiently small perturbation most tori are preserved. The determination of which tori survive perturbation depends on delicate number-theoretic properties of the frequencies characterizing the unperturbed tori. Despite the preservation of (a large measure of) tori, the perturbed system is fundamentally different from its unperturbed counterpart in that it is *nonintegrable*, that is, it no longer possesses a full complement of constants of the motion. It is for this reason that a global solution to the Hamilton–Jacobi solution no longer exists.

The phase-space structure of a nonintegrable system (such as that created by the generic perturbation of an integrable one) is immensely complicated. The tori which are not preserved break up into sets of alternating stable and unstable orbits – a result which can be proved by a theorem due to Poincaré and Birkhoff. In addition it is possible to prove – and demonstrate by numerical computation – that *chaotic* orbits appear. These orbits are characterized by a great sensitivity to small changes in initial conditions and wander over large regions of accessible phase space in an apparently stochastic (but still deterministic!) manner. Typically, though, such orbits do not explore all of the energy shell, and ergodicity on this shell cannot be claimed. Indeed, it can be proved that a generic Hamiltonian system is neither integrable nor ergodic. Nonetheless it is this appearance of deterministic chaos that provides an understanding of the transition from analytical dynamics to statistical mechanics.

See also: Center-of-Mass Systems; Chaos; Hamiltonian Function; Kinematics and Kinetics; Newton's Laws.

Bibliography

V. I. Arnold, *Mathematical Methods of Classical Mechanics*. Springer-Verlag, New York, 1978. (A)

V. I. Arnold and A. Avez, *Ergodic Problems of Classical Mechanics*. Benjamin, New York, 1968. (A)

H. Goldstein, *Classical Mechanics*, 2nd ed. Addison-Wesley, Reading, MA, 1980. (I)

L. D. Landau and E. M. Lifshitz, *Mechanics*, 2nd ed. Pergamon Press, Oxford, 1960. (I)

M. Tabor, *Chaos and Integrability in Nonlinear Dynamics: An Introduction*. John Wiley & Sons, New York, 1989. (E-I)

Eigenfunctions

L. Eisenbud

Eigenfunctions were introduced initially in relation to the mathematics of function spaces or Hilbert spaces. The concept is a natural generalization of the notion of an eigenvector of a linear operator in a finite-dimensional linear vector space. Eigenfunctions find wide physical applications in classical linear field theories but they are probably best known to physicists for their extensive use in Schrödinger's formulation of quantum mechanics.

In pure mathematics the eigenfunction, or characteristic function, is related to a linear operator defined on a given vector space of functions on some specified domain. The operator is a map or rule of transformation that assigns to each function f in the function space a uniquely defined function, say g, in that space. If we denote the operator by Q and if Q maps f into g, we write $Qf = g$. *If a function h* (but not the null function of the function space) *is mapped by Q into a multiple of itself*, i. e., if $Qh = ch$, where c is a scalar, *h is said to be an eigenfunction of Q belonging to the eigenvalue c*. If to c there belong several independent eigenfunctions, c is said to be degenerate. From the linearity of Q it follows that all linear combinations of the set of independent functions also belong to c. Thus the eigenfunctions of Q belonging to its eigenvalues fill out separate subspaces of the function space.

In the theory of quantum mechanics, states of a physical system may be described by suitably normalized functions defined over the domain of possible configurations Y of the system. (This domain requires extension in general to include spin and other nonclassical properties that may be fixed in a particular configuration.) A state function over configuration space is properly normalized if $\int |\psi(C)|^2 \, dC = 1$, where the integration is extended over the range of possible system configurations. These state functions fill out a vector space – the so-called state space of the system. For each observable (e. g., position of a particle, energy, total angular momentum) that can be measured on the physical system, the theory of quantum mechanics defines an associated linear operator on the functions of the state space for the system. For a system consisting of a single particle (to take a simple example) the state space contains all normalizable functions over three-space; the operator for the vector momentum

Encyclopedia of Physics, Third Edition. Edited by George L. Trigg and Rita G. Lerner
Copyright ©2005 WILEY-VCH Verlag GmbH & Co. KGaA, Weinheim
ISBN: 3-527-40554-2

observable $\mathbf{p}$ is given by the theory as $-i\hbar\nabla$ (where $\hbar$ is Planck's constant divided by 2π) and an eigenfunction $\psi(\mathbf{r})$ belonging to the momentum $\mathbf{p}'$ satisfies $-i\hbar\nabla\psi_{p'}(\mathbf{r}) = \mathbf{p}'\psi_{p'}(\mathbf{r})$ so that $\psi_{p'}(\mathbf{r}) = \exp(i\pi\mathbf{p}' \cdot \mathbf{r})$ (not normalized).

The state functions – or, more accurately, the subspaces of state functions (suitably normalized) – that are eigenfunctions of an operator observable for some particular eigenvalue describe states in which a measurement of the observable is certain to find that eigenvalue. (The observable has no definite value in states not described by eigenfunctions of the associated operator.) If, for example, H is the operator for the observable $\mathcal{E}$ – say the energy of the system – then a normalized eigenfunction $\psi_{\mathcal{E}'}$ of H belonging to the eigenvalue $\mathcal{E}'$ (so that $H\psi_{\mathcal{E}'} = \mathcal{E}'\psi_{\mathcal{E}'}$) describes a state of the physical system in which the system *has the energy* $\mathcal{E}'$. The eigenfunction describes the physical system in the following sense: the function $|\psi_{\mathcal{E}'}(C)|^2$ evaluated at a particular configuration C measures the probability of observing configuration C in state $\psi_{\mathcal{E}'}$. More generally from a state function $\psi(C)$ the probability of finding observable q to have the value q' on measurement is given by $|\int \chi_{q'}(C)^* \psi(C)\, dC|^2$, where $\chi_{q'}(C)$ is the eigenfunction (supposed for simplicity to be nondegenerate) of the operator for observable q with the eigenvalue q'.

See also: Operators; Quantum Mechanics.

Bibliography

Mathematical with Applications to Physics

R. Courant and D. Hilbert, *Methods of Mathematical Physics*. Wiley, New York, 1989. (A)

P. M. Morse and H. Feshbach, *Methods of Theoretical Physics*. McGraw–Hill, New York, 1953. (I–A)

F. W. Byron and R. W. Fuller, *Mathematics of Classical and Quantum Physics*. Dover, Mineola, 1992. (I)

Quantum Mechanics

J. von Neumann, *Mathematical Foundations of Quantum Mechanics*. Princeton Univ. Press, Princeton, 1996. (A)

E. Merzbacher, *Quantum Mechanics*. Wiley, New York, 1961. (I)

R. Eisberg and R. Resnick, *Quantum Physics*. Wiley, New York, 1974. (E)

Elasticity

W. E. Bron

Interest in the elastic and inelastic properties of matter dates back to Galileo and has spawned in the interim an overwhelming literature on the mathematical, engineering, and physical aspects of the problem. Much of this body of work ignores the atomic nature of real solids by noting that displacements with wavelengths λ long compared to the interionic spacing a are influenced only weakly by the individual ions and can, accordingly, be treated in the continuum limit. Recently, however, a major interest in the elastic properties of solids has arisen as a result of their applicability to the determination of the interionic forces present in lattice vibrations. It is with the modern goal in mind that this survey is written.

In the long-wavelength limit, $\lambda \gg a$, the equation of motion of a displacement wave can be simply expressed in terms of Hooke's relation between the macroscopic parameters stress and strain. Since most solids are anisotropic, Hooke's law is a tensor relationship of the form

$$\sigma_{ik} = c_{ikjl}\varepsilon_{jl} \tag{1a}$$

in which all indices take on the values x, y, z or alternatively 1, 2, and 3, and summation over repeated indices is implied; σ_{ik} are elements of the stress tensor, ε_{jl} elements of the strain tensor, and c_{ikjl} are variously called the elastic stiffness constants, the elastic moduli, or the elastic constants. (Their dimension is newtons per square meter.) In order to shorten the cumbersome indexes, a convention exists to transform them to matrix notation as follows:

tensor notation (c)	xx	yy	zz	yz	zy	xz	zx	xy	yx
	11	22	33	23	23	13	31	12	21
matrix notation (C)	1	2	3	4		5		6	

such that Hooke's law becomes

$$\sigma_m = C_{mn}\varepsilon_n . \tag{1b}$$

This transformation follows from the symmetries

$$c_{ikjl} = c_{kilj} = c_{iklj} = c_{jlik} ,$$

the first three of which are always present; the last follows from the existence of an elastic potential. In the most general case there remain 21 independent elements of the C tensor.

In the matrix notation Eq. (1) explicitly becomes

$$
\begin{aligned}
\sigma_1^P &= C_{11}\varepsilon_1 + C_{12}\varepsilon_2 + C_{13}\varepsilon_3 + C_{14}\varepsilon_4 + C_{15}\varepsilon_5 + C_{16}\varepsilon_6 , \\
\sigma_2^P &= C_{21}\varepsilon_1 + C_{22}\varepsilon_2 + C_{23}\varepsilon_3 + C_{24}\varepsilon_4 + C_{25}\varepsilon_5 + C_{26}\varepsilon_6 , \\
\sigma_3^P &= C_{31}\varepsilon_1 + C_{32}\varepsilon_2 + C_{33}\varepsilon_3 + C_{34}\varepsilon_4 + C_{35}\varepsilon_5 + C_{36}\varepsilon_6 , \\
\sigma_4^S &= C_{41}\varepsilon_1 + C_{42}\varepsilon_2 + C_{43}\varepsilon_3 + C_{44}\varepsilon_4 + C_{45}\varepsilon_5 + C_{46}\varepsilon_6 , \\
\sigma_5^S &= C_{51}\varepsilon_1 + C_{52}\varepsilon_2 + C_{53}\varepsilon_3 + C_{54}\varepsilon_4 + C_{55}\varepsilon_5 + C_{56}\varepsilon_6 , \\
\sigma_6^S &= C_{61}\varepsilon_1 + C_{62}\varepsilon_2 + C_{63}\varepsilon_3 + C_{64}\varepsilon_4 + C_{65}\varepsilon_5 + C_{66}\varepsilon_6 .
\end{aligned}
\tag{2}
$$

In Eq. (2) the superscripts P and S refer to pressure (stress) and shear stress, respectively; ε_1, ε_2, and ε_3, refer to simple dilational displacements, whereas ε_4, ε_5 and ε_6 refer to shear displacements.

It follows that

$$C_{mn} = \frac{\partial \sigma_m}{\partial \varepsilon_n} . \tag{3}$$

For crystals with cubic lattice symmetry

$$C_{11} = C_{22} = C_{33} \quad C_{44} = C_{55} = C_{66} \quad C_{12} = C_{23} = C_{31} ,$$

and

$$C_{14} = C_{15} = C_{16} = C_{24} = C_{25} = C_{26} = C_{34} = C_{35} = C_{36} = C_{45} = C_{46} = C_{56} = 0 . \quad (4)$$

Young's modulus Y is defined as the longitudinal stress σ_1^P per unit strain ε_1, when the body is not constrained in the transverse directions, i. e., when ε_2 and ε_3 are free to change. Accordingly, for a cubic crystal

$$Y_{cub} = \frac{(C_{11} - C_{12})(C_{11} + 2C_{12})}{C_{11} + C_{12}} .$$

The bulk modulus B, defined in terms of the pressure P and the volume V as

$$B = -V \frac{dV}{dP} ,$$

becomes for a cubic crystal

$$B_{cub} = (C_{11} + 2C_{12}) ,$$

and the compressibility $K = 1/B$. The shear modulus (or modulus of rigidity) S becomes

$$S_{cub} = C_{44} .$$

The corresponding elastic moduli for an isotropic body are

$$Y_{iso} = \frac{C_{44}(3C_{12} + 2C_{44})}{C_{12} + C_{44}} ,$$

$$B_{iso} = C_{12} + C_{44} ,$$

$$S_{iso} = C_{44} .$$

If the forces between lattice ions are central, the elastic constants of isotropic bodies obey the further relations

$$C_{12} = C_{66} , \quad C_{23} = C_{44} , \quad C_{31} = C_{55} ,$$
$$C_{14} = C_{56} , \quad C_{25} = C_{64} , \quad C_{36} = C_{45} . \quad (5)$$

For cubic crystals this becomes the well-known Cauchy relation; i. e., $C_{12} = C_{44}$.

Interest in the elastic constants in regard to lattice dynamics stems from the fact that the propagation of elastic waves in solids can be written in terms of these quantities. We define the density of the medium as ρ, and u, v, and w as the lattice displacements caused by the elastic wave in a small incremental volume Δx, Δy, and Δz in, respectively, the x, y, and z directions of the volume element. The elastic wave is generated by a stress gradient across the incremental volume. The equation of motion of the medium in, say, the x direction is then

$$\rho \frac{\partial^2 u}{\partial t^2} = \frac{\partial \sigma_1}{\partial x} + \frac{\partial \sigma_6}{\partial y} + \frac{\partial \sigma_5}{\partial z} , \quad (6)$$

Table 1: Sound velocities.

[100] direction	[110] direction	[111] direction
$\rho v_l^2 = C_{11}$	$\rho v_l^2 = \frac{1}{2}(C_{11} + C_{12} + 2C_{44})$	$\rho v_l^2 = \frac{1}{3}(C_{11} + 2C_{12} + 4C_{44})$
$\rho v_t^2 = \rho v_{t'}^2 = C_{44}$	$\rho v_t^2 = C_{44}$	$\rho v_t^2 = \rho v_{t'}^2$
	$\rho v_{t'}^2 = \frac{1}{2}(C_{11} - C_{12})$	$= \frac{1}{3}(C_{11} - C_{12} + C_{44})$

and similarly for v and w. From Eq. (3) and the conditions (4) it follows for cubic crystals that

$$\rho \frac{\partial^2 u}{\partial t^2} = C_{11} \frac{\partial \varepsilon_1}{\partial x} + C_{12} \left(\frac{\partial \varepsilon_2}{\partial x} + \frac{\partial \varepsilon_3}{\partial x} \right) + C_{44} \left(\frac{\partial \varepsilon_6}{\partial y} + \frac{\partial \varepsilon_5}{\partial z} \right) . \tag{7}$$

The elements of the strain tensor are, of course, themselves related to the displacements. Specifically

$$\varepsilon_1 = \frac{\partial u}{\partial x}, \quad \varepsilon_2 = \frac{\partial v}{\partial y}, \quad \varepsilon_3 = \frac{\partial w}{\partial z},$$

$$\varepsilon_4 = \frac{\partial v}{\partial z} + \frac{\partial w}{\partial y}, \quad \varepsilon_5 = \frac{\partial u}{\partial z} + \frac{\partial w}{\partial x}, \quad \varepsilon_6 = \frac{\partial u}{\partial y} + \frac{\partial v}{\partial x} .$$

Therefore Eq. (7) can be rewritten as

$$\rho \frac{\partial^2 u}{\partial t^2} = C_{11} \frac{\partial^2 u}{\partial x^2} + C_{44} \left(\frac{\partial^2 u}{\partial y^2} + \frac{\partial^2 u}{\partial x^2} \right) + (C_{12} + C_{44}) \left(\frac{\partial^2 v}{\partial x \partial y} + \frac{\partial^2 w}{\partial x \partial z} \right) . \tag{8}$$

Similar equations of motion exist for $\partial^2 v / \partial t^2$ and $\partial^2 w / \partial t^2$. Solutions to these equations can be written in the form

$$u = u_0 \exp[i(kx - \omega t)] , \tag{9}$$

in which k is the wave vector of the wave and ω its angular frequency. The magnitude of the k vector is related to the wavelength by $k = 2\pi/\lambda$, and ω is related to the linear frequency v by $\omega = 2\pi v$. The velocity of propagation is given by $v = v\lambda = \omega/k$. Solutions exist for both longitudinal (l) and transverse (t) waves – that is, respectively, waves for which the displacement is in the same direction as the propagation direction and (two) waves in which the displacement is perpendicular to the propagating direction. The velocities depend then on the propagation direction. For waves propagating along the [100], [110), and [111] symmetry directions of cubic crystals the relations determined by substituting Eq. (9) into Eq. (8) and displayed in Table 1 can be used to determine the propagation (sound) velocity of the displacement waves.

The relations of Table 1 are, in fact, the ones that permit accurate measurements to be made of the elastic constants. Modern methods of determining the elastic constants proceed by measuring the sound velocity propagating in the body under observation, in various directions and polarizations, until enough relations of the type of Table 1 are obtained to specify the number of independent elastic constants. In these experiments piezoelectric transducers (such as quartz crystals) are excited to produce ultrasonic pulses whose time of flight across a known length of material yields the desired velocity measurement.

In modern (classical) models for the lattice vibrations of real crystals the ions of the lattice are replaced by nondeformable masses ("cores") and the outer electrons are replaced by deformable "shells." The cores and shells are interconnected by springs that simulate the interionic and polarization forces. In lowest order the springs are considered to be harmonic; i. e., the force exerted by the spring is directly proportional to its change in length. The solutions to the equations of motion of such systems of particles and forces also result in displacement waves whose wave vector and frequency vary over a wider range than those of the waves discussed earlier. The form of these waves is, however, related to that of those of Eq. (9). In fact, in the harmonic limit and in the limit of long wavelength (i. e., small k vectors) the solutions obtained from the lattice of discrete masses and springs reduce to the sound waves in elastic media. This means that the elastic constants, as defined earlier, can be used in part to define the properties of the springs in the dynamical model. This procedure is extensively and successfully performed at this time.

Experimentally the elastic constants are found to depend on temperature. In general they tend to decrease in magnitude with increasing temperature. This behavior results in part from the expansion of the crystal lattice with temperature. Clearly, as the lattice expands the interionic forces change. In addition, however, increasing temperature results in an increase in the average amplitude of vibration of the lattice ions. The combined effects have been shown to be the major source of the observed temperature dependence of the elastic constants, at least in simple solids.

So far the discussion has been limited to small lattice displacement and lattice vibrations in the "harmonic" limit for which the elastic energy stored in the lattice is described by terms up to second order in the strain components. [Within the elastic limit the lattice energy is the integral over the product of the stress and the strain, and hence, according to Eq. (1a), is proportional to the square of the strain components.] Effects resulting from even higher-order products of the strain components are described in elasticity theory by higher-order elastic constants. Third-order elastic constants can be measured through sound-velocity measurements on statically stressed solids, and the relationships between these and even higher-order elastic constants can be derived.

See also: Anelasticity; Deformation of Crystalline Materials; Lattice Dynamics; Rheology.

Bibliography
Theory of Elasticity
D. F. Nelson, *Phys. Rev. Lett.* **60**, 608 (1988). (I)
L. D. Landau and E. M. Lifshitz, *Theory of Elasticity*. Butterworth-Heinemann, London, 1986. (A)
H. Leipholz, *Theory of Elasticity*. Nordhoff, Leyden, 1974.
A. E. H. Love, *The Mathematical Theory of Elasticity*, 4th ed. Dover, Mineola, 1972. (A)
S. P. Timoshenko and I. N. Gordier, *Theory of Elasticity*, 3rd ed. McGraw–Hill, New York, 1970. (I)

Application to Solid-State Physics
N. W. Ascroft and N. D. Mermin, *Solid State Physics*. Holt, Rinehart & Winston, New York, 1976. (I)
C. Kittel, *Introduction to Solid State Physics*, 7th ed. Wiley, New York, 1995. (E)
C. Zener, *Elasticity and Anelasticity of Metals*. Univ. of Chicago Press, Chicago, 1948. (E)

Measurement and Values of Elastic Constants

O. L. Anderson, *Physical Acoustics* (W. P. Mason, ed.), Vol. IIIB, pp. 43–95. Academic, New York, 1965. (E)

R. Truell, C. Elbaum, and B. B. Chick, *Ultrasonic Measurements in Solid State Physics*. Academic, New York, 1969. (E)

Higher-Order Elastic Constants

R. A. Cowley, *Adv. Phys.* **12**, 421 (1963). (I)

R. C. Hollinger and G. R. Barsch, *J. Phys. Chem. Solids* **37**, 845 (1976). (I)

J. M. Ziman, *Electrons and Phonons*, Chapter 3. Oxford University Press, Oxford, 2001.

Application to Lattice Dynamics

W. Cochran and R. A. Cowley, *Handb. Phys.* **25/2a**, 59 (1967). (A)

J. A. Reissland, *The Physics of Phonons*. Wiley, New York, 1973. (I)

Electric Charge

M. Phillips[†]

Electric charge is a property of matter, first observed in the behavior of certain materials we now call dielectrics. Specifically, it was noted in ancient Greece that amber, on being rubbed, attracts bits of straw and other light objects. Hence the name, since the Greek word for amber is electron. That electrified bodies repel as well as attract was discovered by Niccolo Cabeo during the first half of the seventeenth century. Nearly a hundred years later Charles-François du Fay demonstrated that there are two kinds of electricity: "Each [body] repels bodies which have contracted an electricity of the same nature as its own, and attracts those whose electricity is of a contrary nature." The two electricities were designated *positive* and *negative* by Benjamin Franklin, a terminology that implies the *conservation of charge*: Charge cannot be created or destroyed, for the algebraic sum of the positive and negative charges in a closed or isolated system does not change in any circumstances. Franklin also established the arbitrary convention as to which is called positive and which negative.

 The force between two charges varies with distance in the same way as the gravitational force between two masses, inversely as the square of the distance. This law was first deduced by Joseph Priestley (discoverer of oxygen) from Franklin's observation that there is no electric force *inside* a metal container, and published (1767) in his history of electricity. The empirical discovery of the inverse square law was made by Charles Augustin Coulomb in 1785. (Henry Cavendish anticipated Coulomb's work, but failed to publish.) The mathematical theory of electrical interactions (*see* Electrostatics; Gauss's Law) is quite analogous to that of gravitational attraction. The necessary modifications and extensions of the celestial mechanics of Lagrange and Laplace were developed by Poisson in 1812–1813.

[†]deceased

In the twentieth century it has become known that charge is intrinsic to the stable components of atoms, electrons and nuclei. This charge is quantized; i. e., it is observed only in multiples of a smallest amount, the electronic charge, usually designated e. In neutral matter the number of elementary charges of one sign is equal to the number of opposite sign. Early in the eighteenth century it was discovered that electricity could be conveyed from one object to another by certain materials, which became known as conductors, and the science of current electricity began. Electric currents result from the motion of charge carriers. In metals the carriers are electrons, each with one elementary charge that is negative in sign. This is essentially because electrons are much more mobile than the massive positive nuclei. In fluids the atomic or molecular charges of both signs may move, and such charge carriers are called ions.

Quantitatively, Coulomb's law for the force F between two point charges q and q' a distance r apart is written

$$F = k \left(\frac{qq'}{r^2} \right) ,$$

where k is a constant that depends on the units. If k is taken to be 1 and dimensionless and both F and r are measured in cgs units, the electrostatic unit of charge (esu), still often used in atomic physics, is defined. In the International System of units (SI) charge is measured in coulombs. The coulomb (C) is derived from the basic SI unit of current, the ampere, and is equal to 1 ampere second (A·s). The standard ampere is determined mechanically from the magnetic force between two current-carrying conductors. In SI units k has a magnitude very nearly equal to 9×10^9, exactly, $k = c^2 \times 10^{-7}$ where c is the speed of light in meters per second. To three significant figures, the elementary charge $e = 1.60 \times 10^{-19}$ C. Charge is always associated with mass, although some unstable massive particles, the neutron and the neutral mesons and baryons, have no net charge. Neutrinos have no charge, and no mass so far as is presently known.

See also: Electron.

Bibliography

Edmund Whittaker, *A History of the Theories of Aether and Electricity*. Vol. I. Dover, New York, 1990.
Richard Becker, *Electromagnetic Fields and Interactions*. Vol. I. Dover, New York, 1982.
J. D. Jackson, *Classical Electrodynamics*, 3rd ed. Wiley, New York, 1998.

Electric Moments

J. Macek

The spatial distribution $\rho(x,y,z)$ of a physical quantity is characterized by its moments m_{ijk} given by

$$m_{ijk} = \int \rho(x,y,z) x^i y^j z^k \, d\tau . \tag{1}$$

Under suitable conditions, $\rho(x,y,z)$ can be reconstructed from a knowledge of its moments, and thus ρ or m offer different but equally complete parametrizations of distributions. It is particularly convenient to parametrize electric charge distributions by their moments since the static electric potential $\Phi(\mathbf{r})$ outside of a source relates directly to multipole moments.

The electric potential $\Phi(\mathbf{r})$ of a charge distribution $\rho(\mathbf{r})$ is given by

$$
\begin{aligned}
\Phi(\mathbf{r}) &= \int \frac{\rho(\mathbf{r}')}{|\mathbf{r}-\mathbf{r}'|}\,d\tau' \\
&= \sum_{lm} \frac{2l+1}{4\pi} Y_{lm}(\hat{\mathbf{r}}) \int \frac{r_<^l}{r_>^{l+1}} \rho(\mathbf{r}) Y_{lm}(\hat{\mathbf{r}}')\,d\tau' ,
\end{aligned}
\tag{2}
$$

where $r_<$ is the smaller of the radii r or r' and $r_>$ is the larger. The spherical harmonics $Y_{lm}(\hat{\mathbf{r}})$ are defined in terms of associated Legendre functions $P_l^m(\cos\theta)$ according to

$$
Y_{lm}(\theta,\phi) = \left(\frac{2l+1}{4\pi} \frac{(l-m)!}{(l+m)!} \right)^{1/2} P_l^m(\cos\theta) e^{im\phi}
\tag{3}
$$

where

$$
P_l^m(x) = \frac{(-1)^m}{2^l l!} (1-x^2)^{m/2} \frac{d^{l+m}}{dx^{l+m}} (x^2-1)^l .
\tag{4}
$$

Outside of the charge distribution where $r_> = r$ and $r_< = r'$, Eq. (2) takes the form

$$
\Phi(\mathbf{r}) = \sum_{lm} \frac{2l+1}{4\pi} Y_{lm}(\hat{\mathbf{r}}) \frac{q_{lm}}{r^{l+1}} ,
\tag{5}
$$

with

$$
q_{lm} = \int \rho(\mathbf{r}') r'^l Y_{lm}(\hat{\mathbf{r}}')\,d\tau' .
\tag{6}
$$

Equation (6) defines the electric multipole moments of the charge distribution $\rho(\mathbf{r})$ expressed in units of charge$\times$(length)^l. In the physics literature these units are used for the monopole and dipole moments, while higher moments are defined in units of (length)^l by dividing q_{lm} of Eq. (6) by the unit of electric charge $e = 4.8 \times 10^{-10}$ esu. Since $r^l Y_{lm}(\mathbf{r})$ is a homogeneous polynomial in x, y, and z, the multipole moments q_{lm} equal certain linear combinations of the moments m_{ijk} with $i+j+k=l$. One cannot use the multipole moments to reconstruct $\rho(\mathbf{r})$, however, since there are fewer moments q_{lm} than moments m_{ijk} with $i+j+k=l$; rather the usefulness of the q_{lm}'s rests on Eq. (5).

In essence, Eq. (5) represents the potential outside of a source in terms of the potentials of point multipoles located at the origin. A macroscopic dielectric medium is made up of microscopic constituents whose static electrical properties are characterized by the multipole moments. The potential of the medium is a linear superposition of potentials due to the idealized point multipoles located at a distribution of points in space. The multipole moments thus represent an essential link between microscopic and macroscopic theories of matter. They are also important for the information that is provided about the internal dynamics of the elementary constituents of matter.

The lowest orders of electric moments are of most significance. The zero'th multiple q_{00} just equals the net charge q. The first-order multipoles q_{1m} relate directly to the dipole moment vector $\mathbf{p} = \int \rho(\mathbf{r})\mathbf{r}\,d\tau$ according to

$$
\begin{aligned}
q_{10} &= (3/4\pi)^{1/2} p_z , \\
q_{1\pm 1} &= \mp(3/4\pi)^{1/2}(p_x \mp i p_y) ,
\end{aligned}
\tag{7}
$$

while the multipoles q_{2m} relate to the components of the quadrupole tensor

$$
Q_{ij} = \int \rho(\mathbf{r})(3x_i x_j - r^2 \delta_{ij})\,d\tau ,
$$

for example,

$$
q_{20} = \frac{1}{2}(5/4\pi)^{1/2} Q_{33} .
\tag{8}
$$

It should be noted that the values of the multipole moments depend, in general, upon the arbitrary origin of coordinates; however, the lowest nonvanishing moment does not. For charged ions and atomic nuclei the lowest-order moment is the total charge q. With the origin located at the center of charge the next-order moment, the dipole moment, vanishes by definition. The center of charge need not coincide with the center of mass, but to the best of our knowledge the center of mass and the center of charge of atomic nuclei and charged elementary particles do indeed coincide. For example, the electric dipole moment of the electron is measured to be $-0.2 \pm 3.2 \times 10^{-26}$ $e{\cdot}cm$, where e is the magnitude of the charge on the electron. The small value, consistent with 0, of this quantity implies implies that the center of mass and center of charge of the electron coincide within a distance of the order of 10^{-26}cm.

For electrically neutral particles the absence of an electric dipole moment is unrelated to the mechanical center of mass, rather it is a property of the electric charge distribution. The dipole moments of such elementary particles relate closely to fundamental theories of elementary particles and have therefore been measured to high accuracy. The present limit of the dipole moment d_n of the neutron, one of the most accurately known, is $|d_n| < 6.3 \times 10^{-26} e{\cdot}cm$.

In addition to a net charge, atomic nuclei frequently exhibit nonzero quadrupole moments. The existence of these moments has provided valuable information on nuclear forces; for example, the quadrupole moment of the deuteron, equal to 0.273×10^{-26} cm^2, implies a non-central character for the force between the constituent proton and neutron.

Atoms and molecules in gases are normally neutral so that $q_{00} = 0$; indeed, atoms are normally in isotropic states so that all multipole moments vanish. Application of an external electric field distorts atomic charge distributions by partially separating the positive and negative charges, thereby inducing a nonzero dipole moment. At low values of the electric field the induced dipole moment is proportional to the external electric field. The constant of proportionality is called the polarizability and its value depends on the dynamics of the particular atomic state.

The permanent moments of molecules are defined as the moments of the molecular charge distribution in a hypothetical coordinate frame where the molecules as a whole do not rotate. In such a frame, molecules may exhibit both induced and permanent dipole moments, quadrupole moments, etc. The order of magnitude of molecular dipole moments equals the product of electronic charge 4.8×10^{-10} esu times an atomic diameter 10^{-8} cm, i.e., 4.8×10^{-18} esu cm. The quantity 10^{-18} esu cm is called a debye. Permanent and induced dipole moments are typically measured indirectly by measuring the macroscopic temperature-dependent dielectric constant and applying the Clausius–Mosotti equation to deduce the microscopic dipole moments.

See also: Electrostatics; Polarizability.

Bibliography

Amos de Shalit and Herman Feshbach, *Theoretical Nuclear Physics*, Vol. 1, pp. 53–58. Wiley, New York, 19XX.

J. D. Jackson, *Classical Electrodynamics*, 2nd ed., pp. 136–167. Wiley, New York, 1975.

Norman F. Ramsey, in *Atomic Physics* (Richard Marrus, Michael Prior, and Howard Shugart, eds.), pp. 453–471. Plenum Press, New York, 1977.

Emilio Segre, *Nuclei and Particles*, pp. 221–244. Benjamin Advanced Book Program, Reading, MA, 1965.

P. G. Harris *et al.*, *Phys. Rev. Lett.*, **82**, pp. 904–07, 1999.

J.J. Hudson, B. E. Saurer, M. R. Tarbutt, and E. A. Hinds, *Phys. Rev. Lett.*, **89**, pp. 023003-1–023002-4, 2002.

Electrochemical Conversion and Storage

G. E. Blomgren

Electrochemical conversion of chemical energy into electrical energy occurs in a Galvanic cell by an oxidation reaction contributing electrons to an external circuit through an electrode called the anode, while a reduction reaction removes electrons from the external circuit through an electrode called the cathode. The active materials which undergo the oxidation and reduction reactions must be in physical (i.e., electronic) contact with the anode and cathode, respectively, so that electronic charge transfer to or from the electrodes can occur. The electrodes must not touch each other, since such contact would cause an internal short circuit, but must each contact a continuous ionically conductive medium such as an electrolyte solution. To ensure that the electrodes do not touch, a porous, insulating material called the separator is often interposed between the electrodes. The separator pores are then filled with electrolyte solution.

An active material which is a sufficiently good electronic conductor can function as its own electrode. Thus, a metal anode often acts as its own electrode and, in the Leclanché dry cell, as the case of the battery as well. An easily understood example of these definitions is the Daniell cell consisting of zinc anode active material and copper sulfate cathode active material. The anodic and cathodic reactions are

$$\text{Zn} \quad \rightarrow \quad \text{Zn}^{2+} + 2\text{e}^- \quad \text{oxidation (anode)}$$
$$\text{CuSO}_4 + 2\text{e}^- \quad \rightarrow \quad \text{Cu} + \text{SO}_4^{2-} \quad \text{reduction (cathode)}$$

When a voltage source of opposite polarity to the Galvanic cell can easily reverse the anode and cathode reactions, a storage cell or secondary battery results. A cell made for discharging the active materials only one time is called a primary battery.

Michael Faraday first described the basis of the operation of a Galvanic cell. These statements, now called "Faraday's laws of electrolysis", are: (1) The amount of chemical change produced by electrolysis is proportional to the total amount of electrical charge passed through the cell; (2) the amount of chemical change produced is proportional to the equivalent weight of the substance undergoing chemical change. Thus, the amount of charge in coulombs (C) passed through the cell is proportional to the number of chemical equivalents undergoing chemical change and the constant of proportionality is F or Faraday's constant (equal to $96\,484.56\,C$/equivalent). Under conditions of maximum efficiency, the Gibbs free energy change, ΔG, of the overall chemical reaction of the Galvanic cell is equal to the negative of the electrical work capable of being obtained. The counter electromotive force (EMF) of the device upon which the electrical work is done must be, under these conditions, the reversible or maximum EMF of the cell, E. The electrical work is given by the amount of electricity flowing through the cell (equal by Faraday's laws to nF, where n is the number of equivalents corresponding to the cell reaction) multiplied by the EMF. Thus

$$\Delta G = -nFE \ . \tag{1}$$

Walther Nernst extended the considerations of Faraday to take into account the dependence of the potential on the concentrations or activities of reactant and product species. The free energy of each species i (reactant or product) depends on the activity a_i and the standard free energy, G_i^0:

$$G_i = G_i^0 + RT \ln a_i \ . \tag{2}$$

The reaction free energy depends on the difference between the sum of product free energies and the sum of reactant free energies. With use of (1) and (2) the EMF is given by the Nernst equation as

$$E = E^0 - RT \ln \frac{\prod_i a_i(\text{products})}{\prod_i a_i(\text{reactants})} \ . \tag{3}$$

The above discussion shows that the energy obtainable from an ideal Galvanic cell (operating under reversible conditions) is equivalent to the free energy inherently available from the chemical system. The Carnot limitation of a heat engine does not apply to Galvanic cells. The ratio of the free energy change to the total energy change is usually close to unity since the entropy effects are small for most practical cells. However, in spite of theoretical superiority, the difficulty of scale-up and problems of presenting a continuous feed of reactants and removal of products in a Galvanic cell has led to almost universal use of heat engines (including wind and water power based turbines) for large-scale production of electrical energy. Thus, the main application for Galvanic cells is the field of isolated or portable power sources.

Losses of energy in a Galvanic cell consist of voltage losses and Coulombic or active material losses. Active material losses can occur from shedding of active material from the battery plate, by electronic isolation as active materials are surrounded by insulating reaction products, or by parasitic chemical or corrosion reactions which convert active materials into inactive products. Voltage losses are far more complicated in origin, and the study of these losses forms the basis of much present work in electrochemistry. There are three main types of voltage loss or polarization: activation, concentration, and Ohmic. All three depend on the current delivered by the cell in a monotonically increasing way. Ohmic polarization arises from the resistance of all of the elements of the cell and follows Ohm's law for the current dependence. Concentration polarization arises from the creation of concentration gradients in the cell and obeys the Nernst equation. Current dependence enters indirectly through the concentration gradient dependence on the current. Activation polarization is a kinetic factor for the charge transfer reaction and is in general different for each electrode in the functional dependence on current and concentrations and in the kinetic parameters describing the charge transfer reaction. The effect of time of discharge or charge is to increase the voltage loss of the cell. All of these losses contribute to lowering the efficiency of the Galvanic cell.

Many hundreds of couples of active materials have been proposed for use as Galvanic cells. However, very few of these have been manufactured successfully. Table 1 lists the main types of primary and secondary cells available as articles of commerce, the anode and cathode active materials, the open circuit voltages, the observed energies per unit weight and volume, and the cell reaction on discharge. The Leclanché and alkaline MnO_2, primary cells and the lead acid secondary cell have experienced the most applications by far, although this picture is rapidly changing. The superior energy densities of the silver oxide and mercuric oxide primary cells and the nickel cadmium secondary cell are balanced by their high cost of material

In addition, the increasing concerns of the public and governments toward discarding devices which may contain toxic materials such as mercury and cadmium has caused many changes in the marketing and availability of certain batteries. For example, the use of mercury batteries has decreased markedly in nearly all countries, even where laws restricting their use are not in place. In part, this is because the silver-oxide cell and the recently developed lithium cells are more than adequate substitutes. Also, the use of mercury in zinc–MnO_2 cells, acid or alkaline, has been greatly reduced or eliminated in commercial batteries (mercury has traditionally been used on zinc as an additive to improve kinetics and reduce corrosion). The high energy density and absence of toxic heavy metals have encouraged the gradual introduction of lithium primary batteries, as noted in Table 1. The high voltage of unit cells in the cases of manganese dioxide and carbon fluoride cells, however, has made the introduction more difficult since devices must be designed either to accommodate a high voltage or battery compartments must be modified to accommodate fewer unit cells. The lithium–iron sulfide system can be a direct replacement for zinc cells and has been so used in miniature batteries. The major use of lithium–sulfur dioxide and lithium–thionyl chloride batteries is for military applications, although thionyl chloride cells are seeing vastly increasing industrial and specialty use because of their remarkable energy coupled to excellent low temperature performance. These batteries are capable of delivering very high currents and have outstanding specific energy and energy density although disposal remains a problem. They are not widely available to consumers for this reason and for reasons of safety. The lithium–iodine battery has been widely used for medical purposes; in fact the widespread use of heart pacemakers relies on the excellent properties of this battery. In spite of its very high energy, however, it is only useful for very small current drains because of its high impedance. For secondary batteries, the use of toxic lead and cadmium continues because there are not sufficiently good

Table 1: Properties of some presently manufactured galvanic cells.

Cell name	Anode active material	Cathode active material	Open circuit voltage	Specific energy ($J/kg \times 10^{-5}$)	Energy density ($J/m^3 \times 10^{-8}$)	Cell reaction
Aqueous primary cells						
Leclanché	Zn	MnO_2	1.58	2.6	5.5	$Zn + 2NH_4Cl + 2MnO_2 \rightarrow Zn(NH_3)_2Cl_2 + 2MnO(OH)$
Alkaline MnO_2	Zn	MnO_2	1.55	3.4	9.0	$Zn + H_2O + 2MnO_2 \rightarrow ZnO + 2MnO(OH)$
Mercury	Zn	HgO	1.35	3.6	21.6	$Zn + HgO \rightarrow ZnO + Hg$
Silver	Zn	Ag_2O	1.60	4.7	18.7	$Zn + Ag_2O \rightarrow ZnO + 2Ag$
Air	Zn	O_2 (air)	1.4	14.4	28.0	$Zn + \frac{1}{2}O_2 \rightarrow ZnO$
Nonaqueous primary cells						
Lithium MnO_2	Li	MnO_2	3.0	7.2	14.4	$Li + MnO_2 \rightarrow LiMnO_2$
Lithium carbon fluoride	Li	CF_x	3.1	7.2	14.4	$xLi + CF_x \rightarrow xLiF + C$
Lithium iron disulfide	Li	FeS_2	1.8	9.0	16.6	$4Li + FeS_2 \rightarrow 2Li_2S + Fe$
Lithium sulfur dioxide	Li	SO_2	3.0	9.0	15.8	$2Li + 2SO_2 \rightarrow Li_2S_2O_4$
Lithium thionyl chloride	Li	$SOCl_2$	3.6	10.8	36.0	$4Li + 2SOCl_2 \rightarrow 4LiCl + S + SO_2$
Lithium iodine	Li	I_2 polymer	2.8	8.6	28.0	$Li + \frac{1}{2}I_2 \rightarrow LiI$
Secondary cells						
Lead acid	Pb	PbO_2	2.2	1.6	2.4	$Pb + 2H_2SO_4 + PbO_2 \rightarrow 2PbSO_4 + 2H_2O$
Nickel cadmium	Cd	$NiO(OH)$	1.35	3.2	5.8	$Cd + 2NiO(OH) + H_2O \rightarrow CdO + Ni(OH)_2$
Nickel metal hydride	AB_5H	$NiO(OH)$	1.35	3.2	10.8	$H(AB)_5 + NiO(OH) \rightarrow (AB)_5 + Ni(OH)_2$
Lithium ion	C_6Li	$Li_{0.5}CoO_2$	3.6	6.8	18.0	$\frac{1}{2}C_6Li + Li_{0.5}CoO_2 \rightarrow \frac{1}{2}C_6 + LiCoO_2$
Lithium ion polymer	C_6Li	$Li_{0.5}CoO_2$	3.6	5.7	11.9	$\frac{1}{2}C_6Li + Li_{0.5}CoO_2 \rightarrow \frac{1}{2}C_6 + LiCoO_2$

alternatives at the present time for certain applications. In the case of large batteries, most of the lead and cadmium is recycled when new batteries are purchased.

The last three systems in Table 1 have been introduced in about the past 12 years. The outstanding energy density and specific energy of these batteries have accounted for their massive acceptance in the consumer market place. Along with the revolution in microelectronics, these batteries have allowed the development of portable computers, communication devices of all kinds, digital photographic devices and many other types of consumer products.

The nickel metal hydride system was the first of the modern rechargeable batteries to be introduced and has developed a major market for itself. The system is not as good as lithium ion in specific energy and energy density, however, so has itself lost market share to this newer battery. The nickel metal hydride cell has the advantage of an aqueous electrolyte (KOH in water), so it has safety advantages over the lithium system. It also has an intrinsic overcharge protection mechanism, like nickel cadmium, since oxygen is evolved at the positive electrode during overcharge, and is subsequently reduced at the negative electrode. The cell, however, has a relatively high rate of self discharge, of the order of a few percent per day, which leads to loss of capacity and consumer complaints. This is another advantage of the lithium ion battery, in that it has only a few percent loss per month by comparison. The main detriment, in addition to safety of lithium ion batteries is the relatively higher cost. This difference is fast disappearing as the product matures, however.

One of the cost improvements resulted in the most recent system, namely the lithium ion polymer battery. The lithium ion battery contains an electrolyte composed of an inorganic salt in a mixture of organic solvents. This electrolyte must be handled carefully since it is flammable and subject to hydrolysis in the presence of water vapor. On rupture of the case due to an external or internal force, the electrolyte can catch on fire with a very hot flame if the lithiated carbon negative electrode is ignited as well. In the lithium ion polymer battery, the safety is improved and manufacturing is simplified if a gel type electrolyte is used with a cross-linked polymer network for the gel forming agent (hence the name lithium ion polymer). This battery has been somewhat slower to gain acceptance, but appears to have a solid niche business in telephone and personal calendar devices.

The lithium ion battery has been widely adapted to space applications, replacing nickel cadmium batteries, and appears to be gaining in all areas. Some of the most recent applications have involved higher power systems than were previously available and this will no doubt diminish the field of nickel cadmium still further. Nickel metal hydride batteries have developed a major application in hybrid vehicles because of their low weight compared to the older systems and safety advantage compared to the lithium systems. This is projected to be a fast growing business area as well. The societal benefits of these new battery systems should not be overlooked. They permit a much more efficient propulsion system for vehicles (the various hybrid systems), they do not contaminate land fills with heavy metals compared with the older rechargeable batteries. The lithium ion systems are beginning to be used in implantable medical applications along with new lithium primary batteries to the benefit of many previously untreatable conditions.

The future of the battery industry seems certain to continue growth, particularly in the new rechargeable systems. It is also anticipated that new positive and negative electrode materials will be introduced. Already, new positive electrode materials such as $LiMn_2O_4$ and $LiFePO_4$ have been introduced in commercial products with improved safety prospects. Many alloying metals offer the potential of higher energy density and specific energy than the lithiated carbon electrode and are under wide study at the research level. Of course, the difficult performance requirements of cycle life, power capability, toxicity and availability of materials, battery safety and cost must all be established before an advance can be made in battery technology. It is a fundamental part of energy technology, however, and will continue to undergo development. One of the many exciting possible applications is providing electrical storage with wind and solar generators for remote sites. Many highway control applications of such devices have already begun. Other energy storage systems such as fuel cells, fly wheels, capacitors, etc. are also under rapid development and will undoubtedly also play a role in future energy systems. The battery is uniquely situated to work in complicated hybridized systems and will likely always be involved in such efforts.

See also: Electrochemistry; Free Energy.

Bibliography

International Power Sources Symposium, IPSS, 1968 *et seq.* Proceedings of the Biennial International Symposium of Batteries at Brighton, England. Presents state of the art information on electrochemical energy conversion devices and the results of studies on battery related problems. (A)

D. Aurbach (ed.), *Nonaqueous Electrochemistry*. Marcel Dekker, New York, 1999. Discusses all aspects of nonaqueous electrolytes used in lithium and lithium ion batteries. (A)

C. A. Hampel (ed.), *The Encyclopedia of Electrochemistry*. Reinhold, New York, 1964. The entries under "Batteries" and "Fuel Cells" are especially useful; as are the entries under specific battery systems. (E)

A. B. Hart and G. J. Womack, *Fuel Cells*. Chapman and Hall, London, 1967. An introduction to and survey of fuel cell systems. (I)

G. W. Heise and N. C. Cahoon (eds.), *The Primary Battery*. Wiley, New York, 1971, Vol. 1; 1976, Vol. 2. Chapter 1, Vol. 1 is a comprehensive history of the development of the science and practice of electrochemical energy conversion. Remainder treats most kinds of primary batteries which have been studied. (A)

Kirk–Othmer Encyclopedia of Chemical Technology, 4th ed. Wiley, New York, 2005. Entries under "Batteries" survey the field of electrochemical energy conversion. (E)

K. V. Kordesch (ed.), *Batteries*. Marcel Dekker, New York, 1974, Vol. 1. Highly detailed discussions of manufacturing aspects and principles of commercial batteries. (A)

D. Linden and T. Reddy, *Handbook of Batteries and Fuel Cells*. McGraw-Hill, New York, 2002. Up-to-date discussion of experimental and production batteries and fuel cells for military as well as consumer and medical applications. (A)

Proceedings of the Power Sources Symposium. PSC Publications Committee, Red Bank, NJ, 1946, *et seq.* Biennial symposium at Atlantic City, New Jersey, on power sources and energy conversion sponsored by several agencies of the U.S. Department of Defense. Presents recent research development results. (A)

W. A. van Schalkwijk and B. Scrosati (eds.), *Advances in Lithium Ion Batteries*. Kluwer Academic/Plenum Publ., New York, 2002. Presents recent work on lithium ion batteries and supercapacitors. (A)

Ullmann's Encyclopedia of Industrial Chemistry, 6th ed. Wiley-VCH, Weinheim, 2000. Entries under "Batteries" give broad view of batteries and electrochemical energy conversion. (E)

W. Vielstich, A. Lamm, H. A. Gasteiger, *Handbook of Fuel Cells: Fundamentals, Technology, and Applications*. Wiley, Hoboken, NJ, 2003. Presents a broad description of the various types of fuel cells and their fundamental principles and technology. (A)

Electrochemistry

D. R. Franceschetti and J. R. Macdonald

Modern electrochemistry is the study of ionic conductors, materials in which ions participate in the flow of electric current, and of interfaces between ionic conductors and materials which conduct current by electron flow, at least in part. It is a truly interdisciplinary field which draws heavily upon many branches of physics and chemistry. Electrochemical phenomena often have analogs in vacuum-tube and semiconductor electronics. Electrochemistry finds numerous applications throughout the natural sciences, the engineering disciplines, and the health-related fields.

Ionics

The study of ionic conductors in themselves has been termed ionics. Ionic conductors are also known as electrolytes. The ionic conductors of interest to electrochemists include (i) liquid solutions of ionic solids (e. g., NaCl in water); (ii) certain covalently bound substances dissolved in polar media, in which ions are formed on solution (e. g., HCl, which is completely dissociated in water, or acetic acid, which is only partially dissociated); (iii) ionic solids containing point defects (e. g., solid AgCl at high temperatures); (iv) ionic solids whose lattice structure allows rapid movement of one subset of ions (e. g., β-Al_2O_3–Na_2O); (v) fused salts (e. g., molten NaCl); and (vi) ionically conducting polymer–salt complexes (e. g., polyethylene oxide–$LiClO_4$).

When an electric field is established in an electrolyte, the migration of positive ions (cations) and negative ions (anions) is observed as a flow of electric current. In defect solid electrolytes, case (iii), the charge carriers are cation and anion interstitials or vacancies (anion vacancies are regarded as positively charged, cation vacancies as negatively charged). Some solid electrolytes are also electronic semiconductors so that the current may also include important contributions from conduction-band electrons and valence-band holes.

The mobility of charge carriers is determined by their interactions with each other and with their environment. Ions in solution often form long-lived aggregates with a characteristic number of solvent molecules. Point defects in solid electrolytes deform the lattice around them. Similarly, the mobility of ions in polymer electrolytes is closely coupled to fluctuations in polymer conformation. A charge carrier is, on the average, surrounded by an "atmosphere" of carriers bearing a net opposite charge. The Debye–Hückel model of dilute electrolytes yields $z_i e_0 \exp(-r/L_D)/\varepsilon r$ as the average electrostatic potential at a distance r from a carrier of charge $z_i e_0$ (e_0 is the proton charge) in a medium of dielectric constant ε. The screening or Debye length, L_D, is determined by the concentrations and charges of the carriers. The flow of current is the result of both carrier drift, in response to the electric field E, and diffusion, which acts to reduce concentration gradients. The current density resulting from charge carriers of species i is given by

$$\mathbf{J}_i = z_i e_0 (\mu_i \mathbf{E} - D_i \nabla) c_i ,$$

where c_i is the concentration of carriers and μ_i is the carrier mobility, related in dilute electrolytes to the diffusion coefficient D_i by the Einstein relationship, $\mu_i = D_i z_i e_0 / kT$. Charge carriers may form pairs, bound by electrostatic or chemical forces, whose members do not readily separate. The equation of continuity for charge carriers of species i,

$$\frac{\partial c_i}{\partial t} = G_i - R_i - (z_i e_0)^{-1} \nabla \cdot \mathbf{J}_i .$$

includes terms for the generation (G_i) and recombination (R_i) of carriers corresponding to the dissociation and formation of bound pairs.

Electrodics

The study of interfaces between ionic conductors and electronic conductors constitutes electrodics, the second major subdivision of electrochemistry. Differences in electrical potential may be determined unambiguously only between electronic conductors; thus electrochemical measurements are usually made on cells with two electrodes. Such cells may be divided into two half-cells, each containing a single electrode, either metal or semiconductor, in contact with an ionic conductor. The half-cell, involving a single interface, is the basic unit studied in electrodics.

The most thoroughly studied electrode material is liquid mercury, for which a clean and atomically smooth interface with electrolyte solutions is readily obtained. Solid metal and semiconductor electrodes are also widely employed, but have not been as extensively characterized. Polymer electrodes, which can be formed electrochemically on metal substrates, have also become a subject of appreciable interest.

Because the potential difference between the electronic and ionic phases of a half-cell is not measurable, a standard half-cell has been chosen and arbitrarily assigned an electrode potential of zero. This cell, called the standard hydrogen electrode, consists of a platinum electrode in contact with hydrogen ions at 1 atm pressure and an aqueous solution containing hydrogen ions at unit mean activity (see below). The electrode potential of other half-cells is defined as the open-circuit potential of the cell which would be formed with the standard hydrogen electrode.

Electrode potentials provide information about the electrochemical reactions by which charge is transferred between electrode and electrolyte. Simple electrode reactions include (i) ionization of the electrode metal (e. g., $Ag \rightleftharpoons Ag^+ + e^-$), (ii) change in the state of an ion (e. g., $Fe^{2+} \rightleftharpoons Fe^{3+} + e^-$), and (iii) ionization of a gas (e. g., $\frac{1}{2}H_2 \rightleftharpoons H^+ + e^-$ in the presence of a nonreactive metal). When electrons are removed from a species, the species is said to be oxidized; when electrons are added, the species is said to be reduced. In a half-cell in equilibrium, oxidation and reduction occur at equal rates. Away from equilibrium, the electrode is termed an anode if oxidation predominates over reduction, and a cathode in the opposite case. Most electrode reactions consist of a sequence of chemical and charge transfer steps, some of which may involve short-lived ionic species not present in the bulk electrolyte.

An example of a simple electrochemical cell is the Daniell cell, which consists of zinc and copper electrodes immersed, respectively, in aqueous solutions of $ZnSO_4$ and $CuSO_4$, the two solutions being separated by a membrane which allows the passage of charge but prevents rapid mixing of the two solutions. When electrons are allowed to flow between the electrodes, the zinc electrode dissolves to form Zn^{2+} ions while Cu^{2+} ions are deposited on the copper electrode, in accord with the overall reaction $Zn + Cu^{2+} \rightarrow Zn^{2+} + Cu$. The open-circuit cell potential E is the energy released by the reaction per unit charge transfer between the electrodes and is related to the Gibbs free energy ΔG of the cell reaction by $\Delta G = 2FE$, where F is Faraday's constant, $96\,485\,\text{C}\,\text{mol}^{-1}$, and two electrons are transferred for each atom of Cu deposited. E is related to the thermodynamic ion activities $a(\text{Cu}^{2+})$ and $a(\text{Zn}^{2+})$ by Nernst's equation, which in this case becomes

$$E = E^0 - \frac{RT}{2F} \ln \left\{ \frac{a(\text{Zn}^{2+})}{a(\text{Cu}^{2+})} \right\} ,$$

where E^0 is the cell potential at unit activity and R is the gas constant, $8.314\,\text{J}\,\text{mol}^{-1}\,\text{K}^{-1}$. Activities are quantities related to the ion concentrations (identical at infinite dilution), which take into account ion–ion interactions in the electrolyte.

Nernst's equation is strictly applicable only to systems in thermodynamic equilibrium. In general, the potential of each half-cell is a function of the cell current, which is determined by the slowest step in the electrode reaction sequence. For many half-cells the current is approximately given by the Butler–Volmer equation, which can be cast in the form

$$i = i_0 [\exp(\alpha_{\text{an}} \eta F / RT) - \exp(-\alpha_\text{c} \eta F / RT)]$$

with

$$\alpha_\text{c} + \alpha_{\text{an}} = n/\nu .$$

Here i_0 is the exchange current, determined by the rate of the electrode reaction at equilibrium; η is the electrode overpotential, the deviation of the half-cell from its equilibrium value; and n is the number of electrons transferred. The parameters α_{an} and α_c are transfer coefficients for the anodic (oxidation) and cathodic (reduction) processes and ν is the stoichiometric coefficient, the number of times the rate-determining step occurs in the overall half-cell reaction.

Electrodes, or more properly half-cells, are classified as polarizable or nonpolarizable depending on the amount of overpotential required for a fixed ion current flow. Limiting cases, which can be closely approximated in practice, include the perfectly polarizable, or blocking, electrode, one in which no ion current flows regardless of the overpotential, and the perfectly nonpolarizable, or reversible, electrode, one in which the electrode potential retains its equilibrium value regardless of the amount of current flow.

By variation of the potential drop across an electrochemical cell, the rates of the electrode reactions may be altered, and one may even reverse the direction of the net cell reaction. If zinc metal is immersed in CuSO_4 solution, copper metal and ZnSO_4 are produced spontaneously. By connecting the electrodes of a Daniell cell to a load, useful work may be obtained from the energy of this spontaneous reaction. Cells operated in an energy-producing manner are termed galvanic cells. By rendering the copper electrode sufficiently positive with respect to the zinc, one may effect dissolution of copper and deposition of zinc. The operation of a cell for the production of substances not obtainable spontaneously from the cell materials is termed electrolysis and the cell so operated, an electrolytic cell.

A fundamental theoretical problem in electrodics is the nature of the electrical "double layer," the region of charge separation formed when an electrode is in contact with an ionic conductor. The double layer formed at a metal electrode in an aqueous electrolyte has received particularly intensive study. The traditional Gouy–Chapman–Stern model involves a (usually charged) idealized metal surface, an adjoining plane of chemisorbed water molecules and (often) ions, and a region of increased concentration of cations or anions, depending on the charge on the electrode. The plane of centers of chemisorbed molecules and ions defines the inner Helmholtz plane (ihp), while the plane of closest approach of solvated ions is the outer Helmholtz plane (ohp). The region from the metal surface to the ohp is termed the compact double layer, characterized by an effective dielectric constant which describes the loss in orientational freedom of the adsorbed molecules. The region of space charge beyond the ohp is the diffuse double layer.

The development of atomic-scale microscopies, such as scanning tunneling electron microscopy and atomic force microscopy, have allowed the *in situ* study of compact layer structures on solid electrodes. New computational methods have provided some insights into the effects of current inhomogeneities on the electrical behavior of real solid–solid and solid–liquid contacts. Microfabrication techniques borrowed from semiconductor technology have also enhanced the level of detail with which at least some semiconductor electrode–electrolyte interfaces can be studied.

Applications

Electrochemical methods are employed widely in quantitative and qualitative chemical analysis. Electrolytic methods are the primary industrial means of purifying many metals, of extracting several metals from their ores or salts, and of producing many nonmetallic substances. Electrolytic separation methods make possible the reclamation of valuable materials from industrial waste and reduction in the quantity of pollutants released into the environment. Much industrial research is directed at retarding the corrosion of metals, a phenomenon involving electrochemical reactions at the surface of the metal. The electroplating of metals with thin layers of inert but costly materials is one of a number of electrochemical remedies to this problem.

Electrochemical cells offer an efficient and often portable source of energy. In fuel cells, the energy of a combustion reaction, such as the combination of hydrogen and oxygen to form water, is converted directly to electrical energy, circumventing the thermodynamic restriction on the efficiency of heat engines. The use of solid electrolytes, particularly those whose crystal structure permits rapid ion movements, is a topic of high current interest and offers new possibilities for high-temperature fuel cells and for high-energy-density storage batteries. The properties of semiconductor electrodes are also of interest in their application to photogalvanic energy conversion.

Electrochemical phenomena are also of considerable importance in biology and medicine. The conduction of nerve impulses depends on the current–voltage relationship for sodium-ion transport across the cell membrane. Much of living matter is colloidal, consisting of small (10–10^4 Å) particles suspended in an aqueous solution. Through adsorption of ions, colloid particles acquire a double-layer structure which determines the stability of the suspension. Advances in the development of miniaturized ion-selective electrodes now offer the prospect of real-time monitoring of some medical conditions, and electrochemical measurements on living single cells are now becoming feasible.

See also: Conduction; Crystal Defects; Diffusion; Electrochemical Conversion and Storage.

Bibliography

J. O'M. Bockris and A. K. N. Reddy, *Modern Electrochemistry*, 2nd ed. Plenum, New York, 1997. (E)

P. G. Bruce, B. Dunn, J. W. Goodby and J. R. West, *Solid State Electrochemistry*. Cambridge University Press, New York, 1997. (A)

A. J. Bard and L. R. Faulkner, *Electrochemical Methods: Fundamentals and Applications*, 2nd ed. Wiley, New York, 2000. (A)

National Materials Advisory Board, *New Horizons in Electrochemical Science and Technology*. National Academy Press, Washington, 1986. (E)

Electrodynamics, Classical

J. D. Jackson

Electrodynamics, a word used by Ampère in his pioneering researches 150 years ago, may properly be used to encompass all electromagnetic phenomena. There is also a more restricted meaning: electromagnetic fields, charged particles, and their mutual interaction at a microscopic level, excluding in practice, if not in principle, phenomena associated with macroscopic aggregates of matter. *Classical* electrodynamics then consists of the regime where (relativistic) classical mechanics applies for the motion of particles, and the photon nature of electromagnetic fields can be ignored. Its quantum generalization, called quantum electrodynamics, is necessarily employed for phenomena without classical basis (e. g., pair production), as well as where quantum effects are significant.

Separate articles exist on many aspects of macroscopic electromagnetism (*see* Electromagnetic Radiation, Electrostatics, Magnets and Magnetostatics, Microwaves and Microwave Circuitry). The emphasis here is on basic principles and selected results of classical electrodynamics in the restricted sense. The Gaussian system of units and dimensions is used. See the Appendix of Ref. [1] for the connection to the SI or mksa units of practical electricity and magnetism.

Maxwell Equations in Vacuum

The differential equations of electromagnetism in vacuum are the Maxwell equations;

$$\nabla \cdot \mathbf{E} = 4\pi\rho , \tag{1a}$$

$$\nabla \times \mathbf{B} - \frac{1}{c}\frac{\partial \mathbf{E}}{\partial t} = \frac{4\pi}{c}\mathbf{J} , \tag{1b}$$

$$\nabla \times \mathbf{E} + \frac{1}{c}\frac{\partial \mathbf{B}}{\partial t} = 0 , \tag{1c}$$

$$\nabla \cdot \mathbf{B} = 0 . \tag{1d}$$

The quantities ρ and $\mathbf{J}$ are the source densities of charge and current, related because of conservation of charge by the differential continuity equation

$$\frac{\partial \rho}{\partial t} + \nabla \cdot \mathbf{J} = 0 . \tag{2}$$

The electromagnetic field quantities $\mathbf{E}$ and $\mathbf{B}$, called respectively the electric field and the magnetic induction, are related to the mechanical force per unit charge according to the Lorentz force equation

$$\mathbf{F} = q \left(\mathbf{E} + \frac{\mathbf{v}}{c} \times \mathbf{B} \right) \tag{3}$$

where $\mathbf{F}$ is the force exerted on a point charge q moving with velocity $\mathbf{v}$ in the presence of external fields $\mathbf{E}$ and $\mathbf{B}$. The parameter c that enters the Maxwell equations and the force equation has the dimensions of a speed (length/time). Solution of the Maxwell equations in free space shows the existence of transverse waves propagating with the speed c. It is thus the speed of light and other electromagnetic radiation. The speed of light is now defined to be $299\,792\,458\,\mathrm{m/s}$ (*see* Metrology).

The four Maxwell equations (1a)–(1d) (actually eight scalar equations) are expressions of the experimental laws of Coulomb, Ampere and Maxwell, Faraday, and the absence of magnetic charges, respectively. This can be seen more clearly in integral form – by integration over a finite volume V, bounded by a closed surface S, and use of Gauss's law, for Eqns. (1a) and (1d); by integration over an open surface S_0, bounded by a closed curve C, and use of Stokes's theorem, for Eqns. (1b) and (1c). *Coulomb's law* then reads

$$\oint_S \mathbf{E} \cdot \mathbf{n} \, da = 4\pi \int_V \rho \, d^3 r \tag{4}$$

where da is an element of area on S and $\mathbf{n}$ is a unit, outwardly directed normal at da. Equation (4) states that the total electric flux out of the volume V is equal to 4π times the total charge inside. It can be shown that this is a consequence of (a) the inverse-square law of force between point charges, (b) the central nature of that force, and (c) linear superposition. These, plus isotropy of the field of a point charge, are the elements of Coulomb's laws of electrostatics. The corresponding integral of Eq. (1d) has zero on the right-hand side – there are (as far as we presently know) no magnetic charges.

Ampère's (and Maxwell's) *law* has the integral form

$$\oint_C \mathbf{B} \, d\mathbf{l} = \int_{S_0} \left(\frac{4\pi}{c} \mathbf{J} + \frac{1}{c} \frac{\partial \mathbf{E}}{\partial t} \right) \cdot \mathbf{n} \, da \tag{5}$$

where $d\mathbf{l}$ is an element of length tangent to the curve C. The sense of the normal $\mathbf{n}$ to the surface S_0 is determined by the right-hand rule with respect to $d\mathbf{l}$. For static fields the second term on the right is absent. Then Eq. (5) is equivalent to Ampère's laws of forces between current loops. The second term on the right shows that time-varying electric fields produce magnetic fields just as do currents. This term is sometimes called Maxwell's displacement current. It is an essential modification of Ampère's laws for rapidly varying fields.

The analogous integral statement of Eq. (1c), *Faraday's law*, is

$$\oint_C \mathbf{E} \, d\mathbf{l} = -\frac{1}{c} \frac{d}{dt} \int_{S_0} \mathbf{B} \cdot \mathbf{n} \, da \,. \tag{6}$$

The line integral of the electric field around the path or circuit C, called the electromotive force, is proportional to the negative (Lenz's law) time rate of change of magnetic flux through that circuit.

The application of Eqns. (4), (5), and (6) to currents and voltages for conductors and circuit elements is discussed in other articles in this volume, *see* Network Theory, Analysis and Synthesis; Resistance and Impedance. When the time variations are rapid enough to make the displacement current term in Eq. (5) important, the ideas of lumped circuits fail. The finite speed of propagation and the wave nature of the phenomena must be taken into account, *see* Transmission Lines and Antennas; Microwaves and Microwave Circuitry. For explicit discussion of the connection between field and circuit points of view, see Refs. [2] and [3].

Conservation Laws

The differential continuity equation that follows from the Maxwell equations and expresses conservation of electromagnetic energy is

$$\frac{\partial u}{\partial t} + \nabla \cdot \mathbf{S} + \mathbf{J} \cdot \mathbf{E} = 0 \tag{7}$$

where u, the energy density, and $\mathbf{S}$, called Poynting's vector and representing energy flux (energy per unit area per unit time), are given by the expressions

$$u = \frac{1}{8\pi} (E^2 + B^2) \,, \quad \mathbf{S} = \frac{c}{4\pi} (\mathbf{E} \times \mathbf{B}) \,. \tag{8}$$

The term $\mathbf{J} \cdot \mathbf{E}$ in Eq. (7) represents the rate of work per unit volume being performed on the sources by the electromagnetic fields; it describes the conversion of electrical into mechanical energy of the charged particles that give rise to ρ and $\mathbf{J}$. To see this explicitly, suppose that the current $\mathbf{J}$ is caused by point particles of charges q_j, positions $\mathbf{r}_j(t)$, and velocities $\mathbf{v}_j(t)$. Then the volume integral of $\mathbf{J} \cdot \mathbf{E}$ can be written

$$\int_V \mathbf{J} \cdot \mathbf{E} \, d^3 r = \sum_{j(V)} q_j \mathbf{v}_j \cdot \mathbf{E}_J = \sum_{j(V)} \mathbf{v}_j \cdot \mathbf{F}_J \,. \tag{9}$$

The second expression on the right follows from use of the Lorentz force equation, Eq. (3). Since $\mathbf{v}_j \cdot \mathbf{F}_J$ is the rate of change of mechanical energy of the jth particle, the sum represents the total rate of change of mechanical energy of all the particles within the volume V. The volume integral of Eq. (7) thus expresses conservation of electromagnetic and mechanical energy, the first term being the rate of change of electromagnetic energy, the last that of mechanical energy, and the middle term being the total flux of electromagnetic energy out of the volume.

Conservation of momentum is expressed in integral form by

$$\frac{d}{dt} (\mathbf{P}_M + \mathbf{P}_e)_i = \oint_S \sum_j T_{ij} n_j \, da \tag{10}$$

where the rate of change of mechanical momentum is the volume integral of the Lorentz force density:

$$\frac{d\mathbf{P}_m}{dt} = \int_V \frac{1}{4\pi c} \mathbf{E} \times \mathbf{B}\, d^3 r = \frac{1}{c^2} \int_V \mathbf{S}\, d^3 r , \tag{11}$$

and the field momentum within the volume V is

$$\mathbf{P}_c = \int_V \frac{1}{4\pi c} \mathbf{E} \times \mathbf{B}\, d^3 r = \frac{1}{c^2} \int_V \mathbf{S}\, d^3 r . \tag{12}$$

The flow of momentum out of V through the surface S is given by the integral over S of the contraction of the components of the unit normal $\mathbf{n}$ with the *Maxwell stress tensor*:

$$T_{ij} = \frac{1}{4\pi} \left[E_i E_j + B_i B_j - \frac{1}{2}(E^2 + B^2)\delta_{ij} \right] . \tag{13}$$

The conservation laws for energy and momentum (and angular momentum) show that electromagnetic fields have energy, momentum, and angular momentum, as much as particles do. This seems obvious at the quantum level of photons, but is just as true classically. Although it may seem counterintuitive, electromagnetic angular momentum, $\mathbf{L}_e$ defined by

$$\mathbf{L}_e = \frac{1}{4\pi c} \int_V \mathbf{r} \times (\mathbf{E} \times \mathbf{B})\, d^3 r \tag{14}$$

can exist even for static fields and must be included in considerations of conservation of angular momentum for electromechanical systems.

Relativistic Transformations of Fields

The equations of classical electrodynamics can be written in a form that exhibits explicitly their covariance under the transformations of *special relativity*. The charge and current densities ρ and $\mathbf{J}$ form the time and space components of a Lorentz four-vector J^μ; the electromagnetic fields are components of an antisymmetric second-rank Lorentz tensor $F^{\nu\mu}$. For the Lorentz-covariant forms of Eqns. (1) and (2) and other details, see Refs. [1, 8, 10, 11]. Of interest are the explicit relations showing how the electric and magnetic fields $(\mathbf{E}', \mathbf{B}')$ in one inertial frame of reference, K', manifest themselves as the fields $(\mathbf{E}, \mathbf{B})$ in a frame K moving uniformly with respect to the first. Let the frame K' have a velocity $\mathbf{v} = \boldsymbol{\beta} c$ with respect to the frame K and let $\gamma = (1 - \beta^2)^{-1/2}$. Then the relations among the fields are

$$\mathbf{E} = \gamma(\mathbf{E}' - \boldsymbol{\beta} \times \mathbf{B}') - \frac{\gamma^2}{\gamma + 1}\boldsymbol{\beta}(\boldsymbol{\beta} \cdot \mathbf{E}') \tag{15a}$$

$$\mathbf{B} = \gamma(\mathbf{B}' + \boldsymbol{\beta} \times \mathbf{E}') - \frac{\gamma^2}{\gamma + 1}\boldsymbol{\beta}(\boldsymbol{\beta} \cdot \mathbf{B}') \tag{15b}$$

The inverse relations can be obtained by interchanging primed and unprimed quantities and reversing the sign of $\boldsymbol{\beta}$. Either set shows that the field components parallel to the velocity are the same in the two frames, but the transverse components of $\mathbf{E}$ and $\mathbf{B}$ become mixed. In particular, a purely electric field in K' appears with both electric and magnetic components in K.

The fields of a point charge q in uniform motion in the frame K provide an important illustration. Let the charge be at rest in frame K'. In K, let the coordinates of the charge be $x_1 = vt$, $x_2 = x_3 = 0$, and let the fields be observed at a point on the x_2 axis with coordinates $(O, b, 0)$. The distance b is the distance of closest approach of the charge to the observer. The purely static Coulombic field in K' has, in the frame K, time-dependent components,

$$E_1 \quad = \quad E_1' = -\frac{q\gamma vt}{(b^2 + \gamma^2 v^2 t^2)^{3/2}} , \tag{16a}$$

$$E_2 \quad = \quad \gamma E_2' = \frac{q\gamma b}{(b^2 + \gamma^2 v^2 t^2)^{3/2}} , \tag{16b}$$

$$B_3 \quad = \quad \gamma\beta E_2' = \beta E_2 . \tag{16c}$$

For highly relativistic particles ($\gamma \gg 1$) the transverse components, E_2 and B_3, dominate and are almost equal in magnitude. The fields are highly compressed in time ($\Delta t \approx b/\gamma v$) and appear like a pulse of electromagnetic radiation with a spectrum of frequencies extending up to $\omega \approx \gamma v/b$. This equivalence between the fields of a relativistic charged particle and a pulse of radiation can be fruitfully exploited to relate charged-particle-induced and photon-induced processes, for example, Compton scattering and bremsstrahlung, or pair production by photons and electrons on nuclei. The basic photon process may be quantum mechanical, but the spectrum of equivalent photons is essentially a classical concept.

Examples of Charged-Particle Motion in External Fields

Many phenomena involving particles and fields fall into one of two classes: (a) fields given, particle motion to be determined; and (b) sources (particle motion) given, fields, including radiation, to be found. This division ignores the mutually reactive effects of fields on sources and vice versa. For class (a), the effects of radiation during the motion can often be included in an approximate way – see Ref. [1], Chapter 16. Some examples of charged-particle motion in external fields are given below. Special relativity effects are included; emission of radiation is ignored.

1. *Particle initially at rest at the origin in a constant uniform electric field.* Let the field **E** point in the x direction, the particle's charge and mass be q and m, respectively. At time t its speed and position are

$$v = \frac{at}{\left(1 + \dfrac{a^2 t^2}{c^2}\right)^{1/2}} , \quad x = \frac{at^2}{1 + \left(1 + \dfrac{a^2 t^2}{c^2}\right)^{1/2}}$$

where $a = qE/m$ is the initial (nonrelativistic) acceleration. The speed as a function of position is $v = (2ax)^{1/2} \cdot (1 + ax/2c^2)^{1/2}/(1 + ax/c^2)$. For short times ($t \ll c/a$), the motion is the familiar nonrelativistic behavior of a particle under constant acceleration. For $t \gtrsim c/a$, the motion becomes relativistic; the particle continues to gain energy at a constant rate but its speed approaches the speed of light and its position is given simply by $x \simeq ct$.

2. *Particle in a constant uniform magnetic field.* The particle is initially at the origin, moving with speed v in the y direction. The magnetic field **B** is in the z direction. At time t the positional coordinates of the particle are $x = (v/\omega)(1 - \cos\omega t)$, $y = (v/\omega)\sin\omega t$, $z = 0$, where ω is the gyration or cyclotron frequency,

$$\omega = \frac{qB}{\gamma mc} = \frac{qcB}{E} \,. \tag{17}$$

Here $\gamma = (1 - v^2/c^2)^{-1/2}$ and $E = \gamma mc^2$ is the total energy of the particle. The particle's trajectory is a circle of radius $R = v/|\omega| = cp/|q|B$. This circular motion in a uniform magnetic field is the basis of operation of the cyclotron. The inverse dependence on γ of the frequency ω is what necessitates modulation of the frequency of the accelerating voltage as the particle gains energy during the acceleration cycle.

3. *Particle drifts in inhomogeneous magnetic fields.* The simple circular motion (or helical, if the particle has a component of velocity parallel to the magnetic field) in a uniform, constant magnetic field is modified if the field varies in strength and/or direction in space. If the variations are slow enough, the short-term motion is circular, but the center of the circle "drifts" slowly in space and perhaps the radius of the orbit changes, too. There are various kinds of drifts, depending on the type of field variation that occurs. These are important for particle motion in the Van Allen radiation belts around the earth and other planets and for particle confinement in plasmas in thermonuclear research. See Chapter 12 of Ref. [1] for some elementary aspects and Ref. [4] for more advanced and detailed discussion. For the planetary particle belts, see Ref. [5].

Radiation by Charged Particles

A charged particle in uniform motion has only localized "velocity" fields associated with it, fields that, in the inertial frame in which the particle is at rest, are just the inverse-square radial electrostatic field of a point charge. When the particle undergoes acceleration it exhibits additional fields **E** and **B**, equal in magnitude, proportional to the acceleration, transverse to the radius vector from the particle, and falling off with distance only with the first inverse power. These "acceleration" fields represent radiation. There are several features of the radiation by particles that are of interest: total radiated power, angular distribution, and frequency distribution. The limiting examples of nonrelativistic motion and extreme relativistic motion are considered here.

Nonrelativistic Motion

For a nonrelativistic particle of charge q, the instantaneous power radiated per unit solid angle is

$$\frac{dP}{d\Omega} = \frac{q^2}{4\pi c^3}|\dot{\mathbf{v}}|^2 \sin^2\theta \tag{18}$$

where $\dot{\mathbf{v}}$ is the acceleration and θ is the polar angle measured relative to the direction of $\dot{\mathbf{v}}$. The total radiated power is given by Larmor's formula:

$$P = \frac{2}{3}\frac{q^2}{c^3}|\dot{\mathbf{v}}|^2 \,. \tag{19}$$

The frequency distribution of radiated energy is proportional to the square of the Fourier transform of $\dot{\mathbf{v}}(t)$. For *periodic* motion, with period $\tau = 2\pi/\omega_0$, the spectrum consists of discrete lines at $\omega = n\omega_0$, $n = 1, 2, 3, \ldots$ For *simple harmonic* motion there is, of course, only the fundamental.

Relativistic Motion, Synchrotron Radiation

When the particle's speed is comparable with the speed of light, new features enter. For the same *acceleration*, much more radiation is emitted. For speeds close to c, it is of course difficult to produce large longitudinal accelerations, but transverse accelerations can still be appreciable. The Larmor power formula has the relativistic generalization

$$P = \frac{2q^2}{3c}\gamma^6[(\dot{\boldsymbol{\beta}}^2 - (\dot{\boldsymbol{\beta}} \times \boldsymbol{\beta})^2]$$

(20)

where $\boldsymbol{\beta} = \mathbf{v}/c$, $\dot{\boldsymbol{\beta}} = \dot{\mathbf{v}}/c$, and $\gamma = (1 - \beta^2)^{-1/2}$. The presence of a high power of γ shows that for extremely relativistic motion ($\gamma \gg 1$) the radiated power can become very large for a given value of $|\dot{\boldsymbol{\beta}}|$.

The angular and frequency distributions are drastically modified for $\gamma \gg 1$. In an inertial frame where the particle is instantaneously at rest, the angular distribution is given by Eq. (15), but the Lorentz transformation to the laboratory causes the radiation to be concentrated almost entirely along the direction of motion, within angles of the order of $\Delta\theta \simeq \gamma^{-1}$. Thus, independent of the details of the acceleration, the particle radiates a narrow "searchlight" beam in its direction of motion. The frequency spectrum, even for simple harmonic motion, contains very many harmonics, up to a maximum frequency of the order of $\omega_{max} = \gamma^3\omega_0$.

A relativistic electron of charge e moving at constant speed βc in a circular orbit of radius R is an idealization of the actual situation in an electron synchrotron or storage ring. The total radiated power per electron is

$$P = \frac{2}{3}\frac{e^2c}{R^2}(\beta\gamma)^4 .$$

(21)

With $\omega_0 \simeq 10^7$ s and $\gamma \simeq 10^4$, the frequency spectrum is a broad distribution extending up to and beyond $\omega_{max} \simeq 10^{19}\,\text{s}^{-1}$, corresponding to x-ray energies of the order of 10 keV. The radiation, called *synchrotron radiation*, provides an intense, wide-band source of photons for solid-state and biophysical researches. The intensity of the radiation can be enhanced and its properties tailored to specific needs by the insertion of "wiggler" magnets into the storage ring. These magnets have strong bending fields, alternating in polarity in a regular way. The periodicity and electron energy can be chosen to enhance the synchrotron radiation in a desired frequency range. Such amounts of radiation are generated that a long wiggler can serve as the engine for a *free-electron laser*. An accessible discussion of synchrotron radiation, wigglers, and undulators appears in Chapter 14 of Ref. [1].

The unique and well-understood properties of synchrotron radiation (polarization as well as frequency spectrum) permit it to be identified in astrophysical circumstances (radio and infrared emission from planets, radio to x-ray emission from supernovas and pulsars) and aid in establishing the conditions of particle motion there.

Other Reading

Classical electrodynamics is a vast subject and has applications in every field of science and engineering. Its richness and beauty can only be appreciated by exploring its extensive literature. Some recent basic texts are Purcell [6] at an elementary level, and Griffiths [7] and Schwartz [8] at an intermediate level. More advanced physics texts include Jackson [1], Landau and Lifshitz [9, 10], and Panofsky and Phillips [11]. The history of the development of electrodynamics can be found in Feather [12] and Vol. 1 of Whittaker [13], while the present experimental status is reviewed in the Introduction to [1].

See also: Conservation Laws; Electromagnetic Radiation; Electrostatics; Faraday's Law of Electromagnetic Induction; Magnets (Permanent) and Magnetostatics; Maxwell's Equations; Microwaves and Microwave Circuitry; Quantum Electrodynamics; Radiation Belts; Relativity, Special Theory; Synchrotron; Synchrotron Radiation; Transmission Lines and Antennas.

References

[1]   J. D. Jackson, *Classical Electrodynamics*, 3rd ed. Wiley, New York, 1998. (A)
[2]   R. B. Adler, L. J. Chu, and R. M. Fano, *Electromagnetic Energy, Transmission and Radiation*. Wiley, New York, 1960. (I)
[3]   R. M. Fano, L. J. Chu, and R. B. Adler, *Electromagnetic Fields, Energy, and Forces*. Wiley, New York, 1960. (I)
[4]   P. C. Clemmow and J. P. Dougherty, *Electrodynamics of Particles and Plasmas*. Addison-Wesley, Reading, Mass., 1969. (A)
[5]   B. Rossi and S. Olbert, *Introduction to the Physics of Space*. McGraw–Hill, New York, 1970. (I–A)
[6]   E. M. Purcell, *Electricity and Magnetism* (Berkeley Physics Course Vol. 2), 2nd ed. McGraw–Hill, New York, 1985. (E)
[7]   D. J. Griffiths, *Introduction to Electrodynamics*. 2nd ed. Prentice-Hall, Englewood Cliffs, N.J., 1989. (I)
[8]   M. Schwartz, *Principles of Electrodynamics*. McGraw–Hill, New York, 1972. (I)
[9]   L. D. Landau and E. M. Lifshitz, *Electrodynamics of Continuous Media*. Addison-Wesley, Reading, Mass., 1960. (A)
[10]   L. D. Landau and E. M. Lifshitz, *The Classical Theory of Fields*, 34th., rev. English ed. Addison-Wesley, Reading, Mass., 1971. (A)
[11]   W. K. H. Panofsky and M. Phillips, *Classical Electricity and Magnetism*, 2nd ed. Addison-Wesley, Reading, Mass., 1962. (A)
[12]   N. Feather, *Electricity and Matter*. University Press, Edinburgh, 1968. (E)
[13]   E. T. Whittaker, *A History of the Theories of Aether and Electricity*, 2 vols. Nelson, reprinted by American Institute of Physics, New York, 1987. (E–I)

Electroluminescence

M. N. Kabler

Electroluminescence is the efficient generation of light in a nonmetallic solid by an applied electric field. Electroluminescence is "cool" light in the sense that the brightness is far above that characteristic of the temperature alone; thus incandescent light is excluded.

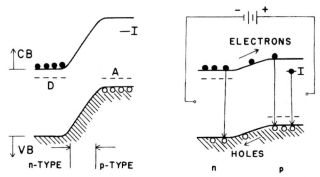

Fig. 1: Schematic representation of electron energies in a *p–n* junction with no bias voltage (left), and with a forward bias in the 1 V range (right). The vertical arrows represent electroluminescent processes, one of which occurs through an impurity level *I*.

The primary electronic states of nonmetallic solids comprise two bands of allowed states separated by a forbidden gap where only states due to impurity atoms or lattice imperfections can exist. At normal temperatures the higher or conduction band (CB) is empty except for a few mobile electrons, while the lower or valence band (VB) is filled with electrons except for a few vacant states called holes, which are also mobile. Because of interactions with thermal vibrations, any electrons in the CB immediately fall to its low-energy edge and any holes in the VB rise to its high-energy edge. When excess electrons and holes are produced and brought into proximity by the action of an applied electric field, the electrons spontaneously fall into or recombine with the holes. A recombination event releases energy comparable to the band gap, which is dissipated as heat or radiated as electroluminescence.

p–n Junction Electroluminescence

The most effective structure for producing electroluminescence is the *p–n* junction. Such a junction is illustrated schematically in Fig. 1, which indicates how electron energy varies as a function of distance perpendicular to the junction both with and without an applied electric field. The device is commonly called a light-emitting diode, or LED. Semiconductors selectively doped with impurity atoms can, at normal temperatures, exhibit high conductivities arising either from extra electrons or extra holes. In the former case the extra electrons come from donor states *D* near the CB edge, and the material is termed *n* type; in the latter case holes are created when VB electrons are trapped in acceptor states *A*, and the material is *p* type. When *n*-type and *p*-type regions are made contiguous with each other, a *p–n* junction is formed. Being nearly void of mobile electrons and holes, the interface region acts as an insulator and is the location of a strong electric field. Under a constant forward bias of a few volts, applied through suitable conducting contacts, electrons and holes are swept in opposite directions across the junction, where they can recombine.

The energy of an emitted photon is within the range of transparency of the material, that is, no higher than the band gap. Although many materials have band gaps in the ultraviolet, efficient LEDs operating beyond the yellow–green range are not yet available. This is largely due to difficulties in producing *p–n* junctions in the wider-band-gap materials.

Luminescent Processes

In general, electrons and holes can recombine near atomic impurities as well as in regions of the unperturbed lattice. The color of the luminescence and the relative probability of producing light instead of heat are characteristic of the particular impurities involved. In many semiconductors, luminescent transitions from the bottom of the CB to the top of the VB cannot take place without the intervention of a lattice vibration or imperfection in order to satisfy the law of conservation of momentum. Such luminescent transitions are termed indirect, and the material is said to have an indirect edge. An important example is GaP, with a band gap in the green spectral region near 2.3 eV (540 nm). In other materials with so-called direct edges this transition can occur without restriction; GaAs is an example, its band gap being in the infrared near 1.5 eV. Because of the momentum-conservation requirement, an indirect luminescent transition is relatively unlikely in pure materials and thus electroluminescence is inefficient. However, a crystal with an indirect edge can be doped with impurity atoms which sequentially trap both electrons and holes, thereby providing a radiative recombination path up to several orders of magnitude more efficient. Such a path is included in Fig. 1, where *I* represents the localized impurity state.

Light-Emitting Diodes

Commercial activity has concentrated on developing efficient electroluminescent devices in the visible spectral range. Most materials with band gaps in this range, particularly toward the violet, cannot be made to incorporate *p–n* junctions, and only compounds and alloys of elements from groups IIIA and VA of the periodic table have thus far yielded LEDs of broad utility.

For the green spectral region, N doping of GaP has proved quite effective in mitigating the inefficiency due to the indirect edge while at the same time retaining an emission wavelength near the band-gap energy. N is isoelectronic with P, for which it substitutes in the lattice. Thus the N normally carries no net charge, and its attractive potential is short range; it traps a conduction electron in a strongly localized orbit and is, for this reason, very effective in providing an interaction through which momentum is conserved during recombination with an approaching hole.

Other dopants, for example, Zn and O incorporated in the lattice as nearest-neighbor pairs, can produce efficient red luminescence in GaP. However, considerable success in the yellow-to-red range has been achieved by N doping of GaAs–GaP alloys. The alloy composition can be chosen to give a particular band-gap energy and emission color. Impurity doping to increase the efficiency of emission is not required for the red and near infrared, since the corresponding alloys have direct edges. The same is true for alloys of AlAs and GaAs.

In many of these *p–n* junction devices the efficiency with which electrical energy is converted to light inside the material can be 10% or greater at room temperature. However, because of geometrical constraints, internal reabsorption, and internal reflection at the semiconductor–air interface, only a small fraction of the photons created actually can emerge into a useful beam. External efficiencies of commercial LEDs generally lie in the 0.1-5% range. The efficiencies of luminescent processes themselves invariably fall at higher temperatures, and thus light output is limited by internal heating due to nonradiative processes. The prominent semiconductors Si and Ge find little utility as LEDs, because their band edges lie in the infrared and are indirect.

Fig. 2: The minute, star-like luminescence from an AlAs–GaAs heterojunction laser glowing beneath the eye of an ordinary sewing needle. This laser is similar to lasers designed for optical communications. (Photograph courtesy of RCA Laboratories.)

Laser Diodes

Electroluminescent lasers can be made by fabricating special p–n junction configurations. Since light amplification requires population inversion, the states near the bottom of the conduction band must be more than half filled with electrons and states near the top of the valence band must be more than half empty, i. e., more than half filled with holes. This requires large hole and electron currents into the recombination region and constriction of this region to a small volume in order to maintain high concentrations. An example of an LED which can be driven into laser oscillations is GaAs heavily doped to increase electron and hole concentrations. However, relatively high threshold current densities are required. The most efficient laser diodes are based on the heterojunction, which is a p–n junction similar to that of Fig. 1 except that there are two different materials with different band gaps on opposite sides of the junction. This effectively restricts recombination to that side of the junction having the smaller band gap, as well as providing a refractive-index gradient which aids in confining the light.

Heterojunctions are formed by depositing thin layers of the different semiconductors on each other, usually from the liquid phase. When contiguous layers differ in lattice parameter, for example, InAs on GaAs, there occurs near the interface an array of defects or dislocations which act as nonradiative recombination centers and thereby drastically lower the luminescent efficiency. Thus heterojunctions are usually constructed from alloys with nearly identical lattice parameters. A prime example is the AlAs–GaAs system, in which the lattice parameter changes by only 0.14% throughout the entire range of composition. A continuous-wave laser comprising two heterojunctions from AlAs–GaAs alloys is depicted alongside an ordinary sewing needle in Fig. 2. This laser is emitting at a wavelength of 750 nm, and the thickness the emitting region is only about 500 nm.

High-Field Electroluminescence

It is generally possible to inject electrons or holes from one material into another if a sufficiently high electric field is applied across the interface. Two characteristic examples of electroluminescence originating in this way are the reverse-biased p–n junction and certain luminescent insulators containing metallic particles.

If a voltage of polarity opposite that shown in Fig. 1 is applied to a p–n junction, the top of the VB on the p side can be raised considerably above the bottom of the CB on the n side. Electrons from the p side can tunnel horizontally through the forbidden gap to the n side and into CB states having high kinetic energies. If this kinetic energy is higher than the band-gap energy, the electron can collide with a normal lattice atom and create a new electron–hole pair which can, in turn, recombine radiatively. In practice a considerable amount of energy is lost as heat, and the reverse-biased p–n junction is therefore not an efficient electroluminescent source.

The second class of high-field process is exemplified by luminescent ZnS in powder form and doped with Cu_2S. Inside the small ZnS crystallites, the Cu_2S precipitates on lattice imperfections to form submicroscopic, needlelike, electrically conducting particles. The external electric field is strongly intensified near the sharp ends of the needles, where electrons or holes are emitted into the ZnS on alternate half cycles of the applied field. The holes are trapped at impurity atoms, and electrons subsequently arrive to produce recombination luminescence. Cu is the principal luminescent impurity, but many others can play various roles in the process, and most visible colors can be produced. Large electroluminescent panels of this type have been constructed which sandwich the luminescent powder between a metallic base and a transparent conducting electrode such as tin oxide, but their efficiencies and lifetimes are too low to be competitive for general space illumination.

Applications

There is now available a wide range of LEDs, which fill an increasingly large fraction of the commercial market for small, low-power sources operating as indicator lights and in symbolic displays. Pocket computers are a conspicuous example. The low voltages required to drive LEDs are compatible with transistor circuitry, a considerable advantage. It is probable that panels of the ZnS type will also gain acceptance for display applications. As the reliability and lifetime of laser diodes improves they will find numerous uses where bright, narrow light beams are required. Optical communications is a particularly promising area, since laser diodes are ideal sources for coupling into fiber-optic waveguides. Their size and operating wavelengths match the diameters and maximum-transmission wavelengths of optical fibers (see Fig. 2), and their output can be modulated at frequencies in the 10^8-Hz range. The less stringent source requirements of many fiber-optic systems can be met by current LEDs. Electroluminescent devices will probably find a role in the interior illumination market, but not before further advances are made in the field of materials science.

See also: Crystal Defects; Electron Energy States in Solids And Liquids; Luminescence; Semiconductors, Crystalline.

Bibliography

H. Kressel, I. Ladany, M. Ettenberg, and H. Lockwood, "Light Sources," *Phys. Today*, p. 38, May 1976.

H. J. Queisser and U. Heim, "Optical Emission from Semiconductors," in *Annual Review of Materials Science* (R. A. Huggins, R. H. Bube, and R. W. Roberts, eds.), Vol. 4, p. 125, 1974.

C. H. Gooch, *Injection Electroluminescent Devices*. Wiley, N.Y., 1973.

R. J. Elliott and A. F. Gibson, *An Introduction to Solid State Physics and Its Applications*. Barnes and Noble, N.Y., 1974.

Electromagnetic Interaction

T. Kinoshita

The electromagnetic interaction is one of the four fundamental interactions known in physics, the other three being the *strong* interaction, which keeps the atomic nucleus together; the *weak* interaction, responsible for spontaneous disintegration (beta decay) of radioactive nuclei; and the *gravitational* interaction, which keeps the stars and galaxies in their orbits. Because of the relative ease with which effects of the electromagnetic interaction can be observed, it is the most thoroughly studied and best understood of the four interactions.

Historically, the structure of the electromagnetic interaction was first determined from the observation of macroscopic electromagnetic phenomena. Its explicit form is given by Maxwell's equations and the Lorentz force acting on the charge, which form the starting point of classical electrodynamics.

Classical electrodynamics fails, however, if it is applied to atomic and subatomic systems; such systems must be treated by quantum mechanics.

Quantization of classical electrodynamics is straightforward if the particles are restricted to nonrelativistic kinematics. In this case, we have only to replace Newton's equation of motion (or, equivalently, Hamilton's equation) by the corresponding Schrödinger equation, and turn the electromagnetic field into a quantized field which means that it can be regarded as a superposition of creation and annihilation operators of photons. (In some applications, the electromagnetic field may be treated as an unquantized field.) The simplest way to introduce the electromagnetic interaction in a quantum-mechanical system is to assume, by invoking the correspondence principle, that it has exactly the same form as in classical electrodynamics. (This does not mean that nothing has changed. In fact, the physics is quite different since the particles now obey the quantum-mechanical equation of motion instead of the classical one.) Such a quantization scheme, if the electron spin is taken into account, is found to be capable of describing practically all properties of atoms and molecules, including emission and absorption of photons.

Quantization of a relativistic particle is considerably more complicated. Straightforward application of the correspondence principle to a relativistic particle leads to the Klein–Gordon equation if the spin of the particle is zero, the Dirac equation if the spin is $\frac{1}{2}$ in units of $\hbar$, etc. However, a satisfactory synthesis of quantum mechanics and relativity requires that not only the electromagnetic field but also the particles be described as quantized fields. The theory thus formulated for the system of interacting electrons and photons is called quantum electrodynamics (QED). This theory can be described by the Lagrangian (or Lagrangian density, to be precise)

$$L(x) = L_e(x) + L_p(x) + L_I(x) \tag{1}$$

where

$$L_e = \bar{\psi}(x)\left(i\sum_{\mu=0}^{3}\gamma^\mu\frac{\partial}{\partial x^\mu} - m\right)\psi(x), \tag{2}$$

$$L_p = -\frac{1}{4}\sum_{\mu=0}^{3}\sum_{\nu=0}^{3}R_{\mu\nu}(x)F^{\mu\nu}(x), \tag{3}$$

$$L_I = -\sum_{\mu=0}^{3}j_\mu(x)A^\mu(x) \tag{4}$$

are the electron, photon, and interaction parts, respectively, of the Lagrangian and

$$F_{\mu\nu}(x) = \frac{\partial A_\nu(x)}{\partial x^\mu} - \frac{\partial A_\mu(x)}{\partial x^\nu}, \tag{5}$$

$$j_\mu(x) = e\psi(x)\gamma_\mu\psi(x). \tag{6}$$

$$\tag{7}$$

The Greek indices μ and ν take the values 0, 1, 2, 3; 0 for the time axis, and 1, 2, 3 for the space axis. Four-vectors with lower and upper indices are related to each other by $A_0 = A^0$, $A_i = -A^i$, $i = 1, 2, 3$. $A_\mu(x)$ is the four-vector potential of the electromagnetic field, γ^μ is the 4×4 Dirac matrix, $\psi(x)$ is a 4×1 matrix representing the "electron field," and $\bar{\psi}(x) = \psi^\dagger(x)\gamma^0$, $\psi^\dagger$ being the Hermitian conjugate of ψ. Finally, e and m are the charge and mass of the electron. (For simplicity we choose the units in which both the velocity of light c and the modified Planck's constant $\hbar$ are equal to one.) The form of the Lagrangian (1) is identical for all observers moving with constant velocity with respect to each other (in other words, invariant under Lorentz transformations). It is also unchanged under another continuous transformation, called the gauge transformation, which is related to the fact that the photon is a massless particle. Finally, it is invariant under discrete transformations C (charge conjugation), P (space reflection), and T (time reversal).

The principle of least action applied to the Lagrangian (1) yields equations of motion for A_μ, ψ, and $\bar{\psi}$. In particular that for A_μ is formally identical (when supplemented by an extra constraint) with the classical Maxwell's equations. The crucial consequence of field quantization is that A_μ is now an operator that creates or annihilates a photon. Similarly, ψ ($\bar{\psi}$) is an operator that annihilates (creates) an electron or creates (annihilates) a positron. The electromagnetic interaction is expressed by L_I, of (4), or equivalently by the interaction energy (i.e., the interaction part of the Hamiltonian density)

$$\mathcal{H}_I(x) = -L_I(x) = j_\mu(x)A^\mu(x) \tag{8}$$

where the summation over the indices μ is suppressed following the convention.

Thus, we finally have a theory of electromagnetic interaction that satisfies the requirements of both relativity and quantum mechanics. Unfortunately, this theory still has a serious fault: Although the lowest-order predictions of the theory in perturbation expansion (which is essentially an expansion of all quantities in powers of the fine-structure constant $\alpha = e^2/4\pi\hbar c = 1/137$) are in good agreement with experiments, the agreement is destroyed completely if we try to improve it by including higher-order terms that are all infinite. (Actually some of these divergence difficulties are inherited from classical electrodynamics.) It took nearly two decades before it was recognized that these infinities can be eliminated by careful examination (called renormalization) of what is meant by the observed mass and charge of an electron.

Renormalized QED as it thus emerged is consistent with all basic principles of physics and is capable of predicting the properties of any process involving electrons and photons as accurately as we might wish. Does this mean that we have finally found the ultimate theory of electromagnetic interaction?

We can answer this question only by performing some critical experiments. There are at least three aspects of the theory that must be examined:

1. How well do higher-order predictions of renormalized QED agree with experiments?

2. Does renormalized QED apply to distances shorter than those already tested?

3. To what extent is the electromagnetic interaction invariant under C, P, T operations?

Some of the most accurate tests of aspect 1 available at present are listed in Table 1. Most theoretical values include correction terms of order α and α^2. In case of the magnetic moment anomaly of the electron, which is a small deviation due to the electromagnetic interaction of the magnetic property of the electron from that predicted by the Dirac theory, correction terms of order α^4 have also been included. In all cases, the agreement of theory and experiment is within the uncertainties of experiment and/or theory. Even more stringent tests will be available in the near future.

It is remarkable that the interaction energy (7), which after all is a straightforward adaptation of classical Maxwell theory, has withstood tests of such high precision. Note, for instance, that from relativistic and gauge invariance alone we cannot exclude the possible existence of an additional interaction term

$$\mathcal{H}_I'(x) = k\frac{e}{2m}j_{\mu\nu}(x)F^{\mu\nu}(x) \quad j_{\mu\nu}(x) = \frac{i}{2}\bar{\psi}(x)(\gamma_\mu\gamma_\nu - \gamma_\nu\gamma_\mu)\psi(x) \ , \tag{9}$$

which would modify the magnetic moment anomaly of the electron by the amount k. Good agreement of theory and experiment implies that the constant k is less than 6×10^{-11}. Actually the presence of such a term would make the theory unrenormalizable. It is reassuring that there is no experimental indication that such a term is present in the Lagrangian $\mathcal{L}$.

The best tests of aspect 2 are provided by electron–positron colliding-beam experiments in which highly accelerated electrons and positrons traveling in opposite directions collide head-on and scatter into large angles,

$$e^+ + e^- \rightarrow e^+ + e^- \tag{10}$$

or annihilate each other and produce a pair of photons (γ) or muons (μ):

$$\begin{aligned} e^+ + e^- &\rightarrow \gamma + \gamma \\ &\rightarrow \mu^+ + \mu^- \ . \end{aligned} \tag{11}$$

Table 1: High precision tests of quantum electrodynamics.

Process	Experiment[a]	Precision ppm[b]	Theory[c]	Precision ppm[b]
Hydrogen fine structure[d]	$10\,969.127(87)$[e]	7.9	$10\,969.034\,8(21)$	0.2
Hydrogen hyperfine structure[f]	$1\,420.405\,751\,766\,7(9)$	6×10^{-7}	$1\,420.405\,3(45)$	3.2
Muonium hyperfine structure[g]	$4\,463.302\,88\,(16)$	0.036	$4\,463.302\,67(186)$	0.42
Helium fine structure[h]	$29\,616.864(36)$	1.2	$29\,616.834(110)$	3.7
Electron magnetic moment anomaly	$0.001\,159\,652\,188\,4(43)$	0.004	$0.001\,159\,652\,133(29)$	0.025

[a] In units of MHz ($10^6\,\mathrm{s}^{-1}$) except for the value of the electron anomaly, which is a pure number.

[b] 1 ppm $= 10^{-6}$.

[c] The theoretical value is calculated using the value of the fine-structure constant $\alpha = 137.035\,997\,9(33)$ obtained by the quantized Hall effect.

[d] Energy interval between the levels $2\mathrm{P}_{3/2}$ and $2\mathrm{P}_{1/2}$ of the hydrogen atom.

[e] The quantity enclosed in parentheses represents the uncertainty in the last digits of a numerical value.

[f] Splitting of the ground-state $1\mathrm{S}_{1/2}$ level of hydrogen due to the interaction of electron spin and proton spin.

[g] Splitting of the ground-state $1\mathrm{S}_{1/2}$ level of muonium (electron–antimuon bound state) due to the spin–spin interaction.

[h] Energy interval between the levels $2\,^3\mathrm{P}_1$ and $2\,^3\mathrm{P}_0$ of the helium atom.

Recent measurements of these processes have produced convincing evidence that QED is valid down to lengths of the order of 10^{-17} m, which is nearly two orders of magnitude smaller than the charge radius of the proton.

In spite of the very impressive experimental confirmation of QED, there are reasons to believe that it is not the ultimate theory and that it will have to undergo some changes sooner or later. For instance, the apparent lack of symmetry between the electric and magnetic fields in Maxwell's equations suggests possible existence of a magnetic monopole, the magnetic counterpart of the electric charge. Experimental confirmation of the existence of a magnetic monopole would certainly affect our understanding of electromagnetic interaction in a very profound way.

Another possibility, which has been pursued with a spectacular success, is that the electromagnetic and weak interactions are nothing but different aspects of the same force. One theoretical scheme for such a unification, the Weinberg–Salam model, has recently been given strong experimental support. In a beautiful experiment in which a tiny parity-nonconservation effect was observed in the inelastic scattering of high-energy electrons off deuterium nuclei, the electromagnetic and weak interactions were found to interfere in a manner predicted by this model. Similar (but larger) interference effects have been observed in processes such as (9) and (10) at high energies. Most importantly, W and Z mesons, which are the main ingredients of the Weinberg–Salam model and about 100 times heavier than proton, have been observed experimentally.

We have thus witnessed an end to "naive" QED. At the same time a way was found to incorporate QED into a deeper law of nature. Note that this does not mean that QED has failed at last. In fact the Weinberg–Salam model can be regarded as a natural generalization of QED.

An even grander scheme of unification, including the strong interaction and possibly the gravitational interaction, is purely speculative at present. However, it is not out of the question that it also may receive experimental support in a not-so-distant future.

As for aspect 3, there is no evidence at present that the electromagnetic interaction has components that violate *C*, *P*, or *T*. However, giving precise upper bounds for such violations is not straightforward.

See also: Electrodynamics, Classical; Gravitation; Maxwell's Equations; Quantum Electrodynamics; Quantum Mechanics; Strong Interactions; Weak Interactions.

Bibliography
L. I. Schiff, *Quantum Mechanics*, 3rd ed. McGraw–Hill, New York, 1968.
J. D. Bjorken and S. D. Drell, *Relativistic Quantum Fields*. McGraw–Hill, New York, 1965.
C. Quigg, *Gauge Theories of the Strong, Weak, and Electromagnetic Interactions*, Addison-Wesley, New York, 1983.

Electromagnetic Radiation
J. C. Herrera

Introduction

The most common form of electromagnetic radiation is visible light. But the fact that this type of emission of energy was electrical in nature was hardly suspected until James Clerk Maxwell (1831–1879) conceived the idea of the electromagnetic theory of light in the year 1864. Prior to that time the phenomenon of light had been studied as a mechanical vibration in a medium called the luminiferous ether, while, instead, the electrical interaction between charged bodies and currents was attributed to a force acting through the intervening empty space similar to the force of universal gravitation discovered by Isaac Newton (1642–1727) about 200 years earlier. Though the mathematical description for the electrical force was more complicated than that for the force of gravity, the interaction was still an instantaneous "action at a distance." Maxwell, guided by the unerring physical intuition of Michael Faraday (1791–1867) and his own genius for constructing mathematical models, was able to develop a set of equations that not only accounted for the known electrical phenomena, but in addition predicted that changing electrical charge distributions would result in electromagnetic waves that traveled at the speed of light. The lines of force that Faraday visualized extending between and around electrical bodies were transformed into the electric and magnetic fields of Maxwell's equations. It was Heinrich Hertz (1857–1894) who verified experimentally in the year 1887 that there existed electromagnetic waves other than light that were indeed generated by oscillating currents in electrical circuits. With the advent at the turn of the century of the theory of relativity formulated by Albert Einstein (1879–1955), the ether medium vanished into space, leaving behind electromagnetic fields. Though the interpretation of Maxwell's theory had changed, the efforts of Einstein and H. A. Lorentz (1853–1928) had served to reinforce the fundamental validity of Maxwell's equations and the reality of electromagnetic waves.

Today the phenomenon of electromagnetic radiation covers the entire spectrum of energies, from very high-frequency cosmic gamma rays to low-frequency, long-wavelength radio waves. Radiant heat, ultraviolet light, and x rays are all electromagnetic waves, distinguishable from each other by their frequency or, equivalently, their wavelength.

Maxwell's Theory

Maxwell's equations are four partial differential equations that relate the electric field vector $\mathbf{E}$ and the magnetic field vector $\mathbf{B}$ at a particular location in space $\mathbf{x}$ and at a time t to the electrical charge density per unit volume ρ and the current density per unit area $\mathbf{J}$ at the same space-time point. Employing the conventional notation of vector analysis, we write them in the classic form

$$\nabla \cdot \mathbf{E} = \frac{\rho}{\varepsilon_0}, \tag{1}$$

$$\nabla \times \mathbf{E} = -\frac{\partial \mathbf{B}}{\partial t}, \tag{2}$$

$$\nabla \cdot \mathbf{B} = 0, \tag{3}$$

and

$$\nabla \times \mathbf{B} = \mu_0 \mathbf{J} + \frac{1}{c^2} \frac{\partial \mathbf{E}}{\partial t}. \tag{4}$$

The units used are SI (Système International). We note that the permittivity of free space, ε_0, and the permeability of free space, μ_0, are dimensional constants that are related to the speed of light according to the expression

$$c = (\varepsilon_0 \mu_0)^{-1/2}. \tag{5}$$

The first of these equations is the differential form of Gauss' law, that is, the equivalent of the statement that the total flux of the electric field vector (the surface integral of $\mathbf{E}$) is a direct measure of the total enclosed electric charge. Faraday's law for the induced electric field due to a magnetic field that changes with time is represented by Eq. (2). We observe that for a stationary condition (one that does not vary with time), the partial derivative of the magnetic field with respect to time vanishes ($\partial \mathbf{B}/\partial t = 0$), and then the first two equations are synonymous with Coulomb's law for the electrostatic force between charges with a fixed separation in space. The well-known fact that there are no free magnetic poles, as compared to the existence of free electric charges [Eq. (1)], gives rise to Eq. (3): The divergence of the magnetic field vector is always equal to zero. Maxwell not only expressed Ampère's law (the line integral of the magnetic field along the closed boundary curve of a surface is a measure of the electric current flowing through the surface) by writing Eq. (4), but he also added the last term on the right-hand side (his so-called displacement current). The four equations as they stand are consistent with the conservation of charge ($\mathbf{V} \cdot \mathbf{J} + \partial \rho / \partial t = 0$), and when considered together, they predict the radiation of electromagnetic waves.

The operational meaning of Maxwell's equations is as follows: Given all the distributions of charges and currents specified by the source functions ($\rho, \mathbf{J}$), we can determine the accompanying electromagnetic field properties in space specified by the vector field functions ($\mathbf{E}, \mathbf{B}$). The subsequent step, one that typifies a field theory, is to associate with these fields an energy and a momentum localized in free space. When the field energy within a given spatial volume changes, it is either dissipated as heat, or motion of charges, within the volume, or instead it passes into the surrounding space. This characterization was derived from Maxwell's theory by John H. Poynting (1852–1914) in 1884. Written in vector notation, the energy balance of the field assumes the compact form

$$-\frac{\partial u}{\partial t} = \mathbf{E} \cdot \mathbf{J} + \mathbf{V} \cdot \mathbf{S} \tag{6}$$

where the electromagnetic energy per unit volume (in joule per cubic meter) is

$$u = (\varepsilon_0/2)(E^2 + c^2 B^2) , \tag{6a}$$

and the electromagnetic power flux density vector (in watt per square meter) is

$$\mathbf{S} = (1/\mu_0)\mathbf{E} \times \mathbf{B} . \tag{6b}$$

In words, the decrease per second in the field energy density (u) is equal to the power per unit volume delivered to the currents, ($\mathbf{E} \cdot \mathbf{J}$), added to the divergence of the field power per unit area ($\mathbf{S}$), the Poynting vector.

Plane Electromagnetic Wave

The radiation concepts that have just been presented can best be illustrated by considering a plane harmonic wave propagating in free space in the direction of the z axis (x, y, and z form a right-handed coordinate system). The electromagnetic fields that satisfy Maxwell's equations (with $\rho = 0$ and $\mathbf{J} = 0$) are, in this instance,

$$\begin{align} E_x &= E \sin 2\pi(\nu t - z/\lambda) , \tag{7a} \\ E_y &= E_z = 0 , \tag{7b} \\ B_y &= (E/c) \sin 2\pi(\nu t - z/\lambda) , \tag{7c} \end{align}$$

and

$$B_x = B_z = 0 . \tag{7d}$$

This transverse wave is plane polarized in the x direction with the wave front in the (x, y) plane and the wave normal in the z direction. The frequency of vibration (ν) and the wavelength (λ) are simply related to the speed of propagation (c), that is,

$$\nu \lambda = c = (\varepsilon_0 \mu_0)^{-1/2} . \tag{8}$$

Table 1: Examples of electromagnetic plane waves.

		Radio signal	Solar radiation
Electric field	E_x	$1 \times 10^{-6}\,\text{V/m}$	$1025\,\text{V/m}$
Magnetic field	B_y	$3.3 \times 10^{-15}\,\text{T}$	$3.4 \times 10^{-6}\,\text{T}$
Poynting vector	$\langle S_z \rangle_{\text{av}}$	$1.3 \times 10^{-15}\,\text{W/m}^2$	$1390\,\text{W/m}^2$
Energy density	$\langle \mu \rangle_{\text{av}}$	$4.4 \times 10^{-24}\,\text{J/m}^3$	$4.6 \times 10^{-6}\,\text{J/m}^3$
Radiation pressure	$\langle \mu \rangle_{\text{av}}$	$4.4 \times 10^{-24}\,\text{N/m}^2$	$4.6 \times 10^{-6}\,\text{N/m}^2$

If we now apply Eqns. (6a) and (6b), we are able to calculate the time average values of the Poynting vector and the electromagnetic energy density. Thus, we obtain

$$\langle S_z \rangle_{\text{av}} = \frac{1}{2} c \varepsilon_0 - E^2 \tag{9}$$

and

$$\langle \mu \rangle_{\text{av}} = \langle S_z \rangle_{\text{av}} / c \,. \tag{10}$$

We can therefore picture an average transfer, or transmission, of energy taking place in the direction of the wave normal and at a speed equal to that of light.

In Table 1 we give two examples of the physical magnitudes involved in the propagation of plane electromagnetic waves. The first column of figures is based on an electric field intensity of $1\,\mu\text{V/m}$, such as might exist at some large distance from a television or radio antenna. The second column of figures has been calculated on the basis of the power arriving at the earth from the sun, that is, corresponding to a solar radiation constant of $20\,\text{kcal/m}^2$ incident every minute $(1390\,\text{W/m}^2)$. In the last row we have also included the magnitude of the radiation pressure that the plane wave would exert on a totally absorbing material surface. Though the idea of radiation pressure was carefully discussed by Maxwell in his *Treatise on Electricity and Magnetism*, published in 1873, the experimental corroboration did not occur until about 1900.

Retarded Potentials

Plane electromagnetic waves, as discussed in the last section, represent an approximation to the observed radiation far away from a localized source. In applying Maxwell's theory to an actual radiating source, it is advantageous to express the two field variables **E** and **B** as functions of the vector and scalar potentials, **A** and ϕ. Thus Eqns. (2) and (3) are satisfied identically if we introduce the two defining relationships

$$\mathbf{B} = \nabla \times \mathbf{A} \tag{11}$$

and

$$\mathbf{E} = -\nabla \phi - \frac{\partial \mathbf{A}}{\partial t} \,. \tag{12}$$

The other Maxwell equations, (1) and (4), containing the source terms, can then be reduced to two similar wave equations

$$\nabla^2 \mathbf{A} - \frac{1}{c^2} \frac{\partial^2 \mathbf{A}}{\partial t^2} = -\mu_0 \mathbf{J} \tag{13}$$

and

$$\nabla^2\phi - \frac{1}{c^2}\frac{\partial^2\phi}{\partial t^2} = -\frac{\rho}{\varepsilon_0} \tag{14}$$

provided that we impose the added stipulation that the potentials satisfy the Lorentz condition

$$\mathbf{\nabla}\cdot\mathbf{A} + \frac{1}{c^2}\frac{\partial\phi}{\partial t} = 0 . \tag{15}$$

It should be emphasized at this point in our discussion that Eqns. (11)–(15) are completely equivalent to Maxwell's equations as far as the determination of the electromagnetic fields produced by a given set of charges and currents is concerned. However, this mathematical representation of the theory does facilitate the actual calculation of the fields in many cases; in addition, it plays an important role in the further understanding of the electromagnetic field. For the case when the electromagnetic radiation is due to localized sources, the physical solutions to the foregoing equations are neatly expressed as the so-called retarded potential functions

$$\mathbf{A}(\mathbf{r},t) = \frac{\mu_0}{4\pi}\int d^3x' \frac{\mathbf{J}(x',t^*)}{|\mathbf{r}-\mathbf{r}'|} \tag{16}$$

and

$$\phi(r,t) = \frac{1}{4\pi\varepsilon_0}\int d^3x' \frac{\rho(x',t^*)}{|\mathbf{r}-\mathbf{r}'|} . \tag{17}$$

These integrals over the space distributions of the charge and current densities are evaluated at the time

$$t^* = t - |\mathbf{r}-\mathbf{r}'|/c , \tag{18}$$

that is, at a time t^*, corresponding to the emission of the radiation, earlier than the observation time t by the time $|\mathbf{r}-\mathbf{r}'|/c$, required for the propagation of the wave from the source point location r' to the field observation point r.

Radiation from a Dipole

A good example of electromagnetic radiation from a localized source is that emitted from a linear electric dipole of moment p (charge multiplied by maximum displacement) oscillating sinusoidally at a frequency ν. The vector potential in the space region far from the dipole, the wave zone, is

$$A_z \simeq \frac{\mu_0\nu p}{2r}\cos 2\pi\nu\left(t - \frac{r}{c}\right) . \tag{19}$$

Here r is the distance from the dipole that is assumed to be located at the origin of the coordinate system and to be oriented along the z direction. The magnetic field vector is transverse to the unit radial vector $\mathbf{n}$ and is given by

$$\mathbf{B} \simeq (\mathbf{n}\times\mathbf{p})\frac{\mu_0\pi\nu^2}{cr}\sin 2\pi\nu\left(r - \frac{r}{c}\right) . \tag{20}$$

As expected, the electric field vector is normal to both the radial vector and the magnetic field. Therefore we write it as

$$\mathbf{E} = -c(\mathbf{n} \times \mathbf{B}) .$$ (21)

The electromagnetic radiation far from an electric dipole can hence be visualized as essentially a plane wave propagating radially outward while the component fields decrease with distance as $1/r$. The power associated with such an outward flow is characterized by the Poynting vector, which has the time-average value of

$$\langle S_r \rangle_{av} = \frac{\pi \mu_0 \nu^4 p^2 \sin^2 \theta}{2cr^2} .$$ (22)

The angle θ is that between the orientation of the source dipole and the direction of the field observation point with respect to the origin. We notice that, according to Eq. (22), the radiation pattern exhibits a maximum in the plane at right angles to the dipole orientation ($\theta = \pi/2$), while the radiation vanishes for field points along the dipole direction ($\theta = 0$). An integration over all directions yields the total radiated power

$$P = \frac{4\pi^3 \mu_0 \nu^4 p^2}{3c} .$$ (23)

We call attention to the dependence of this radiated power on the fourth power of the frequency. It is this characteristic of dipole radiation that basically accounts for the blueness of the sky. Since blue light has a higher frequency than red, it is more effectively reradiated, or scattered, by the bound electrons in the molecules of the atmosphere.

Material Media

In this brief survey we have presented the basic elements of the theory of electromagnetic radiation in free space. As emphasized by Feynman (see the References), it is this aspect of the great synthesis of Maxwell that is of lasting significance. The exact manner of broadening such a development so as to include radiation through material bodies, which have a crystalline, molecular, and atomic structure, is fundamentally a difficult problem. The usual way of doing this is by introducing some phenomenological parameters such as dielectric constant, permeability, and conductivity. We then speak of the effective electric and magnetic polarizations, the electrical displacement ($\mathbf{D}$), and the magnetizing force ($\mathbf{H}$). It should be realized, however, that this approach ultimately requires that these macroscopic material parameters be explained by the interaction of the electromagnetic radiation with matter, that is, by the application of electrodynamics and quantum theory.

See also: Absorption Coefficients; Electrodynamics, Classical; Light; Maxwell's Equations.

References

[1]   E. Whittaker, *A History of the Theories of Aether and Electricity, The Classical Theories* (Reprint of 1954 edition). American Institute of Physics, New York, 1987. (I, A)
[2]   H. H. Skilling, *Fundamentals of Electric Waves*. Krieger, 1974. (E)

[3]   E. M. Purcell, *Electricity and Electromagnetism*. Berkeley Physics Course, Vol. II, McGraw–Hill, New York, 1963. (E, I)

[4]   R. S. Elliott, *Electromagnetics*. McGraw–Hill, New York, 1966.

[5]   R. P. Feynman, R. B. Leighton, and M. Sands, *The Feynman Lectures on Physics, The Electromagnetic Field*, Vol. II. Addison-Wesley, Reading, Mass., 1964. (I, A)

[6]   J. D. Jackson, *Classical Electrodynamics*, 3rd ed. Wiley, New York, 1989. (I, A)

Electromagnets

F. J. Friedlaender

Electromagnets are devices in which magnetic fields are produced by means of current-carrying conductors. Usually the magnetic field in such a device is desired in an air gap or space that is not part of the field-producing structure. Direct-current motors and generators and synchronous motors and generators all require magnetic fields that are produced by either electromagnets or permanent magnets. If time-varying or large fields are required, then the use of permanent magnets is generally ruled out; electromagnets are the only practical means of producing such fields. Lifting magnets, and magnets to produce the fields for (high-gradient) magnetic separators or for MHD generators, are other common applications of electromagnets. Magnetic fields up to over 50 T (500 kG), as required for research purposes, are generally produced by means of electromagnets, and these will be our major concern.

There are, broadly speaking, three classes of electromagnets: (1) those using iron (or a similar ferromagnetic or ferrimagnetic material) in the flux path; (2) those using no iron and having dissipative (normally conducting) coils; (3) those with no iron and nondissipative (i. e., superconducting) coils.

We should also distinguish between pulsed-field or time-varying-field magnets, mostly under classification 2, and the more common steady-field (dc) magnets in all three categories.

Iron in the flux path (category 1) reduces the required coil current greatly, as long as the iron does not saturate (at just over 2.1 T or 21 kG). But iron-clad magnets are useful even at much higher flux densities in the working space (which is always external to the iron). Flux-concentrating means, such as truncated conical iron pole pieces, are commonly used to produce densities of over 10 T (100 kG) in the air gap of iron-clad magnets. As the field is increased in both category 1 and 2 electromagnets, larger currents have to flow in the coils, thus generating increasing thermal losses in the coils requiring adequate cooling to prevent overheating.

There are three major considerations dictating the design of an electromagnet. Of primary concern is the field design, which includes the design of the flux paths for category 1, as well as placement and shapes of coils and current distributions in the coils. Next in importance is thermal design, which also enters into coil design and the provision for adequate cooling, usually by means of water. Somewhat different thermal considerations apply to nondissipative magnets where superconducting coils at cryogenic temperatures are used. Cryogenic fluids (i. e., liquid helium) rather than water are used in such magnets. The third factor is mechanical design: large magnetic stresses occur in high-field electromagnets, and compromises in the design are often necessary in order to achieve a magnet that is mechanically sound.

The current-carrying coils in an electromagnet may have uniform current densities or – to obtain higher efficiencies (i.e., larger fields for the same power) – nonuniform current densities. If the coils have cylindrical symmetry they are called solenoids, but magnets are often designed using rectangular coils. The field H at the center of a field-producing coil can be shown to be related to the applied power P through the equation $H = G(P\lambda/\rho a_1)^{1/2}$ where λ is a winding space factor, ρ the resistivity of the solenoid wire, and a_1 the inside radius of the solenoid. G is a geometrical factor ("G factor," also called *Fabry factor* after Fabry, who first suggested its use).

A major concern of the solenoid designer is the optimization of this G factor by use of an appropriate geometry (i.e., coil dimensions and shape, current distributions, etc.). One such design that has found widespread applications and is named after its inventor, Francis Bitter, uses disks of conductors to form essentially a solenoid with a helical sheet winding. Each disk is a ring with a radial slot. Copper rings are assembled alternately with insulating rings (also with slots), so that a helical conducting path through the copper rings is formed. Axial holes in the disks are provided for water cooling. Gaume modified the Bitter system, in which each disk has the same current density distribution (largest at the inside radius and decreasing with increasing radius), to one in which the thickness of the disk and hence the relative current densities are varied. A Bitter magnet with an inside coil radius of 3.2 cm may require 10 MW to produce 22.5 T (225 kG).

Superconducting magnets can be used to produce fields up to almost 20 T (200 kG). Type II superconductors such as Nb_3Sn with critical fields of over 200 kG are used to make the field coils. Usually it is necessary to stabilize superconducting magnet coils by providing a parallel conducting path of a normal low-resistivity material in good contact with the superconductor. This arrangement allows stable operation of the superconducting magnet and guards against unstable collapse of superconductivity in the entire magnet winding, due to local thermal effects. Superconducting magnets are usually operated at 4.2 K, the temperature provided by the liquid helium that is almost always used as a coolant. Since the coil of a superconducting magnet requires negligible power, superconducting magnets have a considerable economic advantage over other electromagnets for certain applications in which fields of the order of 10 T (100 kG) are required.

In the future, superconductors that operate at liquid-nitrogen or even higher temperatures may become available. Such an event may decrease the cost of the cooling system. But unless superconductors are found that can sustain larger fields and currents, the capabilities of superconducting electromagnets are not likely to change substantially.

In many applications, the only practical method of obtaining the required large fields is by means of pulsed electromagnets. By using a relatively low duty cycle, the high energy required to produce the field can be obtained by means of relatively small average power values. Pulsed magnets provide the only means of obtaining fields over 50 T (500 kG). Usually a bank of capacitors is discharged through the field-producing coil, with suitable electronic control in some recent designs to produce a flat-topped pulse that may last from a fraction of a millisecond to several milliseconds. During the field pulse all the dissipated energy will raise the temperature of the field coil and associated elements, and forced cooling, if any is needed, will depend on the duty cycle (pulse repetition rate) of the magnet. Finally, mechanical considerations play a substantial role in the design of high-field solenoids. The high fields and large currents give rise to forces which can destroy a solenoid that is not designed properly.

See also: Magnetic Fields, High; Magnets (permanent) and Magnetostatics.

Bibliography

D. B. Montgomery, *Solenoid Magnet Design*. Wiley-Interscience, New York, 1969.

H. C. Roters, *Electromagnetic Devices*. Wiley, New York, 1941.

J. Liedl, W. F. Gauster, H. Haslacher, and H. Grossinger "Calculation of the Mechanical Stresses in a High Field Magnet by Means of a Layer Model," *IEEE Trans. Magn.* **MAG-17**, 3256–3258 (1981).

Electron

A. Pais[†]

Intrinsic Properties

Ever since its discovery, the electron has been considered an elementary particle, a fundamental building block of matter that cannot be decomposed into more primary constituents. The electron has the following intrinsic properties (the figures in parentheses denote the 1-standard-deviation uncertainty in the last two digits of the main number):

1. *Stability*: The free electron is generally believed to be absolutely stable. Experimental studies of the absolute validity of electric charge conservation have been made by looking for possible disintegrations of the electron. None has been found. In this way a lower bound of the order of 10^{21} yr for its lifetime was established.

2. *Mass*: $m = 0.5110034(14)\,\mathrm{MeV}/c^2 = 9.109534(47) \times 10^{-28}\,\mathrm{g}$

3. *Charge*: $e = 4.803242(14) \times 10^{-10}\,\mathrm{esu}$.

4. *Spin*: $h/4\pi$ (h is Planck's constant).

5. *Gyromagnetic ratio*: $1.0011596567(35)e/mc$.

6. *Electric-dipole moment*: Not observed; its present experimental upper bound is $3 \times 10^{-24}e\,\mathrm{cm}$.

Discovery

The discovery of the electron finally settled the question, debated for more than a century, whether there exists a quantum of electricity, a smallest unit of electric charge. From Faraday's law (1833), according to which each gram-atom of any univalent electrolyte carries the same charge, it follows that there is such a unit. To see this, however, we should know that the gram-atom of a pure substance consists of a definite number of identical atoms. This "Loschmidt number" was not determined until the 1860s but even then the notion of an elementary unit of

---

[†]deceased

charge seemed to many to be one of terminology rather than of physical reality. For example, in his 1873 treatise on electricity and magnetism Maxwell refers to the electric quantum as "one molecule of electricity. This phrase, gross as it is, and out of harmony with the rest of this treatise …"; and he adds, "It is extremely improbable that when we come to understand the true nature of electrolysis we shall retain in any form the theory of molecular charges …" But proponents of the atomistic view of electricity persisted, among them Helmholtz and the Irishman George Johnstone Stoney. The latter has the distinction of having given the first crude estimate of e as early as 1874, and of having baptized this unit with the name electron in 1891. Thus this term was coined prior to the discovery of the quantum of electricity *and* matter that now goes by that name.

The years of its discovery are 1896 and 1897. In his second paper on the Zeeman effect (28 November 1896), Pieter Zeeman recorded that Lorentz "at once kindly informed me" how the motion of an ion [*sic*] in a magnetic field can be determined. Using an oscillator potential model of the atom, Lorentz interpreted the Zeeman effect in terms of the motion in the atom of a particle with $e/mc \sim 10^7 \, \text{rad s}^{-1} \text{G}^{-1}$. This was the first, albeit indirect, hint of the existence of a particle with a novel low mass.

The direct proof of its actual existence was given by J. J. Thomson and was first communicated on 29 April 1897 in a Friday evening discourse at the Royal Institution. His discovery was that cathode rays are electrons. Such rays had been studied for decades, but their constitution was unclear. Some held them to be "molecular torrents," others "aether disturbances." Thomson determined their e/m, noted that his answer was much like Zeeman's, and observed that "the assumption of a state of matter more finely subdivided than the atom is a somewhat startling one." The first fundamental particle had been isolated. The implicit assumption that the e involved is the same as for univalent electrolytic ions was soon verified, especially by Millikan, who measured e with precision and demonstrated its uniqueness.

The Classical Prerelativistic Electron

"We shall ascribe to each electron certain finite dimensions, however small they may be …, my excuse must be that one can scarcely refrain from doing so if one wishes to have a perfectly definite system of equations …" Thus, with care and caution, Lorentz introduced the classical model of the electron: a charge distribution confined (at rest) to a small sphere of radius r and with the classical mass formula $m = m' + m_e$. Here $m_e = e^2/rc^2$ (up to a number of order unity) is the electromagnetic mass (a concept generally known since 1881) and m', which used to be called the "material mass," is the contribution due to other origins. It was often speculated that $m' = 0$ and hence this particle was ascribed its "classical electron radius": $r = e^2/mc^2 \approx 2.8 \times 10^{-13}$ cm. In fact, in those prerelativistic days it was believed possible to determine experimentally whether or not $m' = 0$ by measuring the electron energy E as a function of velocity v for uniform motion.

For example, if the electron were a rigid sphere (Abraham model), then for $m' = 0$, $E(v)$ is calculable purely as an electromagnetic expression. Progress was not helped by the fact that at first the answer appeared to agree with experiment! For some time this agreement was held as evidence (1) for zero m', (2) for the rigid model; and (3) against the Lorentz model, in which the finite sphere is contracted in its direction of motion.

The Classical Relativistic Electron

With the advent of Einstein's special theory it became clear that the correct $E(v)$ had to be $E(v) = mc^2/(1 - v^2/c^2)^{1/2}$, regardless of what contributes to m, and therefore that the notion of a determination of m' from velocity measurements was illusory. [Not until about 1915 was this form of $E(v)$ verified experimentally.] The focus now shifted to a new problem: None of the existing models, including the one of Lorentz, gave the Einstein $E(v)$. Indeed in all these models the electron is an open system (i.e., there are unbalanced forces on its boundary), whereas Einstein's answer applies to closed systems only. This discrepancy led to conjectures, starting with Poincaré, about nonelectromagnetic cohesive stresses designed to balance the electromagnetic repulsive stresses. Lorentz was not sure of this: "... perhaps we are wholly on the wrong track when we apply to the parts of the electron our ordinary notion of force."

By about 1925 the focus had changed again. At that time the Russian physicist J. Frenkel wrote, "I hold these riddles ... to be a purely scholastic problem ... the electrons are not only physically but also geometrically indivisible. They do not have any extension in space at all ..." But then there appears still another paradox, since m_e becomes infinite for zero radius. We shall return to this shortly.

Electron Spin

A critical examination of the anomalous Zeeman effect led Pauli to propose (December 1924) that the doublet structure of alkali spectra is caused by "a two-valuedness, not describable classically," of the quantum properties of the valence electron. This made him assign four quantum numbers (instead of the customary three) to the electron, and in January 1925 he formulated the exclusion principle: In an atom this set of four numbers cannot be the same for any two electrons (a formulation that was much broadened subsequently). In October 1925, Uhlenbeck and Goudsmit suggested that "it is plausible to assign to the electron with its four quantum numbers also four degrees of freedom ... [one of which] ... can be associated with a proper rotation of the electron" and assigned the value $h/4\pi$ to this intrinsic angular momentum or "spin." The quantitative understanding of the fine structure came only after some struggle. The importance of major relativistic effects (the so-called Thomas factor) had to be recognized. Also, a mysterious value e/mc, twice the value expected classically, had to be assigned to the gyromagnetic ratio of the electron. When in May 1927 Pauli succeeded in describing the spinning electron by a two-component "spinor" wave function he still had to incorporate these effects in an ad hoc way. It was an important advance in description but not yet in understanding.

The Dirac Equation; The Positron

Positron These effects became fully understood when in February 1928 P. A. M. Dirac proposed his relativistic wave equation of the electron, one of the most spectacular advances of modern science. At the same time his equation generated severe new paradoxes.

His equation implies that there are four states for a given momentum p. There is a welcome doubling associated with the spin but an apparently paradoxical further doubling due to the fact that the associated energy takes on the two values $\pm(c^2p^2 + m^2c^4)^{1/2}$. What is the significance of the inevitable negative-energy states? Speculation arose that these might be associated with the proton, the only other fundamental particle then known. However, "one cannot simply assert that a negative energy electron is a proton," Dirac noted. Rather, he argued, we should "assume that there are so many electrons in the world ... that all the states of negative energy are occupied except perhaps a few of small velocity ..." Noting that these holes behave like positively charged particles, "we are therefore led to the assumption that the holes in the distribution of negative energy electrons are protons." This is the earliest version of the "hole theory" (1929).

But this cannot be. Experimentally it would imply that a hydrogen atom would annihilate into photons. Theoretically, as H. Weyl noted, "according to it [the Dirac equation] the mass of the proton should be the same as the mass of the electron ..." The gravity of the situation is illustrated by Weyl's further remark that "the clouds hanging over this part of the subject will roll together to form a new crisis in quantum physics." Once again Dirac found the answer (1931): "a hole, if there were one, would be a new kind of particle, unknown to experimental physics, having the same mass and opposite charge of the electron." This particle, the positron (e^+), was discovered on 2 August 1932 by C. D. Anderson.

The Relativistic Quantum-Mechanical Electron

The revolution of quantum mechanics has erased all pictures of the electron as a tiny sphere. The point model has survived, or (far better stated) the electron is described by a quantized local field. Earlier we noted that for a classical point model $m_e = \lim_{r\to 0} e^2/rc^2$. Also in relativistic field theory the electromagnetic mass (or self-energy) is infinite but the nature of the singularity has changed because of quantum effects. We now have

$$m_e = \lim_{r\to 0}(3\alpha/2\pi)m\ln(h/mcr) + O(e^4)$$

where $\alpha = 1/137$ is the fine-structure constant. Attempts in the 1930s and 1940s to modify the theory such that m_e (and certain other physical quantities) become finite were not successful. However, a major advance has been made since 1947: It is now possible to handle the theory in such a way that in all physical quantities like cross sections or energy levels, only the physical mass m of the electron appears, without hindrance by the infinity in m_e; nor (once again) does the separation into material (or "bare") + electromagnetic mass ever enter. This is the renormalization theory. With most impressive results (e. g., it yields the gyromagnetic ratio given earlier – property 5 in the opening section – to high accuracy) it bypasses the self-energy problem. It does not solve it.

Weak Interactions

In December 1933, Fermi postulated a new kind of electron interaction, the weak interaction. Thereby he reduced all of β radioactivity to the occurrence of the fundamental process neutron → proton + electron (e) + "anti-e-neutrino" ($\bar{\nu}_e$). (The latter particle had been hypothesized by Pauli.) Since that time numerous other "weak processes" have been discovered in which an e and a $\bar{\nu}_e$ or an e^+ and a ν_e (electron neutrino) are created together (*see* Weak Interactions). This has led to the formulation of a new principle, called the conservation of e number: In any reaction the number of $e+\nu_e$ particles minus the number of $e^+ + \bar{\nu}_e$ particles is conserved. This principle ties the mentioned particles together in a "family." In turn, this family is part of a larger family called *leptons* (so named by Møller and Pais). A lepton is a particle that participates in gravitational and weak interactions. It may also participate in electromagnetic interactions (as e and e$^+$); but it does not (at least not directly) participate in strong interactions (which see). Other known leptons (there may be more) are the muon (μ) and (related to these) another brand of neutrinos called ν_μ .

Neutral Currents; Gauge Theories

A search is under way to unify electromagnetic and weak interactions. In such attempts the electron enters into an interconnected set of currents: (I) the electromagnetic current, coupled to a massless neutral vector field (photons); (2) a "weak charged current," coupled to a massive charged vector boson field (*W* bosons), which transmits the Fermi-type weak interactions; (3) a "weak neutral current," coupled to a massive neutral vector boson field (2 bosons). (*see* Gauge Theories; it is not excluded that the set of currents is more complex than sketched here.) This last coupling gives rise to a new class of weak processes in which the electron is *not* accompanied by an e neutrino (but in which e number is conserved). These ideas stimulated a search for such new "neutral-current reactions." Since 1973 there is evidence that these indeed exist.

Theories of this kind are so far speculative (for one thing, *W* and *Z* bosons have not as yet been seen), but they appear to be quite promising. They may yet shed new light on fundamental properties of the electron, such as the origin of its mass and the possibility that it carries a tiny electric-dipole moment. It would appear that the days are numbered both for electromagnetism as a force separate from other forces and for the splendid isolation of the electron as a particle unconnected with other particles.

See also: Beta Decay; Elementary Particles In Physics; Field Theory, Classical; Gauge Theories; Leptons; Quantum Electrodynamics; Weak Interactions.

Bibliography

E. R. Cohen and B. N. Taylor, *J. Phys. Chem. Ref. Data* **2**, 663 (1973). (Intrinsic properties.)
H. A. Lorentz, *The Theory of Electrons*. Dover, New York, 1952.
A. Pais, "The Early History of the Theory of the Electron: 1897–1947," in *Aspects of Quantum Theory* (A. Salam and E. P. Wigner, eds.), p. 79. Cambridge, London and New York, 1972.

Electron and Ion Beams, Intense

J. N. Benford

Theoretical studies of intense relativistic electron beams began in the 1930s with the pioneering works of Bennett on pinched (radially compressed) electron flow and of Alfvén on current limitations in streams of cosmic rays. Practical implementation, however, had to wait until the early 1960s, when J. C. Martin in England succeeded in developing the beginnings of the pulse technology required to deliver the tens of kiloamperes of electron current in the megavolt

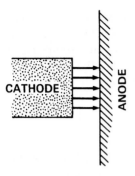

Fig. 1: Flow in unpinched diode.

range that are characteristic of these beams. Since then, vigorous industrial and government laboratory efforts, primarily in the United States, have produced beams with megaampere currents at megavolt energies with pulse durations of $\sim 10^{-7}$ s. In general these beams are generated by employing a voltage multiplier source (Marx generator) to pulse-charge an oil or water dielectric line of 1–50 ohms characteristic impedance. The firing of a gas or liquid output switch allows the voltage pulse to be applied to a field-emission cathode from which a burst of electrons is accelerated toward the anode by the potential difference across the gap.

In a diode with no applied axial magnetic field, uniform current flow occurs (planar Child–Langmuir space-charge-limited flow, Fig. 1) until the turning of electrons by the B_θ self-field causes pinching of the beam toward the axis ($F_r \sim v_z B_\theta$). The critical current level for this pinching to occur is obtained by requiring that electrons emitted at the edge of the cathode reach the anode with a velocity parallel to the surface:

$$I_c = \frac{2\pi mc\beta\gamma}{\mu_0 e}\frac{R}{d} = 8500\,\beta\gamma\frac{R}{d}\,\text{A}$$

where R is the cathode radius, d the anode–cathode gap, γ the relativistic factor $\gamma = (1 - \beta^2)^{-1/2}$, and $\beta = v/c$. For $I > I_c$ the electrons flow toward the axis, and current follows the "parapotential" flow expression, which is derived from a model in which electrons move along conical equipotentials for which $\mathbf{E} - \mathbf{v} \times \mathbf{B} = 0$ (Fig. 2). The current is given by

$$I = 8500g\gamma\ln[\gamma + (\gamma^2 - 1)^{1/2}]\,\text{A}$$

where g is a factor that is geometry dependent, equal to R/d for Fig. 2. Many data agree with this expression, which requires a bias current in the diode as a boundary condition. The bias current is in part provided by ions coming from the anode. This anode plasma seems to be restrained from expanding across the diode gap and shorting the generator by the magnetic pressure of the self-field.

When the anode is a thin metal foil effectively transparent to the relativistic beam electrons, the beam will leave the diode region. Propagation will depend on the magnitudes of the repulsive electrostatic force due to the charge of the beam electrons and the attractive magnetic force that arises from the beam current. In vacuum, the electrostatic force dominates, and the beam rapidly expands radially. Electron repulsion can be alleviated by applying a longitudinal magnetic guide field, by supplying neutralizing ions from preexisting plasmas, or by

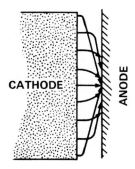

Fig. 2: Parapotential pinched flow.

Fig. 3: Electron trajectories in propagating beam.

direct ionization of gas by beam electrons. In principle, fractional space-charge neutralization sufficient to provide force balance is possible, but in practice propagating beams very quickly become space-charge neutralized.

With the magnetic force determining particle orbits, propagation is characterized by the strength parameter ν/γ, where ν is the number of electrons in a cross-sectional slab of the beam with a thickness of a classical electron radius. This parameter can be expressed as a ratio of currents, $\nu/\gamma = I/I_A$, where I is the beam current and I_A is the Alfvén–Lawson critical current

$$I_A = Ne\beta c = \frac{4\pi mc\beta\gamma}{\mu_0 e} = 1700\beta\gamma\,\text{A}\;.$$

For $\nu/\gamma < 1$, electron orbits are roughly sinusoidal and propagation is possible (Fig. 3). As beam intensity is increased, a point is reached where the gyrodiameter of beam electrons in the B_θ magnetic field of the current is about equal to the beam radius. Beam electrons turn completely around and flow stops. This limit occurs at $I = I_A$ or $\nu/\gamma = 1$ for a uniform radial current distribution. So when $\nu/\gamma - 1$, self-fields strongly influence beam motion.

There are three ways of exceeding this limiting current. The naturally occurring method is "current neutralization," which arises from the tendency of plasma or any conductor to resist sudden changes in magnetic field. A fast-rising beam current induces an opposing axial electric field that drives a partially canceling return current in the plasma. The resulting net current (beam minus plasma current) determines electron orbits, so that ν/γ is replaced by $(\nu/\gamma)_{net}$. For highly conducting plasma, the net current can be reduced to such a low value that beam electrons are not contained and expand to the boundaries of the chamber.

The second method of propagation is to employ a radially nonuniform current distribution. For example, a beam with most of the current flowing in a shell near the edge will propagate with $\nu/\gamma > 1$ by turning beam electrons only in the outer shell. Alternatively, in the Bennett distribution, the current density is peaked on the axis. It can be shown that ν/γ is proportional to the ratio of the magnetic-field energy within the beam to its kinetic energy. Nonuniform beam distributions reduce the magnetic energy, increasing the allowable current that can be propagated.

The third method, application of a longitudinal magnetic guide field, will prevent electrons from turning in their self-field, allowing propagation. The energy density of the field must be at least as great as that of the beam.

Recent experiments have demonstrated an electrostatic propagation aid. A laser ionizes a channel through a background gas. When the head of an electron beam passes, the electrons are expelled, leaving a background of ions. The remainder of the beam has its charge neutralized by the ions, so the beam propagates along the channel.

The basic features of beam propagation away from the diode have been established in efforts to produce copious xray bremsstrahlung from a target and to study the effects of rapid energy deposition in solids. Other applications are the generation of microwaves (1–10 GHz) or infrared to ultraviolet (free-electron laser) by propagation along a periodic magnetic field of suitable choice, and collective acceleration of ions by the space-charge well that exists at the beam front. Recent years have seen rapid development of a wide variety of microwave sources (magnetrons, gyrotrons, backward wave oscillators, virtual cathode oscillators, klystrons) at powers 10^9–10^{10} W for pulse durations $\sim 10^{-7}$ s. Emerging applications for this technology are fusion plasma heating, particle acceleration, and directed energy. Beams are used to excite lasers by exciting electronic and vibrational levels of molecules and excimers, and by initiating chemical reactions.

A major application involves use of the high power and energy of beams for the heating of magnetically confined plasmas in either linear or toroidal geometries. Plasma heating to several-thousand-electron-volt temperatures at particle densities of 5×10^{15} cm^{-3} has been achieved. The heating is due to streaming instabilities. For example, an interaction between beam and plasma electrons causes fluctuating electric fields at the electron plasma frequency. These fields directly heat plasma electrons, and excite oscillations at other frequencies, which then dissipate into particle thermal energy.

Use of beams to create plasma-confining magnetic-field configurations has centered on electron and ion rings, although toroidal systems have shown promise as well. Electron rings are formed by injection of an electron beam transverse to a magnetic field. The electrons gyrate about the field lines and generate a cylindrical current, which in turn produces a magnetic field in its interior in the opposite direction. For a sufficiently large current the direction of field can be reversed, forming a closed set of field lines, a minimum-B magnetic trap. The gyrating electrons radiate their energy rapidly by synchrotron radiation, requiring that ions be used to form the layer in fusion applications.

Pinched electron beams have been applied to inertial-confinement fusion. Self-focused electron beams have been observed with current densities of 10^6–10^7 A/cm^2, resulting in large charge concentration near the diode axis. The electron flow path is long compared with that of ions, which can cross from the anode with little deflection by the magnetic field. Simple arguments show that higher-current generators could have a substantial portion of the current carried by ions that could be focused geometrically for inertial confinement fusion. Ion currents ~ 1 MA have been produced and propagated onto targets from diodes optimized for ion beam production. The basic principle is suppression of electron flow by use of a transverse

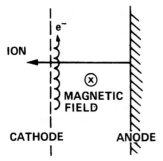

Fig. 4: Ion diode employing external magnetic field for electron current suppression.

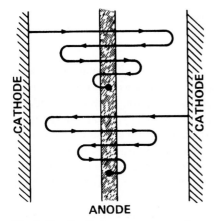

Fig. 5: Double diode arrangement – electron making multiple transits through anode foil.

magnetic field.

$$B_c = \frac{mc}{e} \frac{1}{d} \beta\gamma .$$

Ions cross the gap freely (Fig. 4), undergoing a small deflection, and electrons execute magnetron orbits.

An alternative method of intense ion-beam generation is to prevent pinching with an axial magnetic field and to allow electrons to pass through the anode, losing a small fraction of their energy, and on into another diode. Electrons reflect from the second cathode and return to the first diode. Multiple reflections and the accompanying energy loss result in concentration of the electrons near the anode surface (Fig. 5). The increased space charge there causes vastly enhanced ion emission from both sides of the anode. Currents of $\sim 200\,\text{kA}$ have been generated by this method.

Such intense beam diodes are used in an emerging application, production of very high pulse power ($\sim 10^{13}$–$10^{14}\,\text{W}$). Such diodes serve as opening switches, rapidly diverting energy from high-energy magnetic stores.

The technological developments that have spawned the rapid growth of electron-beam research and have led to the production of intense ion beams are continuing. Higher-current (several megaamperes) and higher-power (10^{13}–10^{14} W) generators are under development.

Bibliography

G. Benford and D. Book, *Adv. Plasma Phys.* **4**, 125 (1971). Reviews beam-equilibrium models. (A)

H. Fleischmann, *Phys. Today* **28**, 34 (1975). Surveys applications of electron beams. (E)

J. Nation, *Particle Accelerators* **8**, (1979). Discusses ion- and electron-beam generation and technology. (A)

Electron and Ion Impact Phenomena

E. W. Thomas

An energetic electron or ion impacting on a surface transfers energy to the lattice of the solid and to the free electrons in the solid by a successive series of collisions. As a result, certain particles are ejected from the surface; familiar examples are electron ejection (secondary-electron emission) and ejection of surface atoms (known as sputtering). The projectile itself may suffer sufficient angular deviation in a collisional scattering event so that it becomes directed out of the surface and may be observed as a reflected primary particle. The deposition of energy into the solid will cause various displacements of the lattice structure, known collectively as radiation damage, and it may excite electrons which subsequently decay with the emission of light (luminescence).

The rates at which these phenomena occur are described by a quantity known as a "coefficient," which is the number of relevant events occurring for every primary projectile incident; thus the secondary-electron emission coefficient is the number of secondary electrons ejected for each primary electron incident. Coefficients are strongly dependent on the condition of the surface and the crystal structure of the substrate. For a reliable measurement that can be related to a theoretical concept it is necessary that the target surface be atomically clean, the substrate must be of known crystallographic structure, and the orientation of the surface to the incident projectile direction must be defined.

Electron Impact Mechanisms

Most often studied is the flux of emergent electrons including both scattered primary electrons and ejected secondary electrons. Figure 1 shows the energy distribution of emergent electrons; region I is due mainly to reflected primaries, region III to secondary electrons, and region II includes components from both sources.

Secondary Electron Ejection

The incident primary electron produces excitation and ionization within the solid. Those excited electrons which diffuse to the surface, overcome the potential barrier, and emerge to be detected are called secondary electrons. Figure 1 indicates that true secondary electrons cannot

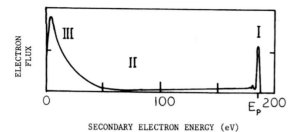

SECONDARY ELECTRON ENERGY (eV)

Fig. 1: A typical energy distribution for electrons emerging from a solid as a result of electron bombardment. E, denotes the energy of the primary electrons which produce the secondary emission.

be distinguished from reflected primary electrons; it is conventional to regard the secondary electrons to be those with energies below 50 eV (excepting Auger electrons discussed below). Secondary electrons arise primarily from excitation of outer-shell electrons of the solid and they have an average energy of a few electron volts. Typical peak secondary-emission coefficients of metals and semiconductors are 1 to 1.5 (secondary electrons out per primary electron in); for insulators and intermetallic compounds yields may be as high as 10 or 20.

Primary electrons of sufficient energy may eject inner-shell electrons from the atoms of the solid, so creating a vacancy which must be subsequently filled by an electron falling from a higher level. The energy liberated as the vacancy is filled may be transferred to some other outer-shell electron which thereby becomes ionized, and escapes from the solid, a process known as Auger-electron emission. Auger electrons are energetic with a relatively small energy spread; they are observed in the secondary-electron energy spectrum as discrete high-energy peaks, superimposed on the general continuous background of other secondaries and reflected primaries. The Auger-electron energy is characteristic of the atom from whence it was ejected so that analysis of the Auger spectrum, using an electron energy analyzer, can indicate which atomic species are present in the target.

Electron Reflection

The high-energy peak of electrons emerging from the solid (I in Fig. 1) is due to primary electrons scattered or reflected from close to the surface and much of the lower energy tail (II in Fig. 2) is due to primary electrons scattered from deeper in the solid. Small subsidiary peaks slightly below the incident energy are due to primary electrons which have lost discrete amounts of energy by exciting plasma oscillations in the electron gas. Reflection coefficients increase with the nuclear charge of the substrate and do not vary strongly with the energy of the projectile.

Electron-Induced Desorption

Because an incident electron has a small mass it cannot transfer appreciable energy to the heavy substrate nuclei by collisions and therefore cannot directly eject atoms. An incident electron may, however, ionize the electrons which bond a surface atom to the substrate; removal of the electron destroys the bond and the atom is ejected. This phenomenon is observed in the removal of weakly bonded adsorbed molecules and is consequently known as electron-induced desorption.

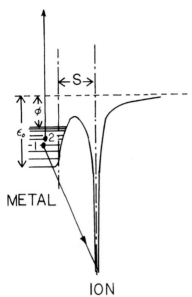

METAL

ION

Fig. 2: A potential energy diagram showing an ion at a distance S from a metal surface; the conduction-band electrons lie at energies between the work function ϕ and the bottom of the conduction band ε_0. Auger neutralization of the ion occurs by electron 1 falling to the bound state in the ion; electron 2 carries away the excess energy.

Luminescence Phenomena

Impact of electrons on a surface gives rise to a variety of light-emission phenomena emanating from the surface, or, if the material is sufficiently transparent, from some distance within.

Broad-band luminescence characteristic of the solid itself is caused by excitation of electrons to a higher level followed by their decay across a forbidden band gap with resulting emission of light. The phenomenon, called cathodoluminescence, is best known for insulators and semiconductors where large gaps exist between bands, but is observed also in metals.

Plasmon radiation is an optical emission from the decay of collective electron oscillations excited in metals; it emanates from the surface. Transition radiation occurs as a result of annihilation of the dipole formed by an electron and its image charge in the metal; this source also is located at the surface. Bremsstrahlung radiation occurs when electrons are decelerated or deflected in the Coulomb field of the atoms and this may emanate from deep in the solid. These three luminescence phenomena are seen only with electrons having energy in excess of tens of keV.

Ion Impact Mechanisms

Impact of an ion or atom on a surface exhibits certain complexities associated with the structure of the projectile. Figure 2 represents a potential energy diagram of an ion at a few Ångstrøms from a metal surface. It is possible for an electron (1) to tunnel through the barrier and enter the ground state of the projectile; the excess energy is given to electron (2) causing its ejection by the Auger mechanisms. Thus as an ion approaches a surface it changes its form to a neutral atom and electrons are ejected before the projectile penetrates the surface. Clearly an incident atom cannot undergo this change. After the projectile enters the solid it may lose some of its electrons by collisional ionization but we have no direct monitor of its structure.

Secondary-Electron Emission

Collisional transfer of kinetic energy from the projectile to electrons in the target will result in a secondary-electron emission spectrum similar to that observed with incident primary electrons. There is a large peak of low-energy emergent electrons and subsidiary peaks at high energies due to Auger transitions. One should note that heavy particles transfer energy inefficiently to light electrons and the projectile energy must exceed some thousands of electron volts before the energy transferred to the electrons is sufficient to overcome the surface potential barrier. Thus low-energy incident ions and atoms cannot eject electrons by "kinetic" mechanisms involving transfer of kinetic energy from the projectile.

Incident ions may also eject electrons by "potential" mechanisms related to their structure; the Auger effect described above and illustrated in Fig. 2 is such a mechanism and will obviously occur at any impact energy however small. Impact of neutral atoms cannot produce secondary electrons by this mechanism.

Reflection of Ions

Projectile ion collisions with electrons cause energy loss but, because of the substantial mass difference between the particles, no appreciable deviation of the heavy projectile. Only collisions of the projectile ion with lattice atoms can produce the deviation necessary to return the projectile toward the surface. For low-energy impact the projectile recoils from the surface with an energy loss related only to the angle of scattering and the mass of the atom from which it recoiled. Measurement of energy loss may be used to determine the mass of the atoms from which scattering occurs and thereby to analyze the composition of the surface. For high-energy impact there is a continuous distribution of recoil energies related to the penetration of projectiles into the solid.

The reflected projectile flux is found to include both ions and atoms. In part the distribution between charge states is related to interaction of the recoiling species with the target surface; Fig. 2 can be used to represent an ion which emerges from a surface and which picks up an electron by the Auger mechanism to become a neutral atom. The distribution of charge states is also partially established while the projectile is in the solid. A small fraction of the reflected projectiles may be excited and will emit normal atomic spectral lines when the projectiles have receded from the surface.

Sputtering

Ion impact on a surface will collisionally transfer energy to the lattice with the result that some lattice atoms may acquire sufficient energy to be ejected or sputtered from the surface. Typically the sputtered particles have an average energy of a few electron volts and are ejected with a coefficient which ranges from 10^{-3} (atoms ejected per ion incident) for a light projectile such as H^+ to 10 or 20 for heavy projectiles such as Pb^+. The ejected particles are found to include both ions and atoms. The distribution between these charge states is related in part to interaction of the emerging particle with the surface. Figure 2 can be used to describe how a sputtered ion can be neutralized as it emerges from the surface. A substantial fraction (up to 10%) of the sputtered particles may emerge as molecules representing the simultaneous ejection of two or more neighboring atoms in the lattice. A fraction of the emerging sputtered particles is found to be excited and will emit atomic line radiation at some distance from the surface.

Luminescence Phenomena

The emission of light from the surface or the bulk occurs through essentially the same mechanisms as for electrons discussed earlier. Intrinsic luminescence is termed ionoluminescence (the counterpart of cathodoluminescence) and is complicated by formation of additional energy levels through radiation damage caused by the ion impact.

Atomic line emission from sputtered or reflected atoms or molecules is not to be termed luminescence as it arises from a region in front of the surface rather than from the bulk material.

Crystallographic Effects

When a simple cubic crystal is "viewed" parallel to the sides of the unit cube it appears highly transparent with the strings of atoms lined up behind one another and large open "channels" exposed between the strings. The majority of projectiles directed along such a channel will lose energy only by collisions with electrons and therefore will penetrate considerable distances. Coefficients for reflection and ejection should be very small. Clearly the magnitude of electron and ion impact coefficients will be closely related to the orientation of the incoming projectile with respect to these crystallographic channels.

Applications

Electron and ion impact phenomena find much application in the analysis and modification of surfaces. The utility of the techniques is related to their performance in the vacuum environment where contaminants may be avoided. Under electron bombardment each element in the near surface region produces Auger electrons of characteristic energy, which may be separately detected by use of an electron energy spectrometer; the analysis provides the atomic composition of the surface and is known as Auger electron spectroscopy (AES). The secondary-electron emission mechanism can be used to provide amplification of weak electron currents using materials where the secondary-emission coefficient exceeds unity; this finds application in photomultipliers. Sputtering of a surface by ion impact provides a method for removal of thin layers of material, one atomic layer at a time. Sputtering is used to remove contaminants from a surface (sputter cleaning), to machine a surface (ion milling), or to remove a layer and expose some underlying surface (sputter etching). Mass analysis of the ions sputtered from a surface can provide a qualitative indication of that surface's composition, a technique known as secondary-ion mass spectrometry (SIMS). Energy analysis of slow ions scattered from the surface gives an indication of surface composition (ion scattering spectrometry or ISS).

See also: Auger Effect; Channeling; Luminescence; Radiation Damage in Solids; Secondary Electron Emission.

Bibliography

J. Schou, in *Physical Processes of the Interaction of Fusion Plasmas with Solids*, W. Hofer and J. Roth (eds.). Academic Press, New York, 1993. A review of electron emission induced by electrons and by ions. (A)

D. Menzel, *Nucl. Instr. Meth. Phys. Res. B* **101**, 1 (1995). Review of electron induced desorption. (I)

R. A. Baragiola , *Phil. Trans. R. Soc. (London)* **A362**, 29 (2004). A useful review of sputtering with references to the major data compendia. (I)

A. Arnau *et al.*, *Surf. Sci. Rep.* **27**, 113 (1997). Discusses all processes induced by multiply charged ions including electron ejection and sputtering. (A)

E. W. Thomas, R. K. Janev and J. Smith, *Nucl. Instr. Meth. Phys. Res. B* **69**, 427 (1992). Compendium of particle reflection data. (A)

Electron Beam Technology

D. P. Kern

Key features that make electron beams attractive for technological applications are that (i) intense electron beams can be easily generated, for example, by using a hot filament, (ii) electron beams can be easily focused by electric and magnetic fields, and (iii) electrons efficiently interact with matter with a range of distances, or effective strengths, which can be controlled by an applied voltage. For example, with modest acceleration potentials of about 20 000 to 200 000 volts and focusing magnetic fields, electron beams can be focused into spots a few Ångstrøms ($1\,\text{Å} = 10^{-8}\,\text{cm}$) in diameter. Thus objects can be imaged with essentially atomic resolution either by magnifying projection or by rapidly scanning such beams by means of time-varying electric or magnetic deflection fields at rates up to hundreds of megahertz. Electrons interact strongly with matter and consequently typical electron beam processes are performed in vacuum of about $10^{-3}\,\text{Pa}$ and better. Electron beam applications can be divided into two categories, those where materials are modified, for fabrication purposes, and the noninvasive techniques used for characterization, such as the various types of electron microscopy, and for display purposes.

Electron Beam Generation and Focusing

An electron optical system typically contains an electron gun and beam forming optics. In the gun, a cathode, mostly a hot tungsten filament, emits electrons which are accelerated toward the anode; the voltage between cathode and anode is in the range 1–200 kV (Fig. 1). The Wehnelt electrode, situated between cathode and anode, is some 100 V more negative than the cathode. It has two purposes: it controls the emission current and it shapes the electric field in such a way that the electrons are collimated to a beam, which passes through an aperture in the anode. Recently also photocathodes and field-emission sources have been used for special applications, particularly in order to increase the brightness (i. e., current density per solid angle) of the electron beam.

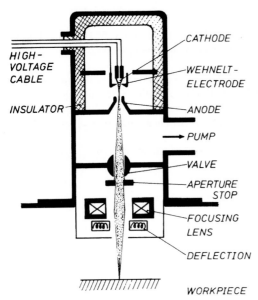

Fig. 1: Schematic diagram of an electron beam gun with a focusing lens and deflection coils.

The beam-forming optics consists of one or more electron lenses. These are typically rotationally symmetric, static electric or magnetic fields, generated by a set of electrodes or by current-carrying coils, mostly surrounded by appropriately shaped ferromagnetic material. With these lenses, the beam can be focused or any plane, e. g., that of a thin film inserted into the beam path, can be imaged. Scanning of the beam can be achieved by time-varying electric or magnetic fields.

Electron Beam Fabrication

Electron beam processes utilize the energy of the electron beam to modify the target material by heating or by altering chemical bonds. In most cases a focused beam is used and depending on the application a wide range of power densities, total beam power, acceleration voltage which determines the penetration depth, and beam diameter is available (see Table 1).

Electron Beam Melting is used for refining of metals. Refractory metals (e. g., with high melting point) and reactive materials can be kept in molten state in vacuum while impurities evaporate.

Electron Beam Evaporation is commonly used to produce metal and insulator coatings, such as mirrors, antireflection coatings on optical lenses, and in the semiconductor industry. Again high-melting-point and reactive materials, including alloys and compounds, can be handled, since the power is directly supplied to the material to be evaporated while the crucible can be kept cool.

Electron Beam Welding relies on the formation of a thin cavity channel by removing material under the impact of a high-power-density beam. The solidifying melting bath around

Table 1: Summary of EB processes and their technical parameters.

Process	Acceleration voltage (kV)	Beam power (kW)	Power density (W m^{-2})	Beam diameter (mm)
Melting	15–35	10–2000	10^8–10^9	1–10
Evaporation	10–35	5–500	10^9–10^{10}	0.5–5
Welding	30–170	0.1–100	10^9–10^{11}	0.2–2
Fusion treatment	20–150	1–10	10^8–10^{10}	0.5–5
Drilling	80–170	1–60	10^{10}–10^{13}	0.05–0.5
Lithography	10–25	10–10	10^5–10^8	10^{-6}–10^{-3}
Polymerization	20–100	1–50	10^4–10^6	10–100

the channel joins the workpieces together. A remarkable depth-to-width ratio of the welding seams (up to 50:1 and several centimeters deep) can be achieved, while only a small volume around the seam is thermally affected, resulting in low material distortions. This has led to a variety of new construction principles, such as welding of precision machined parts and of disparate materials.

Electron Beam Drilling involves pulsed beams with diameters ranging from 0.01–1 mm and power densities up to 10^{13} W/m^{-2} causing explosive removal of the material via holes and cavities of controlled depth can be produced.

Electron Beam Surface Treatment uses the energy of the electron beam to modify thin surface layers, involving melting, often together with previously deposited additives, and subsequent quenching of the liquid, or crosslinking of the molecules of organic surface coatings (paints). In many cases, surfaces with remarkable properties can be attained.

Electron Beam Lithography is the main application of electron beams for microfabrication in semiconductor technology. In electron beam lithography the substrate is first coated with a thin film of radiation sensitive material, commonly called a resist (in most cases an organic polymer). The electron beam irradiation changes the chemical solubility of the resist in the exposed areas. Subsequently, in a development step, the more soluble part of the resist (either the exposed or the unexposed, depending on the polarity of the resist) gets removed and a pattern stays behind. This remaining resist then serves as a mask to transfer the pattern to the substrate, either by etching, deposition, or ion implantation, for example. Two features make electron beam lithography particularly attractive: the high degree of flexibility, accuracy, and speed with which original patterns can be directly generated with a computer controlled beam and the high-resolution capability of electron optics.

The relative ease with which original patterns can be generated by a scanning focused beam under computer control makes electron beam lithography the main technology for fabrication of masks for other types of lithography such as standard optical projection lithography. This will continue to be true in the future for submicrometer optical lithography and for x-ray lithography as the resolution requirement becomes more demanding. Another application

Fig. 2: Electron beam lithography tool for manufacturing, including the electron optical column and the control computer (photo IBM).

is direct-write lithography for manufacturing of relatively low-volume products, application-specific integrated circuits (ASICs), for which the demand is increasing due to the diversification of applications and sophisticated performance requirements.

Electron beam lithography tools today are of the scanning type, in which a fine beam that can be rapidly turned on and off is steered to delineate a pattern serially. Such scanning tools have adequate speeds to prepare optical masks, which are used many times, but are relatively slow and cannot compete with optical lithography tools in processing throughput. State-of-the-art tools (Fig. 2) can scan a beam at pixel rates of several hundred megahertz, and, using complex variable-shaped spots that can expose about 100 pixels in parallel processing rates of $> 10^{10}/\mathrm{s}$ will be achieved (a pixel is the smallest resolution spot in the pattern). The quick turnaround time of direct writing tools together with the high resolution make such tools extremely suitable for development work even when future high-volume manufacturing is performed with optical lithography.

Areas with very little practical alternative to electron beam lithography are the development of submicrometer integrated circuits, research into limits of technologies, and the search for techniques and devices based on novel effects, which requires structures with 100 nm dimensions and below. At this point of time, electron beam lithography is the only proven technology for sub-0.5-μm and even sub-100-nm work. As an example, Fig. 3 shows a scanning electron micrograph of gates from circuits of field effect transistors (FET). These circuits with gates as short as 70 nm have been fabricated with five levels of electron beam lithography, overlaid with an accuracy of better than 30 nm.

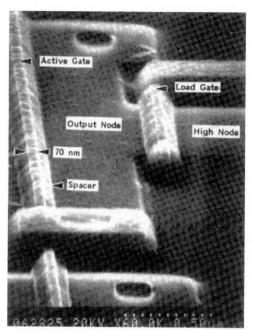

Fig. 3: High-resolution scanning electron micrograph of 70-nm gates in an experimental silicon circuit, fabricated by electron beam lithography.

Fundamental resolution limits involve several factors: (i) the electron optical resolution, which can be a fraction of a nanometer, if current density in the beam and therefore exposure speed is not an issue, (ii) the range of secondary electrons with sufficient energy to expose the resist, (iii) the resolution, or effective size of the basic building blocks of the resist material. For poly-methyl-methacrylate (PMMA), the most widely used resist for high-resolution applications, the latter two factors seem to limit resolution to ~ 10 nm, even if the electron-optical resolution is 10 times better. In inorganic resist materials such as AlF_3, patterns with 2 nm minimum dimension have been produced by electron beam lithography. These materials, however, require rather high exposure doses, so that breaking this apparent resolution limit of ~ 10 nm with highly sensitive material remains one of the challenges for electron beam lithography.

Normal operation of a scanning tunneling microscope (STM) has little or no effect on the sample. But increasing the voltage and distance between tip and substrate so that a strongly confined beam of electrons with sufficient energy to cause chemical reactions is formed by field emission enables the STM to perform nanolithography. Thin layers of electron beam resists have been exposed, including PMMA, Langmuir–Blodgett films, metal halides, and hydrocarbon contamination. Metallic deposits have been formed by decomposing metal-organic adsorbates and bumps have been created on the surface of metallic glasses by thermally and electrostatically enhancing surface diffusion. Atomic scale resolution has been shown in modifications to a single-crystal germanium surface. These concepts are very new and promising for ultrahigh-resolution processing.

Noninvasive Techniques

Noninvasive techniques exploit the various effects that occur when an electron beam interacts with a sample: the electrons get absorbed in thin layers, depending on the local sample properties (in thin films of low atomic number material, essentially the phase of the electron waves gets shifted), they suffer material-characteristic energy losses, they excite secondary radiation in the form of photons (luminescence, x rays) or secondary electrons, again with sample specific energies and yields.

Cathode Ray Tubes

One or more (e. g., three for color applications) finely focused electron beams are scanned across a phosphor coated screen where light is generated at the points of impact. Major applications are video displays (television, terminals) and oscilloscopes.

Electron Beam Characterization

The scanning electron microscope (SEM) in its various forms has come to play a key role in semiconductor technology, in particular because optical microscopy has poor resolution in the submicrometer regime. Sample preparation is simple – even production wafers or packaging modules can be inspected – since a focused electron beam is scanned across the surface, and secondary radiation is collected to form a point-by-point image. Major new developments are advances in the capability to focus significant currents into small spots, leading to high-resolution images with good signal-to-noise ratio. The advent of field emission sources and sophisticated electron optics has led to commercial instruments with better than 1 nm resolution and to high-resolution images with low-energy electron beams (a few hundred electron volts), so that even insulating samples can be investigated without charging. Together with automation and improved metrology by means of laser interferometry, this opens the way to use SEMs for inspection and critical dimension measurements in semiconductor production lines.

The transmission electron microscope (TEM) has been the driving force for the developments in electron optics earlier this century. In the conventional case, an unfocused electron beam is incident on a very thin sample and, by using the transmitted electrons, magnifying projection electron optics form an image of the sample on photographic film with a resolution of a few Ångstrøms. Thus, essentially atomic resolution has been achieved, in particular with the scanning version of the tool. Lattice imaging, where the contrast is improved as a result of the many atoms lined up in a properly oriented lattice plane, has proven to be very powerful in materials research, in particular interface studies. The analytical capabilities by means of energy loss spectroscopy and filtering, and small-area diffraction, for example, are becoming more and more important. A drawback is the rather complicated sample preparation – thinning to a few tens of nanometers is typically required – and the limited sample size. Recent advances in electron holography promise new possibilities for image evaluation and enhancement.

An exciting new form of electron microscopy has become available with the invention of the scanning tunneling microscope (STM). In an STM, a very fine metal tip is scanned across a sample at a height of only a few atomic spacings. At such spacings, a tunneling current occurs when only a few volts are applied between tip and sample. This tunneling current depends sensitively on the tip-to-sample spacing and is used to measure and control the height of the tip. In this way, scanning micrographs of the surface with a height resolution of one-tenth to one-hundredth of an atom and lateral resolution of approximately one to three atomic distances can be directly produced. Also, the electron energy distribution of the tunneling current depends on the electronic properties of the sample and can be used to select or view different types of surface atoms.

Noncontact electrical testing with focused electron beams is another application that will increase in importance as the complexity of circuits and packaging modules increases and critical dimensions further decrease. The current induced by the beam can directly be detected on the substrate and in this way give information on electrical continuity (specimen current) and electronic properties (EBIC). Another technique measures the energy of secondary electrons and thus the voltage of their point of origin on the specimen. Moreover, the electron beam can be chopped into pulses of picosecond duration, and thus fast changes in voltages on IC lines can be sampled at high temporal resolution.

See also: Electron Bombardment of Atoms and Molecules; Electron and Ion Beams, Intense; Electron and Ion Impact Phenomena; Electron Microscopy; Scanning Tunneling Microscope.

Bibliography

A. Septier (ed.), *Focusing of Charged Particles*, 2 vols. Academic Press, New York and London, 1967.
O. Winkler and R. Bakish, *Vacuum Metallurgy*. Elsevier, Amsterdam, London, and New York, 1971.
A. H. Meleka, *Electron Beam Welding*. McGraw–Hill, London, 1971.
G. R. Brewer and J. P. Ballantyne, *Electron Beam Technology in Microelectronic Fabrication*. Academic Press, New York, 1980.
T. H. P. Chang *et al.*, "Nanostructure Technology," *IBM J. Res. Develop.* **32**, 462–493 (1988).
L. Reimer, *Transmission Electron Microscopy*. Springer-Verlag, Berlin, 1984.
O. C. Wells, A. Boyd, E. Lifschin, and A. Rezanowich, *Scanning Electron Microscopy*. McGraw–Hill, New York, 1974.

Electron Bombardment of Atoms and Molecules
E. C. Beaty

Introduction

In scientific research and in technology the bombardment atoms and molecules by electrons occurs under a great variety of physical circumstances. Some are deliberately arranged for a special purpose, such as stripping electrons from gas molecules in order to use a magnetic deflection spectrometer for chemical analysis of the gas, or transferring energy into an atomic or molecular system in order to induce the emission of electromagnetic radiation. For example, the excitation required for the operation of lasers is generally produced directly or indirectly by electron bombardment. For such applications, we seek to understand the effects of electron bombardment in order to obtain the desired result. In other circumstances, the collision of electrons with atoms or molecules is a disruptive influence. For instance, in cathode ray tubes collisions of electrons with residual gas molecules increase the spot size. For these applications, we seek understand electron bombardment effects in order to reduce their influence. For these and other reasons, there has been a great deal of scientific effort directed at understanding the consequences of electron bombardment of atoms and molecules since the electron was first identified by J. I. Thomson in 1897.

The subject of electron bombardment of atoms and molecules is here restricted to the case of atoms and molecules in the gaseous state. Thus, it is adequate to consider a projectile electron as interacting with only one molecule at a time. Even in a gas as dense as the Earth's atmosphere, it is relatively rare for two molecules to be in close enough proximity that a free electron could interact significantly with them simultaneously. Although interesting and important effects follow electron bombardment of solids and liquids, the considerations are very different.

Many of the effects of electron collisions with atoms and molecules are relatively easy to measure and describe. There is an extensive and long-standing literature describing research in this field. Over the years, the pursuit of more accurate and complete descriptions has caused both the experiments and the descriptions of the data to become rather complex. The following paragraphs attempt to provide an elementary description of the phenomena and an introduction to some of the language normally used in the research literature.

General Considerations

Among the early observed phenomena that could not be described in the framework of classical mechanics were some effects of electron bombardment of atoms. While many aspects of an electron beam moving through a rarefied gas can be described adequately using classical mechanics, some features demand description in terms of quantum mechanics.

Quantum mechanics now offers a mathematical framework for describing all collision effects. With a small set of experimentally determined atomic constants (the charge and mass of the electron, Planck's constant, and the speed of light) and a great deal of mathematical work, it is possible to compute both qualitatively and quantitatively all of the electron collision effects to high accuracy. While the preceding sentence is universally believed by atomic physicists as a matter of principle, in practice, the mathematics is so complicated that even experts cannot carry it out accurately except in special circumstances. The prescriptions embodied in quantum mechanics are well-defined but so complex that, except in unusual situations, it is necessary to make approximations which lead to errors of unknown magnitude.

Current research in this field involves using a combination of mathematical analysis and experimental results to work out ways of making the simplifying approximations without introducing unduly large errors. Thus in spite of the acceptance of the quantum-mechanical description as accurate, it is still standard practice to describe electron bombardment effects in the context of classical mechanics with suitable *ad hoc* modifications to account for the quantum or wave effects. In judging the adequacy of a particular classical model, it is necessary to appeal to quantum mechanics or to make a direct check with observations.

It is almost always a good approximation to treat a beam of electrons as a stream of particles. Many years ago the wave effects of diffraction and interference were observed in electron beams, establishing an important principle. However, such observations require rather special circumstances such that the classical approximation is ordinarily used in work on electron motion. Similarly, it is a good approximation to treat the motion of the target molecule from a classical point of view. On the other hand, the electronic structure of atoms and molecules cannot be well described in the context of classical mechanics. It is sometimes convenient to speak of the electrons as being in orbit; however, such descriptions quickly prove inadequate.

In the encounter of a free electron with a free molecule, linear momentum is conserved in the classical sense. The encounter may also cause a change in the wave function of the molecule. In almost all circumstances it is appropriate to regard the molecule as being in one of its energy eigenstates both before and after the encounter. If the final state is different from the initial state, the collision is known as inelastic and the energy difference must come from (or go to) the projectile. Thus energy is conserved in the classical sense; any energy absorbed by the molecule is treated as potential energy.

Generally the mechanics of a collision problem are best described in center-of-mass coordinates. For the case of electrons colliding with molecules, the center of mass is sufficiently close to the molecule that little error is made if one considers the molecule to be a fixed target, with the electron being a projectile hurled at the target. The probability of causing some particular effect upon collision is usually specified by giving the cross section for that effect. More specifically, if a beam of electrons of particle current density J (particles per unit area per unit time) is directed at molecular targets having a density N (molecules per unit volume), the rate R of collisions (a collision being defined by some prescribed measurement of changed trajectory, change of energy state, or other specific effect) will be proportional to both J and N. This may be expressed as

$$R = \sigma J N , \tag{1}$$

where σ is the collision cross section. In this expression, R is in units of events per unit volume per unit time, and σ has units of area per molecule per electron. In giving the cross section for an effect, we specify the equivalent area of the target which the projectile must hit to produce the effect. While it is convenient to think in terms of a single electron being projected toward a single target, such detailed control is not actually possible. Furthermore, as a matter of principle, the effects can only be described and measured in a statistical sense.

Scattered Electrons

One way to observe the effects of electron bombardment of molecules is to pass a beam of electrons through a sample of gas and note the changes in the electron beam caused by the gas. The electron properties capable of being changed by collision with a molecule are momentum and spin. Spin changes are observable only in highly specialized situations and are not considered further here. Instead of working with the vector quantity, momentum, it is frequently more useful to use the equivalent quantities kinetic energy and the polar coordinates giving the direction of motion.

Description of scattering cross sections is assisted by extending the definition to "differential" cross sections. Suppose the collisional effect to be examined is the scattering of an electron into solid angle $\delta\Omega$ with its final energy in the range δE. The cross section for this effect as defined in Eq. (1) will be proportional to $\delta\Omega$ and δE (for appropriately small $\delta\Omega$ and δE). Let

$$S \equiv \lim_{\delta\Omega, \delta E \to 0} \frac{\sigma}{\delta E \, \delta \Omega} . \tag{2}$$

S is called a differential cross section, or sometimes a double differential cross section. The integral of S over either energy or solid angle is useful and is also called a differential cross section.

The double differential cross section embodies a complete description of the electron scattering effects. It is normally defined to include changes in the target as well. Consideration of the energy and momentum of the electron before and after the collision allows deduction of the momentum and energy transferred to the target. However, the target might have more than one way to respond to the changes. A general rule of quantum mechanics is that some energies of excitation cannot be accepted by atoms and molecules with the result that S is zero for some values of the transfer energy. To illustrate this effect let us consider an example. Suppose a narrow beam of 100-eV electrons is passed through helium gas. Examine the electrons which leave the gas sample. (We are imagining that it is possible to restrict the gas to a small volume without having solid objects interfering with the observations.) Most of these electrons will not have been affected, with the result that the original beam will be present with reduced intensity. Electrons will be observed with 100 eV energy but traveling in directions different from the beam. These electrons are said to be elastically scattered. Other electrons will be observed with 80.2 eV energy. These have been inelastically scattered, leaving the helium atom in the first excited state, spectroscopically designated the 2^3S state. No electrons will be found with energies in the range 80.3–99.9 eV. (In helium the energy gap between the ground state and the first excited state is particularly large; however, such a gap exists with all atoms and molecules.) The result is that with enough control over the electron energies it is possible to distinguish elastically scattered electrons from others. In considering elastic scattering it is useful to define a cross section that is differential only in solid angle. Such a cross section is equivalent to integrating S as defined in Eq. (2) over a small energy range which includes the energy of incident electrons. For some purposes it is desirable to know the total scattering cross section a which is related to S by

$$\sigma_T = 2\pi \int_0^\pi \int_0^\pi S(E, \theta) \sin\theta \, dE \, d\theta . \tag{3}$$

To be complete, Eq. (3) should include a sum over all the final states of the target; however, the total cross section is most often used when all the scattering is elastic. Measurements of σ_T are most conveniently made by passing a beam of electrons through a gas with magnetic or electric fields applied in such a way that electrons which lose energy or change direction by collisions are removed from the beam. By measuring the transmitted beam current with and without the gas it is straightforward to deduce σ_T. The total scattering cross section σ_T can be considered the effective area of an atom or molecule for all collisional effects. An important observation is that σ_T is finite.

Target Effects

As noted above, a collision between an electron and a molecule can do nothing to the electron except change its kinetic energy and direction of motion. The molecule, being a much more complex object, is capable of much more extensive changes. It is useful to speak of the cross section for making a particular change in a target with the understanding that what happens to the projectile is of no concern. Thus we might be interested in the cross section of helium atoms for excitation to the $2\,^1P$ state by 100-eV electrons. Such a cross section might be inferred from observation of the photons generated as the excited atoms return to the ground state. For atoms, a change of state means a transition to a different energy eigenstate corresponding to a different electronic configuration. For molecules, vibrational and rotational excitations are also possible. For both atoms and molecules the final electronic state may be high enough that a previously bound electron is set free, leaving the target with a net positive charge (i. e., ionized).

Perhaps the most consequential effect of electron bombardment of atoms and molecules is ionization. A major result is that additional free electrons are produced. Gases at ordinary temperatures and pressures are electrically insulating; however, a few free electrons in an electric field may gain enough energy to have ionizing collisions. The result is an electron avalanche which makes the gas electrically conducting.

Gaseous Conduction

A large array of physical processes is involved in gaseous conduction. A high density of electrons cannot be maintained in the absence of positive ions because the electrons repel each other. But when the source of electrons is the ionization of molecules, the positive ions produced cause an electric field which on the average cancels the electric field due to the electrons. Such a mixture of positive ions and electrons is called a plasma and can be a very good electrical conductor. An applied electric field causes the electrons to move in one direction and the positive ions to move in the other. Since the electrons have much smaller mass they move faster and thus are the primary agent for the movement of charge. The movement of electrons through a gaseous conductor involves a balance of momentum (and energy) gained from the force of the electric field and with that lost from collisions with the gas molecules.

See also: Atoms; Electron Beam Technology; Molecules; Quantum Mechanics; Scattering Theory.

Bibliography

Earl W. McDaniel, *Collision Phenomena in Ionized Gases*. Wiley, New York, 1964. A relatively elementary survey of ionized gas effects. Included are the effects of electron bombardment of atoms and molecules with a brief survey of relevant classical and quantum theory and a substantial compilation of data.

H. S. W. Massey, *Electronic and Ionic Impact Phenomena, Vols. I and II, Electron Collisions with Molecules and Photo-ionization*, 2nd ed. Oxford University Press, London, 1969. The second edition of this monograph is published in four volumes, the first two of which cover electron collisions with atoms and molecules. The phenomena are reviewed with special attention to experimental and theoretical techniques. Included is a good summary of available data.

N. F. Mott and H. S. W. Massey, *The Theory of Atomic Collisions*. Oxford University Press, London, 1965. A thorough account of the application of quantum mechanics to atomic collisions. Includes a thorough discussion of collisions of electrons with atoms and molecules.

Electron Diffraction

M. A. Van Hove, M. A. O'Keefe, and L. Marton[†]

Diffraction, in light optics, has been defined as "Any departure of the actual light path from that prescribed by geometrical optics" [1]. Contrary to light, where the discovery of diffraction preceded the wave theory, electron diffraction was discovered as a consequence of a deliberate attempt to prove the wave nature of the electron.

In 1924 Louis de Broglie advanced the, for the times, revolutionary view that electrons (or any corpuscles) are accompanied by waves [2]. The origins of that view can be traced back to Einstein's 1905 theory about the wave–particle duality of light. Starting with this theory of Einstein, de Broglie came to the conclusion that electrons should exhibit the same kind of dual nature as light, where the relation between a particle's mass m_0, its velocity v, and its wavelength λ could be expressed as

$$\lambda = \frac{h}{m_0 v}$$

h being the constant of Planck. Today we usually write the de Broglie relation as

$$\lambda = h/p \, ,$$

where p is the momentum of the particle; this formulation implies the necessary relativistic correction.

In 1927 came definite evidence for the particle–wave duality through the discovery of electron diffraction by crystal lattices. This discovery was made almost simultaneously by C. J. Davisson and Lester H. Germer, using slow electrons diffracted by a nickel single crystal [3], and by George P. Thomson and A. Reid, using fast electrons passing through a thin celluloid film [4]. Their work was followed by rapid confirmation and extensive further development. Quite early it was shown that the angular dependence of the diffracted beams satisfied Bragg's law (originally established for x-ray diffraction):

$$n\lambda = 2d \sin\theta$$

where d is the distance between atomic planes, θ the diffraction angle and n an integer. The de Broglie wavelength λ, as calculated from the relativistically correct equation

$$\lambda = \frac{h}{[2m_0 eV(1 + eV/2m_0 c^2)]^{1/2}}$$

was confirmed with good accuracy. In this equation e is the charge of the electron, c the light velocity, and V the potential difference applied to accelerate the electron. It was also

[†]deceased

shown that, while the Davisson–Germer experiment corresponded to the Laue diagram of x-ray diffraction, the Thomson–Reid experiment was a counterpart of the Debye–Scherrer diagram of x rays.

Because of easier experimental requirements (lesser vacuum, less stringent control of the accelerating potentials) the high-voltage, transmission method developed much faster than the slow-electron reflection technique. The original diffraction cameras, consisting of an electron source and an aperture, collimating the beam before it passed through the specimen, were soon replaced by the arrangement first suggested by A. A. Lebedev [5]. In this arrangement an axially symmetrical magnetic field, the axis being parallel to the electron beam, is used to focus the beam on the recording medium (usually the photographic emulsion). This procedure has the advantage of replacing a well-collimated beam with a divergent beam and still attaining a higher intensity of the recorded pattern. At the same time this device allows the use of a larger area of the specimen. The accelerating potentials used by different authors were generally in the 10- to 100-kV range, resulting in electron wavelengths corresponding to $(12.2–3.7)\times10^{-12}$ m (0.0122–0.0037 nm).

In the experiments described so far, diffraction was produced by periodic structures, such as crystal lattices. This limitation was lifted by the electron microscope which was developed by using magnetic or electrostatic lenses to bring the diffracted beams together so they could interfere and form an image [6]. By 1940 the electron microscope was sufficiently developed for the observation of diffraction fringes. By that time, all workers in electron microscopy agreed that the resolving power, as in light microscopy, is diffraction limited and obeys the Abbé relation

$$\rho = \frac{0.61\lambda}{n\sin\alpha}$$

where ρ is the least resolved distance, n the index of refraction of the medium surrounding the object, and α the semiangle subtended by the object at the lens. However, spherical aberrations in the electron microscope objective lens limits α to about 0.01, and the resolution of transmission electron microscopy to about 150×10^{-12} m (0.15 nm)

The year 1940 also marks the beginning of another important link between electron diffraction and electron microscopy. Until that time the dominant interpretation of the origin of contrast in electron microscope images was based on mass-thickness scattering by the specimen [7]. B. van Borries and E. Ruska found, however, variations of intensities in chromium oxide smoke images, which could not be interpreted by scattering alone [8]. This finding was followed by the observation of Bragg reflections in images of crystalline specimens [9, 10] and led finally to the theory of diffraction contrast in images of crystalline specimens [11, 12].

Introduced in 1939, convergent-beam electron diffraction started being used only in the 1970s [13, 14]. Instead of illuminating the sample with a parallel beam, the beam is focused as a convergent cone onto a submicron area in the transmission mode. Advantages are the small-scale analysis area, avoiding twinning and other faults, for example, and microstructure fingerprinting of complex materials.

For an understanding of intensity relations in electron diffraction and in electron microscope images of crystalline objects, detailed calculations based on the theory of diffraction phenomena are necessary. There are two theories of electron diffraction: the kinematic (or geometric) theory and the dynamic theory. The kinematic theory, in the ideal case, describes the interaction of the electron with a single atom and derives the intensities of the resulting beams. The dynamic theory takes into account the multiple scattering resulting from the presence of many atoms. This theory requires a wave-mechanical treatment, since the scattering produces a background that very considerably modifies the intensity distributions. X rays, which produce much less scattering than electrons, can usually be treated by means of the kinematic theory. For electrons, even for a thickness of less than 100 atom layers, there is so much scattering that we must have recourse to the dynamic theory [11, 12, 15]. Typical specimen thicknesses employed in high resolution electron microscopy require the application of the dynamic theory in calculating image intensities [16].

With the advent in the 1970s of microscopes capable of producing images showing details smaller then the unit cell size ("lattice images"), dynamic theory [17] was able to compute simulated images that could be used for structure determination when compared with experiment [18].

In the 1980s, electron microscopy started being applied to surfaces, using either samples of submicron thickness with the transmission mode, or submicron-size particles grown on carbon supports in the profile imaging mode [19, 20]. In this manner, deviations from the bulk structure were observed in the first one or two atomic layers of the surface. Obtaining ultrahigh-vacuum conditions is a necessity for control of surface composition [20].

In the 1990s, the "C_s-barrier" limited the resolution of electron microscopes (C_s is the coefficient of spherical aberration). This barrier was overcome by methods of hardware [21] and software correction [22], allowing resolution below the one-Ångström (0.1 nm) mark [23]. The improved resolution allowed the imaging of lighter atoms such as oxygen and even lithium [24].

In 1928, S. Kikuchi, while observing the diffraction of electrons passing through thin mica sheets, discovered a pattern differing widely from the single-crystal Laue spots or the polycrystalline Debye–Scherrer patterns [25]. He found a system of lines and bands, requiring a modification of the dynamic theory. These Kikuchi patterns consist of a continuous angular distribution of inelastically scattered electrons modulated by lines or band edges at the position of the various Bragg reflections. The energy analysis of the electrons diffracted into a Kikuchi line has shown that, for silicon, the average energy loss was about 450 eV, in contrast to the ten-times-lower loss for the ordinary patterns taken in transmission [26].

Similar effects are found with electrons that have energies on the order of 1000 eV and that have either lost energy to an excitation or have themselves been emitted by, for example, a photon in photoelectron emission, or by an Auger decay. Such electrons radiate outward from the point of interaction and are diffracted by the surrounding lattice. They thereby carry structural information which can modulate the excitation process, as in extended x-ray absorption fine structure (EXAFS) [27]. Near a surface, the electrons themselves can escape to a detector and deliver surface structure information, as is the case with photoelectron and Auger electron diffraction [28], since they also undergo diffraction from the surrounding atoms.

For three decades high-energy electron diffraction (HEED) remained the only practical approach to the problems accessible by diffraction methods, although L. H. Germer pointed out quite early [29] the advantages of low-energy electron diffraction (LEED). By 1960, however, two of the experimental obstacles to LEED were removed: high-vacuum technology was radically improved to the extent that the surfaces investigated remained stable during the experiment; and the low intensities of the diffraction patterns, which hampered earlier work, could be enhanced by post-acceleration of the diffracted electrons. The usual energy range of LEED is from 20 to 500 eV, corresponding to a wavelength range of $(274\text{--}55) \times 10^{-12}$ m (0.274–0.055 nm). In the case of LEED, multiple scattering is largely unavoidable and the dynamic theory therefore a necessity [30]. Due to strong inelastic and elastic scattering at LEED energies, the electron mean free path is small, $(500\text{--}1000) \times 10^{-12}$ m (0.5–1 nm), and the electron–atom scattering cross section is large, comparable to the geometrical atomic size. The LEED theory in common use is an adaptation of the theory of electronic band structure, after inclusion of inelastic damping and of x-ray-like Debye–Waller factors to describe vibration effects. The theory is capable of determining structures within atomic and molecular monolayers to an accuracy of $(1\text{--}10) \times 10^{-12}$ m (0.001–0.01 nm). Traditionally a trial-and-error search procedure has been applied to structure determination, but advances such as tensor LEED [31] permit rapid and computationally efficient determinations.

The discovery of electron diffraction was followed, within a remarkably short time, by application of the new effect to many other problems. Within the framework of this short survey it is necessary to limit ourselves to an enumeration of some of these applications, without extensive discussion.

As with x-ray crystallography, structure determinations have been in the forefront. HEED in transmission is very widely used for the study of thin films of monocrystalline, polycrystalline, or amorphous materials. In the gaseous phase HEED is very useful for the study of molecular structures [32]. Within 10 years after the discovery of electron diffraction, results were published on more than 150 substances and by 1953 Pinsker's book [15] tabulated the structures of about 500 molecular species.

Because of the limited penetration of electrons at lower energies, surface structures are much more accessible to electron diffraction investigation than to x-rays. Thus, surface science benefits both from HEED reflection observations and from measurements by means of LEED. For example [30], it was found that many clean metal surfaces exhibit ideal bulk lattice terminations, but show small relaxations in the first few atomic layers. Other metal surfaces and most semiconductor surfaces, by contrast, reconstruct: one or more atomic layers adopt lattices that differ from the bulk lattice. Also, several hundred structures of atomic adsorption have been solved. They often involve simple bonding in high-coordination "hollow" sites of the substrate. Increasingly many cases of adsorbate incorporation into or adsorbate-induced restructuring of the substrate have been uncovered. Such results are of direct value in such diverse fields as microelectronic device properties, catalysis, corrosion, tribology, nanoscience and nanotechnology.

Selected-area diffraction has been extended from the high energies [11] to surface-sensitive energies in the low-energy electron microscope (LEEM) [33] and the photoemission electron microscope (PEEM) [34]. Thus areas of the order of $0.1\text{--}100\,\mu\text{m}^2$ can be selectively examined.

In 1986, Szöke suggested the possibility of performing holography with photoelectrons in the keV range [35]. Barton then showed the feasibility of this approach [36], which subsequently has been much refined and applied to the three-dimensional real-space imaging of the structure of various surfaces with atomic resolution [37].

Electron holography at higher electron energies is possible in the electron microscope by using a bi-prism to combine scattered and direct beams. Holography has been used to improve resolution, for example to image defects in metal crystals at atomic resolution [38]. Information far beyond the point resolution limit of an electron microscope can be retrieved using the holographic off-axis reconstruction technique [39]. High-energy electron holography has been applied to the imaging of magnetic structures [40] and superconductors [41], among other materials.

See also: Crystal Defects; Crystallography, X-Ray; Diffraction; Electron Microscopy; Gratings; Holography; Interferometers and Interferometry; Low-Energy Electron Diffraction (LEED); Quantum Mechanics; Reflection High-Energy Electron Diffraction (RHEED); Surfaces and Interfaces; X-Ray Spectra.

References

[1]   E. U. Condon and H. Odishaw, eds., *Handbook of Physics*, p. 6. McGraw–Hill, New York, 1958.

[2]   L. de Broglie, Thèse, Paris. Masson, Paris, 1924.

[3]   C. J. Davisson and L. H. Germer, *Nature* **119**, 558 (1927); *Phys. Rev.* **30**, 705 (1927).

[4]   G. P. Thomson and A. Reid, *Nature* **119**, 890 (1927).

[5]   A. A. Lebedev, *Nature* **128**, 491 (1931).

[6]   M. Knoll and E.Ruska, *Ann. Physik* 12, 607; 641 (1932).

[7]   L. Marton and L. I. Schiff, *J. Appl. Phys.* **12**, 759 (1941).

[8]   B. v. Borries and E. Ruska, *Naturwiss.* **28**, 366 (1940).

[9]   J. Hillier and R. F. Baker, *Phys. Rev.* **61**, 722 (1942)

[10] R. D. Heidenreich, *Phys. Rev.* **62**, 291 (1942).

[11] R. D. Heidenreich, *Fundamentals of Transmission Electron Microscopy.* Wiley (Interscience), New York, 1964.

[12] P. B. Hirsch, A. Howie, R. B. IVicholson, D. W. Pashley, and M. J. Whelan, *Electron Microscopy of Thin Crystals.* Butterworth, Washington, 1965.

[13] J. W. Steeds, in *Introduction to Analytical Electron Microscopy* (J. J. Hren, J. I. Goldstein, and D. C. Joy, eds.), p. 387. Plenum Press, New York, 1979.

[14] J. M. Cowley and J. C. H. Spence, *Ultramicroscopy* **6**, 359 (1981).

[15] Z. G. Pinsker, *Electron Diffraction* (translated from the Russian by A. A. Spink and E. Feigl). Butterworths, London, 1953.

[16] M. J. Whelan, "Electron Diffraction Theory and Its Application to the Interpretation of Electron Microscope Images of Crystalline Materials," *Adv. Electronics Electron Phys.* **39**, 1 (1975).

[17] J. M. Cowley and A. F. Moodie, *Acta Crys.* **10**, 609 (1957).

[18] M. A. O'Keefe, P. R. Buseck and S. Iijima, *Nature* **274**, 322 (1978).

[19] L. D. Marks and D. J. Smith, *Nature* **303**, 316 (1983).

[20] K. Yagi, K. Takayanagi, and G. Honjo, in *Crystals, Growth, Properties and Applications* (H. C. Freyhardt, ed.), Vol. 7, p. 47. Springer-Verlag, Berlin, Heidelberg, 1982.

[21] M. Haider, G. Braunshausen and E. Schwan, *Optik* **99**, 167 (1995).

[22] W. M. J. Coene, A. Thrust, M. Op de Beeck and D. Van Dyck, *Ultramicroscopy* **64**, 109 (1996).

[23] M. A. O'Keefe, C. J. D. Hetherington, Y. C. Wang, E. C. Nelson, J. H. Turner, C. Kisielowski, J.-O. Malm, R. Mueller, J. Ringnalda, M. Pan and A. Thrust, *Ultramicroscopy* **89**, 215 (2001).

[24] S.-H. Yang, L. Croguennec, C. Delmas, E. C. Nelson and M. A. O'Keefe, *Nature Materials* **2**, 464 (2003).

[25] S. Kikuchi, *Proc. Imp. Acad. Japan*, **4**, 271, 275, 354, 471 (1928); *Jpn. J. Phys.* **5**, 83 (1928).

[26] W. Hartl and H. Raether, *Z. Phys.* **161**, 238 (1961).

[27] P. A. Lee, P. H. Citrin, and B. Kincaid, *Rev. Mod. Phys.* **53**, 769 (1981).

[28] J. J. Barton, S. W. Robey, and D. A. Shirley, *Phys. Rev. B* **34**, 778 (1986).

[29]   L. H. Germer, *Z. Phys.* **54**, 408 (1929); *Bell Syst. Tech. J.* **8**, 591 (1929).

[30]   J. B. Pendry, "Low Energy Electron Diffraction," in *Techniques of Physics* (G. K. T. Conn and K. R. Coleman, eds.), Vol. 2. Academic Press, New York, 1974. M. A. Van Hove, W. H. Weinberg, and C.-M. Chan, *Low-Energy Electron Diffraction: Experiment, Theory, and Surface Structure Determination*. Springer-Verlag, Berlin, 1986.

[31]   J. B. Pendry, K. Heinz, and W. Oed, *Phys. Rev. Lett.* **61**, 2953 (1988).

[32]   L. S. Bartell, "Electron Diffraction by Gases" in *Physical Methods of Chemistry* (A. Weissburger and B. W. Rossiter, eds.), 4th ed., p. 125. Wiley (Interscience), New York, 1972. K. Kuchitsu, "Gas Electron Diffraction" in *Molecular Structure and Properties* (G. Allen, ed.), p. 203. MTP, Oxford, 1972. J. Karle, "Electron Diffraction" in *Determination of Organic Structures by Physical Methods* (F. G. Nachod and J. J. Zuckerman, eds.), p. 1. Academic Press, New York, 1973.

[33]   W. Telieps and E. Bauer, *Ultramicroscopy* **17**, 57 (1985).

[34]   S. Gunther, B. Kaulich, L. Gregoratti and M. Kiskinova, *Progress in Surface Science* **70**, 187 (2002).

[35]   A. Szöke, in *Short Wavelength Coherent Radiation: Generation and Applications*, AIP Conf. Proc. No. 147, edited by D. T. Attwood and J. Boker (AIP, New York, 1986), p. 361.

[36]   J. J. Barton, *Phys. Rev. Lett.* **61**, 1356 (1988).

[37]   C.S. Fadley, S. Thevuthasan, A.P. Kaduwela, C. Westphal, Y.J. Kim, R.X. Ynzunza, P. Len, E. Tober, F. Zhang, Z. Wang, S. Ruebush, A. Budge and M.A. Van Hove, *J. El. Spectrosc. Rel. Phen.* **68**, 19 (1994).

[38]   A. Orchowski and H. Lichte, *Ultramicroscopy* **64**, 199 (1996).

[39]   H. Lichte and W.D. Rau, *Ultramicroscopy* **54**, 310 (1994).

[40]   A. Tonomura, T. Matsuda, J. Endo, T. Arii and K. Mihama, *Phys. Rev. Lett.* **44**, 1430 (1980).

[41]   A. Tonomura, H. Kasai, O. Kamimura, T. Matsuda, K. Harada, Y. Nakayama, J. Shimoyama, K. Kishio, T. Hanaguri, K. Kitazawa, M. Sasae and S. Okayasu, *Nature* **412**, 620 (2001).

Electron Energy States in Solids and Liquids

J. Callaway[†]

Introduction

In an atomic or molecular gas at ordinary temperatures and densities, electrons are bound to specific atoms or molecules and occupy states characteristic of isolated systems. Such states have wave functions which decay exponentially at large distances from the home-base atom or molecule, and the energy levels of the system are reasonably sharp. Interactions between electrons on different atoms are, on the average, quite weak.

In contrast, in condensed matter, interactions cause broadening of the energy levels of electrons in outer atomic shells. It is more correct to describe the electron states as belonging to the system as a whole, rather than to any specific atom or molecule. Such states are called band states, and one says that energy bands are formed. At the normal interatomic distances in solids, only the outer shells have broadened into bands. The energies of tightly bound core states are still quite sharp.

[†] deceased

These ideas are, in a rough way, independent of the particular atoms or molecules involved, and are valid whether the condensed phase is solid or liquid. Closer investigation indicates that in molecular crystals, such as solid hydrogen, or in crystals of rare gases, solid argon for example, the band of valence electron states is quite narrow, and a molecular or atomic description is quite appropriate. Also, if the system is strongly disordered, the disorder itself may lead to electron localization. On the other hand, in metals, the broadening is quite advanced and states originating in different atomic levels may be strongly mixed.

It has proved much easier to develop a detailed understanding of electron states in crystals than in glassy solids or liquids, and in pure materials rather than alloys. This is because the existence of long-range structural order makes possible enormous simplifications in the mathematical description of electron states. Ordered systems are also often easier to characterize from an experimental point of view, and many types of experiments are possible only for ordered systems because of increased electron mean free path. We shall therefore devote most of this article to electron states in perfect crystalline solids, and reserve only a few paragraphs toward the end for disordered materials.

Electron States in Ordered Solids

The essential characteristic which simplifies the description of states in perfect crystals is periodicity. We will ignore here the existence of surfaces. Then there exists an infinite set of direct-lattice vectors $\mathbf{R}_i$,

$$\mathbf{R}_i = n_{i1}\mathbf{a}_1 + n_{i2}\mathbf{a}_2 + n_{i3}\mathbf{a}_3 \ , \tag{1}$$

in which the $\mathbf{a}_i$ are the three primitive lattice vectors and the n_{ij} are integers, such that the crystal is unaltered if all atoms are displaced by any $\mathbf{R}_i$. We shall consider the electron states from a single-particle point of view in which each electron has its own wave function. There are some materials, particularly antiferromagnetic insulators (NiO and CoO are good examples), for which a single-particle description of the electron states is totally inadequate. We will not consider such systems here. The following discussion does apply rather well to metals, common semiconductors (such as Si, Ge, GaAs, etc.), and insulators (such as NaCl) in which the atoms or ions have filled shells of electrons. In these cases, the one-electron wave function is the solution of a Schrödinger equation

$$H\psi = E\psi \ , \tag{2}$$

in which the Hamiltonian is periodic,

$$H(\mathbf{r}+\mathbf{R}_i) = H(\mathbf{r}) \ . \tag{3}$$

It is a fundamental property of solutions of the Schrödinger equation for a periodic Hamilton that (Bloch's theorem)

$$\psi(\mathbf{r}+\mathbf{R}_i) = \exp(i\mathbf{k} \cdot \mathbf{R}_i)\psi(\mathbf{r}) \ . \tag{4}$$

If the coordinate of an electron is changed by a direct-lattice vector, the only change in the wave function is multiplication by a phase factor. The quantity $\mathbf{k}$ is called the wave vector and can be regarded as a quantum number specifying the electron state. It is customary to incorporate **k** into the notation for the wave function so that it is written $\psi_{\mathbf{k}}(\mathbf{r})$ or $\psi(\mathbf{k},\mathbf{r})$. It follows from (4) that the wave function can be written in the form

$$\psi(\mathbf{k},\mathbf{r}) = \exp(i\mathbf{k}\cdot\mathbf{r})u(\mathbf{k},\mathbf{r}) \ , \tag{5}$$

in which $u(\mathbf{k},\mathbf{r})$ has the same periodicity as the Hamiltonian,

$$u(\mathbf{k},\mathbf{r}+\mathbf{R}_i) = u(\mathbf{k},\mathbf{r}) \ . \tag{6}$$

Thus, $u(\mathbf{k},\mathbf{r})$ repeats itself in each unit cell of the crystal, and the wave function ψ has the form of a plane wave modulated in each cell by the periodic function u.

It is useful to define a set of reciprocal-lattice vectors $\mathbf{K}_s$ through

$$\mathbf{K}_s = n_{s1}\mathbf{b}_1 + n_{s2}\mathbf{b}_2 + n_{s3}\mathbf{b}_3 \ , \tag{7}$$

where the n_{si} are integers and the **b**'s are related to the **a**'s by

$$\mathbf{b}_i \cdot \mathbf{a}_i = 2\pi\delta_{ij} \ . \tag{8}$$

It is easy to find **b**'s which obey (8); for example,

$$\mathbf{b}_i = \frac{2\pi(\mathbf{a}_2 \times \mathbf{a}_3)}{\mathbf{a}_1 \cdot (\mathbf{a}_2 \times \mathbf{a}_3)} \ , \tag{9}$$

and the other **b**'s are obtained by cyclic permutation. All direct- and reciprocal-lattice vectors satisfy

$$\mathbf{K}_s \cdot \mathbf{R}_i = 2\pi \times \text{integer} \ . \tag{10}$$

The reciprocal lattice may or may not look like the direct lattice; for example, if the direct lattice is simple cubic, so is the reciprocal lattice, but if the direct lattice is body-centered cubic, the reciprocal lattice is face-centered cubic, and vice versa. Equation (10) has the following implication: If **k** and $\mathbf{k}'$ are two wave vectors which differ by a reciprocal-lattice vector, $\mathbf{k}' = \mathbf{k} + \mathbf{K}_s$ then

$$\exp(i\mathbf{k}\cdot\mathbf{R}_i) = \exp(i\mathbf{k}'\cdot\mathbf{R}_i) \tag{11}$$

for all $\mathbf{R}_i$. Hence $\psi(\mathbf{k},\mathbf{r})$ and $\psi(\mathbf{k}',\mathbf{r})$ behave identically under all lattice translations according to (4) and may be considered to be the same function. For this reason, wave vectors can be restricted to be in a region of momentum space called the Brillouin zone (abbreviated BZ) such that no two interior points differ by a reciprocal-lattice vector. Two examples of BZ's are shown in Fig. 1.

We adopt the convention that wave vectors **k** are restricted to a BZ. When this is done, there will be many solutions of the Schrödinger equation which have the same wave vector but different energies. We designate these different, orthogonal, solutions by an index, usually in the order of increasing energy; thus (2a) is written in a more explicit way as

$$H\psi_n(\mathbf{k},\mathbf{r}) = E_n(\mathbf{k})\psi_n(\mathbf{k},\mathbf{r}) \ . \tag{12}$$

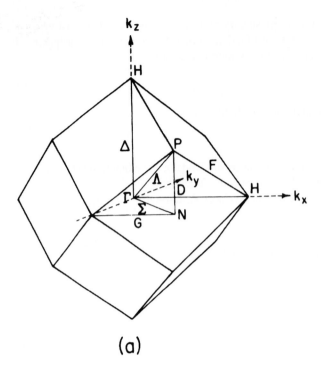

(a)

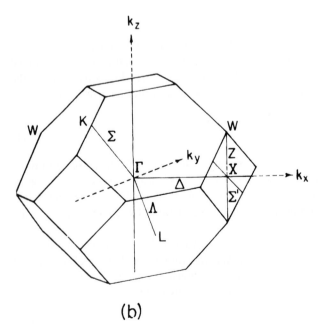

(b)

Fig. 1: Brillouin zones for the body-centered cubic lattice (a), and the face-centered cubic lattice (b). Points and lines of symmetry are indicated according to the accepted notation.

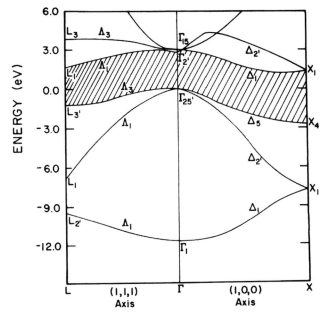

Fig. 2: Energy bands calculated for the semiconductor silicon for two directions of the wave vector **k**. The points Γ, X, L, etc., on the horizontal axis refer to the Brillouin zone of Fig. 1(b). The crosshatched region is the band gap. Bands are labeled according to symmetry properties of the wave functions (see references for details).

The energies of states with a given value of the index n form a continuous manifold when regarded as functions of **k** inside a BZ. This manifold is called an energy band, and n in (2b) is referred to as the band index.

As an example, calculated energy bands for the semiconductor silicon are shown in Fig. 2.

Band states can be regarded as occupied at $T = 0\,\mathrm{K}$ in order of increasing energy according to the rules of Fermi–Dirac statistics, until the required number of electrons per atom has been accommodated. If the energy of the highest occupied state falls inside an energy band, the substance is a metal. The energy of this state is the Fermi energy E_F. In the (excellent) approximation in which the wave vector **k** is treated as continuous, the solutions $\mathbf{k}_F$ of

$$E_n(\mathbf{k}_F) = E_F \tag{12}$$

(for all relevant bands) define the Fermi surface. The Fermi surface can simply be considered to be the surface(s) bounding the occupied region of **k** space. There are several experimental methods which can be used to determine the Fermi surface and so check theoretical calculations. Some of the most important ones include the de Haas–van Alphen effect, the radio-frequency size effect, and the magnetic field dependence of the electrical resistivity and of ultrasonic attenuation. Limitations of space preclude much discussion of these techniques here, but additional information can be found in the references.

If there are no allowed states immediately adjacent in energy to the highest occupied state, the material is a semiconductor or an insulator. The minimum energy separation between the highest occupied state and lowest empty state is the "band gap," which determines the temperature dependence of the electrical conductivity of a pure semiconductor. If the minimum separation between full and empty states is determined for states of the same **k**, one has the "direct" band gap, which is of great importance in the study of optical properties.

It is found in many applications that electrons in solids move in weak applied external electric (**E**) and magnetic (**B**) fields according to the simple equation

$$m^* \left(\frac{d\mathbf{v}}{dt} + \frac{1}{\tau} \right) = -e(\mathbf{E} + \mathbf{v} \times \mathbf{B}) \,, \tag{13}$$

in which e is the magnitude of the electronic charge, τ is a relaxation time which roughly describes the effect of collisions, and m^* is known as the effective mass. The quantity m^* is, in general cases, a tensor; that is, acceleration and force are not necessarily in the same direction. It is possible to determine the effective-mass tensor from the derivatives of the energy band function. Suppose that only a single band need be considered. Then one has

$$\left(\frac{1}{m^*} \right)_{\alpha\beta} = \frac{1}{\hbar^2} \frac{\partial^2 E(\mathbf{k})}{\partial k_\alpha \partial k_\beta} \,, \tag{14}$$

in which α, β are Cartesian indices. A variety of experimental measurements are available to determine m^*, and values smaller than the free electron mass by factors of the order of 100 are found in some materials (e. g., InSb).

It is possible for regions of **k** to be important, where $E(\mathbf{k})$ has negative ("downward") curvature. In this case m^* is negative, but it is customary to consider the mass to be positive and to delete the minus sign in (13) so that one describes the motion of a positive charge. Such effective positive charges are known as holes. The concept of a hole is particularly important in semiconductors, where states near the top of a full valence band are described in this way.

Another important concept is the "density of states." This measures the fractional number of states which have energies in a certain interval. Specifically let $G(E)\,dE$ be the number of states of a given spin direction which have energies between E and $E + dE$. $G(E)$ is given by the expression

$$G(E) = \frac{\Omega}{(2\pi)^3} \sum_n \int d^3k \delta(E - E_n(\mathbf{k})) \,, \tag{15}$$

in which the sum includes all bands and the integral runs over the Brillouin zone. The delta function serves as a counter, registering a unit contribution from those bands (n) and wave vectors (**k**) such that $E_n(\mathbf{k}) = E$. The quantity Ω in the multiplicative constant is the volume of the unit cell. Since the volume of the Brillouin zone is $(2\pi)^3/\Omega$, the constant normalizes the integral so that the contribution from each band to the integral of $G(E)$ over all energies is unity. A very simple case in which (15) may be explicitly evaluated is that of a parabolic band characterized by an effective mass m^* as is found for a nearly free electron system. Suppose $E(\mathbf{k}) = \hbar^2 k^2 / 2m^*$ (a scalar effective mass) and consider only a single band. The result is

$$G(E) = \frac{\Omega}{2\pi^2} \left(\frac{2m^*}{\hbar^2} \right)^{3/2} E^{1/2} \,. \tag{16}$$

This expression will not be valid in a region of energies where an energy band intersects a Brillouin-zone face, for then the expression for $E(\mathbf{k})$ must become more complicated, as some derivatives of the energy must vanish. In fact, it can be shown that the density of states must possess sharp structure at such energies, called "van Hove singularities." Further information can be found in the references.

Electrons in a metal produce a contribution to the specific heat that depends linearly on temperature. This contrasts with the contribution from lattice vibrations, which is proportional to T^3 at low temperatures. At sufficiently low temperatures the linear term dominates the cubic one, and one can determine experimentally the coefficient γ in the relation

$$C_V^{(e)} = \gamma T \ , \tag{17}$$

in which $C_V^{(e)}$ is the electronic contribution to the specific heat per atom. It can be shown that γ is given by

$$\gamma = \frac{1}{3}\pi^2 K^2 G(E_F) \ , \tag{18}$$

in which K is Boltzmann's constant and $G(E_F)$ is the density of states evaluated at the Fermi energy. Hence measurement of γ yields a value for $G(E_F)$ that can be compared with calculations.

Calculation and Measurement of Electronic Energies

There are several rather standard techniques available for the numerical calculation of electronic energies and wave functions from Eq. (2). The most commonly applied methods include (1) the "Green's-function" method, (2) the "augmented-plane-wave" method, (3) the method of "linear combination of atomic orbitals," and linearized versions of the first two (linear muffin tin orbitals method, linearized augmented plane wave method). Space permits only a brief description of some common general features, but details of specific approaches are described in the Bibliography. All of the methods involve the expansion of the unknown $\psi(\mathbf{k},\mathbf{r})$ as a combination of some previously specified functions $\Phi_s(\mathbf{k},\mathbf{r})$, which are considered to obey Bloch's theorem explicitly,

$$\Phi_s(\mathbf{k},\mathbf{r}+\mathbf{R}_i) = \exp(\mathbf{i}\mathbf{k} - \mathbf{R}_i)\Phi_s(\mathbf{k},\mathbf{r}) \ .$$

The simplest set of potentially useful functions are plane waves of the form

$$\exp[\mathbf{i}(\mathbf{k}+\mathbf{K}_s) \cdot \mathbf{r}]$$

in which $\mathbf{K}_s$ is a reciprocal-lattice vector. In general, one writes

$$\psi_n(\mathbf{k},\mathbf{r}) = \sum_s c_{ns}(\mathbf{k})\Phi_s(\mathbf{k},\mathbf{r}) \ . \tag{19}$$

The sum includes, in principle, an infinite number of terms, but of course only a finite subset can be employed in actual calculations. The coefficients $c_{ns}(\mathbf{k})$ have to be determined so that (2) is satisfied. Equation (19) is substituted into (2); the result is multiplied by $\Phi_t^*(\mathbf{k},\mathbf{r})$, where

Φ_t is a general member of the basis set, and integrated over the entire crystal. A set of linear homogeneous equations is obtained:

$$\sum_s H_{ts}(\mathbf{k})c_{ns}(\mathbf{k}) = E_n(\mathbf{k})\sum_s S_{ts}(\mathbf{k})c_{ns}(\mathbf{k}) , \tag{20}$$

in which the H_{ts} are matrix elements of the Hamiltonian on the basis Φ,

$$H_{ts}(\mathbf{k}) = \int \Phi_t^*(\mathbf{k},\mathbf{r})H\Phi_s(\mathbf{k},\mathbf{r})\,\mathrm{d}^3r , \tag{21}$$

and the S_{ts} are overlap matrix elements,

$$S_{ts}(\mathbf{k}) = \int \Phi_t^*(\mathbf{k},\mathbf{r})\Phi_s(\mathbf{k},\mathbf{r})\,\mathrm{d}^3r . \tag{22}$$

The overlap matrix S_{ts} is usually not simply a unit matrix since the functions Φ are not generally orthonormal. However, a variety of standard mathematical techniques exist to enable numerical solution of (20) once the elements of matrices H and S are constructed.

The most important problem encountered in these calculations is the determination of the function $V(\mathbf{r})$, which represents the potential energy of an electron in the crystal. At present most calculations of energy bands are based on density functional theory (see the article by Callaway and March in the Bibliography). This approach gives a procedure by which the contributions from the nuclei of the system, the average electrostatic field of the other electrons, and exchange and correlation interactions can be included in $V(r)$ (the latter only approximately). The fundamental quantity is the electron density which has to be computed self-consistently. That is, one assumes some initial distribution of electrons; computes the corresponding potential function; then solves the Schrödinger equation to find energy bands and wave functions. The new distribution of electrons must now be used and the process repeated until the electron distributions resulting from two successive stages of the procedure agree within some assigned limit of error. This iterative process is quite time consuming. However, reasonably good results are usually obtained in a single-pass, non-self-consistent calculation if the potential energy is computed by solving Poisson's equation for a charge distribution produced by the superposition of neutral-atom charge densities centered on crystal lattice sites.

Many ingenious techniques have been developed to permit the experimental determination of the properties of electron states. The easiest to understand are optical. The absorption of light can produce a transition between an occupied state in band n and an unoccupied state in band l. Because the wave vector of a light wave in the infrared, visible, or near-ultraviolet regions of the spectrum is quite small on a scale appropriate to crystalline lattices, it is an excellent approximation to suppose that the initial and final states have the same wave vector. Such transitions can be represented by a vertical line on an energy band diagram, and are said to be direct. Indirect transitions also occur in which one or more phonons are emitted or absorbed. The interpretation of the spectra of band-to-band transitions is not so simple as in the case of free atoms because both initial and final states belong to continuous bands. Sharp structures that are easily identifiable are associated with the onset of transitions into a band or with the occurrence of van Hove singularities in the density of states. Ultraviolet photoemission spectroscopy, in which the excited electron escapes from the solid, also provides information concerning the distribution of initial states.

Optical investigations can determine the energy differences between states substantially removed from the Fermi energy. States near the conduction-band minimum or valence-band maximum in semiconductors, and states close to the Fermi surface in metals, can be studied experimentally by applying a strong external magnetic field. There are many interesting effects of which cyclotron resonance is the easiest to understand. Consider Eq. (13) in the absence of an electric field and suppose (1) that the relaxation time τ is long enough so that $1/\tau$ can be neglected, and (2) that the effective mass m^* is a scalar rather than a tensor. Then it is easy to verify that an electron will spiral around the direction of the magnetic field with a frequency

$$\omega_c = eB/m^* . \tag{23}$$

The quantity ω_c, is called the cyclotron frequency. Electromagnetic waves of frequency ω_c will be absorbed and in this way m^* can be determined. The discussion is easily extended to allow for a tensor effective mass, and for relaxation. Another effect of great importance is the de Haas–van Alphen effect. The magnetic susceptibility of a metal in a strong magnetic field exhibits oscillations as a function of the field whose period is proportional to $1/B$. The observed constant of proportionality can be used to determine the areas of cross sections of the Fermi surface in a direction perpendicular to the magnetic field.

Electron States in Disordered Materials

Our ability to describe electron states in extended systems rests on the applicability of Bloch's theorem. If the Hamiltonian is not periodic because of disorder on a microscopic scale, the wave vector $\mathbf{k}$ is no longer a good quantum number. We can imagine that the disorder (caused either by random atomic positions in an alloy, or by thermal motions as in a liquid) causes electrons to scatter rapidly between states. Even in pure materials, some such scattering is always present, and gives rise to the electrical resistivity of a metal. In a liquid metal, the scattering is so frequent that the mean free path may be only a few atomic separations. If a state of definite $\mathbf{k}$ were somehow to be prepared initially, it would soon be destroyed.

A straightforward description of the properties of a disordered system in terms of energy bands is therefore not to be expected. The concept of the density of states, however, remains a useful one. The density of states of a disordered system will not exhibit the sharp structure characteristic of ordered systems in which the periodic lattice structure gives rise to van Hove singularities, but need not be a featureless object. An energy gap between filled and empty regions of a density of states may still persist. In some cases, especially highly anisotropic systems such as thin films (effectively two dimensional) or long narrow wires (effectively one dimensional), disorder may lead to complete localization of the electrons, even though bulk, well-ordered samples of the same materials would be normal metals. In certain amorphous solids, particularly semiconductors, disorder may lead to at least partial filling of the energy gap. It has been shown for such materials that states in the vicinity of the gap are localized and therefore will not permit electrical conduction, while states more removed in energy are extended and will conduct, The energy separating conducting and nonconducting states is called the mobility edge.

Much current research is devoted to the development of a quantitative theory of disordered systems. In regard to normal bulk metallic alloys, this work has emphasized the development of approximations in which the actual disordered system is conceptually replaced by an ordered effective medium. The properties of this medium are determined from the scattering properties of the individual atoms by the condition that, on the average, no further scattering should occur (the coherent potential approximation). An effective, non-Hermitian Hamiltonian results, which can be used for the computation of a density of states. Studies of electron states in strongly disordered systems in which electrons are localized, especially those of reduced dimensionality, have to be undertaken by more powerful methods because an effective medium approach is not appropriate. One finds, for example, that the electrical resistance of a disordered one-dimensional system increases exponentially with its length, and at sufficiently low temperatures, the resistance of a given sample may fluctuate as a function of time.

See also: Crystal Symmetry; Cyclotron Resonance; De Haas–van Alphen Effect; Fermi Surface; Hamiltonian Function; Liquid Metals; Liquid Structure; Metals; Quantum Mechanics; Resistance; Schrödinger Equation; Semiconductors, Crystalline; Ultrasonics.

Bibliography
A. Elementary
J. S. Blakemore, *Solid State Physics*, 2nd ed. Cambridge University Press, Cambridge, 1985.

G. Burns, *Solid State Physics*. Academic Press, New York, 1985.

R. E. Hummel, *Electronic Properties of Materials*, 3rd ed. Springer-Verlag, Berlin, Heidelberg, 2000.

C. Kittel, *Elementary Solid State Physics*. Wiley, New York, 1966.

M. N. Rudden and J. Wilson, *Elements of Solid State Physics*. Wiley, New York, 1980.

B. Intermediate
P. W. Anderson, *Concepts in Solids*. World Scientific, Singapore, 1998.

N. W. Ashcroft and N. D. Mermin, *Solid State Physics*. Holt, Rinehart and Winston, New York, 1976.

R. J. Elliott and A. F. Gibson. *An Introduction to Solid State Physics and Its Applications*. Barnes and Noble, New York, 1974.

H. Y. Fan, *Elements of Solid State Physics*. John Wiley, New York, 1987.

C. Kittel, *Introduction to Solid State Physics*, 7th ed. Wiley, New York, 1995.

J. M. Ziman, *Principles of the Theory of Solids*, 2nd ed. Cambridge University Press, London, 1979.

C. Advanced
J. Callaway, *Quantum Theory of the Solid State*. Academic Press, New York, 1974.

J. Callaway and N. H. March, "Density Functional Methods, Theory and Applications," *Solid State Phys.* **38**, 135 (1984).

W. A. Harrison, *Solid State Theory*. Dover, New York, 1980.

W. A. Harrison, *Electronic Structure and the Properties of Solids*. Dover, New York, 1989.

W. Jones and N. H. March, *Theoretical Solid State Physics*. Dover, New York, 1985.

C. Kittel, *Quantum Theory of Solids*, 2nd ed. Wiley, New York, 1987.

P. A. Lee and T. V. Ramakrishnan, "Disordered Electronic Systems," *Rev. Mod. Phys.* **57**, 287 (1985).

O. Madelung, *Introduction to Solid State Theory*. Springer-Verlag, Berlin, Heidelberg, 1995.

G. Mahan, *Many Particle Physics*, 3rd ed. Plenum Press, New York, 2000.

J. C. Slater, *Quantum Theory of Molecules and Solids*, Vols. 2 and 3. McGraw–Hill, New York, 1963.

Electron–Hole Droplets in Semiconductors

J. Shah

A pure semiconductor crystal is devoid of free electrons and holes (vacant electron states) at low temperatures. Excitation of such a semiconductor by photons of appropriate energy creates electrons and holes which interact among themselves via Coulomb interactions. At low densities and temperatures, an electron becomes bound to a hole to form a hydrogen-atom-like complex known as an exciton. In many semiconductors it is possible to create a large density of excitons by intense photoexcitation. In 1968 Keldysh suggested that observations of anomalous photoconductivity in Ge at high excitation intensities might be explained if, at some critical density, the gas of excitons condensed into a metallic liquid phase consisting of electron–hole droplets (EHD). This gas–liquid transition is analogous to condensation of atomic Na vapor into a metallic Na liquid, i. e., an electron is not bound to a particular hole but is free to move independently.

The correctness of this idea was demonstrated by three key experiments. In the first, a rapid increase in the intensity of a luminescence feature of Ge with slight decrease in temperature was interpreted as a first-order phase transition and the line shape of this feature was shown to be that expected for EHD. In the second experiment, small differences in dielectric constants between the metallic droplets and the surrounding exciton gas in an optically excited crystal of Ge scattered a probe laser beam in much the same way as fog scatters light. The angular distribution of the scattered laser light conclusively demonstrated the existence of macroscopic droplets with an average radius of the order of $5\,\mu$m. In the third experiment, discrete giant charge pulses were observed from an optically excited reverse-biased *p–n* junction in Ge. These result from dissociation of EHD in the strong electric field at the junction and provided further evidence for the finite size of EHD.

A detailed theory describing this electron–hole liquid has been developed by considering the ground-state energy $E_G(r_s)$ of an electron–hole pair in a degenerate electron–hole plasma as a function of r_s the interparticle separation in units of the exciton radius. This energy is the sum of the kinetic, exchange, and correlation energies. The kinetic and exchange energies can be calculated analytically, but various approximations have to be used to calculate the correlation energy. Many factors, such as the anisotropy and the multivalley nature of the band structure, affect these energies. However, it has been shown that the sum of the exchange and correlation energies (ε_{xc}) is independent of the band structure of semiconductors. $\varepsilon_{xc}(r_s)$, in the unit of the excitonic Rydberg (ε_x), has been shown to obey the following simple expression for many semiconductors over a large range of r_s: $\varepsilon_{xc}(r_s) = (a + br_s)/(c + dr_s + r_s^2)$, where $a = -4.8316$, $b = -5.0879$, $c = 0.0152$, and $d = 3.0426$. Typically, $E_G(r_s)$ has the shape shown in Fig. 1, with a minimum at r_0. If this minimum energy lies below the exciton binding energy, i. e., if $\psi_0 = |E_G(r_0)| - |\varepsilon_x|$ is positive, then a condensate can form because it is stable against emission of an exciton at absolute zero degrees ($T = 0$); ϕ_0 is then the binding energy of the condensate. As a result of the Coulomb interactions between the carriers the band gap within the EHD is renormalized from E_g to $E_g(r_0)$ as is shown schematically in Fig. 2. The chemical potential $\mu(r_0)$ and $E_g(r_0)$ can be simply calculated from $E_G(r_0)$.

The formation of a metallic state in optically excited semiconductors provides a unique opportunity to explore the properties of quantum metals, and a wide variety of experimental techniques have been used to study this novel state of matter. One technique which stands

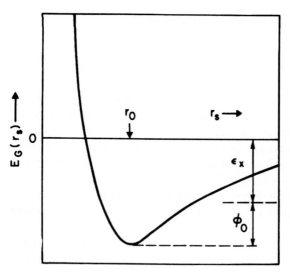

Fig. 1: Schematic representation of ground-state energy per pair versus normalized interparticle separation.

out above all others is the study of photoluminescence spectra of EHD and excitons under a variety of conditions. It is clear from Fig. 2 that at $T = 0$ the energies of the lowest- and the highest-energy emitted photons correspond to $E_g(r_0)$ and $\mu(r_0)$, respectively. Since $E_F^e + E_F^h = -\mu(r_0) - E_g(r_0)$, one can obtain the liquid density n_0 from these measurements. Also, ϕ_0 is obtained by measuring the difference in energy between $\mu(r_0)$ and the low-energy edge of exciton emission. One can also obtain these quantities at $T \neq 0$ by analyzing the line shapes; such measurements show that n_0 and ϕ_0 vary with T in a manner expected for a metal or a degenerate plasma, confirming once again the metallic character of EHD.

When the source of excitation is removed, the EHD luminescence decays with a characteristic decay time. The e–h pairs within a droplet decay by radiative recombination, by nonradiative Auger recombination, and also by thermionic emission into the surrounding gas if the temperature is not too low. The finite lifetime of this metallic liquid phase distinguishes it from other liquids. However, assuming quasi-equilibrium between the gas of excitons and the liquid, one can discuss the gas–liquid coexistence curve in the temperature–density phase diagram of EHD, as for any other gas which undergoes a first-order transition to a liquid. A schematic phase diagram for EHD is shown in Fig. 3. The exciton–EHD system has a critical temperature T_c. above which liquid cannot form. The region under the coexistence curve is the unstable region in that a plasma created at any point (n, T) in this region will spontaneously break up into droplets of density n_0 and gas of density n_{ex}. The coexistence curves for Ge and Si, including the values of T_c, have been determined from luminescence measurements. Phase diagrams have also been calculated theoretically.

At low temperatures and densities, the photoexcited e–h pairs form excitons, corresponding to the lower left region in Fig. 3. For $T < T_c$ as the density of the gas phase is increased beyond the density of gas on the coexistence curve, the gas becomes supersaturated and condensation occurs. Experimentally, there is a well-defined excitation intensity (I_{th}) beyond which the

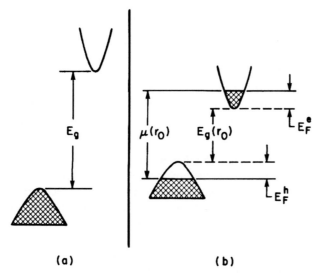

(a) **(b)**

Fig. 2: Schematic energy versus wave-vector representation of the band structure of an indirect-bandgap semiconductor (a) outside and (b) inside EHD. The shaded regions are occupied by electrons. E_F^e and E_F^h are the electron and hole Fermi energies.

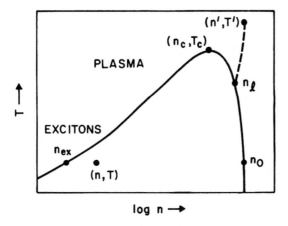

Fig. 3: Schematic phase diagram of EHD.

EHD becomes observable, i. e., the EHD luminescence increases extremely rapidly for $I > I_{th}$. It has been shown that the threshold for monotonically increasing intensity is considerably larger than that for monotonically decreasing intensity. This hysteresis is a result of the surface energy σ of EHD and provides a technique of determining σ.

The theory of nucleation of drops from excitons has been developed by including the effect of finite lifetime of drops in the well-known classical theory of homogeneous nucleation. Density fluctuations of excitons create clusters of excitons or embryos. Embryos with radius R less than a critical radius R_c vanish quickly while embryos with $R > R_c$ continue to grow

Table 1: Binding energy (ϕ_0), liquid density (n_0), critical temperature (T_c), and lifetime (τ) of EHD in various semiconductors

	ϕ_0 (meV)	n_0 (cm^{-3})	T_c (K)	τ (μs)
Ge	1.8	2.4×10^{17}	6.5	40
Si	8.2	3.3×10^{18}	27	0.15
GaP	14	6×10^{18}	> 35	0.035

to a macroscopic size limited by the EHD lifetime. The time to form a macroscopic EHD is determined by the time to form a critical embryo and the time it takes for the critical embryo to collect excitons to grow to its final size. At high supersaturation ($I \gg I_{th}$), the second factor dominates, while at small supersaturation, the first factor dominates. Nucleation theory predicts a very strong dependence of nucleation time on supersaturation close to threshold. Time-resolved luminescence spectroscopy on nanosecond time scales provides a direct confirmation of these predictions. This technique also demonstrates that EHD can form not only by nucleation from a supersaturated gas but also through expansion and cooling of a hot dense plasma, i.e., expansion and cooling of (n', T') first creates a liquid of density n_l (Fig. 3) and then leads to phase separation into EHD and excitons.

The droplets can move through the crystal under exciton or phonon pressure or under external forces such as strain gradients, or electric field, or phonon pressure. Acceleration and deceleration of EHD has been directly observed by Doppler velocimetry. Motion produced by applied electric fields has been used to measure the net charge on EHD in Ge. A remarkable phenomenon has been observed in Ge in the presence of a stress gradient; the EHD accelerate to the point of maximum stress and coalesce into a single large drop whose radius and lifetime are 300 μm and 500 ms, respectively, at high pump intensities. These large drops have also been photographed using infrared image-scanning cameras. Early work on EHD was done exclusively on Ge and Si. The existence of the metallic liquid phase has now also been established in GaP; the polar character of GaP and CdS has been shown to play an important role in the theoretical calculation of the binding energy in these semiconductors. In addition to the thermodynamic considerations, one must also consider whether the plasma lifetime in direct-bandgap semiconductors such as CdS is sufficiently long to allow the formation of the liquid phase. Other subjects which have attracted recent attention are the influence of the exciton–plasma Mott transition on the EHD phase diagram and interactions of phonons with EHD. The question of EHD in quasi-2D systems such as quantum wells has also been considered. Calculations predict liquid formation under certain conditions. However, no experimental results on EHD in quasi-2D semiconductors have been reported.

A study of electron–hole droplets in semiconductors has been a fascinating field of research, and has provided increased understanding not only of semiconductors, but also of phenomena normally associated with other branches of physics of condensed matter. It allows us to investigate, for example, effective ultrahigh densities, quantum and many-body effects in metals, gas–liquid phase transitions, and thermodynamics of multicomponent systems.

See also: Electron Energy States in Solids and Liquids; Excitons; Luminescence; Semiconductors, Crystalline.

Bibliography

V. S. Bagaev, *Springer Tracts Mod. Phys.* Vol. 73, 72 (1985). (I)

C. D. Jeffries, *Science* **189**, 955 (1975). (E)

Ya. E. Pokrovskii, *Phys. Stat. Sol. (a)* **11**, 385 (1972). (I)

G. A. Thomas, *Sci. Am.* **234**, 28 (June 1976). (E)

A two-part review of EHD (Theory by T. M. Rice and Experimental Aspects by J. C. Hensel, T. G. Phillips and G. A. Thomas) can be found in *Solid State Physics*, Vol. 32 (H. Ehrenreich, F. Seitz and D. Turnbull, eds.). Academic Press, New York, 1977 (A); this is an excellent, exhaustive review covering the literature up to mid-1976.

Electron Microscopy

D. J. Smith

The electron microscope uses a finely focused beam of energetic electrons to provide a highly magnified image of the specimen region of interest. Since the electron wavelength is typically much less than 1 Å, for example at 100 kilovolts it is 0.04 Å (where $1\,\text{Å} = 10^{-10}$ m), it might be anticipated that the electron image would show structural details on the same scale. In practice, however, all existing electron lenses suffer from an unavoidable imaging defect, known as spherical aberration, which causes off-axis electrons to be improperly focused. The microscope resolution can then be expressed in the form

$$D = AC_s^{1/4}\lambda^{3/4}$$

where C_s is the spherical aberration coefficient of the microscope objective lens, λ is the electron wavelength, and A is a constant, varying between about 0.43 and 0.7, which depends on the operating conditions. Typical theoretical values for D range from about 5.0 Å for an accelerating voltage of 20 kV to about 1.2 Å at 1000 kV. The quandary for the microscopist is to choose between operation at higher voltages when the resolution is better but the sample structure is likely to be altered by collisions with the energetic electron beam, or to operate at lower voltage and accept that the resolution will be somewhat degraded. The recent trend has been towards operating voltages of 200–400 kV with corresponding image resolutions in the range of 2.5–1.6 Å. Details of atomic configurations are then easily obtained in low-index projections of many types of inorganic materials. Most biological and organic materials can not, however, withstand the effects of the electron beam and resolution figures are typically worse by an order of magnitude or more.

There are two basic modes of operation of the electron microscope, utilizing either fixed or scanning electron beams.

The conventional transmission electron microscope (CTEM) involves a relatively broad beam of electrons, perhaps 0.5–1.0 micron across, which is transmitted through a suitably thinned specimen. A highly magnified image can then be seen on the final viewing screen or on a television monitor via an image pickup system. By judicious choice of the size and position of the so-called objective aperture, which is normally located within the magnetic field of the objective lens pole-pieces, it is possible to highlight significant aspects of the sample morphology, particularly if it has an irregular or defective crystalline structure. Electron diffraction patterns can also be interpreted to provide useful specimen information.

Fig. 1: Atomic-resolution electron micrograph showing a grain boundary
between two nickel oxide crystals. Black spots represent rows of nickel atoms
viewed end-on ($10\text{Å} = 10^{-9}\,\text{m}$).

The scanning electron microscope (SEM) has a finely focused electron beam which is
scanned across the sample in a raster-like fashion. Electrons which are scattered by interaction with the sample are synchronously collected by nearby detectors and, after suitable
amplification, form highly magnified images of the specimen surface on a viewing monitor. Three-dimensional views of the sample can be obtained by recording successive pairs
of (stereo) images which differ only by small changes in the direction of the incident beam.
The image resolution of the SEM is closely related to the minimum probe diameter, which
mainly depends on the type of electron source, Typical resolutions are perhaps 50 Å but with a
field emission electron gun, which has a very sharp metallic tip as its electron source, surface
details as fine as 5 Å across are sometimes discernible.

The scanning transmission electron microscope (STEM) also involves rastering of the focused beam across the sample in synchronization with the detection of electrons which have
passed through the sample, followed by modulation of the image monitor. The sequential nature of the image formation process again lends itself to online manipulation of the electron
signal, in particular to take advantage of the fact that the amount of scattering of the electron
beam depends strongly upon the elemental composition of the sample. Local variations in
composition or structure, sometimes down to the atomic scale, can be differentiated by large
variations in image contrast. Finally, by stopping the probe on a selected region of the sample,
elemental information about the irradiated area can be obtained using either the characteristic
xrays emitted by the sample or by measuring the energy spectrum of the transmitted electrons.

Overall, the electron microscope represents an instrument suitable for probing the *local*
structure and composition of materials at levels which are unapproachable by most other techniques. For example, Fig. 1 shows a grain boundary in a nickel oxide bicrystal. Each black
spot represents a row of nickel atoms viewed end-on, and so it is possible to deduce directly

the arrangement of atoms in the vicinity of the boundary. Since the microscopic properties of most solids depend to a large extent on such local irregularities the electron microscope is a powerful tool for the characterization of advanced materials.

See also: Electron Diffraction; Field Emission.

Bibliography

D. J. Smith, "The Realization of Atomic Resolution with the Electron Microscope", *Rep. Prog. Phys.* **60**, 1513 (1999).
A. V. Crewe, "A High Resolution Scanning Electron Microscope," *Sci. Am.* **224**(4), 1971

Electron Spin Resonance

A. Rassat and S. Gambarelli

A free electron with spin $\hbar\mathbf{s}$ and magnetic moment $\boldsymbol{\mu}_s = -g_e\beta_e\mathbf{s} = -\gamma_s\hbar\mathbf{s}$ ($\hbar$ is Planck's constant/2π, β_e is the Bohr magneton $= \hbar|e|/2m$, γ_s is the electron magnetogyric ratio $= g_e|e|/2m$, g_e is the Landé factor $= 2.0023\ldots$, e is the electron charge, m is the electron rest mass) has in a magnetic field $\mathbf{B}_0$, (here supposed infinitely homogeneous, unless specified) a "Zeeman energy" $W_Z = -\boldsymbol{\mu}_s \cdot \mathbf{B}_0 = g_e\beta_e\mathbf{B}_0 \cdot \mathbf{s} = \gamma_s\hbar\mathbf{B}_0 \cdot \mathbf{s}$ with two stationary states, α ($m_s = +\frac{1}{2}$) and β ($m_s = -\frac{1}{2}$), separated by $\Delta W = g_e\beta_e B_0$. $\mathbf{B}_0$ defines the z axis of a Cartesian coordinate system.

If a small oscillatory magnetic field $\mathbf{B}_1$ [conveniently provided by an electromagnetic radiation of frequency ν_0 ($= \omega_0/2\pi$) and amplitude $B_1 \ll B_0$] is applied in the xy plane so that $h\nu_0 = g_e\beta_e B_0$ (resonance condition), transitions are induced between the α and β levels with absorption or emission of electromagnetic energy: (electron spin) resonance occurs. As induced absorption and emission have equal probability, in a collection of free electrons the populations of each level tend to become equal; there is no net absorption or emission of radiation, and resonance cannot be detected. However, in most systems, *relaxation* occurs: some mechanism ("spin–lattice relaxation") permits thermal equilibrium to be transferred to the spins with a characteristic time constant T_1, the "spin–lattice relaxation time". $\mathbf{B}_0$ also induces precession of the spins at angular velocity $\boldsymbol{\omega}_0 = \gamma_e\mathbf{B}_0$. In a collection of free electrons, spins precess in phase, unless some mechanism produces a fluctuation $\delta\mathbf{B}$ of the magnetic field, which broadens the Zeeman levels by $\delta W_Z = g_e\beta_e|\delta B|/2$, without modifying the level populations. After a time τ, this fluctuation induces a spread in phases, inducing "spin–spin relaxation" with characteristic time constant T_2 (T_2 may be visualized as the time for an average mean square dephasing $\Delta\phi$ between spins to become of the order of $1\,\mathrm{rad}$). Now, if relaxation is not too slow and B_1 is not too strong ($\gamma_e^2 B_1^2 T_1 T_2 \ll 1$), resonance may be detected.

Parallel to the previous quantum model, dealing with energy levels and transition probabilities, these considerations can be conveniently expressed in the so-called gyroscopic model governed by the Bloch equations (*see* Nuclear Magnetic Resonance) relative to the motion of the macroscopic magnetization $\mathbf{M} = \sum_i \boldsymbol{\mu}_{s_i}$ where μ_{s_i} is the magnetic moment of the electron i.

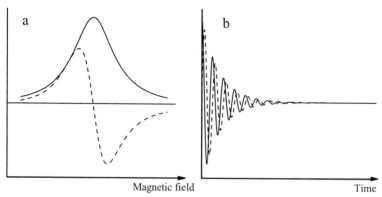

Magnetic field   Time

Fig. 1: (a) CW presentation of the resonance absorption for a collection of electrons resonating at $B = h\nu_0/\beta_e g$; plain line, intensity I of absorption recorded as a function of the applied magnetic field, ν_0 being fixed; dashed line, derivative curve dI/dB, the usual presentation of results in CW ESR. Schematically, a single ESR absorption line is expected to be Lorentzian, $I = 1/[\Delta B^2 + (B - B_0)^2]$, where linewidth $\Delta B \,(= 1/\gamma_e T_2)$ is the damping term of a usual (mechanical) resonance curve (homogeneous broadening), or Gaussian, $I = \exp[-(B - B_0)^2/(\Delta B)^2]$, as in solids, where linewidth ΔB comes from a statistical distribution of local magnetic fields (inhomogeneous broadening). (b) FID detection of the same resonance as in (a) after a $\pi/2$ pulse.

The basic components of an electron-spin-resonance (ESR) *instrument* are a homogeneous magnetic field, a source of monochromatic electromagnetic radiation, and a device to detect its absorption or emission. For technical reasons, the frequency is fixed. Two main different techniques, "continuous wave" (CW) and "pulse ESR" are used to record absorption or emission. For both, various frequencies are available, from 1 GHz (L Band) up to 300 GHz, X band (*ca.* 10 GHz) being of most common use. In CW ESR, absorption I of radiation is recorded as a function of B_0 and is maximum at resonance. In pulse ESR, spins are manipulated according to Bloch equations by turning on and off the electromagnetic radiation with characteristic durations and time intervals, both smaller than T_1 and T_2. For example, after a so-called $\pi/2$ pulse, magnetization **M** originally aligned along $\mathbf{B_0}$ (and so with no apparent precession), precesses with a non-zero component in the *xy* plane, with emission of electromagnetic radiation at the precession frequency. Due to relaxation, M_x and M_y, and so the corresponding emission, undergo "*free induction decay*" (FID) which by Fourier analysis yields the characteristic frequency of this system (Fig. 1). In general, because of magnetic interactions, the spins resonate at different frequencies that are easily detected by this technique. Since T_1 and T_2 are much smaller in ESR than in NMR, pulse-ESR techniques are more sophisticated and of less general use.

Generally, samples are paramagnetic solids or liquids. Let us study first a crystalline sample containing *perfectly oriented* paramagnetic molecules (or ions) having spin $\frac{1}{2}$ ("one unpaired electron"), diluted in a diamagnetic matrix, so that the electronic spins do not interact appreciably. Resonance occurs when $h\nu_0 = g_e \beta_e B$. **B** the local field felt by the electron spin, is now the sum of $\mathbf{B_0}$ and of the magnetic fields due to orbital angular momenta (spin–orbit interactions)

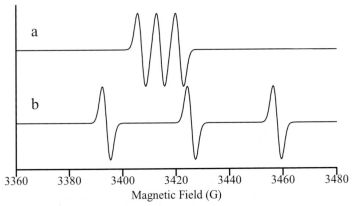

Magnetic Field (G)

Fig. 2: Computed spectra at operating frequency $\nu_0 = 9.6\,\text{GHz}$ for a single crystal containing a diluted and oriented nitroxide free radical at two perpendicular orientations of the magnetic field B_0. In this paramagnetic molecule, the unpaired electron interacts with one nitrogen nucleus (spin $I = 1$). The three lines correspond to $m_I = -1, 0, +1$. $g(\theta, \varphi)$ is measured of the central line $g(\theta, \varphi) = h\nu_0 / \beta_e B_0(\theta, \varphi)$. The hyperfine coupling constant $A(\theta, \varphi)$ is measured as the average hyperfine splitting A' and A'' between the lines. If $g\beta_e B_0 \gg A_N$, $A' = A'' = A_N$. Principal axes of $\mathbf{g}$ and $\mathbf{A}$ tensors coincide. The principal values are $A_{xx} = 7.1\,\text{G}$, $A_{yy} = 5.6\,\text{G}$, $A_{zz} = 32\,\text{G}$, and $g_{xx} = 2.0089$, $g_{yy} = 2.0061$, $g_{zz} = 2.002$.

and to nuclear spins in the molecule (hyperfine interactions). These anisotropic interactions induce an angular dependence of the resonance conditions. If the unpaired electron interacts with one nuclear spin, I_N, the local field depends on the $2I_N + 1$ possible stationary orientations of the I_N spin relative to $\mathbf{B}_0$: $2I_N + 1$ resonance lines are observed. The resonance conditions at one orientation (θ, φ) of $\mathbf{B}_0$ relative to the molecule may be written (if $A_N \ll g_e \beta_e B_0$)

$$h\nu_0 = g(\theta, \varphi)\beta_e B_0 + A_N(\theta, \varphi)m_{I_N} \quad m_{I_N} = -I_N, -I_N + 1, \ldots, +I_N .$$

$g(\theta, \varphi)$ and $A_N(\theta, \varphi)$ are the Landé factor and hyperfine coupling constant of nucleus N with the unpaired electron for direction (θ, φ) of $\mathbf{B}_0$. If there are different interacting nuclei N, resonance conditions are $h\nu_0 = g\beta_e B_0 + \sum_N A_N m_{I_N}$. The number of observed lines is often less than $\sum_N (2I_N + 1)$ because of some coincidences. Figure 2 shows a spectrum obtained at two orientations for one electron interacting with one spin $I = 1$. If resonance is studied at all possible orientations, results may be given in the form of an "effective-spin Hamiltonian," $\mathcal{H}_{\text{spin}} = \beta_e \mathbf{B}_0 \cdot \mathbf{g} \cdot \mathbf{s}' + \sum_N \mathbf{I}_N \cdot \mathbf{A}_N \cdot \mathbf{s}'$ (nuclear terms may be added). $\mathbf{g}$ and $\mathbf{A}_N$ are second rank "tensors" ($\mathbf{B}_0 \cdot \mathbf{g} \cdot \mathbf{s}' = \sum_u \sum_v B_u g_{uv} s'_v$, $u, v = x, y, z$) which by a suitable choice of Cartesian axes ("principal axes") may be written in diagonal form (as a set of three "principal values"). Except for symmetry reasons, both tensors will not be diagonal in the same axes. Diagonalization of the matrix representation of this Hamiltonian in the basis of the effective electron spin $|s' = \frac{1}{2}, m_{s'}\rangle$ and nuclear-spin $\prod_N |I_N, m_{I_N}\rangle$ states gives energy levels reproducing exactly the experimental resonance conditions, if $\mathbf{g}$ and $\mathbf{A}_N$ have correctly been determined. Independently from their experimental determination, the $\mathbf{g}$ and $\mathbf{A}_N$ tensors could in principle be computed from a magnetic Hamiltonian, in the basis of the (electronic) kinetic and (electronic and

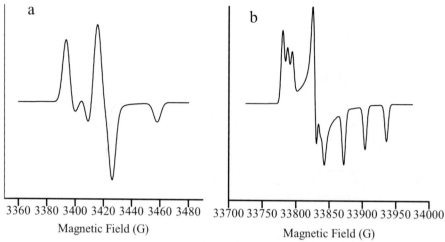

a

b

3360 3380 3400 3420 3440 3460 3480   33700 33750 33800 33850 33900 33950 34000

Magnetic Field (G)   Magnetic Field (G)

Fig. 3: Computed spectra of a frozen solution of the same nitroxide radical as in Figure 2 at two operating frequencies: 9.6 GHz (a) and 95 GHz (b).

nuclear) electrostatic spinless Hamiltonian eigenfunctions. The **g** tensor comes from Zeeman and spin–orbit interactions. The $\mathbf{A}_N$ tensor comes from two interactions: dipolar interaction with nucleus N,

$$H_{\mathrm{dip}_N} = \sum_i^{\text{all electrons}} (\gamma_S \gamma_N \hbar^2 \mu_0 / 4\pi) \times [3(\mathbf{I}_N \cdot \mathbf{r}_{iN})(\mathbf{s}_i \cdot \mathbf{r}_{iN}) r_{iN}^{-2} - \mathbf{I}_N \cdot \mathbf{s}_i] \, r_{iN}^{-3}$$

(γ_N is nuclear magnetogyric ratio of nucleus N, μ_0 is free space permeability, $\mathbf{r}_{iN}$ connects electron i and nucleus N) and (Fermi) contact interaction with nucleus N and all electrons

$$H_{\mathrm{contact}_N} = \sum_i^{\text{all electrons}} (2\mu_0/3)\gamma_S \gamma_N \hbar^2 \delta(\mathbf{r}_{iN}) \mathbf{s}_i \cdot \mathbf{I}_N$$

[$\delta(\mathbf{r}_{iN})$ is the three-dimensional Dirac "function"]. H_{dip} gives rise to a traceless tensor $\mathbf{A}'$ ("purely anisotropic") and H_{contact} to an isotropic term $a_N \cdot \mathbf{1}$ ($\mathbf{A}_N = a_N \cdot \mathbf{1} + \mathbf{A}'_N$).

Let us study a different situation for the same paramagnetic species, still in a "rigid" diamagnetic matrix, but such that the orientations of $\mathbf{B_0}$ relative to the molecule have angular distribution. The observed "immobilized spectrum" $I = F(B_0)$ is the superposition of spectra $I(\theta, \varphi) = f(\theta, \varphi, B_0)$ corresponding to the various orientations: $F(B_0) = \iint f(\theta, \varphi, B_0)\rho(\theta, \varphi)$ $\sin\theta \, d\theta d\varphi / 4\pi$. For an isotropic distribution ("powder spectrum"), the angular distribution $\rho(\theta, \varphi) = 1$. For an anisotropic distribution, $\rho(\theta, \varphi)$ may be estimated from the experimental spectrum. Anisotropies of $\mathbf{g}$ and hyperfine $\mathbf{A}$ terms both affect the "powder spectrum", as shown in Fig. 3. It is difficult to extract all the spin-Hamiltonian parameters from a single spectrum and complementary informations can be obtained by changing the operating frequency (and the corresponding magnetic field) and/or using pulse sequences. As the magnetic field is changed, the relative influence of the field-dependent and field-independent terms of the spin Hamiltonian are changed in the resulting spectra (Fig. 3). In pulse ESR, a convenient

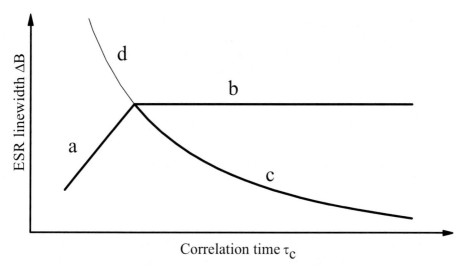

Fig. 4: Schematic ESR linewidth ΔB of a paramagnetic species in solution having an anisotropic interaction $\delta\omega$ as a function of its rotational correlation time τ_c: when $\tau_c \ll 1/\delta\omega$, $\Delta B = \gamma_e^{-1}\delta\omega^2\tau_c$ (a); when $\tau_c \gg 1/\delta\omega$, the previous single line is split into two lines separated by $\Delta B = \gamma_e^{-1}\delta\omega$ (b), each of line width $\Delta B = 1/\gamma_e\tau_c$; (c), part of the life-time broadening curve of which part (d) is counterbalanced by motional narrowing (a).

choice of pulse sequences allows specific detection of a particular term of this Hamiltonian (*see* Nuclear Magnetic Resonance).

Let us consider now the same paramagnetic species in *dilute fluid solution*. Magnetic interactions are anisotropic and the energy levels vary as the molecules rotate. For a spherical molecule of radius a, this random tumbling can be characterized by one rotational correlation time τ_c, related to temperature t and viscosity η of the medium by $\tau_c = 4\pi a^3\eta/3kt$ (k is Boltzmann's constant). This random motion has a component at the resonance frequency and induces transitions, thus modifying the spin–lattice relaxation time T_1. Spin–spin relaxation time T_2 is also influenced by random rotation if anisotropic interactions $\delta\omega$ are present: let us simulate this rotation by sudden random jumps between two orientations of resonance field B_0 and $B_0 + \delta B$ (at fixed ν_0), τ_c being the average lifetime at each orientation (thus simulating the correlation time). For spins precessing initially in phase, each jump introduces a dephasing $\delta\phi = \pm\tau_c\delta\omega = \pm\tau_c\gamma_e\delta B$. If $\tau_c \gg T_2$, the phase difference will be 1 rad after T_2, $1/T_2 \simeq \gamma_e\delta B$. (In the limiting case $\tau_c \to \infty$, the "immobilized spectrum" consists of two lines whose separation is of course δB.) If $\tau_c \ll T_2$, dephasing after n random jumps is given by the accumulation of n random dephasing $\delta\phi$: $\Delta\phi^2 = n\overline{\delta\phi^2} = n\tau_c^2\overline{\delta\omega^2}$. $\Delta\phi \approx 1$ rad is reached after T_2/τ_c jumps: $1/T_2 \approx (\delta\omega)^2\tau_c = \gamma_e^2(\delta B)^2\tau_c$. In this case, the spectrum is averaged out to a single line, of width $1/T_2$, at the average position between B_0 and $B_0 + \delta B$. Spin–spin relaxation may be induced by other mechanisms (such as vibrations, internal motion in a free radical, or chemical exchange of an electron between molecular systems). The same relationships exist between T_2, the difference of resonance frequencies and the characteristic lifetime of the processes (see Fig. 4).

Let us consider now a case of *spin-1* species ("two unpaired electrons") diluted in a diamagnetic environment. There are singlet and triplet states, the lowest having ^1E and (average) ^3E energy. The "exchange integral" J is by definition $2J = {}^1E - {}^3E$. Electron spin–spin dipolar interaction is nonzero in the triplet state: to first order, it arises from the spin–spin magnetic dipolar interaction

$$H_{\text{dip}} = \sum_{i}^{\text{all electrons}} \sum_{j<i} [\gamma_s^2 \hbar^2 \mu_0/4\pi][\mathbf{s}_i \cdot \mathbf{s}_j - 3(\mathbf{r}_{ij} \cdot \mathbf{s}_i)(\mathbf{r}_{ij} \cdot \mathbf{s}_j)r_{ij}^{-2}] r_{ij}^{-3}$$

($\mathbf{r}_{ij}$ connects electrons i and j), and to second order, from spin–orbit interactions,

$$H_{\text{LS}} = \sum_{N}^{\text{nuclei}} \sum_{i}^{\text{electrons}} \zeta(r_{iN}) \mathbf{l}_{iN} \cdot \mathbf{s}_i ,$$

where l_{iN} is the orbital moment of electron i relative to nucleus N and $\zeta(r_{iN})$ is a characteristic function of nucleus N. The influence of singlet and triplet levels and of the dipolar interaction on ESR spectra can be simulated by an effective-spin Hamiltonian $\mathcal{H}'$ similar to the one used for spin-$\frac{1}{2}$ systems, but dealing with two effective spins $\frac{1}{2}$ ($\mathbf{s}_1'$ and $\mathbf{s}_2'$), and having an exchange term $\mathcal{H}'_{\text{spin(exch)}} = -2J\mathbf{s}_1' \cdot \mathbf{s}_2'$ and a dipolar term $\mathcal{H}'_{\text{spin(dip)}} = \mathbf{s}_1' \cdot \mathbf{D} \cdot \mathbf{s}_2'$. $\mathbf{D}$ is a traceless tensor whose principal values are written $(2D, -D+3E, -D-3E)$ and, when spin–orbit can be neglected, $D = g_e^2 \beta_e^2/r^3$, r being an average distance between the two unpaired electrons. In rigid matrix (or in viscous solutions, if $\tau_c \gg \hbar/D$), the dipolar interaction gives "fine structure" and also "half-field transitions" at $g \simeq 4$. As for spin-$\frac{1}{2}$ systems, various situations occur depending on the state (solid or liquid) of the system and of the relative order of magnitude of the Zeeman, exchange, fine, or hyperfine terms. It may happen that the instrument frequency ν_0 is such that $h\nu_0 \ll D$: as the available photons have an energy smaller than the difference between some magnetic levels, they cannot induce resonance transitions between these levels. A higher frequency is then useful to solve this problem. In fluid solutions, the dipolar interaction is averaged out to zero if $\tau_c \ll \hbar/D$, but contributes (very effectively) to line broadening.

The case of two interacting spins may also be found for spin $\frac{1}{2}$ not perfectly isolated from their neighbors, i.e., for concentrated solutions of radicals in a diamagnetic matrix or liquid. Pairwise interactions create a spin-one system, giving rise to exchange J and dipolar D energies. In a pair of spin-coupled electrons, the individual electron spins (as any nonconservative observable in a two-level system) oscillate with a period $\tau = h/2J$. By analogy, we may expect this oscillation ("exchange") to induce a spin–spin relaxation with $1/T_2 = \pi D^2/\hbar J$. This is indeed the case: as the concentration increases, lines broaden (dipolar broadening) and then narrow (exchange narrowing).

Applications of ESR use information related to the spectral moments: total *intensity* is proportional to the number of paramagnetic species. Concentration is measured by comparison with a reference, because absolute determination is difficult. Under favorable conditions 10^{11} spins may be detected. *Position* of spectral lines permits identification of the paramagnetic species, of its surroundings, and of its orientation. *Linewidth*, related to T_2, depends on a characteristic time τ and some characteristic difference $\delta\omega$. For instance, τ may be the lifetime of some process involving a jump between positions having resonance at ω and $\omega + \delta\omega$, or the rotational correlation time in solution. In this case, the linewidth is a measure of the "effective volume" of the species, of the solvent viscosity and temperature, and of anisotropic interactions, among which dipolar interactions ($\simeq r^{-3}$) provide geometrical information. T_1 measurements (saturation, pulses) give information complementary to those obtained from T_2. This information is used, for instance, in chemical (structural and dynamical) analysis of free radicals, of excited triplet molecules, of transition-metal ions, and in biophysics, especially in the "spin-label method", where a free radical reveals information about the biomolecule to which it is bound.

See also: Nuclear Magnetic Resonance; Paramagnetism; Relaxation Phenomena; Resonance Phenomena.

Bibliography

Elementary treatment of ESR may be found in quantum-mechanics textbooks, for instance, in D. Park, *Introduction to the Quantum Theory*, p. 151, McGraw–Hill, New York, 1964 (E), or in C. Cohen-Tannoudji, B. Diu, and F. Laloë, *Quantum Mechanics*, p. 443, Wiley-Hermann, Paris, 1977.(I)

General Texts

N. M. Atherton, *Principles of Electron Spin Resonance*. Ellis Horwood, New York, 1993. (E,I)

A. Carrington, A. D. McLachlan, *Introduction to Magnetic Resonance*. Harper & Row, New York, 1966. (E)

G. E. Pake and T. Estle, *The Physical Principles of Electron Spin Resonance*. Addison-Wesley, Reading, 1973. (I,A)

C. P. Slichter, *Principles of Magnetic Resonance*. Harper & Row, 2nd ed., New York, 1978. (I,A)

J. A. Weil, J. R. Bolton and J. E. Wertz, *Electron Spin Resonance: Elementary Theory and Practical Applications*. Wiley-Interscience, New York, 1994. (E,I)

A. Schweiger and G. Jeschke, *Principles of Pulse Electron Paramagnetic Resonance*. Oxford University Press, Oxford, 2001. (A)

Specialized

A. Abragam and B. Bleaney, *Electron Paramagnetic Resonance of Transition Ions*. Clarendon Press, Oxford, 1970. (A)

L. J. Berliner (ed.) *Spin Labelling: Theory and Applications*, Vol. 1, 1976. Academic Press, New York, Vol. 2, 1979. (E, A)

J. A. Weil, M. K. Bowman, J. R. Morton, and K. F. Preston (ed.), *Electronic Magnetic Resonance of the Solid State*. The Canadian Society of Chemistry, Ottawa, 1987. (I, A)

J. R. Pilbrow, *Transition Ion Electron Paramagnetic Resonance*. Clarendon Press, Oxford, 1990. (E, A)

P. P. Borbat, A. J. Costa-Filho, K. A. Earle, J. K. Moscicki, and J. H. Freed *Science* **291**, 266, (2001)(I)

F. E. Mabbs and D. Collison, *Electron Paramagnetic Resonance of d Transition Metal Compounds*. Elsevier, Amsterdam, 1992.(E, A)

Spin Hamiltonian and Magnetic Interactions

R. McWeeny, *Spins in Chemistry*. Academic Press, New York, 1970. (I)

H. F. Hameka, *Advanced Quantum Chemistry*. Addison-Wesley, Reading, MA, 1965. (A)

J. E. Harriman, *Theoretical Foundations of Electron Spin Resonance*. Academic Press, New York, 1978. (A)

A. Bencini and D. Gatteschi, *EPR of Exchange Coupled Systems*. Springer-Verlag, Berlin, 1990. (I, A)

Relaxation

A. Hudson and G. R. Luckhurst, *Chem. Rev.* **69**, 191, (1969). (E)

R. Lenk, *Fluctuations, Diffusion and Spin Relaxation.* Elsevier, Amsterdam, 1986. (A)

A. Manenkov and R. Orbach, *Spin-Lattice Relaxation in Ionic Solids.* Harper & Row, New York, 1966. (A)

L. T. Muus and P. W. Atkins (eds.), *Electron Spin Relaxation in Liquids.* Plenum Press, New York, 1972. (A)

Web Site of the International EPR (ESR) Society: `www.ieprs.org`.

Electron Tubes

J. D. Ryder[†]

The Edison effect, observed by Thomas A. Edison in 1883 as an electric current between a metal plate and a heated filament in an evacuated bulb, marked the birth of the electron tube. Current passed only with the plate positive to the filament, and the charge carriers were later identified as negative electrons. Edison patented several ideas for application but did not investigate further.

Since this two-element device or diode was unilateral in conduction properties, it was used as a rectifier or detector of radio signals by J. A. Fleming, ca. 1904. In 1906, Lee deForest, following unsuccessful experiments with gas lamps, placed a wire grid as a control element between filament and plate, making a sensitive relay, the three-element or triode tube.

Thermal Emission of Electrons

To achieve emission from a surface, the electron must be given sufficient outward directed energy to overcome surface binding forces. The energy needed is called the work function, E_W at $0\,\mathrm{K}$, $= mv^2/2$. Velocity v is the outward-directed velocity component. This required energy depends on cathode material and surface condition; values for a few common emitting materials are given in Table 1.

In thermionic emitters the work-function energy is supplied by raising the temperature of the emitter. The emitted current is predicted by an equation proposed by S. Dushman in 1923:

$$I = A_0 S T^2 \varepsilon^{-b_0/T} A \,, \tag{1}$$

where A_0 is the surface constant, S is the emitting area in m^2, T is the surface temperature in Kelvin, $b_0 = 11\,600 E_W = e E_W/k$, the voltage equivalent of temperature.

Thermionic cathodes are (1) filamentary or (2) indirectly heated. In a filament, emission occurs from a wire, heated by the passage of a current. In the indirect form, a heater is used inside a metal cylinder to raise the surface to about $1000\,°\mathrm{C}$, sufficient to cause emission from the surface, usually coated with barium and strontium oxides. A short heating time is required for the surface to reach emitting temperature.

[†]deceased

Table 1: Values of the emission constants.

Material	A_0 ($\times 10^4$)	b_0	E_W (eV)[a]
Cesium	16.2	21 000	1.81
Thorium	60.2	39 400	3.4
Tungsten	60.2	52 400	4.52
Thorium on tungsten	3.0	30 500	2.6
Oxide (BaO + SrO)	0.01	11 600	1.0

[a] $1\,\text{eV} = 1.60 \times 10^{-19}\,\text{J}$.

Oxide-coated cathodes were normally used, but high-power tubes employed tungsten with a monatomic thorium layer or pure tungsten.

The Diode

About 1912, independent studies by Langmuir and Child allowed explanation of the diode volt–ampere curve, Fig. 1. The toe from a to b indicates that some electrons have outward velocities and need no accelerating potential to reach the plate. From b to c the current increases with potential but is less than predicted by the Dushman relation. Temperature saturation, as predicted by the Dushman relation, is reached above c.

The lowered current from b to c is caused by the repulsion of the electronic space charge near the cathode, stopping electrons with low outward velocities. If the anode potential is increased, more electrons are attracted from the space cloud to the anode and the repulsion is weakened. Equilibrium is reached at a space charge density at which the number of electrons reaching the anode is balanced by the number leaving the cathode.

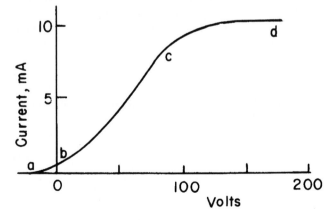

Fig. 1: Volt–Ampere diagram for a vacuum diode.

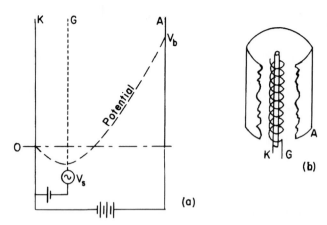

Fig. 2: (a) Space potential with space charge in a triode; (b) typical triode structure.

The equation relating the current collected to the applied voltage and the geometry of the diode is called the three-halves power law, and for large parallel-plane electrodes is

$$J = \frac{4\varepsilon_v}{9} \left(\frac{2e}{m}\right)^{1/2} \frac{V_b^{3/2}}{d^2} = 2.34 \times 10^{-6} \frac{V_b^{3/2}}{d^2} \frac{A}{m^2} \, , \tag{2}$$

where ε_v is the space permittivity $= 10^7/4\pi c^2 = 8.85 \times 10^{-12}$; m is the electron mass $= 9.106 \times 10^{-31}$ kg; d is the cathode–anode spacing, in meters.

The Triode

The potential minimum, established by the space charge near the cathode surface, is illustrated in Fig. 2(a). It acts as a barrier to the lower velocity electrons, and by raising or lowering the potential by the signal v_s the current can be varied or controlled. The average grid potential is maintained negative to cathode, and the grid current is zero in most operating regions.

The anode current is a function of the grid-cathode voltage v_c; if v_b is changed the potential minimum shifts up or down, so we may say

$$i_b = f(v_c, v_b) \tag{3}$$

and the general triode relations are shown in Fig. 3. To show the effect of small changes, as with a grid signal, Δv_c, and Δv_b, we have

$$\Delta i_b = \frac{\partial i_b}{\partial v_c} \Delta v_c + \frac{\partial i_b}{\partial v_b} \Delta v_b \, . \tag{4}$$

We define

$$g_m = \frac{\partial i_b}{\partial v_c} \, , \quad g_p = \frac{\partial i_b}{\partial v_b} \tag{5}$$

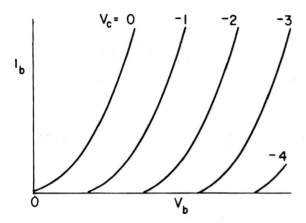

Fig. 3: Voltage–current characteristics for a triode.

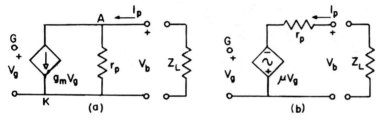

Fig. 4: (a) Current-source equivalent circuit for a triode; (b) voltage-source equivalent circuit. The diamonds are control sources.

and these are the transconductance and the plate conductance, respectively. From Eq. (5) we also have

$$\mu = \frac{g_m}{g_p} = g_m r_p = -\frac{\partial v_b}{\partial v_c}\bigg|_{i_b=K} , \tag{6}$$

where μ is the amplification factor and r_p is known as the plate resistance. Small ac signals may replace the Δ values and Eq. (4) becomes a circuit equation as

$$I_p = g_m V_g + V_b/r_p . \tag{7}$$

This equation represents a current summation at A in Fig. 4(a); this is known as the current-source equivalent circuit for the triode. The circuit dual is shown in Fig. 4(b) as the voltage-source equivalent circuit. Loads may be connected at the output as shown. The voltage gain is $A =$ (output voltage)/(input voltage) and

$$A = -\frac{\mu Z_L}{r_p + Z_L} = -\frac{g_m Z_L}{1 + Z_i/r_p} . \tag{8}$$

The minus sign indicates a 180° shift due to the tube; any additional angle on A is called the circuit phase shift. The grounded-cathode circuit was most often used because of its high gain, but the cathode follower and the grounded-grid circuits were used to take advantage of particular gain and impedance properties. These circuits appear in Fig. 5.

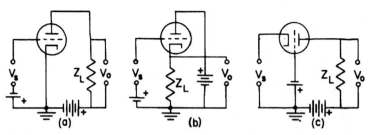

Fig. 5: (a) Grounded-cathode triode circuit; (b) cathode follower; (c) grounded-grid circuit.

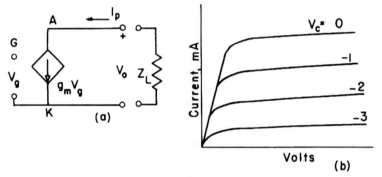

Fig. 6: (a) Equivalent circuit for a pentode: (b) output characteristics of a pentode.

The Pentode

At frequencies of the order of 1 MHz, the triode may become unstable and oscillate in the grounded-cathode circuit. The anode, at a higher signal potential, drives a current to the grid through the grid-plate capacitance, in phase with the signal. This raises the grid voltage, and a cumulative process of regeneration occurs, with instability of gain. Shielding of the grid from the anode by a second and grounded grid reduces this current through C_{gp}. A tube with a screen grid, and often another grid called the suppressor, has a C_{gp} of about 0.004 pF, compared to 2–5 pF for a triode. With five internal elements, the tube is called a pentode.

The first screen is operated at a high dc potential to accelerate the electrons, but is grounded for signal frequencies by a shunt capacitor. The energy of the electrons is sufficient for them to reach the positive anode. The values of μ and r_p are much higher than for a triode, and the low internal capacitance makes the pentode useful to frequencies above 100 MHz.

Output loads of 100–200 kΩ are used, small compared to an r_p of 600–1000 kΩ. Equation (8) with $r_p \gg Z_L$ reduces to

$$A = g_m Z_L . \tag{9}$$

The equivalent circuit for a pentode becomes Fig. 6(a), with characteristics in Fig. 6(b). Voltage gains of several hundred per stage are possible, with g_m values of 0.01 to 0.04 siemens.

Photocells or Photoemissive Tubes

Electrons are emitted when light of appropriate frequency strikes a specially prepared plate. This photoelectric emission differs from thermionic emission only because the work-function energy is derived by the energy of a photon. The Einstein equation explains the phenomena; photon energy hf must supply the work energy, eE_W J. That is

$$hf = eE_W + mv^2/2 . \tag{10}$$

Any excess photon energy appears as kinetic energy of the emitted electron: the photon then disappears. Below some threshold frequency $f_0 = eE_W/h$ the emitted current becomes zero. Emitting surfaces, usually with cesium or rubidium, are available with peak responses from the infrared to the ultraviolet. All emitted electrons are collected at voltages over 25 V. The internal resistance is many megaohms and the cell approximates a current source. Vacuum cells are linear with light intensity, but gas cells are linear only for small light variations. Ionization by collision with gas atoms gives such cells additional sensitivity, gains of 4 to 7 being usual.

The Gas Diode Rectifier

By introduction of a gas – argon, xenon, or mercury vapor – into a thermionic diode, the voltage required for conduction is reduced and power efficiency is raised. When the anode–cathode voltage exceeds the gas-ionization potential, electrons acquire sufficient energy to ionize upon collision with gas atoms. The resulting positive ions neutralize the negative space charge due to the electrons, and more electrons are accelerated in a cumulative action leading to an arc discharge. Upon reversal of the anode polarity the ions migrate to the anode and tube walls, recombining with charges there, and the current goes to zero.

The positive ions are heavy and slow moving and their function is to neutralize the negative space charge. It was possible to secure emission from deep cavities, and heat shielding was added to reduce the heating power needed. With mercury vapor the pressure is dependent on temperature, and below 20 °C insufficient ions are formed. The tube voltage rises and positive ions bombard the cathode, destroying the emitting surface; xenon or argon were often used, but with reduced efficiency. At temperatures above 80 °C the vapor provides insufficient insulation. Such devices were efficient rectifiers operating with forward voltages of 10–15 V with mercury vapor and handling forward currents of 100 A with reverse voltages as high as 15 000 V. The fragility of the oxide-coated cathodes posed maintenance and application problems.

The Thyratron or Gas Triode

The thyratron (door tube) was developed by addition of a grid structure to the gas diode, Fig. 7(a). The grid was a cylinder with baffles pierced by large holes, and the anodes were often graphite blocks.

With the grid negative to cathode, emitted electrons face a repelling field and there is no conduction. With a less negative grid, a few electrons pass through the holes in the baffles, are accelerated toward the anode, and ionize gas atoms. The positive gas ions neutralize the negative space charge, more gas ions are produced, and the discharge proceeds to an arc. The

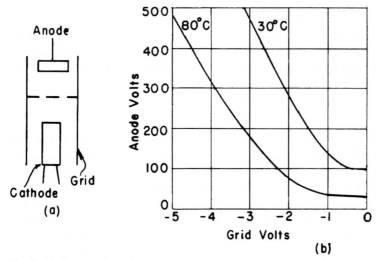

Fig. 7: (a) Cross section of a negative-grid thyratron; (b) critical grid characteristics for a thyratron.

grid potential is the critical factor in starting the discharge, and the relation between critical grid voltage and anode voltage is plotted in Fig. 7(b) for a typical tube.

Positive ions are attracted to the negative grid and build up a positive space charge there. This sheath of ions insulates the grid, and changes in grid potential alter the sheath thickness but have no effect on the current to the anode. Reduction of the anode potential to zero or a negative value must occur to stop the current. Positive ion current to the grid should be limited by resistance.

The time of deionization was typically 10–1000 µs. This is the time for the grid sheath to dissipate and for the grid to regain control; this fixed an upper frequency limit for thyratron operation. Lowest deionization time was obtained with hydrogen as the gas.

See also: Thermionic Emission; Work Function.

Bibliography

W. L. Chaffee, *Theory of Thermionic Vacuum Tubes.* McGraw–Hill, New York, 1933.

S. Dushman, "Thermionic Emission," *Rev. Mod. Phys.* **2**, 381 (1930).

A. L. Hughes and L. A. Dubridge, *Photoelectric Phenomena.* McGraw–Hill, New York, 1932.

A. W. Hull, "Hot Cathode Thyratrons," *Gen. Elec. Rev.* **32**, 213, 390 (1929).

A. W. Hull and N. H. Williams, "Characteristics of Shielded-Grid Pliotrons," *Phys. Rev.* **27**, 432 (1926).

J. D. Ryder, *Engineering Electronics*, 2nd ed. McGraw–Hill, New York, 1967.

F. E. Terman *et al.*, "Calculation and Design of Resistance-Coupled Amplifiers Using Pentode Tubes." *Trans. AIEE.* **59**, 879 (1940).

Electronics

J. A. McCray and W. Nadler

From the physicist's standpoint electronics may be defined as the use of materials and devices which transport or store electrons for the purpose of recording, transmitting, manipulating, or storing physical information. The variables usually monitored as functions of time are voltage, current, and charge. In actual practice an arrangement of passive components, such as resistors, capacitors, or inductors, and active components, such as field effect transistors (FETs), tubes, or bipolar transistors, make up collectively an electronic circuit which is designed to perform a given function. This circuit may be a discrete component circuit, where the components are individually arranged on a circuit board, or it may be an integrated circuit (IC) where thousands of mainly active but also passive components are intimately arranged within a small silicon "chip." Another possibility is a hybrid combination where most of the circuitry is contained in an IC chip with just a few passive components "outboarded" around the chip. Changing these passive components alters the function of the circuit.

It is the responsibility of the physicist to understand the function of the circuit, that is, how the circuit modifies an input voltage or current, so that the physical information represented by these variables is changed in a known manner.

A very important point that must be remembered is that the function of a circuit depends on the relative values of the parameters of the circuit (for example, RC time constants) and the values of the incident waveform (for example, the rise time of a pulse).

Since electronic circuits may be situated at some distance from one another, they must be connected together by transmission lines. Therefore, the properties of the transmission line, such as transit time and characteristic impedance, reflections at its terminations, and voltage and current traveling waves on the line must be understood if the transmitted physical information is not to be distorted.

Electronic circuits may be classified as analog, where for example the magnitude of the information variable is monitored directly, or digital, where the information is contained in a sequence of binary states. A further classification of electronic circuits is to separate them into linear and nonlinear circuits. If the coupled differential equations which describe the variation with time of the relevant charges involved in a circuit are linear in the charges and their time derivatives, then the circuit is described as a linear circuit. If there are any nonlinear terms in the differential equations, then the circuit is called nonlinear.

A given configuration of ideal resistors, capacitors, and inductors in addition to an ideal voltage source constitute an electronic circuit and may be represented by a drawing which is called a circuit diagram. An example of such a diagram that represents an integrating circuit is given in Fig. 1. The physical laws governing the operation of electronic circuits are Maxwell's equations and the equation of continuity. A discussion of these laws and conservation of charge and conservation of energy leads to Kirchhoff's laws. In order to determine the function of each linear circuit, either experimentally or theoretically, the response, i. e., output voltage or current, is obtained for two standard types of input. The steady-state, frequency-dependent properties of a circuit are obtained by using a sinusoidal input voltage or current. For Fig. 1 the input voltage would be $v_0(t) = A \sin(\omega t + \theta)$, where A, ω, and θ are the amplitude, angular frequency, and phase, respectively, of the sinusoidal input. Instead of using Kirchhoff's laws directly and then solving the differential equations it is easier to use the phasor technique. The

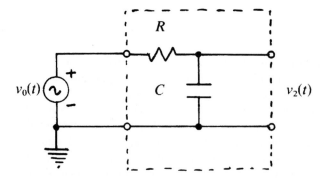

Fig. 1: RC integrating circuit.

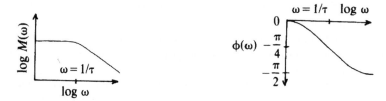

Fig. 2: Amplitude and frequency responses of the RC integrating circuit.

properties of the circuit to be determined are the amplitude frequency response (relative *gain* or *attenuation* as a function of frequency) and the frequency-dependent phase response (*phase shift*). Since only the induced phase shift of the circuit is desired, the circuits are tested with the input $v_0(t) = A\sin(\omega t)$ (which is mathematically equal to the imaginary part of the complex function $\hat{v}_0 = Ae^{i\omega t}$, where $i = \sqrt{-1}$. This latter form is called a *phasor* and, when the phasor form of current is used, $\hat{i} = Be^{i\omega t}$, the voltage drops given above for resistance, capacitance, and inductance become $\hat{v} = R\hat{i}$, $\hat{v} = \hat{i}/i\omega C$, and $\hat{v} = i\omega L\hat{i}$. These all have the form of Ohm's law if R, $1/i\omega C$, and $i\omega L$ are taken as generalized resistances (impedances). Kirchhoff's laws then become just algebraic equations with complex numbers and the analysis reduces to that for dc circuits. For the integrating circuit shown in Fig. 1 the output voltage phasor $\hat{v}_2$ may be easily found by using the voltage divider ratio. Thus $\hat{v}_2 = \{[1/i\omega C]/[R + 1/i\omega C]\}\hat{v}_0$. The ratio of the two complex numbers $\hat{v}_2/\hat{v}_0$ is called the voltage transfer function $\hat{T}$ and can be written in polar form: $\hat{T}(\omega) = M(\omega)e^{i\phi(\omega)}$. $M(\omega)$ is the amplitude frequency response and $\phi(\omega)$ is the phase response of the circuit. For the circuit of Fig. 1 we have $\hat{T}(\omega) = 1/(1 + i\omega\tau)$, where $\tau = RC$ is called the time constant of the circuit. In polar form this complex number is $\hat{T}(\omega) = [1/(1 + \omega^2\tau^2)^{1/2}]\exp(-i\tan^{-1}\omega\tau)$, so that the amplitude response is $M(\omega) = 1/(1 + (\omega^2\tau^2)^{1/2}$ and the phase response is $\phi(\omega) = -\tan^{-1}\omega\tau$. These results are shown in Fig. 2 and illustrate that the circuit can also be called a *low-pass filter* since high frequencies are attenuated, or a lag circuit since the phase shift is negative. When $\omega = 1/\tau$, the amplitude has been decreased by $1/\sqrt{2} = 0.707$ and the power has been decreased by $\frac{1}{2}$. Thus this frequency is a measure of the pass band of the filter and is called the upper half-power point. In terms of decibels the power has fallen off 3 dB.

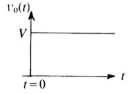

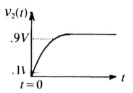

Fig. 3: Response of circuit to step input (rise time).

The transient voltage response of a linear circuit is obtained by using a voltage step input, which is zero up to a specified time origin and then ideally jumps to a value V and remains constant thereafter:

$$v_0(t) = \begin{cases} 0, & t < 0 \\ V, & 0 \le t \end{cases}$$

The transient response may be readily found if Laplace transforms are used. For Fig. 1 the result is $v_2(t) = V(1 - e^{-t/\tau})$, which is an exponential rise. Figure 3 illustrates this basic property of electronic circuits. All circuits have some shunt capacitance to ground which must be charged, and, hence, there will be a certain amount of time necessary for the circuit to respond to a step input and this will depend on the capacitances and resistances involved. A measure of this response time is the 10–90% rise time shown in Fig. 3. The rise time depends on the time constant of the circuit, and it is easily shown that $T_R = 2.2\tau$. This result may be combined with the upper half-power-point frequency for this circuit, $f_2 = 1/2\pi\tau$, to yield the relation $f_2 T_R \simeq \frac{1}{3}$. This expression is particularly useful since most wide-band amplifier circuits have this circuit as a high-frequency approximation. Thus if the frequency band pass of an amplifier, f_2, is given, then the rise time T_R may be easily determined. The frequency spectrum of noise usually encountered is so-called "white noise" and is uniform with frequency. Thus one would like to minimize the electronic bandwidth used. However, if too small a bandwidth is used, then the rise time of the amplifier $T_R \simeq \frac{1}{3} f_2$ will be too long to measure a desired signal rise time or initial round-off of the signal will result. Thus one must adjust the bandwidth in order to optimize the signal-to-noise ratio.

Active Devices

One of the most important features of some electronic circuits is that they are capable of amplification. If an external source of energy is applied to a circuit containing an active device such as a tube, FET, or bipolar transistor, then it is possible to arrange that a low voltage or current control the flow of a large current through the active device. Semiconductor devices are the most important of these. One of the basic features of active semiconductor devices is the p–n junction. This device is produced when a silicon crystal is grown and the doping is changed at a given point from n type to p type, as shown in Fig. 4.

On the n-type side there is effectively an electron "gas" which can diffuse into the-p-type side leaving behind at the junction a layer of bound positive charges. Likewise on the p-type side there is effectively a positive "hole gas" which can diffuse into the n-type region leaving behind a layer of negative charges. The dipole layer electric field created causes Ohmic current in the opposite direction to that of the diffusion current. At steady state the

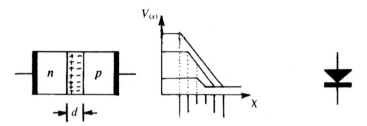

Fig. 4: *p–n* junction diode.

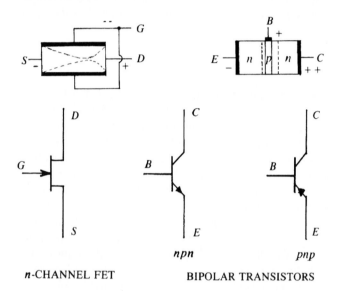

***n*-CHANNEL FET**   **BIPOLAR TRANSISTORS**

Fig. 5: *n*-channel field effect transistor (FET) and bipolar transistor.

currents just balance and a region, called the depletion region, has been established where the potential changes and an electric field exists. This is shown by the middle curve of Fig. 4. If an external voltage is applied across the *p–n* junction so that the positive side is connected to the *n*-type material, then the electric field adds to that already there, resulting in an increased potential barrier and a larger depletion width. It is difficult then for current – conventional positive current – to flow from *p*-type to *n*-type material, and the junction is said to be back-biased. However, if the polarity of the external voltage is reversed, then the additional electric field at the junction subtracts from that already there thus resulting in a lower potential barrier. Current then may more easily flow through the device, and the junction is said to be forward biased. This voltage-dependent directionality of current flow allows rectification of ac currents to dc currents, so the device is called a rectifier or diode and is indicated by the symbol shown in Fig. 4.

p–n junctions may be combined in various ways to form, for example, a voltage-dependent device, the field-effect transistor (FET), or a current-dependent device, the bipolar transistor. These devices are shown in Fig. 5 and both are capable of amplification.

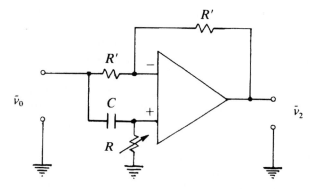

Fig. 6: Variable constant-amplitude op-amp phase shifter.

In the FET case a small voltage change at the gate (G) produces a large change in channel current from source (S) to drain (D) by changing the size of the opposite depletion layers. For the bipolar transistor shown a small change in base (B) current produces a large change in emitter (E) to collector (C) current by varying the base-emitter potential barrier. The emitter and collector region are heavily doped while the base region is very thin and lightly doped. Here, again, one can achieve an amplification, in this case, of current.

Feedback is a very important concept in electronics. A voltage, for example, which is proportional to the output voltage of a circuit, is introduced in either a negative or positive sense into the input circuit of an amplifier. Negative voltage feedback is used for stability and positive voltage feedback is used to make oscillator circuits. In modern circuits an integrated-circuit differential amplifier with high gain, high input impedance, and low output impedance is used to form various functional linear circuits. Various passive components, such as resistors, capacitors, and inductors, are "outboarded" around the so-called operational amplifier (op-amp) and the function of the overall circuit depends on the type and arrangement of these components. For example, application of Kirchhoff's laws to the circuit shown in Fig. 6 indicates that this circuit is a constant-amplitude variable 180-deg phase shifter with voltage transfer function $\hat{T}(\omega) = e^{i(-2\tan^{-1}\omega\tau)}$, where $\tau = RC$. The voltage characteristics of the op-amp (triangular symbol) are represented by $\hat{V}_2 = \hat{A}(\hat{V}_+ - \hat{V}_-)$, where $\hat{V}_+$ and $\hat{V}_-$ are measured with respect to ground. These circuits are called operational amplifier circuits because various analog operations may be performed such as differentiation, integration, summing, etc.

An interesting application of these circuits is to solve nonlinear differential equations, for example, the driven Duffing oscillator equation. Nonlinear effects such as period-doubling leading to chaos can be demonstrated. In fact, it is possible to see and even hear such effects with a simple driven L–R diode circuit where the diode is a nonlinear capacitor.

Digital Electronics

As can be seen from the previous discussion in this article the fidelity of information transfer of analog signals greatly depends on the frequency and phase responses of the circuits and transmission lines through which these signals pass. For many applications this possible distortion of data is intolerable. It is also possible to perform analytical computation with analog circuits and many of the first computers were analog computers, but the accuracy, reliability and possible memory storage were not considered satisfactory.

An alternative approach to data transfer and computation is based upon the binary system of counting. Instead of using the base 10 (decimal system) where a number is expressed in terms of powers of 10 [for example, the number $35486 = N_4N_3N_2N_1N_0$ is $N = N_4 \times 10^4 + N_3 \times 10^3 + N_2 \times 10^2 + N_1 \times 10^1 + N_0 \times 10^0$, where $10^0 = 1$, $10^1 = 10$, $10^2 = 100$, etc.] the number is expressed instead in terms of base 2. As an example, the number $38_{10} = 3 \times 10^1 + 8 \times 10^0$ in base 10 would become $100110_2 = 1 \times 2^5 + 0 \times 2^4 + 0 \times 2^3 + 1 \times 2^2 + 1 \times 2^1 + 0 \times 2^0$ in base 2. The general form is $N = B_5B_4B_3B_2B_1B_0 = \cdots + B_5 2^5 + B_4 2^4 + B_3 2^3 + B_2 2^2 + B_1 2^1 + B_0 2^0$. The reason for the choice of the binary system is clear when it is realized that to represent one binary digit (bit) a device with only two states (0 or 1) is needed while to represent one decimal digit a device with 10 states $(0, 1, 2, 3, 4, 5, 6, 7, 8, 9)$ would be necessary. There are many electrical two-state devices that can be used for binary representation such as switches, relays, bipolar transistor binaries (flip-flops), MOSFET (metal oxide semiconductor field effect transistor) binaries, magnetic cores, magnetic domains on tapes, disks, or bubbles, etc. Binary numbers can then be stored in a linear sequence of such devices, called a register, and can also be added and subtracted, thus making possible all of the higher mathematical operations such as multiplication, division, differentiation, integration, etc. This is done by using repeated addition and data transfer operations which are expressed in a computer program and stored in the memory of a computer.

Since our analytical thinking processes are based on a two-state logic system (true or false), two-state electronic devices are very readily arranged so that various logical operations can be represented by circuits such as the NOT, AND, NAND, OR, NOR, and exclusive OR circuits. These circuits (gates) and various combinations are available in integrated circuit chips. The packing density of gates on a chip is indicated by the terminology SSI (small-scale integration, < 10 gates), MSI (medium-scale integration, < 100 gates), LSI (large-scale integration, ~ 1000 gates) and VLSI (very large-scale integration, > 1000 gates). The type of two-state devices used to make up the logic circuits gives rise to various logic families such as CMOS (complementary metal oxide semiconductor logic – low power consumption), ECL (emitter-coupled logic – high speed), TTL (transistor–transistor logic), and I^2L (integrated injection logic). Computer chip sets made up of the central microprocessor chip and the coprocessor (math processor) chip such as the Motorola 68020/68881, the Intel 80386/80387, and the Digital Equipment Corporation 78032/78132 chip sets form, when combined with ROMs (read-only memory chips) and RAMs (random-access memory chips) along with several other chips necessary for proper matching of circuits, the basis for the Apple MAC II computer, the IBM AT computer, and the DEC Micro-Vax computer, respectively. The new supercomputers connect miniaturized central microprocessor chips in parallel with very short interconnections in order to achieve a higher through-put and greater overall speed of calculation. High density requires special cooling techniques; for example, the Cray-3 supercomputer is cooled by an inert fluorocarbon liquid.

The general philosophy now would be to obtain analog information from a transducer at an experimental site, then convert this information into digital form as soon as possible with an A/D converter (analog-to-digital), store this information in a digital memory, and subsequently process this information in a digital computer. There is, however, one disadvantage in handling information in digital form. If only one bit of information is lost, then the total information may be incorrect. Modern semiconductor and solid-state technology and circuit-design philosophy have been so successful that device and circuit reliability can be excellent even for very large computers. Finally it should be noted that all types of functional circuits such as optical couplers, LEDs (light emitting diodes), PLLs (phase-locked-loops), CCDs (charge-coupled devices), sample and hold circuits, etc. have been designed and produced on single chips. Thus before designing or building any circuit it is advisable to check various manufacturer's data and application manuals (see Bibliography) to see if that particular desired function can be obtained with the use of one or two integrated circuit chips.

See also: Circuits, Integrated; Electron Tubes; Network Theory: Analysis and Synthesis; Semiconductors, Crystalline; Transistors; Transmission Lines and Antennas.

Bibliography
General Electronics
J. J. Brophy, *Basic Electronics for Scientists*, 4th ed. McGraw–Hill, New York. 1983. (E)

L. O. Chua and R. N. Madan, "Sights and Sounds of Chaos," *IEEE Circuits and Devices Magazine*, January 1988, and references therein. (A)

C. F. G. Delaney, *Electronics for the Physicist with Applications*. Halsted Press, New York, 1980. (I)

A. J. Diefenderfer, *Principles of Electronic Instrumentation*. Saunders College Publishing, Orlando, FL, 1979. (E)

T. M. Frederiksen, *Intuitive Operational Amplifiers*. McGraw–Hill, New York, 1988. (E)

P. Horowitz and W. Hill, *The Art of Electronics*. Cambridge University Press, Cambridge, 1987. (I)

W. G. Jung, IC *Op-Amp Cookbook*, 3rd ed. Howard W. Sams & Co., Indianapolis, 1986. (E)

D. Lancaster, *Active-Filter Cookbook*. Howard W. Sams & Co., Inc., New York, 1988. (E)

J. W. Nilsson, *Electronic Circuits*. Addison-Wesley, Reading, MA, 1985. (I)

R. F. Pierret and G. W. Neudeck, eds., *Modular Series on Solid State Devices*. Addison-Wesley, Reading, MA, 1989. (A)

M. J. Sanfilippo, *Solid-State Electronics Theory – with Experiments*. Tab Books Inc., Blue Ridge Summit, PA, (E)

M. R. Spiegel, *Theory and Problems of Laplace Transforms*. McGraw–Hill, New York, 1965. (I)

W. T. Thomson, *Laplace Transformation*. Prentice-Hall, Englewood Cliffs, N.J., 1962. (A)

T. H. Wilmshurst, *Signal Recovery from Noise in Electronic Instrumentation*. Adam Hilger Ltd., Accord, MA, 1985. (I)

Digital Electronics
P. Antognetti and G. Massobrio, *Semiconductor Device Modeling with Spice*. McGraw–Hill, New York, 1988. (A)

M. Bird and R. Schmidt, *Practical Digital Electronics* (laboratory workbook). Hewlett-Packard Co., Santa Clara, CA, 1974. (I)

J. Blukis and M. Baker, *Practical Digital Electronics* (textbook). Hewlett-Packard Co., Santa Clara, CA, 1974. (I)

P. Burger, *Digital Design – A Practical Course*. Wiley, New York, 1988. (E)

L. A. Glaser and D. W. Dobberpuhl, *The Design and Analysis of VLSI Circuits*. Addison-Wesley, Reading, MA, 1985. (I)

R. L. Goodstein, *Boolean Algebra*. Pergamon Press, London, 1963. (A)

Handbooks, Dictionaries and Tables

S. Gibilisco, ed., *Encyclopedia of Electronics*. Tab Books, Blue Ridge Summit, PA, 1985.

R. F. Graf, *Modern Dictionary of Electronics*, 6th ed. Howard W. Sams & Co., New York, 1984.

M. Kaufman and A. A. Seidman, *Handbook for Electronics Engineering Technicians*. McGraw–Hill, New York, 1976.

J. Marcus and C. Weston, *Essential Circuits Reference Guide*. McGraw–Hill, New York, 1988.

P. A. McCollum and B. F. Brown, *Laplace Transform Tables and Theorems*. Holt, Rinehart and Winston, New York, 1965.

The Radio Amateur's Handbook, published annually by the Headquarters Staff of the American Radio Relay League, Newington, CT.

Magazines and Newspapers

BYTE, A McGraw–Hill publication, BYTE Subscriptions, P.O. Box 551, Hightstown, NJ.

EDN, A Cahners Publication, P.O. Box 5563, Denver, CO.

Electronic Design, VNU Business Publications, Inc., 10 Holland Drive, Hasbrouck Heights, NJ.

Electronic Engineering, Morgan-Grampian House, 30 Calderwood St., Woolwich, London.

EE Product News, P.O. Box 12973, Overland Park, KS.

Electronic Products, Hearst Business Communications, Inc./UTP Division, 645 Stewart Ave. Garden City, NY.

Electronics, VNU Business Publications, Inc., 10 Holland Drive, Hasbrouck Heights, NJ.

HP benchbriefs, Hewlett-Packard Co., 1820 Embarcadero Road, Palo Alto, CA.

Hewlett–Packard Journal, 3200 Hillview Avenue, Palo Alto, CA.

Journal of Physics E: Scientific Instruments, IOP Publishing Ltd., 7 Great Western Way, Bristol.

Nuclear Instruments and Methods in Research A, Elsevier Science Publishers B. V., P.O. Box 211, 1000 AE Amsterdam.

Personal Engineering, Box 182, Brookline, MA.

Review of Scientific Instruments, American Institute of Physics, 335 East 45th St., New York, NY.

Manuals

A great deal of use information may also be obtained from manufacturer's data and application manuals and books from companies such as American Telephone and Telegraph, General Electric, General Instruments, Harris, Intel, Motorola, National Semiconductor, Radio Corporation of America, Signetics, Texas Instruments, TRW, and many others. *IC Master*, Hearst Business Communications, 645 Stewart Ave., Garden City, NY 11530, is quite useful. Information about electronics manufacturers and distributors may be found in sources such as *Electronics Buyer's Guide*, published annually by VNU Business Publications, Inc., 10 Holland Drive, Hasbrouck Heights, NJ 07604; *Electronic Design's Gold Book*, published annually by Hayden Publishing Co., Inc., 10 Holland Dr., Hasbrouck Heights, NJ 07604; and *Electronic Engineers Master*, published semiannually by Hearst Business Communications, Inc., 645 Stewart Avenue, Garden City, NY 11530.

Electronic Noses

D. Kohl

The availability of micro controller chips with low prices has stimulated more complex gas sensor systems, usually termed "Electronic Noses". Such noses are based on arrangements or integrated arrays of several gas sensor elements. Commercial electronic noses are more or less compact machines comprising up to 48 sensor elements and electronics recognizing a single gas out of a given set of gases, or complex gas mixtures, e. g., the aroma of a certain type of food.

Signal Generation

Homogenous semiconducting gas sensors are frequently included in electronic noses. Both anorganic, mostly oxidic, and organic layers are in use. The first electronic noses tended to use organic semiconductors, well known is the pioneering and for clinical applications still commercially available Aromascan nose developed at UMIST university. On semiconducting layers operated below 400 °C the adsorbed molecules can form ionic bonds and transfer electric charge into the semiconductor grains of a sensitive layer. For such polycrystalline layers the current transport through the boundaries between neighbored grains dominates the sensor resistance. Depending on temperature the bonds between the gas molecules and the sensor surface are broken at a certain rate. For a given concentration of gas, the molecules leaving the sensor are replaced by new ones, producing a steady state situation. In more complicated cases a sequence of surface reactions is involved. The resistance becomes a reversible function of the gas concentration. Common materials are SnO_2, Ga_2O_3, WO_3, ZnO, In_2O_3, TiO_2 and Fe_2O_3. The sensivity and the specifity (ability to to respond to a certain species in a gas mixture) of the device can be controlled to some extent by selecting the temperature at which the oxide is held. Additions of catalytically active metals enhance the sensitivity. Gas concentrations down to about 1 ppm are detectible. Gas-FETs are devices consisting of a field effect transistor with a polymer covered gate open to atmosphere. The source-drain current at a fixed (regulated) overall gate voltage is taken as signal. This signal is proportional to the work function change caused by the adsorption. Operation temperature is limited to about 300 °C on silicon FETs, and to about 600 °C for silicon carbide FETs. Gas-FET arrays were found in one electronic nose [AppliedSensor Nordic Nose, production discontinued]. Microgravimetric sensors detect a weight change of a polymer layer caused by gas adsorption. The simplest device of that type features the same quartz discs which are used to measure thickness in a thin film deposition apparatus. The added mass lowers the resonance frequency. The frequency decrease serves as sensor signal. The coatings are often the same as in gas chromatography, so the existing knowledge can be used. The sensitivity increases with frequency, typically BAWs with 6 to 50 MHz are realized. Lower sensitivity limits are around 50 ppm at 30 MHz. Higher frequencies are attainable by surface acoustic wave (SAW) devices. In this case one uses the filter devices of cellular phones, consisting of a silicon substrate with piezoelectric (ZnO) layers covered by interdigital structures to generate an acoustic surface wave which travels along the silicon surface (covered by an adsorptive polymer layer) to a detector structure (same structure as the generator). Frequencies up to more than 1 GHz were realized. So sensitivities below 1 ppm were attained. To bring this type of devices closer to micromachining standards, more recently also an eight-cantilever array (50 kHz) with adsorptive coatings has been shown by Gimzewski at IBM. The first electronic nose with BAWs, developed by Dr. Horner at HKW, later produced by Perkin-Elmer, is discontinued. Manufacturers with SAWs followed (Lennartz, Rapp). Microcalorimetric gas sensors burn explosible gases with the surrounding air on the surface of a small ball or film of a catalytically active metal. The catalyst, eg, Pt, Pd or Rh, is kept at 500–600 °C. The heat of combustion in the presence of a gas is balanced by a reduction of electrical heating power. The power consumption serves as signal indicating the concentration of flammable gases. One electronic nose utilizes this sensor principle (Lennartz). Electrochemical cells ionize the gas molecule at a boundary layer (electrode with catalyst in contact with electrolyte). The ion current (e. g., H^+, O^-, Cl^-, F^-) through the electrolyte to the counter electrode serves as signal in amperometric-type cells. One electronic nose utilizes this sensor principle (Lennartz).

Signal Evaluation

If more than two linear sensors are used to characterize a sample, a multivariate evaluation has to be applied to visualize and evaluate the data. There are many multivariate techniques in use. In electronic noses pattern recognition, PARC, techniques are generally applied. Unsupervised PARC techniques try to cluster the results which are usually visualized in a two-dimensional plot. This allows to find off odors of unknown origin for example preceding the refilling of beer bottles, a new application of the WMA Airsense Nose. More often a supervised technique is used to find out if an identification of a set of materials with different smells is possible. During a learning process a set of descriptors is generated for identification of future samples. An important side effect of these models is, that one can find out how much a single sensor element contributes to a reliable identification. Thereby the initial number of sensors can be reduced without significant loss of accuracy. Sensors with linear response, electrochemical cells or microgravimetric sensors, are usually evaluated by Principal Component Analysis (PCA). The linear output of the mass spectromter based SMart Nose, mentioned below, is also evaluated by PCA. The characteristics of semiconducting sensors may be linear, if an initial conductance increase is used for evaluation. If, however, saturation values are taken, nonlinear methods have to be applied. Neural networks and polynome networks can be used, a comparison of both for sensor applications is found in reference.

Applications

Solvent Detection

Already in 1980 Stetter addressed the problem of identification and quantification of compounds for emergency response personnel in field situations. The US Coast Guard, responsible for spills of hazardous chemicals in waterways and on land, ordered the development and delivery of the first commercial Electronic Nose. The chemicals of concern ranged from petrochemicals for plastics and fuels to those used in agriculture and pharmacological processes (e. g., acetic acid, hydrogen sulphide, acrylonitrile, Nitrobenzene, …, more than 20 species). Two filaments were operated as crackers before the gas reached an array of four electrochemical cells. The battery operated sensor system equipped with a micro controller was installed in 1984 by the Coast Guard and performed well. A set of 4 electrochemical cells, made by Stetter, is also available as optional EC-module for MOSES II (Lennartz), a modular electronic nose. This instrument, can be equipped with eight oxide sensors, eight BAW sensors and electrochemical cells. A microcalorimetric module contributed by the University of Freiberg/Germany is also available. For the detection of organic vapors KAMINA (abbreviation for Karlsruhe micro nose, in German *nose = Nase*) is equipped with a semiconductor oxide sensor array. The elements are covered with a thin ceramic membrane (a few nanometers) of selective permeability, either SiO_2 or Al_2O_3, to tune the specifity. The resulting signal patterns allow a sensitive detection of solvents in complex gaseous atmospheres. KAMINA is also offered as a complete system. In practical applications the concentrations of relevant compounds can vary over many orders of magnitude. In such cases an electronic nose with an adaptive gas dilution system is necessary, as offered by the PEN nose of WMA Airsense.

Food Smells

At the Federal Dairy Research Station in Bern (Schaller) the repeatability of three brands of sensor based electronic noses and one mass spectrometer nose were investigated with real food samples (ripening stages (ages) of Emmental cheese). The competing instruments were featuring microgravimetric elements and semiconducting elements (oxidic, organic, and FET-type). Since these electronic noses were not specifically developed or optimized for food control, it may not be surprising that it was resumed "none of the tested sensor technologies was sufficiently stable to obtain well repeatable results over extended periods of time". Nevertheless, it was concluded that metal oxide semiconductor sensors were the most adequate for the measurement. The characteristic flavors of foods, such as bread or coffee, are often generated during roasting by a sequence of chemical reactions from odorless precursors. When the human nose is used as a "sensor array" it has been shown e. g., in aroma extract dilution analysis for a number of foods that of the hundreds of volatiles present, only a small number is needed to generate an overall food aroma. That means the human nose needs only a very limited number of compounds (recombinate) to create an odor impression. Therefore, food processing can be more efficiently controlled by sensor elements optimized to detect the volatiles relevant for the human nose. The ability of single sensor elements to detect food volatiles was evaluated by means of the high resolution gas chromatography/selective odorant measurement multisensor array (SOMSA) technique. The compounds under investigation are separated by high-resolution gas chromatography and the effluent is split into a standard detector and an array of sensors. Key odorants for baking of bread and cake (a pyrroline with a popcorn type note) and for meat roasting (a thiazoline) could be identified and sensor elements responding specifically were developed. A German manufacturer offers a cooking device with fully automatic baking of cake. The smell of the crust detected by a pair of specific sensors developed with the SOMSA technique is used to terminate the baking process.

Fire Detection

Besides electronic noses for general use also dedicated noses are developed and available since some years. For fire detection various concepts were developed with up to eight gas sensors. One brand broadly used in fossile power plants detects carbon monoxide, hydrogen, nitrogen dioxide, and specific hydrocarbons.

Health Control

Universal electronic noses are gradually replaced by specialized ones. An approach, very similar to SOMSA described above, is realized by Osmetech plc using the Aromascan hardware base. Microbiologists detect bacteria in infectious organisms by their smell using gas chromatography with mass spectrometer detection (GC-MS) and investigate the processes involved. They identify key volatiles emitted from target organisms. This information is then fed to the sensor development team where application specific arrays are developed and produced. Already realized point of care devices (results are available after 15 minutes) are screening of urine samples for urinary tract infections, bacterial vaginosis, especially during pregancy and pneumonia of Intensive Care Unit patients.

In general, the electronic noses are on the way to be fitted with the implementation of the corresponding know how into well defined surroundings, e. g.,clinics, food processing facilities or production processes utilizing solvents.

Bibliography
General Surveys
D. Kohl: "Electronic Noses", in *Nanoelectronics and Information Technology*, R. Waser (ed.), pp. 853–863. Wiley-VCH, Weinheim, 2003;
> www.wiley-vch.de/publish/en/books/ISBN3-527-40363-9/.

Handbook of Biosensors and Electronic Noses: Medicine, Food & the Environment, E. Kress-Rogers (ed.). CRC Press, Boca Raton, 1996;
> www.crcpress.com/shopping_cart/products/product_detail.asp?sku=8905

Extended List of Gas Sensor (including Electronic Noses) Textbooks:
> www.chem.utoronto.ca/coursenotes/CHM414/texts1.html.

Special Topics
J. R. Stetter in *Sensors and Sensory Systems for an Electronic Nose*, J. W. Gardner and P. N. Bartlett (eds.), ch. 117, p. 273–301. NATO ASI Series E: Applied Sciences – Vol. 212. Kluwer Academic Publishers, Dordrecht, 1992; www.iit.edu/ jrsteach/enose.html,
electrochem.cwru.edu/ed/encycl/art-n01-nose.htm.

SOMSA technique: C.-D. Kohl, A. Eberheim and P. Schieberle, "Odour Constitutents Detectible by Oxidic Gas Sensors", *tm* **5**, 298–304 (2004);
> www.extenza-eps.com/extenza/loadHTML?objectIDValue=33114&type=abstract.

D. Kohl, J. Kelleter and H. Petig, *Detection of Fires by Gas Sensors*, Sensors Update Vol. 9, pp. 161–223. Wiley-VCH, Weinheim, 2001; www.sensorsupdate.com.

E. Schaller, S. Zenhäusern, T. Zesiger, J.O. Bosse and F. Escher, *Analusis* **28**, 743–749 (2000)

Emmanuelle Schaller, Thesis "Applications and limits of electronic noses in the evaluation of dairy products", ETH Zürich, 2000.
> www.edpsciences.org/articles/analusis/abs/2000/08/an2009/an2009.html.

F. M. Battiston, J.-P. Ramseyer, H.P. Lang, M. K. Baller, Ch. Gerber, J. K. Gimzewski, E. Meyer and H.-J. Güntherodt, *Sensors and Actuators* **B77**, 122–131 (2001).
> www.chem.ucla.edu/dept/Faculty/gimzewski/id11.htm.

Signal Evaluation by Neural Networks
wwwiap.physik.uni-giessen.de/sensorik/grundlagen_teil_2.htm.
www.eng.auburn.edu/department/ece/sensorfusion/docs/vidynn98paper.pdf.

Instruments

E. Vanneste, "Commercial E-nose Instruments", chap. 7 in *Handbook of Machine Olfaction: Electronic Nose Technology*, Tim C. Pearce, Susan S. Schiffman, H. Troy Nagle and Julian W. Gardner (eds.). Wiley-VCH, Weinheim, 2003.

Portable Electronic Nose WMA-Airsense, www.airsense.com/english/index6.html.

Lennartz Moses II, Lennartz electronic GmbH, Bismarckstrasse 136, D-72072 Tübingen, www.lennartz-electronic.de/Pages/Homepage_english.html.

M. Rapp, F. Bender, M. Dirschka, K. H. Lubert, A. Voigt, "PPB-level organic gas detection with surface acoustic wave sensor systems". Sensor 2003 Conference, 13.–15. Mai 2003 in Nuremberg; Conference Proceedings S. 151–156.
www-ifia.fzk.de/IFIA_Webseiten/Webseiten_Rapp/index.php?c_lang=en.

Karlsruhe Micro Nose "KAMINA"; www-ifia.fzk.de.

Osmetech plc, Electra Way, Crewe CW1 6WZ, UK; www.osmetech.plc.uk.

Mass Spectrometer Based Electronic Nose, EM Microelectronic-Marin SA, Laboratory Dr. Zesiger, Fleur-de-Lys 9, CH-2074 Marin-Epagnier; www.smartnose.com/.

Electrophoresis

A. Chrambach

Electrophoresis, the migration of charged particles in an electric field, is a method for separating such particles. In most biological applications, the separation proceeds in aqueous buffers, within 3 to 4 pH units on either side of neutrality, at low (0.01 to 0.03 M) ionic strength and in the presence of an inert polymer network (gel). According to the most common gel types used, the separation method is designated polyacrylamide or agarose gel electrophoresis [1, 2].

For separation of particles differing in size but sharing a common surface net charge density, such as nucleic acids or protein subunits derivatized with sodium dodecyl sulfate (SDS-proteins), separation occurs at gel concentrations at which electrophoretic mobility is inversely related to molecular size (i. e., by "molecular sieving"). At any of those gel concentrations, separated species can be characterized by their migration distances which are related to size, commonly expressed as molecular weights (kDa) or basepairs in the case of DNA and numerically evaluated in reference to size standards [3].

For particles differing from one another both in size and in surface net charge density (related to isoelectric point), electrophoretic mobility is a function of both. Separation between such particles is most efficiently carried out by molecular sieving (as above) if size differences predominate. However, in that case molecular weight is related to the rate of change of electrophoretic mobility with gel concentration, rather than the mobility at any one gel concentration. That rate is expressed as the slope of the plot of log(mobility) versus gel concentration ("Ferguson plot"), designated as the retardation coefficient. The "free electrophoretic mobility," related to net charge density, is given by the intercept of the Ferguson plot with the mobility axis. Gel electrophoresis, therefore, allows one to evaluate separately the two elements of electrophoretic mobility, size and net charge, and thus to define the two most important properties of macromolecules [4].

If, in a particular separation problem, the sizes of all components are similar and differences are preponderantly based on surface net charge densities, the most efficient separation is that on the basis of free mobilities. The gel in those separations serves as an anticonvective medium, not a means of sieving. One applicable technique in such cases is gel electrophoresis in a pH gradient ("isoelectric focusing") [5], provided that particles carry both positively and negatively charged groups and are therefore reduced to zero mobility (and arrest of migration) at a pH at which the two balance one another (isoelectric point). Separation under those circumstances leads to an alignment of particles separated in the order of their isoelectric points.

Two-dimensional electrophoresis of proteins (2-D gel electrophoresis) [6] achieves a consecutive separation based on net charge differences and a separation based on size differences by combining an isoelectric focusing gel in the first dimension with a polyacrylamide gel separation in SDS-containing buffer in the second dimension. This technique is important for the analysis of multicomponent protein systems. Each separated component on a 2-D gel after application of a detection procedure gives rise to a spot defined by coordinates of size and isoelectric pH. On appropriately large gels, thousands of such separated spots can provide a protein map of the system.

Another important application of nonsieving, anticonvective gels is in the pulsed field gel (PFG) electrophoresis of large DNA [7]. DNA larger than 2×10^4 bp is stretched on agarose gel electrophoresis, migrates by "reptation" at a rate independent of size, or is entangled in the gel network and does not migrate at all. Pulsing of the electric field and variation of the direction of the field during electrophoresis restores both the migration of entangled species and the separation by size. Using PFG electrophoresis, chromosomal-size DNA up to sizes higher than 10 megabasepair has been separated.

Electrophoresis can be applied preparatively. At the microgram and milligram scale, gel electrophoretic separation, sectioning of the gel, and electrophoretic extraction are applicable [8]. At a larger scale, free-flow electrophoretic separation techniques [9] are available, but necessarily suffer in resolving power for being free-mobility separations in the absence of molecular sieving effects in those applications where size and shape differences between species predominate, e. g., in the bulk separation of whole cells [10].

See also: Rheology; Electrophotography.

References

[1] A. T. Andrews, *Electrophoresis*. Clarendon Press, Oxford, 1986. (I)

[2] R. C. Allen, C. A. Saravis, and H. R. Maurer, *Gel Electrophoresis and Isoelectric Focusing of Proteins*. de Gruyter, Berlin and New York, 1984. (A)

[3] D. Rodbard, "Estimation of Molecular Weight by Gel Filtration and Gel Electrophoresis," in *Method of Protein Separation*, Vol. 2, N. Catsimpoolas (ed.). Plenum Press, New York, 1976. (A)

[4] A. Chrambach, *The Practice of Quantitative Gel Electrophoresis* (V. Neuhoff and A. Maelicke, eds.). Verlag Chemie, Weinheim, 1985. (I)

[5] P. G. Righetti, *Isoelectric Focusing: Theory, Methodology and Applications*. Elsevier, Amsterdam, 1983. (A)

[6] M. J. Dunn, "Two-dimensional polyacrylamide gel electrophoresis," in *Advances in Electrophoresis, Vol. 1* (A. Chrambach, M. J. Dunn and B. J. Radola, eds.). Verlag Chemie, Weinheim, 1987. (A)

[7] M. V. Olson, "Pulsed-Field Gel Electrophoresis," in *Genetic Engineering* (J. K. Setlow, ed.). Plenum Press, New York, 1989. (A)

[8] R. Horuk, "Preparative polyacrylamide gel electrophoresis of proteins," in *Advances in Electrophoresis*, Vol. 1 (A. Chrambach, M. M. Dunn, and B. J. Radola, eds.). Verlag Chemie, Weinheim, 1987. (A)

[9] M. Bier, "Effective principles for scaling-up of electrophoresis," in *Frontiers in Bioprocessing* (S. K. Sikdar, M. Bier, and P. Todd, eds.). CRC Press, Boca Raton, FL, 1990. (A)

[10] A. W. Preece and K. A. Brown, "Recent Trends in Particle Electrophoresis," in *Advances in Electrophoresis, Vol. 3* (A. Chrambach, M. J. Dunn, and B. J. Radola, eds.). Verlag Chemie, Weinheim, 1989. (I)

Electrophotography

L. B. Schein

Electrophotographic (also called xerographic) printers and copiers are based on two well-known but not well-understood physical phenomena: electrostatic charging and photoconductivity in insulators. It is a tribute to the genius of Chester F. Carlson, the inventor of electrophotography, that he was able in 1938 to combine such little-understood physical phenomena into a process that is now at the heart of a rapidly growing, multi-billion industry. Commercial products span the range from low-cost personal copiers and laser printers that produce six pages per minute to high-speed printers that produce up to 220 pages per minute. Laser printers (and inkjet printers) have completely replaced impact printers. High-speed, high-quality color laser printers are slowly replacing offset presses, and low-speed, medium-quality laser printers are becoming available at low enough prices that they are expected to be as common as black-and-white printers in the near future.

The process of electrophotography is shown schematically in Fig. 1. It is a complex process involving six distinct steps in most cases:

Charge. An electrical corona discharge caused by air breakdown uniformly charges the surface of the photoconductor, which, in the absence of light, is an insulator.

Expose. Light, reflected from the image (in a copier) or produced by a laser (in a printer), discharges the normally insulating photoconductor producing a latent image – a charge pattern on the photoconductor that mirrors the information to be transformed into the real image.

Develop. Electrostatically charged and pigmented polymer particles called toner, ≈ 10 microns in diameter, are brought into the vicinity of the latent image. By virtue of the electric field created by the charges on the photoconductor, the toner adheres to the latent image, transforming it into a real image.

Transfer. The developed toner on the photoconductor is transferred to paper by corona charging the back of the paper with a charge opposite to that of the toner particles.

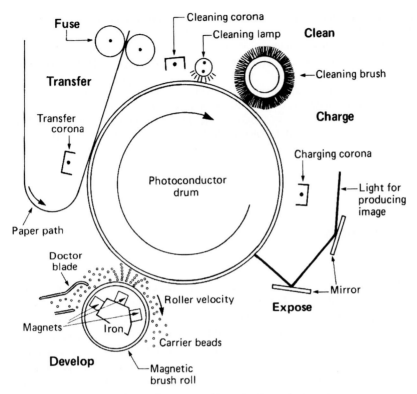

Fig. 1: Electrophotography can be separated into six steps: charge, expose, develop, transfer, fuse, and clean. The diagram locates the steps schematically in a commercial copier.

Fuse. The image is permanently fixed to the paper by melting the toner into the paper surface.

Clean. The photoconductor is discharged and cleaned of any excess toner using coronas, lamps, brushes or scraper blades.

Discussed below are the physical phenomena (1) occurring in the photoconductor (during the expose step), (2) associated with the development step, and (3) related to toner charging.

Much of the scientific literature of electrophotography has focused on the microscopic processes occurring inside photoconductors. All photoconductors, either inorganic (amorphous selenium alloys) or organic (molecularly doped polymers), appear to have similar photogeneration and charge-transport characteristics. Photogeneration is usually modeled as the escape, by random diffusion, of a charge carrier from a Coulomb well, whose source is the countercharge. The escape process is aided by temperature and the electric field, which lowers the well height. Charge transport is usually pictured as a hopping process with the holes (or electrons) hopping among dopant molecules (in the organic photoconductor). However, the electric field independence of the mobility remains puzzling. A remarkable feature of photoconductors involves trapped charges. After repeated charging and discharging, charge may become trapped in the photoconductor which cannot be dissipated by light. The amount of trapped charge must remain small compared to the charges associated with the latent image. It may be shown that very low levels of traps, $5 \times 10^{13}\,\mathrm{cm}^{-3}$, or 1 molecule in 10^8, will produce a useless photoreceptor. That any such material exists in nature is remarkable; that it is a glassy or amorphous material is even more surprising given the discussions of bandtail states that are postulated to exist in amorphous materials.

One important example of a development system is the magnetic brush development system, which is used today in almost all copiers and printers operating above 30 cpm, and is shown schematically in Fig. 1. In this system, the toner (approximately 10 μm in diameter) is mixed with carrier beads (200 μm in diameter). The materials are chosen such that the mixing causes the toner and carrier particles to become oppositely charged, as described below, thereby causing the toner to adhere electrostatically to the carrier particles. The carrier beads are made from a soft magnetic material so that they form magnetic "brushes" on a roller that carries them by frictional and magnetic forces past stationary magnets into the development zone. Here, in response to the electric field of the latent image, the toner transfers to the photoconductor from the carrier particles at the end of the brushes.

The conditions for toner development depend upon the type of image and development system. For solid areas or lines, in the development system shown in Fig. 1, toner develops until the local electric field goes to zero. The local electric field is determined primarily by two terms, one due to the latent image and one due to charge buildup on the carrier particles adjacent to the photoconductor as toner develops.

An evolution of development systems was introduced in 1980. This approach, called monocomponent development, eliminates the need for carrier particles. It is therefore smaller and lower cost, important characteristics for personal copiers and printers. The physical principles behind their development characteristics are similar to those discussed above.

In discussing development, we assumed that the toner is charged. The electrostatic charge exchange between the toner and carrier surfaces (triboelectrification) is key to successful image formation. This occurs by the mixing action in the hopper upstream of the developer. The phenomenon is familiar to anyone who has touched grounded metal after walking across a rug in a dry room or whose hair clings to a comb in the winter.

Improving development requires control of the toner charge. Knowledge of the physics of contact electrification is crucial to this effort, yet the physics of electrification of insulators is very poorly understood. At present, the most is known, both experimentally and theoretically, about contact electrification for metal–metal contacts, and successively less for metal–insulator and insulator–insulator contacts. In electrophotography we are interested primarily in insulator–insulator electrification.

A straightforward model for metal–metal contact electrification involves a simple equilibration of Fermi levels. As the metals are separated, the levels remain in equilibrium out to a distance of 10 Å, where the tunneling currents vanish. Quite reasonable semiquantitative agreement between theory and experiment has been achieved.

For metal–insulator or insulator–insulator triboelectrification, experiments between laboratories do not appear to be reproducible and theoretical uncertainties exist concerning the mobility of charges in the insulators and the applicability of the concept of a Fermi level when excess charge can be trapped in nonequilibrium situations for long times. Attempts to arrange insulators (along with metals) in an empirical triboelectric series such that materials higher up in the series become positively charged on contact with materials lower down in the series have had limited success. Unfortunately, there is not general agreement on the ordering of insulators and in some cases the series appears to be a ring.

The physics of electrostatic charge exchange remains one of the least understood branches of solid-state physics. Part of the reason is the difficulty of performing definitive experiments in this area. When the surfaces of two materials are brought into contact and separated, the actual area which made contact is difficult to determine. Whether pure contact or friction is required has not been determined. The precise nature of the surfaces are usually not well defined: dust particles, surface contaminants, and even water layers may be the "surface." Even for "clean" surfaces the nature of intrinsic and extrinsic surface electron states on insulators is not well understood. The magnitude of return currents during separation remains controversial. Finally, the number of surface molecules involved in the charging process is extremely small, on the order of one molecule in 10^5. Nevertheless, progress in understanding insulator–insulator charging would be highly useful, not only for electrophotography but also for a variety of processes that involve electrostatic charging, such as dust precipitation, spray painting, and reduction of sparking. Recent work has shown that insulator charging can be characterized by an electric field theory of charging that assumes that particles charge until a material-dependent electric field is created at the surface of the particle. But a microscopical theory, which relates toner charging to physical or chemical properties of materials, is not yet available.

See also: Corona Discharge; Electrophoresis; Photoconductivity; Tribology.

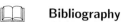

Bibliography
General Sources
L. B. Schein, *Electrophotography and Development Physics*. Laplacian Press, Morgan Hill, CA, 1996.
E. M. Williams, *The Physics and Technology of Xerographic Processes*. Wiley, New York, 1984.
R. M. Schaffert, *Electrophotography*. Focal, New York, 1980.
D. M. Burland and L. B. Schein, *Phys. Today* **39**(5), 46 (1986).
P. M. Borsenber, D. S. Weiss, *Organic Photoreceptors for Xerography*. Marcel Dekker, New York, 1998.

Physics of Electrophotography
J. Mort and D. M. Pai (eds.), *Photoconductivity and Related Phenomena*. Elsevier, Amsterdam, 1976.
J. Mort and G. Pfister (eds.), *Electronic Properties of Polymers*. Wiley, New York, 1982.
L. B. Schein, A. Peled, and D. Glatz, *J. Appl. Phys.* **66**, 686 (1989).
D. M. Pai, *J. Non-Cryst. Solids* **60**, 1255 (1983).
L. B. Schein, K. J. Fowler, G. Marshall, and V. Ting, *J. Imaging Technol.* **13**, 60 (1987).
J. Lowell and A. C. Rose-Inner, *Adv. Phys.* **29**, 1947 (1980).
W. R. Harper, *Contact and Frictional Electrification*. Oxford University Press, Oxford, 1967.
D. A. Sennor, in *Physiochemical Aspect of Polymer Surfaces*, Mittal (ed.), Vol. 1, p. 477, Plenum, New York, 1983.
L. B. Schein, *J. Electrostatics* **46**, 29 (1999).

Electrostatics

R. C. Alig

Electrostatics is concerned with the effects of positive and negative charges. The fundamental charges are the electron and the proton. They are equal in magnitude. The electron is negative, the proton is positive. Like charges repel; unlike charges attract.

The basic electrostatic law is Coulomb's law: Two electric charges attract or repel each other with a force that is proportional to the product of the charges and that varies inversely with the square of the distance between them. The unit charge, the coulomb (C), consists of 0.6242×10^{19} electrons or protons. In a vacuum a charge of 1 C repels a like charge, at a distance of 1 m, with a force of about 9×10^9 newtons, (N) or approximately 1 million tons. In physics and industry the charges encountered are typically vastly smaller than the coulomb.

All the Coulomb forces on a charged object define the electric field. The force is proportional to the charge on the object and to the field intensity at its location. Near the center of a parallel-plate capacitor, the field is uniform. At a separation of 1 cm and with 1 V applied, the central field intensity is $100 \, V/m$.

When water is raised to a higher level against gravitational force, energy is stored as potential energy. Likewise, when a charged object is moved against the Coulomb force, it is raised to a higher electric potential, and energy is stored. The volt (V) is the unit of electric potential, measured in joules per coulomb.

The capacitance measures the capacity of a system to store electric energy. The farad (F) is the unit of capacitance, measured in coulombs per volt. The farad is far greater than the amount of capacitance normally encountered; e. g., the capacitance of a parallel-plate capacitor with area $1 \, m^2$ and separation 1 cm is less than $10^{-9} \, F$.

The electric field is related to the volume charge density $\rho(\mathbf{r})$ by the differential equation $\nabla^2 \phi = \rho(\mathbf{r})$, where $\phi(\mathbf{r})$ is the electric potential and $\rho(\mathbf{r})$ is the volume charge density. This equation is called Poisson's equation, and when $\rho(\mathbf{r}) = 0$, it is called Laplace's equation. The old definition of electrostatics restricted it to the effects of fixed charges. More recently, dynamic phenomena are included in electrostatics, i. e., $\phi(\mathbf{r})$ and $\rho(\mathbf{r})$ are time dependent, provided the time variation is slow enough to ignore the other phenomena of electrodynamics.

Particles with unequal numbers of electrons and protons are said to be charged. Ions are charged atoms or molecules. Larger charged particles are formed when ions or electrons become attached to bits of solid or liquid matter. Most applications of electrostatics can be grouped into phenomena involving (1) the direction of electron and ion beams or (2) the transfer of larger charged particles. Bulk matter may contain an internal fixed charge density. To describe the forces on charges inside the material, the charge density is divided by the dielectric constant of the material. The semiconductor *p–n* junction is an example of an internal fixed charge density.

Sources

Gaseous atoms and molecules are neutral. They can be ionized, i. e., separated into positive ions and electrons, by an intense nonuniform field, such as that near a wire or point. This is called a corona discharge. The electrons formed in this way join neutral molecules to form negative ions. Ions can also be formed by cosmic rays or radioactive sources. Solid and liquid particles can be charged through the exchange of electrons or ions on contact; this is frictional, or tribo, electricity. Historically this was the earliest experience with electricity; frequently the exchange was made by rubbing cat's fur on amber. Solids can emit electrons. Thermionic emission is the release of electrons by heating, and field emission is their release in an intense electric field.

Atmospheric Electrostatics

Cosmic rays and contact electrification give rise to myriad atmospheric electrostatic phenomena. In fair weather the atmosphere is positive, and the earth is negative, resulting in a downward electric field of about $100\,\mathrm{V/m}$. Inside thunderheads the fields can be $10000\,\mathrm{V/m}$. Local corona discharges produce ions, and when a path of dense ion concentration develops, charge is rapidly transported to lower the field. In this way, lightning occurs in the atmosphere. On a smaller scale, sparks discharge the contact electrification that occurs in daily life, e. g., from walking on rugs. While these discharges from the body are generally only physically annoying, in some environments, such as petroleum tanks or grain elevators, such sparks can be quite dangerous. On a yet smaller scale, discharges of voltages 100 times below the threshold for human sensitivity can damage modern integrated circuits (ICs). People working with unshielded ICs must be grounded and exercise extreme caution to avoid sporadic, latent damage to the ICs.

Electron and Ion Beams

A multitude of uses for electron and ion beams exist in research and industry. The most common commercial product is the cathode ray tube, used in television. These beams are focused and directed by electrostatic fields. The design of these fields is called electron optics, because they guide beams much as lenses and prisms guide light. Electron and ion microscopes reveal details extending down to atomic dimensions. Electron-beam lithography is used extensively in manufacturing ICs. In ion implantation, ion beams are used for the detailed control of material composition. Ion propulsion is widely used to guide objects in outer space.

Charged Particles

Many electrostatic applications use the charge on particles to move them. For example, in precipitation, ions from a corona charge airborne particles and move them to precipitator walls to be collected. In this way 60 million tons of fly ash were collected in 2002 from coal-burning power stations. Electrostatic coating saves great quantities of paint: corona from the paint gun charges tiny particles; with the target grounded, the electric field guides them to it. Powder coating is a growing art: a refrigerator can be dry-powder coated and the coat later fused or baked on. Crop spraying often uses electrostatic coating to distribute insecticides and herbicides. Xerography, in copying machines, depends on a corona to charge a coated drum's surface and on the triboelectric effect to charge the toner. Charged-particle movement is basic to the manufacture, electrostatically, of a fifth of a billion dollar's worth of sandpaper and grit cloth per year.

See also: Charged-Particle Optics; Electric Charge; Electrodynamics; Electron and Ion Beams; Electron Beam Technology; Electron Microscopy; Electrophotography; Field Theory, Classical; Electrodynamics, Classical.

Bibliography

A. D. Moore, *Electrostatics: Controlling and Using Static Electricity*, 2nd ed. Laplacian Press, Morgan Hill, 1997.

J. M. Crowley, *Fundamentals of Applied Electrostatics*. Laplacian Press, Morgan Hill, 1999.

Elementary Particles in Physics

S. Gasiorowicz and P. Langacker

Elementary-particle physics deals with the fundamental constituents of matter and their interactions. In the past several decades an enormous amount of experimental information has been accumulated, and many patterns and systematic features have been observed. Highly successful mathematical theories of the electromagnetic, weak, and strong interactions have been devised and tested. These theories, which are collectively known as the standard model, are almost certainly the correct description of Nature, to first approximation, down to a distance scale 1/1000th the size of the atomic nucleus. There are also speculative but encouraging developments in the attempt to unify these interactions into a simple underlying framework, and even to incorporate quantum gravity in a parameter-free "theory of everything." In this article we shall attempt to highlight the ways in which information has been organized, and to sketch the outlines of the standard model and its possible extensions.

Classification of Particles

The particles that have been identified in high-energy experiments fall into distinct classes. There are the *leptons* (*see* Electron; Leptons; Neutrino; Muonium), all of which have spin $\frac{1}{2}$. They may be charged or neutral. The charged leptons have electromagnetic as well as weak interactions; the neutral ones only interact weakly. There are three well-defined lepton pairs, the electron (e^-) and the electron neutrino (ν_e), the muon (μ^-) and the muon neutrino (ν_μ), and the (much heavier) charged lepton, the tau (τ), and its tau neutrino (ν_τ). These particles all have antiparticles, in accordance with the predictions of relativistic quantum mechanics (*see* CPT Theorem). There appear to exist approximate "lepton-type" conservation laws: the number of e^- plus the number of ν_e minus the number of the corresponding antiparticles e^+ and $\bar{\nu}_e$ is conserved in weak reactions, and similarly for the muon and tau-type leptons. These conservation laws would follow automatically in the standard model if the neutrinos are massless. Recently, however, evidence for tiny nonzero neutrino masses and subtle violation of these conservations laws has been observed. There is no understanding of the hierarchy of masses in Table 1 or why the observed neutrinos are so light.

In addition to the leptons there exist *hadrons* (*see* Hadrons; Baryons; Hyperons; Mesons; Nucleon), which have strong interactions as well as the electromagnetic and weak. These particles have a variety of spins, both integral and half-integral, and their masses range from the value of $135\,\mathrm{MeV}/c^2$ for the neutral *pion* π^0 to $11\,020\,\mathrm{MeV}/c^2$ for one of the upsilon (heavy quark) states. The particles with half-integral spin are called *baryons*, and there is clear evidence for baryon conservation: The number of baryons minus the number of antibaryons

Table 1: The leptons. Charges are in units of the positron (e^+) charge
$e = 1.602 \times 10^{-19}$ coulomb. In addition to the upper limits, two of the
neutrinos have masses larger than $0.05\,\mathrm{eV}/c^2$ and $0.005\,\mathrm{eV}/c^2$, respectively.
The ν_e, ν_μ, and ν_τ are mixtures of the states of definite mass.

Particle	Q	Mass
e^-	-1	$0.51\,\mathrm{MeV}/c^2$
μ^-	-1	$105.7\,\mathrm{MeV}/c^2$
τ^-	-1	$1777\,\mathrm{MeV}/c^2$
ν_e	0	$< 0.15\,\mathrm{eV}/c^2$
ν_μ	0	$< 0.15\,\mathrm{eV}/c^2$
ν_τ	0	$< 0.15\,\mathrm{eV}/c^2$

Table 2: The quarks (spin-$\frac{1}{2}$ constituents of hadrons). Each quark carries
baryon number $B = \frac{1}{3}$, while the antiquarks have $B = -\frac{1}{3}$.

Particle	Q	Mass
u (up)	$\frac{2}{3}$	1.5–$5\,\mathrm{MeV}/c^2$
d (down)	$-\frac{1}{3}$	5–$9\,\mathrm{MeV}/c^2$
s (strange)	$-\frac{1}{3}$	80–$155\,\mathrm{MeV}/c^2$
c (charm)	$\frac{2}{3}$	1–$1.4\,\mathrm{GeV}/c^2$
b (bottom)	$-\frac{1}{3}$	4–$4.5\,\mathrm{GeV}/c^2$
t (top)	$\frac{2}{3}$	175–$180\,\mathrm{GeV}/c^2$

is constant in any interaction. The best evidence for this is the stability of the lightest baryon, the proton (if the proton decays, it does so with a lifetime in excess of 10^{33} yr). In contrast to charge conservation, there is no deep principle that makes baryon conservation compelling, and it may turn out that baryon conservation is only approximate. The particles with integer spin are called *mesons*, and they have baryon number $B = 0$. There are hundreds of different kinds of hadrons, some almost stable and some (known as *resonances*) extremely short-lived. The degree of stability depends mainly on the mass of the hadron. If its mass lies above the threshold for an allowed decay channel, it will decay rapidly; if it does not, the decay will proceed through a channel that may have a strongly suppressed rate, e. g., because it can only be driven by the weak or electromagnetic interactions. The large number of hadrons has led to the universal acceptance of the notion that the hadrons, in contrast to the leptons, are composite. In particular, experiments involving lepton–hadron scattering or e^+e^- annihilation into hadrons have established that hadrons are bound states of point-like spin-$\frac{1}{2}$ particles of fractional charge, known as quarks. Six types of quarks have been identified (Table 2). As with the leptons, there is no understanding of the extreme hierarchy of quark masses. For each type of quark there is a corresponding antiquark. Baryons are bound states of three quarks (e. g., proton = uud; neutron = udd), while mesons consist of a quark and an antiquark. Matter and decay processes under normal terrestrial conditions involve only the e^-, ν_e, u, and d. However, from Tables 2 and 3 we see that these four types of fundamental particle are replicated in two heavier *families*, (μ^-, ν_μ, c, s) and (τ^-, ν_τ, t, b). The reason for the existence of these heavier copies is still unclear.

Classification of Interactions

For reasons that are still unclear, the interactions fall into four types, the *electromagnetic*, *weak*, and *strong*, and the *gravitational* interaction. If we take the proton mass as a standard, the last is 10^{-36} times the strength of the electromagnetic interaction, and will mainly be neglected in what follows. (The unification of gravity with the other interactions is one of the major outstanding goals.) The first two interactions were most cleanly explored with the leptons, which do not have strong interactions that mask them. We shall therefore discuss them first in terms of the leptons.

Electromagnetic Interactions

The electromagnetic interactions of charged leptons (electron, muon, and tau) are best described in terms of equations of motion, derived from a Lagrangian function, which are solved in a power series in the *fine-structure constant* $e^2/4\pi\hbar c = \alpha \simeq 1/137$, a small parameter. The Lagrangian density consists of a term that describes the free-photon field,

$$\mathcal{L}_\gamma = -\frac{1}{4} F_{\mu\nu}(x) F^{\mu\nu}(x) , \tag{1}$$

where

$$F_{\mu\nu}(x) = \frac{\partial A_\nu(x)}{\partial x^\mu} - \frac{\partial A_\mu(x)}{\partial x^\nu} \tag{2}$$

is the electromagnetic field tensor. $\mathcal{L}_\gamma$ is just $\frac{1}{2}[\mathbf{E}^2(x) - \mathbf{B}^2(x)]$ in more common notation. It is written in terms of the vector potential $A_\mu(x)$ because the terms that involve the lepton and its interaction with the electromagnetic field are simplest when written in terms of $A_\mu(x)$:

$$\mathcal{L}_l = i\bar{\psi}(x)\gamma^\alpha \left(\frac{\partial}{\partial x^\alpha} - ieA_\alpha(x) \right) \psi(x) - m\bar{\psi}(x)\psi(x) . \tag{3}$$

Here $\psi(x)$ is a four-component spinor representing the electron, muon, or tau, $\bar{\psi}(x) = \psi^\dagger(x)\gamma^0$, the $\gamma^\alpha (\alpha = 0, 1, 2, 3)$ are the Dirac matrices [4×4 matrices that satisfy the conditions $(\gamma^1)^2 = (\gamma^2)^2 = (\gamma^3)^2 = -(\gamma^0)^2 = -1$ and $\gamma^\alpha\gamma^\beta = -\gamma^\beta\gamma^\alpha$ for $\beta \neq \alpha$]; m has the dimensions of a mass in the natural units in which $\hbar = c = 1$. If e were zero, the Lagrangian would describe a free lepton; with $e \neq 0$ the interaction has the form

$$-eA^\alpha(x)j_\alpha(x) , \tag{4}$$

where the current $j_\alpha(x)$ is given by

$$j_\alpha(x) = -\bar{\psi}(x)\gamma_\alpha\psi(x) . \tag{5}$$

The equations of motion show that the current is conserved,

$$\frac{\partial}{\partial x_\alpha} j_\alpha(x) = 0 , \tag{6}$$

so that the *charge*

$$Q = \int d^3\mathbf{r}\, j_0(\mathbf{r}, t) \tag{7}$$

is a constant of the motion.

The form of the interaction is obtained by making the replacement

$$\frac{\partial}{\partial x^\alpha} \longrightarrow \frac{\partial}{\partial x^\alpha} - ieA_\alpha(x) \tag{8}$$

in the Lagrangian for a free lepton. This *minimal coupling* follows from a deep principle, *local gauge invariance*. The requirement that $\psi(x)$ can have its phase changed locally without affecting the physics of the lepton, that is, invariance under

$$\psi(x) \longrightarrow e^{-i\theta(x)}\psi(x)\,, \tag{9}$$

can only be implemented through the introduction of a vector field $A_\alpha(x)$, coupled as in (8), and transforming according to

$$A_\alpha(x) \longrightarrow A_\alpha(x) - \frac{1}{e}\frac{\partial\theta(x)}{\partial x^\alpha}\,. \tag{10}$$

This dictates that the free-photon Lagrangian density contains only the gauge-invariant combination (2), and that terms of the form $M^2 A_\alpha^2(x)$ be absent. Thus local gauge invariance is a very powerful requirement; it implies the existence of a massless vector particle (the photon, γ), which mediates a long-range force [Fig. 1(a)]. It also fixes the form of the coupling and leads to charge conservation, and implies masslessness of the photon. The resulting theory (*see* Quantum Electrodynamics; Compton Effect; Feynman Diagrams; Muonium; Positron) is in extremely good agreement with experiment, as Table 3 shows. In working out the consequences of the equations of motion that follow from (3), infinities appear, and the theory seems not to make sense. The work of S. Tomonaga, J. Schwinger, R. P. Feynman, and F. J.

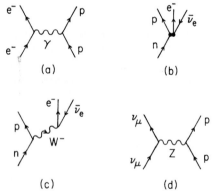

Fig. 1: (a) Long-range force between electron and proton mediated by a photon. (b) Four-fermi (zero-range) description of beta decay (n → pe⁻ν̄ₑ). (c) Beta decay mediated by a W^-. (d) A neutral current process mediated by the Z.

Table 3: Extraction of the (inverse) fine structure constant α^{-1} from various experiments, adapted from T. Kinoshita, *J. Phys.* **G29**, 9 (2003). The consistency of the various determinations tests QED. The numbers in parentheses (square brackets) represent the uncertainty in the last digits (the fractional uncertainty). The last column is the difference from the (most precise) value $\alpha^{-1}(a_e)$ in the first row. A precise measurement of the muon gyromagnetic ratio a_μ is $\sim 2.4\sigma$ above the theoretical prediction, but that quantity is more sensitive to new (TeV-scale) physics.

Experiment	Value of α^{-1}		Difference from $\alpha^{-1}(a_e)$
Deviation from gyromagnetic ratio, $a_e = \frac{1}{2}(g-2)$ for e^-	137.035 999 58(52)	$[3.8\times10^{-9}]$	–
ac Josephson effect	137.035 988 0(51)	$[3.7\times10^{-8}]$	$(0.116\pm0.051)\times10^{-4}$
h/m_n (m_n is the neutron mass) from n beam	137.036 011 9(51)	$[3.7\times10^{-8}]$	$(-0.123\pm0.051)\times10^{-4}$
Hyperfine structure in muonium, μ^+e^-	137.035 993 2(83)	$[6.0\times10^{-8}]$	$(0.064\pm0.083)\times10^{-4}$
Cesium D_1 line	137.035 992 4(41)	$[3.0\times10^{-8}]$	$(0.072\pm0.041)\times10^{-4}$

Dyson in the late 1940s clarified the nature of the problem and showed a way of eliminating the difficulties. In creating *renormalization theory* these authors pointed out that the parameters e and m that appear in (3) can be identified as the charge and the mass of the lepton only in lowest order. When the charge and mass are calculated in higher order, infinite integrals appear. After a rescaling of the lepton fields, it turns out that these are the only infinite integrals in the theory. Thus by absorbing them into the definitions of new quantities, the renormalized (i. e., physically measured) charge and mass, all infinities are removed, and the rest of the theoretically calculated quantities are finite. Gauge invariance ensures that in the renormalized theory the current is still conserved, and the photon remains massless (the experimental upper limit on the photon mass is $6 \times 10^{-17}\,\mathrm{eV}/c^2$).

Subsequent work showed that the possibility of absorbing the divergences of a theory in a finite number of renormalizations of physical quantities is limited to a small class of theories, e. g., those involving the coupling of spin-$\frac{1}{2}$ to spin-0 particles with a very restrictive form of the coupling. Theories involving vector (spin-1) fields are only renormalizable when the couplings are minimal and local gauge invariance holds. Thus gauge-invariant couplings like $\bar{\psi}(x)\gamma^\alpha\gamma^\beta\psi(x)F_{\alpha\beta}(x)$, which are known not to be needed in quantum electrodynamics, are eliminated by the requirement of renormalizability. (The apparent infinities for non-renormalizable theories become finite when the theories are viewed as a low energy approximation to a more fundamental theory. In that case, however, the low energy predictions have a very large sensitivity to the energy scale at which the new physics appears.)

The electrodynamics of hadrons involves a coupling of the form

$$-eA^\alpha(x)\,j_\alpha^{\mathrm{had}}(x) \ . \tag{11}$$

For one-photon processes, such as photoproduction (e. g., $\gamma p \to \pi^0 p$), matrix elements of the conserved current $j_\alpha^{\mathrm{had}}(x)$ are measured to first order in e, while for two-photon processes, such as hadronic Compton scattering ($\gamma p \to \gamma p$), matrix elements of products like

$j_\alpha^{had}(x) j_\beta^{had}(y)$ enter. Within the quark theory one can write an explicit form for the hadronic current:

$$j_\alpha^{had}(x) = \frac{2}{3}\bar{u}\gamma^\alpha u - \frac{1}{3}\bar{d}\gamma^\alpha d - \frac{1}{3}\bar{s}\gamma^\alpha s \ldots , \tag{12}$$

where we use particle labels for the spinor operators (which are evaluated at x), and the co-efficients are just the charges in units of e. The total electromagnetic interaction is therefore $-eA^\alpha j_\alpha^\gamma$, where

$$j_\alpha^\gamma = j_\alpha + j_\alpha^{had} = \sum_i Q_i \bar{\psi}_i \gamma_\alpha \psi_i , \tag{13}$$

and the sum extends over all the leptons and quarks ($\psi_i = e, \mu, \tau, \nu_e, \nu_\mu, \nu_\tau, u, d, c, s, b, t$), and where Q_i is the charge of ψ_i.

Weak Interactions

In contrast to the electromagnetic interaction, whose form was already contained in classical electrodynamics, it took many decades of experimental and theoretical work to arrive at a compact phenomenological Lagrangian density describing the weak interactions. The form

$$\mathcal{L}_W = -\frac{G}{\sqrt{2}} J_\alpha^\dagger(x) J^\alpha(x) \tag{14}$$

involves vectorial quantities, as originally proposed by E. Fermi. The current $J^\alpha(x)$ is known as a charged current since it changes (lowers) the electric charge when it acts on a state. That is, it describes a transition such as $\nu_e \to e^-$ of one particle into another, or the corresponding creation of an $e^-\bar{\nu}_e$ pair. Similarly, $J_\alpha^\dagger$ describes a charge-raising transition such as $n \to p$. Equation (14) describes a zero-range *four-fermi interaction* [Fig. 1(b)], in contrast to electro-dynamics, in which the force is transmitted by the exchange of a photon. An additional class of "neutral-current" terms was discovered in 1973 (*see* Weak Neutral Currents; Currents in Particle Theory). These will be discussed in the next section. $J^\alpha(x)$ consists of leptonic and hadronic parts:

$$J^\alpha(x) = J_{lept}^\alpha(x) + J_{had}^\alpha(x) . \tag{15}$$

Thus, it describes purely leptonic interactions, such as

$$\mu^- \to e^- + \bar{\nu}_e + \nu_\mu ,$$
$$\nu_\mu + e^- \to \nu_e + \mu^- ,$$

through terms quadratic in J_{lept}; semileptonic interactions, most exhaustively studied in decay processes such as

$$n \to p + e^- + \bar{\nu}_e \text{ (beta decay) },$$
$$\pi^+ \to \mu^+ + \nu_\mu ,$$
$$\Lambda^0 \to p + e^- + \bar{\nu}_e ,$$

and more recently in neutrino-scattering reactions such as

$$\nu_\mu + n \quad \rightarrow \quad \mu^- + p \text{ (or } \mu^- + \text{hadrons) },$$
$$\bar{\nu}_\mu + p \quad \rightarrow \quad \mu^+ + n \text{ (or } \mu^+ + \text{hadrons) };$$

and, through terms quadratic in J_{had}^α, purely nonleptonic interactions, such as

$$\Lambda^0 \quad \rightarrow \quad p + \pi^- ,$$
$$K^+ \quad \rightarrow \quad \pi^+ + \pi^+ + \pi^- ,$$

in which only hadrons appear. The coupling is weak in that the natural dimensionless coupling, with the proton mass as standard, is $Gm_p^2 = 1.01 \times 10^{-5}$, where G is the *Fermi constant*.

The leptonic current consists of the terms

$$J_{\text{lept}}^\alpha(x) = \bar{e}\gamma^\alpha(1-\gamma_5)\nu_e + \bar{\mu}\gamma^\alpha(1-\gamma_5)\nu_\mu + \bar{\tau}\gamma^\alpha(1-\gamma_5)\nu_\tau . \tag{16}$$

Both polar and axial vector terms appear ($\gamma_5 = i\gamma^0\gamma^1\gamma^2\gamma^3$ is a pseudoscalar matrix), so that in the quadratic form (14) there will be vector–axial-vector interference terms, indicating *parity nonconservation*. The discovery of this phenomenon, following the suggestion of T. D. Lee and C. N. Yang in 1956 that reflection invariance in the weak interactions could not be taken for granted but had to be tested, played an important role in the determination of the phenomenological Lagrangian (14). The experiments suggested by Lee and Yang all involved looking for a pseudoscalar observable in a weak interaction experiment (*see* Parity), and the first of many experiments (C. S. Wu, E. Ambler, R. W. Hayward, D. D. Hoppes, and R. F. Hudson) measuring the beta decay of polarized nuclei (^{60}Co) showed an angular distribution of the form

$$W(\theta) = A + B\mathbf{p}_e \cdot \langle \mathbf{J} \rangle , \tag{17}$$

where $\mathbf{p}_e$ is the electron momentum and $\langle \mathbf{J} \rangle$ the polarization of the nucleus. The distribution $W(\theta)$ is not invariant under mirror inversion (P) which changes $\mathbf{J} \rightarrow \mathbf{J}$ and $\mathbf{p}_e \rightarrow -\mathbf{p}_e$, so the experimental form (17) directly showed parity nonconservation. Experiments showed that both the hadronic and the leptonic currents had vector and axial-vector parts, and that although invariance under particle–antiparticle (*charge*) conjugation C is also violated, the form (14) maintains invariance under the joint symmetry CP (*see* Conservation Laws) when restricted to the light hadrons (those consisting of u, d, c, and s). There is evidence that CP itself is violated at a much weaker level, of the order of 10^{-5} of the weak interactions. As will be discussed later, this is consistent with second-order weak effects involving the heavy (b, t) quarks, though it is possible that an otherwise undetected superweak interaction also plays a role. The part of J_{had}^α relevant to beta decay is $\sim \bar{u}\gamma_\alpha(1-\gamma_5)d$. The detailed form of the hadronic current will be discussed after the description of the strong interactions.

Even at the leptonic level the theory described by (14) is not renormalizable. This manifests itself in the result that the cross section for neutrino absorption grows with energy:

$$\sigma_\nu = (\text{const})G^2 m_p E_\nu . \tag{18}$$

While this behavior is in accord with observations up to the highest energies studied so far, it signals a breakdown of the theory at higher energies, so that (14) cannot be fundamental. A number of people suggested over the years that the effective Lagrangian is but a phenomeno-logical description of a theory in which the weak current $J^\alpha(x)$ is coupled to a charged *in-termediate vector boson* $W_\alpha^-(x)$, in analogy with quantum electrodynamics. The form (14) emerges from the exchange of a vector meson between the currents (*see* Feynman Diagrams) when the W mass is much larger than the momentum transfer in the process [Fig. 1(c)]. The intermediate vector boson theory leads to a better behaved σ_V at high energies. However, massive vector theories are still not renormalizable, and the cross section for $e^+e^- \rightarrow W^+W^-$ (with longitudinally polarized Ws) grows with energy. Until 1967 there was no theory of the weak interactions in which higher-order corrections, though extraordinarily small because of the weak coupling, could be calculated.

Unified Theories of the Weak and Electromagnetic Interactions

In spite of the large differences between the electromagnetic and weak interactions (massless photon versus massive W, strength of coupling, behavior under P and C), the vectorial form of the interaction hints at a possible common origin. The renormalization barrier seems insur-mountable: A theory involving vector bosons is only renormalizable if it is a gauge theory; a theory in which a charged weak current of the form (16) couples to massive charged vector bosons,

$$\mathcal{L}_W = -g_W[J^{\alpha\dagger}(x)W_\alpha^+(x) + J^\alpha(x)W_\alpha^-(x)] , \qquad (19)$$

does not have that property. Interestingly, a gauge theory involving charged vector mesons, or more generally, vector mesons carrying some internal quantum numbers, had been invented by C. N. Yang and R. L. Mills in 1954. These authors sought to answer the question: Is it possible to construct a theory that is invariant under the transformation

$$\psi(x) \rightarrow \exp[i\mathbf{T}\cdot\boldsymbol{\theta}(x)]\psi(x) , \qquad (20)$$

where $\psi(x)$ is a column vector of fermion fields related by symmetry, the T_i are matrix repre-sentations of a Lie algebra (*see* Lie Groups; Gauge Theories), and the $\theta(x)$ are a set of angles that depend on space and time, generalizing the transformation law (9). It turns out to be pos-sible to construct such a *non-Abelian gauge theory*. The coupling of the spin-$\frac{1}{2}$ field follows the "minimal" form (8) in that

$$\bar{\psi}\gamma^\alpha\frac{\partial}{\partial x^\alpha}\psi \rightarrow \bar{\psi}\gamma^\alpha\left(\frac{\partial}{\partial x^\alpha} + ig T_i W_\alpha^i(x)\right)\psi , \qquad (21)$$

where the W_i are vector (gauge) bosons, and the *gauge coupling constant g* is a measure of the strength of the interaction. The vector meson form is again

$$\mathcal{L}_V = -\frac{1}{4}F_{\mu\nu i}(x)F_i^{\mu\nu}(x) , \qquad (22)$$

but now the structure of the fields is more complicated than in (2):

$$F_{\mu\nu i}(x) = \frac{\partial}{\partial x^\mu}W_\nu^i(x) - \frac{\partial}{\partial x^\nu}W_\mu^i(x) - g f_{ijk}W_\mu^j(x)W_\nu^k(x) , \qquad (23)$$

because the vector fields W_μ^i themselves carry the "charges" (denoted by the label i); thus, they interact with each other (unlike electrodynamics), and their transformation law is more complicated than (10). The numbers f_{ijk} that appear in the additional nonlinear term in (23) are the *structure constants* of the group under consideration, defined by the commutation rules

$$[T_i, T_j] = i f_{ijk} T_k \ . \tag{24}$$

There are as many vector bosons as there are generators of the group. The Abelian group $U(1)$ with only one generator (the electric charge) is the local symmetry group of quantum electrodynamics. For the group $SU(2)$ there are three generators and three vector mesons. Gauge invariance is very restrictive. Once the symmetry group and representations are specified, the only arbitrariness is in g. The existence of the gauge bosons and the form of their interaction with other particles and with each other is determined. Yang–Mills (gauge) theories are renormalizable because the form of the interactions in (21) and (23) leads to cancelations between different contributions to high-energy amplitudes. However, gauge invariance does not allow mass terms for the vector bosons, and it is this feature that was responsible for the general neglect of the Yang–Mills theory for many years.

S. Weinberg (1967) and independently A. Salam (1968) proposed an extremely ingenious theory unifying the weak and electromagnetic interactions by taking advantage of a theoretical development (*see* Symmetry Breaking, Spontaneous) according to which vector mesons in Yang–Mills theories could acquire a mass without its appearing explicitly in the Lagrangian (the theory without the symmetry breaking mechanism had been proposed earlier by S. Glashow). The basic idea is that even though a theory possesses a symmetry, the solutions need not. A familiar example is a ferromagnet: the equations are rotationally invariant, but the spins in a physical ferromagnet point in a definite direction. A loss of symmetry in the solutions manifests itself in the fact that the ground state, the vacuum, is no longer invariant under the transformations of the symmetry group, e. g., because it is a Bose condensate of scalar fields rather than empty space. According to a theorem first proved by J. Goldstone, this implies the existence of massless spin-0 particles; states consisting of these *Goldstone bosons* are related to the original vacuum state by the (spontaneously broken) symmetry generators. If, however, there are gauge bosons in the theory, then as shown by P. Higgs, F. Englert, and R. Brout, and by G. Guralnik, C. Hagen, and T. Kibble, the massless Goldstone bosons can be eliminated by a gauge transformation. They reemerge as the longitudinal (helicity-zero) components of the vector mesons, which have acquired an effective mass by their interaction with the groundstate condensate (the *Higgs mechanism*). Renormalizability depends on the symmetries of the Lagrangian, which is not affected by the symmetry-violating solutions, as was elucidated through the work of B. W. Lee and K. Symanzik and first applied to the gauge theories by G. 't Hooft.

The simplest theory must contain a W^+ and a W^-; since their generators do not commute there must also be at least one neutral vector boson W^0. A scalar (*Higgs*) particle associated with the breaking of the symmetry of the solution is also required. The simplest realistic theory also contains a photon-like object with its own coupling constant [hence the description as $SU(2) \times U(1)$]. The resulting theory incorporates the Fermi theory of charged-current weak interactions and quantum electrodynamics. In particular, the vector boson extension of the Fermi theory in (19) is reproduced with $g_w = g/2\sqrt{2}$, where g is the $SU(2)$ coupling, and

$G \approx \sqrt{2}g^2/8M_W^2$. There are two neutral bosons, the W^0 of $SU(2)$ and B associated with the $U(1)$ group. One combination,

$$A = \cos\theta_W B + \sin\theta_W W^0 , \tag{25}$$

is just the photon of electrodynamics, with $e = g\sin\theta_W$. The weak (or Weinberg) angle θ_W which describes the mixing is defined by $\theta_W \equiv \tan^{-1}(g'/g)$, where g' is the $U(1)$ gauge coupling. In addition, the theory makes the dramatic prediction of the existence of a second (massive) neutral boson orthogonal to A:

$$Z = -\sin\theta_W B + \cos\theta_W W^0 , \tag{26}$$

which couples to the neutral current

$$J_\alpha^Z = \sum_i T_3(i)\bar{\psi}_i\gamma_\alpha(1-\gamma_5)\psi_i - 2\sin^2\theta_W j_\alpha^\gamma , \tag{27}$$

where j_α^γ is the electromagnetic current in (13) and $T_3(i)$ $[+\frac{1}{2}$ for u, v; $-\frac{1}{2}$ for e^-, $d]$ is the eigenvalue of the third generator of $SU(2)$. The Z mediates a new class of weak interactions (*see* Weak Neutral Currents),

$$(v/\bar{v}) + p, n \quad \rightarrow \quad (v/\bar{v}) + \text{hadrons} ,$$
$$(v/\bar{v}) + \text{nucleon} \quad \rightarrow \quad (v/\bar{v}) + \text{nucleon} ,$$
$$v_\mu + e^- \quad \rightarrow \quad v_\mu + e^- ,$$

characterized by a strength comparable to the charged-current interactions [Fig. 1(d)]. Another prediction is that of the existence, in electromagnetic interactions such as

$$e^- + p \rightarrow e^- + \text{hadrons} ,$$

of tiny parity-nonconservation effects that arise from the exchange of the Z between the electron and the hadronic system. Neutral current-induced neutrino processes were observed in 1973, and since then all of the reactions have been studied in detail. In addition, parity violation (and other axial current effects) due to the weak neutral current has been observed in polarized Möller (e^-e^-) scattering and in asymmetries in the scattering of polarized electrons from deuterons, in the induced mixing between S and P states in heavy atoms (*atomic parity violation*), and in asymmetries in electron–positron annihilation into $\mu^+\mu^-$, $\tau^+\tau^-$, and heavy-quark pairs. All of the observations are in excellent agreement with the predictions of the standard $SU(2) \times U(1)$ model and yield values of $\sin^2\theta_W$ consistent with each other. Another prediction is the existence of massive $W^\pm$ and Z bosons (the photon remains massless because the condensate is neutral), with masses

$$M_W^2 = \frac{A^2}{\sin^2\theta_W} , \qquad M_Z^2 = \frac{M_W^2}{\cos^2\theta_W} . \tag{28}$$

where $A \sim \pi\alpha/\sqrt{2}G \sim (37\,\text{GeV})^2$. (In practice, a significant, 7%, higher-order correction must be included.) Using $\sin^2\theta_W$ obtained from neutral current processes, one predicted

Summer 2004

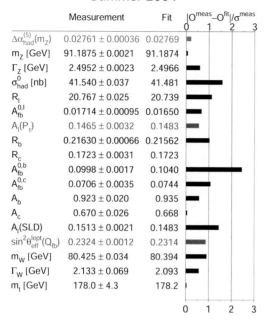

	Measurement	Fit	$|O^{meas}-O^{fit}|/\sigma^{meas}$ 0 1 2 3
$\Delta\alpha_{had}^{(5)}(m_Z)$	0.02761 ± 0.00036	0.02769	
m_Z [GeV]	91.1875 ± 0.0021	91.1874	
Γ_Z [GeV]	2.4952 ± 0.0023	2.4966	
σ_{had}^0 [nb]	41.540 ± 0.037	41.481	
R_l	20.767 ± 0.025	20.739	
$A_{fb}^{0,l}$	0.01714 ± 0.00095	0.01650	
$A_l(P_\tau)$	0.1465 ± 0.0032	0.1483	
R_b	0.21630 ± 0.00066	0.21562	
R_c	0.1723 ± 0.0031	0.1723	
$A_{fb}^{0,b}$	0.0998 ± 0.0017	0.1040	
$A_{fb}^{0,c}$	0.0706 ± 0.0035	0.0744	
A_b	0.923 ± 0.020	0.935	
A_c	0.670 ± 0.026	0.668	
$A_l(SLD)$	0.1513 ± 0.0021	0.1483	
$\sin^2\theta_{eff}^{lept}(Q_{fb})$	0.2324 ± 0.0012	0.2314	
m_W [GeV]	80.425 ± 0.034	80.394	
Γ_W [GeV]	2.133 ± 0.069	2.093	
m_t [GeV]	178.0 ± 4.3	178.2	

0   1   2   3

Fig. 2: Precision observables, compared with their expectations from the best SM fit, from *The LEP Collaborations*, hep-ex/0412015.

$M_W = 80.2 \pm 1.1\,\text{GeV}/c^2$ and $M_Z = 91.6 \pm 0.9\,\text{GeV}/c^2$'. In 1983 the W and Z were discovered at the new $\bar{p}p$ collider at CERN. The current values of their masses, $M_W = 80.425 \pm 0.038\,\text{GeV}/c^2$, $M_Z = 91.1876 \pm 0.0021\,\text{GeV}/c^2$, dramatically confirm the standard model (SM) predictions.

The Z factories LEP and SLC, located respectively at CERN (Switzerland) and SLAC (USA), allowed tests of the standard model at a precision of $\sim 10^{-3}$, much greater than had previously been possible at high energies. The four LEP experiments accumulated some $2 \times 10^7 Z's$ at the Z-pole in the reactions $e^+e^- \rightarrow Z \rightarrow \ell^+\ell^-$ and $q\bar{q}$. The SLC experiment had a smaller number of events, $\sim 5 \times 10^5$, but had the significant advantage of a highly polarized ($\sim 75\%$) e^- beam. The Z pole observables included the Z mass (quoted above), decay rate, and cross section to produce hadrons; and the branching ratios into e^+e^-, $\mu^+\mu^-$, $\tau^+\tau^-$ as well as into $q\bar{q}$, $c\bar{c}$, and $b\bar{b}$. These could be combined to obtain the stringent constraint $N_\nu = 2.9841 \pm 0.0083$ on the number of ordinary neutrinos with $m_\nu < M_Z/2$ (i. e., on the number of families with a light ν). The Z-pole experiments also measured a number of asymmetries, including forward-backward (FB), polarization, the τ polarization, and mixed FB-polarization, which were especially useful in determining $\sin^2\theta_W$. The leptonic branching ratios and asymmetries confirmed the lepton family universality predicted by the SM. The results of many of these observations, as well as some weak neutral current and high energy collider data, are shown in Fig. 2.

The LEP II program above the Z-pole provided a precise determination of M_W (as did experiments at the Fermilab Tevatron $\bar{p}p$ collider (USA)), measured the four-fermion cross sections $e^+e^- \rightarrow f\bar{f}$, and tested the (gauge invariance) predictions of the SM for the gauge boson self-interactions.

The Z-pole, neutral current, and boson mass data together establish that the standard (Weinberg–Salam) electroweak model is correct to first approximation down to a distance scale of 10^{-16} cm (1/1000th the size of the nucleus). In particular, this confirms the concepts of renormalizable field theory and gauge invariance, as well as the SM group and representations. The results yield the precise world average $\sin^2 \theta_W = 0.23149 \pm 0.00015$. (It is hoped that the value of this one arbitrary parameter may emerge from a future unification of the strong and electromagnetic interactions.) The data were precise enough to allow a successful prediction of the top quark mass (which affected higher order corrections) before the t was observed directly, and to strongly constrain the possibilities for new physics that could supersede the SM at shorter distance scales. The major outstanding ingredient is the Higgs boson, which is hard to produce and detect. The precision experiments place an upper limit of around $250\,\text{GeV}/c^2$ on the Higgs mass (which is not predicted by the SM), while direct searches at LEP II imply a lower limit of $114.4\,\text{GeV}/c^2$. Some physicists suspect that the elementary Higgs field may be replaced by a dynamical or bound-state symmetry-breaking mechanism, but the possibilities are strongly constrained by the precision data. Unified theories, such as superstring theories, generally imply an elementary Higgs. It is hoped that the situation will be clarified by the next generation of high energy colliders.

The Strong Interactions

The strength of the coupling that manifests itself in nuclear forces and in the interaction of pions with nucleons is such that perturbation theory, so useful in the electromagnetic interaction, cannot be applied to any field theory of the strong interactions in which the mesons and baryons are the fundamental fields. The large number of hadronic states strongly suggests a composite structure that cannot be viewed as a perturbation about noninteracting systems. In fact, it is now generally believed that the strong interactions are described by a gauge theory, quantum chromodynamics (QCD), in which the basic entities are quarks rather than hadrons. Nevertheless, prior and parallel to the development of the quark theory a wealth of experimental information concerning the hadrons and their interactions was accumulated. In spite of the absence of guidance from field theory, and in spite of the fact that each jump in available accelerator energy brought a shift in the focus of attention, certain simple patterns were identified.

Internal Symmetries

The first hint of a new symmetry can be seen in the remarkable resemblance between neutron and proton. They differ in electromagnetic properties, and, other than that, by effects that are very small; for example, they differ in mass by 1 part in 700. W. Heisenberg conjectured that the neutron and proton are two states of a single entity, the *nucleon* (*see* Nucleon), just as an electron with spin up and an electron with spin down are two states of a single entity, even though in an external magnetic field they have slightly different energies. Pursuing this analogy, Heisenberg and E. U. Condon proposed that the strong interactions are invariant under transformations in an internal space, in which the nucleon is a spinor (*see* Isospin).

Thus, the nucleon is an isospin doublet, with $I_z(p) = \frac{1}{2}$ and $I_z(n) = -\frac{1}{2}$, and isospin (in analogy with angular momentum) is conserved. In the language of group theory, the assertion is that the strong interactions are invariant under the transformations of the group $SU(2)$, and that particles transform as irreducible representations. The electromagnetic and weak interactions violate this invariance. The expression for the charge of the nucleons and antinucleons,

$$Q = I_z + B/2 , \tag{29}$$

shows that the charge picks out a preferred direction in the internal space. (It is now believed that the strong interactions themselves have a small piece which breaks isospin symmetry, in addition to electroweak interactions.)

With the discovery of the three pions (π^+, π^0, π^-) with mass remarkably close to that predicted by H. Yukawa (1935) in his seminal work explaining nuclear forces in terms of an exchange of massive quanta of a mesonic field, the notion of isospin acquired a new significance. It was natural, in view of the small $\pi^\pm$–π^0 mass difference, to assign the pion to the $I = 1$ representation of $SU(2)$. The invariance of the pion–nucleon interaction under isospin transformations led to a number of predictions, all of which were confirmed. In particular, states initiated in pion-nucleon collisions could only have isospin $\frac{1}{2}$ or $\frac{3}{2}$. Early work on pion–nucleon scattering led to the discovery of a resonance with rest mass $1236\,\mathrm{MeV}/c^2$, width $115\,\mathrm{MeV}/c^2$, and angular momentum and parity $J^P = \frac{3}{2}^+$. This resonance occurred in $\pi^+ p$ scattering, so that it had to have $I = \frac{3}{2}$, and its effects seen in $\pi^- p \rightarrow \pi^- p$ and $\pi^- p \rightarrow \pi^0 n$ should be the same as those in $\pi^+ p \rightarrow \pi^+ p$. This prediction was borne out by experiment.

Formally, $SU(2)$ invariance is described by defining generators I_i; ($i = 1, 2, 3$) obeying the Lie algebra

$$[I_i, I_j] = i e_{ijk} I_k , \tag{30}$$

where e_{ijk} is totally antisymmetric in the indices and $e_{123} = 1$. The statement that a pion is an $I = 1$ state then means that the pion field Π_a transforms according to

$$[I_i, \Pi_a] = -(I_i)_{ab} \Pi_b , a = 1, 2, 3 , \tag{31}$$

where the I_i are 3×3 matrices satisfying (30). In relativistic quantum mechanics conservation laws must be local, so the conservation law

$$\frac{dI_i}{dt} = 0 \tag{32}$$

really follows from the local conservation law

$$\frac{\partial}{\partial x^\mu} I_i^\mu(x) = 0 \tag{33}$$

for the isospin-generating currents, for which

$$I_i = \int d^3\mathbf{r}\, I_i^0(\mathbf{r}, t) . \tag{34}$$

Isospin [and $SU(3)$] are global symmetries: The symmetry transformations are the same at all space-time points, as opposed to the local (gauge) transformations in (20). Hence, they are not associated with gauge bosons or a force.

In the early 1950s a number of new particles were discovered. The great confusion generated by the widely differing rates of production and decay was cleared up by M. Gell-Mann and K. Nishijima, who extended the notion of isospin conservation to the strong interactions of the new particles, classified them (and along the way noted "missing" particles that had to exist, and were subsequently found), and discovered that the observed patterns of reactions could be explained by assigning a new quantum number S (strangeness) to each isospin multiplet.

The selection rules were

$$\Delta S = 0 \tag{35}$$

for the strong and electromagnetic interactions, and

$$\Delta S = 0, \pm 1 \tag{36}$$

for the weak interactions. Relation (29) now takes the form

$$Q = I_z + (B + S)/2 . \tag{37}$$

[Equation (37) holds for all hadrons except for those involving heavy (c, b, and t) quarks, discovered in the 1970s and later.]

The success of the strangeness scheme immediately started a search for a higher symmetry that would include isospin and strangeness (or hypercharge, $Y = B + S$), and that would, in some limit, include the nucleons and the newly discovered strange baryons in a supermultiplet. The search ended when M. Gell-Mann and Y. Ne'eman discovered that the Lie group $SU(3)$ was the appropriate (global) symmetry. The group is generated by eight operators F_i ($i = 1, 2, 3, \ldots, 8$), of which the first three may be identified with the isospin generators I_i, and (by convention) F_8 is related to hypercharge. The other four change isospin and strangeness. The nucleons and six other baryons discovered in the 1950s fit into an eight-dimensional (octet) representation containing doublets with $I = \frac{1}{2}$ and $Y = \pm 1$, and $I = 1, 0$ states with $Y = 0$. Similarly, the $I = 1$ pions, the (K^+, K^0) with $I = \frac{1}{2}, Y = 1$, and $(\bar{K}^0, K^-)$ with $I = \frac{1}{2}, Y = -1$, could be fitted into an octet that was soon completed with the discovery of an $I = Y = 0$ pseudoscalar meson, the η (see Table 4). $SU(3)$ is only an approximate symmetry of the strong interactions. Mass splittings within $SU(3)$ multiplets and other breaking effects are typically 20–30%.

Most interesting is that the search for partners of the *resonance* $\Delta(1236)$ with $I = \frac{3}{2}$ led to a dramatic confirmation of $SU(3)$. The simplest representation containing an ($I = \frac{3}{2}, Y = 1$) state is the 10-dimensional representation, which also contains ($I = 1, Y = 0$) and ($I = \frac{1}{2}, Y = -1$) states and an isosinglet $Y = -2$ particle. The symmetry-breaking pattern that explained the mass splittings among the isospin multiplets in the octet predicted equal mass splittings. Thus, when the $I = 1$ $\Sigma(1385)$ was discovered, predictions could be made about the $I = \frac{1}{2}$ Ξ^*, found at mass 1530 MeV/c^2, and the Ω^-, predicted at 1675 MeV/c^2. The latter mass is too low to permit a strangeness-conserving decay to $\Xi^0 K^-$, so the Ω^- had to be long-lived, only decaying

Table 4: Table of low-lying mesons and baryons, grouped according to $SU(3)$ multiplets. There may be considerable mixing between the $SU(3)$ singlets η', φ, and f' and the corresponding octet states η, ω, f.

Particle	B	Q	Y	I	J^P	Mass (GeV/c^2)	Quark content
π	0	$1, 0, -1$	0	1	0^-	0.14	$u\bar{d},\ u\bar{u}-d\bar{d},\ d\bar{u}$
K	0	$1, 0$	1	$\frac{1}{2}$	0^-	0.49	$u\bar{s},\ d\bar{s}$
$\bar{K}$	0	$0, -1$	-1	$\frac{1}{2}$	0^-	0.49	$s\bar{d},\ s\bar{u}$
η	0	0	0	0	0^-	0.55	$u\bar{u}+d\bar{d}-2s\bar{s}$
η'	0	0	0	0	0^-	0.96	$u\bar{u}+d\bar{d}+s\bar{s}$
ρ	0	$1, 0, -1$	0	1	1^-	0.77	$u\bar{d},\ u\bar{u}-d\bar{d},\ d\bar{u}$
K^*	0	$1, 0$	1	$\frac{1}{2}$	1^-	0.89	$u\bar{s},\ d\bar{s}$
$\bar{K}^*$	0	$0, -1$	-1	$\frac{1}{2}$	1^-	0.89	$s\bar{d},\ s\bar{u}$
ω	0	0	0	0	1^-	0.78	$u\bar{u}+d\bar{d}$
ϕ	0	0	0	0	1^-	1.02	$s\bar{s}$
A_2	0	$1, 0, -1$	0	1	2^+	1.32	$u\bar{d},\ u\bar{u}-d\bar{d}\,d\bar{u}$
$K^*(1430)$	0	$1, 0$	1	$\frac{1}{2}$	2^+	1.43	$u\bar{s},\ d\bar{s}$
$\bar{K}^*(1430)$	0	$0, -1$	-1	$\frac{1}{2}$	2^+	1.43	$s\bar{d},\ s\bar{u}$
f	0	0	0	0	2^+	1.28	$u\bar{u}+d\bar{d}$
f'	0	0	0	0	2^+	1.53	$s\bar{s}$
N	1	$1, 0$	1	$\frac{1}{2}$	$\frac{1}{2}^+$	0.94	$uud,\ udd$
Λ	1	0	0	0	$\frac{1}{2}^+$	1.12	$uds-dus$
Σ	1	$1, 0, -1$	0	1	$\frac{1}{2}^+$	1.19	$uus,\ uds+dus,\ dds$
Ξ	1	$0, -1$	-1	$\frac{1}{2}$	$\frac{1}{2}^+$	1.32	$uss,\ dss$
Δ	1	$2, 1, 0, -1$	1	$\frac{3}{2}$	$\frac{3}{2}^+$	1.23	$uuu,\ uud,\ udd,\ ddd$
$\Sigma(1385)$	1	$1, 0, -1$	0	1	$\frac{3}{2}^+$	1.39	$uus,\ uds,\ dds$
$\Xi^*(1530)$	1	$0, -1$	-1	$\frac{1}{2}$	$\frac{3}{2}^+$	1.53	$uss,\ dss$
Ω^-	1	-1	-2	0	$\frac{3}{2}^+$	1.67	sss

by a chain of $\Delta S = 1$ weak interactions with a very clear signature. The dramatic discovery in 1964 of the Ω^- with all the right properties convinced all doubters. [*see* SU(3) and Higher Symmetries; Hyperons; Hypernuclear Physics and Hypernuclear Interactions].

S-Matrix Theory

The construction of higher-energy accelerators, the invention of the bubble chamber by D. Glaser, and the combination of large hydrogen bubble chambers, rapid scanning facilities, and high-speed computers into a massive data production and analysis technology, pioneered by L. Alvarez and collaborators, led to the discovery of many new resonances during the 1950s and 1960s. The basic procedure was to measure charged tracks in bubble-chamber pictures, taken in strong magnetic fields, and to calculate the invariant masses $(\sum E_i)^2 - (\sum \mathbf{p}_i c)^2$ for various particle combinations. Resonances manifest themselves as peaks in mass distributions, and the events in the resonance region may be further analyzed to find out the spin and parity of the resonance. Baryonic resonances were also discovered in phase-shift analyses of angular distributions in pion–nucleon and K–nucleon scattering reactions. The patterns of masses and quantum numbers of the resonances showed that all the mesonic resonances came in $SU(3)$ octets and singlets, and the baryonic ones in $SU(3)$ decuplets, octets, and singlets.

There was good evidence that there was no fundamental distinction between the stable particles and the highly unstable resonances: The Δ and the Ω^-, discussed above, are good examples, and theoretically it was found that both stable (bound) states and resonant ones appeared in scattering amplitudes as pole singularities, differing only in their location. Furthermore, the role assigned by Yukawa to the pion as the nuclear "glue" – it was the particle whose exchange was largely responsible for the nuclear forces – had to be shared with other particles: Various vector and scalar mesons were seen to contribute to the nuclear forces, and G. F. Chew and F. E. Low explained much of low-energy pion physics in terms of nucleon exchange. Chew, in collaboration with S. Mandelstam and S. Frautschi, proposed to do away with the notion of any particles being "fundamental." They hypothesized that the collection of all scattering amplitudes, the scattering matrix, be determined by a set of self-consistency conditions, the *bootstrap* conditions (*see* S-Matrix Theory), according to which, crudely stated, the exchange of all possible particles should yield a "potential" whose bound states and resonances should be identical with the particles inserted into the exchange term.

Much effort was devoted to bootstrap and S-matrix theory during the 1960s and early 1970s. The program had its greatest success in developing phenomenological models for strong interaction scattering amplitudes at high energies and low-momentum transfers, such as elastic scattering and total cross sections. In particular, Mandelstam applied an idea due to T. Regge to relativistic quantum mechanics, which related a number (perhaps infinite) of particles and resonances with the same $SU(3)$ and other internal quantum numbers, but different masses and spins, into a family or *Regge trajectory*. The exchange of this trajectory of particles led to much better behaved high-energy amplitudes than the exchange of one or a small number, in agreement with experiment (*see* Regge Poles). Related models had some success in describing inclusive processes (in which one or a few final particles are observed, with the others summed over) and other highly inelastic processes (*see* Inclusive Reactions).

The more ambitious goal of understanding the strong interactions as a bootstrap (self-consistency) principle met with less success, although a number of models and approximation schemes enjoyed some measure in limited domains. The most successful was the dual resonance model pioneered by G. Veneziano. The dual model was an explicit closed-form expression for strong-interaction scattering amplitudes which properly incorporated poles for the Regge trajectories of bound states and resonances that could be formed in the reaction,

Table 5: The u, d, and s quarks.

	B	Q	Y	I	I_z
u	$\frac{1}{3}$	$\frac{2}{3}$	$\frac{1}{3}$	$\frac{1}{2}$	$\frac{1}{2}$
d	$\frac{1}{3}$	$-\frac{1}{3}$	$\frac{1}{3}$	$\frac{1}{2}$	$-\frac{1}{2}$
s	$\frac{1}{3}$	$-\frac{1}{3}$	$-\frac{2}{3}$	0	0

Regge asymptotic behavior, and duality (the property that an amplitude could be described *either* as a sum of resonances in the direct channel *or* as a sum of Regge exchanges). However, the original simple form did not incorporate unitarity, i. e., the amplitudes did not have branch cuts corresponding to multiparticle intermediate states, and the resonances in the model had zero width (their poles occurred on the real axis in the complex energy plane instead of being displaced by an imaginary term corresponding to the resonance width). Perhaps the most important consequence of dual models was that they were later formulated as *string theories*, in which an infinite trajectory of "elementary particles" could be viewed as different modes of vibration of a one-dimensional string-like object (*see* String Theory). String theories never quite worked out as a model of the strong interactions, but the same mathematical structure reemerged later in "theories of everything."

Many of the S-matrix results are still valid as phenomenological models. However, the bootstrap idea has been superseded by the success of the quark theory and the development of QCD as the probable field theory of the strong interactions.

Quarks as Fundamental Particles

The discovery of $SU(3)$ as the underlying internal symmetry of the hadrons and the classification of the many resonances led to the recognition of two puzzles: Why did mesons come only in octet and singlet states? Why were there no particles that corresponded to the simplest representations of $SU(3)$, the triplet 3 and its antiparticle 3^*? M. Gell-Mann and G. Zweig in 1964 independently proposed that such representations do have particles associated with them (Gell-Mann named them quarks), and that all observed hadrons are made of ($q\bar{q}$) (quark + antiquark) if they have baryon number $B = 0$ and of (qqq) (three quarks) if they have baryon number $B = 1$. They proposed that there exist three different kinds of quarks, labeled u, d, and s. These were assumed to have spin $\frac{1}{2}$ and the internal quantum numbers listed in Table 5. The quark contents of the low-lying hadrons are given in Table 4. The vector meson octet (ρ, K^*, ω) differs from the pseudoscalars (π, K, η) in that the total quark spin is 1 in the former case and zero in the latter. The (A_2, $K^*(1490)$, f) octet are interpreted as an orbital excitation ($^3P_2$). All of the known particles and resonances can be interpreted in terms of quark states, including radial and orbital excitations and spin.

The first question was answered automatically, since products of the simplest representations decompose according to the rules

$$\begin{aligned} \mathbf{3} \times \mathbf{3}^* &= \mathbf{1} + \mathbf{8}\,, \\ \mathbf{3} \times \mathbf{3} \times \mathbf{3} &= \mathbf{1} + \mathbf{8} + \mathbf{8} + \mathbf{10}\,. \end{aligned} \tag{38}$$

A problem immediately arose in that the decuplet to which $\Sigma(1236)$ belongs, being the lowest-energy decuplet, should have its three quarks in relative S states. Thus the Δ^{++}, whose composition is uuu, could not exist, since the spin-statistics connection requires that the wave function be totally antisymmetric, which it manifestly is not when the Δ^{++} is in a $J_z = \frac{3}{2}$ state, with all spins up, for example. The solution to this problem, proposed by O. W. Greenberg, M. Han, and Y. Nambu and further developed by W. A. Bardeen, H. Fritzsch, and M. Gell-Mann, was the suggestion that in addition to having an $SU(3)$ label such as (u, d, s) – named *flavor* by Gell-Mann – and a spin label (up, down), quarks should have an additional three-valued label, named *color*. Thus according to this proposal there are really nine light quarks:

$$
\begin{array}{ccc}
u_R & u_B & u_Y \\
d_R & d_B & d_Y \\
s_R & s_B & s_Y
\end{array}
$$

Hence, the low-lying (qqq) state could be symmetric in the flavor and spin labels, provided it were totally antisymmetric in the color (red, blue, yellow) labels. More colors could be imagined but at least three are needed. Transformations among the color labels lead to another symmetry, $SU(3)_{\text{color}}$. The totally antisymmetric state is a color singlet. The mesons can also be constructed as color singlets, for example,

$$
\pi^+ = \frac{1}{\sqrt{3}} (u_R \bar{d}_R + u_B \bar{d}_B + u_Y \bar{d}_Y) \,.
$$

The existing hadronic spectrum shows no evidence for states that could be color octets, for example, so the present attitude is that either color nonsinglet states are very massive compared with the low-lying hadrons or that it is an intrinsic part of hadron dynamics that *only color singlet states are observable.*

The first evidence that there are *three* (and not more) colors came from the study of $\pi^0 \to 2\gamma$ decay. Using general properties of currents, S. Adler and W. A. Bardeen were able to prove that the π^0 decay rate was uniquely determined by the process in which the π^0 first decays into a $u\bar{u}$ or a $d\bar{d}$ pair, which then annihilates with the emission of two photons. The matrix element depends on the charges of the quarks, and a calculation of the width yields $0.81\,\text{eV}$. With n colors, this is multiplied by n^2, and the observed width of $7.8 \pm 0.6\,\text{eV}$ supports the choice of $n = 3$. Subsequent evidence for three colors was provided by the total cross section for e^+e^- annihilation into hadrons (see below), and by the elevation of the $SU(3)_{\text{color}}$ symmetry to a gauge theory of the strong interactions.

The quark model has been extremely successful in the classification of observed resonances, and even predictions of decay widths work very well, with much data being correlated in terms of a few parameters. The ingredients that go into the calculation are (a) that quarks are light, with the (u, d) doublet almost degenerate, with mass in the 300-MeV$/c^2$ range (one-third of a nucleon), (b) that the s quark is about $150\,\text{MeV}/c^2$ more massive – this explains the pattern of $SU(3)$ symmetry breaking – and (c) that the low-lying hadrons have the simple $q\bar{q}$ or qqq content, without additional $q\bar{q}$ pairs. However, nobody has ever observed an isolated quark (free quarks should be easy to identify because of their fractional charge). It is now generally believed that quarks are confined, i. e., that it is impossible, even in principle, for them to exist as isolated states. However, in the 1960s this led most physicists to doubt the existence of quarks as real particles. That view was shattered by the deep inelastic electron scattering experiments in the late 1960s.

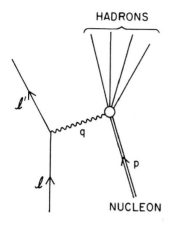

Fig. 3: Kinematics of deep inelastic lepton scattering.

Deep Inelastic Reactions and Asymptotic Freedom

In 1968 the first results of the inelastic electron-scattering experiments (Fig. 3),

$$e + p \rightarrow e' + \text{hadrons}$$

measured at the Stanford Linear Accelerator Center (SLAC), were announced. The experiments were done in a kinematic region that was new. Both the momentum transfer squared (that is, the negative mass squared of the virtual photon exchanged) and the "mass" of the hadronic state produced were large. The cross section could be written as

$$\frac{d^2\sigma}{dE'd\Omega} = \left(\frac{d\sigma}{d\Omega}\right)_{\text{point}} \left(W_2(x, Q^2) + 2W_1(x, Q^2)\tan^2\frac{\theta}{2}\right), \tag{39}$$

where $(d\sigma/d\Omega)_{\text{point}}$ is essentially the cross section for a collision with a free point particle, and the hadronic part of the process was expressed in terms of certain structure functions W_1 and W_2. In (39), E' is the energy of the final electron, θ the scattering angle, $Q^2 = -(p_e - p_{e'})^2$, and the quantity x is $Q^2/2m_p\nu$, where $\nu = p \cdot q/m_p$ is the electron energy loss. No one knew what to expect for the behavior of W_1 and W_2. On the one hand, cross sections for production of definite resonances (*exclusive* reactions rather than *inclusive* ones) fell as powers of $1/Q^2$; on the other hand, J. D. Bjorken had predicted, on simple grounds of the irrelevance of masses when all the variables were large, that the dimensionless functions $F_1(x, Q^2) \equiv m_pW_1$ and $F_2(x, Q^2) \equiv \nu W_2$ should depend on x alone.

The results spectacularly confirmed Bjorken's conjecture of *scaling*. R. P. Feynman interpreted the detailed shapes of the distributions with his parton model, in which the proton, in a frame in which it is moving rapidly, looks like a swarm of independently moving point "parts" without any structure. The shape of F_2 can be interpreted as $\sum_p q_p^2 x f_p(x)$, where $f_p(x)$ is the probability distribution for a parton to carry a fraction x of the proton's momentum and q_p is the parton's charge, while the relation between F_1 and F_2 depends on the parton spin. The observed relation $F_2 \simeq 2xF_1$ establishes that the partons have spin $\frac{1}{2}$. Comparing the structure functions obtained from e and μ scattering (from proton and nuclear targets) with those

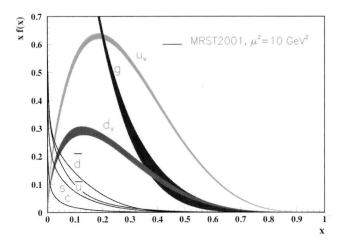

Fig. 4: Distributions $f(x)$ times the fraction x of the proton's momentum carried by valence quarks u_{v}, d_{v}; gluons g; and sea quarks $\bar{d}$, $\bar{u}$, $s = \bar{s}$, and $c = \bar{c}$, from S. Eidelman *et al.* [Particle Data Group], *Phys. Lett.* **B592**, 1 (2004).

obtained by weak reactions such as

$$\nu_\mu(\bar{\nu}\mu)p \to \mu^-(\mu^+) + \text{hadrons} ,$$

one can constrain the parton quantum numbers. They are consistent with the assumption that the partons are quarks, and that the proton consists of the three *valence* quarks assigned to it by the naive quark model, supplemented with a *sea* of quark–antiquark pairs. The relative amount of $q\bar{q}$ sea and its composition (e. g., amount of $s\bar{s}$ relative to $u\bar{u}$) are also determined. The mechanism for deep inelastic scattering is the ejection of a single quark by the virtual photon, or by the weak current in the neutrino reactions. The model assumes that the quarks that make up the proton do not interact, and that seems somewhat mysterious. Furthermore, the model of the mechanism suggests that one quark is strongly deflected from the original path. If that is so, where is it?

The problem of how the quarks appear to be noninteracting is answered by quantum field theory. There we find that the coupling strength is really momentum dependent. For example, in quantum electrodynamics, because of the polarizability of the vacuum, the net charge of an electron seen from afar (low momentum transfer) is smaller than the charge as seen close in (large momentum transfer) where it is not shielded by the positrons produced virtually in the vacuum. Quantum electrodynamics is not the right kind of theory for quarks, since the coupling (charge) increases with momentum transfer. It was pointed out by D. Gross and F. Wilczek, by D. Politzer, and by G. 't Hooft that a theory of quarks coupled via Yang–Mills vector mesons will have the property desired for the quarks probed with high-momentum-transfer currents. The requirement of such a high-momentum-transfer decoupling, named *asymptotic freedom*, thus suggests that the "glue" that binds the quarks together is generated by a non-Abelian gauge theory, which has the attraction of being renormalizable, universal (only one coupling constant), and unique, once the number of "colors," that is, the group

Fig. 5: Fundamental QCD interactions. g_s is the $SU(3)_{\text{color}}$ gauge coupling, which becomes small at large momentum transfers, and G is a gluon.

structure, is determined. The high-energy lepton scattering experiments provide evidence for the existence of some kind of flavor-neutral glue, in that the data are well fitted in terms of quarks, except that only about 50% of the momentum of the initial proton is attributable to quarks. It is now understood that the proton momentum is shared by the valence and sea quarks and electrically neutral *gluons*. Figure 4 shows the current experimental situation.

Quantum Chromodynamics

Quantum chromodynamics (QCD), the modern theory of the strong interactions, is a non-Abelian gauge theory based on the $SU(3)_{\text{color}}$ group of transformations which relate quarks of different colors. (The transformations are carried out simultaneously for each flavor, which does not change. The number of flavors is arbitrary.) The gauge bosons associated with the eight group generators, known as *gluons*, can be emitted or absorbed by quarks in transitions in which the color (but not flavor) can change. Since the gluons themselves carry color they can interact with each other as well (Fig. 5). As long as there are no more than 16 flavors, QCD is weakly coupled at large momentum transfers (asymptotic freedom) and strongly coupled at small momentum transfers, in agreement with observations.

For the theory to be renormalizable the gluons must either acquire mass through a Higgs mechanism (spontaneous symmetry breaking) or remain massless. The first type of theory destroys asymptotic freedom (its *raison d'être*), and the second has the difficulty that no massless vector mesons, aside from the photon, have ever been observed. It has been proposed that the theory has a structure such that only color singlets are observable, so that the vector mesons, like the quarks, are somehow *confined*. It has been speculated that the non-Abelian field lines, in contrast to electric and magnetic field lines, do not fan out all over space, but remain confined to a narrow cylindrical region, which leads to an interaction energy that is proportional to the separation of the sources of the field lines, and thus confinement. The linearly extended structure so envisaged is reminiscent of the string models suggested by duality, and thus may yield the spectrum characteristics of linear Regge trajectories and the associated high-energy behavior. At large separations the potential presumably breaks down, with energy converted into $(q\bar{q})$ pairs, that is, hadrons. This would explain why quarks are never seen in deep inelastic scattering or other processes. This picture of quark and gluon confinement has not been rigorously established in QCD, but is strongly supported by calculations in the most promising approximation scheme for strongly coupled theories: viz., *lattice calculations*, in which the space-time continuum is replaced by a discrete four-dimensional lattice (*see* Lattice Gauge Theory).

QCD is very successful qualitatively, but is hard to test quantitatively. This is partly because the coupling is large for most hadronic processes. Also, QCD brings a subtle change in perspective. The "strong interactions" are those mediated by the color gluons between quarks,

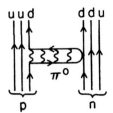

Fig. 6: One-pion exchange in QCD.

and they give rise to the color singlet hadronic bound states. The interaction between these states need not be simple, any more than the interactions between molecules (the van der Waals forces) manifest the simplicity of the underlying Coulomb force in electromagnetism. It is hoped, but not conclusively proved, that successful phenomenological models such as Regge theory or the one-boson-exchange potential emerge as complicated higher-order effects (Fig. 6). Similarly, it has not been possible to fully calculate the hadron spectrum (because of strong coupling, relativistic, and many-body effects), but lattice attempts are promising. *Glueballs* (bound states of gluons) and other nonstandard color singlet states are expected. Candidate mesons exist but have not been unambiguously interpreted. Similar statements apply to *pentaquark* states, such as $uudd\bar{s}$. QCD fairly naturally explains the observed hadronic symmetries. Parity and *CP* invariance (except for possible subtle nonperturbative effects) and the conservation of strangeness and baryon number are automatic, while approximate symmetries such as isospin, $SU(3)_{\text{flavor}}$, and chiral symmetry (see below) can be broken only by quark mass terms.

More quantitative tests of QCD are possible in high-momentum-transfer processes, in which one glimpses the underlying quarks and gluons. To zeroth order in the strong coupling g_s, QCD reproduces the quark–parton model. Higher-order corrections lead to calculable logarithmic variations of F_1 and F_2 with Q^2, in agreement with the data. These experiments have been pushed to much higher Q^2 at the ep collider HERA at DESY in Hamburg, and the results are in excellent agreement with QCD.

Another consequence of the quark–parton picture is the prediction that at high energies the cross section for $e^+ + e^- \to$ hadrons should proceed through the creation (via a virtual photon) of a $q\bar{q}$ pair, which subsequently converts into hadrons through the breakdown mechanism (Fig. 7). Thus the cross section is expected to be point-like, with the modification that the quark charges appear at the production end, so that

$$R \equiv \frac{\sigma(e^+ + e^- \to \text{hadrons})}{\sigma(e^+ + e^- \to \mu^+ + \mu^-)} \Rightarrow \sum_{\text{all quarks}} Q_i^2 \,, \tag{40}$$

where the Q_i are the quark charges. At relatively low energies the (u, d, s) contribution is 2/3 per color, that is, 2. Above the energy (3–4 GeV) needed to produce a charm quark pair $(c\bar{c})$ one expects $R = 10/3$, while above the bottom $(b\bar{b})$ threshold (~ 10 GeV), $R \sim 11/3$. Calculable higher-order corrections in QCD increase these predictions slightly. The new contribution of a virtual Z boson becomes apparent above ~ 30 GeV, with the Z peak dominating at ~ 90 GeV. These predictions are in excellent agreement with the data (Fig. 8), strongly supporting QCD and the existence of color.

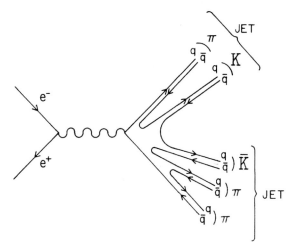

Fig. 7: Schematic picture of hadron production in $e^+ + e^-$ annihilation.

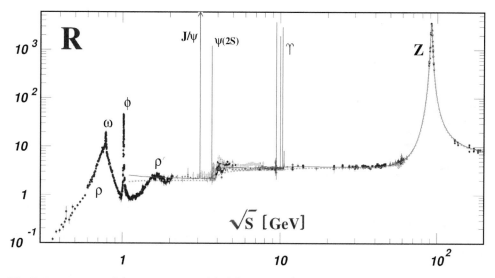

Fig. 8: Data on $R = \sigma(e^+ + e^- \to \text{hadrons})/\sigma(e^+ + e^- \to \mu^+ + \mu^-)$ as a function of the center-of-mass energy $W = \sqrt{s}$, from S. Eidelman *et al.* [Particle Data Group], *Phys. Lett.* **B592**, 1 (2004). The predictions of the quark parton model and a fit to QCD are also shown.

In large-Q^2 processes at sufficiently high energies it is expected (and observed) that the produced hadrons tend to cluster in reasonably well-collimated *jets* of particles following approximately the direction of the final quarks. For example, the angular distribution of the jets observed in e^+e^- annihilation at SLAC and DESY confirms that the quarks have spin $\frac{1}{2}$, as well as the existence of the intermediate quark–antiquark state. Similarly, experiments at DESY have shown the existence of three-jet events whose characteristics are consistent

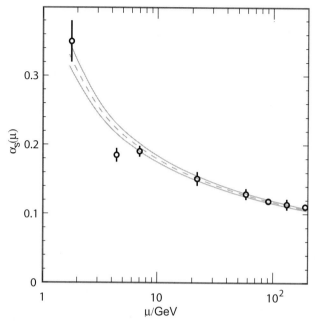

Fig. 9: Values of the strong coupling $\alpha_s = g_s^2/4\pi$ from a variety of determinations as a function of the scale μ at which they are measured, from S. Eidelman *et al.* [Particle Data Group], *Phys. Lett.* **B592**, 1 (2004). The predictions of QCD, with an overall scale determined from the data, are also shown.

with the hadronization of a $q\bar{q}$ pair as well as a gluon (gluon bremsstrahlung, analogous with electron bremsstrahlung). These results give fairly convincing evidence for the existence of gluons, and in particular establish their spin as 1. Finally, jets produced in hadronic processes, especially at high-energy proton–antiproton colliders at CERN and Fermilab, probe the strong interactions at extremely high Q^2 (e. g., $10^4\,\mathrm{GeV}^2$). The observations are all consistent with the QCD predictions of underlying hard quark and gluon-scattering processes.

One test of QCD is to extract the coupling g_s at various scales to see whether it decreases at higher energy scales as predicted. The results, shown in Fig. 9, are in spectacular agreement with the QCD expectations.

It is believed that at high temperatures and densities confinement would no longer be relevant and that a plasma of quarks and gluons should be possible. Presumably, quarks and gluons were unconfined in the very early Universe when the temperature exceeded $\sim 1\,\mathrm{GeV}$. Experiments at the high energy heavy-ion collider RHIC at the Brookhaven National Laboratory are attempting to recreate such a state. They show possible indications, but the signatures are not completely clear.

In view of these various successes, QCD is almost certainly the "correct" theory of the strong interactions, even though there has been no single "gold-plated" test. In fact, QCD is the only realistic candidate within the framework of renormalizable field theories.

Hadronic Weak Interactions, Current Algebra, and Heavy Quarks

The weak interactions, whether in the phenomenological local current–current form or in the Weinberg–Salam [$SU(2) \times U(1)$] form, need a specification of the hadronic currents, both vector and axial. The experimental evidence showed that the phenomenological form for the part of the current involving the proton and neutron was

$$\bar{n}(x)[(0.9744 \pm 0.0010)\gamma^{\alpha} - (1.227 \pm 0.004)\gamma^{\alpha}\gamma_5]\, p(x) \,. \tag{41}$$

This form bears a strong resemblance to the leptonic form in (16), and that, paradoxically, is surprising, since, in general, *form factors* due to strong-interaction corrections should appear. For the vector part, for example, the general expectation would be that for low momentum transfers from proton to neutron (as in beta decay)

$$\langle n|V_{\alpha}|p \rangle = G_V(q^2)\bar{n}\gamma_{\alpha}p \tag{42}$$

and the question is, Why should $G_V(0)$ be equal to unity? (The small deviation of the vector coefficient in (41) from unity is due to the Cabibbo angle factor discussed below.) R. P. Feynman and M. Gell-Mann, as well as S. Gershtein and Y. B. Zel'dovich, pointed out that if the vector current were conserved, and if $\int V_0(x)\, d^3x$ is normalized like the generator of a symmetry group, then the observed result would follow. Feynman and Gell-Mann identified the vector current with the current generating the isospin transformations. This satisfied the conditions that led to $G_V(0) = 1$, and it led to a model-independent characterization of the weak vector current. From this it was possible to predict unambiguously the rate for the decay $\pi^+ \to \pi^0 + e^+ + \nu_e$, and certain "magnetic" corrections to beta-decay spectra, all in excellent agreement with experiment.

In 1963, after the discovery of flavor $SU(3)$, N. Cabibbo generalized this assignment and proposed that the weak current, now also describing the weak interactions of the strange particles, has a form involving the $SU(3)$-generating currents. The next important step was to give a general characterization of the axial current. M. Gell-Mann proposed that there exists an additional global symmetry of the strong interactions, also of the $SU(3)$ type, but generated by eight pseudoscalar charges whose associated currents are axial currents, in fact, the weak axial currents. The difficulty that in the symmetric limit a pseudoscalar charge acting on a proton state, for example, yields another "odd" proton, for which there is no experimental evidence, was resolved by Y. Nambu, who made use of the Goldstone mechanism discussed earlier. The proposal was that in the symmetry limit the axial current is conserved, but that the solutions do not obey the symmetry. This leads to the prediction of massless pseudoscalar particles, which would approximately describe the pions and their $SU(3)$ partners. The symmetry generated by the $SU(3)$ generators F_i and the pseudoscalar F_{5i} is called a *chiral* [$SU(3) \times SU(3)$] symmetry, and through the study of weak interactions it has been established that it is a good symmetry, leading to a number of experimentally verified relations involving matrix elements of the weak hadronic currents. In particular, in 1965 S. Adler and W. Weisberger independently derived a relation between the coefficient $G_A(0)$ of the axial current in (41) in terms of other observables, such as the πp cross section. They obtained $G_A(0) \simeq 1.21$, in excellent agreement with experiment. The algebra $SU(3) \times SU(3)$ emerges quite naturally in a quark model. With minimal couplings, the currents

$$\bar{q}(x)\gamma^{\alpha}\lambda_i q(x) \,, \qquad \bar{q}(x)\gamma^{\alpha}\gamma_5\lambda_i q(x) \,, \tag{43}$$

where the λ_i are the $SU(3)$ generalizations of the $SU(2)$ Pauli matrices, are conserved in the limit that the quark masses vanish. The *current* quark masses, which are the masses appearing in the QCD Lagrangian (they may actually be generated in the electroweak sector by the same Higgs mechanism which yields the W and Z masses), break the symmetries associated with the axial currents explicitly and generate small masses for the π, K, and η. Quark mass differences break the vector symmetries and lead to multiplet mass differences. From these effects one can estimate the current masses given in Table 2. The u and d masses are extremely small. (The much larger *constituent* masses of order $300\,\mathrm{MeV}/c^2$ in the naive quark model are dynamical masses associated with the spontaneous breaking of chiral symmetry.) Experiments indicate $m_u \neq m_d$, implying a breaking of isospin in the strong interactions, which is however no larger than the (separate) electromagnetic breaking because of the small scale of the masses. The much larger m_s leads to a substantial breaking of $SU(3)$.

The quark model gives a simple expression for the electromagnetic current of the hadrons [Eq. (12)], with the coefficients determined by the quark charges. Similarly, the weak neutral current coupling to the Z is given in (27). The weak current that couples to the charged W is given (for three quarks) by

$$
\begin{aligned}
J_W^\alpha &= (\bar{u}\,\bar{d}\,\bar{s})Q_W\gamma^\alpha(1-\gamma_5)\begin{pmatrix} u \\ d \\ s \end{pmatrix} \\[2mm]
&= (\bar{u}\,\bar{d}\,\bar{s})\begin{pmatrix} 0 & 0 & 0 \\ \cos\theta_C & 0 & 0 \\ \sin\theta_C & 0 & 0 \end{pmatrix}\gamma^\alpha(1-\gamma_5)\begin{pmatrix} u \\ d \\ s \end{pmatrix} \\[2mm]
&= (\bar{d}\cos\theta_C + \bar{s}\sin\theta_C)\gamma^\alpha(1-\gamma_5)u\ .
\end{aligned}
\tag{44}
$$

This form leads to the Cabibbo theory, and the angle θ_C, the so-called Cabibbo angle, is of magnitude $0.22\,\mathrm{rad}$. Its origin lies in the difference in the ways in which the strong and weak interactions break $SU(3)$ symmetry. This is associated with the quark masses, which are generated by the Higgs mechanism in the standard model. Their values, as well as θ_C and the other *mixing angles* introduced later, are free parameters that are not understood at present.

The part of the weak current involving s in (44) has the property that the change in strangeness and the change in charge are equal ($\Delta S = \Delta Q$), in agreement with experiment. The Cabibbo theory explains a large number of strange particle decays with a universal choice of θ_C. It has one difficulty: If this current is incorporated into a gauge theory of the weak interactions (e. g., the $SU(2) \times U(1)$ model) in a manner analogous to the leptonic current, then the neutral intermediate W^0 vector meson couples naturally to a neutral current obtained by commuting Q_W with its adjoint,

$$
[Q_W, Q_W^\dagger] = \begin{pmatrix} -1 & 0 & 0 \\ 0 & \cos^2\theta_C & \sin\theta_C\cos\theta_C \\ 0 & \sin\theta_C\cos\theta_C & \sin^2\theta_C \end{pmatrix} .
\tag{45}
$$

Among the neutral currents there will be *strangeness-changing neutral currents* of the type $(\bar{d}s + \bar{s}d)\sin\theta_C\cos\theta_C$ (we ignore the γ matrices for brevity), which give rise to processes such as

$$K^+ \rightarrow \pi^+ + \nu + \bar{\nu} \,,$$
$$K_L^0 \rightarrow \mu^+ + \mu^-$$

$$(46)$$

at rates much larger than experimental limits or observations. Thus, a major modification is needed. The solution had actually been proposed before the Weinberg–Salam theory became popular. In 1970, S. Glashow, J. Iliopoulos, and L. Maiani, building on some earlier work of Glashow and J. D. Bjorken, proposed that the number of quark flavors be extended, with a fourth quark c carrying a new conserved quantum number called charm. The c quark is taken to have charge $\frac{2}{3}$, hypercharge $\frac{1}{3}$, and baryon number $B = \frac{1}{3}$. The weak current, constructed to have the form

$$(\bar{u}\bar{d}\bar{s}\bar{c}) Q_W \gamma^\alpha (1 - \gamma_5) \begin{pmatrix} u \\ d \\ s \\ c \end{pmatrix}$$

$$(47)$$

with

$$Q_W = \begin{pmatrix} 0 & 0 & 0 & 0 \\ \cos\theta_C & 0 & 0 & -\sin\theta_C \\ \sin\theta_C & 0 & 0 & \cos\theta_C \\ 0 & 0 & 0 & 0 \end{pmatrix},$$

$$(48)$$

treats the s quark symmetrically with the d. This implies that the neutral current constructed as in (45) does not have any strangeness-changing (or charm-changing) terms. The smallness of the Cabibbo angle implies that the dominant charged-current transitions are $u \rightarrow d$ and $c \rightarrow s$, which means that mesons involving c quarks are predicted to usually decay into final states with strangeness (e. g., a $\bar{K}$) rather than into nonstrange (e. g., pions) final states.

Notice that if we set $\theta_C = 0$ and replace (u, d, s, c) by $(e^-, \nu_e, \mu^-, \nu_\mu)$, we get the leptonic current. It was this analogy that originally led Bjorken and Glashow to generalize the $SU(3)$ flavor symmetry to the four-flavor $SU(4)$ symmetry. The analogy goes deep. It turns out that renormalizability demands that each lepton pair [e. g., (e, ν_e)] must have a compensating quark pair [e. g., (u, d)], so that the existence of the charmed quark is compelling in gauge theories. In the early 1970s it was also realized that the observed mass difference between the two neutral kaons could be accounted for by a calculable higher-order weak effect (Fig. 10) if the c quark existed and had a mass around $1.5\,\mathrm{GeV}/c^2$.

In 1974 in the experimental study of the ratio R in (40), B. Richter and collaborators at the Stanford Linear Accelerator Center, simultaneously with S. Ting and collaborators at Brookhaven National Laboratory, who were studying the reaction

$$p + Be \rightarrow e^+ + e^- + hadrons,$$

found an extremely sharp resonance at $3097\,\mathrm{MeV}$ and, soon after that, another at $3685\,\mathrm{MeV}$ (Fig. 8). It is interesting that these had been anticipated theoretically by T. Appelquist and D. Politzer, who advanced reasons why the $J = 1$ $c\bar{c}$ (charmed quark–antiquark) states should be quite long-lived. This interpretation of the $J/\psi(3097)$ and $\psi(3685)$ as $1\,^3S_1$ and $2\,^3S_1$

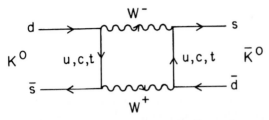

Fig. 10: Higher-order weak diagram leading to the mass difference between $K_s \sim K^0 + \bar{K}^0$ and $K_L \sim K^0 - \bar{K}^0$. Inclusion of the diagram with intermediate t quarks can also account for the observed *CP* violation.

charmonium states was confirmed by the later discovery of $^3P_0$, $^3P_1$, $^3P_2$, and $^2S_0$ states as well as additional $^3S_1$ radial excitations in the vicinity of the ones already discovered. There now exists a complete and well-studied charmonium spectroscopy. The arrangement of levels and the spacing cannot be understood in terms of a simple $1/r$ potential, as for positronium. Rather, a better fit is obtained with a potential of the form $V(r) = A/r + Br$, where the first term represents the short-range (asymptotically weakly coupled) contribution and the second term represents the long-range confining potential. The extreme narrowness of the $^3S_1$ resonances can be understood qualitatively by arguing that a $c\bar{c} \rightarrow u\bar{u}$ (say) transition can only take place through the mediation of three color-carrying gluons, and these couple weakly for large momentum transfers.

Numerous hadrons consisting of a single c quark bound to ordinary quarks or antiquarks have also been identified and their decay modes studied. These include the spin-0 $(c\bar{u})$ D meson, with mass $1865\,\mathrm{MeV}/c^2$, the corresponding spin-1 D^* meson of mass $2007\,\mathrm{MeV}/c^2$, their charged $(D^+ = c\bar{d})$ partners, and the D_s^+ $(c\bar{s})$ at $1968\,\mathrm{MeV}/c^2$, as well as charmed baryons with isospin 0, 1, and $\frac{1}{2}$ at 2285, 2455, and $2470\,\mathrm{MeV}/c^2$, respectively. The spectroscopy of the charmed hadrons, as well as additional evidence from deep inelastic neutrino scattering, have thus clearly established the existence of the c quark as needed for a consistent and realistic gauge theory of the weak interactions.

Similarly, a third (heavy) τ lepton was discovered at SLAC in 1976. The observed weak interactions of the τ left little doubt of the existence of a third (approximately) massless neutrino partner, but the interactions of the ν_τ were not observed directly until 2000 at Fermilab. Renormalizability then implied that another quark pair, which we label (t,b) for *top* and *bottom*, respectively, with charges $\frac{2}{3}$ and $-\frac{1}{3}$, should exist. This notion of lepton–quark symmetry was confirmed in 1977 by the discovery by L. Lederman and collaborators at the Fermi National Accelerator Laboratory of two new narrow resonances, the $\Upsilon(9460\,\mathrm{MeV}/c^2)$ and $\Upsilon'(10023\,\mathrm{MeV}/c^2)$, in the reaction

$$p + Be \rightarrow \Upsilon + hadrons \rightarrow \mu^+ + \mu^- + hadrons.$$

The interpretation of these states as $^3S_1$ states of a new *quarkonium* composed of $b\bar{b}$ quarks with charge $\frac{1}{3}$ was supported by their production in e^+e^- collisions at DESY and at the Cornell Electron Storage Ring (CESR), where additional $^3S_1$ as well as P wave states were also discovered. The bottom mesons $B^+ = u\bar{b}$ and $B^0 = d\bar{b}$, with masses $\sim 5279\,\mathrm{MeV}/c^2$, as well as their antiparticles and many other bottom mesons and baryons, have also been identified and studied in detail at a number of laboratories.

The observed weak interactions of the b quark as well as renormalizability and lepton–quark symmetry indicated that a top quark (t) should exist. Furthermore, by the early 1990's indirect arguments based on the consistency of weak neutral current and Z-pole data, which are sensitive to a heavy t quark mass through higher-order corrections, predicted a mass $m_t = 150 \pm 30\,\mathrm{GeV}/c^2$. The t was finally discovered at the Tevatron in 1994 in the expected range. It is too short-lived to form observable bound states. Rather, pairs of t and $\bar{t}$ are produced through the strong interactions, and the direct decays into b ($\bar{b}$) and additional quark jets or leptons are observed. The current indirect prediction $m_t = 172^{+10}_{-7}\,\mathrm{GeV}/c^2$ is in excellent agreement with the observed value of $178.0 \pm 4.3\,\mathrm{GeV}/c^2$, dramatically confirming the standard electroweak model. Why the t is so much heavier than the other fermions is not understood.

In the four-quark model, Q_W in (47) involves a single mixing angle $\theta_C \simeq 0.22$, and the higher-order diagram in Fig. 10 can account for the K_L–K_S mass difference. The generalization of Q_W to six quarks involves three mixing angles and one complex phase. In addition to θ_C, the dominant b quark transition $b \to c$ is described by a mixing angle $\theta_{bc} \simeq 0.04$ (determined from the relatively long b lifetime). The much weaker $b \to u$ transition is characterized by the angle $\theta_{bu} \simeq 0.004$. These angles suffice to account for the large mixing observed at DESY between the $B^0 = d\bar{b}$ and $\bar{B}^0 = d\bar{b}$ states, by a second-order diagram analogous to Fig. 10 (with the external s replaced by b). Large mixing between the $B_s^0 = s\bar{b}$ and $\bar{B}_s^0 = \bar{s}b$ is predicted.

Even after the discovery that the weak interactions violated parity (P) and charge conjugation (C) invariance, it was generally believed that CP invariance was maintained. In particular, the $CP = +1$ state $K_S \sim K^0 + \bar{K}^0$ should decay rapidly ($\tau \sim 0.89 \times 10^{-10}$ s) into $\pi^+\pi^-$ or $\pi^0\pi^0$, while the $CP = -1$ state $K_L \sim K^0 - \bar{K}^0$ should decay into 3π in $\sim 5.2 \times 10^{-8}$ s. However, in 1964 J. Cronin, V. Fitch, and collaborators working at Brookhaven observed the CP violating decays $K_L \to 2\pi$ with branching ratios of $\sim 10^{-3}$. The results could be accounted for by a small CP violating mixing between the K_L and K_S states. One possibility was that this mixing is generated by an entirely new *superweak* $\Delta S = 2$ interaction. In 1973, M. Kobayashi and M. Maskawa suggested that CP breaking could be generated by the higher-order diagram in Fig. 10 if there were three (or more) fermion families, implying observable phases in Q_W. The observed K_L–K_S mixing could thus be generated either by the three-family standard model or by the superweak model. However, in the 1990's experiments at CERN and Fermilab observed small differences in the CP violating $K_L \to \pi^0\pi^0$ and $\pi^+\pi^-$ rates (relative to K_S,) that required direct CP breaking in the decay amplitude, not just in the state mixing, strongly confirming the standard $SU(2) \times U(1)$ model interpretation.

Subsequently, there have been very detailed studies of CP-violating asymmetries, rare decays, and other aspects of B meson decays at a number of laboratories, especially the "B factories" at CESR, SLAC and KEK (Japan). All data is consistent with the standard model, in particular with the interpretation that the patterns of decays and CP-violation can be accounted for by the unitary Cabibbo–Kobayashi–Maskawa (CKM) quark mixing matrix Q_W. This is illustrated by the *unitarity triangle* shown in Fig. 11. The charged current sector will be tested even more stringently in the future following the (probable) measurement of $B_s^0\bar{B}_s^0$ mixing and more precise measurements of rare decays and CP-violating asymmetries in B meson decays.

Neutrino Mass and Oscillations

For many years all experimental searches for nonzero neutrino masses yielded negative results, implying that they are either massless or much lighter than the charged leptons and

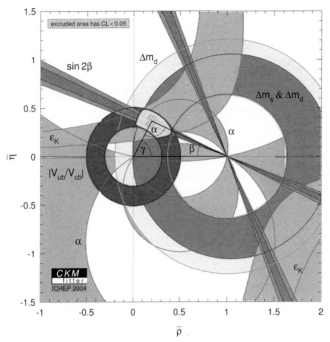

Fig. 11: The current status of the CKM matrix, from J. Charles *et al.*
[CKMfitter Group Collaboration], hep-ph/0406184 and
ckmfitter.in2p3.fr. $\bar{\rho}$ and $\bar{\eta}$ parametrize elements of the matrix. Various
observations from K and B decays determine the shaded bands. They are all
consistent with each other within the region surrounding the vertex (labeled by
angle α) of the unitarity triangle.

quarks. The original $SU(2) \times U(1)$ standard model was constructed to yield massless ν's, but
most theoretical ideas for more complete theories predicted that m_ν would be nonzero at some
level. The first experimental hints of nonzero mass were from the Homestake Solar Neutrino
Experiment in the late 1960's. R. Davis and collaborators observed events in a large underground chlorine experiment that were later confirmed to be induced by electron neutrinos (ν_e)
produced by nuclear reactions in the core of the Sun. However, they only observed about one
third of the rate predicted by J. Bahcall and collaborators based on the observed Solar luminosity and theoretical models of the Sun. This *Solar neutrino problem* was later confirmed by
other experiments using gallium and water-based detectors. One explanation was that the ν_e
were oscillating (converting) into other types of neutrino (ν_μ or ν_τ), to which the experiments
were insensitive, through effects associated with (tiny) neutrino masses and mixings (analogous to the mixings observed for the quarks). However, there was the alternative possibility
that the astrophysical uncertainties in the theoretical calculation had been underestimated. The
situation was resolved in 2002 when the Sudbury Neutrino Observatory (Canada) was able to
measure the ν_e and total fluxes independently using a heavy water detector, with the result that
the ν_e really were converting to ν_μ or ν_τ. This was later confirmed by the observation by the
KamLAND experiment (Japan) of the disappearance of $\bar{\nu}_e$'s produced in power reactors. Prior

to the Sudbury results, a different type of *atmospheric neutrino* oscillation, of ν_μ's produced by cosmic ray interactions in the atmosphere into (presumably) ν_τ's, was established by the Super-Kamiokande experiment (Japan).

There are three types of neutrinos, and *neutrino oscillations* presumably involve three mixing angles and one or more leptonic *CP*-violating phases, similar to the CKM quark mixing. However, a simplified description assumes that only two states are relevant for a given type of experiment. For example, suppose that the ν_e and ν_μ, the states which are associated with the e^- and μ^-, are superpositions of states $\nu_{1,2}$ of definite masses $m_{1,2}$,

$$\begin{aligned} \nu_e &= \nu_1\cos\theta + \nu_2\sin\theta\,, \\ \nu_\mu &= -\nu_1\sin\theta + \nu_2\cos\theta\,. \end{aligned} \tag{49}$$

Then if an initially produced ν_e has energy E large compared to $m_{1,2}$ the mixing effect may cause it to transform into ν_μ after traveling a distance L, with probability

$$P(\nu_e \to \nu_\mu; L) = \sin^2 2\theta \sin^2 \frac{1.27\Delta m^2 L}{E}\,, \tag{50}$$

where $\Delta m^2 = m_2^2 - m_1^2$ is expressed in eV/c^2, L in m, and E in MeV. The observed Solar and atmospheric oscillation parameters (which correspond to different Δm^2 and θ) are shown in Fig. 12. Note that the mass scales are tiny compared to the other fermions, but that the mixing angles are large, unlike the small quark mixings (there is a third neutrino mixing angle that is consistent with zero).

The neutrino masses and mixings are interesting because they are so different, and they may be an indication of new physics underlying the standard model at much shorter distance scales (i. e., much larger mass scales). For example, there are various *seesaw* models which predict very small neutrino masses $m_\nu \sim m_D^2/M \ll m_D$, where m_D is comparable to the quark and charged lepton masses and $M \gg m_D$ is the new physics scale (e. g., 10^{14} GeV/c^2). Another interesting possibility is that the neutrinos are *Majorana*, which means that the mass effects can convert neutrinos into antineutrinos so that the total *lepton number* (number of leptons minus the number of antileptons) is not conserved. This will be probed in *neutrinoless double beta decay* ($\beta\beta_{0\nu}$) experiments searching for rare nuclear decays $(Z,N) \to (Z+2, N-2) + e^- + e^-$, (i. e., without the two antineutrinos expected from two successive β decays). Another outstanding issue is the absolute neutrino mass scale, since oscillations only probe mass-square differences. There is an upper limit of $\sim 0.2\,\mathrm{eV}/c^2$ from cosmology (otherwise, relic neutrinos left over from the big bang would modify the formation of structure in the Universe). This will be refined in the future, and $\beta\beta_{0\nu}$ and kinematic effects in ordinary β decay may also be important. More detailed information on the spectrum, mixings, and possible leptonic *CP* violation are also anticipated, as is clarification of possible indications of additional types of oscillations that cannot be accommodated in the three-neutrino framework.

Problems with the Standard Model

The standard model (QCD plus the Weinberg–Salam electroweak model and general relativity) has been spectacularly successful. Although the elementary Higgs mechanism for symmetry breaking has not yet been tested and may possibly be replaced by a dynamical mechanism, the basic structure of the standard model is almost certainly correct at some level. However,

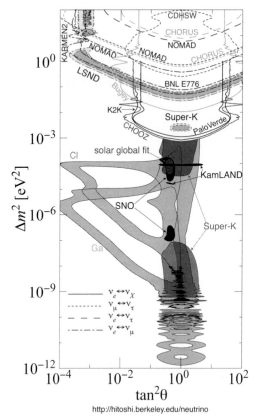

Fig. 12: Current status of neutrino oscillations, from H. Murayama, `hitoshi.berkeley.edu/neutrino`. The regions for $\Delta m^2 \sim 2 \times 10^{-3}\,\mathrm{eV}^2$ and $\tan^2\theta \sim 1$, and for $\Delta m^2 \sim 8 \times 10^{-5}\,\mathrm{eV}^2$ and $\tan^2\theta \sim 0.4$ are indicated by the atmospheric and Solar neutrino oscillations, respectively. (The two regions represent different combinations of neutrino states.)

it contains far too much arbitrariness to be the final story of Nature. One way of seeing this is that the minimal version has 27 free parameters (29 for Majorana neutrinos), *not* including electric charges. In addition, the standard model suffers from:

(a) *The Gauge Problem:* The standard model gauge group is a complicated direct product of three factors with three independent coupling constants. Charge quantization, which refers to the fact that the magnitudes of the proton and of the electron electric charges are the same, is not explained.

(b) *Fermion Problem:* The standard model involves a very complicated reducible representation for the fermions. Ordinary matter can be constructed out of the fermions of the first family (ν_e, e^-, u, d). We have no fundamental understanding of why the additional families (ν_μ, μ^-, c, s), (ν_τ, τ, t, b), which appear to be identical with the first except that they are heavier, exist. In addition, the standard model does not explain or predict the pattern of fermion masses, which are observed to vary over many orders of magnitude, or mixings. Also, the *CP* violation associated with quark mixing is not sufficient to explain the generation of the excess of matter compared to antimatter in the universe (the *baryon asymmetry*), therefore requiring the existence of some additional form of *CP* violation.

(c) *Higgs/Hierarchy Problem:* The spontaneous breakdown of the $SU(2) \times U(1)$ symmetry in the standard model is accomplished by the introduction of a Higgs field. Consistency requires that the mass of the Higgs boson be not too much different from the weak interaction scale; that is, it should be equal to the W mass within one or two orders of magnitude. However, there are higher-order corrections which change (renormalize) the value of the square of the Higgs mass by $\delta m_H^2 \sim m_P^2$, where $m_P = (G_N/\hbar c)^{-1/2} \simeq 10^{19}\,\text{GeV}/c^2$ is the *Planck* (gravity) scale. Therefore, $\delta m_H^2/M_W^2 \geq 10^{34}$, and the bare value of m_H^2 in the original Lagrangian must be adjusted or fine-tuned to 34 decimal places. Such a fine-tuning is possible, but extremely unattractive.

(d) *Strong CP Problem:* It is possible to add an additional term, characterized by a dimensionless parameter θ, to the QCD Lagrangian which breaks *P*, *T*, and *CP* invariance. Limits on the electric dipole moment of the neutron require θ to be less than 10^{-9}. However, weak interaction corrections change or renormalize the lowest-order value of θ by about 10^{-3} – that is, 10^6 times more than the total value is allowed to be. Again, one must fine-tune the bare value against the correction to a high degree of precision.

(e) *Graviton Problem:* The graviton problem has several aspects. First, gravity is not unified with the other interactions in a fundamental way. Second, even though general relativity can be incorporated into the model by hand, we have no idea how to achieve a mathematically consistent theory of quantum gravity: attempts to quantize gravity within the standard model framework lead to horrible divergences and a nonrenormalizable theory. Finally, there is yet another fine-tuning problem associated with the cosmological constant. The vacuum energy density $\langle V \rangle$ associated with the spontaneous symmetry breaking of the $SU(2) \times U(1)$ model generates an effective renormalization of the cosmological constant $\delta \Lambda = 8\pi G_n \langle V \rangle$, which is about 50 orders of magnitude larger than the observed limit (or value, if the observed *dark energy* or *acceleration* of the universe is due to a cosmological constant). One must fine-tune the bare cosmological constant against the correction to this incredible degree of precision.

Extensions of the Standard Model

There must almost certainly be new physics beyond the standard model. Some of the possible types of new physics that have been discussed extensively in recent years are shown in Table 6. Additional gauge bosons, fermions, or Higgs bosons do not by themselves solve any problems, but may exist at accessible energies as remnants of underlying physics such as unified gauge groups or superstrings. The Z-pole data from CERN and SLAC, as well as the experimental evidence from the cosmological abundance of helium (interpreted within the framework of the standard "big-bang" cosmological model), implies that there are no additional fermion families beyond the three already known, unless the associated neutrinos are very heavy (greater than $\simeq 45\,\text{GeV}/c^2$). More likely are *exotic* heavy fermions, predicted by many theories, which are not simply repetitions of the known families. New gauge interactions, especially heavy electrically neutral Z' bosons analogous to the Z in (26) are predicted by many extensions, as are extended Higgs sectors.

Table 6: Some possible extensions of the standard model.

Model	Typical scale (GeV)	Motivation
New Ws, Zs, fermions, Higgs	10^2–10^{19}	Remnant of something else
Family symmetry	10^2–10^{19}	Fermion (No compelling models)
Composite fermions	10^2–10^{19}	Fermion (No compelling models)
Composite Higgs	10^3–10^4	Higgs (No compelling models)
Composite W, Z (G, γ?)	10^3–10^4	Higgs (No compelling models)
Little Higgs	10^3–10^4	Higgs
Large extra dimensions ($d > 4$)	10^3–10^6	Higgs, graviton
New global symmetry	10^8–10^{12}	Strong CP
Kaluza–Klein Higgs (0) $\Leftrightarrow$ gauge (1) $\Leftrightarrow$ Graviton (2) ($d > 4$)	10^{19}	Graviton
Grand unification Strong $\Leftrightarrow$ electroweak	10^{14}–10^{19}	Gauge
Supersymmetry/supergravity Fermion $\Leftrightarrow$ boson	10^2–10^4	Higgs, graviton
Superstrings Strong $\Leftrightarrow$ electroweak $\Leftrightarrow$ gravity Fermion $\Leftrightarrow$ boson ($d > 4$)	10^{19}	All problems!?

Family symmetries are new global or gauge symmetries which relate the fermion families. Another approach to understanding the fermion spectrum is to assume that the quarks and leptons are bound states (*composites*) of still smaller constituents. However, experimental searches for substructure suggest that the underlying particles must be extremely massive ($> 1000\,\mathrm{GeV}/c^2$). Hence, unlike all previous layers of matter, extremely strong binding would be required. Neither of these ideas has led to a particularly attractive model.

The fine-tuning problem associated with the Higgs mechanism could be solved if the elementary Higgs fields were replaced by some sort of bound-state mechanism. However, models which generate fermion as well as W, Z masses are complicated and tend to predict certain unobserved decays at too rapid a rate and lead to unacceptably large corrections to precision electroweak data. *Little Higgs* models are a variant in which the elementary Higgs is replaced by the approximately massless Goldstone boson of a new global symmetry, which ensures cancelation between different higher-order corrections to the square of the Higgs mass. Composite W and Z bosons could be an alternative to spontaneous symmetry breaking, but such theories abandon most of the advantages of gauge theories. The strong *CP* problem could be resolved by the addition of a new global symmetry which ensures that θ is a dynamical variable which is zero in the lowest energy state. Such models imply a weakly coupled Goldstone boson (the *axion*) associated with the symmetry breaking. Constraints from astrophysics and cosmology limit the symmetry breaking scale to the range 10^8–10^{12} GeV.

Kaluza–Klein theories are gravity theories in $d > 4$ dimensions of space and time. It is assumed that four dimensions remain flat while the other $d - 4$ are *compactified* or curled up into a compact manifold with a typical radius of $\hbar/m_P c \sim 10^{-33}$ cm. Gravitational interactions associated with these unobserved compact dimensions would appear as effective gauge interactions and Higgs particles in our four-dimensional world. Although such ideas are extremely attractive for unifying the interactions they have great difficulty in incorporating parity violation and also in achieving a stable configuration for the compact manifold. Many of the aspects have reemerged more successfully within the framework of *superstrings*.

Similarly, theories of *large extra dimensions* postulate the existence of one or more extra dimensions of space that could be as large as a fraction of a mm! There are many variants on these ideas, but one promising version assumes that *gravitons* (the spin-2 quanta of the gravitational field) can propagate freely in the *bulk* of the extra dimensions, while the ordinary particles are somehow confined to the ordinary 3-dimensional space (the *brane*). This eliminates the Higgs/hierarchy problem because the fundamental (largest) mass scale of nature m_f can be far smaller (e. g., 10^5 GeV/c^2) than the Planck scale $m_P \sim 10^{19}$ GeV/c^2, with the apparent weakness of gravity due to the fact that the $1/r^2$ force law is modified for distances smaller than the size of the extra dimension. Such ideas may emerge as a limiting case of superstring theories (which involve extra dimensions), but introduce a new hierarchy problem, viz., why is the large dimension much larger than the expected $\hbar/m_f c$?

Much work has been devoted to *grand unified gauge theories* (GUTs), in which it is proposed that at very high energies (or very short distances) the underlying symmetry is a single gauge group that contains as its subgroups both $SU(3)$ for color and $SU(2) \times U(1)$ for the weak and electromagnetic interactions. The larger symmetry is manifest above a unification scale M_X at which it is spontaneously broken. At lower energies, the symmetry is hidden and the strong and electroweak interactions appear different. The unification scale can be estimated from the observed low-energy coupling constants. The energy-dependent couplings are expected to meet at M_X, as in Fig. 13. Since they vary only logarithmically, M_X is predicted to be extremely large. Two versions of GUTs are typically considered, one in which only the standard model particles exist below M_X and contribute to the variation of the couplings, and one in which a new class of *superpartners* and an extra Higgs multiplet expected in *supersymmetry* (discussed below) are included. The predicted values of M_X are typically around 10^{14} GeV/c^2 and 3×10^{16} GeV/c^2 for these two cases, respectively. (These heady speculations rest on the assumption that nothing fundamentally different occurs in a leap in energy scale from 1 to M_X. This is extremely unlikely, but then who would have thought that Maxwell's equations are good over distances ranging from astronomical down to 10^{-15} cm?) From the strong and electromagnetic couplings one can predict the weak angle $\sin^2 \theta_W$. The simplest non-supersymmetric GUTs predict a value ~ 0.20, well below the present precisely known value, while the supersymmetric ones are in reasonable agreement with experiment, especially when uncertainties from mass differences at low energy and M_X are taken into account.

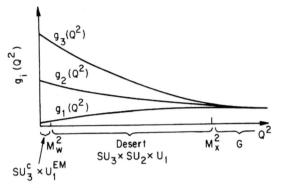

Fig. 13: The momentum dependent normalized coupling constants $g_3 \equiv g_s$, $g_2 \equiv g$, and $g_1 \equiv \sqrt{5/3} g'$.

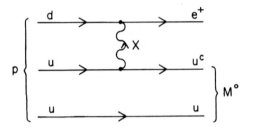

Fig. 14: A typical proton decay diagram.

In addition to unifying the interactions, the additional symmetries within a grand unified theory relate quarks, antiquarks, leptons, and antileptons. For example, the simplest grand unified theory – the 1974 $SU(5)$ model of H. Georgi and S. Glashow – assigns the left-helicity (i. e., spin oriented opposite to momentum) fermions of the first family to a reducible 5*+ 10-dimensional representation:

$$
W^{\pm} \updownarrow \quad
\begin{pmatrix} \nu_e \\ \bar{d} \\ e^- \end{pmatrix}
\overset{5^*}{\underset{\leftarrow X,Y \rightarrow}{}}
\quad
\begin{pmatrix} & u & \\ e^+ & & \bar{u} \\ & d & \end{pmatrix}
\overset{10}{\underset{\leftrightarrow X,Y \leftrightarrow}{}}
\tag{51}
$$

The neutrino and electron can be rotated into the anti-down quark by the new symmetry generators, as can the positron and the up, down, and anti-up quarks. Because of this relation between the different types of fermions, the grand unified theories naturally explain charge quantization. In addition, there are new gauge bosons associated with these extra symmetries. The $SU(5)$ model contains new bosons which carry both color and electric charge, known as the X and Y bosons, with masses $M_X \sim M_Y \sim 10^{14}\,\mathrm{GeV}/c^2$ ($3 \times 10^{16}\,\mathrm{GeV}/c^2$) for the non-supersymmetric (supersymmetric) case. These can mediate proton decay (and also the decay of otherwise stable bound neutrons). A diagram for $p \rightarrow e^+ \pi^0$ is shown in Fig. 14. Motivated by such predictions, a number of experiments have searched for proton decay. No events have been observed, and one finds $\tau(p \rightarrow e^+ \pi^0) > 5 \times 10^{33}\,\mathrm{yr}$, in conflict with the prediction $4 \times 10^{29\pm2}$ yr of the simplest non-supersymmetric GUTs. The predicted lifetime is much longer in the supersymmetric case because of the larger M_X. However, there are other diagrams involving the superpartners in those models that lead to such modes as $p \to K^+\bar{\nu}$, with typical lifetimes around 10^{33} yr. The simplest such models are excluded by the corresponding experimental bounds, but some extended or nonminimal models are still viable.

One of the mysteries of nature is the *cosmological baryon asymmetry*. It is an observational fact that our part of the Universe consists of matter and not antimatter: there is approximately one baryon for every 10^{10} microwave photons, but essentially no antibaryons. This could not be due to initial conditions if the Universe underwent a period of extremely rapid expansion (*inflation*), which many cosmologists believe is the most likely explanation of the almost exact flatness and homogeneity of the Universe. At one time it was believed that in GUTs this asymmetry could be generated dynamically in the first instant (the first 10^{-35} s or so) after the big bang, by baryon number violating interactions related to those which lead to proton decay. This now appears unlikely because the simplest GUTs predict that equal baryon and lepton asymmetries would have been generated, and these would later have been erased due to nonperturbative baryon and lepton number violating processes associated with the electroweak theory prior to the *electroweak phase transition*, i. e., when the temperature was higher than the electroweak scale. (A subsequent period of inflation would also have wiped out the asymmetry.) However, alternative ideas have emerged for dynamically generating the asymmetry. These include the possibility of first generating the lepton asymmetry by heavy particle decays in seesaw models of Majorana neutrino mass (*leptogenesis*), with some of the lepton asymmetry converted to a baryon asymmetry by the nonperturbative effects; or the possibility of actually generating the asymmetry during the electroweak transition, especially in extensions of the supersymmetric standard model.

Grand unified theories predict the existence of superheavy *magnetic monopole* states with masses $M_M \sim M_X/\alpha_G \sim 10^{16} - 10^{18} \, \text{GeV}/c^2$. These may have been produced during phase transitions during the early Universe. If they were still left over as relics today they would contribute far more mass to the present Universe than is allowed by observations. However, they would have been diluted to a negligible density if the Universe underwent a period of inflation.

Grand unified theories offer little or no help with the fermion, Higgs, strong *CP*, or graviton problems, and the simplest non-supersymmetric versions are ruled out by $\sin^2\theta_W$ and proton decay. GUTs are probably not ambitious enough, but it is likely that some ingredients (e. g., related to charge quantization) will survive. It is also possible that there is an underlying grand unification in a theory with extra space dimensions, such as a superstring theory, but the full GUT symmetry is not manifest in the three large dimensions of space.

Supersymmetry (*see* Supersymmetry and Supergravity) is a new kind of symmetry in which fermions are related to bosons. Realistic models are complicated in that they require more than a doubling of the number of fundamental particles. For example, one must introduce a spin-0 superpartner of each known fermion, a spin-$\frac{1}{2}$ partner of each gauge boson or Higgs field, etc. The primary motivation is the Higgs problem: higher-order corrections associated with the new particles cancel the unacceptable renormalization of the Higgs mass that occurs in the standard model. If supersymmetry were exact, the new particles would be degenerate with the ordinary particles. However, there exist experimental lower limits of order $100 - 200 \, \text{GeV}/c^2$ on the masses of most superpartners, so that supersymmetry (if it exists) must be broken. Some models assume that supersymmetry is broken at some very large scale (e. g., 10^{11} GeV) in a sector which is only coupled very weakly (by gravitational-strength interactions) to the ordinary particles and their partners, implying superpartner masses in the range 10^2–10^4 GeV$/c^2$. (Larger masses would fail to solve the Higgs problem.) Variant versions break the supersymmetry at a lower scale (e. g., 10^5 GeV) in a sector that is somewhat more strongly coupled (by gauge interactions) to the ordinary sector, again leading to 10^2–10^4 GeV$/c^2$ superpartner masses. In both cases, the superpartner mass scale actually determines the scale of electroweak symmetry breaking.

Many supersymmetric theories involve a conserved *R-parity*, which distinguishes the ordinary particles from their superpartners. This implies that the lightest superpartner (the LSP) is stable. If it is electrically neutral, such as the spin-$\frac{1}{2}$ partner of a neutral gauge or Higgs field, it would be an excellent candidate for the *cold dark matter* of the Universe. (An alternate possibility if *R-parity* is not conserved would be the axion described above. Massive neutrinos in the eV$/c^2$ range are *not* good candidates because they are too light, and would prevent mass from clustering on small scales.) As described in the GUT section, supersymmetric grand unified theories lead to larger values for $\sin^2 \theta_W$ (in better agreement with experiment), and a larger unification scale, reducing but not necessarily solving the problem that proton decay has not been observed.

Just as the promotion of an ordinary internal symmetry to a gauge symmetry implies the existence of spin-1 gauge bosons, the requirement that supersymmetry transformations can be performed independently at different space-time points implies the existence of spin-2 gravitons; i. e., gauged supersymmetry automatically unifies gravity. However, *supergravity* theories do not solve the problems of quantum gravity. Higher-order corrections are still divergent and nonrenormalizable. There is no experimental evidence for supersymmetry, but if it exists at the scales suggested by the Higgs problem it should be observed in the near future at the LHC *pp* collider being constructed at CERN. Whether Nature chooses to make use of supersymmetry remains to be seen.

Superstrings are a very exciting development which may yield finite theories of all interactions with no free parameters. Superstring theories introduce new structure around the Planck scale. The basic idea is that instead of working with (zero-dimensional) point particles as the basic quantities, one considers one-dimensional objects known as strings, which may be open (e. g., ending on membrane-like objects known as *branes*) or closed. The quantized vibrations of these strings lead to an infinite set of states, the spectrum of which is controlled by the string tension, given by the square of the Planck scale $\tau \sim m_p^2$. When one probes or observes a string at energies much less than the Planck scale, one sees only the "massless" modes, and these represent ordinary particles. The physical size of the string is given by the inverse of the Planck scale $\simeq 10^{-33}$ cm, and at larger scales a string looks like a point particle. There are actually 5 types of consistent string theories, but it is now believed that these (and one 11-dimensional supergravity theory) are limiting cases of an even more fundamental (and mysterious) *M theory*.

The mathematical consistency of superstring theories requires (in most versions) that there are nine dimensions of space and one of time. Presumably, the extra six space dimensions are curled into a compact manifold of radius $\sim 10^{-33}$ cm, reminiscent of Kaluza–Klein theories. (It is possible that one or more or the dimensions have much larger sizes, leading to a realization of the large extra dimension theories described earlier.) In the most realistic closed string case the interactions are due to gauge interactions in the ten-dimensional space. The absence of mathematical pathologies requires an essentially unique gauge group called $E_8 \times E_8$. At energy scales less than the Planck scale an effective particle field theory in four dimensions emerges, with a supersymmetric gauge symmetry based on a subgroup of $E_8 \times E_8$. The effective group, the number of fermions, and their masses, mixings, etc., are all determined by the way in which the extra dimensions are compactified. In the open string theories the gauge and other interactions are determined by the configurations of the branes in the nine space dimensions.

In principle, superstring theories have no arbitrary parameters or other features, and most likely they yield completely finite (not just renormalizable) quantum theories of gravity and all the other interactions. That is, they are candidates for the ultimate "theory of everything." However, there are enormous numbers of possible ways in which the extra dimensions can compactify, and at present we do not possess the principles or mathematical tools to determine which is chosen. It is therefore not clear what the predictions of superstring theories really are (e. g., whether they lead to an effective supergravity GUT) or whether they correspond to the real world. In fact, there have been recent speculations that there is no selection principle, and that the choice of possible compactifications is essentially random out of an enormously large *landscape* of possible vacua. There may be different physics occurring in different regions of a very large chaotic Universe. In that case it is possible that supersymmetry is broken at a high scale, and the Higgs hierarchy problem is apparently solved by a fine-tuning, perhaps related to *anthropic* selection principles, i. e., that life could only develop in the subset of worlds for which the "fine-tuning" occurred.

The Future

The standard model is extremely successful, but there is almost certainly new physics underlying it. There are many theoretical ideas concerning this new physics. Some of the most promising involve energy scales much larger than will ever be probed by direct experimentation, though they may still lead to testable low-energy predictions. Apart from theoretical ideas and new computational techniques (e. g., lattice calculations or efficient ways to calculate Feynman diagrams in QCD emerging from string techniques), many types of experiments will improve the tests of the standard model and search for manifestations of new physics. These include direct searches for new particles at high-energy accelerators, especially the high energy pp LHC collider under construction at CERN, which will have enormous discovery reach, and a proposed e^+e^- International Linear Collider (ILC), which would be complementary to the LHC by being able to carry out more precise measurements. There are also plans for a new generation of neutrino experiments to elucidate the details of the neutrino masses and mixings, and possibilities for searches for rare decays or interactions of muons, kaons, B mesons, etc., that are forbidden or strongly suppressed in the standard model. Other probes include future high precision electroweak measurements and searches for electric dipole moments, magnetic monopoles, and proton decay. Finally, there has been an increasingly close connection between particle physics and cosmology. The dynamics of the early Universe was controlled by the elementary particles and their interactions, and refined studies of the large-scale structure of the Universe and of other relics from the big bang place severe constraints on new physics.

It is impossible to do justice to the subtlety and richness of this field in a brief survey, or to give their due to the thousands of researchers who have painstakingly uncovered the beautiful structure that is emerging. The interplay of imaginative experimentation and daring conjectures has been a source of wonder to all who have witnessed the growth and maturing of elementary-particle physics.

See also: Baryons; Compton Effect; Conservation Laws; Cosmology; Currents in Particle Theory; Electron; Feynman Diagrams; Field Theory, Axiomatic; Gauge Theories; Grand Unified Theories; Gravitation; Hadrons; Hypernuclear Physics and Hypernuclear Interactions; Hyperons; Inclusive Reactions; Isospin; Lattice Gauge Theory; Leptons; Lie Groups; Mesons; Muonium; Neutrino; Nucleon; Partons; Positron–Electron Colliding Beams; Positronium; Quantum Electrodynamics; Quantum Field Theory; Quarks; Regge Poles; Renormalization; *S*-Matrix Theory; Solar Neutrinos; Symmetry Breaking, Spontaneous; String Theory; Strong Interactions; *SU(3)* and Higher Symmetries; Supersymmetry and Supergravity; Weak Interactions; Weak Neutral Currents.

Bibliography

Space limitations preclude an exhaustive bibliography. At best we can provide references to some standard textbooks, monographs, readily available review articles, and popular articles covering some recent developments.

Textbooks
I. Aitchison and A. Hey, *Gauge Theories in Particle Physics: A Practical Introduction*. IOP, Bristol, 2003.
D. Bailin and A. Love, *Supersymmetric Gauge Field Theory and String Theory*. IOP, Bristol, 1994.
T. P. Cheng and L.-F. Li, *Gauge Theory of Elementary Particle Physics*. Clarendon Press, Oxford, 1984.
J. Donoghue, E. Golowich, and B. Holstein, *Dynamics of the Standard Model*. Cambridge University Press, 1992.
S. Gasiorowicz, *Elementary Particle Physics*. Wiley, New York, 1966.
F. Halzen and A. Martin, *Quarks and Leptons: An Introductory Course in Modern Particle Physics*. Wiley, New York, 1984.
D. H. Perkins, *Introduction to High Energy Physics*. Cambridge University Press, 2000.
M. E. Peskin and D. V. Schroeder, *An Introduction to Quantum Field Theory*. Addison-Wesley, 1995.
S. Pokorski, *Gauge Field Theories*. Cambridge University Press, 2000.
P. Ramond, *Field Theory: a Modern Primer*. Addison-Wesley, 1989.
P. Renton, *Electroweak Interactions: an Introduction to the Physics of Quarks and Leptons*. Cambridge University Press, 1990.
L. Ryder, *Quantum Field Theory*. Cambridge University Press, 1996.
S. Weinberg, *The Quantum Theory of Fields*. Cambridge University Press, 1995.

Monographs
V. Barger and R. Phillips, *Collider Physics*. Addison-Wesley, 1996.
I. Bigi and A. Sanda, *CP Violation*. Cambridge University Press, 2000.
G. Branco, L. Lavoura, and J. Silva, *CP Violation*. Oxford University Press, 1999.
S. Coleman, *Aspects of Symmetry*. Cambridge University Press, New York, 1985.
E. D. Commins and P. H. Bucksbaum, *Weak Interactions of Leptons and Quarks*. Cambridge University Press, New York, 1983.

M. Cvetič and P. Langacker, ed., *Testing the Standard Model*. World Scientific, 1991.

H. Georgi, *Lie Algebras in Particle Physics*. Perseus Books, 1999.

M. B. Green, J. H. Schwarz, and E. Witten, *Superstring Theory*. Cambridge UP, New York, 1987.

J. Gunion, H. Haber, G. Kane, and S. Dawson, *The Higgs Hunter's Guide*. Addison-Wesley, 1990.

G. Kane, ed., *Perspectives on Higgs Physics II*. World Scientific, 1997.

G. Kane, ed., *Perspectives on Supersymmetry*. World Scientific, 1998.

B. Kayser, *The Physics of Massive Neutrinos*. World Scientific, 1989.

T. Kinoshita, ed., *Quantum Electrodynamics*. World Scientific, 1990.

K. Kleinknecht, *Uncovering CP Violation*. Springer-Verlag, 2003.

E. Kolb and M. Turner, *The Early Universe*. Addison-Wesley, 1990.

P. Langacker, ed., *Precision Tests of the Standard Electroweak Model*. World Scientific, 1995.

A. Linde, *Particle Physics and Inflationary Cosmology*. Harwood Academic, 1990.

R. N. Mohapatra, *Unification and Supersymmetry*. Springer-Verlag, 2003.

R. N. Mohapatra and P. B. Pal, *Massive Neutrinos in Physics and Astrophysics*. World Scientific, 1998.

J. Polchinski, *String Theory*. Cambridge University Press, 1998.

N. Polonsky, *Supersymmetry: Structure and Phenomena: Extensions of the Standard Model*. Springer-Verlag, 2001.

C. Quigg, *Gauge Theories of the Strong, Weak, and Electromagnetic Interactions*. Benjamin, Reading, MA, 1983.

G. C. Ross, *Grand Unified Theories*. Benjamin, Menlo Park, CA, 1985.

S. Weinberg, *The First Three Minutes, A Modern View of the Origin of the Universe*. Basic, New York, 1977.

S. Weinberg, *The Discovery of Subatomic Particles*. Cambridge University Press, 2003.

J. Wess and J. Bagger, *Supersymmetry and Supergravity*. Princeton University Press, 1992

Advanced Review Articles

E. S. Abers and B. W. Lee, "Gauge Theories," *Phys. Rep.* **9**C, 1 (1973).

G. Altarelli, "A QCD primer," `arXiv:hep-ph/0204179`.

U. Amaldi *et al.*, "A Comprehensive Analysis of Data Pertaining to the Weak Neutral Current and the Intermediate Vector Boson Masses," *Phys. Rev. D* **36**, 1385 (1987).

G. S. Bali, "QCD forces and heavy quark bound states," *Phys. Rept.* **343**, 1 (2001).

M. A. B. Bég and A. Sirlin, "Gauge Theories of Weak Interactions," *Annu. Rev. Nucl. Sci.* **24**, 379 (1974); *Phys. Rep.* **88**, 1 (1982).

J. Bernstein, "Spontaneous Symmetry Breaking, Gauge Theories, the Higgs Mechanism and All That," *Rev. Mod. Phys.* **46**, 7 (1974).

S. Bethke, "QCD studies at LEP," *Phys. Rept.* **403–404**, 203 (2004).

R. Blumenhagen, M. Cvetič, P. Langacker and G. Shiu, "Toward realistic intersecting D-brane models," `arXiv:hep-th/0502005`.

D. J. H. Chung, L. L. Everett, G. L. Kane, S. F. King, J. Lykken and L. T. Wang, "The soft supersymmetry-breaking Lagrangian: Theory and applications," *Phys. Rept.* **407**, 1 (2005).

K. R. Dienes, "String Theory and the Path to Unification: A Review of Recent Developments," *Phys. Rept.* **287**, 447 (1997).

A. D. Dolgov, "Neutrinos in cosmology," *Phys. Rept.* **370**, 333 (2002).

M. K. Gaillard, B. W. Lee, and J. L. Rosner, "Search for Charm," *Rev. Mod. Phys.* **47**, 277 (1975).

J. Gasser and H. Leutwyler, "Quark Masses," *Phys. Rep.* **87**, 77 (1982).

J. S. M. Ginges and V. V. Flambaum, "Violations of fundamental symmetries in atoms and tests of unification theories of elementary particles," *Phys. Rept.* **397**, 63 (2004).

G. F. Giudice and R. Rattazzi, "Theories with gauge-mediated supersymmetry breaking," *Phys. Rept.* **322**, 419 (1999).

M. C. Gonzalez-Garcia and Y. Nir, "Developments in neutrino physics," *Rev. Mod. Phys.* **75**, 345 (2003).

M. W. Grunewald, "Experimental tests of the electroweak standard model at high energies," *Phys. Rept.* **322**, 125 (1999) .

H. E. Haber and G. L. Kane, "The Search for Supersymmetry," *Phys. Rep.* **117**, 75 (1985).

J. L. Hewett and T. G. Rizzo, "Low-Energy Phenomenology Of Superstring Inspired $E(6)$ Models," *Phys. Rept.* **183**, 193 (1989).

J. Hewett and M. Spiropulu, "Particle physics probes of extra spacetime dimensions," *Ann. Rev. Nucl. Part. Sci.* **52**, 397 (2002).

C. T. Hill and E. H. Simmons, "Strong dynamics and electroweak symmetry breaking," *Phys. Rept.* **381**, 235 (2003).

V. W. Hughes and T. Kinoshita, "Anomalous g values of the electron and muon," *Rev. Mod. Phys.* **71** (1999) S133.

G. Jungman, M. Kamionkowski and K. Griest, "Supersymmetric dark matter," *Phys. Rept.* **267**, 195 (1996).

S. F. King, "Neutrino mass models," *Rept. Prog. Phys.* **67**, 107 (2004).

P. Langacker, "Grand Unified Theories and Proton Decay," *Phys. Rep.* **72**, 185 (1981); *Comm. Nucl. Part. Phys.* **15**, 41 (1985).

P. Langacker, M. X. Luo and A. K. Mann, "High precision electroweak experiments: A Global search for new physics beyond the standard model," *Rev. Mod. Phys.* **64**, 87 (1992).

P. Langacker, "Structure of the standard model," `arXiv:hep-ph/0304186`.

A. Leike, "The phenomenology of extra neutral gauge bosons," *Phys. Rept.* **317**, 143 (1999).

W. Marciano and H. Pagels, "Quantum Chromodynamics," *Phys. Rep.* **36**C, 137 (1978).

R. N. Mohapatra *et al.*, "Theory of neutrinos," arXiv:hep-ph/0412099.

M. Neubert, "Heavy quark symmetry," *Phys. Rept.* **245**, 259 (1994).

H. P. Nilles, "Supersymmetry, Supergravity, and Particle Physics," *Phys. Rep.* **110**, 1 (1984).

H. Pagels, "Departures from Chiral Symmetry," *Phys. Rep.* **16**C, 219 (1975).

C. Quigg, "The electroweak theory," `arXiv:hep-ph/0204104`.

G. G. Raffelt, "Astrophysics probes of particle physics," *Phys. Rept.* **333**, 593 (2000).

R. Slansky, "Group Theory For Unified Model Building," *Phys. Rept.* **79**, 1 (1981).

There are also many excellent reviews available in conference and summer school Proceedings.

Scientific American Articles

N. Arkani-Hamed, S. Dimopoulos, G. Dvali, "The Universe's Unseen Dimensions," August 2000.

J. N. Bahcall, "The Solar Neutrino Problem," May 1990.

R. Bousso and J. Polchinski, "The String Theory Landscape," September 2004.

D. B. Cline, "The Search for Dark Matter," March 2003.

F. E. Close and P. R. Page, "Glueballs," November 1998.

M. J. Duff, "The Theory Formerly Known as Strings," February 1998.

G. Feldman and J. Steinberger, "The Number of Families of Matter," February 1991.

A. Guth and P. Steinhardt, "Inflation," May 1984.

G. Kane, "The Dawn of Physics Beyond the Standard Model," June 2003.

E. Kearns, T. Kajita, and Y. Totsuka, "Detecting Massive Neutrinos," August 1999.

L. M. Krauss and M. S. Turner, "A Cosmic Conundrum," September 2004.

L. Lederman, "The Tevatron," March 1991.

T. M. Liss and P. L. Tipton, "The Discovery of the Top Quark," September 1997.

C. Llewellyn Smith, "The Large Hadron Collider," July 2000.

A. B. McDonald, J. R. Klein and D. L. Wark, "Solving the Solar Neutrino Problem," April 2003.

M. Moe and S. P. Rosen, "Double-Beta Decay," November 1989.

M. Mukerjee, "Explaining Everything," January 1996.

S. Myers and E. Picasso, "The LEP Collider," June 1990.

H. R. Quinn and M. S. Witherell, "The Asymmetry between Matter and Antimatter," October 1998.

J. R. Rees, "The Stanford Linear Collider," October 1989.

K. Rith and A. Schäfer, "The Mystery of Nucleon Spin," July 1999.

S. Weinberg, "A Unified Physics by 2050?," December 1999.

D. Weingarten, "Quarks by Computer," February 1996.

Other

The sites *Online Particle Physics*, `www.slac.stanford.edu/library/pdg/hepinfo.html`, and *Interactions*, `www.interactions.org/cms` contain a great deal of information on elementary particle physics.

The following recent studies are very useful:

> *Quantum Universe*, available at `interactions.org/quantumuniverse`;
>
> *From Quarks to the Cosmos*, available at `www.quarkstothecosmos.org`;
>
> *The Neutrino Matrix*, available at `www.aps.org/neutrino`.

Every other April the Particle Data Group's updated "Review of Particle Properties", which has many useful review articles, appears in *Physical Review D* or *Physics Letters B* and online at `pdg.lbl.gov`.

Elements

O. L. Keller, Jr.

The concept that one or a few elementary substances could interact to form all matter was originated by Greek philosophers beginning in the sixth century B.C. The atomic hypothesis, which has proven indispensable in the development of our understanding of chemical elements, originated with the philosopher Leucippus and his follower Democritus in the fifth century B.C. These two interacting concepts of elements and atoms have been unsurpassed in their importance to the development of modern science and technology.

Aristotle accepted from the earlier philosophers that air, water, earth, and fire were elements, and he added a fifth – the ether – representing the heavenly bodies. Although all matter was supposed to be formulated from these elements, in Aristotelian thought the elements themselves represented qualities and were nonmaterial. Centuries later, as the alchemists found new transformations, three more elements were added. These were sulfur, mercury, and salt, thought to represent the quantities of combustibility, volatility, and incombustibility, respectively. Aristotle's view of the elements as nonmaterial qualities thus prevailed for about 2000 years. It was only in the seventeenth century that a scientific atomic theory began to emerge from the philosophical theory of Democritus. In 1661 Robert Boyle, who had developed a chemical atomic theory based on the concepts of Democritus, gave the definition of chemical elements as "certain primitive and simple, or perfectly unmingling bodies, which, not being made of any other bodies or one another" are the constituents of chemical compounds. This physical and rather modern definition did not allow Boyle to conceive how such elements might be distinguished from compounds, however. The experimental determination of which substances were actually elements awaited the development of chemistry as a quantitative science. Over one and one quarter centuries after Boyle had given his definition of an element, Antoine-Laurent Lavoisier was able to determine a list of elements based on an experimentally verifiable definition: A chemical element is a substance that cannot be decomposed further into simpler substances by ordinary chemical means. He had determined his elements (about 30) by making careful quantitative studies of decomposition and recombination reactions. However, his list still included some very stable compounds such as silica and alumina, which he had not been able to break down. Also, from a philosophical point of view, it is interesting that he included heat and light! Apparently, the Greek influence was lingering on.

Work of the sort being carried out by Lavoisier soon led to the development of the law of definite proportions, which stated that in any given compound the elements always occur in the same proportions by weight no matter how the compound is synthesized. This law led to the definition of "equivalent weights" of elements as that weight which will combine with or replace a unit weight of some standard element such as hydrogen. John Dalton, in 1808, was the first to postulate an atomic theory that incorporated atomic weight as distinguished from equivalent weight and was capable of explaining the empirically derived laws of chemical combination. Dalton assumed that (1) all atoms of a given element are identical but are different from the atoms of other elements; (2) compounds are formed from these elemental atoms; (3) chemical reactions result from the atoms becoming rearranged in new ways; and (4) if only one compound can be formed from two elements X and Y, then that compound contains one atom of X and one atom of Y. So, for example, Dalton assumed water was HO instead of H_2O. Dalton needed postulate (4) in order to determine atomic weights experimentally, and he decided that nature is simpler than it is. Really, the only new idea introduced by Dalton was that of atomic weight as the ratio of the mass of a given element's atoms to the mass of some standard atom such as hydrogen. Naturally, Dalton's too-simple postulate (4) led to confusion in the atomic weight field, but a considerable resolution of these difficulties by Joseph-Louis Gay-Lussac, Amedeo Avogadro, and Stanislao Cannizzaro resulted in the publication of useful specific results on atomic weights.

The availability of fairly reliable atomic weights for a number of elements allowed chemists to seek relationships among them on a weight basis. In 1869, Dimitri Mendeleev knew of 65 elements with their atomic weights. While looking for relationships among these elements he made one of the most important discoveries in the history of chemistry: The properties of the elements are periodic functions of their atomic weights. This periodic law allowed the arrangement of the elements in a table in order of increasing atomic weight such that the table contains columns and rows of elements. Elements with similar chemical properties, such as silicon, tin, and lead, were found to fall in the same column. Thus there appeared a regular recurrence of chemical and physical properties of the elements from the top to the bottom of the column even though the elements were widely separated in atomic weight. For the first time in history it was shown that the chemical elements form an entity in their interrelationships, and undiscovered elements with predictable properties could be sought to fill up the holes in the table.

The periodic law proposed by Mendeleev was a daring break with the thought of the scientific community in 1869. In fact, Mendeleev's bold predictions of the chemical and physical properties of still undiscovered elements undoubtedly furnished the touch of drama needed to gain acceptance of his system. The predictions are probably the main reason Mendeleev's name is more firmly associated with the periodic law than that of Lothar Meyer, who discovered it independently and simultaneously. The three most famous predictions by Mendeleev concerned *eka*-aluminum (gallium), *eka*-boron (scandium), and *eka*-silicon (germanium). When the elements themselves were discovered, Mendeleev's detailed predictions were found to be amazingly accurate.

Earlier formulations of somewhat limited periodicities had led to obscurity or ridicule. A particularly brilliant insight is represented by the three-dimensional model presented in 1862 to the French Academy by Alexandre E. Beguyer de Chancourtois. In this model the elements were placed on a helix drawn on a cylinder in such a way that the distances of the elements along the helix were proportional to their atomic weights. This procedure resulted in elements of similar properties falling on parallel vertical lines along the surface of the cylinder. De Chancourtois stated, as a result, that "the properties of substances are properties of numbers." His remarkable insight went unnoticed, however, and the *Comptes Rendus* did not publish his table for nearly 30 years. The limited observation by J. A. R. Newlands in 1864 that a periodicity with atomic weight existed after each eighth element was another important foreshadowing of the periodic law. Unfortunately, his "law of octaves" received only ridicule at the time and the Chemical Society refused to publish his paper, although 23 years later they awarded him the Davy Medal for it.

Mendeleev's periodic table emphasized the value of more accurate atomic weights, partly because some elements were not in the proper order. As more accurate values for atomic weights were obtained, it was found that some of the elements persisted in disobeying the periodic law. Nickel is lighter than cobalt, iodine is lighter than tellurium, and potassium is lighter than argon, yet in each case, it was clear from the chemical properties that the lighter element must follow the heavier in the periodic table. Also, further anomalies in atomic weights developed as measurements became more precise. For example, T. W. Richards obtained the perplexing result that the atomic weight of lead depends on its geological source. Difficulties of this sort were finally resolved through the unexpected discovery by Sir J. J. Thompson that all atoms of an element are not identical, but can differ in their atomic weights. In 1912, Thompson was studying a gas discharge of neon in a mass spectrometer he had constructed. Much to his surprise he found that the neon atoms corresponded to two atomic weights, 20 and 22, rather than one atomic weight of 20.2. Thus, one of the most cherished of the concepts of the atomic hypothesis was reluctantly acknowledged to be wrong – all atoms of a given element are *not* identical. After the neutron was discovered, it was possible to understand that these atoms had the same number of protons in the nucleus but a different number of neutrons. Such atoms are called isotopes. Their chemical properties are virtually identical, although not completely so. The atomic weight of an element was redefined as an average for its naturally occurring isotopes. Also, it became the practice to simply assign a number to each element according to its place in the periodic table, since the order was obviously not strictly by atomic weight. This identifying number from the periodic table came to be called the *atomic number* of the element.

After the discovery of the electron in 1897 by Sir J. J. Thompson following experimentation of Sir William Crookes, several models of atomic structure were proposed, including the nuclear model of Hantaro Nagaoka. Nagaoka viewed the positive charge as concentrated at the center of the atom, with the electrons forming rings (like the rings of Saturn) around the central charge. The essential correctness of the nuclear picture of the atom was established by Ernest Rutherford through studies of the scattering of alpha particles (helium ions) by thin metallic foils. Thus atoms were shown to have structure, and another of the basic concepts of the atomic hypothesis was dethroned. Rutherford's experiments also gave the value of the

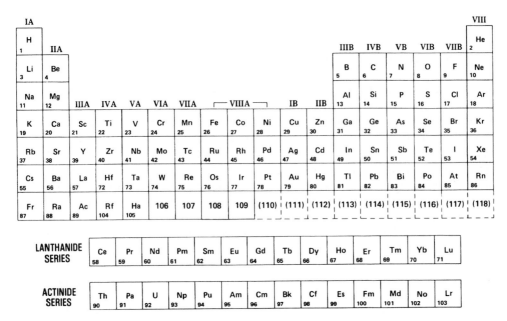

Fig. 1: Periodic table of elements.

charge on the nucleus for the elements he studied. The value was found to be roughly one half the atomic weight. In 1911 A. van den Broek suggested that the atomic number of an element could be identified with the number of protons in the nucleus. The suggestion of van den Broek was confirmed in a brilliant series of experiments carried out by H. G. J. Moseley in 1913. Moseley found that the atomic number (nuclear charge) of an element can be easily identified with the energy of its characteristic K and L x rays. Through the characteristic x rays the atomic number of each element could be positively identified with its position in the periodic table. It was thus found that, indeed, the properties of the elements are correlated in a periodic fashion through their atomic numbers rather than through their atomic weights as originally assumed by Mendeleev.

In 1922, Niels Bohr was able to interpret the periodicity of the chemical and physical properties of the elements by atomic number as a reflection of periodicity in electronic structure. In current periodic tables (Fig. 1) there are three short periods (rows) of 2, 8, and 8 elements; three long periods of 18, 18, and 32 elements; and an incomplete long period of 20 elements – a total of 106 elements in all. The progression of properties in a row occurs as each electron is added in going from each atomic number to the next higher atomic number. The electrons are added to the electronic shells of the atoms in accordance with the laws of quantum mechanics. The development of an understanding of the electronic structure of atoms on the basis of quantum theory began with Bohr's formulation in 1913. The culmination of the basic foundation was achieved in 1925 with the discovery of the electron spin by Uhlenbeck and Goudsmit, the discovery of the exclusion principle by Pauli, and the formulation of quantum mechanics by Heisenberg and Schrödinger. The electronic structure of atoms is described in terms of "orbitals," each of which is able to accommodate up to two electrons. The description of these orbitals is given in terms of so-called quantum numbers, which are designated n, l, and m_l.

The principal quantum number n can assume the values 1, 2, 3, ...; the azimuthal quantum number l can have the values 0, 1, 2, ..., $n-1$; and the orbital magnetic quantum number m_l can be 0 and any positive or negative integer up to and including l. There are therefore $2l+1$ values of m_l for each value of l. Designations that are now arbitrary have been given to the orbitals. For $l=0$ the orbital is given the designation s; for $l=1$, p; for $l=2$, d; for $l=3$, f; and so on. The Pauli exclusion principle states that a maximum of two electrons can occupy one of these orbitals. This is because of the spin magnetic quantum number m_s. which can assume the values $\pm\frac{1}{2}$. In order for two electrons to occupy the same orbital, their spins must be paired.

All electrons with the same value of n appear in the same "shell." The shells are called K, L, M, N, ... for $n=1,2,3,4,...$ It is the orderly filling of these electronic shells that gives the periodicity to the properties of the elements (Fig. 1). The K shell is filled at helium with two electrons ($n=1$; $l=0$). The first short period completes the L shell at neon by filling the 2s and 2p subshells ($n=2$; $l=0,1$). The second short period ends at argon where the 3s and 3p subshells are filled, although the M shell is still incomplete. The filling of the five 3d orbitals in the M shell produces the first transition-element series of 10 elements (the "iron group"), which begins at scandium and ends at zinc. The 4s subshell is filled at calcium and the 4p subshell is filled at the next rare gas, krypton. This leaves the M shell full but the N shell incomplete at the end of the first long period. The second long period, beginning at rubidium and ending at xenon, leaves the N shell complete with an octet of electrons in the O shell. The next long period contains 32 elements rather than 18. This results from the appearance of 14 elements characterized by 4f electrons in addition to the 18 elements corresponding to the filling of the 5d subshell of 10 electrons and the 6s, 6p octet. The 4f or lanthanide elements have extremely similar properties because their valence and outer electronic shells are essentially the same, with the 4f orbitals being deeply buried in the atom where they can have little effect. They thus effectively represent one "place" in the periodic table, and this is represented by placing element 57, lanthanum, in the body of the table and placing the other lanthanides at the bottom of the table. The next period, currently pausing at element 106, will presumably end at element 118, the next rare gas. In this period, the 5f or actinide elements are built up in a fashion analogous to the lanthanides. They are therefore placed beneath the lanthanides to show this similarity, as first suggested by Glenn T. Seaborg.

Discovery of new elements is still an active and challenging area of chemistry and physics. The heaviest element discovered to date (December 1988) is 109. The difficulties in new element discovery are illustrated by the 11-day bombardment time needed by the group at the Gesellschaft für Schwerionenforschung (GSI) in Darmstadt (FRG) to produce one atom of element 109. Another difficulty is posed by 109's short half-life of several milliseconds. Similar characteristics are found for elements 107 and 108.

Although the production cross sections are so small and the half-lives so short, according to the liquid drop model of the nucleus, these elements should not exist at all. It is, indeed, through their production and study that stabilization of matter by nucleon shell effects has been established in this region. Their enhanced stability holds out the possibility of the existence of a predicted "island of stability" for superheavy elements centering on a nucleus of 110 protons and 184 neutrons. The elements 107–109 could be looked upon as a sort of "causeway" from the main group of elements to the superheavy region. Relativistic effects on the electrons in these superheavy elements will be especially important in causing differences to appear between their chemistry and that of the lighter members of their group. The discovery of superheavy elements with long enough half-lives to study their chemistry would therefore open up a whole new area of the periodic table for the chemist as well as the physicist.

The GSI workers have designed and built a powerful accelerator (the UNILAC) capable of delivering the high intensity beams of chromium and iron needed to bombard bismuth and lead targets to produce a few atoms of elements such as 107, 108, and 109. They have also designed and built ultrasensitive, rapid detection and identification methods. Such equipment and methods are also useful for searching for superheavy elements. The search for superheavy elements is, however, requiring the heaviest transuranium target elements including einsteinium-254.

Other groups that are actively searching for new elements are located at the University of California, Berkeley (USA), and the Joint Institute for Nuclear Research, Dubna (USSR).

The techniques for a valid identification of a new element must, by definition, be capable of determining the atomic number unequivocally. Applicable chemical methods must be capable of yielding definitive results using only one atom at a time. Experience shows that ion exchange, solvent extraction, and gas chromatography can be definitive. Applicable physical methods include (1) the detection of characteristic x rays emitted by the daughter of the new element in coincidence with unique alpha particles from the decay of the new element itself and (2) the observation of "mother–daughter" decay chains. In (2) the new element is detected as it decays (for example, by alpha emission) into a known daughter. Then, through a sequence of alpha decays, a series of known nuclides is produced. Sufficient information for identifying the original nuclide in the chain can thus be obtained by recording the *chain of events* set up by the original event that occurs in the decay of the new, unknown nuclide.

It should also be mentioned that the upper limit of 10^{-14} s for a compound nucleus lifetime is considered to be the shortest lifetime of a composite nuclear system that could be thought of as an element.

See also: Atoms; Bohr Theory of Atomic Structure; Isotopes; Rare Earths; Rare Gases and Rare-Gas Compounds; Superheavy Elements; Transuranium Elements.

References

[1]   M. E. Weeks and H. M. Leicester, *Discovery of the Elements*. Kessinger Pub., 2003.
[2]   G. Munzenberg, "Recent Advances in the Discovery of Transuranium Elements," *Rep. Prog. Phys.* **51**, 57–104 (1988).
[3]   E. K. Hyde, D. C. Hoffman, and O. L. Keller, Jr., "A History and Analysis of the Discovery of Elements 104 and 105," *Radiochim. Acta* **42**, 57–102 (1987).

Ellipsometry

H. Lüth

Ellipsometry is an optical reflectance technique in which the change of the state of polarization of light upon reflection on a surface is measured, rather than the change of intensity. Even though the method dates back to Drude (1889) [1], ellipsometry has recently attracted considerable attention because of a wide variety of applications in modern solid-state optics, especially in the physics of surfaces, interfaces and thin films [2].

The change of the state of polarization upon reflection can be expressed in terms of the ratio of the two complex reflection coefficients $r_\parallel$ and $r_\perp$ for light polarized parallel and perpendicular to the plane of incidence, r being the ratio of reflected and incident electric field strength. The complex quantity

$$\rho = \frac{r_\parallel}{r_\perp} = \tan\psi\exp(i\Delta)$$

defines the two ellipsometric angles Δ and ψ that are measured in ellipsometry.

In contrast to a standard reflectivity (intensity) measurement, the two angles Δ and ψ completely determine the two optical constants n (refractive index) and κ (absorption coefficient) of an isotropic reflecting medium. For the mathematical dependence of n and κ on Δ and ψ calculated from Fresnel's formulas, see, e. g., Ref. [3].

The application of ellipsometry in surface physics is favored because in suitable cases an adsorbed species can be detected in a coverage as low as 1/100 of a monolayer. If used at a fixed wavelength λ, measured changes $\delta\Delta$ and $\delta\psi$ due to adsorption of gases can give information about adsorption kinetics (e. g., sticking coefficients) or absolute coverages [4]. The analysis, mostly done by means of computers, is rather involved because the two measured changes $\delta\Delta$ and $\delta\psi$ are not sufficient to determine all parameters (optical constants, thickness) of one or more adsorbed phases. For the study of adsorption processes, therefore, ellipsometry is most effectively used in combination with other methods, such as Auger electron spectroscopy [4].

Surface electronic structure can be successfully investigated by ellipsometric spectroscopy, in which $\delta\Delta$ and $\delta\psi$ caused by surface treatments (e. g., gas adsorption) are taken at various wavelengths over a given spectral range [5]. Transitions between surface states as they can be found by this type of optical spectroscopy are of considerable importance, e. g., for semiconductor device technology and for corrosion and catalysis studies. Another promising development can be expected from an extension of ellipsometric spectroscopy into the infrared in order to study vibration frequencies of adsorbed gases [6]. In the physics of thin films, ellipsometry can be used as a convenient *in situ* technique to follow epitaxial growth.

The conventional experimental setup for ellipsometry consists of two rotatable polarizers and a fixed quarter-wave plate as compensator to compensate the change of polarization of a parallel light beam due to reflection (null ellipsometry) [7]. The parameters Δ and ψ are calculated from component settings at extinction. For spectroscopy or high-speed measurements, automatic ellipsometers are much more convenient: The extinction settings can be done and read out by means of electronically controlled Faraday rotators in a self-compensating ellipsometer [8]. Polarization modulators [9] or rotating analyzers [10] are used in other types of automatic ellipsometers in which the parameter information is contained in the phase and relative amplitude of the ac component of the transmitted light intensity.

See also: Absorption Coefficients; Polarized Light; Reflection; Surfaces and Interfaces.

References

[1]   P. Drude, *Ann. Phys. Chem.* **36**, 532 (1889). (I)
[2]   "Proc. Symp. Recent Developments in Ellipsometry (Nebraska, 1968)," *Surf. Sci.* **16**, (1969); "Proc. 3d Int. Conf. on Ellipsometry (Nebraska, 1975)," *Surf. Sci.* **34**, (1976). (I)

R. M. A. Azzam and N. M. Bashara, *Ellipsometry and Polarized Light*. North-Holland, Amsterdam, 1971. (A)

[3]   G. A. Bootsma and F. Meyer, *Surf. Sci.* **14**, 52 (1969). (A)
[4]   R. Dorn, H. Liith, and G. J. Russell, *Phys. Rev. B* **10**, 5049 (1974). (I)
[5]   H. Lüth, *Appl. Phys.* **8**, 1 (1975). (I)
[6]   M. J. Dignan, B. Rao, M. Moskovits, and R. W. Stobie, *Can. J. Chem.* **49**, 1115 (1971). (A)
[7]   F. L. McCrackin, E. Passaglia, R. R. Stromberg, and H. L. Steinberg, *J. Res. Natl. Bur. Stand.* **A67**, 363 (1963). (A)
[8]   H. J. Mathieu, D. E. McClure, and R. H. Muller, *Rev. Sci. Instrum.* **45**, 798 (1974). (A)
[9]   S. N. Jasperson and S. E. Schnatterly, *Rev. Sci. Instrum.* **40**, 761 (1969). (A)
[10]   D. E. Aspnes and A. A. Studna, *Appl. Opt.* **14**, 220 (1975). (A)

Energy and Work

J. D. Walker

Energy is a certain abstract scalar quantity that an object (either matter or wave) is said to possess. It is not something that is directly observable, although in certain cases, the behavior of the object possessing a particular amount of energy can be observed and the energy inferred. The usefulness of the concept comes from the fact that total energy cannot be eliminated or created in the world, rather energy must be conserved. In predicting the behavior of objects, one uses the conservation of energy to keep track of the total energy and the interchange of energy between its various forms and between objects. The conservation of energy does not account for the method by which energy is transformed between its forms, only that the total amount must remain constant.

Work is the transfer of energy from one object to another by a force from one on the other when that second object is displaced by the force. Often the nature of the first object need not be specified if the force doing the work is known. For example, one can push a box, and the force performing the work is the pushing that the box experiences. The work can be calculated without knowing anything about the person pushing. In general, the work done on an object can be calculated from knowledge of either the applied force or the energy changes involved.

For there to be work in the scientific definition, the object must be displaced, and the displacement must have a component parallel to the force. The work done in such a displacement is the integral

$$W = \int \mathbf{F} \cdot d\mathbf{r} = \int F \cos\theta \, dr ,$$

where θ is the angle between the force and displacement vectors. If the applied force is constant and is at a constant angle with respect to a displacement Δx, then

$$W = F \cos\theta \Delta x .$$

For rotation of a rigid body about an axis by a torque τ, the work is

$$W = \int \boldsymbol{\tau} \cdot d\boldsymbol{\theta} .$$

If the torque is constant, then

$$W = \tau \Delta\theta .$$

Because work is a scalar quantity, the total work done by several forces on an object can be calculated by finding the work done by each separately, and then algebraically adding the work. Alternately, the total work is that work done by the resultant of the several applied forces.

If the dot product is positive (i. e., if the displacement and the force component along the displacement axis are in the same direction), then the work done on the object is positive and energy is added to the object. If the dot product is negative, then the work is negative and energy is removed from the object. These results do not mean that the conservation of energy is violated. If energy is added to the object, that energy came from the second object responsible for the force used in the work calculation. The total energy of the two objects is constant, and the work done is merely the transfer of the energy (equal to the work done) from one object to the other.

An example of a work calculation is the following. A block initially at rest has two forces acting on it: 3N to the right and 2N to the left. It moves to the right and after 2m, the net work done on it is 2J. If there are no other forces on it, then this 2J appears as kinetic energy of the moving block. The right-going force did a positive 6J of work. The left-going force did a negative 4J of work.

Kinetic energy is a form of energy associated with the motion of an object, that is, with the magnitude of its linear velocity v or rotational velocity ω. In classical mechanics, kinetic energy is given by

$$K = \frac{1}{2}mv^2$$

for linear velocity and by

$$K = \frac{1}{2}I\omega^2$$

for rotation of a body with a moment of inertia I about the rotational axis. For an example of the first equation, a mass of 4.0kg moving with a velocity magnitude of 5.0m/s has a kinetic energy of 50J.

To stop an object having a particular kinetic energy, that amount of energy must be removed from the object. In the previous example, the object would come to a stop if an external force such as friction does work on the object such that 50J of energy is removed.

In the special theory of relativity, the kinetic energy is defined by

$$K = E - E_0$$

where E is the total energy and E_0 is the rest energy of the object. For velocities smaller than about 0.1 the speed of light, this expression can be approximated by the classical one.

Potential energy is an energy associated with the relative position of an object. As with kinetic energy, this energy can be removed from the object by the object doing work on another system or this potential energy can be increased by an external force doing positive work on the object. Gravitational potential energy is related to the distance r of a point mass m_2 from another point mass m_2:

$$U = G\frac{m_1 m_2}{r} \, ,$$

where G is the gravitational constant. This classical expression for gravitational potential energy is zero for an infinite separation of the masses and is not defined for zero separation. If an external force is to increase the distance between the objects from r_1 to r_2 that external force must do an amount of work given by

$$W = \Delta U = G m_1 m_2 \left(\frac{1}{r_1} - \frac{1}{r_2} \right) \, .$$

If one of the point masses, say m_1, is released and is allowed to accelerate toward the other mass, its gravitational potential energy is converted to kinetic energy as the gravitational force between the masses does work on the moving mass. The gravitational potential energy and change in that energy near the surface of the earth can be approximated, respectively, by

$$U = mgh \, , \qquad \Delta U = mg\Delta h \, ,$$

where a point mass is a vertical distance h above a reference level, and where g is the local gravitational acceleration, approximately $9.8 \, \text{m/s}^2$. For example, if an object of mass 4.0 kg is raised 3.0 m near the earth's surface, then the external force causing the rise does work of 117.6 J. Thus the object's gravitational potential energy is increased by that amount. If the object is released and allowed to fall through 3.0 m, the gravitational force does work on it during the fall and converts the gravitational potential energy to kinetic energy. At the end of that fall, the object has kinetic energy of 117.6 J.

With such a definition the amount of potential energy a particular object has depends on an arbitrary choice of reference level from which to measure h. The change of vertical height results in the same change of potential energy regardless of the choice of reference level. Thus, the potential energy change of an object moving in a gravitational field is not arbitrary.

Potential energy is also associated with the compression or elongation of an ideal spring:

$$U = \frac{1}{2}kx^2$$

where x is the change in the spring's length and k is the spring constant. This type of potential energy is useful in any situation in which the force on an object obeys Hooke's law, either exactly or approximately.

Electrostatic potential energy between two point charges q_1 and q_2 is given by

$$U = k\frac{q_1 q_2}{r} \, ,$$

where k is a constant, r is the separation distance, and the charges can be either positive or negative. If both charges have the same sign, then an external force would have to do positive work to bring the charges closer. If the charges are already close and are then released, the electrostatic repulsive force would do work on them to accelerate them away from each other and to convert the electrostatic potential energy into their kinetic energy.

If there is no net external force on a system, its total energy must remain constant. The various forms of the energy of the system can change, but the total amount cannot.

If a net external force does work on a system and then the process is reversed, and if the system returns to its initial value of energy, the net work done is zero and the force is said to be conservative. The energy of the system is thus independent of how the work was done and depends only on the initial and final states of the system. An example of a nonconservative force is friction. The amount of energy lost to friction by a moving body depends on the distance over which the body slides on the surface having friction and therefore the choice of path determines the amount of energy lost. If a force F is conservative, a potential energy U can be defined by

$$F = -\nabla U .$$

In the special theory of relativity the total energy E of an object is related to its relativistic mass m,

$$E = mc^2 .$$

Mass and energy are considered to be the different and interchangeable forms of the same thing.

See also: Conservation Laws; Dynamics, Analytical.

Entropy
L. Tisza

The term entropy derives from the Greek expression for "transformation," and was suggested by Rudolf Clausius in 1865. However, the concept was already implicit in his resolution of an impasse in thermodynamics 15 years earlier. The problem was how to account for the fundamental asymmetry in the conversion of work and heat. The dissipation of work into heat is a spontaneous process, whereas the reverse conversion calls for such special arrangements as heat engines, and it is at best of limited efficiency. It was sensed that the underlying factor is the "lower quality" of energy associated with random molecular motion as compared with ordered mechanical energy. However, this contrast was not adequately expressed by the terms of heat and work, since heat was at least partially convertible. The problem was how to express the difference in randomness in terms of measurable quantities.

This complex requirement was filled by the concept of entropy, which was established by Clausius and William Thomson, later Lord Kelvin, by the most careful logic. The resulting practical rules can be stated with relative ease.

Let us inject a small quantity of heat dQ into a system at constant absolute temperature T. The entropy increase of this system is then $dS = dQ/T$.

Suppose that the heat quantity has been extracted from a reservoir of temperature T' which loses the entropy dQ/T', then the total entropy change is

$$dS_{\text{tot}} = dQ(1/T - 1/T') . \tag{1}$$

This quantity is positive for the natural process in which heat flows "downhill" ($T' > T$) and it would be negative for the impossible process in which heat would flow "uphill" ($T' < T$). Of course, such an uphill flow of heat does occur in an air conditioner, but this happens with the intervention of the compressor, and the dissipated work ensures that the total entropy change is positive.

For the limiting case of $T' = T$ the entropy is unchanged; it appears to be conserved. In this ideal case the process can be reversed and we speak of reversible heating.

Let us consider now the situation in which the heat quantity dQ is obtained from the dissipation of mechanical energy. Since the latter is not associated with entropy, we have no compensating entropy decrease, but the entropy increase of the system is the same as before; we say that entropy is a function of the state of the system only, and does not depend on the nature of the process that led up to it. In this respect it differs from the heat quantity dQ, which is not a differential of a state function Q; there is no such thing as a heat content Q of a system. This is sometimes expressed by slashing the symbol of differential:

$$đQ = T \, dS . \tag{2}$$

Note the analogy with the expression of compressional work:

$$đW = -p \, dV . \tag{3}$$

Excess pressure decreases volume and excess temperature increases molecular disorder, as expressed by entropy.

The requirement that the total entropy cannot decrease enables us to distinguish natural irreversible processes, for which the entropy increases, from the impossible reverse processes, for which the entropy would decrease and which do not occur in nature. It is worthwhile to note, however, that the term "irreversible" has a different meaning in thermodynamics from that in the everyday language. The melting of ice by a heating coil is said to be irreversible, but this does not preclude the refreezing of the ice, provided that we supply the necessary entropy to another system.

Although it is remarkable that a simple concept of entropy could be established entirely within classical thermodynamics [1], more elaborate versions emerge in a variety of roles. By merely postulating the entropy, this concern serves to organize the structure of much of thermodynamics. Having been demonstrated in an austere setting by Gibbs, this idea has proved successful on an elementary level [2, 3].

Considered as a measure of molecular disorder, the entropy concept generates a vast range of quantitative calculations within quantum statistics [3–6]. These references also elaborate the conceptual connections with quantum mechanics [6]; more specifically, the "absolute entropy" of the Third Law yields a variety of nontrivial applications if used with sufficient care [6–8].

The connection with information theory [9] and biology [10] marks the extent to which this concept evolved from its original beginnings.

See also: Carnot Cycle; Heat Engines; Heat Transfer; Statistical Mechanics; Thermodynamics, Equilibrium; Thermodynamics, Nonequilibrium.

References

[1]   E. Fermi, *Thermodynamics*. Dover, New York, 1956. (E)

[2]   H. B. Callen, *Thermodynamics*. Wiley, New York, 1960. (I)

[3]   David Chandler, *Introduction to Modern Statistical Mechanics*. Oxford Press, New York, 1987. (E)

[4]   L. D. Landau and E. M. Lifshitz, *Statistical Physics*, 2nd ed. Pergamon, New York, 1969. Third edition with L. P. Pitaevskii. Pergamon, New York, 1980. (I)

[5]   F. Reif, *Fundamentals of Statistical and Thermal Physics*. McGraw–Hill, New York, 1965. (I)

[6]   L. Tisza, *Generalized Thermodynamics*. MIT Press, Cambridge, Mass., 1966. (I)

[7]   J. Wilks, *The Third Law of Thermodynamics*. Oxford, London and New York, 1961. (I)

[8]   L. Tisza, in *Energy Transfer Dynamics. Studies and Essays in Honor of Herbert Fröhlich*, T. W. Barrett and H. A. Pohl (eds.). Springer-Verlag, Berlin, 1987. (A)

[9]   L. E. Brillouin, *Science and Information Theory*, 2nd ed. Academic Press, New York, 1962. (I)

[10] E. Schrödinger, *What Is Life?* Macmillan, New York, 1945. Doubleday Anchor Books, Garden City, NY, 1956. (E)

Equations of State

R. D. Mountain

An equation of state describes the relationship of the directly observable quantities that specify the thermodynamic state of a system. For a fluid the observable quantities are the pressure p, the specific volume v, and the temperature T. The equation of state is then the relationship

$$p = p(v, T) . \tag{1}$$

For a solid, it may be necessary to specify the stress and strain components in addition to the fluid variables. For a ferromagnet the applied field $\mathbf{H}$ and magnetization $\mathbf{M}$ must be included in the set of variables.

The simplest example of an equation of state is the ideal gas equation

$$pV = nRT . \tag{2}$$

This applies to n moles of a gas of noninteracting point molecules occupying a volume V; thus the ideal-gas form of Eq. (1) is

$$p = RT/V \tag{3}$$

with specific volume

$$v = V/n . \tag{4}$$

$R = 8.314471 \pm 0.000014 \, \mathrm{J \, mol^{-1} \, K^{-1}}$ is a universal constant known as the gas constant. Another example of an equation of state is provided by the van der Waals equation

$$p = \frac{RT}{(v - b)} - \frac{a}{v^2} . \tag{5}$$

The constants a and b are intended to take into account the attraction between the molecules and the finite size of the molecules, respectively. With suitable choices for a and b, Eq. (5) provides a qualitative representation of the equation of state for fluids.

Statistical mechanics provides the connection between the thermodynamic properties of a system and the molecular properties of the molecules which make up the system. Given the interaction potential for molecules, the methods of statistical mechanics provide a procedure for calculating the equation of state of a gas. The resulting equation (known as the "virial equation of state") takes the form of an infinite series in powers of v^{-1} with the ideal-gas expression as the first term:

$$p = \frac{RT}{v}\left(1 + \frac{B(T)}{v} + \frac{C(T)}{v^2} + \dots\right) . \tag{6}$$

The temperature-dependent quantities $B(T)$ and $C(T)$ are the second and third virial coefficients, and these objects contain information on the interaction of pairs and triples of molecules. $B(T)$ and $C(T)$ are known both theoretically and experimentally for various types of molecular interactions. The higher-order virial coefficients are difficult to obtain accurately. Theoretically this is so because the computations are complex. Experimentally it is due to the need for very precise p–v–T data over a large range of v if one is to extract the virial coefficients by fitting the data with Eq. (6). Even so, the virial equation of state is an important way of describing the equation of state of gases.

Empirical representations of the equation of state of fluids are frequently developed by fitting expressions of the type

$$p = \sum_{i,j} A_{ij} T^i v^{-j} \tag{7}$$

to experimental data. Such expressions have considerable computational utility even though little physical significance can be attached to the coefficients A_{ij}.

A portion of the p–v projection of an equation of state for a fluid is sketched in Fig. 1. Three isotherms (lines of constant temperature) are shown for the critical temperature T_c for $T_3 > T_c$ and for $T_1 < T_c$. Several universal features for equations of state for fluids should be noted.

For temperatures greater than T_c the pressure is a decreasing function of v and the isotherms are smooth functions of v^{-1}. This changes at the critical point (CP). The coexistence of liquid and gas phases requires nonanalytic regions in the equation of state for temperatures less than the critical temperature. This is because p and T are constant for a range of specific volumes. Analytic equations, such as Eqns. (5) or (7), can be used to describe much of the subcritical region only if they are augmented with the Maxwell construction to determine the location of the phase boundary. Such equations are not suitable in the near vicinity of the critical point as there the p–v–T relationship is known to involve nonintegral powers of $T - T_c$ and $v - v_c$.

Theoretical studies of the equation of state for gases use the ideal gas as a reference system. A different reference is needed for liquid-state studies. We shall arbitrarily suppose that liquids are fluids with $v < v_c$, the value of the specific volume at the critical point. Since the equation of state for a liquid is dominated by the strong repulsion that molecules experience when close together, the hard-sphere fluid provides a useful reference system for liquid-state studies when the attractive part of the potential is treated as a perturbation.

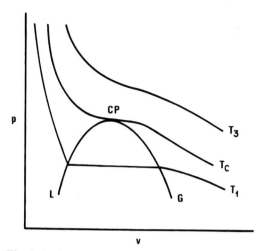

Fig. 1: A sketch of a portion of the p–v projection of the equation of state of a fluid. The curve L–CP–G represents the liquid (L)–gas (G) phase boundary which terminates at the critical point (CP). Three isotherms are indicated for temperatures $T_3 > T_c > T_1$. The critical temperature is T_c.

The hard-sphere fluid has been extensively studied using a combination of statistical-mechanical and computer-simulation methods. An accurate representation of the hard-sphere equation of state is

$$p = \frac{RT}{v} \frac{1 + y + y^2 + y^3}{(1-y)^3} , \tag{8}$$

where $y = \pi N_A d^3 / 6v$, d is the diameter of the sphere, and N_A is Avogadro's constant. A variety of techniques have been devised to obtain numerical equations of state starting with Eq. (8). For example, and diameter d can be used as a variational parameter to minimize the free energy derived by perturbation theory. Such calculations yield values for the equation of state which are accurate to a percent or so over a wide range of liquid-state temperatures and volumes.

Theoretical studies of the equation of state of solids use a collection of coupled harmonic oscillators as the reference system. Debye's theory provides an example which illustrates the concepts involved. In that theory it is assumed that the crystal can be represented as a collection of harmonic oscillators with a distribution of oscillator frequencies which is quadratic in frequency v up to some maximum value μ_{max} . The distribution is identically zero for $v > v_{max}$. The maximum frequency is determined by the number of modes possible for the system. The maximum frequency can be expressed as a temperature, the Debye temperature,

$$\Theta = h v_{max} / k , \tag{9}$$

where h is Planck's constant and $k = R/N_A$, is Boltzmann's constant. The explicit form for the equation of state is too complicated to exhibit here. The Debye temperature is commonly used to characterize a solid since it can be readily inferred from heat-capacity measurements.

If $T < \Theta$ is satisfied, then not all modes of vibration of the lattice are excited and explicitly quantum-mechanical effects are observed in the thermodynamic properties of the solid. If $T > \Theta$ is satisfied, then the solid behaves much like a classical system of oscillators.

More accurate theories of solids modify Debye's theory in two ways. The first modification is to replace the postulated quadratic distribution of frequencies with one based on the lattice structure and interaction parameters of the crystal. The second modification is important at high temperatures and involves the introduction of anharmonic interactions into the equations used to determine the frequencies of the oscillators. These changes are refinements in the theory rather than departures from Debye's theory. One other aspect of equations of state should be mentioned. These equations do not provide a complete, local description of the thermodynamics of a system. The thermodynamic potentials, such as $A(v, T)$, the Helmholtz free energy, do. Thus, while the pressure can be obtained by differentiation

$$P(v, T) = - \left(\frac{\partial A(v, T)}{\partial v} \right)_T , \tag{10}$$

the determination of $A(v, T)$ from $p(v, T)$ requires both a global knowledge of the equation of state and information (the integration constant) not contained in the equation of state. The latter information is obtained from calorimetric studies, a topic outside the domain of this article.

See also: Calorimetry; Phase Transitions; Thermodynamics, Equilibrium.

Bibliography

M. R. Moldover, J. P. M. Trusler, T. J. Edwards, J. B. Mehl, and R. S. Davis, "Measurement of the Universal Gas Constant R Using a Spherical Acoustic Resonator," *J. Res. Natl. Bur. Stand.* **93**, 85–144, 1988. (A)

A. Münster, *Statistical Thermodynamics*, Vol. 2. Academic Press, New York, 1974. (A)

J. O. Hirschfelder, C. F. Curtis and R. B. Bird, *Molecular Theory of Gases and Liquids*. Wiley, New York, 1954. (I)

J. Kestin and J. R. Dorfman, *A Course in Statistical Thermodynamics*. Academic Press, New York, 1971. (I)

John R. Dixon, *Thermodynamics I: An Introduction to Energy*. Prentice Hall, Englewood Cliffs, N.J., 1975. (E)

Ergodic Theory

J. Ford[†]

The ice in a glass of tea placed on a dining room table is always observed to melt and, eventually, the water–tea mixture comes to room temperature. Quite generally on the macroscopic level, a system initiated in a specified disequilibrium state is observed to approach and finally reside in a unique macroscopic equilibrium state having specified values for quantities such

---

[†]deceased

as temperature T, volume V, and pressure P. In its most general physical (but not mathematical) sense, classical ergodic theory is that branch of classical particle mechanics that attempts to rigorously explain and predict the observed macroscopic behavior of matter starting from microscopic particle dynamics, namely, Newton's equations. We immediately note that the macroscopic state of the system is specified by giving only a few quantities such as T, P, and V, while the microscopic state is specified by giving the positions and velocities for each of an enormous number of individual particles. Thus many distinct microstates correspond to each macrostate, whether equilibrium or disequilibrium; however, it may convincingly be argued that, for a given isolated mechanical system, the vast majority of microstates correspond to a single equilibrium macrostate. The central problem of ergodic theory then lies in rigorously proving that almost all microstates corresponding to macroscopic disequilibrium evolve with time into the sea of microstates corresponding to macroscopic equilibrium, and in rigorously computing the times required to reach equilibrium.

Although much work remains to be done in this area, equally, much success has been achieved. The flavor of work in this field may be sampled by considering an extremely simple (but unphysical) model that, nonetheless, exhibits many of the characteristics of more physically realistic systems. Consider a system with a large number of particles in which the microstate for each particle is specified by a position q and a speed p, with both q and p confined to lie in the unit interval $(0, 1)$. Let the equations of motion for each particle (giving the time evolution of microstates at 1-s intervals) be given by

$$q_{n+1} = 2q_n \quad \text{and} \quad p_{n+1} = p_n/2 \tag{1a}$$

when $0 \leq q_n < \frac{1}{2}$, or by

$$q_{n+1} = 2q_n - 1 \quad \text{and} \quad p_{n+1} = (p_n/2) + (1/2) \tag{1b}$$

when $\frac{1}{2} \leq q_n < 1$. Equations of motion (1) generate an area-preserving transformation that first transforms the unit square into a rectangle having twice the original width and one half the original height, and then it slices the rectangle into a left and a right half and places the right half on top of the left to reform the unit square; in addition, each initial microstate evolves via Eqns. (1) through a unique set of microstates (a trajectory). Unique trajectories and the area-preserving property are generic characteristics of dynamical systems.

Macroscopic equilibrium for this system clearly corresponds to a macroscopically uniform distribution of particles over the (q, p) unit square, and here ergodic theory can rigorously prove that almost all initial particle distributions tend to this unique equilibrium state. In order to understand this intuitively, suppose that all the initial-particle microstates are uniformly spread over an arbitrary, small rectangle $(\Delta q \Delta p)$. By Eqns. (1), this small initial rectangle grows exponentially $(\sim 2^n)$ in the q direction and shrinks exponentially $(\sim 2^{-n})$ in the p direction, eventually wrapping many times across the unit square and uniformly covering the square as a set of thin, horizontal ribbons. This exponential (called C-system or K-system) approach to equilibrium is characteristic of many physical systems, most notable the hard-sphere gas. Moreover, for Eqns. (1), since every small rectangle spreads uniformly over the whole unit square, this system is mathematically mixing (analogous to stirring a mixture to uniformity) and hence mathematically ergodic (which means, loosely speaking, the time-evolved iterates of almost all initial-particle microstates densely cover the unit square). Finally, the trajectories of this simple system, like those of certain more physically realistic systems, exhibit striking statistical behavior. Indeed, this system can be proved to be a Bernoulli system, which means that if the unit square is divided into N disjoint regions, successive iterates of most initial-particle microstates generated by the deterministic Eq. (1) sequentially appear in the various regions in a "random" sequence that could equally well have been selected by a roulette wheel.

See also: Statistical Mechanics; Thermodynamics, Nonequilibrium.

Bibliography
V. I. Arnold and A. Avez, *Ergodic Problems of Classical Mechanics*. Benjamin, New York, 1968. (A)
J. Ford, "How Random Is a Coin Toss?," *Phys. Today* 36 (No. 4), 40 (1983). (E)
J. Ford, "What Is Chaos That We Should Be Mindful of It?," in S. Capelin and P. C. W. Davies, eds. *The New Physics*. Cambridge University Press, Cambridge, 1988. (I)
J. Ford, "Chaos: Solving the Unsolvable, Predicting the Unpredictable!," in M. F. Barnsley and S. G. Demko (eds.), *Chaotic Dynamics and Fractals*. Academic Press, New York, 1986. (I)

Error Analysis

P. S. Olmstead[†]

Any discussion of error analysis must make its objective clear. In this article, the objective is a description of how to obtain and present data that are sufficiently "trouble free" to satisfy anyone who would wish to analyze them in a different way. It is recognized that all data are subject to error. However, it is the duty of the experimenter to show what he has done to reduce that error to a point that gives essentially "trouble-free" data. The experimenter's first step toward this goal is to record the data in a form that provides all information necessary for repeating the experiment successfully. The following form is suggested:

$$x_i = f(H_i, C_i, t_i)$$

where i is each successive test; f is the function to be either derived or verified; H_i is the identification of the human factor for the ith test; C_i is the identification of the physical factors for the ith test; t_i is the time that the ith test was made; and x_i is the "error" observed on the ith test. When data are presented in this form, any future analyst with a new hypothesis has an opportunity to test it with these data that is equal to that of the original experimenter.

The simplest set of data in this form is where x is assumed to be a constant. (In the very simplest form, the average error could be zero if certain very stringent conditions were met.) Many experimenters assume that all that is required is to find the average of the x_i's. Experience has shown that this may not be so. Most of us have made experiments where

1. a single observation seems inconsistent with the rest, or
2. several observations, low in value, are followed by a group that are high, or

---

[†] deceased

3. the spread of the values is first low but becomes high, or

4. there is a tendency to increase in value during the experiment, or

5. the data show periodic peaks and valleys, etc.

Usually, we associate these observations with something that we have failed to control, such as

1'. misreading a meter – a human error – or

2'. a shift in the conditions of test, or

3'. change in the sensitivity of a meter, or

4'. drift in calibration, or

5'. picking up a stray cyclical field, etc.

Most experimenters can add other symptoms and explanations to these lists. The purpose here is to point out that the things we have observed are related to patterns in the data that some statisticians have studied. They have been concerned with similar findings in other fields as well as physics. Their approach is to calculate how often a similar "presumed error" would be expected in an experiment with only random effects present. If an "error" as large as that observed could happen half the time, it would be considered a normal occurrence. When, then, should we look for "trouble"? How often should we be willing to look for a false trouble? In other fields, scientists are willing to look 5% of the time for a "nonexistent trouble" particularly if they have found that they can locate a similar "real trouble" at least 50% of the time. This ability to locate "real trouble" is much more important than the inability to locate "nonexistent trouble" indicated by a false clue. This means that scientists have more to gain than to lose by applying statistical criteria to their data.

In considering an outlying observation, for example, one statistical test is the ratio of the difference between the largest and the second largest and the difference between the second largest and the smallest. For a ratio of 1.0 to be meaningful at the 5% level, at least 8 measurements are needed; for a ratio of 0.5, at least 16; for a ratio of 0.4, at least 23; etc., making it clear that it is very dangerous to discard data simply because they "look bad." This is a serious matter in the case of three observations where we find that the ratio must exceed 16 at the 5% level and be at least 8 at the 10% level.

A shift in level has several possibilities for statistical test: a long run of observations on

a. one side of the median,

b. either side of the median,

c. each side of the median, or

d. each side of any cut.

For a sample of size 40, the length criteria for these to be significant are a. 8, b. 9, c. 7 and d. 7; for a sample of 100, they become: a. 10, b. 11, c. 8, and d. 9. The criteria do not change markedly with increase in sample size but experimenters will often find that "real trouble" will be associated with lengths of run that are not marginal as judged by the statistical criteria.

The simplest measure of spread in the observations is the range from maximum to minimum in small samples. To make use of this for detecting trouble, control charts for ranges of consecutive groups of four are recommended. For these, the average range is obtained and any range exceeding $2.28\bar{R}$ is considered as indicating variability worth investigating. If control charts for averages of the groups of four are also made, the expected limits about the grand average are $\pm 0.73\bar{R}$. A necessary condition for data consistency is to find 25 consecutive groups of four with their averages and ranges within control limits.

A slow fluctuation or trend in a set of data may be detected by applying a run test or by looking for a run-up or rundown in either individual observations or averages. Such a run of length 5 has a probability of not over 5% for up to 25 observations; for a run of length 6, the same probability exists to 154 observations.

The best method of identifying a fast fluctuation or cycle is to plot the individual observations in the order in which they were obtained and note any periodicity in the occurrence of peaks and valleys. For more information, this should be followed by lag correlation plots (x_i vs x_{i-L}). If a cyclical effect does exist in the data, the various lag plots will approximate a series of elliptical plots with superimposed random variation. Absence of data near the center of the plot is typical.

Up to this point, it has been assumed that the objective has been to determine a universal constant based on data obtained by several observers at various times and with differing experimental setups. From the results, a standard procedure is to be defined with predictable limits of error. A second type of physical experiment is to establish a curve of relationship for which a theoretical expression,

$$y = f(x) \, ,$$

has been assumed. If this expression is the median curve expected, a check of less than 15 points over the range of interest should not have a run of length 6 on one side of the curve in more than 1% of the experiments. A run-up or rundown of length 6 is even less likely. Failure to meet these criteria suggests need for modification in the expected curve of assumed relationship.

It should be pointed out that use of statistical concepts in various fields differs and the vocabulary used is seldom identical. This paper considers the problem of a physicist examining data that may contain errors worth investigating. A few simple clues to the possible existence of such errors are shown. How to identify the probable cause of the suspected error will depend on the experimental ingenuity of the physicist. Data that have been examined for the types of error discussed here may not be "trouble free" but they will have passed the first requirement for reaching such a condition.

For those interested in reading more about data analysis as it is practiced in industry, a bibliography is appended. Some of these show the fundamental steps outlined here, some show how particular tests have been derived, some show examples of use, and others extend the inquiry into the determination of correlated relationships. No text covers completely the problems of the physicist. However, one very useful text is *Precision Measurement and Calibration, Selected NBS Papers on Statistical Concepts and Procedures*, Harry H. Ku (ed.), National Bureau of Standards Special Publication 300, Volume 1, issued February 1969.

Standardization in this field is important. In the United States, this is being carried on by Committee E-11 of the American Society for Testing and Materials, 1916 Race Street, Philadelphia, PA 19103.

See also: Probability; Statistics.

Bibliography
General
Precision Measurement and Calibration (Harry H. Ku, ed.). NBS Special Publication 300, Vol. 1, 1969, $9.00 (Sup't Doc., U.S. Printing Office, Washington, D.C. 20402), with particular attention to articles by Churchill Eisenhart, W. J. Youden, John Mandel, R. B. Murphy, and Milton E. Terry.

Elementary
American Society for Testing and Materials, Committee E-1 l, *Manual on Quality Control of Materials*, STP No. 15, 1960.

American Society for Testing and Materials, Committee E-1 l, *Use of the Terms Precision and Accuracy as Applied to Measurement of a Property of a Material*, ASTM Standards, Part 41, 1975, pp. 165–182, E 177-71.

Paul S. Olmstead, "How to Detect the Type of an Assignable Cause," *Industrial Quality Control* **9**(3), 32–36 (1952); **9**(4), 22–32 (l953).

Intermediate
Paul S. Olmstead, "Distribution of Sample Arrangements for Runs Up and Down," *Ann. Math. Stat.* **17**, 24–33 (1946).

Paul S. Olmstead, "Runs Determined in a Sample by an Arbitrary Cut," *Bell System Tech. J.* **37**, 55–82 (1958).

Paul S. Olmstead, "Grouping of High (or Low) Values in Observed Data," *Statistica Neerlandica* **26**(3), 29–36 (1972).

Paul S. Olmstead and John W. Tukey, "A Corner Test for Association," *Ann. Math. Stat.* **18**, 495–513 (1947).

Walter A. Shewhart, *Statistical Method from the Viewpoint of Quality Control.* (The Graduate School, Department of Agriculture, Washington, D.C., 1939).

Advanced
ASTM, Committee E-11, "Dealing with Outlying Observations," ASTM Standards, Part 41, 1975, pp. 183–211, E 178-75.

ASTM, Committee E-11, *Manual on Fitting Straight Lines*. STP No. 313, 1962.

ASTM, Committee E-11, *Manual on Conducting an Interlaboratory Study of a Test Method.* STP No. 335, 1963.

Richard L. Anderson, "Serial Correlation," *Ann. Math. Stat.* **13**, 1–33 (1942).

Cuthbert Daniel, "Calibration Designs for Machines with Carry-Over and Drift," *J. Qual. Tech.* **7**, 103–108 (1975).

Wilfrid J. Dixon, "Analysis of Extreme Values," *Ann. Math. Stat.* **21**, 488–506 (1950).

Wilfrid J. Dixon, "Ratios Involving Extreme Values," *Ann. Math. Stat.* **22**, 68–98 (1951).

Jane F. Gentleman and Martin B. Wilk, "Detecting Outliers in a Two-Way Table, l. Statistical Behavior of Residuals," *Technometrics* **17**,1–14 (1975).

Frank E. Grubbs, "Procedures for Detecting Outlying Observations in Samples," *Technometrics* **11**, 1–21 (1969).

Richard E. Lund, "Tables for Approximate Test for Outliers in Linear Models," *Technometrics* **17**, 473–476 (1975).

Excitons

C. Jeffries[†]

The term exciton is broadly used to describe elementary localized excited states in solids, with the characteristic feature that the excitation can propagate through the crystal lattice. Two limiting experimental cases are recognized depending on the type of solid. For cubic band structures these can be shown to form a theoretical continuum.

Frenkel Excitons

In molecular crystals such as anthracene, in rare-gas crystals, and in some alkali halides, a particular molecule, atom, or ion may be initially in an electronically excited state. But since the coupling to the adjacent lattice atoms is strong this excitation can be rapidly transferred from site to site without motion of the atoms themselves. Indeed these Frenkel excitons are best described as propagating waves of electric polarization. The excitons are experimentally manifested as optical absorption bands or characteristic luminescence radiation.

Mott–Wannier Excitons

In semiconducting crystals like Si, Ge, CdS, and Cu_2O, optical excitation promotes an electron from a lower-energy valence band to an upper conduction band, leaving a vacant state, or hole, in the valence band. The electrons and holes are independently mobile and give rise to the electrical conductivity of the semiconductor. However, an electron and a hole have an attractive electrical interaction. They may combine into a hydrogen-like particle – a neutral mobile excited state. The radius of this Mott–Wannier exciton is much larger than the crystal lattice spacing and the binding energy is correspondingly small. In the simplest cases the exciton displays a set of energy levels approximately like $E \propto n^{-2}$ i. e., like atomic hydrogen. As an example the exciton in Ge has a binding energy of 4 MeV and a radius of 115 Å, to be compared to 13.6 eV and 0.5 Å for the H atom. The exciton in Ge may be visualized as a rather large, loosely bound particle (an electron and a hole in orbit about their center of mass) drifting through the crystal lattice; the electron and hole ultimately recombine, giving up characteristic luminescence radiation in the infrared. Being neutral, excitons do not contribute directly to electrical conductivity. They can be trapped on impurity atoms and other lattice defects; the radiation from commercial light-emitting diodes is usually due to the decay of trapped excitons.

Exciton Condensation

At sufficient densities and low temperatures, excitons form macroscopic droplets of a conducting electron–hole liquid in crystals like Ge and Si, much like the condensation of water vapor into fog droplets. In the liquid phase the electrons and holes are no longer bound into excitons but are free to move independently: The medium is a plasma of constant density, displaying both quantum and classical properties. The droplets have surface tension; they move freely through the crystal and can be accelerated to the velocity of sound by crystal strain gradients.

[†]deceased

Bibliography

A. S. Davydov, *Theory of Molecular Excitons*. McGraw–Hill, New York, 1962. (A)

J. J. Hopfield, "Excitons and their Electromagnetic interactions," in *Quantum Optics*, pp. 340–395, R. J. Glauber (ed.). Academic Press, New York, 1969. (I)

C. D. Jeffries, "Electron–Hole Condensation in Semiconductors," *Science* **189**, 955–964 (1975). (E)

R. S. Knox, *Theory of Excitons*. Academic Press, New York, 1963. (A)

D. C. Mattis and J.-P. Gallinar, *Phys. Rev. Lett.* **53**, 1391 (1984).

Far-Infrared Spectra

W. G. Rothschild and K. D. Moeller

The far-infrared (FIR) spectrum is the electromagnetic radiation which begins at wavelengths (λ) longer than can be dispersed by prisms [about 20–50 μm; 1 μm = 1 micrometer (or "micron", μ) $= 10^{-6}$ m], and ends where continuous light sources cease to furnish FIR energy ($\lambda \sim 10\,000$ μm $= 1$ cm). *Far-infrared instruments* use FIR from the electron plasma of mercury lamps; they eliminate unwanted λ with low-pass transmission or reflection filters. Dispersion is by a mirror with ruled grooves which pass a narrow λ range ("grating spectrometer"). Such an instrument successively scans a total of N spectral events during time T, spending time T/N on each event. A more recent technique is *FIR Fourier transform spectroscopy*. It observes all N events during T – which gains $N^{1/2}$ in signal-to-noise ratio ("Fellgett" or "multiplex advantage") – by taking an "interferogram" and transforming it numerically to the ordinary frequency spectrum. Usually a Michelson interferometer is used: a beam splitter generates two FIR beams, one following an optical path of fixed length, the other's travel being varied by distance x ($-L \le x \le L$, $L \sim 10$ cm) through reflection on a mirror which moves, parallel to the FIR beam, back and forth. The two beams are recombined to give the interferogram

$$s(x) = 2 \int_0^\infty A^2(\nu)\{1 + \cos(2\pi\nu x)\}\,d\nu \,,$$

where $A(\nu)$ is beam amplitude at frequency ν, and $2\pi\nu x$ is phase angle difference due to the moving mirror. Fourier transformation gives the spectrum

$$S(\nu) = \int_{-\infty}^{\infty} s(x)\cos(2\pi\nu x)\,dx \,.$$

Much thought has gone into writing programs which tell the detector when to observe $s(x)$ at mirror positions 0, x_1, x_2, ... ("sampling"), to reconcile $|x| \le L$ with infinity integration

Encyclopedia of Physics, Third Edition. Edited by George L. Trigg and Rita G. Lerner
Copyright ©2005 WILEY-VCH Verlag GmbH & Co. KGaA, Weinheim
ISBN: 3-527-40554-2

range ("apodization"), to do the many integrations rapidly, etc. Until continuously tunable narrow-band FIR sources (or detectors) are realized, Fourier-transform spectroscopy remains the choice for the FIR. (It is a classical technique: Michelson only lacked an on-line computer.) A *typical FIR wavelength* of $100\,\mu m$ corresponds to the frequency $2.997 \times 10^{12}\,Hz = 100\,cm^{-1}$ (cm^{-1} = number of waves per cm, imprecisely called "wave number") and to an energy $1.986 \times 10^{-14}\,erg = 285.8\,cal/mol = 0.0124\,eV$, or an energy- equivalent temperature $143.9\,K$ (room temperature $\sim 300\,K$). *Far-infrared phenomena* consequently involve transitions between small energy differences or low temperatures. Most such phemonena are found in crystalline and solid states of matter because the multiplicity of three-dimensional directions of forces, the propagation of waves as patterned through the mobility of many closely spaced particles, the long-range order permitting magnetic and ferroelectric phenomena, etc., yield closely spaced energy levels. The breakdown of long-range order in liquids eliminates most of these cooperative effects, and the chaos of positions and orientations in vapors and gases restricts the application of FIR spectroscopy to weak intramolecular and intermolecular effects.

One-phonon effects: A "primitive unit cell" (i. e., smallest building block of a crystal) containing m atoms or ions possesses $3m$ degrees of freedom ("lattice modes"), which can be described by branches of standing waves ("phonon branches") $\omega = A\exp[i(\mathbf{q}\cdot\mathbf{r} - \omega t)]$ of angular frequency $\omega = 2\pi\nu$ and phonon wave vector $\mathbf{q}$ in crystal direction $\mathbf{r}$. Their spectra are usually in the FIR. For instance, cubic rock salt Na^+Cl^- ($m = 2$, lattice constant $d = 5.6402\,\text{Å}$), has six phonon branches with q between 0 ("center of Brillouin zone") and $\pm\pi/2d = \pm2.79 \times 10^7\,cm^{-1}$ ("boundary"). In contrast, a FIR photon wave vector K is small: For instance, at $100\,\mu m$; $K = 2\pi/\lambda = 628\,cm^{-1}$. Its interaction with the lattice modes ("destruction of a FIR photon and creation of a FIR phonon") therefore involves the $q \sim 0$ phonon ("law of energy and wave vector conservation"). Only a doubly degenerate pair and a single phonon branch of the six are FIR-active ("optical branches"). The dipole moment in the pair is "perpendicularly polarized" (oscillates perpendicularly to phonon propagation direction) and thus can couple to the FIR field (always perpendicularly polarized). This pair is the "transverse optic branch" TO, with resonance frequency ω_{TO}. (Two people wiggling between them a long rope, up- down and left-right, illustrate TO.) The three other branches, the doubly degenerate TA ("transverse acoustic") and the single LA ("longitudinal acoustic"), are FIR inactive since no dipole moment is set up during their motion ("compressional waves"). When the photon–phonon interaction in the TO branch exceeds the resonance energy $\hbar\omega_{TO}$ (the latter depends on the forces, masses, and charge distributions in the crystal), the effective dielectric constant becomes negative. According to Maxwell's equations, this means that the FIR field cannot be sustained within the crystal and is thus strongly reflected ("Reststrahlen") in the range $\omega_{TO} < \omega < \omega_{LO}$ or $\lambda_{TO} > \lambda > \lambda_{LO}$ ("forbidden band"). Examples (λ_{TO} given) are $33\,\mu m$ (NaF), $52\,\mu m$ (NaCl), $61\,\mu m$ (KCl), $77\,\mu m$ (KBr), $150\,\mu m$ (CsI). For more complicated crystals consisting of multiatomic ions (e. g., NO_3^-) or molecules, there are correspondingly more active phonon branches.

Multiphonon effects: For strongly anharmonic crystal forces, the lattice modes are not independent and several phonon branches interact with a FIR photon. For example, two phonons of frequency ω_i, ω_j lead to summation and difference frequencies $\omega_i \pm \omega_j$, ("two-phonon process") and total wave vector $q = q_i \pm q_j \sim 0$. Since $q_i, q_j \neq 0$, they are not at the center of the Brillouin zone but (usually) at the boundary. Optic modes can, in this way, couple with acoustic modes (FIR inactive in the one-phonon approximation; see above). For instance, TO–

LA = 105 cm^{-1} (95.2 μm) for cadmium telluride at 110 K. FIR spectroscopy is thus helpful in characterizing the density of states at critical points of phonon branches, leading to a better understanding of compressibility, ionic charges, heat capacity, thermal expansion, dielectric properties, sound velocity, etc.

Impurity-induced absorption: Crystals with impurity atoms or structure faults show additional FIR absorption or reflection because their translational symmetry is broken, (i) *Local modes*: The vibrational motion of the impurity is concentrated within a few neighboring particles and is strongly influenced by them. Such impurity modes afford a probe for studying vibrational energy transfer and relaxation in solids. (ii) *Gap modes:* The impurity mode moves into the energy gap between the acoustic and optic branches. For instance, the gap mode of a Cl$^-$ (chloride) impurity ($\sim$ 0.4 ppm) in K$^+$I$^-$ (potassium iodide) appears at 77.10 and 76.79 cm^{-1} (λ = 129.7 and 130.2 μm) showing the ^{35}Cl and ^{37}Cl isotope splitting due to natural chlorine. (iii) *Resonance modes*: The impurity mass, driven by the rest of the lattice motions, performs large-amplitude forced vibrations which extend far into the crystal and can give rise to one-phonon absorption in the acoustic (usually FIR-inactive) branches. Study of FIR impurity absorption affords an important probe of lattice dynamics: The introduction of impurities need not change the crystal symmetry but the local disruption of ideality leads to additional absorption and hence to a greater wealth of available crystal parameters.

Soft-mode behavior: On cooling a crystal toward its Curie temperature (below which it becomes ferroelectric), it is found that certain TO branches tend to zero frequency. This means the crystal would disrupt, since $\omega_{LO}^2/\omega_{TO}^2$ = const. ("Lyddane–Sachs–Teller relation"), unless it undergoes a phase transition to another structure (e. g., cubic→tetragonal). This has been studied extensively for perovskite-type crystals (CaTiO$_3$). It offers electro-optic and FIR laser applications (tuning a soft mode with an electric field).

Spin-wave excitation and crystalline-field-effect spectra: The oscillatory motion of the ordered spins in the antiparallel sublattices of antiferromagnetic insulators in an applied magnetic field can couple to the magnetic component of FIR radiation, leading to "single-" ($q = 0$) or "two-magnon" ($q = \pi/2d$) resonance absorption below the ordering (Néel) temperature. For instance, FeF$_2$ has a one-magnon resonance at 52.7 cm^{-1}. In certain rare-earth ion garnets, e. g., 5Fe$_2$O$_3$·3Y$_2$O$_3$ (a mixed iron–ytterbium oxide), temperature-dependent collective $q \sim 0$ spin-wave excitation is observed (e. g., 20 cm^{-1} = 500 μm at 50 K) due to mutual spin precession of the iron and rare-earth sublattices. Temperature-independent FIR single-rare-earth ion transitions also appear, within energy levels determined by the atomic environment ("ligand-field potential") and the iron exchange field.

Semiconductors: Refractive index and absorption coefficients from FIR reflectance data are related to free-carrier concentration and mobility. For instance, in InSb (indium antimonide), free-carrier reflection occurs at $\lambda > 90$ μm, far beyond the *Reststrahlen* band ($\lambda_{TO} = 56$ μm). Of particular interest are ground-state impurity energy levels in regard to photocurrent production for solid-state FIR detectors. Beryllium-doped germanium (Ge) is a good detector for 15–52 μm at 4.2 K, copper-doped Ge is usable to $\sim$ 130 μm. The photon-excited change in the electronic mobility of InSb is utilized for very-long-wavelength detection. Since this is an "intraband effect," there is no long-λ cutoff due to the gap between impurity level and conduction band. Such devices ("Putley detector") are useful to $\lambda > 3000$ μm.

Cyclotron-resonance measurements of free carriers determine many features of band structure in semiconductors. Typically, they are observed for electron carriers, e. g., in indium arsenide and phosphide, at 23–160 μm using strong magnetic fields, and at much longer λ for the heavier holes, e. g., in cadmium-doped *p*-type Ge. On sweeping of the magnetic field, several distinct absorption peaks appear in the FIR which allow distinction between the mass of the different holes involved in the transitions. LO phonons, which do not couple to perpendicularly incident FIR radiation (see above), can do so if a magnetic field is applied to the semiconductor, giving the free electron carriers both longitudinal and transverse components. These couple the longitudinal LO to the transverse FIR photon ("collective plasma cyclotron-LO modes"). Such magnetoplasma effects are useful, e. g., for the generation of tunable coherent radiation sources in the FIR.

High-temperature superconductivity: Far-infrared properties of ceramic high-T_c superconductors, in particular efforts to determine the superconducting energy gap 2Δ, have been actively pursued since the detection of the effect by Bednorz and Müller. Several reflectivity experiments seemed to have verified the Bardeen–Cooper–Schrieffer weak-coupling value $2\Delta = 3.5 k_B T_c$. However, the spectra are complex (sample inhomogeneity, contamination, among others); certainty and consensus are still elusive at this writing.

Astrophysics: Several galactic and interstellar sources emit FIR radiation, for instance the Orion nebula. It is at present believed that the FIR is emitted from dust particles, surrounding the distant source and heated up to ∼ 100 K. Recent far-infrared luminosity observations of nebulae by the IRAS satellite have divulged features that allow tentative attributions as to the chemical nature of the radiating dust particles, such as silicates, silicone carbides, polycyclic hydrocarbons, and ice matrices with chemical inclusions. Airborne far-infrared Fabry–Perot observations have revealed emissions from a large abundance of warm (about 300 K) and dense atomic and molecular (predominantly carbon monoxide) material in galactic star formation regions, in the inner region of our galaxy, in the galactic center, and in external galaxies. It appears that these fine-structure emissions originate from surfaces of molecular clouds excited photoelectrically from 0B stars or the galactic interstellar radiation field. These observations open exciting vistas regarding future developments of theories of star formation and other cosmological phenomena.

Chemical physics: FIR is used to study rotational mobility in liquids, the conformational behavior of flexible vapor molecules, aspects of hydrogen bonding, collision-induced absorption in pressurized rare gases and homodiatomic molecules (hydrogen, oxygen, etc.), and rotational spectra of gaseous small molecules (H_2O, HF). The results of such experiments are of greatest interest to chemists.

See also: Infrared Spectroscopy; Molecular Spectroscopy.

Bibliography

R. J. Bell, *Introductory Fourier Transform Spectroscopy*. Academic Press, New York, 1972. (I/A)

Far Infrared Science and Technology, J. R. Izatt (ed.), SPIE-International Society of Optical Engineering, Bellingham, WA, 1986.

K. D. Moeller and W. G. Rothschild, *Far-Infrared Spectroscopy*. Wiley-Interscience, New York, 1971. (I/A)

W. G. Rothschild and K. D. Moeller, *Phys. Today*, **23**, 44, 1970.

Faraday Effect

L. M. Roth

In 1845, Michael Faraday discovered that when plane-polarized light passed through lead glass in the direction of a magnetic field, the plane of polarization was rotated. The Faraday effect has since been observed in many media and over a wide range of frequencies. A plane-polarized light wave can be resolved into left and right circularly polarized waves, and the rotation arises from a difference in index of refraction between the two. The effect is usually proportional to the magnetic field, and the rotation per unit field per unit path length is given by the Verdet constant V. For positive V, the rotation is in the sense of the current flowing in the coil producing the field. Thus on reflection, the rotation increases, in contrast to the optical activity of quartz and many organic molecules which is due to an intrinsic left or right handedness of the structure, for which the rotation reverses upon reflection. Related effects are the Voigt effect, the Cotton–Moulton effect, and the Hall effect.

In atoms and in impurity ions in solids such as lead glass, the Faraday effect is related to the strong dispersion which occurs at frequencies near an optical absorption line. In a magnetic field the lines undergo Zeeman splittings, with a differential dispersion for the two circularly polarized components. For paramagnetic ions the rotation is enhanced by population effects. In semiconductors, Faraday rotation occurs which is related to electronic transitions across the energy gap. At frequencies above the absorption edge for interband transitions oscillatory effects are seen in large magnetic fields. The Verdet constant of Ge is -0.16 at $1.5\,\mu$m, and of Corning 8363 lead glass is 0.17 at 5000 Å, in min/Oe cm.

In solids, free charge carriers also contribute to the Faraday effect. In a plasma, such as the ionosphere, it is the electrons which contribute at radio frequencies. The rotation due to free carriers depends on their mass and density and is proportional to the square of the wavelength. In solids, the electron and hole effective masses can be measured by infrared Faraday rotation when the damping is too large to observe microwave cyclotron resonance. The density of electrons in intragalactic space is extremely small but the path length is enormous so that Faraday rotation of polarized radio waves, for example from pulsars, has been observed and has been used to estimate galactic magnetic fields. Ionospheric Faraday rotation has been measured by satellite.

In magnetic materials the Faraday rotation is related to the tensor property of the magnetic permeability. For ferromagnets and ferrimagnets the rotation is extremely large and proportional to the magnetization. The corresponding effect in reflection is the magneto-optic Kerr effect. Magnetic semiconductors also have large Faraday rotations. Metallic ferromagnets absorb light, but Faraday rotation has been observed in thin films. Transparent ferromagnets such as yttrium iron garnet (YIG) have large rotations ($100°$/cm at $7.9\,\mu$m for YIG) and are important optical materials. Ferrites and antiferromagnets are good magneto-optic materials at microwave frequencies.

In applications of the Faraday effect the figure of merit is the ratio of rotation to absorption. Faraday rotation is used to observe magnetic domain structure in transparent ferromagnets as well as thin films of metallic ferromagnets. Faraday rotation isolators are used in microwave and optical systems including optical fibers to prevent the highly amplified signal from reflecting back to the source. Other applications are mode converters, magnetic sensors, and optical recording systems.

See also: Hall Effect; Ionosphere; Kerr Effect, Magneto-optical; Magnetic Materials; Optical Activity; Zeeman and Stark Effects.

Bibliography

C. L. Andrews, *Optics of the Electromagnetic Spectrum*, Chap. 19. Prentice-Hall, Englewood Cliffs, NJ, 1960. (E)

E. Scott Barr, "Men and Milestones in Optics V: Michael Faraday," *Appl. Opt.* **6**, 631 (1967). (E)

. K. Furdyna, "Diluted Magnetic Semiconductors," *J. Appl. Phys.* **53**, 7637 (1982). (A)

F. A. Jenkins and H. E. White, *Fundamentals of Optics*, 4th ed. McGraw-Hill, New York, 2001. (E)

M. Lambeck, "Image Formation by Magneto-Optic Effects," *Op. Acta* **24**, 643 (1977). (I)

J. G. Mavroides, "Magneto-Optical Properties," in *Optical Properties of Solids* (F. Abeles, ed.). North-Holland, Amsterdam, 1972. (I)

P. A. H. Seymour, "Faraday Rotation and the Galactic Magnetic Field-A Review," *Quart. J. R. Astron. Soc.* **25**, 293 (1984). (A)

K. Shiraishi, S. Sugaya, and S. Kawakami, "Fiber Optic Rotator," *Appl. Opt.* **23**, 1103 (1984). (E)

S. Wang, M. Shah, and J. Crow, "Studies of the Use of Gyrotropic and Anisotropic Materials for Mode Conversion in Thin Film Optical Wave Guide Application," *J. Appl. Phys.* **43**, 1861 (1972). (A)

K. A. Wickersheim, "Optical and Infrared Properties of Magnetic Materials," and K. J. Button and T. S. Hartwick, "Microwave Devices," in *Magnetism*, Vol. I (G. T. Rado and H. Suhl, eds.). Academic Press, New York, 1963. (I)

H. J. Zeiger and G. W. Pratt, *Magnetic Interactions in Solids*, Secs. 4.10.3 and 6.10. Clarendon, Oxford, 1973. (A)

Faraday's Law of Electromagnetic Induction

L. L. Foldy[†]

The observation that time-varying magnetic fields induce electric fields proportional to the time rate of change of the former was made independently by Michael Faraday and Joseph Henry in the early 1830s. Its quantitative expression is one of the fundamental laws of electromagnetism and is commonly referred to as *Faraday's law of electromagnetic induction*. Its most familiar statement takes the form: *A change, by whatever means, in the magnetic flux linking a closed circuit will result in an electromotive force in the circuit instantaneously proportional to the time rate of change of the linking flux.*

Electromotive force (emf) is here interpreted in the sense that if the circuit is formed of a conducting material obeying Ohm's law, then the emf is the driving force for the electrical current which by Ohm's law is equal to the emf divided by the resistance of the circuit. The change in the linked flux may arise through any one or combination of the following means: (a) motion of a permanent magnet in the vicinity of the circuit, (b) a change of the position of, or the electric current in, a neighboring circuit such as an electromagnet, (c) motion of the original circuit through an externally produced magnetic field, (d) a change in the electric current in the original circuit (self-induction), (e) a change in the shape of the latter. The above

---

[†]deceased

statement of Faraday's law encompasses the explanation of the operation of many familiar electromagnetic devices including inductances (chokes), induction or spark coils, transformers, induction motors, generators of various types, etc.

A complete appreciation of Faraday's law is only possible by introducing the field concepts which so strongly influenced Faraday's thinking about electromagnetism. The electric field **E** and the magnetic induction **B** then exist at each point in space independently of the presence of any material substance at these points. The circuit referred to may then be conceived of as a closed mathematical curve, not necessarily material. The electromotive force in this circuit (whether at rest or changing in time with velocity small compared to vacuum light velocity) is expressed as a line integral about this curve in the form

$$\mathcal{E} \;=\; \text{emf} = \oint_C \mathbf{E}' \cdot \mathbf{l} \,,$$

$$\mathbf{E}' \;=\; \mathbf{E} + \mathbf{v} \times \mathbf{B} \,,$$

where **E** is the electric field and **B** is the magnetic induction at the element d**l** of the circuit and **v** is the velocity of this element. The magnetic flux linking the circuit can be expressed as an integral of the magnetic induction over a surface S bounded by the circuit:

$$\Phi = \text{flux} = \int_S \mathbf{B} \cdot d\mathbf{S} \,.$$

The quantitative expression of Faraday's law then takes the form

$$\mathcal{E} = -\frac{\partial \Phi}{\partial t} \,.$$

The minus sign indicates that the direction of the emf is such as to induce a current in the circuit whose magnetic field has the direction required to oppose the change in flux linking the circuit (Lenz's law). The same result written in differential (purely local) form for a stationary infinitesimal circuit becomes

$$\text{curl}\,\mathbf{E} = -\frac{\partial \mathbf{B}}{\partial t} \,,$$

which is one of Maxwell's equations. This together with another of Maxwell's equations, $\text{div}\,\mathbf{B} = 0$, expressing the nonexistence of magnetic monopoles, can be written as a Lorentz-invariant equation thus demonstrating the consistency of Faraday's law with Einstein's principle of special relativity.

With only minor changes in interpretation, Faraday's law in its differential form has survived the upheavals in physics of the twentieth century: special relativity, general relativity, and quantum mechanics. Recent theoretical attempts to unite weak, electromagnetic, and strong interaction (hadronic) phenomena within a unified theory have seriously raised again the question of the existence of free magnetic monopoles and their associated currents. Experimental discovery of magnetic monopoles would finally herald the necessity of a modification of Faraday's law through the addition of an inhomogeneous term in the mathematical equation expressing it.

The equations above are correct in various unit systems, one of which is: $\mathcal{E}$ in volts, E in volts/meter, Φ in maxwells, B in maxwells/meter^2, distances in meters, and time in seconds.

See also: Electrodynamics, Classical; Magnetic Monopoles; Maxwell's Equations.

Bibliography

E. W. Cowan, *Basic Electromagnetism*. Academic Press, New York, 1968. (I)

R. P. Feynman, R. B. Leighton, and M. Sands, *The Feynman Lectures on Physics*, Vol. II. Addison-Wesley, Reading, Mass., 1964. (E)

J. D. Jackson, *Classical Electrodynamics*. 2nd ed. Wiley, New York, 1975.(A)

Fatigue

J. Weertman

Everyone is familiar with the fact that although a paper clip cannot be broken in two by simply bending it in one direction with the fingers, it is possible to break the clip without much trouble by repeatedly bending it back and forth. This is a simple example of the phenomenon of fatigue – the breaking of material by subjecting it to a cyclic stress whose amplitude is significantly smaller than the magnitude of the stress required to cause failure under static conditions.

In an engineering test of the fatigue properties of a metal the number N of stress cycles that a sample can withstand before failure occurs may be measured as a function of the amplitude S of the cyclic stress. If the experimental data are plotted on a graph of cyclic stress amplitude S versus the log of the number N of cycles to failure to obtain what is called an S–N curve, it is often (especially for steel) found that S decreases monotonically with increasing N from a value equal to the static failure stress (at $N = \frac{1}{4}$) to an apparent limiting value (for N larger than about 10^3–10^5) equal to about one half of the static failure stress. This lower value of S is called the fatigue limit or the endurance limit. A cyclic stress whose amplitude is smaller than the fatigue limit will never produce a fatigue failure, regardless of the number of stress cycles. However, in a number of metals and alloys there is no indication that the S–N curve ever levels off and for these materials a fatigue limit does not exist. Even in a material for which a fatigue limit appears to exist there is no complete assurance that if N is increased to a value larger than any of those used in the tests, fatigue failure may not occur under a cyclic stress whose amplitude is smaller than the apparent fatigue limit. There is virtually no frequency effect in fatigue if the temperature is not high relative to the melting point of the material and an active environment is not present. Whether the cyclic stress is applied at 1 cycle per hour or at 100 cycles per second, failure occurs at the end of the same number of total cycles. Only if the frequency is so very high that heat produced by plastic working cannot be dissipated easily is a frequency effect observed.

In material that contains no preexisting cracks, fatigue failure occurs in two stages. In the first stage small cracks are nucleated. If the material already contains cracks, the first stage of fatigue failure is of course bypassed. Small cracks may be nucleated in the interior of the material by the coalescence of a number of crystal dislocations on intersecting slip planes or, more commonly, at small second-phase particles. It is easier for small cracks to be nucleated at the surface of a metal part or fatigue specimen. The escape at a free surface of the dislocations that create slip bands causes the surface to be roughened. With an electron microscope it can be seen that small notches are produced in the roughened surface that are the equivalent of small surface cracks.

The subsequent growth of a small crack to a large size constitutes the second stage of the fatigue failure process. The rate of growth is given by the Paris equation:

$$\frac{da}{dN} = C(\Delta K)^n$$

where $2a$ is equal to the length of the crack, N now is the number of stress cycles (not the number required to produce failure), C is a constant whose value depends on the material tested, and ΔK is the cyclic stress intensity factor. During each stress cycle a fatigue crack will increase its length by an incremental amount if ΔK is large. If ΔK is small an incremental advance occurs only after many cycles. The term ΔK is a measure of the strength of the stress singularity at the tip of a crack. Depending on the crack geometry and the specimen geometry, the stress intensity factor either is equal to, or is approximately equal to, $\Delta K \cong S(\pi a)^{1/2}$. The exponent n in the Paris equation usually is equal to 4 but it can have values within the range of $n = 2$ to $n = 6$. In some materials, particularly in aluminum, regularly spaced striations can be seen on the fatigue crack surface. The spacing between the striations is such that it is exactly equal to the experimentally measured crack growth rate, proving that one striation is created in each stress cycle.

Complete failure of a part or test specimen occurs when a crack reaches a length sufficiently large that it will propagate catastrophically under a static stress of the same magnitude as S. The fatigue life of a part is thus determined by how long it takes a fatigue crack to be nucleated if no cracks are present initially, and by how long it takes a small crack to grow to a large size. Generally the growth stage occupies the major fraction of the fatigue life of a specimen or metal part when ΔK is large and the nucleation stage is the more important one when ΔK is small.

When the value of the stress intensity factor ΔK approaches the value required for catastrophic crack propagation under a static load, the fatigue crack propagation rate becomes greater than that predicted by the Paris equation. The Paris equation also breaks down when ΔK is so small that the predicted crack growth rate is smaller than about 3–10 atom distances per stress cycle. In these circumstances fatigue cracks apparently do not propagate at all and thus $da/dN \equiv 0$.

In tests in which a specimen is cycled under conditions of a constant plastic strain amplitude ε_p (the plastic strain is equal to the total strain less the elastic strain component ε_e) the number N of cycles required to reach failure is given by the Coffin–Manson law:

$$N^q = \frac{C_p}{\varepsilon_p}$$

if $\varepsilon_p \gg \varepsilon_e$. Here q and C_p are constants ($q \simeq \frac{1}{2}$ and $C_p \simeq \varepsilon_s/2$ where ε_s is the plastic strain at failure under a static load). If the specimen is cycled under a constant elastic strain amplitude and $\varepsilon_e \gg \varepsilon_p$, the failure equation is the Basquin law $N^{q*} = C_e/\varepsilon_p$, where $q^* \simeq \frac{1}{8}$ and C_e is another constant.

The fatigue life is affected by the environment around a specimen. For example, the fatigue crack growth rate of a crack in aluminum is an order of magnitude slower if the specimen is in a high vacuum than if it is in air of 50% humidity. When the temperature is high, diffusion-controlled cavity or crack-like cavity formation generally is the dominant fatigue mechanism. At high temperatures the fatigue life is a strong function of frequency as well as the wave shape of the cyclic stress. Failure paths tend to become intercrystalline rather than transcrystalline.

A phenomenon called static fatigue is observed in glass and in some inorganic crystalline material. No cyclic loading is involved in this phenomenon. A crack will grow under a static load very slowly until its length reaches the size required for catastrophic failure. A continuously occurring corrosive reaction at the crack tip is required to produce static fatigue.

See also: Anelasticity; Crystal Defects; Elasticity; Mechanical Properties of Matter.

Bibliography

N. E. Frost, K. J. Marsh, and L. P. Pook, *Metal Fatigue*. Oxford (Clarendon Press), London and New York, 1974. (E)

W. O. Soboyejo and T. S. Srivatsan (eds.), *High Cycle Fatigue of Structural Materials*. TMS, Warrendale, PA 1997. (A)

S. Suresh, *Fatigue of Materials*, 2nd ed. Cambridge University Press, Cambridge, 1998. (A)

Special Technical Publications on fracture and fatigue, published at irregular intervals by the American Society for Testing and Materials, Philadelphia. (A)

Fermi–Dirac Statistics

P. M. Platzman

In classical mechanics physically similar particles are identifiable. We can number the particles at some instant of time and, in principle, follow each one in its motion during some dynamical process. In quantum mechanics there is, even in principle, no possible way of distinguishing or labeling identical particles. This is true because the system is described by a wave function Ψ satisfying some type of Schrödinger equation which, in turn, implies an uncertainty principle and the impossibility of labeling trajectories.

If we consider a system of two identical particles the states of the system obtained by interchanging the two particles must be equivalent. In particular, this means that the wave function $\Psi(\xi_1, \xi_2)$ can change only by a phase factor when we interchange the two particles, i. e.,

$$\Psi(\xi_1, \xi_2) = e^{i\alpha}\Psi(\xi_2, \xi_1)$$

where α is a real number. By repeating the interchange in three-dimensional space we return to the original state so that $e^{2i\alpha} = 1$, i. e., $e^{i\alpha} = \pm 1$. This argument can be generalized to any number of particles, i. e., the wave function of identical particles is either symmetrical or antisymmetrical under the interchange of any pair of particles. The symmetrical wave function corresponds to bosons, whereas the antisymmetric wave function corresponds to fermions. Fermions are said to obey Fermi–Dirac statistics.

Using some reasonable arguments regarding the general properties of relativistic quantum mechanics, Pauli (1940) constructed an argument which linked the intrinsic spin of the particle to the symmetry or antisymmetry of its wave function. In particular, he showed that half-integer spin particles were fermions, whereas integer spin particles were bosons. Atoms, the building blocks of ordinary matter, are made up of fermions. However, complex particles containing an even number of fermions, for example, He^4, behave as bosons when one considers processes (scattering, statistical mechanics, etc.) at energies small compared to the binding energy of the fermions.

If we consider a system composed of N noninteracting fermions, then the wave function of the various one-particle stationary states ψ_k with energy ε_k which each of the particles may occupy specifies the wave function of the N-particle system. Since the wave function is antisymmetric in the interchange of any pair, it follows that no two fermions may occupy the same one-particle state. This is called the Pauli exclusion principle and was first set forth by Jordan and Wigner (1928).

There are many dramatic experimental consequences of Fermi–Dirac statistics or equivalently, at zero temperature, the Pauli exclusion principle. In particular, all descriptions of atoms, the interpretation of their optical spectra, chemical valence, etc., use the concept of singly occupied one-particle states. Classification of nuclei and the spectra of their excited states are also linked to the exclusion principle.

At a finite temperature T it is easy to show that the thermodynamic potentialof a set of noninteracting fermions is given by

$$\Omega = -k_\mathrm{B}T \sum_k \ln\{1 + \exp[(\mu - \varepsilon_k/k_\mathrm{B}T]\} \ .$$

Here μ is the chemical potential. The average number of particles in each state ε_k is

$$\bar{n}_k = \{1 + \exp[(\mu - \varepsilon_k)/k_\mathrm{B}T]\}^{-1} \ .$$

For free or almost free electrons, $\varepsilon_k = \hbar^2 k^2/2m$. Near $T = 0$, $\mu \cong E_\mathrm{F} = \hbar^2 k_\mathrm{F}^2/2m$, where n is the density.

The low-lying excited states of such a degenerate system are quantitatively described by exciting a particle from an occupied state $\bar{n}_k = 1$ to an empty state $\bar{n}_k = 0$, i. e., creating a particle–hole pair. Landau (1956) pointed out that many interacting-fermion systems (Fermi liquid) should have excitation spectra which could be put in one-to-one correspondence with the noninteracting system. The excitations in this case are called quasiparticle–quasihole pairs. There are many examples in nature of such degenerate Fermi liquids. For example, the Landau theory of Fermi liquids describes most of the low-temperature low-frequency properties of metals: their specific heat, their electrical conductivity, and their magnetic susceptibility. It also describes the properties of such diverse systems as low-temperature liquid He^3, the excited states of heavy nuclei, and the observed properties of white dwarf stars.

Modifications of Fermi liquid theory to include the attractive interactions between quasiparticles leads naturally to the so-called Bardeen–Cooper–Schrieffer theory of superconductivity. This theory describes most known low-temperature superconductors.

See also: Bose–Einstein Statistics; Helium, Liquid; Quantum Fluids; Quantum Statistical Mechanics; Quasiparticles; Superconductivity Theory.

Bibliography

A more complete discussion of this topic and specific references may be found in:

L. D. Landau and E. M. Lifshitz, *Quantum Mechanics*, Non-relativistic Theory. Butterworth-Heinemann, 1981.

L. D. Landau and E. M. Lifshitz, *Statistical Physics*. Butterworth-Heinemann, 1980.

Fermi Surface

B. R. Cooper

The concept of the Fermi surface in metals is intimately related to the existence of lattice periodicity, the most fundamental property of crystalline solids. The existence of the Fermi surface is a consequence of the presence of lattice periodicity and the fact that electrons obey Fermi–Dirac statistics (i. e., obey the Pauli exclusion principle). Interest in the Fermi surface stems from the fact that it is the most important single entity characterizing the electron states in a metal, and also from the fact that macroscopic thermal, electric, magnetic, and optical properties are determined by, or have important contributions from, the behavior of the electrons at and near the Fermi surface. In this article we define the Fermi surface, particularly relating its existence to the lattice periodicity and to the occupation of electron energy bands in solids. We then briefly describe several of the most important experimental techniques for measuring the properties of the Fermi surface.

For a perfect three-dimensional lattice there are three primitive translations, $\mathbf{a}_1$, $\mathbf{a}_2$, $\mathbf{a}_3$, which define the periodicity of the lattice. That is, the crystalline atomic arrangement looks the same in every respect when viewed from any point $\mathbf{r}$ as when viewed from the point

$$\mathbf{r}' = \mathbf{r} + \mathbf{t}_n , \tag{1}$$

where

$$\mathbf{t}_n = n_1\mathbf{a}_1 + n_2\mathbf{a}_2 + n_3\mathbf{a}_3 . \tag{2}$$

Here n_1, n_2, and n_3 are arbitrary integers; and for $\mathbf{a}_1$, $\mathbf{a}_2$, $\mathbf{a}_3$ to be primitive, any two points from which the crystalline atomic arrangement looks the same must be given by (1).

The requirement that the potential acting on an electron moving through a solid is identical at two points r and r' related by (1) leads to the requirement that the electronic wave functions obey Bloch's theorem and are labeled by a crystal-momentum quantum number $\mathbf{k}$. So the most general electron wave function in a three-dimensional periodic lattice has the form

$$\psi_k(\mathbf{r}) = u_k(\mathbf{r})\,\mathrm{e}^{\mathrm{i}\mathbf{k}\cdot\mathbf{r}} , \tag{3}$$

where u_k is periodic so that

$$u_k(\mathbf{r}) = e_k(\mathbf{r} + \mathbf{t}_n) .\tag{4}$$

The crystal momentum $\mathbf{k}$ takes on a discrete set of values determined by periodic boundary conditions. We consider a system in the shape of a parallelepiped of edges $N_1 a_1$, $N_2 a_2$, and $N_3 a_3$; and take the wave function to be identical for locations separated by a translation $\mathbf{t}_N = N_1 \mathbf{a}_1 + N_2 \mathbf{a}_2 + N_3 \mathbf{a}_3$. (That is, we imagine all of space to be filled by identical contiguous parallelepipeds, and require the electronic wave functions to behave identically in each parallelepiped.) Then the discrete values of $\mathbf{k}$ are given by

$$\mathbf{k} = 2\pi \left(\frac{k_1 \mathbf{b}_1}{N_1} + \frac{k_2 \mathbf{b}_2}{N_2} + \frac{k_3 \mathbf{b}_3}{N_3} \right) ,\tag{5}$$

where k_1, k_2, and k_3 are integers; and $\mathbf{b}_1$, $\mathbf{b}_2$, and $\mathbf{b}_3$ are the primitive reciprocal-lattice vectors given by

$$\mathbf{b}_1 = \frac{\mathbf{a}_2 \times \mathbf{a}_3}{\mathbf{a}_1 \cdot (\mathbf{a}_2 \times \mathbf{a}_3)} ,\tag{6}$$

with $\mathbf{b}_2$ and $\mathbf{b}_3$ obtained by cyclic permutation of the indices on the $\mathbf{a}$'s.

Reciprocal space is defined as the space in which the vectors $\mathbf{b}_1$, $\mathbf{b}_2$, and $\mathbf{b}_3$ are primitive vectors in the same way that $\mathbf{a}_1$, $\mathbf{a}_2$, and $\mathbf{a}_3$ are in ordinary space. If we consider a vector

$$\mathbf{K}_m = 2\pi(m_1 \mathbf{b}_1 + m_2 \mathbf{b}_2 + m_3 \mathbf{b}_3) \quad \text{with } m_1, m_2, m_3 \text{ integers},\tag{7}$$

then because the electron wave functions are of the form given by Eqns. (3) and (4) as required by the Bloch theorem for a periodic lattice, it is not possible to distinguish between the wave function with crystal momentum $\mathbf{k}$ and that with $\mathbf{k} + \mathbf{K}_m$. The set $\mathbf{K}_m$ (or strictly speaking $\mathbf{K}_m/2\pi$) are called reciprocal-lattice vectors and define the reciprocal lattice in the same way that the $\mathbf{t}_n$ of Eq. (2) define the real-space crystal lattice. The wave function shows periodicity with $\mathbf{k}$ in reciprocal space (also referred to as $\mathbf{k}$ space) analogous to the periodicity of $u_k(\mathbf{r})$ in real space. Thus all distinct electronic states can be considered by treating only that part of $\mathbf{k}$ space closer to the origin than to any nonzero $\mathbf{K}_m$. This part of $\mathbf{k}$ space is called the first Brillouin zone, and is constructed by setting up planes that perpendicularly bisect the lines connecting the origin to all $\mathbf{K}_m$, and then taking the volume about the origin enclosed by these intersecting planes. Figure 1 shows the first Brillouin zone for the face-centered-cubic (fcc) structure with conventional labeling of points and lines of high symmetry.

There are $N_1 N_2 N_3$ discrete values of $\mathbf{k}$ contained within the first Brillouin zone, i.e., one discrete value per real-space lattice point. For $\mathbf{k} = 0$, there will be a number of discrete electronic energy eigenvalues. For instance, for a noninteracting free electron gas these values are given by $E_m = \hbar^2 K_m^2/2m$ for K_m taking on all values given by Eq. (7), including 0. As $\mathbf{k}$ varies across the Brillouin zone, the values of E emanating from each E_m form a semicontinuous band with band index m. The fact that $\mathbf{k}$ is a good quantum number and that the band index m serves to lump all other quantum numbers (allowing for degeneracies for $\mathbf{k}$ at high-symmetry points and lines), together with the fact that electrons obey Fermi statistics, means that each band can hold two electrons (one with spin up and one with spin down) for a general discrete $\mathbf{k}$

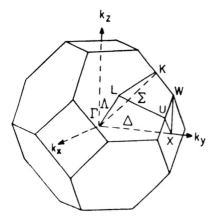

Fig. 1: The first Brillouin zone for the face-centered-cubic structure with points and lines of high symmetry labeled in conventional notation, [after B. Segall, *Phys. Rev.* **125**, 109 (1962)].

in the Brillouin zone. Thus the first Brillouin zone can hold two electrons per real-space crystal lattice site per band. At absolute zero the band energy states will be filled with electrons up to an energy level, the Fermi energy (E_F), such that all the electrons are accounted for. For a metal there are one or more partially filled bands. Thus there is a surface in reciprocal space separating filled from empty states. This surface is the *Fermi surface*.

The simplest example of a Fermi surface is that for a noninteracting electron gas. The Fermi surface in that case is a sphere, and the radius of that sphere is the Fermi radius. For example, if one had an fcc lattice with one free electron per lattice site, the Fermi sphere would have a volume equal to one-half that of the first Brillouin zone shown in Fig. 1. Copper presents a simple and much studied case for a real metal. The 10 3d electrons per copper atom form filled bands in the solid and do not give a Fermi surface. There is one valence electron per atom, corresponding to the atomic 4s state. The band formed from the valence electrons is partially filled and has behavior close to that for free electrons (i. e., the dispersion relationship over much of the Brillouin zone is approximately $E \sim k^2$ with an electron effective mass differing somewhat from the free electron mass); and as shown in Fig. 2, the Fermi surface is basically spherical. However, interaction effects between the s and d electrons give rise to "necks" contacting the hexagonal $\langle 111 \rangle$ faces of the Brillouin zone.

The Fermi surface has a symmetry appropriate to the particular crystal. Depending on the location in the Brillouin zone of the lowest-energy states and the number of electrons available for filling states, the Fermi surface can be quite complicated and bear no resemblance to a sphere. There may be disconnected surfaces in different parts of the Brillouin zone. Especially for metals with several valence electrons there may be several Fermi surfaces corresponding to partial filling of different bands. One can view the surface either from the direction of the filled states or from that of the unfilled states, the holes. If the filled states occupied the outer part of the Brillouin zone in Fig. 1, then the Fermi surface for electrons would be concave. For example, we could interchange the filled and unfilled states for the copper Fermi surface of Fig. 2. In such a case it is simpler to deal with the convex surface enclosing the unfilled volume, and one then treats the situation in terms of a *hole surface*.

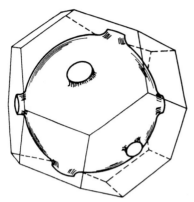

Fig. 2: A sketch of the Fermi surface of copper. The polyhedron represents the Brillouin zone [after B. Segall, *Phys. Rev.* **125**, 109 (1962)].

At finite temperature, following the laws of Fermi–Dirac statistics, the electron states with energy lower than E_F have only a finite probability for being filled, and those above E_F have only a finite probability for being unfilled. However, these probabilities differ significantly from unity only within kT of E_F. Since E_F is typically several electron volts and $1\,eV$ is the energy equivalent of $T = 11.606\,K$, this results in a slight blurring of the Fermi surface at finite temperatures.

Because of its relationship to the filling of the energy band states, the Fermi surface characterizes the electronic structure of a metal. Also the behavior of electrons at and near the Fermi surface determines, or gives important contributions to, many macroscopic properties of metals such as the electrical transport properties, magnetic behavior, and specific heat. For example, under the influence of a dc electric field, the only electrons that can respond are those at the Fermi surface, since they can be displaced into adjacent empty states with different lattice momenta. Thus the behavior of the electrons at the Fermi surface determines the electrical transport properties of metals. Because of this fundamental relationship to the electronic structure and to the macroscopic properties of metals, over a period of about 25 years there has been a great deal of experimental and theoretical work aimed at characterizing the Fermi surfaces of metals.

The experiments most directly characterizing the Fermi surface are those measuring the sizes and shapes of the orbits of electrons moving on the Fermi surface in a magnetic field. Changing the direction of the magnetic field causes the observed behavior to be governed by different parts of the Fermi surface thereby allowing one to map out the Fermi surface. Such experiments are possible only at very low temperature in very pure materials since the electron must be able to complete its orbit without scattering for the necessary observations to be possible.

Among the most valuable techniques for measuring important features of the Fermi surface are the following:

1. *The de Haas–van Alphen effect.* This involves measuring the periodic fluctuation of magnetic susceptibility in a varying magnetic field. The period of oscillation gives a direct measure of extremal cross-sectional areas of the Fermi surface normal to the magnetic

field. By making measurements at all orientations of the crystal relative to the magnetic field, one can reconstruct the Fermi surface almost exactly. The de Haas–van Alphen effect has probably yielded the most precise Fermi-surface measurements.

2. *Magnetoacoustic attenuation.* This involves ultrasonic attenuation in a magnetic field applied perpendicular to the direction of sound propagation. Variation of the attenuation with magnetic field gives extremal diameters of the Fermi surface for sections normal to the applied field.

3. *High-field magnetoresistance.* This involves observing whether the electrical resistance transverse to a high magnetic field saturates or increases indefinitely on increasing H for varying directions of H. Nonsaturation for a given direction implies a multiply connected Fermi surface, i. e., the geometry in directions such as the necks in Fig. 2 where the Fermi surface intersects the Brillouin-zone boundary. Thus one can map out "neck" directions or directions of other multiply connected Fermi surfaces, as opposed to directions of closed Fermi surface such as the spherical portion in Fig. 2.

4. *Azbel'–Kaner cyclotron resonance.* This involves absorption of microwave energy as a function of varying magnetic field applied parallel to the surface of the metal. The period of oscillation in $1/H$ depends on the integral of the inverse of the component of velocity of the electron normal to **H** around the circumference of the orbit. If the shape of the Fermi surface is known by other means, this allows one to measure the electron velocity, i. e., the derivative of electron energy with lattice momentum, at the Fermi energy.

Finally, we should point out that once the Fermi surface is known in detail, as is now the case for most metallic elements, this knowledge can be combined with measurements of macroscopic thermal, electrical, magnetic, and optical properties to obtain information about the interactions of conduction electrons with each other, with collective excitations of various types, and with electromagnetic fields. Thus, once the Fermi surface is known, one can use that knowledge to consider effects beyond those treated by the one-electron theory (i. e., theory based on the model of electrons moving independently in an effective potential) for the electronic structure of metals.

In recent years there has been great interest in qualitative changes of behavior of certain metallic systems from the predictions of one-electron theory as described above. This one-electron theory involves describing the electronic behavior in terms of uncorrelated motion of the itinerant "band" electrons in the solid-state potential. Thus the departures of interest in the electronic behavior are referred to as correlated-electron effects and systems showing such effects are commonly referred to as correlated-electron systems.

The class of correlated-electron system that has gained the most attention are the "heavy-electron metals" often referred to as "heavy-fermion systems." These systems are typically compounds of light rare earths or light actinides (mostly of cerium or uranium) and are primarily characterized by having a low-temperature electronic contribution to the specific heat hundreds of times that corresponding to the band-electron density of states near the Fermi energy as found in one-electron band-theory calculations. There have been a variety of theories trying to explain heavy fermion behavior. These vary from Fermi-liquid-type theories where the Fermi-surface-type description remains essentially intact, to theories which question the existence of a measurable Fermi surface. Furthermore, the nature of the superconductivity in heavy fermion systems is strongly linked to behavior on the Fermi surface, e. g., whether there are anisotropic electron pairing states differing from the usual isotropic BCS pairing state and whether these new states are characterized by the vanishing of the energy gap at points or on lines on the Fermi surface. Thus the experimental observation of a Fermi surface, as for example by the de Haas–van Alphen measurements of Taillefer *et al.* (1987) on UPt$_3$, was very important in restricting and defining allowable mechanisms and theories. For UPt$_3$ the Fermi surface is multisheeted. The complicated and difficult measurements involved have called for refinements and improvements in the de Haas–van Alphen measuring techniques, very high-purity single-crystal samples, high magnetic fields, and very low temperatures involving use of magnetic-dilution refrigeration.

See also: Cyclotron Resonance; de Haas–van Alphen Effect; Electron Energy States in Solids and Liquids; Heavy-Fermion Materials; Magnetoacoustic Effect; Magnetoresistance.

Bibliography

R. J. Elliott and A. F. Gibson, *An Introduction to Solid State Physics*. Harper & Row, New York, 1974. (E)

Charles Kittel, *Introduction to Solid State Physics*, 5th ed. Wiley, New York, 1976. (E)

A. P. Cracknell and K. C. Wong, *The Fermi Surface*. Oxford Univ. Press, Oxford, 1973. (I)

J. M. Ziman, *Electrons in Metals*. Taylor & Francis, London, 1970. (I)

Walter A. Harrison, *Solid State Theory*. McGraw–Hill, New York, 1970. (A)

W. A. Harrison and M. B. Webb (eds.), *The Fermi Surface*. Wiley, New York, 1960. (A)

A. B. Pippard, *The Dynamics of Conduction Electrons*. Gordon and Breach, New York, 1965. (A)

Z. Fisk, D. W. Hess, C. J. Pethick, D. Pines, J. L. Smith, J. D. Thompson, and J. O. Willis, "Heavy-Electron Metals: New Highly Correlated States of Matter," *Science* **239**, 33–42 (1988).

L. Taillefer, R. Newbury, G. G. Lonzarich, Z. Fisk, and J. L. Smith, "Direct Observation of Heavy Quasiparticles in UPt$_3$, via the de H–van A Effect," *J. Magn. Magn. Mater.* **63**, 64, 372–376(1987).

Ferrimagnetism

W. P. Wolf

Ferrimagnets are magnetic materials which exhibit a spontaneous magnetization below a certain temperature, but one which in contrast to ferromagnets arises from atomic moments which are not all parallel. In the simplest ferrimagnets there are two sets of antiparallel moments (generally described as spins) which are unequal either in number or in the magnitudes of their magnetic moments. Thus, a ferrimagnet represents the general intermediate case between a ferromagnet of parallel spins and an antiferromagnet of equal antiparallel spins.

History

The concept of ferrimagnetism was formulated by Néel in 1948 to explain the properties of ferrites with the spinel structure and the general composition Fe_2O_3MO, with $M = Fe$, Mn, Co, Ni, etc. These were known to exhibit a spontaneous magnetization, but the moment extrapolated to $T = 0K$ did not correspond to that which would be expected if all of the spins were parallel. For example, for iron ferrite (magnetite), with $M = Fe^{2+}$, the two Fe^{3+} moments would give $5\mu_B$ each and the Fe^{2+} moment $4\mu_b$, so that parallel spins would give $14\mu_B$ per molecule of Fe_3O_4. The observed moment, on the other hand, was close to only $4\mu_B$-Néel postulated that the spins aligned antiparallel, with the Fe^{2+} and one of the two Fe^{3+} spins pointing in one direction, and the other Fe^{3+} spin in the opposite direction. Such an arrangement immediately explained the observed moment. The same model also explained the observed moments for the ferrites with $M = Ni^{2+}$, Co^{2+}, and Mn^{2+}.

Néel's insight was based on earlier crystallographic analyses, which had shown that the ferrites which showed the spontaneous magnetization had a so-called inverse spinel structure in which half the trivalent Fe^{3+} ions and all of the divalent M ions occupied the same type of crystallographic site (denoted as the B sites) while the remaining trivalent ions occupied the so-called A sites. It was then only necessary to postulate that there exists a dominant interaction between the spins on the A and B sites which favors antiparallel alignment.

Such a mechanism had already been proposed for antiferromagnets by Kramers, in 1934. It is explained by a form of indirect exchange interaction which involves the nonmagnetic O^{2-} ions.

Néel's simple model was subsequently extended in a number of ways, but the fundamental idea of *sublattices* of spins aligned along different directions is common to all ferrimagnets.

Mean Field Theory

Temperature Dependence of Magnetization

Néel's theory also explained quantitatively the temperature dependence of the magnetization. For this he generalized the mean field theory originally due to Weiss to explain ferromagnetism.

If we consider a typical spin on, say, the A sublattice, the effect of the exchange interactions with other spins on the A sublattice and with spins on the B sublattice can be approximated by a molecular or mean field

$$\mathbf{H}_A = -N_{AA}\mathbf{M}_A - N_{AB}\mathbf{M}_B , \tag{1a}$$

where $\mathbf{M}_A$ and $\mathbf{M}_B$ are the average sublattice magnetizations and N_{AA} and N_{AB} are proportional to the appropriate exchange interaction constants. Similarly, for a spin on the B sublattice

$$\mathbf{H}_B = -N_{BA}\mathbf{M}_A - N_{BB}\mathbf{M}_B . \tag{1b}$$

The sublattice magnetizations $\mathbf{M}_A$ and $\mathbf{M}_B$ are determined self-consistently using the relation between magnetization and field:

$$M_A = \sum_i n_i g_i \mu_B S_i B_{S_i}(x_A) , \tag{2}$$

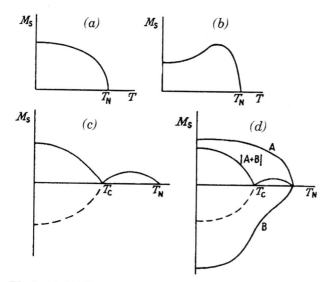

Fig. 1: (a)–(c) Three types of spontaneous magnetization temperature curves predicted by Néel's theory, (d) illustrates the origin of the compensation point T_c arising from the cancelation of two-sublattice magnetizations.

where

$$x_A = \frac{g_i \mu_B S_i}{k_B T} H_A$$

and

$$B_{S_i}(x_A) = \frac{2S_i + 1}{2S_i} \coth\left(\frac{2S_i + 1}{2S_i} x_A\right) - \frac{1}{2S_i} \coth\left(\frac{x_A}{2S_i}\right)$$

is the Brillouin function for spins S_i with spectroscopic splitting factor g_i. Here, μ_B is the Bohr magneton, k_B is Boltzmann's constant, T is the temperature, and n_i is the number of spins per unit volume of species i on the A sublattice. The sublattice magnetization $\mathbf{M}_B$ is given by a similar expression involving $\mathbf{H}_B$.

As in the corresponding case for ferromagnetism, it is found that there are nontrivial solutions with $M_A \neq M_B \neq 0$ for T less than some critical temperature, which is now known as the Néel temperature, T_N.

The temperature dependences of $M_A(T)$ and $M_B(T)$ are generally similar, but not identical, depending on the relative values of the coupling constants N_{AA}, N_{AB}, etc., and the other parameters. Correspondingly, the vector sum $\mathbf{M}_S = |\mathbf{M}_A + \mathbf{M}_B|$ can show a variety of unusual variations. Three of these are shown in Fig. 1. All three were predicted by Néel in his original paper and all were observed subsequently. Néel's theory also predicted some other variations which were later shown to be unphysical and to correspond to more complex situations with more than two sublattices (see below).

Compensation Temperature

The most unusual and interesting variation is case (c), whose origin is explained in more detail in Figure 1(d). This shows the situation when one sublattice has a larger low-temperature magnetization, but slower initial temperature variation. It is then possible for $|M_A|$ to equal $|M_B|$ at some temperature T_c, the so-called *compensation point*. Since experiments often measure the absolute magnitude of $\mathbf{M}_A + \mathbf{M}_B$, the change of sign at T_c may not always be evident, but it can be observed in weak magnetic fields and has given rise to some important practical applications (see below).

High-Temperature Susceptibility

For $T > T_c$, there is no spontaneous magnetization, but an applied field can induce a paramagnetic moment in the usual way. The temperature dependence of the corresponding low-field susceptibility χ is also predicted by the mean field theory. For a simple two sublattice ferrimagnet, χ is given by

$$\frac{1}{\chi} = \frac{T}{C_A + C_B} + \frac{1}{\chi_0} - \frac{\sigma}{T - \theta}, \tag{3}$$

where

$$C_A = \sum_i n_i g_i^2 \mu_b^2 S_i (S_i + 1) / 2k_B$$

is the Curie constant for the A sublattice, and C_B is the same for the B sublattice, and χ_0, σ and θ are constants which depend on the interaction parameters N_{AA}, N_{AB}, etc.

At high temperatures, Eq. (3) tends to the form $1/\chi = T/C + \theta'$, as for an antiferromagnet, consistent with the concept that the dominant interaction favors antiparallel neighboring spins. As the temperature decreases, however, $1/\chi$ will fall below the high-temperature asymptote and at some temperature, T_N, it will reach the value zero. Figure 2 shows experimental results for yttrium iron garnet (YIG) and the fit by an expression of the form of Eq. (3). It can be seen that the agreement is very good, except for temperatures very close to T_N.

Extension of the Two-Sublattice Model

The simple Néel model can be extended in a number of ways. The most obvious is the extension to more than two sublattices. Such a situation arises naturally in materials with a crystal structure with more than two inequivalent sites occupied by magnetic ions.

The best known examples of this situation are the rare-earth iron garnets, which have compositions of the form $5Fe_2O_3 \cdot 3M_2O_3$, where M is a rare earth or yttrium. In these materials, the Fe^{3+} ions occupy two crystallographically inequivalent sites, the octahedrally coordinated [a] sites and the tetrahedrally coordinated (d) sites, which occur in the ratio 4:6. A strong antiferromagnetic a–d coupling aligns the Fe^{3+} spins ferrimagnetically below about 570 K with a spontaneous moment which tends to $(6 - 4) \cdot 5(\mu_B = 10\mu_B$ as $T \to 0$ K. In the yttrium compound (commonly known as YIG for yttrium iron garnet), these are the only magnetic spins and the observed moment indeed tends to $10\mu_B$ at low temperatures, as shown in Fig. 3. The small discrepancy can be ascribed to impurities in these (early) results.

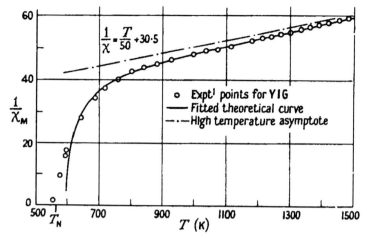

Fig. 2: Temperature variation of the reciprocal of the susceptibility for YIG above its Néel point (after Aléonard and Barbier, 1959).

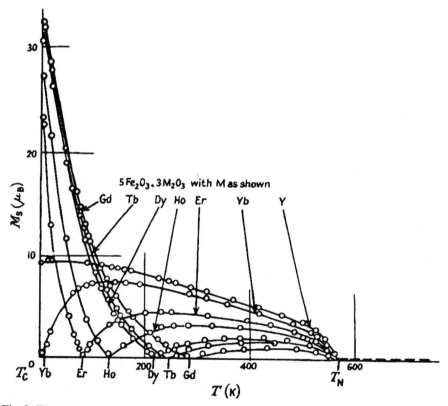

Fig. 3: Temperature variation of the spontaneous magnetization of several rare-earth garnets (after Pauthenet, 1958).

The rare-earth ions occupy the eightfold coordinated $\{c\}$ sites and are relatively weakly coupled to the Fe^{3+} on the $[a]$ and (d) sites. Their presence does little to affect the ordering near T_N but, as the temperature is decreased, the relatively weak exchange field due to the Fe^{3+} sublattices polarizes the rare-earth moments and significantly changes the magnetization, as shown in Figure 3.

The simplest case to understand quantitatively is GdIG ($M = Gd$). If one assumes that the Gd^{3+} moments, with $7\mu_B$ each, are all antiparallel to the net magnetic moment of the Fe^{3+} spins, we would expect a total moment of $[(6) \cdot 7 - (6-4) \cdot 5]\mu_B = 32\mu_B$ in excellent agreement with the observed value. It must be noted that the net moment is larger than that of the Fe^{3+} sublattices only at low temperatures, and that there is thus a compensation point, T_c, at which the two become equal. Above T_c, the Fe^{3+} moment dominates.

For the other rare earths the quantitative moment predictions are complicated by orbital crystal-field effects, but the general behavior is similar. The orbital effects also lead to another complication which was not included in the original Néel theory: the possibility of noncollinear sublattices.

Noncollinear sublattices are generally the result of competing interactions. For rare-earth ions with large angular momenta, the competition is between local anisotropy effects, which favor different directions for different crystallographic sites, and the exchange field which tends to favor a common direction for the spins. For some of the rare-earth ions in the garnets, this leads to six rare-earth sublattices related to one another in threes by cyclic permutation about a [111] axis.

Noncollinear sublattice structures can also arise from competition between *inter*sublattice (N_{AB} and N_{BA}) and *intra*sublattice exchange interactions (N_{AA} and N_{BB}). This possibility was first recognized by Yafet and Kittel, in 1952, who showed that relatively strong antiferromagnetic interactions *within* a given sublattice (say the B sublattice in a two-sublattice ferrimagnet) can actually cause that sublattice to split into two nonparallel sublattices, which then make a common angle with the other (A) sublattice, resulting in a triangular three-sublattice structure.

These ideas can clearly be extended further and in general there is no *a priori* way to determine how many inequivalent sublattices will actually form. However, in practice it is often found that the simple two-sublattice model will adequately describe most ferrimagnets, provided one allows for two conceptual extensions: the possibility of more than one type of spin on a particular sublattice and also the possibility that some of the spins may deviate from the common direction but with an average moment which coincides with that of the rest of the sublattice. The averaging implied by such extensions is readily accommodated in the mean field approximately. Figure 4 summarizes schematically some of the commonly found ferrimagnetic sublattice structures.

More Exact Theories

Of course, the mean field theory is not exact, even for a simple two-sublattice ferrimagnet, and more exact models can be developed. Specifically, for low temperatures and for microwave properties it is appropriate to consider the role of spin waves, quantized excitations with characteristic wavelengths and energies. Mean field theory is also a poor approximation close to T_N, as in other cooperative transitions.

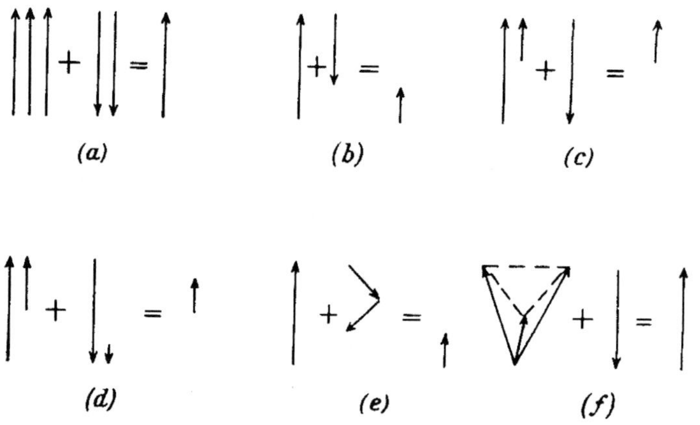

Fig. 4: Illustration of six simple sublattice arrangements which can give rise to a ferrimagnetic moment.

Materials Engineering

Both the spinel and the garnet structures allow wide varieties of substitutions of both magnetic and nonmagnetic ions. Correspondingly, it is possible to synthesize ferrimagnetic materials with specific magnetic properties and to control such parameters as magnetic anisotropy, magnetostriction, magneto-optical parameters, and microwave response, and the temperature dependence of all these properties. This has resulted in a large number of studies of related materials and a large body of experimental data. An extensive compilation may be found in the Bibliography.

Historically, the basic idea of making atomic substitutions to achieve specific properties was closely related to the early development of the theory of ferrimagnetism and it may have served as a useful model for the more general concept of materials engineering which is now widely used.

Technical Applications

Practical applications make use of several distinct properties of ferrimagnetic materials. Some of these have nothing to do with the specifically *ferri*magnetic nature of the spin alignment but rather the almost coincidental fact that most of the earliest ferrimagnets also happened to be good electrical insulators. Consequently, they could be used for magnetic applications involving alternating fields without the complication of eddy current losses associated with most ferromagnetic metals and alloys.

Such applications include radio-frequency transformers and inductances, and microwave devices such as isolators, circulators, and phase shifters. They were also important for early computer memories, using ferrite cores, and for magnetic bubble memories, using thin films of garnets deposited by liquid-phase epitaxy.

Another technical application which remains very important is the use of γ-Fe_2O_3 and its derivatives as the recording medium for audio, video, and computer tapes and for both floppy and hard discs. Although γ-Fe_2O_3 does not have the typical spinel composition, it does, in fact, have a crystal structure closely related to that of the spinels and its formula may be written as $(Fe)[\square_{1/3}Fe_{5/3}]O_4$, where $\square_{1/3}$ denotes a vacancy on one-sixth of the $[B]$ sites in the spinel structure. The corresponding moment should thus be $(5/3 - 1) \cdot 5\mu_B = 3.3\mu_B$, in excellent agreement with the observed value of $3.2\mu_B$ per formula unit.

Other technically important ferrimagnetic materials include a series of complex hexagonal ferrites with compositions such as $BaFe_{12}O_{19}$ and $Ba_3Co_2Fe_{24}O_{41}$, which have high magnetic anisotropies and, correspondingly, applications at high microwave frequencies. The high anisotropy also leads to a large magnetic remanence and these materials are used extensively as permanent magnets for such applications as small electric motors.

All of these applications depend only on the existence of a net spontaneous moment, but not on the presence of more than one sublattice. Recently, however, there have been some important developments which make use of thermomagnetic hysteresis near compensation points for recording digital information. The materials which have been used are amorphous *metallic* alloys with compositions such as $Tb_xFe_yCo_z$, with various x, y, and z, in which the transition-metal moments align antiparallel to those of the rare earths. The readout is by magneto-optical rotation of reflected light (Kerr effect) which depends on the direction of the remanent magnetization.

Concluding Remark

The discovery of ferrimagnetic materials which are metals rather than oxides, and moreover amorphous, underscores the generality of the concept of ferrimagnetism, and it suggests that there can be many different systems which will exhibit this type of cooperative magnetism.

See also: Ferromagnetism; Magnetic Materials; Magnetic Moments; Paramagnetism.

Bibliography

References to the specific papers cited in this article may be found in W. P. Wolf, "Ferrimagnetism," *Rep. Prog. Phys.* **24**, 212 (1961), which also contains a general discussion and an extensive bibliography of work prior to that date.

Extensive summaries of experimental data on ferrimagnetic oxides may be found in *Landolt–Bornstein: Numerical Data and Functional Relationships in Science and Technology*, K.-H. Hellwege and A. M. Hellwege (eds.), Group 3, Vol. 12, Parts a–c. Springer-Verlag, Berlin, 1978, 1980, 1982.

General discussions of ferrimagnetism are given in many textbooks, including:

J. Smit and H. P. J. Wijn, *Ferrites*. Philips Technical Library, Eindhoven, 1959.

S. Chikazumi and S. H. Charap, *Physics of Magnetism*. Wiley, New York, 1964.

A. H. Morrish, *The Physical Principles of Magnetism*. Wiley, New York, 1965.

A. Herpin, *Théorie du Magnétisme*. Presses Universitaires du Magnétisme, Paris, 1968.

D. J. Craik (ed.), *Magnetic Oxides*. Wiley, New York, 1975.

Recent research results may be found in the proceedings of two annual conferences on magnetism, published in the Journal of Applied Physics and the IEEE Transactions on Magnetics.

Ferroelasticity

S. C. Abrahams

Ferroelasticity is a crystal property that was first fully recognized in 1969, although its effects had long been observed previously in the form of mechanical twinning. A crystal is ferroelastic if it can contain two or more equally stable orientational states, in the absence of mechanical stress, that may be transformed reproducibly from one to the other under the application of stress along an appropriate direction. The minimum stress required in order to transform states is the coercive stress E_{ij}, where i, j denote the effective applied stress and transformed strain directions. The magnitude of E_{ij} generally ranges from 10^4 to 10^8 N m^{-2} and is often strongly dependent upon temperature, pressure, and defect distribution within the sample. The spontaneous strain vector $\mathbf{e}_s$ denotes the distortion direction within the unit cell as developed from a higher symmetry cell and also gives a measure of the distortion. The magnitude of $\mathbf{e}_s$ is typically 10^{-3} or less.

Ferroelastic crystal growth is generally accompanied by the formation of domains in which $\mathbf{e}_s$ is distributed over all directions allowed by the higher symmetry. The resulting twinned crystal can, in many cases, be detwinned by the application of compressive stress at room temperature; all ferroelastically twinned crystals may be detwinned by raising the crystal temperature above the transition temperature (T_c) at which the higher-symmetry phase forms, then cooling from this paraelastic phase under compressive stress applied along an appropriate direction. This direction is often one assumed by $\mathbf{e}_s$ in the twinned crystal but analysis is required to determine it in each case. Reorientation of the spontaneous strain direction in large transparent single-domain ferroelastic crystals is often readily observable optically, as the domain wall induced by the application of compressive stress sweeps through the crystal across the direction of view. Ferroelastic switching of $\mathbf{e}_s$ is always detectable by means of direction sensitive properties within the switching plane. Conoscopic examination of a ferroelastic biaxial crystal, for example, may exhibit readily controlled rotation of the plane containing the optic axes as stress is applied appropriately.

Ferroelastic point groups always form as subgroups of higher-temperature paraelastic point groups. The tetragonal point group $4/mmm$, for example, can give rise to the ferroelastic orthorhombic point groups 222, $mm2$, and mmm, with the normal equality of tetragonal a_1 and a_2 axes replaced by unequal orthorhombic a and b axes such that $e_s = 2(a-b)/(a+b)$ with $\mathbf{e}_s$ oriented along either [100] or [010].

The reorientation of $\mathbf{e}_s$ under compressive stress is accompanied by a rearrangement of the atomic distribution. The structure within the ferroelastic unit cell is such that all atoms are related pseudosymmetrically, either in pairs or by the identity operation. Thus, for each ith atom at the position $x_i y_i z_i$ there is a jth atom of the same element at the position $x_j y_j z_j$ with $x_i y_i z_i = f(x_j y_j z_j) + \Delta_{ij}$, where Δ_{ij} is an atomic displacement vector between the ith and jth atoms with magnitude on the order of one ångström and $f(x_i y_i z_i)$ is a transformation that leads to reorientation of the lattice vectors. A typical example is the perovskite-like material SmAlO$_3$, crystallizing in space group *Pbnm*, which has the atomic position relationship

$$x_i y_i z_i = \left(\tfrac{1}{2} - y_j, \ \tfrac{1}{2} - x_j, \ \tfrac{1}{2} + z_j \right) + \Delta_{ij} .$$

At room temperature, Δ_{ij} for Sm is 0.18 Å; for Al, Δ_{ij} is zero, and for O, Δ_{ij} has values 0.40 and 0.64 Å. The supergroup of the paraelastic phase is $\bar{4}2m$, $6/mmm$, $m3m$ or $m3$, since

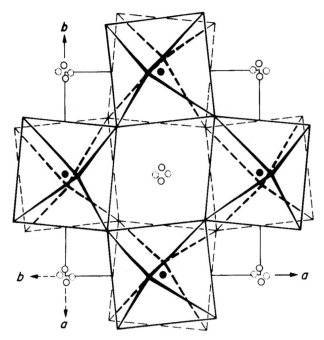

Fig. 1: Unit cell and atomic positions in $SmAlO_3$. Solid lines represent the **abc** orientation and broken lines the $\mathbf{\bar{b}ac}$ orientation following ferroelastic transformation. Octahedra represent O, solid circles Al, and open circles Sm atoms; the lower octahedral edges are omitted for clarity [after S. C. Abrahams, J. L. Bernstein, and J. P. Remeika, *Mat. Res. Bull.* **9**, 1613 (1974)].

the subgroup is *mmm*. With lattice constants at 298 K of $a = 5.29108$, $b = 5.29048$, and $c = 7.47420$ Å, $\mathbf{e}_s = 5.67 \times 10^{-5}$. Compressive stress of $50\,MN\,m^{-2}$ applied along [100] transforms the a axis into b and the b axis into a, as implied by the atomic position relationship which can be expressed in the equivalent form $\mathbf{abc} \rightarrow \mathbf{\bar{b}ac}$, i.e., the **a** and **b** basis vectors exchange and reverse sense under the application of stress as illustrated in Fig. 1.

Reported maximum values of Δ_{ij} in a variety of ferroelastic crystals range from about 0.4 to 2.4 Å. The coercive stress required to reorient $\mathbf{e}_s$ in crystals with large Δ_{ij} may exceed the cohesive strength, as is the case for the langbeinite-type structure $K_2Cd_2(SO_4)_3$ in which the maximum value of Δ_{ij} is 0.91 Å even at 418 K, with $T_c = 432$ K. A crystal that may be ferroelastically reoriented only by heating to T_c and cooling under stress is a *frozen* ferroelastic.

The thermal dependence of $\mathbf{e}_s$, i.e., $(d e_s / e_s dT)$, for a number of ferroelastics has been reported in the range -1 to $-20 \times 10^{-3}\,K^{-1}$. Ferroelasticity is often accompanied by such other properties as ferroelectricity, ferromagnetism, or superconductivity. For materials in which the other property is coupled, ferroelastic reorientation of $\mathbf{e}_s$ may cause reorientation of the second property. A well-known example of coupling between ferroelasticity and ferroelectricity is provided by $Tb_2(MoO_4)_3$ (TMO), where $x_i y_i z_i = (\frac{1}{2} - y_i, x_j, 1 - z_j) + \Delta_{ij}$, so that the a and b axes interchange under the application of compressive stress as the polar c-axis reverses sign simultaneously, with consequent reversal of the spontaneous polarization P_s, i.e., the basis vector transformation is **abc** → **b̄ac̄**. Similarly, reversal of P_s by the application of an electric field results in the interchange of the a and b crystal axes and hence the reorientation of e_s. The normalized thermal dependence of e_s is identical to that of P_s (*see* Pyroelectricity, Fig. 2), confirming experimentally that these two properties are fully coupled in TMO.

In the case of a fully coupled ferroelastic–ferromagnetic crystal such as Fe_3O_4 at room temperature, or Mn_3O_4 below T_c, either field reversal of the spontaneous magnetization or the application of compressive stress will simultaneously reorient both spontaneous magnetization and spontaneous strain.

Among the high-T_c cuprate superconductors are some with a slight orthorhombic distortion from tetragonal symmetry; these are generally ferroelastic and may have enhanced properties if detwinned as outlined above, since conductivity losses at domain wall boundaries are eliminated in single-domain crystals.

Ferroelastic devices, based upon the control of domain wall motion by a small electric field, include optical switches, adjustable optical slits, pattern generators including color modulators, and variable acoustic delay controllers. Micropositioners capable of displacements larger than those possible with piezoelectric positioners have also been designed on the basis of the dimensional changes obtained ferroelastically.

See also: Elasticity; Ferrimagnetism; Ferroelectricity; Ferromagnetism; Phase Transitions; Pyroelectricity; Superconductivity Theory.

References
[1]   S. C. Abrahams, *Mat. Res. Bull.* **6**, 881 (1971).
[2]   K. Aizu, *J. Phys. Soc. Japan* **27**, 387 (1969).
[3]   F. Lissalde, S. C. Abrahams, J. L. Bernstein, and K. Nassau, *J. Appl. Phys.* **50**, 485 (1979).

Ferroelectricity
J. D. Axe and M. E. Lines

Ferroelectrics are materials which possess an electric polarization in the absence of an externally applied electric field, together with the property that the direction of the polarization may be reversed by an electric field. The study of the properties of ferroelectric materials and the attempt to understand the nature of the ferroelectric state constitute the field of ferroelectricity.

Any macroscopic collection of matter is, to a very high degree of accuracy, electrically neutral. However, the positive and negative charges which make up the material are not necessarily distributed in a symmetric manner. If the "centers of gravity" ($\pm d$) of the summed positive and negative charges ($\pm q$) do not coincide (i. e., $d \neq 0$), then the material is said to possess an *electric dipole moment*, which is a vector quantity of magnitude $2dq$ and direction from $-d$ to $+d$. The dipole moment per unit volume is defined as the dielectric polarization. Any material develops a dielectric polarization P when placed in an electric field E, but a substance which has such a natural charge separation even in the absence of a field is called a *polar* material and is said to possess a *spontaneous* polarization P_s. Whether or not a material is polar is determined solely by its crystal structure in the sense that only certain classes of crystal structure are compatible with a polar charge distribution. Crystal structures can be divided into 32 classes, or *point groups*, according to the number of rotational axes and reflection planes they exhibit which leave the structure unchanged. Only those which contain a single axis of rotation symmetry without a reflection plane perpendicular to it are polar. Only 10 of the 32 point groups are polar although they do include examples from all crystal symmetry types lower than cubic (e. g., tetragonal, orthorhombic, hexagonal, trigonal, monoclinic, and triclinic).

Under normal circumstances even polar crystals do not display a net dipole moment. As a consequence there are no electric dipole equivalents of bar magnets. This is because the intrinsic electric dipole moment of a polar material is neutralized by "free" electric charge that builds up on the surface, migrating there from within the crystal (by internal conduction) or from the ambient atmosphere. An equivalent buildup of surface "magnetic charge" on bar magnets cannot take place because the free magnetic monopole (which is the magnetic counterpart of electronic charge) is not known to exist in nature. Polar crystals consequently reveal their polarity only when perturbed in some fashion which momentarily disturbs the balance with compensating surface charge. Since spontaneous polarization is in general temperature dependent, the most universally available such perturbational probe is a change of temperature which induces a flow of charge to and from the surfaces. This is the *pyroelectric* effect. All polar crystals are consequently pyroelectric and, as a result, the 10 polar crystal classes are often referred to as the pyroelectric classes.

Another rather obvious perturbational probe of polarization is the electric field itself. Surprisingly, however, the dielectric response of most pyroelectrics to such a field is small, linear, and not significantly different from that of a nonpolar material. The exceptions are those for which the applied field is sufficient in magnitude to reverse the direction of spontaneous polarization. For these the dielectric response is large and uniquely different, a typical variation of polarization P with applied field E being shown in Fig. 1. The highly nonlinear characteristic shape of this curve, which is called a *hysteresis* loop, is therefore the signature of that small subset of polar crystals which can be dielectrically "switched" by externally applied electric fields. It is this particular subset for which the term *ferroelectric* is reserved. More formally, a crystal is said to be ferroelectric if it possesses two or more orientational polar states and can be shifted from one to another of these states by an electric field. Since crystal perfection, electrical conductivity, temperature, and pressure are all factors which affect the ability to reorient polarization, the question of whether or not a particular material is ferroelectric is not one which can be answered by crystallographic determination alone. Also understood in this definition is the fact that the polar character of the orientational states should be absolutely stable in the absence of a field. Materials which mimic ferroelectrics but for which the polar states are only metastable (i. e., decay slowly with time when the field is removed) are defined separately as *electrets*.

Since ferroelectrics can be "switched", or in a sense structurally reoriented, by relatively modest fields (i. e., smaller than those which would produce crystalline dielectric breakdown), the implication is that they have crystal structures which are only slightly removed from some higher-symmetry (usually nonpolar) parent structure. The highest-symmetry phase in terms of which the ferroelectric phase can be described by small internal atomic rearrangements is called the *prototype* phase. In most instances this prototype phase actually exists as the

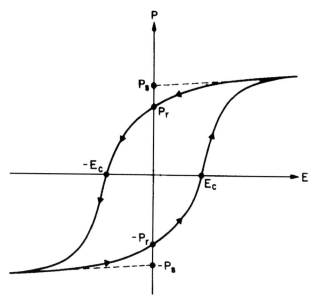

Fig. 1: A typical ferroelectric hysteresis loop showing the variation of polarization P with applied field E as the latter is scanned through a positive and negative cycle in the direction of the arrows. Special values of P and E on the loop, as depicted in the figure, define such ferroelectric parameters as P_s (spontaneous polarization), P_r (remanent polarization) and E_c (coercive field).

highest-temperature phase of the crystal. For such a case the ferroelectric phase develops from the prototype, as the temperature is lowered, by spontaneous displacements of the equilibrium positions of atoms within the unit cell of the crystal. If this takes place in the absence of a biasing applied electric field, then, because of the high symmetry of the prototype, at least two (and often more) polar atomic displacements are equivalent by symmetry and are distinguished only by their direction of polarization. For example, if the prototype phase is cubic and the incipient ferroelectric phase energetically favors a cubic axis for its polar orientation, then six equivalent such directions exist and, in general, all six of these equivalent crystal structures will coexist in the cooled sample in the form of *ferroelectric domains*. A single-domain sample can be induced by *poling* with an external field which energetically favors one directional-domain over the others.

Quite generally, the stable atomic configuration of any crystal at any particular temperature T is that which minimizes the crystal energy at that temperature. This energy results partly from thermally induced atomic vibrational motion and partly from forces of interaction between the atoms.

The interplay of these two energy components as a function of T can often result in a sudden spontaneous displacement of the equilibrium (i. e., vibrationally averaged) positions of the atoms. These sudden displacements are called *structural transitions* and the temperatures T_C at which they occur *Curie temperatures*. Most such transitions do not result in a separation of the "centers of gravity" of the positive and negative charges and are consequently not ferroelectric. Only those that involve the appearance of a spontaneous polarization are ferroelectric.

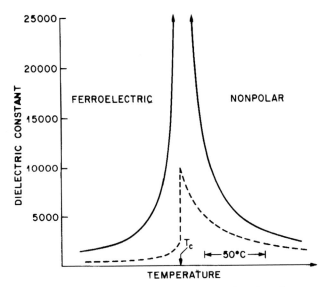

Fig. 2: Typical scans of dielectric constant as a function of temperature near the onset of ferroelectricity at T_C for a second-order (full curve) and first-order (dashed curve) ferroelectric.

It is quite possible for a whole sequence of phases, some ferroelectric, to occur as a function of changing temperature in some systems. It is even possible for one or more ferroelectric phases to occur sandwiched between nonpolar lower- and higher-temperature structures, although such an occurrence is rare. Most often ferroelectrics possess a single low-temperature (polar) phase for which the spontaneous polarization P_s disappears at a single Curie temperature T_C. For most ferroelectrics P_s goes to zero discontinuously at T_C (which is then called a *first-order* transition temperature) although a few examples are known for which P_s goes smoothly to zero at T_C (a *second-order* transition). Second-order ferroelectric transitions are characterized by a linear dielectric response which diverges at T_C while first-order transitions exhibit a marked, but finite, discontinuity at T_C (Fig. 2). Linear dielectric response, which is essentially a measure of the sensitivity of polarization to small applied electric fields, is recorded by a dimensionless quantity called the *dielectric constant*. Most ferroelectrics not only exhibit pronounced peaks in dielectric constant near T_C but tend to maintain large values of this quantity (often in excess of 1000) over wide temperature ranges. Normal nonferroelectric dielectric materials by contrast exhibit dielectric constants which are very much smaller (typically in the range between about 2 and 10). Ferroelectrics are also type characterized as *displacive* or *order–disorder* depending respectively upon whether the prototype phase is nonpolar in a truly microscopic sense (with all atoms vibrating about unique mean positions which together make up a nonpolar configuration) or only in a macroscopic sense (with some atoms hopping between different low-symmetry sites in a manner which averages to a nonpolar configuration only in a statistical sense). Since displacive transitions involve only small oscillations, their physics is simpler, being describable solely in terms of crystal lattice vibrations (or *phonons*).

When atoms vibrate they do not do so independently. Because of the forces between them they move in a collective fashion which defines a *normal mode* of oscillation. A great many normal modes of different spatial symmetries are present in a typical prototype phase. However, only one such normal mode is involved in an essential manner at a particular displacive structural transition. This mode is the one for which the spontaneous atomic displacements that set in below T_C represent a "time-frozen snapshot" of the vibrational displacement pattern of the mode above T_C. The mode is said to "drive" the transition and, becoming anomalously low in frequency on approach to T_C, is called a *soft mode*. The analogous dynamic anomaly driving an order–disorder transition does not directly involve phonons but concerns the much slower large-displacement intersite hopping motions which characterize this type of transition.

If the soft mode has a polar character, then the phase transition is said to be *intrinsically* ferroelectric and to define a *proper* ferroelectric phase. However, if the driving mode is non-polar, it is still sometimes possible for the phase below T_C to be ferroelectric. This can occur if there is an interaction of suitable symmetry between the polarization and the nonpolar driving-mode displacement pattern. Such transitions are said to be *extrinsically* ferroelectric and to define an *improper* ferroelectric phase. Only proper ferroelectrics have a dielectric response which becomes anomalously large near T_C; improper ferroelectrics exhibit only a minor (if any) anomaly in dielectric response at T_C.

In a similar fashion, many other physical properties can couple to a greater or lesser extent with a soft polar (or ferroelectric) mode. As a consequence all of these properties undergo extrinsically induced anomalies at a proper ferroelectric transition, anomalies which may become quite large near T_C itself. One universal example involves the appearance of a spontaneous mechanical strain which, if "switchable" by applied stress, defines the phenomenon of *ferroelasticity*. Accompanying the appearance of spontaneous strain is a linear coupling between polarization and stress (and between applied electric field and strain) referred to as *piezoelectric* response. As a result, all ferroelectrics, and indeed all polar materials, are *piezoelectric*. Another common manifestation of extrinsic coupling is the appearance of spontaneous birefringence (the velocity of light becoming different parallel and perpendicular to the polar axis) and more generally a linear dependence of refractive index on applied field known as the *electro-optic effect*. Even magnetic anomalies, in rare instances, can be extrinsically induced by ferroelectric transitions.

Most of the technical applications of ferroelectrics utilize one or more of these anomalous properties in some way. The high dielectric constants of ferroelectric materials make them useful in fabricating high-energy-storage capacitors.

Materials with large piezoelectric response are useful for sensitive sonar detectors or *transducers* (devices that transform pressure variations into voltage variations or vice versa), whereas those with large pyroelectric response have received application as infrared thermal detectors and, more recently, for the recording of infrared images. At present the fastest growing area of technical application centers on the favorable electro-optic properties. These have been put to use in the rapidly developing field of *photonics*, an information technology in which optics and electronics play complementary roles. In this context, important devices include optic switches and light modulators (for signal processing and interconnection), frequency converters, and light valve arrays (in the development of information storage and optical display technologies). In these various capacities some of the more frequently used ferroelectric materials are barium titanate ($BaTiO_3$), lithium niobate ($LiNbO_3$), lithium tantalate ($LiTaO_3$), potassium dihydrogen phosphate (KH_2PO_4), triglycine sulphate (often abbreviated as TGS), gadolinium molybdate [$Gd_2(MoO_4)_3$], bismuth titanate ($Bi_4Ti_3O_2$), and various ceramics based on mixtures of lead zirconate and lead titanate (PZT). Although, in principle, it is also possible to use the bistable nature of ferroelectric domains as the basis for high-storage-density computer memories, ferroelectrics in this context have not yet been produced with performance characteristics which can compete with the more reliable magnetic and semiconductor memory elements.

The phenomenon of ferroelectricity was first discovered in 1920 in sodium potassium tartrate tetrahydrate (better known as Rochelle salt) long after the related phenomena of piezoelectricity and pyroelectricity had been recognized and studied in the mid-to-late nineteenth century. For many years Rochelle salt remained the sole known example, and ferroelectricity was considered to be a rarity in nature. By contrast, over 500 other examples have now been discovered so that ferroelectrics are in fact far from being uncommon although, for reasons which are still not fully understood, the overwhelming majority are oxides. Perhaps the simplest (and certainly the most studied) high-symmetry prototype crystal structure which "spawns" ferroelectrics is cubic perovskite, with only five atoms per unit cell. Among its better known ferroelectric family members are barium titanate ($BaTiO_3$), lead titanate ($PbTiO_3$), and potassium niobate ($KNbO_3$). The most thoroughly researched improper ferroelectric is gadolinium molybdate ($Gd_2(MoO_4)_3$). Although most of the presently recognized ferroelectric materials are inorganic crystals, an increasing number of examples are now being discovered among organic polymers and even liquid crystals.

See also: Crystal Symmetry; Dielectric Properties; Neutron Spectroscopy; Piezoelectricity; Pyroelectricity.

Bibliography

C. Kittel, *Introduction to Solid State Physics*, 7th ed. Wiley, New York, 1995, Chap. 13. An elementary account of the basic phenomenon.

R. Blinc and B. Zeks, *Soft Modes in Ferroelectrics*. Elsevier, New York, 1974.

M. E. Lines and A. M. Glass, *Principles and Applications of Ferroelectrics and Related Materials*. Oxford University Press, Oxford, 2001.

T. Mitsui, *An Introduction to the Physics of Ferroelectrics*, Gordon and Breach, New York, 1976.

J. C. Burfoot and G. W. Taylor, *Polar Dielectrics and their Applications*. University of California Press, Berkeley, 1979.

Ferromagnetism

A. S. Arrott

Coulomb's law governs the interaction between electrical charges at rest. It is a requirement of the principle of relativity that Coulomb's law be modified when the charges are in motion relative to an observer. Though these modifications are usually very small, depending on the ratio of the products of velocities to the square of the velocity of light, their effects have been observed since antiquity. If the velocity of light were infinity, there would be no magnetism.

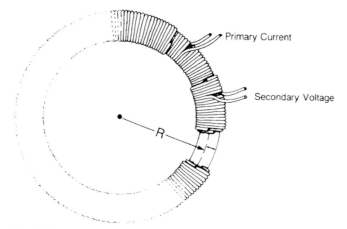

Fig. 1(a): A toroidal transformer. The cross section could be circular, or as assumed in the text a rectangle of area *ab*.

If it were but ten times larger than it is, electrical technology with its generators, transformers, and motors would be most difficult to achieve. Yet the genius, Faraday, was able to create a language of fields and fluxes that describes these workings of magnetic devices without reference to relativity or the velocity of light. Faraday discovered the law of induction that governs the behavior of the transformer. Not only is the transformer a practical device but also it is useful for studying and teaching the properties of magnetic materials.

Consider a magnetic material in the form of a toroid with two coils of wire wound about its surface as shown in Fig. 1(a). The toroid has a mean radius R and rectangular cross-section ab. A current i is passed through the primary coil and a voltage V is measured across the ends of the secondary coil. The voltage is proportional to the rate of change of current according to Faraday's law of induction

$$V = -L_{12}\frac{di}{dt} ,$$
(1)

where the mutual inductance L_{12} depends on the material of the toroid as well as upon its dimensions. In SI units V is in volts, di/dt is in amperes/second, and L_{12} has the units volt-seconds/ampere which is called the henry. For some materials L_{12} can be much larger than for others. One of these is iron. Materials with properties similar to iron are called ferromagnetic. The least ambiguous definition of ferromagnetism is the existence of a spontaneous magnetic flux density within a material. It is possible to deduce this from the behavior of a transformer, but it is more satisfactory to ascertain this by probing the material with neutrons. From the scattering of neutrons one may map out the magnetic flux density on the scale of atomic dimensions. With a transformer one obtains averages of the magnetic flux density over the volume of the material. If the current were changed at a constant rate from large negative to large positive values, the voltage might vary with current as shown in Fig. 1(b). The voltage would be proportional to the mutual inductance. The mutual inductance at large values of the current would approach the value it would have if no iron were present in the core of the toroid. In the absence of iron one could say that the primary current i_p creates a magnetic field

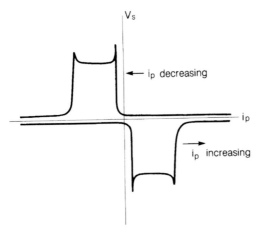

Fig. 1(b): The secondary voltage generated when the primary current increases and decreases linearly in time is plotted against primary current.

strength H given approximately by

$$H = \frac{N_p i_p}{2\pi R} , \tag{2}$$

where N_p is the number of turns in the primary winding. The magnetic field strength would be almost parallel to the circulating axis of the toroid. The units of H are A/m. The magnetic field strength is said to polarize the vacuum and create a magnetic flux density of magnitude

$$B = \mu_0 H , \tag{3}$$

where μ_0 is the permeability of the vacuum, a concept of convenience in establishing the SI units of B and H. The flux density is measured in webers/(meter)^2, which are also called teslas. The permeability of vacuum is defined to be $4\pi \times 10^{-7}\,\mathrm{T\,m\,A^{-1}}$ [or N/A^2 or H/m]. The changing current produces a changing magnetic field strength which produces a changing magnetic flux density. The changing magnetic flux $\Phi = abB$ produces the voltage in the secondary winding according to Faraday's law in the form

$$V = -N_s \frac{d\Phi}{dt} , \tag{4}$$

where N_s is the number of turns in the secondary. Actually there are electrons accelerating in the primary giving a push to the electrons in the secondary according to Coulomb's law with relativistic corrections, but Faraday's view is very helpful. The unit of flux is the volt-second or the weber. Magnetic flux is defined as an integral of the magnetic flux density B over an area, here, the cross section of the secondary coil, $a \times b$. The mutual inductance of the toroidal transformer with no iron core is then approximately

$$L_{12} \cong N_p N_s \mu_0 \left(\frac{ab}{2\pi R} \right) ; \tag{5}$$

when the iron is present,

$$L_{12} \cong N_p N_s \mu_R \mu_0 \left(\frac{ab}{2\pi R} \right) ; \tag{6}$$

where μ_R is the relative permeability, which can be as large as 10^6 for iron. As μ_R depends on how the magnetic flux density B varies throughout the toroid and how B changes in time in response to the current in the primary, it will depend on that current and also on the past history of currents applied. At high currents (large H) B uniformly curls about the toroid and has contributions from the current in the primary and from the material of the toroid. This is expressed by the relation

$$B = \mu_0(H + M_s) , \tag{7}$$

where M_s is called the saturation magnetization, H is the field which would be present if the iron were not there, and M_s approaches an almost constant value at large values of H. If a large current i_1 is reversed, the direction of B, M_s, and H will be reversed. The change in B, ΔB, will be

$$\Delta B = \mu_0(\Delta H + 2M_s) = \mu_0 \left(\frac{N_p}{2\pi R} \right) \cdot 2i_1 + 2\mu_0 M_s .$$

One can determine ΔB experimentally, calculate ΔH from the current, and then obtain the value of M_s. ΔB can be determined from the time integral of the secondary voltage, that is,

$$- \int V \, dt = N_s \int \frac{d\phi}{dt} \, dt = N_s \Delta \phi = N_s ab \Delta B . \tag{8}$$

The voltage in the secondary coil results mainly from the reorientation of the magnetization pattern in the iron as the primary current changes, but also has a direct contribution from the changing current itself. The time integral can be carried out by using an operational amplifier with capacitive feedback as an analog computer or by using a voltage-to-frequency converter and counting. The integration of Fig. 1(b) produces Fig. 1(c), which is called the major or saturation hysteresis loop. The axes are labeled using average values of B ($= \phi/ab$) and H ($= N_p i/2\pi R$). The value of this H for which B is zero is called the coercivity H_c. The value of B for which $H = 0$ is called the remanence B_r. For the particular loop shown in Fig. 1(c), $B_r = \mu_0 M_s$.

The study of ferromagnetism divides neatly into attempts to answer two questions: How does the spontaneous magnetization arise in certain materials such as iron, cobalt, and nickel and several of the rare-earth elements as well as in oxides and other compounds of these and a few other metals? And how does the reversal process take place? The latter question is crucial in understanding the technical use of ferromagnetic materials. Though neither question has been fully answered in any single material, the study of magnetism has contributed significantly to understanding the behavior of electrons in solids and to the development of complex technologies based on the behavior of magnetic materials.

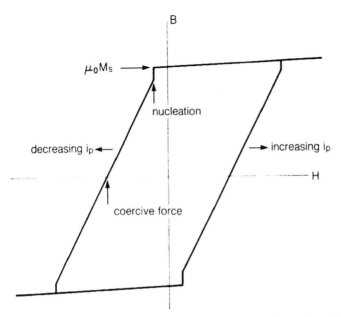

Fig. 1(c): The magnetic flux density, calculated from the integration of Fig. 1(b), plotted against the magnetic field, calculated from the primary current.

The carrier of magnetism is the electron, which can be described as possessing an intrinsic angular momentum (spin) as well as a charge. The intrinsic magnetic moment of an electron (the Bohr magneton, $\mu_B = 9.27 \times 10^{24}$ J/T) is then a relativistic consequence of a spinning charge. Electrons in atoms combine in such a manner that the spin of the atom is generally less than the sum of the spins of the individual electrons. When atoms combine in solids, the spin of the atom is generally less than when the atoms are separated. The quantum-mechanical symmetry effect known as the Pauli exclusion principle makes it impossible for two electrons to have the same spatial coordinates if they have parallel spins (spins are either parallel or antiparallel for two electrons). Thus electrons with parallel spins avoid one another. This decreases the Coulomb repulsion between them below what it would be on the average for two electrons with antiparallel spins. Thus the net electrostatic binding of electrons to the atom, including the Coulomb attraction to the nucleus, would be greater for the parallel spin electrons. Yet to avoid one another the parallel-spin electrons need more kinetic energy. It is the balance between these effects which determines the net spin of an atom and the spontaneous magnetization of a material.

Calculations of the electronic structure of free atoms are sufficiently accurate to account for Hund's rules, which relate the magnetic moments to the number of electrons on the atom.

The moment of an atom is more likely to be preserved in a solid if the electrons responsible for the moment are not appreciably involved in the binding of the solid. This is the case for elements of the first transition series, the lanthanides, and the actinides. The elements of the first transition series keep their atomic moments better in compounds such as oxides and sulfides than they do in metals.

A free manganese atom has five electrons in orbitals of d symmetry, all with their spins parallel. In some solids manganese exhibits a magnetic moment per manganese atom of $5\mu_B$. If this happened in the metal, it would make a most attractive magnetic material with a saturation magnetic flux density equal to 4.5 T. It is technologically unfortunate that this does not happen. The highest spontaneous magnetic flux density at room temperature is found in an alloy of iron and cobalt for which $B_s = 2.4$ T. The spontaneous magnetic flux density for the ferromagnetic elements iron, cobalt, nickel, and gadolinium are 2.20, 1.82, 0.64, and 2.45 T, respectively, in the limit of low temperatures. At room temperatures iron, cobalt, and nickel have B_s values 2.15, 1.80, and 0.61 T, respectively.

In some materials the direction of alignment of the magnetic moment varies from atom to atom, e. g., staggered up–down–up–down. If the net magnetic flux density is zero in any microscopic region, such a material is called antiferromagnetic. If a staggered pattern results in a net magnetic flux density, the material is called ferrimagnetic. The arrangement of magnetic moments in ferrimagnetic and antiferromagnetic materials is studied in detail by neutron diffraction.

At sufficiently high temperatures all of these types of magnetically ordered materials become magnetically disordered and exhibit paramagnetism. The disordering temperature is called the Curie temperature. For gadolinium this Curie temperature is just below room temperature. For iron, cobalt, and nickel the Curie temperatures are ~ 1040 K (~ 770 °C), ~ 1390 K (~ 1120 °C), and ~ 630 K (~ 358 °C), respectively.

Perhaps the behavior of Gd metal is best understood. The seven electrons in the 4f shell of Gd are aligned parallel in accordance with Hund's rule. The f electrons on one atom are aligned with respect to the f electrons on neighboring atoms through a polarization of the itinerant conduction electrons which are also responsible for the binding of the metal. The degree of this polarization ($\sim \frac{1}{2}\mu_B$ per atom) and the Curie temperature are reasonably well accounted for by first-principles theory aided by modern computing techniques.

A general framework for discussing ferromagnetism is provided by the Heisenberg model, which assumes a net spin on each atom and an interaction between spins on neighboring atoms. The energy is written as

$$E_{ex} = -\sum_{i=1}^{N}\sum_{j=1}^{n} J_{ij}\mathbf{S}_i \cdot \mathbf{S}_j , \tag{9}$$

where J_{ij} is called the exchange parameter and E_{ex} is called the exchange energy. The sums are over all pairs of n near neighbors on a lattice of N sites. The detailed calculation of the statistical-mechanical properties of the Heisenberg model has been a major activity in theoretical physics for three-quarters of a century. One attempts to calculate the spectrum of magnetic excitations known as spin waves, the temperature dependence of the spontaneous magnetization, and the nature of magnetic fluctuations near the Curie temperature. Comparisons of the predictions of the model with experimental results are often rather satisfactory for a model with but one parameter $J_{ij} = J$ for the nearest neighbors.

The concept of exchange energy plays an essential role in models of the magnetization reversal process in technical magnetization. In these models the magnetization is taken as a continuum vector field which changes direction by at most a small fraction of a radian from atom to atom and does not change in magnitude. The changing of direction is a magnetic strain which increases the (negative) exchange energy by an amount

$$\Delta E_{\text{ex}} = A \int \frac{1}{M_s^2} \left\{ [\text{curl}\,\mathbf{M}_s]^2 + [\text{div}\,\mathbf{M}_s]^2 \right\} dV + \text{surface terms} . \tag{10}$$

where $A = JS^2/a$ (a is an atomic distance) is an exchange constant. The exchange energy, which is responsible for ferromagnetism, resists rapid changes in the magnetization direction. In the toroidal transformer core the magnetization direction tends to curl about the axis of the toroid and to be everywhere parallel to the surfaces. The reason for this is that in addition to the exchange interaction between the spins on neighboring atoms there is a much weaker magnetostatic interaction among all magnetic moments. The magnetostatic interaction is again a relativistic effect between spinning atoms. This energy is a minimum if $\text{div}\,\mathbf{M}_s = 0$ everywhere in the material and $\hat{n} \cdot \mathbf{M} = 0$ on all surfaces, where $\hat{n}$ is the normal to the surface. If these conditions are not met, each magnetic moment will experience a demagnetizing field H_D which has as its sources $\text{div}\,\mathbf{M}_s$ and $\hat{n} \cdot \mathbf{M}_s$. The curling pattern satisfies $\text{div}\,\mathbf{M}_s = 0$ and $\hat{n} \cdot \mathbf{M}_s = 0$ while at the same time making only a very small increase in the exchange energy ($|\text{curl}\,\mathbf{M}_s| = M_s/R$, where R is the radius of the toroid). To reverse the sense of the curling pattern in response to a reversal of current in the primary of the toroid both the exchange energy and the demagnetizing field energy must increase in the process. In addition the process will pass through configurations which are unstable, irreversible changes will occur, and hysteresis will result. A complete theoretical description of this process does not exist. Observations on an iron transformer show that the crystal structure plays an important role in the reversal process. The magnetic moment is not entirely due to the spin of the electrons, but also includes a contribution from the orbital motion of the electrons. This is the origin of the magnetocrystalline anisotropy, an energy which depends on the orientation of the magnetization with respect to the crystal axes. If the toroidal core itself is wound of a thin metallic tape in which all the crystal grains are oriented along the tape, the crystalline anisotropy as well as the exchange energy and the demagnetizing field energy prefers the magnetization to curl about the toroidal axis. All these energies would be the same if the magnetization pattern were completely reversed. But the energy of interaction of the magnetization with the primary current does change if the primary current is reversed with respect to the magnetization pattern. This is the driving force for reversals. The effect of magnetic anisotropy on the reversal process is to inhibit all processes that take place by a gradual rotation of the magnetization everywhere. What happens is that somehow one or more domain walls nucleate and pass through the cross section of the toroid reversing the magnetization as they go. The walls separate regions aligned in opposite directions. The magnetization in a wall rotates from one direction to the other in a distance of a few hundred atoms for iron at room temperature. The wall has exchange energy because of the rapid rotation, anisotropy energy because the magnetization no longer is confined to preferred directions, and generally some demagnetizing field energy because either $\text{div}\,\mathbf{M} \neq 0$ in the wall or $\hat{n} \cdot \mathbf{M} \neq 0$ where the wall intersects the surfaces. A change in the direction of magnetization can also give rise to mechanical strains through the elongation (or shrinkage) of the atomic lattice in the direction of magnetization. This is called magnetostriction. Magnetoelastic effects are long-ranged, affecting the reversal process in ways that have yet to be fully calculated. Furthermore, domain walls interact with crystal imperfections, grain boundaries, and surfaces. Presumably imperfections are also important for the nucleation of the domain walls. Magnetic vortices can enter at surfaces leading

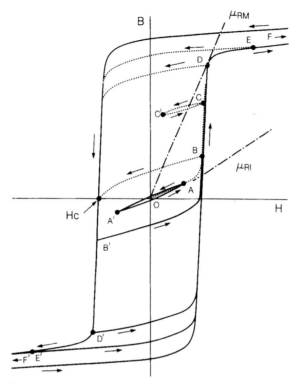

Fig. 2: Hysteresis loops for a commercial transformer material. Portions of the major loop are closed beyond the limits of the diagram (*F* and *F'*).

to generation of domain walls and changing the symmetry of magnetic configurations. If the field from the current in the primary which is sufficient to nucleate the walls is greater than the field necessary to move the walls, the material has a "square" hysteresis loop. If the reverse is the case, the field necessary to move the walls influences the form of the hysteresis loop.

For the loop shown in Fig. 1(c), the nucleation takes place as in a square-loop material, but the walls reach a stable configuration in that nucleation field. Further changes follow the change in the magnetic field, until the walls disappear, leaving the magnetization reversed. In a part of the loop where walls are present, if the direction of field change is reversed, minor hysteresis loops will be followed. Figure 2 shows the behavior of a typical commercial transformer material including a few minor loops. By cycling the magnetic field with decreasing amplitude it is possible to demagnetize the core reaching a state with no net B for zero applied field. The virgin magnetization curve $OABCDEF$ is then traced out by increasing this magnetic field monotonically. The initial permeability μ_{RI}, and the maximum permeability μ_{RM} are indicated in Fig. 2. The loops $DD'D$ or $EE'E$ would correspond to efficient operation of a power transformer, while the loop $AA'A$ might be employed in a communications transformer. In the first case the area of the hysteresis loop contributes to undesirable energy losses, in the second case the area of the loop corresponds to a decrease in the quality of the signal transmission.

The most used magnetic material is polycrystalline motor-grade iron (99.6% Fe) with μ_{RM} about 5×10^3 and a hysteresis loss, ω_H, about 150 joules per cubic meter of core material for one cycle of a loop with a maximum induction of $B_M = 1.2$ tesla. High-grade power transformers use grain-oriented silicon-iron (3.2 wt% Si) with μ_{RM} values over 7×10^4. It is possible to operate with flux densities near 1.9 tesla for fields of 800 A/m ($\mu_0 H = 10^3$ tesla). Core losses for a loop with $B_M = 1.7$ tesla at 60 Hz can be as little as 200 J/(m³ cycle). Only one-quarter of these losses comes from the area of the hysteresis loop as measured slowly.

An alloy with 79Ni-5Mo-16Fe when suitably heat treated has μ_{RM} over 5×10^5 and a hysteresis loss of less than 0.05 J/m³ at $B_M = 0.5$ tesla. As the superior magnetic materials are metallic electrical conductors, it is necessary to use thin laminations with surface insulation to minimize the eddy-current losses. A transformer core as shown in Fig. 1(a) would be a continuous wound tape. At radio frequencies it is necessary to use low-conductivity materials such as nickel-ferrite which has $\mu_R = 2 \times 10^3$. The best materials in each higher frequency range have lower permeabilities, e. g., $\mu_R = 10$ for $3BaO \cdot 2CoO \cdot 12Fe_2O_3$ at 1 GHz. Ferrites are used as components for microwave devices. Amorphous alloys are devoid of crystalline anisotropy, possess enhanced resistivity, and can be engineered to minimize the effects of magnetostriction. By partial recrystalization, small magnetic particles are embedded in an amorphous matrix, combining the low coercivity of amorphous alloys with the higher saturation magnetization of crystalline alloys. These materials are penetrating the markets for transformers.

The preparation of suitable materials for magnetic circuits, motors, generators, transformers, and electromagnets has been an important metallurgical activity for a century. More recently magnetic materials have played a dominant role in the storage of information and are important to the computer revolution. In devices operating at GHz frequencies, magnetic damping becomes important.

For fifty years the most familiar permanent magnets are made of barium ferrite and are found pinning notes to metal cabinets or refrigerators in most households. Barium ferrite is very cheap because it is extracted in an almost ready-to-use form from mines. Permanent magnets made of $Fe_{14}Nd_2B$ have made possible many new applications from disk drives to dentures. These are so strong that they are dangerous to handle.

Permanent magnets are first of all materials for which there is a strong preference for the magnetization to lie in particular directions, but in addition they must be sufficiently heterogeneous to suppress the growth of domains oriented in the directions preferred by the field. A cylinder of such a material with its preferred direction along the axis would retain most of its magnetization parallel to its sides after being magnetized briefly in a strong axial field, despite the strong demagnetizing fields arising from $\hat{n} \cdot M$ at its ends. In modern permanent magnets the magnetocrystalline anisotropy and the impedance to domain wall motion are sufficient that even quite short cylinders will remain magnetized. Though the stronger permanent magnets are much more expensive, even in terms of stored energy, often design considerations favor the additional cost. The stored energy density in a permanent magnet is a maximum if the remanent magnetization $M_r = B_r/\mu_0$ is maintained against a demagnetizing field $H = -\frac{1}{2}M_r$. The maximum energy density $B_r^2/4\mu_0$ is often called the energy product and quoted in megagauss-oersteds which translates into 0.1 MJ/m³. This number has increased by a factor of 10 to 5 MJ/m³ through materials research based on the theory of micromagnetism. That permanent magnets produce and maintain fields without expenditure of energy is of increasing importance in the twenty-first century.

See also: Faraday's Law of Electromagnetic Induction; Ferrimagnetism; Hysteresis; Magnetic Domains and Bubbles; Magnetic Materials; Magnets (Permanent) and Magnetostatics; Maxwell's Equations; Paramagnetism; Spin.

Bibliography

Current research in ferromagnetism is reported and reviewed in the *Journal of Magnetism and Magnetic Materials* and in the proceedings of annual magnetism conferences by that journal, the *Journal of Applied Physics*, and the *IEEE Transactions on Magnetism*.

Feynman Diagrams

T. Fulton

Feynman diagrams (or graphs) are graphical representations of perturbation-theory calculations involving the scattering and propagation of interacting particles. In the case of relativistic quantum field theory, the particles can interact with external fields or with other particles (through the emission and absorption of virtual quanta). In the many-body problem, the particle-particle interaction can also occur through potentials.

The diagrams serve two principal purposes: to guide actual calculations, and to provide an intuitive physical picture of the process being considered. With their aid, we can ascertain that all contributions to a given order in the perturbation expansion have been counted. Further, a specific integral is associated with each diagram. This integral can be obtained with the use of the "Feynman rules" as indicated in what follows.

As an illustration, consider quantum electrodynamics, involving only electrons (e^-), positrons (e^+), and photons (γ), in the absence of external fields. The diagrams representing the scattering amplitude for e^+–e^- scattering (Bhabha scattering) in lowest order are shown in Fig. 1. When the diagrams are interpreted in space-time, Fig. 1 can be thought of as representing the physical processes that take place in such a scattering, with the various lines corresponding to world lines of particles. Thus, in Fig. 1a, reading from the bottom up, an e^+ and an e^- in some initial state (i) scatter each other into a final state (f) through the exchange of a virtual γ, which then propagates to 2 and creates the final e^+–e^- pair. In Fig. 1b the initial e^+ and e^- annihilate at 1, producing a virtual γ that propagates to 2 and there produces the final e^+ and e^-.

For calculational purposes, it is often more convenient to think of Feynman diagrams in momentum rather than coordinate space. In either case, a general Feynman diagram of arbitrary order (involving $e^\pm$ and γ) consists of external lines (corresponding to incoming and outgoing particles), internal lines representing Feynman propagators or Green's functions, and corners or vertices at which two $e^\pm$ lines and one γ line meet. Figure 1 illustrates a case where all the external lines are $e^\pm$ and the internal line is a photon propagator. Figure 2, corresponding

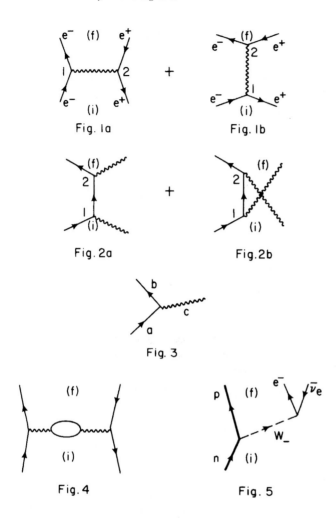

Fig. la

Fig. lb

Fig. 2a

Fig.2b

Fig. 3

Fig. 4

Fig. 5

to lowest-order perturbation theory for Compton scattering, has γ's as incoming and outgoing particles as well, and the internal line is an electron propagator. The Feynman rules associate specific functions with each external line, internal line, and vertex; specify sign conventions; and, in momentum space, require conservation of energy and momentum at each vertex. As a consequence, the propagators are in general "off mass-shell" (i. e., the usual relationship, relativistic or nonrelativistic, between energy and momentum of a free particle does not hold). The arrows in Figs. 1 and 2 represent another convention: electrons propagate "forward" and positrons "backward."

A particular advantage of Feynman diagrams in relativistic field theories (which is absent in the many-body applications) is the way in which a given diagram can represent a number of processes. Consider a basic vertex, as in Fig. 3, which is part of a Feynman diagram. (The lines can be external or parts of internal lines.) Line a in Fig. 3 has alternative interpretations as an incoming electron or outgoing positron, line b as an outgoing electron or incoming positron, and line c as an outgoing or incoming photon. Thus, if we are less explicit about the labels $e^{\pm}$ and i, f, Fig. 1 can also be used to represent $e^- - e^-$ (Møller scattering) as well as $e^+ - e^-$ scattering (with Fig. 1b corresponding to the exchange term) and Fig. 2 can describe $e^+ - e^-$ annihilation into two photons as well as Compton scattering.

Feynman diagrams of two other types are shown in Figs. 4 and 5. Figure 4 represents one of the radiative corrections to Bhabha scattering – that due to the lowest-order vacuum polarization correction to virtual γ exchange. Figure 5 illustrates a Feynman diagram from weak-interaction theory – the decay $n \rightarrow p + e^- + \bar{\nu}_e$ mediated by an intermediate vector boson W^- . Although charge and fermion-number conservation at vertices are implicit in the Feynman rules for quantum electrodynamics, they are much more apparent in this β-decay process.

See also: Quantum Electrodynamics; Quantum Field Theory.

Bibliography

A. A. Abrikosov, L. P. Gorkov, I. E. Dzyaloshinski, *Methods of Quantum Field Theory in Statistical Physics*, Chapters 2 and 3. Prentice-Hall, Englewood Cliffs, NJ, 1963.

J. D. Bjorken and S. D. Drell, *Relativistic Quantum Mechanics*, Appendix B. McGraw–Hill, New York, 1964.

F. J. Dyson, *Phys. Rev.* **75**, 486, 1736 (1949).

R. P. Feynman, *Phys. Rev.* **76**, 769 (1949); **80**, 440 (1950).

J. M. Jauch and F. Rohrlich, *The Theory of Photons and Electrons*, Chapter 8. Addison-Wesley, Reading, Mass., 1955 (2nd ed., Springer-Verlag, Berlin and New York, 1976).

F. Mandl, *Introducton to Quantum Field Theory*, Chapter 14. Wiley-Interscience, New York, 1959.

R. D. Mattuck, *A Guide to Feynman Diagrams in the Many-Body Problem*. McGraw–Hill, New York, 1967.

J. J. Sakurai, *Advanced Quantum Mechanics*, Appendix D. Addison-Wesley, Reading, Mass., 1967.

S. S. Schweber, *An Introduction to Relativistic Quantum Field Theory*, Chapter 14. Harper, New York, 1961.

Fiber Optics

W. P. Siegmund

Fiber optics is the branch of physics pertaining to the passage of light through thin filaments of transparent material and includes transmission theory, applications, and manufacturing technology. Although some references of historical interest go back over 100 years, practical development began in the early 1950s [1, 2]. In particular, the invention of optically "clad" fibers in 1953 [3] placed fiber optics on a firm foundation. Since then the field has grown steadily both in the range of its applications and in commercial value.

The basic phenomenon involved in fiber optics is the channeling of light by total internal reflections from the fiber walls. In conventional (step-index) fibers the reflections are produced by cladding the walls with a material of lower refractive index to isolate the fibers from one another and ensure a clean reflecting interface. Alternatively the refractive index of the fiber may be graded radially outward from a higher to a lower value to achieve this channeling (graded-index fiber).

The principal characteristics of optical fibers are their diameter (in μm), attenuation (in dB/km), and numerical aperture (*NA*). In step-index fibers the *NA* is simply related to the refractive indices of the fiber core ($_1$) and the cladding (n_2) by $NA = (n_1^2 - n_2^2)^{1/2}$.

The *NA* of a fiber is a measure of the cone of light which the fiber can accept and, therefore, contributes to its overall transmission efficiency, analogous to the optical "speed" of a camera lens.

In communications applications optical fibers are generally considered as wave guides having discrete transmission modes. To obtain maximum transmission bandwidth through such fibers it is possible to minimize the differences in group velocities of the various modes by means of a precisely controlled graded-refractive-index profile.

Alternatively the fiber may be designed with a sufficiently small core diameter and/or small numerical aperture such that only a single mode is transmitted (single-mode fiber) thus eliminating differences in mode velocities. Such fibers are also of great interest in very long transmission lines.

Materials used in the manufacture of optical fibers include optical glasses, fused silica, and certain plastics. The glasses used in various forms of fiber-optic devices are chosen for their refractive index, transmittance, thermal, and in some cases chemical characteristics. For long flexible fibers transmittance is the principal factor, while for fused rigid-type fiber components such as intensifier tube faceplates, refractive index is the principal factor. In the latter case a core glass having a high refractive index (typically 1.8) is used to obtain an *NA* value near unity which provides the high diffuse light transmittance required for these applications.

For "low-loss" fibers, used in long distance communication, materials of the highest purity are required. One of the principal methods by which such purity is achieved is by chemically depositing silica (or a silica-based glass whose refractive index is modified by additions of boron or germanium oxides) from the vapor phase onto a pure silica substrate (chemical vapor deposition or CVD process). In this process, impurities such as transition metals which are a major cause of optical absorption are effectively eliminated. Fibers drawn from "preforms" made by this process have exhibited an optical attenuation of less than $1.0\,\mathrm{dB/km}$ in the near ir spectrum ($1.3\,\mu\mathrm{m}$).

In addition to the low attenuation gradient index and single-mode fibers for long-distance communication cables, fused silica fibers are also being designed for special purposes including polarization maintenance in interferometer applications (such as fiber-optic gyroscopes), as well as low-birefringence single-mode fibers for other sensor applications. In general, optical fibers are finding increased application in a variety of sensors involving either their intrinsic characteristics such as micro-bending losses and birefringence, or as low attenuation relays from remote transducers.

Bundles of coordinated optical fibers (i. e., coherent bundles) capable of image transmission have important applications in medical endoscopes and industrial inspection instruments. In such bundles the individual fibers are positioned identically at both ends of the bundle but are free to flex in between. Thus an image projected on one end appears in recognizable form at the end although it consists of a series of discrete dots when viewed under high magnification. Such bundles are usually made by precisely winding layers or "ribbons" of fibers on a drum or spinning frame and then laminating a large number of such ribbons together for a small portion of their circumference into a tightly packed assembly. After bonding, this portion is cut open and the exposed ends are polished to form the "coherent" ends of the bundle. Continuous winding of the entire cross section of the bundle has also been used but generally does not provide as precise fiber alignment.

Small-diameter, precise, coherent image bundles are also made by the leaching process. In this case, the low refractive index cladding on each fiber is surrounded by a second cladding of an acid-soluble glass. The coherent bundle is made in a series of steps starting with the double-clad preform which is drawn into large single fibers. These are assembled into a precise array containing many thousand fibers and redrawn into a rigid bundle of the desired cross section. With its ends protected against leaching, this bundle is immersed in acid to remove the soluble cladding leaving the bundle highly flexible and precisely "coherent."

One measure of the image quality through such a bundle is the total number of fibers contained i. e., essentially equal to the number of image elements transmitted. For medical or industrial instruments this ranges from about 5 000 to 50 000 fibers depending on the size and type of instrument. The individual fiber size may range from about 8 to 20 μm. For medical instruments (gastroscopes, colonoscopes, bronchoscopes, etc.) the length ranges up to about 200 cm while for industrial instruments up to 450 cm.

A small objective lens at the tip projects an image onto the coherent bundle and the transmitted image is viewed by an eyepiece. Means for remotely focusing the objective lens and articulating the end are often provided in such instruments. In such instruments one or more additional fiber bundles are used as flexible light guides to illuminate the object. These transmit light from a remote light source to the distal tip of the instrument. The fibers in these bundles are not normally coordinated (i. e., they are noncoherent bundles) and thus cannot transmit a recognizable image.

Such noncoherent fiber-optic bundles (light guides) are used in a variety of applications for remote illumination and light sensing. Bundle sizes of 1–6 mm diameter and lengths of $\frac{1}{2}$–2 m are commonly available as stock items; units having special diameters, lengths, and end configurations also are manufactured to meet custom requirements. Fiber size typically ranges from 30–80 μm for glass fibers and up to 250 μm for plastic fibers. Very large fibers (clad rods) of almost any diameter are also available in the trade.

Another important form of image-transmitting fiber optics is the class of "fused" fiber bundles. These consist of a stack of clad glass fibers fused together in which the cladding serves not only as the optical insulation but also as the mortar to cement the fibers together. Usually these are made up of assemblies of "multifibers" or even "multi–multi" fibers (rather than single fibers) made by successive steps of drawing, assembly, and redrawing so as to accumulate the large number of small fibers required in such arrays. The final fusing step is carried out under carefully controlled conditions of temperature and pressure to ensure that all the interstices between fibers are completely sealed off. The fused block or "boule" can then be sliced, ground, and polished into faceplates for use as vacuum-tight end windows in electronic image tubes or CRTs. For these applications the fibers are designed for high numerical aperture (up to $NA = 1.0$) and in sizes down to about 4 μm. Such faceplates provide image resolution in excess of 100 line pairs per millimeter. To reduce stray light, which can otherwise degrade image contrast, minute light-absorbing glass fibers are located interstitially among the light-transmitting fibers. These effectively attenuate any stray light moving laterally through the faceplate.

In addition to the faceplate configuration, in which the fibers are precisely parallel, fused bundles may be reshaped with careful heating into "twisters", which rotate the transmitted image about the axis, or into tapers (fiber-optic "cones"), which provide a change in image magnification in the amount of the taper ratio. Such components find important uses in miniature image intensifier tubes for night vision "goggles" and for coupling image intensifier tubes to other electro-optical components such as TV camera tubes or charge-coupled sensor arrays (CCDs).

Bibliography

A. C. S. Van Heel, "A new method of transporting optical images without aberrations." *Nature* **173** (1954).

H. H. Hopkins and N. S. Kapany, "A flexible fiberscope using static scanning." *Nature* **173** (1954).

A. C. S. Van Heel, *De Ingenieur* **65**, 25 (1953).

W. B. Allan, *Fiber Optics Theory and Practice*. Plenum Press, New York, 1973.

M. K. Barnoski, *Fundamentals of Optical Fiber Communications*, 2nd ed. Academic Press, New York, 1981.

The Handbook of Optics, Chap. 13. McGraw–Hill, New York, 1979.

N. S. Kapany, *Fiber Optics*. Academic Press, New York, 1967.

D. A. Krohn, *Fiber Optic Sensors, Fundamentals and Applications*, 3rd ed. Instrumentation Systems, 2000.

D. Marcuse, *Theory of Dielectric Optical Waveguides*, 2nd ed. Academic Press, New York, 1991.

R. Tiedeken, *Fibre-Optics and Its Applications*. Focal Books, 1972.

Field Emission

L. W. Swanson

Electrons may be extracted from cold conductors or semiconductors by application of a strong electric field. The phenomenon, called *field emission*, which was first reported by R. W. Wood in 1897, occurs at fields of the order of 10^9–10^{10} V/m.

Fowler and Nordheim, in 1928, provided the first generally accepted explanation of field emission in terms of the newly developed quantum mechanics which they applied to the Sommerfeld model for the electronic energy levels in a metal. Application of a high field to the metal (see Fig. 1) produces a triangular potential-energy barrier through which electrons, arriving at the metal surface, may quantum mechanically tunnel. By solving the Schrödinger equation, Fowler and Nordheim obtained the barrier penetration probability $D(E_x)$ as a function of electron kinetic energy in a direction perpendicular to the barrier E_x.

By multiplying $D(E_x)$ by the number of electrons arriving at the surface with energy E_x and integrating over all values of E_x, Fowler and Nordheim derived the "Fowler–Nordheim" equation which relates field emission current I to field F and work function ϕ:

$$I = 6.2 \times 10^6 \{E_{\mathrm{F}}^{1/2}/C\phi^{1/2}\} F^2 \exp\left\{-\frac{6.8 \times 10^9 \phi^{3/2}}{F}\right\}, \tag{1}$$

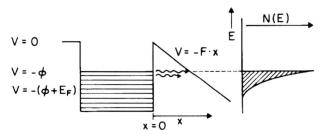

Fig. 1: Sommerfeld model for a metal at $T = 0\,\mathrm{K}$ in the presence of a high field. Electrons are able to tunnel through the potential barrier, especially at the level $V = -\phi$ where the barrier is thinner. The energy distribution curve $N(E)$ is shown. Notice the sharp cutoff at $E = -\phi$ due to the absence of electrons above the Fermi level at $T = 0\,\mathrm{K}$.

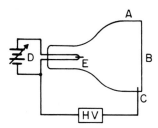

Fig. 2: Simple field-emission microscope. A is an evacuated ($p < 10^{-9}\,\mathrm{Torr}$) glass envelope, B is a conductivized fluorescent screen on the inside of the envelope, C is an external electrical connection to the screen, and D is a low-voltage heating supply for flash cleaning of the field emitter E.

where I is in $\mathrm{A/m^2}$; C, ϕ are in volts; F is in $\mathrm{V/m}$; and $C = \phi + E_\mathrm{F}$. Equation (1) represents the variation of I with field very well except at high current densities ($> 10^{10}\,\mathrm{A/m^2}$), where space-charge effects become noticeable.

In 1937 E. W. Müller invented the field-emission microscope (FEM), which consists of a sharp conducting point directed at a fluorescent screen deposited on conductivized glass that is maintained about 1–$10\,\mathrm{kV}$ positive with respect to the point. The resulting projection microscope (see Fig. 2) has a magnification of upwards of 10^5 and a resolution of $20\,\mathrm{\mathring{A}}$, which is enough to image certain single adsorbed atoms.

From the slope of a plot of $\ln I/F^2$ vs $1/F$, an average work function for the emitter surface may be obtained since, according to Eq. (1), we have $\partial\ln(I/F^2)/\partial(1/F) \propto \phi^{3/2}$. When annealed, field emitters develop hemispherical end forms with large facets of single-crystallographic planes; the work functions of these planes may also be obtained by intercepting the electrons emitted from them in a Faraday cup.

In the hands of Gomer and co-workers, at the University of Chicago, the FEM has been a particularly useful tool for adsorption studies, and it can be used to make measurements of heats of desorption, work function changes, and diffusion energies for adsorbates. For example, in an elegant use of the FEM the relative binding energies of potassium to several planes of tungsten were measured by relating single-plane work functions to surface coverages at different temperatures.

Useful information about emitter band structure, surface states, broadening of adsorbate energy levels, and even vibrational energy levels may be obtained from studies of the total energy distribution (TED) of field-emitted electrons (Fig. 1). An anomalous hump in the energy distribution for the (100) plane of molybdenum and tungsten has been discovered that has been attributed to the existence of a surface state on this plane which disappears when adsorption takes place. A similar structure in the TED from regions containing a single atom of Ca, Sr, or Ba has also been observed. In this case the structure was due to a resonance-enhanced tunneling when the broadened energy levels of the adsorbate atom overlapped the energy levels of the metal just below the Fermi level of the metal. By comparing the TED of hydrogen and deuterium adsorbed on a tungsten field emitter at 78 K, followed by subsequent warming to 300 K, the detection of vibrational energy levels of molecularly adsorbed species has been accomplished.

The small optical source size and very high current densities of field-emission cathodes make them attractive electron sources for microprobe applications because, for focused beam sizes below about 5000 Å, field-emission sources provide higher currents than thermionic cathodes. Thus Crewe and co-workers at the University of Chicago were able to develop a field-emission scanning electron microscope (SEM) with a resolution of 5 Å and a beam current one hundred times larger than the corresponding thermionic one; this allows the SEM to scan at a correspondingly higher rate.

Whereas the field-emission process described by Eq. (1) is valid for room temperature and below, the simultaneous application of an electric field and increased temperature to the cathode results in thermal field emission (TFE) at moderate temperatures and Schottky emission (SE) at high temperatures and lower electric field strength. The use of a $\langle 100 \rangle$-oriented tungsten field emitter with a zirconium oxide coating results in a low-work-function, thermally stable (up to 1900 K) electron source that can be operated with long life in the TFE or SE modes. The advantage of these operational modes [when compared with the lower-temperature cold field emission (CFE) mode] is the relaxation of the residual gas pressure requirements from 10^{-10} to 10^{-8} Torr while still maintaining a high emission intensity. Typically, the current fluctuations are $< 1\%$ and the energy spread of the electron beam $\sim 0.4 \, \text{eV}$ when operated in the SE mode. This results in a high-brightness cathode more practical than the CFE cathode while maintaining emission characteristics favorable to probe forming optics.

The SE cathode is currently finding use in such applications as high-resolution scanning Auger microscopy, high-density information recording, low-voltage scanning electron microscopy, and electron-beam microfabrication of integrated circuits.

In other applications, a CFE cathode array is being used to initiate a field-emission-initiated vacuum arc to produce very large currents in submicrosecond pulses for x-ray generation.

See also: Adsorption; Electron Microscopy; Field-Ion Microscopy; Tunneling; Work Function.

Bibliography

W. P. Dyke, "Advances in Field Emission," *Scientific American* **210** (1), 108 January (1964). (E)

R. Gomer, *Field Emission and Field Ionization.* AIP Press, 1993. (I) An introduction to theory and techniques of field emission.

R. H. Good and E. W. Müller, "Field Emission," *Handbuch der Physik*, Vol. 21, p. 1976, Springer Verlag, Berlin, 1956. (I)

J. W. Gadzuk and E. W. Plummer, "Field Emission Energy Distribution", *Rev. Mod. Phys.* **45**, 487 (1973). (A)

L. W. Swanson and A. E. Bell, "Recent Advances in Field Electron Microscopy of Metals", *Adv. Electr. Electr. Phys.* **32**, 193 (1973). (A)

L. W. Swanson and N. A. Martin, "Field Electron Cathode Stability Studies: Zirconium/Tungsten Thermal Field Cathode", *J. Appl. Phys.* **46**, 2029 (1975).

M. G. R. Thomson, R. Liu, R. J. Collier, H. T. Carroll, E. T. Doherty, and R. G. Murray, "The EBES4 Electron-Beam Column", *J. Vac. Sci. Technol.* **B5**, 53 (1987).

D. W. Tuggle and S. G. Watson, "A Low-Voltage Field-Emission Column with a Schottky Emitter", *Proc. 42nd Electron Microscopy Society of America*, p. 454 (1984). Electron Microscopy Society of America, Woods Hole, MA.

Field-Ion Microscopy

G. L. Kellogg

Introduced by E. W. Müller in 1951, the field-ion microscope is an instrument which can provide direct images of individual atoms on a solid surface. For nearly 20 years field-ion microscopy had the distinction of being the only technique capable of atomic resolution. Although no longer unique in this respect, the field-ion microscopy remains a powerful research tool for the study of problems in materials and surface science.

The field-ion microscope achieves its high magnification from a nearly radial projection of ions from the apex of a sharply pointed needle to a fluorescent screen. The needle is called a "field emitter" or more commonly a "tip" and has a radius of curvature at the end of several tens of nanometers. Ions are generated above the tip surface in a process known as field ionization. The tip is placed in a vacuum chamber and a low pressure (1–100 mPa) of an "imaging gas" such as helium or neon is introduced. A positive voltage of 5–30 kV is applied to the tip which produces electric fields at the apex of the order of $(1-6) \times 10^{10}$ V/m. At this field strength, the imaging gas atoms in the vicinity of the tip are ionized by an electron tunneling from the gas atom to the surface. The positive "field ions" formed in this manner follow the electric field lines away from the tip surface to the fluorescent screen where they produce image spots. Because the field ions are created preferentially above the protruding surface atoms, the image which appears on the screen consists of a pattern of spots, each spot corresponding to an individual surface atom.

The field-ion microscope is very similar in design to the field-emission microscope invented by Müller in 1936. In the field-emission microscope a negative voltage is applied to the tip and the electrons which tunnel from the surface into the vacuum are imaged on a fluorescent screen. Since the tunneling process is very sensitive to the local work function, different surface regions image with different contrast. The field-emission microscope does not require an imaging gas and can be operated at field strength approximately an order of magnitude lower than the field-ion microscope. However, the large lateral velocity component and the large de Broglie wavelength of the emitted electrons limit the resolution of the field-emission

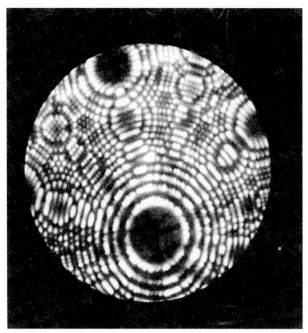

Fig. 1: A field-ion microscope image showing the atomic structure at the apex of a rhodium tip. The image was recorded in 1.3 mPa He at an applied voltage of 6.5 kV. The dark, circular region in the lower central portion of the image corresponds to the flat (111) plane.

microscope to about 2 nm, not sufficient to achieve atomic resolution. As mentioned above, atomic resolution is achieved in the field-ion microscope with the use of positive ions. The lateral velocity component of the ions can be reduced by lowering the emitter temperature (typically to 77 K or below) and the short de Broglie wavelength of the ions poses a negligible diffraction limit.

An example of a field ion microscope image taken from a Rh tip is shown in Fig. 1. The dark, circular regions seen in the image correspond to flat, low-index planes. These regions are dark because the interior atoms do not protrude sufficiently to produce local field enhancements. The rings of spots which surround these dark regions are images of the edge atoms of individual atomic layers, each larger diameter ring corresponding to a successively deeper atomic layer. From the symmetry of the patterns it is possible to determine the Miller indices of the observed planes. For some of the higher-index planes even the interior atoms are resolved.

Since the field-ion microscope is a point-projection microscope, its magnification is given by

$$M = \frac{D}{\beta r},$$

where D is the tip-to-screen distance, r is the tip radius, and β is an image-compression factor which accounts for the fact that the tip is not a true sphere in free space. The value of β

typically lies in the range from 1.5 to 1.8. By preparing a tip which has a radius of curvature at the apex of a few tens of nanometers, a magnification of several million times can be obtained. Such tips are routinely produced by chemical and electrochemical polishing procedures. It is also important to note that, because the ions are generated at the tip surface, there is no relative motion of the surface with respect to the ion beam. As a result, external vibrations are not magnified as they are in a conventional electron microscope.

The resolution of the field-ion microscope is determined by the spatial extent of the region above the protruding surface atom where ionization occurs (the ionization zone), the lateral velocity component of the emerging ions, and diffraction effects due to the finite de Broglie wavelength of the field ions. Compared to the first two effects, the last is negligible and the resolution is given by

$$\delta = \delta_0 + C \left(\frac{rT}{F} \right)^{1/2} ,$$

where δ_0 is the diameter of the ionization zone (ideally the diameter of the imaging gas atom), C is a constant, r is the tip radius, T is the tip temperature, and F is the applied electric field. The above equation implies that for the best resolution the tip radius should be as small as possible, the tip temperature should be as low as possible, and the applied field should be as high as possible. These conditions are best satisfied with helium as the imaging gas, because of its high ionization potential and low condensation temperature. However, as discussed below, many materials cannot be imaged at the field strengths required to ionize helium. For these, other imaging gases such as neon, argon, or hydrogen are commonly used.

The applied electric field used to image the field ion tip can also be used to remove surface atoms in a controlled, layer-by-layer fashion. At a given field strength, which is typically higher than the ion imaging field, surface atoms are ionized and desorb from the surface. This process is known as field evaporation when the removed atoms are substrate surface atoms or field desorption when the removed atoms are adsorbates. It is quite remarkable that field evaporation in the region of a low-index plane removes the substrate atoms from the edge of the plane inward, one layer at a time. The process is very useful as the final step of sample preparation before field-ion imaging because it leaves behind a surface which is smooth on an atomic scale and free of contaminants. Field evaporation may also be used to probe into the near-surface region of a sample to examine defects which do not extend to the initial surface.

The process of field evaporation is understood qualitatively in terms of classical, one-dimensional energy curves. In his original model Müller viewed field evaporation as the escape of an ion of charge n over a potential energy barrier formed by a superposition of the ionic potential energy curve (approximated by an image potential) and the potential due to the applied electric field. This barrier is known as the Schottky barrier and the activation energy for field evaporation is given by the simple expression

$$Q = Q_0 - (n^3 e^3 F)^{1/2} ,$$

where Q_0 is the energy required to remove an ion of charge n from the surface under field-free conditions, e is the elementary charge, and F is the applied field. Corrections which account for the difference in polarizability between the atom and the ion have also been included in subsequent refinements of the model. At a finite temperature the rate of field evaporation is given by an Arrhenius expression of the form

$$k_e = k_0 \exp\left(-\frac{Q}{kT}\right),$$

where k_0 is the Arrhenius prefactor, k is Boltzmann's constant, and T is the temperature. Although it is generally accepted that this "image hump" model is not physically realistic, i. e., there is no classical image hump at such close distances, the model is surprisingly accurate at predicting the evaporation fields of most metals. It does not, however, accurately predict the measured field dependence of the activation energy for W and Rh.

In Gomer's "charge exchange" or "intersection" model for field evaporation, ionization occurs at the intersection of the atomic and ionic potential energy curves. If this point lies beyond the maximum of the classical Schottky barrier, the image hump model does not apply. Although the exact form of the ionic potential-energy curve typically is not known, by making reasonable approximations for the potential-energy curves it is possible to reproduce most of the existing experimental data on field evaporation within the framework of this model. More recent quantum-mechanical treatments of field evaporation have also been able to predict accurately the field dependence of the activation energy for field evaporation.

In order to produce a stable image in a field-ion microscope, the evaporation field of the material must be higher than the applied field required for imaging. However, only the most refractory metals have evaporation fields greater than the 4.5×10^{10} V/m required for optimum imaging in He. Many other metals and alloys can be imaged in Ne and Ar, but these heavier ions have a much lower efficiency for exciting the phosphor on the fluorescent screen. A major breakthrough for field-ion microscopy was the development of channel-plate image intensifiers. These intensifiers can convert an ion image to an electron image with gains exceeding 10^3 for a single plate and 10^6 for stacked plates. With a single channel plate and the various imaging gases available it is now possible to obtain stable images of most metals and alloys and even some semiconductors.

The field-ion microscope is primarily a probe of atomic structure. Chemical identification of selected atoms observed in a field-ion microscope image became a possibility in 1967 with the introduction of the atom-probe field-ion microscope. There are several types of "atom probes" currently in use. The most widely used atom probe, shown schematically in Fig. 2, follows the original design of Müller, Panitz, and McLane. The tip is externally adjusted to align the atom(s) of interest with a probe hole in the viewing screen of the field-ion microscope. The superposition of a short-duration (10–100 ns), high-voltage (0.5–3 kV) electrical pulse onto the imaging voltage removes (field evaporates) atoms as n-fold-charged, positive ions. The ions of interest travel through the probe hole and enter a drift tube. The mass-to-charge ratios of the desorbed ions are determined by a measurement of their flight time from the tip through the probe hole and drift tube to a sensitive ion detector. A dramatic improvement in mass resolution can be achieved with the use of an energy-focusing, curved drift tube instead of the straight tube indicated in Fig. 2. The ability to identify selected atoms associated with structural features (e. g., lattice defects) observed in a field-ion microscope image makes the probe–hole atom probe well suited for investigation of metallurgical problems such as impurity segregation to grain boundaries.

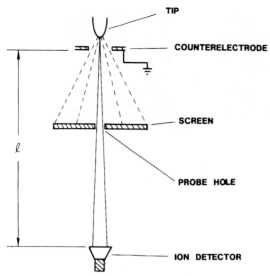

Fig. 2: A schematic representation of the probe–hole, atom-probe field-ion microscope. Ions removed from the tip by a high-voltage pulse are mass analyzed by time-of-flight measurements.

Another type of atom probe, known as the magnetic-sector atom probe, uses a magnetic field to separate and identify field-desorbed or field-evaporated ions. Although this type of atom probe has an inherently high mass resolution, it is not commonly used because it suffers from a limited mass range and requires a large number of ions to obtain a reasonable mass spectrum.

In the imaging atom probe, introduced by Panitz in 1972, the viewing screen of the field-ion microscope is replaced with an imaging detector sensitive to the impact of individual ions. Surface species desorbed from anywhere on the imaged portion of the tip surface are detected and identified by their flight times. The increased number of ions detected per pulse makes the imaging atom probe better suited to the study of surface adsorption and reaction processes than the probe–hole atom probe, but its shorter flight path results in significantly poorer mass resolution. An additional feature of the imaging atom probe is the ability to obtain elemental maps of selected surface species.

Atom-probe investigations of semiconductors and insulators have been greatly facilitated by the introduction of the pulsed-laser atom probe, developed independently by Block and co-workers and by Kellogg and Tsong in 1979. In this version of the atom probe the high-voltage electrical pulse used to stimulate desorption is replaced with a dc voltage and a short-duration laser pulse. The thermal activation provided by the laser pulse initiates the field evaporation or field desorption of surface species. Time-of-flight mass analysis is carried out the same way as in either the probe–hole or imaging atom probe. The use of laser pulses permits analysis of high-resistivity materials which will not transmit short-duration electrical pulses. Elimination of the so-called "energy deficits" associated with high-voltage pulsing also results in a significant improvement in mass resolution in the pulsed-laser atom probe, even better than that obtained with the energy-focusing drift tubes. In addition, the ability to

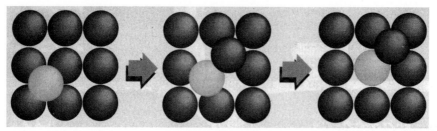

Fig. 3: A schematic representation of the atomic exchange mechanism for single-atom diffusion on fcc(001) single-crystal surfaces. The process is energetically favorable compared to ordinary hopping because fewer chemical bonds are broken through the transition. The mechanism was experimentally verified in studies using the field ion microscope [G. L. Kellogg and P. J. Feibelman, *Phys. Rev. Lett.* **64**, 3143 (1990); C. L. Chen and T. T. Tsong, *Phys. Rev. Lett.* **64**, 3147 (1990)].

vary the electric field in a continuous manner in field desorption studies makes the pulsed-laser atom probe useful for the investigation of a variety of surface chemical reactions and field desorption phenomena.

A more recent development in atom-probe mass spectrometry is the three-dimensional (3D) atom probe. This instrument uses a position-sensitive ion detector to determine both the flight time and the impact position of each evaporated ion. By field evaporating at a low rate (less than one ion/pulse) and pulsing with a high repetition rate, one can produce a three-dimensional atomic map of the element distributions within the analyzed volume. 3D atom probes go by different names depending on the specific type of detector used, including position-sensitive atom probe (POSAP), tomographic atom probe (TAP), and local electrode atom probe (LEAP).

The field-ion microscope and atom-probe mass spectrometer have been applied to a wide range of problems in surface and materials science. One of the most unique applications of the field-ion microscope has been the investigation of the diffusion and interaction of individual surface atoms. Pioneered by Ehrlich in 1966, these studies have provided quantitative diffusion parameters and interaction potentials for a variety of atoms on various single-crystal substrates.

One of the more intriguing insights to come from FIM surface diffusion studies is the discovery that atoms on certain surfaces migrate by an "exchange" or "substitutional" mechanism. Instead of hopping to a neighboring binding site, the diffusing atom finds it energetically favorable to take the place of a surface layer atom leaving the displaced atom to continue the migration process. A schematic illustration of the exchange process for fcc(001) surfaces is shown in Fig. 3. In addition to single-atom diffusion studies, the FIM has been used to investigate the clustering of atoms and the initial stages of epitaxial growth.

The most widespread application of the field-ion microscope and atom probe has been in the area of metallurgy. With the field-ion microscope it is possible to examine the microstructure of various lattice defects with atomic resolution. Even point defects, which are beyond the resolution of conventional electron microscopes, can be observed. Prior to the development of the atom probe, the field-ion microscope had been used to examine the imaging characteristics of various alloys and solid solutions, grain boundaries, and precipitates. Computer simulations aided greatly in the interpretation of the field-ion images. The extended capability for chemical analysis offered by the atom probe has led to more quantitative investigations of impurity and solute segregation to interfacial boundaries, surface segregation, ordering and clustering in alloys, and alloy precipitation.

The field-ion microscope and atom probe have had impact on several other scientific disciplines. Defect structures due to radiation damage have been examined at the atomic level in the field-ion microscope. The morphology of biological materials has been examined with field-ion microscopy and related techniques. Compositional variations at semiconductor–metal interfaces have been determined with the atom probe. Various surface chemical reactions and the initial stages of oxide formation on metals have been investigated with atom-probe techniques. Very recently, it has been established that even composite materials such as ceramic-oxide superconductors can be examined at the atomic level with the field-ion microscope and atom-probe mass spectrometer.

See also: Adsorption; Crystal Defects; Electron Microscopy; Surfaces and Interfaces.

Bibliography

R. Gomer, *Field Emission and Field Ionization.* Harvard University Press, Cambridge, MA, 1961. (I)

J. J. Hren and S. Ranganathan (eds.), *Field Ion Microscopy.* Plenum, New York, 1968. (E)

M. K. Miller and G. D. W. Smith, *Atom Probe Microanalysis: Principles and Applications to Materials Problems.* Materials Research Society, Pittsburgh, PA, 1989.

E. W. Müller and T. T. Tsong, *Field Ion Microscopy, Principles and Applications.* Elsevier, New York, 1969. (I)

E. W. Müller and T. T. Tsong, "Field Ion Microscopy, Field Ionization, and Field Evaporation," in *Progress in Surface Science*, S. Davison (ed.), Vol. 1, part 4. Pergamon Press, New York, 1973. (A)

J. A. Panitz, "Field Ion Microscopy-A Review of Basic Principles and Selected Applications," *J. Phys. E* **15**, 1281–1294 (1982). (I)

G. L. Kellogg, "Pulsed-Laser Atom-Probe Mass Spectroscopy," *J. Phys. E* **20**, 125–136 (1987). (E)

G. L. Kellogg, "Field Ion Microscope Studies of Surface Diffusion and Cluster Nucleation on Metal Surfaces,", *Surf. Sci. Rep.* **21**, 1 (1994).

T. T. Tsong, *Atom-Probe Field Ion Microscopy.* Cambridge University Press, Cambridge, 1990.

Field Theory, Axiomatic

A. S. Wightman

The phrase "axiomatic field theory" is sometimes used to mean "the general theory of quantized fields" and sometimes to mean any treatment of quantum field theory that has some pretensions to mathematical precision. It is a matter of convention whether constructive quantum field theory, which starts from specific Lagrangian models and constructs solutions of them, is regarded as distinct from axiomatic field theory or as a part of it. Here it will be regarded as distinct but nevertheless will be discussed briefly.

Axiomatic field theory was created in the early 1950s in response to a need for clarification in the foundations of quantum field theory. The great advances in renormalization theory in the late 1940s were based on the use of perturbation theory. This progress made it clear that the ideas of relativistic quantum field theory had more coherence and consistency than had been thought in the 1930s, but it shed little light on what a nonperturbative quantum field theory would be. Axiomatic field theory was regarded as novel in those days because it laid down a set of requirements (axioms) on an acceptable quantum field theory and then investigated what followed from those requirements alone. The consistency of the axioms was clear from the start. They were satisfied by theories of noninteracting fields. The fundamental question that axiomatic field theory left open was whether there are theories of fields in nontrivial interaction that satisfy the axioms.

The early development of the theory is summarized in the books of Jost [1] and Streater and Wightman [2]. The book of Bogolyubov, Logunov, and Todorov [3] contains in addition a summary of an alternative approach created by Lehmann, Symanzik, Zimmermann, and Glaser. In the late 1960s the focus in the study of the general theory of quantized fields shifted from axiomatic field theory to two other areas: the theory of local algebras of bounded operators, sometimes called *local quantum theory*, and constructive quantum field theory. The former, initiated by Haag, Araki, and Kastler, uses the deep theory of operator algebras. It is often called the algebraic approach to quantum field theory [4]. On general grounds it is not obvious that local quantum theory is equivalent to axiomatic field theory. However, that turned out to be true for the models in space-times of two and three dimensions whose existence was established by Glimm and Jaffe and others by the early 1970s [5] using the methods of constructive field theory. By the end of the 1980s constructive quantum field theory had established the existence of solutions for some quantum field theory models in the physically interesting case of four space-time dimensions. However, to prove that the requirements of axiomatic quantum field theory and those of local quantum theory are satisfied for these models remains an open problem. It is plausible that all the axiomatic approaches are, in fact, treating the same objects, and that Lagrangian field theories have solutions satisfying their axioms. That is what the above-mentioned results of constructive field theory state for the models treated.

The Axioms for a Quantum Field Theory of a Scalar Field

The basic constituents of a relativistic quantum field theory are, first, a Hilbert space $\mathcal{H}$ whose vectors describe the quantum-mechanical states, and a unitary representation of the inhomogeneous Lorentz (Poincaré) group $\{a, \Lambda\} \rightarrow U(a, \Lambda)$ by unitary operators in $\mathcal{H}$ describing the transformation law of states under Poincaré transformation. Second, there are the field operators themselves. For simplicity of exposition, the axioms will be stated only for a theory of a single neutral scalar field ϕ.

Group-theoretical analysis of U shows that the operators $U(a, 1)$ representing translations in space-time must be of the form

$$U(a, 1) = \exp[i(P \cdot a)]$$

where

$$P \cdot a = P^0 a^0 - \mathbf{p} \cdot \mathbf{a}$$

and the P^0 and $\mathbf{p}$ are self-adjoint operators whose physical significance is as observables of total energy and total momentum, respectively. The first axiom expresses the physical assumption that the energy is bounded below and that a unique state vector Ψ_0 representing the vacuum exists. This is formally expressed as follows.

1. *Spectral Condition.* (a) $P^2 = (P^0)^2 - (\mathbf{P})^2 \geq 0$, $\mathbf{P}^0 \geq 0$.

   (b) There exists a nonvanishing vector Ψ_0 unique up to normalization satisfying

   $$U(a,\Lambda)\Psi_0 = \Psi_0 \,.$$

The next assumption is about the field, ϕ. The assumption must take into account that the field makes sense as an operator in $\mathcal{H}$ only when it is smeared with an appropriate test function f in a manner formally indicated by

$$\phi(f) = \int \mathrm{d}^4 x \, f(x)\phi(x) \,.$$

It turns out that although $\phi(f)$ is an operator for each f, $\phi(f)$ cannot in general be applied to every vector of the Hilbert space $\mathcal{H}$. Thus the assumption on $\phi(f)$ must include some kind of specification of the domain on which the $\phi(f)$ are defined. There are several natural choices for the class of test functions. Here only the most commonly used choice will be mentioned: S, the space of all infinitely differentiate complex-valued functions f on space-time that decrease, together with all their derivatives, faster than any negative power of $R^2 = (ct)^2 + \mathbf{x}^2$ as $R \to \infty$.

2. *The Field and Its Domain.* For each f in $S(\mathbb{R}^4)$ there is an operator $\phi(f)$ acting in $\mathcal{H}$ whose domain includes the dense linear set D; $\phi(f)$ and D satisfy

   $$U(a,\Lambda)D \subset D, \qquad \Psi_0 \in D\,,$$
   $$\phi(f)D \subset D, \qquad \phi(f)^*D \subset D\,,$$

   where $\phi(f)^*$ is the Hermitian adjoint of $\phi(f)$. Further,

   $$\phi(\alpha f) = \alpha\phi(f), \qquad \phi(f+g) = \phi(f) + \phi(g)$$
   $$\phi(f)^* = \phi(\bar{f})\,,$$

   where $\bar{f}$ is the complex conjugate of f. These equations are understood as valid when the operators are applied to vectors in the domain D.

   If Φ and Ψ are any two vectors in D, then

   $$(\Phi, \phi(f)\Psi)$$

   is continuous as a linear functional of the variable f.

   The next assumption connects the transformation law of states, U, with the transformation law of a scalar field: $f \to \{a,\Lambda\}f$ where

   $$(\{a,\Lambda\}f)(x) = f(\Lambda^{-1}(x-a)) \,.$$

3. *Transformation Law of Field.*

$$U(a,\Lambda)\phi(f)U(a,\Lambda)^{-1} = \phi(\{a,\Lambda\}f) \,,$$

again understood as valid when applied to vectors of D.

The following assumption expresses the idea that measurements of the field ϕ taking place at spacelike separated points cannot influence each other.

4. *Local Commutativity.* If the support, $\sup f$, of a function/is defined as the closure of the set of all space-time points where f is nonvanishing, then

$$[\phi(f),\phi(g)] = 0 = [\phi(f),\phi(g)^*]$$

whenever the support of f is spacelike with respect to the support of g, i. e., whenever

$$f(x)g(y) = 0$$

for all x and y such that $(x-y)^2 \geq 0$.

To state the last axiom, it is necessary to discuss the notion of scattering states within the framework of a theory satisfying assumptions 1–4. This was done early in the development of axiomatic field theory by D. Ruelle [6] following ideas of R. Haag, but only under the assumption of a strengthened form of the spectral condition in which 1 is replaced by $1'$.

$1'$. *The Spectral Condition with Mass Gap.*

    (a) $P^2 \geq 0$.

    (b) $P = 0$ is a simple eigenvalue with unique (up to normalization) eigenvector Ψ_0.

    (c) The spectrum of P^2 on $\{\Psi_0\}^{\perp}$, the orthogonal complement of the vacuum state, lies entirely above m^2 (for some $m > 0$) and contains isolated eigenvalues M_1^2, $M_2^2, \ldots$

The Haag–Ruelle procedure consists in constructing many-particle scattering states out of the single-particle states of mass M_1, M_2, … provided by assumption (c). There are two sets of such states, corresponding to ingoing and outgoing wave boundary conditions, respectively; they span subspaces of $\mathcal{H}$, designated $\mathcal{H}^{\mathrm{in}}$ and $\mathcal{H}^{\mathrm{out}}$, respectively. Then the last axiom says that the scattering states span the whole Hilbert space.

5. *Asymptotic Completeness.*

$$\mathcal{H} = \mathcal{H}^{\mathrm{in}} = \mathcal{H}^{\mathrm{out}} \,.$$

In a theory with massless particles, the strengthened spectral condition $1'$ will be violated and the Haag–Ruelle scattering theory has to be generalized. Nevertheless, scattering states for the massless particles themselves can be defined as was shown by Buchholz [7]. However, it appears that there is still no general rigorous definition of the scattering states for those particles (which for brevity we can refer to as the charged particles) capable of emitting and absorbing the massless particles. The technical difficulty is that there is no eigenvalue of P^2 for such a charged particle but only an end point of a stretch of continuous spectrum. It seems likely that when this technical problem has been overcome, it will still be sensible to require asymptotic completeness in the form shown above.

There is an important phenomenon that may lead to a quantum field theory satisfying 1 or 1′, 2, 3, and 4, but not 5. That is the occurrence of superselection sectors other than the vacuum sector. This will be explained after the axioms of local quantum theory have been introduced.

The Axioms for Local Quantum Theory

The basic objects of local quantum theories are algebras $\mathcal{U}(O)$ of bounded operators attached to regions O of space-time. The bounded observables that can be measured in O correspond to self-adjoint elements of $\mathcal{U}(O)$, and $\mathcal{U}(O)$ is supposed to be generated by such observables. More precisely, $\mathcal{U}(O)$ is supposed to be a C^* algebra (*see* Operators).

The first assumption on the $\mathcal{U}(O)$ is

1. *Isotone Property.*

    If $O_1 \subset O_2$, then $\mathcal{U}(O_1) \subset \mathcal{U}(O_2)$.

Using Property 1, the quasilocal algebra $\mathcal{U}$ can be defined:

$$\mathcal{U} = \overline{\cup_O \mathcal{U}(O)} \ .$$

Here the regions admitted to the union comprise all bounded open sets and the closure is understood as closure in the sense of $C^* *$ algebras.

The transformation law is stated algebraically in local quantum theory.

2. *Transformation Law Under Poincaré Group.*

    There is a representation of the Poincaré group, $\{a, \Lambda\} \to \alpha(a, \Lambda)$, by automorphisms of the quasilocal algebra $\mathcal{U}$ such that

    $$\alpha\{a, \Lambda\} \mathcal{U}(O) = \mathcal{U}(\{a, \Lambda\}O) \ .$$

Local commutativity is stated in precise analogy with assumption 3 in axiomatic quantum field theory.

3. *Local Commutativity.* If the points of O_1 are all separated from the points of O_2 by spacelike intervals, then the operators of $\mathcal{U}(O_1)$ commute with those of $\mathcal{U}(O_2)$.

For the finer developments of local quantum theory it is customary to adjoin other axioms, but on the basis of 1–3 we can already discuss a number of significant physical issues. The first concerns the notion of state and, in particular, vacuum state.

A state in the theory of C^* algebras is defined as a complex-valued linear function ω on the algebra $\mathcal{U}$ such that

$$\omega(A^*, A) \geq 0 \quad \text{for all} \quad A \in \mathcal{U}$$

and

$$\omega(1) = 1$$

where 1 is the unit element of the algebra. A state ω on $\mathcal{U}$ determines a Hilbert space $\mathcal{H}_\omega$ and a representation π_ω of $\mathcal{U}$ by operators in $\mathcal{H}$: $A \to \pi_\omega(A)$ such that

$$\omega(A) = (\Omega_\omega, \pi_\omega(A)\Omega_\omega)$$

where Ω_ω is a certain fixed vector of $\mathcal{H}_\omega$ such that $\{\pi_\omega(A)\Omega_\omega\}$ is a dense set in $\mathcal{H}_\omega$. If ω happens to be invariant under a group G of automorphisms α_g of $\mathcal{U}$,

$$\omega(\alpha_g(A)) = \omega(A) \quad \text{for all} \quad g \in G,$$

then there exists a unitary representation of the group, $\alpha_g \to U(\alpha_g)$, such that

$$\pi_\omega(\alpha_g(A)) = U(\alpha_g)\pi_\omega(A)U(\alpha_g)^{-1}$$

and

$$U(\alpha_g)\Omega_\omega = \Omega_\omega.$$

The construction of the Hilbert space $\mathcal{H}_\omega$, the state Ω_ω, and the representations π_ω and U is called the GNS construction after Gelfand, Naimark, and Segal.

The importance of this notion of state and its associated representation is that it makes possible a theory of superselection rules. A superselection rule may be defined in a quantum-mechanical theory by an operator that commutes with all observables. Then the Hilbert space of states breaks up into a sum of subspaces called sectors (or coherent subspaces) within which the superposition principle is valid but between which it is invalid. According to this theory of superselection rules, there are representations of the algebra of observables within the sectors that are unitarily inequivalent. For example, in quantum electrodynamics the algebra of observables has a representation in the charge-1 sector that is inequivalent to the one in the charge-0 (vacuum) sector. In principle, we can find out how many sectors there are by studying the representations of the algebra of observables. In practice, additional assumptions of one kind or another have so far been made [8].

This theory of superselection rules provides an important insight into the axiom of asymptotic completeness. It may happen that the field theory constructed from a given Lagrangian has a Hilbert space $\mathcal{H}_0$ that is the vacuum sector of a larger theory, i. e., a theory with a Hilbert space of states $\mathcal{H}$ that includes $\mathcal{H}_0$ as a proper subspace and that has extra field variables acting in $\mathcal{H}$ but not leaving $\mathcal{H}_0$ invariant. Then we might expect asymptotic completeness to fail in $\mathcal{H}_0$ because there would be multiparticle states with the quantum numbers of the vacuum, some of whose single-particle states are not in the vacuum sector. For example, in a theory of charged particles, the vacuum sector would contain electron–positron pairs even though the states of a single electron or of a single positron are not in the vacuum sector. Thus in order to obtain a Hilbert space in which asymptotic completeness would hold, we would have here to adjoin to $\mathcal{H}_0$ (by direct sum) the Hilbert spaces $\mathcal{H}_{ne}$, describing the other sectors of charge ne, $n = \pm 1, \pm 2, \ldots$ There is convincing evidence that this happens in a variety of model field theories. In particular, it appears that the Sine–Gordon theory of a scalar field describes the vacuum sector of the massive Thirring model [9].

Principal Results

Since the principal objective of axiomatic quantum field theory has been to put the conceptual and mathematical foundations of relativistic quantum field theory in order, it is not surprising that its impact on particle physics has been mild. What it has done is to make clear how general the arguments are for *CPT* symmetry, the connection of spin with statistics, the LSZ reduction formulas, the connection of broken symmetry with Goldstone bosons, and the occurrence of superselection rules. Its offspring, constructive quantum field theory, has shown the internal consistency of nontrivial theories of interacting fields in space-times of dimensions two and three and established a connection, via Euclidean field theory, between statistical mechanics and quantum field theory.

The ultimate objective of axiomatic field theory is a structure theory for the objects satisfying the axioms. Further progress in constructive field theory seems essential before such a goal can be reached.

See also: *CPT* Theorem; Fields; Operators; Quantum Field Theory; Space-Time; Symmetry Breaking, Spontaneous.

References

[1]   R. Jost, *The General Theory of Quantized Fields*. American Mathematical Society, Providence, 1965.
[2]   R. F. Streater and A. S. Wightman, PCT, *Spin and Statistics and All That*. Princeton University Press, Princeton, 2000.
[3]   N. Bogolyubov, A. Logunov, and I. Todorov, *Introduction to Axiomatic Quantum Field Theory*. Benjamin Advanced Book Program, Reading, Mass., 1975.
[4]   G. Emch, *Algebraic Methods in Statistical Mechanics and Quantum Field Theory*. Wiley-Interscience, New York, 1972.
[5]   G. Velo and A. Wightman, eds., *Constructive Quantum Field Theory* (Lecture Notes in Physics No. 25). Plenum Press, New York, 1991.
[6]   D. Ruelle, "On the Asymptotic Condition in Quantum Field Theory", *Helv. Phys. Acta* **35** (1962) 147–174. (See also [1], Chapter VI.)
[7]   D. Buchholz, "Collision Theory for Massless Fermions and Bosons," *Commun. Math. Phys.* **42**, 269–279 (1975); **52**, 147–173 (1977).
[8]   S. Doplicher, R. Haag, and J. Roberts, "Local Observables and Particle Statistics I, II," *Commun. Math. Phys.* **23**, 199–230 (1971); **35**, 49 (1974). S. Doplicher and J. Roberts, "Fields, Statistics, and Non-Abelian Gauge Groups," *Commun. Math. Phys.* **28**, 331–48(1972).
[9]   S. Coleman, "Quantum Sine-Gordon Equation as the Massive Thirring Model," *Phys. Rev. D* **11**, 2088–2097 (1975).

Field Theory, Classical

D. Weingarten

Classical field theory concerns systems whose (measurable) physical properties at each instant of time are given by a collection of real-valued functions on some region of space. These functions are called fields. The set of fields for each system considered in classical field theory is governed by deterministic rules of time evolution that in most cases permit the values of all fields at any time t_a to be deduced from the values of all fields and a sufficient set of time derivatives at any $t_b < t_a$, or in some more complicated systems, from the values of all fields over the time interval preceding any t_b, $t_b < t_a$. Quantum generalizations of classical field theory lacking such deterministic rules of time development are described in separate articles (*see* Quantum Field Theory, Quantum Electrodynamics).

Classical fields may be either structural or phenomenological. A structural field specifies completely a system's actual microscopic configuration. A phenomenological field, on the other hand, represents only a macroscopic property related in some indirect way to the system's microscopic configuration. Classical field theory may be divided into two branches concerned primarily with structural fields – classical electrodynamics and the general relativistic theory of gravitation – and a third branch concerned mainly with phenomenological fields – the mechanics of continuous media. A typical structural field is the electromagnetic field of electrodynamics; an example of a phenomenological field is the velocity field of continuum mechanics, which may be interpreted as a local average of actual microscopic particle velocities.

In most branches of classical field theory it is possible to identify in a natural way local densities and currents of energy and momentum carried by the fields. These quantities can be combined to form the energy–momentum tensor $T^{\mu\nu}$, where μ and ν run from 0 to 3. The component T^{00} gives the field's energy density, T^{0i} gives the density of the i component of momentum, T^{i0} gives the rate per unit area at which energy is being carried by the field in the i direction, and T^{ij} gives the rate per unit area at which the j component of momentum is being carried by the field in the i direction. For fields interacting in the absence of other sources of energy and momentum, the field energy and momentum currents are conserved:

$$\partial_\mu T^{\mu\nu} = 0 .$$

Here ∂_μ is an abbreviation of $\partial/\partial x^\mu$, (x^μ) is the four-component vector of special relativity (t, x^1, x^2, x^3), a system of units has been chosen in which the speed of light c is 1, and a summation is performed over repeated indices. These conventions will also be used throughout the remainder of this article.

The dynamical laws which occur in classical field theory are often conveniently stated as principles of stationary action. For both electrodynamics and general relativity a satisfactory action can be found which is the integral of a Lagrangian density $L(x)$ given by a function of the fields at x and their derivatives of at most second order. For a set of fields φ_i, where i runs from 1 to n, with a Lagrangian density $L(x)$ of the form $L[\varphi_i(x), \partial_\mu\varphi_i(x), \partial_\mu\partial_\nu\varphi_i(x)]$, the principle of stationary action

$$\delta \int d^4x\, L(x) = 0 \tag{1}$$

yields the dynamical equations

$$\partial_\mu\partial_\nu \frac{\partial L}{\partial[\partial_\mu\partial_\nu\varphi_i(x)]} + \partial_\mu \frac{\partial L}{\partial[\partial_\mu\varphi_i(x)]} - \frac{\partial L}{\partial\varphi_i(x)} = 0 . \tag{2}$$

The variation in (1) is required to be zero with respect to each $\varphi_i(x)$ for all i and x. The relations given by (2) are called the Euler–Lagrange equations.

In the remainder of this article we will briefly outline each of the major branches of classical field theory.

Electrodynamics

The electromagnetic field of classical electrodynamics is given, in relativistic notation, by a rank-2 antisymmetric tensor field $F^{\mu\nu}$ with indices μ and ν ranging from 0 to 3. The relation between $F^{\mu\nu}$ and the electric and magnetic fields $\mathbf{E}$ and $\mathbf{B}$, respectively, is

$$|F^{\mu\nu}| = \begin{vmatrix} 0 & E^1 & E^2 & E^3 \\ -E^1 & 0 & B^3 & -B^2 \\ -E^2 & -B^3 & 0 & B^1 \\ -E^3 & B^2 & -B^1 & 0 \end{vmatrix}.$$

These fields can be measured, in principle, by using the Lorentz force law

$$m\frac{\mathrm{d}^2 x^\mu}{\mathrm{d}\tau^2} = qF^\mu_\nu(x)\frac{\mathrm{d}x^\nu}{\mathrm{d}\tau}$$

giving the acceleration of a particle of mass m and charge q moving in an electromagnetic field with no other forces applied. The second index of $F^{\mu\nu}$ in this equation, and the indices of special relativistic tensors in general, are raised and lowered by contraction with the diagonal metric tensors $\eta^{\mu\nu} = \eta_{\mu\nu}$ with $\eta^{00} = -\eta^{11} = -\eta^{22} = -\eta^{33} = 1$. The parameter τ is the particle's proper time, defined by $\mathrm{d}\tau^2 = \mathrm{d}x^\mu \mathrm{d}x^\nu \eta_{\mu\nu}$.

The time development of $F^{\mu\nu}$ is governed by Maxwell's equations, which also restrict the possible field configurations that can occur at a single instant of time. In the presence of an external electric current density J^μ, Maxwell's equations take the form

$$\partial_\mu F^{\mu\nu} = J^\nu, \tag{3}$$
$$\partial_\nu {}^*F^{\mu\nu} = 0. \tag{4}$$

The field ${}^*F_{\mu\nu}$ is the dual of $F^{\mu\nu}$ defined as $-\frac{1}{2}\varepsilon_{\mu\nu\alpha\beta}F^{\alpha\beta}$ where $\varepsilon_{\mu\nu\alpha\beta}$ is the completely anti-symmetric tensor with $\varepsilon_{0123} = 1$.

Maxwell's equations and the Lorentz force law are form invariant under Poincaré transformations of the coordinate frame, $x'^\mu = a^\mu_\nu x^\nu + b^\mu$, where a^μ_ν is a Lorentz transformation defined by the condition $a^\mu_\lambda a^{\nu\lambda} = \eta^{\mu\nu}$ and b^μ is an arbitrary vector. Under Poincaré transformations, J^μ and $F^{\mu\nu}$ transform as relativistic tensors of ranks 1 and 2, respectively, where a general rank-n tensor transforms by the rule

$$T'^{\mu_1 \cdots \mu_n}(x') = a^{\mu_1}_{\nu_1} \cdots a^{\mu_n}_{\nu_n} T^{\nu_1 \cdots \nu_n}(x).$$

Equation (4) implies that $F_{\mu\nu}$ can be obtained from a vector potential field A_μ:

$$F_{\mu\nu} = \partial_\nu A_\mu - \partial_\mu A_\nu. \tag{5}$$

If $F^{\mu\nu}$ is considered a function of A_μ and a Lagrangian density is defined by $\mathcal{L} = -\frac{1}{4}F_{\mu\nu}F^{\mu\nu} - A_\mu J^\mu$, then (3) becomes the Euler–Lagrange equations (2) following from the principle of stationary action (1). The remaining Maxwell equations (4) follow automatically once (5) is taken as the definition of $F_{\mu\nu}$.

Under Poincaré transformations A^μ transforms as a tensor field of rank 1 leaving L form invariant when J_μ is 0. The invariance of it combined with the Euler–Lagrange equations (3) implies, by Noether's theorem, the conservation relations $\partial_\mu T^{\mu\nu} = 0$, $\partial_\mu M^{\mu\nu\lambda} = 0$ for the tensor fields

$$
\begin{aligned}
T^{\mu\nu} &= -F^\mu_\delta F^{\nu\delta} + \frac{1}{4}\eta^{\mu\nu} F_{\alpha\beta}F^{\alpha\beta}\,, \\
M^{\mu\nu\lambda} &= x^\mu T^{\mu\lambda} - x^\lambda T^{\mu\nu}\,.
\end{aligned}
$$

The tensor $T^{\mu\nu}$ may be interpreted as the electromagnetic field's energy–momentum tensor, while $M^{\mu\nu\lambda}$ may be taken to be the relativistic generalization of an angular momentum current. For example, the three-component angular momentum density is $(M^{023}, M^{031}, M^{012})$.

General Relativity

The physical system considered in general relativity is space-time itself. Points in space-time are specified by a set of four coordinates y^μ, and the measurable property of space-time at each point is its metric, given by the symmetric tensor $g_{\mu\nu}$ or equivalently by its inverse $g^{\mu\nu}$, where the indices μ and ν again run from 0 to 3. A measurement of $g_{\mu\nu}$ can be made, in principle, using the information that an ideal clock following a trajectory successively passing through y^μ and $y^\mu + dy^\mu$ will have $dy^\mu dy^\nu g_{\mu\nu}$ positive and show an elapsed time of $(dy^\mu dy^\nu g_{\mu\nu})^{1/2}$ in the course of the displacement dy^μ. At all points in space $g_{\mu\nu}$ has one strictly positive eigenvalue and three strictly negative. Displacements with $dy^\mu dy^\nu g_{\mu\nu}$ positive are called timelike and those with $dy^\mu dy^\nu g_{\mu\nu}$ negative are called spacelike. At any point a small displacement purely in the y^0 direction is timelike and small displacements in the y^1, y^2 or y^3 directions are spacelike.

The dynamical laws governing the behavior of $g_{\mu\nu}$ are most easily stated by defining a sequence of intermediate fields. The affine connection is obtained from the metric tensor by

$$
\Gamma^\alpha_{\mu\nu} = -\frac{1}{2}g^{\alpha\beta}\left(\frac{\partial g_{\beta\mu}}{\partial y^\nu} + \frac{\partial g_{\beta\nu}}{\partial y^\mu} - \frac{\partial g_{\mu\nu}}{\partial y^\beta}\right)\,.
$$

Then the curvature tensor is defined by

$$
R^\alpha_{\beta\mu\nu} = \frac{\partial}{\partial y^\mu}\Gamma^\alpha_{\beta\nu} - \frac{\partial}{\partial y^\nu}\Gamma^\alpha_{\beta\mu} + \Gamma^\alpha_{\nu\lambda}\Gamma^\lambda_{\beta\mu} - \Gamma^\alpha_{\mu\lambda}\Gamma^\lambda_{\beta\nu}
$$

and contracted curvatures are defined by $R_{\beta\nu} = R^\alpha_{\beta\alpha\nu}$, $R = R_{\beta\nu}g^{\beta\nu}$. Geometric interpretations of $\Gamma^\alpha_{\mu\nu}$ and $R^\alpha_{\beta\mu\nu}$ can be found in the 1973 text by Misner, Thorne, and Wheeler listed in the Bibliography. Finally, Einstein's equations for $g_{\mu\nu}$ in the presence of energy and momentum described by the energy–momentum tensor $T_{\mu\nu}$ are

$$
R_{\mu\nu} - \frac{1}{2}g_{\mu\nu}R = 8\pi G T_{\mu\nu} \tag{6}
$$

where G is the Newtonian gravitational constant.

As in the case of Maxwell's equations, Einstein's equations both determine the evolution of the field in a timelike direction and constrain the possible values on spacelike hypersurfaces.

Equation (6) is form invariant under twice-differentiable reparametrizations of space-time by a new set of coordinates $y'^{\mu} = y'^{\mu}(y)$ with nonvanishing Jacobian if $g_{\mu\nu}$ and $T_{\mu\nu}$ are both transformed as rank-2 covariant tensor fields. In general, such a tensor $V_{\mu\nu}$ transforms as

$$V'_{\mu\nu}(x') = \frac{\partial x^{\alpha}}{\partial x'^{\mu}}\frac{\partial x^{\beta}}{\partial x'^{\nu}}V_{\alpha\beta}(x) .$$

Equation (6) in the absence of an energy–momentum tensor $T_{\mu\nu}$ follows from the variational principle Eq. (1) if the Lagrangian density for the metric tensor is taken to be $\sqrt{-g}R/(16\pi G)$, where g is the determinant of $g_{\mu\nu}$. With this choice, the integrated action in Eq. (1) is form invariant under general coordinate transformations. A Lagrangian density yielding (6) in the presence of an energy-momentum tensor can be found by introducing in the Lagrangian for the fields or matter giving rise to $T_{\mu\nu}$ factors of $g_{\mu\nu}$ and $\Gamma^{\alpha}_{\mu\nu}$ in such a way that the resulting action becomes invariant under general coordinate transformations (see Misner, Thorne, and Wheeler for a discussion of how this can be accomplished), then adding to this Lagrangian density the term $\sqrt{-g}R/(16\pi G)$.

Mechanics of Continuous Media

The motion of a solid, liquid, or gas can be specified in one of two equivalent ways. The material description of motion consists of a field $\mathbf{x}(\mathbf{a},t)$ giving the position $\mathbf{x}$ at time t of the material whose position was a at a reference time t_0. The spatial description consists of a field $\mathbf{v}(\mathbf{x},t)$ giving the velocity of the material at position $\mathbf{x}$ at time t. These descriptions are related by the differential equation

$$\frac{\partial}{\partial t}\mathbf{x}(\mathbf{a},t) = \mathbf{v}(\mathbf{x},t) \tag{7}$$

with the boundary condition $\mathbf{x}(\mathbf{a},t_0) = \mathbf{a}$.

The thermodynamic properties of a homogeneous medium can be described by a pair of fields giving two conveniently chosen thermodynamic quantities as functions of position and time, for example, a density field $\rho(\mathbf{x},t)$ and a temperature field $T(\mathbf{x},t)$. Any other thermodynamic fields can be obtained from these using the material's equation of state, which by the assumptions of homogeneity is constant throughout the substance. Inhomogeneous media will not be considered here.

The flow of momentum caused by forces which each region of material exerts on neighboring regions is specified by a stress tensor $T^{ij}(\mathbf{x},t)$, where i and j run from 1 to 3, consisting of 9 of the 16 components of the material's energy–momentum tensor $T^{\mu\nu}(\mathbf{x},t)$ described earlier. The flow of heat caused by thermal conduction is given by a field $\mathbf{q}(\mathbf{x},t)$ where $q^i(\mathbf{x},t)$ is the rate of flow of heat per unit area in the i direction. Additional fields which may enter the dynamical equations governing materials include the body force $\mathbf{f}(\mathbf{x},t)$, giving the external force per unit mass acting on the material, and the macroscopic electromagnetic field $F^{\mu\nu}(\mathbf{x},t)$ and current density $J^{\mu}(\mathbf{x},t)$, given by local averages of the corresponding fields of the section on electrodynamics.

The dynamical equations governing a continuous system can be grouped into general laws fulfilled by all materials and constitutive equations which vary from one substance to another.

Among the first class of equations are conservation of mass,

$$\frac{D\rho}{Dt} + \rho \nabla \cdot \mathbf{v} = 0 , \tag{8}$$

and conservation of momentum and energy, which in the absence of electromagnetic effects take the form

$$\rho \frac{Dv^i}{Dt} - \rho f^i - \frac{\partial T^{ij}}{\partial x^j} = 0 , \tag{9}$$

$$\rho v^i \frac{Dv^i}{Dt} + \rho \frac{DE}{Dt} - \rho v^i f^i - \frac{\partial}{\partial x^j}(v^i T^{ij} - q^j) = 0 . \tag{10}$$

The operator D/Dt in these equations is called the material derivative, $D/Dt = \partial/\partial t + v^i \partial/\partial x^i$, and $E(\mathbf{x}, t)$ is the thermodynamic internal energy per unit mass of the material at $\mathbf{x}$ and t with respect to its rest frame. Without an external body-couple field exerting a torque on small regions proportional to their mass, conservation of angular momentum requires T^{ij} to be a symmetric tensor.

The constitutive equations for a substance may include its thermodynamic equation of state, an equation expressing T^{ij} as a function of either $\mathbf{x}(\mathbf{a}, t)$ or $\mathbf{v}(\mathbf{x}, t)$, and an equation determining $\mathbf{q}$ from the temperature field T. For many materials the constitutive equations for a field $r(\mathbf{x}, t)$ actually require the values of another set of fields $s_1(\mathbf{x}, t'), \ldots, s_n(\mathbf{x}, t')$ not only at $t = t'$ but also over the material's entire preceding history $t' < t$. A satisfactory approximation, however, can often be obtained using only $s_1(\mathbf{x}, t), \ldots, s_n(\mathbf{x}, t)$ and a finite set of their time derivatives.

A typical closed set of conservation laws and constitutive equations are the Navier–Stokes equations for an incompressible Newtonian liquid. These equations apply, for example, to water. Assume ρ and T can be replaced by constants ρ_0 and T_0, respectively, and that T^{ij} obeys the constitutive equation for a Newtonian liquid:

$$T^{ij} = -p\delta^{ij} + \mu \left(\frac{\partial v^i}{\partial x^j} + \frac{\partial v^j}{\partial x^i} \right) . \tag{11}$$

The constant μ is the liquid's viscosity and $p(x, t)$ is the pressure field. Equation (11) combined with (8) and (9) forms a closed system of equations for the time development of $p(x, t)$ and $\mathbf{v}(\mathbf{x}, t)$. The boundary condition used with these is that the fluid at each boundary point must be at rest with respect to the confining material.

See also: Electrodynamics, Classical; Fields; Gravitation; Hydrodynamics; Quantum Electrodynamics; Quantum Field Theory; Relativity, General.

Bibliography

J. D. Jackson, *Classical Electrodynamics*, 3rd ed. Wiley, New York, 1998. (A)

W. M. Lai, D. Rubin, and E. Krempl, *Introduction to Continuum Mechanics*. Butterworth, New York, 1995. (I)

L. D. Landau and E. M. Lifshitz, *The Classical Theory of Fields*, 4th rev. English ed. Butterworth, New York, 1980. (A)

L. D. Landau and E. M. Lifshitz, *Fluid Mechanics*. Butterworth, New York, 1987. (A)

R. E. Meyer, *Introduction to Mathematical Fluid Dynamics*. Dover, New York, 1982. (A)

C. W. Misner, K. S. Thorne, and J. A. Wheeler, *Gravitation*. W. H. Freeman, San Francisco, 1973. (I-A)

E. M. Purcell, *Electricity and Magnetism* (Berkeley Physics Course), Vol. 2. McGraw–Hill, New York, 1963. (E)

M. Schwartz, *Principles of Electrodynamics*. Dover, New York, 1987. (I)

Field Theory, Unified

J. L. Anderson

In the early part of this century Einstein achieved in the general theory of relativity a remarkable unification of geometry and gravity by identifying the gravitational field with the Riemannian metric of space-time. This unification in turn served as the impetus for a number of attempts to achieve a further unification by generalizing the geometrical structure of space-time and identifying the electromagnetic field with the additional geometrical elements needed for this generalization. The names of Weyl, Schrödinger, Kaluza, Klein, and especially Einstein and his co-workers figured prominently in the list of authors of such unified field theories (as they came to be called), and indeed Einstein continued to search for such a theory to the end of his life.

Since no unique generalization of the Riemannian geometry of space-time exists, various authors tried different approaches. Einstein for example at one time considered a geometry that utilized an asymmetric metric tensor instead of the symmetric tensor of Riemannian geometry. Other authors introduced an asymmetric affinity into the geometry and identified the electromagnetic field with the antisymmetric components of this affinity. (An affinity or affine connection is used in geometry to define a notion of local parallelism between vectors.) Kaluza, and later Kaluza and Klein, sought the desired generalization by increasing the dimensionality of space-time from four to five but retained its Riemannian character. The additional components of the five-dimensional metric over the ten of four dimensions were then associated with the electromagnetic field.

In judging the degree of unification achieved by these various theories it is helpful to compare them to theories that already possess some degree of unification. Minimal unification is achieved in the Einstein–Maxwell theory of the gravitational and electromagnetic fields in general relativity. In this theory the two fields are represented by two completely independent geometrical objects that transform independently of one another under an arbitrary coordinate transformation. We are not even compelled to introduce the electromagnetic field at all into the Einstein gravitational theory. The only unification that can be said to exist in this theory is the dependence of the field equations for the electromagnetic field on the metric-gravitational field and vice versa. On the other hand the special relativistic theory of the electric and magnetic fields is a highly unified theory. In special relativity a theory of the electric or magnetic field cannot exist by itself; both fields are needed to construct a complete theory. Furthermore, the electric and magnetic fields do not transform independently of one another under a Lorentz transformation. Finally, the group of invariant transformations of special relativity cannot be decomposed into a product of two or more smaller groups.

A comparison of the foregoing theories suggests two possible criteria for a unified theory:

 (i) The invariance group of a unified theory cannot be decomposed into a direct product of two or more smaller groups.

(ii) The field variables of a unified theory cannot be decomposed into a direct sum of two or more sets of variables that transform independently of one another under the invariance group of the theory.

Taken together these two conditions are almost certainly too restrictive; even the Einstein gravitational theory fails to satisfy (ii). It does, however, satisfy a somewhat weaker condition that is implied by (ii) but not vice versa.

(ii') It is not possible to construct a complete theory using a subset of the field variables of a unified theory.

We do not wish to imply here that these are the only criteria for a unified field theory or that there is even any general agreement on what constitutes such a theory. However, they do serve as a convenient yardstick for judging the degree of unification of a theory and we shall so use them here.

While the asymmetric metric or affine theories mentioned earlier fail to satisfy either condition (ii) or (ii'), the five-dimensional theory of Kaluza and Klein does satisfy the latter as well as condition (i). Furthermore, as in the case of general relativity and in contrast to the asymmetric metric and affine theories, the field equations for the metric tensor in this theory are essentially uniquely determined by the condition that they are of second differential order. Unfortunately, in spite of its attractive features, the Kaluza–Klein theory shares with these theories the defect of having no presently testable consequences beyond those of the Einstein–Maxwell theory itself. This fact, coupled with the lack of any physical motivation for unification such as existed in the case of general relativity, was mainly responsible for its not receiving greater acceptance than it has.

In the years that followed the first attempts to construct a unified theory of the electromagnetic and gravitational fields, several developments occurred in physics that cast serious doubt on the possibility of ever constructing a satisfactory unified field theory. One of these developments was quantum mechanics. It soon became clear, especially after the successes of quantum electrodynamics, that any fundamental field theory would have to be a quantum field theory. Because of the difficulties that were encountered in the many attempts to quantize the gravitational field by itself it appeared likely that a unified theory would be even more difficult to quantize.

The second development that had to be taken into account was the discovery of an ever increasing number of fundamental fields in nature. Even if we restrict our attention to the fields that are now believed to be responsible for the fundamental interactions of matter, we have to add to the electromagnetic and gravitational fields those associated with the so-called strong and weak interactions between elementary particles. Since the properties of these four fields are so different from one another – among other things, the former are long-range fields while the latter are short range – it appeared doubtful that they could ever be unified into a single entity. Yet in spite of these difficulties there is good reason to believe that a unified theory of these fields is not only possible but probably even necessary for the construction of a consistent quantum theory of the fundamental interactions.

The search for a unified theory of the elementary interactions is today a very active field of research not only because of the esthetic appeal of such theories, but more important, because of the limited number of such theories we can construct that satisfy conditions such as those given earlier. Since a final theory of these interactions is still to be achieved, such conditions are extremely useful in limiting the possible candidates for such a theory. The first modern unified theory is the non-Abelian gauge theory, developed during the last decade by Steven Weinberg, Abdus Salam, and John Ward, which succeeded in unifying the weak and electromagnetic interactions in spite of their great dissimilarities. It would take us too far afield to describe the details of this theory here. We can say, though, that it has two virtues not shared by the original unified theories: It has testable consequences; and perhaps even more important is that, as Gerard 't Hooft and Martin J. G. Veltman showed in the early 1970s, it can be quantized in a consistent manner even though weak interaction theory by itself cannot.

Recently a new class of unified theories called supergravity theories has been developed. These theories are also non-Abelian gauge theories that include the gravitational interaction and give hope of uniting it and the other basic interactions. They also give hope of constructing a consistent quantum theory of gravity since, for those calculations that we have been able to carry through so far, they give finite results instead of the infinite ones obtained from a quantized version of the Einstein theory. It is, however, still too early to tell if these hopes can be realized and whether we can ultimately construct a single unified quantum field theory of the fundamental interactions found in nature.

See also: Gauge Theories; Group Theory in Physics; Maxwells Equations; Relativity, General.

Bibliography

An eminently readable account of recent developments in unified field theory is given by Daniel Z. Freedman and Peter van Nieuwenhuizen in "Supergravity and the Unification of the Laws of Physics," *Sci. Am.* **238**, 126 (1978).

A technical introduction to non-Abelian gauge theories can be found in Jeremy Bernstein's "Spontaneous Symmetry Breaking, Gauge Theories, the Higgs Mechanism and All That," *Rev. Mod. Phys.* **46**, 7 (1974).

For a technical account of the various classical field theories, see Peter G. Bergmann; *An Introduction to the Theory of Relativity*. Dover Reprints, New York, 1976.

Fields

C. Taylor

The concept of a "field" has been central to the development of modern theoretical physics. The idea was originally introduced as a conceptual alternative to treating forces in terms of action at a distance. Thus, rather than thinking about the earth instantaneously causing the moon to accelerate toward it, one introduces the earth's gravitational field, which is defined throughout space and tells you how a test particle of infinitesimal mass would move. Formulated thus, the field concept is not of much more than philosophical interest; the power becomes apparent only when one assumes that the field obeys mathematical equations governing its evolution. In this sense, the first field theory was Euler's theory of hydrodynamics. He described the motion of a fluid in terms of the velocity at each point (a velocity field). This motion is then governed by a set of partial differential equations. This procedure of abstracting certain important aspects of physical media, representing them as continuous fields, and assuming a suitable set of equations governing them has been a key element in the development of theories of sound and elasticity, and, in the recent past, in the development of the renormalization group governing aspects of phase transitions in condensed matter systems.

The construction of a satisfactory theory of hydrodynamics, in which momentum transport (and hence, forces) are described by partial differential equations, without the notion of an action at a distance, had a profound impact on the development of other branches of physics. Faraday was the first to argue that conventional action-at-a-distance theories would be inadequate to understand electromagnetic phenomena and pointed the way toward the development of a field theory of electromagnetism. This was brought to fruition by Maxwell, who, using symmetry principles, together with mechanical models for wave propagation, proposed a satisfactory and essentially complete theory of isolated electromagnetic phenomena. At the same time, Maxwell pointed the direction towards a more abstract understanding of fields, arguing that the mechanical models he introduced would aid in the understanding of electrical phenomena, but were designed to be illustrative, not explanatory.

The twentieth century has added two principles which underlie all modern field theories of fundamental processes: relativity and quantum mechanics. Indeed, Einstein's formulation of special relativity is, in one sense, an elucidation of symmetries already present in Maxwell's theory of electromagnetism. In contrast, the rise of quantum theory has resulted in new interpretations of the "meaning" of field theories. This began with Schrödinger, who introduced a nonrelativistic equation which describes the evolution of physical systems. This equation is formulated in terms of a "wave function", a field which has the interpretation of a probability amplitude. That is, quantum mechanically, one cannot think of a classical particle as being defined by its position and momentum, but must instead introduce a wave function $\psi(x)$, such that the probability of finding the particle within a distance dx of the position x is $P(x)\,dx = |\psi(x)|^2\,dx$.

Schrödinger's equation is a useful approximation to much of nonrelativistic quantum physics, but requires modification in order to be consistent with relativity. Such modifications were first carried out by Klein and Gordon; the resulting wave equation governs a field having a single component at each point of space-time. It came as a surprise that the interpretation of Schrödinger was no longer tenable. Instead, the field has to be reinterpreted as defining a charge density for an indefinite number of particles rather than a probability density for a single particle. In attempting to circumvent this problem, Dirac introduced a different field theory, involving a four-component "spinor" field, which can be interpreted as defining a probability density for a single electron at the cost of introducing "antiparticles" which formally have negative energy. This, rather than being a drawback, became a virtue when antiparticles were observed. This, in turn, helped motivate the development of second quantization, in which the fields are quantized, and can be decomposed into operators which create and annihilate particles.

In general, fundamental fields are classified according to their properties under the Lorentz group (the relativistic generalization of rotations), and under internal symmetries. As far as the Lorentz group is concerned, particles are classified by their "spin": the Klein–Gordon equation describes a spin-0 object, the Dirac equation, a spin-$\frac{1}{2}$ object, and Maxwell's equations describe a spin-1 particle. Formally, Einstein's theory of general relativity, which also takes the form of a field theory, describes a spin-2 particle, but no existing theory of gravity is satisfactory at the quantum-mechanical level. (One possible resolution of these problems is string theory, which evades technical problems to which other theories of gravity are subject.)

The internal symmetries of particles can be of two types: global or local. A global invariance of a field theory is a simultaneous modification of the field at all points of space-time which leaves the physical properties of the theory invariant. A local symmetry is a transformation of the fields in an infinitesimal neighborhood which leaves the theory invariant. In the latter case, these local symmetries indicate the presence of a gauge invariance, of which the prototypical example is again Maxwell's theory. Generalizations of the gauge invariance of Maxwell's equations, first proposed by Yang and Mills in 1954, form the basis for our modern theory of the electroweak and strong interactions. In this context, it is important to note that 't Hooft and Veltman showed that these theories are renormalizable; that is, that they retain their predictive power after second quantization. It should also be noted that not all of the symmetries of the fundamental laws of physics are actually realized in the physical world: spontaneous breaking of gauge symmetries provides the basis for models of superconductivity as well as for the generation of masses in the standard model of the electroweak interactions.

Field theory is still a rapidly developing subject. Recent years have seen the intensive study of the nonperturbative structure of quantum field theories, especially those related to string theory, often with surprising results. These advances can be expected to continue. Additional insights are likely to result from the confrontation of theory with experiment as the next generation of particle accelerators is used to test the detailed structure of the standard model and search for new physics.

See also: Electrodynamics, Classical; Elementary Particles in Physics; Field Theory, Classical; Quantum Field Theory; Relativity, General Theory; String Theory.

Bibliography
A nice review of the early history of field theory is: M. B. Hesse, *Forces and Fields*. Philosophical Library, New York, 1961.
A standard textbook on field theory, including applications to both particle physics and critical phenomena is: M. E. Peskin and D. V. Schroeder, *An Introduction to Quantum Field Theory*. Perseus Books, New York, 1995.
A nice treatment of the classical theory of the electromagnetic and gravitational fields is: L. D. Landau and E. M. Lifshitz, *The Classical Theory of Fields*. Pergamon, New York, 1979.

Fine and Hyperfine Spectra and Interactions
G. W. F. Drake

The spectral lines of hydrogen, when observed with high resolution, are found to consist of multiplets which can be ascribed to a doublet splitting of every level except those for which $l = 0$. The multiplet structure, known as the fine structure, is the result of relativistic effects which are neglected in the simple Bohr picture of hydrogen. A yet smaller splitting, the hyperfine structure, arises from the electron interaction with higher multipole moments of the nucleus. In nonhydrogenic atoms, fine structure is often described in terms of various spin–orbit and spin–spin interactions. However, for heavy atoms, these effects become too large to be treated as small perturbations and a complete relativistic calculation of energy levels becomes necessary. The effects also become large in states of high angular momentum, producing a strong mixing of states of the same total angular momentum and parity, but different spin. The hyperfine interaction, in contrast, is generally small enough to be treated as a perturbation irrespective of the atom's complexity.

The fine structure of hydrogen was first described comprehensively by Dirac's relativistic theory of the electron. The major contribution is easily understood by a semiclassical argument. The electron, moving with velocity $\mathbf{v}$, "sees" a motional magnetic field $\mathbf{B}_m = -(\mathbf{v}/c) \times \mathbf{E}$, where $\mathbf{E}$ is the Coulomb field. The motional field gives rise to an interaction $-\boldsymbol{\mu}_e \cdot \mathbf{B}_m$, where $\boldsymbol{\mu}_e$ is the electron's magnetic moment. A straightforward calculation leads to an interaction term of the form

$$\mathcal{H}_{so} = \zeta(r)\mathbf{L} \cdot \mathbf{S}$$

where $\zeta(r) = (\frac{1}{2}m^2c^2r)\,dV/dr$. V is the electrostatic energy. The expression includes a factor of $\frac{1}{2}$, the Thomas factor, arising from the relativistic transformation from the accelerating frame of the electron to an inertial frame. The $\mathbf{L} \cdot \mathbf{S}$ interaction, known as the spin–orbit coupling, is the major contribution to the fine structure.

An additional contribution to atomic energies arises from the variation of the electron's mass with velocity. The total energy of the electron is $W = V + (p^2c^2 + m^2c^4)^{1/2}$. The kinetic energy is $T = W - V - mc^2 \cong p^2/2m - p^4/8mc^2$. The last term,

$$\mathcal{H}_{rel} = -\frac{p^4}{8m^3c^2},$$

is the lowest-order relativistic correction to the classical kinetic energy. It shifts fine structure levels by the same amount, and so does not affect the splitting.

If we treat $\mathcal{H}_{so}$ and $\mathcal{H}_{rel}$ as perturbations to the hydrogenic term energy, we eventually obtain the following expression for the fine structure energy:

$$E_{fs} = \langle \mathcal{H}_{so} + \mathcal{H}_{rel} \rangle = \frac{-hcRZ^2}{n^2}\left[\frac{\alpha^2Z^2}{n}\left(\frac{1}{j+\frac{1}{2}} - \frac{3}{4n}\right)\right],$$

R is the Rydberg constant, n is the principal quantum number, $\hbar j$ is the total electronic angular momentum, Z is the nuclear charge, and $\alpha = e^2/\hbar c \approx 137^{-1}$ is the fine structure constant. The leading factor is the term energy of a hydrogenic atom. The bracketed expression reveals that the fine structure of hydrogen is smaller than the term energy by approximately $\alpha^2Z^2/n \approx 3 \times 10^{-5}/n$, which justifies its treatment as a perturbation. The Dirac theory, when taken to lowest order, gives the identical result.

Fine structure splits each angular momentum state into doublets depending on whether $j = l + \frac{1}{2}$ or $j = l - \frac{1}{2}$. For $l = 0$, only $j = l + \frac{1}{2}$ is possible. Thus the configurations for hydrogen are of the form $^2S_{1/2}$, $^2P_{1/2,3/2}$, $^2D_{3/2,5/2,...}$. An important feature of hydrogenic fine structure is the degeneracy of all states of a given term having the same j, such as the pairs $(^2S_{1/2}, {}^2P_{1/2})$, $(^2P_{3/2}, {}^2D_{3/2})$.

The fine structure of hydrogen has served as a testing ground for relativistic quantum theory and for quantum electrodynamics. The earliest studies involved optical spectroscopy of the Balmer spectrum ($n > 2 \rightarrow n = 2$) of hydrogen and deuterium. The experiments were inconclusive, however, because of the small fine structure splitting of the $n = 2$ state, and the large Doppler width of hydrogen's spectral line. (The fine structure of the $n = 2$ term is only $0.37 \, \text{cm}^{-1} = 11 \, \text{GHz}$; the Doppler width at room temperature of the principal Balmer line is $0.2 \, \text{cm}^{-1}$.)

Starting in the late 1940s Willis E. Lamb and his colleagues carried out a series of experiments in which the fine structure was studied by radio-frequency spectroscopy of the transitions $^2S_{1/2} \rightarrow {}^2P_{1/2}, {}^2P_{3/2}$, with a resolution of $10^{-4} \, \text{GHz}$. He confirmed that the fine structure of the ^2P state is accurately given by the Dirac theory, and in fact used his measurements to obtain a more precise value for α. A more dramatic finding, for which Lamb received the Nobel prize, was that $^2S_{1/2}$ and $^2P_{1/2}$ levels for $n = 2$ are not, in fact, degenerate. The $^2S_{1/2}$ state was found to be shifted upward by 1060 MHz. The displacement, known as the Lamb shift, is dominantly due to radiative coupling of the electron with the vacuum field. The discovery of the Lamb shift played a central role in the development of quantum electrodynamics. Experiments and calculations on the fine structure and the Lamb shift in hydrogen have undergone continuing refinement. Experiment and theory have been pushed to an accuracy of about 11 kHz, or 1 part in 10^5, and are in good agreement.

Lamb shift measurements are also available for the heavier hydrogenic ions up to one-electron U^{91+}. The most accurate result for He^+ is obtained indirectly from the anisotropy in the angular distribution of Ly-α quenching radiation emitted by the metastable $2 \, {}^2S_{1/2}$ state in an electric field. The accuracy of 1.8 parts in 10^5 provides the most sensitive available test of the higher-order quantum electrodynamic contributions to the Lamb shift. Hyperfine interaction arises principally from the coupling of the nuclear magnetic moment $\boldsymbol{\mu}_I$ with the magnetic field produced by the electron, $\mathbf{B}_e$. From symmetry arguments it can be shown that $\boldsymbol{\mu}_I \propto \mathbf{I}$ and that the hyperfine interaction has the general form

$$\mathcal{H}_{\text{hf}} = a\mathbf{I} \cdot \mathbf{J} \, .$$

The quantity a, known as the hyperfine constant, has the value for hydrogen

$$a = hcR \frac{Z^3}{n^3} \frac{m}{M} g_I \frac{1}{(l + \frac{1}{2})(j + 1)} \, .$$

This expression neglects small corrections due to nuclear structure and radiative and relativistic effects. m/M is the electron–nuclear mass ratio and g_I is the nuclear g-factor. In the ground state of hydrogen $I = J = \frac{1}{2}$, and $\mathbf{I} \cdot \mathbf{J}$ has the values $\frac{1}{4}$ and $-\frac{3}{4}$ for the two hyperfine components. Thus the separation between the hyperfine components is simply

$$\Delta E_{\text{hf}} = a \, .$$

The hyperfine constant in frequency units, $a/h \approx 1.420 \, \text{GHz}$, has been measured with the hydrogen maser to a precision of $10^{-3} \, \text{Hz}$; it is probably the best known physical quantity. The hyperfine separation of hydrogen has played a major role in radioastronomy by giving rise to the 21-cm line which is extensively used for mapping hydrogen radio sources.

Table 1: Fine and hyperfine structure of hydrogen and alkali atoms.

| | Fine structure | | Hyperfine structure[a] | |
	State	(cm^{-1})	State	(GHz)
H	2p	0.37	1s	1.42
Li	3p	0.44	2s	0.83
Na	4p	17.2	3s	1.77
K	5p	57.7	4s	0.46
Rb	5p	237.6	5s	3.03
Cs	5p	554.1	6s	9.19

[a] Hyperfine structure for most abundant isotope.

In states for which the nuclear and electronic angular momenta are greater than $\frac{1}{2}$, hyperfine structure can arise from an electrostatic quadrupole interaction between the nucleus and the electron. Higher-order magnetic and electric interactions can also occur, though they are generally minute.

The alkalis, which resemble hydrogen in being essentially single-electron atoms, also display doublet structure for all states except those for which $l = 0$. The closed-shell core has, however, a major effect on the doublet splitting. Inside the core the potential departs radically from the Coulombic form. The spin–orbit coupling, which varies predominantly as dV/dr, is very sensitive to the core structure, and states which appreciably penetrate the core generally have fine structure separations which are large compared with those of hydrogen. This is illustrated in Table 1, which displays the fine structure intervals of the lowest p states for the alkalis and hydrogen. Hyperfine separations of the ground states are also listed for comparison. For s states, hyperfine interactions depend on the density of the electron at the nucleus and on the nuclear moment. As Table 1 shows, although hyperfine separations for the alkalis differ appreciably from that for hydrogen, they are the same order of magnitude.

In certain excited states of the alkalis, both the fine and hyperfine coupling constants are inverted with respect to hydrogen. The mechanism for this involves a spin-dependent exchange interaction between the valence electron and core electrons. For high angular momentum, the interactions assume their normal sign and the systems are well described by hydrogenic theory.

For many-electron atoms, the fine and hyperfine structure can be complex. Systems described by L–S coupling, however, obey a simple rule—the Landé interval rule—which is of great help in identifying spectra. For such systems the spin–orbit interaction can again be written $\zeta \mathbf{L} \cdot \mathbf{S}$, where ζ now represents contributions from each of the electrons. If the total angular momentum can be written in the form $\mathbf{J} = \mathbf{L} + \mathbf{S}$, then simple vector arguments give $\langle \mathbf{L} \cdot \mathbf{S} \rangle = [j(j+1) - l(l+1) - s(s+1)]/2$. The separation between levels with angular momentum j and $j - 1$ is ζj; thus the ratio of separations between adjacent components of a fine structure multiplet, $(j, j-1)$ and $(j-1, j-2)$, is $j/(j-1)$. (This rule also applies to magnetic hyperfine structure with j replaced by the total electronic and nuclear angular momentum, F.)

Table 2 shows intervals between the $j = 2$, 1, and 0 components of the lowest P terms for atoms with two valence electrons. The intermediate-weight elements obey the rule reasonably, but the very light and very heavy elements depart appreciably. In the case of light elements,

Table 2: Fine structure splittings of atoms with two valence electrons[a]

	State	$\Delta E(^3P_2-^3P_1)$	$\Delta E(^3P_1-^3P_0)$	Ratio
He	1s2p	0.076 4261	−0.987 9122	–
Be	2s2p	2.35	0.68	3.5
Mg	3s3p	40.7	20.1	2.03
Ca	4s4p	105.9	52.2	2.03
Sr	5s5p	394.2	186.8	2.11
Ba	6s6p	878.1	370.6	2.37

[a] Energy in cm^{-1}.

fine structure is not well described by a simple $\mathbf{L} \cdot \mathbf{S}$ term; the spin of each electron couples to the spin and orbital moment of the other electron. For helium these effects are so large that the fine structure is actually inverted.

Departures from the interval rule for heavy atoms indicate a breakdown of L–S coupling due to strong configuration mixing by the spin–orbit interaction. In the case of hyperfine structure, departure from the interval rule signals the presence of quadrupole or higher-order interactions.

The theoretical calculation of fine and hyperfine structure from first principles provides a sensitive test of the accuracy of the electronic wave functions used. For helium and the helium-like ions Li^+, Be^{++} ,..., high-precision variational wave functions constructed from linear combinations of correlated functions of the form $r_1^i r_2^j r_{12}^k e^{-\alpha r_1 - \beta r_2}$ are available. Here, $\mathbf{r}_1$ and $\mathbf{r}_2$ are the position vectors of the two electrons and $r_{12} = |\mathbf{r}_1 - \mathbf{r}_2|$ is the interelectronic coordinate. Fine structure then arises from matrix elements of the Breit interaction, which includes spin–other-orbit, spin–spin, and Fermi contact interactions in addition to the spin–orbit interaction mentioned above. The corresponding hyperfine structure Hamiltonian is

$$\mathcal{H}_{hf} = -2\mu_0 \sum_{i=1}^{2} \left[-\frac{1}{r_i^3}\mathbf{l} \cdot \boldsymbol{\mu} + \frac{1}{r_i^3}\left(\frac{\mathbf{s}_i \cdot \boldsymbol{\mu} - 3(\mathbf{s}_i \cdot \mathbf{r}_i)(\boldsymbol{\mu} \cdot \mathbf{r}_i)}{r_i^3} \right) - \frac{8\pi}{3}(\mathbf{s}_i \cdot \boldsymbol{\mu})\delta(\mathbf{r}_i) \right]$$

where $\boldsymbol{\mu} = -g_I(m/M)\mu_0 \mathbf{I}$ is the nuclear magnetic moment and $\mu_0 = e/2mc$. The above three terms correspond to the electron-orbit–nuclear spin, electron-spin–nuclear spin, and contact terms, respectively. There are numerous small corrections for relativistic, radiative, and finite nuclear mass and size effects. When these are included, the theoretical value for the dominant contact term for the 1s2p ^{3}P state of ^{3}He is -4283.8 ± 0.2 MHz, as compared with the experimental value -4282.72 ± 0.04 MHz. The corresponding values for the orbital term are -28.13 ± 0.02 MHz and -29.85 ± 0.14 MHz. The slight differences are due to higher-order terms not included in the calculation.

The fine structure intervals of the 1s2p ^{3}P state of ^{4}He have been extensively studied, both theoretically and experimentally. When higher-order relativistic and quantum electrodynamic effects are included, theory and experiment for the $^3P_1-^3P_0$ interval agree to within 3 kHz or 1 part in 10^8. At this level of accuracy, the comparison can be interpreted as a measurement of the fine structure constant that is comparable in accuracy to other standard measurements such as the ac Josephson and quantum Hall effects from condensed-matter physics.

Interesting studies of fine structure splittings have recently been made in the high angular momentum Rydberg states of helium through the measurement of transitions such as 1s10g–1s10h, 1s10h–1s10i, etc., by fast beam microwave/laser resonance techniques. Here, spin–orbit coupling dominates the small singlet–triplet electrostatic splittings to produce four approximately equally spaced levels for each angular momentum. These splittings have fruitfully been analyzed by asymptotic expansion methods which neglect exchange effects and treat the outer electron as moving in the field of a polarizable core consisting of the nucleus and the inner electron. An important feature of these studies is that they are sensitive to long-range Casimir–Polder retardation effects.

A priori calculations of fine and hyperfine structure for many-electron atoms are much more difficult because the splittings are sensitive to the details of the wave function near the nucleus. Simple Hartree–Fock approximations are generally of low accuracy. Only in the case of three-electron atoms such as Li and Be^+ have high-precision correlated variational wave functions of the type described above for helium been obtained. Results matching experimental accuracies of $0.0000019 \, cm^{-1}$ (6 parts per million) have been achieved for the fine structure splittings of lithium, and similarly for the hyperfine structure. The differences from experiment are due to higher order relativistic and quantum electrodynamic effects not included in the calculations, rather than to errors in the many-electron wave functions. For heavier atoms, considerable progress has been made in the application of many-body perturbation techniques, with uncertainties below $\pm 1\%$ in the hyperfine splittings for atoms as heavy as cesium. This particular case is important as a test of the similar atomic structure calculations required to interpret parity nonconservation measurements in terms of the electroweak interaction and the Weinberg angle.

The term *fine structure* is used in a somewhat different context with respect to molecular spectra. There it refers to the rotational structure of an electronic or vibrational molecular band. The term is descriptive because the structure only becomes apparent when the spectrum is observed with moderate resolution. Molecular spectra can also exhibit hyperfine structure arising from nuclear–electronic interactions. Although magnetic dipole interactions tend to be most important for free atoms, molecular hyperfine structure is often dominated by the electric quadrupole interaction.

Fine structure is generally large enough to be studied by the methods of conventional spectroscopy (an important exception is hydrogen for which most of the precision studies have employed techniques of radio-frequency spectroscopy and atomic beams). Hyperfine structure, on the other hand, is so small that optical observation usually requires high resolution interferometry. Ground-state hyperfine structure is often studied by molecular-beam electric and magnetic resonance methods which can yield precision in the range of parts per million to parts per billion, or better. Hyperfine structure studies can provide information on the spin, magnetic dipole moment, and electrical quadrupole moment of the nucleus, as well as the electronic charge distribution in atoms and molecules. One practical application for hyperfine spectroscopy is to atomic clocks; the second is defined in terms of the hyperfine separation of ^{133}Cs, which is taken to be $9\,192\,631.770 \, Hz$. With progressive improvements, the uncertainty has been steadily decreasing to its current value of less than 1×10^{-15} second.

Interest in fine and hyperfine spectra has been renewed by the advent of laser spectroscopy, which has allowed the study of excited states that were previously inaccessible, and which overcomes the limitations of Doppler broadening, permitting precision far greater than previously possible. The development of ion traps has opened the way to high-precision measurements in ions as well as atoms. By this method, a hyperfine transition in the 2s ground state of Be+ has been measured to precision of about one part in 10^{13}, a precision which is exceeded only by similar measurements in neutral cesium.

See also: Atomic Spectroscopy; Laser Spectroscopy; Quantum Electrodynamics; Quantum Mechanics.

Bibliography

H. A. Bethe and E. S. Salpeter, *Quantum Mechanics of One- and Two Electron Atoms*. Academic Press, New York, 1957.

E. U. Condon and G. H. Shortley, *Theory of Atomic Spectra*. Cambridge University Press, London, 1951.

G. W. F. Drake, "Progress in helium fine structure calculations and the fine structure constant", *Can. J. Phys.* **80**, 1195 (2002).

W. R. Johnson, S. A. Blundell, and J. Sapirstein, in *Atomic Physics 11*, S. Haroche, J. C. Gay and G. Grynberg (eds.). World Scientific, Singapore, 1989.

D. Kleppner, in *Atomic Physics and Astrophysics*, N. Chretien and E. Lipworth (eds.). Gordon and Breach, New York, 1971.

H. G. Kuhn, *Atomic Spectra*, 2nd ed. Longmans, London, 1969.

I. Lindgren, *Rep. Prog. Phys.* **47**, 345 (1984).

M. Mizushima, *Quantum Mechanics of Atomic Spectra and Atomic Structure*, Ch. 9. Benjamin, New York, 1970.

K. Pachucki, "Lithium hyperfine splitting", *Phys. Rev. A* **66**, 062501 (2002).

G. W. Series, in *The Spectrum of Atomic Hydrogen: Advances*, G. W. Series (ed.). World Scientific, Singapore, 1988; and other articles therein.

J. Sapirstein and K.T. Cheng, "Calculation of radiative corrections to hyperfine splittings in the neutral alkali metals", *Phys. Rev. A* **67**, 022512 (1993).

J. C. Slater, *Quantum Theory of Atomic Structure*, Vol. II. McGraw–Hill, New York, 1960.

I. I. Sobel'man, *Introduction to the Theory of Atomic Spectra*. Pergamon Press, Oxford, 1972.

C. H. Townes and A. L. Schawlow, *Microwave Spectroscopy*. McGraw–Hill, New York, 1955.

A. van Wijngaarden, J. Patel, and G. W. F. Drake, in *Atomic Physics 11*, S. Haroche, J. C. Gay and G. Grynberg (eds.). World Scientific, Singapore, 1989.

G. Werth, in *Atomic Physics 9*, R. S. van Dyck, Jr. and E. N. Fortson (eds.). World Scientific, Singapore, 1984.

D. J. Wineland, W. M. Itano, J. G. Bergquist, J. J. Bollinger, and J. D. Prestage, in *Atomic Physics 9*, R. S. van Dyck, Jr. and E. N. Fortson (eds.). World Scientific, Singapore, 1984.

Fluctuation Phenomena

T. J. Greytak

By fluctuation we usually mean the deviation of some quantity from its mean or most probable value. Almost all the quantities that a physicist might be interested in studying exhibit fluctuations, some on quite a microscopic scale (e. g., the motion of an atom about its equilibrium lattice position in a crystalline solid). This discussion, however, will be limited to macroscopic quantities that can be treated as thermodynamic variables. Fluctuations in these quantities manifest themselves in several ways: They may limit the precision of measurements of the mean value of the quantity, as is the case with thermal noise in electronic circuits or photon shot noise in light beams; they are the cause of some of the familiar features of our surroundings, such as the blue sky; or they may cause spectacular and unexpected effects, such as the critical opalescence of an apparently simple and homogeneous fluid.

Fluctuations in a Hydrostatic System

The simplest example of thermodynamic fluctuations occurs in a hydrostatic system such as a gas, a liquid, or a simple solid under hydrostatic pressure. In such a system of a fixed number of particles, the internal energy U is related to the principal state variables of temperature, entropy, pressure, and volume by the first and second laws of thermodynamics:

$$dU = T\,dS - P\,dV \; . \tag{1}$$

Only two of the state variables can be chosen independently; the others are then fixed through the equation of state and an energy equation such as an expression for the heat capacity. Let these variables pertain to a subsystem specified by a certain number N of contiguous particles. N is assumed to be large enough that the definition of local thermodynamic variables makes sense, yet the subsystem must be small enough that its energy is only a very small part of the energy of the total system. In thermal equilibrium the *average* values of the intensive variables T and P are equal to the temperature and pressure of the total system. The average values of S and V will be much smaller than those of the total system and can be related to the average T and P.

At any instant of time there is a finite probability that the subsystem will not be in its most likely state. The probability that a pair of independent variables X and Y will deviate from their mean values by amounts ΔX and ΔY is given by a bivariate Gaussian density that depends only on the three parameters $\langle \Delta X^2 \rangle$, $\langle \Delta Y^2 \rangle$, and $\langle \Delta X \Delta Y \rangle$. These quantities are listed in Table 1. Notice that the cross correlation between ΔT and ΔV and between ΔS and ΔP vanishes. If either of these pairs of variables were chosen to be the independent ones, their joint probability density would factor into the product of two single Gaussian densities, one for each variable. In this case the variables are statistically independent as well as independent in the thermodynamic sense: a knowledge of the fluctuation in one gives no additional information about the fluctuation in the other. It should be pointed out, though, that ΔT and ΔV or ΔS and ΔP are statistically independent random variables only when taken at the same instant of time. Inspection of the table will show that the ratio of the root mean square fluctuation in any variable to its mean value, $\langle \Delta X^2 \rangle^{1/2}/\langle X \rangle$, is proportional to $N^{-1/2}$. This is why conventional thermodynamics need deal only with the mean values of the variables.

Through expansion of any other thermodynamic variable in terms of two of the principal ones, the fluctuations in that quantity can be found. For example, the fluctuations in the number density, $n \equiv N/V$, are given by

$$\langle \Delta n^2 \rangle = \frac{n^3 kT \kappa_T}{N} \tag{2}$$

Table 1: Fluctuations in a hydrostatic system[a].

$\langle \Delta T^2 \rangle = \dfrac{kT^2}{C_V}$	$\langle \Delta V^2 \rangle = kTV\kappa_T$
$\langle \Delta S^2 \rangle = kC_P$	$\langle \Delta P^2 \rangle = \dfrac{kT}{V\kappa_S}$
$\langle \Delta T \Delta V \rangle = 0$	$\langle \Delta S \Delta P \rangle = 0$
$\langle \Delta T \Delta S \rangle = kT$	$\langle \Delta V \Delta P \rangle = -kT$
$\langle \Delta T \Delta P \rangle = \dfrac{\alpha kT^2}{\kappa_T C_V}$	$\langle \Delta V \Delta S \rangle = kTV\alpha$

[a] C_V is the heat capacity at constant volume; C_P the heat capacity at constant pressure. The isothermal compressibility is

$$\kappa_T \equiv -\frac{1}{V}\frac{\partial V}{\partial P}\bigg|_T \; ;$$

the adiabatic compressibility

$$\kappa_S \equiv -\frac{1}{V}\frac{\partial V}{\partial P}\bigg|_S \; ;$$

the expansion coefficient

$$\alpha \equiv \frac{1}{V}\frac{\partial V}{\partial T}\bigg|_P \; .$$

and the fluctuations in the internal energy can be shown to be

$$\langle \Delta U^2 \rangle = kT^2 C_V + kTV\kappa_T\left(P - \frac{\alpha T}{\kappa_T}\right)^2 . \tag{3}$$

In a similar manner all of the entries in Table 1 can be found from the ones pertaining to a single pair of variables.

Fluctuations in Other Systems

The results for the hydrostatic system can be carried over directly to several other systems that are thermodynamically isomorphic to it. To identify the new variables with those of the hydrostatic case we must examine the increment of work done on the system for a given change in the mechanical variables. For the hydrostatic case [Eq. (1)] the work is $(-P)\,dV$; for a magnetic system with magnetization M and an external or applied magnetic field H it is $H\,dM$; for a dielectric body with polarization $\mathcal{P}$ and electric field E it is $E\,d\mathcal{P}$; for an interface of surface tension S and area A it is $S\,dA$; for an elastic rod with tension $\mathcal{T}$ and length L it is $\mathcal{T}\,dL$. For any of these systems we need only make the appropriate substitution of variables in Table 1 in order to find the fluctuations. For example, for the magnetic system $-P \rightarrow H$ and

$V \to M$, so we find that the fluctuations in T and M are statistically independent and that

$$\langle \Delta T^2 \rangle = \frac{kT^2}{C_M}, \quad \langle \Delta M^2 \rangle = kT \left. \frac{\partial M}{\partial H} \right|_T \tag{4}$$

where C_M is the specific heat at constant magnetization and $(\partial M/\partial H)|_T$ is the isothermal susceptibility.

In a more general system with m independent thermodynamic variables the fluctuations in these variables form a set of m jointly Gaussian random variables. The associated probability density function contains as parameters the m mean square fluctuations $\langle \Delta X^2 \rangle$ and the cross correlations of all possible pairs $\langle \Delta X \Delta Y \rangle$. The most important example is probably a binary mixture where $m = 3$. If the concentration of one of the components is $c_a \equiv N_a/N$ and the difference in chemical potential of the two components is $\mu' \equiv \mu_a - \mu_b$, then we can show that

$$\langle \Delta c_a \Delta T \rangle = \langle \Delta c_a \Delta P \rangle = 0 \tag{5}$$

$$\langle \Delta c_a^2 \rangle = \frac{kT}{N} \left. \frac{\partial c_a}{\partial \mu'} \right|_{T,P}$$

It is interesting to note that for this system we cannot choose a set of three conventional variables that are all mutually statistically independent.

Time Evolution of the Fluctuations

So far only static fluctuations – those that are evaluated at one instant of time – have been considered. The fluctuations are actually random processes that evolve in time, $\Delta n(t)$, for example. (Density fluctuations will be used for illustration since they are the ones most often studied experimentally, but the discussion that follows could be applied to any thermodynamic variable.) A great deal of information about the temporal behavior of the fluctuations is contained in the time correlation function, $\langle n(t) n(t + \tau) \rangle$. It is usual to consider each spatial Fourier component of the variable separately. If $n_q(t)$ is the complex amplitude of the qth Fourier component of the density, the relevant correlation function is $\langle n_q(t) n_q^*(t + \tau) \rangle$. In order to calculate such a quantity we use the Onsager "regression of fluctuations" hypothesis, a version of the fluctuation–dissipation theorem , which states that the time dependence of the equilibrium correlation function is the same as the time evolution of an externally induced density disturbance of wave vector $\mathbf{q}$ which is released at $t = 0$. This reduces to a well-posed initial-value problem that can be solved *exactly* once the macroscopic equations of motion are known. The details of this type of calculation for simple fluids have been discussed by Mountain. The general result of such calculations is that the equilibrium correlation function is made up of a sum of contributions, one from each hydrodynamic normal mode of the system, each reflecting the time dependence associated with the macroscopic decay of that mode. In the case of the simple fluid there is a sound-wave contribution to $\langle n_q(t) n_q^*(t + \tau) \rangle$ which oscillates at a frequency $\omega = uq$ (u is the velocity of sound) while it decays toward zero because of processes that damp macroscopic sound waves. There is also a thermal diffusion contribution, which has a simple exponential decay at a rate proportional to the thermal conductivity. For a binary mixture there would be a third contribution to the correlation function, another exponential decay at a rate proportional to the concentration conductivity.

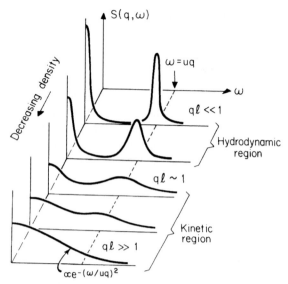

Fig. 1: Dynamic structure factor for density fluctuations in a classical monatomic gas; l is the atomic mean free path and u is the velocity of sound.

This simple picture for the time dependence of the correlation function is based on the assumption that well-defined hydrodynamic normal modes can exist in the system. In a gas when the mean free path l of the molecules becomes comparable to or greater than q^{-1} this assumption is no longer valid. The coherent atomic motion characteristic of such modes is dissipated by the distribution of individual molecular velocities. In this case a kinetic theory, such as the Boltzmann equation, must be used to calculate the correlation function. Such calculations are discussed by Yip.

Measurement of the Fluctuations by Scattering Spectroscopy

A major advance in the study of fluctuations occurred with the realization that all scattering experiments on condensed-matter systems, whether they used light, x rays, neutrons, or electrons, measured essentially the same quantity, the dynamic structure factor $S(\mathbf{q}, \omega)$. $S(\mathbf{q}, \omega)$ is the space and time Fourier transform of the correlation function for the variable that couples to the scattering probe. For the density fluctuations in a classical system it can be written as

$$
\begin{aligned}
S(\mathbf{q}, \omega) &= \int_{-\infty}^{\infty} d\tau \int_{V} d^3 R\, e^{i(\mathbf{q}\cdot\mathbf{R}-\omega\tau)} \langle n(\mathbf{r},t)n(\mathbf{r}+\mathbf{R},t+\tau)\rangle\,, \\
&= \int_{-\infty}^{\infty} d\tau\, e^{-i\omega\tau} \langle n_{\mathbf{q}}(t)n_{\mathbf{q}}^{*}(t+\tau)\rangle\,.
\end{aligned}
\tag{6}
$$

Experimentally the scattering geometry singles out one scattering angle, which, through the Bragg condition, determines the wave vector of the fluctuations being observed. A spectrometer then carries out the frequency analysis of the scattered beam. Figure 1 shows the features of $S(q, \omega)$ for a classical gas. The correlation function for such a system was discussed in the previous section. A spectrum of this type can be obtained by the Brillouin scattering of laser light from a monatomic gas. Notice that in the hydrodynamic region there are distinct contributions from the two normal modes of the system, sound waves and thermal diffusion, but in the kinetic region such a distinction no longer makes sense.

Light scattering has been used to study the dynamics of density fluctuations in gases, liquids, superfluids, and solids. It has also been applied to concentration fluctuations in solutions and orientation fluctuations in liquid crystals. Neutron scattering has studied fluctuations in the magnetization of ferromagnets and antiferromagnets, and fluctuations in the local atomic order associated with structural phase transitions. Inelastic x-ray and electron scattering are also being applied to studies of similar fluctuation phenomena, but often the fluctuations studied are on too microscopic a scale to be classified as thermodynamic fluctuations.

Divergence of the Fluctuations

It was pointed out earlier that usually we need only deal with the mean of a thermodynamic variable when discussing macroscopic phenomena, since the amplitude of the fluctuations relative to the mean is always inversely proportional to the square root of the number of particles involved. However, in some circumstances, usually at a second-order phase transition, the constant of proportionality (some thermodynamic derivative) diverges. For example, the isothermal compressibility [Eq. (2)] diverges at the gas–liquid critical point, the isothermal magnetic susceptibility [Eq. (4)] diverges at the Curie temperature of a ferromagnet, and the osmotic compressibility $(\partial c_a/\partial \mu')|_{T,P}$ [Eq. (5)] diverges at the consolute point of a binary mixture. In such cases the fluctuations themselves dominate the dynamics of the system and the equations of motion dealing only with the mean values are no longer valid. Exactly how the medium does behave in these circumstances is not well understood. It is the subject of a field of research known as dynamic critical phenomena. The clearest example of this fluctuation-dominated behavior is the critical opalescence of a simple fluid at its critical point. The density fluctuations scatter so much light that the fluid becomes translucent or opaque. It is interesting that this particular phenomenon, which first attracted physicists to the study of thermal fluctuations over a hundred years ago, is still the subject of frontier research in physics.

See also: Critical Points; Dynamic Critical Phenomena; Light Scattering; Statistical Mechanics; Thermodynamics, Equilibrium.

Bibliography
Theory of Thermodynamic Fluctuations
L. D. Landau and E. M. Lifshitz, *Statistical Physics, 3rd Edition Part 1*, Chapter 12. Pergamon Press, 1980.

Time Dependence of Fluctuations
E. M. Lifshitz and L. P. Pitaevskii, *Statistical Physics, Part 2*, Chapter 9. Pergamon Press, 1980.
R. D. Mountain, *Rev. Mod. Phys.* **28**, 205 (1966).
S. Yip, *J. Acoust. Soc.* **49**, 941 (1971).

Scattering Spectroscopy

W. Marshall and S. W. Lovesey, *Theory of Thermal Neutron Scattering*. Oxford, London and New York, 1971.

P. A. Fleury and J. P. Boon, "Laser Light Scattering in Fluid Systems", *Adv. Chem. Phys.* **24** , 1–93 (1973).

Critical Phenomena

H. E. Stanley, *Introduction to Phase Transitions and Critical Phenomena*. Oxford, London and New York, 1988.

Fluid Physics

D. Bershader[†]

What is a Fluid?

Matter exists in two forms, solid and fluid, the latter, in turn, being divided into the categories of liquids and gases. None of these distinctions is really sharp, but there is a gradation in both spacing and latitude of motion of the atomic components as one proceeds from solids to gases. Intermolecular forces play a dominant role in the closely spaced atomic arrangements in a solid, a smaller role in a liquid, and are much smaller in a gas. As a consequence, matter in the fluid State will generally deform continuously when subjected to a shear stress, no matter how small; such behavior can, in fact, be taken as a working definition of a fluid. For present purposes we will only mention but otherwise ignore more complex materials which show a dual character, such as jellies or paints; these may behave as elastic solids when allowed to stand, but lose their elasticity and behave as a liquid when strongly disturbed by an outside force.

Although fluid properties and related dynamic behavior are ultimately explained in terms of molecular parameters, the largest segment of fluid physics deals with fluid behavior on a macroscopic scale, i. e., large compared with the average molecular spacing. Stated differently, even "local" fluid behavior from a macroscopic view will involve enough molecules so that fluctuations will be smoothed out with no disturbing effect on the measured average values of physical quantities.

If one notes that gases with significant charged-particle content, i. e., plasmas, are included in the fluids category, then it turns out that more than 99.99% of the matter in the universe is in the fluid state. In our own earthly existence, there are many technical subject areas based on fluid physics. They include aerodynamics; thermal hydraulics of nuclear reactors; aerosols and emulsification for fluid transport purposes; meteorological flows; ballistics and atmospheric penetration by hypervelocity vehicles; flow in pipes, turbomachinery, and open channels; energy transfer processes such as magnetohydrodynamic flows, combustion, and propulsion; cavitation associated with behavior of ship propellers or entrance of a solid body into a fluid; flow through porous media in connection with oil extraction and other geophysical

---

[†]deceased

applications; and biofluid mechanics, to mention just a few. In what follows we will discuss the more important features of fluid physics which are basic to an understanding and utilization of fluid phenomena.

Fluid Properties

Here, we treat briefly a few properties that are of special importance for fluid motion. The two which appear most frequently in flow studies are density and viscosity. The fluid density is basic to the analysis of buoyancy and other static fluid phenomena, and contributes to the inertial behavior of fluids in dynamic situations. Thus, drag forces on solid bodies are proportional to fluid density in certain flow regimes and more complex functions of density in other regimes. Heat exchange between immersed bodies and flowing fluids is also a function of fluid density. In flows of gases with heat exchange or at velocities comparable to or larger than the speed of sound in the fluid, the density changes from a constant to a variable. This fact is of special importance because such density variation or "compressibility" couples the flow dynamics with the flow energetics, resulting in a behavior which is quite different from incompressible, isothermal flows.

A fluid bounded by two parallel planes in relative parallel motion will undergo a shearing strain in response to the applied stress. Propagation of the stress is possible because any real fluid, unless very highly rarefied, "sticks" to the wall and exhibits internal friction as well. Such behavior is termed viscosity and stems from the transfer of molecular momentum among flow regimes of different fluid velocity. For a wide range of parameters, a large variety of fluids will strain at a rate which is proportional to the applied stress. Thus, if the planes just mentioned are parallel to the x and z axes of an orthogonal coordinate system and there is steady relative motion in the x direction, then the strain is expressed as a y gradient of the velocity u. The stress in this case may be denoted by τ_{yx} and it is given by

$$\tau_{yx} = \mu \frac{du}{dy} , \tag{1}$$

where the proportionality factor μ is termed the coefficient of viscosity. A so-called Newtonian fluid is one for which μ is constant, and with which the present article is concerned. There are, however, several important non-Newtonian fluids such as blood, paint, and plastics for which μ is not constant, and for which the above relation must be modified in other ways. Note that the fluid pressure does not appear in the above relation, showing how completely different fluid friction is as compared with solid friction. Note further that the relation is also independent of fluid density, which implies an independence from macroscopic inertial fluid features as well.

Viscosity is a function of temperature, increasing for gases and decreasing for liquids. In the temperature range between $-100\,°C$ and $+300\,°C$, the viscosity of air is well described by

$$\frac{\mu}{\mu_r} = \left(\frac{T}{T_r}\right)^{0.76} , \tag{2}$$

where μ_r and T_r are reference values.

Table 1: Viscosities of some common fluids.

	Temperature (°C)	Viscosity (centipoises)
Liquids		
Water	0	1.798
	20	1.0019
Alcohol	20	1.200
Benzene	20	0.652
Glycerin	20	1.490
Liquid sodium	250	0.381
Gases		
Air	0	0.0171
	20	0.0183
	40	0.0190
Argon	20	0.0222
Ammonia	20	0.0098
Carbon dioxide	20	0.0148
Methane	20	0.0109
Oxygen	19	0.0202

The ratio of viscosity coefficient to fluid density, μ/ρ, is called the *kinematic viscosity*, usually denoted by ν. Its dimensions of (length)^2/time consist, in fact, of kinematic quantities and are those of a diffusion coefficient. It is the kinematic viscosity which determines the diffusive spread of fluid straining behavior originating at a local disturbance in an otherwise uniform flow. Such transport is of the same nature as that of heat conduction or of mass diffusion in an inhomogeneous fluid mixture. The latter two properties play an important role, respectively, in fluid heat transfer behavior and in chemically reactive fluid flows involving mixing. The basic metric unit of ν is the stokes, which is $1\,\mathrm{cm}^2/\mathrm{sec}$, as compared with the metric unit for viscosity μ, the poise, denoting $1\,\mathrm{dyn}\,\mathrm{s}/\mathrm{cm}^2$. Tabulated values are often given in centipoises or centistokes, which are units 100 times smaller than those quoted above (Table 1).

There is a problem in uniquely specifying fluid physical properties associated with viscous, thermal, or mass transport, as just described; it relates to the effect of flow turbulence. The latter, which is discussed more fully later, is characterized by local, rather randomized fluctuations of physical quantities, and includes, as well, a pronounced effect on macroscopic gradients of such quantities as temperature and velocity. The result is that turbulent diffusive transfer may be an order of magnitude greater than so-called transfer of the laminar type, the latter term referring to a flow which involves gradients but which is macroscopically smooth.

An important property of a fluid is its isentropic compressibility, denoted by β_S, a quantity which is essentially a ratio of fractional density change per unit change of applied pressure under conditions of constant entropy:

$$\beta_S = \frac{1}{\rho}\left(\frac{\partial\rho}{\partial p}\right)_S. \tag{3}$$

This quantity determines the speed of small-amplitude disturbances, i. e., sound waves, in the fluid in accordance with the relation

$$a^2 = 1/\rho\beta_S .$$ (4)

The value of β_S for water at $20\,°C$ and a pressure of $1\,atm$ is $46 \times 10^{12}\,cm^2/dyn$, and the corresponding sound velocity is $1470\,m/s$, about 4.3 times the corresponding sonic velocity in air.

For most gases of interest, the density and pressure equation relating two states connected by an isentropic change is

$$\frac{p}{p_1} = \left(\frac{\rho}{\rho_1}\right)^\gamma ,$$ (5)

where γ is the ratio of specific heats at constant pressure and at constant volume, respectively. The equivalent expression for the compressibility is $(\gamma p)^{-1}$ so that the velocity of sound a is now given by

$$a = (\gamma p/\rho)^{1/2} .$$ (6)

Recalling that pp, ρ, and temperature T are related by the equation of state

$$p = \rho R T ,$$ (7)

where R is the gas constant per unit mass, we obtain the sound velocity in a gas as a function of temperature only:

$$a = (\gamma R T)^{1/2} = 20.05 T^{1/2}$$ (8)

for air, with T expressed in kelvin. The sound speed in air at $293\,K$ or $20\,°C$ is $343\,m/s$.

Other principal physical features of fluids which can only be mentioned in the present article, but are adequately treated in the literature, include surface tension and capillarity, thermal conductivity, and additional thermophysical properties such as critical values of pressure, molar volume, and temperature, at which the densities of coexisting liquid and gas phases become identical. At temperatures above the critical temperature a gas cannot be liquified.

Fluid Dynamics

First a few remarks on fluid kinematics. In contrast to the so-called Lagrangian description of motion typical of classical physics, the description of fluid flow is termed Eulerian, i. e., instead of following a particle of fixed identity, one relates the changes experienced by a fluid element to the configuration of the flow field. The laws of motion are expressed in terms of the Eulerian or "material" derivative of physical quantities, D/Dt, given by

$$\frac{D}{Dt} = \frac{\partial}{\partial t} + \mathbf{V} \cdot \nabla ,$$ (9)

i. e., by the sum of the local and convective derivatives.

To express variations in a fluid velocity field, we relate the velocity components (u, v, w) at two nearby points in terms of a Taylor series expansion. Thus, the appropriate first-order terms form a nine-component array $\partial u_i / \partial x_k$.

Application of Newton's second law to a Newtonian fluid, together with certain basic assumptions about the nature of fluid stresses and about viscosity coefficients for shearing and for longitudinal stresses, respectively, yields the following basic relation for the x component of the substantial acceleration:

$$\frac{Du}{Dt} = \frac{\partial u}{\partial t} + \mathbf{V} \cdot \text{grad}\, u , \tag{10}$$

in a flow field where each component of total velocity $\mathbf{V}$ may be a function of all the space coordinates and the time as well:

$$\rho \frac{Du}{Dt} = f_x - \frac{\partial p}{\partial x} + \frac{\partial}{\partial x}\left[\mu\left(2\frac{\partial u}{\partial x} - \frac{2}{3}\,\text{div}\,\mathbf{V}\right)\right] + \frac{\partial}{\partial y}\left[\mu\left(\frac{\partial u}{\partial y} + \frac{\partial v}{\partial x}\right)\right] + \frac{\partial}{\partial z}\left[\mu\left(\frac{\partial w}{\partial x} + \frac{\partial u}{\partial z}\right)\right] \tag{11}$$

with equivalent expressions for the y and z accelerations. Here, (u, v, w) are the velocity components and f_x is the x component of any body force, per unit volume, such as gravity, acting on the fluid. These relations are known as the Navier–Stokes equations. Comparatively few solutions of the general equations for stipulated boundary and thermodynamic conditions are known, but many problems of interest usually permit application of these relations in a simplified form.

Hydrodynamic (liquid) flow and gas flows at speeds considerably below the sonic value and with no heat exchange are essentially incompressible, i. e., the density is constant. In addition, such flows will be isothermal as well in the absence of heat sources or sinks, thus decoupling the dynamic analysis from that of the energetics. Because the temperature does not vary, the viscosity coefficient may then be taken as constant, and the equation for conservation of mass flow,

$$\frac{\partial \rho}{\partial t} + \text{div}(\rho \mathbf{V}) = 0 , \tag{12}$$

requires that the quantity $\text{div}\,\mathbf{V}$ vanish. We can now write the x component of the Navier–Stokes equation as

$$\frac{Du}{Dt} = -\frac{1}{\rho}\frac{\partial p}{\partial x} + \nu\left(\frac{\partial^2 u}{\partial x^2} + \frac{\partial^2 u}{\partial y^2} + \frac{\partial^2 u}{\partial z^2}\right) + f_x . \tag{13}$$

For a nonviscous flow, the second-order terms in the above relation disappear and it reduces to the so-called Eulerian equation of motion. Integration of the Euler equation along a streamline (streamlines are the curves which are everywhere tangent to the direction of the velocity vector) for the case of a conservative body force derivable from a potential energy U per unit mass, i. e., $f = -\text{grad}\,U$, yields

$$\frac{\partial}{\partial t}\int \mathbf{V} \cdot d\mathbf{r} + \frac{V^2}{2} + \int \frac{dp}{\rho} + U = \text{constant along streamline}, \tag{14}$$

where d*r* is the differential path along the streamline. Now if the flow is, in addition, irrotational, i. e., $\mathbf{V} \times \mathbf{V} = 0$, then the velocity is derivable from a potential ϕ in accordance with $\mathbf{V} = \operatorname{grad}\phi$. Then the equation becomes (in three dimensions)

$$\frac{\partial\phi}{\partial t} + \frac{V^2}{2} + \int \frac{dp}{\rho} + U = \text{const.}, \tag{15}$$

the same constant holding everywhere in the flow, rather than varying from streamline to streamline. For steady, incompressible flow, the above relation takes the form

$$\frac{V^2}{2} + \frac{p}{\rho} + U = \text{const.}, \tag{16}$$

which is the familiar form of Bernoulli's equation.

Incompressible, inviscid flow deals, then, with studies of velocity or pressure distributions in a flow field, the two being related by the Bernoulli equation. The use of velocity potential ϕ, defined by

$$\mathbf{V} = \nabla\phi, \tag{17}$$

or of the stream function ψ is a mathematical aid for the analysis. Thus, an irrotational flow is completely specified by the scalar function ϕ instead of the three velocity components.

On the other hand, the stream function also gives the complete velocity distribution in a two-dimensional flow whether irrotational or not. Its relation to the velocity components, namely

$$u = -\frac{\partial\psi}{\partial y}, \qquad v = \frac{\partial\psi}{\partial x}, \tag{18}$$

is a consequence of the equation of continuity, $\operatorname{div}\mathbf{V} = 0$. The value of ψ is constant along streamlines and has physical meaning which relates to the mass flow rate in a stream-tube bounded by two streamlines.

In spite of the implied restrictions, two-dimensional, inviscid, incompressible, and irrotational flow fields represent a central area of fluid dynamics with important applications to aerodynamics. The point is that with no shearing stresses acting on a fluid element, the net pressure force passes through its center of mass, and thus there are no turning moments. Therefore, if the fluid is not rotating initially, it will not begin to do so. A body accelerated from rest to a steady velocity in stationary air will experience irrotational motion. Even in the case of a real fluid such as air, flows at typical aeronautical velocities are such that the rotational behavior due to viscosity is limited to narrow regions near the flow boundaries, the so-called boundary layers and near wakes, while the major parts of the flow field may be treated as irrotational in their contributions to the lift and drag on an immersed body.

The use of the velocity potential has been extended to compressible, irrotational flow; it obeys the relation

$$\frac{\partial^2\phi}{\partial t^2} + \frac{\partial}{\partial t}(\nabla\phi)^2 + \frac{1}{2}\nabla\phi \cdot \nabla(\nabla\phi)^2 - a^2\nabla^2\phi = 0, \tag{19}$$

where a is the local sound velocity. This is a difficult nonlinear equation but there are two important special cases: For incompressible flows $a \to \infty$, and we have Laplace's equation $\nabla^2 \phi = 0$. On the other hand, for small-amplitude motions of a compressible flow, we obtain the wave equation

$$\frac{\partial^2 \phi}{\partial t^2} - c^2 \nabla^2 \phi = 0 \,. \tag{20}$$

Linearization of the ϕ equation has also been utilized to solve for a wide range of small-disturbance, steady supersonic flows, e. g., thin airfoils immersed in a uniform flow field. Assume that the airfoil chord lies along the x axis and that the undisturbed flow far away from the airfoil moves in the x direction at a Mach number M_∞ relative to the airfoil. Then ϕ satisfies the equation

$$(1 - m_\infty^2) \frac{\partial^2 \phi}{\partial x^2} + \frac{\partial^2 \phi}{\partial y^2} + \frac{\partial^2 \phi}{\partial z^2} = 0 \,. \tag{21}$$

For subsonic flow, the compressibility feature is accounted for by an affine transformation, which again reduces the problem to that of solving Laplace's equation. However, when $M_\infty > 1$, the ϕ equation becomes hyperbolic or wavelike. The flow field is now qualitatively different. The disturbance represented by insertion of the airfoil into the flow is propagated along so-called characteristic or Mach lines. The angle μ between the characteristic and flow direction at any point is related to the local flow Mach number M by

$$M = \operatorname{cosec} \mu \,. \tag{22}$$

A "point" disturbance in such a flow would generate a Mach cone comprised of such characteristics. Existence of the disturbance would be known only within the downstream confines of the conical surface. Physically speaking, we are saying that sound signals cannot travel upstream in a supersonic flow.

Fluid Energetics

Application of energy conservation to the flow of a viscous, heat-conducting fluid yields a relation which is conveniently formulated in terms of change of stagnation or total enthalpy per unit mass h_s, defined by $h_s = h + V^2/2$; in turn, h, the specific static enthalpy, is defined as usual in terms of the specific internal energy e, pressure p, and fluid density ρ by $h = e + p/\rho$. Using the summation convention for repeated indices, we may write the energy equation as

$$\rho \frac{Dh_s}{Dt} = \frac{\partial p}{\partial t} + \frac{\partial}{\partial x_k} (\tau_{ik} u_i - q_k) + f_k u_k + Q \,, \tag{23}$$

where

$$q_k = -\kappa \frac{\partial T}{\partial x_k} \text{, the Fourier heat-conduction flux;} \tag{24}$$

f_k = component of body force per unit volume;

Q = power per unit volume exchanged with the outside environment, e. g., radiative energy injected by a laser pulse;

and

τ_{ik} = stress tensor, assumed linearly related to the fluid strain in terms of the viscosity coefficient μ by

$$\tau_{ik} = \mu \left\{ \frac{\partial u_i}{\partial x_k} + \frac{\partial u_k}{\partial x_i} - \frac{2}{3} \delta_{ik} \nabla \cdot \mathbf{V} \right\} , \tag{25}$$

in which $\delta_{ik} = 1$ for $i = k$ and zero otherwise.

Apart from the term $\partial p / \partial t$ which vanishes in steady flow, the energy relation shows the contributions to the change of total enthalpy of the "work" of the viscous stresses, of heat conduction, of the work done by the body forces, as well as energy transfer with outside systems. In most applications over a wide range of flow conditions, the relation can be suitably simplified to calculate heat transfer rates, temperature distributions, and related quantities.

Some manipulation of the equation yields an energy relation in a form which takes us directly from the first to the second law of thermodynamics. We recall that the differential change of specific entropy s associated with changes in enthalpy and pressure is given by

$$ds = dh - dp/\rho . \tag{26}$$

The energy equation reformulation is

$$\frac{Ds}{Dt} = \frac{Dh}{Dt} - \frac{1}{\rho}\frac{Dp}{Dt} = \frac{1}{\rho}\tau_{ik}\frac{\partial u_i}{\partial x_k} - \frac{1}{\rho}\frac{\partial q_k}{\partial x_k} + \frac{Q}{\rho} . \tag{27}$$

Apart from outside energy addition, we see explicitly that viscous stress "work" (really heat production) and heat conduction increase the entropy and are therefore dissipative in nature. Conversely, in adiabatic flows without viscous or heat conduction effects, we have $Ds/Dt = 0$; i. e., the entropy is constant along streamlines and in that sense the flows are reversible.

Shock Waves

A shock wave is a front across which there is a nearly discontinuous, finite jump in pressure, with corresponding jumps in temperature, density, and other fluid properties. Shadow photographs showing such waves ahead of or in the wake of high-speed projectiles have been widely published (see Fig. 1), and they are commonly seen in the neighborhood of supersonic nozzle exhausts such as rocket nozzles during the launching of space vehicles. Such waves may be changing with time as part of an unsteady flow field, but they are also found in steady flow and may be "made stationary" by a suitable Galilean transformation of the fluid flow equations.

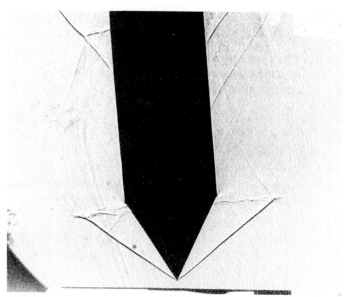

Fig. 1: Shadowgraph showing bow shock wave on a 30° half-angle cone cylinder model immersed in a supersonic flow of Mach number 1.7 at an angle of attack of 5°.

The nonlinear nature of fluid flows is central to the formation of shocks and to the understanding of their behavior. Consider a sinusoidal sound wave of sufficient intensity that (because sound waves propagate adiabatically) the temperature variation across the wave is non-negligible. This means that the higher-temperature portion of the wave will move faster than the lower-temperature portion; thus, the compressive part of the wave will steepen to form a "vertical" pressure front or shock. Shocks, being waves, are formed only in hyperbolic-type, i. e., supersonic, flow fields. In portions of such flows where the pressure is increasing, there is a tendency for the characteristic lines, themselves described by a linear equation, to merge into an envelope, thus creating a shock.

Because shocks are nonlinear, analysis of their behavior is complex, e. g., reflection of a shock wave does not obey any such simple laws as equality of the incident and reflected angles. Indeed, under certain parametric conditions, one finds an unexpected triple shock configuration with a so-called Mach stem. Other types of shock-wave interaction are also more tenuous because linear superposition does not apply. Flows entering a shock at an oblique angle are deflected through a finite angle away from the normal direction. The deflection is due to the decrease of the normal component of the fluid velocity entering the wave. The conservation equations for the normal shock transition of an otherwise steady and uniformly flowing fluid are

$$\text{Mass:} \quad \rho_1 V_1 = \rho_2 V_2 = \text{const.} ;$$
$$\text{Momentum:} \quad p_1 + \rho_1 V_1^2 = p_2 + \rho_2 V_2^2 = \text{const.} ;$$
$$\text{Energy:} \quad h_s = h_1 + V_1^2/2 = h_2 + V_2^2/2 = \text{const.} ;$$

where h_s is the total or stagnation enthalpy. For the case of a thermally perfect ($p = \rho RT$)

and calorically perfect gas flow ($h = c_p T$, where c_p is constant), the above relations can be manipulated to give the Rankine–Hugoniot relation relating the pressure and density ratios across a shock:

$$\frac{p_2}{p_1} = \left(\frac{\gamma+1}{\gamma-1}\frac{\rho_2}{\rho_1} - 1\right) \bigg/ \left(\frac{\gamma+1}{\gamma-1} - \frac{\rho_2}{\rho_1}\right) \tag{28}$$

In the limiting case of very weak shocks with $p_2/p_1 \to 1$, the above relation approaches the isentropic formula [Eq. (5)]. For a shock of nonvanishing strength, the finite jump from state 1 to state 2 is irreversible and the entropy change $s_2 - s_1$ is positive and increases with pressure ratio.

The shock transition for ideal gases increases in strength with the Mach number of the incident flow relative to the shock, M_1; the relation between fractional pressure change and M_1 is

$$\frac{p_2 - p_1}{p_1} = \frac{2\gamma}{\gamma+1}(M_1^2 - 1) . \tag{29}$$

Thus, if $M_1 = 3$, the pressure changes by an order of magnitude through the shock for the case of a diatomic gas such as air. The temperature increases by about 70%, and other quantities change correspondingly. The velocity across a normal shock always changes from supersonic to subsonic. The density jump is limited to $(\rho_2/\rho_1)_{max} = (\gamma+1)/(\gamma-1)$ for the ideal gas case. Thus, for air the density jump can be no larger than 6 according to this formula. In practice, however, gases lose their ideal behavior at the elevated temperature conditions behind very strong shocks, and the density jump for air shocks may approach about 15. Note, by the way, that sonic boom disturbances produced on the ground by supersonic aircraft passing by are rather weak shocks, with changes of pressure typically under 10%.

Many shocks of common experience are unsteady, decaying with time. This is true of thunder associated with electrical storms, blast waves from explosions, and the bursting of a balloon or pressure vessel. The latter type of process is best illustrated with the help of a wave diagram showing the events in a very useful experimental device called a shock tube, following rupture of the diaphragm separating the low-pressure chamber from the high-pressure section. Figure 2 indicates a shock wave advancing into the low-pressure region while a rarefaction wave travels into the high-pressure section, also accelerating that gas in the positive x direction. The rarefaction wave will reflect from the end wall and begin to overtake the shock. The boundary between the gases in the high- and low-pressure chambers, respectively, called a contact surface, travels with the fluid velocity. Across the contact surface, fluid velocity and pressure are equal, but there is a jump in density, temperature, and entropy. Shock tubes are used for basic studies in chemical kinetics, radiative and convective energy transfer, physics of gases, aerodynamics, magnetohydrodynamics, and plasma behavior.

Supersonic and Hypersonic Flows

The existence of shock waves is perhaps the most dramatic feature distinguishing supersonic from subsonic flow. There are other basic physical differences, however, relating especially to the nature of the energy balance at low versus high speeds. In the former case (if we omit the application to heat exchangers), work done on or by the fluid results in a change of velocity,

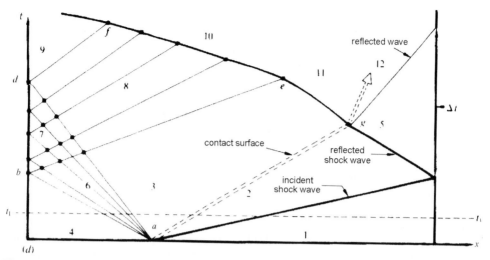

Fig. 2: Wave history in a shock tube given by *x–t* diagram showing wave paths after breakage of diaphragm at point *a* at *t* = 0. Note the reflection of the shock and rarefaction waves [latter shown as a fan, regime (6)] from the right- and left-hand ends of the tube, respectively; and the complex interactions which follow. (Reprinted by permission, from Maurice J. Ducrow and Joe D. Hoffman, *Gas Dynamics*, Vol. 2. Wiley, New York, 1977.]

therefore kinetic energy, of the fluid; this is in accordance with the simple form of the Bernoulli relation, Eq. (16). Changes in temperature as well as density are small throughout the flow field. For the high-speed case, however, the thermal energy of the flowing fluid participates in the energy balance as well; that is, the moving fluid can show substantial variations in temperature. Such behavior has important consequences in applications to internal flows (e. g., nozzles and aircraft engine inlets) and also to external flows around aircraft, wind tunnel models, or turbine components. For example, air accelerating from rest at room temperature, say 77 °F, to Mach 3 cools off to −267 °F. On the other hand, a supersonic plane traveling at Mach 3 in an ambient atmosphere at 77 °F will experience a large rise of temperature to 984 °F across its bow shock wave.

At appreciably higher Mach numbers, these effects are substantially increased. Thus, above Mach 5, say, a nozzle expansion results in a freezing out of the main components of air, namely, nitrogen and oxygen. At suitably high aircraft speeds, a whole host of high-temperature phenomena occur, relating to the fact that now the internal energy states of the molecules and atoms participate in the energy balance. They include vibrational excitation, dissociation, and ionization (the latter is responsible for the communication blackout which occurs during certain phases of astronaut reentry). Radiation takes place as well; that is, in certain velocity and altitude ranges, the shock layer (flow behind the bow shock) may be observed to glow. At high enough speeds, radiative heat transfer to the vehicle may be comparable or even substantially larger than the convective heat transfer more familiar to the aerothermodynamics engineer. For the Apollo project, the spacecraft returning from the moon entered the earth's atmosphere at about 11 km/s; for this case the peak radiative and convective heat transfers were comparable. As a more extreme example, the ultrahigh-speed entry of the Galileo probe into the Jovian atmosphere will result in peak heating conditions where the radiative heating is about four times as large as the convective heating. The phenomena just described are often categorized by the term *aerophysics*. The corresponding vehicle Mach numbers may be very large, approaching Mach 50 in some applications to transatmospheric flight. In this velocity range, the Mach number designation is somewhat less significant than the basic kinetic energy of the flow relative to the vehicle. However, the term *hypersonics* is widely used for high Mach number flows. *Hypervelocity flows* would be better terminology.

There are several special features of supersonic and hypervelocity flows, relating both to their aerodynamic and to their aerophysical behavior. For example, the existence of shock waves introduces the concept of wave drag which has to be included in the determination of the overall drag coefficient of the aircraft. Another practical problem which affects both drag and heat transfer is that of transition from laminar to turbulent flow in the so-called surface boundary layer; the empirical transition data which hold at supersonic speeds do not easily extrapolate to hypersonic case. Indeed, several aerodynamic phenomena need further study to determine behavior in the hypervelocity regime. For example, three-dimensional flows which are accented at high angles of attack show considerable cross flow over aircraft wings, accompanied by vortex bursting, effects which are important in determining aircraft performance features such as stall. The dependence of such flow configurations on Mach number is a subject of continuing research.

Viscous Flow Behavior

The participation of viscous effects in the dynamics and energetics of fluid flows has already been indicated in the previous sections dealing with those topics. The relative importance of viscosity in a particular flow situation stems from the magnitude of the viscous terms as compared to other terms in the equations. Such comparison in the case of flow dynamics leads to a nondimensional term called the Reynolds number, which is (apart from a factor of 2) simply a ratio of dynamic pressure $\frac{1}{2}\rho V^2$ to viscous stress $\mu V/L$, L being a typical length in the flow problem under study. Thus

$$Re = \rho V L/\mu . \tag{30}$$

The Reynolds number is the most important scaling parameter in aerodynamics and a controlling guideline for the overall behavior and analysis of a wide variety of fluid flows. Thus, consider the steady motion of a sphere immersed in a fluid. At low Reynolds numbers, say $Re = 1$ or less, we can ignore the inertial transfer of momentum to the fluid in comparison to the surface shear or fluid "friction". This behavior is known as Stokes flow and is familiar to physicists in connection with the motion of oil drops in Millikan's famous experiment. The drag force F on a sphere of radius a moving with velocity V is, for Stokes flow,

$$F = 6\pi\eta aV . \tag{31}$$

On the other hand, in a Reynolds number range around 10^4, the force is

$$F = c\rho a^2 V^2 , \tag{32}$$

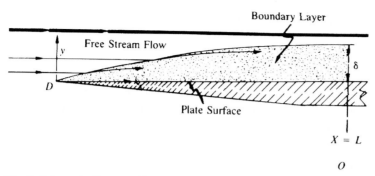

Fig. 3: Schematic diagram of boundary layer development over a flat surface immersed parallel to the flow in a uniform stream. Thickness of layer is much exaggerated for illustrative purposes.

where c is a constant somewhat less than but of order of magnitude of unity. For $Re > 10^5$, the drag force determination is complicated by the transition from laminar viscous to turbulent flow near the spherical surface and to changes in the location of flow separation from the surface. Note that the Stokes drag depends linearly on viscosity μ and velocity V, but not on density, i. e., not on the inertial feature of the flow. On the other hand, the higher Reynolds number formula does show a dependence on density ρ and a quadratic dependence on V but no dependence on viscosity.

Low-Reynolds-number flows, e. g., glycerine or molasses flowing through a pipe, show viscous effects everywhere in the flow field. However, at high Reynolds numbers, say 10^4 or higher, predominant regions of the flow away from boundaries can be treated on an inviscid basis, while near the boundaries relatively thin layers exist in which both viscous and inertial terms must be taken into account. These so-called boundary layers were first referred to by Prandtl in 1904, and by now have been studied in great detail. To illustrate, if x is the direction of a uniform steady flow parallel to a plane surface with leading edge at $x = 0$ (see Fig. 3), then the angular growth of the laminar boundary layer is given by the formula

$$\frac{\delta}{x} = \frac{A}{\sqrt{Re}} , \qquad (33)$$

where δ is the thickness of the layer at a distance x from the leading edge, and A is a constant whose order of magnitude is unity. Since typical Reynolds numbers associated with aircraft and other flight vehicles range around several millions, it is evident that the boundary layers, even when modified by turbulence and other factors, are thin indeed. From a basic conceptual point of view, viscous behavior at the surface generates vorticity which diffuses into the flow at a rate determined by the viscous diffusivity $\nu = \mu/\rho$, while, at the same time, being swept along by the streaming fluid. The equation for δ/x gives the resultant of those two effects.

At high speeds, compressibility effects produce temperature gradients and surface heat transfer to accompany the viscous shear in the boundary layer. The comparison of viscous and thermal behavior is represented by the ratio of viscous diffusivity ν to diffusivity associated with heat conduction, $\kappa/\rho C_p$, κ being the Fourier heat-conduction coefficient and c_p the specific heat at constant pressure. The ratio is known as the Prandtl number, denoted by σ:

$$\sigma = c_p \mu / \kappa . \tag{34}$$

For $\sigma = 1$, viscous and thermal boundary layers are equal in thickness, the viscous production of heat being balanced in a sense by the heat conduction effects. For air $\sigma = 0.74$, close enough to $\sigma = 1$ so that the resultant behavior of air boundary layers is not too different from the simplified case of unit Prandtl number.

Turbulent Fluid Flow

There are several examples of flow fields which represent solutions of the basic fluid equations but which are not stable with respect to small perturbations. This is especially true of viscous shear flows at high Reynolds numbers. The classical experiments on this problem were performed by Reynolds himself, who observed the rapid breakup of dye filaments in liquids flowing through long straight pipes. The phenomenon represented a transition from laminar to turbulent flow. Stability and transition problems have since received much attention, but should be considered distinct from that of the description and understanding of fully developed turbulent flow.

There is no precise definition of turbulent flow. Many of its properties are unmistakable, however. In contrast to the deterministic nature of a laminar flow, a turbulent flow shows random or stochastic behavior. Physical quantities such as the velocity or vorticity show random fluctuations in time and space, but in such a way that meaningful statistical averaging can be applied to provide at least a partial description of the flow. Associated with these fluctuations is a granularity or "lumpiness" pattern of "eddies" whose size provides a scale for the turbulence. Although such eddies vary in size from the time they are energized and agglomerate to when they disintegrate, the scale of turbulence is in any case much larger than molecular dimensions, i. e., it is macroscopic in that sense.

The largest turbulent eddies are determined by the size of the apparatus, say by the pipe diameter in the case of turbulent pipe flow. These relate to energy injection into the turbulent fluctuations (consider a spoon stirring coffee in a cup). Eddy interactions then take place by a complex nonlinear process which ultimately results in much smaller eddies. The latter lose energy by viscous dissipation. The statistical specification of the sizes of both the large and small eddies utilizes the double correlation coefficient for the velocity components at two points in the flow. In turn, the frequency spectrum of the fluctuations is the Fourier transform of the autocorrelation function of the fluctuating velocities.

Following Reynolds, the analysis of turbulent flows is usually performed by separating the mean fluid properties from the fluctuations with respect to the mean values. In such formulations it is assumed that the scales of motion and energy balance are not changing in such a way as to indicate a really unsteady flow; and therefore the flow is quasisteady in the sense that time averages give definable mean values if the averaging time is long enough. We write

$$\bar{U}_i = U_i + u_i ,$$

where $\bar{U}_i$ and U_i are the total and mean values of the ith velocity component, and u_i is its fluctuating value. Substituting into the Navier–Stokes equation [see Eq. (11)], and recalling that averages of products such as $\bar{U}_i\bar{U}_j$ are given by

$$\overline{\bar{U}_i\bar{U}_j} = U_iU_j + \overline{u_iu_j} , \tag{35}$$

we obtain for the momentum equation of ith component of the mean flow (assumed steady and incompressible for simplicity)

$$U_j\frac{\partial U_i}{\partial x_j} = -\frac{\partial P}{\partial x_i} - \frac{\partial}{\partial x_j}(\overline{u_iu_j}) + \nu\nabla^2 U_i , \tag{36}$$

where P is the mean pressure and where the repeated-index summation convention has been used. This differs from the corresponding equation for laminar, steady, incompressible flow by the presence of fluctuation-velocity correlation terms of the type $\overline{u_iu_j}$. It is as if additional stresses $-\overline{u_iu_j}$ were added to those already present in laminar flow. They are indeed termed the Reynolds stresses. The off-diagonal elements of this tensor are shearing stresses which effect transfer of mean momentum by the turbulent motion.

The energy equation for the type of flow just discussed may be written

$$\frac{1}{2}U_i\frac{\partial}{\partial x_i}\overline{q^2} = -\overline{u_iu_j}\frac{\partial U_i}{\partial x_j} - \frac{\partial}{\partial x_k}(\overline{u_kq^2}) - \frac{1}{\rho}\overline{u_k\frac{\partial p}{\partial x_k}} + \overline{\nu u_k\nabla^2 u_k} , \tag{37}$$

in which $q^2 = u_iu_i$ the total turbulent kinetic energy per unit mass; p is the fluctuating static pressure; and ν is the kinematic viscosity. The first term represents the rate of convection of turbulent energy by the mean motion, while the second term expresses the rate of production of turbulent energy from mean motion energy by the Reynolds stresses. The third term gives the rate of convection of turbulent energy by the turbulent motion, and the fourth represents turbulent energy transfer resulting from the work of fluctuating pressure gradients. Finally, the last term is equal to the difference between the rate of diffusive transport of turbulent energy by the viscous forces and the rate of dissipation of turbulent energy to heat.

Any turbulence modeling using extensions of the analysis just given must take into account major physical features of turbulent flows. Apart from the irregularities and granularity, the high level of diffusivity in a turbulent flow is of special importance in many practical applications such as fluid flows in processing plants, and dynamical and thermal design of flight vehicles. Transition from laminar to turbulent flow can increase by an order of magnitude or more the rates of momentum, heat, and mass transfer. For predicting this type of behavior, a purely statistical procedure is less useful than a phenomenological approach in which suitable modeling, based partly on dimensional analysis, is employed, typically including constants which have to be determined by experiment. Thus, to explain the highly increased friction in the case of turbulent as compared to laminar flow along a wall, one introduces an eddy viscosity, which can be visualized as corresponding to the product of a length scale times a velocity scale [units (length)^2/time]. The turbulent spreading of jets, for example, has been analyzed by assuming some scaled but constant value of the eddy viscosity across any plane perpendicular to the flow.

A phenomenological approach to turbulent length scales was made by Prandtl who introduced the concept of mixing length of a turbulent lump of fluid, in parallel with the molecular picture of viscosity resulting from "mean-free-path" exchanges of particles among flow layers of different momenta. It is the difference between the original velocity of the fluid lump and that of its "new" environment which is equated with the transverse velocity fluctuation in the fluid flow. Considering a mean flow U_1 in the x direction and principal gradient in the y direction, the mixing-length formulation leads to the following relation for the shearing stress τ_{xy}:

$$\tau_{xy} = \rho l^2 \left| \frac{dU_1}{dy} \right| \frac{dU_1}{dy} , \tag{38}$$

where ρ is the density and l is the Prandtl mixing length. From dimensional considerations, we can see that the latter is related to the eddy viscosity ε mentioned above by

$$\varepsilon = l^2 \left| \frac{dU_1}{dy} \right| . \tag{39}$$

Prandtl's formulation has been usefully applied to turbulent boundary layers along solid surfaces and also to turbulent mixing of free flows such as the boundaries of a jet.

Currently, there is considerable interest in the technique of modeling turbulent shear flows with the aid of large-scale computers. The approach is to separate the large- from the small-scale structures. The former show a certain degree of coherent structure, and the attempt is made to model that part of the flow by applying the full Navier–Stokes equations to be solved with the aid of a computer. The smaller-scale or higher wave-number phenomena are treated with a more statistical type of formulation, with the necessity of cutting off the calculation at some upper wave number corresponding to the capabilities of the computer being utilized. Interesting results have been obtained for such flows as turbulent boundary layers along solid surfaces. One advantage of the approach is a better ability to handle the inherently three-dimensional nature of turbulent fluctuations with the help of the computer. It bears repeating, however, that no general solution of the physical equations is known for the case of turbulent flow. As of now, the equations simply do not give the entire story. No adequate model of turbulence currently exists.

Somewhat more recently the theory of chaos has been applied to the dynamics of turbulent fluid flow (as well as to several other scientific problems). Turbulence is an example of dynamical chaos, i. e., apparently random or stochastic fluid behavior. As we know, such motion typically develops from an initially laminar or regular flow representing the solution to a deterministic equation of motion with specified initial and/or boundary conditions. The apparent paradox relates to the extreme sensitivity of the motion to certain types of very small disturbances, amounting essentially to correspondingly small changes in, say, the initial conditions; that is, the regular fluid motion can be considered unstable with respect to such changes which, by the way, should be distinguished from random forcing functions. The growth of such instabilities is also intimately related to the underlying nonlinear character of the equations.

The concepts of dynamical chaos have supplied additional tools for the description of fluid turbulence and, hopefully, are providing a deeper perspective as an aid in the still-elusive attempts to understand the phenomenon. The mechanics of chaos analysis investigates the paths of dissipative systems in phase space, under circumstances where the dissipative energy loss is being balanced, in general, by energy gain from other sources. The phase-space trajectories take the form of orbits which are not closed, i. e., not periodic, but which tend to show a layered, ribbon-like structure. In that sense, the motion can be considered repetitive. The orbits occupy a limited region in phase space and show an inclination to converge to, or are "attracted" to, one or more structures within the region. The sets of such structures are referred to as *strange extractors*, or simply *attractors*.

A parallel concept used in the application of chaos theory to turbulence is that of fractional dimensions or fractals, originally introduced to describe the behavior of a very irregular coastline such as that of Great Britain. More recently, the fractal concept has been applied to the analysis of experimental data obtained from fluid-turbulence studies. In order for an irregular physical process to be treated in terms of fractal behavior, it should exhibit scale invariance. The latter is also a fundamental consideration in turbulence theory. It is probably too much to expect that the concepts of strange attractors, fractals, and others associated with chaos theory will fully clarify the underlying physics of turbulence. However, another "dimension" has now been added to our ways of assessing fluid turbulence, one that is providing further insight into the nature of this ever-challenging phenomenon. For further information on chaos theory, please refer to the article on Chaos in this volume.

See also: Boundary Layers; Capillary Flow; Chaos; Critical Points; Equations of State; Fractals; Hydrodynamics; Rheology; Shock Waves and Detonations; Surface Tension; Turbulence; Viscosity.

Formation of Stars and Planets
S. W. Stahler

Stars are self-gravitating balls of gas, supported against inward collapse for long periods by the heat arising from nuclear reactions. They exist in vast numbers throughout the Universe, and in fact constitute most of the visible matter in space. Although some were born eons ago, still others are forming now, and in relatively nearby regions. How these objects originate is one of the fundamental problems of astrophysics. Remarkable progress in both observation and theory has given us fresh insight within the past few decades. Among these results is the realization that many, if not most, young stars are surrounded by thin disks of orbiting gas and dust. It is from these peripheral structures that planets are born. The underlying physics is murkier here, but again there have been exciting discoveries that offer important clues.

Star-forming Clouds

The Orion Nebula is a mottled, glowing patch of light, located near the tip of the Hunter's sword (Fig. 1). Four brilliant stars, known collectively as the Trapezium, provide most of this illumination. These objects, with masses from 10 to 30 times that of the Sun, emit copious ultraviolet radiation that ionizes surrounding gas and causes it to glow in the optical. The dark patches are created by submicron-size dust grains that partially block this emission.

Stars as massive as those in the Trapezium can only survive a few million years. This is the time required for them to fuse much of their internal hydrogen into helium. After exhausting the hydrogen in its central region, a dying, massive star undergoes more accelerated nuclear

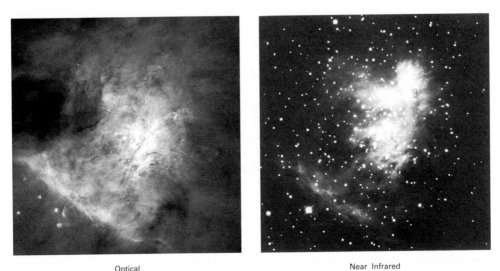

Optical   Near Infrared

Fig. 1: Left: Optical photograph of the Orion Nebula, with the Trapezium stars
at the center. Right: Near-infrared image of the same area.

processing. Helium is fused to carbon, carbon to oxygen, and so on up the periodic table.
Soon the object explodes as a supernova and vanishes altogether. A less massive star, such
as our own Sun, is also less luminous, and requires about ten billion years to consume its
hydrogen fuel. Are there regions like the Orion Nebula, full of dust and gas, that produce only
lower-mass stars?

The answer is yes. Indeed, the majority of stars have even less mass than the Sun, and do
not light up their surroundings like the members of the Trapezium. Discovering the birthplaces
of these common, but less spectacular, objects requires identification of cooler gas. This gas
does not radiate in the optical, but at longer wavelengths. The 1970 discovery of millimeter
emission from carbon monoxide (CO) provided the key to observing cold, star-forming clouds.
Carbon monoxide is a minor constituent of the clouds, which are composed principally of
molecular hydrogen (H_2). But H_2, being a homonuclear molecule, emits weakly in the radio,
while a rotating CO molecule provides radiation in discrete spectral lines that can be detected
even in other galaxies.

Molecular clouds, i. e., those made of molecular, as opposed to atomic, hydrogen, comprise
only a few percent of our Galaxy's total mass. Yet they are the birthplaces of all stars, both
the short-lived and brilliant Trapezium stars and the dimmer, but far more numerous, solar-
type objects. The largest clouds are several hundred light years in extent and contain about a
million solar masses of gas. These giant complexes are concentrated heavily along the spiral
arms of our own Milky Way and similar galaxies. They arise out of more diffuse clouds
of atomic hydrogen. The complexes produce *both* massive stars and many more objects of
solar-like mass.

Other, less extensive, clouds spawn exclusively low-mass stars. In all cases, individual
stars form within compact "cores" embedded within the larger, parent molecular cloud. These
cores, although again composed chiefly of molecular hydrogen, are best probed by species

even rarer than CO, such as ammonia (NH_3) and carbon sulfide (CS). It is found through these tools that each core has a very low temperature, only 10 to 20 degrees above absolute zero. The same is true of the parent cloud. However, the radio emission also reveals that the core's density is several orders of magnitude higher than the background gas. This combination of low temperature and relatively high density enhances the effect of self-gravity, a critical element of the star formation process.

Inside-out Collapse

Although both cloud cores and stars are composed of gas, the former have densities that are larger by twenty orders of magnitude. An essential theoretical issue is how this transition occurs. The driving force is self-gravity, i. e., the mutual attraction of atoms and molecules within the cloud. But identifying the force is only the first step in answering the question. How, in detail, does an individual cloud core, increase its density by such a huge factor?

We first note that self-gravity is not the only force at play. Since the cloud core has a non-zero temperature, there is an outward push associated with its internal gas pressure. There is an analogous pressure arising from the magnetic field, which permeates interstellar space and is known to be present inside molecular clouds. Let us focus, for simplicity, on the thermal pressure. Imagine first a cloud whose self-gravity is substantially larger than the outward pressure force. Such an object immediately goes into free-fall collapse. Since the cloud density rises during collapse, it seems we are on our way to creating a star.

The difficulty is that the cloud has not one center of collapse but many. That is, the object tends to break apart into numerous lumps. Each lump again fragments in an analogous fashion. The process halts when a fragment has so little mass that its self-gravity and pressure forces come into balance. The net result, which has been seen in many computer simulations, is that the original cloud breaks apart into a swarm of clumps in mutual orbit. Although each clump is more compact than the original parent, its density still falls far short of stellar-type values. Worse yet, these clumps are dynamically dead, with no further tendency to evolve.

We could have started with a cloud in which self-gravity is significantly *less* than the pressure force. But this case is also uninteresting, as the object simply disperses into space. The answer must be that star-forming clouds are those in which self-gravity only *slightly exceeds* the internal pressure. In other words the cloud core begins from a state not far from force balance. This inference is bolstered by observations. The temperatures and densities of cloud cores, as obtained from radio emission, are indeed consistent with approximate equality of self-gravity, on the one hand, and the combined thermal and magnetic pressure, on the other.

How does such an object evolve with time? Initially, the pressure of gas near the cloud's center must be relatively high, in order to counteract the full weight of overlying material. By the same token, the pressure diminishes toward the cloud edge. But higher pressure implies higher gas density. Material of high density also experiences a greater force of self-gravity. The collapse process therefore starts in the middle, and only gradually spreads outward. Figure 2 illustrates this *inside-out collapse* in a schematic fashion.

We have not explored how these delicately balanced clouds originate. Perhaps they acquire mass gradually from the surrounding media, until self-gravity just begins to dominate. But whatever their origin, clouds undergoing inside-out collapse are not susceptible to further

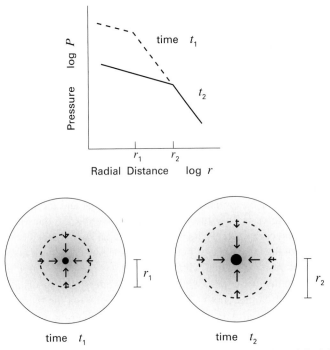

Fig. 2: The process of inside-out collapse. An interior region of diminished pressure advances from radius r_1 at time t_1 to r_2 at t_2. Within this region, gas falls onto the central protostar of growing mass.

fragmentation. Any nascent lump is torn apart by the streaming motion toward the center. Thus, our previous dilemma is resolved. The converging, central flow builds up a *protostar*. Internally, this object is similar to more mature stars, although its central temperature is comparatively low. Its most important distinguishing characteristic is the infalling envelope of gas that surrounds it, cloaks it in obscuring dust grains, and continually adds to its mass.

From Disks To Planets

The central star becomes visible only after the disappearance of this enveloping mantle of dusty gas. Most stars, including our own Sun, emit winds from their surfaces. There is strong evidence that younger objects, including protostars, drive even stronger outflows. It is widely held that such winds dispel the overlying envelope and reveal the star in optical light. However, this idea has yet to be confirmed by direct observations of a cloud in the act of dispersal. Note again that this problem does *not* apply to massive stars, whose devastating effect on surrounding gas is evident.

Once a low-mass star becomes visible, it is said to have entered the *pre-main-sequence* phase. The terminology indicates that the interior stellar temperature is still too low to sustain hydrogen fusion. With time, its self-gravity causes the star to contract, and its central temperature to increase. A star with the mass of the Sun contracts for thirty million years, until hydrogen finally ignites at the center. Paradoxically, the star is more luminous *before* hydrogen ignition. Energy in the young object is produced by self-gravity. The luminosity, or total energy emitted per unit time, is is relatively high because of the object's large surface area. During the subsequent, long main-sequence phase, the star is no longer contracting, and must rely on hydrogen fusion for its energy. Photons from this reaction diffuse outward until they escape from the surface layers.

Pre-main-sequence stars, being both optically visible and relatively bright, are not difficult to identify observationally. Thousands have now been catalogued. Those of roughly solar mass and below are known as *T Tauri stars*. Many of these show evidence for orbiting disks of dust and gas. Light striking such a disk from the central star is absorbed by the dust and reeemitted in the infrared. Thus, stars exhibiting anomalous infrared emission are likely to possess disks. Other objects in the same class have a spatially extended, asymmetric halo when examined in the radio lines of CO. This emission stems from the disk's gaseous component.

Disks arise because the original dense core producing the star has nonzero angular momentum. During inside-out collapse, this quantity is conserved for every fluid element. Those elements with relatively low angular momentum strike the protostellar surface directly. Others spin up so much that they go into orbit around the star. As these fluid elements settle into the equatorial plane, they build up the disk.

It is the solid dust grains within the disk that are critical in the formation of planets. During the pre-main-sequence phase, these grains settle toward the disk midplane. Once the central layer become sufficiently dense, it undergoes clumping. The process is again driven by the ubiquitous force of self-gravity. Clumping proceeds until the accumulated body measures a kilometer or so in size. At this stage, the young star is surrounded by a swarm of *planetesimals*.

What happens next to a group of planetesimals depends on its radial distance from the central star. Closer in, the surrounding gas is dispersed by radiant heating. We are left only with the swarm of planetesimals, which continue to perturb each others' orbits via their mutual gravity. Relatively small bodies tend to pick up speed and have a more difficult time impacting others. In contrast, larger planetesimals have slower velocities, and their greater mass pulls in neighbors more effectively. The next result is that a few bodies exhibit runaway growth, sweeping their neighborhoods clean of the smaller debris. Such was the origin of our own terrestrial (rocky) planets: Mercury, Venus, Earth, and Mars.

The gas giants (Jupiter, Saturn, Uranus, and Neptune in our system) tell another tale, but one which has yet to be fully unraveled. All seem to contain solid cores, which again could have arisen through runaway growth. In addition, they possess extensive mantles of light gas. This volatile component can only persist because the planets reside in the relatively cold, outer regions of the system. Gas was apparently added to the core through the latter's gravitational pull. But such an accretion process may have been quite leisurely in this nether region, where the gas density of the disk is also low. Nature tells us, however, that young stars like the Sun must complete the process within ten million years. Stars older than this show no more evidence of circumstellar disks.

Yet another problem comes from the recent observations of extrasolar planets around solar-type stars. These objects were detected by their gravitational tug on the central object. Not surprisingly, the ones discovered so far are very massive planets relatively close to their stars. Their measured masses (which are actually lower bounds) show that they are gas giants. Thus, they akin to our own Jupiter but often even more massive. But these facts seem to fly in the face of our general picture. For this picture to hold, the "super-Jupiters" must have been born farther out in the disk, and then later spiraled inward. How this orbital migration is accomplished is not at all clear, and is one of several outstanding mysteries in the subject of planetary origins.

Bibliography

I. De Pater and J. Lissauer, *Planetary Sciences*, Cambridge University Press, New York, 1999.
S. W. Stahler and F. Palla, *The Formation of Stars*, Wiley-VCH, Weinheim, 2004.

Fourier Transforms
Y. G. Biraud

In the early part of the nineteenth century the French physicist Joseph Fourier [1] studied the propagation of heat in solids [2]. He realized that the solution of this problem was greatly simplified if he decomposed the physical quantities to be calculated into sums of weighted trigonometric functions which bear his name. But the Fourier series are much more than a powerful mathematical tool. They cover the general notion of spectral analysis and that is why they have been widely studied and extended theoretically by mathematicians and used by physicists.

Fourier series had been devised to represent periodic functions and many restrictions appear on the properties these functions must satisfy to insure the convergence of the series [3, 4, 9, 15]. In order to represent nonperiodic functions a new mathematical entity was necessary. It is the so-called Fourier transform (very often denoted as FT). The FT of $f(x)$ is most often defined as

$$F(u) = \alpha \int_{-\infty}^{+\infty} f(x) e^{i\beta ux} \, dx \,. \tag{1}$$

This formula needs some explanations:

- This integral is taken here in the Riemann sense. Its existence demands several conditions [3, 4].

- The argument of the exponential being dimensionless u and x must have inverse dimensions.

- Most often, too, $\alpha = 1$ but $\alpha = 1/2\pi$ or $\alpha = 1/\sqrt{2}$ are sometimes used.

- Usually $\beta = -2\pi$ but sometimes it is set to ± 1. But the choices of the values α and β and of their signs are of no importance for both theory and practice.

In Eq. (1), $F(u)$ is called the direct Fourier transform of $f(x)$. It is very interesting to note that under precise conditions [3, 4], we may define the inverse FT of $F(u)$ as

$$f(x) = \alpha \int_{-\infty}^{+\infty} F(u) e^{\pm i\beta xu} \, du \,. \tag{2}$$

When this is possible $f(x)$ and $F(u)$ are called a Fourier pair. Many tables of Fourier (and/or of Laplace) pairs have been calculated and are very useful [5–8]. Frequently Eq. (1) is rewritten as

$$F_c(u) + iF_s(u) = \alpha \left(\int_{-\infty}^{+\infty} f(x) \cos(2\pi\beta ux) \, dx \right.$$

$$= \pm i \left. \int_{-\infty}^{+\infty} f(x) \sin(2\pi\beta ux) \, dx \right)$$

where the real and imaginary parts are called Fourier cosine and sine transforms, respectively.

Most often in physics the conditions for the existence of the two integrals in Eqns. (1) and (2) are met. But sometimes they are not (mainly when discontinuities appear, when $f(x)$ is null almost everywhere, or when $f(x) \notin L$ or L_2). In these situations the calculation of the transform as a limit may be quite hazardous. Hopefully the introduction of distributions [4, 9–11] has allowed a generalization of the notion of functions. This solves the problem of Fourier transforming functions like $e^{i2\pi ux}$, $H(x)$ (Heaviside step function), $\delta(x-a)$ (Dirac's "impulse function"), the sampling function $\text{Ш}(x) = \sum_k \delta(x-k)$... In this theory all the quantities appearing in physics now possess FTs.

This new mathematical concept allows us to deal with physical situations which were very difficult to handle otherwise such as:

– point masses in mechanics,

– point charge distributions in electrostatics,

– magnetic doublets,

– atoms or ions in crystalline structures [12],

– point sources or narrow slits in optics and astronomy [13],

– particles in quantum mechanics [14].

Moreover the use of the δ distribution, of the sampling distribution $\text{Ш}(x)$, and of the Heaviside step function (and their FTs) renders the Z- or Laplace transforms quite outdated or at least obsolete (and so unpractical).

The FTs of functions and distributions have been considerably investigated by both mathematicians and physicists who have derived many properties of this transform. Let us denote

$$f(x) \rightleftharpoons F(u) \qquad \text{a Fourier pair } [F(u) = \text{FT}(f(x))]$$

Then if $f(x)$ is real, we may draw up Table 1. This table summarizes the most important properties of the FT. Formula (3) is also known as Parseval's or Plancherel's theorem. In Eq. (4) the derivative of $f(x)$ is taken in the distribution sense. Notice that we have described above the FT of functions or distributions of one variable. It may be easily defined and extended to functions operating on 2-D or 3-D spaces (but notice that some theorems established for 1-D do not apply any longer). Two-dimensional FTs are very frequently used in electromagnetism (antenna patterns, optics, astronomy, ...). Notice the existence of transforms related to the FT: Hankel or Mellin transforms [15].

Table 1: Properties of Fourier transforms.

x-space	$\rightleftharpoons$ *u*-space (FT space)	
$f(x)$	$\rightleftharpoons F(u)$	
Real even	$\Rightarrow$ Real even	
Real odd	$\Rightarrow$ Imaginary odd	
Real (nor even, nor odd)	$\Rightarrow$ Hermitian	
$f(x-a)$	$\rightleftharpoons F(u)e^{-i2\pi au}\,du$	The shift theorem
$f(ax)$	$\rightleftharpoons (1/|a|)F(u/a)$	The similarity theorem
$\int |f(x)|^2\,dx$	$= \int |F(u)|^2\,du$	Rayleigh's theorem (3)
$df(x)/dx$	$\rightleftharpoons i2\pi uF(u)$	(4)
$\delta(x))$	$\rightleftharpoons 1$	$\forall u$ (5)
$e^{i2\pi ax}$	$\rightleftharpoons \delta(u-a)$	(6)
$\text{ш}(x) = \sum_k \delta(x-k)$	$\rightleftharpoons \text{ш}(u) = \sum_k \delta(x-k)$	[see Fig. 1(a) and 1(A)] (7)
$H(x)$ (Heaviside step function)	$\rightleftharpoons \frac{1}{2}[\delta(u) - i(1/\pi u)]$	
$f(x) \cdot g(x)$	$\rightleftharpoons F(u) * G(u)$	Convolution theorem (8)
		[see Eq. (10) below]

A very important notion which is closely related to FT is the convolution. For two functions $f(x))$ and $g(x)$ it is defined by

$$h(x) = \int f(t)g(x-t)\,dt = \int g(t)f(x-t)\,dt \tag{9}$$

and also denoted as

$$h(x) = f(x) * g(x) \tag{10}$$

and may be extended to distributions. Its importance comes from the following statement: the output $h(x)$ of any feasible physical system which is continuous, x-invariant, and linear is related to its input $f(x)$ by a convolution such as in Eq. (9). Because $\delta(x)$ is the convolution unit

$$\phi(x) = \delta(x) * \phi(x) \tag{11}$$

and if in Eq. (10) $f(x) = \delta(x)$, then

$$h(x) = g(x) \ .$$

Hence $g(x)$ which is characteristic of the linear system is simply its impulse response. Now taking the FT of Eq. (10) provides

$$H(u) = F(u) \cdot G(u) \tag{12}$$

[where $H(u)$, $F(u)$, and $G(u)$ are the FTs of $h(x)$, $f(x)$, and $g(x)$, respectively] which shows that a linear system simply multiplies the FT of its input by its complex transfer function $G(u)$.

This may be shown in another way. If the input is $e^{i2\pi vx}$, Eqns. (6) and (12) yield

$$H(u) = G(u) \cdot \delta(u-v) = G(v) \cdot \delta(u-v)$$

which proves that complex exponentials are eigenfunctions of linear systems and may be used to determine the variations of attenuation and dephasing of a filter with frequency.

Equation (12) shows all the interest of the FT. The convolution is an integral relation [Eq. (9)] and the value $h(x_0)$, the output for a fixed x_0, is a weighted sum of all the values of the input. Whereas, if we consider the same problem in the Fourier domain, Eq. (12) shows that the value $H(u_0)$ is simply the multiplication of $F(u_0)$ by the known constant $G(u_0)$. This simplification is one of the reasons of the success of the FT and also allows us to face more easily the deconvolution problem which tries to solve Eq. (9): find the un known $f(x)$ for given $h(x)$ and $g(x)$.

FT and convolution are absolutely essential to understand, model and investigate linear filtering in the wide sense. They apply to functions of time and to functions of space variables as well. The conjugate of time is the usual frequency when in the second case it is space frequency. But the same methods apply to both situations (optical filtering or heterodyne, for example).

Nevertheless it must be noticed that FT is not suitable for the study of time-varying spectra. Other techniques like AR or ARMA methods or Wigner–Ville representations are successful in this situation [16, 17].

As an illustration, we shall emphasize now the efficiency of the use of distributions, of their FTs, and of the convolution, in solving the problem of the sampling of signals (Shannon's/Nyquist's theorem and interpolation). The introduction of the $\text{Ш}(x)$ distribution (see Table 1) simplifies the derivations.

Let s denote the sampling rate on a signal $f(x)$ having a FT $F(u)$. The sampled $f(x)$ is

$$f_s(x) = f(x) \cdot \text{Ш}_s(x) = \sum_k f(ks)\delta(x-ks) .$$

Fourier transforming this equation and using Table 1 yield

$$\begin{aligned}
F_s(u) = F(u) * \frac{1}{s}\text{Ш}_{1/s}(u) &= \frac{1}{s}\sum_k F(u) * \delta\left(u - \frac{k}{s}\right) \\
&= \frac{1}{s}\sum_k F\left(u - \frac{k}{s}\right) .
\end{aligned}$$

$F_s(u)$ is the sum of an infinity of shifted versions of $(1/s)F(u)$. With regard to multiplication the Ш distribution acts as a sampler and to the convolution as a replicating symbol. Two situations may arise:

1. $f(x)$ is such that the definition interval of $F(u)$ is $]-\infty, +\infty[$. See Figs. 1(b) and 1(B).

2. $\phi(x)$ is a so-called band-limited spectrum function:

$$\Phi(u) \equiv 0 \qquad \forall u \notin [-u_M, +u_M] .$$

See Figs. 1(d) and 1(D).

In both situations the FT $F_s(u)$ and $\Phi_s(u)$ are periodical with period $1/s$.

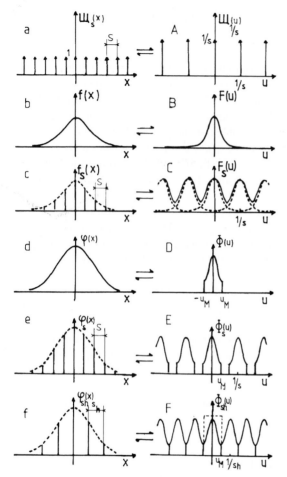

Fig. 1: (a), (A) The sampling function $Ш_s(x)$ and its FT $Ш_{1/s}(u)$. (b), (B) A first signal $f(x)$ and its FT $F(u)$ defined on $]-\infty, +\infty[$. (c), (C) $f_s(x)$, the sampled $f(x)$. [—] Its periodical FT: $F_s(u)$; (---) the different $(1/s)F(u-k/s)$. (d), (D) A second signal $\phi(x)$ and its FT $\Phi(u)$ null outside $[-u_M, u_M]$. (e), (E) A first sampling $\phi_s(x)$ and its FT $\Phi_s(u)$ composed of nonoverlapping $(1/s)\Phi(u-k/s)$. (f), (F) (—) Shannon's sampling $\phi_{s_h}(x)$. Its FT $\Phi_{s_h}(u)$ composed of just touching patterns $(1/s_h)\Phi(u-k/s_h)$; (---) the window function allowing recovery of $\Phi(u)$.

In case 1, the different patterns $F(u-k/s)$ overlap [Fig. 1(C)], whereas in case 2, the $\Phi(u-k/s)$ no longer overlap if $u_M \le 1/2s$ [Fig. 1(E)]. This allows one to sample $\phi(x)$ with a larger period [Fig. 1(f)] which must satisfy

$$s \le s_h = \frac{1}{2u_M}.$$

s_h is often called Shannon's sampling period. If $s = sh$, then the different patterns $\Phi(u-k/s_h)$

are just touching [Fig. 1(F)] and $\Phi(u)$ can be recovered by just multiplying by the window function $(1/2u_M)\Pi(u/2u_M)$ $\{\Pi(\tau) = 1, \forall \tau \in [-\frac{1}{2}, \frac{1}{2}]$, and is null elsewhere$\}$. Hence

$$\Phi(u) = \Phi_{s_h}(u)\frac{1}{2u_M}\Pi\left(\frac{u}{2u_M}\right).$$

Fourier transforming this relation yields the so-called Shannon's interpolation formula,

$$\begin{aligned}
\phi(x) &= \phi_{s_h}(x) * \frac{\sin(\pi x/s_h)}{\pi x/s_h} \\
&= \sum_k \phi(ks_h)\delta(x - ks_h) * \frac{\sin(\pi x/s_h)}{\pi x/s_h}
\end{aligned}$$

or

$$\phi(x) = \sum_k \phi(ks_h)\frac{\sin[\pi(x/s_h - k)]}{\pi(x/s_h - k)}$$

We shall now review different domains which are relevant to the FT. This list does not pretend to be exhaustive but tries to present its most common applications.

First, many situations are concerned with the product of a function $f(x)$ by its shifted version $f(x-a)$. If $f(x)$ is random, the FT of its autocorrelation yields the power spectrum which is of frequent use in noise analyses. In the same field the characteristic function of a random variable may be calculated as the FT of its density probability.

Now if $f(x)$ is no longer random but deterministic, the same process describes the general problem of interferometry. Two different examples:

- In radioastronomy two (or a 2-D array of) antennas measure the visibility function $V(u,v)$ which is Fourier transformed to get the brightness distribution $v(\alpha,\delta)$ of radio sources in the sky (u,v are spatial frequencies, δ and α declination and right ascension).

- In spectroscopy a Michelson interferometer measures the autocorrelation $I(\Delta)$ of the incoming radiation whose spectrum $B(\sigma)$ is the FT of $I(\Delta)$ (Δ: the path difference, σ: the wave number).

Another domain where FT is very useful is the diffraction of electromagnetic waves (radio or optics). Huygens' principle is very easily applied in the Fourier formalism. FT is eventually the tool for a good understanding of image formation, phase contrast, holography, etc.

Many phenomena in physics are naturally periodical and their study is evidently relevant to Fourier analysis:

- in acoustics or vibration mechanics;

- in astrometry: rotation of satellites or planets;

- in crystallography: ions or atoms are positioned at the nodes of a more or less regular array;

- in biology or pharmacology FT allows study of the rhythms

At last FT may be also considered as a tool:

- in pure or applied mathematics [18] and statistics [15],

- in electronics,

- in signal processing for which it represents a day-to-day tool [19],

- in image processing (associated with Hadamard, Walsh, or Haar transforms),

- in communications.

This multipurpose characteristic of the FT explains quite well why so much research on FT has been undertaken. The demand for fast processing was the origin of the success of the FFT (fast FT) algorithms [20, 21] which now flourish and are theoretically investigated and practically widely used. This is perhaps the best sanction of an almost two-century-old discovery.

References

[1]   J. Herivel, *Joseph Fourier*. Oxford (Clarendon Press), London and New York, 1975.

[2]   J. Grattan-Guiness, *Joseph Fourier 1768–1830*. MIT Press, Boston and London, 1972.

[3]   E. C. Titchmarsh, *The Theory of Fourier Integrals*. Oxford (Clarendon Press), London and New York, 1937.

[4]   J. Arsac, *Fourier Transform and the Theory of Distributions*. Prentice Hall, Englewood Cliffs, NJ, 1966.

[5]   A. Erdelyi, *Tables of Integral Transforms*. Vol. I. McGraw–Hill, New York, 1954.

[6]   F. Oberhettinger, *Tabellen zur Fourier Transformation*. Springer-Verlag, Berlin, Göttingen, Heidelberg, 1957.

[7]   J. Lavoine, *Transformation de Fourier des Pseudo-fonctions*. CNRS, Paris, 1963.

[8]   G. Harburn, C. A. Taylor, and T. R. Welberry, *Atlas of Optical Transforms*. Bell, London, 1975; Cornell University Press, Ithaca, NY, 1975.

[9]   L. Schwartz, *Mathematics for the Physical Sciences*. Addison-Wesley, Reading, MA, 1966.

[10] I. M. Gelfand *et al.*, *Generalized Functions*. Academic Press, London, 1964.

[11] M. J. Lighthill, *An Introduction to Fourier Analysis and Generalised Functions*. Cambridge University Press, Cambridge, 1958.

[12] H. Lipson and C. A. Taylor, *Fourier Transforms and X-Ray Diffraction*. Bell, London, 1958.

[13] L. Mertz, *Transforms in Optics*. Wiley, New York, London, 1965.

[14] P. T. Matthews, *Introduction to Quantum Mechanics*. McGraw–Hill, New York, 1963.

[15] R. Bracewell, The Fourier Transform and its Applications. McGraw–Hill, New York, 1965 (with an excellent bibliography).

[16] P. Flandrin, "Some features of time frequency representations of multicomponent signals," IEEE Int. Conf. on Acoustics, Speech and Signal Processing, ICASSP-84, San Diego, 1984, pp. 41B4.1-41B4.4.

[17] T. A. C. M. Claasen and W. F. G. Mecklenbrauker, "On the time-frequency discrimination of energy distributions: can they look sharper than Heisenberg?," IEEE Int. Conf. on Acoustics, Speech and Signal Processing, ICASSP-84, San Diego, 1984, pp. 41B7.1–41B7.4.

[18] B. Davies, *Integral Transforms and their Applications*. Springer-Verlag, New York, Heidelberg, Berlin, 1978.

[19] E. A. Robinson, *Statistical Communication and Detection*. Griffin, England, 1967.

[20] S. W. Cooley and J. W. Tukey, *Math. Comput.* **19**, 297 (1965).

[21] H. J. Nussbaumer, *Fast Fourier Transform and Convolution Algorithms*. Springer-Verlag, Berlin, Heidelberg, New York, 1982.

Bibliography

N. Ahmed and K. R. Rao, *Orthogonal Transforms for Digital Signal Processing*. Springer-Verlag, Berlin, Heidelberg, New York, 1975.

H. S. Carslaw, *Introduction to the Theory of Fourier's Series and Integrals*. Dover, New York, 1930.

D. C. Champney, *Fourier Transforms and Their Physical Applications*. Academic Press, London, 1973.

S. H. Crandall and W. D. Mark, *Random Vibrations in Mechanical Systems*. Academic Press, London, 1963.

J. D. Gaskill, *Linear Systems, Fourier Transforms and Optics*. Wiley, New York, 1978.

J. W. Goodman, *Introduction to Fourier Optics*. McGraw–Hill, New York, 1968.

E. A. Guillemin, *Communication Networks*. Vol. II. Wiley, London, 1947.

H. P. Hsu, *Fourier Analysis*. Simon and Schuster, New York, 1970.

J. Kauppinen, J. Partanen, *Fourier Transforms in Spectroscopy*. Wiley-VCH, Weinheim, 2001.

R.E.A.C. Paley and N. Wiener, *Fourier Transforms in the Complex Domain*. American Mathematical Society, New York, 1934.

A. Papoulis, *The Fourier Integral and Its Applications*. McGraw–Hill, New York, 1962.

K. Ramanohan Rao (ed.), *Discrete Transforms and Their Applications*. Van Nostrand–Reinhold, New York, 1985.

J. P. G. Richards and R. P. Williams, *Waves*. Penguin, England, 1972.

I. N. Sneddon, *Fourier Transforms*. McGraw–Hill, New York, 1951.

Fractals

F. Family

Fractal geometry is a mathematical language and a quantitative approach for describing complexity in shapes, forms, and patterns. The word *fractal*, from the Latin *fractus*, was coined by Benoit B. Mandelbrot to denote highly irregular and fragmented objects or mathematical sets. In recent years fractal concepts have been developed into an effective mathematical tool and have been applied to a wide range of phenomena and processes of scientific and practical importance. Although Euclidean geometry and the theory of smooth functions can describe regular shapes and patterns (e. g., lines, planes, and differentiable functions), the concepts of fractal geometry are needed for understanding irregular shapes and forms as well as the behavior of extremely irregular mathematical functions. A major part of the success in the application of fractals is in describing disparate physical phenomena, which evolve through similar fundamental processes and therefore can be characterized by the same set of fractal attributes.

The most profound property of a fractal is its scale in- variance or self-similarity. A fractal is a shape or pattern made of parts similar to the whole in some way. The reason is that a fractal object has no characteristic length scale. What this implies is that the essential features of a fractal exist at all length scales. Therefore, magnifying a small piece of a fractal results in a larger object; however, the pattern is similar. One of the fundamental quantitative measures of a fractal is its dimension. Unlike regular geometrical objects, fractals can have a dimension that is generally not an integer. For example, whereas a line has a dimension of 1, a tortuous path, like a river network or a long polymer chain in a solution, has a dimension somewhere

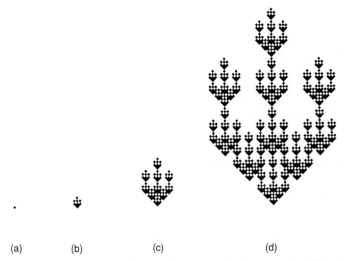

(a)   (b)   (c)   (d)

Fig. 1: This tree-like fractal pattern is an example of an exactly self-similar fractal. It is formed by an iterative process which can continue indefinitely.

between 1 and 2. A square or a triangle has a dimension of 2. But a rough surface, e. g., a mountain or the edge of a broken glass, can have a fractal dimension anywhere from 2 to 3.

The simplest fractals – called self-similar fractals – are invariant under isotropic dilation or contraction. Self-similar fractals can be described by a single fractal dimension. One example is the tree-like pattern shown in Fig. 1. This fractal is made up of squares. One starts with a single square. In the next step, ten squares are put together to construct the square pattern shown in Fig. 1(b). This pattern is the basic building block of the fractal. In each subsequent generation, ten of the objects of the last generation are put together to form a new one having the same overall shape as Fig. 1(b), but with more details. Clearly this process can be continued *ad infinitum* to obtain a very large pattern. Since the same rule is repeated in each iteration, this object is exactly self-similar, having the same shape at any length scale.

In contrast, the shapes and patterns found in nature are usually random fractals. The reason is that they consist of random shapes or patterns that are formed stochastically at any length scale. Because of the randomness, self-similarity of natural fractals is only statistical. Furthermore, natural fractals are self-similar only over a finite range, for example, from atomic sizes to the maximum length in the system. Examples of random fractals includes such objects as the coastlines and mountains, as well as many disordered materials like long polymer chains in solutions, aggregates of colloidal particles, aerosols, and polymeric gels. An example of a random fractal is shown in Fig. 2. This is a viscous fingering pattern formed by injection of a less viscous fluid into a more viscous fluid filling a porous medium. Since the pressure at the interface obeys the Laplace equation, which is a special case of the diffusion equation, this pattern is similar to many other patterns formed in diffusion-limited growth processes.

The fractal dimension of an object or a pattern serves to characterize it and often indicates the manner in which it has been formed. There are several ways to determine the fractal dimension of a fractal, but one of the most widely used methods is to relate the size of the

Fig. 2: An example of a random fractal pattern formed in a viscous fingering experiment where a low-viscosity fluid (air) is pushed into a more viscous fluid (oil) occupying a porous medium.

object to its length, which is an extension of our everyday experience of dimensionality for Euclidean objects. Let us consider the fractal tree shown in Fig. 1. In each generation the length of the new pattern is 4 times the length of the previous one, and its mass is 10 times the previous mass. In contrast, if we construct a nonfractal object, like a regular triangle, then every time its length is doubled, its mass is quadrupled. If the length of a regular triangle is increased by a factor of n, then its mass would be $n^2M(r)$, where r is the length of the original triangle and $M(r)$ is its mass. The number 2 is the dimension of the object; a trivial result for a regular two-dimensional triangle. In general, the fractal dimension is the exponent or power to which n is raised. The fractal dimension D is defined by the relation $M(nr) = n^DM(r)$, where $M(nr)$ is the mass after the length has been scaled by a factor of n. For the fractal tree $M(4r) = 10M(r)$, because every time r is quadrupled, the mass is increased tenfold. Therefore, the fractal dimension D of the fractal tree can be determined from the relation, $10 = 4^D$, which implies that $D = \log 10/\log 4 = 1.66\ldots$ This method can be used to determine the fractal dimension of any object or pattern. In the case of a random fractal, one can simply draw circles of increasing radii about a point on the object and determine the mass $M(r)$ within a circle of radius r and compare it with the mass within a circle of radius nr. The fractal dimension D can be calculated from the relation $M(nr) = n^DM(r)$, as before.

Self-similarity is a consequence of invariance under an isotropic dilation. In general, the transformations that rescale an object may not be isotropic. Objects that are invariant under a more general type of rescaling transformations are called self-affine fractals. This class of fractals must be described by more than one fractal dimension. The reason is that a piece of a self-affine fractal must be enlarged anisotropically in order to match larger parts of the object. Self-affine geometry is frequently encountered in the study of rough surfaces, including mountains, thin films, and fracture surfaces.

The advent of fractal geometry has provided scientists and engineers with a fresh mathematical tool for investigating complex phenomena that had been viewed intractable by generations of scientists. Fractal concepts are rapidly be- coming a standard tool in many scientific investigations, as witnessed by the rich variety of books published in this field in recent years.

Bibliography

R. Devaney, *Chaos, Fractals and Dynamics*. Addison-Wesley, Reading, MA, 1989.
F. Family and D. P. Landau (eds.), *Kinetics of Aggregation and Gelation*. North-Holland, Amsterdam, 1984.
J. Feder, *Fractals*. Plenum, New York, 1988.
B. B. Mandelbrot, *The Fractal Geometry of Nature*. Freeman, San Francisco, 1982.
H. E. Stanley and N. Ostrowsky (eds.), *On Growth and Form: Fractal and Non-fractal Patterns in Physics*. Martinus Nijhoff Publishers, Dordrecht, 1986.
T. Vicsek, *Fractal Growth Phenomena*. World Scientific, Singapore, 1989.

Franck–Condon Principle

B. R. Judd

In the analysis of the photodissociation of diatomic molecules made in 1925, Franck assumed that a transition between two different electronic states takes place so quickly that the positions of the nuclei immediately after the transition are the same as they were just before it. Within a few years, this idea was described in the language of quantum theory by Condon. The variation of the potential energy W of a diatomic molecule with respect to internuclear distance r is illustrated for two electronic states by the curved lines in each of the two parts of Fig. 1. On the left, a molecule in its ground state A absorbs a quantum of radiation and is excited to B, where the two nuclei execute oscillations over those values of r corresponding to the segment BC. If the upper potential curve were to be so flat on its right-hand side that no intersection C occurs, then dissociation would take place. The situation for emission is illustrated in the right part of the figure. A molecule oscillating over a range of r values specified by DE spends most of its time near the extreme positions D and E. The most probable transitions to the lower electronic state correspond to DF and EI, and hence the oscillations corresponding to FG and HI are favored. Quantum mechanics modifies these statements slightly. The most probable values of r lie at points inside the classical ranges of oscillation: the midpoint of the range is the most probable for the lowest vibrational state, as indicated at A. In addition, values of r outside the classical ranges are permitted. Relative transition probabilities between the vibrational states represented by the wavefunctions ψ_v and $\psi_{v'}$ of two different electronic states depend principally on the square of the overlap integral $\int \psi_v \psi_{v'} \, dr$. The essence of the Franck–Condon principle is that the arrowed lines AB, DF, and EI are vertical: each transition occurs for a well-defined nuclear separation r. The principle can be equally well applied to more complicated molecules, including those exhibiting the Jahn–Teller effect.

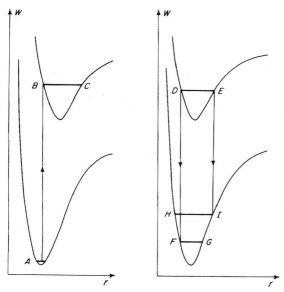

Fig. 1: The Franck–Condon principle illustrated for absorption (on the left) and emission (on the right).

The usefulness of the Franck–Condon principle lies in its capacity to predict the nuclear motion of a molecule after an electronic transition has taken place. If the energy of the transition and also the relevant set of potential-energy curves are known, then the prediction consists of a statement of probabilities for various types of nuclear oscillation (including possibly dissociation). Conversely, of course, an observed nuclear motion can provide information on the shape of the potential-energy curves.

See also: Molecular Spectroscopy.

Bibliography

B. Bak, *Elementary Introduction to Molecular Spectra.* North-Holland, Amsterdam (1962). (E)

G. Herzberg, *Molecular Spectra and Molecular Structure. I. Spectra of Diatomic Molecules.* Van Nostrand, New York (1950). (I)

Fraunhofer Lines

A. Leitner

Fraunhofer lines are the numerous dark lines in the continuous spectrum of sunlight, first noted by W. H. Wollaston in 1802. He reported seeing a few dark lines – i. e., the most prominent ones – but seems to have paid no further attention to them. They were studied in detail between 1806 and 1826 by J. Fraunhofer, who built the first precision spectrometers, slits, quality prisms, and diffraction gratings ruled with up to 300 lines/mm (7500 lines/in). Some of the dark lines in the solar spectrum became his reference marks for charting the refractive indices of optical glasses. These data enabled Fraunhofer to design and construct outstanding achromatic and otherwise corrected optical instruments, including astronomical refracting telescopes. He determined the wavelengths of a number of Fraunhofer lines to the then unheard of precision of four significant figures. It was Fraunhofer who traced the origin of the spectrum to the sun. He reported that the spectra of other stars, though similar, showed different dark line patterns. Around 1890 H. A. Rowland published wavelength tables and a photographic atlas for about 15 000 Fraunhofer lines in the range between 300 and 650 nm. Charting of this spectrum has continued and is even now being extended into the far ultraviolet as a result of observations from satellites.

The Fraunhofer lines became understood as an *absorption* spectrum by the work of G. Kirchhoff and others in the 1860s. Many of them have since been identified with the spectra of known atoms and ions. As a consequence, the constitution of the solar photosphere is quite well known. From line profiles and from the wavelengths at the line centers it is possible to infer much about what is going on in the solar surface. When observations are confined to small portions of the solar disk, much detail can be culled from spectral line data. The Doppler shift gives evidence about the gross rotational motion of the sun and about local velocity fields. The ambient magnetic field at the solar surface has been calculated from line broadening interpreted as unresolved Zeeman splitting. The far stronger magnetic field in sunspots has also been determined from measurements of Zeeman splitting.

The Fraunhofer spectrum observed from below the Earth's atmosphere contains a few lines whose intensity is strongest at the time of sunset or sunrise, and whose wavelengths display no Doppler shifts due to solar motions (for example: the group of lines in the red, called A, which is due to oxygen). These are due to the earth's atmosphere and are called telluric. The Fraunhofer spectrum is also observed as a discrete emission spectrum of bright lines on a dark field of view during the brief flash from the photosphere in the solar limb just before or just after totality in an eclipse.

In the present-day context, the Fraunhofer spectrum of sunlight forms only a very small part of the great role played by spectroscopy in astrophysics. The interested reader is referred to the large textbook literature in astronomy for further information.

See also: Atomic Spectroscopy; Gratings, Diffraction; Sun; Zeeman and Stark Effects.

Bibliography

Dictionary of Scientific Bibliography, Vol. V, pp. 142–144. Charles Scribner's Sons, New York, 1972. (E)

"Prismatic and Diffraction Spectra", in *The Wave Theory of Light and Spectra*. Arno Press, New York, 1981. (E)

H. A. Rowland, *Astrophysical Journal*, **1–6**, 1895–1898 lists series of Fraunhofer lines between 3 000 and 6 500 Å. (I)

Alfred Leitner, The Life and Work of Joseph Fraunhofer (1797–1826), *American Journal of Physics* **43**, 59–68 (1975). (E)

L. Debouille, L. Neven and G. Roland, *Photometric Atlas of the Solar Spectrum from 3 000 to 10 000*, Institut d'Astrophysique, Université Liége, Liége (1973). (I)

Free Energy

G. Weinreich

In terms of the internal energy U, the temperature T, and the entropy S, the free energy of a thermodynamic system is defined by

$$F = U - TS .$$

In a process whose initial and final states are at the same temperature, the change in free energy is given by

$$\Delta F = \Delta U - T\Delta S .$$

If in addition the process is reversible, the last term is equal to the amount of heat transferred, so that the decrease in free energy becomes equal to the work performed by the system. Thus F acts as an "isothermal potential energy."

For irreversible processes, $-\Delta F$ represents an upper limit on the work that the system can perform. The free energy does not, however, measure energy stored within the system. For example, in the reversible isothermal expansion of an ideal gas, none of the work comes from inside the system, since the internal energy does not change.

The functional form of the free energy, when given in terms of the temperature and of the variables whose differentials appear in the element of reversible work, comprises a complete thermodynamic description of the system. This puts F into the category of "thermodynamic potentials". For example, the element of reversible work done by a fluid system is $p\,dV$ where p is the pressure and V the volume; accordingly, all thermodynamic properties of such a system are determined as soon as the functional form $F(T,V)$ is known. Specifically, entropy and pressure are the negative partial derivatives of F with respect to T and V. and other quantities are derivable by further differentiations. A general thermodynamic system, for which the element of reversible work is $\sum_i Y_i\,dX_i$ is similarly completely specified by giving $F(T,X_i)$.

Spontaneous processes at fixed T, X_i always go in a direction that decreases F; hence, any equilibrium state has a minimum value of F when compared to other states of the same T, X_i. This is the sense in which "systems tend toward a state of lowest energy."

In terms of microscopic properties, the free energy is given by

$$F = -kT \log \left[\sum_n \exp(-\varepsilon_n/kT) \right] .$$

where ε_n are the energy levels available to the system and k is Boltzmann's constant. The dependence on the macroscopic coordinates X_i enters through the energy levels. The simplicity of the foregoing relation makes the free energy one of the primary links between microscopic and macroscopic descriptions of a system.

Both the nomenclature and the notation for free energy vary considerably. The quantity we have been discussing is associated with the name of Helmholtz, and is sometimes called "work content" or "work function" and denoted by A instead of F. American chemists almost universally use "free energy" to mean the Gibbs potential $G = F + pV$ (which, however, they denote by F).

See also: Thermodynamics, Equilibrium.

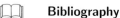

Bibliography

G. N. Lewis and Merle Randall, *Thermodynamics* (revised by K. S. Pitzer and Leo Brewer). McGraw–Hill, New York, 1965. (The chemist's approach.) (I)

Max Planck, *Treatise on Thermodynamics*. Dover, New York. 1945. (Translation from the German treatise whose first edition appeared in 1897, giving the classical view.) (I)

Gregory H. Wannier, *Statistical Physics*. Wiley, New York, 1966. (Connections with statistical mechanics.) (I)

Gabriel Weinreich, *Fundamental Thermodynamics*. Addison-Wesley, Reading, Mass., 1968. (A modern approach that emphasizes thermodynamic potentials.) (I)

Friction

F. E. Kennedy, Jr.

Whenever one body moves tangentially against another, there is resistance to that motion. The resistance is called friction and the resisting force is the friction force. Although there is also friction between a solid surface and a fluid, most of the following discussion will deal with friction between solid bodies. Particular attention will be focused on the contacting surfaces of the solids, since friction is primarily a surface phenomenon.

Coefficient of Friction (Macroscopic)

If two stationary contacting bodies are held together by a normal force N and if a tangential force is applied to one of them, the tangential force can be increased until sliding occurs. The maximum tangential force at the time of incipient sliding is, in most cases, approximately proportional to the normal force. The ratio of the friction force at incipient sliding to the normal force is known as the static coefficient of friction, f_s. After sliding commences, the friction force always acts in the direction opposing motion and the ratio between that friction force and the normal force is the kinetic coefficient of friction, f_k.

Generally f_k is slightly smaller than f_s, and both coefficients are independent of the size or shape of the contact surface under typical conditions. Both coefficients are, however, very much dependent on the materials of the two bodies and on the cleanliness of the contacting surfaces. Typical values of the friction coefficients are given in Table 1. For ordinary metallic surfaces, the friction coefficient is not very sensitive to the surface roughness. For ultrasmooth or very rough surfaces, however, the friction coefficient can be larger for reasons that will be discussed. If a metal is sliding against a very hard or very soft material, the roughness of the harder material can have an important influence on friction.

The kinetic coefficient of friction, f_k, for metallic or ceramic surfaces is relatively independent of sliding velocity at low and moderate velocities, although there is often a slight decrease in f_k with increasing sliding speed at higher velocities. With polymers, elastomers and very soft metals there may be an increase in friction with increasing velocity or temperature until a peak is reached, after which friction decreases with further increases in velocity

Table 1: Some Typical Friction Coefficients[a]

Material pair	f_s in air	f_s in vacuo	f_k dry air	f_k oiled
Mild steel vs. mild steel	0.75		0.57	0.16
Mild steel vs. copper	0.53	0.5 (oxidized) 2.0 (clean)	0.36	0.18
Copper vs. copper	1.3	21.0	0.8	0.1
Tungsten carbide vs. copper	0.35		0.4	
Tungsten carbide vs. tungsten carbide	0.2	0.4	0.15	
Mild steel vs. PTFE	0.04		0.05	0.04

[a] The friction coefficient values were compiled from several of the references listed in the Bibliography.

or temperature. The decrease in kinetic friction with increasing velocity, which may become especially pronounced at higher sliding velocities, can be responsible for friction-induced vibrations (stick-slip oscillations) of the sliding systems. If the sliding velocity is high enough, the amount of frictional heating may be sufficient to cause surface melting of one of the sliding materials.

This can lead to very low friction coefficients. For example, the friction coefficient of ice against itself at a very low velocity (10^{-5} m/s) is about 0.5, whereas at a higher velocity (10^{-2} m/s), f_k is less than 0.05.

Basic Mechanisms of Sliding Friction

Solid surfaces are not perfectly flat and smooth, but have roughness consisting of peaks and valleys produced in the manufacturing process. When two solids are forced against one another, contact occurs not within the entire nominal or apparent contact area, but only at the tips of contacting peaks or asperities on the surfaces. The total real area of contact is generally much smaller than the nominal area, and it is within that small area that friction occurs. Recent advances in surface analysis tools, particularly the atomic force microscope, surface force apparatus, and other scanning probe devices, have enabled measurement of friction at the single asperity level. In addition, theoretical methods including molecular dynamics simulation and phenomenological rate-state models have enabled considerable knowledge to be obtained about friction at the atomic or molecular level. These experimental and analytical advances have led to a better understanding of the basic mechanisms of friction.

Adhesion occurs when the atoms or molecules of the two contacting surfaces approach each other so closely that attractive forces between approaching atoms (or molecules) bond them together. The strength of the bond depends on the size of the atoms, the distance between them, and the presence or absence of contaminant matter (e. g., a lubricant film or adsorbed water molecules) on the surface. In order to separate the materials, the adhesive bonds must be broken. In the presence of an applied tangential force, relative motion between the two surfaces requires that all adhesive junctions formed within the real area of contact must be sheared. The force required to do this can be responsible for irreversible plastic deformation within the near-surface region and this often constitutes the major part of the friction force.

If deformation of the near-surface atoms is elastic instead of plastic, owing to lower contact forces or a very hard surface layer, the friction is considerably lower. Even if there is an interposed film preventing the formation of adhesive bonds between the solid atoms on the two contacting surfaces, there is still friction resulting from plastic shear of the surface films.

Another contribution to the friction force may occur whenever one of the two contacting surfaces is much softer than the other. In such cases, when the surfaces are forced together by a normal force the peaks on the harder surface will indent the softer surface to some extent. Relative tangential motion of the surfaces results in the plowing of the softer surface by the harder one, so the indentations become fine grooves in the softer surface. The deformation in the softer surface could be plastic or viscoelastic in nature, with the former occurring in metallic surfaces and the latter in polymers. The force required to displace the softer material ahead of the moving hard asperities is a second contribution to the friction force, the plowing or deformation contribution. For most dry sliding situations, the deformation contribution is considerably smaller than the contributions related to adhesive surface interactions. If one of the surfaces is both harder and rougher than the other, though, or if a lubricant or contaminant surface film is present on the surfaces, the deformation contribution could become significant.

If one considers only the friction related to adhesive surface interactions, one could conclude that the macroscopic friction force is related to the product of the real area of contact times the shear strength of the adhesive junctions within that area. Large differences in friction coefficient, such as those evident in Table 1, are generally caused by differences in the shear strength term. Ordinary surfaces are not completely clean, but are covered with oxide films, adsorbed gases, or other contaminant films which reduce both interfacial shear strength and adhesion. Thus, friction of sliding solids is considerably reduced by the presence of lubricants, oxides, or other contaminant films. If ultraclean surfaces are placed together in a vacuum, the interfacial adhesion is very strong and the application of a tangential force causes the junctions to grow, resulting in a very high friction force or even complete seizure.

Within adhesive junctions, the strength of the adhesive bonds is determined by the relative sizes of the surface atoms, the number of atoms that can approach each other closely, and the type of bond formed between the atoms (or molecules). The adhesive bonds and the resulting friction coefficients are greatest when the two surfaces are identical metals, slightly lower with dissimilar but mutually soluble metals, still lower for metal versus nonmetal, and lowest for dissimilar nonmetals.

Effect of Lubrication on Friction

When a small amount of lubricant is applied to a surface or when lubricated surfaces slide together at low sliding speeds or with a high applied normal load, the lubricant will likely not be successful in separating the two solid surfaces. The lubricant can still significantly reduce the friction coefficient by reducing the shear strength of adhesive junctions between the solid surfaces. In this situation, called the boundary lubrication regime, the effectiveness of the lubricant can be improved if the lubricant molecules adhere well to the solid surfaces. This is best accomplished by introducing a lubricant or additive that forms a surface film through physical adsorption, chemisorption, or chemical reaction with the surface. Even a molecular monolayer of lubricant on the surface can be quite effective in reducing friction. The reduced shear strength of the surface film can lower the friction coefficient by as much as an order of magnitude from the dry friction value. If the contacting surfaces are ultra-smooth, there may be a significant additional contribution to static friction (or 'stiction') from liquid-mediated adhesive forces, such as meniscus forces, in the presence of water vapor or a very thin lubricant layer on the surface.

When a good supply of a viscous lubricant is available, the separation between the surfaces will increase as the sliding speed increases or the normal load decreases. As the separation increases, the amount of solid/solid contact between the surfaces will decrease, as will the friction coefficient. In this "mixed lubrication" regime, friction is determined by the amount of plowing deformation by the harder asperities and by adhesion within the solid/solid contacts. When the surfaces become completely separated by a self-acting or externally pressurized lubricant film, the lubricant regime is hydrodynamic and friction usually reaches a low value governed by fluid shear of the lubricant. If the lubricant is Newtonian, the local shear stress is equal to the product of the lubricant viscosity times the local shear rate of the lubricant film. The shear rate is approximately equal to the difference in velocity between the two surfaces divided by the lubricant film thickness. The total friction force can be obtained by integrating the shear stress over the nominal contact area. Friction coefficients in such cases can be 0.001 or lower, depending on the surface velocities and the lubricant viscosity.

Most fluid film bearings, either journal or thrust geometry, take advantage of the low friction resulting from hydrodynamic lubrication, and the lubricant in such bearings could be mineral oil, a synthetic lubricant, gas, water, or other liquids. Unless the lubricant contains long-chain polymer additives, the flow in these bearings is generally assumed to be Newtonian. A difficulty arises in a lubricated contact if the sliding velocity goes to zero, because the fluid film separating the surfaces disappears and friction increases to a value determined by boundary lubrication or solid contact conditions.

In concentrated, nonconformal contacts such as in rolling element bearings, gears or cams, the pressures are often high enough to cause deformation of the surfaces and variations in lubricant viscosity. The result is a lubrication mode known as elasto-hydrodynamic lubrication (EHL). In such cases, the laws of Newtonian hydrodynamic lubrication are not necessarily obeyed and friction must generally be determined by empirical methods.

See also: Mechanical Properties of Matter; Rheology; Tribology.

Bibliography

A. D. Berman and J. N. Israelachvili, "Microtribology and Microrheology of Molecularly Thin Liquid Films", in *Modern Tribology Handbook*, B. Bhushan (ed.), pp. 568–615. CRC Press, Boca Raton, FL, 2001.

B. Bhushan, "Friction, Scratching/wear, Indentation, and Lubrication Using Scanning Probe Microscopy", in *Modern Tribology Handbook*, B. Bhushan (ed.), pp. 667–716. CRC Press, Boca Raton, FL, 2001.

F. P. Bowden and D. Tabor, *Friction. An Introduction to Tribology*. Anchor Press/Doubleday, New York, 1973.

D. Dowson, *History of Tribology*. Longman Group Ltd., London, 1979.

K. C. Ludema, "Friction", in *Modern Tribology Handbook*, B. Bhushan (ed.), pp. 205–233. CRC Press, Boca Raton, FL, 2001.

M. O. Robbins and M. H. Muser, "Computer Simulations of Friction, Lubrication, and Wear", in *Modern Tribology Handbook*, B. Bhushan (ed.), pp. 717–765. CRC Press, Boca Raton, FL, 2001.

Fullerenes

W. Krätschmer

Introduction

Fullerenes are all-carbon molecules of closed-cage structure. The most prominent species is the spherical C_{60}, the so-called "Buckminsterfullerene" (named after the American architect Richard Buckminster Fuller, 1895–1983), or "buckyball" for short. It consists of hexagonal and pentagonal carbon rings which together form a truncated icosahedron (or soccer-ball) with one carbon atom at each of its 60 vertices. Other prominent fullerenes, for example the rugby-ball shaped C_{70}, have no special names. The spherical C_{60} with a cage diameter of about 0.7 nm is the smallest stable fullerene which can also be synthesized in large amounts. Recently, plants having production rates of the order of 10 tons/year have been commissioned in Kyushu (Japan). The intention behind this effort is to stimulate the research and commercial application of fullerenes. Related to fullerenes are carbon nanotubes. These are cylindrical structures with diameters of the order of nanometers which can grow to considerable lengths. The ends of the tubes are not open but capped by suitable carbon networks, giving the nanotubes an overall closed-cage structure. Research on tubes began to flourish after single walled carbon nanotubes of a rather narrow diameter distribution became available. One remarkable feature of nanotubes is their extraordinary mechanical strength. Others are their electronic and optical properties, which are entirely determined by their cylindrical structures (that is their tube diameter and their "helicity" – the screw in the arrangement of hexagonal carbon rings along the tube). Besides searching for possible applications in chemistry, medicine, and materials science, fullerenes and nanotubes have also been used as model systems to study fundamental physical effects.

Forms of Carbon

For quite a while graphite and diamond were the only known forms of crystalline carbon. In diamond, each carbon atom is bound to four neighbors in a tetrahedral arrangement; in graphite, each carbon atom has three neighbors and a plane hexagonal network consisting of individual layers is formed. L. Pauling was the first to explain these bonding schemes in terms of the quantum mechanical superposition (i. e., hybridization) of the 2s and 2p electron orbitals in the carbon atom. The carbon in diamond assumes sp^3 hybridization, whereas in graphite it assumes sp^2 hybridization. In the latter case one p electron remains relatively loosely bound giving rise to the electrical conductivity of graphite. No crystalline form seems to exist in which carbon is exclusively in the state of sp hybridization, for which the carbon atoms would be arranged in linear chains. Fullerenes are related to graphite. However, the atomic arrangement is not hexagonal and planar but is curved and bent into a closed cage. In order to assume such a structure it is necessary for topological reasons (Euler's rule) to provide a pattern of carbon atoms which, besides an arbitrary number of hexagonal rings, also must contain twelve pentagonal carbon rings. Like a soccer ball, or the corresponding polyhedron called "truncated icosahedron", C_{60} has 20 hexagonal and 12 pentagonal faces (i. e., rings of carbon atoms) arranged in such a way that each pentagon is completely surrounded by hexagons. Such a configuration yields a uniform spherical curvature and possesses a high symmetry (I_h – which is the highest possible finite point group symmetry). The closed-cage structure also implies that all fullerenes must consist of an even number of atoms. For the cylindrical nanotubes, the twelve pentagons are located at the closing caps and the entire tube section consists of bent hexagons. The bent and curved hexagons can be considered either as mechanically stressed sp^2 bonds or as carbon in an sp^x ($2 < x < 3$) state of hybridization. The mechanical stress and the unfavorable pentagonal carbon rings make fullerenes and nanotubes energetically less stable than graphite, which is the most stable form. Fullerenes and nanotubes are cluster forms of carbon which decay irreversibly under heat and pressure. They cannot be located in the conventional diamond-graphite phase diagram.

Discoveries

In 1985 H. W. Kroto from the University of Sussex (Brighton, England) together with R. E. Smalley, R. F. Curl and their students from Rice University (Houston, Texas) performed mass spectroscopic studies of carbon clusters produced by laser ablation of graphite in a "universal cluster source", which had been developed by Smalley some years before. Under certain conditions they observed mass spectra which showed a very pronounced peak at the positions of C_{60} and a less intense peak at the mass of C_{70}. The researchers speculated that C_{60} might have the structure of a truncated icosahedron and baptized C_{60} as "Buckminsterfullerene". In later experiments the same researchers could confirm the closed-cage structures of both C_{60} and C_{70}. They suggested that these molecules are members of a new family of closed-cage carbon species for which they introduced the name "fullerenes". For their pioneering works Kroto, Curl, and Smalley were awarded the 1996 Nobel Prize for Chemistry. In the years 1982–1990, D. R. Huffman from the University of Arizona (Tucson, Arizona), the author and their students worked on the preparation and spectroscopy of graphitic carbon nanoparticles which might occur in interstellar space. The particles were produced by electric evaporation of graphite rods in a helium atmosphere. The latter enforces re-condensation of carbon vapor into nanometer-sized grains without chemical alteration. The researchers studied the produced particles by optical spectroscopy in the UV and in the IR, observing absorption bands which turned out to belong to C_{60}. The conclusion was that, along with the graphitic particles, fullerenes are formed in yields of order 10%. Until recently, the helium-filled carbon evaporator known as the "fullerene generator", was one of the standard devices for fullerene synthesis. The separation of the fullerenes from the particles can be readily achieved by (a) sublimation (fullerenes evaporate at 600–700 °C) or (b) by solvent extraction (fullerenes are soluble in non-polar solvents such as toluene). The separation of the different fullerene species (the product consists of ca. 80% C_{60}, 20% C_{70}, and 1% higher fullerenes such as C_{76}, C_{78}, C_{82}, C_{84}) can be accomplished by chromatography. With the pure materials at hand, the closed-cage nature of fullerenes could be proven beyond doubt. Fullerenes form crystals, usually of the closest packing type, i. e., face-centered cubic for C_{60} or hexagonal closed packed for C_{70}.

In 1991 S. Iijima from the NEC Corporation in Tsukuba (Japan) studied the graphite cathodes of used fullerene generators by transmission electron microscopy. He discovered a variety of nanoscopic needle-like carbon tubes which had grown from the graphite rod. Through an electron diffraction analysis he demonstrated that the tubes were nested cylinders of hexagonally arranged carbon, which are now called "multi-walled carbon nanotubes" (MWCNs for short). Later it was found that when the graphite rods in fullerene generators are doped with catalytic transition elements, particularly with cobalt, one can obtain single-walled carbon nanotubes (SWCNTs) in high yield. Under certain conditions the SWCNTs can be quite uniform in diameter. Such tubes form "crystals" in which the SWCNTs are entangled and stuck together in bundles like the threads in a rope.

Synthesis Methods

The already mentioned fullerene generator consists of two graphite rods serving as electrodes through which a high current can flow. The electrodes are located inside a vessel filled with helium at a pressure of 100–200 mbar. An arc is struck between the electrodes and the graphite evaporates in a continuously burning DC or AC discharge. When the rods become consumed by evaporation, they can be moved together to keep a constant gap for the arc of about one mm. Electrodes and vessel may be water cooled to allow large power and throughput. The produced fullerene soot is very fluffy and can be easily removed from the inside.

Smalley and co-workers have developed a device in which laser ablation of graphite takes place in a uniform hot (1000–1200 °C) argon atmosphere. For this purpose, the graphite rod is placed inside an argon floated furnace, providing rather uniform conditions for the carbon vapor re-condensation. The fullerene yields are close to those obtained in a conventional generator, with the advantage that the nanostructure of the produced soot is much more uniform. When the graphite is doped by suitable transition metals like cobalt, SWCNTs of very uniform diameters (1.4 nm) are obtained in high yield, along with the fullerenes.

Employing more conventional chemical techniques, fullerenes have been obtained through the pyrolysis of suitable derivates of polycyclic hydrocarbon precursor compounds. The reported yields are rather modest but the importance of this approach rests on the control with which the reaction can be directed towards a specific species. This method may become important in the future, should fullerenes with extraordinary properties be required.

The arc process is very energy consuming. In the early 1990s, the former Hoechst Company optimized this process and produced several kg of fullerenes for scientific application and commercial use. However, soon after their discovery, fullerenes were also detected in sooting flames. A systematic search for optimum fullerene yield conditions in flames was undertaken and the resulting method is known as "incomplete combustion". This method has been recently applied to produce fullerenes in ton/year quantities. For application of this method it is of considerable advantage that soot is already produced on large scales by incomplete combustion, since soot is added to rubber to reduce the wear of car tires. With some modifications in the combustion parameters, fullerene-containing soot can be produced. However, the purification and separation process, which in any kind of production method is very time and money consuming, here requires even more effort, since combustion creates various hydrocarbons, which are rare or absent in the chemical clean arc process. Large-scale production of SWCNTs is attempted by chemical vapor deposition but so far it seems to be extremely difficult to obtain a uniform high quality product.

Properties

Fullerenes are soluble and thus introduce a new chemistry. The closed-cage character allows two kinds of chemistry, namely on the outside (exohedral), and on the inside (endohedral). Some of the most intriguing developments may be highlighted here (see the biography for other details). The C_{60}-based exohedral compounds with the most striking properties are of the composition A_3C_{60} (A stands for alkali element), the "fullerides". They can either be prepared by solid state or solution reactions. Since the LUMO of C_{60} can host six electrons, fullerides exhibit half-filled electron orbitals and are thus conductors. Surprisingly, some of these are superconductors (of type II) with relatively high transition temperatures ($T_c = 32\,K$ for Cs_2RbC_{60}). In contrast to the layered cuprate high-temperature superconductors, the fullerides conduct isotropically. T_c was observed to increase with the fcc lattice constant of the fulleride. Such trend is in accord with BCS theory which relates T_c to the level density at the Fermi-energy. The latter increases with increasing lattice constant. The C_{60} fullerides provide a uniquely simple and highly symmetric system for the study of superconductivity. No C_{70}-based fulleride superconductor has yet been found. In the case of C_{60} fullerides, the buckyballs are electron-doped. The other possibility, namely to prepare hole-doped C_{60} (that is, to remove electrons from the ten electrons hosting C_{60} HOMO) could so far not be realized in practice.

The endohedral compounds are usually denoted by AB, which means molecule A is inside cage B. In case of the C_{60} cage, the guest usually consists of a single atom (or in rare cases of a molecule of two atoms), located in the center of the cage (as for example for noble gas or nitrogen atoms). Metals are usually eccentrically attached at the inside cage wall. Larger fullerenes can host molecules or metallic clusters. In general, the electronic structure of the host cage becomes modified by the guest, for example by charge transfer, leading to a change in the chemical properties of the endohedral cage as compared to the empty cage. The endohedral guest species can also stabilize the outer fullerene cage. An example of this is Sc_3NC_{80} in which the C_{80} cage has I_h symmetry. Such a C_{80} cage can only exist if six electrons are transferred from the guest cluster to the host cage.

So far, most endohedral fullerene compounds have been produced in situ, meaning during the fullerene formation process. This has been achieved by doping the electrodes or adding the desired compounds to the buffer gas. Noble gas atoms containing endohedrals can also be produced *ex situ*, i. e., from fullerene powder exposed to the desired gas under high pressure and temperature. Recently, the fullerene cage could be opened, filled, and then closed again by "chirurgical chemistry" leading to the compound H_2C_{60}. At present, however, the separation of the desired endohedral fullerene from the co-produced empty fullerenes is cumbersome and the obtained yields are quite low.

Nanotubes are too large to be soluble. For SWCNTs which are most attractive for potential applications, various recipes for cleaning and handling exist. Like colloids, tubes can be dispersed in solutions by means of suitable detergents. Alternatively, chemical cutting and functionalization has been attempted. The SWCNT rough product is a felt-like material consisting of numerous entangled ropes of tubes and is called "buckypaper". The preparation of individual SWCNTs requires considerable skill. The electrical and optical properties of such tubes are remarkable. Depending on the structure, the tubes can be metallic or semi-conducting, and changes in tube geometry (for example through bending, or other distortions) result in changes in conductivity, e. g., can lead to diode or transistor characteristics. Since the Raman spectra of SWCNTs can be well understood in terms of structure, Raman spectroscopy has become – besides electron and tunneling microscopy – an important tool for SWCNT characterization. Individual SWCNTs have been used as a model system to study fundamental quantum transport effects in one dimension such as ballistic conduction, fluctuation phenomena, and the effects of magnetic fields on conduction. The mechanical strength (the graphitic carbon bond is stronger than that in diamond) and their elasticity make SWCNTs the "ultimate fiber" considerably superior to any other fiber material. Endohedral nanotube systems, for example SWCNTs filled with metals or with fullerenes (the so-called peapods) have been successfully prepared. The hope is to modify the properties of the host tube, for example to obtain metallic nanowires. Alternatively, nanotubes have been used to study chemical reactions going on in their inside, such as carbide formation or polymerization of fullerenes. Efficient hydrogen storage by nanotubes has also been attempted, but so far the storage capacities required for technical applications could not be reached.

Polymerization

Fullerenes tend to polymerize under the action of pressure and/or heat. Photochemical polymerization has also been studied in detail. The compound KC_{60}, which seem to form a one-dimensional chain polymer, is particularly remarkable. Polymeric fullerenes seem to form ultra hard structures and can serve as precursors for diamond synthesis.

See also: Chemical Bonding; Crystal Symmetry; Mesoscopic Systems; Superconducting Materials; Superconductivity Theory.

Bibliography

J. Baggott, *Perfect Symmetry*. Oxford University Press, Oxford, New York, Tokyo, 1994.

A. Hirsch, *The Chemistry of the Fullerenes*. Thieme Publishers, Stuttgart, New York, 1994

R. Taylor, *Lecture Notes on Fullerene Chemistry*. Imperial College Press, 1999.

P. W. Fowler and D.E. Manolopoulos, *An Atlas of Fullerenes*. Clarendon Press, Oxford, 1995

M. S. Dresselhaus, G. Dresselhaus and P. C. Eklund, *Science of Fullerenes and Carbon Nanotubes*. Academic Press, San Diego, Boston, New York, London, Sydney, Tokyo, Toronto, 1996.

R. Sato, G. Dresselhaus, M. S. Dresselhaus, *Physical Properties of Carbon Nanotubes*. Imperial College Press, 1998.

O. Gunnarsson, *Alkali-Doped Fullerides*. World Scientific Publishers, London, Singapore, Beijing, Hong Kong, 2004.

B. Sundqvist, *Adv. Phys.* **48**, 1–134 (1999).

Galaxies

A. Dressler

In 1924 Edwin Hubble discovered Cepheid variable stars in M33 and M31, the Andromeda nebula. With one stroke he put an end to the debate that had been raging among astronomers for years [1]. The Andromeda nebula was proved to be a giant stellar system - a galaxy – at an immense distance (about 2 million light years) and similar in size to the Milky Way system in which the Sun is situated.

Astronomers now recognize galaxies as the fundamental units in the organization of baryonic matter [2]. A typical galaxy like our own Milky Way is roughly 60 000 light years in diameter. Galaxy sizes range in order of magnitude larger and smaller than this. Galaxy masses, which range from 10 million to 1 trillion times the mass of the sun, are mainly made up of ordinary stars, with smaller contributions from neutral and ionized gas and dust. However, convincing evidence from the motions of stars and extended gas distributions reveals that the luminous structures we recognize as galaxies are condensed regions of ordinary baryonic matter within more extensive mass concentrations containing, typically, 10 times the visible mass. From a variety of evidence, it now seems clear that these so-called "dark matter halos" are made of non-baryonic matter, for example, the mass may be in the form of weakly interacting elementary particles like neutrinos (which would have to have a nonzero rest mass). The exact nature of dark matter remains a crucial, unsolved problem in astronomy.

Galaxies are found to cluster on many scales. Rich clusters include environments several orders of magnitude more dense than the average background, which is mainly composed of loose groups. Such clusters are conspicuous, but contain only 5–10% of all galaxies. A surprisingly complex structure has recently been discovered in the large-scale distribution of galaxies. Giant superclusters, containing tens of thousands of galaxies, have diverse topologies, including flat and filamentary forms in addition to basically spherical shapes. These superclusters are interspersed with immense spherical or tube-like voids in which few galaxies are found.

Encyclopedia of Physics, Third Edition. Edited by George L. Trigg and Rita G. Lerner
Copyright ©2005 WILEY-VCH Verlag GmbH & Co. KGaA, Weinheim
ISBN: 3-527-40554-2

Observations of galaxies now span the electromagnetic spectrum from radio waves to X-rays. Starlight is the primary source of continuum light in the ultraviolet, visible, and near infrared; dust heated by these stars produces most of the far-infrared flux. Optical line emission from ionized regions surrounding hot stars provides important information on the densities, temperatures, and atomic abundances in regions of active star formation. Radio emission is commonly observed as the 21-cm line from cool, neutral hydrogen gas, and even cooler gas in the form of what are called "molecular clouds" is studied in millimeter and sub-millimeter wavelengths. Emission from molecules, most importantly CO, is the dominant tracer of this component of the interstellar medium. Thermal radio emission from recombination of ionized hydrogen in star forming regions is also observed. Continuum radio flux of non-thermal origin arises from synchrotron emission from an "active nucleus" (see below), from the remnants of exploded stars called supernovae, and from the diffuse emission emanating from the interaction of cosmic ray particles with a galaxy's magnetic field.

X-rays from galaxies arise primarily from binary star systems where mass is being transferred to a white dwarf, neutron star, or black hole companion. Thermal X-rays have been observed from hot gas around some galaxies. Both X-ray and radio emission also arise in energetic processes in galactic nuclei, described below. Neutrinos of a bona fide extragalactic origin were first seen with the detection of supernova 1987a in the Large Magellanic Cloud, the closest neighbor galaxy to the Milky Way.

The distribution of galaxy luminosities, and, by inference, galaxy mass, has been found to be remarkably similar over a wide range of environment. This *luminosity function* has a power law form which is truncated exponentially at the bright end. Some outstandingly bright "cD" galaxies at the centers of clusters appear to be star piles formed from the aggregation of many galaxies. On the faint end, the luminosity function continues to rise to systems with 10 million solar luminosities. Though these dwarf galaxies are the most common galaxies in the universe, their contribution by mass is far less than that of luminous systems like the Milky Way.

Galaxies come in a variety of morphological forms which have been described simply by Hubble and more elaborately by Sandage, de Vaucouleurs, and van den Bergh. These classification schemes recognize the basic elements of galaxy form as the disk and the spheroid, structural forms that reflect the specific angular momentum of the system. Elliptical galaxies are ellipsoidal or triaxial distributions of stars whose nearly round shapes are supported against gravity by mainly random motions. Similar spheroidal components are also found in the centers of many disk galaxies. In general, spheroidal components, including the so-called "bulges" of these disk galaxies, are made up of old stars and have little cold gas or dust, the raw materials of star formation.

Disk systems that show a pronounced spiral pattern are called spiral galaxies. The spiral pattern outlines sites of ongoing star formation, usually associated with cold, dense gas clouds rich in molecules, and ionized gas surrounding young, hot stars. The spiral structure is often due to a resonance phenomena wherein a density fluctuation propagates around the galaxy. Morphological classification systems refer to the tightness of the spiral pattern or its coherence. Spiral galaxies with central straight structures called *bars* represent a general class of axisymmetric distortion resulting from dynamical instabilities in stellar systems with high angular momentum. Irregular morphologies, including amorphous and disturbed forms, comprise a few percent of luminous galaxies, but are more common among dwarf galaxies.

Though the connection of morphology to the history of star formation is self-evident, the mechanism that determined whether a galaxy became spiral rather than an elliptical disk, rather than spheroid dominated, is still unclear. An important clue is the strong correlation of the type of galaxy with the crowdedness of its surroundings. Densely-packed clusters of galaxies are mainly composed of spheroidally dominated galaxies like elliptical and S0 galaxies; low-density groups and isolated galaxies are usually disk-dominated spirals. Most galaxies are found in lower-density environments, so, overall, spiral galaxies are the most common. The morphology–density relation suggests that galaxy–galaxy interactions, particularly early in a galaxy's life, may determine its ultimate form. The effect on angular-momentum distributions, perhaps connected to the relationship between the luminous material and the dark halo, could be of crucial importance. Also suggestive are observations of merging galaxies which appear to be in the act of forming spheroidal systems from former disk galaxies. This highlights the possibility that morphology may be drastically altered late in the lives of at least some galaxies.

Although a satisfactory, detailed model of galaxy formation has yet to be formulated, and may, in fact, depend strongly on the nature and distribution of dark matter, the role of gaseous dissipation in achieving the high average density of a typical galaxy seems clear. Some 1–2 billion years after the Big Bang, gas clouds the size of galaxies were cooled by radiation and by inverse Compton scattering of cosmic background photons. In this way they dissipated thermal energy, contracting until they reached densities suitable for copious star formation. Larger structures with the mass of tens or hundreds of galaxies were unable to follow this path because their cooling times scale exceeded the dynamical time.

The exact epoch at which galaxy structure became distinguishable and the first stars formed is still not known. A considerable effort has gone into looking directly for *primeval* galaxies at high redshifts, but only rare, nonrepresentative cases have been identified. However, the common ~ 15 billion year age found for the oldest stars in the Milky Way and its neighboring galaxies, and the approximate coincidence of this time with the present age of the universe, seems to ensure that such a process occurred over a relatively short interval. It is also unknown to what extent the collapse to a galaxy-size structure was a single, coherent process as opposed to the more random coagulation of a number of fragments. In spiral galaxies, at least, the process of turning gas into stars has persisted over the age of the universe, resulting in a steady increase in the abundance of elements heavier than hydrogen. Our own Sun was born in a later generation of star formation some 5 billion years ago, in a spiral arm near the rim of the Milky Way galaxy.

Among the most interesting phenomena in astrophysics are those galaxies which release enormous amounts of energy, often far exceeding the stellar output, from solar-system size regions at their centers. The most luminous examples of these "active galactic nuclei," *radio galaxies* and *quasars*, are important probes of the early universe since their light can be seen over far greater distances, and thus to greater lookback times, than the light of ordinary galaxies. Furthermore, absorption of quasar light by intervening galaxies probes the properties of gas in these objects that are themselves too faint to be observed directly. Successful models of the central energy source of active nuclei invoke accretion onto massive black holes of tens of millions of solar masses. There is now convincing evidence based on the motions of stars and gas in the nuclei of nearby galaxies which indicates that such massive black holes are common, and that the mass of these a central black hole approximately is scaled to large-scale properties of the galaxy, for example, the mass and/or velocity dispersion of the galaxy's bulge. Such correlations suggest a close coupling of the origin of central black holes with the process of galaxy building.

See also: Astrophysics; Cosmology; Interstellar Medium; Milky Way; Quasars; Universe.

References

[1]   E. P. Hubble, *The Realm of the Nebula*, p. 93. Oxford University Press, Oxford and London, 1936.
[2]   D. Mihalas and J. Binney, *Galactic Astronomy*. W. H. Freeman, San Francisco, 1981.

Galvanomagnetic and Related Effects

E. Y. Tsymbal and D. J. Sellmyer

These are transport effects which occur in conductors in the presence of forces produced by electric and magnetic fields, and temperature gradients. The irreversible, cross-coupled processes can be described by the expressions:

$$E_m = \sum_n \rho_{mn}(\mathbf{B})J_n + \sum_n \alpha_{mn}(\mathbf{B})\frac{\partial T}{\partial x_n}\,, \tag{1}$$

$$Q_m = \sum_n \pi_{mn}(\mathbf{B})J_n - \sum_n \kappa_{mn}(\mathbf{B})\frac{\partial T}{\partial x_n}\,, \tag{2}$$

where E_m, J_m, Q_m, and $\partial T/\partial x_m$ represent the mth component of the electric field, electric current density, heat current density, and temperature gradient, respectively. $\mathbf{B}$ is the magnetic flux density and m and $n = 1, 2, 3$ represent x, y, and z Cartesian coordinates, respectively. The transport-coefficient arrays ρ_{mn}, α_{mn}, π_{mn}, and κ_{mn} are known as the *resistivity, thermoelectric power, Peltier coefficient*, and *thermal conductivity*, respectively.

A large number of experimental situations can be described by Eqns. (1) and (2). By convention, three classes of transport effects are defined: (1) *galvanomagnetic* effects in which there are no temperature gradients, (2) *thermoelectric* effects in which there is no magnetic field present, and (3) *thermomagnetic* effects in which all quantities in Eqns. (1) and (2) are allowed to be nonzero. The thermoelectric effects will not be considered further here and some of the more important galvanomagnetic and thermomagnetic effects, with their names and experimental conditions, are defined for an isotropic conductor in Table 1.

The effects are characterized as isothermal or adiabatic depending on whether temperature gradients or heat flows are allowed to exist. In addition, the galvanomagnetic effects are classified according to the three basic relative orientations of applied electric and magnetic fields:

1. *Transverse* effects in a *transverse* magnetic field. $\mathbf{B}$ is perpendicular to the primary current density $\mathbf{J}$ flowing in the sample, and the galvanomagnetic effect is measured in the direction mutually perpendicular to $\mathbf{B}$ and $\mathbf{J}$.

Table 1: Galvanomagnetic and related effects in isotropic media.

Class	Name	Primary fluxes $(\mathbf{B} = \mathbf{B}_z)$	Quantity measured	Defining conditions
1. Transverse effects in a transverse field	Hall (isothermal)	$\mathbf{J} = \mathbf{J}_x$	E_y/J_x	$J_y = \frac{\delta T}{\delta x} = \frac{\delta T}{\delta y} = 0$
	Hall (adiabatic)	$\mathbf{J} = \mathbf{J}_x$	E_y/J_x	$J_y = \frac{\delta T}{\delta x} = Q_y = 0$
	Ettingshausen	$\mathbf{J} = \mathbf{J}_x$	$\frac{1}{J_x}\frac{\delta T}{\delta y}$	$J_y = \frac{\delta T}{\delta x} = Q_y = 0$
2. Longitudinal effects in a transverse field	Electrical transverse magnetoresistivity (isothermal)	$\mathbf{J} = \mathbf{J}_x$	E_x/J_x	$J_y = \frac{\delta T}{\delta x} = \frac{\delta T}{\delta y} = 0$
	Electrical transverse magnetoresistivity (adiabatic)	$\mathbf{J} = \mathbf{J}_x$	E_x/J_x	$J_y = \frac{\delta T}{\delta x} = Q_y = 0$
	Nernst (isothermal)	$\mathbf{J} = \mathbf{J}_x$	$\frac{1}{J_x}\frac{\delta T}{\delta x}$	$J_y = Q_x = \frac{\delta T}{\delta y} = 0$
	Nernst (adiabatic)	$\mathbf{J} = \mathbf{J}_x$	$\frac{1}{J_x}\frac{\delta T}{\delta x}$	$J_y = Q_x = Q_y = 0$
3. Longitudinal effects in a longitudinal field	Electrical longitudinal magnetoresistivity	$\mathbf{J} = \mathbf{J}_z$	E_z/J_z	$\frac{\delta T}{\delta z} = 0$
	Unnamed	$\mathbf{J} = \mathbf{J}_x$	$\frac{1}{J_z}\frac{\delta T}{\delta z}$	$Q_z = 0$

2. *Longitudinal* effects in a *transverse* magnetic field. **B** is again perpendicular to **J**, but the galvanomagnetic effect is measured along the direction of **J**.

3. *Longitudinal* effects in a *longitudinal* electric field. **B** is parallel to **J** and the galvanomagnetic effect is measured along **J**.

Fundamental to an understanding of the galvanomagnetic effects is the Lorentz force equation (in SI units):

$$\mathbf{F} = q(\mathbf{E} + \mathbf{v} \times \mathbf{B}) . \tag{3}$$

Here, q is the charge on a charged particle (such as an electron), **v** is the particle's velocity, and $\mathbf{v} \times \mathbf{B}$ is the vector product of these two quantities (which is perpendicular to both). The galvanomagnetic effects arise from a combination of two circumstances: the deflection of the itinerant electrons in a conductor by the Lorentz force, and the constraint on the electron flow produced by the boundaries of the sample. Consider a conducting sample subjected to a steady electric field **E**. Under the influence of the electrostatic force, each itinerant electron has superimposed upon its random thermal motion an acceleration along the direction exactly opposite to **E** (since an electron carries a negative charge). During its mean free path between randomizing collisions, which occur especially with lattice vibrations or impurities in the substance, each electron contributes to a general current drift along **E**. The Lorentz force produced by a uniform flux density **B**, assumed for generality to be noncollinear with **E**, deflects each moving electron from its unperturbed path. A few will at first be deflected without hindrance, but since electrons are mutually repulsive, and since they are constrained to remain within the bounds of the sample, they eventually pile up against an inside face.

This creates an electric field which opposes the Lorentz force. Ultimately, a new dynamical equilibrium is reached among the electrons in which the current drift is maintained along **E** but is now nonuniform because of the established concentration gradients. Since the electrons possess thermal energy as well as charge, a temperature gradient will be set up concomitantly with the electric field. A third effect is the decrease in the effective electrical conductivity that is implied as electrons are deflected from their otherwise forward motion under the influence of **E**. These three features are typical manifestations of a galvanomagnetic effect.

The most informative and most-often measured of the galvanomagnetic and thermomagnetic effects are two in the former class: *Hall effect* and *magnetoresistance*. For these effects two field regions are defined. In the presence of a magnetic field the Lorentz force causes the electrons to move in circular or helical paths. The angular frequency of this periodic motion is called the *cyclotron frequency* which is given by

$$\omega_c = \frac{qB}{m_c} , \tag{4}$$

where m_c is the cyclotron mass of the carrier. However, the periodic motion is interrupted by collisions of the carriers with impurity atoms or lattice vibrations, with the average time between collisions, or relaxation time, being defined as τ. The two field regions defined above can then be characterized by

$$\omega_c \tau \ll 1 \quad \text{(low-field region)}, \tag{5}$$
$$\omega_c \tau \gg 1 \quad \text{(high-field region)}. \tag{6}$$

Physically, in the low-field region the carriers are prohibited from making a complete cyclotron orbit by collisions. In the high-field region, typically reached in very pure single crystals at liquid-helium temperatures ($T \approx 4\,\text{K}$), many cyclotron orbits can be transversed before scattering takes place. The behavior of the galvanomagnetic effects in the two field regions is markedly different.

In the low-field region, the transverse Hall effect for a conductor with only one type of carrier present (electrons or holes) gives a transverse voltage *inversely* proportional to the carrier density. This permits the measurement of this density for simple monovalent metals like the alkali and noble metals. But the small size of the Hall effect in metals makes it difficult to measure and interpret. For semiconductors, where the carrier density is several orders of magnitude smaller than for metals, the magnitude of the Hall field, E_y (see Table 1), is correspondingly larger. This leads to applications such as magnetometers, susceptibility meters, Hall-effect amplifiers, multiplying elements, displacement transducers, and contactless switches.

In metals and semiconductors where there is more than one type of carrier present – for example, electrons and holes – the theory for the low-field Hall effect becomes more difficult and the information obtained less precise.

In the high-field region the transverse magnetoresistance of certain pure single-current metals can show striking anisotropy as a function of magnetic field direction. These effects have been shown to be related to the topology of the Fermi surface of the metal, the surface bounding the region of occupied electron states in wave vector space. In addition, quantum oscillations in the magnetoresistance are directly related to extremal cross sections of the Fermi surface. Thus these galvanomagnetic effects are important for obtaining fundamental information about the Fermi surfaces of metals.

In ferromagnetic materials galvanomagnetic effects exhibit remarkable features due to the presence of the spontaneous magnetization. This leads to new phenomena, one of them being the *anomalous Hall effect* (sometimes referred to as *extraodinary Hall effect*). In ferromagnetic metals the Hall field per unit current density can be written as

$$E_{\mathrm{H}} = R_0 B + R_1 M \,, \tag{7}$$

where R_0 is the "ordinary" Hall coefficient, R_1 the "anomalous" or "extraordinary" Hall coefficient. The second term stems from the presence of the spontaneous magnetization M in a ferromagnetic metal. In thin-film magnetic structures, measurements of the Hall coefficient have been shown to be a sensitive method for studying the field and temperature dependence of the magnetization.

Magnetoresistive phenomena in magnetic materials include *anisotropic magnetoresistance*, *giant magnetoresistance*, *colossal magnetoresistance*, and *tunneling magnetoresistance*, as described below.

Anisotropic magnetoresistance (AMR) is the dependence of resistivity of a ferromagnetic metal on the relative angle between the electric current and the magnetization direction. This phenomenon was discovered by Thomson in 1857. The importance of this phenomenon was recognized more than a century later in the 1970s when AMR of a few percent at room temperature was found in a number of alloys based on iron, cobalt, and nickel which stimulated the development of AMR sensors for magnetic recording. The AMR originates from the anisotropy of scattering produced by the spin-orbit interaction. Ferromagnetic metals exhibiting a normal AMR effect show maximum resistivity when the current is parallel to the magnetization direction and minimum resistivity when the current is perpendicular to the magnetization direction.

Giant magnetoresistance (GMR) is the large change in electrical resistance in magnetic metallic layered structures when the magnetizations of the ferromagnetic layers are reoriented relative to one another under the application of an external magnetic field. This reorientation of the magnetic moments alters both the electronic structure and the scattering of the conduction electrons in these systems, which causes the change in the resistance. GMR is observed in magnetic multilayers with alternating layers of magnetic and nonmagnetic metals and in granular materials with ferromagnetic clusters imbedded in a non-magnetic matrix. The highest values of a few tens of percent at room temperature are obtained in Fe/Cr and Co/Cu multilayers. The GMR effect is applied in low-field sensors, such as read heads for magnetic recording.

Colossal magnetoresistance (CMR) is an extremely large change of resistance due to an applied magnetic field which was found in doped manganite perovskites such as $La_{3-x}Ca_xMnO_3$. The CMR effect can result in a resistance change of a few orders in magnitude. CMR originates from a metal-insulator transition in the vicinity of the Curie temperature and requires magnetic fields of the order of several Tesla. The latter property makes the applicability of CMR materials fairly limited.

Tunneling magnetoresistance (TMR) is observed in magnetic tunnel junctions, in which ferromagnetic metallic layers are separated by an insulating spacer layer. The insulating layer is so thin (a few nm or less) that electrons can tunnel through the barrier if a bias voltage is applied between the two metal electrodes across the insulator. The tunneling current depends on the relative orientation of the magnetizations of the two ferromagnetic layers, which can be changed by an applied magnetic field. Magnetic tunnel junctions that are based on 3d-metal ferromagnets and Al_2O_3 barriers show TMR values up to 70 percent at room temperature, making them suitable for industrial applications, such as magnetic random access memories.

The Quantum Hall Effect is an interesting galvanometric phenomenon which is present in quasi-two-dimensional semiconducting structures. When the Hall effect is measured at low temperatures in a sample that is so thin that the electrons are confined to move only in a plane, the Hall resistance deviates from the classical behavior. At sufficiently high fields and low temperatures, a series of flat steps appear in the dependence of the Hall resistance versus magnetic field. The Hall resistance steps could take on fractional values to a precision close to one part in a billion, making the Hall effect one of the most precise phenomena known in physics.

The galvanomagnetic effects have been helpful in advancing our knowledge of conduction processes in both semiconductors and metals. Measurements of the low-field Hall and magnetoresistance effects are made routinely to determine the sign, density, and mobility of the charge carriers in semiconductors. Studies of the dependence of these effects on the magnetic field strength lead to information about the nature of the electron scattering processes in both semiconductors and metals, while studies in the high-field condition, particularly in metals, are an important tool for investigating the substance's electronic structure. Recent research on novel nanostructures has led to important new information of great interest in pure and applied physics. In particular, magnetoresistive phenomena in magnetic nanostructures are very interesting from a fundamental physics viewpoint and have a significant potential for applications in electronic devices.

See also: Conduction; Hall Effect; Hall Effect, Quantum; Magnetoresistance; Resistance.

Bibliography

T. Chatterji, *Colossal Magnetoresistive Manganites*. Kluwer Academic Publishers, 2004.

C. L. Chien and C. R. Westgate (eds.), *The Hall Effect and its Applications*. Plenum, New York, 1980.

J. I. Gersten and F. W. Smith, *The Physics and Chemistry of Materials*. Wiley, New York, 2001.

C. M. Hurd, "Galvanomagnetic Effects in Anisotropic Metals," *Adv. Phys.* **23**, 315–433 (1974).

L. Mihály and M. C. Martin, *Solid State Physics*. Wiley, New York, 1996.

R. E. Prange and S. M. Girvin (eds.), *The Quantum Hall Effect*. Springer, New York, 1986.

W. A. Reed, "Experimental Methods of Measuring High-Field Magnetoresistance in Metals", in *Methods of Experimental Physics*, Vol. 11 (R. V. Coleman, ed.), pp. 1–31, 1974.

A. C. Smith, J. F. Janak, and R. B. Adler, *Electronic Conduction in Solids*. McGraw–Hill, New York, 1967.

E. Y. Tsymbal and D. G. Pettifor, "Perspectives of giant magnetoresistance," in *Solid State Physics*, Vol. 56 (H. Ehrenreich and F. Spaepen, eds.). Academic Press, San Diego, pp. 113–237, 2001.

E. Y. Tsymbal, O. N. Mryasov, and P. R. LeClair, "Spin-Dependent Tunneling in Magnetic Tunnel Junctions," *J. Phys. Cond. Matter* **15**, R109–R142 (2003).

M. Ziese and M. J. Thornton (eds.), *Spin Electronics*. Springer, Berlin, 2001.

Gamma Decay

E. der Mateosioan

Gamma decay is a process in which atomic nuclei change their states to states of less energy with the simultaneous emission of electromagnetic radiations, or photons, called *gamma rays*. The probabilities with which these transitions take place may be vastly different and are strongly dependent, in a complicated fashion, on the angular momenta, energies, and parities of the initial and final energy states of the nuclei involved. Other considerations give rise to "selection rules" and perturbations that may prohibit or affect the probability of certain transitions. Since this process involves the electromagnetic field, familiar to physicists through classical events, the study of gamma decay assumed historical importance because it made possible the understanding and determination of nuclear properties (energy, angular momentum, parity) even though the nuclear forces were not known.

The atomic nucleus is a complex system of particles (nucleons) that may absorb or emit energy according to the rules of quantum mechanics (QM). One of the striking restrictions imposed by QM on the nucleus is that it may not absorb or emit any arbitrary amount of energy; rather, the nucleus may exist only in a set of discrete (not continuous) energy states. It is customary to illustrate this situation by means of an energy diagram called a *nuclear level scheme* (Fig. 1). A permissible nuclear energy state for a particular atom is represented by a horizontal line whose vertical position indicates the energy content of the nucleus when in that state. The energy scale is only a relative one based upon the assumption that the lowest

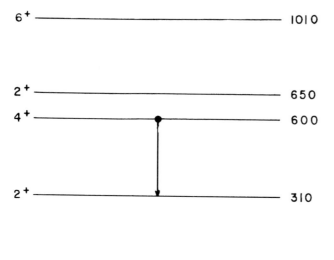

Fig. 1: A nuclear energy level diagram. The horizontal lines represent discrete energy levels in a hypothetical nucleus. To the right of each level appears its energy above the ground state (or lowest level) and to the left, its angular momentum and parity. The vertical line connecting the 4+ and 2+ levels indicates that the nucleus, when in its 4+ state, will make a transition to its 2+ state with the emission of a gamma ray.

energy state (ground state) has zero energy. Properties of the nucleus when in one of these states, such as its energy, angular momentum, and parity, are written beside the level lines. In a quantum-mechanical system such properties are given by *quantum numbers*. When a nucleus spontaneously changes from one state to another with less energy, it emits the difference in energy between these states in the form of a monoenergetic photon, or electromagnetic radiation. If the energy of the photon is $\hbar\omega$ and the energies of the initial and final states are E_a and E_b, respectively, then

$$E_a - E_b = \hbar\omega \tag{1}$$

and energy is conserved. The transition of the nucleus from one energy state to another by means of the emission of a photon is indicated on the energy-level scheme of Fig. 1 by a vertical line between two levels. A dot on the end of this line indicates the initial state; an arrowhead indicates the final state. Thus, Fig. 1 may be taken to mean that the particular nucleus under consideration was in the 600-keV state (which has 4 units of angular momentum) and it decayed to the 310-keV state (2 units of angular momentum). However, Fig. 1 is interpreted to mean more than this. The transition between the two levels does not take place haphazardly but with a probability that can be predicted by various model-dependent theoretical calculations or the systematics of experimentally determined decay probabilities. Figure 1 predicts that if the nucleus is excited to the 600-keV state, it will decay to the 310-keV state (with a certain probability expressed, when known, reciprocally as a half-life). It is thus seen that Fig. 1 gives information not only on the states of the nucleus but on its behavior in these states as well. A nuclear level scheme with all of the known transitions of the nucleus indicated on it is called a decay scheme. The usual decay scheme in practice can be quite complicated. An example of what one may expect is shown in Fig. 2.

Decay schemes of nuclei are sufficiently unique that if the nuclear gamma rays are detected and their energies determined, it is frequently possible to identify the nuclei emitting the radiations; but more than energy characterizes a photon emitted in a nuclear transition between two specific states of a nucleus. For example, in a mechanical system not only energy but angular momentum is conserved, and a transition between two nuclear levels with different angular momenta may take place only if the emitted photon, or gamma ray, carries that amount of angular momentum ℓ such that

$$\mathbf{J}_a = \mathbf{J}_b + \ell , \tag{2}$$

where $\mathbf{J}_a$ and $\mathbf{J}_b$ are the angular momenta of initial and final states of the nucleus. Equation (2) is a vector equation that describes the addition of angular momenta in quantum mechanics and it results in the selection rule that transitions between states with angular momenta $\mathbf{J}_a$ and $\mathbf{J}_b$ can take place only if

$$|J_a - J_b| \leq l \leq J_a + J_b . \tag{3}$$

Another selection rule says

$$M_a - M_b = m \tag{4}$$

where M_a and M_b are the Z components of $\mathbf{J}_a$ and $\mathbf{J}_b$ and m is the Z component of the angular momentum associated with the emitted photon. Photon radiation with $l = 0$ is forbidden and gamma decay between two states with angular momenta zero cannot take place.

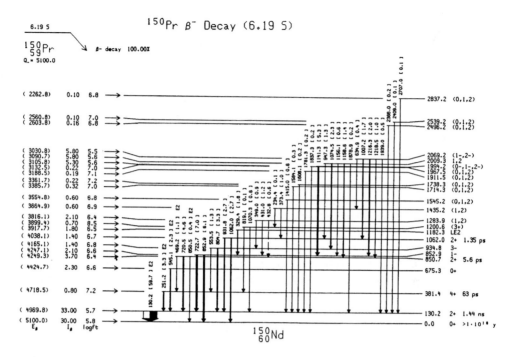

Fig. 2: β decay of ¹⁵⁰Pr to ¹⁵⁰Nd. Gamma decay of excited states of ¹⁵⁰Nd formed in the beta decay of ¹⁵⁰Pr is shown. γ-ray information (energy [intensity] multipolarity) appears above each vertical line representing a specific γ-ray transition. The information to the left of the decay scheme is related to the β decay of ¹⁵⁰Pr. To the right, the three columns give the energy, spin, and parity, and the half-life of the excited energy levels of ¹⁵⁰Nd. In this decay scheme the thickness of a given transition line is related to the relative intensity of that γ ray.

A precise and complete description of the emission (and absorption) of photons by the nucleus is given only in a quantum-mechanical theory of radiation. However, some aspects of the problem may be understood through a comparison of the nucleus to a classical system of charges and currents varying in time and confined to a space small compared to the wavelength of the radiation emitted. The starting point of the classical treatment is Maxwell's equations and it has been found convenient to separate the radiation field of such a classical system into components with definite angular momentum and parity (which see), which process is described as separating the electromagnetic field into its electric and magnetic multipole components. The multipole with $l = 1$ is called dipole radiation; that with $l = 2$, quadrupole radiation. Depending on whether or not parity changes in the transition, the multipole may be electric or magnetic. Multipole radiations of odd l accompanied by a change of parity and of even l with no change of parity are electric. The reverse combinations of angular momentum and parity change give magnetic multipole radiations.

The probability of multipole radiation depends (in the absence of selection rules and other perturbations) on l and the presence of parity change. For example, magnetic dipole radiation ($l = 1$, no parity change) is less probable than electric dipole radiation ($l = 1$, parity change). This is true for all l. Multipole radiation probability rapidly decreases as l increases and increases as the energy of the transition increases. It varies also, but to a lesser degree, with the size of the nucleus. The probability of a transition is frequently given in terms of half-life, which equals 0.693 divided by the probability. The fantastic range of values assumed by the probability for gamma decay is dramatically illustrated by ^{113}Cd, in which one level has a transition probability of $7.7 \times 10^{10} \, \text{s}^{-1}$ and another level has a probability for gamma emission of $3.5 \times 10^{-13} \, \text{s}^{-1}$ (the nucleus in this energy state mostly decays by β-particle emission, the probability for which is $1.6 \times 10^{-9} \, \text{s}^{-1}$). Because of the rapid variation of probability of radiation with l, radiations in gamma decay are most often pure in multipolarity, although radiations that are a mixture of magnetic dipole ($M1$) and electric quadrupole ($E2$) radiations are observed.

In gamma decay the direction in which a photon is emitted is also a function of the multipolarity of the radiation. Normally, an assembly of excited (radioactive) nuclei will be evenly distributed among all directional orientations. If the angular distribution of photons emitted by this array is studied, any inherent anisotropy will be washed out. There are conditions in which the angular dependence of multipole radiations may be observed. For example, nuclei excited through charged-particle bombardment or reactions are often aligned in space, and gamma decay on the part of such an assembly of nuclei can show evidence of a nonisotropic angular distribution. If the gamma-decay scheme of the nuclei under investigation is such that more than one photon is emitted in rapid succession by any member of the assembly, it is not necessary to align the nuclei to see the angular distribution of the emitted photons. If two detectors are used to catch two photons emitted by the same nucleus, the detection of the first photon has the effect of selecting out of the assembly of nuclei those that are aligned in such a fashion as to make the detection of the first photon probable in the first detector. If the second photon is detected simultaneously in the second detector, the probability of observing this event will be a function of the angle between the two detectors. This type of experiment is called a coincidence measurement. These types of experiments have verified the theoretical predictions that radiations accompanying gamma decay have an angular distribution. The angular distribution is the same for electric and magnetic multipole radiation of the same l and m. Other differences, however, do exist that allow one to differentiate between these cases (determination of parity).

See also: Angular Correlation of Nuclear Radiations; Gamma-Ray Spectrometers; Isomeric Nuclei; Multipole Fields; Nuclear Properties; Nuclear States.

Bibliography

R. D. Evans, *The Atomic Nucleus*. Krieger, 1982.(E)
J. M. Blatt and V. F. Weisskopf, *Theoretical Nuclear Physics*. Dover, New York, 1991.(1)

Gamma-Ray Spectrometers

M. P. Carpenter and R. V. F. Janssens

Gamma rays from radioactive nuclei have energies that range from several keV to several MeV. The energies of gamma rays emitted during the course of nuclear reactions can extend up 20 MeV or more, depending on the energies and the properties of the reaction partners involved. Most gamma-ray spectroscopy is focused in this energy range. However, there is a class of spectrometers which measure gamma rays of even higher energy, often to study issues related to astrophysics.

Crystal diffraction gamma-ray spectrometers make use of the wave properties of gamma rays by taking advantage of the constructive interference between amplitudes for elastic scattering from many atoms arranged in a regular crystalline array. Such spectrometers can achieve both high resolution and high precision in the determination of gamma-ray energies. One general drawback to such spectrometers is their low efficiency for detecting gamma rays. As a result, they have been used mostly in instances where the highest energy resolution is of the essence, such as in studies of gamma rays following neutron capture, for example.

All other gamma-ray spectrometers convert the gamma-ray energy into kinetic energy of charged particles, primarily electrons. The principal processes by which this occurs are the photoelectric effect, Compton scattering and pair production. Most spectrometers in use today employ scintillation and/or semiconductor detectors.

In scintillation gamma-ray spectrometers the kinetic energy of electrons is converted into optical and ultraviolet light. The intensity of the scintillation light is proportional to the absorbed electron energy and is measured by the output pulse height of a photomultiplier tube optically connected to the scintillating material. Historically, thallium-doped sodium iodide [NaI(Tl)] is the material most widely used in spectrometers due to its large efficiency for absorbing gamma rays and modest energy resolution (typically 30–40 keV for 662-keV gamma rays). However, newer scintillation materials incorporating high-Z elements are currently in wide use, such as bismuth germanate ($Bi_2Ge_3O_{12}$) and barium fluoride (BaF_2). Bismuth germanate has somewhat poorer energy resolution than sodium iodide, but it is a material of higher density able to stop gamma rays more efficiently. Scintillation from barium fluoride occurs with two distinct time constants, one of which is particularly short (less than 1 ns), making this scintillator especially suitable for applications where gamma rays need to be discriminated from particles through time of flight. Scintillation spectrometers are well suited for routine use where high gamma-ray detection efficiency is preferred over energy resolution, for example in PET scanners used in medical imaging.

In semiconductor spectrometers the kinetic energy of the fast electrons from gamma-ray absorption or scattering is dissipated by the promotion of electrons from below the band to above, and a bias voltage allows the charge in the conduction band to be collected. The energy resolution of semiconductor spectrometers is excellent and on the order of 2 keV for a 1 MeV gamma ray. Both germanium and silicon can be used, but germanium spectrometers are more efficient at detecting gamma rays except for energies below 100 keV. In order to achieve the high energy resolution afforded by germanium detectors, the crystal must be cooled to liquid nitrogen temperature. Progress in growing ultra-pure germanium crystals of large dimensions has been such over the last few decades that it is now possible to manufacture detectors with good efficiency. For example, full absorption efficiencies for modern germanium detectors can now exceed that of a standard 3-in. by 3-in. NaI(Tl) counter.

For many applications in nuclear science research, it is desirable to surround the emitting source with gamma-ray spectrometers covering the entire space (so-called 4p coverage) in order to measure as completely as possible the gamma cascade emitted by the deexciting nucleus. For radioactive decay, the number of emitted gamma rays is typically 1 to 3. For nuclei created in a reaction where an isotopically pure target interacts with an energetic beam of nuclei, the number of emitted gamma rays can range upwards of thirty per reaction. Total absorption spectrometers are typically built out of scintillation detectors and are used to measure the total gamma-ray energy emitted in the decay of a nucleus. In some instances, such spectrometers are segmented into a very large number of scintillation detectors of approximately the same size and efficiency. This segmentation is then used to measure as accurately as possible the total number of gamma rays emitted in each deexcitation, i. e., to determine the gamma-ray multiplicity. Such devices are often referred to as sum-energy-multiplicity spectrometers.

Another example of modern day gamma-ray devices used in nuclear science research is the Compton-suppressed germanium arrays. These arrays employ up to 100 Compton-suppressed germanium spectrometers. Each spectrometer consists of a large germanium detector surrounded by a scintillation counter, usually bismuth germanate. When a gamma ray strikes the germanium detector and a portion of its energy scatters out of the crystal (generally due to Compton scattering) into the scintillator, the event can be vetoed. Such spectrometers greatly reduce the background associated with gamma rays whose total energy is not entirely absorbed by the germanium crystal. Typically, 1 MeV gamma rays deposit their entire energy in a germanium crystal only about 20% of the time. In other words, roughly 80% of the events measured in a germanium detector contribute to the background. In modern day Compton-suppressed spectrometers, the number of 1 MeV events were the full energy is collected is increased to about 65%, thereby enabling the detection of gamma rays with intensities as low as 0.01% that of the strongest photons. These modern arrays of 100 Compton-suppressed detectors have an absolute efficiency for detecting the full energy of 1 MeV photons of about 10%. An example of such an array is the Gammasphere spectrometer which operates at low-energy nuclear accelerator facilities in the United States.

Most recently, there has been interest in electronically segmented germanium detectors. This is accomplished by segmenting the outer electrode of the germanium crystal in the case of cylindrical detectors (the so-called coaxial configuration). In the planar geometry, both the front and back electrodes can be segmented orthogonally to one another producing a double-sided strip detector. The goal of segmentation is to determine the position of the interaction of the photon with the crystal to within a few mm or better. Once this segmentation is accomplished, interactions resulting from Compton scattering of a single gamma ray can be tracked across segmentation or detector boundaries and all interactions can be added together to reconstruct the incident photon energy and its angle of incidence. A spectrometer made out of a large number of segmented germanium detectors could replace the Compton-suppressed spectrometer arrays available today, increasing the efficiency for detecting a 1 MeV gamma ray to around 50%.

For high-energy gamma rays, above a few tens of MeV, the series of events associated with absorption of the energy of a gamma ray in matter necessarily occupies an extended space and becomes a "shower". Large volumes of scintillating material are used in some spectrometers for such high-energy gamma rays. Others use Cerenkov light generated by the electrons and positrons of the shower. In the 100-MeV range, a thin converter followed by magnetic analysis of the generated electron and positron pairs can provide good energy resolution at the cost of high efficiency. An example of a gamma-ray spectrometer for high-energy gamma rays is the EGRET spectrometer which operated in space on the Compton Gamma-Ray Observatory. This spectrometer consisted of a spark chamber for directional information and an array of NaI(Tl) detectors for energy measurements. This spectrometer operated in the energy range 20 MeV to 30 GeV.

In almost all gamma-ray spectrometers, electronic amplification of small signals and ana-log-to-digital conversion play an important role as does the circuitry enabling the high-speed recording and storing of multi-parameter events. In the development of gamma-ray tracking detectors, digitization of the pre-amplifier signals coming directly from the segments of the detector must be undertaken. Low noise, linearity, and efficiency of these electronic systems are crucial for high-quality results.

Bibliography

H. Morinaga and T. Yamasaki, *In-Beam Gamma-Ray Spectroscopy*. North-Holland, Amsterdam, 1976.

P. J. Nolen, F. A. Beck and D. B. Fossan, *Annu. Rev. Nucl. Part. Sci.* **44**, 561–608 (1994).

K. Debertin and R. G. Helmer, *Gamma-and X-ray Spectrometry with Semiconductor Detectors*. North-Holland, Amsterdam, 1988.

G. F. Knoll, *Radiation Detection and Measurement*. Wiley, New York, 1999.

H. Ejiri and M. J. A. de Voigt, *Gamma-Ray and Electron Spectroscopy in Nuclear Physics*. Clarendon Press, Oxford, 1989.

G. Kanbach *et al.*, *Space Sciences Review*, Vol. 49, pp. 69. EGRET Detector, 1988.

Gauge Theories

W. J. Marciano

Gauge theories provide a fundamental description of elementary particles and their interactions. In that framework, forces between particles are mediated by gauge fields associated with an underlying symmetry. A gauge theory based on the symmetry group $SU(3)_C \times SU(2)_L \times U(1)_Y$ accounts for all observed strong, weak, and electromagnetic interaction phenomena. It correctly predicted weak neutral currents as well as the existence and properties of gluons, $W^{\pm}$, and Z gauge bosons. Because of its mathematical elegance, simplicity, and many phenomenological successes, that theory is called the "standard model" of elementary particles, a label that describes its acceptance as a standard against which future discoveries and alternative theories are to be compared.

The concept of local gauge invariance is simply illustrated for a free electrically charged field $\chi(\mathbf{x}, t)$ governed by the (nonrelativistic) Schrödinger equation (with units $\hbar = c = 1$)

$$-\frac{\nabla^2}{2m}\chi(\mathbf{x}, t) = -\mathrm{i}\frac{\partial}{\partial t}\chi(\mathbf{x}, t) \ . \tag{1}$$

The overall phase of a complex field such as χ is not measurable; so Eq. (1) must be and is invariant under global phase transformations of the form $\chi \to e^{ie\theta}$, $0 \le \theta < 2\pi$. (The unit of electric charge, e, is introduced in the exponent, so the phase symmetry can be identified with electric charge conservation.) The set of all such phase transformations form a $U(1)$ group. They are global in that the same phase change is made on $\chi(\mathbf{x},t)$ at each space-time point. If one generalizes that symmetry and requires invariance under the set of local gauge transformations,

$$\chi(\mathbf{x},t) \to e^{ie\theta(\mathbf{x},t)}\chi(\mathbf{x},t) , \tag{2}$$

where $\chi(\mathbf{x},t)$ can vary over space-time points, then Eq. (1) is no longer adequate. It must be modified by the introduction of an electromagnetic potential (gauge field) $(\mathbf{A}(\mathbf{x},t),\phi(\mathbf{x},t))$ such that

$$-\frac{1}{2m}(\nabla - ie\mathbf{A})^2\chi = i\left(\frac{\partial}{\partial t} - ie\phi\right)\chi , \tag{3}$$

where under the local gauge transformation

$$\begin{aligned}
\mathbf{A}(\mathbf{x},t) &\to \mathbf{A}(\mathbf{x},t) + \nabla\theta(\mathbf{x},t) \\
\phi(\mathbf{x},t) &\to \phi(\mathbf{x},t) + \frac{\partial}{\partial t}\theta(\mathbf{x},t) .
\end{aligned} \tag{4}$$

Equation (3) is invariant under the combined transformations in Eqns. (2) and (4). The specific manner in which the electromagnetic potential is introduced into Schrödinger's equation is called minimal coupling. The terminology gauge field and gauge transformation correspond to the freedom of changing one's standard or measurement gauge for the electromagnetic potential via Eqns. (2) and (4). The gauge field $A_\mu \equiv (\mathbf{A},\phi)$ itself satisfies Maxwell's equations which are also invariant under Eq. (4).

The principle of local gauge invariance is one of the fundamental precepts of modern physics. It elevates ordinary global symmetries to space-time-dependent gauge symmetries via the introduction of a gauge field. The validity of that approach is borne out by the existence of 12 known gauge fields with the properties specified by the local gauge invariance prescription of the standard model.

Quantum electrodynamics (QED), the theory of interacting electrons and photons, is based on the relativistic version of the principle of local $U(1)$ gauge invariance. The electron field, $\psi(x)$, $x = (\mathbf{x},t)$, is a four-component Dirac spinor which satisfies the Dirac equation ($\partial_\mu \equiv \partial/\partial x^\mu$)

$$i[\partial_\mu - ieA_\mu(x)]\gamma^\mu\psi(x) = m\psi(x) , \tag{5}$$

where m is the electron mass; γ^μ, $\mu = 0,1,2,3$, are 4×4 Dirac matrices; and the repeated index μ is summed over. Each spinor component of ψ transforms according to Eq. (2) while the gauge field $A_\mu(x)$ satisfies Eq. (4) under local $U(1)$ gauge transformations, thus rendering Eq. (5) invariant.

Second quantizing the $\psi(x)$ and $A_\mu(x)$ fields, they become operators on a Hilbert space and represent spin-$\frac{1}{2}$ electrons and spin-1 photons, respectively. In that full framework, QED provides a relativistic quantum-mechanical description of interacting electrons and photons with the interaction specified by the coupling between $\psi(x)$ and $A_\mu(x)$ in Eq. (5). Despite its simplicity and elegance, there is a problem with QED. Products of operators become ill defined when evaluated at the same space-time point. That results in short-distance (or ultraviolet) divergences in perturbative expansions of QED. Fortunately, QED is renormalizable quantum field theory which means that ultraviolet divergences can be consistently absorbed into the difference between bare (i. e., before turning on interactions) and physical parameters (in this case the electron's mass and charge). After renormalization, high-order perturbative calculations are finite and unambiguously given in terms of e and m. Comparison of such predictions with very high-precision experiments confirms the validity of QED at a level better than 1 part in 10^{11}, making it the best tested theory in physics.

The $U(1)$ symmetry underlying QED is an Abelian (commuting) group. Any two transformations commute, $e^{ie\theta_1}e^{ie\theta_2} = e^{ie\theta_2}e^{ie\theta_1} = e^{ie(\theta_1+\theta_2)}$. As shown by C. N. Yang and R. L. Mills in 1954, the concept of local gauge invariance can be extended to non-Abelian (i. e., noncommuting) symmetry groups such as $SU(N)$, the group of unitary $N \times N$ matrices with determinant 1. For example, the $SU(2)$ isodoublet

$$\psi(x) = \begin{pmatrix} \psi_1(x) \\ \psi_2(x) \end{pmatrix}, \tag{6}$$

where each ψ_i, $i = 1, 2$, is a four-component Dirac spinor and satisfies the free Dirac equation

$$i\partial_\mu\gamma^\mu\psi_i(x) = m\psi_i(x), \tag{7}$$

is invariant under the general set of global $SU(2)$ transformations

$$\begin{aligned} \psi(x) &\rightarrow U\psi(x) \\ U &= \exp(-i\tau \cdot \boldsymbol{\alpha}), \end{aligned} \tag{8}$$

where the a_i, $i = 1, 2, 3$, are real parameters and τ_i are 2×2 Pauli matrices. That symmetry can be (gauged) expanded to local $SU(2)$ gauge invariance with $\boldsymbol{\alpha} \rightarrow \boldsymbol{\alpha}(x)$ by introducing three gauge fields $W_\mu^i(x)$, $i = 1, 2, 3$ [i. e., one for each generator of $SU(2)$], which transform as an $SU(2)$ triplet. They are introduced into Eq. (7), analogous to minimal coupling, by replacing the derivative ∂_μ with a covariant derivative operator

$$\partial_\mu \rightarrow D_\mu = \partial_\mu + ig\frac{\tau}{2} \cdot \mathbf{W}_\mu, \tag{9}$$

where, under $U(x)$, $\mathbf{W}_\mu^i$ transforms as

$$\frac{\tau}{2} \cdot \mathbf{W}_\mu \rightarrow U(x)\frac{\tau}{2} \cdot \mathbf{W}_\mu U^{-1} - \frac{i}{g}U^{-1}(x)\partial_\mu U(x). \tag{10}$$

The coupling g characterizes the strength of the interaction of $\psi(x)$ with the gauge field in analogy with the electric charge, e, of QED. In non-Abelian gauge theories, the fields themselves carry $SU(2)$ charge or are self-coupled because they transform as a nontrivial isovector multiplet. That is to be contrasted with QED where the photon carries no charge and is, therefore, not self-interacting. The generalization to $SU(N)$ symmetries, $N = 3, 4, \ldots$, is straightforward.

$N^2 - 1$ gauge bosons (corresponding to the number of group generators) are introduced via a covariant derivative analogous to Eq. (9).

Like QED, non-Abelian gauge theories are renormalizable; so unambiguous finite perturbative calculations are possible. However, the fact that non-Abelian gauge fields (unlike the photon) are self-interacting makes their dynamical properties very different. The effective coupling g of non-Abelian gauge theories grows at low energies (large distances) and decreases at high energies (short distances). Those two properties find their natural application in quantum chromodynamics (QCD), an $SU(3)$ gauge theory of strong interactions. In that theory, quarks interact by exchanging gluons [non-Abelian $SU(3)$ gauge quanta]. Because their coupling to gluons grows at long distances, quarks are confined within hadrons (infrared slavery). At very short distances, quark couplings decrease and they have much less interaction with one another. The latter property known as asymptotic freedom is well established by high-energy scattering processes.

The standard model of strong, weak, and electromagnetic interactions is based on gauging the symmetry group $SU(3)_C \times SU(2)_L \times U(1)_Y$ where the subscripts denote special features of a given symmetry. The symmetries act on quark and lepton fields. (Leptons are electrons, muons, neutrinos, etc.) The C in $SU(3)_C$ stands for color. Each quark has three color components and $SU(3)_C$ transforms them into one another. $SU(3)_C$, the basis of quantum chromodynamics, is an exact symmetry of nature. There are eight massless gluons which correspond to the eight gauge fields of $SU(3)_C$.

The L on $SU(2)_L$, the weak isospin group, denotes the fact that only left-handed components, ψ_L, of spinor fields,

$$\psi_L \equiv \frac{1 - \gamma_5}{2} \psi , \tag{11}$$

transform as doublets under that group. Right-handed spinor components $\psi_R \equiv \frac{1}{2}(1 + \gamma_5)\psi$ are isosinglets under $SU(2)_L$, i.e., they are unchanged under $SU(2)_L$ transformations and therefore do not couple to its three gauge fields which we denote by $W_\mu^\pm \equiv (W_\mu^1 \mp W_\mu^2)/\sqrt{2}$ and W_μ^3. The fact that only left-handed quarks and leptons couple to those gauge fields makes their (weak) interactions maximally parity violating.

The Y on $U(1)_Y$ stands for weak hypercharge, the charge associated with that Abelian group. That gauge group has one gauge field B_μ which couples to quarks and leptons via their weak hypercharge Y. It couples to left- and right-handed components of those particles differently and therefore also violates parity, but not maximally.

The $SU(2)_L \times U(1)_Y$ part of the standard model is not an exact symmetry. If it were, the $W^\pm$, W^3, and B would all be massless gauge bosons. That is not the case. To accommodate electroweak phenomenology, a scalar (spin 0) field is introduced which breaks the symmetry $SU(2)_L \times U(1)_Y$ down to the $U(1)$ symmetry of QED. That breaking gives mass to the $W^\pm$ and the combination of fields

$$Z_\mu = W_\mu^3 \cos\theta_W - B_\mu \sin\theta_W \tag{12}$$

which is called the Z boson. The orthogonal combination

$$A_\mu = B_\mu \cos\theta_W + W_\mu^3 \sin\theta_W \tag{13}$$

remains massless and is identified as the photon. The angle θ_W, called the weak mixing

Table 1: Elementary particles.

Particle	Symbol	Spin	Charge	Color	Mass (GeV)	
Electron neutrino	ν_e	$\frac{1}{2}$	0	0	$< 1 \times 10^{-9}$	
Electron	e	$\frac{1}{2}$	-1	0	0.51×10^{-3}	1st
Up quark	u	$\frac{1}{2}$	$\frac{2}{3}$	3	5×10^{-3}	generation
Down quark	d	$\frac{1}{2}$	$-\frac{1}{3}$	3	9×10^{-3}	
Muon neutrino	ν_μ	$\frac{1}{2}$	0	0	$< 0.25 \times 10^{-3}$	
Muon	μ	$\frac{1}{2}$	-1	0	0.106	2nd
Charm quark	c	$\frac{1}{2}$	$\frac{2}{3}$	3	1.25	generation
Strange quark	s	$\frac{1}{2}$	$-\frac{1}{3}$	3	0.175	
Tau neutrino	ν_τ	$\frac{1}{2}$	0	0	< 0.035	
Tau	τ	$\frac{1}{2}$	-1	0	1.78	3rd
Top quark	t	$\frac{1}{2}$	$\frac{2}{3}$	3	178	generation
Bottom quark	b	$\frac{1}{2}$	$-\frac{1}{3}$	3	4.5	
Photon	γ	1	0	0	0	
W boson	$W^\pm$	1	± 1	0	80.43	Gauge
Z boson	Z	1	0	0	91.1875	bosons
Gluon	g	1	0	8	0	
Higgs scalar	H	0	0	0	$114 < m_H < 250$	

angle, is experimentally found to be $\sin^2 \theta_W = 0.23$. That leads to the standard model's predictions $m_W \simeq 80\,\text{GeV}$ and $m_Z \simeq 91\,\text{GeV}$ for $W^\pm$ and Z boson masses. Those values have been confirmed by measurements at high-energy accelerators.

The full particle spectrum of the standard model is illustrated in Table 1. All 12 gauge bosons as well as all quarks and leptons have been observed. Only the Higgs scalar, a remnant of the symmetry breakdown, remains to be discovered. Its mass is presently unknown, but some theoretical and experimental bounds do exist (see Table 1).

Attempts have been made to embed the standard model into a larger compact simple symmetry group such as $SU(5)$, $SO(10)$, and E_6. Such models are called grand unified theories (GUTs). The idea of a simple unified gauge theory is mathematically appealing; but, so far, there is no direct experimental support for any GUT. A fairly generic prediction of GUTs is that the proton should be unstable and decay, albeit with a very long lifetime. Experiments have searched for proton decay with negative findings and now give a bound on its lifetime $\tau_p > 10^{34}$ year.

See also: Electromagnetic Interaction; Elementary Particles in Physics; Field Theory, Unified; Grand Unified Theories; Quantum Field Theory; Quantum Mechanics; Weak Interactions.

Bibliography
E. S. Abers and B. W. Lee, "Gauge Theories", *Phys. Rep.* **9**, 1 (1973).
W. Marciano and H. Pagels, "Quantum Chromodynamics", *Phys. Rep.* **36C**, 137 (1978).
J. C. Taylor, *Gauge Theories of Weak Interactions.* Cambridge, London and New York, 1976.

Gauss's Law

R. W. Brown

In its narrowest definition, Gauss's law refers to the integral form of a Maxwell equation in electrodynamics: The electric flux through a closed surface S of any volume V is proportional to the electric charge contained in that volume. Generalizations include a result from macroscopic averaging, an "electroweak" extension, and "non-Abelian" gauge theory analysis.

The law in electrodynamics is

$$\int_S \mathbf{E} \cdot \hat{\mathbf{n}} \, dS = K \int_V \rho \, dV \ . \tag{1}$$

In this equation, $\mathbf{E}$ is the electric field, ρ is the charge density, and $\hat{\mathbf{n}}$ is a unit vector normal to the surface at the point of integration, pointing away from the enclosed volume. The proportionality constant K depends on the system of electromagnetic units. In Gaussian units, $K = 4\pi$. In another popular system, rationalized mks units, $K = \varepsilon_0^{-1}$, where $(4\pi\varepsilon_0)^{-1} = 10^{-7} c^2$ (c is the speed of light).

The left-hand side of Eq. (1) can be taken as a definition of the flux through S associated with a vector field (a vector whose direction and magnitude may change in space). If we imagine drawing lines in space denoting the direction of $\mathbf{E}$ at any point with the density of such lines proportional to the magnitude $|E|$, then the flux through S is proportional to the number of lines crossing S. Historically, one thought of the flow of something through S, a picture which helps us to introduce similar statements about other physical phenomena. Equations analogous to Eq. (1) can be developed for heat conduction, gravity, water flow, particle diffusion, quantum-mechanical probability, electric current conservation, and so forth. Aside from sources or sinks (charges, heat generators, masses, etc.), the common denominator is a conservation law prohibiting the disappearance of the "stuff" flowing out or in.

The point is that Eq. (1) implies that static electric field lines cannot end outside of the source region and will spread according to the familar inverse-square law, which brings to mind the usual derivation of Eq. (1) for electrostatics. Consider the Coulomb field of an electric point charge q at rest:

$$\mathbf{E} = \frac{Kq}{4\pi} \frac{\hat{\mathbf{r}}}{r^2} \ . \tag{2}$$

The fact that this field is radial (central) and drops off like r^{-2} leads to

$$\int_S \mathbf{E} \cdot \hat{\mathbf{n}} \, dS = 0 \tag{3}$$

if the charge is outside V. That is, contributions to the integral from opposite areas on S subtending the same solid angle at the point charge cancel. (Observe that $dS = r^2 d\Omega / |\cos\theta|$, where θ is the angle between $\hat{\mathbf{n}}$ and $\mathbf{E}$.) Now suppose S consists of disconnected inner and outer surfaces with V the volume between them and the charge inside the inner surface [still outside V so Eq. (3) is still applicable]. If the inner surface is a sphere centered around the charge, the corresponding inner contribution to Eq. (3) is $-Kq$ [see Eq. (2)]. By linear superposition of the fields due to an assemblage of charges, Gauss's law (1) follows.

We can convert Gauss's law into differential form. What is needed is the divergence theorem, sometimes called Gauss's theorem, which reads

$$\int_S \mathbf{F} \cdot \hat{\mathbf{n}} \, dS = \int_V \mathbf{\nabla} \cdot \mathbf{F} \, dV \tag{4}$$

for any suitable vector field $\mathbf{F}$. Since this is an integral identity, the class of vectors $\mathbf{F}$ can be generalized to include those whose derivatives are distributions or generalized functions, which actually arise in $\mathbf{\nabla} \cdot \mathbf{E}$ for a point charge. In fact, Eqns. (2) and (4) imply the identification

$$\mathbf{\nabla} \cdot \mathbf{E} = Kq\delta(\mathbf{r}) \tag{5}$$

in terms of the Dirac delta function. The derivation of Eq. (4), in which partial integration of the right-hand side is the principal step, shows that this theorem applies to an arbitrary number of dimensions with S a hypersurface of V; this is relevant to conservation laws and the investigation of conserved "charges" in four-dimensional space-time. Getting back to Gauss's law, Eqns. (1) and (4) can be combined to give

$$\mathbf{\nabla} \cdot \mathbf{E} = K\rho \tag{6}$$

by considering arbitrarily small volumes. This is the promised differential form.

Equation (6) is an equivalent formulation of Gauss's law and one of the Maxwell differential equations. It must be emphasized that no change is required when time-dependent fields and sources are considered. We can use Eq. (1) even if the charges are moving relativistically; the field is to be evaluated at all points on S and the density at all points inside V at the same time in a given Lorentz frame of reference.

Furthermore, the full quantum theory of electrodynamics assumes the correctness of Gauss's law where the electric field and the charge density are now operators in a Hilbert space. Therefore this law is a basic "equation of motion" which governs electromagnetism from microscopic distances (checked down to 10^{-17} cm for electrons and photons) to intergalactic plasmas. It is interesting that Gauss's law is a statement about the masslessness of photons, the electromagnetic quanta. If the photon had a mass λ, the Coulomb potential would be modified by a factor $e^{-\lambda cr/\hbar}$, where $\hbar$ is Planck's constant (essentially a Yukawa potential). Flux lines could disappear as we go away from the sources and Eq. (1) would not be valid.

In fact, generalizations are characterized by nonvanishing divergence. In electroweak theory, there is a partner to the photon, the neutral Z boson, whose huge mass most definitely changes things. In non-Abelian gauge theories, the fields carry "charge". In macroscopic media, the polarization charge is absorbed into a displacement vector $\mathbf{D}$, hiding the fact that $\mathbf{\nabla} \cdot \mathbf{E}$ may not be zero.

See also: Electrodynamics, Classical; Maxwell's Equations.

Bibliography
R. P. Feynman, R. B. Leighton, and M. Sands, *The Feynman Lectures on Physics*, Vol. II. Addison-Wesley, Reading, Mass., 1964.(E)
C. Itzykson and J.-B. Zuber, *Quantum Field Theory*. McGraw–Hill, New York, 1980. (A)
J. D. Jackson, *Classical Electrodynamics*, 2nd ed. Wiley, New York, 1975. (I)

Geochronology
M. A. Lanphere

Introduction

Geochronology is the study of time in relation to the history of the earth. The term includes *relative* dating systems based on fossils and *physical* (sometimes called absolute) dating systems based on radioactive decay. These physical dating methods, collectively called radiometric dating, are based on decay of naturally occurring radioactive nuclides in rocks and minerals.

In 1896 the French physicist Henri Becquerel discovered that uranium salts emitted radiation that had properties similar to x rays. This spontaneous emission of radiation was termed "radioactivity", and this property is the basis for all radiometric dating methods. In the next few years many investigators provided evidence on the manner in which one element is produced from another by radioactive decay. In 1905 Ernest Rutherford suggested that the rate of transformation of radioactive nuclides might be used to determine the age of geologic materials. Rutherford in 1906 determined the first mineral ages using the uranium and helium contents of minerals and the disintegration rate of radium, a decay product of uranium. Bertram Boltwood suggested that lead was the final product of the radioactive decay of uranium, and in 1907 he published uranium-lead ages for several minerals. These ages were calculated before the disintegration rate of uranium was accurately known, before isotopes were discovered, and before it was discovered that lead also is a product of the radioactive decay of thorium. But it was clear even then that the phenomenon of natural radioactivity offered great potential in constructing a quantitative scale of geologic time.

Principles of Radiometric Dating

When a radioactive parent atom decays to produce a stable daughter atom, each disintegration results in one more atom of daughter and one less atom of parent. Several workers in the early 1900s conducted experiments on uranium and thorium salts that led to construction of a general theory to explain radioactivity. It was suggested that atoms of radioactive elements are unstable and a fixed proportion of the atoms spontaneously disintegrate in a given period of time to form atoms of a new element. It was further proposed that the activity or intensity of radiation is proportional to the number of atoms that disintegrate per unit time. The probability of a parent atom decaying in a given period of time is the same regardless of temperature, pressure, or chemical conditions; this probability of decay is called the decay constant. Thus,

Table 1: Parent and daughter isotopes used in radiometric dating.

Parent isotope	Daughter isotope	Half-life/a
Carbon-14 (^{14}C)	Nitrogen-14 (^{14}N)	5.73×10^3
Potassium-40 (^{40}K)	Argon-40 (^{40}Ar)	1.25×10^9
Rubidium-87 (^{87}Rb)	Strontium-87 (^{87}Sr)	4.88×10^{10}
Samarium-147 (^{147}Sm)	Neodymium-143 (^{143}Nd)	1.06×10^{11}
Lutetium-176 (^{176}Lu)	Hafnium-176 (^{176}Hf)	3.59×10^{10}
Rhenium-187 (^{187}Re)	Osmium-187 (^{187}Os)	4.3×10^{10}
Thorium-232 (^{232}Th)	Lead-208 (^{208}Pb)	1.40×10^{10}
Uranium-235 (^{235}U)	Lead-207 (^{207}Pb)	7.04×10^8
Uranium-238 (^{238}U)	Lead-206 (^{206}Pb)	4.47×10^9

if one knows the decay constant and can measure the amounts of parent and daughter isotopes in a mineral, then the time since the mineral formed can be calculated. Another measure of the rate of radioactive decay is the half-life which is inversely proportional to the decay constant; the half-life is the time required for one-half of the number of radioactive parent atoms to decay.

A number of long-lived radioactive isotopes have been used in radiometric dating; some of the parent and daughter nuclides and half-lives are shown in Table 1. Some radioactive parent nuclides decay directly to stable daughter nuclides in a process known as simple decay. The radioactive uranium and thorium parents decay to radioactive daughter products that also decay, and so forth, until stable lead isotopes are produced in a process known as chain decay. ^{40}K decays to two different daughter elements in a process known as branching decay. All of the decay schemes in Table 1, except radiocarbon, are based on accumulation of daughter nuclides. The radiocarbon (^{14}C) method is based on disappearance of the parent nuclide by radioactive decay.

Key elements in radiometric dating are knowing the decay constants accurately and demonstrating that the rate of decay is constant. The decay constants for radioactive elements listed in Table 1 are known with an accuracy of a few percent from laboratory experiments involving counting the number of particles emitted per unit time by a known quantity of parent nuclide. Because radioactive decay occurs spontaneously in the nucleus of an atom, the decay rate should not be affected by physical or chemical conditions. Many attempts have been made to change radioactive decay rates, but these have not produced any significant changes. A type of decay known as electron capture, in which an orbital electron falls into the nucleus, is affected very slightly by external conditions because the process involves a particle outside the nucleus. Measurements have shown, however, that physical and chemical conditions have no significant effects on the decay constant of isotopes which decay by electron capture.

There are several fundamental conditions that must be met before a radioactive decay scheme can be used to successfully measure ages of geologic material. These are:

1. The decay of a radioactive parent takes place at a constant rate that is accurately known.

2. The present-day proportion of a radioactive parent nuclide to the total quantity of the same element is the same in all materials. There is no natural fractionation of isotopes of the same element.

3. A rock or mineral may contain nonradiogenic nuclides of the same mass as the radiogenic daughter; in this case a correction must be made for the nonradiogenic nuclides present when the rock or mineral formed.

4. The rock or mineral has been a closed system since the time of formation. That is, there has been no gain or loss of radioactive parent or radiogenic daughter.

Radiometric Dating Methods

This review is too short to describe all of the dating methods in use. Only the three most commonly used methods will be described. These are the potassium–argon (K–Ar), uranium–lead (U–Pb), and radiocarbon (^{14}C) methods. These methods have been applied to problems covering the entire span of geological time from the present to the formation of Earth, moon, and meteorites.

The K–Ar Method

The K–Ar method is the radiometric dating technique used most widely by geologists. Potassium is a common element found in many minerals and the half-life of ^{40}K has a value that permits application to very young as well as ancient rocks. ^{40}K has a branching decay to produce the daughter products ^{40}Ca and ^{40}Ar. ^{40}Ca is the most abundant isotope of calcium which also is very abundant in the Earth's crust. Thus, it usually is not possible to correct for the ^{40}Ca initially present and the ^{40}K–^{40}Ca decay is rarely used for dating, except with potassium-rich, calcium-poor salts. ^{40}K also decays to ^{40}Ar, a noble gas, and a K–Ar age commonly is referred to as a gas-retention age. The K–Ar method is used mostly on igneous rocks, such as lava or granite, where generally there is no initial radiogenic ^{40}Ar. While a rock is molten, any radiogenic ^{40}Ar will escape from the liquid, but as the rock solidifies and cools the radiogenic ^{40}Ar is trapped within mineral grains and accumulates as time passes. However, argon makes up approximately 1% of the Earth's atmosphere, and this nonradiogenic argon is present in most minerals and the laboratory apparatus in which samples are processed. The correction for this nonradiogenic argon is easily made and, unless the atmospheric argon is a large proportion of the total argon, the calculated age is not adversely affected.

For the conventional K–Ar method, the quantities of potassium and argon in a sample are measured in separate experiments by different techniques. In a variant of K–Ar dating known as the ^{40}Ar/^{39}Ar method the sample is irradiated in a nuclear reactor in order to convert part of the ^{39}K, which is the most abundant isotope of potassium, to ^{39}Ar. Then the ratio of daughter to parent is given by the ratio of ^{40}Ar to ^{39}Ar. Corrections must be made for atmospheric argon and interfering isotopes produced by undesirable neutron reactions with calcium and potassium. If the irradiated sample is totally melted in one experiment, the method gives ages with precisions comparable to conventional K–Ar ages.

An irradiated sample also can be heated in increments to progressively higher temperature and the gas released at each temperature treated as a separate experiment. Data from these experiments are plotted as an age spectrum (apparent age versus percentage of ^{39}Ar released). From the shape of the spectrum one can infer whether the sample is undisturbed, has suffered argon loss since crystallization, or contains excess (initial) argon. Although many age spectra from rocks with complex histories cannot yet be interpreted, this new technique has greatly expanded the usefulness of the K–Ar method.

Conventional K–Ar ages younger than five thousand years have been measured on lavas; these ages overlap the range covered by ^{14}C ages. ^{40}Ar/^{39}Ar ages older than four billion years have been measured on lunar rocks and meteorites. The ability to apply the method to the entire span of geologic time underscores the importance of the K–Ar dating method.

The U–Pb Method

The radioactive decays of ^{238}U to ^{206}Pb and ^{235}U to ^{207}Pb yield two independent ages. The decay of ^{232}Th to ^{208}Pb gives a third independent age. An age also can be calculated from the ^{207}Pb/^{206}Pb ratio since this ratio changes as a function of time because the half-lives of the two parent uranium isotopes differ by a factor of about 6. If these four ages agree, then this represents the age of the rock. However, the ages often do not agree because lead is often lost from minerals whose crystal structures have been damaged by radioactive decay. Some graphical and numerical procedures have been developed that permit interpretation of discordant U–Pb ages.

On a U–Pb parent-daughter diagram the locus of concordant ^{206}Pb/^{238}U and ^{207}Pb/^{235}U ages is a curve called "concordia". However, most rock-forming minerals yield U–Pb ages that are discordant to some degree and do not plot on concordia. This is true for zircon, an accessory mineral in igneous rocks which contains uranium when it crystallizes but very little lead. The ages for a suite of zircon samples which have lost different amounts of lead generally form a linear array, called discordia, below the concordia curve. A chord drawn through this array will intersect concordia in two places. The older intercept is considered the original age of the rock. The younger intercept may be interpreted in several ways, but usually it indicates the time at which lead loss occurred. The U–Pb concordia–discordia system is quite resistant to heating and is the dating method of choice in rocks with complex histories.

The Radiocarbon Method

Radiocarbon (^{14}C) dating is quite different from other radiometric dating methods because it depends on disappearance of the parent isotope rather than accumulation of the daughter isotope. The ^{14}C method is used on organic material, and its half-life of 5730 years restricts the method's use to about 50 000 years.

^{14}C is produced in the upper atmosphere by a number of nuclear reactions, the most important of which is the interaction of cosmic ray neutrons and ^{14}N. ^{14}C is radioactive and decays back to ^{14}N. The ^{14}C produced in the atmosphere is incorporated into organic material, and as long as the organism lives, the ^{14}C in the organism and the atmospheric reservoir remain in equilibrium. When the organism dies, the ^{14}C begins to disappear with time because it ceases to be replenished. The ^{14}C age of a sample is given by the amount of ^{14}C left compared to the amount of ^{14}C in a modern sample of known age.

For the ^{14}C dating method to work requires a constant inventory of ^{14}C in the atmosphere. It is known, however, that the production rate of ^{14}C has not been constant with time because of variations in the cosmic ray flux in the upper atmosphere. Deviations from a constant production rate have been documented by dating tree rings from certain long-lived species. This detailed dendrochronology has been determined back to 5400 b.c. This calibration curve is used to adjust measured ages for variations in the ^{14}C production rate. The accuracy of ages older than 5400 b.c. may have uncertainties of $\pm 10\%$ or so. However, the relative differences between older ^{14}C ages are not affected by production rate variation.

Conclusions

Radiometric dating now is a standard technique in most geological investigations. The use of radioactive decay to measure elapsed time has been applied to problems in the earth sciences since early in the century. However, accurate measurement of decay constants, development of analytical techniques, and availability of adequate instrumentation dates mostly from about 1950. During the past four decades technique development has permitted the measurement of ages with analytical precisions of better than ±0.5% for several of the dating methods. During the past 10 years techniques for measuring K–Ar and U–Pb ages on single mineral grains have been developed. The future promises continued improvements and extension of radiometric dating to a broad range of geological and geophysical problems.

See also: Cosmic Rays, Solar System Effects; Geophysics; Radioactivity.

Bibliography

G. B. Dalrymple and M. A. Lanphere, *Potassium–Argon Dating*. Freeman, San Francisco, 1969. (I)
G. Faure, *Principles of Isotope Geology*. John Wiley & Sons, New York, 1986. (A)
D. York and R. M. Farquhar, *The Earth's Age and Geochronology*. Pergamon Press, London, New York, 1972. (E)

Geomagnetism

J. A. Jacobs[†] and D. Gubbins

At its strongest near the poles, the Earth's magnetic field is several hundred times weaker than between the poles of a toy horseshoe magnet – being less than $100\,\mu T$. Thus in geomagnetism we are measuring extremely small magnetic fields; a convenient unit is the nanoTesla (nT).

In a magnetic compass the needle is weighted so that it will swing in a horizontal plane, and its deviation from geographic north is called the declination D. Over most of the northern hemisphere the north-seeking end of a magnetized needle balanced horizontally on a pivot will dip downward, the angle it makes with the horizontal being called the magnetic dip or inclination I. Over most of the southern hemisphere, the north-seeking end of the needle points upward.

At points on the Earth's surface where the horizontal component of the magnetic field vanishes, a dip needle will rest with its axis vertical. Such points are called dip poles. Two principal poles of this kind are situated near the north and south geographic poles and are called the magnetic north and south poles. Their positions in 1980 were 77.3° N, 258.2° E and 65.6° S, 139.4° E. They are thus not diametrically opposite, each being about 2500 km from the point antipodal to the other. The total intensity F of the Earth's magnetic field is a maximum near the magnetic poles – its value is just over $60\,\mu T$ near the northern dip pole and just over $70\,\mu T$ near the southern dip pole. In some areas such as Kursk, south of Moscow, and Berggiesshubel in Germany, the magnitude of F may exceed $300\,\mu T$, but this is entirely due to local concentrations of magnetic ore bodies.

[†]deceased

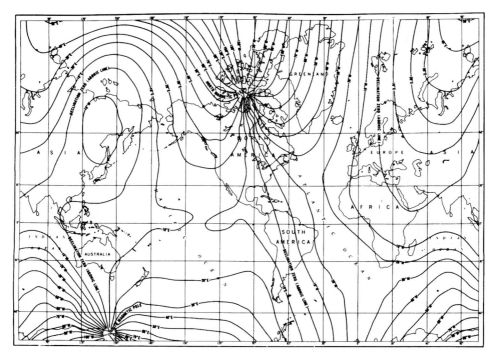

Fig. 1: Simplified isogonic chart of the entire world except polar regions, for 1955. Mercator projection. (Derived from U.S. Navy Hydrographic Office Chart 1706.)

The variation of the magnetic field over the Earth's surface is best illustrated by isomagnetic charts, i. e., maps on which lines are drawn through points at which a given magnetic element has the same value. Figures 1 and 2 are world maps showing lines of equal declination (isogonics) and lines of equal inclination (isoclinics) for the year 1955.

In 1839 Gauss showed that the field of a uniformly magnetized sphere (which is the same as that of a geocentric dipole) is an excellent first approximation to the Earth's magnetic field. This was nearly 250 years after William Gilbert had reached the same conclusion from experimental studies on the variation in direction of the magnetic force over the surface of a piece of the naturally magnetized mineral lodestone which he had cut in the shape of a sphere. The points where the axis of the geocentric dipole which best approximates the Earth's field meet the surface of the Earth are called the geomagnetic poles and are situated approximately at 79° N, 289° E and 79° S, 109° E. The geomagnetic axis is thus inclined at about 11° to the Earth's geographic axis.

In 1635 Gellibrand discovered that the magnetic declination changed with time. This change in the magnetic field with time is called the secular variation and is observed in all magnetic elements. A spherical harmonic analysis of the Earth's surface magnetic field shows that its source is predominantly internal. Superimposed on this field is a rapidly varying external field giving rise to transient fluctuations. Unlike the secular variation, which is also of internal origin, these transient fluctuations produce no large or enduring changes in the

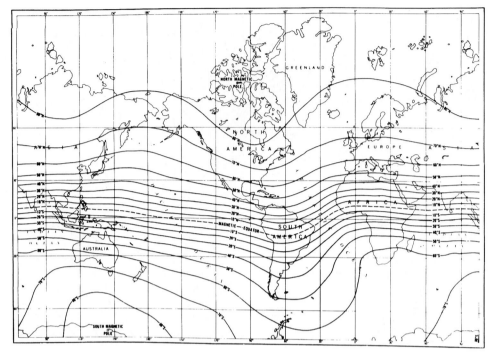

Fig. 2: Simplified isoclinic chart of the entire world except polar regions, for
1955. Mercator projection. (Derived from U.S. Navy Hydrographic Office
Chart 1700.)

Earth's field. They are mostly due to solar effects which disturb the ionosphere and give rise
to a number of related upper atmospheric phenomena such as magnetic storms and aurora.

The Earth's main field is generated by a dynamo acting in the liquid iron core. Rotation,
through the Coriolis force, provides the organisation to the flow required for it to behave in
this way, as reflected in the alignment of the magnetic field with the Earth's spin axis. Early
theoretical results, such as Cowling's theorem of 1933 that prohibited dynamo generation of
a magnetic field symmetric about an axis, cast doubt on the theory but it is now thoroughly
established that most sufficiently vigorous and sufficiently complex flows are capable of generating magnetic fields through a dynamo instability. It has recently become possible to solve
the equations of magnetohydrodynamics in a sphere to simulate the Earth's dynamo. These
numerical dynamos, while falling short of matching all the parameters of the geodynamo, do
reproduce many aspects of the Earth's magnetic field, including polarity reversals and secular
variation.

The energy needed to drive the geodynamo originates in a variety of forms (gravitational,
chemical, and thermal); ultimately it is converted into heat and flows into the mantle. The
efficiency of any heat-driven dynamo is only 5–10%. Latent heat is released as the Earth cools
and the liquid outer core freezes to enlarge the solid inner core. Light elements in the outer
core separate during this process and release gravitational energy, producing a large boost
to the power supply to the dynamo. Favoured light elements are S, Si, O. Recent quantum

mechanical calculations on liquid iron mixtures suggest S and Si form solid solutions in the inner core whereas O does not form solid solution and remains in the liquid.

Frequent polarity reversal is one of the most interesting results of paleomagnetic studies. Reverse magnetization was first discovered in 1906 by Brunhes in France and examples have since been found in every part of the world. Over the last 45 Myr the field has reversed at least 150 times. Furthermore, there have been many excursions or failed reversals since the last full reversal 750 thousand years ago; it seems the geodynamo has attempted to reverse 10 or 20 times in the present interval of normal polarity, achieving large movements of the magnetic poles and reduction of field strength before reverting to the present configuration.

The frequency of reversals has however changed throughout geologic time, and there were times when there were almost no reversals; e. g., during the Cretaceous (from about 107 to 85 Myr ago) the field remained normal and during the upper Carboniferous and Permian (from about 290 to 235 Myr ago) the field remained reversed with but few changes in polarity. These long-term changes in reversal frequency are attributed to changes in the solid overlying mantle: nothing in the core possesses such a long timescale.

Reversals of the Earth's field have played a key role in the development of plate tectonics, which has revolutionized geologic thinking. As new oceanic crust forms at the center of an oceanic ridge and cools through its Curie temperature, the permanent component of its magnetization will assume the ambient direction of the Earth's magnetic field. The striped pattern of magnetic anomalies observed in the ocean basins is thus due to seafloor spreading as the upwelling mantle material is carried away from the ridge in both directions.

See also: Geophysics.

Bibliography

D. Gubbins and E. Herrero-Bervera, *Encyclopedia of Geomagnetism and Paleomagnetism*. Springer, Dordrecht, to appear 2006. This reference work is dedicated to the late John Arthur Jacobs, the original author of the present article, who passed away after shortly contributing two articles for it.

J. A. Jacobs, *Reversals of the Earth's magnetic field*, 2nd ed. Cambridge University Press, 1994.

R. T. Merrill, M. W. McElhinny and P. L. McFadden, *The Magnetic Field of the Earth*. Academic Press, New York, 1996.

W. D. Parkinson, *Introduction to Geomagnetism*. Scottish Academic Press, 1983.

Geometric Quantum Phase

J. H. Hannay

The geometric quantum phase, or "Berry phase", results from a fundamental feature of quantum states: while the phase of $\langle a|b\rangle\langle b|a\rangle$ is zero, the phase of $\langle a|c\rangle\langle c|b\rangle\langle b|a\rangle$ is *not* zero. Moreover, because each of the states occurs twice, it is independent of their individual, arbitrary, phases. The Berry phase is defined not for this discrete chain of states, but for any *continuous* cyclic chain of states $|n(\mathbf{R})\rangle$, labeled by their position $\mathbf{R}$ in parameter space, as the phase of the product $\langle n(\mathbf{R}_0)|n(\mathbf{R}_\infty)\rangle \ldots \langle n(\mathbf{R}_2)|n(\mathbf{R}_1)\rangle\langle n(\mathbf{R}_1)|n(\mathbf{R}_0)\rangle$.

Each term is of the form $\langle n+dn|n\rangle = 1 + \langle dn|n\rangle = \exp(id\gamma)$ in the limit $|dn\rangle \to 0$, where $d\gamma = -i\langle dn|n\rangle = i\langle n|dn\rangle$ (using $d\langle n|n\rangle = 0$). So the Berry phase is $\gamma = \int d\gamma = i\int\langle n|dn\rangle$. Here $\langle n|dn\rangle$ means $\langle n|\nabla n\rangle \cdot d\mathbf{R}$, or it can be read directly as a 1-form: the "phase 1-form" in parameter space.

Schematically, if the parameter space is thought of as the surface of a sphere, the cycle of states could be represented as a loop of points on it, with the (arbitrary) phase of each state indicated by the direction of a little arrow in the surface attached to its representative point. The change of direction between neighboring arrows (the angle between one and its *parallel transported* neighbor) could be likened to the phase of $\langle n(\mathbf{R}+d\mathbf{R})|n(\mathbf{R})\rangle$. Just as the sum of these direction increments around the loop is nonzero for a curved surface, so also the phases add up around the cycle to give a nonzero sum. In both these "holonomy" effects, the sum captures a fundamental quantity enclosed by the loop. On the one hand, the direction change captures the integrated curvature of the surface (the integrated flux of the curvature 2-form). On the other hand, the Berry phase captures, by Stokes's theorem, the integrated flux of a "curl": $\nabla \wedge \langle n|\nabla n\rangle = \langle \nabla n| \wedge |\nabla n\rangle$, or more correctly, of a 2-form $d\langle n|dn\rangle = \langle dn| \wedge |dn\rangle$. Unlike the phase 1-form, the phase 2-form does *not* depend on the arbitrary choice of phases of the states $|n\rangle$. For the simplest circumstance of a quantum system with only two basis states, the sphere construction above is not merely a schematic description but a quantitative one (the sphere being the "Bloch sphere", or "Poincaré sphere" in optics, whose geometric phase was anticipated by Pancharatnam).

The Berry phase is realized physically by adiabatic (i.e., infinitely slow) cyclic change of a quantum Hamiltonian $H(\mathbf{R}(t))$. The quantum adiabatic theorem guarantees that if a system starts in an eigenstate of this Hamiltonian, the nth say, then it will stay in the nth, making no transitions to others on the way. The phases of the initial and final states, however, are not the same. During the excursion the phase changes at a rate $-\hbar^{-1}$ times the energy, $E_n(\mathbf{R}(t))$, of the nth eigenstate of $H(\mathbf{R}(t))$ so there is a "dynamical" phase change given by the integral of this over the duration of the excursion: $-\hbar^{-1}\int E_n(\mathbf{R}(t))\,dt$. If the Hamiltonian, and therefore the eigenstates, are purely real, this is the only phase change apart from a possible change of sign (see next paragraph). However for a general (Hermitian) Hamiltonian there is an extra phase change, the Berry phase, due to the cycle that $|n(R(t))\rangle$ has executed. The derivation is straightforward: the instantaneous state, $|\psi(t)\rangle$ is given by $e^{i\chi(t)}|n(\mathbf{R}(t))\rangle$, where $\chi(t)$ is to be determined. Substitution into the Schrödinger equation $H|\psi\rangle = i\hbar|\dot\psi\rangle$ and use of $H|n\rangle = E_n|n\rangle$ yields $E_n e^{i\chi}|n\rangle = i\hbar[i\dot\chi e^{i\chi}|n\rangle + e^{i\chi}|\nabla n\rangle\cdot\dot{\mathbf{R}}]$. Premultiplying by $\langle n|$ gives $\dot\chi = -E_n/\hbar + i\langle n|\nabla n\rangle\cdot\dot{\mathbf{R}}$. The time integral of the second term is recognized as the Berry phase; "geometrical" because it only depends on the sequence of parameters passed through, not on the duration of each.

In this adiabatic realization of the geometrical phase, the phase 2-form $\langle dn| \wedge |dn\rangle$ can be expressed as a sum over states

$$\langle dn| \wedge |dn\rangle = \sum \langle dn|m\rangle \wedge \langle m|dn\rangle = \sum \langle n|dH|m\rangle \wedge \langle m|dH|n\rangle/(E_n - E_m)^2\,,$$

as can be verified by differentiating the relations $\langle n|H|m\rangle = 0$, $\langle n|n\rangle = 1$, and using $H|n\rangle = E_n|n\rangle$. This expression shows that the 2-form is singular near "accidental" degeneracy points $\mathbf{R}$, generic in parameter space of 3-D (or more), where $E_m(\mathbf{R})$ contacts a neighboring energy $E_m(\mathbf{R})$ ($m = n \pm 1$). Local analysis shows it to take an inverse square form (like the magnetic field a magnetic monopole would produce). If we imagine a succession of loops in parameter space, enclosing the degeneracy point like the lines of latitude of a globe, the Berry phase factor $e^{i\gamma}$ changes from loop to loop. It starts at unity for the little loop near the north pole and traces out the unit circle in the complex plane, ending at unity again at the south pole. The integrated flux of the phase 2-form through the enclosing surface is the change of γ: $\pm 2\pi$ depending on the direction of the tracing. (The integrated flux for the partner state m is the opposite: $\mp 2\pi$.) Degeneracies thus bring topology into the geometric phase. Indeed if the Hamiltonian is purely real, only topology survives: the geometric phase is zero for any loop except those enclosing degeneracy lines (the reality restriction makes degeneracies lines, not points, in 3-D). For enclosing loops the phase turns out to be π, so the Berry phase factor is a sign change.

One manifestation of the Berry phase arises in the Born–Oppenheimer theory of molecules in which the nuclear coordinates $\mathbf{R}$ are adiabatically changing parameters as far as the (nondegenerate) electron state $|n(\mathbf{R})\rangle$ is concerned. This provides a standard potential energy function $V_n(\mathbf{R}) = \langle n(\mathbf{R})|H_{\text{elec}}|n(\mathbf{R})\rangle$ felt by the nuclei which affects their motion by adding a term $V_n(\mathbf{R})|N\rangle$ to the Schrödinger equation for the state $|N\rangle$ of the nuclear motion. However the kinetic energy form for the nuclei is also influenced, being not merely $\nabla^2|N\rangle$ but $\langle n|\nabla^2|n\rangle|N\rangle$. This evaluates to $(\nabla - \mathbf{A})(\nabla - \mathbf{A})|N\rangle + \phi|N\rangle$ where the vector $\mathbf{A}(\mathbf{R}) \equiv \langle n|\nabla n\rangle$ acts like a magnetic vector potential corresponding to the phase 1-form, and $\phi(\mathbf{R})$ is a certain additional potential energy. The derived magnetic field $\nabla \wedge \mathbf{A}$ corresponds to the phase 2-form. A similar effect is present in the theory of gauge anomalies in quantum field theory. Other phenomena in which the Berry phase plays a role include magnetic resonance and the quantum Hall effect.

A geometrical holonomy effect analogous to the Berry phase is present in the classical mechanics of integrable (i. e., chaos free) systems. Their "action" variable I is adiabatically conserved, while the change in the "angle" variable under cyclic excursion of the classical Hamiltonian is a dynamical part plus a geometric angle θ. In the semiclassical limit ($\hbar \to 0$) the geometric phase γ is related to θ by $\theta = -\hbar d\gamma/dI$.

See also: Hall Effect, Quantum; Hamiltonian Function; Resonance Phenomena.

Bibliography
M. V. Berry, *Proc. Roy. Soc.* **A392**, 45 (1984).
M. V. Berry, *Sci. Am.* **259** No. 6, 46, December (1988).
A. Shapere, and F. Wilczek, "Geometric Phases in Physics", *Advanced Series in Mathematical Physics,* Vol. 5. World Scientific (1989).
D. Chruscinski and A. Jamiolkowski, *Geometric phases in classical and quantum mechanics,* Birkhäuser (2004).
A. Bohm, A. Mostafazadeh, H. Koizumi, Q. Niu, and J. Zwanziger, *The geometric phase in quantum systems.* Springer (2003).

Geophysics

T. J. Ahrens

Geophysics in its broadest sense includes the study of the physical processes and properties of the atmosphere and hydrosphere as well as those of the solid earth. Hence atmospheric physics, meteorology, oceanography, and hydrology may be considered as branches of geophysics, although the term more commonly refers to the study of the solid earth (and planets) alone. The major subdivisions of solid-earth geophysics are seismology, geodesy and the earth's gravitational field, geomagnetism and electricity, thermal properties of the earth, rheology, and mineral geophysics. Applied geophysics describes the techniques and theory of geophysical exploration and prospecting; these include seismic methods, gravity and magnetic surveys, electrical and combined electromagnetic measurements, and investigation of surface heat flow. Marine geophysics is a broad subdivision concerned with the properties of the earth beneath the oceans. In the study of geophysics much use is made of knowledge from other branches of science, principally from physics (e. g., elasticity, electromagnetic theory, fluid dynamics, thermodynamics and solid-state physics), but also from geology, chemistry and geochemistry, crystallography, and mineralogy.

The theories of sea-floor spreading and plate tectonics have been the most significant unifying hypotheses in the earth sciences, and have provided explanations for, and relationships between, many geophysical, geochemical, geological, and paleoecological observations. The scientific study of the physics of plate tectonics has been termed "geodynamics". The basic tenet of plate tectonics is that the entire surface of the earth consists of a small number of rigid lithospheric plates, $\sim 150\,$km thick, that move continuously relative to one another and with respect to the underlying, more plastic, asthenosphere by rotations about axes through the center of the earth. The plate boundaries are defined by narrow belts of seismicity and volcanism. The plates may contain both continental and oceanic regions. The continental component of plates appears not to be destroyed, and has generally grown throughout earth history; however, the configuration of a plate may be changed by collision with another continent (which causes the formation of mountains like the Himalayas) or by rifting to form small continents and new intervening oceans. Oceanic plates are continuously created at mid-ocean ridges and reabsorbed (subducted) at destructive boundaries (trenches). Conservative boundaries, where the plates simply slip past one another horizontally, are termed transform faults. The subduction process only involves the oceanic portion of plates. When a plate sinks in the earth, the intermittent sliding motion of the downgoing plate against the overriding plate gives rise to major earthquakes. Upon bending downwards, the fact that $150\,$km thick plate is cooler and undergoes phase changes as higher pressures are encountered gives rise to earthquakes within the plate as the plate sinks down to depths in some places of $650\,$km. Whether subducting plates go deeper than $650\,$km has been subject to much controversy and is at present unclear. Plate motions are driven by mantle convection, which is part of the process of cooling of the earth. The energy for convection is provided both by radioactivity of the mantle and by the initial gravitation energy of the earth which is released by several processes including contributions from the latent heat of freezing of the inner core of the earth. The rise of molten material from the upper mantle at mid-ocean ridges is also thought to be an intrinsic driving process of plate tectonics.

The original paradigm of plate tectonics became widely accepted in the 1960s after it was recognized that all the continents were part of a super continent which broke some $200\,$M years ago (Fig. 1), forming the present distribution of plates (Fig. 2). Two additional concepts, which have been   accepted, are the presence of mantle plumes and the movement of oceanic and continental terranes by plate tectonics. The study of the geology of China and the easternmost portion of the USSR and Alaska indicates that most of the continental areas of these regions stems from the succession of on the order of 10 oceanic and continental small plates, only

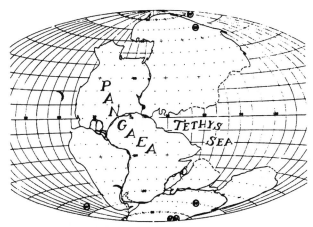

Fig. 1: Reconstruction of the supercontinent Pangea as it may have looked 200 million years ago. The relative positions of the continents are based upon numerical best fits (R. S. Dietz and J. G. Holden, *J. Geophys. Res.* **75**, 4939 (1970)).

100 km in extent, which have been rafted into their present position by ancient plate tectonic motion.

Mantle plumes appear to be fixed, virtually cylindrical regions of the mantle which are continual sources of molten rock for periods approaching 80 My. The plume sources appear to be deep seated, at least below a depth of 150 km, and a number appear to be associated with mid-ocean ridges. The minimum depth of plume sources is inferred from the observation of the resulting molten material which pierces the oversliding 150 km thick plate. Some geodynamicists have suggested that the ultimate source of the heat which is required from mantle plumes come from a proposed thermal boundary layer at the core mantle boundary (Fig. 3). In the case of, for example, the Hawaiian hotspot, a chain of submerged volcanic islands with successively older rocks, extends east northeast 3000 km from the present active volcanic Kilauea to Mawaukee sea mount (submerged, submarine volcano). From this point the chain continues north northeast 2000 km to the intersection of the Kurile and Alaskan trenches (Fig. 4). Mantle plumes are apparently a major source of molten rocks of the earth's interior (largely basalt) and taken with the other two sources of basalt – midocean ridges and subduction volcanoes – provide a framework for understanding the origin of the earth's volcanic rocks. This can be seen in Fig. 5. The geoid, upon removal of the effect of the earth's ellipticity, is dominated by highs (around the Pacific) whose origin can be accounted for with the excess density associated with the subducting oceanic plates.

Seismology is the study of earthquakes, their mechanisms and near source effects, and the propagation of seismic waves through the earth. As predicted by the theory of plate tectonics, the propagating fractures (earthquakes) that produce seismic waves occur principally along zones of preexisting weakness around plate boundaries. The propagation of waves generated by earthquakes can be described using elasticity theory. Three types of waves are generated: *P* (longitudinal), *S* (transverse), and surface waves; in addition, large earthquakes may excite whole-earth vibrations termed free oscillations.

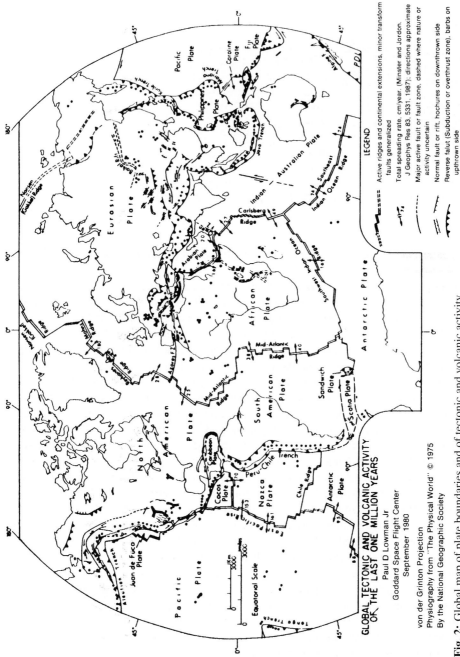

Fig. 2: Global map of plate boundaries and of tectonic and volcanic activity. (P. D. Lowman, Jr., *Bull. Int. Assoc. Eng. Geol.* **23**, 37 (1981)).

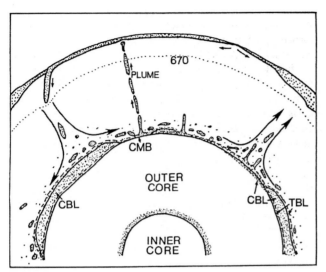

Fig. 3: A schematic model of the core–mantle transition zone. Geophysical measurements suggest a heterogeneous chemical boundary layer (CBL) is embedded in a thermal boundary layer (TBL). Thermal plumes caused by boundary layer instabilities ascend from the core-mantle boundary (CMB) (T. Lay, EOS, *Trans. Amer. Geophys. U.* **70**, 49 (1979)).

Earthquakes are of major public concern if they occur in populous areas, and seismic risk evaluation and the siting and vibration-resistant design of manmade structures have given rise to the field of earthquake engineering. Although there has been much research conducted on testing a wide range of proposed techniques for predicting the time and place of the occurrence of earthquakes, no reliable technology exists for earthquake prediction. However, it is known that the largest earthquakes are associated with either the subduction process (thrust faulting) or motion on transform faults. It is empirically observed that due to the relative motion on a fault, similar lengths of faults rupture on a quasiperiodic basis. The historical record of periodicity of rupture has recently been augmented backwards in time by as much as 50 000 years using ^{14}C dating methods in many places. What is measured is displacement of ancient soil layers disturbed by faulting. These data have led to the gap theory of earthquake prediction. The theory assigns characteristic earthquakes to sectors of characteristic length on a major fault. Each length can be assigned a repeat rupture period. Typical values range from 30 to 15 000 years, and are the basis of assigning earthquake probabilities to a particular region along a plate boundary. Although the gap theory predicts where earthquakes occur and what magnitude is expected, it cannot predict exactly when an earthquake will occur. Thus the gap theory is most useful in providing information on the degree of shaking which can be expected to be experienced by critical structures, such as bridges, dams, nuclear reactors, and public buildings.

Seismic methods are used to derive both shallow and whole-earth structure. The latter is determined from the travel times and wave forms of seismic waves, and the periods of free oscillations; resultant models give longitudinal velocity (α), transverse velocity (β), and density (ρ) as a function of depth (from which may be derived bulk and shear moduli). These

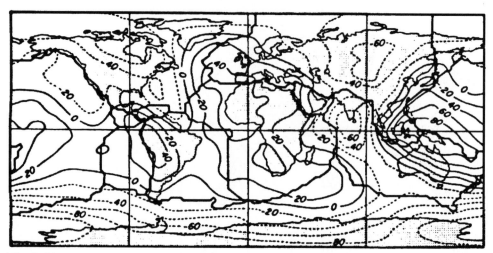

Fig. 4: Global distribution of hotspots (dots) and hotspot traces (lines). (S. T. Crough, *Annu. Rev. Earth Planet. Sci.* **11**, 165 (1983)).

show that the earth consists of a number of layers (Table 1).

Geodesy is concerned with the shape of the earth, and classically has been divided into two parts: determination of the spheroid (the ellipsoid giving the best fit to the sea-level surface) and the measurement of the deviation of the actual sea-level surface (geoid) from the spheroid (Fig. 5). The shape of the earth is basically that of an ellipsoid of revolution slightly flattened at the poles; accurate measurements indicate that it is in fact slightly nonhydrostatic. Determinations of the geoid and spheroid are currently made by a combination of precise surveying and Doppler tracking of artificial satellite orbits. Currently, geologists are attempting to measure directly the absolute distance changes that accompany plate motion by using radio interferometry in combination with the global positioning satellite system. The gravitational potential field for the earth is described via expansion in a series of spherical harmonics (associated Legendre polynomials). The higher-degree harmonics ($n \approx 10$), and to some extent the lower ones, are caused by lateral variations in the upper mantle and transition zone; these may be due to chemical differences, or thermal fluctuations, or may possibly be the direct consequence of convective motion in the mantle. As can be seen in Fig. 5, the geoid, upon removal of the effect of the earth's ellipticity, is dominated by highs whose origin can be accounted for by the density which is associated with subduction of plates.

Local fluctuations in the gravitational field are caused by density variations within the crust, and are used in geophysical exploration to locate structures such as dense ore bodies. Spatial gravity anomalies extending over 10^2–10^3 km do not correlate well with topography because surface features and density variations appear to be compensated at depth (isostasy).

The earth's magnetic field, its history and origin, and the behavior of naturally occurring magnetic minerals are studied in geomagnetism. The field is commonly believed to be caused by electrical currents in the fluid, conducting, outer core of the earth, and undergoes secular variations in strength, quasiperiodic fluctuations on a time scale of 10^2–10^4 yr, and reversals of polarity. Measurements of the intensity and direction of magnetism in rocks of known ages

Table 1: Major layers of the earth and their approximate characteristics.[a]

Layer	Approx. depth km	α (km s⁻¹)	β (km s⁻¹)	ρ (g cm⁻³)	σ (Ω⁻¹ m⁻¹)	η P	T °C	
Crust	0–*ca.* 35[b] 0–*ca.* 10[c]	*ca.* 2–7	*ca.* 1.2–3.8	*ca.* 2–3.0				Laterally heterogeneous, layered; complex velocity, density and conductivity variations
Mohorovicic (M-) discontinuity								Usually known as the "Moho"
Upper mantle	*ca.* 35–400[b] *ca.* 10–400[c]	*Jumps* 8–9	*Jumps* 4.3–4.7	*Jumps* 3.3–3.6	*ca.* 10⁻²	*ca.* 10²¹	*ca.* 400–1500	Laterally inhomogeneous, possibly convecting; principal minerals probably magnesium-rich olivine and pyroxene
Mantle transition zone	400–1000	9–11.5	4.7–6.4	3.6–4.7	10⁻²–10²	*ca.* 10²² at 1000 km	*ca.* 50–2000	Mixed-phase region with complex velocity and density gradients. Two major discontinuities: ~400 km, olivine → spinel; ~600 km, spinel → postspinel
Lower mantle	1000–2900	11.5–13.8	6.4–7.5	4.7–5.7	*ca.* 10²	*ca.* 10²³ or[d] *ca.* 10²⁶	2000–*ca.* 3500 (±500)	Fairly homogeneous; possibly chemically similar to upper mantle
Core–Mantle Boundary (Gutenberg discontinuity)[e]								
Outer core	2900–4980	*ca.* 8–10	0	9.4–12	*ca.* 5 × 10⁵		3800±500	Liquid; alloy of iron–nickel with lighter elements, probably sulfur, carbon, or silicon; probably convecting and source of earth's magnetic field
Transition zone[f]	4980–5120							
Inner core	5120–6370	10–11.7	3.4–3.6	15.8–17.2			6300±1000	Solid, probably similar to outer core in composition; pressure at center of earth *ca.* 3.1 Mbar.

[a] Key: α, *P* velocity; β, shear velocity; ρ, density; σ, electrical conductivity; η, viscosity; *T*, temperature. α, β, ρ, and *T* generally increase with depth within each layer.
[b] Continents.
[c] Oceans.
[d] Depends on method.
[e] Chemical boundary; pressure ~ 1.4 Mbar.
[f] Probably phase boundary between liquid outer and solid inner core.

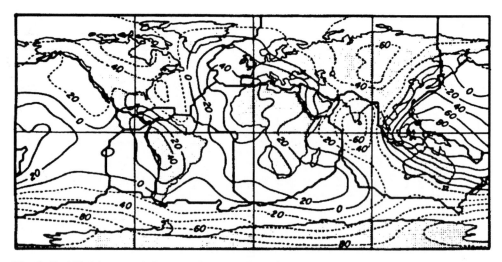

Fig. 5: Geoid height anomaly in meters from the GEM 10B method of Lerch *et al.* (1979) corrected by removal of the dynamically inferred flattening of Nakiboglu (1980). Solid contours are positive anomalies, dashed contours are negative anomalies. Contour interval is 20 m.

enable study of the history of the magnetic field; these have also provided key evidence in favor of the theories of plate tectonics and continental drift.

The electrical conductivity of the earth as a function of depth is determined from damping of magnetic-storm-induced fluctuations in the magnetic field, and temporal and spatial variation of magnetic variations induced by air tides. Magnetotelluric methods utilize the interaction of magnetic and electrical field fluctuations, generated by naturally or artificially induced currents, to determine shallow conductivity structure for prospecting purposes or for studies of the lower crust and upper mantle, depending on frequency.

Thermal properties of the earth are generally investigated by measuring heat flow and the thermal diffusivity and radioactivity near the surface, and then extrapolating to greater depths using constraints provided by the measured rock-melting intervals at high pressure and the position of the known phase changes in the earth's upper mantle (Table 1).

Solar heating has a negligible effect on the interior of the earth, and the energy released by earthquakes and tidal friction is also small compared with geothermal heat losses. The major sources of heat within the earth, at present, are thought to be the radioactive decay of long-lived isotopes and, to a lesser and more uncertain degree, the release of gravitational energy and the decay of short-lived isotopes, such as ^{26}Al.

The present heat flow out of the earth is highly variable (on land an average of $57\,\mathrm{mW/m^2}$), and correlates strongly with the radioactivity and hence heat production of continental rocks. The contribution to land heat flow from the mantle accounts for approximately 50% of the heat flow. In contrast, the average heat flow measured on the oceans is $100\,\mathrm{mW/m^2}$, most of which results from convection in the mantle below.

Below about 100 km the distribution of temperature and heat sources and the mechanism of heat transfer are uncertain. Lattice thermal conductivity mechanisms are thought to dominate throughout the entire mantle as the iron-bearing silicates of mantle become opaque at high pressure. Convection probably occurs in the upper mantle and outer core. To what degree convection now takes place in the lower mantle of the earth is unclear. It appears, however, that thermal convection was more vigorous during the early history of the earth than at present. Higher heat flow is found in regions of volcanism such as mid-ocean ridges and geothermal activity, and geothermal energy may be a source of power in the future.

Rheology is the study of the deformation and flow of matter. When applied to the earth, this generally means the magnitude and time dependence of viscosity within the earth. The study of fracture processes in rock, creep processes, and the motion along faults at depth, as well as the sliding of the lithosphere over the asthenosphere and the subduction of the lithosphere, may also be considered geophysical applications of rheology. Direct measurements of the earth's viscosity are not possible, but estimates can be made by studying the rate of uplift of areas such as the ancient terranes of Canada and Fennoscandia, earlier subjected to down-warping by ice-sheet loading, and provides a determination of the upper-mantle viscosity in the range from 10^{20} to 10^{22} P. Estimates of the lower-mantle viscosity vary from 10^{22} to 10^{26} P.

Mineral geophysics includes the study of mineral and rock properties at high pressures, and the interpretation of velocity and density models of the earth in terms of possible mineral assemblages. (Estimates of the bulk composition of the earth are based on solar and meteoritic abundances and on the composition of crustal and upper-mantle rocks.) The development of theories concerning high-pressure behavior of properties such as density, crystal structure, conductivity, and melting point utilizes results from thermodynamics, elasticity, and solid-state physics. Experimental methods include ultrasonic measurements of velocity as a function of pressure (limited to ~ 30 kbar), and high-pressure techniques such as the static compression of samples in diamond anvil cells are currently being carried out to pressures of at least 1000 kbar. Using dynamic compression via shock waves and diamond anvil cells, pressures in excess of those at the center of the earth (~ 3600 kbar) can be achieved.

Ideally, measurements of compressional and shear velocities, the elastic moduli and their pressure derivatives, density, thermal expansion, specific heat, melting points, viscosity, and electrical and thermal conductivity, all as functions of pressure, are needed for a complete understanding of processes occurring at depth in the earth. However, most of these data are not available, and much present knowledge is in the form of equations of state for density as a function of pressure. Often it is necessary to extrapolate experimentally determined data to higher pressures using theoretical equations of state. Lattice-dynamical, quantum-mechanical, and finite-strain theories are used to construct equations of state.

Experimentally determined empirical relationships such as those between velocity, density, and mean atomic weight have played an important role in estimating properties of materials whose elastic properties are as yet unmeasured, and in interpreting the seismic data for the earth in terms of composition and phase. All earth models show discontinuities in density and velocity at various depths; these are due to changes in chemical composition, but some have been identified with phase changes or rearrangements of crystal structure in the minerals present. Experimental observations of phase changes have been made using x-ray diffraction methods; these studies allow identification of discontinuities in the earth with phase changes.

Phase changes in the earth occur over a band of pressure, and are also "broadened" by the simultaneous transformation of several coexisting minerals. The transition zone between depths of 400 and 1000 km includes two major discontinuities at depths of ~ 400 km and ~ 670 km, which have been identified as being olivine [$(Fe, Mg)_2 SiO_4$] to spinel and spinel related structures and spinel to a denser assemblage of perovskite plus magnesiowüstite. This latter transition is very sharp and takes place in the earth in a depth interval range of less than about 4 km and thus could be expected to be a good seismic reflector.

Recently, using tomographic methods, lateral variations in seismic velocity of $\pm 2\%$ have been mapped in the earth's mantle. The velocity structure of the upper mantle, especially the outer 100 km, can be related to many surface features and to heat flow. For example, the mid-ocean ridge system and the stable continental interiors indicate hotter and cooler shear wave velocities than does the average earth. The lower mantle also demonstrates similar magnitude compressional and shear wave velocity variability except that it is not correlated with surface features.

The core-mantle boundary is a chemical one; the outer core is thought to consist largely of liquid iron-nickel with a lighter alloying element, probably sulfur, oxygen, or silicon.

The most accurate methods of determining the ages of rocks (*see* Geochronology) involve radioactive decay processes within the rock, and are studied in isotope geochronology, which forms a link between geophysics, geochemistry, and nuclear physics. Radioactive dating is accomplished by measuring the proportions of parent and daughter isotopes present and, allowing for the initial concentration of daughter isotope, using the measured half-life of the parent to calculate the age. Originally, the uranium and thorium decay series were widely used, but now a wide range of methods have found more general applicability. Recently the samarium method has found wide use to infer the petrogenetic history of igneous and metamorphic rocks. The age of the earth is inferred from the ages of primitive meteorites to be $\sim 4.65 \times 10^9$ yr. However, the oldest dated terrestrial rocks are only $\sim 4.0 \times 10^9$ yr old.

See also: Atmospheric Physics; Geochronology; Geomagnetism; Gravity, Earth's; Meteorology; Rheology; Seismology.

Bibliography
D. L. Anderson, *Theory of the Earth*. Blackwell, Boston, 1989.
J. M. Bird, *Plate Tectonics*, 2d ed., Am. Geophys. U., Washington, D.C., 1980.
W. R. Peltier, *Mantle convection*. Gordon and Breach, New York, 1989.
G. Ranalli, *Rheology of the Earth*. Allen and Unwin, London, 1986.
F. D. Stacey, *Physics of the Earth*, 2d ed. Wiley & Sons, New York, 1979.

Glass

B. Golding

Glasses and amorphous solids comprise a class of materials (i) whose microscopic atomic arrangement exhibits no periodicity or long-range order, in contrast to crystals; and (ii) whose shear viscosity is sufficiently large that macroscopic shapes are maintained for relatively long times, in contrast to fluids. A substance in its amorphous phase is metastable, as its atoms or molecules exist in a higher configurational energy state than in its crystalline phase, but are unable to reach that state owing to substantial energy barriers that separate the two phases. An amorphous solid state can be achieved (1) by rapid cooling of a viscous fluid, the traditional preparation method for inorganic glasses such as SiO_2 (vitreous silica), As_2S_3, sulfur, organics such as polystyrene and polymethyl methacrylate, and for glassy metal alloys ; or (2) through circumvention of the liquid state by preparation from the vapor phase, as in vacuum deposition, sputtering, gaseous decomposition, etc., as used in the preparation of amorphous silicon and germanium. The term *glass* is often restricted to specify materials formed by process (1). Alternatively, the term *glassy* is frequently applied to physical systems with periodic structures that lack static long-range order in auxiliary degrees of freedom, e. g., magnetic or electric dipoles, molecular orientation.

If, upon cooling, a liquid successfully avoids crystallization, i. e., becomes supercooled, it passes through the glass transition temperature T_g, defined as the temperature at which the substance's shear viscosity exceeds approximately 10^{14} poise. On cooling through T_g, the specific volume and entropy of a glass decrease abruptly but continuously from values characteristic of the liquid toward values approaching, but not reaching, those of a crystalline state. The rate of cooling through the glass transition region determines the total change in the substance's extensive thermodynamics properties by influencing the overall degree of structural rearrangement.

The structures of many inorganic glasses can be visualized as made up of small molecular groups (e. g., SiO_4 tetrahedra in silica glass), joined together to satisfy all, or nearly all, bonding requirements. This continuous random network differs from a crystalline network in that the bonding angles linking the groups are not fixed but have distributions of values about the crystalline angle. Nevertheless, the short-range order of the nearest atomic coordination shells about a central atom in a crystal is largely retained in the amorphous phase.

The glassy network of vitreous silica can be modified and the softening temperature lowered for glasses of commercial and technological importance by incorporating other oxides (e. g., Na_2O and B_2O_3) into the structure to form the soda silicates and borosilicates. Such glasses are transparent to visible light because optical excitation of electronic transitions begins in the ultraviolet part of the spectrum, whereas excitation of vibrational modes occurs in the infrared. The minimum optical loss due to absorption and scattering in silica glass fibers occurs in the near infrared at a wavelength of approximately 1.5 μm.

A glass is an elastically isotropic solid, characterized by two independent elastic constants, and supports a long-wavelength longitudinal and a doubly degenerate transverse acoustic mode. Oxide glasses are generally brittle, exhibit little plastic deformation, and fracture by crack propagation, whereas many organic polymeric glasses are capable of withstanding strains of several hundred percent before breaking. Heat is carried by sound waves but because of the disordered structure the thermal vibrations undergo strong scattering, resulting in low thermal conductivities.

A glass may be regarded as a collection of intrinsic structural defects. When it is subjected to external stimuli such as time-dependent electric or stress fields, the induced rearrangement of defects creates frequency- and temperature-dependent features in the dielectric and elastic responses, i. e., large dielectric and elastic relaxational loss and dispersion characteristic of a particular glass. In addition to a primary relaxational process occurring near T_g, a secondary structural relaxation appears at temperatures below T_g. The specific heat of glasses reflects the extra atomic degrees of freedom by an enhancement of the T^3 Debye specific heat at low temperatures and the appearance of a quasilinear temperature-dependent defect contribution below 1 K.

See also: Crystal Defects; Elasticity; Glassy Metals; Liquid Structure; Semiconductors, Amorphous

Bibliography
R. H. Doremus, *Glass Science*. Wiley, New York, 1994.
S.R. Elliott, *Physics of Amorphous Materials*. Wiley, New York, 1990.
A. Paul, *Chemistry of Glasses*. Chapman and Hall, London, 1990.
R.C. Ropp, *Inorganic Polymeric Glasses*. Elsevier, Amsterdam, 1992.
J. Wong and C. A. Angell, *Glass*. Marcel Dekker, New York, 1976.
R. Zallen, *The Physics of Amorphous Solids*. Wiley-Interscience, New York, 1983.

Glassy Metals

P. Duwez and T. Egami

The crystalline state is characterized by a periodic repetition of a unit cell which contains a certain number of atoms. In the amorphous, or glassy, state the atoms are arranged rather randomly without any periodicity, as in the liquid state. The transition from the liquid to the crystalline state is very sharp, with various properties changing discontinuously at the transition. By contrast the transition from the liquid to the glassy state is gradual and continuous. When a liquid is cooled, the viscosity of the liquid changes rapidly but continuously, until it becomes so high that the material behaves like a solid. During this transition, or the glass transition, the atoms remain in approximately the same configuration. Therefore the glass transition is not really a phase transition in a rigorous sense, and glassy metals can be considered as supercooled liquid metals.

Fused silica and mixtures of silica with other oxides of Al, Na, Ca, etc., constitute the oldest and most commonly seen class of glassy inorganic solids. Other glassy solids include chalcogenide glasses containing elements of group Va, VIa, and VIIa, and amorphous polymers. Glassy metals belong to the same category of materials as these more traditional glasses and share many properties with them. On the other hand many of them require much higher cooling rates to form from the liquid state. Around 1960 new methods were developed to cool liquid metallic alloys at very high rates – up to 1 million degrees per second or more – and under these extreme conditions crystallization can be avoided and glassy metals can be formed with some metallic alloys such as Fe–B and Ni–P, and metallic alloys such as Cu–Zr, Fe–Tb, Au–La, and Al–Ca.

Glassy metals obtained with very high cooling rates are thin foils or ribbons, since the rate of heat extraction is governed largely by thermal conductivity of the sample. However, in the 1990s glassy alloys with much lower critical cooling rates were discovered. Unlike the early glassy metals that require special equipment to attain rapid cooling, these alloys can be cast just like ordinary metals into thicker rods or plates (Fig. 1), or any shape as long as the cooling rate does not fall below a certain critical value at any part of the cast. For this reason they are called bulk metallic glasses. Alloy compositions which produce bulk metallic glasses are quite complex, usually containing more than six elements, for instance, Zr-Ti–Cu-

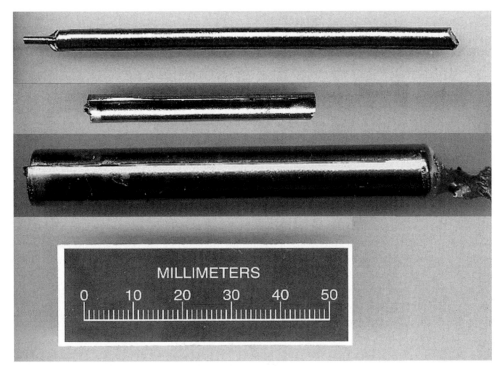

Fig. 1: Ingot of iron based bulk metallic glass, courtesy of C. T. Liu, Oak Ridge National Laboratory.

Ni–Be or Fe–Mn–Mo–Y–C–B. They show mechanical strength far superior to the crystalline counterpart, making them extremely attractive for application as structural materials. They are already used for golf clubs, phone cases and other special applications.

Glassy metals can be produced by other methods as well. The earliest reports of amorphous phase formation are by electrochemical deposition and vapor deposition at low temperatures. The samples obtained by the latter, however, are usually unstable at room temperature. In the earlier studies the researchers really could not distinguish amorphous phases from microcrystalline phases. It took the advent of rapid cooling methods to initiate extensive studies of this new phase. Amorphous thin films can be produced by sputtering deposition and are beginning to be used for magnetic applications.

It is also possible to produce an amorphous phase directly from the crystalline state, without going through a gas or liquid phase. Irradiating a crystal with electrons and ions produces lattice defects, and eventually leads to a collapse of the whole lattice structure into the amorphous state. More interestingly, it was discovered that when a mixture of two elements is annealed at a temperature just high enough to promote sufficient atomic diffusion, an amorphous phase can result due to interdiffusion. If powders of two elements are mechanically mixed by a ball-milling machine, the product can again be amorphous. Thus glassy metals can be produced by many thermal or athermal processes.

At least theoretically, everything becomes glassy when it is quenched from the liquid state with a sufficiently high cooling rate. In reality, however, the critical cooling rate necessary for some materials to become glassy is technically too high to attain. For instance, it is generally recognized that pure metals cannot be quenched from the liquid state into a glassy solid, at least not with the presently achieved rates of cooling which are in the range of 10^5–$10^7\,^\circ\mathrm{C/s}$. So all glassy metals actually are alloys containing two or more elements. Over the last four decades a very large number of alloy systems have been found to form glasses, including bulk metallic glasses. It is now clear that glass formation is not limited to particular classes of alloy systems, but is a phenomenon widely found in metallic alloys.

Various criteria have been proposed in attempting to predict glass-forming combinations of elements. However, there is yet no perfectly reliable method of predicting the glass formability. A large variety of chemical interactions found in glass-forming alloy systems imply that chemistry has little to do with glass formation, except that a strongly positive heat of mixing would lead to phase segregation and prevent glass formation. A factor which is definitely important in glass formation is the atomic radii of the elements. A large difference in the atomic size makes the solid solution unstable and makes the glass formation easier. In binary systems (A–B) the minimum composition (in atomic percent) of the second (B) element necessary for glass formation, $c_B^{\min}$, is usually related to the atomic volume of each, v_A and v_B, by $c_B^{\min} \sim 10 v_a / |v_A - v_B|$.

In addition, in order for the glass to be stable, the rate of crystallization at room temperature has to be sufficiently low. Glassy metals are always unstable against crystallization when the atomic diffusivity is high enough, and they usually crystallize when they are heated to near the glass-transition temperature, T_g. For many transition-metal–metalloid glasses such as $Fe_{80}B_{20}$, T_g is around $400\,^\circ\mathrm{C}$; thus they crystallize at around this temperature. Since T_g is roughly equal to $0.4\bar{T}_m$, where $\bar{T}_m$ is the compositionally averaged melting temperature of the constituent elements, $\bar{T}_m$ has to be above $1000\,\mathrm{K}$ for the glass to be stable. The length of time required for crystallization to occur, τ_c, changes exponentially with temperature approximately obeying the Arrhenius law, $\tau_c = \tau_0 \exp(E_c/kT)$, with the apparent activation energy of 2–5 eV. Thus sufficiently below T_g, glassy metals are stable for quite a long time and can be regarded stable in practice.

The atomic structure of glassy metals can be best studied by diffraction experiments, using x-rays, neutrons, or electrons. A typical diffraction pattern consists of several broad peaks, rather than a number of sharp Bragg peaks as for crystalline solids. The diffraction peak width, if interpreted in terms of microcrystalline aggregates, corresponds to the grain size of 15–20 Å. It is now well established, however, that the structure of glassy metals cannot be described as a collection of small crystalline grains. The simplest and yet physically most sensible model of liquid and glassy metals, the dense random packing (DRP) model, was proposed by Bernal and Scott. The model just consists of a large number of steel balls randomly packed in a bag. Today the model can be built by a computer, using more realistic interatomic potentials or even quantum mechanically from the first principles. The calculated distribution of the atomic distances in the model compares very well with those of the glassy metals directly obtained from the diffraction intensities by the Fourier transformation, as shown in Fig. 2.

Because the transition from liquid to glass is a process of kinetic arrest, the structure of glass is not quite settled as produced, and relaxes when they are annealed at temperatures below T_g. This process, known as structural relaxation, occurs both for oxide glasses as well as metallic glasses. Various properties change during this process, some drastically, and others only slightly. For instance diffusivity changes by many orders of magnitude, while the changes

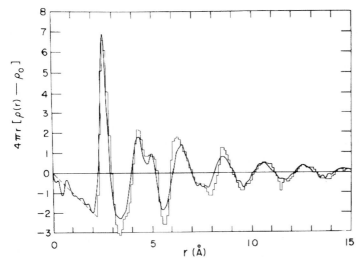

Fig. 2: The distribution of interatomic distances presented in terms of
$G(r) = 4\pi r[\rho(r) - \rho_o]$, where $\rho(r)$ is the atomic pair distribution function
which describes the probability of finding two atoms at a distance r and ρ_0 is
the average number density of atoms in the alloy. The solid line is $G(r)$
determined by x-ray diffraction experiment for amorphous $Fe_{40}Ni_{40}P_{14}B_6$,
while the histogram describes that of a DRP model.

in volume and structure are only barely observable. For many applications pre-annealing
treatment should be applied to finish off this relaxation before use.

Since glassy metals are isotropic, their elastic properties can be described by only two elastic constants, namely, bulk and shear constants. The bulk modulus of glassy alloys is similar to that of crystalline alloys of the same composition, while the shear modulus is significantly (by about 30%) lower than in the corresponding crystalline materials. Mechanical deformation of crystalline solids occurs via the motion of lattice dislocations. In glassy metals the dislocations cannot be defined in the absence of a lattice; thus they show very high mechanical strength, often of the order of $300 \, kg/mm^2$ or more, which is comparable to or higher than that of the strongest steel (piano) wires. At least in the as-quenched state the glassy metals exhibit considerable ductility in compression. However, the ductility is lost when they are annealed, and the glasses become brittle. The degree of embrittlement depends on composition, but it poses a challenge for structural application of bilk metallic glasses.

Glassy metals show relatively high electrical resistivity, ranging from 50 to $200 \, \mu\Omega \, cm$. This high resistivity is due to structural disorder and is only weakly dependent upon temperature, with the temperature coefficient being either positive or negative. The resistivity and its temperature dependence of some alloys, particularly the simple metal alloys, can be understood well in terms of the scattering theory by Ziman which was successfully applied to many liquid alloys. At low temperatures the resistivity sometimes shows a minimum similar to that of Kondo alloys, but the details of its mechanism are not well understood. Some of the glassy metals containing Zr, Mo, and La are superconducting at low temperatures. The critical temperature is somewhat lower than in crystalline alloys of a similar composition, but the critical field, H_{c2}, and the critical current density are higher. The superconducting properties of glassy metals are resistant against radiation damage, and thus they may be useful for applications in nuclear reactors or in accelerators.

Ferromagnetism is not limited to crystalline solids, and many glassy alloys containing Fe, Co, and Ni are ferromagnetic at room temperature and above. The saturation magnetic moment, M_s, and the Curie temperature, T_C, of Co-rich alloys are very similar to the crystalline counterparts, while those of Fe-rich alloys are usually smaller than in the crystalline alloys. This contrast reflects the difference in the electronic structures of these two. In Co-rich alloys the majority-spin d band is full (strong magnetism), so that the magnitude of the magnetic moment depends only upon the electron density, while in Fe-rich alloys both the majority and minority bands are partially full (weak magnetism), and the spin polarization is influenced by structure and other factors.

Their magnetic transition at T_C is well defined and sharp, unless they contain significant compositional fluctuations. The critical behavior in the vicinity of T_C is very similar to the corresponding crystalline alloys, since the atomic-level disorder does not influence the long-range magnetic fluctuations. Away from T_C, disorder has some effects on magnetism, but the compositional disorder which can occur in crystalline solids as well has more significant effects than topological disorder. Changes in the compositional short-range order can affect magnetic properties, such as T_C and field-induced magnetic anisotropy.

Many ferromagnetic glassy alloys show very low coercivity and therefore high permeability, except when they contain rare-earth elements, because they do not have grain boundaries and lattice defects like dislocations. Their permeability after annealing can be equal to or better than that of the crystalline alloy with the highest permeability, Supermalloy. Ferromagnetic glassy alloys are already used in a large number of devices. Their use in distribution power transformers results in significant energy savings, and, in switching regulators which supply stable dc power to computers and other devices, they save much space and weight. In antitheft devices the mechanical toughness is an important asset, because the magnetic signals do not become deteriorated by handling. On the other hand those containing rare-earth elements show high coercivities, particularly in thin films. Sputter-deposited Fe–Co–Tb thin films are used for magneto-optical data storage in which information is written by a laser in the form of tiny magnetic bubbles.

Glassy metals are a new class of materials rich in prospect of leading to interesting and important science and applications, some of which have already materialized. They posed a challenge to the conventional theories of solid-state physics which are largely based upon the periodicity and symmetry of crystalline lattices, and are stimulating the growth of new and more general theory of solids. Even though they are not thermodynamically stable, they can be used in many applications just as a number of metastable crystalline solids which are currently in use. The glassy metallic state is not an unusual state which is an object of scientific curiosity alone, but represents a general state of matter which can occur in a large number of metallic alloy systems.

See also: Alloys; Conduction; Crystal Growth; Glass; Liquid Metals; Liquid Structure; Viscosity.

Bibliography
Pol Duwez, *Annu. Rev. Mater. Sci.* **6**, 83–117 (1976).
H.-J. Gunterodt and H. Beck (eds.), *Glassy Metals.* Vols. 1 and 2. Springer, Berlin, 1981, 1984.
F. E. Luborsky (ed.), *Amorphous Metallic Alloys.* Butterworth, London, 1983.
T. Egami, *Rep. Prog. Phys.* **47**, 1601–1725 (1984).

Grand Unified Theories

S. M. Barr

Historical Background

The first example of a unification of forces was Maxwell's electromagnetic theory. This was not simply a matter of the electric and magnetic fields appearing together in the same equations but, more profoundly, involved a new symmetry principle: Lorentz invariance. (A symmetry is a transformation which leaves the form of something unchanged; and the set of such transformations form a mathematical structure called a "group". For example, rotations by multiples of $2\pi/3$ leave an equilateral triangle unchanged and so make up the group of symmetries of the triangle.) Lorentz transformations leave the form of Maxwell's Equations unchanged. Since certain Lorentz transformations rotate electric and magnetic fields into each other, one can only speak in an absolute sense of a unified entity called the "electromagnetic field".

A point of cardinal importance is that a deep connection exists between forces or "interactions" and symmetries that act locally in space-time. For example, underlying the electromagnetic force is a local symmetry called electromagnetic gauge invariance (Weyl 1929). Einstein's theory of gravity is based on a local symmetry called general coordinate invariance. Other similarities between these two forces are their long-range character (inverse square law) and their coupling to conserved charges (electric charge and energy–momentum, respectively). The other two interactions, the "strong" and "weak" nuclear forces, seemed not to share any of these features. Thus the first attempts were to find a "unified field theory" of electromagnetism and gravity with a larger local symmetry containing both electromagnetic gauge invariance and general coordinate invariance. Such theories can be found (Kaluza 1921, Klein 1927) but suffer from severe difficulties. In the late 1960s and early 1970s it was understood that the strong and weak forces, like the electromagnetic, are based on "gauge" symmetries. This had been obscured, in the case of the weak interactions by "spontaneous symmetry breaking", and in the case of the strong interactions by "confinement". The local symmetry of gravity is of a somewhat different kind and the early attempts of Einstein and others to unify gravity with other forces are now seen as premature. Modern grand unified theories unify the three nongravitational forces.

The electroweak theory of Glashow, Salam, and Weinberg (1967) is based on the gauge group $SU(2) \times U(1)$. [$SU(N)$ is the group whose elements can be represented by the $N \times N$ unitary matrices with unit determinant. $U(1)$ is the group whose elements can be represented by complex numbers of unit magnitude.] $SU(2)$ is the "weak isospin" symmetry; $U(1)$ is the "weak hypercharge" symmetry. The vacuum state (i. e., ground state) does not respect this full

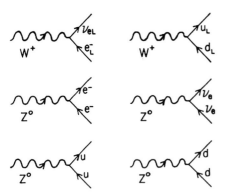

Fig. 1: The absorption or emission of the weak gauge bosons $W^{\pm}$ causes transitions within particle multiplets.

symmetry, however, but only a $U(1)$ subgroup which is the electromagnetic gauge group. The rest of the symmetries (associated with the weak interactions) are said to be spontaneously broken. In this theory the left-handed components of the electron and electron neutrino form a multiplet of weak isospin as do the left-handed u and d quarks:

$$\left[\begin{array}{c} \nu_{eL} \\ e_L \end{array}\right], \left[\begin{array}{c} u_L \\ d_L \end{array}\right].$$

This means there are $SU(2)$ transformations that rotate $\nu_{eL} \leftrightarrow e_L^-$ and $u_L \leftrightarrow d_L$. A remarkable feature of gauge theory is that there corresponds to every type of gauge symmetry transformation a particle – a "gauge boson" – which by emission or absorption actually performs it on the particles. For the weak interactions these are the $W^{\pm}$ and Z^0 (Fig. 1). To broken gauge symmetries correspond massive gauge bosons ($M_W \cong 80\,\text{GeV}$, $M_Z \cong 90\,\text{GeV}$), to unbroken ones massless gauge bosons ($M_{\text{photon}} = 0$). Symmetry breaking accounts for the nonconserved charges and the short range that distinguish the weak from the electromagnetic forces. Spontaneous symmetry breaking is associated with a mass or energy scale; that of electroweak breaking is about $10^2\,\text{GeV}$ ($\sim M_W$, M_Z).

The strongly interacting particles (hadrons, e. g., proton, neutron, pion) are made up of more elementary constituents called quarks and gluons. It is believed that only hadrons and not free quarks or gluons can be directly observed experimentally (confinement). The theory describing the interactions of quarks and gluons is called quantum chromodynamics (QCD) and is based on the gauge group $SU(3)$. The gauge bosons are called gluons and couple to a conserved charge called "color". Each quark is really a multiplet of $SU(3)$ whose components have different color. For example "the u quark" is actually (u^r, u^g, u^b). ($r =$ "red", $g =$ "green", $b =$ "blue"). $SU(3)$ transformations rotate these components among themselves, and, correspondingly, gluon emission or absorption physically accomplishes this.

Together the electroweak theory and QCD constitute the standard model. The elementary spin-$\frac{1}{2}$ particles of this theory are the quarks (which have color) and the leptons (which do not). These come in three sets called families identical except for their masses. The first family consists of the following multiplets:

multiplet		$SU(3) \times SU(2) \times U(1)$	electric charge
leptons	$\left\{ \begin{bmatrix} \nu_{eL} \\ e_L^- \end{bmatrix} \right.$	$(1, 2, -\tfrac{1}{2})$	$\begin{bmatrix} 0 \\ -1 \end{bmatrix}$
	$\left. e_R^- \right.$	$(1, 1, -1)$	(-1)
quarks	$\left\{ \begin{bmatrix} u_L \\ d_L \end{bmatrix} \right.$	$(3, 2, \tfrac{1}{6})$	$\begin{bmatrix} \tfrac{2}{3} \\ -\tfrac{1}{3} \end{bmatrix}$
	$\left. u_R \right.$	$(3, 1, \tfrac{2}{3})$	$(\tfrac{2}{3})$
	$\left. d_R \right.$	$(3, 1, -\tfrac{1}{3})$	$(-\tfrac{1}{3})$

This assortment of particles is strange and complicated, yet also exhibits certain mysterious regularities. We note three: (1) Left-handed $SU(2)$ doublets, right-handed singlets. (2) Quarks have three colors and $\tfrac{1}{3}$-integral charges. (3) Electric charge is quantized. Also note that since quarks and leptons come in separate multiplets, gauge interactions (except for "anomalies") conserve both lepton number ($L \equiv N_{\text{leptons}} - N_{\text{antileptons}}$) and baryon number ($B \equiv [N_{\text{quark}} - N_{\text{antiquark}}]/3$). Protons as the lightest baryons ($B = 1$) cannot decay.

Grand Unified Theories

The technical definition of a grand unified theory (GUT) is a gauge theory of elementary-particle interactions in which the gauge symmetries of the strong, weak, and electromagnetic forces are contained within a single simple Lie group. ("Simple" is a group-theoretical term.) An apparent problem is that the three standard-model forces have different strengths given by the gauge coupling constants g_1, g_2, and g_3. However, in quantum field theory, coupling "constants" are really functions of the energy scale at which interactions are taking place. This is described by the renormalization group equations, which predict a slow (logarithmic) dependence of g_i ($i = 1, 2, 3$) on energy. Georgi, Quinn, and Weinberg (1974) found that these three couplings if extrapolated to very high-energy scales appear to converge to a common "unified coupling" g_{GUT} at an energy near 10^{15} GeV. This is the unification scale, M_{GUT}.

In the simplest (and first) grand unified theory, that of Georgi and Glashow (1974), the unification group is $SU(5)$ which undergoes two stages of spontaneous symmetry breaking. At the unification scale, M_{GUT}, $SU(5)$ breaks down to the standard model group $SU(3) \times SU(2) \times U(1)$, which itself undergoes the electroweak breaking at the scale M_W. The ratio of these two energy scales, M_W/M_{GUT}, is roughly 10^{-13} and its smallness is very hard to explain satisfactorily in the context of quantum field theory where it receives contributions that are of order $g_i^2/4\pi$ ($\approx 10^{-2}$) from higher-order quantum effects. Different contributions must therefore conspire to cancel to many decimal places which involves a highly unnatural "fine tuning" of parameters. This is called the "gauge hierarchy problem" (Gildener 1976) and is the key theoretical difficulty afflicting unified theories. The fermions of the first family form only two multiplets of $SU(5)$, an enormous simplification (Fig. 2). (Instead of the right-handed particles, e_R^-, u_R, and d_R we display their *CP* conjugates, the left-handed antiparticles, e_L^+, u_L^c, d_L^c.) Importantly, the quarks and leptons appear together in the same multiplets, so that there are $SU(5)$ gauge transformations, and hence also gauge bosons – called usually X and Y – which convert leptons into quarks and vice versa (Fig. 3). Thus baryon and lepton numbers are violated and the proton decays, for example into $\pi^0 e^+$, $\pi^+ \bar{\nu}$, and $K^0 \mu^+$. As $M_X \cong M_Y \approx M_{\text{GUT}}$ the proton decay rate is suppressed by $(m_p/M_{\text{GUT}})^4$ relative to typical

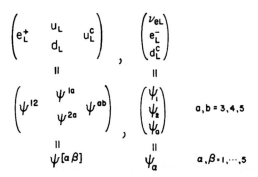

Fig. 2: The two $SU(5)$ multiplets containing the quarks and leptons of the first family.

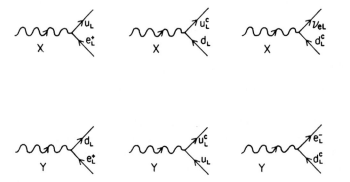

Fig. 3: The superheavy ($\approx 10^{15}$ GeV) gauge bosons of the $SU(5)$ unified theory, X and Y, can cause transitions between leptons and quarks.

nuclear interaction rates (10^{24} s^{-1}). So $\tau_p = 2 \times 10^{29 \pm 1.7}$ yr in the minimal $SU(5)$ model, with $\tau(p \to e^+ \pi^0)$ about three times that.

Since, in the minimal $SU(5)$ model, the unification of the three standard model gauge couplings g_i allows one to compute them in terms of only g_{GUT} and M_{GUT}, there is a prediction which is that $\sin^2 \theta_W [\equiv g_1^2 / (g_1^2 + g_2^2)] = 0.214 \pm 0.004$ (at M_W). This agrees roughly with the experimental value of 0.231 ± 0.00016. However, the minimal $SU(5)$ model was definitively ruled out when the Irvine–Michigan–Brookhaven (IMB) experiment set a limit of $\tau(p \to e^+ \pi^0) > 2.5 \times 10^{32}$ yr.

In models with unification groups other than $SU(5)$ at least two stages of symmetry breaking are required to reach the standard model: $G \longrightarrow G' \longrightarrow \ldots \longrightarrow SU(3) \times SU(2) \times U(1)$. With more unknown parameters there is no definite prediction for τ_p or $\sin^2 \theta_W$. Since the unknown scales of breaking enter as the fourth power in τ_p and only logarithmically in $\sin^2 \theta_W$, the proton decay rate is much more sensitively model-dependent. That is why the failure so far to see proton decay, though disappointing, is not regarded as a fatal problem for the idea of grand unification.

Another remarkable prediction of grand unification is the existence of magnetic monopoles with mass of order M_{GUT}. These are too heavy to be produced terrestrially, but primordial monopoles produced in the Big Bang have been searched for. Given various astrophysical bounds on such particles, finding them would seem a long shot. There are other possibilities for confirming grand unification in the laboratory. One would be the discovery of so-called extra Z bosons (which can arise in GUTs with group of rank > 4) and a measurement of their properties. Another would be relations coming from grand unified symmetries among the masses of new particles (such as "gauginos") that are predicted to exist in models with supersymmetry (SUSY).

More precise measurements of the gauge couplings g_i showed in 1991 that they did not unify when extrapolated to high energy in the minimal *non-supersymmetric SU(5)* theory, but did in the minimal supersymmetric $SU(5)$ theory. In fact, in supersymmetric $SU(5)$ the value of $\sin^2 \theta_W$ is predicted accurately to within about one percent. This is taken by many theorists to be strong evidence in favor of both grand unification and supersymmetry, and led to a strong revival of theoretical work on "SUSY GUTs" in the 1990s.

Aside from unification of couplings, grand unification has other impressive explanatory successes. In particular, it provides simple group-theoretical explanations of the "mysterious regularities" of the quark and lepton quantum numbers enumerated above. Many theorists believe that the exact group-theoretical "fit" of the Standard Model within unified models based on $SU(5)$ and related groups is too impressive to be dismissed as coincidence. Grand unified theories also relate the masses of the quarks and leptons. The fact that the b-quark mass and τ-lepton mass when extrapolated to the unification scale are approximately equal is considered evidence in favor of unification. There are many interesting proposals for explaining various features of the mass spectrum of the quarks and leptons on the basis of grand unified symmetries.

Grand Unified Groups

A unification scheme just slightly older than $SU(5)$ is the $SU(4) \times SU(2) \times SU(2)$ model of Pati and Salam (1974). This is not truly a grand unified model since the group is not "simple" (it has three factors). Nonetheless a very appealing unification of quarks and leptons is achieved. The first family looks like

$$\begin{bmatrix} \nu_{eL} & u_L^r & u_L^g & u_L^b \\ e_L^- & d_L^r & d_L^g & d_L^b \end{bmatrix} \quad \begin{bmatrix} N_{eR} & u_R^r & u_R^g & u_R^b \\ e_R^- & d_R^r & d_R^g & d_R^b \end{bmatrix}$$

The left-handed particles are doublets under one $SU(2)$ group and the right-handed particles are doublets under the other – a pleasing left–right symmetry (which can be made mathematically exact). Moreover one can see that leptons become the "fourth color" of quark. Thus, as great a degree of quark–lepton unification is achieved in the Pati–Salam model as in $SU(5)$. (Notice that a right-handed neutrino N_R must be present in each family.)

While $SU(5)$ may be the smallest and simplest possibility, $SO(10)$ is in many ways a privileged unification group. It is the smallest which allows the quarks and leptons of one family to be unified into a single multiplet – namely, the 16-component spinor. $SO(10)$ also mathematically contains the Georgi–Glashow and Pati–Salam groups as subgroups – and thus in a sense as special cases. One can make a chart of the most interesting groups for unification (Fig. 4). Of course, ultimately only experiment can tell us which, if any, unified group is correct. But theoretical (or aesthetic) considerations have focused greater attention on certain ones. One

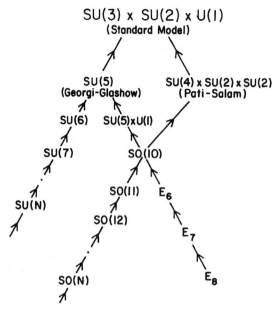

Fig. 4: Groups of most interest for unification of forces. $G \to H$ means here that H is a subgroup of G.

approach to understanding why there are families is the idea that all of them are contained in a single irreducible multiplet. For three (or more) families this can be achieved with the groups $SO(N)$, $N > 14$, E_7, and E_8 (which all contain $SO(10)$). The exceptional series (E_N) is interesting for other reasons both mathematical and physical. E_8 is unique as the largest exceptional group (with finite-dimensional representations), whereas there is no largest $SO(N)$ or $SU(N)$ group. E_6 allows the fermions (the quarks and leptons) and the scalar bosons (the Higgs fields required for spontaneous symmetry breaking) of the theory to appear in multiplets of the same size, which is significant from the viewpoint of supersymmetry where a fermion–boson unification occurs. E_6 and E_8 aroused increased attention as unification groups once it was found that they arise naturally in the context of heterotic superstring theory.

Unification and Neutrinos

The large underground detectors that were built to search for proton decay, though failing of their original purpose, have made dramatic discoveries about neutrinos. The Kamiokande detector in Japan and the IMB detector observed neutrinos emitted from supernova 1987a, which was of great importance to the theory of supernovas, as well as placing bounds on neutrino masses. In 1998, the super-Kamiokande detector found conclusive evidence of $\nu_\mu \to \nu_\tau$ "flavor oscillations" occurring in neutrinos in cosmic ray air showers ("atmospheric neutrinos"). In subsequent years, super-Kamiokande, the Solar Neutrino Observatory (SNO) in Sudbury, Canada, and other underground laboratories showed definitively that the long-known deficit of neutrinos from the Sun was due to oscillations of ν_e into heavier flavors of neutrino. The experimental evidence strongly supports the conclusion that both the solar and atmospheric neutrino oscillation effects are due to non-vanishing neutrino masses in the range 10^{-1} to 10^{-2} eV. These neutrino oscillation effects tend to support the idea of grand unification in two ways. First, grand unified theories based on $SO(10)$ and related groups predict the existence of right-handed neutrinos and thus typically predict that the observed left-handed neutrinos should have non-vanishing masses. Second, the masses predicted for left-handed neutrinos in unified theories are typically of order $M_W^2/M_{GUT} \sim 10^{-2}$ eV, which is in remarkable qualitative agreement with the data.

Cosmological Implications

The ideas of unification have had an enormous impact on cosmology. It is now believed that the predominance of matter over antimatter in our universe is due to baryon-number-violating processes that occurred in the early universe. This idea (Sakharov 1967) is a powerful argument in favor of grand unification. Spontaneous symmetry breaking associated with grand unified theories would have had dramatic consequences in the early universe when phase transitions connected with such breakings would have occurred. The study of such transitions led (Guth 1981) to the important idea of inflationary cosmology [which was also motivated by attempts to solve the "primordial magnetic monopole problem" (Preskill 1979) that arises in unified theories].

Future

Theory has gone beyond the original grand unified theories in two main respects. First, the attempt to solve the "gauge hierarchy problem" and to achieve a unification of bosons with fermions led to models with low-energy supersymmetry. The attempt to include gravity led to supergravity and superstring unification. Whether any of these beautiful, important and theoretically fruitful ideas describe physical reality awaits experimental confirmation.

See also: Electromagnetic Interaction; Elementary Particles in Physics; Gauge Theories; Group Theory in Physics; Invariance Principles; Neutrino; Proton; Strong Interactions; $SU(3)$ and Higher Symmetries; Symmetry Breaking, Spontaneous; Weak Interactions.

Bibliography

H. Georgi, "A Unified Theory of Elementary Particles and Forces", *Sci. Am.* 48–61 (1981). (E)

A. Zee, *Fearful Symmetry, The Search for Beauty in Modern Physics.* Macmillan Publishing Co., London, 1986. (E)

T.-P. Cheng and L.-F. Li, *Gauge Theory of Elementary Particle Physics*, pp. 428–453. Oxford University Press, Oxford, 1984. (A)

P. Langacker, "Grand Unified Theories and Proton Decay", *Phys. Rep.* **74**, 185–385 (1981). (A)

A. Zee, *The Unity of Forces in the Universe* (2 vols.). World Scientific, Singapore, 1982.

Gratings, Diffraction

E. G. Loewen

Introduction

Diffraction gratings are the most widely used elements for dispersing light into its spectral components. They are typically in the form of plane or spherically concave mirror surfaces that contain a large number of straight, parallel grooves. Groove frequencies vary from 20 to 6000 per millimeter. Incident light is diffracted by each of the grooves, but for a given wavelength is visible only in that direction in which light from each groove interferes constructively with that of the others. Depending on the ratio of wavelength to groove spacing, there may be one or more of such sharply defined directions, the *grating orders*. As reflection devices they can be used over the entire electromagnetic domain from 0.1 nm to 1 mm. In the form of transmission gratings, their use is restricted by the transparency of available materials to a band from the near ultraviolet to the near infrared.

Since gratings can be considered a special type of interferometer, it has always been clear that to obtain clean separation of wavelengths requires the grooves to be located with a uniformity of 0.1–0.001 of the wavelength of the light being analyzed. To achieve accuracy of such levels while mechanically impressing the grooves on the grating surface is the ultimate challenge to contemporary machinery.

History

Joseph Fraunhofer was the first (1823) to rule precision diffraction gratings and use them to isolate solar spectral lines. He derived the basic grating equation:

$$m\lambda = d(\sin\alpha + \sin\beta)$$

where m is the spectral order of wavelength λ, d the groove spacing, and α and β are the angles of incidence and diffraction, respectively, with respect to the normal of the grating surface.

The next major step was taken by Henry Rowland, who built the first ruling engine that produced large gratings of high resolving power (1885), with the grooves diamond burnished onto blanks made of speculum metal. He also invented the concave grating that performed both the imaging and the dispersing functions, eliminating the need for collimating optics required for plane gratings. With that invention he opened up spectroscopy in the entire vacuum ultraviolet.

Albert Michelson devoted decades to the building of large ruling engines, recognizing in 1915 the desirability for long (250-mm) travels to be monitored interferometrically, instead of relying on even the best and most carefully corrected lead screws. George Harrison took advantage of technological advances in isotope lamps, electronics, and servos to bring this concept into being in 1950. At the same time he introduced the idea of the echelle grating, used in high orders at high angles of incidence. With this grating, resolution of over 10^6 is attainable.

R. W. Wood was responsible (1920) for showing how the diffraction efficiency could be greatly improved through the shaping of individual grooves. John Strong introduced the concept of ruling in vacuum-deposited aluminum layers on glass substrates, an idea crucial to the ruling of larger and finer pitch gratings.

Michelson indicated in 1927 the possibility of generating high-precision gratings by recording in a suitable medium the highly uniform interference fringe field generated when two monochromatic collimated beams intersect. He lacked an intense source (i. e., a laser) as well as a grainless photographic recording medium. Labeyree in France, and Rudolph and Schmahl in Germany, began in 1967 to use photoresists and argon ion lasers to make such gratings, usually termed holographic. For some applications these gratings have the advantage of eliminating the effects of residual mechanical errors, but they have restrictions in groove spacing (no coarser than 600 per millimeter) and in diffraction efficiency.

From a user's point of view a very important step forward was the development by 1950 of a plastic casting process with which it was possible to make replicas of master gratings that were optically identical. For the first time precision gratings became available in quantity at low cost, and quickly displaced prisms from most spectroscopic instruments. Compared to prisms, they are cheaper, easier to mount, identical from one to the other, mathematically predictable, and not limited by transparency of substrates. Resolving powers such as 10^5, beyond the reach of prisms, became routine.

Monochromators and Spectrographs

Spectrometry requires not only dispersion but also imaging of an entrance slit onto a receiving surface. The latter may be a fixed exit aperture, in which case we speak of monochromators, because only one wavelength is transmitted at a time; or it may be a receiving surface, in which case we speak of spectrographs.

Most monochromator designs are based on collimating the entrance beam so that the collimated light diffracted by a plane grating can be focused stigmatically onto the exit slit. All elements are fixed except the grating, whose angular setting with respect to the incident beam determines the wavelength transmitted. It is possible to use concave gratings instead, which can image without the use of separate collimating optics, but at the expense of nonstigmatic imaging.

Spectrographs are frequently designed around concave gratings because the optics achieves the ultimate in simplicity, i. e., just the grating. The traditional photographic film plate recording of spectra has been largely displaced by electronic array detectors; sensitivity is increased and direct computer data handling becomes straightforward. The grating' s usefulness is greatly enhanced by the development of holographic-generating methods that reduce aberrations and flatten the image field. One important consequence is that much more compact systems become possible.

A high-dispersion, high-resolution system with an unusually compact display is possible by combining a plane echelle-type grating with a cross-dispersing grating or prism. The latter conveniently separates the otherwise overlapping high orders and leads to a rectangular display.

Largely due to lasers, a number of nonspectrometric uses have been found for gratings. Examples are beam splitters and recombiners, beam-scanning systems, and pulse compression.

See also: Fraunhofer Lines; Holography; Interferometers and Interferometry; Visible and Ultraviolet Spectroscopy.

Bibliography

R. M. Barnes and R. F. Jarrell, "Gratings and Grating Instruments", in *Analytical Emission Spectroscopy* (E. Grove, ed.), Ch. 4. Dekker, New York, 1968.

G. R. Harrison, R. C. Lord, and J. R. Loofbourow, *Practical Spectroscopy*. Prentice-Hall, Englewood Cliffs, N.J., 1948.

M. C. Hutley, *Diffraction Gratings*. Academic Press, London, New York, 1982.

E. W. Palmer, M. C. Hutley, A. Franks, J. F. Verrill, and B. Gale, "Diffraction Gratings", *Rep. Prog. Phys.* **38**, 975–1048 (1975).

R. W. Sawyer, *Experimental Spectroscopy*. Dover, New York, 1963.

Gravitation

J. Friedman

Introduction

Early History

Our knowledge of gravity is rooted in astronomy and geometry. Like astronomy, geometry has written records spanning four millenia, and, like astronomy, it began as an empirical science. Greek geometers learned their ropes in apprenticeship to Egyptians, whose stretched cords had served as both ruler and compass. From direct measurement, the Egyptians had developed a set of geometric laws that included special cases of the Pythagorean theorem and approximate areas of circles and spheres. Over five centuries the Greeks gradually built their theory of geometry on an experimental, initially Egyptian, foundation. Plato, recognizing the parallel between astronomy and geometry, hoped similarly to unify the rather complicated Babylonian rules governing the recurrence of eclipses and planetary positions: to his students he set the problem

> *to find the uniform and ordered movements by the assumption of which the motion of the planets can be accounted for.*

Solving Plato's problem required a quantitative understanding of gravity, and it ultimately led, early in the present century, to general relativity, a subtle unification of astronomy and geometry.

The parallel between the two sciences can, in fact, be drawn quite closely if one regards early geometry as a study of space via objects that show extreme spatial regularity, for early astronomy can be called a study of time via objects whose dynamics (behavior in time) show extreme temporal regularity. The importance of astronomy to agricultural societies, in whose religions it played a usually dominant role, was related to the accurate timekeeping the stars provided. In fact, until the development of atomic clocks stellar positions provided the most accurate measure of elapsed time.

Once the Greeks had recognized that the earth was spherical (by 200 b. c., Eratosthenes had measured its circumference to better than 10%), it was possible to explain the apparent daily revolution of the stars from east to west as the result of a real daily rotation of the earth from west to east, and at least a minority of Greek astronomers adopted this view. Although accounting for the simple apparent motion of the stars was thus a straightforward step from a round earth, solving Plato's problem – explaining the apparent motion of the planets – was much more difficult. The key obstacle was that, seen from the earth, the planets move against the background stars in paths that are only roughly repetitive and in which an average west-to-east motion is periodically interrupted by a retrograde motion from east to west. The Greek models of planetary motion reproduce retrograde motion by supposing that the planets move in epicycles (small circles) whose centers themselves move in circles about the earth. This description is in fact correct for a frame of reference centered at the earth: Seen from earth, the sun appears to move yearly in a circle, and the nearly circular motion of each planet about the sun is then approximated by an epicycle whose center (the sun) itself appears to circle the earth. Because planetary orbits are not exactly circles, the most accurate Greek model (Ptolemy's) was burdened by additional small epicycles for each planet.

Copernican Revolution

A simple description of planetary motion requires one to choose a frame of reference centered at the sun – to abandon thereby the apparently obvious fact that the earth is at rest at the center of the celestial sphere. A sun-centered system brings a dramatic conceptual simplification: To an observer at rest relative to the sun, there is no retrograde motion – the need for major epicycles disappears for all five visible planets at once. Although Aristarchos had, by 250 b.c., suggested such a system, subsequent Greek models failed to exploit his idea, and nearly two millenia passed before Copernicus finally contrived a model in which planets moved about the sun (Fig. 1).

The subsequent logical development has been rapid. Within a hundred years it was clear that if the earth did not define an absolute rest frame, neither did the sun: After Galileo and Newton it was clear that there was no natural candidate for an object at rest, and so no natural way to decide what it meant to be at the same place at two different times. One lost absolute space, the idea that one can pick out the same point in space at two different times. When the earth is a natural rest frame, one seems to need a force to make an object move. With no natural frame, there is no way to know if an object is moving or at rest. Thus, a force is needed not to make something move – with no force, a puck slides forever on perfect ice – but to change its motion.

With force needed only for acceleration, the force that acts on a planet in circular orbit about the sun must point to the sun, because the acceleration, the change in velocity per unit time, is an arrow pointing to the center of a circular orbit. This was the insight that Kepler, still tied to the idea of absolute space, had missed: from the fact that all objects fell to a round earth, he guessed that gravity was universal, that

> If two stones were placed anywhere in space near to each other and outside the reach
> of a third like body, then they would come together, after the manner of magnetic
> bodies, at an intermediate point, each approaching the other in proportion to the
> other's mass.

Failing to understand Galileo, Kepler still thought a force was needed to keep a planet moving. His spectacular success in deducing the elliptical shape of planetary orbits and the speed of a planet along any orbit was accompanied by a failed attempt to explain these simple motions in terms of the sun's magnetic field. To the next generation, knowing both Kepler and Galileo, it

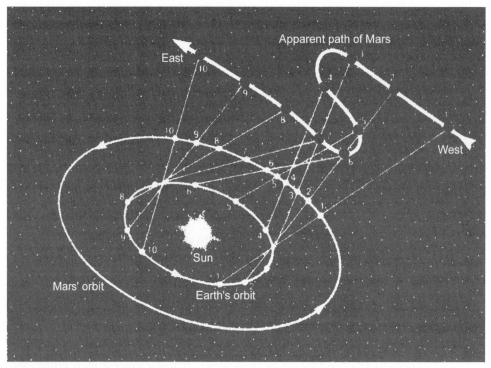

Fig. 1: The motion of planets, complex as seen from earth, is simple in a frame of reference at rest relative to the sun. The motion of Mars seen from earth and sun is depicted above: The earth moves more quickly in its orbit than Mars does, and as we pass Mars, it appears to reverse its direction of motion relative to the background stars.

was clear that gravity pointed in the right direction to bend a planet's path, and both Newton and Hooke guessed that the force between two point masses decreased with distance as $1/r^2$ (r the distance between the masses). Armed with his calculus, Newton proved that the single new axiom of a $1/r^2$ force (and Galileo's law that force was proportional to acceleration) implied Kepler's three laws and hence the motion of planets.

Newtonian Gravity

Newton's law governs the acceleration of one particle from the gravity of a second. If the second particle is at a distance r from the first and has mass M, the acceleration of the first particle given by the equation,

$$\frac{d^2\mathbf{r}}{dt^2} = -\frac{GM}{r^2}\hat{\mathbf{r}}, \tag{1}$$

where the constant G has the value $6.67 \times 10^8\,\mathrm{kg^{-1}m^3s^{-2}}$. This relation describes the geometry of each particle's path in spacetime: its path in space as a function of time. For mass distributed through space with density ρ, Eq. (1) takes the form

$$\frac{d^2\mathbf{r}}{dt^2} = -\nabla\Phi \,, \tag{2}$$

where

$$\nabla^2\Phi = 4\pi G\rho \,. \tag{3}$$

Using Eq. (2) one can easily prove a result that Newton struggled with for some years, that the acceleration of a ball (a spherically symmetric distribution of mass m) due to the gravity of a second ball of mass M is again given by Eq. (1), with r now the distance between the centers of the balls. The corresponding mutual attraction of two spherically symmetric objects is described by the force $F = GMm/r^2$.

The value of G was not known to Newton. To find it, one must measure the gravitational force between two known masses, and the gravitational forces that could be measured were between planets and the Sun or between the Earth and ordinary objects. John Michell, in 1795, designed an experiment in which a quartz fiber holds up a a horizontal dumbbell, a rod with lead balls on either end. Another lead ball is brought close enough to one of the suspended balls that its gravitational attraction twists the quartz fiber, and measuring the angular twist allows one to measure G. Michell died before doing the experiment, but he gave his torsion balance to Cavendish, who successfully measured G to about 1% accuracy. Using the acceleration $g = \frac{GM}{R^2} = 9.8 \,\mathrm{m/s}^2$ of falling objects, and the known radius of the Earth, Cavendish then famously found the Earth's mass.

At the close of the Copernican revolution, marked by Newton's law of gravity, all but one of the ancient problems of astronomy were explained – the apparent motion of sun, moon, stars, and planets, the seasons, tides, and eclipse cycles, the precession of the equinoxes. Only the "guest stars", apparent new stars whose sightings were recorded once every few centuries, remained mysterious.

The Geometry of Spacetime

Space and time are to some extent already blended in the Newtonian framework. By adjoining time to the three spatial coordinates of Descartes, Newtonian dynamics acquired at least the formal trappings of spacetime, the set of points or "events", $(t, \mathbf{x})$. More striking, however, is the mixing of space and time that Newton's laws imply: The path of an observer "at rest" and one moving at constant speed are both straight lines in spacetime, pointing to the future, but not parallel. Because there is no empirical way to decide who is at rest, one has no unique way of splitting spacetime into space + time. If we suppress a dimension and think of space as two dimensional, a plane with two-dimensional objects moving about on it, then spacetime can be regarded as a set of photographs of the plane, placed one on top of the other. An observer stacking the photographs aligns them so that successive pictures of an object at rest relative to the observer are directly over one another, forming a vertical line. Thus an observer at rest relative to the earth would place the earth on the $t = 1$ second photograph directly above the point that marks the earth on the $t = 0$ photograph. On the other hand, for an observer at rest relative to the sun, the earth moves at 30 km/s, forming a helix in spacetime. Any straight line directed toward the future is a path of an unaccelerated observer, and the fact that there is no absolute space means that there is no preferred direction for such a line.

By the middle of the last century, however, it appeared that absolute space was reestablished, when Maxwell found that light was an electromagnetic wave moving across space with finite speed c. For you can claim to be at rest relative to space – now conceived of as the "aether" through which light propagates – if, when you set off a firecracker, its light at all later times forms a sphere with you at the center. Instead of resurrecting absolute space, however, Maxwell's discovery turned out to destroy absolute time: In 1887, Michelson and Morley found experimentally that the speed of light was, in fact, independent of the motion of its observer (and of its source)! Thus, the new prescription for determining when you are at rest fails: Suppose, for example, you are in the middle of a train car and you try to decide whether you and the train are moving left or right by setting off a firecracker and observing whether you first receive the light reflected from the right wall or the left wall. Then regardless of the (uniform) speed of the train, both reflected signals will hit you at once. What is more, you will thereby deduce (by symmetry) that the light hit left and right walls simultaneously. But if the train is moving to the right, say, relative to an observer on the earth, the observer will report that the light hit the left wall before it hit the right wall. Thus, the two observers disagree on which events took place at the same time, and absolute time has been lost.

What remains? The absolute speed of light and the fact that no information can travel faster than light allows causality to survive. Only if two events occur so close together in time that information does not have time to travel from one to another can different observers disagree on which was first. The expanding sphere of light from a firecracker set off at a point $P = (t, x)$ expands to form a cone in spacetime, the "future light cone" of P. Anything that starts at P and travels more slowly than light remains inside this light cone, and the event P can thus affect only future events inside the light cone. (See Fig. 2.)

Observers who move along straight, parallel lines in spacetime are again at rest relative to each other, and they agree on what events occurred at the same time. These sets of simultaneous events are their surfaces of simultaneity, slicing spacetime into space + time. In the absence of gravity, these spatial slices are flat: they continue to obey the laws of Euclidean geometry. For each choice of time-like direction, one can label the surfaces of simultaneity by a coordinate $x^0 = ct$ for which the light cone through a point P, at $t = 0$, is the set of points $(\mathbf{x}, t)$, for which

$$\eta_{\alpha\beta} x^\alpha x^\beta \equiv c^2 t^2 - \sum (x^i)^2 = 0 , \qquad \alpha = 0, 1, 2, 3 ,$$

where $\|\eta_{\alpha\beta}\| = \mathrm{diag}(-1, 1, 1, 1)$ and x^i, $i = 1$–3, are Cartesian coordinates of the surface of simultaneity. The matrix η is the flat metric of this "Minkowski" spacetime, giving the distance between any two events, x^α and x^β.

If, for two events, the squared interval is positive, then there is an observer for which the events are simultaneous, and their spatial distance is $|x - x'| \equiv |\eta_{\alpha\beta}(x^\alpha - x'^\alpha)(x^\beta - x'^\beta)|^{1/2}$. If $\eta_{\alpha\beta} x^\alpha x^\beta$ is negative, then a clock moving along the straight line joining the events will measure the time interval $|x - x'|$.

This is the framework of Einstein's special relativity, but Newton's law of gravity is not consistent with it. When Eq. (1) is satisfied on the surfaces of simultaneity of one observer, it fails on those of any other. The problem is that according to Eq. (1), the position of the

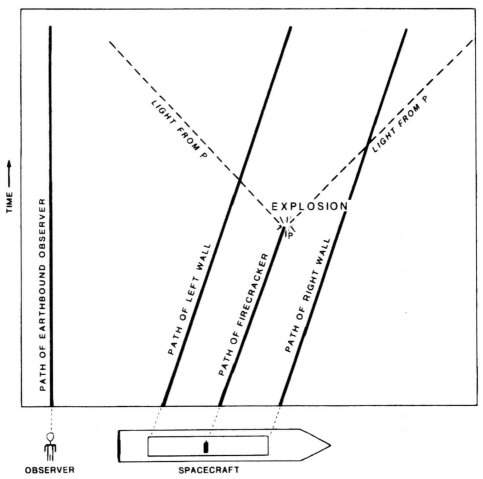

Fig. 2: Event *P* is the explosion of a firecracker set off midway between the left and right walls of a spacecraft. Seen by an observer on the spacecraft, the light hits the two walls simultaneously: A line joining events L and R is "space", the set of events that is simultaneous for this observer. But to an earthbound observer the light hits the left wall first: L occurs before R. The fact that nothing can move faster than light limits the tilt of line LR relative to the horizontal space of the earthbound observer, and it implies that events can be seen as simultaneous only if light does not have time to travel from one to another.

second particle (*M*) instantaneously determines the acceleration of the first, contradicting the requirement that no information can travel faster than light. Instead, to find a theory of gravity that agrees with Newton's for small relative velocities and is consistent with special relativity, one must abandon the assumption that space is flat. In its place Einstein realized in part the dream of the mathematician Riemann that particles curve owing not to forces but to the curvature of space.

The idea seems on the face of it absurd. Balls thrown on earth deviate obviously and macroscopically from straight-line motion, but space is experimentally flat near the earth to better than one part in a million. The key, however, is simple. In measuring intervals between events in spacetime, one uses the speed of light c to compare times to distances. A time of 1 s has the same magnitude as a distance of one light-second (300000 km), the distance light travels in 1 s.

Thus, in falling for 1 s, a ball drops less than 10^{-7} light-seconds, and its path in spacetime barely deviates from the 1-s-long straight line it would have described in the absence of gravity. Similarly, because the earth's speed in its orbit about the sun is $10^{-4}c$, in one year it traces out a nearly straight line in spacetime, a helix whose radius is 10^{-4} times smaller than its height.

The advent of general relativity at once destroyed Euclidian geometry and Newtonian astronomy, but it showed that a unification of space and time implied a unification of the two sciences as well. Newton's law (1) is replaced by the statement that particles move on geodesies, paths of shortest length on a curved spacetime. Equation (2), relating the gravitational potential Φ to the distribution of mass, becomes a limiting case of the equation relating the curvature of spacetime to the matter density. The flat metric of Minkowski space is replaced by a metric describing the curved spacetime geometry. The way in which mass curves space is given by the equation

$$R_{\alpha\beta} - \tfrac{1}{2}g_{\alpha\beta}R = 8\pi G T_{\alpha\beta} , \tag{4}$$

where $R_{\alpha\beta}$ is the Ricci curvature tensor ($R \equiv g^{\alpha\beta}R_{\alpha\beta}$), constructed from the metric and its first two derivatives. For a collection of particles with mass density ρ moving along paths with unit tangent u^{α}, the matter tensor $T^{\alpha\beta}$ is $\rho u^{\alpha}u^{\beta}$.

Newtonian Limit: Gravity in the Solar System

The sun, planets, and moons that comprise the solar system move with relative velocities small compared to c. In the frame of the solar system, the components of u^{α} are then approximately given by $(1,0,0,0)$. The metric has the form,

$$||g_{\alpha\beta}|| = \left\| \begin{matrix} 1+\dfrac{2\phi}{c^2} & & & \\ & 1+\dfrac{2\phi}{c^2} & & \\ & & 1+\dfrac{2\phi}{c^2} & \\ & & & 1+\dfrac{2\phi}{c^2} \end{matrix} \right\| \tag{5}$$

and Eq. (4) becomes, in this Newtonian limit, Eq. (2). The solar system's metric deviates from the flat metric η by at most one part in 10^6, and the corrections provided by general relativity to Newtonian gravity are similarly small. At the time it was formulated, the only discrepancy in Newtonian gravity was an inaccurate prediction of the precession of Mercury's elliptical orbit that would arise from the other planets' gravity. The predicted precession disagreed with observations by $43''$ of arc/century, and the first triumph of general relativity was to account for that difference.

Not until the past two decades was gravity systematically measured to a precision high enough that one could verify the geometry predicted by relativity. Among the most accurate of these are the Eötvös experiments, tests of the "principle of equivalence", the statement that particles with the same initial velocities move on identical trajectories, independent of the mass or composition of the particles. Galileo's demonstration at the tower of Pisa is the precursor to experimental work of Eötvös, and a number more recent investigators, searching for a dependence of the gravitational acceleration on the internal structure of two suspended masses. Lack of such a dependence is currently verified at a level of 1 part in 10^{13}. Suggestions of a "fifth force" that modifies the $1/r^2$ law at short distances is now ruled out at an accuracy of about 1 part in 10^4 at lengths above a meter, and at much higher accuracy, from orbital tracking of satellites, the Moon, and planets, at distances above 10^3 km.

The implication of these experiments is that, in a gravitational field that is nearly uniform over the size of a laboratory, one cannot distinguish the gravitational field from an accelerating laboratory. In a uniformly accelerating laboratory, or in one at rest in the earth's nearly uniform field, free particles will fall with constant acceleration relative to the laboratory. To discover that he is not in uniform acceleration, an observer confined to a closed laboratory must make measurements accurate enough to see a change in the gravitational field from one side of the room to the other, or to detect the energy lost in radiation by a particle at rest relative to the room. The approximate inability to distinguish a slowly varying gravitational field from uniform acceleration is known as the *principle of equivalence*. Historically one of the guiding principles leading to relativity, its shadow remains in the theory as the statement that any smooth geometry is approximately flat when measured over intervals short compared to its radius of curvature.

A second, related, class of experiments measure the gravitational redshift, the change in frequency of light as it climbs higher in a gravitational field. To conserve energy, the work done by gravity on a photon of energy E that travels upward a height h in the gravitational field of the earth must be the work done on a mass of magnitude E/c^2, namely, $(E/c^2)gh$, where g is the acceleration due to the earth's gravity, $g = 9.8 \, \mathrm{m/s^2}$. Because the energy of a photon is related to its frequency ω by $E = \omega$), a photon climbing up a height h is redshifted, changing its frequency by $\Delta\omega/\omega = -gh/c^2$. The first successful, high-precision redshift measurements (due to Pound, Rebka, and Snider) were made in 1960–1965. The redshift measures the part of the metric that determines rates of clocks at different heights in a static gravitational field, and hydrogen maser clocks on orbiting satellites have restricted the deviation from general relativity to a part in 10^4.

An additional set of tests involves the bending of light by the sun's gravity and delays in light travel time due to the comparably large curvature of the spacetime near the planets. The bending of light by the sun, the most famous of the theory's early predictions, was first observed during the 1919 solar eclipse. Current striking examples of light deflection are gravitational double images of distant quasars, for whom intervening galaxies act as gravitational lenses. Light deflection and light travel times depend on a parameter independent of the redshift, and measurements by Viking spacecraft and satellites have limited its departure from the prediction of relativity to 0.001%.

The predicted precession of the orbit of Mercury has been measured to a precision of about 0.1%, by using radar determination of the planet's position: By bouncing radio waves off an object and carefully timing their round-trip time, one obtains solar system positions to within a few meters. The test of precession, however, is limited not only by planetary position, but by our knowledge of the sun's shape: rotation of the solar interior could make the sun slightly oblate, and a departure from spherical symmetry also implies a rate of planetary precession.

Finally, comparison of orbital precession of the earth, and measurements of solid tides of the earth due to the combined gravity of sun, moon, and earth, restrict each of two additional independent parameters to within 0.1% of its predicted value. In summary, the past two decades of solar system tests have left little doubt that general relativity correctly describes gravity for objects that are large compared to atoms and small enough that a particle can escape their gravity with a speed small compared to light.

Gravitational Collapse and Black Holes

Black Holes The most dramatic predictions of general relativity are in exactly the cases where it is poorly tested, where objects are very small compared to atoms or have a large enough concentration of mass that one must travel at speeds approaching light to escape. In the Newtonian approximation, a particle of mass m can escape the gravitational field of an object if its kinetic energy, $\frac{1}{2}mv^2$, is larger than its gravitational binding energy, $m\phi$. For a spherical star, $\phi = -GM/R$, where R is the distance from the object's center. This means that a particle can escape only when its velocity is given by

$$v > v_{esc} \equiv \sqrt{2GM/R} \, . \tag{6}$$

At the surface of the earth, $v_{esc} = 11\,km/s$. From Eq. (6), the escape velocity from the surface of an object exceeds the speed of light when its radius shrinks to within the "Schwarzschild radius",

$$R_S = 2GM/c^2 \, . \tag{7}$$

This turns out to be the correct result in the exact theory as well, if R is defined as the length for which $2\pi R$ is the circumference of the object (the fact that space is curved then implies a larger distance from the object's center to its edge). For the sun, the Schwarzschild radius is 3 km, vastly smaller than the star's 700 000-km radius. Remarkably, however, stars with masses larger than 10 or 20 times that of the sun seem fated to end their lives by collapsing to within their Schwarzschild radius – to form black holes.

As a star evolves, the atomic nuclei in its core fuse to build up a sequence of heavier elements. The free protons and electrons (hydrogen ions) that largely comprise the initial star gradually fuse together to create the elements from helium through iron that populate the universe. Iron is the most stable nucleus, and in the most massive stars, an iron core gradually builds up as lighter nuclei fuse together. Because iron nuclei cannot fuse, the core supports itself by the pressure of its electrons, each restricted to a distinct state.

There is, however, a maximum mass that such matter can support, related to the finite speed of light. For an ideal gas, the pressure per unit mass (p/ρ) is equal to the kinetic energy per unit mass, equal to v^2, where v is the average speed of a gas molecule. The restriction $v < c$ suggests a limit on the pressure per unit mass of c^2, and this limit is, in fact, implied for any matter by the causality requirement that the speed of sound be less than the speed of light.

But there is no limit on the gravitational attraction per unit mass. As a result, there is a maximum mass that can be supported by electron pressure, the Chandrasekhar limit of $1.4M_\odot$, where $M_\odot$ is the mass of the sun. Shortly before the mass of a star's core reaches this limit, as the speed of the electrons approaches the speed of light, the core collapses, with the electrons pushed by gravity onto the iron nuclei. The entire core is thereby converted to neutrons,

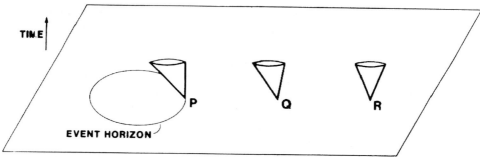

Fig. 3: The diagram above shows the path followed by light emitted at points outside the event horizon of a black hole. One spatial dimension is suppressed, so the horizon of the black hole is a circle. A flash of light emitted at time 0 expands to form a widening sphere, depicted in the diagram as a widening circle. Far from the event horizon (point R), the circle is centered about the point where the light was emitted; the flash from Q is pulled somewhat inward; and the flash at P, on the event horizon, is pulled inward so strongly that even the outward directed light cannot escape – rays directed radially outward remain forever at the same radius.

becoming a neutron star whose mass of nearly $1\frac{1}{2}$ times the mass of the sun is compressed to a radius of about 10 km. The gravitational energy of the collapse blows off the rest of the star in an explosion called a supernova. A closely similar collapse is thought to occur when white dwarfs accrete enough matter from a companion star that they exceed the Chandrasekhar limit. The "guest stars" of ancient astronomy were supernovae, their explanation an achievement of the last half-century.

Like the iron core or white dwarf from which it forms, a neutron star has a maximum mass set by causality – by the finite speed of light. Our lack of knowledge of the behavior of matter at the extreme densities of neutron stars, 10^{15} g/cm^3, means that the largest neutron star mass is known only roughly, but general relativity implies that no spherical configuration of matter at neutron density or higher can have a mass greater than $5M_\odot$. For dense matter with mass greater than this, the escape velocity exceeds the speed of light, and the star collapses to form a black hole.

In the collapse of a star, the escape velocity at the star's surface increases as the star contracts. When the surface passes through the Schwarzschild radius, R_S, the escape velocity exceeds that of light, and no light (or matter) can escape from the region inside a sphere of circumference $2\pi R_S$. In fact, within the classical theory, the matter inside is drawn to a single point. The gravitational field, however, persists and a region of empty space outside the tiny collapsed star is cut off from the surrounding universe: the light cones at the Schwarzschild radius tip inward, preventing anything from emerging (Fig. 3). Events inside the "event horizon" at $r = R_S$ are inaccessible to external observers. When the collapse is not spherical, as in the collapse of a rapidly rotating star, the event horizon is oblate, but a theorem of Penrose again implies, within classical gravity, a singular evolution inside the black hole.

The experimental evidence for black holes has become extremely strong. Black holes can in principle be observed in binary systems (systems of two stars orbiting their common center of mass) when mass from the ordinary star falls onto the black hole. (The expansion of the ordinary star to form a red giant late in its evolution is commonly the event that triggers the accretion: matter at the outer edge of the giant is closer to the black hole than to the center of its parent star.) The accreting matter moves at speeds comparable to c, and emits a rapidly varying spectrum of x rays. While matter falling onto a neutron star will also emit x rays, observations of the binary system can set a lower limit on the mass of the dense member. In cases where this mass is above $5M_\odot$ there seems to be no way to avoid the conclusion that one is observing a black hole. The best studied black hole candidates of this kind are in the binary systems Cygnus X-1, LMC X-3, and Cir X-l, all x-ray sources for which the candidate is a dense star with mass greater than $8M_\odot$.

Even more remarkable is a second class of black hole candidates in the centers of large galaxies. Observations of the central regions of galaxies show a sharp increase in brightness near the center. To account for the emission with a cluster of stars, the density of stars would have to be exceptionally high, high enough that it seems quite difficult to avoid rapid coalescence to giant stars, leading to gravitational collapse to large black holes. The resulting prediction of sharp central concentrations of mass has been dramatically verified in recent years by radio telescopes electronically joined across the globe (railed very long baseline interferometers). The resolution of these telescopes is comparable to a single telescope the size of the earth, and by looking at the velocity of clouds orbiting the centers of massive galaxies, they reveal masses as large as $5 \times 10^9 M_\odot$ in a region the size of our solar system. Observations appear to show a central black hole in every large galaxy. In the center of own galaxy, the Milky Way, stars a a few solar-system distances from the core orbit at $1/100$ the speed of light, implying an object of 2.6 million solar masses in a volume less than 10^{-6} the average galactic volume per star.

Gravitational Radiation

A second implication of the finite speed of information is the prediction of gravitational waves, waves of curvature that travel at the speed c. For nearly flat spacetimes, with $h_{\alpha\beta} \equiv g_{\alpha\beta} - \eta_{\alpha\beta}$ small, Eq. (4) becomes the wave equation for $h_{\alpha\beta}$:

$$h_{\alpha\beta} \equiv \left[\frac{\partial^2}{\partial x^{0^2}} - \frac{\partial^2}{\partial x^{1^2}} + \frac{\partial^2}{\partial x^{2^2}} + \frac{\partial^2}{\partial x^{3^2}} \right] h_{\alpha\beta} = 0 \,.$$

The first evidence for these waves came from observations observations of a binary system of two neutron stars, discovered in 1974 by Hulse and Taylor. The two identical stars orbit their common center of mass, in an orbit that slowly shrinks as the system radiates energy in gravitational waves at the rate corresponding to that predicted by general relativity. Thirty years of accumulated data verify the predicted energy loss to an accuracy of about 0.1%, and more recently discovered neutron star binaries will significantly improve the precision.

Ultimately pairs of orbiting neutron stars and pairs of orbiting black holes lose enough energy to coalesce. In the last few seconds of their life, they move at speeds up to $1/3$ the speed of light, orbiting hundreds of times each second and emitting gravitational waves strong enough to be detected. The exciting prospect of a gravitational wave astronomy that could see the coalescence of otherwise invisible black holes and reveal the structure of neutron stars has spurred the US, Europe and Japan to build large gravitational wave detectors. These *laser interferometers* measure minute changes in the distance between a pair of mirrors a few km apart by laser beam that bounces more than 10 000 times between them. Gravitational waves passing between the mirrors slightly distort the geometry, periodically shortening or lengthening the distance between the mirrors, with a period half that of the orbiting stars that emitted them. By monitoring the shift in interference fringes between beams bouncing between mirror pairs that lie along perpendicular directions, one can detect changes in distance smaller than the size of an atomic nucleus. And that is the sensitivity required to see the strongest sources of gravitational waves.

Within the next decade, the most sensitive earthbound detectors are expected to observe several events each year, and may possibly see hundreds of coalescing neutron-star and black-hole binaries. The supermassive black holes in the centers of galaxies may have grown by a series of coalescences, but the waves emitted in events involving giant black holes are too long to be seen by earthbound detectors. Their gravitational waves are the target of a space-based laser interferometer (LISA) planned for about 2015 by the European and US space agencies (ESA and NASA).

Cosmology

On the largest scale, the visible universe appears to be uniform in its distribution of matter, which on smaller scales is clumped into stars, galaxies, and a hierarchy of clusters of galaxies. Models for the universe, satisfying the field equations (4), are either expanding or contracting, and the discovery by Slipher, Hubble, and Humason of a uniform expansion traced by galactic redshifts should have been an early triumph of the theory. It was not only because Einstein had by then added an additional term $\Lambda g_{\alpha\beta}$ to the equation to allow a static universe. If one follows an expanding universe backward in time, one encounters ever-increasing densities, and a second theorem of Penrose, this time in collaboration with Hawking, again implies a singularity, marking the beginning of the universe, or at least of the part of the universe in which we reside. The explosive expansion immediately after the singularity is termed the "big bang".

There is little doubt that the big bang in fact occurred. The most remote galaxies, over 1×10^{10} light-years distant, are seen as they were over 1×10^{10} years in the past, and as one looks back in time toward the big bang the character of galaxies changes. Finally at redshifts greater than about 10, one sees no galaxies at all: They could not have formed until after the big bang. The age of the universe found from the big-bang expansion is consistent with the independently found age of its oldest stars; and the abundances of the lightest elements is consistent with the nuclear reaction-rates in big-bang models. But the key piece of evidence concerns the cosmic microwave background radiation predicted by Alpher, Bethe, and Gamow in 1948. In the hot, dense, early universe, light and matter are in equilibrium, but as the universe expands, the time between successive collisions of a photon with the surrounding matter grows longer, until, after about 1×10^5 years, the density is low enough that the time for a photon to hit a particle is longer than the time to cross the universe. Light from that time should now be uniformly distributed and should have been redshifted in tandem with the expansion, its wavelength proportional to the radius of the universe. In 1965, the prediction was confirmed, with the spectacular discovery by Penzias and Wilson of isotropically distributed microwave radiation, and its identification with the predicted background by Dicke *et al*. Subsequent satellite observations depict a nearly exact blackbody spectrum at 2.73 K.

Beginning in 1996, data from the COBE satellite showed a slight deviation from perfect uniformity in the microwave background. These fluctuations show temperature differences in the big bang of about one part in 10^5 and the corresponding density fluctuations are thought to be the seeds of the large-scale structure of the present universe. Subsequent, more sensitive observations showed a universe whose nearly flat large-scale spatial geometry supports an early "inflation," in which an exponential growth of the universe in the first 10^{-32} s after the big bang imitates the behavior induced by the cosmological constant Einstein introduced and then discarded. Remarkably, recent improvements in the accuracy of distance determinations to the most distant galaxies apparently reveal a universe that is accelerating in the way a different, mysteriously small, but nonzero Λ would predict.

Quantum Gravity

On scales larger than the size of an atom, gravity appears to be accurately described as a deterministic theory of the geometry of spacetime. The theory, however, is not consistent with quantum mechanics, whose probabilities characterize matter on small scales, as well as large. The logical inconsistency of the two theories is exacerbated by the singularity theorems, which show that within black holes, and at early times in the history of our universe, the density of matter is so high that one can no longer do physics without a way to unify the theories.

While a quantum-mechanical geometry must follow the quantum-mechanical probabilities for atoms on an atomic scale, gravitational fields on such scales are far too small to be detected. Only when one looks at distances of the order of the Planck length,

$$ l = \left(\frac{c}{G^3} \right)^{1/2} = 1.6 \times 10^{-33} \, \text{cm} , $$

do quantum fluctuations in the geometry become noticeable. At such lengths, even the topology of spacetime may fluctuate. Spacetime might have more than three spatial dimensions, with compact topology, lengths too small too small, and a geometry that reveals itself as the non-gravitational forces. Child universes might branch off our own, or the concept of spacetime as smooth geometry might simply fail to describe reality on so small a scale. For somewhat larger distances, however, fluctuations of the geometry can be ignored and a convincing approximate theory has been developed. Among its more striking implications is a predicted evaporation of black holes: The quantum nature of matter alters gravity, allowing the horizon of an isolated black hole gradually to shrink as it radiates its mass in a thermal spectrum of particles.

See also: Mach's Principle; Newton's Laws; Relativity, General.

Bibliography

J. B. Hartle, *Gravity, An Introduction to Einstein's General Relativity*. Addison-Wesley, San Francisco, 2003.

S. W. Hawking, *A Brief History of Time*. Bantam, New York, 1988.

C. W. Misner, K. S. Thorne, and J. A. Wheeler, *Gravitation*. W. H. Freeman, San Francisco, 1973.

R. M. Wald, *Space, Time, and Gravity*. Chicago, 1977.

C. M. Will, "The Confrontation Between General Relativity and Experiment," *Living Reviews in Relativity*, **4**, 2001; Online at www.livingreviews.org/lrr-2001-4.

Gravitational Lenses

J. N. Hewitt

Gravitational lensing occurs when the ray paths of electromagnetic radiation are altered as they pass through a gravitational field. The first measurements of the influence of a gravitational field on the propagation of light took place during the solar eclipse of 1919, when the deflection due to the gravitational field of the sun was measured. After this measurement, there were many articles in the literature of physics and astronomy that discussed the expected properties of images of objects viewed through a gravitational lens. In particular, it was recognized that a gravitational lens is likely to produce distorted, multiple images of the background source. Two short articles that have received considerable attention were those of Einstein [1], in which he showed that perfect alignment of the source and a circularly symmetric lens would produce a circular ring image (now popularly referred to as an "Einstein ring"), and of Zwicky [2], in which he pointed out that the large masses and distances of galaxies and clusters of galaxies implied that gravitational lenses would be observed in extragalactic sources of radiation. The predictions of both articles have been realized: the first gravitational lens, consisting of a distant quasar lensed by a large galaxy and its associated cluster of galaxies, was discovered in 1979 [3], and the first Einstein ring was discovered in 1988 [4]. There are now some 20 gravitational lens systems known, all involving propagation of light across intergalactic distances. They include multiply imaged quasars and galaxies, two Einstein rings, and the luminous arcs which surround many dense clusters of galaxies.

The deflection of a ray of light in the gravitational field of a spherically symmetric mass distribution is given by the following formula, due to general relativity, for the angle θ between the perturbed and unperturbed light ray:

$$\theta = \frac{4GM}{c^2 b} \, ,$$

where G is the gravitational constant, M is the mass, c is the speed of light, and b is the impact parameter of the light ray with respect to the mass. In the limit of a weak gravitational field, one can sum the contributions to the deflection angle of a collection of point masses; therefore, the lensing properties of an arbitrary mass distribution can be calculated. Most of the qualitative features of the known gravitational lenses can be reproduced by calculations that invoke a gravitational potential with elliptical symmetry.

Gravitational lenses provide a new laboratory in which to address problems in astrophysics. They are unique in that they signal the presence of concentrated matter, regardless of whether the matter emits electromagnetic radiation. There is evidence for dark matter in the universe in the dynamics of galaxies and clusters of galaxies. The presence of dark matter on several scales should manifest itself in gravitational lenses. On the smallest scales, from planet-sized bodies to masses of approximately 100 solar masses, the lenses within galaxies act as "microlenses", causing apparent fluctuations in the brightness of images. The time scale of these fluctuations depends on the amount of matter present on these scales, as well as on the geometry of the lens, the relative velocities of the lens and the source, and the size of the source [5]. On intermediate scales, consisting of galaxies and clusters as a whole acting as lenses, the geometric properties of the images give independent measures of the ratio of mass to light in the systems. On the largest scale, that of the universe itself, the frequency of occurrence of

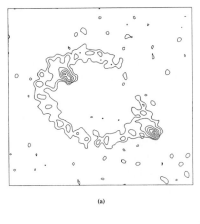

(a)

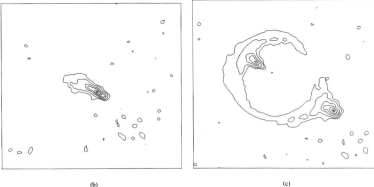

(b)   (c)

Fig. 1: (a) A radio image of an "Einstein ring" gravitational lens as observed with the National Radio Astronomy Observatory's Very Large Array radio telescope. The contours represent loci of equal surface brightness, (b) The hypothesized background source responsible for the Einstein ring, viewed without the gravitational lens in the foreground, (c) The calculated appearance of the source in (b) if it were viewed through a gravitational lens characterized by an elliptical potential. The model of (b) and (c) reproduces the major features of the object in (a). Figures (b) and (c) represent calculations presented by C. Kochanek *et al.*, *Monthly Notices of the Royal Astronomical Society*, Volume 238, page 43.

gravitational lenses places limits on the quantity of cosmologically distributed dark matter in the form of concentrated matter.

A second application of gravitational lensing is in the determination of Hubble's constant. The different images of a gravitational lens travel along different paths from source to observer, and in general have different propagation times. Therefore, a fluctuation in the brightness, or other property, of the source will be observed at different times in the different images, giving a measure of the time delay between the images. The angle between the images is measured, and this is related to the time delay by the distances between the source, the lens, and the observer. For example, for a point mass lens, the expression relating the angular image separation, $\Delta\theta$, and the time delay Δt, is

$$\Delta t = \frac{(1+z_L)}{2c} \frac{D_{OL}D_{OS}}{D_{LS}} \left[\beta\Delta\theta + 2\ln\frac{\Delta\theta + \beta}{\Delta\theta - \beta} \right] ,$$

where z_L is the redshift of the lens, D_{OL} is the observer–lens distance, D_{OS} is the observer–source distance, D_{LS} is the lens–source distance, β is the angle between the (unlensed) source and the optic axis of the lens measured in units of the Einstein ring radius of the lens, and c is the speed of light. The distances are inferred from measurements of the lens and source redshifts, and scale inversely with Hubble's constant. Therefore, measurements of the image separation, time delay, and redshifts in principle give a measurement of Hubble's constant. In practice, the relationship between the time delay and the image separation may not be clearly determined because of uncertainties in the mass model for the lens. To date, measurements of the time delay have been reported in only one system [6], and the mass model in that system is not well constrained. However, the inferred value for Hubble's constant is consistent with estimates made using other methods.

A third application is simply to use the gravitational lens as a magnifying glass, giving astronomers the means to view distant objects in more detail than otherwise possible. Of particular interest would be the observation of a magnified quasar or active galactic nucleus. The degree of magnification, however, is a very strong function of the alignment of the source and the lens. While several known lens systems appear to offer moderate magnification of the background object, a highly magnified quasar or active galactic nucleus remains to be discovered.

The study of gravitational lenses is relatively young, and it is most likely that their potential has yet to be realized. Gravitational lenses can be applied to many problems in astrophysics, and observational programs aimed at exploiting their properties are under way. New systems are being discovered at a rate of several per year, and many with remarkable symmetry, such as the Einstein rings, are among them.

See also: Astrophysics; Cosmology; Gravitation; Relativity, General Theory.

References
[1]   A. Einstein, *Science* **84**, 506 (1936).
[2]   F. Zwicky, *Phys. Rev.* **51**, 290 (1937).
[3]   D. Walsh, R. F. Carswell, and R. J. Weymann, *Nature* **279**, 381 (1979).
[4]   J. N. Hewitt, E. L. Turner, D. P. Schneider, B. F. Burke, G. I. Langston, and C. R. Lawrence, *Nature* **333**, 537 (1988).
[5]   R. Kayser, S. Refsdal, and R. Stabell, *Astron. Astrophys.* **166**, 36
[6]   C. Vanderriest, J. Schneider, G. Herpe, M. Chevreton, M. Moles, and G. Wlérick, *Astron. Astrophys.* **215**, 1 (1989).

Bibliography
R. D. Blandford, C. S. Kochanek, I. Kovner, and R. Narayan, "Gravitational Lens Optics", *Science* **245**, 797 (1988).
E. L. Turner, "Gravitational Lenses", *The Fourteenth Texas Symposium on Relativistic Astrophysics* (E. J. Fenyves, ed.). Acdemy of Sciences, New York (1989).
E. L. Turner, "Gravitational Lenses", *Sci. Am.* **259**, 54 (July 1988).

Gravitational Waves

P. R. Saulson and J. A. Tyson

Einstein's field equations relating space-time curvature to energy-momentum,

$$R_{\mu\nu} - \frac{1}{2} g_{\mu\nu} R = \frac{8\pi G}{c^4} T_{\mu\nu} ,$$

have radiative solutions: gravitational waves. These waves in the curvature of space-time exist embody the requirement that gravity obey the key principle of relativity, that no signal can travel faster than the speed of light. The meaning of these solutions is clearest in the weak field limit, i. e., when the metric can be written as $g_{\mu\nu} = \eta_{\mu\nu} + h_{\mu\nu}$, with h a small perturbation. In an appropriate coordinate system (the transverse-traceless gauge), the wavelike perturbations have a particularly simple form. Consider a wave that propagates in the z direction. The wavelike perturbations can be a superposition of two polarizations,

$$h_+ = \begin{pmatrix} 0 & 0 & 0 & 0 \\ 0 & a & 0 & 0 \\ 0 & 0 & -a & 0 \\ 0 & 0 & 0 & 0 \end{pmatrix}$$

and

$$h_\times = \begin{pmatrix} 0 & 0 & 0 & 0 \\ 0 & 0 & b & 0 \\ 0 & b & 0 & 0 \\ 0 & 0 & 0 & 0 \end{pmatrix}.$$

Gravitational waves are produced by accelerated mass-energy, analogous to the production of electromagnetic waves by accelerated charges. However, there is no gravitational dipole radiation, since linear momentum and angular momentum are conserved. The first nonvanishing radiative multipole is quadrupole, with an amplitude of

$$h_{\mu\nu} = \frac{2G}{Rc^4} \ddot{I}_{\mu\nu}, \tag{1}$$

where $I_{\mu\nu}$ is the quadrupole moment of the source,

$$I_{\mu\nu} \equiv \int dV \left(x_\mu x_\nu - \frac{1}{3} \delta_{\mu\nu} r^2 \right) \rho(\mathbf{r}) .$$

J. Taylor *et al.* have presented compelling evidence that the binary pulsar 1913+16 is losing orbital energy and angular momentum at approximately the rate expected from gravitational radiation due to the 7.8-hour orbital motion of the two neutron stars. The observed derivative of the orbital period, $(-2.1 \pm 0.4) \times 10^{-12}\,\mathrm{s\,s^{-1}}$ is in excellent agreement with theory $(-2.38 \pm 0.02) \times 10^{-12}\,\mathrm{s\,s^{-1}}$, using the separately determined masses, orbital period, and eccentricity. In the absence of direct detection of gravitational waves, pulsar 1913+16 remains our only experimental check of the general relativistic quadrupole radiation formula. It is an ideal relativistic laboratory for testing theories of gravity.

For a physical interpretation of a weak gravitational wave (all that we are likely to ever encounter!), imagine three freely falling astronaut-scientists arranged like the letter "L", in which the two astronauts at the ends of the "L" are a distance l from the corner. Chart their progress by plotting their world lines: If space-time were flat, their world lines would be straight. If a gravitational wave passed by in a direction perpendicular to the plane of the L, their world lines would become wavy. The quadrupolar character of wave emission is mirrored in the quadrupolar character of wave reception; the separation change between the astronaut at the vertex and the one at one end of the L will be equal but opposite to the separation change along the other arm. (This is the physical meaning of the form of h_+ and $h_\times$.) When the wave and the L are optimally aligned, the fractional separation change $\Delta l/l$ is given by $1/2$ of the amplitude h of the wave. The astronauts could detect the presence of this wave by exchanging bursts of light and measuring their separation (now oscillating) as a function of time. An interferometric detector works in much the same way. A Michelson interferometer is a device that compares the travel time of a light beam along one of its arms to that in the other, giving a response when the difference between them changes. It can be turned into a gravitational wave detector by making the mirrors (nearly) freely falling, by suspending them as pendulums. The shift in arm lengths causes a phase difference between the arms that is revealed as a change in the brightness of the superposed light beams returning from the two arms. Another actual implementation is the Doppler measurement of the variations in the velocities of one or more interplanetary spacecraft with respect to a tracking station on the earth.

In a frame of reference such as one used to describe ordinary laboratory experiments (in particular, with distances marked by rigid rods), the effect of a gravitational wave appears as equal and opposite forces on the two masses. Joseph Weber's pioneering experiments to search for astrophysical bursts of gravitational waves basically used this idea, with an elastic bar (made of aluminum) playing the role of two separated masses (the ends of the bar) connected by a spring (the middle of the bar.) As a gravitational wave passes by, the wave does work on the bar's elastic forces, imparting energy to the fundamental longitudinal vibrational mode of the rod. After the wave has passed, the bar still rings; the persistence of the vibration until the elastic energy has been dissipated is a key part of Weber's detection strategy. By contrast, the separations of freely-falling masses (in an interferometer) track the instantaneous value of $h(t)$. This makes it more straightforward to extract the waveform, but puts extreme demands on the noise in the system used for reading out the test mass separations.

By 1969 Weber had assembled two aluminum bar detectors at 1660 Hz and was claiming coincident detection of bursts apparently in the direction of the center of our Galaxy. Estimates of source luminosity ranged between 10^3 and $10^5 M_\odot c^2 \, \text{yr}^{-1}$. Subsequently, however, several other groups using detectors of greater sensitivity searched for kilohertz-band bursts and found none. Undaunted, the search for gravitational waves goes on, with detectors of ever increasing sensitivity.

Weber's room temperature bars were read out by dozens of piezoeletric strain sensors glued to the bar's midsection. More recent examples cool the bars to the temperature of liquid helium (or below) to reduce Brownian motion of the bar's modes, and read out the vibration of the fundamental longitudinal mode with ultra-sensitive resonantly-tuned accelerometers coupled to superconducting (SQUID) amplifiers (also made possible by the low temperature of the bar.) The best of the latest generation are now sensitive to waves whose amplitude h is below 10^{-18}, for signals with power near the resonance frequencies of about 900 Hz.

Interferometric detectors use a very different strategy to achieve good sensitivity. They operate at room temperature, but minimize the impact of mechanical noises such as Brownian motion of the mirrors by placing the mirrors very far apart (up to 4 km in the interferometers of the Laser Interferometer Gravitational Wave Observatory, or LIGO.) Precision of the arm length difference measurement itself is limited by shot noise in the photon flux in the interferometers; this is made small by using powerful lasers and optical resonance techniques.

How weak are the signals now being searched for? For an order of magnitude estimate, consider a binary star system, say a double neutron star system like PSR 1913+16. Calculation of the second time derivative of its moment of inertial yield the following approximate equality: $h \approx r_{S1} r_{S2} / r_0 R$, where r_{S1} is the Schwarzschild radius $2GM/c^2$ of star 1, $2r_0$ is the separation of the stars, and R is the distance of the system from the detector. If we consider the radiation shortly before the neutron star binary system coalesces, then some representative numbers would be: $M_1 = M_2 = 1.4 M_\odot$, $r_{S1} = r_{S2} = 4.15$ km, $r_0 = 20$ km. If we imagine the system at the distance of the Virgo Cluster, $R = 16$ Mpc $= 4.9 \times 10^{20}$ km. Then, the instantaneous amplitude of the gravitational wave at the earth would be about 2×10^{-21}.

Detector technology is advancing rapidly, and is expected soon to lead to successful detections of astronomical signals. The LIGO and Virgo multi-km interferometers will be sensitive to brief bursts with amplitudes around 10^{-21} in their present data runs, for signals from around 10 Hz to a few kHz. GEO's (0.6 km) and TAMA's (0.3 km) are not far behind, focused on the upper part of the interferometer signal band. Thus these detectors have the sensitivity to detect plausible signals like the one in the previous paragraph. Conservative astronomical estimates suggest that up to an order of magnitude better sensitivity might be required to see frequent signals. Upgrades envisioned for the next decade will in fact improve sensitivity by more than one order of magnitude in h. In addition to neutron star binary coalescences, signals from stellar core collapse, non-symmetric pulsars, cosmic strings, black holes, gamma ray bursts, and various early universe processes are also being sought.

Also in the mid-2010s, the Laser Interferometer Space Antenna (or LISA) should be launched. It will place 3 spacecraft into solar orbit in the form of an equilateral triangle of arm length 5×10^6 km. Using drag-free technology to minimize mechanical noise and phaselocked laser transponders to link the spacecraft into a set of interferometers, LISA will have its best sensitivity in the band from 1 to 100 mHz. LISA is expected to be able to detect the formation of supermassive black holes at the centers of galaxies at cosmological distances, as well as the signals from many white dwarf binaries on our Galaxy.

Low-frequency signals (in the vicinity of 10^{-8} Hz) have long been sought in fluctuations in the arrival times of ultra-stable pulsars. Gravitational waves of horizon-scale wavelengths in the early Universe may be detected as distinctive patterns in maps of the polarization of the cosmic microwave background radiation.

See also: Black Holes; Pulsars; Relativity, General; Spacetime.

Bibliography

M. Bartusiak, *Einstein's Unfinished Symphony*, Joseph Henry Press, Washington, D.C., 2000. (E)

P. R. Saulson, *American Journal of Physics* **65** (6), 501 (1997). (E)

H. M. Collins, *Gravity's Shadow*, University of Chicago Press, Chicago, 2004. (E, I)

P. F. Michelson, J. C. Price, and R. C. Taber, *Science* **237**, 150 (1987). (I)

B. C. Barish and R. Weiss, *Physics Today* **52** (10), 44 (1999). (I)

P. R. Saulson, *Fundamentals of Interferometric Gravitational Wave Detectors*, World Scientific, Singapore, 1994. (I)

D. G. Blair, *The Detection of Gravitational Waves*, Cambridge University Press, Cambridge, 1991. (I)

C. W. Misner, K. S. Thorne, and J. A. Wheeler, *Gravitation*, pp. 943–1044. W. H. Freeman, San Francisco, 1973. (I,A)

J. A. Tyson and R. P. Giffard, *Annu. Rev. Astron. Astrophys.* **16**, 521 (1978). (I,A)

J. M. Weisberg and J. H. Taylor, *Phys. Rev. Lett.* **52**, 1348 (1984). (A)

Gravity, Earth's

J. C. Harrison

Newton's law of gravitation states that particles attract each other with a force proportional to the product of their masses and inversely proportional to the square of the distance between them. Gravity at a point on the earth's surface, defined as force per unit mass or equivalently as acceleration of a freely falling body, consists of the net gravitational attraction of the earth's mass together with the centrifugal force of rotation. The latter (about 0.3% at the equator) is not required in space applications where an inertial reference frame is employed. The unit the Gal (after Galileo), equal to $0.01\,\mathrm{m/s^2}$, is frequently used. Other related quantities include astronomical latitude and longitude – the direction of the gravity vector; and the geoid – the equipotential or level surface which, over the oceans, coincides with mean sea level. Gravity varies predominantly with latitude (from 983.2 Gal at the poles to 978.0 Gal at the equator), because of the rotation and nonspherical shape of the earth, and with height above sea level (0.3086 mGal/m). The remaining variations, 400 mGal extreme, 34 mGal root mean square, are due to topography and density variations within the earth.

On land, gravity can be measured by a transportable free-fall apparatus with an absolute accuracy of about 3 µGal [1]. However, more commonly and easily, gravity meters [2] (sensitive spring balances) capable of 30-µGal accuracy worldwide and 3-µGal accuracy locally under carefully controlled conditions are used to measure gravity differences. Heavily damped gravity meters and high-quality accelerometers operating on gyroscopically stabilized platforms are used on ships with an accuracy which can be as high as 1 mGal but is often degraded by navigational uncertainties [3]. These instruments can also be used on helicopters and fixed-wing aircraft with some loss of accuracy and resolution [4, 5]. Absolute levels and scale factors of gravimetric surveys are based on the International Gravity Standardization Net, 1971 [6], which replaced the Potsdam System in use from 1909 to 1971. This network of nearly 500 stations is based on a reduction of many pendulum and gravity-meter measurements, with the pendulum measurements controlling the scale factors of the meters employed and 11 absolute determinations by free-fall methods providing the absolute level. Its worldwide accuracy is 0.1 mGal, although much higher accuracies are attained in some national gravity base networks as a result of great improvements in the accuracy and transportability of free-fall apparatus since 1971. Gravity at a given site varies periodically with time as a result of the tidal attractions of the sun and the moon, of the ocean tides, and of the associated deformation of the Earth. This variation, less than $200\,\mu$Gal in amplitude, is predictable at most places to within a few tenths of a μGal [7]. Nonperiodic changes due to vertical ground motion, subterranean motion of fluids, and igneous activity have been reported from some sites, usually at the sub-mGal level.

The correlations with latitude and elevation account for much of the variation of gravity, and so it is convenient to work with gravity anomalies, the amount by which gravity differs from a normal (ellipsoidal) model in which these factors are taken into account. This model compares gravity with that outside a body whose surface is (1) an exact ellipsoid of revolution and (2) a gravity equipotential, and (3) whose equatorial radius, geocentric gravitational constant (GM), rate of rotation, and dynamical form factor J_2 (see below) have specified values close to those observed for the actual earth [8]. Gravity γ on this surface is given by

$$\gamma(\phi) = 978032.677(1 + 0.001931851 \sin^2 \phi)/(1 - 0.006694380 \sin^2 \phi)^{1/2} ,$$

where ϕ is geodetic latitude. This formula replaces earlier versions adopted in 1930 and 1967. The vertical gradient of normal gravity is close to $-0.3086\,$mGal$/$m and the difference

$$g_{meas} - \{\gamma(\phi) - 0.3086h\} ,$$

where h is the station elevation above sea level (the geoid), the anomaly after correction for height is the *free air anomaly*. (A similar expression but using height above the reference ellipsoid gives the *gravity disturbance*.) The *Bouguer anomaly* is obtained by subtracting the attraction of the topography above sea level from the free air anomaly.

The free air anomaly is strongly correlated with topography on a local scale but not regionally (continents to oceans, mountains to plains, etc.). The local topographic effects are eliminated from the Bouguer anomalies, which are therefore the anomalies normally used in geophysical prospecting. These anomalies are, however, strongly correlated with regional topography because this is largely compensated – that is, mountains are underlain by less dense rocks than oceans, so that the obvious mass excesses and deficits due to the topography are normally canceled out (isostasy). There are many specific hypotheses for the distribution of this compensating mass, and for each of these an isostatic gravity anomaly can be computed. The isostatic effects are often represented spectrally in the form of a response function giving the mGal of free air gravity anomaly associated with 1 m of topographic amplitude as a function of wavelength. This response tends to zero at long wavelengths, tends to $2\pi G\rho$ (where ρ is the rock density) at short wavelengths, and its behavior between these extremes is related to the ability of the lithosphere to support topographic loads.

The gravity field in space is important because of its influence on the orbits of artificial satellites. Accurate prediction of these orbits requires that the gravity field be well known; conversely, gravity field models can be improved by adjusting gravity field parameters and tracking station coordinates to minimize the sum of the squares of the residuals between satellite positions and velocities as predicted from the model and as obtained from tracking data. In geophysics the potential is defined at a point in space as

$$V = G\sum_i \frac{m_i}{r_i} ,$$

where masses m_i are at distances r_i from the point. Force per unit mass is given by grad$V$. (This sign convention is opposite to that adopted in physics.) In free space, V obeys Laplace's equation and in the vicinity of the earth V can be written in a series expansion in spherical harmonics. A frequently used form is

$$V_{\text{space}} = \frac{GM}{r} \left[1 + \sum_{n=2}^{n_{\max}} \sum_{m=0}^{n} \left(\frac{a_e}{r}\right)^n \bar{P}_{nm}(\sin\phi) \times \{\bar{C}_{nm}\cos m\lambda + \bar{S}_{nm}\sin m\lambda\} \right] ,$$

where GM is the product of the Newtonian gravitational constant and the mass of the earth, a_e is its equatorial radius, and r, ϕ, and λ are radial distance from the earth's center of gravity and geocentric latitude and longitude, respectively.

The $\bar{C}_{nm}$ and $\bar{S}_{nm}$ are coefficients and the $\bar{P}_{nm}(\sin\phi)$ are normalized associated Legendre polynomials with the property that the mean square values of $\bar{P}_{nm}(\sin\phi)\cos m\lambda$ and $\bar{P}_{nm}(\sin\phi)$ $\sin m\lambda$ over a sphere are 1 [9]. Traditionally the zonal terms ($m=0$) were written in an expansion

$$V_{\text{zonal}} = \frac{GM}{r} \left[1 - \sum_{n=2}^{n_{\max}} J_n \left(\frac{a_e}{r}\right)^n P_n(\sin\phi) \right] ,$$

where the P_n are conventional (not normalized) Legendre polynomials and the J_n are coefficients related to the $\bar{C}_{n0}$ by

$$J_n = -(2n+1)^{1/2}\bar{C}_{n0} .$$

These expansions link the field in space to that on the earth's surface; the earthbound observer adds a centrifugal term $\frac{1}{2}p^2\omega^2$, where p is the perpendicular distance from the axis of rotation and ω the earth's angular velocity.

An anomalous potential T is defined as $V - U$, where U is the potential of the normal (ellipsoidal) model. The geoid height is given by T/g, where g is the mean value of gravity, and the gravity anomaly field by $-(\partial T/\partial r + 2T/r)$ [10]. The Goddard Earth Model GEM-T1 [11], complete to degree and order 36, is the most recent gravity field model based entirely on satellite tracking.

A number of satellites carrying radar altimeters to measure the distance from the satellite to the sea surface have been launched since 1975. If the satellite orbit is known, these measurements, after a number of corrections including one for ocean tides, allow the height of the sea surface relative to the reference ellipsoid to be determined. This surface is almost a gravity equipotential (the geoid) but includes a small ($\sim 2\,\text{m}$ maximum, $\sim 70\,\text{cm}\,\text{rms}$) signal due to ocean dynamics. These effects cannot be entirely separated but sea surface altimetry has vastly extended our knowledge of the marine gravity field and will be important in oceanography. The radial orbital errors, initially of the order of 5 m and reduced to 70 cm by the early 1980s, are larger than those in the altimeter ($\sim 10\,\text{cm}$ for SEASAT, $\sim 3\,\text{cm}$ for GEOSAT). However, they are dominantly of long wavelength and a strategy of removing biases on the basis of discrepancies at the crossovers of ascending and descending tracks has been successful in reducing their influence to about 25 cm. SEASAT was launched in June 1978 into a repeat orbit which gave a spacing of about 160 km between ascending tracks at the equator but it unfortunately failed after 3 months of operation. GEOSAT has been repeating the SEASAT orbit since late 1987, yielding a more precise and complete data set and a better understanding of the temporal variations in sea surface height. Mean gravity anomalies over $0.5°\times 0.5°$ blocks have been recovered with a precision of about 3.5 mGal for most of the world's ocean from the SEASAT and earlier GEOS 3 altimetry. In addition some important relative information can be extracted at finer detail. The GEOSAT results are expected to be even better. A useful summary is given in [12] and an attractive map of the gravity field of the world's oceans has been published [13].

Spherical harmonic expansions of the gravity have been made to degree 180 ($1° \times 1°$ resolution) and higher by combining data from satellite tracking, altimetry, and terrestrial measurements [14]. The long-wavelength information (degree less than 25) is derived from satellite tracking whereas that at shorter wavelengths comes largely from altimetry over the oceans and from terrestrial measurements on land. Of the continents, only Europe, North America, and Australia are adequately surveyed.

The interpretation of gravity anomalies in terms of the causative density variations is not unique and there generally exist an infinite number of possibilities that must be narrowed using other geological or geophysical information. Features less than a few hundred kilometers in size usually correlate well with geological structure and are very helpful in modeling this structure. The longer-wavelength features originate in the mantle and at the core-mantle boundary, and show little correlation with surface features; their origin is not understood, although they are of great interest in connection with convective and other geodynamic models.

See also: Geophysics.

References

[1]   G. Peter, R. E. Moose, C. W. Wessells, J. E. Faller, and T. M. Niebauer, "High-Precision Absolute Gravity Observations in the United States", *J. Geophys. Res.* **94**, 5659–5674 (1989).

[2]   J. C. Harrison, "Gravity Sensors", in *Geoscience Instrumentation* (E. A. Wolff and E. P. Mercanti, eds.). Wiley, New York, 1974.

[3]   L. J. B. LaCoste, "Measurement of Gravity at Sea and in the Air", *Rev. Geophys.* **5**, 447–526 (1967).

[4]   S. Hammer, "Airborne Gravity Is Here", *Geophysics* **48**, 213–223 (1983).

[5]   J. M. Brozena and M. F. Peters, "An Airborne Gravity Study of Eastern North Carolina", *Geophysics* **53**, 245–253 (1988).

[6]   *The International Gravity Standardization Net, 1971.* Int. Assoc. Geodesy Publ. No. 4, 1974.

[7]   T. F. Baker, R. J. Edge, and G. Jeffries, "European Tidal Gravity: An Improved Agreement Between Observations and Models", *Geophys. Res. Lett.* **16**, 1109–1112 (1989).

[8]   H. Moritz, "Geodetic Reference System 1980", *Bull. Geodesique* **62**, 348–358 (1988).

[9]   W. M. Kaula, "Determination of the Earth's Gravitational Field", *Rev. Geophys.* **1**, 507–551 (1963).

[10]   H. Moritz, *Physical Geodesy*, Chap. 2.14. Freeman, San Francisco, 1967.

[11]   J. G. Marsh *et al.*, "A New Gravitational Model for the Earth from Satellite Tracking Data: GEM-T1", *J. Geophys. Res.* **93**, 6169–6215 (1988).

[12]   B. C. Douglas, D. C. McAdoo, and R. E. Cheney, "Oceanographic and Geophysical Applications of Satellite Altimetry", *Rev. Geophys.* **25**, 875–880 (1987).

[13]   W. F. Haxby, *Gravity Field of the World's Oceans.* National Geophysical Data Center, NOAA, Boulder, Colorado, 1987.

[14]   R. H. Rapp and J. Y. Cruz, Spherical harmonic expansions of the Earth's gravitational potential to degree 360 using 30′ mean anomalies. Report no. 376, Dept. of Geodetic Science and Surveying, The Ohio State University, 1986.

Group Theory in Physics

J.-P. Antoine

Origins

As far back as we know, people have been fascinated by symmetry. From megalithic monuments to Mycenian jewelry, from Egyptian pyramids to Greek temples, symmetrical forms were clearly perceived and intuitively understood as manifestations of harmony and perfection. When modern science developed, this innate inclination became incorporated into the work of the great pioneers of physics, whose beautiful theories and equations were thought of as reflecting the harmony of the world (Kepler). However, the technical study of symmetry properties requires a level of mathematical sophistication not available before the late nineteenth century when, following E. Galois, the abstract theory of groups was developed by mathematicians like G. Frobenius, I. Schur, W. Burnside, E. Cartan, and H. Weyl. Classical physics had not created much motivation for it; from Newton to Lagrange or Maxwell, the aim was to derive and solve the partial differential equations describing the structure of matter. Only in some works of Euler do we see a first approach to the idea of a group and its role in physics. Thus physics followed (and in some cases preceded) analysis rather than algebra.

Group theory first entered physics with crystallography. Given an arbitrary crystal, it is an easy exercise to figure out all symmetry operations that leave it unaffected: reflection through certain planes or inversion with respect to the center, rotations around given axes through the center (only the angles $2\pi/n$, with $n = 2, 3, 4$, or 6, will be compatible with the periodicity of the crystalline lattice), or any combination of these. A systematic investigation shows there exist exactly 32 different combinations of symmetry properties, and accordingly crystals are subdivided into 32 *crystal classes*. For a given class, any two symmetry operations can be combined to give a third one (and this product is associative), there is one that does nothing at all (the identity), and every operation has an inverse (the combination of the two being the identity); in other words, the symmetry operations form a *group* with a *finite* number of elements.

Each crystal class corresponds to such a group, which is called a *point group*, since it consists of symmetry operations around a fixed point; thus there are 32 different point groups. Combining these with the lattice translations for each of the 14 different types of lattices (the Bravais lattices), we obtain, after some hard work, 230 possible invariance groups for crystal lattices, the so-called *space groups*. This remarkable achievement, due to E. V. Fedorov (1885) and A. Schönflies (1891), illustrates the primordial role of group theory in physics, namely, to organize data in a rational fashion.

Groups and Representations

Let there be given an arbitrary physical system, for which we know a symmetry group G. In order to exploit this information we need another concept, that of a (linear) group representation. An element $g \in G$ transforms a given configuration of the system into another one: The symmetry group is realized in the space V of configurations. Quite often, the latter has or can be given the structure of a vector space. Every $g \in G$ will then be represented by a nonsingular transformation or operator $L(g)$, acting on V in such a way that the operators $\{L(g) : g \in G\}$ obey the same algebraic relations as the group elements themselves: $L(g_1)L(g_2) = L(g_1g_2)$,

$L(g^{-1}) = [L(g)]^{-1}, L(e) = 1$ (e is the neutral element of G, 1 is the identity operator). We say in this case that the map $L: g \mapsto L(g)$ is an n-dimensional (linear) representation of G, where n is the dimension of V ($1 \leqslant n \leqslant \infty$).

If the space V possesses an inner product and the latter is invariant under G, the representation L is called *unitary* (if V is a complex vector space) or *orthogonal* (if V is real). So the representation L describes completely the action of the symmetry group on the system. For instance, the configurations may be decomposed into subsets V_j (vector subspaces of V), each of which is transformed into itself by all the operations of G; the representation is then said to be *reducible*. If L is unitary or orthogonal, it can be decomposed into a direct sum of subrepresentations L_j: $L = \oplus_j L_j$, corresponding to $V = \oplus_j V_j$; if a given V_k no longer contains an (nontrivial) invariant subspace, then L_k is called *irreducible*. With this machinery, we can describe as follows the common pattern of most applications of group theory in physics. First, identify the symmetry group G of the system under study; next determine the representation L of G in the space of configurations V, and decompose L into a direct sum of irreducible representations L_j. At this stage, in general, the analysis of the system will be enormously simplified, since the number of configurations available in V_j is much smaller than the one in V. Accordingly, we can evaluate the quantities of interest, such as normal frequencies or transition probabilities, rather easily. A typical example of this procedure is the study of the normal modes of vibration of a mechanical system, that is, a system composed of coupled harmonic oscillators; this is the lowest approximation of the classical description of a molecule or a crystal. In all such cases the symmetry group of an object is an essential element for simplifying its investigation.

The Principle of Relativity and the Associated Groups

Geometrical symmetries, as discussed earlier, are very useful, but group theory enters physics in a much more fundamental way through the *principle of relativity*. The latter asserts that two observers will describe a physical process by the same equations whenever they are *equivalent*, by which one usually means that they are at rest or in uniform linear motion with respect to each other (other notions of equivalence are possible, as we shall see below). "Observer" here means simply a reference frame in space-time. (Notice that the existence of inertial frames is an independent assumption.) In other words, a transformation of space-time that maps any observer into an equivalent one cannot affect the description of physical processes. These transformations obviously form a group, called the *relativity group* G^{rel} of the theory. Thus, technically, space-time is a homogeneous manifold on which the group G^{rel} acts. The importance of the principle of relativity is that it guarantees (because of the homogeneity of space-time) that experiments can be repeated whenever and wherever we wish, i. e., that science is possible.

However, the principle of relativity does not determine G^{rel} uniquely; an additional postulate is needed. Three possibilities have been used. We can say that two observers are equivalent only if they are at rest with respect to each other; G^{rel} is then the Euclidean group $\mathcal{E}(3)$, which consists of all rotations and translations in three-dimensional space as met in ordinary vector calculus. If, in addition, equivalent observers are allowed to be in relative uniform motion, but if time remains absolute, we have the Galilei group $\mathcal{G}$, which is the relativity group of Newtonian mechanics. If, instead, we require that light propagate with the same speed for every (inertial) observer (which is incompatible with an absolute time), we obtain the inhomogeneous Lorentz group, also called the Poincaré group $\mathcal{P}$. This is the relativity group of Einstein's theory of special relativity, which adapts classical mechanics to the symmetry properties inherent in electromagnetism. The principle of relativity can be extended to observers in gravitational fields, but the interpretation of general relativity in group-theoretical terms is no longer straightforward. Indeed, the equivalence principle forces one to consider only local frames, so that Poincaré invariance is only valid locally.

Once the relativity group of the theory G^{rel} has been determined, the principle of relativity must be put into action: It asserts that all laws of physics are relations between tensors associated to the chosen G^{rel}. Tensors are usually introduced through multilinear algebra, but they can also be defined in a purely group-theoretical language. Each of the relativity groups $\mathcal{E}(3)$, $\mathcal{G}$, $\mathcal{P}$, has a defining matrix representation M; a tensor of order (p,q) may then be defined as an element of the vector space of the representation $M^{(p,q)} = M \otimes \cdots \otimes M \otimes M^* \otimes \cdots \otimes M^* = M^{\otimes p} \otimes M^{*\otimes q}$, the tensor product of p factors M and q factors M^* (M^* is the contragredient representation defined by the relation $\langle M^*(g)\phi|\psi\rangle = \langle\phi|M(g)\psi\rangle$ for any vectors ϕ, ψ in the representation space). Thus the classification of irreducible tensors of order (p,q) amounts to the decomposition of $M^{(p,q)}$ into its irreducible constituents. At this point, however, a new element enters, for this reduction involves the study not only of representations of G^{rel}, but also of those of the group S_n of permutations, which act on the indices of the tensors. As a result, we can write explicitly, in principle, all possible forms for the laws of physics in a theory governed by a given relativity group. This is, of course, an enormous simplification, which is achieved by postulating the principle of relativity from the outset. In practice, further restrictions (based on simplicity, for instance) are used.

One example of such an enterprise is the derivation of all covariant wave equations for a given relativity group. For instance, the Dirac equation is the unique first-order covariant equation describing a relativistic ($G^{\text{rel}} = \mathcal{P}$) spin-$\frac{1}{2}$ particle of mass $m > 0$. It should be mentioned that all these equations are classical, although historically they have been discovered in a quantum context. (Second) quantization is the reinterpretation of their solutions as operators acting on the Hilbert space of the theory.

Invariance and Conservation Laws

The crystallographic groups are very special; they have either a finite or a countably infinite number of elements; such groups are called *discrete*. In contrast, the three relativity groups $\mathcal{E}(3)$, $\mathcal{G}$, and $\mathcal{P}$ depend *continuously* on some parameters (e. g., rotation angles or translations). Such groups are called *Lie groups*, after the Norwegian mathematician Sophus Lie, who first investigated them systematically. In a Lie group, there exist infinitesimal transformations, arbitrarily close to the identity. The law of group multiplication can be linearized in terms of their coefficients, called *infinitesimal generators*. The latter form a vector space known as the *Lie algebra*. Conversely (and this was Lie's main achievement), the group can be reconstructed from its algebra. The latter plays an important role in physics, for it contains the essential information that can be extracted from the symmetry of a system, namely, the conservation laws that it satisfies. This crucial connection was formulated in the following fundamental theorem by the German mathematicians David Hilbert (1916) and Emmy Noether (1918): If a Lagrangian theory is invariant under an N-parameter Lie group of transformations (in the sense that the Lagrange function $\mathcal{L}$, or, more generally, the action integral $I = \int \mathcal{L} \, dt$, is invariant), then the theory possesses N conserved quantities. This theorem, which is a consequence of a variational principle, holds in any Lagrangian theory, whether the Lagrange function depends only on finitely many coordinates (point mechanics), or is a functional of fields (classical field theory). It is one of the cornerstones of classical physics and, by the correspondence principle, of quantum physics as well.

As an example, consider the relativity groups introduced earlier. The Euclidean group $\mathcal{E}$ is a six-parameter Lie group. Accordingly, any Lagrangian system invariant under $\mathcal{E}$ possesses six constants of the motion, which span the Lie algebra, namely, the components of the total momentum and those of the total angular momentum. Both G and P have 10 parameters and contain $\mathcal{E}$ as a subgroup. Hence systems invariant under either G or P have four additional constants of the motion, namely, the total energy (the Hamiltonian) and three others that guarantee that the center of mass moves as a *free* particle.

It should be noted that those conservation laws, although fundamental, are not sufficient in general for a complete integration of a classical mechanical problem. There are, however, some problems that can be integrated in an elementary way, without solving any differential equations of motion, thanks to additional symmetry properties (called *hidden dynamical symmetry*). The two main cases are the n-dimensional oscillator, which has a hidden $\mathrm{SU}(n)$ symmetry [$\mathrm{SU}(n)$ is the group of all $n \times n$ unitary matrices of determinant 1]; and Kepler's problem, which is invariant under a hidden four-dimensional rotation group $\mathrm{SO}(4)$. In the latter case, the three additional constants of motion have an immediate physical interpretation, namely, the components of the major semiaxis of the trajectory; hence, not only does the orbit lie in a fixed plane (angular momentum conservation), but its position in the plane does not change in time. This fact has an obvious importance for astronomy!

Symmetries in Quantum Mechanics

So far we have discussed symmetry principles of classical physics only, but it is in quantum theory that group-theoretical ideas have found their most fertile ground, with a spectacular development as the result.

The basic reason is the *linearity* of the theory; indeed, the first axiom of quantum mechanics asserts that the states of a system are represented by unit rays in a Hilbert space $\mathcal{H}$. Thus, contrary to classical physics, quantum theory has a built-in linear structure through which symmetries are automatically realized by group representations, thanks to fundamental theorems of E. P. Wigner and V. Bargmann. More precisely, if a symmetry is defined as a map of the states into themselves that preserves all transition probabilities, i.e., the modulus of the inner product between any two states, then every group G of symmetries of the system is realized, up to phase factors, by a unitary representation U of G into $\mathcal{H}$.

For exploiting this theorem, we proceed along the general lines described earlier. The unitary representation U can be decomposed into irreducible constituents U_j, corresponding to subspaces $\mathcal{H}_j$. Most quantities of interest are given by matrix elements of certain observables that have a simple behavior under G (e.g., vector and tensor operators, when G is a rotation group). If A is one of these observables, a matrix element $\langle \phi | A | \psi \rangle$, with $\phi \in \mathcal{H}_j$, $\psi \in \mathcal{H}_k$ can be evaluated using only properties of the symmetry group; either it will vanish (selection rule), or it will depend to a large extent on the subrepresentations U_j and U_k

only, but not on the individual states ϕ, ψ (this is the idea of the famous Wigner–Eckart theorem).

Furthermore, many observables themselves can be derived from the symmetry group via Noether's theorem and the correspondence principle. For instance, invariance under $\mathcal{E}$ (translations and rotations) yields total momentum and total angular momentum; time translation gives the Hamiltonian; Galilei invariance yields position observables. In all cases, these observables belong to the Lie algebra of the symmetry group and are realized in the state space by self-adjoint operators.

Approximate Symmetries

So far we have discussed only *exact* symmetries, but we can go beyond that and, in the spirit of perturbation theory, consider *approximate* symmetries. This concept is useful whenever the Hamiltonian can be split into two terms, where the first is invariant under a given group and the second is a small correction, invariant only under a subgroup; in addition, the symmetry-breaking term is assumed to have well-defined transformation properties under the full approximate symmetry group. Then we can resort to the general procedure described earlier for computing matrix elements. The procedure can be repeated, leading to a hierarchy of approximate symmetries, more and more badly broken, corresponding to a descending chain of subgroups. This idea, which might be traced back, for instance, to the analysis of the Zeeman effect, has been remarkably successful.

In summary, besides implying fundamental conservation laws, the symmetries of a system, exact or approximate, provide an extremely powerful tool for computing physical quantities. This explains the considerable development of group-theoretical methods in the various domains of physics, starting in the late twenties under the impulsion of such great physicists or mathematicians as Heisenberg, Pauli, Weyl, van der Waerden, Wigner, and Bargmann.

Applications to Atoms, Molecules, and Solids

The first applications of group theory were to atomic physics, where in most cases the bewildering complexity of spectroscopic data resisted analysis. The basic fact is the rotational symmetry of a free atom: In the center-of-mass frame, the Hamiltonian is invariant under all rotations around the origin. This SO(3) symmetry guarantees that the total angular momentum of the atom is conserved.

Consider first a simplified hydrogen atom: a spinless electron bound in a Coulomb potential (i. e., to a spinless proton). The Schrödinger equation can then be solved exactly in polar coordinates; the energy levels obtained, indexed by an integer $n = 1, 2, \ldots$, are highly degenerate: the nth level accommodates angular momenta $l = 0, 1, 2, \ldots, n-1$, and for each l there are still $(2l+1)$ different states indexed by $m_l = -l, -l+1, \ldots, l$. This so-called accidental degeneracy of the Coulomb potential has a group-theoretical explanation, exactly as in the classical case: for a given level n, the irreducible representations $D^{(l)}$ ($l = 0, 1, \ldots, n-1$) of SO(3) can be grouped to form a single irreducible representation of SO(4). This fact was recognized by W. Pauli in 1926 and exploited by him to solve the Coulomb problem in a purely algebraic way. If the central potential is not exactly Coulombic, the energy levels depend on both n and l. Following Pauli, we now take into account the electron spin: According to the exclusion principle, each state $|n, l, m\rangle$ can accommodate two, and only two, electrons. The total angular momentum of the atom is now the (vector) sum of the orbital angular momentum and the spin, and can take values $y = l \pm \frac{1}{2}$; this law of addition is nothing but the decomposition into irreducible constituents of the *tensor product* of two SU(2) representations: $D^{(l)} \otimes D^{(1/2)} = D^{l+1/2} \oplus D^{l-1/2}$.

Notice that the usual rotation group SO(3) describes properly integer angular momenta only; for describing half-integer ones, we need SU(2), which is closely related to it [technically, SO(3) is the quotient of SU(2) by a two-element subgroup].

Consider now a many-electron atom. The Schrödinger equation can no longer be solved exactly, and one must have recourse to approximation. The crudest one is the so-called central field approximation, in which each electron moves in an average, central potential created by all the other electrons. This allows one to find an energy spectrum $E(n, l)$ for each electron, and the global spectrum by addition. The result is the description of an atom called the *shell model*. All electrons with the same values of n and l are said to form a "shell", and we can list all possible configurations within each shell. This model explains the periodic classification of the elements (Mendeleev table). Group theory enters this process in two ways. First, the addition of all individual angular momenta and spins is simply, as before, the decomposition of a tensor product of many representations $D^{(l_i)}$ or $D^{(1/2)}$ of SU(2). Then, also, the permutation groups enter, through the Pauli exclusion principle. A further step was taken by G. Racah, who introduced in this context an *approximate symmetry*. The idea is to consider the single-electron eigenfunctions in a given shell (n, l) as the basis of a representation of the much larger SO($8l + 5$). A classification of all configurations is then obtained by studying this representation and its restriction to a chain of smaller and smaller subgroups. At the same time, transformation properties under these larger groups may be exploited for computing various matrix elements, and this is useful for the next approximation, namely, when electron spins and spin–orbit coupling (which is a truly relativistic effect) are introduced as a perturbation. By this method we can obtain a good picture of the complete energy spectrum of the heaviest atoms. It can be fairly said that without the enormous simplifying power of group theory, the unraveling of atomic spectra would be all but impossible.

A different extension of the notion of symmetry has been considered, first in the case of the hydrogen atom. Some operators that describe transitions between different levels, such as the dipole operator, generate, if combined with the generators of SO(4), two still larger groups, SO(4,1) and SO(4,2). A single irreducible representation of each of these contains *all* the representations of SO(4) corresponding to all energy levels. These groups are variously called *dynamical symmetry groups, noninvariance groups,* or *spectrum generating groups*; they are not invariance groups (although they contain the true symmetry group as a subgroup), but can describe transitions as well. The use of such groups has proved very fruitful.

Going one step higher, to molecules, a similar story can be repeated. The rotation group, however, is restricted to the geometrical symmetry group of the molecule (known from chemistry). The simplest case is a diatomic molecule; it retains an axis of symmetry, and is accordingly invariant under SO(2). More complicated molecules are invariant only under a (finite) crystallographic group: e. g., the benzene ring (hexagonal symmetry) or the ammonia molecule. Here too group theory provides an efficient tool, first for classifying the stable configurations and energy levels of the molecule, then also for analyzing interactions with radiation (such as Raman scattering).

Then, of course, we may return to crystals. The quantum theory of solids encompasses all the knowledge accumulated from classical crystallography and applies it to dynamical problems: energy bands in solids, and the behavior in a perfect solid of foreign particles, such as electrons (semiconductors and metals), phonons or photons (optical or electromagnetic properties of solids), or ions (impurities), including their movement under the influence of external electric or magnetic fields. In each case, the use of crystal symmetry is crucial for solving the problem.

Coherent States and Wavelets

The usefulness of group theory does not stop at isolated atoms, but can be extended to their interaction with radiation (emission, absorption, scattering). The key remark here is that creation and annihilation operators of the radiation field obey commutation relations identical to those of the Lie algebra of the Weyl–Heisenberg group, generated by the familiar position and momentum operators of quantum mechanics. This fact can be used for solving all problems with a Hamiltonian at most quadratic in these operators. These Hamiltonians cover a large part of quantum optics, notably lasers and other coherent phenomena. Quite naturally, the states describing such systems are called *(canonical) coherent states*. These states, which have remarkable properties, were initially introduced by Schrödinger in 1926, as those states which describe best the classical limit of quantum mechanics. They were, however, quickly forgotten, until the early sixties, when they were rediscovered by J. R. Klauder, R. J. Glauber and E. C. G. Sudarshan in the context of a quantum optical description of coherent light beams emitted by lasers. This approach seemed to have little to do with group theory, until the mathematician A. M. Perelomov and the physicist R. Gilmore, independently, pointed out in 1972 that coherent states are precisely obtained by letting a unitary representation of a Lie group act on a fixed vector (ground state or fiducial vector) – and this applies not only to the Weyl–Heisenberg group, but to an arbitrary Lie group – and even to Lie superalgebras that will be mentioned below. Since then, these (generalized) coherent states have pervaded nearly all branches of quantum physics – quantum optics, of course, but also nuclear, atomic and solid state physics, quantum electrodynamics (the infrared problem), quantization and dequantization problems and path integrals, to mention just a few.

A particularly spectacular development of coherent states is *wavelet analysis*, nowadays ubiquitous in physics, mathematics, and engineering. Actually, wavelets are in fact nothing but the coherent states associated to the affine group of the line, consisting of translations and dilations. This was the crucial discovery made in 1984 by A. Grossmann and J. Morlet [they were analyzing the mathematical structure of a technique of signal processing that had been designed empirically by J. Morlet in the context of oil prospecting with microseismology]. From there, the general theory of coherent states allowed an immediate generalization to the analysis of two-dimensional signals, i. e., images, the relevant group here being the similitude group of the plane (translations, dilations and rotations). It is true that the real take-off of wavelet analysis was the conjunction of the ideas above with established techniques of signal processing (filter banks), yet group theory was the necessary starting point and the key to all subsequent generalizations, e. g., to three-dimensional signals, to signals on the two-sphere, or to time-dependent signals (such as video sequences).

Nuclear Physics

A free nucleus can be viewed as a bound state of N particles (protons and neutrons) interacting through two-body forces; its geometrical invariance group is again $SO(3)$. To understand its structure better, we use exactly the same method as for atoms, which leads to the shell model of nuclei. To a first approximation, each of the N particles may be visualized as moving in the average field of the others. But here this average field is approximately a harmonic potential [which has an $SU(3)$ invariance]. We then build an approximate symmetry group again, this time a unitary group of large dimensions, and use it to estimate the energy levels of the nucleus.

A New Kind of Symmetry

All the symmetries discussed so far are geometrical in origin; but in nuclear physics, and even more so in elementary-particle physics, internal symmetries play an essential role. The simplest case is *charge conjugation*, which exchanges particle and antiparticle – this is still a discrete symmetry. Then, as Heisenberg was the first to realize in 1932, protons and neutrons may be considered as two different states of the same entity, the *nucleon*. They differ only through electromagnetic, but *not* through strong (i. e., nuclear), interactions. The typical quantum system with only two possible states is a spin-$\frac{1}{2}$ particle; hence Heisenberg attributed to the nucleon a new *internal* degree of freedom, later called *isospin*. Mathematically it is isomorphic to ordinary spin: The nucleon is a particle with spin $\frac{1}{2}$ and isospin $\frac{1}{2}$. Hence the Hamiltonian of a nucleus must be invariant under the isospin $SU(2)$ group. If only Wigner and Majorana forces are present, it will also be invariant, to a good approximation, under the ordinary spin $SU(2)$, hence under the enveloping group, $SU(4)$. This approximate symmetry, proposed by Wigner in 1937, has given good clues for understanding nuclear ground states.

The Classification of Elementary Particles

Nucleons are not the only particles subject to strong interactions. Since the discovery of the π meson in 1947, the list of strongly interacting particles, or hadrons, has increased steadily, to more than 400 entries at present! Here again group theory has provided a unique tool for organizing such overwhelming experimental data.

First, all the hadrons can be grouped in isospin multiplets. Then another internal degree of freedom, called the *hypercharge* Y (or the closely related *strangeness*) was introduced by M. Gell-Mann; it satisfies the famous relation $Q = T_3 + \frac{1}{2}Y$, where Q is the electric charge and T_3 the third component of the isospin. After many unsuccessful attempts, Gell-Mann (and others independently) discovered in 1962 how to merge the isospin and the hypercharge into a new internal symmetry group $SU(3)$. The latter is an approximate symmetry that permits the classification of all known hadrons in multiplets corresponding to irreducible representations and thereby predicts many new particles. Here again a use of the Wigner–Eckart theorem has led to correct predictions of many physical properties, such as mass differences, magnetic moments, or branching ratios for various decay modes. The most spectacular success was the discovery in 1964 of the quasistable particle Ω, whose mass was correctly predicted by Gell-Mann. Various extensions of this model have been proposed, but none of them proved really satisfactory.

Instead the classification problem took a new direction. Shortly after $SU(3)$ was introduced, Gell-Mann (and, independently, G. Zweig) suggested that all known hadrons could be thought of as bound states of three elementary building blocks, the so-called *quarks*, and their antiparticles. This model, naive at first but steadily refined, has been remarkably successful for describing the dynamical properties of hadrons (although the quarks themselves have never been isolated – this is the famous *confinement* phenomenon). More recently, the discovery of totally new particles has forced the theorists to enlarge their model. First, in 1974–1975, came the charmonium family $(J/\psi, \psi', \psi'', \dots)$, whose properties can be best understood by the existence of a new quantum number called *charm*; this demands a fourth quark. Similarly, the upsilon family, discovered in 1977, requires a fifth quark, the so-called *bottom* quark. More important, the phenomenological quark model is now incorporated, and understood, within a genuine theory, the so-called *Standard Model* (see below). The latter actually predicts the existence of six quarks, and the last of them, the *top*, has finally been discovered at Fermilab in 1993.

Understanding Dynamical Properties

However, group-theoretical ideas were not confined to classification. As far back as 1958, R. P. Feynman and M. Gell-Mann had proposed that part of the (conserved) electromagnetic current and the (almost conserved) weak current form an isospin triplet (the CVC, or conserved vector current, hypothesis). A corresponding but weaker hypothesis (PCAC) was soon made for the axial currents. These assumptions were extended to $SU(3)$ by N. Cabibbo, who thus obtained very good predictions of weak decay rates. These ideas finally led to Gell-Mann's *algebra of charges*. To each current there corresponds a charge (space integral of the zeroth component of the current); the vector charges transform like the adjoint representation of $SU(2)$, or $SU(3)$ if hypercharge-changing currents are included; this implies that the commutator of a vector charge with an axial charge is again an axial charge. Then Gell-Mann postulated that the commutator of two axial charges be a vector charge, i. e., the algebra of all charges closes under commutation to the Lie algebra of $SU(2) \otimes SU(2)$, or $SU(3) \otimes SU(3)$. From this so-called *chiral symmetry* a large number of predictions were obtained, simply by taking matrix elements of the commutation relations between adequate states and using the Wigner–Eckart theorem again. Finally, going one step further, Gell-Mann postulated that the currents themselves satisfy a local $SU(3) \otimes SU(3)$ algebra: this is the famous *current algebra*. These developments represent a remarkable evolution since the early applications of group theory. The precise structure of the various hadronic currents is unknown, but only their symmetry properties are important: the line of thought is exactly opposite to the one originally used, e. g., in atomic physics!

Gauge Theories and All That

By far the most promising development, however, is the emergence of the so-called *gauge theories*, which are based on an extension of the concept of symmetry. Within a field theory, an internal symmetry is said to be *global* if the action of the symmetry group on the field $\phi(x)$ is independent of the space-time point x; this is the concept used so far. The symmetry is called *local* if the action is, in addition, allowed to vary from point to point; such a theory is called a *gauge field theory*, and the local symmetry group is called a *gauge* group. The idea goes back to H. Weyl in 1918, who treated electromagnetism as a gauge theory based on the commutative gauge group $U(1)$. A noncommutative theory, based on $SU(2)$, was proposed by C. N. Yang and R. L. Mills in 1954, but the gauge concept did not become popular until the Dutch physicist G. t'Hooft proved in 1971 that a noncommutative gauge field theory is renormalizable (i. e., susceptible of giving consistently finite predictions). The key point in his proof was that a clever use of group identities produces sufficiently many cancelations between potentially divergent terms. Since then, gauge theories (and with them, differential geometry) have invaded the whole field of particle physics, with rather remarkable results.

The most important aspect is that a gauge symmetry must necessarily be exact; this eliminates lots of arbitrary parameters and gives the theory a much greater coherence (and elegance, too). In particular the form of the interaction Lagrangian is uniquely determined. On the other hand, the interaction in a gauge theory is mediated by massless particles; the canonical example is the photon, corresponding to the fact that electromagnetism is a $U(1)$ gauge theory.

This mechanism extends to the other interactions. On the one hand, the model of S. Weinberg, A. Salam, and S. Glashow, based on the gauge group $SU(2) \otimes U(1)$ (which requires the sixth quark), gives a unified description of weak and electromagnetic interactions; it has accumulated excellent experimental support and is by now almost universally accepted (although a key ingredient – a particle known as the Higgs boson – has not been found so far). On the other hand, a new theory of strong interactions, called *quantum chromodynamics* (QCD), has emerged. It is also a gauge theory, with gauge group $SU(3)$, that generalizes the old quark model by assuming that every quark comes in three types (or colors). Notice that, in both cases, all the gauge particles, except the photon and the gluons, acquire nonzero masses by a subtle mechanism based on the notion of *spontaneously broken symmetry* (this term describes the situation where the ground state of a system has *less* symmetry that the Hamiltonian – the canonical example is the Heisenberg ferromagnet).

Taken together, these two models constitute what is now known as the *Standard Model*, covering simultaneously all three types of interactions. In fact it may fairly be called a *theory* instead of a mere model, although it contains a large number of free parameters, notably the masses of the fundamental fermions (quarks and leptons). This implies that, in the standard model, the isospin symmetry and even more the flavor $SU(3)$ symmetry are in fact *accidental*. On the other hand, the color $SU(3)$ is unbroken and truly fundamental, because it reflects dynamical properties, as did already current algebra and chiral symmetry. The fact that the gluons remain massless is a manifestation of this state of affairs, which in turn suggests the confinement of quarks the latter is so far purely empirical, no real proof has been given).

Recent Developments

Where do we go now? Although the answer to that question is certainly confused (things become clear only with hindsight, of course), one aspect is certain: The role of group theory is more central than ever!

A first direction to be mentioned is that of the so-called *grand unified theories* (GUTs), which try to unify all interactions, except gravity. Various schemes have been proposed, based on groups like $SU(5)$, $SO(10)$, etc., but all predict new particles and the decay of the proton – both unseen so far.

Another interesting development is *supersymmetry* (which leads to *supergravity* when gravitation is included), a theory that seeks to unify bosons and fermions in a common framework. This idea has opened a new branch of (super) mathematics, namely, analysis (including group theory) with *anticommuting* variables. As a physical model, supersymmetry is very elegant (although it is badly broken in Nature), but it has received so far no experimental confirmation whatsoever; in particular, it predicts the existence of many new particles with fancy names (photinos, gluinos, etc.), none of which has been seen. So the question remains totally open. On the other hand, supersymmetry has led to very interesting mathematical developments, both in the realm of groups, namely, supergroups and Lie superalgebras, and in analysis, for instance analysis with anticommuting variables.

A totally different idea yet has emerged from gauge theory, and that one is much more promising. The original idea of gauge invariance introduced by Weyl was that the theory should be insensitive to a redefinition of length standards – in other words, that it should be invariant under reparametrization. Although not correct in Einstein's general relativity (the coupling constant is not dimensionless, hence it fixes a mass scale), this idea has proven extremely successful (note that general relativity is also a gauge theory, albeit of a very special type). On the other hand, two-dimensional models have been popular among field theorists for a long time: they are easier to solve than their 4-dimensional, real world, counterparts, and they present striking (and probably deep) similarities with various models in statistical physics. Applying the idea of reparametrization in a 2-dimensional world leads to *conformal* invariance, which has become an extremely successful concept in high-energy physics and statistical mechanics as well. First of all, it lies at the basis of the theory of strings or superstrings, according to which the basic constituents of matter are no longer pointlike, but 1-dimensional, string-like, objects. Various models have been proposed, and they are all based on heavy use of (unexpected) groups like $SO(32)$ or the exceptional $E(6)$, $E(7)$, $E(8)$. Second, the conformal group in two dimensions is no longer a Lie group, since it has infinitely many parameters. Thus *infinite-dimensional Lie algebras* have entered physics. First there is the Virasoro algebra, closely related to the conformal algebra. Next, combining the idea of reparametrization invariance with the formulation of classical string theory, one is led to a whole class of simple, infinite-dimensional Lie algebras, the so-called Kac–Moody algebras (discovered, independently, by the mathematicians V. Kac and R. V. Moody). The representation theory of those algebras is by now well under control, and they play a central role both in quantum string theory and in various models of classical statistical mechanics (conformal invariance is the link between the two). It is interesting to notice that Kac–Moody algebras were encountered previously as invariance algebras of some nonlinear differential equations giving rise to soliton solutions, such as the famous Korteweg–de Vries equation describing waves in shallow water.

Another recent development is that of the so-called *quantum groups*, which are in fact not groups, but Hopf algebras. This is an upshot of noncommutative geometry, a framework supposedly better adapted to quantum theories than the usual, commutative, geometry. Although mathematical progress has been spectacular, such constructions have had so far rather little impact on physics. Only future will tell if they are really relevant.

Conclusion

Clearly our present understanding of elementary particles is sketchy and the whole picture is confusing. Yet, whatever theory finally emerges, it seems fair to say that group theory has grown into one of the essential tools of contemporary physics. Besides its fundamental role in relativity, it has provided physicists with a remarkable analyzing power for exploiting known symmetries, and thereby with a considerable predictive capability, precisely in cases where the basic physical laws are unknown. One of the striking aspects is its versatility: Going from the rather restrictive study of exact symmetries to that of approximate ones, more and more badly broken, group theory has pervaded all fields of physics, often in a fundamental way. Except for calculus and linear algebra, no mathematical technique has been so successful.

See also: Elementary Particles In Physics; Gauge Theories; Grand Unified Theories; Invariance Principles; Isospin; Lie Groups; Relativity, Special; String Theory; *SU*(3) And Higher Symmetries; Supersymmetry and Supergravity.

Bibliography

H. Weyl, *The Theory of Groups and Quantum Mechanics*. Dover, New York. 1950 (original, 1st German ed., 1928).

E. P. Wigner, *Group Theory and Its Application to the Quantum Mechanics of Atomic Spectra*. Academic Press, New York, 1959 (original, 1st German ed., 1931).

F. J. Dyson, *Symmetry Groups in Nuclear and Particle Physics* (a lecture note and reprint volume). Benjamin, New York, 1966.

E. M. Loebl (ed.), *Group Theory and Its Applications*. Vols. 1–3. Academic Press, New York, 1968-1975.

M. B. Green, J. H. Schwarz, and E. Witten, *Superstring Theory*. Vols. I–II. Cambridge University Press, Cambridge, 1987.

A. Bohm, Y. Ne'eman, and A. O. Barut (eds.), *Dynamical Groups and Spectrum Generating Algebras*. Vols. I–II. World Scientific, Singapore, 1988.

S. T. Ali, J-P. Antoine, and J-P. Gazeau, *Coherent States, Wavelets and Their Generalizations*, Graduate Texts in Contemporary Physics, Springer-Verlag, New York, Berlin, Heidelberg, 2000.

Gyromagnetic Ratio

J. H. Freed

The gyromagnetic (or magnetogyric) ratio γ is defined as the ratio of the magnetic moment $\boldsymbol{\mu}$ to the angular momentum $\mathbf{J}$ for any system. Specifically, one introduces for electrons with electron spin angular momentum $\mathbf{J} = \hbar\mathbf{S}$ (where $\hbar = h/2\pi$ and h is Planck's constant) the electron gyromagnetic ratio γ_e by $\boldsymbol{\mu} = -\gamma_e\hbar\mathbf{S}$, where the negative sign represents the fact that the spin and moment are oppositely directed. For a nucleus, which is a composite particle with total spin $\mathbf{I}$, one introduces the nuclear gyromagnetic ratio γ_I by $\boldsymbol{\mu}_I = \gamma_I\hbar\mathbf{I}$. Both $\boldsymbol{\mu}_I$ and $\mathbf{I}$ (or ($\boldsymbol{\mu}_e$ and $\mathbf{S}$) must be regarded as quantum-mechanical operators. In the case of nuclear moments, the defining equation for γ_I must be considered as implying that the expectation values of $\boldsymbol{\mu}_I$ and $\mathbf{I}$ are taken for the given state (usually the ground state) of the nucleus. It then follows from symmetry considerations expressed in the Wigner–Eckart theorem that $\boldsymbol{\mu}_I$ and $\mathbf{I}$ may be taken as collinear, with $\gamma_I \hbar$ the proportionality constant.

A classical spinning spherical particle with mass m and charge e can be shown to give rise to a magnetic moment $e\hbar/2mc$, where c is the velocity of light. For an electron, this moment is known as the Bohr magneton, β. But one has $\gamma_e \hbar = g_s \beta$, where g_s is the anomalous g value of the electron spin, which was first derived from the relativistic Dirac equation to be exactly 2. Schwinger showed how to correct this for quantum-electrodynamic effects to first order in $\alpha = e^2/\hbar c$ to give $g_s = 2(1 + \alpha/2\pi) = 2.0023$.

The most accurate theoretical calculation, due to Hughes and Kinoshita, is in excellent agreement with the experimental value obtained by van Dyck and collaborators by microwave-induced transitions between Landau–Rabi levels of an electron in a magnetic field. The NIST value of g_s is 2.0023193043718. Earlier atomic-beam measurements by Kusch on the hydrogen atom, for which corrections must be made for the relativistic mass change due to binding, yielded $g_s = 2.002292$. The NIST value of $\gamma_e = 1.76085974 \times 10^{11} \, \mathrm{s^{-1} \, T^{-1}}$.

The nuclear gyromagnetic ratio, representing a composite nuclear property, may be measured very accurately by molecular-beam techniques. Also useful are nuclear magnetic resonance and optical, microwave, electron paramagnetic, Mössbauer, and electric quadrupole resonances. One typically introduces the nuclear magneton β_N and the nuclear g value; thus $\gamma_I \hbar = g_I \beta_N$. One finds for the proton $g_p = 5.585694701$ with $\gamma_p = 2.67522205 \times 10^8 \, \mathrm{s^{-1} \, T^{-1}}$ (NIST).

See also: Magnetic Moments; Nuclear Moments.

Bibliography

V. W. Hughes, and T. Kinoshita, *Rev. Mod. Phys.* **71**, 5133 (1999).

H. Kopfermann, *Nuclear Moments*. Academic Press. New York 1958.

C. M. Lederer, J. M. Hollander, and I. Perlman, *Tables of Isotopes*. Wiley, New York, 1967.

N. F. Ramsey, *Molecular Beams*. Oxford, London, 1956.

B. N. Taylor, W. H. Parker, and D. N. Langenberg, *Rev. Mod. Phys.* **41**, 375 (1969).

R. S. van Dyck, Jr., in *Quantum Electrodynamics*, T. Kinoshita (ed.), p. 322. World Scientific, Singapore, 1990.

NIST Web Page: `physics.nist/cuu/Constants/`.

H Theorem

M. Kac[†] and G. W. Ford

Originally the H theorem, or more precisely the Boltzmann H theorem, referred to the following statement first enunciated by L. Boltzmann in a famed memoir in 1872:

   In a spatially homogeneous dilute (in the sense that only binary collisions need be considered) gas of molecules interacting through central forces, let $f(\mathbf{v},t)\,\mathrm{d}^3v$ denote the number of molecules having velocity $\mathbf{v}$ within the (three-dimensional) volume element d^3v; then setting

$$H = \int \mathrm{d}^3 v\, f(\mathbf{v},t) \log f(\mathbf{v},t) \,,$$

one has that for any initial distribution H decreases monotonically with increasing time, with H constant only for the Maxwellian equilibrium distribution

$$f(\mathbf{v}) = \left(\frac{m}{2\pi kT} \right)^{3/2} \exp\left(-\frac{mv^2}{2kT} \right) \,.$$

This is a rigorous consequence of the Boltzmann equation (first derived in the aforementioned memoir of 1872):

$$\frac{\partial f}{\partial t} = \int \mathrm{d}^3 v_1 \int \mathrm{d}\Omega\, gI(g,\Omega)\{f(\mathbf{v}',t)f(\mathbf{v}_1',t) - f(\mathbf{v},t)f(\mathbf{v}_1',t)\} \,,$$

where the velocity variables refer to the four velocities of a binary collision $(\mathbf{v},\mathbf{v}_1) \rightleftarrows (\mathbf{v}'\mathbf{v}_1')$, $g = |\mathbf{v}_1 - \mathbf{v}| = |\mathbf{v}_1' - \mathbf{v}'|$ is the magnitude of the relative velocity which in an elastic collision is invariant, and $I(g,\Omega)$ is the differential cross section for a collision in which the relative velocity turns through angle θ into the element of solid angle $\mathrm{d}\Omega$.

   Since in equilibrium $-H$ is proportional to the thermodynamic entropy and $-H$ increases in time, Boltzmann saw his theorem as a mechanistic derivation (at least for dilute gases)

[†] deceased

Encyclopedia of Physics, Third Edition. Edited by George L. Trigg and Rita G. Lerner
Copyright ©2005 WILEY-VCH Verlag GmbH & Co. KGaA, Weinheim
ISBN: 3-527-40554-2

of the second law of thermodynamics for irreversible processes. Subsequent critical analyses showed that Boltzmann's derivation of his equation did not follow from the mechanics of collision processes alone but required in addition a nonmechanical statistical assumption equating the actual number of collisions of a certain type with its average (Stosszahlansatz). This objection was answered by the statistical method of Boltzmann and the Boltzmann–Gibbs statistical mechanics that is at the heart of much of present-day physics. The modern view is that Boltzmann's *H* theorem embodies the essential features of the irreversible approach to equilibrium.

Since Boltzmann's time the *H* theorem has been generalized to a wide variety of mechanical systems, both classical and quantum mechanical. Gradually the term *H* theorem has acquired a much broader meaning, referring essentially to any statement about the approach to equilibrium which asserts that an appropriate quantity decreases (or increases) with time.

See also: Entropy; Statistical Mechanics; Thermodynamics, Equilibrium; Thermodynamics, Nonequilibrium.

Bibliography

S. G. Brush, "Kinetic Theory. Vol. 2. Irreversible Processes", in *Selected Readings in Physics*. Pergamon Press, New York. 1966. Here one finds an English translation of the 1872 Boltzmann memoir.

J. R. Dorfman and H. van Beijeren, "The Kinetic Theory of Gases", in *Statistical Mechanics. Part B: Time-Dependent Processes* (B. J. Berne, ed.). Plenum Press, New York, 1977. An account of more recent developments.

Martin J. Klein, *Paul Ehrenfest*. North-Holland, Amsterdam and American Elsevier, New York, 1970. Contains the best critical and historical treatment of the *H* theorem and its role in elucidating the statistical nature of the second law of thermodynamics.

George E. Uhlenbeck and George W. Ford, *Lectures in Statistical Mechanics*. American Mathematical Society, Providence, RI, 1963. Chapter IV contains a succinct derivation of the Boltzmann equation and the *H* theorem.

Hadrons

M. L. Perl

Hadrons such as protons, neutrons, and mesons are the largest family of subnuclear particles. Hadrons have three related properties:

1. They interact with each other through the strong force or strong interaction. Since the strong force also holds protons and neutrons together in a nucleus, its older name is the nuclear force. The simplest measure of the strong force is that when two hadrons collide at high energy (above several GeV) the total cross section is 10–60 mb (1 mb is $10^{-27}\,\mathrm{cm}^2$).

2. In the collision of two hadrons at high energy, additional hadrons are usually produced, up to 50 additional hadrons when the energy is in the thousand GeV range.

3. Hadrons are roughly spherical, with radii of the order of $10^{-13}\,\mathrm{cm}$.

Table 1: Properties of some hadrons. The quark content is in the order of the electric charge, for example the π^+ contains the $u\bar{d}$ quark pair and π^- contains the $\bar{u}d$ pair. The quark symbols are u = up, d = down, s = strange, c = charm, and b = beauty, also called bottom. The bar means antiquark. The K^0 has two different lifetimes according to how it decays.

Name	Symbol	Mass (GeV/c^2)	Electric charge	Quarks in hadron	Spin	Lifetime (seconds)
Charged pion	$\pi^{\pm}$	0.140	$+1, -1$	$u\bar{d}, \bar{u}d$	0	2.6×10^{-8}
Neutral pion	π^0	0.135	0	$u\bar{u} + d\bar{d}$	0	8.4×10^{-17}
Charged kaon	$K^{\pm}$	0.494	$+1, -1$	$\bar{s}u, s\bar{u}$	0	1.2×10^{-8}
Neutral kaon	K^0	0.498	0	$\bar{s}d$	0	5.2×10^{-8} and 8.9×10^{-11}
Rho meson	ρ	0.771	$+1, 0, -1$	$u\bar{d}, u\bar{u} + d\bar{d}, \bar{u}d$	1	4.4×10^{-24}
Proton	p	0.938	$+1$	uud	$\frac{1}{2}$	Stable
Neutron	n	0.940	0	udd	$\frac{1}{2}$	886
Charged D meson	$D^{\pm}$	1.869	$+1, -1$	$c\bar{d}, \bar{c}d$	0	11×10^{-13}
Neutral D meson	D^0	1.865	0	$c\bar{u}$	0	4.1×10^{-13}
Psi	ψ	3.097	0	$c\bar{c}$	1	7.6×10^{-21}
Charged B meson	$B^{\pm}$	5.279	$+1, -1$	$\bar{b}u, b\bar{u}$	0	1.5×10^{-12}
Neutral B meson	B^0	5.279	0	$\bar{b}d$	0	1.7×10^{-12}
Upsilon (1S)	Υ (1S)	9.460	0	$b\bar{b}$	1	1.3×10^{-20}

Each hadron is composed of elementary particles called quarks and gluons. The quarks contribute to the mass of the hadron and determine other hadron properties such as electric charge and spin. Depending on the types of quarks, the hadron may also possess the properties called strangeness, charm, or beauty. The gluons carry the strong force which holds the quarks together inside the hadron, in the same quantum-mechanical manner that photons carry the electromagnetic force. Both quarks and gluons are less than 10^{-16} cm in size. Thus the hadron volume of 10^{-13} cm radius is mostly empty space containing a few quarks and gluons. Yet the strong force makes it difficult for hadrons to penetrate each other.

There are two types of hadrons. The baryons such as protons and neutrons contain three quarks or three antiquarks. The mesons contain one quark and one antiquark.

More than 100 different kinds of hadrons have been found. A few prominent examples and some of their properties are listed in Table 1. All hadrons, except for the proton, are unstable. Some hadrons decay into other hadrons through the strong interaction, for example, $\rho^0 \rightarrow \pi^+ + \pi^-$. Other hadrons decay through the electromagnetic interaction, $\pi^0 \rightarrow \gamma + \gamma$, or through the weak interaction $\pi^+ \rightarrow e^+ + \nu_e$. There is a rough rule that the larger the mass, the shorter the lifetime and hence the more unstable the particle. However, when hadrons decay through the weak interaction their lifetimes are relatively longer. Examples in Table 1 are the charged pions, K mesons, neutron, D mesons and B mesons.

The hadron mass depends partly on the mass of the quarks that compose the hadron and partly on the other partons. When the quarks are relatively light – u with mass of 1.5 to 4.5 MeV/c^2 , d with mass of 2.5 to 5.5 MeV/c^2 , and s with mass of 30–50 MeV/c^2 – it is primarily the other factors that determine the hadron mass. But for the more massive quarks – c with mass 1.0 to 1.4 GeV/c^2 and b with mass 4.0 to 4.5 GeV/c^2 – the hadron mass comes mostly from the quark mass.

There is a sixth very heavy quark not listed in Table 1, it is the $t =$ top quark with a mass of about 178 GeV/c^2. The t quark has a very short lifetime, so short that there is no time for a hadron containing a t quark to form. Therefore the most massive hadrons are made up of b quarks not t quarks.

See also: Baryon; Cross Section; Electromagnetic Interation; Elementary Particles in Physics; Gluon; Hyperons; Mesons; Neutron; Nucleon; Nucleus; Partons; Proton; Quarkonium; Quarkonium; Quarks; Strong Interaction; Weak Interaction.

Bibliography
F. Close, M. Marten, and C. Sutton, *The Particle Explosion*. Oxford University Press, New York, 1987.
Donald H. Perkins, *Introduction to High Energy Physics*. Cambridge University Press, Cambridge, 2000.
Martinus J. G. Veltman, *Facts and Mysteries in Elementary Particle Physics*. World Scientific, New Jersey, 2003.

Hadron Colliders at High Energy

R. R. Rau and T. Satogata

Introduction

Interactions from high-energy colliding beams were first observed in 1963. They were produced by counter-rotating beams of 250 MeV electrons and positrons in a small ring called AdA, built by Italian scientists at the Frascati Laboratory. In succeeding years, approximately 30 colliders at a dozen locations have been built and used for experiments, and results from some of these colliders have revolutionized the understanding of high-energy particle physics. Only four of these (ISR, Spp̄S, Tevatron, and RHIC) have produced hadron-hadron collisions, usually either proton–proton (pp) or proton–antiproton (pp̄), while another (HERA) produces electron–proton (ep) collisions.

Both hadron and lepton colliders are important for high-energy and nuclear physics research, though they are usually considered complementary. Hadron colliders provide a broader spectrum of collision data at higher energies, so they tend to excel at making new discoveries in particle physics; lepton colliders, with cleaner beams and smaller backgrounds, tend to excel at precise measurements of phenomena first observed in hadron collisions. The energy reach of hadron colliders is 10–20 times that of lepton colliders, since they can accelerate particles in rings with minimal synchrotron radiation energy loss.

The first hadron collider was the ISR (Intersecting Storage Rings) pp collider at CERN. The ISR reached the energy of 31 GeV per beam, or 62 GeV center of mass (c.m.) energy, in 1971. This state of the art collider demonstrated the importance of hadron colliders to elementary particle physics research. The ISR was decommissioned in December 1983 after producing collisions between beams of alpha particles as well as protons.

In the late 1970s, the major challenge for construction of a pp̄ collider was the production of enough antiprotons with enough phase space density to attain a large interaction rate between the beams of protons and antiprotons. The problem was solved at CERN with the invention of "stochastic cooling", a process that reduces the relative momentum between antiprotons and increases their phase space density. With high-density bunches of p̄ available, counter-rotating beams of protons and antiprotons could be stored in an existing accelerator ring and produce large numbers of interactions. This approach, providing pp̄ collisions using one ring by modifying an existing accelerator complex, provided a minimal cost route to reaching very high c.m. energy hadron–hadron collisions. At CERN, the SpS accelerator ($\sim 250\,\text{GeV}$) was converted to the Spp̄S, which accelerated and collided protons and antiprotons in the original accelerator tunnel.

Two hadron colliders are in operation as of this writing; these accelerators are the forefront machines for experiments with hadrons at very high energies. The Relativistic Heavy Ion Collider (RHIC), located at Brookhaven National Laboratory (BNL), is an ion and polarized pp collider that began operating for experiments in 2000, with c.m. energies ranging from $200\,\text{GeV}/\text{u}$ for gold ions to $500\,\text{GeV}$ for polarized protons. RHIC consists of two intersecting 3.8 km rings of magnets, with a magnetic field of up to 3.5 T (Tesla). These rings intersect in six locations where beam collisions can occur; two of these locations contain the large nuclear physics experiments STAR and PHENIX, while two more contain smaller experiments, BRAHMS and PHOBOS.

The Tevatron, located at the Fermi National Accelerator Laboratory (FNAL), is a pp̄ collider that began operating for experiments in 1987; its c.m. energy is $1960\,\text{GeV}$ using superconducting 4.5 T dipole magnets. The top quark was discovered at this collider in 1995. The Tevatron collider uses a 6.9 km ring of superconducting magnets installed in the original accelerator tunnel, and stores counter-rotating bunches of protons and antiprotons. The energy of each beam during collisions is $980\,\text{GeV}$. There are provisions for four collision areas, but only two are currently used for the large detectors CDF and D0.

A third collider, HERA, is a hybrid ep collider. It produce collisions of $28\,\text{GeV}$ electrons with $920\,\text{GeV}$ protons, with a c.m. energy of $320\,\text{GeV}$. HERA is located at DESY Laboratory in Hamburg, Germany, and has been in operation since 1995 with peak luminosities ranging from $4\text{--}15 \times 10^{32}\,\text{cm}^{-2}\,\text{s}^{-1}$. The 6.3 km HERA proton ring has superconducting 4.7 T dipole magnets.

Operation of pp̄ colliders is quite complex, as the example of the Tevatron shows. To produce p̄'s, protons are accelerated in the original accelerator ring to $120\,\text{GeV}$. The beam is then extracted, striking an external target. p̄'s are collected by a magnet focusing system and injected into a debunching ring, where the bunch structure is removed and the p̄'s are cooled with stochastic cooling, as mentioned earlier. They are then transferred to a concentric accumulating ring, in which p̄'s are collected from many repetitions of this operation, lasting several hours. After sufficient p̄'s are collected, they are injected into the main injector, orbiting in a sense opposite to protons, accelerated to $150\,\text{GeV}$, then transferred into the Tevatron. Thirty-six bunches of p̄'s are stored, followed by thirty-six bunches of protons circulating opposite to the antiprotons. Both beams are then accelerated to $980\,\text{GeV}$ and brought into collision in the experimental regions, providing collisions at a total c.m. energy of $1960\,\text{GeV}$. Figure 1 is a schematic showing the machine elements of the complete Tevatron collider, and Table 1 lists some of the relevant parameters for various existing and planned hadron colliders.

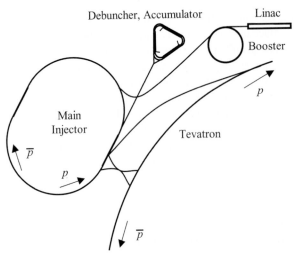

Fig. 1: This schematic drawing shows the primary elements and accelerators for the injection chain of the FNAL Tevatron p$\bar{\text{p}}$ collider.

Why Colliding Beams?

The key parameter in particle physics research has always been the available c.m. energy. As higher-energy accelerators were constructed, startling discoveries were made. For example, in 1952, at the Chicago synchrotron with its 1 GeV proton beam, the π^+p cross-section was observed to increase unexpectedly and dramatically up to the maximum π^+ energy available of $\sim$ 140 MeV. A short time later the π^+p I (isotropic spin)$= 3/2$ resonance at a mass of 1232 MeV was discovered. In 1953 at the Brookhaven National Laboratory 3 GeV Cosmotron, the experimental proof was obtained for the associated production of newly discovered "strange" particles. In 1955, the antiproton was produced at the 6 GeV Bevatron at Lawrence Berkeley Laboratory. The 30 GeV Brookhaven AGS provided several surprises: 1962, the discovery of the muon neutrino; 1964, CP (C = charge conjugation, P = parity) violation in neutral kaon decay; 1974, discovery with SLAC of the fourth quark, "charm". New phenomena have been observed with each new higher-energy accelerator.

These discoveries, however, were made in fixed-target collisions. At relativistic energies, the c.m. energy, E_{cm}, increases only with the square root of the projectile energy,

$$E_{\text{cm}} \cong \sqrt{2m_{\text{p}}E_{\text{i}}} \,, \tag{1}$$

where m_{p} is the mass of the target proton, and E_{i} ($\gg m_{\text{p}}$) is the incident projectile energy in the laboratory frame of reference.

On the other hand, when particles of equal and opposite momenta collide in the laboratory, the c.m. energy increases linearly with the energy of the particles. It therefore follows that with current accelerator technology, colliding beams offer the only practical method of achieving c.m. energies of hundreds or thousands of GeV in hadron collisions. The example of the discovery of the weak intermediate vector bosons ($W^{\pm}$, mass 81 GeV, and Z^0, mass 92.4 GeV) at CERN in 1983, illustrates this argument. In the late 1970s, the SpS at CERN accelerated protons to 270 GeV, and provided 23 GeV c.m. energy when delivering this beam to fixed-

Table 1: Parameters of Tevatron/RHIC/LHC colliders

Parameter/Accelerator	Tevatron	RHIC	LHC
Injection Energy [GeV]	150	24.3	450
Max Energy/beam [TeV]	0.98	0.25	7.0
Circumference [km]	6.3	3.8	26.7
Dipole Field [T]	4.5	3.5	8.4
No. bunches/beam	36	112	2800
Particles/bunch [$\times 10^{11}$]	2 p, 0.3 p̄	2	1
Peak luminosity [$\times 10^{32}\,\mathrm{cm}^{-2}\,\mathrm{s}^{-1}$]	1	1	100
Luminosity lifetime [h]	10–15	12	10

target experiments. For a fixed-target proton accelerator to achieve 92 GeV c.m. energy, a beam of 4300 GeV would be required. However, by modifying the 270 GeV SpS to become the SppS pp̄ collider, more than enough c.m. energy was available to produce the Z^0, and even the $W^\pm$ which must be produced in pairs.

Parameters and Technical Considerations

In comparison to liquid hydrogen or solid targets used in fixed-target accelerators, the colliding beams of a hadron collider are extremely low-density. It is therefore natural to question whether useful interaction rates can be achieved with colliding hadron beams. Luminosity, L (cm^{-2} s^{-1}), is the parameter that connects the interaction rate, R, and the relevant cross section, σ_{cs}, through the relation

$$R = \sigma_{cs} L . \tag{2}$$

At the Tevatron collider energy, the cross section of pp̄ interactions is $\sigma_{cs} \sim 50\,\mathrm{mb}$ ($50 \times 10^{-27}\,\mathrm{cm}^2$). With $L = 10^{32}\,\mathrm{cm}^{-2}\,\mathrm{s}^{-1}$, $R = 5 \times 10^6$ interactions per second, which is indeed a useful event rate. For bunched beams, assuming N_b particles in each bunch and head-on collisions, L is given by

$$L = \frac{N_b^2 k f}{4\pi\sigma^2} , \tag{3}$$

where k is the number of bunches in the beam, f is the revolution frequency, and σ is the root mean square beam size for round beams. Equation (3) suggests how to maximize the luminosity: decrease the beam size σ, increase the number of particles per bunch N_b, and increase the number k of bunches in each beam. f is fixed by the collider's circumference. Naturally there are limitations on the values for these parameters.

There are two components contributing to beam size. One arises from the intrinsic nature of the beam itself, the other from the linear focusing properties of the collider ring magnet system. The beam size σ in one dimension can be written

$$\sigma = \sqrt{\varepsilon \beta^*} , \tag{4}$$

corresponding to one root mean square beam height or width. ε is the emittance of the beam, in units of mm mrad. β^* is the value of the betatron amplitude function (or beta function) $\beta(s)$ at the collision point, and s is the longitudinal coordinate around the ring.

$\beta(s)$ is characteristic of the linear focusing properties of the magnet structure around the ring (the "lattice") and is independent of the beam itself. The luminosity is maximized by minimizing β^*. However, decreasing β^* requires that $\beta(s)$ at some other point (such as at a quadrupole focusing magnet) increases. The aperture of magnets near the collision point thus usually become a limitation on β^*. In hadron machines, it is possible to achieve β^* values as small as 0.5–1.0 m.

For a linearly-focusing lattice at a fixed central momentum, the emittance, ε, is an invariant characteristic of the beam. Even as momentum changes there is an invariant of the motion, termed the normalized emittance,

$$\varepsilon_N = \beta\gamma\varepsilon , \tag{5}$$

where β and γ are the usual relativistic variables. Combining Eqns. (4) and (5) we see that the beam size shrinks as $\sqrt{\beta\gamma}$ or $\sqrt{\text{momentum}}$. Although the emittance ε_N is an invariant for an accelerator where motion is purely linear, there are effects in a real accelerator or collider that tend to make the emittance grow. Sources of some of these effects are (1) non-linearities present in the magnet system, (2) electromagnetic fields generated by the "other'" beam (beam–beam effects), and (3) scattering processes: intrabeam scattering, i. e., multiple Coulomb scattering of protons (hadrons) within bunches in the circulating beam, or beam–gas multiple Coulomb scattering. The cumulative effect of these processes is a kind of "second law", embodied in Liouville's theorem, which states that for a given current the real beam emittance, at any energy, cannot be less than the initial emittance. Any collider design must ensure that the growth time of the emittance is large compared to the time needed for the performance of useful physics. That this was indeed possible was first demonstrated at the ISR where useful circulating beams, lasting 24 hours or more, were routinely available for physics experiments. ε_N is initially determined by the injection system and the source of particles for the collider; hence these sources must be designed to minimize ε_N to achieve high luminosity.

The normalized emittance ε_N can be decreased by cooling processes which seem to violate Liouville's theorem, but in actuality do not. In stochastic cooling of $\bar{p}$'s, mentioned earlier, the minute statistical variations of beam centroid position are fed back to a low-noise high-bandwidth beam kicker, gradually reducing the beam emittance in timescales of minutes to hours. Hadron beams can also be cooled with momentum-matched beams of electrons with very low transverse momentum spread (or temperature). Here higher-energy hadrons preferentially transfer momentum to the electron beam, and the hadron beam cools in a timescale of minutes to hours.

There are also limitations on N_b, the number of particles in each bunch. For example, when two beams collide head on, the particles in a bunch moving clockwise experience a defocusing by the electromagnetic field of the colliding bunch moving counterclockwise, and vice versa. This is expressed by a tune shift ΔQ,

$$\Delta Q = \frac{N_b \beta r_A}{4\pi\varepsilon_N} , \tag{6}$$

where $\beta = v/c$, and $r_A = (Z^2 e^2 / 4\pi\varepsilon_0 A m_p c^2)$ is the classical radius of the hadron where Ze is the hadron charge, $A m_p$ is the hadron mass, and ε_0 is the permittivity of free space. If ΔQ

becomes too large, individual beam particles experience forces that eventually remove them from the beam. From experience, a maximum value of $\Delta Q \sim 0.004$ for each collision point, in hadron colliders, appears to be a safe limit.

Similarly, k, the number of bunches in the ring, is limited. Two different effects illustrate limitations. For head-on collisions, if the distance between bunches, d_b, is less than the field-free space on either side of the collision point, then bunches in one beam have close encounters with several bunches in the oppositely moving beam and this acts to increase the effective beam-beam tune shift. Second, if the luminosity is high so that the effective number of interactions per beam crossing is ≥ 1, then when d_b is short, a particle detector could record several interactions from succeeding beam crossings. These multiple interactions would be difficult to separate in the subsequent analysis of data. This is not a problem with current colliders ($d_b \sim 1000\,\text{m}$), but it is an important consideration for the LHC, where $d_b \sim 5\text{--}10\,\text{m}$.

Future Facilities

Under Construction: the LHC

The LHC (Large Hadron Collider) is the next generation of hadron collider, currently under construction at CERN in the 27 km circumference former LEP (large electron–positron collider) tunnel. This collider has a c.m. energy of 14 TeV, using two intersecting rings of 8.4 T superconducting magnets to collide protons as well as lead ions. As of 2004, LHC is expected to complete construction in 2008, with beam tests in portions of the ring in upcoming years. The total cost of LHC is comparatively low because its construction leverages the injectors and existing LEP tunnel at CERN. Parameters for LHC are included in Table 1, compared to the present operating hadron colliders Tevatron and RHIC.

Proposed Facilities

With the LHC nearing completion, design work is already underway to study the feasibility of an LHC upgrade based on a future hypothesized improvement in superconducting magnet strength by a factor of two, to nearly 17 Tesla. This upgrade, called the "Super" LHC or SLHC, would double the number of bunches in each beam, and separate beams near the interaction points with angle crossings to avoid long-range parasitic beam-beam collisions. These upgrades would provide a peak luminosity improvement of a factor of ten, to $L = 10^{35}\,\text{cm}^{-2}\,\text{s}^{-1}$, and would extend the energy reach of the LHC to nearly 30 TeV to characterize the Higgs boson and explore for new phenomena, such as supersymmetric particle production. There are also two design concepts for a Very Large Hadron Collider (VLHC), designed to reach the 100 TeV energy scale using either aggressive high-field (HF) 12.5 T magnets and a modest 89 km circumference collider ring, or conservative low-field (LF) 2 T magnets and a large 520 km collider ring. Parameters for these proposed facilities are summarized in Table 2.

To complement and extend nuclear structure data from ep collisions at HERA, another proposed electron-hadron collider is eRHIC, an extension of RHIC at Brookhaven. This collider is designed to provide 10 GeV polarized electrons or positrons colliding with various hadrons, including 50–250 GeV polarized protons and 100 GeV/u gold ions, with luminosities ranging from $10^{30}\text{--}10^{33}\,\text{cm}^{-2}\,\text{s}^{-1}$. Collisions would be produced for one experiment in about 2014 by adding an 1.3 km circumference electron ring and 10 GeV electron linac.

Table 2: Parameters for proposed pp colliders, LF/HF VLHC and SLHC.

Parameter/Accelerator	LF VLHC	HF VLHC	SLHC
Injection Energy [GeV]	2500	2500	450
Max Energy/beam [TeV]	50	50	14.0
Circumference [km]	520	89	26.7
Dipole Field [T]	2	12.5	16.8
No. bunches/beam	21000	20000	5600
Particles/bunch [$\times 10^{11}$]	0.22	0.125	1
Peak luminosity [$10^{32}\,\mathrm{cm}^{-2}\,\mathrm{s}^{-1}$]	1	1	1000
Luminosity lifetime [h]	25	13	~ 3
Number of experiments	2	2	2(4)

Interaction of Collider and Experiments

Colliders differ from other large accelerators such as synchrotrons in significant respects. The need for high-field DC magnet operation and longtime stability of the beams are examples which are briefly touched upon elsewhere in this article. However, the crucial difference is the close interrelationship between the design and operations of each experiment and the collider.

Since collisions between the circulating beams occur within the vacuum chamber of the collider, experiments become an integral part of the machine and must be designed about the specific properties of the beam and with consideration for the geometry of the collider and the stability of the beam.

The vacuum system provides a clear example of how the experimental needs affect the machine design. The beams must be able to circulate for many hours with minimal loss of beam quality or luminosity. Protons scattering from the residual gas is one such loss mechanism and implies the need for a far better vacuum than in conventional accelerators. For example, a vacuum of 10^{-10} Torr or better is required to minimize background particles, against which the experimental equipment must discriminate, and to preserve the beam lifetime over many billions of revolutions around the collider. The collider design must also include long field-free regions on either side of the beam-collision points to accommodate the particle detectors that are often the size of a three-story building.

Conclusion

For producing interactions above several hundred GeV c.m. energy, only pp colliders currently have the required high luminosity, $L > 10^{32}\,\mathrm{cm}^{-2}\,\mathrm{s}^{-1}$. p$\bar{\mathrm{p}}$ and heavy-ion colliders have significantly lower maximum luminosity. New technology will be needed beyond the next generation of high-energy colliders to keep costs within reason, perhaps in high-field magnet technology. In the past, new ideas and technologies permitted higher energies to be obtained, from the synchrotron principle in 1944, to the alternating gradient or strong focusing principle in 1952, to the use of superconducting magnets in 1985. Will the cycle repeat?

Bibliography

General

E. D. Courant and H. S. Snyder, "Theory of the Alternating Gradient Synchrotron", *Ann. Phys. (New York)* **3**, 1 (1958).

M. S. Livingston and J. P. Blewett, *Particle Accelerators*. McGraw-Hill, New York, 1962.

C. Pelligrini, "Colliding-Beam Accelerators", *Ann. Rev. Nucl. Sci.* **20**, 1 (1972).

J. D. Lawson and M. Tigner, "The Physics of Particle Accelerators", *Ann. Rev. Nucl. Part. Sci.* **34**, 29 (1984).

W. Scharf, *Particle Accelerators and Their Uses*. Harwood Academic Publishers, 1986 (2 volumes).

A. W. Chao and M. Tigner (eds.), *Handbook of Accelerator Physics and Engineering*. World Scientific, Singapore, 1999.

F. Zimmermann, "Luminosity Limitations at Hadron Colliders", in *Proc. 18th International Conference on High Energy Accelerators* (HEACC2001), Tsukuba, Japan, CERN-SL-2001-009 AP, 2001.

Specific Accelerator Components

F. T. Cole and F. E. Mills, "Increasing the Phase-Space Density of High-Energy Particle Beams", *Ann. Rev. Nucl. Part. Sci.* **31**, 295 (1981).

R. Palmer and A. V. Tollestrup, "Superconducting Magnet Technology", *Annu. Rev. Nucl. Part. Sci.* **34**, 247 (1984).

The Fermilab Antiproton Source Design Report. Fermi National Accelerator Laboratory, Batavia, IL, February 1982.

Proceedings

The Proceedings of various conferences and schools contain the current state of the art on accelerator theory and practice. A few are provided here.

Proceedings of the U.S. Particle Accelerator School.

Proceedings of the CERN Accelerator School.

Proceedings of the International Conference on Magnet Technology.

M. Month, P. Dahl, and M. Dienes (eds.), *The Physics of High Energy Particle Accelerators*. AIP Conference Proceedings No. 105 and 127. American Institute of Physics, New York, 1983 and 1985.

Hall Effect

P. R. Emtage

In 1879, E. H. Hall found that when a metal strip which bore a current was placed in a magnetic field, as in Fig. 1a, a voltage was produced across the strip. This transverse voltage is the simplest and most widely useful of the galvanomagnetic effects, since the number and nature of the current carriers can often be found from it.

The Hall voltage comes from a transverse electric field that cancels the sideways deflection of the current carriers by the magnetic field. Suppose the current is carried by electrons of charge $-e$ and drift velocity v_x in the direction of the current. In a magnetic field B_z they are acted on by a transverse force f_y – the Lorentz force – which causes the cyclotron motion and

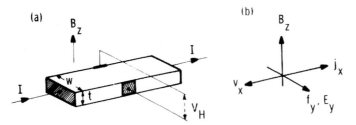

Fig. 1: (a) Configuration in which Hall effect is found, (b) Fields, etc., in the system.

is at right angles to both field and velocity,

$$f_y = ev_x B_z .$$

The electron trajectories are therefore bent toward one side of the conductor; a surface charge forms on the sides and causes an electric field E_y that opposes further deflection – see Fig. 1b. Equilibrium is reached when the total force is zero,

$$ev_x B_z - eE_y = 0 . \tag{1}$$

The transverse field E_y should be expressed in terms of measurable quantities, such as electric current, rather than electron velocity. If there are n electrons per unit volume, the current density is

$$j_x = -nev_x ,$$

the negative sign being inserted because the charge on the electron is negative. Accordingly from Eq. (1) we find

$$E_y = Rj_x B_z , \tag{2}$$

wherein R is defined as the Hall coefficient, and from the above argument is given by

$$R = -\frac{1}{ne} . \tag{3a}$$

Note that, had we supposed the current to be carried by p positive carriers of charge $+e$, we should have found

$$R = \frac{1}{pe} . \tag{3b}$$

The sign of the Hall coefficient therefore says whether the current is borne by positive or by negative charges, and its magnitude yields the density of current carriers.

The above results are so simple that they should be regarded with the gravest suspicion. Equations (3) hold only in very high magnetic fields, when each electron makes several cyclotron revolutions between collisions. In low fields no unique electron drift velocity v_x exists, and therefore Eq. (1) cannot be satisfied for all groups of electrons simultaneously. The total

Table 1: Hall coefficients R (at room temperature) and apparent number of
free electrons per atom, n_H/n_a, for some elemental conductors.

Element	Group^a	$10^{10}R/(\text{m}^3\text{C}^{-1})$	n_H/n_a
Na	Ia (M)	-2.1	1.17
Cu	Ib (M)	-0.536	1.38
Be	IIa (M)	$+2.44$	-0.21
Al	IIIa (M)	-3.0	0.31
Cr	IVb (TM)	$+3.55$	-0.21
Pt	VIII (TM)	-0.244	3.87
As	Va (SM)	-70	1.9×10^{-2}
Sb	Va (SM)	$+250$	-7.6×10^{-3}
Ge	IVa (SC)	-8×10^{8}	1.7×10^{-9}

^{a} In this column M = metal, TM = transition metal, SM = semimetal,
  SC = semiconductor.

transverse current must, however, be zero, and therefore a well-defined low-field Hall coefficient does exist, and in simple cases has the form

$$R = -\frac{r}{ne},$$

r being a factor of order unity. It is common to define a "Hall density of carriers" through

$$n_H \equiv -\frac{1}{Re}. \tag{4}$$

Where other evidence suggests that the electron structure is in fact simple, this definition is a useful guide to the electron density.

Measurements

The Hall coefficient R is defined from Eq. (2), and measurements of it are carried out on bar-shaped samples such as that in Fig. 1a. Let the bar have width w and thickness t; if the total current is I, the current density is $j_x = I/wt$. The Hall voltage V_H is measured between electrodes on the sides, and is related to the Hall field through $E_y = V_H/w$. Therefore,

$$R = \frac{E_y}{j_x B_z} = \frac{t V_H}{I B_z}.$$

In practice it is always necessary to find the mean of V_H from four measurements with field and current both forward and reversed, so as to eliminate errors resulting from misalignment of the electrodes, thermoelectric effects, and other galvanomagnetic effects, as well as inhomogeneities or anisotropy in the sample.

    Values of the Hall coefficient at room temperature for a variety of elemental conductors are shown in Table 1; the last column shows the apparent number of free electrons per atom, n_H being defined by Eq. (4). The results for the polyvalent metals are distressingly irregular, but there is some order amidst the chaos: all the metals have high carrier densities, and the monovalent metals have nearly one free electron per atom; the semimetals show much lower

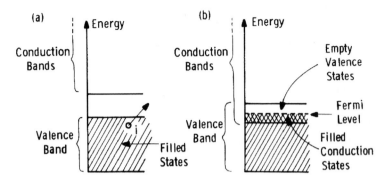

Fig. 2: Electron bands in (a) a semiconductor or insulator; (b) a semimetal.

carrier densities, as expected; while Ge, a semiconductor with no free electrons at all at low temperatures, is nearly destitute of current carriers.

The outstanding oddity of Table 1 is that the Hall coefficients are in many cases positive. It was found in Eqns. (3) that a positive Hall coefficient should correspond to positively charged current carriers; the results therefore conflict with the electron theory of conduction, and for a resolution of this enigma we must turn to the band theory of electrons.

Band Theory

There is not space here for more than the briefest description of the band theory. In a semiconductor the valence electrons occupy a complete set or band of states that spans a finite range of energies; no electron in this set can be moved to any other valence state without violating the Pauli exclusion principle, so the valence electrons are "rigid" and can carry no current. There must, however, be states of higher energy, the conduction bands, within which an electron can move – otherwise, the material would be impervious to high-energy electrons.

In a semiconductor the conduction band is separated from the valence band and is normally empty, as in Fig. 2a. In many semimetals the conduction and valence bands overlap somewhat, as in Fig. 2b; to minimize energy, electrons in the upper part of the valence band "fall" into the lower part of the conduction band. In most polyvalent metals the overlap of bands is large, and no clear distinction between conduction and valence bands can be drawn.

The peculiarities of the band picture can be seen if we imagine an electron in state i removed from a filled valence band, as in Fig. 2a; conduction can now occur since other valence electrons can move into the vacant state. Denote by $\sum$ a sum over all valence states, and by $\sum'$ a sum over all states except i. The charge on each valence electron is $-e$; let the energy of each electron relative to the band maximum be $-\varepsilon_v$. The energy ε_h and charge q_h of the "hole" in the band are

$$\varepsilon_h = \sum{}'(-\varepsilon_v) - \sum(-\varepsilon_v) = \varepsilon_i ,$$
$$q_h = \sum{}'(-e) - \sum(-e) = +e .$$

The hole in the valence band therefore acts as a "particle" of positive energy and positive charge, and the Hall coefficient due to its motion is positive. It is from the motion of electrons in incompletely filled bands that so many materials derive positive Hall coefficients.

See also: Conduction; Electron Energy States in Solids and Liquids; Galvanomagnetic and Related Effects.

Bibliography
A. A. Abrikosov, *Introduction to the Theory of Normal Metals*. Academic Press, New York, 1972. (A)
C. Kittel, *Introduction to Solid State Physics*, 7th ed. Wiley, New York, 1995.(E)
E. H. Putley, *The Hall Effect and Related Phenomena*. Butterworth, London, 1960. (I)

Hall Effect, Quantum

K. von Klitzing

The most exciting result of the quantum Hall effect (QHE) is the fact that from measurements on microelectronic devices a new type of electrical resistance can be deduced [quantized Hall resistance (QHR)] with a value which is independent of the material and microscopic details of the conductor. High-precision measurements in different countries demonstrated that within the experimental uncertainty of about 1×10^{-9} the same value R_K is found for the quantized Hall resistance, and on the basis of a recommendation of the Comité Consultatif d'Electricite (September 9, 1988) the best value for R_K is given as $R_K = 25\,812.807\,\Omega$ with an uncertainty of $\Delta R_K = \pm 0.005\,\Omega$. Up to now, all theories of the QHE show that the quantized Hall resistance is identical with the fundamental constant h/e^2 (h = Planck constant, e = elementary charge) and the table of recommended values of fundamental constants (2002) uses the expression von Klitzing constant for the unit h/e^2 and gives the value $25\,812.807449 \pm 0.000086\,\Omega$. The quantity h/e^2 is directly proportional to the Sommerfeld fine-structure constant (the proportionality constant is a fixed number which is known without any uncertainty) and therefore the title of the first publication in 1980 about the quantized Hall effect was "New method for high precision determination of the fine structure constant...".

Since January 1, 1990 all calibrations of resistance are based on a fixed value of the quantized Hall resistance with $R_{K-90} = 25\,812.807\,\Omega$ (conventional von Klitzing constant). Experimentally, the quantized Hall resistance is observed in Hall-effect measurements at low temperature T and high magnetic fields B (typically $T = 2\,K$ and $B = 10\,T$) on two-dimensional electronic systems. Two dimensional means that the electrons are free to move within a plane but have a fixed energy for the motion in the direction perpendicular to the plane.

Normally silicon field-effect transistors or GaAs–AlGaAs heterostructures are used as two-dimensional electronic systems for the investigation of the QHE. The electrons in these devices are confined in such a thin layer close to the interface Si–SiO$_2$ or GaAs–AlGaAs that the energy of the electrons for the motion perpendicular to the interface becomes quantized into well-separated electric subbands E_i. At low temperatures only the lowest electric subband E_0 is occupied with electrons and the energy of the electrons can be written

$$E = E_0 + \frac{\hbar^2 k_\parallel^2}{2m} . \tag{1}$$

The second term in this equation characterizes the free motion of the electrons (effective mass m and momentum $\hbar k_\parallel$) within the plane parallel to the Si–SiO$_2$ or GaAs–AlGaAs interface (x-y plane). The free motion of the electrons within the x-y plane is drastically changed if a strong magnetic field B_z is applied perpendicular to the two-dimensional electronic systems. The motion of the electrons on closed cyclotron orbits leads to a quantization in the energy (Landau levels) comparable with the discrete energies of electrons in the hydrogen atom. Under this condition the energy spectrum of the electrons becomes discrete,

$$E_n = E_0 + \left(n + \frac{1}{2}\right)\hbar\omega_c, \quad n = 0, 1, 2\ldots \tag{2}$$

The cyclotron energy $\hbar\omega_c = eB_z/m$ is typically of the order of 10 meV at $B_z = 10$ T. Each cyclotron orbit occupies an area of h/eB_z within the x-y plane which leads to a degeneracy factor per unit area for each Landau level of $N = eB_z/h$. Since the area of a cyclotron orbit is relatively small (the cyclotron radius at $B_z = 10$ T is only 8 nm), a very large number of electrons (2.4×10^{11} cm^{-2} at 10 T) can occupy each Landau level.

The classical Hall effect which relates the Hall voltage U_H (measured perpendicular to the magnetic field B_z and the current direction I) to the carrier density n becomes independent of the magnetic field and the dimension of the sample if for a two-dimensional electron gas, the carrier density agrees exactly with an integer number i of fully occupied Landau levels. Under this condition the Hall voltage has the value

$$U_H = \frac{h}{ie^2}I, \quad i = 1, 2, 3\ldots \tag{3}$$

where the proportionality constant between the Hall voltage and the current through the sample is the quantized Hall resistance R_H.

Figure 1 shows a typical result of Hall effect and resistivity measurements on a two-dimensional electron gas (2DEG). A GaAs–Al$_{0.3}$Ga$_{0.7}$As heterostructure was used in this experiment, because a fully quantized energy spectrum, necessary for the observation of the quantized Hall resistance, is already obtained at a relatively low magnetic field: the cyclotron energy for electrons in GaAs is much larger than in silicon for the same magnetic field.

The Hall voltage U_H and the voltage drop U_p between the potential probes are measured under constant current conditions ($I =$ const) as a function of the magnetic field applied perpendicular to the plane of the 2DEG. The oscillation in $R_x = U_p/I$ originates from variations in the filling factor of the energy levels E_n in a magnetic field. Whenever R_x becomes zero, an integer number i of energy levels E_n is occupied and the Hall resistance adopts the value $R_H = h/e^2 i$. Maxima in R_x are observed at approximately half-filled energy levels.

The simple theory discussed before cannot explain the flat regions in the $U_H(B)$ measurements (Hall plateaus). For a fixed surface carrier density n only at well-defined magnetic field values $B_i = hn/ei$ should the Hall resistance be expressed by $R_H = h/ie^2$. A number of theoretical papers have discussed the phenomena of Hall *plateaus* and the authors conclude that potential fluctuations within the area of the sample (for example due to ionized impurities) lead to a localization of electrons which stabilize the Hall resistance at the quantized values. The plateaus should disappear for an ideal system. Experimentally the width of the plateaus becomes smaller if the number of impurities is reduced (higher mobility of the electrons). Simultaneously, new phenomena related to the fractional quantum Hall effect (FQHE) become

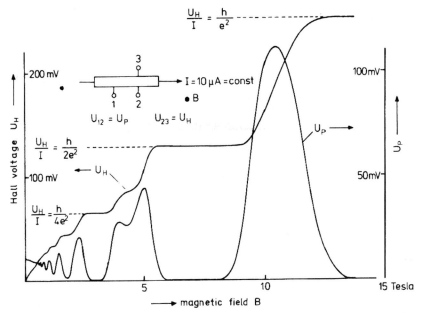

Fig. 1: Hall voltage (U_H) and resistivity ($\rho_{xx} \sim U_p$) data for a GaAs–Al$_{0.3}$Ga$_{0.7}$As heterostructure as a function of the magnetic field B at a temperature of $T = 1.6\,\mathrm{K}$. Steps in the Hall voltage are visible with resistance values $R_H = h/e^2 i$ at magnetic fields where U_p becomes zero.

more pronounced which lead to Hall plateaus not only at integer values of i in the equation $R_H = h/ie^2$ but also at "fractional values" of i like $i = \frac{1}{3}, \frac{2}{3}, \frac{3}{5} \dots$

The origin of these additional Hall plateaus are new gaps in the energy spectrum due to the electron–electron interactions. Very often the FQHE is discussed as the normal QHE of Composite Fermions, where one electron plus two flux quanta form a new quasi particle which feels an effective magnetic field $B_{eff} = B - B^*$ where B^* is the magnetic field of half filled Landau level $i = 1/2$. The appearance of (integer or fractional) Hall plateaus is always connected with a vanishing energy dissipation (vanishing voltage drop U_p in the current direction I as shown in Fig. 1). The existence of such supercurrents forms the basis of different theories which are able to explain the quantum Hall effect in a more general way. The experimental finding that a quantized longitudinal resistance $R = h/ie^2$ ($i = 2, 4, 6, \dots$) is also observed *without* magnetic field if the width of the two-dimensional system is so small ($\approx 100\,\mathrm{nm}$) that an additional quantization in the energy spectrum of the electrons becomes effective (one-dimensional channel for the electron motion) favors an interpretation of the quantum Hall effect on the basis of one-dimensional transport without backscattering. In this picture the interior of the two-dimensional sample is only a reservoir of localized electronic states. It seems that the edges of the sample are extremely important for an explanation of the quantum Hall effect since they consist of metallic (compressible) and insulating (incompressible) stripes origination from the reduction of the carrier density to zero within the depletion of about 100 nm close to the edge of the sample.

See also: Constants, Fundamental; Hall Effect.

Bibliography
K. von Klitzing, "The quantized Hall effect" (Nobel Lectures in Physics 1985), *Rev. Mod. Phys.* **58**, 519 (1986).
D. Yoshioka, *The Quantum Hall Effect*, Springer Series in Solid State Sciences **133** (2002).
R. Haug and D. Weiss (eds.) Proc. of the International Symposium "Quantum Hall Effect: Past, Present and Future", *Physica E* **20**/1–2 (2003).

Hamiltonian Function
R. H. Good, Jr.

The Hamiltonian is the function that, through the way that it depends on its arguments, specifies the time development of a system.

Consider first a classical (nonquantum) system with coordinates q_i and momenta p_i. The Hamiltonian equations of motion are

$$\frac{dq_i}{dt} = \frac{\partial H}{\partial p_i}, \quad \frac{dp_i}{dt} = -\frac{\partial H}{\partial q_i}. \tag{1}$$

As we know the functional dependence of the Hamiltonian on coordinates and momenta, $H(q_i, p_j)$, we have here a set of differential equations that determine the time dependence of the coordinates and momenta. An example is a system with one coordinate, one momentum, and

$$H(q, p) = (p^2/2m) + V(q). \tag{2}$$

The Hamiltonian equations of motion lead to

$$\frac{dq}{dt} = \frac{p}{m}, \quad \frac{dp}{dt} = -\frac{\partial V}{\partial q}. \tag{3}$$

The first equation serves to define the momentum in terms of the velocity; if the momentum is eliminated, then

$$m\frac{d^2q}{dt^2} = -\frac{\partial V}{\partial q}, \tag{4}$$

which is Newton's second law for the motion of a particle with coordinate q in a potential field $V(q)$. In general the Hamiltonian equations of motion are equivalent to Newton's equations. The Hamiltonian equations apply to a variety of problems and they permit general discussions of equations of motion and transformation between systems, the detailed descriptions of the systems being relegated to the functional dependence of the Hamiltonian. Because of the formal elegance of the equations and the discussions they lead to, they are often called the canonical equations of motion. Another aspect of the role of the Hamiltonian as determining the evolution in time is the theorem

$$\frac{du}{dt} = [u, H] . \tag{5}$$

Here $u(q_i, p_j)$ is any function of the coordinates and momenta and the right-hand side is the Poisson bracket of the two functions, defined by

$$[u, H] = \sum_i \left(\frac{\partial u}{\partial q_i} \frac{\partial H}{\partial p_i} - \frac{\partial u}{\partial p_i} \frac{\partial H}{\partial q_i} \right) . \tag{6}$$

Next consider a quantum-mechanical system. Here also there are classical-type coordinates q_i, momenta p_i, and a Hamiltonian $H(q_i, p_j)$ that describes the system to be studied. Now, however, these are all operators, the coordinates and momenta satisfying

$$[p_i, q_i] = -i\hbar\delta_{ij} \tag{7}$$

where the brackets indicate the commutator $[p_i, q_j] = p_i q_j - q_j p_i$. In a quantum-mechanical system there may also be nonclassical coordinates, as for the spin of a particle. In the Schrödinger picture the state of the system is described by a time-dependent wave function $\psi(q, t)$. The operators act on the wave function, and the commutation rules, Eqns. (7), are realized by identifying p_i as $-i\hbar\partial/\partial q_i$. The Hamiltonian again determines the time evolution of the system, here through the Schrödinger equation

$$H\psi = i\hbar\frac{\partial\psi}{\partial t} . \tag{8}$$

For example, if the system has the Hamiltonian of Eq. (2), the equation for the time development of the wave function is

$$-\frac{\hbar^2}{2m}\frac{\partial^2\psi}{\partial q^2} + V\psi = i\hbar\frac{\partial\psi}{\partial t} . \tag{9}$$

The Hamiltonian will be a Hermitian operator in some sense, so that the wave function can be normalized. In the example above H is Hermitian in the sense that

$$\int \psi_1^* H\psi_2 \, dq = \int (H\psi_1)^* \psi_2 \, dq \tag{10}$$

where ψ_1 and ψ_2 are any two wave functions that go to zero at the limits of the integration, and then it follows from Eq. (8) that

$$\frac{d}{dt}\int \psi^*\psi \, dq = 0 , \tag{11}$$

so that the normalization condition

$$\int \psi^*\psi \, dq = 1 \tag{12}$$

can be assigned independent of time. In the Heisenberg picture the time dependence is moved from the wave function to the operators. For any operator $u_S(q_i, p_j)$ in the Schrödinger picture the Heisenberg operator is defined by

$$u_H(t) = e^{-Ht/i\hbar} u_S e^{Ht/i\hbar} \tag{13}$$

and the Heisenberg operator has time dependence such that

$$\frac{du_H}{dt} = (i\hbar)^{-1}[u_H, H] . \tag{14}$$

Here again the Hamiltonian determines the time development. There is a close parallel between classical mechanics, as expressed by Eq. (5), and quantum mechanics in the Heisenberg picture, as expressed by Eq. (14). The Poisson bracket and $(i\hbar)^{-1}$ times the commutator play corresponding roles.

In classical mechanics and nonrelativistic quantum mechanics the Hamiltonian is the total energy of the system. For example, when Eq. (2) applies, H is the kinetic energy plus the potential energy. In relativistic quantum mechanics this interpretation cannot be applied straightforwardly. In Dirac's theory of the electron–positron, for example, the Hamiltonian for a free particle is

$$H = a\boldsymbol{\alpha} \cdot \mathbf{p} + mc^2\beta , \tag{15}$$

where $\boldsymbol{\alpha}$ and β are 4×4 matrices satisfying

$$\alpha_i\alpha_j + \alpha_j\alpha_i = 2\delta_{ij}$$
$$\alpha_i\beta + \beta\alpha_i = 0 , \qquad \beta^2 = 1 .$$

Consider eigenstates of the operator $\mathbf{p}$ and use the same symbol for the eigenvalue. The eigenvalues of H are found to be $\pm(p^2c^2 + m^2c^4)^{1/2}$. We cannot identify H as the energy operator without further discussion, in view of the negative eigenvalues. One way out of the difficulty is Dirac's hole theory. He suggested that H be identified as the energy operator and that the vacuum consists of all the positive-energy states empty, all the negative-energy states filled. A physical electron is the vacuum plus a particle in a positive-energy state; a physical positron is the vacuum minus a particle in a negative-energy state. Both electron and positron then have positive energy. Another way out of the difficulty is to abandon the interpretation of H as the energy operator. We can define $|H|$ as the operator $(p^2c^2 + m^2c^4)^{1/2}$, having always positive eigenvalues, and identify it as the energy operator. The operator $H/|H|$ has eigenvalues $+1$ and -1, corresponding to electron states and positron states.

See also: Dynamics, Analytical; Kinematics and Kinetics; Quantum Mechanics; Schrödinger Equation.

Bibliography

H. C. Corben and P. Stehle, *Classical Mechanics*, 2nd ed. Wiley. New York, 1960.
P. A. M. Dirac, *The Principles of Quantum Mechanics*, 4th ed. Oxford, London and New York, 1958.
H. Goldstein, *Classical Mechanics*, 2nd ed. Addison-Wesley, Reading, Mass., 1980.

Heat

J. W. Morris, Jr.

In thermodynamics the term heat is used to denote the quantity of energy exchanged through thermal interaction between the system of interest and its environment. If, for example, a flame is applied to a cool body of water, the energy content of the water increases, as evidenced by its increased temperature, and we say that heat has passed from the flame to the water. If energy losses from the water to the atmosphere may be ignored, the heat transferred is numerically equal to the energy gained by the water.

In more complex processes, which may involve mechanical as well as thermal interaction, the heat transferred is more difficult to identify. In fact, much of the history of thermodynamics concerns the slow evolution of clear concepts of heat, energy, and entropy, and of the distinctions between them. In the resulting science of thermodynamics heat is defined and measured indirectly in terms of the objectively measurable quantities energy and mechanical work.

Specifically, thermodynamics postulates the conservation of energy (the first law of thermodynamics) and distinguishes two general ways in which the conserved quantity, energy, may be transferred from one system to another. First, the interacting systems may do work on one another through the action of the forces imposed by their mechanical, electromagnetic, or chemical interaction. Second, the energy may be exchanged without sensible work and is in this case called heat. The change in the energy (ΔE) of a system is objectively measurable; since energy is a state function (i. e., is uniquely determined by the thermodynamic state of the system) the net change in energy may be found by identifying the initial and final states of the system and computing the difference between their associated energies. The net work (W) done on a system is also objectively measurable through the rules supplied by the science of mechanics. The heat, in contrast, is not directly measurable. Thermodynamics rather defines the heat supplied to a system as the difference between its energy change and the work done on it:

$$Q = \Delta E - W .\tag{1}$$

The concept of a thermal interaction is then defined, in an equally indirect way, as a nonmechanical interaction that results in the transfer of heat.

Since heat and work are forms of the same physical quantity, energy, and since heat is not directly measurable, we might wonder why it is important to retain the concept of heat in modern physics. The answer is contained in the second law of thermodynamics, which defines a new state function, the entropy (S) of the system. The change in the entropy of a system is sensitive to the manner in which energy is supplied; it does not necessarily change when work is done on the system, but necessarily increases when heat is added. The entropy change governs the reversibility of thermodynamic processes. When two systems exchange energy solely in the form of work the thermodynamic process may, in theory, be reversed to reestablish their initial states. However, when the thermodynamic process involves an exchange of heat or the conversion of work into heat via friction, the process generally cannot be reversed without causing a corresponding irreversible change in some third system.

See also: Energy and Work; Entropy; Thermodynamics, Equilibrium; Thermodynamics, Nonequilibrium.

Heat Capacity

N. Pearlman

An intrinsically positive, extensive property of matter, defined as the limiting ratio of heat input ΔQ to temperature increment ΔT, is the heat capacity

$$c = \lim_{\Delta T \to 0} \Delta Q / \Delta T \ .$$

Constraints, such as constant volume or pressure, are usually indicated by subscripts. Tabulated values are ordinarily heat capacity per mole. The quantity normally measured is c_p, which is related to c_V by

$$c_p - c_V = VTB\beta^2$$

(V is volume, β the thermal expansion coefficient, B the bulk modulus). This is related to energy differences,

$$\Delta E = \int_{T_1}^{T_2} c_V(T) \, dT \ ,$$

and, according to the constraint, to entropy differences,

$$\Delta S_{p,V} = \int_{T_1}^{T_2} \frac{c_{p,V}(T)}{T} \, dT \ ,$$

thereby yielding information on the excitations concerned. For instance, the temperature-independent molar heat capacity

$$c_V = \frac{\nu}{2} R$$

(ν is an integer R the gas constant per mole, $R = N_A k_B$, with N_A Avogadro's number and k_B Boltzmann's constant) observed for many gases and solids at room temperature corresponds to the classical equipartition energy $k_B T / 2$ per degree of freedom, of which there are ν per molecule. In solids, $\nu = 6$ accounts for lattice energy in terms of normal-mode harmonic oscillators. Their energy, and hence also their heat capacity, vanishes exponentially as T approaches zero, as required by the third law of thermodynamics. For the quadratic oscillator frequency distribution assumed in Debye's theory,

$$\left(\frac{c_V}{R} \right)_{\text{lattice}} = \left(\frac{12\pi^4}{5} \right) \left(\frac{T}{\theta_D} \right)^3 \ , \quad T \ll \theta_D$$

(θ_D, the Debye temperature, is $\hbar\omega_m / k_B$, with ω_m the maximum oscillator frequency). It is possible to relate θ_D to lattice elastic properties. The electrons in normal metals form a degenerate Fermi gas, and so their heat capacity contribution can be observed only at low temperatures, where $(c_V)_{\text{lattice}}$ has become negligible. The electronic contribution, $(c_V)_{\text{electronic}}$ is found to be linear in T, as calculated in free-electron theory,

$$\left(\frac{c_V}{R} \right)_{\text{electronic}} = \gamma \frac{T}{R} = \left(\frac{\pi^2}{2} \right) \left(\frac{T}{T_F} \right) \ .$$

The Fermi degeneracy temperature T_F is related to the Fermi level E_F by $T_F = E_F/k_B$, and thereby to the density of electron states per electron at the Fermi level,

$$D(E_F) = \frac{3}{2}E_F$$

and hence also to the electron effective mass, which is proportional to $D(E_F)$. In superconductors, below the transition temperature T_c the ground state is separated from excited states by an energy gap E_g which, according to BCS theory, is zero at T_c and increases to about $3.5 k_B T_c$ at $0\,\mathrm{K}$. In consequence, there is a jump in $(c_V)_{\text{electronic}}$ at T_c:

$$\frac{(c_V)_{\text{electronic}}^{\text{supercond}}(T_c)}{\gamma T_c} = 2.43$$

according to BCS theory. Also, the heat capacity vanishes exponentially rather than linearly as T approaches zero:

$$\frac{(c_V)_{\text{electronic}}^{\text{supercond}}(T)}{\gamma T_c} = a\,e^{-bT_c/T}$$

where both a and b vary with T. While in a superconductor the ground state and the energy gap are collective properties of all the electrons due to their interaction with the lattice, a number of interactions (magnetic, crystal field, etc.) involving isolated degrees of freedom also produce a gap, ΔE, between the ground state and the first excited state. In terms of the temperature parameter $T_S = \Delta E/k_B$, the heat capacity of such a mode (called a "Schottky anomaly") is

$$\frac{c}{k_B} = \frac{\left(\frac{T_S}{T}\right)^2 e^{T_S/T}}{\left(1 + e^{T_S/T}\right)^2} .$$

Unlike the contributions described earlier, this is not monotonic. It has a peak at $T = 0.416 T_S$, decreasing, as $(T_S/T) \times e^{T_S/T}$ for $T < T_S$, and as $(T_S/2T)^2$ for $T > T_S$. Since the excited state is empty at $T = 0\,\mathrm{K}$ while its occupancy equals that of the ground state for $T > T_S$, the total area under c/T is the entropy difference corresponding to this rearrangement, or $k_B \ln 2$.

See also: Fermi–Dirac Statistics; Lattice Dynamics; Thermodynamics, Equilibrium.

Bibliography

C. Kittel, *Introduction to Solid State Physics*, 7th ed. Wiley, New York, 1996.
M. W. Zemansky and R. H. Dittman, *Heat and Thermodynamics*. McGraw–Hill, New York, 1997.

Heat Engines

M. Garbuny[†]

Systems that receive heat and convert it into mechanical work in a periodic process are called heat engines. The heat is supplied to a working substance externally, as in the steam turbine, or internally, as in the gasoline engine. At constant volume the working substance, which may be either a gas or a liquid, experiences an increase of its internal energy U, where U represents the sum of the molecular kinetic energies due to thermal motion and of the potential energies caused by molecular fields. However, if the working substance is allowed to expand its volume V against a moving piston or turbine blade at a pressure P, it will convert part of the received heat into work $W = \int P(V) \, dV$. The extent of this conversion is governed by the first and second laws of thermodynamics.

The first law, a manifestation of the energy conservation theorem, asserts that heat δQ_{in} supplied to a system is, in the limit of lossless operation, just equal to the sum of the internal energy increase in the system and the work performed by the system or, in differential form, $\delta Q_{in} = dU + P \, dV$. In executing a cycle, a system returns to its starting point, i.e., to the original values of pressure, volume, and temperature T. Therefore the contribution of heat to the internal energy after a complete cycle is zero, and the first law then yields the relation $W = Q_{in} - Q_{out}$, where Q_{out} is the heat not converted into work. A *thermal efficiency* η can now be defined as the ratio W/Q_{in}, i.e., $\eta = 1 - Q_{out}/Q_{in}$.

The existence of residual heat Q_{out} after the completion of a thermal engine cycle is a requirement imposed by the second law. The mathematical formulation of this law introduces the *entropy* S defined by the differential $dS = \delta Q_r/T$, where δQ_r is an amount of heat converted *reversibly*, so that reversing the heat transfer reestablishes conditions inside and outside the system to those before the conversion. The second law asserts that dS is an exact differential, so that after a cycle completed by any thermal system with any working substance, $\oint \delta Q_r/T = 0$. Thus an engine cannot completely convert a heat input Q_{in} into work W. If Q_{in} is obtained from a heat source at a temperature T_H, then for reversible conversion into work, the system requires a heat sink at a lower temperature T_L, so that $\oint \delta Q_r/T = (Q_{in}/T_H) - (Q_{out}/T_L) = 0$. Thus $Q_{in}/T_H = Q_{out}/T_L$. With the first law, this yields the Carnot thermal efficiency

$$\eta_r = 1 - T_L/T_H . \tag{1}$$

According to Eq. (1), the thermal efficiency of a reversible cycle depends only on the temperatures T_H of the source and T_L of the sink. An ideal cycle operating reversibly between two heat reservoirs at T_H and T_L was first proposed in 1824 by Sadi Carnot. The operation of the cycle is shown by the pressure–volume (P–V) diagram in Fig. 1a and is most easily visualized as applied to a gas passing through appropriate engine components such as cylinders with pistons and heat exchangers. Starting at point 1, the gas is compressed *adiabatically*, i.e., isolated from heat exchange with the surroundings (constant entropy), until its temperature has increased from T_L at point 1 to T_H at point 2 as the result of converting compressional work into heat. In the process 2–3, the gas is brought into thermal contact with a heat reservoir at temperature T_H and expands isothermally, i.e., without change of temperature. At point 3, the gas continues its expansion adiabatically until, because of the partial conversion of heat into

---

[†]deceased

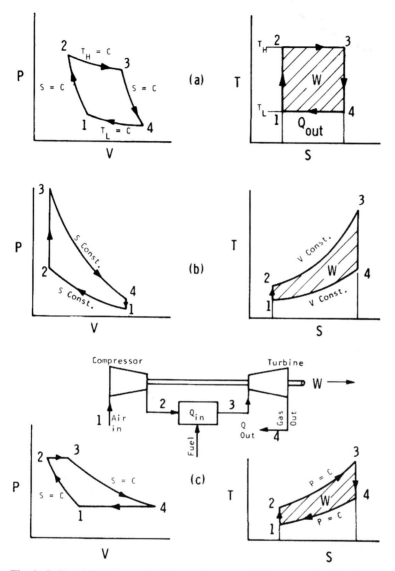

Fig. 1: *P–V* and *T–S* diagrams of thermal engines: (a) Carnot cycle; (b) Otto
cycle (gasoline engine); (c) Brayton cycle (gas turbine).

work, it has cooled to the temperature T_L at point 4. In the final process 4–1, the gas is brought
into thermal contact with a heat reservoir at temperature T_L and compressed isothermally, so
that at point 1 it assumes the values of pressure, volume, and temperature it had at the start of
the cycle. If the four steps are performed reversibly, the gas receives heat Q_{in} isothermally at
T_H and rejects heat Q_{out} isothermally at T_L, while the two adiabatic processes contribute nothing to a heat exchange with the environment of the gas. The conditions required for achieving
the Carnot efficiency η_r according to Eq. (1) are therefore fulfilled. The area enclosed by

the *P–V* diagram of steps 1–2–3–4–1 is representative of the work $\oint P \, dV$ performed by the system, but since the diagram yields no information on the isothermal heat exchange, it provides no indication of the thermal efficiency. That information is obtainable, however, from the temperature–entropy (*T–S*) diagram shown at the right in Fig. 1a. Since, by definition, $\delta Q_r = T \, dS$, the area under a curve $T(S)$ for which dS increases indicates heat Q_{in} received, while that under a curve for which dS decreases indicates heat gout rejected. Therefore the area $\oint T \, dS$ enclosed by 1–2–3–4–1 represents the work W produced reversibly, and the ratio of this area to that under $T(S)$ for $dS > 0$ (i. e., 2–3) is equal to the thermal efficiency. For the Carnot cycle the efficiency is read directly as $(T_H - T_L)/T_H$.

Since the efficiency of a reversibly operating thermal engine depends only on T_H and T_L, that of the Carnot cycle cannot be exceeded. In fact, however, reversibility of the various processes can only be approached by infinitely slow and frictionless operation. In actual operation, the upper temperature of the gas must be somewhat lower than T_H to produce the required heat transfer. To invert the process by isothermal compression, the gas temperature must be somewhat higher than T_H. Thus somewhat less work is gained than must be applied to reverse the process, but complete reversibility is then not achieved. Similar considerations apply to the other three processes of the Carnot cycle, with the effect that the areas enclosed by indicator diagrams are larger than those utilized by the gas, so that the actual efficiencies are smaller than the ideal η_r. The susceptibility of an engine to losses incurred by irreversible processes is large for small values of a merit factor called the *work ratio* r_w,

$$r_w = \frac{W_{out} - \sum W_{in}}{W_{out}} = \frac{W}{W_{out}} \, . \tag{2}$$

The work $W = \oint \delta Q_r$ ideally produced is the difference of the work output W_{out} and the sum $\sum W_{in}$ of energies required for isothermal and adiabatic compression. W_{out} and W_{in} are the work contributions expected from the ideal Carnot cycle, but because of losses, the actual work required is always larger than W_{in} and the actual output is smaller than W_{out}. Therefore, if r_w is small, the practically achieved $(r_w)_{pract}$ may be negative. The work ratio of a Carnot engine using a gas is quite small, since the work contributions of adiabatic and isothermal compression and expansion are large compared to the net work generated. Despite the thermally ideal cycle, Carnot gas engines are not practical. An improvement results in a Carnot engine using steam. In this case the steps 2–3 and 4–1 in the *T–S* diagram of Fig. 1a represent, respectively, transitions from the liquid to the vapor phase in a boiler and from the vapor to the liquid phase in a condenser. Both processes occur at constant temperature and pressure. Step 1–2 is adiabatic compression of the liquid requiring relatively little work and step 3–4 represents adiabatic expansion of the vapor, which performs work on a piston or turbine. Practical implementation, however, is difficult for a variety of reasons. Therefore, practical steam turbines (and the older steam engines) use the *Rankine* cycle, which differs from the Carnot steam cycle mainly in step 1–2 of the *T–S* diagram. This step consists here of compressing the liquid to the pressure at point 2 and heating it at that pressure to the saturation (boiling) temperature T_H at point 2. Important alternatives of the Rankine cycle include the use of "superheated" steam and operation above the triple point. Although the thermal efficiency of the Rankine cycle is less than that of the Carnot cycle, practical efficiencies of 40–45% result from the much larger work ratio.

In addition to steam (or vapor) power cycles, in which the working substance alternates between the liquid and vapor states, there exists the general class of gas power cycles, in which the working substance, although usually injected as a fluid, remains in the gaseous state throughout the cycle. These cycles are particularly suitable for engines of relatively small weight, and for powers below about 10^4 kW, such as those used for ground and air transportation. Most of these engines operate by internal combustion, and their design must be primarily aimed at optimizing the conversion of the available chemical energy rather than at approaching a Carnot cycle. Nevertheless, combustion should begin at the highest, and heat rejection at the lowest, possible temperature. The conventional thermodynamic engineering approach to an analysis of these cycles is the air-standard approximation, which treats the working substance as an ideal gas and combustion as simple heat addition.

The Otto cycle, represented by the P–V and T–S diagrams of Fig. 1b, is the model for the operation of the gasoline engine. After an intake stroke (not shown), the gas is adiabatically compressed in step 1–2, receives heat (by spark-initiated explosion) at constant volume in step 2–3, expands adiabatically against a piston (power stroke) in step 3–4, and rejects heat (after passing through the exhaust valve) at constant volume in step 4–1. The cycle is simple because two steps involve only heat, but not work, and the two other steps involve work, but not heat. The air-standard analysis (assuming temperature-independent specific heats, c_p and c_v) yields $T_2/T_1 = T_3/T_4$ and $\eta_r = 1 - r^{1-\kappa}$ for a compression ratio r and $\kappa = c_p/c_v$.

The Brayton cycle represents the model for the operation of the basic gas turbine process as shown in Fig. 1c. In step 1–2, air drawn into the compressor is compressed adiabatically. In step 2–3, the air is mixed with fuel in the combustion chamber and receives heat at constant pressure. Step 3–4 represents adiabatic expansion, which drives the turbine. In step 4–1, heat is rejected at constant pressure. The air-standard analysis yield for the thermal efficiency, at a pressure ratio r_p, $\eta_r = 1 - r_p^{(1-\kappa)/\kappa}$.

The diesel engine is in many respects similar to the gasoline engine, but operates without spark ignition at very high compression ratios. After the intake, air is compressed adiabatically and reaches a temperature exceeding the ignition point of the fuel that is injected when the piston begins to reverse its motion. Fuel is injected at such a rate that its consumption and the expansion of the volume occur at approximately constant pressure. The rest of the cycle proceeds as in steps 3–4 and 4–1 of the Otto cycle. The T–S diagram is a hybrid of the Otto and Brayton cycles in that heat is received at constant pressure and rejected at constant volume.

See also: Carnot Cycle; Thermodynamics, Equilibrium.

Bibliography

J. B. Jones and G. A. Hawkins, *Engineering Thermodynamics*, 2nd ed., Wiley, New York, 1986. (I)

E. F. Obert and R. L. Young, *Elements of Thermodynamics and Heat Transfer*, 2nd ed. (reprint of 1962 ed.). Krieger, Melbourne, FL, 1980. (I)

M. W. Zemansky and R. Dittman, *Heat and Thermodynamics*, 7th ed. McGraw–Hill, New York, 1996. (I)

R. Becker, *Theory of Heat*, 2nd ed. (revised by G. Leibfried), Springer, Berlin, 1967. (A)

Heat Transfer

A. J. Chapman

Introduction

Heat is defined as that energy exchange which occurs as the result of a temperature difference or a temperature gradient. The term heat transfer is used to denote that body of knowledge which attempts to describe, physically and mathematically, the rate of exchange of heat between bodies, various parts of bodies, fluids, etc., when the appropriate physical conditions are specified. The two fundamental mechanisms of heat transfer are known as *conduction* and *radiation*. However, a third mechanism is customarily identified when the conduction process is coupled with the macroscopic motion of a fluid medium; it is termed *convection*. Although problems of physical interest exist in which only one of these mechanisms is present, there are many for which a complex simultaneous accounting for two or more must be made. Perhaps the most complex heat transfer problems are those in which the situation is further complicated by the simultaneous occurrence of a phase change process such as condensing, boiling, solidification, or sublimation.

Conduction

Heat conduction is the exchange of heat from one body at a given temperature to another body at a lower temperature with which it is in contact, or between portions of a body at different temperatures, by the transfer of kinetic energy of motion through molecular impact. The transfer of energy takes place from the more energetic (higher-temperature) molecules to the less energetic (lower-temperature) molecules. In a gas the molecules are spaced relatively far apart and their motion is random, leading to an energy transfer that is less than that observed in a liquid, in which the molecules are more closely packed. Amorphous solids exhibit conduction much like liquids; however, in dielectric solids, which form a crystal lattice, the transfer of energy by conduction is enhanced by the vibratory motion of the lattice structure itself. In solids that are also electrical conductors, heat conduction is further enhanced by the drift of free electrons within the lattice structure.

Macroscopically, the above-described transfer of heat by conduction is hypothesized to be described by *Fourier's conduction law*:

$$\frac{q}{A} = -k\frac{\partial T}{\partial n} , \tag{1}$$

in which the heat transfer rate q per unit of area A is presumed to be proportional to the gradient of the temperature T in the direction normal to the area n. The proportionality constant k is identified as the *thermal conductivity*. The thermal conductivity is a physical property of the matter composing the body under consideration and is a function of the composition of the matter, the phase in which it exists, and such state variables as pressure and temperature.

In general, the solution of a conduction problem consists of finding, in a body of given geometry and initial temperature distribution, the temperature at any point at any subsequent time. This solution is obtained from the *heat conduction equation*, which results from the combination of Fourier's law and the energy conservation principle:

$$\nabla \cdot (k\nabla T) + q^* = \rho C_p \frac{\partial T}{\partial t} \tag{2}$$

where q^* represents the volumetric internal production of heat, if any; ρ the material density; C_p the specific heat of the material; and t the time. When internal heat generation is absent and the thermal conductivity is taken as constant we have

$$\nabla^2 T = \frac{\rho C_p}{k} \frac{\partial T}{\partial t} . \tag{3}$$

A special case of common interest is that of the *steady state*, in which the temperatures are independent of time. In this instance, if internal heat generation is absent and the material properties are constant, the heat conduction equation reduces to the familiar Laplace equation:

$$\nabla^2 T = 0 . \tag{4}$$

A complete solution to a heat conduction problem involves the solution of Eq. (2), or (4), subject to an initial known temperature distribution (if nonsteady) and whatever boundary conditions apply. Typically, the boundary condition commonly encountered is that of either a specified temperature on the body surface or convection from the surface into a bounding moving fluid of known temperature. The latter condition is usually represented by *Newton's law of cooling*, which states

$$\frac{q}{A} = -k \left(\frac{\partial T}{\partial n} \right)_s = -h(T_s - T_\infty) \tag{5}$$

in which the subscript "s" denotes the body surface, n the coordinate direction normal to the surface, and T_∞ the temperature of the fluid far away from the surface. The quantity h is known as the surface heat transfer coefficient and is the primary quantity sought in the analysis of convection.

Convection

Heat convection is the term applied to the heat transfer mechanism that occurs when a fluid of some known temperature flows past a solid surface at some different temperature. The convection is termed *natural*, or *free*, when the fluid motion results from buoyant forces created by temperature (and, hence, density) differences in the fluid. It is called *forced* convection when the fluid motion is produced by imposed pressure differences or by motion of the surface itself. The flow of heat between the surface and the fluid is usually expressed in terms of the heat transfer coefficient h defined in Eq. (5).

The determination of the heat transfer coefficient necessitates the solution of the governing equations of a viscous, heat-conducting fluid. These equations in their general form are not solvable analytically, and the problem is usually simplified by introduction of the simplifications provided by boundary layer theory. In this simplified form, the fundamental equations that need to be solved near the body surface are noted in the following equations (for an incompressible fluid), where the coordinate x is measured parallel to the body surface, y is measured normal to x, and the symbols v_x and v_y denote the fluid velocities in those directions.

Conservation of mass:

$$\frac{\partial v_x}{\partial x} + \frac{\partial v_y}{\partial y} = 0 \ . \tag{6}$$

Equation of motion:

$$v_x \frac{\partial v_x}{\partial x} + v_y \frac{\partial v_x}{\partial y} = \frac{1}{\rho} \frac{dp}{dx} + \frac{1}{\rho} \frac{\partial \tau}{\partial y} + g\beta(T_f - T_\infty) \ . \tag{7}$$

Conservation of energy:

$$v_x \frac{\partial T_f}{\partial x} + v_y \frac{\partial T_f}{\partial y} = \frac{1}{\rho C_p} \left(\tau \frac{\partial v_x}{\partial y} \right) - \frac{1}{\rho C_p} \frac{\partial}{\partial y} \left(\frac{q}{A} \right) \ . \tag{8}$$

In these equations, ρ represents the fluid density, p the pressure, C_p the specific heat, and β the coefficient of volume expansion. The symbol T_f is used to denote the fluid temperature, to distinguish it from temperatures within the solid surface, and it is generally a function of position within the fluid (i.e., a function of x and y). In the momentum equation and energy equation, the quantities τ and q/A are used to denote the shear stress and heat flux, respectively, within the fluid domain. For *laminar* flow these are usually expressed in terms of the gradients of velocity and temperature within the fluid by introduction of the following hypothesized rate laws:

$$\tau = \rho v \frac{\partial v_x}{\partial y} \ , \tag{9}$$

$$\frac{q}{A} = -\rho C_p \alpha \frac{\partial T_f}{\partial y} \ , \tag{10}$$

where v and α represent the kinematic viscosity and thermal diffusivity, respectively. In the instance when *turbulent* flow exists in the fluid, it is customary to augment these rate coefficients with their turbulent counterparts, the so-called eddy viscosity and eddy diffusivity.

In the equation of motion, Eq. (7), the last term represents buoyant forces arising from temperature gradients in the fluid and may be absent in cases of pure forced convection. Likewise, in the energy equation, Eq. (8), the first term on the right-hand side represents the rate of dissipation of kinetic energy into thermal energy through viscous action, and it may in certain cases be negligible.

In general, the solution to a particular convection problem consists of the simultaneous solution of Eq. (6), (7), and (8), under the appropriate boundary conditions, for the fluid temperature $T_f(x,y)$ from which we eventually ascertain the sought-for heat transfer coefficient by application of Fourier's law at the surface:

$$h = -\frac{k(\partial T_f/\partial y)_s}{(T_f - T_\infty)} \ , \tag{11}$$

the subscript "s" denoting the surface.

Usually, the solutions just referred to are expressed in dimensionless form. If Eqns. (6)–(11) are nondimensionalized by introducing a characteristic velocity U and a characteristic dimension L, we can deduce that the above-implied dependence of h on the other parameters may be expressed as

$$N_{NU} = f_n(N_{RE}, N_{GR}, N_{EC}, N_{PR}) . \qquad (12)$$

In Eq. (12) the quantities shown are dimensionless groupings of variables that bear the following designations:

Nusselt number:	$N_{NU} = hL/k$;	
Reynolds number:	$N_{RE} = UL/v$;	
Grashoff number:	$N_{GR} = \dfrac{L^3 g \beta (T_s - T_\infty)}{v^2}$;	
Eckert number:	$N_{EC} = \dfrac{U^2}{C_p(T_s - T_\infty)}$;	
Prandtl number:	$N_{PR} = v/\alpha$.	

A vast number of solutions for convection problems comprising a wide variety of surface geometries and hydrodynamic boundary conditions have been carried out and are reported in the literature. An equally vast number of empirical convection solutions have been deduced in the form of Eq. (12) by conducting appropriate experiments.

The foregoing discussion applies to an incompressible fluid for which the density is taken to be constant. When fluid compressibility is important, as might be the case in aerodynamic applications, the governing equations become more complex. However, the same dimensionless correlation implied in Eq. (12) still holds.

Radiation

Heat transfer by thermal radiation constitutes a completely different mechanism from those discussed earlier for convection and conduction. The basic mechanism is that of electromagnetic radiation in a wavelength band usually taken to extend from 0.1 to 100 μm. Thermal radiation emitted by a heated surface is found to be a complex function of the surface temperature, the nature of the surface, the wavelength of the emission, and the particular direction of concern. If the "hemispherical" emission is considered, wherein the radiation emitted in all directions is summed together, the *monochromatic emissive power* W_λ is defined as the total energy emitted, per unit time and area, at the wavelength in question. The symbol λ is used to denote the wavelength. If the surface in question is a perfect *blackbody* (i. e., absorbs all radiant energy incident upon it), the monochromatic emissive power $W_{b\lambda}$ is known to be a function of wavelength and temperature as described by Planck's radiation law. When integrated over all wavelengths to obtain the *total* emissive power W_b, Planck's equation yields the Stefan–Boltzmann law:

$$W_b = \sigma T^4 . \qquad (13)$$

For nonblack surfaces, the monochromatic emissive power is found to be less than that for a blackbody at the same temperature. The ratio of these emissive powers is defined as the *monochromatic emissivity* ε_λ:

$$\varepsilon_\lambda = W_\lambda / W_{b\lambda} . \qquad (14)$$

We may also define a *monochromatic absorptivity* α_λ as the fraction of an incident radiation at a given wavelength that is absorbed by a nonblack surface. Kirchhoff's law states that, at equal surface temperatures, the monochromatic emissivity and monochromatic absorptivity are equal:

$$\varepsilon_\lambda = \alpha_\lambda \; . \tag{15}$$

Generally, these quantities are functions of the wavelength λ. However, if a surface exhibits the characteristic by which $\varepsilon_\lambda = \alpha_\lambda$ is constant, the surface is termed *gray*.

In a similar fashion, the total emissivity ε of a nonblack surface is defined as the ratio of the energy emitted, per unit time and area, at all wavelengths to the same quantity for a black surface at the same temperature. Thus, for the nonblack surface

$$W = \varepsilon \alpha T^4 \; . \tag{16}$$

Similarly, there is defined a total absorptivity α, which represents the fraction of incident radiation absorbed by a nonblack surface at all wavelengths. Kirchhoff's law in the total sense,

$$\varepsilon = \alpha \; , \tag{17}$$

will follow from Eq. (15) only if the surface is gray, or if the incident radiation is from a black surface at the same temperature as the receiving surface.

A frequently encountered problem of practical interest is that of determining the radiant exchange between a collection of finite surfaces, black or gray, maintained at different temperatures. In such instances it is necessary to know the fraction of the radiation leaving one surface that is intercepted by another. This fraction is expressed by the *shape factor* F_{1-2}, in which 1 denotes the originating surface and 2 the receiving surface. If we presume that Lambert's law of diffuse radiation governs the spatial distribution of the emitted energy – that is, the radiant energy density varies as the inverse square of the distance from the source and as the cosine of the angle between the surface normal and the direction in question – we may deduce that the shape factor between two surfaces A_1 and A_2 of given shape and orientation is

$$F_{1-2} = \frac{1}{A_1} \int_{A_1} \int_{A_2} \frac{\cos\theta_1 \cos\theta_2 \, dA_1 \, dA_2}{\pi r_{12}^2} \; , \tag{18}$$

in which θ_1 and θ_2 represent the angles between the surface normals to dA_1 and dA_2 and the line connecting them, r_{12}. The shape factor is a geometric factor that depends only on the shape and orientation of A_1 and A_2. It is generally very difficult to determine analytically because of the twofold double integrals in Eq. (18). Values for many surface combinations are reported in the literature, and computer techniques are available for complex geometries.

For an enclosure consisting of a number of finite black surfaces, maintained at specified temperatures, the heat exchanged between any pair of the surfaces (say A_i and A_j maintained at T_i and T_j) is

$$q_{i-j} = A_i F_{i-j} \sigma (T_i^4 - T_j^4) \tag{19}$$

and the total energy given up by one surface is then

$$q_i = \sum A_i F_{i-j} \sigma (T_i^4 - T_j^4) \,, \tag{20}$$

in which the summation is taken over all the surfaces in the enclosure. (In the event that a true enclosure does not exist, the vacant space may be represented by a surface at zero temperature to simulate an enclosure.) The use of Eqns. (19) and (20) requires the prior determination of the shape factors of all the surface-pair combinations in the enclosure. This may be very time consuming.

   If the surfaces of an enclosure are gray instead of black, the effect of multiple reflections must be included. Likewise, if the heat flux of a surface is specified (such as zero) instead of the temperatures, the problem becomes much more complex. In such instances highly developed techniques exist that yield formulations similar to those in Eqns. (19) and (20) where F_{i-j} is replaced with complex exchange coefficients ϕ_{i-j} involving the F's, the surface emissivities, etc.

See also: Absorption Coefficients; Blackbody Radiation; Boundary Layers; Lattice
   Dynamics; Turbulence.

Bibliography
All three modes of heat transfer; applications
Alan J. Chapman, *Heat Transfer*, 4th ed. Macmillan, New York, 1984. (I)
E. R. G. Eckert and R. M. Drake, *Analysis of Heat and Mass Transfer*. McGraw–Hill, New York, 1972.
   (A)
J. P. Holman, *Heat Transfer*, 9th ed. McGraw–Hill, New York, 2001. (E)
F. P. Incropera and D. P. Dewitt, *Fundamentals of Heat and Mass Transfer*, 5th ed. Wiley, New York,
   2001. (I)
M. Jakob, *Heat Transfer*. Wiley, New York, 1949 (Vol 1), 1957 (Vol. 2). (A)

Conduction
V. S. Arpaci, *Conduction Heat Transfer*. Addison-Wesley, Reading, Mass., 1966. (A)
H. S. Carslaw and J. C. Jaeger, *Conduction of Heat in Solids*, 2nd ed. Oxford, London and New York,
   1959. (A)
P. J. Schneider, *Conduction Heat Transfer*. Addison-Wesley, Reading, Mass., 1955. (I)

Convection
W. M. Kays and M. E. Crawford, *Convective Heat and Mass Transfer*. 3rd ed. McGraw–Hill, New York,
   1980. (A)
H. Schlichting, *Boundary Layer Theory*, 7th ed. McGraw–Hill, New York, 1979. (A)

Radiation
H. C. Hottel and A. F. Sarofim, *Radiative Transfer*. McGraw–Hill, New York, 1967. (A)
R. Siegel and J. R. Howell, *Thermal Radiation Heat Transfer*. 4th ed. Taylor and Francis, New York,
   2001. (A)

Heavy-Fermion Materials

D. E. MacLaughlin

Heavy-fermion or heavy-electron materials are metals that exhibit enormously enhanced effective conduction electron masses [1]. The classical heavy-fermion metals contain 4f or 5f elements. These are usually Ce or U, e. g., $CeAl_3$, UBe_{13}, but can also be Yb ($YbRh_2Si_2$) or Pr ($PrAg_2In$). The heavy-electron mass is inferred from the large enhancement of the electronic specific heat C_e; more specifically, the Sommerfeld coefficient γ of the linear term in C_e is a factor of 100–1000 larger than in ordinary metals. (Observed values of γ for rare-earth intermetallic compounds range between ~ 10 and $\sim 1000\,\mathrm{mJ\,K^{-2}\,mol^{-1}}$; a value greater than $\sim 200\text{–}400\,\mathrm{mJ\,K^{-2}\,mol^{-1}}$ is usually arbitrarily taken to define a heavy-fermion system.)

This mass enhancement (which reflects the strongly-correlated nature of the band electrons and is not a "real" change of the electron mass) is associated with strong hybridization between conduction and f electrons. At high temperatures the effect of hybridization is weak, and the properties of heavy-fermion systems are those of a collection of weakly-interacting localized f ions embedded in a normal metal, but at low temperatures heavy-fermion materials exhibit a rich variety of ground states and unusual phenomena [1, 2]. Heavy-fermion systems have been discovered that are superconducting, that order magnetically, and that do not undergo a phase transition but appear to approach a metallic "normal" ground state at zero temperature. Experimental and theoretical interest has been particularly attracted to heavy-fermion superconductivity, which offers the possibility of exotic Cooper pairing mechanisms and previously unknown pairing symmetries [3], and to the effect of extremely strong correlations between conduction electrons characteristic of these materials.

Over the past decade evidence has been found that "normal" (nonmagnetic, non-superconducting) heavy-fermion ground states are not always described by the Landau Fermi-liquid theory [4, 5]. Non-Fermi-liquid (NFL) systems exhibit temperature dependences, some weakly singular (logarithms or small powers), of quantities, such as the Sommerfeld specific heat coefficient and the Pauli susceptibility, that are constant for Fermi liquids. Such behavior signals a breakdown in the Fermi liquid picture for these systems, which is surprising because of the belief that in the absence of a phase transition (magnetic, superconducting, ...) this picture should be very robust. An interesting feature of some NFL systems ($CeIn_3$, $CePd_2Si_2$) is that the depression of a magnetic ordering temperature T_m by applied pressure leads to NFL behavior above a small "dome" of superconductivity around the "quantum critical point" $T_m = 0$. This is very reminiscent of the situation in the high-T_c cuprates.

The characteristic energy scales of all of the heavy-fermion ground states are very low (0.1–10 MeV, or 1–100 K in temperature units) compared to typical Fermi energies of ordinary metals (1–10 eV, 10^4–10^5 K). (NFL systems may have no energy scale at all.) In Fermi liquids one expects on very general grounds an effective characteristic or degeneracy energy ε_0 (Fermi energy for a weakly interacting Fermi gas) to be given by $\varepsilon_0 = p_0^2/2m^*$, where m^* is the effective mass and p_0, an effective quasiparticle momentum at the Fermi surface, is determined by the electron concentration (Luttinger theorem): $p_0/\hbar$ is typically 1–10 Å^{-1} independent of m^*. Therefore, an enhanced electron mass and a reduced energy scale are equivalent descriptions of heavy-fermion behavior. This characteristic energy (or equivalent

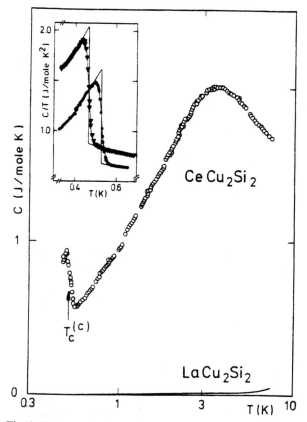

Fig. 1: Temperature dependence of the molar specific heat C of $CeCu_2Si_2$ in zero magnetic field. The arrow gives the superconducting transition temperature $T_c = 0.51 \pm 0.04\,K$. Inset: C/T vs. T near T_c for two other $CeCu_2Si_2$ samples, showing the large jump. From Ref. 6, with permission.

characteristic temperature $T_0 = \varepsilon_0/k_B$) has been associated with the Kondo temperature T_K of a single magnetic impurity in a normal metal, in part because heavy-fermion metals behave approximately as "concentrated Kondo systems."

It is hard to understand the stability of heavy Fermi liquids even in the normal state, because the narrow bandwidths characteristic of nearly localized electronic states should be costly in energy of localization. The various ground states of heavy-fermion systems are in fact not very stable and are easily modified, e. g., by application of pressure or magnetic field, or by doping with impurities. Such modification can sometimes provoke NFL behavior.

It was realized some time ago [6] that the observed large discontinuity of the specific heat in heavy-fermion superconductors at the superconducting critical temperature T_c, shown in Fig. 1 for $CeCu_2Si_2$, implies that the heavy electrons themselves are involved in the superconducting Cooper pairing. (Figure 1 also shows the extremely enhanced linear specific heat in the normal state of $CeCu_2Si_2$, compared to the isostructural compound $LaCu_2Si_2$ which has no $4f$ electrons.) The fact that nearly localized electrons experience strong Coulomb repulsion when on the same site makes a conventional (BCS) phonon-mediated attraction suspect, since the latter is a local interaction and would be seriously weakened by the tendency of the electrons to avoid each other. Instead, Cooper pairing might take advantage of an interaction that is attractive at intermediate electron separations. Such an interaction, together with the on-site repulsion, would favor a non-s-like Cooper pair state with a node at the origin of the relative coordinate.

Experimental evidence on the nature of the pairing interaction is difficult to obtain. Consequences of unconventional Cooper pair states are somewhat less obscure, and the experimental evidence by and large supports claims for unconventional superconductivity in particular systems. This evidence includes [2, 7] (a) signs of an extremely anisotropic superconducting energy gap, (b) rapid depression of T_c by nonmagnetic impurities, and (c) evidence for multiple superconducting phases in a number of systems (UPt$_3$, U$_{1-x}$Th$_x$Be$_{13}$, PrOs$_4$Sb$_{12}$, ...). All of these phenomena are predicted for unconventional Cooper pairing [3, 7]. Group-theoretical considerations put the various kinds of energy-gap anisotropy into two general classes: point nodes and line nodes (zeros) of the gap function $\Delta(\mathbf{k})$ on the Fermi surface. At low temperatures each of these node geometries gives rise to characteristic power-law temperature dependences of quantities, such as the ultrasonic attenuation coefficient and the nuclear spin–lattice relaxation rate, which depend on the number of thermal excitations across the gap. These power laws are more or less easily distinguished from the activated temperature dependence $\exp(-\Delta/k_BT)$ expected for a nonzero gap, and have been verified in a number of experiments.

The recent discovery of strong antiferromagnetic (AF) spin fluctuations in several heavy-fermion superconductors is consistent with unconventional pairing mediated by such spin fluctuations, as is the observation in UPt$_3$ of an unusual phase boundary in the $H-T$ plane. A case has been made, however, for conventional but highly anisotropic pairing [1]. Some authors [8] have speculated that a relation exists between superconductivity in heavy-fermion metals and in high-T_c cuprates, involving spin fluctuations of rare-earth ions in the former and of Cu^{2+} ions in the latter.

The other heavy-fermion ground states are scarcely less puzzling. Thermodynamic measurements in "normal" heavy-fermion materials suggest extremely low characteristic energies, as noted above. In particular the magnetic susceptibility tends toward a constant Pauli-type value $\chi(0)$ at low temperatures that is also enhanced compared to normal metals. Surprisingly, the so-called "Wilson ratio" $\chi(0)/\gamma(0)$ is seldom far from the value for free-electron metals. Various electron spectroscopies [photoemission spectroscopy (PES), bremsstrahlung isochromat spectroscopy (BIS), and their variants] yield information on the single-particle Green's function, which is in turn affected by interactions between the added hole or electron and the rest of the system. Another recently-developed "facility-based" experimental technique, muon spin rotation (μSR), is conceptually similar to nuclear magnetic resonance (NMR) and, like NMR, probes magnetism on the local (atomic) scale with great sensitivity [9].

There is evidence for the onset of "coherence" effects (consequences of well-defined quasiparticles with long mean free paths) at a temperature T_{coh} well below the characteristic or "Kondo" temperature T_0. Coherence effects are most notable in transport properties, but also appear to influence the neutron scattering form factor and NMR properties at low temperatures. The nature of the coherence is not well understood. The best evidence for the Fermi-liquid nature of the normal ground state is the observation of de Haas–van Alphen (dHvA) oscillations in the magnetic susceptibility of heavy-fermion metals at extremely low temperatures [2]. The dHvA effect is due to quantization of quasiparticle orbits in a magnetic field and its effect on the population of orbits at the Fermi surface, and its very existence indicates the presence of a Fermi surface and long-lived charged fermion quasiparticles in these systems. For UPt_3 the dHvA experiments yield a Fermi surface geometry in good agreement with band structure calculations, but the measured cyclotron masses are about an order of magnitude larger than the calculated values.

Some heavy-fermion compounds undergo a transition to a magnetically ordered state. This magnetism has some experimental features in common with the itinerant AF (spin density wave) states of 3d metals, but differs in that the antiferromagnetism is usually commensurate and of a simple structural symmetry. In addition, the ordered magnetic moment per f ion is often much smaller than a Bohr magneton, and the entropy associated with the transition is often much smaller than $k_B \ln(2J+1)$. Both of these features can be understood as consequences of strong but incomplete Kondo screening.

Magnetic order and superconductivity appear to coexist in a number of heavy-fermion compounds, although in some cases the magnetism may originate from the superconducting Cooper pairs rather than a different set of electrons. In $U_{1-x}Th_xBe_{13}$, $0.018 < x < 0.045$, μSR experiments show the onset of a small spontaneous magnetic field below a critical temperature T_{c2} less than the (maximum) superconducting transition temperature T_{c1}. Since a number of experiments suggest that T_{c2} is a transition between superconducting states, rather than the onset of magnetism in a separate set of electrons, the spontaneous field revealed by μSR is evidence that this low-temperature superconducting state in $U_{1-x}Th_xBe_{13}$ breaks time-reversal symmetry: the magnetic order is a feature of the superconducting state itself. Similar behavior has been observed in the heavy-fermion superconductor $PrOs_4Sb_{12}$.

In URu_2Si_2 (with moderately heavy band electrons) a phase transition with a sizable specific-heat anomaly is observed at 17 K, with no evidence for an order parameter, magnetic or otherwise, associated with this transition. The origin of this "hidden order" is a subject of considerable research, in part because there is also evidence for hidden order in the normal state of some copper oxide systems.

A great deal of effort has been put into understanding NFL behavior in heavy-fermion metals, both because of its intrinsic interest and because the phenomenon is found in an even wider variety of settings [4, 5]. Theoretical efforts can be grouped into three rough and overlapping classes. NFL behavior is often found in the neighborhood of magnetic order in an appropriate phase diagram, e. g., temperature vs. a "control parameter" δ (pressure, magnetic field, chemical composition, . . .), where the magnetic ordering temperature T_m is suppressed to zero at a critical value of δ. For a second-order transition this defines a quantum critical point in the phase diagram, because the critical fluctuations, which are thermal excitations for $T_m > 0$, become quantum fluctuations at $T_m = 0$. This quantum criticality leads to NFL behavior. Another approach invokes the multichannel Kondo effect, in which more than one "channel" (band index) of otherwise identical conduction electrons leads to additional degrees of freedom in the Kondo problem. Then the local moment can be overscreened, leading to a multiplet ground state and NFL behavior. Lastly, it has been noted that disorder in a heavy-fermion system and the resultant distribution of hybridization strengths can lead to NFL behavior, either on its own or in concert with other mechanisms. At this writing the situation is far from clear, either experimentally or theoretically, and NFL behavior in heavy-fermion materials remains a topic of intense interest.

See also: de Haas–van Alphen Effect; Kondo Effect; Metals; Non-Fermi Liquid Behavior; Quantum Criticality; Quasiparticles; Superconducting Materials; Transition Elements.

References

[1]   For a review see P. Fulde, J. Keller, and G. Zwicknagl, in *Solid State Physics* (H. Ehrenreich and D. Turnbull, eds.), vol. 41. Academic Press, New York, 1988. (I-A)

[2]   N. Grewe and F. Steglich, in *Handbook on the Physics and Chemistry of Rare Earths* (K. A. Gschneidner, Jr., and L. Eyring, eds.) vol. 14, p. 343. North-Holland, Amsterdam, 1991. (I-A)

[3]   M. Sigrist and K. Ueda, *Rev. Mod. Phys.* **63**, 239 (1991) (I-A)

[4]   G. R. Stewart, *Rev. Mod. Phys.* **73**, 797 (2001). (I)

[5]   C. M. Varma, Z. Nussinov, and W. van Saarloos, *Phys. Reports* **361**, 267 (2002). (I-A)

[6]   F. Steglich, J. Aarts, C. D. Bredl, W. Lieke, D. Meschede, W. Franz, and J. Schäfer, *Phys. Rev. Lett.* **43**, 1892 (1979). (E-I)

[7]   See P. A. Lee, T. M. Rice, J. W. Serene, L. J. Sham, and J. W. Wilkins, *Comments Cond. Mat. Phys.* **12**, 99 (1986). (I-A)

[8]   Z. Fisk, D. W. Hess, C. J. Pethick, D. Pines, J. L. Smith, J. D. Thompson, and J. O. Willis, *Science* **239**, 33 (1988). (E-I)

[9]   A. Amato, *Rev. Mod. Phys.* **69**, 1119 (1997). (I)

Helium, Liquid

R. A. Guyer

The liquid-helium systems, liquid ^{4}He, liquid ^{3}He, and liquid ^{3}He–^{4}He mixtures, are among the most extensively studied systems in physics. The primary reason for this is that the helium systems do not solidify down to arbitrarily low temperature unless pressures in excess of about 25 atm are applied. Liquid ^{3}He and ^{4}He can be carried to low temperatures and for any kind of interaction can be studied where the thermal energy, $k_B T$, is comparable to or less than the interaction energy; e. g., for liquid ^{4}He at $T \approx T_\lambda = 2.17$ K, where the normal-liquid to Bose superfluid transition occurs, or for liquid ^{3}He at $T \approx T_A = 2.1$ mK, where the Fermi-liquid to A-phase superfluid transition occurs, etc. The only other laboratory system which remains liquid (or gaseous) to arbitrarily low temperature is the electron gas that inhabits the ion lattice of a metal, but the properties of this gas are enormously complicated by its environment.

The helium systems are so exceptional because the helium–helium interaction (which is the same between 4–4 pairs, 3–3 pairs, and 3–4 pairs) is very weak, of order 10 K, and the mass of the atoms is very small. Thus the force tending to bond the atoms is relatively weak and the zero-point energy, proportional to $\hbar^2/m$, which is inimical to bonding and localization, is relatively large. So in the absence of a substantial external pressure the systems are liquid.

The most important temperatures encountered in describing these liquids are the temperatures at which quantum-statistical effects set in for it is below these temperatures that these liquids exhibit properties not seen in classical liquids. Liquid ^{4}He is composed of composite particles which are bosons; the Bose character of this system becomes important when the

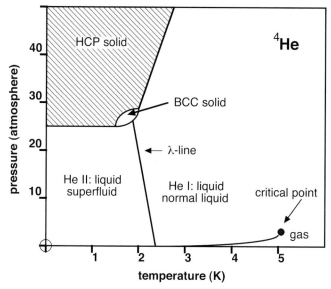

Fig. 1: Phase diagram of ^{4}He. ^{4}He remains a liquid as $T \to 0\,$K unless pressures in excess of 25 atm are applied. At temperatures less than about 2 K, $P < 25\,$atm, the liquid changes from a normal liquid, He I, to a superfluid, He II. Below $T_\lambda(P)$ the fluid is described as a combination of "normal" and superfluid components; it exhibits many unusual properties.

thermal de Broglie wavelength $\lambda(T) = h/p$ $(p^2/2m = k_B T)$ becomes comparable to the inter-particle spacing $\rho^{-1/3}$, where ρ is the number density. Liquid ^{3}He is composed of particles that are fermions; it is for $k_B T \leq k_B T_F$ $[\lambda_T(T_F) \approx \rho^{-1/3}]$ that the Fermi character of the constituents leads to liquid behavior unlike that of classical liquids. For both liquid ^{4}He and liquid ^{3}He, $T_B \approx T_F \approx 1\,$K.

Helium-four was first liquefied by Kamerlingh Onnes in 1908. This liquid and liquid ^{3}He, which is relatively difficult to obtain in large quantities, have been studied intensively for the past 60 years. With the discovery of superfluidity in ^{3}He at $T \leq 3\,$mK, by Osheroff, Richardson, and Lee in 1972, ^{4}He was left behind, while during the next decade attention turned to the ^{3}He superfluid. Liquid ^{3}He and ^{4}He, of interest because of their unusual low-temperature properties, are currently regarded as well understood. These fluids continue to be the focus of serious research activity that more and more makes use of their properties to probe other systems.

Liquid ^{4}He

The phase diagram of ^{4}He is shown in Fig. 1. At suitably high temperature and low pressure ^{4}He is a gas; as the temperature is lowered (at, e. g., 1 atm) a gas–liquid phase transition (of the normal variety) occurs at about 4 K and a more or less conventional liquid results. As the temperature is lowered further the liquid passes through $T_\lambda(P)$ at about 2.2 K and becomes liquid helium II, the superfluid. Above $T_\lambda(P)$ the liquid has properties similar to those of a normal liquid; below T_λ it has many unusual properties. For example, liquid ^{4}He in a bucket

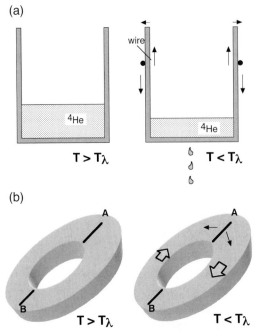

(a)

(b)

Fig. 2: Superfluid phenomena. Among the many superfluid properties of the liquid are those illustrated here. Experiment 1: liquid helium is placed in a bucket that contains an unglazed ceramic plug in the bottom. With the system (bucket, liquid, etc.) at $T > T_\lambda$ the liquid helium remains in the bucket. When the system is carried to $T < T_\lambda$ the fluid pours through the bottom of the bucket. Experiment 2: a heat pulse is propagated from a heater (a piece of carbon resistor board) through liquid helium to a detector of the same material. The temperature at the detector is recorded as a function of time from initiation of the heater pulse.

at the same temperature as the bath and with the bottom of the bucket having an unglazed ceramic plug will remain in the bucket at $T > T_\lambda(P)$ but pour through the bottom of the bucket as T is lowered below $T_\lambda(P)$; and at $T < T_\lambda(P)$ a temperature pulse will propagate through the liquid as a temperature wave (see Fig. 2). These phenomena are understood in terms of a phenomenologic model, the two-fluid model first introduced by Tisza, which has been very successful in describing macroscopic ^{4}He phenomena. The essential element of this model is that below T_λ the fluid, which has been unexceptional at $T > T_\lambda$, can be described as being made up of two components, a normal component with density $\rho_n(T)$ and a superfluid component with density $\rho_s(T)$ (Fig. 3):

$$\rho = \rho_n(T) + \rho_s(T) . \tag{1}$$

ρ is only weakly temperature dependent because of thermal expansion, but $\rho_s(T)$ and $\rho_n(T)$ have striking temperature dependences. The two components of the density have different properties: the normal component is like a normal liquid in that it carries entropy and has viscosity; the superfluid component carries no entropy and flows without viscosity. Superfluid

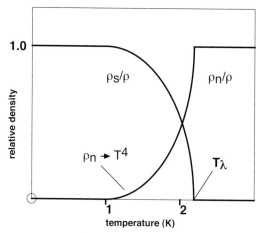

Fig. 3: Superfluid density. The superfluid component and normal fluid
component of the total density are strong functions of T for $T < T_\lambda$. As
$T \to 0\,\mathrm{K}$, $\rho_s \to \rho$; as $T \to T_\lambda$, $\rho_n \to \rho$. The normal fluid density is a measure of
the degree of thermal excitation of the system above the ground state.

^{4}He phenomena are described by the linearized two-fluid equations (ignoring viscosity and
nonlinear terms)

$$\rho = \rho_n + \rho_s , \tag{2}$$

$$\mathbf{j} = \rho_n\mathbf{v}_n + \rho_s\mathbf{v}_s , \tag{3}$$

$$\frac{\partial\rho}{\partial t} + \mathbf{\nabla}\cdot\mathbf{J} = 0 , \tag{4}$$

$$\frac{\partial\mathbf{j}}{\partial t} = -\nabla P , \tag{5}$$

$$\frac{\partial\mathbf{V}_s}{\partial t} = -\nabla G , \tag{6}$$

where $\mathbf{v}_n$ and $\mathbf{v}_s$ are the velocities of the two components, $\mathbf{j}$ is the momentum density, and P
and G are the pressure and Gibbs free energy. The physical content of these equations is that
the total fluid density is driven by the forces that normally drive a fluid but that the superfluid
component obeys a special equation of motion, Eq. (6). Application of the two-fluid model
with many elaborations (beyond the linear regime, with viscosity) leads to a remarkably good
description of the properties of liquid ^{4}He.

In 1938 London suggested that the basic physical event that presages the superfluidity is
the occurrence of Bose condensation in the liquid. Below T_λ the system goes into a single
quantum-mechanical state, the Bose condensed ground state, with excitations above the Bose
condensed ground state corresponding to the normal fluid. Although this picture is strictly correct only for the dilute Bose gas, it is believed that the essential elements remain unchanged
at the densities of the real liquid. London's suggestion gives the two-fluid model some microscopic underpinnings. A complete understanding of the model comes from the work of
Landau, who emphasized the special character of the ground state and excitations above it.
Excitations above the ground state are identified with the normal component of the fluid and

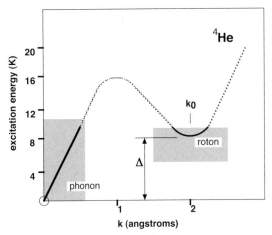

Fig. 4: Excitation spectrum of ^{4}He. The excitation spectrum of ^{4}He is directly observable in neutron scattering experiments. The excitation spectrum is found to contain two branches: a phonon branch as $k \to 0$, $\varepsilon_k \propto k$, and a roton branch at $k \approx k_0$, $\varepsilon_k \propto (k - k_0)^2$. There are also single-particle-like excitations at $k > k_0$, $\varepsilon_k \propto k^2$. At $T \neq 0$ these excitations are present in the system in numbers that depend on the temperature; they are the normal component.

are called the phonons and rotons; they have been carefully studied in neutron scattering experiments. (See Fig. 4.) As $k \to 0$ the excitations are long-wavelength longitudinal phonons,

$$\varepsilon(k) \to \hbar C_l |k| \; ; \tag{7}$$

at $k \approx k_0$ the excitations are single-particle-like and are called rotons,

$$\varepsilon(k) \to (\hbar^2/2\mu)(k - k_0)^2 \; . \tag{8}$$

The specific character of the excitation is observable in thermodynamic measurements at $T <$ T_λ, e. g., in the specific heat, and in two-fluid phenomena since $\rho_n(T)$, $S_n(T)$, etc., as well as the transport coefficients that enter the two-fluid equations, are given in terms of them. The normal density $\rho_n(T)$ is a measure of the density of excitations; at $T \ll 0.5\,\mathrm{K}$,

$$\rho_n(T) = \frac{2\pi^2}{45} \frac{k_B^4 T^4}{\hbar^3 C^5} \tag{9}$$

due to the phonons.

^{3}He

The ^{3}He phase diagram is shown in Fig. 5. Like ^{4}He, the liquid ^{3}He system can be carried to arbitrarily low temperature at pressures less than 30 atm. In the temperature range $T \approx 1\,\mathrm{K}$ the liquid phase diagram has none of the exotic features of the ^{4}He phase diagram. When the system is carried to $T \lesssim 1\,\mathrm{K}$, just as in ^{4}He, it is approaching the temperatures at which quantum statistics become important. Unlike ^{4}He, this does not bring on superfluidity but rather the continuous evolution of a nondegenerate Fermi system to a degenerate Fermi system.

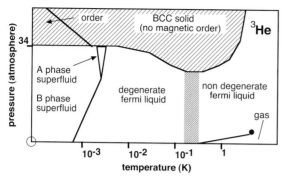

Fig. 5: Phase diagram of ^{3}He. ^{3}He remains a liquid as $T \to 0\,$K unless pressures in excess of 30 atm are applied. At temperatures less than about 1 K the liquid goes from a nondegenerate Fermi system to a degenerate Fermi system (this "transition" is not sharp). The liquid is a Fermi liquid for $3\,$mK $\lesssim T < 1\,$K. At temperatures below about 3 mK the Fermi liquid undergoes a transition to one of the two superfluid ^{3}He states. The details of phase diagram in this temperature range are a sensitive function of magnetic field. An interesting and not well-understood magnetic transition also occurs in solid ^{3}He in this temperature range.

Although in principle the ^{3}He liquid is a "can of worms", it was suggested by Landau that quite simple behavior might occur. The ^{3}He system at $T \lesssim 1\,$K, like the ^{4}He system, is strongly interacting in the sense that the strength of the pair interaction, $\varepsilon = 10\,$K, is large compared to the temperature (this interaction has a very repulsive short-range core) and the system is dense so that the typical 3He atom is in repeated strong interaction with many other particles. Landau suggested that almost all of the repeated strong interaction among particles goes into determining the ground state. The little bit that is left over is seen in weak interactions of quasiparticle excitations of the system above the ground state. Specifically, the excitation properties of the system can be described by an aggregate of fermion quasiparticles whose energy relative to the ground state is

$$\varepsilon(\mathbf{p}) = \varepsilon_0(\mathbf{p}) + \int f(\mathbf{p}\mathbf{p}') \delta n(\mathbf{p}') \, d\mathbf{p}' \,, \tag{10}$$

where

$$\varepsilon_0(p^0) = \mu + \left(\frac{\partial \varepsilon}{\partial p}\right)_0 (p - p_0) \,. \tag{11}$$

The quasiparticles are single particles as, in leading approximation, the term $\varepsilon_0(\mathbf{p})$ of Eq. (10). They have a finite lifetime by virtue of their interactions through $f(\mathbf{p}\mathbf{p}')$ with other quasiparticles that occupy the system with density $\delta n(\mathbf{p})$. As the quasiparticles are fermions their mutual interactions go to zero as the temperature is lowered as $(T/T_F)^2$ so that they are long lived as $T \to 0\,$K. Lifetimes and mean-field effects for the quasiparticles depend upon the Fermi-liquid parameters which are employed to parametrize the phenomenologic interaction $f(\mathbf{p}\mathbf{p}')$; these parameters and the other characteristics of the quasiparticle excitation spectrum are determined by the many experiments on ^{3}He. For example, the specific heat as a function

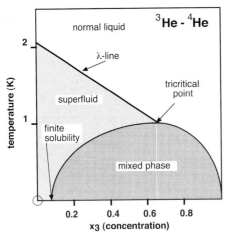

Fig. 6: The phase diagram of ^{3}He–^{4}He mixtures. The ^{3}He–^{4}He mixtures are phase separated at temperatures less than about 1 K. But the unique feature of the mixture phase diagram is the finite solubility of ^{3}He in ^{4}He as $T \to 0\,\mathrm{K}$ up to concentrations of order 6%. This feature makes it possible to study dilute ^{3}He in ^{4}He, i.e. dilute Fermi liquids, to arbitrarily low temperature.

of temperature varies as T, $C = \gamma T$, like a degenerate ideal Fermi gas (the magnitude of the constant γ determines one of the "Fermi-liquid" parameters); the magnetic susceptibility behaves like that of a Pauli paramagnet but with a magnitude 10 to 20 times as great as an ideal Fermi gas (this enhancement of the susceptibility is understood in terms of the behavior of the Fermi-liquid parameter which measures the degree to which the polarization of one atom affects another). The Landau theory of a Fermi liquid provides a scheme for correlating the many data that are available on liquid ^{3}He. Furthermore, the structure of the physical ideas in the theory leads to the prediction of new phenomena, e. g., zero sound, the observation of which phenomena was a crowning achievement of the extensive theoretical and experimental attack on liquid ^{3}He.

^{3}He–^{4}He Mixtures

The phase diagram of liquid ^{3}He–^{4}He mixtures at 1 atm is shown in Fig. 6. When ^{3}He is mixed into ^{4}He at concentrations less than that at the tricritical point, x_t, the resulting solution is a normal liquid at $T > T_\lambda(x_3, P)$ and undergoes a normal liquid to superfluid transition at $T = T_\lambda(x_3, P)$. At temperatures below $T_\lambda(x_3, P)$ the liquid behaves like a superfluid and is describable by the two-fluid model. The ^{3}He atoms in the fluid join the usual normal component and the Bose condensed ground state continues to be identified with the superfluid. At suitably low 3He concentrations ^{3}He atoms see the ^{4}He as an inert medium and behave like single particles with dispersion relation

$$\varepsilon_k = \varepsilon_0(x_3, P) + \hbar^2 k^2 / 2m^* , \tag{12}$$

where m^* is an effective mass that measures the departure in the behavior of the ^{3}He atom in ^{4}He from the behavior of a free ^{3}He atom, and $\varepsilon_0(x_3, P)$ is an energy which measures the bonding of the ^{3}He atom to the ^{4}He medium. The normal component of the superfluid is then made up of three parts, phonons, rotons, and ^{3}He atoms:

$$\rho_n = \rho_p(T) + \rho_r(T) + \rho_3(T) , \qquad (13)$$

with $\rho_3(T) \approx x_3 m^* N/V$. As $T \to 0\,$K, $\rho_p(T)$ and $\rho_r(T) \to 0$ and the normal component is dominated by the ^{3}He excitations; this occurs at $T \lesssim 0.5\,$K. Then, in conventional two-fluid-model experiments on the mixtures, e. g., second sound, osmotic pressure, etc., one learns about the properties of the ^{3}He normal component, e. g., m^*, the ^{3}He–^{4}He interactions, etc.

   In 1962 Edwards showed, from an analysis of scant thermodynamic data, that one should expect that as $x_3 \to 0$, $\varepsilon_0(x_3, P) \to 0$; i.e. a single ^{3}He atom would prefer to be in a ^{4}He medium than in a ^{3}He medium. There is a natural tendency for mixing of ^{3}He in ^{4}He at low ^{3}He concentrations. This is seen in Fig. 6 as the finite solubility of ^{3}He in ^{4}He at $T = 0\,$K for concentrations less than about 6%. Because of this finite solubility it is possible to carry low-concentration ^{3}He in ^{4}He mixtures to very low temperature; the ^{3}He phase does not separate out (compare to solid ^{3}He–^{4}He mixture diagram for a contrast) but the ^{4}He phonon–roton excitations disappear completely. The dilute gas of fermions that now inhabit the ^{4}He medium have a degeneracy temperature of

$$
\begin{aligned}
k_B(T_F(x_3)) &= k_B T_F x_3^{2/3} \ll 1\,\text{K} ; \\
k_B T_F &= (\hbar^2/2m^*)(3\pi^2\rho)^{2/3} ; \\
k_B T_F(5\%) &\approx 70\,\text{mK} .
\end{aligned}
$$

At $T < T_F(x_3)$ the dilute mixtures are systems of degenerate fermions. As such these systems constitute a second class of Fermi systems which have been explored extensively by the full range of experimental techniques and against which the Landau theory of Fermi liquids can be tested. Furthermore, the dilute mixtures can be employed to operate a "dilution refrigerator" that makes the temperature range $10 \le T \le 100\,$mK readily accessible.

^{3}He Below 3 mK

In 1972 Osheroff, Richardson, and Lee encountered a "glitch" in P vs T on the melting curve of ^{3}He near 2.7 mK during an experiment in which they were trying to elucidate the properties of solid ^{3}He. Within a few months Osheroff, Richardson, and Lee had identified the glitch with a modification of the behavior of the liquid ^{3}He in the sample chamber and had identified several important characteristics of liquid ^{3}He below 2.7 mK, i.e., superfluid ^{3}He. A ^{3}He superfluid had been expected on theoretical grounds (by analogy with the electron superfluid in metals and the hadron superfluids in nuclei and dense matter) but the experimental search for this phenomenon had driven the theory before it to lower and lower temperatures.

   The phase diagram of liquid ^{3}He in the millidegree range is shown in Fig. 5. There are two ^{3}He superfluids, the A superfluid ^{3}He-A at $2.2 \le T \le 2.1\,$mK on the melting curve and the B superfluid ^{3}He-B at $T < 2.2\,$mK on the melting curve. Also shown is the solid ^{3}He magnetic transition at 1 mK. Unlike ^{4}He which is composed of bosons, the ^{3}He liquid is composed of fermions and the occurrence of superfluidity requires a coupling that will tend to bond ^{3}He pairs. The ^{3}He–^{3}He interaction is strongly repulsive at short distances so that one expects an effective attractive interaction to occur in an $l \neq 0$ relative angular momentum state for a pair.

Experimental evidence points to an $l = 1$ pairing state. With $l \neq 0$ the ^{3}He pairs may have a variety of spin states, unlike the $l = 0$ pairing in superconductors that have antiparallel spins only. The A phase is composed of $l = 1$, with $\uparrow\uparrow$ pairs and $\downarrow\downarrow$ pairs behaving as two weakly coupled spin fluids; the B phase has all three $l = 1$ spin states in equal number. Because of the nature of the microscopic pairing a correct characterization of the ^{3}He superfluids is substantially more complex than characterization of ^{4}He superfluids; e. g., a piece of the fluid is described by $\mathbf{I}$, a measure of the angular momentum of the pairs in the fluid. Similarly a piece of the fluid can be described by $\mathbf{S}$, a measure of its spin, and $\mathbf{d}$, the order parameter. Because of the weak dipolar interaction between ^{3}He spins the energy of the system depends on the relative orientation of $\mathbf{S}$, $\mathbf{I}$, and $\mathbf{d}$. Furthermore, $\mathbf{S}$ couples directly to a magnetic field; $\mathbf{I}$ is influenced by walls, etc. As a consequence a great deal about the ^{3}He superfluids has been learned from experiments that probe the relationship of $\mathbf{S}$, $\mathbf{I}$, and $\mathbf{d}$ and the influence of $\mathbf{H}$, walls, etc. Experiments on superfluid ^{3}He have explored these relationships and have provided strong evidence in support of an $l = 1$ ground state and in support of the Leggett theory of the dynamics of $\mathbf{S}$, $\mathbf{I}$, and $\mathbf{d}$. The understanding of superfluid ^{3}He advanced in a few years (1972–1982) to a level comparable to that of our understanding of superfluid ^{4}He. Superfluid ^{3}He and ^{4}He are currently regarded as well understood.

Uses of Liquid Helium

Having a well-understood physical system in hand permits the possibility of employing this system as a probe.

1. As a consequence of the superfluidity of ^{4}He, thin films (thickness of order $10\,\text{Å}$) of ^{4}He support vortices having quantized circulation, $+$ or $-$, that interact with one another through a $\log r$ potential. Thus, thin films of ^{4}He have behavior that is due to the $d = 2$ neutral Coulomb gas of vortices that resides in them. It was in experimental studies of thin ^{4}He films (Bishop and Reppy, 1978) that compelling evidence for the Kosterlitz–Thouless scenario for $d = 2$ phase transitions (a scenario built on the neutral Coulomb gas model) was found.

2. The fourth sound mode of ^{4}He in packed powders (a mode made possible because the normal fluid is held immobile in the confined space available to it between the powder particles) can be used to study the properties of an elastic fluid, the superfluid component of ^{4}He, in a porous medium. Experimental and theoretical studies of the ^{4}He system, ^{4}He in Vycor glass, have been important in elucidating the properties of a correct description of complex fluid–matrix systems; e. g., ^{4}He in Vycor glass is analogous to oil in a rock.

3. The single-particle states appropriate to describing particles in a thin film, e. g., a thin film of ^{3}He, are strongly influenced by the film's size (thickness). The magnetic properties of the film depend on the occupation of these states by the ^{3}He particles, à la Pauli. Thus the magnetization of a thin ^{3}He film is a sensitive probe of the geometry of the space in which it resides. Should this space be fractal the magnetization of the ^{3}He will show it.

The quantum fluids, well known, are valuable probes of the unknown.

See also: Bose–Einstein Statistics; Fermi–Dirac Statistics; Quantum Fluids; Quasiparticles; Second Sound.

Bibliography

E. E. Keller, *Helium 3 and Helium 4*. Plenum Press, London, 1969.

J. C. Wheatley, *Am. J. Phys.* **36**, 181 (1968).

J. C. Wheatley, *Physics Today* **29**, No. 2, 32 (1976).

J. Wilks, *The Properties of Liquid and Solid Helium*. Clarendon Press, Oxford, 1967.

Helium, Solid

H. R. Glyde

Introduction

Helium was first solidified at the famous Kamerlingh Onnes low-temperature physics laboratories in Leiden by W. H. Keesom [1] on June 25, 1926. The initial experiments by Sir Francis Simon at Oxford University and by Keesom and their collaborators focused on the melting curve, the specific heat, and the thermal conductivity of solid helium as a test of our early understanding of solids. These measurements showed, for example, that the Lindemann criterion of melting does not hold in solid helium. This pioneering work up to 1957 is elegantly and beautifully reviewed by Domb and Dugdale [2], a review that stands today as an excellent introduction to solid helium along with the books and reviews by Wilks [3], Keller [4], Wilks and Betts [5], Glyde [6], Dobbs [7] and Roger *et al.* [8].

The pair potential $v(r)$ between helium atoms is precisely known [9, 10]. It is weakly attractive at large separation, $r \gtrsim 3\,\text{Å}$, with a maximum well depth $\varepsilon = 10.95\,\text{K}$. At close approach, $r \leq \sigma = 2.63\,\text{Å}$, where the hard-core radius σ, defined by $v(\sigma) = 0$, $v(r)$ becomes steeply repulsive. The potential parameters σ and ε of the rare gases are compared in Table 1. The potential seen by a helium atom lying between two atoms in a linear lattice is depicted in Fig. 1. The well shape, which is wide and anharmonic, is clearly dominated by the repulsive core of $v(r)$.

Table 1: Comparison of solid ^{3}He (at $V = 24\,\text{cm}^3/\text{mol}$) and solid ^{4}He (at $V = 21.1\,\text{cm}^3/\text{mol}$) with the heavier rare-gas crystals.[a]

Rare-gas crystal	Debye temp. θ_D (K)	Melting temp. T_M (K)	Debye zero point energy $E_{ZD} = \frac{9}{8}\theta_D$	Lindemann parameter $\delta = \frac{\langle u^2 \rangle^{1/2}}{R}$	Potential parameters σ (Å)	ε (K)	de Boer parameter Λ
^{3}He(bcc)	19	0.65	21	0.368	2.637	10.95	0.325
^{4}He(bcc)	25	1.6	28	0.292	2.637	10.95	0.282
Ne	66	24.6	74	0.091	2.758	42.25	0.061
Ar	84	83.8	95	0.048	3.357	143.22	0.019
Kr	64	161.4	72	0.036	3.579	199.9	0.011
Xe	55	202.0	62	0.028	3.892	282.35	0.0065

[a] The interatomic potential parameters are the core radius σ [$v(\sigma) = 0$] and the well depth ε for the following potentials: He [9, 10]; Ne, HFD-C2 [9]; Ar, HFD-C [9]; Kr, HFD-C (HFGKK) [9]. For Xe we quote σ and ε from Barker *et al.* [11].

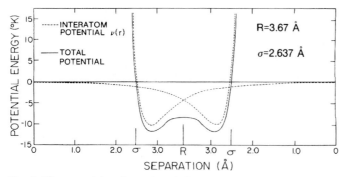

Fig. 1: The potential well seen by a helium atom in a linear solid arising from its nearest neighbors (solid line). R is the interatom space of the linear lattice. Dashed line is the interatomic potential, $v(r)$. Note that the second derivative of the potential well is negative at the lattice point R.

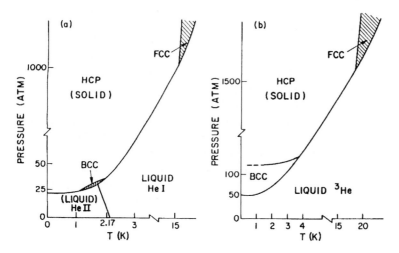

Fig. 2: Phase diagram of helium.

Since helium is light, its thermal wavelength, λ_T, is long, e. g., at $T = 1.0\,\mathrm{K}$, $\lambda_T \sim 10\,\text{Å}$ for ^{4}He. Helium is therefore difficult to localize. Attempts to localize it lead to a high kinetic or zero point energy. Since $v(r)$ is weak, helium does not solidify under attraction via $v(r)$. Rather, it solidifies only under pressure and then solidifies because of the hard core of $v(r)$, much as billiard balls form a lattice under pressure. The total energy E is small and positive except at the lowest pressures.

The phases of helium are sketched in Fig. 2. At $p \approx 35\,\mathrm{bar}$, the lighter isotope ^{3}He solidifies into an expanded body-centered-cubic (bcc) structure, having $V = 24.4\,\mathrm{cm}^3/\mathrm{mol}$ with interatom spacing $R = 3.75\,\text{Å}$. At higher pressure, both ^{3}He and ^{4}He are compressed into the close-packed (fcc and hcp) phases (e. g., at $p = 4.9\,\mathrm{kbar}$, fcc ^{4}He has $V = 9.03\,\mathrm{cm}^3/\mathrm{mol}$, $R = 2.11\,\text{Å}$). A discussion of phases can be found in the books by Keller [4], Wilks [3], and Wilks and Betts [5].

Helium, the Quantum Solid

The basic character of solid helium is revealed in the specific heat, C_v^p, arising from vibration of the atoms about their lattice points, the phonons (p). At low temperature (T), C_v^p is well described in all phases by the traditional Debye expression

$$C_v^p = 3R \left(\frac{4\pi^4}{5} \right) \left(\frac{T}{\theta_D} \right)^3 . \tag{1}$$

Here R is the gas constant and θ_D is a free parameter, the Debye temperature, obtained by fitting to the observed C_v^p. θ_D is the characteristic energy of the phonons. From Table 1, θ_D is clearly large compared to the melting temperature (T_M) in solid helium.

In Debye's model the zero point vibrational energy is

$$(E_Z)_D = 2\langle KE \rangle_D = \frac{9}{8}\theta_D . \tag{2}$$

This energy, a purely quantum effect, clearly dominates thermal energies (T_M) so that solid helium may be called a quantum solid. From θ_D we may also evaluate the mean square vibrational amplitude $\langle u^2 \rangle$ of the atoms about their lattice points. At $T = 0\,\text{K}$ this is

$$\langle u^2 \rangle = \sum_{qj} \left(\frac{\hbar}{2M\omega_{qj}} \right) = 109.2 \left(\frac{1}{M\theta_D} \right) \text{Å}^2 , \tag{3}$$

where $\hbar$ is Planck's constant, M is the atomic mass and ω_{qj} are the phonon frequencies. From Table 1 we see that $\langle u^2 \rangle$ and the Lindemann [12] ratio $\delta \equiv \langle u^2 \rangle^{1/2}/R$ are large in solid helium.

Simply from θ_D we see that the atoms are not well localized. There is a trade-off between reducing the (energy), e. g., E_Z in Eq. (2) and localizing the particle $\langle u^2 \rangle$ in Eq. (3). To keep the zero point energy manageable each nucleus has a large vibrational distribution about its lattice point. The atoms therefore explore a wide range of the potential well depicted in Fig. 1. For the dynamics, the hard core is the important part of the potential and all force constants will be the derivatives of $v(r)$ averaged over large-amplitude vibration.

The vibrational distribution may be described by a nuclear wave function with wavelength given by the famous de Broglie wavelength, λ. The degree of quantumness of a solid can be characterized by the de Boer [13] parameter,

$$\Lambda \equiv \frac{1}{2\pi} \frac{\lambda}{\sigma} = \frac{h}{\sigma(2m\varepsilon)^{1/2}} \tag{4}$$

which is the ratio of the de Broglie wavelength, $\lambda = h/p$ [with momentum $p = (2/m\varepsilon)^{1/2}$], to the minimum separation of atoms in the crystal, σ. If λ is comparable to σ, the solid is highly quantum (see Table 1). Also, Λ is clearly mimicked by the Lindemann ratio δ (see Table 1). A large Λ therefore means quantum and very anharmonic vibration.

The large nuclear wave function of atoms means that wave functions of atoms on adjacent lattice sites overlap. This leads to direct nuclear exchange integrals and tunneling between sites. In solid ^{3}He, which has a nuclear spin $\frac{1}{2}$, this exchange can be traced in nuclear magnetic resonance experiments and in thermodynamic properties. The characteristic energies in helium are the phonon energies (θ_D) or kinetic energy $\langle KE \rangle \gtrsim 25\,\text{K}$, the total energy $E \sim 1\,\text{K}$, and the exchange energy, $J \sim 10^{-3}\,\text{K}$.

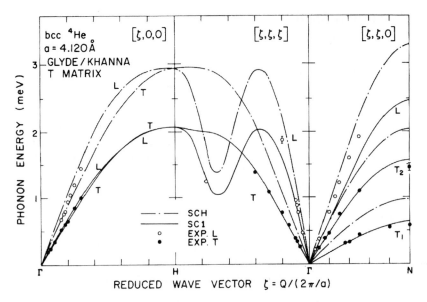

Fig. 3: The phonon-frequency dispersion curves observed in bcc ^{4}He by Minkiewicz *et al.* [18] and calculated by Glyde and Goldman [19] in the self-consistent harmonic (SCH) and complete first-order self-consistent (SC1) approximations.

Ground State and Dynamics

A key to understanding helium is finding a suitable wave function $\Psi(r_1 \ldots r_N)$ to describe the vibrational distribution of the atomic nuclei. Nobel laureate Max Born and his student D. J. Hooten first recognized that this vibrational distribution and the phonon dynamics must be determined consistently [2]. Nosanow [14] proposed a Gaussian function, the wave function for a harmonic crystal, for the distribution with corrections to account for short-range correlations between pairs of atoms induced by the hard core of $v(r)$. The short-range correlations can be described by a Jastrow pair function or by use of a Brueckner T-matrix method, both developed initially to describe short-range correlations between nucleons in nuclei. The self-consistent determination of the phonon frequencies and lifetimes and the vibrational distribution constitutes the self-consistent phonon (SCP) theory. This theory is exhaustively reviewed [6, 15–17].

Phonons in solid helium are most directly observed by inelastic neutron scattering. Measurements on all phases show that, although solid helium is highly anharmonic, most phonons are well defined and have long lifetimes [6]. The expanded bcc ^{4}He phase is so anharmonic that phonon energies can be determined at low wave vector (Q) only (see Fig. 3). At higher Q, the scattering intensity contains fascinating interference effects between the one-phonon and multiphonon scattering contributions that broadens the response and can be well described using the SCP theory [6]. The more compressed fcc phase of helium is significantly less anharmonic and can be quite well described without short-range corrections. Recent measurements [20] identify an anomaly in the intensity suggesting a new mode.

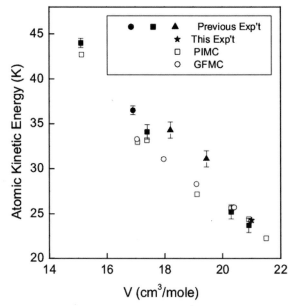

Fig. 4: Atomic kinetic energy in solid ^{4}He. The solid symbols are experimental values. The open symbols are GFMC [22] and PIMC [26] calculations (from Ref. [24]).

The static properties such as the ground state energy, $E = \langle KE \rangle + \langle V \rangle$, and the atomic momentum distribution, $n(\mathbf{k})$, are most accurately calculated using Monte Carlo (MC) methods [21], Diffusion Monte Carlo (DMC) [22, 23] at $T = 0$ K and Path Integral Monte Carlo (PIMC) [21] at $T > 0$ K. Typically $\langle KE \rangle \simeq 25$–50 K depending on the density. The $\langle KE \rangle$ and $\langle V \rangle$ nearly cancel to give $E = \pm 1$ K in solid ^{3}He. In spite of this, DMC values of E for both solid ^{3}He and ^{4}He agree remarkably well with experiments over a wide density range [23].

The atomic kinetic energy $\langle KE \rangle$ and the atomic momentum distribution $n(\mathbf{k})$ can be measured directly by high energy transfer neutron scattering [6]. As seen from Fig. 4, the observed $\langle KE \rangle$ and those calculated by Monte Carlo methods agree well for solid ^{4}He. It is interesting that the observed and MC $\langle KE \rangle_{\mathrm{MC}} = 25$ K at $V = 21$ cm^3/mol in Fig. 4 is twice the Debye model value $\langle KE \rangle_D = (9/16)\theta_D = 14$ K obtained using the observed θ_D in Table 1. This difference is a direct manifestation of the anharmonic hard core of $v(r)$. Interactions via the hard core introduce high-energy anharmonic tails in the phonon response functions. These high-energy components raise the average $\langle KE \rangle$ above that predictable by an adjusted Debye model which has no tails. High-energy tails are common to all phases of helium. Recent measurements Diallo [24] show that $n(\mathbf{k})$ in solid ^{4}He differs significantly from a Gaussian (see Fig. 5). There are more low-momemtum (low $\mathbf{k}$) atoms than in a classical, Maxwell–Boltzmann $n(\mathbf{k})$. It is not clear whether this arises from anharmonic efforts or from the onset of Bose statistics at low temperature since the shape of $n(\mathbf{k})$ is similar [24, 25] in liquid and solid ^{4}He. This deviation from a Gaussian $n(\mathbf{k})$ is also found in PIMC calculations [21, 26] in both liquid and solid ^{4}He. In addition, the Debye–Waller factor is found to deviate from a Gaussian in PIMC calculations [27].

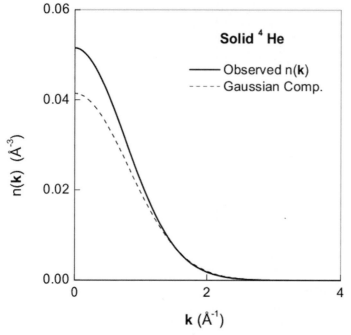

Fig. 5: Momentum distribution of solid ^{4}He at $T = 1.6\,$K and $V = 20.87\,\text{cm}^3/\text{mol}$ compared with its Gaussian component. The momentum distribution $n(\mathbf{k})$ is observed to be non-Gaussian and is characterized by a larger occupation of the low-momentum states [24].

Exchange and Magnetic Properties

Solid ^{3}He, a nuclear spin-$\frac{1}{2}$ Fermi solid, displays a rich spectrum of nuclear spin phases at low temperature ($T \lesssim 10^{-3}\,\text{K} = 1\,\text{mK}$). This is in both bulk, 3D bcc solid ^{3}He close to the melting line [8, 28] and in 2D triangular lattice layers that form on grafoil surfaces. In bcc ^{3}He there is a transition from the paramagnetic to an antiferromagnetic phase at $T = 1\,$mK at low magnetic field ($B \leq 0.46\,$T). Nuclear magnetic resonance [29] and neutron-diffraction data [30] are consistent with this being a tetragonal [29] U2D2 ordering displayed in Fig. 6. The antiferromagnetic resonant frequency, Ω_0, of the U2D2 ordering agrees better with experiment [31] than do other orderings not excluded by neutron and NMR data. At higher field, $B \gtrsim 0.46\,$T, there is a transition to a second ordered phase believed to be a canted normal antiferromagnetic (CNAF) phase (see Fig. 6). As B is increased, the canting angle decreases until it goes to zero at an upper critical field [31], $B_{C2} = 21.7 \pm 1\,$T. For $B \geq 10\,$T, the phase diagram is not well known.

The physical origin of the rich magnetic behavior is the large vibrational amplitudes of the nuclei noted above. The nuclear wave functions are broad and the wave functions of nuclei on adjacent sites overlap leading to large nuclear exchange integrals, $J \sim 1\,$mK. This is true only at large molar volumes where $\langle u^2 \rangle$ is large. J decreases exponentially with decreasing molar volume [32,33]. Also, since the pair potential has a repulsive hard core, simple pair exchange is suppressed relative to exchange integrals involving ring exchanges of three or

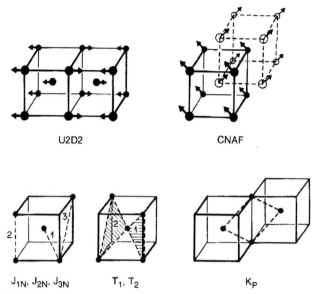

U2D2   CNAF

J_{1N}, J_{2N}, J_{3N} T_1, T_2 K_P

Fig. 6: The tetragonal up–up down–down (U2D2) and cubic canted normal antiferromagnetic (CNAF) structures. Exchange processes involving two, three, and four atoms in H_{MSE}, $J_2 = -J_{1N}$, $J_3 = -T_1$, $J_4 = -K_p$.

more atoms [34]. Indeed in 2D layers the magnitudes [33] of the J_n involving exchange of n atoms are in the order $n = 3, 2, 4, 6, 5$. Also exchanges of an even number of atoms (odd-parity exchanges) favor antiferromagnetism while exchanges of an odd number (even parity) favors ferromagnetism [34]. This leads to a multiple spin exchange (MSE) Hamilitonian [35, 36] with competing antiferromagnetic and ferromagnetic terms,

$$
\begin{aligned}
H_{\mathrm{MSE}} &= \sum_n (-1)^n J_n \sum_{(n)} P_n \\
&= J_2 \underset{(2)}{\sum P_2} - J_3 \underset{(3)}{\sum P_3} + J_4 \underset{(4)}{\sum P_4} - J_5 \underset{(5)}{\sum P_5} + J_6 \underset{(6)}{\sum P_6} ,
\end{aligned}
\tag{5}
$$

where P_n is a permutation involving n atoms. The sign convention in H_{MSE} is chosen so the exchange energies J_n are positive. The first term in H_{MSE} is the Heisenberg Hamiltonian, $H_H = J_2 \sum_{\langle ij \rangle} P_2(ij)$ with $P_2(ij) = 2S_i \cdot S_j + \frac{1}{2}$.

In pioneering work, Roger, Hetherington and Delrieu [8] and Roger [28] were able to reproduce the magnetic phase diagram of bulk bcc ^{3}He as well as the magnetic specific heat keeping up to four particle exchanges ($n = 2 - 4$) in H_{MSE}. The exchange constants $J_2 = -J_{1N}$, $J_3 = -T_1$ and $J_4 = K_p$ were treated as fitting parameters where J_{1N}, T_1 and K_p is the notation used by Roger *et al.* [8]. Ceperley and Jacucci [32] in a remarkable application of path integrals have evaluated the J_n. They confirm that the first four terms in H_{MSE} are the most important and find $J_2 = 0.46\,\mathrm{mK}$, $J_3 = 0.19\,\mathrm{mK}$ and $J_4 = 0.27\,\mathrm{mK}$ at $V = 24.12\,\mathrm{cm}^3/\mathrm{mol}$, within 10% of the Roger *et al.* [8] values. Exchanges $n > 4$ are non-negligible and, including these, Godfrin and [37] obtain impressive agreement with experiment, especially the Curie–Weiss constant.

The magnetization and specific heat of 2D solid ^{3}He (on grafoil) also displays remarkable behavior [38, 39]. For example, there is a crossover from antiferromagnetic to ferromagnetic response in the second 2D layer on grafoil with increasing density. This magnetic behavior and the crossover can be reproduced using H_{MSE} in (5) keeping terms up to $n = 6$ with the antiferromagnetic (J_4, J_6) and the ferromagnetic (J_3) terms having different density dependencies [35, 40]. Spin liquids are suggested [41]. Indeed exotic spin states having magnetic short-range order, a spin gap but no Néel long-range order are predicted [36] using H_{MSE}. These novel spin orderings serve as realizations of spin states, such as valence-bond spin liquids, of broad interest in condensed matter [36].

The direct energies such as $\langle \mathrm{KE} \rangle$, $\langle V \rangle$ and the ground state E are 10^4 times larger than the exchange energies, J_n. Thus statistics (Fermi or Bose) play no role in determining the the the ground static wave function. The magnitude of the exchange integrals will therefore be similar in solid ^{3}He and ^{4}He.

Supersolid Helium

Kim and Chan [42–44]have observed a nonclassical, superfluid component in solid ^{4}He. This is in both bulk [42] solid ^{4}He and in solid ^{4}He immersed in porous media [43, 44]. The moment of inertia of solid or liquid ^{4}He can be accurately measured using a torsional oscillator. In a normal solid or liquid, the full moment of the solid is observed. When solid ^{4}He is cooled below 0.23 K, the moment of inertia is observed to decrease indicating that a fraction of the solid has decoupled from the walls of the oscillator exactly as if it were a superfluid. The "superfluid" fraction, ρ_s/ρ, of solid ^{4}He is approximately 2% at low temperature (as against 100% in liquid ^{4}He). The temperature dependence of $\rho_s(T)/\rho$ is very different in solid and liquid ^{4}He. Superfluidity can be suppressed by adding a few hundred parts per million of ^{3}He. The "supersolid" phase extends up to pressures over 100 bar (compressed ^{4}He) which is surprising. Possible mechanisms of superflow are based on the large exchange integrals in quantum solids (tunneling) and often involve a quantum defect (e. g., vacancies) as the superfluid component. This observation of "supersolidity" opens up an exciting new dimension to quantum solids.

See also: Crystal Binding; Crystal Defects; Interatomic and Intermolecular Forces; Rare Gases and Rare-Gas Compounds; Tunneling.

References

[1] W. H. Keesom, *Comm. Kamerlingh Onnes Lab., Leiden* No. **184b** (1926); *Helium*, p. 180. Elsevier, Amsterdam, 1942.

[2] C. Domb and J. S. Dugdale, *Prog. Low Temp. Phys.* **2**, 338 (1957).

[3] J. Wilks, *The Properties of Liquid and Solid Helium*. Clarendon Press, Oxford, 1967.

[4] W. E. Keller, *Helium-3 and Helium-4*. Plenum, New York. 1969.

[5] J. Wilks and D. S. Betts, *Liquid and Solid Helium*. Clarendon Press, Oxford, 1987.

[6] H. R. Glyde, *Excitations in Liquid and Solid Helium*. Clarendon Press, Oxford, 1994.

[7] E. R. Dobbs, *Solid Helium Three*. Clarendon Press, Oxford, 1994.

[8] M. Roger, J. H. Hetherington, and J. M. Delrieu, *Rev. Mod. Phys.* **55**, 1 (1983).

[9] R. A. Aziz, in *Inert Gases* (M. L. Klein, ed.). Springer-Verlag, Berlin, Heidelberg, 1984.

[10] T. Korona, H. L. Williams, R. Bukowski, B. Jeziorski and K. Szalewicz, *J. Chem. Phys.* **106**, 5109 (1997).

[11] J. A. Barker, M. L. Klein, and M. V. Bobetic, *IBM J. Res. Dev.* **20**, 222 (1976).

[12] F. A. Lindemann, *Phys. Z.* **11**, 609 (1911).

[13] J. de Boer, in *Progress in Low Temperature Physics*, Vol. 2 (J. C. Gorter, ed.). North-Holland, Amsterdam, 1957. We use the definition introduced by Newton Bernardes, *Phys. Rev.* **120**, 807 (1960), which is $(2\pi\sqrt{2})^{-1}$ times de Boer's definition.

[14] L. H. Nosanow, *Phys. Rev.* **146**, 120 (1966).

[15] T. R. Koehler, in *Dynamical Properties of Solids*, Vol. II (G. K. Horton and A. A. Maradudin, eds.). North-Holland, Amsterdam, 1975.

[16] H. R. Glyde, in *Rare Gas Solids*, Vol. I (M. L. Klein and J. A. Venables, eds.). Academic Press, New York, 1976.

[17] H. Horner, in *Dynamical Properties of Solids*, Vol. I (G. K. Horton and A. A. Maradudin, eds.). North-Holland, Amsterdam, 1974.

[18] V. J. Minkiewicz, T. A. Kitchens, G. Shirane, and E. B. Osgood, *Phys. Rev. A* **8**, 1513 (1973).

[19] H. R. Glyde and V. V. Goldman, *J. Low Temp. Phys.* **25**, 601 (1976).

[20] T. Markovich, E. Polturak, J. Bossy, and E. Farhi, *Phys. Rev. Lett.* **88**, 195301 (2002).

[21] D. M. Ceperley, *Rev. Mod. Phys.* **67**, 279 (1995).

[22] P. A. Whitlock and R. M. Panoff, *Can. J. Phys.* **65**, 1409 (1987).

[23] S. Moroni, F. Pederiva, S. Fantoni, and M. Boninsegni, *Phys. Rev. Lett.* **84**, 2650 (2000).

[24] S. O. Diallo, J. V. Pearce, R. T. Azuah and H. R. Glyde, *Phys. Rev. Lett.* **93**, 075301 (2004).

[25] H. R. Glyde, R. T. Azuah and W. G. Stirling, *Phys. Rev. B* **62**, 14337 (2000).

[26] D. M. Ceperley, in *Momentum Distributions*, (R. N. Silver and P. E. Sokol, eds). Plenum, New York, 1989.

[27] E. W. Draeger and D. M. Ceperley, *Phys. Rev. B* **61**, 12094 (2000).

[28] M. Roger, *Phys. Rev. B* **30**, 6432 (1984).

[29] D. D. Osheroff, M. C. Cross, and D. S. Fisher, *Phys. Rev. Lett.* **44**, 792 (1980).

[30] A. Benoit, J. Bossy, J. Flouquet, and J. Schweizer, *J. Phys. Lett.* **46**, L923 (1985).

[31] D. D. Osheroff, H. Godfrin, and R. R. Ruel, *Phys. Rev. Lett.* **58**, 2458 (1987).

[32] D. M. Ceperley and G. Jacucci, *Phys. Rev. Lett.* **58**, 1648 (1987).

[33] B. Bernu and D. M. Ceperley, *J. Phys. Condens. Mat.* **14**, 9099 (2002).

[34] D. J. Thouless, *Proc. Phys. Soc.* **86**, 893 (1965).

[35] M. Roger, *Phys. Rev. B* **56**, R2928 (1997).

[36] G. Misguich, C. Lhuillier, B. Bernu, and C. Waldmann, *Phys. Rev. B* **60**, 1064 (1999).

[37] H. Godfrin and D. D. Osheroff, *Phys. Rev. B* **38**, 4492 (1988).

[38] H. Godfrin and R. E. Rapp, *Adv. Phys.* **44**, 113 (1995).

[39] E. Collin and Yu. M. Bunkov and H. Godfrin, *J. Phys. Condens. Mat.* **16**, S691 (2004).

[40] M. Roger, *Phys. Rev. Lett.* **64**, 297 (1990).

[41] K. Ishida, M. Morishita, K. Yawata and H. Fukuyama, *Phys. Rev. Lett.* **79**, 3451 (1997).

[42] E. Kim and M. H. W. Chan, *Science* **305**, 1941 (2004).

[43] E. Kim and M. H. W. Chan, *Nature* **427**, 227 (2004).

[44] E. Kim and M. H. W. Chan, *J. Low Temp. Phys.* **138**, 859 (2005).

Hidden Variables

M. Horne

The quantum-mechanical state of a system provides, in general, statistical predictions on the results of measurements. A controversy long existed as to whether the dispersions predicted by quantum mechanics indicate that the theory is incomplete and whether it is possible to reproduce the quantum-mechanical statistics by suitable ensembles of dispersion-free states for which the results would be determined individually. The dispersion-free states are commonly referred to as hidden variables, and a theory which uses dispersion-free states to explain quantum-mechanical statistics is called a hidden variable theory. If dispersion-free states could actually be prepared in the laboratory, quantum mechanics would be observably inadequate. However, the controversy focused not on whether the hidden variables should be observable but on whether the mere hypothesis of their existence is compatible with the structure of quantum mechanics. Another facet of the controversy focused on whether hidden variables are needed, i. e., whether their existence solves any pressing dilemmas in physics.

To describe the results of research on the existence question, it is helpful to recognize, within the general family of hidden variable theories, two distinct classes which may be called *contextual* theories and *noncontextual* theories. A noncontextual hidden variable theory would be related to quantum mechanics in much the same way that classical mechanics is related to classical statistical mechanics. Each individual system is specified by a state λ from a phase space Γ of hidden states. Each quantum-mechanical state ψ is associated with a probability density $\rho_\psi(\lambda)$ on the space Γ; that is, the state ψ is associated with an ensemble of the λ states. The value of each observable A is specified by a function $f_A(\lambda)$ mapping the hidden states λ into the real line; that is, the result of measuring the observable A is predetermined by the hidden state of the individual system. Clearly, agreement with quantum mechanics requires that the range of the function $f_A(\lambda)$ be the set of eigenvalues of the operator $\hat{A}$ associated with the observable A. And finally, the statistical predictions of quantum mechanics are reproduced: for each observable A and state ψ

$$|\langle \psi | a_i \rangle|^2 = \int_{\Gamma(a_i)} \rho_\psi(\lambda) \, d\lambda \, ,$$

where $|a_i\rangle$ is the eigenvector of $\hat{A}$ with eigenvalue a_i and $\Gamma(a_i)$ denotes the subspace in Γ where $f_A(\lambda) = a_i$. The central feature of this scheme is quite simple: each individual system in effect possesses a definite value for every observable.

It is now known that noncontextual hidden variable states are incompatible with the formal structure of quantum mechanics. For many years it was widely believed that von Neumann had already demonstrated the incompatibility in 1932, but in 1966 his proof was examined by Bell and shown to be inadequate. Bell and others noted, however, that incompatibility does follow as a corollary to a 1957 theorem of Gleason. The proof presupposes a Hilbert space of dimension greater than two, which permits the existence of noncommuting operators $\hat{B}$ and $\hat{C}$, each commuting with a third operator $\hat{A}$, such that one can in principle simultaneously observe either A and B or A and C. The proof then shows from quantum mechanics that there exists no function $f_A(\lambda)$ which specifies the value of the observable A and which is independent of whether B or C is measured simultaneously with A. Thus the above noncontextual hidden variable scheme is impossible. However, the proof does not exclude contextual hidden variable theories: a necessary condition for a hidden variable interpretation of quantum mechanics is that the function specifying the value of an observable A depends not only on the hidden state A but also on the context of the whole apparatus, e. g., on whether the apparatus measures A and B or A and C.

Contextual theories do not suffice to salvage the hidden variable program because it turned out that any such theory which agrees with all the statistical predictions of quantum mechanics must have an undesirable property of nonlocality. This fact was shown by Bell, who worked out the implications of an argument initiated in 1935 by Einstein, Podolsky, and Rosen (EPR). In their argument, EPR consider a composite system consisting of two spatially separated, yet correlated, subsystems and argue persuasively that the quantum-mechanical description is incomplete unless hidden variables exist. Consider, for example, a pair of spin-4 particles produced somehow with total spin angular momentum zero (singlet spin state) and moving freely in opposite directions. Measurements can be made, say with Stern–Gerlach magnets, on selected components of the spins of the two particles. If measurement of some component of the spin of particle one yields the result "up" or $+1$, then, according to quantum mechanics, measurement of the same component of the spin of particle two must yield the result "down" or -1, and vice versa. Now assume the two measurements are made at widely separated places and consider the following premises of EPR:

1. Since at the time of measurement the two systems no longer interact, no real change can take place in the second system in consequence of anything that may be done to the first system.

2. If, without in any way disturbing a system, we can predict with certainty (i. e., with probability equal to unity) the value of a physical quantity, then there exists an element of physical reality corresponding to 'this physical quantity.

3. Every element of the physical reality must have a counterpart in the [complete] physical theory.

Since, in our example, we can predict in advance the result of measuring any component of the spin of one particle by previously measuring the same component of spin of the other remotely located particle, it follows from premises 1 and 2 that the value of any spin component of either particle is an element of physical reality. That is, the value of every spin component of either particle must actually be predetermined. And reflection will convince the reader that the only way premise 3 may be fulfilled is through the existence of a hidden variable theory. Because of premise 1, the hidden variable theories required by EPR have the property called *locality*.

In 1964 Bell investigated the local hidden variable theories and found that the whole class is incompatible with certain statistical predictions of quantum mechanics. It is significant that he was able to expose an *observable* discrepancy between quantum mechanics and the local hidden variable theories. Consider, as does Bell, the pair of spin-$\frac{1}{2}$ particles in the singlet state discussed above. Measurements can be made on the **a** component of the spin of particle one and on the **b** component of the spin of particle two. If the unit vectors **a** and **b** are not parallel, all combinations of results are possible: $++$, $+-$, $-+$, and $--$. For given directions **a** and **b**, quantum mechanics predicts a probability for the occurrence of each of these outcomes: $P_{++}(\mathbf{a},\mathbf{b})$, $P_{+-}(\mathbf{a},\mathbf{b})$, $P_{-+}(\mathbf{a},\mathbf{b})$, and $P_{--}(\mathbf{a},\mathbf{b})$, where the first and second subscripts refer to the results for the first and second particle. The specific statistical quantity considered by Bell is the expectation value of the product of the two results:

$$E(\mathbf{a},\mathbf{b}) = P_{++}(\mathbf{a},\mathbf{b}) - P_{+-}(\mathbf{a},\mathbf{b}) - P_{-+}(\mathbf{a},\mathbf{b}) + P_{--}(\mathbf{a},\mathbf{b}) .$$

The quantum mechanical prediction for this is

$$E_{qm}(\mathbf{a},\mathbf{b}) = -\mathbf{a} \cdot \mathbf{b} .$$

Bell proves that for any local hidden variable theory the expectation value satisfies the inequality

$$|E_{hv}(\mathbf{a},\mathbf{b}) - E_{hv}(\mathbf{a},\mathbf{c})| - E_{hv}(\mathbf{b},\mathbf{c}) - 1 \leq 0 ,$$

where $\mathbf{a}$, $\mathbf{b}$, and $\mathbf{c}$ are any three directions. The quantum-mechanical function E_{qm} does not, in general, satisfy the inequality. For example, choosing $\mathbf{a}$, $\mathbf{b}$, and $\mathbf{c}$ coplanar, letting the angle between $\mathbf{a}$ and $\mathbf{b}$ be 60°, the angle between $\mathbf{b}$ and $\mathbf{c}$ be 60°, and the angle between $\mathbf{a}$ and $\mathbf{c}$ be 120°, insertion of E_{qm} into the hidden variable inequality yields the contradiction $\frac{1}{2} \leq 0$.

Bell's theorem concerning the local hidden variable theories is a profound development in the history of hidden variable discussions. First, the theorem proves that the EPR premises are untenable if quantum mechanics is correct. Second, the theorem makes it possible to design experiments for testing the entire family of local hidden variable theories, on the chance that quantum mechanics may break down for widely separated, yet correlated, pairs of systems. Of course, the suggestion of such a breakdown is a radical proposal in view of the success of quantum mechanics. But on the other hand, the premises of EPR are plausible and some physicists may prefer experimental evidence before giving them up. Many experiments have been designed and carried out, the majority yielding results in good agreement with quantum mechanics and in violation of local hidden variable predictions. An important feature of one recent experiment is the use of time-varying polarization analyzers. In this experiment, the settings of each analyzer are effectively changed during the flight of the particles, thereby substantially reducing the possibility that one particle or its analyzing apparatus "knows" the setting of the distant analyzing apparatus. Although the experiments so far are not sufficiently close to ideal to be conclusive, experimental evidence is accumulating that the EPR premises and the hidden variables they imply are untenable.

See also: Quantum Mechanics; Quantum Statistical Mechanics.

Bibliography

For the EPR argument that local hidden variables are needed, see *Phys. Rev.* **47**, 777 (1935); for Bohr's reply, see *Phys. Rev.* **48**, 696 (1935).

For an explicit hidden variable model of elementary wave mechanics, see D. Bohm, *Phys. Rev.* **85**, 166, 180 (1952).

For criticism of early noncontextual incompatibility proofs and for proof based on Gleason's work, see J. Bell, *Rev. Mod. Phys.* **38**, 447 (1966); for simpler proofs, see F. J. Belinfante, *A Survey of Hidden-Variables Theories*. Pergamon Press, New York 1973.

For Bell's theorem concerning local theories, see *Physics (Long Island City)* **1**, 195 (1964); for proposal of an experimental test, see J. Clauser *et al.*, *Phys. Rev. Lett.* **23**, 880 (1969); for the experiment, see S. Freedman and J. Clauser, *Phys. Rev. Lett.* **28**, 938 (1972).

For general discussion of experimental tests, see J. Clauser and M. Home, *Phys. Rev. D* **10**, 526 (1974).

For extensive discussion of the implications of Bell's work, see B. d'Espagnat, *Conceptual Foundations of Quantum Mechanics*, 2nd ed. Benjamin, New York 1976; M. Redhead, *Incompleteness, Nonlocality and Realism*. Clarendon, Oxford, 1987.

For a comprehensive review of experimental tests through 1978 and their implications, see J. Clauser and A. Shimony, *Rep. Prog. Phys.* **41**, 1881–1927 (1978).

For the experiment with time-varing analyzers, see A. Aspect *et al.*, *Phys. Rev. Lett.* **49**, 1804 (1982).

For collections of reprints, including many of the papers listed in this bibliography, see J. A. Wheeler and W. H. Zurek (eds.), *Quantum Theory and Measurement*. Princeton University Press, Princeton, NJ, 1983; J. S. Bell, *Speakable and Unspeakable in Quantum Mechanics*. Cambridge University Press, Cambridge, 1987; and L. E. Ballentine (ed.), *Foundations of Quantum Mechanics Since the Bell Inequalities*. American Association of Physics Teachers, College Park, MD, 1988.

High-Field Atomic States

J. E. Bayfield

An atom placed in a sufficiently strong electric, magnetic, or electromagnetic field is in some high-field atomic quantum state. The strength of the field must be high enough for the alteration of the energy of the atom not to be a small perturbation. Thus by definition, high-field states are quite different from those of isolated atoms. The electron density distribution within the atom is controlled by both the nuclear Coulomb field and the applied field. While some high-field states can be considered to be stationary, others involve an explicit time evolution of the electron's energy.

Atoms in Strong Static Fields

One early interest in high-field states was in astrophysics, as atoms can be in the intense static magnetic fields of neutron stars. An analysis of the spectra of such atoms yields the value of the magnetic field. The interaction Hamiltonian

$$\frac{e}{mc}\mathbf{A}\cdot\mathbf{p}+\frac{e^2}{2mc^2}\mathbf{A}^2 ,$$

where $\mathbf{p}$ is the electron's momentum and the quantity $\mathbf{A}$ is related to the position $\mathbf{r}$ of the electron relative to the nucleus and to the magnetic field strength $\mathbf{B}$ by

$$\mathbf{A}(\mathbf{r}) = \frac{1}{2}\mathbf{B}\times\mathbf{r} .$$

The Zeeman effect occurs at low fields $\mathbf{B}$ and is a perturbative effect arising from the first term in the interaction energy. High-field states are formed when the second term exceeds the first. Whereas the classical physics of the Zeeman effect involves a field-induced precession of the electron's orbit plane about the magnetic field direction, the high-field states classically can involve a keen competition between the electron orbiting around the nucleus and orbiting around the magnetic field direction. This competition produces a mixture of possible orbits in the classical limit, with some being periodic, some quasiperiodic, and some chaotic. The periodic classical orbits play a major role in determining the stationary high-field quantum states that produce the spectrum of the atom in the field. On the other hand, one quantum manifestation of the presence of chaotic orbits is an alteration of the distribution of quantum energy level spacings in a certain characteristic way.

Whereas a free electron classically orbits in a circle when in a static magnetic field, a static electric field accelerates such an electron along the field direction. Thus magnetic high-field states tend to have the electron's localization near the nucleus maintained, whereas electric high-field states tend to involve ultimately a removal of the electron from the region of the nucleus, a process called field ionization. When both electric and magnetic high fields are present, some high-field states are slowly ionizing while others ionize rapidly.

Atoms in High-Strength Electromagnetic Waves

When an atom is exposed to a monochromatic electromagnetic wave of high-field strength and of frequency comparable to the classical electron orbit frequency of the unperturbed atom, a large number of photons can be absorbed from the wave before the electron leaves the vicinity of the nucleus. More photons may be absorbed than the minimum energetically required for photoionization, a phenomenon called above-threshold ionization. The many-photon absorption has been studied for the case of highly excited atoms, where the classical electron orbit is very large and the electron orbit frequency is in the microwave region, much lower than the optical frequencies corresponding to lowest quantum atomic energy levels. In the classical limit, the possible electron dynamics again spans the range from periodic to chaotic. Although the high-field states cannot be stationary since the applied electromagnetic field is time-varying, in the absence of field ionization there exist "quasienergy" states that display the time periodicity of the field. When the electric field strength of the wave reaches a frequency-dependent threshold value of the order of 10% of the mean atomic Coulomb field strength, a many-photon field ionization process does rapidly occur that in the classical limit involves the chaotic electron orbits only. During the ionization process the time evolution of electron energy exhibits a near-classical growth characteristic of a diffusion process. As a result the ionization threshold fields are near-classical. At ionizing field strengths the quasienergy level separations again exhibit the distribution function characteristic of chaos in the classical limit. This situation is altered when the frequency of the wave is higher than twice the initial electron orbit frequency, where destructive quantum wave interference suppresses the near-classical time evolution and forces the ionization threshold field to higher values.

See also: Chaos; Dynamics, Analytical; Order-Disorder Phenomena; Photoionization; Quantum Mechanics; Zeeman and Stark Effects.

Bibliography

H. Hasegawa, M. Robnik, and G. Wunner, in "New Trends in Chaotic Dynamics of Hamiltonian Systems", *Prog. Theoret. Phys. Supplement*, 1989. (A)

J. E. Bayfield, *Comments At. Mol. Phys.* **20**, 245 (1987). (A)

G. Casati, B. V. Chirikov, D. L. Shepelyansky, and I. Guarneri, *Phys. Rpt.* **154**, 77–123 (1987). (A)

High Temperature

J. W. Hastie, D. W. Bonnell, and J. B. Berkowitz

High temperature is a pervasive condition, occurring as a natural intra- and extraterrestrial phenomenon and in many aspects of ancient and modern science and technology. The human sense of temperature derives from the fact that biological processes (as we know them) are confined to a very narrow temperature range. A person coming into contact with an object at a much higher (or lower) temperature than that of his own body will describe the object as "hot" (or "cold"). In a thermodynamic sense, an object will feel hot when the direction of heat flow is from the object to the individual. The observation that heat will only flow spontaneously (i. e., without expenditure of work) from a hot object to a cold object, and never in the reverse, is in essence one statement of the second law of thermodynamics. In general, if two bodies are brought into contact and there is no heat flow from one to the other, they are said to be at the same temperature. This forms the basis for more formal definitions of temperature, and for the establishment of a scale by which to measure temperature quantitatively. The realization of the measurement of temperature is the purpose of the International Temperature Scale of 1990 (ITS-90), which replaces IPTS-68(75), the previous standard temperature scale. The four temperature scales in use are Kelvin (K), Celsius (°C), Fahrenheit (°F), and Rankine (°R). They are interrelated as: $t(°C) = T(K) - 273.15$; $t(°F) \cong 9/5t(°C) + 32$; $t(°R) \cong 9/5T(K)$. The Kelvin and Rankine temperature scales are referred to as *absolute* thermodynamic scales, because their lower reference point is absolute zero. The Kelvin scale is currently realized in practice through the ITS-90, and the other scales are derivative. The Rankine scale is of historic interest because many tabulated engineering data are based on that scale. The Fahrenheit scale is used mainly in the United States, and for nonscientific purposes. Implicit in the concept of temperature is the concept of equilibrium. A system not at equilibrium cannot be said to have a temperature. However, in many globally nonequilibrium situations, if one carries out measurements on an appropriate scale, either in space or time, then a local equilibrium can be identified with which a temperature can be associated. This concept provides the basis for dealing with the real world of matter and energy flows. For example, in a system with a temperature gradient and with mass transport occurring, there is no three-dimensional volume at constant temperature. However, if the energy and mass transport are occurring under equilibrium conditions, a temperature still exists at every point in the system.

From a human physiological perspective, high temperatures are those in excess of the range of stability of living systems. From a materials or chemical perspective, the influence of high temperatures can often be defined from the thermodynamic relationship for the Gibbs energy change for a physicochemical process, $\Delta G = \Delta H - T\Delta S$, where ΔH is the heat energy, or *enthalpy*, change, ΔS is the entropy change, and T is the absolute temperature. In many practical situations, ΔG provides a measure of the thermal stability of a system. The more negative ΔG is, the more stable the products of the process are. Often, and particularly in the range from 1000 to 5000 K, both the ΔH and $T\Delta S$ terms are influential in determining the sign and magnitude of ΔG. This interplay between the opposing stability factors of bond formation (negative ΔH) and bond breaking (positive ΔS) leads to complex, often unpredictable, behavior as a function of temperature. Thus, the usual insight at ambient temperature regarding heat energy as the driving force can be counterintuitive in understanding high-temperature processes. From a scientific viewpoint, high-temperature phenomena are often associated with entropy-dominated processes.

From an engineering and technology point of view, high temperatures are usually those temperatures sufficiently in excess of normal ambient temperatures to change significantly the properties of physical and chemical systems. While ambient temperatures are usually considered to fall around room temperature, 298 K, the recent development of "high-temperature" superconductors, which exhibit zero electrical resistance at temperatures reported as high as about 100 K, is an example of the nebulous and changing nature of our concept of high temperature.

The properties of greatest interest at high temperatures are either those obtained under equilibrium conditions, such as thermodynamic stability and physical, molecular, and mechanical properties, or under kinetic (dynamic or nonglobal equilibrium) conditions. The significant changes in conditions of equilibrium include changes in phase (solid, liquid, gas), and changes in molecular composition. Physical properties include density, surface tension, viscosity, thermal and electrical conductivity, and the like. Molecular properties include lattice structure and bond energies, and spectroscopic parameters such as vibrational frequencies, electronic states, and so on. Basic thermodynamic properties include heat of formation, entropy, heat capacity and equation of state. Mechanical properties of solids and glasses include a variety of correlative properties (creep, deformation, strain, fracture resistance) which are all measures of cohesive strength, i. e., surface and grain-boundary energy. Microstructure is also important to understanding chemical, mechanical, and physical properties of solids. The significant changes in kinetic interactions include changes in the rates at which thermodynamically favorable processes occur and changes in the mechanisms (i. e., the ways) by which chemical reactions take place. In addition to chemical kinetic interactions, rates of energy transfer, particularly among rotational, vibrational, and electronic quantum states, and their conversion of kinetic or translational energy, are of key significance to the concepts of equilibrium and temperature. In a gas, translational energy and thermodynamic temperature, T, are related by $\frac{1}{2}m\bar{v}^2 = 3/2kT$, where m is molecular mass, $\bar{v}$ is average velocity, and k is the Boltzmann constant, 1.38×10^{-16} erg/K.

Significant changes in properties are generally observed at temperatures above 1000 K. At temperatures above 3000 K, the only metals that are still stable in the solid phase at ordinary pressures are rhenium, tantalum, and tungsten. The normal boiling points of some common metals are: aluminum, 2720 K; copper, 2855 K; gold, 2980 K; silver, 2450 K; tin, 2960 K; and zinc, 1181 K. Above these temperatures the only stable phase of the metal at atmospheric pressure will be gaseous. Temperatures above 3000 K are reached in both industrial and natural processes. In the manufacture of silicon carbide (one of the common abrasives), temperatures in the center of the fabricating furnace are above 3600 K. Temperatures within the earth range from about 1000 K in the mantle regions, 100 km below the surface, to possibly 6900 K in the metal core. The core temperature of Jupiter is estimated to be 10 000–20 000 K. Some stars have surface temperatures as high as 100 000 K. Our own sun has a surface temperature of about 6000 K, with temperatures of its coronal gases reported as high as 2 000 000 K. In thermonuclear reactions (e. g., nuclear fusion and stellar cores), temperatures can be more than 100 000 000 K. In astrophysically catastrophic events, such as supernovas, violent galaxies, quasars, and the standard model of the big bang, or the bizarre objects of gravitational collapse, neutron stars and black holes, temperatures are often measured in units of 10^9 K. The maximum realizable temperature in nature, limited by the onset of hadron creation from the available energy, have been estimated to be of the order of 2×10^{12} K. Experimental nuclear temperatures from nucleus–nucleus collisions in accelerators of over 1×10^{12} K have been reported.

The remainder of the article is devoted to a number of illustrative examples of the properties of physical and chemical systems at temperatures higher than 1000 K. Since the dominant physicochemical properties, and the methods of temperature generation and measurement, vary considerably between 1000 and 10^{12} K, the temperature range is subdivided for purposes of discussion as follows: 1000–5000 K; 5000–10000 K; 10 000–50000 K; and more than 50000 K.

The Temperature Range From 1000 to 5000 K

In the 1000 to 5000 K temperature range, at a pressure of one bar (= 0.1 MPa, ≈ one atmosphere), all condensed phases (solids and liquids) become unstable against a gaseous phase. There are no known solids stable at one bar above 4300 K, the approximate melting point of a mixture of tantalum carbide and hafnium carbide. There are stable liquids over the entire range, although they have not been studied extensively. The normal boiling point (1 bar) of tungsten for example, is about 6150 K. This temperature range is one of the most significant for high-temperature science and technology. It is particularly important to the processing and performance of present day high-performance materials, such as high strength alloys, advanced ceramics, high-temperature structural composites, semi- and superconductor materials – as well as traditional materials, such as steels, aluminum, and other metals and refractories. Modern technologies, such as chemical vapor deposition used in semiconductor production and the relatively new area of diamond films, rely heavily on processes occurring in this temperature range.

Although the intermolecular forces responsible for the stability of solids and liquids begin to weaken as the temperature is increased from 1000 to 5000 K, chemical valence forces that bind atoms into molecules are still of considerable importance in the gas phase. In fact, complex molecular species that are unstable at room temperature have been found under conditions of equilibrium in high-temperature vapors.

In the late 1940s and early 1950s, there was considerable controversy over the correct value for the heat of sublimation of graphite to gaseous atoms; i. e., the heat of the reaction

$$C \, (\text{graphite}) \rightarrow C(g) \, .$$

The sublimation of graphite had been studied experimentally by a number of investigators, and the heats of sublimation obtained from seemingly good data tended to cluster around one of two values – 140 kcal/mol or 170 kcal/mol. The controversy was settled in favor of the higher value when the composition of the equilibrium vapor over graphite at temperatures between 2400 and 2700 K was analyzed directly with a mass spectrometer. The vapor was found to contain not only gaseous carbon atoms, $C(g)$, but also $C_2(g)$ and $C_3(g)$ gaseous molecules. In fact, for this temperature range, the concentration of $C_3(g)$ molecules in the vapor is roughly six times that of carbon atoms (and continues to increase up to the graphite triple point, where solid, liquid, and gas are all in equilibrium, at a pressure of about 100 bar and a temperature of approximately 4500 K). The low value for the heat of sublimation had been calculated from experimental data, with the assumption that the vapor consisted almost entirely of atomic carbon.

Through the use of the high-temperature mass spectrometer, many complex gas-phase molecules have been discovered. For example, about one-third of the vapor molecules in equilibrium with molten table salt at 1200 K are in the form of the dimer, Na_2Cl_2, with the remainder in the form of the monomer, NaCl. Electron diffraction studies in the vapor phase and other spectroscopic evidence show that the dimer molecules are planar tetragons of the form

$$Na^+$$

$$Cl^- \qquad\qquad Cl^-$$

$$Na^+$$

The first electron diffraction studies that were carried out were interpreted on the assumption that the vapor was entirely monomeric. The calculated Na–Cl bond distance was significantly larger than that obtained from later microwave spectra. The discrepancy was resolved based on the fraction of dimer in the vapor, as determined by mass spectrometry. These types of complex molecules have the ability to combine with other metal halide, oxide, sulfide, etc. species to form molecules of even greater complexity, e. g., $MnFe_2Cl_8$. The exceptional thermodynamic stability of these complexes allows their use as vapor transport agents in modern metallurgical processes. Similar schemes are typically used in various forms of chemical vapor deposition (CVD) in the burgeoning semiconductor industry. CVD generally exploits the differing stabilities/reactivities of molecules over a thermal gradient to provide highly controlled film deposition. A relatively recent discovery of unusually stable clusters of atoms or molecules (e. g., C_{60}) has been made, using techniques similar to those applied to complex species of the type just mentioned. These clusters represent an intermediate state of matter between the gas and condensed phase. The formation of smoke from a fuel-rich hydrocarbon flame is a common example, where high-temperature cluster formation occurs as an intermediate step. The research area of cluster physics and chemistry is currently developing at a rapid pace.

There are many ways to achieve temperatures in the 1000 to 5000 K range. The one most commonly known is electrical resistance heating (used in both the electric toaster and the tungsten-filament light bulb). Other traditional techniques include combustion processes, induction heating, electron bombardment, and radiant heating, including solar furnaces and arc imaging furnaces. In recent years, high-power laser beams have been used as heat sources, particularly where localized heating and short heating times are desired. A serious difficulty in high-temperature research is the chemical reactivity of container materials with the system of interest. These reactions are generally prevalent at high temperatures, and have increasing rates (i. e., speed) as the temperature is raised. Various methods of "containerless" heating have been used in recent years. Electromagnetic levitation/heating and electrical pulse heating of conductors have been used to generate temperatures to above the melting temperature of tungsten (3695 K). Extending these methods using microgravity levitation and auxiliary heating is a promising future technique. The use of pulsed laser heating has also been used to alleviate the containment difficulty by providing heat only to a localized hot spot on the surface of the sample. The use of thermocouples for high-temperature measurement is also complicated by the materials interaction problem, as well as the tendency of insulating sleeves to be conductive, particularly above 2500 K. The most common temperature measurement method for the 1000 to 5000 K range is optical pyrometry, calibrated against a tungsten strip lamp. This method is the basis for the high temperature region above 1234.93 K of the ITS-90.

The Temperature Range From 5000 to 10 000 K

In the temperature range from 1000 to 5000 K, stable molecules exist in the gas phase, and for some systems, larger and more complex vapor molecules appear as the temperature is increased. However, as temperatures are raised above 5000 K, a point is reached where no molecules at all can exist. This temperature has been estimated to lie somewhere between 8000 and 12 000 K.

Consider nitrogen, for example. At 5000 K and a pressure of one bar, N_2 molecules predominate in the gas phase, much as they do at room temperature, but with a higher average translational energy, and more importantly, higher vibrational energies; about 1% of the molecules will have dissociated into nitrogen atoms because their internal vibrational energy exceeded the N–N bond strength. At 6000 K, more than 10% of the original nitrogen (N_2) molecules will be split into atoms. At 10 000 K, nitrogen atoms will predominate in the vapor phase, with molecules accounting for less than 1% of the particles present.

Temperatures above 5000 K are both difficult to achieve and difficult to measure. Temperature is generally defined by the equilibrium established among the accessible states of the system under consideration and the resulting average of the distribution of energy (*equipartition* of energy). However, the methods available for bringing systems to temperatures in excess of 5000 K often do not result in thermal equilibrium; i. e., evidence exists that all parts of the system, on the atomic and molecular level, may not be at the same temperature.

Flames have been used to achieve temperatures in the lower part of the range of interest. A one-bar carbon subnitride (C_4N_2)–oxygen flame, burning to carbon monoxide and nitrogen, yields temperatures around 5260 K. Theoretical flame temperatures are calculated from tabulated thermodynamic data by assuming that the heat released in the combustion reaction goes into raising the temperature of the combustion products. The calculations assume that equilibrium is achieved, and depend on a knowledge of the nature of the product species. The reaction zone of aflame is usually only a steady state region, and can often be far from equilibrium. The simplest frequently used method for measuring flame temperatures is the "spectral-line reversal" method, which makes use of the fact that certain chemical species at high temperatures emit visible light at well-defined wavelengths. If light from a calibrated continuum source (preferably a thermal equilibrium light emitter) is passed through a flame containing sodium vapor, the two sodium spectral lines (589.0 and 589.6 nm) will appear dark in absorption if the background source is at a higher temperature than the flame, and bright in emission if the background source is at a lower temperature than the flame. When the (known) source temperature matches the flame temperature, the lines disappear. For the highest-temperature flames, however, well-calibrated sources are not generally available.

Spectral methods to measure gas temperatures that do not require a calibrated source have been developed, but the measurements often reflect the temperature of abnormally excited

(i. e., having nonequilibrium energy distribution) species rather than mean average temperatures. Observed spectral lines reflect the transition of a chemical species from one allowed energy level to another. If two spectral lines of the same species (molecule, radical, atom, ion, etc.) are excited at different energy levels, E_1, and E_2, and if the respective transition probabilities are P_1 and P_2, then the equilibrium ratio of the intensities of the two lines will be given by the ratio of Boltzmann factors, $P_1 e^{-E_1/kT}/P_2 e^{-E_2/kT}$, defining the distribution between energy levels 1 and 2, where k is the Boltzmann constant, T is the thermodynamic temperature (K), and e is the natural logarithm base. On a quantum scale, the overall Boltzmann distribution, which defines the equilibrium populations among system energy levels, is a definition of temperature. Those systems which do not obey the distribution are considered to be nonequilibrium by definition. The use of the Boltzmann distribution to derive temperatures for distributions, such as the state population inversions characteristic of lasing, have led to the concept of negative temperatures. To avoid infinite energies being associated with creating such a condition, the state spectrum must have an upper limit, the states at negative temperature must be inaccessible to the rest of the system states which are at a positive temperature, *and* the states at negative temperature must be in thermal equilibrium with the appropriate Boltzmann distribution for the assigned negative temperature. Often, when different degrees of freedom of a system have different equilibrium distributions, distinct temperatures can be assigned to the various distributions, particularly if they form isolated subsystems, as in the above example. If the transition probabilities are determined by measuring spectral-line intensities in sources of known temperature, then the ratio of the spectral-line intensities in a source of unknown temperature can be used to calculate the temperature. Use has been made of spectra arising from electronic transitions in atoms (such as iron) and from rotational and vibrational transitions in diatomic molecules (such as OH or CN). Measured rotational temperatures of OH in the reaction zone of flames have ranged from 5400 K in a one-bar oxyacetylene flame to 10000 K in a low-pressure oxyhydrogen flame containing a trace of acetylene. These temperatures are much higher than the theoretical flame temperatures, and probably reflect the fact that the OH radicals can be formed in a highly excited nonequilibrium rotational distribution.

Temperatures up to 10000 K (and even higher in some cases) can be achieved, for times on the order of microseconds, in shock waves. A shock wave can be produced in a so-called shock tube by the sudden bursting of a diaphragm separating a gas at low pressure from one at high pressure. A compression wave, which rapidly steepens to form a shock front, is generated in the low-pressure gas, and temperatures at the front rise abruptly. Shock waves are always produced in explosions or detonations. The energy released at the shock front goes immediately into increasing the kinetic energy (or mean square speed) of the gas molecules, which is equivalent to increasing translational temperatures.

For a gas at equilibrium, almost by definition, translational, rotational, vibrational, and electronic temperatures will all be equal. At the shock front, however, there may be some delay in achieving equilibrium between the translational energies of the molecules and their internal energies. Equilibrium is brought about by the interconversion of energy from translational to internal modes during molecular collisions. This process of energy equilibration is known as "relaxation", and has been studied for a number of gases in shock tubes. Generally, transfer of vibrational and electronic energy is relatively slow, which can lead to these internal state systems becoming isolated. Dissociation of molecules to equilibrium concentrations of atoms and ions has also been studied in shock tubes and is mechanistically complex, depending not only on collision frequency, but also on the energy of the colliding particles and the distribution of that energy between translational and internal modes. Equilibrium temperatures are, of course, lower than initial translational temperatures.

Direct laser heating of surfaces is used for industrial processes such as welding and cutting, yielding industrial process temperatures estimated to be above 5000 K. Surface interaction with very high power infrared or visible lasers ($> 10^7\,\mathrm{W/cm^2}$), or with near-UV lasers, produces a luminous plume which contains, in addition to atoms and molecules, an abundance of ions having translational energies which correspond to temperatures of 8000 K and above. Use of lower power levels, with visible or infrared lasers, can produce approximately equilibrium high temperatures. This appears to be true even for short pulse-length ($\approx 10\,\mathrm{ns}$) lasers. The bulk temperature of the surface can vary from near ambient for very low power levels to values above 5000 K. The majority of reported measurements have used spectroscopic probes, which are generally only sensitive to excited species; mass spectrometric determinations have found that vaporization of neutral species with apparent thermal equilibrium among emitted species can be the major process.

Gas discharges and electric arcs have been used for studies between 5000 and 10 000 K, but since they are also capable of achieving higher temperatures, they are discussed in the next section.

The Temperature Range From 10 000 to 50 000 K

At 10 000 K, atoms will predominate over molecules in most gases at equilibrium, and concentrations of ions can be appreciable. As temperatures are increased further, ions and free electrons become the dominant species, and gases that were electrically insulating at lower temperatures become electrically conducting. In nitrogen at one bar pressure, for example, the concentrations of nitrogen atoms, singly ionized nitrogen ions (N^+), and free electrons become equal at approximately 14 000 K. Doubly ionized nitrogen (N^{2+}) begins to form at about 20 000 K, and N^{3+} becomes significant at about 34 000 K. By 30 000 K, neutral nitrogen atoms have virtually disappeared.

The devices most commonly used to produce temperatures in the 10 000 to 50 000 K range are known generically as electric discharge devices. Examples include dc glow discharges, induction and microwave plasmas, electric arcs, and plasma jets or torches. The devices differ in many ways, but they all have a source of electrical power and a coupling mechanism for delivering the power to a plasma environment (i. e., a glowing environment characterized by the presence of positive ions and free electrons). The nature of the plasma produced (e. g., electron and gas temperatures, degree of gas ionization, deviations from equilibrium among the species present, electric field strengths, gas density, ratio of plasma volume to device volume) is highly sensitive to the particular device and its mode of operation. The basic principle of the electric discharge, however, is that electrons accelerated by large electric fields transfer energy from the power supply to the gas under study. Collisions between accelerated electrons and gas molecules result in the production of ions and dissociated free radicals or atoms. Laser-based welding systems produce a laser-opaque plasma by direct vaporization/ablation of the surface, followed by further heating of the plasma by direct transfer of laser energy to the plasma.

The mean kinetic energies, and hence the temperatures, of species present in glow discharge and laser-generated plasmas can be very different. Electrons can reach 30000 K while neutral gas molecules in the same discharge can be present effectively at 300 K. In a hydrogen discharge, both atoms and molecules can have velocity distributions characteristic of room temperature, while the fraction of atoms present is that which would be expected for hydrogen at equilibrium at 4000 K. It is possible, however, to generate plasmas in which there is at least local thermal equilibrium – i. e., regions where the concentrations of atoms, ions, and electrons, and the distribution of energy among particle vibrations, rotations, and electronic excitations, can be described by a single temperature. Such equilibrium plasmas are generally formed at higher pressures (one bar and above) than for plasmas where there is a wide divergence in particle temperatures.

Temperatures Over 50 000 K

As temperatures are increased beyond 50000 K, atoms lose more and more electrons in a stepwise fashion until a stable gas at 10 million to about 10 billion kelvins consists of bare nuclei and free electrons. It is this type of plasma environment that exists in the cores of "normal" stars and within the fireball of a nuclear explosion. It is also the environment of the nuclear fusion reactors that are looked upon as possible power sources of the future.

The stability of atomic nuclei increases rapidly with atom mass from hydrogen (whose nucleus is a proton) to neon (atomic mass 20), peaking at iron (atomic mass 56). The fusion of lighter nuclei to form heavier nuclei therefore results in a high release of energy. For technical and economic reasons, the combination of deuterium and tritium nuclei (heavy hydrogen nuclei with atomic masses of 2 and 3, respectively) to form a 3.5-MeV α particle (a helium nucleus, atomic mass 4) and a 14-MeV neutron (atomic mass 1) is still considered to be the best reaction for generation of fusion power. The deuterium and tritium (D–T) have to be heated to a temperature between about 50000000 K and 1000000000 K (depending on the plasma density), which requires a large energy input; clearly if a fusion reactor is to be practical, there must be an energy payback (i. e., the energy output must exceed the energy required to build and operate the system). The deuterium and tritium nuclei formed in the plasma will of course be positively charged, and therefore mutually repulsive. Both high temperatures and plasma confinement schemes are needed to promote sufficient closeness of approach between nuclei so that fusion can occur.

Current research centers around two quite different techniques. The more traditional is electromagnetic/electrostatic confinement, in which strong magnetic and/or electrical fields are used to confine and heat a fusion plasma. Recent successes with tokamak-type devices have brought this method close to "break-even", as evidenced by the detection of very high neutron fluxes characteristic of fusion reactions. The other method is inertial confinement, in which a very high-energy laser beam is divided, and the resulting beams are all directed at a tiny sphere containing a D–T mixture. Laser impact simultaneously implodes and heats the pellet to (near) fusion temperatures. The NOVA laser system at Lawrence Livermore Laboratory is a major U.S. facility of this type. Currently, both methods are within about a factor of two of the 5×10^{14} keV cm^{-3} s temperature × density × confinement-time product, which represents the break-even fusion energy objective. The D–T ignition point is about an order of magnitude higher. Both techniques are reaching the point of "scientific feasibility" – where fusion plasmas can be expected to be achieved in the laboratory with realistic apparatus. The question of economic viability of fusion power still hinges on a variety of materials science issues, such as development of an acceptable "first wall" material and reactor construction materials to withstand the expected mechanical, thermal, and radiation fatigue problems. A variety of approaches are being considered, representing a rapidly growing area of high-temperature research and development.

Temperatures even hotter than the several billion kelvin fusion reaction temperatures of highly evolved class O stars are the province of the interiors of supernova explosions, violent galaxies, quasars, and gravitationally collapsed objects such as "White Dwarf" stars and neutron stars. Above 10 billion kelvins, the plasma requires a relativistic treatment to describe it. At the pressures and temperatures present inside neutron stars, the "electron pressure" limit is exceeded, and electrons are forced into nuclei to create an object of neutrons. These conditions are of keen interest to theoreticians, and are experimentally accessible in high energy nucleon-colliding particle accelerators. Temperatures in collider events are now nearing nature's limits, where additional energy goes into creating matter (hadrons), rather than increasing the temperature of the nuclear fluid.

Although many scientific and technological areas have benefitted greatly from high temperature research, even more interesting and challenging problems still remain in each of the temperature regions.

See also: Alloys; Combustion and Flames; Geophysics; Levitation, Electromagnetic; Maxwell–Boltzmann Statistics; Shock Waves and Detonations; Stellar Energy Sources and Evolution; Superconductors; Thermodynamics; Thermometry; Ultrahigh-Pressure Techniques; Vapor Pressure.

Bibliography

Bibliography on the High Temperature Chemistry and Physics of Materials (M. G. Hocking and V. Vasantasree, current eds.) Vol. 33. Imperial College of Science and Technology, London, 1989.

M. Bass, "Laser Heating of Solids", in *Physical Processes in Laser–Materials Interactions* (M. Bertolotti, ed.), pp. 77–115. Plenum, New York, 1983.

B. D. Blaustein (ed.), *Chemical Reactions in Electrical Discharges*. American Chemical Society, Washington, DC, 1969.

D. W. Bonnell, R. L. Montgomery, B. Stephenson, P. C. Sundareswaran, and J. L. Margrave, "Levitation Calorimetry", in *Specific Heat of Solids* (C. Y. Ho and A. Cezairliyan, eds.), pp. 265–298. Hemisphere Publishing Corp., Washington, DC, 1988.

L. E. Brus, R. W. Siegal, *et al.* "Research Opportunities on Clusters and Cluster-Assembled Materials: A Department of Energy, Council on Materials Science Panel Report", *J. Mater. Res.* **4**(3), 704–736 (1989). Brus and Siegal were panel cochairmen.

G. Chaudron and F. Trombe (eds.), *Les Hautes Températures et leurs utilisations en physique et en chimie*, Vols. 1–2. Masson et Cie Editeurs, Paris, 1973.

W. A. Chupka and M. G. Inghram, "The Thermodynamics of Carbon Molecules as Determined in the Mass Spectrometer". *Mem. Soc. Roy. Set. Liege, Quatrieme Ser.*, **15**, 373–377 (1954).

Committee on High Temperature Science and Technology, *High Temperature Science: Future Needs and Anticipated Developments*. National Academy of Sciences, Washington, DC, 1979.

H. P. Furth, "High-Temperature Plasma Physics", in *Physics in a Technological World. XIX General Assembly International Union of Pure and Applied Physics*, pp. 315–345. American Institute of Physics, New York, 1988.

I. Glassman, *Combustion*. Academic Press, New York, 1977.

R. Goulard (ed.), *Combustion Measurements: Modern Techniques and Instrumentation*. Academic Press, New York, 1976.

W. Greinerand H. Stocker, "Hot Nuclear Matter", *Sci. Am.* **252**(1) 76–87 (1985).

J. W. Hastie (ed.), *Characterization of High Temperature Vapors and Gases*. Government Printing Office, Washington, DC, 1979.

J. W. Hastie, *High Temperature Vapors: Science and Technology*. Academic Press, New York, 1975.

J. W. Hastie (ed.), *Sixth International Conference on High Temperatures: Chemistry of Inorganic Materials*. Humana Press, Clifton, NH, 1990. The conference was held in 1989.

High Temperature Technology (ILJPAC International Symposium on High-Temperature Technology). Butterworth, Washington, DC, 1964.

C. Kittel, *Thermal Physics*. Wiley, New York, 1969.

G. N. Lewis, M. Randall, K. S. Pitzer, and L. Brewer, *Thermodynamics*. McGraw–Hill, New York, 1961.

B. W. Mangum, "Special Report on the International Temperature Scale of 1990: Report of the 17th Session of the Consultative Committee on Thermometry", *J. Res. Natl. Inst. Std. Technol.* **95**, 69–77 (1990).

J. L. Margrave, ed., The Characterization of High Temperature Vapors. Wiley, New York, 1967.

G. de Maria and G. Balducci (eds.), *Fifth International Conference on High Temperature and Energy-Related Materals*. Pion Ltd., London, 1989. The conference was held in 1987.

Physics through the 1990's, Vols. 1–9. National Academy Press, Washington, DC, 1986

H. Preston-Thomas, "The International Temperature Scale of 1990 (ITS-90)", *Metrologia* **27**, 3–10 (1990).

Symposium on High Temperature and Materials Chemistry, LBL-27905. Lawrence Berkeley Laboratory Materials and Chemical Science Division, Berkeley, CA, 1989.

Temperature: Its Measurement and Control in Science and Industry. American Institute of Physics, New York, 1941–1989. The volumes for 1941–1971 are proceedings of the Symposium on Temperature; the volumes for 1982–1989 are proceedings of the International Temperature Symposium.

E. T. Turkdogen, *Physical Chemistry of High Temperature Technology*. Academic Press, New York, 1980.

S. Woosley and T. Weaver, "The Great Supernova of 1987", *Sci. Am.* **261**(2), 32–40 (1989).

K. M. Young, "Summary Abstract: New Diagnostic Approaches for High-Temperature Plasmas", *J. Vac. Sci. Technol. A* **6**(3), 2061–2062.

History of Physics

L. Pyenson

Physics in the sense of Aristotle, the study of the material representation of natural phenomena, may be identified in all civilizations and cultures. If the historian were to write a history of physics based on Aristotle's definition, he would reveal many unfamiliar cosmologies that provide unusual taxonomies for natural knowledge. This essay has a more modest focus. It concerns physics understood as a scientific discipline with a recognizably modern syllabus and institutional locus. Beyond preliminary remarks the following text does not address the history of physical thought before the crystallization of the modern discipline of physics, around 1830. The restriction may be deceiving, for even over the past six generations physics has been expressed through ideologies and institutions that no longer exist. Physics in the world we have just lost is a mixture of recognizable equations, slightly unusual life-styles, and unfamiliar methodological preoccupations. In focusing on modern physics it must be remembered that many of the social forms and philosophical prejudices associated with physics during the past 180 years have roots in previous centuries; indeed, one of the most exciting problems in the history of science has long been the transformation in physical world views between 1600 and 1800. Notwithstanding the traditional focus of interest of historians of physics, reconstructing the recent past provides a much-needed perspective on the decisions and pressures that confront physics today, and in part for this reason the discussion here will concentrate on developments in the late nineteenth and the twentieth centuries.

Given the persistence of classical learning in medieval Europe and the continual stimulation provided by Islamic institutional developments, it is difficult to speak of a renaissance in natural knowledge in the same way that the term has come to be applied in literature and art. To a considerable extent natural philosophy in the sixteenth and seventeenth centuries was a reaction against classical wisdom and the established institutions – the universities – that consecrated it. The vehicle for circulating new ideas beyond the university was the printed book. By 1500 the book had ceased to be an innovation concerned only with disseminating the classics, and it interacted with the nonverbal tradition of the mechanical arts. Engineers, navigators, cartographers, engravers, watchmakers, surveyors, and architects availed themselves of the new medium and contributed to it in increasing numbers. The printing press was also open to the hermetic tradition of magic and occult learning which influenced the labors of Johannes Kepler, Robert Hooke, and Isaac Newton. Finally, the press promoted new learned disciplines, such as chemistry, that were in part based on a pedagogical tradition beyond the universities.

Geometrical astronomy was a high art when the posthumously published work of Nicholas Copernicus suggested that a heliocentric astronomical model might provide a simpler picture of the heavens, a point of view that by 1600 was widely discussed throughout northern Europe. When in that year Giordano Bruno died at the stake for his Copernican heresies, however, only about a dozen writers had come out in print against geocentrism. It was left to Galileo, at the beginning of the seventeenth century, to invent modern mechanics and physical astronomy. Although he taught the traditional wisdom at several universities. Galileo limited his own classical inspiration to Archimedes. Instead of following any master, he urged reasoning through experimentation. Galileo was best at synthesizing numerous unrelated phenomena, and for this task his superb physical intuition allowed him to set aside irrelevant, but persistent, experimental irregularities. Galileo's work is usually designated as the beginning of the "scientific revolution". During the seventeenth century intense concern with creating a unified picture of the physical world was reflected in the algebraic interpretation of mechanics and the final acceptance of the differential calculus. At the same time, new experimental and inductivist methods were explored by the earliest scientific societies. Conventional disciplinary lines became blurred as new mathematical and experimental approaches slowly penetrated universities at Leiden, Jena, and Edinburgh. The seventeenth-century natural philosopher was different from a scientist. He sought to investigate the entire natural world. At the same time, he navigated currents that we would call nonscientific. The natural philosopher might equally be a magus casting horoscopes, a hermeticist believing in occult traditions, or a divine obsessed with biblical prophecy.

Isaac Newton was an exemplary seventeenth-century natural philosopher. Frank Manuel has argued that Newton acquired Puritan religious sympathies during Oliver Cromwell's Protectorate and that his later religious convictions, personal temperament, and scientific prejudices were conditioned by an early fixation on his mother. Newton entered Trinity College, Cambridge, as a poor subsizar who was required to perform menial tasks. It appears as if throughout his life, even after having been knighted and having served as Master of the Mint and President of the Royal Society, he felt insecure about his low birth. At the urging of his friends, Newton set down his three laws of motion and derived Johannes Kepler's three laws of planetary motion from the gravitational attraction of the sun. The publication of these results in 1685 provided conclusive evidence against Cartesian celestial vortex mechanics, which had been unable to account for Kepler's laws. The mature Newton saw his youthful work as the prelude to more important labors in theology, alchemy, and speculation about the ultimate composition of matter. He fretted over biblical numerology and believed that he had discovered a way to produce "philosophical mercury".

Newton's endeavors did not directly result in the creation of the modern *physics* discipline. At the end of the eighteenth century, mathematicians transformed Newton's geometrical representation of mechanics into more familiar expressions in differential and integral calculus. The experimental study of matter was refined by mineralogists, apothecaries, and especially professional chemists. Observation of nature was the business of anatomists, botanists, geologists, and astronomers. Engineering was the product of a largely independent tradition of mathematical practitioners – those who had for many generations designed and built optical systems, navigational equipment, and chronometers. Distinct from these activities, Newtonian natural philosophy around 1800 constituted speculative preoccupation with the essence of physical substances such as heat and light. For example, the followers of Pierre Simon de Laplace attempted to interpret physical processes through Newtonian mechanics by the hypothesis of short-range forces. Late-eighteenth-century natural philosophers who called themselves physicists did study mechanics, light, heat, acoustics, and occasionally electricity, but few systematic treatments of physics as a unified field of knowledge existed before the second third of the nineteenth century.

Modern physics may be traced directly to curriculum innovations at several German universities around 1830. The most influential program was founded at the University of Königsberg by Carl Gustav Jacobi in collaboration with Carl Neumann and the astronomer Wilhelm Friedrich Bessel. The Königsberg physics seminar included training in both theory and experiment. Partial differential equations from Joseph Fourier's new science of heat provided the theoretical language of physics; precision measurement, itself not a critical innovation, spanned a wide range of physical phenomena and for the first time took account of significant figures. Within a generation the Königsberg innovation had spread to universities throughout Prussia and the other German states. The high value accorded all branches of scholarly research in Germany provided incentives for talented youth to pursue physical science despite the disappointingly low salaries often paid university professors. At mid-century, education ministries of the German states were recognized leaders in processing large numbers of students. However autocratic these authorities may seem to us now, at the time they made competent and even brilliant choices for professorships in physics and mathematics, and they managed to provide at least some funds for the earliest laboratories devoted entirely to physics.

After mid-century the concept of a physics discipline was conveyed from Germany to France and Great Britain, where it interacted with national traditions. French and British investment in physics manpower and fixed assets became comparable with that of Germany by 1900, but the theoretical part of the discipline assumed a distinct character in each of the three countries. Since the eighteenth century France had maintained a strong tradition of mechanics, a subject that united astronomy, analysis, and engineering. A good deal of theoretical physics in France was treated by mathematicians in the discipline of mechanics. In Britain university physics was not clearly distinguished from either mathematics or chemistry until the end of the century.

At the beginning of the twentieth century neither France nor Great Britain had a system of higher education to match that of Germany, and it was above all in universities that physics flourished. Because physics was seen as abstract learning, it occupied a natural place in the curricula of elite institutions of higher education. In nineteenth-century Britain and Germany (although not in France), the ordinary practitioners of physics came from the upper classes in proportionately greater numbers than did other physical scientists. Physicists were educated in the best secondary schools and were under less pressure than chemists or geologists to practice their profession in industry or commerce. It might even be maintained that the life-style of nineteenth-century physicists resembled that of classical philologists more closely than that of chemists. The great democratization of higher education in the United States and the Soviet Union during the first part of the twentieth century significantly transformed the class structure of the physics profession. Vestiges of the former elite status of physicists may be found in the disproportionate influence that physicists until recently exerted on university affairs and in the carefully cultivated image of the physicist as an omnivorous intellectual.

It would be a mistake to imagine that elite nineteenth-century physicists devoted themselves exclusively to a search for synthetic representations of fundamental physical laws. The research of some was directed to a deeper understanding of properties of certain chemical elements or minerals, to routine collection of meteorological data, or to technical improvements in electrical or mechanical devices. The published work of other physicists was indistinguishable from that of physical chemists, applied mathematicians, or engineers. At least in terms of the problems that came to dominate physics in the early years of the twentieth century, however, three threads may be identified in nineteenth-century physical discourse: electrodynamics, thermodynamics, and kinetic theory. During the last half of the nineteenth century, each comprised a circumscribed, though not entirely self-contained, set of theories and experiments.

Although various theories of electric current had been explored by physical scientists of the Laplacian school in the period around 1830, electrodynamics emerged as a science when William Thomson (Lord Kelvin) and James Clerk Maxwell succeeded in expressing Michael Faraday's intuitive picture of an electromagnetic field in mathematical language. Maxwell formulated a set of partial differential equations to describe the electromagnetic field. For over a generation his achievement was not clearly recognized by physicists on the continent, who continued to use action-at-a-distance formulations of electromagnetism in the tradition of André-Marie Ampère and Wilhelm Weber. Heinrich Hertz, who first demonstrated the existence of electromagnetic waves in 1887, was instrumental in gaining acceptance for Maxwellian electrodynamics in Germany. Hertz argued that the mathematical form of physical laws, in particular those of mechanics and electrodynamics, did not have to be based on physical models. In the middle 1890s Hendrik Antoon Lorentz reduced Hertz's formulation of Maxwell's equations to the now-familiar expressions.

Thermodynamics, a second important area of nineteenth-century physics, emerged from the disintegration of the caloric theory (where heat was viewed as a fluid) in part through the work of Joseph Fourier and Sadi Carnot. The conservation of energy, a principle independently announced at nearly the same time by four scientists, formed the basis of the first law of thermodynamics. Beginning around 1850, Rudolf Clausius and others elaborated a general criterion for physical processes in the second law of thermodynamics. To express the second law succinctly Clausius formulated a new quantity, *entropy*. The second law of thermodynamics was of great practical advantage for chemists. The change in entropy for any conceivable chemical reaction could be calculated easily once the heats of formation of the initial and final products were known. During the last third of the nineteenth century, some chemists such as Marcellin Berthelot refused to acknowledge the validity of the second law, and they proposed a variety of alternative rules for explaining reactions. Nevertheless, by 1900 the two laws of thermodynamics seemed as secure to most physicists as Newton's laws and Maxwell's equations.

Kinetic theory was the third important development in nineteenth-century physics. Around the middle of the century many physicists believed that if heat were equivalent to mechanical motion, as the first law of thermodynamics requires, then heat might reasonably be produced by the vibrations and collisions of molecules, the microscopic building blocks of matter that some physicists and chemists assumed to exist but about which few detailed investigations had been carried out. From data on the specific heats of gases and the gas laws, Clausius, Maxwell, and others were able to calculate molecular dimensions. For Maxwell and those who followed, the temperature and heat of a macroscopic quantity of gas was related to the *average* velocity of gas molecules. At the end of the century, Ludwig Boltzmann was able to interpret the laws of thermodynamics through an elastic-sphere model of gas molecules, and Josiah Willard Gibbs formulated a statistical mechanics.

Around 1900 serious doubts arose about the completeness of classical mechanics, and partisans of each of the three threads of nineteenth-century physics offered their specialty as the basis for a new picture of the world. Lorentz, Emil Wiechert, and Wilhelm Wien thought that some modification of Maxwell's equations might provide the basis for all physical laws. Their optimism stimulated much work on the theory of the recently discovered electron, then believed to be the smallest physical particle. Although the so-called electromagnetic view of nature was short-lived, intense and mathematically exacting work on electron dynamics during the period around 1900 sharpened the analytical tools of atomic theorists and conditioned contemporary physicists within the intellectual orbit of German-speaking Europe for the reception of Einstein's special theory of relativity. Indeed, until around 1911 many physicists and mathematicians thought that special relativity was equivalent to the electron theory of Lorentz. The electron theory provided inspiration for Hermann Minkowski's ideas of four-dimensional space-time and David Hilbert's version of the covariant field equations of general relativity. The laws of thermodynamics, too, furnished several physical scientists with a world view called energeticism. According to Georg Helm, Pierre Duhem, and Wilhelm Ostwald, all the laws of mechanics and electrodynamics would emerge from thermodynamics. At the end of the century atomism was held in disrepute by some physicists, for it postulated physical objects that could not be observed; most continued to make use of the atomic hypothesis, nevertheless, in their daily research. Marian Smoluchowski and Jean Perrin then gave reason to current belief.

Albert Einstein accomplished more than any other physicist attempting to unify these three threads of nineteenth-century physics. It is ironic that Einstein, who in some ways had little in common with the professional scientists of his day, has come to be seen as the greatest representative of theoretical physics. He held teaching professorships only at Zurich and Prague between 1909 and 1913. Many physicists felt that Einstein wasted the last 40 years of his life in a search for a field theory to unify electromagnetism and gravitation. Unlike most physicists, he never accepted Niels Bohr's Copenhagen interpretation of indeterminism in quantum mechanics. Einstein reported to his collaborator Leopold Infeld that he was considered an old fool at the Institute for Advanced Study in Princeton, where he worked from 1933 to his death in 1955. Einstein's unpretentious and direct lifestyle contrasted with the formality, extravagance, and chauvinism of powerful professors and laboratory directors. He never sanctioned the use of physics for national ends. In December 1914 he wrote to his good friend Paul Ehrenfest about the war: "The international catastrophe has imposed a heavy burden upon me as an internationalist. In living through this 'great epoch', it is difficult to reconcile oneself to the fact that one belongs to that idiotic, rotten species which boasts of its freedom of will. How I wish that somewhere there existed an island for those who are wise and of good will! In such a place even I should be an ardent patriot". Notwithstanding his "convinced" pacifism, Einstein vigorously opposed the rise of fascism in Europe.

Einstein was born and raised in southern Germany. He dropped out of school at age 15 to join his parents in Milan when his father moved there from Munich. Within a year he sat for entrance examinations in the engineering faculty at a school near Milan that offered advanced instruction in German, the Swiss Federal Institute of Technology in Zurich. He failed the nonscientific part of the examination and was advised to spend a year at a secondary school in nearby Aarau. One year later he entered the Zurich Polytechnic, this time in the section for preparing secondary-school teachers. He studied in Zurich for four years, although upon receiving his teaching diploma in 1900 he could not find a university assistantship in physics. After a lean two years he was hired as a scientific examiner in the Patent Office at Berne. During the years at Berne, Einstein developed the special theory of relativity, as well as his ideas on molecular motion and the quantum theory of light. Soon after he became associate professor of theoretical physics at the University of Zurich in 1909, Einstein was recognized as one of the leading theoretical physicists in German-speaking Europe.

Along with other theoretical physicists Einstein believed that mathematics and experiment provided tools for constructing theories in physics, but that the theoretician required above all physical insight. Einstein's early master in thermodynamical reasoning was Boltzmann, and the laws of thermodynamics provided him with a logical model in much of his early work. The young Einstein felt himself attracted to Ernst Mach's critical approach to physical theories, but he was not a consistent follower of Mach's philosophy. Later, in the period around 1920, he came to appreciate the ideas of Immanuel Kant. From his youth (his family ran various electrotechnical enterprises), Einstein had special affinity for scientific instruments and apparatus, but contemporary experiments seem not to have influenced the course of his theoretical work. In formulating the special theory of relativity, Einstein sought to eliminate the apparent contradiction between classical electromagnetic theory and classical mechanics. He argued that the notions of absolute motion and ether were not essential elements of physical

Fig. 1: Theoreticians. Albert Einstein and Niels Bohr, photographed by Paul Ehrenfest.

reality. Before he settled on the covariant field equations late in 1915, Einstein occasionally wrote that observation might disprove the general theory of relativity. Nevertheless, by the period around 1914 his faith in general relativity became fixed. He wrote to his good friend Michele Angelo Besso in 1914 about an attempt to verify the gravitational deflection of light during a solar eclipse: "Now I am fully satisfied, and I no longer doubt the correctness of the whole system, whether the observation of the eclipse will succeed or not. The reasonableness of the matter is too evident". As he grew older aesthetic criteria from pure mathematics played an ever more important role in his formulation of physical laws.

Einstein's contemporary Ernest Rutherford has epitomized the modern experimental physicist. To our sophisticated eyes his ingenious equipment may seem small and improvised, but using it, Rutherford had a hand in revealing more about the nature of submicroscopic physical

Fig. 2: A Couple of Experimentalists. Jacob Clay and Tettje Jolles, 1906, in Leiden. Both studied with Heike Kamerlingh Onnes. After more than a decade of work on low-temperature physics and the philosophy of science, he became the first physics professor at the Bandung Institute of Technology, on Java. There, in the 1920s, the Clays carried out research into cosmic rays and solar radiation. (Courtesy of Mevr. Nelke van Osselen-Clay.)

reality than perhaps any other experimentalist of the twentieth century. He was an expansive and commanding man, gifted with penetrating physical insight and the talent for demonstrating his intuitions. Around his laboratories in Montreal, Manchester, and Cambridge gathered a brilliant constellation of twentieth-century physical scientists.

Rutherford received early university physics training in his native New Zealand. He went to Cambridge in 1895 when the first choice for a scholarship declined to accept the award. Working under J. J. Thomson at the Cavendish Laboratory, he examined the ionizing radiation

Fig. 3: Master and Disciple. Ernest Rutherford and J. J. Thomson, June 1934.
(Bainbridge Collection, Niels Bohr Library, American Institute of Physics).

produced by x rays, which had just been discovered by Wilhelm Conrad Röntgen at Würzburg. Rutherford quickly showed that the spontaneous radiation discovered in uranium by Henri Becquerel was the same as the ionizing radiation that he had studied, and that the uranium radiation had two components, which he called alpha and beta rays. In 1898 the 27-year-old Rutherford was appointed to a research chair endowed by the Macdonald tobacco fortune at McGill University in Montreal. There he showed that radioactivity was solely a property of atoms, discovered atomic transmutation through radioactive decay, and identified alpha rays as helium atoms with a positive charge twice that carried by an electron. In 1907 he went to fill Arthur Schuster's chair at Manchester University, where in 1911 he demonstrated the existence of atomic nuclei by observing the recoil of alpha particles that had been directed at a thin gold foil. In 1919 Rutherford succeeded his old teacher J. J. Thomson at the Cavendish. During the interwar period Cambridge was radioactive with Rutherford-dominated illuminati.

By the eve of the first World War quantum physics and relativity together with new ray and radioactivity phenomena were widely discussed in German periodicals. The theoretical side of these developments only slowly penetrated to Great Britain, France, and the United States. The acceptance of the relativity theories and quantum physics cannot be summarized easily, but the most important factor in their reception may be that theoretical physics was institutionalized principally in greater German-speaking Europe. Theoretical physicists often sought to synthesize all fundamental laws in what they called a physical world picture. Used by Max Planck, Einstein, and others, the term world picture – Weltbild – indicated a comprehensive structure of the physical universe that was based on a number of fundamental physical principles or prejudices. A physical world picture was considerably more than a model useful for explaining a limited range of phenomena. Theoretical physics was framed in the language of mathematics and it explained or addressed experimental propositions, but many theoretical physicists saw a world picture as the necessary basis of physical theories.

Theoretical physics beyond German-speaking Europe was carried out by and addressed to a broad community of physical scientists. Theoretically-minded physicists in Britain studied to succeed in the Cambridge mathematics tripos, examinations that emphasized mathematics as well as mathematical problems in mechanics. The British physics community showed little interest in quantum theory or the theoretical side of atomic physics. Furthermore, distinguished physicists like Oliver Lodge and Joseph Larmor were hostile to relativity, and they refused to abandon the ether. Many others thought that the principle of relativity itself was nonsense. French theoreticians were trained as mathematicians, lesser copies of Henri Poincaré. They focused for the most part on problems in classical mechanics. Young theoreticians who studied quantum theory and relativity, such as Perrin, Paul Langevin, and Edmond Bauer, were exceptional.

The first World War disrupted physics throughout Europe. No different from men in other professions, H. G. J. Moseley, Karl Schwarzschild, and Friedrich Hasenöhrl marched to their death. Funds and personnel bled away from major centers of research at Manchester, Göttingen, and Leipzig. Dozens of young French physicists, graduates of the Polytechnical School of Paris and junior officers in the army, led hopeless charges against German machine-gun emplacements. German physicists traveling in Allied lands – Erich Hupka passing through Ceylon, Peter Pringsheim visiting Australia – were interned. When Allied forces overran German institutions, Bruno Meyermann and Gustav Angenheister spent the duration under various forms of confinement. Laboratories and observatories – in Lebanon's Beqa'a Valley, on Tahiti, at Tsingtao, in Belgium – suffered violation and ruin. Loyal British subjects – Australia's T. H. Laby and Canada's J. C. MacLennan – volunteered their talents to professional killers. And through it all came brilliant elaborations of general relativity, atomic physics, and meteorology by physicists whom fortune or conscience had granted respite from the slaughter. Only a small flow of information passed between belligerent powers through Niels Bohr's Copenhagen, Paul Ehrenfest's Leiden, and Svante Arrhenius's Stockholm.

The greatest part of quantum mechanics was forged by German-speaking physicists during the years after the war. In 1913 Bohr's atomic model had combined quantum constraints with classical mechanics and succeeded in interpreting the gross features of atomic spectra. Nevertheless, ten subsequent years of atomic model building failed to explain spectral fine

Fig. 4: Astrophysicist in Paradise. Milan R. Štefánik on Tahiti, 1910, for observing Halley's comet. A Slovak educated in Prague, Štefánik moved to France and across the decade before the First World War traveled widely on French astrophysical missions. During the war he commanded Czechoslovak troops under French colors; then he became the first minister of war in Czechoslovakia. (Emil Purghart, ed., *Štefánik vo fotografii. 2: Vydanie* [Orbis, Prague, 1938].)

structure. Arnold Sommerfeld and his students at Munich formulated complicated, although never completely satisfying, atomic models by quantizing the action principle to describe electrons orbiting a central nucleus in precise paths. What we know today as matrix mechanics emerged when Werner Heisenberg abandoned classical models in favor of formalism from a quantum theory of radiation that had been used by Rudolf Ladenburg to explain dispersion of light in gases. By the spring of 1926, P. A. M. Dirac, Max Born and Norbert Wiener, and Erwin Schrödinger had proposed alternative formulations of quantum mechanics, although all were soon demonstrated to be equivalent to matrix mechanics.

In 1927 Heisenberg announced the principle of uncertainty, according to which the product of certain pairs of operators that correspond to physical quantities has as a lower limit a fixed value determined by Planck's constant. This principle later gave rise to a large philosophical literature about modern physics, causality, free will, and human destiny, but at the time many physicists quickly assimilated it as a fundamental limitation on physical knowledge. Paul Forman suggested in 1971 that physicists embraced indeterminism in response to a hostile environment in Weimar Germany. As the nation careened through political and economic crises, many nonscientists developed a strong antipathy to logic, reason, and deterministic science. These highly valued attributes of culture in Imperial Germany were placed at the root of Weimar Germany's troubles. German physicists had long sought to precipitate crises in physics, and during the early 1920s increasing numbers of them hoped for a clean break with the past. A flurry of papers asserted the necessity of a quantum mechanics predicated on abandoning classical principles. It was even argued that energy might not rigorously be conserved in all physical processes. German physicists looked to the uncertainty principle in a perhaps unconscious attempt to appease critics of physical science and to rehabilitate their own image as important contributors to Weimar culture.

If the theoretical underpinning of physics today is a product of nineteenth- and early twentieth-century German scientific culture, contemporary experimental physics has been most strongly conditioned by events in the United States and to a lesser extent in the Soviet Union. Teams were able to mount expensive experimental research with Rutherford at Cambridge, with Enrico Fermi at Rome, and with Frédéric Joliot-Curie at Paris, but it was in the United States that experimental physics emerged on a massive scale. Requiring many experts working for years to construct enormous experimental installations, big physics in America was guided by an elite circle of science brokers who channeled money from many sources into selected research programs. European physicists watched while their American colleagues helped themselves from a groaning table.

Big physics depended on funding from three sources. The first source, industry, had furnished university laboratories in physical chemistry and mechanics in Germany and France since the end of the nineteenth century, and industrial laboratories such as those at Eindhoven and Schenectady cautiously began to fund basic research. Private foundations were a second source of support. Some of these, such as the Teyler's Stichting in Haarlem, the Royal Institution in London, and the dozens of Jesuit congregations, had long played a key role in financing physical research, but the generation of enormous private fortunes late in the nineteenth century gave new meaning to scientific philanthropy. Empires in physics arose from the generosity of plutocrats and robber barons, although it is an open question if results at the most lavishly appointed institutions – the Carnegie Institution of Washington, the Nice Observatory in France, the Nobel Institute in Stockholm – came close to satisfying expectations. A third source of funding was government largesse distributed outside the usual channels that supported higher education. Governments had long favored research at their own institutions, such as the Imperial Institute of Physics and Technology at Berlin and the Naval Observatory in Washington. By the 1920s national supervisory agencies were established in Canada and much of Europe to direct government funds to authors of the cleverest grant proposals. The rise of accelerator physics depended on consummate skill in obtaining money from all three sources, and during the lean 1930s enterprising physicists exploited lucrative technologies generated by their research. Ernest O. Lawrence financed part of his cyclotron research at Berkeley by dispensing cancer radiation therapy.

The second World War and its aftermath produced an awesome respect for the knowledge wielded by physicists. The period 1939–1945 could be called, with some accuracy, the "physicists' war". Although the world pictures of physicists could not directly power the engines of industry or the wheels of commerce, physicists were held to guide basic research that yielded useful gadgets, and they possessed the key – the "secret", many believed – to the enormous energies of the atom. Respect was tempered with official sanctions, and none of the great powers hesitated to discipline physicist heroes for entertaining politically subversive notions. Nevertheless, by the 1950s most reasonable funding requests from physicists were honored. Driven forward by military and space research, physics flourished in the lavish environment of the late 1940s and 1950s.

About mid-twentieth century, national military establishments funded the lion's share of physics-related work. Armies and navies, long interested in purchasing or pioneering new technologies, supported a great deal of pure research. The integrity of pure learning – the proud cornerstone of science in the nineteenth-century German universities – fell to crass expediency. In the United States, the Soviet Union, and elsewhere, expediency has been a way of life for several generations, for what escaped the military's grasp has been caught up by governmental regulation and control. The pattern is familiar to students of physics in France, where for nearly 200 years generals and admirals, sporting academic laurels, have captained major research boards and institutions, and where physics has long tacked to political winds.

The postwar development of big physics resulted in a new kind of physics laboratory. Only during the last third of the nineteenth century had laboratories especially designed for physics research and teaching first been erected. The principal desideratum in their construction was an attempt to provide space for undisturbed measurement: Whole wings were built without iron, and massive piles were fixed to resist vibrations. By around 1910 these laboratories had outlived their design. Vibrations from new trolley lines made precision mechanical adjustments almost impossible. Furthermore, many measurements of spectra, radioactivity, and phenomena using cathode-ray tubes did not require vibration-free, magnetic isolation. Accelerators brought an end to the self-contained physics laboratory. Just as geophysics and astrophysics had left the physics laboratory by 1914, so the apparatus of atomic and nuclear physics moved to new quarters during the 1930s.

In the period after the second World War the accelerator laboratory was much larger than even the biggest prewar government or industrial research installations in physics. Big physics centered around big machines, much as astronomy had long depended on powerful telescopes. Large teams of specialists conducted experiments conceived by several principal investigators. It became common to see papers authored by tens of physicists who were affiliated with a handful of institutions. In addition to resident and visiting physicists, accelerator laboratories employed many service and auxiliary personnel. Rutherford's talented mechanics —Kay, Baumbach, and Niedergesass – were reflected in scores of engineers who built and serviced high-energy accelerators. Routine analysis of experimental data, often recorded in cloud-chamber photographs, was performed by nonphysicists. Housewives and bohemians, modern counterparts of Justus von Liebig's laboratory janitor Aubel, scanned hundreds of thousands of photographs for evidence of elementary-particle interactions. New elementary particles were discovered by nonphysicists who stood near the bottom of the social hierarchy in large national laboratories.

Big physics contributed to increasing estrangement between theoreticians and experimentalists. In the 1950s Enrico Fermi was sometimes held to be the last physicist at ease equally with experiment and theory. During the course of the twentieth century increasingly fewer physicists mastered both the yin and the yang of their discipline. Twentieth-century theoretical physicists found that they could speak more easily with mathematicians and theoretically minded chemists and astronomers than with other physicists interested in experimental

Fig. 5: A Modern Coeducational Teaching Laboratory, 1893–1894. West Physics
Building, University of Michigan. Built with high ceilings and adequate plumbing,
outfitted with a spectroscope and variable-resistance box, such a laboratory would not
have seemed out of place for elementary undergraduate instruction as late as the 1960s.
(Physics Department, University of Michigan.)

problems. Experimentalists drew closer to nuclear chemists and those scientists interested in
designing equipment using optical, electrical, and vacuum technologies. The unity of experiment and theory continued to supply the foundation for physics education after the war. The
two were indeed combined by many physicists who undertook research in physical optics,
solid-state electronics, and low-temperature phenomena, but the physics profession nevertheless continued to drift even farther into theoretical and experimental camps.

Modern physics has so far been described within the developed countries of Western Europe and North America. This emphasis should not be surprising, for English, German, and
French cultures ruled over intellectual activity through the first third of the twentieth century, and their impact in science has survived even when the physical presence of military
garrisons and cultural missions has disappeared. At the same time, by the beginning of the
twentieth century indigenous physics had developed in perhaps a dozen regions previously
dominated by a foreign scientific culture. By around 1920 the discipline of physics was taught
and successfully practiced in popular languages in Poland, Finland, Norway, Hungary, and
Flanders. Among those who issued from emerging twentieth-century physics communities in
Europe were Leopold Infeld, Gunnar Nordström, Vilhelm Bjerknes, Leo Szilard, and Marcel
Minnaert.

Fig. 6: Physicists Preparing the Big Eye. Enrique Gaviola (with beret) and
John Strong inspecting the aluminum coating on the 100-inch mirror for Mt.
Wilson in California, June 1935. Gaviola, a native Argentinian who had
obtained a doctorate at Berlin in 1926, was on leave from the University of
Buenos Aires; in 1940 he became director of the Córdoba Observatory
(Argentina). Strong, at the California Institute of Technology, was the
supervisor of Gaviola's Guggenheim fellowship. (Courtesy of Dr. Gaviola.)

By the eve of the first World War physics had been established in several non-European
cultures. The most unusual implantation occurred in Japan. The Meiji regime that came to
power in 1868 formulated a national policy of adopting Western learning. Japanese science
students went abroad for extended periods of study and, upon returning to Japan, taught their
specialty in a Western language. The first generation of Japanese physicists was predisposed
to acquire learning from a foreign culture, for all had been trained in Chinese as Confucian
scholars. Some Japanese physicists sought to preserve their culture from destruction at the
hands of the Western barbarians. In 1888 the young physicist Nagaoka Hantarõ wrote in
English to Tanakadate Aikitsu, then studying in Glasgow: "We must work actively with an
open eye, keen sense, and ready understanding, indefaticably [*sic*] and not a moment stopping
...There is no reason why the whites shall be so supreme in everything, and as you say,
I hope we shall be able to beat those *yattya hottya* [pompous] people in the course of 10
or 20 years: I think there is no use of observing the victory of our descendents over the
whites with the telescope from *jigoku* [hell]". Nagaoka need not have been alarmed. Japanese
physics had grown to maturity by 1922 when Einstein spent ten weeks in Japan visiting with

Fig. 7: Van de Graaff Generator Constructed by the Department of Terrestrial Magnetism of the Carnegie Institution of Washington. By the end of the 1930s projects were underway to produce particle accelerators many orders of magnitude greater in cost and energy output. This photograph shows the Van de Graaff on 18 May 1935. (Courtesy of the Department of Terrestrial Magnetism of the Carnegie Institution of Washington.)

his colleague and translator Ishihara Jun. Within twenty years the Japanese had research and teaching institutions in physics throughout their Asian and Pacific empire.

When the Japanese conquered European colonies in Asia, they found themselves in possession of laboratories and observatories that in some cases surpassed the standards of establishments in Japan. Particularly impressive were the French Jesuit observatory at Zikawei (Xujiahui) near Shanghai, directed by physicist Pierre Lejay, and the Dutch observatories and faculties on Java, which had been home to physicists Willem van Bemmelen and Jacob Clay, among others. After China and Indonesia freed themselves from colonial rule, each country slowly built upon a heritage of excellence in physical research; significant programs, in place during the 1960s, have since translated into vital enterprises in nuclear, astrophysical, and geophysical specialties.

Physics also came to countries with established, national scientific communities, such as Argentina. Eager to create a technical elite, in 1904 the Argentine educator Joaquín V. González, renovated his moribund provincial University of La Plata. Gonzalez advertised throughout Germany the availability of well-paying faculty positions. He succeeded in attracting Emil Bose, editor of the physics journal *Physikalische Zeitschrift*; Jakob Johann Laub, Einstein's first scientific collaborator; and Richard Gans, a recognized authority on magnetism. The La Plata program declined after Bose died and Laub and other German scientists left for positions elsewhere. Before the first World War there were enough physicists for critical discussions in the Buenos Aires region, and distinguished scientists such as Walther Nernst came for short lecture tours. The foreigners began to publish in Spanish, but they continued to measure professional aspirations by circumstances in Germany. When Einstein lectured at La Plata in 1925 he was well received by a German community whose numbers had been augmented after the war, but Argentine physics was a pale reflection of German forms. The wholesale transplantation of foreign physicists has been repeated many times throughout Africa and Asia among former colonies governed by unenlightened despots and dominated by ravaged economies and oppressive social structures.

The latest discoveries in physics have been made at a time when many physicists in the industrial Western countries are adjusting their professional and personal aspirations to new economic and social conditions. Research in all fields of science and scholarship suffered a major decline in financial support when the postwar economic expansion foundered in the middle 1960s. Fueled by American investment in exotic military and aerospace hardware, inflation and economic stagnation wrought havoc with research budgets. Physics continued to receive relatively greater support than other fields because it could claim to be directly relevant to military and technological goals. Even though only a fraction of young physicists found permanent academic or research positions in their own specialties, unemployment among American physicists remained considerably lower than the national average. Though many universities instituted programs cultivating narrow specialties in the physical sciences, well-trained physicists still commanded the respect of industry, government, and neighboring disciplines, a favored status that they had enjoyed for nearly a century.

Changes in the physics discipline also stemmed from the growth of disillusionment among young people with physics and with scientific world views. In the late 1960s physics acquired the image of an activity estranged from human values. Many critics observed that physicists were working with extravagant or destructive technologies. Others argued that abstract physics contributed to the anomie of industrial societies and to the grinding poverty of former colonies. As evidence that physics was suffering from a profound malaise, critics cited the absence of major synthetic formulations over the preceding generation. They argued that some problems in applied physics – for example, research in nuclear fusion power – were being suppressed by powerful interest groups in favor of other, less promising investigations. An undercurrent of discontent with both the Copenhagen interpretation of quantum mechanics and mechanical explanation in general, blossomed into alternative visions of physical reality. By the 1970s prophets of a new holism – David Bohm, René Thom, Ilya Prigogine, James

Fig. 8: Danish Feminist in Argentina. Margrete Bose (*née* Heiberg), recipient of an advanced degree in chemistry from the University of Copenhagen (*Magistra Scientiarum i Kemi*, 1901) and an alumna of Walther Nernst's Göttingen, here instructing students as a professor of physics at the National University of La Plata, circa 1914. (Courtesy of Walter B. L. Bose.)

Lovelock – contributed to a physics counterculture. The holists married physics to biology in an ecological vision of natural processes: Molecular biology (which by this time had come to rival physics for popular attention), subatomic physics, and the vast cosmos came under the rule of indeterminism, nonlinearity, and intangible emotion.

The holists emboldened nonphysicists in the 1970s to question the cost of big physics, believing that it was the same sort of cultural pursuit as music or art. High-energy physics seemed to confirm an increasingly popular notion that the discipline related less to seeking physical truths than to formulating an aesthetic view of the world. This understanding received encouragement from Thomas Pynchon and Stanislaw Lem, whose metaphorical use of physics continued a tradition upheld earlier by Paul Valéry, Henry Adams, and Robert Musil. Other authors were stimulated by the discovery of new astrophysical objects, and artists and musicians found new means of expression in laser optics and electronic data processing. Physicists tacked to the winds of public opinion. Departing from the moral justifications of Wilhelmian physicists and the arguments of technical utility used by early twentieth-century American physicists, in the late 1960s and early 1970s many physicists pointed to the aesthetic value of their research. Aesthetes like Hubert Reeves, Carl Sagan, and Stephen Hawking retained a large following during the 1970s and 1980s, and their paeans have contributed to public support for large projects – colossal particle accelerators, enormous telescopes, and ambitious space missions.

The holists and the aesthetes contended for hegemony beyond the confines of nation and language. In the last decade of the twentieth century, progress in physics has ceased to be practised only within national institutions financed by the caprice of national resolve. International laboratories and research efforts – CERN, Ariane, and extensive mountaintop observatories – configure the cutting edge of research. Research teams increasingly have partners or entire laboratories in foreign lands, just as professors sport permanent overseas appointments.

International activity in physics has long served national ends. The nineteenth-century commissions on physical standards, for example, helped furnish extensive laboratories in national sectors; a nod from internationalist initiatives around 1900 such as the Nobel, Solvay, and Carnegie foundations and institutes (all of them endowed by bachelor industrialists with extensive experience in more than one nation-state) propelled physicists to the top of a national research infrastructure. International visibility has provided a stern correction to the pretensions of unaccomplished but well-connected laboratory administrators.

For a generation, a picture of a researcher's international productivity and visibility has been available in tabulations of publications and citations published by the Institute for Scientific Information in Philadelphia, a resource now available over the internet. This tool, along with a search engine like Google, allows isolated researchers to keep up with the latest advances in their field. The internet has opened science to all interested parties. From a remote location, one may now conduct experiments and observations using sophisticated and expensive equipment, like particle accelerators and large telescopes.

The dawn of the Information Age, as the computer revolution is called, has provided wide access to knowledge and has facilitated contact between like-minded people. A reaction to this democratic ideal of science has occurred with a renewed emphasis on oral exchanges and tacit understanding. Over the past fifteen years, postmodernist commentators at elite institutions have insisted that all knowledge derives entirely from social circumstances and political imperatives. It is not sufficient to read the literature and see the pictures, they contend; one must be initiated into the culture that produced the words and the images; or as physicist David Brewster's faithful contributor to the *Edinburgh Encyclopaedia*, the nineteenth-century historian Thomas Carlyle wrote, one can always sniff out class. With its thick prose and its rejection of clear and causal reasoning, the postmodernist offensive against a dispassionate evaluation of natural knowledge reinforces the privileged status of exclusive institutions. It recalls how theology allied with science in early nineteenth-century Cambridge, and how class and race entered into early twentieth-century scientific methodology at European and American universities.

Following the innovations of scientific academies in the seventeenth century, popular associations and institutes in the eighteenth century, disciplinary societies in the nineteenth century, and democratic universities in the twentieth century, the new age will generate new kinds of institution. A revolution in access to information may erode the pertinence of lecture courses and subvert the value of both orthodox curricula and university diplomas. Advanced learning may once more become a personal relationship between master and disciple, calling to mind Hermann Hesse's Castalia in his novel, *Das Glasperlenspiel*, depicting a future where the Western world is an economic backwater. But these new congregations for exploring the regularities of the physical world are still indistinct. Their form depends on circumstances dimly perceived.

Note: Quotations in the text from: Otto Nathan and Heinz Norden (eds.), *Einstein on Peace*, Schocken, New York, 1960; Lewis Pyenson, "Einstein's Early Scientific Collaboration", *Historical Studies in the Physical Sciences*, **7**, 83–124 (1976); Kenkichiro Koizumi, "The Emergence of Japan's First Physicists, 1868–1900", *ibid*, **6**, 3–108 (1975).

See also: Accelerators, Linear; Accelerators, Potential-Drop Linear; Atomic Spectroscopy; Bohr Theory of Atomic Structure; Cloud and Bubble Chambers; Dynamics, Analytical; Electrodynamics, Classical; Electromagnetic Interaction; Electromagnetic Radiation; Elementary Particles in Physics; Entropy; Heat; Kepler's Laws; Kinetic Theory; Light; Lorentz Transformations; Maxwell's Equations; Michelson–Morley Experiment; Newton's Laws; Philosophy of Physics; Quantum Mechanics; Radioactivity; Relativity, General; Relativity, Special Theory; Statistical Mechanics; Thermodynamics, Equilibrium; Uncertainty Principle.

Bibliography

Useful, recent encyclopedias include: Roshdi Rashed and Régis Morelon (eds.), *Encyclopedia of the History of Arabic Science*, 3 vols., (London, 1996); Helaine Selin (ed.), *Encyclopedia of the History of Science, Technology, and Medicine in Non-Western Cultures*, (Dordrecht, 1997); Arne Hessenbruch (ed.), *Reader's Guide to the History of Science*, (London, 2000), and John Heilbron (ed.), *Oxford Companion to the History of Modern Science*, (Oxford, 2003).

Readers seeking the technical side of recent physics may look to Laurie Brown, Abraham Pais, Brian Pippard (eds.), *Twentieth Century Physics*, 3 vols., (London, 1995); a more wieldy narrative is found in Helge Kragh, *Quantum Generations: A History of Physics in the Twentieth Century*, (Princeton, 1999).

Physics in several national settings is treated in Daniel J. Kevles, *The Physicists: The History of a Scientific Community in Modern America*, (New York, 1977); Dominique Pestre, *Physique et physiciens en France 1918–1940*, (Paris, 1984); and Paul Josephson, *Physics and Politics in Revolutionary Russia*, (Berkeley, 1991).

Some of the themes in the present article are elaborated in Lewis Pyenson and Susan Sheets-Pyenson, *Servants of Nature: A History of Scientific Institutions, Enterprises and Sensibilities*, (New York, 1999).

The premier journal for the history of science in the modern period, *Historical Studies in the Physical and Biological Sciences* (founded in 1969 by Russell McCormmach and continued by John Heilbron), contains much that is relevant to the present topic. For Latin America, *Quipu* (Mexico

City) and *Saber y tiempo* (Buenos Aires) have published significant studies; for Japan, *Historia scientiarum* (Tokyo) is indispensable as a means of accessing Japanese scholarship. An overview of European expansion may be found in the pages of *Itinerario* (Leiden). Historiography is a central theme of *History of Science* (Cambridge). Not all of these journals are indexed in the standard bibliographies published by the History of Science Society and the bibliographies available over the Internet from the Institute for Scientific Information in Philadelphia, but all contain scintillating contributions.

Holography

P. Hariharan

Holography is a technique that makes it possible to store and reproduce three-dimensional images. Holographic images can be produced, in principle, with any type of wave. However, this article will only deal with optical holography, excluding related techniques such as acoustical and microwave holography, whose applications have not developed to the same extent.

Holography was invented by Dennis Gabor in 1947 in an attempt to improve the resolution of images obtained with an electron microscope. Within a few years of its invention, it appeared destined for obscurity because of two serious problems: the desired image was overlaid by a spurious image in line with it and there was no suitable source of coherent light for making holograms. However in the early 1960s, Leith and Upatnieks demonstrated a new approach to holography based on communications theory that made it possible to obtain images of good quality. At about the same time the invention of lasers provided an intense source of coherent light. These developments triggered a major research effort in holography.

While Gabor's original goal has not yet been achieved, the last two decades have seen several breakthroughs which have established holography as a practical technique with a remarkably wide range of applications.

Basic Concepts

The unique characteristic of holographic imaging is the idea of recording both the amplitude and the phase of the light waves coming from the object. Since all available recording media respond only to light intensity, it is necessary to convert the phase information into variations of intensity. This is done by using a coherent, monochromatic source and adding to the wave front from the object a reference wave front of known amplitude and phase derived from the same source.

To record a hologram, the object is illuminated with light from a laser as shown in Fig. 1, and a photographic film is placed so that it receives light reflected from the object. The first difference from ordinary photography is that no lens is needed. Each point on the object reflects light to the entire photographic film. The second difference is that a portion of the light from the laser is incident directly on the film. This is called the reference beam. The film records the interference pattern between the light waves reflected from the object and the reference beam. This record is called the hologram of the object.

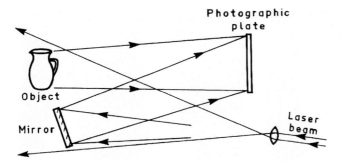

Fig. 1: Schematic of the optical setup used to record a hologram.

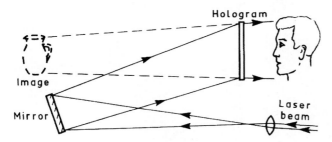

Fig. 2: Viewing the holographic image.

Since the light waves reflected from the object have a very complex form, the interference pattern making up the hologram is quite irregular. It is also so fine that its details can only be seen under a microscope. To the naked eye the hologram looks like a grey blur which bears little or no resemblance to the object. However, all the details of the light waves reflected by the object are recorded in it.

To view the image, the hologram is illuminated with a beam of monochromatic light as shown in Fig. 2. The light waves diffracted by the hologram are a replica of the light waves that came originally from the object. An observer looking through the hologram sees a lifelike image of the object in the same position in which it was when the hologram was made.

This reconstruction presents a natural 3-D appearance and has all the visual properties of the original object. The hologram looks like a window through which the image can be seen. When the observer moves his head, the perspective of the image changes, as shown in Fig. 3. If an object in the foreground masks something, the observer can look around it.

If we try to photograph the image, we find that it also exhibits depth-of-focus effects. With the camera lens wide open, it is possible to focus on only one plane in the image; other planes in front of this plane and behind it are out of focus. To bring the whole image into focus, it is necessary to stop down the camera lens just as with the original object.

An interesting feature of the hologram is that if it is broken, each part still reproduces the entire image. This is because each point on the hologram has received light from all parts of the object. The only change is that the viewing window through which the image is seen is smaller. For the same reason, holograms can be scratched quite badly with little effect on the image, which appears some distance behind the hologram.

Fig. 3: Views from two positions of the image reconstructed by a hologram, showing different perspectives.

Several practical problems have to be faced when making holograms. In the first place, the spacing of the interference fringes recorded in making a hologram is extremely small-usually less than a micrometer. Special ultrafine-grain films are needed, which are correspondingly slow. In addition, it is necessary to see that the object and, in fact, the whole setup, does not move by more than $0.1\,\mu m$ during the exposure. For these reasons, holograms are normally recorded on a massive steel or granite table floated on an antivibration system. Alternatively, it is possible to use a pulsed ruby laser which produces a flash of light lasting only a few nanoseconds.

Applications

Holography has many fascinating applications. Current trends suggest that the areas of greatest interest are art and advertising, high-resolution imaging, information storage and security coding, holographic optical elements, and holographic interferometry.

Art and Advertising

Some very spectacular applications of holography as a medium for art and advertising were demonstrated at quite an early stage. These included the production of a hologram of the Venus de Milo and portraits of living human beings with pulsed lasers. However, these early holograms had several drawbacks which stood in the way of their wider acceptance. In the first place, they needed a laser or a bright point source of monochromatic light to illuminate them. Even then, the reconstructed image was dim and could be viewed only in subdued light. Finally, the image was reconstructed in a single color, that of the source used to illuminate the hologram.

Some progress was made toward the solution of these problems in the early 1960s. This period saw the development of white-light imaging using reflection holograms, the development of new recording materials such as photopolymers and dichromated gelatin, and improved processing techniques for commercial photographic materials.

A major advance was the invention in 1969 of the rainbow hologram by Benton. This is what may be called a second-generation hologram. A hologram of a solid object is recorded initially in the conventional way. When this primary hologram is properly illuminated, it projects a real image of the object. The aperture of this primary hologram is then deliberately limited by a horizontal slit and a second hologram is made of the projected real image.

When this second hologram is illuminated with monochromatic light, it forms, in addition to an image of the object, an image of the slit which is projected out into the viewing space. With white light, the slit image is spread out into a spectrum in the vertical plane. The observer can position his eyes at any point on this spectrum (hence the name "rainbow" hologram); he then sees a bright, sharp image in the color corresponding to that part of the spectrum.

Two further advances followed. One was the white-light holographic stereogram of Cross. This was a composite rainbow hologram built up from a number of views of the subject from different angles in the horizontal plane. These views can be recorded with a motion picture camera in white light and subject movement presents no problems. Another was the production of multicolor rainbow holograms. In this case, the final hologram is built up by successive exposures to the real images projected by three primary holograms, made either with different laser wavelengths or with the reference beam incident at different angles. When this multiply-exposed hologram is illuminated with a white-light source, the observer sees a full-color three-dimensional image reconstructed from the three color separations. These techniques have made it possible to produce multicolor holographic stereograms up to 2 m wide and 1 m high.

Another valuable technique which was developed during the 1970s was the production of copies of holograms by embossing on a thin sheet of plastic backed with an evaporated metal coating. Embossed holograms can be produced cheaply in large quantities and give a bright image when viewed by reflection. They have opened up a rapidly expanding field of applications ranging from novelties and greeting cards to book and record covers.

Holographic movies have also attracted considerable attention. The major problem here is that the holograms have to be small for economy, while the image has to be large enough for viewing by a number of people. The most advanced system so far is that developed by Komar in the USSR, in which a series of holograms are recorded with a pulsed laser on 70-mm film using a lens with an aperture of about 200 mm. The same lens is then used to project the reconstructed images on a holographic screen which forms a number of images of the pupil of the lens in the viewing space. Each of these imaged pupils constitutes a viewing zone within which an observer can see a three-dimensional image.

High-Resolution Imaging

One of the interesting features of a hologram is that it can reconstruct an image with very high resolution over a great depth. Because of this, one of the first successful applications of holography was in the study of aerosols. Measurements on moving microscopic particles distributed over an appreciable volume are not possible with a conventional microscope because of its very limited depth of field. Pulsed laser holography makes it possible to store an image of the whole particle field at a given instant. The stationary image reconstructed by the hologram can then be studied in detail with a microscope.

Holography can also be used to obtain an image which is unaffected by lens aberrations and even by the presence of aberrating media in the optical path. The object is illuminated with coherent light and a hologram is recorded of the aberrated image wave using a collimated reference beam. When the hologram is illuminated by the same collimated reference beam reflected back along its original path, it reconstructs the conjugate to the object wave, which has the same phase errors as the original object wave but with the opposite sign. Accordingly, when this conjugate wave propagates back through the optical system, these phase errors cancel out exactly, resulting in an aberration-free image. Such an image can be produced in real time if a photorefractive crystal (e. g., bismuth silicon oxide) is used as the recording material.

One possible application of holographic high-resolution imaging is in the production of photolithographic masks for semiconductor devices. Holographic systems could overcome the problem of designing and producing lenses which can give very high resolution over a large field.

Information Storage

The fact that information stored in a hologram is not easily corrupted by local surface damage led at an early stage to research on holographic memories. Due to the lack of a suitable recording material, this application now appears to have been overtaken by other techniques. However, holographic information storage still appears to have advantages for some specialized applications. One is the reduction of the space required for archival storage of documents, far below that possible with microfilm. Others are for storage of multicolor material, such as motion pictures, and for storing three-dimensional information.

An application with considerable commercial importance is security coding for credit and identity cards. A simple method adopted by some of the major credit cards involves an embossed hologram incorporated in the card to provide an additional safeguard against forgery.

In the cardphone, now widely used in Britain, the credit card has a number of holograms imprinted on its surface. These are illuminated by a light source in the cardphone and counted to work out how many units the user has left. As the call progresses, the holographic patterns are erased one by one by a focused heat source in the cardphone.

Holographic Optical Elements

Diffraction gratings formed by recording an interference pattern in a photoresist layer are commonly known as holographic gratings. Holographic gratings are free from ghosts and exhibit very low levels of scattered light. They have replaced conventional ruled gratings for many spectroscopic applications.

Holographic scanners are now widely used in point-of-sale terminals. Typically, the scanner consists of a disc with a number of holograms recorded on it. Rotating the disc about an axis perpendicular to its surface causes the reconstructed image spot to scan the image plane in the desired pattern.

Holographic optical elements (HOEs) are much lighter than lenses and mirrors, since they can be produced on thin substrates. Another advantage is that several holograms can be recorded in the same layer to produce multifunction elements. In addition, holograms can correct system aberrations. One of the most successful applications of HOEs has been in head-up displays for high-performance aircraft. The optical system in such a head-up display projects an image of the instruments along the pilot's line of vision so that he can monitor critical functions while looking straight ahead through the windscreen.

An attractive possibility is the use of computer-generated holograms as HOEs. With these it should be possible to realize quite unusual optical systems. However, the main application of computer-generated holograms so far has been in interferometric tests of aspheric optical surfaces. The hologram corresponds to the interferogram that would be obtained if the wave front from the desired aspheric surface were to interfere with a tilted plane wave front .This hologram is placed in the plane in which the surface under test is imaged. The superposition of the actual interference pattern and the computer-generated hologram then produces a moiré pattern which gives the deviation of the actual wave front from the ideal wave front.

A potentially very interesting application of HOEs is to provide optical interconnections between integrated circuits; these could overcome many of the limitations of conventional wire interconnections.

Holographic Interferometry

Holographic interferometry makes use of the fact that a hologram can store an accurate, three-dimensional image of an object. This makes it possible to compare the shape of the object when it is stressed with a hologram image of its shape in its normal condition.

To do this, the hologram is replaced, after processing, in exactly the same position in which it was recorded and is illuminated with the original laser beam, so that it reconstructs an image which coincides with the object. If, then, a stress is applied to the object, so that its shape changes very slightly, the observer receives two sets of light waves which have traversed slightly different paths. These waves reinforce each other at points in the field of view where the separation of the surfaces of the object and the image corresponds to an optical path difference of zero or a whole number of wavelengths and cancel each other at some points in between. As a result, the image is covered with a pattern of bright and dark fringes which, as shown in Fig. 4, map in real time the displacements of the surface of the object. Very accurate measurements can be made of these changes since the fringes are not affected by surface roughness, and the contour interval corresponds to a movement of the surface of approximately half a wavelength (a fraction of a micrometer).

Alternatively, two holograms can be recorded on the same photographic plate, one of the object in its normal state and the other while it is under stress. In this case, interference fringes are generated by the two reconstructed images, so that the hologram is a permanent record of the changes in the shape of the object. With a pulsed laser it is possible to study transient deformations and rotating objects.

Holographic interferometry has found wide application in nondestructive testing to detect cracks and areas of poor bonding and is now an important part of the quality-control process in the aerospace industry. Turbine blades, load-bearing structures in aircraft, and even tires are routinely subjected to holographic inspection. The automobile industry is also using this technique to improve engine performance and reduce noise from transmissions and door

Fig. 4: A hologram reconstructs an image of the unstressed object. The
interference fringes map the changes in the shape of the object.

panels. Other applications are in studies of plasmas and in medical research to measure the
deformations of implants under stress.

Holographic interferometry can also be used to produce an image modulated by a fringe
pattern corresponding to contours of constant elevation with respect to a reference plane.
Three basic methods of holographic contouring are available; these involve changing either
the wavelength used or the angle of illumination or, alternatively, the refractive index of a
medium in which the object is immersed.

Yet another application of holographic interferometry has been in studies of vibrating surfaces. One way is to use stroboscopic illumination, but a widely used alternative is time-
average holographic interferometry. In this technique a hologram of the vibrating object is
recorded with an exposure which is long compared to the period of vibration.

The scattered light from the object now exhibits a time-varying phase modulation, and
hence contains frequency-shifted components. Only the component at the original frequency
can interfere with the reference wave. As a result, the brightness at the image at any point is
multiplied by a factor $J_0^2(\xi)$, where ξ is proportional to the amplitude of vibration. The image
appears overlaid with dark fringes corresponding to the zeros of this function which can be
used to map the vibration amplitude.

Figure 5 shows a set of time-average holograms of some of the modes of vibration of the
soundboard of an acoustic guitar. Such studies have led to instruments with improved acoustic
performance.

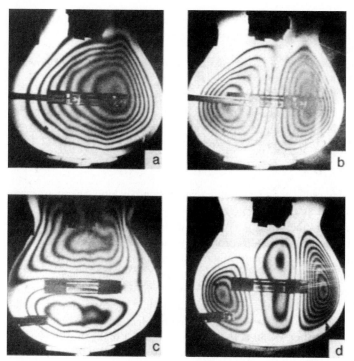

Fig. 5: Time-average holograms showing the vibration modes of the soundboard of an acoustic guitar at frequencies of (a) 195, (b) 292, (c) 385, and (d) 537 Hz.

Digital phase-stepping techniques now make possible very accurate measurements of the surface displacements at a uniformly spaced network of points covering the object. Because of the speed with which data can be acquired and manipulated, digital phase-stepping techniques open up many new applications for holographic interferometry in quantitative stress analysis.

Conclusion

It is clear from this survey that holography has already established itself in several rapidly expanding areas. We can look forward confidently to other completely new applications in the next few years.

See also: Gratings, Diffraction; Interferometers and Interferometry; Lasers; Light; Optics, Physical; Waves.

Bibliography

E. N. Leith and J. Upatnieks, *Sci. Am.* **212**, 24 (June 1965). (E)

R. J. Collier, C. B. Burckhardt, and L. H. Lin, *Optical Holography*. Academic Press, New York, 1971. (A)

H. M. Smith, *Principles of Holography*. Wiley, New York, 1976. (A)

E. N. Leith, *Sci. Am.* **235**, 80 (October 1976). (E)

H. J. Caulfield (ed.), *Handbook of Optical Holography*. Academic Press, New York, 1979. (A)

C. M. Vest, *Holographic Interferometry*. Wiley, New York, 1979. (A)

G. Saxby, *Holograms*. Focal Press, London, 1980. (I)

P. Hariharan, *Optical Holography: Principles, Techniques and Applications*. Cambridge University Press, Cambridge, 1984. (A)

Hot Atom Chemistry

P. P. Gaspar

Hot atom chemistry is the study of the chemical events, such as bond making, energy transfer, and electron exchange, that occur after excitation of an atom by a nuclear transformation.

The initial excitation leading to the formation of a hot atom can be induced by bombardment with high-energy particles such as protons or neutrons, and such transformations often result in high recoil energies that lead to bond rupture. Other processes, such as thermal neutron activation, and radioactive decay by beta emission, electron capture, or internal conversion, often endow daughter atoms with low recoil energies. Nevertheless, bond rupture can occur in these processes, despite their low recoil energy, as a result of the formation of vacancies in the inner shells of electrons. The vacancies are filled by electrons from outer shells, but these nonradiative atomic events, known as Auger processes, result in the loss of as many as 10 additional electrons, leading to highly positively charged atoms in excited electronic states. If these atoms are bound in molecules, the electrons will be redistributed, leading to several positively charged atoms in the same molecule, which immediately flies apart due to electrostatic repulsion. Photoionization with monochromatic x rays can also bring about inner electron-shell vacancy cascades by a non-nuclear process.

Thus, hot atoms are generally formed as ions carrying multiple positive charges and a recoil energy of from tens to millions of electron volts. The recoiling atom will lose energy to its surroundings through collisions with atoms and molecules in which electron exchange can also occur. When the hot atom is sufficiently deexcited so that its collisions can lead to the formation of new chemical bonds, a variety of chemical reactions can take place. At this stage the hot atom is typically electrically neutral, but monopositive ions can also reach the epithermal energy range, from 1 to 100 eV.

The production of hot atoms by nuclear recoil allows the study of chemical reactions of atoms and molecules possessing far more energy than the threshold for thermally induced processes. It is virtually the only technique known for producing superexcited molecules containing tens of eV of internal excitation in a bath of room-temperature molecules. Through the study of the decomposition of molecules formed with high internal energies through reactions of hot atoms, the distribution of recoil energies of the hot atoms undergoing reaction can be deduced.

One limitation on the nuclear recoil technique is the small number of hot atoms produced. In a typical experiment employing an accelerator or a nuclear reactor, 10^6 to 10^{10} hot atoms are produced over a 1-h period, each with a chemical lifetime of only on the order of 1 μs. The instantaneous concentrations of hot atoms are therefore so low, perhaps one hot atom among 10^{20} normal molecules, that it is difficult to detect a hot atom until it has incorporated itself via its reactions in a chemically stable molecule. Detection of hot atoms via their radioactive decay is facilitated by the production of short-lived isotopes.

Nevertheless, much has been learned about the chemistry of free atoms from the end products of hot atom reactions. Since new bonds tend to be formed one or two at a time in the collisions of hot atoms with molecules, novel molecules in which the valence of the hot atom is not saturated are produced as reactive intermediates in the reactions of hot atoms that must make more than two bonds to satisfy their bonding capability. Many new chemical reactions of free atoms and reactive intermediates have been discovered in hot atom experiments. Nuclear techniques for producing hot atoms complement such instrumental methods as molecular and ion beam experiments that can provide more direct information about reaction kinetics and dynamics.

When hot atoms with high recoil energies are produced by nuclear transformations in the gas phase, they can escape the molecule debris of atoms, radicals, and ions formed in the collisional energy loss cascade that brings the hot atom down to the energy range of chemical reactions. The much shorter range of recoiling atoms in the liquid and solid state sometimes results in chemical reactions of hot atoms with their own debris.

From the earliest days of hot atom chemistry in the 1930s, practical use has been made of the chemical consequences of nuclear transformations. Recoil processes facilitate the isolation of radioisotopes in high specific activities by delivering the product of a nuclear transformation in a different chemical form from that of the target nucleus. Thus, an organic halide could be irradiated with neutrons, and the radioactive halide atoms freed by nuclear recoil dissolved in aqueous solution without dissolving the stable halide bound in the parent molecules, an effect known as the Szilard–Chalmers reaction.

Molecules can be labeled directly by the reactions of hot atoms, but since these reactions can be quite unselective, the use of hot atom reactions to provide labeled molecules has been limited to such simple systems that the number of reaction channels is small. This has led to the practice of employing recoil reactions to form small molecules such as $^{11}CO_2$ and $H^{11}CN$ which are converted by chemical synthesis into larger molecules, no mean feat with half-lives as short as 20 min for ^{11}C. Hot atom chemistry has made important contributions to nuclear medicine by providing biomolecules containing short-lived isotopes for physiological and diagnostic studies.

The processes following nuclear recoil in solids are of interest because of the short range of the hot atom in the solid state, and have important implications for materials science and reactor technology. Solid-state hot atom chemistry can lead to the synthesis of simple inorganic and organic compounds that can be converted into a variety of useful molecules.

in situ methods have been directed at answering the most basic questions in solid-state hot atom chemistry: What fraction of the recoiling atoms remains in the chemical form of the target atom? What is the nature of the recoil species produced? What reactions do the recoiling atoms undergo?

Mössbauer emission spectroscopy has revealed changes in charge state, ligands, and ligand geometry of atoms following nuclear recoil on the time scale of 10^{-9}–10^{-7} s. Measurements of the perturbed angular correlation of emitted γ-rays caused by the coupling of nuclear moments with extranuclear electromagnetic fields can provide information about the local environment of a recoiling atom on a nanosecond time scale following a nuclear transformation.

See also: Atoms; Radioactivity.

Bibliography

T. Tominaga and E. Tachikawa, *Modern Hot-Atom Chemistry and Its Applications.* Springer-Verlag, Berlin, 1981.

T. Matsuura, ed., *Hot Atom Chemistry: Recent Trends and Applications in the Physical and Life Sciences and Technology.* Kodansha, Tokyo and Elsevier, Amsterdam, 1984.

"Hot Atom Chemistry" (special issue), *Radiochim. Acta* **43** (1988).

Hot Cells and Remote Handling Equipment

J. M. Davis

In nuclear terminology, "hot" means radioactive, and a "cell" is an enclosure that provides containment for the radioactive material and shielding from penetrating radiation. Remote handling equipment includes mechanical, electrical, and hydraulic devices with which workers outside the hot cell can handle, process, and examine radioactive material inside the cell.

Hot Cells

The design of hot cells and remote handling equipment varies greatly, depending on the kinds and amounts of radioactive materials to be handled, the processes or examinations to be conducted, and the production efficiency desired. A small steel box with a window, one ball-and-socket tong, and a syringe-actuated pipette is a hot cell for dispensing medical isotopes. A 450-m-long "canyon" with 1.2-m-thick concrete walls and inside bridge crane is a hot cell (see Fig. 1) for a reactor fuel reprocessing plant.

Design bases common to all hot cells, regardless of size or purpose, follow:

1. *Containment Integrity.* The radioactive paniculate or gaseous effluent(s) from the hot cell must be ALARA (*as low as reasonably achievable*), and in no case greater than allowed by Federal regulation. This is usually achieved by tight construction, operation at negative pressure, and highly efficient air filtration.

2. *Radiation Protection.* The hot cell itself must provide sufficient shielding to reduce radiation exposure of workers to ALARA, and never more than permitted by Federal regulation. This requires correct construction materials. These can vary from thin sheet-metal walls and plate glass or plastic windows for alpha and weak beta emitters to thick concrete, (①, Fig. 1) steel, or lead walls and shielding glass windows (②, Fig. 1) for gamma and neutron emitters.

3. *Consequences of Malfunction and Accidents.* The designer of hot cells and associated remote handling equipment must be able to predict normally expected malfunctions and accidents, and provide for recovery without breaching containment or significantly adding to occupational radiation exposure. He must also predict unexpected and more serious accidents, as well as the maximum credible accident, and provide for protecting workers

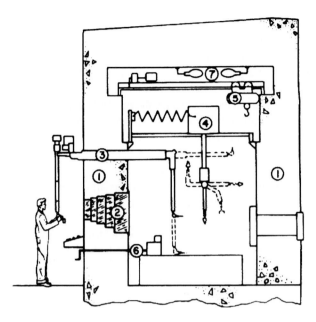

Fig. 1: A typical hot cell.

and the general public against the consequences. Since some malfunction or accident scenarios may involve the possibility of breaching hot cell containment, the room containing the hot cell must also be at negative pressure and have exhaust air filters. The room thus becomes an emergency secondary containment system.

Remote Handling Equipment

The type, design, and use of remote handling equipment varies greatly with the nature and quantities of radioactive materials being handled, and with the processes or examinations being conducted. The major basis for decision is the ALARA principle of controlling occupational radiation exposure. If the materials to be handled emit little or no penetrating radiation, and if there is insignificant hazard from heat, corrosion, or machinery, then manipulations can be safely accomplished by hand through long rubber gloves affixed to openings in a thin-walled containment cell (glove box). Otherwise, remote handling becomes necessary.

General Purpose
During the growth of nuclear technology, a few general-purpose remote handling tools were developed to a high degree of excellence and now can be purchased from several vendors. Perhaps the most generally useful of these tools is the master–slave manipulator. Through an ingenious system of cables, steel tapes, gears, and electric drives operating through a penetration in a hot cell wall of any thickness, this manipulator (③ in Fig. 1) duplicates the movements and forces of the human hand inside the hot cell. Given sufficient light (⑦ in Fig. 1) and vision, a skilled operator can thread a needle or operate an engine lathe with a pair of these manipulators.

Somewhat less dexterous, but far more powerful, is the rectilinear power manipulator (④, Fig. 1). When rail-mounted on a trolley and bridge inside the hot cell, it can reach and manipulate within nearly the entire volume. With interchangeable jaws, special attachments, and power tools, it is capable of performing a great variety of work that is beyond the reach or capacity of master–slaves. Control is by console.

Perhaps the oldest and simplest of the remote handling tools is the in-cell bridge crane (⑤, Fig. 1). Equipped with inching controls and variable speeds, and operated by a person skilled in its use, it becomes a marvelously versatile manipulator.

Special Purpose

Remote handling equipment for special purposes in the hot cell performs all of the required functions that cannot or should not be done with general-purpose equipment, for reasons of inaccessibility, overstress, economy of operator time, etc. (⑥, Fig. 1). Material transfer can often be better done by conveyor, pneumatic tube, or even toy electric train, than by crane or master-slave. Most repetitive motions can easily be automated or computer controlled.

Requirements for all special-purpose equipment are reliability, simplicity, accessibility, visibility, and ease of repair or replacement by remote handling.

Other Considerations

1. *Vision.* Vision is usually afforded by shielding glass windows (②, Fig. 1) designed and furnished by vendors who are specialists in this field. Closed-circuit television is also proving increasingly useful.

2. *Maintenance, Repair, Replacement.* Careful analysis and preplanned solutions to anticipated problems will pay off handsomely in reduced downtime, process interruption, and occupational radiation exposure.

3. *Quality.* The best quality of equipment, materials, and workmanship that money can buy will prove to be the least expensive approach to building, equipping, and operating a hot cell.

4. *Decommissioning.* Cleanup and disposition of the hot cell, when obsolete or not needed, must be considered in the initial design.

See also: Radiation Interaction with Matter; Radioactivity; Radiological Physics.

Bibliography

Title 10, Code of Federal Regulations, Parts 20, 30, 40, 50, 70, 71.
Proc. 9th-26th Conf. Remote Systems Technology. American Nuclear Society, 1961–1978.
Proc. Int. Symposium High Activity Hot Laboratories Working Methods, Grenoble, France, June 1965, published by ENEA/OECD.
Remote Handling of Mobile Nuclear Systems. U.S. Atomic Energy Commission, 1966.

Hubble Effect

K. R. Lang

As early as 1917 Vesto M. Slipher of the Lowell Observatory had accumulated the spectra of 25 spiral nebulae, using the Doppler effect to show that they are receding from our Galaxy with tremendous velocities as large as 1 800 kilometers per second. Slipher had inferred these large velocities by observing the difference between the wavelengths, λ_0, of the spectral lines of the nebulae and the wavelengths, λ_L, of the same lines observed in terrestrial laboratories. According to the Doppler effect, the radial velocity of the nebula, V, is given by

$$V = \frac{c(\lambda_0 - \lambda_L)}{\lambda_L} = cz \, ,$$

where $c = 2.997\,924\,56 \times 10^5$ kilometers per second is the velocity of light, and the parameter z is called the redshift of the nebula. Six years after Slipher's pioneering work, Carl Wirtz looked for correlations between these velocities and other observable properties of the nebulae. Wirtz found that when suitable averages of the available data were taken an approximate linear dependence of velocity and apparent magnitude was visible. Because the observed light intensity of an object falls off as the square of its distance, this meant that the more distant objects had larger outward velocities. Two years after Wirtz argued for a velocity–distance relation, Arthur Eddington showed that when matter is placed in the de Sitter solution for Einstein's field equations it will expand; and Hermann Weyl showed that the redshift of this matter will, to first order, increase linearly with distance. By 1923 it was clear that Einstein's field equations suggested an expansion of the Universe, and that the Wirtz interpretation of Slipher's observations suggested that the spiral nebulae participate in this expansion.

In 1926, the Belgian astrophysicist and Catholic priest, Georges Edouard Lemaître, visited Slipher at the Lowell Observatory, Flagstaff, Arizona, and Edwin Hubble at the Mount Wilson Observatory, California, learning of the latest redshift measurements of the radial velocities of spiral nebulae. In a prophetic article, published in 1927, Lemaître interpreted the receding velocities of extra-galactic nebulae as a cosmic effect of the expansion of the Universe in accordance with Albert Einstein's General Theory of Relativity.

It was not until 1929 that Edwin Hubble put the expected linear dependence of velocity on distance in a quantitative observational framework. When the observed radial velocities, V, were compared with the estimated distances, D, of the spiral nebulae, the now famous Hubble law

$$V = H_0 \times D$$

was found to hold for each spiral nebula. Here the parameter H_0 is called Hubble's constant. Hubble's original velocity–distance plot, from which the Hubble law was inferred, is given in Fig. 1. Such a plot is now known as the *Hubble diagram*.

Hubble's constant essentially measures the rate of recession of the nebulae, and can therefore be used to infer an age parameter of the Universe through the relation age $= 1/H_0$. Hubble obtained a value of $H_0 = 550$ kilometers per second per megaparsec for his recession constant, and this value of the constant gives an age parameter of 2×10^9 years.

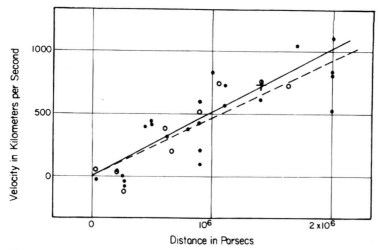

Fig. 1: The velocity–distance relationship for galaxies, as plotted by E. P. Hubble in 1929. The solid line represents the relation from individual galaxies (dots), the dashed line the relation when the galaxies are combined into groups (circles). The cross indicates the mean for 22 galaxies of uncertain distance.

Then in 1931, Lemaître suggested that the systematic outward motions observed for the distant galaxies could be extrapolated backwards in time, attributing them to the explosion of an incredibly dense "primeval atom" or "cosmic egg", sending its contents out in all direction. Today we call this event the Big Bang.

We now know that Hubble's basic ideas were correct, but that his estimates of distance were too small. Many of the distance calibrations were based on Cepheid variable stars, and involved the assumption that all Cepheids obeyed the same relationship between period of light variation and luminosity. But, as Walter Baade showed in 1952, the period–luminosity relation for Cepheids in spiral arms is quite different from that for Cepheids in galactic nuclei or in globular clusters. When this difference was taken into account, the extragalactic distance scale was corrected, and Hubble's constant was revised downward to about 180 kilometers per second per megaparsec. During the next 15 years Allan Sandage and his colleagues at the Mount Wilson and Palomar Observatories have published at least five successively smaller values for Hubble's constant.

In order to test the Hubble effect for the more distant objects, Milton Humason, Nicholas Mayall, and Allan Sandage reformulated Hubble's law in the form

$$\log(V) = 0.2m + B ,$$

where m is the apparent magnitude of the object and the constant B depends on the absolute magnitude of the object and Hubble's constant. They then examined the log velocity–apparent magnitude relation for 474 nebulae using data obtained during a 25-year period with the Mt. Wilson, Palomar, and Lick observatories. In order to test the velocity–distance relation, redshifts were corrected for the solar motion and the observed magnitudes were corrected for absorption within our Galaxy and for the effects of Doppler shifts on the radiation spectrum. The linearity of the corrected log velocity–apparent magnitude relation was then tested

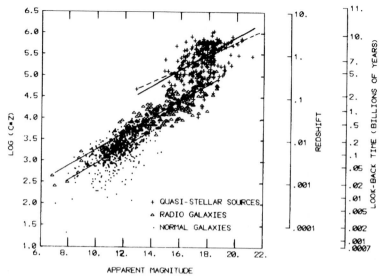

Fig. 2: The author's composite Hubble diagram, in which are plotted 663 normal galaxies, 230 radio galaxies, and 265 quasistellar objects. The vertical axis at right gives the look-back time in billions of years, calculated on the assumption of a Hubble constant of 50 kilometers per second per megaparsec. For normal galaxies and radio galaxies, the solid line denotes the least-squares fit to the data. For the quasistellar objects, the dashed line denotes the least-squares fit, while the solid line represents the theoretical Hubble relation that would hold for a homogeneous, isotropic, expanding Universe.

by a linear least-squares fit. Within the accuracies of the data, the relation is linear and the slope is 0.2. It has since been shown that this slope is exactly the expected result for a homogeneous, isotropic, expanding Universe that obeys the laws of Einstein's general theory of relativity.

As illustrated in Fig. 2, the author and his colleagues have shown that the Hubble-diagram plots for normal galaxies, radio galaxies, and quasistellar objects are all consistent, within the errors, with that expected for a homogeneous, isotropic, expanding Universe. The presently available data suggest that all extragalactic objects with redshifts ranging from 0.001 to about 6 participate in the expansion of the Universe. Unhappily, the great dispersion in the observed data makes it impossible to determine the details of the expansion. The data suggest instead that the distant quasistellar objects are about a factor of 10 brighter than the nearer radio galaxies, which are in turn about a factor of 10 brighter than the very nearby normal galaxies. Because the light with which we view distant objects left them some time ago, this change in brightness suggests that younger extragalactic objects are brighter than older ones.

In more recent times, Wendy L. Freedman and their colleagues have used the Hubble Space Telescope, or HST for short, to refine estimates for the Hubble constant and to infer a more accurate age of the Universe since the Big Bang. They have measured the distances of Cepheid variable stars in 23 spiral nebulae out to a distance of 65 million light-years. When these results are extended to larger distances by observations of supernovae of type Ia, and compared with other diverse cosmic measurements, including those from the Wilkinson Microwave Anisotropy Probe, abbreviated WMAP, an expansion rate of $H_0 = 71$ kilometers per second per megaparsec is obtained, with a margin of error of about 5%, and the age of the Universe of 13.7 billion years since the Big Bang is obtained, with a margin of error close to 1%.

See also: Astrophysics; Cosmology; Galaxies; Universe.

Bibliography

W. Baade, *Pub. Astron. Soc. Pacific* **68**, 5 (1956). (I)

A. S. Eddington, *The Mathematical Theory of Relativity*. Cambridge University Press, London, 1923. (I)

E. P. Hubble, *Proc. Natl. Acad. Sci.* **15**, 168 (1929). (I)

E. P. Hubble, *The Realm of the Nebulae*. Yale University Press, New Haven, 1936. Reprinted by Dover Publications, New York, 1958. (E)

M. L. Humason, N. U. Mayall, and A. R. Sandage, *Astron. J.* **61**, 57 (1956). (I)

K. R. Lang, *Astrophysical Formulae*, 3rd ed. Springer-Verlag, New York, Berlin-Heidelberg, 1999. (A)

K. R. Lang and O. Gingerich, *Source Book in Astronomy and Astrophysics 1900–1975*. Harvard University Press, Cambridge, Mass., 1979. (I)

K. R. Lang, S. D. Lord, J. M. Johanson, and P. D. Savage, *Astrophys. J.* **202**, 583 (1975). (I)

K. R. Lang and G. S. Mumford, *Sky and Telescope* **51**, 83 (1976). (E)

A. R. Sandage, *Astrophys. J.* **178**, 25 (1972). (I)

A. R. Sandage and G. A. Tamman, *Astrophys. J.* **196**, 313 (1975). (I)

V. M. Slipher, *Proc. Amer. Philos. Soc.* **56**, 403 (1917). (I)

S. Weinberg, *Gravitation and Cosmology: Principles and Applications of the General Theory of Relativity*. Wiley, New York, 1972. (A)

H. Weyl, *Phys. Zeit.* **24**, 230 (1923). (A)

C. Wirtz, *Astron. Nach.* **215**, 349 (1921). (I)

Hydrodynamics

M. Nelkin

Hydrodynamics is the study of the mechanics of fluid flow. More recently called fluid mechanics or fluid dynamics, it has become an extensively developed discipline of applied science. This is hardly surprising since fluid flows are such an important part of our natural and man-made environment. In this article, however, we focus on the growing interest in hydrodynamics as an active research subject in physics. We will see that there are many problems for which a basic physical understanding is still lacking.

The equations of motion of a fluid, the Navier–Stokes equations, are relatively simple partial differential equations. The molecular properties of the fluid enter only through the values of certain transport coefficients, such as the viscosity. When properly scaled, these equations describe the same motions in air that they do in water. This is well confirmed experimentally, so that we can be confident that the phenomena we wish to study are consequences of the underlying hydrodynamic equations. These phenomena, however, occur in bewildering variety, and there are very few of them that we can calculate directly. This is a consequence of the nonlinearity of the underlying equations and of our inability to predict, from nonlinear equations of motion, even the qualitative features of the resulting motion. This same situation exists in many other parts of physics, but seldom do we have so much confidence that we have the correct equations of motion. Thus hydrodynamics becomes a very important testing ground for our developing understanding of nonlinear phenomena. (For an elementary but elegant expansion of this viewpoint see Ref. 1.)

We turn now to a more quantitative description. The velocity of a fluid can be described by a vector function of space ($\mathbf{x}$) and time t, written as $\mathbf{u}(\mathbf{x},t)$. The density of the fluid is given by a scalar function $\rho(\mathbf{x},t)$ of the same variables. The conservation of mass is expressed by the equation of continuity,

$$\frac{\partial \rho}{\partial t} + \nabla \cdot (\rho \mathbf{u}) = 0 \ . \tag{1}$$

For many fluid flows of interest, typical flow speeds are very small compared to the speed of sound in the fluid. Under these conditions the density of the fluid is nearly constant, and Eq. (1) can be replaced by

$$\nabla \cdot \mathbf{u} = 0 \ . \tag{2}$$

Such fluid flows are called incompressible, and Eq. (2) is sometimes called the incompressibility condition. It is important to recognize that Eq. (2) expresses a property of a fluid flow, and not of a fluid.

The momentum balance for a moving fluid element is given by

$$\rho \left[\frac{\partial \mathbf{u}}{\partial t} + (\mathbf{u} \cdot \nabla) \mathbf{u} \right] = -\nabla p + \mathbf{f}_{\text{ext}} + \mathbf{f}_{\text{vis}} \tag{3}$$

where $p(\mathbf{x},t)$ is the pressure, $\mathbf{f}_{\text{ext}}$ is the force per unit volume due to external forces such as gravity, and $\mathbf{f}_{\text{vis}}$ is the force per unit volume due to internal viscous friction. Of all the terms in Eq. (3) only $\mathbf{f}_{\text{vis}}$ depends on the molecular nature of the fluid. For an ordinary isotropic classical fluid, $\mathbf{f}_{\text{vis}}$ can be simply expressed in the form

$$\mathbf{f}_{\text{vis}} = \eta \nabla^2 \mathbf{u} + \left(\frac{1}{3} \eta + \eta_v \right) \nabla (\nabla \cdot \mathbf{u}) \tag{4}$$

where η is the ordinary shear viscosity and η_v is a volume coefficient of viscosity. These are to be taken as empirical properties of the fluid that depend on temperature and density. For gases they can be calculated from kinetic theory. For liquids theory is more complicated and less reliable.

Fluids that obey the viscous friction law of Eq. (4) are called Newtonian. Air and water and most other gases and liquids are Newtonian under usual conditions. At high enough frequencies, however, all fluids exhibit viscoelastic effects that make them non-Newtonian. Other fluids, such as polymer solutions, may exhibit non-Newtonian effects even at ordinary frequencies. Anisotropic fluids such as liquid crystals have a more complicated behavior. Finally, superfluids such as helium II below 2.17 K have a totally different hydrodynamics.

To close our system of equations we need an equation for the energy balance of a moving fluid element. This will contain the temperature $T(\mathbf{x},t)$. We also need the equation of state $\rho(p,T)$. The problem simplifies greatly, however, if we make the approximation of incompressible flow, $\rho = $ constant. By use of Eq. (2), Eq. (4) becomes

$$\frac{\partial \mathbf{u}}{\partial t} + (\mathbf{u} \cdot \nabla)\mathbf{u} = -\frac{1}{\rho}\nabla p + \frac{1}{\rho}\mathbf{f}_{\text{ext}} + \nu \nabla^2 \mathbf{u} , \tag{5}$$

where we have introduced the kinematic viscosity $\nu = \eta/\rho$. For air under standard conditions $\nu = 0.15\,\text{cm}^2/\text{s}$, and for water $\nu = 0.01\,\text{cm}^2/\text{s}$. For incompressible flow this is the only property of the fluid that enters.

Equation (5) is the Navier–Stokes equation of hydrodynamics. It is to be supplemented by the incompressibility condition of Eq. (2). Taking the divergence of Eq. (5) and using Eq. (2), we have

$$\nabla \cdot (\mathbf{u} \cdot \nabla)\mathbf{u} = -\frac{1}{\rho}\nabla^2 p + \frac{1}{\rho}\nabla \cdot \mathbf{f}_{\text{ext}} . \tag{6}$$

The pressure $p(\mathbf{x},t)$ can be obtained by integrating this equation over space subject to the appropriate boundary conditions. Thus the pressure term in the Navier–Stokes equation is a quadratically nonlinear contribution that is spatially nonlocal. The subtle interplay of the two nonlinear terms with the linear viscous term gives the Navier–Stokes equation very special and interesting properties.

Under almost all conditions the appropriate boundary condition is that the velocity vanish at a solid surface. The fluid sticks to the solid. As stated nicely by Feynman, "You will have noticed, no doubt, that the blade of a fan will collect a thin layer of dust – and that it is still there after the fan has been churning up the air. In spite of the fact that the fan blade is moving at high speed through the air, the speed of the air relative to the fan blade goes to zero right at the surface. So the very smallest dust particles are not disturbed". Exceptions to this boundary condition can occur in highly rarified gases, and in attempts to extend hydrodynamic ideas to motion at the molecular level, but these are unusual cases. In the two-fluid model of liquid helium, the normal-fluid component obeys the usual boundary condition, but the superfluid slips freely at a surface.

In most flows there is a well-defined largest speed U and a well-defined largest length L. It is then convenient to rewrite the Navier–Stokes equations in dimensionless form. We measure velocities in units of U, distances in units of L, times in units of L/U, and pressures in units of ρU^2. Leaving out any external forces for simplicity, the dimensionless equation of motion becomes

$$\frac{\partial \mathbf{u}}{\partial t} + (\mathbf{u} \cdot \nabla)\mathbf{u} = -\nabla p + \frac{1}{R}\nabla^2 \mathbf{u} . \tag{7}$$

The parameter

$$R = UL/\nu \tag{8}$$

is called the Reynolds number, and completely defines the characteristics of the flow. The length scale L, velocity scale U, and fluid viscosity v can be varied, but if the Reynolds number is the same, the flow will be the same. This is well confirmed experimentally even for very complicated turbulent flows.

When the Reynolds number is small, corresponding to large viscosity or low speed or small length scales, the linear viscous term dominates the nonlinear terms, and the problem is straightforward and well understood. Consider for example the flow past a sphere of radius a. At small Reynolds numbers the drag force on the sphere is given by the famous Stokes formula

$$F = 6\pi\eta aU \tag{9}$$

where U is the free-stream speed far from the sphere. The physics of this situation is best seen by looking at the rate of energy dissipation

$$P = FU = 6\pi\eta aU^2 . \tag{10}$$

The motion of the sphere through the fluid creates a velocity field extending a considerable distance from the sphere. The energy dissipation is accounted for by the decay of the fluid kinetic energy into heat through action of the viscosity.

For another simple example consider a cylindrical pipe of diameter D whose axis is in the z direction. In cylindrical coordinates (r, θ, z) the flow field is given by

$$u_z(r) = \frac{1}{16\eta} \left(\frac{dp}{dz} \right) (D^2 - 4r^2) , \tag{11}$$

where dp/dz is the pressure gradient along the pipe.

As a third simple example, consider two concentric cylinders with the outer cylinder stationary, and the inner cylinder rotating at angular frequency ω (Couette flow). The flow field, again in cylindrical coordinates, is given by

$$u_r = u_z = 0 , \qquad u_\theta(r) = \frac{\omega R_1^2}{r} \frac{(R_2^2 - r^2)}{(R_2^2 - R_1^2)} \tag{12}$$

where R_1 and R_2 are the radii of the inner and outer cylinders, respectively.

These three examples are illustrative in their differences as well as in their similarities. The theoretical results of Eqns. (9), (11), and (12) all agree with experiment at sufficiently low Reynolds numbers. They all disagree with experiment at high Reynolds numbers, but in different ways and for different reasons. For the flow past a sphere, Eq. (9) is an approximate result in which the nonlinear terms in the equation of motion have been explicitly neglected. Corrections can be calculated as a power series in the Reynolds number, but this perturbation expansion is itself valid only for small Reynolds numbers. At high Reynolds number, the flow becomes more complicated, and eventually becomes turbulent. For the flow in a pipe or between rotating cylinders, the solutions we have given are exact steady solutions of the Navier–Stokes equations at all Reynolds numbers. Whether these solutions are observed is a question of hydrodynamic stability.

For pipe flow the Poiseuille profile of Eq. (11) is the only known steady solution, but for large enough Reynolds numbers this solution is unstable. The observed flow when the stability of the Poiseuille flow is lost is fully developed turbulence. Under some conditions the laminar Poiseuille flow may be metastable, much as a supercooled liquid can be metastable against solidification. The transition to turbulence and the detailed nature of turbulent pipe flow have been extensively studied experimentally. The transition is reasonably well understood, but the fully turbulent state is very complicated. A fundamental theoretical understanding of fully developed turbulence is still lacking. There is, however, no reason to believe that turbulent pipe flow cannot be understood in principle starting from the Navier–Stokes equations. It is just an extremely difficult problem, both mathematically and computationally. Qualitatively it is not surprising that high Reynolds number flow should be turbulent. The Navier–Stokes equations can be thought of as a nonlinear dissipative dynamical system with a very large number of degrees of freedom. When the dissipation is weak, such systems generically exhibit chaotic time dependence. The essential feature is sensitivity to initial conditions. Two solutions that are close together in the appropriate phase space at some initial time will be far apart at much later times. In such a situation, a statistical description is the natural one. Turbulent flows exhibit a subtle mixture of randomness and regularity that is beyond our present ability to calculate, but turbulence is an active field of research, and considerable progress can be expected. For more about turbulence, see the article in this volume.

The Couette flow problem exhibits a more interesting structure before the onset of turbulence. At a critical Reynolds number the steady flow of Eq. (12) becomes unstable. The stable flow is still steady, but has a complex spatial structure consisting of toroidal rolls, the so-called Taylor vortices discovered by G. I. Taylor in 1923. At a second critical Reynolds number this structured steady flow becomes unstable to a transverse wave motion of the Taylor vortices. The stable flow is then still laminar, but is now structured in space and periodic in time. The behavior up to this point can be understood by hydrodynamic stability theory. With a further increase in Reynolds number, the flow becomes more complicated, and appears to show a sudden transition to turbulence. This transition is quite different from that in pipe flow, and is currently the subject of active experimental and theoretical research. At much higher Reynolds numbers the turbulence becomes stronger. The fully developed turbulence is quite similar in Couette and in pipe flow, but the pattern of transitions to reach it is quite different.

The flow past an obstacle such as a sphere or a cylinder is again of a different character. In contrast to Couette and Poiseuille flow, in this case the flow is perturbed in a negligible fraction of the total space. At sufficiently large distances cross-stream or upstream, the velocity field will approach its uniform unperturbed value. Downstream perturbations may extend farther, but these also will eventually die out. At large distances upstream or cross-stream the flow will be slowly varying and streamlined. A variety of elegant mathematical methods of classical hydrodynamics are applicable in this region, but the problem of how to join on to the rapidly varying flow near the obstacle makes the application of these methods to real problems quite subtle. Near the obstacle the flow varies rapidly, and the singular perturbation of the viscous term is important. In a thin layer near the obstacle useful approximate solutions including this singular perturbation can be obtained by boundary layer theory. The asymptotic solution obtained by boundary layer theory near the obstacle, and from potential flow theory far away, must be joined together using the method of matched asymptotic expansions. All of this works in precise form when the flow is laminar, but useful empirical methods exist also for the case when the boundary layer and wake are turbulent.

There is an immense amount known about fluid mechanics, and there are a tremendous variety of qualitatively interesting phenomena. The spirit of the subject as a part of physics is perhaps best given by Landau and Lifshitz in their book on fluid mechanics [2], in almost every chapter of which are several examples of phenomena whose understanding and application are of interest in modern physics.

See also: Boundary Layers; Chaos; Fluid Physics; Nonlinear Wave Propagation; Turbulence; Viscosity.

References

[1]   *The Feynman Lectures on Physics*, Vol. II, Chapters 40 and 41. Addison-Wesley, Reading, Mass., 1964. (E)

[2]   L. D. Landau and E. M. Lifshitz, *Fluid Mechanics (Course of Theoretical Physics*, Vol. 6, translated from the Russian by J. B. Sykes and W. H. Reid). Pergamon, New York, 1987. (I)

Bibliography

G. K. Batchelor, *An Introduction to Fluid Dynamics*. Cambridge, London and New York, 1970. (E)
P. G. Drazin and W. H. Reid, *Hydrodynamic Stability*. Cambridge, London and New York, 1980. (I)
E. Guyon, J. P. Hulin, L. Petit and C. D. Mitescu *Physical Hydrodynamics*. Oxford, Oxford, 2001. (E)
Advanced work appears principally in *J. Fluid Mech.*, *Phys. Fluids*, and *Phys. Rev. E.* (A)

Hydrogen Bond

G. C. Pimentel[†]

Introduction

In the preponderant majority of its compounds, the element hydrogen is found attached to one atom, whether in a diatomic or polyatomic molecule. The bond energies (from 40 kcal/mol in RbH to 134 kcal/mol in HF) identify the attachments as chemical bonds and form the basis for the classical bonding rule that hydrogen has a "valence" of one, i. e., it can bond to one other atom.

In many molecules, however, there is evidence of additional attractive interaction that specifically links the hydrogen atom in a group (or molecule) A–H to another atom B in the same or a different molecule. In most cases, this additional interaction involves bond energies of a few kilocalories per mole, well above van der Waals interactions but well below chemical-bond energies. Since the hydrogen atom has already expended its classical bonding capacity, this interaction is given a special name; it is called a hydrogen bond (hereafter, called H bond). It is usually designated $A–H \cdots B$. H bonds are formed by the hydrohalides and by molecules that contain hydrogen that is bonded to oxygen or to nitrogen but not to carbon (examples

---

[†] deceased

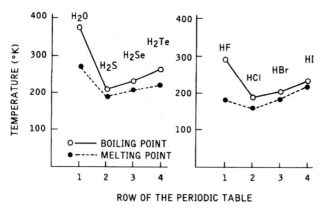

ROW OF THE PERIODIC TABLE

Fig. 1: H-bond formation by H_2O and HF revealed by melting and boiling point trends.

are alcohols, phenols, carboxylic acids, amides, and amines). Such an A–H group can form H bonds to any other molecule that contains an effective electron donor, B. (Examples are oxygen- and nitrogen-containing compounds including ethers, ketones, amines, etc.).

Macroscopic Manifestations of Hydrogen Bonding

Two readily observed properties that can reveal H bonding are the melting and boiling points. Figure 1 shows the trends in these properties for the sixth and seventh column hydrides. The anomalous values for ice and solid HF are symptomatic of specially strong intermolecular forces in these condensed phases. It is now recognized that H_2O and HF form relatively strong H bonds.

The entropy of vaporization demonstrates another facet of this interaction. Most substances have about the same value of $\Delta S_{vap} = \Delta H_{vap}/T$ (Trouton's rule). In contrast, hydrogen-bonded substances all display ΔS_{vap} values higher than 21 cal/degmol (H_2O, 26.1; CH_3OH, 25.0; NH_3, 23.3 cal/deg mol). The excess over the Trouton-rule norm indicates that there is special order in the liquid state when hydrogen bonding occurs. This special order is associated with the fixed, linear orientation in the A–H$\cdots B$ link.

This fixed orientation gives H-bonding liquids specially high dielectric constants. These high dielectric constants reflect the ability of the H bonds to reorient in response to an electric field.

The specially strong intermolecular attractions caused by H bonding are evidenced in the gas phase through the PVT behavior. Specially high deviations from the ideal gas law are always observed when H bonding can be expected. Thus, the experimental second virial coefficients of methanol and of water(-1220 and -976, respectively) are each over twice as negative as those calculated from the Berthelot equation (-550 and -421, respectively). The Berthelot-equation estimate does not include the H-bond part because it is based on the critical constants, and H bonds do not persist at very high temperatures.

The viscosities of H-bonding substances tend to be higher than those of other substances of comparable molecular weight. This is particularly true if the molecule has two H-bonding sites, as for water and for polyhydroxy alcohols. For example, the viscosities of diethyl ether

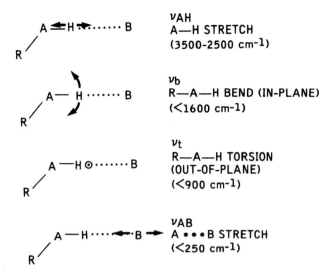

Fig. 2: Vibrational modes of a hydrogen bonded system A–$H\cdots B$.

and n-octane are 0.23 and 0.54 centipoise, respectively. In contrast, the hydrogen-bonding liquids n-octanol, ethylene glycol, and glycerol have viscosities of 10.6, 19.9, and 1490 centipoise, respectively.

This catalogue of observable properties that are affected by H bonding could be extended to include density, surface tension, heat of solution, solubility, conductance, ferroelectric behavior, refractive index, thermal conductivity, acoustic conductivity, and adiabatic compressibility.

Spectroscopic Manifestations

Spectroscopic techniques furnish effective means for detecting and studying H bonds. Infrared absorption spectra are most used because there are quite characteristic spectral changes caused by H bonding. There are four characteristic vibrational modes of the A–$H\cdots B$ group that can be examined, as pictured in Fig. 2.

The A–H stretching mode ν_{AH} has been most studied. For example, the vibration ν_{OH} of monomeric CH_3OH (in dilute CCl_4 solution) absorbs at $3642\,\mathrm{cm}^{-1}$ with a bandwidth near $20\,\mathrm{cm}^{-1}$. At higher concentrations, H bonding causes the formation of dimers, $(CH_3OH)_2$, trimers, $(CH_3OH)_3$, and higher polymers, $(CH_3OH)_n$. The involvement of the O–H group in the H bond causes its absorption band to shift to lower frequencies, to broaden, and to become much more intense (see Fig. 3). The extent of each of these changes is found to increase as the H-bond strength increases, so these spectral properties, $\Delta\nu$, $\nu_{1/2}$, and B/B_1 (B_1 being the integrated absorption of the monomer) furnish useful criteria for ordering substances according to hydrogen bonding strength. Furthermore, systematic concentration studies in an inert solvent (such as CCl_4 or n-hexane) furnish equilibrium data. Such measurements carried out at various temperatures lead to ΔH^0 and ΔS^0 for H bond formation. All of these valuable results make the infrared study of H bonds one of the most fruitful techniques available.

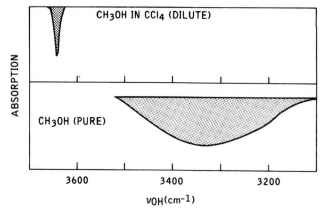

Fig. 3: The O–H stretching vibration in the infrared spectrum of methanol: (a) monomeric CH_3OH in carbon tetrachloride; (b) H-bonded CH_3OH.

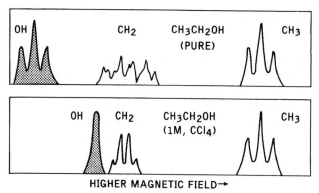

Fig. 4: The effect of H bonding on the NMR resonance of the OH proton in ethanol.

The other vibrations, v_b, v_t, and v_{AB}, are less studied and they are somewhat less useful diagnostically. The bending mode v_b shifts upward slightly ($< 0.2\%$) and is not intensified. The torsional mode v_t can exhibit as spectacular shifts (to higher frequencies), broadening, and intensification as those of v_{AH}. Rather less is known about v_{AB} because the low frequencies ($< 250 \, cm^{-1}$) have made them less accessible, but there is significant potential for their future study and application.

The nuclear-magnetic-resonance (NMR) spectrum also furnishes a rather definitive criterion for detecting H bonding and for classifying it as to strength. The characteristic NMR proton frequency shifts significantly toward lower fields upon H-bond formation. Figure 4 shows how distinctive the bonding perturbation of the NMR resonance can be. In pure ethanol, the OH proton NMR resonance is at much lower magnetic field than that of the CH_2 protons. Dilution to $1.0 \, M$ in CCl_4 causes the OH proton resonance to move over close to the CH_2 signal. Further dilution to less than $0.1 \, M$ reveals the monomeric OH proton resonance to be at higher field than that of the CH_3 protons!

Thermodynamic Properties

The energy and entropy changes (ΔH^0 and ΔS^0) that accompany H-bond formation can be deduced from almost any of the manifestations mentioned above. Some of the most reliable values come from the spectroscopic measurements. In general, H-bond energies are in the range 2–7 kcal/mol, with the notable exception of H bonds to negative ions, as in the bihalide ions, HF_2^- (-37 kcal/mol) and HCl_2^- (-14 kcal/mol).

Entropy changes ΔS^0 on hydrogen-bond formation correlate with the ΔH^0 values. The reason is obvious: As ΔH^0 becomes more negative, the H bond becomes more rigid, and hence its positional randomness decreases.

Structural Properties

X-ray and neutron-diffraction studies of crystals and electron-diffraction studies of gases provide us with information about the geometry of H bonds. Data of this type, which are voluminous, lead to some important generalizations:

(i) In crystals, the A–H bond is almost always within $15°$ of collinearity with $A\cdots B$.

(ii) The $A\cdots B$ distance R_{AB} is less than the sum of the van der Waals radii of A and B, usually by about 0.1 to 0.2 Å. As the H-bond strength increases, R_{AB} decreases.

(iii) The A–H distance R_{AH} is always longer in A–H$\cdots B$ than in the non-H-bonded A–H parent molecule. For the very strongest H bonds, as in HF_2, the proton is symmetrically placed between the A and B atoms.

(iv) H bonds figure prominently in the crystal lattices of all molecules that have the capability of forming intermolecular H bonds to other, like molecules.

These generalizations are of great value in predicting and deducing the molecular structures of biologically important molecules (proteins, DNA, RNA, etc.) in which H bonding is a structural element.

Theory

Early explanations of the H bond were essentially electrostatic in character, and these persist today. However, no correlation is found between the H-bond energy and the dipole moments of either acid or base. There are other deficiencies, such as inability to explain the intensity effects in the A–H stretching and bending modes or the zigzag arrangement of hydrogen halide molecules in their crystal lattices in preference to the parallel orientation preferred by a dipolar array. Any calculational successes obtained using the electrostatic theory must be attributed, at least in part, to the parametric freedom associated with charge placement.

The molecular-orbital description gives a satisfactory qualitative explanation of the existence of the H bond. Axial molecular orbitals are constructed from the hydrogen 1s orbital and the terminal, axially directed p orbitals. If the hydrogen atom is symmetrically placed (as in HF_2^-), there results one bonding, one nonbonding, and one antibonding molecular orbital. With four electrons to be accommodated, only the first two orbitals need to be occupied, so two half-order bonds result. The nonbonding orbital places the charge on the terminal atoms, those with high electronegativity.

The most extensive *ab initio* calculations on H bonds utilize the Hartree–Fock approximation. These calculations depend upon cancelation of electron-correlation effects since configuration interaction has not been included. Yet the completeness of the basis sets has approached the point that useful calculations of the rather small H-bond energies can be made. These studies seem to corroborate the charge distribution implied by the molecular orbital approach just described.

Incorporation of configuration interaction in calculations of H-bond energies and geometries must be high on the agenda of today's theorists because of the crucial role played by hydrogen bonding in biological molecules such as the proteins and DNA. Both intramolecular H bonds within the DNA molecule and intermolecular H bonds involving the aqueous environment play a role in determining primary, secondary, and tertiary structures of these macromolecules. Accurate theory is needed to predict such structures because these structures determine biological function.

Significance

The mere existence of the oceans on earth is, of course, attributable to the H bonding that elevates the boiling point of water (see Fig. 1). Thus we are afforded the vast thermostat that moderates the climate and that provided the primordial brew in which life first appeared. Water's ionizing solvent properties, too, can now be seen to involve H bonding. The H-bond energies of bifluoride and bichloride ions imply an $HOH \cdots Cl^-$ bond energy near 20 kcal. Four such H bonds to Cl^- account nicely for the heat of aquation of a chloride ion, a heat that is essential to the stability of electrolyte solutions. Finally, all of the sugars, carbohydrates, cellulose, proteins, and other stuff of which plants and animals are made are generously sprinkled with ether linkages, hydroxyl groups, amides, and esters – groups that raise aqueous solubility and that provide compatibility with the aqueous milieu in which they must function.

The structure of a protein even more intimately depends on H bonding. Its skeleton is a long polymer of amide linkages formed from α-amino acids. This skeleton coils into a specially configured helical structure, called the α-helix, which is maintained by H bonds between successive coils (see Fig. 5a). Even more ornate in their dependence on H bonding are the structures of nucleic acids, including DNA. Figure 5b pictures diagrammatically the double-stranded helix that represents the basic makeup of DNA. The spiraling ribbons denote sugar-phosphate chains and the bars represent matched pairs of bases that bond the chains together by H bonds. Each base pair includes a purine molecule linked to a pyrimidine partner that matches in a lock and key H-bonded configuration. Despite the size of a DNA molecule (molecular weight can be as high as 10^9), it contains only four different amine bases, adenine, thymine, cytosine, and guanine (abbreviated A, T, C, and G). Adenine and thymine have geometrically fixed capacities to form hydrogen bonds that match each other so well that an adenine base can recognize a thymine molecule and bond to it in strong preference to any other bases present. Cytosine and guanine match in a similar way.

This recognition capability is incorporated when two sugar-phosphate strings twist into the double-helix structure. Because of the specificity of the A–T and C–G hydrogen bonds, the helix can form only if the sequence of bases on the first string is matched in perfect complementarity to the sequence on the second string. The order in which these base pairs occur creates an information code in the molecule. This information can be copied (replicated)

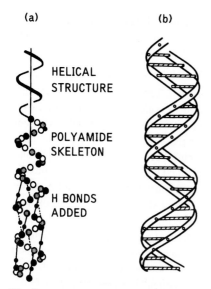

(a) **(b)**

HELICAL
STRUCTURE

POLYAMIDE
SKELETON

H BONDS
ADDED

Fig. 5: (a) A schematic representation of the a-helix in protein: the helical configuration is determined and maintained by H bonds, (b) A schematic representation of DNA: the spiraling ribbons denote sugar-phosphate chains and the bars represent matched pairs of bases linked by H bonds.

to produce duplicate DNA molecules through enzymatic synthesis. The double strands unzip their hydrogen bonds to expose a single strand that serves as a sequence guide in enzymatic synthesis of a complementary copy. This process involves making and breaking of relatively weak H bonds which can be carried out under conditions that do not break the much stronger covalent bonds that carry and preserve the molecular information. Thus, the genetic coding in DNA and its replication are accomplished through an elegant orchestration of weak H bonds and strong covalent bonds.

See also: Chemical Bonding; Interatomic and Intermolecular Forces; Molecular Structure Calculations.

Bibliography

S. Bratoz, *Adv. Quantum Chem.* **3**, 209 (1966).

D. Hadzi (ed.), *Hydrogen Bonding, Symposium on Hydrogen Bonding.* Pergamon Press, New York, 1957.

G. C. Pimentel and A. L. McClellan, *The Hydrogen Bond.* Freeman and Co., San Francisco, Calif., 1960.

G. C. Pimentel and A. L. McClellan, "Hydrogen Bonding", *Ann. Rev. Phys. Chem.* **22**, 347 (1971).

J. A. Pople, W. G. Schneider, and H. J. Bernstein, *High Resolution Nuclear Magnetic Resonance,* Chap. 15. McGraw–Hill, New York, 1959.

N. D. Sokolov, *Ann. Chim. (Paris)* **10**, 497 (1965).

Hypernuclear Physics and Hypernuclear Interactions

A. R. Bodmer

Introduction

Hypernuclear physics deals with the interactions of hyperons. These are "strange baryons" which are heavier than neutrons (n) or protons (p) referred to collectively as nucleons (N). Strangeness, represented by the quantum number S, is conserved in strong interactions but is violated in weak interactions which are responsible for the slow decay of hyperons with a lifetime of about 10^{-10} s. The lightest hyperon is the neutral Λ with $S = -1$, spin $\frac{1}{2}$, and a mass of 1116 MeV, i.e., about 1.2 times that of a nucleon. The free Λ decays are mostly $\Lambda \to p + \pi^-$ or $n + \pi^0$ where π denotes a pion (the lightest meson). The next heavier hyperon is the Σ also with $S = -1$ and spin $\frac{1}{2}$; it exists in three charge states Σ^+, Σ^-, Σ^0 (positive, negative, and neutral) and is about 80 MeV heavier than the Λ.

In the quark model, baryons are composites of three (valence) quarks (plus "sea" quarks). Nucleons are composites of the lightest up (u) and down (d) quarks, whereas hyperons contain one or more strange (s) quarks which carry strangeness $S = -1$ and which are about 150 MeV heavier than the u, d quarks. Thus a proton $p \equiv (uud)$ is a composite of two u quarks and a d quark whereas a $\Lambda \equiv (uds)$ is a composite of a u, d, and s quark.

Hypernuclei are "strange" nuclei with one or more hyperons bound to an (ordinary) nucleus. Most observed hypernuclei are Λ hypernuclei with one Λ, but Σ hypernuclei have also been observed and there is evidence for $\Lambda\Lambda$ hypernuclei with two Λs. The ground states of Λ hypernuclei decay via the weak interaction with lifetimes about the same as that of a free Λ, i.e., about 2×10^{-10} s. Thus for most purposes the low-lying hypernuclear states may be considered as completely stable, just as are the β-decaying states of ordinary nuclei. Striking differences between ordinary and hypernuclei arise because the Λ is distinct from the nucleons, with no Pauli principle between it and these. The Λ can then go into deeply bound orbits already occupied by nucleons. In particular for the ground state the Λ occupies the 1s orbit which has zero orbital angular momentum.

The Strong Hyperon–Nucleon Force

Because Λs and Σs are short-lived one cannot produce beams of these at low energies. The information about the Λ–N and Σ–N forces is thus much less than about the N–N force and has been obtained from the scattering of Λs and Σs by protons in a hydrogen bubble chamber. Nevertheless, these data show quite clearly that the Λ–N nuclear force is strong and about half as attractive as the N–N force. A distinctive feature of the ΛN, ΣN systems is their strong coupling to each other because the Λ and Σ may be rapidly converted into each other since they have the same strangeness. Thus above the threshold energy of about 80 MeV one has the nuclear reaction $\Lambda + N \to \Sigma + N$. However, even for the ΛN system below the ΣN threshold, where no Σ can be produced, this "$\Lambda\Sigma$" conversion can strongly influence the net ΛN force because of the rather small $\Lambda\Sigma$ mass difference. Thus the time–energy uncertainty principle allows the Λ to turn into a Σ for a few percent of the time, proportional to $(M_\Sigma - M_\Lambda)^{-1}$.

Theoretical models of these forces invoke the exchange of mesons for larger separations, greater than about 0.5×10^{-13} cm, and the interaction of quarks for shorter distances. For a meson of mass m, the force range is about $\hbar/mc$. Because there is only a single neutral λ (its isospin is zero), the exchange of a pion – the lightest meson – between Λ and N, or another Λ, is strongly inhibited. Thus this long-range pion-exchange force is strongly suppressed in the Λ–N (and Λ–Λ) system and mostly only heavier mesons contribute. The inner part of the N–N force is known to be strongly repulsive, and can be explained as mostly due to the interaction of the quarks in the two nucleons when these overlap at small separations. The short-range character of the hyperon–nucleon forces is not established, although quark models suggest a repulsion for the Λ–N and Σ–N forces. Since strange quarks are now involved, knowledge about this part of the force could give important insight into the strong interactions of strange quarks. It is clearly important to obtain more data about the hyperon–nucleon forces. This could be done by studying the interactions of a nucleon and a hyperon produced, for example, by the inelastic scattering of energetic electrons by nuclei using the CEBAF electron accelerator projected to come on-line in a few years time.

Hypernuclei and Strong Interactions

The earlier data (1950s and 1960s) are from emulsion and bubble chamber experiments in which hypernuclei are an end product of negative kaon–nucleus interactions, the negative kaon (K^- meson) with $S = -1$ transmitting its strangeness to a nucleon to produce a Λ or Σ (e. g., $K^- + n \to \pi^- + \Lambda$). Especially through observation of the decay products of a hypernucleus one has determined accurate ground-state binding energies, the spin of very light hypernuclei, and certain decay characteristics for individually identified hypernuclei up to a mass number (total number of nucleons plus Λ) of $A = 14$ corresponding to hypernuclear nitrogen, $^{14}_{\Lambda}$N.

More recently, extensive experiments with K^- beams of well-defined low energy have produced many excited states of hypernuclei by means of the strangeness-exchange (K^-, π^-) reaction $K^- + {}^A Z \to \pi^- + {}^A_{\Lambda} Z$, where $^A Z$ is the target nucleus and $^A_{\Lambda} Z$ is the final hypernucleus. The fundamental process is $K^- + n \to \pi^- + \Lambda$. The Λ may be produced with a controllable and small momentum and thus various states of the final hypernucleus may be produced. In particular, the Λ can be substituted with large probability in place of a loosely bound neutron in the original nucleus, producing a "strangeness-analog" or "substitutional" state where the Λ is in the same higher angular momentum orbit as the replaced neutron, whereas for the ground state the Λ has to occupy the 1s orbit. In this way many excited states of relatively light (p-shell) hypernuclei have been identified. In some cases γ rays (photons) emitted by the decay of an excited state to the ground state have also been observed. Σ hypernuclei have also been observed in (K^-, π^-) reactions.

Very recently beams of positive pions have been used to produce hypernuclei through the (π^+, K^+) reaction: $\pi^+ + {}^A Z \to {}^A_{\Lambda} Z + K^+$ with detection of the K^+ meson (strangeness $= +1$), the basic process being $\pi^+ + n \to K^+ + \Lambda$. In this reaction the Λ is in general produced with a larger momentum than in the (K^-, π^-) reaction and is thus less favorable for the production of substitutional-type states. However, this larger momentum makes the (π^+, K^+) reaction much more favorable for the production of hypernuclei with the Λ in deeply bound states, in particular the 1s state. Also since the K^+ is absorbed much less inside a nucleus than the K^-, the (π^+, K^+) reaction allows the production and observation of much heavier hypernuclei than the (K^-, π^-) reaction, with the heaviest hypernucleus studied so far being $^{89}_{\Lambda}$Yb with $A = 89$.

The following are some of the principal results and problems of (strong) hypernuclear physics.

The lightest Λ hypernuclei are more weakly bound than ordinary nuclei because the Λ–N force is less attractive than the N–N force. There is in fact no two-body bound ΛN (or ΣN) system, whereas a neutron and proton are bound together as a deuteron. The lightest hypernucleus is the hypertriton $^3_\Lambda H$ (bound state of npΛ) with the Λ only just bound. For heavier hypernuclei a dramatic effect is that the energy needed to separate a Λ from a hypernucleus in its ground state is much greater than that ($\simeq 8\,\text{MeV}$) needed to separate a nucleon from a nucleus. This nucleon, because of the Pauli principle, must go into the last unfilled shell and thus has a large (Fermi) kinetic energy which largely compensates the attractive potential energy due to its interactions with the other nucleons. The Λ, however, can go into deeply bound orbits already occupied by nucleons; in particular it can go into the $1s$ orbit to produce the hypernuclear ground state. For this orbit the Λ kinetic energy can be quite small and the resulting binding energy quite large, even though the potential energy of the Λ is less than that for a nucleon. For very heavy hypernuclei the $1s$ state of the Λ is bound by about $30\,\text{MeV}$ which is then also the ground-state Λ separation energy. The recent (π^+, K^+) experiments, made for a large range of nuclei, show in fact a very well-defined sequence of deeply bound single-particle-like states which are very well explained as a result of the Λ moving in a potential well generated by the nucleons of the hypernucleus. In a given hypernucleus these states correspond to increasing orbital angular momentum of the Λ, with the $1s$ state being the most deeply bound. These results also show very clearly that the Λ maintains its identity inside a nucleus and does not dissociate into its constituent quarks.

More detailed calculations of the ground-state energies of Λ hypernuclei which use Λ–N forces consistent with Λp scattering show the need for strongly repulsive many-body, in particular three-body, forces involving the Λ and at least two nucleons. Thus the free Λ–N force is effectively weakened in a nucleus. One possible mechanism for such a weakening is the inhibition of the $\Lambda\Sigma$ conversion process by the presence of other nucleons. Recent calculations have shown that three-body forces between nucleons are important also for ordinary nuclei.

A noteworthy result obtained from the excited states of hypernuclei (in particular of $^9_\Lambda Be$ and $^{16}_\Lambda O$) observed in the (K^-, π^-) experiments is that the Λ–nucleus spin–orbit force is much weaker than the nucleon–nucleus one which plays such a vital role for the nuclear shell model. The (K^-, π^-) experiments have also led to the observation of excited Σ hypernuclear states with remarkably narrow widths in view of the strength of the (strong) $\Sigma + n \to \Lambda + n$ reaction by which these states can decay. These narrow widths have so far received no satisfactory explanation.

Recently an observation of the $A = 4$ helium hypernucleus $^4_\Sigma He$ ($\Sigma^+ p 2n$) has been reported. There is now a whole family of $A = 4$ nuclei: the ordinary ^4He nucleus (the alpha particle), the Λ hypernuclei $^4_\Lambda H$ ($\Lambda 2np$) and $^4_\Lambda He$ ($\Lambda n2p$) in both the ground and an excited state, and now, if confirmed, also $t_\Sigma He$, with all the hypernuclei being much less bound than ^4He. The excited states of $^4_\Lambda H$ and $^4_\Lambda He$ are observed by their γ decay to their respective ground states, with the Λ flipping its spin relative to the spin of the nucleons. Analysis of the ground-state and excited-state energies of both $^4_\Lambda H$ and $^4_\Lambda He$ give quite detailed information about the charge-symmetry breaking of the Λ–N force, i.e., about the difference between the Λ–p and Λ–n forces. The Λ–p force is found to be slightly more attractive than the Λ–n force. The lack of any spin dependence in the difference is in strong disagreement with meson-exchange models, suggesting the possible importance of quark effects.

Two emulsion events of $\Lambda\Lambda$ hypernuclei have been reported. The binding energy of such a $\Lambda\Lambda$ hypernucleus, where the two Λs are bound to an ordinary nucleus, can then give information about the Λ–Λ force. Analysis of the $^{10}_{\Lambda\Lambda}$Be event (two alpha particles + two Λs) indicate that the Λ–Λ force is about as attractive as the corresponding Λ–N force. Proposed experiments to produce $\Lambda\Lambda$ hypernuclei are clearly very much needed.

An exciting and fundamental question is the existence and properties of bound or resonant states of six quarks with a baryon number of 2. In particular, theoretical models suggest that the most strongly bound such state is the so-called H dibaryon$\equiv (uuddss)$ with the same quantum numbers as two Λs, namely, $S = -2$ (and zero spin). The H could possibly be quite strongly bound with respect to two Λs, and could then decay only by weak interactions and would be long-lived. Experiments to search for the H have been proposed and are under way. An exciting possibility is the existence of superdense hadronic "nuclei" of very large strangeness which are more strongly bound than ordinary nuclei. Such "strangelets" would be multiquark objects with roughly equal numbers of u, d, s quarks, and would thus have a large negative strangeness equal to the number of s quarks and a small (positive) charge (Bodmer, Witten). Their properties would be quite unusual, e. g., the ability to grow by "eating up" nucleons. It has been suggested that very rapidly rotating very compact pulsars may then be strange quark stars rather than ordinary neutron stars.

An intriguing situation where hypernuclear interactions may play an important role is in the deeper interior of neutron stars (pulsars). For fairly massive neutron stars (about the mass of the sun or greater) the matter not too deep in the interior is expected to be mostly in the form of neutrons. However, deeper in the interior the pressure due to gravitation may be sufficient to compress the matter so much beyond ordinary nuclear densities that it may become energetically favorable for some or even most of the neutrons to transform into hyperons. The interior of massive neutron stars may thus consist of a baryonic "soup" containing many species of hyperons whose effect would be to soften the equation of state with possible observational consequences.

Weak Interactions

The weak decays of the ground states of Λ hypernuclei are a unique source of information about the weak Λ–N force. The free pionic decay $\Lambda \rightarrow N + \pi$ leaves the final nucleon with only a small kinetic energy ($\simeq 5\,\mathrm{MeV}$). This decay will however be strongly suppressed inside a hypernucleus since most of the states for the final nucleon are now unavailable being occupied by the nucleons of the hypernucleus as a result of the Pauli principle. The decay of the Λ then proceeds mainly by the nonmesonic strangeness-changing weak interaction: $\Lambda + N \rightarrow N + N$ whereby the Λ loses its strangeness turning into a nucleon, the net result being two energetic nucleons for which there is little reduction in the available final states. The large energy release is a signature of the nonmesonic decay. Thus recent measurements on $^{12}_{\Lambda}$C show the nonmesonic decay rate to be about 20 times the pionic rate. In spite of the very different decay mechanism from that of a free Λ the lifetime of $^{12}_{\Lambda}$C is observed to be about the same as the free Λ decay time, i. e., about $2.5 \times 10^{-10}\,\mathrm{s}$. For even heavier hypernuclei the nonmesonic rate is about 100 times the pionic rate, dramatic evidence for Pauli suppression of the latter. Experimentally, the neutron nonmesonic rate ($\Lambda + \mathrm{n} \rightarrow \mathrm{n} + \mathrm{n}$) is about the same or even greater than the proton rate ($\Lambda + \mathrm{p} \rightarrow \mathrm{n} + \mathrm{p}$). A puzzle is that theoretical models predict the neutron rate to be much smaller than the proton rate.

See also: Hyperons; Neutron Stars; Nuclear Forces; Nuclear Structure; Nucleon; Pulsars; Weak Interaction.

Bibliography

R. H. Dalitz, *Nuclear Interactions of the Hyperons.* Oxford University Press, London and New York, 1965.

A. Gal, "Strong Interactions in Λ-Hypernuclei", in *Advances in Nuclear Physics*, Vol. 8 (Michel Baranger and Erich Vogt, eds.). Plenum Press, New York and London, 1975.

B. Povh, *Annual Review of Nuclear Science*, Vol. 28, p. 1. Annual Reviews, Inc., Palo Alto, CA, 1978.

Proceedings of the 1985 International Symposium on Hypernuclear and Kaon Physics, *Nucl. Phys.* **A450** (1986).

Proceedings of the 1988 International Symposium on Hypernuclear and Low-Energy Kaon Physics, *Nuovo Cimento* (1989).

A. R. Bodmer, *Phys. Rev. D* **4** (1971) 1601; E. Witten, *ibid.* **D4** (1984) 272.

Hyperons

D. O. Caldwell

Hyperons are particles that are baryons, like the proton, but they are heavier than the proton and possess a different property, designated by the quantum number of "strangeness". Baryons are fermions (i.e., they obey Fermi–Dirac statistics and have half-integer intrinsic angular momentum, or spin) and hadrons, meaning that they interact via the strong interactions.

Of the first two "strange" particles observed in cosmic rays in 1947, one was a hyperon. The particles were considered strange because they were produced copiously (i.e., via a strong interaction) and decayed slowly (i.e., via a weak interaction), and yet the two processes appeared to be the inverse of each other and both should have been strong. That is, the decay should have taken about 10^{-23} s, instead of the observed 10^{-10} s, a decided anomaly. As was proved in 1954 when strange particles could be produced in large numbers at the Cosmotron accelerator, their apparently strange behavior was due to their possessing a new quantum number, which has been called strangeness or hypercharge. Since this quantum number is conserved by the strong interactions, nonstrange particles (like nucleons – such as the proton, p – and π mesons) can produce strange particles only in pairs having opposite values of the quantum number. Thus the Λ^0 hyperon (the first observed) has strangeness $S = -1$, whereas the K^0 meson (the other particle seen first in 1947) has $S = +1$, making the sum $S = 0$ for both sides of the production equation, $\pi^- + p \rightarrow \Lambda^0 + K^0$. On the other hand, the decay $\Lambda^0 \rightarrow p + \pi^-$ can occur only by the weak interaction because the new quantum number is not conserved by the process (i.e., $S = -1 \rightarrow S = 0$). In the cosmic-ray observations the Λ^0 and K^0 were not seen in the same interaction because they were observable only by their short-lived charged-particle decays (e.g., $K^0 \rightarrow \pi^+ + \pi^-$) and not when $K^0 \rightarrow \pi^0 + \pi^0$ or $\Lambda^0 \rightarrow n + \pi^0$. In addition, half the time the K^0 has a much longer-lived decay, which was not then observable. Thus the apparently strange behavior is not really strange, but was the first observation of the effects of a new quantum number.

Table 1: Properties of the "Stable" Hyperons

Particle	Mass (MeV)	Mean life (s)	Strangeness	Spin
Λ^0	1115.7	2.63×10^{-10}	-1	$\frac{1}{2}$
Σ^+	1189.4	0.80×10^{-10}		
Σ^0	1192.6	7×10^{-20}	-1	$\frac{1}{2}$
Σ^-	1197.4	1.48×10^{-10}		
Ξ^0	1314.8	2.9×10^{-10}	-2	$\frac{1}{2}$
Ξ^-	1321.3	1.64×10^{-10}		
Ω^-	1672.4	0.82×10^{-10}	-3	$\frac{3}{2}$

The hyperons differ from the strange mesons not only in having the opposite sign (for particles, as opposed to antiparticles) of their new quantum number S, but also in possessing another property called baryon number, B. Baryons are part of the wider class of fermions (unlike the mesons, which are bosons), but both protons and electrons are fermions. Since the proton is not observed to decay into the much less massive positive electron, or positron, the difference between the particles is characterized by assigning $B = 1$ to the proton, whereas an electron has an analogous lepton number, describing its membership in its own family. The baryons, as opposed to the leptons (electron, muons, and tauons, and their neutrinos), are strongly interacting particles, or hadrons. It is sometimes convenient to utilize B and replace S with the hypercharge quantum number, referred to above, which is $Y = S + B$.

Because their decays could be observed directly, the first hyperons discovered were those that decayed by the weak or electromagnetic (e. g., $\Sigma^0 \rightarrow \Lambda^0 + \gamma$) interactions. Some of the properties of these so-called "stable" hyperons are given in Table 1. Note that all the hyperons have a larger mass than that of the proton (938.3 MeV), a fact that was the original source of their name. Note also that hyperons exist with larger values of the strangeness quantum number, the Ξ having $S = -2$ and the Ω^- having $S = -3$.

All hadrons are made up of quarks, mesons being a quark–antiquark combination and baryons having three quark constituents. One type of quark has $S = -1$, so the Λ and Σ possess one of these, the Ξ has two, and the Ω^- has all three quarks of the strange variety. The strange quark has charge $-\frac{1}{3}$, and the remainder of the quarks in the hyperons are the charge $+\frac{2}{3}$ up quark and the charge $-\frac{1}{3}$ down quark. Thus the Λ^0 hyperon consists of one strange, one up, and one down quark. While baryons with charm or bottom quarks can have strange quarks as well, these particles are not usually called hyperons.

There are a large number of hyperons which decay via the strong interaction. These particles are massive enough so that they have available other particles into which they can decay while still conserving the strangeness quantum number. At present, about 14 Λ-like, 9 Σ-like, 6 Ξ-like, and 1 Ω-like heavier particles are well established. For the particles in Table 1, as well as many of these heavier resonant states, their quantum numbers, masses, and partial decay rates have been determined. There are so many of these particles that the term "hyperon" is now less used, and instead members of each group are referred to as "Λ baryons," "Σ baryons," etc.

Except for the Σ^0, the hypersons of Table 1 have been produced at very high energy accelerators in beams of sufficiently long effective life that quite accurate magnetic moments have been measured. Even the transition magnetic moment of the Σ^0 has been determined by a different technique. A good first approximation to these magnetic moments is provided by the simplest quark model, using the p, n, and Λ moments as inputs, providing strong support for the quark structure.

See also: Baryons; Elementary Particles in Physics; Strong Interactions; *SU*(3) and Higher Symmetries; Weak Interactions.

Bibliography
R. H. Dalitz, *Rep. Prog. Phys.* **20**, 163 (1957). (On early history.)
Particle Data Group, *Phys. Lett.* **592B**, 1 (2004), or later versions, which appear every two years, alternately in *Reviews of Modern Physics*. (On hyperon properties.)

Hysteresis

R. A. Dunlap

Hysteresis is a phenomenon in which the state of a system does not reversibly follow changes in an external parameter. Examples of hysteresis include the dependence of strain or stress for many materials and the resonant response of nonlinear systems. Most commonly, however, the term hysteresis refers to the behavior of the magnetic induction B or, analogously, the magnetization M as a function of the applied magnetic field intensity H in ferromagnetic and ferrimagnetic materials, or the behavior of the polarization as a function of applied electric field in ferroelectric materials.

For a typical ferromagnetic material, the hysteresis in B as a function of H is illustrated in Fig. 1. This hysteresis loop arises principally from irreversible domain wall motion in the material which results from structural imperfections such as grain boundaries, dislocations, etc. The hysteresis loop is measured by applying a cyclic magnetic field. Beginning with an initially unmagnetized sample in zero field, the B–H behavior will follow the curve *oa* as H is increased. This portion of the curve is the initial magnetization curve. In sufficiently large H, B reaches its saturation value, B_s. As H is reduced along the portion of the curve *ab*, the magnetic induction decreases to a remanent value, B_r, at $H = 0$. The direction of the applied field is then reversed and the magnetic induction follows the portion of the curve *bc*, reaching zero at $H = H_c$, the coercive field or coercivity. Reverse saturation is obtained at point *d* on the curve. The portion of the curve *defa* results from applying progressively more positive H and in most magnetic materials is symmetric with the *abcd* portion of the curve. Repeated cycling of the applied magnetic field will repeatedly trace out the hysteresis loop *abcdefabc*.... The so-called minor hysteresis loop *a'b'c'd'e'f'a'b'c'*... is obtained by cycling the applied field with an amplitude which is insufficient to attain magnetic saturation. A family of such curves exists for different driving field amplitudes.

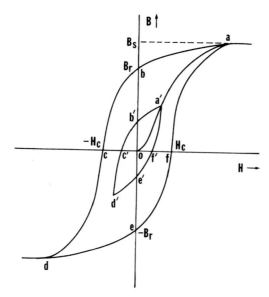

Fig. 1: Hysteresis loop of a ferromagnetic material.

The area enclosed by the hysteresis loop is the energy loss W per cycle of the applied field and is an integral over the loop of the form

$$W = \oint H \, dB \tag{1}$$

where H is in A/m, B is in Wb/m^2 (or tesla), and W is in J/m^3. Materials with small H_c and B_r have correspondingly small W and are referred to as soft magnetic materials. Soft magnetic materials with sufficiently large B_s have applications as transformer cores, electric motor stators and rotors, tape recorder heads, etc. Commercial applications, such as power distribution transformer cores, frequently use Fe–Si alloys with room-temperature $W \approx 70 \, \text{J/m}^3$. Newly developed low-loss amorphous and nanocrystalline alloys typically have $W < 1 \, \text{J/m}^3$.

The so-called "hard" magnetic materials have large H_c and B_r, and a correspondingly large value of W, and form the class of permanent magnets. The maximum of the product BH in the second quadrant of the hysteresis loop (i. e., the portion of the curve bc) is a convenient measure of the energy density of a permanent magnet. Commercially available Co–Sm and Nd–Fe–B permanent magnet materials have room-temperature BH_{max} up to $\sim 200 \, \text{kJ/m}^3$ and $\sim 400 \, \text{kJ/m}^3$, respectively.

See also: Ferroelectricity; Ferromagnetism; Magnetic Domains and Bubbles; Magnetic Materials; Magnets (Permanent) and Magnetostatics.

Bibliography
R. M. Bozorth, *Ferromagnetism*. Van Nostrand, New York, 1951.
B. D. Cullity, *Introduction to Magnetic Materials*. Addison-Wesley, Reading, MA, 1972.
A. H. Morrish, *The Physical Principles of Magnetism*. Wiley, New York, 1965.

Ice

H. Engelhardt

Ice, the frozen form of liquid water, is abundant on the earth's surface, in the planetary system, and in interstellar space. If all the ice presently existing on earth melted, the sea level would rise about 66 m. In some planets and in most moons, ice is the major constituent. Pluto is 80% ice; Jupiter's moons Europa, Ganymede and Callisto and Saturn's Titan contain up to 40% ice. Ice is also present in many other moons, in the planetary ring systems, and in comets. Planet Mars has significant polar ice caps and apparently frozen oceans. The mass of ice stored in the terrestrial ice sheets of Antarctica (91%) and Greenland (8%) and in alpine glaciers (1%) is closely related to climate.

All of the natural ice on earth is hexagonal ice, ice Ih, as manifested in six-cornered snow flakes (see Fig. 1). At lower temperatures and at pressures above 0.2 GPa at least 12 other ice phases with different crystalline structures exist. No other known substance exhibits such a variety of forms. The phase diagram of ice shows the conditions of stability for the ice phases (Fig. 2; Table 1). The equilibrium line between water and ice Ih has negative slope, which is a consequence of the peculiar ice structure with large hexagonal channels resulting in a lower density than the liquid, unlike most other substances and the high-pressure ice phases. The equilibrium lines extend as metastable phase boundaries into the area of stability of other ice phases. Ice IV exists only as a metastable phase within the stability fields of ice III, V, and VI. After forming by nucleation from supercooled water, it quickly transforms into one of the other phases unless it is quenched to low temperatures.

The individual water molecule remains intact as the fundamental building unit in all the ice phases except ice X. Since there are equal numbers of protons and of lone electron pairs in the water molecule, each proton can match up with an electron pair of a neighboring molecule forming a hydrogen bond (H-bond). Each water molecule forms H-bonds to four nearest neighbors in a tetrahedral arrangement with one proton occupying each H-bond. The tetrahedral bond geometry explains the openness and relatively low density of the ice Ih structure. In ice Ih the O–O–O angles are nearly the same as the perfect tetrahedral angle (109.5°), which

Encyclopedia of Physics, Third Edition. Edited by George L. Trigg and Rita G. Lerner
Copyright © 2005 WILEY-VCH Verlag GmbH & Co. KGaA, Weinheim
ISBN: 3-527-40554-2

Table 1: Structural data on the phases of ice.

Ice phase	Crystal system	Space group	Cell dimensions[a] (pm, deg)	No. of molecules in a unit cell	No. of nearest neighb.	Distance of nearest neighbors (pm)	O-O-O angles[a] (deg)	Hydrogen positions	Density[a] (Mg m^{-3})
Ih	Hexagonal	$P6_3/mmc$	$a = 450; c = 732$	4	4	275	109.3–109.6	Disordered	0.93
Ic	Cubic	$Fd3m$	$a = 635$	8	4	275	109.6	Disordered	0.93
a$_I$	Amorphous	–	–	–	4	277		Disordered	0.94
a$_{II}$	Amorphous	–	–	–	4	280		Disordered	1.31
II	Rhombohedral	$R\bar{3}$	$a = 778; \alpha = 113.1$	12	4	275–284	80–128	Ordered	1.18
III	Tetragonal	$P4_12_12$	$a = 673; c = 683$	12	4	276–280	87–141	Disordered	1.15
IV	Rhombohedral	$R\bar{3}c$	$a = 760; \alpha = 70.1$	16	4	279–292	87.7–127.8	Disordered	1.27
V	Monoclinic	$A2/a$	$a = 922; b = 754;$ $c = 1035; \beta = 109.2$	28	4	276–287	84–128	Disordered[b]	1.24
VI	Tetragonal	$P4_2/nmc$	$a = 627; c = 579$	10	4	280–282	76–128	Disordered[c]	1.33
VII	Cubic	$Pn3m$	$a = 341$	2	8	295	109.5	Disordered	1.66
VIII	Tetragonal	$I4_1/amd$	$a = 480; c = 699$	8	8	296 and 280[d]	109.5	Ordered	1.66
IX	Tetragonal	$P4_12_12_1$	$a = 673; c = 683$	12	4	276–280	87–140	Ordered	1.16
X	Cubic	$Pn3m$	$a = 283$	2	8	245	109.5	Symmetric	2.51

[a] Cell dimensions and density are at 110 K and atmospheric pressure. Ice X at 44 GPa and 25 °C.

[b] Becomes partially ordered when quenched to low temperatures. A more fully ordered phase V has space group $P1/a$.

[c] When cooled slowly to −150 °C, ice VI' is partly ordered with orthorhombic space group $Pmmn$. A fully ordered ice VI' has space group Pn.

[d] Four of the nearest neighbors are hydrogen-bonded to central molecule at a distance of 296 pm; four are nonbonded at 280 pm.

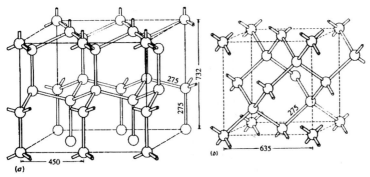

Fig. 1: The structure of (a) ice Ih and (b) ice Ic. Note the tetrahedral coordination and the openness of the structure. On each hydrogen bond, shown by a rod joining the oxygen atoms, lies one proton in an asymmetric position (not shown). Bond lengths, 275 pm, are indicated. The hexagonal *c* axis is labeled 732 pm, and one of the hexagonal a axes is labeled 450 pm. In (b), the cubic unit cell is outlined with dashed lines; dimensions are in pm at 110 K.

matches fairly well the H–O–H angle of the water molecule, little changed from the angle in the free molecule (104.4°). In the higher-pressure phases the tetrahedral bonding geometry is distorted; the bond angles are bent from perfect tetrahedral and the H-bonds are stretched. The distances to nonbonded near neighbors become shorter the higher the pressure. In ice VII the nonbonded neighbors come as close as the nearest neighbors. Ice VII and ice VIII can be visualized as two ice Ic structures completely intertwined with one another. A general feature repeated in all of the ice structures is H-bonded rings. In ice I and ice II the smallest rings are of six molecules, but in higher-pressure phases rings of five and four molecules occur.

Since the H-bond lengths are in the range of 274 pm (ice Ih) to 296 pm (ice VIII) and the O–H distances are only 98.5 pm (ice Ih), the position of the protons in the H-bonds is asymmetric. There are six different orientations possible for a water molecule in its tetrahedral bonding environment, each corresponding to a different arrangement of protons in its four H-bonds. If all the possible orientations of the water molecule at each lattice site are equally realized, the phase is proton-disordered; if one orientation is preferred the phase is proton-ordered. Most ice phases have a high- and low-temperature modification which are distinguished by their degree of proton order. Ice VIII is the proton-ordered version of ice VII and ice IX is the proton-ordered version of ice III. Partially proton-ordered modifications exist for ice V and ice VI. Ice II is a completely proton-ordered phase for which the proton-disordered version is unstable and is not observed. In most cases the existence of proton order or disorder affects the crystallographic symmetry of the ice phase. The cubic symmetry of fully disordered ice VII changes to tetragonal symmetry on transformation to ordered ice VIII.

If water vapor is condensed on a cold substrate between −80 °C and −130 °C, a cubic modification, ice Ic, is formed. Ice Ic is related to the structure of ice Ih in the same way that cubic diamond is related to hexagonal diamond, the cubic and hexagonal forms having almost the same density. Below −130 °C a noncrystalline amorphous solid known as low-density amorphous ice, ice a_I, appears. A high-density amorphous phase, ice a_{II}, with a density of 1.31 Mg/m^3 can be made at 77 K by compressing ice Ih to 1 GPa.

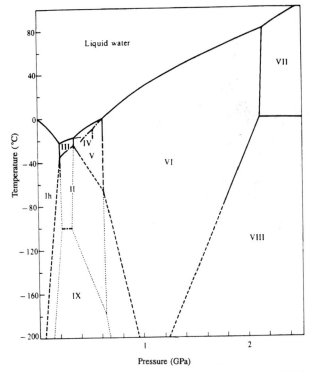

Fig. 2: The phase diagram of ice: —, measured stable equilibrium lines;
·—·—·—·, measured metastable lines; – – –, extrapolated or estimated stable
lines; ···, extrapolated or estimated metastable lines. Ice IV is metastable in
the region of stability of ice V. The line above ice IV is the equilibrium line
between liquid water and ice IV. Not shown are ice Ic, the amorphous phases,
and the ordered versions of ice V and ice VI. Ice X exists above 44 GPa.

At a pressure of 44 GPa, the H-bonds in ice VII are so shortened by compression to 245 pm
that the protons leave their asymmetric positions in the bonds and move to the centers, forming
the structure of ice X, in which, because of the symmetric H-bonding, discrete H_2O molecules
are no longer identifiable.

Many of the physical properties of ice are unique and relate to the special features of structure, especially the H-bonding (Table 2). Certain physical properties arise from defects in the
ideal ice structure. Important among the point defects are the H_3O^+ and OH^- ions and the D
and L bond defects – sites where H-bonds are broken by violation of the rule that there be one
and only one proton along each O–O bond line. The violation is caused when a water molecule
undergoes rotation from one of its six possible orientations to another without compensating
adjustments by the adjacent molecules to which it is H-bonded. Such a rotation interchanges
some of the molecule's protons and electron pairs, with the result that some of the H-bonds are
converted to nonbonded O–O contacts with no protons (L defect) or with two associated protons (D defect). The ions and orientational defects, together with substitutional or interstitial
impurity atoms or ions, have important influence on the electrical and mechanical properties

Table 2: Physical Properties of Ice Ih at 0 °C.

Density	$0.917\,\mathrm{Mg\,m^{-3}}$
H-bond length	$276.5\,\mathrm{pm}$
Adiabatic compressibility	$0.119\,\mathrm{GPa^{-1}}$
Isothermal compressibility	$0.33\,\mathrm{GPa^{-1}}$
Melting point	$273.15\,\mathrm{K}$
Melting point depression	$-74\,\mathrm{K\,GPa^{-1}}$
Specific heat	$2.01\,\mathrm{kJ\,kg^{-1}\,K^{-1}}$
Heat of melting	$334\,\mathrm{kJ\,kg^{-1}}$
Heat of sublimation	$2.84\,\mathrm{MJ\,kg^{-1}}$
Thermal conductivity	$2.2\,\mathrm{W\,m^{-1}\,K^{-1}}$
Linear expansion coefficient	$55 \times 10^{-6}\,\mathrm{K^{-1}}$
Cubical expansion coefficient	$166 \times 10^{-6}\,\mathrm{K^{-1}}$
Vapor pressure	$610.7\,\mathrm{Pa}$
Static dielectric constant	96.5
High-frequency dielectric constant	3.2
Dielectric relaxation time	$20\,\mathrm{\mu s}$
Activation energy for dielectric relaxation	$55\,\mathrm{kJ\,mol^{-1}}$
Refractive index	1.31
Electric dc conductivity	
of ice single crystals at $-10\,°\mathrm{C}$	$1.1 \times 10^{-8}\,\mathrm{\Omega^{-1}\,m^{-1}}$
of polycrystalline glacier ice	$10^{-5} - 10^{-6}\,\mathrm{\Omega^{-1}\,m^{-1}}$
Proton mobility	$0.8 \times 10^{-4}\,\mathrm{m^2\,V^{-1}\,s^{-1}}$
Acoustic velocity	
longitudinal wave	$3828\,\mathrm{m\,s^{-1}}$
transverse wave	$1951\,\mathrm{m\,s^{-1}}$
Velocity of radio waves	$170\,\mathrm{m\,\mu s^{-1}}$

of ice and in some cases completely dominate them. Also important are crystal dislocations (line defects) and subgrain or grain boundaries (surface defects), the latter in polycrystalline ice.

The electrical conduction in ice is carried by the protons, whose motion is closely tied to the motion of ionic defects. An ionic defect can move from one lattice site to the next by a small jump of a proton from one off-center (asymmetric) position to the other in an H-bond that connects the two lattice sites. Since the small proton jumps occur collectively along a favorably oriented chain of H-bonds, the ionic defects and the effective charge carried by them have a very high mobility in the ice structure, 10 times greater than the mobility of ions in normal ionic conductors. Protonic conduction plays a major role in biological charge transfer processes across membranes. Proton conduction in ice and H-bonded materials is analogous to electron conduction in semiconductors. Ice Ih has an unusual high static dielectric constant (96.6 at 0 °C compared to 88 for liquid water). The reorientation of the molecular dipoles in ice is facilitated by the mobility of orientational L and D defects, which are generated by thermal activation. The dielectric relaxation time is extremely slow compared to water ($20\,\mathrm{\mu s}$ in ice and $10\,\mathrm{ps}$ in water at 0 °C).

Ice is transparent to visible light. It has the lowest index of refraction for the sodium D line of any known crystalline material. It is doubly refracting, uniaxial, optically positive with very small birefringence. The proton-disordered phases have a broad infrared absorption band for the fundamental intramolecular bending and stretching vibrations (near $3220\,cm^{-1}$ and $1650\,cm^{-1}$ for ice Ih). The infrared band for hindered rotations of the water molecules in ice Ih is centered around $840\,cm^{-1}$. The translational lattice vibrations absorb in the range from 50 to $350\,cm^{-1}$ (with peak for ice Ih at $229\,cm^{-1}$). The proton-ordered phases show distinct narrow peaks in their infrared and Raman spectra. From infrared and Raman spectroscopy on the ice phases, much has been learned about the intermolecular coupling mechanism, lattice dynamics, and the properties of H-bonds.

For electromagnetic radar waves with frequencies from 5 to 300 MHz the loss of energy by absorption in ice is sufficiently small that they can penetrate large ice masses great distances. Radio waves are reflected by inhomogeneities in the ice and at material boundaries, especially at the ice–water and ice–rock interfaces, and these waves can therefore be utilized to examine the internal structure of glaciers and to determine the depth and bottom topography of large polar ice sheets and ice shelves.

Ice is a viscoelastic material with a nonlinear flow law. When shear stress is applied to a single crystal of ice, it undergoes plastic shear strain easiest parallel to the basal plane, which is perpendicular to the hexagonal c axis. In other directions the stress needed to produce plastic shear deformation is much higher. When polycrystalline ice is subjected to stress, it immediately deforms elastically, followed by transient creep, and finally steady viscous flow called secondary creep is reached. For high stresses in excess of 400 kPa the creep curve accelerates which is called tertiary creep. Several physical processes are responsible for these deformations: movement of dislocations, sliding along grain boundaries, and recrystallization. The steady-state secondary creep rate $\dot{\varepsilon}$ in secondary creep is related to the stress σ by the flow law

$$\dot{\varepsilon} = A\sigma^n \exp\left(-\frac{Q}{kT}\right) ,$$

where A is a temperature-independent constant, Q is the activation energy for creep, n is the nonlinear exponent, k is the Boltzmann constant, and T is the absolute temperature. Values of A, Q, and n depend to some extent on the grain size, grain orientation distribution, and impurity content of the polycrystalline material. At temperatures between 0 and $-10\,°C$ and for stresses between 100 and 250 kPa, $A = 5 \times 10^{-15}\,s^{-1}\,kPa^{-3}$, $Q = 139\,kJ\,mol^{-1}$, $n = 3$. At stresses lower than 100 kPa, $n = 2$.

The surface of ice Ih shows unique properties. Near the melting point, the surface contains many dangling broken bonds that promote the existence of a liquid-like layer. Consequences of the surface properties are sintering of snow to cohesive snowballs, recrystallization of snow to firn and its transformation to glacier ice, and the low friction of many materials on ice, which is useful for sled riding, skiing, and skating. Regelation is a unique ice property: melting of ice under pressure, coupled with adjacent refreezing of meltwater at lower pressure and transfer of heat from one side to the other. It is the mechanism by which a loop of wire can be pulled slowly through an ice block without cutting the block in two.

There is a whole class of solids, clathrates, where the ice forms a H-bonded host lattice that encages a great variety of small guest atoms or molecules like argon or methane. Ice VI, VII, and VIII can be viewed as self-clathrates, where two equal ice lattices interpenetrate each other but are not H-bonded to each other. Ice-like structures or structurally arranged water molecules are encountered in hydrates and hydrated compounds including all macromolecules that would not be biologically active without their structured water or ice-like shells. Even water can be viewed as a partially broken down structure of ice.

H-bonds, so pervasive in the crystalline ice phases, play a major role in many substances from glues and mortar to the life-supporting structure of proteins. The properties of ice, and H-bonds, are recognized in many aspects of physics, chemistry and biology, in several branches of geophysics, including glaciology (dynamics of large ice masses), meteorology (cloud physics), and in oceanography (sea ice), in planetary sciences, and in astronomy.

See also: Hydrogen Bond; Lattice Dynamics; Phase Transitions; Water.

Bibliography

N. H. Fletcher, *The Chemical Physics of Ice*. Cambridge University Press, Cambridge, 1970. (An intermediate and advanced text that illustrates many aspects of solid-state physics using ice as an example.)

P. V. Hobbs, *Ice Physics*. Clarendon Press, Oxford, 1974. (A comprehensive treatment and documentation of ice at an intermediate and advanced level.)

B. Kamb, "Crystallography of Ice", in *Physics and Chemistry of Ice*, E. Whalley, S. J. Jones, and L. W. Gold (eds.). Royal Society of Canada, Ottawa, 1973. (An advanced text.)

E. Whalley, "The Hydrogen Bond in Ice", in *The Hydrogen Bond*, P. Schuster, G. Zundel, and C. Sandorfy (eds.). North-Holland, Amsterdam, 1976. (An advanced text.)

W. S. B. Paterson, *The Physics of Glaciers*. Pergamon Press, Oxford, 1994. (An intermediate text on applied ice physics.)

Journal of Glaciology, published by the International Glaciological Society, Cambridge. (A professional journal.)

Annals of Glaciology. (Proceedings of International Conferences on all aspects of Glaciology).

Inclusive Reactions

T. Ferbel

Collisions at high energies generally involve the production of many elementary particles. To establish which features of multiparticle production reflect the basic dynamics requires a detailed study of such production processes. One approach is to examine as many complete reactions as are accessible to experimental investigation. For example, in a collision between particle A and particle B we might study the specific channels $A + B \rightarrow A + B + C$, $A + B \rightarrow C + D$, $A + B \rightarrow A + C + D + G$, or the like (where C, D, and G represent produced particles). Reactions such as these, in which all final-state particles are measured, are termed *exclusive* reactions. A complementary approach to the problem is based on a form of statistical sampling of all exclusive channels using *inclusive* reactions. An inclusive experiment measures the probability for the production of a specified configuration of particles in a collision, independent of whatever else might be produced. Thus the simplest (albeit trivial) inclusive measurement, yielding what might be termed a zero-particle inclusive cross section, is one in which no particles at all are specified, but rather a sum is taken over all possible final states. The rate for the zero-particle inclusive reaction A + B → *anything* is consequently determined by the total cross section (a Lorentz-invariant quantity, σ_t) for the interaction of particle A with particle B.

A more informative inclusive reaction is a single-particle inclusive process of the kind A + B → C + *anything*, in which the cross section for the production of particle C is measured regardless of whatever is produced along with C. This type of reaction is characterized by a function, $d\sigma/(d^3 p/E)$, proportional to the probability for emitting particle C, of momentum p and energy E, into a Lorentz-invariant differential element of momentum space $d^3 p/E$. Further information pertaining to multiparticle production can be obtained by examining a two-particle inclusive reaction of the kind A + B → C + D + *anything*; in this case the production of C and D is characterized by a Lorentz-invariant two-particle differential cross section, $d\sigma/[(d^3 p_C/E_C)(d^3 p_D/E_D)]$. Although we might expect that additional dynamic information could be obtained from studies of higher-order inclusive reactions, this does not seem to be the case; it appears that inclusive cross sections of order greater than 2 may be calculable from lower-order data for total collision energies in the center-of-mass (c.m.) system (E_T^*) of at least up to ~ 50 GeV.

Inclusive production cross sections depend in general on the nature of the colliding particles, on their states of polarization, on the energy available in the center-of-mass frame, and on the momentum vectors of the specified final-state particles. In the absence of polarization, the cross section for any specific inclusive reaction can depend only on E_T^* and on the longitudinal and transverse momenta of the specified particles (longitudinal being along the collision axis). Transverse momenta (p_T) of most particles produced in hadronic reactions tend to be small and their distribution only weakly dependent on E_T^*. (The average p_T value for pions is ~ 0.4 GeV/c.) Longitudinal momenta (p_L) can be large, and are often comparable to incident momenta. Two p_L-like variables have been found to be particularly useful in describing inclusive reactions. These are (1) the Feynman x variable, $x = 2p_L^*/E_T^*$, where the asterisks signify c.m. quantities; and (2) the rapidity variable $y = \frac{1}{2}\ln[(E + p_L)/(E - p_L)]$. The range of x is limited between −1 and +1, while the allowed range of y values grows logarithmically with E_T^*.

It is often convenient to distinguish two regimes of particle production in "soft" high-energy collisions (i.e., collisions that do not ostensibly involve constituents). One of these is the "fragmentation region": here, the produced particles have momenta comparable to those of the incident particles (i.e., large $|x|$ values) and can be thought of as fragments from the breakup of the target or projectile. The other regime is the "central region"; here particles are produced with comparatively small momenta in the c.m. frame and therefore do not appear to be associated with either the target or the projectile particle.

Total cross sections for hadrons interacting with other hadrons vary substantially at low energies ($E_T^* < 3$ GeV), but at intermediate incident momenta ($E_T^* < 100$ GeV) the dependence is much weaker. For $E_T^* > 100$ GeV, σ_T appears to grow logarithmically (almost doubling every decade). It has been suggested that a similar simplification may obtain in the case of inclusive reactions for large E_T^*. In particular, the hypothesis of limiting fragmentation (HLF), motivated in part by a diffraction picture of particle production, asserts that at fixed p_T the invariant cross section (or this cross section divided by σ_T) for producing a particle of any finite p_L, as measured in the rest frame of the target or the projectile particle, approaches a finite, energy-independent limit as E_T^* grows.

An alternative statement of limiting behavior in inclusive reactions is the scaling hypothesis, which is based on an analogy between hadron emission in high-energy collisions and the radiation of photons by an accelerating charge. This hypothesis states that for fixed p_T and large E_T^*, the invariant cross section for any x value approaches a limit that is E_T^* independent. Yet another formulation of asymptotic behavior is available in a macroscopic picture of high-energy collisions that involves a short-range order (SRO) hypothesis. This view of particle production suggests that among the particles involved in a scattering event only those with similar rapidities are correlated. Thus at asymptotic energies, particle production and correlations among produced particles should depend not on rapidity but rather on differences in the y values of produced particles. At asymptotic E_T^*, HLF and scaling yield the same predictions in the fragmentation region. The scaling and the SRO hypotheses also provide predictions for limiting behavior in the central production domain.

A comparison of the x dependence of invariant inclusive cross sections for pion production at fixed p_T and varying E_T^* indicates that up to $E_T^* \sim 900 \, \text{GeV}$, scaling appears to hold surprisingly well at small p_T, particularly at large $|x|$ values (fragmentation region). However, small but significant deviations are observed near $x = 0$ (central region). If inclusive cross sections were to start scaling at some E_T^*, then the invariant cross sections would develop plateaus in the rapidity variable and, as E_T^* is increased, those plateaus would grow in extent as $\ln E_T^*$. It is easy to show that as a result of this, the average number of particles of any particular species produced in an inelastic collision would rise linearly with $\ln E_T^*$. This is only approximately what happens for $E_T^* \gtrsim 10 \, \text{GeV}$. In fact, the average number of charged particles ($\langle n \rangle$) produced in inelastic p–p collisions for $E_T^* \gtrsim 10 \, \text{GeV}$ is well represented by $\langle n \rangle = 0.5 \ln^2 E_T^* + 0.9 \ln E_T^* + 1$. (Pi mesons account for about 90% of all produced particles, and K mesons for about 10%.) The values of $\langle n \rangle$ for other types of incident particles differ only somewhat from that in p–p reactions. In fact, except for an overall normalization (given approximately by σ_T), inclusive particle production in the central region appears to be independent of the incident channel.

At small p_T, where there is no fundamental theory of particle production (such as, e. g., QCD at large p_T), the rate at which inclusive cross sections (or their normalized values) might approach their asymptotic forms and relationships between different reactions can be examined by means of a generalization of the optical theorem. The usual optical theorem relates the amplitude for the two-body elastic-scattering process $A + B \to A + B$ to the total cross section (zero-particle inclusive reaction $A + B \to anything$); there is an analogous relationship connecting cross sections for higher-order inclusive reactions to multiparticle elastic-scattering amplitudes ($A + B \to C + anything$ is related to $A + B + \bar{C} \to A + B + \bar{C}$ scattering, where $\bar{C}$ is the antiparticle of C). Although experiments on multiparticle scattering are not feasible, models can be used to calculate these "elastic"-scattering amplitudes and such calculations provide the phenomenology for relating different inclusive reactions. There has been substantial success in this regard in the application of Regge models to inclusive reactions.

High-energy collisions typically produce a great number of particles, but the number of exclusive channels that are simple enough to be amenable to experimental probing at high energies is rather small. It therefore appears that, of the two approaches available for investigation of high-energy phenomena, the inclusive technique may be the more valuable one for gaining an understanding of the dynamics of multiparticle production.

See also: Elementary Particles in Physics; Hadrons; Strong Interactions.

Bibliography
H. Bøggild and T. Ferbel, *Annu. Rev. Nucl. Sci.* **24**, 451 (1974). (I-A)
D. Horn and F. Zachariasen, *Hadron Physics at High Energies.* Benjamin, Advanced Book Program, Reading, Mass., 1973. (I-A)
J. G. Rushbrooke, in *Proceedings of the International Symposium on Multiparticle Dynamics*, J. Granhans (ed.). World Scientific Publisher, Singapore, 1985.

Inertial Fusion

S. W. Haan

Inertial fusion (IF) is one of the two principal schemes proposed for producing technologically useful energy from nuclear fusion. In IF the burning plasma is confined by its own inertia, the burn time being limited by hydrodynamic disassembly. For electromagnetically confined fusion *see* Plasma Confinement Devices.

In IF, beams of radiation or ions (designated driver beams) are used to implode targets containing fusionable material, in order to compress and heat them sufficiently so that thermonuclear reactions ensue. Generation of electric power from such a process is potentially attractive, as large amounts of energy could be produced from readily available fuel with relatively little radioactive waste. Also, an IF facility could be of value to defense research, both for studies of effects of the explosions' output on military hardware and for basic studies of plasma processes under conditions relevant to thermonuclear weapons.

IF was proposed in the early 1960s, very quickly following the invention of the laser in 1960. Work by Stirling A. Colgate, Ray E. Kidder, John H. Nuckolls, Ronald F. Zabawski, and Edward Teller in that period already defined many of the crucial concepts as currently understood. Most of their work was initially classified and was not published until the early 1970s. A succession of more and more powerful laser has explored the physics of IF, culminating in a 1.8 MJ laser called the National Ignition Facility with which researchers plan to produce thermonuclear ignition in approximately 2010.

In the following we consider the requirements for an implosion intended to produce significant yield efficiently. Basic features of this implosion determine many characteristics of the drivers, described in subsequent sections here. They also determine the overall scale for high-yield IF, which requires drivers somewhat larger than any currently existing.

IF targets can be "direct drive," in which the driver energy is deposited directly in an ablator, or "indirect drive," in which symmetry is enhanced by converting the driver energy to thermal x rays that drive the ablator. Details regarding indirect drive were classified until about 1990. Generally, the implosion requirements are the same for direct and indirect drive.

Implosion Characteristics

A target generally includes two nested spherical shells (see Fig. 1). The outer shell, the ablator, is heated by energy from the driver-either directly or by use of a *hohlraum* (Fig. 1). Ablation creates pressure at the outer surface, which causes inward acceleration, and ultimately com-

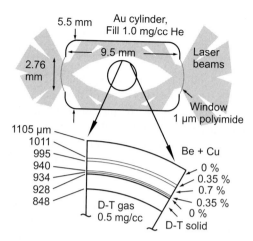

Fig. 1: Sectional view of an *indirect drive* IF target. Energy from a powerful laser irradiates the inside of a cylindrical cavity, called a "hohlraum", filling it with nearly Planckian x rays. These irradiate and implode the spherical central capsule consisting of a low atomic number ablator enclosing a shell of fusible deuterium–tritium fuel. The specific target shown here uses a beryllium ablator doped with copper, with dimensions appropriate for the National Ignition Facility, currently under construction. The x-ray drive on the spherical capsule is smooth at small length scales; its symmetry over longer scales is controlled by the geometry of the hohlraum and beam placement. In *direct drive*, there is no cavity and the laser irradiates the ablator. In that case about 60 individual beams, each focused to a spot size about equal to the capsule diameter, are required to produce sufficiently symmetric illumination.

pression, of the remainder of the pellet. The interior shell is fusionable material, usually equimolar deuterium and tritium (D–T). Schemes have been proposed using other fusion fuels such as D–^{3}He, but such fuels require higher temperatures for the reactions.

For fusion to occur, the thermal kinetic energy of the D and T ions must be high enough to overcome the Coulomb barrier. The Maxwellian-averaged D–T reaction rate, $\overline{\sigma v}$, is therefore strongly temperature dependent. Significant energy is produced only when temperatures of a few keV are reached. Each reaction produces a 14-MeV neutron and a 3.5-MeV α particle. The neutrons generally escape. (In a reactor, these would be absorbed in a heat-transfer medium, and in a lithium supply for tritium reproduction.) The hot D–T in the center of the fuel (the hot spot) must have column density (ρr) at least 0.3 g/cm^2, so that the αs are stopped in the hot spot providing further heating. The compressed main shell of D–T, cooler and at higher density, must have (ρr) greater than about 1.0 g/cm^2 in order to confine the fuel inertially while it ignites and burns.

Self-sustaining burn can ensue, provided there is a positive energy balance of four processes: α deposition and $P\,dV$ work from the implosion, which deposit energy into the fuel, and bremsstrahlung radiation and thermal conduction, which remove energy. The balance of these four processes depends on the implosion velocity of the igniting material, the tempera-

ture, and the ρr of the hot spot. Ignition requires temperatures of about 5 keV, implosion velocities of about 3×10^7 cm/s, and ρr as described above. Once such conditions are achieved,
the burn rate increases further, and the ignited fuel "bootstraps" via accelerating α deposition
up to temperatures of tens of keV.

Subsequent dynamics are controlled by the disassembly. The net energy produced per mass
of fuel depends on the density, since the reaction rate is proportional to ρ^2, on the temperature
via the thermally averaged cross section velocity product $\overline{\sigma v}$, and on the confinement time,
which is essentially the time it takes a rarefaction wave to travel in from the perimeter. These
variables enter in such a way that, once ignition occurs, the burn-up fraction depends only on
the total ρr. This is equivalent to the density-confinement time product needed for magnetic
fusion.

Significant burn-up (30%) requires $n\tau$ of about 2×10^{15} cm^{-3} s, or ρr of at least 2 g/cm^2.
Clearly there is advantage to having ρr larger than the 0.3 g/cm^2 required for ignition, but
heating additional fuel to ignition temperature wastes driver energy. So the optimal imploded
configuration has two distinct fuel regions: a hot spot with $\rho r \cong 0.3$ g/cm^2 and $T \cong 5$ keV to
produce ignition, and a surrounding "main fuel" layer, at lower temperature so that it can be
efficiently compressed to increase the total ρr to about 3 g/cm^2.

For a given mass of D–T in radius R, $\rho r \propto 1/R^2$. So the required ρr values could in principle
be achieved by compression of an arbitrarily small mass of D–T and correspondingly small
energy investment. However, while the energy produced by burning a mass of D–T at a given
ρr is proportional to the mass, the energy required to compress a dense Fermi gas to density
ρ is proportional to $m\rho^{2/3}$. So the gain (energy out/energy in) drops for smaller masses,
which require proportionately more energy for compression. Target gains of order 100 are
required to produce a significant net system gain given a driver efficiency of order 10–30%.
The mass required to produce such a target gain depends on how efficiently the driver energy
is coupled to compression, how much energy is used to heat the hot spot, and other factors,
but for most configurations the requirement that the gain be about 100 sets the main fuel mass
at about 10^{-3} g. Another factor limiting the mass is the compression that can be realistically
achieved. This is limited by the symmetry of the implosion. Plausible convergence ratios
(initial outer radius/final hotspot radius) are in the range 30–50. The main fuel itself then
converges in linear dimensions by a factor of about 10, corresponding to compressions of
order 1000. Given an initial density of 0.25 g/cm^3 for cryogenic D–T, a mass of 2 mg would
produce a ρr of 3 g/cm^2. This corresponds to about the same mass as is set by the requirement
of gain 100. These two issues, fortunately compatible, set the overall scale for high-gain IF:
the target contains a few mg of D–T, at initial radius R of a few mm. The energy produced is
between 100 MJ and 1 GJ. In a reactor designed to produce about 1 GW of electricity, targets
of this scale would have to be shot at a rate of 2 to 30 per second, depending on target yield
and reactor efficiency. The energy in the assembled DT is about 100 kJ; depending on the
coupling efficiency, the driver energy E is required to be in the range 1–10 MJ. The pulse
length, $t \cong R/(3 \times 10^7$ cm/s$)$, is about 10 ns, so that the power E/t is a few times 10^{14} W, and
the intensity $E/4\pi r^2 t$ is a few times 10^{14} W/cm^2. These numbers are characteristic of both
direct and indirect drive, which have coupling efficiencies similar enough not to change these
estimates.

Additional requirements are determined by considering how the ignition configuration can
be assembled. The main fuel must be compressed and accelerated by the ablation pressure;

the kinetic energy thereby achieved is then converted into further compression of the shell, as well as compression and $P\,dV$ heating of the hot spot. If the hot-spot mass is a few percent of the main fuel mass, the kinetic energy in the latter at the velocity required for ignition $(3 \times 10^7\,\text{cm/s})$ is also adequate to produce heating of the hot spot to several keV. Thus the object of the first part of the implosion is to accelerate the main fuel to about $3 \times 10^7\,\text{cm/s}$, while keeping its entropy very low so that its pressure is close to the minimum imposed by Fermi degeneracy. Minimal entropy requires that the fuel be initially liquid or solid. A series of shocks may pass through the fuel, but they must be relatively weak. Other sources of energy deposition in the main fuel must be minimized.

Achieving the needed velocity requires high ablation pressure, low mass per unit shell area, and/or large distance over which the acceleration can act. The aspect ratio of the compressed shell in flight is limited by hydrodynamic stability considerations to be in the range of 30 to perhaps 60, depending on how well the implosion can be designed to limit instability growth. This requires that the ablation pressure be in the range 50–200 Mbar. Such a large pressure cannot be applied immediately to cold D–T, but four or five shocks in succession can produce adequate pressure while maintaining sufficiently low entropy. Each shock is launched by an increase in driver intensity, and the increases are timed so that the shocks coalesce just inside of the high-density fuel layer.

Another important issue is the efficiency with which the incoming driver energy is converted to ablation pressure and then to kinetic energy. The ablation efficiency depends on the driver type used. With direct drive, the driver energy is absorbed somewhere outside the actual ablation surface, and then conducted to the ablation front. Differences in the radii of absorption and ablation help to improve symmetry, but reduce the efficiency. X-ray drive is absorbed at the ablation surface, and has a higher mass ablation rate which reduces the hydrodynamic instability growth. This advantage compensates for the reduction in efficiency that results from heating the hohlraum.

Other ignition configurations have been considered that are more complicated than the hot-spot/main-fuel configuration described here. If the igniting hot spot is surrounded by a shell of material with high atomic number, and hence high density and high opacity, the energy balance for ignition improves. Such a shell can also be driven to very high velocity by collision with another, heavier, shell converging from a larger radius. However, these more complicated configurations generally involve interfaces that are hydrodynamically unstable. They also risk contamination of the fuel with high atomic number material, which can increase the energy losses dramatically. Another important alternative configuration is the "exploding pusher", in which the entire shell is heated rapidly and about half of it implodes while the other half explodes. Such targets have low gains, but have been very useful for physics experiments since they produce high temperatures, albeit at very low ρr.

Potential Drivers and Related Trade-offs

Four driver systems are being actively considered at the time of this writing: two short-wavelength laser systems, heavy ions, and x rays produced by Z-pinches.

High-energy lasers are most readily available in the infrared, but research in the 1970s found that laser–plasma interactions generate energetic electrons which produce copious preheat and seriously degrade the implosion. The most important such process is stimulated

Raman scattering, in which the incident infrared resonantly decays into a scattered electromagnetic wave and an electron plasma wave, which produces the energetic electrons. A similar process, stimulated Brillouin scattering, is the decay into an electromagnetic wave and an ion acoustic wave; this process can degrade the absorption or change its location, since nearly all the incident energy can be transferred to the scattered light wave. In more recent experiments, these deleterious plasma processes have been sharply reduced by using laser light with wavelengths 0.35 μm or shorter. The short-wavelength laser light is absorbed primarily by inverse bremsstrahlung (collisional heating). In all cases, the interactions are concentrated near the surface of so-called critical density, at which the incident frequency equals the plasma frequency and inside of which the light cannot propagate. To a large degree, the advantage of shorter-wavelength illumination is the higher critical density, so that the plasma is more collisional where the light is being absorbed. In indirect drive the laser is incident on high atomic number material, and the efficiency of its conversion to thermal x rays is a key issue.

Of the two candidate short-wavelength laser systems, the most thoroughly investigated is doped Nd glass, pumped with optical flashlamps. This system emits 1.06-μm infrared, which is frequency-tripled in optically nonlinear crystals before hitting the target. In most current glass lasers the infrared pulse is generated by passing through a series of slabs, individually pumped, which serve as amplifiers. Several systems of this type have been built, the largest of which could produce 120 kJ. The Omega laser currently operating at the University of Rochester is of this type, with an active experimental program utilizing about 10 shots per day. This type of laser, at larger scale, would adequately function as a driver for a very low repetition-rate high-gain experimental facility. Such a facility is being constructed at the time of this writing, the National Ignition Facility, with scheduled completion in 2009. One of 48 beamlines is operative. It is designed to produce 1.8 MJ of frequency-tripled light, which is expected to be adequate to achieve the burn conditions described above and produce 10–30 MJ of thermonuclear yield. With respect to electricity production, it has yet to be demonstrated that glass lasers can be sufficiently efficient and that they can be built to sustain the thermal loads resulting from a repetition rate of order $10 \, \mathrm{s}^{-1}$.

The other candidate laser system is the KrF gas laser. A gaseous mixture of Ar–Kr–F_2 is pumped either by electronic discharge or by an electron beam, and the system lases at a wavelength of 0.28 μm wavelength via decay of Kr^+F^- ion pairs to a ground state of dissociated covalent Kr and F. Efficient amplification requires a pulse length of $> 100 \, \mathrm{ns}$, and so the short pulse required for IF must be created by a pulse-compression scheme. For example, a synthetic pulse can be created from a sequence of short pulses which propagate through the gain medium in slightly different directions. After amplification, the pulses are individually temporally delayed by passing them through different path lengths before they are overlapped on the target.

X rays produced by a high current Z-pinch are also being investigated as driver candidates. Z-pinch drivers appear to be relatively inexpensive, but it is difficult to produce sufficiently high power, and to provide the symmetry and pulse-shaping needed for implosions. There is an active experimental program on a facility named Z at Sandia Laboratory in Albuquerque, NM, using hohlraums heated to temperatures up to $\sim 150 \, \mathrm{eV}$.

Finally, heavy-ion beams could become the most attractive driver option. Singly to triply charged ions with atomic mass 130 or higher would be accelerated to about 10 GeV. The target coupling physics appears to be much simpler than with lasers and has desirable scaling with intensity. The system appears to be efficient, and a $10\,s^{-1}$ repetition rate is feasible. Although cost scaling to a large facility is very attractive, it is also the principal difficulty at this time: a small, "experimental" heavy ion driver is much more expensive than a comparable laser or light-ion machine.

Current Research

Most experimental facilities are generally doped Nd glass lasers, frequency multiplied. Experiments, details of which are classified, have also been done using nuclear explosions in underground tests by the United States and possibly other countries.

The basic physics of the implosions is being tested at various laboratories, with both direct and indirect drive. High-density compressions and high temperatures have been generated, and it is generally agreed that the spherically symmetric implosion hydrodynamics is well understood. Detailed simulations with large computer codes are used to model the experiments.

The effects of deviations from spherical symmetry are the topic of much current research, both theoretical and experimental. It is important to estimate the effect of irradiation asymmetry and hydrodynamic instabilities, because the shell as described above is hydrodynamically unstable on the outside while being accelerated, and on the inside while being decelerated. Instability growth determines the acceptable values for the aspect and convergence ratios, and thus is related to the required size and other properties of the driver. Also, since inhomogeneities in driver intensity can couple to the instabilities, smooth beams are very important and work is being done to improve beam quality and test the consequences.

Laser–matter interactions are another important area of current research. Hot electron production has been reduced with short-wavelength lasers, but questions remain regarding backscattered light, beam-to-beam energy transfer in the plasma, and how the phenomena may change in plasmas with larger length scales and the densities and temperatures characteristic of high-gain designs. The efficiency of absorption, and for indirect drive, conversion to x rays, are very important in establishing the size and properties of a high-gain driver.

Inertial fusion experimental and theoretical programs are closely coupled to a broader research field of high energy density science. The facilities described above are used to study hydrodynamics, equation of state and opacity physics, and other areas of research accessible with these facilities, with applications in astrophysics, planetary science, and other areas.

Experiments to date indicate that high-gain targets would work within the parameter range described above; the major research efforts now are related to achieving ignition on the National Ignition Facility as soon as possible after the facility's completion, and to directing the program thereafter towards efficient, high-gain targets and applications to energy production and research using high energy density plasmas.

Acknowledgement
This research was supported by the U.S. Department of Energy, Lawrence Livermore National Laboratory under Contract No. W-7405-ENG-48. Much of this article is based on an original article by Heiner Meldner.

See also: Bremsstrahlung; Fermi–Dirac Statistics; Fluid Physics; Lasers; Nuclear Fusion; Plasmas; Plasma Waves; Shock Waves and Detonations.

Bibliography

S. Atzeni and J. Meyer-ter-Vehn, *The Physics of Inertial Fusion*. Clarendon Press, Oxford, 2004.

R. S. Craxton, R. L. McCrory, and J. M. Soures, *Sci. Am.* **255**, No. 2, 68 (1986).

J. F. Holzrichter, E. M. Campbell, J. D. Lindl, and E. Storm, *Science* **229**, 1045 (1985).

J. D. Lindl, *Inertial Confinement Fusion*. Springer, New York, 1998.

J. D. Lindl *et al.*, *Physics of Plasmas* **11**, 339 (2004).

J. H. Nuckolls, L. L. Wood, A,. R. Thiessen, and G. B. Zimmerman, *Nature* **239**, 139 (1972).

Infrared Spectroscopy

D. A. Dows

Infrared spectroscopy is primarily a method for the study of the vibrations of atomic nuclei in molecules and crystals. When electromagnetic radiation in the infrared region of the spectrum falls on a sample, energy may be transferred to or from the moving nuclei. The resulting removal or addition of electromagnetic energy is measured to record the spectrum. The infrared region of the electromagnetic spectrum lies at longer wavelengths, or lower energy, than the visible light region. It is common to refer to the near infrared (about 0.7–2 μm), the middle infrared (about 2–50 μm), and the far infrared (greater than 50 μm) wavelength regions.

Energy level spacings for molecular and crystal vibrations lie in the 0–0.5 eV energy range, and transitions between them which can absorb or emit light lie in the infrared. These transitions can also be observed in light scattering, where a monochromatic incident light beam, usually from a visible laser, gives rise to frequency-shifted lines in the scattered light (the Raman spectrum); the shift is just the vibrational energy difference. They can also be observed as vibronic structure in the electronic spectrum (in the visible and ultraviolet regions of the spectrum).

It is usual in infrared spectroscopy to describe transition energies by giving the wave number of the transition, obtained from the Bohr frequency condition $\bar{v} = (E_1 - E_2)/hc$, where E_1 and E_2 are the upper and lower state energies, h is Planck's constant, and c is the velocity of light. The units of $\bar{v}$ are cm^{-1}, but $\bar{v}$ is often called the "frequency" of the transition (the frequency, measured in hertz, is really equal to $c\bar{v}$). The vibrational "frequency" ranges from about 10 to 4000 cm^{-1}, and the corresponding wavelength of infrared light from about 1000 to 2.5 μm.

Since the frequencies are determined by the molecular or crystal potential function, or by its parameters, the "force constants", study of the infrared or Raman spectrum most immediately yields information about the strengths and stiffness of chemical bonds and valence angles or about the binding of atoms in crystals. In addition, due to the characteristics of the physical processes leading to interaction with light and thus to spectroscopic phenomena, there exist "selection rules" limiting the number of transitions which can be observed. These selection rules (which may be different for the infrared and the Raman spectra) depend on the symmetry of the molecule or crystal, and their existence often adds to the determination of molecular structure or crystal symmetry.

Infrared Spectroscopic Method

Infrared spectroscopy, first developed in the early years of the last century by Coblentz, experienced a long period of dormancy followed by a dramatic resurgence at the end of the Second World War due to development of new technologies. In an ideal spectrometer, strong radiation from a perfectly monochromatic, tunable light source would pass through the sample to a sensitive, noise-free detector. In reality, a typical spectrometer used continuous radiation, from a heated source, which passes through the sample, is dispersed by a monochromator into narrow frequency bands, and is measured by a thermal or photoconductive detector. The wartime development of electronic systems, particularly the phase-sensitive detector and low-noise amplifier, together with more recent photoconductive detectors and bolometers brought infrared spectroscopy to the status of a routine tool in physics and chemistry. An alternative to the monochromator has now become the standard; the light from the sample is passed through a Michelson interferometer, which records the Fourier transform of the spectrum as its movable mirror is scanned. The Fourier-transform (FTIR) method of spectroscopy has a large advantage in light-gathering power over the monochromator method, but requires a small computer to make the Fourier analysis and cannot as directly study a single infrared wavelength as a function of time.

When a narrow band of radiation selected by the spectrometer is of a frequency given by the Bohr condition, absorption of the light by the sample may take place. For interaction with the radiation field the sample must have an electrical polarity change resonating with the field frequency. In practice it is necessary that the electric dipole of the molecule (or the electric polarization of the crystal) change during the course of the vibration in order that the transition between energy levels of the vibration appear in the infrared spectrum.

The requirement of an oscillating dipole leads to the conclusion that homopolar molecules (e. g., H_2, Cl_2) and crystals (e. g., diamond) will not have infrared spectra. Heteropolar diatomics (e. g., HCl, CO) have infrared spectra, though the strengths of the characteristic absorptions vary greatly. Polyatomic molecules and almost all crystals absorb infrared radiation, though their various vibrations vary in absorption strength. If symmetry is present in the absorber, some vibrations may not carry an oscillating dipole. In crystals, second-order effects are much more important than in molecules, and cause extensive complex absorptions not simply given by the basic selection rules.

Gases, liquids, and solutions to be studied in the infrared are ordinarily contained in cells ranging from 10 or more to 0.1 or less millimeters in thickness. The cell windows are usually cut from crystals of salts (e. g., KBr, NaCl) which do not absorb near- or middle-infrared radiation. In the case of crystalline materials, where it is difficult to prepare samples of the requisite thinness (*ca.* 0.001 cm), most routine analytical work is done in mixtures where a small amount of finely powdered sample is intimately mixed with an oil (a "mull" spectrum) or mixed with a powdered alkali halide and pressed into a transparent pellet. It is also possible to study the spectrum in the reflected light from the sample surface. And the emission of infrared radiation from heated samples (flames, shock waves) is an important technique.

The advent of the computer and of affordable integrated circuit and chip technology has profoundly modified modern infrared instrumentation. In turn, this has opened new areas of exploration. Primarily, the computer has pushed Fourier-transform spectroscopy to the fore to such extent that the purchase of new grating spectrometers for infrared studies is no longer an option for many industrial and university laboratories – save for specific purposes such as infrared observations from satellites and research on phenomena that stretch over narrow- band regions. The advantage of Fourier-transform infrared spectroscopy over a grating instrument ("Fellgett" or "multiplex" advantage) not only yields a gain in signal-to- noise level normalized to the observation time, but also an increase in resolution by a mere increase of the time of observation if the original signal strength is demanded. Further, the ease of data handling (smoothing, transforming, storing and retrieving, analyzing, and plotting) is appreciable since sampling is, *a priori*, done at precise discrete intervals.

New technology of lens construction in combination with the intensity-gathering advantages of infrared Fourier-transform spectroscopy have spawned a method of observing the diffusely scattered infrared radiation ("DRIFT"). The technique has been very useful in exploring vibrational modes of samples, such as powders and entire flat panels, which do not lend themselves to being pressed into pellets (possible modification by the applied high pressures) or otherwise being decimated (art objects). It is also very promising for *in situ* adsorption/desorption studies on amorphous materials.

Vibrational Motions Studied by Infrared Spectroscopy

The molecular motions studied by the infrared technique are understood by considering well-established theories of classical and quantum mechanics [1, 2] as are those of atoms in crystals [3].

In a molecule containing N atoms, $3N - 6$ linearly independent coordinates ($3N - 5$ in linear molecules) describe the relative positions of the atoms with respect to one another. The displacements of these coordinates from their equilibrium values form the basis for description of the molecular vibrations. When an atom is displaced from equilibrium, a force, approximately described by Hooke's law, resists the motion, and a vibration results. A typical molecular vibration frequency is $1000 \, cm^{-1}$ ($3 \times 10^{13} \, Hz$), and with a typical atomic mass of $10^{-26} \, kg$ the Hooke's-law force constants are found to be of the order of $10^5 \, dyn/cm$ ($100 \, newtons/m$). This number is characteristic of the magnitude of force constants opposing the stretching of a chemical bond.

Polyatomic molecules require a large number of Hooke's law force constants to represent their force fields; the theoretical treatment of the vibrational problem, called a "normal coordinate analysis" is straightforward though tedious. Seldom is there enough experimental data to determine all of the force constants in a general force field for a polyatomic molecule, and a number of approximate fields (the "valence force field", the "Urey–Bradley field", etc.) have been adopted by different investigators. Though a degree of arbitrariness is involved, the force fields and vibrational spectra of many polyatomic molecules have come to be quite well understood, and this understanding has contributed greatly to modern pictures of chemical bonding and molecular structure. It is now possible, using *ab initio* quantum mechanics programs such as Gaussian, to compute vibrational frequencies and force constants for quite large molecules.

In a crystalline solid, the translational symmetry, based on the crystalline unit cell, greatly simplifies what would appear to be an intractable theoretical problem. Description of the crystalline vibrational motions is reduced to a separate description of the motions of atoms in one unit cell and a statement of the phase relationship between unit cells. A given possible phase relationship corresponds to a point in the "Brillouin zone". For a crystal containing N atoms in its unit cell, there are $3N$ vibrational coordinates describing each unit cell, and therefore $3N$ vibrational motions at any single point in the Brillouin zone.

The requirement of an oscillating polarization for interaction of a sample with infrared radiation takes on a special importance for crystals. If a vibrational motion of a single unit cell generates an oscillating polarization, then the motion can in principle be observed as a transition in the infrared spectrum. But the wavelength of the infrared radiation is very long (of the order of micrometers) compared to the dimensions of the unit cell (nanometers); thus for a particular vibrational motion to absorb radiation, a very large number of unit cells must oscillate in phase with one another to avoid cancelation of the interaction by interference. The phase relationship in which all unit cells oscillate exactly in phase with one another corresponds to the "center of the Brillouin zone". Thus the crystalline modes which can cause infrared transitions are limited to the $3N$ at the zone center (since the frequencies of the modes change slowly with change of location in the zone, the limitation to the zone center, though not rigorous, is in practice correct).

Of the zone-center modes, three correspond to motions where all atoms in the cell have the same displacement, i. e., to a shift of the whole cell; since all cells move in phase, this motion corresponds to a translation of the whole crystal, and has zero frequency. Thus the infrared spectrum can, in the simplest approximation, contain $3N - 3$ frequencies. This number itself can be further curtailed if the crystal has symmetry beyond the translational.

Quantum mechanics tells us that the various normal vibrational modes can be treated independently to a good approximation, and that the energy of the molecule or crystal is a sum of terms, one for each mode. If the normal-mode frequencies are ν_i then the energy of vibrational motion is $\sum_i (n_i + \frac{1}{2})hc\bar{\nu}_i$, where n_i is the quantum number for the i'th mode. Light-induced transitions in the infrared (or Raman) spectrum are most commonly those where only one of the n_i changes by one unit. Thus the various normal frequencies appear directly in the infrared spectrum. Breakdown of the independence of the normal vibrations, and of the simple rule governing changes in n_i, can come as a result of anharmonic terms in the force law governing the displacements of atoms from equilibrium. The effects of anharmonicity in gaseous molecules are generally small, but in crystals quite strong effects occur. Of particular interest is the observation of "second-order effects" including spectroscopic transitions in which two or more n_i change simultaneously. In these transitions the two vibrational modes involved need not be zone-center modes, and as a result it is possible to observe indications of all the vibrations in the Brillouin zone in some crystals.

Applications of Infrared Spectroscopy

Infrared spectra of gases are due to simultaneous transitions involving both vibrational and rotational energy levels of the molecules. By study of these spectra it is possible to locate the band origin, and thus determine the pure vibrational frequency, and also to determine the moments of inertia of the molecule, which control the rotational line spacings. From the moments of inertia, and with spectra of different isotopic species of the molecule, it is possible to determine the structure of the molecule, as in microwave spectroscopy.

In order to make a full analysis of the rotational spectrum high resolution is needed; for small, light molecules a great deal of detail may be obtained with good grating or Fourier transform spectrometers. At lower resolution this fine rotational structure is blurred, leaving only a "rotational envelope" which may have a number of maxima. Some information on moments of inertia may still be derived from the rotational contour, but the multiple maxima may cause confusion if several vibrational transitions are closely spaced. Much higher resolution is obtainable using diode lasers, and there has been extensive work on molecules in high vacuum jet expansions. In particular, studies are made of molecular dimers formed in jets, and of molecules dispersed in small helium droplets. Since these studies are done at low temperatures, the rotational structure of the spectrum is simplified, aiding in analysis.

An upsurge of interest in irreversible mechanisms of the condensed phase has created an entire field of experimental and theoretical research on the dynamics of molecular motion by applying the principles of linear response theory and of the fluctuation–dissipation theorem to vibrational–rotational spectra of condensed-phase molecules. The principle, which relates the frequency distribution of a spectral profile to the time-dependent correlation function of the observed transition moment tensor has generated a large base of quantitative knowledge on the nature of rotational and vibrational motion ("relaxation", "fluctuations") of molecules that are constantly perturbed by their neighbors. Rather precise notions on the complicated aspects of "hindered" and "free" rotation, the statistics and dynamics of environments local to the observed oscillator, the clustering of molecules, the persistence of local order within a (macroscopically) isotropic medium were among the fruits of these studies. In addition, new theories on vibrational resonant energy transfer, collision-induced effects, conformation changes, short-lived phonon modes, etc. were engendered by these principles. Finally, experimental data from this kind of analysis has spurred the use of computer simulations or "experiments" for the solutions of Newton's equations of motion (molecular dynamics calculations). These, in turn, have led to workable and transferable intermolecular potential functions, thereby allowing the separation of a particular dynamic process from the panoply of concurring mechanisms.

Infrared spectroscopy has contributed heavily to solid-state physics as well as to the chemical analysis of solid materials. In general the forces between atoms in crystals are somewhat weaker than chemical bonds, and as a result the vibrational transitions of solid materials occur at lower frequencies than those involving distortion of the chemical bonds within molecules. For this reason much solid-state information is obtained in the far-infrared spectrum, and many details of the spectroscopic study of solid-state phenomena are given in the article on far-infrared spectra in this encyclopedia. Molecular crystals are characterized by containing strongly bonded molecular groups of atoms which interact with weaker forces to form the crystal. In this case the internal or molecular vibrations cause transitions in the middle infrared while the external motions of the molecules, derived from rotation and translation, give rise to lattice vibrations in the far infrared.

To obtain a crystalline sample thin enough (from 1 to $10\,\mu m$ in thickness) for infrared transmission studies is quite difficult. In some cases thin crystals can be grown, but a common and powerful method of preparing solid samples involves mixing the sample with a larger amount of powdered potassium bromide, KBr, which is a transparent crystal often used for windows in the mid-infrared. The mixture is then pressed into a transparent disk which is studied in the normal way. In addition to allowing a convenient control of the amount of the sample, this method reduces the anomalous reflection and scattering effects which occur in solid-state spectra by enclosing the sample particles in a medium (the KBr) which has about the same index of refraction. The KBr pellet technique is of great value in the qualitative analysis of synthetic mixtures or of unknown solids. The study of well-characterized surfaces has benefitted from the use of infrared spectroscopy to define the bonding and configuration of adsorbed molecules.

Methods of analysis based on vibrational spectra involve both the frequency specificity and the intensity of vibrational transitions. Their use for qualitative characterization and for quantitative measurement of the amounts of materials is widespread.

Qualitative and quantitative study of a sample for chemical content, as in air pollution, characterization of mineral samples, detection of impurities in solid samples, or study of planetary atmospheres, often relies on the presence of characteristic group frequencies. Because of the similarities in strengths of bonds which are similar chemically (e. g., carbon–hydrogen bonds, carbonate groupings), there are found to exist quite definite "group frequencies" which are present in the spectrum whenever the chemical group is present in the sample. Extensive tables of such group frequencies are available.

Molecular vibration frequencies depend primarily on *intra*molecular force constants, and vibrational spectra are commonly rather independent of the state of the sample, with gases, liquids, solids, or solutions yielding quite similar frequencies. The analyst takes advantage of this fact to adapt his techniques to the other problems which face him, and he can often combine data taken in two or more phases to identify a molecule. However, the effects of *inter*molecular forces are appreciable, and these forces may be studied by infrared spectroscopy. Of particular chemical interest are studies of association and hydrogen bonding, while in solid materials the small shifts in frequency of an impurity or dopant may give a clue to understanding multiple sites available to the guest particle.

Quantitative analysis by infrared absorption spectroscopy uses of the absorption law common to all absorption spectroscopic techniques, $\log(I_0/I) = \alpha c l$, where the logarithm of the ratio of incident to the transmitted light is dependent on an absorption coefficient (α, a property of the sample and of the frequency of the light), the sample concentration (c), and cell length (l). The basic law is not always followed exactly, but corrections derived from measurements on standard samples make the analytical procedure an accurate one. Use of highly monochromatic light, for example of laser origin, to single out a particular absorption feature of a sample (e. g., a single rotational line of the molecular of interest in a gas mixture) can give a highly sensitive, specific analytical method for a wide variety of problems. Pollution control, for example, relies heavily on infrared measurements. The concentrations of reaction intermediates and products in such inhospitable environments as flames and plasmas can be followed by observations of the infrared spectrum, often in emission; the dependence of line intensities in the gas phase on temperature makes the careful quantitative measurement of line intensities a useful thermometer for high-temperature systems.

Lasers have been constructed in which the energy of a chemical reaction, deposited in excited vibrational levels of the reaction products, emerges as infrared radiation. The best-known example is the powerful hydrogen fluoride laser, deriving its energy from the gas phase reaction $H_2 + F_2 = 2HF$. The reaction product, HF, is produced in excited vibrational levels, and laser action can be stimulated. Conversely, the study of infrared radiation, often in laser-amplifier configurations, gives detailed evidence of the exact states which result in chemical reactions. The information on microscopic chemical reaction kinetics which emerges is revolutionizing the study of kinetic processes.

See also: Far-Infrared Spectra; Interatomic and Intermolecular Forces; Microwave Spectroscopy; Molecular Spectroscopy; Molecules; Raman Spectroscopy.

Bibliography

G. Herzberg, *Molecular Spectra and Molecular Structure II. Infrared and Raman Spectra of Polyatomic Molecules*. D. Van Nostrand, New York, 1945.

E. B. Wilson, J. C. Decius, and P. C. Cross, *Molecular Vibrations*. McGraw–Hill, New York, 1955.

M. Born and K. Huang, *Dynamical Theory of Crystal Lattices*. Oxford, London, 1954.

R. W. Ramirez, *FFT, Fundamentals and Concepts*. Prentice-Hall, Englewood Cliffs, NJ, 1985. (I)

R. G. Gordon, *Adv. Magn. Resonance* **3**, 1 (1968). (A)

B. J. Berne and G. D. Harp, Adv. *Chem. Phys.* **17**, 63 (1970). (A)

W. G. Rothschild, *Dynamics of Molecular Liquids*. Wiley, New York, 1984. (A)

Insulators

R. A. Anderson

Introduction

An insulator, or dielectric, could be conveniently defined as a material that does not conduct electricity, if not for the fact that all known materials conduct a measurable amount of electricity. Nevertheless, insulators are readily distinguished from conductors. At room temperature, the conductivity of a good insulator can be more than 20 orders of magnitude smaller than that of a good conductor, and the difference becomes larger at lower temperatures.

The physics that underlies the vast difference between the electrical conductivity of insulators and conductors is the first topic considered. Following that, discussions are devoted to properties of insulators, including the importance of the insulator–electrode contacts in determining conduction processes. This article concludes with a brief description of some of the ways an insulator loses its ability to insulate when electrically stressed to the point of breakdown.

Band Structure and Conductivity

A medium consisting of isolated atoms or molecules cannot conduct electricity because no charged particles are free to flow in response to an electric field. However, if atoms are brought close together, as in a crystalline solid, electrons can move from one atom to another. It would appear, then, that all crystalline solids should conduct electricity.

The distinction between insulators and conductors originates in the quantum-mechanical behavior of electrons. As a consequence of the periodic structure of a crystalline solid, electron energies are restricted to well-defined energy bands, normally separated by bands of forbidden energies. Each band of allowed energies consists of a very large number of closely spaced allowed momentum values, each of which can be occupied by, at most, two electrons having opposite spins. For this reason, there is a finite number of electrons that can be included within an energy band, and the lower-lying energy bands are completely or partially filled with electrons.

Electrical conductivity is crucially dependent on the degree to which the energy bands are filled with electrons. A completely empty band obviously cannot contribute to electrical conductivity. On the other hand, no charge can be transported by a completely filled band: Because all available momentum values are occupied, the net momentum must remain zero in the presence of an electric field. In a partially filled band, however, the boundary between occupied and unoccupied allowed momenta can become displaced in response to an electric field. Charge is transported because the net momentum is no longer zero. The charge carrier in a nearly filled band is the absence of an electron, called a hole. A hole behaves very much as though it were an electron of reversed charge in a nearly empty band.

In most cases for a material to be an insulator, all energy bands must be either full or empty. If one or more energy bands are partially filled, the material usually is a metallic conductor. A further characteristic of insulators is that the gap between filled (valence) and empty (conduction) bands is large enough (usually $> 2\,\text{eV}$) that few electrons are thermally excited across this band gap. If, instead of a band gap, there is a small overlap between these bands, electrons at the top of the valence band spill over into the bottom of the conduction band. Such materials, of which antimony is an example, are known as semimetals. In the case of a pure semiconductor, such as germanium, the band gap is so small that a portion of the valence-band electrons are thermally excited into the conduction band, thereby creating electrons and holes that can conduct electricity. Such a material is an insulator at very low temperatures. Other semiconductors owe their conductivity to impurities that either remove electrons from the valence band or add them to the conduction band. While there is no natural demarcation between insulators and semiconductors, it is convenient to place the boundary at a conductivity of $10^{-6}\,\text{S/m}$. At the other extreme, fluorinated ethylene–propylene copolymer, a very good insulator, has a room-temperature conductivity less than $10^{-17}\,\text{S/m}$.

At this point the reader may be wondering how the unlikely situation of exactly enough electrons to fill the valence bands can occur at all. The reason it occurs frequently is as follows. Each repeating unit of the crystal structure contributes one allowed momentum value, which can accommodate two electrons, to each energy band. Therefore, bands can be filled without leftover electrons if each repeating unit contains an even number of valence electrons, a situation that is not at all unlikely.

In the preceding discussion, we have considered only crystalline solids. However, the short-range atomic order in amorphous materials is sufficient to establish energy bands similar to those found in solids having highly periodic structure. Examples of insulators, semiconductors, and conductors are found in both crystalline and amorphous solids, as well as in liquids.

Metal–Insulator Transitions

The electrical conductivity of certain materials changes many orders of magnitude with a small change in some parameter value. These complicated effects are collectively referred to as metal–insulator transitions. As an example, the conductivity of vanadium sesquioxide

(V_2O_3) abruptly falls by more than seven orders of magnitude when the temperature drops below approximately 150 K. The low-temperature phase then reveals typical insulator behavior, a vanishing conductivity as the temperature approaches zero. Depending on the material, transitions between metallic or semi-metallic conductivity and an insulating state can be caused by changes in temperature, pressure, or applied magnetic field, and metal-insulator transitions can also result from changes in the composition, the degree of disorder, or the dopant concentration in semiconductors.

Insulator–Electrode Contacts

The conductivity of an insulator is strongly influenced by interactions between the insulator and the electrodes attached to it. When an insulator and a conductor are brought into contact, electrons diffuse either into or out of the insulator, depending on the particular materials involved, to establish thermodynamic equilibrium.

If electrons diffuse into the insulator, the insulator-electrode contact is called Ohmic, or injecting. Because an accumulation of electrons injected into the conduction band at the insulator surface is established, which can supply current to the bulk, conduction is limited by the bulk conductivity of the insulator. The bulk conductivity itself depends on whether or not the injected charge extends throughout the thickness of the insulator. If this charge is everywhere larger than the intrinsic charge in the conduction band, then the conductivity reflects the properties of the insulator-electrode contacts rather than the intrinsic conductivity, and the conduction process is referred to as space-charge limited.

A blocking contact occurs when electrons diffuse out of the insulator to establish thermodynamic equilibrium. In this case, conduction is limited by the rate at which electrons can cross the interface, and the conduction process is referred to as electrode limited. Current–voltage characteristics depend on whether electrons are thermally excited over, or tunnel through, the potential barrier at the insulator surface.

Contacts also may be either blocking or injecting for holes, which introduces the complexity of recombination of electrons and holes if both charge carriers are flowing simultaneously.

Traps

All known insulators contain localized energy levels in the band gap called traps, that tend to capture and immobilize the few charge carriers present in the valence and conduction bands. Traps originate from impurity atoms as well as naturally occurring defects. Because trapping affects recombination as well as the flow of carriers, essentially no conduction process escapes being influenced.

Photoconductivity

The absorption of light by an insulator may be accompanied by an increase in the electrical conductivity during illumination. Some insulators also display an increased conductivity in the dark as the result of previous exposure to light. These phenomena are called primary and secondary photoconductivity, respectively. Primary photoconductivity occurs when absorbed photons have sufficient energy to excite electrons into the conduction band from traps or from the valence band. Secondary photoconductivity is ascribed to light-induced changes in the occupation of traps near the electrodes, giving rise to local electric fields that assist the injection of electrons into the insulator. A well-known application of photoconductivity is the xerographic duplicating process.

Photoconductivity provides a means of investigating the mobility and lifetime of charge carriers in insulators. For example, a thin layer of carriers can be generated near an electrode by strongly absorbed light, so that the conductivity due to either electrons or holes can be measured.

Alternating-Current Conductivity

The amount of alternating current flowing through a capacitor depends on the electrical polarizability of the insulator between the capacitor electrodes. Polarization, however, does not respond instantaneously to changes in applied voltage. Rather, several separate processes are involved, each of which proceeds at a different rate and contributes a frequency dependence to the polarizability of the insulator. Alternating-current measurements are a useful probe of slow processes such as chain motion in polymers and drift of mobile ions, as well as faster electronic processes, and can provide a wealth of information when performed at a series of different temperatures.

Electrical Breakdown

The most important application of insulators is electrical insulation, without which electricity would be of little use. Failure of insulators under electrical stress, therefore, remains a topic of research interest.

Ideally, electrical breakdown can be classified as either thermal or electronic. Thermal breakdown occurs when the conductivity increases rapidly enough with increasing temperature to permit thermal runaway. Electronic breakdown involves an instability of the conduction electrons, such as a runaway drift velocity which leads to an ionization avalanche, or enhanced electron injection due to trapped charge near an electrode, as in secondary photoconductivity. The ultimate stage of a destructive electronic breakdown event is probably thermal.

Some insulators are able to withstand very high electric fields. Thin sheets of polyethylene terephthalate, a widely used capacitor dielectric, have a dielectric strength near $4 \times 10^8 \, \text{V/m}$. The highest known values exceed $10^9 \, \text{V/m}$, such as reported for layers of silicon dioxide grown on silicon during microelectronic device fabrication.

In practice, breakdown often occurs at stresses well below the intrinsic dielectric strength of an insulator, and the mechanisms leading to breakdown are complex and poorly understood. For example, insulation on ac power transmission cables is susceptible to breakdown phenomena collectively called treeing, in which branching networks of tubular voids resembling trees slowly develop. Eventually the insulation is penetrated and a damaging discharge occurs. Trees tend to start growing wherever the electric field is enhanced: at cracks or voids in which gas discharges occur, or at metallic inclusions or electrode irregularities. Cable insulation in a moist environment often develops trees that are filled with water.

We consider, finally, another class of breakdown phenomena called surface flashover, in which discharges follow the insulator surface rather than penetrating the bulk. Insulator surfaces exposed to vacuum are particularly susceptible. Although vacuum surface flashover often appears to be the consequence of an avalanche of electrons which multiply on the insulator surface by secondary emission, mechanisms responsible for flashover in other cases may involve bulk breakdown processes in a shallow layer near the insulator surface.

See also: Conduction; Dielectric Properties; Electron Energy States in Solids and Liquids; Photoconductivity.

Bibliography

A. R. Blythe, *Electrical Properties of Polymers*. Cambridge University Press, Cambridge, 1979. (I)

H. Fröhlich, *Theory of Dielectrics*. Oxford, London and New York, 1958.(A)

T. J. Gallagher, *Simple Dielectric Liquids*. Oxford, London and New York, 1975. (A)

N. E. Hill, W. E. Vaughan, A. H. Price, and M. Davies, *Dielectric Properties and Molecular Behavior*. Van Nostrand, Princeton, NJ, 1969. (A)

R. E. Hummel, *Electrical Properties of Materials*, 2nd ed. Springer-Verlag, Berlin, 1993. (I)

C. Kittel, *Introduction to Solid State Physics*, 7th ed. Wiley, New York, 1996. (E)

M. A. Lampert and P. Mark, *Current Injection in Solids*. Academic, New York, 1970. (A)

N. F. Mott, *Metal–Insulator Transitions*, 2nd ed. Taylor and Francis, London and New York, 1990. (A)

N. F. Mott and R. W. Gurney, *Electronic Processes in Ionic Crystals*, 2nd ed. Oxford, London and New York, 1948. (I)

E. H. Nicollian and J. R. Brews, *MOS (Metal Oxide Semiconductor) Physics and Technology*, Chap. 11. Wiley, New York, 1982. (A)

J. J. O'Dwyer, *The Theory of Electrical Conduction and Breakdown in Solid Dielectrics*. Oxford, London and New York, 1973. (A)

T. M. Rice, "Metal–Insulator Transitions", in G. L. Trigg (ed.) *Encyclopedia of Applied Physics*, Vol. 10, pp. 113–127. VCH Publishers, Weinheim and New York, 1994. (I)

J. G. Simmons, "Conduction in Thin Dielectric Films", *J. Phys. D: Appl. Phys.* **4**, 613 (1971). (A)

L. Solymar and D. Walsh, *Lectures on the Electrical Properties of Materials*, 5th ed. Oxford, New York, 1993. (I)

W. A. Stygar *et al.*, "Flashover of a Vacuum–Insulator Interface: A Statistical Model", *Phys. Rev. ST Accel. Beams* **7**, 070401 (2004), Sec. II D. (A)

J. M. Ziman, *Principles of the Theory of Solids*, 2nd ed. Cambridge University Press, Cambridge, 1972. (I)

Interatomic and Intermolecular Forces

J. A. Barker and J. P. McTague

Molecules, and atoms which are not chemically bonded to one another, exert forces on one another which are attractive at large distances and repulsive at short distances. This is what is generally meant by the phrases "intermolecular forces" and "interatomic forces", which are therefore not normally used to describe the "chemical" or "valence" forces acting between atoms bonded together in the same molecule. It is the existence of these intermolecular forces which explains the existence of matter in solid, liquid, and gaseous states, and the stability of different states of matter in particular regimes of temperature and pressure (phase diagrams and equations of state). Modern statistical mechanics has made it possible to predict accurately the properties of solids, liquids, and gases starting from information on intermolecular forces.

The repulsive forces are dominant at distances comparable to and smaller than the size of the molecules, a few Å($1 \text{ Å} = 10^{-10}$ m) for nonpolymeric molecules. In fact, it is the repulsive forces which determine the size and shape of the molecule. The attractive forces, which hold together the molecules in solids and liquids, are of longer range and fall off proportionally to R^{-7} for nonpolar molecules, where R is the distance between the molecules. The corresponding potential energy of interaction is proportional to $-R^{-6}$ (the force is the derivative of the potential energy with respect to distance). These forces are consequences of the electrostatic interactions between the electrons and nuclei and are described satisfactorily by quantum mechanics (before the advent of quantum mechanics, classical mechanics and electrodynamics were unable to describe these phenomena since they predicted that both static and dynamic charge distributions were unstable). The nonelectrostatic interactions between nuclei arising from the "strong" and "weak" nuclear forces are of much shorter range than the electrostatic and electrodynamic interactions, and completely negligible at distances comparable with molecular dimensions. The gravitational interaction is also negligible compared with the electrostatic interactions at such distances, though it falls off much more slowly with distance (potential energy proportional to $-R^{-1}$), and provides the dominant interaction between electrically neutral and nonmagnetic bodies at macroscopic distances. Magnetic contributions to intermolecular forces are usually unimportant.

It is almost universal to consider the interaction between pairs of molecules as a sum of separate elementary interactions. This is strictly valid only when the molecules are sufficiently far apart that their electronic charge do not overlap. From a practical point of view, this requires intermolecular distances R about 1 Å greater than the nearest-neighbor spacings in solids. In this range the permanent electrostatic, induction, and dispersion interactions can be related to the properties of the individual molecules.

The permanent electrostatic interactions reflect the more or less asymmetrical average distribution of the electronic charge within the molecule, which can be described in terms of the charge (zero for neutral molecules) and the dipole, quadrupole, and higher permanent multipole moments of the molecule. The dipole moment describes the separation of positive and negative charge in the molecule and the quadrupole and higher moments describe in successively higher detail the average distribution of the charge. The dipole–dipole interaction, and all higher multipole interactions, give an average of zero when averaged over orientations of the molecules. Both dipole and quadrupole forces can play a significant role in determining the relative orientation of molecules in molecular crystals, but do not play a predominant role in the total binding energy of the solid.

Induction forces arise because a permanent molecular dipole or quadrupole moment produces electric fields which induce charge distortions, and therefore dipole and higher moments, in neighboring molecules. Those forces play only a small role in molecular orientation and binding energies.

None of the above-named forces can exist for spherical systems, such as rare gases, for which all multipole moments are zero. Nevertheless their binding energies can be as large as those for polar molecules. F. London (1930) showed that, owing to the constant motion of the electrons in an atom or molecule, there are instantaneous electrical moments, which can set up transient electrical fields at neighboring molecules. These electric fields cause the electrons

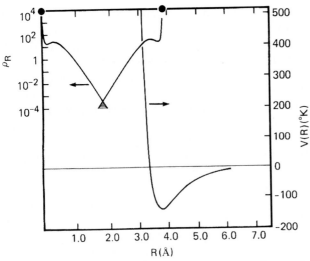

Fig. 1: Potential energy of a pair of argon atoms versus internuclear separation R. The electron density ρ (electrons/Å^3) is also shown for the equilibrium internuclear distance.

in the neighbor to correlate their motion with those of the first molecule, leading to dynamic dipole–dipole, dipole–quadrupole, quadrupole–quadrupole, etc., attractive interactions. From the point of view of molecular quantum mechanics these are "electron correlation" effects, which are not included in, for example, a Hartree–Fock self-consistent field calculation. The leading (dipole–dipole) contribution to the energy varies as $-R^{-6}$, with higher-order terms in $-R^{-8}$, $-R^{-10}$, etc. (At very long range they are modified by relativistic effects, becoming proportional to $-R^{-7}$, etc.) These forces, variously called London, van der Waals, or dispersion forces, are universal and account for the major part of the binding energy of molecular solids, liquids, and even such weakly bound dimers as Ar_2, which cannot have ordinary chemical bonds.

The above-mentioned long-range forces, which are all directly related to properties of the isolated molecules, do not explain the repulsion at very short distances. This additional force comes into play when the electronic charge distributions of different molecules overlap. They have their origin in the quantum-mechanical exchange interaction, which is also involved in chemical bonding. In fact, the repulsive interaction may be looked upon as "antibonding". Since these energies are of the order of chemical interactions it is not surprising that only a very small amount of electron overlap is required to cancel out completely the weak long-range attractive forces, as illustrated in Fig. 1.

The short-range repulsive energy varies approximately as $\exp(-R/R_0)$, where R_0 depends on the pair of atoms involved, but is typically of the order of 0.1–0.2 Å. The extremely steep repulsion suggests that a useful approximation is to consider the atomic cores to be impenetrable spheres, whereas the effect of the attractive forces can be approximated as a uniform background potential. These approximations are remarkably successful in correlating the thermodynamic properties and geometrical arrangement of molecules in both liquid and solid phases. The energies of collections of molecules are approximately, but not exactly, pairwise additive. For liquid and solid densities the additional term is of the order of 5–10% of the pair energy, where it is known.

There is detailed information on the nature of the intermolecular potential energy function for only a few relatively simple atomic and molecular pairs, in particular the rare gases. For the simplest case, that of two helium atoms, the interatomic potential has been derived with impressive accuracy from *ab initio* quantum-mechanical calculations, and the results agree very closely with empirical estimates. This provides a convincing validation both of our understanding of the origin of the intermolecular forces and of the empirical methods used to study them.

For systems involving more electrons accurate *ab initio* calculations are so far prohibitively time-consuming. However for heavier rare-gas pairs the interatomic potentials are known very accurately from empirical studies which use theoretically motivated mathematical forms together with a wide range of experimental data to determine the values of a number of adjustable parameters. The experimental data which are useful include the deviation of real gases from ideal gas behavior ("virial coefficients"); the viscosity, thermal conductivity, and diffusion coefficients of dilute gases; spectroscopic data on energy levels of bound pairs; and measurements of scattering cross sections for pairs of atoms and molecules. Earlier potentials which were derived from measurements of only one property were much less satisfactory.

For more complicated cases, such as small organic molecules, mathematically simple models involving a minimum of adjustable parameters, often two, are commonly used. The model most commonly used in calculations of bulk properties is the Lennard-Jones 6-12 potential:

$$V(R) = 4\varepsilon \left[(\sigma/R)^{12} - (\sigma/R)^{6} \right] \, ,$$

where ε is the attractive energy at the potential minimum and σ the intermolecular distance where $V = 0$. The parameters are fixed to reproduce selected bulk properties. These phenomenological potentials have often been used as effective potentials for approximate calculations even for systems where the potential is more accurately known. This is because of their simplicity and the scaling properties of V/ε and R/σ which enable the properties of one substance to be related to those of another at a different temperature and density (principle of corresponding states). In quantitative studies where more accurate potentials are available this is no longer necessary.

The interactions of large molecules are often modeled as a sum of atom–atom interactions between the atoms of one molecule and those of the other, with the atom–atom interactions described by the Lennard-Jones potential. Several sets of "transferable" Lennard-Jones parameters for atom pairs relevant to organic materials are available in the literature. These atom–atom potentials are useful in predicting both thermodynamic and structural properties of organic solids and liquids and of polymeric materials including those of biological importance such as proteins. For flexible polymeric molecules it is necessary to include atom–atom potentials for nonbonded atoms in the same molecule, since these intramolecular interactions are important in determining the structure adopted by the molecule.

See also: Electric Moments; Equations of State; Magnetic Moments; Molecular Structure Calculations; Molecules; Polarizability; Statistical Mechanics.

Bibliography

G. C. Maitland, M. Rigby, E. B. Smith, and W. A. Wakeham, *Intermolecular Forces, Their Origin and Determination*. Clarendon Press, Oxford, 1981. (I)

M. L. Klein and J. A. Venables (eds.), *Rare Gas Solids*, Vol. 1. Academic Press, New York, 1976. (I)

J. O. Hirschfelder, C. F. Curtiss, and R. B. Bird, *The Molecular Theory of Gases and Liquids*. Wiley, New York (1964 corrected printing). (I)

J. A. Barker and D. Henderson, "What Is 'Liquid'? Understanding the States of Matter", *Rev. Mod. Phys.* **48**, 587 (1976). (I)

H. Margenau and N. R. Kestner, *Theory of Intermolecular Forces*, Vol. 18. Pergamon Press, New York, 1969. (A)

J. N. Israelachvili, *Intermolecular and Surface Forces with Application to Colloidal and Biological Systems*, Academic Press, New York, 1985. (I)

Califano (ed.), *International School of Physics "Enrico Fermi", Course LV*. Academic Press, New York, 1975. (A)

Interferometers and Interferometry

K. M. Baird

Light is propagated according to laws of wave motion, and when two or more beams of light are combined under appropriate conditions, they will sometimes reinforce, sometimes tend to nullify, each other, like waves in and out of phase. This can cause cyclical intensity changes, corresponding to changes in the difference in the phases of the light waves in the respective beams, giving rise to an "interference" pattern of light and dark contours, usually called fringes, whose interval corresponds to a change of one cycle in the phase difference – i. e., a change of one wavelength (λ) in the relative distance traveled by the beams. This effect is the basis of interferometry, widely used to obtain information about the path traversed by a light beam and to study the characteristics of the light itself.

With natural light the phase due to different atoms varies independently and very rapidly, so that interference cannot be observed on the combination of beams that have come from different parts of a light source. Observable interference fringes are produced by the superposition of a very large number of nearly identical patterns, each due to a pair or set of beams that have come from the same atom by different paths; an interferometer provides the means for doing this by division of light into separate beams, which are then recombined under suitable conditions regarding polarization imaging, and relative path lengths. For laser-produced light, atoms are stimulated to emit in phase with one another and the restrictions are relaxed by many orders of magnitude; for example, the paths traveled by the interfering beams can differ in length by hundreds of meters, as compared with the few decimeters possible with natural sources.

Since the wavelength of visible light is about $0.5\,\mu m$ and it is quite possible to observe changes as small as 10^{-3} fringe, interferometry provides an extremely sensitive means for observing small displacements, microtopography of surfaces, etc. Furthermore, even natural light will produce fringes over a length of up to 10^6 wavelengths, so that physical lengths

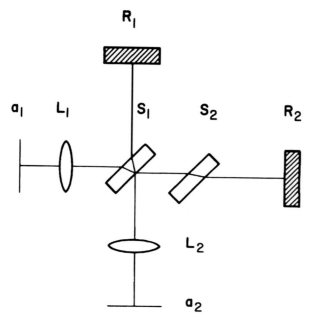

Fig. 1: Twyman–Green form of the Michelson interferometer.

can be determined in terms of wavelengths with a precision approaching $1 : 10^9$. This fact and the very precise reproducibility of some wavelengths accounts for the great importance of interferometry in metrology, as exemplified by the definition of the international meter in terms of a wavelength of light. Similarly, interferometry makes possible great precision and resolution in spectroscopic measurements.

There exist a very large number of types of interferometer, differing in the arrangement of reflectors, lenses, etc. necessary for the separation and recombination of the light beams in a manner convenient for specific applications. They can be classified in a number of ways – according to whether two or a multiplicity of beams interfere, whether the beams are separated by division of the area of a wave front or by division of its amplitude at a partially transmitting reflector, etc. These will be illustrated in the following description of the most widely used types of interferometer that can be considered as basic examples.

Two-Beam Interferometers

One of the most important interferometers is the Michelson, which is described in some detail with reference to later modifications that included lenses and apertures; this type employs two beams and division of amplitude. Referring to Fig. 1, we see that light from given points at a_1 is rendered parallel by the lens, L_1, and split into two beams at the partially transmitting reflector, S_1. The beams travel to mirrors R_1 and R_2, respectively, and thence back to S_1, where they recombine to form a beam that is focused at a_2 where an aperture can be used to block unwanted light such as that reflected by the second surface of S_1. The optional plate S_2 is to provide symmetry in the light paths.

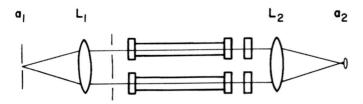

Fig. 2: Rayleigh interferometer.

The recombining beams produce interference fringes that can be observed in three ways:

1. If the reflectors are plane and adjusted so that the recombined beams are collinear, the phase difference between wave fronts from a given point in the plane of a, will be constant across the whole wave front and the intensity of the recombined beam is given by

$$I = I_0 \cos^2\left[\frac{2\pi}{\lambda}(S_1R_1 - S_1R_2)\cos\theta\right]$$

   where θ is the angle a beam makes with the mirror normal. Beams perpendicular to the mirrors, which become focused at the center of a_2, have the greatest phase difference; obliquely traveling beams, which are focused at points removed from the center of a_2, have lesser phase differences and there results a pattern of circular fringes having a sinusoidal intensity distribution.

2. If under the foregoing conditions apertures are placed at the centers of a_1 and a_2 and are made small compared with the central circular fringe, the total light passed will vary sinusoidally with displacement of one of the reflectors, provided the light is of a sufficiently narrow spectral band. This type of observation is particularly suited to photoelectric detection and is useful for length metrology and for Fourier analysis of spectra.

3. Viewed through a small aperture at a_2, the reflectors will appear uniformly illuminated because of the constant phase difference. If now a variation of phase difference across the wave front is introduced, say, by an inclination of one of the reflectors, or by a distortion resulting from reflection or transmission by an object placed in one of the beams, a pattern of interference fringes showing contours of equal optical path length will be seen "localized" near the plane of R_1. This mode of observation, devised by Twyman and Green, is applied to the testing of optical components, microtopography of surfaces, refractive index studies, length metrology, etc.

The basic features described above are common to a large number of similar interferometers modified for special applications by the use of arrangements such as a prism beam splitter, separated beam splitter and combiner, retro reflectors (R_1 and R_2), curved wave fronts, unsymmetrical beam sizes, and a shear displacement of the wave fronts.

Another class of two-beam interferometer separates the beams by division of the wavefront area. Important examples are Michelson's stellar interferometer and the Rayleigh, shown in Fig. 2, which is used for refractive index measurement. Light scattering, diffraction, and polarization are other methods used for separation of the beams.

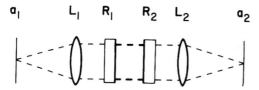

Fig. 3: Fabry–Perot interferometer.

Multiple-Beam Interferometers

This important class is typified by the Fabry–Perot interferometer or "etalon", of great importance in spectroscopy and, in modified forms, in lasers. In Fig. 3, light from given points in the plane a_1, rendered parallel by passage through L_1 is partially transmitted by the plane-parallel reflecting surfaces, R_1 and R_2, which may have reflectances as high as 99%. Light in the space between R_1 and R_2 is multiply reflected back and forth, resulting in a series of wave fronts leaving R_2 with successive retardations of twice the separation of the mirrors. The light is brought to a focus at the plane a_2, where interference can be observed in much the same manner as described in connection with Fig. 1. An essential difference in the form of the fringes arises, however, because of the multiple reflections. These create a standing-wave system between R_1 and R_2 that is relatively powerful when the retardation of successive wave fronts is an exact multiple of λ, under which conditions the combination of reflectors has a transmittance of nearly 100%. The standing-wave power, and consequently the transmittance, falls off rapidly with departure from this condition as given by the Airy function $A(n) = T^2/[(1-R)^2 + 4R\sin^2(\pi n)]$ where R is the reflectance, T the mean transmittance of the reflectors, and $n = 2(\mu t/\lambda)\cos\theta$, the order of interference, μt being the optical separation of the reflectors. The very sharp "resonance" effect in this class of interferometer produces fringes that are very narrow compared to the spacing between them.

In the Fabry–Perot, the fringes are observed either as circles formed at a_2 or as a change in flux passing through a small aperture at a_2 while the optical separation μt or the wavelength λ is varied. This form has had its greatest use in spectroscopy, where it has been used for wavelength measurement and very high-resolution spectrometry; it is also used in a form, originated by Connes, having curved reflectors.

The Fizeau form of interferometer makes use of the fringes localized at the reflectors observed through the small aperture at a_2. It has had its greatest application in microtopography, sometimes in conjunction with a microscope to provide great resolution in three dimensions.

Multiple-beam interferometry using division of amplitude is also the basis of the Lummer–Gehrcke plate. The echelon of Michelson formerly used in spectroscopy and the diffraction grating can be considered multiple-beam interferometers displaying division of aperture.

Miscellaneous

The fringes observed in the interferometers described are formed by the use of single very narrow bands of the spectrum and represent contours of either equal path length or equal angle. Another type of fringe can be observed when light in a broad band of the spectrum passes successively through an interferometer and a spectroscope, in which case they are said to represent contours of equal chromatic order.

Interferometry has a history of well over a hundred years but continues to be actively developed by the application of newly available sources and detection and recording techniques.

It is applied throughout the spectrum from radio waves (for astronomy) to x rays (for measuring fundamental constants) and it has been enormously influenced by the development of the laser, which has made possible a wide variety of new techniques, such as optical heterodyning, holography, and laser speckle, that themselves justify book-length treatment.

See also: Gratings, Diffraction; Holography; Waves.

Bibliography

K. M. Baird and G. R. Hanes, *Applied Optics and Optical Engineering*, Vol. 4, pp. 309–361. Academic Press, New York, 1967. (I)

J. C. Dainty, *Laser Speckle and Related Phenomena*, 2nd ed. Springer-Verlag, Berlin and New York, 1984. (A)

Encyclopaedic Dictionary of Physics, Vol. 3, pp. 873–893. Pergamon Press, New York, 1961. (E)

P. Hariharan, *Optical Interferometry*, 2nd ed. Academic Press, Sydney and Orlando, 2003. (A)

G. Hernandez, *Fabry–Perot Interferometers*. Cambridge University Press, London and New York, 1988. (A)

Elliott R. Robertson and James M. Harvey, *The Engineering Uses of Holography*. Cambridge University Press, London and New York, 1970. (I)

Current issues of journals of the Optical Society of America. (A)

Intermediate Valence Compounds

J. M. Lawrence

Intermediate valence (IV) [1] occurs in certain rare-earth compounds such as $CePd_3$, SmS, $EuPd_2Si_2$ and $YbAl_3$ where the rare-earth valence is fractional, or intermediate between two integral values. In these compounds the valence has the same fractional value at each rare-earth site; this is referred to as homogeneous mixed valence to distinguish it from inhomogeneous mixed valence where the rare-earth element has different values of integral valence at different lattice sites. (This latter case occurs in compounds such as Sm_3S_4.) A key characteristic of these solids is a *crossover* from local moment, integral-valence behavior at high temperatures to nonmagnetic, moderately heavy-mass Fermi liquid behavior at low temperatures. The heavy-fermion compounds [2] are a related class of materials.

Rare-earth elements have an electronic configuration consisting of a filled xenon core, 2–4 bonding (valence) electrons in 5d and 6s states, and 0–14 4f electrons. The configuration can be expressed as $(5d6s)^z 4f^n$, where z is the valence and n the 4f count. In the solid state the bonding electrons form energy bands. The 4f orbital has a small radius and does not contribute directly to the bonding. In most rare-earth compounds the 4f behaves as a localized orbital, with a well-defined magnetic moment. Ordinarily the energy separation $E = E_n - E_{n-1}$ between configurations $(5d6s)^z 4f^n$ and $(5d6s)^{z+1} 4f^{n-1}$ is large (5–10 eV) but for compounds of the elements Ce, Sm, Eu, Tm, and Yb the separation can be much smaller (0–2 eV). Under these circumstances the two configurations hybridize via a process where the conduction electrons hop on and off the rare-earth sites, causing them to fluctuate between the $4f^n$ and $4f^{n-1}$ configurations. In Ce the nearly degenerate states are $4f^0(5d6s)^4$ and $4f^1(5d6s)^3$; for Yb they are $4f^{14}(5d6s)^2$ and $4f^{13}(5d6s)^3$. The valence fluctuations between these states lead to a quantum-mechanical ground state which contains an admixture of both configurations. The strength of this mixing interaction is given by $\Gamma = V^2 N(E_n)$, where V is the matrix element of the interaction and $N(E_n)$ is the number of conduction electron states per unit energy interval. The degree of admixture depends on the ratio E/Γ; when this is small, both states contribute equally and the valence is nonintegral.

In a band of uncorrelated (non-interacting) 4f electrons, charge fluctuations into all fourteen $4f^n$ configurations would be allowed, but in IV metals the hybridization V is small compared to the Coulomb energy $U = E_{n+1} - E_n$ required to excite the next highest energy 4f orbital, so that occupancy of the $4f^{n+1}$ configuration is suppressed. The electrons cannot move freely on and off the f-orbital, but are restricted to hopping on to a $4f^{n-1}$ orbital or off a $4f^n$ orbital. The resulting slow hopping, or valence fluctuations , between two nearly-localized 4f configurations represents an archetypal example of correlated electron behavior in solids.

At high temperatures the behavior of IV compounds approaches that of a conventional rare earth: integral valence, with a well-defined local moment with total angular momentum J. The susceptibility obeys a Curie law $\chi = C_J/T$ and the entropy approaches $R\log(2J+1)$. For Ce the high temperature state is trivalent $[4f^1(5d6s)^3]$ with $n_f = 1$, $z = 4 - n_f = 3$ and $J = 5/2$. For Yb, the high temperature state is again trivalent $[4f^{13}(5d6s)^3]$ with $n_f = 1$, $z = 2 + n_f$ and $J = 7/2$. (Yb is the hole-analogue of Ce, and the convention for the 4f occupancy is that n_f is taken as the number of holes in the $4f^{14}$ shell.)

As the temperature is lowered through a characteristic temperature T^*, the behavior crosses over to that of a Fermi liquid. The 4f occupation (and hence the valence) approaches a fractional value. The susceptibility is finite as $T \to 0$ (Fig. 1). Finite susceptibility (Pauli paramagnetism) is characteristic of nonmagnetic metals; hence for IV compounds the local moment is *quenched*. In addition, the low-temperature specific heat has a linear temperature dependence $C = \gamma T$. Both $\chi(0)$ and γ are enhanced, by factors of order 10–50, over the values observed in normal metals. These facts taken together imply that the low-temperature behavior is that of a heavy mass *Fermi liquid* where the enhancement of $\chi(0)$ and γ are proportional to the effective mass m^*. That is, despite the extremely complicated correlations between the conduction electrons and the 4f ions, in the ground state the conduction electrons behave as though they are noninteracting, albeit with a moderately large effective mass m^*.

Such crossover behavior is exhibited in Fig. 1 for the IV compound $YbAl_3$. In this figure, the temperature dependent occupation number is determined from x-ray absorption. The L_{III} absorption edge occurs when the photon energy equals the value necessary to excite a 2p deep core electron to the valence band. For a rare-earth ion the L_{III} edge occurs at different energies for the $4f^{n-1}$ and $4f^n$ configurations and in mixed valence compounds *two* absorption edges are observed, at the values expected for the two integral valence cases. The ratio of the intensity of the two absorption peaks is used to estimate the valence or 4f occupation number n_f.

The theory which describes mixed valence compounds is based on the Anderson impurity model (AIM) [3]. This describes a single 4f impurity embedded in a nonmagnetic host with conduction bandwidth W. The energy separation $E = E_n - E_{n-1}$, the hybridization V, and the density of states $N(E_n)$, as well as the Coulomb energy $U = E_{n+1} - E_n$, are taken as model

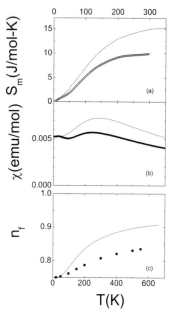

Fig. 1: Experimental data (symbols) for YbAl$_3$ for the (a) 4f entropy,
(b) susceptibility and (c) 4f count n_f (defined for Yb compounds as the number
of holes in the 4f^{14} shell). The lines are the predictions of the Anderson
impurity model.

parameters. For the rare earths U is typically 6 eV; V is of order 0.1 eV and E varies from close
to zero for Sm, Eu, Tm, and Yb mixed valence compounds to 2 eV for cerium. In the AIM
the valence fluctuations give rise to a *Kondo resonance* which enhances the density of states
(DOS) of the conduction electrons over an energy scale T_K about the Fermi energy. Here T_K
is the *Kondo temperature* which in the model varies as $T_K \sim W e^{-E/\Gamma}$.

The model predicts the basic features of IV compounds, in particular the crossover from
high temperature local moment behavior to low temperature Fermi liquid behavior. The characteristic temperature T^* for the crossover is the Kondo temperature T_K. A key feature of the
model (as well as of the actual compounds) is *universality*: the susceptibility, specific heat and
4f count are functions of a scaled variable T/T_K.

The Kondo energy $k_B T_K$ is the energy scale for fluctuations of the 4f spin. The local moment is quenched by these spin fluctuations whose ultimate origin lies in the fluctuations
between the two integral-valent configurations. The enhanced specific heat coefficient arises
from thermal excitation of spin fluctuations. The enhancement of χ and γ is proportional to
$1/T_K$. The energy spectrum $S_{mag}(\Delta E)$ of the spin fluctuations can be measured using inelastic
neutron scattering which measures the energy required to flip a 4f spin. The spin-fluctuation
spectrum for a typical IV compound is a broadened Lorentzian, with a maximum at an energy
$k_B T_K$. This is shown in Fig. 2 for YbAl$_3$, a compound with T_K of order 600 K; the additional
feature seen at 6 K will be discussed further below. The neutron spectra show very little dispersion (momentum dependence). The magnetic excitation is thus a highly localized and highly
damped spin fluctuation (oscillation) at a characteristic energy $k_B T_K$.

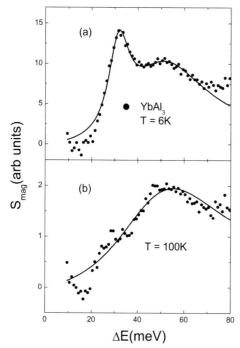

Fig. 2: Magnetic neutron scattering spectra (symbols) for YbAl$_3$ at (a) 6 K and (b) 100 K. The lines are Lorentzian fits, with the main component centered near 50 meV. An additional peak centered at 32 meV occurs at low temperatures.

The characteristic temperature T_K takes on values between 100 and 2000 K for different IV compounds and can be varied in a given compound by pressure or alloying. There is a correlation between T_K and valence; e. g., in cerium compounds, large T_K corresponds to smaller 4f count.

For some IV compounds, such as YbAgCu$_4$, the Anderson impurity model provides excellent fits to the experimentally measured susceptibility, specific heat and 4f count over a broad temperature range and to the energy dependence of the spin-fluctuation spectrum [4]. Such agreement with the impurity model is perhaps expected at high temperatures $T > T_K$ where the 4f ions behave as a set of noninteracting impurities. At low temperatures, however, such agreement is surprising, because the Anderson model describes an impurity, whereas mixed valence compounds consist of a periodic lattice of 4f sites for which interactions among the 4f ions should affect the behavior. Such agreement with the impurity model is a consequence of the fact that the above mentioned properties (χ, γ, n_f and $S_{mag}(\Delta E)$) are dominated by the spin fluctuations which are essentially localized at a single 4f site.

However, for most IV compounds, agreement with the impurity model is merely semi-quantitative. This can be seen for YbAl$_3$ in Fig. 1. A characteristic deviation from the predictions of the AIM , observed for some compounds (including YbAl$_3$), is that the data approaches the high temperature limit more slowly than the model. This *slow crossover* [4] is a consequence of the fact that the change of 4f count with temperature (e. g., Fig. 1c) for an Anderson impurity affects only one degree of freedom in the whole conduction band, but for an periodic lattice of 4f sites, a large redistribution must occur in the vicinity of the Fermi surface to accommodate the valence change.

The effects of the 4f interactions are seen most graphically in the transport behavior, which is extremely sensitive to the degree of lattice periodicity. The resistivity, which is large and finite for an Anderson impurity, vanishes as $T \rightarrow 0$; this is a manifestation of Bloch's Law. At low temperatures, the resistivity varies as $[T/T_K]^2$, which is Fermi liquid behavior. This buildup of interactions, the vanishing of the resistivity, and the onset of Fermi liquid behavior is referred to as *coherence*. The temperature scale T_{coh} for the onset of coherence is typically an order of magnitude smaller than the Kondo temperature; e. g., in YbAl$_3$, where $T_K > 600$ K, $T_{coh} \approx 50$ K.

Coherence is observed most convincingly in de Haas–van Alphen (dHvA) experiments, which show that the wave functions in the ground state are not random (as they would be for impurity scattering) but are fully periodic band states with a high degree of 4f character. Indeed, calculations [5] for mixed valence compounds using standard band theoretic methods adequately describe the geometry (in particular, the extremal areas) of the Fermi surface as measured in the dHvA experiments. However, such calculations fail to predict the large effective masses observed in dHvA and specific heat experiments, essentially because the band theories are not currently capable of handling the highly localized, correlated character of the 4f electrons.

The description of coherence on the theoretical level has been studied using the Anderson *lattice* model [3, 6]. While the theory of the Anderson lattice is currently not as well-understood as that of an Anderson impurity, important conclusions can be drawn from approximate treatments of the theory. A key prediction is that the hybridization between the lattice of 4f states and the conduction electrons results at low temperatures in a renormalized band structure with a hybridization gap [Fig. 3, inset]. For intermediate valence metals, the Fermi level lies in the vicinity of the gap, in a region of a high density of states of 4f character; the large DOS is responsible for the enhanced effective masses. For a small number of intermediate valence compounds, including Ce$_3$Bi$_4$Pt$_3$ and YbB$_{12}$, the Fermi level lies in the gap; such compounds, known as Kondo insulators [6, 7], behave as small ($\sim$ 1–10 meV) gap semiconductors. At higher temperatures, for both the IV metals and the Kondo insulators, the electronic structure crosses over to that of the background conduction band (with no gap) with the 4f spins serving as incoherent scattering centers.

The best experimental evidence for such a gap is found in optical conductivity experiments [8], which at low temperatures $T < T_{coh}$ show a very narrow Drude resonance (representing excitations very close to the Fermi surface) separated by a very deep minimum from a near-infrared feature (representing excitations across the gap). This is shown schematically in Fig. 3. At high temperatures, as the hybridization gap disappears, the behavior crosses over to a normal broad Drude response reflecting incoherent scattering from the localized 4f spins.

The onset of the hybridization gap at low temperatures also affects the Hall coefficient and in some compounds causes anomalies in the susceptibility (e. g., the small feature seen in YbAl$_3$ below $T_{coh} = 50$ K in Fig. 1b) and specific heat [9]. Additional peaks are observed in the low temperature spin-fluctuation spectra [10] which appear to reflect scattering across the (indirect) hybridization gap. For YbAl$_3$ when $T < T_{coh}$ a new peak is observed at 32 meV (Fig. 2a); since the energy of this peak corresponds to the energy of the minimum in the optical conductivity, it must represent such hybridization gap scattering.

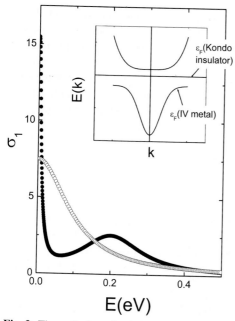

Fig. 3: The optical conductivity of IV compounds versus energy at low temperature (closed circles) and room temperature (open circles). The inset shows schematically the prediction of the Anderson lattice model for the renormalized low temperature band structure.

A significant characteristic of IV compounds is that the crystal density is unusually sensitive to changes in temperature and/or pressure [11]. The radius of the 4f ion in the $4f^n$ state is larger than that of the $4f^{n+1}$ state so that the cell volume varies proportionally to $n_f(T)$. As the temperature increases through T_K the increase of n_f leads to a thermal expansion anomaly. In a few cases this leads to an unusual kind of phase transition, namely, an *isomorphic* transition where at a transition temperature the lattice constant undergoes a large (5%) discontinuous increase without any change in the crystal symmetry. These are valence transitions where the 4f count increases discontinuously. A famous example is the $\alpha - \gamma$ transition in elemental cerium metal.

The renormalization of the electronic spectral density with temperature and the concomitant enhanced low temperature effective masses are consequences of the large Coulomb energy U which prevents uncorrelated hopping; as such these features are generic to correlated electron systems. The intermediate valence compounds thus serve as archetypal materials for studying correlated electron behavior. The heavy-fermion (HF) compounds [2] are a related class of materials. The key feature distinguishing HF from IV materials is that in the former, the Kondo energy $k_B T_K$ is small compared to the crystal field splitting energies $k_B T_{cf}$ of the 4f state. Because the effective mass is inversely proportional to the Kondo temperature, the IV compounds have mass enhancements ($\sim$20–50) which are much smaller than those of the HF compounds where $m^*/m \sim 100$–1000. The main properties of IV metals (the crossover from high temperature local moment behavior to coherent low temperature heavy mass Fermi liquid behavior, where the behavior is dominated by localized spin fluctuations) also hold for HF metals. The key difference is that the heavy-fermion compounds are essentially trivalent and are nearly magnetic. They reside in close proximity to a $T = 0$ quantum critical point for a transition to antiferromagnetism and hence antiferromagnetic fluctuations contribute strongly to the low temperature behavior of the truly heavy-fermion compounds. This is not the case for IV compounds, where antiferromagnetic interactions are relatively insignificant. Rather than being nearly magnetic, they reflect more strongly the effects of hybridization gap, and proximity to the Kondo insulator state [7].

See also: Heavy-Fermion Materials; Phase Transitions; Rare Earths.

References
[1]   J. M. Lawrence, P. S. Riseborough and R. D. Parks, *Rep. Prog. Phys.* **44** 1, (1981).
[2]   N. Grewe and F. Steglich, in *Handbook on the Physics and Chemistry of Rare Earths* Vol. 44, p. 343, (K. A. Gschneidner, Jr., and L. Eyring, eds.). Elsevier, Amsterdam, 1991.
[3]   A. C. Hewson, *The Kondo Problem to Heavy Fermions*, Cambridge University Press, Cambridge, 1993.
[4]   J. M. Lawrence *et al.*, *Phys. Rev. B* **63**, 054427 (2001).
[5]   A. Hasegawa and H. Yamagami, *Prog. Theor. Phys.* **108**, 27 (1992). For YbAl$_3$, see T. Ebihara *et al.*, *Phys. Rev. Lett.* **90**, 166404 (2003).
[6]   P. S. Riseborough, *Adv. Phys.* **49**, 257 (2000).
[7]   Z. Fisk and J. L. Sarrao, in *Physical Phenomena at High Magnetic Fields II*, p. 139, (Z. Fisk, L. Gorkov, D. Meltzer and R. Schrieffer, eds.). World Scientific, Singapore, 1996.
[8]   L. Degiorgi, *Rev. Mod. Phys.* **71**, 687 (1999). For YbAl$_3$ see H. Okamura *et al.*, *Acta Phys. Polon. B* **34**, 1075 (2003).
[9]   A. L. Cornelius *et al.*, *Phys. Rev. Lett.* **88**, 117201 (2002).
[10]   P. A. Alekseev *et al.*, *Physica B* **281–282**, 34 (2000).
[11]   J. D. Thompson and J. M. Lawrence, in *Handbook on the Physics and Chemistry of Rare Earths*, Vol. 19, p. 383, (K. A. Gschneidner, Jr., L. Eyring, G. H. Lander and G. R. Choppin, eds.). Elsevier, Amsterdam, 1994.

Internal Friction in Crystals

A. V. Granato

A description of the mechanical properties of crystals is formally similar to that for electrical or magnetic properties, with the elastic constant playing a role analogous to that of the dielectric constant or the permeability.

The mechanical behavior (response to applied forces, or strain produced by a given stress) of a crystal is determined by the interatomic forces between the atoms as well as by the motion of defects in the crystal. The interatomic forces act quickly and may be taken to be instantaneous for mechanical tests, whereas the displacements or strain from defect motion requires a finite time. A perfect or ideal crystal can be defined as one for which there are no defects providing energy-loss mechanisms, or internal friction, for mechanical excitations. The elastic strain response ε_{el}, of an ideal elastic crystal to an applied stress σ is given by elasticity theory as $\varepsilon_{el} = \sigma/G$, where G is the elastic modulus of the crystal. When this stress–strain relation is combined with Newton's laws for a volume element of the crystal, plane-wave solutions without attenuation, or energy loss, of the equations of motion are found.

Real crystals contain defects. When the defect contribution to the strain is linear with a unique equilibrium value for a given stress, but is achieved only after the passage of a characteristic time, the crystal is called anelastic. For a sinusoidal applied external stress, the in-phase (real) part of the delayed response (ε_R) leads to an effective modulus change ΔG, while the out-of-phase (imaginary) part (ε_I) leads to an energy loss, or internal friction Q^{-1}. These are given by

$$\frac{\Delta G}{G} = \frac{G\varepsilon_R}{\sigma_0} \quad \text{and} \quad Q^{-1} = \frac{G\varepsilon_I}{\sigma_0}, \tag{1}$$

respectively. The response is often of the Debye type for which

$$\frac{\Delta G}{G} = \frac{\Delta_R}{1+(\omega\tau)^2} \quad \text{and} \quad q^{-1} = \frac{\Delta_R \omega\tau}{1+(\omega\tau)^2} \tag{2}$$

where Δ_R is the relaxation strength, ω is the angular frequency of the applied stress, and τ is the relaxation time for the delayed anelastic component of the strain.

The classic example of an internal friction peak, which occurs when $\omega\tau = 1$, is the Snoek relaxation of carbon in bcc iron. In bcc crystals interstitial impurities normally occupy octahedral sites and have tetragonal symmetry. Depending on the particular site chosen, the tetragonal axis may lie along any one of the cube axes x, y, or z. When a stress is applied along one of the Cartesian x, y, or z directions, the equivalence of the sites is lost, and the carbon atoms will jump by diffusion, preferentially occupying the lowest energy sites. This changes the length of the crystal in the direction of the stress, leading to an anelastic strain with a characteristic delay determined by the time of diffusion of the carbon atom between the sites. When a stress is applied along the $\langle 111 \rangle$ direction, no effect is observed, since this stress deforms the x, y, and z directions by the same amount. The relaxation strength is determined by the number of interstitials and the direction of the applied stress.

The overall behavior can be classified conveniently with the help of simple models, as illustrated in Fig. 1. For a crystal containing point defects and dislocations, there is a point-defect strain ε_p and a dislocation strain ε_d in addition to the elastic strain ε_{el}. The point-defect strain may be paraelastic (stress-induced rotation between equivalent positions of permanent dipoles with the help of thermal fluctuations, as in the Snoek effect), or diaelastic (induced dipole, which produces an internal strain, as with interstitials in metals). Also, the effects of phonons and electrons can be included within the formalism as point defects.

The dislocation strain is a diaelastic effect requiring a mass term in the general case and is given by

$$\varepsilon_d = \Lambda b y \tag{3}$$

where Λ is the dislocation density, b the Burgers vector, and y the average dislocation displacement. The displacement y is given by the dislocation equation of motion

$$M\ddot{y} + B\dot{y} + Ky = b\sigma \tag{4}$$

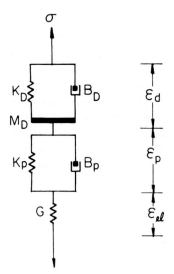

Fig. 1: Schematic representation of strain in a specimen containing point defects and dislocations.

where $M \simeq \rho b^2$ is the dislocation mass per unit length, B is the dislocation viscosity, and K is a restoring-force constant for the dislocation. Comparing Eqns. (3) and (4), we obtain the model values M_d, B_d, and K_d from M, B, and K by dividing the latter values by Λb^2. For small enough B, the response ε_d has a resonance character, but for large B, inertial effects can be ignored and we have a relaxation with

$$\Delta_R = \frac{\Lambda G b^2}{K_d} \quad \text{and} \quad \tau = \frac{B_d}{K_d}. \tag{5}$$

The coefficients B and K are phenomenological, can arise from many sources representing the dislocation interactions with other defects, and are related to the corresponding quantities appearing for the point-defect strain ε_p. In general, any mechanism that damps ultrasonic shear waves in a crystal containing no dislocations will also provide a viscous drag on dislocations, since a moving dislocation generates a shear strain rate in the crystal. Sources of drag B are phonons, electrons, and point defects, while sources of K are the dislocation tension and immobile aligned dipolar point defects. Dislocation internal-friction effects have been detected over a frequency range from 10^{-2} to 10^{12} Hz, and over a point-defect concentration range from 10^{12} to 10^{-2}.

See also: Anelasticity; Crystal Defects; Elasticity.

Bibliography

R. DeBatiste, *Defects in Crystalline Solids.* North-Holland. Amsterdam, 1972.

A. S. Nowick and B. S. Berry, *Anelastic Relaxation in Crystalline Solids.* Academic, New York, 1972.

R. Truell, C. Elbaum, and B. B. Chick. *Ultrasonic Methods in Solid State Physics.* Academic, New York, 1969.

Interstellar Medium

D. J. Hegyi

The interstellar medium refers to the matter and radiation lying between the stars within the region of space associated with a galaxy. Most interstellar matter is found in the planes of spiral galaxies; observations of elliptical galaxies have revealed relatively little interstellar material. Until quite recently, almost all data on the interstellar medium have been obtained from optical and radio studies of our own galaxy, a spiral, and, in fact, even more locally, from observations in the solar neighborhood. However, new observations are just beginning to reveal information about the interstellar medium in nearby galaxies. Using mostly data from our galaxy, an overview of the interstellar medium will be presented by discussing the following: the hierarchy of structure, chemical composition, cosmic rays, the galactic magnetic field, and the interaction between the components.

Hydrogen is the main constituent of both the interstellar medium and the stars. Hydrogen's widespread abundance, coupled with the ability of radio telescopes to detect interstellar hydrogen in absorption and emission using the 21-cm hyperfine transition [1], has made it possible to trace out the distribution of hydrogen over most of the galaxy. Beginning with the grossest features and proceeding to those on smaller scales, the hydrogen is confined to a thin flat disk, the galactic plane, whose thickness [2] inward of the sun (solar galactic radius $\simeq 10\,\mathrm{kpc}$; $1\,\mathrm{pc} = 3.26\,\mathrm{light\text{-}years} = 3.09 \times 10^{18}\,\mathrm{cm}$) is about 230 pc and decreases in thickness to less than 100 pc at a galactic radius of less than 300 pc. Beyond the galactic radius of the sun, the galactic disk is known to be warped, possibly because of tidal distortion by a nearby galaxy, the Large Magellanic Cloud [3].

The observations show that neutral hydrogen (H_I) in the disk has spiral-like structure. This pattern is superimposed on the overall differential rotation of the galaxy. To account for the persistence of this structure, which is not smeared by differential rotation, it has been suggested [4, 5] that the spiral features are density waves propagating in a uniform-density disk. This is consistent with observations showing the interarm regions to be only a factor of 2 less dense than the spiral arms.

The predominant motion of the hydrogen in the galactic plane is differential rotation, for which the angular velocity is a decreasing function of radius. The linear velocity reached at the solar position is approximately $225\,\mathrm{km\,s^{-1}}$ [6, 7] and increases slightly at larger galactic radii as a consequence of the massive dark halo inside of which the disk is located. Deviations from circular motion in the plane are typically less than $10\,\mathrm{km\,s^{-1}}$. Away from the galactic plane, H_I clouds have been observed with velocity components ranging from 100 to $200\,\mathrm{km\,s^{-1}}$ moving toward the plane [8]. It appears that these clouds are from the galaxy-Magellanic Cloud system and have reached these high velocities by having been gravitationally accelerated by the galaxy from distances larger than 20 kpc from the galactic disk [9].

The structure of the interstellar gas in the galactic plane is usually described as cloud-like. The reference to clouds is meant to invoke a picture not of matter in a void, but rather of a density enhancement in a more tenuous background. Since the clouds are not gravitationally stable, they are expected to dissipate due to internal thermal pressure on a time scale of about $10^7\,\mathrm{yr}$. To stabilize the clouds, Spitzer [10] proposed a hot ($\sim 10^6\,\mathrm{K}$) tenuous intercloud medium in pressure equilibrium with the clouds. The existence of the hot component of the interstellar medium is now well established from observations of the diffuse x-ray background and observations of O_{VI} absorption lines [11]. The present picture of the interstellar medium [12] consists of cool clouds in a mixed intercloud medium consisting of both warm and hot gas in pressure equilibrium. The energy needed to heat and compress the interstellar medium comes from a combination of stellar winds and supernovae while the advance of shock fronts forms the clouds by compressing the initially cool material.

Parameters describing individual H_I clouds lie in the following ranges: size, 0.01–50 pc; density, $10–10^7$ particles cm^{-3}; mass, $(0.1 - 10^6)M_\odot$; and kinetic temperature, 10–100 K. Taken collectively, approximately 5–10% of the mass of the galactic plane resides in neutral hydrogen. The distribution function for interstellar cloud masses is consistent with a single power-law slope [14] extending from cloud masses of less than $0.01M_\odot$ to $10^6 M_\odot$. For cloud masses higher than $(10^2 - 10^3)M_\odot$, the distribution shifts from atomic to molecular clouds because the more massive clouds can more effectively shield the molecules from photodissociation.

Hydrogen is also found in an ionized state (H_{II}) in the radiation fields surrounding hot stars ($T_{star} > 20000$ K) (see, e. g., Spitzer [15]). These emission nebulae (which are among the most photogenic astronomical objects, containing very delicate shadings of light and dark) sometimes exhibit a sharp outer boundary beyond which is H_I. The abrupt transition between H_{II} and H_I occurs when the ionizing flux due to the hot central star has dropped to a low enough level so that hydrogen is mostly in the ground state. Because H_{II} requires an ionizing source, it is not distributed as widely nor as uniformly throughout the galaxy as H_I [16].

Not only do H_I and H_{II} exist in interstellar clouds, but a significant amount of H_2 has been found, and it may be extremely abundant in the interior of dark clouds where it is difficult to observe. Molecular hydrogen is believed to form on the surfaces of grains. At low temperatures an appreciable fraction of the hydrogen atoms that collide with a grain adhere, then move about the grain surface occasionally finding another hydrogen, and coalesce to form H_2. Because H_2 is only weakly bound to a grain, it can escape, leaving the grain as gaseous H_2. Molecular hydrogen does not exhibit the 21-cm hyperfine transition and must be detected in absorption in the far ultraviolet from above the atmosphere.

It is generally believed that the present chemical composition of the interstellar medium is relatively similar to the solar composition, since the sun formed from interstellar material only 4.6 billion years ago, a short time ($\sim$25%) relative to the age of the galaxy. Nevertheless, this point of view does require some explanation when confronted with data that ostensibly show 25 elements to be depleted [17] by factors as high as 10^{-4}. It is argued that a significant amount, if not all, of the depleted material must be condensed out onto dust grains [18]. These are particles $\sim 0.1\,\mu$m in size that account for the obscuration in certain directions in space. They are most evident as the cause of the dark band in the center of the Milky Way visible to the unaided eye on a dark summer night. Supporting condensation onto dust grains is an observation [19] that shows that at the higher temperatures encountered in the intercloud medium, about 5000 K in this case, the calcium-hydrogen ratio is an order of magnitude larger than the value obtained from the cooler interstellar clouds. Presumably, the elevated temperature reduces the probability of a calcium atom sticking to a grain. Also supporting grain condensation is the observed correlation between the amount of gas and dust in a particular direction in space [13]. Since grains contribute about 1% of the mass of the interstellar medium, they must be quite rich in heavy elements.

In addition to H_2, the interstellar medium contains other molecules, which were detected first in the interstellar medium. The first molecules to be discovered, CN, CH, and CH^+, were found as optical interstellar absorption lines by Adams and Dunham in 1937 [20]. The next interstellar molecule to be discovered, the OH radical, detected in the radiofrequency spectrum in 1963, precipitated an avalanche of other detections. Now, well over 50 molecules have been discovered, including molecules as complex as $(CH_3)_2O$, dimethyl ether, and C_2H_5OH, ethanol. Almost all of these molecules have been found at millimeter wavelengths. The principal reason for the productivity of the millimeter techniques for detecting molecules is that most molecules have millimeter rotational transitions, and these are about the only lines that can be excited in the cold interior of interstellar clouds. For molecules other than hydrogen, there is no agreement about the formation mechanism. They might form in two-body reactions in the gas phase [21] or on the surfaces of grains [22].

The isotopic abundance ratios of the elements in the interstellar medium can be interpreted to provide a variety of different types of information. For example, the interstellar deuterium to hydrogen ratio is sensitive to the density in baryons at the epoch of primordial nucleosynthesis, when the universe was less than 3 minutes old. Observations of the deuterium to hydrogen abundance ratio have been obtained from Ly-α absorption features in the spectrum of nearby stars using the International Ultraviolet Explorer satellite [23] and yield a range of values consistent with cosmological expectations. Also, isotopic abundance ratios can be used to reveal information about the cycling of interstellar material through stars and galactic chemical evolution. Using both radio and optical techniques many workers have collected data on the $^{12}C/^{13}C$ ratio, for example [24], finding galactic gradients such that the ratio varies from about 25 at the galactic center to about 50 [25] in the local interstellar medium with variations to above 100 that complicate the interpretation. The corresponding solar value for the $^{12}C/^{13}C$ ratio is 89 and probably reflects the interstellar value when the sun was formed 4.6 billion years ago.

The interstellar medium is also filled with cosmic rays. These are energetic particles consisting mostly of protons and heavier nuclei with an electron component amounting to about 1% of the proton flux. Cosmic rays ranging in energy from less than 10^6 eV to more than 10^{20} eV are observed. However, it is only at energies from 10^9 eV to about 10^{17} eV that the cosmic ray data is believed to be characteristic of the galactic cosmic radiation. At lower energies the solar wind influences fluxes, whereas at higher energies extragalactic cosmic rays can penetrate the galactic magnetic field and, conversely, galactic cosmic rays can escape. The solar wind, the plasma flowing from the solar corona, carries with it frozen-in magnetic field lines that affect the diffusion of cosmic rays in the solar cavity [26, 27]. The cosmic ray element abundances are somewhat similar to the solar abundances except for Li, Be, and B. The abundances of these elements can be accounted for by the fragmentation of heavier nuclei as they spiral around galactic magnetic field lines, a path traversing about $5\,\mathrm{g\,cm^{-2}}$ of matter.

The origin of cosmic rays is a topic that has been discussed for years. At present, the most common viewpoint is that low-energy cosmic rays are accelerated in supernova shocks. For energies above 10^4 eV, the acceleration is believed to occur in pulsar magnetospheres. There, the energy density in cosmic rays, about $10^{-12}\,\mathrm{erg\,cm^{-3}}$, is significant and comparable to the energy density of the 2.8 K cosmic blackbody radiation, of starlight in the galactic plane, of turbulence in the interstellar medium, and of the galactic magnetic field. Whether this is a coincidence or has fundamental significance is an unanswered question.

The existence of the galactic magnetic field has been deduced from observation of the polarization of starlight [28, 29]. To explain the consistency of magnitude and direction of the polarization of light from stars viewed through the same interstellar cloud, a galactic magnetic field has been postulated that aligns elongated grains in the cloud, establishing a preferred scattering direction. The value of the galactic magnetic field, approximately $1-3\,\mu G$ [30], may be obtained most directly by use of the Faraday effect, the rotation of the plane of polarization of radiation, and the Zeeman splitting of the 21-cm line. Somewhat less directly, magnetic field values may be obtained from synchrotron radiation due to electrons spiraling in the galactic magnetic field. The topology of the galactic magnetic field is of interest to understand the origin of the field. If the field is generated by a self-starting dynamo [31], the magnetic field lines in the disk are expected to be closed and circular, but if they are primordial they are expected to have a bisymmetric configuration open to intergalactic space [32]. The observations appear to favor a bisymmetric spiral pattern [33].

After enumerating the various constituents of the interstellar medium, it is not yet possible to proceed significantly further by discussing the interactions among the components. Any complete theory must begin with the formation of our galaxy, approximately 10^5 yr after the "big bang" when the universe consisted of a homogeneous gas containing only hydrogen and helium. It must follow the matter through cycles of star formation and nuclear burning to account for the formation of heavy elements, and stellar mass ejections, including supernovae, which disperse these elements throughout the interstellar medium. Since, for example, the star formation rate is dependent upon the composition of the matter, magnetic field strength, and cosmic ray flux, no individual aspect of the system can be treated without taking all the other components into account. Consequently, any advance in understanding of the interstellar medium is related to general progress in astrophysics.

See also: Cosmic Rays, Astrophysical Effects; Cosmic Rays, Solar System Effects; Cosmology; Galaxies; Milky Way; Solar Wind.

References

[1]   H. I. Ewen and E. M. Purcell, *Nature* **168**, 356 (1951).
[2]   W. B. Burton, *Galactic and Extra-Galactic Radio Astronomy* (G. L. Verschuur and K. I. Kellerman, eds.). p. 82. Springer Verlag, Berlin and New York, 1974.
[3]   C. Hunter and A. Toomre, *Astrophys. J.* **155**, 747 (1969).
[4]   B. Linblad, *Stockholm Obs. Ann.* **22**, 5 (1963).
[5]   C. C. Lin, C. Yuan, and F. H. Shu, *Astrophys. J.* **155**,721 (1969).
[6]   L. Blitz, *Ap. J.* **231**, 115 (1979).
[7]   J. E. Gunn, G. R. Knapp, and S. D. Tremaine, *Astrophys. J.* **84** (1979).
[8]   G. L. Verschuur, *Annu. Rev. Astron. Astrophys.* **13**, 257 (1975).
[9]   I. F. Mirabel, *Astrophys. J.* **256** (1982).
[10]   L. Spitzer, *Astrophys. J.* **120**, 1 (1954).
[11]   E. B. Jenkins, *IAU Colloquium 81, The Local Interstellar Medium* (Y. Kondo, F. C. Bruhweiler, and B. D. Savage eds.), p. 155. NASA Conf. Pub. 2345, 1984.
[12]   R. M. Cretcher, *Astrophys. J.* **254**, 82 (1982).
[13]   L. Spitzer and E. B. Jenkins, *Annu. Rev. Astron. Astrophys.* **13**, 133 (1975).
[14]   J. M. Dickey and R. W. Garwood, *Astrophys. J.* **341**,201 (1989).

[15]   L. Spitzer, *Diffuse Matter in Space*, p. 112. Wiley-Interscience, New York, 1968.

[16]   F. J. Lockman, *Astrophys. J.* **209**, 429 (1976).

[17]   D. C. Morton, *Astrophys. J. Lett.* **193**, L35 (1974).

[18]   G. B. Field, *Astrophys. J.* **187**, 453 (1974).

[19]   L. M. Hobbs, *Astrophys. J. Lett.* **206**, L117 (1976).

[20]   W. S. Adams and T. Dunham, *Pub. Astrom. Astrophys. Soc.* **9**, 5 (1937).

[21]   E. Herbst and W. Klemperer, *Astrophys. J.* **185**, 505 (1973).

[22]   W. D. Watson and E. Salpeter, *Astrophys. J.* **175**, 659 (1972).

[23]   W. B. Landsman, R. C. Henry, H. Moos, and J. L. Lansky, *Astrophys. J.* **285**, 801 (1984).

[24]   P. G. Wannier, *Annu. Rev. Astron. Astrophys.* **18**, 399 (1980).

[25]   P. Crane and D. J. Hegyi, *Astrophys. J. Lett.* **326**, L35 (1988).

[26]   J. R. Jokipii, *Rev. Geophys. Space Phys.* **9**, 27 (1971).

[27]   H. K. Volk, *Rev. Geophys. Space Phys.* **13**, 547 (1975).

[28]   W. A. Hiltner, *Science* **109**, 165 (1949).

[29]   J. S. Hall, *Science* **109**, 166 (1949).

[30]   G. L. Verschuur, in *Galactic and Extragalactic Radio Astronomy* (G. L. Verschuur and K. I. Kellerman, eds.), p. 179. Springer-Verlag, Berlin and New York, 1974.

[31]   M. Stix, *Astron. Astrophys.* **47**, 243 (1976).

[32]   J. H. Piddington, *Astrophys. Space Sci.* **59**, 237 (1978).

[33]   Y. Sofue, M. Fujimoto, and R. Wielebinski, *Annu. Rev. Astron. Astrophys.* **24**, 459 (1986).

Invariance Principles

F. Gürsey[†]

Invariance principles came to occupy their preeminent position in contemporary physics chiefly through the impact of three successive waves. The first was relativity, which emphasized the invariance of the fundamental laws of physics under the Poincaré transformations relating equivalent inertial systems and the invariance of the laws of gravitation under Einstein's general coordinate transformations (*see* Relativity, Special Theory; Relativity, General) in spacetime. The second was the study of the invariance properties of the Hamiltonian in quantum mechanics, which led to an explanation of degeneracies in energy of the different states of a physical system. The pioneering work of Weyl, Wigner, and Fock and the extensive use of Noether's theorem showed the importance and usefulness of symmetry considerations. The third wave came with particle physics, where invariance principles turned out to be indispensable tools for the understanding of selection rules in particle reactions, for the classification of particle states, and for the restriction of the possible forms of the interaction Hamiltonian or the scattering matrix.

Various Kinds of Invariance

Invariance principles may be classified in different ways: (a) Those related to *spacetime* (like parity, time reversal, Poincaré invariance, conformal invariance, or general relativistic invariance) and those related to *internal* coordinates or internal quantum numbers (like charge con-

---

[†]deceased

jugation, gauge invariance, charge independence, etc.). (b) Those associated with *discrete* or *continuous* groups of transformations. For instance T (time reversal), P (parity), C (charge conjugation), and their combinations, as well as invariance under discrete translations in crystals, fall in the first category, while translations, rotations, scaling transformations, isospin, or unitary transformations are examples in the second category, (c) Those which are *exact* and those which are *approximate*. Poincaré invariance, TCP invariance, invariance under fermion -number gauge transformations (and perhaps separate baryon- and lepton-number transformations), gauge invariance related to electric charge are all thought to be exact. On the other hand, invariance principles related to C, P, and T separately, charge independence, unitary symmetry are approximate in varying degrees, (d) Those corresponding to a *local* (x-dependent) or *global* (x-independent) continuous transformations. For example, the fermion conservation law is a global invariance principle, whereas invariance of electromagnetism is local, (e) Those which are *manifest*, being associated with an explicit symmetry of the Lagrangian of the system, and those which are *hidden* as in the case of spontaneous symmetry breaking where the Lagrangian is expressed in terms of physical fields which are not covariant with respect to the invariance group. Topological quantum numbers that arise in special solutions like monopole or soliton solutions to field equations provide another example of hidden symmetry not contained explicitly in the Lagrangian.

Invariance Groups and Conserved Quantities

In order to illustrate the strong links between invariance principles and symmetry groups, let us characterize a system by a Lagrangian $L(q_i, \dot{q}_i)$ where q_i are the generalized coordinates and $\dot{q}_i$ (their time derivatives) the generalized velocities. The equations of motion are given by the variational principle

$$\delta \int_{t_1}^{t_2} L \, dt = 0 \tag{1}$$

which tells us that the action is minimized (or extremized) by the actual motion. The Hamiltonian is $H(p, q) = p_i \dot{q}_i - L$ where the momenta p_i are given by $\partial L / \partial \dot{q}_i$. If we consider a physical observable $\Omega = \Omega(p, q)$ as a function of the canonical variables p_i and q_i the time variation of Ω is given by

$$\frac{d\Omega}{dt} = \frac{\partial \Omega}{\partial q_i} \frac{dq_i}{dt} + \frac{\partial \Omega}{\partial p_i} \frac{dp_i}{dt}. \tag{2}$$

Now the equations of motion derived from Eq. (1) can be written in the Hamiltonian form

$$\frac{dq_i}{dt} = \frac{\partial H}{\partial p_i}, \quad \frac{dp_i}{dt} = -\frac{\partial H}{\partial q_i} \tag{3}$$

so that we can write

$$\frac{d\Omega}{dt} = \{\Omega, H\} = \frac{\partial \Omega}{\partial q_i} \frac{\partial H}{\partial p_i} - \frac{\partial H}{\partial q_i} \frac{\partial \Omega}{\partial p_i} \tag{4}$$

by means of a Poisson bracket.

The quantum-mechanical case is similar. Ω and H are then Hermitian operators that are functions of the Hermitian operators p_i and q_i that obey Heisenberg's commutation relations. In Dirac's formulation of quantum mechanics, the Poisson bracket is replaced by i times the commutator, i. e.,

$$\frac{d\Omega}{dt} = i[\Omega, H] = i(\Omega H - H\Omega) .\tag{5}$$

Conservation in time of the observable Ω implies the vanishing of its bracket (Poisson or Dirac) with the Hamiltonian. In particular, the Hamiltonian H is a constant of motion if L does not depend on t explicitly. Since H is the sum of kinetic and potential energies both in classical and quantum mechanics, this means conservation of total energy when L is invariant under time translations. This is an example of *Noether's theorem* according to which invariance of the Lagrangian under a one-parameter transformation implies the existence of a conserved quantity associated with the generator of the transformation. Another example is provided by invariance under space translations. Let q_i represent the six coordinates $\mathbf{r}_1$ and $\mathbf{r}_2$ of a two-point system. If the potential energy is invariant under space translations $\mathbf{r} \to \mathbf{r} + \mathbf{a}$, it must be a function of $\mathbf{r}_1 - \mathbf{r}_2$ only. Using for Ω in Eq. (4) or (5) the three components of the total momentum $\mathbf{p} = \mathbf{p}_1 + \mathbf{p}_2$, we find $d\mathbf{p}/dt = 0$, so that the total momentum is conserved. Thus momentum conservation is associated with the principle of invariance under the translation group. Using the commutators of $\mathbf{r}_1$, $\mathbf{r}_2$ with $\mathbf{p}_1$, $\mathbf{p}_2$ in the quantum-mechanical case we can show readily that

$$\exp(i\mathbf{p} \cdot \mathbf{a}) f(\mathbf{x}_1, \mathbf{x}_2) \exp(-i\mathbf{p} \cdot \mathbf{a}) = f(\mathbf{x}_1 + \mathbf{a}, \mathbf{x}_2 + \mathbf{a}) .\tag{6}$$

This shows that the conserved quantity $\mathbf{p}$ is also the generator of space translations.

In a similar way, conservation of angular momentum follows from the isotropy (invariance under rotations) of the Lagrangian.

According to Poisson's theorem, if $\mathbf{H}$ has vanishing brackets with quantities C_r, it also has vanishing brackets with $\{C_r, C_s\}$ (or $[C_r, C_s]$ in the quantum-mechanical case). Thus the conserved quantities C_r have a Lie algebraic structure, so that they can be associated with the generators of a Lie group. There are N conserved quantities corresponding to the dimension N of the group. For example, there are eight conserved quantities (three of them being the angular momenta) in the case of the three-dimensional harmonic oscillator, the Lagrangian of which is invariant under the group $SU(3)$. In the case of the hydrogen atom (or the Kepler) problem, the invariance group is $SO(4)$ leading to six conserved quantities, namely, the three angular momenta and the three components of the Runge–Lenz vector.

If we now turn to a discrete invariance principle, we can consider a system invariant under the parity operator with the two eigenvalues ± 1. A parity noninvariant term in the Lagrangian will induce transitions among states $P = 1$ and $P = -1$ causing a parity-violating decay of the system in the course of time. This is what happens in a weak interaction such as beta decay. Strong and electromagnetic interactions, insofar as they can be separated from weak interactions, obey P (and C) invariance. Again, while the usual weak interactions are invariant under CP (and T) but not under C and P separately, a special kind of weak decay associated with neutral kaons exhibits a small violation of CP (or T) invariance. Such a violation of time reversal is ascribed to superweak forces.

Gauge Invariance and Internal Symmetries

In the domain of internal symmetries, the best known (and oldest) invariance principle is associated with the conservation of electric charge. This exact conservation law and the vanishing of the photon mass are also related to invariance under a local gauge transformation which affects simultaneously the electromagnetic potential A_ν and the wave function (or field operator) ψ of the charged particle (e. g., the electron) so that the Lagrangian remains invariant under the transformation

$$A_\nu \to A_\nu - \frac{1}{e}\frac{\partial \Lambda}{\partial x^\nu}\,, \quad \psi \to \exp(i\Lambda)\psi\,, \tag{7}$$

where Λ is an arbitrary function of the space-time variables x^ν and e is the charge of the particle. The electromagnetic field tensor $F_{\mu\nu}$ and the charge current j_ν are gauge invariant. Then the charge $Q = \int j^0 \, d^3x$ has a vanishing bracket with the Hamiltonian and represents an exactly conserved quantity. The photon field associated with the charge remains massless in virtue of Eq. (7). The fact that the field ψ is charged is expressed by the quantum-mechanical relation

$$[Q,\psi] = e\psi\,. \tag{8}$$

The principle of local gauge invariance associated with electric charge has been generalized by C. N. Yang and R. L. Mills in 1954 to charges Q_i that are the generators of a Lie group G [say $SU(2)$]. The G invariance is supposed to hold independently at every space-time point x. In the color gauge theory of strong interactions, G is taken to be the color group $SU(3)$. The electron field is then replaced by colored quark fields and the photon is replaced by eight massless gluons. In the $SU(2)$ example, ψ is a doublet and we can use the 2×2 Pauli matrices τ to write a generalization of Eq. (7), i. e.,

$$\tau \cdot \mathbf{A}_\nu \to S\tau \cdot \mathbf{A}_\nu S^{-1} - \frac{1}{g}S\left(\frac{\partial S^{-1}}{\partial x^\nu}\right)\,, \quad \psi \to S\psi\,, \tag{9}$$

where $S = \exp[\frac{1}{2}\tau \cdot \boldsymbol{\omega}(x)]$. In the case of global gauge transformations the parameters Λ or $\boldsymbol{\omega}$ become constant and the inhomogeneous terms in the transformation laws of A_ν or $\mathbf{A}_\nu$ drop out. Established field theories such as gravitation, electromagnetism, and quantum chromodynamics for the quark–gluon system are all based on exact local gauge invariance principles and involve massless boson fields whose interaction with other fields is universal and determined by the gauge principle.

Invariance Principles Associated with Spontaneously Broken Symmetries

Systems can be constructed in such a way that the Lagrangian is invariant under a group G, but the ground state (vacuum) which minimizes the energy is not invariant under G. Then the excited states will not be G covariant. The action of G on such states will transform them into other degenerate states. Then we speak of a spontaneous symmetry breakdown. If the ground state is still invariant under a subgroup H of G, the generators of H are associated with massless boson fields, whereas the other fields corresponding to the remaining generators of G acquire a mass. When the Hamiltonian is reexpressed in terms of physical noncovariant fields the H symmetry is manifest, whereas the remainder of the G symmetry is hidden. A prototype of spontaneous symmetry breakdown is the ferromagnet that has a ground state with all its atomic spins aligned in one direction although interatomic forces are rotationally invariant. In this case, rotational invariance is restored by noting that the magnet pointing in an arbitrary direction through rotation of its initial position provides an equally good ground state. The totality of all rotated magnets forms an isotropic set. Another example is the generalization of electrodynamics based on a local $U(1)$ group to a unified gauge field theory of weak and electromagnetic interactions based on the Weinberg–Salam group $SU(2) \times U(1)$. A realistic theory is obtained by allowing the group to be spontaneously broken so that out of the four generators of the group one remains associated with a massless boson (the photon), while the other three correspond to massive weak bosons that mediate charged and neutral weak interactions.

Usefulness and Power of Invariance Principles

Invariance principles are of great practical value in determining the general form of physical observables. Wave functions and transition amplitudes related to a Hamiltonian invariant under G can be decomposed into irreducible representations of G with coefficients that are functions of invariants only. For example, a rotationally and translationally invariant quantity depending on $\mathbf{r}_1$ and $\mathbf{r}_2$ can only be a function of $|\mathbf{r}_1 - \mathbf{r}_2|^2$. Projection operators for spin 0 and 1 in a system composed of two spinors can only be a function of the invariant combination $\boldsymbol{\sigma}^{(1)} \cdot \boldsymbol{\sigma}^{(2)}$ and because of the algebraic properties of Pauli matrices they must be linear in this expression. Invariants of discrete translations in a lattice are periodic functions. When a scaling invariance principle holds at high energy and momentum transfer, the structure functions occurring in the cross sections can only be functions of dimensionless combinations of kinematic variables called scaling variables. Homogeneity and isotropy of the universe at large determine the form of the cosmological metric (the Robertson–Walker metric).

Some Recent Developments

In two-dimensional dynamical systems (like spin models) scaling invariance near the critical point leads to conformal invariance, hence to a new invariance principle related to the behavior of the system under holomorphic mappings. This also holds for the system represented by a two-dimensional field theory. Such conformal field theories have a very high symmetry described by infinite algebras like the Virasoro algebra and Kac–Moody algebras. The invariants of these algebras are in turn related to critical exponents and other characteristics of the dynamical systems they represent. Since string theory also makes use of conformal field theories, the principle of two-dimensional conformal invariances has become a keystone of modern mathematical physics.

Another development concerns a possible symmetry (broken in nature) between fermions obeying the Fermi–Dirac statistics and bosons obeying the Bose–Einstein statistics. It is called supersymmetry. If a supersymmetric invariance principle were fundamental, then the Poincaré group would be extended to the super Poincaré group that would imply the existence of new particles in the role of superpartners of known fundamental particles. Although there is no experimental evidence for such an extension of the particle spectrum, supersymmetric theories are attractive because of their improved convergence properties and the possibilities they offer for the unification of all fundamental forces.

Following the lead of Einstein, physicists would like to derive the fundamental dynamical laws for all fundamental interactions from general invariance principles comprising and generalizing relativity and gauge principles and perhaps including symmetries between bosons and fermions (supersymmetry) and the conformal invariance of a world sheet substructure. If the dream comes true, all of fundamental physics could be reduced to invariance principles.

See also: Conservation Laws; *CPT* Theorem; Quantum Mechanics; Relativity, General; Relativity, Special Theory; String Theory; Supersymmetry and Supergravity.

Bibliography

A. Belavin, A. Polyakov, and A. Zamolodchikov, "Infinite Conformal Symmetry in Two-dimensional Quantum Field Theory". *Nucl. Phys. B* **241**, 333 (1984).

J. P. Elliott and P. G. Dawber, *Symmetry in Physics*, Vols. I and II, Oxford University Press, New York, 1979.

D. Friedan, Z. Qiu, and S. Shenker, "Conformal Invariance, Unitarity and Critical Exponents in Two Dimensions", *Phys. Rev. Lett.* **52**, 1575 (1984).

T. D. Lee, Symmetries, *Asymmetries and the World of Particles*. University of Washington Press, Seattle, 1988.

C. Quigg, *Gauge Theories of the Strong, Weak and Electromagnetic Interactions*. Benjamin Cummings, New York, 1983.

J. J. Sakurai, *Invariance Principles and Elementary Particles*. Princeton University Press, Princeton, NJ, 1964.

J. Wess and J. Bagger, *Introduction to Supersymmetry*. Princeton University Press, Princeton, NJ, 1983.

H. Weyl, *Space – Time – Matter*. Dover, New York, 1951.

E. P. Wigner, *Symmetries and Reflections*. MIT Press, Cambridge, MA, 1970.

Inversion and Internal Rotation

L. C. Krisher

Both inversion and internal rotation, despite their names, belong to the category of molecular dynamics called molecular vibrations. In molecular spectroscopy we find an energy hierarchy by which general molecular dynamics can be usefully described in decreasing magnitudes, as electronic transitions, molecular vibrations, and molecular rotations. Vibrational transitions ordinarily fall in the infrared region of the spectrum around $1000\,\mathrm{cm}^{-1}$ and rotations in the microwave region near $1\,\mathrm{cm}^{-1}$. Incongruously enough, both inversion phenomena and internal rotations have been extensively studied from observations made in the microwave region, since they often represent relatively low-energy (low-frequency) vibratory motions which can be strongly coupled to the overall molecular rotations.

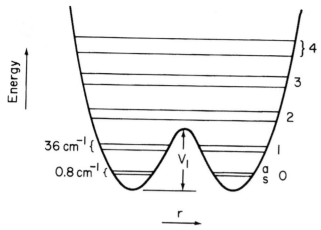

Fig. 1: Potential-energy curve for the ammonia molecule (NH_3) showing the allowed inversion energy levels, labeled in pairs with the appropriate vibrational quantum number. The energies are not to scale; the doublet separations are greatly exaggerated. The molecular coordinate r indicates the distance of the nitrogen atom from the center of the triangle formed by the hydrogen atoms.

Inversion is a vibration characterized by a potential function which has a double minimum, idealized as a parabola with a small bump at the bottom. Because of the rather atypical nature of the allowed energy levels for these potential functions, a number of common molecules, the most well-known perhaps being ammonia, exhibit pure vibrational transitions at microwave frequencies. For ammonia, NH_3, the classical motion involves the passing back and forth of the nitrogen atom through the plane formed by the three equidistant hydrogen atoms. The vibrational levels are shown in Fig. 1, and for high values of the quantum number v can be seen to exist more or less like harmonic oscillator levels, with a dissociation limit at the top, not indicated in the diagram. The bump at the bottom is sufficiently high, for molecules such as ammonia, to stabilize quantum states, in pairs, below the top of the classical barrier. The "inversion doublet" spectrum involves jumps between members of these pairs, and corresponds quantum mechanically to the nitrogen atom tunneling through the barrier. The lowest inversion doublet, labeled $0.8\,\mathrm{cm}^{-1}$ in Fig. 1, actually represents a large set of microwave transitions near that energy, the detailed frequencies of which depend on the rotational J and K quantum numbers for the molecule, although the latter do not change during the transition. In effect, the rotational state, specified by J and K, indicates the height on the barrier at which the tunneling is to take place. At ordinary temperatures the most intense member of this set is the 3,3 transition ($J = 3$, $K = 3$) at $23870\,\mathrm{MHz}$, which was used for the first maser oscillator and time standard.

The concept of internal rotation involves motion around the type of chemical bond known as a single bond, such as that between the carbon atoms in the ethane molecule, CH_3–CH_3. The idea of internal rotation, the twisting of one of the methyl groups with respect to the other, when hindered by a potential-energy function such as the threefold barrier shown in Fig. 2, actually represents a vibrational degree of freedom, the torsional vibration. This type of mo-

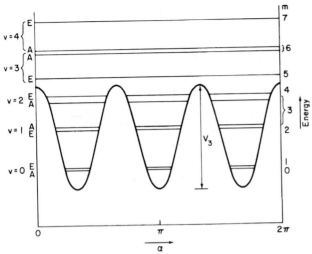

Fig. 2: Potential-energy diagram for a threefold symmetric internal rotation, such as that presented by a methyl group (CH_3-), showing the allowed vibrational (torsional) levels. The vibrational quantum numbers are indicated on the left with the corresponding "free rotor" or m quantum numbers at the right. The molecular coordinate α is an angle describing the degree of internal rotation; for a methyl group it indicates the position of the three hydrogen atoms with respect to the remaining part (framework) of the molecule.

tion for ethane and similar organic chemicals, long thought to be virtually free, is hindered by potential barrier heights of about $1000\,cm^{-1}$, sufficient to stabilize several vibrational states. For more complex hydrocarbons, this type of barrier also stabilizes the definite molecular geometries which are known as the conformations of the carbon chain, and which give rise to rotational isomers, or *rotamers*.

The finite threefold barrier for ethane-type molecules also imposes a pairing of the torsional vibration levels, with nondegenerate A and doubly degenerate E symmetry, as shown in Fig. 2. Transitions can occur between states with like symmetry, and for the lower levels are typically in the far infrared, at $100-300\,cm^{-1}$. The height of the torsional barrier can also be derived very accurately from a complex quantum -mechanical treatment of the coupling between the torsional vibration and the pure rotational motion of the molecule. For reasonably high internal barriers such as those found for the ethanes, the expected "pure" rotational transitions appear in the microwave region as close-spaced doublets. In effect, a molecule in a specific low torsional state, say $v = 0$, and of symmetry species A will have its own slightly different rotational constants and rotational spectrum from one in the $v = 0$ state but of symmetry E. The doublet splittings (frequency differences) afford a very accurate, if indirect, measurement of the barrier height. It should be carefully noted that the microwave energies do not effect transitions from one torsional state to another, nor are they energetic enough to allow jumps over the barrier to yield new conformations. The microwave measurements indicate the two different rotational behaviors of the A and E vibrational species, which exist because of the presence of the barrier.

Inversion Spectra

Ammonia is usually singled out because of the intensity of the microwave transitions and its historical importance. The first microwave transition, observed by Cleeton and Williams in 1934, involved the lowest inversion levels of this molecule. Following the development of radar in the 1940s, as microwave technology became available to spectroscopists, ammonia was one of the earliest molecules studied in detail, and in the 1950s was used by C. H. Townes and co-workers as the working substance for the first molecular oscillator, known as the *maser*.

The presence of the bump at the bottom of the potential, shown in Fig. 1, gives rise to a pairing of the inversion energy levels, with wave functions of alternating symmetry. The molecular symmetry allows direct transitions between these pairs, only two of which are bound by V_1, the lowest at about $0.8\,cm^{-1}$ accounting for the microwave spectrum. The inversion frequency is very sensitive to the height of the barrier, as is found for most tunneling phenomena, and several semiempirical approaches lead to a value for V_1 of about $2080\,cm^{-1}$.

The NH_3 molecule can also rotate, of course, but because of its very small moment of inertia, allowed transitions between rotational levels occur at much higher energies, in the far infrared. The inversion energies add to and subtract from these to form a doublet structure, and thus a rather special case of a vibration-rotation band. In addition, the "pure" inversion transition frequency shows a strong dependence on the particular rotational state in which the molecule is found. This gives rise to the large number of intense microwave inversion transitions, with frequencies ranging, from 17 000 to 40 000 MHz, dependent on the values of the rotational quantum numbers J and K, which remain unchanged for the transition. From the form of the dependence of the inversion frequency on the potential, V_1, Costain has developed an exponential power series expansion which fits these transitions very accurately through $J = 16$, as

$$\nu(\text{MHz}) = 23\,785.88\exp\{-6.36996\times10^{-3}J(J+1)+8.88986\times10^{-3}K^2$$
$$+8.6922\times10^{-7}J^2(J+1)^2-1.7845\times10^{-6}J(J+1)K^2+5.3075\times10^{-7}K^4\}\,.$$

Quantum-mechanical considerations prevent the number $K = 0$ from being used in this formula.

Many other trigonal-pyramidal molecules should in principle exhibit this "umbrella inversion", but only the deuterated ammonias such as ND_3 and the phosphine molecule PH_3 appear to have realistically observable frequencies. Such molecules as $AsCl_3$, because of mass effects and much higher values for V_1, have the inversion doublets collapsed to essentially zero frequency, as do more complex trigonal molecules such as CH_3Cl.

Beyond these symmetric umbrella inversions, there are many other vibrational phenomena which have double-minimum potentials. In larger molecules, a trivalent nitrogen atom may often be regarded as tunneling back and forth through a molecular plane, which motion is compensated in a complex way by the remainder of the molecule. Another example is provided by "ring-puckering" in molecules such as cyclopentene and trimethylene oxide. In cyclopentene, for example, the carbon atom furthest from the double bond is not coplanar with the other four, and the pertinent vibration allows it to move to an equivalent position on the opposite side of the plane. In more complex molecules, however, other low-frequency vibrational motions are often possible which, along with the molecular rotations, may be coupled to the inversion in such a way as to prevent a separable quantum mechanical description.

Internal Rotation Spectra

The ethane molecule, CH_3–CH_3, is often regarded as the prototype exhibiting the property of hindered internal rotation, largely because of interest in the thermodynamic properties of hydrocarbons. The bond between the two carbon atoms represents the chemist's classic "single bond", of Pauling hybridization type sp^3, as contrasted to "double-bond" molecules of the ethene series, with Pauling hybridization sp^2, which afford separable isomers (the *cis–trans* species) for many substituents.

Careful thermodynamic work in the late 1930s, along with developing spectroscopic techniques, indicated that internal rotatory motion in the ethane case was also hindered by a periodic energy barrier of about $1000\,cm^{-1}$, with bound torsional vibration states as shown in Fig. 2. Quantum-mechanical considerations for this case forbid transitions between the "doublets" (the selection rules are $A \leftrightarrow A$, $E \leftrightarrow E$), and torsional transitions, such as between the $v = 0$ and the $v = 1$ levels, are found in the far infrared at about $290\,cm^{-1}$.

Interestingly enough, the barrier heights hindering such motion can often be determined very accurately by a somewhat indirect method, which essentially involves only the rotational spectrum in the microwave region near $1\,cm^{-1}$. This requires analysis of the coupling between the internal and overall rotations, actually a vibration-rotation effect, in which the quantity V_3 appears as a parameter.

Analogues of ethane, and, more generally, series with threefold symmetric internal rotors, have been extensively studied. A methyl group or any other group with C_3 rotational symmetry provides a simplification of the dynamics, in that the overall moments of inertia of a molecule remain unchanged regardless of the detailed orientation of the internal rotor coordinate, labeled α in Fig. 2.

The model for a molecule such as acetaldehyde, CH_3CHO, assumes that the molecule is composed of a rigid threefold symmetric internal rotor (the CH_3– group) connected to a rigid framework (the CHO group) and is able to rotate about the intervening bond. All vibrations other than the torsion are thus assumed to be uncoupled and to provide a rotational average set of moments of inertia. The shape of the internal potential is assumed to be sinusoidal, and a small number of other parameters appear which can in principle be determined from the molecular structure.

The Hamiltonian for this model is expressed by Herschbach as

$$\mathcal{H} = \mathcal{H}_r + F(p - \mathcal{P})^2 + (V_3/2)(1 - \cos 3\alpha) \, ,$$

where $\mathcal{H}_r$ is the usual rigid-rotor expression involving the three rotational constants, F is a derived "internal rotational constant", and the operator $p - \mathcal{P}$ represents the relative angular momentum of the top (CH_3– group) and the framework. Expansion of the second term gives Fp^2 which can be combined with the V_3 term to give Mathieu's equation, and $F\mathcal{P}^2$ which, since $\mathcal{P}$ is a linear function of the overall molecular angular momenta, can be combined with $\mathcal{H}_r$ to give the new "effective rigid-rotor" constants. The cross term $-2F\mathcal{P}p$ is treated by perturbation theory in a basis set composed of the modified rigid-rotor functions times Mathieu functions. Contributions from the cross term lift the A, E degeneracy of a given torsional level and, for a certain range of barrier height, give rise to two sets of slightly different effective rotational constants. Because of the selection rule $A \leftrightarrow A$, $E \leftrightarrow E$, the rotational spectrum thus appears as close-spaced doublets, with frequency splittings $(\nu_A - \nu_E)$ which are a sensitive function of the barrier height V_3.

   This PAM (for "principal-axis method") treatment is especially useful for high barriers and for those cases where $A–E$ splittings in the ground torsional state, $v = 0$, are too small to observe. One must then measure the pure rotational spectrum of molecules in a higher torsional state, and if there are $A–E$ splittings, the barrier can be calculated. An alternative treatment called the IAM (for "internal-axis method") is often used for molecules which have lower barriers or which have other low-frequency vibrational modes besides the torsion. For very low barriers it is more convenient to treat the barrier energy itself as a perturbation to an otherwise "free internal rotor", using a model which employs the m quantization scheme indicated in Fig. 2 at the right.

Internal Barriers and Conformations

Molecules for which barriers have been determined are chemically quite diverse but tend to fall into series, the members of which differ by the progressive change of one substituent. Substituents with widely different electronegativities appear to have remarkably little effect on the barrier values through a given series. Thus the ethanes, when progressively fluorinated to give CH_3CH_2F, CH_3CHF_2, continue to have barriers near $1000 cm^{-1}$. Analogous methyl silane molecules, CH_3SiH_3, CH_3SiH_2F, etc., all have barriers near $500 cm^{-1}$, and the methyl germanes at about $350 cm^{-1}$. For the "acetyl series" in which the methyl group rotates against a double–single bond structure, substitution of any halogen for the acetaldehyde hydrogen leads to very little change in the barrier from $380 cm^{-1}$. Acetic acid appears to be an anomaly with an intermediate barrier value of about $170 cm^{-1}$. Much lower barriers are found for molecules such as $CH_3C=CSiH_3$ in which the "central bond" is a linear group, and for pure sixfold barrier cases such as CH_3NO_2, which has a barrier of about $2 cm^{-1}$.

   For internal rotors which lack threefold symmetry, the presence of the barrier allows one to describe in detail the preferred molecular conformations, indicated by the coordinate α. The overall moments of inertia and thus the pure rotational spectra of such molecules are now strongly dependent on α, and species at potential minima (which no longer need be threefold symmetric) are called *rotamers*. As shown in Fig. 3, the *cis–trans* terminology is no longer sufficient, and the term *gauche* is applied to the asymmetric forms which occur as a result of the threefold bonding.

   It is interesting to compare these barrier energies with those of conventional chemical bonds. The energy of formation of a chemical single bond, such as that between the two carbon atoms in the ethane molecule, is about $30000 cm^{-1}$. Also, the energy of an ethene type *cis–trans* conversion, which can be thought of as an internal rotation about a chemical double bond, is about $10000 cm^{-1}$. The rotamer barriers, typically less than $1000 cm^{-1}$, are thus a fraction of a conventional bond energy. The stable species shown in Fig. 3 can exist, however, through many hundreds of overall rotations, and be observable with distinct values for I_a, I_b, and I_c. These single-bond conformations when stabilized in condensed phases are also of great interest to molecular biologists. An entire issue of the *Journal of Physical Chemistry* (**91**, No. 21, October 8, 1987), is concerned with "dynamical stereochemistry", including these conformational effects in the synthesis of complex organic compounds.

   From the study of the conformations of small molecules, two powerful generalizations have emerged. When one threefold group, such as a methyl group or any sp^3 hybridized carbon, is bonded to another, the stable internal conformation is "staggered", as shown in Fig. 3 for the

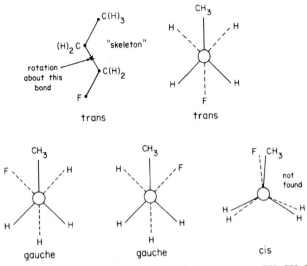

Fig. 3: Conformations of the molecule 1-fluoropropane $CH_3CH_2CH_2F$. The first sketch shows the carbon-fluorine chain or "skeleton" in the *trans* conformation, accompanied by the appropriate Newman diagram. The second row shows the two inertially equivalent *gauche* forms, and the unstable *cis* form which represents the "top of the hill" for the internal rotation potential. Also note that both the *trans* and *gauche* forms are "staggered" whereas the *cis* form is "eclipsed".

trans and the two *gauche* forms of 1-fluoropropane. The *cis* form represents a conformation corresponding to a maximum in V, and is said to be "eclipsed". Secondly, when a threefold sp^3 group is bonded to an sp^2 carbon, such as in the molecule propylene or in the acetyl series, the energy minima occur for those conformations in which a methyl hydrogen eclipses the double bond of the sp^2 group.

The internal-rotation coupling phenomenon is an important example from a larger area of vibration–rotation interactions, which serves to indicate the difficulties sometimes encountered when one tries to employ the often convenient and useful idea of the separation of molecular dynamics.

See also: Microwave Spectroscopy; Molecules; Tunneling.

Bibliography
Inversion
T. M. Sugden and C. N. Kenney, *Microwave Spectroscopy of Gases*. Van Nostrand, London, 1965 (I)
C. H. Townes and A. L. Schawlow, *Microwave Spectroscopy*. Dover, New York, 1975. (A)

Internal Rotation
C. C. Lin and J. D. Swalen, *Rev. Mod. Phys.* **31**, 841 (1959). (A)
E. B. Wilson, Jr., *Advances in Chemical Physics*, Vol. II, p. 367. Wiley, New York, 1959. (I)
J. E. Wollrab, *Rotational Spectra and Molecular Structure*. Academic Press, New York, 1967. (I)
R. B. Bernstein, D. R. Herschbach, and R. D. Levine, *J. Phys. Chem.* **91**, No. 21 (1987), see page 5366ff.

Ionization

M. E. Rudd

Ionization is the process in which a neutral atom or molecule is given a net electrical charge. Some atoms can accept an electron to form a negative ion in the process called *attachment*. One or more electrons can be removed from an atom or molecule forming a positive ion.

Of all the basic atomic processes, ionization requires the greatest energy. The amount of energy required to remove the least tightly bound electron from a neutral atom or molecule is called the *first ionization potential I* and is usually measured in electron volts (eV). The additional energy needed to remove the next electron is the *second ionization potential*, etc. The energy by which an excess electron is bound in a negative ion is known as the *electron affinity*. Tables of electron affinities, ionization potentials, and binding energies of inner-shell electrons are available for most atoms and many molecules.

Because ions and charged radicals are chemically highly reactive, much of the radiation damage to matter caused by high-energy photons or particles is a result of ionization, caused both by the primary radiation and by secondary products.

It is an empirical fact that approximately one ion-electron pair is formed for every 30 eV of energy lost by a fast charged particle traversing nearly any material. This makes it possible to determine the approximate energy of a particle by counting the ions formed in a detector which is sufficiently thick to stop the incoming particle.

Photoionization

Electromagnetic radiation of sufficiently short wavelength will cause ionization. The energy of an individual photon, $h\nu$, must equal or exceed the first ionization potential of the atom or molecule for this to take place. For most atoms the ionization threshold is in the ultraviolet region of the spectrum. The cross section (which is a measure of the probability that a given process takes place) for photoionization rises abruptly at the threshold and for most atoms falls off at higher energies as shown in Fig. 1. Photoionization in the upper atmosphere caused by ultraviolet radiation from the sun is largely responsible for the presence of the ionosphere. Two-photon ionization can also occur in which a single photon has insufficient energy to ionize an atom but enough to raise it to a real or virtual excited state. If a second photon arrives before the atom decays back to the ground state, the additional energy may then be enough to cause ionization. Three-photon and higher multiphoton ionizations have also been observed. For these processes to take place the radiation field must be sufficiently intense so that successive photons interact with a single atom within a time span shorter than the decay time. Such radiation fields are usually obtained with lasers.

Electron Collisions

If an electron makes a collision with an atom or molecule, enough energy may be transferred to one of the electrons to eject it. To do this, the incident electron must have an energy at least equal to the first ionization potential *I*. As shown in Fig. 1, the cross section for this process increases rapidly above threshold to a maximum and then falls off monotonically at high energies. In atoms with more than one subshell, additional thresholds occur at higher energies yielding structure in the ionization function.

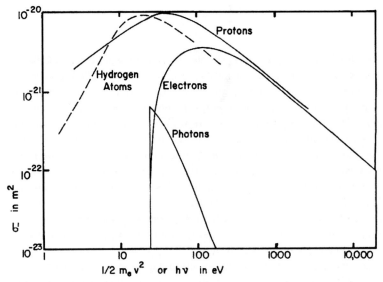

Fig. 1: Cross section for ejection of electrons from a helium atom by various incident particles. The abscissa is the incident energy for photons or electrons; for heavy particles it is the energy of an electron of equal velocity.

The cross section varies with impact energy E as $E^{-1} \log E$ at high energies. This prediction of the Bethe–Born approximation has been well verified experimentally. The dependence of the cross section on energy near threshold is of the form $(E-I)^n$. The semiclassical treatment by Wannier yields $n = 1.127$, a value verified by experiment.

Heavy-Particle Collisions

Because of the added possibilities of electron removal from the projectile (called *stripping* or *electron loss*) and the transfer of electrons (variously called *electron transfer*, *charge transfer*, and *charge exchange*), ionization by heavy atomic particles is a more complex process than photon or electron ionization. For example, the following are possible final charge states when a proton collides with a helium atom:

$$H^+ + He \quad \rightarrow \quad H^+ + He^0 , \tag{1}$$
$$H^+ + He^+ + e^- , \tag{2}$$
$$H^0 + He^+ , \tag{3}$$
$$H^+ + He^{2+} + 2e^- , \tag{4}$$
$$H^0 + He^{2+} + e^- , \tag{5}$$
$$H^- + He^{2+} . \tag{6}$$

Of these six processes, 2, 4, and 5 contribute to the production of free electrons; 2, 3, 4, 5, and 6 contribute to the production of positive ions; and 3, 5, and 6 are electron-transfer reactions. Note that while reactions 2 and 4 are the only ones that might fairly be called simple ionization, other reactions contribute to the charging of the target or to the release of free electrons.

In most experiments involving the collision of a beam of particles with a stationary gas target, what is measured are the positive and negative charges left behind or the change of charge state of the projectile. If we call the negative charge production cross section σ_-, the cross section for production of slow positive ions σ_+, and the electron-transfer cross section σ_c, then one can readily show from the basic reactions listed above that $\sigma_- = \sigma_+ + \sigma_c$.

For collisions of fast neutral atoms with other neutral atoms a similar analysis can be made. If the cross section for producing negative ions is small compared to the other cross sections (as it usually is), then the approximation $\sigma_- = \sigma_+ + \sigma_s$ results where σ_s is the electron-loss cross section.

Figure 1 shows cross sections for ejection of electrons from collisions of protons or neutral hydrogen atoms with helium atoms. Note that at high energies, equal-velocity protons and electrons yield the same ionization cross sections.

Cross sections for ionization from various electronic shells of large numbers of targets have been and are being measured as a function of many parameters. These include the collision energy, the excitation and charge state of the target (and the projectile in the case of heavy-particle impact), the impact parameter (or distance of closest approach), and the energy and angular distributions of the ejected electrons and of the scattered and recoil particles.

As a result of such measurements and the concomitant theoretical investigations, a number of different mechanisms of ionization have been identified. Some of these are as follows.

Direct Coulomb Ionization

If a charged particle such as an electron or proton makes a sufficiently close collision with an atom, enough energy may be transferred to one (or more) of the atomic electrons to remove it from the atom. This transfer of energy is through the Coulomb interaction between the electron and the incident charged particle and can occur for inner-shell as well as outer-shell electrons.

This mechanism of ionization may be described classically either in a binary-encounter model (taking account only of the interaction between the charged projectile and one electron of the target) or by the use of Monte Carlo methods to solve the three-body problem. Quantum mechanically, calculations may be made using the Born approximation or the Continuum-Distorted Wave Eikonal-Initial State (CDW-EIS) method.

In the case of a collision between two neutral atoms, the electric fields of the electrons and the nucleus tend to screen each other for large distances of separation, but at close approaches, the Coulomb field of the nucleus and the electrons separately may cause ionization.

Electron Promotion

A pair of colliding atoms can be thought of as a temporary molecule with an internuclear distance which decreases, reaches a minimum, and then increases to an infinite separation. The molecular energy levels must change accordingly, and if the collision is slow enough, the electronic wave functions adjust adiabatically except near crossings of the energy levels. At such crossings electrons readily transfer from one level to another and upon separation electrons may be "promoted" to higher excited levels or into the continuum. Provided there are crossings at reasonably large internuclear separations, this mechanism is very efficient in producing inner-shell vacancies. This picture is called the *molecular-orbital promotion model* or the *Fano–Lichten model*.

Autoionization and the Auger Effect

When an atom is put into an excited state with an energy above the ionization potential, instead of decaying by the emission of a photon it is usually more likely to make a transition to a continuum state, i. e., a state of the ion plus a free electron with kinetic energy. This spontaneous ejection of an electron is known as *autoionization*. Electrons ejected by this mechanism have discrete energies characteristic of the initial and final states of the atom.

Autoionizing states can be produced by photons, electrons, or heavy particles. Excited states with energies above the ionization potential may result, e. g., from the simultaneous excitation of two electrons. The excitation may also be caused by the removal of an inner-shell electron. In this case the resulting transition involving the filling of the vacancy and the ejection of an electron is known as the *Auger effect*. This competes with x-ray emission, the branching ratio in favor of the latter being called the *fluorescence yield*. Successive Auger transitions in a multishell atom can leave the outer shell completely or nearly completely stripped of electrons.

Electron Transfer

A collision between a fast positive ion and a neutral atom or molecule may result in a transfer of an electron to the ion. Even though no electron is released, this reaction leaves a positive ion behind and therefore can be thought of as a mechanism for producing ionization.

It has been found that the transferred electron may end up not only in the ground state of its new host atom or in an excited state, but can also be transferred to a continuum state of the projectile. Electrons ejected by this mechanism are distinguishable from those in continuum states of the target atom only insofar as their angular and energy distribution is characteristic of the moving frame of reference rather than that of the stationary target.

Electron Shakeoff

When an electron is ejected from a multielectron atom, the remaining electrons suddenly find themselves in a different central potential. After the ensuing readjustment there is a probability that a given electron will find itself in a new state. If this is a higher excited state, the process is called *shakeup*, but if it is a continuum state it is known as *shakeoff*.

Theoretically, the sudden approximation says that the probability of a transition is the square of the overlap integral of the wave functions for the initial and final states. The outermost electrons are the ones most likely to be ejected by shakeoff.

Penning Ionization

If one atom has a greater excitation energy than the ionization potential of a second atom with which it makes a collision, the energy released by the deexcitation of the first atom can go to ionize the second. This process, known as *Penning ionization*, normally requires that the excited state be a metastable (i. e., long lived) one so that the atom can retain its excitation energy long enough to make a collision.

Dissociative Ionization

Photon, electron, or heavy-particle impact on a molecule can cause it to break up or dissociate into fragments which may be charged. Dissociative ionization also occurs when ionic solids such as salts are dissolved in certain solvents. This occurs when the ions of the solute are more strongly attracted to the dipoles of the solvent molecules than to the oppositely charged ions and therefore are separated from them.

See also: Auger Effect; Collisions, Atomic and Molecular; Photoionization.

Bibliography

J. W. Gallagher, C. E. Brion, J. A. R. Samson, and P. W. Langhoff, "Absolute Cross Sections for Molecular Photoabsorption, Partial Photoionization, and Ionic Photofragmentation Processes", *J. Phys. Chem. Ref. Data* **17**, 9–153 (1988). (I)

J. B. Hasted, *Physics of Atomic Collisions*, 2nd ed. American Elsevier, New York, 1972. (I)

Y.-K. Kim, K.K. Irikuna, M. E. Rudd, D. S. Zucker, J. S. Coursey, K. J. Olsen, and G. G. Wiersma, "Electron Impact Ionization Cross Sections."
http://physics.nist.gov/PhysRefData/Ionization/Xsection.html.

Wolfgang Lotz, "Electron Binding Energies in Free Atoms", *J. Opt. Soc. Am.* **60**, 206–210 (1970). (E)

T. D. Märk and G. H. Dunn, eds, *Electron Impact Ionization*. Springer, Wien and New York, 1985. (I)

H. S. W. Massey, E. H. S. Burhop, and H. B. Gilbody, *Electronic and Atomic Impact Phenomena*, 2nd ed. Oxford, London, 1969–1975 (5 volumes). (I)

E. W. McDaniel, *Collision Phenomena in Ionized Gases*. Wiley, New York, 1964. (I)

N. F. Mott and H. S. W. Massey, *Theory of Atomic Collisions*, 3rd ed. Oxford, London, 1965. (A)

M. E. Rudd and J. H. Macek, "Mechanisms of Electron Production in Ion-Atom Collisions", *Case Studies At. Phys.* **3**, 47–136 (1972). (I)

M. E. Rudd, Y.-K. Kim, D. H. Madison, and J. W. Gallagher, "Electron Production in Proton Collisions: Total Cross Sections", *Rev. Mod. Phys.* **57**, 965–994 (1985). (I)

M. E. Rudd, Y.-K. Kim, D. H. Madison, and T. J. Gay, "Electron Production in Proton Collisions with atoms and molecules: Energy Distributions," *Rev. Mod. Phys.* **64**, 441–490 (1992).

H. Tawara and, T. Kato, "Total and Partial Ionization Cross Sections of Atoms and Ions by Electron Impact", *At. Data Nucl. Data Tables* **36**, 167–353 (1987). (E)

U. Wille and R. Hippler, "Mechanisms of Inner-Shell Vacancy Production in Slow Ion–Atom Collisions", *Phys. Rep.* **132**, 129–260 (1986). (I)

Ionosphere

J. Weinstock

The ionosphere is an upper region of the atmosphere where gas particles are ionized, thereby forming a plasma that contains free electrons and positive ions. The presence of free electrons enables the ionosphere to reflect high-frequency waves and control radio communication around the earth. The study of this region, whose altitude extends from approximately 50 km to several hundred km, is called aeronomy.

The electrons and ions of the ionosphere are produced by shortwave solar radiation and cosmic particles that penetrate the upper atmosphere and ionize molecules and atoms. The dominant ionization process can be described by the photoionization formula

$$A + h\nu \rightarrow A^+ + e^- \ ,$$

where A is an atmospheric molecule or atom, ν is the frequency of a photon of solar ultraviolet radiation, h is Planck's constant, A^+ is the positive ion produced, and e^- is the free electron. With a knowledge of the atmospheric composition of the various particles A, the solar radiation flux, absorption cross sections, and ionization efficiencies, it is possible to compute the ion production rate q at each altitude. When q is balanced against electron loss processes, such as recombination, there results a steady-state value of electron number density n_e. Since the atmospheric composition varies with altitude, it follows that n_e varies with altitude.

Large values of n_e occur from roughly 50 km to several hundred km. This (ionospheric) range is divided into three principal regions, or layers: D, below about 90 km; E, between 90 and 150 km; and F, above 150 km. Each region has either a local maximum of n_e with altitude or a slow variation of n_e with altitude. The slow variation is referred to as a ledge. The F region has both an absolute maximum of n_e, the F_2 layer, and a bottom-side ledge, the F_1 layer. Typical maximum (noontime) values of n_e are: D region, 10^3; E region, 10^5; and F region, 10^6, in units of electrons/cm^3.

These values of n_e vary with the intensity of incident solar radiation and, consequently, vary with time of day and of year. There are also less predictable variations due to solar disturbances. Superimposed on the smooth variations of ne are small-scale irregularities that are caused by atmospheric waves and plasma waves and are somewhat random. These have been studied intensively because they are difficult to predict.

The F region makes possible communication by high-frequency radio waves since it reflects (back down to earth) radio waves below a critical frequency. This critical frequency ω^*, the plasma frequency, is given by

$$\omega^* = (4\pi n_e^* e^2/m)^{1/2} ,$$

where n_e^* is the maximum value of n_e in the ionosphere, e is the electronic charge, and m is the electron mass. Radio waves above this frequency propagate through the ionosphere into outer space. Additionally, there is a lower bound to radio wave frequencies because of absorption in the D region. High-frequency radio communications are also influenced by the mentioned small-scale irregularities in n_e which can cause static. Very-short-scale irregularities can sometimes be utilized for communication in a mode called *scatter propagation*, by scattering VHF or UHF radio waves back down to earth. An advantage of VHF is that it is less susceptible to interference than ordinary radio waves. In the past two decades, it has become possible to modify (100-km) sections of the ionosphere with powerful (e. g., 2 MW) ground-based radio wave transmitters. The intense radio waves heat electrons and excite short-scale plasma turbulence. This turbulence has been used for scatter propagation to transmit photographs from Texas to California.

Most recently, the general circulation of the lower ionosphere (also referred to as the middle atmosphere) has been studied intensively. The circulation pertains to motions of neutral particles. The coupling of these neutral motions to charged particle motions and emissions is presently being investigated with renewed interest.

Aeronomy has advanced greatly in the past three decades by the use of rockets, satellites, and radar, in addition to classic radio reflection and chemiluminescent emission techniques. The ionosphere remains an active field of study.

See also: Atmospheric Physics; Electromagnetic Radiation; Ionization; Photoionization; Plasmas.

Bibliography
J. A. Ratcliffe, *Sun, Earth, and Radio*. Weidenfeld and Nicolson, London, 1970. (E)
H. Risbeth and O. K. Garriot, *Introduction to Ionospheric Physics*. Academic Press, New York, 1969.
(I)

Ising Model

F. Y. Wu

It is an empirical fact that the magnetization of a permanent magnet diminishes in strength as the magnet is heated, disappearing completely above a certain temperature, called the Curie point. The Ising model is a simple physical model for explaining this phenomenon from a microscopic point of view. Historically, the model was first proposed by Lenz in 1920 and later studied by his student Ernst Ising. For this reason, it is also referred to in the literature as the Lenz–Ising model.

In the microscopic picture, the atoms of a magnetic substance are themselves tiny magnets. The existence of a spontaneous magnetization in the bulk is then explained as a result of an alignment of these tiny magnets, or "spins", due to their mutual interactions. To construct a simple model which simulates this situation, Lenz assumed that (1) the spins are arranged on a regular lattice, (2) each spin can point in only one of two directions, "up" or "down", and (3) there exists an interaction energy $-J$ between two neighboring spins which point in the same direction, and an energy J if they point in opposite directions. This describes the Ising model. Since a physical system attains its lowest energy at absolute zero temperature, for positive J the Ising model will be in a configuration with all spins aligned at absolute zero. It is then hoped that the spin alignment would persist, at least partially, up to some nonzero but finite temperature identified as the Curie point.

The mathematical formulation of the model is as follows: associate a two-valued variable σ_i to the ith spin such that $\sigma_i = +1$ (or -1) denotes an up (or down) spin. The interaction energy between two spins i and j can then be written as $-J\sigma_i\sigma_j$, and the total energy of the system is

$$E = -J \sum_{\langle i,j \rangle} \sigma_i\sigma_j ,$$

where the summation is taken over all neighboring pairs $\langle ij \rangle$. Then, according to the principles of statistical mechanics, the thermodynamics of the system can be derived from the partition function

$$Z = \sum_{\text{configurations}} e^{-E/kT}$$

where k is the Boltzmann constant and T is the absolute temperature. The summation is taken over all possible spin configurations. The mathematical problem is to find a closed-form expression for Z. Once this is done, the Curie point, if any, will manifest itself in this expression as a point of mathematical nonanalyticity in the variable T.

In his dissertation summarized in a 1925 paper, Ising solved this problem for a one-dimensional system (spins arranged on a chain) and found no Curie point. The existence of a spontaneous magnetization, which is related to the occurrence of a Curie point, in two- and three-dimensional systems was established by Peierls in 1936, and in 1944 Onsager published the exact solution of a two-dimensional model. For a square lattice of N spins, Onsager obtained the following expression for the partition function Z in the limit of infinite N:

$$\lim_{N \to \infty} \frac{1}{N} \ln Z = \ln 2 + \frac{1}{2\pi^2} \int_0^\pi d\theta \int_0^\pi d\phi \, \ln \left[\cosh^2 \left(\frac{2J}{kT} \right) - \sinh \left(\frac{2J}{kT} \right) (\cos\theta + \sin\phi) \right] .$$

Analysis of this expression indicates that the two-dimensional Ising system possesses a unique Curie point T_C located at

$$\sinh(2J/kT_C) = 1$$

near which the specific heat diverges as $-\ln|1 - T/T_C|$. The spontaneous magnetization of this Ising model, whose derivation was first given by Yang in 1952, is

$$\begin{aligned} M &= [1 - \sinh^{-4}(2J/kT)]^{1/8}, & T < T_C \\ &= 0, & T > T_C . \end{aligned}$$

While the problems of the three-dimensional Ising model and the two-dimensional model in a magnetic field remain unsolved to this date, their thermodynamic properties have been studied extensively by numerical analyses and computer simulations. The subject of the Ising model has found applications in many diverse areas of research including chemistry, biology, and materials science, and has over years developed into a major field of research.

See also: Ferromagnetism.

Bibliography

S. G. Brush, "History of the Lenz–Ising Model", *Rev. Mod. Phys.* **39**, 883 (1967). (E)

C. Domb, "On the Theory of Cooperative Phenomena in Crystals", *Adv. Phys.* **9**, 149 (1960). (I)

C. Domb, "Ising Model", in *Phase Transitions and Critical Phenomena*, Vol III (C. Domb and M. S. Green, eds.). Academic Press, New York, 1974. (I)

E. Ising, "Beitrag zur Theorie des Ferromagnetismus", *Z. Physik* **31**, 253 (1925). (I)

B. M. McCoy and T. T. Wu, *The Two-Dimensional Ising Model*. Harvard University Press, Cambridge, MA, 1973. (A)

L. Onsager, "Crystal Statistics, I. A Two-Dimensional Model with an Order-Disorder Transition", *Phys. Rev.* **65**, 117 (1944). (I)

C. N. Yang, "The Spontaneous Magnetization of a Two-Dimensional Ising Model", *Phys. Rev.* **85**, 809 (1952). (I)

Isobaric Analog States

H. L. Harney

The term "isobaric analog state" (IAS) refers to a multiplicity occurring in nuclear states. States are said to be isobaric analogs of each other if they differ only by their number Z of protons but have the same intrinsic configuration otherwise. Especially their mass number A is the same – as the qualification "isobaric" says.

The basic example is given by the proton-neutron pair. Protons and neutrons can be treated as the same object – the nucleon – charged differently. According to Heisenberg (1932) this alternative is described by the isobaric spin $\frac{1}{2}$ (shortly isospin) of the nucleon, allowing two charge states. This concept is identical to that of ordinary spin which allows quantized orientations in ordinary space. Isospin has been widely used to classify the states of light nuclei since the pioneering work of Wigner in 1937 (see Wilkinson in *Isospin in Nuclear Physics*, 1969), but its full impact on nuclear science was only recognized with the discovery of isobaric analog states in heavy nuclei.

As an example from light nuclei, Fig. 1 displays the lowest energy levels of the mirror nuclei $^{11}_{5}\text{B}_6$ and $^{11}_{6}\text{C}_5$. Their ground states have been placed at the same energy, i.e. their Coulomb energy difference has been disregarded. This is the energy required to add one unit of charge to the ^{11}B-nucleus. One observes a one-to-one correspondence in quantum numbers and a close correspondence in the spacings between the states of these two nuclei. The states in ^{11}C and ^{11}B are isobaric analogs of each other. The small differences in the energy spacings indicate a small breaking of isospin. It is due to the Coulomb force which affects only the protons not the neutrons.

The strong interaction between the nucleons can be assumed to conserve isospin. This explains the close correspondence between isobarically analogous spectra in light nuclei. As long as the Coulomb energy is less than the binding energy of the last nucleon, the nuclear forces dominate and preserve isospin symmetry. How about heavy nuclei where the Coulomb energy significantly exceeds the nucleon binding energy?

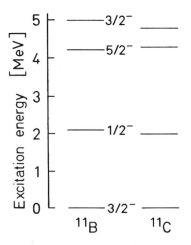

Fig. 1: Level diagrams for ^{11}B and ^{11}C.

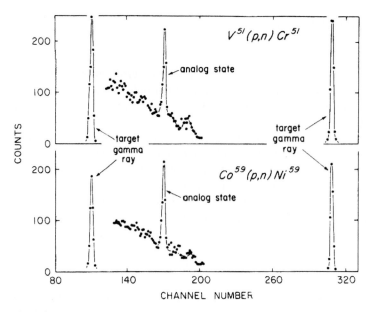

Fig. 2: Neutron spectra showing IASs. From Anderson *et al.*, 1962.

In 1961, Anderson and Wong published experiments that led to the unexpected conclusion that isospin is a valid quantum number even for heavy nuclei. A study of the (p, n) reaction on target nuclei with mass numbers A between 48 and 93 revealed an intense neutron group in each spectrum as shown in Fig. 2. Conceptually this (p, n) reaction is very simple: The incident proton exchanges its charge with one of the (excess) neutrons of the target, leaving the remaining characteristics of the target unchanged. Thus an IAS of the target is formed. The resultant neutron is emitted with an energy loss (the Q-value of the reaction) essentially equal to the Coulomb energy. The surprise was that the line width (the natural energy spread) of the IAS is narrow. Let its isospin be $T_>$. The IAS is are located in a region of excitation energy (indeed above nuclear binding energy) where the density of states with isospin $T_< = T_> - 1$ is large. Any isospin violating force spreads the IAS over the $T_<$ states within some energy range $\Gamma^\downarrow$. Hence, although an IAS may be strongly mixed into the local $T_<$ states, this mixing occurs only within a small energy range $\Gamma^\downarrow$ and therefore the state stands out from the background of $T_<$ states. In this sense isospin is a useful quantum number even for heavy nuclei. A compilation of experimental spreading widths $\Gamma^\downarrow$ is given in Fig. 15 of Harney *et al.*, 1985. Up to the heaviest nuclei, $\Gamma^\downarrow$ does not exceed $\approx 100\,\mathrm{keV}$.

From some IASs populated in the last-mentioned experiment, proton emission is energetically possible. It is compatible with isospin conservation. One could therefore excite IASs in heavy nuclei via proton scattering. They should show up as resonances in the compound nucleus. This was indeed discovered by Fox *et al.* in 1964. Figure 3 shows a later example which is especially beautiful. The resonances occur in the compound system $(^{208}\mathrm{Pb} + \mathrm{p}) \rightarrow {}^{209}\mathrm{Bi}$ and are isobaric analogs of the lowest states of $^{209}\mathrm{Pb}$. If one removes the Coulomb energy difference between $^{209}\mathrm{Bi}$ and $^{209}\mathrm{Pb}$, much as in Fig. 1, the spectrum of resonances agrees (according to spins, parities, and energy spacings) with the spectrum of $^{209}\mathrm{Pb}$ sketched at the top of Fig. 3.

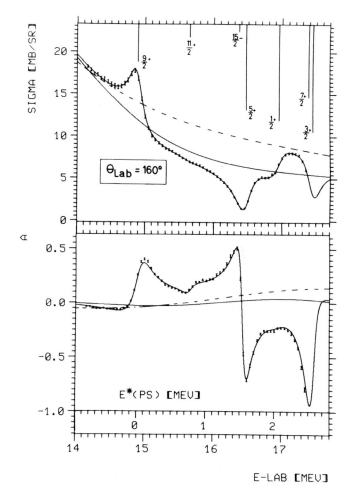

Fig. 3: Excitation functions of cross section and analyzing power for elastic proton scattering on ^{208}Pb, showing resonances in the compound system ^{209}Bi. They are the isobaric analogs of low-lying states in ^{209}Pb as indicated by the vertical lines with the spin-parity assignments. From Melzer *et al.*, 1985.

The spread of the isobaric analog configuration over the fine structure of nearby $T_<$ states has been impressively demonstrated by the high-resolution proton scattering experiments of Bilpuch *et al.* (1974, 1976). The isobaric analog is not a single state, its configuration is completely dissolved into the $T_<$ states. In this sense, isospin is strongly broken. However, said configuration is coupled to the $T_<$ states only in a small energy region, see Åberg *et al.* (2004), of width $\Gamma^\downarrow \approx 17$ keV. In this sense, isospin is a useful quantum number: Fig. 4 shows the distribution of the coupling strengths, i.e., the partial widths of the fine structure resonances. The IAS appears as a doorway state to the fine structure. This phenomenon is similar to a giant resonance.

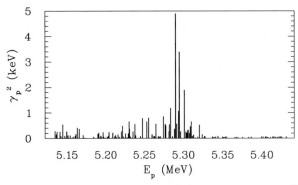

Fig. 4: Strength distribution of the fine structure of an IAS in ^{93}Tc. The experiment is due to Bilpuch *et al.* (1974), the figure is taken from Åberg *et al.* (2004) who analyzed the chaotic aspects of the fine structure.

Bibliography

S. Åberg, A. Heine, G. E. Mitchell, and A. Richter, *Phys. Lett.* B **598**, 42 (2004).

F. D. Anderson and C. Wong, *Phys. Rev. Lett.* **7**, 250 (1961) (A); *Phys. Rev. Lett.* **8**, 442 (1962). (A)

E. G. Bilpuch, J. D. Moses, F. O. Purser, W. H. Newson, G. E. Mitchell, R. O. Nelson, and D. A. Outlaw, *Phys. Rev.* C **9**, 1589 (1974) (A); E. G. Bilpuch, A. M. Lane, G. E. Mitchell, and J. D. Moses, *Phys. Rep.* **28**C, 145 (1976). (A)

Joseph Cerny (ed.), *Nuclear Spectroscopy and Reactions*. Academic Press, New York, 1974. See the articles by G. M. Temmer, Part B, p. 61 (E, I), P. von Brentano, John G. Cramer, Part B, p. 89 (E, I), and D. Robson, Part D, p. 179 (E, I).

W. R. Coker and C. F. Moore, *Phys. Today* **22** (No. 1), 53 (1969). (E)

J. D. Fox, C. F. Moore, and D. Robson, *Phys. Rev. Lett.* **12**, 198 (1964). (A)

H. L. Harney, A. Richter, and H. A. Weidenmüller, *Rev. Mod. Phys.* **58**, 607 (1986). (A)

W. Heisenberg, *Z. Physik* **77**, 1 (1932). (A)

R. Melzer, P. von Brentano, and H. Paetz gen. Schieck, *Nucl. Phys.* **A432**, 363 (1985). (A)

D. H. Wilkinson (ed.), *Isospin in Nuclear Physics*. North-Holland, Amsterdam, 1969. (E,I,A)

Isomeric Nuclei

G. Scharff-Goldhaber[†]

The term isomers [from the Greek ισο (same) and μερος (part, share)] refers to two or more aggregates consisting of the same components, but in different arrangements. It was introduced into nuclear physics in analogy to its use in molecular physics. Whereas in molecules the difference in arrangement has to do with the structure and/or spatial distribution of the atoms within a molecule (e. g., two isomeric molecules may be mirror images of each other), the most noticeable distinction between two or more nuclear isomers [i. e., nuclei with the same atomic number (Z) and mass number (A)] is the half-life ($\tau_{1/2}$). One of the isomers,

---

[†]deceased

Table 1: Electric and magnetic multipoles.

	Dipole		Quadrupole		Octupole		Hexadecapole		32-pole	
l	1		2		3		4		5	
Δπ	no	yes	no	yes	no	yes	no	yes	no	yes
	*M*1	*E*1	*E*2	*M*2	*M*3	*E*3	*E*4	*M*4	*M*5	*E*5

representing the ground state of a given nuclide (N, Z), may be stable, i. e., $\tau_{1/2} = \infty$. The other isomer or isomers are metastable states that decay to one or more lower-lying states by "isomeric transitions", emitting energy quanta in the form of γ rays, or internal-conversion electrons from the various atomic shells. The percentages of conversion electrons from each shell and subshell are accurately described by electromagnetic theory. The ratio of conversion electrons to γ rays is referred to as the internal-conversion coefficient α. Competition with the isomeric transition by β radiation, electron capture, α radiation, etc., may occur. The half-life of an isomeric transition increases by a large factor for each unit of the difference in angular momentum of the isomeric state and the final state. Although in principle all excited states of a given nuclide might be considered isomers, in practice only states with "measurable" half-lives are given this name. An isomer whose half-life $\tau_{1/2} \gtrsim 1$ s is frequently referred to as a "long-lived" isomer. The first, and for a long time only, pair of isomers was found by Otto Hahn in 1921 among the decay products of ^{238}U in the element protactinium $(Z = 91)$. They are ^{234m}Pa (then named UX$_2$) with $\tau_{1/2} = 1.17$ min, and ^{234}Pa (the ground state of the nuclide, then named UZ) with $\tau_{1/2} = 6.75$ h. Although both isomers decay mainly by β emission into ^{234}U, ^{234}Ta decays in 0.1% of the cases to the ground state via a "two-step isomeric transition". The energy quantum released in the first of the two transitions, which determines the half-life of the state, can so far only be estimated (~ 10 keV). It is followed by a 73.9-keV transition (Fig. 1, part a).

In the early 1950s the study of isomeric transitions contributed importantly to the establishment and detailed testing of the shell model of the atomic nucleus. While this model, conceived in close analogy to the atomic shell model, had been shown to describe the angular momenta (or spins) I of the nuclear *ground states* of odd-A nuclei, the study of the "multipole order" l of isomeric transitions as a function of N and Z proved that the shell model was valid also for a large number of *excited states*. Here $l \geq I_{init} - I_{final}$, where the I's are the spins of the initial and final states of the transition. In addition, it is of importance whether the wave functions of the initial and final states have the same parity (π), which may be even $(+)$ or odd $(-)$, or whether the parity changes. If $l = 1$, and no parity change takes place, the transition is said to be a magnetic dipole ($M1$); if $l = 1$ and the parity changes, it is an electric dipole ($E1$). The lowest magnetic and electric multipoles – the most common ones found – are shown in Table 1. The multipole order of an isomeric transition can be determined in various ways, e. g., by determining α, or ratios of the numbers of conversion electrons from the subshells of an atomic shell, or by determining the angular correlation of the gamma rays with those of a transition following it in the decay. For the isomeric transition in the "odd–odd" nucleus ^{214}Pa referred to earlier, it is believed that $I = 3$ and Δπ yes, i. e., that the transition is $E3$ (Fig. 1, part a).

An interesting long-lived triplet of isomers is known in ^{192}Ir $(Z = 77)$. Here the ground state, which decays mainly by β decay to ^{192}Pt $(Z = 78)$ with a 74.2-day half-life, has spin 4.

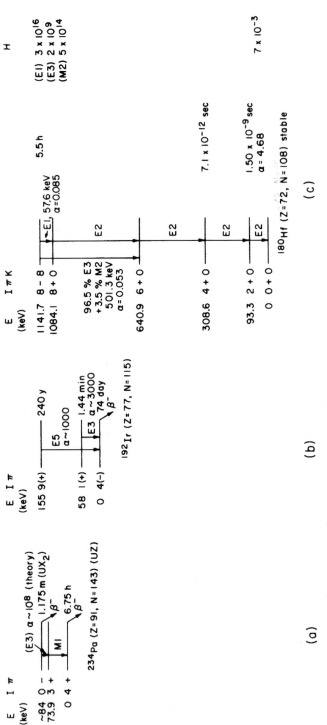

(a)

(b)

(c)

Fig. 1: Three examples of isomerism. At the left of each diagram displaying the level scheme of the three nuclides are shown the energy E in keV, the spin I, and the parity π. (Since for ^{192}Ir only the parity *changes* for the two transitions have been determined, the probable parities are given in parentheses.) The arrows representing isomeric transitions originating from metastable states are labeled with the corresponding multipole order ($E3$, etc.). At the right of each level scheme the half-lives of the states and the conversion coefficients α are stated.

(a) The first pair of isomers, discovered in nature in the decay of uranium (^{238}U), occurs in protactinium (^{234}Pa). The ~10-keV energy of the isomeric transition is based on indirect evidence. Theory predicts $\alpha \sim 10^8$, i. e., γ rays are too weak to be observed.

(b) Triple isomerism in ^{192}Ir: The 74-day ground-state activity, which is efficiently produced by neutron capture in natural iridium, plays an important role in industry and medicine. The 240-year isomer is $\sim 10^{11}$ times weaker than the ground state at the time of production.

(c) K isomerism in ^{180}Hf: The 5.5-h $E1$ transition of 57 keV is exceedingly retarded, as indicated by the hindrance factor H in the last column, whereas the $E2$ transitions within the rotational band, especially the 93.3-keV transition, are strongly enhanced. Since the 501-keV transition is less K-forbidden than the 57-keV transition, it can compete, in spite of the higher multipole orders.

Table 2: Electric and magnetic reduced transition probabilities.

Multipolarity	$B(El)$	Multipolarity	$B(Ml)$
$E1$	1.0×10^{14}	$M1$	2.9×10^{13}
$E2$	7.4×10^{7}	$M2$	8.4×10^{7}
$E3$	3.4×10^{1}	$M3$	8.7×10^{1}
$E4$	1.1×10^{-5}	$M4$	4.8×10^{-5}
$E5$	2.5×10^{-12}	$M5$	1.7×10^{-11}

The 58-keV first excited state, with $I = 1$, lives 1.44 min. It decays to the ground state by an $E3$ transition, i. e., the parity changes. The second excited state, at 155 keV, with $I = 9$, decays to the ground state by an $E5$ transition. This state has a half-life of approximately 240 yr (Fig. 1, part b). In nature, there exists an isomer, ^{180m}Ta, with an abundance of 0.012% and a half-life $> 10^{17}$ yr.

The shell model, which is based on the assumption that the nucleus has a spherical shape, makes predictions for the half-life of an isomeric transition of a single proton, depending on the transition energy E (in MeV), the multipole order l, the absence or presence of parity change, and the mass number A: for electric transitions.

$$[\tau_e(1+\alpha)]^{-1} = B(El)E^{2l+1}A^{2/3}$$

and for magnetic transitions,

$$[\tau_m(1+\alpha)]^{-1} = B(Ml)E^{2l+1}A^{(2l-2)/3}$$

where $B(El)$ and $B(Ml)$, the electric and magnetic reduced transition probabilities, are estimated to be as in Table 2. Whereas these predictions were found to be in amazingly good agreement with observed $M4$ transition half-lives (within a factor of 2), considerable discrepancies were found for most other multipoles. Some of these have been attributed to the presence of large numbers of nucleons outside closed shells, which bring about a change of the nuclear shape from spherical to spheroidal [usually prolate (like a football), sometimes also oblate (like a pancake)]. The difference in the length of the axis of symmetry from that of the two other axes of the spheroid is expressed by the deformation parameter ((3). The following deviations from the foregoing shell-model predictions for lifetimes of isomeric transitions are noteworthy.

(i) The $E2$ transitions taking place within a rotational band, especially that built on the ground state, are strongly enhanced. For the largest values of β, their half-lives are about 200 times shorter than the individual-particle shell model predicts. An example of an enhanced $E2$ transition to the ground state of the even–even nucleus ^{180}Hf is given in Fig. 1c.

(ii) Many transitions, in particular $E1$, $M1$, and $M2$, are greatly retarded because of "K isomerism". This name refers to the quantum number K, which measures the component of the nuclear spin along the axis of symmetry. For a rotational band built on the ground state of a deformed even–even nucleus, $K^\pi = 0^+$. If the transition takes place from a state with a high spin I whose direction coincides with the axis of symmetry, $K = I$, to a

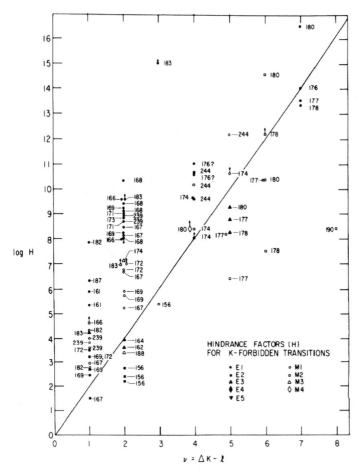

Fig. 2: $\log H$ versus $\nu = \Delta K - l$, where H is the hindrance factor and ν denotes the degree of K forbiddenness. The symbols indicate the multipole order of the transition in a K isomer characterized by its mass number. The points with arrows show lower limits for $\log H$. (Prepared by G. T. Emery and G. Scharff-Goldhaber.)

state in the ground-state band, the retardation (compared with shell-model predictions), measured by the *hindrance factor H*, follows (within very wide limits) the rule

$$\log H \approx 2(\Delta K - l) .$$

Let us take as an example the 5.5-h isomer of ^{180}Hf (Fig. 1, part c) with $K = I = 8^-$, which decays to the state $K = 0, 8^+$ by a 57.5-keV $E1$ transition. Hence, according to the rule, $\log H \approx 2(8 - 1) = 14$; i. e., $\tau_{1/2}$ is expected to be $\sim 10^{14}$ times longer than the individual-particle model predicts. As seen from Fig. 2, which displays the observed distribution of $\log H$ versus $\Delta K - l$, the experimental value of H for this isomer (labeled 180) is even somewhat larger, namely, 3.1×10^{16}.

(iii) A second type of retarding effect may be due to a shape change of the nucleus, e. g., from oblate to prolate, or to a change representing a jump in deformation to almost twice that of the ground state, as has been observed in the so-called fission isomers occurring in the heaviest elements known. A fission isomer is a metastable state of a heavy nucleus that decays largely by fission. Little is known about the isomeric transition lifetimes from these states.

See also: Gamma Decay; Nuclear Properties; Nuclear Structure.

Bibliography
D. E. Alburger, Encyclopaedic *Dictionary of Physics*, Vol. 4, pp. 101–102. Pergamon, New York, 1961.
Y. Y. Chu and G. Scharff-Goldhaber, "Decay of ^{234}Th to the ^{234}Pa Isomers", *Phys. Rev. C* **17**, 1507 (1978).
J. Godart and A. Gizon, "Niveaux de ^{234}Pa atteints par la disintegration de ^{234}Th", *Nucl. Phys.* **A217**, 159 (1973).
M. Goldhaber and R. D. Hill, "Nuclear Isomerism and Shell Structure", *Rev. Mod. Phys.* **24**, 179-239 (1952).
M. Goldhaber and A. W. Sunyar, "Classification of Nuclear Transition Rates", in *Alpha-, Beta- and Gamma-Ray Spectroscopy* (K. Siegbahn, ed.), pp. 931–949. North-Holland, Amsterdam, 1965.
S. A. Moszkowski, "Theory of Multipole Radiation", in *Alpha-, Beta- and Gamma-Ray Spectroscopy*, (K. Siegbahn, ed.), pp. 863–886. North-Holland, Amsterdam, 1965.
H. C. Pauli. K. Alder, and R. M. Steffen. "The Theory of Internal Conversion". *The Electromagnetic Interaction in Nuclear Spectroscopy*, pp. 341–440. North-Holland. Amsterdam. 1975.
G. Scharff-Goldhaber. "Multipole Order and Enhancement Factor of Long-Lived Isomeric Transition in ^{192}Ir.", *Bull. Am. Phys. Soc.* **22**, 545 (1977).
G. Scharff-Goldhaber. C. B. Dover and A. L. Goodman, "The Variable Moment of Inertia Model and Nuclear Collective Motion". *Annu. Rev. Nucl. Sci.* **26**, Section 3 (1976).
G. Scharff-Goldhaber and M. McKeown, "Triple Isomerism in ^{192}Ir". *Phys. Rev. Lett.* **3**, 47 (1959).
R. Vandenbosch and J. R. Huizenga, *Nuclear Fission*, pp. 59–76. Academic. New York. 1973.

Isospin
D. Wilkinson

Isospin, also called isotopic or isobaric spin, is a concept that arises only for systems of particles, hadrons, that interact with each other through the nuclear force. The most familiar of such hadronic systems are the atomic nuclei, comprised of neutrons (n) and protons (p) – together called nucleons (*N*)– but hadrons include the mesons whose exchange between the nucleons generates the *N–N* force and also include the excited isobaric states or resonances into which nucleons can be raised under appropriate bombardment.

The essence of the matter is that hadrons exist in families, of varying numbers of members, called multiplets, all members having identical quantum numbers (spin, parity, charge conjugation, *G*-parity, etc., as appropriate) and closely similar masses; the various members are distinguished essentially only by their electric charges. Thus the neutron and proton form a family of two and have charges of 0, +1 but differ in mass by only 0.14%; the lightest of the mesons, the pion (π), forms a family of three members of charge −1, 0, +1 whose masses span a range of only 3.3%; the lightest of the nucleon resonances, the Δ, forms a family of four of charge −1, 0, +1, +2 whose masses differ over all by only about 1%, and so on. These mass splittings, of order 1%, within the multiplets, separating the members of different electric charge, are of the same order as the factor by which the distinguishing electrical force is weaker than the nuclear force that gives the family its definition in terms of its common quantum numbers. This suggests that from the point of view of the dominating nuclear force that defines the particles the multiplet members are identical each to the others and that the charge acts merely as a kind of label to distinguish the members; since electrical charge carries electrostatic energy and energy by $E = mc^2$ implies mass, the different charge states therefore also have slightly different masses.

This idea of identity from the viewpoint of the nuclear force with distinction only by the electric force is strongly reinforced by the observation that the interaction between the members of a multiplet (called an isobaric or charge multiplet because of the near identity of the masses of the various charge states that we have noted) does not depend on the charges of the multiplet members involved after allowance for the direct effect of the electric charges, if any, upon each other and for other electromagnetic interactions such as those associated with the magnetic moments, if any. In other words, the nuclear force that defines the particles does not depend on the charge state of the members. Thus the np force between a neutron and a proton in the $^1S_0$ state is the same to within about 2% as that, pp, between two protons in the same spectroscopic state after subtraction of the electromagnetic effects, while the similar force between two neutrons and between two protons is the same to within 1%: nn $\simeq$ np $\simeq$ pp. We can even understand the fact that the np strong force is slightly different (2%) from the pp strong force: The strong force is generated by the exchange of mesons between the nucleons; the allowed charges for the exchanged mesons depend on the charges of the nucleons that are interacting – thus two protons can exchange only neutral mesons if they are to remain nucleons because the exchange of a meson of charge ±1 would involve one of the nucleons becoming of charge +2 which does not exist for the nucleon, but a neutron and a proton can exchange either neutral mesons or mesons of charge ±1, the latter simply effecting an interchange of the charge label between the nucleons; but since the charged and neutral mesons themselves have slightly different masses the forces that they engender by their exchange will also be slightly different. It is quite possible that if we could correct for this slight charge difference *within* the meson multiplets, itself presumably due to the electric force, then the np and pp strong forces would be found to be identical.

This indifference of the nuclear force to the charge states is called the principle of charge independence. It extends to forces *between* isobaric multiplets as well as *within* multiplets; thus in the interaction between pions and nucleons we have only one strength, the πN force, which is experimentally practically the same for all charge combinations: $\pi^+ p \simeq \pi^+ n \simeq \pi^0 p \simeq \pi^0 n \simeq \pi^- p \simeq \pi^- n$.

This idea of charge independence of the nuclear force, the charge merely acting as a label, leads us to ask how we might most conveniently define that label in mathematical terms. Since the multiplet families are of different sizes, two members for *N*, three for π, four for Δ, etc., and some with only a single member such as the η and ω mesons, we naturally look for a labeling that will not only tell us the charge state but also how many members there are in the family. A ready-made analogy is to hand in angular momentum. A particle of intrinsic spin s (e. g., $s = 0$ for an α particle, $s = \frac{1}{2}$ for an electron, $s = 1$ for a deuteron, $s = \frac{3}{2}$ for the Δ, etc.) is allowed by the ordinary rules of quantum mechanics to assume $2s + 1$ alternative orientations in space distinguished by the projections s_z of the angular momentum along whatever axis, z, we choose for our consideration. The particle is identical in its properties in each of its $2s + 1$ possible states but the states are distinguishable physically. Thus a proton with its spin "up", $s_z = +\frac{1}{2}$, is no different intrinsically when it is turned upside down, $s_z = -\frac{1}{2}$, but we can tell the difference by an appropriate test such as passing it through an inhomogeneous magnetic field which will bend the trajectory up or down depending on s_z. So we invent an analogous quantity, t, for the hadrons and call it isospin by analogy with the ordinary spin s. We imagine it to exist in its own, totally fictitious, isospace within which it can take up its $2t + 1$ alternative orientations each of which, t_z, labels the charge state of the particle. t tells us the number, $2t + 1$, of members in the family and t_z tells us the charge of the individual member. Thus for the nucleon, with two members to the family, $t = \frac{1}{2}$; conventionally in nuclear physics, $t_z = +\frac{1}{2}$ represents the neutron and $t_z = -\frac{1}{2}$ represents the proton (to emphasize that it is only a convention, high-energy physicists have it the other way around). Because of charge independence a nucleon with its isospin "up", a neutron, is not essentially changed when it is turned upside down in isospace, then merely changing its charge label and becoming a proton distinguished not by any "nuclear" property but only by its charge which totally changes its chemistry but does not affect its physics. So for the other families so far mentioned: η and ω have $t = 0$; π has $t = 1$; Δ has $t = \frac{3}{2}$.

But now comes the critical point. Ordinary angular momentum is a conserved quantity: In the absence of external torques the overall angular momentum of a mechanical system remains constant no matter how violent the forces at play *within* that system. This principle is congruent with the remark that the mechanical properties of the system are independent of its orientation in space: a proton with its spin up, $s_z = +\frac{1}{2}$, is identical to a proton with its spin down, $s_z = -\frac{1}{2}$, so that an *ensemble* of protons of total spin $\mathbf{S} = \Sigma \mathbf{s}$ may exist in $2S + 1$ different orientations $S_z = \Sigma s_z$ in space, with identical properties, no matter how violent the interactions between the protons that may result in a constant turning of each other upside down, i. e., a constant interchange of their individual s_z values. S is then a constant of the motion, a conserved quantity, a "good quantum number". (This illustration assumes no orbital angular momentum to be present and so ignores the Pauli principle, but this is irrelevant to our illustration.) Consider now an *ensemble* of, say, nucleons. Because of charge independence a nucleon in its neutron state, $t_z = +\frac{1}{2}$, is not changed in its interactions when in its proton state, $t_z = -\frac{1}{2}$, so the *ensemble* with $\mathbf{T} = \Sigma \mathbf{t}$ may exist in $2T + 1$ different orientations $T_z = \Sigma t_z$ in isospace with identical properties no matter how violent the interactions that turn the nucleons upside down in isospace, viz., that interconvert neutrons and protons, and T, as S, will be a good quantum number. So for N neutrons and Z protons we have $T_z = \frac{1}{2}(N - Z)$, but this must belong to a multiplet of $2T + 1$ nuclear states, identical to each other in physical properties but being found in $2T + 1$ different chemical elements, the only restriction being $\frac{1}{2}(N + Z) \geq T \geq |T_z|$. The actual value of T for a given multiplet is determined by the Pauli principle, viz., the antisymmetrization of the overall wave function in the three variables of space, spin, and

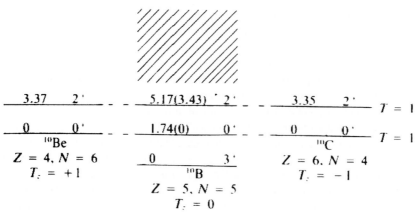

Fig. 1: Some low-lying levels of $A = 10$. The numbers on the left of the lines are the excitation energies in MeV; on the right of the lines are the J^π values. In brackets for ^{10}B are the excitation energies relative to the lowest $T = 1$ state. The unlabeled states of ^{10}B (and the ground state) are of $T = 0$. $T = 0$ states are dense in the hatched region (the next $T = 1$ states lie about 2 MeV higher up). The lowest $T = 1$ states have been drawn at the same level to display the symmetry; in practice, because of the extra Coulomb energy in the more highly charged nuclei, the ground state of ^{10}C is some 1.4 MeV heavier than its companion state at 1.74 MeV in ^{10}B, which in turn is some 0.7 MeV heavier than the ground state of ^{10}Be.

isospin, and, crudely speaking, is a measure of the number of successive conversions of, say, neutrons into protons that can be made before two neutrons or two protons are placed into identical quantum states. We might offhand expect the most symmetrical systems, of lowest T, to lie lowest in energy, and although this is usually the case it does not necessarily have to be so and sometimes is not. So any nuclear state must be accompanied by $2T$ other nuclear states of identical physical properties in $2T$ other chemical elements. Illustrations of this are given in the energy level diagrams of Figs. 1 and 2 for both odd and even mass systems. Note particularly that all members of an isospin multiplet must have the same ordinary spin J and parity π labeled J^π.

With perfect overall charge independence the masses of the members of an isomultiplet would be identical. But we do not have perfect charge independence because of the Coulomb force, the effect of which differs for differing members of the multiplet. This has been subtracted out in the figures by drawing the members of the lowest multiplet all at the same level so that the figures effectively show relative excitation energies within the different chemical elements.

The figures demonstrate the utility of isospin as a label for nuclear states. The other aspect of isospin, namely, its conservation, is also of great importance. Nothing in our argument says that the N neutrons and Z protons have to constitute or continue to constitute a single nucleus. Thus if two nuclei A and B engage in a nuclear reaction: $A + B \rightarrow C + D$, then, because of charge independence and exactly as with ordinary angular momentum, the overall isospin T must be conserved: $\mathbf{T} = \mathbf{T}_A + \mathbf{T}_B = \mathbf{T}_C + \mathbf{T}_D$. As an illustration consider the reaction

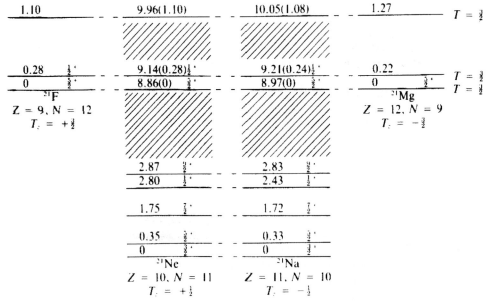

Fig. 2: Energy levels of $A = 21$ with labeling conventions as for $A = 10$. The states without a T label, dense within the hatched regions, are of $T = \frac{1}{2}$. Where J^{π} is not shown it is unsure. (Note the energy gap, not to scale, between the $T = \frac{1}{2}$ states at about 2.8 MeV and the onset of the $T = \frac{3}{2}$ states; it contains at least 40 $T = \frac{1}{2}$ states and probably many more.)

$^{12}\text{C} + \text{d} \rightarrow {}^{10}\text{B} + \alpha$. The target ^{12}C which is in its ground state has $T = 0$ as also do the incident deuteron and the product a particle. So the left-hand side of the equation has $T = 0$ and so must the right-hand side; since $T_{\alpha} = 0$ we can only form those states of ^{10}B that have $T = 0$. By reference to Fig. 1 we should therefore expect to be able to form ^{10}B in, for example, its ground state but not in those at excitation energies of 1.74 and 5.16 MeV: It is indeed found that production of these latter states is very considerably suppressed.

Similarly in the decay of excited nuclear states. Consider the lowest $T = \frac{3}{2}$ state of ^{13}N at 15.06 MeV. Energetically this can emit a proton (of isospin $\frac{1}{2}$) leaving the ground state of ^{12}C which, however, has $T = 0$ so that this is forbidden and indeed this emission takes place about 1000 times more slowly than we might otherwise have expected.

The fact that the isospin selection rules, which also exist in a somewhat subtler form for gamma-ray emission and β decay, are well, but not perfectly, obeyed reflects the role of the Coulomb force, which prevents *de facto* overall charge independence from being perfect; the energy of the T multiplet *does* depend slightly on its T_z value, its orientation in isospace, because, for example, the Coulomb energy of the ground state of ^{10}C is bigger than that of the 1.74-MeV state of ^{10}B which is bigger than that of the ground state of ^{10}Be (see Fig. 1 where these differences have been subtracted out). T, unlike ordinary angular momentum, cannot therefore be a perfect quantum number but it is good enough to be of very great utility not only in nuclear physics, from which most of the present illustrations have been drawn, but also for high-energy or particle physics.

This discussion has been conducted in terms of the analogy between isospin and the rotational invariance of ordinary space. In an alternative symmetry language the multiplets that have been discussed here are $SU(2)$ multiplets: The algebras of three-dimensional rotation and $SU(2)$ symmetry are identical so the discussions are completely equivalent.

See also: Electromagnetic Interaction; Elementary Particles in Physics; Hadrons; Nuclear Properties; Nuclear Reactions; Nuclear States; Rotation and Angular Momentum; Strong Interactions.

Isotope Effects
W. A. Van Hook

The term isotope effect (IE) denotes a difference in some molecular or atomic property consequent to a change of mass or mass distribution caused by isotopic substitution. Radioactive and other specific nuclear effects are excluded. IEs are of theoretical interest because within the framework of the Born–Oppenheimer approximation the electronic properties of a system of atoms or molecules can be separated from those involving nuclear motion. Solution of the electronic part of the problem determines an electronic potential surface on which nuclear motion occurs. The electronic structure determines most of the chemistry and much of the physics of the molecule or process under description. Investigation of isotopic substitution amounts to a study of how mass and/or mass distribution affects motion on the (isotope-independent) electronic potential surface. Alternatively, experimental data on IEs can be regarded as a tool to gain information about the potential-energy surface. We note that isotope effects are expected to be small; some representative values are given in Table 1. IEs on equilibrium constants or vapor pressure amount to several percent or more for H/D substitution (around room temperature) but are smaller for more massive isotopic isomers ($^{12}C/^{13}C$, for example).

Table 1: Some representative isotope effects.

System	Process	Temperature (K)	Light/Heavy
$H_2 + 2DI \rightleftharpoons D_2 + 2HI$	Equilibrium (gas)	600	$1.21 = \bar{K}_H/\bar{K}_D$
$^{15}NH_3 + {}^{14}NH_4^+ \rightleftharpoons {}^{14}NH_3 + {}^{15}NH_4^+$	Equilibrium (solution)	298	$1.034 = \bar{K}_{14}/\bar{K}_{15}$
$^{10}BF_3\,(g) + donor\cdot{}^{11}BF_3\,(l) \rightleftharpoons {}^{11}BF_3\,(g)$	Equilibrium	303	$1.026 = \bar{K}_{10}/\bar{K}_{11}$
$+ donor\cdot{}^{10}BF_3\,(l)\;[donor = (CH_3)_2O]$			
$^{36}Ar{-}^{40}Ar$	Vapor pressure	87	$1.006 = P_{36}/P_{40}$
$H_2O{-}D_2O$	Vapor pressure	373	$1.054 = \bar{P}_{H_2O}/P_{D_2O}$
$C_6D_6{-}C_6H_6$	Vapor pressure	353	$0.976 = P_H/P_D$
$H_2 + Cl\cdot \rightarrow HCl + H\cdot$			
$D_2 + Cl\cdot \rightarrow DCl + H\cdot$	Reaction rate	298	$9.6\ \ = K_H/K_D$
$(CH_3)_3CCl \rightarrow (CH_3)_3C^+ + Cl^-$			
$(CD_3)_3CCl \rightarrow (CD_3)_3C^+ + Cl^-$	Reaction rate	298	$2.4\ \ = K_H/K_D$

IEs on the rate constants of chemical reactions are typically larger. Isotope effects on a given property are normally much much smaller than the corresponding differences between chemically distinct systems. Part of the practical interest in isotope effects stems from the World War II and postwar efforts at preparing sizable quantities of separated isotopes.

A convenient starting point for a theoretical understanding of isotope effects is to consider isolated molecules in the gas phase. Nuclear motion can be described in terms of small displacements from the equilibrium nuclear positions. The potential energy is approximated with an expansion in terms of displacement coordinates where the leading (quadratic) term is the largest, but cubic and higher-order terms can also be important. Even so, the cubic and higher terms are often neglected and this leads to a description of the molecule in terms of a set of $3n - j$ harmonic oscillator frequencies ($j = 5$ for a linear, 6 for a nonlinear, or 3 for a monatomic molecule). These frequencies are often described as the set of solutions which follow from diagonalization of a matrix, $\bar{F}\bar{G}$, where $\bar{F}$ is a $(3n - j)$-dimensional potential-energy (force constant) matrix and $\bar{G}$ a kinetic-energy matrix which takes due account of mass and mass distribution. From the paragraph above we recognize $\bar{F}$ as isotope independent whereas $\bar{G}$ reflects the known mass differences. Given $\bar{F}$ and $\bar{G}$ for a set of isotopic molecules one can readily calculate a complete set of frequencies and isotopic frequency shifts which can then be employed to evaluate partition functions, partition function ratios, and, therefore, thermodynamic properties via the methods of a statistical thermodynamics. In the calculation of the thermodynamic properties it is necessary to properly account for the j zero-frequency rotations and translations, which are usefully described in the classical limit in the gas, but which take on nonzero values in the condensed phase and are sometimes treated there in an harmonic approximation. It is to be noted that the isotopic frequency shifts are small, at least for heavier isotopes, and it therefore becomes convenient to formulate the isotopic partition function ratios (which are near unity) in powers of $1/T$. In this connection it has been long established that IEs vanish in the classical limit, and Jacob Bigeleisen and Maria G. Mayer in 1947 pointed out the convenience of defining a reduced partition function ratio: q/q(classical). Their approach and notation has since been widely employed.

The methodology outlined above has been applied to a wide variety of special cases including gas-phase chemical equilibrium constants (Table 1), isotope effects on rate constants in chemical reactions, vapor pressure, molar volume, and other condensed phase isotope effects, etc. In suitable cases, corrections for such effects as anharmonicity, rotation–vibration interaction, centrifugal distortion, the effect of anharmonicity on the zero-point energy, and corrections for deviations from the Born–Oppenheimer approximation have been made. Most analyses to date have considered systems in thermal equilibrium so that Maxwell–Boltzmann distribution functions are appropriate for use in the calculations. Isotope effects for systems not in thermal equilibrium are also of considerable interest although not discussed in this article. Non-Maxwellian distributions are important in the treatment of laser methods of isotope separation, IEs on fast unimolecular reactions, etc.

The theory outlined above for equilibrium readily lends itself to the interpretation of data on isotope effects on rate constants of chemical reactions. The analysis is based on the absolute rate (activated complex) approach of Eyring and others. The theoretical construct readily allows assumptions about the nature of the transition state and reaction path to be tested in a reasonably straightforward way. This area has been treated in detail with extensive computer model calculations. It has been demonstrated that isotope effects essentially probe force-constant changes at the position of isotopic substitution, transition state minus reactant. The effects are normal (light faster than heavy) if force constants in the "transition state" (read "product" if isotope effects on equilibria are being considered) are smaller than in the reactant, and inverse in the opposite case. Unusual temperature dependencies can occur and corrections for quantum-mechanical tunneling can be important.

The IE on equilibrium between condensed and vapor-phase species as expressed by the vapor-pressure isotope effect (VPIE) has been measured and interpreted for a wide variety of compounds. The VPIE is directly related to the mean square force or Laplacian of the intermolecular potential. The available data on argon, for example, are in excellent agreement with recent sophisticated calculations. For polyatomic molecules both normal and inverse IEs are observed together with complicated temperature dependencies. The inverse effects (commonly seen for protio-deuterocarbons) are a consequence of the net red shift in internal frequencies, gas to liquid, due to the influence of the intermolecular van der Waals forces. In many cases detailed correlation with spectroscopic information has been possible. Differences in vapor pressure between equivalent isomers such as *ortho*, *meta*, and *para* dideuterobenzene or *cis*, *trans*, and *gem* dideuteroethylene have allowed the importance of rotation–vibration interaction in the liquid phase to be assessed. Studies on solutions of isotopic isomers (C_6H_6 in C_6D_6, H_2 in D_2, deutero/protio polystyrene, etc.) have been made demonstrating significant nonideality. Such effects owe their origin to isotopic differences in vibrational amplitudes (in turn described in terms of an isotope-independent intramolecular potential function). In cases where the temperature can be forced low enough, i. e., solutions of He^3/He^4 or H_2/D_2, or the cumulative effect per substituted atom is large enough, i. e., mixtures of H/D isomers of certain polymers, the excess free energy from this effect can be large enough to cause phase separation.

In addition to the effects described above, IEs on density, molar volume, transition temperatures, superconductivity, crystal structure, and gas-phase virial coefficients have been investigated. IEs on molecular properties like polarizability, dipole moments, chemical shifts in magnetic resonance, etc., are understandable in terms of amplitude differences. These arise from the fact that the mean square displacement of any internal coordinate (even in the harmonic approximation) is nonzero and is temperature and mass dependent. The effects are important for those molecular properties which must be averaged over vibrational motion.

See also: Kinetics, Chemical; Molecular Structure Calculations.

Bibliography

J. Bigeleisen, *Science* **147**, 463 (1965); J. Bigeleisen, M. W. Lee, and F. Mandel, *Annu. Rev. Phys. Chem.* **24**, 407 (1973); *Acc. Chem. Res.* **8**, 179 (1975).

E. Buncel and C. C. Lee (eds.) *Isotopes in Organic Chemistry*. Elsevier, Amsterdam, Vols. 1 (1975)–7 (1987).

C. J. Collins and N. S. Bowman (eds.), *Isotope Effects in Chemical Reactions*. Van Nostrand-Reinhold, New York, 1970.

A. Imre, W. A. Van Hook, *Recent Res. Devel. Polymer Sci. Transworld, Trivanaddam* **2**, 539 (1998).

T. Ishida, W. A. Van Hook, and M. Wolfsberg (eds.), "Festschrift in honor of J. Bigeleisen". *Z. Naturforschung A*, **44** (1989).

G. Jancso in *Handbook of Nuclear Chemistry*, A. Vertes, S. Nagy and Z. Klencsar (eds.), Vol. 4, p. 85, Kluwer, 2004.

G. Jancso, L. P. N. Rebelo and W. A. Van Hook, *Chem. Rev.* **93**, 2645 (1993); *Chem. Soc. Rev.* 257–264 (1994).

G. Jancso and W. A. Van Hook, *Chem. Rev.* **74**, 689 (1974): W. A. Van Hook, *J. Chem. Phys.* **83**, 4107 (1985): R. R. Singh and W. A. Van Hook, *J. Chem. Phys.* **88**, 2969 (1987); *J. Chem. Phys.* **87**, 6088, 6097 (1987); *Macromolecules* **20**, 1855 (1987).

L. Melander and W. H. Saunders, *Reaction Rates of Isotopic Molecules*. John Wiley and Sons, New York, 1980.

I. B. Rabinovitch, *Effect of Isotopy on Physico-Chemical Properties of Liquids*. Consultants Bureau, New York, 1970.

W. A. Van Hook, in *Isotope Effects in Chemistry and Biology*, A. Kohen and H. H. Limbach (eds.). Marcel Dekker, New York, 2005.

M. Wolfsberg, *Annu. Rev. Phys. Chem.* **20**, 449 (1969); *Acc. Chem. Res.* **5**, 225 (1972).

Isotope Separation

T. Ishida

Most elements exist in nature as a mixture of isotopes differing in mass as a result of differences in the number of neutrons in their atomic nuclei. Isotopes of an element have very similar physical and chemical properties that make their separation difficult, but some possess distinctly different nuclear properties that make those isotopes particularly useful.

Enrichment plants for isotopes such as uranium-235 (^{235}U) and deuterium are of industrial scale yielding, e. g., hundreds of tons per year; others yield product at rates ranging only from grams per day to grams per year [1]. However, the scale of production does not necessarily reflect the relative importance of the separated isotopes. Uranium-235 and deuterium are essential materials for the nuclear power industry. Deuterium oxide (heavy water) is a coolant and moderator in heavy-water reactors, which use natural abundance uranium rather than enriched ^{235}U as fuel. Deuterium is also an essential ingredient of fuels for all types of nuclear fusion schemes currently under development. Lithium-6 may be a useful fuel in fusion reactors and is also a starting material for producing tritium. Boron-10 is used in control rods for fission reactors.

Carbon-13, nitrogen-15, and oxygen-17 are produced in plants whose typical capacities are grams per day. They are powerful tracers for various fields of research in the physical and, particularly, the life sciences because they are nonradioactive and thus physiologically harmless. They occur in nature at low abundances. For this reason relatively low enrichments yield significant signals, and their nuclear magnetic moments make it possible to use them to determine molecular environments surrounding the atoms by means of nuclear magnetic resonance spectroscopy.

Critical needs exist for enriched isotopes of virtually every element in the periodic table, although the required quantities of some may only be on the order of grams per year [2]. Some highly enriched stable isotopes are vital to fundamental research in nuclear physics and chemistry, solid state physics, geoscience, and biology and medicine. Many separated stable isotopes are used as nuclear targets in research reactors or particle accelerators to produce particular radioisotopes, which are then used as radiotracers or radiopharmaceuticals. High isotopic enrichment of the target is the key to minimizing interference in measurements and possible physiological side effects because of the presence of other radioisotopes, which may result from isotopic impurities in the target.

A variety of methods has been developed especially adapted to separating isotopes. Some methods, such as gaseous diffusion, gas centrifuge, separation nozzle, and electromagnetic separation, are obviously based directly on mass differences; others, such as thermal diffusion, chemical exchange, chromatographic and ion exchange, distillation, and photochemical methods, result from less obvious mass-dependent changes in atomic and molecular properties.

General Comments

Although the common principles underlying isotope separation are relatively few, a large variety of processes has been considered. Examination of processes for separating deuterium turned up 98 potential candidates, and one study group evaluated at least 25 known processes other than gaseous diffusion and gas centrifuges [3] for separating uranium isotopes. All individual processes have some merit for the separation of isotopes, but evaluating them depends on several factors: whether isotopes of light-weight (e. g., deuterium), intermediate-weight (e. g., nitrogen), or heavy-weight elements (e. g., uranium) are to be separated; and whether the quantities needed are grams (as in research) or tons (as for use in power reactors). The choice depends on the properties of the element, the degree of separation needed, and the scale and continuity of the demand; even for a given isotope, there is no one best method. For large-scale applications availability of feed materials, capital costs, and power requirements may be the overriding considerations, while for laboratory needs simplicity of operation or versatility may be primary.

More than 30 years ago Manson Benedict and Thomas Pigford (in the first edition of [1]) drew general conclusions that are still valid regarding isotope separation methods:

1. the most versatile means for the production of small quantities of isotopes is the electromagnetic method;

2. the simplest and most inexpensive means for small-scale separation of many isotopes is thermal diffusion;

3. distillation and chemical exchange are the most economical methods for large-scale separation of the lighter elements; and

4. gaseous diffusion and the gas centrifuge are most economical for large-scale separation of the heaviest elements.

In addition, laser-induced isotope separation (LIS) has become an extremely attractive method for separating isotopes of many elements, as a result of recent advances in laser technology [4], and probably the most economical option for future large-scale production of isotopes of heavy elements such as Pu and U.

Since isotopic molecules have very similar properties, the degree of separation achieved in a single separating unit device (e. g., for distillation, a vessel containing a multicomponent

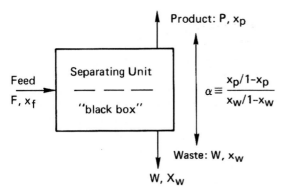

Fig. 1: An elementary separating stage.

liquid and its vapor phase in equilibrium with each other) is extremely small. The degree of separation achieved in such a unit is measured by a property of the separating process called the elementary separation factor. If one envisions a separating unit for any process as a "black box" (Fig. 1) that separates a single feed stream into a product stream–somewhat enriched in the desired isotope – and a waste stream–somewhat depleted in that component – the elementary separation factor, a, is defined as the abundance ratio of the isotopes in the product stream divided by the same ratio for the waste stream:

$$\text{elementary separation factor: } \alpha = \frac{[x/(1-x)]_{\text{product}}}{[x/(1-x)]_{\text{waste}}}$$

Here, x is the mole fraction of the desired isotope. If α is unity, no separation occurs. For typical (other than electromagnetic and laser isotope separation) processes and most elements other than hydrogen, α is at most only several percent different from unity, and often only tenths of a percent away from unity. A single separating stage can thus increase the enrichment by only a few percent or less, relative to the enrichment level of its feed.

To achieve a useful enrichment it is thus necessary to construct a stack of separating stages in which the product stream from a lower stage (processing less enriched material) is fed to the next higher stage, whose product is in turn fed to the stage above, and so on. In addition, the waste streams from upper stages usually contain desired isotope concentrations higher than the natural abundance, so they are recycled as part of the feed to an appropriate intermediate stage. Such a stack of interconnected stages is called a separation cascade. Furthermore, when, as is often the case, the desired isotope has a low relative abundance in nature, it is a minor component of the natural-abundance feed material going into the separation plant. Consequently, the section of such an enriching cascade near the feed-point must handle considerably higher material flow rates than those close to the product-end. The higher-flow section should therefore contain larger flow-capacity separative elements or a number of elements connected in parallel.

Material conservation (see Fig. 1) requires that the total amount of material fed into the stage be balanced by the amounts leaving in the product and waste streams, and further that the quantities of the desired isotope in the two exit streams equal the amount of the isotope that entered the stage:

total material balance: $F = P + W$;

isotopic material balance: $F x_f = P x_p + W x_w$.

Here F, P, and W symbolize the molar flows in the feed, product, and waste streams, and x again represents the mole fraction of the desired isotope. These two material-balance conditions, together with the elementary separation factor, govern the steady-state operation of a separating unit. In fact, a similar set of three equations can be applied to either an element or a stage, and the performance of an entire cascade is governed by the overall material balance equations and the cumulative effect of the separative capability of the individual elements.

More detailed theoretical treatments of cascade parameters will be found in the references (see particularly [1, 3, 5–7]). Now to illustrate the principles just outlined, the article will provide brief descriptions of several isotope-separating processes used (or proposed) on the industrial or laboratory scales.

Separation Processes

Gaseous Diffusion [7, 6–9]

Most of the current ^{235}U enrichment capacity of the world is still provided by gaseous diffusion plants. In each stage of the process, gaseous uranium hexafluoride (UF_6, the only volatile compound of uranium at ordinary temperature) flows through a diffusion barrier with very fine holes from a high-pressure chamber into a lower-pressure region. This phenomenon is gaseous effusion, although the process is usually termed gaseous diffusion. The lighter $^{235}UF_6$ molecules have a slightly higher mean speed than their ^{238}U counterparts; therefore, the gas passing through the barrier is slightly richer in ^{235}U than the portion remaining behind. The mean speeds of the molecules are in inverse ratio to the square roots of the molecular weights of the isotopic molecules: $(^{238}UF_6/^{235}UF_6)^{1/2} = (352/349)^{1/2} = 1.0043$, the elementary separation factor for $^{235}U/^{238}U$ enrichment by gaseous diffusion.

The low separation factor per stage requires the use of many enriching stages in a countercurrent cascade to produce a useful degree of enrichment. Figure 2 depicts schematically the operation of a stage, and Fig. 3 shows the arrangement of three stages in a cascade. The gas in the enriched stream is at a lower pressure, so it must be recompressed before it is fed to the next higher stage, accounting for the high electrical power usage of such enrichment plants; about 70% of the operating expenses of a gaseous diffusion plant (exclusive of feed material) is for electricity. An "ideal cascade" to separate natural uranium feed containing 0.71% ^{235}U into a product containing 3% and waste containing 0.2% requires 1272 separating stages of varying size and capacity arranged in series as in Fig. 3. (To produce product containing 90% ^{235}U, at the same feed and waste compositions, would require 3919 stages.) Figure 4 shows the arrangement of stages in an "ideal" cascade, one in which the compositions of streams from the upper and lower stages at each enrichment level are equal (Fig. 3).

Many of the plant parameters that make large contributions to the cost of enriched product (other than the cost of feed material) are proportional to a simple mathematical function of the isotopic abundances and the quantities (moles or kilograms) of the feed, product, and waste streams entering and leaving the plant. This function – applicable to an ideal separating cascade and called the separative work [1, 6, 7] – is a measure of the value added by the plant to the material processed by the cascade. It is proportional to the quantity of the enriched product, and it increases with the change in the enrichment levels of product and waste from

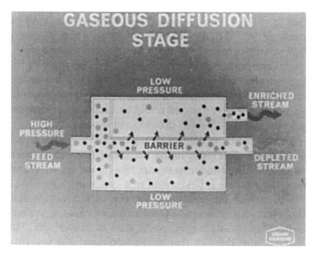

Fig. 2: A gaseous diffusion stage.

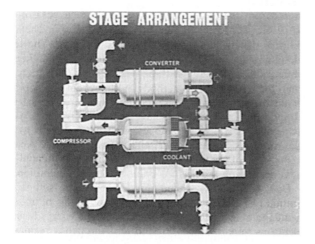

Fig. 3: Arrangement of gaseous diffusion stages.

that of the feed. It does not, however, depend on the stage separation factor. The cost of an enriched product is proportional to this separative work, which is a measure of the quantity of separation achieved, and an additional factor-a function of the separation factor and thus dependent on the efficiency of the separation process. Thus the relative merits of various separation processes can be discussed in terms of their costs per separative work unit (SWU). The kilogram of uranium is the conventional unit for the SWU for ^{235}U.

For a plant with a single waste, product, and feed stream, there are three mole fraction variables and three quantity variables, of which only four can be specified independently because of the two material balance requirements. Thus, for instance, two plants each taking the same fixed feed concentration and yielding the same amount of product enriched to the same isotopic concentration and the same waste assay have the same SWUs, no matter what separation

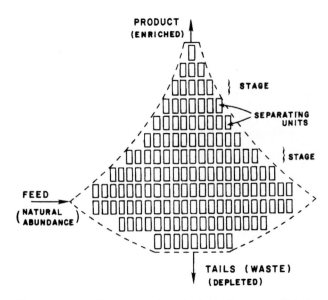

Fig. 4: A schematic representation of relative flow rates of ideal cascade.

principle and engineering design they operate on. A plant of a given SWU capacity may be run (a) to produce either a small quantity of material at high enrichment or a larger quantity at lower enrichment; and (b) to maintain a given quantity and quality of product by extracting more ^{235}U from the feed (lower waste assay) when natural uranium is in short supply and discarding wastes of higher assay when feed material is plentiful.

The gaseous diffusion plants in the world have a combined annual capacity of about 50 million SWUs, or 50 thousand metric tons, of uranium enrichment, of which the United States and the Soviet Union have, respectively, capacities of 27 million SWUs/yr and about 10 million SWUs/yr [9].

Overall, the advantages of the gaseous diffusion process are its proven reliability and demonstrated ability (over more than 40 years) to operate at a capacity factor of over 99%. Among its disadvantages are its high energy consumption-about 2200 kwh/SWU – an inevitable consequence of the need to recompress the gas many times – and its need for large installations to make economical operation possible.

Gas Centrifugation [1, 6–11]

When gaseous uranium hexafluoride spins in a centrifuge at a high rotational speed, the heavier ^{238}U molecules move preferentially toward the periphery, leaving the inner zone enriched in ^{235}U. Modern developments in this method (see e. g., Fig. 5) are all variants of the countercurrent gas centrifuge design of Gernot Zippe. UF$_6$ gas enclosed within a rapidly rotating cylinder is subject to centrifugal acceleration thousands of times greater than gravity. At peripheral speeds above 500 m/s, made possible in current centrifuges using rotors composed of high-performance carbon-fiber- or aramide-reinforced polymers, the ^{235}U content at the center of the cylinder can be as much as 18% greater than at the periphery. A system of rotating baffles and stationary scoops induces longitudinal countercurrent gas flow with light gas (rich

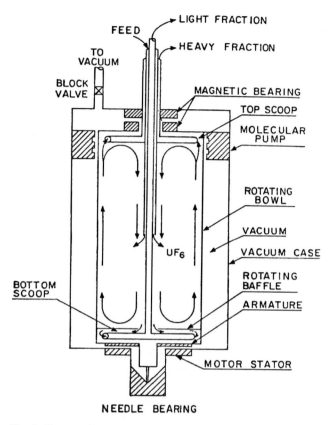

Fig. 5: Gas centrifuge.

in $^{235}UF_6$) flowing upward near the axis and heavy gas (depleted in ^{235}U) flowing downward near the periphery. This sets up a cascade of multiple separative elements in a single centrifuge and, in a sufficiently long centrifuge, a ^{235}U concentration ratio as high as 2 between the top and bottom; the enriched product is then withdrawn from the top of the cylinder. Only seven stages would be needed in an ideal cascade to produce 3% ^{235}U product and waste at 0.25%.

The intrinsic separation factor in centrifuges is proportional to the difference in the masses of isotopic molecules rather than the ratio (or its square root) of the masses. Therefore, this process is best suited to a system such as $^{235}UF_6$–$^{238}UF_6$, which has a mass difference of three mass units. Unfortunately, uranium hexafluoride condenses at pressures higher than one-sixth atmosphere at room temperature. This, coupled with the fact that at the periphery of a centrifuge the pressure is millions of times the pressure near the axis, necessarily makes the throughput of gas (the amount of material processable in a given time) by an individual centrifuge quite low. Thus, a centrifuge plant with a capacity of 3×10^6 kgSWU/yr producing 3% ^{235}U would need 6×10^4 centrifuges, each 10 m long, to do the separative work produced by 1200 gaseous diffusion stages. On the other hand, the electrical energy consumed would be only 90 kwh/SWU as opposed to 2200 kwh/SWU for gaseous diffusion.

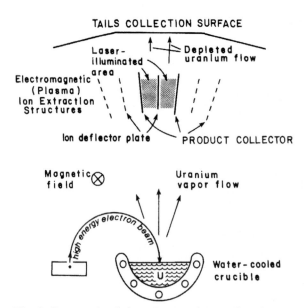

Fig. 6: Cross-sectional view of an atomic vapor laser isotope separation module.

Among centrifuge plants existing or designed around the world are a plant with a capacity of about 1×10^6 SWUs per year built and operated by a joint British, Dutch, West German venture, a plant with an annual capacity of 2.2×10^6 SWUs per year designed by the U.S. Department of Energy at Portsmouth, Ohio, and in Japan a prototype plant with 2×10^5 SWUs per year capacity built and operating, and another with a capacity of 1.5×10^6 SWUs per year scheduled for completion in 1991.

Laser Isotope Separation [1, 4, 6–9, 12, 13]

This process takes advantage of the differences in absorption spectra of different isotopic species. By using sufficiently monochromatic light of an appropriate wavelength, it preferentially excites a particular isotopic species to an upper energy level. The excited species must then be separated from its isotopic partners before it exchanges isotopes or before it loses excitation by some physical or chemical process that need not be isotopically selective. The technique was successfully used to separate small quantities of mercury isotopes almost three decades ago. Attention has been focused on these methods by the recent development of narrow-band, tunable lasers of high intensity and high repetition rate. Laser methods have been notably successful in achieving separation of isotopes of elements such as hydrogen, boron, chlorine, sulfur, and bromine, but the greatest importance of the method is likely to lie in its potential as a next generation process for uranium isotope separation.

One such process, called atomic vapor laser isotope separation (AVLIS), is illustrated in Fig. 6. Atoms of uranium heated in a high-vacuum chamber by a high-energy beam of electrons vaporize and flow upward through a region between a pair of negatively charged plates. The space between the plates is illuminated by light from a system of lasers passing through in the direction perpendicular to the page. Laser photons with an appropriate wavelength selectively excite ^{235}U atoms but leave ^{238}U atoms unexcited. At least one additional laser beam is used to ionize the excited ^{235}U atoms. The ionizing photons have energies sufficiently high to ionize the already excited ^{235}U atoms but insufficient to ionize unexcited ^{238}U atoms. The ^{235}U ions are then collected by negatively charged plates, while ^{238}U atoms continue their upward flow until condensed on the ceiling surface. It is a major attraction of the method that under ideal conditions, a complete separation of ^{235}U and ^{238}U in a single LIS stage is theoretically possible.

The future of the LIS process depends on successful development of laser systems. Since even the largest energy difference between the atoms of ^{235}U and ^{238}U corresponds only to a change on the order of 1 in 5×10^4, the excitation laser photon energy must be tuned more closely than this to avoid accidental excitation of ^{238}U atoms. Lasers generate pulses (or packets) of photons at a fixed repetition rate. The lasers for the LIS must have a sufficiently rapid repetition rate to catch nearly all the ^{235}U atoms passing between the collection plates; ^{235}U atoms left un-ionized by the lasers will not be collected by the plates, thus reducing the effective separation factor. Only a fraction of laser photons passing through an LIS chamber is actually absorbed by uranium, and it takes one mole of absorbed photons to excite one mole of uranium atoms. Thus, LIS requires a high-power (high photon production rate), rapid-repetition, finely tunable laser. In addition, laser photons of high energy are expensive to produce. Since a higher density of uranium atoms in the illuminated region increases the likelihood of collision between excited ^{235}U atoms and ^{238}U atoms, which can produce excited ^{238}U atoms at the expense of de-excited ^{235}U, the throughput of LIS processes is necessarily low.

In spite of these technical difficulties, LIS for ^{235}U [13] holds high promise because of the theoretical possibility of its extremely high separation factor. The laser isotope separation processes other than the AVLIS for ^{235}U that have been developed include the molecular LIS for ^{235}U, in which UF_6 is working material, and "Special Isotope Separation" for converting fuel-grade [7–19% ^{240}Pu] into weapon-grade [less than 7% ^{240}Pu] by removing unwanted isotopes such as ^{240}Pu from ^{239}Pu.

Separation Nozzle [1, 3, 6–9]

This aerodynamic process achieves partial separation of isotopes in a flowing gas stream subjected to high linear or centrifugal acceleration, or both. A cross section of an improved separation nozzle developed by E. W. Becker and his co-workers in Germany is shown in Fig. 7. A feed gas, typically consisting of a mixture of about 5 vol% UF_6 and 95% H_2 at a pressure of about 1 atm, passes through a nozzle with first a convergent, then a divergent cross section into a lower-pressure region. The change in cross section accelerates the mixture to supersonic speed, and the curved groove produces centrifugal acceleration. The gas adjacent to the curved wall is preferentially enriched in ^{238}U, and the knife edge downstream divides the stream into a (waste) portion depleted in ^{235}U (region C) and a portion enriched in the light isotope (region B).

The separation factor increases with the mean speed of the gas molecules and with their centrifugal acceleration. Therefore, a larger pressure ratio between the feed and product gases, a narrower throat width at the entrance, and a smaller radius of curvature of the curved groove all contribute to increasing the separation factor. But the resulting miniaturization of nozzle elements reduces the throughput and makes fabrication of the elements a formidable task.

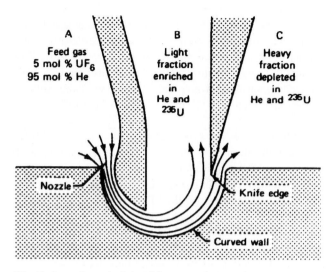

Fig. 7: Operating principle of the separating nozzle.

A separation factor of 1.015 has been achieved with a mixture of 5% UF_6–95% H_2 flowing through a pressure ratio of 3.5. The dilution of UF_6 with a gas of low molecular weight yields (1) a higher sonic velocity and thus a higher separation factor, and (2) higher diffusion rates of UF_6 molecules, which allow operation at higher product pressures and correspondingly increased uranium throughput. The Becker nozzle process offers a higher separation factor than gaseous diffusion, but the 20-fold dilution of UF_6 also makes power consumption (about 3×10^3 kwh/SWU) higher than for gaseous diffusion. A demonstration cascade of about 0.3×10^6 SWU/yr has been operated in Brazil with technical assistance from West Germany.

Electromagnetic Separation [1-3, 14]
In this process, the principle is essentially that of a large-scale mass spectrometer, and its use dates back to the World War II Manhattan District Project when such machines were used to separate ^{235}U in kilogram quantities. Since the end of World War II, the electromagnetic separators (called Calutrons for *Cal*ifornia *U*niversity Cyclo*tron* because they were originally developed at the University of California by E. O. Lawrence) have been used for separating an amazing variety of isotopes, both stable and radioactive. A retrospective paper by L. O. Love [14] describes activities at Oak Ridge National Laboratory (ORNL) in this area over three decades. See Ref. [2] for a discussion of needs for various isotopes of practically every element that can be separated by this versatile method, and a description of the current ORNL electromagnetic separation program.

Thermal Diffusion [1, 3, 6]
The thermal diffusion effect – namely, that a concentration gradient is produced in a gaseous mixture subjected to a temperature gradient (e. g., in a vessel with walls at different temperatures) – was transformed from a laboratory curiosity into a useful and simple method for separating fluid mixtures (including mixtures of isotopes) by the invention of the thermal diffusion column by Clusius and Dickel. The operating principle of such a column is illustrated

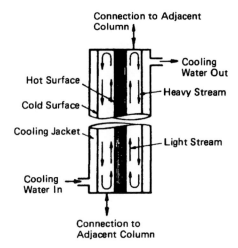

Fig. 8: Thermal diffusion column.

in Fig. 8. A fluid is confined in a long vertical cylinder between an inner wall, kept at a higher temperature, and an outer wall, kept at a lower temperature. A countercurrent convective flow is set up in the fluid, as indicated in Fig. 8, descending along the cool wall and rising adjacent to the heated wall. Since horizontal slices of the column at different levels along the column length act as unit separation stages, the countercurrent vertical flows multiply the small (less than 1%) isotopic concentration difference between the two walls, so that a single column contains a large number of separating stages.

Laboratory-scale cascades suitable for daily separation of milligram quantities of isotopes in gaseous form are remarkably simple to construct and operate. Notably, the isotopes of the noble gases are separated by this method.

Chemical Exchange [1, 3, 5–6, 15–16]

Isotopic exchange equilibria, which lead to nonequipartition of isotopes between different chemical species, result from differences in internal vibrational frequencies of the isotopic forms. The magnitude of the separative effect depends on the isotopic mass difference between the isotopes ($\Delta m/m$), and on the difference in free energy (produced by the mass difference) in a pair of different chemical species [3, 5, 15]. The first criterion leads to the conclusion that chemical exchange is most effective for separating isotopes of the lighter elements and becomes increasingly difficult as one proceeds to heavier elements ($>$ sulfur). The second criterion leads one to expect a large separation effect between two molecules, one in which the isotopic atom is unbonded or weakly bonded, and another in which it is connected by a maximum number of strong bonds. Unfortunately, such compounds do not necessarily undergo isotope exchange rapidly, and exchange processes often require a catalyst to promote the reaction at an effective rate.

Reactions that have been used to produce quantities of separated isotopes include the exchanges: H_2O (liq)/HDS (gas); H_2O (liq)/HD (gas); NH_3 (liq)/HD (gas); $H^{14}NO_3$ (aq)/ ^{15}NO (gas); $^{12}CN^-$ (aq)/$H^{13}CN$ (gas); $^{10}BF_3$–ether (complex)/$^{11}BF_3$ (gas), and $H^{32}SO_3^-$ (aq)/ $^{34}SO_2$ (gas).

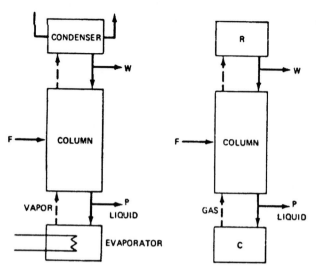

Fig. 9: Analogy between distillation and gas–liquid chemical exchange.
R = reflux converter; C = chemical reactor or electrolytic cell; F = feed;
P = product; W = waste.

The elementary exchange effects are multiplied in long exchange columns similar to those used for distillation. In contrast to distillation, where reflux at the ends is effected by a boiler (at the bottom) to evaporate the liquid, and a condenser (at the top) to liquefy the vapor, exchange columns use chemical reactors at each end to interconvert the exchanging species. Figure 9 illustrates the analogy between the gas–liquid exchange process and distillation.

Among the above listed exchanges, those involving exchange of hydrogen and deuterium have been and are being used in plants producing thousands of tons of D_2O annually [16]; the exchange between nitric acid and nitric oxide (termed the Nitrox process) has been widely used throughout the world in simple laboratory scale cascades to produce grams per day of nitrogen-15 enriched to purities up to 99.8from the natural abundance of 0.37%. Operating characteristics for the systems are described in the references, particularly [1, 3, 5–6, 16].

Distillation [1, 3, 5–6]

H. C. Urey discovered deuterium by concentrating it sufficiently through distillation of liquid hydrogen and, for the first time, detecting it spectroscopically. The method, which is one of the common chemical engineering processes for separating chemical substances, has been used for producing deuterium by the distillation of liquid hydrogen or water. Isotopes of carbon, nitrogen, and oxygen have been produced by distillation of their compounds, carbon monoxide, nitric oxide, and water.

The energy consumed by distilling water to produce heavy water is relatively high, due partly to the fact that heavy water is less volatile than light water. This makes distillation less competitive than chemical exchange processes for primary enrichment of heavy water. However, its high throughput (because the working material is liquid), reasonably high separation factor (1.027 at 100 °C), design simplicity, and operational reliability make it the preferred process for final enrichment – from several percent deuterium to 99.9% D. This illustrates the comment made earlier in this article that, even for a given isotope, there is no one best separation method.

Acknowledgement

This work is based on the article "Isotope Separation" coauthored by late Dr. William Spindel and the present author (T. I.) in the second edition of this Encyclopedia (VCH Publishers, New York, 1991).

See also: Isotope Effects.

Bibliography

M. Benedict, T. H. Pigford, and H. W. Levi, *Nuclear Chemical Engineering*, 2nd Edition. McGraw–Hill, New York, 1981. (A)

Separated Isotopes: Vital Tools for Science and Medicine. Office of Chemistry and Chemical Technology, National Research Council, Washington, DC, 1982. (E)

W. Spindel, in *Isotopes and Chemical Principles*, ACS Symposium Series, Vol. 11, Chap. 5. American Chemical Society, Washington, DC, 1975. (E)

V. S. Letokhov and C. Bradley Moore, "Laser Isotope Separation", in *Chemical and Biochemical Applications of Lasers* (C. Bradley Moore, ed.). Academic Press, New York, San Francisco, London, 1977. (I)

J. Bigeleisen, in *Isotope Effects in Chemical Processes*, Advances in Chemistry Series, Vol. 89, Chap. 1. American Chemical Society, Washington, DC, 1969. (I).

S. Villani, *Isotope Separation*. American Nuclear Society, Hinsdale, IL, 1976. (A)

S. Villani (ed.), *Uranium Enrichment*. Vol. 35 of Topics in Applied Phys., Springer-Verlag, Berlin, Heidelberg, New York, 1979. (A)

M. Benedict (ed.), *Developments in Uranium Enrichment*, AIChE Symp. Ser. 169, Vol. 73. American Institute of Chemical Engineers, New York, 1977. (I)

M. Benedict, in AIChE Symp. Ser. 235, Vol. 80, pp. 149–156. American Institute of Chemical Engineers, New York, 1983. (E)

D. R. Olander, *Sci. Am.* **239**, 37 (August 1978). (E)

S. Whitley, *Rev. Mod. Phys.* **56**, 41 (1984); **56**, 67 (1984). (A)

R. N. Zare, *Sci. Am.* **236**, 86 (February, 1977). (E)

J. A. Paisner and R. W. Solarz, "Resonance Photoionization Spectroscopy" (pp. 175–260), and J. L. Lyman, "Laser-Induced Molecular Dissociation: Applications in Isotope Separation and Related Processes" (pp. 417–506), in *Laser Spectroscopy and Its Applications* (L. J. Radziemski, R. W. Solarz, and J. A. Paisner, eds.). Marcel Dekker, New York, 1987.

L. O. Love, *Science* **182**, 343 (1973). (E)

J. Bigeleisen, *Science* **147**, 463 (1965). (E)

H. K. Rae (ed.), *Separation of Hydrogen Isotopes*, ACS Symposium Series, Vol. 68. American Chemical Society, Washington, DC, 1978. (I)

Isotopes

R. C. Barber

Introduction

An isotope is one of two or more different species of the same chemical element, having the same atomic number Z but different mass number A. Thus, for a given element the nuclei of isotopes possess the same number of protons but differ in the number of neutrons. The term, proposed by Frederick Soddy (1913), derives from the Greek *isos* (equal) and *topos* (place) and reflects the fact that such species occupy the same position in the periodic table of the elements. For this reason the term isotope implies the existence of at least two different species of a given element. The general term for a particular atomic species characterized by an atomic number Z and mass number A is nuclide, although the term isotope has frequently been used loosely with this meaning.

Within each of the known elements at least three nuclides have been observed, some of which may be stable with the remainder being radioactive. However, studies of the isotopic nature of elements are normally confined to those nuclides that are either stable or sufficiently long-lived to exist in significant amounts in nature. The occurrence and relative abundances of such species have been studied by mass spectrometry.

The largest number of stable isotopes, ten, occurs in tin ($Z = 50$, $A = 112$, 114, 115, 116, 117, 118, 119, 120, 122, 124). Xenon possesses nine isotopes; tellurium and cadmium each have eight. A total of 62 elements possess naturally occurring isotopes; a further 20 elements are mononuclidic, that is, in each case only one nuclide of the element is stable. In more precise terms, the Commission on Atomic Weights and Isotopic Abundances of IUPAC has defined an element to be mononuclidic if it has one and only one nuclide that is stable or has a half-life greater than 3×10^{10} years. Therefore the Commission considers Be, F, Na, Al, P, Sc, Mn, Co, As, Y, Nb, Rh, I, Cs, Pr, Tb, Ho, Tm, Au, and Bi to be mononuclidic, while Pa is not. Among the elements below bismuth ($Z = 83$) only two, technetium and promethium, do not have any stable nuclide.

For those elements that possess naturally occurring isotopes, the relative abundances of these isotopes, in most cases, are remarkably independent of the source of the material. Accordingly, the chemical atomic weight, which is the average value of the masses of all of the stable isotopes, weighted according to the relative abundances of each, is found to be a constant for most materials. Situations in which relative abundances do vary are then of special interest; some of these are discussed later.

Discovery

Against the background of the idea, generally accepted in the latter part of the nineteenth century, that a given number of atoms of a particular element (Avogadro's number) had a mass characteristic of the element (the atomic weight), the first speculation recognizably consistent with the modern idea of isotopes was made by Sir William Crookes. Addressing the Chemical Section of the British Association at Birmingham in 1886 he suggested that "when we say the atomic weight of, for instance, calcium is 40, we really express the fact that, while the majority of calcium atoms have an actual atomic weight of 40, there are not a few which are represented by 39 or 41, a less number by 38 or 42 and so on". He further considered the possibility that "these heavier and lighter atoms may have been in some cases subsequently sorted out by a process resembling chemical fractionation. This sorting out may have taken place in part while atomic matter was condensing from the primal state of intense ignition, but also it may have been partly effected in geological ages by successive solutions and precipitations of the various earths".

Subsequently he developed this idea in connection with his pioneer work with the rare earths, subdividing a sample of "yttria" into a number of components by fractional precipitation. These components, which he called "meta-elements", possessed different phosphorescent spectra and differed a little in atomic weight but closely resembled each other in chemical properties. With advances in chemical techniques, however, one after another of these components was identified as a distinct element, each having its characteristic atomic spectrum and atomic weight. Thus the yttria studied by Crookes proved to be a mixture of elements and the idea of meta-elements was abandoned.

By the end of the nineteenth century, the increased precision in the determinations of atomic weights, coupled with the lack of variations in these values, had further reinforced the general assumption that the atomic weight of an element was a fundamental property. Two problems remained unsolved: Why were so many atomic weights nearly integers on the scale $O = 16$, and why were there a few notable cases of fractional atomic weights (e. g., $Cl = 35.5$)?

The first experimental evidence that chemically identical substances might possess different physical properties emerged from the studies of radioactivity in the elements heavier than bismuth (i. e., $Z > 83$). In particular these studies exploited the fact that minute quantities of a particular substance could be detected and identified by observing the type of radiation and the half-life. While studying the decay chains of uranium, B. B. Boltwood (1906) identified a new element, which he called ionium and which, when mixed with thorium, could not be separated by any known chemical process. Shortly thereafter, similar behavior was observed in mixtures of radiothorium (a disintegration product of thorium) and thorium by H. N. McCoy and W. H. Ross, and in mixtures of mesothorium 1 (also in the thorium series) and radium by W. Marckwald and by F. Soddy. Further evidence of the chemical identity of ionium and thorium came from the examination of the visible emission spectrum of a mixture of the two by A. S. Russell and R. Rossi (1912). No new lines attributable to ionium were observed when compared to the spectrum of pure thorium.

Soddy, convinced that in spite of the unique radioactive decay properties and distinct atomic weights, the chemically inseparable species were in fact the same elements, undertook a survey of the radiochemistry of the heavy elements and proposed the existence of isotopes, along with a general scheme that reconciled the chemical properties with the decay data and notably with the recently established identity of the a particle. Moreover he saw no reason to restrict the incidence of isotopes to the radioactive elements and recognized that it might account for fractional atomic weights.

Clear evidence of the existence of stable isotopes emerged from the investigations by J. J. Thomson of the nature of positive rays in gas discharge tubes. The positive rays from a gas discharge were collimated, passed through an arrangement of parallel and coterminous electric and magnetic fields, and allowed to fall on a photographic plate. In such apparatus all particles having the same value of e/M, regardless of energy, strike the plate at some point on a particular parabola. Thomson observed that, in addition to the intense line for neon at atomic weight 20, there was a much fainter line that accompanied it at mass 22. This result led Thomson to encourage F. W. Aston, then a research student at Cambridge, to pursue the study of stable isotopes.

Aston initially attempted to enrich neon in its suspected isotopes but then turned his attention to the construction of a "mass spectrograph" (1919), which both focused and analyzed positive-ion beams. With this instrument he demonstrated conclusively that neon consists of a mixture of isotopes and shortly thereafter confirmed the isotopic nature of chlorine, mercury, nitrogen, and the noble gases.

Independently and simultaneously, A. J. Dempster, working at the University of Chicago, devised a "mass spectrometer" (1918), which he used to discover, and make accurate abundance determinations for, the isotopes of magnesium and, shortly thereafter, lithium, potassium, calcium, and zinc.

Thus with the work of Aston and Dempster the isotopic nature of the stable elements was unequivocally demonstrated and an important instrument in the study of isotopes, the mass spectrometer, was introduced.

Isotopes and Nuclear Structure

The work of Aston indicated early that the atomic masses of the various isotopes are close to integers. This is confirmed in the modern values for atomic masses, including many nuclides not known to Aston, all of which lie within $0.1\,u$ of the value given by the mass number.

Initial attempts to account for this behavior in terms of nuclear electrons were recognized to be unsatisfactory and were discarded when the neutron, a neutral particle with a mass of $\sim 1\,u$, was discovered by Chadwick in 1932 and identified as a nuclear particle. Differing numbers of neutrons were seen to have no effect on the electronic structure, which determined chemical properties, but altered the mass in steps of $\sim 1\,u$ and led to very different nuclear structure and properties.

Aston also realized that the study of divergences from "whole numbers" could be highly rewarding inasmuch as it was a measure of nuclear stability. In this connection, he realized that the definition of atomic weights used in chemistry at the time, wherein oxygen in its "natural" abundance of isotopes was taken to be 16 amu, was imprecise and adopted instead $^{16}O = 16$ amu exactly as his standard.

This double standard persisted until 1960 when the standard for the "chemical" scale was found to be insufficiently precise for modern chemical work, while the standard for the "physical" scale was generally inconvenient for use. Accordingly, in a coordinated action, both the International Union of Pure and Applied Chemistry and the International Union of Pure and Applied Physics adopted in that year the "unified" scale of atomic mass, with the symbol u, where $^{12}C = 12\,u$ exactly. Thus for this scale, 12 g of ^{12}C contains Avogadro's number of atoms, $6.022\,136\,7\,(36) \times 10^{23}$ atoms.

All atoms are stable with respect to spontaneous decomposition into the constituent particles. Thus, as required by special relativity, the mass of the aggregation of protons and neutrons in the form of a tightly bound nucleus, with the relatively loosely bound electrons surrounding it, must be less than the sum of the masses of the dissociated, free particles. That is, we may express the mass of an atom, $M(A,Z)$, as

$$M(A,Z) = A[1 + f(A,Z)] \qquad (1)$$

where $f(A,Z)$ is the packing fraction, a quantity that for stable nuclides lies in the range 8×10^{-3} to -8×10^{-4}. This approach to the description of the average binding energy of the nucleus was first used by Aston, who defined the packing fraction and presented sufficient experimental data to determine the general outline of the way that $f(A,Z)$ varies with mass number. In hindsight, this remarkable work indicated the large changes in energy to be expected from the fusion of light nuclei or from the fission of very heavy nuclei following neutron capture.

In more recent work, the mass of an atom is usually expressed in some variation of the form

$$M(N,Z) = Z[m_p + m_e] + Nm_n - (\text{BE})_{\text{nucleus}} - (\text{BE})_{\text{electron}} \qquad (2)$$

where N is the number of neutrons; mp, m_e, and m_n are the masses of the proton, electron, and neutron, respectively; and $(\text{BE})_{\text{nucleus}}$ and $(\text{BE})_{\text{electron}}$ represent the total binding energies of the nuclear particles and of the atomic electrons, respectively, with the former being much larger than the latter. The variation of average nuclear stability with nuclear size is usually expressed in terms of the binding energy per nucleon, $(\text{BE})_{\text{nucleus}}/A$, and given as a function of the mass number. This quantity increases quickly from 1.1 MeV for ^{2}H to -6.5 MeV at $A = 10$ and increases thereafter to a maximum value of 8.793 MeV for ^{58}Fe. From this value it declines to 7.6 MeV in uranium.

Evidence of the enhanced stability associated with closed nuclear shells may be seen in the fine detail of the variation of the binding energy per nucleon. It may be seen more clearly in a comparison, among isotopes of a given element, of the systematic changes in the energy required to remove the last pair of neutrons. Abrupt and dramatic decreases in this energy are seen as the "magic" numbers ($N = 28, 50, 82, 126$) are exceeded. In a similar way it can be demonstrated that the same magic numbers of protons are also configurations of enhanced stability.

Current values for atomic masses are based on data that derive from two major sources: precise values for mass differences between the members of doublets in mass spectra obtained with large, high-resolution mass spectrometers; and precise determinations of the energies of particles taking part in nuclear reactions, i. e., in a reaction

$$a + X \rightarrow Y + b + Q$$

where Q is the net change in rest mass, converted to energy. Data of both types may be regarded as determinations of the mass-difference connections between the various nuclides. Such values outnumber the actual mass differences, and so constitute an overdetermined set for which the method of least squares is appropriate for deducing "best" values for the atomic masses and the mass differences. The values of the atomic masses given in Table 1 have been derived in this manner.

The distribution and relative abundances of the naturally occurring nuclides also reflect certain features of the nuclear force.

The 287 nuclear species that occur in nature in appreciable quantities (Table 1) may be classified as follows: 168 with even Z and even N; 57 with even Z and odd N; 53 with odd Z and even N; and 9 with odd Z and odd N. From this it is seen that there is a strong tendency for protons to pair with protons and neutrons to pair with neutrons. The distribution also

Table 1: Naturally occurring nuclides.

Atomic no. Z	Element	Atomic weight[a] (u)		Mass no. A	Atomic mass[b] (u)		isotopic composition[c] (%)	Notes*
1	H	1.00794	(7)	1	1.0078250319	(6)	99.9850	d, e, f
				2	2.0141017779	(6)	0.0115	
2	He	4.002602	(2)	3	3.0160293094	(12)	0.000137	d, e
				4	4.0026032497	(15)	99.999863	
3	Li	6.941	(2)	6	6.0151223	(5)	7.59	d, e, f
				7	7.0160041	(5)	92.41	
4	Be	9.012182	(3)	9	9.0121822	(4)	100	
5	B	10.811	(7)	10	10.0129371	(3)	19.9	d, e, f
				11	11.0093055	(4)	80.1	
6	C	12.0107	(8)	12	12.0000000	(0)	98.93	d, e
				13	13.003354838	(5)	1.07	
7	N	14.0067	(2)	14	14.0030740074	(18)	99.632	d, e
				15	15.000108973	(12)	0.368	
8	O	15.9994	(3)	16	15.9949146223	(25)	99.757	d, e
				17	16.99913150	(22)	0.038	
				18	17.9991604	(9)	0.205	
9	F	18.9984032	(5)	19	18.99840320	(7)	100	
10	Ne	20.1797	(6)	20	19.992440176	(3)	90.48	e, f
				21	20.99384674	(4)	0.27	
				22	21.99138550	(25)	9.25	
11	Na	22.989770	(2)	23	22.98976966	(26)	100	
12	Mg	24.3050	(6)	24	23.98504187	(26)	78.99	
				25	24.98583700	(26)	10.00	
				26	25.98259300	(26)	11.01	
13	Al	26.981538	(2)	27	26.98153841	(24)	100	
14	Si	28.0855	(3)	28	27.97692649	(22)	92.2297	d

Table 1 (*continued*): Naturally occurring nuclides.

Atomic no. Z Element	Atomic weight^a (u)	Mass no. A	Atomic mass^b (u)	isotopic composition^c (%)	Notes*
14 Si (*continued…*)	29	29	28.97649468 (29)	4.6832 (22) *d*	
		30	29.97377018 (22)	3.0872	
15 P	30.973761 (2)	31	30.97376149 (27)	100	*f*
16 S	32.065 (5)	32	31.97207073 (15)	94.93 (15)	*d, e*
		33	32.97145854 (15)	0.76	
		34	33.96786687 (14)	4.29	
		36	35.96708088 (25)	0.02	
17 Cl	35.453 (2)	35	34.96885271 (4)	75.78 (4)	*f*
		37	36.96590260 (5)	24.22 (5)	
18 Ar	39.948 (1)	36	35.96754626 (27)	0.3365 (27)	*d, e*
		38	37.9627322 (5)	0.0632 (5)	
		40	39.962383124 (5)	99.6003 (5)	
19 K	39.0983 (1)	39	38.9637069 (3)	93.2581 (3)	*e*
		40	39.96399867 (29)	0.0117 (29)	*g* 1.28 × 10⁹ a
		41	40.96182597 (28)	6.7302 (28)	
20 Ca	40.078 (4)	40	39.9625912 (3)	96.941 (3)	*e*
		42	41.9586183 (4)	0.647 (4)	
		43	42.9587668 (5)	0.135 (5)	
		44	43.9554811 (9)	2.086 (9)	
		46	45.9536927 (25)	0.004 (25)	
		48	47.952533 (4)	0.187 (4)	
21 Sc	44.955910 (8)	45	44.9559102 (12)	100 (12)	
22 Ti	47.867 (1)	46	45.9526295 (12)	8.25 (12)	
		47	46.9517637 (10)	7.44 (10)	
		48	47.9479470 (10)	73.72 (10)	
		49	48.9478707 (10)	5.41 (10)	
		50	49.9447920 (11)	5.18 (11)	

Table 1 (*continued*): Naturally occurring nuclides.

Atomic no. Z	Element	Atomic weight[a] (u)		Mass no. A	Atomic mass[b] (u)		isotopic composition[c] (%)	Notes*
23	V	50.9415	(1)	50	49.947 1627	(14)	0.250	$g > 4 \times 10^{16}$ a
				51	50.943 9635	(14)	99.750	
24	Cr	51.9961	(6)	50	49.946 0495	(14)	4.345	
				52	51.940 5115	(15)	83.789	
				53	52.940 6534	(15)	9.501	
				54	53.938 8846	(15)	2.365	
25	Mn	54.938049	(9)	55	54.938 0493	(15)	100	
26	Fe	55.845	(2)	54	53.939 6147	(14)	5.845	
				56	55.934 9418	(15)	91.754	
				57	56.935 3983	(15)	2.119	
				58	57.933 2801	(15)	0.282	
27	Co	58.933200	(9)	59	58.933 1999	(15)	100	
28	Ni	58.6934	(2)	58	57.935 3477	(16)	68.0769	
				60	59.930 7903	(15)	26.2231	
				61	60.931 0601	(15)	1.1399	
				62	61.928 3484	(15)	3.6345	
				64	63.927 9692	(16)	0.9256	
29	Cu	63.546	(3)	63	62.929 6007	(15)	69.17	
				65	64.927 7938	(19)	30.83	
30	Zn	65.409	(4)	64	63.929 1461	(18)	48.63	*d*
				66	65.926 0364	(17)	27.90	
				67	66.927 1305	(17)	4.10	
				68	67.924 8473	(17)	18.75	
				70	69.925 325	(4)	0.62	
31	Ga	69.723	(1)	69	68.925 581	(3)	60.108	
				71	70.924 7073	(20)	39.892	

Table 1 (continued): Naturally occurring nuclides.

Atomic no. Z	Element	Atomic weight[a] (u)	Mass no. A	Atomic mass[b] (u)	isotopic composition[c] (%)	Notes*
32	Ge	72.64 (1)	70	69.9242500 (19)	20.84	
			72	71.9220763 (16)	27.54	
			73	72.9234595 (16)	7.73	
			74	73.9211784 (16)	36.28	
			76	75.9214029 (16)	7.61	
33	As	74.92160 (2)	75	74.9215966 (18)	100	
34	Se	78.96 (3)	74	73.9224767 (16)	0.89	
			76	75.9192143 (16)	9.37	
			77	76.9199148 (16)	7.63	
			78	77.9173097 (16)	23.77	
			80	79.9165221 (20)	49.61	
			82	81.9167003 (22)	8.73	
35	Br	79.904 (1)	79	78.9183379 (20)	50.69	
			81	80.916291 (3)	49.31	
36	Kr	83.798 (2)	78	77.920388 (7)	0.35	e, f
			80	79.916379 (4)	2.28	
			82	81.9134850 (28)	11.58	
			83	82.914137 (4)	11.49	
			84	83.911508 (3)	57.00	
			86	85.910615 (5)	17.30	
37	Rb	85.4678 (3)	85	84.9117924 (27)	72.17	e
			87	86.9091858 (28)	27.83	g 4.7×10^{10} a
38	Sr	87.62 (1)	84	83.913426 (4)	0.56	d, e
			86	85.9092647 (25)	9.86	
			87	86.9088816 (25)	7.00	
			88	87.9056167 (25)	82.58	

Table 1 (*continued*): Naturally occurring nuclides.

Atomic no. Z	Element	Atomic weight[a] (u)		Mass no. A	Atomic mass[b] (u)		isotopic composition[c] (%)	Notes*
39	Y	88.90585	(2)	89	88.9058485	(26)	100	
40	Zr	91.224	(2)	90	89.9047022	(24)	51.45	e
				91	90.9056434	(23)	11.22	
				92	91.9050386	(23)	17.15	
				94	93.9063144	(26)	17.38	
				96	95.908275	(3)	2.80	
41	Nb	92.90638	(2)	93	92.9063762	(24)	100	
42	Mo	95.94	(2)	92	91.906810	(4)	14.84	e
				94	93.9050867	(20)	9.25	
				95	94.9058406	(20)	15.92	
				96	95.9046780	(20)	16.68	
				97	96.9060201	(20)	9.55	
				98	97.9054069	(20)	24.13	
				100	99.907476	(6)	9.63	
43	Tc							
44	Ru	101.07	(2)	96	95.907604	(9)	5.54	e
				98	97.905287	(7)	1.87	
				99	98.9059385	(22)	12.76	
				100	99.9042189	(22)	12.60	
				101	100.9055815	(22)	17.06	
				102	101.9043488	(22)	31.55	
				104	103.905430	(4)	18.62	
45	Rh	102.90550	(2)	103	102.905504	(3)	100	
46	Pd	106.42	(1)	102	101.905607	(3)	1.02	e
				104	103.904034	(5)	11.14	
				105	104.905083	(5)	22.33	
				106	105.903484	(5)	27.33	

Table 1 (*continued*): Naturally occurring nuclides.

Atomic no. Z	Element	Atomic weight^a (u)	Mass no. A	Atomic mass^b (u)	isotopic composition^c (%)	Notes*
46	Pd (*continued…*)		108	107.903895 (4)	26.46	
			110	109.905153 (12)	11.72	
47	Ag	107.8682 (2)	107	106.905093 (6)	51.839	e
			109	108.904756 (3)	48.161	
48	Cd	112.411 (8)	106	105.906450 (6)	1.25	e
			108	107.904183 (6)	0.89	
			110	109.903006 (3)	12.49	
			111	110.904182 (3)	12.80	
			112	111.9027577 (30)	24.13	
			113	112.9044014 (30)	12.22	g 9 × 10^5 a
			114	113.9033586 (30)	28.73	
			116	115.904756 (3)	7.49	
49	In	114.818 (3)	113	112.904062 (4)	4.29	
			115	114.903879 (4)	95.71	g 5 × 10^{14} a
50	Sn	118.710 (7)	112	111.904822 (5)	0.97	e
			114	113.902783 (3)	0.66	
			115	114.903347 (3)	0.34	
			116	115.901745 (3)	14.54	
			117	116.902955 (3)	7.68	
			118	117.901608 (3)	24.22	
			119	118.903311 (3)	8.59	
			120	119.9021985 (27)	32.58	
			122	121.9034411 (29)	4.63	
			124	123.9052745 (15)	5.79	
51	Sb	121.760 (1)	121	120.9038222 (26)	57.21	e
			123	122.9042160 (22)	42.79	

Table 1 (*continued*): Naturally occurring nuclides.

Atomic no. Z	Element	Atomic weight[a] (u)	Mass no. A	Atomic mass[b] (u)	isotopic composition[c] (%)	Notes*
52	Te	127.60 (3)	120	119.904026 (11)	0.09	e
			122	121.9030558 (29)	2.55	
			123	122.9042711 (20)	0.89	g 1.2 × 10^13 a
			124	123.9028188 (16)	4.74	
			125	124.9044241 (20)	7.07	
			126	125.9033049 (20)	18.84	
			128	127.9044615 (19)	31.74	
			130	129.9062229 (21)	34.08	
53	I	126.90447 (3)	127	126.904468 (4)	100	
54	Xe	131.293 (6)	124	123.9058954 (21)	0.09	e,f
			126	125.904268 (7)	0.09	
			128	127.9035305 (15)	1.92	
			129	128.9047799 (9)	26.44	
			130	129.9035089 (11)	4.08	
			131	130.9050828 (18)	21.18	
			132	131.9041546 (15)	26.89	
			134	133.9053945 (9)	10.44	
			136	135.907220 (8)	8.87	
55	Cs	132.90545 (2)	133	132.905477 (3)	100	
56	Ba	137.327 (7)	130	129.906311 (7)	0.106	
			132	131.905056 (3)	0.101	
			134	133.904504 (3)	2.417	
			135	134.905684 (3)	6.592	
			136	135.904571 (3)	7.854	
			137	136.905822 (3)	11.232	
			138	137.905242 (3)	71.698	

Table 1 (*continued*): Naturally occurring nuclides.

Atomic no. Z	Element	Atomic weight[a] (u)		Mass no. A	Atomic mass[b] (u)		isotopic composition[c] (%)	Notes*
57	La	138.9055	(2)	138	137.907108	(4)	0.090	e, g 1.05×10^{11} a
				139	138.906349	(4)	99.910	
58	Ce	140.116	(1)	136	135.907140	(50)	0.185	e
				138	137.905986	(11)	0.251	
				140	139.905435	(3)	88.450	
				142	141.909241	(4)	11.114	$g > 5 \times 10^{16}$ a
59	Pr	140.90765	(2)	141	140.907648	(3)	100	
60	Nd	144.24	(3)	142	141.907719	(3)	27.2	e
				143	142.909810	(3)	12.2	
				144	143.910083	(3)	23.8	g 2.1×10^{15} a
				145	144.912569	(3)	8.3	
				146	145.913113	(3)	17.2	
				148	147.916889	(4)	5.7	
				150	149.920887	(4)	5.6	
61	Pm							
62	Sm	150.36	(3)	144	143.911996	(4)	3.07	e
				147	146.914894	(3)	14.99	g 1.0×10^{11} a
				148	147.914818	(3)	11.24	g 8×10^{15} a
				149	148.917180	(3)	13.82	$g > 1 \times 10^{16}$ a
				150	149.917272	(3)	7.38	
				152	151.919729	(3)	26.75	
				154	153.922206	(3)	22.75	
63	Eu	151.964	(1)	151	150.919846	(3)	47.81	e
				153	152.921227	(3)	52.19	
64	Gd	157.25	(3)	152	151.919789	(3)	0.20	e, g 1.1×0^{14} a
				154	153.920862	(3)	2.18	

Table 1 (*continued*): Naturally occurring nuclides.

Atomic no. Z	Element	Atomic weight[a] (u)	Mass no. A	Atomic mass[b] (u)	isotopic composition[c] (%)	Notes*
64	Gd (*continued…*)		155	154.922619 (3)	14.80	
			156	155.922120 (3)	20.47	
			157	156.923957 (3)	15.65	
			158	157.924101 (3)	24.84	
			160	159.927051 (3)	21.86	
65	Tb	158.92534 (2)	159	158.925343 (3)	100	
66	Dy	162.500 (1)	156	155.924278 (7)	0.06	e, g 2 × 10^14 a
			158	157.924405 (4)	0.10	
			160	159.925194 (3)	2.34	
			161	160.926930 (3)	18.91	
			162	161.926795 (3)	25.51	
			163	162.928728 (3)	24.90	
			164	163.929171 (3)	28.18	
67	Ho	164.93032 (2)	165	164.930319 (3)	100	
68	Er	167.259 (3)	162	161.928775 (4)	0.14	e
			164	163.929197 (4)	1.61	
			166	165.930290 (3)	33.61	
			167	166.932046 (3)	22.93	
			168	167.932368 (3)	26.78	
			170	169.935461 (3)	14.93	
69	Tm	168.93421 (2)	169	168.934211 (3)	100	
70	Yb	173.04 (3)	168	167.933895 (5)	0.13	e
			170	169.934759 (3)	3.04	
			171	170.936323 (3)	14.28	
			172	171.936378 (3)	21.83	
			173	172.938207 (3)	16.13	

Table 1 (*continued*): Naturally occurring nuclides.

Atomic no. Z	Element	Atomic weight[a] (u)		Mass no. A	Atomic mass[b] (u)		isotopic composition[c] (%)	Notes*
70	Yb (*continued...*)			174	173.938858	(3)	31.83	
				176	175.942569	(3)	12.76	
71	Lu	174.967	(1)	175	174.9407682	(28)	97.41	e
				176	175.9426827	(28)	2.59	g 2.7×10^{10} a
72	Hf	178.49	(2)	174	173.940042	(4)	0.16	g 2×10^{13} a
				176	175.941403	(3)	5.26	
				177	176.9432204	(27)	18.60	
				178	177.9436981	(27)	27.28	
				179	178.9458154	(27)	13.62	
				180	179.9465488	(27)	35.08	
73	Ta	180.9479	(1)	180	179.947466	(3)	0.012	$g > 1.6 \times 10^{13}$ a
				181	180.947996	(3)	99.988	
74	W	183.84	(1)	180	179.946706	(5)	0.12	
				182	181.948205	(3)	26.50	
				183	182.9502242	(30)	14.31	
				184	183.9509323	(30)	30.64	
				186	185.954362	(3)	28.43	
75	Re	186.207	(1)	185	184.952955	(3)	37.40	
				187	186.9557505	(30)	62.60	g 5×10^{10} a
76	Os	190.23	(3)	184	183.952491	(3)	0.02	e
				186	185.953838	(3)	1.59	
				187	186.9557476	(30)	1.96	
				188	187.9558357	(30)	13.24	
				189	188.958145	(3)	16.15	
				190	189.958445	(3)	26.26	
				192	191.961479	(4)	40.78	

Table 1 (*continued*): Naturally occurring nuclides.

Atomic no. Z	Element	Atomic weight^a (u)		Mass no. A	Atomic mass^b (u)		isotopic composition^c (%)	Notes*
77	Ir	192.217	(3)	191	190.960591	(3)	37.3	
				193	192.962923	(3)	62.7	
78	Pt	195.078	(2)	190	189.959930	(7)	0.014	$g\ 7 \times 10^{11}$ a
				192	191.961035	(4)	0.782	
				194	193.962663	(3)	32.967	
				195	194.964774	(3)	33.832	
				196	195.964934	(3)	25.242	
				198	197.967875	(5)	7.163	
79	Au	196.96655	(2)	197	196.966551	(3)	100	
80	Hg	200.59	(2)	196	195.965814	(4)	0.15	
				198	197.966752	(3)	9.97	
				199	198.968262	(3)	16.87	
				200	199.968309	(3)	23.10	
				201	200.970285	(3)	13.18	
				202	201.970625	(3)	29.86	
				204	203.973475	(3)	6.87	
81	Tl	204.3833	(2)	203	202.972329	(3)	29.524	
				205	204.974412	(3)	70.476	
82	Pb	207.2	(1)	204	203.973028	(3)	1.4	$d, e, g\ 1.4 \times 10^{17}$ a
				206	205.974449	(3)	24.1	
				207	206.975880	(3)	22.1	
				208	207.976636	(3)	52.4	
83	Bi	208.98038	(2)	209	208.980384	(3)	100	
84	Po							
85	At							

Table 1 *(continued)*: Naturally occurring nuclides.

Atomic no.		Atomic weight[a] (u)		Mass no. A	Atomic mass[b] (u)		isotopic composition[c] (%)	Notes*
Z	Element							
86	Rn							
87	Fr							
88	Ra							
89	Ac							
90	Th	232.0381	(1)	232	232.0380495	(22)	100	e, g 1.4 × 10^{10} a
91	Pa							
92	U	238.02891	(3)	234	234.0409447	(22)	0.0055	e, f, g 2.44 × 10^5 a
				235	235.0439222	(21)	0.7200	g 7.04 × 10^8 a
				238	238.0507835	(22)	99.2745	g 4.47 × 10^9 a

* The general symbol for the year is a.

[a] Values are those recommended by the Commission on Atomic Weights and Isotopic Abundances, Inorganic Chemistry Division, IUPAC, in *Pure Appl. Chem.* **75**, 1107–1122 (2003). Uncertainty is in parentheses.

[b] Values are given by A. H. Wapstra and G. Audi, *Nucl. Phys.* **A565**, 1 (1993). Uncertainty is in parentheses.

[c] Values are those recommended by the Commission on Atomic Weights and Isotopic Abundances, IUPAC, in *Pure Appl. Chem.* **70**, 217–236 (1998).

[d] Variations in isotopic composition prevent a more precise atomic weight being given.

[e] Specimens having anomalous isotopic composition are known in which the atomic weight may exceed the uncertainties given.

[f] Commercially available element may vary in isotopic composition for inadvertent or undisclosed reasons.

[g] Radioactive isotope; half-life is given.

suggests that odd-neutron configurations are similar to odd-proton ones and that the nuclear force is charge independent. This is further supported by the tendency toward equal numbers of neutrons and protons in light nuclei.

Additionally, the shell nature of the nuclear force is reflected in the incidence of stable nuclides. The existence of a large number of stable isotopes is consistent with a particularly stable arrangement of protons. For example, tin, with $Z = 50$, has the largest number of stable isotopes. Moreover, for a particular isotope having a magic number of neutrons, the isotopic abundance will be larger than would otherwise be expected. In barium, which has seven stable isotopes, the heaviest one, ^{138}Ba, for which $N = 82$, has a relative abundance of 71.9%.

Examination of Table 1 shows that there is a stable nuclide for every value of $Z \leq 83$, except for $Z = 43$ (Tc) and for $Z = 61$ (Pm). At least one stable nuclide occurs for each value of $A \leq 209$ except for $A = 5$ and $A = 8$.

In general, even-Z elements have many more isotopes than their nearby odd-Z neighbors. This is especially evident above $A = 16$. Only in one case ($Z = 19$, K) throughout the entire table does an odd-Z element have more than two isotopes. By contrast, among the even-Z elements, only Be ($Z = 4$) exists in a single stable form, and all elements from oxygen and above ($Z \geq 8$) have at least three stable isotopes except the heavy radioactive element Th ($Z = 90$).

Mattauch's rule, reflecting the properties of β decay, excludes the existence of stable isobars (nuclides with the same value of A) whose atomic numbers differ by unity. In each of the ten groups of naturally occurring adjacent isobars, one has been found to be radioactive, as indicated in Table 1.

Of the 287 naturally occurring species appearing in the table, 23 are now known to be radioactive, decaying with long half-lives, comparable at least to the age of the earth (4.5×10^9 years). Of these, seven have half-lives between 10^{11} years and 10^{15} years, whereas seven have half-lives from 10^{15} years to 10^{17} years.

Natural Isotopic Abundances

The relative amounts of various elements have been seen to vary widely, depending on the location in the solar system. However, for many elements the relative isotopic abundances seem to be remarkably constant. Moreover, for these elements the same abundances are found in terrestrial, lunar, and meteoritic samples. Accordingly, it is presumed that the observed relative abundances reflect the process by which the elements were originally formed in the solar system. For this reason the reproduction of the observed isotopic abundances by any proposed mechanism of nucleosynthesis constitutes an important requirement of the theory.

Elements in which variations in isotopic abundances do occur are therefore of special interest. Such changes may take place as a result of the very small differences in the physical properties of isotopes that are reflected in physical and chemical processes. Fractionation, or the alteration of relative isotopic abundances in this way, is invariably small for each such process. By contrast, nuclear transformations involving particular nuclides may result in dramatically altered abundances. The latter form the basis of what is sometimes called nuclear geology.

As indicated in Table I, the half-lives for the decays 40K to ^{40}Ar (and ^{40}Ca), ^{87}Rb to ^{87}Sr, ^{232}Th to ^{208}Pb, ^{235}U to ^{207}Pb, and ^{238}U to ^{206}Pb are well known and are appropriate to the geological time scale. In each case the parent nuclide is present in rocks of geological interest and decays to form a particular stable daughter isotope whose abundance may be determined relative to other isotopes of that element. We can thus identify the amount of the daughter isotope that is a product of the decay and deduce from it the time interval since the mineralization of the sample.

The formation of ^{3}He, the lighter and very rare isotope of helium, is believed to take place in rocks by the nuclear reactions

$$^6\text{Li}(n,\alpha)^3\text{H}\,, \qquad ^3\text{H}(\beta^-)^3\text{He}\,,$$

and in the atmosphere by

$$^{14}\text{N}(n,^{12}\text{C})^3\text{H or } ^{14}\text{N}(n,3\alpha)^3\text{H and } ^3\text{H}\,(\beta^-)^3\text{He}\,.$$

In rocks the energetic neutrons come from the spontaneous fission of ^{238}U and from a variety of (α,n) reactions on various elements; in the atmosphere the neutrons are present in the cosmic rays. Variations in the relative abundance of 3He reflects both the formation process and the retention of the helium. Values range widely from the one shown in Table 1 with observations as large as 7.7%.

A third example, in which natural abundances of isotopes have been drastically altered by nuclear reactions, is the natural nuclear reactor at the Oklo quarry in Gabon. The structure of the geological formation was such that natural uranium was sustained in a critical reaction for a significant length of time. The ore is correspondingly found to be anomalously depleted in ^{235}U, the isotope that is effectively responsible for slow-neutron-induced fission. Also, as a result of fission, anomalous isotopic abundances have been observed for Kr, Zr, Mo, Ru, Pd, Ag, Sn, Sb, Te, Xe, Ce, Nd, Sm, Eu, and Gd.

Similarly, variations in the abundances of Kr and Xe can be related to the accumulation of fission-product gases in uranium or thorium minerals arising from both neutron-induced and spontaneous fission.

Among the processes based on the differences in physical properties between isotopes (apart from nuclear properties) that may lead to isotopic fractionation, the most significant are diffusion, evaporation, and chemical exchange reactions. These three processes, which depend primarily on the relative differences in mass, are most pronounced among the light elements, especially hydrogen, carbon, nitrogen, and oxygen.

The rate of diffusion of a gas through a porous membrane is inversely proportional to the square root of the molecular weight (Graham's law). In a more elaborate geological setting, considerable variation in the ^{14}N/^{15}N ratio has been observed in natural gas fields and related to crude oils as a result of molecular flow through porous rock and surface diffusion.

In evaporation, the vapor is enriched in lighter isotopes whereas a corresponding enrichment in heavier isotopes occurs in the residual liquid. This has been observed in the ^{2}H/^{1}H and ^{18}O/^{16}O ratios for precipitation compared to ocean water and to isolated bodies of water from which extensive evaporation has taken place.

In chemical exchange reactions, the differences in molecular vibrational energies, moments of inertia, etc., of structurally similar molecules under equilibrium conditions lead to fractionation. For example, carbon is involved in the carbonate–atmospheric-CO_2 exchange reaction

$$^{13}CO_2 + {}^{12}CO_3^{2-} \rightleftharpoons {}^{12}CO_2 + {}^{13}CO_3^{2-}\,,$$

for which the equilibrium constant differs somewhat from unity and is temperature dependent. By making use of the empirical values of the equilibrium constant as a function of temperature and of the measured $^{13}C/^{12}C$ ratios from $CaCO_3$ in the shells of marine organisms, it is possible to determine paleo-temperatures.

Both carbon and sulfur participate in reaction cycles in which biological processes are involved, and the isotopic fractionation in these processes has been studied very extensively.

For carbon the $^{13}C/^{12}C$ ratio can be used to distinguish materials originating as limestone carbon, organic carbon of marine origin, and carbon from land plants. These variations are presumed to be determined by the influence of the carbon isotopic exchange reaction for the limestone, by the preferential assimilation of ^{12}C in photosynthesis by both marine and land plants, and by kinetic effects for the various reactions involving carbon in the atmosphere.

The $^{32}S/^{34}S$ ratio determined in meteoritic samples has been found to be remarkably invariant and is presumed to represent an average value for all terrestrial material. However, individual terrestrial deposits are found to vary widely, primarily as a result of the action of bacteria that reduce sulfate to produce H_2S or free sulfur, either of which is depleted in ^{34}S with a corresponding enrichment of ^{34}S in the sulfate.

Separated Isotopes

Inasmuch as nuclear structure and properties are unique to particular nuclides, there are many experiments in nuclear physics where it is desirable or essential to use material enriched in, or consisting solely of, a particular isotope. Such a case would be the studies of a reaction on a particular nucleus. Similarly, nuclear properties determine the use of certain isotopes in reactor applications where large quantities of the material are required, e. g., deuterium in heavy-water-moderated natural uranium reactors, or fuel rods enriched in ^{235}U for light-water-moderated reactors.

Stable isotopes may be used to examine biochemical reactions by the introduction of material with an abnormal isotopic composition that can be monitored at subsequent stages. The method is suitable for reactions involving oxygen and nitrogen where radioactive isotopes suitable for tracer work do not exist.

A variety of techniques by which differing degrees of isotope separation are achieved are summarized here. The processes of gaseous diffusion, evaporation, and chemical exchange, mentioned in the previous section, have been exploited on a large scale. Other processes in which separation takes place include the following:

1. *Centrifuging*: The sample in a centrifuge experiences an effective force that depends on the mass; hence the separation factor depends on the ratio of the masses.

2. *Electrolysis*: When a water solution of NaOH has been electrolyzed, the residual electrolyte is enriched in ^2H.

3. *Electromagnetic separation*: Mass spectrometers with ion sources capable of producing large elemental ion beams may be used to achieve high degrees of separation for small quantities of virtually any isotope.

4. *Laser separation*: The optical spectra of atoms and molecules depend on the mass, volume, and shape of the nuclei involved. Laser light may be used to excite and then ionize a given isotope. The ions are then separated from residual material by an electric field.

Atomic Weight

Current values of the atomic weights of the elements as recommended by the Commission on Atomic Weights of the Inorganic Chemistry Division of IUPAC are given in Table 1. The determination of isotopic abundances by mass spectroscopic methods combined with the very precisely known atomic masses usually yields the most precise values of atomic weights. In certain unfavorable cases, however, chemical methods are of comparable precision, as described in the detailed IUPAC report on atomic weights of the elements (1987).

In Table 1, attention is drawn to the fact that there are commercially available materials whose isotopic composition has been altered from that of the natural material. Materials that vary in isotopic abundances according to the source are so indicated in the table.

See also: Atomic Spectroscopy; Atoms; Elements; Geochronology; Isotope Effects; Isotope Separation; Nuclear Properties; Radioactivity.

Bibliography

F. W. Aston, *Mass Spectra and Isotopes*. Edward Arnold, London, 1942. (I)

H. E. Duckworth, R. C. Barber, and V. S. Venkatasubramanian, *Mass Spectroscopy*, 2nd ed. Cambridge University Press, Cambridge, 1986. (A)

IUPAC Inorganic Division, Commission on Atomic Weights and Isotopic Abundances, "Element by Element Review of Their Atomic Weights", *Pure Appl. Chem.* **56**, 695 (1984). (A)

IUPAC Inorganic Division, Commission on Atomic Weights and Isotopic Abundances, "Atomic Weights of the Elements 1987", *Pure Appl. Chem.* **60**, 841 (1988). (I)

K. Rankama, *Isotope Geology*. McGraw–Hill, New York, 1954. (I)

A. Romer, *Radiochemistry and the Discovery of Isotopes*. Dover, New York, 1970. (E)

M. A. Preston, *Physics of the Nucleus*. Addison-Wesley, Reading, MA, 1962. (A)

F. W. Walker, D. G. Miller, and F. Feiner, *Chart of the Nuclides*, 13th ed. Available from General Electric Company, Nuclear Energy Operations, 175 Curtner Ave., M/C 684, San Jose, CA 95125. (E, I)

F. A. White and G. M. Wood, *Mass Spectrometry, Applications in Science and Engineering*. Wiley, New York, 1986. (A)

Jahn–Teller Effect

B. R. Judd

The Jahn–Teller effect refers to the tendency of molecular systems to distort when electronic degeneracy is present. The electronic energy of such a system almost always depends linearly on a displacement coordinate d of some kind, while the elastic energy resisting the distortion is invariably a quadratic function of d. Consequently, the total energy of the system can be lowered if a distortion takes place. There are two exceptions to this general rule. Linear molecules (like CO_2) do not exhibit the required linearity of the electronic energy with respect to d, and may or may not distort. In addition, all molecular systems possessing an odd number of electrons necessarily exhibit Kramers degeneracy, a phenomenon that pairs every quantum-mechanical state with its time-reversed companion. Such twofold degeneracy cannot be split by distortions of the nuclear frame, and thus the inducement to undergo a distortion is absent. The possible existence of the Jahn–Teller effect was pointed out to Teller by Landau in 1934, and a systematic study of the symmetry types for which the effect can be expected was carried out by Jahn (in collaboration with Teller) and presented in 1936. In that form it is referred to today as the "static" Jahn–Teller effect. It is considered responsible for the comparatively low symmetry of the sites of many transition-metal ions in crystals. The paradoxical result that a Hamiltonian with a definite symmetry may lead to systems with lower symmetry is resolved by noting that, in principle, the many equivalent systems of lower symmetry are connected in virtue of quantum-mechanical tunneling. Given enough time (perhaps an eternity), all systems will be represented in a way consistent with the original symmetry. The apparent preference of Nature for broken symmetries has been commented on in recent years by Teller in the wider context of elementary particles.

Because the electronic energy depends on d, there is a coupling between the electronic states and vibrations. Thus the absorption line of an electron trapped in an oxygen vacancy in CaO, which corresponds to an excitation from an s state to a p state, exhibits a broad absorption band on its low-wavelength side. Some of the structure can be associated with the excitation of vibrational modes of the octahedron of calcium ions surrounding the defect site.

Encyclopedia of Physics, Third Edition. Edited by George L. Trigg and Rita G. Lerner
Copyright © 2005 WILEY-VCH Verlag GmbH & Co. KGaA, Weinheim
ISBN: 3-527-40554-2

This is referred to as the "dynamic" Jahn–Teller effect. It is often accompanied by a partial quenching of electronic properties such as the spin–orbit coupling, a phenomenon known as the Ham effect after its discoverer.

The static Jahn–Teller effect provides good examples of phase differences between physically identical states of the same system (referred to as Berry's phase). In the theoretical limit of strong coupling between the p orbital and the displacements of the calcium ions in the example mentioned above, it can be shown that a rotation by 180° of the p orbital (which interchanges the positive and negative lobes) brings the calcium ions back to their original positions. The system is physically unchanged, but the rotation produces a phase reversal in the wave function of the entire system.

See also: Geometric Quantum Phase; Symmetry Breaking, Spontaneous; Tunneling.

Bibliography

I. B. Bersuker, *The Jahn–Teller Effect and Vibronic Interactions in Modern Chemistry.* Plenum, New York, 1984. (I)

C. C. Chancey and M. C. M. O'Brien, *The Jahn–Teller Effect in C_{60} and other Icosahedral Complexes.* Princeton University Press, Princeton, New Jersey, 1997. (A)

Yu. E. Perlin and M. Wagner (eds.), *The Dynamical Jahn–Teller Effect in Localized Systems.* North-Holland, Amsterdam, 1984. (A)

E. Teller, "The Jahn–Teller Effect – Its History and Applicability", in *Group-Theoretical Methods in Physics* (L. L. Boyle and A. P. Cracknell, eds.). North-Holland, Amsterdam, 1982. (E)

Josephson Effects

D. J. Scalapino

Brian Josephson won the Nobel Prize in 1973 for theoretical work he carried out on the properties of two superconducting metals separated by a thin insulating oxide layer. This structure, a sandwich in which the two superconducting films are the bread with the oxide analogous to a very thin slice of cheese separating the metal films, is now often called a Josephson tunnel junction, and the phenomena associated with the superconducting current flow between the metals are called the Josephson effects. Figure 1 shows a typical junction configuration. Here two cross strips of Pb are separated by a thin oxide layer 10–20 Å in thickness.

In order to understand the Josephson phenomena it is essential to realize that the superconducting state exhibits quantum-mechanical properties on a macroscopic scale. In a superconducting metal, pairs of electrons are bound together by an attractive interaction mediated by the motion of the ionic lattice. According to the basic laws of quantum mechanics, the motion of the center of mass of the pairs is described by a wave. Now, ordinarily the quantum-mechanical waves of the many electrons which make up a metal differ from one another, so that on a macroscopic scale we are not directly aware of the phase properties of the electron wave functions. However, in the superconducting state, the state of lowest free energy cor-

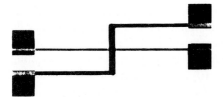

Fig. 1: A typical Josephson tunnel junction formed by two overlapping Pb strips separated by an oxide layer. The glass slide on which this junction is mounted is about an inch square.

responds to a situation in which all of the pairs have the same center-of-mass wave function. With this vast number of electron pairs having the same wave structure, it becomes possible to observe the relative change in phase $\Delta\varphi$ of the center-of-mass wave function.

The first Josephson equation states that if there is a phase difference $\Delta\varphi$ between the pair waves in the superconductors on the two sides of a Josephson junction, pairs will tunnel through the oxide giving rise to a supercurrent

$$I = I_1 \sin\Delta\varphi \,. \tag{1}$$

This is the dc Josephson effect. It predicts, for example, that if a phase difference $\Delta\varphi = \pi/2$ exists between the superconductors on either side of a Josephson junction, then a current I_1 will flow across the junction in the absence of a voltage difference. Naturally, one needs leads on the junction to supply and remove this current, or excess charge will build up giving rise to a voltage which produces further effects which we will discuss below.

Suppose leads from a current source are attached across the junction shown in Fig. 1 and another set of high-impedance leads are used to monitor the voltage V across the junction. The low-temperature $I(V)$ characteristic of the junction will look like that shown in Fig. 2. At low temperatures, the current I passing through the junction can be increased at zero voltage to a critical value beyond which the junction will switch (dashed line of Fig. 2) to a state

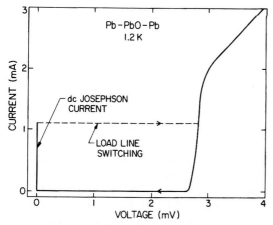

Fig. 2: The $I(V)$ characteristic of a Pb–PbO–Pb Josephson junction showing the dc Josephson current and the quasiparticle current onsetting at a voltage $2\Delta/e$.

with a finite voltage. Then as the current is reduced, the voltage will decrease toward zero as shown by the arrow on the solid line in Fig. 2. The zero-voltage current corresponds to the dc Josephson effect and was first reported by P. W. Anderson and J. M. Rowell. The magnitude of the critical current $I_1(T)$ is temperature dependent, vanishing above the superconducting transition temperature T_c and approaching a constant $I_1(0)$ at low temperatures. For a junction composed of the same type of superconducting metal on both sides, $I_1(0)$ is equal to the current which is carried by the junction in the normal (nonsuperconducting) state at a voltage $\pi\Delta/2e$, where Δ is the superconducting energy gap and $I_1(T)/I_1(0) = \tanh(\Delta/2kT)$. A typical Josephson junction will have a current density of order amperes per square centimeter while a very strongly coupled junction can have a current density several hundred times larger.

If a voltage difference V exists across the junction, quantum mechanics implies that the relative difference $\Delta\varphi$ in the pair phase between the two superconductors changes at a rate set by $2eV/\hbar$:

$$\frac{\partial}{\partial t}\Delta\varphi = \frac{2eV}{\hbar} \, . \tag{2}$$

Here e is the electron charge, $\hbar$ is Planck's constant divided by 2π, and the 2 arises from the fact that we are dealing with electron pairs. If the voltage is constant at a value V_0 then the phase difference will increase with time, $\Delta\varphi(t) = 2eV_0t/\hbar + \Delta\varphi(0)$, and a current $I = I_1\sin(2eV_0t/\hbar + \Delta\varphi)$ will oscillate back and forth across the oxide at a frequency $f = 2eV_0/h$. Using the values of e and h, one finds that this frequency is approximately $500\,\text{MHz}/\mu\text{V}$. This is the ac Josephson effect.

In order to observe the ac Josephson effect, Josephson suggested applying an rf field to the junction and looking for current steps in the dc $I(V)$ characteristic. With both a dc bias voltage V_0 and an rf bias voltage $V_1\cos\omega t$ across the junction,

$$\frac{\partial\Delta\varphi}{\partial t} = \frac{2e}{\hbar}(V_0 + V_1\cos\omega t)$$

and

$$I = I_1\sin\left(\frac{2e}{\hbar}V_0t + \frac{2eV_1}{\hbar\omega}\sin\omega t + \Delta\varphi(0)\right) \, .$$

This current has a dc step whenever $2eV_0 = n\hbar\omega$. Figure 3 shows this step structure for a Josephson junction irradiated by a microwave field with frequency $f \sim 12\,\text{GHz}$. With the approximate value of $500\,\text{MHz}/\mu\text{V}$, this implies a step spacing of order $25\,\mu\text{V}$. The ac Josephson effect was first observed in this way by S. Shapiro.

The frequency of the ac Josephson current can be measured with high precision. If, in addition, a careful measurement of the applied voltage V_0, is made, the ratio of the fundamental constants $2e/h$ can be obtained. Measurements of this type were carried out by Taylor, Parker, and Langenberg, who found that $2e/h = 4.835976(12) \times 10^{14}\,\text{Hz}/\text{V}_{\text{NBS}}$. Here V_{NBS} is the NBS volt. Given a value of $2e/h$ and the ability to measure frequency, the Josephson relation $f = 2eV_0/h$ can be used to determine V_0. This type of measurement is, in fact, used to define the present NBS voltage standard.

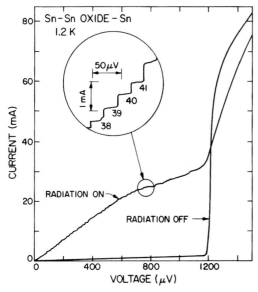

Fig. 3: The $I(V)$ characteristic of a Josephson junction in the presence of a microwave field of frequency f shows vertical steps separated by $hf/2e$.

The final Josephson relation concerns the behavior of a junction in a magnetic field. If a magnetic field H is applied parallel to the oxide layer, it again follows from quantum mechanics that the relative pair phase difference between the superconducting film on opposite sides of the junction varies with the position in the plane of the junction. Specifically, the change of $\Delta\varphi$ with distance is greatest as one moves in the plane of the junction perpendicular to the magnetic field. For identical superconducting films which are thick compared to the penetration depth λ of the magnetic field into the superconducting film, the change of $\Delta\varphi$ over a distance Δx perpendicular to the field H is $(2e/\hbar c)H(2\lambda+l)\Delta x$. Here l is the thickness of the oxide and $\hbar$ is Planck's constant divided by 2π. This means that the current density varies with position. For a junction lying in the $x-y$ plane with a uniform field H in the oxide parallel to the y axis, the current density is

$$ j(x) = j_1 \sin\left(\frac{2eH(2\lambda+l)x}{\hbar c} + \Delta\varphi(0) \right) , $$

with $j_1 = I_1/A$, where A is the area of the junction.

This effect of a magnetic field H on the current density is illustrated in Fig. 4. In part a the magnetic field is zero, and the current density is uniform. As the field is increased, parts b–d, the current density oscillates in a direction perpendicular to H. The value H_0 is just such that $2eH_0(2\lambda+l)\cdot L/\hbar c = 2\pi$, where L is the length of the junction perpendicular to the field direction. This corresponds to one flux unit $hc/2e$ passing through the junction. When H is equal to H_0, Fig. 4c, the current density makes one complete oscillation and no net current flows across the junction. In Fig. 4d the field has been increased to $3H_0/2$ so that $3/2$ wavelengths fit into the junction. In this case a net current smaller than that shown for Figures a or b can flow across the junction.

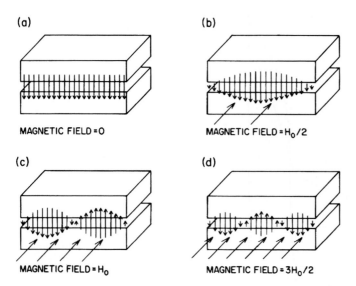

Fig. 4: Effect of a magnetic field on the current density in a Josephson junction.

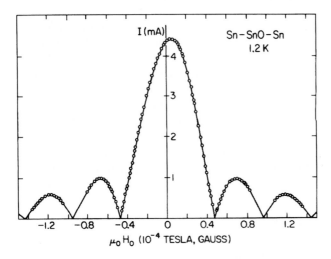

Fig. 5: The critical Josephson dc current versus magnetic field.

A plot showing how the critical junction dc current depends on H is shown in Fig. 5. This single-slit-like interference pattern, first observed by J. Rowell, shows in a vivid manner the de Broglie wave properties of electron pairs in superconductors. Thus, by monitoring the Josephson current, it is possible to tell when $(2\lambda + l)LH/(hc/2e) = 2\pi n$. The flux quantum $hc/2e$ corresponds to approximately 2×10^{-7} G/cm^2. Sensitive instruments can measure changes of parts in a thousand in the maximum Josephson current. Thus, Josephson junctions provide the means for making flux detectors with sensitivities of 10^{-10} G/cm^2.

Besides their use as magnetometers and voltage standards, Josephson junctions are now used in a variety of devices such as picovoltmeters and high-frequency electromagnetic detectors and mixers. Perhaps their most important applications will be as switching elements in which the junction switches from a zero-voltage current-carrying state to a finite-voltage current-carrying state (see the dashed line of Fig. 2) as the critical current I_1 is exceeded. This switching can occur in times of order tens of picoseconds and with power dissipation less than microwatts. While equivalently fast semiconductor switching elements can achieve similar switching times, their power dissipation is many orders of magnitude larger. Thus the density with which one can pack Josephson junctions can be much higher, reducing the time for electromagnetic signals to pass between them. At present, work on high-speed microcircuit technology using Josephson junctions is an active area of research.

Acknowledgement
The author thanks the University of Pennsylvania tunneling group for the experimental figures used. Support from the Office of Naval Research and the J. S. Guggenheim Foundation is gratefully acknowledged.

See also: Superconductive Devices; Superconductivity Theory.

Bibliography
P. W. Anderson, in *Lectures on the Many-Body Problems* (E. R. Caianello, ed.). Academic Press, New York, 1964.
P. W. Anderson, *Physics Today*, (November, 1970).
B. D. Josephson, *Phys. Lett.* **1**, 251 (1962).
D. N. Langenberg, D. J. Scalapino, and B. N. Taylor, *Scientific American*, **214**(5), 30 (May, 1966).

Kepler's Laws

E. H. Kerner[†]

One of the giant decisive strides on the way toward that watershed of physics that is the Newtonian world view was taken by Johannes Kepler (1571–1630) in his *Astronomia Nova* of 1609, containing (besides much else) his first two laws:

I. the planets circulate around the sun in elliptical orbits with the sun at one focus;

II. the radius vector from the sun to a planet sweeps out equal areas in equal times;

and in his *Harmonices Mundi* of 1619, stating the third law,

III. "The ratio which exists between the periodic times of any two planets is precisely the ratio of the $\frac{3}{2}$th power of the mean distances"

(i. e., the ratio of the $\frac{3}{2}$ power of the two semimajor elliptical axes).

The laws were, with great and skillful computational labor, drawn out of the large body of pretelescopic planetary observations of the master observer Tycho Brahe (1546–1601) with whom the theoretically bent Kepler worked in Prague for a time prior to Tycho's death, continuing thereafter for years to mine Tycho's data. It is remarkable that Tycho right off gave Kepler the problem of Mars' motion, notoriously intractable since Hipparchus (and thought at first by Kepler to be manageable in two weeks work), but, as it happens, with an orbit far enough from circular to be clearly discernable as such from the accuracy (by eye observation) then available. Though the Copernican controversy was in full furore, Kepler adopted the heliocentric position definitively. However, the Platonic conception of the "perfectness" of the sphere and circle completely dominated all thinking, including Kepler's; to break this domination was to be a principal achievement. After some 70 attempts to fit Mars' orbit to a circle, employing Tychonian observations-in-opposition (for which Sun–Earth–Mars are well aligned and for which uncertainties of Earth's orbit are then minimized), Kepler's best fit to selected observations was within 2 min of arc (or about a sixteenth of the moon's angular breadth) of them,

---

[†]deceased

Encyclopedia of Physics, Third Edition. Edited by George L. Trigg and Rita G. Lerner
Copyright © 2005 WILEY-VCH Verlag GmbH & Co. KGaA, Weinheim
ISBN: 3-527-40554-2

viz., within the limit of direct eye judgment. But then still other observations-in-opposition were off by as much as 8 min. These were fateful minutes: for Kepler was sure that Tycho, "that most diligent observer", could not have been that much in error: and he had perforce to give up circles, later saying "... These eight minutes alone have pointed the way to the reformation of the whole of astronomy".

It took another year for Kepler to realize that the Martian egg in the sky was indeed no mere oval but exactly an ellipse; he had gotten the equation of his oval but did not recognize it for what it was (the times were before analytic geometry); then he cast it aside, thought independently about ellipses after trials with other figures, and thereto reinvented his self-same oval equation! With charming candor he wrote, "Why should I mince my words? The truth of Nature, which I had rejected and chased away, returned by stealth through the back door ...".

As regards the third law, which within 70 years came to be the critical linkup to Newton's law of gravitation and equations of motion, it was a dream of Kepler's for 25 years to pierce through the (to him) tantalizing data on orbit sizes and periods to the simple rule he felt with intuitive certainty must be controlling them.

The state of Tycho's art may have been providential in another way. Had Tycho had a telescope, his observations might have been too good, in that deviations from Keplerian ellipses due to perturbations from other planets might well have come into range, blurring and bothering possibly fatally the simplicity of the coarser view. Similar good fortune perhaps attended the study centuries later of another Kepler-system-in-little, that of a planetary electron wheeling about a heavy attracting proton, when Balmer in 1885 discerned from the hydrogen-spectrum data of Ångström and of Huggins – nicely set between coarseness and fineness – the simple rule for the spectral wavelengths that, stamped decisively with integers, foreshadowed quantum physics.

While uncovering the geometrodynamical peaks of the world-machine, Kepler was also acutely sensitive all along to the question of how the machine could operate as pure clockwork from fundamental universal law applying both terrestrially and cosmically, recognizing the need for a "true doctrine concerning gravity" and for a comprehensive theory of motion. But, wholly in thrall to the Aristotelian (mis-) conception about inertia, by which a body had to be shoved along to sustain even uniform motion (otherwise coasting to rest), his quest after the clockwork had to go unfulfilled. Even so, Kepler had the brilliant physical intuition that the Sun as center of both light and motion produced its "moving effect" on planets so as to be weakened "through spreading from the Sun in the same manner as light"; but then, in thought of spreading in a plane, Kepler adduced an inverse first-power law that was convenient to his Aristotelian theorizing.

On the Newtonian base, the laws of Kepler fall into place swiftly and simply. If we assume for simplicity that the Sun's mass M is so great that it may be taken to be at the center of an inertial frame, a planet of mass m at radius vector r moves according to the equation (G is the universal gravitational constant)

$$m\ddot{\mathbf{r}} = -\frac{GMm}{r^2}\hat{r}$$

(note too the powerful assertion that inertial mass on the left is equal to gravitational mass on the right, exemplifying the principle of equivalence that undergirds the later Einsteinian gravitation). Two conservation laws follow readily, reflecting the rotational invariance and time-translational invariance of the equation of motion:

$$\mathbf{r} \times m\dot{\mathbf{r}} = \mathbf{L} = \text{conserved vector angular momentum,}$$

$$\frac{1}{2}m(\dot{r}^2 + r^2\dot{\theta}^2) - \frac{GMm}{r} = E = \text{conserved energy}$$

(plane polar coordinates r, θ for the position of m). The scalar content of the first is $mr^2\dot{\theta} = L$ or $\frac{1}{2}r(rd\theta) = dS$ (the area swept out in time dt) $= (L/2m)dt$. Hence Kepler's second law (which was actually discovered first) with its physically vivid portrayal that m at close approach to M must be speeding along, while at the far reach of orbit it must be relatively lolling. In the statement of energy conservation, one may introduce $\dot{\theta} = L/mr^2$ and then go to the geometry of the orbit by $dr/dt = (dr/d\theta)\dot{\theta} = (dr/d\theta)(L/mr^2)$ whose differential equation then integrates to the polar equation of an ellipse with focus at M,

$$\frac{1}{r} = \frac{GMm^2}{L^2} + \left(\frac{2mE}{L^2} + \frac{G^2m^4M^2}{L^4}\right)^{1/2} \cos\theta$$

provided $E < 0$. The third law follows from integration of the first, $S = \pi ab = (L/2m)T$ (T is the period and a and b are semimajor and semiminor axes), upon using the geometry of the preceding ellipse for $b = b(a)$, giving $T = 2\pi(GM)^{-1/2}a^{3/2}$.

Keplerian "harmony" in more recent times has been embraced in the perhaps wider term "symmetry", and in the Kepler problem a further symmetry beyond those of space rotations and time translations shows itself in the conserved Runge–Lenz vector

$$\mathbf{M} = \dot{\mathbf{r}} \times \mathbf{L} - GMm\hat{r}.$$

It is orthogonal to $\mathbf{L}$, lying therefore in the plane of motion, and it is an expression of the fact that the Kepler orbits are closed. The higher symmetry of the Kepler problem, with all of E, $\mathbf{L}$, $\mathbf{M}$ conserved, is that of the four-dimensional orthogonal group usually designated as $O(4)$.

The Kepler problem has continued to course like some deep inevitable stream in Nature's ocean or man's mind down through the centuries from Kepler and Newton on celestial planes, to Balmer, Bohr, and Schrödinger at the atomic level, to Einstein and Schwarzschild again cosmically but on a grander geometrorelativistic base for gravitation, to Dirac and Lamb in our own day once again atomistically. Through all the refinements and revolutions of later times, Kepler's laws and Newton's synthesis stand steadfastly as foundation stones.

See also: Conservation Laws; Group Theory in Physics; Newton's Laws.

Bibliography

Alexander L. Fetter and John D. Walecka, *Theoretical Mechanics of Particles and Continua*. McGraw–Hill, New York, 1980.

Arthur Koestler, *The Watershed: A Biography of Johannes Kepler*. Anchor Books, Garden City, New York, 1960.

L. Pars, *Treatise on Analytical Dynamics*. Wiley, New York, 1965.

Kerr Effect, Electro-Optical

R. Hebner

The electro-optic Kerr effect is the name given to a modification of the index of refraction of a material by an electric field E. In general, an electric field induces a difference between two components of the index of refraction, Δn, which can be expressed as

$$\Delta n = \alpha + \beta E + \gamma E^2 + \Lambda . \tag{1}$$

For simplicity, the fact that the elements of this equation are tensors has been neglected and only scalar quantities are considered. If only the quadratic term is nonzero, Eq. (1) describes the Kerr effect. Similarly, if the first-order term is the only one which is nonzero, Eq. (1) describes the Pockels effect. There is not complete symmetry in naming between these electro-optical effects and magneto-optic effects. Analogous to the Pockels effect, the Faraday effect refers to a change in polarization for transmission through a material aligned by a first-order magnetic field. The magneto-optical Kerr effect, however, does not refer to transmission through a material. Rather it refers to a change of polarization upon reflection from a magnetic material in a magnetic field.

For the electro-optic Kerr effect, Eq. (1) can be written as

$$n_{\parallel} - n_{\perp} = \lambda B E^2 , \tag{2}$$

where $n_{\parallel}$ and $n_{\perp}$ are the indices of refraction of the electro-optic material for light polarized parallel and perpendicular to the electric field, respectively; λ is the wavelength of the transmitted light; B is the Kerr coefficient of the material; and E is the applied electric field.

Equation 2 is a convenient way to describe the electro-optical Kerr effect when the light source is a laser, i. e. a monochromatic light source. In this case, λ can usually be represented with sufficient precision by a single number.

The Kerr effect can be modeled as dealing with two different electric fields: one called the orienting field and the other the sensing field. To appreciate the distinction between the two, it is useful to consider, qualitatively, the behavior of a system on a molecular scale. Assume that the molecules in question have some electrical anisotropy, e. g., a dipole moment. If a low-intensity, linearly polarized light beam is passed through the material, no net effect of the anisotropy is observed because the dipoles are randomly oriented. The electric field associated with this light beam is the sensing field. If a second field, with sufficiently high intensity, is applied, a detectable alignment of the dipoles results. This second field is the orienting field. The distinction between the orienting and sensing field is based solely on intensity.

The frequency dependence of the Kerr effect results, at least in part, from the fact that different electrical anisotropies dominate in different frequency ranges. The effect is primarily due to molecular reorientation for orienting fields with temporal variations slow compared to molecular relaxation time and to molecular deformation for higher frequency orienting fields, e. g., high-power laser beams.

Values for Kerr coefficients range from $10^{-21}\,\mathrm{m/V^2}$ for helium gas, to $10^{-12}\,\mathrm{m/V^2}$ for nitrobenzene, to $10^{-7}\,\mathrm{m/V^2}$ for bentonite in water. These depend on temperature, wavelength of the sensing light, and frequency of the orienting field.

The Kerr effect has been used in a variety of applications. It is the basis for a very accurate measurement technique for high-voltage pulses. The Kerr effect has been used to measure the electric field and space-charge distributions in insulating materials. Measurements of the Kerr effect have provided insight into molecular structure and into molecule-molecule interactions in the liquid and gaseous states. Optical shutters and digital light beam deflectors based on the Kerr effect have been constructed using a variety of electro-optic materials.

Bibliography

Elementary
F. A. Jenkins and H. E. White, *Fundamentals of Optics*, 3rd ed. McGraw-Hill, New York, 1976. R. W. Wood, *Physical Optics*. Dover Publications, New York, 1967.

Intermediate
E. Fredericq and C. Houssier, *Electric Dichroism and Electric Birefringence*. Clarendon Press, Oxford, 1973.
R. E. Hebner *et al.*, *Proc. IEEE* **65**, 1524 (1977).
C. G. LeFevre and R. J. W. LeFevre, *Rev. Pure Appl. Chem.* **5**, 261 (1955).
C. T. O'Konski, *Molecular Electro-Optics: Part 1 – Theory and Methods*. Marcel Dekker, New York, 1976.

Kerr Effect, Magneto-Optical

J. F. Dillon, Jr., and R. Hebner

When polarized light is reflected from the surface of a magnetic material, there are changes in polarization state attributable to the magnetization. This phenomenon is called the magneto-optical Kerr effect after its discoverer, the Scottish physicist John Kerr (1824–1907). The magneto-optical Kerr effect refers to a polarization change for a light beam reflected from the surface of a material in a magnetic field, while the electro-optical Kerr effect refers to a similar polarization change, but for a light beam transmitted through a material in an electric field.

Though usually thought of relative to visible light, the magneto-optical Kerr effect may be seen with radiation anywhere in the electromagnetic spectrum. This effect is closely related to the magneto-optical rotation of the axis of linear polarization seen on transmission through a magnetized material (Faraday rotation) and to first-order Raman scattering from magnetic excitations. All three can be attributed to off-diagonal terms in the dielectric or magnetic susceptibility tensors. For instance in the dielectric case, which is dominant for visible light, the tensor ε^0 in the relation $D = \varepsilon_0 \varepsilon^0 E$ has the following form for cubic and isotropic materials (to first order in **M**):

$$\varepsilon = \varepsilon_0 \begin{bmatrix} 1 & -iQM_z & iQM_y \\ iQM_z & 1 & -iQM_x \\ -iQM_y & iQM_x & 1 \end{bmatrix} .$$

In this equation, ε_0 is the dielectric constant in the absence of magnetization and M_i are the components of the magnetization. The complex quantity Q is a magneto-optical parameter

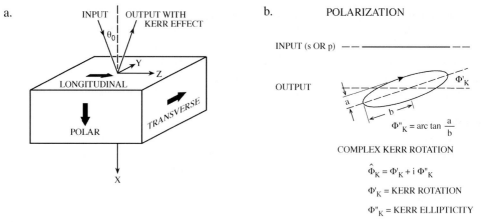

Fig. 1: (a) Magnetization orientations for the three magneto-optical Kerr effects. (b) For longitudinal and polar cases, incident light is linearly polarized with **E** parallel (p) or perpendicular (s) to the plane of incidence. The polarization of the reflected light may be an ellipse with its major axis slightly rotated from the original s- or p-axis. The sketch shows the definitions of the complex Kerr rotation.

which varies with material, frequency, and temperature. A quantum-mechanical theory for Q is based on the differing probabilities for transitions between ground and excited states for the two signs of circular polarization.

The off-diagonal terms are first order in **M**, and thus the effects that arise from them change sign when **M** is reversed. This tensor expresses the fact that, in magnetized materials, a radiation field **E** not parallel to the magnetization can give rise to dielectric polarization perpendicular to both **E** and the static **M**. This new time-varying polarization then radiates, and the net polarization of the radiation field is altered. These first-order magneto-optical effects may be derived by solving the appropriate boundary-value problem using these dielectric and magnetic susceptibility tensors in Maxwell's equations.

Since reflection from a surface generally produces changes in polarization state even without the presence of a magnetic field, the magneto-optical Kerr effect is traditionally defined for three geometries in which nonmagnetic effects play no part. All three are illustrated in Fig. 1a. Light with an angle of incidence Θ_0 from the normal is reflected at an equal angle. These two beams define the plane of incidence. In the absence of a magnetization, light which is linearly polarized with **E** in this plane (p-polarization or direction) or normal to it (s-polarization) is reflected without change of polarization. On the other hand, if the material has a magnetization, the output state may be elliptical, as shown exaggerated in Fig. 1b. The angle between the major axis of this ellipse and the original s- or p-directions is called the Kerr rotation Φ'_K. The shape of the ellipse is given by the Kerr ellipticity Φ''_K, as defined in Fig. 1b. Φ'_K is taken as positive if the major axis of the ellipse appears to have rotated in a counterclockwise sense to an observer looking into the beam. Φ''_K is positive if the tip of optical frequency **E** is seen to go around the ellipse in a counterclockwise sense by such an observer. However, different sign conventions have been used. Together, this rotation and ellipticity are taken to be the real and imaginary components of a complex Kerr rotation, i. e., $\hat{\Phi}_K = \Phi'_K + i\Phi''_K$.

Table 1: Selected values for the magneto-optical Kerr effect.

Material	T/K	λ/nm	Longitudinal[a] p-polarization Φ_p^l/°	Φ_p^u/°	Polar[b] Φ^l/°	Φ^u/°	Transverse[a] $10^2\Delta R_{pp}$
Fe[c]	300	633	0.14	0.07	0.63	−0.47	0.54
EuO[c]	4.2	550	2.35	0.00	1.60	2.01	0.39
Mn Bi[c]	300	633	0.95	0.23	2.47	0.00	2.00
$Y_3Fe_5O_{12}$[c]	300	430	0.001	0.02	0.002	0.05	0.001
U_3As_4[d]	15	2250	–	–	5.90	4.00	–
Amorphous $Tb_{0.21}Fe_{0.79}$[e]	300	600	–	–	0.37	0.30	–

[a] Incident from air at $\theta = 45°$.
[b] Incident from air at $\theta = 0°$.
[c] J. M. Judy, *Ann. NY Acad. Sci.* **189**, 239 (1972).
[d] J. Schoenes and W. Reim, *J. Magn. Mat.* **54–57**, 1371 (1986).
[e] R. Allen and G. A. N. Connell, *J. Appl. Phys.* **54**, 2353 (1982).

The longitudinal (meridional) Kerr effect pertains to the geometry in which the static **M** lies both in the surface and in the plane of incidence. In the dielectric tensor, this situation is described by $M_x = M_y = 0; M_z \neq 0$. Both rotations and ellipticities are different in magnitude and often in sign for incident p- and s-polarizations.

The polar Kerr effect is seen when the magnetization lies normal to the surface. In this case, only M_x is nonzero. Measurements are usually made with light incident normally, so there is no distinction between s- and p-polarizations.

In the transverse (equatorial) geometry (only M_y nonzero), it is found that the reflection of s-polarized light is not changed at all on reversing M_y, but that there are changes in the reflectivity of p-polarized light. The effect can be seen even with unpolarized incident light.

Table 1 shows representative values of the longitudinal $\Phi_K'(p)$ and $\Phi_K''(p)$ for p-polarized light and of the polar Φ_K' and Φ_K''. The last column gives ΔR_{pp}, the fractional change in the intensity of reflected p-polarized light on the reversal of magnetization in the transverse case. Conditions must be specified for every value in Table 1 since all of these quantities vary with the refractive index of the incident medium, the angle of incidence, the temperature, and the wavelength. They can be drastically modified or enhanced by thin covering layers of other materials. Though polar Kerr rotation and ellipticity for Fe are about 0.5°, there are a few materials with much larger values. Usually the polar rotation is much larger than the longitudinal effect.

In magnetics research, the various manifestations of the magneto-optical Kerr effect are used to visualize the distribution of magnetization, i. e., to see magnetic domains, at the surface of a sample. It is often valuable to construct magneto-optical hysteresis loops by measuring rotation or ellipticity as the field is cycled. The effects are used to measure the off-diagonal elements of the dielectric tensor, and thus to obtain valuable spectroscopic information on energy levels associated with magnetization. An exceedingly important technological application of the polar Kerr effect has been developed in recent years, the erasable magneto-optical memory. In these computer memories, information is stored in the orientation of the magnetization in small spots on a magnetic film. These bits of information may be read optically using the polar Kerr effect. The materials now used in this technology are amorphous alloys of transition metals and the rare earths such as the amorphous TbFe show in Table 1. In addition, the magneto-optical Kerr effect has been incorporated as the active element in sensors. These can be used for magnetic field measurement or to non-invasively measure electric current in a circuit by measuring the magnetic field it produces.

Bibliography

R. Carey and E. E. Isaac, *Magnetic Domains and Techniques for Their Observation*. Academic Press, New York, 1966.

R. P. Hunt, *IEEE Trans. Magn.* **MAG-5**, 700 (1969). Theory.

J. H. Judy, *Ann. NY Acad. Sci.* **189**, 239 (1972). Phenomenological theory and Kerr effects of many materials.

M. H. Kryder, *J. Appl. Phys.* **57**, 3913 (1985). Magneto-optical recording technology.

M. R. Parker, *Physica* **86–88B**, 1171 (1977). A review article on magneto-optical Kerr effects.

Kinematics and Kinetics

E. C. G. Sudarshan

Physics is the study of changes and of invariance: In the stability of matter it sees the dynamics of atomic systems, while in the movement of a stream it sees the unfolding of a system by a family of canonical transformations. The simplest changes pertain to *motion*, to changes of configuration. The study of motion, without reference to the causes or effects of motion or the nature of the object moving, is *kinematics*.

When the moving entity has no structure, we refer to a particle, a position as a function of time, a *trajectory*. In general, the position is a vector **r** in three dimensions. The time rate of change of position is the *velocity* **v** and its rate of change is the *acceleration* **a**.

$$\mathbf{a} = \frac{\mathrm{d}}{\mathrm{d}t}\mathbf{v} \; ; \qquad \mathbf{v} = \frac{\mathrm{d}}{\mathrm{d}t}\mathbf{r} \; .$$

For constant acceleration

$$\mathbf{v}(t) \;=\; \mathbf{v}(0) + t\mathbf{a} \; ,$$

$$\mathbf{r}(t) \;=\; \mathbf{r}(0) + t\mathbf{v}_0 + \frac{1}{2}t^2\mathbf{a} \; .$$

The trajectory is a parabola with its axis parallel to the direction of acceleration, suitable for describing terrestrial projectiles and electrically charged particles in constant electric fields.

Motion along a circle of constant radius and with constant speed is accelerated since the direction of velocity changes. If the radius of the circle is R and the speed v, the acceleration is v^2/R directed toward the center. Such orbits describe possible motions of planets and satellites, and of electrically charged particles in constant magnetic fields and in the field of a central electric charge.

A *rigid body* may be pictured as a collection of particles with invariable mutual distances. The motion of a rigid body can then be obtained as the motion of the particles under these constraints. Alternatively, we may use the position of the center of mass and the orientation angles of the rigid body as coordinates. The center of mass moves like a mass point; the changes of the orientation involve *angular velocities*. When the center of mass remains fixed, we have the rigid rotator. More generally, any system of particles subject to any number of constraints between their coordinates can be dealt with in terms of *generalized coordinates*, not all necessarily Cartesian position variables.

So far motion has been treated as secondary to configuration, as a derived quantity. It is possible to view motion as a quality in the same sense as position; and profitable to do so when we recognize that motion is seen against (or from) a *reference frame* and that when the frame is changed, motion is also changed along with position. Dynamical quantities undergo well-defined transformations under changes of reference frames. Since the composition of frame changes is a frame change, they can be seen to form a group; the transformation of the dynamical quantities furnish realizations of this group. But to see it thus requires that positions and motions are to be treated on par. Another reason so to treat them is the change in motion produced by collisions or more generally under interaction. We have to take account of the quantity of motion or momentum. It is traditional to denote generalized coordinates by q_j and generalized momenta by p_j. The number of distinct values that j takes is called the number of degrees of freedom.

The trajectory of a particle may be seen either as a particular solution to a differential equation of motion or as an abstract general solution, which is mapped into a particular trajectory by specifying the initial state (coordinates and momenta). The latter view leads to the generalization, to a statistical description, as well as the passage to quantum physics. In quantum physics coordinates and momenta cease to be numerical variables, but become noncommuting *operators* with characteristic commutation relations. For Cartesian coordinates and momenta these are

$$q_j p_k - q_k p_j = (ih/2\pi)1$$

where h is Planck's constant (6.5×10^{-26} erg s). For a rigid rotator the angular momenta J_1, J_2, J_3 satisfy the commutation relations

$$J_1 J_2 - J_2 J_1 = (ih/2\pi)J_3 \quad \text{cyclically.}$$

Such commutation relations together with equations of motion constitute *quantum kinematics*.

The causes, effects, and interplay of motions constitute *kinetics*. The traditional starting point of mechanics is the set of Newton's three *laws of motion*: (i) A body continues in its state of rest or of uniform motion unless acted upon by an external force, (ii) Force equals product of acceleration induced times the mass of the body, (iii) Action and reaction are equal and opposite. The first law is a characterization of free motion; the second a quantitative definition of force. The third law states that the total momentum of a system under interaction is additive and invariable.

The energy of an isolated system is a scalar invariant consisting of an additive *kinetic energy* (quadratic in the momentum) and a nonadditive *potential energy*. The specification of these energy functions is the seed which generates the complete kinetics given the kinematics.

In *Lagrangian mechanics* one starts with a function $L(q,\dot{q})$ of the *generalized coordinates* and velocities and defines the generalized momenta p_j by

$$p_j = \frac{\partial L}{\partial q_j} \,.$$

These equations may be derived as the Euler–Lagrange variational equations for the action principle

$$\delta \int L(q,\dot{q})\mathrm{d}t = 0 \,.$$

For a system where the potential energy is independent of the generalized velocities the Lagrangian is the difference between the kinetic and potential energies.

In the Lagrangian formulation in terms of the action principle the consequences of *invariance properties* of the system are seen immediately. If we have invariance of the action with respect to, say, the position of the system, when we choose these quantities as generalized coordinates, the Lagrangian is independent of them; and by virtue of the equations of motion the corresponding generalized momentum is invariant. An invariance becomes translated into an invariant dynamical quantity: Symmetry and conservation become related. Invariance with respect to time origin corresponds to conservation of energy; indifference to the choice of space origin implies constancy of the space momentum (Newton's third law!); irrelevance of spatial orientation is equivalent to a conserved angular momentum (Kepler's second law). Conversely, conserved dynamical variables entail symmetries of the dynamical system: For motion under the inverse square law the axes of the ellipse remaining fixed implies the existence of a symmetry group changing the eccentricity of the ellipse without changing the energy. These ideas have important consequences for modern particle physics.

The clear separation between kinematics and dynamics is already blurred by including the "boost" transformation to moving frames. In this process the kinematic quantity of momentum and the dynamic quantity of energy undergo transformations into linear combinations of themselves. However, if a system obeyed Newton's First Law of Motion, it will continue to do so. There are other transformations to noninertial frames which affect the dynamical picture more drastically. We now turn to two different cases of this.

First is the transformation to rotating coordinates. A particle at rest in the original inertial frame has an apparent motion along a circle as seen from the rotating coordinate system. This is no longer an inertial motion. If the angular velocity is given by the (axial) vector $\boldsymbol{\omega}$ any vector dynamical quantity $\mathbf{A}$ has time derivatives $\mathrm{d}\mathbf{A}/\mathrm{d}t$ and $\mathrm{d}'\mathbf{A}/\mathrm{d}t$ with respect to the inertial and rotating coordinate system related by

$$\frac{\mathrm{d}\mathbf{A}}{\mathrm{d}t} = \frac{\mathrm{d}'\mathbf{A}}{\mathrm{d}t} + \boldsymbol{\omega}\times\mathbf{A} \,; \qquad \frac{\mathrm{d}\boldsymbol{\omega}}{\mathrm{d}t} = \frac{\mathrm{d}'\boldsymbol{\omega}}{\mathrm{d}t} \,.$$

Specializing to the radius vector $\mathbf{r}$ we relate the velocities in the two frames

$$\mathbf{v} = \frac{\mathrm{d}\mathbf{r}}{\mathrm{d}t} \,, \qquad \mathbf{v}' = \frac{\mathrm{d}'\mathbf{r}}{\mathrm{d}t}$$

by

$$\mathbf{v} = \mathbf{v}' + \boldsymbol{\omega}\times\mathbf{r} \,.$$

But for accelerations we have a more complicated relation:

$$\mathbf{a} = \mathbf{a}' + \boldsymbol{\omega} \times (\boldsymbol{\omega} \times \mathbf{r}) + 2\boldsymbol{\omega} \times \mathbf{v}' + \frac{d\boldsymbol{\omega}}{dt} \times \mathbf{r} \,.$$

The last term vanishes for steady rotation. The second term,

$$\boldsymbol{\omega} \times (\boldsymbol{\omega} \times \mathbf{r}) = \omega^2 \mathbf{r} - (\mathbf{r} \cdot \boldsymbol{\omega})\boldsymbol{\omega} \,,$$

is the "centrifugal" acceleration which could be pictured as if there is a centrifugal force on *all* particles *proportional to their mass*. The term $2\boldsymbol{\omega} \times \mathbf{v}'$ is the Coriolis acceleration, which is perpendicular to both the axis of rotation and the velocity as measured in the rotating system. It vanishes for particles at rest or seen to be moving parallel to the axis of rotation. The corresponding apparent Coriolis force on a particle is *velocity dependent* and *proportional to its mass*.

These noninertial frame-dependent forces are thus proportional to the mass of a particle betraying their kinematic origin. Viewed in this light, the fact of gravitational force on a particle being proportional to its mass suggests the natural way of understanding *gravitation is as a kinematically induced force*. This idea is implemented in the General Theory of Relativity in which a gravitating mass causes a curvature in the space around it; more precisely, the *stress tensor* composed of the momentum and energy flows and their densities is the *source of space-time curvature*.

Hamiltonian mechanics starts with a collection of phase-space variables q, p. The equations of motion are now

$$\dot{p}_j = -\frac{\partial H}{\partial q_j} \,,$$

$$\dot{q}_j = +\frac{\partial H}{\partial p_j} \,,$$

where $H(q, p)$ is the Hamiltonian (energy) function. Hamiltonian mechanics can be extended to systems with more general phase-space variables ω, like the rigid rotator. For any two phase-space functions $F(\omega)$ and $G(\omega)$ a new function $E(\omega)$ is defined called the *Poisson bracket* of F and G:

$$E = \{F, G\}$$

with the properties

$$\{F, G\} = -\{G, F\} \,,$$

$$\{c_1 F_1 + c_2 F_2, G\} = c_1 \{F_1, G\} + c_2 \{F_2, G\} \,,$$

$$\{\{F_1, F_2\}, F_3\} + \{\{F_2, F_3\}, F_1\} + \{\{F_3, F_1\}, F_2\} = 0 \,,$$

where c_1 and c_2 are any constants. The last property is called the Jacobi identity. With this structure the phase-space functions constitute a *Lie algebra*, which is associated with the Lie group of canonical transformations of phase-space functions and dynamical variables:

$$\phi(\omega) \to \phi(\omega; \tau) \,,$$

with

$$\frac{\partial \phi(\omega; \tau)}{\partial \tau} = \{\phi(\omega; \tau), F(\omega)\} .$$

By virtue of the Jacobi identity, we have

$$\{F(\omega; \tau), G(\omega; \tau)\} = E(\omega; \tau) .$$

The dynamical structure is therefore preserved under canonical transformations; the *transformation is a change of perception.*

We may adjoin the derivation property of the Poisson brackets,

$$\{F_1 \cdot F_2, G\} = F_1 \cdot \{F_2, G\} + \{F_1, G\} \cdot F_2 ,$$

to construct all Poisson brackets from the fundamental brackets:

$$\{\omega^\mu, \omega^\nu\} = \varepsilon^{\mu\nu}(\omega) .$$

For canonical variables

$$\begin{aligned} \{q_j, q_k\} &= \{p_j, p_k\} = 0 , \\ \{q_j, p_k\} &= \delta_{jk} , \end{aligned}$$

so that

$$\{F(q, p), G(q, p)\} = \sum_j \left(\frac{\partial F}{\partial q_j} \frac{\partial G}{\partial p_j} - \frac{\partial F}{\partial p_j} \frac{\partial G}{\partial q_j} \right) .$$

For the rotator

$$\{J_1, J_2\} = J_3 \text{ (cyclically)}$$

so that

$$\{F(J), G(J)\} = \sum_{jkl} \frac{\partial F}{\partial J_j} \frac{\partial G}{\partial J_k} \varepsilon_{jkl} J_l .$$

Quantum dynamics is obtained by identifying the Poisson bracket with a multiple of the commutator,

$$\{F, G\} = (i\hbar)^{-1}(FG - GF) ,$$

and identifying the Hamiltonian operator of the system.

Recognition that canonical transformations are changes of perception entails the dynamical transformations due to change of reference frames forming a canonical transformation group. This is the *relativity group* built up from the following:

1. change of space origin: displacement generator **P**;

2. change of time origin: time evolution generator H;

3. change of space orientation: rotation generator **J**;

4. change to moving frames: boost generator **K**.

These generators are dynamical variables, which have, respectively, the physical significance of linear momentum **P**, energy (Hamiltonian) H, angular momentum **J**, and moment of energy **K**. The Poisson bracket relations deduced from the structure of the frame transformations are

$$\{J_j, P_k\} = \sum_l \varepsilon_{jkl} P_l \, ,$$

$$\{J_j, J_k\} = \sum_l \varepsilon_{jkl} J_l \, ,$$

$$\{J_j, K_k\} = \sum_l \varepsilon_{jkl} K_l \, ,$$

$$\{K_j, P_k\} = \delta_{jk} H \, ,$$

$$\{K_j, H\} = P_j \, ,$$

$$\{K_j, K_k\} = \sum_l \varepsilon_{jkl} J_l \, ,$$

with all other Poisson brackets vanishing. All relativistic systems are realizations of this *Poincaré group*; the simplest realization is by a particle with $H = (m^2 + p^2)^{1/2}$, **K** = qH, **J** = **q** × **p**, where m is the mass of the particle. In this case we have a simple geometrical interpretation of the free-particle trajectory as a *world-line*: the quantities $\mathbf{q}(t), t$ transform as the components of a four-vector. A collection of interacting particles which have both the canonical structure and world-line trajectories cannot have any interaction; this is the *no-interaction theorem* of classical dynamics.

For many purposes the velocity of light is very large compared with the velocities of interest. The *Galilean group* of transformations is adequate to describe them. The Lie algebra consists of **P**, H, **J**, **G**, with the new generator **G** (the moment of mass) with the Poisson bracket relations

$$\{J_j, G_k\} = \sum_l \varepsilon_{jkl} G_l \, ,$$

$$\{G_j, G_k\} = 0 \, ,$$

$$\{G_j, H\} = P_j \, ,$$

$$\{G_j, P_k\} = M \delta_{jk} \, ,$$

where M has vanishing brackets with all generators (and may be identified with the constant total mass of the system). The generator **G** is the moment of mass $m\mathbf{q}$ for a particle.

The dynamical description by Newtonian/Lagrangian or Hamiltonian equations has three limitations: first, it is a short-time description and does not deal with *asymptotic behavior*; second, the equations of motion may become inapplicable or indeterminate for certain configurations causing *catastrophes*; third, it deals only with a finite number of degrees of freedom.

The initial configuration can be determined only with a finite accuracy; the predictions of the dynamical equations ought to be such that neighboring initial configurations should lead to neighboring final configurations. Dynamics should not only map configuration space points into points but also map *"neighborhoods" into neighborhoods*. For most dynamical systems this property is true only for short times; for sufficiently long times two arbitrarily near configurations can become separated widely. A simple example is provided by a cluster of particles of varying speeds moving along (1) a straight line and (2) a circle. In the first case we have a Hubble law: proportionality between velocity and distance from origin. In the second case although such correlations obtain for short times, if we wait long enough there is no correlation between the (angular) positions of the particles and their speeds. This simple example also illustrates the importance of global (topological) properties on apparently local correlations for the asymptotic development of the system.

The inertia of a particle participating in a nonlinear interaction is a dynamical variable and it could vanish for certain dynamical configurations. This in turn leads to unpredictable behavior leading to catastrophes.

The limitation to systems with a finite number of degrees of freedom must be removed if the dynamics of fluids, elastic bodies, and fields is to be considered. Natural generalizations of the Lagrangian, the action principle, the Hamiltonian, and the transformation group obtain; but since there are an infinity of fundamental dynamical variables, their algebra is not locally compact and considerable care is needed in handling realizations. One important consequence is the existence of states of the system which do not admit the original relativity transformations meaningfully. The restriction to finite time behavior has to be removed if we have to study the longtime asymptotic behavior of dynamical systems, especially in the context of the foundations of statistical mechanics. A discussion of these fascinating topics is beyond the scope of this article.

See also: Catastrophe Theory; Center-of-Mass System; Chaos; Conservation Laws; Dynamics, Analytical; Energy and Work; Gravitation; Group Theory in Physics; Hamiltonian Function; Invariance Principles; Lie Groups; Newton's Laws; Relativity, General; Relativity, Special Theory.

Kinetic Theory
R. N. Varney

Kinetic theory is a term applied to the study of gases by assuming that gases are composed of molecules subject to rapid random motions colliding with one another and with the container walls, analyzing these motions, and deducing various properties of the gases such as their molal heat capacities, the pressure they exert, their viscosity, coefficient of diffusion, and rate of effusion through apertures. The adjective *kinetic* refers to the fact that the properties of the gases are deduced from the *motions* of the molecules. The term *theory* refers to the *deduction* of the properties of the gases *without* reference to *any* known experimental properties, not even the ideal gas law (q.v.).

The atomic nature of matter, as a hypothesis, dates back to antiquity. It is generally accepted that the concept of *molecules* as distinct from atoms and their role in the behavior of gases dates from 1811 with the formulation of Avogadro's hypothesis which states that equal *volumes* of gases at the same temperature and pressure contain equal numbers of *molecules*. The key feature of the hypothesis lies in Avogadro's recognition of molecules as distinct from atoms.

The great and key developments of the kinetic theory of gases were made largely during the next 100 years despite the fact that some highly respected scientists (Ostwald, Mach) denied the existence of molecules. The reality of their existence, as opposed to the hypothesis, was finally proved to the satisfaction of even the disbelievers by the experimental work on Brownian motion by Jean Perrin in 1908 coupled with the analysis of it by A. Einstein.

In 1843, Joule applied the hypothesis of the intense motions of molecules in gases to explain the pressure exerted by a gas. His simplified assumptions were that (i) the molecules all had the same mass, m; (ii) they all had the same speed, designated by c; (iii) an experimental volume of gas V contained n molecules; (iv) the molecules bombarded the walls of the container with speed c and momentum mc, always striking the walls normally and rebounding elastically, hence with the same speed but reversed direction, and traveled unimpeded to the opposite wall of the rectangular container where the bombardment and rebounding occurred again. Momentum $2mc$ was transferred to the wall at each impact, the factor of 2 arising because the momentum changed from $+me$ to $-me$ upon impact. The time to cross the box of length x from one wall across and back was $t = 2x/c$, so that the transfer of momentum per *unit time* was $2mc/t = 2mc/(2x/c) = mc^2/x$. By Newton's second law, the rate of change of momentum equals the force/exerted by one molecule on one wall, (v) Joule assumed that of the n molecules in the box, $\frac{1}{3}n$ at each instant traveled in the x direction so that the force of all of these molecules on one wall was $(n/3)(mc^2/x)$. If the wall had dimensions y by z and area yz, then the pressure on the wall, $p = f/A$, became

$$p = \frac{1}{3}\frac{nmc^2}{xyz} .$$

Since the product xyz is the volume of the box V, the result is

$$pV = \frac{1}{3}nmc^2 .$$

Before commenting about this result of Joule's it may be noted that when the derivation was later redone with far more rigorous and realistic assumptions to include the collisions of molecules with one another, the random direction of hitting the walls, a wide distribution of values of the velocity c, and even mixtures of molecules with various mass values, the result was unchanged except for replacing c^2 by $\bar{c}^2$, the average squared speed, and the replacing of p by the partial pressure of each homogeneous constituent gas. This feature of kinetic theory of giving a correct or nearly correct result with highly simplified assumptions recurs in many other derivations.

While neither n nor m could be known to Joule, the product nm was the mass of gas in the volume V and was subject to experimental measurement. With a known pressure p, known volume V, and known mass of gas, the value of c or at least a mean value of c could be found. For N_2 gas at room temperature it proved to be about $5 \times 10^4 \, \mathrm{cm\,s^{-1}}$.

A series of deductions was at once made from Joule's equation.

(i) Since Boyle's law states that at constant temperature T, pV is constant, it follows that the mean speed c is not a function of p.

(ii) Since the ideal gas law gives $pV = RT$, it follows that $hMc^2 = RT$, where M is the molecular weight of the gas, in kg if joules and n s^{-1} are the units used.

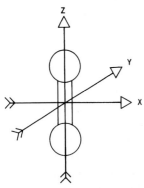

Fig. 1: A diatomic molecule oriented along the z direction has equal moments of inertia about axes through the center of mass parallel to the x and the y axes. The z-axis moment of inertia is negligible.

(iii) The kinetic energy of all the molecules of total mass M is $\frac{1}{2}Mc^2$ and hence this kinetic energy is equal to $\frac{3}{2}RT$. The quantity $\frac{3}{2}RT$ has been known from thermodynamics to be the internal energy of one mole of any monatomic gas. Its derivative $(\mathrm{d}(\frac{3}{2}RT)/\mathrm{d}T)_\mathrm{v}$ is the molal heat capacity at constant volume, C_v, and is equal to $\frac{3}{2}R$, a result widely verified for monatomic gases such as He, Ar, and Ne.

A new concept now enters the subject of kinetic theory from the domain of mechanics, called the *equipartition of energy*, one that has been more disputed than any other in kinetic theory. In a simple form it may be stated that each degree of freedom of the gas possesses the same amount of energy. This idea was tacitly assumed by Joule when he pictured one-third of all the molecules to be moving in the x direction, one-third in the y direction, and one-third in the z direction, all with the same speed c. He thus assumed that the same amount of kinetic energy was to be found in each of these components of motion or *degrees of freedom*. There would seem to be little reason to doubt this assumption; if it were false, it would seem that a wind must be blowing in a favored direction to cause the unbalance. Thus $\frac{1}{3}(\frac{1}{2}Mc^2) = \frac{1}{2}RT$ must be the kinetic energy to be ascribed to each degree of translatory motion.

The concept, as a hypothesis, was immediately extended in two respects: The first was that if the gas consisted of a mixture of molecules having different masses, then each molecular type would have the same kinetic energy per degree of freedom as each other type. It followed at once that the more massive molecules, having larger M, must have slower speeds of agitation c in order that the kinetic energies would all be the same. The second extension was that if the molecules of a gas were not simply single pointlike atoms but had molecular structure so as to have moments of inertia about principal axes and kinetic energies of rotation $\frac{1}{2}I\omega^2$ about each axis, where I is the moment of inertia about that axis, then each degree of rotational freedom would also possess the same kinetic energy $\frac{1}{2}RT$. A simple diatomic molecule composed of two identical atoms, as shown in Fig. 1, has the same moment of inertia about an axis in the x direction and the y direction, and essentially no moment of inertia at all about an axis in the z direction. The total kinetic energy of a mole of such molecules may then be written

$$E_k = \frac{1}{2}nm(c_x^2 + c_y^2 + c_z^2) + \frac{1}{2}nI(\omega_x^2 + \omega_y^2) \; ;$$

but by the equipartition of energy theorem, each of the five terms is the same, each is equal to $\frac{1}{2}RT$, and the total kinetic energy becomes $\frac{5}{2}RT$. The molal heat capacity at constant volume C_v should then be $\frac{5}{2}R$, and this has also been abundantly demonstrated to be true. However, it definitely fails to be true when quantum limitations become applicable. Thus when hydrogen gas is cooled to nearly its boiling temperature, its heat capacity at constant volume drops from $\frac{5}{2}R$ to $\frac{3}{2}R$ because the equipartition energy is less than one quantum of rotational energy of the hydrogen molecule.

The result most commonly associated with the term kinetic theory is the law of distribution of molecular velocities, a law that describes how many molecules out of each mole of gas have molecular velocities in any selected range of values. It may equally well be described as telling the probability that one molecule will have a velocity in any defined range of values. Perhaps the most remarkable feature of the distribution is that it exists, i. e., that the random motions settle down to a distinct pattern whereby only a few molecules have very high or very low speeds and the number in any range in between remains essentially constant.

The expression *Maxwell distribution* is often used for the distribution law, as J. C. Maxwell gave one of the earliest derivations of it. A derivation leading to the same result by L. Boltzmann is generally regarded as more rigorous but both derivations are criticized on the grounds of containing unproved, if experimentally valid, assumptions. (The basic issue has been whether the law can be derived solely from considerations of Newtonian mechanics or whether a theorem of probability is required in addition. The latter has proved to be the case.) The criticisms do not alter the validity of the law nor its enormous importance in physics but only the philosophical perfection of kinetic theory in reaching an important result entirely free of experimental aspects.

The Maxwell distribution law is illustrated graphically in Fig. 2. In this form, the x axis shows the speed (regardless of direction) of molecules, and the y axis shows the probability that a molecule will have this speed within a narrow but specified range Δc. (If there were an infinite number of molecules, it would be possible to speak of a probability of an exact speed, but since there is only a finite number, it is not possible to speak of an exact speed or the probability of it.) The curve is represented by the equation,

$$P_c = \frac{4}{\alpha^3 \pi^{1/2}} c^2 e^{-c^2/\alpha^2} \Delta c \; .$$

Here P_c is plotted on the y axis and is the probability of a speed c, shown on the x axis, in a range of speeds Δc. The symbol e is the base of natural logarithms, and α, a constant, proves to be the value of c at the highest point of the curve, i. e., at the maximum value of P_c. It is thus correct to say that α is the most probable speed of a molecule.

Some important features of the equation and its graph may be noted. The curve is *not* symmetric around the most probable speed α, since it terminates at zero on the low side but extends to infinity on the high side. By use of the mean-value theorem of integral calculus, it is possible to determine the average speed of molecules, $\bar{c}$. It is shown in Fig. 2 and has the value $\bar{c} = 2\pi^{-1/2}\alpha$, some 12.8% greater than α. In addition, because of the appearance of c^2

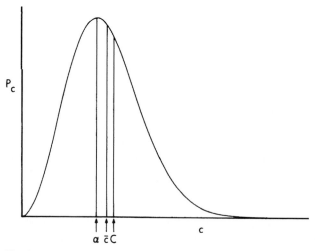

Fig. 2: Probability P_c that a molecule has a speed between c and $c + \Delta c$. Shown in the figure are the most probable speed α, the average speed c, and the root-mean-square speed $C = (\bar{c}^2)^{1/2}$.

in the Joule derivation, there is interest in the average value of c^2, written $\bar{c}^2$. Its square root is called the root-mean-square speed, often abbreviated rms speed, and is commonly designated by C. The value of C is $C = \frac{3}{2}^{1/2}\alpha$ or some 22.5% greater than α. It is also marked on the curve of Fig. 2.

Some caution is necessary in using the distribution law. Thus if it is desired to find the probability that a molecule has a certain component of velocity *in a certain direction*, say, for example, parallel to the x axis, the equation is

$$P_x = \frac{1}{\alpha\pi^{1/2}}e^{-c^2/\alpha^2}\Delta c \ ,$$

and is pictured in Fig. 3 for positive directions only. Since it is equally likely that a molecule should have a negative as well as a positive component of velocity in any given direction, the symmetric negative half of the curve is not drawn. The most probable component of velocity *in a given direction* is zero.

Finally, it is possible to calculate the probability that a molecule has a given energy, within a narrow band of energies, and this variation of the distribution law is often called the *Maxwell–Boltzmann law*. It is usually written in a somewhat different form from the law of distribution of speeds, and the symbols used are presented first.

P_E is the probability of a molecular kinetic energy between E and $E + \Delta E$. If the molecules are in a potential-energy field such as the earth's gravitational field, this potential energy is introduced as E_p. It is usually set equal to zero for a standard condition such as sea level in the gravitational case. Finally, recourse to Joule's derivation is taken wherein the mean molecular kinetic energy arises from

$$\frac{1}{2}nm\bar{c}^2 = \frac{3}{2}RT$$

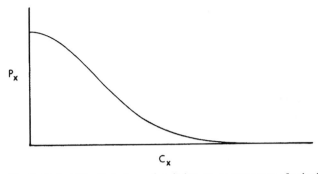

Fig. 3: Probability P_x that a molecule has an x component of velocity between c_x and $c_x + \Delta c_x$. Only positive values of c_x are shown, but the curve extends symmetrically to negative values of c_x. The most probable x component of velocity is seen to be zero.

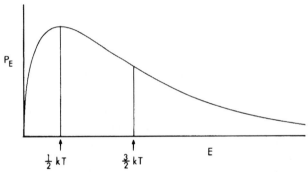

Fig. 4: Probability P_E that a molecule has kinetic energy between E and $E + \Delta E$. The most probable energy is $\frac{1}{2}kT$ and the average kinetic energy is $\frac{3}{2}kT$.

or

$$\frac{1}{2}m\overline{c^2} = \frac{3}{2}(R/n)T = \frac{3}{2}kT \ .$$

The molal gas constant R, divided by Avogadro's number n, is called Boltzmann's constant and is represented by k. The best value of k is $1.3806 \times 10^{23}\,\mathrm{J\,K^{-1}}$. Since $\frac{3}{2}kT$ the average kinetic energy of a molecule, this quantity is used instead of a or a most probable kinetic energy. The Maxwell–Boltzmann law then is

$$P_E = \frac{2}{\pi^{1/2}(kT)^{3/2}}E^{1/2}\mathrm{e}^{-(E+E_p)/kT}\,\Delta E$$

and it is plotted in Fig. 4.

The curve has several distinctive characteristics. Thus its slope at the origin is infinite whereas the speed-law curve has zero slope at the origin. The curve has a maximum, and the most probable kinetic energy proves to be $\frac{1}{2}kT$. At first glance this is surprising as a

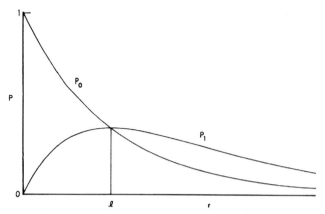

Fig. 5: Two curves, P_0 and P_1 are shown. P_0 is the probability that a molecule travels a distance r without a collision, and P_1 that the molecule having traveled a distance r has had exactly one collision – no more, no less. The value of $r = l$ is the mean free path. P_0 has the value unity at $r = 0$ and the value e^{-1} or 0.368 at $r = l$. P_1 has its peak value at $r = l$.

calculation of $\frac{1}{2}m\alpha^2$ with α the most probable speed gives kT and not $\frac{1}{2}kT$. The average energy $\frac{3}{2}kT$ is also indicated on the graph.

Direct experimental verification of the distribution law of molecular velocities was achieved in the late 1920s and early 1930s when the techniques of molecular beams (q.v.) had been developed. Indirect verification of a sort was achieved earlier by studies of the Doppler broadening of spectroscopic lines occurring from the speed of the radiating atoms toward or away from the spectroscope.

One of the great achievements of kinetic theory has been the analysis of interdiffusion of gases. The experimental problem in its simplest form is to expose some ammonia in one corner of a draft-free room and determine the time interval until the odor is detectable several meters away. The molecular velocity of $5 \times 10^2 \, \mathrm{m\,s^{-1}}$ obtained from application of the Joule derivation might suggest that the odor should be noted 2 m away in 4 ms, a time much too short to correspond with observations that are in fact measurable in minutes. The first analysis at the hands of R. Clausius generated the concept of the *mean free path* whereby the ammonia molecules travel only some 10^{-5} cm between collisions with gas molecules at atmospheric pressure, in time intervals of about 10^{-10} s between collisions, and describe a long and intricate path in the process of acquiring a displacement of several meters.

Assuming that a molecule has just had a collision at the origin of coordinates, the probability that at a distance r from the origin it still has not had a new collision is given by $P_0 = e^{-r/l}$. The symbol l, a constant in the equation, proves to be the mean value of r and hence is called the mean free path, commonly abbreviated mfp. The curve of P_0 vs r is shown in Fig. 5. Somewhat more enlightening is the curve marked P_1 in the same figure, which is the probability that the molecule *will* have collided *once* after traveling the distance r. Note that P_0 is the probability that the molecule *will not* have collided after traveling a distance r, and P_1 is the probability that it *will* have collided *just once*. The curve for P_1 drops at distances $r > l$, since the likelihood of two or more collisions rises. The value of P_1 is given by $P_1 = (r/l)e^{-(r/l)}$.

From the mean free path l and the average molecular speed c, J. C. Maxwell deduced expressions for the coefficients of viscosity, of thermal conduction, and of diffusion in gases. The derivations required calculation, respectively, of the transfer of momentum, of energy, and of mass through the gases, and the results proved to be

$$\text{Coefficient of viscosity} \quad = \quad \eta = \frac{1}{3}\rho\bar{c}l \, ,$$

$$\text{Thermal conductivity} \quad = \quad \kappa = \frac{1}{3}C_v\rho\bar{c}l = \eta C_v \, ,$$

$$\text{Coefficient of diffusion} \quad = \quad D = \frac{1}{3}\bar{c}l = \eta/\rho \, .$$

In these equations, ρ is the gas density, C_v is the molal heat capacity at constant volume, and the other symbols have been defined previously. The equations agree with experimental measurements to 5% or better over a large range of the variables, limited only by the requirement that the mean free path l be short compared with the main apparatus dimensions. Greater refinement is obtained by replacing the product $\bar{c}l$ by the average $\langle cl(c)\rangle$, that is, by multiplying the velocity by the path length for that velocity and then averaging.

Derivation of the value of the coefficient of viscosity is presented here in simplified form by way of illustrating kinetic-theory analysis. It is assume that by some mechanism of moving belts, a velocity gradient $\partial u/\partial z$ is created across a gas. Here u refers to a flow velocity in the x direction. Then across a mean free path l in the z direction, there is a difference of x velocity $l\partial u/\partial z$, and $ml\partial u/\partial z$ is the difference of momentum across such a free path. If it is now assumed that there are n molecules per cm^3 and that $\frac{1}{3}n$ are moving at any instant in the z direction, with velocity $\bar{c}$, it follows that $\frac{1}{3}n\bar{c}$ molecules per cm^2 per second travel across a plane at constant z. Hence the product $\frac{1}{3}n\bar{c}$ times $ml\partial u/\partial z$ is the momentum transferred per cm^2 per second across such a plane, and by Newton's second law, this must be the force per cm^2 acting across a plane at constant z. The coefficient of viscosity is defined by the relation

$$\eta = -\frac{f/A}{\partial u/\partial z} \, ,$$

so that it emerges that $\eta = \frac{1}{3}nm\bar{c}l$. Since nm is the mass of gas per cm^3, or ρ, Maxwell's result above is obtained.

The value derived for η, the coefficient of viscosity, depends on the product ρl, and these two quantities change oppositely with change of gas pressure, annulling any change in η with p. The original prediction of this fact, made by J. C. Maxwell, seemed unbelievable at first but has been verified over a large range of pressures. The reason for the surprise is traceable to an attempt to compare the viscosity of gases with that of liquids, a comparison that fails because in liquids, the viscosity arises from intermolecular forces, while in gases it arises from the collisions and the transfer of momentum. This difference causes η to rise with temperature in gases but fall in liquids, an experimentally substantiated distinction.

The kinetic theory of gases would probably not have a place in this encyclopedia if its major underlying hypothesis, the existence of molecules, had not been proved beyond any lingering doubts in 1908 by the work of Perrin and of Einstein on Brownian motion. A by-product of the 1908 work was the first reliable evaluation of Avogadro's number, the number of molecules in a mole. There are today no less than 20 methods for evaluating this number (see L. B. Loeb in the bibliography). The best current value is believed to be $6.022045(31) \times 10^{26}$ molecules per kg mole.

A precise distinction between kinetic theory and statistical mechanics cannot be made. The latter is usually more highly formalized but utilizes the same basic laws of mechanics and collisions. Statistical mechanics is generally regarded as the aspect of kinetic theory that deals with interpreting thermodynamic quantities like entropy and free energy by molecular collision analysis.

See also: Equations of State; Statistical Mechanics; Viscosity.

Bibliography

L. B. Loeb, *Kinetic Theory of Gases*. Dover, New York, 2004. This reference is highly readable, contains many applications, and is notable for balancing out excessive mathematical rigor in favor of physical comprehensibility.

A. I. Khinchin, *Mathematical Foundations of Statistical Mechanics*. Dover, New York, 1960. This book, translated from Russian, is essentially mathematical and contains the proof that the Maxwell–Boltzmann distribution law cannot be derived solely from Newton's second law without the aid of probability theory. It is not easily readable below advanced levels.

C. Schaefer, *Einführung in die theoretische Physik*, Vol. 2, part 1. Akad. Verlagsges, Leipzig, 1937. This reference is cited for the benefit of readers who prefer German to English. It is comparable with Loeb's book, going somewhat further in statistical mechanics.

Kinetics, Chemical

R. M. Noyes[†]

Chemical kinetics is the study of the rates of those processes by which systems move toward equilibrium states determined by chemical thermodynamics. The importance of the subject is its use to elucidate the mechanism by which chemical change takes place in any system.

A system is *homogeneous* if it consists of a single uniform phase; the rate of chemical change per unit volume is then a function only of temperature, pressure, and the concentrations of various component species. The system is *heterogeneous* if at least one additional phase is present; the rate may then depend on factors such as distribution and extent of subdivision of that phase.

Let us postulate a homogeneous uniform system in which extent of chemical change can be described in terms of a single balanced equation such as

$$aA + bB + \cdots \rightleftarrows cC + dD + \dots , \tag{1}$$

where capital letters are chemical species and lowercase letters are rational numbers. The *rate*

[†]deceased

of this reaction v is defined by

$$v = -\frac{1}{a}\frac{d[A]}{dt} = -\frac{1}{b}\frac{d[B]}{dt} = \frac{1}{c}\frac{d[C]}{dt} . \tag{2}$$

where $[A]$ denotes concentration of A in units such as $mol\,dm^{-3}$.

If such a system is far from thermodynamic equilibrium, an equation of the form

$$v = k[A]^\alpha[B]^\beta[C]^\gamma[D]^\delta[P]^\pi \tag{3}$$

may be valid over a wide range of conditions if temperature and either pressure or volume are maintained constant. The coefficients α, β, ..., may be positive, negative, or zero, and the concentration of species P may affect the rate even though it does not appear in Eq. (1) for net chemical change. Such a reaction is said to be α *order* in A, β order in B, and $(\alpha + \beta + \gamma + \delta + \pi)$ order overall. Often orders are small integers and the equation can be integrated in closed form. Thus, if the rate depends only on $[A]$ and if $a = \alpha = 1$, the solution becomes $\ln([A]_0/[A]) = kt$.

The *reaction rate constant* k is a function of temperature but not of species concentrations. It can often be fitted empirically to the Arrhenius equation

$$k = Ae^{-E/RT} , \tag{4}$$

where the *pre-exponential factor* A and *activation energy* E are disposable parameters, while R and T are the gas constant and absolute temperature, respectively.

It is generally believed that net chemical change in any system is the consequence of a number of *elementary processes* each of which involves simultaneous interaction of one, two, or at most three molecules. If an elementary process involves single isolated molecules of a species, the rate is first order in that species. If the process involves collisions between pairs of molecules, it is second order. The order of an elementary process is thus determined by its molecularity. However, the empirical observation of a simple reaction order does not prove the reaction takes place in a single step with the indicated molecularity.

At chemical equilibrium, $v = 0$ and all reactants and products are at finite concentrations. Therefore, Eq. (3) cannot remain valid as the system approaches equilibrium. In general, the exact expression for v contains another term of opposite sign that is of negligible magnitude when the system is far enough from equilibrium. Because the equation must reduce to $v = 0$ for all conditions of thermodynamic equilibrium, the allowable coefficients in the second term are strongly restricted. However, they are not uniquely determined except that $[P]^\pi$ must appear in both terms or neither because this species does not affect the position of equilibrium.

For many chemical reactions, an equation of the form of (3) can not describe the data over a significant range of concentrations no matter how far the system is from equilibrium. Such reactions must involve more than a single elementary process to accomplish net chemical change. Detailed measurements of the dependence of v on various concentrations provide insight to the molecular *mechanism* of the overall reaction. Such measurements can never unequivocally establish the way in which elementary processes accomplish net chemical change, but they can permit many otherwise plausible mechanisms to be rejected. If a comparatively simple explanation remains consistent with all measurements designed to test it, we develop an ever increasing confidence that it does describe true molecular behavior. When hypothetical mechanisms are being tested in this way, it is helpful to recognize that elementary processes may be *consecutive* involving formation of intermediates that subsequently react by unique paths, or they may be *competitive* so that species have different probable fates depending on experimental conditions.

The subject of chemical kinetics cannot be covered in detail here. Reference 1 illustrates how theories of elementary processes can be related to rates of collisions of gas molecules, and Ref. 2 illustrates an alternative viewpoint based on statistical thermodynamic partition functions. Reference 3 is an excellent presentation of the overall subject. Reference 4 illustrates how kinetic measurements are used to elucidate chemical mechanism, and Ref. 5 develops the importance of energetic considerations. Reference 6 develops the concepts of consecutive and competitive processes and is found in one of two volumes that offer detailed information about kinetic techniques.

See also: Thermodynamics, Equilibrium.

References

[1]   L. S. Kassel, *The Kinetics of Homogeneous Gas Reactions*. Chemical Catalog Company, New York, 1932.
[2]   S. Glasstone, K. J. Laidler, and H. Eyring, *The Theory of Rate Processes*. McGraw–Hill, New York, 1941.
[3]   S. W. Benson, *The Foundations of Chemical Kinetics*. Krieger, 1980.
[4]   A. A. Frost and R. G. Pearson, in *Kinetics and Mechanism*, 2nd ed., Chap. 12, pp. 285–387. Wiley, New York, 1976. Unfortunately, this material is omitted in a later edition.
[5]   S. W. Benson, *Thermochemical Kinetics*, 2nd ed. Wiley, New York, 1976.
[6]   R. M. Noyes, in *Investigations of Rates and Mechanisms of Reactions*, 4th ed., Part I, Chap. V, pp. 373–423. C. F. Bernasconi (ed.) of *Techniques of Chemistry*. Vol. VI, A. Weissberger (ed). Wiley, New York. 1986.

Klystrons and Traveling-Wave Tubes

J. A. Arnaud

Klystrons and traveling-wave tubes (TWT) are electron tubes that can amplify (or generate with the help of a suitable feedback mechanism) electromagnetic waves in the microwave range of frequency, from about 1 to 200 GHz. The operation of these devices can be understood on the basis of a classical theory, the electrons being viewed as point charges subjected to the electric force, and the electromagnetic waves as solutions of the Maxwell equations satisfying the boundary conditions imposed by the conductors.

The klystron tube was first demonstrated by the Varian brothers in 1937. As shown schematically in Fig. 1a, a klystron consists of an electron gun; a modulating cavity, called the buncher; and a collecting cavity, called the catcher. The electrons originating from the gun are accelerated by a dc field. When they reach the buncher they have almost all the same velocity, except for a small spread due to the cathode temperature. When a particular electron of the

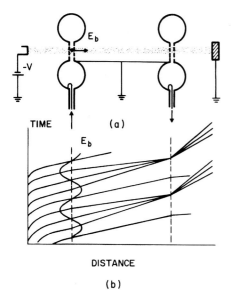

TIME   (a)

DISTANCE

(b)

Fig. 1: Schematic view of a two-cavity klystron amplifier.

beam traverses the buncher, it is accelerated or decelerated, depending on whether the axial electric field in the buncher is negative or positive, respectively. Some electrons traverse the buncher at a time when the field is equal to zero. Their trajectory, then, is unaffected. Because the buncher modulates the velocity of the passing electrons, the klystron is called a velocity-modulation device. In Fig. 1b the positions of a few electrons are shown as a function of time. It is easily seen that the electrons emitted within a given period of field oscillation tend to meet at a distance from the modulating cavity that is inversely proportional to the strength of the modulating field. The velocity-modulated electrons are said to form "bunches". At the particular location where the electrons are most concentrated the electron beam carries a large alternating current whose fundamental frequency is equal to that of the modulating field. This alternating current, flowing through the catcher, generates a power that may be considerably larger than the power used to modulate the beam. In fact, according to the mechanism of operation just described, electron bunches could be obtained with arbitrarily small modulating fields. The gain of klystrons could be increased to arbitrarily large values merely by increasing the distance between the catcher and the buncher. This is not the case, however, because electrons repel each other. In order to obtain a large gain, it is necessary to introduce additional cavities between the buncher and the catcher. These cavities are not coupled directly to each other, but only to the electron beam.

Klystrons generate electromagnetic waves when a suitable feedback mechanism is provided. In the "reflex" klystron, the modulated electron beam is reflected back into the buncher with the help of a negatively biased electrode called the reflector. The reflex klystron therefore incorporates a single cavity, which plays the role of both the buncher and the catcher. The frequency of oscillation is primarily determined by the resonant frequency of that cavity, but it can be changed slightly by changing the reflector voltage. In spite of increasing competition from solid-state sources, particularly Gunn and Impatt devices, reflex klystrons remain

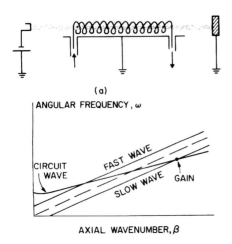

Fig. 2: Schematic view of a helix-type traveling-wave amplifier.

invaluable as low-noise local oscillators, particularly in the 30 to 170-GHz frequency range. High-power multicavity klystrons, which can generate peak powers of tens of megawatts, are commonly used as sources of electromagnetic energy in linear accelerators.

The traveling-wave tube, invented and demonstrated during the second World War by Kompfner, has some similarity to the multicavity klystron described earlier. However, the traveling-wave tube makes use of a metallic structure that can guide electromagnetic waves independently of the electron beam. The traveling-wave tube is inherently a broadband device because of the absence of any resonant structure required for its basic operation. The guiding structure that was first considered and that remains the most useful is the helix, shown schematically in Fig. 2a. The phase velocity of waves guided by the helix is approximately equal to the velocity of light in free space, c, times the ratio of the helix period to its perimeter. A typical value for the phase velocity is $c/10$. The wave guided by the helix is amplified when it interacts with an electron beam whose velocity exceeds slightly the phase velocity of the guided wave. When this near-synchronism condition is met, the electrons are submitted to an almost constant axial field. On the average, they supply energy to the electromagnetic wave. In a precise theory, the fact that electrons repel each other must be taken into account. When waves are excited on electron beams (in the absence of any circuit), the electronic fluid tends to oscillate at the so-called plasma angular frequency ω_p as a result of the electrostatic forces. If the angular frequency of the electron wave in the laboratory frame of reference is ω, the apparent wave number can be either $\beta_+ = (\omega + \omega_p)/v_e$ (slow wave) or $\beta_- = (\omega - \omega_p)/v_e$ (fast wave), where v_e denotes the average electron velocity. From the fact that the electron-wave energy in the electron-beam frame of reference is positive, and that the group velocity $d\omega/d\beta_+ = v_e$ in the laboratory frame exceeds the phase velocity ω/β_+ of the slow wave (see Fig. 2b), it can be shown quite generally that the energy of the slow wave in the laboratory frame is negative. Approximately, this means that as the amplitude of the slow wave grows, the total energy of the electron beam decreases. Thus, when the slow electron wave is synchronous with the circuit wave, the electron beam supplies energy to the circuit wave, both the circuit and the slow electron waves growing exponentially as a function of distance. This is the condition for maximum gain. If, however, the circuit wave is synchronous with the fast wave, the circuit wave supplies energy to the electron beam and is attenuated. The stability of traveling-wave tubes with respect to reflections from the load is considerably improved when a properly located attenuation is introduced on the circuit: Reflected waves are attenuated but the gain is not very much reduced.

Traveling-wave tubes are commonly used to amplify weak signals up to a level of a few tens of watts in the 3- to 30-GHz range of frequency, for example in communication satellites. Below 10 GHz and for powers less than a few watts, field-effect transistors can successfully compete with TWT. Gigawatts of peak power have been reported in TWT having beam voltages of a few megavolts.

Two important modifications of the conventional TWT should be mentioned. In this analysis it has been implicitly assumed that the circuit group velocity, as well as the circuit phase velocity, is in the forward direction. However, some circuits, particularly periodic circuits such as the inderdigital line, may in fact support waves whose phase and group velocities are in opposite directions. When such circuits are made to interact with electron beams, a continuous feedback takes place that results in an oscillation of the device. The electromagnetic energy flows in the direction opposite to the electron flow. This is the principle of the backward-wave oscillators. These devices are commonly used as electronically tunable sources of radiation. They provide milliwatts of power up to 1000 GHz.

The second modification concerns the electron beam. Electron beams can be kept confined in static electric and magnetic fields perpendicular to each other and to the direction of propagation. Crossed-field tubes, such as the magnetron, usually exhibit higher efficiencies than conventional traveling-wave tubes, but they are more noisy. Some magnetrons operate at voltages up to 1 MV.

A recently developed tube is the gyrotron which operates in the relativistic regime and can generate up to 100 kW, cw, at 80 GHz.

The free-electron laser is a related device whose frequency range extends to optics.

See also: Microwaves and Microwave Circuitry.

Bibliography
General
C. Susskind, *The Encyclopedia of Electronics*, Reinhold, New York, 1962. (E)

History
E. L. Ginzton, "The $100 Idea; How Russel and Sigurd Varian, with the Help of William Hansen and a $100 Appropriation, Invented the Klystron", *IEEE Spectrum* **12** (2) (1975). (E)
R. Kompfner, *The Invention of the Traveling-Wave Tube*, San Francisco Press, San Francisco, 1964. (E)

Technical
J. F. Gittins, *Power Traveling-Wave Tubes*, Elsevier, New York, 1965. (I)
V. Granatstein and I. Alexeff, *High Power Microwave Sources*, Artech House, 1987.
E. Okress, ed., *Crossed-Field Microwave Devices*, Academic Press, New York, 1961. (I)

Theory
J. R. Pierce, *Traveling-Wave Tubes*, Van Nostrand, Princeton, NJ, 1950. (A)
J. E. Rowe, *Nonlinear Electron–Wave Interaction Phenomena*, Academic Press, New York, 1965. (A)

Kondo Effect

A. S. Edelstein

The Kondo effect is the name given originally to describe the temperature dependence of the resistivity in certain alloys containing a small concentration of magnetic impurities. The term has now been generalized to cover other cases of the interaction of localized electron spins with conduction electrons. Understanding these interactions is central to answering many important problems in solid state physics. More recent cases of the Kondo effect are *single-electron transistors* (SET) and *quantum dots*.

The resistivity of some alloys as a function of decreasing temperature exhibits a minimum, followed by a logarithmic increase at lower temperatures and finally becomes temperature independent at still lower temperatures. The effect was named after J. Kondo who in 1964 performed a perturbation-theory calculation which correctly predicted the logarithmic behavior. Though first observed in 1939 in "pure" Cu, the effect was later understood to be associated with residual impurities in the Cu. These investigations started a research area that is still active and that has evolved into several closely related research areas that include valence fluctuations or mixed valence, heavy fermions, and strongly correlated electron systems. These areas are discussed separately in this volume. It will be pointed out below that heavy-fermion systems are concentrated Kondo systems, i. e., concentrated systems that exhibit the Kondo effect.

The reasons for the interest in and longevity of the field are threefold. First, since the resistivity in Kondo's calculation increased as $\ln T$, it would become infinite at low temperatures and violate the unitarity bound for scattering. Instead of this behavior, experiments on dilute alloys containing magnetic impurities showed that the low-temperature resistivity is temperature independent and relatively small. The qualitative difference between the prediction of perturbation theory and experiment is an indication that the low-temperature properties and the ground state of the system cannot be obtained from perturbation theory. New theoretical techniques had to be developed to treat this problem. Second, as discussed below, the magnetic moments of the impurities in Kondo systems become unstable at low temperatures. Thus, it is necessary to understand the Kondo effect in order to understand magnetic moment formation in metals. Third, the discovery of a continuing variety of new phenomena, such as f-band superconductivity, has served to sustain the field.

It is useful to contrast the Kondo effect with the usual temperature dependence of a metal. Usually the resistivity of a metal, $\rho(T)$, monotonically decreases with decreasing temperature to a temperature-independent value, ρ_0, at low temperatures. The resistivity ρ_0 is due to nonmagnetic impurities and other scattering centers that break the perfect periodicity of the lattice. The resistivity increases with increasing temperature since there are increasingly larger thermal vibrations which also disrupt the lattice periodicity. Based upon calculations using first-order perturbation theory, it was thought that magnetic impurities would only increase ρ and thus shift $\rho(T)$ upward. Instead of this behavior, Kondo systems show a logarithmic increase in their resistance with decreasing temperature in a temperature region which is approximately a decade wide. The temperature where this increase occurs, which is system dependent, often is below $10\,\mathrm{K}$. This behavior is illustrated in Fig. 1 for LaB_6 containing various percentages of Ce as the impurity.

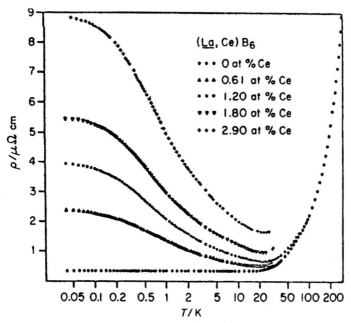

Fig. 1: Electrical resistivity of LaB$_6$ and four (La,Ce)B$_6$ samples as a function of temperature (after Samwer and Winzer 1976) (from "Exact Results in the Theory of Magnetic Alloys", by A. M. Tsvelick and P. B. Wiegmann, from *Advances in Physics*, Vol. 32, published by Taylor & Francis Ltd, copyright 1983. Used by permission).

Let us consider the scattering of conduction electrons off a magnetic impurity to see how the temperature dependence of $\rho(T)$ arises. Scattering can alter an electron's momentum (both its direction and magnitude) and can cause the conduction electron's spin to flip. Since scattering from nonmagnetic impurities can alter a conduction electron's momentum, clearly spin-flip scattering must be essential in giving rise to the Kondo effect. In a perturbation theory treatment of scattering, one attempts to express the state of the system after the scattering in terms of the initial state. It turns out that this is accomplished by considering repeated scattering off the impurity. Because of the Heisenberg uncertainty principle, the uncertainty of the lifetime of the intermediate states between successive scatterings gives rise to an uncertainty in the electron's energy when it is in these intermediate states. Pauli's exclusion principle, which states that there can only be one electron in a given state, places an important constraint on the occupancy of intermediate states.

In metals at zero temperature, all electronic states below the Fermi energy, E_F, are occupied, and all the higher-energy states are empty. At an absolute temperature T, there is a band of states of width $k_B T$ (where k_B is Boltzmann's constant) centered at E_F where there are empty states below E_F and occupied states above E_F. For nonspin-flip scattering, when one sums over the allowed intermediate states, there are two competing effects which cancel the temperature dependence of the scattering. Kondo's calculation showed that these two competing effects no longer cancel in the case of spin-flip scattering.

In his calculation Kondo assumed that the interaction between a conduction electron and the impurity has the form

$$E_{int} = -2J\mathbf{S} \cdot \mathbf{s} \,, \tag{1}$$

where J is an energy and is called the exchange integral and $\mathbf{S}$ and $\mathbf{s}$ are the spins of the magnetic impurity and the conduction electron, respectively. Using the second Born approximation in a perturbation calculation, Kondo found that scattering off magnetic impurities increases the resistivity by an amount $\Delta\rho$ given by

$$\Delta\rho = c\rho_m \left[1 + N(E_F)J \ln\left(\frac{k_B T}{D}\right) \right] \,, \tag{2}$$

where c is the impurity concentration, ρ_m is the contribution to the resistivity of a single impurity calculated using the first Born approximation, and $N(E_F)$ is the density of states of the conduction electrons evaluated at the Fermi energy. For simplicity, Kondo assumed that the conduction-electron density of states is independent of E and of width $2D$. One sees from Eq. (2) that $J < 0$ is necessary for the Kondo effect to occur, i. e., for the resistivity to increase at low temperatures. One also sees that $\rho \to \infty$ for $T \ll D/k_B$ and $J < 0$. This is an unphysical result. There is a limit, the unitary limit, which sets an upper bound on the amount of scattering from a single impurity. The fact that Eq. (2) is incorrect at low temperatures indicates that the method used in its derivation, i. e., perturbation theory, is not applicable.

Thus, there is special interest in the low-temperature properties. There is a temperature, called the Kondo temperature, below which the system gradually condenses into its ground state. No sharp phase change occurs at T_K, since an isolated impurity does not have enough degrees of freedom to permit a sharp phase transition. The value of T_K is given by

$$k_B T_K = \tilde{D} [|J|N(E_F)]^{1/2} \exp\left[-\frac{1}{|J|N(E_F)} \right] \,, \tag{3}$$

where $\tilde{D}$ is an effective bandwidth which is of order D. The value of T_K is important in that it determines the energy and temperature scale; i. e., $T/T_K > 1$ and $E/k_B T_K > 1$ correspond to high temperatures and energies, respectively. Much work has been done to develop a theory that is valid at all temperatures. Sophisticated methods, including renormalization-group and Bethe-*Ansatz* techniques, have been used to derive the low-temperature properties correctly. Wilson, who used a renormalization-group approach, was the first person to solve the spin-$\frac{1}{2}$ Kondo problem. The Ansatz techniques were used to obtain formulas describing the behavior in both weak and strong magnetic fields, for the specific heat at all temperatures, for treating the s–d model for spins greater than 1/2 and for including degeneracy and crystalline fields.

Since the Kondo temperature T_K sets the energy scale, it also determines the temperature scale for experimental properties. Thus, one expects these properties to be approximately functions of just T/T_K. Experimentally the value of T_K for a given system is determined by fitting experimental data to theoretical predictions. The experimental values for T_K obtained in this way can vary drastically depending on the impurity and host. For dilute Mn in Cu, $T_K < 0.01\,\text{K}$, while for dilute Fe in Cu, $T_K \approx 30\,\text{K}$.

The exchange form of the interaction energy, Eq. (1), can, in a certain limit, be derived from the more fundamental formulation of P. W. Anderson. Anderson was interested in magnetic moment formation of dilute $3d$ transition-element impurities in a nonmagnetic host. In the dilute limit one need only calculate the contribution of a single, isolated impurity. He considered the energy of the total system to consist of a conduction-electron contribution, an impurity contribution, and an interaction term between the impurities and the conduction electrons. The impurity contribution was taken to be a sum over σ of $\varepsilon_d n_{d\sigma} + U n_{d\sigma} n_{d-\sigma}$, where ε is the energy of the d state, $n_{d\sigma}$ is the number of d electrons of spin σ occupying the impurity site, and U, which is positive, is the Coulomb repulsion between d electrons of opposite spin. For a nondegenerate d orbital, double occupancy by electrons with the same spin is not allowed because of the Pauli exclusion principle. The interaction term allows an electron to hop on or off the impurity site without changing its spin. A spin flip occurs if an electron with spin σ hops off the site and is replaced by an electron with spin $-\sigma$. This interaction causes the original sharp impurity level to have a width Δ proportional to $|V_{dk}|^2$, where $|V_{dk}|$ is the magnitude of the matrix element involved in the hopping. For $\Delta \ll U$ one can transform the interaction energy in Anderson's model to the form of Eq. (1). The exchange integral that then appears in Eq. (1) is $J = J_d + J_{\text{eff}}$ where J_d is a positive quantity called the direct exchange integral and J_{eff} is given by

$$J_{\text{eff}} = \frac{|V_{dk}|^2 U}{E_{df}(E_{df} + U)} , \qquad E_{df} = E_d - E_f < 0 . \tag{4}$$

In Anderson's model, the conditions that most favor magnetic moment formation are $\Delta \ll U$ and $E_{df}/U = -\frac{1}{2}$. As mentioned earlier, the total exchange integral must be negative in order for the Kondo effect to occur. Since J_d is positive, this requires that $E_{df}(E_{df} + U) < 0$ and sufficiently small in magnitude that $-J_{\text{eff}} > J_d$. Notice that the conditions required for the Kondo effect to occur are different from those favoring moment formation. Thus it is not surprising that the ground state of Kondo systems is a nonmagnetic singlet.

For dilute, stable magnetic moments the susceptibility χ per impurity of a substance having a magnetic moment μ per impurity is given by Curie's law, $\chi = \mu^2/3k_B T$. In Kondo systems for $T < T_K$, the interactions between the local moment and the conduction electrons reduce the value of χ below the Curie's law prediction. One can define an effective moment $\mu_{\text{eff}}^2 = 3kT\chi$ as a measure of how much the magnetic response of the impurity is reduced. This reduction might be anticipated from Eq. (1). Since $J < 0$, the total energy of the system is minimized if the conduction electrons' spins are aligned opposite to the impurity spins. This antiparallel arrangement reduces the effective moment. In fact, for Kondo systems for $T \ll T_K$, χ approaches a constant χ_0 and hence $\mu_{\text{eff}}^2 \to 0$. Numerical calculations using the renormalization group show that the distribution of allowed energy states in the limit $T \ll T_K$ is similar to that which would exist if $|J|$ in Eq. (1) were to become infinite. Since $T \to 0$ is like $|J| \to \infty$ in Eq. (1), the coupling between a spin-$\frac{1}{2}$ impurity and the conduction electrons becomes very strong. This implies for $T \ll T_K$ that there is always a conduction electron paired with its spin antiparallel to that of the impurity. Thus, the ground state is a nonmagnetic singlet. The spin of the impurity is compensated by the spins of the conduction electrons in a dynamic way. There is not a static compensating conduction electron cloud.

Evidence against the existence of a static compensating conduction electron cloud was provided by neutron-scattering measurements on Y:Ce alloys and nuclear-magnetic-resonance measurements on Cu:Fe alloys. The measurements on Y:Ce alloys showed that the spatial distribution of the Ce moment was the same as that of Ce^{3+} and did not appear to be modified by the conduction electrons for $T < T_K$. The nuclear-magnetic-resonance measurements were

Table 1: Temperature Dependence of Kondo Impurity Contributions to Some Macroscopic Properties[a]

	Resistivity	Susceptibility	Specific heat
$T \gg T_K$	See Eq. (2)	$\mu^2/3k_B(T + T_K)$	$\propto 1/T^2$
$T \ll T_K$	$\propto (1 - bT12/T_K^2)$	$\chi_0(1 - dT^2/T_K^2)$	$\propto cT/T_K$

[a] b and d are constants.

performed on the Cu nuclei, and it was found possible to detect the separate resonance signals from Cu nuclei which were different distances from the Fe impurities. Signals from the first through fourth nearest neighbors of the Fe impurities have been separated from the main resonance of most of the Cu nuclei. From the temperature dependence of these signals it is possible to conclude that the magnetization or spin density has the form $\sigma(r,T) = f(r)\chi(T)$, i.e., it does not change its shape for $T < T_K$ at the position of these neighbors.

For $0 < T \ll T_K$ the renormalized J is large but finite. In this case real transitions out of the ground state are impossible since they require a finite amount of energy. Virtual transitions are possible and the impurity is polarizable. This leads to a repulsive interaction between the conduction electrons and to a finite value for the susceptibility for $T \ll T_K$. Hence $\mu_{eff} \to 0$ as $T \to 0$.

One can also understand why $\mu_{eff} \to 0$ as $T \to 0$ by considering the time that an impurity spin will remain correlated with itself, which is called the intrinsic spin correlation time, τ. For Kondo impurities it turns out that $\tau \approx h/k_B T_K$, where h is Planck's constant. The susceptibility of a dilute impurity is the time average of $\{S(0)S(t)\}$ from 0 to a time $t = h/k_B T$. If a spin only stays correlated with itself a time t, then for $T \gg T_K$ the susceptibility is approximately given by Curie's law, while for $T \ll T_K$ the susceptibility is given by $\chi \approx \mu^2\tau/3h$. Hence, $\mu_{eff} \propto (T\tau)^{1/2}$ for $T \ll T_K$.

A variety of both microscopic and macroscopic experiments have been performed on Kondo systems in which the impurity, the host, the concentration of impurities, and parameters such as the temperature and the magnetic field were varied. For most properties, the single-impurity contribution must be proportional to the impurity concentration. For many systems one must employ concentrations less than 100 ppm to observe the single-impurity contribution. Early experimental results were often complicated by interaction effects.

Table 1 shows the qualitative behavior for the resistivity, susceptibility, and specific heat for Kondo systems for $T \gg T_K$ and $T \ll T_K$. The temperature dependence of these properties gradually changes from its high- to its low-temperature behavior. There is a broad specific-heat anomaly, which occurs over several decades in T/T_K, that is centered at approximately $T_K/3$. The entropy change derived from the area under a plot of C/T vs. T, where C is the impurity specific heat, is approximately equal to $R\ln(2S+1)$. This is consistent with the ground state being a singlet. Note that the specific heat is proportional to cT/T_K, where c is the impurity concentration for $T \ll T_K$. Thus, if c/T_K is large, then the low-temperature specific heat will have a large term which is linear in T. Heavy-fermion systems are defined as systems which have such a large linear term. The mechanism which gives rise to this term in heavy-fermion systems is believed to be large values of c/T_K. Theoretical calculations predict that χ is a universal function of T/T_K. It is interesting that the low-temperature susceptibility of several Kondo systems can be fitted by this function of T/T_K.

Magnetic impurities drastically affect the properties of superconductors. The electrons in superconductors condense into states in which pairs of electrons are bound and have their spins oppositely directed. Magnetic impurities break this symmetry property and cause a large depression in the superconducting transition temperature T_c. Kondo impurities cause an especially large depression in T_c and gapless superconductivity at very low impurity concentrations. The impurities' tendency to inhibit superconductivity is greatest for $T \approx T_K$. Thus, if T_K is smaller than the superconducting transition temperature of the pure system, it is possible for the system to undergo a transition into the superconducting state at a temperature T_c, above which the impurity has its full effect and then undergo a second transition out of the superconducting state at a temperature $T_{c_2} \approx T_K$. These effects have been observed in (La,Ce)Al$_2$ and (La,Th):Ce alloys.

The Kondo effect remains a very active area of research. Though there continues to be research on the original problem of dilute magnetic impurities in metals, other systems that are discussed below also exhibit the Kondo effect. A common feature of all these cases is an internal degree of freedom of an unpaired electron spin and non-communicative terms in the expressions for the scattering amplitude.

In recent times, much of the focus of the research on the Kondo effect has been devoted to studying systems in which electrons are confined in some way such as in lower dimension systems. Some of these studies provide unique information about a single scattering center and, thus, avoid the averaging over many scattering centers that occurs in bulk sample. For example, an elliptical quantum corral was created by placing atoms on the surface of a single crystal of copper. A cobalt atom was placed at one of the foci. Spectroscopic evidence of a Kondo resonance was obtained both at the position of the Co atom and at the position of the other foci. Studies of the Kondo effect have also been made in single electron transistors (SET). A cloud or droplet of electrons is established in a small channel in the SET. If the number of electrons in the droplet is odd there is an unpaired electron with a free spin that can form a singlet with the electrons at the Fermi level in the leads. At low temperatures, this gives rise to an increased density of states of the electrons in the leads that increases the conductance of the device. In SET the Kondo effect gives rise to a zero voltage peak in the conductance at $T < T_K$. Another example of the Kondo effect in limited dimensions was the study of single-walled carbon nanotubes with gold source and drain contacts and substrate gate electrodes. This arrangement forms a one-dimensional quantum dot. In these systems, besides varying the temperature and magnetic field, one can study also the effect of varying the gate voltage and the voltage between the source and the drain. In some of these systems, one can control whether the system acts as a Kondo system or a mixed valence system by varying the gate voltage. Kondo behavior in C$_{60}$ coupled to gold electrodes has been deduced from the observed zero bias conductance peak. A low temperature magnetic sensor has been fabricated by placing C$_{60}$ between two nickel electrodes. In this case, the field dependence of the Kondo-assisted tunneling is enhanced by the interactions with the nickel ferromagnetic electrodes. Another case of Kondo behavior in a restricted geometry has been observed in point contacts containing a single charge trap. A zero-bias anomaly was observed whose magnitude increased logarithmically with decreasing temperature.

Low-temperature scanning tunneling spectroscopy has been employed to study the Kondo effect. The spectroscopy was used to observe the Kondo resonance and to measure its special extent as a function of its distance from a single magnetic impurity on a metallic substrate. As expected, the resonance is centered at the Fermi energy.

In summary, the term Kondo effect is used to describe the interaction of localized spins with conduction electrons that gives rise to many body bound states below some characteristic temperature, T_K. Increasing the interaction increases the characteristic temperature. Similar to what occurs in alloy systems, the effective coupling strength is strongly renormalized at low temperatures. The localized electrons can reside on impurity spins, on spins in compounds, in quantum dots, on C_{60}, on carbon nanotubes, and in single electron transistors. The conduction electrons can be those of a host alloy, compound, or those of the contact electrodes of quantum dots and single electron transistors.

See also: Conduction; Diamagnetism and Superconductivity; Heavy-Fermion Materials; Intermediate Valence Compounds.

Bibliography

G. A. Fiete and E. J. Heller, *Rev. Mod. Phys.* **75**, 933 (2003); Theory and experimental results on quantum corrals. (A)

D. Goldhaber-Gordon *et al.*, *Nature* **391**, 156 (1998); Kondo effect in single-electron transistor. (I)

G. Gruner and A. Zawadowski, *Rep. Prog. Phys.* **37**, 1497 (1974); theoretical review. (A)

A. J. Heeger, *Solid State Physics*, (F. Seitz, D. Turnbull, and H. Ehrenreich, eds.), Vol. 23, p. 283. Academic Press, New York, 1969; experimental review. (I)

C. Kittel, *Introduction to Solid State Physics*, 4th ed., p. 660. Wiley, New York, 1971. (E)

J. Kondo, *Solid State Physics* (F. Seitz, D. Turnbull, and H. Ehrenreich, eds.), Vol. 23, p. 183. Academic Press, New York, 1969; theoretical review. (I)

P. Nozieres, *J. Low Temp. Phys.* **17**, 31 (1974); Fermi-liquid description of the problem. (I)

C. Rizzuto, *Rep. Prog. Phys.* **37**, 147 (1974); experimental review. (I)

P. L. Rossiter, *Aust. J. Phys.* **39**, 529 (1986). (I)

K. Samwer and K. Winzer, *Z. Phys. B* **25**, 269 (1976).

P. Schlottmann, *Phys. Rep.* **181**, 1 (1989) (A). Review of theory that includes Coqblin–Schrieffer model and the (compensated) multichannel Kondo problem.

H. Suhl (ed.). *Magnetism*, Vol. V. Academic Press, New York, 1973; both theoretical and experimental reviews which include the effect on superconductivity. (I–A)

A. M. Tsvelick and P. B. Wiegmann, *Adv. Phys.* **32**, 453 (1983). (A)

K. G. Wilson, *Rev. Mod. Phys.* **47**, 773 (1975); review of renormalization theory. (A)

Laser Spectroscopy

S. Chu and F. Cataliotti

The interaction of laser light with matter has dramatically extended the capabilities of optical spectroscopy. These developments are due to the synergism between technological advances in lasers, a deeper understanding of how light interacts with matter, and new spectroscopic techniques that have emerged from this understanding. Laser spectroscopy has a profound relevance in numerous applications in physics, chemistry, biology and medicine.

Some of the technological capabilities of lasers are summarized below. The frequency range of lasers has been extended from the far-infrared to the x-ray region of the electromagnetic spectrum range. There is almost complete coverage of the ultraviolet to mid-infrared portion of the spectrum with broadly tunable light sources. The frequency control of lasers now exceeds the precision of the best microwave sources: stability of the most precisely controlled lasers is better than one part in 10^{15}. If linked to suitable atomic references, they have the potential to be stable to better than one part in 10^{18}. Lasers can deliver high energy pulses in excess of 10^{12} watts, and because photons interact extremely weakly with each other without the aid of intervening matter, this power can be concentrated into volumes of less than $100\,\mu m$ in diameter. Even with small laser systems, electric fields can be made far stronger than atomic electric fields. Pulses as short as a few femtoseconds (10^{-15} seconds) can be generated in precisely regular pulse trains or as a single high intensity burst. Miniature solid state diode lasers are now commonly used in laser spectroscopy as well as in optical communications, laser printing, and information retrieval.

Advantages of Lasers in Spectroscopy

In classical absorption spectroscopy the radiation source has a broad emission continuum spectrum, this means that, in order to obtain a spectrum, one is forced to use a dispersive spectrometer which limits the spectral resolution to $\Delta v/v \geq 10^{-7}$ which can be obtained only with large and expensive instruments. Furthermore the detection sensitivity is limited by intensity fluctuations of the radiation source and by the relatively short absorption path achievable due to the limited collimation of classical sources. On the contrary tunable lasers offer several advantages:

Encyclopedia of Physics, Third Edition. Edited by George L. Trigg and Rita G. Lerner
Copyright ©2005 WILEY-VCH Verlag GmbH & Co. KGaA, Weinheim
ISBN: 3-527-40554-2

Because of the monochromaticity of the source no dispersive element is needed and the spectral resolution is limited only by the line-widths of the observed samples.

Intensity fluctuations of most lasers can be essentially suppressed by active intensity stabilization increasing the detection sensitivity.

Because of the good collimation of a laser beam, long absorption paths can be realized by multiple reflections back and forth through the sample allowing the detection of tiny traces of gases.

The coherence and high spectral power density of single mode lasers allows to easily access nonlinear effects in the dependence of the optical properties of materials on the incident power. These has opened the way for the development of nonlinear spectroscopy.

High-Sensitivity Spectroscopy

In propagating through a medium light is absorbed with an exponential law. For small absorptions this law reduces to linear, therefore an intensity measurement directly yields the absorption coefficient once the length of the light path is known. From the absorption coefficient it is possible to extract many useful information such as the density of absorbing matter or the light absorption cross section. Various techniques have been developed to increase the sensitivity and accuracy of absorption measurements pushing the limit to relative absorptions of 10^{-17} and beyond.

The first objective to increase the sensitivity is to increase the absorption path length, which can be achieved by reflecting a laser beam back and forth through the sample either realizing an open path (multipass cell) or a closed path (optical cavity). The second objective is to reduce the effect of noise sources, the most important of which are often intensity fluctuations of the laser itself. This is achieved with frequency modulation techniques where either the laser intensity or the laser frequency is modulated and the detection is synchronized to the modulation. This effectively realizes a subtraction of the laser noise and eliminates all the other noise sources outside the detection bandwidth.

When the absorption is really low it can be useful not to detect the laser light at all but to detect only the absorbed radiation. This can be achieved in many different ways. For visible and ultraviolet light, the laser excites the electronic states of atoms or molecules, therefore excitation spectroscopy, where one detects fluorescence photons coming out of the material, is generally the most suitable technique. For more energetic photons, or for the spectroscopy of states lying very close to the ionization limit or else for two-photon absorption, one can monitor the produced ions in ionization spectroscopy. This technique is the most sensitive due to the very high collection efficiency that can be achieved with ions. For less energetic radiation, deep in the infrared region, no ions are produced and excitation spectroscopy is not very sensitive. In this case one has to resort to photoacoustic spectroscopy which uses the collision induced transfer of excitation energy into thermal energy. In special environments, such as gas discharges, optogalvanic spectroscopy, where one measures the change in the discharge resistance in presence of a laser beam, is a very sensitive technique. Other more specific techniques can be used for special atoms or molecules with large permanent dipole moments.

Overall all these techniques can be combined with intracavity absorption and frequency modulation to achieve the ultimate sensitivity limit.

Nonlinear Spectroscopy: Double-Resonance Techniques

The interaction of two beams of monochromatic light with atoms or molecules, satisfying a double resonance condition, can be used to eliminate Doppler broadening and allow the much narrower homogeneous linewidth to be resolved. When only a single transition is involved, the fields can be of the same frequency, but must propagate in opposite directions (saturation spectroscopy, Lamb-dip effect). When two transitions sharing a common level are excited, the two fields may be of different frequencies and can propagate either in the same direction (resonant Raman effect) or in opposite directions (two-photon absorption). Energy and momentum conservation or perturbation theory show that the Doppler shift $k_{1,2}v_z$ is eliminated for the Lamb-dip case when $\omega_1 = \omega_2 = \Omega_1 = \Omega_2$, for the resonant Raman case when $\Omega_1 - \Omega_2 = \omega_1 - \omega_2$, and for the two-photon absorption case when $\Omega_1 + \Omega_2 = \omega_1 + \omega_2$. Here, $\omega_{1,2}$ is the transition frequency of laser beams 1 and 2, $\Omega_{1,2}$ is the laser frequency, $\mathbf{k}_1 - \mathbf{k}_2$ is the propagation vector of light, and v_z is the molecular velocity component along the laser beam direction. To lowest order, the intensity of the double-resonance signal depends on the product of intensities of the two applied fields (nonlinearly) and exhibits a sharp Doppler-free tuning behavior.

Applications of Double-Resonance Techniques

One significant application of the Lamb dip has been the precise measurement of the speed of light. As a result, the speed of light is now a defined quantity, $c = 299\,792\,458\,\mathrm{m/s}$, that ties the cesium time standard to the length standard. The frequency of a continuous wave (cw) 3.39-μm He–Ne laser has been frequency-locked ($\Delta v/v = 10^{-13}$) to the Lamb dip of a CH_4 sample located inside the laser cavity. Figure 1 shows the tuning behavior of this laser, a nonlinear resonance 400 kHz wide riding on top of a 260-MHz-wide Doppler profile. The 88 THz optical frequency corresponding to the 3.39 μm wavelength was directly tied to the cesium microwave standard by generating a chain of microwave and laser oscillators linked

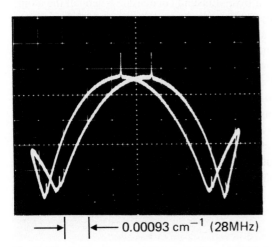

$\longrightarrow$ |$\longleftarrow$ 0.00093 cm^{-1} (28MHz)

Fig. 1: The tuning characteristic of a 3.39-μm He–Ne laser that contains an intracavity CH_4 sample (by permission of The American Physical Society).

by ultrahigh-frequency mixing diodes. Lamb-dip measurements of the hydrogen 2S–nD transitions in atomic hydrogen using a cw tunable dye laser have produced a new Rydberg value, $R_\infty = 109\,737.315\,714\,(19)\,\mathrm{cm}^{-1}$.

Two-photon spectroscopy has been used to measure the 1S→2S transitions in fundamental atoms such as hydrogen, positronium (e^+-e^-) and muonium (μ^+-e^-). The precise energy measurements of these intervals provide some of the best tests of quantum electrodynamics for a bound system.

Laser Cooling of Trapped Ions and Atoms

Neutral atoms and trapped ions can be cooled to temperatures on the order of ten microkelvins with laser light. At those temperatures, the atoms have velocity spreads on the order of 10 cm/s. Both the first and second order Doppler shifts of the particles are greatly reduced and the available measurement time is greatly increased. These techniques have yielded linewidths in the microwave regime of less than 10 mHz, and optical regime of less than 30 Hz, and will probably form the basis of our next generation time standards. Tunable lasers can be phase-locked to a reference Fabry–Pérot cavity, and the length of the cavity can then be stabilized by signals derived from laser cooled samples of ions or atoms. Laser cooling techniques are also broadly applicable to spectroscopic measurements where the signal appears as a small frequency shift. Tests of mass anisotropy, nonlinearity in quantum mechanics, time reversal invariance, and charge neutrality of atoms are examples of work where laser cooling is playing a role. Laser spectroscopy of cooled ions and atoms has also opened up the study of single and few ion systems, ultracold plasmas, ultracold collisions, and the formation of dilute quantum gases.

Laser cooled samples are also the starting point for other cooling methods, such as evaporative cooling, that are capable of taking the temperature of a sample of atoms in the picokelvin regime. At those temperatures the quantum statistical nature of the atoms plays a crucial role. Bosonic atoms perform a phase transition to a Bose–Einstein condensate where all the atoms in the sample occupy the same energy state. This is a new state of matter which is, to some extent, the analogous of laser light. The same laser light can be used to manipulate atoms at these temperatures with a completely coherent control. This has allowed the observation of a new class of phase transitions driven only by quantum fluctuations. Fermionic atoms, on the contrary, reach a degeneracy limit where atoms occupy all the available energy levels similar to what electrons do in a solid. This new quantum gases can be used to study, with a perfect control, many solid-state phenomena such as superconductivity or electron transport in metals. Furthermore atoms at these temperatures can be manipulated with potentials created with integrated electronic circuits opening the way to the development of a totally new generation of quantum devices where one could exploit the quantum correlations between the ultracold atoms to realize quantum logic circuits or quantum limited sensors with unprecedented capabilities.

Coherent Transients

Nonlinear optical processes of a coherent nature can be observed in the time domain and are capable of yielding detailed dynamic information about relaxation processes as well as

Doppler-free spectra. Coherent transient phenomena, such as optical nutation, free induction decay, and photon echoes arise in a sample excited by a sequence of resonant laser pulses. These phenomena are completely analogous to effects in pulsed NMR spectroscopy. Each transient effect may be used to select and examine a specific dephasing process. The laser field prepares the sample by placing the transition levels in quantum-mechanical superposition, and the resulting dipoles radiate in accordance with the coupled Schrödinger–Maxwell wave equations. The detection of the emitted radiation then yields information about the environment of the oscillators.

Picosecond and Femtosecond Spectroscopy

Extremely short pulses of light are used to examine events such as vibrational and rotational relaxation of molecules, the thermalization of photo-excited electron plasmas in a semiconductor, and nonthermal processes such as desorption of adsorbates from surfaces. Subpicosecond snapshots of chemical reactions, studies of the primary photochemical behavior of retinal in rhodopsin, and ultrafast electrical pulses are now possible with the short pulse technology. Most experiments use some variation of the "pump-probe" technique, where the first optical pulse is used to trigger an event with the absorption of one or more photons. A second pulse optically delayed with respect to the first pulse then probes the system spectroscopically. Figure 2 shows the induced transmittance change for the molecule Nile Blue using a 6-femtosecond pump and probe. Coherent transient techniques such as photon echoes or four-wave mixing are also applicable in time domain spectroscopy. The short laser pulses can also be used to generate very short bursts of x rays, phonons, electrons, and electrical pulses.

Applications of Laser Spectroscopy Outside Physics

Laser spectroscopy is virtually ubiquitous in natural sciences but it is also extensively used in many other fields such as engineering, e. g., for combustion monitoring or art restoration.

In analytical chemistry it is used for ultrasensitive detection of small concentration of pollutants, tracer elements or short lived intermediate species in chemical reactions. It can be used for the investigation of collision-induced energy-transfer processes or to induce chemical reactions by excitation of one or more of the reactants. Ultrashort laser pulses have led to the birth of real time femtochemistry where the formation or dissociation of molecules can be observed as it is happening. Laser spectroscopy can also be used for isotope identification and separation.

In biology laser spectroscopy has been fundamental for the determination of the structure of biological molecules or of energy transfer in DNA, while the extreme time resolution allowed by ultrafast lasers allows the study of fast dynamical processes such as isomerization during photosynthesis.

In medicine, besides the many successful therapeutic applications of lasers, laser spectroscopy is now extensively considered as a diagnostic tool, e.g. in breath analysis or in cancer management.

Geophysics and environmental research benefit from laser spectroscopy in the detection of natural gas emissions or in pollution monitoring. Of particular interest is the LIDAR (*l*ight *d*etection and *r*anging) technique, where a laser beam is sent into the atmosphere through an

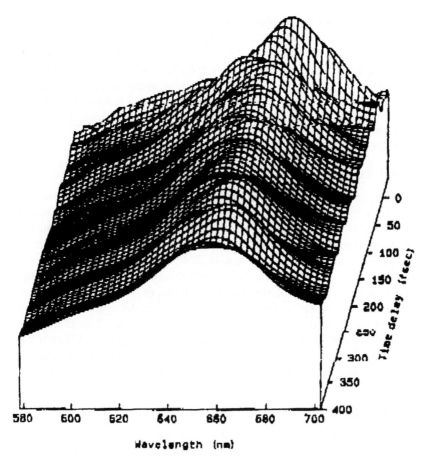

Fig. 2: Three-dimensional plots of the differential absorption spectrum as a function of time delay and wavelength for the dye Nile Blue. The amplitude and peak of the induced absorption oscillates with a period of 60 femtoseconds. (From C. V. Shank, *et al.*, p. 302 of *Laser Spectroscopy IX*, M. S. Feld, J. E. Thomas, and A. Mooradian (eds.), Academic Press, Boston, 1989.)

expanding telescope and backscattered off a distant object, e. g., a volcanic cloud of ashes or a polluted church facade. By measuring the backscattered radiation with spectral and time resolution it is possible to gain information on the object distance, extension and composition.

Other examples of laser spectroscopy applications are the aforementioned combustion monitoring which can be performed in situ and in real time, i. e., inside a functioning engine; or the measurement of flow velocities and turbulence.

Raman spectroscopy is widely used for the non invasive determination of the chemical composition of paint in frescos and paintings as well as stone analysis in sculpture.

Laser spectroscopy can also provide an answer to questions about the purity of material, its composition and the quality of production processes in the fabrication of electronic circuits.

Optical Frequency Synthesizers

One of the most recent breakthroughs in laser spectroscopy has been the development of the optical frequency synthesizer, which has completely revolutionized the world of time and frequency standards allowing the connection of the optical frequency domain with the radio frequency domain. Optical frequency synthesizers are the most precise measurement tool now available with an inherent stability that has been tested beyond 10^{-18}.

An optical frequency synthesizer is based on a mode–locked pulsed laser capable of emitting pulses that can be made as short as a few femtoseconds. The frequency spectrum of a train of pulses of emitted by such a laser is a comb of equally spaced modes (the frequency spacing is just the laser repetition rate) that can span over one octave. It is possible to compare the two ends of the spectrum thus comparing the modes at a given low frequency with the modes at twice that frequency. This can be achieved by frequency doubling the low frequency modes in a nonlinear crystal and observing on a fast photodiode the beatnote between the doubled modes and the high frequency modes. This procedure allows one to know the frequency of each comb mode but for an offset frequency. The offset frequency can be measured by comparing a suitably chosen mode of the comb with a laser beam stabilized on a stable frequency reference. Indeed the modes of the comb can be actively stabilized to the reference, this procedure effectively transfers the stability of the reference to each of the comb modes, which cover the entire visible range, and to the comb frequency spacing, which is in the radiofrequency domain. The intrinsic stability of optical frequency synthesizers is such that at the moment we are lacking a sufficiently stable reference to take full advantage of the synthesizer. This is one of the main reasons forcing us to reconsider the definition of the second abandoning the Caesium microwave standard in favor of an optical transition either of a laser cooled atom or of a single trapped ion.

See also: Lasers; Nonlinear Optics; Clocks.

Bibliography

L. Allen and J. H. Eberly, *Optical Resonance and Two-Level Atoms.* Wiley, New York, 1975.

S. Chu and C. Wieman (eds.), *J. Opt. Soc. Am. B* **6** (1989), Special Issue "Laser Cooling and Trapping of Atoms".

H. Figger, D. Meschede, and C. Zimmermann (eds.), *Laserphysics at the Limit.* Springer, Heidelberg-Berlin-New York, 2001.

M. Inguscio, S. Stringari, C. E. Wieman (eds.), *Bose–Einstein Condensation in Atomic Gases.* IOS Press, Amsterdam, 1999.

F. deMartini, C. Monroe (eds.), *Experimental Quantum Computation and Information.* IOS Press, Amsterdam, 2003.

W. Demtröder, *Laser Spectroscopy.* Springer, Berlin, 1998.

M. Feld, J. E. Thomas, and A. Mooradian (eds.), *Laser Spectroscopy IX.* Academic Press, Boston, 1989.

G. R. Fleming and A. E. Siegman (eds.), *Ultrafast Phenomena V.* Springer, Berlin, 1986.

S. Haroche, J. C. Gay, and G. Grynberg, *Atomic Physics 11.* World Scientific, Singapore, 1988.

Y. Prior, A. Ben-Reuven, and M. Rosenbluth (eds.), *Methods of Laser Spectroscopy.* Plenum Press, New York, 1986.

Y. R. Shen, *Principles of Nonlinear Optics.* Wiley, New York, 1984.

A. Siegman, *Lasers.* University Science Books, Mill Valley, 1986.

A. Yariv, *Quantum Electronics*, 3rd ed. Wiley, 1988.

Laser Cooling

A. Sasso

Introduction

When an atom interacts with a light beam, the emitted and absorbed photons carry much information about the atomic structure, this being the essence of spectroscopy. But the interaction of photons with atoms can also be used to manipulate the kinetic status of the atoms, i. e., their velocity. This phenomenon, usually named *laser cooling*, was first suggested independently by Hänsch and Schawlow [1] for neutral species and by Wineland and Dehmelt [2] for ions.

The production of ultra-cold atomic samples, having temperatures below microKelvin, has opened up a rich area for experiments in atomic and molecular physics. One of the main applications is in ultra-high-resolution spectroscopy, since for very slow atoms the first- and second-order Doppler shifts and pressure shift are eliminated. Moreover, Bose–Einstein condensation (BEC) was observed in 1995 [3] in a remarkable series of experiments on vapors of rubidium in which the atoms were confined in magnetic traps and cooled down to extremely low temperatures, of the order of fractions of microkelvins.

Basic Principles

The principle of laser cooling is best illustrated by a two-level atomic system. We consider atom of mass M and moving along the z-axis with a velocity v and we describe it in terms of a simplified two-level scheme: $|1\rangle$–$|2\rangle$. A photon of frequency ν traveling in the opposite direction can be absorbed if ν is appropriately Doppler-shifted with respect to the resonance frequency $\nu_0 = (E_2 - E_1)/h$. The photon momentum $h\nu/c$ changes the velocity of the atom by $\Delta v = h\nu/Mc$. We suppose that the excited atom can decay radiatively only back to the initial state. During this process, the atom also receives a velocity kick when it radiates a photon by spontaneous emission (we neglect stimulated emission). Since spontaneous photons are emitted isotropically the average recoil momentum transfer is zero, and there is thus overall a net momentum transfer from the radiation to the atoms. After de-excitation the atom is ready to absorb another photon, and so on. The cooling transition $|1\rangle$–$|2\rangle$ must therefore be closed, in the sense that atoms in upper level $|2\rangle$ may decay radiatively only to the initial state $|1\rangle$. For instance, for a sodium atom ($M = 23$ amu) which absorbs photons at $\lambda = 589$ nm ($3\,S_{1/2}$–$3\,P_{3/2}$ transition), the velocity change per photon absorbed is about 3 cm/s. If the initial atomic velocity is typical of room temperature distributions (about 10^5 cm/s), the number of scattered photons required to stop the atoms will be $n_{phot} = v/\Delta v = 3 \times 10^4$. If τ is the lifetime of the excited level, atoms undergo a maximum deceleration given by $a = h\nu/2Mc\tau$ during the laser cooling process. Again for sodium, $a \approx 10^6$ m s^{-2} which corresponds to about 10^5 times the gravitational acceleration g. This strong acceleration is sufficient to bring a thermal sodium atom to rest in 1 ms and over a distance of 0.5 m, which is quite reasonable on the laboratory scale.

Of course this is only a very simple description of the phenomenon. Many other effect should be considered. For instance, the sodium transition $3\,S \to 3\,P$, or similar alkaline atom transitions, is not a true two-level system but a multi-level system because of the hyperfine structure (hfs). After optical pumping on the hfs transition $3\,^2S_{1/2}(F'' = 2) \to 3\,^2P_{3/2}(F' = 3)$, the fluorescence can only reach the initial lower level $F'' = 2$ (see Fig. 1). A true two-level

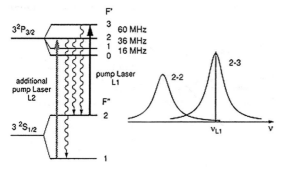

Fig. 1: Level diagram of the Na $3\,{}^2S_{1/2} \to 3\,{}^2P_{3/2}$ transition with hyperfine splittings. Optical pumping on the hfs component $F'' = 2 \to F' = 3$ represents the cooling transition while the transition $F'' = 1 \to F' = 2$ is used to refill atoms in the lower level $F'' = 2$.

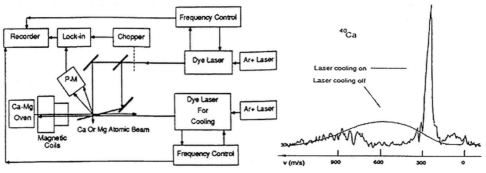

Fig. 2: Left: Simplified experimental setup for the deceleration of atoms in a collimated beam by photon recoil. Right: Laser cooling of calcium atoms calcium using the transition ${}^1S_0$–${}^1P_1$ at 422 nm (from [4]).

system would be realized if any overlap of the pump transition with other hfs components could be avoided. Nevertheless, since the lines 2–2 and 2–3 are partially overlapping, this causes also a pumping of the level $3\,{}^2P_{3/2}(F' = 2)$. As a consequence, some of the atoms are lost through the radiative decay toward the level $3\,{}^2S_{1/2}(F'' = 1)$. To overcome this problem, a second laser tuned on the transition $3\,{}^2S_{1/2}(F'' = 1) \to 3{}^2P_{3/2}(F' = 2)$ is used to refill atoms in the level $3\,{}^2S_{1/2}(F'' = 2)$.

Moreover, owing to the Doppler effect, the laser frequency ν has to be properly detuned in order to be resonant with the cooling transition and hence to avoid interruption of photon absorption.

An experimental apparatus suitable for laser cooling is shown in Fig. 2. A laser beam, appropriately detuned with respect to the atomic resonance, travels against an effusive atomic beam. A second laser beam, acting as an analysis beam, is roughly collinear with the cooling laser. When the laser frequency is tuned onto resonance, atoms absorb and re-emit photons which can be detected by a photomultiplier. The analysis beam is swept in order to probe the velocity distribution modified by the cooling beam. Part of the analysis beam is sent

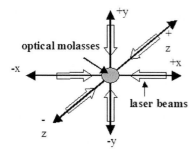

Fig. 3: Optical molasses with six pairwise counter-propagating laser beams.

perpendicular to the atomic beam to provide a zero-velocity marker. The magnetic coils in Fig. 2 produce a varying magnetic field along the atomic beam direction which shifts the Zeeman sub-levels so that the transition frequency between two Zeeman levels changes with position.

For cooling of atoms in a thermal gas where all three velocity components $\pm v_x, \pm v_y, \pm v_z$ have to be reduced, six laser beams propagating into the $\pm x, \pm y, \pm z$ directions are required (*optical molasses*). All six beams are generated by splitting a single laser beam (Fig. 3). If the laser frequency is tuned to the red side of the atomic resonance ($\Delta\omega < 0$), a repulsive force is always acting on the atoms, because for atoms moving toward the laser wave ($\mathbf{k} \cdot \mathbf{v} < 0$) the Doppler-shifted absorption frequency is shifted toward the resonance frequency ω_0, whereas for the counter-propagating wave ($\mathbf{k} \cdot \mathbf{v} > 0$) it is shifted away from resonance.

It can be shown that the net force resulting from the interaction with two counter-propagating beams is, for low velocity regime, viscous-like: $F_i = -a v_i$ ($i = x, y, z$) with a being a constant. The effectiveness of the optical molasses for cooling atoms anticipates that the atoms are trapped within the overlap region of the six laser beams for a sufficiently long time. This demands that the potential energy of the atoms shows a sufficiently deep minimum at the center of the trapping volume, that is, restoring forces must be present that will bring escaping atoms back to the center of the trapping volume.

We will briefly discuss the two most commonly used trapping arrangements. The first is based on induced electric dipole forces in inhomogeneous electric fields and the second on magnetic dipole forces in magnetic quadrupole fields. Letokhov [5] proposed to use the potential minima of a three-dimensional standing optical field composed by the superposition of three perpendicular standing waves for spatial trapping of cooled atoms, whereas Ashkin and Gordon [6] calculated that the dispersion forces in focused Gaussian beams could be employed for trapping atoms.

Induced Dipole Forces in a Radiation Field

When an atom with the polarizability α is brought into an inhomogeneous electric field $\mathbf{E}$, a dipole moment $\mathbf{p} = \alpha\mathbf{E}$ is induced and the force

$$\mathbf{F} = -(\mathbf{p} \cdot \nabla)\mathbf{E} = -\alpha(\mathbf{E} \cdot \nabla)\mathbf{E} = -\alpha\left[\left(\frac{1}{2}\nabla E^2\right) - \mathbf{E} \times (\nabla \times \mathbf{E})\right] \tag{1}$$

acts onto the induced dipole. The same relation holds for an atom in an optical field. However, when averaging over a cycle of the optical oscillation the last term in Eq. (1) vanishes, and we obtain for the mean dipole force (or gradient force)

$$\langle F_D \rangle = -\frac{1}{2} \alpha \nabla (E^2) = -a \Delta \omega \nabla I , \tag{2}$$

where $\Delta \omega = \omega - (\omega_0 - kv)$ is the detuning of the laser frequency against the Doppler-shifted resonance frequency of the atom, $I = \varepsilon_0 c E^2$, and $a = e^2/(m \varepsilon_0 c \gamma \omega_0 S)$ (with S the saturation parameter). Equation (2) reveals that in a homogeneous optical field, as a plane wave, the gradient force is zero because $\nabla I = 0$. For a Gaussian beam with the beam waist w propagating in the z-direction, the intensity $I(r)$ in the x–y-plane is:

$$I(r) = I_0 e^{-2r^2/w} \quad \text{with} \quad r^2 = x^2 + y^2 , \tag{3}$$

and the intensity gradient results:

$$\nabla I = -\frac{4}{w^2} I(r) r . \tag{4}$$

Therefore the dipole force F_D into the radial direction is directed toward the axis $r = 0$ for $\Delta \omega < 0$ (red traps) or radially outwards for $\Delta \omega > 0$ (blue traps).

For $\Delta \omega < 0$ the z-axis of an intense Gaussian laser beam with $I(r = 0)$ represents a minimum of the potential energy where atoms with sufficiently low radial kinetic energy may be trapped.

If a Gaussian laser beam is strongly focused then an intensity gradient is created in the r direction as well as in the z-direction. If the two forces are sufficiently strong, atoms can be trapped in the focal region. Nevertheless we have to remember that besides this dipole force in the r and z directions the recoil force (or scattering force) tends to push the atom in the $+z$ direction. In a standing wave configuration in $\pm z$ direction the scattering force along $+z$ and $-z$ are balanced and the gradient force may be sufficiently strong to trap atoms in all directions.

Magneto-optical Trap

Optical molasses can keep cold atoms in the region where the six laser beam intersect. In order to get high atomic density the tom should be concentrated in a small volume. A very elegant experimental realization for cooling and trapping of atoms is the magneto-optical trap (MOT), which is based on a combination of optical molasses and an inhomogeneous magnetic quadrupole field (Fig. 4). Its principle can be understood as follows.

In a magnetic field the atomic energy levels E_i experience Zeeman shifts

$$\Delta E_i = -\boldsymbol{\mu} \cdot \mathbf{B} = -\mu_B g_F m_F B , \tag{5}$$

which depend on the Landè g-factor g_F, Bohr's magneton μ_B, the quantum number m_F of the projection of the total angular momentum F onto the field direction, and on the magnetic field B.

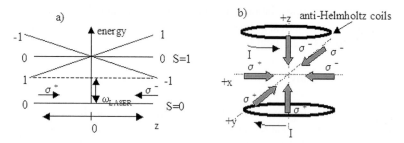

Fig. 4: Arrangement for a magneto-optical trap. (a) The energy-level scheme for an atom having spin $S = 0$ in the ground state and $S = 1$ in the excited state. (b) Experimental arrangement of a three-dimensional trap (from [8]).

In the MOT the inhomogeneous field is produced by two equal electric currents flowing into opposite directions through two coils with radius R and distance $D = R$ (anti-Helmholtz arrangement). If we choose the z direction as the symmetry axis through the center of the coils, the magnetic field around $z = 0$ in the middle of the arrangement can be described by the linear dependence $B = bz$, where the constant b depends on the size of the anti-Helmholtz coils. The Zeeman splittings of the transition from $F = 0$ to $F = 1$ are shown in upper part of Fig. 4. Atoms in the center of this MOT are exposed to the six red-tuned laser beams of the optical molasses. Let us at first only consider the two beams in the $\pm z$ direction, where the laser beam in $+z$ direction is σ^+ polarized. Then the reflected beam in the $-z$ direction is σ^- polarized. For an atom at $z = 0$ where the magnetic field is zero, the absorption rates are equal for both laser beams, which means that the average momentum transferred to the atom is zero. For an atom at $z > 0$, however, the σ^- beam is preferentially absorbed because here the frequency difference $\omega_L - \omega_0$ is smaller than for the σ^+ beam. This means that the atom experiences a net momentum transfer into the $-z$ direction, back to the center. In a similar way an atom at $z < 0$ shows a preferential absorption of σ^+ light and acquires a net momentum in the $+z$ direction. This shows that the atoms in the MOT are compressed toward the trap center.

Generally, the MOT is filled by slowing down atoms in an atomic beam. Spin-polarized cold atoms can also be produced by optical pumping in a normal vapor cell and trapped in a magneto-optic trap. This was demonstrated by Wieman and coworkers [7], who captured and cooled 10^7 Cs atoms in a low-pressure vapor cell by six orthogonal, intersecting laser beams. A weak magnetic field gradient regulates the light pressure in conjunction with the detuned laser frequency to produce a damped harmonic motion of the atoms around the potential minimum. This arrangement is far simpler than atomic beam. Effective kinetic temperatures of 1 mK have been achieved for Cs atoms.

The lowest achievable temperatures of the trapped atoms can be estimated as follows: because of the recoil effect during the absorption and emission of photons, each atom performs a statistical movement comparable to the Brownian motion. If the laser frequency ω_L is tuned to the resonance frequency ω_0 of the atomic transition, the net damping force becomes zero. Although the time average $\langle v \rangle$ of the atomic velocity approaches zero, the mean value of $\langle v^2 \rangle$ increases, analogous to the random-walk problem. The optical cooling for $\omega - \omega_0 < 0$ must compensate this "statistical heating" caused by the statistical photon scattering. If the velocity of the atoms has decreased to $v < \gamma/k$, the detuning $\omega - \omega_0$ of the laser frequency must be smaller than the homogeneous linewidth γ of the atomic transition in order to stay in resonance. This yields a lower limit of $\eta\Delta\omega \leq k_B T_{min}$, or with $\Delta\omega = \gamma/2$

$$T_{min} = \frac{\eta\gamma}{2k_B}.$$ (6)

In the meanwhile, experiments have shown that in fact temperatures lower than this calculated Doppler limit can be reached. The experimental results can be explained by a model which takes into account the effect of the optical pumping and level shifts caused by the intensity and polarization gradient (polarization-gradient cooling, Cohen-Tannoudji *et al.* [9]), Chu *et al.* [10]).

Applications of Cooled Atoms

Optical cooling and deflection of atoms open new areas of atomic and molecular physics. Collisions at very small relative velocities, where the de Broglie wavelength $\lambda_{dB} = h/mv$ is large, can now be studied. They provide information about the long-range part of the interaction potential, where new phenomena arise, such as retardation effects and magnetic interactions from electron or nuclear spins. One example is the study of collisions between Na atoms in their $3\,^2S_{1/2}$ ground state. The interaction energy depends on the relative orientation of the two electron spins $S = \frac{1}{2}$. The atoms with parallel spins form a Na_2 molecule in a $^3\Sigma_u$ state, while atoms with antiparallel spins form a Na_2 ($^1\Sigma_g^+$) molecule. At large internuclear distances ($r > 1.5\,nm$) the energy differences between the $^3\Sigma_u$ and $^1\Sigma_g^+$ potentials become comparable to the hyperfine splitting of the $Na(3\,^2S_{1/2})$ atoms. The interaction between the nuclear spins and the electron spins leads to a mixing of the $^3\Sigma_u$ and $^1\Sigma_g^+$ states, which corresponds in the atomic model of colliding atoms to *spin-flip* collisions (Fig. 5).

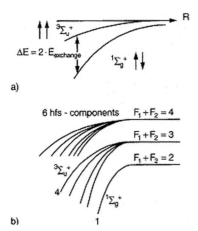

Fig. 5: Interaction between two Na atoms at large internuclear distances R for different spin orientations: (a) without hyperfine structure and (b) including the nuclear spins $I = \frac{3}{2}$, which gives three dissociation limits.

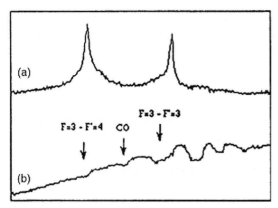

Fig. 6: (a) Resonant fluorescence from cold atomic cesium produced in a magneto-optic trap. (b) The saturated absorption spectrum for comparison.

Collisions between cold atoms in a trap can be studied experimentally by measuring the loss rate of trapped atoms under various trap conditions (temperature, magnetic-field gradients, light intensity, etc.). It turns out that the density of excited atoms cannot be neglected compared with the density of ground-state atoms, and the interaction between excited- and ground-state atoms plays an essential role. For collisions at very low temperatures the absorption and emission of photons during the collisions is important, because the collision time $\tau_c = R_c/v$ becomes very long at low relative atomic velocities v. The two dominant energy-transfer processes are collision-induced fine-structure transitions in the excited state and radiative redistribution, where a photon is absorbed by an atom at the position $\mathbf{r}_1$ in the trap potential $V(r)$ and another photon with a slightly different energy is re-emitted after the atom has moved to another position $\mathbf{r}_2$.

Another application is the deflection of atoms by photon recoil. For sufficiently good beam collimation, the deflection from single photons can be detected. The distribution of the transverse-velocity components contains information about the statistics of photon absorption. Such experiments have successfully demonstrated the antibunching characteristics of photon absorption [11]. The photon statistic is directly manifest in the momentum distribution of the deflected atoms [12]. Optical collimation by radial recoil can considerably decrease the divergence of atomic beams and thus the beam intensity. This allows experiments in crossed beams that could not be performed before because of a lack of intensity.

A very interesting application of cold trapped atoms is their use for an optical frequency standard [13]). They offer two major advantages: reduction of the Doppler effect and prolonged interaction times on the order of 1 s or more. Optical frequency standards may be realized either by atoms in optical traps or by atomic fountains. For the realization of an atomic fountain, cold atoms are released in the vertical direction out of an atomic trap. They are decelerated by the gravitational field and return back after having passed the culmination point with $v_z = 0$.

Since cold atoms move with velocities of a few centimeters per second, that allows direct Doppler-free spectroscopic measurements. Figure 6 shows a spectrum of the cesium $6\,S_{1/2}(F = 3,4)$–$6\,P_{3/2}(F = 3,4)$ transition obtained by laser irradiation of a cloud of trapped

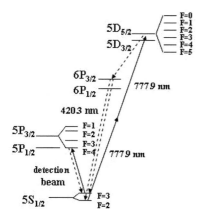

Fig. 7: "Spin shelving" two-photon spectroscopy scheme in ^{85}Rb.

atoms (Fig. 6a) and a conventional saturated absorption spectrum (Fig. 6b). The linewidth observed with the cold atoms is about 30 MHz, somewhat wider than the natural width of the transition (5.3 MHz). This is essentially due to the saturating effect of the trapping laser. In order to improve the resolution, the atoms should be measured after switching off the trapping and repumping laser beams.

Laser cooling techniques allow to achieve strong confinement together with low interaction between the atoms and the trapping light. A two-photon spectroscopy measurement of the transition $5\,S_{1/2} \rightarrow 5\,D_{5/2}$ in Rb (see Fig. 7), has been performed using extremely weak laser probe power by employing so-called "spin shelving" in analogy to electron-shelving spectroscopy. In this scheme, atoms are first loaded into the trap and optically pumped to a specific hyperfine level of the ground state ($F = 2$ in our case). Then, the probe two-photon laser is applied for a certain amount of time, exciting atoms to the excited state ($5\,D_{5/2}$) at some low rate. The excited atoms decay spontaneously back to the ground state, either to the initial hyperfine level, from which they can be excited again, or to the other hyperfine level ($F = 3$). In this level, the atoms can be stored or "shelved" for a long time, since they do not interact with either the probe laser or the trapping laser (dark trap, large detuning). The fraction of atoms transferred to this hyperfine ground level is the signal of the measurement. It can be accumulated for about a second, and then detected using a detection beam in resonance with a cycling transition. This scheme yields a quantum amplification of $\sim 10^7$ in photon rate as compared to measuring the spontaneous emission directly. In this experiment, it is possible to measure the above transition with only 25 mW laser power, about 100 times lower (10^4 times lower transition rate) then in previous experiments.

There are many possible applications of cold trapped molecules. One example is the spectroscopy of highly forbidden transitions, which becomes possible because of the long interaction time. Another aspect is a closer look at the chemistry of cold trapped molecules, where the reaction rates and the molecular dynamics are dominated by tunneling and a manipulation of molecular trajectories seems possible. Experiments on testing time-reversal symmetry via a search for a possible electric dipole moment of the proton or the electron are more sensitive when cold molecules are used .

References

[1] T. W. Hänsch and A. Schawlow, *Opt. Commun.* **13**, 68–69 (1975).

[2] D. Wineland and H. Dehmelt, *Bull. Am. Phys. Soc.* **20**, 637 (1975).

[3] M. H. Anderson, J. R. Ensher, M. R. Matthews, C. E. Wieman, and E. A. Cornell, in *Laser Spectroscopy XII*. World Scientific, Singapore, 1995.

[4] N. Beverini, F. Giammanco, E. Maccioni, F. Strumia, and G. Vissani, *J. Opt. Soc. Am. B* **6**, 2188–2193 (1989).

[5] V. S. Letokhov and B. D. Pavlik, *Appl. Phys.* **9**, 229 (1976).

[6] A. Ashkin and J. P. Gordon, *Opt. Lett.* **4**, 161 (1979).

[7] C. Wieman *et al.*, *Phys. Rev. Lett.* **65**, 1571 (1990).

[8] E. Raab, M. Prentiss, A. Cable, S. Chu, and D. Pritchard, *Phys. Rev. Lett.* **59**, 2631–2634 (1986).

[9] C. Cohen-Tannoudji *et al.*, *J. Opt. Soc. Am. B* **6**, 2023 (1989).

[10] S. Chu and C. Wieman (eds.), *J. Opt. Soc. Am. B* **6**, 2018–2278 (1989); Special issue on Laser Cooling and Trapping of Atoms.

[11] Y. Z. Wang *et al.*, in *Laser Spectroscopy VII*, T. Hänsch and Y. R. Shen (eds.), p. 238. Springer Series in Optical Science, Vol. 49, Heidelberg, 1985.

[12] Akulin *et al.*, *Phys. Rev. A* **44**, R1462 (1991).

[13] Ertmer *et al.*, *Metrologia* **22**, 195 (1986).

Lasers

J. J. Zayhowski and P. L. Kelley

Introduction

The laser is a device that generates coherent, highly directional electromagnetic radiation somewhere in the wavelength range from submillimeter through x-ray. Lasers can operate at a single wavelength (and frequency) or, when mode locked, on a large number of frequencies. The word *laser* is an acronym for "*l*ight *a*mplification by *s*timulated *e*mission of *r*adiation." The principle of operation of the lasers is similar to that of the maser, which is somewhat arbitrarily defined as a device operating in the range from the radio or microwave region down to millimeter wavelengths. Since the first laser was operated in 1960, the laser has come to play an important role through its revolutionary impact on applied optical technology, including fiber-optical communications and optical data storage.

Basic Theory of Operation

Population Inversion and Stimulated Emission

Quantum theory shows that matter can exist only in certain allowed energy levels or states. In thermal equilibrium, lower-energy states of matter are preferentially populated, with an occupation probability proportional to $e^{-E/kT}$, where E is the state energy, T the temperature, and k the Boltzmann constant. An excited state can decay spontaneously (i. e., with only zero-point electromagnetic radiation present) to a lower-energy state, emitting a quantum or wave packet of electromagnetic radiation (photon) with transition frequency $\nu = \Delta E/h$, where ΔE is the energy difference between the two states and h is Planck's constant. In the presence of

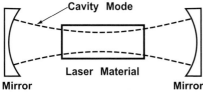

Fig. 1: Simplified schematic of a laser oscillator. The mirrors at the ends of the laser form an open resonator. Stable modes that consist of electromagnetic waves that travel back and forth in the resonator are amplified by the active laser material. In the radiative steady state, the gain due to amplification balances the loss due to intracavity absorption, mirror reflection losses, and diffraction beyond the edges of the mirrors. The pumping system is not shown, nor are ancillary intracavity elements that are often used for temporal (including frequency selection) and spatial control of the laser output. Usually, one of the mirrors is partially transmitting so that some of the highly directional radiation leaves the cavity through the mirror. The dashed lines are approximately characteristic of the transverse extent of the lowest-order transverse mode.

radiation at frequency ν, a transition from the upper state to the lower state can be induced, with the simultaneous emission of a photon in phase (coherent) with the stimulating radiation. This stimulated emission process is the inverse of the absorption process. If matter can be forced out of thermal equilibrium to a sufficient degree, so that the upper state has a higher population than the lower state (population inversion), then more stimulated emission than absorption occurs, leading to coherent growth (amplification or gain) of the electromagnetic wave at the transition frequency.

Laser transitions in the optical region are most often electric dipole in character. In the dipole approximation to the Hamiltonian, the transitions arise from a term of the form $er\mathcal{E}$, where e is the electronic charge, r is the quantum mechanical coordinate operator defined relative to the center of coordinates of the material system (such as an atom or molecule), and $\mathcal{E}$ is the electric field of the optical wave at the center of coordinates. The transition rate and gain cross section are proportional to the square of this interaction term. The transition matrix element of the coordinate operator between upper and lower laser levels ranges from about one hundredth of a Bohr radius ($\approx 0.5 \times 10^{-8}$ cm), for vibrational transitions in molecules and for local-field-induced transitions of rare earths in solids, to several hundreds of Bohr radii, for highly excited Rydberg atoms.

A laser generally consists of three components: (i) an active medium with energy levels that can be selectively populated, (ii) a pump to produce population inversion between some of these energy levels, and (usually) (iii) a resonant electromagnetic cavity that contains the active medium and provides feedback to maintain the coherence of the electromagnetic field (see Fig. 1). In a continuously operating laser, coherent radiation will build up in the cavity to the level required to balance the stimulated emission and cavity losses (see the discussion of *Threshold* below). The system is then said to be lasing, and radiation is emitted in a direction defined by the cavity.

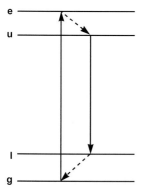

Fig. 2: Schematic representation of a four-level system. Population is pumped from g to e and laser operation occurs on the transition between u and l.

Pumping and Relaxation Processes

A material system can become excited and displaced from normal thermal equilibrium when driven by processes such as chemical reactions or under sufficiently strong external influence. External influences include electron beams and optical fields that selectively excite energy levels of the material. Applied voltages can create electrical currents, also resulting in disequilibrium. Disequilibration, if by a sufficiently selective process, can result in population inversion and laser operation. This "pumping" can be carried out continuously, with single pulses, or with multiple pulses of excitation. The inversion and its duration depend on the relaxation rates for the different energy levels and the degrees of freedom of the system, as well as on the rate of stimulated emission. The energy-level scheme of the laser plays an important role in obtaining inversion; in the section *Rate-Equation Model* we will discuss the difference in the operation of three-level and four-level lasers.

Let us try to understand pumping and relaxation in an "ideal" four-level laser with the aid of Fig. 2. The pumping process, indicated by the upward arrow, is assumed to excite the system from the lowest energy level, denoted by g for ground state, to the highest level, denoted by e for excited state. Pumping might occur in a variety of ways, one of which could be through radiative excitation using light whose frequency coincides with the transition frequency between g and e. The state e is assumed to relax to the upper laser level u. The population of the upper laser level is radiatively transferred, either through spontaneous or stimulated emission, to the lower laser level l. Finally, the lower laser level can either relax to the ground state or absorb the laser radiation and repopulate the upper laser level. Several conclusions concerning optimal operation can be made from this model. First, the relaxation rates from e to u and from l to g should be as rapid as possible in order to maintain the maximum population inversion between u and l. Second, the pumping rate between g and e should be sufficiently rapid to overcome the spontaneous emission from u to l. Third, the thermal equilibrium population of l should be as small as possible. Fourth, decay of e to any level other than u should be as slow as possible (for optical pumping, e can decay radiatively to g) and the nonradiative decay of u should be slow. For radiative pumping it is advantageous to have e be distinct from u and not to have radiative decay of u to g; for other types of pumping this advantage is not obvious.

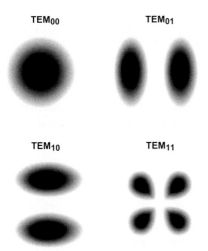

Fig. 3: Lowest-order transverse modes for a "stable" resonator with square symmetry. The TEM$_{mn}$ notation indicates that the modes have nearly transverse electric and magnetic fields with m nodes vertically and n nodes horizontally.

Not every practical laser satisfies this "ideal" model. The Cu-vapor laser violates the first conclusion listed in the last paragraph, since the lower laser level has a very slow decay rate, and the laser self-terminates due to filling of this level with concomitant reduction in the population inversion. Nevertheless, the Cu-vapor laser operates with fair efficiency as a powerful repetitively pulsed laser.

Resonators and Cavity Modes

An important aspect of the laser involves the design of resonators to accommodate the characteristics of the active medium and the diffractive properties of radiation, and at the same time meet requirements such as low angular beam divergence and high efficiency. The electromagnetic field in a resonator has well-defined modes that have patterns both transverse to and along the cavity axis. Waveguiding with index-of-refraction profiles or reflecting walls can be usefully employed in some cases (most notably in semiconductor and fiber lasers), not only to confine the radiation to the amplifying medium, but also to force the laser to operate in a single transverse mode. More frequently, however, laser resonators are open in the sense that the transverse structure is defined only by axial mirrors or lenses (see Fig. 1). Open resonators formed with convergent optics ("stable" resonators) generally have the lowest diffraction losses, while planar resonators have higher losses, and resonators formed with divergent optics ("unstable" resonators) have the highest losses. Figure 3 shows a few of the lowest-order transverse mode distributions for a stable resonator.

There is generally one transverse mode of a cavity that has the largest net gain (product of the amplifier gain and the transmittance of the remainder of the cavity). This is the transverse mode that oscillates first, and is typically the lowest-order (TEM$_{00}$) mode (see Fig. 3). Single-transverse-mode lasers, in particular lasers operating in the TEM$_{00}$ mode, have nearly optimal "spatial brightness" in the sense that the beam divergence is near the minimum value for the spot size of the laser on the output mirror; such a laser is said to have diffraction-limited output. The optimal resonator for a particular laser (e. g., "stable" or "unstable") is determined by the geometry of the gain medium, the desired cavity length, and the single-pass gain.

Cavity modes have an axial periodicity that is determined by the cavity length. The frequency spacing between the axial (longitudinal) modes is the inverse of the round-trip time for radiation in the cavity. The gain in a laser is peaked at a transition frequency determined by the energy levels of the active medium, and laser operation tends to occur at the axial-mode frequency (or frequencies) closest to the gain peak.

Important Characteristics of Laser Radiation

Linewidth – Spectral Brightness

Lasers can have very narrow linewidths and, therefore, very high spectral brightness (power per unit spectral interval). Linewidths as narrow as a fraction of a hertz have been obtained (one-hertz stability corresponds to a fractional stability of about 2×10^{-15} for visible lasers).

The first step in achieving narrow-line operation is to design the laser to operate in a single longitudinal and transverse mode (see the above section on *Resonators and Cavity Modes*). Once single-mode operation has been obtained, the linewidth of the output is determined by technical and fundamental noise. Technical noise arises from sources that can be controlled, such as power-supply fluctuations, variations in the thermal environment, environmental vibrations, etc. Fundamental noise arises from sources that cannot be eliminated, such as spontaneous emission and fundamental thermal fluctuations. The random effect of noise causes the laser frequency to drift. The influence of noise can often be reduced by adjusting controllable parameters such as cavity lifetime. In addition, the frequency drift can be reduced by measuring the frequency of the laser and providing feedback to adjust the cavity-mode frequency to compensate for the drift. The frequency can be measured in several ways: by using a stable Fabry–Pérot etalon, by comparing the laser frequency with a narrow resonance in an atom or molecule, or by comparing the frequency with a nearby frequency reference.

Intensity and Directionality (Angular Confinement) – Spatial Brightness

As discussed in the above section on *Resonators and Cavity Modes*, lasers operating solely in the lowest-order transverse mode (TEM_{00}) have optimal spatial brightness and are described as operating in the diffraction limit. In this case, for propagation of initially collimated radiation in a nonconfining medium, the product $A\Omega$, where A is the laser output aperture and Ω is the solid angle into which the power is confined at long distances, has a minimum value of order λ^2, where $\lambda = c/\nu$ is the wavelength of the radiation and c is the velocity of light. This aperture–solid-angle product is invariant to *ideal* transformations of the laser beam such as beam expansion. The invariance goes under various names: conservation of radiance or brightness, brightness theorem, and antenna theorem. Because of their diffraction-limited character, single-mode laser beams can maintain high intensities over long distances and have a high degree of directionality (small solid angle). The "times-diffraction-limit" or XDL of an actual beam is a (usually one dimensional) measure of how much the beam exceeds the ideal single-mode diffraction limit.

One utility of using diffraction limited laser beams can be seen by comparing their focusing characteristics to that of sunlight. A 20 μW diffraction limited laser beam expanded to fill a 3 cm diameter lens will produce the same intensity at the focus of the lens as the approximately 1.5 W of direct sunlight falling on the same lens.

Beam outputs can also be obtained through amplified spontaneous emission (ASE) without the directional feedback provided by a cavity simply by choosing an appropriate shape and timing of the excitation in the amplifiying structure. Typically, this does not lead to diffraction limited operation but can be important for short wavelength systems where fast spontaneous decay limits the usefulness of cavity feedback.

Short Pulses – Temporal Brightness

Short pulses of laser radiation can be made in a variety of ways including: gain switching, Q switching, cavity dumping, mode locking, and nonlinear frequency broadening and pulse compression. These techniques are described in following sections. To obtain the shortest possible pulses, mode locking is used to obtain pulses in the picosecond (1 ps $= 10^{-12}$ s) to femtosecond (1 fs $= 10^{-15}$ s) range, often followed by frequency broadening and pulse compression. Pulses as short as 4.5 fs have been obtained at a center wavelength of 800 nm, corresponding to less than 2 cycles of light, and very high peak powers are possible. To obtain even shorter pulses, nonlinear optical techniques are used to frequency convert the output of mode locked lasers into the VUV and soft x-ray spectral regions, where pulses as short as 100 attoseconds (1 as $= 10^{-18}$ s) have been reported.

High temporal brightness (high peak power) competes with high spectral brightness in the sense that the shortness of a pulse is limited by the bandwidth of the radiation. Said another way, for a fixed average power in a continuous train of bandwidth-limited pulses, peak power cannot be increased without reducing spectral brightness.

Types of Lasers

A rich variety of physical systems have been exploited to produce laser radiation over a five-decade range of wavelengths. We can only briefly describe here some of the most significant of these.

Solid-State Lasers

The first laser used a rod of ruby (single-crystal Al_2O_3 with the ion Cr^{3+} substituted for a small fraction of the Al^{3+} ions) as the active medium. In this laser, a xenon flashlamp pumps the chromium ions from their ground state to a broad band of states, from which they rapidly decay nonradiatively to a long-lived state at an energy of about 14422 cm^{-1} (693.4 nm) above the ground state. (In spectroscopy and laser physics, energy E and frequency v are often given in terms of wave number $1/\lambda$, where $E = hv = hc/\lambda$.) It is the narrow (≈ 11 cm^{-1}) emission line from this level back to the ground state that gives rise to ruby laser emission. When lasing occurs, the emission narrows in frequency width to less than 1 cm^{-1}, and leaves the ruby rod with an angular spread of a few milliradians.

The ruby laser is a three-level system (see Fig. 4 and the section on the *Rate-Equation Model* below), and therefore requires depopulation of the ground state by more than 50% to obtain population inversion. For this reason, pumping of the laser requires a high-intensity source, and continuous-wave (cw) operation of the system is difficult to achieve. In the free-

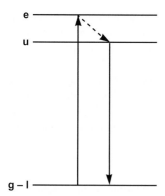

Fig. 4: Schematic representation of a three-level system. Note that g and l are now the same level.

running mode, as opposed to the *Q*-switched mode (see later in this section and the section on *Q-Switched Operation*), the output radiation of the ruby laser fluctuates rapidly over a time of about 1 ms and, for a 1-cm-diameter × 10-cm-long laser rod, is emitted in pulses of about 1 J.

The high threshold pump-power requirement of a three-level laser is greatly relaxed in a four-level system. In the four-level system, radiative transitions do not terminate on the ground state (see Fig. 2). If the final state is not significantly populated at the operating temperature, then population inversion can be maintained with only moderate pump power, and such lasers can usually be operated on a continuous basis.

One type of four-level solid-state laser is based solely on the electronic levels of ions in a crystal or other solid-state host, with the laser transition terminating on an excited electronic level of the ion. The rare-earth ions Nd^{3+}, Ho^{3+}, Er^{3+}, and Tm^{3+}, which can also be operated as quasi-three-level systems (with the lower laser level partially occupied), are the most frequently used active ions in such lasers. Extensively developed lasers of this type are the Nd-glass laser and the Nd:YAG ($Y_3Al_5O_{12}$) laser, both of which have their most important transitions in the vicinity of 1.06 μm. Nd:YAG lasers have produced in excess of 5 kW in continuous operation, several joules in low-repetition-rate (10 Hz) pulsed operation, and about 10 mJ in high-repetition-rate (1 kHz) pulsed operation. In order to increase efficiency and decrease heat loading in the solid-state laser medium, Nd:YAG lasers have been pumped with arrays of semiconductor diode lasers. Large Nd-glass lasers have produced tens of joules at repetition rates of several Hz, and several kilojoules at very low repetition rates ($\approx$ 1 per hour).

Rare-earth-doped glass fiber lasers are of utility in optical communications. Single-mode Er^{3+} fiber lasers at 1.55 μm (a quasi-three-level transition) have given output powers of the order of 10 W. At power levels of the order of 100 mW, they are in use as amplifiers in long-distance fiber transmission.

The availability of high-brightness diode lasers has made some quasi-three-level solid-state systems very attractive for continuous, as well as pulsed, operation. The rare-earth ion Yb^{3+}, which operates as a quasi-three-level system at 1.03 μm, has been found to be very suitable for diode laser pumping because of its relatively broad pump band. Yb^{3+} also has a paucity of low-lying electronic states, resulting in no absorption of pump light by ions in the excited state. The small energy difference between the excitation photon and the emitted photon leads to low thermal loading, but comes with the disadvantage of significant thermal occupation of the lower laser level at room temperature, and the Yb-ions must be pumped sufficiently hard to overcome the consequent absorption. The relatively small amount of heat generated in Yb-doped materials is attractive for scaling to high average powers, and Yb:YAG lasers have been operated with more than 1 kW of continuous output, making them competitive with more extensively developed Nd:YAG systems. A nitrogen temperature Yb:YAG laser has produced 165 W of diffraction limited output. Double-clad fiber lasers using Yb have also been operated with output powers over 1.3 kW. In the double-clad structure, the laser radiation is confined to a single-mode inner core while the multimode pump radiation travels through both the inner core and a larger multimode region. The shape of the multimode region can be tailored to the shape of the pump beam. For instance, a rectangular region can be used to accommodate the shape of a pump-diode array.

A second type of four-level solid-state laser involves fewer electronic levels of the active ion. Instead, laser transitions are employed that terminate on an excited vibrational level of the ion in the host lattice. The basic optical pumping–emission cycle is similar to that of a molecular dye laser (see the section on *Dye Lasers* below). One important class of these solid-state lasers uses transition-metal ions such as Cr^{3+}, Cr^{4+}, Ti^{3+}, Ni^{2+}, and Co^{2+}. As with the ruby (Al_2O_3) laser, the alexandrite ($BeAl_2O_4$) laser employs the Cr^{3+} ion. In the latter case, however, the laser transition can terminate on a variety of final vibrational states. As a result, this laser is tunable from 700 to 818 nm, and has pulsed output energies similar to the ruby laser. The Ti:sapphire (Ti:Al_2O_3) laser has even broader tunability, covering a range from 660 nm to beyond 1.1 μm. It has operated continuously at power levels up to 17 W with Ar-laser pumping, and pulsed at energies of hundreds of millijoules when pumped by doubled Nd:YAG laser radiation.

Lasers based on color centers in alkali-halide crystals also operate on a principle similar to dye lasers (see the section on *Dye Lasers* below) and transition-metal lasers. Using different types of *F* centers in various alkali halides, wavelength coverage over a range from 0.82 to 3.3 μm can be obtained, with continuous output powers ranging from tens of milliwatts to over a watt. Stability of *F* centers can be a problem, and low-temperature storage is required for several of these lasers.

The energy stored in a pumped solid-state laser medium can be delivered as a giant pulse in a time much shorter than the spontaneous lifetime of the upper laser energy level through the use of *Q* switching (quality-factor switching). This technique (see the section on *Q-Switched Operation* below), which was first used in the ruby laser, has been applied to many laser systems. In the case of the Nd:YAG laser, *Q*-switched pulses are typically of the order of 10 ns in duration and can be obtained with the use of a saturable absorber, an electro-optic element, or an acousto-optic element as the switch in the laser cavity. Miniature, or "microchip" *Q*-switched Nd:YAG lasers produce pulses as short as 100 μs.

Because solid-state lasers have spectrally broad gain regions, many equally spaced longitudinal cavity modes can lie within the gain bandwidth when cavity lengths are of the order of tens of centimeters. As a result, mode locking (see the section on *Mode-Locked Operation* below) can be used to obtain a train of high-intensity, short pulses. Mode locking has been applied to many solid-state laser systems such as Nd:YAG (20-ps pulses) and Ti:Al_2O_3 (5-fs pulses). Mode locking has also been used in many other types of laser (see, e. g., the section on *Dye Lasers* below).

Gas Lasers

A number of methods can be used to produce population inversion in gaseous media. Inversion can exist between some of the energy levels of the constituents in a gas discharge (electrical discharge in a gas). The first such system, demonstrated not long after the announcement of the ruby laser, was the He–Ne laser, now a standard item in optics laboratories. This system makes use of a discharge in He at a pressure of about 1 Torr, with an admixture of Ne at about 0.1 Torr. The discharge excites He atoms to their first excited level, about $160000 \, \mathrm{cm}^{-1}$ above their ground state. This excitation is readily transferred by collisions to a Ne atomic level with nearly the same excitation energy (resonant transfer). These excited states decay radiatively to lower-energy Ne states, giving rise to continuous laser emission in the red at $15820 \, \mathrm{cm}^{-1}$ ($632.8 \, \mathrm{nm}$) with an output power in the range of $10^{-2} \, \mathrm{W}$. Other transitions produce strong emission at $8680 \, \mathrm{cm}^{-1}$ ($1.15 \, \mu\mathrm{m}$) and $2957 \, \mathrm{cm}^{-1}$ ($3.39 \, \mu\mathrm{m}$). In a pure Ne discharge, excitation would occur to many Ne levels, and population inversion would not occur as effectively.

An important gas-discharge laser is based on the energy levels of the argon ion (Ar^+). Through a complex series of steps, argon-ion–electron collisions in the discharge lead to population inversion and lasing at a number of frequencies near $20500 \, \mathrm{cm}^{-1}$ ($488 \, \mathrm{nm}$). Continuous output at power levels of tens of watts in the blue-green make this device especially useful as a spectroscopic source in Raman scattering, and for pumping continuously operating tunable dye and $Ti:Al_2O_3$ lasers.

Other intense laser sources arise from atomic transitions of metal ions in a pulsed He discharge (the Cu- and Cd-vapor lasers, for example). Gas-discharge lasers have also been made to operate in the UV, but special problems arise in this frequency range; these are discussed below in the section *UV and X-ray Lasers*.

Gas lasers with output at longer wavelengths make use of the vibration-rotational energy levels of molecules. In addition to the electronic-state energy-level structure characteristic of atoms, there is vibration-rotational structure associated with the relative motion of the nuclei. The spacing of vibrational energy levels corresponds to frequencies in the infrared. It is in this region of the electromagnetic spectrum that molecular gas-discharge lasers are especially important.

The most efficient and powerful of the molecular gas-discharge lasers is the CO_2 laser. One version of this laser makes use of a dilute mixture of CO_2 in an N_2 discharge. The N_2 molecules are excited by collisions with electrons to their first excited vibrational state, from which the excitation is resonantly transferred by molecular collisions to preferentially excite CO_2 molecules to a particular vibrational state. These molecules in turn undergo radiative transitions to lower vibrational levels. The presence of the rotational structure gives rise to a cluster of many lines that can lase, grouped near frequencies of $944 \, \mathrm{cm}^{-1}$ ($10.6 \, \mu\mathrm{m}$) and $1042 \, \mathrm{cm}^{-1}$ ($9.6 \, \mu\mathrm{m}$). The electric-discharge CO_2 laser is quite efficient (better than 10% electrical power converted to laser power) and is capable of producing continuous output powers of greater than $1 \, \mathrm{kW}$. Other important molecular gas-discharge lasers make use of vibration-rotational or pure rotational transitions of H_2O, CO, and HCN, and produce emission at ($78 \, \mu\mathrm{m}$, $119 \, \mu\mathrm{m}$), $5.3 \, \mu\mathrm{m}$, and ($337 \, \mu\mathrm{m}$, $311 \, \mu\mathrm{m}$), respectively.

A number of other methods of generating laser radiation using molecular or atomic energy levels in gases have been devised. Powerful pulses of laser radiation at $1.315 \, \mu\mathrm{m}$ from excited iodine atoms have been produced by flash photolysis (UV photodecomposition) of CH_3I

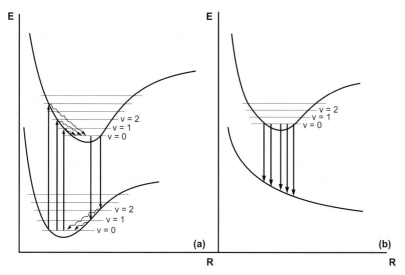

Fig. 5: Light emission from molecular systems: a) bound–bound system;
b) bound–free system.

(methyl iodide) and, using energy transfer from chemically generated $O_2(^1\Delta)$, kilowatts of continuous power have been obtained. In the gas-dynamic laser, a nonthermal distribution of molecular vibrational energy levels is produced by the rapid expansion of a hot gas through a nozzle. This method has produced continuous emission of tens of kilowatts at 10.6 μm from CO_2 gas. In the chemical laser, two reacting molecular species in a gas produce a product that is left in an excited vibration-rotation state and returns to the ground state radiatively. An example is the HF (DF) laser, which lases in the 2.5- to 3.5-μm region when H_2 (D_2) and F_2 gases combine chemically. Other types of lasers produced by excitation of gases are TEA (transversely excited atmosphere) lasers, e-beam (electron-beam excited) lasers, and UV-preionized electric-discharge lasers.

CO$_2$ laser radiation has been used to pump other gases, yielding far-infrared emission. If there is a coincidence between a CO_2 laser line and a vibration-rotational transition in another gas, an excited level can be populated directly or by collisional transfer of excitation. Pure rotational transitions can take place radiatively to an unpopulated level, producing far-infrared laser radiation. Such gases as NH_3 (output at 291 μm) and CH_3OH (output at 164 and 205.3 μm) have been made to lase by this method.

Dye Lasers

Many organic dyes, when illuminated with visible or UV radiation, fluoresce strongly at lower frequencies. This so-called Stokes shift of the fluorescence can be understood in terms of the change of equilibrium internuclear position with electronic excitation, the rapid vibrational relaxation within an electronic state, and the Franck–Condon principle, which states that an electronic transition in a molecule takes place so rapidly that the nuclear coordinates can be regarded as nearly fixed. An optical pumping-fluorescence cycle for a molecular system is indicated schematically in Fig. 5a. After absorption of a photon and nonradiative vibrational cascading to the lowest vibrational level, the excited electronic state can decay by fluorescence at a lower frequency, and the molecule returns to the ground state by nonradiative vibrational cascading. Since the final state in the radiative transition process is unoccupied, the pumping requirements for such a laser system need not be too severe. Discrete vibrational structure is washed out in a liquid, but the general outline of the cycle indicated in Fig. 5a is still preserved in dye fluorescence in a liquid. This process is virtually identical to that occurring in tunable transition-metal solid-state lasers such as $Ti^{3+}:Al_2O_3$, $Cr^{3+}:BeAl_2O_3$, and $Cr^{4+}:Mg_2SiO_4$. In the solid, the substitutional ion's coordinates with respect to neighboring host ions play the same role as the internuclear coordinates in the dye molecules.

A problem that arises with the use of dyes in a laser is the existence of long-lived electron spin-triplet states into which the usual excited electron spin-singlet state can relax nonradiatively. (For rhodamine 6G, a commonly used dye, this nonradiative relaxation time is $\approx 10^{-7}$ s.) This process interferes with laser action, but can be largely circumvented by circulating the dye solution through the laser cavity. Dye lasers can be flashlamp pumped, or pumped with radiation from Ar-ion, frequency-doubled Nd:YAG, or N_2 lasers. Dye systems fluoresce over a wide band of frequencies from the near-infrared through the visible, and are well adapted for use in a cw laser, with wavelength tuning over as much as 40 nm for a single dye. A battery of dyes placed in optical cavity structures gives tunable laser radiation over a range from roughly $10\,000\,cm^{-1}$ (1 μm) to $25\,000\,cm^{-1}$ (400 nm).

Because of the broad bandwidth of dye fluorescence lines and, in the case of pulsed dye lasers, the low Q of the resonant structure, the dye laser emission line is fairly broad. It can, however, be greatly narrowed without much loss in output power through the use of a diffraction grating in place of one of the mirrors in the cavity. Even narrower linewidths may be obtained by using intracavity frequency-selective elements (etalons). Single-longitudinal-mode operation can be obtained in both pulsed and cw dye lasers. Dye-jet fluctuations normally limit the stability of cw dye lasers, although stabilities of hundreds of kilohertz are readily obtained, and stabilities of a fraction of a hertz have been achieved. Thus, the dye laser became an important source of radiation for spectroscopy, but recently the tunable solid-state laser has proven more practical.

Since the dye laser has a spectrally broad gain region and many equally spaced longitudinal cavity modes falling within the gain bandwidth, it is well suited for mode-locked operation (see the discussion of *Mode-Locked Operation* below). Using continuous mode-locked dye lasers, pulse trains having pulses of less than 10 fs duration have been obtained. These short pulses have been used for studies of fast electronic processes in solids and organic molecules.

Semiconductor Lasers

In 1962, it was reported that a forward-biased semiconductor diode of GaAs radiated efficiently at about $11\,800\,cm^{-1}$ (850 nm). In the following year, a number of groups reported the observation of laser emission in this frequency region from suitably prepared diode structures. While these semiconductor devices are solids, they are differentiated from those lasers involving optically active ions in ionic hosts on account of their markedly different physical and technological characteristics.

To understand how semiconductor lasers function, it is necessary to consider the nature of the electronic energy states in a semiconductor. A periodic crystal has bands of allowed energy levels separated by forbidden energy gaps. In an intrinsic semiconductor at low temperatures,

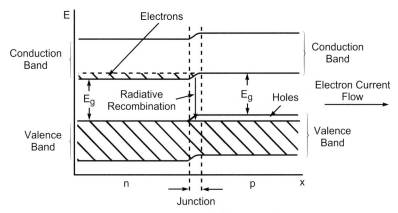

Fig. 6: Schematic showing the emission of radiation from a forward-biased light-emitting diode.

there are just enough electrons present to fill the uppermost occupied energy band (valence band), leaving the next higher band (conduction band) empty. In an n-type semiconductor, impurity atoms (donors) are present that contribute electrons to the conduction band; in a p-type semiconductor, there are impurity atoms (acceptors) present that can bind electrons, leaving behind missing electrons (holes) in the valence band.

Figure 6 shows a schematic of a p–n junction, fabricated by forming p- and n-type semiconductor layers in intimate contact. When a voltage is applied in the forward direction, electrons are injected from the n region into the depletion region of the junction (a region of the junction about $1\,\mu m$ thick). At the same time, holes are injected from the p region. As electrons drop into hole states they may emit radiation (electron-hole recombination radiation) at a frequency in the vicinity of the energy gap ($\nu = E_g/h$). When the injection current density is sufficiently high, population inversion and gain will be induced.

A typical semiconductor diode-laser heterostructure is shown schematically in Fig. 7. The dielectric film serves to guide the current into a narrow stripe region in order to concentrate the electron-hole-pair density. The x coordinate shown in Fig. 6 corresponds to the vertical direction in Fig. 7.

In many diode (or injection) lasers, a cavity structure (typically a fraction of a millimeter in dimensions) is provided by plane-parallel, cleaved facets at right angles to the junction plane. Laser emission is perpendicular to the cleaved facets, once the diode injection current reaches a threshold value. Multimode behavior often occurs in the plane of the junction. Since stimulated emission occurs in a narrow area near the junction (a few to several hundred micrometers in the plane of the junction and a fraction of a micrometer perpendicular to the junction), the angular spread of emitted radiation is fairly large, as expected from consideration of diffraction.

The technology of diode lasers has undergone considerable development, with the primary goals of achieving room-temperature operation, low thresholds, high output powers, improved mode quality, wavelength diversity, and long lifetimes. Progress has included improvements in electrical and optical confinement, and closer coupling of the gain region to the heat sink. Much of the improvement has involved advances in materials growth, including molecular-

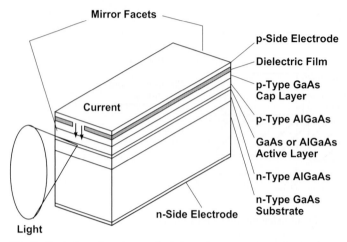

Mirror Facets

p-Side Electrode

Dielectric Film

p-Type GaAs
Cap Layer

Current

p-Type AlGaAs

GaAs or AlGaAs
Active Layer

n-Type AlGaAs

n-Type GaAs
Substrate

n-Side Electrode

Light

Fig. 7: Typical semiconductor diode-laser heterostructure. The material compositions shown produce lasers operating in the 750–850-nm wavelength range. The device length is typically 250 μm, the width 50–100 μm, and the height of the order of 50 μm. The epitaxial layers (i. e., excluding the substrate and the contacts) contribute only a few micrometers to the height. For thermal and electrical connection, the device is soldered to an electrically conducting heat sink, and an electrical lead is attached to the top side of the device. The width of the emitting region is 10–20 μm, while the height is approximately 0.5 μm; this causes the beam to diverge rapidly in the vertical direction.

beam epitaxy (MBE) and metal-organic chemical vapor deposition (MOCVD). These growth techniques have resulted in material of exceptionally high quality as well as structures with quantum confinement of carriers (electrons and holes).

Quantum-well diode lasers have exceptional optical and electrical properties, including very low laser thresholds. By reducing the volume of the active region through the use of quantum wells, the amount of current required to bleach the absorption of the semiconductor (the transparency current) is reduced, with a commensurate reduction in threshold. In addition, the quantum confinement of carriers in the active region increases their radiative gain cross section. Other quantum confinement structures, such as quantum rods and quantum dots, are being studied in an effort to further improve semiconductor lasers. In strained quantum well devices the density of states in the valence band is reduced, leading to a further reduction in the transparency current and reducing parasitic effects such as Auger recombination, which is important at long wavelengths.

In 1969 (before the advent of quantum-well devices), room-temperature continuous operation was achieved in a GaAlAs double-heterostructure laser. This laser structure consists of a small region of GaAs sandwiched between p- and n-type layers of the wider-bandgap alloy $Al_xGa_{1-x}As$ ($x < 1$). With further development, device lifetimes of tens of years were obtained, with output powers in the tens of milliwatts. These improvements opened up applications of considerable significance – in particular, fiber-optical communication and optical data storage. GaAlAs diode lasers have also been operated continuously at room temperature with output powers in the range of several watts. Linear arrays of diodes, in the form of multiple stripes in a 1-cm-wide bar, have given a total output of over 50 W. In addition, electrical-to-optical power conversion efficiencies of greater than 50% have been obtained.

Continuous laser operation at room temperature has also been obtained in the quaternary alloy system InGaAsP. Because laser operation is much less sensitive to dislocation effects in this system, long-lifetime devices were readily achieved. When operating at 1.3 and 1.55 µm, these lasers are matched to low-loss ($< 1\,db\,km^{-1}$), low-dispersion fiber optics, and are currently used in high-data-rate, long-distance communication.

InGaAsSb/AlGaAsSb quantum-well lasers have produced 1 W of continuous output (5 W pulsed) at wavelengths as long as 2.5 µm, at near room temperature. Cooled to liquid-nitrogen temperature, this material system can be used to produce laser output at wavelengths a long as 4 µm.

The compositionally tuned lead-salt lasers ($Pb_xSn_{1-x}Te$, PbS_xSe_{1-x}) operate at cryogenic temperatures. As with other semiconductor lasers, changes in composition change E_g and, therefore, the frequency of laser emission. Since E_g and hence the dielectric constant at near-bandgap wavelengths are sensitive functions of temperature in these small-gap semiconductor systems, cavity-mode frequencies and gain peaks are tunable, giving rise to temperature-tunable laser output. These tunable sources have been used extensively for high-resolution infrared spectroscopy in the 5- to 20-μm region.

Quantum-cascade lasers offer an alternative to lead-salt lasers for operation at mid-infrared wavelengths and longer. The active region of a quantum-cascade laser features several epitaxially grown layers of semiconductor. The device generates radiation based on electronic transitions that occur in the stacked layers. The thickness of the layers, rather than the fundamental bandgap E_g of the materials, determines the frequency of the emitted radiation. Quantum-cascade lasers have been operated at wavelengths from 3.4 to 67 µm, with the potential to extend the range from 1 to 100 µm. Continuous operation with output powers in excess of 0.5 W has been obtained at liquid-nitrogen temperature, and pulsed operation with peak powers approaching 1 W has been demonstrated at room temperature. These systems have also demonstrated tunability in excess of 150 nm.

In the visible region, using the quaternary alloy AlGaInP, continuous room-temperature operation has been obtained at 635 nm and pulsed operation at 603 nm; continuous liquid-nitrogen-temperature operation has been achieved at 583.6 nm, in the yellow region of the spectrum. Initial attempts to obtain operation in the blue-green spectral region focused on $Zn_{1x}Cd_xSe/ZnS_ySe_{1-y}$ quantum-well heterostructure diodes. These lasers have been operated continuously at room temperature, but have not yet demonstrated acceptable lifetimes for commercial applications. Operation of SiC diode lasers at 403 nm, in the blue-violet spectral region, has faced a similar problem. More recently, heterostructure diode lasers based on InAlGaN alloys have demonstrated in excess of 10,000 h lifetime for continuous room-temperature operation at wavelengths as short as 375 nm, and this system has the potential to be tailored for outputs throughout the visible and near-ultraviolet. Operating at an output power of a few milliwatts, InAlGaN diode lasers are now in use in CD/DVD recorders.

VCSELs (vertical-cavity surface-emitting lasers) have become an important class of semiconductor lasers. As with cleaved-cavity semiconductor lasers, these devices are constructed in layers using MBE or MOCVD. The peculiarity of the VCSEL is that some of the layers are used to form distributed Bragg mirrors on either side of the junction region. Radiation builds up in the direction normal to the junction instead of in the plane of the junction. Careful cleaving does not have to be done and the output beams are circularly symmetric. A high operating temperature (90 °C) VCSEL has been operated single mode at 1530 nm with 1.5 mW output power.

Semiconductor lasers with one of the radiation feedback elements external to the semiconductor material are also technologically significant. Wavelength-tuned external-cavity lasers are flexible sources for optical communications and other applications. Both conventional cleaved-cavity diodes and VCSELs have been used to make external-cavity lasers. In order to reduce the effect of the internal cavity (formed by the interface of the semiconductor to air), very good antireflection coatings must be used, having reflectivities of the order 0.0001. A passively mode-locked external cavity VCSEL has generated 1.42 W of average power at 960 nm.

Semiconductor optical amplifiers (SOAs) are being developed to provide high-power outputs and booster amplification in fiber-communication systems. Tapered-waveguide designs are used to prevent filamentation or "hot spots" at the end of the amplifier.

In addition to injection pumping of semiconductor lasers, a number of other methods not requiring the fabrication of a junction have been used successfully. These include electron-beam pumping and optical pumping.

UV and X-ray Lasers

To produce stimulated emission in the UV and x-ray regions, special problems must be addressed. Short lifetimes of inverted populations become important, since spontaneous radiative lifetime varies with frequency as v^{-3}. For example, the nitrogen gas-discharge laser, which radiates in the near UV (337 nm), can only operate in a pulsed mode, and must be pumped by a powerful intermittent source. Another problem is associated with the difficulty of devising resonant structures, since the reflectivity of materials becomes very small in the vacuum-UV and x-ray regions. A different conceptual scheme must frequently be used in these spectral regions, involving directionally amplified spontaneous emission (ASE) from an inverted population (superradiance), where the directional amplification is achieved by the geometry of the pumped region.

The nitrogen and hydrogen discharge lasers make use of radiative transitions between two bound electronic levels (bound-bound transitions). On the other hand, there are molecular systems (excimers) in which radiative transitions occur between a bound excited state and a free or very weakly bound ground state (bound-free transitions, see Fig. 5b). Xe and Kr form the excited molecular states Xe_2^* and Kr_2^*, although the diatomic molecules are unstable in their ground states. At high pressure these gases, when pumped by powerful electron-beam sources, emit superradiantly at 172 and 145.7 nm, respectively. At somewhat longer wavelengths, rare-gas–halide excimers, such as ArF* at 248 nm, KrF* at 193 nm, and XeCl* at 308 nm, have been operated both by electron-beam pumping and by transversely excited discharge. These lasers can produce multijoule pulses at 100-Hz repetition rates and are commercially available.

X-ray lasers at wavelengths shorter than a few tens of nanometers present a more difficult challenge. Nevertheless, several groups have achieved laser operation in this region. A multijoule visible laser can directly pump the gain medium through inversion produced in a laser-generated plasma, or indirectly by using the x-rays from a laser plasma to pump a separate x-ray laser medium. The first approach has been used to achieve operation near 21 nm in Se XXV, which has a $1s^2 2s^2 2p^6$ Ne-like electron configuration. Current research involves efforts to improve the very low efficiency of x-ray lasers as well as studies of potential new x-ray laser systems.

Free-Electron Lasers

Relativistic electrons traveling in a periodically alternating transverse magnetic field (wiggler) can be stimulated to give up radiation to a copropagating electromagnetic field of wavelength $\lambda = \lambda_w/2\gamma^2$, where λ_w is the wiggler period and γ is the ratio of the electron energy to its rest-mass energy. This phenomenon can be pictured as stimulated inverse Compton scattering from the electromagnetic field of the wiggler seen in the electron rest frame. To obtain efficient conversion over reasonable distances with high output power, the electron beam must be very monoenergetic, constitute a high current, and have very low angular divergence. Efficient operation also requires either the use of an electron storage ring or a tapered-period wiggler. Free-electron lasers have been operated at several wavelengths ranging from the visible to millimeter waves. While free-electron lasers do not have the wavelength restrictions imposed on other lasers, and do not have the inhomogeneities characteristic of many laser media, they require electron beams of a current level and quality that are difficult to achieve simultaneously. Further, the electron-beam and wiggler requirements tend to become more demanding as wavelength decreases. These systems have the large sizes that are associated with relativistic electron-beam sources.

Laser Dynamics

Rate-Equation Model

Many of the properties of the laser can be determined from a rate-equation model for the population of the laser levels and the photon number in the laser cavity. The rate equations provide a simple and intuitive, yet accurate, picture of the behavior of many lasers. In the most simplified form, the increase in photon number within the laser cavity is balanced by the decrease in the population difference between the upper and lower laser levels. In addition, the population difference increases on account of the pumping process, while the photon number decreases because of absorption, diffraction of the beam out of the cavity, and transmission through the mirrors.

The rate-equation model can be derived as an approximation to the fundamental equations relating the electromagnetic field, the material polarization, and the populations. The validity of the rate equations requires that the polarization can be accurately approximated by assuming that it instantaneously follows the field; this is a situation that applies to most lasers. In order to describe the problem in terms of total population and total photon number within the laser cavity, it is necessary that the gain of the laser be small during one pass through the cavity and that the laser operate in a single longitudinal mode (see the discussion of *Multimode Operation* below).

As stated above, the number of photons q within the laser cavity is affected by two types of events: the emission of a photon by the gain medium ($\dot{q}_{em}$) and the escape of a photon from the cavity or absorption by unpumped transitions ($\dot{q}_{dec}$). Photon emission can be either stimulated ($\dot{q}_{stim}$) or spontaneous ($\dot{q}_{spon}$). Once a laser is above threshold, the stimulated emission rate is much greater than the spontaneous rate and, to first order, spontaneous emission can be ignored. We will return to the issue of spontaneous emission in the next section when we discuss the buildup of a laser from noise.

The stimulated-emission rate is proportional to the number of photons within the cavity, the total population inversion N, and the probability per unit time B that a given photon will interact with a given inverted site. The interaction probability B is the product of the probability that a photon will pass within the gain cross section σ of a given inverted site as it traverses the laser cavity and the number of times per second the cavity is traversed. Mathematically, this reduces to $B = \sigma/A \times c/l = \sigma c/V$, where A is the cross-sectional area of the laser mode, l is the cavity length, and V is the volume of the lasing mode. The stimulated emission rate is, therefore, $\dot{q}_{\text{stim}} = qN\sigma c/V$.

The escape of photons from the laser cavity and their absorption within the cavity are characterized by the cold-cavity lifetime τ_c (cavity lifetime in the absence of any inversion) and corresponds to $\dot{q}_{\text{dec}} = q/\tau_c$. The total rate of change of the number of photons within the laser cavity is thus

$$\dot{q} = \dot{q}_{\text{stim}} - \dot{q}_{\text{dec}} = \frac{qN\sigma c}{V} - \frac{q}{\tau_c} \, . \tag{1}$$

We will now derive the rate equations for the population inversion of both a four-level and a three-level laser. The energy-level diagrams of a four- and a three-level laser are shown in Figs. 2 and 4. In both cases, the pump excites the active medium from the ground level (with population N_g) to level e (N_e). It is then assumed that the excited state quickly decays to the upper laser level, level u (N_u), so that $N_e \approx 0$. Lasing occurs between levels u and l (N_l). The difference between a four- and a three-level laser is that for the three-level laser the lower laser level is also the ground state.

To derive the rate equation for the population inversion $N = N_u - N_l$, we start by considering the population of the upper laser level N_u. The population of the upper level is affected by pumping P_r, stimulated emission $\dot{N}_{u,\text{stim}}$, and spontaneous decay $\dot{N}_{u,\text{spon}}$. For most pumping schemes, the pump rate is proportional to the number of ions in the ground state and can be written as $P_r = W_p N_g$. The stimulated-emission process decreases the population of the upper laser level by one for every photon created, so that $\dot{N}_{u,\text{stim}} = -\dot{q}_{\text{stim}}$. Spontaneous decay is characterized by the spontaneous lifetime τ, corresponding to $\dot{N}_{u,\text{spon}} = -N_u/\tau$. The rate equation for the upper-level population is thus

$$\dot{N}_u = P_r + \dot{N}_{u,\text{stim}} + \dot{N}_{u,\text{spon}} = W_p N_g - \frac{qN\sigma c}{V} - \frac{N_u}{\tau} \, . \tag{2}$$

In an ideal four-level laser there is a very rapid decay of the lower laser level to the ground state, so that N_l and $\dot{N}_l$ are approximately equal to zero. Since the total number of active ions N_t is constant, $N_t \approx N_g + N_u$. The rate equations for a four-level laser are, therefore,

$$\dot{q} = \left(\frac{N\sigma c}{V} - \frac{1}{\tau_c} \right) q \, , \tag{3}$$

$$\dot{N} = W_p(N_t - N) - \frac{qN\sigma c}{V} - \frac{N}{\tau} \, . \tag{4}$$

Since, in an ideal three-level laser, the lower laser level is the ground state, $N_t \approx N_1 + N_u$ and $\dot{N_1} \approx -\dot{N_u}$. As a result, the rate equations for a three-level system reduce to

$$\dot{q} = \left(\frac{N\sigma c}{V} - \frac{1}{\tau_c} \right) q , \tag{5}$$

$$\dot{N} = W_p(N_t - N) - \frac{2qN\sigma c}{V} - \frac{(N_t + N)}{\tau} . \tag{6}$$

The photon rate equations [Eqns. (3) and (5)] for the four-level and three-level lasers are the same. The rate equations for the population inversion are slightly different. In particular, the stimulated emission term for a three-level laser is twice that of a four-level laser.

Buildup from Noise

In the photon rate equation derived in the previous section, the term corresponding to spontaneous emission was left out. Laser action is initiated by spontaneous emission, or noise. As a result, these rate equations cannot account for the onset of lasing, as is seen by setting $q = 0$ at time $t = 0$. When spontaneous emission is properly taken into account, the photon rate equation becomes

$$\dot{q} = \frac{(qN + N_u)\sigma c}{V} - \frac{q}{\tau_c} . \tag{7}$$

The net effect is as if there were initially one photon in the cavity. This one photon of noise stimulates optical transitions and initiates lasing.

Threshold

The threshold inversion required for lasing is derived by requiring that the photon rate equation have a nontrivial solution in steady state. In steady state $\dot{q} = 0$, resulting in the condition $qN\sigma c/V - q/\tau_c = 0$. Physically, this states that the number of photons leaving the cavity must be balanced by the number of photons generated though the stimulated-emission process. The threshold inversion is, therefore,

$$N_{\text{thresh}} = \frac{V}{\sigma c \tau_c} . \tag{8}$$

The pump rate W_p required to reach threshold is obtained by setting $\dot{N} = 0$, $\dot{q} = 0$, and $N = N_{\text{thresh}}$. For a four-level laser

$$W_{p,\text{thresh}} = \frac{N_{\text{thresh}}}{(N_t - N_{\text{thresh}})\tau} \approx \frac{V}{N_t \sigma c \tau_c \tau} , \tag{9}$$

where we have assumed that $N_{\text{thresh}} \ll N_t$. For a three-level laser

$$W_{p,\text{thresh}} = \frac{(N_t + N_{\text{thresh}})}{(N_t - N_{\text{thresh}})\tau} \approx \frac{1}{\tau} . \tag{10}$$

For the same value of τ, the threshold pump rate for a four-level laser is smaller than the threshold pump rate for a three-level system by a factor of $V/N_t\sigma c \tau_c$, which is usually quite large. This is the basis of the superior performance of a four-level system over a three-level system in cw operation.

Gain Saturation

The gain g of an active medium is defined as the fractional change in optical intensity per unit length as a light beam passes through. From the above discussion of the *Rate-Equation Model*, it follows that $g = N\sigma/V$. In the presence of a strong optical field the population inversion is reduced and the gain is saturated. The rate equation for the population inversion (of a three- or four-level gain medium) in steady state can be rewritten in the form

$$g = \frac{g_0}{(1 + I/I_s)},\tag{11}$$

where g_0 is the unsaturated gain (the gain in the absence of an optical field), $I = h\nu qc/V$ is the optical intensity ($h\nu$ is the energy of one photon), and I_s is known as the saturation intensity.

The steady-state photon rate equation ($\dot{q} = 0$) predicts that above threshold ($q \neq 0$) the inversion density and the gain of a laser are clamped at their threshold values; the round-trip gain of the cavity is equal to the round-trip loss. With increased pumping, the gain remains fixed while the photon number and the output of the laser increase.

Laser Efficiency

The efficiency of a laser is often discussed in terms of the slope efficiency η_s. Slope efficiency is defined as the ratio of the change in output power to the change in pump power of a laser once it has reached threshold, and is determined by four factors: the pump efficiency η_p, the area efficiency η_a, the intrinsic efficiency η_i, and the output-coupling efficiency η_o. Mathematically,

$$\eta_s = \eta_p \eta_a \eta_i \eta_o.\tag{12}$$

The pump efficiency η_p is the ratio of the energy absorbed by the gain medium to the energy of the pump source. In an optically pumped laser, part of the incident optical energy may be reflected by the gain medium and part may be transmitted. Both of these effects decrease the pump efficiency. The area efficiency η_a is a measure of how well the pumped volume is used by the oscillating mode. If the cross section of the pumped volume is much larger than the cross section of the lasing mode, only a small portion of the pumped volume contributes to the gain of the system, and the area efficiency will be low. The intrinsic efficiency η_i is simply the ratio of the energy of the photon created during lasing to the energy required to create one excitation. In an optically pumped system, the intrinsic efficiency is the ratio of the energy of a photon at the oscillating frequency to the energy of an absorbed pump photon. Finally, the output-coupling efficiency η_o is the ratio of the output coupling to the total round-trip loss of the laser cavity.

The total efficiency η of a laser (power out divided by power in) is dependent on the slope efficiency and the laser threshold. Using the rate equation model, the slope efficiency is constant and the total efficiency is given by

$$\eta = \eta_s \left(1 - \frac{P_{thresh}}{P}\right),\tag{13}$$

where P is the total pump power and P_{thresh} is the pump power required to reach threshold.

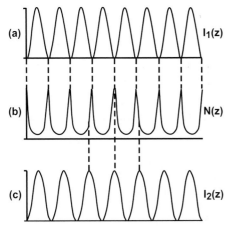

(a) $I_1(z)$

(b) $N(z)$

(c) $I_2(z)$

Fig. 8: Illustration of spatial hole burning, showing (a) the intensity profile of the first longitudinal mode to lase, $I_1(z)$; (b) the population inversion in the presence of the first oscillating mode, $N(z)$; and (c) the intensity profile of the second longitudinal mode to lase, $I_2(z)$.

Multimode Operation

For most lasers, the frequency spacing between adjacent longitudinal modes is much less than the gain bandwidth. As a result, lasers tend to oscillate at several frequencies simultaneously.

Although the above statement is true, the reasons are subtler than they may initially seem. In the early days of lasers, it was believed that lasers with homogeneously broadened gain spectra should operate in a single longitudinal mode. The reasoning behind this can be understood from the rate equations. If we assume a uniform optical intensity within the laser cavity, the steady-state solution to the photon rate equation fixes the inversion density at its threshold value. The first cavity mode to lase (the one with the highest net gain) clamps the inversion density and no other mode can reach threshold. The flaw in this reasoning lies in the assumption of uniform optical intensity.

Experimentally, lasers with both homogeneously and inhomogeneously broadened gain media tend to oscillate in several longitudinal modes as a result of spatial and spectral hole burning.

Spatial Hole Burning   In standing-wave laser cavities, the coherent superposition of the optical fields traveling in two directions within the cavity results in a sinusoidal intensity distribution. At positions were the intensity distribution is at its maximum, there is strong gain saturation and the population inversion is depleted. However, at nulls in the optical field the oscillating mode is unable to deplete the inversion. As a result, the inversion density is no longer uniform, but has "holes" at the positions corresponding to the peaks in the optical intensity. This phenomenon is known as spatial hole burning. The gain at the nulls in the optical field will continue to increase as the gain medium is pumped harder. Since other cavity modes have a different spatial profile than the first mode, and can use the population inversion at these positions, this will lead to the onset of multimode operation. These ideas are illustrated in Fig. 8.

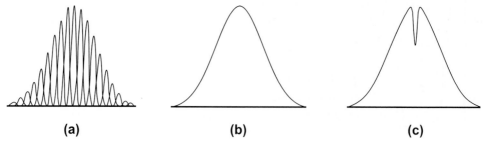

(a)	**(b)**	**(c)**

Fig. 9: Illustration of spectral hole burning, showing (a) several closely spaced homogeneously broadened spectra; (b) the inhomogeneously broadened spectrum resulting from the sum of several homogeneously broadened spectra; and (c) the inhomogeneously broadened spectrum with a spectral hole burnt in the center as a result of the saturation of one of the homogeneously broadened components.

Spectral Hole Burning   Homogeneous gain broadening occurs when each excited state of the gain medium has exactly the same energy distribution. This is often the case in crystalline solid-state gain media at room temperature. In materials such as glasses, each excited ion sees a slightly different environment, resulting in a slightly different energy spectrum. The ensemble effect is inhomogeneous gain broadening. Another example of an inhomogeneous system is a gas, where each atom or molecule is moving at a slightly different speed and has its energy spectrum Doppler shifted by a different amount.

In an inhomogeneous system, only those excitations with gain at the lasing frequency are able to participate in the stimulated emission process. As a result, only those excitations become depleted, producing a gain spectrum that has a dip at the lasing frequency. This is known as spectral hole burning, and is illustrated in Fig. 9. Excitations that cannot contribute to the lasing process for the first mode can contribute to the onset of lasing for other modes, resulting in multimode oscillation.

Single-Frequency Operation   There are many techniques for obtaining single-frequency operation from a laser. Several of these involve introducing an element into the cavity such that the cavity sees a frequency-dependent loss, thereby decreasing the net gain bandwidth and selecting an individual longitudinal cavity mode. Examples of such elements are a prism, a grating, a Fabry–Pérot etalon, and the combination of a birefringent filter and a polarizing element. In some cases, a cavity may require more than one device in order to obtain enough frequency selectivity to ensure single-frequency operation.

Single-mode operation has also been obtained by reducing the length of the cavity so that the longitudinal mode spacing is comparable to, or less than, the gain bandwidth. This is most easily done with gas lasers, which have a narrow gain bandwidth, but has also been achieved in very short solid-state lasers.

Alternatively, single-mode operation can be achieved in a laser with a homogeneously broadened gain medium by reducing or eliminating the effect of spatial hole burning. A unidirectional ring cavity has a uniform optical intensity within the cavity, rather than the sinusoidal intensity distribution of a standing-wave cavity. Spatial hole burning is therefore eliminated, and such a laser may operate at a single frequency well above threshold. A variation of the same idea involves placing a quarter-wave plate on either side of the gain medium. As a result, the optical fields traveling in opposite directions in the gain medium are orthogonally polarized and do not interact coherently. The optical intensity within the gain medium is, therefore, uniform and there is no spatial hole burning.

In some gain media there is a large amount of energy diffusion. Energy diffusion moves some of the excited states away from the peaks in the population inversion, toward the minima. This smoothes out the population-inversion profile, reducing the effect of spatial hole burning. One important example of such a gain medium is a semiconductor.

Finally, the effect of spatial hole burning is reduced if the gain medium is located very close to a cavity mirror. At a mirror, the phase of the optical intensity distribution is pinned at zero for all of the cavity modes. The phase difference between longitudinal modes increases gradually as one moves toward the center of the cavity. Quite close to the mirror, the peaks and nulls of the optical intensities for each of the modes occur in approximately the same place, and that portion of the gain that is not depleted by one mode will not be in a good position to contribute to any other.

Types of Pulsed Operation

Long-Pulse Operation

Long-pulse, or quasi-cw, operation refers to a pulsed laser with a pulse duration long enough for all relevant parameters within the system to come to their steady-state value. Although the behavior of the system is cw-like at the end of the pulse, it will, in general, be quite different at the beginning of the pulse.

Let us consider a laser with a step-function pump source. The pump may quickly create a population inversion. It will take some time, however, for a lasing mode to buildup from spontaneous emission. During this time, the inversion density may greatly exceed threshold. The large inversion density eventually results in a large optical intensity, well in excess of the cw value. This optical intensity, in turn, drives the inversion density below threshold, substantially reducing the laser intensity. The entire process then starts again. For a single-mode laser, this often leads to regular spiking at the beginning of the pulse. The process is damped, however, and with time the intensity of the spikes decreases. Spiking eventually gives way to damped oscillations (known as relaxation oscillations) in the optical intensity and finally cw-like behavior.

Figure 10 shows the computer solutions to the rate equations for a laser under the conditions just described. In this computer simulation, the spiking was heavily damped and cw-like behavior was quickly obtained. In a multimode laser the interaction between modes often leads to mode hopping, mode beating, and very irregular spiking, which may never damp out.

Relaxation Oscillations   Relaxation oscillations occur whenever the population inversion of a laser is disturbed from its steady-state value. It is a result of the coupling between the population inversion and the photon density within the laser cavity, as described above. From the rate equations, it can be shown that for a single-mode laser a small disturbance in the

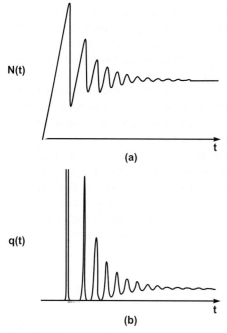

N(t)

(a)

q(t)

(b)

Fig. 10: Computer solutions to the rate equations for a laser with a step-function pump source, showing (a) the population inversion $N(t)$, and (b) the photon number $q(t)$.

inversion density results in damped oscillations with an oscillation frequency

$$\omega = \left[\frac{(N/N_{\text{thresh}} - 1)}{\tau_c \tau} \right]^{1/2} \tag{14}$$

and a damping constant

$$t_0 = \frac{2\tau N_{\text{thresh}}}{N} . \tag{15}$$

Note that if $1/t_0 > \omega$ the oscillations are overdamped and spiking will not occur. Although this condition is not satisfied in solid-state lasers, it is common in gas lasers.

Q-Switched Operation

For many applications it is desirable to obtain short, high-peak-power pulses from a laser. This can be achieved by creating a large population inversion and then quickly decreasing the cavity loss so that the inversion density is well in excess of its new threshold value. The large inversion density allows an intracavity optical field to rapidly develop. This field then depletes the population inversion and turns itself off. The cavity loss is subsequently increased to prevent the development of a second pulse. This technique is known as Q switching, since the quality factor, or Q, of the optical cavity is changed.

Q switching relies on the fact that the lifetime of the population inversion is much longer than the output pulse. The gain medium is therefore able to store energy, which can be quickly released in the form of a short output pulse.

The length of the Q-switched output pulse is dependent on several factors. In order to obtain a rapid buildup of the optical pulse, it is desirable to have a large gain cross section and a large population inversion when the Q of the cavity is switched. The rapid buildup of the output pulse also relies on the presence of a strong intracavity optical field, which argues in favor of a high cavity Q. In order to obtain a short output pulse, however, the intracavity intensity must also decay quickly after the peak of the pulse. Since the Q of the cavity is constant for the duration of the pulse (it is difficult to change the Q of the cavity significantly during the duration of a short output pulse, although it can be changed in the relatively long pulse buildup time preceding the pulse), this would argue for a low cavity Q. Solutions to the rate equations show that the minimum-width output pulse is obtained when the total round-trip loss of the cavity during the output pulse is adjusted so that the initial inversion density is about three times the threshold value.

Methods for Q switching a laser include the use of electro-optic shutters, acousto-optic Q switches, and mechanical devices. Passive Q switching can be obtained through the use of an intracavity saturable absorber with a long recovery time.

Gain-Switched Operation

Gain switching is another way to obtain short, high-peak-power pulses from a laser. The idea is to rapidly increase the pump power so that the population inversion of the laser is well in excess of the threshold value by the time the first output spike develops. The optical pulse then drives the population inversion below its threshold value. The pump power is subsequently reduced so that the inversion remains below threshold and only a single pulse is obtained.

The factors that are important for short gain-switched pulses are the same as for short Q-switched pulses.

Cavity-Dumped Operation

Cavity dumping allows the energy in a laser cavity to be output in a time comparable to the cavity round-trip time. The concept is to rapidly (within a cavity round-trip time) introduce a large output coupling (nearly 100%) into a cavity that previously had no output coupling. Methods for cavity dumping include the use of an electro-optic Pockels cell and polarizing beam splitter, and acousto-optic devices.

Mode-Locked Operation

Mode locking refers to the situation when the phases of several cavity modes are fixed (or locked) with respect to each other such that the electric fields add coherently and constructively for a short period of time. This allows the generation of a train of high-intensity, ultrashort pulses. To understand this, consider the case of $2n + 1$ equally spaced longitudinal modes oscillating with the same amplitude E_0. Assume that the phases ϕ_m of the modes are locked according to $\phi_m - \phi_{m-1} = \phi$, where ϕ is a constant. The total electric field is the sum

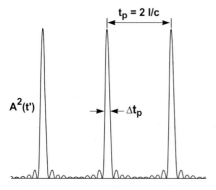

Fig. 11: Train of mode-locked pulses, made up of 11 modes of equal intensity.

of all of these modes:

$$E(t) = \sum_{m=-n}^{n} E_0 \exp[i(\omega_0 - m\Delta\omega)t + m\phi] = A(t')\exp(i\omega_0 t) , \tag{16}$$

where ω_0 is the frequency of the center mode, $\Delta\omega$ is the frequency difference between two adjacent modes, $t' = t + \phi/\Delta\omega$, and

$$A(t') = E_0 \frac{\sin[(2n+1)\Delta\omega t'/2]}{\sin[\Delta\omega t'/2]} . \tag{17}$$

Equation (16) shows that $E(t)$ can be represented in terms of a wave with a carrier frequency ω_0 whose amplitude $A(t')$ is time dependent. The intensity of this wave is given by $A^2(t')$, which consists of a train of pulses whose peak intensity is $A^2(0) = (2n+1)^2 E_0^2$, pulse width is $\Delta\tau_p = 2\pi/(2n+1)\Delta\omega$, and separation between pulses is $\tau_p = 2\pi/\Delta\omega$. Since the total oscillating bandwidth $\Delta\nu_{osc}$ is given by $(2n+1)\Delta\omega/2\pi$, the pulse width can also by written as $\Delta\tau_p = 1/\Delta\nu_{osc}$. Using the relationship $\Delta\omega = \pi c/l$, where l is the length of the laser cavity, the time between pulses is $\tau_p = 2l/c$. In words, a mode-locked laser produces a train of output pulses whose pulse width is given by the inverse of the oscillating bandwidth and whose separation between pulses is equal to the round-trip time of the laser cavity. Fig. 11 shows $A^2(t')$ for $n = 5$.

In the above example, the phases of all of the modes were locked so that the output pulse had its minimum possible duration. Such a pulse is referred to as a transform-limited pulse, since its temporal profile is the Fourier transform of its spectral profile. This need not be the case – it is possible to obtain longer pulses, but not shorter pulses. Also, it is worth noting that, unlike the other pulsed schemes described in this section, a mode-locked laser is a cw device and there is phase coherence between pulses.

The above discussion tells us what mode locking is, but sheds little light on why or how it occurs. In general, mode locking will occur if the net gain for a mode-locked train of pulses is greater than the net gain of any other combination of cavity modes.

Active Mode Locking   Mode locking can be obtained actively or passively. Active mode locking can be broken into two categories, "AM mode locking" (produced using an amplitude modulator within the laser) and "FM mode locking" (produced using a frequency or phase modulator within the laser). In AM mode locking, the loss of some element in the laser cavity is modulated at the round-trip cavity frequency. Light circulating in the cavity will see less loss, and therefore more net gain, when it is incident at the loss element during the time of minimum loss. This encourages short-pulsed operation and mode locking can be induced. The same result occurs if the gain of the cavity is modulated. Gain modulation through modulation of the pump source is known a synchronous pumping.

In FM mode locking, the optical length of the laser cavity (length or refractive index) is modulated at the round-trip cavity frequency. For simplicity, let us consider the case where one of the mirrors is moved sinusoidally along the direction of the cavity axis. Light incident on the mirror during its motion will be Doppler shifted. As a result, it will not reproduce itself after one round trip, and will not produce a coherent oscillating mode. Light incident on the mirror at its turning points (maximum or minimum cavity length) sees a stationary mirror and will not experience a Doppler shift. The net result is that mode-locked pulses will tend to form. Modulation of the refractive index at some point in the cavity has the same effect.

Passive Mode Locking   Passive mode locking can occur when a laser cavity contains a nonlinear optical element, such as a saturable absorber. In this case, the more intense the light incident on the saturable absorber, the less the total absorption. The total loss of the cavity is therefore minimized by putting all of the energy into short pulses. This is essentially self-induced AM mode locking. A similar effect is obtained by putting a Kerr lens and an aperture within the cavity. Other techniques include the use of interferometric elements containing nonlinear media.

Passive mode locking must be initiated by the presence of a pulse within the cavity. If the optical intensity within the cavity is uniform in time, there is no loss or gain element that is modulated at the round-trip cavity frequency in order to induce mode locking. Noise, however, is capable of introducing a small amplitude modulation on the optical field. In some lasers, this small modulation is sufficient to start the mode-locking process. Such lasers are referred to as self starting. In other systems a pulse (or AM modulation) must be intentionally introduced into the cavity to start the mode-locking process. Once started, however, mode locking can persist for a long time.

The introduction of an appropriate nonlinear optical element into a laser cavity is not sufficient to guarantee mode locking. In order for passive mode locking to work, the relative phases of all of the longitudinal modes must remain constant. One effect that can destroy the phase relationship, and therefore prevent mode locking, is dispersion. Passively mode-locked lasers must be dispersion free if mode locking is to occur. To accomplish this, prisms are often introduced into the laser cavity to compensate the dispersion of other intracavity elements, such as the gain medium. In a properly compensated cavity, non–mode-locked operation can be unstable.

The other effect that can destroy the phase relationship between the spectral components in a mode-locked pulse train is spontaneous emission, or noise. The phase of the noise is unrelated to the phase of the oscillating mode. The net phase will be shifted when the two are combined.

Control of Laser Output

Frequency Tuning

Frequency tuning of a laser can occur in one of two ways. If the longitudinal mode spacing of the laser cavity is much less than the gain bandwidth, the cavity is capable of supporting several modes, each at a different frequency. A single frequency is then selected through the insertion of an element into the cavity such that the cavity sees a frequency-dependent loss, as discussed in the above section on *Single-Frequency Operation.* In most of the examples listed in that section, a small repositioning of the frequency-selective element would result in a new longitudinal mode (and hence a new operating frequency) being selected. The frequency-selective element is used to select one of the several cavity modes, and discrete tuning is obtained. Fast tuning can be obtained through the use of electro-optic frequency-dependent components.

The other way a laser can be tuned is to change the frequency of a given cavity mode, by changing the optical length of the cavity. Since the cavity length can be changed continuously, this leads to continuous tuning. This type of tuning is often limited by the free spectral range of the cavity. Once the cavity modes are shifted by a full free spectral range, an adjacent cavity mode is positioned at the frequency where the initial mode started. For the same reasons that the initial mode was originally favored, the adjacent mode is now favored, and the laser will have a tendency to mode-hop back to the original frequency.

Amplitude Modulation

The output power of a laser can be controlled by changing the pump power, the output coupling, or the intracavity loss. This type of amplitude modulation is usually limited to frequencies below the frequency of the relaxation oscillations. The relaxation frequency characterizes the response time of the cavity. Near the relaxation frequency there is resonant enhancement of the modulation response; above the relaxation frequency the response rolls off.

Methods used for direct amplitude modulation of a laser may have the side effect of introducing frequency modulation as well. For example, changing the pump power affects the thermal load on the gain medium, and therefore the temperature. This, in turn, affects the refractive index, changing the optical length of the cavity and the oscillating frequency. For amplitude-modulation applications where frequency stability is critical, it is often better to modulate the laser power external to the cavity.

Oscillator–Amplifier Systems

As the required output power of the laser increases, the need for amplification stages becomes more apparent. Master oscillators (lasers) that operate at low powers can be easily controlled to produce a desired output. Successive amplification stages can be designed to provide increasingly higher output powers by increasing the pumping power and the amplifier aperture. Limiting the gain of each stage can eliminate parasitic oscillation. Isolators are used to prevent feedback from subsequent amplification stages. The design of each stage can also be optimized for heat removal and low optical distortion. Beam cleanup between stages can be used to suppress unwanted spatial frequencies and, thereby, maintain single-transverse-mode operation. Amplifiers can also be paralleled by splitting the master oscillator beam; if a coherent output is desired, the outputs of the amplifiers must be phased.

There are two basic types of amplifier, regenerative and traveling wave. A regenerative amplifier provides a feedback loop and can oscillate without the laser (master oscillator) input. Control of the regenerative amplifier is achieved by injection seeding it with the master oscillator signal. In a traveling-wave configuration, the amplifier simply boosts the signal injected by the master oscillator. Regenerative amplifiers are, by nature, multi-pass. Traveling-wave amplifiers can be designed so that the radiation travels through the amplifying medium more than once without retracing its path.

Laboratory-scale oscillator-amplifier systems can be designed to produce petawatt (10^{15} W) pulses. Very short (femtosecond) pulses cannot be directly amplified to very high powers without damage to the amplifier material. Instead, a dispersive element can be used to chirp and stretch the pulse in time. This lowers the peak power so that the pulse can be passed through a broadband amplifier, such as $Ti:Al_2O_3$, and later through another dispersive element that un-chirps and compresses the amplified pulse.

Issues in Laser Design

The design of a laser is dependent on many interdependent factors, including the requirements placed on the output beam (wavelength, spectral purity, tunability, divergence, polarization, power, and power stability), the operating environment (temperature, humidity, vibration, acceleration, and externally applied forces), and practical considerations (size, cost, and available power). There is an increasingly large number of gain media, cavity designs, and pump configurations that have been employed in lasers, and several texts have been written on the subject of laser design. No one design is well suited for all applications; every laser is optimized for operation at one point in the multidimensional parameter space outlined above.

A very important issue in the design of many lasers is the extraction of heat from the gain medium. In the process of pumping the gain medium, heat is generated. As the temperature of the gain medium changes, so too do its physical length and refractive index. Each of these contributes to a change in the optical length and resonant frequencies of the laser cavity. Nonuniform heating results in thermal lensing and internal stress. Thermal lensing changes the confocal parameters of the laser cavity and can destabilize an otherwise stable cavity (or visa versa). Internal stress leads to stress birefringence and, eventually, stress fracture.

Other issues that must be considered in high-power lasers are nonlinear optical effects and optical damage. The electrical field within the optical beam of a high-power laser can be large enough to damage optical components. This is particularly important in high-power pulsed lasers. At optical intensities below the optical damage level, deleterious nonlinear optical interactions can still degrade the performance of the laser, and even destroy the device. One example is stimulated Brillouin (acoustic wave) scattering in fiber lasers. In this case, nonlinear interactions create acoustic waves that can blow off the ends of the fiber.

Frequency Conversion and Nonlinear Control of Laser Radiation

Nonlinear optical techniques can be used to extend the frequency coverage of lasers as well as to modify other characteristics of laser radiation. Frequency conversion is a very important adjunct, converting the output of practical lasers to regions where primary laser sources may not exist or may not be very practical. Harmonic generation, frequency mixing, optical parametric oscillation, and stimulated Raman scattering have been used for frequency conversion.

Nonlinear processes have also been used to produce mode locking (see the above section on *Passive Mode Locking*), to improve transverse beam quality (i. e. to produce output closer to the diffraction limit), and to dampen relaxation oscillations in pulsed lasers. The article Nonlinear Optics describes the nonlinear processes in detail; here, we will only discuss their general significance for lasers. Finally, it should be mentioned that laser oscillators and amplifiers are nonlinear optical devices in that there is a reduction in optical gain caused by partial depletion of the population inversion by the laser radiation. This partial depletion, or saturation, stabilizes the laser output amplitude in cw lasers and plays a vital role in determining the operating characteristics of pulsed lasers, as we have seen in the section on *Types of Pulsed Operation*.

The most frequently used frequency-conversion techniques are second-harmonic and sum-frequency generation. Typically, the output of infrared lasers in the 1-μm region, such as Nd:YAG or Nd:glass, is doubled into the green. Depending on the application, shorter wavelengths may be obtained (for example, by summing the green radiation with the infrared). There are a variety of uses for this short-wavelength output, including pumping short-wavelength lasers and optical parametric oscillators (OPOs). For second-harmonic generation to the green, conversion efficiencies as high as 90% have been reported and average powers of greater than 50 W have been obtained. Materials such as $KTiOPO_4$ (KTP), periodically polled $LiNbO_3$ (PPLN), LiB_3O_5 (LBO), and β-BaB_2O_4 (BBO) are used. In one example, a pulsed Yb double-clad glass fiber laser and an LBO doubler were used to obtain 60 W of average green power.

Difference-frequency generation and OPOs have been used as tunable sources of radiation for various spectroscopic applications, primarily in the infrared. Continuous-output difference-frequency generation, while producing very low average power with low efficiency, has been used for ultrahigh-resolution molecular spectroscopy in the mid-infrared. OPOs are useful sources of tunable pulsed output with high peak power and high efficiency.

Stimulated Raman scattering has been used to generate large pulse energies with essentially unity quantum efficiency at a variety of wavelengths. However, because energy is deposited in the Raman process, moving Raman media are often required to obtain high-average-power conversion. For high energies and powers, gasses such as H_2 have been used. With a very-high-finesse cavity, a diode-laser-pumped hydrogen Raman laser has been operated continuously. Fiber Raman lasers have produced tens of watts of continuous power, while fiber Raman amplifiers have been used to increase the useful extent of the fiber-communications spectrum. Stimulated Raman scattering has also been observed in quantum-cascade semiconductor lasers.

The optical Kerr effect, which is the change in refractive index proportional to the optical intensity, can cause short pulses, typically 0.1 to 10 ps in duration, to acquire a frequency sweep, or chirp. By sending the chirped pulse through a frequency-dispersive delay line, which might be a grating or prism pair, even shorter pulses, typically 5 to 50 fs, can result.

Nonlinear phase conjugation can reverse the phase distortion acquired from the active medium in a laser (for example, from an optically imperfect laser crystal or a laser material that has large thermal gradients due to high-average-power operation). On a second pass through the distorting medium, the reversed phase distortion is canceled. Most often, backward stimulated Brillouin scattering is used to reverse the phase; the small acoustic frequency shift produces an optical wave that remains within the gain bandwidth of most lasers and has linear propagation characteristics that match the input to the Brillouin cell. Other techniques used take advantage of the photorefractive effect or absorptively induced nonlinearities.

Harmonic generation has been used to moderate the amplitude of the relaxation oscillations that often occur in long-pulse lasers. Relaxation oscillations (see the above section on *Relaxation Oscillations*) involve the flow of energy back and forth between the population inversion and the radiation field. Conversion to the second harmonic can clip the high initial peaks in the laser power, an action that reduces the amount of depletion in the population inversion, thereby damping the oscillations.

Bibliography

Popular discussions of lasers and their applications and history
J. Hecht, *Laser Pioneers*. Academic Press, Boston, 1992.
J. Hecht and D. Teresi, *Laser: Light of a Million Uses*. Dover, Mineola, 1998.

General discussions of laser science and technology
D. A. Eastham, *Atomic Physics of Lasers*. Taylor & Francis, London, 1986.
I. P. Kaminow and A. E. Siegman, *Laser Devices and Applications*. IEEE Publications, New York, 1973.
K. Shimoda, *Introduction to Laser Physics*, 2nd ed. Springer, New York, 1991.
A. E. Siegman, *Lasers*. University Science, Mill Valley, 1986.
W. T. Silfvast, *Laser Fundamentals*, 2nd ed. Cambridge University Press, Cambridge, 2004.
O. Svelto, *Principles of Lasers*, 4th ed. Plenum Press, New York, 1998.
J. T. Verdeyen, *Laser Electronics*, 3rd ed. Prentice Hall, Englewood Cliffs, 1995.
A. Yariv, *Quantum Electronics*, 3rd ed. Wiley, New York, 1989.

Detailed theoretical discussions of lasers
H. Haken, *Laser Theory*. Springer, New York, 1986.
M. Sargent III, M. O. Scully, and W. E. Lamb, Jr., *Laser Physics*. Addison-Wesley, Reading, 1974.

Physics of laser resonators
D. R. Hall and P. E. Jackson (eds.), *Physics and Technology of Laser Resonators*. Adam Hilgar, Bristol, 1989.

Specific types of lasers
D. C. Brown, *High Peak Power Nd:Glass Laser Systems*. Springer, New York, 1981.
H. C. Casey, Jr., and M. B. Panish, *Heterostructure Lasers*, Parts A and B. Academic Press, New York, 1978.
L. Coldren and S. W. Corzine, *Diode Lasers and Photonic Integrated Circuits*. Wiley, New York, 1995.
F. J. Duarte (ed.), *Tunable Lasers Handbook*. Academic Pres, San Diego, 1995.
F. J. Duarte and L. W. Hillman (eds.), *Dye Laser Principles*. Academic Press, Boston, 1990.
C. G. B. Garrett, *Gas Lasers*. McGraw–Hill, New York, 1967.
A. A. Kaminskii, *Laser Crystals*, 2nd ed. Springer, New York, 1990.
E. Kapon (ed.), *Semiconductor Lasers I: Fundamentals*. Academic Press, San Diego, 1999.
W. Koechner, *Solid-State Laser Engineering*, 5th ed. Springer, New York, 1999.
W. Koechner, M. Bass, and H. Roth, *Solid State Lasers: A Graduate Text*. Springer, New York, 2003.
L. F. Mollenauer, J. C. White, and C. R. Pollock (eds.), *Tunable Lasers*, 2nd ed. Springer, New York, 1992.
F. P. Schäfer (ed.), *Dye Lasers*, 3rd ed. Springer, New York, 1990.
C. S. Willett, *Introduction to Gas Lasers: Population Inversion Mechanisms*. Pergamon Press, Oxford, 1974.

Data sources

M. J. Weber (ed.), *Handbook of Laser Science and Technology*, Vols. I and II, 1982; Vols. III and IV, 1986; Vol. V, 1987; Suppl. 1, 1991; Suppl. 2, 1995. CRC Press, Boca Raton.

M. J. Weber (ed.), *Handbook of Lasers*. CRC Press, Boca Raton, 2003.

Lattice Dynamics

R. F. Wallis

Introduction

Lattice dynamics concerns the vibrations of atoms of a solid about their equilibrium positions and the effect of these vibrations on the properties of the solid. In the adiabatic approximation [1] the total wave function of the solid, $\Psi(\mathbf{r}, \mathbf{R})$, is written as the product of the electronic wave function, $\psi(\mathbf{r}, \mathbf{R})$, and the nuclear wave function, $\chi(\mathbf{R})$, where $\mathbf{r}$ represents all the electronic coordinates and $\mathbf{R}$ all the nuclear coordinates. The term adiabatic signifies that the nuclear motion does not induce electronic transitions. The vibrational motion of the atoms is determined by the nuclear potential energy, which is the electronic energy eigenvalue. The latter is a function of the nuclear coordinates $\mathbf{R}$.

Lattice-Dynamical Coupling Constants

We assume that the crystal is in a nondegenerate electronic ground state and that the atoms execute small vibrations about their equilibrium positions. The nuclear potential energy, $\Phi(\mathbf{R})$, may then be expanded in powers of the displacement components of the nuclei from their equilibrium positions:

$$
\begin{aligned}
\Phi(\mathbf{R}) = \; & \Phi(\mathbf{R}^{(0)}) + \sum_{l\kappa\alpha} \Phi_\alpha(l\kappa) u_\alpha(l\kappa) \\
& + \frac{1}{2} \sum_{l\kappa\alpha l'\kappa'\beta} \sum \Phi_{\alpha\beta}(ll', \kappa\kappa') u_\alpha(l\kappa) u_\beta(l'\kappa') \\
& + \frac{1}{6} \sum_{l\kappa\alpha l'\kappa'\beta l''\kappa''\gamma} \sum \sum \Phi_{\alpha\beta\gamma}(ll'l'', \kappa\kappa'\kappa'') \\
& \times u_\alpha(l\kappa) u_\beta(l'\kappa') u_\gamma(l''\kappa'')
\end{aligned}
\tag{1}
$$

In Eq. (1), $u_\alpha(l\kappa) = R_\alpha(l\kappa) - R_\alpha^{(0)}(l\kappa)$ is the αth Cartesian component of the displacement from equilibrium of the κth nucleus in the lth primitive unit cell and the superscript zero denotes the equilibrium configuration. The equilibrium condition requires that the quantities $\Phi_\alpha(l\kappa) = 0$. The $\Phi_{\alpha\beta}(ll', \kappa\kappa')$ are the *harmonic* coupling constants and the $\Phi_{\alpha\beta\gamma}(ll'l'', \kappa\kappa'\kappa'')$, ..., are the *anharmonic* coupling constants. The coupling constants are partial derivatives of the potential energy and therefore satisfy symmetry conditions such as $\Phi_{\alpha\beta}(ll', \kappa\kappa') = \Phi_{\beta\alpha}(l'l, \kappa'\kappa)$. Other conditions [1, 2] satisfied by the coupling constants are those imposed by infinitesimal translational invariance and infinitesimal rotational in variance. For the harmonic coupling constants, these conditions take the respective forms

$$\sum_{l'\kappa'} \Phi_{\alpha\beta}(ll', \kappa\kappa') = 0 \tag{2}$$

and

$$\sum_{l'\kappa'} \{\Phi_{\alpha\beta}(ll', \kappa\kappa')X_{\mu}(l'l, \kappa'\kappa) - \Phi_{\alpha\mu}(ll', \kappa\kappa')X_{\beta}(l'l, \kappa'\kappa)\} = 0, \tag{3}$$

where $X_{\mu}(l'l, \kappa'\kappa) = R_{\mu}^{(0)}(l'\kappa') - R_{\mu}^{(0)}(l\kappa)$. Analogous equations are satisfied by the anharmonic coupling constants. Additional conditions on the coupling constants are provided by utilizing the point symmetry and translational symmetry of the crystal [2].

Typically, the nuclear potential energy of a crystal has several physical origins. These include the Coulomb interaction between electrically charged ions, van der Waals interactions, interactions associated with chemical bonds, and the short-range quantum-mechanical repulsion between atoms due to the Pauli principle.

Equations of Motion and Normal Coordinate Transformation

The classical equations of motion for the nuclei can be written as

$$\partial_{\kappa}\ddot{u}_{\alpha}(l\kappa) = -\frac{\partial \Phi(\mathbf{R})}{\partial u_{\alpha}(l\kappa)}, \tag{4}$$

where the double dot above the u represents a second time derivative. Making use of Eq. (1) and retaining only the harmonic terms, we obtain

$$M_{\kappa}\ddot{u}_{\alpha}(l\kappa) = \sum_{l'\kappa'\beta} \Phi_{\alpha\beta}(ll', \kappa\kappa')u_{\beta}(l'\kappa'). \tag{5}$$

The set of equations (5) corresponding to the various values of l, κ, and α describe a system of coupled harmonic oscillators. Their solution is facilitated by the substitution $u_{\alpha}(l\kappa) = M_{\kappa}^{-1/2}w_{\alpha}(l\kappa)\exp(i\omega t)$, which leads to

$$\omega^2 w_{\alpha}(l\kappa) = \sum_{l'\kappa'\beta} D_{\alpha\beta}(ll', \kappa\kappa')w_{\beta}(l'\kappa'), \tag{6}$$

where the quantities $D_{\alpha\beta}(ll', \kappa\kappa') = (M_{\kappa}M_{\kappa'})^{-1/2}\Phi_{\alpha\beta}(ll', \kappa\kappa')$ are the elements of the *dynamical matrix*. For a nontrivial solution, the determinant of coefficients of the $w_{\alpha}(l\kappa)$ in Eq. (6) must be zero:

$$|D_{\alpha\beta}(ll', \kappa\kappa') - \omega^2 \delta_{ll'}\delta_{\kappa\kappa'}\delta_{\alpha\beta}| = 0, \tag{7}$$

where the δ_{ij}'s are Kronecker deltas. Equation (7) is applicable to any crystal, perfect or imperfect, and specifies the normal-mode frequencies of the crystal. If the crystal contains N atoms, there are $3N$ normal modes and hence $3N$ solutions for ω to Eq. (7). We shall distinguish the various normal modes by the index s, $1 \le s \le 3N$. Typically, the normal-mode frequencies range from zero up to the infrared region.

For a macroscopic crystal, Eq. (7) is a polynomial equation in ω^2 of very high order. Calculations are simplified for a perfect crystal if we introduce periodic boundary conditions, justified by Ledermann's theorem [2], and write

$$w_\alpha(l\kappa) = W_{\alpha\kappa}(\mathbf{k}) \exp[i\mathbf{k} \cdot \mathbf{R}^{(0)}(l)] , \tag{8}$$

where $\mathbf{k}$ is the wave vector. Periodic boundary conditions require that the atoms at corresponding points on opposite faces of the crystal have the same displacements. There are $3N$ values of $\mathbf{k}$ that occupy the first Brillouin zone and that specify the independent normal modes [2], The analog of Eq. (7) becomes

$$|D_{\alpha\beta}(\mathbf{k},\kappa\kappa') - \omega^2 \delta_{\alpha\beta}\delta_{\kappa\kappa'}| = 0 , \tag{9}$$

where

$$D_{\alpha\beta}(\mathbf{k},\kappa\kappa') = \sum_{l'} D_{\alpha\beta}(ll',\kappa\kappa') \exp\{i\mathbf{k} \cdot [\mathbf{R}^{(0)}(l') - \mathbf{R}^{(0)}(l)]\}$$

is independent of l. If there are r atoms per unit cell, the determinant in Eq. (9) is $3r \times 3r$ in size, far smaller than that in Eq. (7) and therefore more suitable for computation. Let us denote the eigenvectors and eigenvalues of $D(\mathbf{k},\kappa\kappa')$ by $e_\kappa(\mathbf{k}j)$ and $\omega^2(\mathbf{k}j)$, respectively, where j is the branch index. The normal coordinates, $Q(\mathbf{k}j)$, can be introduced through the transformation

$$w_\alpha(l\kappa)N^{-1/2} \sum_{\mathbf{k}j} e_{\alpha\kappa}(\mathbf{k}j)Q(\mathbf{k}j) \exp\left[i\mathbf{k} \cdot \mathbf{R}^{(0)}(l)\right] . \tag{10}$$

The Hamiltonian for the harmonic crystal can be written as

$$H = \frac{1}{2} \sum_{\mathbf{k}j} \left\{ |\dot{Q}(\mathbf{k}j)|^2 + \omega^2(\mathbf{k}j)|Q(\mathbf{k}j)|^2 \right\} , \tag{11}$$

which consists of the sum of contributions from independent harmonic oscillators. In the quantum-mechanical theory in the absence of anharmonicity, the nuclear wave function is a product of harmonic oscillator functions. The energy eigenvalue can be written as

$$E = \sum_{\mathbf{k}j} \hbar\omega(\mathbf{k}j) \left[n(\mathbf{k}j) + \frac{1}{2} \right] , \tag{12}$$

where $n(\mathbf{k}j)$ is the harmonic oscillator quantum number for the mode $\mathbf{k}j$. The quantum of excitation energy for a normal vibrational mode is called a phonon.

Phonon Dispersion Curves

The solutions of Eq. (9) determine the phonon dispersion curves, i.e., the normal-mode frequencies $\omega(\mathbf{k}j)$ as functions of the wave vector $\mathbf{k}$ for the various branches j. As a simple example, we give the normal-mode frequencies for a monatomic linear chain with atomic mass M and nearest-neighbor interactions characterized by the coupling constants

$$\Phi_{xx}(l,l+1) = \Phi_{xx}(l,l-1) = \frac{1}{2}\Phi_{xx}(l,l) = -\sigma ; \quad \omega(k) = (4\sigma/M)^{1/2} \sin(ak/2) , \tag{13}$$

where a is the lattice constant. In this case we have only one atom per primitive unit cell and only one branch. For a crystal with r atoms per unit cell, there are $3r$ branches of which three are acoustical branches with $\omega(\mathbf{k}j) \to 0$ as $|\mathbf{k}| \to 0$ and $3r - 3$ are optical branches. In crystals of high symmetry with $\mathbf{k}$ in a high-symmetry direction, the branches can be classified as longitudinal (displacements $\mathbf{u}$ parallel to $\mathbf{k}$) or transverse ($\mathbf{u}$ perpendicular to $\mathbf{k}$). The phonon frequencies associated with a given branch are restricted to a finite range or "band". There may be gaps between the various phonon bands.

Phonon dispersion curves can be determined experimentally by means of the coherent inelastic scattering of cold neutrons or by the thermal diffuse scattering of x rays. Inelastic neutron scattering [3] is the more widely used method. If the incident and scattered neutron energies are E_0 and E_s, respectively, the energy, $\hbar\omega(\mathbf{q}j)$, of the phonon created or destroyed in the scattering process is given by conservation of energy as

$$E_0 - E_s = \pm\hbar\omega(\mathbf{q}j) , \tag{14}$$

where the plus sign refers to the creation and the minus sign to the destruction of the phonon. In addition, conservation of momentum requires that

$$\mathbf{k}_0 - \mathbf{k}_s = \mathbf{q} + \mathbf{G} , \tag{15}$$

where the wave vectors of the incident and scattered neutrons are $\mathbf{k}_0$ and $\mathbf{k}_s$, respectively, and $\mathbf{G}$ is a reciprocal lattice vector. From experimental values of E_0, E_s, $\mathbf{k}_0$, and $\mathbf{k}_s$, we can determine $\omega(\mathbf{q}j)$ and $\mathbf{q}$ and, hence, the phonon dispersion curves. Results for NaBr [4] are shown in Fig. 1. The gap between the acoustical and optical branches is evident.

Specific Heat

The specific heat associated with the lattice vibrations of a crystal can be calculated in the harmonic approximation by taking the thermal average, $\langle E \rangle$, of the energy given by Eq. (12). Replacing the sums over $\mathbf{k}$ and j by an integral over frequency ω, we obtain for the specific heat

$$
\begin{aligned}
C_{\mathrm{v}}(T) &= \left(\frac{\partial \langle E \rangle}{\partial T}\right)_V \\
&= k_{\mathrm{B}}\beta^2 \int_0^{\omega_{\mathrm{m}}} d\omega\, g(\omega) \hbar^2 \omega^2 \frac{e^{\beta\hbar\omega}}{(e^{\beta\hbar\omega} - 1)^2}
\end{aligned}
\tag{16}
$$

where $\beta = 1/k_{\mathrm{B}}T$, k_{B} is Boltzmann's constant, ω_{m} is the maximum phonon frequency, and $g(\omega)$ is the phonon frequency distribution. In general, $g(\omega)$ must be evaluated numerically using a computer. However, for the case of an isotropic elastic continuum, $g(\omega)$ takes the simple form

$$
\begin{aligned}
g(\omega) &= \frac{\Omega}{2\pi}\left(\frac{2}{c_{\mathrm{t}}^3} + \frac{1}{c_{\mathrm{l}}^3}\right)\omega^2 , \quad 0 \le \omega \le \omega_{\mathrm{m}} , \\
&= 0 , \qquad\qquad\qquad\quad \omega > \omega_{\mathrm{m}} ,
\end{aligned}
\tag{17}
$$

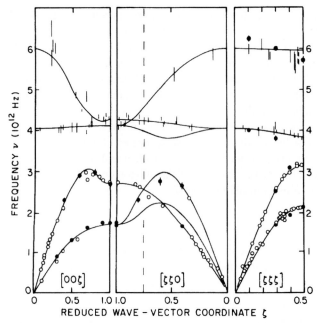

Fig. 1: Phonon dispersion curves for NaBr in the [001], [110], and [111] directions (from Ref. 4). The experimental data are indicated by bars and circles. The solid curves are calculated from a lattice-dynamical model and $\nu = \omega/2\pi$.

where c_t and c_l are the speeds of transverse and longitudinal sound waves and Ω is the volume of the crystal. Substitution of Eq. (17) into Eq. (16) gives

$$C_v(T) = 9Nk_B \left(\frac{T}{\Theta}\right)^3 \int_0^{\Theta/T} dx \, \frac{x^4 e^x}{(e^x - 1)^2} , \qquad (18)$$

where $\Theta = \hbar\omega/k_B$ is the Debye temperature [5]. At low temperatures, $T \ll \Theta$, we obtain

$$C_v(T) \cong (12Nk_B\pi^4/5)(T/\Theta)^3 , \qquad (19)$$

which is the "Debye T^3 law". A tabulation of some Debye temperatures is given in Table 1. The T^3 law is observed experimentally at sufficiently low temperatures; at higher temperatures, however, deviations from Eq. (18) appear because a real crystal is not an elastic continuum. The correct frequency distribution is in general a complicated function of ω and exhibits Van Hove singularities [6], which arise from maxima, minima, or saddle points on the constant-frequency surfaces in **k** space. The frequency distribution for copper [7] is shown in Fig. 2.

Optical Properties

An ionic crystal can interact with the electromagnetic field through the vibrations of its ions, since the latter are electrically charged and undergo acceleration during a lattice vibration.

Table 1: Debye Temperatures of Some Cubic Crystals.[a]

Solid	Debye temperature (K)
Sodium	157
Copper	342
Lead	102
Nickel	427
Silicon	647
NaCl	321
LiF	732

[a] From J. S. Blakemore, *Solid State Physics*, 2nd ed., p. 130. Saunders, Philadelphia, PA, 1974.

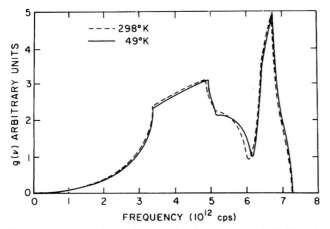

Fig. 2: Frequency distribution function for copper (Ref. 7). Note that $\nu = \omega/2\pi$.

Optical phenomena are simplest in cubic crystals, which are optically isotropic and can be described by the complex dielectric constant $\varepsilon(\omega) = \varepsilon_1(\omega) + i\varepsilon_2(\omega)$. For a crystal such as NaCl, with two atoms per primitive cell, we can show that [1, 8]

$$\varepsilon(\omega) = \varepsilon_\infty + \frac{4\pi e^{*2}}{\mu\Omega_0} \left[\frac{\omega_t^2 - \omega^2 + i\omega\gamma}{(\omega_t^2 - \omega^2)^2 + \omega^2\gamma^2} \right] , \tag{20}$$

where ω_t is the transverse optical phonon frequency of long wavelength, γ is the damping constant, e^* is the transverse effective charge, μ is the reduced mass of the two ions in the unit cell, Ω_0 is the volume of the unit cell, and ε_∞ is the limiting value of $\varepsilon(\omega)$ for $\omega \gg \omega_t$.

The real part of the dielectric constant, $\varepsilon_1(\omega)$, can be written in the absence of damping in the simple form

$$\varepsilon_1(\omega) = \frac{\varepsilon_\infty(\omega_l^2 - \omega^2)}{(\omega_t^2 - \omega^2)} , \tag{21}$$

Table 2: Lattice-dynamical parameters of some cubic crystals.

Solid	ε_∞	$\varepsilon(0)$	ω_t (cm^{-1})	ω_l (cm^{-1})
LiF[a]	1.9	8.9	307	662
NaCl[a]	2.25	5.9	164	264
KI[a]	2.7	5.1	101	139
ZnS[b]	5.14	8.67	271	352
GaAs[b]	10.9	12.9	269	292
InSb[b]	15.6	17.7	185	197
PbS[b]	17.2	202	65	223

[a] Ref. 8.
[b] Ref. 9.

where $\omega_l = [\omega_t^2 + (4\pi e^{*2}/\mu\Omega_0\varepsilon_\infty)]^{1/2}$ is the longitudinal optical phonon frequency of long wavelength. Without damping, the dielectric constant has a pole at ω_t and a zero at ω_l. If we set $\omega = 0$ in Eq. (20), we obtain

$$\frac{\varepsilon(0)}{\varepsilon_\infty} = \frac{\omega_l^2}{\omega_t^2}, \tag{22}$$

which is the Lyddane–Sachs–Teller relation. Values of the lattice-dynamical parameters for several crystals are given in Table 2.

The complex index of refraction, $n - i\kappa$, is defined by $(n - i\kappa)^2 = \varepsilon(\omega)$. The reflectivity at normal incidence, R, is given by

$$R = \frac{(n-1)^2 + \kappa^2}{(n+1)^2 + \kappa^2} \tag{23}$$

and is large between ω_t and ω_l, the region of *Reststrahlen*. The absorption coefficient, $\eta = 2\omega\kappa/c$, is strongly peaked close to ω_t if γ/ω_t is small. The coupling between the electromagnetic field and the transverse optical phonon of long wavelength leads to coupled modes called polaritons whose dispersion relation is given by $c^2k^2 = \omega^2\varepsilon(\omega)$. Experimental information about the polariton dispersion relation can be obtained by Raman scattering [10]. The dispersion curves for polaritons in GaAs are shown in Fig. 3. Note that there are two branches, one above and one below the *Reststrahlen* region.

Impurity Modes

Associated with an impurity atom in a crystal are changes in atomic mass and coupling constants compared to the perfect crystal. These changes lead to modifications of the normal-mode eigenvectors and frequencies. In accordance with Rayleigh's theorem [2], a decrease (increase) in an atomic mass or increase (decrease) in a coupling constant produces either no change or an increase (decrease) in the normal-mode frequencies. Of particular importance are modes whose frequencies lie in regions that are forbidden for the frequencies of the perfect crystal. Such modes are termed localized modes because the atomic displacement amplitudes decrease exponentially from the impurity site. For an isotopic impurity of mass M'

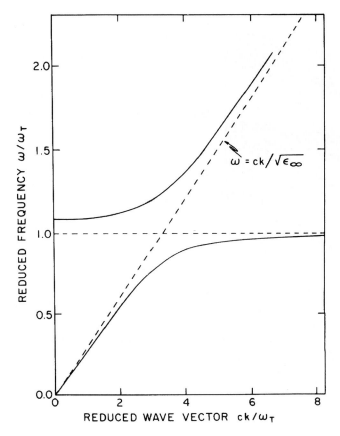

Fig. 3: Polariton dispersion curves for GaAs.

replacing a host crystal atom of mass M, the localized-mode frequency, ω_i, is specified by the equation [11]

$$\varepsilon\omega_i^2 \int_0^{\omega_m} \frac{g(\omega)\,d\omega}{\omega_i^2 - \omega^2} = 1 , \qquad \omega_i > \omega_m , \tag{24}$$

where $\varepsilon = (M - M')/M$ and $g(\omega)$ is the frequency distribution of the unperturbed crystal normalized to unity. Typically, a localized mode appears if M' is sufficiently light compared to M.

For a crystal with two atoms per unit cell there are optical modes of vibration whose frequencies can be separated from those of the acoustical modes by a gap (Fig. 1). If an atom of the host crystal is replaced by an impurity atom of mass M', a localized impurity mode can appear with frequency above the maximum frequency of the optical modes or in the gap between the optical and acoustical modes. Whether an impurity mode exists and where its frequency lies depends on which host atom is replaced and the relation of the impurity mass M' to the host atom masses M_1 and M_2 [12].

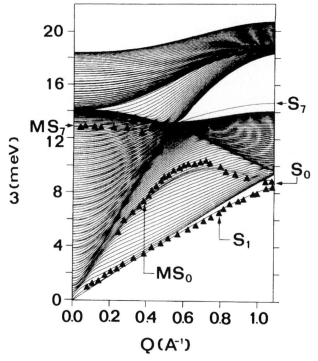

Fig. 4: Dispersion curves for surface and bulk modes of silver with a (110) surface. The continuous curves are theoretical results (Ref. 15) and the triangles are experimental results (Ref. 16).

Surface Modes

A crystal surface may be regarded as a defect in which certain coupling constants are set equal to zero. Modes localized at a surface may be derived from both acoustical and optical branches. Long-wavelength surface acoustic modes are known as Rayleigh waves [12], whose speed, c_R, in an isotropic medium is given by

$$16\left(1 - \frac{c_R^2}{c_l^2}\right)\left(1 - \frac{c_R^2}{c_t^2}\right) = \left(2 - \frac{c_R^2}{c_t^2}\right)^4 . \tag{25}$$

The speed of Rayleigh waves is always less than c_l and c_t.

In recent years complete surface-phonon dispersion curves have been measured for many crystal surfaces using the techniques of electron energy loss spectroscopy [13] and inelastic helium atom scattering [14]. The surface-mode frequencies appear below the bulk continuum or in gaps within the bulk continuum. An example is shown in Fig. 4 for the silver (110) surface [15] where the Rayleigh mode is indicated by S_1, the gap surface mode by S_7, and so-called resonance modes by MS_0 and MS_7.

Anharmonicity

The cubic, quartic, and higher-order terms in the displacements in Eq. (1) give rise to anharmonic effects such as thermal conductivity, thermal expansion, broadening of optical absorption lines, and ultrasonic attenuation. The coefficient of thermal conductivity, K, for an isotropic crystal can be shown [17] to have the form

$$K = \frac{1}{3} C s \lambda , \qquad (26)$$

where C is the specific heat per unit volume, s is the average speed of sound, and λ is the phonon mean free path. The coefficient K increases as T^3 at very low temperatures, reaches a maximum, and then decreases as $1/T$ at higher temperatures.

Anharmonicity also leads to a volume dependence of the Debye temperature described by the Grüneisen constant, γ, defined by

$$\gamma = -\frac{V}{\Theta} \frac{d\Theta}{dV} . \qquad (27)$$

The thermal expansion coefficient, α, can be expressed in terms of γ,

$$\alpha = \frac{1}{V} \left(\frac{\partial V}{\partial T} \right)_p = \frac{\kappa \gamma C_v}{V} , \qquad (28)$$

where κ is the compressibility and C_v is the specific heat at constant volume V.

In certain circumstances it is possible to have vibrational waves propagating in an anharmonic lattice without damping. Such waves are known as solitons [18].

See also: Crystal Defects; Crystal Symmetry; Nonlinear Wave Propagation; Phonons; Quasiparticles; Solitons.

References

[1]   M. Born and K. Huang, *Dynamical Theory of Crystal Lattices*. Oxford, London and New York, 1998.

[2]   A. A. Maradudin, E. W. Montroll, G. H. Weiss, and I. P. Ipatova, *Theory of Lattice Dynamics in the Harmonic Approximation*. Academic Press, New York, 1971.

[3]   B. N. Brockhouse, in *Phonons and Phonon Interactions* (T. A. Bak, ed.), p. 221, Benjamin, New York, 1964.

[4]   J. S. Reid, T. Smith, and W. J. L. Buyers, *Phys. Rev. B* **1**, 1833 (1970).

[5]   P. Debye, *Ann. Phys. (Leipzig)* **39**, 789 (1912).

[6]   L. Van Hove, *Phys. Rev.* **89**, 1189 (1953).

[7]   R. M. Nicklow, G. Gilat, H. G. Smith, L. J. Raubenheimer, and M. K. Wilkinson, *Phys. Rev.* **164**, 922 (1967).

[8]   E. Burstein, in *Phonons and Phonon Interactions* (T. A. Bak, ed.), p. 276. Benjamin, New York. 1964.

[9]   E. Burstein, A. Pinczuk, and R. F. Wallis, *The Physics of Semimetals and Narrow Gap Semiconductors* (D. Carter and R. Bate, eds.), p. 251. Pergamon, Oxford, 1971.

[10]   C. H. Henry and J. J. Hopfield, *Phys. Rev. Lett.* **15**, 964 (1965).

[11]   P. G. Dawber and R. J. Elliott, *Proc. Roy. Soc. (London)* **A273**, 222 (1963).

[12]   Lord Rayleigh, *Proc. London Math. Soc.* **17**, 4 (1887).

[13]   S. Lehwald, J. M. Szeftel, H. Ibach, T. S. Rahman, and D. L. Mills, *Phys. Rev. Lett.* **50**, 518 (1983).

[14]   R. B. Doak, U. Harten, and J. P. Toennies, *Phys. Rev. Lett.* **51**, 578 (1983).

[15]   A. Franchini, G. Santoro, V. Bortolani, and R. F. Wallis, *Phys. Rev. B* **38**, 12139 (1988).

[16]   G. Bracco, R. Tatarek, F. Tommasini, V. Linke, and M. Persson, *Phys. Rev. B* **36**, 2928 (1987).

[17]   C. Kittel, *Introduction to Solid State Physics*, 7th ed. Wiley, New York, 1995.

[18]   M. Toda and M. Wadati, *J. Phys. Soc. Japan* **34**, 18 (1973).

Lattice Gauge Theory

M. Creutz

Supercomputers have recently become a crucial tool for the quantum field theorist. Applying the formalism of lattice gauge theory, numerical simulations are providing fundamental quantitative information about the interactions of quarks, the fundamental constituents of those particles which experience nuclear interactions. Perhaps most strikingly, these simulations have provided convincing evidence that the interquark forces can prevent the isolation of these constituents.

Quarks are the primary constituents of particles subject to the strong nuclear force. Their basic interactions are believed to follow from a generalization of the gauge theory of electromagnetism. Instead of a single photon, this theory involves eight spin-1 quanta, referred to as gluons. Furthermore, these eight gluons are themselves charged with respect to one another. This introduces subtle nonlinear effects which appear even in the pure glue theory.

One particularly important consequence of these nonlinearities is that the quark interactions weaken at small separations. This phenomenon, known as "asymptotic freedom", is essential to many of the successes of the simple quark model. As long as the quarks remain near each other, their interactions are small.

In contrast, the behavior of the gauge fields changes dramatically as the quarks are pulled apart. The experimental nonobservance of free quarks has led to the conjecture of the phenomenon of "confinement", wherein interquark forces increase and remain strong as quarks are pulled apart to arbitrary separations. In this picture, it requires an infinite amount of energy to separate a single quark from the other constituents of a physical particle. This explains why free quarks are not produced in nature.

Standard field-theoretical tools are severely hampered in the regime of large distances where these effects come into play. Perturbation theory, the historic mainstay of quantum field theory, begins with free particles and then treats their interaction as a small correction. With confinement, however, the fundamental constituents become increasingly strongly interacting as their separation is increased. In this domain the conventional perturbative approach fails totally.

Lattice gauge theory, originally formulated by K. Wilson, provides a novel framework for calculations in this regime. This approach replaces the relativistic continuum of space and time with a discrete space-time lattice. The quarks move through this scaffolding by a sequence of discrete hops between nearest-neighbor sites. The gluon fields lie on the bonds connecting these sites.

This lattice is a mathematical trick, introduced for calculational purposes only. It should not be taken as requiring a crystalline basis for physical space. At the end of any calculation, one should consider a continuum limit, wherein the lattice spacing is extrapolated to zero. In this limit observable quantities, such as the masses of particles and the forces between them, should approach their physical values.

The lattice artifice, however, has several advantages. First, by replacing an infinite number of space-time points in any given volume by a finite number, the field-theoretical system becomes mathematically considerably simpler and better defined. Continuum quantum field theories are notorious for the appearance of formally infinite quantities. These divergences involve short-distance singularities and must be exorcised by a renormalization procedure. A space-time lattice provides a particularly convenient regulator of such divergences. Indeed, the lattice spacing represents a minimum length and singularities arising from wavelengths shorter than this distance are automatically excluded.

Second, this formulation makes no assumptions on calculational schemes to be applied. Other techniques for controlling the singular behavior of a field theory are usually formulated directly in terms of some calculational method. For example, conventional discussions of renormalization regulate the divergences only after they are encountered in the perturbative expansion. On the lattice the theory is mathematically well defined at the outset.

Finally, the lattice formulation is particularly well suited to numerical simulation. While there are several analytic techniques which have been applied to the strongly interacting lattice gauge problem, numerical simulations by Monte Carlo techniques currently dominate the field. These simulations have given compelling evidence that the confinement phenomenon does indeed occur in the standard gauge theory of the nuclear force. In addition, the approach is now giving quantitative predictions for long-range hadronic properties not accessible to more traditional theoretical methods.

In the Wilson approach, the gauge degrees of freedom are represented by matrices, one of which is associated with each lattice bond. To describe the physical theory of quarks, these are 3×3 unitary matrices with determinant 1; thus, they are elements of the group $SU(3)$. The number three is dictated by the phenomenological fact that protons and neutrons are each made up of three quarks. The interactions of these degrees of freedom are most concisely summarized in the "action"

$$S = -\frac{1}{3} \sum_p \operatorname{Re} \operatorname{Tr} U_p .$$

Here the sum is over all elementary squares, or "plaquettes", p, and U_p is the product of the link variables around the plaquette in question. The latter represents the flux of the gauge fields through the corresponding tile. For slowly varying fields, the above sum reduces to the conventional gauge-theory action as the lattice spacing goes to zero.

The numerical techniques used for lattice gauge simulations are borrowed directly from statistical mechanics. Indeed, there is a deep mathematical relationship between quantum field theory and classical statistical mechanics in four dimensions. In this relationship, the strength of the quark coupling to the gauge fields corresponds directly to temperature and the action S corresponds to the classical energy. Thus, a study of confinement and long-distance quark interactions is equivalent to a study of a high-temperature statistical model.

In a Monte Carlo simulation of such a lattice system, one attempts to create an ensemble of field configurations with a Boltzmann-like distribution where the probability for any given configuration C takes the form

$$P(C) \propto e^{-\beta S(C)} \ .$$

Here β is proportional to the inverse of the gauge coupling squared. Thus one wants configurations typical of "thermal equilibrium". The procedure begins with the storage of some initial values for all the lattice fields in the computer memory. These are then updated with pseudorandom changes on the field variables, thus mimicking thermal fluctuations. The structure of such a program is quite simple. On the outside is a set of nested loops overall the system variables. These loops surround calls to the random number generator, so as to simulate a thermal coupling to these degrees of freedom. The field changes are constructed with a bias toward lower values of S so as to obtain the appropriate thermal weighting of configurations.

Having the values of all fields at his disposal, the physicist is free to calculate any quantity of interest. There will, of course, be statistical errors coming from the thermal fluctuations. In addition, there will be errors coming from the requisite extrapolation to the continuum limit and from the practical requirement of working with a finite volume. It is attempts to reduce these uncertainties that have driven the theorists to the most powerful computers available.

Despite the inevitable uncertainties, several important results have been extracted. Perhaps the most dramatic of these is the measurement of the confinement force and how it relates to the weaker interactions of the quarks at short distances. Then there are successful studies of the mass spectra of the bound states of quarks. These calculations are being refined to give information on the distributions of the quarks and on the strong-interaction effects on other processes, such as weak decays.

In addition, there have been quantitative studies of physics at temperatures sufficiently high that strongly interacting particles are created by thermal fluctuations. Here lattice gauge calculations have provided strong evidence that the vacuum undergoes a phase transition at a temperature of $kT \sim 170\,\text{MeV}$. This transition is from a phase of ordinary matter, made up of quarks bound into the familiar nuclear particles, to a new quark-gluon plasma phase where the quarks and their attendant gluon fields form a thermal gas. Indeed, the lattice approach gives the best estimates for the temperature of the transition to this phase, which will be looked for in future accelerator experiments.

Until recently, the bulk of the numerical simulations in lattice gauge theory considered the full dynamics of the gauge fields but only included the quarks as fixed sources. In this way the confinement potential has been mapped out. A more refined approach allows the primary "valence" quarks to carry kinetic energy, but still ignores the creation of matter–antimatter quark pairs by quantum fluctuations. Most of the hadronic spectrum calculations have been done in this approximation.

Current computers have now reached the stage that new results are being obtained beyond this valence approximation. To proceed in this direction, one must allow for the possibility of unlimited numbers of virtual quarks being created by quantum fluctuations. The most difficult part of the problem is the inclusion of the effects of the Pauli exclusion principle in the simulations. Algorithms to treat the dynamical quarks appropriately remain rather awkward, and do not perform well when the quark masses are light. Improvements of these algorithms represent an area of intense current research.

In conclusion, computer simulations of lattice gauge theory provide a powerful tool for the study of nonperturbative phenomena. The technique provides a first-principles approach to calculating particle properties as well as details of the phase transition to a quark gluon plasma.

See also: Gauge Theories; Hadrons; Monte Carlo Techniques; Quantum Field Theory; Quarks; Strong Interactions.

Bibliography
Elementary
M. Creutz, "Simulating quarks," *Computers in Science & Engineering*, March/April 2004, p. 80 (IEEE CS and AIP, 2004).
M. Creutz, "Lattice gauge theory: A retrospective," *Nucl. Phys. Proc. Suppl.* **94**, 219 (2001) [arXiv:hep-lat/0010047].

Intermediate
C. Rebbi, *Sci. Am.* **248**, 54 (1983).
D. H. Weingarten, *Sci. Am.* **274**, 104 (1996).

Advanced
K. Wilson, *Phys. Rev. D* **10**, 2445 (1974).
M. Creutz, *Quarks, Gluons, and Lattices*. Cambridge University Press, Cambridge, 1983.
I. Montvay, G. Munster, *Quantum Fields on a Lattice*, Cambridge University Press, Cambridge, 1994.

Leptons
M. L. Perl

The known leptons are a class of elementary particles having two defining characteristics:

1. They are fermions. That is, they have spin $\frac{1}{2}$ and obey Fermi–Dirac statistics.

2. They have no strong interactions among themselves or with any other known particles. Table 1 lists the known leptons and their properties. The name lepton, meaning small or light in Greek, was coined when the heaviest known lepton was the muon which is lighter in mass than all hadrons. However, the tau discovered in 1975 is heavier than many hadrons.

The electron is, of course, the best known of the leptons, being a constituent of all ordinary matter. An atom consists of electrons moving in orbits around a nucleus. The symbol for the electron which has one unit of negative electric charge is e^-. Associated with the electron is its antiparticle, the positron with one unit of positive charge, denoted e^+. The e^+ has the same mass and spin as the e^-.

Table 1: Properties of the known leptons. Only the particles are listed. The antiparticles e^+, $\bar{\nu}_e$, μ^+, $\bar{\nu}_\mu$, τ^+ and $\bar{\nu}_\tau$ have the same mass and lifetime as their corresponding particles e^-, ν_e, μ^-, ν_μ, τ^-, ν_τ.

Generation	1	2	3
Charged lepton name	Electron	Muon	Tau
Charged lepton symbol	e^-	μ^-	τ^-
Charged lepton mass (MeV/c^2)	0.511	105.7	1777.0
Charged lepton lifetime (seconds)	Stable	2.197×10^{-6}	2.906×10^{-13}
Neutrino name	Electron neutrino	Muon neutrino	Tau neutrino
Neutrino symbol	ν_e	ν_μ	ν_τ
Neutrino mass upper limit	$< 3\,\text{eV}/c^2$	$< 0.19\,\text{MeV}/c^2$	$< 18.2\,\text{MeV}/c^2$

Also associated with the electron is the electron-neutrino, ν_e, and the electron-antineutrino, $\bar{\nu}_e$. These neutrinos are neutral, that is have zero electric charge, and they have a mass small compared to the electron mass. The e^-, e^+, ν_e, and $\bar{\nu}_e$ constitute the electron family of leptons. For convenience only the e^- and ν_e are listed in Table 1. The tabulated properties of the antiparticles e^+ and $\bar{\nu}_e$ are the same except that the e^+ has opposite charge to the e^-.

Two more lepton families have been discovered. The muon family consists of the negative muon, μ^-, and its associated neutrino, ν_μ, plus the associated antiparticles, μ^+ and $\bar{\nu}_\mu$. Similarly, the tau family consists of τ^-, ν_τ and the antiparticles τ^+ and $\bar{\nu}_\tau$. In Table 1 the three families are listed in order of increasing mass of the charged lepton, which is also the historical order in which the families were discovered.

The families are also called generations, 1st, 2nd, and 3rd for convenience. There are believed to be connections between these lepton generations and the three quark generations: e^-, ν_e corresponding to the u, d quark pair, μ^-, ν_μ corresponding to the c, s quark pair and τ, ν_τ corresponding to the t, b quark pair.

The three lepton generations are separated from each other and from all quarks in collision and decay processes by a well-established but unexplained lepton conservation law. For example, a positron, e^+, can annihilate an electron, e^-, producing two photons

$$e^+ + e^- \rightarrow \text{photon} + \text{photon} .$$

But reactions read as

$$e^+ + \mu^- \rightarrow \text{photon} + \text{photon} ,$$

or

$$e^+ + \tau^- \rightarrow \text{photon} + \text{photon} ,$$

have never been detected. The e, μ, and τ have different intrinsic properties.

The electron and positron are stable, that is, they do not decay. The muon and tau are unstable, decaying through the weak interaction. The negative muon has only one major decay mode,

$$\mu^- \rightarrow \nu_\mu + e^- + \bar{\nu}_e .$$

The corresponding decay mode for the positive muon is

$$\mu^+ \to \bar{\nu}_\mu + e^+ + \nu_e \ .$$

For another example of the lepton conservation law note that in the decay of the muon

$$\mu^- \to \nu_\mu + e^- + \bar{\nu}_e \ ,$$

the intrinsic muon property is preserved when the muon decays by the appearance of a muon neutrino. But the simpler decay process

$$\mu^- \to e^- + \text{photon},$$

in which a muon with its intrinsic property just disappears has never been detected. The experimental upper limit on its probability is 10^{-11}.

The tau because of its greater mass has many decay modes such as

$$
\begin{aligned}
\tau^- &\to \nu_\tau + e^- + \bar{\nu}_e \ , \\
\tau^- &\to \nu_\tau + \mu^- + \bar{\nu}_\mu \ , \\
\tau^- &\to \nu_\tau + \pi^- \ , \\
\tau^- &\to \nu_\tau + \pi^- + \pi^0 \ , \\
\tau^- &\to \nu_\tau + \pi^- + \pi^+ + \pi^- \ ,
\end{aligned}
$$

with corresponding modes for the τ^+. The symbol π stands for a pion.

Neutrinos do not seem to decay, but one type of neutrino can change into another type of neutrino by a process called neutrino oscillation.

All of the leptons have sizes less than 10^{-16} cm; they may be much smaller. There is no known natural law which limits the number of different leptons. Table 1 lists all leptons discovered as of 2004. It is also possible that other kinds of leptons may exist, leptons with zero spin for example, but no other kinds have been found.

See also: Conservation Laws; Electron; Elementary Particles in Physics; Muon; Neutrinos; Positron; Positron–Electron Colliding Beams; Tau; Weak Interactions.

Bibliography
F. Close, M. Marten, and C. Sutton, *The Particle Explosion*. Oxford University Press, New York, 1987.
Donald H. Perkins, *Introduction to High Energy Physics*. Cambridge University Press, Cambridge, 2000.
Martinus J. G. Veltman, *Facts and Mysteries in Elementary Particle Physics*. World Scientific, New Jersey, 2003.

Levitation, Electromagnetic
G. Wouch and A. E. Lord, Jr.

Levitation is the act of holding up an object with no visible support. The generation of eddy currents in an electrical conductor by a time-varying magnetic field is the basis of electromagnetic levitation. (It is also the basis of induction heating, most metal detectors at the airport and beach, some geological surveys, and eddy-current nondestructive testing of metal parts. Transformer cores are laminated in order to break up the eddy currents and hence reduce loss.)

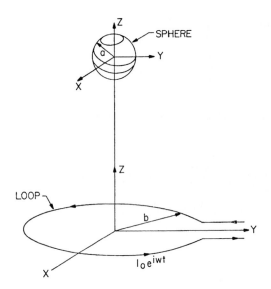

Fig. 1: Sphere positioned on the axis of a single current loop.

Electromagnetic levitation can be described in very simple terms. A time-varying magnetic flux generates eddy currents (in a conductor) and Lenz's law says that the magnetic field of these circulating currents must be in such a direction as to oppose the flux. If the flux varies spatially, then there is a force on the object

$$\mathbf{F} = \mathbf{m} \cdot \nabla \mathbf{B} \tag{1}$$

where $\mathbf{m}$ is the total magnetic moment included in the metal object and $\nabla \mathbf{B}$ is effectively the gradient of the applied magnetic field (actually $\nabla \mathbf{B}$ is a dyadic). Thus, for example, if at a particular instant $\mathbf{B}$ is in, say, the Z direction, m will be in the $-Z$ direction and if $dB/dZ < 0$, Eq. (1) gives a positive or lifting force.

Now let us be a bit more analytical [1] with a very simple model, but one that is quite realistic. Consider the situation shown in Fig. 1. A small metal sphere of radius a is located on the axis of a circular loop of wire of radius b. The wire is carrying a current $I_0 e^{i\omega t}$. A most important quantity in the problem is the ratio of the skin depth δ to the radius a. The skin depth (here taken as the effective depth of penetration of the magnetic field) is given by [2]

$$\delta = (2/\omega\sigma\mu_0)^{1/2} , \tag{2}$$

where ω is the circular frequency, σ is the electrical conductivity, and μ_0 is the permeability of free space ($= 4\pi \times 10^{-7}$ in SI units). A typical induction heating unit also used for levitation studies operates at 400 kHz and a typical metallic conductivity is $10^7 \, \Omega^{-1} \, \mathrm{m}^{-1}$. This gives $\delta = 4 \times 10^{-4} \, \mathrm{m} = 0.4 \, \mathrm{mm}$. Usual metal spheres being levitated are at least 1 cm in diameter. Thus $\delta/a \ll 1$; this also applies to most metal-detector situations where the frequency may be lower but the objects to be detected are usually somewhat larger. Thus the calculation here will be limited to the regime of $\delta/a \ll 1$. No essential physics is lost in this way.

In the extreme case of $\delta/a \ll 1$ the eddy currents induced on the surface of the sphere completely shield the interior from magnetic flux. Hence

$$\mathbf{B} = \mu_0(\mathbf{H} + \mathbf{M}) , \tag{3}$$

where $\mathbf{B}$ is the magnetic induction, $\mathbf{H}$ is the total magnetic field, and $\mathbf{M}$ is the magnetization in the sphere. Now

$$\mathbf{H} = (H_0 e^{i\omega t} - H_{\text{demag}})\hat{k} , \tag{4}$$

where for a sphere the demagnetizing field is given by [3]

$$H_{\text{demag}} = \frac{1}{3}\mathbf{M} . \tag{5}$$

Thus setting $\mathbf{B} = 0$ we obtain

$$0 = \mu_0(H_0 e^{i\omega t} + \frac{2}{3}M) , \tag{6}$$

and so

$$\mathbf{M} = -\frac{3}{2}H_0 e^{i\omega t}\hat{k} . \tag{7}$$

The magnetic moment $\mathbf{m}$ of the sphere is given by

$$\begin{aligned}\mathbf{m} &= \left(\frac{4}{3}\pi a^3\right)\left(-\frac{3}{2}H_0 e^{i\omega t}\hat{k}\right) \\ &= -2\pi a^3 H_0 e^{i\omega t}\hat{k} .\end{aligned} \tag{8}$$

Levitation Force

The force on a dipole in a nonuniform field is given by [4]

$$\mathbf{F} = \mathbf{m} \cdot \nabla\mathbf{B} . \tag{9}$$

The axial magnetic field of a single loop is the well-known result [5]

$$\mathbf{B} = (I_0/2)[\mu_0 b^2/(b^2 + z^2)^{3/2}] e^{i\omega t}\hat{k} \tag{10}$$

and

$$\nabla\mathbf{B} = -[3b^2 I_0\mu_0 z/(b^2 + z^2)^{3/2}] e^{i\omega t}\hat{k},\hat{k} . \tag{11}$$

From Eqns. (8), (9), and (11), the levitation force is

$$\mathbf{F} = [3\pi a^3 b^4 z/2(b^2 + z^2)^4]\mu_0 I_0^2 e^{2i\omega t}\hat{k} . \tag{12}$$

Taking the real part, averaging over a period, and using $I_{\text{rms}} = I/\sqrt{2}$ gives the average force as

$$\langle\mathbf{F}\rangle = [3\pi a^3 b^4 z/2(b^2 + z^2)^4]\mu_0 I_{\text{rms}}^2\hat{k} . \tag{13}$$

where I_{rms} is the root-mean-square current. This result agrees exactly with the general treatment of levitation forces given by Okress *et al.* [6] and Fromm and Jehn [7] in the case where

$\delta/a \ll 1$. A typical levitation situation might be

$$a = 0.5\,\text{cm} = 5 \times 10^{-3}\,\text{m}\,,$$
$$b = 2.5\,\text{cm} = 2.5 \times 10^{-2}\,\text{m}\,,$$
$$z = 0.5\,\text{cm} = 5 \times 10^{-3}\,\text{m}\,;$$

this yields the numerical result

$$\langle F \rangle = 8.12 \times 10^{-8} I_{\text{rms}}^2 \,. \tag{14}$$

A current of $I_{\text{rms}} = 1000\,\text{A}$ will give a levitation force of $0.08\,\text{N}$ or about $7.2\,\text{g}$, which is about twice the weight of the above sphere assuming a density of $8\,\text{g/cm}^3$. A current of $707\,\text{A}$ is sufficient to levitate the sphere of radius a and density $8\,\text{g/cm}^3$ at a distance of $0.5\,\text{cm}$ above the plane of the loop. Normally a coil of n turns is used so that I_{rms} to first order can be replaced by NI_{rms} in the equation. For a three-turn coil, then the I_{rms} required would be $236\,\text{A}$ to levitate the above sphere at that distance. Those numbers are quite realistic. The change in inductance of the coil due to the sphere and the power absorption in the sphere is also worked out in Ref. 1.

The eddy currents induced in the conductor by the time-varying electromagnetic field heat it and may melt it. It can be shown [7] that for $a/\delta \gg 1$, the average power absorbed by the sphere producing Ohmic heating is given by

$$P = \frac{4}{3}\pi\sigma\mu_0^2\omega^2\delta^3 a^2 H_0^2 \,.$$

Research work conducted jointly by the authors has resulted in some new accomplishments of electromagnetic levitation. This work has primarily been directed at obtaining independent control over levitation force and heating. The achievement of this has made possible some interesting and perhaps useful laboratory investigations of solidification phenomena [8–12]. Containerless solidification of melts, which avoids contamination by crucibles, has been accomplished by utilizing an auxiliary heating source, i.e., electron beam, laser, or thermal imaging source, to obtain independent control of levitation force and heating power [8–10]. Containerless solidification of melts has also been achieved by utilizing the microgravity environment of space, where objects are naturally levitated [8, 11–13]. In that environment only very weak electromagnetic fields are required to confine objects, and specimens can be melted and solidified as slowly or as rapidly as desired while containerless. The utilization of the space environment for containerless experiments represents a definite advance in this field of research. By the use of weak fields to confine melts (either electromagnetic or acoustic), physical and chemical phenomena can be studied with negligible disturbance of the melt. This is in contrast to terrestrial levitation, where the large forces required to levitate appreciably disturb the melt, i.e., electromagnetic stirring and agitation. The microgravity environment of space thus allows almost complete independence of control over all of the parameters, i.e., heating, confinement, stirring, for containerless experiments. This, coupled with the microgravity environment itself, has permitted new studies of phenomena in melts and during solidification. As an example of this, in the microgravity environment, the rates of rising and settling of constituents of different densities in a melt and gravity-driven convection are greatly reduced. Hence the time for agglomeration and separation from the melt of particles or a dispersed phase in a melt is greatly extended, if other agitations are minimized (i.e., electromagnetic

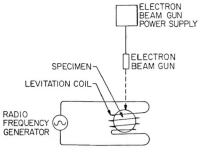

Fig. 2: Schematic of terrestrial containerless melting and solidification apparatus.

stirring). This has been observed both in containerless and noncontainerless experiments in space [8, 11–13].

Figure 2 is a schematic of an advanced levitation apparatus combining electromagnetic levitation with electron beam heating [14]. Figure 3 is a sketch of an electromagnetic containerless processing payload, flown twice on sounding rockets [10], using electromagnetic fields to confine the specimen and inductively melt it while floating in the microgravity environment. Independent control over confinement and heating is achieved here by the use of the space environment and adjustment of the strength of the electromagnetic field. The apparatus has a high-power mode for heating and a low-power mode for containerless solidification. Figure 4 shows the levitation coil assembly used in the terrestrial levitation apparatus shown in Fig. 2. A shaped molybdenum ring shown in Fig. 5 above the coil is used to achieve stable levitation of molten metals. Eddy currents induced in this ring by the alternating magnetic field of the levitation coil produce a magnetic field that presses down on the levitated specimen, preventing it from jumping out of the coil. Active rings carrying current or passive rings carrying only the induced currents are used to obtain stable levitation. The thick molybdenum ring above the stabilizing ring is used to collimate the electron beam and prevent it from striking the coil.

Figure 6 shows a levitated solid tungsten sphere. Figure 7 shows a sequence where tungsten is melted while levitated, using an electron beam [8].

Electromagnetic levitation is used in laboratory investigations of the physical and chemical properties of materials. Examples of these investigations are the measurement of the thermodynamic properties of metals [15–18], the study of gas-metal interactions [19, 20], observations of supercooling and nucleation phenomena in liquid metals [21, 22], determination of the densities and surface tensions of liquid metals [23], and the study of evaporative purification of liquid metals [24].

It must be mentioned that objects can also be levitated by intense acoustic fields [25], electrostatic fields [26], magnetostatic fields [27], and gas streams [28]. Other forms of electromagnetic levitation include utilizing the radiation pressure of laser light waves [29] and microwave confinement [29]. Many ingenious schemes have also been utilized to neutralize the effects of gravity such as obtaining up to 3 s of free fall in the NASA drop towers [30], up to 20 s of free fall during KC135 airplane ballistic trajectory flights [31], and sounding rocket flights [8, 13].

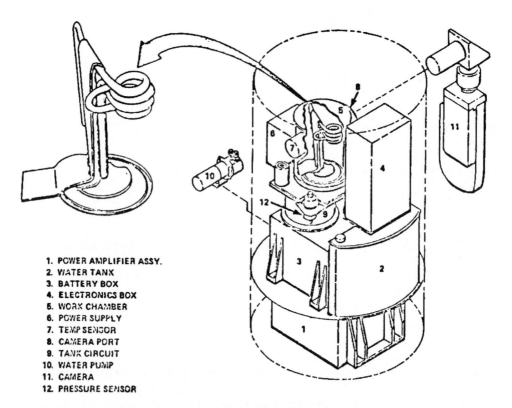

1. POWER AMPLIFIER ASSY.
2. WATER TANK
3. BATTERY BOX
4. ELECTRONICS BOX
5. WORK CHAMBER
6. POWER SUPPLY
7. TEMP SENSOR
8. CAMERA PORT
9. TANK CIRCUIT
10. WATER PUMP
11. CAMERA
12. PRESSURE SENSOR

Fig. 3: Electromagnetic containerless processing payload schematic.

Fig. 4: Levitation coil assembly with insertion pedestal extended and specimen positioned in the coil for levitation.

Fig. 5: Levitation coil with passive stabilizing ring and electron beam collimator above it.

Fig. 6: Levitated solid tungsten ball at a temperature of 2873 K.

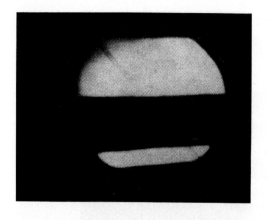

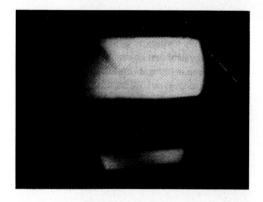

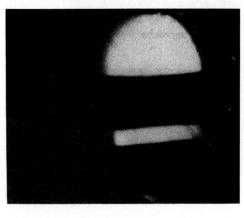

Fig. 7: Levitation melting sequence. Topmost view is solid tungsten sphere levitated at a temperature of 2873 K. The dark band is a turn of the levitation coil; center view is when melting is begun by allowing electron beam to strike the specimen from above; bottom view is completely molten, levitated tungsten specimen at a temperature of about 3773 K.

See also: Magnets (Permanent) and Magnetostatics.

References

[1] G. Wouch and A. E. Lord, Jr., "Eddy Currents: Levitation, Metal Detectors and Induction Heating", *Am. J. Phys.* **46**, 464 (1978).

[2] J. R. Reitz and F. J. Milford, *Foundations of Electromagnetic Theory*, p. 305. Addison-Wesley, Reading, Mass. 1967.

[3] J. R. Reitz and F. J. Milford, in Ref. 2, p. 213.

[4] J. R. Reitz and F. J. Milford, in Ref. 2, p. 243.

[5] J. R. Reitz and F. J. Milford, in Ref. 2, p. 156.

[6] E. C. Okress, D. M. Wroughton, G. Comenetz, P. H. Brace, and J. C. R. Kelly, "Electromagnetic Levitation of Solid and Molten Metals", *J. Appl. Phys.* **23**, 545 (1952).

[7] E. Fromm and H. Jehn, "Electromagnetic Forces and Power Absorption in Levitation Melting", *Br. J. Appl. Phys.* **16**, 653 (1965).

[8] G. Wouch, "Containerless Melting and Solidification of Metals and Alloys in the Terrestrial and Space Environments", Ph.D. Thesis, (Drexel University, Copyright, 1978).

[9] G. Wouch *et al.*, *Rev. Sci. Instrum.* **46**(8), 1122 (1975).

[10] G. Wouch, R. T. Frost, and A. E. Lord, Jr., *J. Crystal Growth* **37**, 181 (1977).

[11] G. Wouch *et al.*, *Nature* **274**, 235 (1978).

[12] G. Wouch, NBS Sponsored Conference on Applications of Space Flight in Materials Science and Technology, at NBS, Gaithersburg, Maryland, April 20–21, 1977.

[13] G. Wouch *et al.*, General Electric Company, Final Report, NAS8-31963, June, 1977.

[14] L. S. Nelson, *Nature* **210**, 410 (1966).

[15] A. K. Chadhuri, D. W. Bonnell, L. A. Ford, and J. L. Margrave, *High Temperature Sci.* **2**, 203 (1970).

[16] J. A. Treverton and J. L. Margrave, *J. Phys. Chem.* **75**, 3737 (1971).

[17] J. A. Treverton and J. L. Margrave, *J. Chem. Thermodyn.* **3**, 473 (1971).

[18] L. A. Stretz and R. G. Bautista, *Metall. Trans.* **5**, 921 (1974).

[19] O. C. Roberts, D. G. C. Robertson, and A. E. Jenkins, *Trans. Metall. Soc. AIME* **245**, 2413 (1969).

[20] LL. A. Baker, N. A. Warner, and A. E. Jenkins, *Trans. Metall. Soc. AIME* **230**, 1228 (1964).

[21] E. T. Turkdogan, *Trans. Metall. Soc. AIME* **230**, 740 (1964).

[22] E. T. Turkdogan and K. C. Mills, *Trans. Metall. Soc. AIME* **230**, 750 (1964).

[23] S. Y. Shiraishi and R. G. Ward, *Canadian Metall. Quart.* **3**, 118 (1964).

[24] B. F. Oliver, *Trans. Metall. Soc. AIME* **227**, 996 (1963).

[25] R. R. Whymark, "Acoustic Positioning for Containerless Processing", *Proc. Third Space Proc. Symp.*, Skylab Results, Vol. II, April 30–May 1, 1974, NASA M-74-5.

[26] A. T. Nordsieck, "Free-gyro systems for navigation or the like", U.S. Patent 3 003 356 (A. Nov. 1, 1954: P. Oct. 10, 1961).

[27] F. T. Backers, "A Magnetic Journal Bearing", *Philips Tech. Rev.* **22**(7), 232 (April 5, 1961).

[28] R. A. Happe, "Oxide Glass Processing", in Ref. 25.

[29] A. H. Boerdijk, "Technical Aspects of Levitation", *Philips Res. Rep.* **11**(1), 45 (Feb., 1956).

[30] H. F. Wuenscher, "Space Processing Experiments on Sounding Rockets", in Ref. 25.

[31] R. T. Frost *et al.*, Free Suspension Processing Systems for Space Manufacturing, General Electric Company, Final Report Contract NAS8-26157, June 15, 1971.

Lie Groups

M. Hamermesh[†]

A group is an algebraic structure with the following properties: The group consists of a set of elements $g, h, k, \ldots$, together with an operation of binary combination such that to each ordered pair of elements g and h of the set there is associated a unique element of the set, p, which is often called the *product* of g and h, so that we write $gh = p$. This multiplication is associative. There is an element e in the set such that its product with any element of the set in either order simply reproduces that element: $eg = ge = g$ for any g in G. Also, for each g in G there exists an element g' that is the *inverse* of g: $gg' = g'g = e$. The inverse of g is usually denoted by g^{-1}. Using these algebraic properties we can deduce a rich structure of theorems concerning groups with a finite or countable number of elements. Among these groups are the crystal symmetry groups that have important applications to the physics of the solid state.

  If we add to the algebraic structure some notion of *nearness* or *continuity* of the elements of the group, we obtain a *continuous* group. The most general form of this extension is the *topological* group, where the elements of the group G form a topological space. A special case of this extension, which includes practically all the groups that have physical applications, is the case where the elements of the group G can be labeled by a finite number r of continuously varying parameters $a_1, \ldots, a_r$, so that two elements of the group $R(a) = R(a_1, \ldots a_r)$ and $R(b)$ are near if the Euclidean distance $[\sum_{i=1}^{r}(a_i - b_i)^2]^{1/2}$ between the points a and b is small. If we take the product of the elements $R(a)R(b)$ we get an element $R(c)$ of the group, where the parameters $c: (c_1, \ldots c_r)$ are functions of the parameters a and b:

$$c_i = \varphi_i(a_1, \ldots, a_r; b_1, \ldots, b_r) = \varphi_i(a; b) \qquad (i = 1, \ldots, r) . \tag{1}$$

Similarly, the parameters of the inverse of the group element $R(a)$ will be functions of the parameters a. If all these functions are analytic, the group is said to be an *r-parameter Lie group*.

  Of particular interest in physics are Lie groups of transformations. For example, the assumption that space is homogeneous means that the transformations $x' = x + a$, $y' = y + b$, $z' = z + c$ must leave all physical statements unchanged for all real a, b, c. Similarly, isotropy of space implies the equivalence of descriptions in rotated systems. In classical mechanics such symmetry statements lead to Noether's theorem: Invariance of the Hamiltonian under a group of transformations implies the existence of corresponding constants of the motion. For example, translation invariance implies conservation of linear momentum of an isolated system.

  In quantum mechanics similar results hold, but because we have a superposition of states in quantum mechanics the results of symmetries are much more profound. If the Hamiltonian H of a system is invariant under a group G of transformations g, there exists a set of unitary operators $U(g)$ that commute with H: $UHU^{-1} = H$. As a result, if ψ is an eigenstate of H belonging to eigenvalue E, then $U(g)\psi$ is also an eigenstate with the same eigenvalue E. Unless there is some accidental degeneracy, the states belonging to a given eigenvalue form a subspace in Hilbert space and are transformed among themselves by the operators $U(g)$. We are then led to the notions of irreducibility, the splitting of multiplets when a perturbation

---

[†]deceased

is applied (the lowering of the symmetry means that the group G' is a subgroup of G and the states that previously were degenerate may no longer transform into one another), and selection rules on transitions.

For most of the problems of physics we need not deal with the Lie group as a whole. If we consider the transformations of the group G that are near the identity element of the group, we can, because of analyticity, regard them as being generated by integration of infinitesimal transformations, $U(g) = 1 + \varepsilon \hat{g}$, where 1 is the identity element and ε is a small parameter. For an r-parameter Lie group we can construct r independent basis operators $\hat{g}_i$, $(i = 1, \ldots r)$. By the fundamental theorem of Lie these operators form an algebra: They can be added, and multiplied by real constants, and the product $\hat{g}_i \hat{g}_j - \hat{g}_j \hat{g}_i$, the commutator $[\hat{g}_i, \hat{g}_j]$ of any two elements, is a linear combination of the basis operators:

$$[\hat{g}_i, \hat{g}_j] = \sum_{k=1}^{r} c_{ij}^{k} \hat{g}_k \, , \tag{2}$$

where the c_{ij}^{k} are the structure constants of the Lie group. We thus arrive at a new algebraic structure, the Lie algebra corresponding to the original Lie group. For the translation group the Lie algebra consists of three basis elements, p_x, p_y, p_z, that commute with one another. The p_i are simply the momentum operators along the three axes. For the three-dimensional rotation group, the Lie algebra has three basis elements, the angular momentum operators for x, y, z. Thus we can, instead of working with the continuous manifold of the Lie group, work with the finite-dimensional Lie algebra. We can associate with each element of the Lie algebra $\hat{g}$ a unitary operator $U(\hat{g})$ that acts on the states of the system. In this way we recover the simple theory of representations.

See also: Group Theory in Physics; Invariance Principles; Operators; Quantum Mechanics; Rotation and Angular Momentum.

Bibliography

M. Hamermesh, *Group Theory and Its Applications to Physical Problems*. Addison-Wesley, Reading, Mass., 1962.

M. Tinkham, *Group Theory and Quantum Mechanics*. McGraw–Hill, New York, 1964.

H. Lipkin, *Lie Groups for Pedestrians*. North-Holland, Amsterdam, 1965.

R. Gilmore, *Lie Groups, Lie Algebras, and Some of Their Applications*. Wiley (Interscience), New York, 1974.

B. Wybourne, *Classical Groups for Physicists*. Wiley, New York, 1974.

M. Gourdin, *Basics of Lie Groups*. Editions Frontieres, Paris, 1982.

J. F. Cornwell, *Group Theory in Physics*, Vol. 2. Academic Press, New York, 1984.

D. H. Sattinger and O. L. Weaver, *Lie Groups and Algebras with Applications to Physics*, Geometry, and Mechanics. Springer-Verlag, New York, 1986.

P. Olver, *Applications of Groups to Differential Equations*. Springer-Verlag, New York, 1986.

Light

V. L. Granatstein

Nature of Light

Light is energy in the form of radiation that is preferentially received by the human eye and causes the sensation of vision. It can be regarded as composed either of photons in the energy range $(2.5$ to $5.2) \times 10^{-19}$ J or equivalently of electromagnetic waves in the wavelength range 800 nanometers (nm) down to 380 nm. Undoubtedly, it is no coincidence that the spectrum of solar radiation reaching the earth's surface is peaked in this same spectral range where humans and other earth animals possess sensitive receptors.

The foregoing definition speaks of both photons and electromagnetic waves and thus brings to mind the question whether light is composed of particles or of waves. The answer to this question is that, strictly speaking, light is like neither the waves nor the particles that we are familiar with from everyday experience. Rather, light is something more subtle that displays both wavelike and particlelike properties. The behavior of light is adequately described by the theory of quantum electrodynamics. This theory uses Maxwell's equations to determine the pattern of fields in an electromagnetic wave and interprets the square of these fields as a probability density function for the photons that compose the light radiation. The photons themselves have particlelike characteristics (e. g., each photon has associated with it a discrete value of energy); however, they also have properties that are very different from the particles of normal experience (e. g., it is postulated that photons have zero rest mass, which implies that they always travel at a constant speed independent of their energy).

Electromagnetic Waves

Light forms a small part of the continuous spectrum of electromagnetic radiation, which encompasses radio waves and infrared radiation at wavelengths longer than light, and ultraviolet, x-ray, gamma-ray, and cosmic-ray radiation at progressively shorter wavelengths. At all wavelengths, electromagnetic radiation is made up of a magnetic field **H**, which exerts a force on moving electric charges, and an electric field **E**, which exerts a force on charges even when they are stationary.

Maxwell's equations may be combined to form a wave equation for the fields. In a medium that is homogeneous, stationary, and free of electric charges, this equation takes the form

$$\nabla^2 \mathbf{E} - \mu\varepsilon \frac{\partial^2 \mathbf{E}}{\partial t^2} - \mu\sigma \frac{\partial \mathbf{E}}{\partial t} = 0 \tag{1}$$

where ∇^2 is the Laplacian operator ($\nabla^2 = \partial^2/\partial x^2 + \partial^2/\partial y^2 + \partial^2/\partial z^2$ in Cartesian coordinates) and t denotes time. The properties of the medium through which the wave propagates are given by three parameters: (1) the magnetic permeability μ; (2) the dielectric constant ε; and (3) the conductivity σ.

In free space, $\sigma = 0$, while $\varepsilon_0 = 10^7/4\pi c^2$ F/m and $\mu_0 = 4\pi \times 10^{-7}$ H/m (mks units). Then, Eq. (1) takes the form

$$\nabla^2 \mathbf{E} - \frac{1}{c^2} \frac{\partial^2 \mathbf{E}}{\partial t^2} = 0 . \tag{2}$$

A simple solution is obtained by restricting spatial variation to the z direction and specifying **E** to point in the y direction. Then Eq. (2) yields

$$E_y = e_0 \sin 2\pi[\nu t - (z/\lambda) + \phi]$$

where $\lambda\nu = c$, and e_0 (the wave amplitude) and ϕ (the phase angle) are constants. This expression describes a propagating plane wave that is sinusoidal in both space and time. The spatial periodicity is characterized by the wavelength λ while the time variation has a frequency ν. Any one crest of the wave moves forward a distance λ in a time $1/\nu$, so that the speed of wave propagation is $\lambda\nu = c$.

The electric field in our example is transverse to the direction of propagation. This is a general characteristic of electromagnetic waves. The time-varying electric field is accompanied by a time-varying magnetic field **H**, which is oriented so that the direction of wave propagation is given by $\mathbf{E} \times \mathbf{H}$. In free space, the amplitude of the magnetic field is related to the amplitude of the electric field by $H = (\mu_0/\varepsilon_0)^{1/2}E$.

The Speed of Light

In vacuum, light waves travel at a constant speed c. The value of the constant c has been determined by measuring the time taken by a modulated light beam to cover a known distance. The presently accepted value of c is $2.99792458 \times 10^8 \, \mathrm{m/s} \pm 400 \, \mathrm{m/s}$.

A photon always travels at the speed c. In a material medium, however, the photons are absorbed and reemitted by molecules, a process that slows down the effective rate of travel of a light wave. Macroscopically, this retardation is represented in Maxwell's equations by having the parameters μ, ε, and σ depart from their free-space values. We can define a refractive index $n = (\varepsilon\mu/\varepsilon_0\mu_0)^{1/2}$, and in a weakly conducting medium the light wave propagates with phase velocity c/n, and has its fields exponentially attenuated through absorption as $\exp(-k_I z)$ where $k_I = (\mu/\varepsilon)^{1/2}\sigma/2$.

In general, the refractive index n is a function of wavelength, so that two light beams of different colors (different frequencies) propagate through materials at two different speeds. This phenomenon is known as dispersion and it is important to distinguish between phase velocity and group velocity in a dispersive medium. Any realizable measurement of the speed of light involves measurement on a wave packet, i.e., a wave that persists only for a finite time T. Such a wave packet is not monochromatic but covers a range of frequencies $\Delta\nu \approx 2/T$. As a wave packet travels through a dispersive medium the phase relation between the different frequency components changes. However, this change is relatively slow for an almost monochromatic wave in which $\Delta\nu \ll \nu$, and the wave packet progresses as a recognizable entity. The measured speed of the packet will be the group velocity $-\lambda^2 \partial\nu/\partial\lambda$ rather than the phase velocity $\lambda\nu$. Information is transmitted at the group velocity, which will always be $\leq c$.

Refraction and Reflection

When an electromagnetic wave in one medium impinges on a second medium it will be partly reflected and only a portion of the wave will be transmitted into the second medium. Moreover, the direction of wave propagation will be changed in the second medium; i.e., the light "rays" will be bent or refracted. This bending is the basis for simple optical devices such as lenses.

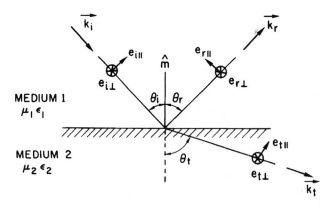

Fig. 1: Reflection and refraction at a planar interface.

The rules for reflection and transmission may be derived by a straightforward application of electromagnetic theory. The results are conveniently stated in terms of the configuration shown in Fig. 1. A plane electromagnetic wave is incident at an angle θ_i on a plane interface between two transparent media. The propagation vector $\mathbf{k}_i$ of this incident wave and the normal $\hat{m}$ to the interface lie in and define the plane of incidence. The propagation vectors of the reflected and transmitted waves (denoted by $\mathbf{k}_r$ and $\mathbf{k}_t$, respectively) lie in this plane of incidence.

The direction of $\mathbf{k}_r$ is given by Snell's law of reflection, viz., $\theta_r = \theta_i$. The direction of $\mathbf{k}_t$ is given by Snell's law of refraction,

$$\sin\theta_t = [(\mu_1\varepsilon_1)/(\mu_2\varepsilon_2)]^{1/2}\sin\theta_i = n_{12}\sin\theta_i$$

where n_{12} is the relative refractive index of the two media.

When the refractive index in medium 1 is greater than that in medium 2 (i. e., $n_{12} > 1$), as happens when light is refracted as it passes from glass into air, the refracted "ray" grazes the surface if $\sin\theta_i = 1/n_{12}$. For $\sin\theta_i > 1/n_{12}$ the incident wave is totally reflected. This phenomenon is exploited to confine and guide light inside optical fibers.

The magnitudes of the transmitted and reflected electric fields are given by the Fresnel coefficients. For the incident wave polarized with its *electric field in the plane of incidence* the field magnitudes are

$$e_{r\|} = -e_{i\|}\tan(\theta_i - \theta_t)/\tan(\theta_i + \theta_t)$$

and

$$e_{t\|} = e_{i\|}2\sin\theta_t\cos\theta_i/\sin(\theta_i + \theta_t)\cos(\theta_i - \theta_t)$$

while for the incident *electric field perpendicular to the plane of incidence*

$$e_{r\perp} = -e_{i\perp}\sin(\theta_i - \theta_t)/\sin(\theta_i + \theta_t)$$

and

$$e_{t\perp} = e_{i\perp}2\sin\theta_t\cos\theta_i/\sin(\theta_i + \theta_t) .$$

For the case of normal incidence ($\theta_i = 0$), $e_r = e_i(n_{12} - 1)/(n_{12} + 1)$ and $e_t = e_i 2n_{12}/(n_{12} + 1)$ independent of polarization. From the Fresnel coefficients given above, it is clear that $e_{r\parallel}$ vanishes when $\theta_i + \theta_t = \pi/2$. The angle of incidence at which this occurs is known as Brewster's angle and is given by $\tan\theta_i = (n_{12})^{-1}$. When randomly polarized light is incident on an interface at the Brewster angle, the reflected light will be linearly polarized with the electric field perpendicular to the plane of incidence; also, light polarized with the electric field in the plane of incidence will be totally transmitted. Typically, output windows on gas lasers are mounted at the Brewster's angle to produce a polarized light output.

Dispersion, Anisotropy, and Nonlinearity

In the preceding discussion of refraction, the refractive index was assumed to be constant for the sake of simplicity. In general, the refractive index will be a function of wavelength (dispersion), of polarization and direction of wave propagation (anisotropy), and of light intensity (nonlinearity).

Dispersion is easily observed. In most "transparent" media, the refractive index increases regularly toward the blue end of the spectrum. Thus white light passing through an air–glass interface at some oblique angle will be dispersed into a spectrum of colors, with the rays of blue light ($\lambda = 400\,\text{nm}$) being bent more than the rays of red light ($\lambda = 640\,\text{nm}$). A glass prism is useful in displaying the spectral content of white light or in measuring the wavelength of monochromatic light as in a prism spectrometer. The Kramers–Kronig relation allows us to calculate the light absorption properties of a medium when its dispersion is known.

Anisotropy is displayed strongly by certain crystalline materials such as calcite. A beam of linearly polarized light normally incident on a calcite plate may be resolved into two components with mutually perpendicular electric fields that will propagate through the calcite with different phase velocities. If the plate is of such a thickness that the phase difference between the two components is $90°$ after traversing the plate, the linearly polarized wave will have been converted into an elliptically polarized wave. Such a quarter-wave plate is a useful device for studying the polarization of light.

The application of a strong steady electric or magnetic field to a medium can significantly affect the interaction of a light wave with the molecules composing the medium. For example, Kerr demonstrated that strong anisotropy could be induced in glass by applying an external electric field. Of course, if external electric and magnetic fields can affect the propagation of light through matter, then the self fields of the light beam can have a similar effect if they are sufficiently strong. In recent years, with the development of lasers, light beams have been created with sufficiently great intensity to produce nonlinear effects such as harmonic generation.

Diffraction and Interference

Usually, nonlinear effects can be neglected. In that case, when two or more light waves are present at the same place and at the same time, their fields add algebraically and in a linear way, as given by the principle of superposition,

$$\mathbf{E} = \mathbf{E}_1 + \mathbf{E}_2 + \mathbf{E}_3 + \cdots + \mathbf{E}_N . \tag{3}$$

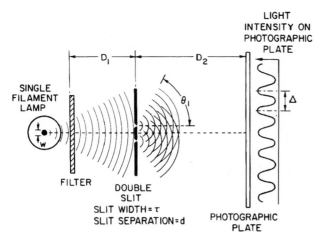

Fig. 2: Experimental arrangement for observing interference fringes (not to scale).

This implies the possibility of forming interference fringes, which are an observable confirmation of the wavelike properties of light.

A simple experimental demonstration of interference fringes can be made using the apparatus in Fig. 2, which constitutes one form of Young's double-slit experiment. A single-filament lamp is used as the light source. A filter (e. g., a piece of colored glass) is placed in front of the lamp to make the light more nearly monochromatic. This quasimonochromatic light is now passed through two closely spaced slits in an opaque plate.

Semicylindrical waves spread out from each slit. The spreading of a light beam that is restricted by a narrow slit is a manifestation of the phenomenon of diffraction; the trajectories of photons passing through a slit spread over an angle ϕ given by $\sin\phi \approx \lambda/w$ where w is the slit width. Thus for narrow slits the spread angle is considerable and the light beams from the two slits overlap.

The circular lines in Fig. 2 represent the crests of the waves. The intersection of two lines indicates the arrival at that point of two waves with the same phase or with phase difference equal to an integral multiple of 2π. Here the fields add according to Eq. (3) to yield maximum brightness. On the other hand, when a trough of one wave coincides with a crest of the second wave the fields cancel, resulting in darkness.

Through this interference between the two waves, a pattern of bright and dark bands or fringes is formed that may be recorded on a photographic plate placed far from the slits ($D_2 \gg d$). Bright fringes occur for displacement from the centerline of $x = p\lambda D_2/d$ and dark fringes for $(p + \frac{1}{2})\lambda D_2/d$ where p is an integer.

In a typical experiment, the double slits might be separated by $d = 0.2$ mm with the distance from the slits to the photographic plate $D_2 = 1$ m. Then, the spacing between adjacent bright fringes is $\Delta = 5000\lambda \approx 3$ mm, and the fringes are readily observable. Thus, although the spatial period of light waves is too small to be seen directly, interference techniques can effectively magnify the periodicity so that it becomes observable.

Coherence and Lasers

The production of fringes indicates that light waves from the two slits have a high degree of spatial coherence (i. e., their phase difference remains constant during the period of observation.) However, the light waves will be highly coherent and fringes will be produced only when d, the slit separation, is small. For strong fringe formation $d < \lambda D_1/\tau$ where τ is the thickness of the filamentary source. When d becomes large, the light waves will not interfere and their intensities will add (law of photometric summation) instead of their fields, producing a low-level illumination without fringes. The angular width of an inaccessible source can be determined by studying fringe visibility as d is varied; Michelson used this method to obtain the angular diameter of a star.

In addition to measuring coherence between light at two points separated transversely to the direction of wave propagation, measurements can also be made of the mutual coherence between light at points separated along the direction of propagation. This allows us to define a coherence length l_c such that strong coherence is measured when the separation between the two sampling points is less than l_c. A related coherence time can be defined as $t_c = l_c/c$. Measurement of l_c or t_c yields information on the frequency spread of the wave Δv, viz., $\Delta v \approx t_c^{-1}$.

Light beams produced by lasers are highly coherent and monochromatic. Strong spatial coherence extends over the beam cross section, and the coherence length along the direction of propagation may be many meters long. In conventional light sources even when filters are employed as in Fig. 2, coherence lengths are at best on the order of centimeters.

A description of laser physics is beyond the scope of this article. To be very brief, lasers are sources that are based on in-phase light emission from an ensemble of excited atoms or molecules at a resonant frequency characteristic of the particular atomic or molecular species. Because of the excellent coherence properties of laser beams they are being employed in myriad unique and important applications in such diverse fields as communications, weaponry, physics research, and corneal surgery.

Finally, we take note of the recent advent of free-electron lasers which are coherent light sources based on passing a stream of relativistic electrons with injection velocity v through a spatially periodic transverse magnetic field with period l_w and magnetic flux density $\mathbf{B}_\perp$. Such a laser amplifies light at a wavelength

$$\lambda = l_w(1 + a_w^2)(1 - v/c)^{-1}, \tag{4}$$

where $a_w = e\mathbf{B}_\perp l_w/2\pi mc$ in MKS units. Thus, the wavelength is continuously tunable by varying the electron velocity. Free-electron lasers are also characterized by unusually high power and efficiency.

This article has attempted to treat a number of important topics in sufficient detail so as to be meaningful to a wide readership. It has not been possible to give a comprehensive treatment of the subject of light; for a more detailed treatment of both topics surveyed in this article and of the many topics that were omitted, see the Bibliography.

See also: Diffraction; Electromagnetic Radiation; Faraday Effect; Kerr Effect, Electro-Optical; Lasers; Maxwell's Equations; Photons; Polarized Light; Reflection; Refraction.

Bibliography

G. Birnbaum, *Optical Masers*. Academic Press, New York, 1964. (I)

M. Born and E. Wolf, *Principles of Optics*, 4th ed. Pergamon, New York, 1970. (A)

C. A. Brau, *Free-Electron Lasers*. Academic Press, New York, 1990.

R. Ditchburn, *Light*, 2nd ed. Wiley (Interscience), New York, 1963. (I)

H. P. Freund and T. M. Antonsen, Jr., *Principles of Free-Electron Lasers*, 2nd ed. Kluwer Academic Publishers, Dordrecht, 1996.

F. A. Jenkins and H. E. White, *Fundamentals of Optics*. McGraw–Hill, New York, 1957. (I)

M. Kerker, *The Scattering of Light and Other Electromagnetic Radiation*. Academic Press, New York, 1969. (A)

M. G. J. Minnaert, *The Nature of Light and Color in the Open Air*. Dover, New York, 1954. (E)

J. W. Simmons and M. J. Guttman, *States, Waves and Photons; a Modern Introduction to Light*. Addison-Wesley, Reading, Mass., 1970. (A)

S. Tolansky, *Curiosities of Light Rays and Light Waves*. American Elsevier, New York, 1965. *Revolution in Optics*. Pelican Books, Harmondsworth, England, 1968. (E)

Light Scattering

R. W. Detenbeck

The scientific study of light scattering began in 1869 with the experiments of Tyndall, who sent a beam of white light through a suspension of fine particles. The scattered light was bluish and, observed at right angles to the beam, was highly polarized. These effects suggested to Tyndall that the blue color and polarization of light from the sky are produced in the scattering of sunlight by atmospheric dust. In a subsequent, theoretical treatment Rayleigh calculated the scattering from a random collection of small (compared with a wavelength), widely separated spheres. He predicted the observed polarization phenomena, as well as the inverse-fourth-power dependence of scattered intensity on wavelength which gives the sky its blue color. That the blue color and polarization were most striking in the clearest skies led Rayleigh to conclude that the molecules of air itself, not added dust particles, are responsible for the blue sky.

Wavefronts propagate straightforward within a homogeneous medium, and the effect of the medium is described in terms of its average electromagnetic properties. Light scatters when it passes through a medium that contains inhomogeneities on the scale of a wavelength. These may be embedded particles, as in Tyndall's scattering experiments, or the molecules of the medium itself, as in Rayleigh's theory of the blue sky. However, Rayleigh's molecular theory applies only to gases, where it is proper to add the *intensities* of light scattered by individual molecules. In condensed matter the medium is quite homogeneous on the scale of optical wavelengths, and the coherent addition of scattered *amplitudes* cancels radiation in all but the forward direction. Using a different model suggested by Smoluchowski, Einstein in 1910 described the weak light scattering from liquids in terms of thermodynamic density fluctuations within small volume elements of a continuous medium.

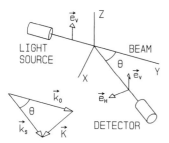

Fig. 1: Light scattering from a single scattering center located at the origin of the coordinate system.

Descriptions of light-scattering phenomena can be classified as static or dynamic. Static descriptions include time-averaged angular distributions of scattered intensity and polarization. One interesting case is the scattering of linearly polarized light from a single particle, illustrated in Fig. 1. If the scattering takes place within a surrounding medium of index n, and the vacuum wavelength of the incident light beam is λ_0, the incident wave vector has magnitude $k_0 = 2\pi n/\lambda_0$. The scattered light has wavelength λ_s, and the magnitude of its wave vector is $k_s = 2\pi n/\lambda_s$. The incident beam propagates in the direction of the vector $\mathbf{k}_0$, and the scattered light travels from the particle to the detector in the direction of $\mathbf{k}_s$. The two wave vectors determine the scattering plane, in which the scattering angle between them is denoted by θ. In a common arrangement the incident light is polarized with its electric field aligned along the normal to the scattering plane, often called the vertical direction and denoted here by the unit vector $\mathbf{e}_v$. The detector may select either the vertically polarized component or the horizontal component, with electric field aligned along $\mathbf{e}_h$.

Electromagnetic scattering from a particle remains an interesting theoretical problem in itself. The exact solution for the scattering from an isolated sphere was obtained by Mie in 1908, but the only other exact solution for a finite particle, an ellipsoid of revolution, awaited the work of Asano and Yamamoto in 1975. Numerical approximation techniques, which can calculate the scattering in almost any particular case, have been developed to run on modern computers. Moreover, they can be tested experimentally by scaling the electromagnetic wavelength and a model of the scatterer to centimeter sizes. However, many of the most important applications of light scattering to other fields (e. g., astronomy, atmospheric and ocean physics) deal with naturally occurring collections of particles differing in size, shape, orientation, and composition. In such cases one cannot afford a simple sum over all parameters, and theoretical ingenuity and intuition are required. In more complicated situations, such as dense atmospheric clouds, where the light is multiply scattered, even more powerful approximation techniques are needed.

Light carries momentum and energy. Therefore, when it is scattered or absorbed, there is a momentum transfer to the scatterer. A small particle can be levitated in a focused laser beam. Even single atoms can be manipulated by radiation forces. Laser cooling of atoms exploits the enhancement and strong frequency dependence of the scattering of light when its frequency is near that of an atomic resonance. The Doppler shift of near-resonance radiation can be exploited to favor scatterings which slow moving atoms in a viscous manner. Temperatures less than 1 mK for small numbers of atoms have been obtained.

Frequency shifts in the scattered light carry information about dynamic processes involving the scatterers. Brillouin in 1922 predicted a doublet in the frequency distribution of the light scattered from thermally excited sound waves in solids. Although experiments were very difficult with the light sources available at the time, Gross observed the Brillouin doublet in liquids in the early 1930s. Meanwhile, Raman and others had observed large frequency shifts associated with light scattered by molecular excitations in condensed matter.

The intense, coherent light from laser sources developed in recent decades is ideally suited to light-scattering studies of time-dependent phenomena. The spatial scale of accessible phenomena is determined by the wavelength of light. The scattering geometry is that of Fig. 1. The scattering vector $\mathbf{K} = \mathbf{k}_0 - \mathbf{k}_s$ is the difference between the incident and scattered wave vectors. In Rayleigh and Brillouin scattering the fields may be treated classically. The scattering process selects spatial Fourier components of the scattering fluctuations with a wave vector $\mathbf{K}$. Each temporal frequency component to ω_s of the scattered light is produced by a corresponding frequency component Ω of the fluctuations, where $\omega_s = \omega_0 \pm \Omega$. One associates Brillouin scattering in condensed matter with sound waves of gigahertz frequencies, inaccessible to acoustic measurements. Frequency shifts in Rayleigh scattering are caused by translational and rotational motions of particulate scatterers or by nonpropagating collective modes in condensed matter. In Rayleigh/Brillouin scattering the frequency shift of the scattered light is often such a small fraction of the incident frequency that k_s is nearly equal to k_0. Then the vector-addition triangle of Fig. 1 is isosceles, and the direction of $\mathbf{K}$ is along the bisectrix of $-\mathbf{k}_0$ and $\mathbf{k}_s$. The magnitude is $K = (4\pi n/\lambda_0)\sin(\frac{1}{2}\theta)$; the scattering angle θ selects a particular value of K. The maximum value of K occurs for backward scattering, where $K_{max} = 2k_0$. For light from a He–Ne laser, k_0 is about $10\,\mu m^{-1}$, and K_{max} corresponds to a spatial resolution of the order of $0.1\,\mu m$.

When the frequency shifts in dynamic light scattering are very small, the scattering process is called quasielastic. Light scattering from swimming microorganisms, diffusing macromolecules, and critical opalescence falls into this category. The narrow bandwidth of the incident laser beam permits measurement of these very small frequency shifts, but in some cases they are beyond the resolution capabilities of tuned optical elements. Then the frequency spectrum can be measured by beating the scattered light against a sample of the incident laser light, by beating the various Fourier components against each other, or by equivalent intensity-autocorrelation techniques.

Raman scattering is discussed in terms of quantum excitations. Momentum conservation requires that the scattering excitation accept a momentum equal to $\hbar K$. The corresponding frequency ω_s of the outgoing photon, for a change of excitation energy $\pm \Delta E$, is determined by energy conservation: $\hbar \omega_s = \hbar \omega_0 \pm \Delta R$. Typically, Raman scattering involves optical phonons of wavelength $2\pi/K$ in solids, or molecular excitations in fluids, but it is also valuable for the study of other, more exotic excitations in condensed matter.

The intense optical fields of high-power lasers produce nonlinear scattering effects, such as stimulated Raman and Brillouin scattering, in addition to the linear phenomena discussed above.

See also: Aerosols; Atmospheric Physics; Brillouin Scattering; Dynamic Critical Phenomena; Fluctuation Phenomena; Light; Nonlinear Optics; Polarized Light; Raman Spectroscopy; Rayleigh Scattering.

Bibliography

B. Berne and R. Pecora, *Dynamic Light Scattering*. Dover, Mineola, 2000. (A)

C. F. Bohren and F. R. Huffman, *Absorption and Scattering of Light by Small Particles*. John Wiley, New York, 1983. (A)

W. Bragg, *The Universe of Light*. Dover, New York, 1959. (E)

B. Chu, *Laser Light Scattering*. Academic Press, New York, 1974. (A)

I. L. Fabelinskii, *Molecular Light Scattering*. Plenum Press, New York, 1968. (A)

M. Kerker, *The Scattering of Light*. Academic Press, New York, 1997. (A)

M. Minnaert, *Light and Colour in the Open Air*. G. Bell and Sons, London, 1940 (reprinted by Dover, New York, 1954). (E)

D. W. Schuerman, (ed.), *Light Scattering by Irregularly Shaped Particles*. Plenum Press, New York, 1980. (A)

D. J. Wineland and W. M. Itano, "Laser Cooling", *Phys. Today* **40**(6), 34 (1987). (I)

H. C. van de Hulst, *Light Scattering by Small Particles*. John Wiley, New York, 1957 (reprinted by Dover, New York, 1982). (I)

H. C. van de Hulst, *Multiple Light Scattering*, Vols. 1 and 2. Academic Press, New York, 1980. (A)

Light-Sensitive Materials

H. E. Spencer and O. Horváth

Numerous materials display sensitivity to visible and ultraviolet radiation. They include gases, liquids, and solids, and exhibit a wide variety of properties. Their common feature is, however, that absorption of visible and ultraviolet photons generates excited electronic states. The fate of these states determines the behavior of the irradiated materials, which may undergo physical and/or chemical change. These excited states can best be discussed in terms of two somewhat different types of materials. In molecular materials, both the ground-state and the excited-state properties are characteristic of individual molecules. In many solids, however, individual atoms, molecules, or ions interact strongly and both the ground and excited electronic states are determined by the collection of these in the solid. In molecular organic materials both the ground and excited states, in spectroscopic notation, are usually singlets. Deactivation of the excited singlet back to the ground state, mostly by a radiationless path but sometimes accompanied by fluorescence, usually dominates unless internal conversion to a triplet state is energetically and kinetically possible. The long lifetime of the triplet state coupled with its high energy relative to the ground state results in a high probability of its reaction with another species.

A specific example of a molecular organic material that has been studied extensively is benzophenone, $C_6H_5COC_6H_5$, in solution at room temperature. It absorbs weakly in the violet and near-ultraviolet region. The excited singlet converts with near unit efficiency to a triplet state that can react with numerous compounds. For instance, if isopropyl alcohol is present in solution, the benzophenone triplet abstracts hydrogen atoms from the alcohol. This is but an initial step; the final products are acetone and benzopinacol. If isobutylene is present instead of isopropyl alcohol, a triplet benzophenone molecule and an isobutylene molecule combine to form a new molecule, an oxetane. Such photochemical reactions often provide the easiest means of synthesizing certain complex molecules. Isomerization is also an important photoinduced reaction of organic molecules. For example, stilbenes can easily undergo photochemical *cis–trans* isomerization. Photochemical sigmatropic shift, i.e., switching of double and single bonds is another type of isomerization. Photoisomerization of fulgids results in special photochromic systems because they require irradiation (generally of longer wavelength) for the reverse reaction as well. Upon photoexcitation polyenes can undergo cyclization or polymerization.

Numerous molecular inorganic materials are photosensitive as well. For instance, the photochemistry of NO_2 in the troposphere is important in the chain of events whereby the atmospheric pollutant ozone, O_3, one of the main components of photochemical smog, is formed. NO_2 is generated as a by-product of high-temperature combustion such as occurs in automobile engines. Absorption in NO_2 of ultraviolet radiation of 300 to 400 nm wavelengths from the sun produces an excited state that rapidly decomposes into NO and O. The latter reacts with O_2 to form O_3. Moreover, ozone itself is light-sensitive absorbing sunlight in the stratosphere in the 200–300 nm range, resulting in its dissociation to O and O_2. This absorption by atmospheric ozone protects us from damage by the short-wavelength solar ultraviolet radiation.

Other examples of photosensitive inorganic materials are found among coordination compounds of transition metals. For example, $K_3Fe(C_2O_4)_3 \cdot 3H_2O$, both as a solid or in solution, is decomposed by visible and ultraviolet radiation. In solution, products consisting of Fe(II) complexes and CO_2 form with high quantum efficiency. One of the most commonly employed actinometers is an acidic solution of $K_3Fe(C_2O_4)_3 \cdot 3H_2O$. The direction of the photoinduced charge transfer can be ligand to metal (LMCT) as in the previous example and metal to ligand (MLCT) as in the photochemistry of tris(bipyridyl)ruthenium(II) utilized in solar energy conversion systems. Besides photoredox reactions, metal complexes can undergo photoinduced substitution and isomerization as well. One of the most typical examples for the first process is the aquation of the Reineckate ion, *trans*-$Cr(NH_3)_2(NCS)_4^-$, upon excitation in the 315–600 nm range, producing $Cr(NH_3)_2(NCS)_3(H_2O)$ and free NCS^-. This reaction is particularly useful for actinometry in the visible region. A well-known photoisomerization is displayed by mercury(II) dithizonate (diphenylthiocarbazone) in organic solvents. In this compound, a Hg(II) center is coordinated to nitrogen and sulfur donors. Visible irradiation promotes a transform of the orange isomer to the blue one via transfer of a proton and a double bond. Interestingly, the blue isomer converts back to the orange starting complex in the dark, realizing a photochromic system.

In many solids, absorbed ultraviolet and visible radiation energetically raises electrons from the valence band to the conduction band, leaving behind mobile holes. Often these electronic carriers are detected in photoconductive or photovoltaic experiments. The electrons and holes commonly recombine, often in a radiationless manner, but sometimes they are accompanied by luminescence, particularly at low temperatures. Photochemically generated electrons and holes that avoid recombination can be utilized for catalytic processes involving simultaneous oxidation and reduction, as in the application of titanium dioxide semiconductor dispersion for wastewater treatment.

Silver halides exemplify photosensitive solids of scientific and commercial importance. The chloride, bromide, and iodide salts, and mixtures thereof form the basis of many photographic systems. After absorption, the hole–electron pairs that do not recombine reduce silver ions to metallic silver and oxidize halide ions to halogen. For photographic applications of a silver salt, microcrystals of micron or submicron linear dimensions are usually dispersed in a gelatin coating on an inert polymeric substrate. Silver halides vary in their sensitivity to the different wavelengths of light; all are sensitive to ultraviolet, but the sensitivity to visible wavelength drops off from iodide to bromide to chloride. The spectral sensitivity can be enhanced by the addition to the emulsion of photosensitizers. A color film is made made by sandwiching three layers of silver emulsions containing sensitizers, separated by suitable filters if necessary, such that each layer is sensitive to only one of the three primary colors, red, blue, and green.

In color-reversal film, during the developing and reversing processes, the dye formed becomes opaque to the primary color to which the emulsion layer was sensitive; therefore after all of the developed silver is dissolved from the film there remain three layers of subtractive filters that recreate the color and light intensity information of the original image when illuminated.

Silver chloride can also be used for photochromic lenses which darken when exposed to bright light due to the formation of elemental silver clusters. In the presence of copper(I) chloride, when the lenses are removed from light, photo-generated chlorine atoms oxidize Cu^+ to Cu^{2+}, regenerating Cl^- ions, and the Cu^{2+} formed oxidizes the silver atoms. The net effect of these reactions is that the lenses become transparent again as the silver and chloride atoms are converted to their original oxidized and reduced states.

See also: Photoionization.

Bibliography

R. P. Wayne, *Principles and Applications of Photochemistry*. Oxford University Press, Oxford, 1988.

O. Horváth and K. L. Stevenson, *Charge Transfer Photochemistry of Coordination Compounds*. VCH Publishers, New York, 1993.

N. J. Turro, *Modern Molecular Photochemistry*. University Science Books, Sausalito, 1991.

M. Klessinger and J. Michl, *Excited States and Photochemistry of Organic Molecules*. VCH Publishers, New York, 1995.

P. Boule (ed.), *Environmental Photochemistry*. Springer, Berlin, 1999.

Lightning

M. A. Uman

Lightning is a transient, high-current electric discharge whose path length is generally measured in kilometers. The source of most lightning is the electric charge separated in the common thunderstorm, although forms of lightning occur due to charge separation in snowstorms, in sandstorms, in the clouds generated by some volcanoes, and near thermonuclear explosions. Most thunderstorm-produced lightning occurs within the cloud (intracloud discharges). Cloud-to-ground lightning (sometimes called streaked or forked lightning) has been studied

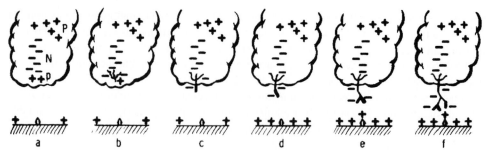

Fig. 1: Stepped-leader initiation and propagation, (a) Cloud charge distribution prior to lightning; (b) discharge in lower cloud; (c)–(f) stepped-leader progression toward ground. Scale of drawing is distorted for illustrative purposes. (Adapted from M. A. Uman, *All About Lightning*, Dover, New York, 1986.)

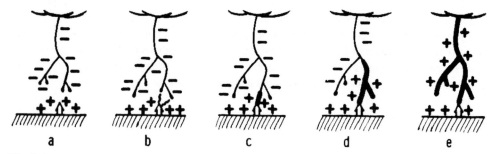

Fig. 2: Return-stroke initiation and propagation, (a) Final stage of stepped-leader descent; (b) initiation of upward-moving discharges; (c)–(e) return-stroke propagation from ground to cloud. Scale of drawing is distorted for illustrative purposes. (Adapted from M. A. Uman, *All About Lightning*, Dover, New York, 1986.)

more extensively than other forms of lightning because of its practical interest and because it is easily photographed. Cloud-to-cloud and cloud-to-air lightning are less common than intracloud or cloud-to-ground discharges.

The typical lightning between cloud and ground is initiated in the cloud and results in the lowering of tens of coulombs of negative charge to ground in about 0.2 s. The total discharge is called a *flash* and is composed typically of three or four component *strokes*, each lasting about 1 ms with a separation time of roughly 50 ms. Lightning often appears to flicker because the eye resolves the individual luminous stroke pulses. The process that initiates the first stroke in a flash, the *stepped leader*, is sketched in Figs. 1 and 2. In the idealized model of cloud charge shown in Fig. 1a, P and N are of the order of many tens of coulombs of positive and negative charge, respectively, and p is a few coulombs of positive charge. The negative charge N is located at heights where the temperature is between $-10\,°C$ and $-20\,°C$, about 6 to 8 km above sea level in temperate regions, and is more horizontally extensive than illustrated. The local discharge in the cloud base (Fig. 1b) allows negative charge (electrons) to be funneled toward the ground in a series of luminous steps of typically 1 µs duration and 50 m length with

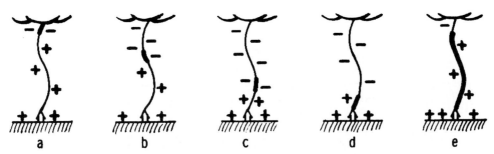

Fig. 3: Dart leader and subsequent return stroke, (a)–(c) Dart leader deposits negative charge on defunct first-stroke channel; (d)–(e) return stroke propagates from ground to cloud. Scale of drawing is distorted for illustrative purposes. (Adapted from M. A. Uman, *All About Lightning*, Dover, New York, 1986.)

a pause time between steps of about 50 µs (Fig. 1c–f, Fig. 2a). A fully developed stepped leader has about 5 C of negative charge on it, has traveled to ground in about 20 ms with an average speed of the order of 10^5 m/s and an average current of about 100 A, and has an electric potential with respect to ground of about -10^8 V. The intermittent leader steps have a pulse current of about 1 kA and are brighter than the channel above. The stepped leader branches in a downward direction during its trip to ground.

When the stepped leader is near ground, its electric field causes upward-moving discharges to be launched from the ground (Fig. 2b). When one of these discharges contacts the leader some tens of meters above the ground, the leader bottom is connected to ground potential. The leader is discharged by virtue of a ground potential wave, the return stroke, which propagates continuously up the leader channel at a velocity of typically one third the speed of light (Fig. 2c–e), the trip taking less than 100 µs. The return-stroke channel carries a peak current of typically 30 kA with a time to peak of a few microseconds. Currents measured at the channel base fall to half of peak value in about 50 µs, and currents of the order of hundreds of amperes often flow for milliseconds and may flow for many tens of milliseconds. The return stroke energy input heats the leader channel to near 30 000 K with the result that the tortuous high-pressure channel expands, creating shock waves that become the thunder we hear.

After the stroke current has ceased to flow, the flash may end. On the other hand, if additional charge is made available to the decaying channel top in a time less than about 100 ms, a continuous or *dart leader* (Fig. 3) will traverse the defunct return-stroke channel at about 2×10^6 m/s, depositing charge along the channel by virtue of its 1-kA current and carrying cloud potential earthward once more. The dart leader thus sets the stage for the second (or any subsequent) return stroke. Dart leaders and strokes subsequent to the first are generally not branched. Some leaders begin as dart leaders and end their trip to ground as stepped leaders. A drawing of a streak photograph and a still photograph of a lightning flash are given in Fig. 4.

While most lightning to ground lowers negative charge, about 10% of the worldwide ground flashes are initiated by downward-moving positive leaders, and the resultant flash lowers positive charge. Positive ground discharges are relatively more common at the higher latitudes, in winter, in the dissipating stage of thunderstorms, and in certain shallow or strati-

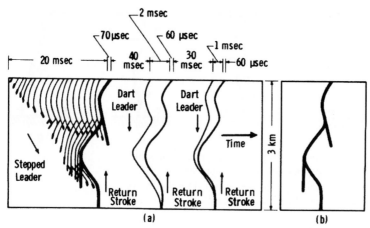

Fig. 4: (a) The luminous features of a typical lightning flash as would be recorded by a camera with relative motion (horizontal and continuous) between lens and film, a so-called streak camera. Scale of drawing is distorted for illustrative purposes, (b) The same lightning flash as recorded by an ordinary camera. (Adapted from M. A. Uman, *Lightning*, Dover, New York, 1984.)

form clouds. They generally have only one return stroke followed by a relatively long period of continuous current flow between cloud and ground, and have produced the largest measured peak currents, near 300 kA.

In addition to being initiated by downward leaders, lightning can occasionally be initiated from high structures by upward-going stepped leaders which can be either positively or negatively charged. The upward-going leaders branch in an upward direction. Upward-going leaders can be artificially initiated by firing from ground under a charged cloud small rockets trailing a few hundred meters of wire, either connected or unconnected to the earth. The wire serves to concentrate and enhance the electric field of the cloud charge to the point of lightning initiation. For both natural and artificially initiated upward lightning, the initial upward discharge may be followed by dart leader-return stroke sequences similar to those in normal downward-initiated lightning.

Intracloud discharges take place between two charge centers in the cloud, have a total duration about equal to that of ground discharges, and appear to the camera as a continuous luminosity with brighter pulses superimposed. It is thought that a leader moves through the cloud generating weak return strokes (called K-changes) when contacting pockets of opposite charge. The charge neutralized in a cloud discharge is about the same as in a ground discharge.

Heat and *sheet* lightning are due to cloud illumination by normal lightning in situations where the storm is too distant for thunder to be heard (generally over 25 km). Rocket lightning or spider lightning is the name given to the very long air discharges that often travel along the bases of the clouds. *Ribbon* lightning occurs when the cloud-to-ground discharge channel is shifted horizontally by the wind in the time between strokes. *Bead* lightning is the name given to the form of lightning in which the channel to ground breaks up into a "string of pearls" which persists in luminosity for a longer-than-normal flash duration. *Ball* lightning is the name given to the mobile luminous spheres that have been observed during thunderstorms, often in the vicinity of strokes to ground. A typical ball lightning is the size of a grapefruit and has a lifetime of a few seconds. At present there is no completely satisfactory explanation for ball lightning.

Lightning has been identified on the planet Jupiter from spacecraft television pictures and has been postulated to exist on Venus, Saturn, Uranus, and Neptun, less than convincingly, primarily from radio noise measurements.

See also: Arcs and Sparks; Atmospheric Physics; Corona Discharge.

Bibliography
R. H. Golde, *Lightning Protection*. Edward Arnold, London, 1973. (A)
R. H. Golde, (ed.), *Lightning*, Vol. 1: *Physics of Lightning*; Vol. 2: *Lightning Protection*. Academic Press, New York, 1977. (A)
D. J. Malan, *Physics of Lightning*. English Universities Press, London, 1963.(A)
B. Schonland, *The Flight of Thunderbolts*, 2nd ed. Oxford (Clarendon Press), London and New York, 1964.
M. A. Uman, *Lightning*. Dover, New York, 1984. (A)
M. A. Uman, *All About Lightning*. Dover, New York, 1986. (E)
M. A. Uman, *The Lightning Discharge*. Dover, New York, 2001. (A)
P. E. Viemeister, *The Lightning Book*. Doubleday, Garden City, NY, 1961; MIT Press (paperback), Cambridge, Mass., 1972. (E)

Liquid Crystals
P. G. de Gennes and J. Prost

Classification and Examples

In a crystal the building blocks (atoms or molecules) are regularly arranged in a periodic lattice. In a liquid they are completely disordered. With suitably chosen molecules, we can find other states of organization, with less symmetry than in a crystal but more symmetry than in a liquid: they are called *mesomorphic phases* or, more loosely, liquid crystals. The main types known at present are the following [1].

1. *Nematics* [2] are fluids of rodlike molecules with their centers disordered but with a common axis of alignment for the rods (Fig. 1a). A typical molecule giving rise to a nematic phase is methoxybenzilidenebutylaniline (MBBA):

The sequence of phases for MBBA is

$$\text{solid} \xrightarrow{10\,°C} \text{nematic} \xrightarrow{43\,°C} \text{isotropic liquid.}$$

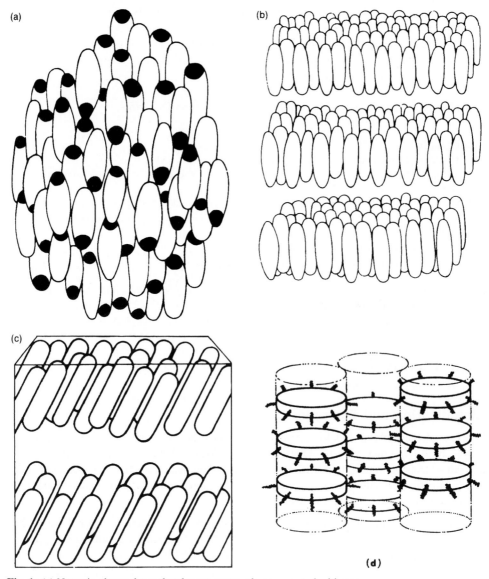

Fig. 1: (a) Nematic phase: the molecules are purposely represented with one black end; note that as many molecules are "up" as are "down" ; even if the molecules carry a permanent dipole, the phase is not ferroelectric, (b) Smectic-A phase, (c) Smectic-C phase. Note that in parts (b) and (c) each layer is disordered (liquid-like), (d) Thermotropic "columnar" phase: the disks represent the rigid part of a "plate-like" molecule and the wavy lines alkyl chains. Note the absence of periodicity along a column.

2. *Smectics A* and *smectics C* [3] are layered systems, each layer behaving like a two-dimensional liquid. The A type is optically uniaxial (Fig. 1b), whereas the C type is biaxial (Fig. 1c). Both types can be obtained with certain long organic molecules, such as terephthal-ifs-p-butylaniline (TBBA):

The sequence of phases of TBBA is complex: the main phases are

$$\text{crystal} \longrightarrow B \longrightarrow C \longrightarrow A \longrightarrow N \longrightarrow I$$
$$113\,^{\circ}\text{C} \quad 144\,^{\circ}\text{C} \quad 172\,^{\circ}\text{C} \quad 200\,^{\circ}\text{C} \quad 236\,^{\circ}\text{C}$$

(where I stands for isotropic and N for nematic, and B is another smectic type, to be discussed later). This shows that one compound may have many states in between the crystal and the liquid! This situation is not exceptional; the number of known molecules giving rise to mesomorphic phases is of the order of 2000.

In pure materials like TBBA, the main physical variable that allows us to explore the sequence of phases is the temperature; for this reason the resulting liquid crystals are called *thermotropic*. Another group is obtained with mixtures, where the phases can be obtained by changes in concentration (lyotropic liquid crystals). A major example of lyotropic smectics is found with mixtures of lipid and water [4]. A lipid molecule contains a polar head (hydrophilic) and a hydrocarbon tail (hydrophobic). Many lipid-water systems show a lamellar phase (Fig. 2a) that has the symmetry of a smectic A.

What is the origin of the smectic layers? For lyotropic systems they appear as a natural way of having the heads close to the water and the tails far from it. For thermotropic systems the situation is less clear, but something similar may happen: the molecules usually have an aromatic – rigid – part (Ar) plus terminal chains that are aliphatic (Al). It may be that (Ar)–(Ar) attractions dominate the system's behavior, and they favor a layered system.

3. Smectics A and C are not the only smectic phases. Hexatic smectics are also stacks of liquid layers. Like the A type, they are optically uniaxial, but the pair correlation function exhibits a sixfold macroscopic modulation. Locally, there is a triangular order like in a crystal, but a high enough defect density destroys the periodicity without decorrelating the triangles directions.

Smectics I, F, and K are hexatics in which molecules are tilted with respect to the layers. In terms of symmetry, they are not different from smectics C.

More recently, an "incommensurate" smectic has been reported. A real-space picture is not easy to give: imagine two conventional smectic A systems with incommensurate layer spacing; mix them together so that the two percolate through each other, this gives an approximate image of an incommensurate smectic.

4. Many other phases have been called smectics historically, smectics B in particular. The transition from normal solid to smectic B is often associated with the "melting" of aliphatic chains at the end of the molecule while other parts (the aromatic region in thermotropic systems, or the polar heads in lipid water) remain well ordered: this gives a structure in which successive layers are nearly uncoupled. X rays show that they should be classified as crystals. The same remark holds for smectics D, E, and G. The most interesting is the D phase which is a cubic gel with several hundred molecules per unit cell.

5. "Canonic" phases [5], also called columnar phases, have periodicity in two directions and not in the third; they are found with lipid water and also with thermotropic systems. In the latter case, disc-like molecules pack on top of each other, forming columns; however, the lack of a fixed repeat distance between the centers of gravity of the molecules gives a liquid-like order along the column axis (Fig. 1d). In the former case, one component is associated in rods, each rod being a one-dimensional fluid tube (Fig. 2b ,c). The macroscopic symmetry is the same. There are many possible variations of this basic arrangement among which about a dozen are currently known.

The physics of all these liquid crystals has expanded rapidly during the last twenty years, and cannot be summarized easily. We have chosen here to omit all further discussion of properties at the molecular level (for which there exist many reviews [9, 10] and to focus on certain macroscopic features, which are most unusual.

Optical Properties of Nematics

Nematics are optically uniaxial, but they flow like liquids. Their optical properties are interesting mainly because the optical axis may be *rotated by weak external agents*: the effects of the walls of the container, of magnetic and electric fields, and of flows are most important in practice.

1. *Alignment by walls*: It is possible to fix the orientation of the optical axis in a nematic slab by suitable treatment of the container surfaces: rubbing in one direction, or covering the surface with suitable detergents, or evaporating metallic films at an oblique angle. It is thus possible to prepare single-domain samples with well-defined boundary conditions.

2. *Alignment by magnetic fields H*: Nematic fluids have a weak diamagnetic susceptibility χ, and χ is anisotropic. Usually, this anisotropy favors an optical axis parallel to the magnetic field H. With suitable orientation of H, there may be an interesting competition between wall alignment and field alignment, first studied long ago by Freedericksz [6]. Typically, for a slab 50 μm thick, the fields involved are of the order of $10^3 - 10^4$ Oe. The field effects can be monitored very directly by optical observations.

3. *Flows* are strongly coupled to the molecular alignment [7]. In a simple shear flow the long axis of the molecules tends usually to become nearly parallel to the lines of flow. In many cases, a competition can be set up between wall alignment and flow alignment, leading to unusual mechano-optic effects and to instabilities even at extremely low Reynolds numbers.

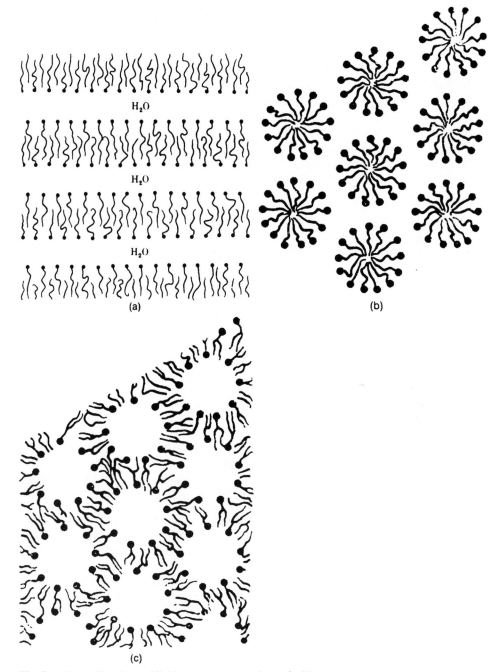

Fig. 2: (a) Lamellar phase of lipid + water systems (smectic A).
(b), (c) Hexagonal phases of lipid–water systems (canonic).

4. *Electric fields E* have remarkable effects on nematics. Alignments similar to the *H* effect discussed above provide display devices which are sensitive to voltages as low as a few volts. Optical anisotropy or linear dichroism provide the necessary coupling to light. Multiplexing requires the use of active addressing schemes. More interesting from a fundamental standpoint is the instability obtained with slightly conducting fluids ($\sigma \approx$ 10^{-9} mho): (a) The molecular alignment is slightly deformed by a thermal fluctuation; (b) this reacts on the conduction-current pattern *J*, and makes *J* inhomogeneous; (c) the inhomogeneous *J* builds up a space charge *Q*; (d) the field *E* acting on *Q* gives a bulk force in the liquid; (e) the result is a flow pattern, which distorts the alignment as explained in item (3) and *enhances* the original deformation – hence the instability [8]. When *V* is high enough, a state of turbulent flow is reached, the alignment is strongly perturbed, and the system scatters light very strongly.

These examples show that the optical axis can be tilted easily by external perturbations. As usual in physics, this also implies that, in the absence of any perturbation, the *spontaneous fluctuations* of the axis are important: they give rise to a scattering of light (in the absence of any field *E*) that is much higher than for normal liquids: nematics are *turbid* when viewed in thicknesses $\gtrsim 1$ mm.

Smectics A

Smectics A are the simplest example of a system with liquid layers, and they have unusual mechanical properties, intermediate between those of crystals and liquids.

Deformability of the Layer Structure

The ideal arrangement is a succession of flat sheets. But the sheets can be deformed very easily. The distance between them is essentially fixed, but they can be moved over large regions since successive layers can slip freely over one another. This shows up in various problems.

Textures: If the boundary conditions do not allow for the ideal flat arrangement, the layers build an interesting deformed texture. The simplest example is a jelly-roll shape (with a singular line on the axis). But there are more complex patterns, where the singular lines become ellipses and hyperbolas ("focal conies") [11].

Undulation modes: An oscillation with a wave vector parallel to the layers does not alter the interlayer distance; thus, the restoring force for these particular modes is extremely weak, and they have a large thermal amplitude, giving rise to a special type of light scattering.

Instability under traction: A smectic slab (typical thickness 100 μm) with the layers parallel to the slab plane is rapidly pulled along the normal to the layers; as soon as the relative displacement of the limiting plates exceeds a few hundred angstroms, the smectic layers fold to fill the available space without changing their thickness (Fig. 3).

Flows in a Smectic Structure

At first sight, a smectic appears very viscous – very much like a soap. But the real situation is more subtle. Flow *parallel* to the layers involves only weak friction, very much like in a conventional fluid. Flow *through* the layers requires a pressure gradient, just like the flow of

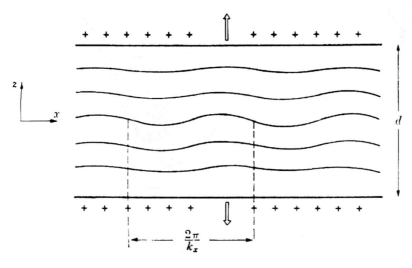

Fig. 3: Instability of smectic-A layers under traction.

water through a filter. But here the filter and the flowing fluid are one and the same species! The hydrodynamic consequences of these strange features have now been observed in certain weak flows. Strong flows usually imply the creation and motion of many defects (dislocations, focal conies) and are just beginning to be understood.

Fluctuations play an unusual role in smectics: elastic moduli and viscosities become wave-vector and frequency dependent in the long-wavelength long-time limit.

Canonic Phases

They again have unusual mechanical properties bearing some similarities with those of smectics.

Deformability of the Structure The ideal arrangement is that of straight columns. However, bending the columns, keeping the distance between them constant, is possible: this costs only nematic energy and thus occurs easily.

Textures The equivalent of the smectic "focal conies" are developable domains. The simplest example is obtained by wrapping the columns around a cylinder of revolution perpendicularly to its axis.

Instabilities A slab of canonic sample with the columns parallel to the slab behaves pretty much like the smectic of Fig. 3. The novelty is now that one can get a buckling instability when the columns are perpendicular to the slab and under compressive stress.

Flows have not been studied, but pulling on a columnar material produces thin strands (a few microns diameter, a few millimeters long), almost perfectly oriented.

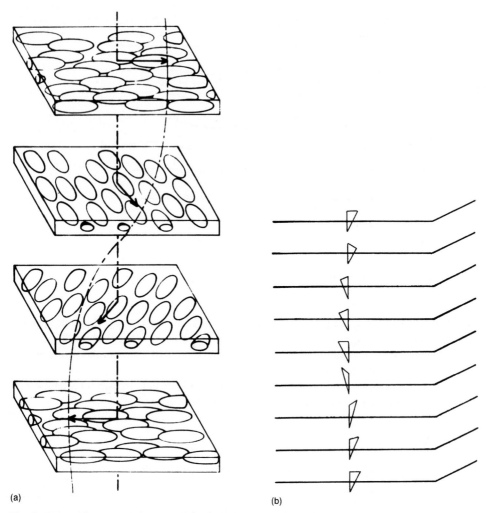

(a)   (b)

Fig. 4: Twisted liquid crystals: (a) cholesteric N* (twisted nematic); (b) twisted smectic C (C*). Here for simplicity each molecule is represented as a thin rod, and only one molecule per layer is shown.

Chiralization

Another remarkable example of the deformability of liquid-crystal structures is connected with the effect of solute molecules that are not identical to their mirror image (chiral molecules). A nematic N doped with such molecules becomes twisted into a helical form N* (Fig. 4a), usually known as a *cholesteric* phase [12]. The pitch of the helix is typically in the micron range and thus gives rise to Bragg scattering of optical light. For this reason many cholesterics show beautiful iridescent colors. Also, the pitch (and thus the color) can be very sensitive to various external agents (temperature, fields, chemical contaminants, etc.) and this has been the basis of certain applications [13]. When added to a smectic A, chiral molecules cannot induce a macroscopic twist (because this twist is incompatible with the condition of constant layer thickness). But in a smectic C, chiralization does give a twisted phase C* (Fig. 4b). The C* systems have Bragg reflections and optical rotations very much like N*. They have one further property that was recognized recently [14]: each C* layer is *ferroelectric*, with a permanent moment parallel to the layers and normal to the tilt direction. The ferroelectric moment is usually rather small (of the order of 10^{-2} debye per mole), but it can reach important values (almost one debye per mole) for well-chosen molecules. The existence of a moment in a system that can flow leads to remarkable cross effects. In particular, a thin slab ($\approx 1\,\mu m$) of C*, in which the layers are perpendicular to the slab, provides a novel, highly multiplexable and fast display. In this geometry, the helical precession is unwound by the boundaries, and only two states of polarization (up and down) are possible. A modest electric field commutes the states.

On the whole, we see that liquid cystals seem to have a relatively broad spectrum of future applications.

See also: Liquid Structure; Viscosity.

References

[1]   The classification of liquid crystals has been constructed mainly by G. Friedel [*Ann. Phys. (Leipzig)* **18**, 273 (1922)] using optical observations, and by H. Sackmann and D. Demus [*Mol. Cryst.* **2**, 81 (1966)] using miscibility criteria and textural features.

[2]   From the greek *nema*, thread; this refers to certain threadlike defects (disclinations) that are found in poorly oriented nematics (see Ref. 6).

[3]   From the greek *smegma*, soap; certain lamellar phases of the soaps (salts of fatty acids) are smectics.

[4]   For a review of the polymorphism of lipid–water systems, see V. Luzzati in *Biological Membranes* (H. Chapman, ed.), Academic Press, New York, 1968.

[5]   From the greek *canon*, rod; this terminology was introduced by F. C. Frank.

[6]   The continuum elastic theory on which the interpretation of phase effects is based is due to C. Oseen, H. Zocher, and F. C. Frank; see F. C. Frank, *Disc. Faraday Soc.* **25**, 19 (1958).

[7]   The equations of "nematodynamics" were constructed by J. L. Ericksen and F. M. Leslie. A more general formulation covering the smectics as well can be found in P. Martin, O. Parodi, and P. Pershan, *Phys. Rev. A* **6**, 2401 (1972).

[8]   The complete explanation is due to W. Helfrich, *J. Chem. Phys.* **51**, 4092 (1969).

[9]   G. Gray, *Molecular Structure and the Properties of Liquid Crystals.* Academic Press, New York, 1962. A. Saupe, *Angew. Chem. Int. Ed. Engl.* **7**, 97 (1968). G. H. Brown, W. Doane, and V. Neff, *C.R.C. Crit. Rev. Solid State Sci.* **1**, 303 (1970). P. G. de Gennes, *The Physics of Liquid Crystals*, Oxford, London and New York, 1974.

[10]   For a recent perspective on the field, see the *Proceedings* of the 5th International Conference on Liquid Crystals (Stockholm, 1974) published by *J. Phys. (Paris), Colloq.* **C1**, (1975).

[11]   The focal conies are always with sets of equidistant surfaces, such as wave fronts in optics. G. Friedel realized this very early, and from observations of focal conies under the microscope he was able to infer – long before any x-ray work (see Ref. 1) – that smectics A were made up of equidistant fluid layers.

[12]   Because it was found first with cholesterol esters.

[13]   J. Fergason, *Sci. Am.* **211**, 77 (1964).

[14]   R. B. Meyer, L. Liébert, L. Strzelecki, and P. Keller, *J. Phys. (Paris) Lett.* **36**, L69(1975).

Liquid Metals

N. E. Cusack

Among the new insights into liquid-metal physics which began to emerge in the 1960s was the realization that several strands of theory were converging. Simple nonmetallic liquids like argon had become well understood through steady progress in liquid-state theory. This theory was ready for use for liquid metals if their interatomic forces resembled, at least in some ways, those of nonmetals. Striking developments in the physics of crystalline metals were showing that this was indeed conceivable, so a fusion of metal and liquid-state theories began, and continues still. In another perspective, liquid metals can be seen as one class of structurally disordered, or noncrystalline, materials which also includes all other liquids and glasses and amorphous semiconductors. Thus, liquid metals pose problems which arise, *mutatis mutandis*, in crystalline metals, metallic glasses, and nonmetallic liquids. To see what these problems are let us describe a metal in general terms.

The general character of metals results from the dense fluid of free electrons that permeates them. Indeed a pure liquid metal can be seen as a mixture of (i) a fluid of positively charged ions and (ii) a gas of negative electrons containing one electron per ion in sodium, two in calcium, and so on according to valency. There will, therefore, be about 10^{28} electrons/m^3, and these are loosely called "free electrons" because they can move through the entire system. The free-electron gas conducts electricity well and is also responsible for thermal conduction, metallic reflectivity, and all the other typically metallic properties.

To understand a liquid metal we must tackle three problems which are not really separate. They concern the forces between the ions and the electrons, the forces between one ion and another, and the way electrons behave in a disordered (noncrystalline) environment. The following three sections touch on these.

Forces Between Ions and Electrons

The ions contain electrons tightly bound to nuclei. The free electrons flow through the ions under the control of two major influences: (a) the electrostatic attraction of the positive ions and (b) a complicated response to those electrons bound to the nuclei. A powerful development of the modern wave-mechanical theory of solids made it possible to treat influence (b) mathematically as if it were an electrostatic influence like (a) except that it is a repulsion, which partially cancels the attraction (a). The net effect of the two is quite small and is called a *weak pseudopotential*.

An electron passing through an ion may therefore be thought of as deviated by the appropriate pseudopotential. However, the ion does not suddenly exert its influence as the electron crosses a well-defined ionic boundary. The ions are positively charged and the electron gas is mobile, and therefore the ions exert a long-range influence on the gas by attracting electrons to themselves. Meanwhile the free electrons are also repelling each other, and a complex balance of forces sets in whose ultimate effect is to envelop each ion in an electron cloud. This cloud, being negative, neutralizes the effect of the positive ion rather completely at large distances (say, several atomic diameters) and less completely at smaller distances.

The cloud of electrons piled up near an ion is called the ion *screening cloud* and the total effect on an approaching electron of the pseudopotential and the screening cloud is called the *screened pseudopotential*.

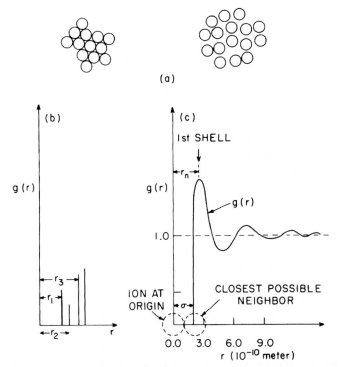

Fig. 1: (a) Schematic regular and irregular arrays of atoms shown in two dimensions, (b) Radial distribution function for a perfect regular or crystalline array. Atoms occur only at definite distances r_1, r_2, etc. (c) Radial distribution function $g(r)$ for a liquid metal. Atoms occur at any distance greater than σ (the effective hard-sphere diameter) with varying probability; r_n is the mean distance of nearest-neighbor shell, which typically accommodates 11–12 neighbors on average.

In summary we may therefore say that from the point of view of one electron passing through a liquid metal it encounters the screened potentials of the ions (the nearest ones being felt more strongly) and is scattered by them. The total effect of all the ions acting through their weak screened pseudopotentials on all the electrons is to produce the very electrons clouds that screen the ions. This idea underlies most current thinking about liquid-metal physics.

Pseudopotentials can be calculated from first principles, i. e., from the known nature of the ions, and they enable a good account to be given of the conductivity and other electronic properties in the following way.

Perfectly regular arrays of atoms in space are known theoretically not to have any electrical resistance. Cold crystalline metals are good approximations to this and their resistance is indeed very low. As they are heated, their electrical resistance rises because the ions vibrate about their ordered positions, thus acquiring *thermal disorder*. On melting, the array of ordered positions dissolves and *structural disorder* sets in as illustrated in Fig. 1. This causes a further rise of resistance by a factor of about 2 at the melting point. Further heating increases the resistance of liquid metals still more. Thus, although the resistance of liquid metals is higher than that of the corresponding crystals, it is not catastrophically so and liquid metals are good conductors. They also have thermoelectric powers and thermal conductivities comparable with those of solid metals.

These properties are now well understood on the basis of wave mechanics. As explained earlier, the electrons are deflected when they collide with the ions because the latter have weak screened pseudopotentials. This is in terms of particle language. In the wave language of wave mechanics, the electron waves are scattered by the ions to a degree depending on the pseudopotential, and the scattered wavelets from the different ions may be in or out of phase, thus reinforcing or canceling one another. To find the net effect of all the wavelets is a typical problem in diffraction theory, well understood in, for example, x-ray crystallography. The combination of pseudopotential and diffraction theory, employing the structural knowledge of Fig. 1c, leads to expressions for electrical resistance, thermoelectric power, and thermal conductivity in good agreement with observations on many pure liquid metals and alloys.

Forces Between Ions

Any two positive ions repel one another electrostatically. But in a metal they have another, indirect, mutual influence because they both exert force on the same electron gas. The latter contribution must depend on the density of the system because this determines the density of the electron gas. This contrasts with the intermolecular forces in nonmetals which are independent of density to a good approximation. However, for a *fixed density*, it turns out that the total ion–ion interaction in a metal can be treated just like the forces in nonmetallic liquids. In particular, the ions have pairwise interactions. The exact form of the ion–ion potential function will vary from metal to metal and differ from those of nonmetals; it depends on the pseudopotential. But once it is known, it can be inserted into liquid-state theory to calculate thermodynamic properties such as free energy and entropy. This is technically complicated but has been done with some success for pure liquid metals and some binary alloys, thus bringing these materials into the theory of liquids.

It is even possible to use liquid-state theory and the ion–ion potentials to say something about structure. But generally the details of the interaction are not well enough known to calculate structure accurately from first principles. The structure of crystals is of course revealed in full by x-ray crystallography. This is not so for liquids. What x-ray and neutron diffraction studies can do for liquids is to provide a quantity known as the radial distribution function, denoted by $g(r)$ and defined as follows: choose any one ion as origin; then in a spherical shell of radius r, the average number density of the ions is $g(r)$ times the average number density in the whole liquid. Figure 1 contrasts the structures of liquid and crystalline metals. The function $g(r)$ exceeds unity at the first closely packed shell of neighbors and is zero at smaller radii because the neighbors cannot penetrate the origin atom. Note that $g(r)$ is an average. The ions are constantly moving about and interchanging neighbors like a milling crowd; $g(r)$ represents the average appearance of numerous instantaneous snapshots of the crowd. To think of the arrangement of ions in a liquid metal as a jumble of billiard balls in a bag is not far from the truth, and a hypothetical system of hard spherical atoms is called the "hard-sphere model" of a liquid. Because hard spheres are comparatively simple objects, it is possible to treat the hard-sphere model mathematically and to derive its thermodynamic properties by statistical mechanics and computing.

It is a somewhat remarkable fact that the structure and some of the thermodynamic properties of liquid metals can be represented quite well by those of the hard-sphere model provided that the diameter of the hypothetical spheres is selected by a theoretical process that takes the electron–ion and ion–ion interactions into account.

Electrons in Disordered Matter

The free or conduction electrons in metals are of course not perfectly free – they interact with the ions. But when, as assumed in the preceding two sections, the electron–ion pseudopotential is weak, the electrons are nearly free, and to treat them as wholly so between collisions with the ions is an acceptable approximation. This is valid for many metals for which the division of their electrons into those bound in the ions and those that are free in the electron gas is clear; examples are the alkali metals, Ca, Al, Pb, and Bi. In transition metals of the iron group the situation is more obscure for the 3d shell of electrons; these interact strongly with the ions but are partially free as well. There are other circumstances, such as particular compositions of binary liquid alloys, where anomalous electrical and thermodynamical properties show clearly that strong electron–ion interactions render the free-electron concept dubious.

The obvious question then is: how do electrons behave under strong forces from the ions? For crystals the answer is available in the Bloch or band theory of solids (q.v.) and the regular order of the crystal structure is essential for the solution. In disordered liquids or glasses, the problem is much more difficult and apt to lead quite rapidly to mathematical technicalities and unanswered questions. The following difficulties are characteristic: (i) the actual ionic arrangement in liquids is unknown – the radial distribution function gives only limited information; (ii) even if a typical configuration were wholly known, it would remain difficult to solve Schrödinger's equation for the electron motion and to average the solution over all possible configurations of the moving ions; (iii) the multiple collisions of an electron with the ions make the simple electron diffraction theory of electrical resistance untenable and more powerful, but very complex theories have to be tried instead. By now there are several theoretical and computational attempts to deal with these problems and it is fairly clear that the electron energy spectrum in a molten metal will normally be similar to that of its crystal but with the sharpness removed from peaks and discontinuities, a general smoothing out of features, and a removal of any anisotropy.

In alloys, strong electron–ion interactions are an expression of binding forces and a conceptual approach that emphasizes chemical bonds and valency can be helpful in the absence of complete descriptions of the electron motion from physical first principles. For example, the extraordinarily low conductivity of the liquid alloys of special compositions Li_4Pb and CeAu – the latter actually being nonmetallic! – testify to the enduring challenge of dealing fully with the physics of liquid metals outside the weak pseudopotential regime. The same applies to liquid metals of which the density is greatly reduced by heating under pressure to high temperatures near the liquid–vapor critical point. This process induces a gradual disappearance of metallic properties.

In the late 1980s the phenomena mentioned in the preceding paragraph are at the frontiers of the subject. The emphasis now is less on describing liquid metals as such and more on understanding what conditions of density, temperature, and composition make a disordered material metallic, semiconducting, or insulating and how its electrons are behaving to cause these properties.

See also: Conduction; Electron Energy States in Solids and Liquids; Glassy Metals; Liquid Structure; Metals; Molten Salts; Semiconductors, Amorphous.

Bibliography

N. E. Cusack, *The Physics of Structurally Disordered Matter*. Adam Hilger, Bristol, 1987. (I, A)
T. E. Faber, *Theory of Liquid Metals*. Cambridge University Press, Cambridge, 1972. (I, A)
M. Shimoji, *Liquid Metals*, Academic Press, London, New York, San Francisco, 1977. (I, A)

Liquid Structure

B. Alder

Van der Waals Equation

Over 100 years ago van der Waals proposed that a liquid could be distinguished from a gas by a simple modification of its equation of state. The equation of state constitutes the functional relation between the pressure p, volume v, and temperature T of a material, which for a perfect gas was then known to be $pv = RT$, where R is a universal constant. The modification that was required came about because the volume occupied by a liquid is considerably less than that of a gas. Hence the volume b taken up by the molecules themselves could no longer be ignored, and furthermore, when the molecules on the average are close together, they attract each other, so that the pressure required to contain the liquid is reduced. Thus, he derived

$$\left(p + \frac{a}{v^2}\right)(v - b) = RT \ ,$$

where the constant a is a measure of the cohesive force between the molecules of the particular gas.

One of the great achievements of this formulation is that it led to what was then by no means generally accepted, namely, the continuity of the states of matter, which conceives of the liquid and gas phases as being part of a single model. The van der Waals equation is formally a cubic equation in the volume with three real roots below a certain temperature and only one above that temperature, called the critical temperature. The largest and smallest of the three volume roots at a given pressure and temperature describe the liquid in equilibrium with its vapor (gas), the coexistence curve, while the middle root has no physical meaning. Above the critical temperature the liquid cannot exist.

What van der Waals did not realize at the time was that his concept of the continuity of states of matter could be extended using his model to include the solid state. This can be demonstrated by rearranging his equation to

$$\frac{pv}{RT} = \frac{v}{v - b} - \frac{a}{vRT}$$

so that the $v/(v - b)$ term can be recognized as a purely geometrical factor, describing the packing of the particles without consideration of their attractive force. Such particles can be idealized as hard spheres and the $v/(v-b)$ term then represents a crude approximation to their equation of state. To this day there exists no adequate theory for the equation of state of hard spheres that accounts for their complex arrangements at all possible volumes; however, numerical simulation computer results have been obtained for this system. These results convincingly demonstrate that hard spheres for a range of volumes beyond close-packing arrange themselves in an ordered solid structure, beyond which they undergo melting. Thus the van der Waals equation, modified by the correct hard-sphere properties, also describes the melting curve, including the temperature where all three phases of matter coexist, called the triple point.

Virial Expansion

Subsequent to the van der Waals model there have been many attempts to describe the liquid structure in terms of either a modified gas or a modified solid. None of these have been entirely successful, because both of these states of matter are too different from a liquid. The motivation for such models is primarily the availability of simple theories for the perfect gas and the harmonic solid, and the unavailability of any equivalent idealization of a liquid that leads to a simple, manageable mathematical theory. Thus the perfect-gas model was modified to account for the clustering of molecules leading to the so-called virial expansion. Each successive term in this expansion accounts for the properties of a cluster larger by one molecule than the previous one, the liquid being thought of as a gigantic cluster. The hope, not yet realized, was that one could derive some general properties of the infinite cluster relevant to liquids, or could at least prove the divergence of the virial series at the condensation point of the gas.

Lattice Models

On the other hand, it was thought that an adequate description of a liquid could be obtained by introducing holes (missing molecules at lattice sites) into a perfect crystal to account for the larger volume and increased mobility of molecules in a liquid. There has never been the slightest bit of encouragement from either experimental or quantitative considerations that such holes exist. The larger volume available to the molecules in a liquid, over that in a solid, seems to be democratically shared by all the molecules. For example, the entropy of a liquid, which is a measure of its order, is such that it requires an unreasonable number of holes in view of the energetic requirements to create such holes. The mechanism of mobility in a liquid is not by a large jump of a molecule adjacent to a vacancy into the hole, as in a solid, but overwhelmingly by a large number of much smaller steps. Such lattice models impose both a local geometrical arrangement and a long-range order, known, from x-ray diffraction studies, to be absent from a liquid.

Melting

Similarly, there is at present no clear evidence in the solid phase just prior to melting of any liquid-like behavior. The search for such premelting phenomena in the solid phase is motivated by an effort to try to understand the universal mechanism by which all solids melt or reach their stability limit. The only generally valid observation on when melting occurs was made long ago by Lindemann, who observed that solids melt at a temperature at which they are about 30% expanded from their volume at absolute zero or, equivalently, when the root mean square excursion of a particle from its lattice site exceeds about 10% of the interatomic distance. This geometrical criterion is quantitatively confirmed by the hard-sphere simulations. The universality of melting is then ascribed to the general similar nature of the steep repulsive forces between all atoms separated by close distances, which can be well approximated by hard spheres.

This leaves open, however, the question of the mechanism of instability leading to melting. A number of instability criteria for the solid phase, such as the disappearance of shear stress, have been investigated, but none has been shown to coincide with melting. In two-dimensional systems, constructed by polystyrene spheres confined to a plane, melting has been associated with the dissociation of dislocation pairs. By the same mechanism, helium films have been proposed to lose their superfluidity when vortex pairs start to dissociate. However, computer simulation of hard disks does not confirm such an instability criteria; it is rather a metastability criterion very close to melting. It is, of course, evident that at best only the least stable mode will coincide with melting, while all others will lead to metastable extensions of the solid phase into the liquid. The most conservative view is simply that melting occurs whenever the chemical potentials or Gibbs free energy of the solid and liquid are the same, and this need not be associated with an instability mechanism.

The melting curve exhibits a maximum whenever the liquid has a smaller volume than the solid, as the thermodynamic relation known as the Clausius–Clapeyron equation quantitatively demonstrates. The cause of the denser liquid is associated with the fewer constraints on the configurational arrangement imposed by the liquid structure than by the solid. The ice–water transition is a case in point; the ice crystal is constrained to a low-coordination-number (4), tetrahedral, regular structure by the highly directional interaction between the molecules, whereas in water the coordination number does not have to be exactly 4, and is in fact larger, and hence the fluid phase is denser.

Because the entropy of melting, $\Delta S/R$, has been observed to be of the order of 1 for many simple fluids, the notion that this entropy could be identified with the communal entropy, which is exactly 1, seemed attractive. The communal entropy derives from the difference between N particles, each confined to a volume per particle around a lattice site, and hence distinguishable, and these same N particles permitted to roam over the entire volume of the system, and hence indistinguishable, which leads to $N!$ additional identical configurations. A more detailed examination of the problem, however, leads to the conclusion that the entropy of melting is primarily due to the expansion of the system and only a small fraction can be identified as communal entropy.

Critical Point

The melting curve at high temperature and pressure can be predicted on quite general grounds to continue indefinitely; that is, there is no solid–fluid critical point, contrary to the gas–liquid coexistence curve. The argument, attributed to Landau, depends on the observation that if there were such a critical point, one could find paths in the phase diagram by which a substance could go continuously from a state of long-range order (solid) to one of short-range order (liquid), which is not allowable. On the other hand, this argument does not rule out the existence of a gas–liquid critical point, since only the degree of short-range order varies across it. This same argument eliminates critical points for phase transitions within the solid phase unless the two solids in equilibrium happen to have the same crystal structure. There is at least one such known case, cerium, where the atom in the two different phases has a different internal electronic structure.

The liquid–gas critical point has been found experimentally for all liquids, including liquid metals, not to be of the van der Waals type, that is the coexistence curve is not quadratic but closer to cubic. The universality of this behavior has been ingeniously and quantitatively described by the renormalization theory. For cesium the critical point coincides with the metal–insulator transition.

Radial Distribution Function

Because the liquid does not have a predominant structure but a wide range of possible arrangements of the atoms, the only practical description is in terms of statistically averaged distributions. There have been many attempts, particularly for water, to pick out a few of the more stable configurations and represent the structure as an average over them weighted by their relative energy. Even for water, however, there are so many configurations that have nearly the same energy or probability of occurrence that the task becomes hopeless. The x-ray or neutron diffraction pictures obtained from liquids give no hint of any definite arrangements of atoms but yield a set of smeared-out bands of rings. From the Fourier transform of this intensity pattern the radial distribution function is experimentally determined. It quantitatively describes the structure of a liquid in terms of the probability of finding another particle at a given distance from a central particle. The rapidly damped oscillations of this function with the period of about a lattice spacing show that the short-range order in a liquid extends to about three neighboring shells, beyond which deviations from the average density are no longer discernible. The accurate calculation of this function is the central problem in the theory of liquids, since using this weighting function for the appropriate pairwise property allows the determination of any thermodynamic property. For example, the potential part of the internal energy is calculated as an average of the pair intermolecular potential. For the calculation of the pressure of a liquid under normal conditions this radial distribution function must be exceedingly accurately known, since the pressure is the result of the near cancelation of the kinetic pressure with the potential part of the pressure determined from the average of the derivative of the potential (virial).

Theoretical expressions for the radial distribution function describing the structure of a liquid are derived from the partition function. Mathematical transformation of this partition function leads to a hierarchy of equations that express the pair radial distribution function in terms of triplet distribution functions, which in turn are expressed in quadruplet distribution functions, etc., *ad infinitum*. To close this hierarchy, the higher-order distribution functions must be expressed in terms of lower-order ones. For example, the triplet distribution function, describing the probability of three atoms being separated by three given distances, is expressed as a triple product of pair distribution functions, each for a separation by one of the three distances. This neglect of triplet correlations is called the pairwise superposition approximation and leads to a semiquantitative theory of the pair radial distribution function. Different closure approximations, but physically less transparent, have led to more quantitative predictions.

Most of the foregoing theories for the radial distribution are mathematically quite complex and of uncertain accuracy, so that the need for an alternative method led to the development of a computer simulation method – the so-called Monte Carlo method. As applied to hard spheres, for example, the particles are placed initially in some nonoverlapping configuration in a box and then randomly displaced, one at a time. Upon displacement, a check is made whether in the new position the sphere is closer than a diameter to any other sphere (overlaps). If yes, the displacement is rejected and the particle is placed in its former position. If no overlap occurs, a new configuration has been generated and the process is repeated. Averaging over a sufficient number of such configurations leads, via the evaluation of the radial distribution function, to an exact calculation of any thermodynamically required average.

Perturbation On Structure

These computer calculations, and particularly those for hard spheres, have led to a great deal of insight into the structure of simple liquids and a greater understanding of the validity of the van der Waals equation. It can be shown that the addition of a weak and long-range attractive potential to the hard-sphere potential leads rigorously to the van der Waals equation. Since this attractive potential is unrealistic, another way to obtain the van der Waals equation is to add a short-range attractive potential to the hard-sphere potential as a perturbation. But then it is necessary to show that the higher-order perturbation terms are negligible. The first-order perturbation term represents the mean number of hard-sphere particles within the range of the attractive forces and leads rigorously to the van der Waals equation, while the higher-order perturbation terms represent higher-order fluctuations in the number of particles about the mean.

If the added attractive potential is sufficiently long range or so weak that first-order perturbation suffices, the attractive potential does not change the structure of the fluid from that of pure hard spheres. This is a good approximation for a normal liquid, since the particles are so closely packed that they rarely escape outside the range of their attractive forces. The validity of the van der Waals equation at much lower density can be traced to the structural observation that the fluctuations of the number of particles within the range of the attractive forces does not change significantly with density over the entire fluid density range. For this reason the higher-order perturbation terms can be neglected.

Complex Liquid Mixtures

Under this topic quantum fluids such as helium, metals, liquid crystals, fluids made up of molecules as big as polymers and microemulsions are considered. All these topics are under active investigation but much progress has been made already, primarily via computer simulations. For helium a Monte Carlo path integral simulation has succeeded in describing the thermodynamic properties of the so-called λ transition between the normal and superfluid phase. For metals, the electrons must be dealt with quantum mechanically as well; however, since they are fermions instead of bosons, as in helium, a stable algorithm has not as yet been developed. Good results are, however, obtained by treating the electrons approximately by the density functional theory, used so successfully in solid-state band theory, and by treating the ion cores classically.

The notion that the formation of liquid crystals required anisotropic attractive potentials between the molecules was dispelled by a computer simulation of hard ellipsoids. In that simulation a phase transition from an isotropic liquid phase to a nematic phase in which the ellipsoids are only rotationally but not translationally ordered was found. It is believed that most of the many possible other rotationally ordered phases, such as a smectic phase, can be found for particles with other purely repulsive anisotropic potentials. On the other hand, the details of the attractive forces appear to be essential to the formation of lamellar phases from microemulsions in such mixtures as oil, water, and soaps. Such systems by suitably adjusting concentrations, temperature, and attractive potentials assume layer-like structures with alternate layers of water and oil with soap layers sandwiched between each interface. They are modeled by a soft ellipsoid (the soap molecules) which attracts water molecules on one end and oil molecules on the other, each represented by soft spheres. Lattice model versions of this system have been able to analytically predict this complex phase diagram qualitatively. Simpler mixtures of molecules have been successfully described by the van der Waals theory with appropriate values of a and b for each species. Statistical mechanical theories of simpler mixtures more realistically represent each molecule by a spherical soft potential originating at the site of each atom in the molecule, leading to what is called the interaction site potential. When dealing with polymers, that is, a highly viscous system of entangled very long molecules, the simulation procedure uses the reptation process to sample more efficiently the possible configurations of the polymer in the melt. Reptation, as the name implies, is a process by which a snake moves, namely, the head moves forward as the tail contracts.

Transport Properties

Computer simulation of hard spheres has shown that the fundamental assumption of molecular chaos by which collision becomes eventually uncorrelated is not valid at any finite density. Thus, strictly speaking, the Boltzmann transport equation is not valid although the quantitative corrections are exceedingly small. The persistence of velocity between successive collisions is caused by the generation of hydrodynamic vortex modes in which the momentum of a particle is fed back into its motion by the vortex it created in the medium. The phenomenon is well understood by graph theoretical means as well as mode-mode coupling (hydrodynamic) arguments. The latter lead to the unexpected conclusion that hydrodynamics is quantitatively valid on a microscopic scale, that is, on a few-molecular-diameters and a few-collision-times scale. The slow decay of the stress is not of the same origin as that of the velocity. That slow decay is not, at present, well understood, but leads to the rapid increase of the viscosity near melting and is the key to the understanding of glass formation.

The structure and dynamics of liquids is hence primarily determined by the steep repulsive forces. The attractive forces, whose range is typically about half that of the repulsive forces, do not alter the structure significantly. The hard-core computer simulation results, unfortunately in numerical form, are hence accurate idealizations of the liquid structure, even for metals and liquid crystals.

See also: Equations of State; Fluid Physics; Monte Carlo Techniques; Order–Disorder Phenomena.

Bibliography
J. P. Hansen and I. R. McDonald, *Theory of Simple Liquids*. Academic Press, London, 1986.
J. S. Rowlinsen and F. L. Swinton, *Liquids and Liquid Mixtures*. Butterworth, London, 1982.

Lorentz Transformations

A. E. Everett

The Lorentz transformation equations give the relations between the space and time coordinates of a single event as measured by observers in two different inertial reference frames in motion relative to one another. Let S and S' be the two reference frames, and v be their relative speed. We can think of a "reference frame" as being a rigid framework of measuring rods (e. g., meter sticks), which allow us to determine the spatial coordinates of any point in space relative to some origin, together with a set of identical clocks distributed throughout space and attached to the measuring rods. The clocks must be properly synchronized so that observers moving with the reference frame agree that all of the clocks show the same time, even though they are at different locations. An inertial frame may be defined, somewhat loosely, as one in which the stars have, on the average, a constant velocity. Hence, given one such frame, the set of all inertial frames consists of all possible reference frames in uniform linear motion relative to the original one. For simplicity we will suppose that the coordinate axes in S and S' are chosen to be parallel to one another, and that the velocity of the origin of S' as seen from S is along the positive x axis. Furthermore we suppose that the observers in the two reference frames agree to set their clocks in such a way that, at the instant the origins of the frames coincide, the clocks located at their origins both read zero. Let us suppose that some event, for example the detection of some particle or of a light pulse, occurs at the point with spatial coordinates x, y, z in S at a time t as measured on the clock fixed in S at the point where the event occurs. The Lorentz transformation equations giving the coordinates x', y', z', and t' of the same event in S' are

$$x' = (x - vt)(1 - v^2/c^2)^{-1/2} \tag{1a}$$
$$y' = y \tag{1b}$$
$$z' = z \tag{1c}$$
$$t' = (t - vx/c^2)(1 - v^2/c^2)^{-1/2} \tag{1d}$$

where c is the speed of light. If we solve this set of equations for the unprimed quantities in terms of the primed, we obtain the identical set of equations except that v is replaced by $-v$. Hence, S and S' enter the transformation equations on an equal footing, the only difference arising from the fact that the velocities of S' relative to S and S relative to S' are the negatives of one another.

These transformation equations have a crucial property. It follows from them by simple algebra that

$$x'^2 + y'^2 + z'^2 - c^2 t'^2 = x^2 + y^2 + z^2 - c^2 t^2 \ . \tag{2}$$

Suppose that the event in question is the detection of a light pulse that was emitted at the origin at $t = t' = 0$. Then, if the speed of light in S is c, we will have $x^2 + y^2 + z^2 - c^2 t^2 = 0$. But it then follows from Eq. (2) that the light pulse will also be observed to have speed c in S'. Thus the Lorentz transformations imply that the speed of light is the same in all inertial frames, in accordance with the experimental results of the Michelson–Morley experiment and the basic assumptions of the special theory of relativity. The Lorentz transformation equations (1) in fact follow from the requirement that Eq. (2) be satisfied and the assumptions that the form of the transformations is independent of the choice of origin of the coordinate system, and that all directions in space are equivalent.

The quantities x, y, z, and t are said to be the components of a 4-vector. There are other examples of 4-vectors, whose components by definition transform according to Eqns. (1). An important case is the 4-momentum, with x, y, z, and t replaced by p_x, p_y, p_z, and E/c^2, where p and E are momentum and energy. The form of the relativistic expressions for E and p is in fact determined by the requirement that the components of the 4-momentum transform between inertial frames by Eqns. (1).

The unusual and unintuitive properties of the Lorentz transformations, such as that the time of an event is different in different reference frames, result from the experimentally based requirement that the transformations leave the speed of light unchanged. These properties become important only for speeds comparable to c. For $v \ll c$, Eqns. (1a) and (1d) reduce to the simpler equations $x' = x - vt$, $t' = t$, which are called the Galilean transformations and which yield predictions in agreement with everyday experience, based on the behavior of objects with speeds small compared to c.

See also: Michelson–Morley Experiment; Relativity, Special Theory; Twin Paradox.

Bibliography

P. G. Bergmann, *Introduction to the Theory of Relativity*. Prentice-Hall, Englewood Cliffs, NJ, 1942. (1)
M. Born, *Einstein's Theory of Relativity*. Dover, New York, 1962. (E)
A. Einstein, *Relativity, the Special and the General Theory* (translated by R. W. Lawson). Crown, New York, 1961. (E)
A. P. French, *Special Relativity*. Norton, New York, 1968. (E)
H. Goldstein, *Classical Mechanics*. 2nd ed. Addison-Wesley, Reading, Mass., 1980. (I)
W. Pauli, *Theory of Relativity* (translated by G. Field). Pergamon, New York, 1958. (A)
J. L. Synge, *Relativity: The Special Theory*. North-Holland, Amsterdam, 1965. (A)

Low-Energy Electron Diffraction (LEED)

C. B. Duke

Low-energy electron diffraction (LEED) is the coherent reflection of electrons in the energy range $5 \lesssim E \lesssim 500\,\mathrm{eV}$ from the uppermost few atomic layers of crystalline solids. A schematic diagram of an apparatus for measuring the configuration and intensities of the diffracted beams

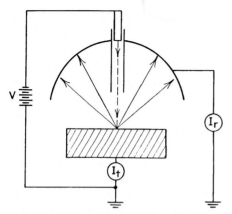

Fig. 1: Schematic diagram of a low-energy electron diffraction apparatus (i. e., $V \sim 100\,\mathrm{V}$) illustrating that the current drawn from the gun, $I = I_t + I_r$, either passes through the target to ground (I_t) or is reflected back from the target (I_r). [After C. B. Duke, *Adv. Chem. Phys.* **27**, 1 (1974).]

is presented in Fig. 1. Diffraction occurs because the wavelength λ (in angstroms) is related to the accelerating voltage V (in volts) via

$$\lambda = 12.2638/V^{1/2} \tag{1}$$

leading to λ of the order of atomic dimensions for V in the "low-energy" range. The surface sensitivity of the diffraction process arises from the short inelastic-collision mean free paths, $5 \lesssim \lambda_{ee} \lesssim 10\,\text{Å}$, for substantial energy losses ($w \geq 5\,\text{eV}$) by these low-energy electrons. Since most electron detectors exhibit an energy resolution of $\Delta E \leq 0.5\,\text{eV}$, the detection only of elastically scattered electrons effectively discriminates against reflected electrons emanating from depths greater than λ_{ee}. The detection of electrons that have experienced only a single energy-loss process also measures the consequences of scattering events occurring within distances of about λ_{ee} of the surface.

The nature of electron diffraction from crystals can be understood most simply by regarding a crystal as composed of geometrically equivalent layers of atoms parallel to a given surface. An incident electron of wave vector **k** (i. e., momentum $\mathbf{p} = \hbar\mathbf{k}$) is diffracted from the array of these two-dimensional gratings stacked together to form the crystal. The periodic translational symmetry of these diffraction gratings is manifested in the electron-solid scattering cross sections by virtue of the reflected electrons emerging from the solid in a series of diffracted beams as indicated in Fig. 2. The configuration of the diffracted beams relative to the beam of incident electrons may be predicted from the translational symmetry of the system, which leads to the momentum conservation law

$$\mathbf{K}'_\parallel = \mathbf{k}_\parallel + \mathbf{g}, \tag{2a}$$

$$k_\parallel = (2mE/\hbar^2)^{1/2}\sin\theta, \tag{2b}$$

for electrons of energy E incident on a planar surface at an angle θ relative to its exterior normal. The vectors **g** are the reciprocal lattice vectors of the two-dimensional atomic diffraction gratings, as indicated in Fig. 2. The only information conveyed by the configuration of the

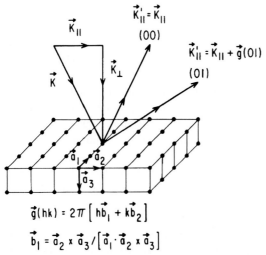

$$\vec{g}(hk) = 2\pi \left[h\vec{b}_1 + k\vec{b}_2 \right]$$

$$\vec{b}_1 = \vec{a}_2 \times \vec{a}_3 / \left[\vec{a}_1 \cdot \vec{a}_2 \times \vec{a}_3 \right]$$

Fig. 2: Schematic illustration of an incident electron beam of wave vector $\mathbf{k} = \mathbf{k}_\perp + \mathbf{k}_\parallel$, scattered elastically from a single crystal into a state characterized by the wave vector $\mathbf{k}' = \mathbf{k}_\parallel + \mathbf{k}'_\perp$. The construction of the reciprocal lattice associated with the single-crystal surface is also shown. The vectors $\mathbf{g}(hk)$ designate the reciprocal lattice vectors associated with the lowest-symmetry Bravais net parallel to the surface. [After C. B. Duke, *Adv. Chem. Phys.* **27**, 1 (1974).]

diffracted beams is the translational symmetry parallel to the surface of the two-dimensional atomic diffraction gratings that make up the solid. Determination of the unit-cell structure within each of these gratings and of the packing sequence of the gratings relative to each other requires analysis of the diffracted intensities.

The observation in 1927 of diffracted beams corresponding to the parallel-momentum conservation law (Fig. 2) was one of the original demonstrations of the wave nature of the electron, a fact that led to the recognition in 1937 of Clinton Davisson as a Nobel laureate. The potential for the use of LEED as a technique for surface crystallography was recognized in Davisson and Germer's original 1927 paper. This application of LEED was not achieved, however, until the early 1970s. The reason for this 40-year delay between recognition and achievement lies in the nature of the electron–solid interaction. The fact that it is strong (i. e., $\lambda_{ee} \sim 5\,\text{Å}$) is necessary for LEED to be a surface-sensitive spectroscopy. Because it is strong, however, a single-scattering (Born-approximation) calculation, patterned on the analysis of x-ray diffraction from bulk solids, does not suffice to describe the scattered intensities. Since analysis of these intensities is required for surface crystallography, the accomplishment of the latter awaited the construction of an adequate renormalized quantum field theory of the electron-diffraction process that properly describes the occurrence of multiple elastic and inelastic electron-scattering events. Such theories were forthcoming during 1969–1971. Surface crystallography via LEED intensity analysis subsequently became routine for low-index faces of clean metals and semiconductors, as well as for ordered adsorbed overlayers, when computer programs based on these theories became fast and inexpensive on modern computers.

The structures thereby attained have led to many new results in surface physics including the discoveries that the top layer spacing on most metal surfaces is contracted, that semiconductor surfaces exhibit new types of chemical bonding, and that surface relaxations of atomic layer spacings exhibit an oscillatory behavior extending several layers inward for both metals and semiconductors.

Measurements of the configuration and intensity of LEED beams are carried out under ultrahigh-vacuum conditions ($p \leq 10^{-8}\,\mathrm{N/m^2}$) in order to prevent contamination of the surface during the experiment. A typical apparatus consists of the vacuum chamber and pumps, electron gun, electron detector, and goniometer (i. e., a sample holder permitting precise control of the orientation and temperature of the specimen). Most LEED measurements are performed, however, in conjunction with other experiments which determine the elemental composition and electronic structure of the surface under examination. Thus, LEED is best viewed as one of a number of spectroscopic techniques utilized in concert to characterize solid surfaces. Specifically, LEED, ion scattering, and scanning tunneling microscopy are the primary techniques for the determination of the atomic geometries of crystalline surfaces and ordered overlayers thereon. ILEED is only one of many techniques for the assessment of the electronic structure of surfaces, the most prominent of which is angle-resolved photoemission spectroscopy.

The theory of LEED has been extended both to inelastic low-energy electron diffraction (ILEED) and to low-energy positron diffraction (LEPD). ILEED has been used to measure the energy–momentum relations of collective surface excitations in metals. LEPD has become an even more precise technique for surface structure determination than LEED because the interaction of an incident positron with the valence electrons of the sample is known more accurately than in the case of an incident electron.

See also: Crystallography, X-Ray; Diffraction; Electron Diffraction; Scanning Tunneling Microscopy.

Bibliography

History and General Usage

J. B. Pendry, *Low-Energy Electron Diffraction Theory and Its Application to Determination of Surface Structure*. Academic Press, London, 1974 (A)

M. A. Van Hove, W. H. Weinberg and C. M. Chan, *Low-Energy Electron Diffraction*. Springer, Berlin, 1986.

C. B. Duke (ed.), "Surface Science: The First Thirty Years", *Surf. Sci.* **299/300** (1994).

W. N. Unertl (ed.) *Physical Structure: Handbook of Surface Science*, Vol. 1. North Holland, Amsterdam, 1996.

LEED Theory and Applications to Surface Crystallography

C. B. Duke, Adv. *Chem. Phys.* **27**, 1 (1974). (I)

A. Kahn, *Surf. Sci. Rept.* **3**, 193 (1983). (I)

M. A. Van Hove and S. Y. Tong, *Surface Crystallography by LEED*. Springer-Verlag, Berlin, 1979. (A)

D. P. Woodruff, *Surf. Sci.* **500**, 147 (2002).

Role of LEED in Surface Science

F. W. de Wette (ed.), *Solvay Conference on Surface Science*. Springer-Verlag, Berlin, 1988. (A)

M. A. Van Hove and S. Y. Tong (eds.), *The Structure of Surfaces*. Springer-Verlag, Berlin, 1985. (A)

C. B. Duke, *J. Vac. Sci. Technol. A* **21**, S36 (2003).

Luminescence (Fluorescence and Phosphorescence)

A. M. Halpern

Luminescence is the emission of optical radiation (infrared, visible, or ultraviolet light) by matter. It represents the conversion of energy from one form to another, in this case, light. This phenomenon is to be distinguished from incandescence, which is the emission of radiation by a substance by virtue of its being at a high temperature ($> 500\,°C$) (*see* Blackbody Radiation). Luminescence can occur in a wide variety of substances and under many different circumstances. Thus, atoms; organic, inorganic, or organometallic molecules; polymers; organic or inorganic crystals; and amorphous materials luminesce under appropriate conditions. The most straightforward way in which luminescence can be produced is by exposure of nonionizing radiation (visible or ultraviolet light) to some form of matter. For example, a solution of the organic dye rhodamine-B emits orange light if it is irradiated with green light (or shorter-wavelength light). Many papers, fabrics, and laundry detergents contain "optical brighteners," which are organic compounds that emit blue or purple light when exposed to ultraviolet radiation. The production of light in this manner is called photoluminescence.

Other common types of luminescence and their modes of production are as follows:

Radioluminescence (or scintillation) is produced by ionizing radiation. Some polymers contain organic molecules that emit visible light when exposed to such radiation as x rays, γ rays, or cosmic rays, and thus act as detectors for high-energy radiation. Scintillators are organic molecules that emit visible light when they are excited by the decay products of certain radioactive isotopes such as ^3H (tritium), ^{14}C, and ^{18}O, and are used in many biomedical applications.

Electroluminescence is produced by the exposure of atoms or molecules to an electric field or plasma. An example is the gas-discharge tube. These are lamps that are filled with a gas such as neon, argon, xenon, nitrogen, etc., or the vapor of metals such as mercury or sodium. The lamps contain electrodes, which are connected to an electrical power supply. The applied voltage is high enough to ionize the atoms and produce a low-pressure plasma, or ionic (current-carrying) gas. The recombination of metal ions and electrons in the plasma leads to the production of electronically excited states of the atoms and hence the emission of light, usually in the form of very narrow lines throughout the electromagnetic spectrum. The common fluorescence lamp is a discharge tube in which electronically excited mercury atoms transfer this energy to a material called a phosphor, which coats the inside surface of the tube. The phosphor, after being excited by collisions with electrons or light produced by atoms or molecules, emits most of the observed light. Another example is lightning, in which the static electricity generated by a moving air mass ionizes (removes electrons from) the nitrogen and oxygen molecules. Charge recombination produces nitrogen molecules in an excited electronic state, which release this energy by emitting radiation as visible light. The Northern Lights (aurora borealis) also represents an example of this type of luminescence. Electroluminescence is produced by the familiar light-emitting diode, which is constructed from a pair of aluminum-gallium-arsenide (AlGaAs) sheets in contact with each other. One contains impurities that add electrons, while the other has deficiencies of electrons. When an electric current flows between the sheets, electrons are promoted to a higher energy level (the conduction band), and the subsequent transition of these electrons to a lower level (the charge-depleted sites) results in the emission of light, often in the red region of the spectrum.

Chemiluminescence is produced as a result of a chemical reaction usually involving an oxidation–reduction process, in which, simply viewed, electronic charge is transferred from one species to another with the resultant formation of an excited state. One common example is the luminol reaction. When luminol, an amino-substituted phthalazine derivative, is treated with base and an oxidizing agent such as hydrogen peroxide, nitrogen molecules are produced along with the emission of blue light. This light comes from the excited state of a product, the aminophthalate molecule. In this example, the light emitted represents the conversion of chemical energy to optical energy.

Bioluminescence is the result of certain oxidation processes (usually enzyme-catalyzed) in biological systems. The firefly and its larva, the glowworm, are well-studied examples; light emission for the firefly serves as a mating signal. These organisms produce a molecule called luciferin and an enzyme known as luciferase. When these molecules are in the presence of oxygen and adenosine phosphate, yellow-green light is produced with very high efficiency. Many other organisms such as bacteria, fungi, and certain marine species are bioluminescent. It is thought that chemiluminescence is a now-defunct evolutionary mechanism that was developed by some anaerobic organisms for the disposal of oxygen, which can be toxic to organisms.

Thermoluminescence (TL) is light produced when a substance, previously exposed to ionizing radiation, is heated. Many natural substances, such as quartz and feldspar, are exposed to ionizing radiation produced by the radioactive decay of the isotopes of elements such as potassium, thorium and uranium that are present in their immediate environment (and also from cosmic rays). As a result of this exposure, electrical charges are gradually placed in low-energy sites in the solid called traps. When this substance is heated, these (opposite) charges recombine, producing luminescence whose intensity reflects the number of stored-up charges. If in some past event the material became heated, either through geological processes or anthropological activities, the accumulated charge pairs are depleted (the "clock is reset"), and their population again increases with time. Deliberate exposure of the sample to heat at some later time produces TL, which can be used to date it. Alternatively, this light can be released by exposing the sample to light in a technique called Optically Stimulated Luminescence.

Triboluminescence is light emitted when a solid is subjected to friction, such as when it is scratched, ground, or rubbed. The luminescence produced by some minerals in this way is akin to thermoluminescence, except that a mechanical stimulus, rather than bulk heating, causes the effect. In other cases involving organic solids, such as sucrose, the frictional forces result in charge separation, which, in turn, leads to the production of electronically excited nitrogen molecules that emit light (similar to what is seen in lightening). Crushing candy products that contain wintergreen flavor (salicylate), causes a bright blue triboluminescent response. In this case the excited nitrogen (or other species) transfers its energy to the salicylate causing it to fluoresce.

Although there is a very wide range of circumstances in which luminescence is observed, there is one common aspect to this phenomenon. For solid state materials charge recombination produces the excited states of the emitting species. In the case of atoms or molecules, luminescence results from the transition between discrete energy levels. These levels arise when electrons are confined to a region of space that is defined by the positions of the positively charged nuclei. The way in which the molecule's (or atom's) electrons fill up, or occupy, these levels (which also have spatial notations called orbitals) determines the physical and chemical properties of the system.

The *ground state* of a system corresponds to the situation in which the electrons occupy the lowest possible energy levels. The Pauli exclusion principle requires that no more than two electrons occupy the same orbital. Electronic excited states refer to other arrangements in which the electrons occupy the orbitals in such a way that the total energy is higher. In this context, then, luminescence is the release of energy via photon emission, and is a consequence of a transition from an electronically excited state to some other, lower-energy state, usually the ground state.

Because the electron possesses intrinsic spin angular momentum, the picture describing electronic states is more complicated. It is found experimentally that when an electron is confined to an atom or molecule (or when it is under the influence of a magnetic field), its spin angular momentum is quantized and can thus have only one of *two* possible values; e. g., + or $-h/2$ (where h is Planck's constant, a universal quantity). The + or − value indicates the sense of electron spin, or rotation around its axis, i. e., clockwise or counterclockwise. If two electrons populate the same orbital, they must have opposite spin values; in such a case, the electrons are said to be *paired* (Pauli principle). In an *excited state*, one of the electrons is "promoted" to occupy a higher, vacant orbital. Thus the excited state is depicted as having two half-filled orbitals: the half-populated lower one, and the half-filled upper one. The electrons in these two half-filled orbitals can have opposite values (or spin quantum numbers), or the same ones. The former situation is referred to as a *singlet* state, while the latter denotes a *triplet* state. These terms stem from the observed behavior of atomic or molecular spectra under the influence of a magnetic field.

It is observed experimentally that a triplet state has a lower total energy relative to the corresponding singlet state. The reason for this state ordering is that in a triplet state, the two unpaired electrons are prevented from being too close to each other, and this space restriction reduces the electrostatic repulsion between them. Moreover, it is known that in the presence of a magnetic field, the triplet state splits into three sublevels whose separation is proportional to the strength of the magnetic field; hence the term "triplet" state.

Thus in cases where spin quantum numbers play an important role in determining state energies, one may distinguish between two different types of luminescence: *fluorescence*, in which emission occurs with no net change in spin quantum number, and *phosphorescence*,

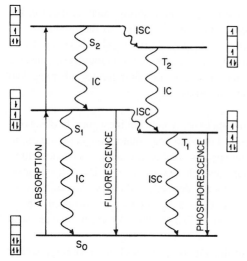

Fig. 1: Schematic electronic-state diagram for a four-electron system. S_0, S_1, and S_2 refer to the ground, first, and second excited singlet states, respectively, while T_1 and T_2 denote the first and second excited triplet states. Alongside each of the electronic state positions is an orbital diagram in which the direction of the arrow symbolizes the electron's spin value. The straight arrows connecting states describe radiative transitions. For example, S_1 or S_2 can be reached from S_0 via photon absorption, and the ground state is produced from S_1 or T_1 by photon emission. The wavy arrows portray radiationless transitions: internal conversion (IC), or intersystem crossing (ISC).

in which emission is accompanied by a change in spin quantum number. This distinction is valid for species possessing "light" atoms, i.e., atoms derived from the first or second row of the periodic table. For "heavy" atoms (e.g., $\geq$ third row), or molecules containing such atoms, one may often speak of emission in a general sense as "luminescence" because heavy atoms may promote changes in electron spin. The relative energies of the ground state and various excited singlet and triplet states are schematically portrayed in Fig. 1. In this diagram, each state is represented by an orbital-energy-level scheme, which is filled with four electrons. The upward and downward arrows symbolize the two spin quantum numbers of the electron.

Because most organic molecules are ground-state singlet systems (all electrons paired), the direct production of an excited singlet state (via photon absorption) a spin-allowed transition. In contrast, the direct production of a triplet state is a spin-forbidden transition because during such a process there would have to be a net change in the spin angular momentum of the system. This has the consequence that singlet-singlet transitions occur with much higher probability relative to singlet-triplet transitions. For organic molecules containing "light" atoms, the ratio of a singlet-singlet *vis-à-vis* singlet–triplet transition probability is typically 10^8. This has the consequence that fluorescence is much shorter-lived (ca. 10^{-9}–10^{-7} s) as compared with phosphorescence (ca. 10^{-3}–10 s).

An excited singlet state (e. g., S_1 or S_2 in Fig. 1) can be directly produced by photon absorption. This excited state can then undergo what are called *radiationless transitions*. Through these processes, energy initially deposited in the molecule is dissipated into vibrational (i. e., thermal) energy, or sometimes chemical energy (if bonds are broken or rearranged). The spin-forbidden radiationless transition (i. e., singlet → triplet or triplet → singlet) is called intersystem crossing (ISC), and the spin-allowed transition (singlet → singlet or triplet → triplet) is referred to as internal conversion (IC) (see Fig. 1). It is generally observed that the rate of a radiationless process decreases sharply as the energy gap between the two electronic states increases. Because for most molecules, the energy differences between the excited states are much smaller than the gap between the lowest excited state and the ground state, the lowest excited state is usually formed with very high efficiency after optical excitation into any of the higher states. Thus, with very few exceptions, fluorescence and phosphorescence take place from the lowest excited singlet and triplet states (S_1 and T_1), respectively.

For most organic molecules phosphorescence lifetimes are generally much longer than fluorescence lifetimes (see above). Because the triplet state is so long-lived, it has a high probability of undergoing a reaction by which excitation energy is dissipated by some means. If this process occurs as a result of an interaction (e. g., via a collision) with another molecule, it is called *quenching*. One interesting and important consequence of the quenching of a triplet state by another molecule is energy transfer in which the excitation energy is "picked up" by the quencher (the energy acceptor), The quencher, now enriched, can undergo some chemical or physical reaction just as it would if it were directly excited. If energy transfer takes place in a system, the donor luminescence is diminished (because it is quenched) and in some cases the acceptor emits light. This process is called *sensitized* luminescence and can be observed in singlet–singlet or in triplet–triplet energy transfer (i. e., sensitized fluorescence or phosphorescence, respectively). In certain systems where there is a high concentration of molecules, such as in crystals, excitation energy can "migrate" over fairly large distances before it is dissipated. If the crystal contains an impurity (or a site defect) having a lower excited-state energy, the impurity acts as an energy trap, and it may luminesce with its own characteristics.

Triplet states also have a high reactivity toward other molecules that might be present in very small amounts (such as impurities). For this reason, phosphorescence in fluid media is usually very weak because of the high mobility of the molecules and the correspondingly high probability of an encounter between the triplet state and a quencher. The phosphorescence intensity of a system can be significantly enhanced by cooling the system so that the bulk solvent viscosity increases. This results in a decrease in the encounter probability, the (excited) triplet state and a potential quencher. Alternatively, phosphorescence may be observed at room temperature if the phosphorescent molecules are dissolved in an immobilizing medium, such as a rigid polymeric "glass", a micelle, or some other type of isolating matrix.

See also: Atomic Spectroscopy; Chemiluminescence; Electroluminescence; Thermoluminescence.

Bibliography

R. S. Becker, *Theory and Interpretation of Fluorescence and Phosphorescence.* Wiley-Interscience, New York, 1969. (I)

J. B. Birks, *Photophysics of Aromatic Molecules.* Wiley-Interscience, London, 1970. (A)

M. D. Lumb (ed.), *Luminescence Spectroscopy.* Academic Press, London, 1978. (A)

C. A. Parker, *Photoluminescence in Solutions.* Elsevier, Amsterdam, 1968. (I)

P. G. Seybold, "Luminescence", *Chemistry* **42**, 6 (1973). (E)

G. H. Schenk, *Absorption of Light and Ultraviolet Radiation.* Allyn & Bacon, Boston, 1973. (E)

N. J. Turro, "The Triplet State", *J. Chem. Educ.* **46**, 2 (1969). (E)

F. Williams (ed.), *Luminescence of Crystals, Molecules, and Solutions.* Plenum Press, New York, 1973. (A)

M. J. Aitken, *An Introduction to Optical Dating.* Oxford University Press, New York, 1998. (A)

M. J. Aitken, *Thermoluminescence Dating* Academic Press, London, 1985. (A)